The New American Machinist's Handbook

OTHER McGRAW-HILL HANDBOOKS OF INTEREST

AMERICAN SOCIETY OF MECHANICAL ENGINEERS · ASME Handbooks:
- Engineering Tables
- Metals Engineering—Design
- Metals Engineering—Processes
- Metals Properties

BAUMEISTER AND MARKS · Standard Handbook for Mechanical Engineers
BEEMAN · Industrial Power Systems Handbook
BRADY · Materials Handbook
BURINGTON AND MAY · Handbook of Probability and Statistics with Tables
CARRIER AIR CONDITIONING COMPANY · Handbook of Air Conditioning System Design
CARROLL · Industrial Instrument Servicing Handbook
CONSIDINE · Process Instruments and Controls Handbook
CONSIDINE AND ROSS · Handbook of Applied Instrumentation
CROCKER AND KING · Piping Handbook
DE CHIARA AND CALLENDER · Time-Saver Standards for Building Types
DUDLEY · Gear Handbook
EMERICK · Handbook of Mechanical Specifications for Buildings and Plants
EMERICK · Heating Handbook
EMERICK · Troubleshooters' Handbook for Mechanical Systems
FACTORY MUTUAL ENGINEERING DIVISION · Handbook of Industrial Loss Prevention
FINK AND CARROLL · Standard Handbook for Electrical Engineers
FLÜGGE · Handbook of Engineering Mechanics
GARTMANN · DeLaval Engineering Handbook
HARRIS · Handbook of Noise Control
HARRIS AND CREDE · Shock and Vibration Handbook
HEYEL · The Foreman's Handbook
KALLEN · Handbook of Instrumentation and Controls
KING AND BRATER · Handbook of Hydraulics
KLERER AND KORN · Digital Computer User's Handbook
KOELLE · Handbook of Astronautical Engineering
KORN AND KORN · Mathematical Handbook for Scientists and Engineers
MACHOL · System Engineering Handbook
MAGILL, HOLDEN, AND ACKLEY · Air Pollution Handbook
MANAS · National Plumbing Code Handbook
MANTELL · Engineering Materials Handbook
MAYNARD · Industrial Engineering Handbook
MERRITT · Building Construction Handbook
MORROW · Maintenance Engineering Handbook
PERRY · Chemical Engineers' Handbook
PERRY · Engineering Manual
ROSSNAGEL · Handbook of Rigging
ROTHBART · Mechanical Design and Systems Handbook
SHAND · Glass Engineering Handbook
SOCIETY OF MANUFACTURING ENGINEERS:
- Die Design Handbook
- Handbook of Fixture Design
- Manufacturing Planning and Estimating Handbook
- Tool Engineers Handbook

STANIAR · Plant Engineering Handbook
STREETER · Handbook of Fluid Dynamics
TOULOUKIAN · Retrieval Guide to Thermophysical Properties Research Literature
TRUXAL · Control Engineers' Handbook

THE NEW
AMERICAN MACHINIST'S
HANDBOOK

Edited by **RUPERT LE GRAND**
Senior Associate Editor, American Machinist

Based upon earlier editions of "American Machinists' Handbook"

edited by **FRED H. COLVIN** *and* **FRANK A. STANLEY**

McGraw-Hill Book Company

NEW YORK ST. LOUIS SAN FRANCISCO DÜSSELDORF JOHANNESBURG
KUALA LUMPUR LONDON MEXICO MONTREAL NEW DELHI
PANAMA RIO DE JANEIRO SINGAPORE SYDNEY TORONTO

THE NEW AMERICAN MACHINIST'S HANDBOOK

ISBN 07-037065-6

Library of Congress Catalog and Card Number: 55-8908

242526 MUBP 09

Preface

No handbook serving metalworking can be considered up to date unless it has been completely rewritten and revised since 1950.

This is true because of the exceptionally rapid advancement since the last war in the methods for machining, forming, and joining of metals; the improvement in cutting materials; the new constructional steels and tool steels; and the refinements in heat-treatment and finishing. Further, there have been many important developments in gearing, splines, threads, fastening devices, drafting practice, machine-tool components, and power-transmission equipment. In short, the technology, standards, and practices of metalworking have grown and broadened in keeping with the industry's expansion to the number 1 position in the American economy.

These considerations led to the development of "The New American Machinist's Handbook," as successor to that great work, "The American Machinists' Handbook," or "*Colvin & Stanley's*," as it was popularly known Founded in 1908 by Fred H. Colvin and carried on by him through eight editions, with assistance by Frank A. Stanley, the original "American Machinists' Handbook" was considered a bible by hundreds of thousands of shop men and engineers all over the world.

Like its predecessor, the guiding principle in writing "The New American Machinist's Handbook" has been practicality of subject matter. Information has been drawn from hundreds, perhaps thousands, of sources, then condensed and rewritten into a logical and coherent whole. The purpose has been to create an encyclopedic treatment that would serve both as a reference work and as a text to broaden the knowledge of the reader in the many branches of metalworking activity by means of self-study.

An up-to-date handbook compresses into small space the essentials of important articles, technical papers, engineering and product standards, and proved shop and engineering practice. Needed information is conveniently available, in contrast to the task of hunting through a mass of books, papers, or back issues of magazines, only to find that the wanted material has been mislaid or thrown out. Then, too, the professional handbook writer has many sources of information unknown to the average reader.

In addition to being up-to-date, practical, and comprehensive, a handbook must be written to serve a broad audience. In this case, "The New American Machinist's Handbook" is intended to serve as a supplementary instruction manual for the vocational student and apprentice, and particularly as a reference work and text for machinists, toolmakers, machine-repair men, inspectors, foremen, superintendents, managers, estimators, process engineers, production and manufacturing engineers, tool engineers, product designers, machine designers, draftsmen, purchasing agents, and general executives in the metalworking industry.

A modern format has been adopted for "The New American Machinist's Handbook." Subject matter is divided into 45 sections grouped in 11 parts. For example, Part 1, Machining Methods, consists of 614 pages on 13 machining subjects: broaching; drilling; files and burrs; gears and gear cutting, splines and serrations; grinding processes; milling; planing and shaping; reaming; sawing; threading and thread systems; tapping; turning and boring; and screw-machine work. Part 2, Metal-forming Methods, logically groups spinning; pressworking and cold-roll forming; forging, upsetting, and cold heading; cold working of metals; die casting; and babbitting of bearings The remaining 26 sections are likewise appropriately grouped.

An important feature of this book is the care with which all information on a single subject is combined in one section. For example, if you are interested in how to drill a specific material, it will be found in Section 2, Drills and Drilling, along with drill types, grinds, and selection, and not scattered through the sections on materials, which are reserved for properties, analyses, and forms.

Acknowledgements. Compilation of a handbook can be achieved only with the active support of many scores of individuals and organizations. In this book I have had the active help and wise counsel of Fred H. Colvin. He prepared certain material and made the index. The *American Machinist* and *Product Engineering* were the sources of most of the material derived from magazine articles. Then, too, much help was given by the AISI, ASA, ASME, SAE, AWS, the Metal Cutting Tool Institute, the Fasteners Institute, and many other associations, private companies, and individuals.

Rupert Le Grand

Contents

Part 1

MACHINING METHODS

Section 1

BROACHES AND BROACHING

Consultant: C. Morawski, Chief Tool Designer,
Detroit Broach Company

SECTION 1

BROACHES AND BROACHING

Definition. Broaching is a generating process, whereby metal is removed with a multiple-point tool, usually a bar, with tooth height increasing from the starting end. When the broach is pulled or pushed through or over the work, each tooth removes a chip of uniform thickness, in contrast to a milling cutter tooth, which removes a wedge-shaped chip. The chip thickness normally ranges from 0.007 to 0.001 in., depending on whether cutting is being done by the roughing, semifinishing, or finishing teeth. Usually one pass completes the hole or surface.

Applications. Broaching is firmly established as a mass-production process but is used also for short-run jobs. Some broaching machines are specially built to machine one product, like a cylinder block; others permit rapid changes of broaches and work-holding fixtures for a variety of pieces in small lots. And, of course, there are simple machines for occasional keyway-cutting jobs and the like.

Broaching is economical because only a single cut is usually required, and subsequent finishing operations are not needed. The process is also economical because of the number of elements of a surface, external or internal, that can be cut simultaneously. Only limitations are: all elements of a broached surface must be parallel with the broach axis; there must be no obstructions in the plane of the broached surface; and depth of cut is governed by the stroke and tonnage of the machine.

Kinds of Broaches. A pull broach cuts when it is pulled through a hole or over a surface. A push broach cuts when it is pushed through a hole or over a surface.

Broach classification:

1. Method of operation—pull or push.
2. Type of operation—internal or external.
3. Construction—solid or built-up.
4. Function—keyway, surface, round hole, serration, combination round and spline, spline, helical spline, rifling, and burnishing. In helical spline cutting and rifling, the broach may be rotated by a special head, lead bar, and puller adapter.

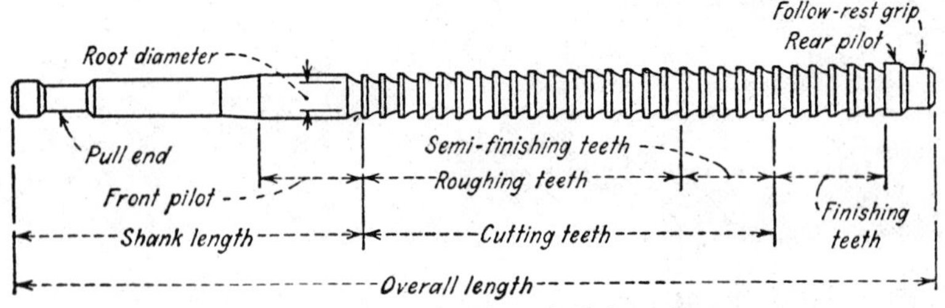

Fig. 1. Typical internal broach of the pull type.

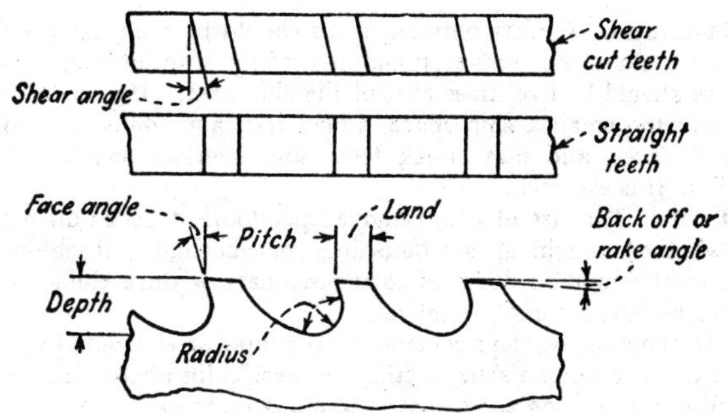

FIG. 2. Broach nomenclature.

BROACH DESIGN

The broach user is interested in broach design as an aid to select the proper tool for a job. Broaches are ordinarily designed by specialists because of problems in correct design for specific circumstances and difficulty of manufacture.

Broach Materials. Steels used in manufacture of broaches are tungsten and moly high-speed steel. Carbide cutting sections have been brazed or inserted in broaches. Some of the most modern broaches use a series of inserts like single-point tools, which can easily be removed, sharpened, and replaced.

Pitch of Teeth. Pitch, or spacing, of teeth (Fig. 2) affects chip space and strength of the tooth and controls the number of teeth in contact with the work and alignment while in the cut. Pitch is determined by length of cut, chip thickness, and material being broached. Cast iron does not require the chip space necessary for steel as cast-iron chips crumble while steel chips curl up and need more room.

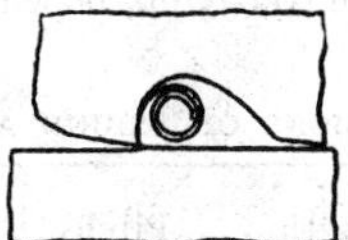

FIG. 3. Chip space depends on the material broached.

The Broaching Tool Institute suggest the formula

$$\text{Pitch} = 0.35\sqrt{\text{length of cut}}$$

EXAMPLE: With a cut 4 in. long, the pitch is $0.35 \times 2 = 0.70$ in. For a 9-in. cut in the same material, the pitch is $0.35\sqrt{9} = 1.05$ in. This gives more total room for chips but not in proportion to the length of the cut. Table 1 gives dimensions of round broaches for various cut lengths.

For successful broaching, at least two teeth should always be engaged with the work, and three is better. This holds true for both horizontal and vertical broaching. It is better to reduce the chip per tooth than to use too few teeth, where the power of the broaching machine is limited.

Differential Tooth Spacing. If a broach vibrates and leaves tooth marks on the work it may be advisable to vary the tooth spacing in groups of two or three teeth. This prevents chatter in internal broaching. On surface broaching, chatter may be caused by contact with more than one surface at once.

Gullet Dimensions. Gullets between teeth should be round at the bottom and have an area ten times the cross-sectional area of the chip for steel. For cast iron the gullet area should be five times that of the chip area. If a spiral chip, as from steel, curls into too small a chip space, added resistance dulls the broach, causes rough work surfaces, and may break teeth and overload machine. Good chip formation (Fig. 3) is essential.

Cut per Tooth. The size of chip removed per tooth depends upon stage of the operation—whether roughing, semifinishing, or finishing. Roughing teeth on broaches for steel should be designed to remove narrow thick chips by use of chip breakers. Such chips are easily removed.

STEELS. In general, the chips removed in roughing steel should be from 0.005 to 0.010 in. thick. The semifinishing section removes chips about 0.005 in. thick; the finishing section takes chips about 0.0005 to 0.002 in. thick.

For free cutting steel use a cut per tooth of 0.004 to 0.006 in. on dia for splines and 0.0015 to 0.003 in. on dia for round holes. For keyway and surface broaching use 0.003 to 0.006 in. per tooth.

NICKEL-ALLOY STEELS. For broaching spline holes, cut per tooth can be 0.004 in. on dia for splines, 0.002 in. on dia for round broaches, 0.004 in. for keyway and surface broaches. Shear angles may be 5 to 20°.

Face angle varies between 8 and 20°, decreasing with hardness. Back-off or rake angle varies between ½ and 2° for internal broaches, up to 3½° for surface broaches.

NITRIDING STEELS. If treated to obtain correct machinability, these steels may be broached with tools used for other alloy steels, but broaching speed may have to be reduced to improve finish and increase broach life. For internal broaching, a cut per tooth of 0.004 to 0.005 in. is recommended. On surface broaching the cut may vary from 0.0025 to 0.0035 in., depending on length, shape, and size of part.

TABLE 1. DIMENSIONS OF ROUND AND SPLINE BROACHES FOR STEEL

Cut Length	Pitch	Land	Depth	Radius
3/16	1/8	3/64	3/64	1/32
1/4	3/16	1/16	1/16	1/32
3/8	7/32	1/16	5/64	3/64
1/2	1/4	1/16	3/32	1/16
3/4	5/16	3/32	1/8	5/64
1	11/32	3/32	1/8	5/64
1 1/8	3/8	7/64	9/64	3/32
1 1/4 to 1 5/8	7/16	1/8	5/32	7/64
1 3/4 to 2	1/2	1/8	3/16	1/8
2 1/4 to 2 1/2	9/16	9/64	7/32	9/64
2 3/4 to 3 1/4	5/8	5/32	15/64	5/32
3 1/2 to 4	11/16	5/32	9/32	3/16
4 1/4 to 4 3/4	3/4	5/32	19/64	3/16
5 to 5 1/2	13/16	5/32	5/16	13/64
5 3/4 to 6	7/8	3/16	11/32	7/32
6 1/2 to 7	15/16	3/16	23/64	7/32
7 1/2 to 8	1	7/32	3/8	1/4
8 1/2 to 10 1/2	1 1/8	1/4	25/64	1/4
11 to 12	1 1/4	9/32	13/32	7/64

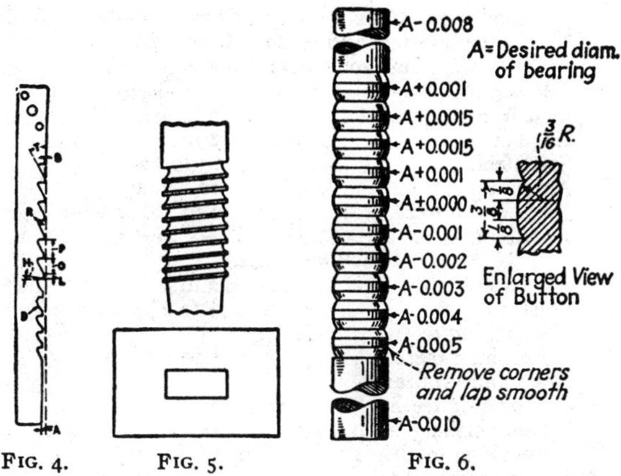

FIG. 4. Broach for cutting keyways or slots. FIG. 5. A shear angle may be used on a square-hole broach to avoid vibration. FIG. 6. Burnishing bars are frequently used after broaching of holes. Size of steps depends on the work material.

STAINLESS STEELS. Cut per tooth, 0.001 to 0.005 in. per tooth for round broaches. Speeds, between 8 and 20 fpm. Face angle, 12 to 18°. Back-off, a minimum, or about 2°. Use chip breakers.

NICKEL ALLOYS. For spline broaches the step per tooth should not exceed 0.003 in. for monel and inconel. Depth of cut for round broaches used on material harder than Rockwell 14 C should not exceed 0.0015 in., and for spline broaches 0.002 in. Chip breakers are recommended.

CAST IRONS. A greater cut per tooth can be used than for free-cutting steel. Less chip room is required.

BRASSES AND BRONZES. A somewhat greater cut per tooth than for steel is permissible, and a chip breaker is valuable on roughing teeth.

ALUMINUM AND MAGNESIUM. Larger chip spaces than provided for other metals are desirable. Standard broaches can be used. To overcome trouble in maintaining tolerances, increase the finishing cut to 0.002 in. per tooth.

Face and Back-off Angles. See Table 2. The back-off angle is frequently varied from a maximum at the starting end of the broach to ½° at the finishing end.

Land. Strength of a broach tooth depends on the land, which must be greater for heavy than for light cuts. The teeth are backed off clear to the edge for roughing cuts, varying from ½ to 3° for internal broaches and up to 3½° for surface broaches.

Maintenance of size is helped by not backing off the entire length of the land on the finishing teeth, increasing the straight land from the first finishing tooth. This helps to give longer life to the broach. Too much straight land increases friction, which may expand the work and gall, or abrade, the finished surface.

Shear Angle. A shear angle can be used on a surface broach to give better finish and eliminate vibration. For cast iron, use a shear angle of 20°, for steel forgings

TABLE 2. FACE AND BACK-OFF ANGLES
(Cylindrical and Surface Broaches)

Face, or Hook, Angle	
Material Broached	*Degrees*
Cast iron	6 to 8
Hard steel	8 to 12
Soft steel	15 to 20
Aluminum	10, or more
Brass and bronze	0 to 10, or more
Brittle brass	−5 to +5
Back-off Angle	
Cast iron	2 to 5
Steel:	½ to 2*
Roughing teeth	1½
Finishing teeth	½ to 1†
Brass and bronze:	
Roughing teeth	2
Semifinishing teeth	1
Finishing teeth	½†
Spline broaches:	
Roughing teeth	3
Finishing teeth	1½†
Surface broaches	Up to 3½

* Back-off angle may vary from 2° at the beginning end of broach to ½ at the finishing end. Holding back-off angle to a minimum reduces size loss when the broach is sharpened.

† Part of the land of finishing teeth may be straight and may be graduated from first to last finishing tooth. Size of land determines number of resharpenings possible before the broach is ground under size. Too much land increases cutting friction, causing expansion and galling of broached surfaces on some material.

about 10 to 15°. In slotting, however, a shear cut forces the chips against one side to roughen the surface.

Chip Breakers. Tools that broach tough material and form wide chips should have chip breakers. These are nicks on the roughing-section teeth but are seldom used on the semifinishing section and never on the finishing section of the broach. Chip breakers produce grooves that must be removed by succeeding teeth. As a rule, chip breakers are not used in broaching cast iron, except on extra heavy roughing cuts.

Broaches are generally made of 18-4-1 or 18-4-2 tungsten high-speed steels or from molybdenum high-speed steels, but carbides are also used. For round-hole and rifling broaches, the carbide is shaped in the form of rings and brazed to a bar. But surface broaches are being made of a series of toolholders incorporating removable carbide inserts especially for heavy cuts on cast irons.

BURNISHING BARS

Bearings and bushings are frequently burnished after broaching by a bar with polished buttons (Fig. 6). This tool has ten buttons which increase the diameter by a total of only 0.001 in. The lower or entering end is 0.010 in. below size, the button diameters increasing by 0.001 in. until 0.001 in. oversize is reached. The eighth and ninth buttons are 0.0015 in. oversize, whereas the last is 0.001 in. and the upper end has a clearance of 0.002 in. The diameter of the final button insures a bearing clearance of 0.001 in., even though the metal may close in after the 0.0015-in.

buttons pass. Burnishing buttons are sometimes included in the broach following the cutting teeth.

ESTIMATING PRODUCTION

Normal broaching speed for many types of steel has been set at 30 fpm for the usual hydraulic broaching machine up to 20 tons capacity. Small parts have been broached at more than 40 fpm, but the hydraulic equipment must be increased beyond the economical limit. The higher the tonnage capacity of the machine the slower its economical speed. The range is usually from 4 to 30 fpm, with 18 to 24 fpm usual for average work.

Production depends on the speed of cutting and return, starting and stopping, and the handling of the work in and out of fixtures. Starting and stopping is usually figured at 2 sec and loading at 5 sec. An efficiency of 85% is considered good.

EXAMPLE:

Cutting speed = 24 fpm = 288 ipm
Return speed = 34 fpm = 408 ipm
Stroke = 40 in.
Starting and stopping time = 2 sec
Loading time = 5 sec

$$\text{Cutting time} = \frac{40 \times 60}{288} = 8.33 \text{ sec}$$

$$\text{Return time} = \frac{40 \times 60}{408} = 5.9 \text{ sec}$$

$$\text{Starting and stopping} = 2.0 \text{ sec}$$

$$\text{Loading time} = 5.0 \text{ sec}$$

$$\text{Complete cycle} = 21.23 \text{ sec}$$

$$\text{Predicted output} = \frac{60 \times 60 \times 85\%}{21.23} = 144 \text{ pcs per hr}$$

SHARPENING BROACHES

It is not necessary to grind all finishing teeth each time a broach is sharpened. Grinding the first one or two teeth is usually sufficient until they have worn under size. Suggestions for sharpening internal and external broaches are given in Fig. 7. The grinding cut on the face of the tooth should blend into the radius as shown. Round broaches can be ground as shown in Fig. 8.

POINTS IN BROACH SHARPENING

1. Maintain original tooth form because design characteristics affect operating efficiency (Fig. 9).
2. Maintain original chip space to permit smooth chip flow.
3. Remove just enough stock to sharpen tooth. Grinding away more material shortens broach life.

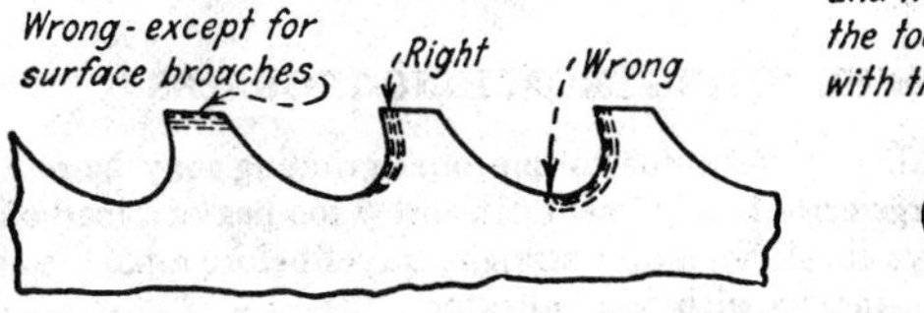

FIG. 7. Incorrect and correct methods of sharpening broach teeth. The gullet must be a smooth curve.

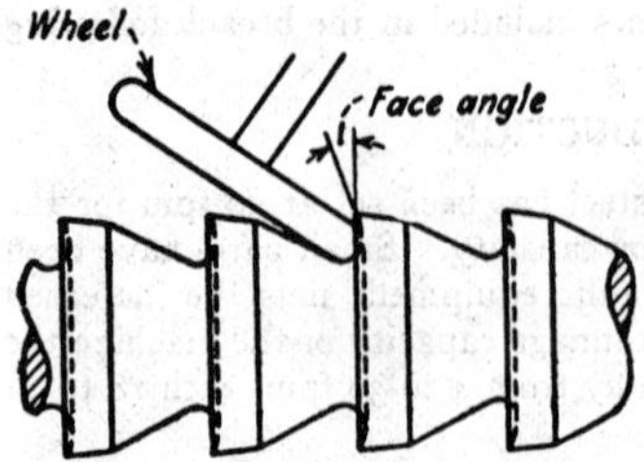

FIG. 8. Wheel plane is set a greater angle than the face angle of the broach tooth, in order to avoid reducing the face angle when sharpening.

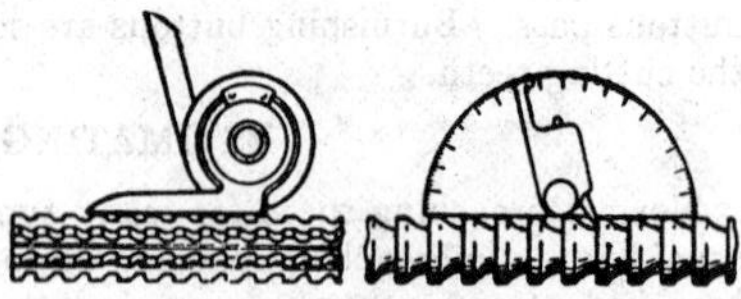

FIG. 9. Face angle of a new broach should be measured before use and checked after sharpening. A combination hook and radius gage, right, determines face angle and proper curvature of chip space.

4. Use blueprint as guide to face angle, back-off angle, tooth depth, radius, and land width.

5. Remove galls and nicks on OD and also on spline sides before sharpening.

6. If you have no broach sharpener, mount a high-speed grinder on the cross-slide of a lathe for sharpening cylindrical broaches on centers. Wheel-head angle = face angle × wheel dia ÷ broach root dia.

7. Do not throw away a broach with a broken tooth. Remove tooth and restep three or four following teeth to distribute load.

8. Back off cylindrical broach teeth only when absolutely necessary because of extremely poor condition. Use a broach sharpener or a good cylindrical grinder to increase back-off angle without changing tooth diameter at cutting edge.

9. Use a steadyrest on a long broach to prevent sag, vibration, and chatter.

10. Measure tooth height carefully when grinding surface-broach lands.

11. Demagnetize the broach if sharpened on a magnetic chuck.

12. Be sure, when setting up broach inserts in a holder, that all contacting surfaces are clean and free from chips, that the teeth blend to give proper chip space where two sections meet, that inserts are not tightened too hard as they are likely to break.

13. Grind dry or wet. In dry grinding, be careful to avoid burning the cutting edges and letting the wheel "spark out."

14. Do not let broach teeth strike any metal surface, as teeth are extremely hard and easily damaged. Store them in wood or lined racks with individual compartments.

15. Use the right type of wheel—Recommended: face grinding—vitrified aluminum oxide disk wheel of 46 to 80 grit with soft or medium bond for roughing and 100 grit for finishing; backing off—vitrified aluminum oxide cup wheel of 60 grit with medium bond. For extremely smooth finish, use finer grain up to 400.

16. Stone cutting edges lightly to remove burrs and gain smoother surfaces, but do not remove enough material to form a negative land.

TROUBLE-SHOOTING BROACHING TROUBLES

Broken Teeth. Packing of chips due to improper grinding may be one cause. On surface broaching a large error in alignment can throw too heavy a load on teeth, causing breakage. Always check holder for straight travel before a part is actually broached. Check steps of inserts with dial indicator.

Spoiled Work or Broken Insert. Check insert assembly in holder to see if screws are too long or too short. If the insert is loose, the screws are too long. If screws are too short and are pulled up with force, the screw hole becomes weak and eventually pulls out.

POOR FINISH AND VARIATION IN SIZE. Look for loose clamps. Check loading fixture and seating of pieces. Improper loading and chip accumulation are causes.

BREAKAGE OF INTERNAL BROACH. Check alignment. See that direction of pull is at right angles to faceplate. Check center axis of broach with axis of faceplate.

DRIFTING. Check the center of the starting hole. It probably is not centralized with broach center.

ROUND OR SPLINE BROACHES CUT OFF CENTER. This is caused by "drifting." On round holes one side does not clean up. On spline broaches the splines will be eccentric. See above recommendation to eliminate drifting.

EXCESSIVE WEAR AND DULLING OF TEETH. Again, this is usually the result of drifting. Also check lubricant. If too rich in sulfur, cut back with paraffin oil.

CHATTER. Inserts may have featheredge and require stoning. Parts not held tight enough. Part vibrates from forcing the cut. Chatter can also develop from using too light a machine. Check hydraulic system.

PARTS WILL NOT HOLD SIZE. Look for something loose while broach is cutting. Part may be springing due to cutting force. Check clamps. Are they strong enough? Check for deflection in machine.

TEARING AND/OR HEAVY BURRS. Dead soft steel is draggy and can be the cause of this condition. Material should be about 28 to 36 Rockwell.

TABLE 3. CUTTING FLUIDS FOR BROACHING VARIOUS MATERIALS

Material Group*	Cutting Fluids†	Material Group*	Cutting Fluids†
1	90 K + 10 M 70 K + 50 M SM-SML	5	10–20 W + 1 SO SM-SML M + (10–15) L LM
2	20–25 W + 1 SO LM	6	10–20 W + 1 SO SM-SML M + (10–15) L LM
3	5–15 W + 1 SO SM-SML M + (10–20) L	7	10–15 W + 1 SO SM-SML M + (10–20) L LM
4	5–15 W + 1 SO SM-SML	8	5–10 W + 1 SO SM-SML M + (10–20) L LM

* Material groups:
1. Aluminum and alloys; al. and zinc die castings
2. Brass
3. Bronze
4. Copper, Everdur, inconel, monel, nickel
5. Wrought and malleable iron
6. Low-carbon and free-cutting steels
7. Medium-carbon and tough low-alloy steel
8. High-carbon high-alloy steels, including stainless

† Cutting fluids:
W = water
K = kerosene
M = mineral oil
LM = straight mineral oil
SM = sulfurized mineral oil
SML = sulfurized mineral lard oil

REASONS FOR PART GALLING AND PICKUP

1. Broach Teeth Damaged

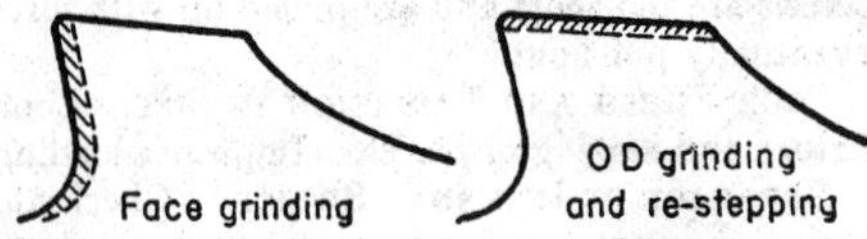

Repair mutilated teeth by: (1) heavy face grind, or (2) send to supplier for OD grind and restepping. Check handling and setup practices. Use follower supports if needed.

2. Hard Fixtures or Liner Bushings

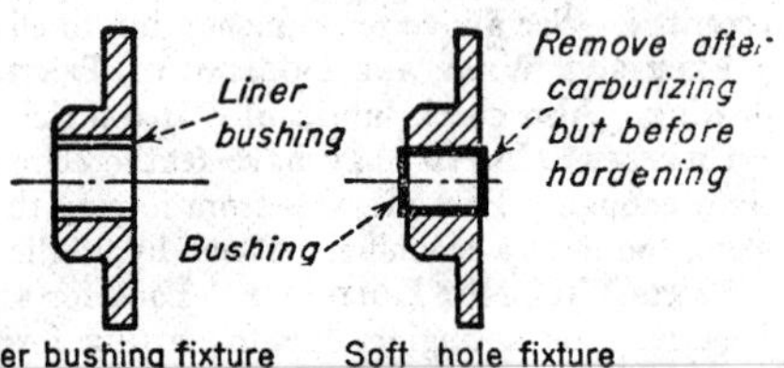

Use soft (Rockwell C 30-35) liner bushings or soft-hole fixtures. On a carburized and hardened bushing or fixture, the hole can be bored after carburizing to remove the hardening agent before hardening the piece, and thus produce the desired "soft hole" in the hardened part.

3. Improper Face Angle

4. Negative Rake

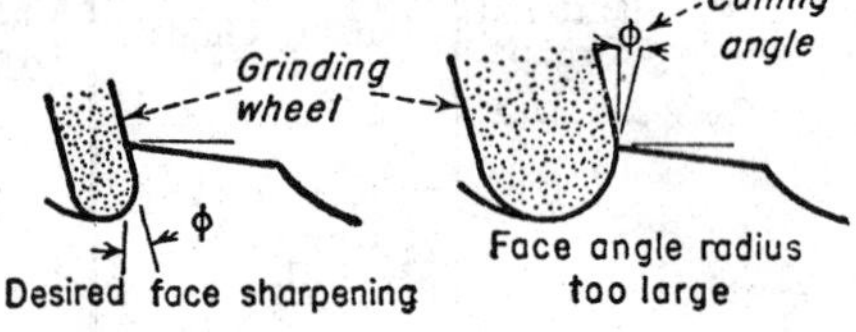

Check sharpening practice, broach print, and broach design. Check face-angle radius on grinding wheel.

5. Deep or Shallow Face-angle Radius

Check sharpening practice and broach print.

6. Rounded Cutting Edges

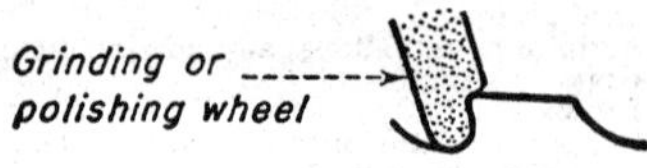

Check grinding-wheel dressing method. Use bottom of chip space instead of broach OD for steadyrest support.

Section 2

DRILLS AND DRILLING

SECTION 2

DRILLS AND DRILLING

Drills are probably more widely employed than any other tool in the shop, but their selection, grinding, and conditions of use often receive the least care. Cost per hole is frequently higher than it should be, because a drilled hole is considered either a clearance hole for fastening purposes or the starting point for tapping or production of an accurate hole by second operations like reaming or boring.

If accuracy and smoothness are not important in a drilled hole, at least the cost is. Too often a drill is selected merely for size and length. Much drilling can be done satisfactorily with the standard point angle of 118°. Where quantities of holes must be produced, however, refinements in practice are desirable. Factors that bear watching are: (1) machine sharpening that achieves good geometry of the drill point, such as cutting edges of the same length and at equal angles with the drill axis; (2) point angle, lip clearance, and web thickness related to the material being drilled; (3) proper support in a close-fitting drill bushing of suitable length; and (4) machine, chuck or driver, and fixture in good condition. Sometimes special drills are justified.

TWIST DRILLS

The broad definition for a twist drill is "an end-cutting tool having one or more cutting edges, and having helical or straight flutes for the passage of chips." In general practice, a standard twist drill (Fig. 2) has two cutting edges, two helical

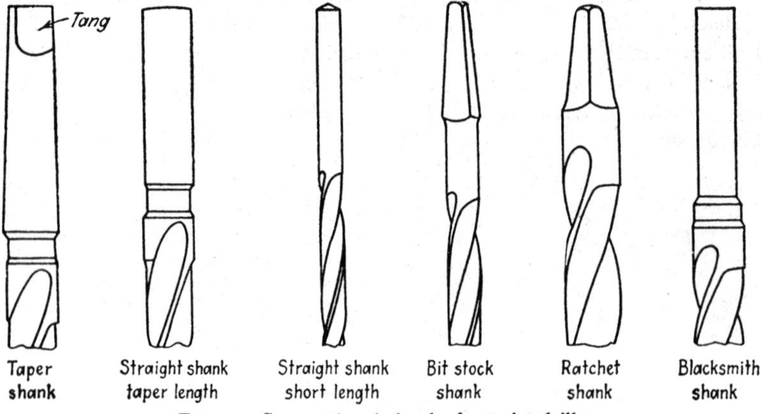

Fig. 1. Conventional shanks for twist drills.

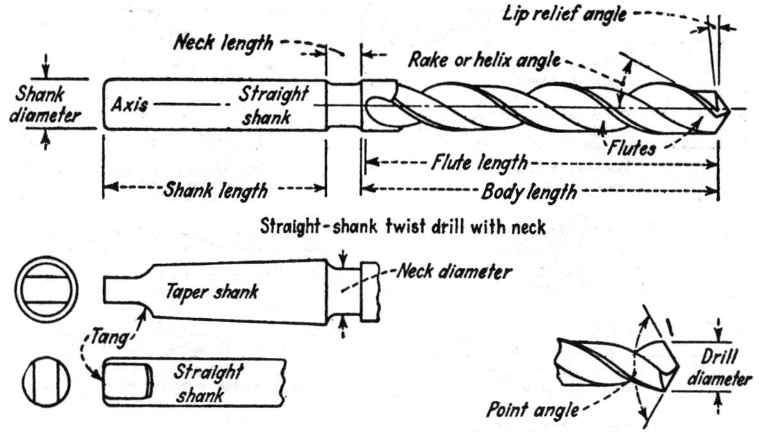

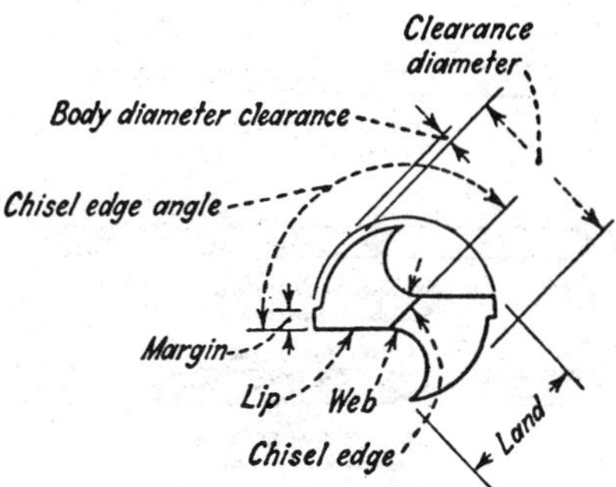

FIG. 2. Terminology for straight-shank and taper-shank standard twist drills. Tolerances on various features of standard drills now make them interchangeable in the user's shop. Straight-shank drills in number, letter, and fractional sizes have corresponding lengths.

flutes, and a straight or taper shank. See standard sizes of twist drills, Tables 1 to 4. Such drills are designed for cutting holes from solid metal but can be used to enlarge holes too. Three- or four-fluted core drills are intended for enlarging cored holes in castings, drilled holes in various materials, and punched holes.

Standard twist drills are available in various grades of high-speed steel and carbon steel. Carbide-tipped or solid carbide drills have not yet been standardized.

Drill Sizes. Three designations are currently used for English-measure drills: numbers—No. 80 to No. 1 (0.0135 to 0.228 in.), letters—A to Z (0.234 to 0.413 in.), and fractions—$\frac{1}{64}$ to $3\frac{1}{2}$ in. Not listed are millimeter sizes, but these are available.

Nomenclature for Twist Drills

According to American Standard B5.12-1950 published by the ASME, the parts of standard twist drills with straight or tapered shanks are defined as follows:

Point Angle. The angle included between the lips projected upon a plane parallel to the drill axis and parallel to the two cutting lips is known as the point angle.

Lip Relief Angle. This is the angle measured between a tangent on the surface back of the cutting edge at the periphery, and a plane at right angles to the axis of the drill.

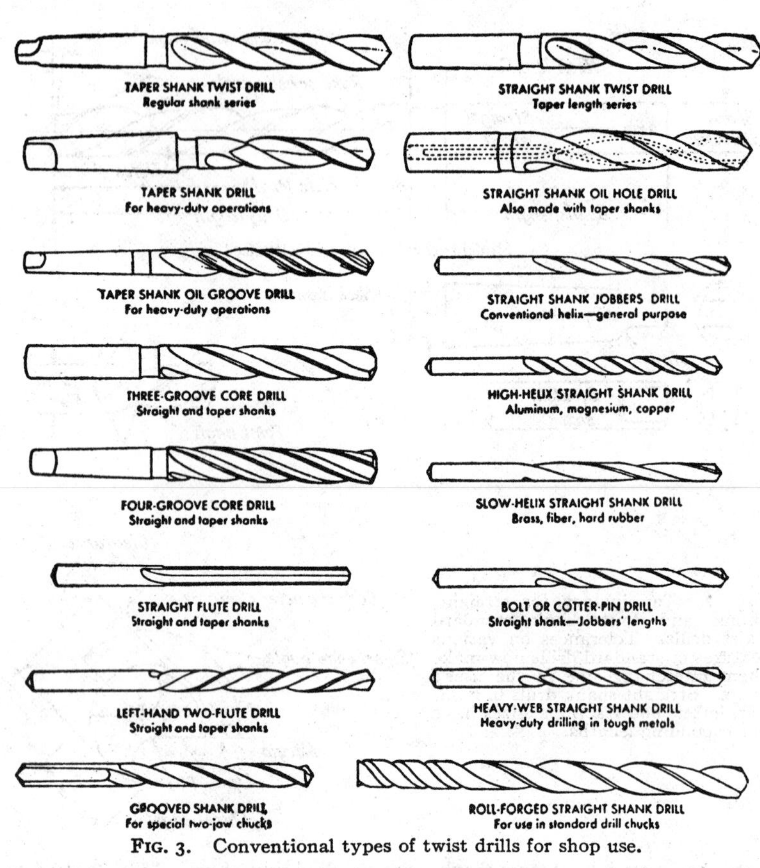

FIG. 3. Conventional types of twist drills for shop use.

Chisel Edge Angle (Center Angle). The angle included between the chisel edge and the cutting edge as seen from the end of the drill is called the chisel edge angle.

Back Taper (Longitudinal Relief). Drills are usually made slightly smaller in diameter at the shank end than at the point. This is known as back taper.

The **axis** is the longitudinal center line through the drill.

Classification, Based on Kind of Shank

Straight-shank Drills. Drills having cylindrical shanks. The shank may be of the same or of a different diameter from that of the body of the drill. It may be made with or without driving flats, tang, or grooves.

Taper-shank Drills. Drills having conical shanks suitable for direct fitting into tapered holes in drilling-machine spindles or driving sockets. The tapered shanks generally have a driving tang.

Taper Square-shank Drills. Drills having tapered shanks with four flat sides for fitting ratchets and braces.

Classification, Based on Number of Flutes

Two-flute Drills. These are the conventional drills used for originating holes.

Single-flute Drills. Drills used principally for drilling wood and other soft substances.

Three-flute Drills (Core Drills). These drills are used for enlarging and finishing holes. They will not originate holes because they do not have any cutting edges at the center.

Four-flute Drills. These drills are used interchangeably with three-flute drills. They are of the same construction except for the number of flutes.

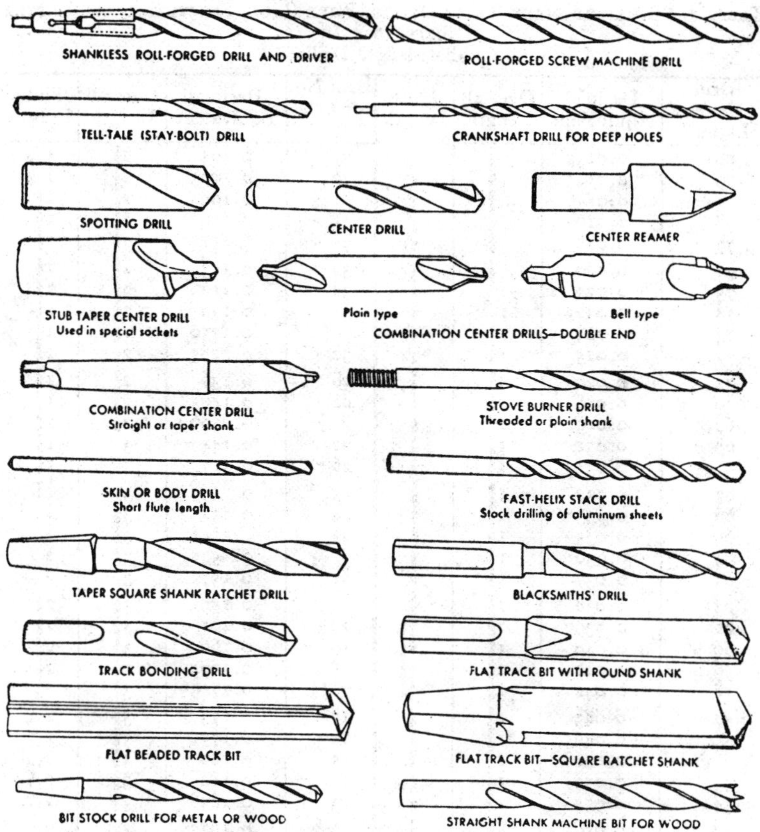

FIG. 4. Conventional drills and special-purpose drills.

Classification, Based on "Hand of Rotation"

Right-hand Drills. The great majority of drills are made "right-hand"; that is, looking toward the point of these drills with the shank extending away, they must be rotated in a counterclockwise direction in order to cut.

Left-hand Drills. These drills are made to cut when rotated in a clockwise direction. They are not used extensively.

Classification, Based on Material

All drills covered by this standard are available made from high-speed steel. Carbon-steel drills also are available in many of the types covered by this standard.

Drills are also made from other materials for special applications, but such special drills do not necessarily follow this standard for dimensions or tolerances.

Classification, Based on Standard or Special

Standard drills are carried in stock. Drills are available made to a great number of variations in form, dimensions, and tolerances. Such drills are not usually carried in stock but are made to specifications and are called special.

Drills for Reamed Holes. The user should specify drills to give desired reaming allowance. Normally $\frac{1}{64}$ in. is sufficient for reaming and is supplied by some makers when the size is not specified.

TABLE 1. STRAIGHT-SHANK TWIST DRILLS, WIRE GAGE, FRACTIONAL AND LETTER SIZES TO 1/2 IN.
Short-length Drills

Drill No. or Letter*	Decimal Equivalent	Over-all Length	Flute Length	Drill No. or Letter*	Decimal Equivalent	Over-all Length	Flute Length
80	0.0135	3/4	3/16	37	0.104	2 1/2	1 7/16
79	0.0145	3/4	3/16	36	0.1065	2 1/2	1 7/16
1/64	0.0156	3/4	3/16	7/64	0.1094	2 5/8	1 1/2
78	0.016	7/8	3/16	35	0.110	2 5/8	1 1/2
77	0.018	7/8	3/16	34	0.111	2 5/8	1 1/2
76	0.020	7/8	3/16	33	0.113	2 5/8	1 1/2
75	0.021	1	1/4	32	0.116	2 3/4	1 5/8
74	0.0225	1	1/4	31	0.120	2 3/4	1 5/8
73	0.024	1 1/8	5/16	1/8	0.1250	2 3/4	1 5/8
72	0.025	1 1/8	5/16	30	0.1285	2 3/4	1 5/8
71	0.026	1 1/4	3/8	29	0.136	2 7/8	1 3/4
70	0.028	1 1/4	3/8	28	0.1405	2 7/8	1 3/4
69	0.0292	1 3/8	1/2	9/64	0.1406	2 7/8	1 3/4
68	0.031	1 3/8	1/2	27	0.144	3	1 7/8
1/32	0.0312	1 3/8	1/2	26	0.147	3	1 7/8
67	0.032	1 3/8	1/2	25	0.1495	3	1 7/8
66	0.033	1 3/8	1/2	24	0.152	3 1/8	2
65	0.035	1 1/2	5/8	23	0.154	3 1/8	2
64	0.036	1 1/2	5/8	5/32	0.1562	3 1/8	2
63	0.037	1 1/2	5/8	22	0.157	3 1/8	2
62	0.038	1 1/2	5/8	21	0.159	3 1/4	2 1/8
61	0.039	1 5/8	11/16	20	0.161	3 1/4	2 1/8
60	0.040	1 5/8	11/16	19	0.166	3 1/4	2 1/8
59	0.041	1 5/8	11/16	18	0.1695	3 1/4	2 1/8
58	0.042	1 5/8	11/16	11/64	0.1719	3 1/4	2 1/8
57	0.043	1 3/4	3/4	17	0.173	3 3/8	2 3/16
56	0.0465	1 3/4	3/4	16	0.177	3 3/8	2 3/16
3/64	0.0468	1 3/4	3/4	15	0.180	3 3/8	2 3/16
55	0.052	1 7/8	7/8	14	0.182	3 3/8	2 3/16
54	0.055	1 7/8	7/8	13	0.185	3 1/2	2 5/16
53	0.0595	1 7/8	7/8	3/16	0.1875	3 1/2	2 5/16
1/16	0.0625	1 7/8	7/8	12	0.189	3 1/2	2 5/16
52	0.0635	1 7/8	7/8	11	0.191	3 1/2	2 5/16
51	0.067	2	1	10	0.1935	3 5/8	2 7/16
50	0.070	2	1	9	0.196	3 5/8	2 7/16
49	0.073	2	1	8	0.199	3 5/8	2 7/16
48	0.076	2	1	7	0.201	3 5/8	2 7/16
5/64	0.0781	2	1	13/64	0.2031	3 5/8	2 7/16
47	0.0785	2	1	6	0.204	3 3/4	2 1/2
46	0.081	2 1/8	1 1/8	5	0.2055	3 3/4	2 1/2
45	0.082	2 1/8	1 1/8	4	0.209	3 3/4	2 1/2
44	0.086	2 1/8	1 1/8	3	0.213	3 3/4	2 1/2
43	0.089	2 1/4	1 1/4	7/32	0.2187	3 3/4	2 1/2
42	0.0935	2 1/4	1 1/4	2	0.221	3 7/8	2 5/8
3/32	0.0937	2 1/4	1 1/4	1	0.228	3 7/8	2 5/8
41	0.096	2 3/8	1 3/8	A	0.234	3 7/8	2 5/8
40	0.098	2 3/8	1 3/8	15/64	0.2344	3 7/8	2 5/8
39	0.0995	2 3/8	1 3/8	B	0.238	4	2 3/4
38	0.1015	2 1/2	1 7/16	C	0.242	4	2 3/4

TABLE 1. STRAIGHT-SHANK TWIST DRILLS, WIRE GAGE, FRACTIONAL AND LETTER SIZES TO $\frac{1}{2}$ IN. (*Continued*)

Drill No. or Letter*	Decimal Equivalent	Over-all Length	Flute Length	Drill No. or Letter*	Decimal Equivalent	Over-all Length	Flute Length
D	0.246	4	$2\frac{3}{4}$	$\frac{11}{32}$	0.3437	$4\frac{3}{4}$	$3\frac{7}{16}$
E & $\frac{1}{4}$	0.250	4	$2\frac{3}{4}$	S	0.348	$4\frac{7}{8}$	$3\frac{1}{2}$
F	0.257	$4\frac{1}{8}$	$2\frac{7}{8}$	T	0.358	$4\frac{7}{8}$	$3\frac{1}{2}$
G	0.261	$4\frac{1}{8}$	$2\frac{7}{8}$	$\frac{23}{64}$	0.3594	$4\frac{7}{8}$	$3\frac{1}{2}$
$\frac{17}{64}$	0.2656	$4\frac{1}{8}$	$2\frac{7}{8}$	U	0.368	5	$3\frac{5}{8}$
H	0.266	$4\frac{1}{8}$	$2\frac{7}{8}$	$\frac{3}{8}$	0.375	5	$3\frac{5}{8}$
I	0.272	$4\frac{1}{8}$	$2\frac{7}{8}$	V	0.377	5	$3\frac{5}{8}$
J	0.277	$4\frac{1}{8}$	$2\frac{7}{8}$	W	0.386	$5\frac{1}{8}$	$3\frac{3}{4}$
K	0.281	$4\frac{1}{4}$	$2\frac{15}{16}$	$\frac{25}{64}$	0.3906	$5\frac{1}{8}$	$3\frac{3}{4}$
$\frac{9}{32}$	0.2812	$4\frac{1}{4}$	$2\frac{15}{16}$	X	0.397	$5\frac{1}{8}$	$3\frac{3}{4}$
L	0.290	$4\frac{1}{4}$	$2\frac{15}{16}$	Y	0.404	$5\frac{1}{4}$	$3\frac{7}{8}$
M	0.295	$4\frac{3}{8}$	$3\frac{1}{16}$	$\frac{13}{32}$	0.4062	$5\frac{1}{4}$	$3\frac{7}{8}$
$\frac{19}{64}$	0.2969	$4\frac{3}{8}$	$3\frac{1}{16}$	Z	0.413	$5\frac{1}{4}$	$3\frac{7}{8}$
N	0.302	$4\frac{3}{8}$	$3\frac{1}{16}$	$\frac{27}{64}$	0.4219	$5\frac{3}{8}$	$3\frac{15}{16}$
$\frac{5}{16}$	0.3125	$4\frac{1}{2}$	$3\frac{3}{16}$	$\frac{7}{16}$	0.4375	$5\frac{1}{2}$	$4\frac{1}{16}$
O	0.316	$4\frac{1}{2}$	$3\frac{3}{16}$	$\frac{29}{64}$	0.4531	$5\frac{5}{8}$	$4\frac{3}{16}$
P	0.323	$4\frac{5}{8}$	$3\frac{5}{16}$	$\frac{15}{32}$	0.4687	$5\frac{3}{4}$	$4\frac{5}{16}$
$\frac{11}{64}$	0.3281	$4\frac{5}{8}$	$3\frac{5}{16}$	$\frac{31}{64}$	0.4844	$5\frac{7}{8}$	$4\frac{3}{8}$
Q	0.332	$4\frac{3}{4}$	$3\frac{7}{16}$	$\frac{1}{2}$	0.5000	6	$4\frac{1}{2}$
R	0.339	$4\frac{3}{4}$	$3\frac{7}{16}$				

Long-length Drills

Drill Dia	Decimal Equivalent	Over-all Length	Flute Length	Drill Dia	Decimal Equivalent	Over-all Length	Flute Length
$\frac{1}{8}$	0.1250	$5\frac{1}{8}$	$2\frac{3}{4}$	$\frac{11}{64}$	0.3281	$6\frac{1}{2}$	$4\frac{1}{8}$
$\frac{9}{64}$	0.1406	$5\frac{3}{8}$	3	$\frac{11}{32}$	0.3437	$6\frac{1}{2}$	$4\frac{1}{8}$
$\frac{5}{32}$	0.1562	$5\frac{3}{8}$	3	$\frac{23}{64}$	0.3594	$6\frac{3}{4}$	$4\frac{1}{4}$
$\frac{11}{64}$	0.1719	$5\frac{3}{4}$	$3\frac{3}{8}$	$\frac{3}{8}$	0.375	$6\frac{3}{4}$	$4\frac{1}{4}$
$\frac{3}{16}$	0.1875	$5\frac{3}{4}$	$3\frac{3}{8}$	$\frac{25}{64}$	0.3906	7	$4\frac{3}{8}$
$\frac{13}{64}$	0.2031	6	$3\frac{5}{8}$	$\frac{13}{32}$	0.4062	7	$4\frac{3}{8}$
$\frac{7}{32}$	0.2187	6	$3\frac{5}{8}$	$\frac{27}{64}$	0.4219	$7\frac{1}{4}$	$4\frac{5}{8}$
$\frac{15}{64}$	0.2344	$6\frac{1}{8}$	$3\frac{3}{4}$	$\frac{7}{16}$	0.4375	$7\frac{1}{4}$	$4\frac{5}{8}$
E & $\frac{1}{4}$	0.250	$6\frac{1}{8}$	$3\frac{3}{4}$	$\frac{29}{64}$	0.4531	$7\frac{1}{2}$	$4\frac{3}{4}$
$\frac{17}{64}$	0.2656	$6\frac{1}{4}$	$3\frac{7}{8}$	$\frac{15}{32}$	0.4687	$7\frac{1}{2}$	$4\frac{3}{4}$
$\frac{9}{32}$	0.2812	$6\frac{1}{4}$	$3\frac{7}{8}$	$\frac{31}{64}$	0.4844	$7\frac{3}{4}$	$4\frac{3}{4}$
$\frac{19}{64}$	0.2969	$6\frac{3}{8}$	4	$\frac{1}{2}$	0.5000	$7\frac{3}{4}$	$4\frac{3}{4}$
$\frac{5}{16}$	0.3125	$6\frac{3}{8}$	4				

Tolerances

Element	Range	Direction and tolerance
Drill dia........	Up to $\frac{3}{64}$ incl.	+0.0000 to −0.0006
	Over $\frac{3}{64}$ to $\frac{1}{8}$ incl.	+0.0000 to −0.0008
	Over $\frac{1}{8}$ to $\frac{1}{4}$ incl.	+0.0000 to −0.0010
	Over $\frac{1}{4}$ to $\frac{1}{2}$ incl.	+0.0000 to −0.0015
Back taper......	Up to $\frac{5}{32}$ incl.	0.0000 to 0.0008 in. per in.
	Over $\frac{5}{32}$ to $\frac{21}{64}$ incl.	0.0002 to 0.0008 in. per in.
	Over $\frac{21}{64}$ to $\frac{1}{2}$ incl.	0.0002 to 0.0009 in. per in.

* Fractions are drill diameters.

TABLE 2. STRAIGHT-SHANK, LONG-LENGTH DRILLS IN SIZES FROM $\frac{33}{64}$ TO 2.000 IN.

Drill Dia	Over-all Length	Flute Length	Drill Dia	Over-all Length	Flute Length	Drill Dia	Over-all Length	Flute Length
$\frac{33}{64}$	8	$4\frac{3}{4}$	$\frac{27}{32}$	10	$6\frac{1}{8}$	$1\frac{11}{64}$	12	$7\frac{3}{8}$
$\frac{17}{32}$	8	$4\frac{3}{4}$	$\frac{55}{64}$	10	$6\frac{1}{8}$	$1\frac{3}{16}$	12	$7\frac{3}{8}$
$\frac{35}{64}$	$8\frac{1}{4}$	$4\frac{7}{8}$	$\frac{7}{8}$	10	$6\frac{1}{8}$	$1\frac{13}{64}$	$12\frac{1}{8}$	$7\frac{1}{2}$
$\frac{9}{16}$	$8\frac{1}{4}$	$4\frac{7}{8}$	$\frac{57}{64}$	10	$6\frac{1}{8}$	$1\frac{7}{32}$	$12\frac{1}{8}$	$7\frac{1}{2}$
$\frac{37}{64}$	$8\frac{3}{4}$	$4\frac{7}{8}$	$\frac{29}{32}$	10	$6\frac{1}{8}$	$1\frac{15}{64}$	$12\frac{1}{2}$	$7\frac{7}{8}$
$\frac{19}{32}$	$8\frac{3}{4}$	$4\frac{7}{8}$	$\frac{59}{64}$	$10\frac{3}{4}$	$6\frac{1}{8}$	$1\frac{1}{4}$	$12\frac{1}{2}$	$7\frac{7}{8}$
$\frac{39}{64}$	$8\frac{3}{4}$	$4\frac{7}{8}$	$\frac{15}{16}$	$10\frac{3}{4}$	$6\frac{1}{8}$	$1\frac{9}{32}$	$14\frac{1}{8}$	$8\frac{1}{2}$
$\frac{5}{8}$	$8\frac{3}{4}$	$4\frac{7}{8}$	$\frac{61}{64}$	11	$6\frac{3}{8}$	$1\frac{5}{16}$	$14\frac{1}{4}$	$8\frac{5}{8}$
$\frac{41}{64}$	9	$5\frac{1}{8}$	$\frac{31}{32}$	11	$6\frac{3}{8}$	$1\frac{11}{32}$	$14\frac{3}{8}$	$8\frac{3}{4}$
$\frac{21}{32}$	9	$5\frac{1}{8}$	$\frac{63}{64}$	11	$6\frac{3}{8}$	$1\frac{3}{8}$	$14\frac{1}{2}$	$8\frac{7}{8}$
$\frac{43}{64}$	$9\frac{1}{4}$	$5\frac{3}{8}$	1	11	$6\frac{3}{8}$	$1\frac{13}{32}$	$14\frac{5}{8}$	9
$\frac{11}{16}$	$9\frac{1}{4}$	$5\frac{3}{8}$	$1\frac{1}{64}$	$11\frac{1}{8}$	$6\frac{1}{2}$	$1\frac{7}{16}$	$14\frac{3}{4}$	$9\frac{1}{8}$
$\frac{45}{64}$	$9\frac{1}{2}$	$5\frac{5}{8}$	$1\frac{1}{32}$	$11\frac{1}{8}$	$6\frac{1}{2}$	$1\frac{15}{32}$	$14\frac{7}{8}$	$9\frac{1}{4}$
$\frac{23}{32}$	$9\frac{1}{2}$	$5\frac{5}{8}$	$1\frac{3}{64}$	$11\frac{1}{4}$	$6\frac{5}{8}$	$1\frac{1}{2}$	15	$9\frac{3}{8}$
$\frac{47}{64}$	$9\frac{3}{4}$	$5\frac{7}{8}$	$1\frac{1}{16}$	$11\frac{1}{4}$	$6\frac{5}{8}$	$1\frac{9}{16}$	$15\frac{1}{4}$	$9\frac{5}{8}$
$\frac{3}{4}$	$9\frac{3}{4}$	$5\frac{7}{8}$	$1\frac{5}{64}$	$11\frac{1}{2}$	$6\frac{7}{8}$	$1\frac{5}{8}$	$15\frac{5}{8}$	$9\frac{7}{8}$
$\frac{49}{64}$	$9\frac{7}{8}$	6	$1\frac{3}{32}$	$11\frac{1}{2}$	$6\frac{7}{8}$	$1\frac{11}{16}$	$15\frac{3}{4}$	10
$\frac{25}{32}$	$9\frac{7}{8}$	6	$1\frac{7}{64}$	$11\frac{3}{4}$	$7\frac{1}{8}$	$1\frac{3}{4}$	$16\frac{1}{4}$	$10\frac{1}{2}$
$\frac{51}{64}$	10	$6\frac{1}{8}$	$1\frac{1}{8}$	$11\frac{3}{4}$	$7\frac{1}{8}$	$1\frac{13}{16}$	$16\frac{1}{4}$	$10\frac{1}{2}$
$\frac{13}{16}$	10	$6\frac{1}{8}$	$1\frac{9}{64}$	$11\frac{7}{8}$	$7\frac{1}{4}$	$1\frac{7}{8}$	$16\frac{1}{2}$	$10\frac{3}{4}$
$\frac{53}{64}$	10	$6\frac{1}{8}$	$1\frac{5}{32}$	$11\frac{7}{8}$	$7\frac{1}{4}$	$1\frac{15}{16}$	$16\frac{5}{8}$	$10\frac{3}{4}$
						2	$16\frac{5}{8}$	$10\frac{3}{4}$

Tolerances for Long-length Drills

Element	Range	Direction and Tolerance
Drill dia........	Over $\frac{1}{2}$ to $\frac{3}{4}$ incl.	+0.0000 to −0.0015
	Over $\frac{3}{4}$ to $1\frac{1}{2}$ incl.	+0.0000 to −0.0020
	Over $1\frac{1}{2}$ to 2 incl.	+0.0000 to −0.0025
Back taper......	Over $\frac{1}{2}$ to $\frac{3}{4}$ incl.	0.0002 to 0.0011 in. per in.
	Over $\frac{3}{4}$ to $\frac{31}{32}$ incl.	0.0002 to 0.0012 in. per in.
	Over $\frac{31}{32}$ to $1\frac{1}{2}$ incl.	0.0002 to 0.0015 in. per in.
	Over $1\frac{1}{2}$ to 2 incl.	0.0002 to 0.0020 in. per in.

Tolerances for Taper-shank Drills

Element	Range	Direction and Tolerance
Drill dia........	$\frac{1}{8}$	+0.0000 to −0.0008
	Over $\frac{1}{8}$ to $\frac{1}{4}$ incl.	+0.0000 to −0.0010
	Over $\frac{1}{4}$ to $\frac{3}{4}$ incl.	+0.0000 to −0.0015
	Over $\frac{3}{4}$ to $1\frac{1}{2}$ incl.	+0.0000 to −0.0020
	Over $1\frac{1}{2}$ to $3\frac{1}{2}$ incl.	+0.0000 to −0.0025
Back taper......	Up to $\frac{5}{32}$ incl.	0.0000 to 0.0008 in. per in.
	Over $\frac{5}{32}$ to $\frac{21}{64}$ incl.	0.0002 to 0.0008 in. per in.
	Over $\frac{21}{64}$ to $\frac{1}{2}$ incl.	0.0002 to 0.0009 in. per in.
	Over $\frac{1}{2}$ to $\frac{3}{4}$ incl.	0.0002 to 0.0011 in. per in.
	Over $\frac{3}{4}$ to $\frac{31}{32}$ incl.	0.0002 to 0.0012 in. per in.
	Over $\frac{31}{32}$ to $1\frac{1}{2}$ incl.	0.0002 to 0.0015 in. per in.
	Over $1\frac{1}{2}$ to $3\frac{1}{2}$ incl.	0.0002 to 0.0020 in. per in.

TABLE 3. AUTOMOTIVE-SERIES STRAIGHT-SHANK DRILLS

Dia Drill	Comparable Letter, Fraction, No., or MM Drill	Shank			Short Length		Long Length		Dia Drill	Comparable Letter, Fraction, No., or MM Drill	Shank			Short Length		Long Length	
		Dia Max	Dia Min	Ground Length	Over-all Length	Flute Length	Over-all Length	Flute Length			Dia Max	Dia Min	Ground Length	Over-all Length	Flute Length	Over-all Length	Flute Length
0.2500	1/4	0.2485	0.2475	1 1/8	4 1/2	3 1/8	6 1/2	4 5/16	0.3750	3/8	0.3730	0.3720	1 3/8	5 5/8	4 1/8	7 5/8	5 1/8
0.2520	6.40MM	0.2505	0.2495	1 1/4	4 1/2	3 1/8	6 1/2	4 5/16	0.3860	W	0.3840	0.3830	1 1/2	5 5/8	4 1/8	7 5/8	5 1/8
0.2570	F	0.2550	0.2540	1 1/4	4 1/2	3 1/8	6 1/2	4 5/16	0.3906	25/64	0.3886	0.3876	1 1/2	5 5/8	4 1/8	7 5/8	5 1/8
0.2610	G	0.2590	0.2580	1 1/4	4 1/2	3 1/8	6 1/2	4 5/16	0.3870	X	0.3950	0.3940	1 1/2	5 5/8	4 1/8	7 5/8	5 1/8
0.2656	17/64	0.2636	0.2626	1 1/4	4 1/2	3 1/8	6 1/2	4 5/16	0.4062	13/32	0.4042	0.4032	1 1/2	5 5/8	4 1/8	7 5/8	5 1/8
0.2720	I	0.2700	0.2690	1 1/4	4 1/2	3 1/8	6 3/4	4 1/2	0.4219	27/64	0.4199	0.4189	1 1/2	5 5/8	4 1/8	7 5/8	5 1/8
0.2770	J	0.2750	0.2740	1 1/4	4 1/2	3 1/8	6 3/4	4 1/2	0.4375	7/16	0.4355	0.4345	1 1/2	6 1/8	4 7/16	8 1/8	5 7/16
0.2812	9/32	0.2792	0.2782	1 1/4	4 3/4	3 3/8	6 3/4	4 1/2	0.4531	29/64	0.4511	0.4501	1 1/2	6 1/8	4 7/16	8 1/8	5 7/16
0.2854	7.25MM	0.2834	0.2824	1 1/4	4 3/4	3 3/8	6 3/4	4 1/2	0.4687	15/32	0.4667	0.4657	1 1/2	6 1/8	4 7/16	8 1/8	5 7/16
0.2913	7.40MM	0.2893	0.2883	1 1/4	4 3/4	3 3/8	6 3/4	4 1/2	0.4844	31/64	0.4824	0.4814	1 9/16	6 1/8	4 7/16	8 1/8	5 7/16
0.2969	19/64	0.2949	0.2939	1 1/4	4 3/4	3 3/8	6 3/4	4 1/2	0.5000	1/2	0.4980	0.4970	1 9/16	6 1/8	4 7/16	8 1/8	5 7/16
0.3020	N	0.3000	0.2990	1 1/4	4 3/4	3 3/8	7	4 11/16	0.5156	33/64	0.5131	0.5121	1 9/16	6 5/8	4 13/16	8 5/8	5 3/4
0.3071	7.80MM	0.3051	0.3041	1 1/4	4 3/4	3 3/8	7	4 11/16	0.5312	17/32	0.5287	0.5277	1 9/16	6 5/8	4 13/16	8 5/8	5 3/4
0.3125	5/16	0.3105	0.3095	1 1/4	5	3 5/8	7	4 11/16	0.5469	35/64	0.5444	0.5434	1 9/16	6 5/8	4 13/16	8 5/8	5 3/4
0.3160	O	0.3140	0.3130	1 3/8	5	3 5/8	7	4 11/16	0.5625	9/16	0.5600	0.5590	1 9/16	6 5/8	4 13/16	8 5/8	5 3/4
0.3230	P	0.3210	0.3200	1 3/8	5	3 5/8	7	4 11/16	0.5781	37/64	0.5756	0.5746	1 23/32	6 5/8	4 13/16	8 5/8	5 3/4
0.3281	21/64	0.3261	0.3251	1 3/8	5	3 5/8	7	4 11/16	0.5937	19/32	0.5912	0.5902	1 23/32	7 1/8	5 3/16	9 1/8	6 1/8
0.3320	Q	0.3300	0.3290	1 3/8	5	3 5/8	7 1/4	4 15/16	0.6094	39/64	0.6069	0.6059	1 23/32	7 1/8	5 3/16	9 1/8	6 1/8
0.3390	R	0.3370	0.3360	1 3/8	5	3 5/8	7 1/4	4 15/16	0.6250	5/8	0.6225	0.6215	1 23/32	7 1/8	5 3/16	9 1/8	6 1/8
0.3437	11/32	0.3417	0.3407	1 3/8	5 1/4	3 7/8	7 1/4	4 15/16	0.6406	41/64	0.6381	0.6371	1 23/32	7 1/8	5 3/16	9 1/8	6 1/8
0.3480	S	0.3460	0.3450	1 3/8	5 1/4	3 7/8	7 1/4	4 15/16	0.6562	21/32	0.6537	0.6527	1 23/32	7 1/8	5 3/16	9 1/8	6 1/8
0.3543	9MM	0.3523	0.3513	1 3/8	5 1/4	3 7/8	7 1/4	4 15/16	0.6719	43/64	0.6694	0.6684	1 7/8	7 5/8	5 5/8	9 5/8	6 1/2
0.3594	21/64	0.3574	0.3564	1 3/8	5 1/4	3 7/8	7 1/4	4 15/16	0.6875	11/16	0.6850	0.6840	1 7/8	7 5/8	5 5/8	9 5/8	6 1/2
0.3680	U	0.3660	0.3650	1 3/8	5 1/4	3 7/8	7 5/8	5 1/8									

Tolerances

Element	Range	Direction and Tolerance
Drill dia.	1/4	+0.000 to −0.0010
	Over 1/4 to 11/16 incl.	+0.0000 to −0.0015
Back taper	From 1/4 to 21/64 incl.	0.0002 to 0.0008 in. per in.
	Over 21/64 to 1/2 incl.	0.0002 to 0.0009 in. per in.
	Over 1/2 to 11/16 incl.	0.0002 to 0.0011 in. per in.

TABLE 4. TAPER-SHANK TWIST DRILLS, REGULAR SHANK LENGTH

Size Drill	ASA Taper (Morse)	Over-all Length	Flute Length	Size Drill	ASA Taper (Morse)	Over-all Length	Flute Length	Size Drill	ASA Taper (Morse)	Over-all Length	Flute Length
1/8	1	5 1/8	1 7/8	55/64	3	10 3/4	6 1/8	1 19/32	5	16 7/8	9 7/8
9/64	1	5 3/8	2 1/8	7/8	3	10 3/4	6 1/8	1 39/64	5	17	10
5/32	1	5 3/8	2 1/8	57/64	3	10 3/4	6 1/8	1 5/8	5	17	10
11/64	1	5 3/4	2 1/2	29/32	3	10 3/4	6 1/8	1 41/64	5	17 1/8	10 1/8
3/16	1	5 3/4	2 1/2	59/64	3	10 3/4	6 1/8	1 21/32	5	17 1/8	10 1/8
13/64	1	6	2 3/4	15/16	3	10 3/4	6 1/8	1 43/64	5	17 1/8	10 1/8
7/32	1	6	2 3/4	61/64	3	11	6 3/8	1 11/16	5	17 1/8	10 1/8
15/64	1	6 1/8	2 7/8	31/32	3	11	6 3/8	1 45/64	5	17 1/8	10 1/8
1/4	1	6 1/8	2 7/8	63/64	3	11	6 3/8	1 23/32	5	17 1/8	10 1/8
17/64	1	6 1/4	3	1	3	11	6 3/8	1 47/64	5	17 1/8	10 1/8
9/32	1	6 1/4	3	1 1/64	3	11 1/8	6 1/2	1 3/4	5	17 1/8	10 1/8
19/64	1	6 3/8	3 1/8	1 1/32	3	11 1/8	6 1/2	1 25/32	5	17 1/8	10 1/8
5/16	1	6 3/8	3 1/8	1 3/64	3	11 1/4	6 5/8	1 13/16	5	17 1/8	10 1/8
21/64	1	6 1/2	3 1/4	1 1/16	3	11 1/4	6 5/8	1 27/32	5	17 1/8	10 1/8
11/32	1	6 1/2	3 1/4	1 5/64	4	12 1/2	6 7/8	1 7/8	5	17 3/8	10 3/8
23/64	1	6 3/4	3 1/2	1 3/32	4	12 1/2	6 7/8	1 29/32	5	17 3/8	10 3/8
3/8	1	6 3/4	3 1/2	1 7/64	4	12 3/4	7 1/8	1 15/16	5	17 3/8	10 3/8
25/64	1	7	3 5/8	1 1/8	4	12 3/4	7 1/8	1 31/32	5	17 3/8	10 3/8
13/32	1	7	3 5/8	1 9/64	4	12 7/8	7 1/4	2	5	17 3/8	10 3/8
27/64	1	7 1/4	3 7/8	1 5/32	4	12 7/8	7 1/4	2 1/32	5	17 3/8	10 3/8
7/16	1	7 1/4	3 7/8	1 11/64	4	13	7 3/8	2 1/16	5	17 3/8	10 1/4
29/64	1	7 1/2	4 1/8	1 3/16	4	13	7 3/8	2 3/32	5	17 3/8	10 1/4
15/32	1	7 1/2	4 1/8	1 13/64	4	13 1/8	7 1/2	2 1/8	5	17 3/8	10 1/4
31/64	2	8 1/4	4 3/8	1 7/32	4	13 1/8	7 1/2	2 5/32	5	17 3/8	10 1/4
1/2	2	8 1/4	4 3/8	1 15/64	4	13 1/2	7 7/8	2 3/16	5	17 3/8	10 1/4
33/64	2	8 1/2	4 5/8	1 1/4	4	13 1/2	7 7/8	2 7/32	5	17 3/8	10 1/8
17/32	2	8 1/2	4 5/8	1 17/64	4	14 1/8	8 1/2	2 1/4	5	17 3/8	10 1/8
35/64	2	8 3/4	4 7/8	1 9/32	4	14 1/8	8 1/2	2 5/16	5	17 3/8	10 1/8
9/16	2	8 3/4	4 7/8	1 19/64	4	14 1/4	8 5/8	2 3/8	5	17 3/8	10 1/8
37/64	2	8 3/4	4 7/8	1 5/16	4	14 1/4	8 5/8	2 7/16	5	18 3/4	11 1/4
19/32	2	8 3/4	4 7/8	1 21/64	4	14 3/8	8 3/4	2 1/2	5	18 3/4	11 1/4
39/64	2	8 3/4	4 7/8	1 11/32	4	14 3/8	8 3/4	2 9/16	5	19 1/2	11 7/8
5/8	2	8 3/4	4 7/8	1 23/64	4	14 1/2	8 7/8	2 5/8	5	19 1/2	11 7/8
41/64	2	9	5 1/8	1 3/8	4	14 1/2	8 7/8	2 11/16	5	20 3/8	12 3/4
21/32	2	9	5 1/8	1 25/64	4	14 5/8	9	2 3/4	5	20 3/8	12 3/4
43/64	2	9 1/4	5 3/8	1 13/32	4	14 5/8	9	2 13/16	5	21 1/8	13 3/8
11/16	2	9 1/4	5 3/8	1 27/64	4	14 3/4	9 1/8	2 7/8	5	21 1/8	13 3/8
45/64	2	9 1/2	5 5/8	1 7/16	4	14 3/4	9 1/8	2 15/16	5	21 3/4	14
23/32	2	9 1/2	5 5/8	1 29/64	4	14 7/8	9 1/4	3	5	21 3/4	14
47/64	2	9 3/4	5 7/8	1 15/32	4	14 7/8	9 1/4	3 1/16	6	24 1/2	14 5/8
3/4	2	9 3/4	5 7/8	1 31/64	4	15	9 3/8	3 1/8	6	24 1/2	14 5/8
49/64	2	9 7/8	6	1 1/2	4	15	9 3/8	3 3/16	6	24 1/2	14 5/8
25/32	2	9 7/8	6	1 33/64	5	16 3/8	9 3/8	3 1/4	6	25 1/2	15 1/2
51/64	3	10 3/4	6 1/8	1 17/32	5	16 3/8	9 3/8	3 5/16	6	25 1/2	15 1/2
13/16	3	10 3/4	6 1/8	1 35/64	5	16 5/8	9 5/8	3 3/8	6	25 1/2	15 1/2
53/64	3	10 3/4	6 1/8	1 9/16	5	16 5/8	9 5/8	3 7/16	6	25 1/2	15 1/2
27/32	3	10 3/4	6 1/8	1 37/64	5	16 7/8	9 7/8	3 1/2	6	26 1/2	16 3/8

Drills for Bolt-clearance Holes. Clearance between bolt body or threads and the work varies with the nature of the work: $\frac{1}{64}$ in. is considered *fine*, $\frac{1}{32}$ in. *moderate*, and $\frac{1}{16}$ in. *coarse* or *heavy*. Drills should be specified to the clearance desired by adding it to the nominal size. A $\frac{5}{8}$ in. clearance drill should be ordered $\frac{41}{64}$, $\frac{21}{32}$, or $\frac{11}{16}$ in., depending on the grade of work.

Drills for Dowel-pin Holes. Sizes of drills to produce drive fits and slight clearance for dowel pins are given in Table 5.

TABLE 5. DRILLS AND REAMERS FOR DOWEL PINS

Sizes of Rod		Drills and Reamers for Drive Fits			Drills for Clearance	
No. of Gage (Stubbs Steel Wire)	Dia, In.	Size of Drill	Dia of Drill, In.	Dia of Reamer, In.	Size of Drill	Dia of Drill, In.
54	0.055	No. 55	0.052		No. 54	0.055
45	0.081	No. 47	0.0785		No. 46	0.081
33	0.112	No. 36	0.1065	0.110	No. 33	0.113
30	0.127	No. 31	0.120	0.125	No. 30	0.1285
21	0.157	No. 24	0.152	0.155	No. 22	0.157
10	0.191	No. 13	0.185	0.189	No. 11	0.191
	0.252	C	0.242	0.250 –0.2505	F	0.257
	0.315	5/16 Reamer	0.307	0.3125–0.313	O	0.316
V	0.377	3/8 Drill	0.366	0.375 –0.3755	V	0.377
	0.439	7/16 Drill	0.427	0.4375–0.438		
	0.503	1/2 Drill	0.489	0.500 –0.5005		
	0.628	5/8 Drill	0.616	0.625 –0.6255		
	0.753	3/4 Drill	0.734(47/64)	0.750 –0.7505		

Drills for Tapped Holes. Except in special cases, tap drills should always be larger than the root diameter of the tap. In average practice, the drill gives about three-quarters thread depth. Some metals flow enough to make nearly a full thread, especially with a dull tap.

The simple rule of subtracting the pitch of one thread from the tap diameter is accurate enough for most work. Following this:

A $\frac{3}{8}$-in. tap, 16-thread, would be $\frac{3}{8}$ minus $\frac{1}{16} = \frac{5}{16}$ drill; a $\frac{3}{4}$-in. tap, 10-thread, would be $\frac{3}{4}$ minus $\frac{1}{10} = \frac{75}{100} - \frac{10}{100}$ or $0.75 - 0.10 = \frac{65}{100}$ or 0.65, or a little over $\frac{5}{8}$ in.; so a $\frac{5}{8}$-in. drill will do nicely. With a 1-in. tap we have $1 - \frac{1}{8} = \frac{7}{8}$-in. drill, which is a little large but leaves enough thread for most cases.

There are few cases in which it is advisable to use a tap drill small enough to give a completely full thread in the tapped hole. A full-depth thread takes three times the power needed to tap a 75% depth thread and is only 5% stronger. A nut with 50%, or half-depth, thread, will break the bolt before the thread will strip. The tougher and harder the material or the deeper the threaded hole, the less the thread depth can safely be. Good manufacturing practice uses from 62 to 75% of thread depth and never more than $83\frac{1}{3}$%.

RECOMMENDED TAP DRILL SIZES. The Bureau of Standards "Handbook H28" contains the accompanying Tables 6 and 7 for tap drills for American National coarse and fine threads, showing the percentage of thread depth given by the use of standard drills. These tables show the basic, maximum, and minimum diameters of the nut.

It must be remembered that the diameter of the hole made by any drill depends to some extent on the way it is ground, on the material drilled, and on the lubricant used. Holes that have been drilled before heat-treatment sometimes go out of round and cause tap breakage. The drills shown are carried in stock, some being in metric dimensions.

The essential requirement of a tap drill is that the hole produced by it shall be such that, when tapped with a screw thread, the minor diameter of the tapped hole

TABLE 6. TAP DRILLS FOR AMERICAN NATIONAL COARSE THREADS

Size of Thread	Threads per In.	Minor Dia of Nut, In.			Stock Drills and Corresponding Percentage of Basic Thread Depth		
		Basic	Max	Min	Nominal Size	Dia, In.	Percentage of Depth of Basic Thread
1	64	0.0527	0.0623	0.0561	1.45 mm	0.0571	78
					1.50 mm	0.0591	68
					1.55 mm	0.0610	59
2	56	0.0628	0.0737	0.0667	No. 51	0.0670	82
					No. 50	0.0700	69
					No. 49	0.0730	56
3	48	0.0719	0.0841	0.0764	5/64 in.	0.0781	77
					No. 46	0.0810	67
					2.10 mm	0.0827	60
4	40	0.0795	0.0938	0.0849	No. 44	0.0860	80
					No. 43	0.0890	71
					2.30 mm	0.0906	66
					3/32 in.	0.0937	56
5	40	0.0925	0.1062	0.0979	No. 39	0.0995	79
					2.60 mm	0.1024	70
					No. 37	0.1040	65
6	32	0.0974	0.1145	0.1042	No. 36	0.1065	78
					7/64 in.	0.1094	70
					No. 33	0.1130	62
8	32	0.1234	0.1384	0.1302	3.40 mm	0.1339	74
					No. 29	0.1360	69
					3.50 mm	0.1378	65
10	24	0.1359	0.1559	0.1449	No. 26	0.1470	79
					No. 24	0.1520	70
12	24	0.1619	0.1801	0.1709	No. 17	0.1730	79
					No. 16	0.1770	72
					No. 15	0.1800	67
1/4	20	0.1850	0.2060	0.1959	No. 8	0.1990	79
					13/64 in.	0.2031	72
5/16	18	0.2403	0.2630	0.2524	F	0.2570	77
					G	0.2610	71
3/8	16	0.2938	0.3184	0.3073	5/16 in.	0.3125	77
					O	0.3160	73
7/16	14	0.3447	0.3721	0.3602	U	0.3680	75
1/2	13	0.4001	0.4290	0.4167	27/64 in.	0.4219	78
9/16	12	0.4542	0.4850	0.4723	31/64 in.	0.4844	72

TABLE 6. TAP DRILLS FOR AMERICAN NATIONAL COARSE THREADS (*Continued*)

Size of Thread	Threads per In.	Minor Dia of Nut, In.			Stock Drills and Corresponding Percentage of Basic Thread Depth		
		Basic	Max	Min	Nominal Size	Dia, In.	Percentage of Depth of Basic Thread
5/8	11	0.5069	0.5397	0.5266	17/32 in.	0.5312	79
3/4	10	0.6201	0.6553	0.6417	16.5 mm	0.6496	77
7/8	9	0.7307	0.7689	0.7547	49/64 in. 19.5 mm	0.7656 0.7677	76 74
1	8	0.8376	0.8795	0.8647	22 mm 7/8 in.	0.8661 0.8750	82 77
1 1/8	7	0.9394	0.9858	0.9704	25 mm 63/64 in.	0.9842 0.9844	76 76
1 1/4	7	1.0644	1.1108	1.0954	28 mm 1 7/64 in.	1.1024 1.1094	80 76
1 3/8	6	1.1585	1.2126	1.1916	30.5 mm 1 13/64 in.	1.2008 1.2031	80 79
1 1/2	6	1.2835	1.3376	1.3196	1 21/64 in.	1.3281	79
1 3/4	5	1.4902	1.5551	1.5335	39 mm 1 35/64 in. 39.5 mm	1.5354 1.5469 1.5551	83 78 75
2	4 1/2	1.7113	1.7835	1.7594	45 mm 1 25/32 in.	1.7716 1.7812	79 76
2 1/4	4 1/2	1.9613	2.0335	2.0094	51.5 mm 2 1/32 in.	2.0276 2.0312	77 76
2 1/2	4	2.1752	2.2564	2.2294	57 mm 2 1/4 in.	2.2441 2.2500	79 77
2 3/4	4	2.4252	2.5064	2.4794	2 31/64 in. 2 1/2 in.	2.4844 2.5000	82 77
3	4	2.6752	2.7564	2.7294	2 47/64 in. 2 3/4 in. 70 mm	2.7344 2.7500 2.7559	82 77 75
3 1/4	4	2.9252	3.0064	2.9794	2 63/64 in. 76 mm 3	2.9844 2.9921 3.0000	82 79 77
3 1/2	4	3.1752	3.2564	3.2294	3 1/4 in.	3.2500	77
3 3/4	4	3.4252	3.5064	3.4794	3 1/2 in.	3.5000	77

Table 7. Tap Drills for American National Fine Threads

Size of Thread	Threads per In.	Minor Dia of Nut, In.			Stock Drills and Corresponding Percentage of Basic Thread Depth		
		Basic	Max	Min	Nominal Size	Dia, In.	Percentage of Depth of Basic Thread
0	80	0.0438	0.0514	0.0465	3/64 in.	0.0469	81
					1.25 mm	0.0492	67
1	72	0.0550	0.0634	0.0580	1.50 mm	0.0591	77
					1.55 mm	0.0610	67
2	64	0.0657	0.0746	0.0691	No. 50	0.0700	79
					No. 49	0.0730	64
3	56	0.0758	0.0856	0.0797	No. 46	0.0810	78
					2.10 mm	0.0827	70
					No. 44	0.0860	56
4	48	0.0849	0.0960	0.0894	2.30 mm	0.0906	79
					3/32 in.	0.0937	68
					No. 41	0.0960	59
5	44	0.0955	0.1068	0.1004	2.60 mm	0.1024	77
					No. 37	0.1040	71
					No. 36	0.1065	63
6	40	0.1055	0.1179	0.1109	No. 33	0.1130	77
					No. 32	0.1160	68
8	36	0.1279	0.1402	0.1339	3.40 mm	0.1339	83
					No. 29	0.1360	78
					3.50 mm	0.1378	73
					9/64 in.	0.1406	65
10	32	0.1494	0.1624	0.1562	5/32 in.	0.1562	83
					No. 21[3]	0.1590	76
					No. 20	0.1610	71
					No. 19	0.1660	59
12	28	0.1696	0.1835	0.1773	No. 15	0.1800	78
					4.70 mm No. 13	0.1850	67
					3/16 in.	0.1875	61
1/4	28	0.2036	0.2173	0.2113	No. 3	0.2130	80
5/16	24	0.2584	0.2739	0.2674	17/64 in.	0.2656	87
					I	0.2720	75
3/8	24	0.3209	0.3364	0.3299	Q	0.3320	79
7/16	20	0.3725	0.3906	0.3834	W	0.3860	79
					25/64 in.	0.3906	72
1/2	20	0.4350	0.4531	0.4459	29/64 in.	0.4531	72
9/16	18	0.4903	0.5100	0.5024	0.5062	0.5062	78
5/8	18	0.5528	0.5725	0.5649	14.5 mm	0.5709	75
3/4	16	0.6688	0.6903	0.6823	11/16 in.	0.6875	77
					17.5 mm	0.6890	75
7/8	14	0.7822	0.8062	0.7977	51/64 in.	0.7969	84
					20.5 mm	0.8071	73
1	14	0.9072	0.9312	0.9227	23.5 mm	0.9252	81
1 1/8	12	1.0167	1.0438	1.0348	26.5 mm	1.0433	75
1 1/4	12	1.1417	1.1688	1.1598	29.5 mm	1.1614	82
1 3/8	12	1.2667	1.2938	1.2848	1 9/32 in	1.2812	87
					1 19/64 in.	1.2969	72
1 1/2	12	1.3917	1.4188	1.4098	36 mm	1.4173	76

shall be within the specified limits. It should be noted that the minor diameters of the tapped holes are the same for classes 1 to 4.

If the drill is too large, the minor diameter of the tapped hole will also be too large and the thread in the nut will be too shallow, that is, too small a percentage of a full thread.

If, on the other hand, the tap drill is too small, the tap will be forced to cut a thread of full depth, and in the extreme case to act as a reamer also. This will result in excessive power consumption and tap breakage and will also make the minor diameter of the tapped hole dependent upon the minor diameter of the tap. This is undesirable since the minor diameter of the tap is not, in general, held to the same close limits as the other tap elements. As a result the minor diameter of a hole tapped under these conditions may be in error.

TABLE 8. TAP DRILLS FOR INSTRUMENT THREADS*

Size of Screw and Threads	Tap Drill Sizes			Clearance Drill
	75 to 80% Full Thread Wrought Brass Nickel Babbitt Wrought Alum. Alloy Fiber White Metal Hard Rubber	70 to 75% Full Thread Mild Steel Cast Aluminum Cast Iron Cast Brass	65 to 75% Full Thread Bronze Tool Steel Drop Forging Stainless Steel Cast Steel Nickel Copper	
00-96	64 (0.036)	63 (0.037)	63 (0.037)	55 (0.052)
0-80	55 (0.052) 56 (0.0465)	55 (0.052)	55 (0.052)	51 (0.067)
1-72	53 (0.0595)	53 (0.0595)	52 (0.0635)	47 (0.0785)
2-64	50 (0.070)	49 (0.073)	48 (0.076)	42 (0.0935)
3-56	46 (0.081)	45 (0.082)	44 (0.086)	36 (0.065)
4-48	43 (0.089)	42 (0.0935)	41 (0.096)	31 (0.120)
5-44	38 (0.1015)	37 (0.104)	35 (0.110)	29 (0.136)
6-40	33 (0.113)	32 (0.116)	31 (0.120)	26 (0.147)
8-36	29 (0.136)	29 (0.136)	28 (0.1405)	17 (0.173)
10-32	21 (0.159)	20 (0.161)	19 (0.166)	7 (0.201)

* Kollsman Instrument Co. compiled these data to provide accuracy required for aviation instruments.

Drills for Taper-pin Holes. Wear on taper-pin reamers is reduced by step-drilling the hole first. Table 9 gives the sizes of the several drills that may be required and the depths to drill with each one in order to step-drill holes for stand-

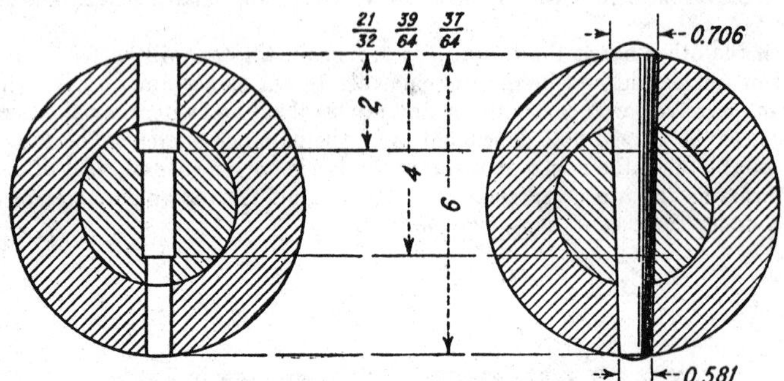

Fig. 5. To produce a taper-pin hole for a No. 10 taper pin 6 in. long, the two component parts are step-drilled with three drills and the hole is then taper-reamed.

Table 9. Size and Depth When Step-drilling Taper-pin Holes

Size Pin	First Drill through Size	Second Drill		Third Drill		Fourth Drill		Fifth Drill	
		Size	Depth	Size	Depth	Size	Depth	Size	Depth
7/0	3/64								
6/0	3/64								
5/0	1/16								
4/0	5/64								
3/0	3/32								
2/0	3/32	7/64	1 1/4						
0	3/32	1/8	1 1/2						
1	7/64	9/64	1 1/2						
2	7/64	9/64	1 3/4						
3	9/64	11/64	1 3/4						
4	11/64	13/64	1 3/4						
5	3/16	15/64	2						
6	15/64	17/64	3 1/4	19/64	1 3/4				
7	19/64	21/64	3 1/4	3/8	1 5/8				
8	25/64	27/64	3 1/4	29/64	1 5/8				
9	15/32	1/2	4	35/64	2				
10	37/64	39/64	4	21/32	2				
11	43/64	23/32	6 3/4	49/64	4 1/2	13/16	2 1/4		
12	27/32	57/64	6 3/4	15/16	4 1/2	63/64	2 1/4		
13	1	1 3/64	8 3/4	1 3/32	6 3/4	1 9/64	4 1/2	1 3/16	2 1/4
14	1 1/4	1 19/64	10 1/2	1 23/64	7 3/4	1 13/32	5 1/4	1 15/32	2 3/4

ard-length taper pins in sizes from No. 7/0 to 10, and for taper pins Nos. 11 to 14, inclusive. These last sizes are made in the following maximum lengths: Nos. 11 and 12—9 in.; No. 13—11 in.; and No. 14—13 in. Figure 5 gives a case example of drill selection and shows that three sizes are required to step-drill for a No. 10 taper pin 6 in. long.

Tap Drills for Plastics. The Boonton Plastics Co. suggests the selection of tap drills for plastics to give the following percentages of full depth of thread:

Screw Size	Percent Thread Depth	Screw Size	Percent Thread Depth
Up to No. 6	50	1/4 to 1/2 NC	70
Nos. 6 to 12	60	9/16 to 1 NC	75
		1/4 to 1 NF	70

The formula for finding the tap-drill diameter is

$$D = T - N \times 2h$$

where D = drill diameter, T = major diameter of tap, N = percent of thread, and h = depth of thread.

SPECIAL TWIST DRILLS AND THEIR USES

While the ordinary twist drill can handle the great majority of drilling jobs, special circumstances may require the designing of a new type of drill. Some of these special tools are listed below and shown in Fig. 4.

Core Drills. Not adapted to drill holes from the solid but used to enlarge cored, punched, or previously drilled holes. Sometimes used in place of roughing reamers.

Combination Tools. Any combination of rotary cutting tools, such as drill and reamer; drill and countersink; step drill of several diameters; drill, reamer, countersink, etc.

Shell Drills. Used for the same purpose as three or four-fluted drills. Made with tapered hole and fit on an arbor. One size of arbor will hold a large range of shell drills.

Oil-hole Drills. Used mostly in screw-machine or turret-lathe work for drilling deep holes. Have one or two oil holes running from the shank to the cutting point, and oil may be forced through these holes for lubrication.

Oil-tube Drills. Have oil tubes sunk in grooves cut in the lands of the drill instead of oil holes.

Straight Fluted Drill. These have two straight flutes running parallel to the axis. They are well adapted for brass, copper, or other soft metals, as they will not run ahead or grab. Formerly called Farmer drill.

Dual-cut Drills. Two- or three-step combination drills with the lands of the small diameter ground to size for the full length of the flutes.

High-helix Drills. Has a helix angle of about 40°. Developed for the drilling of slate and marble. Also useful in drilling deep holes in aluminum, magnesium, wood, copper, and fiber.

Bakelite Drills. Have a wide, polished flute adapted for use in bakelite, fiber, and hard rubber. Generally made of high-speed steel to resist abrasion.

Stove-burner Drill. Has a very short flute for great strength. Well adapted for drilling short holes in quantity.

Brass Drills. Usually carbon steel with special shape. Superior to regular drills for brass work. Also satisfactory for magnesium alloys.

Crankshaft Drills. Specially designed for oil-hole drilling in crankshafts and connecting rods. Do not often exceed 60 diameters in length. Have heavy webs. Drill must be withdrawn frequently in deep-hole drilling.

Flat-track Drills. Forged flat and have a special point milled on the end.

Manganese Drills. Developed for drilling work-hardening manganese steels. They are short and stubby with a heavy cross-section and thick web. Slow speed and a heavy power feed are recommended.

Bonding Drills. Designed and tempered for drilling holes for bonding wires in track-circuit signal work. Shorter and heavier than regular drills.

Bobbin Bit. Developed for the drilling of deep holes in wood. Its chip-clearing ability makes it useful in drilling celluloid.

Spoon Bit. For drilling stacked layers of paper, cardboard, or thin wood. Has a crescent-moon section with sharp edges. They remove a central core of the material.

Tube Drill. For drilling holes in paper or cardboard. Similar to spoon bit in action. The hollow tube with sharpened ends removes a solid core.

Router Bits. Made in an endless variety of styles and shapes. Widely used in wood-carving and engraving machines.

Glass Drills. At least three ways of drilling holes in glass are in use. In one method a three-cornered file has its point ground off smooth and is rotated in a drill press. Another method is to use a copper tube of the correct diameter and apply plenty of abrasive powder to its cutting end. A solid copper rod may be substituted for the tube but is much slower. Liberal application of turpentine must be made in all cases.

Flat Drills. The flat drill is well adapted for very small sizes. They are sold in diameters as small as 0.002 in.

Hollow Drill. Especially adapted for horizontal screw machine use in drilling deep holes, such as gun barrels. The drill is provided with a threaded shank. A hole through this shank runs into the flutes. The drill is screwed to the end of a substantial pipe. The work is generally rotated and lubricant is forced through the pipe to the cutting lips.

Gun-drill Tip. The gun-drill tip is used almost exclusively for drilling long small-diameter holes, such as rifle barrels. The tool is fastened to the end of a long hollow rod, and the work is rotated. Lubricant is forced through the rod to the drill tip. The tip is of round section with a deep V groove milled parallel to its axis.

Deep-hole and Gun Drills. Special, thick-web twist drills (Fig. 6) are employed for deep-hole drilling, as in oil holes for crankshafts. But deep holes of $\frac{3}{8}$ in. and larger are often drilled with single- and two-lip gun drills. Made of high-speed steel (HSS) or tipped with carbide, these drills are normally held stationary while

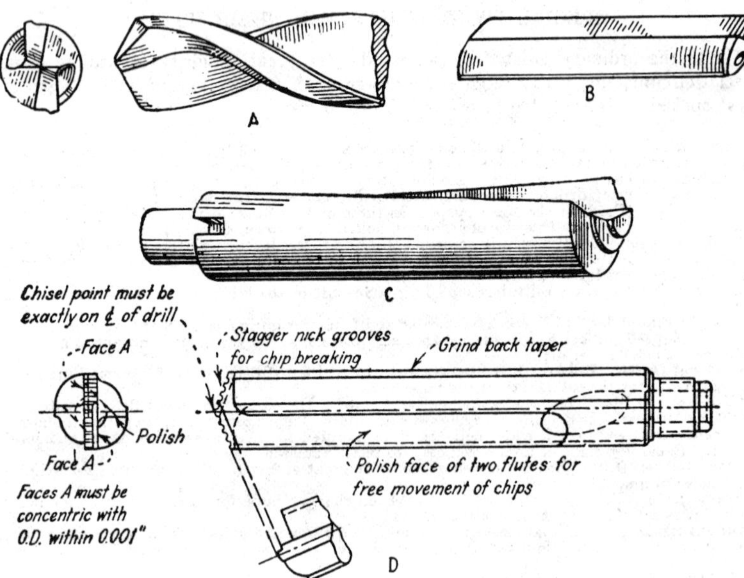

FIG. 6. Drills for deep holes: *A*, twist drill with thick web; *B*, single-lip or gun drill with oil hole; *C*, larger drill with stepped cutting edge; and *D*, two-lipped gun drill with nicked cutting edges to break up chips.

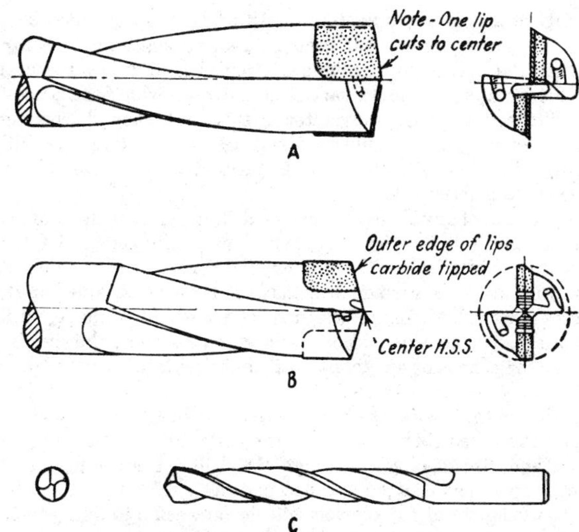

FIG. 7. Three forms of carbide twist drills: *A*, tipped drill with one carbide lip that cuts to center; *B*, chisel point of high-speed steel at area of least wear and carbide tips on the outer edges; and *C*, a solid carbide drill.

the work revolves, in an effort to get the hole to follow the axis of rotation. However, sometimes both work and tool are caused to revolve.

CARBIDE DRILLS

No definite standards for carbide drills have been established, but most large drill manufacturers now furnish carbide drills.

Types of carbide drills (solid and tipped) which have been used to drill brass, bronze, aluminum, pastics, cast irons, and low-carbon free-machining steels such as SAE 1020, SAE 1113, and X1112 are shown in Fig. 7.

Carbide drills should be run about three times faster than HSS drills. Feed per revolution should be from 0.004 to 0.010 ipr, depending on drill size. Less than 0.004 in. usually causes rapid cutting-edge wear with a build-up on carbide.

For maximum rigidity, the drill should be as short as possible. This does not sacrifice drill life, since sharpening does not shorten a carbide drill to the same extent as HSS drills.

It is preferable to spot-drill first when drilling steel. "Oil-hole" type drills (such as the "all-depth" drills) should be used for holes ½ in. dia and larger.

Carbide drills of the "all-depth" type resemble gun drills but are useful for shallow holes too. The solid-carbide type (Fig. 8) is made in sizes from ¼ to ⅝ in.; whereas the wear-strip types (Figs. 9 and 10) are produced in size ranges of 0.2905 (30 cal) to 1 in., and 1 to 3 in., respectively. All three types are designed for connection to a drill tube or drill bar, as the case requires, so that oil can be fed to the cutting edges at high pressure to wash out chips.

Grind the OD of all carbide-tipped all-depth drills to an included back taper of 0.0006 in. per in. After the OD has been ground, a longitudinal clearance should be ground along the cutting tip, leaving a circular land below the cutting edge. This land *J* (Figs. 9 and 10) should vary from 0.030 in. wide for a 0.290-in. drill to 0.050 in. wide for a 3-in. drill. Never touch this land when sharpening the tool.

The OD of the steel bodies should be about 2% smaller than the drill size, and the cutting edge of the tool should always be on center to 0.003 in. below center to produce a free cutting action.

Successful operation of any all-depth carbide drill depends on use and maintenance of correct cutting and relief angles (Fig. 11). Sharpening should be done only on the end of the drill. The face of the cutting tip should never be touched.

Best results with all-depth carbide drills have so far been obtained on regular gun-drilling machines, because these are designed for the use of stationary drills. Ordinarily, drills of the type considered here cannot be rotated at the speeds proper for carbide, because long drills of unsymmetrical shape tend to result in an unbalanced condition.

Practically all gun-drill work has been carried out in the feed range of 0.0006 to 0.0036 ipr, the lighter feeds being used for the smaller drills. But feed rate depends on oil pressure and size of oilhole through the drill. Depending on the job, the oil pressure will range between 300 and 800 psi.

The correct cutting speed for the job will lie between 130 and 350 sfpm, when

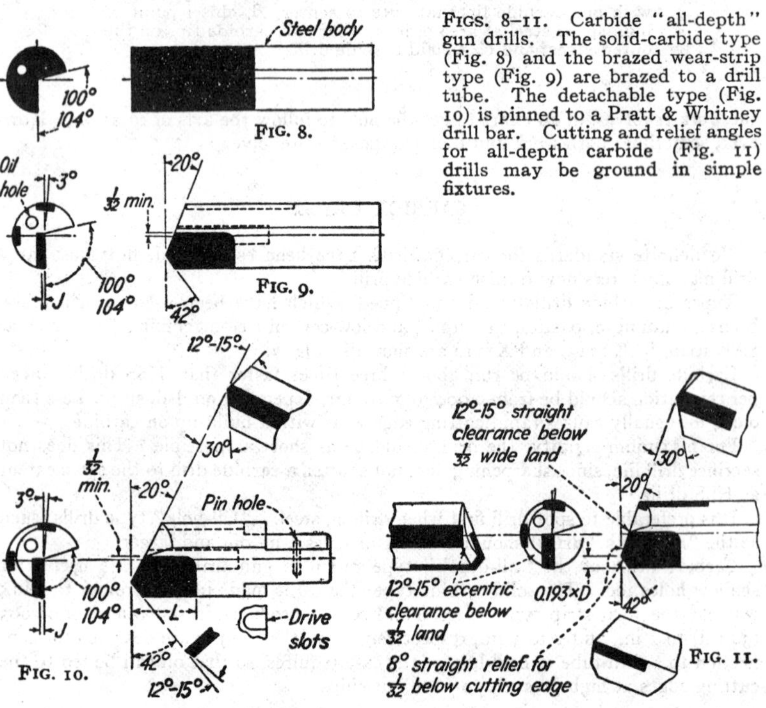

FIGS. 8–11. Carbide "all-depth" gun drills. The solid-carbide type (Fig. 8) and the brazed wear-strip type (Fig. 9) are brazed to a drill tube. The detachable type (Fig. 10) is pinned to a Pratt & Whitney drill bar. Cutting and relief angles for all-depth carbide (Fig. 11) drills may be ground in simple fixtures.

drilling steel. Start a new job at 200 to 250 sfpm, and vary the speed in either direction until a satisfactory balance is achieved between drill life and feed rate.

When all-depth drills approach $2\frac{1}{2}$ in. dia, break the broad chip, which comes from the 20 and 42° cutting angles by steps ground in the cutting edges.

While the preceding practice has been successful, it is evident that carbide gun drilling can be done at far higher cutting speeds and feeds per minute than often thought possible. Nickel cast iron has been drilled at 22 ipm, and B1113 steel at 17 ipm. Feeds per revolution have ranged between 0.002 and 0.005 in. Steel has been drilled at an actual cutting speed of approximately 1200 sfpm. Coolant is very important to machining rate and finish.

Carbide-drill Troubles, Causes, and Remedies*

Trouble	Cause	Remedy
Drill sticks in hole	1. Clearance on margin gone	*A*. Sharpen back to point where original clearances are unaffected *B*. Reduce rpm
Cutting edges chip	1. Excessive lip clearance	*A*. Resharpen with less clearance
	2. Vibration	*A*. Shorten length drill projecting from chuck
	3. Excessive feed	*A*. Reduce feed
	4. Rough handling	*A*. Caution operator
Pickup on margin	1. Excessive heat	*A*. Reduce rpm *B*. Use cutting compound
	2. Clearance gone	*A*. Sharpen back to point where original clearances are unaffected *B*. Reduce rpm
Retarded drill penetration	1. Dull point	*A*. Resharpen
	2. Lip clearance is too small	*A*. Resharpen with greater clearance
	3. Web too heavy	*A*. Thin out web at point and resharpen
Chips weld together	1. Drill not cleared from hole frequently enough	*A*. Retract drill at more frequent intervals
	2. Excessive heat due to lack of cutting compound	*A*. Use generous supply of cutting fluids
Drill breaks	1. Dull point	*A*. Resharpen
	2. Feed too heavy	*A*. Reduce feed
	3. Faulty setup	*A*. Check alignment of jig bushing
	4. Point improperly ground	*A*. Resharpen

SMALL-HOLE DRILLS

For holes from 0.001 to 0.020 in., flat drills as in Fig. 12 are best. Above this diameter both flat and twist drills are used. Standard "Najet" sizes of flat drills are given in Table 10. Point angles from 90 to 135° are used, depending on work material. Tolerances are held to minus 0.0001 in. plus nothing. The larger sizes may be ground in fixtures; smaller drills are ground freehand on copper disks charged with diamond dust and oil.

Tolerances of ±0.0002 in. on holes from 0.004 to 0.008 in. dia and ±0.0003 in. on holes from 0.008 to 0.012 in. dia can be maintained. Above 0.012 in. twist drills will hold tolerances of ±0.0004 in.

Rules for Drilling of Microscopic Holes

1. According to J. A. Cupler, II, general manager of the National Jet Co., drilling speeds in excess of 4000 rpm for drills as small as 0.020 in. dia are impracticable. Speeds not in excess of 2800 rpm will give the most satisfactory results as to hole quality and drill life, regardless of the material drilled, except in exceptional cases. If any rule should be kept in mind, it should be this: As the drill decreases in diameter below 0.020 in., the rpm should be decreased rather than increased, and the feed per revolution of the drill should be compatible with the ability of the drill to withstand the torque. As the diameter of the drill is decreased one-half, the area is four times less, and consequently the torque resistance is very appreciably less.

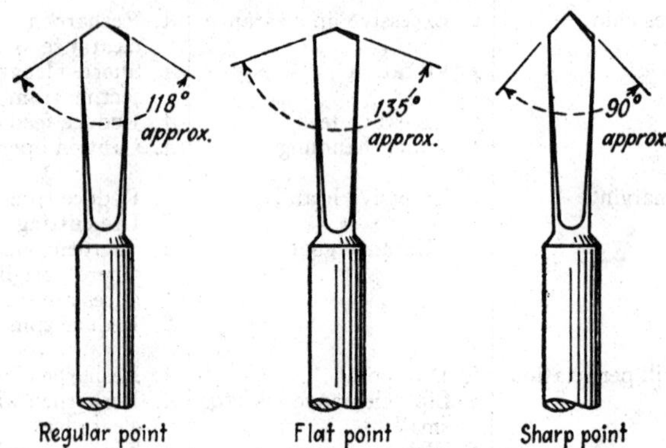

FIG. 12. Flat drills for microscopic holes. The regular point is for carbon and alloy steels, the flat point for hard and tough metals, and the sharp point for plastics.

2. Drill diameter governs hole size. Smaller drills drill holes closer to the drill's own size than larger drills.

3. The drill must be chucked absolutely true, preferably under a 20-power binocular. The drill should be tightly chucked to resist flexing in the collet.

4. Web thickness determines hole size by governing necessary drill pressure.

5. Rake angle further determines drilling pressure beyond that for the original web thickness of the drill.

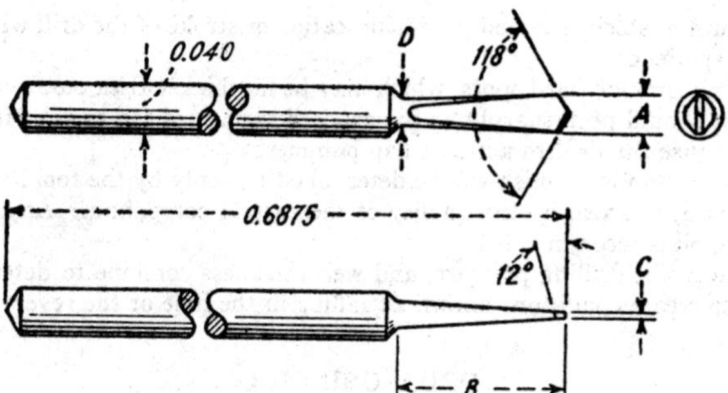

TABLE 10. "NAJET" STANDARDS FOR SMALL AND MICROSCOPIC DRILLS*

Dia *A*	Length *B*	Web Thickness *C*	Back Taper *D*
0.0010	0.0070	0.0005–0.00075	0.0001
0.0020	0.0140	0.0011–0.0015	0.0001
0.0030	0.0210	0.0016–0.002	0.0002
0.0040	0.0280	0.0018–0.0022	0.0002
0.0050	0.0350	0.002 –0.0024	0.0003
0.0060	0.0420	0.0022–0.0026	0.0003
0.0070	0.0490	0.0024–0.0028	0.0004
0.0080	0.0560	0.0026–0.0030	0.0004
0.0090	0.0630	0.0028–0.0032	0.0005
0.0100	0.0700	0.0030–0.0034	0.0005
0.0110	0.0770	0.0032–0.0036	0.0006
0.0120	0.0840	0.0038–0.0042	0.0006
0.0130	0.0910	0.0042–0.0046	0.0007
0.0140	0.0980	0.0044–0.0048	0.0007
0.0150	0.1050	0.0046–0.0050	0.0008
0.0160	0.1120	0.0048–0.0052	0.0008
0.0170	0.1190	0.0050–0.0054	0.0009
0.0180	0.1260	0.0053–0.0058	0.0009
0.0190	0.1330	0.0056–0.0060	0.0010
0.0200	0.1400	0.0060–0.0064	0.0010

National Jet Co.

6. Point angle affects hole size, because the longer the point angle, the more the centering effect for the drill. The blunter the point angle, the greater care and skill must be exercised to control hole size.

7. Since drilling pressure affects hole size, the operator must develop a sensitive touch to duplicate results over many holes.

8. For metals having a tendency to work harden or glaze, never allow the drill to dwell without being progressively fed. To avoid torque on the tiny drill, reciprocate it.

9. Amount of stock removed per reciprocation or stroke of the drill will be determined by experience.

In breaking through hard spots, which may be no thicker than 0.001 or 0.002 in., the operator should peck sharply at the exposed surface of the metal with the drill in order to cause the drill to act as a trip hammer.

10. Hole size and wall finish will be determined not only by the tool itself and the technique used, but also by the quality of the cutting compound. Any light lard-base cutting oil is recommended.

11. Cutting oils, drilling pressure, and web thickness combine to determine wall finish and to create conditions known as galling in the hole or the reverse of it.

DRILL GRINDING

Fully 95% of drilling troubles can be traced to improper grinding of the point. The point must have (1) both lips of same length and at the same angle to the drill axis (Fig. 13), (2) correct lip clearance (Fig. 14), and (3) correct angle between the lips and the chisel edge (Fig. 15).

The cutting edges must be at equal angles and of equal length. When the point is central but the angles of the cutting edges are different, the drill will bind on the side of the hole opposite to the lip which is cutting. It will drill too large a hole, and all the work will fall on the one cutting edge. Figure 16 illustrates this condition at *A*.

When the point is ground with equal angles but with cutting edges of different lengths, the point will no longer be central, and the condition shown at *B* will result.

When both angle and length of cutting edges are wrong, the drill will be laboring under the severe conditions shown at *C*.

Drill-point Angles. Standard drills have points with an included angle of 118°. Some materials are drilled more easily if this angle is varied. See Fig. 17, Table 11, and Notes on Drilling Various Materials, page 2-29.

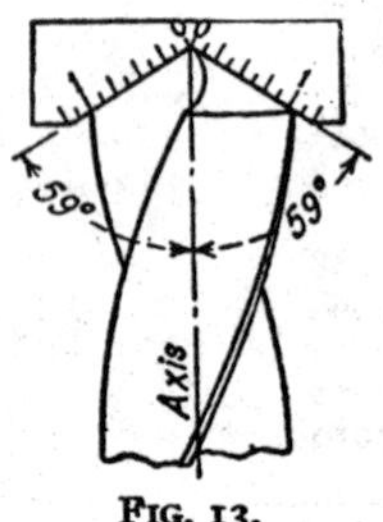

Fig. 13.

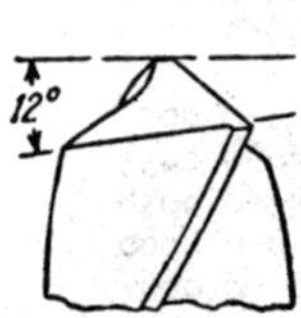

Fig. 14.

Figs. 13–16. Correct geometry of the drill point is essential to avoid poor holes and lost production. Shown are relationships for the standard point. Fig. 13. 118° included angle for lips. Fig. 14. 12° lip clearance. Fig. 15. 135° chisel-edge angle. Fig. 16. Badly ground drills will cut in one of these ways.

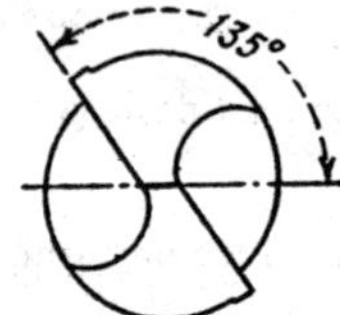

Fig. 15.

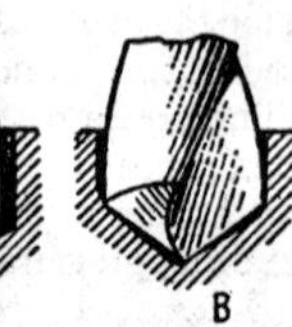

Fig. 16.

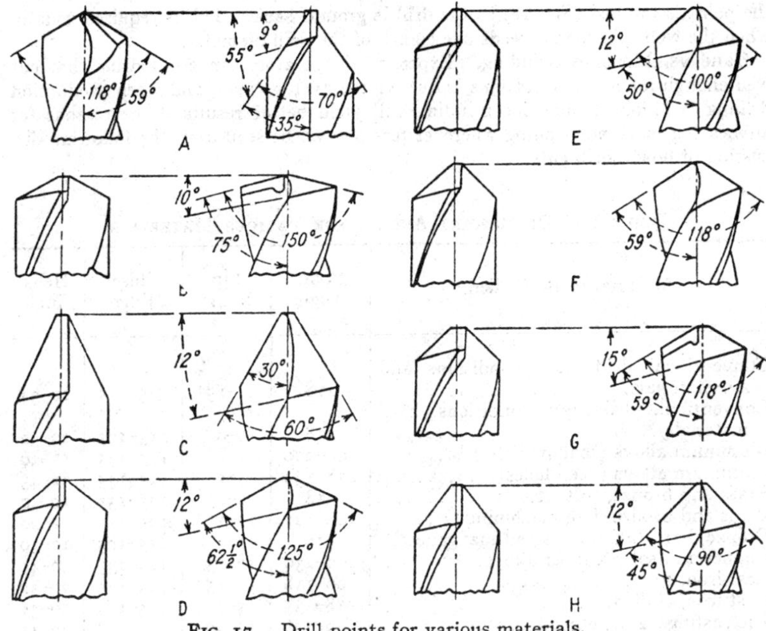

FIG. 17. Drill points for various materials.

The drill points shown in Fig. 17 are recommended by the Chicago-Latrobe Twist Drill Works for these purposes:

A—Crankshaft and deep-hole drilling.
B—Manganese steel and hard materials.
C—Wood, fiber, hard rubber, and aluminum.
D—Heat-treated steels and drop forgings.
E—Copper and some copper alloys.
F—Molded materials.
G—Brass and soft bronze.
H—An alternative for *C*, and used for cast iron and die castings.

Points as blunt at 170° have been used successfully in production drilling, where the tool is guided by a bushing.

Lip Clearance. Experience shows that 12° is the best angle of lip clearance, at the periphery of the drill, for general applications. This angle should be increased gradually as the center of the drill is approached; and, when the point is correctly ground, the line across the center of the web stands at an angle of approximately 135° with the cutting edges. Failure to give sufficient angle of lip clearance at the center of the drill is the principal cause of splitting drills up the web.

Lip Shape. In order to break up chips, it is sometimes necessary to alter the shape of the lips by grinding their faces. If it is desired to curl the chips more, this can be done by grinding more rake off on the faces of the lips. By decreasing the rake, the chips will tend to break up in smaller pieces.

Web Thinning. The central web of the drill increases in thickness toward the shank. This decreases chip clearance and tends to build up end pressure. Hence, the point is thinned (Fig. 18) as the drill is ground back. This is required usually when the web thickness exceeds one-eighth of the drill diameter.

Hand vs. Machine Grinding. Experienced operators can grind drills that cut well enough for many operations. But for the average man, and when production drilling is involved, machine grinding will yield better results. The reasons for advocating machine grinding wherever feasible can be seen from the following discussion of how a drill cuts.

TABLE 11. DRILL-POINT ANGLES FOR VARIOUS MATERIALS

Material to Be Drilled	Point Angle	Lip Relief	Chisel Edge	Helix Angle
Conventional drills (av. conditions and av. materials)	118	12–15	125–135	20–32
Conventional drills (av. conditions, hard materials)	118	6–9	115–125	20–32
Aluminum alloys (shallow holes)	90–120	12	125–135	17–20
Aluminum alloys (deep holes)	118–130	12	125–135	32–45
Brass and bronze, soft	118	12–15*	125–135	10–30
Brass and bronze, free machining	118–125	12–15*	125–135	0–20
Bronze, hard (aluminum and manganese)	118	5–7	115–125	10–30
Copper and some copper alloys	100–130	10–15	125–135	30–40
Cast iron, soft	90–118	12–15	125–135	20–32
Cast iron, chilled	118–135	5–7	115–125	20–32
Die castings, zinc, etc	60–136	12–20	125–135	32–45
Fiber	60–90	12–15	125–135	17–20
Magnesium alloys (shallow holes)	70–118	12–15	120–135	10–20
Magnesium alloys (deep holes)	118	12–15	135–150	40–45
Marble	90	12	125–135	20–32
Nickel alloys, monel, nickel, etc	118	12–15	125–135	20–32
Nickel alloys, hard, inconel	135–140	5–7	115–125	20–32
Plastics, laminated (Bakelite, etc.)	90–118	12–15	125–135	10–20
Plastics, molded	60–90	12–15	125–135	10–20
Rubber, hard	60–90	12–15	125–135	10–20
Slate	118	12–15	125–135	20–32
Steel, soft, low-carbon	118	12–15	125–135	20–32
Steel, forged, annealed	118–125	12–15	125–135	20–32
Steel, cast	118	12–15	125–135	20–32
Steel, alloy forged (crankshafts)	118–140	9 and 55	100	25–35
Steel, manganese (7–13% manganese)	150	10*	115–125	20–32
Steel, high-speed	135	5–7	115–125	20–32
Steel, nickel (up to 200 brinell)	118	10	115–125	20–32
Steel, nickel (about 250 brinell)	130–140	5–7	115–125	20–32
Steel, nickel (about 300 brinell)	130	5–7	115–125	20–32
Steel, nickel (about 350 brinell)	130–150	5–7	115–125	20–32
Steel, nickel (about 400 brinell)	150	5–7	115–125	20–32
Steel, nitriding (about 250 brinell)	130–140	5–7	115–125	20–32
Steel, nitriding (free machining)	118	10	125–135	20–32
Steel, stainless	118–140	5–7	115–125	20–32
Wood	70	12	125–135	30–40

NOTE: In general, the helix angle for conventional drills varies with the diameter of the drill.
* Lips flattened to reduce rake angle.

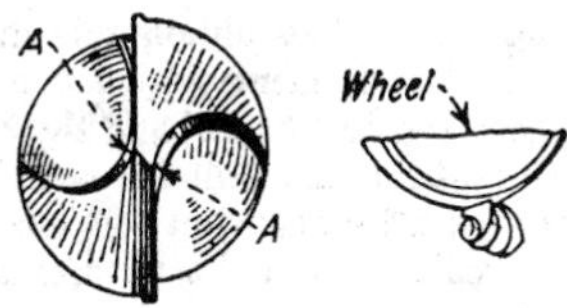

FIG. 18.

FIGS. 18, 19. Thinning of the web, as at *A-A* (Fig. 18) is necessary as the drill is ground back, because the thickness of the web increases toward the shank. Lip clearance (Fig. 19) should be greater at the center than at the periphery.

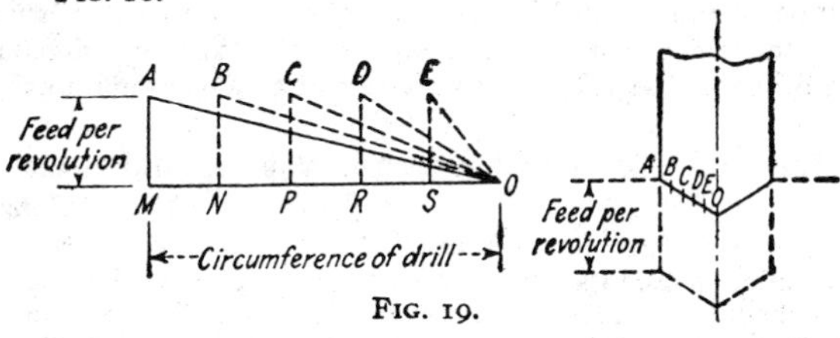

FIG. 19.

Hollow Grinding. Every point of a drill lip when at work travels in a helix. Each point travels in a helix of the same lead but with a different diameter, depending on its distance from the axis. A point near the axis travels on a helix of larger angle than one near the periphery; hence the clearance should be greater nearer the center than at the periphery.

In Fig. 19, points *A*, *B*, *C*, *D*, *E*, and *O* lie on the cutting edge of a drill. As the drill makes one revolution, the cutting edge moves forward the depth of feed. Line *AM* represents the feed, and *MO* the circumference of the drill. Angle *AOM* represents the helix angle of the helix over which point *A* travels and also represents the minimum lip-clearance angle at point *A* for this feed.

For point *E*, the feed is still the same, but the circumference *SO* is much smaller, and the helix angle *EOS* is much greater. Thus, the clearance angle ground on the lip should be greater nearer the center than at the periphery.

The theoretical modification of the cutting edge is approached by a "hollow-grinding" process, which has shown marked improvement in tool life between grinds in production work.

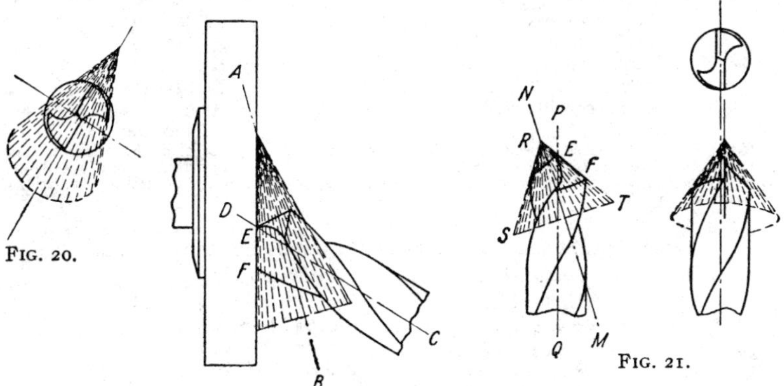

FIG. 20.

FIG. 21.

FIGS. 20, 21. Two methods of producing approximately correct lip clearance by grinding a segment of a cone.

Approximate Grinding Method. The most practical way to obtain this drill-point clearance is to use a type of machine which grinds a segment of a cone, as in Fig. 20, about an axis *AB* which is at an angle to both the drill lip *EF* and the axis of the drill *CD*. Axis *AB* is also a slight amount above the drill axis. When one lip is ground, the drill must be removed and the other lip placed in position for grinding. Here points near the center of the drill oscillate about shorter arcs and receive more clearance than points near the periphery.

A similar type of point formed in a slightly different manner is shown in Fig. 21. Here the lip clearance is generated as a segment of a cone in a different way by revolving the drill about axis *MN*. This axis is at an angle to drill axis *PQ* and is

TABLE 12. DRILLING SPEEDS FOR VARIOUS MATERIALS

Material to Be Drilled	*Cutting Speed, Fpm*
Aluminum and its alloys	200–300
Bakelite	100–150
Brass and bronze, soft	200–300
Bronze, high tensile	70–100
Carbon, pure (carbide drills)	100
Cast iron, soft	100–150
Cast iron, hard	70–100
Cast iron, chilled	30–40
Copper graphite alloy (carbide drills)	60–70
Glass (carbide drills)	20–30
Magnesium and its alloys	250–400
Malleable iron	80–90
Marble	15–25
Marble (carbide drills)	60–80
Nickel and monel	40–60
Slate	15–25
Slate (carbide drills)	40
Steel, machinery (0.2–0.3 C)	80–110
Steel, annealed (0.4–0.5 C)	60–70
Steel, tool (1.2 C)	50–60
Steel, forged	50–60
Steel, alloy	50–70
Steel, stainless, free machining	60–70
Steel, stainless, hard	30–40
Steel, manganese	15
Stone	15–25
Stone (carbide drills)	30
Wood	300–400

NOTE: Except as noted, speeds given above are for high-speed-steel drills. Carbon-steel drills should be run at from 40 to 50% of those for high-speed-steel drills. These speeds are given as starting points; the best speed in each case must be based on the prevailing conditions, material, setup, etc.

TABLE 13. SUGGESTED FEEDS FOR VARIOUS SIZES OF DRILLS

Dia of Drill, In.	*Feed, Ipr*
Under $\frac{1}{8}$	0.001 to 0.002
$\frac{1}{8}$ to $\frac{1}{4}$	0.002 to 0.004
$\frac{1}{4}$ to $\frac{1}{2}$	0.004 to 0.007
$\frac{1}{2}$ to 1	0.007 to 0.015
1 in. and over	0.015 to 0.025

NOTE: It is best to start with a moderate speed and feed, increasing either one, or both, after observing the action and condition of the drill.

also offset below the drill axis. Rotation about axis *MN*, while the point is held against a grinding wheel, causes the lip clearance to assume the shape of the cone *RST*.

Estimating Speeds and Feeds. Feeds are governed by the use of the drill and the material drilled. The general rule is to use a feed of 0.001 to 0.002 ipr for drills smaller than ⅛ in.; 0.002 to 0.004 ipr for drills ⅛ to ¼ in.; 0.004 to 0.007 ipr for drills ¼ to ½ in.; 0.007 to 0.015 ipr for drills ½ to 1 in.; and 0.015 to 0.025 ipr for drills larger than 1 in. Alloy and hard steels should generally be drilled at a lighter feed than given above, while cast iron, brass, and aluminum may usually be drilled with a heavier feed. Tables 12, 13, and 14 give speeds and feeds for drills. To find the rpm of small drills to obtain desired cutting speeds, refer to Tables 15 and 16. Coolants for various metals are given in Table 17.

NOTES ON DRILLING VARIOUS MATERIALS

Tables 11, 12, and 13 summarize recommendations for speeds, feeds, and drill-point angles in drilling various materials. These data may vary somewhat from the following details, because they were derived at different times and under circumstances that probably were not entirely the same. However, both sets of data, when taken as a whole, do provide a guide to drilling problems.

Aluminum. On drillpress work, the 28° helix drill is satisfactory on holes of medium depth. Deep or large holes are better done with drills of increased helix. Standard point angle of 118° is satisfactory but may be increased to 130 to 140° to facilitate chip removal and minimize burning. For high-silicon alloys the drill should have a sharper point. Use lip clearance up to 20°, when feed is heavy or in drilling soft alloys.

Use small drills at maximum machine speed; larger drills, 500 fpm for carbon-steel drills, 600 fpm for high-speed drills, and 2000 fpm for carbide-tipped drills.

For HSS drills up to ⅜ in., feeds may be 0.004 to 0.012 ipr; for drills ⅜ to 1¼ in., 0.006 to 0.020 ipr.; and for drills over 1¼ in. dia, 0.016 to 0.035 ipr.

Magnesium. Shallow holes up to five times the diameter can be drilled with few difficulties. The point angle may vary from 70 to 118° with chisel-edge angles 120 to 135°. Lip relief should be about 12°. A 25° helix angle can be used, or the helix may vary from 10 to 30°. Flutes should be highly polished to facilitate chip flow. Corners of cutting edges should be rounded.

Feeds for drilling magnesium are:

Drill Dia	Feed, Ipr		
	Sheet	Shallow Holes	Deep Holes
¼	0.005–0.030	0.004–0.030	0.004–0.008
½	0.010–0.030	0.015–0.040	0.012–0.020
1	0.010–0.030	0.020–0.050	0.015–0.03c

Deep holes require special drills. A high helix, 40 to 45°, should be used. An improved version has a 10° helix at the tip, changing to a 40 to 45° helix for the shank. Flutes should be open and polished for large chip spaces, and the web should have a constant thickness so that there will be as much chip space at the shank end as at the point end. The standard 118° point angle is the most satisfactory, but a spur or pilot point should be ground at the chisel edge to reduce spiraling. Chisel-edge angles should be 135 to 150° to secure a good surface and minimize spiraling.

Use the maximum speed obtainable when drilling. Speeds of 75 to 400 fpm are in general use, but speeds up to 2000 fpm are quite satisfactory.

Copper Alloys. Flat and straight-flute drills, having a natural 0° rake angle, are widely accepted for the alloys in group 1 (free-cutting) and 2 (readily machinable), particularly in screw-machine work.

The slow-helix drill with 10 to 22° helix angle and with wide polished flutes and a thin web provides large chip clearance and is often used for deep-hole drilling on alloys in groups 1 and 2.

For copper and most of the alloys in group 3 (machinability rating of 20 or under), drills having a helix angle of around 40° are used successfully.

TABLE 14. SPEEDS AND FEEDS FOR HIGH-SPEED DRILLS*

Drill Size In.	Aluminum and Copper		Cast Iron		Brass or Bronze		Steel Cstgs, Monel Metal, Stainless Steel, Molybdenum Drop-Forged Tool Steel		Drop Forgings, Alloy Steel Heat-Treated, Also SAE 4140		Mild Steel	
	Feed	Speed	Feed	Speed	Feed	Speed	Feed	Speed	Feed	Speed	Feed	Speed
1/16	.002 to .005	15000 to 18000	.002 to .004	4575 to 6700	.002 to .004	9150 to 12000	.002 to .003	3650 to 4550	.002 to .003	2750 to 3650	.002 to .003	4275 to 5500
1/8	.002 to .003	7600 to 9100	.002 to .004	2300 to 3350	.002 to .004	4575 to 6100	.002 to .003	1800 to 2250	.002 to .003	1375 to 1800	.002 to .003	2100 to 2800
1/4	.003 to .005	3800 to 4600	.004 to .006	1150 to 1675	.004 to .007	2300 to 3000	.003 to .004	925 to 1150	.003 to .004	700 to 900	.003 to .005	1050 to 1375
3/8	.003 to .005	2500 to 3000	.006 to .009	750 to 1125	.007 to .010	1525 to 2025	.004 to .006	600 to 750	.004 to .005	450 to 600	.005 to .007	700 to 925
1/2	.006 to .008	1900 to 2280	.008 to .012	575 to 850	.010 to 0.014	1150 to 1525	.006 to .009	450 to 575	.005 to .006	350 to 450	.006 to .010	525 to 700
5/8	.008 to .010	1500 to 1800	.012 to .016	450 to 675	.014 to .018	900 to 1200	.008 to .012	350 to 450	.007 to .010	275 to 375	.010 to .014	425 to 550
3/4	.008 to .010	1250 to 1500	.012 to .016	375 to 550	.014 to .018	750 to 1000	.008 to .012	300 to 375	.007 to .010	225 to 300	.010 to .014	350 to 450
7/8	.010 to .013	1095 to 1300	.014 to .020	325 to 475	.016 to .022	650 to 875	.010 to .014	260 to 325	.009 to .012	180 to 250	.014 to .016	300 to 400
1	.010 to .013	950 to 1125	.014 to .020	280 to 425	.016 to .022	575 to 750	.010 to .014	225 to 280	.009 to .012	150 to 200	.014 to .016	265 to 340

* Chicago-Latrobe Twist Drill Works

Dry drilling of copper alloys is common, but suitable lubricant gives better results. On bar stock in screw machines, a lubricant is used almost universally.

DRILLING SPEEDS

Alloy	*Fpm*
Group 1	200–500
Group 2	75–250
Group 3	50–125

For carbon drills, reduce speeds approximately 50%.

Cast and Malleable Iron. For drilling cast iron with an analysis of 3.30 to 3.60 carbon, 2.20 to 2.60 silicon, 0.60 to 0.90 manganese, 0.15 phosphorus max, and 0.12 sulfur max, the Ford Motor Co. uses a HSS drill with these features: the point angle

TABLE 15. RPM OF NUMBER-SIZE DRILLS

Ft. per Minute	30	40	50	60	70	80	90	100	110	120
No. Size	Revolutions per Minute									
1	503	670	838	1,005	1,173	1,340	1,508	1,675	1,843	2,010
2	518	691	864	1,037	1,210	1,382	1,555	1,728	1,901	2,074
3	538	717	897	1,076	1,255	1,434	1,614	1,793	1,974	2,152
4	548	731	914	1,097	1,280	1,462	1,645	1,828	2,010	2,193
5	558	744	930	1,115	1,301	1,487	1,673	1,859	2,045	2,230
6	562	749	936	1,123	1,310	1,498	1,685	1,872	2,060	2,247
7	570	760	950	1,140	1,330	1,520	1,710	1,900	2,090	2,281
8	576	768	960	1,151	1,343	1,535	1,727	1,919	2,111	2,303
9	585	780	975	1,169	1,364	1,559	1,754	1,949	2,144	2,339
10	592	790	987	1,184	1,382	1,579	1,777	1,974	2,171	2,369
12	606	808	1,010	1,213	1,415	1,617	1,819	2,021	2,223	2,425
14	630	840	1,050	1,259	1,469	1,679	1,889	2,099	2,309	2,518
16	647	863	1,079	1,295	1,511	1,726	1,942	2,158	2,374	2,590
18	678	904	1,130	1,356	1,582	1,808	2,034	2,260	2,479	2,704
20	712	949	1,186	1,423	1,660	1,898	2,135	2,372	2,610	2,847
22	730	973	1,217	1,460	1,703	1,946	2,190	2,433	2,676	2,920
24	754	1,005	1,257	1,508	1,759	2,010	2,262	2,513	2,764	3,016
26	779	1,039	1,299	1,559	1,819	2,078	2,338	2,598	2,858	3,118
28	816	1,088	1,360	1,631	1,903	2,175	2,447	2,719	2,990	3,262
30	892	1,189	1,487	1,784	2,081	2,378	2,676	2,973	3,270	3,567
32	988	1,317	1,647	1,976	2,305	2,634	2,964	3,293	3,622	3,951
34	1,032	1,376	1,721	2,065	2,409	2,753	3,097	3,442	3,785	4,129
36	1,076	1,435	1,794	2,152	2,511	2,870	3,228	3,587	3,945	4,304
38	1,129	1,505	1,882	2,258	2,634	3,010	3,387	3,763	4,140	4,516
40	1,169	1,559	1,949	2,339	2,729	3,118	3,508	3,898	4,287	4,677
42	1,226	1,634	2,043	2,451	2,860	3,268	3,677	4,085	4,494	4,902
44	1,333	1,777	2,221	2,665	3,109	3,554	3,999	4,442	4,886	5,330
46	1,415	1,886	2,358	2,830	3,301	3,773	4,244	4,716	5,187	5,659
48	1,508	2,010	2,513	3,016	3,518	4,021	4,523	5,026	5,528	6,031
50	1,637	2,183	2,729	3,274	3,820	4,366	4,911	5,457	6,002	6,548
52	1,805	2,406	3,008	3,609	4,211	4,812	5,414	6,015	6,619	7,218
54	2,084	2,778	3,473	4,167	4,862	5,556	6,251	6,945	7,639	8,334
56	2,465	3,286	4,108	4,929	5,751	6,572	7,394	8,215	9,036	9,857
58	2,729	3,637	4,547	5,456	6,367	7,275	8,186	9,095	10,004	10,913
60	2,865	3,820	4,775	5,729	6,684	7,639	8,594	9,549	10,504	11,459
62	3,015	4,020	5,025	6,030	7,035	8,040	9,045	10,050	11,057	12,060
64	3,183	4,244	5,305	6,366	7,427	8,488	9,549	10,610	11,671	12,732
66	3,474	4,632	5,790	6,948	8,106	9,264	10,422	11,580	12,732	13,890
68	3,696	4,928	6,160	7,392	8,624	9,856	11,088	12,320	13,554	14,786
70	4,091	5,456	6,820	8,184	9,548	10,912	12,276	13,640	15,006	16,370
72	4,584	6,112	7,640	9,168	10,696	12,224	13,752	15,280	16,807	18,335
74	5,106	6,808	8,510	10,212	11,914	13,616	15,318	17,020	18,674	20,372
76	5,730	7,640	9,550	11,460	13,370	15,280	17,190	19,100	21,008	22,918
78	7,101	9,548	11,935	14,322	16,709	19,096	21,483	23,870	26,260	28,648
80	8,490	11,320	14,150	16,980	19,810	22,640	25,470	28,300	31,123	33,953

TABLE 16. RPM FOR LETTER AND FRACTIONAL DRILLS

Ft. per Min.	30	40	50	60	70	80	90	100	110	120	130	140	150
Diameter, In.	Revolutions per Minute												
1/16	1,833	2,445	3,056	3,667	4,278	4,889	5,500	6,111	6,722	7,334	7,945	8,556	9,167
1/8	917	1,222	1,528	1,833	2,139	2,445	2,750	3,056	3,361	3,667	3,973	4,278	4,584
3/16	611	815	1,019	1,222	1,426	1,630	1,833	2,037	2,241	2,445	2,648	2,852	3,056
1/4	458	611	764	917	1,070	1,222	1,375	1,528	1,681	1,833	1,986	2,139	2,292
5/16	367	489	611	733	856	978	1,100	1,222	1,345	1,467	1,589	1,711	1,833
3/8	306	407	509	611	713	815	917	1,019	1,120	1,222	1,324	1,426	1,528
7/16	262	349	437	524	611	698	786	873	960	1,048	1,135	1,222	1,310
1/2	229	306	382	458	535	611	688	764	840	917	993	1,070	1,146
5/8	183	244	306	367	428	489	550	611	672	733	794	856	917
3/4	153	203	255	306	357	407	458	509	560	611	662	713	764
7/8	131	175	218	262	306	349	393	436	480	524	568	611	655
1	115	153	191	229	267	306	344	382	420	458	497	535	573
1 1/8	102	136	170	204	238	272	306	340	373	407	441	475	509
1 1/4	92	122	153	183	214	244	275	306	336	367	397	428	458
1 3/8	83	111	139	167	194	222	250	278	306	333	361	389	417
1 1/2	76	102	127	153	178	204	229	255	280	306	331	357	382
1 5/8	70	94	117	141	165	188	212	235	259	282	306	329	353
1 3/4	65	87	109	131	153	175	196	218	240	262	284	306	327
1 7/8	61	81	102	122	143	163	183	204	224	244	265	285	306
2	57	76	95	115	134	153	172	191	210	229	248	267	287
2 1/4	51	68	85	102	119	136	153	170	187	204	221	238	255
2 1/2	46	61	76	92	107	122	137	153	168	183	199	214	229
2 3/4	42	56	69	83	97	111	125	139	153	167	181	194	208
3	38	51	64	70	89	102	115	127	140	153	166	178	191
A	491	654	818	982	1,145	1,309	1,472	1,636	1,796	1,959	2,122	2,285	2,448
B	484	642	803	963	1,124	1,284	1,445	1,605	1,765	1,926	2,086	2,247	2,407
C	473	631	789	947	1,105	1,262	1,420	1,578	1,736	1,894	2,052	2,210	2,368
D	467	622	778	934	1,089	1,245	1,400	1,556	1,708	1,863	2,018	2,174	2,329
E	458	611	764	917	1,070	1,222	1,375	1,528	1,681	1,834	1,968	2,139	2,292
F	446	594	743	892	1,040	1,189	1,337	1,486	1,635	1,784	1,932	2,081	2,229
G	440	585	732	878	1,024	1,170	1,317	1,463	1,610	1,756	1,903	2,049	2,195
H	430	574	718	862	1,005	1,149	1,292	1,436	1,580	1,723	1,867	2,010	2,154
I	421	562	702	842	983	1,123	1,264	1,404	1,545	1,685	1,826	1,966	2,106
J	414	552	690	827	965	1,103	1,241	1,379	1,517	1,655	1,793	1,930	2,068
K	408	544	680	815	951	1,087	1,223	1,359	1,495	1,631	1,767	1,903	2,039
L	395	527	659	790	922	1,054	1,185	1,317	1,449	1,581	1,712	1,844	1,976
M	389	518	648	777	907	1,036	1,166	1,295	1,424	1,554	1,683	1,813	1,942
N	380	506	633	759	886	1,012	1,139	1,265	1,391	1,518	1,644	1,771	1,897
O	363	484	605	725	846	967	1,088	1,209	1,330	1,450	1,571	1,692	1,813
P	355	473	592	710	828	946	1,065	1,183	1,301	1,419	1,537	1,657	1,774
Q	345	460	575	690	805	920	1,035	1,150	1,266	1,384	1,496	1,611	1,726
R	338	451	564	676	789	902	1,014	1,127	1,239	1,355	1,465	1,577	1,690
S	329	439	549	659	769	878	988	1,098	1,207	1,317	1,427	1,537	1,646
T	320	426	533	640	746	853	959	1,066	1,173	1,280	1,387	1,494	1,600
U	311	415	519	623	727	830	934	1,038	1,142	1,246	1,349	1,453	1,557
V	304	405	507	608	709	810	912	1,013	1,114	1,219	1,317	1,418	1,520
W	297	396	495	594	693	792	891	989	1,088	1,188	1,286	1,385	1,484
X	289	385	481	576	672	769	865	962	1,058	1,155	1,251	1,347	1,443
Y	284	378	473	567	662	756	851	945	1,040	1,135	1,229	1,324	1,418
Z	277	370	462	555	647	740	832	925	1,017	1,110	1,202	1,295	1,387

is 100°, the lips are chamfered 0.040 to 0.050 in., and the outer ends of the lips are ground to an included angle of 60°. With these drills, the optimum cutting speed is 85 sfpm and feed 0.007 ipr to produce a tool life of 16 hr between grinds on the stated material. At this tool life, the wear land reaches 0.020 in. and is considered the proper point for regrinding by machine.

The following data on drilling irons with HSS drills are supplied by the Malleable Founders' Society.

	Cutting Speed, Sfpm
Malleable iron:	
Standard	70–90
Pearlitic—180 to 200 brinell	60–80
Pearlitic—200 to 240 brinell	50–70
Cast iron:	
Soft—160 to 193 brinell	80–100
Semisteel—193 to 220 brinell	70–90
Hard—220 to 240 brinell	60–80

Feeds for these materials are:

Drill dia.	1/16	1/8	3/16	1/4	5/16	3/8	7/16	1/2
Malleable iron	0.003	0.0045	0.0058	0.0075	0.009	0.010	0.011	0.0125
Cast iron	0.003	0.0045	0.006	0.008	0.009	0.011	0.012	0.013

Nickel Alloys. Drills ground with standard 118° point angle and clearance angles of 12 to 15° are satisfactory for most of the high-nickel alloys. But provide a point angle of 135 to 140° with just enough clearance to clear the work when drilling the

Table 17. Coolants for Drilling

Material	Coolant	Material	Coolant
Aluminum	Kerosene Kerosene and lard oil Soluble oil	Malleable iron	Dry Soda water
Brass	Dry Soluble oil Kerosene and lard oil	Monel metal	Lard oil Soluble oil
Bronze	Soluble oil Lard oil Mineral oil Dry	Steel, mild	Soluble oil Mineral lard oil Sulfurized oil Lard oil
Cast iron	Air jet Dry Soluble oil	Steel alloys Forgings Tool steel	Soluble oil Mineral lard oil Sulfurized oil
Cast steel	Soluble oil Mineral lard oil Sulfurized oil	Steel, manganese 12 to 15%	Dry
Copper	Soluble oil Dry Mineral lard oil Kerosene	Wrought iron	Soluble oil Mineral lard oil Sulfurized oil

harder alloys. The web of the drill should be thinned out at the point, especially when drilling the harder alloys. Drilling speed ranges from 40 to 60 sfpm for nickel and monel, 20 to 25 sfpm for unhardened K monel, and 30 to 40 sfpm for inconel.

When deep holes $\frac{3}{4}$ in. dia and larger are drilled, oilhole drills are recommended, with the cutting fluid supplied under pressure. These drills should have polished flutes and will operate most efficiently when chip breakers are used.

When any of the high-nickel alloys are drilled, it is essential that the feeding pressure be steady and be sufficiently high to keep the drill cutting at all times. Otherwise the material will polish and work harden quickly. Feeds range from 0.002 to 0.008 ipr for R monel.

Armor Plate. Use included point angle 135 to 140°; clearance angle of 6 to 9° at periphery; chisel point angle of 115 to 125°. For shallow holes use 40 to 50 sfpm, for deeper holes reduce speeds. Use feeds about 15 to 25% lower than ordinary.

EXAMPLE. $\frac{15}{16}$-in. drill at 40 sfpm with feed of 0.010 ipr. This drill should run 163 rpm. For 50 fpm, the speed would be 203 rpm. For coolant use rich mixture (8 or 10 to 1) of soluble oil or good sulfurized cutting oil. If one side is harder than the other, drill the hard side first.

Hardened Steel. Die and tool steels with maximum hardness can be drilled with "Hardsteel" drills. These tools do not cut but really anneal the hardened metal and wipe off the softened steel. There are no cutting edges. The end of the drill has small grooves ground into the flat sides of the triangular point, providing a negative rake and space for the escape of chips. Constant heavy pressure is required and should be run at the speeds shown in the table. If a coolant is used, it should have a rust inhibitor such as soda water or soluble oil, but no lubricating oil.

DRILLING SPEEDS FOR "HARDSTEEL" DRILLS

Drill Size	Without Coolant, Rpm	With Coolant, Rpm
$\frac{1}{8}$	2500 to 3000	3300 to 4000
$\frac{3}{16}$	2000	2700
$\frac{1}{4}$	1600	2100
$\frac{5}{16}$	1500	2000
$\frac{3}{8}$	1200	1600
$\frac{7}{16}$	1000	1300
$\frac{1}{2}$	950	1250
$\frac{5}{8}$	900	1200
$\frac{3}{4}$	800	1000
$\frac{7}{8}$	700	950
1	600	800

Carbon-steel Sheet. In the drilling of $\frac{1}{16}$- to $\frac{5}{16}$-in. holes in carbon sheet steel, high-speed drills, designed with a shorter helix and heavier web than used on a regular drill, are recommended to avoid breakage. Point angle for small drills can be 118° with 10 to 12° lip clearance.

In the drilling of large holes, point angles of 135 to 160° are recommended to prevent causing a burr on the back side. Some drill manufacturers recommend thinning the web on large drills for sheet-metal work, to reduce the pressure needed to force the drill through the work. In order to curtail "hogging in" when breaking through, grind a flat along the cutting edge at the drill point to reduce the rake.

It is often difficult to drill a round hole in sheet steel. This can be partially or entirely eliminated by using a modified spur or pilot point, which should be exactly in the center. With such drills, the web at the point should be thinned to about half the regular thickness. Another method for drilling round holes and eliminating burrs is use of a combination drill having two diameters. The larger diameter tends to remove any burr which the small diameter may cause.

Drilling speeds in low-carbon sheet steel range from 60 to 100 sfpm. Soluble oil is the most generally recommended lubricant.

Nickel Steels. Approximate feeds and speeds for drilling nickel-alloy steels with HSS drills are listed conservatively in Table 18. Peripheral speed and penetration rate decrease with work hardness but are not affected by change in tool size if the same hardness is being cut. Point angle is 118°, lip clearance 10° for hardness to 200 brinell. At 250 brinell, increase the point angle to 135 to 140° and decrease the lip clearance to 5 to 7°. As material hardness increases to 400 brinell, the point angle is increased to 150°.

TABLE 18. APPROXIMATE FEEDS AND SPEEDS FOR DRILLING NICKEL-ALLOY STEELS*

Brinell Hardness	Feed, Ipm	Cutting Speed, Fpm	Drill, Rpm (upper figures), and Feed, Ipr, for Given Drill Size				
			⅛ In.	¼ In.	½ In.	¾ In.	1 In.
Up to 163	4½	75	2300 0.0020	1150 0.0040	575 0.0080	385 0.0120	285 0.0160
163–192	4	70	2140 0.0020	1070 0.0040	535 0.0075	355 0.0110	265 0.0150
192–223	3	60	1830 0.0015	915 0.0035	460 0.0065	305 0.0100	230 0.0130
223–255	2¼	50	1530 0.0015	765 0.0030	380 0.0060	255 0.0090	190 0.0120
255–285	1¾	45	1370 0.0015	685 0.0025	340 0.0050	230 0.0075	170 0.0010
285–321	1¼	40	1220 0.0010	610 0.0020	305 0.0040	200 0.0060	150 0.0085
321–352	1	30	920 0.0010	460 0.0020	230 0.0040	150 0.0060	115 0.0080
352–388	¾	25	760 0.0010	380 0.0020	190 0.0040	125 0.0060	95 0.0080
388–415	½	15	460 0.0010	230 0.0020	115 0.0040	75 0.0060	55 0.0080

NOTE: This table applies when high-speed-steel drills are used with a cutting fluid.
* International Nickel Co.

Nitriding Steels. Cutting speeds range from 40 to 50 fpm, when a sulfur-base cutting oil is used. Feeds range from 0.004 to 0.016 ipr for drilling holes from ⅛ to 1 in. dia, in 225 to 250 brinell material.

Stainless Steels. Point angle for stainless steels is 140°, lip clearance 6 to 8°, and sometimes a secondary clearance of 7 to 9° to prevent digging in. Thin the web to about one-sixth the drill diameter, and down to one-eighth as the drill is ground back.

HSS drills of high helix angle are normally used on stainless. The drill should not dwell without cutting, particularly with the work-hardening chromium-nickel grades. Drill-jig bushings should be short to accommodate the recommended short drills and should be removable if chip congestion is a problem.

Speeds and feeds, particularly with small drills, should be controlled carefully. If the drill diameter approaches or exceeds three-quarters of stock thickness, speeds should be reduced. If the drill is less than one-quarter stock thickness in diameter, speeds should be increased. A "cotter-pin" drill works well on small holes, with proper speed and low feed.

Feeds are:

Drills ⅛ in. and smaller	0.002 ipr
Drills ⅛ to ¼ in.	0.002 to 0.004 ipr
Drills ½ in. and up	0.007 to 0.015 ipr

Cutting speeds when drilling various grades of stainless are:

Stainless No.	Cutting Speed, Sfpm	Stainless No.	Cutting Speed, Sfpm
403, 405, 410	35–75	414, 431	20–40
416	70–110	440	50–70
420	30–60	446	40–60
420 F	70–90	301, 302, 304	15–40
430	35–75	303	35–85
430 F	70–115		

Plastics. For best results in drilling plastics, use twist drills of special design, having highly polished flutes and maximum chip clearance. Carbide drills resist abrasion and minimize down time for grinding during production drilling jobs.

On many plastics, cutting speeds as high as 350 sfpm are possible with HSS drills, if heat generation is closely watched. Fast feeds are desirable, but do not force the tool. Ideal coolant is compressed air, but soluble oil can be employed on production drilling of cellulose acetates. Soap applied to the drill helps on small-lot production.

Drills sometimes cut undersize on plastics, and this situation can be overcome by ordering them 0.002 to 0.003 in. oversize.

Point and clearance angles have pronounced effect on success of drilling. Thermoplastic materials like cellulose acetate should be drilled with tools having a point angle of about 90°, a helix angle of 17°. On thin stock and tubing, use a point angle down to 60° to prevent chipping and grabbing.

Laminates and glass-reinforced plastics, also vulcanized fiber, require a point angle of 55 to 90°, increasing with stock thickness, and a slight negative rake.

Titanium. HSS drills, preferably with a short twist, should be used for titanium. Drills should be ground very sharp (approximately 87° point angle). The drill should be as stiff as possible, and as short as permissible for the particular job.

Section 3

FILES AND BURRS

SECTION 3

FILES AND BURRS

HAND FILES

By Grant Loader, *Plant Superintendent, Heller Bros. Company*

Metalworking hand files are divided in three categories: machinists' files of the American or regular pattern, Swiss pattern files, and special-purpose files. Correct selection and use of files mark the craftsman.

Cutting efficiency of a file is a function of tooth design, construction, and pattern. Important factors are: the sharpness, as indicated by the rake on the tooth face and the clearance on the back face; angle of cut relative to the file axis, coarseness of the cut, and ratio of the pitch of the upcut to that of the overcut in double-cut files.

File Selection. Choice of a file depends on:

Kind of job (flat or curved surface, edge or aperture, circular or square hole).

Work properties (size, shape, and location of work surface, nature of material and its hardness).

Manner of filing (stock to be removed, depth of cut, speed, accuracy or smoothness, rough filing or smooth finishing).

Standard Cuts. Standard cuts in a file are either single-cut or double-cut, although some special and patented cuts such as wavy-cut and curved-cut are available.

A single-cut file has a single set of diagonal rows of teeth, parallel to one another and extending across the face of the file. A double-cut file has two sets of diagonal rows. The first set is called the overcut. On top of these, a second set, cut at a different angle to the file axis, is known as the upcut. These rows are finer and deeper than the overcut.

The tooth spacing, or number of teeth per inch, varies slightly with the make of file and increases in proportion as the length of file increases. For example, a 12-in. mill smooth file has a coarser cut than a 6-in. mill smooth file. Bastard cut is coarse; second-cut is medium coarse; smooth-cut is fine.

Machinists' Files. Fast-working machinists' files remove metal rapidly and are generally used wherever smooth finishing is not the primary objective. They are available in nine shapes to cover almost every application, and vary in length from 4 to 18 in., depending on type and use. In general, sizes move up by 1-in. steps from 4 to 8 in., and by 2-in. steps from 8 to 18 in. They are usually made in double-cut, except small rounds, and the backs of fine-cut half-rounds which are made single-cut.

Flat files are of rectangular cross-section, tapering in both width and thickness. Hand files are somewhat thicker in relation to their width, and taper in thickness but are uniform in width. Pillar files are almost square and are also uniform in width but tapered in thickness. Warding files are thin and taper sharply in width, but are uniform in thickness.

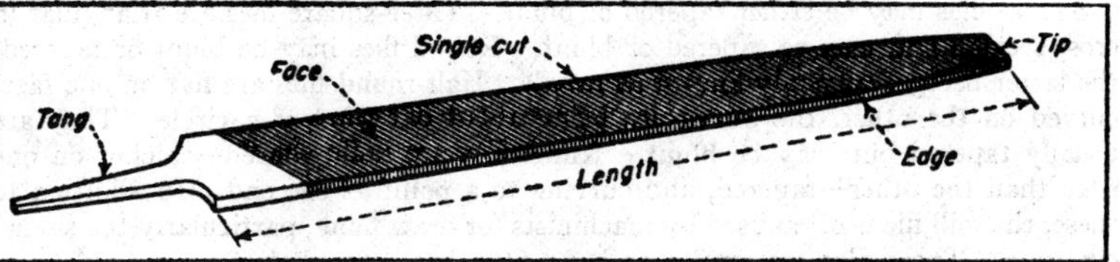

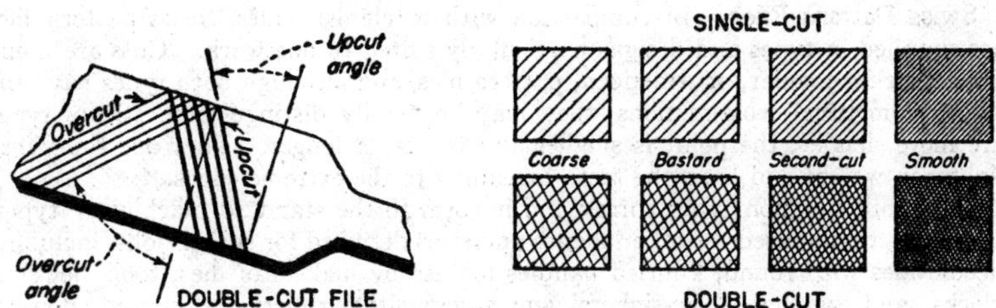

FIG. 1. Selection guide for machinists' files.

File Finder for Machinists' Files

CROSS-SECTION	NAME	CHARACTER OF TEETH	GENERAL USES
	Flat	Usually bastard. Also second-cut and smooth.	A general-purpose file
	Hand	One edge safe. Bastard, second-cut and smooth.	Finishing flat surfaces.
	Pillar	One edge safe. Bastard, second-cut and smooth.	Keyways, slots, narrow work.
	Warding	Usually bastard. Also second-cut and smooth.	Filing ward notches in keys. Narrow work.
	Square	Bastard, second-cut and smooth.	Enlarging holes or recesses. Mortises, keyways and splines.
	Three-Square	Sharp edges. Bastard, second-cut and smooth.	Filing acute angles, corners, grooves, notches.
	Round	Usually bastard. Also second-cut and smooth.	Enlarging holes; shaping curved surfaces.
	Half-Round	Usually bastard. Also second-cut and smooth.	Concave corners, crevices, round holes.
	Knife	Usually bastard. Also second-cut and smooth.	Cleaning out acute angles, corners, slots.

Square files may be either tapered or blunt. Three-square files are triangular in cross-section and may be tapered or blunt. Round files may be blunt or tapered, the latter being commonly known as rattail. Half-round files are flat on one face, curved on the other, the curvature being about one-third of a circle. They are usually tapered but may be blunt. Knife files are knife shaped—thicker on one edge than the other—tapered, and curving to a point at the end. In addition to these, the mill file is often used by machinists for draw filing, particularly the single-cut type. These files are similar to hand files, but somewhat slimmer, and taper slightly in width and thickness.

Swiss Pattern Files. In comparison with machinists' files, Swiss pattern files are supplied in types and designs particularly suited to fine work. Cuts are much finer than for regular, or American, pattern files, and although both types have the same geometrical cross-sections, they may be readily distinguished. Swiss types are more slender; the point is smaller and the taper longer. They are somewhat lighter in weight and have the teeth extending to the extreme edges.

The more common types correspond in form to the standard machinists' types, but there are hundreds of other shapes and sizes designed for special jobs, including needle files with round, knurled handles for use by makers of dies, tools, jewelry, clocks, and watches. Die-sinkers' and silversmiths' rifflers are curved, double-ended files curved upwards at the ends, and are used for getting into the corners, crevices, and holes of intricate dies and for other delicate work.

Special-purpose Files. In addition to general-purpose files such as machinists' files, there are a number of so-called special-purpose files. These are specially designed for a particular kind of work such as lathe filing, or for a particular material such as aluminum, stainless steel, or plastics.

The brass file has a tooth design that incorporates sturdiness and sharpness. It is nonclogging, cuts rapidly, yet gives a smooth finish without chattering or scratching, and without grooving or running off the work.

The aluminum file has a special tooth design, with a fine light overcut, and a coarse, deep upcut, so it will not clog, chatter, or channel under average working conditions. It cuts rapidly, even with moderate pressure. At the same time it leaves a good, smooth finish.

The lead float file has a short angle and coarse, single-cut, ridgelike teeth that shear metal in thin shavings.

The foundry file is intended to file rough iron castings, is equipped with sturdy, strong teeth that will stand the punishment of core sand, sharp edges, fins, sprues, hard projections, and shock without shelling or breaking out.

The special-purpose, long-angle lathe file has teeth cut at a much longer angle than the teeth of a standard mill file. This makes possible a shearing, clean cutting action without dragging, tearing, or chattering. Teeth are on the flat sides only to permit the file to be used in proximity to the chuck or any shoulder on the work.

Filing Various Materials

Cast Iron, Soft Steel. These are the metals on which files are most frequently used. For removing stock rapidly, select a flat bastard. If the work is of thin gage, a second-cut or even a smooth-cut file should be employed. For smooth finishing, choose a second-cut or smooth-cut. On thin or hard metals, use a smooth-cut only.

Copper, Brass, Bronze. On these metals a sharper file is required than for steel and wrought iron. This means that, if the same file is to be applied on a piece of

File Finder for Swiss Pattern Files

CROSS-SECTION	NAME	CHARACTER OF TEETH	GENERAL USES
	Hand	Double-cut on two flat faces and one edge. Other edge safe or uncut.	Flat surfaces.
	Pillar	Double-cut on two flat faces. Both edges safe.	Flat surfaces, slots.
	Warding	Double-cut on two flat faces. Single-cut on two edges.	Slots
	Square	Double-cut.	Corners, holes.
	Three-Square	Double-cut on three faces. Single-cut on edges.	Corners, holes.
	Round	Double-cut.	Corners, holes.
	Half-Round	Double-cut.	Corners, holes.
	Knife	Double-cut on flat faces. Single-cut on edges.	Slots
	Crossing	Double-cut.	Corners, holes.
	Equalling	Double-cut on flat faces. Single-cut on edges.	Slots, corners.
	Barrette	Cut only on wide flat face. Other faces safe.	Córners, flat surfaces, burring gear teeth.
	Crochet	Double-cut.	Slots, flat surfaces, rounded corners.
	Cant	Double-cut on three faces. Single-cut on two sharp edges.	Corners
	Slitting	Double-cut on four faces. Single-cut on two sharp edges.	Slots, corners.
	Pippin	Double-cut.	Rounded corners, holes.

brass and a piece of iron, use it on the brass first. Then, after the file has become too dull for further work on brass, it can be used on cast iron, wrought iron, or steel.

Aluminum. The special-purpose aluminum file with its coarse, deep upcut tooth construction will be found satisfactory for removing stock rapidly and leaving a smooth finish. The special tooth and open gullet design prevents clogging and chattering.

Lead, Babbitt, Other Soft Metals. These metals are satisfactorily filed with the lead float file. Since metals such as lead and babbitt are much softer than copper

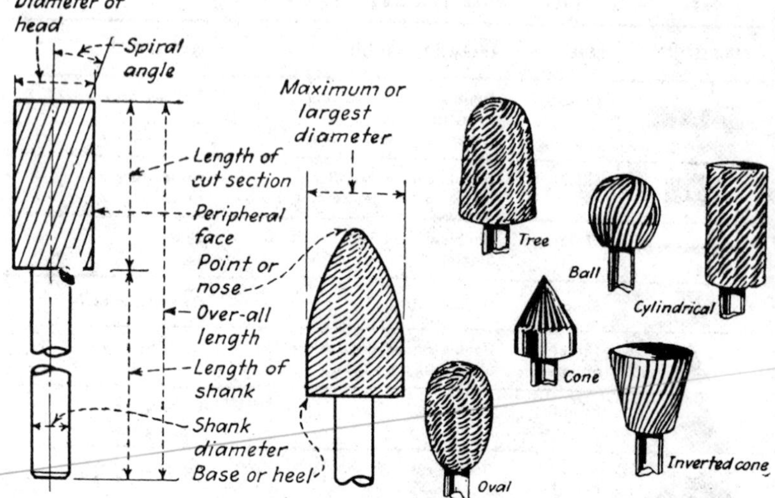

FIG. 2. Designation of style and various parts of rotary files and burrs will avoid confusion. See Simplified Practice Recommendation R233-48, issued by the National Bureau of Standards, Washington, to avoid abnormal stock requirements.

and aluminum, the lead float file has a very sharp, open-tooth construction that cuts rapidly, yet finishes smoothly without clogging.

Castings, Forgings. The scale customarily encountered on this work is very hard. It will dull the teeth of a new file after merely a few strokes. For this reason, a new file should never be employed on rough castings or forgings. Use only files that have become too dull for efficient cutting on other metals.

Filing Various Surfaces

Sheet-metal Work. Several file teeth should always be in contact with the metal being cut. If the cut of the file is so coarse that the metal gets between the teeth, there is a good chance of the teeth being broken. A finer cut file should be used for cutting thin stock—a coarser cut file for relatively heavier stock.

Narrow Surfaces. A file bites more freely on a narrow surface than on a wide one, since fewer teeth are in contact with the surface at any point in the stroke. Also, less pressure is required to make the file bite. It is generally better, on very narrow work, to use a file of fine cut and long tooth ridge. Such a file gives greater strength, and the shear of its cut is smoother—one tooth commencing to take hold as another leaves off. When an ordinary double-cut file is used, it is likely to wear excessively or break.

Broad Surfaces. A broad surface requires a sharper file than a narrow one. If the same file is to be used on both a broad surface and a narrow surface, use it on the broad surface first. For this work, a file with pointed teeth has the advantage.

Flat Surfaces. For filing a flat surface, a square file will be found to require less pressure than a flat or a pillar file. It is under better control and responds more effectively to the touch and feel of the operator.

Concave Surfaces. For rough filing, an ordinary half-round file will be found satisfactory. But a square file is to be preferred, since it requires less effort to shape the surface. After the piece has been brought to the desired dimensions by cross-filing, it can be finished by draw-filing with a half-round or crossing file that has a radius of curvature less than that of the concave surface.

Band Filing

Continuous filing with short file sections fastened to a steel band has largely replaced the reciprocating filing machine to save the time lost on the return stroke. File sections are mounted on a steel band and are readily removable when worn out.

Bastard-cut files with a short angle are generally used. They are listed as extra coarse—12 teeth per in.; coarse—14 teeth; medium coarse—16 teeth; medium—20 teeth; and fine—24 teeth. Files with 16 teeth per in. are used on mild steel and 20-tooth files on tool steel and the harder grades of materials.

Special files with 10 teeth per in. are used on cast iron and on nonferrous metals and cut very fast. The shapes include flats, half round, oval, and specials. Widths vary from ¼ to ½ in. as a rule. Filing speeds rarely exceed 125 fpm on soft metals and from 50 to 60 fpm on steel.

TABLE 1. NUMBER OF TEETH IN ROTARY FILES AND BURRS

Rotary Files, Teeth per inch			Dia. of hd. in.	Burs, Teeth per inch		
Rough cut	Std. cut	Fine cut		Coarse cut	Std. cut	Fine cut
14	20	30	⅛	25	32	36
14	20	30	3/16	20	25	36
14	20	30	¼	16	25	32
12	18	27	5/16	16	20	32
12	18	27	⅜	13	20	25
12	18	27	7/16	13	20	25
12	18	27	½	11	16	25
12	18	27	9/16	11	16	25
12	18	27	⅝	11	16	25
12	18	27	¾	9	13	20
12	18	27	⅞	9	13	20
10	15	24	1	8	13	20
10	15	24	1⅛	8	11	20
10	15	24	1¼	8	11	20
10	15	24	1½			

ROTARY FILES AND BURRS

Rotary power filing is a fast method of abrading or smoothing metal and other materials. Both rotary files and burrs are widely used for chamfering corners, forming fillets, removing fins and burrs, and elongating holes and slots. Both types of tool are available in a wide variety of shapes and sizes.

Tool Definitions. A rotary file is a tool with teeth raised by hammer and chisel. A rotary burr is a tool with the teeth ground from a hardened blank or milled from a soft blank and hardened.

Materials. Hand-cut files and milled burrs are made of high-speed steel; ground-from-the-solid burrs of high-speed steel, carbides, and cast alloys.

Shank Diameters. High-speed-steel tools to 1-in. head size, ¼ in. shank; over 1-in. head, ⅜-in. shank. Threaded shanks—NF thread.

Cut of Teeth. Right-hand helix, right-hand cut.

Number of Teeth. Files—teeth are counted in each row parallel to the axis; burrs—counted at largest tool diameter and at right angles to helix.

Selection of Files and Burrs. Because hand-cut teeth are broken up, they seem to dissipate heat created by friction and are therefore recommended for tough die steels, forgings, welds, and scaly material. Ground burrs are more efficient than rotary files on nonferrous materials, because they free chips more readily.

Cutting Speeds. There is no agreement in the file and burr industry on cutting speeds for these tools. For this reason, cutting speeds are given from two sources.

Use of Rotary Files and Burrs

Proper speed is usually a compromise between the finish required and the amount of material to be removed. Higher speeds give better finish but remove less material.

Effective speed at a constant rpm can be varied by changing the tool diameter or the number of teeth. Similar finish is achieved by using fine-cut tools at the low side of the speed range and coarse-cut tools at the high side. More material can be removed with coarse-cut tools.

File-cut tools should be operated toward the high side of the speed range and, in general, are recommended for iron and steel.

Ground-tooth tools, or burrs, are recommended for nonferrous material but can be used on iron and steel. They are more economical from a resharpening standpoint.

Carbide tools are recommended for high-production jobs on any material.

Table 2. Rpm for Standard-cut Rotary Files and Burrs*

	High-speed-steel Tools					Carbide
Dia	Mild Steel	Cast Iron	Bronze	Aluminum	Magnesium	All Materials
1/8	4,600	7,000	15,000	20,000	30,000	45,000
1/4	3,450	5,250	11,250	15,000	22,500	30,000
3/8	2,750	4,200	9,000	12,000	18,000	24,000
1/2	2,300	3,500	7,500	10,000	15,000	20,000
5/8	2,000	3,100	6,650	8,900	13,350	18,000
3/4	1,900	2,900	6,200	8,300	12,400	16,000
7/8	1,700	2,600	5,600	7,500	11,250	14,500
1	1,600	2,400	5,150	6,850	10,300	13,000
1 1/8	1,500	2,300	4,850	6,500	9,750	
1 1/4	1,400	2,100	4,500	6,000	9,000	

* Nicholson File Co.

Table 3. Sfpm for Rotary Files and Burrs*

Material	High-speed-steel Files and Burrs, Sfpm	Tungsten Carbide Burrs, Sfpm
Cast iron	200–500	1500–4000
Steel—soft or annealed	200–400	1500–4000
Steel—hardened	Not recommended	2000–5000
Copper, bronze, brass	500–1500	1500–4000
Aluminum	500–2000	1000–5000
Magnesium	1000–3000	1000–5000

* M. A. Ford Mfg. Co., Inc.

Section 4

GEARS, SPLINES, AND SERRATIONS

SECTION 4

GEARS, SPLINES, AND SERRATIONS

GEARS

General Designations[1]

Gears are machine elements that transmit motion by means of successively engaging teeth. Of two gears that run together, the one with the larger number of teeth is called the gear; the one with the smaller number of teeth is called the pinion.

A **rack** is a gear with teeth spaced along a straight line. It may be thought of as a gear of infinite diameter.

A **worm** is a gear with one or more teeth in the form of screw threads. Unless otherwise specified, it is cylindrical in form and has helical threads.

Kinds of Gears (See Figs. 1 and 2)

Spur gears are cylindrical in form and operate on parallel axes. The teeth are straight and parallel to the axis.

A **spur rack** has straight teeth that are at right angles to the direction of motion.

A **helical gear** is cylindrical in form and has helical teeth. (The term "spiral gear" is not recommended.)

Parallel helical gears operate on parallel axes.

Crossed helical gears operate on crossed axes and may have teeth of the same or of opposite hand.

Single-helical gears have teeth of only one hand on each gear.

Double-helical (herringbone) gears each have both right-hand and left-hand helical teeth, and operate on parallel axes.

A **helical rack** has straight teeth that are oblique to the direction of motion.

Worm gears include worms and their mating gears. The axes are usually at right angles.

A **worm gear** is the mate to a worm. A worm gear that is completely conjugate to its worm has line contact, and usually is cut by a counterpart of the worm. Some forms of hourglass worms and gears are called double-enveloping. A spur gear or helical gear used with a cylindrical worm has only point contact.

An **hourglass worm** has one or more threads increasing in diameter from its middle portion toward both ends, conforming to the curvature of the gear.

Bevel gears are conical in form and operate on intersecting axes which are usually at right angles.

Miter gears are a pair of bevel gears with equal numbers of teeth and with axes at right angles.

Angular bevel gears are bevel gears in which the axes are not at right angles.

A **crown gear** is a bevel gear in the form of a disk, having a plane pitch surface. The crown gear corresponds in bevel gears to the rack in spur gears.

Straight bevel gears have straight teeth which, if extended, would meet their respective axes.

Spiral bevel gears have teeth that are curved and oblique.

Zerol bevel gears have teeth that are curved but in the same general direction as straight teeth. They are spiral bevel gears of zero spiral angle.

[1] American Gear Manufacturers Association Tentative Standard 112.01.

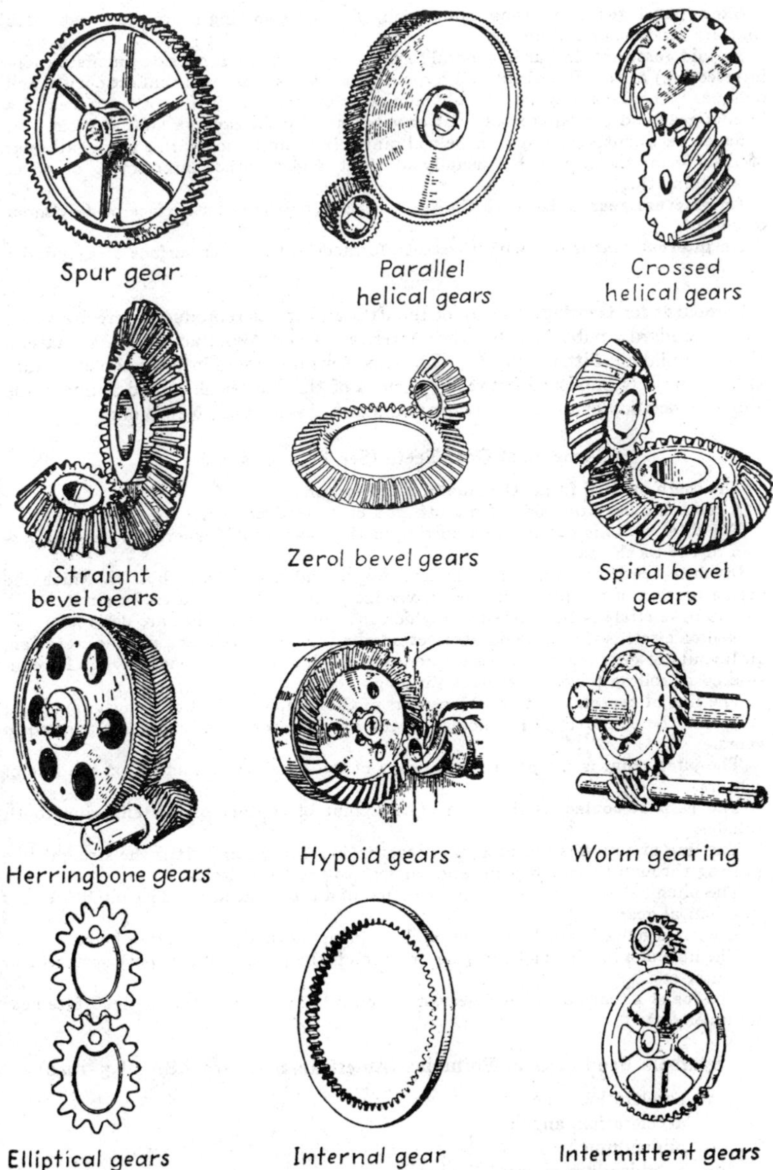

FIG. 1. Types of gears. Most of these have broad application.

Skew bevel gears are those for which the corresponding crown gear has teeth that are straight and oblique.

Hypoid gears are similar in general form to bevel gears, but operate on offset (nonintersecting) axes. Practically all hypoid gears have spiral teeth that are curved and oblique. The axes may be at right angles or otherwise. The tooth surfaces of a hypoid gear and pinion are both cut or generated by the same or similar tools.

Face gears consist of a spur or helical pinion in combination with a conjugate gear of disk form, the axes being usually at right angles, either intersecting or offset (nonintersecting).

An **external gear** is one with the teeth formed on the outer surface of a cylinder or cone.

An **internal gear** is one with the teeth formed on the inner surface of a cylinder or cone.

Standards for Gearing. Many of the data and tables reproduced here are taken from standards published by the American Gear Manufacturers Association, Empire Building, Pittsburgh, Pa. Those persons interested in the complete standards or subjects published between revisions of this handbook should apply to the above association or to the American Standards Association, New York.

Elements of Gear Teeth (See Figs. 3*a* and *b*)

The **tooth surface** forms the side of a gear tooth.

A **tooth profile** is one side of a tooth in a cross-section.

The **fillet curve** is the concave portion of the tooth profile where it joins the bottom of the tooth space.

Involute teeth of spur gears, helical gears, and worms are those in which the active portion of the profile in the transverse plane is the involute of a circle.

The **base circle** is the circle from which involute tooth profiles are derived.

A **pitch circle** is the curve of intersection of a pitch surface and a plane of rotation. In layouts of gear teeth in cross-section, the pitch circles are imaginary circles considered as rolling together without slipping.

The **line of centers** connects the centers of the pitch circles of two engaging gears; it is also the common perpendicular of the axes in crossed helical gears and worm gears.

The **pitch point** is the point of tangency of two pitch circles and is on the line of centers.

The **path of contact** is the path of the point of contact of two engaging tooth profiles.

The **line of action** is the path of contact in involute gears. It is the straight line passing through the pitch point and tangent to the base circles.

The **pitch helix** is the curve of intersection of a tooth surface and its pitch cylinder in a helical gear.

A **normal helix** is a helix on the pitch cylinder, normal to the pitch helix.

The **heel** of a tooth on a bevel gear or pinion is the portion of the tooth surface near its outer end.

The **toe** of a tooth on a bevel gear or pinion is the portion of the tooth surface near its inner end.

Symbols Used in Gear Formulas (American Standard ASA B6.5-1949)

a acceleration, linear

α acceleration, angular

a addendum

- a_G addendum of gear
- a_P addendum of pinion
- a_c chordal addendum (for tooth calipers)

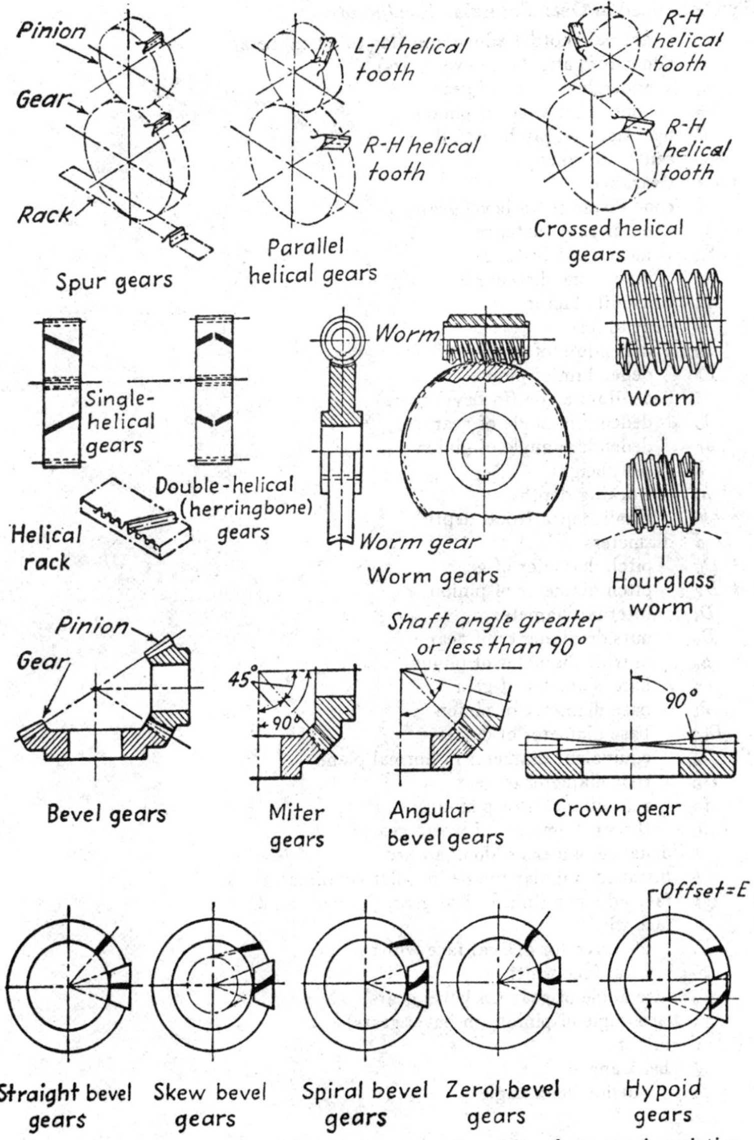

FIG. 2. Types of gears as defined by the American Gear Manufacturers Association.

Symbols Used in Gear Formulas (*Continued*)

- a_{nc} normal chordal addendum (for tooth calipers)
- α addendum angle (in bevel gears)
  - α_G addendum angle of gear
  - α_P addendum angle of pinion
- B backlash (on pitch circle)[1]
- C center distance
- c clearance
- A cone distance (in bevel gears)
  - A_o outer cone distance
  - A_m mean cone distance
  - A_i inner cone distance
- K constant (factor)
- b dedendum
  - b_G dedendum of gear
  - b_P dedendum of pinion
- δ dedendum angle (in bevel gears)
  - δ_G dedendum angle of gear
  - δ_P dedendum angle of pinion
- h depth; height
  - h_k working depth
  - h_t whole depth (total depth)
- D, d diameters
  - D, D_G pitch diameter of gear
  - d, D_P pitch diameter of pinion
  - D_i internal diameter
  - D_o outside diameter of gear
  - d_o outside diameter of pinion
  - D_b base diameter of gear
  - d_b base diameter of pinion
  - D_{bc} base diameter of cutter
  - D_n equivalent diameter in normal plane
  - D_R root diameter of gear
  - d_R root diameter of pinion
  - D_t throat diameter (of worm gear)
- s distance, linear or along an arc
- θ distance, angular (angle in polar coordinates)
- Q_F face advance (in a helical gear, $Q_F = F \tan \psi$)
- F face width
  - F_e effective (or active) face width
  - F_t total face width
- Γ_o face angle of gear (in bevel gears)
- γ_o face angle of pinion (in bevel gears)
- K factor
- ψ helix angle[2]
  - ψ_o outside helix angle

[1] Backlash measured at any given place or in any given direction should be converted to a value on the pitch circle for comparison with a general standard.

[2] In gearing, the helix angle is the acute angle between a tooth and the axis of the gear.

ψ_b base helix angle
ψ_p pitch helix angle
l lead (of worm or helical gear)
λ lead angle ($\lambda = 90° - \psi$, $\tan \lambda = l/\pi D$)
L, l lengths
l length of line of contact
L total length of lines of contact
Z length of line of action
Z_a approach portion of line of action
Z_r recess portion of line of action
W load (total)
W_t transverse, or tangential, component of load
W_n load component normal to helix (in pitch plane)
W_N load normal to surface
W_r radial component of load
w load per unit length[3]
M moment, bending
m module ($m = D/N$), inches or millimeters
N number (of teeth)
N_G number of teeth in gear
N_P number of teeth in pinion
N_C number of teeth in cutter
N_c number of teeth in crown gear
N_W number of threads in worm
E offset (of axes, in hypoid gears and face gears)
p pitch, circular
p_b base pitch (normal to involute)
p, p_t transverse pitch[4]
p_n normal pitch (normal to helix)[4]
p_x axial pitch (preferred to linear pitch)[4]
p_N normal base pitch (normal to surface)
p_{xW} axial pitch of worm
p_X axial base pitch
P, P_d pitch, diametral[5]
P_{nd} normal diametral pitch[5]
Γ pitch angle of gear (in bevel gears)
γ pitch angle of pinion (in bevel gears)
P power[6]
ϕ pressure angle
ϕ, ϕ_t transverse pressure angle[7]
ϕ, ϕ_n normal pressure angle[7]
ϕ_x axial pressure angle[7]
ϕ_C pressure angle of cutter
ϕ_{xW} axial pressure angle of worm

[3] Unit loads frequently referred to in gears are pounds per inch of face width and pounds per inch of line of contact.
[4] Pitch dimensions in the three principal directions.
[5] Use the forms P_d and P_{nd} when necessary to avoid confusion when P must be used for "power."
[6] Note that power is a correct general term and that horsepower is properly a unit of measurement.
[7] Pressure angles in the three principal directions.

Symbols Used in Gear Formulas (*Continued*)

R, r radii

- R, R_G pitch radius of gear
- r, R_P pitch radius of pinion
- R_O outside radius of gear
- r_O outside radius of pinion
- R_C pitch radius of cutter
- r_T edge radius of tool

ρ radius of curvature

- ρ_G profile radius of curvature in gear
- ρ_P profile radius of curvature in pinion
- ρ_n radius of curvature in normal plane
- ρ_o relative radius of curvature ($1/\rho_o = 1/\rho_1 + 1/\rho_2$)
- ρ_f radius of curvature of fillet

m_G gear ratio ($m_G = N_G/N_P$)

m_ω angular velocity ratio

m_p contact ratio (of profiles) ($m_p = Z/p_b$)[8]

m_F face contact ratio (in axial plane, $m_F = F/p_x$)[8]

m_t total contact ratio ($m_t = m_p + m_F$)[8]

n revolutions per unit of time

Γ_R root angle of gear (in bevel gears)

γ_R root angle of pinion (in bevel gears)

Σ shaft angle

ψ spiral angle (in bevel gears)[9]

- ψ_G spiral angle of gear
- ψ_P spiral angle of pinion
- ψ_o spiral angle at outer cone distance
- ψ_m spiral angle at mean cone distance
- ψ_i spiral angle at inner cone distance

X strength factor ($X = t^2/4h$)[10]

s stress

K stress concentration factor

t tooth thickness

- t_G circular thickness of gear
- t_P circular thickness of pinion
- t_n normal circular thickness
- t_b tooth thickness on base circle
- t_c chordal thickness
- t_{nc} normal chordal thickness
- t_x axial thickness

y tooth-form factor for circular pitch

Y tooth-form factor for diametral pitch

Y_K tooth-form factor including stress concentration

T torque

v velocity, linear

- v_t velocity in transverse direction tangential to pitch circle[11]

[8] Contact ratio is the preferred term for what has also been called "average number of teeth in contact."

[9] Spiral angle in bevel gears is understood to be given at the mean cone distance unless otherwise stated.

[10] This strength factor is usually obtained in the graphical determination of tooth strength.

[11] Linear velocities in the three principal directions. Compare with footnotes 4 and 7, p. 4-7.

v_n velocity in normal direction (in pitch plane)[11]
v_x velocity in axial direction[11]
v_N velocity normal to surface
ω velocity, angular
ω_G angular velocity of gear
ω_P angular velocity of pinion

Abbreviations for Terms in Gearing

Symbols for mathematical equations and formulas are to be distinguished from abbreviations which are shortened forms of words suitable for use in correspondence, records, and tables, and in some cases on gear drawings. The use of abbreviations is advocated only when there can be no possible misunderstanding as to their meanings.

When abbreviations for terms in gearing are required, the following are recommended:

AP axial or linear pitch
APA axial pressure angle
BD base diameter
BHA base helix angle
BP base pitch
CD center distance
CP circular pitch
DP diametral pitch
FW face width
HA helix angle
ID internal diameter; inside diameter
L lead
LA lead angle
LH left-hand
NCP normal circular pitch
NDP normal diametral pitch
NPA normal pressure angle
OD outside diameter
OHA outside helix angle
PA pressure angle
PD pitch diameter
RH right-hand
T number of teeth or threads

Dimensions of Gear Teeth

Linear, Circular, and Angular (See Figs. 3*a* and *b*)

Symbols for Dimensions. Symbols for these dimensions conform with AGMA Standard 111.01 (ASA B6.5-1943).

Center distance C is the distance between the parallel axes of spur gears and parallel helical gears, or the crossed axes of crossed helical gears and worm gears. Also, it is the distance between the centers of the pitch circles.

Offset E is the perpendicular distance between the axes of hypoid gears and offset face gears.

Pitch p is the distance between similar equally spaced tooth surfaces, in a given direction and along a given line or curve. The single word "pitch" without qualification has been used to designate circular pitch, axial pitch, and diametral pitch, but such confusing usage should be avoided.

Circular pitch p is the distance along the pitch circle or pitch line between corresponding profiles of adjacent teeth.

Transverse circular pitch p_t is the circular pitch in the transverse plane.

Normal circular pitch p_n is the circular pitch in the normal plane, and also the length of the arc along the normal helix between helical teeth or threads.

Axial pitch p_x is the circular pitch in the axial plane and in the pitch surface between corresponding sides of adjacent teeth, in helical gears and worms. The term axial pitch is preferred to the term linear pitch.

Addendum a is the height by which a tooth projects beyond the pitch circle; also, the radial distance between the pitch circle and the addendum circle.

Dedendum b is the depth of a tooth space below the pitch circle; also, the radial distance between the pitch circle and the root circle.

Clearance c is the amount by which the dedendum in a given gear exceeds the addendum of its mating gear.

Working depth h_k is the depth of engagement of two gears, that is, the sum of their addendums.

Whole depth h_t is the total depth of a tooth space, equal to addendum plus dedendum, also equal to working depth plus clearance.

Pitch diameter D, d is the diameter of the pitch circle. Unless otherwise specified it is *standard pitch diameter* obtained by dividing the number of teeth by the diametrical pitch or by multiplying the circular pitch by the number of teeth, and dividing by π. In parallel-shaft gears, the pitch diameters can be determined directly from the numbers of teeth and the center distance. *Operating pitch diameter* is the pitch diameter at which the gear operates. *Generating pitch diameter* is the pitch diameter at which a gear is generated. In bevel gears, the pitch diameter is understood to be at the outer ends of the teeth, unless otherwise specified.

Outside diameter D_o, d_o is the diameter of the addendum (outside) circle. In a bevel gear it is the diameter at the crown circle. In a throated worm gear it is the maximum diameter of the blank.

Root diameter D_R, d_R is the diameter of the root circle.

Internal diameter D_i is the diameter of the addendum circle of an internal gear.

Throat diameter D_t is the diameter of the addendum circle at the middle of a worm gear or a concave worm.

Pitch radius R, r is the radius of the pitch circle.

Circular thickness t_G, t_P is the length of arc between the two sides of a gear tooth on the pitch circle unless otherwise specified.

Transverse circular thickness t_t is the circular thickness in the transverse plane.

Normal circular thickness t_n is the circular thickness in the normal plane. In helical gears, it is an arc of the normal helix.

Axial thickness t_x in helical gears and worms is the tooth thickness in the axial cross-section at the pitch line.

Chordal thickness t_c is the length of the chord subtending a circular-thickness arc.

Normal chordal thickness t_{nc} is the chordal thickness in the plane normal to the pitch helix or the tooth spiral at the center of the tooth.

Chordal addendum a_c is the height from the top of the tooth to the chord subtending the circular-thickness arc.

Normal chordal addendum a_{nc} is the chordal addendum in the plane normal to the pitch helix or the tooth spiral at the center of the tooth.

Backlash B is the amount by which the width of a tooth space exceeds the thickness of the engaging tooth on the pitch circles. As actually indicated by measuring devices, backlash may be determined variously in the transverse, normal, or axial planes, and either in the direction of the pitch circles, or on the line of action. Such measurements should be corrected to corresponding values on transverse pitch circles for general comparisons.

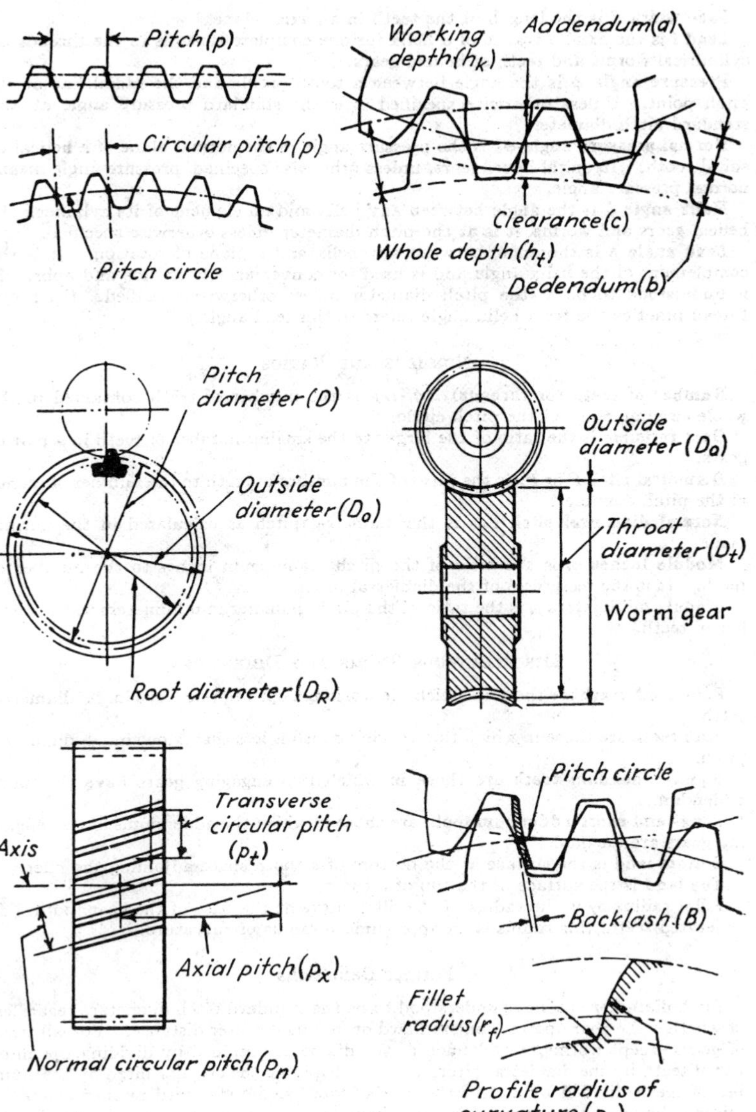

FIG. 3a. Graphical representation of gear dimensions.

Face width F is the length of the teeth in an axial plane.

Lead l is the axial advance of a helix for one complete turn, as in the threads of cylindrical worms and teeth of helical gears.

Pressure angle ϕ is the angle between a tooth profile and the radial line at its pitch point. Unless otherwise specified it is the standard pressure angle at the standard pitch diameter.

Normal pressure angle ϕ_n is the pressure angle in the normal plane of a helical or spiral tooth. In spiral bevel gears, unless otherwise specified, pressure angle means normal pressure angle.

Helix angle ψ is the angle between any helix and an element of its cylinder. In helical gears and worms, it is at the pitch diameter unless otherwise specified.

Lead angle λ is the angle between any helix and a plane of rotation. It is the complement of the helix angle and is used for convenience in worms and hobs. It is understood to be at the pitch diameter unless otherwise specified. (In screw thread practice the term helix angle refers to this lead angle.)

Numbers and Ratios

Number of teeth (or threads) N_G, N_P is the number of teeth contained in the whole circumference of the pitch circle.

Gear ratio m_G is the ratio of the larger to the smaller number of teeth in a pair of gears.

Diametral pitch P or P_d is the ratio of the number of teeth to the number of inches in the pitch diameter.

Normal diametral pitch P_{nd} is the diametral pitch as calculated in the normal plane.

Module inches m is the ratio of the pitch diameter in inches to the number of teeth. It is the reciprocal of the diametral pitch.

Module millimeters $_m$ is the ratio of the pitch diameter in millimeters to the number of teeth.

Miscellaneous Terms and Dimensions

Full-depth teeth are those in which the working depth equals 2.000 in. ÷ diametral pitch.

Stub teeth are those in which the working depth is less than 2.000 in. ÷ diametral pitch.

Equal-addendum teeth are those in which two engaging gears have the same addendums.

Long- and short-addendum teeth are those in which the addendums of two engaging gears are unequal.

Bottom land is the surface at the bottom of a tooth space adjoining the fillet.

Top land is the surface of the top of a tooth.

Fillet radius ρ_f is the radius of the fillet curve at the base of the gear tooth. In generated teeth, this radius is an approximate radius of curvature.

Further Definitions

Pitch diameter is always understood to be the standard pitch diameter, regardless of whether the gear operates on standard or modified center distance. For all types of gears except worms, it is defined as the diameter obtained by dividing the number of teeth by the diametral pitch, or by multiplying the circular pitch by the number of teeth divided by π. In the case of worms, the standard pitch diameter is midway between the standard outside diameter and the working-depth cylinder.

Addendum is the distance between the pitch diameter and outside diameter, regardless of whether the outside diameter is enlarged, decreased, or topped.

Diameter enlargement or decrease is the amount by which a standard spur, helical, or worm gear is modified to avoid undercut or attain a nonstandard center distance.

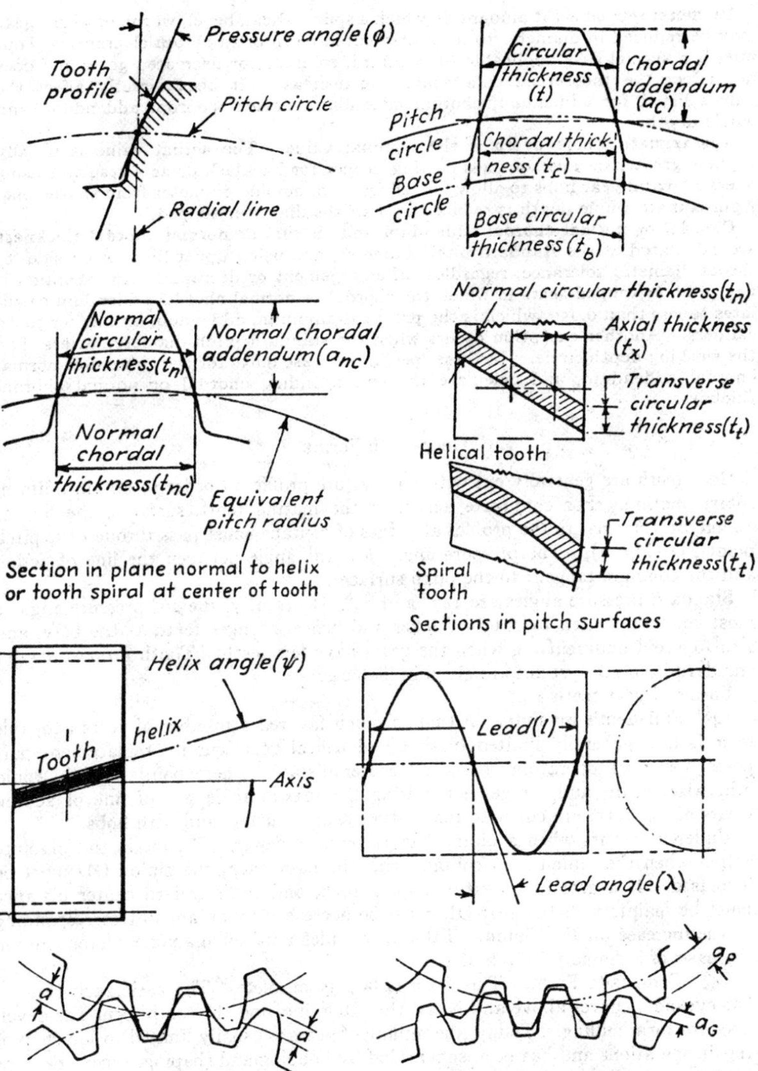

FIG. 3b. Graphical representation of gear dimensions (*continued*).

Diameter topped is the amount by which a spur, helical, bevel, worm, or worm gear may be reduced in diameter to prevent pointed teeth or to obtain clearance. Topping is a special case applicable to standard, enlarged, or decreased gears and does not affect the theoretical enlargement or decrease. It does directly affect the values listed for addendum, chordal addendum, normal chordal addendum, and whole depth.

Approximate whole depth is the nominal value. The actual value is usually slightly greater because of tools cutting deeper for backlash or as a result of using oversize worm-gear hobs to allow for reduction of outside diameter from sharpening. Approximate whole depth is reduced by half the diameter topped.

Chordal or **normal chordal addendum** and **chordal** or **normal chordal thickness** are calculated at the standard pitch diameter, and using upper limit of outside- or throat-diameter tolerance, regardless of enlargement or decrease. An exception is fine-pitch or topped gears in which the chordal or normal chordal addendum calculates to less than 0.050, which is the practical minimum addendum setting for tooth calipers. Another exception occurs when the standard pitch diameter is less than the working-depth circle. In these special cases use 0.050 for the chordal or normal chordal addendum, and calculate the corresponding chordal or normal chordal thickness.

Gear-tooth Forms

Gear teeth are generally cut with an involute profile, in order to obtain uniform rotary motion. For conjugate action of the mating tooth surfaces, the line of action, or a normal to the profiles at points of contact, must pass through the pitch points, O (Fig. 4). The pressure angle ϕ is the angle between the line of action and the common tangent to the pitch surfaces.

Standard pressure angles are $14\frac{1}{2}°$ and 20°. Currently, the 20° pressure angle is most commonly used to obtain thicker and hence stronger teeth at the base, and also to avoid undercutting when the gears have few teeth. Tooth proportions for standard involute systems are given in Table 1.

Forms of gear teeth are:

$14\frac{1}{2}°$ Full-depth Involute. Although much favored a number of years ago, this form is now generally limited to spur and helical gears cut for replacement purposes, for gears operating on spread center distance where pointed teeth would otherwise result, and for gears requiring the maximum length of line of action. Gears of this form are cut with pinion-type shaper cutters and with hobs.

Undercut occurs when a pinion has 31 teeth or fewer. To obtain full involute action when the pinion is in contact with the basic rack, the pinion OD must be increased. If a gear is substituted for a rack, and the standard center distance must be maintained, the gear OD must be decreased by an amount corresponding to the increase on the pinion. Table 8 provides data for diameter increments on the basis of a diametral pitch of 1.

$14\frac{1}{2}°$ Composite Form. This form is largely obsolete. The basic rack (Fig. 5) has cycloidal curves above and below the pitch line, and the tooth form was developed for form milling. Today, the milling of gears is usually limited to job-shop or repair operations and has been superseded by hobbing and shaping, except perhaps for gears of 2 diametral pitch or coarser.

20° Full-depth Involute. Modern gears for general power-transmission purposes are based on this system. The tooth form (Fig. 6) has the same parts as the $14\frac{1}{2}°$ involute but is thicker at the base and hence stronger. Undercutting does not occur until pinions have 17 teeth or fewer. Diameter corrections are given in Table 9. The diametral pitches range from $\frac{1}{2}$ to 18. See Table 2 for tooth parts.

20° Fine-pitch Involute. Fine-pitch gears range from 20 to 200 diametral pitch

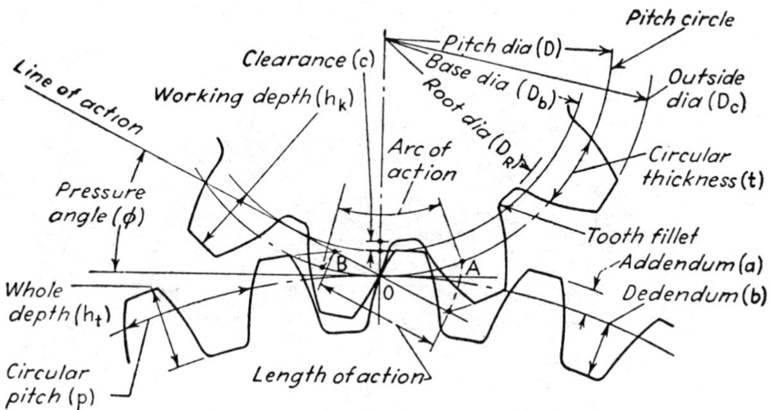

FIG. 4. Nomenclature for spur and helical gears.

TABLE 1. TOOTH PROPORTIONS FOR STANDARD INVOLUTE SYSTEMS

Item	14½° and 20° Full Depth; 14½° Composite	20° Fine Pitch	20° Stub
Addendum	1.000/P	1.000/P	0.8/P
Dedendum	1.157/P	1.200/P + 0.002	1/P
Working depth	2/P	2.000/P	1.6/P
Whole depth*	2.157/P	(2.200/P) + 0.002	1.8/P
Clearance*	0.157/P	(0.200/P) + 0.002	0.2/P
Tooth thickness on pitch dia	1.5708/P	1.5708/P	1.5708/P
Pitch dia	N/P	N/P; n/P	N/P
OD†	$(N + 2)/P$	$(N + 2)/P$, $(n + 2)/P$	$(N + 1.6)/P$
Center distance		$(N + n)/2P$	

N = number of teeth in gear.
n = number of teeth in pinion.
P = diametral pitch.

* Clearance is a minimum for all forms except the 20° fine pitch.
† Pinions having 17 or fewer teeth should be enlarged.

and are applied for transmission of motion more than for power. While there is no mathematical difference between fine- and coarse-pitch gears, fine-pitch gears require special consideration for the following reasons: greater clearances are required for greater wear on point of generating tool; consideration must be given to point widths of gear-cutting tools and top lands of gears; and fine-pitch gears require different production and inspection techniques, except in the coarser ranges. Tooth parts for 20° fine-pitch spur and helical gears are given in Table 3.

Pinions having 17 or fewer teeth are enlarged in accordance with Table 9, and pinions with fewer than ten teeth are not recommended.

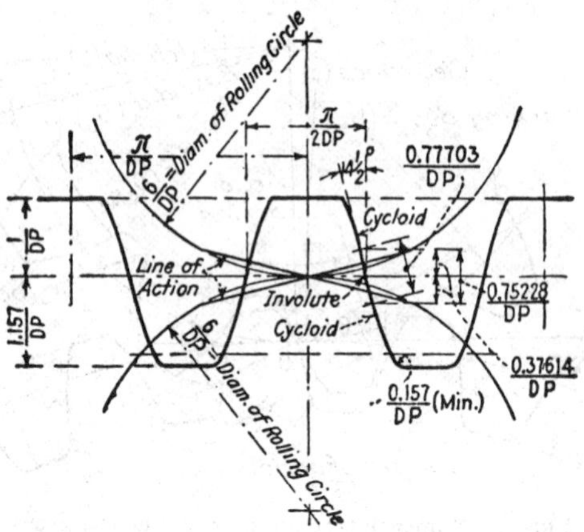

FIG. 5. Basic rack for 14½° composite system (full-depth tooth).

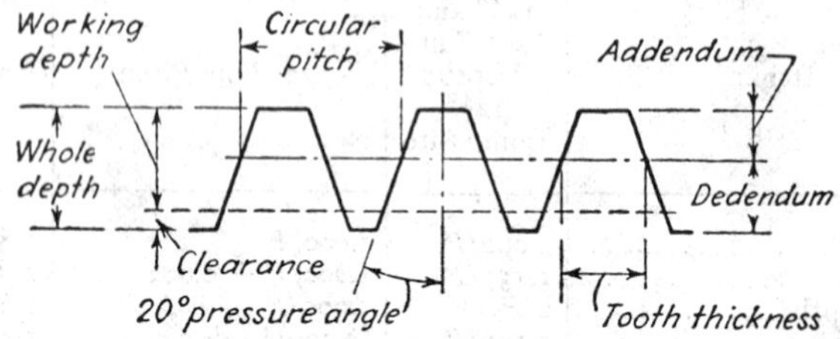

FIG. 6. Basic rack for 20° fine-pitch system. Terminology also applies to 14½° full depth, 20° full depth, and stub forms.

Nonstandard Tooth Forms

20° Fellows Stub Involute. In this system pitches are expressed as fractions like $\frac{10}{21}$. The numerator P_1 determines the pitch diameter from the formula N/P and the tooth thickness from the formula $\pi/2P$ as in other involute gearing systems. The denominator P_2 determines the addendum from the formula $1/P_2$, the dedendum from the formula $1.25/P_2$, and also the whole depth and clearance. Note: Whole depth and clearance of American Standard fine-pitch 20° involute gears and Fellows stub-tooth gears are the same for pitches from 20 to 54.

20° Fellows Full-depth Involute. Revisions in this system make it correspond closely with the American Standard 20° coarse and fine pitches. Pitch diameter, circular thickness, and addendum are identical for pitches from 1 to 200. For diametral pitches from 1 to 18, the dedendum = $1.25/P$, the whole depth = $2.25/P$, and the clearance = $0.25/P$, but for pitches from 20 to 200 these tooth parts are identical with those for the 20° fine-pitch system.

Module System. Gearing produced in the United States is commonly based on the diametral pitch, which equals the number of teeth divided by the pitch diameter. The module system is employed in metric countries, the module being

TABLE 2. TOOTH PARTS* FOR DIAMETRAL (OR NORMAL DIAMETRAL) PITCHES FROM 1 TO 18

Applies to spur, helical, straight bevel, and worm gearing. Parenthetical words may be omitted for spur and straight bevel gearing. Addendum and dedenum proportions apply only to full-depth teeth in a normal section, and to equal-addendum gearing. For straight bevel gearing, dedendum and whole-depth columns do not apply. Refer to bevel-gear formulas, page 4-67. All dimensions in inches.

(Normal) Diametral Pitch	(Normal) Circular Pitch	(Normal) Circular Thickness	Addendum	Dedendum, Approx	Whole Depth, Approx
18	0.174533	0.0873	0.0556	0.0643	0.1198
16	0.196350	0.0982	0.0625	0.0723	0.1348
14	0.224400	0.1122	0.0714	0.0826	0.1541
13	0.241661	0.1208	0.0769	0.0890	0.1659
12	0.261799	0.1309	0.0833	0.0964	0.1798
11	0.285599	0.1428	0.0909	0.1052	0.1961
10	0.314159	0.1571	0.1000	0.1157	0.2157
9	0.349066	0.1745	0.1111	0.1286	0.2397
8	0.392699	0.1963	0.1250	0.1446	0.2696
7	0.448799	0.2244	0.1429	0.1653	0.3081
6	0.523599	0.2618	0.1667	0.1928	0.3595
5	0.628319	0.3142	0.2000	0.2314	0.4314
4	0.785398	0.3927	0.2500	0.2893	0.5393
3.5	0.897598	0.4488	0.2857	0.3306	0.6163
3	1.047198	0.5236	0.3333	0.3857	0.7190
2.75	1.142397	0.5712	0.3636	0.4208	0.7844
2.5	1.256637	0.6283	0.4000	0.4628	0.8628
2.25	1.396264	0.6981	0.4444	0.5143	0.9587
2	1.570797	0.7854	0.5000	0.5785	1.0785
1.75	1.795196	0.8976	0.5714	0.6612	1.2326
1.5	2.094395	1.0472	0.6667	0.7714	1.4381
1.25	2.513274	1.2566	0.8000	0.9257	1.7257
1	3.141593	1.5708	1.0000	1.1571	2.1571
0.75	4.188791	2.0944	1.3333	1.5428	2.8761
0.5	6.283186	3.1416	2.0000	2.3142	4.3142

* Gould & Eberhardt, Inc.

the pitch diameter divided by the number of teeth. The *metric module* equals the pitch diameter in millimeters divided by the number of teeth, and the *English module* equals the pitch diameter in inches divided by the number of teeth. In the German DIN standard tooth form for spur and bevel gears the addendum equals the module and the diametral pitch equals π times the module.

Rack Teeth

Rack teeth are generally cut with a gang of circular gear cutters, or by the Fellows gear-shaping method, in which the rack is fed past the reciprocating cutter at the

TABLE 3. TOOTH PARTS FOR 20° FINE-PITCH SPUR AND HELICAL GEARS

Diametral Pitch	Circular Pitch	Standard Circular Thickness	Working Depth	Whole Depth	Clearance	Standard Addendum	Standard Dedendum
20	0.15708	0.07854	0.1000	0.1120	0.0120	0.0500	0.0620
21	0.14960	0.07480	0.0952	0.1068	0.0116	0.0476	0.0591
22	0.14280	0.07140	0.0909	0.1020	0.0111	0.0455	0.0565
23	0.13659	0.06830	0.0870	0.0977	0.0107	0.0435	0.0542
24	0.13090	0.06545	0.0833	0.0937	0.0104	0.0417	0.0520
25	0.12566	0.06283	0.0800	0.0900	0.0100	0.0400	0.0500
26	0.12083	0.06042	0.0769	0.0866	0.0097	0.0385	0.0482
27	0.11636	0.05818	0.0741	0.0835	0.0094	0.0370	0.0464
28	0.11220	0.05610	0.0714	0.0806	0.0092	0.0357	0.0449
29	0.10833	0.05417	0.0690	0.0779	0.0089	0.0345	0.0434
30	0.10472	0.05236	0.0667	0.0753	0.0087	0.0333	0.0420
31	0.10134	0.05067	0.0645	0.0730	0.0085	0.0323	0.0407
32	0.09818	0.04909	0.0625	0.0708	0.0083	0.0313	0.0395
33	0.09520	0.04760	0.0606	0.0687	0.0081	0.0303	0.0384
34	0.09240	0.04620	0.0588	0.0667	0.0079	0.0294	0.0373
35	0.08976	0.04488	0.0571	0.0649	0.0077	0.0286	0.0363
36	0.08727	0.04363	0.0556	0.0631	0.0076	0.0278	0.0353
37	0.08491	0.04245	0.0541	0.0615	0.0074	0.0270	0.0344
38	0.08267	0.04134	0.0526	0.0599	0.0073	0.0263	0.0336
39	0.08055	0.04028	0.0513	0.0584	0.0071	0.0256	0.0328
40	0.07854	0.03927	0.0500	0.0570	0.0070	0.0250	0.0320
41	0.07662	0.03831	0.0488	0.0557	0.0069	0.0244	0.0313
42	0.07480	0.03740	0.0476	0.0544	0.0068	0.0238	0.0306
43	0.07306	0.03653	0.0465	0.0532	0.0067	0.0233	0.0299
44	0.07140	0.03570	0.0455	0.0520	0.0065	0.0227	0.0293
45	0.06981	0.03491	0.0444	0.0509	0.0064	0.0222	0.0287
46	0.06830	0.03415	0.0435	0.0498	0.0063	0.0217	0.0281
47	0.06684	0.03342	0.0426	0.0488	0.0063	0.0213	0.0275
48	0.06545	0.03272	0.0417	0.0478	0.0062	0.0208	0.0270
49	0.06411	0.03206	0.0408	0.0469	0.0061	0.0204	0.0265
50	0.06283	0.03142	0.0400	0.0460	0.0060	0.0200	0.0260
51	0.06160	0.03080	0.0392	0.0451	0.0059	0.0196	0.0255
52	0.06042	0.03021	0.0385	0.0443	0.0058	0.0192	0.0251
53	0.05928	0.02964	0.0377	0.0435	0.0058	0.0189	0.0246
54	0.05818	0.02909	0.0370	0.0427	0.0057	0.0185	0.0242
55	0.05712	0.02856	0.0364	0.0420	0.0056	0.0182	0.0238
56	0.05610	0.02805	0.0357	0.0413	0.0056	0.0179	0.0234
57	0.05512	0.02756	0.0351	0.0406	0.0055	0.0175	0.0231
58	0.05417	0.02708	0.0345	0.0399	0.0054	0.0172	0.0227
59	0.05325	0.02662	0.0339	0.0393	0.0054	0.0169	0.0223

TABLE 3. TOOTH PARTS FOR 20° FINE-PITCH SPUR AND HELICAL GEARS (*Continued*)

Diametral Pitch	Circular Pitch	Standard Circular Thickness	Working Depth	Whole Depth	Clearance	Standard Addendum	Standard Dedendum
60	0.05236	0.02618	0.0333	0.0387	0.0053	0.0167	0.0220
61	0.05150	0.02575	0.0328	0.0381	0.0053	0.0164	0.0217
62	0.05067	0.02534	0.0323	0.0375	0.0052	0.0161	0.0214
63	0.04987	0.02493	0.0317	0.0369	0.0052	0.0159	0.0210
64	0.04909	0.02454	0.0312	0.0364	0.0051	0.0156	0.0208
65	0.04833	0.02417	0.0308	0.0358	0.0051	0.0154	0.0205
66	0.04760	0.02380	0.0303	0.0353	0.0050	0.0152	0.0202
67	0.04689	0.02344	0.0299	0.0348	0.0050	0.0149	0.0199
68	0.04620	0.02310	0.0294	0.0344	0.0049	0.0147	0.0196
69	0.04553	0.02277	0.0290	0.0339	0.0049	0.0145	0.0194
70	0.04488	0.02244	0.0286	0.0334	0.0049	0.0143	0.0191
71	0.04425	0.02212	0.0282	0.0330	0.0048	0.0141	0.0189
72	0.04363	0.02182	0.0278	0.0326	0.0048	0.0139	0.0187
73	0.04304	0.02152	0.0274	0.0321	0.0047	0.0137	0.0184
74	0.04245	0.02123	0.0270	0.0317	0.0047	0.0135	0.0182
75	0.04189	0.02094	0.0267	0.0313	0.0047	0.0133	0.0180
76	0.04134	0.02067	0.0263	0.0309	0.0046	0.0132	0.0178
77	0.04080	0.02040	0.0260	0.0306	0.0046	0.0130	0.0176
78	0.04028	0.02014	0.0256	0.0302	0.0046	0.0128	0.0174
79	0.03977	0.01988	0.0253	0.0298	0.0045	0.0127	0.0172
80	0.03927	0.01964	0.0250	0.0295	0.0045	0.0125	0.0170
81	0.03879	0.01939	0.0247	0.0292	0.0045	0.0123	0.0168
82	0.03831	0.01916	0.0244	0.0288	0.0044	0.0122	0.0166
83	0.03785	0.01893	0.0241	0.0285	0.0044	0.0120	0.0165
84	0.03740	0.01870	0.0238	0.0282	0.0044	0.0119	0.0163
85	0.03696	0.01848	0.0235	0.0279	010044	0.0118	0.0161
86	0.03653	0.01827	0.0233	0.0276	0.0043	0.0116	0.0160
87	0.03611	0.01806	0.0230	0.0273	0.0043	0.0115	0.0158
88	0.03570	0.01785	0.0227	0.0270	0.0043	0.0114	0.0156
89	0.03530	0.01765	0.0225	0.0267	0.0042	0.0112	0.0155
90	0.03491	0.01745	0.0222	0.0264	0.0042	0.0111	0.0153
91	0.03452	0.01726	0.0220	0.0262	0.0042	0.0110	0.0152
92	0.03415	0.01707	0.0217	0.0259	0.0042	0.0109	0.0150
93	0.03378	0.01689	0.0215	0.0257	0.0042	0.0108	0.0149
94	0.03342	0.01671	0.0213	0.0254	0.0041	0.0106	0.0148
95	0.03307	0.01653	0.0211	0.0252	0.0041	0.0105	0.0146
96	0.03272	0.01636	0.0208	0.0249	0.0041	0.0104	0.0145
97	0.03239	0.01619	0.0206	0.0247	0.0041	0.0103	0.0144
98	0.03206	0.01603	0.0204	0.0244	0.0040	0.0102	0.0142
99	0.03173	0.01587	0.0202	0.0242	0.0040	0.0101	0.0141

TABLE 3. TOOTH PARTS FOR 20° FINE-PITCH SPUR AND HELICAL GEARS (*Continued*)

Diametral Pitch	Circular Pitch	Standard Circular Thickness	Working Depth	Whole Depth	Clearance	Standard Addendum	Standard Dedendum
100	0.03142	0.01571	0.0200	0.0240	0.0040	0.0100	0.0140
101	0.03110	0.01555	0.0198	0.0238	0.0040	0.0099	0.0139
102	0.03080	0.01540	0.0196	0.0236	0.0040	0.0098	0.0138
103	0.03050	0.01525	0.0194	0.0234	0.0039	0.0097	0.0137
104	0.03021	0.01510	0.0192	0.0232	0.0039	0.0096	0.0135
105	0.02992	0.01496	0.0190	0.0230	0.0039	0.0095	0.0134
106	0.02964	0.01482	0.0189	0.0228	0.0039	0.0094	0.0133
107	0.02936	0.01468	0.0187	0.0226	0.0039	0.0093	0.0132
108	0.02909	0.01454	0.0185	0.0224	0.0039	0.0093	0.0131
109	0.02882	0.01441	0.0183	0.0222	0.0038	0.0092	0.0130
110	0.02856	0.01428	0.0182	0.0220	0.0038	0.0091	0.0129
111	0.02830	0.01415	0.0180	0.0218	0.0038	0.0090	0.0128
112	0.02805	0.01402	0.0179	0.0216	0.0038	0.0089	0.0127
113	0.02780	0.01390	0.0177	0.0215	0.0038	0.0088	0.0126
114	0.02756	0.01378	0.0175	0.0213	0.0038	0.0088	0.0125
115	0.02732	0.01366	0.0174	0.0211	0.0037	0.0087	0.0124
116	0.02708	0.01354	0.0172	0.0210	0.0037	0.0086	0.0123
117	0.02685	0.01343	0.0171	0.0208	0.0037	0.0085	0.0123
118	0.02662	0.01331	0.0169	0.0206	0.0037	0.0085	0.0122
119	0.02640	0.01320	0.0168	0.0205	0.0037	0.0084	0.0121
120	0.02618	0.01309	0.0167	0.0203	0.0037	0.0083	0.0120
121	0.02596	0.01298	0.0165	0.0202	0.0037	0.0083	0.0119
122	0.02575	0.01288	0.0164	0.0200	0.0036	0.0082	0.0118
123	0.02554	0.01277	0.0163	0.0199	0.0036	0.0081	0.0118
124	0.02534	0.01267	0.0161	0.0197	0.0036	0.0081	0.0117
125	0.02513	0.01257	0.0160	0.0196	0.0036	0.0080	0.0116
126	0.02493	0.01247	0.0159	0.0195	0.0036	0.0079	0.0115
127	0.02474	0.01237	0.0157	0.0193	0.0036	0.0079	0.0114
128	0.02454	0.01227	0.0156	0.0192	0.0036	0.0078	0.0114
129	0.02435	0.01218	0.0155	0.0191	0.0036	0.0078	0.0113
130	0.02417	0.01208	0.0154	0.0189	0.0035	0.0077	0.0112
131	0.02398	0.01199	0.0153	0.0188	0.0035	0.0076	0.0112
132	0.02380	0.01190	0.0152	0.0187	0.0035	0.0076	0.0111
133	0.02362	0.01181	0.0150	0.0185	0.0035	0.0075	0.0110
134	0.02344	0.01172	0.0149	0.0184	0.0035	0.0075	0.0110
135	0.02327	0.01164	0.0148	0.0183	0.0035	0.0074	0.0109
136	0.02310	0.01155	0.0147	0.0182	0.0035	0.0074	0.0108
137	0.02293	0.01147	0.0146	0.0181	0.0035	0.0073	0.0108
138	0.02277	0.01138	0.0145	0.0179	0.0034	0.0072	0.0107
139	0.02260	0.01130	0.0144	0.0178	0.0034	0.0072	0.0106

TABLE 3. TOOTH PARTS FOR 20° FINE-PITCH SPUR AND HELICAL GEARS (*Continued*)

Diametral Pitch	Circular Pitch	Standard Circular Thickness	Working Depth	Whole Depth	Clearance	Standard Addendum	Standard Dedendum
140	0.02244	0.01122	0.0143	0.0177	0.0034	0.0071	0.0106
141	0.02228	0.01114	0.0142	0.0176	0.0034	0.0071	0.0105
142	0.02212	0.01106	0.0141	0.0175	0.0034	0.0070	0.0105
143	0.02197	0.01098	0.0140	0.0174	0.0034	0.0070	0.0104
144	0.02182	0.01091	0.0139	0.0173	0.0034	0.0069	0.0103
145	0.02167	0.01083	0.0138	0.0172	0.0034	0.0069	0.0103
146	0.02152	0.01076	0.0137	0.0171	0.0034	0.0068	0.0102
147	0.02137	0.01069	0.0136	0.0170	0.0034	0.0068	0.0102
148	0.02123	0.01061	0.0135	0.0169	0.0034	0.0068	0.0101
149	0.02108	0.01054	0.0134	0.0168	0.0033	0.0067	0.0101
150	0.02094	0.01047	0.0133	0.0167	0.0033	0.0067	0.0100
151	0.02081	0.01040	0.0132	0.0166	0.0033	0.0066	0.0099
152	0.02067	0.01033	0.0132	0.0165	0.0033	0.0066	0.0099
153	0.02053	0.01027	0.0131	0.0164	0.0033	0.0065	0.0098
154	0.02040	0.01020	0.0130	0.0163	0.0033	0.0065	0.0098
155	0.02027	0.01013	0.0129	0.0162	0.0033	0.0065	0.0097
156	0.02014	0.01007	0.0128	0.0161	0.0033	0.0064	0.0097
157	0.02001	0.01001	0.0127	0.0160	0.0033	0.0064	0.0096
158	0.01988	0.00994	0.0127	0.0159	0.0033	0.0063	0.0096
159	0.01976	0.00988	0.0126	0.0158	0.0033	0.0063	0.0095
160	0.01964	0.00982	0.0125	0.0158	0.0032	0.0063	0.0095
161	0.01951	0.00976	0.0124	0.0157	0.0032	0.0062	0.0095
162	0.01939	0.00970	0.0123	0.0156	0.0032	0.0062	0.0094
163	0.01927	0.00964	0.0123	0.0155	0.0032	0.0061	0.0094
164	0.01916	0.00958	0.0122	0.0154	0.0032	0.0061	0.0093
165	0.01904	0.00952	0.0121	0.0153	0.0032	0.0061	0.0093
166	0.01893	0.00946	0.0120	0.0153	0.0032	0.0060	0.0092
167	0.01881	0.00941	0.0120	0.0152	0.0032	0.0060	0.0092
168	0.01870	0.00935	0.0119	0.0151	0.0032	0.0060	0.0091
169	0.01859	0.00929	0.0118	0.0150	0.0032	0.0059	0.0091
170	0.01848	0.00924	0.0118	0.0149	0.0032	0.0059	0.0091
171	0.01837	0.00919	0.0117	0.0149	0.0032	0.0058	0.0090
172	0.01827	0.00913	0.0116	0.0148	0.0032	0.0058	0.0090
173	0.01816	0.00908	0.0116	0.0147	0.0032	0.0058	0.0089
174	0.01806	0.00903	0.0115	0.0146	0.0031	0.0057	0.0089
175	0.01795	0.00898	0.0114	0.0146	0.0031	0.0057	0.0089
176	0.01785	0.00893	0.0114	0.0145	0.0031	0.0057	0.0088
177	0.01775	0.00887	0.0113	0.0144	0.0031	0.0056	0.0088
178	0.01765	0.00882	0.0112	0.0144	0.0031	0.0056	0.0087
179	0.01755	0.00878	0.0112	0.0143	0.0031	0.0056	0.0087

TABLE 3. TOOTH PARTS FOR 20° FINE-PITCH SPUR AND HELICAL GEARS (*Continued*)

Diametral Pitch	Circular Pitch	Standard Circular Thickness	Working Depth	Whole Depth	Clearance	Standard Addendum	Standard Dedendum
180	0.01745	0.00873	0.0111	0.0142	0.0031	0.0056	0.0087
181	0.01736	0.00868	0.0110	0.0142	0.0031	0.0055	0.0086
182	0.01726	0.00863	0.0110	0.0141	0.0031	0.0055	0.0086
183	0.01717	0.00858	0.0109	0.0140	0.0031	0.0055	0.0086
184	0.01707	0.00854	0.0109	0.0140	0.0031	0.0054	0.0085
185	0.01698	0.00849	0.0108	0.0139	0.0031	0.0054	0.0085
186	0.01689	0.00845	0.0108	0.0138	0.0031	0.0054	0.0085
187	0.01680	0.00840	0.0107	0.0138	0.0031	0.0053	0.0084
188	0.01671	0.00836	0.0106	0.0137	0.0031	0.0053	0.0084
189	0.01662	0.00831	0.0106	0.0136	0.0031	0.0053	0.0083
190	0.01653	0.00827	0.0105	0.0136	0.0031	0.0053	0.0083
191	0.01645	0.00822	0.0105	0.0135	0.0030	0.0052	0.0083
192	0.01636	0.00818	0.0104	0.0135	0.0030	0.0052	0.0082
193	0.01628	0.00814	0.0104	0.0134	0.0030	0.0052	0.0082
194	0.01619	0.00810	0.0103	0.0133	0.0030	0.0052	0.0082
195	0.01611	0.00806	0.0103	0.0133	0.0030	0.0051	0.0082
196	0.01603	0.00801	0.0102	0.0132	0.0030	0.0051	0.0081
197	0.01595	0.00797	0.0102	0.0132	0.0030	0.0051	0.0081
198	0.01587	0.00793	0.0101	0.0131	0.0030	0.0051	0.0081
199	0.01579	0.00789	0.0101	0.0131	0.0030	0.0050	0.0080
200	0.01571	0.00785	0.0100	0.0130	0.0030	0.0050	0.0080

proper rate to generate the teeth, as with a gear. Racks are sometimes cut in a regular shaper, using a form tool and suitable means for spacing the distance between the teeth.

Measurement of Racks

In measuring a rack, the distance that a 1.728 in. dia wire projects above the pitch line of the teeth for a 1 diametral pitch rack is as follows:

Degrees	$14\frac{1}{2}$	$17\frac{1}{2}$	20	25	30
Inches.	1.2779	1.2463	1.2323	1.2241	1.2316

For any other diametral pitch p divide the values by p and use wires $1.728/p$ in. in diameter.

To find the distance the wire projects above the tops of the teeth of a $1p$ rack subtract 1 from the above values.

TABLE 4. TOOTH PARTS FOR CIRCULAR (OR NORMAL CIRCULAR) PITCH*

Applies to spur, helical, straight bevel, and worm gearing. Parenthetical words may be omitted for spur and straight bevel gearing.

Addendum and dedendum proportions apply only to full-depth teeth in a normal section, and to equal-addendum gearing.

For straight bevel gearing, dedendum and whole-depth columns do not apply. Refer to page 4-67. For AGMA and ASA dedendum and whole depth formulas, refer to ASA B6.1-1932. Proportions for 20 DP and finer have been omitted; refer to AGMA 207.02. All dimensions in inches.

(Normal) Circular Pitch	(Normal) Diametral Pitch	(Normal) Circular Thickness	Addendum	Dedendum, min	Whole Depth, Min
0.1875	16.755163	0.0938	0.0597	0.0690	0.1287
0.25	12.566372	0.1250	0.0796	0.0921	0.1716
0.3125	10.053098	0.1562	0.0995	0.1151	0.2146
0.375	8.377581	0.1875	0.1194	0.1381	0.2575
0.5	6.283186	0.2500	0.1592	0.1842	0.3433
0.625	5.026549	0.3125	0.1989	0.2301	0.4291
0.75	4.188791	0.3750	0.2387	0.2762	0.5150
0.875	3.590392	0.4375	0.2785	0.3223	0.6007
1	3.141593	0.5000	0.3183	0.3683	0.6866
1.125	2.792527	0.5625	0.3581	0.4143	0.7724
1.25	2.513274	0.6250	0.3979	0.4604	0.8583
1.375	2.284795	0.6875	0.4377	0.5064	0.9441
1.5	2.094395	0.7500	0.4775	0.5525	1.0299
1.625	1.933288	0.8125	0.5173	0.5985	1.1158
1.75	1.795196	0.8750	0.5570	0.6445	1.2016
1.875	1.675516	0.9375	0.5968	0.6906	1.2874
2	1.570796	1.0000	0.6366	0.7366	1.3732
2.5	1.256637	1.2500	0.7957	0.9208	1.7165
3	1.047198	1.5000	0.9549	1.1049	2.0598
3.5	0.897598	1.7500	1.1440	1.2591	2.4031
4	0.785398	2.0000	1.2733	1.4731	2.7464
4.5	0.698132	2.2500	1.4324	1.6674	3.0997
5	0.628319	2.5000	1.5915	1.8415	3.4330
5.5	0.571199	2.7500	1.7507	2.0257	3.7763
6	0.523599	3.0000	1.9098	2.2098	4.1196

* Gould & Eberhardt, Inc.

TABLE 5. TOOTH PARTS FOR MODULE, MM (OR NORMAL MODULE, MM)*

Applies to spur, helical, straight bevel, and worm gearing. Parenthetical words may be omitted for spur and straight bevel gearing.

Addendum and dedendum proportions apply only to full-depth teeth in a normal section, and to equal-addendum gearing.

This table does not include module in inches, which is the pitch diameter in inches, divided by the number of teeth.

Dedendum and whole depth may vary from tabulated values; refer to German Standard DIN 867.

Equivalent (Normal) Diametral Pitch, In. Basis	Mm					
	(Normal) Module	(Normal) Circular Pitch	(Normal) Circular Thickness	Addendum	Dedendum	Whole Depth
25.40	1	3.14159	1.57	1.00	1.167	2.167
20.32	1.25	3.92699	1.96	1.25	1.46	2.71
16.93	1.5	4.71239	2.36	1.50	1.75	3.25
14.51	1.75	5.49779	2.75	1.75	2.04	3.79
12.70	2	6.28319	3.14	2.00	2.33	4.33
11.29	2.25	7.06858	3.53	2.25	2.63	4.88
10.16	2.5	7.85398	3.93	2.50	2.92	5.42
9.24	2.75	8.63938	4.32	2.75	3 21	5.96
8.47	3	9.42478	4.71	3.00	3.50	6.50
7.82	3.25	10.21018	5.11	3.25	3.79	7.04
7.25	3.5	10.99558	5.50	3.50	4.08	7.58
6.77	3.75	11.78097	5.89	3.75	4.38	8.13
6.35	4	12.56637	6.28	4.00	4.67	8.67
5.64	4.5	14.13717	7.07	4.50	5.25	9.75
5.08	5	15.70796	7.85	5.00	5.84	10.84
4.62	5.5	17.27876	8.64	5.50	6.42	11.92
4.23	6	18.84956	9.42	6.00	7.00	13.00
3.91	6.5	20.42035	10.21	6.50	7.59	14.09
3.63	7	21.99115	11.00	7.00	8.17	15.17
3.17	8	25.13274	12.57	8.00	9.34	17.34
2.82	9	28.27434	14.14	9.00	10.50	19.50
2.54	10	31.41593	15.71	10.00	11.67	21.67
2.31	11	34.55752	17.28	11.00	12.84	23.84
2.12	12	37.69912	18.85	12.00	14.00	26.00
1.95	13	40.84071	20.42	13.00	15.17	28.17
1.81	14	43.98230	21.99	14.00	16.34	30.34
1.69	15	47.12390	23.56	15.00	17.50	32.50
1.59	16	50.26549	25.13	16.00	18.67	34.67
1.41	18	56.54867	28.27	18.00	21.01	39.01
1.27	20	62.83186	31.42	20.00	23.34	43.34
1.15	22	69.11505	34.56	22.00	25.67	47.67
1.06	24	75.39823	37.70	24.00	28.01	52.01

These formulas apply to module in millimeters or inches:

$$\text{Module} = \frac{\text{pitch dia}}{\text{No. of teeth}}$$

Addendum = normal module

Clearance = 0.167 × normal module

Normal circular pitch = 3.141593 × normal module

Normal circular thickness = 1.570797 × normal module

Spur pitch dia = No. of teeth × module

$$\text{Helical pitch dia} = \frac{\text{No. of teeth} \times \text{normal module}}{\cos \text{helix angle}}$$

* Gould & Eberhardt, Inc.

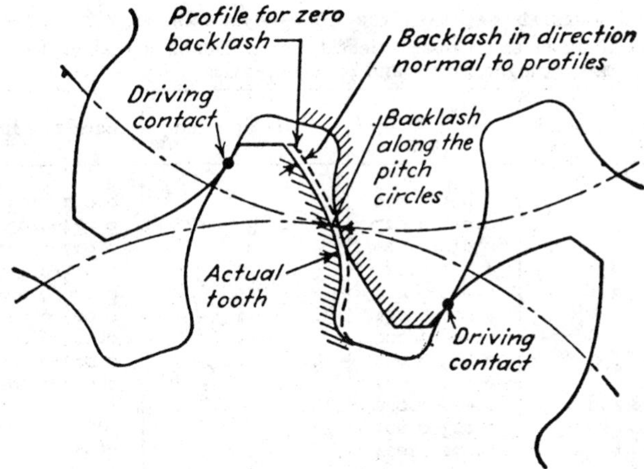

FIG. 7. Backlash between tooth profiles.

Backlash in Gears

In general, backlash in gears is play between mating teeth (Fig. 7). For purpose of measurement and calculation, backlash is defined as the amount by which a tooth space exceeds the thickness of an engaging tooth. When not otherwise specified, numerical values of backlash are understood to be given on the pitch circles.

Purpose of Backlash. The general purpose of backlash is to prevent gears from jamming together and making contact on both sides of their teeth simultaneously.

Lack of backlash may cause noise, overloading, and overheating, even seizing and failure. Excessive backlash is objectionable, particularly if the drive is frequently reversing, or if there is an overrunning load as in cam drives.

Specification of an unnecessarily small amount of backlash allowance will increase the cost of gears, because errors in runout, pitch, profile, and mounting must be held correspondingly smaller. On the other hand, provision of reasonable backlash permits greater leeway in manufacturing and usually is not detrimental to proper gear action.

Proper Amount of Backlash. In setting up proper backlash amounts and tolerances for a pair of gears, many factors should be given consideration. The most important factor is probably the maximum amount of runout expected in both gear and pinion. Next are the errors in profile, pitch, tooth thickness, and helix angle.

Tolerances for tooth errors for various types and classes of gears are given (page 4-45). Obviously the backlash between a pair of gears will vary as successive teeth make contact, because of the effect of tooth errors, particularly runout. The tables of backlash represent minimum backlash at the tightest point between pinion and gear.

The higher the helix angle or spiral angle, the more transverse backlash is required for a given normal backlash. The transverse backlash multiplied by the cosine of the helix angle gives the normal backlash.

Another factor is the pressure angle. For higher pressure angles, more backlash is required on the pitch circles to obtain a given amount of backlash in a direction normal to the tooth profiles.

TABLE 6. SUGGESTED BACKLASH FOR COARSE-PITCH GEARS WHEN ASSEMBLED*

Spur Gears, Parallel and Crossed Helical Gears, Double-helical or Herringbone Gears, Straight, Spiral, and Zerol Bevel Gears, Hypoid Gears

Diametral Pitch	Backlash, In.	Circular Pitch	Backlash, In.
1	0.025–0.040	4	0.032–0.050
$1\frac{1}{2}$	0.018–0.027	3	0.024–0.038
2	0.014–0.020	2	0.017–0.025
$2\frac{1}{2}$	0.011–0.016	$1\frac{1}{2}$	0.013–0.019
3	0.009–0.014	1	0.009–0.014
4	0.007–0.011	$\frac{3}{4}$	0.007–0.011
5	0.006–0.009	$\frac{1}{2}$	0.005–0.007
6	0.005–0.008	$\frac{1}{4}$	0.003–0.005
7	0.004–0.007	$\frac{1}{8}$	0.002–0.004
8 and 9	0.004–0.006		
10 to 13	0.003–0.005		
14 to 32	0.002–0.004		

* AGMA Standard 233.01 and ASA Standard B6.6-1946.

Errors in boring the gear housings, in both center distance and misalignment, are of extreme importance in determining allowances to obtain the backlash desired. The same is true in the mounting of the gears, which is affected by the type and adjustment of bearings, etc.

Other influences in backlash specification are heat-treatment subsequent to cutting the teeth, lapping operations, the possible necessity for recut for any reason, and reduction of tooth thickness from normal wear.

Minimum backlash is necessary for timing, indexing, and certain instrument trains. However, the specification of "zero backlash," so commonly stipulated for gears of this nature, usually involves special and expensive technique and is difficult to obtain.

Providing Backlash. In order to obtain the amount of backlash desired, it is necessary to decrease tooth thickness. The allowances made on tooth thickness, however, almost always must exceed the amount of backlash because of manufacturing and assembling inaccuracies not only in the gears but also in other parts.

It is customary to make half the allowance for backlash on the tooth thickness of each gear of a pair, although there are exceptions. For example, on pinions having very low numbers of teeth it is desirable to provide all the allowance on the mating gear, so as not to weaken the pinion teeth.

In spur and helical gearing, backlash allowance is usually obtained by sinking the cutter deeper into the blank than the theoretically standard depth. In some instances the allowance is provided in the cutter instead, and the cutter is then operated at the standard tooth depth.

Measurement. Backlash is commonly measured by holding one gear of a pair stationary and rocking the other back and forth. The movement is registered by a dial indicator having its pointer or finger in a plane of rotation at or near the pitch diameter and in a direction parallel to a tangent to the pitch circle of the moving gear.

If the direction of measurement is normal to the teeth, or other than as specified above, it is recommended that readings be converted to the plane of rotation and in

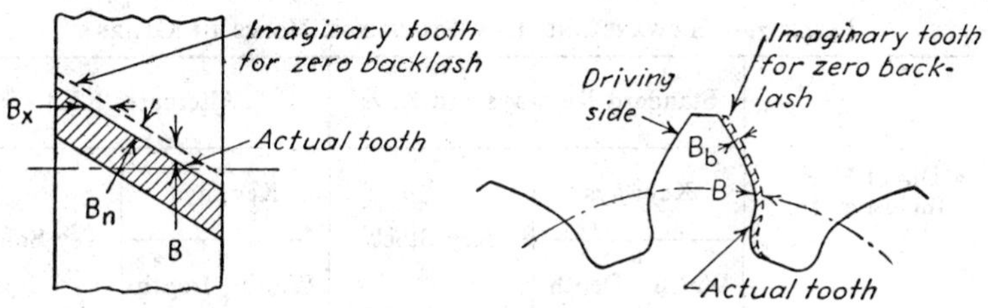

B = Backlash on pitch circle (in transverse direction)
B_n = Backlash in direction normal to helix
B_x = Backlash in axial direction

FIG. 8. Backlash in pitch plane.

B = Backlash on pitch circle
B_b = Backlash on base circle, also in direction normal to profiles

FIG. 9. Backlash in transverse plane.

a tangent direction, for purposes of standardization and comparison (see Figs. 8 and 9).

In spur gears, parallel helical gears, and bevel gears, it is immaterial whether the pinion or gear is held stationary for the test. In crossed helical and hypoid gears, it is customary to hold the pinion stationary and measure on the gear.

In some instances backlash is measured by thickness gages or feelers. A similar method utilizes lead wire inserted between the teeth as they pass through mesh.

Keyways for Holes in Gears

The American Gear Manufacturers Association has revised its standard 261.01 "Keyways for Holes in Gears for General Industrial Practice." This revision is in agreement with the American Standard Association's publication B17.1-1943, entitled "Shafting and Stock Keys." Details of the standard are:

Depth of Keyways. The depth of keyways shall be one-half the key height measured at the edge, according to Fig. 10.

Figure 10 and Table 7 are also used for plain and gib head taper keys with the standard taper of $\frac{1}{8}$ in. per ft where the depth shown is the deep end of the keyway.

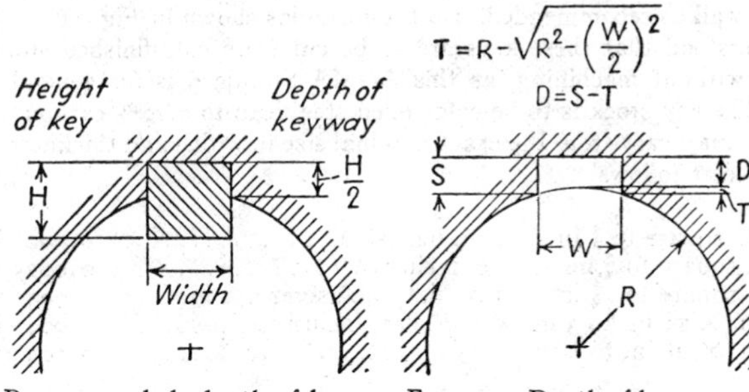

FIG. 10. Recommended: depth of keyway measured on vertical wall.

FIG. 11. Depth of keyway measured on centerline of the key.

TABLE 7. KEYWAYS AND KEY STOCK FOR HOLES IN GEARS

Dia of Holes Inclusive, In.	Standard Keyways and Keys			Alternates XX†		
	Keyways		Key Stock	Keyways		Key Stock
	Width	Depth		Width	Depth	
5/16 to 7/16	3/32	3/64	3/32 × 3/32*			
1/2 to 9/16	1/8	1/16	1/8 × 1/8	1/8	3/64	1/8 × 3/32
5/8 to 7/8	3/16	3/32	3/16 × 3/16	3/16	1/16	3/16 × 1/8
15/16 to 1 1/4	1/4	1/8	1/4 × 1/4	1/4	3/32	1/4 × 3/16
1 5/16 to 1 3/8	5/16	5/32	5/16 × 5/16	5/16	1/8	5/16 × 1/4
1 7/16 to 1 3/4	3/8	3/16	3/8 × 3/8	3/8	1/8	3/8 × 1/4
1 13/16 to 2 1/4	1/2	1/4	1/2 × 1/2	1/2	3/16	1/2 × 3/8
2 5/16 to 2 3/4	5/8	5/16	5/8 × 5/8	5/8	7/32	5/8 × 7/16
2 13/16 to 3 1/4	3/4	3/8	3/4 × 3/4	3/4	1/4	3/4 × 1/2
3 5/16 to 3 3/4	7/8	7/16	7/8 × 7/8	7/8	5/16	7/8 × 5/8
3 13/16 to 4 1/2	1	1/2	1 × 1	1	3/8	1 × 3/4
4 9/16 to 5 1/2	1 1/4	7/16	1 1/4 × 7/8			
5 9/16 to 6 1/2	1 1/2	1/2	1 1/2 × 1			
6 9/16 to 7 1/2	1 3/4	5/8	1 3/4 × 1 1/4*			
7 9/16 to 8 15/16	2	3/4	2 × 1 1/2*			
9 to 10 15/16	2 1/2	7/8	2 1/2 × 1 3/4*			
11 to 12 15/16	3	1	3 × 2*			
13 to 14 15/16	3 1/2	1 1/4	3 1/2 × 2 1/2*			
15 to 17 15/16	4	1 1/2	4 × 3*			
18 to 21	5	1 3/4	5 × 3 1/2*			

* Shaft sizes for these keys are not listed in American Standard B17.1-1943.
† It is recommended that these alternates be used only when conditions make it undesirable to use the standard sizes in the above table.

Where the depth of keyway is measured on the centerline of the key instead of the vertical wall as recommended, use the formulas shown in Fig. 11.

It is understood that these keys are to be cut from cold-finished stock and are to be used without machining, as this AGMA Standard is for general industrial practice. The key stock is to be cold-rolled steel 0.10 to 0.20% carbon.

Key stock may vary from the exact nominal size in width and thickness to a negative tolerance as follows:

3/32 in. square to 3/8 in. square, inclusive	0.002 in.
1/2 in. square to 3/4 in. square, inclusive	0.0025 in.
7/8 in. square to 1 1/2 in. × 1 in. flat, inclusive	0.003 in.
1 3/4 in. × 1 1/4 in. to 3 in. × 2 in. flat, inclusive	0.004 in.
3 1/2 in. × 2 1/2 in. to 5 in. × 3 1/2 in. flat, inclusive	0.005 in.

Keyways to be cut from exact nominal size to plus 0.002 in. in width, and depth shall be nominal to plus $\frac{1}{64}$ in. for straight keys. For taper keys depth shall be nominal to $\frac{1}{64}$ in. minus.

For heat-treated pinions the depth shall be $\frac{1}{32}$ to $\frac{3}{64}$ in. over nominal size with a minimum radius of $\frac{1}{32}$ in. in corners of keyways.

For highly stressed or alternating loads encountered, it is recommended that the corners of the keyway be rounded to a minimum of $\frac{1}{32}$ in. radius and not over a maximum of $\frac{1}{5}$ of the keyway depth. The edges of the key stock are to be rounded to correspond.

Spur Gears

Spur gears are cylindrical in form and operate on parallel axes. The teeth have a constant section throughout their length, unless modified by shaving to a barrel shape, and are parallel to the axes. There is a basic rack form for any spur-gear tooth-form system, said rack meshing with the given tooth form.

Interference. There are two types of interference in involute gearing: *involute* and *fillet*. Involute interference is the term used to indicate that contact between mating teeth takes place at some point other than along the line of action, *before* contact reaches the natural interference point. On the other hand, fillet interference refers to contact at some point other than along the line of action, *after* contact has passed the natural interference point.

The smallest number of teeth to avoid involute interference with an unmodified straight-sided rack tooth is: 32 teeth for $14\frac{1}{2}°$ full-depth teeth; 18 teeth for 20° full-depth teeth, and 14 teeth for stub-tooth gears—AGMA standard with $\frac{8}{10}$ addendum. See Tables 8 and 9 for diameter corrections to avoid undercutting gears with fewer teeth than the above values.

Fillet interference is generally a matter of tool design and can be corrected in cutting.

These diameter corrections, involving lengthening the addendum of the pinion teeth and correspondingly shortening the gear teeth, constitute the "long and short addendum" method of designing gears for avoidance of interference and undercutting. When using these diameter corrections, apply them to cases where standard center distance is to be maintained. If a pinion is to be meshed with a gear that may not have the addendum reduced, the center distance must be spread and special calculations are used. Changes in center distance and backlash are covered in AGMA 207.1 and 207.01.

Spur-gear Formulas,* Equal Addendum

Number of teeth	$N = DP_d$
Diametral pitch	$P_d = N/D$
Circular pitch	$p = 3.141593D/N$
Pressure angle	$\phi = 20°$ or $14.5°$
Pitch dia	$D = N/P_d$
OD	$D_o = D + 2a$
Base dia	$D_b = D \cos \phi$
Center distance	$C = (D_G + D_P)/2$
Addendum	$a = 1/P_d$
Approx whole depth	$h = 2.157/P_d$
Dedendum	$b = h - a$
Clearance	$c = 0.157/P_d$
Circular thickness	$t = p/2$
Chordal addendum	$a_c = a + (t^2/4D)$
Max chordal thickness	$t_c = t - (t^3/6D^2) - (B/2)$
Backlash assembled	B

* Gould & Eberhardt, Inc. Symbols conform with ASA B6.5-1949.

TABLE 8. DIAMETER CORRECTIONS TO AVOID UNDERCUT OF 14½° FULL-DEPTH TEETH. BASED ON 1DP

When Pinion Has 31 or Fewer Teeth. All Dimensions in Inches

No. of Teeth in Pinion	Dia* Increment	Pinion Cir Tooth Thickness	Mating Gear Cir Tooth Thickness	Min No.† of Teeth in Mating Gear Avoiding Undercut	Min No.† of Teeth in Mating Gear for Full Involute Action
10	1.3731	1.9259	1.2157	54	27
11	1.3104	1.9097	1.2319	53	27
12	1.2477	1.8935	1.2481	52	28
13	1.1850	1.8773	1.2643	51	28
14	1.1223	1.8611	1.2805	50	28
15	1.0597	1.8449	1.2967	49	28
16	0.9970	1.8286	1.3130	48	28
17	0.9343	1.8124	1.3292	47	28
18	0.8716	1.7962	1.3454	46	28
19	0.8089	1.7800	1.3616	45	28
20	0.7462	1.7638	1.3778	44	28
21	0.6835	1.7476	1.3940	43	28
22	0.6208	1.7314	1.4102	42	27
23	0.5581	1.7151	1.4265	41	27
24	0.4954	1.6989	1.4427	40	27
25	0.4328	1.6827	1.4589	39	26
26	0.3701	1.6665	1.4751	38	26
27	0.3074	1.6503	1.4913	37	26
28	0.2447	1.6341	1.5075	36	25
29	0.1820	1.6179	1.5237	35	25
30	0.1193	1.6017	1.5399	34	24
31	0.0566	1.5854	1.5562	33	24

NOTE: Table is based on 1 diametral pitch. For other pitches divide values given by desired diametral pitch.

* Dia increment is the amount by which the OD of the pinion must be increased over the standard OD and also the amount by which the OD of the mating gear must be decreased below standard to maintain standard center distances.

† The number of gear teeth, in each case, shown in this column or any larger number of teeth, up to that specified in the preceding column, will produce undercutting of the gear teeth in varying amounts with consequent reduction in strength, the maximum undercutting in each case occurring with the minimum number of gear teeth.

Fine-pitch Gears

Fine-pitch gears have been defined as gears of 20 diametral pitch and finer by the American Gear Manufacturers Association. While there is no mathematical difference between fine- and coarse-pitch gears of involute form, fine-pitch gears require special considerations for the following reasons: Greater clearances than for coarse pitch are required; considerations must be given to point widths of gear cutting tools and to top lands of gears, especially to top lands of enlarged heat-treated

TABLE 9. DIAMETER CORRECTIONS TO AVOID UNDERCUT OF 20° FULL-DEPTH TEETH

No. of Teeth in Pinion	Dia* Increment	Pinion Cir Tooth Thickness	Mating Gear Cir Tooth Thickness	Min No. of Teeth in Mating Gear Avoiding Undercut	Min No.† of Teeth in Mating Gear for Full Involute Action
8	1.0642	1.9581	1.1835	26	16
9	0.9472	1.9156	1.2260	25	16
10	0.8302	1.8730	1.2686	24	16
11	0.7132	1.8304	1.3112	23	16
12	0.5963	1.7878	1.3538	22	15
13	0.4793	1.7453	1.3963	21	14
14	0.3623	1.7027	1.4389	20	14
15	0.2453	1.6601	1.4815	19	14
16	0.1284	1.6175	1.5241	18	13
17	0.0114	1.5749	1.5667	17	13

* Dia increment is the amount by which the OD of the pinion must be increased over the standard OD and also the amount by which the OD of the mating gear must be decreased below standard to maintain standard center distances.

† The number of gear teeth, in each case, shown in this column or any larger number of teeth, up to that specified in the preceding column, will produce undercutting of the gear teeth in varying amounts with consequent reduction in strength, the maximum undercutting in each case occurring with the minimum number of gear teeth.

All dimensions in inches.

SPUR-GEAR SPECIFICATIONS ON DRAWINGS*

Example		Instructions to Draftsmen
Number of teeth	37	
Diametral pitch	12	
Pressure angle	14.50	
Pitch dia	3.0833	4 places
Base dia	2.9851	4 places
Addendum	0.0833	4 places
Dia enlargement	....	4 places. Omit if standard
Dia decrease	....	4 places. Omit if standard
Dia topped	....	4 places. Special case
Approx whole depth	0.1875	4 places
Chordal addendum	0.0847	4 places
Max chordal thickness	0.1289	4 places
180° error	0.0025	
Pitch variation	0.00035	Omit if hand-operated
Profile error	0.0006	Omit if hand-operated
Backlash assembled	0.003–0.005	
Center distance	....	Usually 4 places

* Gould & Eberhardt, Inc.

Table 10. Dimensions Required for 20° Fine-pitch Gears When Using Enlarged Pinions. AGMA Standard 207.02

The OD's of Small Pinions Are Enlarged to Avoid Undercut. All Dimensions Are Given in Inches

	Pinion Dimensions		Standard Center-distance System (Long and Short Addendum) Gear Dimensions*				Enlarged Center-distance System. Standard Mating Gear Diameter†	
No. of Teeth n	OD	Cir Tooth Thickness at Standard Pitch Dia	Decrease in Standard OD	Cir Tooth Thickness at Standard Pitch Dia $\Delta t = \Delta d \tan \phi$	Recommended Min No. of Teeth (N)	Contact Ratio n Mating with N	Increase over Standard Center Distance	Contact Ratio Two Equal Pinions
10	12.8302	1.8730	0.8302	1.2686	33	1.419	0.4151	1.135
11	13.7132	1.8304	0.7132	1.3112	30	1.450	0.3566	1.186
12	14.5963	1.7878	0.5963	1.3538	27	1.473	0.2982	1.238
13	15.4793	1.7452	0.4793	1.3964	25	1.493	0.2397	1.290
14	16.3623	1.7027	0.3623	1.4389	23	1.508	0.1812	1.344
15	17.2453	1.6601	0.2453	1.4815	21	1.516	0.1227	1.398
16	18.1284	1.6175	0.1284	1.5241	19	1.519	0.0642	1.436
17	19.0114	1.5749	0.0114	1.5667	18	1.522	0.0057	1.511

Note: Tabular values are in inches for 1 diametral pitch. For other pitches divide tabular values by the diametral pitch.

* To maintain standard center distances when using enlarged pinions, the mating gear diameters must be decreased by the amount of the pinion enlargement.

† If mating gears are made with standard tooth proportions, the center distances must be increased as shown.

Table 11. 20° Pressure Angle Involute Fine-pitch System for Spur Gears

Calculated Center Distance at Which Enlarged Pinions Will Engage with No Backlash

No. of Teeth	10	11	12	13	14	15	16	17
10	10.6846							
11	11.1461	11.6060						
12	11.6060	12.0644	12.5214					
13	12.0644	12.5214	12.9768	13.4307				
14	12.5214	12.9768	13.4307	13.8833	14.3344			
15	12.9768	13.4307	13.8833	14.3344	14.7840	15.2323		
16	13.4307	13.8833	14.3344	14.7840	15.2323	15.6793	16.1247	
17	13.8833	14.3344	14.7840	15.2323	15.6793	16.1247	16.5687	17.0114

Dimensions are in inches for 1 DP. For other pitches divide tabular values by diametral pitch.

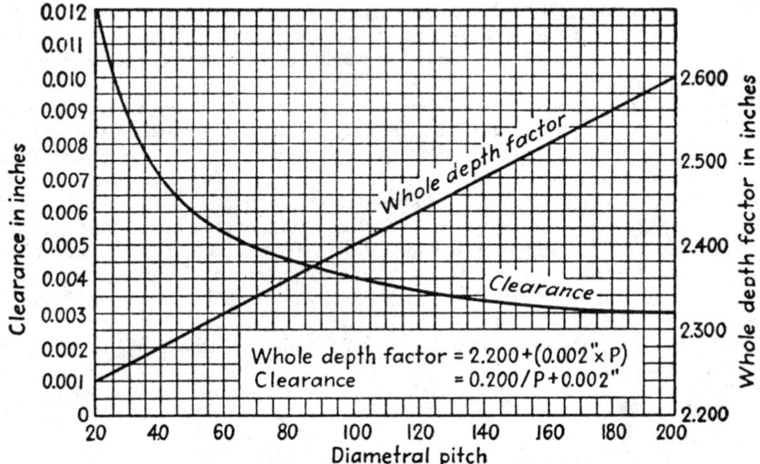

FIG. 12. Clearances and whole-depth factor for fine-pitch gears.

TABLE 12. 20° PRESSURE ANGLE FINE-PITCH SYSTEM FOR SPUR GEARS
Theoretical Backlash Obtained When Meshing Two Enlarged Pinions

No. of Teeth	10	11	12	13	14	15	16	17
10	0.163							
11	0.136	0.112						
12	0.112	0.092	0.074					
13	0.092	0.074	0.058	0.045				
14	0.074	0.058	0.045	0.034	0.024			
15	0.058	0.045	0.034	0.024	0.017	0.011		
16	0.045	0.034	0.024	0.017	0.011	0.006	0.003	
17	0.034	0.024	0.017	0.011	0.006	0.003	0.001	0

NOTE: Enlarged center-distance system. Dimensions are in inches for 1 DP. For other pitches divide tabular values by the diametral pitch.

pinions; fine-pitch gears require different production and inspection techniques than coarse-pitch gears, except in the coarser range of fine-pitch gears, because their use is generally for the transmission of motion rather than power. Surface durability is of far greater importance than beam strength.

The AGMA has developed a number of standards to meet the requirements of fine-pitch gears. Standard 470.01 established clearances for all 20° pressure angle fine-pitch gears of 20 DP and finer, except bevel gears. Greater clearances than for coarse-pitch gears are required to allow for: (1) greater wear on point of generating tool; (2) fillet radius produced by dull generating tool is greater than derived from formula 0.157/DP; (3) accumulation of foreign matter at bottom of teeth. Proportions are: working depth = 2.000 in./P; whole depth + 2.2000 in./P + 0.002 in.; clearance = 0.200 in./P + 0.002 in.; whole depth factor = 2.200 in. + (0.002 in. $\times P$). See Fig. 12.

AGMA Standard 207.02 established the tooth form for fine-pitch involute spur and helical gears of 20° normal pressure angle. Figure 6 gives the basic rack for this system. Table 1 gives the standard tooth proportions. Pinions having 17 or fewer teeth are enlarged in accordance with Table 10. Two options are given in this standard: (1) standard center-distance system (long and short addendum) and (2) enlarged center-distance system. Each system has some advantages and disadvantages over the other. The standard does not recommend the use of pinions having fewer than ten teeth. Almost all the advantages of the involute curve are lost when pinions having fewer than ten teeth are enlarged to avoid objectionable undercut.

Table 11 gives the center distance at which enlarged pinions will engage with no backlash.

Table 12 gives the theoretical backlash obtained when two enlarged pinions are meshed.

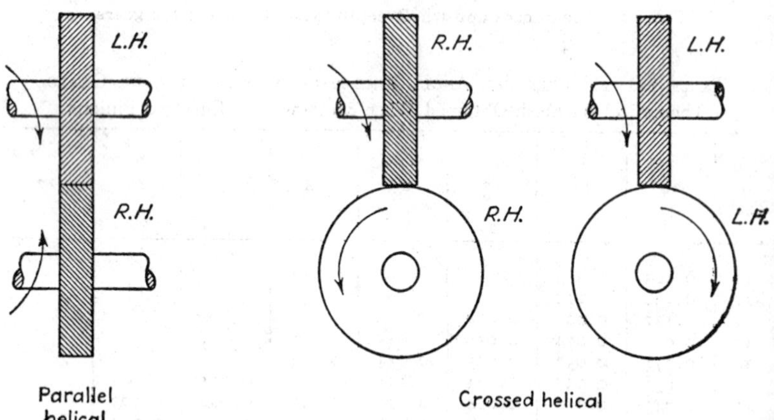

FIG. 13. Types of helical gears.

Helical Gears

Like spur gears, helical gears are cylindrical in form, and the teeth have a constant section, but the teeth are oblique to the axes of the gear cylinders. In other words, the teeth are cut along a helix on the pitch cylinder. Helical gears tend to run smoothly and quietly, but produce end thrust unless double-helical or herringbone teeth are employed.

When helical gears are cut with a spur milling cutter, of either the involute or composite form, the actual number of teeth in the gear cannot be used as a guide in selecting the cutter. The reason is that the tooth form lies in a *normal* section, and hence the equivalent number of teeth must be found.

Helical gears are of two basic types: (1) parallel axis, and (2) crossed axis (Fig. 13). The former type is preferred for power transmission because the teeth have line contact, whereas the crossed-axes type has point contact. Hence the trend in terminology is to use "helical gears" for parallel-axes gears to distinguish them from "crossed helical gears," meaning crossed-axes gears. Furthermore, the term "helical gears" has superseded the old term "spiral gears," which was applied to all gears in which the teeth are not parallel to the axis. The term "spiral gears" is applied properly only to gears with teeth in one plane and to spiral bevel gears.

PARALLEL-HELICAL-GEAR FORMULAS,* FULL-DEPTH TEETH IN NORMAL PLANE, EQUAL ADDENDUM

Number of teeth	$N = P_{nd} D \cos\psi$
Equivalent number of teeth	$N_e = N/\cos^3\psi$
Normal diametral pitch	$P_{nd} = 3.141593/p_n$
Diametral pitch	$P_d = P_{nd}\cos\psi$
Normal circular pitch	$p_n = 3.141593/P_{nd}$
Circular pitch	$p = 3.141593D/N$
Axial pitch	$p_x = 3.141593/P_{nd}\sin\psi$
Normal pressure angle	$\phi_n = 20°$ or $14.5°$
Transverse pressure angle	$\phi_t = \tan^{-1}\tan\phi_n/\cos\psi$
Pitch dia	$D = N/P_{nd}\cos\psi$
OD	$D_o = D + 2a$
Base dia	$D_b = D\cos\phi_t$
Center distance	$C = (D_G + D_P)/2$
Helix angle	$\psi = \cos^{-1} N/P_{nd}D = \cos^{-1}(N_G + N_P)/2CP_{nd}$
Lead	$l = Np_x$
Addendum	$a = 1/P_{nd}$
Approx whole depth	$h = 2.157/P_{nd}$
Dedendum	$b = h - a$
Clearance	$c = 0.157/P_{nd}$
Normal circular thickness	$t_n = P_n/2$
Normal chordal addendum	$a_{nc} = a + t_n^2\cos^2\psi/4D$
Max normal chordal thickness	$t_{nc} = t_n - (t_n^3\cos^4\psi/6D^2) - (B\cos\psi/2)$
Backlash assembled	B

* Gould & Eberhardt, Inc.

Speed Ratios. In considering speed ratios for helical gears the driving gear can be taken as a worm having as many threads as there are teeth and the driven as the worm gear with its number of teeth, so that one revolution of the driver will turn a point on the pitch circle of the driven gear as many inches as the lead of the teeth of the driver. Divide this by the circumference of the pitch circle of the driven gear to get the revolutions of the driven.

It is possible to operate helical gears on any center distance by choosing the correct helix angle ψ.

While the subject of helical gears is rather complex if considered broadly, most of the difficulties disappear when they have a tooth angle of 45°. It is perhaps for this reason that from 75 to 90% of the helical gears used are made with this angle.

This has the added advantage of being the most durable, although there is but a trifling increase in wear down to 30°, and the wear at 20° is not serious. In cases of necessity, even 12° can be used without destructive wear.

Where higher speed ratios than can be had with a 45° angle tooth are necessary, they can be laid out as will be shown later and can be cut on most milling machines. The usual change gears allow about two thousand different helices to be cut.

Helical-gear Specifications for Drawings*

Example		Instructions to Draftsmen
Finishing process	Lap	Omit if not lapped
Number of teeth	56	
Normal diametral pitch	7	
Axial pitch	1.201670	6 places
Normal pressure angle	14.5°	
Pitch dia	8.624044	6 places
Base dia	8.3072	4 places
RH helix angle	21°55′49″	1″
Addendum	0.1429	4 places
Dia enlargement	....	4 places. Omit if standard
Dia decrease	....	4 places. Omit if standard
Dia topped	....	4 places. Special case
Approx whole depth	0.3214	4 places
Normal chordal addendum	0.1441	4 places
Max normal chordal thickness	0.2214	4 places
Shaft angle	....	1″. Omit if parallel axes
180° error	0.0025	
Pitch variation	0.00035	Omit if hand-operated
Profile error	0.0006	Omit if hand-operated
Circumferential lead error:		
In any 1 in. of face	0.0003	Omit if hand-operated
In any 2 in. of face	0.0004	Omit if 1 in. or less
In any 3 in. of face	....	Omit if 2 in. or less
Backlash assembled	0.004–0.007	
Center distance	....	4 places

* Gould & Eberhardt, Inc.

Where the angles are not 45°, the gear with the greatest angle must always be the driver.

All the tooth parts are derived from the normal pitch, whereas the pitch diameters are derived from the circular pitch. These are never the same in two gears of a pair except when both are 45°.

As the diameter of a helical gear does not indicate its speed ratio, the terms *driver* and *follower* are used in place of gear and pinion.

45° Crossed Helical Gears

These gears are the simplest of all helices to lay out and to make, the required speed ratios being obtained by varying the diameters, precisely as with spur or bevel gears, the rules for the speed ratio being the same in both cases. Moreover, the various factors required in laying out and making such gears can be reduced to the simple terms in Table 13. With it anyone can quickly make the few calculations connected with any pair of 45° gears having teeth between 2 and 48 diametral pitch.

Example: Let it be desired to construct a pair of crossed helical gears with 35 teeth in the driver and 16 teeth in the follower, using a 10-pitch cutter. Using Table 13 we have for a 10-pitch cutter, under the proper columns, the following:

PD = 0.14143 × 35 = 4.950
OD = 4.950 + 0.200 = 5.150
Pitch in inches to one turn of helix = 0.44431 × 35 = 15.550

NOTE: A slight variation in one turn makes no practical difference; hence, the ordinary change gears furnished with a universal miller will usually be found sufficient.

Number of teeth in spur with same curvature = 2.828 × 35 = 98.980

Looking at Brown & Sharpe spur-gear-cutter list, we see that 99 is between 55 and 134; therefore, we select a No. 2 cutter.

In a similar manner, using 16 as a multiplier, we obtain the data for the follower. This gives 2.262 as pitch diameter, so that the center distance = (4.950 + 2.262)/2 = 3.606.

The various dimensions follow:

	Driver	Follower
Number of teeth	35	16
Pitch diameter	4.950	2.262
OD	5.150	2.462
Pitch in inches to one turn	15.550	7.108
Angle of helix	45°	45°
Pitch of cutter	10	10
Number of cutter	2	3
Whole depth of tooth	0.216	0.216
Angle of shafts	90°	
Center distance of shafts	3.606	

Figuring Crossed Helical Gears (Angle Other Than 45°)

As there is no direct solution for a pair of helical gears, their calculation is a tedious process and the result must be found by trial.

As numerous calculations are absolutely necessary, this formula should not involve division by large or fractional numbers and should contain the fewest possible operations. Such formulas are:

Let C = center distance.
P = diametral pitch.
N_1 = number of teeth in the driver.
N_2 = number of teeth in the follower.
S_1 = helix angle of driver.
S_2 = helix angle of follower.

Then

$$2C = \frac{(\sec S_1\, N_1) + (\sec S_2\, N_2)}{P}$$

That is, the sum of the secant of the driving angle times the number of teeth in the driver, and the secant of the follower angle times the number of teeth in it, divided by the diametral pitch, equals twice the center distance. This formula is derived as follows: The secant of the helix angle times the pitch diameter of a spur gear of the same number of teeth and pitch equals the pitch diameter of a helical gear of that angle, the pitch of the spur gear being the same as the normal pitch of the helical gear. Now for a spur gear, the number of teeth divided by the diametral pitch equals the pitch diameter. Therefore, *the secant of the helix angle* $\times N/P =$

TABLE 13. CALCULATION OF 45° CROSSED HELICAL GEARS

Pitch of Cutter	Pitch Dia	Pitch of Helical In. to One Turn	No. of Teeth in Spur Same Curvature	OD	Thickness of Tooth at Pitch Line (Normal)	Depth of Tooth	Clearance	Cir Pitch (Normal)
	Multiply by No. of Teeth in Helical Gear			Add to PD				
2	0.70710	2.22142	2.828	1.0000	0.7854	1.0785	0.0785	1.5708
2¼	0.62855	1.97464	2.828	0.8888	0.6981	0.9587	0.0699	1.3963
2½	0.56566	1.77707	2.828	0.8000	0.6283	0.8628	0.0628	1.2566
2¾	0.51425	1.61556	2.828	0.7273	0.5712	0.7844	0.0572	1.1424
3	0.47140	1.48094	2.828	0.6666	0.5236	0.7190	0.0524	1.0472
3½	0.40406	1.26939	2.828	0.5714	0.4488	0.6163	0.0449	0.8976
4	0.35355	1.11071	2.828	0.5000	0.3927	0.5393	0.0393	0.7854
5	0.28283	0.88853	2.828	0.4000	0.3142	0.4314	0.0314	0.6283
6	0.23570	0.74047	2.828	0.3333	0.0618	0.3595	0.0262	0.5236
7	0.20203	0.63469	2.828	0.2857	0.2244	0.3081	0.0224	0.4488
8	0.17677	0.55534	2.828	0.2500	0.1965	0.2696	0.0196	0.3927
9	0.15714	0.49367	2.828	0.2222	0.1745	0.2397	0.0175	0.3491
10	0.14143	0.44431	2.828	0.2000	0.1571	0.2157	0.0157	0.3142
11	0.12856	0.40388	2.828	0.1818	0.1428	0.1961	0.0143	0.2856
12	0.11785	0.37024	2.828	0.1666	0.1309	0.1798	0.0131	0.2618
14	0.10101	0.31733	2.828	0.1429	0.1122	0.1541	0.0112	0.2244
16	0.08836	0.27759	2.828	0.1250	0.0982	0.1348	0.0098	0.1963
18	0.07855	0.24677	2.828	0.1111	0.0873	0.1198	0.0088	0.1745
20	0.07071	0.22214	2.828	0.1000	0.0785	0.1079	0.0079	0.1571
22	0.06428	0.20194	2.828	0.0909	0.0714	0.0980	0.0071	0.1428
24	0.05892	0.18510	2.828	0.0833	0.0654	0.0898	0.0065	0.1309
26	0.05437	0.17081	2.828	0.0769	0.0604	0.0829	0.0060	0.1208
28	0.05050	0.15865	2.828	0.0714	0.0561	0.0770	0.0056	0.1122
30	0.04713	0.14806	2.828	0.0666	0.0524	0.0719	0.0053	0.1047
32	0.04425	0.13901	2.828	0.0625	0.0491	0.0674	0.0050	0.0982
36	0.03929	0.12343	2.828	0.0555	0.0436	0.0599	0.0043	0.0785
40	0.03533	0.11099	2.828	0.0500	0.0393	0.0539	0.0039	0.0873
48	0.02944	0.09249	2.828	0.0417	0.0327	0.0449	0.0033	0.0654

the pitch diameter of a helical gear. The combined pitch diameters times the center distance are equal to

$$\left(\sec S_1 \times \frac{N_1}{P}\right) + \left(\sec S_2 \times \frac{N_2}{P}\right)$$

or $(\sec S_1 N_1) + (\sec S_2 N_2)$ for one diametral pitch.

The quantity sec S_1N_1 is the pitch diameter for the driver, and sec S_2N_2 is the pitch diameter of the follower. To obtain the center distance for any other pitch, it is simply necessary to divide this last result by that pitch.

A table of secants will furnish constants covering the entire range of angles and, therefore, all possible solutions for a pair of gears.

Selecting Secants and Trial Numbers of Teeth

To calculate a pair of crossed helical gears, select secants for the desired angles, assuming the normal pitch, try out the value of $2C$ with trial numbers of teeth for driver and follower.

If the value $2C$ is too small, increase the number of teeth and try again. A very few calculations will show the number of teeth to secure the closest result.

If the center distance thus found is not as desired the angles must be shifted, keeping in mind the general laws governing the change of the center distance with the angle.

It is often found that when the desired center distance is reached the driving angle is too large to be desirable. The only alternative is to change the normal pitch and try again. A slide rule will give approximate results.

When there are limitations placed on the diameter of one or both of the gears, the following formula is of value. It may also serve as a check on the above calculations. The pitch diameters are assumed.

$$\tan S_1 = \frac{(\text{PD of driver} \times \text{number of revolutions of driver})}{(\text{PD of follower} \times \text{number of revolutions of follower})}$$

This will set a limit on the driving angle S_1, to exceed which means that the gear will be too large.

Real Pitches for Circular-pitch Helical Gears

Table 14 will be found convenient in figuring particulars for helical gearing, because it eliminates much of the work by shortening the process, thus making it quite an easy and simple matter to find the dimensions for either helical gears with axes parallel to each other or for gears with right-angle drive.

Formulas for use with the table are as follows: Circumference on pitch line = real pitch multiplied by number of teeth.

Lead of helix = circumference on pitch line divided by the tangent.

Pitch diameter = circumference divided by 3.1416.

For whole diameter add the same amount above pitch line as for spur wheels of the same pitch as the normal pitch.

The following is an example of the use of the table: A pair of wheels is required to be: ratio, 6 to 1; normal pitch, 1 in.; driver, 6 teeth; follower, 36 teeth; angle for driver, 66°; angle for follower, 24°.

Referring to the table, we find that the real pitch for the driver is 2.4585.

2.4585 × 6 (teeth) = 14.751 (circumference on pitch line).

Cir 14.751 ÷ 2.246 (tangent) = 6.567 (lead of helix).

Cir 14.751 ÷ 3.1416 = 4.695 (pitch diameter).

For the follower the real pitch is 1.0946.

1.0946 × 36 = 39.4056 (circumference).

Cir 39.4056 ÷ 0.4452 (tangent) = 88.512 (lead of helix).

Cir 39.4056 ÷ 3.1416 = 12.543 (pitch diameter).

Table 14. Real Pitches for Circular-pitch Helical Gears

Angles from Axis		Normal Pitches													
		1/4″		5/16″		3/8″		7/16″		1/2″		5/8″		3/4″	
Driver	Follower	Driver	Follower	Driver	Follower	Driver	Follower	Driver	Follower	Driver	Follower	Driver	Follower	Driver	Follower
80°	10°	1.4396	0.2538	1.7996	0.3173	2.1595	0.3808	2.5194	0.4442	2.8793	0.5077	3.5992	0.6346	4.3190	0.7616
79°	11°	1.3102	0.2546	1.6377	0.3183	1.9653	0.3820	2.2928	0.4457	2.6304	0.5093	3.2755	0.6366	3.9306	0.7640
78°	12°	1.2024	0.2556	1.5030	0.3195	1.8036	0.3833	2.1042	0.4473	2.4049	0.5112	3.0060	0.6390	3.6072	0.7666
77°	13°	1.1113	0.2565	1.3892	0.3207	1.6669	0.3848	1.9449	0.4490	2.2226	0.5131	2.7784	0.6414	3.3338	0.7696
76°	14°	1.0334	0.2576	1.2917	0.3220	1.5500	0.3865	1.8084	0.4509	2.0668	0.5153	2.5834	0.6440	3.1000	0.7730
75°	15°	0.9659	0.2591	1.2074	0.3238	1.4489	0.3885	1.6903	0.4529	1.9318	0.5181	2.4148	0.6476	2.8978	0.7772
74°	16°	0.9069	0.2600	1.1337	0.3250	1.3605	0.9301	1.5872	0.4551	1.8139	0.5201	2.2674	0.6502	2.7210	0.7802
73°	17°	0.8550	0.2614	1.0688	0.3268	1.2826	0.3931	1.4964	0.4575	1.7101	0.5228	2.1377	0.6536	2.5652	0.7842
72°	18°	0.8090	0.2628	1.0112	0.3286	1.2135	0.3943	1.4157	0.4600	1.6180	0.5257	2.0225	0.6572	2.4270	0.7886
71°	19°	0.7678	0.2644	0.9598	0.3305	1.1518	0.3966	1.3438	0.4627	1.5357	0.5288	1.9196	0.6610	2.3036	0.7932
70°	20°	0.7309	0.2660	0.9137	0.3325	1.0964	0.3990	1.2791	0.4656	1.4619	0.5320	1.8274	0.6651	2.1928	0.7981
69°	21°	0.6976	0.2678	0.8720	0.3347	1.0464	0.4017	1.2208	0.4686	1.3952	0.5356	1.7440	0.6694	2.0928	0.8034
68°	22°	0.6673	0.2696	0.8342	0.3370	1.0010	0.4044	1.1679	0.4718	1.3346	0.5392	1.6684	0.6740	2.0020	0.8088
67°	23°	0.6398	0.2716	0.7998	0.3395	0.9597	0.4074	1.1120	0.4752	1.2796	0.5432	1.5996	0.6790	1.9194	0.8148
66°	24°	0.6146	0.2736	0.7683	0.3420	0.9220	0.4105	1.0756	0.4789	1.2292	0.5472	1.5366	0.6841	1.8440	0.8210
65°	25°	0.5915	0.2758	0.7394	0.3448	0.8873	0.4137	1.0352	0.4827	1.1830	0.5516	1.4788	0.6896	1.7746	0.8274
64°	26°	0.5702	0.2781	0.7129	0.3477	0.8554	0.4172	0.9980	0.4868	1.1406	0.5563	1.4258	0.6954	1.7108	0.8344
63°	27°	0.5507	0.2806	0.6883	0.3507	0.8260	0.4209	0.9636	0.4910	1.1014	0.5612	1.3766	0.7014	1.6520	0.8418
62°	28°	0.5325	0.2831	0.6656	0.3539	0.7988	0.4247	0.9319	0.4955	1.0650	0.5662	1.3312	0.7078	1.5976	0.8949
61°	29°	0.5157	0.2858	0.6446	0.3573	0.7735	0.4287	0.9024	0.5002	1.0314	0.5716	1.2892	0.7146	1.5470	0.8574
60°	30°	0.5000	0.2886	0.6250	0.3608	0.7500	0.4330	0.8750	0.5051	1.0000	0.5773	1.2500	0.7216	1.5000	0.8660
59°	31°	0.4853	0.2916	0.6067	0.3646	0.7281	0.4375	0.8494	0.5104	0.9708	0.5833	1.2134	0.7292	1.4562	0.8750
58°	32°	0.4717	0.2948	0.5897	0.3685	0.7076	0.4422	0.8256	0.5159	0.9435	0.5896	1.1794	0.7370	1.4152	0.8844
57°	33°	0.4590	0.2981	0.5738	0.3726	0.6885	0.4471	0.8033	0.5216	0.9180	0.5962	1.1476	0.7452	1.3770	0.8942
56°	34°	0.4470	0.3015	0.5588	0.3769	0.6706	0.4523	0.7824	0.5277	0.8941	0.6031	1.1176	0.7538	1.3412	0.9046
55°	35°	0.4358	0.3052	0.5448	0.3815	0.6538	0.4578	0.7627	0.5341	0.8717	0.6104	1.0896	0.7630	1.3076	0.9156
54°	36°	0.4253	0.3090	0.5316	0.3862	0.6380	0.4635	0.7443	0.5408	0.8506	0.6180	1.0632	0.7724	1.2760	0.9270
53°	37°	0.4154	0.3130	0.5192	0.3913	0.6231	0.4695	0.7269	0.5478	0.8308	0.6260	1.0384	0.7826	1.2462	0.9390
52°	38°	0.4060	0.317[illegible]	0.5076	0.3965	0.6091	0.4759	0.7161	0.5552	0.8121	0.6345	0.0152	0.7930	1.2182	0.9518
51°	39°	0.3972	0.3217	0.4965	0.4021	0.5959	0.4825	0.6952	0.5628	0.7945	0.6434	0.9930	0.8042	1.1918	0.9650
50°	40°	0.3889	0.3264	0.4861	0.4079	0.5834	0.4895	0.6806	0.5711	0.7778	0.6527	0.9722	0.8159	1.1668	0.9790
49°	41°	0.3810	0.3312	0.4763	0.4140	0.5716	0.4969	0.6668	0.5797	0.7621	0.6625	0.9526	0.8280	1.1432	0.9938
48°	42°	0.3736	0.3364	0.4670	0.4205	0.5604	0.5046	0.6538	0.5887	0.7472	0.6728	0.9340	0.8410	1.1208	1.0092
47°	43°	0.3665	0.34[illegible]8	0.4582	0.4272	0.5498	0.5127	0.6415	0.5982	0.7331	0.6836	0.9164	0.8544	1.0996	1.0254
46°	44°	0.3598	0.3475	0.4498	0.4344	0.5398	0.5213	0.6298	0.6082	0.7197	0.6950	0.8996	0.8688	1.0796	1.0426
45°	45°	0.3536	0.3536	0.4419	0.4419	0.5303	0.5303	0.6187	0.6187	0.7071	0.7071	0.8839	0.8839	1.0606	1.0606

Angles from Axis		Normal Pitches												Tangent of Angle	
		7/8"		1"		1 1/8"		1 1/4"		1 3/8"		1 1/2"			
Driver	Follower	Driver	Follower	Driver	Follower	Driver	Follower	Driver	Follower	Driver	Follower	Driver	Follower	Driver	Follower
80°	10°	5.0388	0.8884	5.7587	1.0154	6.4785	1.1423	7.1983	1.2692	7.8192	1.3962	8.6380	1.5230	5.6713	0.1763
79°	11°	4.5856	0.8914	5.2407	1.0187	5.8961	1.1460	6.5510	1.2732	7.2061	1.4007	7.8612	1.5280	5.1446	0.1944
78°	12°	4.2084	0.8946	4.8098	1.0223	5.4109	1.1501	6.0122	1.2780	6.6134	1.4057	7.2146	1.5335	4.7046	0.2126
77°	13°	3.8898	0.8980	4.4454	1.0263	5.0001	1.1546	5.5568	1.2828	6.1125	1.4111	6.6681	1.5394	4.3315	0.2309
76°	14°	3.6168	0.9018	4.1336	1.0306	4.6500	1.1594	5.1669	1.2882	5.6837	1.4171	6.2004	1.5459	4.0108	0.2493
75°	15°	3.3806	0.9058	3.8637	1.0353	4.3466	1.1658	4.8296	1.2952	5.3126	1.4248	5.7955	1.5543	3.7321	0.2679
74°	16°	3.1744	0.9102	3.6279	1.0403	4.0814	1.1703	4.5348	1.3004	4.9884	1.4304	5.4419	1.5604	3.4874	0.2867
73°	17°	2.9928	0.9150	3.4203	1.0457	3.8478	1.1764	4.2754	1.3071	3.7029	1.4378	5.1304	1.5685	3.2709	0.3057
72°	18°	2.8314	0.9200	3.2360	1.0515	3.6405	1.1829	4.0450	1.3143	4.4495	4.4457	4.8540	1.5771	3.0777	0.3249
71°	19°	2.6876	0.9254	3.0715	1.0576	3.4554	1.1898	3.8394	1.3220	4.2233	1.4542	4.6072	1.5864	2.9042	0.3443
70°	20°	2.5582	0.9312	2.9238	1.0641	3.2892	1.1972	3.6548	1.3302	4.0202	1.4632	4.3857	1.5962	2.7475	0.3640
69°	21°	2.4416	0.9372	2.7904	1.0711	3.1391	1.2050	3.4880	1.3389	3.8368	1.4728	4.1856	1.6067	2.6051	0.3839
68°	22°	2.3358	0.9436	2.6694	1.0785	3.0030	1.2133	3.3369	1.3480	3.6705	1.4829	4.0041	1.6178	2.4751	0.4040
67°	23°	2.2240	0.9504	2.5593	1.0863	2.8792	1.2221	3.1991	1.3580	3.5190	1.4937	3.8389	1.6295	2.3559	0.4245
66°	24°	2.1512	0.9578	2.4585	1.0946	2.7659	1.2314	3.0732	1.3682	3.3805	1.5051	3.6876	1.6418	2.2460	0.4452
65°	25°	2.0704	0.9654	2.3662	1.1033	2.6619	1.2412	2.9577	1.3792	3.2535	1.5171	3.5493	1.6550	2.1455	0.4663
64°	26°	1.9960	0.9736	2.2811	1.1127	2.5663	1.2517	2.8515	1.3908	3.1366	1.5298	3.4218	1.6689	2.0503	0.4877
63°	27°	1.9272	0.9820	2.2026	1.1223	2.4780	1.2626	2.7533	1.4029	3.0287	1.5431	3.3040	1.6835	1.9626	0.5095
62°	28°	1.8638	0.9910	2.1300	1.1326	2.3963	1.2741	2.6625	1.4157	2.9288	1.5572	3.1950	1.6988	1.8807	0.5317
61°	29°	1.8048	1.0004	2.0626	1.1433	2.3205	1.2862	2.5783	1.4292	2.8361	1.5721	3.0940	1.7150	1.8040	0.5543
60°	30°	1.7500	1.0102	2.0000	1.1548	2.2500	1.2990	2.5000	1.4433	2.7500	1.5877	3.0000	1.7320	1.7321	0.5774
59°	31°	1.6988	1.0208	1.9415	1.1667	2.1843	1.3125	2.4269	1.4583	2.6696	1.6042	2.9134	1.7499	1.6643	0.6009
58°	32°	1.6512	1.0318	1.8870	1.1790	2.1229	1.3266	2.3588	1.4740	2.5947	1.6213	2.8306	1.7687	1.6003	0.6249
57°	33°	1.6066	1.0432	1.8360	1.1923	2.0655	1.3414	2.2951	1.4904	2.5246	1.6395	2.7451	1.7885	1.5399	0.6494
56°	34°	1.5648	1.0554	1.7883	1.2062	2.0118	1.3569	2.2353	1.5077	2.4589	1.6585	2.6824	1.8093	1.4826	0.6745
55°	35°	1.5254	1.0682	1.7434	1.2207	1.9614	1.3734	2.1793	1.5260	2.3972	1.6786	2.6151	1.8311	1.4281	0.7002
54°	36°	1.4886	1.0816	1.7012	1.2360	1.9139	1.3905	2.1266	1.5449	2.3392	1.6997	2.5519	1.8541	1.3764	0.7265
53°	37°	1.4538	1.0956	1.6614	1.2521	1.8693	1.4086	2.0070	1.5652	2.2847	1.7217	2.4924	1.8781	1.3270	0.7536
52°	38°	1.4322	1.1104	1.6242	1.2690	1.8273	1.4276	2.0303	1.5862	2.2333	1.7449	2.4364	1.9035	2.2799	0.7813
51°	39°	1.3904	1.1256	1.5890	1.2867	1.7876	1.4476	1.9862	1.6084	2.1847	1.7693	2.3835	1.9301	1.2349	0.8098
50°	40°	1.3612	1.1422	1.5557	1.3054	1.7502	1.4686	1.9446	1.6317	2.1391	1.7949	2.3336	1.9581	1.1918	0.8391
49°	41°	1.3336	1.1594	1.5240	1.3250	1.7148	1.4906	1.9054	1.6562	2.0958	1.8219	2.2863	1.9875	1.1504	0.8693
48°	42°	1.3076	1.1774	1.4944	1.3456	1.6812	1.5138	1.8681	1.6820	2.0549	1.8502	2.2417	2.0184	1.1106	0.9004
47°	43°	1.2830	1.1964	1.4662	1.3673	1.6495	1.5382	1.8328	1.7089	2.0161	1.8001	2.1994	2.0510	1.0724	0.9325
46°	44°	1.2596	1.2164	1.4395	1.3901	1.6194	1.5639	1.7994	1.7377	1.9792	1.9114	2.1593	2.0852	1.0355	0.9657
45°	45°	1.2374	1.2374	1.4142	1.4142	1.5910	1.5910	1.7678	1.7678	1.9445	1.9445	2.1213	2.1213	1.0000	1.0000

Table 15. Helical-gear Table
Shaft Angles 90° for One Diametral Pitch

Angle of Helix,°	To obtain the circular pitch for one tooth, divide by the required diametral pitch	To obtain the PD, divide by the required diametral pitch and multiply the quotient by the required number of teeth	To obtain the lead of helix, divide by the required diametral pitch and multiply quotient by the required number of teeth		To obtain the PD, divide by the required diametral pitch and multiply the quotient by the required number of teeth	To obtain the circular pitch for one tooth, divide by the required diametral pitch	Angle of Helix,°
	Cir Pitch	One Tooth or Addendum	Lead of Helix		One Tooth or Addendum	Cir Pitch	
Small Wheel	Small Wheel	Small Wheel	Small Wheel	Large Wheel	Large Wheel	Large Wheel	Large Wheel
1	3.1419	1.0001	180.05	3.1420	57.298	180.01	89
2	3.1435	1.0006	90.020	3.1435	28.653	90.016	88
3	3.1457	1.0013	60.032	3.1458	19.107	60.026	87
4	3.1491	1.0024	45.038	3.1492	14.335	45.035	86
5	3.1535	1.0038	37.077	3.1527	11.473	36.044	85
6	3.1589	1.0055	30.056	3.1589	9.5667	30.055	84
7	3.1652	1.0075	25.728	3.1651	8.2055	25.778	83
8	3.1724	1.0098	22.573	3.1724	7.1852	22.573	82
9	3.1806	1.0124	20.082	3.1807	6.3924	20.082	81
10	3.1900	1.0154	18.092	3.1901	5.7587	18.092	80
11	3.2003	1.0187	16.464	3.2003	5.2408	16.464	79
12	3.2145	1.0232	15.076	3.2105	4.8097	15.104	78
13	3.2242	1.0263	13.966	3.2294	4.4454	13.988	77
14	3.2377	1.0306	12.986	3.2378	4.1335	12.986	76
15	3.2522	1.0352	12.138	3.2524	3.8637	12.138	75
16	3.2679	1.0402	11.393	3.2678	3.6279	11.397	74
17	3.2848	1.0456	10.417	3.2821	3.4203	10.745	73
18	3.3116	1.0514	10.192	3.3032	3.2360	10.166	72
19	3.3225	1.0576	9.6494	3.3225	3.0715	9.6494	71
20	3.3430	1.0641	9.1848	3.3433	2.9238	9.1854	70
21	3.3650	1.0711	8.7662	3.3652	2.7904	8.7663	69
22	3.3882	1.0785	8.3862	3.3833	2.6694	8.3862	68
23	3.4127	1.0863	8.0399	3.4129	2.5593	8.0403	67
24	3.4451	1.0946	7.7379	3.4391	2.4585	7.7242	66
25	3.4661	1.1033	7.4332	3.4663	2.3662	7.4336	65
26	3.4953	1.1126	7.1664	3.4952	2.2811	7.1663	64
27	3.5258	1.1223	6.9198	3.5257	2.2026	6.9197	63
28	3.5579	1.1325	6.6912	3.5575	2.1300	6.6916	62
29	3.5918	1.1433	6.4799	3.5919	2.0626	6.4799	61
30	3.6276	1.1547	6.2778	3.6277	2.0000	6.2832	60
31	3.6650	1.1666	6.0979	3.6652	1.9416	6.0997	59
32	3.7043	1.1791	5.9282	3.7044	1.8870	5.9282	58
33	3.7457	1.1923	5.7710	3.7459	1.8360	5.7680	57
34	3.7894	1.2062	5.6181	3.7826	1.7882	5.6178	56
35	3.8349	1.2207	5.4754	3.8351	1.7434	5.4770	55
36	3.8830	1.2360	5.3431	3.8834	1.7013	5.3448	54
37	3.9336	1.2521	5.2201	3.9261	1.6616	5.2200	53
38	3.9867	1.2690	5.1028	3.9921	1.6242	5.1026	52
39	4.0482	1.2867	4.9866	4.0416	1.5890	4.9920	51
40	4.1010	1.3054	4.8873	4.1012	1.5557	4.8874	50
41	4.1626	1.3250	4.7885	4.1540	4.5242	4.7884	49
42	4.2273	1.3456	4.6949	4.2272	1.4944	4.6948	48
43	4.2956	1.3673	4.6065	4.2956	1.4662	4.6062	47
44	4.3671	1.3901	4.5223	4.3675	1.4395	4.5225	46
45	4.4428	1.4142	4.4428	4.4428	1.4142	4.4428	45

Another method of finding the lead of helix is to multiply the real pitch by the number of teeth, but for this purpose take the real pitch of the mating wheel.

In the above example we should have

Real pitch of follower, 1.0946 × 6 = 6.5676 (lead of helix).

Real pitch of driver, 2.4585 × 36 = 88.506.

It will be noticed that there is a slight difference in the result, but this is unimportant, as it is only brought about by the dropping of a few decimal points in the tangent.

Helical-gear Table

Table 15 gives the circular pitch and addendum or the diametral pitch and lead of helices for one diametral pitch and with teeth having angles of 1 and 89° to 45 and 45°. For other pitches, divide the addendum given and the helix number by the required pitch, and multiply the results by the required number of teeth. This will give the pitch diameter and lead of helix for each wheel. For the outside diameter add two diametral pitches as in spur gearing.

Suppose we want a pair of helical gears with 10 and 80° angles, 8-diametral-pitch cutter, with 16 teeth in the small gear, having a 10° angle and 10 teeth in the large gear with its 80° angle.

Find the 10° angle of helix and in the third column find 1.0154. Divide by pitch, 8, and get 0.1269. Multiply this by number of teeth — 0.1269 × 16 = 2.030 = pitch diameter. Add 2 pitches — two × $\frac{1}{8}$ = $\frac{1}{4}$. Thus, 2.030 + 0.25 = 2.28 in. outside diameter.

The lead of helix for 10° for small wheel is 18.092. Divide by pitch = 18.092 ÷ 8 = 2.2615. Multiply by number of teeth, 2.2615 × 16 = 36.18, the lead of helix, which means that it makes one turn in 36.18 in.

For the other gear with its 80° angle, find the addendum, 5.7587. Divide by pitch, 8, obtaining 0.7198. Multiply by number of teeth, 10, obtaining 7.198. Add two pitches, or 0.25, obtaining 7.448 as outside diameter.

The lead of helix is 3.1901. Divide by pitch, 8, obtaining 0.3988. Multiply by number of teeth to obtain 3.988 the lead of helix.

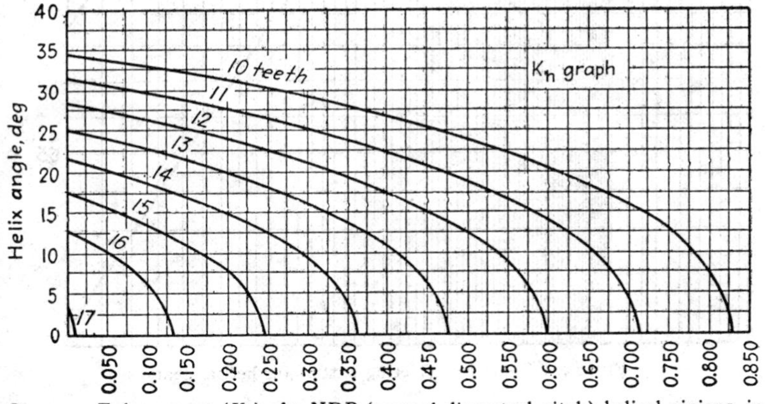

FIG. 14. Enlargement (K_h) of 1-NDP (normal diametral pitch) helical pinions, in inches.

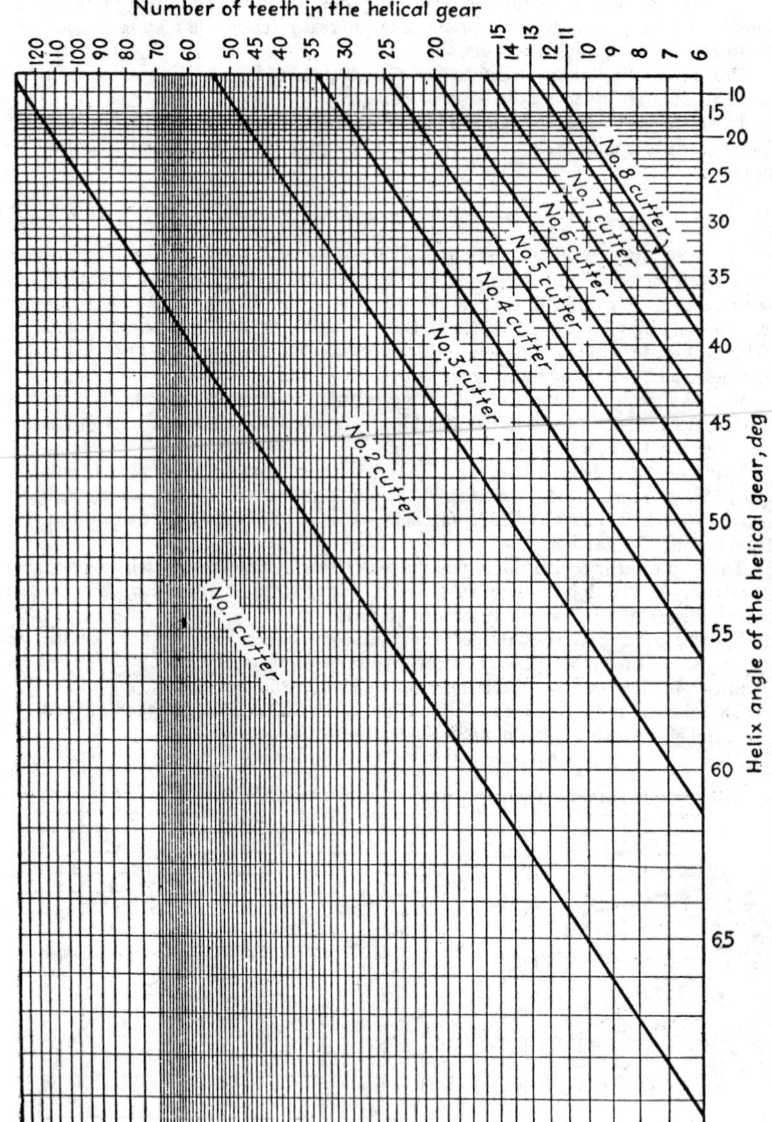

FIG. 15. Chart for selecting cutters for helical gears.

When racks are to mesh with helical gears, divide the number in the circular pitch columns for the given angle by the required diametral pitch to get the corresponding circular pitch.

If we want to make a rack to mesh with a 40° helical gear of 8 pitch: Look for circular pitch opposite 40 and find 4.101. Divide by 8, obtaining 0.512 as the circular pitch for this angle. The greater the angle, the greater the circular or linear pitch, as can be seen by trying an 80° angle. Here the circular pitch is 2.261 in.

Enlargement of Helical Pinions 20° Normal Pressure Angle

The enlargement of helical pinions to avoid undercut cannot be tabulated as simply as for spur pinions because of the wide range of helix angles used. It is recommended that helical pinions be enlarged in accordance with Fig. 14 and formulas.

When using an enlarged helical pinion, either the mating gear must be reduced in diameter or the center distance must be increased to correspond with the enlargement, as in spur gears.

Formulas

$$P_t = P_n \cos \psi \quad (1) \qquad d = \frac{n}{P_t} \quad (2) \qquad d_o = d + \frac{2 + K_h}{P_n} \quad (3)$$

Example: 12 teeth, 32 normal diametral pitch, 20° normal pressure angle, and 18° helix angle.

$$P_t = 32 \times 0.95106 = 30.4339 \quad (1)$$

$$d = \frac{12}{30.4339} = 0.3943 \quad (2)$$

$$K_h = 0.388 \text{ (from graph, Fig. 14)}$$

$$d_o = 0.3943 + \frac{2 + 0.388}{32} = 0.468 \quad (3)$$

Spur-gear Cutters for Helical Gears

To find the number of a spur-gear cutter to be used in cutting a given helical gear, locate the intersection of lines traced from the points representing the number of teeth and the helix angle on the two scales (Fig. 15). The number in the area on the chart within which the intersection falls is the cutter number required in Brown & Sharpe's involute cutter system.

Tolerances for Spur and Helical Gears

Coarse-pitch Gears

Tolerances for spur and helical gears, the latter including parallel and crossed-helical gears and double-helical or herringbone gears, are given in AGMA Standards 231.01, and ASA Standard B6.6-1946. The data in Table 16 do not form a rigid specification but are intended to form a practical basis upon which gearing can be manufactured, inspected, and sold. Tooth errors can be readily measured, except in the case of high-speed large gears, where it is often preferable to obtain desired gear accuracy through control of the machine.

Diameter Range. The diameters considered are from $\frac{3}{4}$ to 100 in., arranged in an approximate geometric progression, namely: $\frac{3}{4}$, $1\frac{1}{2}$, 3, 6, 12, 25, 50, and 100 in. Tolerances for other diameters may be interpolated.

Pitch Range. The pitches considered are from 1 to 32 diametral pitch, arranged in a geometric progression, namely: 1, 2, 4, 8, 16, and 32. Tolerances for other pitches may be interpolated.

Classes and Speeds. Four classes of spur and helical gears are listed according to accuracy. These are:

Class 1.......... Up to 80 fpm
Class 2.......... Up to 400 fpm
Class 3.......... Up to 2000 fpm
Class 4.......... Over 2000 fpm

These classes of gears and their definitions in terms of speed are purely arbitrary. There are cases when tolerances for class 4 gears are applicable for lower speeds, for example, timing mechanisms, dividing gears, and instruments.

Types of Errors. Fundamental errors for which tolerances are given are:

Runout.

Pitch error.

Accumulated error.

Profile error.

Lead error.

Allowances and tolerances for backlash are treated separately on page 4-25. Tooth-thickness and pitch-diameter tolerances are related to backlash and are therefore omitted.

Runout of a gear is the total difference between high and low readings of a dial indicator suitably arranged to denote the off-center relation of the axis of the tooth profiles with respect to the gear journals or the axis about which the gear rotates. It is twice the eccentricity. It includes the effect of side runout or wobble.

Runout and eccentricity can be held to close limits only by accurate setup of workpiece on machine; this calls for good accurate fits in journals and bores of blanks, shafts, arbors, and chucks, and squareness of clamping surfaces.

Runout may be measured by an indicator applied to the bottom lands of tooth spaces, provided the bottom lands have been machined simultaneously with the profiles.

Runout of teeth roughed deeper than the finish cut may be checked by a pin or V-block placed in the tooth spaces and actuating a dial indicator. Indicator readings so obtained may be affected by pitch errors.

It may be measured by an indicator applied at the outside diameter of the teeth if a topping cutter has been used.

It may also be measured by observing or graphically recording the center-distance variation when the gear to be inspected is rotated in mesh under spring pressure with a master gear, excluding the effect on such center-distance variation caused by profile errors, errors in spacing, interference, etc.

Runout is quite commonly measured by using a testing fixture arranged with two fingers, one fixed and the other actuating a dial indicator, both of which are mounted on movable blocks 180° apart, with the fingers making contact on the same set of profiles at the midface. The difference between the highest and lowest indicator readings around the gear represents twice the runout or four times the eccentricity. Obviously readings obtained are influenced by pitch and profile errors, but the two latter are in most instances relatively small as compared with the runout error.

TABLE 16. TOLERANCES FOR SPUR AND HELICAL GEARS
All Readings in 0.0001 In.

Class	Diametral Pitch	Runout of Pitch Dia, Total Indicator Reading								Pitch Error Measured on Pitch Circle in Plane of Rotation							
Pitch Dia		¾	1½	3	6	12	25	50	100	¾	1½	3	6	12	25	50	100
Class 1 (up to 80 fpm)	1					70	90	90	100					50	75	125	180
	2				60	60	80	80	90				40	50	70	100	150
	4			50	60	60	80	80	90			30	40	50	60	90	120
	8		30	50	60	60	80	80			15	20	25	30	35	35	
	16	30	30	50	60	60	80			15	17	20	25	25			
Class 2 (up to 400 fpm)	1					30	35	40	45					25	35	45	60
	2				20	20	25	30	35				10	15	20	25	30
	4			20	20	20	25	30	35			5	6	7	8	9	10
	8		15	15	20	20	25	30			5	5½	6	6½	7	7	
	16	15	15	15	20	20	25	30		3½	3½	4	4	4½	5	5½	
Class 3 (up to 2000 fpm)	2				20	20	25	25	30				5	5	5½	6	6
	4			10	20	20	25	25	30			3½	3½	4	4	5	5
	8		10	10	15	20	25	25	30		3½	3½	3½	4	4	5	5
	16	10	10	10	15	20	25	25		2½	2½	3	3	3½	3½	4	
	32	10	10	10	15	20	25			2	2	2½	2½	3	3		
Class 4 (over 2000 fpm)	4				10	10	12	14					3½	4	4	4	
	8			10	10	10	12	14	16			2½	2½	3	3	3	3
	16		10	10	10	10	12	14	16		2	2½	2½	2½	3	3	3
	32	10	10	10	10	10	12			2	2	2	2	2½	2½		

Class	Diametral Pitch	Accumulated Error between Any Two Teeth Exclusive of Runout Effect								Profile Error, Exclusive of Tip Modification. Total Variation, Not Plus and Minus							
Pitch Dia		¾	1½	3	6	12	25	50	100	¾	1½	3	6	12	25	50	100
Class 1 (up to 80 fpm)	1													50	60	70	80
	2												30	35	40	45	50
	4											25	30	35	40	40	40
	8										20	20	25	25	30	30	
	16									15	15	15	20	20	25		
Class 2 (up to 400 fpm)	1													30	35	40	40
	2												20	20	25	30	30
	4											10	10	11	12	12	12
	8										9	9	9	9	9	10	
	16									7	7	7	7	7	7	8	
Class 3 (up to 2000 fpm)	2				20	20	27	43	70				11	11	11	11	11
	4			15	15	15	21	33	53			8	8	8	8	8	8
	8		15	15	15	15	15	30	45		6	6	6	6	6	6	6
	16	15	15	15	15	15	15	24		5	5	5	5	5	5	5	
	32	15	15	15	15	15	15			5	5	5	5	5	5		
Class 4 (over 2000 fpm)	4				10	10	12	20					5	5	5	5	
	8			10	10	10	12	20	30			4	4	4	4	4	4
	16		10	10	10	10	12	20	30		3	3	3	3	3	3	3
	32	10	10	10	10	10	12			3	3	3	3	3	3		

Class	Diametral Pitch	Circumferential Lead Error per Unit Width of Face*			
Pitch Dia		1 In.	6 In.	12 In.	18 In.
Class 1 (up to 80 fpm)	1				
	2				
	4				
	8				
	16				
Class 2 (up to 400 fpm)	1	6	11		
	2	5	10		
	4	5	9		
	8	4	8		
	16	4	8		
Class 3 (up to 2000 fpm)	2	4	8		
	4	4	7		
	8	3	6		
	16	3	6		
	32	3	6		
Class 4 (over 2000 fpm)	4	3	6	8	10
	8	2	4	6	8
	16	2	4	6	8
	32	2	4	6	8

ASA B6.6—1946.

* See page 4-40.

On wide-face gears, the runout should be measured near both ends of the teeth. Appreciable difference in runout near the two faces of a gear is an indication of wobble, which is more serious than uniform runout.

Pitch error of a gear is the maximum difference between any two successive tooth-to-tooth readings taken at the pitch circle between corresponding sides of adjacent teeth. Either set of profiles may be used for tooth-to-tooth readings, but preferably the driving sides if the gear operates in one direction only.

Tooth-to-tooth readings may be taken by using a device consisting of a pair of fingers contacting adjacent teeth, one fixed and the other actuating a dial indicator, both of which are mounted on a movable block or slide that has some means to return it always to the same position. The *differences* between successive pairs of readings denote pitch variations.

In the case of spur gears or helical gears having low helix angles the tooth-to-tooth readings may be readily taken in the plane of rotation. However, at higher angles the indicator finger may give false readings from the side-cramping effect of the helical teeth. In such cases, readings may be taken in a plane normal to the helix angle, and converted to the plane of rotation for purposes of comparison and standardization by dividing by the cosine of the helix angle.

Accumulated error of a gear is the maximum accumulation of pitch errors obtained by algebraic addition of pitch errors.

Unless otherwise specified, accumulated error is checked over the entire 360° and expressed as the maximum error between any two teeth. In some cases it is measured only for one-half or one-third of a circle and occasionally tolerances are set for maximum accumulation of pitch errors over the number of teeth in contact at any instant. In deriving the accumulated error, the pitch errors must be corrected with reference to the true pitch.

For low- and medium-speed gears, accumulated error is of minor importance unless the gears are to be used for timing mechanisms, dividing gears, gun-sighting devices, or the like.

Profile error of a gear tooth is the difference between the highest and lowest readings of a dial indicator moving along the true involute curve in the plane of rotation and having a finger contacting the active tooth profile.

On spur gears or helical gears of low helix angles, the contacting finger may be set in the plane of rotation, but at higher angles it sometimes tends to give false readings because of the side-cramping effect of the helical teeth.

Tip relief or other departure from a true involute should be taken into consideration in interpreting profile-error readings.

Checking of the involute should be confined to that portion of the tooth comprising the active profile. Usually the inactive portion at or near the base circle is of no interest.

Active involute profile can be predicted when data on the mating gear are available. It may be expressed as a radius below which no involute action occurs, or in degrees of roll.

A considerable error in pressure angle may be acceptable on a gear provided the same error is present on the mating pinion.

Lead error of a helical gear is expressed as the circumferential tooth error per unit of face width.

It may be measured by traversing an indicator along a tooth parallel to the axis while the gear rotates in a timed relation according to the lead. Readings are taken in a plane normal to the helix to avoid side-cramping effect, and converted to the plane of rotation by dividing by the cosine of the helix angle.

The higher the helix angle, the more closely the *axial* lead error must be held to have satisfactory tooth contact between gears. The axial lead error is obtained by dividing the circumferential error by the tangent of the helix angle.

The lead error is usually considered to be independent of gear diameter or pitch.

Lead-error readings are greatly influenced by runout and therefore will be false unless the gear is mounted perfectly true for testing. The lead tolerances herewith do not include the effect of runout or wobble.

The values of circumferential lead error in Table 16 are not to be interpreted as plus and minus. It is assumed that the mating gear either has zero lead error or that its error is *in the same direction.* In other words, the tabulated values represent the maximum acceptable difference in circumferential lead error between a gear and its mate.

In determining whether the lead errors of a given gear are acceptable, it is intended that all columns of errors for the units of face width apply, up to the face width of the gear. For instance, for a precision high-speed gear of 4 pitch 20-in. face width, the maximum error in *any* 1 in. must not exceed 0.0003, in *any* 6 in. it must not exceed 0.0006, in *any* 12 in. it must not exceed 0.0008, and in *any* 18 in. it must not exceed 0.0010 in. Thus, if the error in this gear were 0.0010 in the first 5 in., it would not meet the tolerances, even though the error for the remaining 15 in. were zero.

The term "lead error" is not commonly associated with spur gears, but it is recommended that tolerances for circumferential error per unit of face width be applied to spur as well as helical gears.

Fine-pitch Gears

AGMA Standard 236.02 covers the inspection and tolerances of fine-pitch gears. This standard was developed to meet the diverse applications of fine-pitch gears and safeguard the interests of quality control. Fine-pitch gears lend themselves readily and economically to inspection methods in which measurements are made under conditions closely approaching those of their operation. The gear is rotated in intimate contact with a master, both mounted on a variable-center-distance fixture. The displacements or variations in center distance are measured or charted by suitable means, giving a composite check (see Fig. 16).

Quantity specifications are made on the basis of the results of the composite check, which do not involve backlash since backlash does not directly affect the accuracy of the gear operation and may be produced or eliminated in the mountings. A choice of backlash to suit the particular application is provided in this standard by giving each class group several ranges of backlash. Note that backlash is specified by a letter suffix to the class number, such as Commercial 1B or Precision 3C.

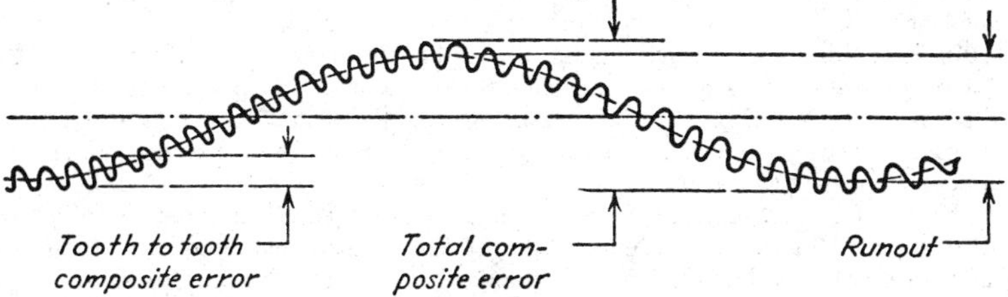

FIG. 16. By rolling a fine-pitch gear with a master in a variable-center-distance fixture, a composite check of errors is obtained.

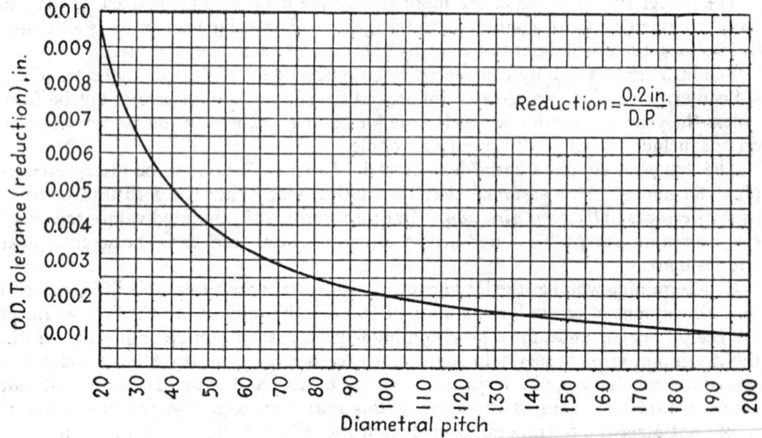

FIG. 17. Permissable reduction of OD tolerance of fine-pitch gears decreases rapidly with the diametral pitch.

OD Tolerance for Fine-pitch Gears

Figure 17 shows the permissible reduction in outside diameter of gears from 20 to 200 DP. This reduction is established as 0.200 in. per DP.

In order to avoid the danger of causing interference, the outside diameter should not exceed $N/DP + 2a$. The outside diameter may be reduced with safety, provided a satisfactory contact ratio is maintained. In fine-pitch gears the minimum satisfactory contact ratio is considered to be 1.200.

While this information is designed to be used in connection with the 20° involute fine-pitch standard gears, it may be used satisfactorily for involute gears of other pressure angles, provided a minimum contact ratio of 1.200 is maintained. Do not use the data for gears whose outside diameter is used for chucking or assembly purposes. In such cases, tolerances must be held closer.

Inspection Methods for Spur and Helical Gears

Measurement of tooth thickness can be done by several methods: (1) over accurately ground wires or pins of proper diameter, (2) by gear-tooth verniers and micrometers, (3) with dial-indicator setups, and (4) with optical comparators. The last method is particularly suitable for fine-pitch gears.

Pin Method. Tooth thickness at the pitch diameter is quickly checked by placing two accurately ground pins of proper diameter in opposite tooth spaces of even-tooth gears (Fig. 18) and measuring over the pins with a micrometer. In the case of gears with an odd number of teeth, the pins are placed in tooth spaces as nearly opposite as possible.

The resultant measurement is compared with the theoretical value in Tables 18 to 21 supplied by the Van Keuren Co. If the measurement is greater than the listed value, the tooth is thicker than standard; if it is less, the tooth is thin.

Pins or wires of the proper diameter must be used for each diametral pitch (see Table 22). The theoretical value of the measurement over wires is obtained by

TABLE 17. TOLERANCES AND BACKLASH FOR FINE-PITCH GEARS

Tolerances for Commercial Fine-pitch Gears

Class	Total Composite Error, In.	Tooth-to-tooth Composite Error, In.
Commercial 1	0.006	0.002
Commercial 2	0.004	0.0015
Commercial 3	0.002	0.001

Tolerances for Precision Fine-pitch Gears

Class	Total Composite Error, In.	Tooth-to-tooth Composite Error, In.
Precision 1	0.001	0.0004
Precision 2	0.005	0.0003
Precision 3	0.00025*	0.0002

Backlash for Different Classes of Gears

Diametral Pitch	*Backlash, In.*
Class A:	
20 to 45	0.004 to 0.006
46 to 70	0.003 to 0.005
71 to 90	0.002 to 0.0035
Class B:	
20 to 60	0.002 to 0.004
61 to 120	0.0015 to 0.003
121 and finer	0.001 to 0.002
Class C:	
20 to 60	0.001 to 0.002
61 to 120	0.0007 to 0.0015
121 and finer	0.0005 to 0.001
Class D:	
No measurable backlash at any pitch	

* This possibly would be the result of selection and segregation.

dividing values tabulated in Tables 18 to 21 by the diametral pitch of the gear being checked.

EXAMPLE: An even-toothed external gear of 26 teeth, 10 diametral pitch, $14\frac{1}{2}°$ pressure angle is to be checked. The correct wire size from Table 22 is 0.1728 in. From Table 18, the measurement over wires of a $1P$ gear is 28.4315 in. For a $10P$ gear, the measurement should be $28.4315 \div 10 = 2.8431$ in.

The measurements over wires, as given in Tables 18 to 21, are for gears of theoretical standard tooth thickness and do not consider backlash or inaccuracies in gear cutting. Because the wires contact the teeth on opposite sides of the gear, and also because the teeth are curved, the change in measurement over wires for thick or thin teeth may amount to $1\frac{1}{2}$ to 4 times the change in tooth thickness. To make corrections in the measurement over wires, when backlash is desired, or to find how much the teeth are thin, use the change factors K in Table 24.

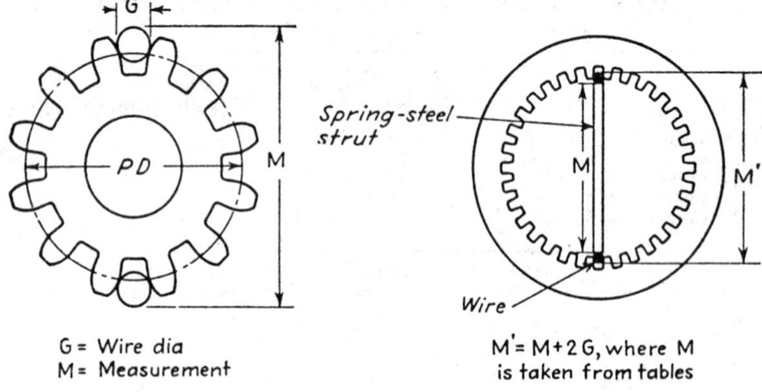

FIG. 18. Measurement over wires for external and internal gears.

How to Make Teeth Thin. EXAMPLE. On a 50-tooth, $14\frac{1}{2}°$ external gear, it is desired to make the teeth 0.016 in. thin. Reduction in measurement equals the change factor for a 50-tooth gear times $0.016 = 3.22 \times 0.016 = 0.0515$ in.

NOTE: The amount of reduction in measurement over wires is the same for all diametral pitches for a given amount of reduction of tooth thickness. In other words, the change factor K is the same for all diametral pitches.

To Obtain Backlash. If a generating gear cutter is fed into the blank an amount exactly equal to the whole depth of the tooth, the gear is cut without backlash. To provide backlash B, the teeth must be made thinner by an equal amount Δt (increment of t). The formulas for finding the excess depth of cut E from Δt and vice versa are:

Pressure Angle	E Extra Depth of Cut to Thin Teeth Δt	Δt Amount Teeth Are Cut Too Thin Because of Excess Depth of Cut E
$14\frac{1}{2}°$	$E = 1.93 \times \Delta t$	$\Delta t = E \div 1.93 = 0.52E$
$20°$	$E = 1.37 \times \Delta t$	$\Delta t = E \div 1.37 = 0.73E$

EXAMPLE: A total backlash of 0.016 in. is desired between a 30-tooth, $14\frac{1}{2}°$ pinion and a 100-tooth gear, the backlash to be taken equally on pinion and gear. From Table 24, the wire measurement for the pinion should be $2.98 \times 0.008 = 0.0238$ in. under the computed value, and $3.48 \times 0.008 = 0.0278$ in. undersize for the gear. If after a trial cut, the wire measurement on the gear is only 0.010 in. undersize, the backlash provided is $0.010 \div 3.48 = 0.0029$ in., or 0.0051 too little. Now from the formula for extra depth of cut, $E = 1.93 \times 0.0051 = 0.0098$ in.

TABLE 18. EXTERNAL GEARS. EVEN TEETH*

Measurements over Wires for 1 DP External Spur Gears with Wires 1.728 In. in Diameter

For Any Other Diametral Pitch, Divide Measurements Given in the Tables by the Diametral Pitch. All Measurements in Inches

No. of Teeth	14½°	20°	25°	No. of Teeth	14½°	20°	25°
6	8.2847	8.3020	8.3298	90	92.5076	92.4438	92.4356
8	10.3160	10.3266	10.3505	92	94.5085	94.4442	94.4359
				94	96.5094	96.4446	96.4362
10	12.3400	12.3445	12.3658	96	98.5102	98.4450	98.4364
12	14.3590	14.3578	14.3768	98	100.5110	100.4454	100.4366
14	16.3746	16.3680	16.3847				
16	18.3877	18.3765	18.3908	100	102.5118	102.4457	102.4368
18	20.3989	20.3837	20.3958	102	104.5126	104.4460	104.4370
				104	106.5134	106.4463	106.4372
20	22.4086	22.3899	22.4000	106	108.5142	108.4466	108.4374
22	24.4171	24.3953	24.4036	108	110.5149	110.4469	110.4376
24	26.4247	26.4000	26.4067				
26	28.4315	28.4040	28.4094	110	112.5156	112.4471	112.4378
28	30.4376	30.4074	30.4118	112	114.5162	114.4474	114.4380
				114	116.5167	116.4477	116.4382
30	32.4431	32.4103	32.4139	116	118.5172	118.4480	118.4384
32	34.4481	34.4128	34.4157	118	120.5177	120.4483	120.4386
34	36.4526	36.4150	36.4173				
36	38.4567	38.4171	38.4187	120	122.5182	122.4486	122.4388
38	40.4604	40.4191	40.4200	122	124.5187	124.4489	124.4390
				124	126.5192	126.4492	126.4391
40	42.4638	42.4211	42.4212	126	128.5197	128.4494	128.4392
42	44.4669	44.4231	44.4223	128	130.5202	130.4496	130.4393
44	46.4698	46.4250	46.4233				
46	48.4726	48.4268	48.4243	130	132.5207	132.4498	132.4394
48	50.4753	50.4284	50.4252	132	134.5212	134.4500	134.4395
				134	136.5217	136.4502	136.4396
50	53.4779	52.4298	52.4261	136	138.5222	138.4504	138.4397
52	54.4804	54.4310	54.4269	138	140.5226	140.4506	140.4398
54	56.4827	56.4320	56.4277				
56	58.4848	58.4329	58.4284	140	142.5230	142.4508	142.4399
58	60.4867	60.4337	60.4291	142	144.5234	144.4510	144.4400
				144	146.5238	146.4512	146.4401
60	62.4885	62.4345	62.4297	146	148.5242	148.4514	148.4402
62	64.4902	64.4353	64.4303	148	150.5246	150.4516	150.4403
64	66.4918	66.4361	66.4309				
66	68.4933	68.4369	68.4314	150	152.5250	152.4518	152.4404
68	70.4948	70.4377	70.4319	152	154.5254	154.4520	154.4405
				154	156.5258	156.4522	156.4406
70	72.4963	72.4384	72.4324	156	158.5262	158.4524	158.4407
72	74.4977	74.4391	74.4328	158	160.5265	160.4525	160.4408
74	76.4990	76.4397	76.4332				
76	78.5002	78.4403	78.4335	160	162.5268	162.4526	162.4409
78	80.5014	80.4409	80.4338	162	164.5271	164.4527	164.4410
				164	166.5274	166.4528	166.4411
80	82.5026	82.4414	82.4341	166	168.5277	168.4529	168.4412
82	84.5037	84.4419	84.4344	168	170.5280	170.4530	170.4413
84	86.5047	86.4424	86.4347				
86	88.5057	88.4429	88.4350	170	172.5283	172.4531	172.4414
88	90.5067	90.4434	90.4353	180	182.5298	182.4536	182.4419
				190	192.5310	192.4541	192.4423
				200	202.5321	202.4545	202.4426

Above 170 teeth, where the gear to be measured is not included in the table, use the number of teeth plus 2 for the figure to the left of the decimal point, and for the figure to the right of the decimal point interpolate between the table values.

* The Van Keuren Co.

Table 19. External Gears. Odd Teeth

Measurements over Wires for 1 DP External Spur Gears with Wires 1.728 In. in Diameter

For Any Other Diametral Pitch, Divide Measurements Given in the Tables by the Diametral Pitch. All Measurements in Inches

No. of Teeth	14½°	20°	25°	No. of Teeth	14½°	20°	25°
5	6.9933	7.0153	7.0472	91	93.4945	93.4304	93.4221
7	9.1121	9.1260	9.1536	93	95.4957	95.4311	95.4227
9	11.1827	11.1905	11.2142	95	97.4969	97.4318	97.4233
				97	99.4980	99.4324	99.4239
11	13.2317	13.2332	13.2536	99	101.4990	101.4330	101.4244
13	15.2677	15.2660	15.2826				
15	17.2957	17.2870	17.3031	101	103.4999	103.4336	103.4249
17	19.3181	19.3046	19.3181	103	105.5008	105.4342	105.4253
19	21.3365	21.3194	21.3299	105	107.5017	107.4348	107.4257
				107	109.5026	109.4354	109.4261
21	23.3521	23.3318	23.3402	109	111.5035	111.4360	111.4265
23	25.3656	25.3423	25.3492				
25	27.3773	27.3512	27.3570	111	113.5044	113.4365	113.4269
27	29.3875	29.3586	29.3637	113	115.5053	115.4370	115.4273
29	31.3965	31.3653	31.3695	115	117.5062	117.4375	117.4277
				117	119.5070	119.4380	119.4280
31	33.4046	33.3709	33.3746	119	121.5078	121.4385	121.4283
33	35.4119	35.3758	35.3791				
35	37.4185	37.3802	37.3831	121	123.5086	123.4390	123.4286
37	39.4245	39.3843	39.3867	123	125.5093	125.4394	125.4289
39	41.4299	41.3881	41.3899	125	127.5100	127.4398	127.4292
				127	129.5106	129.4402	129.4295
41	43.4348	43.3916	43.3928	129	131.5112	131.4406	131.4298
43	45.4394	45.3948	45.3954				
45	47.4437	47.3977	47.3977	131	133.5118	133.4409	133.4301
47	49.4477	49.4004	49.3997	133	135.5124	135.4412	135.4304
49	51.4514	51.4029	51.4015	135	137.5130	137.4415	137.4307
				137	139.5136	139.4418	139.4310
51	53.4548	53.4052	53.4031	139	141.5142	141.4421	141.4313
53	55.4579	55.4074	55.4045				
55	57.4608	57.4094	57.4058	141	143.5147	143.4424	143.4316
57	59.4636	59.4112	59.4070	143	145.5152	145.4427	145.4319
59	61.4663	61.4129	61.4082	145	147.5156	147.4430	147.4322
				147	149.5161	149.4433	149.4325
61	63.4689	63.4145	63.4094	149	151.5165	151.4436	151.4328
63	65.4714	65.4160	65.4105				
65	67.4737	67.4174	67.4115	151	153.5170	153.4439	153.4331
67	69.4758	69.4188	69.4125	153	155.5174	155.4442	155.4334
69	71.4778	71.4201	71.4135	155	157.5179	157.4445	157.4337
				157	159.5183	159.4448	159.4340
71	73.4797	73.4213	73.4144	159	161.5188	161.4451	161.4343
73	75.4815	75.4224	75.4153				
75	77.4832	77.4234	77.4162	161	163.5192	163.4453	163.4346
77	79.4848	79.4243	79.4171	163	165.5196	165.4455	165.4349
79	81.4863	81.4252	81.4180	165	167.5200	167.4457	167.4351
				167	169.5204	169.4459	169.4353
81	83.4877	83.4261	83.4188	169	171.5208	171.4461	171.4355
83	85.4891	85.4270	85.4196				
85	87.4905	87.4279	87.4203	171	173.5212	173.4463	173.4357
87	89.4919	89.4288	89.4209	181	183.5232	183.4473	183.4368
89	91.4932	91.4296	91.4215	191	193.5251	193.4483	193.4378
				201	203.5269	203.4493	203.4388

Table 20. Internal Gears. Even Teeth

Measurements between Wires for 1 DP Internal Spur Gears with Wires 1.44 In. in Diameter

For Any Other Diametral Pitch, Divide Measurements Given in the Tables by the Diametral Pitch. All Measurements in Inches

No. of Teeth	14½°	20°	25°
6	4.8173	4.6595	4.5206
8	6.8270	6.6609	6.5208
10	8.8337	8.6617	8.5209
12	10.8389	10.6623	10.5210
14	12.8433	12.6627	12.5210
16	14.8470	14.6630	14.5210
18	16.8501	16.6633	16.5210
20	18.8528	18.6635	18.5210
22	20.8551	20.6636	20.5211
24	22.8570	22.6638	22.5211
26	24.8586	24.6639	24.5211
28	26.8600	26.6640	26.5211
30	28.8613	28.6641	28.5211
32	30.8624	30.6642	30.5211
34	32.8634	32.6642	32.5211
36	34.8642	34.6643	34.5211
38	36.8650	36.6643	36.5211
40	38.8657	38.6644	38.5211
42	40.8664	40.6644	40.5211
44	42.8670	42.6645	42.5212
46	44.8676	44.6645	44.5212
48	46.8682	46.6646	46.5212
50	48.8687	48.6646	48.5212
52	50.8692	50.6646	50.5212
54	52.8696	52.6647	52.5212
56	54.8700	54.6647	54.5212
58	56.8704	56.6648	56.2212
60	58.8707	58.6648	58.5212
62	60.8711	60.6648	60.5212
64	62.8714	62.6648	62.5212
66	64.8717	64.6649	64.5212
68	66.8720	66.6649	66.5212
70	68.8723	68.6649	68.5212
72	70.8726	70.6649	70.5212
74	72.8729	72.6649	72.5212
76	74.8731	74.6649	74.5212
78	76.8733	76.6649	76.5212
80	78.8735	78.6649	78.5212
82	80.8737	80.6649	80.5213
84	82.8739	82.6649	82.5213
86	84.8741	84.6650	84.5213
88	86.8742	86.6650	86.5213
90	88.8744	88.6650	88.5213
92	90.8746	90.6650	90.5213
94	92.8747	92.6650	92.5213
96	94.8749	94.6650	94.5213
98	96.8750	96.6650	96.5213
100	98.8752	98.6650	98.5213
102	100.8753	100.6650	100.5213
104	102.8755	102.6650	102.5213
106	104.8756	104.6650	104.5213
108	106.8758	106.6650	106.5213
110	108.8759	108.6651	108.5213
112	110.8760	110.6651	110.5213
114	112.8761	112.6651	112.5213
116	114.8762	114.6651	114.5213
118	116.8763	116.6651	116.5213
120	118.8764	118.6651	118.5213
122	120.8765	120.6651	120.5213
124	122.8766	122.6651	122.5213
126	124.8767	124.6651	124.5213
128	126.8767	126.6651	126.5213
130	128.8768	128.6652	128.5213
132	130.8769	130.6652	130.5213
134	132.8770	132.6652	132.5213
136	134.8771	134.6652	134.5213
138	136.8772	136.6652	136.5213
140	138.8773	138.6652	138.5214
142	140.8773	140.6652	149.5214
144	142.8774	142.6652	142.5214
146	144.8774	144.6652	144.5214
148	146.8775	146.6652	146.5214
150	148.8775	148.6652	148.5214
152	150.8776	150.6652	150.5214
154	152.8776	152.6652	152.5214
156	154.8777	154.6652	154.5214
158	156.8778	156.6652	156.5214
160	158.8778	158.6652	158.5214
162	160.8779	160.6652	160.5214
164	162.8779	162.6652	162.5214
166	164.8780	164.6652	164.5214
168	166.8780	166.6652	166.5214
170	168.8781	168.6652	168.5214
180	178.8783	178.6653	178.5214
190	188.8785	188.6653	188.5214
200	198.8787	198.6653	198.5214

Above 170 teeth, where the gear to be measured is not included in the table, use the number of teeth minus 2 for the figure to the left of the decimal point and for the figure to the right of the decimal point interpolate between table values.

Table 21. Internal Gears. Odd Teeth

Measurements between Wires for 1 DP Internal Spur Gears with Wires 1.44 In. in Diameter

For Any Other Diametral Pitch, Divide Measurements Given in the Tables by the Diametral Pitch. All Measurements in Inches

No. of Teeth	14½°	20°	25°	No. of Teeth	14½°	20°	25°
5	3.5517	3.4090	3.2778	95	93.8618	93.6520	93.5081
7	5.6394	5.4823	5.3462	97	95.8623	95.6523	95.5084
9	7.6894	7.5231	7.3847	99	97.8628	97.6526	97.5087
11	9.7219	9.5490	9.4086	101	99.8632	99.6529	99.5090
13	11.7449	11.5670	11.4265	103	101.8636	101.6532	101.5093
15	13.7620	13.5801	13.4390	105	103.8639	103.6534	103.5096
17	15.7752	15.5901	15.4486	107	105.8642	105.6537	105.5098
19	17.7857	17.5980	17.4562	109	107.8645	107.6539	107.5100
21	19.7944	19.6046	19.4624	111	109.8648	109.6542	109.5102
23	21.8018	21.6099	21.4676	113	111.8651	111.6544	111.5104
25	23.8080	23.6144	23.4719	115	113.8653	113.6547	113.5106
27	25.8133	25.6182	25.4756	117	115.8656	115.6549	115.5108
29	27.8177	27.6213	27.4788	119	117.8659	117.6550	117.5110
31	29.8216	29.6240	29.4815	121	119.8662	119.6552	119.5112
33	31.8251	31.6264	31.4841	123	121.8664	121.6553	121.5113
35	33.8283	33.6286	33.4863	125	123.8667	123.6555	123.5114
37	35.8312	35.6306	35.4882	127	125.8669	125.6556	125.5115
39	37.8338	37.6325	37.4898	129	127.8671	127.6558	127.5116
41	39.8361	39.6343	39.4912	131	129.8674	129.6559	129.5117
43	41.8382	41.6360	41.4925	133	131.8676	131.6561	131.5118
45	43.8401	43.6375	43.4937	135	133.8678	133.6562	133.5119
47	45.8418	45.6388	45.4948	137	135.8680	135.6564	135.5120
49	47.8434	47.6399	47.4958	139	137.8682	137.6566	137.5121
51	49.8449	49.6409	49.4968	141	139.8684	139.6568	139.5122
53	51.8463	51.6418	51.4978	143	141.8686	141.6569	141.5123
55	53.8476	53.6426	53.4987	145	143.8688	143.6571	143.5124
57	55.8488	55.6433	55.4995	147	145.8690	145.6573	145.5125
59	57.8499	57.6440	57.5003	149	147.8692	147.6574	147.5126
61	59.8509	59.6447	59.5010	151	149.8694	149.6576	149.5127
63	61.8519	61.6454	61.5016	153	151.8696	151.6577	151.5128
65	63.8528	63.6461	63.5022	155	153.8698	153.6578	153.5129
67	65.8536	65.6467	65.5028	157	155.8700	155.6578	155.5130
69	67.8544	67.6472	67.5033	159	157.8702	157.6579	157.5131
71	69.8551	69.6477	69.5037	161	159.8704	159.6580	159.5132
73	71.8558	71.6481	71.5041	163	161.8706	161.6581	161.5133
75	73.8564	73.6485	73.5045	165	163.8708	163.6582	163.5133
77	75.8570	75.6489	75.5049	167	165.8710	165.6582	165.5133
79	77.8576	77.6493	77.5053	169	167.8712	167.6583	167.5133
81	79.8582	79.6497	79.5057	171	169.8713	169.6584	169.5133
83	81.8587	81.6501	81.5061	181	179.8722	179.6587	179.5135
85	83.8593	83.6505	83.5065	191	189.8730	189.6589	189.5137
87	85.8598	85.6508	85.5069	201	199.8737	199.6591	199.5139
89	87.8603	87.6511	87.5072				
91	89.8608	89.6514	89.5075				
93	91.8613	91.6517	91.5078				

Above 171 teeth, where the gear to be measured is not included in the table, use the number of teeth minus 2 for the figure to the left of the decimal point and for the figure to the right of the decimal point interpolate between table values.

TABLE 22. GEAR WIRE SIZES, IN.*

Diametral Pitch P	For External Gears $G = \frac{1.728}{P}$ In.	For Internal Gears $G = \frac{1.44}{P}$ In.
2	0.864	0.720
2½	0.6912	0.576
3	0.576	0.480
4	0.432	0.360
6	0.288	0.240
7	0.24686	0.20571
8	0.216	0.180
9	0.192	0.160
10	0.1728	0.144
11	0.15709	0.13091
12	0.144	0.120
14	0.12343	0.10286
16	0.108	0.090
18	0.096	0.080
20	0.0864	0.072
22	0.07855	0.06545
24	0.072	0.060
28	0.06171	0.05143
32	0.054	0.045
36	0.048	0.040
40	0.0432	0.036
48	0.036	0.030
64	0.027	0.0225
72	0.024	0.020
80	0.0216	0.018

* These are Van Keuren wires which are 1⅛ in. long and accurate within 0.000025 for size and roundness. Standardized at 1 lb pressure between flat contacts.

TABLE 23. SPECIAL SIZES OF GEAR WIRES

Size Range, In.	*Size Range, In.*
0.002–0.005	0.180–0.250
0.005–0.007	0.250–0.500
0.007–0.010	0.500–0.750
0.010–0.016	0.750–1.000
0.016–0.100	1.000–1.5708
0.100–0.180	

These special sizes are used in measuring helical gears and splines. They are made up in sets of two or in pairs.

TABLE 24. CHANGE FACTORS K FOR 1.728-IN. WIRES USED ON EXTERNAL AND 1.44-IN. WIRES USED ON INTERNAL SPUR GEARS

$$K = \frac{\cos \phi}{\sin \phi w}$$

Reduction in measurement = $K \times$ amount teeth are to be thin (external)
Increase in measurement = $K \times$ amount teeth are to be thin (internal)

No. of Teeth	14½°		20°		25°		No. of Teeth	14½°		20°		25°	
	Ext.	Int.	Ext.	Int.	Ext.	Int.		Ext.	Int.	Ext.	Int.	Ext.	Int.
5	2.00	2.51	1.76	2.42	1.55	2.23	35	3.06	3.45	2.42	2.69	1.98	2.15
6	2.09	2.61	1.83	2.46	1.60	2.21	36	3.07	3.46	2.43	2.69	1.98	2.15
7	2.17	2.70	1.88	2.50	1.64	2.20	37	3.08	3.47	2.43	2.69	1.98	2.15
8	2.24	2.77	1.94	2.52	1.67	2.19	38	3.10	3.47	2.44	2.69	1.99	2.15
9	2.31	2.84	1.98	2.54	1.70	2.19	39	3.11	3.48	2.45	2.69	1.99	2.15
10	2.36	2.89	2.01	2.56	1.73	2.18	40	3.12	3.49	2.45	2.70	2.00	2.15
11	2.42	2.94	2.05	2.58	1.75	2.18	41	3.13	3.50	2.46	2.70	2.00	2.15
12	2.47	2.99	2.09	2.59	1.77	2.18	42	3.14	3.51	2.46	2.70	2.00	2.15
13	2.51	3.03	2.12	2.60	1.79	2.18	43	3.15	3.51	2.47	2.70	2.00	2.15
14	2.55	3.07	2.14	2.61	1.81	2.18	44	3.17	3.52	2.47	2.70	2.01	2.15
15	2.59	3.10	2.17	2.62	1.83	2.17	45	3.18	3.53	2.48	2.70	2.01	2.15
16	2.63	3.13	2.19	2.63	1.84	2.17	46	3.19	3.53	2.48	2.70	2.01	2.15
17	2.66	3.16	2.21	2.63	1.85	2.17	47	3.20	3.54	2.49	2.70	2.01	2.15
18	2.70	3.19	2.23	2.64	1.86	2.16	48	3.21	3.54	2.49	2.70	2.02	2.15
19	2.73	3.21	2.25	2.64	1.88	2.16	49	3.21	3.55	2.50	2.70	2.02	2.15
20	2.76	3.24	2.26	2.65	1.88	2.16	50	3.22	3.56	2.50	2.71	2.02	2.15
21	2.78	3.26	2.28	2.65	1.89	2.16	60	3.30	3.60	2.54	2.71	2.04	2.15
22	2.81	3.27	2.29	2.66	1.90	2.16	70	3.36	3.63	2.56	2.72	2.05	2.15
23	2.83	3.29	2.30	2.66	1.91	2.16	80	3.41	3.66	2.58	2.72	2.06	2.15
24	2.86	3.31	2.32	2.67	1.92	2.16	90	3.45	3.68	2.60	2.72	2.07	2.15
25	2.88	3.33	2.33	2.67	1.93	2.16	100	3.48	3.70	2.61	2.73	2.08	2.15
26	2.90	3.34	2.34	2.67	1.93	2.16	110	3.51	3.71	2.62	2.73	2.08	2.15
27	2.92	3.36	2.35	2.67	1.94	2.16	120	3.54	3.72	2.63	2.73	2.09	2.15
28	2.94	3.37	2.36	2.67	1.94	2.16	130	3.56	3.73	2.64	2.73	2.09	2.15
29	2.96	3.38	2.37	2.68	1.95	2.16	140	3.58	3.74	2.65	2.73	2.10	2.15
30	2.98	3.40	2.38	2.68	1.95	2.16	150	3.59	3.75	2.65	2.73	2.10	2.15
31	2.99	3.41	2.39	2.68	1.96	2.16	160	3.61	3.75	2.66	2.73	2.10	2.15
32	3.01	3.42	2.40	2.68	1.96	2.16	170	3.62	3.76	2.66	2.73	2.10	2.15
33	3.03	3.43	2.41	2.68	1.97	2.16	180	3.63	3.77	2.67	2.73	2.11	2.15
34	3.04	3.44	2.41	2.69	1.97	2.16	190	3.64	3.77	2.67	2.73	2.11	2.15

FIG. 19. Method for holding and aligning measuring wires on a helical gear.

Helical Gears. Special wires $1.728/P_n$ are required, where P_n is the normal diametral pitch. The spur gear tables are used (even teeth) and for odd-tooth gears the decimal is interpolated for the nearest whole number of N. Thus, for 37 teeth the decimal will be the average of the values for 36 and 38 teeth.

EXAMPLE

Given	*Mathematical Data*
12 teeth = N	$\cos 45° = 0.70711$
$6P$	$\dfrac{1}{\cos 45°} = 1.4142$
2 in. pitch diameter	
45° helix angle	$\dfrac{1}{(\cos 45°)^3} = 2.8284$
$14\frac{1}{2}°$ pressure angle	

1. Find normal diametral pitch $P_n = \dfrac{6}{\cos 45°} = 6 \times 1.4142 = 8.4852$

2. Find wire diameter $= \dfrac{1.728 \text{ in.}}{P_n} = \dfrac{1.728 \text{ in.}}{8.4852} = 0.20365$ in.

3. Find number of teeth in equivalent spur gear. $N' = \dfrac{12}{\cos 45°} = 12 \times 1.4142 = 16.9704$

4. Find number of teeth for which cutter would be selected. $n = \dfrac{12}{(\cos 45°)^3} = 12 \times 2.8284 = 33.94$

5. Find measurement for $1P = M$. Use decimal only from table for n teeth (33.94).

$$\begin{aligned} \text{Use nearest whole number } 34 &= 0.4523 \text{ in.} \\ \text{Add } N' + 2 &= 18.9704 \text{ in.} \\ M &= 19.4227 \text{ in.} \end{aligned}$$

6. M for given helical gear $= \dfrac{M}{P_n} = \dfrac{19.4227}{8.4852} = 2.2890$ in.

Table 25 provides values of $1/\cos \psi$ and $1/(\cos \psi)^3$ for helix angles of 5 to 60°.

If special wires are not available, the method reported by Dr. Adam Zahorski in *American Machinist*, July 12, 1939, can be used.

Block-gaging Method. Standard vernier calipers, even inside micrometers, may be used to check gears by the block-gaging method. In this procedure, the jaws are caused to contact the pitch circle over two or more teeth and the chordal

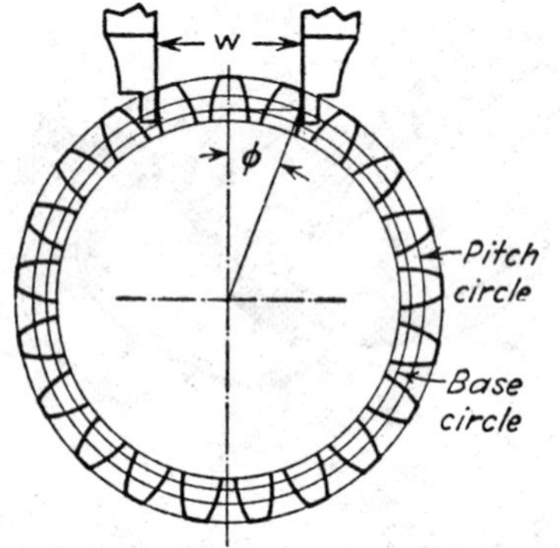

$$w = r\left(\frac{\pi(2t-1)}{N} + 2 \text{ inv } \phi\right)$$

Fig. 20. Block-gaging method of measuring gear teeth.

Table 25. Functions for Even-tooth Helical Gears

Helix Angle ψ	$\frac{1}{\cos\psi}$	$\frac{1}{(\cos\psi)^3}$	Helix Angle ψ	$\frac{1}{\cos\psi}$	$\frac{1}{(\cos\psi)^3}$
5°	1.0038	1.011	37°	1.2521	1.963
10	1.0154	1.047	38	1.2690	2.044
12	1.0223	1.068	39	1.2867	2.130
14	1.0306	1.095	40	1.3054	2.225
16	1.0403	1.126	41	1.3250	2.326
17	1.0457	1.143	42	1.3456	2.437
18	1.0515	1.163	43	1.3673	2.556
19	1.0576	1.183	44	1.3902	2.687
20	1.0642	1.205	45	1.4142	2.828
21	1.0711	1.229	46	1.4395	2.983
22	1.0785	1.254	47	1.4663	3.152
23	1.0864	1.280	48	1.4945	3.338
24	1.0946	1.309	49	1.5242	3.541
25	1.1034	1.343	50	1.5557	3.765
26	1.1126	1.377	51	1.5890	4.012
27	1.1223	1.414	52	1.6243	4.285
28	1.1326	1.453	53	1.6616	4.588
29	1.1433	1.495	54	1.7013	4.924
30	1.1547	1.540	55	1.7434	5.297
31	1.1666	1.588	56	1.7883	5.719
32	1.1792	1.640	57	1.8361	6.190
33	1.1924	1.695	58	1.8871	6.720
34	1.2062	1.755	59	1.9416	7.319
35	1.2208	1.819	60	2.0000	8.000
36	1.2361	1.889			

TABLE 26. SPUR-GEAR BLOCK-GAGING MEASUREMENTS

Pressure Angle $14\frac{1}{2}°$			20°			25°		
I = 0.00537			I = 0.01400			I = 0.02717		
t	N	W	t	N	W	t	N	W
2	12 25	4.6268	2	12 18	4.5962	2	12 14	4.5970
3	26 37	7.7435	3	19 27	7.6464	3	15 21	7.5257
4	38 50	10.8497	4	28 36	10.7245	4	22 29	10.5632
5	51 62	13.9610	5	37 45	13.8026	5	30 36	13.6279
6	63 75	17.0669	6	46 54	16.8808	6	37 43	16.6653
7	76 87	20.1784	7	55 63	19.9589	7	44 51	19.7027
8	88 100	23.2844	8	64 72	23.0370	8	52 58	22.7674
			9	73 81	26.1154	9	59 65	25.8049

t = number of teeth in the block
I = increment per tooth
N = number of teeth in the gear
W = caliper setting for 1 DP gears
W' = caliper setting for finer than 1 P gears = $W \div$ diametral pitch

dimension W (Fig. 20) is ascertained and compared with the value calculated with the aid of Table 26.

When the measurement, or caliper setting, W is obtained from the first number of teeth in the series (12 to 25, 25 to 37, etc.), an increment per tooth I is added for each additional tooth above the least number of teeth in the series. And W' for the diametral pitch involved is found by dividing W (for $1P$) by the diametral pitch.

EXAMPLE: 21 teeth, 20° pressure angle, 10 diametral pitch.
Solution: For 19 teeth, $W = 7.6464$. $I = 0.014$.
For 21 teeth, $W = 7.6464 + 2 \times 0.014 = 7.6744$. Then $W' = 7.6744 \div 10 = 0.7674$.
The measurement is made over three teeth.

Chordal-thickness Method. When cutting gears, if the blank OD has been turned to the correct figure and the cutter is sunk to the required depth for the given pitch and backlash, the tooth thickness should also be correct. Inspection is commonly made with vernier gear-tooth caliper (Fig. 21).

To measure gear teeth with this device, the correct chordal addendum a_c and the correct chordal thickness t_c must first be known. These values are readily solved through use of Table 27. Given there are the basic formulas and values of constants K_a and K_t for gears of 1 diametral pitch and with teeth from 6 to 200.

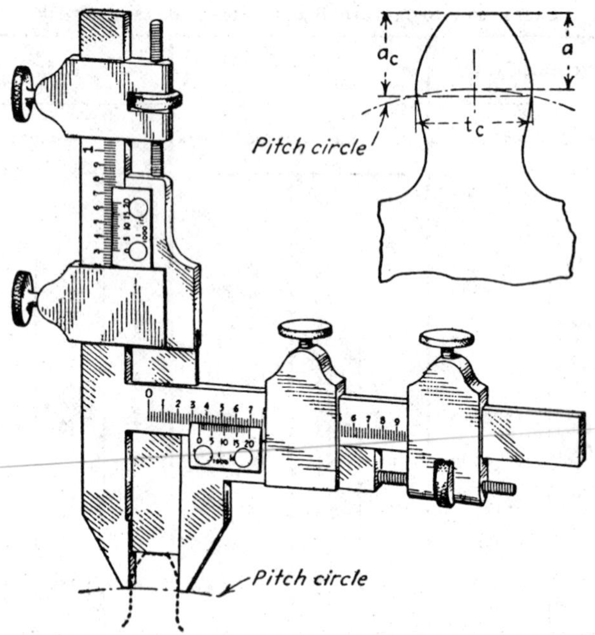

FIG. 21. Gear-tooth vernier caliper.

EXAMPLE: To check a 5-diametral pitch gear having 50 teeth. From tooth parts for spur gears (Table 2) the value of a = 0.2000 in., and the value of t = 0.3142 in. Assume backlash B = 0.016 in. Then

$$a_c = 0.2000 + (0.01233 \div 10) = 0.20123$$
$$t_c = 0.3142 - (0.00026 \div 10) - (0.016 \div 2) = 0.3062$$

With precise values of a_c and t_c known, the vertical scale of the vernier gear-tooth caliper is set to the chordal addendum, so that the jaws are located at the pitch circle. Now, with the sliding jaw and its scale, the tooth thickness is checked against the calculated value.

Master Gears

Specifications for master gears are given in AGMA Standard 235.01 and in Table 28. A master gear is one of known accuracy in every respect. It is usually hardened and ground, and by measurement proved to be highly accurate with respect to runout, pitch error, accumulated error, profile error, and lead or spiral-angle error.

Master gears are used primarily on gear-rolling fixtures for a running check of gears to determine the cumulative effect of errors in individual elements, and provide the means of determining whether these errors compensate for each other or build up in excess of specifications. They determine backlash, runout, and smoothness of operation.

TABLE 27. CHORDAL ADDENDUM AND CHORDAL THICKNESS*

Exact for Equal-addendum Gears. Satisfactorily Approximate for Long- and Short-addendum Gears

Spur: $N_e = N$ $a_c = a + \frac{K_a}{P_d}$ $t_c = t - \frac{K_t}{P_d} - \frac{B}{2}$

Helical: $N_e = \frac{N}{\cos^3 \psi}$ $a_{nc} = a + \frac{K_a}{P_{nd}}$ $t_{nc} = t_n - \frac{K_t}{P_{nd}} - \frac{B}{2}$

Bevel: $N_e = N \sec \Gamma$ $a_c = a + \frac{K_a}{P_d}$ $t_c = t - \frac{K_t}{P_d} - \frac{B}{2}$

Worm: $N_e = \frac{N_w}{\sin^3 \lambda}$ $a_{nc} = a + \frac{K_a}{P_{nd}}$ $t_{nc} = t_n - \frac{K_t}{P_{nd}} - B$

N_e	K_a	K_t	N_e	K_a	K_t	N_e	K_a	K_t
6	0.10222	0.01789	36	0.01714	0.00052	66	0.00933	0.00015
7	0.08777	0.01316	37	0.01667	0.00048	67	0.00920	0.00015
8	0.07686	0.01008	38	0.01623	0.00045	68	0.00907	0.00014
9	0.06836	0.00797	39	0.01582	0.00043	69	0.00893	0.00014
10	0.06156	0.00645	40	0.01542	0.00041	70	0.00880	0.00013
11	0.05598	0.00547	41	0.01504	0.00039	71	0.00867	0.00013
12	0.05133	0.00449	42	0.01471	0.00037	72	0.00855	0.00013
13	0.04739	0.00382	43	0.01437	0.00035	73	0.00843	0.00012
14	0.04401	0.00328	44	0.01404	0.00033	74	0.00832	0.00012
15	0.04109	0.00286	45	0.01370	0.00032	75	0.00821	0.00012
16	0.03852	0.00273	46	0.01336	0.00030	76	0.00810	0.00011
17	0.03625	0.00224	47	0.01311	0.00029	77	0.00799	0.00011
18	0.03425	0.00200	48	0.01285	0.00028	78	0.00789	0.00011
19	0.03244	0.00181	49	0.01258	0.00027	79	0.00780	0.00011
20	0.03083	0.00162	50	0.01233	0.00026	80	0.00772	0.00010
21	0.02936	0.00147	51	0.01209	0.00025	81	0.00762	0.00010
22	0.02803	0.00132	52	0.01187	0.00024	82	0.00752	0.00010
23	0.02681	0.00124	53	0.01165	0.00023	83	0.00743	0.00010
24	0.02569	0.00113	54	0.01143	0.00022	84	0.00734	0.00009
25	0.02466	0.00103	55	0.01121	0.00022	85	0.00725	0.00009
26	0.02371	0.00094	56	0.01102	0.00021	86	0.00716	0.00009
27	0.02234	0.00086	57	0.01083	0.00020	87	0.00708	0.00009
28	0.02194	0.00082	58	0.01064	0.00019	88	0.00700	0.00009
29	0.02121	0.00077	59	0.01046	0.00019	89	0.00693	0.00008
30	0.02055	0.00072	60	0.01029	0.00018	90	0.00686	0.00008
31	0.01990	0.00068	61	0.01011	0.00018	91	0.00679	0.00008
32	0.01926	0.00064	62	0.00994	0.00017	92	0.00672	0.00008
33	0.01869	0.00061	63	0.00978	0.00017	93	0.00665	0.00008
34	0.01813	0.00059	64	0.00963	0.00016	94	0.00658	0.00008
35	0.01762	0.00055	65	0.00947	0.00016	95	0.00651	0.00007

* Gould & Eberhardt, Inc.

TABLE 27. CHORDAL ADDENDUM AND CHORDAL THICKNESS (*Continued*)

N_e	K_a	K_t	N_e	K_a	K_t	N_e	K_a	K_t
96	0.00644	0.00007	131	0.00472	0.00004	166	0.00370	0.00003
97	0.00637	0.00007	132	0.00469	0.00004	167	0.00368	0.00003
98	0.00630	0.00007	133	0.00466	0.00004	168	0.00366	0.00003
99	0.00623	0.00007	134	0.00462	0.00004	169	0.00364	0.00003
100	0.00617	0.00007	135	0.00457	0.00004	170	0.00362	0.00003
101	0.00611	0.00006	136	0.00454	0.00004	171	0.00359	0.00003
102	0.00605	0.00006	137	0.00451	0.00004	172	0.00357	0.00003
103	0.00599	0.00006	138	0.00447	0.00004	173	0.00355	0.00003
104	0.00593	0.00006	139	0.00444	0.00004	174	0.00353	0.00003
105	0.00587	0.00006	140	0.00441	0.00004	175	0.00351	0.00003
106	0.00581	0.00006	141	0.00439	0.00004	176	0.00349	0.00003
107	0.00575	0.00006	142	0.00435	0.00004	177	0.00347	0.00003
108	0.00570	0.00006	143	0.00432	0.00004	178	0.00345	0.00003
109	0.00565	0.00005	144	0.00429	0.00004	179	0.00343	0.00002
110	0.00560	0.00005	145	0.00425	0.00003	180	0.00342	0.00002
111	0.00556	0.00005	146	0.00422	0.00003	181	0.00340	0.00002
112	0.00551	0.00005	147	0.00419	0.00003	182	0.00339	0.00002
113	0.00546	0.00005	148	0.00416	0.00003	183	0.00337	0.00002
114	0.00541	0.00005	149	0.00413	0.00003	184	0.00335	0.00002
115	0.00537	0.00005	150	0.00411	0.00003	185	0.00333	0.00002
116	0.00533	0.00005	151	0.00409	0.00003	186	0.00331	0.00002
117	0.00529	0.00005	152	0.00407	0.00003	187	0.00329	0.00002
118	0.00524	0.00005	153	0.00405	0.00003	188	0.00327	0.00002
119	0.00519	0.00005	154	0.00402	0.00003	189	0.00325	0.00002
120	0.00515	0.00005	155	0.00400	0.00003	190	0.00324	0.00002
121	0.00511	0.00005	156	0.00397	0.00003	191	0.00322	0.00002
122	0.00507	0.00005	157	0.00394	0.00003	192	0.00321	0.00002
123	0.00503	0.00004	158	0.00391	0.00003	193	0.00319	0.00002
124	0.00499	0.00004	159	0.00389	0.00003	194	0.00318	0.00002
125	0.00495	0.00004	160	0.00386	0.00003	195	0.00317	0.00002
126	0.00491	0.00004	161	0.00383	0.00003	196	0.00315	0.00002
127	0.00487	0.00004	162	0.00380	0.00003	197	0.00314	0.00002
128	0.00483	0.00004	163	0.00378	0.00003	198	0.00313	0.00002
129	0.00479	0.00004	164	0.00376	0.00003	199	0.00311	0.00002
130	0.00475	0.00004	165	0.00373	0.00003	200	0.00309	0.00002

Backlash may or may not be provided in a master gear. When it is, a master is usually sized for the average backlash desired.

This standard is based on the use of master gears in variable-center-distance devices. Fixed-center-distance fixtures are not recommended because of the expense of making the necessary master gears to the extreme accuracy of both runout and size over pins.

TABLE 28. SPUR AND HELICAL GROUND MASTER GEAR TOLERANCES
All Dimensions in Inches

	All Classes		
Hole size within	+0.0002		
(Preferred hole sizes $\frac{3}{4}$, $1\frac{1}{4}$, $1\frac{3}{4}$)	−0.0000		
Face runout of locating surfaces with respect to hole, total indicator reading per in. of dia, not to exceed	0.0001		
OD size within	±0.0010		
Runout of OD, with respect to hole	0.0010		
	Class		
Runout (as measured by pin, cone, or ball)	1	2	3
4 dia and smaller	0.0005	0.0004	0.0003
4 + to 8 dia	0.0009	0.0007	0.0005
8 + to 12 dia	0.0013	0.0010	0.0007
	All Classes		
Gear-tooth measurement over pins	+0.020		
(See Note 1)	−0.000		
Pitch error (tooth-to-tooth variation, not over pins)			
4 dia and smaller	0.0002		
4 + to 8 dia	0.0003		
8 + to 12 dia	0.0004		
	Spur	*Helical*	
Profile error (total variation, not plus and minus)			
4 dia and smaller	0.0002	0.0003	
4 + to 8 dia	0.00025	0.00035	
8 + to 12 dia	0.0003	0.0004	
	All Classes		
Circumferential lead error per in. of face (or parallelism of spur teeth with axis)	0.0002		

NOTE 1: Circular tooth thickness at standard pitch diameter to be marked on each master gear.
NOTE 2: Tolerances for root diameter are purposely omitted because they affect only the clearance at the root of the teeth, and therefore are considered unessential.
NOTE 3: Material recommended is high-speed steel hardened to a minimum of Rockwell C63.

Bevel and Hypoid Gears

Straight bevel, spiral bevel, and angular bevel gears have pitch surfaces that are cones, and they operate on intersecting axes. Hypoid gears are similar in form to bevel gears but operate on offset, or nonintersecting, axes.

Gleason 20° Straight-bevel-gear System. Principal changes in this system, as made by the Gleason Works in 1949, cover:

1. Slight increases in clearance and whole depth, which are proportionately greater in the case of fine-pitch teeth.

2. Adoption of 20° pressure angle, to replace the $14\frac{1}{2}$° pressure angle previously used.

3. Somewhat different proportional tooth thicknesses of pinion and gear for approximately equal fatigue life.

The general basis of the system is as follows:

Ratios. These include ratios in common use.

Working Depth. The working depth is 2.000 in. ÷ DP. Use of stub teeth is not recommended because the reduction in contact increases noise and decreases wear resistance.

Clearance. The clearance at the large end of the teeth is 0.188 in. ÷ DP + 0.002 in.

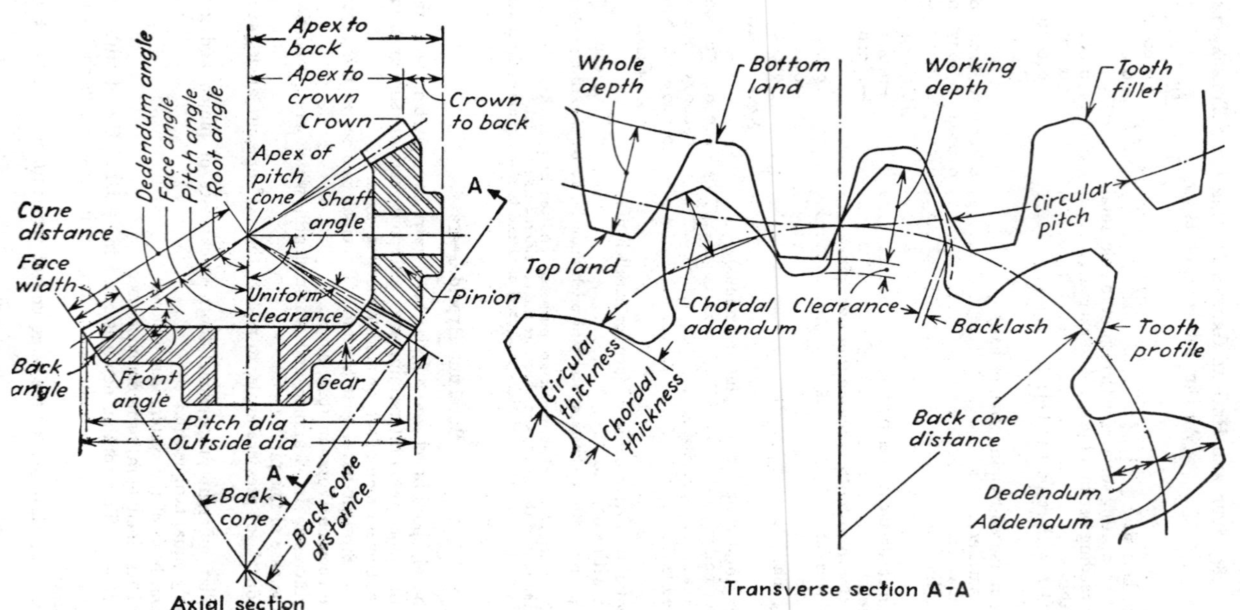

FIG. 22. Nomenclature for bevel gears.

STRAIGHT-BEVEL-GEAR FORMULAS,* AXES AT RIGHT ANGLES, LONG AND SHORT ADDENDUM
Conforms to Gleason 20° Straight-bevel-gear System, 1949

	Pinion, Lesser No. of Teeth		Gear
Number of teeth	$n = dP_d$		$N = DP_d$
Diametral pitch		$P_d = \dfrac{N}{D}$	
Circular pitch		$p = \dfrac{3.141593}{P_d}$	
Pressure angle		$\phi = 20°$	
Pitch dia	$d = \dfrac{n}{P_d}$		$D = \dfrac{N}{P_d}$
Pitch angle	$\gamma = \tan^{-1} \dfrac{n}{N}$		$\Gamma = 90° - \gamma$
Cone distance		$A_o = \dfrac{D}{2 \sin \Gamma}$	
Addendum	a_P		a_G
Approx whole depth		$h = \dfrac{2.188}{P_d} + 0.002$	
Dedendum	$b_P = h - a_P$		$b_G = h - a_G$
Clearance		$C = \dfrac{0.188}{P_d} + 0.002$	
Root angle	$\gamma_R = \gamma - \tan^{-1} \dfrac{b_P - 0.002}{A_o}$		$\Gamma_R = \Gamma - \tan^{-1} \dfrac{b_G - 0.002}{A_o}$
Face angle	$\gamma_o = 90° - \Gamma_R$		$\Gamma_o = 90° - \gamma_R$
OD	$d_o = d + 2a_P \cos \gamma$		$D_o = D + 2a_G \cos \Gamma$
Pitch apex to crown	$x_o = \dfrac{D}{2} - a_P \sin \gamma$		$X_o = \dfrac{d}{2} - a_G \sin \Gamma$
Circular thickness	t_P		t_G
Tooth angle, minutes	$\theta_{tP} = \dfrac{3438}{A_o} \left[\dfrac{t_P}{2} + 0.364(b_P - 0.002) \right]$		$\theta_{tG} = \dfrac{3438}{A_o} \left[\dfrac{t_G}{2} + 0.364(b_G - 0.002) \right]$
Equivalent number of teeth	$n_e = n \sec \gamma$		$N_e = N \sec \Gamma$
Backlash assembled		B	
Chordal addendum	a_{cP}		a_{cG}
Max chordal thickness	t_{cP}		t_{cG}
Face width		$F = 0.25$ to $0.3A_o$	

* Gould & Eberhardt, Inc.

TABLE 29. ADDENDUMS AND CIRCULAR THICKNESS K FACTORS FOR 1 DP STRAIGHT BEVEL GEARS (LONG AND SHORT ADDENDUMS)*

Addendums: For Actual Values Divide by the Diametral Pitch†

Ratio	Addendum		Ratio	Addendum		Ratio	Addendum		Ratio	Addendum	
	Pinion	Gear		Pinion	Gear		Pinion	Gear		Pinion	Gear
1.00–1.00	1.000	1.000	1.15–1.17	1.120	0.880	1.42–1.45	1.240	0.760	2.06–2.16	1.360	0.640
1.00–1.02	1.010	0.990	1.17–1.19	1.130	0.870	1.45–1.48	1.250	0.750	2.16–2.27	1.370	0.630
1.02–1.03	1.020	0.980	1.19–1.21	1.140	0.860	1.48–1.52	1.260	0.740	2.27–2.41	1.380	0.620
1.03–1.04	1.030	0.970	1.21–1.23	1.150	0.850	1.52–1.56	1.270	0.730	2.41–2.58	1.390	0.610
1.04–1.05	1.040	0.960	1.23–1.25	1.160	0.840	1.56–1.60	1.280	0.720	2.58–2.78	1.400	0.600
1.05–1.06	1.050	0.950	1.25–1.27	1.170	0.830	1.60–1.65	1.290	0.710	2.78–3.05	1.410	0.590
1.06–1.08	1.060	0.940	1.27–1.29	1.180	0.820	1.65–1.70	1.300	0.700	3.05–3.41	1.420	0.580
1.08–1.09	1.070	0.930	1.29–1.31	1.190	0.810	1.70–1.76	1.310	0.690	3.41–3.94	1.430	0.570
1.09–1.11	1.080	0.920	1.31–1.33	1.200	0.800	1.76–1.82	1.320	0.680	3.94–4.82	1.440	0.560
1.11–1.12	1.090	0.910	1.33–1.36	1.210	0.790	1.82–1.89	1.330	0.670	4.82–6.81	1.450	0.550
1.12–1.14	1.100	0.900	1.36–1.39	1.220	0.780	1.89–1.97	1.340	0.660	6.81	1.460	0.540
1.14–1.15	1.110	0.890	1.39–1.42	1.230	0.770	1.97–2.06	1.350	0.650			

Circular Thickness: Values for K $\quad t_G = \frac{p}{2} + \frac{K}{p_d} - 0.364(a_P - a_G) \qquad t_P = p - t_G$

N_P	Ratio										
	1.000+ to 1.020	1.020+ to 1.075	1.075+ to 1.140	1.140+ to 1.260	1.260+ to 1.855	1.855+ to 2.250	2.250+ to 2.645	2.645+ to 3.105	3.105+ to 3.650	3.65+ to 4.35	4.35+ to 5.21
13 to 14					+0.075	+0.060	+0.040	+0.020	+0.005	−0.015	−0.035
15 to 16	+0.000	+0.020	+0.035	+0.055	+0.070	+0.060	+0.045	+0.025	+0.010	−0.010	−0.030
17 to 21	+0.000	+0.030	+0.035	+0.050	+0.070	+0.060	+0.045	+0.025	+0.010	−0.005	−0.025
22 to 26	+0.000	+0.015	+0.030	+0.045	+0.060	+0.050	+0.035	+0.020	+0.005	−0.005	−0.025
27 to 35	+0.000	+0.015	+0.025	+0.040	+0.050	+0.040	+0.030	+0.015	+0.000	−0.010	−0.030
36 to 45	+0.000	+0.010	+0.020	+0.030	+0.040	+0.030					

* Gould & Eberhardt, Inc.

† In case of choice, use the smaller pinion addendum.

TABLE 30. TOOTH PROPORTIONS FOR 1 DP FINE-PITCH STRAIGHT BEVEL GEARS
90° Shaft Angle

Ratios	Ratio = $\frac{\text{No. of gear teeth}}{\text{No. of pinion teeth}}$			
	Pinion Addendum	Gear Addendum	Pinion Circular Thickness	Gear Circular Thickness
1.00 to 1.00	1.000	1.000	1.5708	1.5708
1.00 to 1.02	1.010	0.990	1.5781	1.5635
1.02 to 1.03	1.020	0.980	1.5854	1.5562
1.03 to 1.04	1.030	0.970	1.5926	1.5490
1.04 to 1.05	1.040	0.960	1.5999	1.5417
1.05 to 1.06	1.050	0.950	1.6072	1.5344
1.06 to 1.08	1.060	0.940	1.6145	1.5271
1.08 to 1.09	1.070	0.930	1.6218	1.5198
1.09 to 1.11	1.080	0.920	1.6290	1.5126
1.11 to 1.12	1.090	0.910	1.6363	1.5053
1.12 to 1.14	1.100	0.900	1.6436	1.4980
1.14 to 1.15	1.110	0.890	1.6509	1.4907
1.15 to 1.17	1.120	0.880	1.6582	1.4834
1.17 to 1.19	1.130	0.870	1.6654	1.4762
1.19 to 1.21	1.140	0.860	1.6727	1.4689
1.21 to 1.23	1.150	0.850	1.6800	1.4616
1.23 to 1.25	1.160	0.840	1.6873	1.4543
1.25 to 1.27	1.170	0.830	1.6945	1.4471
1.27 to 1.29	1.180	0.820	1.7018	1.4398
1.29 to 1.31	1.190	0.810	1.7091	1.4325
1.31 to 1.33	1.200	0.800	1.7164	1.4252
1.33 to 1.36	1.210	0.790	1.7237	1.4179
1.36 to 1.39	1.220	0.780	1.7309	1.4107
1.39 to 1.42	1.230	0.770	1.7382	1.4034
1.42 to 1.45	1.240	0.760	1.7455	1.3961
1.45 to 1.48	1.250	0.750	1.7528	1.3888
1.48 to 1.52	1.260	0.740	1.7601	1.3815
1.52 to 1.56	1.270	0.730	1.7673	1.3743
1.56 to 1.60	1.280	0.720	1.7746	1.3670
1.60 to 1.65	1.290	0.710	1.7819	1.3597
1.65 to 1.70	1.300	0.700	1.7892	1.3524
1.70 to 1.76	1.310	0.690	1.7965	1.3451
1.76 to 1.82	1.320	0.680	1.8037	1.3379
1.82 to 1.89	1.330	0.670	1.8110	1.3306
1.89 to 1.97	1.340	0.660	1.8183	1.3233
1.97 to 2.06	1.350	0.650	1.8256	1.3160
2.06 to 2.16	1.360	0.640	1.8329	1.3087
2.16 to 2.27	1.370	0.630	1.8401	1.3015
2.27 to 2.41	1.380	0.620	1.8474	1.2942
2.41 to 2.58	1.390	0.610	1.8547	1.2869
2.58 to 2.78	1.400	0.600	1.8620	1.2796
2.78 to 3.05	1.410	0.590	1.8693	1.2723
3.05 to 3.41	1.420	0.580	1.8765	1.2651
3.41 to 3.94	1.430	0.570	1.8838	1.2578
3.94 to 4.82	1.440	0.560	1.8911	1.2505
4.82 to 6.81	1.450	0.550	1.8984	1.2432
6.81	1.460	0.540	1.9057	1.2359

Addendums. Except where the numbers of teeth are equal, the pinion has a long addendum and the gear a short addendum. The amount of departure from equal addendums varies with the ratio. Long addendums are used on the pinion principally to avoid undercut and to increase tooth strength. Tabulated values are for straight bevel gears with generated teeth, axes at 90° and with 13 or more pinion teeth. Numbers of teeth that apply are: ratios with 16 or more teeth in the pinion, $\frac{15}{17}$ and higher, $\frac{14}{20}$ and higher, and $\frac{13}{30}$ and higher.

Face Angles. The face cone of the blank is made parallel to the root cone of the mating gear. This gives constant clearance along the tooth and allows the use of larger edge radiuses on generating tools without fillet interference on the small end, thus increasing tooth strength.

Backlash. The values below for backlash apply to gears assembled ready to run. Because of manufacturing tolerances and changes resulting from heat-treatment, it is frequently necessary to subtract more than one-half the recommended value from the theoretical tooth thickness of each member in order to obtain the correct backlash at assembly.

DP	Backlash	DP	Backlash
1.00–1.25	0.020–0.030	3.50–4.00	0.007–0.009
1.25–1.50	0.018–0.026	4–5	0.006–0.008
1.50–1.75	0.016–0.022	5–6	0.005–0.007
1.75–2.00	0.014–0.018	6–8	0.004–0.006
2.00–2.50	0.012–0.016	8–10	0.003–0.005
2.50–3.00	0.010–0.013	10–20	0.002–0.004
3.00–3.50	0.008–0.011	20 and finer	0.001–0.003

Coniflex Bevel Gears. Straight bevel gears with localized tooth bearing are termed "Coniflex" gears by the Gleason Works. They have teeth that are straight and that, if extended inward, would intersect at the axes.

Spiral Bevel Gears. These gears have teeth that are curved and oblique. The general form of the blank is the same as for straight bevels, but the design and manufacture are highly specialized. Operating advantages are: smoothness, quietness, high efficiency, and durability. Practically any shaft-intersecting angle can be used and almost any operating speed. Spiral bevel gears are particularly recommended where the peripheral speed is in excess of 1000 fpm. Also, they are used advantageously to provide high load capacity and high ratios when the size of the drive must be kept to a minimum.

Zerol Bevel Gears. Trade-named "Zerol" by the Gleason Works, these gears are curved-tooth bevel gears with zero-degree spiral angle. They combined the localized tooth contact of spiral bevel gears with the low axial thrust of straight bevel gears. They are used to replace straight bevel gears in pairs, where the additional accuracy of ground gears is desired and where thrust limitations and other factors do not permit mounting provision for spiral bevels or hypoids.

Hypoid Gears. Hypoids resemble spiral bevel gears in that they have conical shape and curved oblique teeth and are used for transmitting power and motion around a corner. They differ from spiral bevels in that the pinion axis is offset above or below the gear axis. The fact that the gear and pinion shafts can pass each other is of advantage in industrial drives with continuous shafts. And, of course, their use in automobile rear-axle drives is well known. The gears operate with extreme smoothness and quietness. Over-all efficiencies of 95 to 98% are common.

Fine-pitch Straight Bevel Gears. Straight bevel gears of fine pitch are designed in accordance with AGMA Standard 206.02.

The features of this standard are:

1. Clearance increases proportionately as the pitch becomes finer, as in fine-pitch spur and helical gears. While in other types of fine-pitch gears the whole depth is $2.200/P + 0.002$, in order to conform with well-established practices in the bevel-gear field, the whole depth for bevel gears is $2.188/P + 0.002$. This difference is very small and of no practical consequence.

2. Tooth thicknesses correspond to those generated by a crown gear in which the tooth thickness and space width are equal.

3. The maximum face width is limited to three-tenths of the cone distance, or $8/P$, whichever is smaller.

This standard covers generated straight bevel gears:

1. Of 20 diametral pitch and finer. In practice most of the fine-pitch straight bevel gears lie between 20 and 64 DP. The point widths of the cutting tools become extremely small for teeth finer than 64 DP.

2. For all shaft angles.

3. With the numbers of teeth equal to or greater than $\frac{16}{16}$, $\frac{15}{17}$, $\frac{14}{20}$, $\frac{13}{30}$—90° shaft angle only. However, gears with numbers of teeth equal to or greater than $\frac{12}{13}$ may be designed in accordance with the principles of this standard without reduction in contact ratio due to undercut. The theoretical contact ratio is not less than 1.45.

General Dimensions

Pitch P = 20 diametral pitch and finer.

Pressure angle $\phi = 20°$.

Smallest pinion, 13 teeth and smallest miter, 16 teeth, 90° shaft angle.

Working depth $h_k = 2.000/P$.

Whole depth $h_t = 2.188/P + 0.002$.

Maximum face width $F = 3/10A_o$ or $8/P$, whichever is smaller, where A_o = cone distance.

Gear-blank Dimensions and Tolerances. In dimensioning bevel gear blanks, it is necessary to specify properly the items important to the functioning of the teeth. These are:

1. Outside diameter.
2. Crown to back (or mounting surface).
3. Face angle.
4. Mounting distance.

Face widths and back angles are usually of secondary importance.

Recommended tolerances are listed for outside diameter, crown to back, face angle, and back angle. The first three directly affect the addendum height and therefore the contact ratio.

a. Outside Diameter and Crown to Back. These tolerances apply to the theoretical crown point. The outside diameter cannot, therefore, be measured directly if the blank has been rounded or flattened.

b. Face Angles. The face angle is made so that the face-cone element is parallel to the root-cone element of the mate. This produces constant clearance along the tooth.

In establishing these tolerances, consideration was given to the decrease in the tooth height and the corresponding reduction in the contact ratio. The face-angle

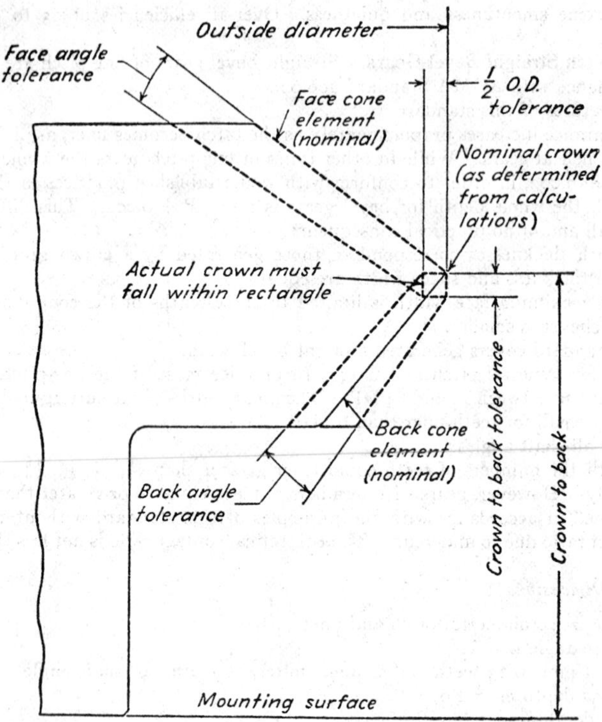

FIG. 23. Application of tolerances to bevel-gear blanks. Heavy dotted lines represent blank outline limits. When located with respect to bore centerline and mounting surface, a blank profile that falls within the dotted lines is acceptable.

tolerances are based on a decrease in the working depth not to exceed 0.0015 in. at the inner ends of the teeth.

c. Mounting Distance. Bevel gears are cut and inspected to the mounting distance. In order to obtain the running characteristics machined into the gears, they should be assembled to the same dimension. The results of errors become more pronounced in low ratio combinations. Allowable tolerances for mounting distance will therefore depend on the design and accuracy required of the application.

Exact position for assembling bevel gears should be determined by the mounting distance. It is not good practice to depend on setting the back cones flush.

d. Back Angle. A tolerance of plus or minus 1° will generally be satisfactory.

NOTE: The design of fine-pitch spiral bevel and hypoid gears has not been standardized, because the tooling is tied in so closely with the physical dimensions of the finished part.

TOLERANCES FOR BEVEL AND HYPOID GEARS

These tolerances apply to straight, spiral and Zerol bevel, and hypoid gears.

Diameter Range. The diameters considered are from $\frac{3}{4}$ to 100 in., arranged in

TABLE 31. TOLERANCES FOR FINE-PITCH STRAIGHT BEVEL GEARS

OD Tolerance		Crown-to-back Tolerance		Face-angle Tolerance	
Diametral Pitch	Tolerance, In.	Diametral Pitch	Tolerance, In.	Face Width, In.	Tolerance, Min
20 to 30	+0.000 −0.005	20 to 47	+0.000 −0.002	$\frac{1}{2}$	+10 − 0
31 to 40	+0.000 −0.004	47 and finer	+0.000 −0.001	$\frac{1}{4}$	+20 − 0
41 to 56	+0.000 −0.003			$\frac{3}{16}$	+30 − 0
57 to 94	+0.000 −0.002			$\frac{1}{8}$	+40 − 0
95 and finer	+0.000 −0.001				

approximate geometric progression, namely: $\frac{3}{4}$, $1\frac{1}{2}$, 3, 6, 12, 25, 50, and 100 in. Tolerances for other diameters may be interpolated.

Pitch Range. The pitches considered are from 1 to 32 diametral pitch, arranged in a geometric progression, namely: 1, 2, 4, 8, 16, and 32.

Classes and Speeds. Three classes of bevel gears are listed according to accuracy. These are:

Class 2 .... Up to 400 fpm
Class 3* .... Up to 2000 fpm
Class 4 .... Over 2000 fpm

* Straight-bevel limitation up to 1000 fpm.

Types of Errors. The fundamental errors for which tolerances are given are: runout, pitch error, accumulated error, and required tooth contact.

Allowances and tolerances for backlash are treated separately on page 4-25. Tooth-thickness tolerances are related to backlash and are therefore omitted.

Tolerances for whole depth are omitted for several reasons. Teeth are often roughed deeper than the finishing depth. A considerable variation in whole depth is acceptable without affecting gear performance in any way.

Tolerances for various dimensions of bevel-gear blanks, such as hole size, outside diameter, face width, and face angle, are not yet established but may ultimately be included because of their importance to gear accuracy.

Pitch Error. Pitch error of a bevel or hypoid gear is the maximum difference between any two successive tooth-to-tooth readings taken in a plane of rotation at the midface between corresponding sides of adjacent teeth. Either set of profiles may be used for tooth-to-tooth readings, but preferably the driving sides if the gear operates in one direction only.

In the case of straight bevel gears, or gears having a low spiral angle, the tooth-to-tooth readings may be taken with a dial indicator in the direction of the pitch circle.

TABLE 32. TOLERANCES FOR STRAIGHT, SPIRAL AND ZEROL BEVEL, AND HYPOID GEARS
All Readings in 0.0001 In.

Class	Diametral Pitch	Runout of Pitch Dia, Total Indicator Reading								Pitch Error Measured at Midface in Plane of Rotation								Accumulated Error between Any Two Teeth Exclusive of Runout Effect								Required Min Initial Area of Contact, %
Pitch Dia		¾	1½	3	6	12	25	50	100	¾	1½	3	6	12	25	50	100	¾	1½	3	6	12	25	50	100	
Class 2 (up to 400 fpm)	1					20	24	32	40					8	9	9	10									
	2					20	24	28						8	9	9										50
	4				16	20	24						7	8	9											
	8			16	16	20	24					7	8	8	9											
	16		16	16	16	20	24				7	7	8	8	9											
Class 3 (up to 1000 fpm)*	2					16	20	24	28				5	6	7	7	8									
	4				14	16	20	24				5	5	6	7	7										
	8			14	14	16	20				5	5	5	6	7	7										60
	16		12	14	14	16	20				5	5	5	6	7											
	32	10	12	14						5	5	5	5	6												
Class 4 (over 2000 fpm)	4				8	10	14	18	24				3	4	5	5	6				10	10	12	20	30	
	8			8	8	10	14	18				3	3	4	5	5				10	10	10	12	20		70
	16		8	8	8	10	14				3	3	3	4	5				10	10	10	10	12			
	32	6	8	8	8					3	3	3	3					10	10	10	10					

* Straight bevels limited to 1000 fpm.

However, at higher angles the indicator may give false readings from the side-cramping effect of the angle. In such cases the indicator may be set normal to the tooth surface, and readings converted to the direction of the pitch circle for purposes of comparison and standardization by dividing by the cosine of the spiral angle and by the cosine of the pressure angle at that diameter.

In the case of spiral bevel and hypoid gears with unusually high spiral angles, these tolerances, which are stated in the direction of the pitch circle, should be increased in the approximate ratio of

$$\frac{\cos 35^\circ}{\text{cos gear spiral angle}}$$

Accumulated Error. Accumulated error of a gear is the maximum accumulation of pitch errors obtained by algebraic addition of pitch errors.

Unless otherwise specified, accumulated error is checked over the entire 360° and expressed as the maximum error between any two teeth.

For low- and medium-speed gears, accumulated error is of minor importance unless the gears are to be used for timing mechanisms, dividing gears, instruments, or the like.

Required Tooth Contact. Determination of the proper tooth contact is best accomplished by means of a testing machine. Proper contact involves three factors, namely: position, shape, and area of contact.

Worms and Worm Gears

Many authorities still consider the design and manufacture of worms and worm gears an art rather than a science. Coarse-pitch worm gearing, which is used to transmit power, has reached the stage of tentative standards for design and inspection, whereas fine-pitch worm gearing, which is used to translate motion, has been standardized as to design.

This situation arises because a worm-gear drive may be likened in simple terms to a screw driving a nut. In other words, the worm is a threaded member and the worm gear is a throated member that is fully conjugate to the mating worm.

Tooth Form. The tooth form is the crux of the problem with coarse-pitch worm gears. It may be based upon the axial pitch of the worm, which is equal to the circular pitch of the worm gear, or upon diametral pitch of the worm gear, or the normal diametral pitch of the worm gear. The choice is dictated by tools available. If there are no tools to be considered, the axial-pitch basis is to be preferred.

Cutting Tools. Information on tools, cutters, and grinding wheels may be summarized:

Machine Operations

1. Chasing.
   a. Cutting edges of tool in axial plane.
   b. Cutting edges of tool in plane normal to helix at middle of thread space.
   c. Various other tool settings.
2. Milling.
   a. "Standard" thread-milling cutter, double-conical, straight-sided, of nominal pressure angle, usually 4 to 7 in. diameter.
   b. Curved-edge (form relieved) milling cutter.
3. Hobbing.
   a. Hob similar to gear-cutting hob, with helical-thread surfaces.
   b. Single-convolution worm-generating hob.

4. Shaping.
   a. Generating cutter with involute profiles to generate straight-sided axial worm threads.
5. Grinding.
   a. Double-conical wheel, straight-sided, of nominal pressure angles, usually 12 to 20 in. diameter.
   b. Wheels similar to above, but diamond-dressed to curved forms.
   c. Plane-sided wheel set at nominal pressure angle, to grind involute helical-thread surface.
   d. Miscellaneous wheel types.

For proper tooth contact, the worm-gear hob must correspond to the worm in thread form and in general size.

About 75% of worms produced in this country are made with double-conical, straight-sided, thread-milling cutters tilted normal to the worm thread. The cutter is usually 4 in. dia. The fact that the tapered flank of the cutter does not reproduce itself because of interference causes tooth curvature, and this becomes greater with the lead angle of the worm. Involute thread forms are produced in England, but the production rates are lower than those achieved in American practice.

If the worm is hardened and then ground, usually with a 16-in. straight-sided wheel, of double-conical form, a curved form is produced in the normal and axial sections.

Lead Angle. There has been much confusion in respect to the term "lead angle," as applied to a worm thread. The lead angle of a worm or hob (Fig. 3*b*) is the angle between any helix and a plane of rotation. It is the complement of the helix angle. Note: In screw-thread practice, the term "helix angle" refers to this lead angle. For maximum efficiency as a power drive, the lead angle of worms ranges between 25 and 50°.

Pressure Angle. Years ago the $14\frac{1}{2}°$ pressure angle was popular because the worm was usually produced in a lathe, and it was single-threaded. Currently the trend is to 20° and higher pressure angles. In general, worms with one or two threads have 20° normal pressure angle; 25° NPA when over two threads. Index worms have 14.5° NPA when one or two threads; 20° NPA when over two threads. Worm-gear normal pressure angle, like tooth form, is not readily defined except in terms of the mating worm.

Enlargement. Worms and worm gears are made with equal addendums wherever possible. An exception is made in large dual-lead index worms, where addendums are varied to shift the zones of contact, in order to minimize the length of the worm.

Worm Diameter. For maximum efficiency as a power drive, it is desirable to have the smallest possible diameter consistent with satisfactory shaft deflection and ample gear-face width: (1) to minimize sliding friction and resulting heat; (2) to reduce oil churning and resulting heat; (3) to attain an efficient lead angle; and (4) to minimize size of bearings, housings, and related parts.

Minimum work diameter is attained by: (1) increasing lead angle; (2) decreasing tooth height; (3) increasing axial pitch; (4) increasing lead; and (5) increasing number of threads.

Worm-gear Outside Diameter. Worm gears ordinarily have cylindrical outside diameters. In special cases, as for fine- or extra-fine-pitch index worm gears, they are partly throated to increase tooth contact, with chamfers to blend OD into faces on the larger sizes.

Worm and Worm-gear Formulas,* Equal Addendum

Number of teeth in gear	$N_G = \frac{3.141593 D_G}{p_x}$
Number of threads in worm	$N_w = \frac{1}{p_x}$
Equivalent number of threads	$N_e = \frac{N_w}{\sin^3 \lambda}$
Circular pitch of gear Axial pitch of worm	$\left.\begin{matrix} p = \\ p_x = \end{matrix}\right\} \frac{3.141593 D_G}{N_G}$
Normal circular pitch	$p_n = p_x \cos \lambda$
Diametral pitch	$P_d = \frac{3.141593}{p}$
Normal diametral pitch	$P_{nd} = \frac{3.141593}{p_n}$
Normal pressure angle	$\phi_n = \begin{cases} 20° \text{ if 1 or 2 thread} \\ 25° \text{ if over 2 thread} \end{cases}$
Axial pressure angle	$\phi_x = \tan^{-1} \frac{\tan \phi_n}{\cos \lambda}$
Pitch dia of gear	$D_G = \frac{N_G p}{3.141593}$
Pitch dia of worm	$D_w = 2C - D_G$
OD	$D_o = D_G + 2a,\ D_w + 2a$
Center distance	$C = \frac{D_G + D_w}{2}$
Lead angle	$\lambda = \tan^{-1} \frac{l}{3.141593 D_w}$
Helix angle with axis (worm)	$\psi = 90° - \lambda$
Lead	$l = p_x N_w$
Addendum	$a = \frac{1}{P_{nd}}$
Approx whole depth	$h = \frac{2.157}{P_{nd}}$
Dedendum	$b = h - a$
Clearance	$c = \frac{0.157}{P_{nd}}$
Normal circular thickness	$t_n = \frac{p_n}{2}$
Normal chordal addendum (worm)	$a_{nc} = a + \frac{t_n^2 \sin^2 \lambda}{4 D_w}$
Max normal chordal thickness (worm)	$t_{nc} = t_n - \frac{t_n^3 \sin^4 \lambda}{6 D_w^2} - B \cos \lambda$
Backlash assembled	B

* Gould & Eberhardt, Inc.

Worm-gear Face Width. The face width equals the square root of the difference between the squares of the worm outside and pitch diameters.

Worm-gear Rim Thickness. The thickness of rim beneath root of tooth is made to the nearest convenient fractional size larger than the tooth depth (ignoring oversize hobs).

Tolerances for Coarse-pitch Gears. Suggested tolerances are given in Tables 33 and 34 for coarse-pitch worms and worm gears of three classes. Worm diameters range from $\frac{3}{4}$ to 12 in. dia in geometric progression: $\frac{3}{4}$, $1\frac{1}{2}$, 3, 6, and 12 in. The worm-gear diameters range from $\frac{3}{4}$ to 100 in. dia in a similar manner. Pitches are from $\frac{1}{4}$ to 4 axial pitch, namely: $\frac{1}{4}$, $\frac{1}{2}$, 2, and 4 in. Tolerances for other diameters and axial pitches may be interpolated.

Classes of Gears.

Class 1. Light-duty worm gearing such as that used in manually operated adjusting devices, or on extremely large low-speed power-driven gearing, gearing not generally requiring ground worms.

Class 2.[1] Worm gearing employed for power-transmission work generally operated in an enclosing housing up to 4000 fpm sliding velocity at the nominal pitch line, and generally employing hardened and ground steel worms.

Class 3.[1] Precision worm gearing, for the transmission of power above 4000 fpm sliding velocity at the nominal pitch line or for the transmission of motion where high accuracy is required, as in indexing gearing.

Application of Tolerances. Tables 33 and 34 show, first, the tolerances that may be allowed in manufacture under ordinary conditions and, second, that tolerances must vary reasonably with respect to pitch and diameter of the gears.

These tables do not in any way represent "standard specifications" for design or manufacture. A manufacturer is expected to supply a worm and worm gear as a pair that will operate satisfactorily under usual conditions and that represent good workmanship.

Interchangeability of worms and worm gears manufactured in different shops is usually not to be expected and can be obtained satisfactorily only by special control and gaging which must be very carefully planned.

When a user expects that he will need special tolerances, he should discuss his requirements with the manufacturer to arrive at a mutually satisfactory understanding.

Recommended Backlash. The following formulas for backlash assume average conditions for general-purpose worm gearing. In many instances, the values obtained require modification for the various reasons stated, and it is therefore emphasized that judgment must be used when specifying backlash for a particular application.

The amounts of backlash as determined by the formulas are for worms and worm gears in actual running position. Minimum values represent backlash at the tightest point. The allowances to be made on tooth thickness and the control of other factors necessary to obtain the backlash specified are the joint responsibility of the manufacturer of the gears and those who assemble the gears.

[1] Formula for sliding velocity:

$$v = \frac{\pi d n_w}{12 \cos \lambda}$$

where v = sliding velocity at the nominal pitch diameter of the worm, fpm; d = worm diameter at center of working depth, in.; n_w = number of worm rpm; λ = lead angle of worm.

Table 33. Tolerances for Worms
In 0.0001 In.

Class	Axial Pitch, In.	Runout Measurement on Pitch Circle, Total Indicator Reading					Pitch Error in Axial Plane Multiple Threads Only					Profile Error, Total Variation, Not Plus and Minus Lead Angle, °			Lead Error, Total Variation per Convolution Lead Angle, °		Lead Error, Total Variation per Active Length of Worm Lead Angle, °		
Pitch Dia, In.		¾	1½	3	6	12	¾	1½	3	6	12	0–15	15–30	30–45	0–15	15–30	0–15	15–30	30–45
Class 1. Light duty	4					80					50	30	40	45	40	80	80	100	120
	2				50	60				35	40	13	15	17	25	35	35	40	45
	1			40	50	60			35	35	40	11	13	15	15	22	28	32	35
	½		30	40	50	60		25	35	35	40	10	12	14	15	20	25	28	32
	¼	30	30	40			20	25	25			10	11	12	12	18	22	25	28
Class 2. Power transmission	2				30	30				20	25	10	12	14	17	22	23	25	30
	1			23	23	30			18	20	25	8	10	12	9	15	19	21	23
	½		18	23	23	30		12	18	20	25	7	9	11	9	14	17	19	21
	¼	15	18	23			12	12	15			7	8	9	8	12	15	17	19
Class 3. Precision	2				15					15		8	10	12	15	20	18	20	24
	1			12	12				12	12		6	8	10	7	12	14	16	18
	½		8	12	12			10	12	12		5	7	9	7	11	12	14	16
	¼	8	8	12			10	10	12			5	6	7	6	9	10	12	14

TABLE 34. TOLERANCES FOR WORM GEARS

In 0.0001 In.

Class	Cir Pitch, In.	Runout Measurement on Pitch Circle, Total Indicator Reading								Pitch Variation Measured on Pitch Circle in Plane of Rotation (Increase Tolerances 50% for Nonhunting Ratios)								Accumulated Error between Any Two Teeth Exclusive of Runout Effect								Required Min Initial Area of Contact, %
Pitch Dia, In.		¾	1½	3	6	12	25	50	100	¾	1½	3	6	12	25	50	100	¾	1½	3	6	12	25	50	100	
Class 1. Light duty	4						90	90	100						60	70	80									20
	2					60	80	80	90					35	40	50	60									20
	1				60	60	80	80	90				30	35	35	40	50									20
	½			50	60	60	80	80				25	25	30	30	35										20
	¼		30	50	60	60	80				20	20	20	25	25											20
Class 2. Power transmission	2					25	30	35	40					15	20	25	30									30
	1				20	25	30	35	40				10	15	15	15	20									30
	½			20	20	25	30	35				10	10	10	12	15										30
	¼		15	20	20	25	30				9	9	9	10	10											30
Class 3. Precision	2					20	25							9	9							20	27			40
	1				20	20	25	25	30				7	7	7	8	8				15	15	15	35	55	40
	½			10	15	20	25	25	30			6	6	6	6	7	7			15	15	15	15	30	45	40
	¼		10	10	15	20	25				3½	4	4	4	4				15	15	15	15	15			40

SUGGESTED BACKLASH FORMULAS

Class	Max	Min
1	$0.0007D + 0.007p + 0.0035$	$0.0004D + 0.0045p + 0.001$
2	$0.0005D + 0.005p + 0.0025$	$0.0003D + 0.0035p + 0.001$
3	$0.00035D + 0.0035p + 0.002$	$0.00025D + 0.0025p + 0.001$

D = pitch diameter of worm gear. p = circular pitch.

Cone-drive Double-enveloping Gears

"Cone-drive" is the trade name for a double-enveloping, right-angle drive produced by Cone Drive Gears Division, Michigan Tool Co., and others. Basically, Cone-drive in its simplest form is a double-enveloping gear and pinion, the pinion threads and gear teeth of which are generated on operating centers in geared time relation. The hob or cutter is in the same relative position to the part being generated as the mating part will occupy when the set is mounted in operating position. The hob and cutter, however, have teeth thinner than the space to be generated but have identical lead, pressure angle, and tooth spacing. While maintaining geared time relation, side feed also rotates the parts about their axis to increase the width of space cut by the thin teeth to that actually desired.

Worm-gear Types. Worm gearing may be divided into three classes or types, (Fig. 24) as follows:

1. Those having neither element throated.
2. Those having one element throated.
3. Those having both elements throated.

When neither element is throated, as in the case of right-angle spiral gears, the nature of the contact is a theoretical point (Fig. 25). When a load is applied, deflection takes place in the materials and, consequently, the point becomes a small spot having some area of contact. This type has the least load-carrying capacity of any of the "worm-gear" types.

In the "conventional" worm gear, one element is throated. Contact is a theoretical line, varying in length with different designs. One member is involute in form, the other straight-sided. Under load, this line widens out to a thin band or zone of contact.

In the double-enveloping, or "Cone-drive," type of worm gear—in addition to having each element envelop the other—all teeth contact full depth, being straight-sided and tangent to a common circle. As a result of this principle, we have, in addition to the contact across the face, as found in single-enveloping types, a line contact extending full depth of the teeth. Thus, double enveloping results in area contact.

The nature of the contact is such that, in the passage of the pinion through the gear during one convolution, every portion of the pinion thread and tooth flank is in contact. For a considerable distance, above and below and at the midplane of the worm wheel, the contact is continuous. The remaining portions are in periodic engagement. The length of the pinion and width of the gear face are in no sense restricted, when the Cone design is employed. The gears may be of any desired ratio and may be cut with any desired amount of backlash, or with no backlash in cases where none is desirable.

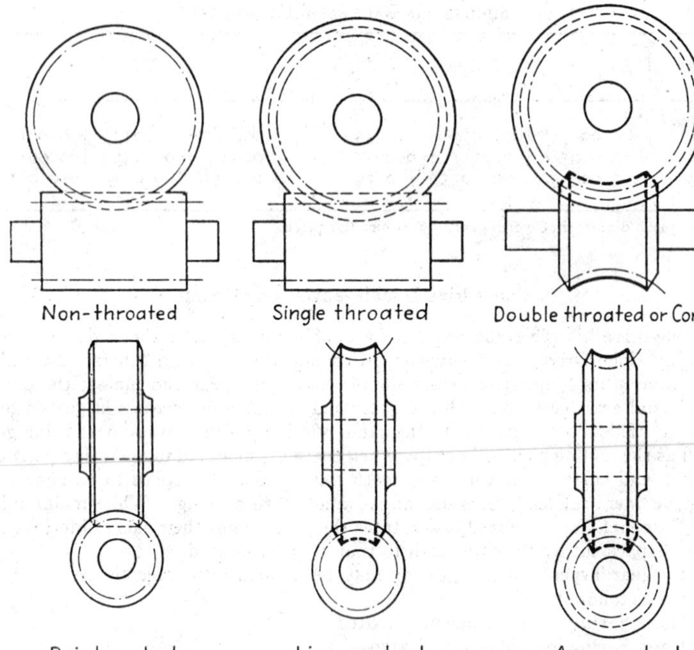

Fig. 24. Comparison of the double enveloping action of Cone-drive gearing with other classes of worm gearing.

Thus it will be seen that "throating" or "enveloping" the gear element or elements always results in an increase in tooth contact, and thereby load-carrying capacity. Furthermore, only throated elements regenerate or reproduce themselves when wear occurs. Such elements also must be form-generated on operating centers and must be located at assembly within reasonable limits.

Cone gears are no more sensitive to misalignment than conventional worm gearing. Any center-distance assembly errors affect the capacity and efficiency of Cone-drive gears to the same extent that they affect conventional worm gearing. Since the gears of both types are throated, it is necessary to hold their side positions within reasonable limits. And since the Cone pinion is also throated, it must similarly be held within the same reasonable limits as for the gears of either type. However, Cone-drives have the advantage that both pinion and gear are regenerative and, therefore, tend to correct themselves if assembled in a misaligned position.

Cone-drive gear sets are not lapped, nor do they require lapping where requirements in lead, pressure angle, and tooth spacing are within commercial limits. Lapping is used only when high-precision drives are needed, and here the purpose of lapping is to reduce the run-in period or to secure "zero backlash." Lapping compounds that do not charge the blanks and readily break down in the lapping process are used, such as powdered turkey bone and oil. The lapping may be done in a lapping machine or in the unit in which the gears will operate.

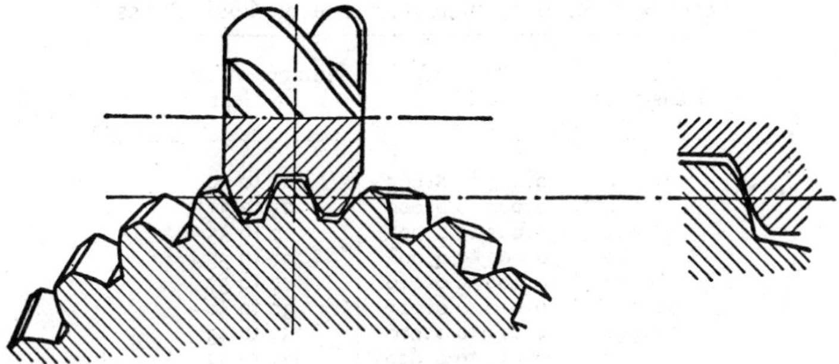

RIGHT ANGLE HELICAL. This type of gearing has theoretical point contact in either plane

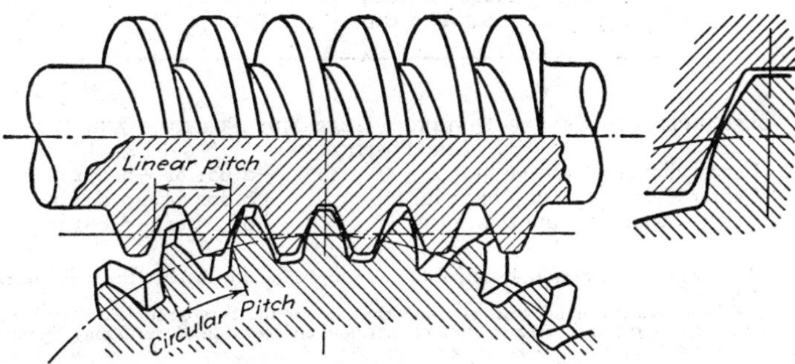

CONVENTIONAL WORM GEAR. This single enveloping type has theoretical line contact, with point contact in section. Under load there is some area contact

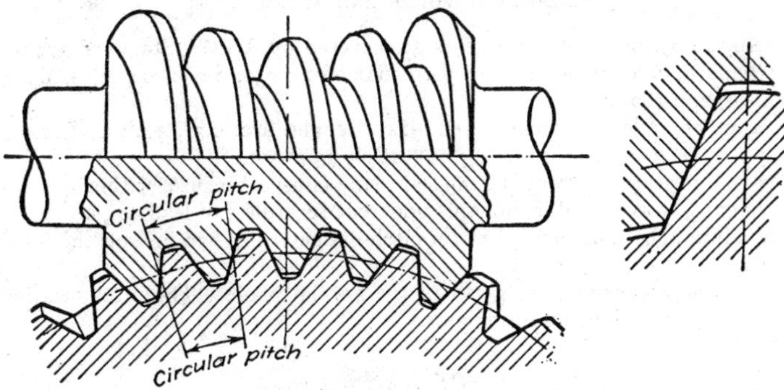

CONE-DRIVE. In addition to line contact along the gear tooth, this type also provides full depth contact, resulting in large area contact and more teeth in contact

FIG. 25. Cone-drive gears have larger area contact and more teeth in contact than right-angle helical and conventional worm gearing.

TABLE 35. RECOMMENDED BACKLASH OF CONE GEARS

Center Distance	Standard Backlash	Number of Teeth in Gear
3.000	0.003 to 0.005	24 to 30
4.000	0.004 to 0.006	25 to 31
5.000	0.005 to 0.007	27 to 33
6.000	0.006 to 0.008	28 to 35
8.000	0.008 to 0.010	29 to 37
10.000	0.010 to 0.015	30 to 40
12.000	0.015 to 0.020	32 to 45
13.500	0.015 to 0.020	36 to 52
15.000	0.015 to 0.020	40 to 55
18.000	0.020 to 0.025	43 to 60
20.000	0.020 to 0.025	45 to 66
22.000	0.025 to 0.030	48 to 70

TOLERANCES FOR CENTER DISTANCE, GEAR SIDE POSITION, AND WORM END POSITION

Center distances up to 6 in. . . . . +0.001 or −0.001 in.
Center distances from 6 to 15 in. . . . . +0.002 or −0.002 in.
Center distances more than 15 in. . . . . +0.003 or −0.003 in.

The side position of the gear and end position of worm should be held to the same tolerances as for center distance.

All Cone-drives are cut on accurate adapters that are ground to close limits. The blueprint dimension for center distance, side position of gear, and end position of worm may therefore be regarded as exactly correct.

FINE-PITCH WORMS AND WORM GEARS

Fine-pitch worms and worm gears are covered by AGMA Standard 374.01. In this standard, 8 axial pitches are used, namely: 0.030, 0.040, 0.050, 0.065, 0.080, 0.100, 0.130, 0.160 in.

A range of 15 lead angles is used. Lead angles are: 0.5°, 1.0°, 1.5°, 2.0°, 3.0°, 4.0°, 5.0°, 7.0°, 9.0°, 11.0°, 14.0°, 17.0°, 21.0°, 25.0°, 30.0°.

The 20° pressure angle in the normal plane is used because, within the range of the standard, it avoids objectionable undercut regardless of lead angle.

The minimum recommended worm pitch diameter is 0.250 in. and the maximum is 2.000 in.

The tooth proportions are given in the normal plane. The pitch relations are expressed by the following formulas:

$$p_n = p_x \cos \lambda = p \cos \psi$$

where p = circular pitch of worm gear, p_x = axial pitch of worm, p_n = normal circular pitch of worm and worm gear, λ = lead angle of worm, and ψ = helix angle of worm gear.

Table 36 gives the worm and worm-gear dimensions.

TABLE 36. PROPORTIONS OF FINE-PITCH WORMS AND WORM GEARS

p = circular pitch of worm gear
p_x = axial pitch of worm
p_n = normal circular pitch of worm and worm gear
n = number of threads in worm
N = number of teeth in worm gear

Term	Symbol	Formula, In.
Worm Dimensions		
Lead	l	np_x
Pitch dia	d	$l \div (\pi \tan \lambda)$
OD	d_o	$d + 2a$
Safe minimum length of threaded portion of worm	F_W	$\sqrt{D_o^2 - D^2}$*
Worm-gear Dimensions		
Pitch dia	D	$Np \div \pi$
Throat dia	D_t	$D + 2a$
OD	D_o	See Figs. 26 and 27
Data Relating to Worm and Worm Gear		
Addendum	a	$0.3183p_n$
Whole depth	h_t	$0.7003p_n + 0.002$
Working depth	h_k	$0.6366p_n$
Clearance	c	$h_t - h_k$
Tooth thickness	t	$0.5p_n$
Approx normal pressure angle (see p. 4-84)	ϕ_n	$20°$
Center distance	C	$0.5(d + D)$

* This formula allows a sufficient length for fine-pitch worms.

Table 37 gives the pitch diameter for each combination of lead and lead angle, together with the number of threads for a particular lead and diameter.

Table 38 gives the tooth proportions based on normal pitch for all combinations of standard axial pitches and lead angles.

Figures 26 and 27 show blank recommendations for fine-pitch worm gears. These two types have been standardized and are tied in with the worm-length formulas for each type of blank design.

Backlash in fine-pitch worm gearing is produced by thinning the worm threads by a suitable amount. The worm gear is cut so that its tooth thickness and space width are equal at the standard pitch diameter.

This standard recognizes the fact that worm profiles are dependent on their method of manufacture and profile departures from the calculated shape occur when production methods, for the same worm, differ. Thus, a worm of a given pitch and lead angle produced with a 2-in. milling cutter will have a different profile and pressure angle than the same worm ground with a 20-in. dia grinding wheel. This standard emphasizes the fact that, in worm gearing, the tools and work are inti-

TABLE 37. DIAMETER RANGE OF WORK FOR FINE-PITCH WORMS

n	Lead, In.	Lead Angles, °														
		0.5	1.0	1.5	2.0	3.0	4.0	5.0	7.0	9.0	11.0	14.0	17.0	21.0	25.0	30.0
1	0.030	1.0937	0.5472	0.3647	0.2735											
1	0.040	1.4583	0.7297	0.4863	0.3646	0.2429										
1	0.050	1.8228	0.9121	0.6079	0.4558	0.3037	0.2276									
2	0.060	2.1874	1.0945	0.7295	0.5469	0.3644	0.2731									
1	0.065		1.1857	0.7903	0.5925	0.3948	0.2959	0.2365								
1, 2	0.080		1.4593	0.9726	0.7293	0.4859	0.3641	0.2911								
3	0.090		1.6417	1.0942	0.8204	0.5466	0.4097	0.3274	0.2333							
1, 2	0.100		1.8242	1.2158	0.9116	0.6073	0.4552	0.3638	0.2592							
3, 4	0.120		2.1890	1.4590	1.0939	0.7288	0.5462	0.4366	0.3111	0.2412						
1, 2	0.130			1.5805	1.1851	0.7896	0.5917	0.4730	0.3370	0.2613						
3, 5	0.150			1.8237	1.3674	0.9110	0.6828	0.5457	0.3889	0.3015	0.2456					
1, 2, 4	0.160			1.9453	1.4585	0.9718	0.7283	0.5821	0.4148	0.3216	0.2620					
6	0.180			2.1884	1.6408	1.0932	0.8193	0.6549	0.4667	0.3618	0.2948					
3	0.195				1.7776	1.1843	0.8876	0.7095	0.5055	0.3919	0.3193	0.2490				
2, 4, 5	0.200				1.8232	1.2147	0.9104	0.7276	0.5185	0.4020	0.3275	0.2553				
7	0.210				1.9143	1.2754	0.9559	0.7640	0.5444	0.4221	0.3439	0.2681				
3, 6, 8	0.240				2.1878	1.4576	1.0924	0.8732	0.6222	0.4823	0.3930	0.3064	0.2499			
5	0.250					1.5184	1.1380	0.9096	0.6481	0.5024	0.4094	0.3192	0.2603			
2, 4	0.260					1.5791	1.1835	0.9459	0.6741	0.5225	0.4258	0.3319	0.2707			
9	0.270					1.6398	1.2290	0.9823	0.7000	0.5426	0.4421	0.3447	0.2811			
7	0.280					1.7006	1.2745	1.0187	0.7259	0.5627	0.4585	0.3575	0.2915			
3, 6, 10	0.300					1.8220	1.3656	1.0915	0.7778	0.6029	0.4913	0.3830	0.3123	0.2488		
2, 4, 8	0.320					1.9435	1.4566	1.1642	0.8296	0.6431	0.5240	0.4085	0.3332	0.2654		
5	0.325					1.9739	1.4794	1.1824	0.8426	0.6532	0.5322	0.4149	0.3384	0.2695		
7	0.350					2.1257	1.5932	1.2734	0.9074	0.7034	0.5731	0.4468	0.3644	0.2902		

This table gives the pitch diameter for each combination of lead and lead angle, together with the number of threads for a particular lead and diameter.

9	0.360						1.6387	1.3098	0.9333	0.7235	0.5895	0.4596	0.3748	0.2985	0.2457	
3, 6	0.390						1.7752	1.4189	1.0111	0.7838	0.6387	0.4979	0.4060	0.3234	0.2662	
4, 5, 8, 10	0.400						1.8207	1.4553	1,0370	0.8039	0.6550	0.5107	0.4165	0.3317	0.2730	
9	0.450						2.0483	1.6372	1.1666	0.9044	0.7369	0.5745	0.4685	0.3732	0.3072	0.2481
7	0.455							1.6554	1.1796	0.9144	0.7451	0.5809	0.4737	0.3773	0.3106	0.2509
3, 6	0.480							1.7463	1.2444	0.9647	0.7860	0.6128	0.4998	0.3980	0.3277	0.2646
5, 10	0.500							1.8191	1.2963	1.0049	0.8188	0.6383	0.5206	0.4146	0.3413	0.2757
4, 8	0.520							1.8919	1.3481	1.0451	0.8515	0.6639	0.5414	0.4312	0.3550	0.2867
7	0.560							2.0374	1.4518	1.1255	0.9170	0.7149	0.5830	0.4644	0.3823	0.3087
9	0.585								1.5166	1.1757	0.9580	0.7469	0.6091	0.4851	0.3993	0.3225
6	0.600								1.5555	1.2059	0.9825	0.7660	0.6247	0.4975	0.4096	0.3308
4, 8	0.640								1.6592	1.2863	1.0480	0.8171	0.6663	0.5307	0.4369	0.3529
5, 10	0.650								1.6852	1.3064	1.0644	0.8298	0.6767	0.5390	0.4437	0.3584
7	0.700								1.8148	1.4068	1.1463	0.8937	0.7288	0.5805	0.4778	0.3859
9	0.720								1.8666	1.4470	1.1791	0.9192	0.7496	0.5971	0.4915	0.3970
6	0.780								2.0222	1.5676	1.2773	0.9958	0.8121	0.6468	0.5324	0.4300
5, 8, 10	0.800									1.6078	1.3101	1.0213	0.8329	0.6634	0.5461	0.4411
9	0.900									1.8088	1.4738	1.1490	0.9370	0.7463	0.6144	0.4962
7	0.910									1.8289	1.4902	1.1618	0.9474	0.7546	0.6212	0.5017
6	0.960									1.9294	1.5721	1.2256	0.9995	0.7961	0.6553	0.5293
10	1.000									2.0098	1.6376	1.2767	1.0412	0.8292	0.6826	0.5513
8	1.040										1.7031	1.3277	1.0828	0.8624	0.7099	0.5734
7	1.120										1.8341	1.4299	1.1661	0.9287	0.7645	0.6175
9	1.170										1.9160	1.4937	1.2181	0.9720	0.7987	0.6451
8	1.280										2.0961	1.6341	1.3327	1.0614	0.8738	0.7057
10	1.300											1.6597	1.3535	1.0780	0.8874	0.7167
9	1.440											1.8384	1.4993	1.1941	0.9830	0.7939
10	1.600											2.0427	1.6658	1.3268	1.0922	0.8821

TABLE 38. TOOTH PROPORTIONS OF FINE-PITCH WORMS AND WORM GEARS BASED ON NORMAL PITCH FOR ALL COMBINATIONS OF STANDARD AXIAL PITCHES AND LEAD ANGLES

Standard Axial Pitch, In.	Tooth Proportions	Lead Angle,°														
		.5	1	1.5	2	3	4	5	7	9	11	14	17	21	25	30
.030	a	.0095	.0095	.0095	.0095	.0095	.0095	.0095	.0095	.0094	.0094	.0093	.0091	.0089		
	h_t	.0229	.0229	.0229	.0229	.0229	.0229	.0229	.0229	.0227	.0227	.0225	.0220	.0216		
	p_n	.0300	.0300	.0300	.0300	.0300	.0299	.0299	.0298	.0296	.0294	.0291	.0287	.0280		
.040	a	.0127	.0127	.0127	.0127	.0129	.0127	.0127	.0126	.0126	.0125	.0124	.0122	.0119	.0115	
	h_t	.0299	.0299	.0299	.0299	.0299	.0299	.0299	.0297	.0297	.0295	.0293	.0288	.0282	.0273	
	p_n	.0400	.0400	.0400	.0400	.0399	.0399	.0398	.0397	.0395	.0393	.0388	.0383	.0373	.0363	
.050	a	.0159	.0159	.0159	.0159	.0159	.0159	.0159	.0158	.0157	.0156	.0154	.0152	.0149	.0144	.0138
	h_t	.0370	.0370	.0370	.0370	.0370	.0370	.0370	.0368	.0365	.0363	.0359	.0354	.0348	.0337	.0324
	p_n	.0400	.0500	.0500	.0500	.0499	.0499	.0498	.0496	.0494	.0491	.0485	.0478	.0467	.0453	.0433
.065	a		.0207	.0207	.0207	.0207	.0206	.0206	.0205	.0204	.0203	.0201	.0198	.0193	.0188	.0179
	h_t		.0475	.0475	.0475	.0475	.0473	.0473	.0471	.0469	.0467	.0462	.0456	.0445	.0434	.0414
	p_n		.0650	.0650	.0650	.0649	.0648	.0648	.0645	.0642	.0638	.0631	.0622	.0607	.0589	.0563
.080	a		.0255	.0255	.0254	.0254	.0254	.0254	.0253	.0252	.0250	.0247	.0244	.0238	.0231	.0221
	h_t		.0581	.0581	.0579	.0579	.0579	.0579	.0577	.0574	.0570	.0563	.0557	.0544	.0528	.0506
	p_n		.0800	.0800	.0800	.0799	.0798	.0797	.0794	.0790	.0785	.0776	.0765	.0747	.0725	.0693
.100	a		.0318	.0318	.0318	.0318	.0318	.0317	.0316	.0314	.0312	.0309	.0304	.0297	.0288	.0276
	h_t		.0720	.0720	.0720	.0720	.0720	.0717	.0716	.0711	.0706	.0700	.0689	.0673	.0654	.0627
	p_n		.1000	.1000	.0999	.0999	.0998	.0996	.0993	.0988	.0982	.0970	.0956	.0934	.0906	.0866
.130	a			.0414	.0414	.0413	.0413	.0412	.0411	.0409	.0406	.0402	.0396	.0386	.0375	.0358
	h_t			.0931	.0931	.0929	.0929	.0926	.0924	.0920	.0913	.0904	.0891	.0869	.0845	.0808
	p_n			.1300	.1299	.1298	.1297	.1295	.1290	.1284	.1276	.1261	.1243	.1214	.1178	.1126
.160	a			.0509	.0509	.0509	.0508	.0507	.0506	.0503	.0500	.0494	.0487	.0475	.0462	.0441
	h_t			.1140	.1140	.1140	.1138	.1135	.1133	.1127	.1120	.1107	.1091	.1065	.1036	.0990
	p_n			.1599	.1599	.1598	.1596	.1594	.1588	.1580	.1571	.1552	.1530	.1494	.1450	.1386

Recommended Design of Blanks for Fine-pitch Worm Gears

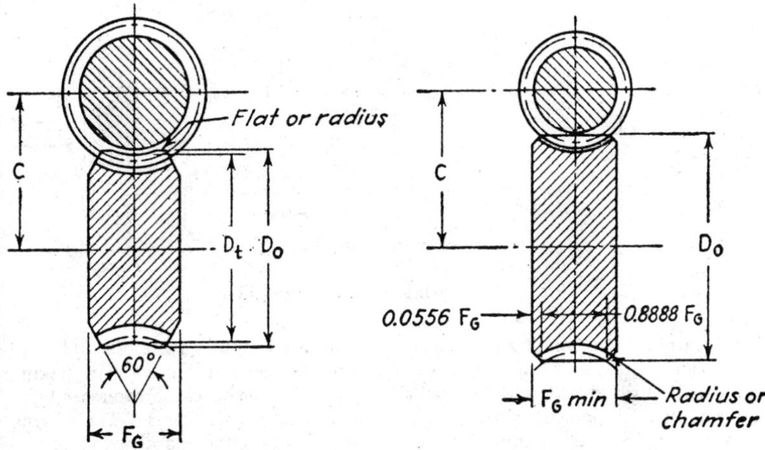

FIG. 26. Throated blank (for power drives).

$$D_o = 2C - 0.891d + 1.782a$$
$$F_G = 0.57735d_o$$
$$D_t = D + 2a \text{ (See Table 36)}$$

where
D = pitch dia of worm gear
D_o = outside dia of worm gear
d_o = outside dia of worm
d = pitch dia of worm
a = addendum
F_G = face width of worm gear
C = center distance

FIG. 27. Nonthroated blank (for transmission of motion).

$$D_o = 2C - d + 2a$$
$$F_{G\,\min} = 1.125\sqrt{(d_o + 2c)^2 - (d - 4a)^2}$$

where
D_o = outside dia of worm gear
C = center distance
d = pitch dia of worm
a = addendum
$F_{G\,\min}$ = minimum face width of worm gear
d_o = outside dia of worm
c = clearance

mately connected and manufacturing information of the work must be furnished when ordering gear-cutting tools in order to obtain gears which are conjugate to worms. A table of profile deviations and pressure-angle changes covering the entire range of the standard is given as part of the standard.

Gear-cutting Methods

Hobbing Method of Cutting Gear Teeth

By Granger Davenport,

Chief Engineer, Gould & Eberhardt, Inc.

Hobbing is a machine process for generating gear teeth. All methods of cutting gear teeth fall into either of two general classes—the forming method or the generating method. In the forming method, the tool is given a shape matching that of the desired tooth space; in the generating method, the tool has a shape conjugate to the form of the tooth when rolled in contact with it and frequently can be substantially straight, as in hobbing.

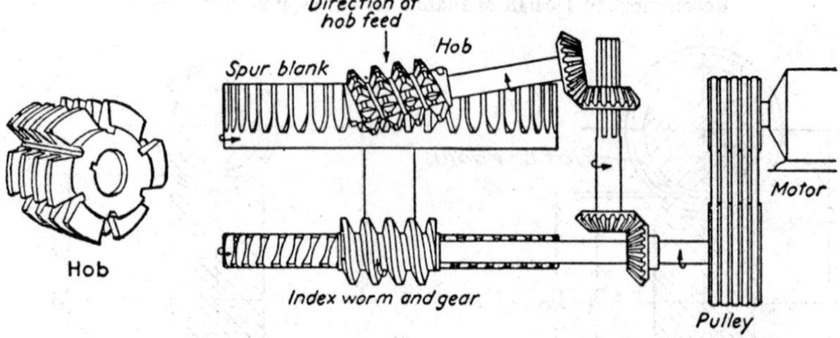

FIG. 28. Gearing in hobbing machine.

The cutter in the hobbing process is called a "hob." The nature of a hob may be best understood by comparing it with a worm, as shown in Fig. 28. In making a hob, a cylindrical blank is first turned and bored; then the helical thread resembling a worm thread is milled; in this state the hob is actually a worm. The cutting teeth are produced by milling flutes across the thread, either parallel to the hob axis or at right angles to the thread. The sides and tops of the individual hob teeth are relieved uniformly to provide and maintain the cutting clearances and precise form throughout the sharpening life of the tool similar to milling cutters. After the hardening operation, the hole, proof diameters, and usually the sides and tops of the teeth are all ground. The hob is sharpened by grinding the flutes.

A hob and blank being cut may be likened to a worm and worm gear in mesh as in Fig. 28. Hobbing is a continuous milling process in which the hob and blank both rotate in timed relation to each other. In addition to the rotary motion, the hob and gear blank are fed relatively to each other to produce the spur, helical, or worm gear. In the hobbing process the cutting action is continuous in one direction until the blank is finished. Modern hobbing machines have a high degree of accuracy built in that can be maintained over a long period of years by adjusting means incorporated therein. Hobbing is therefore widely used for cutting precision gears at a high rate of production.

The hob for spur or helical gears is provided with a tooth form identical with that of an involute tooth from a spur or helical rack of the same pitch, tooth depth, and pressure angle. This tooth form is theoretically curved to mesh properly with the curved involutes generated on the blank. However, the hob form so nearly approaches a straight-line rack-tooth shape that the difference is extremely slight. This is a fundamental advantage of the hobbing process, because hob teeth being straight-sided, or nearly so, are readily made to a very high degree of accuracy and are easily measured, as compared with curved cutter teeth used in some methods of gear cutting.

Sections through a hob and gear blank at right angles to a pair of teeth in contact are represented in Fig. 29. The teeth in the hob section actually form the path of an ellipse, but, practically speaking, they are equivalent to a rack, because the degree of curvature of the ellipse within the arc of tooth contact is almost nil.

Figure 29 also illustrates how successive hob teeth coming into contact with each tooth in the blank generate gear teeth of involute form. Hobbing is not limited to the involute form, although for spur and helical gears this form is used almost exclusively at the present time. In the involute system a straight-sided rack will

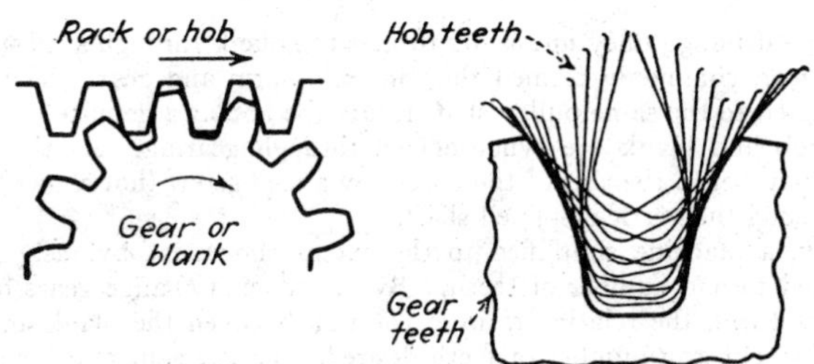

FIG. 29. How teeth are generated.

mesh properly with a gear or pinion regardless of the number of teeth. Consequently, one hob can be used to generate gears of any number of teeth, all of which will mesh properly with each other and with a rack.

Angular Setting of the Hob. The axis of the hob for cutting spur gears is tilted according to the hob thread angle to have the hob teeth in line with the teeth of the blank, as in Fig. 30. In cutting helical gears, the hob is tilted an additional amount so that the teeth coincide with the inclination of the teeth being produced on the blank, also illustrated in Fig. 30. In this case the angle of tilt equals the helix angle of the gear minus the thread angle of the hob, when both are of the same hand. When they are of opposite hands, the angle of tilt is the sum of the two angles.

A slight variation in setting the hob angle of tilt affects neither the tooth form nor the helix angle. An angular error of as much as several minutes can be tolerated with no noticeable effect on the product other than slightly thinner teeth and increased backlash. This is an inherent advantage of the hobbing method.

The Gearing in a Hobbing Machine. The arrangement of a hobbing machine may be readily understood by referring to the simplified gearing diagram shown in

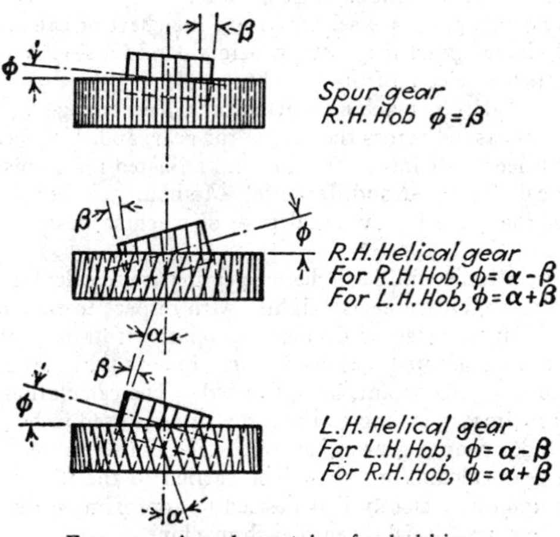

FIG. 30. Angular setting for hobbing.

Fig. 28. The driving pulley uniformly rotates the blank through a master dividing worm and gear, commonly termed the "index" worm and gear. The four bevel gears connected to the same pulley shaft rotate the hob at a definite relative speed. Thus the hob and blank are synchronized through gearing. As the hob is fed downward, parallel to the axis of the blank, by a feed screw (not shown), one bevel gear slides along the vertical splined shaft.

Ordinarily, a machine simplified to the extent shown is obviously impractical unless limited to one number of teeth. By introducing change gears in the drive to the index worm, the relative ratio of rotation between the blank and hob may be changed to cut any number of teeth desired. To cut spur and helical gears of various pitches, additional mechanism must be introduced in the hob drive to permit tilting to the required angle. A feed screw and related mechanism driven from the index worm shaft feed the hob at a definite rate with respect to the rotation of the blank. Suitable change gearing is employed to vary the feed.

The hob speed is changed to suit the cutting conditions by varying the speed of the main drive shaft. This also is accomplished by change gears, which in this case alter the speed of the entire mechanism proportionately. For instance, if the speed of the hob is doubled, it is obvious from Fig. 28 that the speeds of the blank and feed are increased to the same extent and the production is doubled.

Hobbing Spur Gears. The first step in the preparation for cutting a gear is to mount a fixture on the work table or spindle. Second, a blank is accurately clamped to the fixture and indicated for running true. The hob is then clamped upon its arbor and tilted to the correct angle. Next, the proper change gears are selected and placed upon the machine according to the number of teeth to be cut and for the hob speed and feed required.

The ratio of the index change gears depends upon the total ratio of the index and cutter gear trains. For example, if the machine ratio is 30, then the index formula is $30T/N$, in which N equals the number to be cut and T equals the number of threads in the hob. Usually a chart is provided showing index change-gear combinations for a selected range of teeth. In other cases, the formula is applied by substituting values for N and T, and splitting the result into two fractions suitable for change gears. The calculations must be exact.

Next the feed change gears are selected from the chart or calculated by the feed formula. These change gears may be approximate. In a similar manner, the hob-speed change gears are selected from the chart.

The hob is set to depth, machine started, and feed engaged. Cutting is completed after the hob has fed across the face of the gear, and then the gear is removed or the hob is returned to its initial position and adjusted for a finishing cut.

Hobbing Helical Gears—Nondifferential Method. In hobbing helical gears nondifferentially, the same fundamental type of machine described for spur gears is used.

In this method the index and feed change gearing are so calculated that the blank and cutter are advanced or retarded slightly with respect to each other to produce the helix angle. The increase or decrease in relative rotation must be definitely correlated with the feed selected. Consequently, in both the index and feed formulas a factor, in the form of a constant, is introduced. The calculation of this constant may be very approximate, as long as the same value is used in both index and feed formulas. The index change gearing must be exact, as in spur-gear cutting, whereas the feed change-gear calculations should be carried to the fourth or fifth decimal place depending upon how closely it is desired to approximate the theoretical helix angle. Instructions are furnished with each machine.

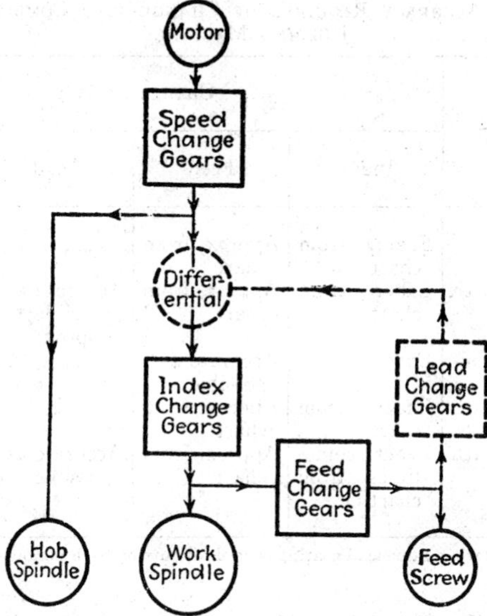

FIG. 31. Diagram of hobbing-machine layout.

Hobbing Helical Gears—Differential Method. In the differential method, additional mechanism is required to obtain the increase or decrease of the relative rotation of the hob and blank to produce the helix angle. This mechanism usually comprises a bevel-gear type of differential connected to effect the rotation of the blank and correlated to the feed motion through an extra set of change gears for the lead. The differential functions by imparting a supplemental motion to the existing direct drive to the work table.

A schematic diagram of a hobbing machine, including a differential and lead gears, is shown in Fig. 31. The differential and lead gears are shown in dotted lines to indicate the mechanism added to a nondifferential machine for cutting helical gears by the differential method. The power originating at the motor passes through speed change gears and branches to the hob and work spindles. The differential and index change gears are located in the branch drive to the work spindle. A branch to the feed screw is taken off between the index change gears and work spindle. The feed change gears are provided in this branch line, and drive both the feed screw and the differential. The lead change gears are located in the drive to the differential.

In the differential method the index change gears are selected from a chart, the same as for spur gears. The feed change gears are also selected directly from a chart. The rate of feed can be changed at will without affecting the index or lead gearing, resulting in a greater flexibility in this respect than the nondifferential method. The lead change gears connecting the feed to the differential determine the helix angle. Therefore, the lead change-gear calculations must be correct to the same degree of accuracy as for *feed* change gears in nondifferential hobbing in order to approximate the theoretical helix angle.

TABLE 39. ACCURACY REQUIRED OF CHANGE-GEAR COMBINATIONS ON HOBBING MACHINES

Types of Gears Being Hobbed	Change Gearing			
	Index	Feed	Lead	Speed
Spur gears..........	Exact, from chart	Approx, from chart	....	Approx, from chart
Helical gears cut with differential	Exact, from chart	Approx, from chart	Accurate as possible;* approx	Approx, from chart
Helical gears cut non-differentially	Exact	Accurate as possible	....	Approx, from chart
Worm gears cut with infeed	Exact, from chart	Approx, from chart	....	Approx, from chart
Worm gears cut with tangential feed	Exact, sometimes from chart	Approx, from chart	Accurate as possible	Approx, from chart

* Permissible only when same lead gearing is used for cutting both pinion and gear.

Differential vs. Nondifferential Hobbing. When a helical gear is cut differentially and a second cut is required, it is merely necessary to disconnect the feed, move the cutter slide to the initial position, and proceed with the recut. In other words, the relation of the hob to the teeth of the blank being cut is always maintained.

On the other hand, if a gear has been cut nondifferentially and the feed is disconnected for the purpose of returning the slide to the initial position, the lead relation of the hob teeth to the blank teeth is destroyed, making it necessary to reset the cutter and blank very carefully to have the second cut clean up evenly on both sides. If there is a reasonable amount of stock left on the teeth for the finishing cut, this operation is comparatively simple. But if the finishing cut is to remove very little stock, the proper setting of the cutter becomes quite difficult. Various devices are employed to aid in resetting the hob for the second cut on nondifferential machines. One is a clutch arranged to temporarily disengage the feed-screw drive until the hob teeth are seen to track exactly in the gear-tooth spaces. Another comprises a clutch in the index-worm drive and a set of three pointers and zero marks, enabling the work table, cutter spindle, and cutter slide to be returned to their initial zero positions for the recut. However, none of these methods is as reliable as the differential method for very light recuts. Thus, it is evident that, while a differential complicates the mechanism, it is valuable if helical gears are to be hobbed in more than one cut.

There are still other advantages of the differential method. For instance, if mating gears are to be cut on the same machine or with the same lead change gearing, a considerable departure from the theoretical helix angle may be tolerated. Under these circumstances the change-gear calculations need not be carried out to the high degree of accuracy required for nondifferential hobbing of helical gears. However, if pinion and gear are cut on different machines with unlike lead formulas, these approximate calculations will not suffice, and the same accuracy is required as for nondifferential feed-gear calculations.

The nondifferential machine is simpler and less expensive than one equipped with a differential. This is an important consideration for mass-production shops.

Other advantages of the nondifferential method are for certain combinations of pitch and helix angle beyond the range of a differential machine and for helical gears with large prime numbers of teeth. In special cases, it is possible to cut nondifferentially and to engage the differential only while returning the slide for the recut, thus maintaining the lead relation.

In general, to obtain the benefits of both differential and nondifferential hobbing, the machine should be equipped with a differential which may be locked out.

Hobs for Spur and Helical Gears. The same hob used for spur gears is suitable for cutting helical gears of either hand and of any helix angle. If the helix angle is more than 10 or 12° and the quantity of helical gears to be cut is large, it will be found advantageous to use two hobs, the hand of each corresponding to the hand of the gear being cut. When the helix angle is more than 20 or 25°, it is preferable to taper the hob to relieve the load on the end teeth.

A ground hob is one having the sides and tops of its teeth ground after hardening. It is more accurate and more expensive than unground hobs.

Four classes of hobs are recognized. Class A is precision ground and is used for finishing gears of the highest accuracy. Class B is commercial ground and is satisfactory for finishing many types of work. Class C is accurate unground, which is frequently suitable for finishing, particularly for the finer pitches. Class D is commercial unground utilized for roughing cuts.

For fine pitches and small lots of medium- and coarse-pitch gears the same hob may be used for both roughing and finishing.

For medium or large lots if the pitch is 12 DP or coarser, it is good practice to rough with a class D hob and finish with a class A, B, or C hob, depending upon the accuracy required. For very coarse pitches it is sometimes necessary to take two roughing cuts and one finishing cut.

Unless otherwise specified, finishing hobs are given a slight modification in form for producing tip relief on the gear teeth. This eases the gears into mesh and helps to quiet the gears during their initial running-in period.

One of the most important factors in the use of hobs is proper sharpening. In sharpening, it is essential to mount the hobs on arbors running exactly true, to maintain exact spacing of the flutes, and to maintain the faces of the teeth exactly radial or to the original hook angle. If hobs are properly sharpened, their initial accuracy of lead and form will remain throughout their life. It is a characteristic of the hobbing process that the tooth forms generated are unchanged by sharpening the hob.

In mounting hobs on the hobbing machine, they must run true within 0.00025 in. (a quarter of a thousandth) indicator reading at each end; otherwise inaccurate tooth forms and noisy gears will result.

The cutting action of a hob is distributed over many teeth, producing a large number of tooth inches before sharpening is required. When a hob shows signs of becoming dull, it should be shifted axially a short distance to utilize sharp teeth in a new zone, the slightly dull teeth continuing to be used in the roughing zone. This should be repeated until the entire hob is dull, before sharpening. The hob may be shifted manually, or automatically on high production runs. The amount of shift may have to be an exact multiple of the pitch, if hob centralizing is necessary for the job being cut. The special tapered-end hobs for high-angle helical gears can be shifted only a limited amount; hence they are made much shorter than standard spur-gear hobs and require more frequent sharpening.

Multiple-thread Hobs. Multiple-thread hobs should be used for roughing and semifinish hobbing prior to shaving, to save time. They cannot be made as accurately as single-thread hobs, and they are not ordinarily practical for pitches coarser than 3 DP. For very low numbers of teeth the number of threads in the hob may not exceed two without causing excessive speed of the index worm in the machine. The cutting action of roughing hobs is improved by providing about 5° positive rake on the teeth, instead of radial flutes.

Shear-type roughing hobs are particularly advantageous in cutting the higher carbon steels at fast feeds and speeds. The shearing action is obtained by making the hand of flute opposite to the hand of the thread, with positive rake on the teeth. In this type of hob the lead is modified to distribute the cutting action among more cutting teeth; this, however, restricts the range of teeth that may be cut by any one shear-cut hob.

Hob-setting Gages. For the general run of hobbed spur and helical gears, and infed worm gears, there is no need for accurately centralizing on either a tooth or a space of the hob. This is particularly true of fine and medium pitches. In general, an axial section through a hob may be compared with a rack in mesh with the gear being cut, and as new rows of cutting teeth come into action the successive sidewise displacement of the hob teeth gives the effect of rolling a rack with the gear. There are usually a dozen or more finishing teeth that form the final involute profile and which are ample to generate smooth and accurate involute forms without traces of flats. Since each tooth generates a small portion of the involute, there is little to be gained by centralizing on any hob or space.

In hobbing coarse pitches, the combination of a low number of teeth being cut and relatively few flutes in the hob will generate visible flats on the work. These flats may not be symmetrical on the two sets of profiles unless a hob-setting gage is used to centralize a hob tooth or space. Balancing the flats in this manner will often improve involute profile of pinions, and definitely affect spline shapes.

On low numbers of teeth the root fillets and chamfers at the tops of the teeth (if used) are likewise sensitive to off-balance conditions which can be readily rectified by the use of a hob-setting gage.

Multiple-thread finishing hobs necessarily have relatively few teeth for generating the involute profile, and consequently centralizing on a tooth or a space is highly desirable when cutting pinions. This is true even if the gears are to be subsequently shaved, because off balance of the profiles may prevent uniform cutting action of the shaving cutter.

Increasing the number of flutes in pinion hobs is sometimes successful in generating flats so close together that they merge into smooth curves. Obviously there are practical limitations to the number of flutes that can be used for any given pitch and hob diameter. This remedy should not be overlooked when production warrants purchase of special hobs.

If the number of teeth to be cut is extremely low, as in automotive starter pinions having five or six teeth, it may not be practical to increase the number of hob flutes sufficiently to eliminate all visible flats, but such pinions are not used at high speeds and therefore accuracy of form is not so important.

When hobbing with a single-position hob as for certain splines, or in the production of worms by a single-convolution hob, centralizing with a gage is essential. The single-convolution worm-generating hob has three finishing teeth, of which the middle one must be centralized.

Speeds and Feeds for Spur Gears. So many variables enter into the selection of speeds and feeds that only general recommendations can be mentioned here.

The speeds for cast iron vary from 40 to 100 fpm, usually being about 80. For steel, the speed is usually between 60 and 150 fpm, averaging approximately 100. On specially designed high-speed hobbing machines, the speed for hobbing steel with ordinary high-speed steel hobs may approach 500 fpm. Cemented-carbide hobs are successfully used on nonmetallic materials at still higher speeds.

The feed per revolution of blank is 0.030 to 0.050 in. for a single-cut job. For roughing, it ranges from 0.060 to 0.200 in. with special roughing hobs. In selecting a roughing feed, both the pitch and the feed-mark pattern must be taken into account. A limiting factor in feed selection is the amount of metal that each hob tooth can safely remove, particularly in the coarser pitches. Another factor is the acceptable feed pattern or distance between feed marks, especially if the cut is semifinishing prior to shaving.

For double-thread hobs, the feed may be the same as for single-thread roughing hobs or reduced up to one-fourth, with the index gearing arranged for half the number of teeth, practically doubling the production. For triple-thread hobs, the feed is somewhat less than for single-thread roughing feeds, sometimes reduced up to one-third, the index gearing being arranged for one-third the number of teeth being cut; in this case the production rate is still higher than for double-thread hobs.

In finishing after roughing cuts, the feed ranges from 0.040 to 0.100 in. according to the finish desired.

Speeds and Feeds for Helical Gears. The speeds for hobbing helical gears are the same as for spur gears. The feeds per revolution of blank given for spur gears should be reduced as the helix angle increases, to maintain the same quality of finish. In selecting a feed for a helical gear, the appropriate spur-gear feed is multiplied by the cosine of the helix angle.

Climb vs. Conventional Hobbing. In "climb hobbing" the directions of hob rotation and feed are so related that the hob tends to pull itself into the work, resulting in a chip section that is maximum at the start, tapering to nothing at the finish. It is comparable to "climb" or "down" milling, as opposed to "up" or "conventional" milling.

It is often advantageous to climb hob, depending upon material to be cut and amount of stock to be removed. The pitch, helix angle, and setup, and especially the condition of the machine, are factors that must be taken into consideration when deciding on whether to climb or conventional cut.

Climb hobbing is recommended for steel, either in a soft state or heat-treated. Decidedly better hob life and surface finish result. The speeds and feeds can be increased materially when climb hobbing heat-treated steel.

In some instances climb hobbing of soft steel has permitted finishing in one cut, where conventional hobbing required two cuts; the speed and feed for the single pass are identical with the *finish* cut by conventional method, thereby effecting a considerable reduction in time.

For very coarse-pitch steel gears, in the order of 1 DP, conventional cutting has remained preferable to climb cutting, because there is less tendency toward chatter.

In hobbing cast iron there is little choice as to whether to climb or conventional hob, from the standpoint of feed, speed, finish, or hob life.

It is good practice to try both climb and conventional cutting on the first few blanks cut on any job involving a considerable number of pieces. This is particularly true for materials other than steel or cast iron, where it is not always possible to foretell which method is preferable.

In conventional cutting it is preferable to have the force of the cut against the work table or spindle, and the same is true when climb cutting. This should be taken into account when determining direction of feed and rotation of hob when climb hobbing. Occasionally, because of limitations of the design of the part or the fixture, it is necessary to depart from the practice of having the cutting force toward the work table, but ordinarily this is not recommended.

Special hobs are not necessary for climb-hobbing spur gears, or helical gears with helix angles up to about 25°. For larger helix angles, where tapered hobs are essential, the hob must be designed with the taper at the opposite end as compared with hobs for conventional cutting.

In general, maximum results from the climb-hobbing method will not be obtained unless the hobbing machine is in good condition with minimum play in the moving parts, particularly between feed screw and nut.

Coolant. Sulfurized mineral oil in combination with lard oil is preferable for hobbing steel to obtain maximum cutter life, improved finish, equalized temperature, and high production. While soluble oil in water is an excellent coolant, particularly for brass and bronze, it is not recommended for use on precision machine tools because sooner or later it will work its way into bearings and gear housings and cause damage. No coolant is necessary for cast iron, but in special cases such as hobbing ductile iron, it is effectively used.

Hobbing Worm Gears—Infeed Method. Worm gears may be produced by the infeed method. The hob is fed in a radial direction toward the center of the blank instead of across the face of the blank as in spur-gear hobbing. When the proper center distance has been reached, as indicated by a scale and vernier on the machine, the feed is disengaged and the blank permitted to rotate at least one extra revolution until the hob stops cutting and the operation is completed.

The index change gears for the infeed method are the same as for spur gears. The feed change gears may be approximate and are selected from a chart.

The hob is usually set at exactly right angles to the axis of the worm-gear blank. Sometimes the hob is tilted angularly when either over- or undersized, to compensate for the slight change in thread angle. In any case, the angular setting of the hob arbor must be accurate if cross-cornered tooth contact is to be avoided. This is an important difference in setup as compared with spur or helical gear hobbing.

An infeed hob should be an exact duplicate in tooth form and lead of the worm that is to mesh with the gear, except that the tooth height is increased for clearance and the tooth thickness increased for the necessary backlash.

Infeeding is suitable only when the thread angle of the worm does not exceed 6 or 7°. On larger angles, infeeding destroys part of the useful tooth surfaces, and the tangential method is therefore preferable.

Hobbing Worm Gears—Tangential Method. In hobbing worm gears tangentially, the direction of feed is along a line tangent to the blank and parallel with the axis of the hob. In this method the hob is set at the exact center distance and to one side of the center line to clear the blank entirely. The gear teeth are first roughed out by the leading end of the hob and finish cut as the hob passes through the blank. The index change gears and angular setting of the hob are the same as in infeeding.

Feeding the hob tangentially requires a supplemental rotative motion to the blank to synchronize the relation between the teeth of the hob and the blank. This is accomplished by a differential and lead change gearing calculated from the lead of the job. These calculations must be very accurate.

The tangential method is inherently much slower than the infeed method; hence, it is common practice to precede the tangential cut by a roughing infeed cut to save time. For this combined cutting, separate hobs may be used, mounted in the same arbor, or one hob may serve both operations.

Hobs for Worm Gears. Infeed hobs are cylindrical in shape and in general appearance resemble spur-gear hobs.

Hobs for tangential feeding usually have the teeth at the entering end tapered in width and height to divide the load among several teeth and ease the cutting action.

Roughing hobs and the roughing teeth on finishing hobs may be unground, but the finishing teeth or tooth should always be ground if satisfactory tooth forms are to be produced.

Hobs of small diameters and coarse pitches are made with solid shanks to avoid small cutter arbors. This condition frequently occurs because, from a design standpoint, worms are made relatively small in diameter. Small worm diameters result in steep thread angles and consequently high worm and gear efficiency.

When infeed hobbing a worm gear of a low number of teeth, where frequently a small worm diameter limits the number of flutes that it is practical to provide in the hob, appreciable flats may be present. These can be eliminated by tangential feeding rather than infeeding.

Flytools tangentially fed provide a means to cut a single gear or limited quantities without an expensive hob. The flytool consists of a single finishing hob tooth clamped in the cutter arbor and fed across the blank like a tangential hob but at a very much reduced rate of feed. Cemented-carbide tips on flytools are used to good advantage on chilled bronze worm-gear blanks; speed of cutting is greatly increased, often to the maximum available in the hobbing machine.

The tooth forms of worm-gear hobs and flytools may be involute or straight-sided in an axial or a normal plane, or they may be given special shapes corresponding to wormthreads generated by straight-sided milling cutters or grinding wheels of definite diameters. The choice of form depends upon the equipment available for making the worms.

The successful operation of a worm and gear depends a great deal upon the care exercised in the shop to produce good contact between the worm and worm-gear teeth. Proper tooth contact is dependent particularly upon the precision of the worm, also upon the condition and setup of the hobbing machine and the accuracy of the worm-gear hob. The precautions in handling hobs, as outlined under spur and helical gears, are equally important for worm gears.

Other Applications of the Hobbing Method. Hobbing is by no means limited to the three general types of gears mentioned. It is also widely used for double-helical, or herringbone, gears, helical gears for operation on crossed axes (sometimes termed "spiral gears"), pump rotors, single- and multiple-thread worms, sprockets, ratchet gears, straight-sided and involute splines, serrations, etc. Special worms and gears, such as the Hindley, hourglass, and Cone types, are also manufactured on hobbing machines.

Importance of Frequent Hob Shifting. In cutting both coarse and fine-pitch gears, hobs should be shifted frequently and in short amounts in order to obtain the best possible cutting conditions and longest hob life. After a hob has been used in one position for any length of time, the operator can tell how many teeth have been in use and how much wear each one shows. From the amount of wear, the operator can calculate the distance the hob should be shifted to get maximum life and uniform wear from all teeth. Many shops establish a maximum amount of wear to be permitted, and when the hob reaches this point it should be shifted to another position. A *right* hand hob, *top coming*, should always be shifted to the *right;* a *left* hand hob to the *left.*

All teeth used will not show the same amount of wear. Select a tooth around the lead which shows maximum wear. Consider this wear in terms of amount of stock which has to be removed to restore the hob to a good keen cutting edge. This

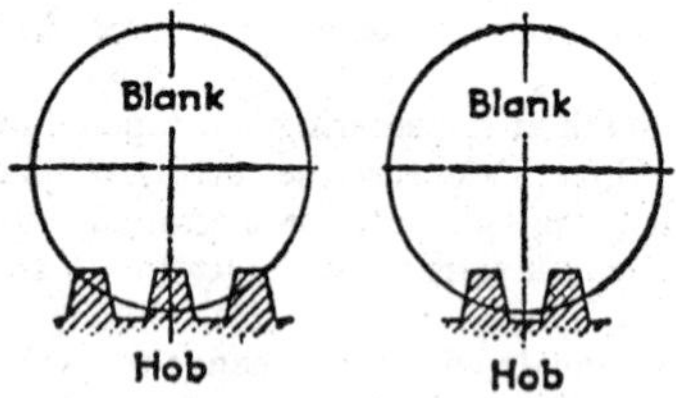

FIG. 32. For proper hob setting, one hob tooth or one tooth space must be centered on the center line of the blank.

should not exceed 0.003 to 0.006 in., depending on the pitch and accuracy of profile. Shift the hob far enough so that the tooth with maximum wear will clear the cut. It is not necessary to shift so far that all teeth used will clear in the new position; those showing minimum wear may be used again cutting in a different position.

To set the hob properly be sure that either one hob tooth or one tooth space is centered on the center line of the gear blank, as in Fig. 32. This position will produce symmetrical generating flats and is usually accomplished by trial moving the hob endwise until a symmetrical form has been obtained, or with the aid of a gage. After this position has been obtained the hob should be shifted *one full circular pitch* each time or fraction of a pitch corresponding to number of flutes in hob. This is important and should be carefully observed if good pinions are to be obtained.

Milling Straight-tooth Bevel Gears

Toolrooms, experimental shops, and maintenance departments often find it necessary to cut straight-tooth bevel gears on a milling machine. This work is done with involute gear cutters and use of the dividing head.

In some instances, it may be required to draw up a pair of gears. This is the procedure. Decide the pitch. Draw center lines *BB* and *CC* intersecting at *A* (Fig. 33). Then draw lines *DD* and *EE* the same distance from *BB* and parallel to it. If the gears are 8 diametral pitch, the distance from *DD* to *EE* will be as many eighths as there are teeth in the gear. Assume a 24-tooth gear and a 16-tooth pinion. Therefore, the distance between *DD* and *EE* is $\frac{24}{8} = 3$ in. Lines *KK* and *LL* are similarly drawn with respect to *CC* and the distance between them is $\frac{16}{8} = 2$ in.

Then through the intersections of *DD* and *LL*, *EE* and *LL*, and *EE* and *KK*, draw the diagonals *FA*. These are the pitch lines. Through the same point draw lines as *GG* at right angles to the pitch lines, forming the backs of the teeth. On these lines lay off $\frac{1}{8}$ in. each side of the pitch lines, and draw *MA* and *NA*, forming the faces and bottoms of the teeth. The lines *HH* are drawn parallel to *GG*, the distance between them being the width of the face.

The face of the larger gear should be turned to the lines *MA*, and the small gear to *NA*. For other diametral pitches, the same rules apply. If 4 diametral pitch, use fourths, instead of eighths; if 3 pitch, thirds, etc.

Bevel bears should always be turned to the exact diameters and angles of the drawings and the teeth cut at the correct angle.

Proportions of Miter and Bevel Gears. To find the pitch angle:

Divide the number of teeth in the gear by the number of teeth in the pinion. This gives the tangent of the pitch angle of the gear. Or divide the number of teeth in the pinion by the teeth in the gear, and get the tangent of the pitch angle

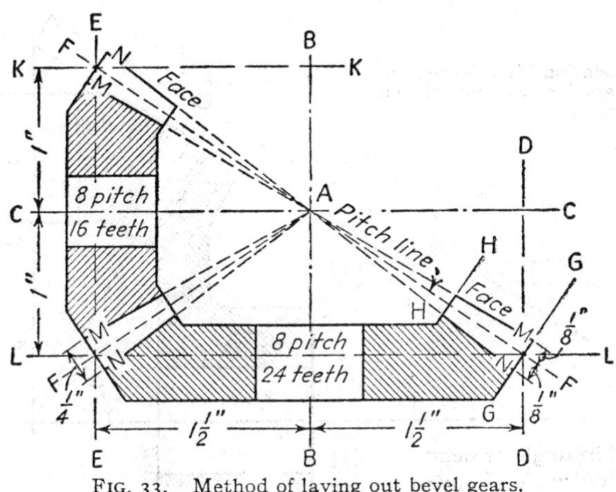

FIG. 33. Method of laying out bevel gears.

of the pinion. Subtracting either pitch angle from 90 gives the pitch angle of the other.

To find the outside diameter:

Multiply the cosine of the pitch angle by twice the addendum, and add the pitch diameter.

To find the cone distance:

Multiply the secant of the pitch angle of the pinion by one-half the pitch diameter of the gear.

Another useful rule is:

The cone distance of a bevel gear or pinion equals its pitch diameter divided by twice the sine of its pitch angle.

To find the face and cutting angles:

Divide the addendum by the cone distance. This gives the tangent of the addendum or outside angle. Subtract this angle from the pitch angle of the pinion to obtain the cutting angle of the pinion and the face angle of the gear. Subtract the same addendum angle from the face angle of the gear to obtain the cutting angle of the gear and the face angle of the pinion. This gives a uniform clearance and is especially for use with rotary cutters.

To find height of addendum at small end of tooth:

Divide the addendum at the large end of the tooth by the cone distance. This gives the decrease in height of the addendum for each inch of gear face. Multiply this by the length of the gear face, and subtract the result from the addendum of the large end of the tooth. The difference is the height of the addendum at the small end of the tooth.

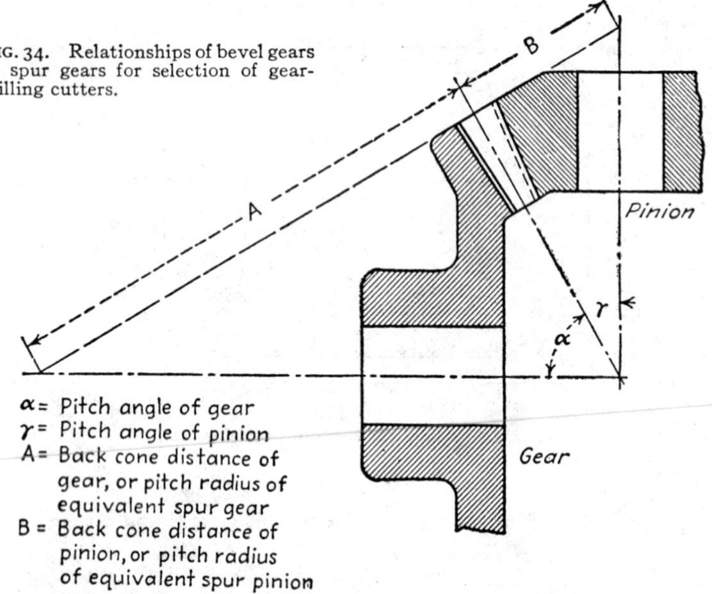

FIG. 34. Relationships of bevel gears to spur gears for selection of gear-milling cutters.

Involute Gear Cutters. Standard involute gear cutters are made for cutting gears of 12 or more teeth. These tools are stamped with a number from 1 to 8 and also the diametral pitch. Example: No. 1, 4P. The numbers indicate:

No. 1 will cut wheels from 135 teeth to a rack, inclusive.
No. 2—wheels from 55 to 134 teeth.
No. 3—wheels from 35 to 54 teeth.
No. 4—wheels from 26 to 34 teeth.
No. 5—wheels from 21 to 25 teeth.
No. 6—wheels from 17 to 20 teeth.
No. 7—wheels from 14 to 16 teeth.
No. 8—wheels from 12 to 13 teeth.

Thus, for any one diametral pitch, it takes 8 cutters to cut all gears having 12 teeth or over.

Cutter Selection for Bevel Gears. The forms of the teeth in a bevel pinion and a bevel gear will differ in such degree, in many cases, that it is necessary to select two cutters to cut them. Thickness of the teeth is greater at the outer ends (the pitch circle) than it is at the inner ends. But the cutter can be no thicker than the space at the small ends of the teeth. For this reason, after the teeth are gashed they must be rolled and set over so that an additional wedge-shaped section can be removed from each tooth flank, in order to obtain proper meshing and correct thickness at the pitch line.

The correct size of cutter for the bevel gear is one that will pass through the small end of the teeth and cut the desired depth at the large end of the teeth. This cutter is one that is suitable for cutting a spur gear having a pitch-circle radius equal to the back-cone distance *A* (Fig. 34). Similarly, the correct cutter for the bevel pinion is one that is suitable for cutting a spur pinion having a pitch-circle radius *B*.

By mathematics, it can be shown that:

Number of teeth in spur gear
= number of teeth in bevel gear divided by the cosine of the pitch angle α
Number of teeth in the spur pinion
= number of teeth in the bevel pinion divided by the cosine of pitch angle r

With the numbers of teeth so ascertained, one selects the proper number of cutter from the involute gear cutters previously listed.

EXAMPLE: A bevel gear with 30 teeth, 6 DP, 45° pitch angle, and 14½° pressure angle. Teeth in spur gear = 30/0.707 = 42+. Therefore, a No. 3, 6P cutter would be selected.

Roll and Setover

To eliminate cut-and-try methods, the engineers of the Cincinnati Milling Machine Co. have devised the following procedure and formulas to calculate roll and setover.

Application of the formulas is described for the gear mentioned above. Full details are given with Fig. 35.

In the first operation, the teeth are gashed to the full depth at the large end of the gear, in this case 0.3595 in. A standard 10-in. dividing head is used. A 30-hole circle in the index plate would be satisfactory, but the 54-hole plate is selected, because it is useful later in obtaining the correct amount of roll.

In the first operation, the width of gashes produced at the pitch line is 0.1745 in. and 0.150 in. at the large and small ends of the gear, respectively. But the correct dimensions for the finished teeth are 0.2618 in. and 0.1878 in., respectively (Fig. 36). In order to obtain these dimensions, an additional amount of stock must be removed on each side of the tooth space as indicated by the shading in Fig. 36. Thus, it is necessary to determine the angle of roll and the amount of setover.

Determining Angle of Roll. The gear blank must be rolled or rotated on its axis through an angle, so that either line *AB* or line *CD* (view 1, Fig. 37) is placed in a direction parallel to the line *EF*. The latter connects the points corresponding to the dimensions of the gashes (Fig. 36) at the pitch line at large and small ends of the gear.

Then point *A*, for example, will have moved to a distance *G* from center line *M* (view 2, Fig. 37), and the distance d_1 between points *A* and *B* (view 1, Fig. 37) will change to the distance *d* between points *E* and *F*.

But *d* is half the difference between the chordal thicknesses T_L and T_S of the gear cutter (Fig. 38) corresponding to the pitch line at the large and small ends of the gear, respectively. From the geometry of view 2, Fig. 37, the distance *G* can be expressed by formula *X*:

$$G = C_r \frac{T_L - T_S}{2w}$$

where G = distance between center line of gear blank and point at pitch line of gear at large end, in.; C_r = pitch cone radius, in., T_L = chordal thickness of gear cutter tooth at pitch line at large end of gear, in.; T_S = chordal thickness of gear cutter tooth at pitch line at small end of gear, in.; w = width of gear-tooth face, in.

The amount of gear-blank roll, from the position in view 1, Fig. 37, to that shown in view 2, is the difference between the circular distances d_2 and *G* of point *A* from

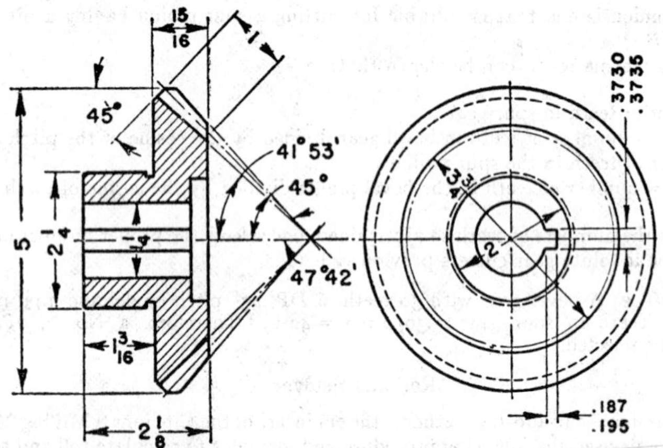

FIG. 35. To avoid cut and try when establishing roll and setover for milling bevel gears, the dimensions of the gear in the example on page 4–103 are calculated:

30 tooth, 6 diametral pitch, 14½° pressure angle
Pitch cone radius—3.535 in.
Pitch diameter at large end—5.000 in.
Pitch diameter at small end—3.585 in.
Circular pitch at large end—0.5236 in.
Circular pitch at small end—0.3756 in.
Tooth thickness and tooth space at large end—0.2618 in.
Tooth thickness and tooth space at small end—0.1878 in.
Whole depth of tooth at large end—0.3595 in.
Whole depth of tooth at small end—0.2588 in.
Addendum at large end of gear—0.166 in.
Addendum at small end of gear—0.1195 in.
Dedendum + clearance at large end of gear—0.193 in.
Dedendum + clearance at small end of gear—0.139 in.
Gear cutter to be used: No. 3, 6 diametral pitch.

the center line M of the blank. The distance d_2 is one-half of 0.2618 in., or one-quarter of the circular pitch C_p and the distance G is obtained from the formula X.

The corresponding angle in degrees is obtained by dividing this difference by the pitch radius of the large end of the gear, and multiplying the result by 57.3, which is the degrees of an arc corresponding to 1 radian. This is expressed by formula Y:

$$C = \frac{57.3}{P_d}\left[\frac{C_p}{2} - \frac{C_r}{w}(T_L - T_S)\right]$$

where C = angle of roll, °; P_d = pitch diameter at large end of gear, in.; C_p = circular pitch at large end of gear, in.; C_r = pitch cone radius at large end of gear, in.; T_S, T_L = chordal thickness of gear cutter tooth pitch line at small and large ends of gear respectively, in.; 57.3 = degrees per radian; w = width of gear tooth face, in.

All the values to be used in formula Y are known with the exception of the quantities T_S and T_L. These are obtained by direct measurement of the gear cutter tooth employed. Dimensions for the present example are given in Fig. 38.

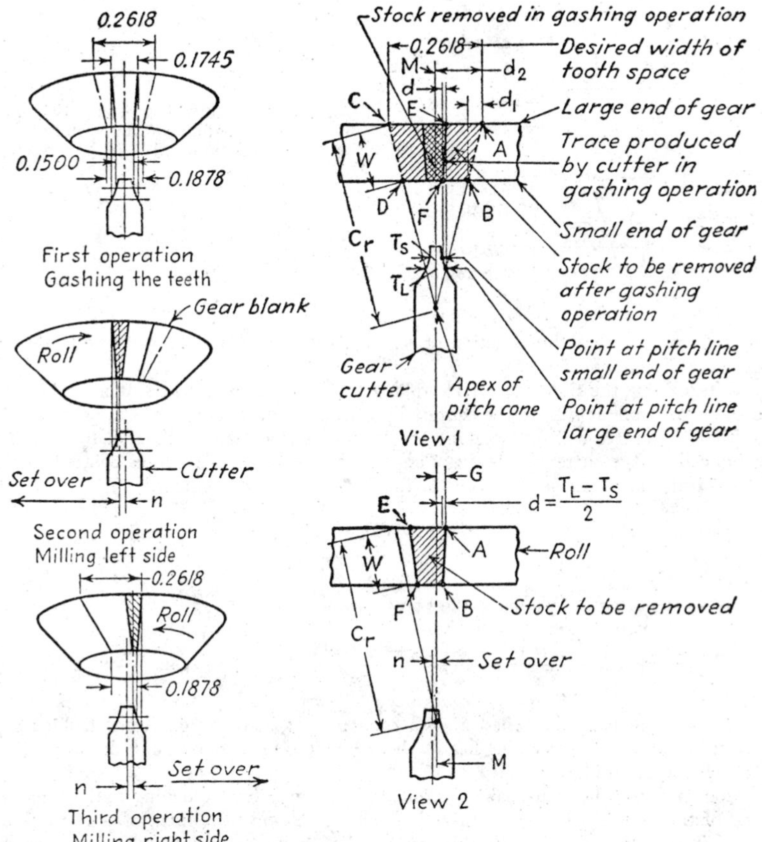

FIG. 36. Three operations are required to mill a tooth space properly in a bevel gear.

FIG. 37. Dimensions for calculating the angle of roll.

Substituting the known values in formula *Y*,

$$C = \frac{57.3}{5}\left[\frac{0.5236}{2} - \frac{3.535}{1}(0.1745 - 0.1500)\right] = 2.007°$$

The angle of roll of 2° is obtained by indexing 12 spaces on the 54-hole circle. The subsequent teeth are, of course, indexed from this new position by turning the index crank one full turn and 18 spaces on the 54-hole circle, as in the case of the gashing operation.

The direction of roll is not important, since the roll is reversed for milling the opposite sides of the teeth after all the teeth have been milled on one side. The only

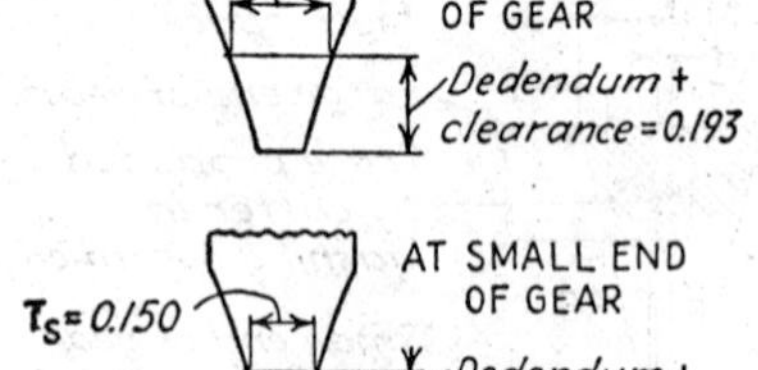

FIG. 38. Chordal thickness of the gear cutter, corresponding to the pitch line at the small and large ends of bevel-gear teeth.

consideration is that the direction of roll and the setover must always be made in opposite directions (Fig. 36).

Determining the Setover. After rolling the blank 2° on its axis, it must be set over by an amount n from the centered position. This is done to locate the blank so that the cutter will follow along the line AB, which is now parallel to line EF produced by the cutter at view 2, Fig. 37. The setover n is also calculated from the dimensions of the gear and cutter tooth, by means of formula Z:

$$n = \frac{T_L}{2} - \frac{(T_L - T_S)C_r}{2w}$$

By substituting in formula Z the known values:

$$n = \frac{0.1745}{2} - \frac{(0.1745 - 0.1500)3.535}{2} = 0.044 \text{ in.}$$

If the blank has been rolled 2° in a *counterclockwise* direction (when looking at the spindle end of the dividing head), the machine table is moved out, or *away from the column*, 0.044 in.

Conversely, if the blank has been rolled 2° *clockwise* (when looking at the spindle end of the dividing head), the table is moved in *toward the column* of the machine 0.044 in., to offset the work by this amount with respect to the center position used in the gashing operation.

To set for milling the opposite side of the teeth after one side has been completed, the table is moved twice the amount of the setover, or 0.088 in., and the blank is rolled twice the angle of roll, or 4°.

Cutting Worms

Lead Gears for Diametral-pitch Worms. C. A. Johnson, Stahl Gear & Machine Co., presents this method. Practically all spur, helical, and bevel gears are now made to the diametral-pitch system because of the simplicity of its calculations. Worm gearing has been made to the circular-pitch system because of the ease of gearing a machine to produce worms with leads which are simple fractions. This practice is now being gradually superseded by the diametral-pitch system. The one disadvantage, that of calculating change gears, is obviated by Table 40, giving the four gears necessary to produce the correct lead on a worm of any diametral pitch from 1 to 60 and any number of threads from 1 to 8.

TABLE 40. LEAD GEARS FOR DIAMETRAL PITCH WORMS

Diametral Pitch	Number of Starts or Leads: 1				2				3				4				5				6				7				8				Machine Constant
	Driver	Follower	Driver	Follower	Driver	Follower	Driver	Follower	Driver	Follower	Driver	Follower	Driver	Follower	Driver	Follower	Driver	Follower	Driver	Follower	Driver	Follower	Driver	Follower	Driver	Follower	Driver	Follower	Driver	Follower	Driver	Follower	
1	68	50	77	100	68	50	77	50	51	25	77	50	68	25	77	50	85	25	77	50	51	25	77	25	85	24	89	43	68	25	77	25	
1¼	68	84	89	86	68	84	89	43	85	70	89	43	68	42	89	43	68	25	77	50	85	35	89	43	85	30	89	43	98	47	90	28	
1½	68	75	77	100	68	75	77	50	68	50	77	50	68	25	77	75	85	25	77	75	68	25	77	50	85	36	89	43	98	47	75	28	
1¾	68	100	88	100	68	100	88	50	68	50	99	75	68	50	88	50	85	50	88	50	68	25	99	75	68	25	77	50	68	25	88	50	
2	68	100	77	100	68	100	77	50	51	50	77	50	68	50	77	50	85	50	77	50	51	25	77	50	85	48	89	43	68	25	77	50	3
2½	34	84	89	86	68	84	89	86	62	55	68	61	68	84	89	43	68	50	77	50	85	70	89	43	85	60	89	43	68	42	89	43	
3	34	75	77	100	68	75	77	100	68	50	77	100	68	75	77	50	85	75	77	50	68	50	77	50	85	72	89	43	68	75	77	25	
3½	68	100	44	100	68	100	88	100	68	100	99	75	68	100	88	50	85	100	88	50	51	50	88	50	68	50	77	50	68	50	88	50	
4	34	100	77	100	68	100	77	100	51	50	77	100	68	100	77	50	85	100	77	50	51	50	77	50	85	48	89	86	68	50	77	50	
5	62	55	34	61	62	55	68	61	93	55	68	61	68	56	89	43	51	25	77	50	51	28	89	43	85	40	89	43	85	35	89	43	
6	68	100	77	100	68	100	77	50	51	50	77	50	68	50	77	50	85	50	77	50	51	25	77	50	85	48	89	43	68	25	77	50	
7	68	100	66	100	68	100	99	75	51	50	99	75	68	50	99	75	85	50	99	75	68	50	99	50	51	25	77	50	68	25	66	50	
8	51	100	77	100	51	100	77	50	66	58	88	85	51	50	77	50	51	40	77	50	66	29	88	85	77	29	88	85	51	25	77	50	
9	34	75	77	100	68	75	77	100	68	50	77	100	68	75	77	50	85	75	77	50	68	50	77	50	85	72	89	43	68	25	77	75	
10	31	55	34	61	62	55	34	61	62	55	51	61	62	55	68	61	51	50	77	50	93	55	68	61	85	80	89	43	85	70	89	43	1
11	51	100	56	100	68	100	84	100	68	100	63	50	68	100	84	50	51	50	98	70	68	50	63	50	51	50	98	50	68	50	84	50	
12	34	100	77	100	68	100	77	100	51	50	77	100	68	100	77	50	85	50	77	100	51	50	77	50	85	48	89	86	68	50	77	50	
13	32	73	43	78	41	65	59	77	48	73	86	78	85	91	89	86	51	50	77	65	96	73	86	78	85	52	89	86	85	91	89	43	
14	51	100	44	100	68	100	66	100	51	50	66	100	68	100	99	75	85	100	99	75	51	50	99	75	51	50	77	50	66	50	68	50	
15	42	94	45	96	34	84	89	86	62	55	34	61	68	84	89	86	68	50	77	100	62	55	68	61	68	96	89	43	68	42	89	86	
16	33	85	44	87	51	100	77	100	66	87	66	85	51	50	77	100	51	50	77	80	99	87	88	85	77	58	88	85	51	50	77	50	

TABLE 40. LEAD GEARS FOR DIAMETRAL PITCH WORMS (*Continued*)

Diametral Pitch	1				2				3				4				5				6				7				8				Machine Constant
	Driver	Follower	Driver	Follower	Driver	Follower	Driver	Follower	Driver	Follower	Driver	Follower	Driver	Follower	Driver	Follower	Driver	Follower	Driver	Follower	Driver	Follower	Driver	Follower	Driver	Follower	Driver	Follower	Driver	Follower	Driver	Follower	
17	42	100	44	100	56	100	66	100	84	100	66	100	84	100	88	100	84	100	99	90	84	100	99	75	98	100	99	75	84	75	99	75	
18	31	99	34	61	34	75	77	100	68	100	77	100	68	75	77	100	51	50	77	90	68	50	77	100	85	72	89	86	68	75	77	50	1
19	32	70	34	94	41	77	59	95	43	73	80	95	63	76	75	94	82	77	59	76	86	73	80	95	85	76	89	86	63	76	75	47	
20	27	64	35	94	31	55	34	61	55	95	70	86	62	55	34	61	51	50	77	100	62	55	51	61	85	80	89	86	62	55	68	61	
22	51	100	56	100	68	100	84	100	68	100	63	50	68	100	84	50	51	50	98	70	68	50	63	50	51	50	98	50	68	50	84	50	
24	34	100	77	100	68	100	77	100	51	50	77	100	68	100	77	50	85	50	77	100	51	50	77	50	85	48	89	86	68	50	77	50	
26	32	73	43	78	41	65	59	77	48	73	86	78	85	91	89	86	51	50	77	65	96	73	86	78	85	52	89	86	85	91	89	43	
28	51	100	44	100	68	100	66	100	51	50	66	100	68	100	99	75	85	100	99	75	51	50	99	75	51	50	77	50	66	50	68	50	
30	42	94	45	96	34	84	89	86	62	55	34	61	68	84	89	86	68	50	77	100	62	55	68	61	68	96	89	43	68	42	89	86	
32	33	85	44	87	51	100	77	100	66	87	66	85	51	50	77	100	51	50	77	80	99	87	88	85	77	58	88	85	51	50	77	50	
34	42	100	44	100	56	100	66	100	84	100	66	100	84	100	88	100	84	100	99	90	84	100	99	75	98	100	99	75	84	75	99	75	
36	31	99	34	61	34	75	77	100	68	100	77	100	68	75	77	100	51	50	77	90	68	50	77	100	85	72	89	86	68	75	77	50	
38	32	70	34	94	41	77	59	95	43	73	80	95	63	76	75	94	82	77	59	76	86	73	80	95	85	76	89	86	63	76	75	47	½
40	27	64	35	94	31	55	34	61	55	95	70	86	62	55	34	61	51	50	77	100	62	55	51	61	85	80	89	86	62	55	68	61	
42	34	100	44	100	68	100	44	100	68	100	66	100	68	100	88	100	85	100	88	100	68	100	99	75	68	100	77	50	68	100	88	50	
44	34	100	42	100	51	100	56	100	68	100	63	100	68	100	84	100	85	100	84	100	68	100	63	50	51	50	98	100	68	100	84	50	
46	26	69	29	80	32	69	43	73	64	92	43	73	63	92	75	94	82	92	59	77	64	92	86	73	85	92	89	86	63	46	75	94	
48	26	79	35	88	34	100	77	100	51	100	77	100	68	100	77	100	85	100	77	100	51	50	77	100	85	96	89	86	68	100	77	50	
50	27	94	42	96	28	57	44	86	56	95	55	86	56	57	44	86	62	55	34	61	56	57	66	86	85	100	89	86	56	57	88	86	
56	34	100	33	100	51	100	44	100	51	100	66	100	68	100	66	100	85	100	66	100	51	50	66	100	51	50	77	100	68	100	99	75	
60	27	94	35	96	42	94	45	96	31	55	34	61	34	84	89	86	68	100	77	100	62	55	34	61	70	82	79	92	68	84	89	86	

The diametral pitch is found at the left-hand side of the table, and in the column under the desired number of threads is found the change gears which will produce the correct lead. The table is divided horizontally into three sections. Gearing for the coarse pitches (1 to 4) is calculated for a machine with a 3-in. constant, that is, a machine which *will produce a lead of 3 in. with equal gears.* The medium pitches (5 to 20) are calculated for a machine with a 1-in. constant, and the fine pitches (22 to 60) are calculated for a machine with a ½-in. constant. In every case the largest gears to produce the desired ratio are given. For some ratios these may be factored if smaller gears are desired. All the gear ratios in the table are the closest to the desired ratio which may be obtained using gears with tooth numbers from 24 to 100. The required gears are as follows:

3-in. machine:
1 each—24, 25, 28, 30, 34, 35, 36, 42, 43, 44, 47, 48, 51, 55, 60, 61, 62, 68, 70, 72, 75, 77, 84, 86, 88, 89, 90, 95, 98, 99.
2 each—25, 50, 100.

1-in. machine:
1 each—25, 27, 28, 29, 31, 32, 33, 34, 35, 40, 41, 42, 43, 44, 45, 47, 48, 51, 52, 55, 56, 58, 59, 61, 62, 63, 64, 65, 66, 68, 70, 72, 73, 76, 77, 78, 80, 81, 82, 84, 85, 86, 87, 88, 89, 90, 91, 93, 94, 95, 96, 98, 99.
2 each—50, 75, 100.

½-in. machine:
1 each—26, 27, 28, 29, 31, 32, 33, 34, 35, 41, 42, 43, 44, 45, 46, 47, 48, 51, 52, 55, 56, 57, 58, 59, 61, 62, 63, 64, 65, 68, 69, 70, 73, 76, 77, 78, 79, 80, 81, 82, 84, 85, 86, 87, 88, 89, 90, 91, 92, 94, 95, 96, 98, 99.
2 each—50, 66, 75, 100.

It may be necessary to cut a worm on a machine of different constant than that for which the change gears are tabulated. The procedure for finding the correct set of gears is as follows:

Suppose that a 3 DP gear is to be cut on a 2-constant machine. The gears for a 3-constant machine are of 34, 75, 77, and 100 teeth. To maintain the same lead on the worm, however, it will be necessary to increase the ratio of the change gears in the ratio of 3 to 2. Thus, $\frac{3 \times 34 \times 77}{2 \times 75 \times 100}$, which by factoring becomes $\frac{68 \times 77}{100 \times 100}$.

Gear Shaving

Gear-tooth shaving is a process in which a small amount of material is cut or shaved from the working tooth surfaces of a gear to overcome errors in index, helical angle, tooth profile, and eccentricity—all of which cause noisy operation and reduce service life. Shaving corrects these deficiencies before hardening.

Various widely differing methods of shaving have been developed. However, shaving with a rotary tool or cutter on an axis skewed with that of the work is the most frequently used method.

Rotary Crossed-axes Shaving. In this process (Fig. 39) a gashed rotary cutter in the form of a helical gear having a different helical angle than the work gear (a spur is one of zero helix angle) is rotated in tight mesh with the work gear. The cutter removes fine hairlike shavings.

The shaving tool is made to have the same normal diametral pitch and pressure angle as the gear it shaves. Given a tool of 15° right-hand helix angle, 20 normal

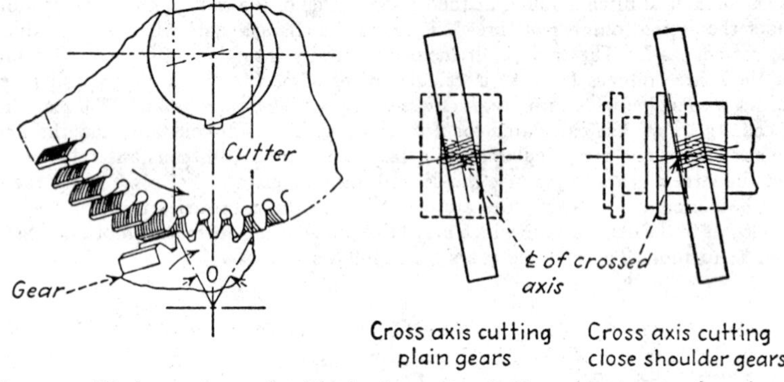

FIG. 39. Work gear in mesh with shaving cutter (*left*), and how axes of work and cutter gears are set (*right*) for plain and shoulder gears.

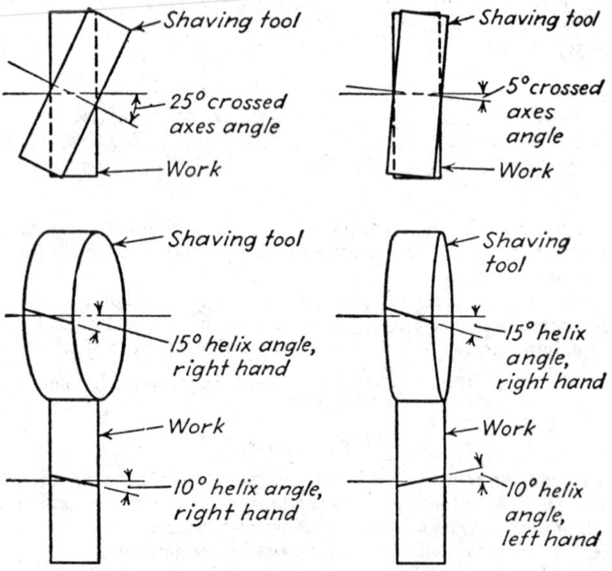

FIG. 40. Relationship between crossed-axes angle and the "hand" of the helix angle of tool and work.

DP, 20° normal pressure angle, it is possible to shave a spur gear at a crossed-axes angle of 15°. The same tool can likewise be employed to shave either right- or left-hand helical gears of the same normal pitch and pressure angle. When the hand of the tool and the hand of the work are the same, the machine setting is equal to the sum of their helix angles. When the helix angle of the tool and the work are of

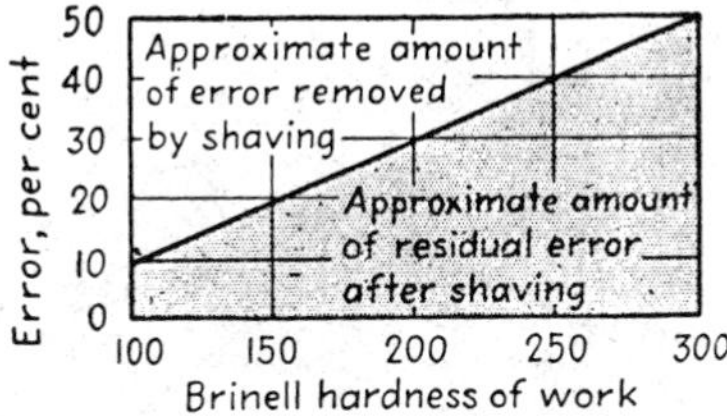

FIG. 41. Approximate relationship between material hardness and the effectiveness of shaving.

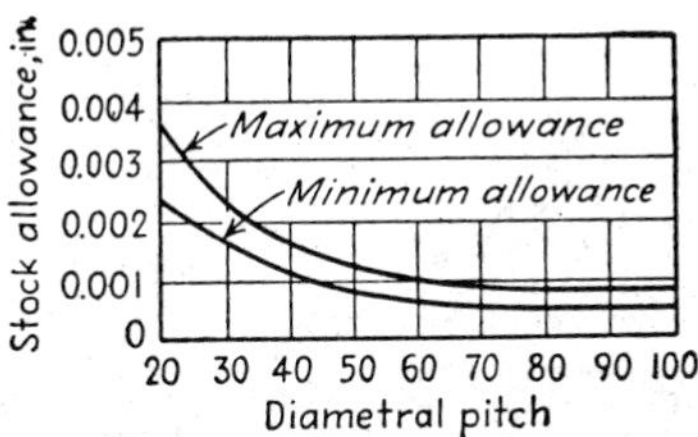

FIG. 42. Recommended stock allowances for shaving gears of various pitches, in terms of tooth thickness.

TABLE 41. STOCK ALLOWANCE FOR SHAVING*

Normal Diametral Pitch	*Stock on Tooth Thickness, In.*
2–4	0.003–0.004
5–6	0.0025–0.0035
7–10	0.002–0.003
11–14	0 0015–0.0020
16–18	0.001–0.002
20–48	0.0005–0.0015
52–72	0.0003–0.0007

These are total tooth-thickness allowances. One-half the amount shown is the stock on each side of the tooth.

* National Broach & Machine Co.

opposite hands, however, the crossed-axes angle is the *difference* between their helix angles (Fig. 40).

In conventional crossed-axes shaving, the tool must be reciprocated for a distance somewhat greater than the face width of the gear. Also the line of tool travel is parallel to the axis of the gear. In the diagonal or full-tooth shaving process the axis of the work is set at an angle relative to the tool traverse and the tool has to travel only a fraction of the distance to carry the "crossed-axes point" across the face of the gear teeth.

For best results, the shaving tool should have a prime number of teeth, or at least be prime to the number of teeth in the gear it shaves. This principle of gear refinement is based on the "hunting-tooth effect" brought about by the continual orientation of the teeth in the cutter with those in the work.

Shaving Teeth in Elliptoid Form. This tooth form is produced in the rotary shaving operation by rocking the inner work-carrying table of the shaving machine as it is reciprocated. Rocking the table causes the cutter to sink deeper at the ends of the work gear tooth than it does at the center, thus thinning the tooth progressively toward the ends. This is also known as "crown shaving."

By positioning the work on the table, the location of maximum tooth thickness or crown is controlled. By the cam setting at the end of the table, the amount of crowning is controlled. The crowning operation requires no more machine time than conventional shaving. With the cam disengaged and the inner table locked, the machine produces conventional (straight) gear teeth—no crown.

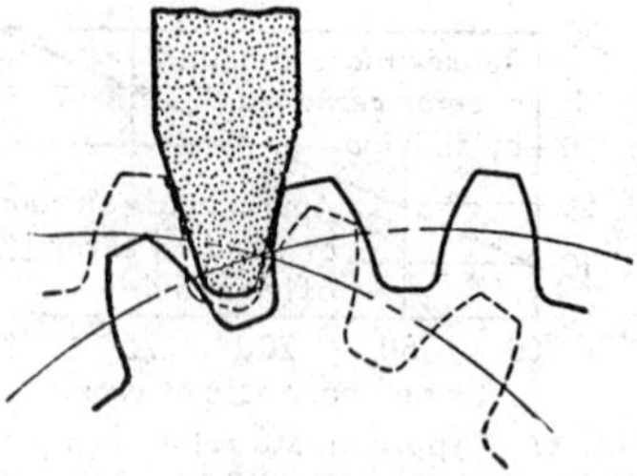

FIG. 44. In a single-wheel generating grinder, the wheel is beveled to the form of a master rack and is traversed across the work.

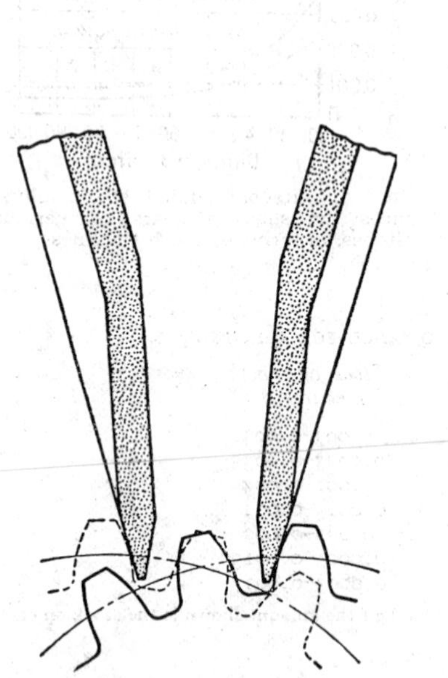

FIG. 43. In a gear grinder for generating relatively short teeth, two large dish wheels grind the sides of two separate teeth simultaneously.

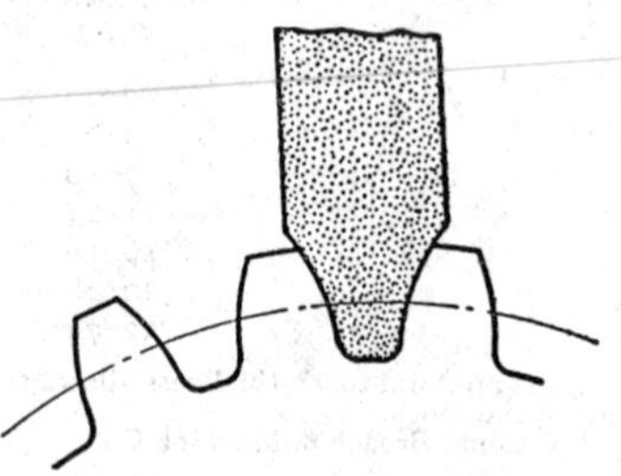

FIG. 45. In form grinding a spur tooth, the wheel is dressed to a negative form of the desired tooth and grinds two adjacent tooth flanks simultaneously.

Gear-tooth Grinding

Quiet smooth-running gears, which can transmit heavy loads uniformly and without excessive stresses in any of the teeth, must be very accurately finished as to both tooth form and tooth spacing according to the Norton Co. and others. Errors as small as 0.0005 in. may under certain conditions increase stresses in a single tooth as much as 100%, and noisy gearing can be caused by even smaller errors.

Most ground gears are used for aircraft purposes. For best results, errors in tooth form or spacing should not exceed 0.0002 in. There is a pronounced difference in the running qualities of gears accurate to 0.0002 in. and those accurate to 0.0005 in.

This degree of accuracy depends on the steel and heat-treatment. Where hardened steels are involved, grinding seems to be the only practical method of finishing gears to the required degree of accuracy.

Grinding Methods. There are two basic methods of grinding gears: generating and form grinding. Both methods are used for grinding spur gears and pinions. Spiral bevel and hypoid gears may be ground by either method, but the pinions are always generated.

Spur Generators. Machines of the generating type can be divided into two subgroups:

In the first of these, the grinding surface of the wheel is a flat plane against which the gear tooth is "rolled" mechanically, thus generating the desired tooth form. The flat grinding surface of the wheel theoretically represents one surface of the flat-sided basic rack. The wheel can, therefore, grind only one tooth surface at a time, but machines employing this method of grinding are usually equipped with two grinding wheels, each tilted at the correct angle so that one surface each on two separate teeth can be ground at one "rolling" (see Fig. 43).

Root of the tooth is an arc corresponding to the periphery of the wheel. This type of grinder is not often used for gears thicker than 1¼ in.

In the second type of generating grinder, the gear tooth is "rolled" against a straight line on the conical surface of a beveled wheel. This method necessitates constant traversing of this straight generating line back and forth along the length of the gear tooth, while the tooth is being "rolled." The periphery of the grinding wheel is beveled on both sides so that a radial cross section of the grinding face resembles the complete tooth of a master rack and grinding is thus done on both the conical surfaces, generating one side each of two adjacent teeth at once (see Fig. 44). There is no practical limitation to the face width of the gears which can be ground within the capacity of the machine.

Spur Form Grinders. Negative form of the tooth is accurately trued in the periphery of the grinding wheel by means of diamond fixtures controlled by accurately formed cams and guides. The work table reciprocates longitudinally under the wheel which grinds the left side of one tooth and the right side of the adjacent tooth simultaneously with the root. The gear is, of course, mechanically indexed between strokes (see Fig. 45). By utilizing small wheels it is possible to grind a gear which may be located close to another gear of larger diameter or to a large shoulder on the same shaft. Also by utilizing small-diameter wheels, it is practical to grind internal gears.

The form grinder is also adaptable with slight modifications to the grinding of splines, both external and internal.

Bevel Generators. This machine is designed to grind spiral bevel gears and pinions with a cup-shaped wheel dressed on the face with beveled grinding surfaces, the angles of which suit the pressure angle required by the work.

This type of machine employs a familiar generating principle in which a relative rolling motion of the work and of the wheel-carrying cradle automatically generates the correct tooth form. The motion corresponds to that of a gear rolling with a crown gear of which the grinding wheel represents a tooth. A single pass of the wheel generates one tooth surface.

Spiral Bevel Grinders. This machine is designed to grind Formate spiral bevel and hypoid gears only. As there is no generating motion the sides of the teeth ground are straight, but a correction for tooth shape is made in the pinion which is always ground by the generating method.

The wheel used is a flaring-cup type and is carried from end to end of the teeth by an oscillating motion. The grinding surfaces of the wheel are dressed to suit the pressure angles required by the gear. The wheel grinds the left side of one tooth and the right side of the adjacent tooth simultaneously. The wheel makes from 10 to 20 passes for each tooth and is fed into depth by a cam.

Although the grinding action is different in each of the grinding methods, the basic requirements are the same. The wheel must cut clean the amount of stock fed to it without measurable wear per revolution of the gear, and without building up excessive pressure. The ability of the wheel to retain form determines the accuracy of the tooth form in the work, but too hard a wheel will burn the work.

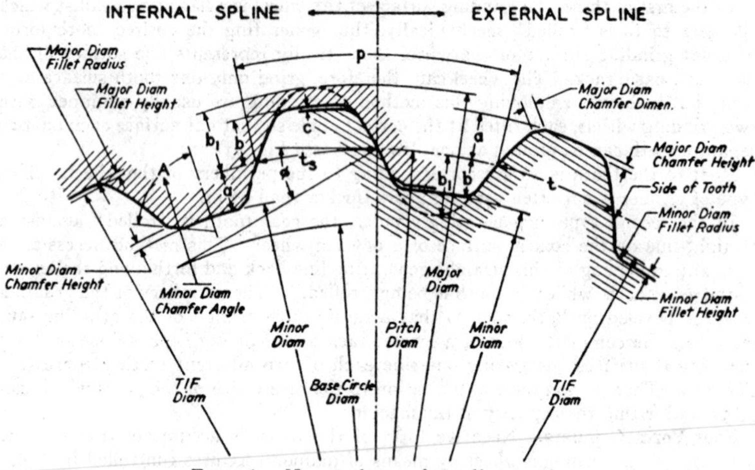

FIG. 46. Nomenclature for spline tooth.

- P Diametral pitch
- p Circular pitch
- a Addendum
- b Dedendum—flat root
- b_1 Dedendum—fillet root
- c Clearance
- t Circular tooth thickness
- t_s Circular space width
- ϕ Pressure angle
- ϕ_e Pressure angle at pin center—external
- ϕ_i Pressure angle at pin center—internal
- M_e Measurement over pins—external
- M_i Measurement between pins—internal
- N Number of teeth
- D Pitch diameter
- D_b Base circle diameter
- D_o Major diameter
- D_R Minor diameter
- TIF Diameter at junction of involute form with fillet
- d_e Measuring pin diameter—external
- d_i Measuring pin diameter—internal

SPLINES AND SERRATIONS

Definitions

Splines are multiple keys cut in a shaft and related parts to prevent relative rotation. With a suitable fit, sliding of one member relative to the other is possible, as in a sliding-gear transmission. There are two types of splines: the original straight-sided tooth form, which is tending to go out of use, except where parts must be ground after hardening, and the involute-tooth form which has been standardized in ASA B5.15-1950 (see Fig. 46).

Serrations have different tooth proportions and are intended primarily for parts permanently fitted together. They are made for use with either uniform or tapered diameters and are well adapted for use on thin-wall tubing. In the internal form, the serrations may be shaped or cold forged, and the external serrations may be shaped or hobbed.

Involute Splines

Involute splines (Fig. 46) have a pressure angle of 30°. They are standardized on a stub-tooth basis, where the depth of the spline is one-half the depth of a standard

Basic Formulas for Involute Splines

Flat and Fillet Root		Fillet Root Only		
		Diametral pitch	1/2–12/24	16/32–48/96
Pitch dia $D = \frac{N}{P}$	Major dia (external) $D_o = \frac{N+1}{P}$	Major dia (internal)	$D_o = \frac{N+1.8}{P}$	$D_o = \frac{N+1.8}{P}$
Circular pitch $p = \frac{3.141593}{P}$	TIF dia (internal) $= \frac{N+1}{P}$	Minor dia (external)	$D_R = \frac{N-1.8}{P}$	$D_R = \frac{N-2}{P}$
Circular tooth thickness $t = \frac{1.570796}{P}$	Minor dia (minor dia fits only) $D_R = \frac{N-1}{P}$			
Addendum $a = \frac{0.500}{P}$	TIF dia (external) $= \frac{N-1}{P}$	Dedendum (internal)	$b_1 = \frac{0.900}{P}$	$b_1 = \frac{0.900}{P}$
Dedendum $b = \frac{0.600}{P} + 0.002$		Dedendum (external)	$b_1 = \frac{0.900}{P}$	$b_1 = \frac{1.000}{P}$

For major-diameter fits the dedendum of the internal spline is the same as the addendum, and for minor-diameter fits the dedendum of the external spline is the same as the addendum. Basic data for involute splines are given in Table 42.

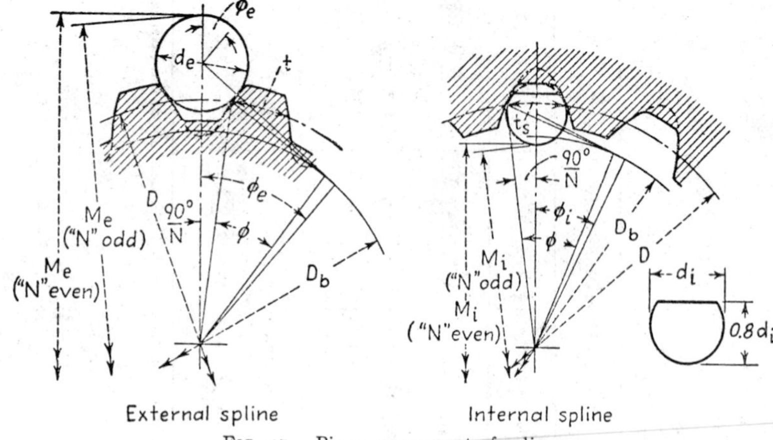

FIG. 47. Pin measurement of splines.

gear tooth. Primary diametral pitch (the numerator in the designation of standardized pitches) controls the pitch diameter and the tooth thickness, while the secondary diametral pitch (denominator) controls the addendum and dedendum. True involute form (TIF) diameter is the junction of the involute form with the fillet.

Standardized Pitches. The ASA standard pitches for involute splines are:

1/2	4/8	8/16	32/64
2.5/5	5/10	10/20	40/80
3/6	6/12	12/24	48/96

Teeth range from 6 to 50 for both the flat-root and fillet-root types. The flat root is used wherever moderate stresses are involved and where thin walls in the splined part dictate the use of shallower teeth. The fillet root is selected for highly stressed parts.

Basic Tooth Proportions. Basic dimensions for both internal and external splines are: pitch diameter, circular pitch, circular tooth thickness, circular space width, addendum, dedendum, major diameter, and minor diameter. Because of the almost universal practice of broaching the internal spline, it has been held to basic dimensions, and the external spline dimensions varied to control the fit. This corresponds to the general practice of fitting to a drilled or reamed hole, because of the fixed condition of the finishing tools and gages. Formulas are given on page 4-115.

Calculation of Any Pitch

Tabulated Data. Tables 43 and 44 are for one/two diametral (1/2) pitch and are the masters for all other tables in ASA B5.15-1950. Tables given here permit complete calculations, but persons having much work to do with splines are advised to obtain a copy of the standard for convenience.

Dimensions for any diametral pitch can be calculated with the formulas in Table 45

(Continued on page 4–129)

Formulas for Pin Measurements for Splines

External Splines

ϕ = pressure angle at pitch dia = 30°
ϕ_e = pressure angle at pin center

$$\text{inv}\ \phi_e = \text{inv}\ 30^\circ + \frac{d_e}{D_b} + \frac{t}{D^{(1)}} - \frac{\pi}{N}$$

From inv ϕ_e find ϕ_e in degrees
When N is even,

$$M_e = D_b^{(1)} \sec \phi_e + d_e \qquad \text{or} \qquad M_e = \frac{D_b^{(1)}}{\cos \phi_e} + d_e$$

When N is odd,

$$M_e = D_b^{(1)} \sec \phi_e \cos \left(\frac{90^\circ}{N}\right) + d_e \qquad \text{or} \qquad M_e = \frac{D_b^{(1)} \cos (90^\circ/N)}{\cos \phi_e} + d_e$$

Example: Standard dimensions for $N = 20$, $P = 10/20$, class B fit.

$$d_e = 0.1920$$

Basic t (tooth thickness) = 0.1571
Allowable errors × 60% + mach. tol. (Table 45) = 0.0001(0.054N + 19) = 0.0020
Min tooth thickness for measurement t = 0.1551

$$\text{inv}\ \phi_e = 0.053751 + 0.110851 + \frac{0.1551}{2.0000} - 0.157080 = 0.085072$$

$$\phi_e = 34.491456^\circ$$
$$M_e = 1.732051 \sec 34.491456^\circ + 0.1920 = 2.2935$$

Internal Splines

ϕ = pressure angle at pitch dia = 30°
ϕ_i = pressure angle at pin center

$$\text{inv}\ \phi_i = \text{inv}\ 30^\circ + \frac{t_s}{D} - \frac{d_i}{D_b}$$

From inv ϕ_i find ϕ_i in degrees
When N is even,

$$M_i = D_b^{(1)} \sec \phi_i - d_i \qquad \text{or} \qquad M_i = \frac{D_b^{(1)}}{\cos \phi_i} - d_i$$

When N is odd,

$$M_i = D_b^{(1)} \sec \phi_i \cos \left(\frac{90^\circ}{N}\right) - d_i \qquad \text{or} \qquad M_i = \frac{D_b^{(1)} \cos (90^\circ/N)}{\cos \phi_i} - d_i$$

Example: Standard dimensions for $N = 20$, $P = 10/20$, internal spline.

$$d_i = 0.1440$$

Basic t_s (space width) = 0.1571
Allowable errors × 60% + mach. tol. (Table 45) = 0.0001(0.072N + 24) = 0.0025
Maximum space width for measurement t_s = 0.1596

$$\text{inv}\ \phi_i = 0.053751 + 0.079800 - 0.083138 = 0.050413 \qquad \phi_i = 29.412396^\circ$$
$$M_i = 1.732051 \sec 29.412396^\circ - 0.1440 = 1.8443$$

Note: inv ϕ is the involute function of ϕ. Involute functions may be found in mechanical engineering handbooks and books on gearing.

TABLE 42. BASIC DATA FOR INVOLUTE SPLINES (ASA B5.15-1950)

	1/2 Diametral Pitch			All Pitches						
	Internal and External		External	Internal			Internal and External	External		
N	D_b	$D_b \cos \frac{90°}{N}$	Measurement over Pins	$\frac{t_s}{D}$	$\frac{d_i}{D_b}$	Factor F	$\cos \frac{90°}{N}$	$\frac{d_e}{D_b}$	$\frac{\pi}{N}$	Factor E
1	2	3	4	5	6	7	8	9	10	11
6	5.196152		8.8660	0.261799	0.277128	1.91		0.369504	0.523599	1.305
7	6.062178	5.910186	9.6819	0.224399	0.237538	1.83	0.974928	0.316718	0.448799	1.302
8	6.928203		10.8944	0.196350	0.207846	1.86		0.277128	0.392699	1.362
9	7.794229	7.675817	11.7535	0.174533	0.184752	1.81	0.984808	0.246336	0.349066	1.364
10	8.660254		12.9144	0.157080	0.166277	1.83		0.221703	0.314159	1.406
11	9.526279	9.429316	13.8003	0.142800	0.151161	1.80	0.989821	0.201548	0.285599	1.409
12	10.392305		14.9295	0.130900	0.138564	1.81		0.184752	0.261799	1.440
13	11.258330	11.176244	15.8335	0.120830	0.127905	1.79	0.992709	0.170540	0.241661	1.443
14	12.124356		16.9412	0.112200	0.118769	1.80		0.158359	0.224399	1.467
15	12.990381	12.919218	17.8584	0.104720	0.110851	1.78	0.994522	0.147802	0.209440	1.471
16	13.856406		18.9507	0.098175	0.103923	1.79		0.138564	0.196350	1.490
17	14.722432	14.659629	19.8778	0.092400	0.097810	1.78	0.995734	0.130413	0.184800	1.493
18	15.588457		20.9584	0.087266	0.092376	1.78		0.123168	0.174533	1.509
19	16.454483	16.398282	21.8934	0.082673	0.087514	1.78	0.996584	0.116686	0.165347	1.512
20	17.320508		22.9649	0.078540	0.083138	1.78		0.110851	0.157080	1.525
21	18.186533	18.135680	23.9062	0.074800	0.079179	1.77	0.997204	0.105573	0.149600	1.528
22	19.052559		24.9704	0.071400	0.075580	1.77		0.100774	0.142800	1.539
23	19.918584	19.872149	25.9168	0.068295	0.072294	1.77	0.997669	0.096392	0.136591	1.541
24	20.784610		26.9752	0.065450	0.069282	1.77		0.092376	0.130900	1.551
25	21.650635	21.607912	27.9259	0.062832	0.066511	1.77	0.998027	0.088681	0.125664	1.553

26	22.516660		28.9793	0.060415	0.063953	1.76		0.085270	0.120830	1.562
27	23.382686	23.343126	29.9337	0.058178	0.061584	1.76	0.998308	0.082112	0.116355	1.564
28	24.248711		30.9829	0.056100	0.059385	1.76		0.079179	0.112200	1.571
29	25.114737	25.077904	31.9405	0.054165	0.057337	1.76	0.998533	0.076449	0.108331	1.573
30	25.980762		32.9862	0.052360	0.055426	1.76		0.073901	0.104720	1.580
31	26.846787	26.812337	33.9465	0.050671	0.053638	1.76	0.998717	0.071517	0.101342	1.581
32	27.712813		34.9890	0.049087	0.051962	1.76		0.069282	0.098175	1.587
33	28.578838	28.546468	35.9517	0.047600	0.050387	1.76	0.998867	0.067183	0.095200	1.589
34	29.444864		36.9916	0.046200	0.048905	1.76		0.065207	0.092400	1.594
35	30.310889	30.280368	37.9565	0.044880	0.047508	1.76	0.998993	0.063344	0.089760	1.596
36	31.176914		38.9939	0.043633	0.046188	1.76		0.061584	0.087266	1.600
37	32.042940	32.014068	39.9607	0.042454	0.044940	1.76	0.999100	0.059920	0.084908	1.602
38	32.908965		40.9960	0.041337	0.043757	1.75		0.058343	0.082673	1.606
39	33.774991	33.747599	41.9645	0.040277	0.042635	1.75	0.999189	0.056847	0.080554	1.608
40	34.641016		42.9980	0.039270	0.041569	1.75		0.055426	0.078540	1.611
41	35.507041	35.480986	43.9680	0.038312	0.040555	1.75	0.999266	0.054074	0.076624	1.613
42	36.373067		44.9996	0.037400	0.039590	1.75		0.052786	0.074800	1.616
43	37.239092	37.214248	45.9711	0.036530	0.038669	1.75	0.999333	0.051559	0.073060	1.617
44	38.105118		47.0013	0.035700	0.037790	1.75		0.050387	0.071400	1.621
45	38.971143	38.947403	47.9740	0.034907	0.036950	1.75	0.999391	0.049267	0.069813	1.622
46	39.837168		49.0028	0.034148	0.036147	1.75		0.048196	0.068295	1.625
47	40.703194	40.680464	49.9766	0.033421	0.035378	1.75	0.999442	0.047171	0.066842	1.626
48	41.569219		51.0041	0.032725	0.034641	1.75		0.046188	0.065450	1.629
49	42.435245	42.413442	51.9790	0.032057	0.033934	1.75	0.999486	0.045245	0.064114	1.630
50	43.301270		53.0055	0.031416	0.033255	1.75		0.044341	0.062832	1.632

Constants $\pi = 3.141593$ $\quad \cos 30° = 0.866025 \quad$ inv $30° = 0.053751$

$$E = \frac{\cos 30°}{\sin \phi_e} = \frac{\text{rate of change of } M_e}{\text{rate of change of } t} \left(\text{for } N \text{ odd, } E = \frac{\cos 30°}{\sin \phi_e} \cos \frac{90°}{N}\right)$$

$$F = \frac{\cos 30°}{\sin \phi_i} = \frac{\text{rate of change of } M_i}{\text{rate of change of } t_e} \left(\text{for } N \text{ odd, } F = \frac{\cos 30°}{\sin \phi_i} \cos \frac{90°}{N}\right)$$

TABLE 43. 1/2 DIAMETRAL PITCH INTERNAL INVOLUTE SPLINES

Pressure Angle 30° | Addendum (Basic) 0.5000 | Circular Pitch 3.1416, Measuring Pin Diam 1.4400

Internal and External			Internal												
			Major-dia Fit		Flat Root Side Fit					Fillet Root Side Fit				All Fits	
						Major-dia fillet		Major Dia							
N	Pitch Dia Ref	Base Circle Dia Ref	Major Dia Basic	TIF Dia	Minor Dia	Radius	Height	Full* Dedenum	Short† Dedenum	TIF Dia	Minor Dia	Major Dia	Fillet Radius	Measurement between Pins	Space Width Min Effective 1.5708
1	2	3	4	5	6	7	8	9A	9B	10	11	12	13	14	15
Recommended Tolerance.........			+0.0007 + L‡ −0.0000	Min	+0.0050 −0.0000	Approx	Max	+0.1050 −0.0000	+0.0007 + L‡ −0.0007 + L	Min	+0.0050 −0.0000	+0.1050 −0.0000	Min §	Max	Dimensional Max¶
6	6.0000	5.1962	7.0000	6.897	5.2062	0.145	0.052	7.2040	7.0000	7.0000	5.2062	7.8000	0.166	4.4022	1.5755
7	7.0000	6.0622	8.0000	7.881	6.0722	0.160	0.060	8.2040	8.0000	8.0000	6.0722	8.8000	0.194	5.2320	1.5756
8	8.0000	6.9282	9.0000	8.870	7.0400	0.170	0.066	9.2040	9.0000	9.0000	7.0400	9.8000	0.208	6.4043	1.5756
9	9.0000	7.7942	10.0000	9.860	8.0000	0.178	0.071	10.2040	10.0000	10.0000	8.0000	10.8000	0.220	7.2707	1.5757
10	10.0000	8.6603	11.0000	10.852	9.0000	0.184	0.075	11.2040	11.0000	11.0000	9.0000	11.8000	0.230	8.4055	1.5757
11	11.0000	9.5263	12.0000	11.845	10.0000	0.189	0.079	12.2040	12.0000	12.0000	10.0000	12.8000	0.238	9.2955	1.5757
12	12.0000	10.3923	13.0000	12.839	11.0000	0.193	0.081	13.2040	13.0000	13.0000	11.0000	13.8000	0.245	10.4063	1.5758
13	13.0000	11.2583	14.0000	13.834	12.0000	0.196	0.084	14.2040	14.0000	14.0000	12.0000	14.8000	0.249	11.3129	1.5758
14	14.0000	12.1244	15.0000	14.830	13.0000	0.199	0.086	15.2040	15.0000	15.0000	13.0000	15.8000	0.253	12.4070	1.5759
15	15.0000	12.9904	16.0000	15.826	14.0000	0.201	0.088	16.2040	16.0000	16.0000	14.0000	16.8000	0.257	13.3258	1.5759
16	16.0000	13.8564	17.0000	16.823	15.0000	0.203	0.090	17.2040	17.0000	17.0000	15.0000	17.8000	0.260	14.4073	1.5759
17	17.0000	14.7224	18.0000	17.820	16.0000	0.205	0.091	18.2040	18.0000	18.0000	16.0000	18.8000	0.262	15.3357	1.5760
18	18.0000	15.5885	19.0000	18.818	17.0000	0.207	0.093	19.2040	19.0000	19.0000	17.0000	19.8000	0.265	16.4078	1.5760
19	19.0000	16.4545	20.0000	19.815	18.0000	0.208	0.094	20.2040	20.0000	20.0000	18.0000	20.8000	0.267	17.3437	1.5761
20	20.0000	17.3205	21.0000	20.813	19.0000	0.209	0.095	21.2040	21.0000	21.0000	19.0000	21.8000	0.269	18.4081	1.5761

21	21.0000	18.1865	22.0000	21.811	20.0000	0.210	0.096	22.2040	22.0000	22.0000	20.0000	22.8000	0.270	19.3499	1.5761
22	22.0000	19.0526	23.0000	22.810	21.0000	0.212	0.097	23.2040	23.0000	23.0000	21.0000	23.8000	0.272	20.4085	1.5762
23	23.0000	19.9186	24.0000	23.808	22.0000	0.212	0.097	24.2040	24.0000	24.0000	22.0000	24.8000	0.273	21.3552	1.5762
24	24.0000	20.7846	25.0000	24.807	23.0000	0.213	0.098	25.2040	25.0000	25.0000	23.0000	25.8000	0.274	22.4088	1.5763
25	25.0000	21.6506	26.0000	25.805	24.0000	0.214	0.099	26.2040	26.0000	26.0000	24.0000	26.8000	0.275	23.3598	1.5763
26	26.0000	22.5167	27.0000	26.804	25.0000	0.215	0.100	27.2040	27.0000	27.0000	25.0000	27.8000	0.276	24.4089	1.5763
27	27.0000	23.3827	28.0000	27.803	26.0000	0.215	0.100	28.2040	28.0000	28.0000	26.0000	28.8000	0.277	25.3637	1.5764
28	28.0000	24.2487	29.0000	28.802	27.0000	0.216	0.101	29.2040	29.0000	29.0000	27.0000	29.8000	0.278	26.4092	1.5764
29	29.0000	25.1147	30.0000	29.801	28.0000	0.217	0.101	30.2040	30.0000	30.0000	28.0000	30.8000	0.279	27.3669	1.5764
30	30.0000	25.9808	31.0000	30.801	29.0000	0.217	0.102	31.2040	31.0000	31.0000	29.0000	31.8000	0.280	28.4094	1.5765
31	31.0000	26.8468	32.0000	31.800	30.0000	0.218	0.102	32.2040	32.0000	32.0000	30.0000	32.8000	0.281	29.3699	1.5765
32	32.0000	27.7128	33.0000	32.799	31.0000	0.218	0.103	33.2040	33.0000	33.0000	31.0000	33.8000	0.282	30.4097	1.5766
33	33.0000	28.5788	34.0600	33.798	32.0000	0.219	0.103	34.2040	34.0000	34.0000	32.0000	34.8000	0.282	31.3725	1.5766
34	34.0000	29.4449	35.0000	34.798	33.0000	0.219	0.103	35.2040	35.0000	35.0000	33.0000	35.8000	0.283	32.4097	1.5766
35	35.0000	30.3109	36.0000	35.797	34.0000	0.219	0.104	36.2040	36.0000	36.0000	34.0000	36.8000	0.284	33.3748	1.5767
36	36.0000	31.1769	37.0000	36.796	35.0000	0.220	0.104	37.2040	37.0000	37.0000	35.0000	37.8000	0.284	34.4100	1.5767
37	37.0000	32.0429	38.0000	37.796	36.0000	0.220	0.104	38.2040	38.0000	38.0000	36.0000	38.8000	0.285	35.3770	1.5768
38	38.0000	32.9090	39.0000	38.795	37.0000	0.220	0.105	39.2040	39.0000	39.0000	37.0000	39.8000	0.286	36.4101	1.5768
39	39.0000	33.7750	40.0000	39.795	38.0000	0.221	0.105	40.2040	40.0000	40.0000	38.0000	40.8000	0.286	37.3787	1.5768
40	40.0000	34.6410	41.0000	40.794	39.0000	0.221	0.105	41.2040	41.0000	41.0000	39.0000	41.8000	0.286	38.4104	1.5769
41	41.0000	35.5070	42.0000	41.794	40.0000	0.221	0.106	42.2040	42.0000	42.0000	40.0000	42.8000	0.287	39.3805	1.5769
42	42.0000	36.3731	43.0000	42.794	41.0000	0.222	0.106	43.2040	43.0000	43.0000	41.0000	43.8000	0.287	40.4106	1.5770
43	43.0000	37.2391	44.0000	43.793	42.0000	0.222	0.106	44.2040	44.0000	44.0000	42.0000	44.8000	0.287	41.3821	1.5770
44	44.0000	38.1051	45.0000	44.793	43.0000	0.222	0.106	45.2040	45.0000	45.0000	43.0000	45.8000	0.288	42.4107	1.5770
45	45.0000	38.9711	46.0000	45.793	44.0000	0.222	0.106	46.2040	46.0000	46.0000	44.0000	46.8000	0.288	43.3835	1.5771
46	46.0000	39.8372	47.0000	46.792	45.0000	0.222	0.107	47.2040	47.0000	47.0000	45.0000	47.8000	0.288	44.4108	1.5771
47	47.0000	40.7032	48.0000	47.792	46.0000	0.223	0.107	48.2040	48.0000	48.0000	46.0000	48.8000	0.288	45.3849	1.5772
48	48.0000	41.5692	49.0000	48.792	47.0000	0.223	0.107	49.2040	49.0000	49.0000	47.0000	49.8000	0.288	46.4111	1.5772
49	49.0000	42.4352	50.0000	49.791	48.0000	0.223	0.107	50.2040	50.0000	50.0000	48.0000	50.8000	0.289	47.3860	1.5772
50	50.0000	43.3013	51.0000	50.791	49.0900	0.223	0.107	51.2040	51.0000	51.0000	49.0000	51.8000	0.289	48.4113	1.5773

* Intended for cutting by a generating process.

† If this dimension is used, the dimension in column 24, Table 44, should be decreased by twice the amount of maximum dimensional tooth clearance and the chamfer applied.

‡ L = 0.0001 × dia (column 4).

§ Represents minimum allowable radius of curvature and is based on 75% of the full tangent radius for maximum depth.

¶ Allowable errors (Table 48), except lead, have been added to the machining tolerance in computing the maximum space width. When allowances for lead errors must be made, add 60% of the lead error to this dimension (see Table 48).

TABLE 44. 1/2 DIAMETRAL PITCH EXTERNAL INVOLUTE SPLINES

1/2 Diametral Pitch, Pressure Angle 30°	Addendum (Basic) 0.5000	Circular Pitch 3.1416, Measuring Pin Dia 1.9200

External

N	Major-dia Fit*					Flat Roof Side Fit			Fillet Root†			Dimensions for all Fits						
	Major Dia			Major-dia Chamfer		Minor Dia	Minor-dia Fillet		Major‡ Dia	Minor Dia	Fillet Rad	TIF Dia	Measurement over Pins			Tooth Thickness Max Effective		
	Class I	Class II	Class III	Dim	Ht		Rad	Ht					Class A	Class B	Class C	Class A 1.5703	Class B 1.5723	Class C 1.5753
1	16	17	18	19	20	21	22	23	24	25	26	27	28	29	30	31	32	33
Recommended Tolerance	+0.0000 −0.0019 + *J*§	+0.0000 −0.0006 + *J*§	+0.0009 + *J*§ −0.0000	Approx	Min	+0.0000 −0.1050	Approx	Max	+0.0000 −0.0100	+0.0000 −0.1050	Min¶	Max	Min	Min	Min	Min Dimensional‖		
6	6.9985	6.9999	7.0015	0.228	0.091	4.7960	0.170	0.168	7.0000	4.2000		5.2062	8.8590	8.8622	8.8661	1.5654	1.5679	1.5709
7	7.9985	7.9999	8.0016	0.230	0.094	5.7960	0.150	0.150	8.0000	5.2000		6.0722	9.6747	9.6780	9.6819	1.5653	1.5678	1.5708
8	8.9985	8.9999	9.0017	0.230	0.095	6.7960	0.140	0.137	9.0000	6.2000	0.379	7.0400	10.8869	10.8903	10.8914	1.5653	1.5678	1.5708
9	9.9985	9.9999	10.0018	0.232	0.097	7.7960	0.130	0.127	10.0000	7.2000	0.373	8.0000	11.7460	11.7494	11.7535	1.5653	1.5678	1.5708
10	10.9985	10.9999	11.0019	0.233	0.098	8.7960	0.120	0.119	11.0000	8.2000	0.367	9.0000	12.9065	12.9100	12.9143	1.5652	1.5677	1.5707
11	11.9985	11.9999	12.0020	0.234	0.099	9.7960	0.116	0.113	12.0000	9.2000	0.362	10.0000	13.7924	13.7959	13.8002	1.5652	1.5677	1.5707
12	12.9985	12.9999	13.0021	0.234	0.100	10.7960	0.110	0.107	13.0000	10.2000	0.357	11.0000	14.9215	14.9250	14.9294	1.5652	1.5677	1.5707
13	13.9985	13.9999	14.0022	0.235	0.100	11.7960	0.108	0.103	14.0000	11.2000	0.353	12.0000	15.8253	15.8289	15.8332	1.5651	1.5676	1.5706
14	14.9985	14.9999	15.0023	0.235	0.101	12.7960	0.105	0.099	15.0000	12.2000	0.350	13.0000	16.9328	16.9365	16.9409	1.5651	1.5676	1.5706
15	15.9985	15.9999	16.0024	0.236	0.102	13.7960	0.102	0.096	16.0000	13.2000	0.348	14.0000	17.8500	17.8537	17.8581	1.5651	1.5676	1.5706
16	16.9985	16.9999	17.0025	0.236	0.102	14.7960	0.100	0.093	17.0000	14.2000	0.345	15.0000	18.9421	18.9458	18.9503	1.5650	1.5675	1.5705
17	17.9985	17.9999	18.0026	0.236	0.103	15.7960	0.100	0.091	18.0000	15.2000	0.343	16.0000	19.8691	19.8729	19.8774	1.5650	1.5675	1.5705
18	18.9985	18.9999	19.0027	0.237	0.103	16.7960	0.100	0.089	19.0000	16.2000	0.341	17.0000	20.9497	20.9534	20.9579	1.5650	1.5675	1.5705
19	19.9985	19.9999	20.0028	0.237	0.103	17.7960	0.100	0.087	20.0000	17.2000	0.339	18.0000	21.8845	21.8883	21.8928	1.5649	1.5674	1.5704
20	20.9985	20.9999	21.0029	0.237	0.104	18.7960	0.100	0.085	21.0000	18.2000	0.338	19.0000	22.9559	22.9597	22.9643	1.5649	1.5674	1.5704

21	21.9985	21.9999	22.0030	0.237	0.104	19.7960	0.100	0.083	22.0000	19.2000	0.336	20.0000	23.8970	23.9009	23.9054	1.5648	1.5673	1.5703
22	22.9985	22.9999	23.0031	0.237	0.104	20.7960	0.100	0.082	23.0000	20.2000	0.334	21.0000	24.9612	24.9650	24.9696	1.5648	1.5673	1.5703
23	23.9985	23.9999	24.0032	0.238	0.104	21.7960	0.100	0.081	24.0000	21.2000	0.333	22.0000	25.9075	25.9114	25.9160	1.5648	1.5673	1.5703
24	24.9985	24.9999	25.0033	0.238	0.105	22.7960	0.100	0.080	25.0000	22.2000	0.332	23.0000	26.9657	26.9696	26.9743	1.5647	1.5672	1.5702
25	25.9985	25.9999	26.0034	0.238	0.105	23.7960	0.100	0.079	26.0000	23.2000	0.331	24.0000	27.9164	27.9203	27.9250	1.5647	1.5672	1.5702
26	26.9985	26.9999	27.0035	0.238	0.105	24.7960	0.100	0.078	27.0000	24.2000	0.330	25.0000	28.9698	28.9737	28.9784	1.5647	1.5672	1.5702
27	27.9985	27.9999	28.0036	0.239	0.105	25.7960	0.100	0.077	28.0000	25.2000	0.330	26.0000	29.9240	29.9279	29.9326	1.5646	1.5671	1.5701
28	28.9985	28.9999	29.0037	0.239	0.105	26.7960	0.100	0.076	29.0000	26.2000	0.329	27.0000	30.9732	30.9771	30.9818	1.5646	1.5671	1.5701
29	29.9985	29.9999	30.0038	0.239	0.106	27.7960	0.100	0.075	30.0000	27.2000	0.328	28.0000	31.9307	31.9347	31.9394	1.5646	1.5671	1.5701
30	30.9985	30.9999	31.0039	0.239	0.106	28.7960	0.100	0.074	31.0000	28.2000	0.327	29.0000	32.9763	32.9802	32.9849	1.5645	1.5670	1.5700
31	31.9985	31.9999	32.0040	0.239	0.106	29.7960	0.100	0.074	32.0000	29.2000	0.327	30.0000	33.9365	33.9405	33.9452	1.5645	1.5670	1.5700
32	32.9985	32.9999	33.0041	0.239	0.106	30.7960	0.100	0.073	33.0000	30.2000	0.326	31.0000	34.9789	34.9828	34.9876	1.5644	1.5669	1.5699
33	33.9985	33.9999	34.0042	0.239	0.106	31.7960	0.100	0.072	34.0000	31.2000	0.326	32.0000	35.9415	35.9455	35.9503	1.5644	1.5669	1.5699
34	34.9985	34.9999	35.0043	0.239	0.106	32.7960	0.100	0.072	35.0000	32.2000	0.325	33.0000	36.9814	36.9854	36.9902	1.5644	1.5669	1.5699
35	35.9985	35.9999	36.0044	0.239	0.106	33.7960	0.100	0.071	36.0000	33.2000	0.325	34.0000	37.9461	37.9501	37.9549	1.5643	1.5668	1.5698
36	36.9985	36.9999	37.0045	0.240	0.106	34.7960	0.100	0.071	37.0000	34.2000	0.324	35.0000	38.9835	38.9875	38.9923	1.5643	1.5668	1.5698
37	37.9985	37.9999	38.0046	0.240	0.107	35.7960	0.100	0.070	38.0000	35.2000	0.324	36.0000	39.9503	39.9543	39.9591	1.5643	1.5668	1.5698
38	38.9985	38.9999	39.0047	0.240	0.107	36.7960	0.100	0.070	39.0000	36.2000	0.323	37.0000	40.9854	40.9894	40.9942	1.5642	1.5667	1.5697
39	39.9985	39.9999	40.0048	0.240	0.107	37.7960	0.100	0.069	40.0000	37.2000	0.323	38.0000	41.9539	41.9579	41.9627	1.5642	1.5667	1.5697
40	40.9985	40.9999	41.0049	0.240	0.107	38.7960	0.100	0.069	41.0000	38.2000	0.323	39.0000	42.9874	42.9914	42.9962	1.5642	1.5667	1.5697
41	41.9985	41.9999	42.0050	0.240	0.107	39.7960	0.100	0.069	42.0000	39.2000	0.322	40.0000	43.9572	43.9612	43.9661	1.5641	1.5666	1.5696
42	42.9985	42.9999	43.0051	0.240	0.107	40.7960	0.100	0.068	43.0000	40.2000	0.322	41.0000	44.9888	44.9928	44.9927	1.5641	1.5666	1.5696
43	43.9985	43.9999	44.0052	0.240	0.107	41.7960	0.100	0.068	44.0000	41.2000	0.322	42.0000	45.9603	45.9643	45.9692	1.5641	1.5666	1.5696
44	44.9985	44.9999	45.0053	0.240	0.107	42.7960	0.100	0.068	45.0000	42.2000	0.322	43.0000	46.9903	46.9943	46.9992	1.5640	1.5665	1.5695
45	45.9985	45.9999	46.0054	0.240	0.107	43.7960	0.100	0.067	46.0000	43.2000	0.321	44.0000	47.9630	47.9670	47.9719	1.5640	1.5665	1.5695
46	46.9985	46.9999	47.0055	0.240	0.107	44.7960	0.100	0.067	47.0000	44.2000	0.321	45.0000	48.9916	48.9957	49.0005	1.5639	1.5664	1.5694
47	47.9985	47.9999	48.0056	0.240	0.107	45.7960	0.100	0.067	48.0000	45.2000	0.320	46.0000	49.9654	49.9694	49.9743	1.5639	1.5664	1.5694
48	48.9985	48.9999	49.0057	0.240	0.107	46.7960	0.100	0.066	49.0000	46.2000	0.320	47.0000	50.9929	50.9969	51.0018	1.5639	1.5664	1.5694
49	49.9985	49.9999	50.0058	0.240	0.107	47.7960	0.100	0.066	50.0000	47.2000	0.320	48.0000	51.9676	51.9717	51.9766	1.5638	1.5663	1.5693
50	50.9985	50.9999	51.0059	0.240	0.107	48.7960	0.100	0.066	51.0000	48.2000	0.319	49.0000	52.9940	52.9981	53.0030	1.5638	1.5663	1.5693

* Measurement over pins for class A is recommended (column 28), but if tighter fits are required, class B (column 29) may be used.

† This may be used for a major-diameter fit by using dimension in column 16, 17, or 18 instead of that in column 24.

‡ When column 9B, Table 43, is used for the internal spline, reduce this dimension as covered in Table 43, note †.

‖ Allowable errors (Table 48), except lead, have been added to the machining tolerance in computing the minimum tooth thickness. When allowances for lead errors must be made, subtract 60% of the lead error from this dimension (see Table 48).

¶ Represents minimum allowable radius of curvature, and is based on 75% of the full tangent radius for maximum depth.

§ $J = 0.0002 \times$ dia (column 24).

TABLE 45. COLUMN FORMULAS AND TOLERANCES FOR INVOLUTE SPLINES

Column No.	Class Fit	Diametral Pitch	Formula for Column Dimension	Machining Tolerance
2		All pitches	$D = N/P$	
3		All pitches	$D_b = D \cos 30°$	
4		All pitches	$D_o = \dfrac{N+1}{P}$	$+(0.0007 + 0.0001\,D_o)$ -0.0000
5			column 4 − 2 (column 8)	Min
6, 11, 27		1/2–4/8 5/10–8/16 10/20–20/40 24/48–48/96	$D_R = \dfrac{N-1}{P}$ (for 1/2 to 20/40, 6 and 7 teeth, $D_R = D_b + 0.010$) [for 24/48 to 48/96, 6 and 7 teeth, $D_R = D_b + 0.25(D - D_B)$]. (Arbitrary correction on 8 tooth for 2.5/5 to 12/24)	$+0.0050$ -0.0000 $+0.0035$ -0.0000 $+0.0025$ -0.0000 $+0.0015$ -0.0000
7, 8, 13, 19, 20, 22, 23, 26, 35		All pitches	Determined by layout for 1/2 diametral pitch. Divide dimensions for 1/2 diametral pitch by that used	
9A		All pitches	$D_o = \dfrac{N+1.2}{P} + 0.0040$ $\left(\text{clearance, } c = \dfrac{0.1}{P} + 0.002\right)$	$+\left(\dfrac{0.2}{P} + 0.005\right)$ -0
9B		All pitches	$D_o = \dfrac{N+1}{P}$	$+(0.0007 + 0.0001D_o)$ $-(0.0007 + 0.0001D_o)$
10		All pitches	$\text{TIF} = \dfrac{N+1}{P}$	Min
12		All pitches	$D_o = \dfrac{N+1.8}{P}$	$+\left(\dfrac{0.1}{P} + 0.005\right)$ -0
14		1/2 2.5/5 3/6 4/8 and 5/10 6/12 and 8/16 10/20 to 48/96	Calculate from formulas, page 4-117 Measurement = $\begin{cases} \text{Basic } M_i + 0.0001F(0.396N + 45) \\ \text{Basic } M_i + 0.0001F(0.156N + 42) \\ \text{Basic } M_i + 0.0001F(0.132N + 40) \\ \text{Basic } M_i + 0.0001F(0.096N + 37) \\ \text{Basic } M_i + 0.0001F(0.072N + 30) \\ \text{Basic } M_i + 0.0001F(0.072N + 24) \end{cases}$	Max toleranee is included in formulas For lead allowances see Table **48**.

15		1/2 2.5/5 3/6 4/8 and 5/10 6/12 and 8/16 10/20 to 48/96	Min is basic Max is basic plus (this includes machining tolerance) $\begin{cases} 0.0001(0.396N + 45) \\ 0.0001(0.156N + 42) \\ 0.0001(0.132N + 40) \\ 0.0001(0.096N + 37) \\ 0.0001(0.072N + 30) \\ 0.0001(0.072N + 24) \end{cases}$	For reference only, included in max* +0.0020 −0.0000 +0.0020 −0.0000 +0.0020 −0.0000 +0.0020 −0.0000 +0.0015 −0.0000 +0.0010 −0.0000
16	I	All pitches	$D_o - 0.0015$ or column 4 − 0.0015	$+0.0000 - (0.0019 + 0.0002D_o)$
17	II	All pitches	$D_o - 0.0001$ or column 4 − 0.0001	$+0.0000 - (0.0006 + 0.0002D_o)$
18	III	All pitches	$D_o + 0.0008 + 0.0001D_o$	$+(0.0009 + 0.0002D_o) - 0.0000$
21		All pitches	$D_R = \frac{N - 1.2}{P} - 0.0040$	$+0$ $-\left(\frac{0.1}{P} + 0.005\right)$
24		1/2 to 4/8 5/10 to 8/16 10/20 to 20/40 24/48 to 48/96	$D_o = \frac{N + 1}{P}$	+0.0000 −0.0100 +0.0000 −0.0070 +0.0000 −0.0050 +0.0000 −0.0030
25		1/2 to 12/24 16/32 to 48/96	$D_R = N - 1.8/P$ $D_R = N - 2.0/P$	$+0$ $-\left(\frac{0.1}{P} + 0.005\right)$
28	A	1/2 2.5/5 3/6 4/8 and 5/10 6/12 and 8/16 10/20 to 20/40 24/48 to 48/96	Calculate M_e from formulas, page 4-117 Measurement = $\begin{cases} \text{Basic } M_e - 0.0001E(0.360N + 52) \\ \text{Basic } M_e - 0.0001E(0.120N + 48) \\ \text{Basic } M_e - 0.0001E(0.090N + 47) \\ \text{Basic } M_e - 0.0001E(0.072N + 43) \\ \text{Basic } M_e - 0.0001E(0.054N + 36) \\ \text{Basic } M_e - 0.0001E(0.054N + 35) \\ \text{Basic } M_e - 0.0001E(0.054N + 30) \end{cases}$	For reference only, included in min* +0.0000 −0.0020 +0.0000 −0.0020 +0.0000 −0.0020 +0.0000 −0.0020 +0.0000 −0.0015 +0.0000 −0.0010 +0.0000 −0.0010
29	B	1/2 2.5/5 3/6 4/8 and 5/10 6/12 and 8/16 10/20 to 20/40 24/48 to 48/96	Calculate M_e from formulas, page 4-117 Measurement = $\begin{cases} \text{Basic } M_e - 0.0001E(0.360N + 27) \\ \text{Basic } M_e - 0.0001E(0.120N + 23) \\ \text{Basic } M_e - 0.0001E(0.090N + 22) \\ \text{Basic } M_e - 0.0001E(0.072N + 18) \\ \text{Basic } M_e - 0.0001E(0.054N + 18) \\ \text{Basic } M_e - 0.0001E(0.054N + 19) \\ \text{Basic } M_e - 0.0001E(0.054N + 16) \end{cases}$	For reference only, included in min* +0.0000 −0.0015 +0.0000 −0.0015 +0.0000 −0.0015 +0.0000 −0.0015 +0.0000 −0.0015 +0.0000 −0.0015 +0.0000 −0.0010
30	C	1/2 2.5/5 3/6 4/8 and 5/10 6/12 and 8/16 10/20 to 20/40 24/48 to 48/96	Calculate M_e from formulas, page 4-117 Measurement = $\begin{cases} \text{Basic } M_e - 0.0001E(0.360N - 3) \\ \text{Basic } M_e - 0.0001E(0.120N - 4) \\ \text{Basic } M_e - 0.0001E(0.090N - 5) \\ \text{Basic } M_e - 0.0001E(0.072N - 3) \\ \text{Basic } M_e - 0.0001E(0.054N + 1) \\ \text{Basic } M_e - 0.0001E(0.054N) \\ \text{Basic } M_e - 0.0001E(0.054N - 5) \end{cases}$	For reference only, included in min* +0.0000 −0.0015 +0.0000 −0.0015 +0.0000 −0.0015 +0.0000 −0.0015 +0.0000 −0.0015 +0.0000 −0.0015 +0.0000 −0.0010

TABLE 45. COLUMN FORMULAS AND TOLERANCES FOR INVOLUTE SPLINES (*Continued*)

Column No.	Class Fit	Diametral Pitch	Formula for Column Dimension: Max	Formula for Column Dimension: Min	Machining Tolerance	
31	A	1/2	Basic − 0.0005	Basic − 0.0001(0.360N + 52)	For reference only, included in min*	+0.0000 −0.0020
		2.5/5	Basic − 0.0005	Basic − 0.0001(0.120N + 48)		+0.0000 −0.0020
		3/6	Basic − 0.0005	Basic − 0.0001(0.090N + 47)		+0.0000 −0.0020
		4/8 to 5/10	Basic − 0.0005	Basic − 0.0001(0.072N + 43)		+0.0000 −0.0020
		6/12 to 8/16	Basic − 0.0005	Basic − 0.0001(0.054N + 36)		+0.0000 −0.0015
		10/20 to 20/40	Basic − 0.0005	Basic − 0.0001(0.054N + 35)		+0.0000 −0.0015
		24/48 to 48/96	Basic − 0.0005	Basic − 0.0001(0.054N + 30)		+0.0000 −0.0010
32	B	1/2	Basic + 0.0015	Basic − 0.0001(0.360N + 27)	For reference only, included in min*	+0.0000 −0.0015
		2.5/5	Basic + 0.0015	Basic − 0.0001(0.120N + 23)		+0.0000 −0.0015
		3/6	Basic + 0.0015	Basic − 0.0001(0.090N + 22)		+0.0000 −0.0015
		4/8 to 5/10	Basic + 0.0015	Basic − 0.0001(0.072N + 18)		+0.0000 −0.0015
		6/12 to 8/16	Basic + 0.0013	Basic − 0.0001(0.054N + 18)		+0.0000 −0.0015
		10/20 to 20/40	Basic + 0.0011	Basic − 0.0001(0.054N + 19)		+0.0000 −0.0015
		24/48 to 48/96	Basic + 0.0009	Basic − 0.0001(0.054N + 16)		+0.0000 −0.0010
33	C	1/2	Basic + 0.0045	Basic − 0.0001(0.360N + 3)	For reference only, included in min*	+0.0000 −0.0015
		2.5/5	Basic + 0.0042	Basic − 0.0001(0.120N − 4)		+0.0000 −0.0015
		3/6	Basic + 0.0042	Basic − 0.0001(0.090N − 5)		+0.0000 −0.0015
		4/8 to 5/10	Basic + 0.0036	Basic − 0.0001(0.072N − 3)		+0.0000 −0.0015
		6/12 to 8/16	Basic + 0.0030	Basic − 0.0001(0.054N + 1)		+0.0000 −0.0015
		10/20 to 20/40	Basic + 0.0030	Basic − 0.0001(0.054N)		+0.0000 −0.0015
		24/48 to 48/96	Basic + 0.0030	Basic − 0.0001(0.054N − 5)		+0.0000 −0.0010
34	X, Y, Z	All	Determined by layout			
36	X	All	$D_R = \frac{N-1}{P} + 0.0015$		$+(0.0019 + 0.0001D_R)$ -0.0000	
37	Y	All	$D_R = \frac{N-1}{P} + 0.0001$		$+(0.0006 + 0.0001D_R)$ -0.0000	
38	Z	All	$D_R = \frac{N-1}{P} - (0.0007 + 0.0002D_R)$		$+0.0000$ $-(0.0009 + 0.0001D_R)$	
39	X, Y, Z	All	$D_R = \frac{N-1}{P}$ same as column 6 except for N, 6 and 7		$+0.0000$ $-(0.0006 + 0.0002D_R)$	
40	X, Y, Z	All	$\text{TIF} = \frac{N-1}{P} + 2$ (column 23)			

* These tolerances are included in the formulas given and are shown for reference only.

TABLE 46. KEY TO TABLES AND COLUMN NUMBERS WHEN APPLYING FITS

Type of Fit	Class of Fit	Internal External		Internal							External								
		Pitch Dia	Base Cir Dia	Major Dia	Minor Dia	TIF Dia	Major-dia Fillet		Meas be-tween Pins	Space* Width	Major Dia	Minor Dia	TIF Dia	Minor-dia Fillet		Major-dia Chamfer		Meas over Pins	Tooth Thick-ness†
							Rad	Ht						Rad	Ht	Dim	Ht		
Major dia	Sliding I......	2	3	4	6	5	7	8	14	15	16	21	27	22	23	19	20	28	31
	Close II.....	2	3	4	6	5	7	8	14	15	17	21	27	22	23	19	20	28	31
	Press III....	2	3	4	6	5	7	8	14	15	18	21	27	22	23	19	20	28	31
Sides of teeth, flat root	Sliding A.....	2	3	9‡	6	10	7	8	14	15	24	21	27	22	23	19§	20	28	31
	Close B.....	2	3	9‡	6	10	7	8	14	15	24	21	27	22	23	19§	20	29	32
	Press C.....	2	3	9‡	6	10	7	8	14	15	24	21	27	22	23	19§	20	30	33
Sides of teeth, fillet root	Sliding A....	2	3	12	11	10	13	..	14	15	24	25	27	26	..	19§	20	28	31
	Close B.....	2	3	12	11	10	13	..	14	15	24	25	27	26	..	19§	20	29	32
	Press C.....	2	3	12	11	10	13	..	14	15	24	25	27	26	..	19§	20	30	33

Numbers refer to column numbers, Tables 43 and 44.

* Use minimum dimension given over column heading, and maximum from column 15.

† Use maximum dimension given over column heading, and minimum from columns 31, 32, or 33.

‡ Column 9A is to be used where full dedendum is desired, with clearance over external major diameter. Column 9B may be used when full dedendum is not required, and broaching of the internal spline is desired.

§ Chamfering which is specified above is required for a major-diameter fit. In some manufacturing a chamfer is desirable because it removes burrs which are a hazard in assembly processes and also saves injuries to those handling the parts. These chamfers are optional and can cause only slight loss of tooth contact.

TABLE 47. DIMENSIONAL AND EFFECTIVE CLEARANCE FOR INVOLUTE SPLINES*

Diametral Pitch	Class A				Class B				Class C			
	Dimensional Clearance†		Effective Clearance		Dimensional Clearance†		Effective Clearance		Dimensional Clearance†		Effective Clearance	
	Min	Max	Min	Max	Min	Max	Min	Max	Min	Max	Min	Max
1/2	+0.0076	+0.0116	+0.0005	+0.0045	+0.0056	+0.0091	−0.0015	+0.0020	+0.0026	+0.0061	−0.0045	−0.0010
2.5/5	+0.0057	+0.0097	+0.0005	+0.0045	+0.0037	+0.0072	−0.0015	+0.0020	+0.0010	+0.0045	−0.0042	−0.0007
3/6	+0.0052	+0.0092	+0.0005	+0.0045	+0.0032	+0.0067	−0.0015	+0.0020	+0.0005	+0.0040	−0.0042	−0.0007
4/8 and 5/10	+0.0044	+0.0084	+0.0005	+0.0045	+0.0024	+0.0059	−0.0015	+0.0020	+0.0003	+0.0038	−0.0036	−0.0001
6/12 and 8/16	+0.0039	+0.0069	+0.0005	+0.0035	+0.0021	+0.0051	−0.0013	+0.0017	+0.0004	+0.0034	−0.0030	−0.0000
10/20 to 20/40	+0.0037	+0.0062	+0.0005	+0.0030	+0.0021	+0.0046	−0.0011	+0.0014	+0.0002	+0.0027	−0.0030	−0.0005
24/48 to 48/96	+0.0037	+0.0057	+0.0005	+0.0025	+0.0023	+0.0043	−0.0009	+0.0011	+0.0002	+0.0022	−0.0030	−0.0010

* Based on using machining tolerances plus 60 % of all allowable errors cumulatively, not including lead.
† For 25 teeth.

See Table 45.

The maximum dimensional clearance is obtained by subtracting column 31, 32, or 33 dimension from that in column 15.

The minimum dimensional clearance is obtained by subtracting the sum of the machining tolerances for column 15 and that for either column 31, 32, or 33 from the maximum dimensional clearance.

The minimum effective clearance is obtained by subtracting the dimension (maximum effective) in the heading over column 31, 32, or 33 from the dimension (minimum effective) in the heading over column 15.

The maximum effective clearance is obtained by adding to the minimum effective clearance the machining tolerances for column 15 and that for either column 31, 32, or 33.

used in conjunction with Tables 42, 43, and 44. Specifically, circular-tooth thickness, addendum, and dedendum can be calculated from the basic formulas. For diametral pitches other than 1/2, some dimensions are inversely proportional to data in Tables 43 and 44, specifically in the case of columns 2, 3, 4, 7, 8, 9B, 10, 19, 20, 22, 23, 24, 35, and 39. Tolerances are determined by manufacturing experience, not by proportions.

Measurement of Tooth Thickness. Tooth thickness is determined by measuring over pins for the external spline, and between pins for the internal spline. The

TABLE 48. ALLOWABLE ERRORS FOR INVOLUTE SPLINES

Diametral Pitch	Involute Profile Errors	Accumulated Pitch Error between Any Two Teeth													Lead Errors					
		Use as Given or Interpolate as Required													Length of Spline					
		No. of Teeth													0 to 0.49	0.50 to 1.24	1.25 to 2.49	2.50 to 3.99	4.00 to 5.24	5.25 to 6.50
		6	8	10	12	14	16	20	25	30	35	40	45	50						
1/2	15	22	22	22	22	23	24	27	30	33	36	39	42	45	0	4	5	6	7	8
2.5/5	11	20	20	20	20	20	20	20	20	20	21	22	23	24	0	4	5	6	7	8
3/6	10	18	18	18	18	18	18	18	18	18	18	19	20	20	0	4	5	6	7	8
4/8 and 5/10	8	15	15	15	15	15	15	15	15	15	15	15	15	15	0	4	5	6	7	8
6/12 and 8/16	6	15	15	15	15	15	15	15	15	15	15	15	15	15	0	3	3	4	5	6
10/20 up	5	15	15	15	15	15	15	15	15	15	15	15	15	15	0	3	3	4	5	6

Out of Roundness

Diametral Pitch	Internal Spline													External Spline												
	No. of Teeth													No. of Teeth												
	6	8	10	12	14	16	20	25	30	35	40	45	50	6	8	10	12	14	16	20	25	30	35	40	45	50
1/2	17	19	21	24	23	23	24	24	24	25	25	26	26	16	18	20	23	23	24	24	24	25	25	25	25	55
2.5/5	13	13	14	15	16	17	18	20	23	23	23	23	24	12	13	14	15	16	17	18	20	22	22	22	22	22
3/6	12	13	13	14	14	15	17	19	20	22	23	24	24	12	12	14	14	15	15	17	18	20	21	21	21	21
4/8 and 5/10	11	12	12	13	13	13	15	16	17	18	20	21	22	10	11	11	12	13	13	14	15	16	18	19	20	21
6/12 to 48/96	9	9	10	10	10	11	11	12	13	14	15	16	17	8	9	10	10	10	11	12	12	14	14	15	16	17

NOTE 1: All dimensions in table are 0.0001 in.

NOTE 2: Allowable errors in the above table are based on class 3 of American Standard Gear Tolerances and Inspection (ASA B6.6-1946). The basis of that standard is experience with gear generating machines. When the internal spline is broached, smaller allowances may be used for mass production, with the consent of the broach maker. If less accuracy can be permitted use 133 % of the errors in the table. In all cases, use 60 % of the total for calculating effective fits.

NOTE 3: To obtain total allowable errors: (*a*) Add the out of roundness divided by factor F or factor E, to the profile error and the accumulated pitch error. (*b*) Multiply the total by 60 %. The result is the total of all allowable errors to be used.

NOTE 4: The formulas in Table 45 are based on 60 % of the total allowable errors except lead. The machining tolerances are also included in these formulas. If allowance for lead must be included, add 60 % of the error to the space width (column 15 max) for the internal, or deduct from the tooth thickness, (column 31, 32, or 33, min) for the external spline.

NOTE 5: Pin measurements, column 14, Table 43, must be increased (when allowance for a lead error is added) by adding the lead error × 60 % × F (Table 42). Likewise, pin measurements, column 28, 29, or 30, must be decreased by the amount of the allowance for lead error × 60 % × factor E.

diameter of the measuring pin is optional within a narrow range. It is not necessary that a pin contact exactly at the pitch circle. For the sake of uniformity and to follow a satisfactory practice, this standard incorporates a recommended diameter indicated at the top of each table. Measurements with pins will not check spacing errors and they must be determined by other means. Pins for measuring flat root internal splines may be flattened as shown in Fig. 47, but for external splines those specified will clear the minor diameter.

Tooth Dimensions. Because of allowable errors which must be provided for in determining fits, space width must be increased and tooth thickness decreased in relation to the diameter of the part, and therefore, in relation to the number of teeth. The minimum space width and the maximum tooth thickness are constant for each type of fit, but the maximum space width must be selected from column 15, and the minimum tooth thickness from column 31, 32, or 33 for the number of teeth used. Table 45 shows the formulas used to determine these dimensions.

Key to Tables with Column Numbers. Because of the different types of splines included in this standard, it is desirable to use a key, so that the proper dimensions will be selected. This key is incorporated in Table 46.

Cutting Tools. Tables 49A and 49B show, for internal splines, the minimum difference of teeth between internal spline and cutter, and a suggested cutter design for general use. Table 49C shows a design for general use with external splines, and Table 49D shows preferred cutters which may be used for either. For the fillet-root type the addendum of the shaper cutter, hob, or broach may be checked by layout or other means to assure the required involute clearance. The addendum is also influenced by the lack of tangency between the tooth profile and the tip radius of the tool. The part drawing should show a tolerance for the dedendum (applied to major or minor diameter) because of the tool tolerances and allowances for wear as well as the depth of cut to obtain the desired fit.

Figure 48 shows the dimensions of a 1/2 diametral pitch hob tooth for flat and fillet-root types. Table 50 lists the recommended over-all dimensions for finishing hobs. Figure 49 shows a 1/2 diametral pitch broach tooth.

Types of Fits. There are three methods of fitting splined members together:

Major Diameter. On the major diameter, where fit is controlled by varying the major diameter of the external spline.

Sides of Teeth. On the sides of the teeth, where fit is controlled by varying the tooth thickness. It is customary to use this type fit for the fillet root type spline.

Minor Diameter. On the minor diameter, where fit is controlled by varying the minor diameter of the internal spline.

Each of the three types of fits listed is divided into three classes: *sliding fits*, which must have clearance at all points; *close fits*, which must be close on either the major diameter, sides of teeth, or minor diameter; *press fits*, which must have an interference on either the major diameter, sides of teeth, or minor diameter.

For all types and classes of fits the internal pin measurements are the same and are to be found in column 14 of Table 43. Table 46 shows the correct combinations of dimensions for any desired fit.

Special Fits. The fits provided for in Table 43 cover usual practice and will be acceptable in a majority of installations. If it is necessary to use a special fit of the side-bearing type, the basic data in Table 42 together with the formulas in Fig. 47 may be used for its calculation. For further suggestions relative to special fits, see Table 48.

Tolerances. In the cutting of a splined part, a machine-tool adjustment may vary the depth of cut. The tolerances suggested are listed in Table 45.

Effective Fits. The effective clearance (positive or negative) between two splined parts is equal to the effective space width of the internal minus the effective tooth thickness of the external spline (Table 47).

Control of Fits

To control fitting of parts, the internal spline is basically constant, and the external spline is varied to obtain the required fit. However, for the minor-diameter

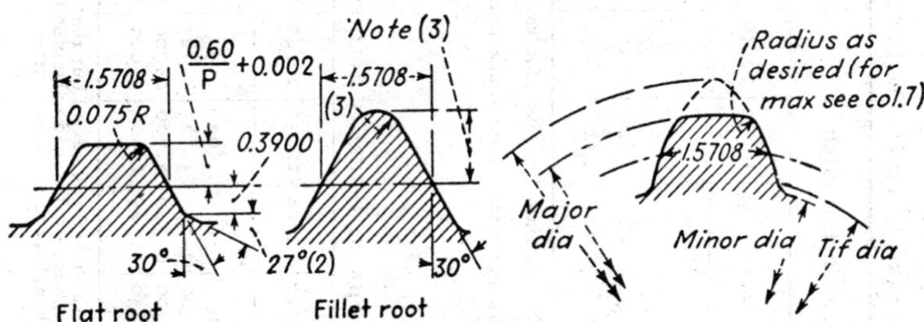

FIG. 48. One/two diametral (1/2) pitch hob tooth.

FIG. 49. 1/2 pitch broach tooth.

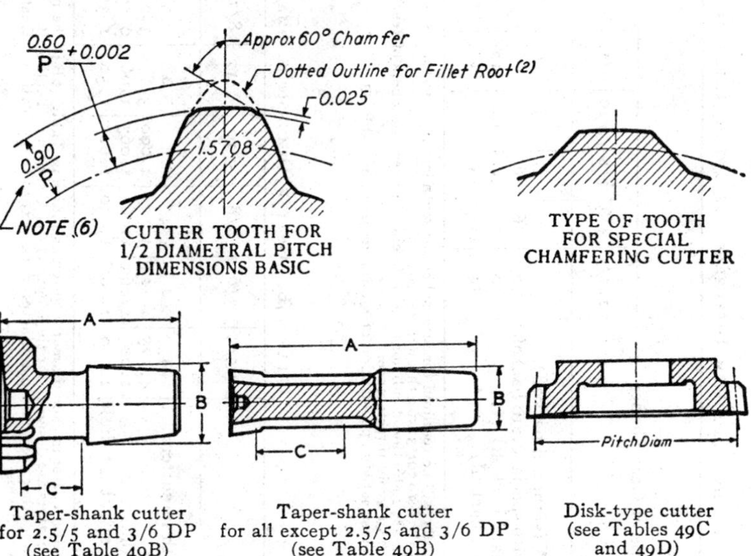

Taper-shank cutter for 2.5/5 and 3/6 DP (see Table 49B)

Taper-shank cutter for all except 2.5/5 and 3/6 DP (see Table 49B)

Disk-type cutter (see Tables 49C and 49D)

Illustrations for Table 49 (notes on pages 1-132 and 1-133)

TABLE 49A. MINIMUM DIFFERENCE OF TEETH, INTERNAL SPLINE AND CUTTER[1] (See Illustrations on Page 4-131)

Type of Gear Shaper	Diametral Pitch													
	2.5/5	3/6	4/8	5/10	6/12	8/16	10/20	12/24	16/32	20/40	24/48	32/64	40/80	48/96
72 or 725A (standard backoff)	..	..	..	6	6	6	6	6	6	6	6	6	6	6
6 or 6A (with special 0.010 backoff)	6	6	6	6	6	6	6	6	6	6	6	6	6	6
6A (standard backoff)	10	10	10	10	10	9	8	7	7	7	7	7	7	7
6 (standard backoff)	14	13	12	11	11	11	11	11	11	11	11	11	11	11

Special study will be required for internal spline teeth less than 10 for 2.5/5 through 6/12 diametral pitch, for less than 12 for 8/16 through 24/48, and for less than 16 for 32/64 through 48/96.

[1] For normal production conditions the difference between numbers of teeth in work and cutter on internal splines should be not less than shown in Table 49A to avoid interference.

[2] Radius must satisfy specification of minimum given in Table 43.

TABLE 49B. CUTTER DIMENSIONS, SMALLEST SHANK, INTERNAL SPLINES[3]

	2.5/5	3/6	4/8	5/10	6/12	8/16	10/20	12/24	16/32	20/40	24/48	32/64	40/80	48/96
No. of teeth in cutter	4	4	4	4	4	5	6	6	6	7	7	9	10	12
A. Length	4.125	4.125	4.125	4.000	3.875	3.625	3.500	3.375	3.125	3.125	3.000	3.000	3.000	3.000
B. Taper dia.	1.063	1.063	1.063	1.063	1.063	0.700	0.700	0.700	0.700	0.700	0.700	0.700	0.700	0.700
C. Length of cut (at life)	2.000	2.000	1.750	1.625	1.500	1.375	1.250	1.125	1.000	1.000	0.875	0.875	0.750	0.750

[3] The smallest cutter in each diametral pitch as shown in Table 49B is the most versatile and will cut all but the smallest numbers of teeth. The taper shank cutters are designed for the length of cut C shown in Table 49B. Where splines are designed to lengths not exceeding C, the availability of cutters from other users and manufacturers can expedite fabrication. Customer should specify whether cutters are to be made to these tables, or for specific conditions.

TABLE 49C. CUTTER DIMENSIONS (DISK TYPE), EXTERNAL OR INTERNAL SPLINES[4,5]

Diametral Pitch	2.5/5	3/6	4/8	5/10	6/12	8/16	10/20	12/24	16/32	20/40	24/48	32/64	40/80	48/96
No. of teeth in cutter....	10	12	16	18	22	28	36	44	56	72	72	96	120	120
Pitch dia of cutter.......	4.000	4.000	4.000	3.600	3.667	3.500	3.600	3.667	3.500	3.600	3.000	3.000	3.000	2.500
Approx min No. of teeth in external if cut on shaper 6 or 6A........	6	6	6	6	6	6	7	9	11	14	14	19	24	24
Approx min No. of teeth in external if cut on shaper 72 or 725A.....	..	..	..	6	6	6	6	6	8	10	10	14	17	17

TABLE 49D. PREFERRED CUTTERS, INTERNAL SPLINES[4,5]

Diametral Pitch	2.5/5	3/6	4/8	5/10	6/12	8/16	10/20	12/24	16/32	20/40	24/48	32/64	40/80	48/96
No. of teeth in cutters...	6 8(*d*)	6 9(*d*)	6 8 13(*d*)	6 10 16(*d*)	6 9 12 18(*d*)	8 12 16 24(*d*)	10 15 20 30(*d*)	12 18 24 36(*d*)	12 18 24	12 24	12 24	14 24	16 24	18 24

[4] Tables 49C and 49D show numbers of teeth in preferred cutters for external and internal splines, respectively, based on Table 49A.

[5] When the 2.5- and 3.0-in. cutters are used (Tables 49C and 49D) on machines with cutter spindles larger than cutter root (minor) diameter, the maximum face width will be limited by positioning of cutter on spindle. For 5/10 and finer diametral pitches, if type 62, 62A, or 7A machines are used, larger cutters with 1.75-in. hole are required.

[6] For external splines, in diametral pitches of 16/32 through 48/96, the addendum is $1.00/P$.

d = disk-type cutters

fit, the minor diameter of the internal spline is varied to control the fit, but the involute teeth are not otherwise varied on the internal spline.

The major-diameter fit is the most practical for maintaining accuracy because the major diameter of the broached or shaped internal spline is controlled by the outside diameter of the broach or cutter during its cutting or generating action. Fitting on the sides of the teeth permits most accurate control of backlash but is affected by the pitch or spacing error, involute profile error, out of roundness, and errors of lead. The minor diameter fit depends on the accuracy of the minor diameter of the external spline and special tooling is required.

Combinations of Types. A fillet root external spline may be used with a flat root internal spline with either a major-diameter or side-bearing type of fit. A flat root external spline may be used with a fillet root internal spline in either a side-bearing or minor-diameter fit. Where both the internal and external members are the fillet type, only a side-bearing type of fit can be used.

Relation to Ball-bearing Size. Where a shaft is splined near a bearing, a major diameter below the bearing size should be selected. The standard includes a table to show the proper selection.

Basic Depth. The basic working or contact depths of involute splines for either flat or fillet root types are shown below.

Diametral Pitch	Depth	Diametral Pitch	Depth	Diametral Pitch	Depth
1/2	1.0000	6/12	0.1667	20/40	0.0500
2.5/5	0.4000	8/16	0.1250	24/48	0.0417
3/6	0.3333	10/20	0.1000	32/64	0.0312
4/8	0.2500	12/24	0.0833	40/80	0.0250
5/10	0.2000	16/32	0.0625	48/96	0.0208

True Involute Form Modifications. For major-diameter fits, the TIF of the internal spline must be reduced from the basic, by twice the height of the major-diameter fillet (column 8). Likewise for minor-diameter fits the TIF of the external spline must be increased by twice the height of the minor-diameter fillet (column 23).

External Spline Fillets and Chamfers. The fillet at the minor diameter is formed by the generating action of the hob or shaper cutter. Since the shaping operation involves two circular parts, the tool and the work, the approximate radius formed is greater than in hobbing, and the height needed for shaping is shown in column 23. Column 22 is derived from the same data. The major-diameter chamfer is obtained by the generating action of the cutter or hob, or can be made by special machining or hand operation. The dimension given in column 19 (length of the approximate flat of the chamfer) and height in column 20 are affected by the tool with depth of cut as controlled by columns 28, 29, and 30. Fillet radius, column 22, is based on a 20-tooth cutter design.

Internal Spline Fillets and Chamfers. The fillet at the major diameter may be constant for broached parts, but, if cut with a shaper cutter, the generating action forms a fillet, the radial height of which is shown in column 8. The minor-diameter chamfer may be cut as a flat by a broach but will be slightly curved if produced by a shaper cutter. The standard allows sufficient height for shaper operation, which is greater than needed for any other method. In this case, an angle of 53 to 54° is

TABLE 50. RECOMMENDED DIMENSIONS FOR HOBS FOR SPLINES

Diametral Pitch	OD	Width	Bore
1/2	7.00	9.00	1.50
2.5/5	4.00	4.50	1.25
3/6	3.75	4.00	1.25
4/8	3.50	3.50	1.25
5/10	3.00	3.00	1.25
6/12	3.00	3.00	1.25
8/16	2.75	2.75	1.25
10/20	2.50	2.50	1.25
12/24	2.50	2.00	1.25
16/32	2.50	2.00	1.25
20/40	2.50	2.00	1.25
24/48	2.50	2.00	1.25
32/64	2.50	2.00	1.25
40/80	2.50	2.00	1.25
48/96	2.50	2.00	1.25

Hob shown in Fig. 52 should not be used for minor-diameter fits.

Ramp is required for major-diameter fits (see Table 46) also for the special combination covered by note † under Table 43.

The addendum of the hob is $0.90/P$ for diametral pitches from 1/2 through 12/24, and $1.00/P$ for diametral pitches from 16/32 through 48/96. The radius must be such as to satisfy specification of fillet radius in the tables.

Depth of cut will vary as controlled to obtain the desired fit (see Table 43).

The broach design will be in accordance with the specifications given in the tables for internal splines.

The length of the broach is determined by the length of the part, the hardness, the type of material to be broached, and the broaching equipment available.

Dotted outline, Fig. 49, shows the fillet-root type.

created between the tooth center and the flat of the chamfer produced by the special chamfer cutter shown on page 4-131. Fillet radius (column 7, Table 43) is based on a four-tooth cutter design

Allowable Errors (see Table 48)

Accumulated pitch error (*spacing*) is the greatest difference in any two teeth between the actual and theoretical tooth spacing on the same circle. Measurements are taken from one pitch point selected as reference, to the corresponding points on all other teeth, and will be affected by involute error and out of roundness.

Profile error is the difference between the highest and lowest readings of a dial indicator moving along the true involute curve in the plane of rotation and having a finger contacting the active tooth profile.

Lead error is a deviation of the teeth from the dimensional lead of the spline. Lead error is usually measured by traversing a dial indicator along the tooth face normal to the pitch line and parallel to the axis of the spline. This error may be disregarded unless experience shows otherwise.

Out of roundness is the difference between the maximum and minimum measurements over or between pins.

Application of errors. Experience has shown that not all the possible extremes will occur at one time, so that, instead of using the total of all errors, 60% has been adopted as a fair maximum for computing effective fits.

Tools for Cutting Splines

External Splines on Shafts. Hobs and gear shaper cutters are preferred.
Internal Splines. Shaper cutters and broaches are employed.

Involute Serrations

Definition and Scope. Involute serrations are multiple keys in the general form of internal and external involute gear teeth, as used for permanent fits between shafts and parts mounted thereon. The purpose of this standard is to provide a uniform, easily fabricated set of serrations that can be made by several manufacturing processes.

The pitches included are 10/20, 16/32, 24/48, 32/64, 40/80, 48/96, 64/128, 80/160, and 128/256, complete from 6 to 100 teeth only for the first three. The scope is from 0.10 diameter to 10.00 diameter.

Nomenclature. The nomenclature agrees as nearly as possible with that adopted by the American Gear Manufacturers Association and the American Standards Association, and corresponds exactly with the Standard for Involute Splines (B5.15-1949).

Pressure Angle. The pressure angle for all serrations is 45°.

Basic Tooth Form. All serrations are based upon involute form as generated with a straight-sided hob of the form included in this standard.[1]

[1] "In internal serrations for fine pitches, particularly with a large number of teeth, the involute curve will approach a straight line, and a straight profile is acceptable in this standard."

TABLE 51. TOOTH DIMENSIONS OF INVOLUTE SERRATIONS

Diametral Pitch	External Addendum Dedendum	Internal Serration		Circular Pitch	Effective Space Min	Effective Tooth Thickness Max			Pin Dia
		Addendum	Dedendum			Class of Fit			
P	$a = b$	a_1	b_1	p	t_s	A	B	C	d
1/2	0.5000	0.3000	0.7000	3.1416	1.7708	1.7703	1.7718	1.7738	1.9200
10/20	0.0500	0.0300	0.0700	0.3142	0.1771	0.1766	0.1781	0.1801	0.1920
16/32	0.0313	0.0188	0.0438	0.1963	0.1107	0.1102	0.1117	0.1137	0.1200
24/48	0.0208	0.0125	0.0292	0.1309	0.0738	0.0733	0.0748	0.0768	0.0800
32/64	0.0156	0.0094	0.0218	0.0982	0.0553	0.0548	0.0563	0.0583	0.0600
40/80	0.0125	0.0075	0.0175	0.0785	0.0443	0.0438	0.0453	0.0473	0.0480
48/96	0.0104	0.0063	0.0145	0.0654	0.0369	0.0364	0.0379	0.0399	0.0400
64/128	0.0078	0.0047	0.0109	0.0491	0.0277	0.0272	0.0287	0.0307	0.0300
80/160	0.0063	0.0038	0.0088	0.0393	0.0221	0.0216	0.0231	0.0251	0.0240
128/256	0.0039	0.0023	0.0055	0.0245	0.0138	0.0133	0.0148	0.0168	0.0150

INTERNAL SERRATION

EXTERNAL SERRATION

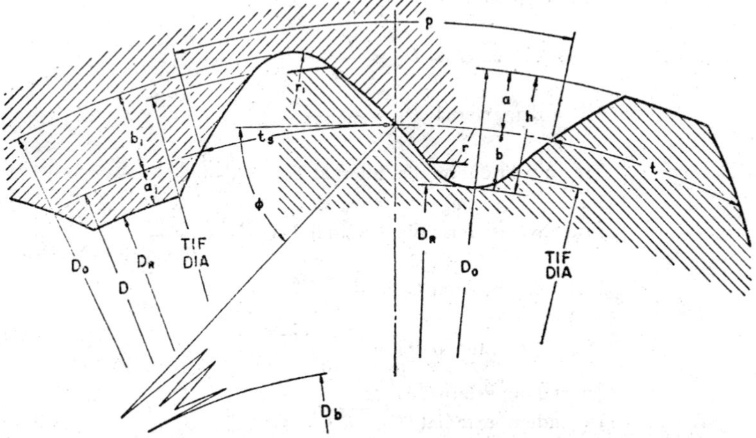

FIG. 50. Nomenclature for involute serrations.

P	Diametral pitch	N	Number of teeth
p	Circular pitch	D	Pitch diameter
a	Addendum, external	D_b	Base circle diameter
a_1	Addendum, internal	D_o	Major diameter
b	Dedendum, external	D_R	Minor diameter
b_1	Dedendum, internal	d	Measuring pin diameter
h	Total depth	TIF	Diameter at junction of involute form with fillet
t	Circular tooth thickness		
t_s	Circular space width	r, r_1	Approximate radius of fillet
ϕ	Pressure angle at pitch line (45°)		

M_e Measurement over pins—external serration
M_i Measurement between pins—internal serration

Tooth Dimensions. The basic dimensions are diametral pitch, pitch diameter, circular pitch, major diameter (of external serration), minor diameter (of external serration) and addendum (of external serration).

The special dimensions are circular tooth thickness and width of space, and both major and minor diameters of the internal serration.

The dimensions for the internal serration are held the same for all fits, and the external serration is varied to obtain the desired fit.

Modifications are made to obtain equally strong teeth in both members and at the same time retain even steps of major diameter in ordinary fractions.

Tables. Table 53 shows basic dimensions from which the dimensions for any serration may be calculated. Some of the data apply equally to all diametral pitches, and some must be divided by the diametral pitch for direct use.

Table 54 shows all dimensions for three classes of fits for 10/20 serrations. All

BASIC FORMULAS FOR INVOLUTE SERRATIONS

Pitch dia $D = \frac{N}{P}$

Circular pitch $p = \frac{\pi}{P}$

Major dia, external $D_O = \frac{N + 1}{P}$

Minor dia, external $D_R = \frac{N - 1}{P}$

True involute form dia, internal TIF $= \frac{N + 1.10}{P}$

True involute form dia, external TIF $= \frac{N - 0.70}{P}$

Major dia, internal $D_O = \frac{N + 1.40}{P}$

Minor dia, internal $D_R = \frac{N - 0.60}{P}$

Addendum, external a
Dedendum, external b $= \frac{0.500}{P}$

Addendum, internal $a_1 = \frac{0.300}{P}$

Dedendum, internal $b_1 = \frac{0.700}{P}$

Circular space width, internal $t_s = \frac{0.5\pi + 0.20}{P}$

dimensions are inversely proportional to diametral pitch, except measurement over pins for the external serrations, and tolerances given at the heading of the columns.

Table dimensions have been rounded off by using the last figure (fourth place) retained unchanged if that which follows is less than 5, and increasing the last figure (fourth place) to the next larger when the following figure is 5 or more.

All dimensions are in inches.

Measurement of Tooth Thickness. Tooth thickness is determined by measuring over pins for the external serrations and between pins for the internal serrations. The diameter of the pins is optional within a narrow range. It is not necessary that a pin contact exactly at the pitch circle. For the sake of uniformity and to follow a satisfactory practice, this standard incorporates a recommended pin diameter indicated at the top of each table. Measurements with pins will not check spacing errors, and these must be determined by other means.

Classes of Fits. There are three classes of fits, *loose*, *close*, and *press*. These are designated in the standard as classes A, B, and C, respectively.

Fillets, *External Serrations.* The fillet for the minor diameter is formed by the generating action of the hob or shaper cutter. The resulting fillet will not be a true radius and will always be larger than that of the tool.

TABLE 52. ALLOWABLE ERRORS FOR INVOLUTE SERRATIONS (IN 0.0001 IN.)

Diametral Pitch	Involute Profile	Accumulated Pitch Error between Any Two Teeth	Lead Errors: Length of Serration 0 to 0.49	0.50 to 1.24	1.25 to 2.49	2.50 to 3.99	4.00 to 5.24	5.25 to 6.50
10/20	5	15	0	3	3	4	5	6
16/32	5	15	0	3	3	4	5	6
24/48	5	15	0	3	3	4	5	6
32/64 up	5	15	0	3	3	4	5	6

Diametral Pitch	Out of Roundness, Internal — No. of Teeth: 6	8	10	12	14	16	20	25	30	35	40	45	50	60	70	80	90	100	Machining Tolerances, Internal
10/20	11	10	9	9	9	10	10	10	11	12	12	13	13	14	16	17	18	19	10
16/32	10	9	9	9	9	9	9	9	10	10	10	11	11	12	13	13	14	15	10
24/48	10	9	9	9	9	9	9	9	9	9	10	10	10	11	11	12	13	13	10
32/64 up	10	9	8	8	8	8	8	8	8	9	9	9	9	9	10	10	11	11	10

Diametral Pitch	Out of Roundness, External — No. of Teeth: 6	8	10	12	14	16	20	25	30	35	40	45	50	60	70	80	90	100	Machining Tolerances, External
10/20	8	9	9	9	10	10	10	11	12	12	13	13	14	15	16	17	18	19	15
16/32	8	8	9	9	9	9	10	10	10	11	11	11	12	12	13	14	14	15	15
24/48	8	8	8	9	9	9	9	9	10	10	10	10	10	11	11	11	12	12	10
32/64 up	8	8	8	8	8	8	9	9	9	9	9	10	10	10	10	10	11	11	10

NOTES: All table figures are 0.0001 in.
Add the profile error, accumulated pitch error, and the out of roundness divided by F or E.
If serration is longer than its pitch diameter, a lead error may be added.

TABLE 53. BASIC DATA FOR INVOLUTE SERRATIONS

1/2P			All Pitches				1/2P			All Pitches			
Internal and External		External	Internal and External		Internal	External	Internal and External		External	Internal and External		Internal	External
N	D_b	Measure over Pins	$\cos \frac{90°}{N}$	$\frac{d}{D_b}$	$\frac{t_s}{D}$	$\frac{\pi}{N}$	N	D_b	Measure over Pins	$\cos \frac{90°}{N}$	$\frac{d}{D_b}$	$\frac{t_s}{D}$	$\frac{\pi}{N}$
1	2	3	4	5	6	7	1	2	3	4	5	6	7
6	4.242641	9.1631		0.452548	0.295133	0.523599	31	21.920310	34.1965	0.998717	0.087590	0.057123	0.101342
7	4.949747	9.9666	0.974928	0.387899	0.252971	0.448799	32	22.627417	35.2387		0.084853	0.055338	0.098175
8	5.656854	11.1820		0.339411	0.221350	0.392699	33	23.334524	36.2006	0.998867	0.082282	0.053661	0.095200
9	6.363961	12.0330	0.984808	0.301699	0.196756	0.349066	34	24.041631	37.2401		0.079861	0.052082	0.092400
10	7.071068	13.1949		0.271529	0.177080	0.314159	35	24.748737	38.2042	0.998993	0.077580	0.050594	0.089760
11	7.778175	14.0749	0.989821	0.246845	0.160982	0.285599	36	25.455844	39.2414		0.075425	0.049189	0.087266
12	8.485281	15.2042		0.226274	0.147567	0.261799	37	26.162951	40.2074	0.999099	0.073386	0.047859	0.084908
13	9.192388	16.1038	0.992709	0.208868	0.136215	0.241661	38	26.870058	41.2425		0.071455	0.046600	0.082673
14	9.899495	17.2114		0.193949	0.126486	0.224399	39	27.577164	42.2103	0.999189	0.069623	0.045405	0.080554
15	10.606602	18.1251	0.994522	0.181019	0.118053	0.209440	40	28.284271	43.2435		0.067882	0.044270	0.078540
16	11.313708	19.2171		0.169706	0.110675	0.196350	41	28.991380	44.2129	0.999266	0.066227	0.043190	0.076624
17	12.020815	20.1413	0.995734	0.159723	0.104165	0.184800	42	29.698485	45.2444		0.064650	0.042162	0.074800
18	12.727922	21.2215		0.150849	0.098378	0.174533	43	30.405592	46.2153	0.999333	0.063146	0.041181	0.073060
19	13.435029	22.1541	0.996584	0.142910	0.093200	0.165347	44	31.112698	47.2453		0.061711	0.040245	0.071400
20	14.142136	23.2253		0.135764	0.088540	0.157080	45	31.819805	48.2175	0.999391	0.060340	0.039351	0.069813
21	14.849242	24.1645	0.997204	0.129300	0.084324	0.149600	46	32.526912	49.2461		0.059028	0.038496	0.068295
22	15.556349	25.2284		0.123422	0.080491	0.142800	47	33.234019	50.2195	0.999442	0.057772	0.037677	0.066842
23	16.263456	26.1731	0.997669	0.118056	0.076991	0.136591	48	33.941125	51.2468		0.056569	0.036892	0.065450
24	16.970563	27.2311		0.113137	0.073783	0.130900	49	34.648232	52.2213	0.999486	0.055414	0.036139	0.064114
25	17.677670	28.1804	0.998027	0.108612	0.070832	0.125664	50	35.355339	53.2474		0.054306	0.035416	0.062832
26	18.384776	29.2334		0.104434	0.068108	0.120830	51	36.062446	54.2230	0.999526	0.053241	0.034722	0.061600
27	19.091883	30.1865	0.998308	0.100566	0.065585	0.116355	52	36.769553	55.2481		0.052217	0.034054	0.060415
28	19.798990	31.2354		0.096975	0.063243	0.112200	53	37.476659	56.2245	0.999561	0.051232	0.033411	0.059275
29	20.506097	32.1918	0.998533	0.093631	0.061062	0.108331	54	38.183766	57.2487		0.050283	0.032793	0.058178
30	21.213203	33.2372		0.090510	0.059027	0.104720	55	38.890873	58.2260	0.999592	0.049369	0.032196	0.057120

1/2P			All Pitches			
Internal and External		External	Internal and External		Internal	External
N	D_b	Measure over Pins	$\cos \frac{90°}{N}$	$\frac{d}{D_b}$	$\frac{t_s}{D}$	$\frac{\pi}{N}$
1	2	3	4	5	6	7
56	39.597980	59.2492		0.048487	0.031621	0.056100
57	40.305086	60.2273	0.999620	0.047637	0.031067	0.055116
58	41.012193	61.2497		0.046815	0.030531	0.054165
59	41.719300	62.2286	0.999646	0.046022	0.030014	0.053247
60	42.426407	63.2502		0.045255	0.029513	0.052360
61	43.133514	64.2297	0.999668	0.044513	0.029030	0.051502
62	43.840620	65.2506		0.043795	0.028561	0.050671
63	44.547727	66.2308	0.999689	0.043100	0.028108	0.049867
64	45.254834	67.2510		0.042426	0.027669	0.049087
65	45.961941	68.2319	0.999708	0.041774	0.027243	0.048332
66	46.669047	69.2514		0.041141	0.026830	0.047600
67	47.376154	70.2328	0.999725	0.040527	0.026430	0.046889
68	48.083261	71.2518		0.039931	0.026041	0.046200
69	48.790368	72.2337	0.999741	0.039352	0.025664	0.045530
70	49.497475	73.2521		0.038790	0.025297	0.044880
71	50.204581	74.2346	0.999755	0.038244	0.024941	0.044248
72	50.911688	75.2525		0.037712	0.024594	0.043633
73	51.618795	76.2354	0.999768	0.037196	0.024258	0.043036
74	52.325902	77.2528		0.036693	0.023930	0.042454
75	53.033009	78.2362	0.999781	0.036204	0.023611	0.041888
76	53.740115	79.2531		0.035728	0.023300	0.041337
77	54.447222	80.2369	0.999792	0.035264	0.022997	0.040800
78	55.154329	81.2534		0.034811	0.022703	0.040277
79	55.861436	82.2376	0.999802	0.034371	0.022415	0.039767
80	56.568542	83.2536		0.033941	0.022135	0.039270
81	57.275649	84.2383	0.999812	0.033522	0.021862	0.038785
82	57.982756	85.2539		0.033113	0.021595	0.038312
83	58.689863	86.2389	0.999821	0.032714	0.021335	0.037851
84	59.396970	87.2541		0.032325	0.021081	0.037400
85	60.104076	88.2395	0.999829	0.031945	0.020833	0.036960
86	60.811183	89.2544		0.031573	0.020591	0.036530
87	61.518290	90.2401	0.999837	0.031210	0.020354	0.036110
88	62.225397	91.2546		0.030856	0.020123	0.035700
89	62.932503	92.2406	0.999844	0.030509	0.019897	0.035299
90	63.639610	93.2548		0.030170	0.019676	0.034907
91	64.346717	94.2411	0.999851	0.029838	0.019459	0.034523
92	65.053824	95.2550		0.029514	0.019248	0.034148
93	65.760931	96.2416	0.999857	0.029197	0.019041	0.033781
94	66.468037	97.2552		0.028886	0.018838	0.033421
95	67.175144	98.2421	0.999863	0.028582	0.018640	0.033069
96	67.882251	99.2554		0.028284	0.018446	0.032725
97	68.589358	100.2426	0.999869	0.027993	0.018256	0.032388
98	69.296464	101.2556		0.027707	0.018069	0.032057
99	70.003571	102.2430	0.999874	0.027427	0.017887	0.031733
100	70.710678	103.2557		0.027153	0.017708	0.031416

Constants

$\pi = 3.141593$

$\cos 45° = 0.707107$

$\text{inv } 45° = 0.214602$

TABLE 54. 10/20 DIAMETRAL PITCH INVOLUTE SERRATIONS

10/20 Diametral Pitch Pressure Angle 45° | Cir Pitch 0.3142, Cir Tooth Thickness 0.1771 | Addendum, External 0.0500, Addendum, Internal 0.0300

Cutting Tool Radius 0.045 | Measuring Pin Diameter, Internal and External 0.1920

Internal and External			Internal				External						
N	Pitch Dia	Base Cir Dia	Major Dia	Minor Dia	TIF Dia	Meas between Pins	Major Dia	Minor Dia	TIF Dia	Measurement over Pins			Ratio E
	Ref	Ref	Min	Min	Min		Basic	Max	Max	Class A	Class B	Class C	
1	2	3	4	5	6	7	8	9	10	11	12	13	14
Recommended Tolerance..			+0.0100 −0.0000	+0.0100 −0.0000		Max	+0.0000 −0.0100	+0.0000 −0.0100		Min	Min	Min	
6	0.6000	0.4243	0.7400	0.5400	0.7100	0.3044	0.7000	0.5000	0.5300	0.9131	0.9144	0.9161	0.872
7	0.7000	0.4950	0.8400	0.6400	0.8100	0.3923	0.8000	0.6000	0.6300	0.9934	0.9948	0.9965	0.884
8	0.8000	0.5657	0.9400	0.7400	0.9100	0.5088	0.9000	0.7000	0.7300	1.1149	1.1162	1.1180	0.893
9	0.9000	0.6364	1.0400	0.8400	1.0100	0.5978	1.0000	0.8000	0.8300	1.2000	1.2013	1.2031	0.901
10	1.0000	0.7071	1.1400	0.9400	1.1100	0.7108	1.1000	0.9000	0.9300	1.3161	1.3174	1.3192	0.908
11	1.1000	0.7778	1.2400	1.0400	1.2100	0.8012	1.2000	1.0000	1.0300	1.4040	1.4054	1.4072	0.914
12	1.2000	0.8485	1.3400	1.1400	1.3100	0.9120	1.3000	1.1000	1.1300	1.5169	1.5183	1.5201	0.919
13	1.3000	0.9192	1.4400	1.2400	1.4100	1.0035	1.4000	1.2000	1.2300	1.6069	1.6083	1.6101	0.924
14	1.4000	0.9900	1.5400	1.3400	1.5100	1.1126	1.5000	1.3000	1.3300	1.7176	1.7190	1.7208	0.928
15	1.5000	1.0607	1.6400	1.4400	1.6100	1.2053	1.6000	1.4000	1.4300	1.8090	1.8104	1.8122	0.931
16	1.6000	1.1314	1.7400	1.5400	1.7100	1.3132	1.7000	1.5000	1.5300	1.9182	1.9196	1.9214	0.935
17	1.7000	1.2021	1.8400	1.6400	1.8100	1.4066	1.8000	1.6000	1.6300	2.0105	2.0119	2.0138	0.938
18	1.8000	1.2728	1.9400	1.7400	1.9100	1.5136	1.9000	1.7000	1.7300	2.1186	2.1200	2.1219	0.941
19	1.9000	1.3435	2.0400	1.8400	2.0100	1.6076	2.0000	1.8000	1.8300	2.2118	2.2132	2.2151	0.943
20	2.0000	1.4142	2.1400	1.9400	2.1100	1.7139	2.1000	1.9000	1.9300	2.3189	2.3203	2.3222	0.945

21	2.1000	1.4849	2.2400	2.0400	2.2100	1.8086	2.2000	2.0000	2.0300	2.4129	2.4143	2.4162	0.948
22	2.2000	1.5556	2.3400	2.1400	2.3100	1.9143	2.3000	2.1000	2.1300	2.5192	2.5206	2.5225	0.950
23	2.3000	1.6263	2.4400	2.2400	2.4100	2.0092	2.4000	2.2000	2.2300	2.6137	2.6151	2.6170	0.951
24	2.4000	1.6971	2.5400	2.3400	2.5100	2.1144	2.5000	2.3000	2.3300	2.7195	2.7209	2.7228	0.953
25	2.5000	1.7678	2.6400	2.4400	2.6100	2.2098	2.6000	2.4000	2.4300	2.8143	2.8157	2.8176	0.955
26	2.6000	1.8385	2.7400	2.5400	2.7100	2.3146	2.7000	2.5000	2.5300	2.9195	2.9210	2.9229	0.956
27	2.7000	1.9092	2.8400	2.6400	2.8100	2.4102	2.8000	2.6000	2.6300	3.0150	3.0154	3.0183	0.958
28	2.8000	1.9799	2.9400	2.7400	2.9100	2.5147	2.9000	2.7000	2.7300	3.1198	3.1212	3.1231	0.959
29	2.9000	2.0506	3.0400	2.8400	3.0100	2.6107	3.0000	2.8000	2.8300	3.2155	3.2169	3.2188	0.960
30	3.0000	2.1213	3.1400	2.9400	3.1100	2.7148	3.1000	2.9000	2.9300	3.3200	3.3214	3.3233	0.961
31	3.1000	2.1920	3.2400	3.0400	3.2100	2.8110	3.2000	3.0000	3.0300	3.4159	3.4174	3.4193	0.962
32	3.2000	2.2627	3.3400	3.1400	3.3100	2.9150	3.3000	3.1000	3.1300	3.5201	3.5216	3.5235	0.963
33	3.3000	2.3335	3.4400	3.2400	3.4100	3.0114	3.4000	3.2000	3.2300	3.6163	3.6178	3.6197	0.964
34	3.4000	2.4042	3.5400	3.3400	3.5100	3.1151	3.5000	3.3000	3.3300	3.7202	3.7217	3.7236	0.965
35	3.5000	2.4749	3.6400	3.4400	3.6100	3.2118	3.6000	3.4000	3.4300	3.8166	3.8181	3.8200	0.966
36	3.6000	2.5456	3.7400	3.5400	3.7100	3.3152	3.7000	3.5000	3.5300	3.9203	3.9218	3.9237	0.967
37	3.7000	2.6163	3.8400	3.6400	3.8100	3.4120	3.8000	3.6000	3.6300	4.0170	4.0185	4.0204	0.968
38	3.8000	2.6870	3.9400	3.7400	3.9100	3.5153	3.9000	3.7000	3.7300	4.1205	4.1220	4.1239	0.968
39	3.9000	2.7577	4.0400	3.8400	4.0100	3.6123	4.0000	3.8000	3.8300	4.2172	4.2187	4.2206	0.969
40	4.0000	2.8284	4.1400	3.9400	4.1100	3.7154	4.1000	3.9000	3.9300	4.3206	4.3221	4.3240	0.970
41	4.1000	2.8991	4.2400	4.0400	4.2100	3.8125	4.2000	4.0000	4.0300	4.4175	4.4190	4.4209	0.971
42	4.2000	2.9698	4.3400	4.1400	4.3100	3.9154	4.3000	4.1000	4.1300	4.5205	4.5220	4.5239	0.971
43	4.3000	3.0406	4.4400	4.2400	4.4100	4.0127	4.4000	4.2000	4.2300	4.6176	4.6191	4.6210	0.972
44	4.4000	3.1113	4.5400	4.3400	4.5100	4.1155	4.5000	4.3000	4.3300	4.7206	4.7221	4.7240	0.972
45	4.5000	3.1820	4.6400	4.4400	4.6100	4.2128	4.6000	4.4000	4.4300	4.8179	4.8194	4.8213	0.973
46	4.6000	3.2527	4.7400	4.5400	4.7100	4.3155	4.7000	4.5000	4.5300	4.9207	4.9222	4.9241	0.973
47	4.7000	3.3234	4.8400	4.6400	4.8100	4.4130	4.8000	4.6000	4.6300	5.0181	5.0196	5.0215	0.974
48	4.8000	3.3941	4.9400	4.7400	4.9100	4.5156	4.9000	4.7000	4.7300	5.1208	5.1223	5.1242	0.974
49	4.9000	3.4648	5.0400	4.8400	5.0100	4.6132	5.0000	4.8000	4.8300	5.2182	5.2197	5.2216	0.975
50	5.0000	3.5355	5.1400	4.9400	5.1100	4.7157	5.1000	4.9000	4.9300	5.3209	5.3224	5.3243	0.975
51	5.1000	3.6062	5.2400	5.0400	5.2100	4.8134	5.2000	5.0000	5.0300	5.4184	5.4199	5.4218	0.976
52	5.2000	3.6770	5.3400	5.1400	5.3100	4.9158	5.3000	5.1000	5.1300	5.5209	5.5224	5.5243	0.976
53	5.3000	3.7477	5.4400	5.2400	5.4100	5.0135	5.4000	5.2000	5.2300	5.6186	5.6201	5.6220	0.977
54	5.4000	3.8184	5.5400	5.3400	5.5100	5.1158	5.5000	5.3000	5.3300	5.7210	5.7225	5.7244	0.977
55	5.5000	3.8891	5.6400	5.4400	5.6100	5.2136	5.6000	5.4000	5.4300	5.8187	5.8202	5.8221	0.977
56	5.6000	3.9598	5.7400	5.5400	5.7100	5.3158	5.7000	5.5000	5.5300	5.9210	5.9225	5.9244	0.978
57	5.7000	4.0305	5.8400	5.6400	5.8100	5.4137	5.8000	5.6000	5.6300	6.0188	6.0203	6.0222	0.978
58	5.8000	4.1012	5.9400	5.7400	5.9100	5.5159	5.9000	5.7000	5.7300	6.1211	6.1226	6.1245	0.979
59	5.9000	4.1719	6.0400	5.8400	6.0100	5.6138	6.0000	5.8000	5.8300	6.2189	6.2204	6.2223	0.979
60	6.0000	4.2426	6.1400	5.9400	6.1100	5.7159	6.1000	5.9000	5.9300	6.3210	6.3225	6.3244	0.979

TABLE 54. 10/20 DIAMETRAL PITCH INVOLUTE SERRATIONS (*Continued*)

10/20 Diametral Pitch Pressure Angle 45° | Cir Pitch 0.3142, Cir Tooth Thickness 0.1771 | Addendum, External 0.0500, Addendum, Internal 0.0300

Cutting Tool Radius 0.045 | Measuring Pin Diameter, Internal and External 0.1920

Internal and External			Internal				External						
N	Pitch Dia	Base Cir Dia	Major Dia	Minor Dia	TIF Dia	Meas between Pins	Major Dia	Minor Dia	TIF Dia	Measurement over Pins			Ratio *E*
	Ref	Ref	Min	Min	Min		Basic	Max	Max	Class A	Class B	Class C	
1	2	3	4	5	6	7	8	9	10	11	12	13	14
Recommended Tolerance..			+0.0100 −0.0000	+0.0100 −0.0000		Max	+0.0000 −0.0100	+0.0000 −0.0100		Min	Min	Min	
61	6.1000	4.3134	6.2400	6.0400	6.2100	5.8139	6.2000	6.0000	6.0300	6.4190	6.4205	6.4224	0.980
62	6.2000	4.3841	6.3400	6.1400	6.3100	5.9159	6.3000	6.1000	6.1300	6.5211	6.5226	6.5245	0.980
63	6.3000	4.4548	6.4400	6.2400	6.4100	6.0140	6.4000	6.2000	6.2300	6.6191	6.6206	6.6225	0.980
64	6.4000	4.5255	6.5400	6.3400	6.5100	6.1159	6.5000	6.3000	6.3300	6.7211	6.7226	6.7245	0.980
65	6.5000	4.5962	6.6400	6.4400	6.6100	6.2141	6.6000	6.4000	6.4300	6.8192	6.8207	6.8226	0.981
66	6.6000	4.6669	6.7400	6.5400	6.7100	6.3160	6.7000	6.5000	6.5300	6.9211	6.9226	6.9245	0.981
67	6.7000	4.7376	6.8400	6.6400	6.8100	6.4142	6.8000	6.6000	6.6300	7.0193	7.0208	7.0227	0.981
68	6.8000	4.8083	6.9400	6.7400	6.9100	6.5160	6.9000	6.7000	6.7300	7.1212	7.1227	7.1246	0.982
69	6.9000	4.8790	7.0400	6.8400	7.0100	6.6142	7.0000	6.8000	6.8300	7.2194	7.2209	7.2228	0.982
70	7.0000	4.9497	7.1400	6.9400	7.1100	6.7160	7.1000	6.9000	6.9300	7.3212	7.3226	7.3246	0.982
71	7.1000	5.0205	7.2400	7.0400	7.2100	6.8143	7.2000	7.0000	7.0300	7.4195	7.4209	7.4229	0.982
72	7.2000	5.0912	7.3400	7.1400	7.3100	6.9160	7.3000	7.1000	7.1300	7.5213	7.5227	7.5247	0.982
73	7.3000	5.1619	7.4400	7.2400	7.4100	7.0144	7.4000	7.2000	7.2300	7.6195	7.6209	7.6229	0.983
74	7.4000	5.2326	7.5400	7.3400	7.5100	7.1160	7.5000	7.3000	7.3300	7.7213	7.7227	7.7247	0.983
75	7.5000	5.3033	7.6400	7.4400	7.6100	7.2144	7.6000	7.4000	7.4300	7.8195	7.8209	7.8229	0.983
76	7.6000	5.3740	7.7400	7.5400	7.7100	7.3161	7.7000	7.5000	7.5300	7.9212	7.9226	7.9246	0.983
77	7.7000	5.4447	7.8400	7.6400	7.8100	7.4146	7.8000	7.6000	7.6300	8.0196	8.0210	8.0230	0.984
78	7.8000	5.5154	7.9400	7.7400	7.9100	7.5162	7.9000	7.7000	7.7300	8.1212	8.1226	8.1246	0.984
79	7.9000	5.5861	8.0400	7.8400	8.0100	7.6146	8.0000	7.8000	7.8300	8.2197	8.2211	8.2231	0.984
80	8.0000	5.6569	8.1400	7.9400	8.1100	7.7162	8.1000	7.9000	7.9300	8.3213	8.3227	8.3247	0.984

10/20 Diametral Pitch Pressure Angle 45°				Cir Pitch 0.3142, Cir Tooth Thickness 0.1771					Addendum, External 0.0500, Addendum, Internal 0.0300				
Cutting Tool Radius 0.045							Measuring Pin Diameter, Internal and External 0.1920						
Internal and External			Internal				External						
N	Pitch Dia	Base Cir Dia	Major Dia	Minor Dia	TIF Dia	Meas between Pins	Major Dia	Minor Dia	TIF Dia	Measurement over Pins			Ratio E
	Ref	Ref	Min	Min	Min		Basic	Max	Max	Class A	Class B	Class C	
1	2	3	4	5	6	7	8	9	10	11	12	13	14
Recommended Tolerance..			+0.0100 −0.0000	+0.0100 −0.0000		Max	+0.0000 −0.0100	+0.0000 −0.0100		Min	Min	Min	
81	8.1000	5.7276	8.2400	8.0400	8.2100	7.8147	8.2000	8.0000	8.0300	8.4197	8.4211	8.4231	0.984
82	8.2000	5.7983	8.3400	8.1400	8.3100	7.9162	8.3000	8.1000	8.1300	8.5213	8.5227	8.5247	0.985
83	8.3000	5.8690	8.4400	8.2400	8.4100	8.0147	8.4000	8.2000	8.2300	8.6198	8.6212	8.6232	0.985
84	8.4000	5.9397	8.5400	8.3400	8.5100	8.1162	8.5000	8.3000	8.3300	8.7213	8.7227	8.7247	0.985
85	8.5000	6.0104	8.6400	8.4400	8.6100	8.2148	8.6000	8.4000	8.4300	8.8199	8.8213	8.8233	0.985
86	8.6000	6.0811	8.7400	8.5400	8.7100	8.3162	8.7000	8.5000	8.5300	8.9213	8.9227	8.9247	0.985
87	8.7000	6.1518	8.8400	8.6400	8.8100	8.4148	8.8000	8.6000	8.6300	9.0199	9.0213	9.0233	0.985
88	8.8000	6.2225	8.9400	8.7400	8.9100	8.5162	8.9000	8.7000	8.7300	9.1214	9.1228	9.1248	0.985
89	8.9000	6.2933	9.0400	8.8400	9.0100	8.6149	9.0000	8.8000	8.8300	9.2200	9.2214	9.2234	0.986
90	9.0000	6.3640	9.1400	8.9400	9.1100	8.7162	9.1000	8.9000	8.9300	9.3214	9.3228	9.3248	0.986
91	9.1000	6.4347	9.2400	9.0400	9.2100	8.8150	9.2000	9.0000	9.0300	9.4200	9.4214	9.4234	0.986
92	9.2000	6.5054	9.3400	9.1400	9.3100	8.9164	9.3000	9.1000	9.1300	9.5213	9.5227	9.5247	0.986
93	9.3000	6.5761	9.4400	9.2400	9.4100	9.0151	9.4000	9.2000	9.2300	9.6200	9.6214	9.6234	0.986
94	9.4000	6.6468	9.5400	9.3400	9.5100	9.1164	9.5000	9.3000	9.3300	9.7213	9.7227	9.7247	0.986
95	9.5000	6.7175	9.6400	9.4400	9.6100	9.2151	9.6000	9.4000	9.4300	9.8200	9.8214	9.8234	0.986
96	9.6000	6.7882	9.7400	9.5400	9.7100	9.3164	9.7000	9.5000	9.5300	9.9213	9.9227	9.9247	0.987
97	9.7000	6.8589	9.8400	9.6400	9.8100	9.4151	9.8000	9.6000	9.6300	10.0201	10.0215	10.0235	0.987
98	9.8000	6.9296	9.9400	9.7400	9.9100	9.5164	9.9000	9.7000	9.7300	10.1214	10.1228	10.1248	0.987
99	9.9000	7.0004	10.0400	9.8400	10.0100	9.6152	10.0000	9.8000	9.8300	10.2201	10.2215	10.2235	0.987
100	10.0000	7.0711	10.1400	9.9400	10.1100	9.7164	10.1000	9.9000	9.9300	10.3214	10.3228	10.3248	0.987

Internal Serrations. The radius of tool given in Table 54 is recommended for use on the broach. The same radius, if used on a shaper cutter, will form a fillet which will not be a true radius and will always be larger than that of the cutter.

Standard Fits (Effective). The effective space width of the internal serration is equal to the dimensional space width minus a portion of the total of all the allowable errors, and likewise the effective tooth thickness of the external serration is equal to the dimensional tooth thickness plus a portion of the total of all the allowable errors.

The effective clearance (positive or negative) between the two serrated parts is equal to the effective space width of the internal serration minus the effective tooth thickness of the external serration. The sign (−) indicates an interference.

DIMENSIONAL AND EFFECTIVE CLEARANCES FOR 25 TEETH*

Diametral Pitch	Class A			Class B			Class C		
	Dimensional Max	Effective Clearance		Dimensional Max	Effective Clearance		Dimensional Max	Effective Clearance	
		Min	Max		Min	Max		Min	Max
10/20	0.0067	0.0005	0.0030	0.0052	−0.0010	0.0015	0.0032	−0.0030	−0.0005
16/32	0.0065	0.0005	0.0030	0.0050	−0.0010	0.0015	0.0030	−0.0030	−0.0005
24/48 Up	0.0059	0.0005	0.0025	0.0044	−0.0010	0.0010	0.0024	−0.0030	−0.0010

* Based on using 60 % of all allowable errors cumulatively, not including lead.

Tolerances. In the cutting of a serrated part, a machine-tool adjustment may vary the depth of cut and therefore the tooth thickness. For any fit, the tolerances are added to the allowable-error total in the calculation of space width and tooth thickness. The total of tolerance and errors provide the range for the manufacture of tools and gages.

The fits and tolerances suggested in this standard are to be considered binding on a manufacturer or seller only when specifically agreed to in writing.

Tapered Serrations. The dimensions in this standard should apply to the big end of the external, or shaft, and the big end of the internal, or fitting. The degree of taper recommended is 0.750 in. per ft included angle, measured on the major diameter (see Fig. 53).

Pin Measurements. For internal serrations minimum pin measurements are not specified because they should not be the basis for rejection as long as effective size, controlled by gages, is not below the specified minimum. When it is not feasible to check effective sizes a minimum pin measurement may be used equal to the maximum pin measurement minus the machining tolerance (Table 52) times factor F.[1] This minimum pin measurement may, however, be reduced as allowable errors are reduced.

[1] The factors E and F refer to the ratio of change of over pin measurement to the change of arc tooth thickness, and is $\cos \phi / \sin \phi_e$ for the external serration, and $\cos \phi / \sin \phi_i$ for the internal serration (see Fig. 51).

For external serrations, maximum pin measurements are not specified for the same reasons as above. A maximum pin measurement may be used equal to the minimum pin measurement plus the machining tolerance times factor E.

Formulas for Pin Measurement of Serrations

External Serration

ϕ = pressure angle at D (45°)
ϕ_e = pressure angle at pin center,°
d = pin dia
D_b = base circle dia ($D \cos 45°$)
t = circular tooth thickness

$$\text{inv}\ \phi_e = \text{inv}\ 45° + \frac{d}{D_b} + \frac{t}{D^{(1)}} - \frac{\pi}{N}$$

When N is even, use

$$M_e = D_b^{(1)} \sec \phi_e + d \quad \text{or} \quad \frac{D_b^{(1)}}{\cos \phi_e} + d$$

When N is odd, use

$$M_e = D_b^{(1)} \sec \phi_e \cos \frac{90°}{N} + d \quad \text{or} \quad \frac{D_b^{(1)} \cos 90°/N}{\cos \phi_e} + d$$

Example: Standard dimensions for $N = 20$, $P = 10/20$, class B fit.

$$d = 0.1920$$

Basic tooth thickness	=	0.1771
Allowable errors (See page 4-150)	=	−0.0023
Min tooth thickness	=	0.1748

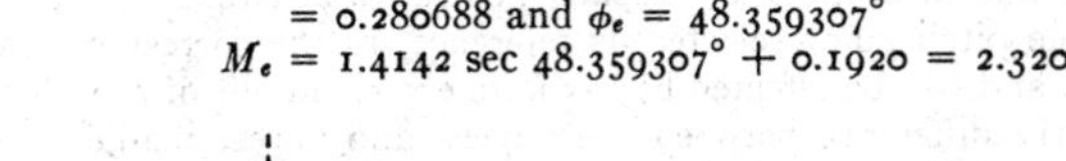

$$\text{inv}\ \phi_e = 0.214602 + \frac{0.1920}{1.4142} + \frac{0.1748}{2.0000} - 0.157080$$

$$= 0.280688 \text{ and } \phi_e = 48.359307°$$

$$M_e = 1.4142 \sec 48.359307° + 0.1920 = 2.3203$$

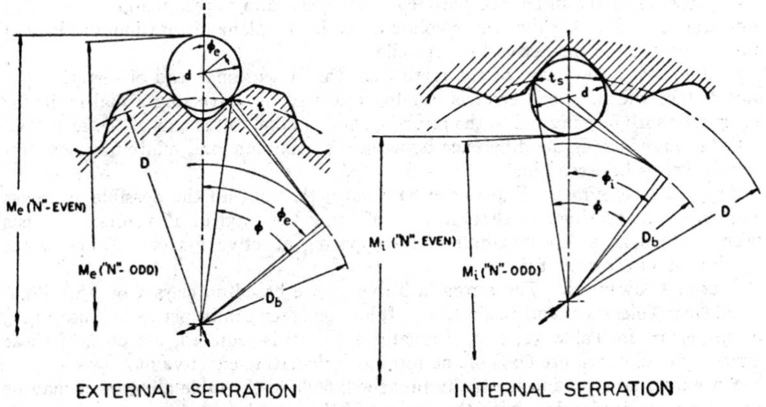

Fig. 51. Pin measurement of involute serrations.

INTERNAL SERRATION

ϕ_i = pressure angle at pin center,°
t_s = circular space width, basic
d = pin dia

$$\text{inv } \phi_i = \text{inv } 45^\circ + \frac{t_s}{D} - \frac{d}{D_b}$$

When N is even, use

$$M_i = D_b^{(1)} \sec \phi_i - d \quad \text{or} \quad \frac{D_b^{(1)}}{\cos \phi_i} - d$$

When N is odd, use

$$M_i = D_b^{(1)} \sec \phi_i \cos \frac{90^\circ}{N} - d \quad \text{or} \quad \frac{D_b^{(1)} \cos 90^\circ/N}{\cos \phi_i} - d$$

EXAMPLE: Standard dimensions for $N = 20$, $P = 10/20$, internal.

$$d = 0.1920$$

Basic space width	=	0.1771
Allowable errors (See page 4-150)	=	+0.0027
Max space width t_s	=	0.1798

$$\text{inv } \phi_i = 0.214602 + \frac{0.1798}{2.0000} - \frac{0.1920}{1.4142}$$
$$= 0.168736 \text{ and } \phi_i = 42.096509^\circ$$
$$M_i = 1.4142 \sec 42.096509^\circ - 0.1920 = 1.7139$$

Allowable Errors

Accumulated pitch error (spacing) is the greatest difference in any two teeth between the actual and theoretical tooth spacing on the same circle. Measurements are taken from one pitch point selected as reference, to the corresponding points on all other teeth, and will be affected by involute error and out of roundness.

Profile error is the difference between the highest and lowest readings of a dial indicator moving along the true involute curve in the plane of rotation and having a finger contacting the active tooth profile.

Lead error is a deviation of the teeth from the dimensional lead of serration per unit of face width. Lead error is usually measured by traversing a dial indicator along the tooth face normal to the pitch line and parallel to the axis of the serration.

Out of roundness is the difference between the maximum and minimum measurements over or between pins.

Application of errors. Experience has shown that not all the possible extremes will occur at one time, so that, instead of using the total of all errors, 60% has been adopted as a fair maximum for computing effective fits (see Table 52 for application of errors to fits).

Special Allowances. The errors in Table 52 are based on class 3 of ASA B6.6-1946, Gear Tolerances and Inspection. If less accuracy can be permitted, use 133% of the errors in Table 52; and, if greater accuracy is required, use 66% of those errors. In all cases, use 60% of the total for calculating effective fits.

Since internal serration is usually broached, 66% of the table allowances may be used for mass production, with the consent of the broach maker.

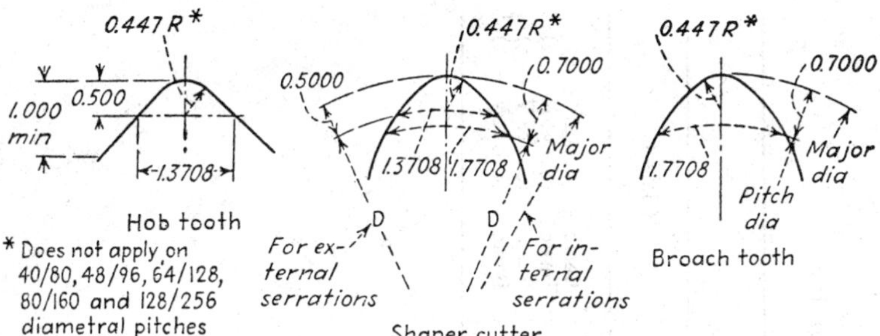

FIG. 52. Form of one/two (1/2) diametral pitch hob, shaper cutter, and broach tooth.

TABLE 55. RECOMMENDED DIMENSIONS OF HOBS FOR INVOLUTE SERRATIONS

Bore	1.2500		0.7500		0.5000		0.3150	
Diametral Pitch	OD	Width	OD	Width	OD	Width	OD	Width
10/20	2.500	2.500	1.750	1.750	1.250	1.250		
16/32	2.500	2.500	1.750	1.750	1.250	1.250		
24/48	2.500	2.500	1.750	1.750	1.250	1.250		
32/64			1.500	1.500	1.125	1.125	0.750	0.500
40/80			1.500	1.500	1.125	1.125	0.750	0.500
48/96			1.500	1.500	1.125	1.125	0.750	0.500
64/128			1.500	1.500	1.125	1.125	0.750	0.500
80/160			1.500	1.500	1.125	1.125	0.750	0.500
128/256			1.500	1.500	1.125	1.125	0.750	0.500

Because of manufacturing restrictions on the total number of teeth in a hob, the length of the hubs may be increased in some cases to effect a reduction in the width of cutting face.

TABLE 56. TAPER SERRATIONS ANGLE A

N	A	N	A	N	A	N	A
6	1°17′	15	1°34′	30	1°40′	50	1°43′
8	1°26′	20	1°37′	35	1°41′	70	1°44′
10	1°28′	25	1°39′	40	1°42′	100	1°45′

Space-tooth Thickness, Formulas. When the three allowable errors of profile, pitch, and out-of-roundness effect on t are added, the result is a gradually increasing amount approximated by the formulas of Table 57, which incorporate 60% of the total amount, excepting lead.

TABLE 57. FORMULAS FOR ERRORS IN TOOTH-SPACE THICKNESS IN 0.0001 IN.
Deduct Calculated Error from Basic Tooth Thickness

Diametral Pitch	Internal	External		
		Class A	Class B	Class C
10/20	Basic + $(0.072N + 26)$	Basic − $(0.06N + 37)$	Basic − $(0.06N + 22)$	Basic − $(0.06N + 2)$
16/32	Basic + $(0.048N + 26)$	Basic − $(0.036N + 37)$	Basic − $(0.036N + 22)$	Basic − $(0.036N + 2)$
24/48	Basic + $(0.036N + 26)$	Basic − $(0.018N + 32)$	Basic − $(0.018N + 17)$	Basic − $(0.018N - 3)$
32/64–128/256	Basic + $(0.024N + 26)$	Basic − $(0.012N + 32)$	Basic − $(0.012N + 17)$	Basic − $(0.012N - 3)$

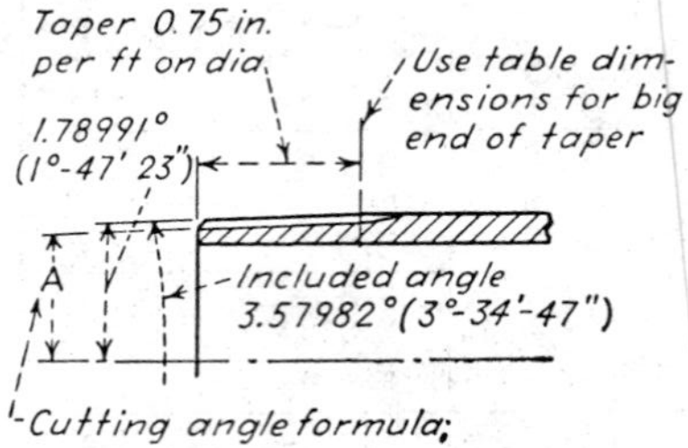

FIG. 53. Tapered serration.

Section 5

GRINDING, HONING, LAPPING, AND SUPERFINISHING

Consultant: E. T. LARSON, Norton Company
Contributors: W. A. CORSE, Carborundum Company
ARTHUR A. CRAFTS CO., INC.
J. C. WILSON and FREDERICK KRAFFT, Thompson Grinder Company

SECTION 5

GRINDING, HONING, LAPPING, AND SUPERFINISHING

GRINDING

Why Grinding Results Vary

Even though grinding is one of the most important metalworking processes, practice has not been reduced to an exact science. It is commonly found that different plants grinding the same or similar parts are forced to use wheels of different specifications. The difference is traceable to the type and condition of the machines employed in the two shops, possibly to some difference in the amount of stock removed, to use of unlike work speeds, and finally to variations in operator skill. In precision grinding, certainly, the operator's training and experience have a very definite effect on the results obtained.

Despite the importance of variables in grinding-wheel selection, the choice of the proper wheel for a given operation is not a matter of guesswork. An understanding of the factors involved will prove helpful to the shop superintendent, methods engineer, foremen, and operator. Grinding-wheel selection is governed by these factors:

Factors in Grinding

Given conditions (see Table 1):

1. Material to be ground.
2. Amount of material to be removed and finish desired.
3. Arc of contact.
4. Type of grinding machine.

Influential variable factors:

1. Wheel speed.
2. Work speed (or pressure if hand grinding).
3. Condition of machine.
4. Skill of the workman.

1. Material to Be Ground. The composite effect of several physical properties of a material to be ground largely governs the selection of the type of abrasive to be used. For materials that are neither brittle nor easily penetrated, aluminum oxide abrasive wheels have been found most suitable. Most materials for which this abrasive is adapted are comparatively high in tensile strength. Therefore, beginning with the hardest and toughest of alloy steels, down to and including the tougher grades of bronze, *aluminum oxide* grinding wheels should be used.

For materials that are hard but very brittle, such as cemented carbides and stone, silicon carbide wheels are more suitable. Cast and chilled irons offer moderate resistance to penetration and are sufficiently brittle compared with hardened steel to make silicon carbide the more effective abrasive for them. Materials of low

TABLE 1. FACTORS THAT INFLUENCE GRINDING-WHEEL SELECTION

Factor	Recommendation
1. Physical properties of material:	
a. Use aluminum oxide grinding wheels for materials of high tensile strength	Carbon steels Alloy steels High-speed steels Annealed malleable iron Wrought iron Tough bronzes Tungsten, etc.
b. Use silicon carbide grinding wheels for materials of low tensile strength	Gray iron Chilled iron Brass and bronze Aluminum and copper Marble Granite Pearl Rubber Leather, etc.
2. Factors affecting the selection of the grit:	
a. Amount of material to be removed	Use coarse wheels for fast removal
b. Finish desired	Use fine grain for fine finish
c. Physical properties of material to be ground	Use coarse grain for ductile materials and finer grain for hard, dense or brittle materials
3. Factors affecting the selection of the grade (degree of hardness):	
a. Physical properties of the material to be ground	Use hard wheels on soft materials and vice versa
b. Arc of contact	The shorter the contact, the harder the wheel should be
c. Wheel speed and work speed	The higher the ratio of wheel speed to work speed the softer the grade should be and vice versa
d. Condition of grinding machine	Machines in poor condition require harder wheels
e. Skill of operator	Skillful operator can use softer wheels than unskilled man Piece-work grinding usually calls for harder wheels than day work
4. Factors affecting the selection of the process:	
a. Dimensions of wheel	Wheels subjected to bending strains should be made by elastic or rubber process Extremely thin abrasive saws must be made by the elastic or rubber process Wheels over 36 in. dia are usually made by the silicate process
b. Rate of cutting	Use vitrified wheels for most rapid cutting at speeds under 6500 sfpm; rubber wheels at higher speed.
c. Finish desired	Use elastic or rubber wheels for highest finish, where rapid production is not a factor Use silicate wheels to replace sandstones on cutlery, etc.

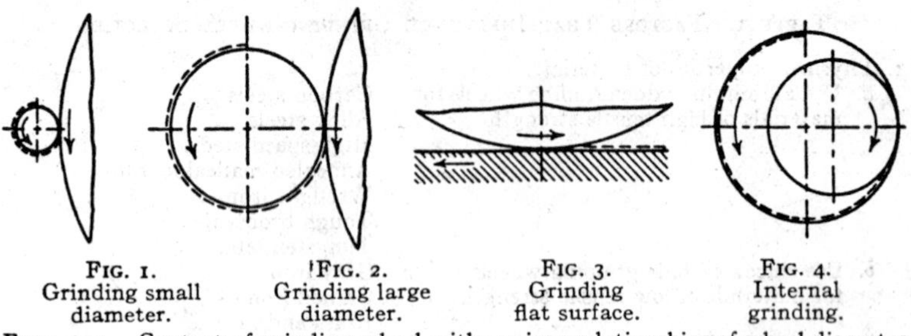

Fig. 1. Grinding small diameter.

Fig. 2. Grinding large diameter.

Fig. 3. Grinding flat surface.

Fig. 4. Internal grinding.

Figs. 1–4. Contact of grinding wheel with various relationships of wheel diameter to work diameter.

tensile strength, such as soft brasses and bronzes, aluminum, and copper, are efficiently ground with *silicon carbide* grinding wheels.

2. Amount of Material to Be Removed. The amount of material to be removed by the grinding wheel and the finish desired influence the selection of the grit size. Where comparatively large amounts of material have to be removed by grinding, the coarser grain combinations should be employed to facilitate rapid removal of stock. This applies particularly to hand grinding operations like snagging castings, where finish is no consideration.

Finish is dependent to a considerable extent on grit size used, except on machine grinding operations. Here, proper truing and dressing afford fine finish without the sacrifice of production usually accompanying the use of fine grit wheels.

3. Arc of Contact. The arc of contact, or the area of contact, has a very important bearing on the grain and grade selection. By the "arc of contact" is meant the length of the arc (measured along the periphery of the wheel) that is in contact with the work. The two extremes of the length of this arc are illustrated by Figs. 1 and 4. The difference in the length of the arc of contact in these two extremes is actually rather small, but the effect on the grinding wheel is great. In the first case, a medium-hard wheel is required, whereas only the softest grades will grind satisfactorily in the second example.

In the case of cup or cylinder wheels, where the grinding is done on the rim, the arc of contact develops into an area of contact and the effect of a change from a small area to a large area is even more marked than with a straight wheel. The extremes here range from the grinding of steel balls, where practically "point contact" exists, to grinding of broad, flat surfaces. To select a wheel suitable to variation in the area of contact, grades F to Z have to be employed.

The "arc of contact" and "area of contact" should not be confused with the "width of contact." By "width of contact" is meant the width of the wheel in contact with the work in the case of cylindrical grinding, or the width of the rim in the case of cup and cylinder wheels. Width of contact has practically no effect on the cutting action of a wheel, provided, that the pressure per unit area remains constant and there is sufficient power in the drive. In other words, there should be no difference in the cutting action of a wheel 1 in. wide and another 4 in. wide, or a cup with 1-in. rim or one with a 2-in. rim, provided that the pressure per unit area remains the same.

4. Type of Grinding Machine. Heavy, rigidly constructed machines can use slightly softer wheels than lighter, more flexible types. Some machines set up greater vibrations than others, thus calling for finer and harder wheels. The combination of speeds and feeds on some precision machines makes the use of different kinds of wheels necessary. Plane-surface grinding machines making use of the rim of a cup or cylinder wheel require much softer wheels than plane-surface machines using the periphery of a straight wheel.

Influential Variable Factors

1. Wheel Speed. Too slow a speed means waste of abrasive without getting much useful work in return, and an excessive speed may retard cutting and if carried too far is dangerous. Run a grinding wheel near the speed recommended by the maker. The grit sizes and grades usually recommended for certain grinding operations are based on the assumption that approximately the recommended speeds will be employed. If, for some reason, these speeds cannot be used, then the grade at least must be changed to suit this condition.

RECOMMENDED WHEEL SPEEDS

	Sfpm
Cylindrical grinding	5,500 to 6,500
Internal grinding	2,000 to 6,000
Snagging, offhand grinding (vitrified)	5,000 to 6,000
Snagging, resinoid wheels	7,500 to 9,500
Surface grinding	4,000 to 5,000
Machine-knife grinding	3,500 to 4,000
Hemming cylinders	2,100 to 5,000
Wet-tool grinding	5,000 to 6,000
Cutlery wheels	4,000 to 5,000
Rubber, shellac, and resinoid cutting-off wheels	9,000 to 16,000*

* The higher speed is recommended only where suitable bearings are employed.

Speed Tables, Rules for Surface Speeds. Table 2 gives the number of rpm at which grinding wheels from 1 to 72 in. dia must be operated to secure peripheral speeds of 4000 to 16,000 fpm.

The exact speed at which any specified wheel should be run depends upon several conditions, such as the type of machine, character of work and wheel, quality of finish desired, and various other factors referred to at other places in this book. Wheels are ordinarily run in practice from about 4000 to 6000 fpm, though in some cases a speed as high as 7500 fpm has been employed. An average speed recommended by most wheelmakers is 5000 fpm. To allow an ample margin of safety, it is recommended that wheel speeds should not exceed 6500 fpm.

Table 3 gives safe speeds for various types of grinding wheels and various types of bonds. These speeds are recommended by the Grinding Wheel Institute and appear in the ASA Safety Code on the subject of grinding wheels.

Table 4 gives minimum spindle diameters for wheels of various sizes operating at speeds up to 7000 fpm.

2. Work Speed. In machine grinding, speed of work exerts a pronounced effect on the grinding wheel. In general, fast work speed, whether the operation be cylindrical, surface, or internal grinding, tends to wear the wheel faster than slower work speed. This is not necessarily a drawback to good grinding practice, but it must be understood to be properly controlled.

Wheel wear is dependent on the ratio of the surface wheel speed to the surface work speed The higher the ratio the less work the wheel is required to do in a given amount of time, hence the wheel naturally wears at a slower rate. If the ratio is

Table 2. Rpm of Grinding Wheels to Secure 4000 to 16,000 Sfpm

Diameter of Wheel in Inches	Peripheral Speed in Feet per Minute																Diameter of Wheel in Inches
	4,000	4,500	5,000	5,500	6,000	6,500	7,000	7,500	8,000	8,500	9,000	9,500	10,000	12,000	14,000	16,000	
	Revolutions per Minute																
1	15,279	17,189	19,098	21,008	22,918	24,828	26,737	28,647	30,558	32,467	34,377	36,287	38,196	45,836	53,474	61,116	1
2	7,639	8,594	9,549	10,504	11,459	12,414	13,368	14,328	15,278	16,238	17,188	18,143	19,098	22,918	26,737	30,558	2
3	5,093	5,729	6,366	7,003	7,639	8,276	8,913	9,549	10,186	10,822	11,459	12,115	12,732	15,278	17,826	20,372	3
4	3,820	4,297	4,775	5,252	5,729	6,207	6,685	7,162	7,640	8,116	8,595	9,072	9,549	11,459	13,368	15,278	4
5	3,056	3,438	3,820	4,202	4,584	4,966	5,348	5,730	6,112	6,494	6,876	7,258	7,640	9,168	10,696	12,224	5
6	2,546	2,865	3,183	3,501	3,820	4,138	4,456	4,775	5,092	5,411	5,729	6,048	6,366	7,639	8,913	10,186	6
7	2,183	2,455	2,728	3,001	3,274	3,547	3,820	4,092	4,366	4,538	4,911	5,183	5,456	6,548	7,640	8,732	7
8	1,910	2,148	2,387	2,626	2,865	3,103	3,342	3,580	3,820	4,058	4,297	4,535	4,775	5,729	6,685	7,640	8
10	1,528	1,719	1,910	2,101	2,292	2,483	2,674	2,865	3,056	3,247	3,438	3,629	3,820	4,584	5,348	6,112	10
12	1,273	1,432	1,591	1,751	1,910	2,069	2,228	2,386	2,546	2,705	2,864	3,023	3,183	3,820	4,456	5,092	12
14	1,091	1,228	1,364	1,500	1,637	1,773	1,910	2,046	2,182	2,319	2,455	2,592	2,728	3,274	3,820	4,366	14
16	955	1,074	1,194	1,313	1,432	1,552	1,672	1,791	1,910	2,029	2,149	2,268	2,387	2,865	3,342	3,820	16
18	849	955	1,061	1,167	1,273	1,379	1,485	1,591	1,698	1,803	1,910	2,016	2,122	2,546	2,970	3,396	18
20	764	859	955	1,050	1,146	1,241	1,337	1,432	1,528	1,623	1,719	1,814	1,910	2,292	2,674	3,056	20
22	694	781	868	955	1,042	1,128	1,215	1,302	1,388	1,476	1,562	1,649	1,736	2,084	2,430	2,776	22
24	637	716	796	875	955	1,034	1,115	1,194	1,274	1,353	1,433	1,512	1,591	1,910	2,228	2,546	24
26	588	661	734	808	881	955	1,028	1,101	1,176	1,248	1,322	1,395	1,468	1,762	2,056	2,352	26
28	546	614	682	750	818	887	955	1,023	1,092	1,159	1,228	1,296	1,364	1,637	1,910	2,182	28
30	509	573	637	700	764	828	891	955	1,018	1,082	1,146	1,210	1,274	1,528	1,782	2,036	30
32	477	537	597	656	716	776	836	895	954	1,014	1,074	1,134	1,194	1,432	1,672	1,910	32
34	449	505	562	618	674	730	786	843	898	955	1,011	1,067	1,124	1,348	1,572	1,796	34
36	424	477	530	583	637	690	742	795	848	902	954	1,007	1,061	1,273	1,484	1,698	36
38	402	452	503	553	603	653	704	754	804	854	904	955	1,006	1,206	1,408	1,608	38
40	382	430	478	523	573	620	669	716	764	812	860	908	956	1,146	1,338	1,528	40
42	366	409	454	500	545	591	636	682	732	775	818	863	908	1,090	1,272	1,464	42
44	347	390	434	478	521	564	608	651	694	737	780	824	868	1,042	1,216	1,388	44
46	333	375	416	458	500	541	582	624	666	708	750	791	832	1,000	1,164	1,332	46
48	318	358	398	438	478	517	558	597	636	676	716	756	796	956	1,116	1,272	48
53	288	324	360	395	432	468	503	539	576	612	648	683	720	864	1,006	1,152	53
60	255	287	319	350	387	414	446	478	510	542	574	606	638	774	892	1,020	60
72	212	239	265	291	318	345	371	398	424	451	477	504	530	637	742	849	72

Note: Centrifugal force, which is the force that tends to rupture a given wheel when overspeeding, increases as the square of the velocity of that wheel. For example, the centrifugal force in a wheel running at 5,500 sfpm is 49% greater than in the same wheel running at 4,500 sfpm, although the speed is actually only 22% greater.

TABLE 3. "SAFE" SPEEDS FOR GRINDING WHEELS
From American Standards Association Safety Code

Classification Number	Types of Wheels	Vitrified and Silicate Bonds			Organic Bonds		
		Low Strength	Medium Strength	High Strength	Low Strength	Medium Strength	High Strength
		FPM	FPM	FPM	FPM	FPM	FPM
1	Type 1—Straight Wheels (Including plate mounted and inserted nut wheels) Type 4—Taper Wheels	5,500	6,000	6,500	6,500	8,000	9,500
2	Types 5 and 7—Recessed Wheels	5,500	6,000	6,500	6,500	8,000	9,500
3	Type 2—Cylinder Wheels (Including plate mounted and inserted nut wheels)	4,500	5,500	6,000	6,000	8,000	9,500
4	Dovetail Wheels Types 11 and 12—Dish and Flaring Cup Wheels Type 13—Saucer Wheels	4,500	5,500	6,000	6,000	8,000	9,500
5	Type 6—Deep Recessed Cup Wheels	4,500	5,000	5,500	6,000	7,500	9,000
6	Cutting Wheels Larger than 16" diameter						7,500 to 14,000
7	Cutting Wheels 16" and smaller						10,000 to 16,000‡
8	Thread Grinding Wheels	5,500 to 8,000	6,000 to 10,000	6,500 to 12,000			9,500 to 12,000
9	Automotive and Aircraft Crank Grinding	5,500	6,000 to 7,300	6,500 to 8,500			
10	Automotive and Aircraft Cam Grinding	5,500	6,000 to 8,000	6,500 to 8,500			
11	Diamond Wheels (all types)	Any Bond—Maximum 6,500 fpm					

NOTE: When wheels of unusual and extreme shapes such as deep cups with thin walls or backs, long drums, or wheels with large center holes are required, consult wheel manufacturer for speeds recommended.

Maximum speeds indicated are based on the strength of the wheels and not on their cutting efficiency. Best speeds may sometimes be considerably lower.

decreased by increasing the work speed, the wheel will be required to do more work in a given time and will wear faster.

In general, the longer the arc of contact in precision grinding operations, the faster should be the speed of the work, in order to have the wheel cut properly.

In hand grinding, the rate at which the work is forced against the wheel or the method of applying the work exerts an influence on grade selection. The harder the grinding wheel is forced, the harder should be the wheel if abrasive economy is to be expected.

3. **Condition of the Grinding Machine.** A grinding wheel is a refined cutting tool and cannot work to good advantage on a machine in poor repair or not properly set up. Spindles loose in their bearings necessitate the use of very much harder wheels than would be used under normal conditions. Insecure or shaky foundations cause no end of trouble in grinding, because hard grades must be employed to overcome the tendency of the wheel to wear rapidly. Thus cutting qualities are sacrificed.

4. **Skill of the Workman.** It has been found on hand grinding that costs vary widely on the same work and with the same kind of a machine in the same factory, owing to the difference in the men's methods of applying the work to the wheel. Even in machine grinding, the operator can vary the results and grinding costs by his methods of handling the machine.

TABLE 4. MINIMUM SPINDLE DIAMETERS FOR WHEEL SPEEDS TO 7000 SFPM

Diameter of Wheel, Inches	Thickness of Wheel, Inches																		
	1/4	3/8	1/2	5/8	3/4	1	1 1/4	1 1/2	1 3/4	2	2 1/4	2 1/2	2 3/4	3	3 1/4	3 1/2	4	4 1/2	5
	Diameter of Spindle, Inches																		
6	1/2	1/2	1/2	1/2	1/2	1/2	5/8	5/8	3/4	3/4	3/4	3/4	3/4	3/4	3/4	3/4	1	1	1
7	1/2	1/2	1/2	1/2	5/8	5/8	5/8	3/4	3/4	3/4	3/4	3/4	3/4	1	1	1	1	1	1
8	5/8	5/8	5/8	5/8	5/8	5/8	3/4	3/4	3/4	1	1	1	1	1	1	1	1 1/4	1 1/4	1 1/4
9	5/8	5/8	5/8	5/8	3/4	3/4	3/4	3/4	1	1	1	1	1	1 1/4	1 1/4	1 1/4	1 1/4	1 1/4	1 1/4
10	3/4	3/4	3/4	3/4	3/4	3/4	3/4	3/4	1	1	1	1 1/4	1 1/4	1 1/4	1 1/4	1 1/4	1 1/4	1 1/2	1 1/2
12	3/4	3/4	3/4	3/4	3/4	1	1	1	1	1	1	1 1/4	1 1/4	1 1/4	1 1/4	1 1/4	1 1/2	1 1/2	1 1/2
14	7/8	7/8	7/8	7/8	1	1	1 1/4	1 1/4	1 1/4	1 1/4	1 1/4	1 1/4	1 1/4	1 1/2	1 1/2	1 1/2	1 1/2	1 1/2	1 1/2
16	..	..	..	..	1 1/4	1 1/4	1 1/4	1 1/4	1 1/4	1 1/4	1 1/2	1 1/2	1 1/2	1 1/2	1 1/2	1 3/4	1 3/4	1 3/4	1 3/4
18	..	..	..	..	1 1/4	1 1/4	1 1/4	1 1/2	1 1/2	1 1/2	1 1/2	1 1/2	1 1/2	1 1/2	1 3/4	1 3/4	1 3/4	1 7/8	1 7/8
20	..	..	..	..	...	1 1/2	1 1/2	1 1/2	1 1/2	1 1/2	1 1/2	1 1/2	1 3/4	1 3/4	1 3/4	1 7/8	1 7/8	1 7/8	1 7/8
24	..	..	..	..	...	1 1/2	1 1/2	1 1/2	1 3/4	1 3/4	1 3/4	1 3/4	1 3/4	1 3/4	2	2	2	2	2
26	..	..	..	..	...	...	1 1/2	1 1/2	1 3/4	1 3/4	1 3/4	1 3/4	2	2	2	2	2 1/4	2 1/4	2 1/4
30	..	..	..	..	...	...	...	1 3/4	1 3/4	2	2	2	2	2	2 1/4	2 1/4	2 1/2	2 1/2	2 1/2
36	..	..	..	..	...	...	...	...	2	2 1/4	2 1/4	2 1/4	2 1/2	2 1/2	2 1/2	2 3/4	2 3/4	3	3

NOTE: For speeds exceeding 7000 sfpm, the spindle sizes shown are usually not adequate. Wheels larger than specified by the machine manufacturer should not be used on a given machine.

Abrasives in Use

Slight variation in grain or grade of the wheel may affect the grinding results to a pronounced degree. Therefore, data pertaining to abrasives and wheels are of first importance in dealing with the subject of grinding.

Commercial Abrasives. Emery, corundum, silicon carbide, aluminum oxide, crushed diamond, and boron carbide are the commercial abrasive materials commonly used today. They vary in hardness and toughness. It does not follow that the hardest abrasive is always the best for a job; the shape and fracture characteristics of the abrasive must also be taken into consideration.

Hardness of an abrasive may be thought of as scratch hardness. Toughness, as used here, is the ability to resist impact. By fracture characteristics are meant the condition of the surface and points after breaking off particles, as well as resistance to fracture.

Emery is a form of corundum found in nature with a variable percentage of impurities. Its use today is confined largely to abrasive paper and cloth.

Corundum is an oxide of aluminum of a somewhat variable purity, according to the locality in which it is mined; its fracture is conchoidal and generally crystalline.

Silicon carbide abrasive is a product of the electric furnace; it breaks with a sharp crystalline fracture. Carborundum and Crystolon are two well-known trade names for this abrasive.

Alundum and Aloxite are also electric-furnace abrasives, being fused oxide of

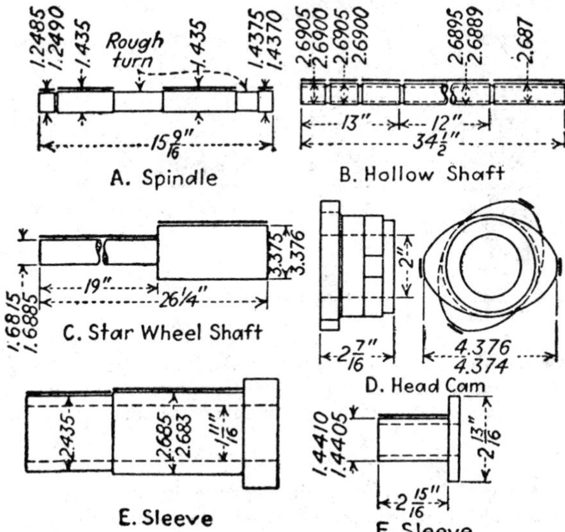

FIG. 5. Examples of commercial grinding. Stock removal ranges from 0.062 to 0.015 in., depending on the part. The depth of cut per pass for hardened work is about 0.002 in., and 0.003 to 0.004 in. for soft work. Light cuts and fast table traverse constitute good grinding practice.

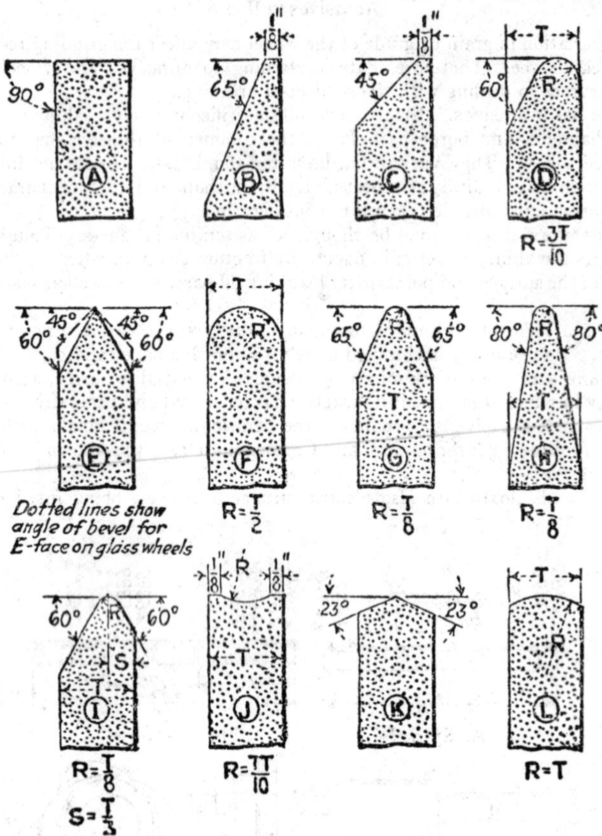

FIG. 6. Straight-wheel types can be furnished with any of the above standard wheel faces.

aluminum. They are of uniform quality, of about 98% purity, breaking with a sharp, crystalline fracture and having unusual toughness or point endurance.

Grinding wheels are usually made of aluminum oxide or silicon carbide.

Selecting the Abrasive

Choosing the proper abrasive for a grinding wheel is not always easy. Both the silicon carbide and aluminum oxide abrasives have their own fields of application, and there are some jobs where choice is a matter of personal opinion. The characteristics of the two types of abrasives vary somewhat with respect to hardness and toughness.

Silicon carbide is both harder and tougher (higher impact strength) than aluminum oxide. It is better adapted for grinding very hard materials, such as cemented carbides, stone, and ceramic materials, which will dull aluminum oxide abrasives.

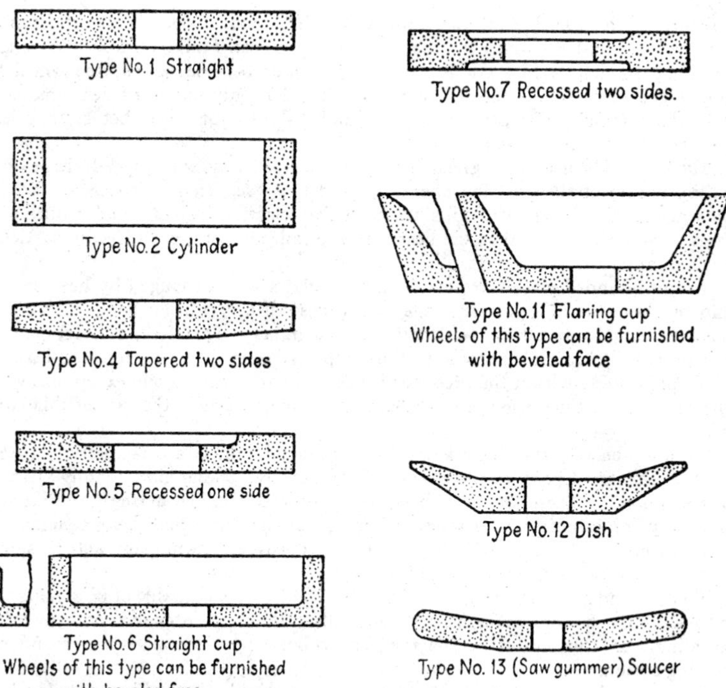

FIG. 7. Standard types of grinding wheels.

Metals and other materials that are of low tensile strength are best ground with silicon-carbide abrasive wheels.

A list of materials suggested for grinding with the two types of abrasives are:

Silicon Carbide	*Aluminum Oxide*
Gray and chilled iron	Carbon steels
Brass and soft bronze	Alloy steels
Aluminum and copper	High-speed steels
Marble and other stone	Annealed malleable iron
Rubber and leather	Wrought iron
Very hard alloys	Hard bronzes
Cemented carbides	
Unannealed malleable iron	

Functions of the Bond

Every grinding wheel has two components: abrasive that does the actual cutting, and bond that holds the abrasive grains while they cut. Cutting efficiency of a wheel depends largely upon the type of abrasive. The grade or hardness of a wheel depends on the relative percentage of bond used.

Functions of the bond are: (1) to hold the abrasive grains together, (2) to provide the proper factor of safety at running speed, and (3) to modify the hardness or strength of the wheel to fit the work it is called upon to do. We often hear the operator say that his wheel is "too hard" or "too soft." He actually means this: the bond retains cutting grains so long they have become dulled; the wheel has stopped cutting. Or, in the case of a soft wheel, the bond is of insufficient amount to hold the cutting grains firmly; they are pulled out of the wheel before they have performed the required work.

The bond to be used for a given operation depends on the wheel speed, shape of the wheel, nature of the grinding operation, and finish required. Wheels are bonded by what are known as vitrified, silicate, shellac (elastic), resinoid, and rubber processes. No one bond makes the best wheel for all purposes; each has its particular fields of application.

Vitrified-bonded wheels are made of fused clays, are unchanged by heat or cold, and can be made in a greater range of hardness than any other bond. This bond does not completely fill the voids between the grains. Therefore, a wheel made by this process, being more porous than any other type, is adaptable for practically all kinds of grinding within the prescribed speed range. There is one exception: where the wheel is not thick enough to withstand side pressure as in the case of thin cutoff wheels.

Silicate wheels derive their name from the bond, which is principally silicate of soda. This bond permits the release of the abrasive grains more readily than the vitrified bond. Therefore, the wheels are considered "milder acting" and are used primarily for grinding edge tools. They are important where heat generated in grinding must be kept at a minimum. For practical reasons in manufacture, silicate bond is used in making very large solid wheels.

Shellac bond is one of the so-called organic bonds. It is capable of producing high finishes on such work as camshafts and rolls. It is not intended for heavy-duty grinding. In thin cutoff wheels the shellac bond provides a very fast and cool cutting action.

To make rubber-bonded wheels, pure rubber is mixed with sulfur as a vulcanizing agent. Rubber-bonded cutoff wheels can be made as thin as 0.006 in. for slitting pen-point nibs.

Resinoid-bonded snagging wheels, operating at 9500 sfpm, are being used on a wide scale for high-speed snagging of steel castings and for billet and weld grinding. Resinoid saw-gumming and cutting-off wheels are also widely used.

Grain and Grade. Grinding wheels are made in various combinations of coarseness and hardness to meet a variety of requirements. The abrasive material is crushed and screened from coarse to fine in many sizes designated by numbers. For example, No. 20 grain means a size that will just pass through a screen having 20 meshes to the linear inch.

The term "grade" refers to the relative hardness of the wheel, or the resistance of the cutting grains to tearing loose under grinding pressure. A wheel from which the grains are easily torn loose, causing it to wear rapidly, is called "soft," while one which retains the grains longer is called "hard." In general, with a given type of bond, it is the *amount* of bond that determines the wheel's hardness or grade.

Grinding-wheel Markings

A standard system for marking grinding wheels and other bonded abrasives (except diamond wheels and sharpening stones) is followed by grinding-wheel manufacturers (see Fig. 8). The marking consists of six parts: (1) abrasive type; (2) grain size; (3) grade of hardness or strength of bond; (4) structure, or spacing of abrasive grains; (5) kind of bond; and (6) manufacturer's record. Details are:

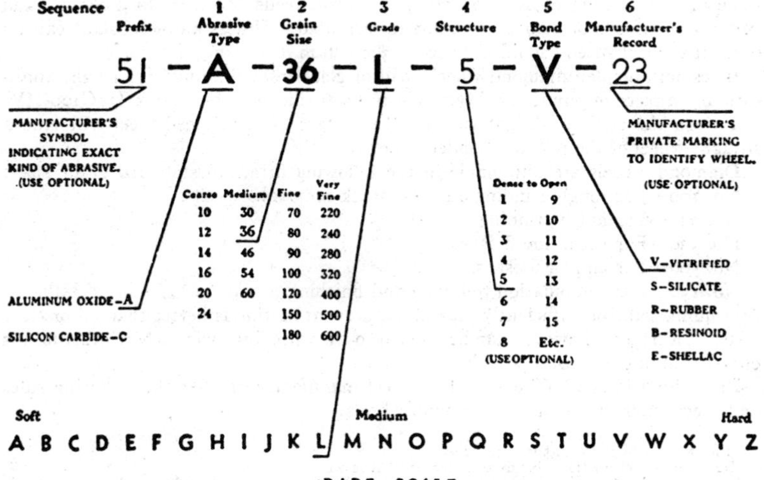

FIG. 8. Standard marking system for grinding wheels.

Type of Abrasive. Abrasives naturally fall into two groups: the aluminum oxide group and the silicon carbide group. Letter symbols are used to identify these two groups, as follows:

A—Aluminum oxide.

C—Silicon carbide.

Grain Size. Grain size is indicated by a number. The following list (from coarse to fine) includes all the ordinary grain sizes commonly used: 10, 12, 14, 16, 20, 24, 30, 36, 46, 54, 60, 70, 80, 90, 100, 120, 150, 180, 220.

The following additional sizes are occasionally used: 240, 280, 320, 400, 500, 600.

Grade. The grade is indicated by a letter of the alphabet, A to Z, soft to hard, in all bonds or processes.

Structure. Numbers from 1 to 15 indicate progressively wider grain spacing (sometimes called "more open" structure).

Bond or Process. The bond or process is designated by the following letters:

V—Vitrified.

S—Silicate.

E—Shellac or elastic.

R—Rubber.

B—Resinoid (synthetic resins).

Manufacturer's Record. This is usually a symbol indicating a modification or a particular type of a standard bond. For example, in a standard Norton Co. wheel marking, the symbol BE in the sixth position denotes a Norton BE type of vitrified bond.

Diamond Abrasive Wheels

Diamond wheels are made of diamond (bort) particles held in either a resinoid, vitrified, or metallic bond. They are recommended for cemented carbides, glass, ceramics, and stone. Surface speeds of 4500 to 6000 fpm are recommended with table or work travel of 100 to 500 ipm. Cross-feeds of 0.030 to 0.060 in. and vertical or infeeds of 0.00025 to 0.002 in. are used. Unless ample coolant can be used, the wheel speed should not exceed 3000 sfpm.

If resinoid-bonded diamond wheels load or glaze from running dry, a light application of a piece of pumice or a very fine and soft silicon carbide stick (37C400-HV) will clean the surface. Light cuts on the surface of high-speed steel will true a straight wheel of the resinoid-bonded type.

Diamond wheels are obtainable in the following principal standard grain sizes:

No. 100—For rough grinding and fast stock removal.
No. 220—For finish grinding.
No. 320—For extra-fine finishes.
No. 500—For lapped finishes.
No. 150—For combination roughing and finishing.

A grinding solution sufficiently alkaline to prevent rusting is better than oil or clear water. Kerosene, however, can be applied to the wheel by a wick where other means of wet grinding are not available.

Rules for Diamond Wheels. The wheel manufacturers offer the following rules and precautions in the use of diamond wheels:

Mount the wheel so that it runs true.
Keep the wheel on its individual collet until it is worn out.
Undercut the steel shank on carbide-tipped tools with silicon carbide or aluminum oxide abrasive wheels before grinding the carbide with a diamond wheel.
Be sure the grinding face of the wheel is lubricated at all times.
Dress the wheel if it should become loaded or glazed.
Do not attempt to true the wheel with a diamond.
Do not operate the wheel dry unless unavoidable.
Do not grind steel or other relatively soft, tough materials except for a light finishing cut.

Vitrified and metal-bonded diamond wheels should be dressed, whenever necessary, with silicon carbide sticks like the sample packed with the wheel.

One method of truing a peripheral type of diamond wheel is by cylindrically grinding the periphery with a silicon carbide, vitrified-bonded abrasive wheel of soft grade and about 80 or 90 grit size. Cup and dish wheels can be trued and dressed at the same time by hand lapping the diamond face on a cast-iron lap, using about 100-mesh silicon carbide with water as a vehicle. Truing of diamond wheels should always be done with a copius flow of water as coolant. In use, a diamond wheel may become glazed or loaded with shank steel. If this occurs, a piece of pumice or a very fine and soft silicon carbide stick may be used to dress or clean the wheel.

Grinding-wheel Recommendations

It is impossible to recommend grinding wheels for all possible operating conditions, because an encyclopedia would be required. The grit size and the grade or hardness of the wheel are the two elements of the wheel specification that may be changed somewhat to suit specific work and machine conditions. When a new job is involved, the operator, using the accompanying Table 5, pages 5-15 to 5-30, draws upon his experience and is generally able to narrow down his selection to one or two specifications.

Then, if the wheel is not exactly right, he copes with the situation by changing the coarseness of diamond dressing, the work speed, rate of table travel, etc., until best grinding conditions are achieved. Table 5 should be adequate as a guide in selecting a wheel where the machine and work conditions represent the average. In fact, they are based upon successful production grinding in industry.

(Continued on page 5-30)

TABLE 5. NORTON GRINDING-WHEEL RECOMMENDATIONS

Work—Material —Operation	Abrasive	Grain Size	Grade	Structure	Bond	Bonding Variation	Abrasive (Trade Mark)	Bonding* Process
Alnico:								
Offhand	57A	36	N	5	V	BE	Alun.	Vit.
Cylindrical	57A	54	L	5	V	BE	Alun.	Vit.
Surfacing	32A	60	H	8	V	BE	Alun.	Vit.
Cutting-off (wet) up to 7500 sfpm	A	80	K	0	R	30	Alun.	Rub.
Aluminum:								
Cylindrical	37C	364	K	5	V	....	Cryst.	Vit.
Surfacing (cups and cylinders)	37C No. 8 treated	24	J	8	V	....	Cryst.	Vit.
Surfacing (disks)	37C	20	K	4	B	....	Cryst.	Res.
Internal	37C	36	K	5	V	....	Cryst.	Vit.
Floor stands 5000–6500 sfpm	37C No. 12 treated	24	O	5	V	....	Cryst.	Vit.
Floor stands 7000–9500 sfpm	37C No. 12 treated	24	P	4	B	5	Cryst.	Res.
Portable grinders 5000–6500 sfpm	37C No. 12 treated	24	O	5	V	....	Cryst.	Vit.
Portable grinders 7000–9500 sfpm	37C No. 12 treated	30	O	4	B	5	Cryst.	Res.
Cutting-off (dry) 9000–16,000 sfpm	A	245	Q	8	B	....	Alun.	Res.
Cutting-off (wet) 7500–12,000 sfpm	A	46	V	8	R	30	Alun.	Rub.
Armatures (laminations):								
Cylindrical	A	36	L	5	V	BE	Alun.	Vit.
Internal	A	36	J	5	V	BE	Alun.	Vit.
	57A	46	J	12	V	BEP	Alun.	Vit.
Asbestos:								
Cutting-off	37C	24	M	8	B	....	Cryst.	Res.
Ax forgings:								
Siding	A	36	T	5	V	BE	Alun.	Vit.
Edging (floor stands) 5000–6500 sfpm	A	24	R	5	V	....	Alun.	Vit.
Edging (floor stands) 9500 sfpm	A	30	Q	5	B	H	Alun.	Res.
Axles (auto):								
Centerless	A	46	M	5	V	BE	Alun.	Vit.
Cylindrical	A	54	L	5	V	BE	Alun.	Vit.
Axles (railway):								
Cylindrical	A	46	L	5	V	BE	Alun.	Vit.
Balls steel (soft—large):								
Roughing	A	46	Z1	7	V	....	Alun.	Vit.
Balls (soft—small):								
Roughing	A	60	Z2	8	V	....	Alun.	Vit.
Balls (soft—large):								
Semifinishing	A	80	Z2	9	V	....	Alun.	Vit.
Balls (soft—small):								
Semifinishing	A	180	Z6	13	V	....	Alun.	Vit.
Balls (hard—large):								
Final finishing	37C	240F	X	10	V	....	Cryst.	Vit.

* Abbreviations:
Vit.—vitrified Res.—resinoid Met.—metal Alun.—Alundum
Rub.—rubber Sil.—silicate Shel.—shellac Cryst.—Crystolon
Diam.—Diamond

TABLE 5. NORTON GRINDING-WHEEL RECOMMENDATIONS (*Continued*)

Work—Material —Operation	Abrasive	Grain Size	Grade	Structure	Bond	Bonding Variation	Abrasive (Trade Mark)	Bonding* Process
Balls (hard—small):								
Final finishing	37C	XF	Z4	11	V	....	Cryst.	Vit.
Ball bearings:								
Surfacing cups and cones —soft	32A	46	I	8	V	BE	Alun.	Vit.
Surfacing cups and cones —hard	32A	80	F		S	....	Alun.	Sil.
Surfacing, segments	32A	70	F	12	V	BEP	Alun.	Vit.
Grind OD cups—roughing and finishing	A	80	M	5	V	BE	Alun.	Vit.
Grind outer race:								
Wheel dia ¼–⅞ in	A	901	T	14	R	6	Alun.	Rub.
	A	150	R	12	R	3	Alun.	Rub.
Wheel dia 1–3 in	1	1001	R	10	R	4	Alun.	Rub.
	A	1001	R	6	R	10	Alun.	Rub.
3 in. up	A	1001	R	2	R	3	Alun.	Rub.
Grind inner race	A	1201	R	6	R	6	Alun.	Rub.
	A	1801	R	6	R	3	Alun.	Rub.
Internal grind bore	A	60	M	5	V	BE	Alun.	Vit.
Billets (alloy and high-speed steels):								
(Swing frames) 7000–9500 sfpm	A	123	R	4	B	H	Alun.	Res.
(Swing frames) 5000–6500 sfpm	A	16	Q	5	V	BE	Alun.	Vit.
Billets:								
Portable grinders 5000–6500 sfpm	A	20	Q	5	V	BE	Alun.	Vit.
Portable grinders 7000–9500 sfpm	A	20	Q	4	B	H	Alun.	Res.
	A	20	T	6	R	30	Alun.	Rub.
Billets and slabs (stainless steels):								
(Swing frames) 7000–9500 sfpm	A	14	V	8	R	16	Alun.	Rub.
	A	14	S	4	B	H	Alun.	Res.
(Portable grinders) 7000–9500 sfpm	A	20	T	2	R	23	Alun.	Rub.
	A	20	Q	4	B	H	Alun.	Res.
Bits (auger):								
Grinding throats	A	36	R	6	R	2	Alun.	Rub.
Fluting	A	70	R	6	R	2	Alun.	Rub.
Bolts (casehardened steel):								
Cylindrical	A	60	L	5	V	BE	Alun.	Vit.
Centerless	A	60	N	5	V	BE	Alun.	Vit.
Boron carbide (norbide):								
Cylindrical:								
Roughing	D	100	L	100	V	....	Diam.	Vit.
Semifinishing	D	220	J	100	B	....	Diam.	Res.
Finishing	D	320	J	100	B	....	Diam.	Res.
	D	400	J	100	B	....	Diam.	Res.
Cutting-off	D	100	N	100	B	....	Diam.	Res.
Internal grinding:								
Roughing	D	100S	L	100	V	....	Diam.	Vit.
Finishing	D	320	N	100	V	....	Diam.	Vit.
Offhand grinding:								
Roughing	D	100	J	100	B	....	Diam.	Res.
Finishing	D	320	J	100	B	....	Diam.	Res.
Surface grinding:								
Roughing	D	100	J	100	B	....	Diam.	Res.
Finishing	D	320	J	100	B	....	Diam.	Res.
Brake drums (automotive):								
Regrinding	19A	46	K		V	....	Alun.	Vit.
Brake-lining (woven):								
Surfacing—(disks)	37C	20	J	4	B	L	Cryst.	Res.

TABLE 5. NORTON GRINDING-WHEEL RECOMMENDATIONS (*Continued*)

Work—Material —Operation	Abrasive	Grain Size	Grade	Structure	Bond	Bonding Variation	Abrasive (Trade Mark)	Bonding Process
Brake lining (molded):								
Surfacing—(disks) roughing	37C	16	M	4	B	....	Cryst.	Res.
Surfacing—(disks) finishing	37C	46	L	5	B	....	Cryst.	Res.
Cutting-off (dry)	37C	30	O	8	B	....	Cryst.	Res.
Brake shoes (railroad) chilled iron:								
Floor stands 5000–6500 sfpm	37C	20	R	5	V	....	Cryst.	Vit.
Floor stands 9000–9500 sfpm	57A	16	Q	4	B	5	Alun.	Res.
Brass:								
Centerless	37C	36	N	5	V	....	Cryst.	Vit.
Cylindrical	37C	36	K	5	V	....	Cryst.	Vit.
Internal	37C	36	J	5	V	....	Cryst.	Vit.
Surfacing (cups and cylinders)	37C	24	H	8	V	....	Cryst.	Vit.
Surfacing—(disks)	37C	20	M	4	B	....	Cryst.	Res.
Snagging (floor stands) 5000–6500 sfpm	37C No. 12 treated	24	P	5	V	....	Cryst.	Vit.
Snagging (floor stands) 7000–9500 sfpm	37C No. 12 treated	24	P	4	B	5	Cryst.	Res.
Cutting-off, gates and risers (dry, Tabor mach. 16,000 sfpm)	37C	245	T	6	R	29	Cryst.	Rub.
Cutting-off rod (dry) 9000–16,000 sfpm	A	30	U	7	B	....	Alun.	Res.
Cutting-off rod (wet) 7500–12,000 sfpm	A	60	V	8	R	30	Alun.	Rub.
Cutting-off tubing (wet) 9000–16,000 sfpm	A	120	V	8	R	30	Alun.	Rub.
Cutting-off tubing (dry) 7500–12,000 sfpm	A	80	V	8	R	30	Alun.	Rub.
Broaches:								
Sharpening	32A	46	K	5	V	BE	Alun.	Vit.
	32A	46	I	12	V	BEP	Alun.	Vit.
Bronze (soft) use same wheels as for brass								
Bronze (hard):								
Centerless	A	60	M	5	V	BE	Alun.	Vit.
Cylindrical	A	46	L	5	V	BE	Alun.	Vit.
Internal	A	60	L	5	V	BE	Alun.	Vit.
Snagging (floor stands) 5000–6500 sfpm	A	20	Q	8	V	BE	Alun.	Vit.
Snagging (floor stands) 7000–9500 sfpm	57A	20	O	4	B	5	Alun.	Res.
Cutting-off (dry) 9000–16,000 sfpm	A	30	O	7	B	....	Alun.	Res.
Cutting-off, gates and risers (dry, Tabor mach. 16,000 sfpm)	37C	245	T	6	R	29	Cryst.	Rub.
Surfacing (straight wheels)	32A	60	K	5	V	BE	Alun.	Vit.
	57A	60	J	5	V	BE	Alun.	Vit.
Surfacing—(disks)	37C	16	L	4	B	L	Cryst.	Res.
Bushings (hardened steel):								
Cylindrical	A	60	L	5	V	BE	Alun.	Vit.
Internal	32A	60	K	5	V	BE	Alun.	Vit.
	32A	80	I	12	V	BEP	Alun.	Vit.
Centerless	A	60	M	5	V	BE	Alun.	Vit.

TABLE 5. NORTON GRINDING-WHEEL RECOMMENDATIONS (*Continued*)

Work—Material —Operation	Abrasive	Grain Size	Grade	Structure	Bond	Bonding Variation	Abrasive (Trade Mark)	Bonding Process
Bushings (cast iron):								
Cylindrical	37C	46	K	5	V	....	Cryst.	Vit.
Internal	37C	46	J	5	V	....	Cryst.	Vit.
	37C	80	I	12	V	P	Cryst.	Vit.
Cams (automotive) (hardened steel):								
Roughing	A	60	L	5	V	BE	Alun.	Vit.
Finishing	A	90	R	2	R	17	Alun.	Rub.
Roughing and finishing								
Hand machines	A	70	P	8	B	2	Alun.	Res.
Automatic	A	70	O	8	B	2	Alun.	Res.
Cams (automotive) (cast alloys):								
Roughing	A	46	M	5	V	BE	Alun.	Vit.
Roughing and finishing (hand machines)	A	70	P	8	B	2	Alun.	Res.
Roughing and finishing (auto. machines)	A	70	O	8	B	2	Alun.	Res.
Camshaft bearings (automotive):								
Cylindrical	A	46	N	5	V	BE	Alun.	Vit.
Car wheels (chilled iron):								
Cylindrical—regrinding	37C	24	Q	4	B	H	Cryst.	Res.
Car wheels (steel):								
Cylindrical—regrinding	A	20	P	5	V	BE	Alun.	Vit.
Car wheels (manganese steel):								
Cylindrical—regrinding	A	16	Q	8	V	BE	Alun.	Vit.
Carboloy (see cemented carbide)								
Carbon:								
Surfacing	37C	16	J	8	V	....	Cryst.	Vit.
Surfacing (disks)	37C	36	M	5	B	....	Cryst.	Res.
Carbon (soft):								
Cutting-off (dry)	37C	36	N	8	B D	.... Sides	Cryst.	Res.
Cutting-off (wet)	37C	54	L	8	B D	.... Sides	Cryst.	Res.
Cutting-off	D	46	N	25	M	....	Diam.	Met.
Carbon (hard):								
Centerless	37C	36	N	5	V	....	Cryst.	Vit.
Cutting-off (dry)	37C	46	N	8	B D	.... Sides	Cryst.	Res.
Cutting-off (wet)	37C	54	K	8	B	....	Cryst.	Res.
Cutting-off	D	60	L	50	M	....	Diam.	Met.
Carbon (hard-plate):								
Stripping	37C	30	O	5	B	....	Cryst.	Res.
Carbon (metallic):								
Cutting-off	37C	36	K	8	B D	.... Sides	Cryst.	Res.
Cast iron:								
Centerless	37C	46	L	5	V	....	Cryst.	Vit.
Cutting-off (dry) 9000–16,000 sfpm	A	36	P	8	B	....	Alun.	Res.
Cylindrical	37C	36	J	5	V	....	Cryst.	Vit.
Internal	37C	46	J	5	V	....	Cryst.	Vit.
	37C	46	I	12	V	P	Cryst.	Vit.
Surfacing (cups and cylinders)	37C	16	H	8	V	....	Cryst.	Vit.
	32A	601	F	12	V	BEP	Alun.	Vit.
Surfacing (segments)	32A	601	E+	12	V	BEP	Alun.	Vit.
Surfacing (disks)	37C	16	M	5	B	L	Cryst.	Res.
	37C	30	K	5	V	....	Cryst.	Vit.
Surfacing (straight wheels)	37C	30	J	8	V	....	Cryst.	Vit.

TABLE 5. NORTON GRINDING-WHEEL RECOMMENDATIONS (*Continued*)

Work—Material —Operation	Abrasive	Grain Size	Grade	Structure	Bond	Bonding Variation	Abrasive (Trade Mark)	Bonding* Process
Cast iron: (*Continued*)								
Snagging (floor stands) 5000–6500 sfpm	37C	20	R	5	V	....	Cryst.	Vit.
Snagging (floor stands) 7000–9500 sfpm	37C	16	P	4	5	....	Cryst.	Res.
Snagging (swing frame) 5000–6500 sfpm	37C	16	S	5	V	....	Cryst.	Vit.
Snagging (swing frame) 7000–9500 sfpm	37C	14	Q	4	B	H	Cryst.	Res.
Snagging (portable grinder) 5000–6500 sfpm	37C	24	S	5	V	....	Cryst.	Vit.
Snagging (portable grinder) 7000–9500 sfpm	37C	20	Q	4	B	H	Cryst.	Res.
Catalin:								
Cutting-off (dry)	37C	90	P	..	B	....	Cryst.	Res.
Cutting-off (wet)	37C	100	R	0	R	30	Cryst.	Rub.
Cylindrical	37C	46	J	5	V	....	Cryst.	Vit.
Cylindrical (form grinding)	37C	100	M	6	V	....	Cryst.	Vit.
Surfacing	37C	36	H	8	V	....	Cryst.	Vit.
Cemented carbides:								
Single-point tools, offhand:								
Cup or plate mounted wheels:								
Roughing (dry) 10 and 14 in. dia.	39C	60	F	12	V	P	Cryst.	Vit.
Less than 10 in. dia	39C	60	I	7	V	....	Cryst.	Vit.
Roughing (wet) 10 and 14 in. dia	39C	60	G	12	V	P	Cryst.	Vit.
	D	100	L	100	V	....	Diam.	Vit.
Semifinishing (dry)	39C	100	H	7	V	....	Cryst.	Vit.
Semifinishing (wet)	39C	100	I	7	V	....	Cryst.	Vit.
Finishing (wet)	D	320	P	100	V	....	Diam.	Vit.
Combination roughing and finishing (wet)	D	150	P	100	V	....	Diam.	Vit.
Straight wheels:								
Roughing (dry)	39C	60	I	7	V	....	Cryst.	Vit.
Roughing (wet)	39C	60	J	7	V	....	Cryst.	Vit.
Semifinishing (dry)	39C	100	H	7	V	....	Cryst.	Vit.
Semifinishing (wet)	39C	100	I	7	V	....	Cryst.	Vit.
Single-point tools, machine grinding								
Straight wheels:								
Roughing and finishing (wet)	39C	80	J	7	V	....	Cryst.	Vit.
Chip breaker grinding								
Straight wheels:								
Less than ¼ in. thick	D	100S	N	100	BX	....	Diam.	Res.
¼ in. and thicker	D	150	N	100	V	....	Diam.	Vit.
Milling cutters, reamers, etc.:								
Backing-off, cup wheel:								
Roughing (type D11B)	39C	60	I	7	V	....	Cryst.	Vit.
	D	100	L	100	B	....	Diam.	Res.
Finishing (type D11B)	D	180	L	100	B	....	Diam.	Res.
Surface grinding straight wheels:								
Roughing (dry)	39C	60	J	7	V	....	Cryst.	Vit.
Roughing (wet)	39C	60	J	7	V	....	Cryst.	Vit.
	D	100	J	100	B	....	Diam.	Res.

TABLE 5. NORTON GRINDING-WHEEL RECOMMENDATIONS (*Continued*)

Work—Material —Operation	Abrasive	Grain Size	Grade	Structure	Bond	Bonding Variation	Abrasive (Trade Mark)	Bonding* Process
Cemented carbides: (*Continued*)								
Finishing (dry)	39C	100	H	7	V	....	Cryst.	Vit.
Finishing (wet)	39C	100	I	7	V	....	Cryst.	Vit.
	D	220	J	100	B	....	Diam.	Res.
Cylindrical grinding:								
Roughing (wet)	39C	60	K	7	V	....	Cryst.	Vit.
	D	100	L	100	B	....	Diam.	Res.
Finishing (wet)	39C	100	J	7	V	....	Cryst.	Vit.
	D	220	J	100	B	....	Diam.	Res.
Internal grinding	D	150	N	100	V	....	Diam.	Vit.
Hand honing or stoning:								
Rectangular hone	D	320	....	..	V	....	Diam.	Hone
	37C	280	N	..	V	....	Cryst.	Vit.
Lapping:								
Cup wheel (wet)	D	400	J	50	B	....	Diam.	Res.
Abrasive grain and cast-iron disk	....	320	F	..	....	....	Norbide	
Cutting-off	D	100S	N	100	B	....	Diam.	Res.
Chain links (annealed malleable iron and steel)								
Snagging (floor stands) 5000–6500 sfpm	57A	20	R	5	V	BE	Alun.	Vit.
Snagging (floor stands) 7000–9500 sfpm	A	16	Q	4	B	5	Alun.	Res.
Chain links (unannealed malleable iron):								
Snagging (floor stands) 5000–6500 sfpm	37C	16	R	5	V	....	Cryst.	Vit.
Snagging (floor stands) 7000–9500 sfpm	37C	16	Q	2	B	H	Cryst.	Res.
	37C	16	T	6	R	30	Cryst.	Rub.
Chasers (thread):								
Surfacing (cup wheels)	32A	36	I	8	V	BE	Alun.	Vit.
Surfacing (straight wheels)	32A	60	K	5	V	BE	Alun.	Vit.
	32A	60	J	12	V	BEP	Alun.	Vit.
Grinding throats	A	60	M	4	E	....	Alun.	Shel.
Chilled iron:								
Snagging (floor stands) 5000–6500 sfpm	37C	16	R	5	V	....	Cryst.	Vit.
Snagging (floor stands) 7000–9500 sfpm	37C	16	P	2	B	5	Cryst.	Res.
Surfacing (cups and cylinders)	37C	24	H	8	V	....	Cryst.	Vit.
Surfacing (straight wheels)	37C	36	I	8	V	....	Cryst.	Vit.
Cylindrical (see rolls)								
Chisels (woodworking):								
Sharpening	57A	60	K	5	V	BE	Alun.	Vit.
Chromium plating (cylindrical):								
Commercial finish	A	100	J	7	V	BE	Alun.	Vit.
Good commercial finish	A	150	K	5	E	....	Alun.	Shel.
High finish	A	500	I	9	E	....	Alun.	Shel.
Commutators:								
Roughing and finishing (wheel)	37C	60	M	4	E	....	Cryst.	Shel.
Connecting rods (automotive):								
Internal	32A	60	K	..	V	....	Alun.	Vit.
Surfacing (cups and cylinders)	19A	24	J	8	V	BE	Alun.	Vit.
Surfacing (disks)	A	36	K	5	B	....	Alun.	Res.

TABLE 5. NORTON GRINDING-WHEEL RECOMMENDATIONS (*Continued*)

Work—Material —Operation	Abrasive	Grain Size	Grade	Structure	Bond	Bonding Variation	Abrasive (Trade Mark)	Bonding* Process
Copper:								
Cylindrical (also see rolls)	37C	100	I	9	V	....	Cryst.	Vit.
Surfacing (cups and cylinders)	37C	14	I	8	V	....	Cryst.	Vit.
Surfacing (disks)	37C	36	M	5	B	....	Cryst.	Res.
Cutting-off:								
Rod (dry)	A	60	V	8	R	30	Alun.	Rub.
Rod (wet)	A	80	V	8	R	30	Alun.	Rub.
Tubing (dry)	A	80	V	8	R	30	Alun.	Rub.
Tubing (wet)	A	120	V	8	R	30	Alun.	Rub.
Gates and risers (dry, Tabor mach. 16,000 sfpm)	37C	24	T	6	R	29	Cryst.	Rub.
Couplers and bolsters:								
Snagging:								
Swing frames 5000–6500 sfpm	57A	10	U	8	V	BE	Alun.	Vit.
Swing frames 7000–9500 sfpm	A	12	R	4	B	H	Alun.	Res.
Floor stands 5000–6500 sfpm	57A	12	S	8	V	BE	Alun.	Vit.
Floor stands 7000–9500 sfpm	A	14	Q	4	B	5	Alun.	Res.
Crankshafts:								
Airplane	38A	60	K	5	V	BE	Alun.	Vit.
Diesel	A	46	L	5	V	BE	Alun.	Vit.
Automotive (pins and bearings):								
Roughing—heavy side removal	A	36	R	4	V	A	Alun.	Vit.
Light side removal	A	46	Q	5	V	A	Alun.	Vit.
Finishing	A	54	O	5	V	A	Alun.	Vit.
Roughing and finishing	A	54	P	5	V	A	Alun.	Vit.
Snagging (for balancing) (floor stands) 5000–6500 sfpm	A	20	O	5	V	BE	Alun.	Vit.
Snagging (for balancing) (floor stands) 7000–9500 sfpm	A	14	O	5	B	5	Alun.	Res.
Automotive (regrinding)	A	46	L	5	V	BA	Alun.	Vit.
Cutlery:								
Knives—butcher:								
Hemming and Klotz machines:								
Carbon steel and stainless steel	A	60	L	7	B	4	Alun.	Res.
	57A No. 22 treated	801	D	12	V	BEP	Alun.	Vit.
Offhand:								
Surfacing sides	A	150	I	..	V	BA	Alun.	Vit.
Sharpening (production)	19A	100	N	..	V	BA	Alun.	Vit.
Knives—paring:								
Hemming:								
Carbon steel and stainless steel	A	100	P	7	E	V	Alun.	Shel.
Knives—pocket:								
Hemming:								
Carbon steel and stainless steel	A	100	P	7	E	V	Alun.	Shel.
Offhand:								
Roughing	A	100	N	..	V	BA	Alun.	Vit.
Finishing	A	220	O	..	V	BA	Alun.	Vit.
Swaging	A	60	Q	5	V	BE	Alun.	Vit.
Backs	A	46	P	5	V	BE	Alun.	Vit.
Tangs	A	60	P	5	V	BE	Alun.	Vit.

TABLE 5. NORTON GRINDING-WHEEL RECOMMENDATIONS (*Continued*)

Work—Material—Operation	Abrasive	Grain Size	Grade	Structure	Bond	Bonding Variation	Abrasive (Trade Mark)	Bonding* Process
Cutlery: (*Continued*)								
Knives—table:								
Hemming:								
Carbon steel.......	57A No. 22 treated	80	D	12	V	BEP	Alun.	Vit.
	A	60	M	4	E	....	Alun.	Shel.
Stainless steel......	A	60	L	4	E	....	Alun.	Shel.
Cutters:								
Sharpening (machine)..	32A	46	K	5	V	BE	Alun.	Vit.
Cylinders, automotive (cast iron)								
Regrinding (wheels)....	37C	36	H	8	V	....	Cryst.	Vit.
	37C	46	G	12	V	P	Cryst.	Vit.
Honing (new cylinders, sticks):								
Commercial finish....	37C	120	P	8	V	....	Cryst.	Vit.
Mirror finish.........	37C	500	I	..	V	....	Cryst.	Vit.
Cylinders (aircraft):								
Molybdenum steel:								
Roughing............	A	36	J	5	V	....	Alun.	Vit.
Finishing............	38A	60	I	8	V	BE	Alun.	Vit.
Regrinding..........	38A	46	K	5	V	BE	Alun.	Vit.
Nitrided:								
Before nitriding......	19A	54	J	7	V	....	Alun.	Vit.
After nitriding.......	37C	60	I	8	V	....	Cryst.	Vit.
	39C	90	J	12	V	P	Cryst.	Vit.
Regrinding..........	37C	60	J	5	V	....	Cryst.	Vit.
Dies (forging):								
Offhand—portable grinding:								
Mounted points and wheels (coarse).....	38A	60	O	..	V	....	Alun.	Vit.
Mounted points and wheels (medium)...	38A	90	M	..	V	....	Alun.	Vit.
Mounted points and wheels (fine).......	38A	120	M	..	V	....	Alun.	Vit.
Straight wheels, roughing 5000–6500 sfpm.	A	46	Q	5	V	BE	Alun.	Vit.
Straight wheels, roughing 7000–9500 sfpm.	A	36	Q	5	B	H	Alun.	Vit.
Dies (drawing):								
Surfacing—hardened:								
Straight wheels (dry).	32A	46	G	12	V	BEP	Alun.	Vit.
	32A	46	H	8	V	BE	Alun.	Vit.
Straight wheels (fast traverse, wet)......	32A	46	J	5	V	BE	Alun.	Vit.
Cup wheels (wet).....	32A	30	H	8	V	BE	Alun.	Vit.
Disks...............	19A	46	J	7	V	BE	Alun.	Vit.
	A	46	J	5	B	....	Alun.	Res.
Segments............	32A	30	E	12	V	BEP	Alun.	Vit.
Surfacing—annealed:								
Straight wheels (dry).	32A	46	I	8	V	BE	Alun.	Vit.
Cup wheels (wet).....	32A	24	I	8	V	BE	Alun.	Vit.
Disks...............	19A	24	I	8	V	BE	Alun.	Vit.
Segments............	32A	30	G	12	V	BEP	Alun.	Vit.
Drills (resharpening):								
¼ in. and smaller—machine..............	19A	100	I	8	V	BE	Alun.	Vit.
¼ in. and smaller—offhand................	A	80	M	5	V	BE	Alun.	Vit.
¼ to 1 in.—machine.....	19A	46	K	..	V	....	Alun.	Vit.
¼ to 1 in.—offhand......	A	46	M	5	V	BE	Alun.	Vit.
Ebonite:								
Cutting-off (dry).......	37C	30	M	8	B	....	Cryst.	Res.

TABLE 5. NORTON GRINDING-WHEEL RECOMMENDATIONS (*Continued*)

Work—Material —Operation	Abrasive	Grain Size	Grade	Structure	Bond	Bonding Variation	Abrasive (Trade Mark)	Bonding* Process
Fiber:								
Surfacing—disks	37C	14	J	4	B	....	Cryst.	Res.
Cutting-off (board)	37C	36	L	8	B	....	Cryst.	Res.
Forgings:								
Centerless	A	60	M	5	V	BE	Alun.	Vit.
Cylindrical	A	46	M	5	V	BE	Alun.	Vit.
Snagging (floor stands) 7000–9500 sfpm	57A	16	Q	2	B	5	Alun.	Res.
(Floor stands) 5000–6500 sfpm	57A	16	R	8	V	BE	Alun.	Vit.
Surfacing—disks	No. 6 treated A	14	L	4	B	....	Alun.	Res.
Gages (plug):								
Cylindrical	32A	80	K	5	V	BE	Alun.	Vit.
Cylindrical, mirror finish	37C	500	J	9	E	....	Cryst.	Shel.
Gages (threads precut):								
Grinding threads 12 pitch and coarser	38A	120	J	7	V	BE	Alun.	Vit.
Grinding threads 13–20 pitch	38A	150	K	8	V	BE	Alun.	Vit.
Grinding threads 24 pitch and finer	38A	220	K	9	V	BE	Alun.	Vit.
Gears (cast iron):								
Cleaning between teeth (offhand)	37C	24	T	6	R	23	Cryst.	Rub.
Gears (hardened steel):								
Teeth—form precision grinding	32A	60	K	5	V	BE	Alun.	Vit.
	32A	80	J	12	V	BEP	Alun.	Vit.
Teeth—generative precision grinding	32A	60	J	5	V	BE	Alun.	Vit.
	32A	80	J	12	V	BEP	Alun.	Vit.
Internal	A	46	L	5	V	BE	Alun.	Vit.
Surfacing (cups and cylinders)	32A	30	I	8	V	BE	Alun.	Vit.
Surfacing (segments)	32A	30	J	8	V	BE	Alun.	Vit.
Surfacing (disks) (remove burrs)	A	24	L	5	B	....	Alun.	Res.
Surfacing (straight wheels)	32A	46	J	5	V	BE	Alun.	Vit.
Lucite:								
Cutting-off (dry)	37C	46	N	8	B	....	Cryst.	Res.
Cutting-off (wet)	37C	902	P	..	B	....	Cryst.	Res.
Magnesium:								
Snagging:								
5000–6500 sfpm	37C	24	N	5	V	....	Cryst.	Vit.
7000–9500 sfpm	37C	20	N	5	B	5	Cryst.	Res.
Malleable castings (annealed):								
Floor stands 5000–6500 sfpm	57A	16	R	5	V	BE	Alun.	Vit.
Floor stands 7000–9500 sfpm	57A	16	P	4	B	5	Alun.	Res.
	A	16	T	6	R	23	Alun.	Rub.
Swing frames 5000–6500 sfpm	57A	14	R	5	V	BE	Alun.	Vit.
Swing frames 7000–9500 sfpm	A	14	P	4	B	5	Alun.	Res.
Portable grinders 5000–6500 sfpm	A	20	Q	5	V	BE	Alun.	Vit.
Portable grinders 7000–9500 sfpm	A	20	Q	4	B	H	Alun.	Res.
Malleable castings (unannealed):								
Floor stands 5000–6500 sfpm	37C	20	R	5	V	....	Cryst.	Vit.

TABLE 5. NORTON GRINDING-WHEEL RECOMMENDATIONS (*Continued*)

Work—Material—Operation	Abrasive	Grain Size	Grade	Structure	Bond	Bonding Variation	Abrasive (Trade Mark)	Bonding* Process
Malleable castings (unannealed): (*Continued*)								
Floor stands 7000–9500 sfpm	37C	20	P	2	B	5	Cryst.	Res.
	A	20	T	6	R	23	Alun.	Rub.
Surfacing—disks	A	20	L	4	B	5	Alun.	Res.
Monel metal:								
Cutting-off (dry)	A	36	N	8	B	....	Alun.	Res.
Cutting-off (wet)	A	60	T	6	R	I	Alun.	Rub.
Floor stands 5000–6500 sfpm	A	24	Q	5	V	BE	Alun.	Vit.
Floor stands 7000–9500 sfpm	57A	20	P	4	B	5	Alun.	Res.
Cylindrical	37C	60	J	8	V	....	Cryst.	Vit.
Needles:								
Pointing	A	70	S	8	V	....	Alun.	Vit.
Nickel:								
Cutting-off (dry)	A	46	Q	8	B	....	Alun.	Res.
Cutting-off (wet)	A	60	T	6	R	1	Alun.	Rub.
Nitralloy:								
Before nitriding	32A	60	I	12	V	BEP	Alun.	Vit.
	A	46	K	5	V	BE	Alun.	Vit.
After nitriding	32A	60	H	8	V	BE	Alun.	Vit.
	37C	100	I	8	V	....	Cryst.	Vit.
Pistons (aluminum):								
Cylindrical	39C	46	I	12	V	P	Cryst.	Vit.
	37C	364	K	5	V	....	Cryst.	Vit.
Centerless	37C	46	I	5	V	....	Cryst.	Vit.
Regrinding	A	46	I	8	V	BE	Alun.	Vit.
Pistons (cast iron):								
Cylindrical	39C	46	I	12	V	P	Cryst	Vit
	37C	36	K	5	V	....	Cryst.	Vit.
Centerless	37C	46	L	5	V	....	Cryst.	Vit.
Regrinding	A	46	I	8	V	BE	Alun.	Vit.
Piston pins:								
Centerless machine:								
Roughing	57A	60	M	5	V	BE	Alun.	Vit.
Semifinishing	57A	80	M	5	V	BE	Alun.	Vit.
Finishing	37C	320	N	8	E	....	Cryst.	Shel.
Surfacing ends (disks)	A	60	L	4	B	....	Alun.	Res.
Lapping (Norton Hyprolap)	37C	XF	K	6	E	....	Cryst.	Shel.
Piston rings (cast iron or semisteel):								
Surfacing rough (cylinders)	32A	30	H	8	V	BE	Alun.	Vit.
Surfacing (disks):								
Roughing	37C	16	M	4	B	....	Cryst.	Res.
Semifinishing	37C	36	L	7	B	....	Cryst.	Res.
Finishing	37C	60	M	5	B	....	Cryst.	Res.
Surfacing (straight wheels)	A	70	L	5	V	BE	Alun.	Vit.
Lapping (Norton Hyprolap)	39C	400	J	9	V	....	Cryst.	Vit.
Internal (snagging)	37C	36	S	5	V	....	Cryst.	Vit.
Piston rods (locomotive):								
Cylindrical	A	46	M	5	V	BE	Alun.	Vit.
Plastics:								
Cutting-off (dry)	37C	46	K	8	B	....	Cryst.	Res.
Cutting-off (wet)	37C	902	P	..	B	....	Cryst.	Res.
Plows (steel):								
Stubbing—landsides	A	16	R	5	V	BE	Alun.	Vit.
Surfacing shares	57A	16	T	5	V	BE	Alun.	Vit.
Edging shares	A	16	R	5	V	BE	Alun.	Vit.
Jointing—mold boards	A	16	R	5	V	....	Alun.	Vit.
Filling—plow bottom	A	30	Q	5	V	....	Alun.	Vit.
	No. 0.0115 treated							
Landsides—plow bottom	A	16	R	5	V	BE	Alun.	Vit.

TABLE 5. NORTON GRINDING-WHEEL RECOMMENDATIONS (*Continued*)

Work—Material —Operation	Abrasive	Grain Size	Grade	Structure	Bond	Bonding Variation	Abrasive (Trade Mark)	Bonding* Process
Plows (chilled iron):								
All operations, wet	A	20	R	7	V	BE	Alun.	Vit.
All operations, dry	A	20	N	5	V	BE	Alun.	Vit.
Pulleys (cast iron):								
Cylindrical	37C	30	J	5	V	....	Cryst.	Vit.
Roughing with pulley grinder (cylindrical wheel)	37C	24	I	8	V	....	Cryst.	Vit.
Finishing with pulley grinder (cylindrical wheel)	A	54	P	0	R	30	Alun.	Rub.
Reamers:								
Backing-off	32A	80	H	12	V	BEP	Alun.	Vit.
	32A	46	K	5	V	BE	Alun.	Vit.
Cylindrical	A	54	L	5	V	BE	Alun.	Vit.
Rifle barrels:								
Cylindrical	A	46	M	5	V	BE	Alun.	Vit.
Rims (automobile):								
Grinding welds 5000–6000 sfpm	A	24	Q	5	V	BE	Alun.	Vit.
Grinding welds 7000–9500 sfpm	A	20	T	..	R	23	Alun.	Rub.
Roller-bearing cups:								
Centerless OD	A	80	L	5	V	BE	Alun.	Vit.
Internal	A	70	N	5	V	BE	Alun.	Vit.
Segments	57A	70	F	12	V	BEP	Alun.	Vit.
Rollers for bearings:								
Centerless—roughing	A	80	M	5	V	BE	Alun.	Vit.
Finishing	A	90	R	2	R	3	Alun.	Rub.
Surfacing ends (disks)	19A	90	O	8	V	....	Alun.	Vit.
Rolls (brass or copper):								
Cylindrical—roughing	37C	46	L	3	E	....	Cryst.	Shel.
Finishing	37C	100	I	8	V	....	Cryst	Vit.
Rolls (cast iron):								
Cylindrical (roughing)	37C	30	K	5	V	....	Cryst.	Vit.
Cylindrical (finishing)	37C	80	J	8	V	....	Cryst.	Vit.
Rolls, hot mill (chilled iron):								
Regrinding:								
Roughing	37C	36	M	7	B	....	Cryst.	Res.
Roughing (on lighter type machines)	37C	30	L	5	E	....	Cryst.	Shel
	37C	30	L	7	B	....	Cryst.	Res
Cutting down fire-cracked rolls	37C	24	O	5	B	....	Cryst.	Res
New rolls:								
Roughing (standard machines)	37C	20	R	4	B	H	Cryst.	Res.
Roughing (heavy and amply powered machines)	A	30	U	4	B	H	Cryst.	Res.
Finishing	37C	36	M	7	B	....	Cryst.	Res
Cutting-off wobblers	37C	24	S	2	B	H	Cryst.	Res
Rolls, cold mill (hardened steel):								
Regrinding:								
Fast cutting action and satin finish	37C	150	L	7	B	....	Cryst.	Res.
	A	80	K	7	B	....	Alun.	Res
	A	100	I	6	V	BE	Alun.	Vit.
On lighter type machines	A	80	M	5	E	....	Alun.	Shel.
Extra-fine finish	37C	320	I	8	E	....	Cryst.	Shel.
	37C	500	G	7	B	4	Cryst.	Res.
New rolls:								
Roughing after hardening	38A	46	L	5	V	BE	Alun.	Vit.
Finishing	A	100	I	6	V	BE	Alun.	Vit.

TABLE 5. NORTON GRINDING-WHEEL RECOMMENDATIONS (*Continued*)

Work—Material —Operation	Abrasive	Grain Size	Grade	Structure	Bond	Bonding Variation	Abrasive (Trade Mark)	Bonding* Process
Rolls, cold mill (chilled iron):								
Regrinding:								
Fast cutting action and satin finish....	37C	100	J	7	B	4	Cryst.	Res.
On lighter type machines............	37C	80	M	5	E	....	Cryst.	Shel.
(Extra-fine finish)....	37C	320	I	8	E	....	Cryst.	Shel.
New rolls:								
Roughing............	37C	20	R	4	B	H	Cryst.	Res.
Finishing............	37C	100	J	7	B	H	Cryst.	Res.
Rolls, Steckel mill (high-speed tool steel):								
Regrinding—excellent finish..............	38A	100	J	6	V	BE	Alun.	Vit.
	A	100	I	6	V	BE	Alun.	Vit.
Rolls, backup (cast steel):								
Regrinding or new:								
Roughing and finishing...............	19A	36	J	5	V	BE	Alun.	Vit.
	A	24	L	5	B	....	Alun.	Res.
Rolls, backup (chilled iron):								
Regrinding or new:								
Roughing and finishing...............	37C	24	N	5	B	....	Cryst.	Res.
Rolls, tinning (heat-treated steel):								
Regrinding, commercial finish..............	A	46	L	5	V	BE	Alun.	Vit.
Rubber (hard):								
Cutting-off............	37C	30	M	8	B	....	Cryst.	Res.
Cylindrical............	37C	30	K	5	B	....	Cryst.	Res.
Saws (band):								
Gumming............	57A	46	N	5	V	....	Alun.	Vit.
	19A	46	M	..	V	BE	Alun.	Vit.
	57A	361	N	7	B	5	Alun.	Res.
	A	30	N	3	E	....	Alun.	Shel.
Saws (circular):								
Gumming............	19A	46	M	..	V	....	Alun.	Vit.
	A	46	M	5	V	BE	Alun.	Vit.
Saws (metal cutting):								
Gumming............	A	60	P	5	V	BE	Alun.	Vit.
	A	80	P	8	B	....	Alun.	Res.
Shear blades (power metal shears):								
Sharpening (segments)..	A	30	G	8	V	BE	Alun.	Vit.
Sharpening (cylinders)..	32A	30	G	8	V	BE	Alun.	Vit.
Spline shafts:								
Centerless............	A	60	O	5	V	BE	Alun.	Vit.
Cylindrical............	A	54	N	5	V	BE	Alun.	Vit.
Grinding splines........	A	54	L	5	V	BE	Alun.	Vit.
Sprayed metal:								
Cylindrical............	37C	60	J	5	V	....	Cryst.	Vit.
Cutting-off (wet) 9000–12,000 sfpm..........	A	46	U	8	R	30	Alun.	Rub.
Springs (coil):								
Squaring ends (disks):								
Small gage wire......	19A	60	M	5	V	BE	Alun.	Vit.
Medium gage wire....	57A	36	M	5	V	BE	Alun.	Vit.
Large gage wire......	A	20	P	5	V	BE	Alun.	Vit.
Steel castings (low-carbon):								
Swing frames—5000–6500 sfpm...........	57A	14	R	5	V	BE	Alun.	Vit.
Swing frames—7000–9500 sfpm...........	A	14	P	4	B	5	Alun.	Res.
Floor stands—5000–6500 sfpm...........	57A No. 0.0115 treated	16	P	5	V	BE	Alun.	Vit.

TABLE 5. NORTON GRINDING-WHEEL RECOMMENDATIONS (*Continued*)

Work—Material —Operation	Abrasive	Grain Size	Grade	Structure	Bond	Bonding Variation	Abrasive (Trade Mark)	Bonding* Process
Steel castings (low-carbon) (*Continued*)								
Floor stands—7000–9500 sfpm..........	A No. 0.0115 treated	16	P	4	B	5	Alun.	Res.
Portable grinders—5000–6500 sfpm..........	A	20	Q	5	V	BE	Alun.	Vit.
Portable grinders—7000–9500 sfpm..........	A	20	Q	4	B	H	Alun.	Res.
	A	20	T	6	R	30	Alun.	Rub.
Steel castings (manganese):								
Swing frames—5000–6500 sfpm..........	A	20	Q	5	V	BE	Alun.	Vit.
Swing frames—7000–9500 sfpm..........	A	14	Q	4	B	5	Alun.	Res.
Floor stands—5000–6500 sfpm..........	A	20	P	5	V	BE	Alun.	Vit.
Floor stands—7000–9500 sfpm..........	A	16	P	4	B	7	Alun.	Res.
Portable grinders—5000 6500 sfpm..........	A	24	Q	5	V	BE	Alun.	Vit.
Portable grinders—7000 9500 sfpm..........	A	20	Q	4	B	5	Alun.	Res.
Surfacing—planer type —5000–6500 sfpm....	57A	14	P	5	V	BE	Alun.	Vit.
Surfacing—planer type —7000–9500 sfpm....	A	16	O	4	B	....	Alun.	Res.
Surfacing—lathes and boring mills—5000–6500 sfpm..........	A	16	P	8	V	BE	Alun.	Vit.
Surfacing—lathes and boring mills—7000–9500 sfpm..........	A	20	O	4	B	....	Alun.	Res.
Internal—rough grinding—5000–6500 sfpm.	A	20	O	5	V	BE	Alun.	Vit.
Internal—rough grinding—7000–9500 sfpm.	A	24	Q	4	B	H	Alun.	Res.
Steel (hardened):								
Centerless............	57A	60	M	5	V	BE	Alun.	Vit.
Centerless (fine finish)..	A	80	T	8	R	23	Alun.	Rub.
	A	80	P	4	R	30	Alun.	Rub.
Cylindrical............	57A	60	L	5	V	BE	Alun.	Vit.
Surfacing (segments)...	32A	30	E	12	V	BEP	Alun.	Vit.
Surfacing (cups and cylinders)..............	32A	30	G	8	V	BE	Alun.	Vit.
Surfacing—(disks) roughing...........	A	16	J	4	B	....	Alun.	Res.
Surfacing—(disks) finishing...........	A	46	J	5	B	....	Alun.	Res.
Surfacing (straight wheels)............	32A	46	H	8	V	BE	Alun.	Vit.
	32A	60	G	12	V	BEP	Alun.	Vit.
Internal...............	32A	60	J	5	V	BE	Alun.	Vit.
Cutting-off (wet) 9000–12,000 sfpm..........	A	60	P	0	R	30	Alun.	Rub.
Cutting-off (dry) 9000–12,000 sfpm..........	A	46	M	8	B	....	Alun.	Res.
Steel (soft):								
Centerless.............	57A	60	N	5	V	BE	Alun.	Vit.
Cylindrical............	A	54	N	5	V	BE	Alun.	Vit.
Surfacing (segments)...	32A	30	E	12	V	BEP	Alun.	Vit.
Surfacing (cups and cylinders)............	32A	24	J	8	V	BE	Alun.	Vit.
Surfacing (disks) roughing...........	A	16	L	4	B	....	Alun.	Res.
Surfacing (disks) finishing................	A	46	K	5	B	....	Alun.	Res.

TABLE 5. NORTON GRINDING-WHEEL RECOMMENDATIONS (*Continued*)

Work—Material—Operation	Abrasive	Grain Size	Grade	Structure	Bond	Bonding Variation	Abrasive (Trade Mark)	Bonding* Process
Steel (soft): (*Continued*)								
Surfacing (straight wheels) narrow.......	32A	46	J	5	V	BE	Alun.	Vit.
Surfacing (straight wheels) wide.........	32A	36	L	8	V	BE	Alun.	Vit.
Internal grinding........	A	46	M	5	V	BE	Alun.	Vit.
Cutting-off (dry) 12,000–16,000 sfpm..........	A	245	Q	8	B	....	Alun.	Res.
Cutting-off (wet) 9000–12,000 sfpm..........	A	46	V	8	R	30	Alun.	Rub.
Steel (high-speed):								
Centerless.............	57A	60	L	5	V	BE	Alun.	Vit.
Cylindrical............	32A	60	K	5	V	BE	Alun.	Vit.
Surfacing (segments)...	32A	30	E	12	V	BEP	Alun.	Vit.
Surfacing (cups and cylinders).............	32A	46	G	8	V	BE	Alun.	Vit.
Surfacing (straight wheels).............	32A	46	H	8	V	BE	Alun.	Vit.
	32A	60	F	12	V	BEP	Alun.	Vit.
Cutting-off (soft, dry) 12,000–16,000 sfpm...	A	24	Q	8	B	....	Alun.	Res.
Cutting-off (sott, wet) 9000–12,000 sfpm....	A	46	V	8	R	30	Alun.	Rub.
Cutting-off (hard, dry) 9000–12,000 sfpm....	A	46	M	8	B	....	Alun.	Res.
Cutting-off (hard, wet) 9000–12,000 sfpm....	A	60	P	0	R	30	Alun.	Rub.
Internal.............	32A	46	J	5	V	BE	Alun.	Vit.
Steel (stainless, free machining):								
Centerless.............	37C	46	M	5	V	....	Cryst.	Vit.
Cylindrical............	37C	46	M	5	V	....	Cryst.	Vit.
Surfacing (cups and cylinders).............	32A	36	G	8	V	BE	Alun.	Vit.
Surfacing (segments)...	32A	601	E+	12	V	BEP	Alun.	Vit.
Surfacing—straight wheels..........	32A	46	H	8	V	BE	Alun.	Vit.
Cutting-off (dry) 12,000–16,000 sfpm..........	A	46	Q	8	B	....	Alun.	Res.
Cutting-off (wet) 9000–12,000 sfpm..........	A	46	V	8	R	30	Alun.	Rub.
Tubing—Cutting-off								
Dry 12,000–16,000 sfpm..............	A	80	V	10	R	29	Alun.	Rub.
Wet 9000–12,000 sfpm..............	A	80	U	10	R	29	Alun.	Rub.
Billets and slabs:								
Swing frames—7000–9500 sfpm..........	A	123	R	4	B	H	Alun.	Res.
	A	16	V	8	R	30	Alun.	Rub.
Portable grinders—7000–9500 sfpm....	A	20	Q	4	B	H	Alun.	Res.
	A	24	T	6	R	30	Alun.	Rub.
Steel (stainless, hardened):								
Centerless.............	57A	60	L	5	V	BE	Alun.	Vit.
Cylindrical............	A	60	L	5	V	BE	Alun.	Vit.
Cutting-off (dry)........	A	46	M	8	B	....	Alun.	Res.
Cutting-off (wet) 9000–12,000 sfpm..........	A	60	P	0	R	30	Alun.	Rub.
Hemming & Klotz								
Cutlery, small:								
Roughing..........	A	60	L	7	B	4	Alun.	Res.
Finishing..........	A	100	P	7	E	V	Alun.	Shel.
Surfacing (cylinders and cups).............	32A	24	G	8	V	BE	Alun.	Vit.
(Segments)...........	32A	301	E	12	V	BEP	Alun.	Vit.
(Straight wheels).....	32A	46	H	8	V	BE	Alun.	Vit.
Internal...............	A	60	L	5	V	BE	Alun.	Vit.

TABLE 5. NORTON GRINDING-WHEEL RECOMMENDATIONS (*Continued*)

Work—Material —Operation	Abrasive	Grain Size	Grade	Structure	Bond	Bonding Variation	Abrasive (Trade Mark)	Bonding* Process
Stellite:								
Cylindrical	A	46	M	5	V	BE	Alun.	Vit.
Cutter grinding	32A	46	J	5	V	BE	Alun.	Vit.
Surfacing (cups and cylinders)	32A	46	G	..	S	....	Alun.	Sil.
Surfacing (disks)	A	36	L	5	V	BE	Alun.	Vit.
Surfacing (straight wheels)	32A	46	H	8	V	BE	Alun.	Vit.
Cutting-off (thin wheel $\frac{1}{16}$ in. thick)	A	60	O	8	B	....	Alun.	Res.
Tools:								
Offhand	57A	46	N	5	V	BE	Alun.	Vit.
Machine	32A	46	L	5	V	BE	Alun.	Vit.
Stoodite:								
Offhand grinding	A	30	P	5	V	BE	Alun.	Vit.
Taps:								
Squaring ends	A	60	N	5	V	BE	Alun.	Vit.
Grinding relief	32A	80	J	5	V	BE	Alun.	Vit.
Fluting (small taps)	A	60	R	2	R	3	Alun.	Rub.
(Large taps)	32A	46	L	5	V	BE	Alun.	Vit.
Threading:								
12 pitch and coarser (precut)	38A	120	M	9	V	BE	Alun.	Vit.
13 pitch to 24 pitch (precut)	38A	180	M	10	V	BE	Alun.	Vit.
13 pitch to 24 pitch (solid)	A	180	S	9	B	H	Alun.	Res.
32 pitch and finer (precut)	38A	220	M	10	V	BE	Alun.	Vit.
Shanks (cylindrical)	38A	80	M	6	V	BE	Alun.	Vit.
Threads (see gages and taps):								
Screws and studs—12 pitch to 16 pitch (solid)	A	100	R	9	B	KH	Alun.	Res.
Screws and studs—18 pitch to 24 pitch (solid)	A	120	S	9	B	KH	Alun.	Res.
Screws and studs—32 pitch and finer (solid)	A	180	T	9	B	KH	Alun.	Res.
Worms (precut)	38A	80	I	8	V	BE	Alun.	Vit.
	38A	100	I	8	V	BE	Alun.	Vit.
	38A	120	J	7	V	BE	Alun.	Vit.
Tools (lathe and planer):								
Carbon and high-speed steel								
Offhand grinding:								
Bench and pedestal grinders								
Coarse	57A	36	O	5	V	BE	Alun.	Vit.
Fine	57A	60	M	5	V	BE	Alun.	Vit.
Wet-tool grinders:								
12 to 24 in. dia wheels	57A	36	O	5	V	BE	Alun.	Vit.
Over 24 in. dia wheels	57A	24	P	..	S	....	Alun.	Sil.
	57A	24	M	5	V	BE	Alun.	Vit.
Machine grinding:								
Straight wheels:								
15 in. dia wheels	57A	36	L	5	V	BE	Alun.	Vit.
24 in. dia wheels	57A	24	M	5	V	BE	Alun.	Vit.
Cup or cylinder wheels	57A	24	L	8	V	BE	Alun.	Vit.
Cutting-off toolbit stock (hard)	57A	46	M	8	B	....	Alun.	Res.

TABLE 5. NORTON GRINDING-WHEEL RECOMMENDATIONS (*Continued*)

Work—Material —Operation	Abrasive	Grain Size	Grade	Structure	Bond	Bonding Variation	Abrasive (Trade Mark)	Bonding* Process
Tubing:								
Centerless	A	60	M	5	V	BE	Alun.	Vit.
Cutting-off (dry) 12,000–16,000 sfpm								
Steel, finish unimportant	A	30	U	7	B	....	Alun.	Res.
Steel	A	80	V	10	R	29	Alun.	Rub.
Stainless steel	A	80	V	10	R	29	Alun.	Rub.
Chrome—molybdenum	A	80	V	10	R	29	Alun.	Rub.
Cutting-off (wet) 9000–12,000 sfpm								
Steel	A	80	V	10	R	29	Alun.	Rub.
Stainless steel	A	80	U	10	R	29	Alun.	Rub.
Chrome—molybdenum	A	80	U	10	R	29	Alun.	Rub.
Valves (automotive):								
Refacing	A	80	L	5	V	BE	Alun.	Vit.
	37C	80	L	5	V	....	Cryst.	Vit.
Stems								
Cylindrical	A	54	N	5	V	BE	Alun.	Vit.
Centerless	A	60	O	5	V	BE	Alun.	Vit.
Cutting-off	A	30	U	7	B	....	Alun.	Res.
Surfacing ends—disks	A	36	J	8	V	BE	Alun.	Vit.
Valve-seat inserts—regrinding:								
Roughing:								
Cast iron	37C	46	M	5	V	....	Cryst.	Vit.
Alloy steel	38A	60	N	6	V	BE	Alun.	Vit.
Stellite	38A	80	I	8	V	BE	Alun.	Vit.
	38A	80	H	12	V	BEP	Alun.	Vit.
Finishing:								
All seats	37C	150	L	..	V	....	Cryst.	Vit.
Valve tappets:								
Centerless	A	80	P	7	V	BE	Alun.	Vit.
Cylindrical	A	46	M	5	V	BE	Alun.	Vit.
Welds:								
Carbon-alloy steels:								
Portable grinders 5000–6500 sfpm	A	24	Q	5	V	BE	Alun.	Vit.
Portable grinders 7000–9500 sfpm	A	24	Q	4	B	H	Alun.	Res.
Stainless steel:								
Portable grinders 7000–9500 sfpm	A	24	T	6	R	30	Alun.	Rub.
Worms (see Threads)								
Wrought iron:								
Floor stands 5000–6000 sfpm	A	16	R	7	V	BE	Alun.	Vit.

Care of Grinding Wheels

Storage. Most grinding wheels are relatively fragile. Therefore, appropriate care should be exercised in handling and storing them. Suitable racks should be provided to accommodate the various types and sizes of wheels carried in stock.

Most straight and tapered wheels are best supported on edge in racks.

Thin rubber, shellac, and resinoid-bonded wheels should be laid flat on a plane surface to prevent warpage.

Cylinder wheels and large cup wheels should be stacked on their flat sides with corrugated paper or other cushioning material between them.

Small cup and other shaped wheels and small internal grinding wheels may be stored in boxes, bins, or drawers.

Inspection. Immediately upon receipt, all wheels should be closely inspected to make sure that they have not been injured in transit or otherwise. As an added precaution, wheels should be tapped gently (while suspended) with a light implement, such as the handle of a screwdriver for light wheels, or a wooden mallet for heavier wheels. If they sound cracked, they should not be used. Wheels must be dry and free from sawdust when the test is applied; otherwise the sound will be deadened. It should also be noted that organic-bonded wheels do not emit the same clear metallic ring as do vitrified and silicate wheels.

Mounting of Grinding Wheels

Diameter of Spindle. No wheel of larger diameter or greater thickness than specified in Table 4 shall be used on any machine of given spindle diameter.

Direction of Spindle Thread. Ends of spindles should be so threaded that the nuts on both ends will tend to tighten as the spindles revolve. Care should be taken in setting up machines that the spindles are mounted so that they will revolve in the proper direction, otherwise the nuts on the ends will loosen.

Note. To remove the nuts, they should both be turned in the direction that the spindle revolves when the wheel is in operation.

Work-rest Adjustment. The work rest should be kept adjusted close to the wheel, with a maximum distance of $\frac{1}{8}$ in., to prevent the work from being caught between the wheel and rest and should be securely clamped after each adjustment. This adjustment shall not be made while the wheel is in motion. The rest should be maintained in good condition.

Flanges. Flanges shall be recessed at least $\frac{1}{16}$ in. on the side next to the wheel. See standard flanges, Table 6.

Fit. The driving flange shall be keyed, screwed, shrunk, or pressed onto the spindle, and the bearing surface shall run true and at right angles with the spindle.

Surface Condition. All surfaces of wheels, washers, and flanges in contact with each other should be free from foreign material.

Bushing. The soft metal bushing shall not extend beyond the sides of the wheel. The hole in the wheel bushing should be 0.005 in. larger than standard size spindles. This permits the wheel to slide on the spindle without cramping and insures a good fit not only on the spindle but against the inside flange, which is essential.

Washers. Washers or flange facings of compressible material shall be fitted between the wheel and its flanges. If blotting paper is used, it should not be thicker than 0.025 in. If rubber or leather is used, it should not be thicker than 1 in. If flanges with babbitt or lead facings are used, the thickness of the babbitt or lead should not exceed $\frac{1}{8}$ in. The diameter of the washers shall not be smaller than the diameter of the flanges. See Table 6.

Tightening of Nut. When spindle-end nuts are tightened, care should be taken to tighten them only enough to hold the wheel firmly; otherwise, the clamping strain is likely to damage the wheel or associated parts.

Guards for Abrasive Wheels. The safety code prescribes guards for abrasive wheels as shown in Fig. 9. The diagrams give the amount and the position of the openings for grinding machines of different types.

General Suggestions. Competent men shall be assigned to the mounting, care, and inspection of grinding wheels and machines.

After a new wheel is mounted, care should be taken to see that the hood is properly replaced.

TABLE 6. STANDARD DIMENSIONS OF PROTECTION FLANGES
All Dimensions in Inches

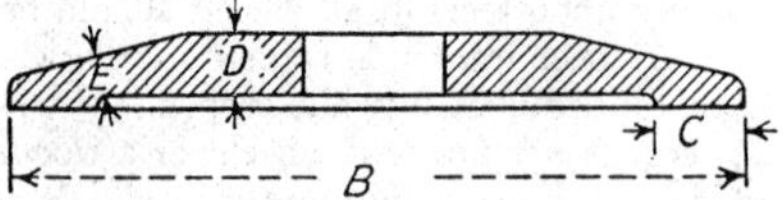

A Diameter of Wheel	B Minimum Outside Diameter of Flanges	C Radial Width of Bearing Surface		D Minimum Thickness of Flange at Bore	E Minimum Thickness of Flange at Edge of Recess
		Minimum	Maximum		
1	3/8	1/16	1/8	1/16	1/16
2	3/4	1/8	3/16	1/8	3/32
3	1	1/8	1/4	3/16	3/32
4	1 1/4	3/16	3/8	3/16	1/8
5	1 1/2	3/16	3/8	1/4	1/8
6	2	1/4	1/2	3/8	3/16
8	3	1/4	1/2	3/8	3/16
10	3 1/2	5/16	5/8	3/8	1/4
12	4	5/16	5/8	1/2	5/16
14	4 1/2	3/8	3/4	1/2	5/16
16	5 1/2	1/2	1	1/2	5/16
18	6	1/2	1	5/8	3/8
20	7	5/8	1 1/4	5/8	3/8
22	7 1/2	5/8	1 1/4	5/8	7/16
24	8	3/4	1 1/4	5/8	7/16
26	8 1/2	3/4	1 1/4	5/8	1/2
28	10	7/8	1 1/2	3/4	1/2
30	10	7/8	1 1/2	3/4	5/8
36	12	1	2	7/8	3/4

Tapered protection flanges shall always be used with tapered wheels having the same degree of taper. The taper should be at least 3/4 in. per ft for each flange.

All new wheels shall be run at full operating speed for at least one minute before applying work, during which time the operator shall stand at one side.

Work should not be forced against a cold wheel but applied gradually, giving the wheel an opportunity to warm and thereby minimize the chance of breakage. This applies to starting work in the morning in cold rooms and to new wheels which have been stored in a cold place.

Grinding on the flat sides of straight wheels is often hazardous and should not be allowed when the sides of the wheels are appreciably worn or when any considerable or sudden pressure is brought to bear against the sides.

DUST-EXHAUST PROVISION. Hoods on machines used for dry grinding and other

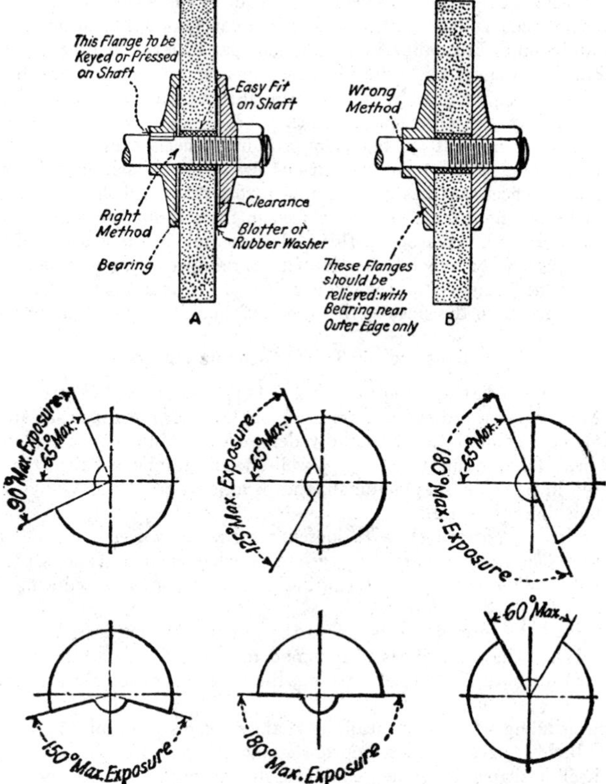

FIG. 9. Right and wrong methods of mounting wheels, and positioning of wheel guards.

operations where dust is produced should have provision made for connection to an exhaust system.

The size of such connections should be in conformity with the following specified dimensions:

Wheel Dia, In.	*Min Dia Branch Pipe, In.*
6 or less..........	3
7 to 16...........	4
17 to 24..........	5
25 to 30..........	6

A modification of the above requirements will be allowed in the case of narrow wheels used for light work where very little dust is generated and where a smaller pipe will satisfactorily remove it.

Balancing of Grinding Wheels. Grinding wheels change in size and weight while in use. This change in mass of the wheel sometimes causes variation in the balance of the rotating-wheel spindle unit, resulting in a varying quality of finish.

To maintain uniform quality in production, particularly in precision grinding, wheels should be balanced as the need becomes apparent. An operator can generally tell if a wheel is out of balance by the vibration it sets up in adjacent machine parts, such as the wheel guard and coolant nozzle.

The standard equipment for balancing grinding wheels consists of a balancing stand and balancing arbors to fit the various sizes of wheel sleeves. Before a new wheel or a used wheel that has been removed from the wheel sleeve is balanced, it should first be mounted on its sleeve, placed in the grinding machine, and rough-trued on the face. (The reason for this is that the hole in the wheel is of necessity larger than the sleeve on which it is mounted. Truing out the resulting eccentricity usually affects the "balance" of the whole unit.) If the wheel is to be used for work requiring grinding with the sides of the wheel, it should also be trued on its sides.

Dressing and Truing Grinding Wheels

Two basic uses of diamond tools are those of (1) dressing and (2) truing grinding wheels. More industrial diamonds are used for these purposes than for any others. Aside from these common uses, specially designed diamond tools are employed in shaping wheels for grinding to form. A well-known example is called the "phonopoint," used in pairs for shaping small, thin wheels to grind the V-angles of taps. This is a finishing operation.

The function of "dressing" is to loosen particles of wheel bond and abrasive material, as well as granular particles of ground stock (steel, metal alloys, etc.), which in time will load or even glaze the grinding surface, greatly reducing abrasive efficiency.

"Truing," as the term suggests, keeps the grinding surface even and uniformly effective. While the same types of diamond tools for given requirements usually are employed for both dressing and truing, the latter operation requires more time and care.

Forming grinding wheels by means of diamond tools calls for special skill and experience besides fixture design appropriate to the purpose. In some cases the diamond itself is shaped for some definite result contingent upon the grinding operation. Common examples are diamonds shaped to a conical point or to a chisel edge. Wheels frequently are fluted, scored, or grooved with diamond tools.

Industrial diamonds have pronounced limitations as to size and fractural propensity, yet their universal employment in conditioning grinding units amply reveals their industrial importance. Although accepted as the hardest material, diamond is relatively brittle. It requires solid mounting in a tool and reasonable care in application (see Fig. 10). The most successful methods of mounting are brazing and sintering. Each method has specific applications, but in any case the diamond must be correctly centered and firmly held. Tendency to lateral vibration can be overcome by correct mounting methods. Diamonds in the largest industrial sizes are comparatively rare and expensive, but those suitable for the most rigorous demands of grinding are stronger because of mass, and more resistant to fracture and wear than smaller stones.

Excessive pressure against the grinding wheel or careless handling will cause the diamond to wear rapidly, which means that minute crystals break away from the parent body. Because of its naturally fragile character the diamond can be chipped or shattered by being thrust violently against a rotating wheel. A loose setting or steadyrest or fixture can produce these results, which are expensive.

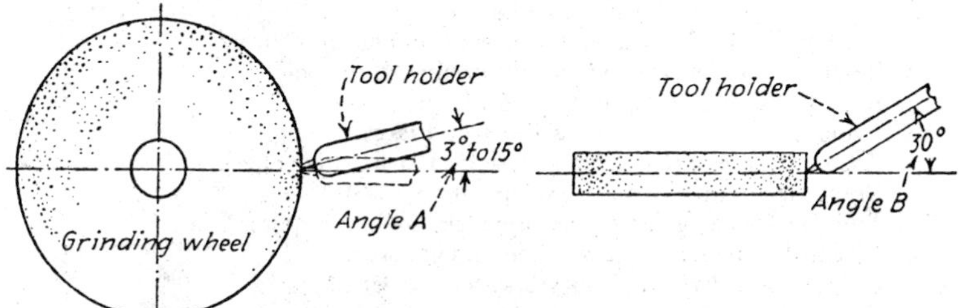

FIG. 10. For truing and dressing, set the diamond holder at angle A with respect to a radial line passing to wheel center, and cant the holder to angle B to make the stone self-sharpening.

Many efforts have been made to classify diamonds for all grinding-wheel applications in respect of definite carat weights. Such standardization, while desirable, has not proved successful because of the great variety of wheel compositions, sizes, speeds, frequency of dressings and truings, and the limitless conditions and requirements involved. A general and simple rule is to use a weight of diamond that is consistent with the work; a large stone for heavy duty, and a smaller stone for light duty. Table 7 will serve as an approximate guide to carat size.

TABLE 7. DIAMOND WEIGHTS FOR WHEEL DRESSING*

	Carats
Wheel dia under 8 in	0.300 to 0.750
Wheel dia under 12 in	1.000 to 1.250
Wheel dia under 24 in	1.500 to 3.000
Wheel dia over 24 in	3.500 to 10.000

* Depending on conditions, type of wheel, grain, etc.

Diamond truing and dressing tools are made in a wide range of sizes and holder design to meet the specific requirements of grinding machines. Single-stone tools are more commonly used than multiple types, because they can be reset and adapted to increasingly lighter work. Multiple-stone dressers are set with three or more diamonds in a single line, or in geometrical grouping. These types are not suitable for resetting and are recommended for frequent light dressing.

The following rules governing the use of diamond tools for dressing and truing are derived from long experience and collective opinion.

1. Take light cuts from 0.001 to 0.003 in.
2. Machine should be level and rigid.
3. When used, water flow should cover full width of wheel face.
4. Coolant should be turned on before truing cuts are made.
5. Feed tool slowly and with minimum pressure.
6. In average cases, or whenever possible, diamond should travel evenly, lightly, and slowly.
7. Relative coarseness of wheel grain and hardness of wheel should be considered in taking cuts.
8. Use the right weight of diamond for wheel size and grain type; do not use a small diamond on a large wheel.
9. Avoid impact of diamond against edge of wheel.

10. Avoid grinding of dressing-tool holder.
11. Turn tool often from (about) one-sixth to one-third turn.
12. Reset diamond when worn nearly level with the holder.
13. Keep transverses to limits of wheel face.
14. Avoid impact of diamond against wheel face.
15. Keep even pressure and transverse feed at finishing edges.
16. Do not attempt to rush truing operation.
17. True as often as grinding performance of wheel requires.
18. A heavy diamond is not required on light work.
19. Do not use a diamond that vibrates in its setting.
20. Do not use a diamond that seems even slightly loose in its setting.

Common sense and reasonable care are the first essentials in dressing and truing. An operator quickly gets the "feel" of the work, with the result that he can make a suitable selection of diamond dresser for the operation in hand.

Magnetic Chucks

Fixed magnetic chucks of rectangular pattern and rotating circular chucks provide means for securely holding flat, thin, or irregular parts without extensive setup. Body of the chuck contains coils which, when energized, set up a strong magnetic field in the steel core. Work laid on the faceplate of the chuck is attracted in relation to its magnetic properties. The attraction or pull may equal 150 psi or more, and and work down to approximately $\frac{1}{32}$ to $\frac{1}{16}$ in. can be held securely. Permanent-magnet chucks are also used.

Hints for Using Magnetic Chucks

Nothing but iron or steel can be held on the chucks.

The holding power depends on the amount of work surface in contact with the chuck.

Work can be held on edge by using adjustable back rest.

Very thin work can be held for grinding on the edges by laying it against the back rest and backing it up with a parallel strip.

Thin work will not hold so well as thick work.

In packing a number of small pieces on a chuck at one setting, it is better to separate them a little with strips of nonmagnetic material.

Do not plug up the vent holes in the chuck.

Keep water away from the switch, the brushes, and the interior of the chuck.

Do not use water on chucks except where they are made for it.

Chucks are usually wound for 110 or 220 v for dc only.

Use of Coolant

In grinding metals and many nonmetallic substances, it is essential to use a coolant to dissipate the heat generated by the abrading action and friction. It is particularly true in precision grinding that an equable, moderate temperature be maintained in the work to obviate distortion and to make accurate size taking possible during or immediately following the grinding operation.

Solutions containing soapy emulsions or soluble oil compounds are valuable as rust preventives and their cooling quality tends to reduce the heat generated and to improve the finish produced by the grinding wheel. Kerosene or other light oil solutions are used for aluminum and its alloys, and these permit settling of the light chips and aid effective grinding action.

Use of a grinding fluid tends to keep the wheel face from loading or filling with particles of the material being ground, thus improving its cutting action. It also settles the abrasive dust, preventing contamination of exposed bearing surfaces and adding to the hygienic conditions of the workman. The coolant should be applied in ample volume directly to the grinding point.

Correct proportioning of coolant compound to water is essential to good grinding. With soft water, less compound can be used; hard water requires more compound.

Correcting Taper in Ground Shafts

By Ralph Waech

Alignment of the swivel table on grinding machines can be done quickly with an indicator and tabulated figures which give the table movements necessary to remove taper from the work. Table 8 is based on a distance of 40 in. from the indicator to

Table 8. Swivel-table Adjustment to Overcome Shaft Taper

Taper per In., In.	Taper in 40 In., In.	Adjustment, In.
0.00001	0.0004	0.0002
0.00002	0.0008	0.0004
0.00003	0.0012	0.0006
0.00004	0.0016	0.0008
0.00005	0.0020	0.0010
0.00006	0.0024	0.0012
0.00007	0.0028	0.0014
0.00008	0.0032	0.0016
0.00009	0.0036	0.0018
0.0001	0.004	0.002
0.0002	0.008	0.004
0.0003	0.012	0.006
0.0004	0.016	0.008
0.0005	0.020	0.010
0.0006	0.024	0.012
0.0007	0.028	0.014
0.0008	0.032	0.016
0.0009	0.036	0.018
0.0010	0.040	0.020
0.0011	0.044	0.022
0.0012	0.048	0.024
0.0013	0.052	0.026
0.0014	0.056	0.028
0.0015	0.060	0.030
0.0016	0.064	0.032
0.0017	0.068	0.034
0.0018	0.072	0.036
0.0019	0.076	0.038
0.0020	0.080	0.040

the pivot point on the swivel table of the grinding machine. Column 1 lists the taper per inch found in the work, column 2 the corresponding taper in 40 in. When taper has to be eliminated from the part being ground, one-half of the taper in 40 in. will have to be taken up at the indicator. These adjustments are listed in column 3.

If micrometer readings at each end of a 5.4-in. shaft show it to have 0.0013-in. taper, the taper per inch is 0.0013/5.4, or 0.00024 in. The table shows that a taper of 0.0002 in. requires a movement of the swivel table of 0.004 in. while a taper of 0.00004 in. requires a movement of 0.0008; therefore, the swivel table must be moved 0.0048 in. at the indicator.

Grinding Processes

Centerless Grinding. The centerless-grinding machine consists of a conventional-speed grinding wheel and an opposed slowly moving regulating wheel, forming a grinding throat. A work rest suitably supports the work in this throat and the several parts are relatively adjustable for different sizes and varieties of work.

The action of the grinding wheel forces the work against the work rest (Fig. 11) owing to the cutting pressure and also against the regulating wheel by virtue of what may be called the "cutting contact" pressure. This pressure, aided by the gravity component of the work, keeps the piece being ground in contact with the regulating wheel. This wheel provides a continuously advancing frictional surface insuring constant and uniform rotation of the work at the same peripheral velocity as that of the regulating wheel.

Average roughing cuts are from 0.005 to 0.008 in. per pass, and finishing cuts range from 0.003 to 0.0015 in. per pass. When work is relatively straight but out of round, the regulating wheel should be set at a slight angle, giving a small lap per revolution. When pieces are warped, as from hardening, an angle of 5 or 6° will give a wide lap per revolution and straighten the work. With the proper angle for best results, the speed of the regulating or feed wheels should be increased to the limit established by the requirements of accuracy and production.

There are three main classes of centerless grinding, namely, through-feed, infeed, end-feed.

THROUGH-FEED GRINDING. Through-feed grinding is accomplished, as the name implies, by passing the work through or between the grinding and regulating wheels. Grinding takes place as the work passes from one side of the wheels to the other.

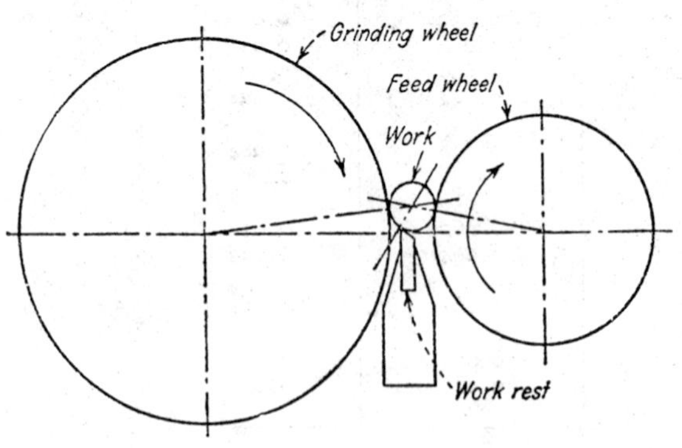

FIG. 11. Setup for centerless grinding.

Obviously, since all points on the work pass all contact points between the w
only straight cylindrical surfaces without interfering shoulders can be ground by t
method.

The axial movement of the work past the grinding wheel is imparted by the regulating wheel.

Machine is arranged in such a manner that the regulating wheel can be swung about a horizontal axis from zero to 7 or 10° relative to the axis of the grinding-wheel spindle.

The speed of the regulating wheel and the diameter of the regulating wheel also influence the feeding rate of the work. It is often necessary to pass work between the wheels more than once. The number of passes is determined by the amount of stock to be removed, the condition of the work as to roundness and straightness, the quality of the material, and the limits of accuracy required.

With this method there is a fixed relation between the grinding wheel, regulating wheel, and the work-supporting blade. The wheels are adjusted so that the distance between their active surfaces, together with the height of the work blade, determines the diameter of the ground piece. The centers of the wheels are stationary during grinding operations and require slight readjustments from time to time to compensate for the wear of the grinding wheel.

A work rest or fixture, which provides means for holding the blade, incorporates adjustable guides, to both the front and rear of the wheels. These guides must be accurately aligned with the regulating wheel face to insure that the work travels in a straight line.

Infeed Grinding. Infeed method is usually employed when grinding work that has a shoulder, head, or some portion larger than the ground diameter. The same method is used for the simultaneous grinding of several diameters of the work as well as for finishing pieces with taper, spherical, or any other irregular profile.

In general, this method corresponds to the plunge-cut or form grinding on the center-type grinder.

The length of the section or sections to be ground in any one operation is limited by the width of the grinding wheel.

As there is no relative axial movement of the work, the regulating wheel is set with its axis approximately parallel to that of the grinding wheel. A slight angle is maintained to keep work tight against an end stop.

End-feed Grinding. The end-feed method is used only on taper work. The grinding wheel, the regulating wheel, and the blade are set in a fixed relation to each other, and the work is fed in from the front, manually or mechanically, to a fixed-end stop. Either the grinding or the regulating wheel, or both, are dressed to the proper taper.

Regardless of the method employed, the grinding of the smaller size work can be assisted by use of a magazine, gravity chute, or hopper feeds, provided that the shape of the piece will permit the use of such devices.

The increase in the application of the centerless method of grinding to precision work is directly due to the following factors:

1. Saving in cost effected by the elimination of the centering operation.
2. Owing to the complete support of the work, the grinding wheel can be made to remove metal at maximum efficiency.
3. Material reduction of idle or nongrinding time in the grinding cycle.
4. The centerless-grinding method permits work to be rounded up with approximately half the surplus stock necessary for other types of cylindrical grinding, which materially reduces the wheel cost per piece ground.

ˏenterless-grinding machine sizes on the diameter and not on the ˏy automatically doubling the factor of safety in securing precision. ˏ plunge-cut work, the control of size is obtained by the utilization of the of leverage applied at a considerable radius terminating in a dead stop. design of the machine and the class of work on which it is generally employed themselves favorably to the application of automatic-hopper feeds for both ˏrough-feed and infeed work. Another factor is the automatic compensation for grinding-wheel wear, by which means the size of work produced is automatically maintained within extremely close limits of accuracy. Centerless-grinding machines are also made for internal work, which is revolved between two wheels while an internal spindle drives the abrasive wheel.

Contour Grinding with Crush-dressed Wheels

By J. C. Wilson, *Chief Engineer*, and Frederick G. Krafft, *Laboratory Head*
The Thompson Grinder Company

Primary Purpose. Contour grinding combines many operations into one to produce contours that would otherwise be difficult to achieve. On contours consisting of tangent radii or blending curves, contour grinding is also particularly useful. The process is highly economical in production of parts such as the fir-tree sections of turbine blades, thread forms in rolling dies and chasers, serrations in milling cutter blades and knurling dies, and punch and die sections for lamination dies. Where two or more identical parts are required or where a blend of several surfaces in one part is required the toolmaker will find the method suitable.

Crushing Rolls. These are generally 3 to 4 in. in dia and $\frac{1}{8}$ in. wider than the grinding wheel. Bore and OD of roll must be concentric; sides of roll must be rigidly mounted so that exact duplication is accomplished. An 18-4-1 high speed steel hardened to Rockwell 62 to 64 C is the most economical material for crushing rolls (see Table 9).

The advantage of additional life, obtained by the use of highly abrasion-resistant roll materials, is more than offset by increased difficulty of grinding. On forms where extreme sharpness of corners is not required, free-machining steel or close-

Table 9. Rate of Crushing-roll Wear

Depth of Contour, In.	Rate of Roll Wear for Different Materials (0.001 In. Wear per Wheel Dressing)			
	18-4-1 HSS	W 2 HSS M 2 HSS	Tungsten Carbide	Meehanite or Soft Steel
0.000	0.0005	0.0004	0.0002	0.001
0.025	0.005	0.0046	0.0025	0.008
0.050	0.010	0.0095	0.005	0.016
0.100	0.020	0.019	0.010	0.032
0.200	0.040	0.038	0.020	0.060
0.400	0.080	0.076	0.040	0.125
0.800	0.160	0.150	0.080	0.260

TABLE 10. SELECTION OF GRINDING WHEELS FOR CRUSH DRESSING

Grain Size

GRAIN SIZE	SIZE OF RADIUS AT ROOT OF FORM	THDS PER INCH
500	0.001	100
320	0.002	36
280	0.003	24
220.	0.004	16
180	0.005	10
120	0.007	6
80	0.010	

Grain Type Silicon carbide is used for molybdenum high-speed steels; high-carbon, high-chrome steels and most grades of Stellite and Tantung. Aluminum oxide is used for non-ferrous metals, and for steels up to including 18-4-1.

Grade—Wheel hardness depends on table feed. Grade "G" is considered best for heavy cuts at 2 to 10 in. feeds, but "H" or "F" may be used. For light cuts at 50 to 100 fpm, Grade "J" is general purpose.

Structure—Most open standard structure is desirable, but not extra porosity secured by use of fillers.

Bond—Vitrified bond is suitable; resinoid bond is unsuitable.

Grit Size—This depends on sharpness of corner at root of form. Finish is controlled by surface finish of crushing roll and coolant cleanliness.

---

grain cast iron may be used. Since these soft materials wear much more rapidly, it is necessary to recondition the roll more frequently. A shaving tool in a slide built into the crushing-roll fixture is a practical method of reconditioning soft rolls.

Use of Two Rolls. When hard material is used for the crushing rolls, the form life is prolonged by the use of two rolls. A work roll is used for the initial crushing of the grinding wheel and for subsequent truing or dressings between grinds. When the work roll loses form beyond tolerance, the grinding wheel is trued or dressed on a reference or master roll. The work roll is then reground to its initial form, and production is continued.

Suitable Grinding Wheels. See Table 10 for grinding-wheel selection. For crush dressing, the speed of the grinding wheel should be reduced to 250 to 300 sfpm for economy, although the rate of crushing-roll wear increases very little up to three times this speed.

Crush Dressing. The grinding wheel should be brought into driving contact with the crushing roll with both members at rest. A preload of 0.002 to 0.020 in. will develop between wheel and roll as crush dressing proceeds. This will vary with the grade and width of the grinding wheel, the rate of crushing, and the rigidity of

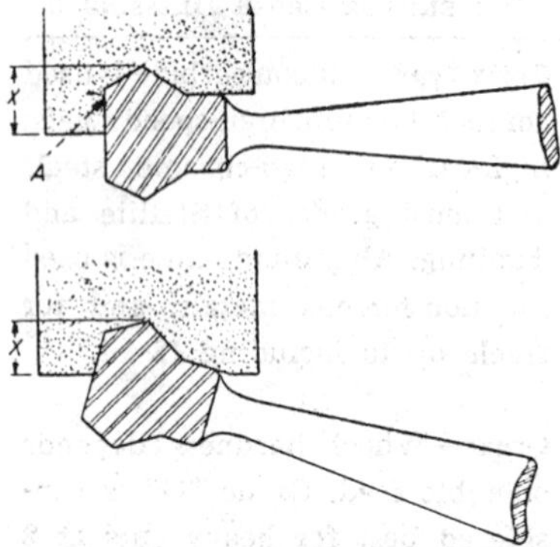

FIG. 12. Grinding the form as in the upper view is difficult because of the vertical side *A* and the additional depth of *X*. A setup as in lower view is preferable because it eliminates all vertical sides and reduces depth *X*.

the machine. This preload should be maintained constant for any one wheel to avoid concentration of stress which would result in breaking or spalling of the wheel.

Life of Rolls. The life of crushing rolls is proportional to the depth of form. A cylindrical roll having no contour will last ten times as long as one having 0.025-in. depth of contour (see Table 9). Very deep forms are impractical to crush-dress. With a HSS crushing roll equivalent of 20-pitch thread form, 0.001 in. of roll wear occurs after approximately 0.175 in. of wheel dressing.

Checking Contact. The crushing roll can be marked with layout fluid, crayon pencil, or other suitable means to show if all surfaces are in contact. The rate of crushing feed is 0.001 in. for two to five revolutions of the grinding wheel, depending on wheel hardness. All unnecessary rolling contact between wheel and roll should be avoided to result in optimum roll life.

Coolant. Best results are obtained in crush dressing by using an ample supply of good grinding oil directed to the point of contact between wheel and work (5 to 8 gpm per in. width of wheel). Very often open forms without sharp corners can be crush-dressed using a soluble oil. Where coolant is not permissible, a copious blast of clean dry air can be used. Since the coolant acts as a lubricant in crush dressing, the roll life is reduced when the process is performed dry.

Width of Form. There is no limit to width of grinding wheel that can be crushed other than the limits of the machine. In grinding extremely wide parts, care must be taken that the temperature rise produced does not distort the parts beyond tolerance.

Pitch of Serrations. Fineness of serrations that can be contour-ground is limited only by fineness of crushing roll. Thread forms as fine as 100 pitch or 90° included angle serrations as fine as 400 pitch are practical.

Slot Grinding. In crush dressing a wheel for grinding vertical-sided slots, it is best to build up the crushing roll from a series of disks or washers. By this method, sides can be maintained within 0.002 in. per in. (7′) of vertical. Maximum ratio of slot depth to width is approximately 10:1. Slots as narrow as 0.010 in. can be ground.

Balance the Form. Where a contour has straight and curved or angular sides, the setup should be made at such angle as will balance the form (see Fig. 12).

Cutting-off Operations

Abrasive cutting-off wheels are used extensively for practically all kinds of metals and for such materials as plastics, glass, porcelain, tile, and asbestos. Shellac, rubber, and resinoid bonds are used for these wheels. Some consider 2 in. dia as the maximum steel bar to be cut by the chopper method, but on special machines with oscillating wheel motion, bar stock up to 6 in. dia is being cut. Tubing is cut efficiently in sizes up to 3½ in. Shellac bonds are used in cutting tool steels. Wheels as thin as 0.005 in. are made with rubber bond for slotting pen nibs. Wheels 0.02 to 0.025 in. thick are used for cutting tungsten rods, and glass tubing is cut with wheels 0.030 to 0.062 in. thick. Resinoid and rubber bonds permit cutting at speeds up to 16,000 sfpm.

Makers of cutting-off wheels should be consulted as to specifications of wheels and the best speed for special materials and all unusual conditions.

Internal Grinding

Recommended Stock Conditions for Internal Grinding. There are many factors which govern conditions of holes, such as straightness, out-of-roundness, eccentricity with the locating points, and condition of the surface in the rough hole. However, it is felt that the following will be helpful for use under general conditions, but, where pieces are ground in large quantities and stock and heat-treating conditions are carefully controlled, less stock may be left in order to reduce the cost of grinding. The figures are for the stock left in the holes as they come to the grinder, whether soft or heat-treated, and for lengths two to three times the diameter. Longer holes usually require more stock, as do out-of-round holes and those that are not true with the finished hole to be produced.

Hole Dia, In.	*Recommended Stock Left for Grinding, In.*
¼ to 7/16	0.005 to 0.007
7/16 to ⅝	0.006 to 0.008
⅝ to 1	0.008 to 0.010
1 to 1½	0.010 to 0.012
1½ to 2½	0.012 to 0.015
2½ to 4	0.015 to 0.018
4 to 6	0.020 to 0.025
6 to 8½	0.025 to 0.030
8½ to 10	0.030 to 0.035

Table 11 gives stock allowances for hardened-steel parts.

Wheel Width and Diameter. It is general practice in internal grinding to use as short stubby wheel spindles or quills as possible, and, where narrow surfaces are being ground, the wheel is usually permitted to uncover half the width of the work. However, where the width of the work is wider than the width of the wheel, the wheel usually withdraws from the hole half the width of the wheel. In this way, straight holes are obtained without bellmouthing. To aid in choosing the width of wheel for a given size of hole, the widths of wheel shown in the following table are recommended practice:

Hole Dia, In.	*Suggested Width of Wheel, In.*
½ to 1½	Use the nearest standard wheel shape that has a width approximately equal to the wheel dia
1¼ to 5	1 to 1½
5 to 10	1½ to 2

When there are short ports or openings in the hole being ground, wheels wide enough to span the openings will prevent grinding low places adjacent to the openings. Where the openings are too long to span with the wheel, narrow wheels will

TABLE 11. STOCK ALLOWANCE FOR INTERNAL GRINDING*

Diam. of Hole in Inches	*LENGTH OF HOLE (in inches)*													
	5/16	½	¾	1	1½	2	2½	3	3½	4	5	6	7	8
⅛	.004 .005	.004 .005	....	....	....	....	....	....	....	....	....	....	....	....
¼	.005 .006	.005 .006	.006 .008	.006 .008	....	....	....	....	....	....	....	....	....	....
½	.005 .006	.005 .006	.006 .008	.006 .008	.008 .010	.008 .010	....	....	....	....	....	....	....	....
¾	.006 .008	.006 .008	.008 .010	.008 .010	.010 .012	.010 .012	.010 .012	.010 .012	....	....	....	....	....	....
1	.008 .010	.008 .010	.008 .010	.008 .010	.010 .012	.010 .012	.010 .012	.010 .012	.010 .012	.010 .012	....	....	....	....
1½	.008 .010	.008 .010	.010 .012	.010 .012	.010 .012	.012 .015	.012 .015	.012 .015	.012 .015	.012 .015	.012 .015	....	....	....
2	.010 .012	.010 .012	.010 .012	.012 .015	.012 .015	.012 .015	.015 .018	.015 .018	.015 .018	.015 .018	.015 .018	.015 .018	.015 .018	....
2½	.012 .015	.012 .015	.012 .015	.012 .015	.015 .018	.015 .018	.015 .018	.015 .018	.018 .020	.018 .020	018 .020	.018 .020	.018 .020	.018 .020
3	.012 .015	.012 .015	.012 .015	.015 .018	.015 .018	.015 .018	.015 .018	.018 .020	.018 .020	.018 .020	.018 .020	.018 .020	.018 .020	.018 .020
4	.015 .018	.015 .018	.015 .018	.015 .018	.015 .018	.018 .020	.018 .020	.018 .020	.018 .020	.020 .025	.020 .025	.020 .025	.020 .025	.020 .025
5	.018 .020	.018 .020	.018 .020	.018 .020	.018 .020	.018 .020	.020 .025	.020 .025	.020 .025	.020 .025	.020 .025	.025 .030	.025 .030	.025 .030
6	.020 .025	.020 .025	.020 .025	.020 .025	.020 .025	.020 .025	.020 .025	.025 .030	.025 .030	.025 .030	.025 .030	.025 .030	.030 .035	.030 .035
7	.025 .030	.025 .030	.025 .030	.025 .030	.025 .030	.025 .030	.025 .030	.025 .030	.025 .030	.030 .035	.030 .035	.030 .035	.030 .035	.030 .035
8	.025 .030	.025 .030	.025 .030	.025 .030	.025 .030	.025 .030	.025 .030	.025 .030	.025 .030	.030 .035	.030 .035	.030 .035	.030 .035	.030 .035

* Norton Company. These allowances are suitable for hardened steel parts with a rigid wall and under average conditions. Stock is given for the *diameter* of the hole and excludes the runout arising from improper centralization of the work in the fixture. Long holes and parts with thin walls require a greater stock allowance to compensate for heat-treating distortion.

reduce the wheel pressure and produce more accurate surfaces. There is no definite rule which can be made to cover all conditions, as experience will determine the best wheel for a job.

Table 12 will aid in obtaining the proper diameter of wheel for a given size of hole. It will be noted that there are two columns given to wheel diameters. One is for the diameter of wheel to be purchased, and the other is the diameter to dress a new wheel to make it suitable for grinding. The diameter to which the new wheel is dressed is he dimension of the wheel after the first dressing and is a suitable size to commence grinding with. However, when purchasing, it is necessary to secure the next largest standard wheel for holes of larger diameter than the desired wheel. When the diameter of the wheel given is the size of the wheel before dressing, it is only necessary merely to true up the wheel in such cases.

Polishing and Buffing

"Polishing" designates that branch of grinding whereby surfaces are smoothed by abrasive cemented or adhering to flexible belts or resilient wheels.

It is not possible to classify all work that should be done by flexible grinding in preference to solid-wheel grinding.

TABLE 12. RECOMMENDED SIZES OF WHEELS FOR INTERNAL GRINDING*
(In Inches)

Hole Dia to Be Ground	Grinding Wheel Dia Purchased	Grinding Wheel Dia Trued before Using	Grinding Wheel Width Av	Grinding Wheel Width Standard Range	Hole Dia to Be Ground	Grinding Wheel Dia Purchased	Grinding Wheel Dia Trued before Using	Grinding Wheel Width Av	Grinding Wheel Width Standard Range
$\frac{1}{8}$	$\frac{1}{8}$	$\frac{7}{64}$	$\frac{3}{16}$	$\frac{1}{8}$ – $\frac{3}{16}$	$1\frac{3}{4}$	$1\frac{5}{8}$	$1\frac{5}{8}$	$1\frac{1}{2}$	1 –$1\frac{1}{2}$
$\frac{3}{16}$	$\frac{3}{16}$	$\frac{11}{64}$	$\frac{1}{4}$	$\frac{3}{16}$ – $\frac{1}{4}$	$1\frac{7}{8}$	$1\frac{5}{8}$	$1\frac{5}{8}$	$1\frac{1}{2}$	1 –$1\frac{1}{2}$
$\frac{1}{4}$	$\frac{1}{4}$	$\frac{15}{64}$	$\frac{5}{16}$	$\frac{1}{4}$ – $\frac{5}{16}$	2	$1\frac{3}{4}$	$1\frac{3}{4}$	$1\frac{1}{2}$	1 –$1\frac{1}{2}$
$\frac{5}{16}$	$\frac{5}{16}$	$\frac{19}{64}$	$\frac{3}{8}$	$\frac{5}{16}$ – $\frac{3}{8}$	$2\frac{1}{4}$	2	2	$1\frac{1}{2}$	1 –$1\frac{1}{2}$
$\frac{3}{8}$	$\frac{3}{8}$	$\frac{11}{32}$	$\frac{7}{16}$	$\frac{5}{16}$ – $\frac{7}{16}$	$2\frac{1}{2}$	$2\frac{1}{4}$	$2\frac{1}{4}$	$1\frac{1}{2}$	1 –$1\frac{1}{2}$
$\frac{7}{16}$	$\frac{7}{16}$	$\frac{13}{32}$	$\frac{1}{2}$	$\frac{3}{8}$ – $\frac{1}{2}$	$2\frac{3}{4}$	$2\frac{1}{2}$	$2\frac{1}{2}$	$1\frac{1}{2}$	1 –$1\frac{1}{2}$
$\frac{1}{2}$	$\frac{1}{2}$	$\frac{15}{32}$	$\frac{1}{2}$	$\frac{3}{8}$ – $\frac{1}{2}$	3	$2\frac{1}{2}$	$2\frac{1}{2}$	$1\frac{1}{2}$	1 –$1\frac{1}{2}$
$\frac{9}{16}$	$\frac{9}{16}$	$\frac{17}{32}$	$\frac{1}{2}$	$\frac{1}{2}$ – $\frac{5}{8}$	$3\frac{1}{4}$	$2\frac{3}{4}$	$2\frac{3}{4}$	$1\frac{1}{4}$	$\frac{3}{4}$ –$1\frac{1}{2}$
$\frac{5}{8}$	$\frac{5}{8}$	$\frac{19}{32}$	$\frac{5}{8}$	$\frac{1}{2}$ – $\frac{5}{8}$	$3\frac{1}{2}$	3	3	$1\frac{1}{4}$	$\frac{3}{4}$ –$1\frac{1}{2}$
$\frac{11}{16}$	$\frac{11}{16}$	$\frac{21}{32}$	$\frac{5}{8}$	$\frac{1}{2}$ – $\frac{5}{8}$	$3\frac{3}{4}$	$3\frac{1}{4}$	$3\frac{1}{4}$	$1\frac{1}{4}$	$\frac{3}{4}$ –$1\frac{1}{2}$
$\frac{3}{4}$	$\frac{3}{4}$	$\frac{23}{32}$	$\frac{3}{4}$	$\frac{5}{8}$ –1	4	$3\frac{1}{2}$	$3\frac{1}{2}$	$1\frac{1}{4}$	$\frac{3}{4}$ –$1\frac{1}{2}$
$\frac{13}{16}$	$\frac{3}{4}$	$\frac{3}{4}$	$\frac{3}{4}$	$\frac{5}{8}$ –1	$4\frac{1}{2}$	$3\frac{1}{2}$	$3\frac{1}{2}$	$1\frac{1}{4}$	$\frac{3}{4}$ –$1\frac{1}{2}$
$\frac{7}{8}$	$\frac{7}{8}$	$\frac{13}{16}$	$\frac{7}{8}$	$\frac{3}{4}$ –1	5	4	4	1	$\frac{3}{4}$ –$1\frac{1}{2}$
$\frac{15}{16}$	$\frac{7}{8}$	$\frac{7}{8}$	$\frac{7}{8}$	$\frac{3}{4}$ –1	$5\frac{1}{2}$	$4\frac{1}{2}$	$4\frac{1}{2}$	1	$\frac{3}{4}$ –$1\frac{1}{2}$
1	$\frac{15}{16}$	$\frac{15}{16}$	1	$\frac{3}{4}$ –1	6	5	5	1	$\frac{3}{4}$ –$1\frac{1}{2}$
$1\frac{1}{8}$	$1\frac{1}{16}$	$1\frac{1}{16}$	$1\frac{1}{8}$	$\frac{3}{4}$ –$1\frac{1}{4}$	$6\frac{1}{2}$	5	5	1	$\frac{3}{4}$ –2
$1\frac{1}{4}$	$1\frac{1}{8}$	$1\frac{1}{8}$	$1\frac{1}{4}$	1 –$1\frac{1}{4}$	7	$5\frac{1}{2}$	$5\frac{1}{2}$	1	$\frac{3}{4}$ –2
$1\frac{3}{8}$	$1\frac{1}{4}$	$1\frac{1}{4}$	$1\frac{1}{4}$	1 –$1\frac{1}{4}$	$7\frac{1}{2}$	6	6	1	1 –2
$1\frac{1}{2}$	$1\frac{3}{8}$	$1\frac{3}{8}$	$1\frac{3}{8}$	1 –$1\frac{1}{2}$	8	6	6	1	1 –2
$1\frac{5}{8}$	$1\frac{1}{2}$	$1\frac{1}{2}$	$1\frac{1}{2}$	1 –$1\frac{1}{2}$	$8\frac{1}{2}$	$6\frac{1}{2}$	$6\frac{1}{2}$	$1\frac{1}{2}$	1 –2

* Norton Co. Internal grinding on a mass-production basis requires correct relationship of the wheel diameter and thickness to the given diameter of hole.

Parts or surfaces of irregular contour which require only reduction of size or shape to approximate dimensions can usually be ground economically on flexible wheels provided that the amount of material to be removed is not too great.

As a rule, flexible wheels tend to follow the original contour of the surface being ground; and, if it is necessary materially to change the shape of a part, it is better to use solid wheels. Work which has been ground with solid wheels is frequently polished or buffed afterward; and, if the grinding has been accurately done, the polished surface will possess accuracy as well as luster.

"Buffing" further smooths a surface and heightens the luster. The work is held against a wheel of fabric, felt, leather, or wood to which bar compound containing fine abrasive or a solution containing abrasive is applied. The abrasive-spray method allows the operator to concentrate on his work and attain higher outputs.

Types of Polishing Wheels and Belts

There are many types of wheels and belts used for flexible grinding and polishing each of which has certain characteristics which adapt it to certain kinds of work.

Cloth polishing wheels are made by gluing together sections of sewed pieced buffs. They are used for much the same class of work as the disk canvas wheels but are softer and more flexible.

Wool-felt wheels are manufactured from disks of woven felt from $\frac{3}{16}$ to $\frac{1}{4}$ in. thick. They provide a stiffer wheel than the cloth polishing wheel, owing to the thickness of the felt disks. They hold a flatter face and have less tendency to mush or round over on the corners.

Solid-felt wheels are made from a fine grade of felt in several qualities—fine or coarse. When coated with abrasives, they provide a wheel with a yielding character but with a face which is solid or unbroken as differentiated from wheels built up of disks.

Walrus hide, because of its great thickness and open porous character, is very valuable for making polishing wheels for certain fine finishes. It is extensively used in the silver and jewelry trades with the finer grades of powdered abrasives applied by moistening the abrasives and simply forcing them into the pores or fibers of the leather.

Sheepskin wheels are made by cementing or sewing together disks of sheepskin. The cemented wheels are quite soft, but the sewed ones are even more so. These wheels are suitable for fine finishing and buffing of soft metals. The hand-sewed wheel is used extensively in the jewelry trade.

Leather-covered wheels are made of laminated wood centers covered with a strip of "back" leather, the best quality of the hide used in belting. These are in common use and are particularly adapted for flat work and for work where it is necessary to maintain sharp corner outlines.

Bull-neck wheels are made by cementing together disks of bull-neck leather of fairly uniform thickness. Bull-neck leather, as differentiated from back leather, is softer, more spongy, and has a more open grain. The hardness and quality of the wheels can be varied by using leathers of different thicknesses, thick disks making softer wheels than thin disks.

Disk canvas wheels are usually of two general kinds of construction, the glued wheels and hand-sewn wheels. The hardest wheels are those made by gluing together the individual disks of canvas into a solid wheel. More flexible wheels are manufactured by sewing together several layers of the canvas into sections and then gluing the sections together into the solid wheel. The flexibility of the face may be increased by not gluing the sections all the way out to the periphery.

Compress polishing wheels are made from rectangular pieces of leather, canvas, felt, paper, or other material arranged radially and compressed to form a ring or cushion. This ring is assembled between two metal side plates which engage in annular grooves in the sides of the ring and hold the section intact.

Buffs. Buffs are manufactured in two general forms—loose or open buffs and sewed pieced buffs. The loose buffs are made from closely woven firm cotton fabrics of various grades and weaves as may be required. Each layer is a full-size disk. These are usually held together with one row of sewing near the center hole.

Sewed pieced buffs are manufactured with the disks comprised of several strips or pieces with outside full disks as covers, the whole assembly held together with rows of machine sewing.

Both these buffs are used with tripoli, crocus, levigated alumina, rouge, and other buffing compositions. Sewed pieced buffs are extensively used in the manufacture of the cloth polishing wheels. Buffs made of canton flannel or wool are used on such materials as hard rubber, bakelite, and precious metals.

Polishing belts are endless and are usually made of canvas. These are sometimes supported on leather belts and sometimes cushion wheels are used as a backing for the belt. On some classes of work, these belts are very efficient.

Importance of Proper Grain. The abrasive grain used in connection with flexible grinding or polishing is of great importance because it is the cutting or abrading medium which does the actual work. Natural abrasives were formerly used almost exclusively, but artificial abrasives of the aluminous type have very largely displaced them. Silicon carbide abrasives are seldom used for polishing. The quality of natural abrasives cannot be controlled.

The abrasives used are designated by number as in grinding wheels, running from 8 to 600 for this work.

Selection of Proper Size. Some work involves the removal of very little material and can be done with a single "polishing" operation. Other jobs require two, three, or even more operations starting with coarser and finishing with finer grain. If coarse grain (say, No. 36) is used in the first operation, this should not be followed by a very fine grain, because a great deal of time would be required to remove the deep scratches made by the coarse grain. It would be better to follow the No. 36 with No. 60 and then use No. 120 grain if a finish as fine as this is required. Still finer grains could follow the No. 120 if necessary to produce the desired luster. When changing from one size to the next finer, some experts advocate changing the direction of the strokes, if practical, in order to facilitate the removal of scratches.

Glue for Setup Wheels. Too much emphasis cannot be placed on the proper selection and application of glue. Unsatisfactory glue conditions are actually responsible for most of the common troubles in the polishing room such as poor work, variable abrasive costs, irregular rate of production, difficulty in maintaining standard piece rates, and many other things of this nature.

Abrasive-belt Grinding

By W. A. Corse, *Abrasive Engineer, Carborundum Company*

The finishing of round and flat work and of sheets has been speeded up remarkably by grinding and polishing with abrasive-coated belts. The "backstand" method, which is very common, employs a contact wheel supporting a long horizontal belt tensioned and tracked by a backstand. Platen-type machines are also much used, and here the working surface of the belt is normally in the vertical plane. Work like shafts or pins can be arranged to feed itself. Numerous special belt machines, such as contour grinders, swing-frame grinders, roll grinders, and deburring machines, are available.

Abrasives Used. Both silicon carbide and aluminum oxide abrasives in grits from 24 to 500 are used. Silicon carbide in wide paper-belt form polishes stainless steel. Grit sizes from 100 to 400 are used, depending on color and finish desired. Nonferrous metals are rough-ground with 24- to 36-grit silicon carbide; intermediate ground with 50 to 80 grit and finished with 100 to 240 grit.

Aluminum oxide is most commonly used for ferrous metals, occasionally on nonferrous materials. For offhand grinding and polishing of steel, roughing is done with 24- to 60-grit cloth belts; intermediate grinding with 80, 100, or 120 grit; and "oiling out," or final polishing, with 180 to 240 grit. Flat steel is polished to eliminate expensive polishing of formed work by rough-grinding with 60- to 100-grit paper-backed, glue-bonded belts and finishing with 150- to 220-grit belts.

Grinding Speeds. In general, 7000 sfpm is satisfactory for backstand work; sometimes 10,000 sfpm is used for rough grinding. With soft buff-type contact wheels, speeds of 3500 to 5000 sfpm avoid destroying the soft action of such wheels. Platen machines and wide-belt polishers run at 3500 to 5000 sfpm.

Coolants Used. Water with a rust inhibitor and oils and greases are used. Water produces the best finish, keeps the work cooler, avoids dust, prevents work distortion, flushes chips away, but tends to stain or rust the work unless soluble oil or inhibitors are added. Also, a more expensive resin-bond belt is required.

Oils are used in polishing stainless steel; kerosene is a good coolant for aluminum.

Greases when used in final polishing operations, soften cutting action of the belt, produce a high finish. A grease stick of the low-melting-point type is required.

Power Required. An average figure for motor horsepower required is 1 hp per in. of belt width, especially when high production is necessary and the machine is working at or near capacity.

HONING

Honing, whether internal or external, is a process wherein bonded abrasive sticks are applied to a surface under controlled pressure, and with a combination of rotary and reciprocating motions. Its function is the production of geometrically round and straight surfaces with a minimum of heat or disturbance of the crystalline structure of the material, and the production of accurate size. Honing also produces a type of surface finish unobtainable by any other method, and to any desired degree of smoothness.

Almost any material used in industry can be honed successfully, including glass and some plastics, but lead, babbitt, and some of the softer brasses may give trouble with loading of the stones. The design of the honing head imposes a certain limitation on size, and the smallest hole known to have been honed is 0.100 in. in dia. At the other end of the scale, maximum bore size to date has been over 40 in., but this is not the limit, and hones can be built to suit any desired size. As regards length, there is virtually no limit, and machines have been built with a stroke of 75 ft and 75-hp motor for driving the hone.

Preparation for Honing

Generally, honing follows fine boring, reaming, or grinding, when it is to be used merely for fine finishing and sizing. The stones remove the tiny ridges and irregularities left by the preparatory operations, and bring the whole surface down to the base metal. Because the stones are forced radially outward from the center of the tool with a constant pressure in all directions, the abrasive grits penetrate deeper into the tight, or out-of-round, spots and more stock is removed from that area. In this way, the bore is made round before any stock is removed from the large segments of diameter.

The hone or the work must be free to float so the tool can follow the original bore The surface quality produced by the preceding operation does not affect the degree of finish developed by honing, but the finer it is, the shorter will be the honing time. Stock allowance in such cases usually varies between 0.0002 and 0.001 in.

Increasing advantage, however, is now being taken of the rapid cutting action to remove large amounts of material, and in these cases a rough drilling or boring operation is all that is needed. It is not unusual to remove as much as $\frac{1}{16}$ in. at a rate of $1\frac{1}{2}$ to 2 cu in. per min, but if a fine finish is also required, the operation is preferably performed in two steps—roughing and finishing—with two different grades of stone.

Honing Practice

In conventional practice, the stones pass over the entire length of the work at each stroke and should project about one-quarter of their length at each end. This will insure enough overrun to produce straight, round, parallel walls without bellmouth. Sometimes, however, blind holes must be honed, and it is then desirable to have a recess at the bottom. This need be only a few thousandths larger than the finished diameter of the bore, but should be as long as possible. Where the recess is small, or nonexistent, the hone must be allowed to dwell for a few revolutions. If this is not done, the hole will be tapered at the closed end.

Rate of stock removal varies with the material and with the type of finish required (which is a function of the grade of stone), but on soft materials it is not unusual to remove 0.020 in. per min from the diameter, 0.015 to 0.018 in. on cast iron, and 0.006 to 0.012 in. on hard steel up to $65R_c$. These figures are based on work three to four diameters in length. On long steel tubing, as much as 200 cu in. may be removed per hr.

Cutting Speed

It is highly important that the stones be kept cutting at all times to insure continuous breakdown of the grits and the bond, and to prevent glazing. The aim should be to obtain free cutting with a minimum of pressure and heat generation. Rotary speed, therefore, is somewhat critical. An accompanying table gives speeds recommended by one manufacturer, but it must be understood that these are only approximate and are given as a general guide. It is, however, simple to select the correct speed at which the stones will cut freely without glazing, and at the same time give a reasonable stone life. The squeal produced by the operation is the clue; too high a speed is indicated by a high-pitched squeak, accompanied by heating. Under such conditions the stones

APPROXIMATE ROTARY SPEEDS FOR HONING

Diameter, in.		Rotary speed, rpm	
Hard steel	Gray iron	Roughing	Finishing
$\frac{5}{16}$	1	400	500
$\frac{5}{8}$	2	280	360
1	3	200	280
$1\frac{1}{2}$	4	160	220
$1\frac{3}{4}$	5	130	180
2	6	100	140
$2\frac{3}{4}$	8	65	100
3	10	50	65
4	12	40	50
$5\frac{1}{2}$	20	20	35

glaze and require more pressure to remove stock, and in extreme cases may cease to cut, especially on hard or close-ground materials. Too low a speed, which causes high stone wear, is indicated by a soft swishing sound, quite readily distinguished from the harsh cutting noise produced at the correct speed.

Linear speed is less critical, but best results are obtained when stroke speed is so related to rotation that the stones move with about a 45° helix angle. This must, however, depend upon the conditions of each particular job, and in some cases a 20 or 30° helix may be preferred. Experience is the best guide.

For holes which are short in relation to their diameter, stone length must be reduced to obtain automatic dressing of the surface and maintain truly parallel bores. Stone length should not exceed one-half the length of the hole.

Abrasive Sticks

The stones commonly employed are molded in stick form from either aluminum oxide or silicon carbide, with a resinoid or vitrified bond. For extremely hard materials, particularly for sintered carbides, diamond sticks may also be used. Grit sizes generally run from 36 to 600, and bonds may be soft, medium, or hard.

Quality of the surface obtained depends chiefly on the size of the abrasive, but the normal surface produced by a given grit can usually be made finer by changing the ratio of reciprocation to rotation so more multidirection action is obtained. In general, however, it is better to use a finer abrasive to get a finer finish. Hardness of the bond is also an important factor, and the harder the material, the softer should be the bond.

Choice of Abrasives

Selection of the abrasive depends chiefly on the material being finished. For cast iron, silicon carbide is generally used; for steel, aluminum oxide is preferred. Diamond sticks are costly but produce excellent results on the hardest materials, such as sintered carbides.

Coolants Are Important

All machine honing is done with the aid of a coolant, and satisfactory results can usually be obtained with a sulfurized mineral base or lard oil mixed with kerosene. For cast iron, straight kerosene or mineral-seal oil is often recommended, and for steel tubing a mixture of lard, oil, kerosene and sulfur base. For hard steel a light machine oil may replace the lard oil in the above mixture. For glass and similar brittle materials turpentine is advisable, and for bronze and other soft metals straight lard oil is preferred, although some bronzes require a sulfur-base oil to prevent loading of the stone. Water-soluble coolants have not proved satisfactory.

The right coolant is important as regards both production and finish, and it must be supplied in liberal quantities; 40 gpm is not unusual. Of equal, or even greater, importance is cleanliness, and every care must be taken to prevent chips and particles of abrasive from recirculating to the work surface. Any of the industrial filters will be satisfactory, provided they are cleaned regularly, but magnetic separators are generally preferred when honing ferrous material. These will operate continuously without attention.

LAPPING

Lapping may be defined as "precision finishing with abrasives." Specifically, the process, which is performed by hand or machine, improves the geometrical accuracy of flat surfaces and holes and refines the surface finish. Lapped surfaces are often required to obtain longer wear of moving parts, to obtain better seals, or to obtain longer life of cutting edges, as in dies.

Abrasives Used

Hand lapping is done with diamond powder or abrasive flours; machine lapping with loose abrasive, bonded abrasives, or cloth- or paper-backed abrasive belts.

Graded diamond powder in oil can be purchased from a number of suppliers, or it can be made in the shop from the small debris from diamond cutting.

Assume that the process is started with 25 carats of debris: Into a mortar place about 5 carats, using an 8-oz hammer to crush it. It takes from 3 to 4 min steady pounding to reduce it to a good average. Scrape the powder free from the bottom and the sides and empty into ½ pt of the best olive oil held in a 1½-pt cup-shaped receptacle. The 25 carats being reduced to powder and in the oil, stir until thoroughly mixed and allow to stand 5 min; then pour off to another dish.

The diamond that remains in the dish is coarse and should be washed in benzine and allowed to dry, and then be repounded, unless extremely coarse diamond is desired. In that case label it No. 0. Now stir that which has been poured from No. 0, and allow to stand 10 min. Then pour off into another dish. The residue will be No. 1. Repeat the operation, following the table below. The settlings can be put into small bottles for convenient use.

Settling Time for Diamond Powder

To obtain No. 0—5 min	To obtain No. 3—1 hr
To obtain No. 1—10 min	To obtain No. 4—2 hr
To obtain No. 2—30 min	To obtain No. 5—10 hr
To obtain No. 6—until oil is clear	

Diamond is seldom hammered into the lap material; it is generally rolled into the metal. For instance, several pieces of wire of various diameters charged with diamond may be desired for use in die work. Place the wire and a small portion of the diamond between two hardened surfaces, and under pressure roll them back and forth until thoroughly charged. In this case No. 2 diamond is generally used. Or one can form the metal to any desired shape and apply the diamond and use a roll to force the diamond into the metal. This is then a file which will work hard steel. But the moment this diamond file, or lap, is crowded, it is stripped of the diamond and is consequently of no use. It is to be used with comparatively light pressure.

Lap Materials

Cast iron, cold-rolled steel, copper, brass, and hardwood are the materials most commonly used for lapping operations. Cast iron is used mostly for machine lapping of flat or other shaped work, and also for hand lapping of flat surfaces and carbide-die openings. Steel and lead are employed in hole lapping. Copper readily takes diamond and is used for shaped laps, also for formed-wheel finishing of hardened steel or carbide on the surface grinder. Hardwood laps are restricted largely to fine finishing operations on die openings.

Hand Lapping

Hand lapping of flat pieces is ordinarily performed by rubbing the parts to be lapped over the accurately finished flat surface of a master lap, the abrading action being accomplished by a very fine abrasive powder mixed with a lubricant. One principal requirement of this operation is the movement of the work relative to the master lap. When properly performed, the piece is passed along an ever-changing path to insure uniform abrasion of both the work and the lap and to eliminate, as far as possible, the production of parallel grain marks.

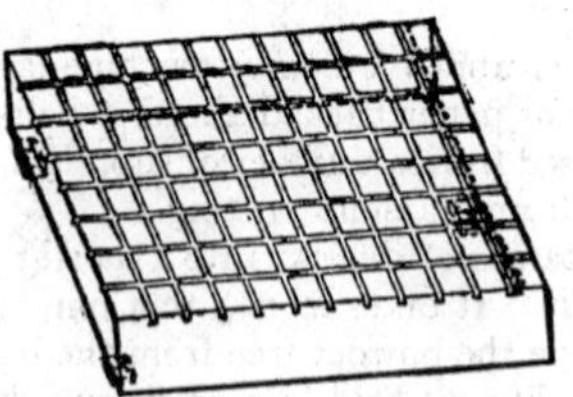

FIG. 13. Cast-iron lapping plates are scored to trap abrasives.

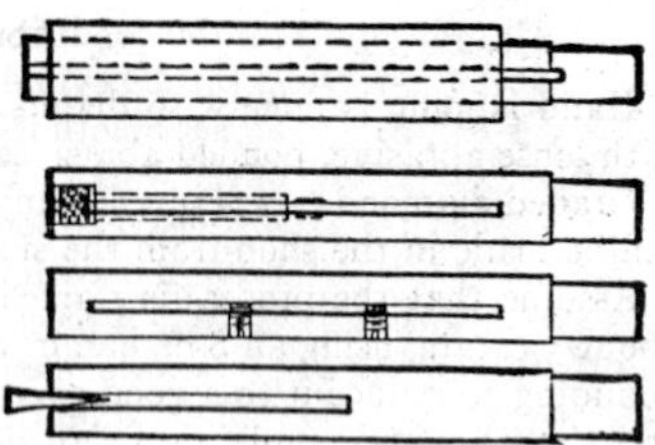

FIG. 14. Laps for holes.

Lapping by means of this ever-changing path of abrasion results in what is known as "matte" surface. It is not always a highly reflective surface.

Laps are roughly divided into three classes. First are those where the lap makes line contact with the work. If the work is cylindrical it is revolved to develop the cylindrical form, but if the work is straight in one direction it is moved back and forth under the lap. Second, full-contact laps may be used for straight surfaces. Third, full-contact laps are used for male and female cylindrical surfaces. In all cases the lap material must be softer than the work. If this is not so, the abrasive will charge the work and cut the lap, instead of the lap cutting the work.

Lapping Flat Surfaces. In lapping flat surfaces, which are usually on hardened steel, a cast-iron plate is used as a lap, with a good abrasive (Fig. 13). For rough work or "blocking down," the lap works better if scored with narrow grooves, about ½ in. apart, both lengthwise and crosswise, thus dividing the plate into small squares, as in Fig. 13. The abrasive is sprinkled loosely on the block, wet with lard oil and the work rubbed on it; care is taken to press hardest on the highest spots. The abrasive and oil get in the grooves and are continually rolling in and out, getting between the plate and the work, and are crushed into the cast iron, thus charging it thoroughly in a short time. About No. 100 or 120 abrasive is best for this purpose.

After blocking down, or if the work has first been ground on a surface grinder, the process is different. A plain plate is used with the best quality of flour of emery as an abrasive. Instead of oil, benzine is used as a lubricant, and the lap should be cleaned off and fresh benzine and emery applied as often as it becomes sticky. The work should be tried from time to time with a straightedge and care taken not to let the emery run in and out from under the work, as this will cause the edges to abrade more than the center and will especially mar the corners. After getting a good surface, the plate and work should be cleaned perfectly dry and then rubbed. The charging in the plate will cut just enough to remove whatever emery may have become charged in the work, take away the dull surface and leave it as smooth as glass and as accurate as it is possible to produce.

Some shops grind flat work to within 0.0005 in. of size and lap to within 0.0002 in. on a type-metal lap charged with flour emery. The lap is used dry and cleaned with benzine. Crocus powder on a cast-iron lap gives a high polish.

In using the hand method on outside cylindrical surfaces, the work to be lapped is held in a chuck mounted on a lathe spindle and rotated, while the operator holds a split lap over the work surface by means of a clamp or holder. Loose abrasive grain is employed with a metal lap of cast iron, lead, brass, or other metal. Measurements are taken periodically to check the work diameter until it is lapped to the required size.

Laps for Holes. In lapping holes, various kinds of laps are used, according to the accuracy required and the conditions under which the work is done. The simplest is a piece of hardwood turned cylindrical with a longitudinal groove or split in which the edge of a piece of emery cloth is inserted. This cloth is wound around the wood until it fills the hole in the work. This method is fit only for smoothing or enlarging rough holes and usually leaves them more out of round and bellmouthed than they were at first. Another lap used for the same purpose—and which produces better results—is made by turning a piece of copper, brass, or cast iron to fit the hole and splitting it longitudinally for some distance from the end. Loose abrasive is sprinkled over it, with lard oil for a lubricant, and a taper wedge is driven into the end for adjustment as the lap wears.

For lapping common drill bushings, cam rolls, etc., in large quantities, where a little bellmouthing can be allowed, and yet a reasonably good hole is required, a great many shops use adjustable copper laps made with more care than the above. One way of making them is to split the lap nearly the whole length, but leaving both ends solid. One side is drilled and tapped for expanding screws for adjustment. Either one screw halfway down the split may be used or two screws dividing the split into thirds. Another and better means of adjustment is to drill a small longitudinal hole a little over half the length of the lap, enlarge it for half its length, and tap the large end for some distance. This is done before splitting. Into this hole a long screw with a taper point is fitted so that when tightened it tries to force itself into a small hole, thus spreading the lap.

For quality work there is nothing better than a lead lap. Lead charges easily, holds the abrasive firmly, and does not scratch or score the work. It is easy to fit to the work and holds its shape well for light cuts. Under hard usage, however, it wears easily. For this reason, while laps for a single hole or a special job are sometimes cast on straight arbors, where much lapping is done it is customary to mold the laps to taper arbors with means for a slight adjustment. After any extensive adjustment the lap will be out of true and must be turned off. All these laps, as shown in Fig. 14, are to be held by one end in a lathe chuck, and the work run back and forth on them by hand, or by means of a clamp held in the hand. If a clamp is used, care should be taken not to spring the work.

How to Lap Good Holes. Several points must be taken into consideration in order to get good results in lapping holes. The most important is that the lap shall always fill the hole. Next in importance to getting a round hole is to have it straight. To attain this, the lap should be a little longer than the work, so that it will lap the whole length of the hole at once, and not have a tendency to follow any curvature there may be in it. What is known as bellmouthing, or lapping large on the ends, is hard to prevent, especially if the emery is sprinkled on the lap and the work shoved on it while it is running. The best way to avoid this condition when using cast-iron or copper laps, which do not charge easily, is to put the abrasive in the slot, near the center of the lap, and after the work is shoved on squirt oil in the slot to float the abrasive. Then, when the lathe is started the abrasive will carry around and gradually work out to the ends, lapping as it goes. Where lead is used, the abrasive can be put on where it is desired to have the lap cut and rolled in with a flat strip of iron. It will not come out easily, so will not spread to any extent, and it is possible with a lap charged in this manner to avoid cutting the ends of the hole at all. The work should always be kept in motion back and forth to avoid lumping of the abrasive and cuttings which will score grooves in the work.

Ring Gage and Other Work. Ring gages are lapped with a lead lap. They are first ground straight and smooth to within 0.0005 in. of size. When lapped, they are cooled to room temperature in benzine before the plug is tried. Some shops leave a

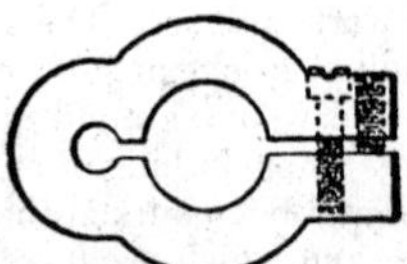

FIG. 15. Lap for plugs.

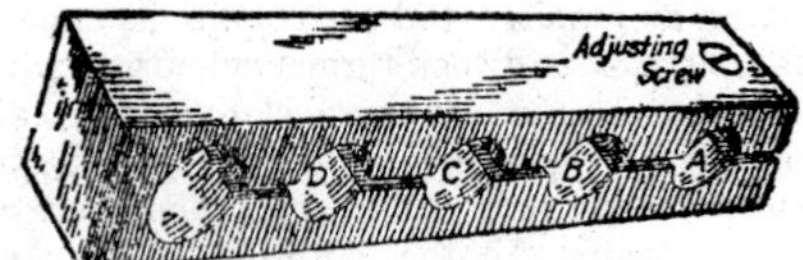

FIG. 16. Step lap for plug gages.

thin collar projecting from each side around the hole, so that, if there is any bell-mouthing, it will be in these collars, which are ground off after the lapping is done.

Soft metals and cast iron will charge. To some extent charging can be taken out without changing the work materially by rubbing by hand with "flour" abrasive cloth. In lapping bronze or brass, crocus and vienna lime are used. Crocus is used with a cast-iron or lead lap, and the charging is removed by running the work for a few seconds on a hardwood stick which fits the hole. Unslaked vienna lime, freshly crushed, is used with a lead or hardwood lap and does not charge. It does a nice job, but is very slow.

For lapping cylindrical articles, a cast-iron lap is usually used, split and fitted with a closing and a spreading screw, as shown in Fig. 15. Sometimes, where a very fine finish is required, or where the work is not hardened, the hole is made larger than the work, and a lead ring cast into it.

Adjustable Step Lap for Plug Gages. The step lap (Fig. 16) can be used to advantage for plug gages. It is of gray cast iron, and the end hole *A* adjusted to size. The lap is used at this hole until it refuses to cut. The plug gage is then entered into hole *B* without changing the adjustment. This hole is used up, and the plug gage is passed to holes *C* and *D*.

As the holes are all made one size and as the plug gage becomes smaller at each hole, there is less wear on the lap as the plug is passed from hole to hole. The holes retain about the same proportions and save measuring for sizes very often, when rough-lapping the plugs to within 0.0001 in. of finish size. As a finishing lap for size, it requires very few trial measurements with the micrometer or ring gage.

There is a tendency among workmen to use an excessive rotary speed when lapping a plug or ring. A speed that will make the work just comfortably warm to the hand is about right. Speed above this rate has a tendency to bake the oil and abrasive, which is a detriment to smooth and fast cutting.

Lapping Cemented Carbide. Lapping with diamond powder is essential for finishing die openings and pilot-pin holes in blanking and piercing dies, in which cemented carbide components are used for the cutting members. Also the approach angle and the bearing angle of draw dies must be lapped and polished.

Two kinds of abrasive are used—boron carbide (Norbide) in 240 grit for roughing out of small holes, and diamond powder dispersed in olive or castor oil. The No. 3 diamond powder (see Table 13) is most commonly used for rough lapping; Nos. 4, 5, and 6 for finishing.

Lap Materials. Cast iron (like Meehanite), cold-rolled steel, and copper are used for laps in the order mentioned.

Lapping Operations. The lapping of four types of holes is considered: straight tapered, shaped, and stepped.

STRAIGHT HOLES. Small holes are line-lapped with tapered (0.001 in. per in.) drill-rod laps (Fig. 17). Larger holes are lapped with split laps (Fig. 18). The face of the lap is not crisscrossed to trap abrasive, and the length of the lapping section is approximately two-thirds the depth of the hole.

TABLE 13. DIAMOND POWDER GRADES FOR LAPPING CARBIDES

Comparative Designations		Grain Size			
		Min		Max	
In Text	Bureau of Standards	Microns*	In.	Microns*	In.
No. 3	40	20	0.00078	60	0.00236
No. 4	25	13	0.00051	37	0.00145
	14	8	0.00051	20	0.00078
No. 5	8X	4	0.00015	12	0.00047
No. 6	3X	0	0	6	0.00023

NOTE: Some diemakers use a grading system for diamond as in column 1; others use the Bureau of Standards designations.
* One micron = 0.000039 in.

Always lap from the same side of the nib, and leave stock on the top face to be ground off to remove bellmouth.

TAPERED HOLES. Considerable skill is required to lap a tapered hole. If the taper is small, say 0.002 in. per side, a solid taper lap may be considered. The rule is: make the lap to one-half the desired hole taper. In use, the lap will wear in.

SHAPED HOLES. Wire-drawing die nib is mounted face up in a die-shaping machine, and the lap to shape the approach angle is reciprocated vertically through a $\frac{1}{16}$- to $\frac{1}{4}$-in. stroke. Boron carbide in 220 grit is used for the abrasive. The bearing and back relief are handled in similar setups. Subsequently, these three laps are mounted in a portable tool of the reciprocating type and are used to finish-lap the surfaces involved with No. 3 diamond powder.

STEPPED HOLES. Grind the holes to 0.005 in. of size. Then a series of laps, decreasing 0.001 in. dia, and all kept within 0.0005 in. for concentricity of steps, are used with No. 3 dust. The final lap is made within 0.0005 in. of size.

Polishing. When small holes must be polished, wrap cotton on a wood dowel and add a few drops of No. 5 diamond powder in oil. Chuck the nib in a speed lathe and apply the dowel by hand.

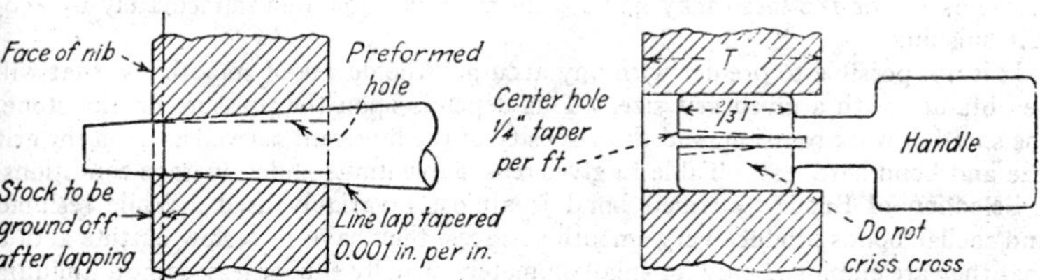

FIG. 17. A tapered lap is used for opening out cored openings under 3/16 in. in carbide die parts.

FIG. 18. For holes over 3/16 in. dia, a split lap is used in finishing holes in carbide die parts.

Machine Lapping

Modern lapping machines have made possible the quantity production of parts, the cost of which would have been prohibitive by the old hand-lapping methods.

Industrial requirements demand lapping machines arranged to use three types of lapping mediums. One uses metal laps and loose abrasive grain mixed with a lubricant and will always find a place in the field of gage manufacture, or for other operations where extreme accuracy is required.

Another type of machine uses bonded abrasives for commercial production work.

A third and more recent type of machine employs abrasive paper or cloth instead of cast-iron or bonded abrasive sticks. A very bright finish is obtained and a genuine lapped surface is produced.

Preparation of Parts for Lapping

From a production standpoint, lapping is not a stock removal operation. In no case should more than 0.0005-in. material be left for the lapping operation. Lapping is not intended to replace grinding, but further to refine pieces beyond that point of finish and accuracy produced by the grinding operation.

Parts should always be ground to close limits for size, parallelism, roundness, or flatness before lapping, although the ground finish is not so important.

SUPERFINISHING

Superfinishing is a method promoted by Gisholt Machine Co. to secure an exceptionally smooth, wear-resistant surface on wearing parts. The machines give a combination of motions to the abrasive materials, mostly at right angles to the work. Only one movement, and that at low velocity, is in the direction of the tooling marks. Oscillating movements are quite rapid, but the actual surface speed is only 12 to 1[illegible] fpm because of the narrow range of the stone oscillations.

Choice of Abrasives. Aluminum oxide is standard for superfinishing steels, and silicon carbide is generally best for cast iron and nonferrous metals.

While slower cutting, silicon carbide is often employed on steels softer than about Rockwell 30 C, and for diameters under about ⅜ in. Also, as silicon carbide can be made in fine grits, it can produce ultrafine surfaces of 1½ μ in.[1] or less on hard steels.

Grit Size. Most superfinishing is preceded by a grind finer than 30 μ in. rms when the desired final finish is 3 μ in. or less. For these average operations, a 600 mesh abrasive has generally proved most satisfactory, in either kind of grit. When the final grind is rougher than 30 μ in. rms, it may be desirable to employ a 500-grit stone in the interest of speed. Grit sizes of 400 or 320 mesh may be used when a definite scratch pattern is desired for some special purpose. In some machines where high rates of production are required—often from a turned surface—grits as coarse as 120 or 180 mesh may be used for roughing, followed immediately by 400 grit finishing.

It is not possible to predict with any accuracy the degree of smoothness that will be obtained with a given grit size. This depends upon the pressure on the stone, the speed of work rotation, and the viscosity of the lubricant, as well as upon the grit size and bond hardness. Table 14 gives reasonable limits under average conditions.

Selection of Bond. Vitrified bond is almost invariably used. While resinoid and shellac bonds produce very smooth surfaces, they have such slow cutting action that they are employed only for small diameters, usually ¼ in. or less; some aluminum alloys; or second operations on a few difficult metals, such as brasses and soft stainless steels.

[1] μ in. = microinches.

Table 16 gives recommended Rockwell H hardnesses for vitrified bond stones to be used on steel parts ½ in. dia and over. Necessary deviations are:

1. Diameters of less than ½ in. require progressively harder bonds as they become smaller.

2. Interrupted surfaces require a bond hardness increased in accordance with the degree of the interruption.

3. Very roughly ground surfaces, which would wear a stone face rapidly, require a harder bond—perhaps 5 points Rockwell H harder than otherwise would be applied.

4. Traversing requires a slightly harder bond than when the work is so short that traversing is not employed.

5. Conical shapes must be superfinished with a bond-hardness of Rockwell 65 to 80 H.

When employing silicon carbide stones on steels, approximately the same bond hardnesses should be used as for aluminum oxide stones.

It has not been found possible to correlate the bond hardnesses of silicon carbide stones with the hardnesses of cast iron and nonferrous metals, as has been done with steels. As a starting point, it is recommended that a stone of Rockwell 50 to 60 H be selected for trial. This range is suitable for steel hardened to Rockwell 40 to 50 C. If steels are Rockwell 60 to 65 C, use stones of Rockwell 20 to 30 H.

TABLE 14. ANTICIPATED FINISHES FOR DIFFERENT GRIT SIZES

GRIT SIZE	FINISH USUALLY OBTAINED
600	1½ to 4 mu in. rms
500	3 to 7 " "
400	4 to 10 " "
320	6 to 20 " "

TABLE 15. STANDARD ABRASIVE STICKS AND CORRESPONDING WORKPIECE SIZES

SIZE of ABRASIVE STICK	DIAMETER OF WORKPIECE
¼ × ⅜ in.	0 to 7/16 in.
⅜ × ½ in.	7/16 to 9/16 in.
½ × ⅝ in.	9/16 to 7/8 in.
¾ × ¾ in.	7/8 to 1¼ in.
1 × 1 in.	1¼ to 3 in.

TABLE 16. RECOMMENDED HARDNESS FOR VITRIFIED SUPERFINISHING STONES FOR STEEL ½ IN. DIA AND OVER

ROCKWELL C OF STEEL	ROCKWELL H OF STONE
Chromium Plate	0 to 15
65 plus	15 to 30
65 to 58	30 to 40
58 to 50	40 to 50
50 to 35	50 to 60
35 to 15	60 to 70
15 and less	70 to 80

Shapes of Superfinishing Stones. Cylindrical and conical shapes are finished with square or rectangular stick-type stones.

Superfinishing sticks are often altered from the standard shape to meet special or high-production requirements. There are two reasons for this practice:

1. To counteract dressing of an area of the stone face by its passage over a hole or other interruption in the work.

2. To maintain straightness in a diameter, most often when there are shoulders at one or both ends to restrict the length of the stone.

Lubricant. The usual lubricant is compounded from a light-mineral-oil base, such as kerosene or mineral seal oil, together with 5 to 15% of a heavier cutting oil to regulate the viscosity—and thus the cutting action of the stone. The addition of a heavy straight mineral oil for this purpose is not satisfactory, and a water-base compound has never been found at all successful.

OILSTONES AND THEIR USES

Natural Stones

The following particulars regarding the well-known Arkansas and Washita stones are given by the Behr-Manning Div., Norton Co.

Arkansas stones are prepared in two grades, hard and soft.

Hard Arkansas stone is composed of pure silica, and its sharpening qualities are due to small, sharp-pointed grains, or crystals, of hexagonal shape, which are much harder than steel and will, therefore, cut away and sharpen steel tools. The extreme fineness of texture makes this stone, of necessity, a slow cutter, but in the very density of the crystals of which it is composed lies its virtue as a sharpener.

Soft Arkansas stone is not quite so fine-grained and hard as the hard Arkansas, but it cuts faster and is better for some kinds of mechanical work. It is especially adapted for sharpening the tools used by wood carvers, filemakers, patternmakers, and all workers in hardwood.

Washita stone is also found in Arkansas and is similar to the Arkansas stone, being composed of nearly pure silica, but is much more porous. It is known as the best natural stone for sharpening carpenters' and general woodworkers' tools.

Artificial Oilstones

Artificial oilstones are manufactured in a multitude of shapes and sizes and are adapted for sharpening all kinds of tools. Such stones are made of Alundum and Crystolon, by the Norton Co., the former being known as India oilstones, the latter as Crystolon sharpening stones. Similar shapes are manufactured by the Carborundum Co. and others.

The stones are made in three grades or grits—coarse, medium, and fine. The coarse stones are used in machine shops for sharpening very dull or nicked tools and machine knives and for general use where fast cutting is desired.

Medium stones are for sharpening mechanics' tools in general, more particularly those used by carpenters and in woodworking shops.

Fine stones are adapted for engravers, die workers, cabinet makers, and other users of tools requiring a very fine, keen-cutting edge.

How to Care for Oilstones

To retain the original freshness of the stone, it should be kept clean and moist. To let an oilstone remain dry a long time or to expose it to the air tends to harden it. A new natural stone should be soaked in oil for several days before it is used.

To keep the surface of an oilstone flat simply requires care in using it. Tools should be sharpened on the edge of a stone as well as in the middle.

To restore an even, flat surface, grind the oilstone on the side of a grindstone or rub it down with sandstone or an emery brick.

An oilstone can be prevented from glazing by use of oil or water.

Water—and plenty of it—should be used on all coarse-grained natural stones.

On medium- and fine-grained natural stones and in all artificial stones, oil should be used always, as water is not thick enough to keep the steel out of the pores.

Further to prevent glazing, the dirty oil should *always be wiped off the stone thoroughly* as soon as possible after using it. Cotton waste is good for this.

If the stone does become glazed or gummed up, a good cleaning with gasoline or ammonia will usually restore its cutting qualities, but if it does not, then scour the stone with loose emery or sandpaper fastened to a flat board.

Never use turpentine on an oilstone for any purpose.

Section 6

MILLING AND MILLING CUTTERS

SECTION 6

MILLING AND MILLING CUTTERS

Milling Process. Milling is a machining process whereby a surface is generated with a rotating toothed cutter. Each tooth takes an individual chip. The surface may be plane or of regular or irregular profile; it may be a thread, a cam profile, etc.

In order to provide feed, relative motion between cutter and workpiece must be secured. In most applications of milling, the cutter revolves about a stationary axis at relatively high speed, and the work is moved past the cutter (with suitable depth of cut) at a comparatively low rate of feed. In other cases, the work may revolve slowly, as in thread milling, while the cutter revolves at a higher speed.

Types of Milling. In *peripheral* milling, the surface is generated by teeth located on the periphery of the cutter body. The surface may be flat or profiled. In *face* milling, the milled surface is generally at right angles to the cutter axis, is flat, and is produced by the combined action of cutting edges on the face of the cutter as well as its periphery.

Methods of Milling. In peripheral milling, the workpiece can be fed either with or against the direction of cutter rotation. When the cutter rotates in the feed direction, the method is termed "in-cut milling" (also climb milling or down milling), because each tooth cuts inward from the work surface and finishes its cut at the machined surface (Fig. 1). When the cutter rotates against the direction of feed, the method is called "out-cut milling" (also conventional or up milling) because each tooth cuts outward from the machined surface and finishes its cut at the work surface (Fig. 2). With the in-cut method, the chip is thickest at the beginning of the cut, whereas with the out-cut method the chip is thinnest at the beginning of the cut.

In-cut, or down, milling is best performed on machines in good condition, and when the work-moving slide is steadily fed; otherwise yielding of the work may cause a broken cutter, as well as damaged work and machine.

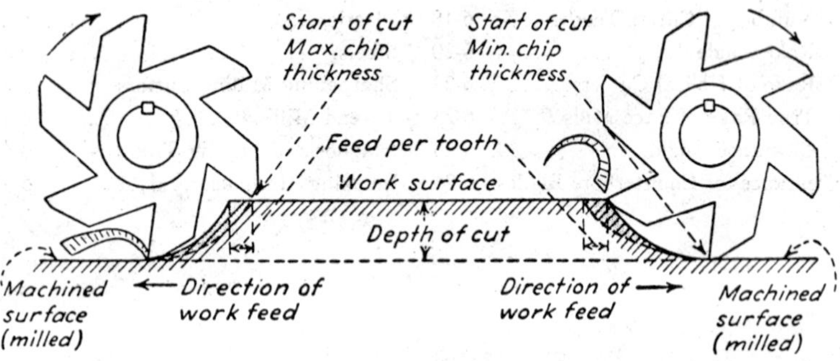

Fig. 1. In-cut milling (or climb milling, down milling).

Fig. 2. Out-cut milling (or conventional milling, up milling).

TYPES OF MILLING CUTTERS

American Standard B5.3-1950

Definition. A milling cutter is a rotary cutting tool provided with one or more cutting elements called teeth, which intermittently engage the workpiece and remove material by relative movement of a workpiece and cutter. Typical milling cutters and their applications are shown in Figs. 3 and 4.

General Classifications

Classification Based on Construction. Milling cutters are often described by using terms which refer to their construction characteristics, such as:

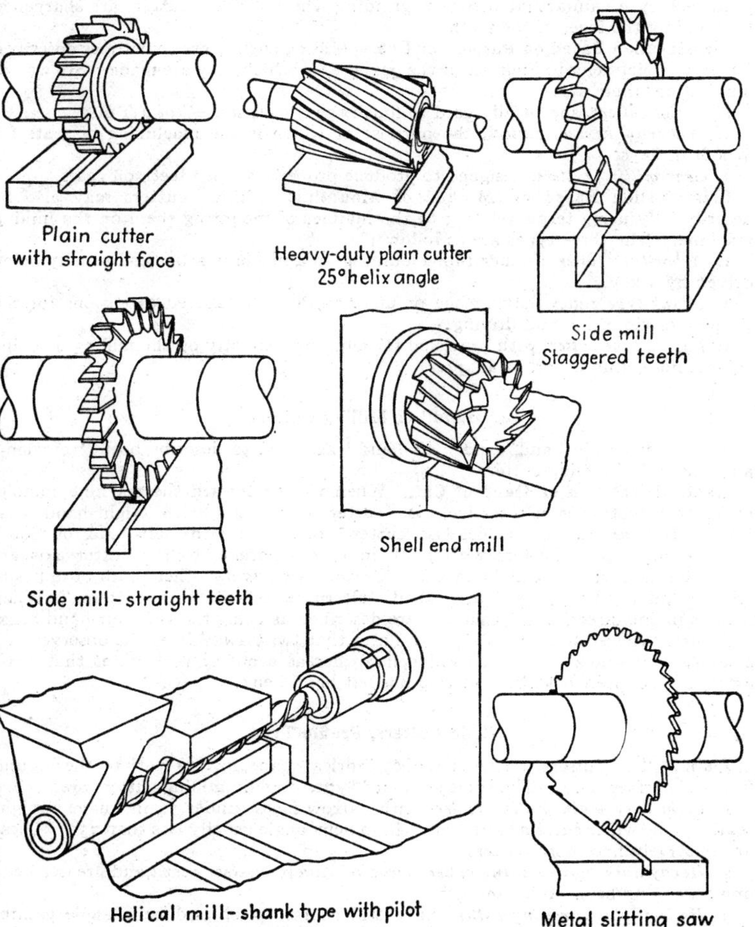

FIG. 3. Typical milling cutters and their applications.

1. *Solid cutters*, those made of one piece of material; for example, high-speed steel.

2. *Tipped solid cutters*, those similar to solid cutters, except that such materials as cast alloys or carbides are joined to them to provide the cutting edges of the teeth.

3. *Inserted-tooth cutters*, those which have mechanically retained, and/or adjustable, solid or hard alloy-tipped teeth or blades.

Classification Based on Relief of Teeth. Milling cutters may also be described on the basis of one of two methods of providing relief for the cutting edges. These methods determine the manner of sharpening.

1. *Profile-relieved cutters*, those on which the relief is obtained and sharpening done by grinding a narrow land back of the cutting edge. Profile-relieved cutters may produce flat, curved, or irregular surfaces, the latter being called profile-type form cutters.

2. *Form-relieved cutters*, those on which a curved relief back of the cutting edge is produced by a cam-actuated tool or grinding wheel. These cutters are sharpened by grinding the faces of the teeth.

Classification Based on Purpose or Use. Milling cutters are sometimes described by terms which refer to their use or the purpose for which they are made. By way of illustration, there are:

1. *T-slot cutters*, specifically used to finish the "head space" of a T-slot.

2. *Woodruff keyseat cutters*, the purpose of which is the machining of seats for Woodruff keys.

3. *Gear milling cutters*, designed to produce properly shaped teeth on gears.

Classification Based on Method of Mounting. Milling cutters may also be described by using terms relating to the manner of mounting them on the milling machine. The three types are as follows:

1. *Arbor-type cutter*, intended for mounting on a machine arbor and usually being driven by a key.

2. *Shank-type cutter*, having one or more straight or tapered extensions for the purpose of mounting and driving.

3. *Facing-type cutter*, with provision for mounting directly on the milling machine spindle nose, stub arbor, or adapter.

"Hand" of Milling Cutters

The terms "right-hand" and "left-hand" are used to describe both "rotation" and "helix" of milling cutters.

Hand of Rotation or Hand of Cut. When viewed toward the machine spindle, and rotary motion is counterclockwise, cutter is operating with "right-hand rotation." If rotary motion is clockwise, cutter is operating with "left-hand rotation."

Hand of Helix. Cutters with teeth in planes parallel to the cutter axis are described as having "straight teeth." Cutters with every other tooth of opposite (right- and left-hand) helix are called "alternate helical tooth cutters." When helix is in one direction only, cutters are described as being right- or left-hand helix.

A cutter viewed from one end, with flutes that twist away from the observer in a clockwise direction, has right-hand helix; whereas a cutter with flutes that twist away in a counterclockwise direction has left-hand helix.

Solid Cutters, Profile Type

Plain Milling Cutters. These are of cylindrical shape, having teeth on the circumferential surface only. The three standard types of plain milling cutters are:

1. *Light-duty plain milling cutters* with narrow faces (under ¾ in.) have straight teeth. The wider cutters have teeth with a helix angle usually less than 25°. These are relatively fine tooth cutters.

2. *Heavy-duty plain milling cutters* have relatively coarser teeth and greater helix angle, ranging between 25 and 45°.

3. *Helical plain milling cutters* have still coarser pitch and helix angle greater than 45°.

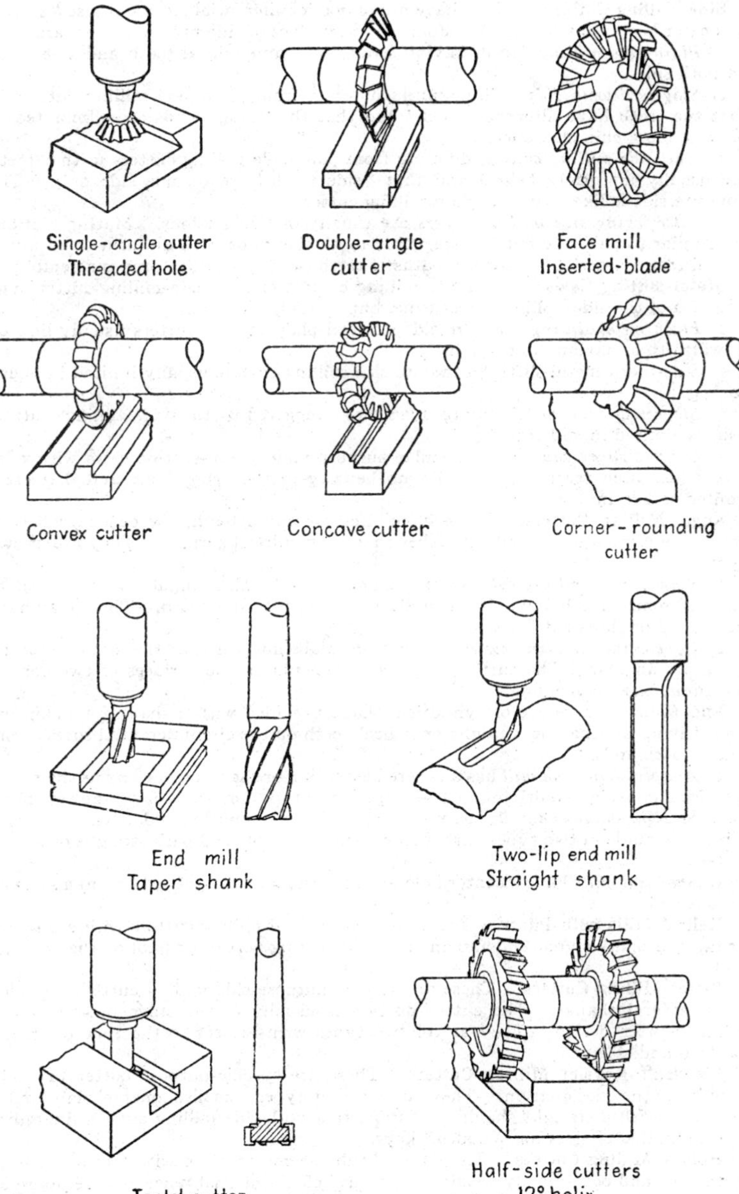

FIG. 4. Typical milling cutters and their applications.

Side Milling Cutters. These have not only circumferential teeth but also "teeth" on one or both sides (or ends). Four common types of side milling cutters are:

1. *Plain side milling cutters* having straight circumferential teeth and side teeth on both sides.

2. *Staggered-tooth side milling cutters* differing from plain side milling cutters in that the teeth have alternating helix and that the "drag" ends are eliminated to provide more chip clearance.

3. *Half side milling cutters*, differing from plain side milling cutters in that teeth are usually uniformly helical and that "side teeth" are on one side only. The opposite side is like that of a plain milling cutter.

4. *Interlocking side milling cutters* are usually of two sections. Mating sections are similar to half side mills or staggered-tooth side mills with uniform or alternate helical teeth so designed that the paths of teeth overlap when in proper assembly.

Metal-slitting Saws. These are milling cutters of the plain-milling-cutter type and also of the side-milling-cutter type, but limited in width.

1. *Plain metal-slitting saws* are side-relieved plain milling cutters usually limited in width to a maximum of 3⁄16 in.

2. Side-tooth metal-slitting saws are side milling cutters usually limited in width to 3⁄16 in.

3. *Staggered-tooth metal-slitting saws* are staggered-tooth side milling cutters usually limited in width to 1⁄4 in.

4. *Screw slotting saws* are particular-purpose plain metal-slitting saws with relatively fine-pitch teeth, occasionally no keyway in the arbor hole, and diameters limited to 2¾ in.

Angle Milling Cutters. These have circumferential teeth, the cutting edges of which are elements of a conical rather than a cylindrical surface. They are of two types:

1. *Single-angle milling cutters* are angular half side mills, angular with respect to circumferential teeth and half side mills with respect to side teeth. These may have keywayed or threaded holes.

2. *Double-angle milling cutters* differ from plain mills in that the circumferential teeth are angular. The cutting edges are elements on the surfaces of two conical frustums, bases coinciding.

End Mills. These are of cylindrical shape provided with a shank for mounting and driving and having straight or helical teeth on the circumferential surface and one or both ends.

1. *A taper-shank end mill* has a tapered extension or shank on one end as the provision for mounting and driving. These are of two types: multiple-flute and two-flute.

2. *Straight-shank end mills* have cylindrical-shaped mounting and driving portions. These are made in two types, either single-end or double-end with straight or helical flutes.

3. *Shell end mills* have a center hole for mounting upon a stub arbor and a keyway on one end.

Helical Mills with Pilots. These are plain helical milling cutters with a straight or tapered shank on one end and an outboard bearing portion or pilot on the opposite end.

T-slot Milling Cutters. These are specific-purpose side milling cutters provided with integral shanks. The cutter portion is usually of the staggered-tooth side milling cutter design. They are of two types with respect to the type of shank, tapered and straight.

Woodruff Keyslot Milling Cutters. These are specific-purpose cutters as indicated in the designation. They are of two types: small-diameter plain mills, provided with a straight shank, and staggered-tooth side milling cutters limited in size to milling of slots for Woodruff keys.

Hollow Milling Cutters. These are of tubular construction with teeth on one or both ends and occasionally within the center hole. Internal relief may be obtained through a plain tapered hole or internal flutes with back taper.

Solid Cutters, Form-relieved

A **convex milling cutter** is one with teeth curved outward on the circumferential surface to form the contour of a semicircle. These are designed to mill a concave surface of circular contour equal to half a circle or less.

A **concave milling cutter** is one with teeth curved inward on the circumferential surface to form the contour of a semicircle. These are designed to mill a convex surface of circular contour equal to half a circle or less.

A **corner-rounding milling cutter** is one with teeth curved inward on the circumferential surface adjacent to one side to form the contour of a quarter circle. These are designed to mill a convex surface of circular contour equal to a quarter circle or less.

Involute gear milling cutters are form cutters designed to mill the spaces between involute form gear teeth.

Roughing (stocking) gear milling cutters are made in several styles for "roughing out" spaces between involute form gear teeth. Various styles are differentiated chiefly by a number of types of chip-breaker design. All styles are sized to leave stock for finishing both sides of gear teeth.

Single-roughing milling cutters mill one undersized space at a time.

Multiple-roughing cutters mill two or more undersized spaces at one time.

Finishing gear milling cutters mill finished spaces between involute form gear teeth.

Sprocket-wheel milling cutters are form cutters designed to mill the spaces between sprocket wheel teeth. Single-sprocket milling cutters mill, one at a time, the spaces between sprocket-wheel teeth.

Straddle-sprocket milling cutters mill the complete profile of one sprocket wheel tooth at one time.

Spline milling cutters are form cutters designed to mill splines on a workpiece.

Solid Cutters, Profile and Form-relieved

Thread milling cutters are designed to mill threads of a specific form on a workpiece. These are of two general types:

1. *Single-thread milling cutters* which are also of two types with respect to the method of producing the relief:

Single-thread profile milling cutters mill one finished thread on a workpiece at one time (generally worm or Acme thread) when used in a specifically designed thread milling machine. These are of the "full-tooth" and the "interrupted-tooth" design. The full-tooth type is generally used to mill fine-pitch worms. The interrupted-tooth type has alternate sides of every other tooth cut away for chip clearance with only one tooth full for sizing. These are used for milling coarser pitch worms.

Single-thread form-relieved milling cutters are also made to do thread milling. These may be the "topping" or "nontopping" types.

2. *Multiple-thread milling cutters*, which mill in one revolution of the work a number of threads, have no "lead." They are regularly made for all standard types of thread and are used for both external and internal thread milling, either straight or tapered. They have form-relieved teeth and are of two types with respect to provision made for mounting and driving.

Inserted-tooth cutters can be made to duplicate practically all common types of solid cutters. One type of inserted-tooth cutter, the "face" mill, however, has no counterpart among solid cutters. The teeth of all inserted tooth cutters may be either of solid cutting material, such as high-speed steel, cast alloy, or sintered carbide, or tipped with these materials. Three classifications of inserted-tooth cutters are possible based on methods of mounting. Popular types of inserted-tooth cutters are:

Tipped solid cutters are available in all solid-style cutters. The distinction between these cutters and solid cutters consists in the tipped cutters having hard-alloy material joined to a cutter body to form the faces and cutting edges of the teeth. The distinction between hard-alloy-tipped cutters and inserted-tooth hard-alloy-tipped cutters consists in the latter having tipped teeth or blades which are retained mechanically in the cutter body.

"Step" cutters are face mills or end mills cutting various planes to produce one plane surface.

TOOTH PARTS AND ANGLES

Plain Milling Cutters

A plain milling cutter (Fig. 3) has teeth with cutting edges on the peripheral or cylindrical surface of the cutter body only. When teeth are inserted in a cutter body, as in the large face mills, they are called blades.

The distance between the two faces or ends of plain, helical, and side milling cutters, or the length of the outside diameter cylinder, is the *cutter width* if small or *cutter length* if large with respect to the diameter.

Land. The land width (Fig. 5) varies from about 1/64 in. for small-diameter cutters to 1/16 in. or more for large-diameter cutters. Sometimes a cylindrical margin from 0.005 to 0.010 in. wide is left back of the cutting edge. The relieved surface is then provided back of the margin.

Relief Angle. The relief angle (peripheral) is the angle between the surface formed by the land and a tangent to the cutter outside circle passing through the cutting edge in a diametral plane.

Relief and clearance are measured in degrees or radial fall in inches at a certain specific distance back of the cutting edge on the land. A dial indicator may be used to measure the radial fall in thousandths.

Clearance Angle. The clearance angle (peripheral) is the angle between the surface back of the land and a tangent passing through the cutting edge. The back of the tooth may be a smooth curve in which the back of the land is variable from point to point or it may be made up of the land and one line (Fig. 6), or two lines in which the smallest clearance angle back of the land is called the first clearance angle and the next smallest is called the second clearance angle.

Radial Rake Angle. The angle formed in a diametral plane between the face of the tooth and a radial line passing through the cutting edge.

Axial Rake Angle (or Helical Rake). When a plain milling cutter has helical teeth, the resulting rake is called helical rake. If the cutting edge is straight, its rake is axial rake.

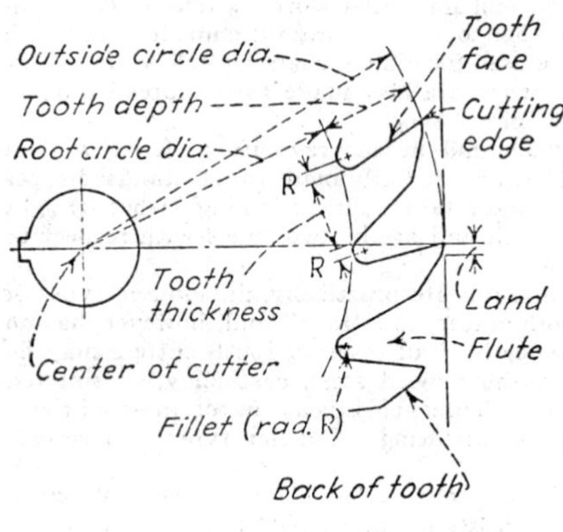

FIG. 5. Tooth parts of a plain milling cutter.

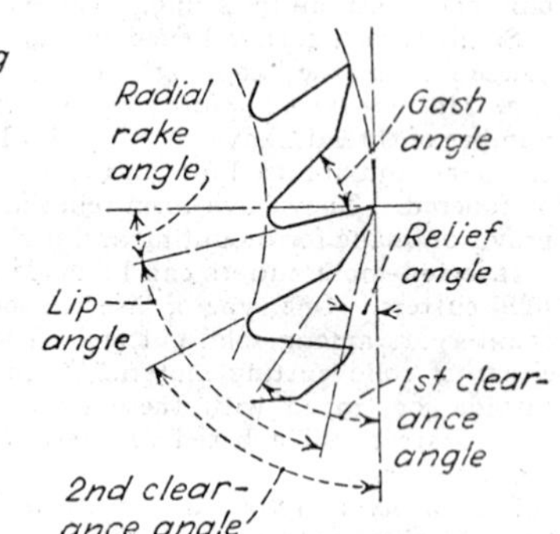

FIG. 6. Tooth angles of a plain milling cutter.

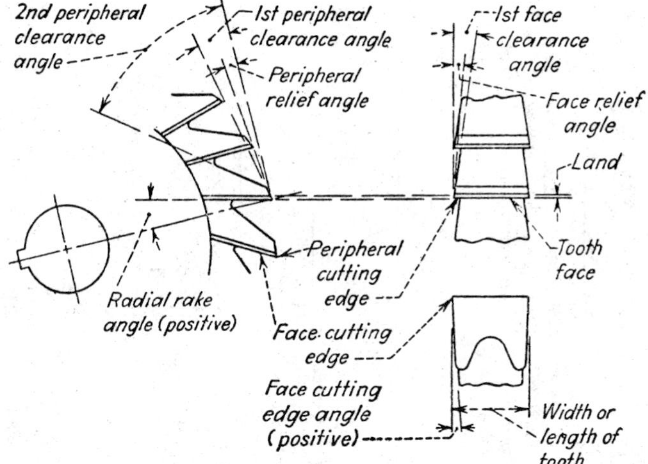

FIG. 7. Nomenclature of side milling cutter teeth.

Normal Rake Angle. The rake angle of the tooth face at right angles to the cutting edge. The true rake is the actual working rake as determined by the direction of the flow of chip. If the chip flows over the tool face at right angles to the cutting edge, the true rake equals the normal rake.

Lip Angle. The included angle of metal between the face of the tooth and the surface of the land. These two surfaces intersect to form the cutting edge.

Side Milling Cutters

Side milling cutters (Fig. 7) have cutting edges on the periphery (or cylindrical surface of the cutter) and on the face (or end) of the cutter. The cutting edges on the periphery are called peripheral cutting edges and those on the face (or end) of the cutter are called face cutting edges.

In the case of side milling cutters, there are relief angles, clearance angles, and rake angles on the periphery and on the face of the cutter.

Radial Rake Angle. The angle measured in the diametral plane between the face of the tooth and a radial line passing through the tooth cutting edge. If the radial line and tooth face coincide in the diametral plane, the angle is zero. If the tooth face is tilted so as to reduce the peripheral lip angle, the rake angle is positive. If the tooth face is tilted from the zero line to increase the lip angle, the rake angle is negative.

Axial Rake Angle. The angle formed between the peripheral cutting edge and a line passing through the nose parallel to the cutter axis. The axial rake is zero if the peripheral cutting edge is parallel to this line, positive if it tends to reduce the face lip angle, and negative if it increases the face lip angle.

Land. The relieved surface back of the cutting edge is known as the peripheral land and the face land, depending on whether it is back of the peripheral cutting edge or face cutting edge.

Face Milling Cutters

By face milling with end mills and face mills is meant the machining of a surface parallel to the face of the cutter (or at right angles to the axis of the cutter), in which each cutter tooth moves across the machined surface as the work is fed past the

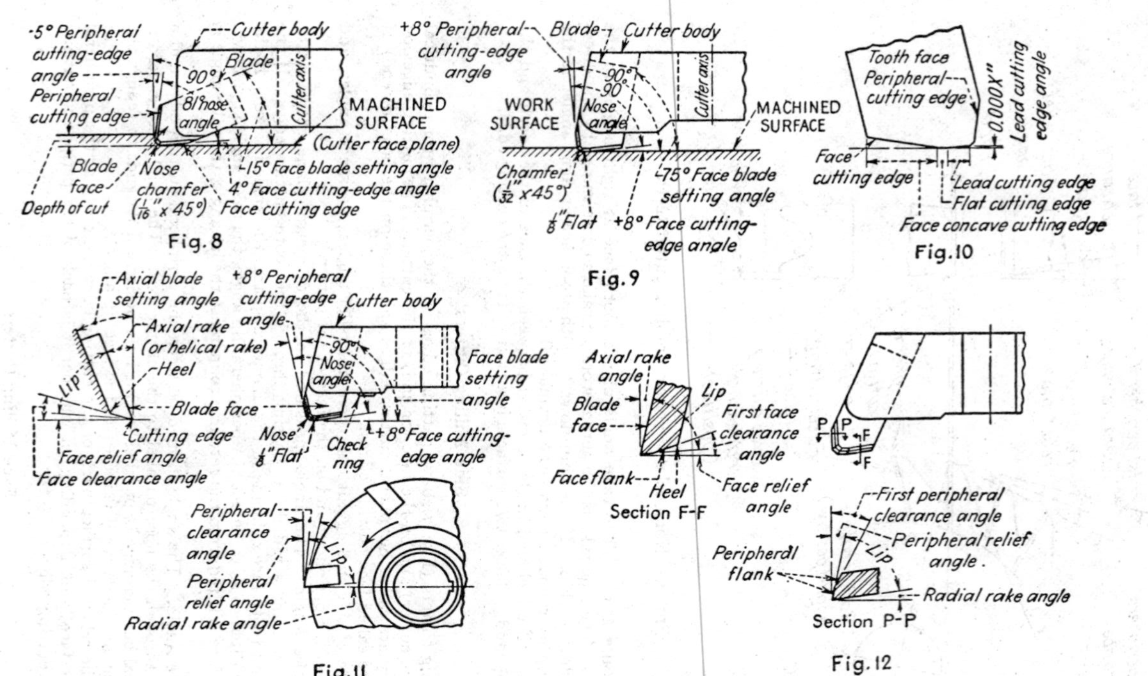

FIG. 8. A low face-blade setting angle is used for roughing face mills. FIG. 9. For finishing face mills, a high face-blade setting angle is used. FIG. 10. Face cutting edge is made up of a lead cutting edge, a flat, and a concave cutting edge. Lead angle may be designated in thousandths or degrees. FIG. 11. A face mill with inserted blades, showing the nomenclature of axial and radial rake. FIG. 12. Besides showing the axial rake, this drawing illustrates peripheral face relief and clearance.

cutter. The machined surface per pass of the cutter may equal the whole width of the face of the cutter (cutter diameter) or be of any fraction of the cutter diameter.

A face milling cutter may have cutting edges on one face of the cutter as well as on the periphery. It may be mounted on an arbor which passes completely through the cutter, or have its face completely free by being mounted on the end of the spindle.

A face milling cutter may be of the solid type, having teeth integral with or brazed to the body, or it may have a body provided with slots in which blades are held.

Face mills may have the blades mounted in two distinct positions. If blades are mounted nearly radially in the body (Fig. 8) so as to have a face blade setting angle less than 45°, the blade adjustment is greater in a radial direction than in an axial direction. This mounting is used advantageously when deep cuts are taken. Most dulling occurs on the peripheral cutting edge, which has to be sharpened more extensively and oftener than the face cutting edge. For light finishing cuts, wear occurs to greater extent on the face-cutting edge, so that it is advantageous to have a large face-blade setting angle (Fig. 9). This provides an axial adjustment of the blade greater than the radial adjustment.

Right-cut Face Mill. A blade of right-cut face mill (turning counterclockwise when viewed from the face) is similar to a right-cut single-point tool. Names of angles are related as follows (those in parentheses are for single-point tools):

Peripheral cutting edge (side cutting edge).
Peripheral cutting edge angle (side cutting edge angle).
Face cutting edge (end cutting edge).
Face cutting edge angle (end cutting edge angle).
Nose (nose).
Peripheral or face relief (side or end relief, respectively).
Axial rake (back rake).
Radial rake (side rake).

Blade Setting Angle. This refers to the position of the blade in the holder with respect to the axis of the cutter and the cutter face. The angle formed between the side edge of the blade (or the side of the slot in the body) and the cutter face plane is the face-blade setting angle. The angle formed between the face of the blade (or the blade-supporting surface in the body) and a line through the nose of the blade parallel to the cutter axis is the axial-blade setting angle.

Peripheral Cutting Edge. That part of the cutting edge of the blade which is on the periphery of the cutter on the outside of the nose, chamfer, or flat, whichever forms the machined surface.

Peripheral Cutting Edge Angle (sometimes called corner angle). The angle between the peripheral cutting edge and a line passing through the nose parallel to the axis of the cutter body. It is negative if it tends to reduce the value of the nose angle, and positive if it tends to increase the nose angle.

Face-cutting Edges. Those cutting edges of the tooth (or blade) on the face of the cutter, inside the nose or chamfer, which produce the machined surface (see Figs. 8 and 9).

There may be the following:

1. A sharp nose or corner and one straight concave cutting edge.
2. A chamfer (or nose radius) and one concave cutting edge.
3. A chamfer (or nose radius), a flat, and a concave cutting edge.
4. A chamfer, a lead cutting edge, a flat, and a concave cutting edge.

Face-cutting Edge Angle. The angle between the face-cutting edge and the cutter face plane. It prevents the face-cutting edge from scraping the cut surface except at the nose or flat. These angles on side mills may be right and left, as being on the right and left side of the tooth when looking toward the tooth face of a right-hand cutter.

If the face-cutting edge angle tends to force the cutter from the work, the angle is negative.

Flat. That part of the face-cutting edge which is in the cutter face plane (see Fig. 10). It is intended to produce a smooth machined surface and its length is

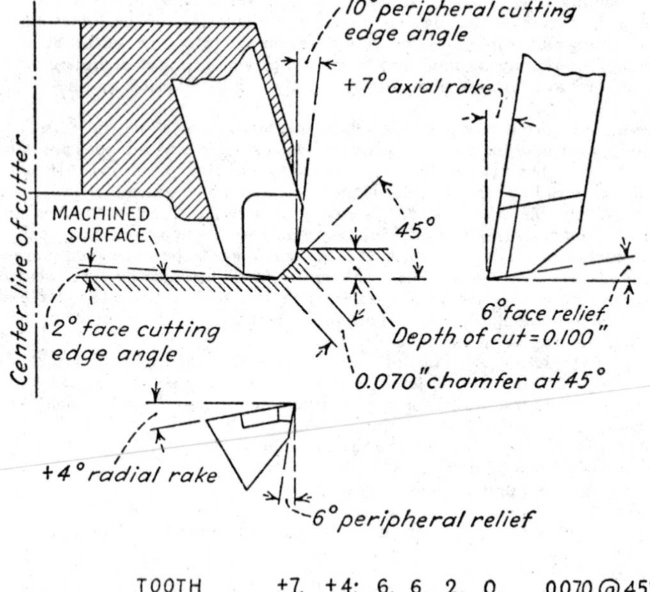

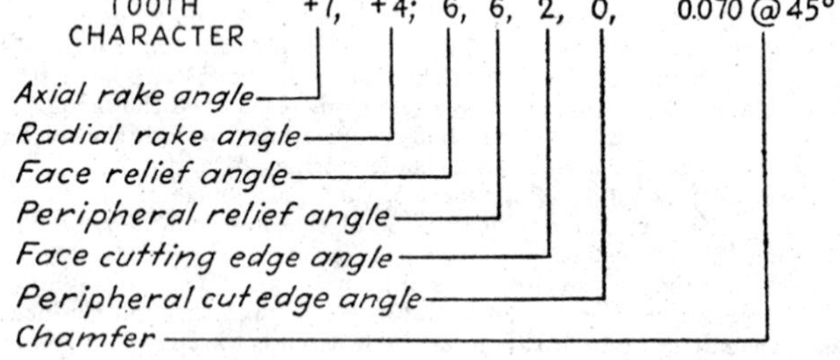

FIG. 13. Standard method of designating the shape of a single tooth of a face milling cutter. (*Courtesy of Cincinnati Milling Machine Co.*)

usually slightly more than the feed per revolution of the cutter. When the flat is in the face plane of the cutter, the face-cutting edge angle of the flat is zero. It may have a slight face-cutting edge angle (or concave) of 0.75 to 2°, in which case it is designated as 2° flat, ⅛ in. long (first concave) while the remainder of the face-cutting edge may have a 4 or 5° face-cutting edge (second concave).

Nose, Corner, or Chamfer. The cutting edge at the nose connecting the peripheral cutting edge with the face-cutting edge. The nose may be a sharp corner or made up of a chamfer consisting of a curved line, a straight line (as 1⁄16 in. wide at an angle of 45° from the cutter face plane), or a series of short lines at different angles with the face. The chamfer is intended to be less than the depth of cut. The nose angle is the angle between the peripheral and face-cutting edges.

Face Relief Angle. The angle between the land or relieved flank of the tool immediately back of the cutting edge and the cutter face (see Fig. 11).

The angle is *normal* face relief if measured in a plane at right angles to the cutting edge.

Face Clearance Angle. The angle between the cleared flank of the tooth back of the relieved surface and the cutter face. The face clearance is larger than the face relief to give additional clearance to the heel of the tooth.

Peripheral Relief Angle. The angle between the relieved flank of the tooth or blade and a tangent to the periphery in a diametral plane passing through the cutting edge. This angle is normal face relief if measured in a plane at right angles to the cutting edge. The peripheral cutting edge may have peripheral and normal relief.

Peripheral Clearance Angle. The angle between the cleared flank of the blade and a tangent to the periphery in a diametral plane passing through the cutting edge.

If there is a chamfered nose, the corresponding relief and clearance are known as peripheral or normal chamfer relief and chamfer clearance, respectively, or nose relief and nose clearance, respectively.

Radial Rake. The angle in the face plane between the blade face and a radial line or plane passing through the cutter axis and blade nose.

Axial Rake. The angle between the face of the blade and a line passing through the nose parallel to the cutter axis or a plane formed by a radial line through the nose and the cutter axis. When the peripheral cutting edge forms a helix, the helical rake and axial rake are equal (see Fig. 12).

As with lathe tools, there is a standard method of showing the shape of a tooth in a face mill. See Fig. 13 for typical drawing and method of designating the tooth character.

METHODS OF MOUNTING CUTTERS

Plain, side, angle, and formed milling cutters and metal-slitting saws are made with a standard center hole and keyway for mounting between an arbor held on the milling-machine spindle and the outer support. The arbor is provided with a full-length keyway to accommodate keys for driving the cutters and the spacing and bearing collars. Table 1 gives the dimensions for keys and keyways for milling cutters and arbors.

Large face mills, usually 7 in. dia and over, are mounted directly on the spindle nose and are held in position by four cap screws. The face mill may be centered by the outer diameter of the standard spindle (see Sec. 41 for standard spindle noses) or by a plug in the spindle. Large face mills may also be held on a stub arbor or adapter fitting the spindle-nose taper, when a quick-change adapter is required for multiple tooling without changing the setup of the workpiece.

The self-releasing taper standardized for milling-machine spindles and the shank of accessories to be mounted thereon is 3½ in. per ft. Standard spindle noses have

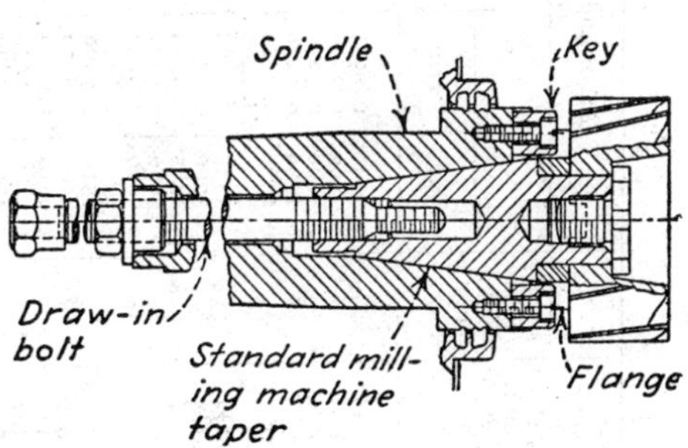

FIG. 14. Method of mounting a shell end mill to a milling-machine spindle, using a stub arbor.

TABLE 1. DIMENSIONS OF KEYS AND KEYWAYS FOR MILLING CUTTERS* AND ARBORS

Nominal Arbor and Cutter Bore Dia	Nominal Size Key, Square	Arbor and Keyseat				Bore and Keyway						Arbor and Key			
		A		B		C		D		H Nom.	Corner Radius	E		F	
		Max	Min	Max	Min	Max	Min	Max	Min			Max	Min	Max	Min
1/2	3/32	0.0947	0.0937	0.4531	0.4481	0.106	0.099	0.5678	0.5578	3/64	0.020	0.0932	0.0927	0.5468	0.5408
5/8	1/8	0.1260	0.1250	0.5625	0.5575	0.137	0.130	0.7085	0.6985	1/16	1/32	0.1245	0.1240	0.6875	0.6815
3/4	1/8	0.1260	0.1250	0.6875	0.6825	0.137	0.130	0.8325	0.8225	1/16	1/32	0.1245	0.1240	0.8125	0.8065
7/8	1/8	0.1260	0.1250	0.8125	0.8075	0.137	0.130	0.9575	0.9475	1/16	1/32	0.1245	0.1240	0.9375	0.9315
1	1/4	0.2510	0.2500	0.8438	0.8388	0.262	0.255	1.1140	1.1040	3/32	3/64	0.2495	0.2490	1.0940	1.0880
1 1/4	5/16	0.3135	0.3125	1.0630	1.0580	0.325	0.318	1.3950	1.3850	1/8	1/16	0.3120	0.3115	1.3750	1.3690
1 1/2	3/8	0.3760	0.3750	1.2810	1.2760	0.410	0.385	1.6760	1.6660	5/32	1/16	0.3745	0.3740	1.6560	1.6500
1 3/4	7/16	0.4385	0.4375	1.5000	1.4950	0.473	0.448	1.9580	1.9480	3/16	1/16	0.4370	0.4365	1.9380	1.9320
2	1/2	0.5010	0.5000	1.6870	1.6820	0.535	0.510	2.2080	2.1980	3/16	1/16	0.4995	0.4990	2.1880	2.1820
2 1/2	5/8	0.6260	0.6250	2.0940	2.0890	0.660	0.635	2.7430	2.7330	7/32	1/16	0.6245	0.6240	2.7180	2.7120
3	3/4	0.7510	0.7500	2.5000	2.4950	0.785	0.760	3.2750	3.2650	1/4	3/32	0.7495	0.7490	3.2500	3.2440
3 1/2	7/8	0.8760	0.8750	3.0000	2.9950	0.910	0.885	3.9000	3.8900	3/8	3/32	0.8745	0.8740	3.8750	3.8690
4	1	1.0010	1.0000	3.3750	3.3700	1.035	1.010	4.4000	4.3900	3/8	3/32	0.9995	0.9990	4.3750	4.3690
4 1/2	1 1/8	1.1260	1.1250	3.8130	3.8080	1.160	1.135	4.9630	4.9530	7/16	1/8	1.1245	1.1240	4.9380	4.9320
5	1 1/4	1.2510	1.2500	4.2500	4.2450	1.285	1.260	5.5250	5.5150	1/2	1/8	1.2495	1.2490	5.5000	5.4940

* Keys and keyways for hobs on the basis of hole size are not the same as for milling cutters in all cases. Tolerances for keys and keyways for hobs and hobbing machine arbors, however, are the same on the basis of square key size.

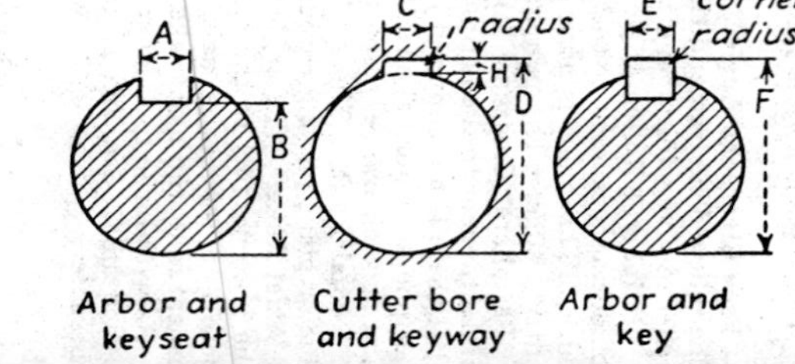

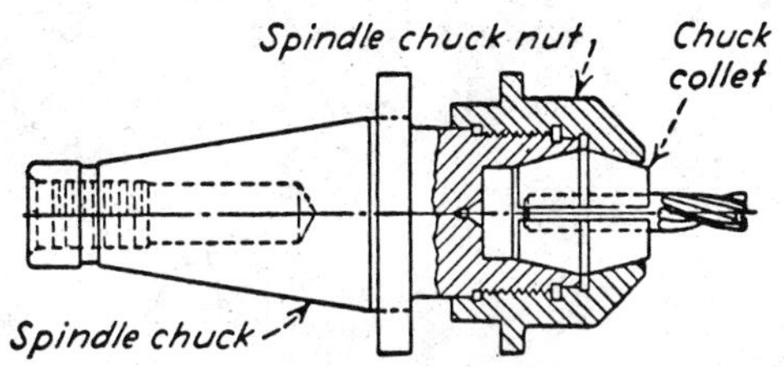

FIG. 15. Single-end straight-shank cutters are held by a split collet. A spring chuck adapter is suitable for a double-end cutter or drill.

tapers Nos. 30, 40, 50, and 60. To extend the range, so that tools with adapters or stub arbors can be utilized on several types of machines, the Joint Industry Committee uses tapers Nos. 5, 10, 15, 20, 25, 35, 42½ (special-purpose for boring machines), 45, and 55 in addition to the four tapers standardized by the ASA.

Shell-end mills are mounted on the spindle by means of "stub arbors" or adapters (see Sec. 41 and Fig. 14). The arbor has keys for driving the cutter and a retaining screw for the tool. The arbor, in turn, is driven by keys on the spindle nose and is retained in the spindle by a draw-in bolt.

Integral-shank cutters are held by several means: (1) in a collet adapter with a tang drive, (2) in a two-tapered sleeve that fits the machine spindle and takes a tool with a slow taper, (3) in a split collet inside a stub arbor, and (4) in a collet chuck held by an adapter or arbor, as in Fig. 15. Through the use of collets and adapters, frequent changes in tooling or setup are accommodated, as in toolroom or jobbing work. Expensive tools can be shifted from machine to machine readily.

HOW TO MAKE A MILLING SETUP

Ten Basic Steps

For successful milling, especially with carbide, there are 10 basic steps in attaining a successful setup. In order of application, these are:

1. Identify and classify material.
2. Select surface speed.
3. Select cutter type and diameter.
4. Determine metal-removal rate of the workpiece.
5. Correlate metal-removal rate with horsepower available in the milling machine designated for the job.
6. Establish feed rate from the foregoing data.
7. Check tooth load.
8. Specify milling method (conventional or climb), holding method or fixture, chip disposal, etc.
9. Establish cutter care and grinding procedure.
10. Check surface and tolerance specifications and their influence on metal-removal rate.

Formulas for Relationships in Milling

TERMINOLOGY

Cutter dia, in	D	Feed per tooth, ipt	f
Cutting speed, fpm	V	Feed per rev	f'
Cutter rpm	N	Feed rate, ipm	F
Number of cutter teeth	n	Depth of cut, in	d
Tooth contacts per min	$n \times N$	Width of cut, in	W
Horsepower at spindle	hp_c	Machinability constant	K

To Calculate	*Known*	*Formula*
fpm = V	D and N	$V = \dfrac{3.1416 \times D \times N}{12}$
rpm = N	V and D	$N = \dfrac{12 \times V}{3.1416 \times D}$
f	F, N, and n	$f = F/N \times n$
F	f, N, and n	$F = f \times N \times n$
f'	F and N	$f' = F/N$
K	d, W, F, and hp_c	$K = \dfrac{d \times W \times F}{hp_c}$

Selection of Cutting Speed

The cutting speed will vary between wide limits, as determined by the work material, the cutter material, and specific job conditions. See Tables 2 and 3 for suggested cutting speeds for a number of materials, when using HSS, carbide, or cast-alloy cutters. Also see Milling Characteristics of Various Materials, page 6-31.

Tensile strength is one of the properties of a work material that has been found to give a practical indication of the cutting speed to use. For the same steel alloy, for example, the tensile strength increases with the hardness. Thus, the brinell hardness number is frequently taken as an indication of the value of the cutting speed to select, when milling ferrous and nonferrous materials, with the possible exception of stainless steels. Figure 16 gives the relationship between the brinell hardness value of the work and cutting speed when milling various materials with common cutting materials.

In milling with carbide-tipped cutters, there will be considerable variation in cutting speed, according to carbide grade, cutter design, (including rake angles), the material, and the nature of the operation. In practice, the selected cutting speed may have to be adjusted in order to obtain satisfactory tool life, good finish of the milled surface, and operation of the machine with available power. For example, aluminum and magnesium alloys are milled with carbide cutters at cutting speeds within the wide range of 3500 and 12,000 fpm, and higher.

Cutter Selection

Cutter Diameter. The size of the job is known; the faces to be machined are dimensioned. From this information, the cutter size is selected. Where clearance presents no limitations, the ratio of cutter diameter to face dimension should be approximately 5:3, according to one authority. Thus, if a piece 3 in. wide is to be milled, a 5-in. cutter is selected.

A cutter larger in diameter than the largest effective gear in the milling machine should not be chosen (cutter will seldom be larger than 12 in. except on bed and planer-type millers), because the flywheel effect of the cutter will cause backlash in the gear train. This will occur with each entrance and exit of the teeth.

A cutter that can be assembled directly on the machine spindle, rather than a setup that requires an arbor, is recommended. Arbors deflect, their setup is time-consuming, and the cost of grinding the cutter assembly is greater than for a face mill mounted on the spindle.

Selection of straddle mills is not recommended, unless there is no other way to perform the operation. Carbide milling has so reduced cutting time that it is preferable to mill one face at a time and turn the piece manually or mechanically in the fixture. Two cutters in a straddle-mill setup never dull equally. Considering teardown and

TABLE 2. MILLING SPEEDS FOR VARIOUS MATERIALS, SFPM
HSS and Carbide Cutters

Material	High-speed Steel		Carbide-tipped		Coolant
	Rough	Finish	Rough	Finish	
Cast iron	50–60	80–110	180–200	350–400	Dry
Semisteel	40–50	65–90	140–160	250–300	Dry
Malleable iron	80–100	100–130	250–300	400–500	Soluble, sulfurized or mineral oil
Cast steel	45–60	70–90	150–180	200–250	Soluble, sulfurized, mineral or mineral lard oil
Copper	100–150	150–200	600	1000	Soluble, sulfurized or mineral lard oil
Brass	200–300	200–300	600–1000	600–1000	Dry
Bronze	100–150	150–180	600	1000	Soluble, sulfurized or mineral lard oil
Aluminum	400	700	800	1000	Soluble or sulfurized oil, mineral oil, and kerosene
Magnesium	600–800	1000–1500	1000–1500	1000–1500	Dry, kerosene, mineral lard oil
SAE steels:					
1020 (coarse feed)	60–80	60–80	300	300	Soluble, sulfurized, mineral or mineral lard oil
1020 (fine feed)	100–120	100–120	450	450	Soluble, sulfurized, mineral or mineral lard oil
1035	75–90	90–120	250	250	Soluble, sulfurized, mineral or mineral lard oil
X-1315	175–200	175–200	400–500	400–500	Soluble, sulfurized, mineral or mineral lard oil
1050	60–80	100	200	200	Soluble, sulfurized, mineral or mineral lard oil
2315	90–110	90–110	300	300	Soluble, sulfurized, mineral or mineral lard oil
3150	50–60	70–90	200	200	Soluble, sulfurized, mineral or mineral lard oil
4340	40–50	60–70	200	200	Sulfurized and mineral oils
Stainless steel*	60–80	100–120	240–300	240–300	Sulfurized and mineral oils

NOTE: Feeds should be as much as the work and equipment will stand, provided a satisfactory surface finish is obtained.

* See Stainless Steels under Milling Characteristics of Various Materials, page 6-35.

TABLE 3. CUTTING SPEEDS AND FEEDS FOR CAST-ALLOY (TANTUNG) MILLING CUTTERS

Material to Be Cut	Cutting Speed, Fpm		Feed per Tooth, In.	
	Roughing	Finishing	Roughing	Finishing
Cast iron	100–150	150–300	0.010–0.015	0.005–0.010
Malleable iron	125–150	150–300	0.010–0.015	0.005–0.010
Steel:				
Soft	200–300	250–350	0.004–0.015	0.004–0.008
Medium	150–225	175–250	0.004–0.015	0.002–0.008
Hard	80–125	185–150	0.004–0.015	0.002–0.008
Cast steel	80–100	100–200	0.008–0.015	0.004–0.010
Aluminum	800–1000	1000–1200	0.005–0.015	0.002–0.015
Brass	500–600	500–1000	0.008–0.030	0.004–0.015
Bronze	175–250	250–300	0.008–0.030	0.004–0.015
Magnesium alloys	800–1000	1000–2000	0.004–0.015	0.002–0.015
Zinc alloys	400–800	700–1500	0.004–0.015	0.004–0.015
Hard rubber	600–900	800–1000	0.004–0.015	0.004–0.015
Plastics	700–800	800–1000	0.005–0.030	0.004–0.015

reassembly time for such a setup, a simple face mill mounted on the spindle will almost always give less floor-to-floor time per piece.

Number of Cutter Teeth. For any cutter diameter, the number of teeth and the size of the chip space are directly related. Large chip space is helpful, particularly when a continuous curly chip is produced, as when milling ductile materials.

HSS CUTTERS. No standard rule exists. The number of teeth in standard cutters has not proved as satisfactory for many, if not most, purposes as the so-called heavy-duty, or coarse-tooth, cutters. The latter type of cutter allows each tooth to take a substantial chip instead of skipping over the work. Sometimes, to improve performance, every other tooth of standard cutters is ground low, in order to reduce the number of teeth that actually cut.

To improve performance of cutters, some shops make their own cutter bodies of malleable iron, cast iron, Meehanite, or mild steel, and braze HSS or carbide blades to them, or any of the cast-alloy materials that will not have their temper drawn, or even deposit HSS from a special welding rod.

CAST NONFERROUS CUTTER MATERIALS. Selection of the number of teeth closely follows that for HSS cutters.

CARBIDE CUTTERS. The maximum number of teeth in a carbide cutter is limited primarily by horsepower available at the machine spindle, and secondarily by chip space.

The formula for number of teeth is

$$n = \frac{K \times \text{hp}_c}{d \times f \times W \times \text{rpm}}$$

where n = number of teeth; K = machinability constant, or K-factor, cu in. per min/hp_c; d = depth of cut; f = feed per tooth, in.; W = width of cut, in.;

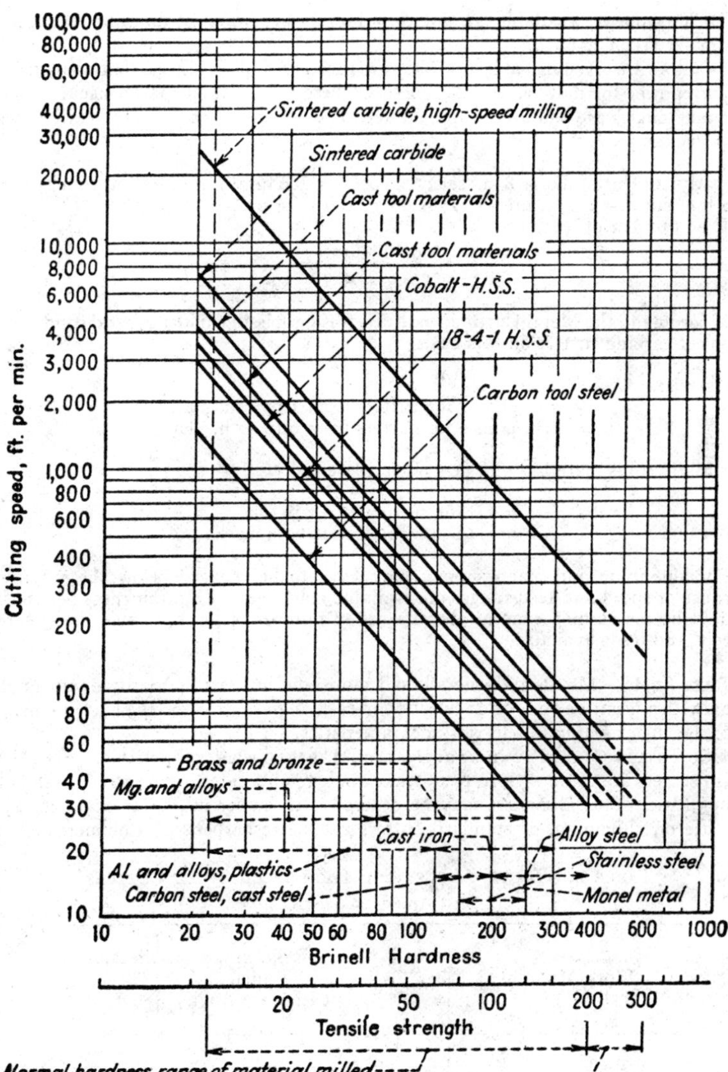

FIG. 16. Relationship between cutting speed and brinell hardness of the work when milling various materials with the several types of tool materials.

hp_e = horsepower available at cutter = horsepower input × machine efficiency (normally taken as 60%).

K-factors are given in Table 7. They are based on dull cutters, so that no increase in horsepower for dulling need be made. These factors apply when usual speeds and rake angles are employed, and the feed is 0.010 in. The K-factor varies with chip thickness.

EXAMPLE: Cutter dia = 8 in.; selected speed = 400 fpm; f = 0.011 ipt; rpm = 192; depth of cut d = 0.075; hp_e = 12; W = 5; and K = 0.66.

Then number of teeth equals

$$n = \frac{0.66 \times 12}{0.075 \times 0.011 \times 5 \times 192} = 10$$

According to the Metal Cutting Institute, it is permissible under general conditions to select carbide cutters on this basis:

For steels:

$n = D + 1$ for cutters from 1 to 5 in. dia
$n = D + 2$ for cutters from 6 to 12 in. dia

For cast ferrous materials and general-purpose use:

$n = 2D$ for cutters 1 to 3 in. dia
$n = 1\frac{1}{2}D$ for cutters above 3 in. dia

For aluminum and magnesium n may have to be reduced below these figures because of speed and feed ranges on available machines. Chip thickness per tooth should be maintained even at the expense of number of teeth, to suit the spindle speeds and feeds available.

Rake Angle. Metals are generally cut more efficiently with a positive rake angle than with a negative one. But selection of rake angle is based on the material being cut, the cutter material, and work characteristics.

HSS CUTTERS. Positive rake angles should be used wherever possible. Standard plain cutters are made with 10 to 15° rake angle to obtain maximum strength for all-around use. Special cutters may be ordered with higher rakes, for example, 25°, for use on aluminum, or standard cutters can be resharpened. Commercial face

TABLE 4. TOOTH ANGLES FOR HIGH-SPEED STEEL CUTTERS
All Angles in Degrees

Work Material	Rake Angle	Relief Angle (Primary) Clearance Angle)	Secondary Clearance Angle
Alloy steel	10–15	2–3	6–10
Mild steel	10–15	3–4	6–10
Cast iron	5–10	3–5	6–10
Brass	10–20	3–4	6–10
Bronze (hard)	10–15	2–4	6–10
Aluminum	10–35	6–10	6–10

NOTE: Maximum width of land is 1/32 to 1/16 in.

TABLE 5. CUTTING ANGLES AND CLEARANCE ANGLES FOR CAST-ALLOY (TANTUNG) MILLING CUTTERS

Work Material	Radial Rake Angle	Axial Rake Angle	Clearance or Relief Angle
Cast iron	0	5–6	4–6*
Malleable iron	0	5–6	4–6*
Mild steel	10	10	4–6
Cast steel	0	5–6	4–6*
Aluminum	20	20	8–10
Brass	0	5–6	4–6*
Bronze	0	5–6	4–6*
Magnesium alloys	20	20	8–10

* Primary clearance or land to support cutting edge. Secondary clearance of approximately 10° allows for heel clearance of tool.

mills usually have positive 10° radial and axial rakes for milling cast iron, forged and alloy steel, and brass and bronze. Table 4 gives conventional rake angles for HSS cutters.

HIGH-RAKE-ANGLE MILLING. HSS cutters with rake angles as high as 35° have given superior performance, as compared with conventionally ground cutters, in respect to cutting speed, tool life, and surface finish produced on the workpiece. Some of the results obtained are listed below.

High-speed Steel. Failure of carbide cutters when milling HSS of the 18-4-1, 18-4-2, M-1, and M-2 grades, because of chromium carbide in the steel, led to adoption of M-2 cutters with a 25° positive rake, at 126 fpm and with soluble oil. Cutting speed is two-thirds faster than with cutters ground with zero radial rake. Less stock removal is required when regrinding the cutter.

Stainless Steel. On 300-series stainless, HSS staggered-tooth side milling cutters operated successfully at 123 fpm, feed of 8 ipm and depth of cut of ⅞ in. Potential cutter life between grinds when rake of 34° is used is 10 to 20 times, and the metal removal rate is 88% greater, as compared with cutters with rake of 4 to 8°.

Aluminum Alloys. 61S and 52 can be cut at 1118 fpm, 63 ipm, and 0.3 in. depth of cut with a 34° rake cutter, dry, without generating heat in the cutter.

CAST-ALLOY CUTTERS. Positive rake angles are generally provided (see Table 5). In case of shock, however, negative rake angles may be used; also to eliminate vibration when milling thin sections.

CARBIDE CUTTERS. Peripheral milling cutters such as slab mills, slotting cutters, and saws tipped with carbide usually have negative radial rake of −5° for soft low-carbon steel to −10° or more for alloy steel, and positive axial rake of +5 to +10° (0° in slotting cutters and saws). On soft materials like aluminum, positive rake angles of +10 to +20° may be used. See Table 6 for rake and clearance angles.

Face milling cutters used on cast iron usually have positive radial and axial rake angles, if the tools have a zero corner angle and a small chamfer or radius. With a large corner angle, positive axial rake with either positive or negative radial rake is used with a 60° corner angle or a double corner angle of 60 and 45°. This is done to increase cutter life and to reduce work breakout.

In face milling of steel, with 0° corner-angle cutters, negative radial and axial rakes are used. If cutters have a large corner angle, a combination of negative radial and

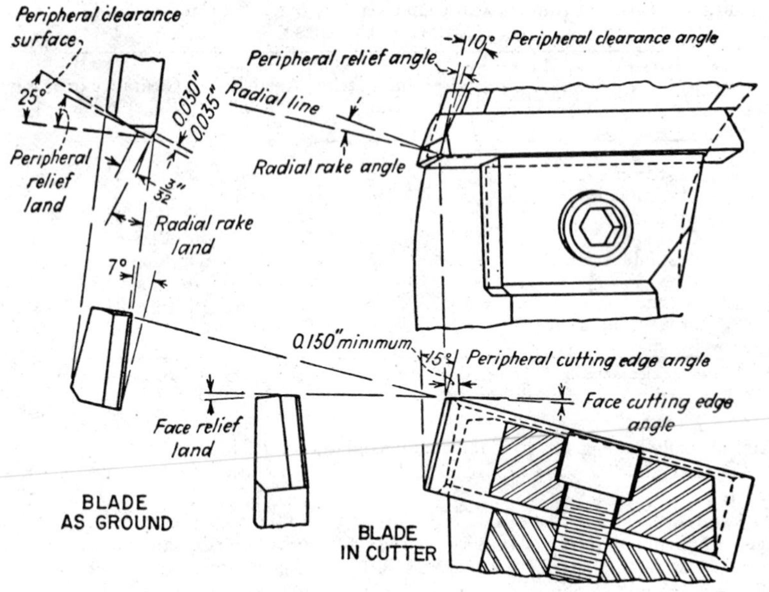

TABLE 6. CUTTING AND CLEARANCE ANGLES FOR CARBIDE CUTTERS

Material	Cutting Angles, Deg	Clearance Angles, Deg
Wrought iron	10 PRR–7 NAR 8 PTR	7 OD–0 FCE 15 FR
Low-carbon steel	10 PRR–7 NAR 8 PTR	7 OD–0 FCE 15 FR
Stainless steel	10 PRR–7 NAR 8 PTR	7 OD–0 FCE 15 FR
Standard steel	7 NRR–7 NAR 7 NTR	7 OD–0 FCE 7 FR
Unmodified cast iron	10 PRR–7 NAR 8 PTR	7 OD–0 FCE 7 FR
Modified cast iron	7 NRR–7 NAR 7 NTR	7 OD–0 FCE 7 FR
All ferrous materials harder than 300 brinell	10 NRR–7 NAR 8 NTR	5 OD–0 FCE 5 FR
Nonferrous materials	15 PRR–7 NAR 12 PTR	7 OD–0 FCE 15 FR

NOTE: PRR = positive radial rake. NRR = negative radial rake. NAR = negative axial rake. PTR = positive true rake. NTR = negative true rake. OD = outside dia, or peripheral clearance. FR = face relief, or clearance. FCE = face cutting edge, or dish.

either positive or negative axial rake is employed. Under good setup conditions, the negative-radial–positive-axial-rake combination is preferred.

Rake of Helical Cutters. The radial rake angle r in the plane normal to the cutter axis is commonly used in describing helical milling cutters. Actually, the chip flows approximately normal to the cutting edge, and the oblique radial rake angle w in that plane is the one that should be considered in selecting the proper angle for the job. For a known helix angle a, the formula is $\tan w = \tan r \times \cos a$.

True Rake of Face Mills. With cutters having a square nose (1/16 in. chamfer or less), the oblique rake angle is measured in a plane normal to the cutting edge at the *periphery* of the cutter, as in the case of helical mills. The axial rake angle then corresponds to the helix angle in helical mills, and the above formula can be used to find the true rake.

If face mills have a chamfer exceeding 1/16 in., the oblique rake angle w is measured in a plane normal to the cutting edge on the *chamfer*, since this approximates the direction of chip flow. Figure 17 can be used to determine the true rake angle.

EXAMPLES: 1. Given: a radial rake of +5° and an axial rake of +10°. Required: the corner angle which will give a true rake of +10°. Solution: The point of intersection between the dash line connecting these points cuts the +10° true rake curve at a corner angle of about 45°.

2. Given: a radial rake of +5° and an axial rake of −10°. What corner angle will give 0° true rake? Solution: The dash line cuts the 0° true rake curve at 25° corner angle.

Clearance Angle. Chief function of the clearance angle is to prevent interference between the land and the surface being generated. The clearance angle is sometimes referred to as the "primary clearance," since a secondary clearance is provided when resharpening cutters. Width of the land is increased by repeated resharpen-

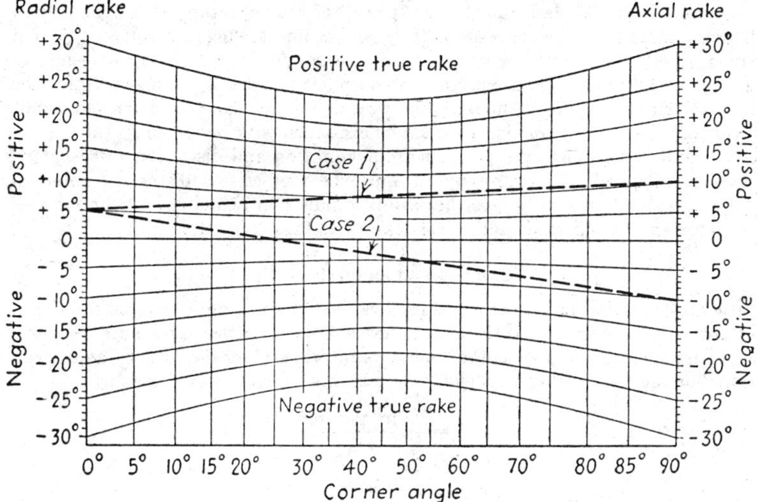

FIG. 17. Values of true rake angle w in face mills are determined with this chart.

ing, and interference might develop. To maintain the desired width of land without weakening the tooth, secondary clearance is ground.

On face mills, the clearance angle must be ground along the complete contour of the cutting edge—the periphery, the face, and the chamfer or round nose. The term "relief" refers to the angle provided radially on the sides or face of the cutter to reduce or eliminate binding or rubbing. Common values of relief are 3 to 5° for side mills and face mills and 1 to 2° for saws.

For general-purpose work, milling cutters from ⅛ to 3 in. dia have clearance angles of 13 to 5°, respectively. In cutters over 3 in. dia, manufacturers ordinarily provide a clearance of 4 to 5°. Width of land is about 1/64, 1/32, and 1/16 in. for small, medium, and large cutters, in that order.

Values of primary clearance angles are changed according to work material (see Tables 4, 5, and 6).

Clearance for Counterbore Blades

Clearance angles on counterbore blades should be as small as possible, to provide maximum blade strength, but at the same time must be large enough to clear the counterbore diameter. The accompanying chart (Fig. 18) can be used to determine the smallest clearance angle that can be used for blades of ¼, 5/16, ⅜, and ½ in. thicknesses.

Locate the diameter to be cut on the left-hand scale, then move across to the blade thickness. The intersection point should be well above the diagonal representing the angle. For example, to cut a 2.5000-in. counterbore with a ½-in. blade, it will be necessary to use a 15° clearance angle, but if a ⅜-in. blade is used, a 10° angle will be satisfactory.

Machining Limitations of Workpiece

Every job has a practical maximum metal-removal rate. Only from experience, preferably firsthand, can one predict the amount of metal that can be removed in unit time. Material, design of the workpiece, the kind of setup that can be made—all these affect the problem, especially when carbide milling is involved. A thin-walled, small and complex casting, for example, may be able to withstand milling 7 cu in. of material per min, whereas an enormous and thick casting will probably allow 100 cu in. per min, or more, probably much more than the cutter and spindle can take. Thus, each component must be examined with regard to its thickness of section and buttressing, and freedom from vibration and distortion when cutting forces are applied. Clamping points must be studied for distortion potential. These considerations apply most heavily in production planning, but the job-shop estimator and the operator, too, must take them into account.

Power Consumption in Milling

For any material the rate of metal removal is a function of power consumption at the cutter. Ratios of metal removal in terms of cubic inches per minute per horsepower are variously known as the "metal removal rate" or the "machinability constant" or the "K-constant." Thus, the relationship can be expressed as

$$K = \frac{\text{cu in. per min}}{\text{hp}_c}$$

or

$$K = \frac{d \times W \times F}{\text{hp}_c}$$

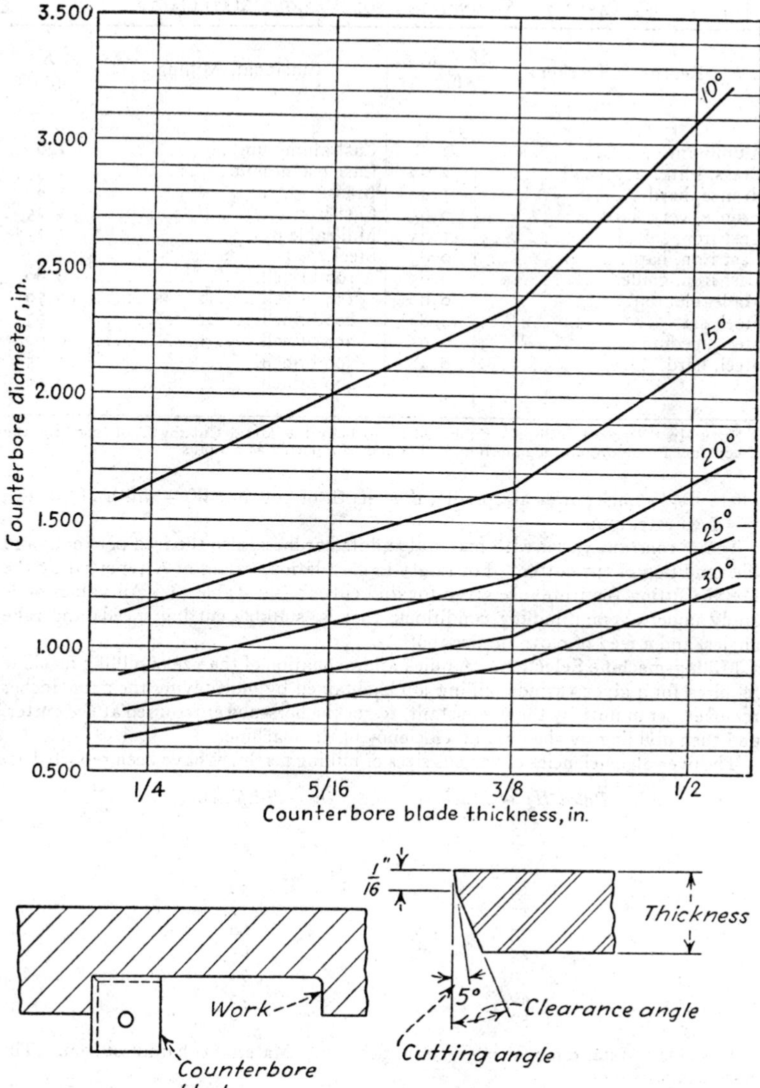

FIG. 18. Chart for clearance angles on counterbore blades.

TABLE 7. *K*-CONSTANTS FOR VARIOUS MATERIALS

Kearny & Trecker	*K* Figures	Cincinnati Milling	*K* Figures
Aluminum	2.28	Cast aluminum	2.0
Brass, soft	2.00	Cast magnesium	2.5
Bronze, hard	1.40	Brass	1.8
Bronze, very hard	0.65	Cast iron	1.25
Cast iron, soft	1.35	Malleable iron	1.25
Cast iron, hard	0.85	Steel:	
Cast iron, chilled	0.65	100 brinell	0.80
Malleable iron	0.90	150 brinell	0.70
Steel, soft	0.85	200 brinell	0.65
Steel, medium	0.65	250 brinell	0.60
Steel, hard	0.48	300 brinell	0.55
		400 brinell	0.50

NOTE: In "Efficient Milling," published in 1950 by the Metal Cutting Tool Institute, the *K*-constant for aluminum, magnesium, and copper is reported as 4.0 plus.

where hp_c = horsepower at spindle; d = depth of cut, in.; W = width of cut in.; F = feed rate, ipm.

The *K*-constant varies with material and its hardness, also the feed per tooth and the condition of the cutter. For safety in calculation, values of *K* reported by the Metal Cutting Institute[1] are given for dull cutters (see Table 7). All values of *K* apply under average milling conditions, that is, suitable cutting speeds and rake angles, and a feed of 0.010 in. per tooth.

Milling-machine Selection. A quick approximation of the size of milling machine required for a given carbide milling job is obtained by multiplying the cubic inches removed per minute by the *K*-constant, to get the horsepower required at the cutter, and then dividing by the over-all efficiency of the machine.

The over-all efficiencies of various sizes of milling machines have been reported as:

Rated Hp of Machine	*Over-all Efficiency, %*
3	40
5	48
7.5	52
10	52
15	52
20	60
25	65
30	70
40	75
50	80

EXAMPLE: Metal removal = 10 cu in. per min. Material is hard cast iron. The *K*-constant = 0.85.

Then $hp_c = 10 \times 0.85 = 8.5$ hp. Assume that a 15-hp milling machine is available. Required horsepower = 8.5 divided by the over-all efficiency (52%) = 16.3 hp. It will probably be all right to use the 15-hp machine for the job, because under normal operating conditions maximum horsepower is not required continuously. For this reason, temporary overloads of 25% or more are being used, in order to

[1] "Milling with Carbides," 1947.

TABLE 8. MILLING-MACHINE SELECTOR TABLE

Material to Be Milled	Rated Hp of Machine									
	3	5	7.5	10	15	20	25	30	40	50
	Max Metal Removal, Cu In. per Min									
Aluminum	2.7	5.5	8.7	12	18	27	37	48	69	91
Brass, soft	2.4	4.7	7.5	10	16	24	32	41	60	79
Bronze, hard	1.7	3.3	5.3	7.3	11	17	23	30	43	56
Bronze, very hard	0.78	1.6	2.5	3.4	5.3	7.8	11	15	20	26
Cast iron, soft	1.6	3.2	5.2	7.1	11	16	22	28	41	54
Cast iron, hard	1	2	3.3	4.6	7	10	14	18	26	35
Cast iron, chilled	0.78	1.6	2.5	3.4	5.3	7.8	10	13	19	26
Malleable iron	1	2.1	3.4	4.7	7.3	11	14	18	26	36
Steel, soft	1	2	3.3	4.6	7	10	14	18	26	35
Steel, medium	0.78	1.6	2.5	3.4	5.3	7.8	10	13	19	26
Steel, hard	0.56	1.1	1.8	2.5	3.9	5.7	7.7	10	14	19

increase the feed rate. If temporary overloads are permitted, the idle time between overloads, or cuts, should theoretically equal the overload time, so that the motor is not excessively overheated. Machine life and motor life may suffer from the practice of "crowding on" the feed rate, but in production use it may pay to wear out the machine faster in order to increase production.

A quick selector for milling-machine size is given in Table 8. This table is based on the number of cubic inches per minute that can be milled with different sizes of machines, and at the over-all efficiencies stated above, and not under overload conditions.

Establish Feed Rate

First ascertain the amount of power available at the cutter. Then calculate the maximum feed rate F in inches per minute from the formula

$$F = \frac{K \times \text{hp}_c}{d \times W}$$

EXAMPLE: Available milling machine has 20-hp motor. Material is soft cast iron. Depth of cut is ¼ in. Width of cut is 5 in. From Table 8, the value of $K \times \text{hp}_c$ for a 20-hp machine when milling soft cast iron is 16 cu in. per min, the maximum metal-removal rate. Then

$$F = \frac{16}{\frac{1}{4} \times 5} = 12.8 \text{ in. per min}$$

Selecting Feed per Tooth

The feed per tooth indicates the amount of material removed by each tooth from the workpiece, and for a given width and depth of cut it is a measure of the load on the cutter tooth. In practice, feed per tooth must be established in relation to type of cutter, the materials in work and cutter, rigidity of the workpiece, rigidity of machine and fixture, and power available. The range of values given in Table 9 must be scaled down where the depth of cut is great, the overhang is considerable, and the cutter or work is fragile. However, from the standpoints of production efficiency and freedom from chatter, the feed per tooth should be as great as possible.

Proper cutter design and correct tooth angle will permit increasing the feed per tooth over the values cited in Table 9. But under average operating conditions, and especially when taking finishing cuts, the values selected are usually lower than those listed—often being one-third to one-half as much.

For cutters made of carbon steel, the feed per tooth is one-half the values given for HSS cutters, while the feed per tooth for cutters with cast nonferrous teeth is the same as for carbide cutters.

Check the Tooth Load

In setting up a milling operation, especially when milling with carbide, it is essential to check the tooth load, also called the chip load or chip thickness. Operation above or below recognized chip loads (Table 9) is not good for the cutter and can be wasteful practice.

The feed per tooth f in inches is found by the formula

$$f = \frac{F}{N \times n}$$

where F = feed rate, in. per min; N = cutter rpm; and n = number of cutter teeth.

EXAMPLE: For the case cited in the example under Establish Feed Rate, we shall have to select a suitable cutting speed V from Table 2. For soft cast iron, V = 180 to 200 fpm for roughing cuts. Actually "cast iron" is an omnibus term. From identification of the material we find that it has a brinell hardness of 150. From Fig. 16, we see that a cast iron of this hardness can be milled with carbide at 300 fpm. So we shall let V = 300 fpm.

For a surface 5 in. wide, we would use a face mill at least 8 in. dia. For this size of cutter, as used with carbide teeth, the number of teeth $n = 1\frac{1}{2}D = 12$.

We are now in a position to calculate the tooth load:

$$N = \frac{12 \times V}{3.1416 \times D} = \frac{12 \times 300}{3.1416 \times 8} = 144 \text{ rpm}$$

$$f = \frac{F(\text{from previous example})}{N \times n} = \frac{12.8}{144 \times 12} = 0.0074 \text{ ipt}$$

According to Table 9, the suggested feed when face milling 150 brinell cast iron with carbide is 0.020 ipt. The tooth load for our setup is far too low.

There are several ways of raising the chip load to an appropriate value. From inspection of the above equation, we see that F can be increased, which will then raise f. To raise the feed rate, we must decrease the depth of cut. As an alternate, we can decrease the number of teeth n in the cutter to possibly six or even four teeth. A considerable amount of milling is currently being done with coarse-pitch cutters, many of them made in the shop by brazing teeth to cast cutter bodies.

A third method of employing proper feed rate and chip load would be to put the job on a smaller machine.

Remember that feed rate is the element in milling that pays off in terms of work done.

Maximum Chip Thickness

It is common practice to use a rule of thumb for feed or to compute the feed per

TABLE 9. SUGGESTED FEED PER TOOTH FOR MILLING VARIOUS MATERIALS*

Material	Face Mills	Helical Mills	Slotting and Side Mills	End Mills	Form-relieved Cutters	Circular Saws
With High-speed-steel Cutters						
Plastics	0.013	0.010	0.008	0.007	0.004	0.003
Magnesium and alloys	0.022	0.018	0.013	0.011	0.007	0.005
Aluminum and alloys	0.022	0.018	0.013	0.011	0.007	0.005
Free-cutting brasses and bronzes	0.022	0.018	0.013	0.011	0.007	0.005
Medium brasses and bronzes	0.014	0.011	0.008	0.007	0.004	0.003
Hard brasses and bronzes	0.009	0.007	0.006	0.005	0.003	0.002
Copper	0.012	0.010	0.007	0.006	0.004	0.003
Cast iron, soft (150–180 brinell)	0.016	0.013	0.009	0.008	0.005	0.004
Cast iron, medium (180–220 brinell)	0.013	0.010	0.007	0.007	0.004	0.003
Cast iron, hard (220–300 brinell)	0.011	0.008	0.006	0.006	0.003	0.003
Malleable iron	0.012	0.010	0.007	0.006	0.004	0.003
Cast steel	0.012	0.010	0.007	0.006	0.004	0.003
Low-carbon steel, free-machining	0.012	0.010	0.007	0.006	0.004	0.003
Low-carbon steel	0.010	0.008	0.006	0.005	0.003	0.003
Medium-carbon steel	0.010	0.008	0.006	0.005	0.003	0.003
Alloy steel, annealed (180–220 brinell)	0.008	0.007	0.005	0.004	0.003	0.002
Alloy steel, tough (220–300 brinell)	0.006	0.005	0.004	0.003	0.002	0.002
Alloy steel, hard (300–400 brinell)	0.004	0.003	0.003	0.002	0.002	0.001
Stainless steels, free-machining	0.010	0.008	0.006	0.005	0.003	0.002
Stainless steels	0.006	0.005	0.004	0.003	0.002	0.002
Monel metals	0.008	0.007	0.005	0.004	0.003	0.002
With Carbide-tipped Cutters						
Plastics	0.015	0.012	0.009	0.007	0.005	0.004
Magnesium and alloys	0.020	0.016	0.012	0.010	0.006	0.005
Aluminum and alloys	0.020	0.016	0.012	0.010	0.006	0.005
Free-cutting brasses and bronzes	0.020	0.016	0.012	0.010	0.006	0.005
Medium brasses and bronzes	0.012	0.010	0.007	0.006	0.004	0.003
Hard brasses and bronzes	0.010	0.008	0.006	0.005	0.003	0.003
Copper	0.012	0.009	0.007	0.006	0.004	0.003
Cast iron, soft (150–180 brinell)	0.020	0.016	0.012	0.010	0.006	0.005
Cast iron, medium (180–220 brinell)	0.016	0.013	0.010	0.008	0.005	0.004
Cast iron, hard (220–300 brinell)	0.012	0.010	0.007	0.006	0.004	0.003
Malleable iron	0.014	0.011	0.008	0.007	0.004	0.004
Cast steel	0.014	0.011	0.008	0.007	0.005	0.004
Low-carbon steel, free-machining	0.016	0.013	0.009	0.008	0.005	0.004
Low-carbon steel	0.014	0.011	0.008	0.007	0.004	0.004
Medium-carbon steel	0.014	0.011	0.008	0.007	0.004	0.004
Alloy steel, annealed (180–220 brinell)	0.014	0.011	0.008	0.007	0.004	0.004
Alloy steel, tough (220–300 brinell)	0.012	0.010	0.007	0.006	0.004	0.003
Alloy steel, hard (300–400 brinell)	0.010	0.008	0.006	0.005	0.003	0.003
Stainless steels, free-machining	0.014	0.011	0.008	0.007	0.004	0.004
Stainless steels	0.010	0.008	0.006	0.005	0.003	0.003
Monel metals	0.010	0.008	0.006	0.005	0.003	0.003

* Cincinnati Milling Machine Co.

tooth. The latter is correct only when the radius of the cutter equals the depth of cut. A more suitable term is "basic table advance per tooth."

The maximum chip thickness (at the point where the tooth leaves the work) is

$$\text{Max chip} = \text{approach} \times \text{basic advance per tooth} \div \text{cutter radius}$$
$$\text{Approach} = \sqrt{d(D-d)}$$

where D = diameter of cutter and d = depth of cut.

Table 10 lists the approaches of various cutters from 1¼ to 12 in. dia.

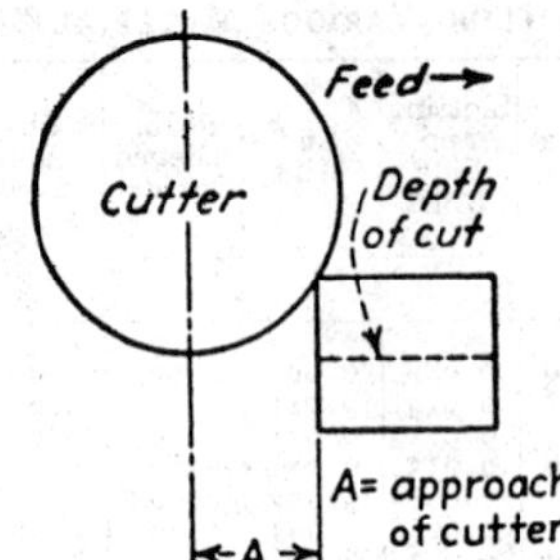

TABLE 10. APPROACH OF CUTTER FOR SPIRAL MILLS, KEYWAY CUTTERS, AND SAWS

Dia of Cutter	Depth of Cut, In.											
	1/16	1/8	3/16	1/4	3/8	1/2	3/4	1	1½	2	3	4
1¼	0.272	0.370	0.440	0.500	0.570	0.612						
1½	0.299	0.410	0.490	0.560	0.650	0.710	0.750					
1¾	0.320	0.450	0.540	0.610	0.720	0.790	0.870					
2	0.350	0.480	0.580	0.660	0.780	0.870	0.970	1.000				
2¼	0.370	0.520	0.620	0.710	0.840	0.930	1.060	1.120				
2½	0.390	0.540	0.660	0.750	0.890	1.000	1.150	1.220				
2¾	0.410	0.570	0.690	0.790	0.940	1.060	1.220	1.320				
3	0.430	0.600	0.730	0.830	0.990	1.120	1.300	1.410	1.500			
3¼	0.450	0.630	0.760	0.870	1.040	1.170	1.370	1.500	1.620			
3½	0.460	0.650	0.790	0.910	1.080	1.220	1.440	1.580	1.730			
3¾	0.480	0.670	0.820	0.930	1.130	1.280	1.500	1.660	1.840			
4	0.490	0.700	0.850	0.970	1.170	1.320	1.560	1.730	1.940	2.000		
4¼	0.511	0.720	0.870	1.000	1.210	1.370	1.620	1.800	2.030	2.120		
4½	0.530	0.740	0.900	1.030	1.240	1.410	1.670	1.870	2.120	2.240		
4¾	0.540	0.760	0.920	1.060	1.280	1.460	1.730	1.930	2.210	2.350		
5	0.560	0.780	0.950	1.090	1.320	1.500	1.790	2.000	2.290	2.450		
5½	0.580	0.820	1.000	1.140	1.390	1.580	1.890	2.120	2.450	2.640		
6	0.610	0.860	1.040	1.200	1.450	1.660	1.980	2.240	2.600	2.830	3.000	
6½	0.634	0.890	1.090	1.250	1.510	1.730	2.080	2.350	2.740	3.000	3.240	
7	0.658	0.930	1.130	1.300	1.580	1.800	2.170	2.450	2.870	3.160	3.460	
7½	0.680	0.950	1.170	1.350	1.630	1.870	2.250	2.550	3.000	3.220	3.670	
8	0.710	0.990	1.210	1.390	1.690	1.940	2.330	2.650	3.120	3.460	3.870	4.000
8½	0.730	1.030	1.250	1.440	1.740	2.000	2.410	2.740	3.240	3.610	4.060	4.240
9	0.750	1.050	1.280	1.480	1.800	2.060	2.490	2.830	3.360	3.740	4.240	4.470
9½	0.770	1.080	1.320	1.520	1.850	2.120	2.560	2.920	3.470	3.880	4.420	4.690
10	0.790	1.110	1.350	1.560	1.900	2.180	2.630	3.000	3.570	4.000	4.580	4.900
10½	0.810	1.140	1.390	1.600	1.950	2.240	2.710	3.080	3.670	4.120	4.750	5.100
11	0.830	1.170	1.420	1.640	2.000	2.290	2.770	3.160	3.770	4.290	4.840	5.290
11½	0.850	1.190	1.460	1.680	2.040	2.340	2.840	3.240	3.870	4.360	5.040	5.480
12	0.860	1.220	1.490	1.720	2.090	2.400	2.910	3.320	3.970	4.470	5.190	5.650

EXAMPLE: Assume the feed per tooth is said to be 0.008 in. What is the actual maximum chip thickness for these conditions?

Dia cutter	5
Number of teeth	30
Spindle rpm	52
Surface fpm	68
Table feed per min	3.4
Feed per rev	0.065
Basic advance per tooth	0.00215
Depth of cut	0.500

$$\text{Approach} = \sqrt{0.5(5 - 0.5)} = 1.5$$

$$\text{Max chip} = \frac{1.5(0.00215)}{2.5} = 0.00128$$

In plants where shopmen are allowed to use their discretion or methods engineers assume a certain feed rate, or "basic table advance per tooth," the actual maximum chip thickness will vary from 0.0002 to 0.007 in. for operations on identical material. Cycle times under these conditions are usually excessive. By using the actual maximum chip thickness, and adjusting the spindle speed and table feed, the cutting speeds will be increased and the milling time reduced, often as much as 50%.

MILLING CHARACTERISTICS OF VARIOUS MATERIALS

Light Metals

Aluminum. Milling is probably the most efficient method of machining aluminum alloys. They have been milled at speeds up to 25,000 fpm. Relatively short chips are produced. Vertical milling may be best for chip disposal, but climb milling results in more efficient cutting and less heating of tool and work.

Inserted-tooth cutters are preferred, and with relatively few teeth. Tooth design follows that of single-point tools. A helix on teeth of plain cutters is desirable, up to 50° if there is considerable top rake. Nicking of teeth is sometimes necessary to give shorter chips.

For good finish, which is usually a prime requirement, face relief angle should be 15°, peripheral clearance or relief angle 7° as a minimum, and radial rake 15° positive. A slippery steel grade of carbide is generally recommended. Table 11 gives details of speeds, feeds, and tool angles.

A conventional side milling cutter reground to high rake (34°) gave excellent performance at 1118 fpm and 63 in. per min feed on aluminum 61S and 52S alloys at ⅜ in. depth of cut.

Magnesium. Surface speeds as high as 9000 fpm are used, but standard practice probably does not exceed 5000 fpm. Cutters of standard pattern can be used if they have plenty of chip space, but half as many teeth as for steel gives better results. Rake angles range from 5 to 20°, relief angles from 7 to 10° for 1⁄16 in., followed by clearance of 20 to 30°. Face mills usually have axial rake of 5 to 15° but sometimes have negative axial rake to throw chips out of the cutter. Helix angle of end and slab mills is customarily 10 to 25°, and on helical mills about 45°.

Feed ranges from 0.004 to 0.025 in. per tooth. Finish cuts are normally 0.003 to 0.004 in. Milling is generally done dry, except that coolant may be employed on thin sections to avoid overheating.

Margins on counterbores should be small, about 0.015; clearance large to prevent rubbing and provide good chip space. Relief should be 6 to 10°, clearance 15 to 25°.

TABLE 11. MILLING PRACTICE WITH ALUMINUM ALLOYS

Alloy Type	High carbon or High-speed Steel				Cemented-carbide Tools			
	Max Cut, In.	Speed, Fpm	Feed		Max Cut, In.	Speed, Fpm	Feed	
			Max Fpm	In. per Tooth			Max Fpm	In. per Tooth
Roughing:								
Soft	0.250	700–2000	10	0.005–0.025	0.300	3000–15,000	20	0.004–0.002
Hard	0.200	500–1500	10	0.005–0.025	0.250	3000–15,000	20	0.004–0.020
Finishing:								
Soft	0.020	Up to 5000	10	0.005–0.025	0.020	3000–15,000	20	0.004–0.020
Hard	0.020	Up to 4000	10	0.005–0.025	0.020	3000–15,000	20	0.004–0.020

Tool Details

Cutting angle	48 to 67°	Cutting angle	68 to 97°
Top rake	20 to 35°	Top rake	−10 to 15°
Clearance:		Clearance:	
Primary	3 to 7°	Primary	3 to 7°
Secondary	7 to 12°	Secondary	7 to 12°
Helix	10 to 50°	Helix	−10 to 20°
Tooth spacing	Coarse	Tooth spacing	Approx one tooth per in. dia or fewer

Titanium. Milling operations on titanium present no difficulty. Conventional roughing speeds are 45 to 60 fpm and finishing speeds 55 to 65 fpm. At these speeds, the cutter teeth should have a 10° rake. With a Douglas Hi-Helix cutter, satisfactory cutting action is secured at 210 fpm and 0.002 in. feed per tooth. As with conventional cutters, climb milling gives best results. The Douglas HSS cutter was 2⅝ in. dia, with 20 teeth, 72° spiral helix, 0° radial rake. Coolant was heavy soluble oil.

Copper Alloys

Copper alloys fall into three machinability groups (see Sec. 26). Milling speeds and tool angles are:

Group	Fpm	Rake	Clearance
1	200–250	0–10°	10–15°
2	150–200	0–10°	6–12°
3	50–150	0–15°	See text

In addition, cutter teeth should have a narrow land of 0.015 to 0.030 in. and secondary clearance.

The group 3 alloys, the nonleaded nickel silvers, phosphor bronzes, supernickel, and copper-silicon alloys have worked well with a 5 to 10° clearance. For nonleaded copper, a clearance of 12 to 15° has been found more effective.

Coarse-tooth cutters permit higher speeds, and the additional space between teeth provides better chip removal.

High-rake milling offers good promise on materials like phosphor bronzes. A 12-tooth HSS cutter 6 in. dia and ground to 35° rake and 10° helix gave free cutting action on ¼- and ½-in. cuts at 240 fpm and 10 in. per min feed.

Irons

Normal Cast Iron. This is a metal with normal carbon content of 2.5 to 3.5%, plus silicon, manganese, sulfur, and phosphorus and having a tensile strength of around 25,000 psi. Cutting speed with carbide is 300 to 450 fpm. If the workpiece is thin or has narrow cross-sections, the 450 fpm speed should be tried to reduce breakout at the leaving edge. On operations where machine down time is costly and the above considerations do not apply, the lower range of speed may be more suitable to reduce cutter wear.

Modified Cast Iron. Controlled or modified cast irons, such as semisteel, Meehanite, and Linite with tensile strengths of approximately 50,000 psi or higher, call for surface speeds no higher than 300 fpm when machined with carbide. These materials usually brinell at 180 to 200. Cutting angles are the same as for standard steel.

Special Cast Irons. Chilled or white irons are milled with carbide at speeds from 100 to 250 fpm. Brinell hardness ranges from 500 to 700. The lower range of cutting speed is used for the harder material.

Malleable Iron. Choose the same cutting angles as for modified cast iron and machine at 350 to 450 fpm. A grade of carbide intended for cast steel will be suitable.

Wrought Iron. It is important first to identify the material as having less than 5 points carbon. Brinell hardness does not indicate machinability of wrought iron. If the material does have less than 5 points carbon, it should be milled with carbide at speeds of 1000 fpm or higher.

Nickel Alloys

Nickel, monel, and inconel produce curling chips, and cutters should have ample space between teeth for chip disposal. Except where indicated, the following recommendations apply to HSS cutters: Positive rake angle of 10 to 15° with plain or slab milling cutters of the heavy-duty helical type. Face-milling cutters should have teeth set into the body to achieve a positive radial rake of 15° and a negative axial rake of 7°. Tool nose should have a double bevel of 22½° and 45°.

General practice in milling monel, KR monel, and nickel with HSS cutters is to use 50 to 65 fpm and a feed per tooth of 0.005 to 0.010 in. With inconel and non-heat-treated K monel, use 40 fpm and a feed per tooth of 0.003 to 0.006 in. Depth per cut should rarely exceed ¼ in.; the average should be about ⅛ in.

K monel has been milled at 425 fpm, using Kennametal KM-tipped cutters.

For Nimonic 80 and Inconel X, see high-rake milling under Titanium.

Plastics

Carbon-steel cutters can be used at speeds as for brass, and HSS cutters at speeds three to four times higher. Carbide-tipped cutters with a slight rake give good results on glass- and asbestos-base laminates at speeds from 600 to 1000 fpm. When milling laminated plastics, never mill in a direction that will tend to separate the laminations. Some consider helical cutters better than straight-tooth types. Single- and double-bladed fly cutters are sometimes used at high speeds and fine feeds.

Steels

Cast Steel. For cast steel, the cutting speed with carbide is critical, usually 300 fpm and not higher than 350 fpm. Surface condition is sandy, making it essential to restrict the cutting speed to a lower rate than for a wrought steel of the same composition and brinell hardness. Tool angles are the same as for standard steel. Carbide grade must be carefully selected for the job. Resistance to abrasion and friction is vitally important. A 45° nose angle, instead of 15°, will prove helpful to resisting shock on contact with large irregular castings.

Low-carbon Steel. Material with from 5 to 15 points carbon content should be milled at speeds from 600 to 900 fpm. Applying speeds recommended for "standard" steels invites failure. Be certain to identify the material from specifications. A low hardness reading of 135 to 140 brinell is not a safe criterion.

SAE Steels. Cutting speeds for SAE steels beginning with 1015 or 1020 are generally selected on the basis of brinell hardness. On mass-production operations, and where the facilities exist for checking microstructure, the speeds will be selected according to machinability studies.

Approximate cutting speeds, based on brinell hardness, are:

Hardness	Cutter Material	
	High-speed Steel	Sintered Carbides
Up to 163	133	615
163–192	103	543
192–223	81	484
223–255	67	438
255–285	55	398
285–321	46	367
321–352	39	340
352–388	34	318
388–415	30	300

Approximate feeds for SAE steels in inches per tooth are:

Brinell Hardness	Cutter Material	Type of Milling Cutter				
		Face Mills	Helical Mills	Slotting and Side Mills	End Mills	Form Cutters
Up to 192	HSS	0.010	0.008	0.006	0.005	0.003
	Carbides	0.012	.....	0.008	0.007	0.004
192–233	HSS	0.008	0.007	0.005	0.004	0.003
	Carbides	0.012	.....	0.008	0.007	0.004
223–285	HSS	0.006	0.005	0.004	0.003	0.002
	Carbides	0.011	.....	0.007	0.006	0.004
285–415	HSS	0.004	0.003	0.003	0.002	0.002
	Carbides	0.010	.....	0.006	0.005	0.003

In face milling SAE 1020 with 30° rake, cutter life at 164 fpm equaled that of a conventional cutter and gave double the production rate. In slab milling with a 30° rake, 45° helix cutter worked well at 152 fpm.

High-speed Steel. Climb milling is preferred. Negative-rake carbide cutters have not proved suitable, because no available grade of carbide will mill the chromium carbide constituent in high-speed steel. Therefore, HSS cutters are used on this material. Conventional grind is 8 to 10° positive rake, 7 to 8° clearance, and no land. Speeds, feeds, and depths of cut follow practice on nickel steel.

With high-rake cutters (25° rake) speed may be up to 130 fpm and feed 4½ to 7¾ in. per min. Climb milling is used.

Hardened Steel. End mills for repair of die and tool parts hardened to Rockwell 62 C have carbide teeth mounted on a heavy cushion of silver solder. Cutters 0.250 in. dia have no rake on the teeth but have negative rake up to 5° on the 1-in. size. Tools have a slight back taper to avoid drag. Cutting-edge relief is 0.005 in. in 1/16 in., then secondary relief of 14°. Two-lip tool is used for roughing; three-lip tool for finishing. Cutting speed is about 210 fpm, feed about 0.015 to 0.030 in. per revolution.

Counterbores are the same except for four teeth and a pilot. Tips are mounted to grind a small hook (the outside edge being 0.015 in. ahead of center). Each lip is ground with a flat on the end and a 14° clearance.

Other materials machined with these tools are Nihard and Stellite.

Die blocks of 300 to 400 brinell, either new or undergoing repair, can be milled with carbide tools. Using 8-in. dia 16-blade cutters, average material removed was 390 cu in. per grind at a wear land of 1/32 in. Cutting speed was 175 fpm, feed rate 10½ in. per min, depth of cut 3/16 in., width of cut 5 in., and length of cut 44 in.

Stainless Steels. Milling speeds for stainless steels, using HSS cutters, are:

Type	Carpenter	Armco
403, 405, 410	70/105	40/60
416	100/125	50/80
420	35/70	35/50
420 F	70/100	35/55
430	70/105	40/60
430 F	110/135	50/80
414, 431	....	40/60
440 ABC	35/80	30/50
440 F	65/80	35/55
446	....	40/60
301, 302, 304	35/65	40/60
303	75/110	40/60
309	....	30/50
316, 317	....	30/50
321, 347	....	40/60

Minimum feed should be 0.005 in. per tooth for roughing and 0.003 in. for finishing. Rake angle is 4 to 10° positive, primary clearance likewise, and width of land 1/32 to 3/16 in.

High-rake milling, with HSS cutters ground to 34°, and primary and secondary clearance angles unchanged, removed nearly twice as much material as a conventional grind. Tool life may be ten to twenty times as great. Cut was made at 123 fpm, feed 8 in. per min, and depth of cut 7/8 in., using a 9/16- by 4½-in. staggered-tooth side milling cutter.

When using carbides on stainless steels, type 304 (without sulfur) calls for a cutting speed of 900 fpm. All other stainless steels should follow the recommendations for standard SAE steels.

The positive radial rake can be changed to a negative radial rake of 7°, when the workpiece or machine-tool conditions adversely affect cutting with a positive rake.

SHARPENING MILLING CUTTERS

Cutter sharpening has two purposes: (1) to regrind a new commercially ground cutter to the rake and clearance angles required for a specific material or operation, and (2) to restore worn cutting edges. Worn cutters should be resharpened when 0.010 to 0.035 in. of wear is evident. Frequent regrinding, at specified intervals, is best, so that not more than 0.010 in. need be removed from HSS and cast-alloy blades and 0.006 in. from carbide blades. Angles of principal concern to the tool sharpener are shown in Fig. 19.

Clearance Angle. Insufficient clearance angle adversely affects cutter life. Grinding increases the land width and decreases the clearance. Therefore, check the clearance after each sharpening. Primary clearance angle is measured easily with a dial indicator (Fig. 20). For each 1/16 in. width of land, each degree of clearance is approximately equivalent to 0.001 in. on the dial indicator. For any width of

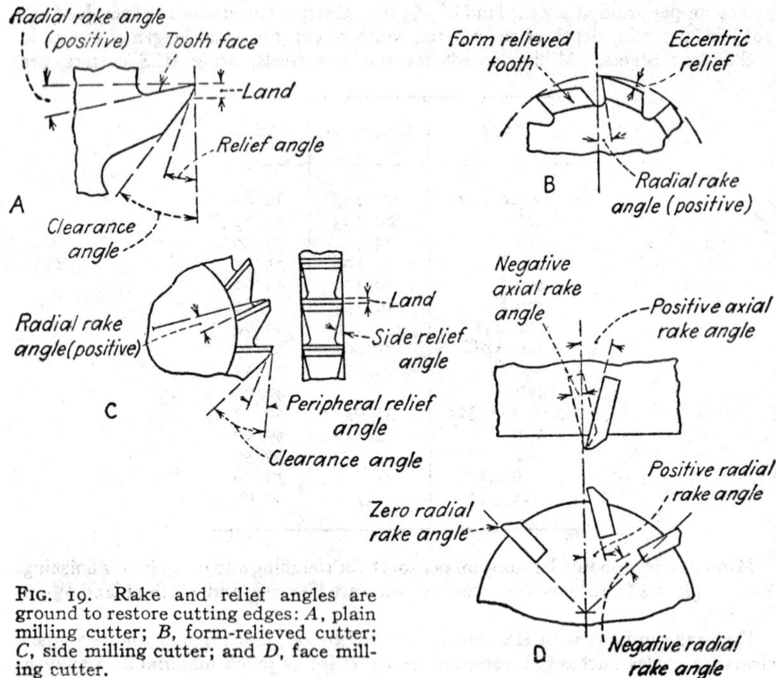

FIG. 19. Rake and relief angles are ground to restore cutting edges: *A*, plain milling cutter; *B*, form-relieved cutter; *C*, side milling cutter; and *D*, face milling cutter.

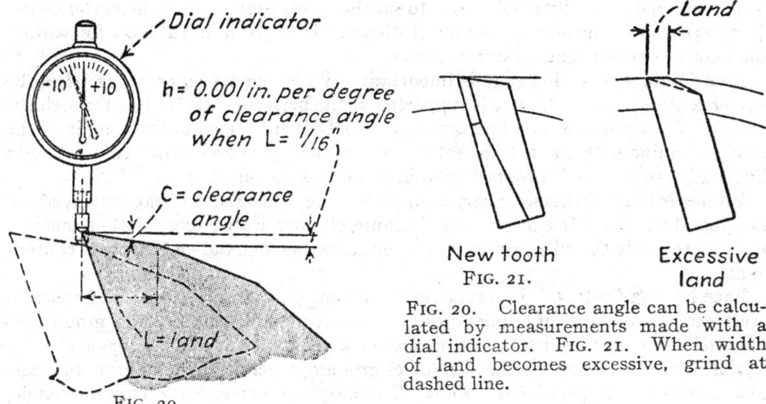

FIG. 20.

FIG. 21.

FIG. 20. Clearance angle can be calculated by measurements made with a dial indicator. FIG. 21. When width of land becomes excessive, grind at dashed line.

land, the clearance C in degrees $= 57.32h/L$, where h = indicator reading and L = width of land, in.

When the width of land is excessive and causes interference, grind secondary clearance from tip of heel of tool, equal to primary clearance plus 3°. This step will remove the high spot (Fig. 21).

Peripheral Cutters. The narrow land on the periphery of the teeth can be ground with a cup wheel, producing a straight surface, or with a disk wheel, leaving a slightly concave surface. In the latter case, a wheel of 6 to 8 in. dia minimizes the dishing effect.

With either disk or cup wheel, the direction of rotation is optional. Grinding away from the cutting edge is safer but leaves a burr. Grinding toward the cutting edge tends to lift the cutter from the tooth rest, so that it must be held by hand.

Clearance is established as follows:

Grinding with Cup Wheel. With wheel center and cutter center in line, raise or lower the tooth rest so that the cutter tooth is turned from horizontal by an amount equal to the clearance angle. The vertical adjustment equals 0.0087 × cutter dia × clearance angle. Conversely, raise or lower the cutter center by the stated offset, and keep the cutter edge at the wheel center line (see Fig. 22).

Grinding with Disk Wheel. For offset centers, the vertical adjustment should be as follows:

Cutter center above wheel center:

$$\text{Offset} = 0.0087 \times \text{cutter dia} \times \text{clearance angle}$$

Cutter center below wheel center:

$$\text{Offset} = 0.0087 \times \text{wheel dia} \times \text{clearance angle}$$

NOTE: For practical purposes, the offset is equal for the above two cases.

Cutter center in line with wheel center:

$$\text{Offset} = \frac{0.0087 \times \text{cutter dia} \times \text{wheel dia} \times \text{clearance angle}}{\text{wheel dia} + \text{cutter dia}}$$

The method generally employed is to set the wheel center below the cutter center (Fig. 23). For convenience, the offset distance H is given in Table 12 for various diameters of cutters and clearance angles.

Location of the cutting edge is important. When the wheel center is *below* the cutter center, the cutting edge is supported at the horizontal center line through the wheel. When the wheel is *above* center, the tooth is supported so that the tip of the tooth is in line with the cutter center. When the wheel and cutter centers are in line, the cutting edge is dropped by the amount of the offset.

Profile Milling Cutters. Sharpening procedure is similar to that employed for peripheral cutters. If trouble arises because of complicated tooth outline, make a templet to guide the wheel or cutter during resharpening, and check with a contour gage.

Face Milling Cutters. Two types are common: (1) with solid, brazed, or inserted teeth sharpened on cutter grinders; and (2) with removable teeth that are ground offhand or in a multiface fixture that incorporates the necessary angles for relief.

Sharpening on a cutter grinder involves grinding relief on the periphery, the face, and the chamfer or round nose. Offset $H = 0.0087 \times$ cutter dia $\times$ clearance angle, or may be taken from Table 12.

Form-relieved Cutters. These are face-sharpened only (Fig. 24). Zero-rake cutters are sharpened radially by centering the wheel face over the cutter center and adjusting the tooth rest against the back of the tooth to be ground. Positive-rake cutters are offset from the wheel face by an amount (usually marked in thousandths

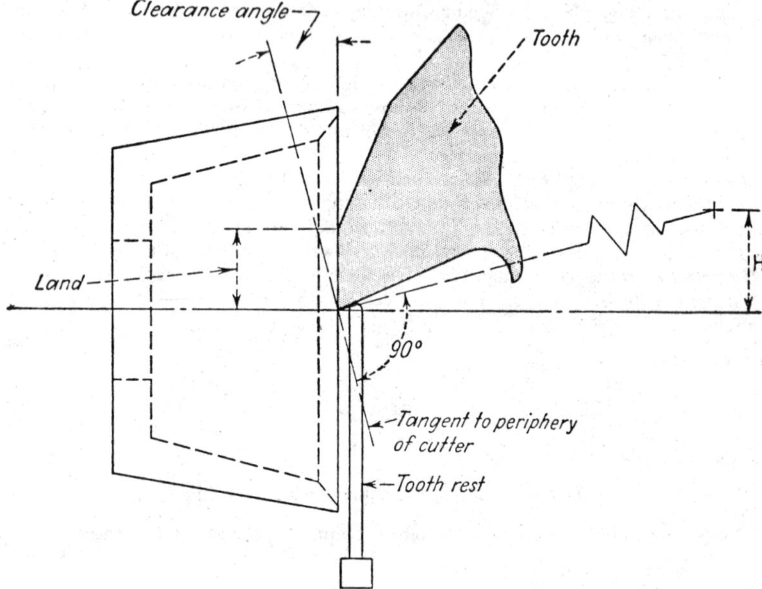

FIG. 22. A straight land is produced when grinding clearance on a peripheral cutter with a cup wheel. (*Courtesy of Cincinnati Milling Machine Co.*)

of an inch on the cutter) to provide required rake. If the rake is in degrees, offset d (in inches) = $R \sin r$. Grinding is done by passing the cutter under the wheel and indexing after each pass. To insure uniform tooth height, each tooth should be first spot-ground on the back for the tooth rest. After grinding, rake and helix angles should be checked, because any change in them affects cutter and work contour and cannot be tolerated on closely held work. To restore the form, an increase in rake angle will result in a narrower form; a decrease in rake will widen the form. Use gages to check rake and helix angles.

Helical End Mills. When a truncated conical wheel is rotated in contact with a tooth of a helical-flute end mill (Fig. 25), varying depths will be ground along the axial width of the cutter tooth. Eccentric relief is produced along the entire length of the tooth by traversing the end mill along its axis. It is necessary to rotate the end mill with the tooth helix, as it moves by the wheel, employing a tooth-rest finger for the purpose. Wheel width, or portion of wheel width to be used, is determined by the helix angle and the dimensions of the cutter. Minimum width must be equal to the width of land measured parallel to the cutter axis. Cone angle on the grinding wheel is obtained by dressing the wheel to angle Φ, where $\tan \Phi = D/W$. Here D is the radial clearance (desired indicator drop over width of primary clearance) and W is the axial width of land to be cleared.

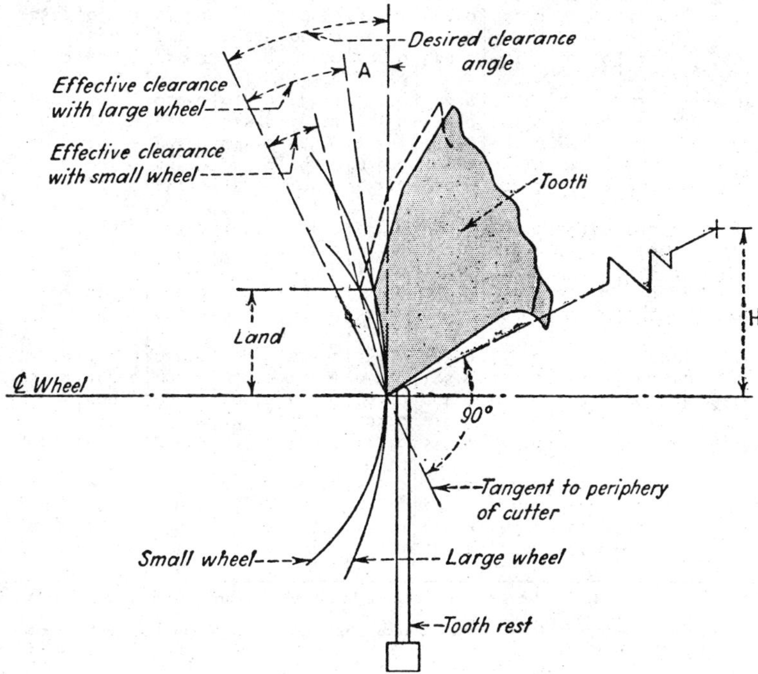

FIG. 23. A concave land is produced when clearance angle is ground with a disk wheel.

TABLE 12. OFFSET *H* WHEN DISK-WHEEL CENTER IS SET ABOVE OR BELOW CUTTER CENTER

Wheel* or Cutter† Dia	Clearance Angle, Deg									
	1°	2°	3°	4°	5°	6°	7°	8°	9°	10°
3	0.026	0.052	0.078	0.104	0.130	0.156	0.182	0.208	0.235	0.261
3¼	0.028	0.056	0.084	0.113	0.141	0.169	0.198	0.226	0.254	0.283
3½	0.030	0.061	0.091	0.122	0.152	0.182	0.213	0.243	0.274	0.304
3¾	0.032	0.065	0.097	0.131	0.163	0.195	0.228	0.261	0.293	0.326
4	0.034	0.069	0.104	0.139	0.174	0.208	0.243	0.278	0.313	0.348
4¼	0.037	0.074	0.111	0.148	0.185	0.222	0.258	0.295	0.333	0.370
4½	0.039	0.078	0.117	0.156	0.196	0.235	0.273	0.312	0.352	0.391
4¾	0.041	0.082	0.124	0.165	0.206	0.248	0.288	0.330	0.372	0.413
5	0.043	0.087	0.130	0.174	0.217	0.261	0.304	0.347	0.391	0.435
5¼	0.045	0.091	0.137	0.183	0.228	0.274	0.319	0.365	0.411	0.457
5½	0.047	0.095	0.144	0.191	0.239	0.287	0.335	0.382	0.430	0.478
5¾	0.050	0.100	0.150	0.200	0.250	0.300	0.350	0.400	0.450	0.500
6	0.052	0.104	0.156	0.208	0.261	0.313	0.365	0.417	0.470	0.521
6¼	0.054	0.109	0.163	0.217	0.272	0.326	0.380	0.435	0.489	0.543
6½	0.056	0.113	0.169	0.226	0.283	0.339	0.395	0.452	0.509	0.565
6¾	0.058	0.117	0.176	0.235	0.294	0.352	0.410	0.470	0.529	0.587
7	0.061	0.122	0.182	0.243	0.304	0.365	0.425	0.487	0.549	0.609
7¼	0.063	0.126	0.189	0.252	0.315	0.378	0.440	0.504	0.568	0.630
7½	0.065	0.130	0.195	0.261	0.326	0.391	0.456	0.522	0.588	0.651
7¾	0.067	0.135	0.202	0.270	0.337	0.404	0.471	0.540	0.607	0.673
8	0.069	0.139	0.208	0.278	0.347	0.417	0.487	0.557	0.626	0.695
8½	0.074	0.148	0.222	0.296	0.370	0.444	0.518	0.592	0.666	0.740
9	0.078	0.157	0.235	0.313	0.391	0.470	0.548	0.626	0.705	0.783
9½	0.083	0.165	0.248	0.331	0.413	0.496	0.579	0.664	0.747	0.827
10	0.087	0.174	0.261	0.348	0.435	0.522	0.609	0.696	0.783	0.870
10½	0.091	0.183	0.274	0.365	0.455	0.547	0.639	0.730	0.822	0.914
11	0.096	0.191	0.287	0.383	0.479	0.574	0.670	0.766	0.861	0.957
11½	0.100	0.200	0.300	0.400	0.500	0.600	0.700	0.800	0.900	1.000
12	0.104	0.209	0.313	0.418	0.522	0.626	0.731	0.835	0.940	1.044
12½	0.108	0.218	0.326	0.435	0.544	0.653	0.761	0.870	0.979	1.088
13	0.113	0.226	0.339	0.452	0.565	0.679	0.792	0.905	1.018	1.131
14	0.122	0.244	0.366	0.487	0.609	0.731	0.853	0.974	1.096	1.218
15	0.131	0.261	0.392	0.522	0.653	0.783	0.914	1.044	1.175	1.305
16	0.139	0.278	0.418	0.557	0.696	0.835	0.974	1.114	1.253	1.392

* Wheel diameter applies for selecting distance to set wheel center above or below cutter center when using a straight wheel.

† Cutter diameter applies for selecting distance to set tooth rest above or below cutter center when using a cup wheel.

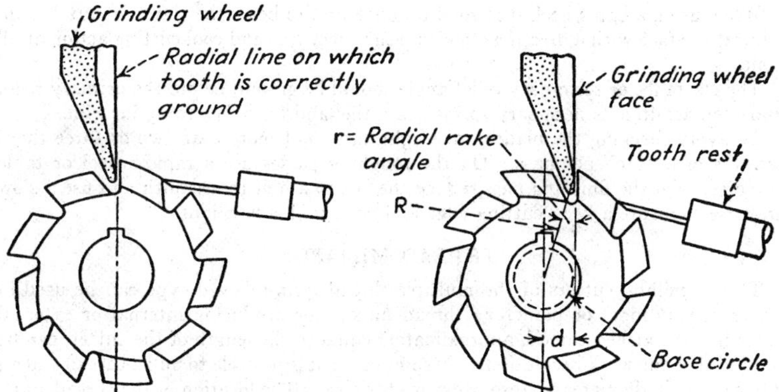

FIG. 24. Form-relieved cutters are face-sharpened only.

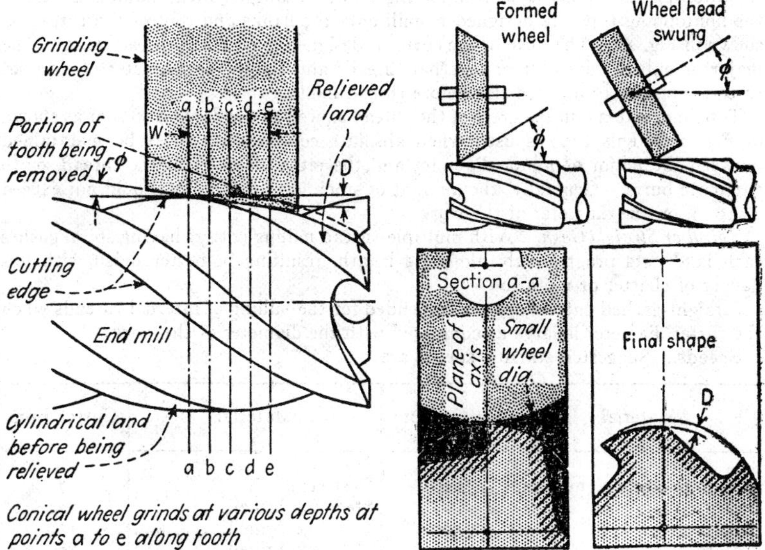

FIG. 25. Conical wheel grinds eccentric relief on helical end mill. Wheel is dressed or wheel head swung to angle Φ.

Cast-alloy Cutters. Face mills equipped with inserted cast-alloy blades and shell end mills with brazed-in blades should be resharpened as soon as a land about 0.010 in. to 1/64 in. wide has been worn on the peripheral cutting edges. Operating the cutter beyond this point causes rubbing on the land and poor finish on the work and will lead to chipping and complete failure of the cutting edge.

When using a cup wheel, it is good practice first to bevel the rim to about 1/16 in. wide at the face with a dressing stick to insure very free and cool cutting action at all times.

The clearance or secondary relief angle should be ground before the primary relief and then as often as necessary to maintain the land from 1/16 to 1/8 in. wide.

To avoid checking the blades, take light cuts—not more than two or three thousandths of an inch per pass. On the roughing passes use a rapid wheel or table traverse. On the finishing pass reduce the cut to a half thousandth and use a slow traverse to insure a keen cutting edge and uniform blade height.

THREAD MILLING[1]

Thread milling cutters of the multiple-thread form-relieved type can be used in either conventional or planetary thread millers for producing internal or external threads. A finished thread, approximately equal to the length of the cutter, can be produced in one work revolution. In some cases, it is possible to mill a thread and an adjacent plain diameter to absolute concentricity. The location of the thread start, or other reference point on the thread, can be maintained accurately in relation to a reference point on the part.

Cutter Types. *Topping or Nontopping Type.* Multiple-thread milling cutters of the nontopping type are designed to mill only the flanks and root of the thread as shown in Fig. 26. The form of the cutter is designed to clear the thread crest. The major or outside diameter of external threads and the minor diameter of internal threads must be to finished size before thread milling.

Topping cutters mill the crest of the thread as well as the flanks and root as shown in Fig. 27. This type is used when absolute concentricity must be maintained between the major or minor diameter and the pitch diameter of the thread, or to eliminate burrs. Generally, the amount of stock left for topping should not exceed 0.010 in. on the diameter of the work.

Spiral or Straight Gash. With multiple-thread milling cutters having spiral gashes each land cuts progressively along its length, resulting in better finish with less danger of chatter or vibration.

Straight-gashed cutters are recommended for the milling of internal threads when the cutter diameter is large as compared with the diameter of the work.

Speeds. Suggested starting speeds are:

Material	Cutter Fpm	Material	Cutter Fpm
Aluminum and magnesium	600	Cast steel	70
Brass:		Steel 100 brinell	120
Soft	500	150 brinell	110
Hard	200	200 brinell	100
Bronze:		250 brinell	90
Soft	200	300 brinell	70
Hard	60 to 110	350 brinell	55
Malleable iron	80	400 brinell	45
Cast iron:		Stainless steel	50 to 100
Soft	150		
Medium	125		
Hard	100		

[1] Elmer Zook, Chief Engineer, Detroit Tap & Tool Co.

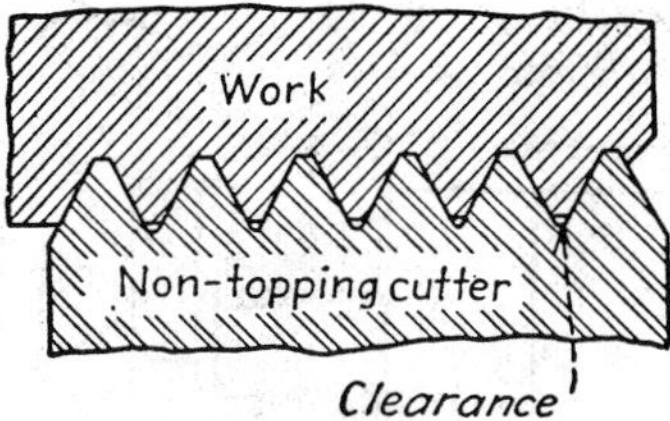

FIG. 26. Nontopping cutters mill only the flanks and root of the thread.

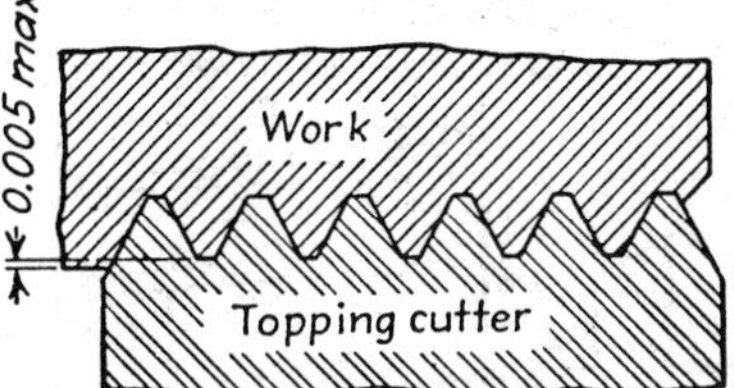

FIG. 27. A topping cutter mills the entire thread, including the crest.

Feeds. Practical feed rates for thread milling range from 0.001 to 0.005 in. per tooth per revolution, with 0.002 in. being a good average rate of feed.

Cutter Maintenance. Grind the cutting face of each gash to remove the worn portion and maintain the original form and accuracy. The rake angle must be maintained at its original value and should always be marked on the cutter in terms of offset distance. This dimension and the relative position of the grinding wheel and cutter are shown in Fig. 28.

All thread milling cutters should be sharpened by using the angle side of a dish wheel, preferably on an automatic cutter-sharpening machine. Indexing for straight-gashed cutters is best accomplished by use of a master index plate or by using an index finger against the back of the tooth. In using the latter method, the cutter must first be accurately index ground on the back of each tooth. This can be done by using the front face as a master.

Spiral-gashed cutters may be sharpened by swinging the wheelhead or the table (depending upon the type of cutter grinder) to the gash and using a follower against a master guide to the spiral lead marked on the cutter (Fig. 29).

The best method of inspecting multiple-thread milling cutters is by inspection of the actual parts produced.

What to Do about Chatter

The following procedure is recommended when chatter is encountered:

1. Examine the work-holding fixture. Unless the work is securely held, chatter is almost sure to develop.

2. Vary the feed and speed rate both up and down. Chatter is often eliminated in this manner. Usually it is better to increase the feed per tooth.

3. Check rigidity of the machine. Loose gibs are often responsible for chatter.

4. Try different cutters. Sometimes a cutter with fewer blades will work per-

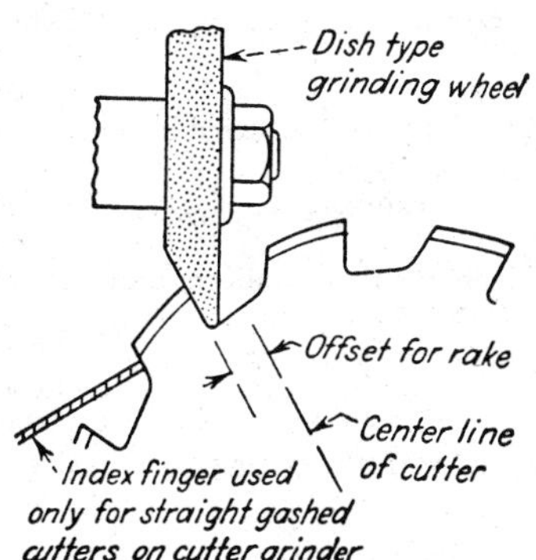

FIG. 28. When sharpening straight-gashed thread milling cutters, index with an index finger against the back of the tooth or use a master index plate.

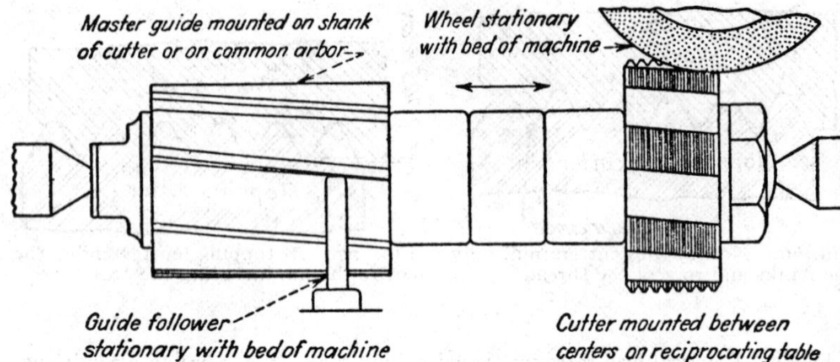

FIG. 29. When sharpening spiral-gashed thread-milling cutters, use a follower agains a master guide milled accurately to the lead marked on the cutter.

fectly. Often changing the number of teeth in contact with the work will eliminat chatter.

5. Vary the cutter grind. Increasing the chamfer, reducing the clearance angle circle grinding to leave a true land of 0.005 to 0.007 in., or reducing the width of fla on the face of the blades are some of the variations which may eliminate the difficulty

DIAGNOSIS OF MILLING TROUBLES[1]

The following suggestions apply to milling cutters and other multiple-point rotat ing tools. Often more than one of the troubles cited will appear at the same time In this case, careful study of the tool in operation will help to isolate and correct th troubles encountered.

Probable Cause	*Suggestions for Correction*
	Breakage
Excessive feed	Reduce feed to a point where weakest cross-sectior of tool will withstand the required torque an thrust
Too many teeth engaged in cut at same time	Reduce number of teeth in cutter or strengthe cross-section to carry load
Chip packing	Reduce number of teeth in cutter or increase chi wells by revising tooth shape
Feeding without uniform rotation of tool	On arbor cutters, use well-fitting key extending int collars on both sides. On shank-type cutters make sure driving mechanism is adequate
Dull cutter	Do not run tool past the time when it needs resharp ening
Improper resharpening	Check resharpening methods to make sure that too is not cracked by improper resharpening

[1] Gorham Tool Co.

Probable Cause	*Suggestions for Correction*
Chipping of Cutting Edge	
Excessive feed	Reduce feed rate or redesign tooth shape to strengthen cutting edge. Possibly use a tougher tool material
Excessive vibration	Keep machine tool, arbors, supports, work fixtures, etc., in first-class condition. Use an arbor of sufficient size to overcome torsional vibration
Excessive relief	Do not exceed recommended amounts of relief when resharpening
Rough Finish	
Too high feed rate	Reduce feed rate
Too low surface speed	Increase surface speed but avoid excessive speed
Dull tool	Do not run tool past the time at which it needs resharpening
Improper cutting fluid	Change to cutting fluid recommended for finish on particular work material
Milling against the feed	If conditions warrant, use in-cut or climb milling
Chip Clogging	
Insufficient chip room	Reduce number of teeth or revise shape of tooth to provide ample chip room
Insufficient cutting fluid	Increase pressure of cutting fluid or so direct flow that chips are washed out of teeth or flutes
Tool magnetized	Make sure that the tool is demagnetized after being held on a magnetic chuck
Rubbing	
Insufficient concave	Increase concave to avoid rubbing
Work closing in on cutter	Keep workpiece cool or clamp so that it cannot shift into cutter
Axis of tool out of relation to feed travel	Check machine tool to make sure that it is properly lined up
Burning of Cutting Edge	
Excessive speed	Reduce surface speed to a point at which the tool will stand up. Possibly use a tool material of higher red hardness
Hard work material	Check hardness of work material to keep it within range of machinability under the operating conditions
Inadequate cutting fluid	Increase flow of cutting fluid or improve direction of flow to cool cutting edge in the cut. Consider use of cutting fluid with better cooling properties
Cratering	
Excessive speed or feed or both eroding face of tooth	Reduce speed or feed to eliminate excessive cratering. Possibly use a more abrasion-resistant tool material or apply a case treatment to present tools
Insufficient rake angles	Increase rake angles to reduce friction of chip against tooth face

INDEXING METHODS

Indexing is the process employed to move and accurately locate a workpiece in a series of positions. The spacing may be circumferential, as in milling of gear teeth, or linear, as in milling of rack teeth. Normally, the spacing between consecutive indexings is uniform, but it may be varied to suit job requirements.

Linear Indexing. Two methods are available: (1) to position the workpiece by making a number of equal table adjustments with the table micrometer dial, or (2) to use a dividing head geared to the lead screw. Change gears are selected for a low lead, so that the lead screw can be rotated by rotating the index plate and index crank.

Circumferential Indexing. Various types of dividing heads are employed, according to accuracy requirements. The ordinary dividing head will accurately rotate the workpiece a full turn or part of a turn, determined by the number of divisions or angular space distance between divisions. Error accumulated in indexing from one hole to the next for a complete circle on a 12 in. dia circle is ordinarily within 0.015 in.

Indexing Centers. Also known as a direct-indexing head, or a single plate head, this milling-machine accessory consists of a tailstock and a headstock. The latter is provided with an index plate or disk that has either notches or one or more concentric circles of holes—of different numbers of holes uniformly spaced. The work spindle is rotated through the desired number of notches or holes and locked while the part is machined. Thus, the circle can be divided into an even number of parts only. Indexing centers are particularly valuable for repetitive work, as in milling reamer flutes, hexagons, gear teeth, keys, etc.

Plain and Universal Dividing Heads. Much greater versatility is obtained with these attachments, especially for the indexing of low leads beyond the capacity of the indexing centers, or the selection of a much broader number of divisions of the circle. The work spindle has a taper-bored hole to receive work mandrels or stub arbors, and usually mounts an index plate for direct-indexing purposes. The plain head is fixed, whereas the universal head is arranged to swing the work spindle from below the horizontal to past the vertical, according to a graduated scale.

These heads normally have a 40-tooth worm wheel on the work spindle and a single-thread worm on a worm shaft. One turn of the worm shaft rotates the work spindle one-fortieth of a revolution, or 9°. For direct indexing, the worm can be dropped out of engagement and the front plate used. But for low leads, an index plate mounted on a shaft, and connected to the worm shaft by idler gears or change gears, is used.

The several methods of indexing are:

Rapid Indexing. If the dividing head is used, the worm and worm wheel are disengaged and the spindle is moved by hand. Only numbers that will divide evenly into the number of holes or notches in the index plate can be indexed. If the plate has 24 holes, for example, the number of divisions of a circle that can be indexed are 2, 3, 4, 6, 8, 12, and 24.

Rule: Divide 24 by the number of divisions required and the result equals the number of holes to move in the 24-hole rapid index plate. Thus, number of holes to move $= 24/D$, where D is the number of divisions.

Plain or Simple Indexing. By this method, numbers beyond the range of rapid indexing are secured. In this operation, the work spindle is moved by turning an index crank attached to the worm shaft.

The general equation for relationship of the motion of the index crank and of the work spindle is $R/D = s/C$, where R is the ratio of the worm wheel to the worm, D

is the number of divisions of the circle, s is the number of spaces to be indexed, and C is the hole circle selected. Since R is normally 40, this equation simplifies to $40/D = s/C$.

We can also let $s/C = T$, the number of turns of the index crank. Then $T = 40/D$.

EXAMPLE: Divisions wanted, $D = 5$. Then $T = 40/5 = 8$ turns. In this case, spacing of holes in the index plate is not required.

But suppose that by inspection of the indexing problem, T is seen to be a fraction or a whole number plus a fraction.

EXAMPLE: $D = 54$. Then $T = 40/54$. If your dividing head is equipped with a plate with a 54-hole circle, you can index 40 spaces. If you do not have this plate, divide both terms in the fraction by a common divisor that will permit you to choose an available plate. If 2 is selected as the divisor, $40/54 = 20/27$. Now, index 20 holes on the 27-hole circle.

If you have a problem in indexing where T is a whole number plus a fraction, and the denominator of the fraction is smaller than any hole circle, proceed thus: Multiply both terms of the fraction by a number that will give a denominator corresponding to an available hole circle.

EXAMPLE: Divisions $D = 9$. Then $T = 40/9$, or 4 turns plus 4/9 turn. Multiply the fraction by a suitable number, say 2. Then 4/9 becomes 8/18. Since you have an 18-hole circle, you will index 4 turns plus 8 spaces on the 18-hole circle.

Angular Indexing. Often a job will come to you with an angular distance specified rather than a number of divisions. The angular distance must be translated into fractions of a turn or complete turns and fractions of a turn of the index crank.

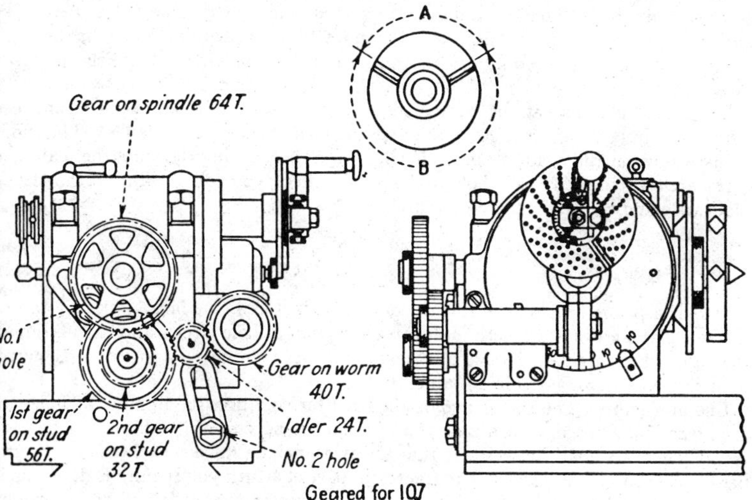

FIG. 30. Brown & Sharpe dividing head set up with compound gearing and one idler to index 107 divisions.

Because the usual dividing head is geared in a 40:1 ratio, one turn of the crank indexes the spindle 360/40 = 9°. If the angular distance is in degrees only, $T = A/9$, where A is expressed in degrees.

If the angular spacing is expressed in degrees and minutes, reduce it to minutes. Then $T = A'/540$.

If the angular spacing is expressed in degrees, minutes, and seconds, then $T = A''/32400$.

EXAMPLE: Suppose four notches are to be milled 50°47′20″ apart. Then A'' = 182,840 (number of seconds in above angular dimension). And T = 182,840/32,400 = 5 and 521/810 turns. Since there is no index plate with a 810-hole circle, we have two choices: (1) to reduce the fraction 521/810 to get the nearest "standard" plate, that is, say ratios of 18/28 or 27/42. With the latter, the total displacement of 5 turns plus 27 holes in the 42-hole plate is 11.43″ less than required. If we have high-number index plates, we might index 5 turns plus 128 holes on the 199-hole circle, and achieve an error of only 0.2″.

If the number of divisions is large, the error per division may accumulate to a sizable amount. Compensation for accumulated error is made by adding or subtracting one space, after making a number of divisions for which the aggregate error totals one or nearly one space in the circle of holes used. In the case of a standard plate, the angular displacement of one space can be considerably greater than the correction needed.

Compound Indexing. When divisions are required beyond the range of simple indexing, compound indexing is infrequently used. In effect, two plain indexings are made. The method is not much used today, because differential indexing is preferred. In brief, for the first indexing, the index crank is rotated in the usual manner. This is followed, after disengagement of the plate stop, by rotation of the index plate with the index crank engaged. Indexing is done with respect to the fixed index pin, for a number of holes on a different circle of holes in the same index plate. To *add* to the previous amount indexed, the index plate is made to rotate in the same direction as the index crank; to *subtract*, the plate movement is in the opposite direction.

These instructions can be expressed mathematically: The crankpin is engaged with hole circle C_1 of an index plate and the lockpin in hole circle C_2. The index crank is now moved s_1 spaces in circle C_1. After disengaging the stop, the plate and crank are rotated s_2 spaces in hole circle C_2. In effect, you add or subtract the second plain indexing to satisfy the equation:

$$T = \frac{40}{D} = \frac{s_1}{C_1} \pm \frac{s_2}{C_2}$$

If D is a prime number, and hence cannot be factored, the hole circles C_1 and C_2 will not give an exact answer. Repeated selection of hole circles and calculation of the value of the right-hand side of the equation will be necessary to reduce the error with respect to $40/D$ within suitable limits.

Use of Sector. The sector avoids the need for counting the number of holes for each fractional part of a turn that you will make in succession. This sector has two radial arms. Adjust the sector so that the arms include *one more hole* than the number of spaces to be indexed. The fractional part of a turn you require is determined by spaces, not holes. After indexing, move the sector so that it will be in position for the next indexing operation.

Differential Indexing. By this technique a wider range of divisions can be

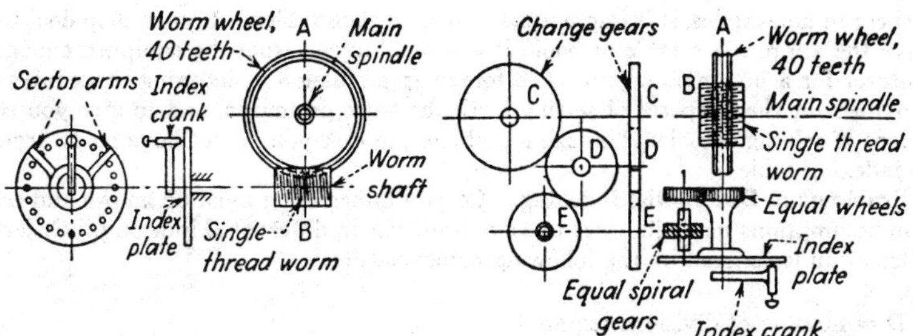

FIG. 31. Spindle drive.

FIG. 32. Application of change gears.

secured than with plain indexing. In differential indexing (Fig. 30), the dividing operation is performed the same as in simple indexing, but the index plate instead of remaining stationary is made to move relative to the index crank. A set of change gears is placed between the index plate and the work spindle. Through this train of gears, the index plate is made to rotate in proper relationship to the spindle, either faster or slower, and in the same direction (positive) or in the opposite direction (negative). The differential in movements of the index plate and the index crank causes the work to turn exactly the amount desired. Expressed mathematically, differential indexing modifies the spindle-drive ratio R by $\pm x$.

For differential indexing, the general formula for plain indexing $R/D = s/C$ is modified to

$$\frac{R \pm x}{D} = \frac{s}{C}$$

With the Brown & Sharpe dividing heads arranged for differential indexing, these change gears are provided: 24 (2 gears), 28, 32, 40, 44, 48, 56, 64, 72, 86, and 100 teeth. These are the change gears used to make a train with a value of x.

If gear S (gear on spindle) and gear W (gear on worm) (see Fig. 30) are equal, and two idlers are between them, the index plate will move counterclockwise when the index crank is moved clockwise. This assumes that the index pin is pulled out. Forty turns of the crank will turn the plate once, but the index pin will have caught up with the original hole 41 times. Thus, the circle has been divided into 41 parts, and you could cut 41 notches or gear teeth by stopping each time the index pin comes back to the original hole.

If one idler is used, the plate and crank will move in the same direction. For 40 turns of the crank the pin will catch up with the original hole 39 times.

Unequal gears S and W change the spindle-to-worm ratio. For example, if S is 48 and W is 24, the index plate will revolve twice while the spindle revolves once.

The Brown & Sharpe index plates have the following hole circles:

Plate 1: 15-16-17-18-19-20.

Plate 2: 21-23-27-29-31-33.

Plate 3: 37-39-41-43-47-49.

Other manufacturers supply different plates.

Indexing Tables. Manufacturers of indexing equipment supply books with each dividing head, showing the number of divisions that can be indexed, the index plate to use, number of turns of index, plus any required change gears and idlers. For space reasons, and because the index plates supplied by various companies vary in respect to hole circles, it is impractical to reprint such tables. If your shop does not have the appropriate table or book, it can be obtained from the equipment manufacturer for a nominal charge. The foregoing discussion of indexing procedures is provided for the purposes of setting forth the basic principles, and to give you the means of solving occasional indexing problems, in case you do not have ready access to indexing tables.

Problems in Differential Indexing. Certain information must be known and certain assumptions made in order to solve problems in differential indexing. Orderly calculation is assisted by the following nomenclature:

D = number of divisions required
C = number of holes in index plate
s = number of spaces taken at each indexing
R = ratio of gearing spindle-worm ratio, normally 40:1
x = ratio of train of gearing between spindle and index plate
S = gear on spindle
G_1 = first gear on stud
G_2 = second gear on stud
W = gear on worm

If $C \times R$ is greater than $D \times s$, then $x = \dfrac{C \times R - D \times s}{C}$

If $C \times R$ is less than $D \times s$, then $x = \dfrac{D \times s - C \times R}{C}$

For simple gearing, $x = \dfrac{S}{W}$ For compound gearing, $x = \dfrac{SG_1}{G_2W}$

The need for idler gears can be determined by inspection of $C \times R$ and $D \times s$:

If $C \times R$ is greater than $D \times s$ and gearing is simple, use one idler.
If $C \times R$ is greater than $D \times s$ and gearing is compound, use no idlers.
If $C \times R$ is less than $D \times s$ and gearing is simple, use two idlers.
If $C \times R$ is less than $D \times s$ and gearing is compound, use one idler.

Select s so that the ratio of gearing is less than 6:1 to avoid excessive stress on the gears.

EXAMPLE 1: $D = 39$. Required C, s, and x. Assume $C = 33$, $s = 22$. Then

$$x = \frac{(33 \times 40) - (59 \times 22)}{33} = \frac{22}{33} = \frac{2}{3}$$

Now select S as 32 and W as 48. Because $C \times R$ is greater than $D \times s$, simple gearing with one idler will be required.

EXAMPLE 2: $D = 319$. Required C, s, and x. Assume $C = 29$, $s = 4$. In this case $C \times R$ is less than $D \times s$. Then

$$x = \frac{(319 \times 4) - (29 \times 40)}{29}$$

$$= \frac{1160 - 1276}{29} = -\frac{116}{29} = -4$$

Therefore compound gearing is required. Because $C \times R$ is less than $D \times s$, one idler is required.

Section 7

PLANING AND SHAPING

SECTION 7

PLANING AND SHAPING

PLANING

When properly used, planers provide an economical means of machining narrow or broad surfaces in the horizontal, vertical, and diagonal planes, and likewise also producing radiuses and forms. The words "properly used" indicate certain reservations. Planers are likely to be equipped with three or four heads. Economical use of the planer requires that as many heads be used as possible. Therefore, how the work is set up is important. Possibly, it can be set up in two rows, or strings, so that all heads can be used at once. Gang tools consisting of three or four tools set in a holder, so that each tool takes its proportional share of the total feed, are sometimes used.

Work Setup. Fixtures are used to good advantage on some kinds of repetitive work, especially when clamping can be done rapidly and effectively. However, a great many jobs must be set up individually. Before pieces are brought to the planer, they should be laid out on a layout or surface plate. This practice saves idle time at the machine, and the layout man develops a higher degree of skill than the machine operator is likely to attain.

A number of factors affect planer output, but skillful work setup is one of the most important. It is said that setting up work on a planer is more difficult than on any other type of machine tool. This is particularly so when a number of small- or medium-sized pieces of irregular shape are planed at one time. The reason for this is that the work is subjected to the intermittent cutting pressures of several tools as well as the inertia forces developed at table reversal. Also, care must be taken to avoid bending and warping of the work when finishing cuts are taken.

A good practice is to sketch the various clamps, screw jacks, end stops, etc., used to clamp the work, all in proper relation to the job. A sketch allows the operator to go ahead with the job without working up a proper setup. This practice avoids the possibility of a poor setup, as when made by a newly trained man; saves machine time while figuring out a setup on the job; and serves as a reference the next time the job is run.

Keep on hand a plentiful supply of clamps, bolts, stop pins, etc., as shown in Fig. 1. These devices will enable many different kinds of jobs to be set up and clamped properly.

Uses of some of these devices are shown in Fig. 2, for making a setup to plane the top and right side of the piece. The part is end-stopped by stop pins *A* and the angle bracket *B*. Side stop pins *C* prevent the work from shifting to the left. Screw jack *D* supports the overhang of the piece. Offset clamps *F* on the left side and pin clamps *G* on the ends clamp the work to the table. Clamp studs *H* screw into the T-slot nuts *J*.

Planer Efficiency. Most planers are sizable machines, they represent a considerable investment, and their hourly machine rate is often much higher than that of many other classes of machine tools. Prewar planers have been made obsolete by

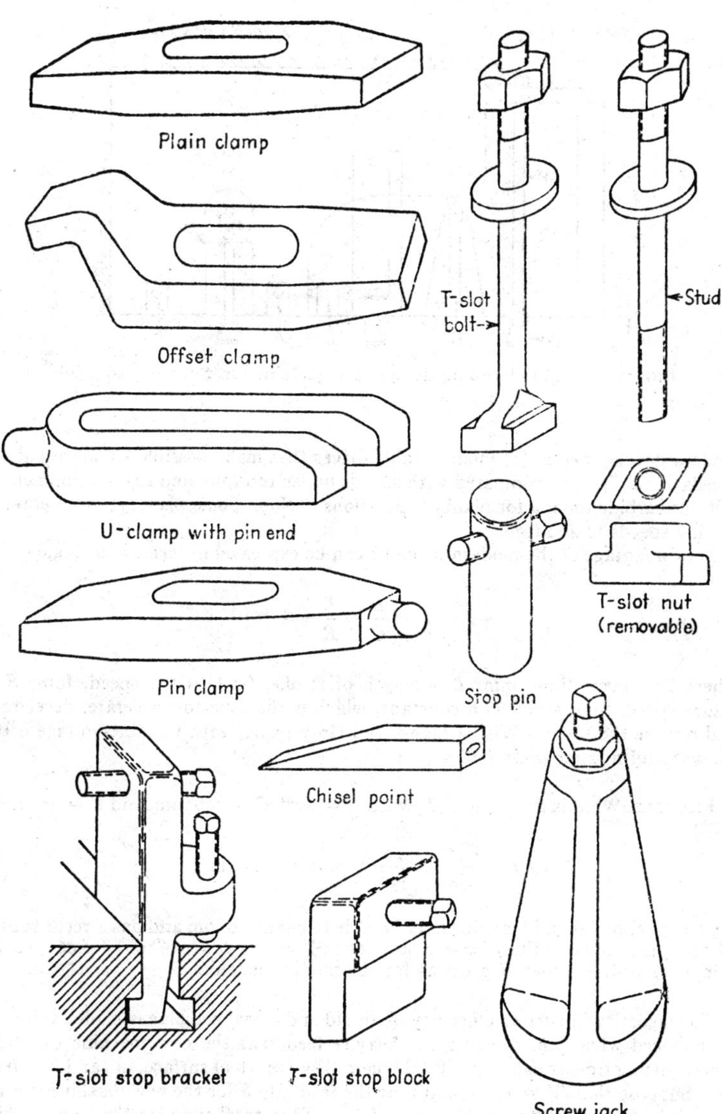

FIG. 1. Various clamps, stops, bolts, screw jacks, and T-slot nuts should be kept on hand for planer setup.

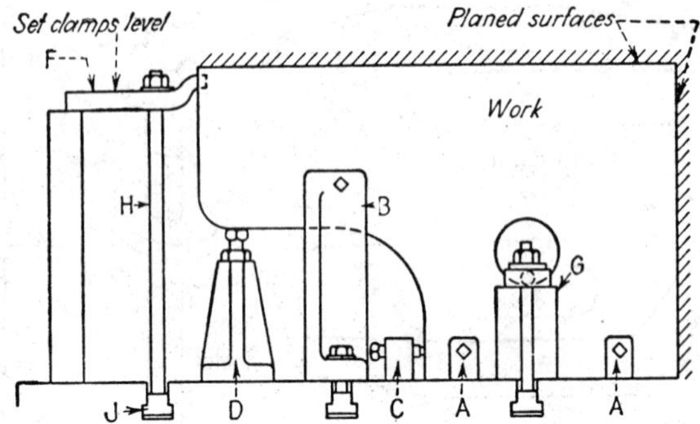

FIG. 2. Uses of clamping devices are explained in reference to a job.

modern developments: (1) modern table drives that make possible standard cutting speeds of 240 fpm, as compared with speeds under 100 fpm, and (2) development of suitable carbide tooling for planing operations. Note: Some planers now operate at cutting speeds to 400 fpm.

The advantage of the modern machine can be expressed in terms of a single cycle:

$$T = \frac{L}{C} + \frac{L}{R} + k$$

where T = cycle time, min; L = length of stroke, ft; C = cut speed, fpm; R = return speed, fpm; and k = a constant, which is the time to accelerate, decelerate, and reverse the table. With a high-speed Gray planer with variable-voltage drive, the value of k = 0.03 min.

EXAMPLE: What is the value of T, when L = 10 ft, C = 200 fpm, and R = 300 fpm?

$$T = \frac{10}{200} + \frac{10}{300} + 0.03 = 0.113 \text{ min}$$

By comparison, an old machine may be able to cut at 30 fpm and has a return speed of 100 fpm, and the value of k = 0.06. Then T = 10/30 + 10/100 + 0.06 = 0.493 min, or over four times as great as for the modern machine.

The disparity in cutting efficiency of an old and a new machine is made even more pronounced when you consider the heavier feed that the new machine can take. Total cutting time equals $T \times W/F$, where W = width of surface, in., and F = feed, in. Suppose that W = 10 in. and that the feed rate F for the new machine is 0.100 in., while F for the old machine is 0.06 in. Then total time for the new machine = 0.113 × 10/0.100 = 11.3 min, whereas for the old machine total time = 0.493 × 10/0.06 = 81.8 min.

Speeds and Feeds. Applicable speeds and feeds depend on the size of the work and its rigidity, the tool material, how the work can be clamped to the planer table or held in a fixture, and the power of the planer motor when heavy cuts are taken with several tools. In general, the best speed and feed combination will be determined by trial and experience. However, Tables 1 and 2 will prove helpful.

So far as tool materials are concerned, experience has shown that:

For heavy roughing cuts, high-speed steel or cast-alloy tipped tools are used. Second roughing cuts on such jobs are often performed with carbide tools.

Finishing cuts on cast iron, bronze, and aluminum are made with tools 1 to 1½

TABLE 1. PLANING SPEEDS WITH VARIOUS TOOL MATERIALS*

Cutting Speeds in Fpm

	High-speed Steel Tools				Cast-alloy Tools				Carbide-tipped Tools				
Depth of cut	1/8	1/4	1/2	1	1/8	1/4	1/2	1	1/16	3/16	3/8	3/4	1
Feed	1/32	1/16	3/32	1/8	1/32	1/16	3/32	1/8	1/32	1/32	1/16	1/16	3/32
Cast iron:													
Soft	95	75	60	50	160	135	110	95	225	200	175	150	150
Medium	70	55	45	35	125	105	90	75	180	160	150	130	130
Hard	45	35	25	...	95	80	65	...	180	160	150	130	130
Steel:													
Free-cutting	90	70	55	40	140	105	85	65	315	245	190	140	
Average	70	55	40	30	105	80	60	45	270	205	160	120	
Low machinability	40	30	25	...	65	50	40	...	195	145	115		
Bronze	150	150	125	...	†	†	†	†	†	†	†	†	
Aluminum	200	200	150	...	†	†	†	†	†	†	†	†	

* G. A. Gray Co.
† Maximum table speed.

TABLE 2. PLANING SPEEDS WITH CARBIDES*

	Roughing				Finishing
Depth of cut, in.	0.500 to 1.000	0.250 to 0.500	0.125 to 0.250	0.015 to 0.125	0.003 to 0.015
Feed per stroke, in.	0.010 to 1.00	0.015 to 0.125	0.015 to 0.175	0.015 to 0.200	0.015 to 1.000
CI hard and Meehanite	150	175	185	200	200
CI soft	175	200	225	250	250
Semisteel CI	150	175	200	210	210
Steel carbon	....	....	250	300	300
Free-cutting steel	225	250	300	300	300
Stainless steel	....	....	....	250	300
Cast steel	150	180	200	250	250
Brass—hard	150	200	225	250	300
Brass—soft	250	275	300	350	400
Aluminum	300	350	400	450	500
Plastics	....	....	450	600	750

For interrupted cuts on steel, reduce speed 25%.
No coolant.
* Cincinnati Planer Co.

TABLE 3. FACTORS FOR CALCULATING HORSEPOWER TO PLANE CAST IRON AND STEEL*

Depth of Cut, In.	Amount of Feed per Stroke, In.										
	1/32	1/16	3/32	1/8	5/32	3/16	1/4	5/16	3/8	7/16	1/2
1/8	0.0115	0.0235	0.036	0.047	0.059	0.071	0.094	0.118	0.142	0.165	0.189
3/16	0.0172	0.0352	0.054	0.070	0.088	0.106	0.141	0.177	0.213	0.248	0.283
1/4	0.023	0.047	0.073	0.094	0.118	0.142	0.189	0.236	0.284	0.331	0.378
5/16	0.029	0.058	0.091	0.117	0.147	0.177	0.236	0.295	0.355	0.414	0.472
3/8	0.035	0.070	0.110	0.141	0.177	0.213	0.283	0.354	0.426	0.497	0.567
7/16	0.041	0.082	0.128	0.165	0.206	0.248	0.330	0.413	0.497	0.580	0.662
1/2	0.047	0.094	0.147	0.189	0.236	0.284	0.378	0.473	0.568	0.663	0.757
7/16	0.055	0.106	0.165	0.212	0.265	0.319	0.425	0.532	0.639	0.745	0.851
5/8	0.063	0.118	0.184	0.236	0.295	0.355	0.473	0.591	0.710	0.828	0.946
11/16	0.071	0.130	0.202	0.260	0.325	0.390	0.520	0.650	0.781	0.911	1.041
3/4	0.080	0.142	0.221	0.284	0.355	0.426	0.568	0.709	0.852	0.994	1.136
11/16	0.083	0.153	0.239	0.307	0.384	0.461	0.615	0.768	0.923	1.076	1.230
7/8	0.087	0.165	0.257	0.331	0.414	0.497	0.662	0.828	0.994	1.159	1.325
15/16	0.090	0.177	0.275	0.354	0.443	0.532	0.704	0.887	1.065	1.242	1.420
1	0.094	0.189	0.294	0.378	0.473	0.568	0.757	0.947	1.136	1.325	1.515
1 1/16	0.100	0.200	0.312	0.401	0.502	0.605	0.804	1.006	1.207	1.407	1.609
1 1/8	0.106	0.212	0.331	0.425	0.532	0.639	0.851	1.065	1.278	1.49	1.704
1 3/16	0.112	0.224	0.349	0.449	0.562	0.674	0.898	1.124	1.349	1.573	1.798
1 1/4	0.118	0.236	0.368	0.473	0.592	0.710	0.946	1.183	1.420	1.656	1.893
1 5/16	0.125	0.250	0.391	0.502	0.628	0.745	0.993	1.241	1.491	1.739	1.987
1 3/8	0.133	0.265	0.414	0.532	0.665	0.781	1.04	1.30	1.562	1.822	2.082
1 7/16	0.140	0.280	0.437	0.561	0.702	0.816	1.087	1.360	1.633	1.905	2.177
1 1/2	0.148	0.295	0.460	0.591	0.739	0.852	1.135	1.420	1.704	1.988	2.272

For a given cut find the factor for desired depth of cut and feed. Multiply this value by the cutting speed in sfpm. The result will be the horsepower required to plane cast iron with one head. If steel up to 40 points carbon is to be machined, multiply the above result by 2; if steel is above 40 points carbon, multiply by 2¼.

* Cincinnati Planer Co.

in. wide, used at ¼ to 1 in. feed. Depth of cut for a good-quality finish is 0.001 to 0.002 in.; for less important finish, depth of cut may be 0.005 in.

A planing speed of 100 fpm minimum is required to warrant the extra expense of carbide tooling. Depth of cut is the most important factor in planing cast iron and steel; hardness is secondary. The advantage of carbides is usually obtained by increasing the cutting speed, not the feed. A deep cut with a light feed is better than a light cut and a heavy feed, because the chip pressure is distributed over a longer cutting edge. As depth of cut is increased, the work speed is decreased.

Power Requirements. Horsepower consumed in planing is directly proportional to the cutting speed. Table 3 can be used to figure the power under ordinary conditions, when planing cast iron and steel.

Carbide-tool Design. Tools with brazed-on tips are generally restricted to relatively fine cuts and light-duty work (Fig. 3). For deep cuts, interrupted cuts, and large-area cuts, the clamped-on tool is preferred (see Figs. 4 to 6). Rules for grinding carbide tools, in reference to Fig. 7, are:

1. For roughing tools, the included cutting-edge angle *A* should increase with material hardness. In fact, on many carbide planing tools, *A* has been increased to the point where the primary side-rake angle *B* is zero for cast iron and up to 20° negative for steel.

2. As mentioned, the primary side-rake angle *B*, shown on land *K*, should be decreased to increase *A*. Width of land *K* should be slightly greater than the feed per stroke.

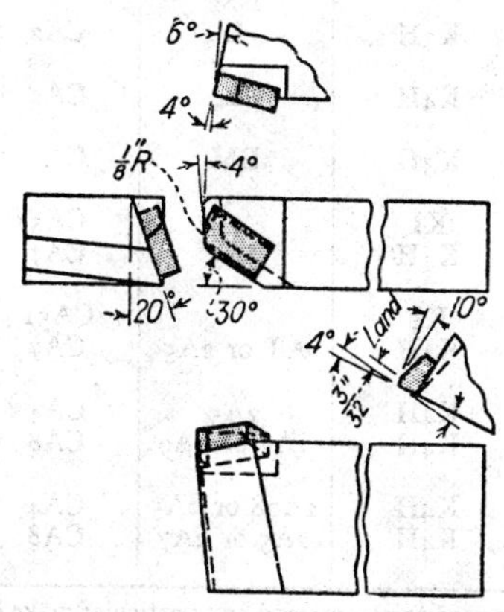

FIG. 3. Carbide-tipped planer tool for horizontal surfaces in steel.

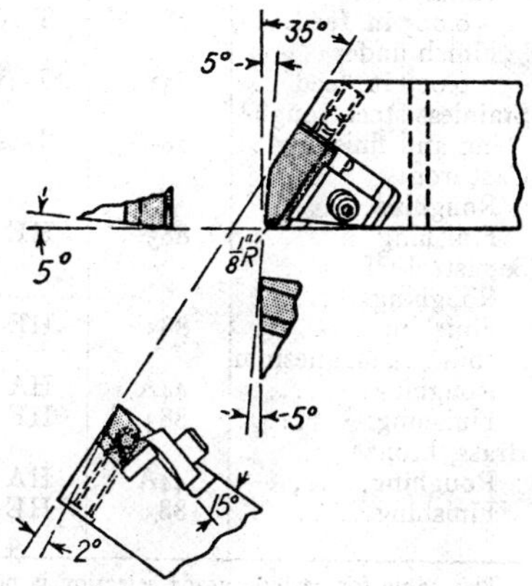

FIG. 4. Clamped-on tool for vertical surfaces in cast iron.

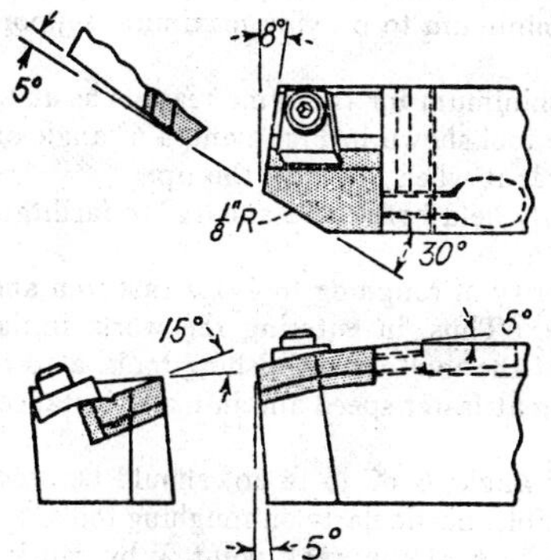

FIG. 5. Clamped-on tool for horizontal surfaces in cast iron.

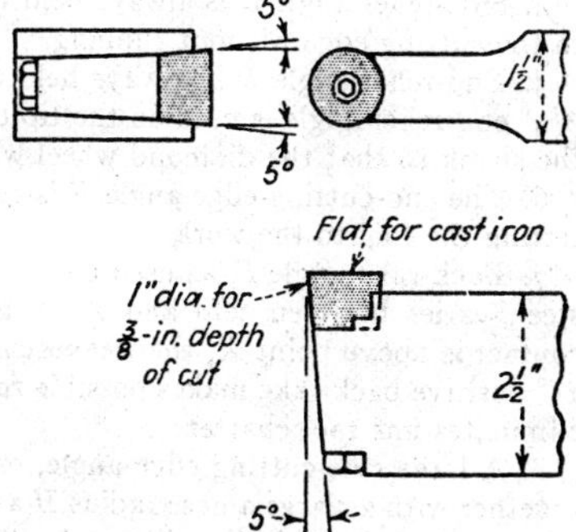

FIG. 6. Roughing tool for horizontal or vertical cuts on cast iron. When the carbide cone is worn, it is rotated to provide a new cutting edge. (*G. A. Gray Co.*)

TABLE 4. CARBIDE GRADES FOR PLANER TOOLS

Material and Operation	Carboloy	Firthite	Kennametal	Vascoloy Ramet	Allegheny Ludlum
Steel:					
Rough forgings and bar stock	78-B	T-04 T-A	KM	EE	CA5
Rough—sandy castings	78-B	T-04 T-A	K2S	EE	CA5
Semifinishing	78	T-16	KH or K3H	EM	CA1
Finish over 0.005 in. feed	78	T-16	K3H	EM E	CA2
Finish under 0.005 in. feed	831	T-31	K4H	E	CA2
Stainless steel, roughing and finishing	78-B	T-04	K3H	EM	CA5
Cast iron:					
Roughing			KI		CA51
Finishing	883	HE	K4H	2A5	CA7
Semisteel CI:					
Roughing			KI		CA51
Finishing	883	HE	K4H	AT or 2A5	CA7
Aluminum, magnesium					
Roughing	44A	HA	K4H	2A5	CA4
Finishing	883	HF	K4H	2A7 or 2A9	CA8
Brass, bronze					
Roughing	44A	HA	K4H	2A68 or 2A5	CA4
Finishing	883	HE	K4H	2A5 or 2A7	CA8

This table for carbide grade selection is not intended to recommend any particular make of carbide but only to show the various types of carbides that can be selected.

3. Secondary side-rake angle *B′* should be sufficient to allow proper chip flow.

4. Side-relief angle *C* is always held to a minimum to provide maximum support to the cutting edge without rubbing.

5. End-relief angle *D* is always held to a minimum for the same reasons as at *C*. A 4° end-relief angle is used on the tip of the tool shown in Fig. 3 and a 6° angle on the shank so that the diamond wheel will clear it when grinding the tip.

6. The end-cutting-edge angle *E* is generally held between 8 and 12° to facilitate setting the tool to the work.

7. Back-rake angle *F*, as used on the majority of roughing tools for cast iron and steel, varies between zero and 20° negative. Thus, in entering the work, initial contact is above point *X*, the weakest part of the tool. For finishing tools, a 10 tc 15° positive back rake makes possible running at faster speed and in many instances eliminates fine tool chatter.

8. A large side-cutting-edge angle, or lead angle, *G* of 30 to 40° should be used, together with as large a nose radius *H* as possible, particularly on roughing tools. A large lead angle thins the chip and relieves shock at the weak point *X* by causing initial contact above that point. But nose radius *H* must not be so large that chip flow is impeded. A nose radius of ⅛ in. has been found satisfactory on tools for roughing both cast iron and steel.

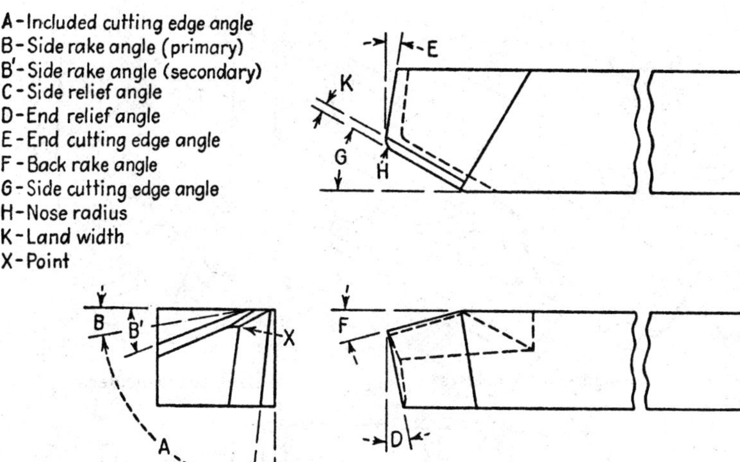

FIG. 7. Text references to preferred planer-tool angles for roughing and finishing cuts follow the above nomenclature.

SHAPING[1]

The shaper resembles the planer in that it produces flat surfaces in the horizontal, vertical, or angular planes and can be arranged to machine radiuses and irregular outlines, using a single-point tool. In the shaper, however, the work is stationary—being clamped to a box or universal table, or to a vise or fixture fastened thereto—and the tool is reciprocated by a ram. The work is fed by cross-feeding the table. The ram may be driven by a crank motion or by means of a hydraulic system. Because the shaper is simple to set up, it is an essential machine tool in general manufacturing plants, tool and die rooms, and railroad shops.

Types of Shapers. The shaper with a horizontal push-cut ram is most common. Draw-cut shapers, used mostly in railroad shops, cut on the return stroke of the ram. Vertical shapers, more commonly known as slotters, have a vertical ram and a rotary table. This type of shaper is extremely useful in die shops and toolrooms for making die components, cams, etc., and is frequently found in very large sizes in railroad shops, turbine plants, and heavy-machinery plants.

Shaper Speeds. On a crank shaper, the term *speed* has two distinct connotations: (1) speed of the machine itself and (2) the average rate of cutting speed during the cutting stroke. Speed of the machine is the number of strokes per minute and remains constant for a given speed of the driving gear, whether the stroke is long or short. Cutting speed is determined by the tool travel during the cutting stroke and the ratio of the cutting-stroke time to the return-stroke time. Obviously, cutting speed changes whenever the stroke is made longer or shorter, and the ram speed is not changed.

To apply the correct cutting speed for a given material and length of stroke, follow these steps.

[1] "Suggested Unit Course in Shaper Work," Delmar Publishers, Albany, N.Y.

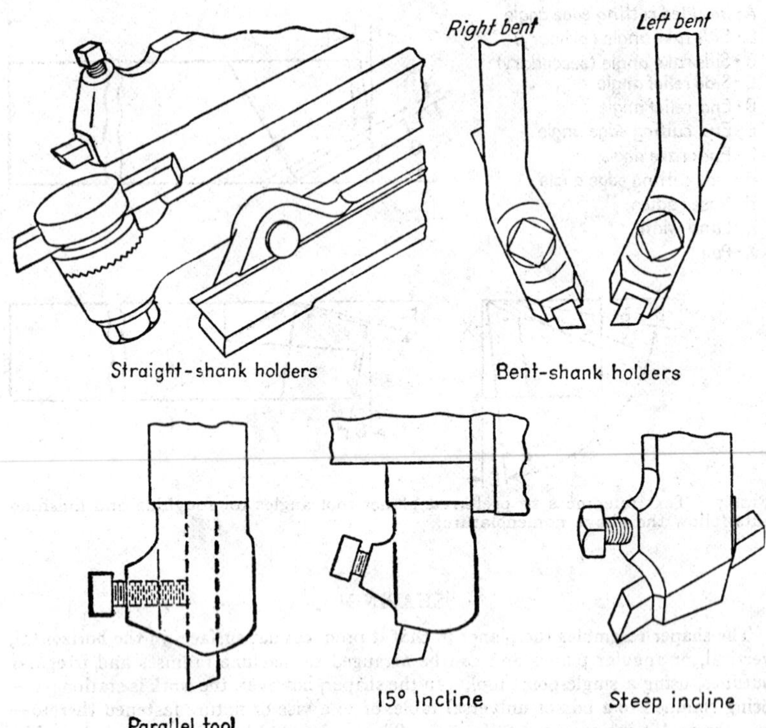

FIG. 8. Shaper toolholders are made in a number of styles.

1. Ascertain the ratio of the cutting-stroke time to the return-stroke time. Suppose the crankpin travels 220° during the cutting stroke and 140° during the return stroke. Then their ratio X is found from $220–140 = X–1$. Then $X = 1.57$. For practical purposes a ratio of 1.57 equals a ratio of 3:2. Then the cutting stroke takes three-fifths the time of the cutting cycle and the return stroke requires two-fifths.

2. The number of inches cut per minute $= N \times L/12$, where N = number of strokes per min, and L = length of stroke, in.

3. The cutting speed is found by dividing the above result by the fraction C of the cutting-cycle time represented by the cutting stroke. Thus $S = N \times L/12C$. In this case $C = 3/5$. Hence $S = N \times L \div 12 \times 3/5 = 0.14\ N \times L$ in fpm.

4. With the length of stroke and the cutting speed known, the number of ram strokes N per minute is found from the formula $N = S/0.14L = 7S/L$.

These calculations are not required when a hydraulic shaper is used. The cutting speed is always shown by an indicator.

The time to take a cut on a shaper is computed from the formula

$$T = \frac{W}{F} \times N$$

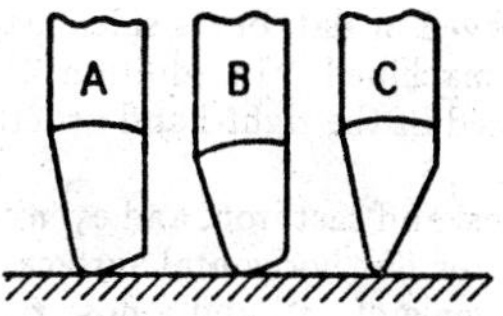

FIG. 9. Round-nosed shaper tools for horizontal surfaces.

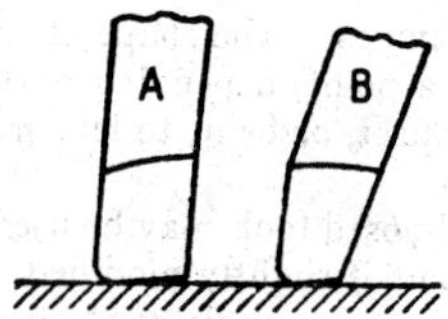

FIG. 10. Tools for vertical surfaces.

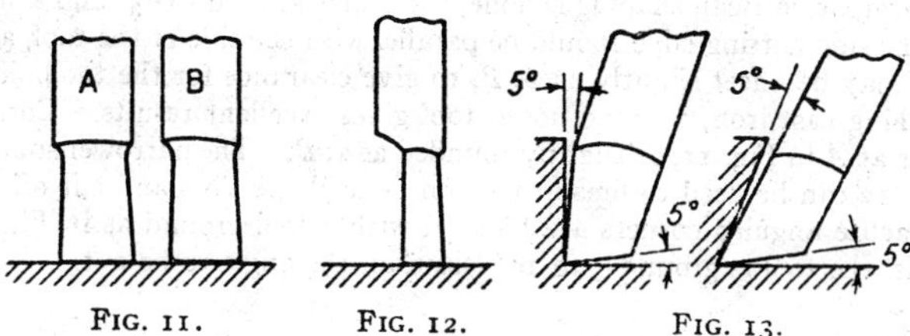

FIG. 11. FIG. 12. FIG. 13.

FIG. 11. Finishing tools for cast iron. FIG. 12. Finishing and slotting tools. FIG. 13. Finishing tools for corners.

where T = time, min; W = width of surface, in.; F = feed, in. per stroke; and N = number of strokes per min.

EXAMPLE: The work is a cast-iron plate 7 in. wide, the stroke is 12 in., and the feed is 0.020 in. per stroke. The cutting speed for cast iron is 60 fpm. Then $N = 7 \times 60/12 = 35$. And $T = 7 \div 0.2 \times 35 = 7 \div 0.7 = 10$ mim.

Feeds. A heavy cut with a fine feed is preferred to a light cut and a coarse feed, in order to produce a good finish. The rate of feed is also influenced by the angle the cutting edge makes with the surface being machined. Chip thickness is reduced as the angle is increased. Best results in roughing ferrous metals have been obtained when the cutting edge of the tool assumes an angle of about 20° with the work's surface. Angular approach of the cutting edge to the work can be obtained in two ways: (1) by the preferred method of setting the toolholder vertically in the toolpost and grinding the tool to the desired angle, or (2) by changing the position of the toolholder in the toolpost.

Toolholders. Shaper toolholders (Fig. 8) are made in a variety of styles: for example, (1) straight-shank, which holds the tool parallel with the sides of the holder, and (2) bent or offset to the left or right. When the tool is held parallel with the toolholder shank, the clearances and angles are easily determined and ground. However, two further methods of holding the tool may be encountered: (1) inclined at a slight angle, and (2) inclined at a steep angle. If the toolholder angle is 15°, most grinding of the top of the tool will be eliminated, and only the front and side clearance need be ground. If the toolholder angle is steep, as sometimes used with form tools, the most satisfactory angle is the tool's front-clearance angle. Then the tool is ground on top and the contour is not changed by repeated grinding.

Tool Shapes. The shape of the tool may be curved or flat, or its sides may converge to a point, depending on the surface being machined. In addition, the tool may be offset, or bent, to left or right, and may feed in the right-hand or left-hand direction.

Round-nosed tools may be used to rough both steel and cast iron, and even to take a finish cut if slightly modified. Tool *A* (Fig. 9) for flat horizontal surfaces has a side-cutting-edge angle of 8°, an end-cutting-edge angle of 15°, and a nose radius of 1⁄16 in. On some roughing cuts a side-cutting-edge angle of 20° on tool *B* gives better results. If this tool is used for roughing out radiuses, for fine cuts, and for other work where the broad surface created by the large radius induces chatter, tool *C* with a smaller radius is employed.

When down, or vertical, shaping is done, the tool is ground to 85° and a small nose radius. The side cutting edge should be parallel with the side of the tool, as at *A* in Fig. 10, or may be offset slightly as at *B*, to give clearance for the toolholder.

For finishing cast iron, a broad-nosed tool gives excellent results. Corners may be sharp, as at *A* in Fig. 11, or slightly rounded as at *B*. The narrower square-nosed tool in Fig. 12 can be used to finish cast iron or to shape slots and cut off material. Square or acute angular corners are finished with a tool ground as in Fig. 13. In other words, the tool is ground 5 to 10° less than the angle to be cut.

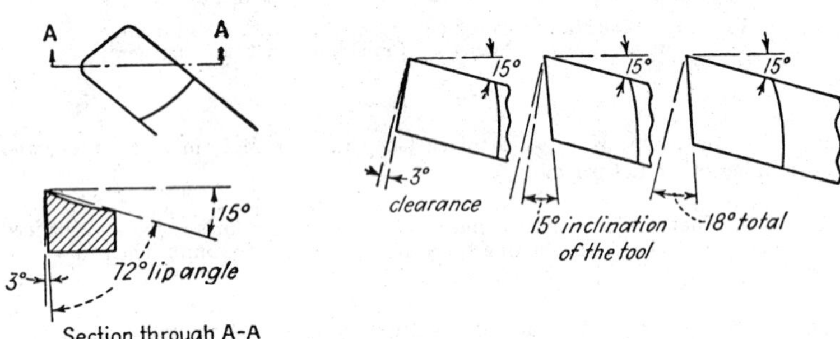

FIG. 14. True lip angle is influenced by the side and back rakes.

FIG. 15. If the tool is held inclined 15° in the holder, the front must be ground at 18° to provide necessary 3° clearance.

Tool Angles. A shaper tool does not feed sideways into the work during cutting. Hence about 4° end and side clearance will be sufficient. Side rake and back rake are more subject to variation. Satisfactory results have been obtained with these rakes: for steel—side rake of 10 to 20°, back rake of 2 to 8°; for cast iron—side rake of 3 to 10°, back rake of 0 to 3°.

The side rake and back rake influence the true lip angle on the tool (Fig. 14). If the tool has a 3° side clearance and a 15° side rake, the lip angle will be 72° (when the toolholder holds the tool parallel with the shank).

In the case where the tool is inclined 15° in the holder, and 3° front clearance is required, an 18° clearance must be ground on the end of the tool to compensate for the tool slope and provide the required 3° front clearance (Fig. 15).

Section 8

REAMERS AND REAMING

SECTION 8

REAMERS AND REAMING[1]

A reamer is a cutting tool for enlarging a previously formed hole to the required diameter as closely as possible. There is a common misconception that a reamer somehow gets into a hole and then scrapes it to size by means of its flutes. Actually, the cutting action resembles in many ways the action of boring tools and twist drills, because in a machine reamer the entering teeth cut their way into the hole and then produce a smoothly finished hole, round, straight, and as near to size as possible.

The standard reamer is an important feature of the standard-hole system of allowances and tolerances for metal fits, and the tool provides the most economical means of securing precision fits and universal interchangeability.

MACHINE AND HAND REAMERS

Reamer Clearances. A reamer without clearance cannot cut, because its cutting edges cannot get under the surface of the metal to lift a chip. Therefore, three kinds of clearance are required on reamers:

1. "Point" clearance, or chamfer relief, on the entering ends of the teeth (Fig. 1). The entering edges of chucking reamers are usually beveled at 40 to 50°. This cone shape helps to keep the reamer sharp and to center it in the work. Fluted chucking reamers intended for very accurate, smooth holes will have a second slight bevel that acts like the taper on a hand reamer.

Hand reamers have a point or cutting edge that is much longer than that on machine reamers, usually tapered about 0.015 in. per in., to allow a scraping cut and production of a smooth hole. Taper edges are ground sharp with clearance. A bevel of about 45° is put at the end of the tapering point to aid in entering the hole, but there should be so little stock left in the hole that this bevel does not cut.

2. "Radial relief," or clearance, along the lands of peripheral part of the reamer. Side clearance, or relief, is necessary for accurate reaming, but a slight margin must be left to aid the tool to maintain correct size and be sharpened without losing size. Wear on the margin cannot be prevented, and hence commercial reamers are usually made a few tenths oversize.

The standard tolerances are:

Up to ¼ in., inclusive	+0.0001 to 0.0004 in.
Over ¼ to 1 in., inclusive	+0.0001 to 0.0005 in.
Over 1 in.	+0.0002 to 0.0006 in.

3. "Longitudinal relief," or "back taper," of about 0.0001 in., making the reamer smaller toward the shank, to prevent the back end from enlarging the hole or dragging and roughening the finish.

Rake Angles. The "rake angle" of a tool is the angle between the top cutting surface of a tool and a plane which is perpendicular to the surface of the work and to the direction of motion of the tool with respect to the work.

[1] T. F. Githens, Mechanical Engineer, Cleveland Twist Drill Co.

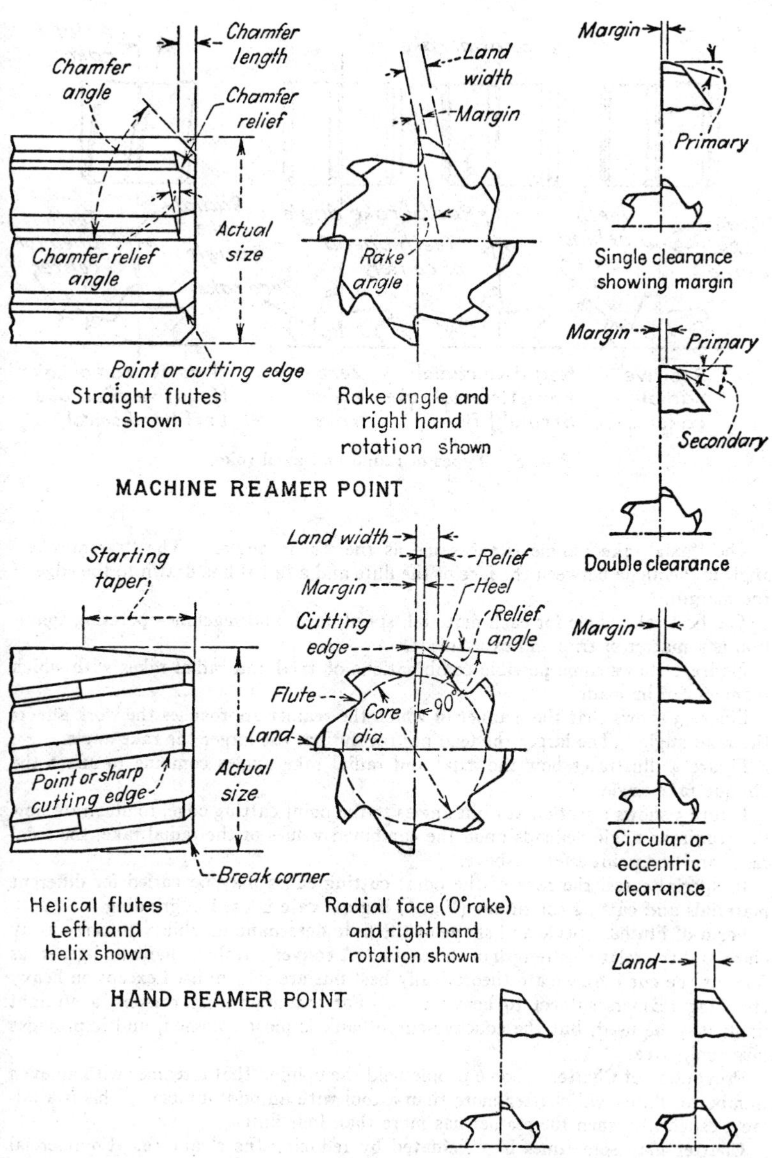

FIG. 1. Machine reamers have 40 to 50° cutting ends or points; hand reamers have more tapered cutting edges. Small sketches show types of side clearance or radial relief.

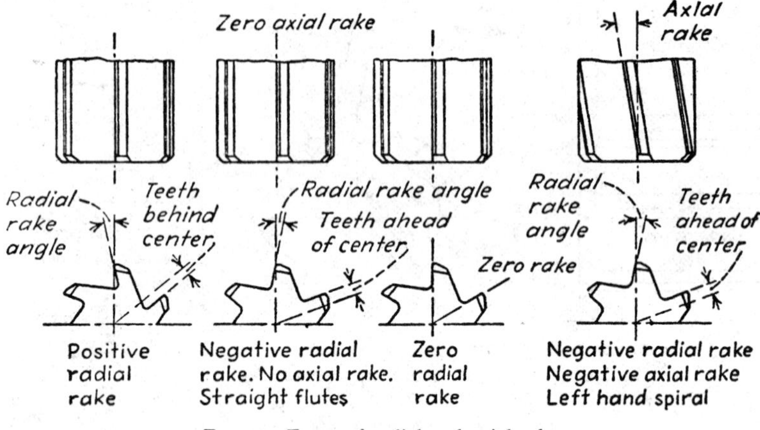

FIG. 2. Types of radial and axial rake.

The "axial-rake" angle is the same as the "helix angle." The "radial-rake" angle is the angle between the face of the flute and a radial line drawn to the edge of the margin.

The best rake angle for each material, speed, feed, and machine-operating condition is a matter of trial and experiment.

Figure 2 shows some possible combinations of axial and radial rakes with which reamers can be made.

Figure 3 shows that the manner in which the reamer approaches the work affects the rake angle. The larger the feed per revolution, the larger the rake angle.

Figure 4 illustrates how the axial and radial rake angles combine to affect the oblique rake angle.

Figure 5 shows a section at right angles to the point cutting edge, to illustrate how the true rake angle depends upon the combined values of the radial rake, the axial rake, and the point angle or bevel.

In special cases, the rake of the point cutting edges may be varied for different materials and cutting conditions (Fig. 6) if great care is used in grinding.

Form of Flutes. Style and shape of the flute determine its ability to carry away chips and the relative strength of the tooth. A convex parabolic flank and generous fillet at the cutting face are theoretically best but are seldom used except on heavy chucking reamers subject to heavy cuts. For manufacturing reasons, a straight flank may be used, but the concave-curve flank is most common, and it provides large chip area.

Prevention of Chatter. Some people hold the opinion that a reamer with an even number of flutes will chatter more than a tool with an odd number. This is a fallacy, especially when the reamer has more than four flutes.

Chatter may sometimes be eliminated by reducing the clearance. Commercial reamers ordinarily have a considerable amount of clearance to make them suitable for use on a wide variety of materials, and the tool may bite in to produce chatter. This effect may often be reduced by making the setup rigid, and using pilots and guide bushings, reducing the speed, or increasing the feed if the hole tends to glaze.

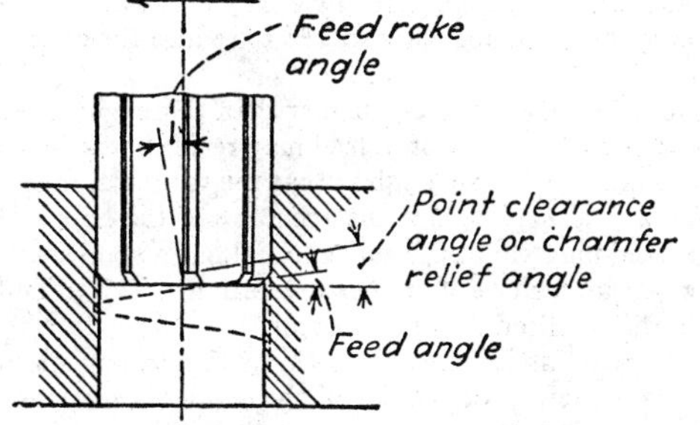

FIG. 3. Action of the feed increases the rake angle of the reamer by the amount of the "feed rake angle." Point-clearance angle or chamfer-relief angle should be greater than the feed angle shown.

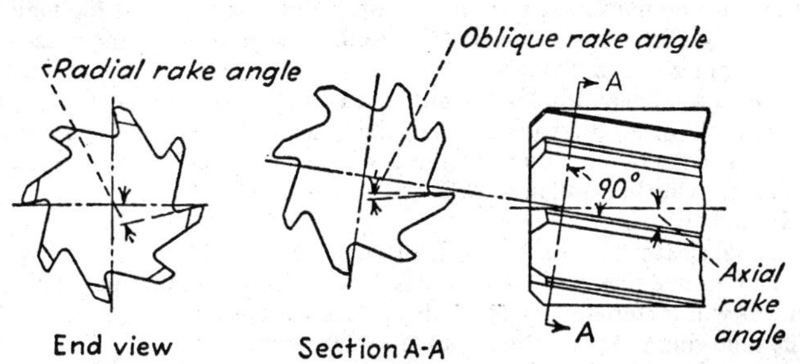

FIG. 4. Rake angle due to the helix angle or oblique rake angle.

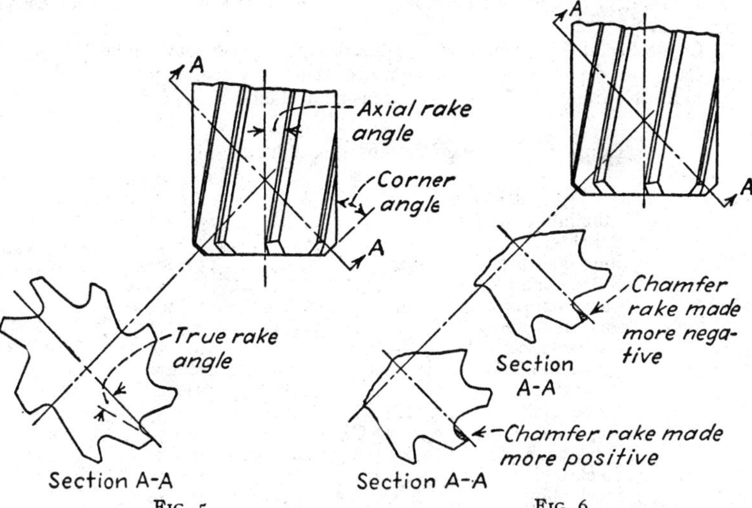

FIG. 5. True rake angle, normal to the cutting edge of the chamfer, as it results from a given combination of radial rake angle, axial rake angle, and corner angle or point chamfer. FIG. 6. Positive or negative rake may be varied on cutting edges of chamfer by careful grinding.

Too much positive rake angle or too much negative rake angle may also cause chatter. Most commercial reamers have the teeth unevenly spaced to reduce the chance of chatter.

Reaming Feeds. In reaming, feeds are usually much higher than those employed for drilling, often being 200 to 300% greater. Too low a feed may result in excessive reamer wear. At all times, the feed must be high enough so that the tool cuts, rather than rubs or burnishes. Too high a feed may reduce the accuracy of the hole and impair the finish. The amount of feed may vary with the material but a good starting point is between 0.0015 and 0.004 in. feed per flute per revolution. Try to find the highest feed that will produce the required finish and accuracy.

Reaming Speeds. Most machine reaming is done at two-thirds the drilling speed for a given material. If the speed is too low, productivity of the reamer is adversely affected with no particular gain in tool life. But if the speed is too high, premature dulling, rough holes, and chatter may result. Suggested reaming speeds for various materials are given in Table 1.

Lubricants. Several functions of lubricants are:

1. To cool both the cutting edges of the tool and the work being machined. This can be best done by directing as large a volume of the coolant as possible on the cutting edges. On thin-walled work it often helps to allow a large volume of flow onto and around the piece.

2. To lubricate the chips; this aids in chip clearance.

3. To improve the finish of the work. The selection and proper application of the lubricant will materially influence the machined finish.

The following lubricants may prove satisfactory:

1. Aluminum and its alloys: soluble oil, kerosene, and lard-oil compounds, light nonviscous neutral oil, kerosene, and soluble-oil mixtures.

2. Brass: dry, soluble oil, kerosene, and lard-oil compounds, light nonviscous neutral oil.

3. Copper: soluble oil, winter-strained lard oil, oleic acid compounds.

4. Cast iron: dry or with a jet of compressed air for a cooling medium.

5. Malleable iron: soluble oil, nonviscous neutral oil.

TABLE 1. REAMIMG SPEEDS FOR VARIOUS MATERIALS

Material	*Cutting Speed, Fpm*
Aluminum and its alloys	130–200
Bakelite	70–100
Brass and bronze, ordinary	130–200
Bronze, high-tensile	50– 70
Cast iron, soft	70–100
Cast iron, hard	50– 70
Cast iron, chilled	20– 30
Magnesium and its alloys	170–270
Malleable iron	50– 60
Monel metal	25– 35
Steel, machinery, 0.2 to 0.3 C	50– 70
Steel, annealed, 0.4 to 0.5 C	40– 50
Steel, tool, 1.2 C	35– 40
Steel, alloy	35– 40
Steel, automotive forgings	35– 40
Steel, alloy, 300 to 400 brinell	20– 30
Steel, free-machining stainless	40– 50
Steel, hard stainless	20– 30

6. Monel metal: soluble oil, sulfurized mineral oil.
7. Steel, ordinary: soluble oil, sulfurized oil, high E.P. value mineral oil.
8. Steel, very hard and refractory: soluble oil, sulfurized oil, turpentine.
9. Steel, stainless: soluble oil, sulfurized mineral oil.
10. Wrought iron: soluble oil, sulfurized oil, high animal-oil-content mineral-oil compound.

Stock Allowances for Reaming. Speeds, feeds, and stock allowances are dependent upon each other, as well as upon type of material, rigidity of setup, lubricant, depth of hole, design and condition of reamer, condition of the machine, and required quality of the product. Therefore, a certain amount of experimenting may be necessary to find the best reaming speed, feed, and stock allowance for a particular setup. However, the following stock allowances may be used as a starting point:

STOCK ALLOWANCES FOR MACHINE REAMING

0.010 in. on a ¼-in. hole
0.015 in. on a ½-in. hole
0.020 in. on a 1-in. hole
Up to 0.025 in. on a 1½-in. hole
0.030 in. on a 2-in. hole
Up to 0.045 in. on a 3-in. hole

STOCK ALLOWANCE FOR HAND REAMING

0.001 to 0.003 in. is the common allowance

Standard-hole Tolerances. Under ideal conditions, a standard commercial reamer will produce holes that are slightly above the minimum basic hole size, but the fact that the reamer is accurately sharpened to size is no guarantee that the hole will be as accurate as the reamer. However, with the standard size of reamer all classes of fit can be produced, according to tolerances that can be held in the shop or are desirable. The main thing is that the tool will not cut below basic hole size, unless excessively worn.

Rules for Reaming. The following elementary rules hold true for all types of reamers:

1. Helical fluted reamers must be used when reaming holes with keyways or oil grooves, or which have an otherwise interrupted surface.
2. Straight-fluted reamers are preferred where extreme accuracy is required.

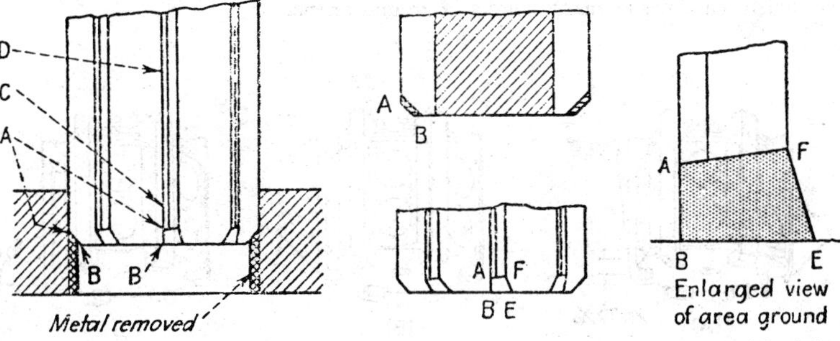

FIG. 7. Wear of a machine reamer takes place along the cutting edge AB (left). To sharpen, grind only the cutting edge, or point, back as shown (center) to proper clearance.

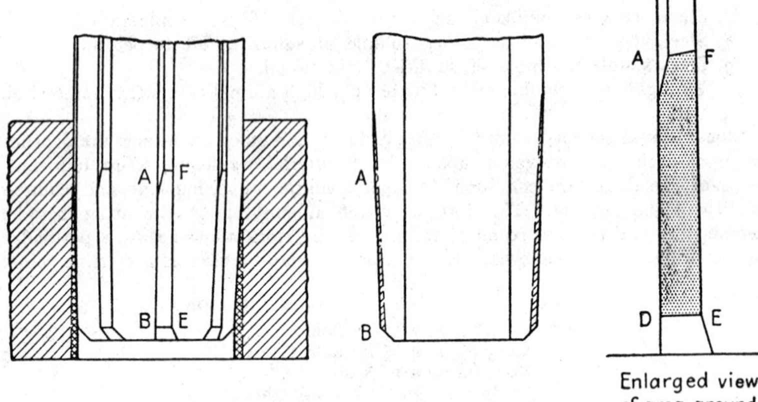

FIG. 8. Wear of a hand reamer is distributed over a longer cutting edge AB (left). Again, the cutting edge is ground back at the correct clearance angle.

3. Reamers always should be turned only in the direction of cutting, even when withdrawing from a hole. Never turn a reamer backward.

4. Rough and finish reamed holes have a surface finish superior to that of holes reamed in a single operation.

5. Better surface finish may be expected whenever provision is made for giving the reamer a uniform feed.

6. Some materials show an improved surface finish if a lubricant is used when reaming. The compound varies with the kind of material.

7. A cutting lubricant also may affect the extent to which a reamer will cut the hole greater than its own diameter.

8. High-speed-steel blades lose their original keenness faster than carbon-steel blades, but last longer.

9. A refined cutting edge has improved wearing qualities. Often a lapped reamer will finish ten times as many holes as a ground reamer.

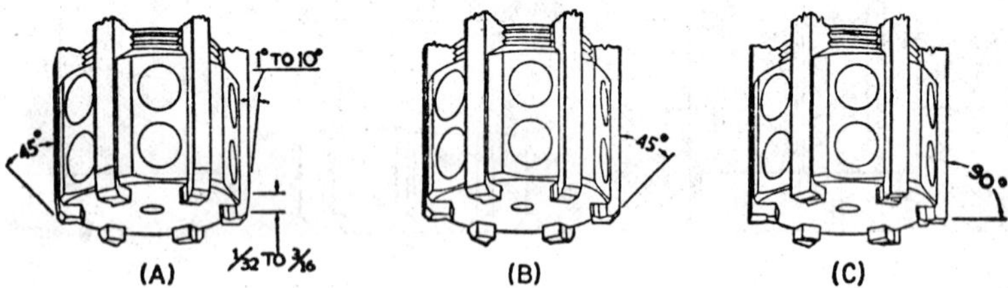

FIG. 9. Three types of grind for HSS adjustable reamers are: A, double bevel for fine finish; B, the common grind; and C, square-end grind that acts like an end mill to straighten misaligned holes.

STANDARD REAMERS

Standards for reamers are developed by Technical Committee 20 of the ASME. Tables 2 to 17 give standard sizes and important dimensions of reamers, as reported in the revised standard ASA B5.14-1949. Dimensions and tolerances are unchanged from the 1941 standard, but obsolete sizes have been removed and new sizes added, to keep pace with industrial needs. For example, some types of carbon-steel reamers are no longer in demand, while new sizes and types of HSS reamers are required.

Among the new types of reamers that have come into use are: fluted chucking reamers with spiral flutes and tapered shank, Brown & Sharpe taper reamers for reaming out B & S taper sockets and sleeves, plus data on center reamers and machine countersinks.

TABLE 2.- HAND REAMERS

Reamer Dia		Over-all Length	Flute Length	Reamer Dia		Over-all Length	Flute Length	Reamer Dia		Over-all Length	Flute Length
Carbon Steel	HSS			Carbon Steel	HSS			Carbon Steel	HSS		
Straight Flutes and Squared Shank											
1/8	1/8	3	1 1/2	3/8	3/8	5	2 1/2	3/4	3/4	8 3/8	4 3/16
9/64	....	3 1/4	1 5/8	25/64	....	5 1/4	2 5/8	25/32	25/32	8 3/4	4 3/8
5/32	5/32	3 1/4	1 5/8	13/32	13/32	5 1/4	2 5/8	13/16	13/16	9 1/8	4 9/16
11/64	....	3 1/2	1 3/4	27/64	....	5 1/2	2 3/4	27/32	27/32	9 3/8	4 11/16
3/16	3/16	3 1/2	1 3/4	7/16	7/16	5 1/2	2 3/4	7/8	7/8	9 3/4	4 7/8
13/64	....	3 3/4	1 7/8	29/64	....	5 3/4	2 7/8	29/32	29/32	10	5
7/32	7/32	3 3/4	1 7/8	15/32	15/32	5 3/4	2 7/8	15/16	15/16	10 1/4	5 1/8
15/64	....	4	2	31/64	....	6	3	31/32	31/32	10 5/8	5 5/16
1/4	1/4	4	2	1/2	1/2	6	3	1	1	10 7/8	5 7/16
17/64	....	4 1/4	2 1/8	17/32	17/32	6 1/4	3 1/8	1 1/16	1 1/16	11 1/4	5 5/8
9/32	9/32	4 1/4	2 1/8	9/16	9/16	6 1/2	3 1/4	1 1/8	1 1/8	11 5/8	5 13/16
19/64	....	4 1/2	2 1/4	19/32	19/32	6 3/4	3 3/8	1 3/16	1 3/16	12	6
5/16	5/16	4 1/2	2 1/4	5/8	5/8	7	3 1/2	1 1/4	1 1/4	12 1/4	6 1/8
21/64	....	4 3/4	2 3/8	21/32	21/32	7 3/8	3 11/16	1 5/16	1 5/16	12 1/2	6 1/4
11/32	11/32	4 3/4	2 3/8	11/16	11/16	7 3/4	3 7/8	1 3/8	1 3/8	12 5/8	6 5/16
23/64	....	5	2 1/2	23/32	23/32	8 1/8	4 1/16	1 7/16	1 7/16	12 7/8	6 7/16
								1 1/2	1 1/2	13	6 1/2
Spiral Flutes and Squared Shank											
1/8	....	3	1 1/2	1/2	1/2	6	3	7/8	7/8	9 3/4	4 7/8
5/32	....	3 1/4	1 5/8	17/32	17/32	6 1/4	3 1/8	.....	29/32	10	5
3/16	....	3 1/2	1 3/4	9/16	9/16	6 1/2	3 1/4	15/16	15/16	10 1/4	5 1/8
7/32	....	3 3/4	1 7/8	....	19/32	6 3/4	3 3/8	.....	31/32	10 5/8	5 5/16
1/4	1/4	4	2	5/8	5/8	7	3 1/2	1	1	10 7/8	5 7/16
9/32	9/32	4 1/4	2 1/8	....	21/32	7 3/8	3 11/16	.....	1 1/16	11 1/4	5 5/8
5/16	5/16	4 1/2	2 1/4	11/16	11/16	7 3/4	3 7/8	.....	1 1/8	11 5/8	5 13/16
11/32	11/32	4 3/4	2 3/8	....	23/32	8 1/8	4 1/16	.....	1 3/16	12	6
3/8	3/8	5	2 1/2	3/4	3/4	8 3/8	4 3/16	.....	1 1/4	12 1/4	6 1/8
13/32	13/32	5 1/4	2 5/8	....	25/32	8 3/4	4 3/8	.....	1 5/16	12 1/2	6 1/4
7/16	7/16	5 1/2	2 3/4	13/16	13/16	9 1/8	4 9/16	.....	1 3/8	12 5/8	6 5/16
15/32	15/32	5 3/4	2 7/8	....	27/32	9 3/8	4 11/16	.....	1 7/16	12 7/8	6 7/16
								.....	1 1/2	13	6 1/2

TABLE 3. EXPANSION HAND REAMERS

Reamer Dia	Over-all Length	Flute Length	Reamer Dia	Over-all Length	Flute Length	Reamer Dia	Over-all Length	Flute Length
Straight Flutes and Squared Shank—Carbon Steel								
1/4	4 3/8	1 3/4	5/8	7	3	1	10	4 1/2
9/32	4 3/8	1 3/4	21/32	7 5/8	3	1 1/16	10	4 1/2
5/16	4 3/8	1 7/8	11/16	7 5/8	3	1 1/8	10 1/2	4 3/4
11/32	5 3/8	1 7/8	23/32	7 5/8	3	1 3/16	10 1/2	4 3/4
3/8	5 3/8	2	3/4	8	3 1/2	1 1/4	11	5
13/32	5 3/8	2	25/32	8	3 1/2	1 5/16	11 3/4	5 1/4
7/16	5 3/8	2	13/16	8	3 1/2	1 3/8	11 3/4	5 1/2
15/32	6 1/2	2 1/4	27/32	8 5/8	3 1/2	1 7/16	11 3/4	5 3/4
1/2	6 1/2	2 1/2	7/8	9	4	1 1/2	12	6
17/32	6 1/2	2 1/2	29/32	9	4			
9/16	6 1/2	2 1/2	15/16	9	4			
19/32	6 1/2	2 1/2	31/32	9	4			
L. H. Spiral Flutes and Squared Shank—Carbon Steel								
1/4	4 3/8	1 3/4	3/4	8 5/8	3 1/2	1 1/4	11 3/8	5
5/16	4 3/8	1 3/4	13/16	9	3 1/2	1 5/16	11 3/4	5 1/4
3/8	6 1/8	2	7/8	9 3/8	4	1 3/8	11 3/4	5 1/2
7/16	6 1/4	2	15/16	9 7/8	4	1 7/16	12	5 3/4
1/2	6 1/2	2 1/2	1	10 1/4	4 1/2	1 1/2	12 1/8	6
9/16	6 7/8	2 1/2	1 1/16	10 3/4	4 1/2			
5/8	8	3	1 1/8	10 7/8	4 3/4			
11/16	8 3/8	3	1 3/16	11 1/8	4 3/4			

Over-all lengths and flute lengths are maximums.
The maximum expansion on these reamers is as follows: 1/4 to 15/32 inclusive, 0.006 in.; 1/2 to 31/32 inclusive, 0.010 in.; and 1 to 1 1/2 inclusive, 0.012 in.
The guides on these reamers are ground slightly undersize.

REAMER SHARPENING

A reamer requires resharpening when it dulls along the cutting edge *AB* (Figs. 7 and 8). Greatest wear occurs at corner *A*. After the corner becomes worn, wear progresses along the land *AC* and eventually to *AD*, and the tool will be ruined.

The first result of a dull reamer will probably be undersize holes, because the tool is forced through the hole without cutting its true size. If at the first sign of holes coming smaller in size the reamer is sharpened, holes up to size will again be produced.

To sharpen a reamer it is necessary only to grind the chamfer at the proper relief angle. To do this effectively, it is desirable to cut off the end of the reamer, if it has no pilot. The cutting edges, in any case, are ground back until the corner and marginal dullness is removed.

Three common types of grind for high-speed-steel adjustable reamers are shown in Fig. 9. The type shown at *A* has a double bevel to obtain better finish or closer

TABLE 4. MACHINE (JOBBERS) REAMERS, STRAIGHT FLUTES AND TAPERED SHANK, HSS

Reamer Dia	Over-all Length	Flute Length	Taper Shank No.	Reamer Dia	Over-all Length	Flute Length	Taper Shank No.
1/4	5 3/16	2	1	11/16	8	3 7/8	2
9/32	5 3/16	2	1	23/32	8	3 7/8	2
5/16	5 1/2	2 1/4	1	3/4	8 3/8	4 3/16	2
11/32	5 1/2	2 1/4	1	13/16	8 13/16	4 9/16	2
3/8	5 13/16	2 1/2	1	7/8	9 3/16	4 7/8	2
13/32	5 13/16	2 1/2	1	15/16	10	5 1/8	3
7/16	6 1/8	2 3/4	1	1	10 3/8	5 7/16	3
15/32	6 1/8	2 3/4	1	1 1/16	10 5/8	5 5/8	3
1/2	6 7/16	3	1	1 1/8	10 7/8	5 13/16	3
17/32	6 7/16	3	1	1 3/16	11 1/8	6	3
9/16	6 3/4	3 1/4	1	1 1/4	12 9/16	6 1/8	4
19/32	6 3/4	3 1/4	1	1 5/16	12 11/16	6 1/4	4
5/8	7 9/16	3 1/2	2	1 3/8	12 13/16	6 5/16	4
21/32	7 9/16	3 1/2	2	1 7/16	13	6 7/16	4
				1 1/2	13 1/8	6 1/2	4

Taper shank machine (jobbers) reamers have approximately the same flute length as hand reamers but are designed for machine use.

tolerance; at *B* is the common grind, and at *C* is a square-end grind that acts like an end mill to straighten out bent or misaligned holes. If a change in diameter is made by resetting the blades, it is sometimes necessary to give the blades a cylindrical grind to insure roundness and correct size. After this step, it is usually necessary to relieve or clear the margins. These relationships may prove of value:

Reamer Dia, In.	Margin Width, In.	Primary Clearance, Deg
1	0.013	9
1 1/2	0.016	7
2 in. and up	0.023	7

Chamfer on carbide reamers should be reground long before the tool shows excessive wear. Occasionally the face of the tips will require polishing slightly to remove rounded portions or built-up edge. A successful method of regrinding the chamfer is to insert the tool in a bushing with only the end of the tool protruding slightly. Press the reamer lightly against the tailstock center, and grind from the face of the tip toward the back of the margin. Keep the setup as compact as possible. This setup achieves a concentric chamfer in relation to the tool OD.

Cylindrical grinding of carbide reamers should be done only when absolutely necessary. Average figures for relieving the margins are given on page 8-15.

TABLE 5. SHELL REAMERS—STRAIGHT AND SPIRAL FLUTES, HIGH-SPEED STEEL

Dia	Length *A*	Hole *H**	Fitting Arbor No.	Dia	Length *A*	Hole *H**	Fitting Arbor No.
3/4	2 1/4	3/8	4	1 7/8	3 1/2	1	8
13/16	2 1/2	1/2	5	1 15/16	3 1/2	1	8
7/8	2 1/2	1/2	5	2	3 1/2	1	8
15/16	2 1/2	1/2	5	2 1/16	3 3/4	1 1/4	9
1	2 1/2	1/2	5	2 1/8	3 3/4	1 1/4	9
1 1/16	2 3/4	5/8	6	2 3/16	3 3/4	1 1/4	9
1 1/8	2 3/4	5/8	6	2 1/4	3 3/4	1 1/4	9
1 3/16	2 3/4	5/8	6	2 5/16	3 3/4	1 1/4	9
1 1/4	2 3/4	5/8	6	2 3/8	3 3/4	1 1/4	9
1 5/16	3	3/4	7	2 7/16	3 3/4	1 1/4	9
1 3/8	3	3/4	7	2 1/2	3 3/4	1 1/4	9
1 7/16	3	3/4	7	2 9/16	4	1 1/2	10
1 1/2	3	3/4	7	2 5/8	4	1 1/2	10
1 9/16	3	3/4	7	2 11/16	4	1 1/2	10
1 5/8	3	3/4	7	2 3/4	4	1 1/2	10
1 11/16	3 1/2	1	8	2 13/16	4	1 1/2	10
1 3/4	3 1/2	1	8	2 7/8	4	1 1/2	10
1 13/16	3 1/2	1	8	2 15/16	4	1 1/2	10
				3	4	1 1/2	10

Spiral Flute

Arbor Dimensions

Arbor No.	Length *A*	Shank *D*	Shank Taper
4	9	1/2	2
5	9 1/2	5/8	2
6	10	3/4	3
7	11	7/8	3
8	12	1 1/8	4
9	13	1 3/8	4
10	14	1 5/8	5

Shell reamers are designed as a sizing or finishing reamer and are held on an arbor provided with driving lugs.

The holes in these reamers are ground with a taper of 1/8 in. per ft.

These reamers are regularly furnished with left-hand spiral flutes.

* Large end of hole, which is tapered 1/8 in. per ft.

TABLE 6. EXPANSION CHUCKING REAMERS, STRAIGHT FLUTES—STRAIGHT SHANK AND STRAIGHT FLUTES—TAPER SHANK

Dia	Length	Flute Length	Straight Shank Dia	Taper Shank No.	Dia	Length	Flute Length	Straight Shank Dia	Taper Shank No.
3/8	7	3/4	5/16	1	1 1/4	11 1/2	1 7/8	1	4
13/32	7	3/4	5/16	1	1 5/16	11 1/2	1 7/8	1	4
7/16	7	7/8	3/8	1	1 3/8	12	2	1	4
15/32	7	7/8	3/8	1	1 7/16	12	2	1 1/4	4
1/2	8	1	7/16	1	1 1/2	12 1/2	2 1/8	1 1/4	4
17/32	8	1	7/16	1	1 9/16	12 1/2	2 1/8	1 1/4	4
9/16	8	1 1/8	7/16	1	1 5/8	13	2 1/4	1 1/4	4
19/32	8	1 1/8	7/16	1	1 11/16	13	2 1/4	1 1/4	4
5/8	9	1 1/4	9/16	2	1 3/4	13 1/2	2 3/8	1 1/4	5
21/32	9	1 1/4	9/16	2	1 13/16	13 1/2	2 3/8	1 1/2	5
11/16	9	1 1/4	9/16	2	1 7/8	14	2 1/2	1 1/2	5
23/32	9	1 1/4	9/16	2	1 15/16	14	2 1/2	1 1/2	5
3/4	9 1/2	1 3/8	5/8	2	2	14	2 1/2	1 1/2	5
25/32	9 1/2	1 3/8	5/8	2	2 1/16	14 1/2	2 3/4	1 1/2	5
13/16	9 1/2	1 3/8	5/8	2	2 1/8	14 1/2	2 3/4	1 1/2	5
27/32	9 1/2	1 3/8	5/8	2	2 3/16	14 1/2	2 3/4	1 3/4	5
7/8	10	1 1/2	3/4	2	2 1/4	14 1/2	2 3/4	1 3/4	5
29/32	10	1 1/2	3/4	2	2 5/16	15	3	1 3/4	5
15/16	10	1 1/2	3/4	3	2 3/8	15	3	1 3/4	5
31/32	10	1 1/2	3/4	3	2 7/16	15	3	1 3/4	5
1	10 1/2	1 5/8	7/8	3	2 1/2	15	3	1 3/4	5
1 1/32	10 1/2	1 5/8	7/8	3	2 9/16	15 1/2	3 1/4	2	5
1 1/16	10 1/2	1 5/8	7/8	3	2 5/8	15 1/2	3 1/4	2	5
1 3/32	10 1/2	1 5/8	7/8	3	2 11/16	15 1/2	3 1/4	2	5
1 1/8	11	1 3/4	7/8	3	2 3/4	15 1/2	3 1/4	2	5
1 5/32	11	1 3/4	7/8	3	2 13/16	16	3 1/2	2	5
1 3/16	11	1 3/4	1	3	2 7/8	16	3 1/2	2	5
1 7/32	11	1 3/4	1	3	2 15/16	16	3 1/2	2	5
					3	16	3 1/2	2	5

ASA B5.14-1949.
High-speed steel reamers are standard.

Reamer Dia	*No. Flutes*
3/8– 15/32	4–6
1/2– 31/32	6–8
1 –1 11/16	8–10
1 3/4–1 15/16	8–12
2 –2 5/16	10–12
2 3/8–2 11/16	10–14
2 3/4–3	10–15

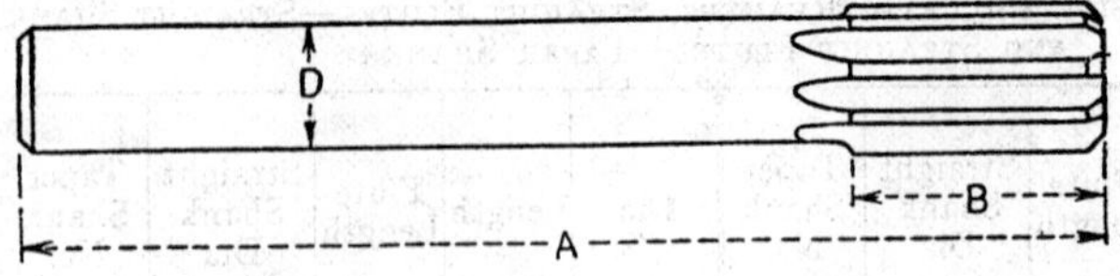

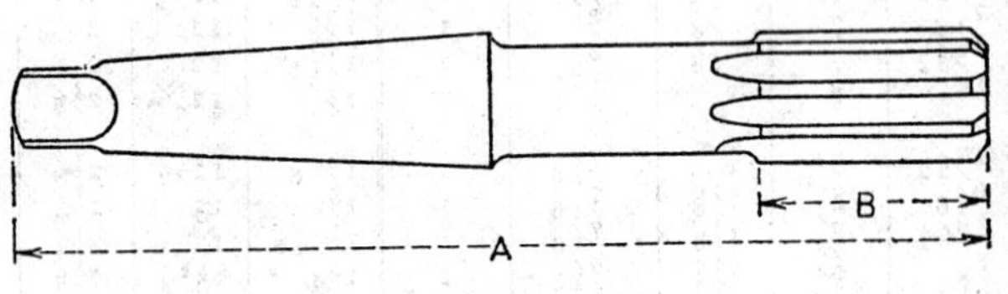

TABLE 7. FLUTED CHUCKING REAMERS (HIGH-SPEED STEEL)

Dia	Length *A*	Flute Length *B*	Straight Shank Dia *D*	Taper Shank No.	Dia	Length *A*	Flute Length *B*	Straight Shank Dia *D*	Taper Shank No.
1/8	3 1/2	7/8	7/64		11/16	9	2 1/4	9/16	2
5/32	4	1	9/64		23/32	9	2 1/4	9/16	2
3/16	4 1/2	1 1/8	11/64		3/4	9 1/2	2 1/2	5/8	2
7/32	5	1 1/4	13/64		25/32	9 1/2	2 1/2	5/8	2
1/4	6	1 1/2	15/64	1	13/16	9 1/2	2 1/2	5/8	2
9/32	6	1 1/2	15/64	1	27/32	9 1/2	2 1/2	5/8	2
5/16	6	1 1/2	9/32	1	7/8	10	2 5/8	3/4	2
11/32	6	1 1/2	9/32	1	29/32	10	2 5/8	3/4	2
3/8	7	1 3/4	5/16	1	15/16	10	2 5/8	3/4	3
13/32	7	1 3/4	5/16	1	31/32	10	2 5/8	3/4	3
7/16	7	1 3/4	3/8	1	1	10 1/2	2 3/4	7/8	3
15/32	7	1 3/4	3/8	1	1 1/16	10 1/2	2 3/4	7/8	3
1/2	8	2	7/16	1	1 1/8	11	2 7/8	7/8	3
17/32	8	2	7/16	1	1 3/16	11	2 7/8	1	3
9/16	8	2	7/16	1	1 1/4	11 1/2	3	1	4
19/32	8	2	7/16	1	1 5/16	11 1/2	3	1	4
5/8	9	2 1/4	9/16	2	1 3/8	12	3 1/4	1	4
21/32	9	2 1/4	9/16	2	1 7/16	12	3 1/4	1 1/4	4
					1 1/2	12 1/2	3 1/2	1 1/4	4

Fluted chucking reamers are made with straight and tapered shanks and with straight or spiral flutes, for use on turret lathes and screw machines. They have a slight chamfer for end cutting. Relief is ground back of the cutting edge for the full length of the land.

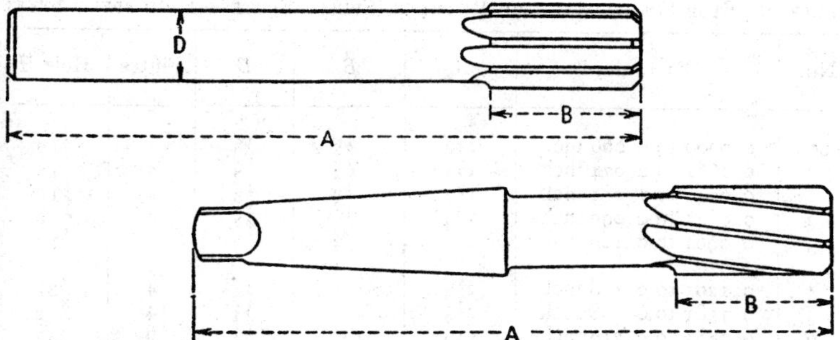

TABLE 8. ROSE CHUCKING REAMERS (HIGH-SPEED STEEL)

Dia	Length *A*	Shank *D*	Taper No.	No. Flutes	Dia	Length *A*	Shank *D*	Taper No.	No. Flutes
1/8	3 1/2	7/8	..	4 to 6 incl.	5/8	9	9/16	2	6 to 8 incl.
5/32	4	1	..	4 to 6 incl.	11/16	9	9/16	2	6 to 8 incl.
3/16	4 1/2	1 1/8	..	4 to 6 incl.	3/4	9 1/2	5/8	2	6 to 10 incl.
7/32	5	1 1/4	..	4 to 6 incl.	13/16	9 1/2	5/8	2	6 to 10 incl.
1/4	6	1 1/2	1	4 to 6 incl.	7/8	10	3/4	2	6 to 10 incl.
9/32	6	1 1/2	1	4 to 6 incl.	15/16	10	3/4	3	6 to 10 incl.
5/16	6	1 1/2	1	4 to 6 incl.	1	10 1/2	7/8	5	6 to 10 incl.
11/32	6	1 1/2	1	4 to 6 incl.	1 1/16	10 1/2	7/8	3	6 to 10 incl.
3/8	7	1 3/4	1	4 to 6 incl.	1 1/8	11	7/8	3	6 to 10 incl.
13/32	7	1 3/4	1	4 to 6 incl.	1 3/16	11	1	3	8 to 12 incl.
7/16	7	1 3/4	1	6 to 8 incl.	1 1/4	11 1/2	1	4	8 to 12 incl.
15/32	7	1 3/4	1	6 to 8 incl.	1 5/16	11 1/2	1	4	8 to 12 incl.
1/2	8	2	1	6 to 8 incl.	1 3/8	12	1	4	8 to 12 incl.
9/16	8	2	1	6 to 8 incl.	1 7/16	12	1 1/4	4	8 to 12 incl.
					1 1/2	12 1/2	1 1/4	4	8 to 12 incl.

Rose chucking reamers are primarily designed for use in turret lathes and screw machines. The lands are ground cylindrically, without radial relief, and are designed to cut on the end.

GRINDS FOR CARBIDE REAMERS

Reamer Dia, In.	Margin Width, In.	Primary Clearance, Deg
1/4	0.007	14
1/2	0.009	11
1	0.013	9
1 1/2	0.016	7
2	0.023	7

TABLE 9. STUB SCREW MACHINE REAMERS, SPIRAL FLUTES—HIGH-SPEED STEEL

No.	Dia Range	*A*	*B*	*D*	Flutes	Hole *H*
00	0.0600 to 0.066 incl.	1¾	½	⅛	4	1/16
0	0.0661 to 0.074 incl.	1¾	½	⅛	4	1/16
1	0.0741 to 0.084 incl.	1¾	½	⅛	4	1/16
2	0.0841 to 0.096 incl.	1¾	½	⅛	4	1/16
3	0.0961 to 0.126 incl.	2	¾	⅛	4	1/16
4	0.1261 to 0.158 incl.	2¼	1	¼	4	3/32
5	0.1581 to 0.188 incl.	2¼	1	¼	4	3/32
6	0.1881 to 0.219 incl.	2¼	1	¼	6	3/32
7	0.2191 to 0.251 incl.	2¼	1	¼	6	3/32
8	0.2511 to 0.282 incl.	2¼	1	⅜	6	⅛
9	0.2821 to 0.313 incl.	2¼	1	⅜	6	⅛
10	0.3131 to 0.344 incl.	2½	1¼	⅜	6	⅛
11	0.3441 to 0.376 incl.	2½	1¼	⅜	6	⅛
12	0.3761 to 0.407 incl.	2½	1¼	½	6	3/16
13	0.4071 to 0.439 incl.	2½	1¼	½	6	3/16
14	0.4391 to 0.470 incl.	2½	1¼	½	6	3/16
15	0.4701 to 0.505 incl.	2½	1¼	½	6	3/16
16	0.5051 to 0.567 incl.	3	1½	⅝	6	¼
17	0.5671 to 0.630 incl.	3	1½	⅝	6	¼
18	0.6301 to 0.692 incl.	3	1½	⅝	6	¼
19	0.6921 to 0.755 incl.	3	1½	¾	8	5/16
20	0.7551 to 0.817 incl.	3	1½	¾	8	5/16
21	0.8171 to 0.880 incl.	3	1½	¾	8	5/16
22	0.8801 to 0.942 incl.	3	1½	¾	8	5/16
23	0.9421 to 1.010 incl.	3	1½	¾	8	5/16

ASA B5.14-1949.

Stub screw machine reamers are particularly adapted for use in floating holders, and the shank is provided with a pinhole for this purpose. These reamers are regularly furnished with right-hand cut and left-hand spiral flutes in any size within the range shown. Left-hand reamers and reamers with right-hand spiral flutes are special.

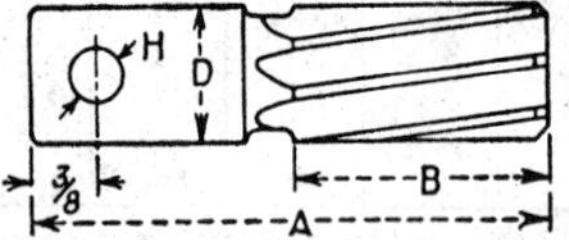

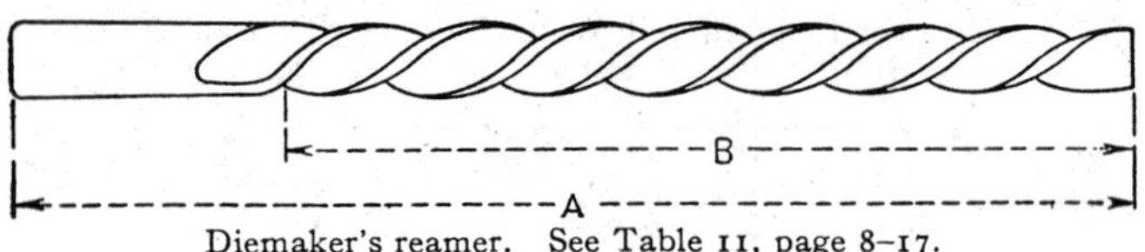

Diemaker's reamer. See Table 11, page 8-17.

TABLE 10. TAPER PIN REAMERS (TAPER 1/4 IN. PER FT), STRAIGHT, SPIRAL, AND HELICAL FLUTES

Reamer	Dia		Over-all Length	Length of Flute	Length of Square	Shank Dia	Size of Square	Flutes	
	Small End	Large End						Straight and Spiral	Helical
7/0	0.0497	0.0666	1 13/16	13/16	5/32	5/64	0.060	3 to 4 incl.	2 to 3 incl.
6/0	0.0611	0.0806	1 15/16	15/16	5/32	3/32	0.070	3 to 4 incl.	2 to 3 incl.
5/0	0.0719	0.0966	2 3/16	1 3/16	5/32	7/64	0.080	3 to 4 incl.	2 to 3 incl.
4/0	0.0869	0.1142	2 5/16	1 5/16	5/32	1/8	0.095	3 to 4 incl.	2 to 3 incl.
3/0	0.1029	0.1302	2 5/16	1 5/16	5/32	9/64	0.105	4 to 6 incl.	2 to 3 incl.
2/0	0.1137	0.1462	2 9/16	1 9/16	7/32	5/32	0.115	4 to 6 incl.	2 to 3 incl.
0	0.1287	0.1638	2 15/16	1 11/16	7/32	11/64	0.130	4 to 6 incl.	2 to 3 incl.
1	0.1447	0.1798	2 15/16	1 11/16	7/32	3/16	0.140	5 to 6 incl.	2 to 3 incl.
2	0.1605	1.2008	3 3/16	1 15/16	1/4	13/64	0.150	5 to 6 incl.	2 to 3 incl.
3	0.1813	0.2294	3 11/16	2 5/16	1/4	15/64	0.175	5 to 6 incl.	2 to 3 incl.
4	0.2071	0.2004	4 1/16	2 9/16	1/4	17/64	0.200	5 to 6 incl.	2 to 3 incl.
5	0.2409	0.2994	4 5/16	2 13/16	5/16	5/16	0.235	5 to 6 incl.	2 to 3 incl.
6	0.2773	0.354	5 7/16	3 11/16	3/8	23/16	0.270	6 to 8 incl.	2 to 3 incl.
7	0.3297	0.422	6 5/16	4 7/16	3/8	13/32	0.305	6 to 8 incl.	2 to 3 incl.
8	0.3971	0.505	7 3/16	5 3/16	7/16	7/16	0.330	6 to 8 incl.	2 to 3 incl.
9	0.4805	0.6066	8 5/16	6 1/16	9/16	9/16	0.420	6 to 8 incl.	2 to 4 incl.
10	0.5799	0.7216	9 5/16	6 13/16	5/8	5/8	0.470	7 to 8 incl.	2 to 4 incl.

ASA B5.14-1949.

Taper pin reamers with straight flutes are made in carbon steel and high-speed steel. The point of each reamer will enter the hole reamed by the next size smaller. Spiral-flute reamers have left-hand spiral flutes. They, like straight-flute reamers, have a squared shank.

Helical taper pin reamers are intended for machine reaming, and hence have a round shank. They are regularly furnished with left-hand helical flutes and in high-speed steel.

TABLE 11. DIEMAKER'S REAMERS, HELICAL TYPE—HIGH-SPEED STEEL

Letter Size	Dia Small End	Over-all Length *A*	Length of Flute *B*	Letter Size	Dia Small End	Over-all Length *A*	Length of Flute *B*
AAA	0.055	2 1/4	1 1/8	K	0.190	3 1/2	2 1/4
AA	0.065	2 1/4	1 1/8	L	0.205	3 1/2	2 1/4
A	0.075	2 1/4	1 1/8	M	0.220	4	2 1/2
B	0.085	2 3/8	1 3/8	N	0.235	4 1/2	3
C	0.095	2 1/2	1 3/8	O	0.250	5	3 1/2
D	0.105	2 5/8	1 5/8	P	0.275	5 1/2	4
E	0.115	2 3/4	1 5/8	Q	0.300	6	4 1/2
F	0.125	3	1 3/4	R	0.335	6 1/2	4 3/4
G	0.135	3	1 3/4	S	0.370	6 3/4	5
H	0.145	3 1/4	1 7/8	T	0.405	7	5 1/4
I	0.160	3 1/4	1 7/8	U	0.440	7 1/4	5 1/2
J	0.175	3 1/4	1 7/8				

ASA B5.14-1949.

All sizes have two or three flutes. These reamers have a taper of 3/4° included angle or 0.013 in. per in.

TABLE 12. MORSE TAPER REAMERS WITH SQUARED AND TAPER SHANKS

Taper No.	Finishing Reamer Dia		Squared Shank					Taper Shank			No. Flutes	
	Small End	Large End	Over-all Length A	Length of Flute* B	Length of Square C	Shank Dia D	Size of Square	Over-all Length A	Length of Flute B	Taper No.	Roughing	Finishing
0	0.2503	0.3674	3¾	2¼	5/16	5/16	0.235	5 11/32	2¼	0	4 to 6 incl.	4 to 6 incl.
1	0.3674	0.5170	5	3	7/16	7/16	0.330	6 5/16	3	1	4 to 6 incl.	6 to 8 incl.
2	0.5696	0.7444	6	3½	5/8	5/8	0.470	7⅜	3½	2	4 to 6 incl.	6 to 8 incl.
3	0.7748	0.9881	7¼	4¼	7/8	7/8	0.655	8⅞	4¼	3	4 to 6 incl.	8 to 10 incl.
4	1.0167	1.2893	8½	5¼	1	1⅛	0.845	10⅞	5¼	4	4 to 8 incl.	8 to 10 incl.
5	1.4717	1.8005	9¾	6¼	1⅛	1½	1.125	13⅛	6¼	5	6 to 10 incl.	10 to 12 incl.
6	2.1119	2.5550	12¼	8½	1½	2	1.500	17 13/16	8½	6	6 to 12 incl.	12 to 14 incl.

ASA B5.14-1949.

* Taper shank reamers have the same length of flute. These reamers are used for reaming out Morse Taper sockets.
Squared-shank reamers are made in:
Roughing type, straight flute, in HSS.
Finishing type, straight flute, carbon or HSS.
Finishing, spiral flute, in HSS.
Taper-shank reamers have ASA standard taper for sizes 1 to 5. They are made in three styles above, in HSS only.

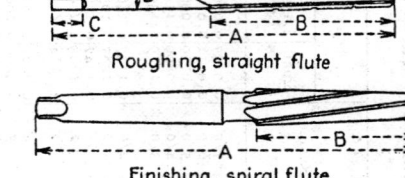

Roughing, straight flute

Finishing, spiral flute

TABLE 13. BROWN & SHARPE TAPER REAMERS WITH SQUARED SHANK

Taper No.	Dia Small End	Dia Large End	Over-all Length	Flute Length	Length of Square	Shank Dia	Size of Square	No. Flutes
1	0.1974	0.3176	4 3/4	2 7/8	1/4	9/32	7/32	4 to 6
2	0.2474	0.3781	5 1/8	3 1/8	5/16	11/32	1/4	4 to 6
3	0.3099	0.4510	5 1/2	3 3/8	3/8	13/32	5/16	4 to 6
4	0.3474	0.5017	5 7/8	3 11/16	7/16	7/16	11/32	4 to 6
5	0.4474	0.6145	6 3/8	4	1/2	9/16	7/16	4 to 6
6	0.4974	0.6808	6 7/8	4 3/8	5/8	5/8	15/32	4 to 6
7	0.5974	0.8011	7 1/2	4 7/8	3/4	3/4	9/16	6 to 8
8	0.7474	0.9770	8 1/8	5 1/2	13/16	13/16	5/8	6 to 8
9	0.8974	1.1530	8 7/8	6 1/8	7/8	1	3/4	6 to 8
10	1.0420	1.3376	9 3/4	6 7/8	1	1 1/8	27/32	6 to 8
11	1.2474	1.5657	10 5/8	7 5/8	1 1/16	1 1/4	15/16	6 to 8
12	1.4974	1.8409	11 3/8	8 1/4	1 1/8	1 1/2	1 1/8	8 to 10

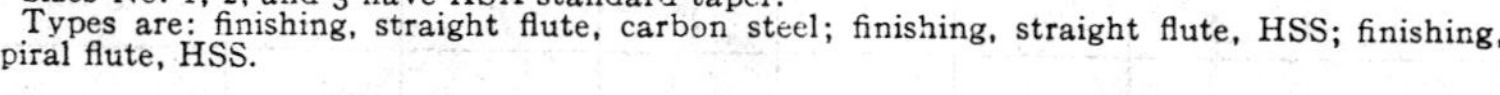
These reamers are designed for use in reaming out Brown & Sharpe standard taper sockets. Sizes No. 1, 2, and 3 have ASA standard taper.

Types are: finishing, straight flute, carbon steel; finishing, straight flute, HSS; finishing, spiral flute, HSS.

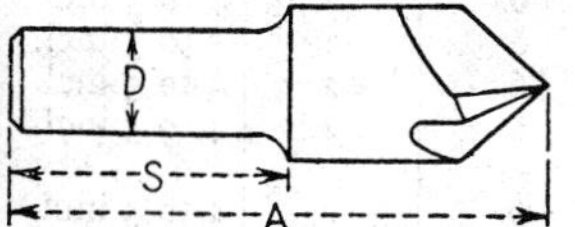

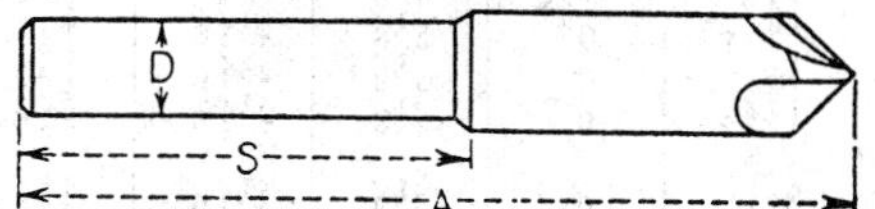

TABLE 14. CENTER REAMERS AND MACHINE COUNTERSINKS, STRAIGHT SHANK—CARBON OR HIGH-SPEED STEEL

Center Reamers				Machine Countersinks				
Size Cut	Length *A*	Shank Dia *D*	Shank Length *S*	Size Cut	Length *A*	Shank Dia *D*	Shank Length *S*	Flutes
1/4	1 1/2	3/16	3/4	1/2	3 7/8	1/2	2 1/4	3 to 4 incl.
3/8	1 3/4	1/4	7/8	5/8	4	1/2	2 1/4	3 to 4 incl.
1/2	2	3/8	1	3/4	4 1/8	1/2	2 1/4	3 to 4 incl.
5/8	2 1/4	3/8	1	7/8	4 1/4	1/2	2 1/4	3 to 4 incl.
3/4	2 5/8	1/2	1 1/4	1	4 3/8	1/2	2 1/4	3 to 4 incl.

Center reamers are used for centering lathe work and also for countersinking for screw heads. They are regularly furnished with either 60 or 82° included angle.

Machine countersinks are regularly furnished with either 60 or 82° included angle.

Both types of tools are made with three or four flutes.

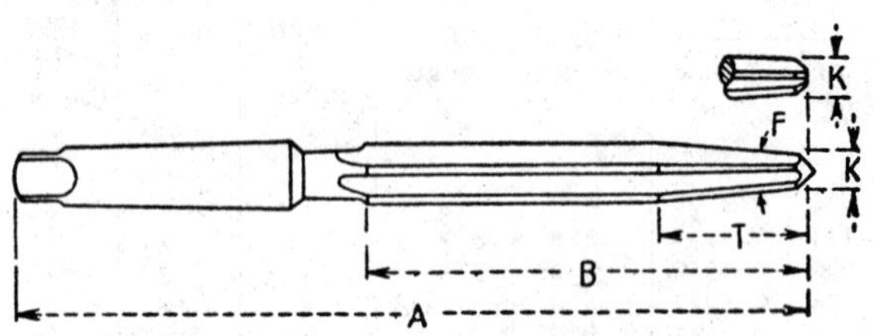

TABLE 15. TAPER BRIDGE REAMERS, STRAIGHT OR SPIRAL FLUTES—TAPER SHANK

Reamer Dia	Length	Length Flute		Dia	Included Angle	Taper Shank No.	No. Flutes
		Total	Taper				
	A	*B*	*T*	*K*	*F*		
13/32	8 1/4	4 3/8	1 3/4	7/32	6	2	4 to 5 incl.
7/16	8 1/4	4 3/8	1 3/4	1/4	6	2	4 to 5 incl.
15/32	9	5 1/8	2 1/16	1/4	6	2	4 to 5 incl.
1/2	9	5 1/8	2 1/16	9/32	6	2	4 to 5 incl.
17/32	9	5 1/8	2 1/16	5/16	6	2	4 to 5 incl.
9/16	9	5 1/8	2 1/16	11/32	6	2	4 to 5 incl.
5/8	10	6 1/8	2 7/16	3/8	6	2	4 to 5 incl.
11/16	11 3/4	7 1/8	2 7/8	25/64	6	3	4 to 5 incl.
3/4	12	7 3/8	2 15/16	7/16	6	3	4 to 5 incl.
13/16	12	7 3/8	2 15/16	1/2	6	3	4 to 5 incl.
7/8	12	7 3/8	2 15/16	9/16	6	3	4 to 5 incl.
15/16	12	7 3/8	2 15/16	5/8	6	3	4 to 5 incl.
1	12	7 3/8	2 15/16	11/16	6	3	4 to 6 incl.
1 1/16	12	7 3/8	2 15/16	3/4	6	3	4 to 6 incl.
1 1/8	12	7 3/8	2 15/16	13/16	6	3	4 to 6 incl.
1 3/16	12	7 3/8	2 15/16	7/8	6	3	4 to 6 incl.
1 1/4	13	7 3/8	2 15/16	15/16	6	4	4 to 6 incl.
1 5/16	13	7 3/8	2 15/16	1	6	4	4 to 7 incl.*
1 3/8	13	7 3/8	2 15/16	1 1/16	6	4	4 to 7 incl.*
1 7/16	13	7 3/8	2 15/16	1 1/8	6	4	4 to 7 incl.*
1 1/2	13	7 3/8	2 15/16	1 3/16	6	4	4 to 7 incl.*

ASA B5.14-1949.

Taper bridge reamers are particularly adapted for reaming rivet and bolt holes in structural iron and steel, boiler plate, etc. They are tapered at the point to facilitate entering holes which are out of alignment.

High-speed-steel reamers are standard; carbon steel, special.

* Four to six flutes for spiral-flute reamers.

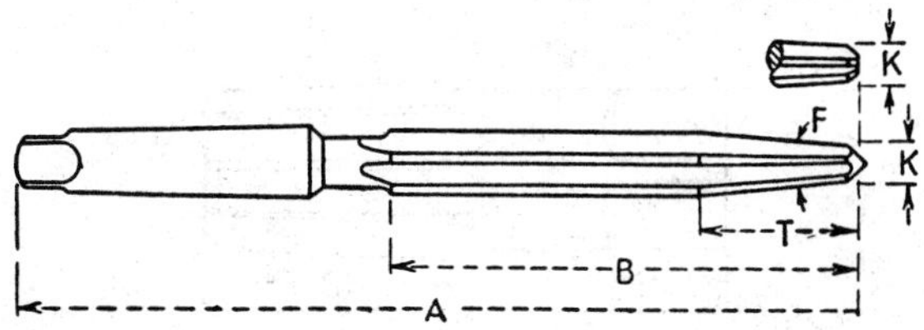

Table 16. Taper Car Reamers—Short Length, Straight or Spiral Flutes—High-speed Steel

Reamer Dia	Length	Length Flute		Dia	Included Angle	Taper Shank No.	No. Flutes
		Total	Taper				
	A	*B*	*T*	*K*	*F*		
1/4	5 7/16	2 1/2	1 1/4	1/8	6	1	4 to 5 incl.
9/32	5 7/16	2 1/2	1 1/4	5/32	6	1	4 to 5 incl.
5/16	5 11/16	2 3/4	1 3/8	11/64	6	1	4 to 5 incl.
11/32	5 11/16	2 3/4	1 3/8	13/64	6	1	4 to 5 incl.
3/8	5 11/16	2 3/4	1 3/8	15/64	6	1	4 to 5 incl.
13/32	6 3/16	2 3/4	1 3/8	17/64	6	2	4 to 5 incl.
7/16	6 15/16	3 1/2	1 3/4	1/4	6	2	4 to 5 incl.
15/32	7 1/16	3 1/2	1 3/4	9/32	6	2	4 to 5 incl.
1/2	7 9/16	4	2	19/64	6	2	4 to 5 incl.
17/32	7 9/16	4	2	1/4	8	2	4 to 5 incl.
9/16	7 9/16	4	2	9/32	8	2	4 to 5 incl.
5/8	8 1/16	4 1/2	2 1/4	5/16	8	2	4 to 5 incl.
11/16	8 13/1	4 1/2	2 1/4	3/8	8	3	4 to 5 incl.
3/4	9 1/2	5	2 1/2	13/32	8	3	4 to 5 incl.
13/16	9 1/2	5	2 1/2	15/32	8	3	4 to 5 incl.
7/8	9 1/2	5	2 1/2	17/32	8	3	4 to 5 incl.
15/16	9 1/2	5	2 1/2	19/32	8	3	4 to 5 incl.
1	9 1/2	5	2 1/2	21/32	8	3	4 to 6 incl.
1 1/16	9 1/2	5	2 1/2	23/32	8	3	4 to 6 incl.
1 1/8	9 1/2	5	2 1/2	25/32	8	3	4 to 6 incl.
1 3/16	9 1/2	5	2 1/2	27/32	8	3	4 to 6 incl.
1 1/4	9 1/2	5	2 1/2	29/32	8	3	4 to 6 incl.

ASA B5.14-1949.

Short length taper car reamers are similar in construction to taper bridge reamers. They are especially adapted for reaming rivet and bolt holes in thin structural sections.

They are tapered at the point to facilitate entering holes which are out of alignment.

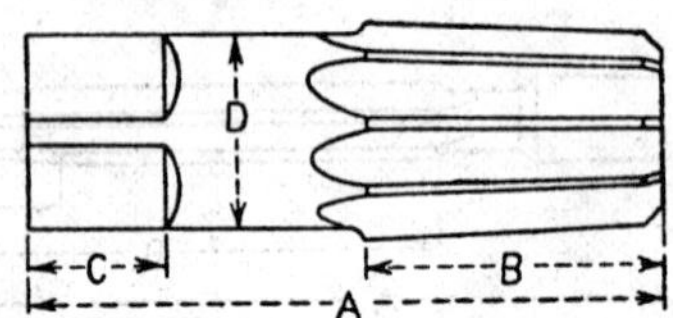

TABLE 17. TAPER PIPE REAMERS, STRAIGHT FLUTES—CARBON OR HIGH-SPEED STEEL

Nominal Size	Dia		Over-all Length	Length of Flute	Length of Square	Dia of Shank	Size of Square
	Large End	Small End	*A*	*B*	*C*	*D*	
1/8	0.362	0.316	2 1/8	3/4	3/8	0.4375	0.328
1/4	0.472	0.406	2 7/16	1 1/16	7/16	0.5625	0.421
3/8	0.606	0.540	2 9/16	1 1/16	1/2	0.7000	0.531
1/2	0.751	0.665	3 1/8	1 3/8	5/8	0.6875	0.515
3/4	0.962	0.876	3 1/4	1 3/8	11/16	0.9063	0.679
1	1.212	1.103	3 3/4	1 3/4	13/16	1.1250	0.843
1 1/4	1.553	1.444	4	1 3/4	15/16	1.3125	0.984
1 1/2	1.793	1.684	4 1/4	1 3/4	1	1.5000	1.125
2	2.268	2.159	4 1/2	1 3/4	1 1/8	1.8750	1.406

ASA B5.14-1949.

These reamers are tapered 3/4 in. to the foot and are intended for reaming holes to be tapped with American Standard Taper Pipe Thread taps.

Section 9

SAWS AND SAWING

SECTION 9

SAWS AND SAWING

HAND HACK-SAW BLADES

Standard hand hack-saw blades are made in 10- and 12-in. lengths, a width of ½ in., and a thickness of 0.025 in. These blades are available in four pitches of teeth: 14, 18, 24, and 32 teeth per inch. Select the blade for the kind of work you have:

No. of Teeth per In.	*Kind and Type of Material*
14	For materials equivalent to 1 in. round or more—aluminum, brass, bronze, cast iron, copper, cold-rolled steel, structural steel, rails, etc.
18	For materials ¼ to 1 in. in dia. Also tool steels, drill rod, cold-rolled steel, and medium-weight structural shapes
24	For materials ⅛ to ¼ in. in thickness. Also pipe and tubing, BX cable, heavy sheet metal, moldings, etc.
32	For materials less than ⅛ in. in thickness. Also tubing, BX cable, sheet metal, moldings, etc.

NOTE: It is suggested that blades of the flexible type be used where the operator is required to assume an awkward position.

The general rule is: make sure that at least three consecutive teeth are in contact with the stock at all times. In starting a cut, keep off sharp angles. Thus you will use an 18-tooth blade for general work, and shift to a 32-tooth blade for thin-wall tubing.

In using a hack saw, employ a long steady stroke, applying pressure on the forward stroke, and raising the blade on the return stroke to avoid dulling the teeth. Do not use enough pressure to cause a crooked cut, and work at about 40 to 50 strokes per minute.

Blade materials are: carbon steel, tungsten alloy steel, molybdenum steel, molybdenum high-speed steel, and tungsten high-speed steel, the last being considered as providing the greatest service, particularly on alloy steels.

POWER HACK-SAW BLADES

Single-edge power hack-saw blades are made in lengths from 14½ to 36 in. in a coarse pitch of 2½ teeth per inch. Double-edge blades range from 12½ to 36 in. long. Blades up to 14⅜ in. long

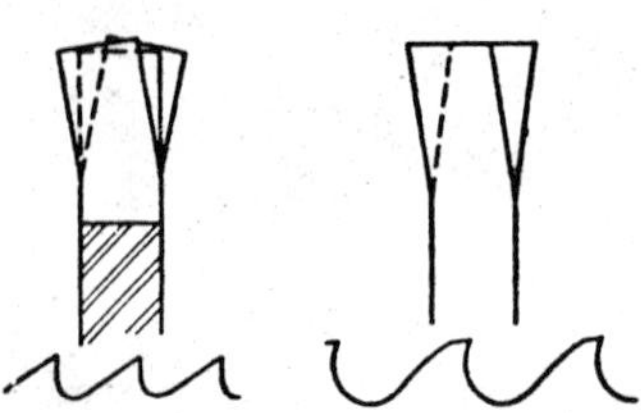

FIG. 1. When every other tooth is set right or left, as shown at left, only the outside points cut. With the straight set, as at right, the entire point of each tooth shares in cutting.

TABLE 1. APPLICATIONS OF POWER HACK-SAW BLADES

Material	Teeth per In.	Strokes per Min	Feed Pressure, Lb
Aluminum	4 to 6	135 to 150	60
Brass, cast:			
Soft	6 to 10	135 to 150	60
Hard	6 to 10	135	60
Cast Iron	6 to 10	135	120
Copper	6 to 10	135	120
Tool steel	4 to 6 to 10	90	120
Cold-rolled steel	4 to 6	135	150
High-speed steel	6 to 10	90	120
Machine steel	4 to 6	135	150
Iron pipe	10 to 14	135	120
Structural steel	6 to 10	135	120
Tubing:			
Steel	14	135	60
Brass	14	135	60

TABLE 2. POWER HACK-SAWING TROUBLES AND REMEDIES

Cause	*Correction*
Pulling Out at Pinhole	
Blade too tight, or blade twisted	Reduce tension on blade
Premature Wear	
Feed too heavy	Reduce
Speed too great	Reduce to recommended speed
Incorrect tooth spacing	Use number of teeth recommended for material
Insufficient feed	Increase feed as recommended
Dry cutting	Use coolant
Stripping Teeth	
Tooth spacing too coarse, or teeth too fine for material	Use number of teeth recommended for material
Crooked Cuts	
Worn frame, or frame out of line with vise	Check machine for wear and adjustment
Blade loose in frame	Adjust
Stock not tight in vise	Inspect clamps and tighten
Feed too heavy	Reduce
Worn-out blade	Install new blade
Hard spot in material	Start new cut. May require new blade
Blades Breaking	
Insufficient tension	Make adjustment
Tooth spacing too coarse	Use number of teeth recommended for material
Feed too heavy	Reduce
New blade in unfinished cut	Start new cut
Side strain on blade	Worn out. Change blade

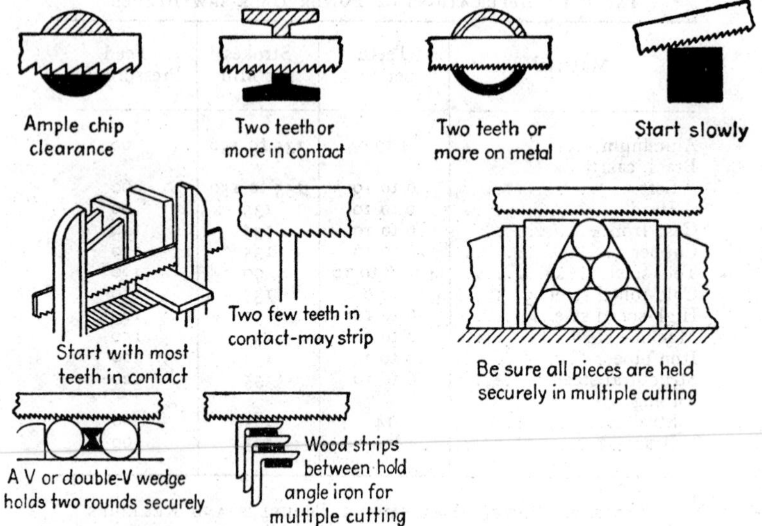

FIG. 2. Two points should be kept in mind in every power hack-sawing setup: (1) equal and adequate clamping of each piece in a multiple setup and (2) proper number of teeth in contact with the work.

may have 10, 14, or 18 teeth depending on size; longer blades have 4 to 6 teeth per inch, sometimes 10. Blade thicknesses range from 0.032 to 0.100 in.

Types of Saw Teeth. The more common type of tooth has a straight face (Fig. 1). Since cutting is done on the outside points, the chip is pushed against the cutting face and is broken into short chips. However, this type of saw blade can be operated only at light feeds. A second type of tooth form, made only in the coarse pitch of 2½ teeth per inch, has a 10° hook, a 15° clearance, and a rounded gullet to curl the chips. With this type of tooth, cutting rates as high as 16 sq in. per min on 1020 CR steel have been attained.

Selection of Blade. Use a blade as short as the work will permit, and thick and wide enough to withstand feed pressure. Select a blade that will have three teeth in the cut at all times. Finer teeth are used on hard stock, coarser teeth on soft stock for better chip removal.

Speeds and Feed Pressure. Cutting speeds range up to 150 strokes per minute on softer material, down to 90 on high-speed steel. Feed pressures should be light for thin materials, tubing, and soft metals, and heavier for large sections and harder materials (see Table 1). Coolants are soluble oil in 1:50 or sulfur-base cutting oil diluted with kerosene.

Work Clamping. Rigid clamping of the work is essential to avoid saw breakage (see Fig. 2). In general, these rules apply: flat stock should be laid flat in the vise. Round bars and tubing may be retained with a V or double-V wedge placed between two bars, whereas multiple bars should be held in a square or rectangular bundle by a holddown clamp, or grouped in a pyramid between two wedges. The main thing in multiple cutting is to be certain that all pieces will be held securely.

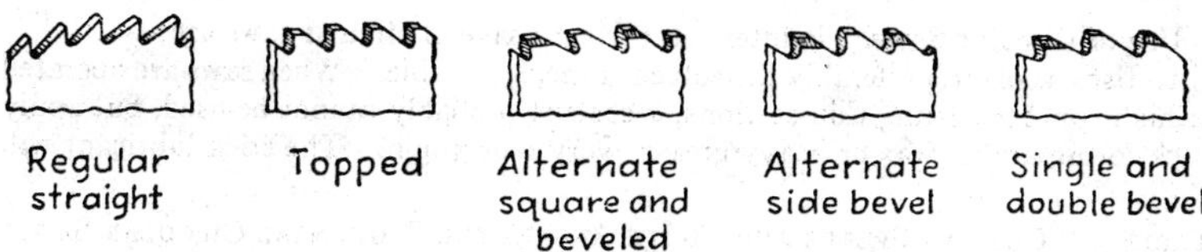

FIG. 3. Common tooth profiles of circular saws.

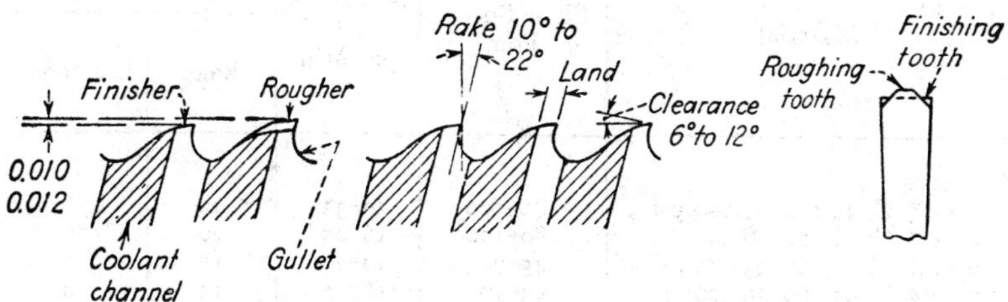

FIG. 4. "Curled-chip" circular saws have ample gullet space and a rake angle to suit the material.

CIRCULAR SAWS

In circular, or cold, sawing of metals, a thin saw blade is fed through the work at milling speeds and feeds. The teeth, therefore, have both rake and clearance.

Circular-saw blades are made in three types: solid, segmental, and inserted-tooth. Solid-blade saws of the metal-slitting type are made in diameters up to 8 in. or more for cutting hard materials at speeds from 50 to 75 fpm. Teeth range from 28 for a 2½-in. saw to 60 for an 8-in. saw. Cup-ground solid blades are made in diameters of say 6 to 16 in. for cutting ferrous and nonferrous metals at a speed range of 50 to 75 fpm. Teeth range from say 50 for a 6-in. saw to 96 for a 16-in. saw.

For cutting tubing and thin extrusions of ferrous or nonferrous materials, hollow-ground saws may operate at speeds up to 18,000 fpm. Fine-pitched teeth are used: 150 or 250 teeth for a 6-in. saw and 300 teeth for a 16-in. saw. Other hollow-ground saws running at 10,000 to 12,000 fpm on similar materials range from 8 in. dia by 200 teeth to 12 in. dia by 84 teeth.

Segmental-type blades consist of segments containing the teeth, the segments being riveted to a disk. When worn teeth are worn or broken, segments are replaced. In contrast to inserted-tooth blades, the segmental type can be provided in the desired tooth pitch but often is ⅜ to ½ in. On the other hand, a single tooth of an inserted-tooth saw can be quickly replaced.

Tooth Profile. Various tooth profiles (Fig. 3) have been used over the years. The "curled-chip" or "cam-generated" contour shown in Fig. 4 provides smooth sawing action, because chips curl up within the gullets and free themselves readily when the teeth pass out of the work. Teeth are ground alternately high and low; that is, the roughing tooth beveled at 45° on both sides is 0.010 to 0.012 in. higher than the square finishing tooth which removes two chips from the corners of the cut. By this arrangement of leader and follower teeth, the cutting load is divided, and the roughing teeth are not subjected to overload.

Teeth of circular-saw blades should have positive rake and clearance angles that will vary with the material cut (see Table 3).

Use of Circular Saws. Pointers on correct usage of circular saws are:

1. Use a coolant, preferably the solvent type, if possible. When saws are operated at high rpm for cutting thin sections, a coolant probably cannot be used, but apply stick-form paraffin wax or heavy grease every 2 or 3 min. The stick lubricant will

TABLE 3. CUTTING SPEEDS AND TOOTH ANGLES FOR SEGMENTAL CIRCULAR SAWS*

Material	Blade Speed, Fpm	Production, Sq In. per Min	Tooth Angles, Deg	
			Rake	Clearance
Steels:				
0.12% C, 42,000–64,000 psi	80–100	19–31	20	6
0.25% C, 64,000 psi	80–100	16–25	20	6
0.40% C, 71,000–85,000 psi	45–70	12–21	15	6
0.60% C, 85,000–100,000 psi	45–70	16–23	15	6
0.90% C, 100,000–115,000 psi	45–55	11–19	15	6
Nonrusting	25–35	3–9	10	6
Tool steel, annealed	35–45	6–12	10	6
Steel tubes, thick-wall	80–100	6–12	20	6
Tubing, thin-wall	1300–8000	62–124	20	6
Structural iron	55–100	16–25	20	6
Rails, hard	25–55	8–12	10	6
Castings, steel	25–55	6–12	10	6
Cast iron	35–45	6–16	15	6
Bronze	350–2400	47–93	20	10
Cast brass	350–2400	47–93	20	10
Rolled brass	500–1400	93–186	20	10
Hard copper	150–850	25–60	15–20	6
Aluminum	1200–5000	62–124	25–30	12
German silver	100–350	16–31	20	10
Pure nickel	25–55	3–6	15	6

* E. C. Atkins Saw Co.

TABLE 4. TOOTH PITCHES FOR SEGMENTAL CIRCULAR SAWS*

Material	*Tooth Pitch, In.*
Round and square bars:	
½–1½ in.	5/16
1½–3 in.	½–¾
3–5	¾–1
5–8	⅞–1⅛
8–12	1–1¼
12–20	1–1¾
Round and square bars in bundles	5/16–¾
Rolled sections in singles and bundles	⅝–¾
Pipes, thin-wall	¼–¾
Pipes, thick-wall	⅜–1
Brass, copper (solid material)	¾–1½
Aluminum (solid material)	1¼–2½
Aluminum sections	⅛–¾

* E. C. Atkins Saw Co.

help the saw and will not impose an additional cleaning operation.

2. Excess saw speed will dull the saw quickly (see Table 3), and a dull saw creates extreme cutting pressure and causes the saw to dish and wander in the cut.

3. Segmental saws are made for maximum feed performance. The chip load should be 0.001 to 0.008 in. per tooth, depending on materials cut. Too light a feed produces a scraping action and dulls the saw.

4. Rigid clamping of the work is essential. Slight movement of the work will cause the saw to grab and shell out the teeth.

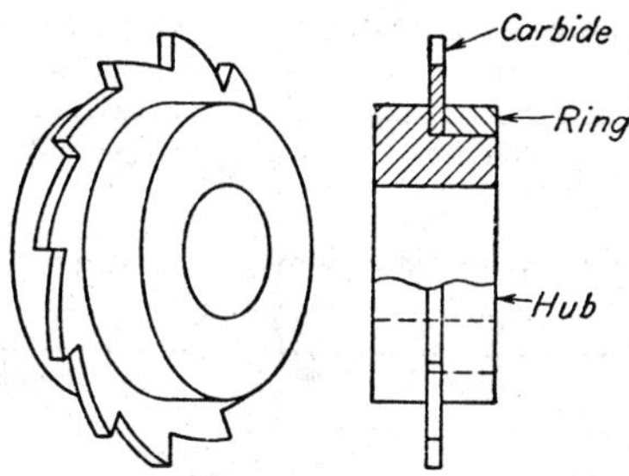

FIG. 5. Carbide ring brazed to a hub provides a high-speed cutting tool for many metals and plastics.

5. The size, shape, and kind of material control the correct number of teeth in the saw (Table 4).

6. When mitering thin angle sections at high saw rpm, notch the work at light feed for 1/16 to 1/8 in. deep, then feed the saw at constantly increasing feed until the cut is completed.

7. Idling the saw in the work will heat the rim, causing the blade to dull, get coated with material, wander in the cut, and develop cracks.

Carbide Saws. Since carbide is difficult to braze to saw teeth, especially in small sizes, the Gay-Lee Co. of Ferndale, Mich., developed saws with a disk of carbide brazed to a hub (Fig. 5). These saws range in from 1/4 to 4 in. dia. Feeds and speeds depend greatly on the application. Saws have been put to work at speeds from 150 to 5000 fpm, and feeds from 2 to 300 in. per min. Chip loads per tooth have ranged from a few tenths to 0.010 in. per tooth. These saws have been used on steel, brass, copper, plastics, mica undercutting, etc. Saws as thin as 0.0067 in. have been used for slotting needle bars with a straight perpendicular cut. Production life, as would be expected of carbide, has been up to 75 times that of HSS saws.

Large carbide-tipped saws have been used in a gang setup for sawing apart multiple castings. The castings, four or five parts en bloc, are machined at once, reducing the machining time per piece, and then cut apart.

BAND SAWS AND BAND SAWING

Many modern shops find that the power band saw is virtually a jack-of-all-trades machine. It serves for job-shop work in infinite variety, and likewise is a high-production tool. Metal-cutting power band saws have speeds ranging from say 50 to 1500 fpm, a range that is more than ample for ferrous materials (see Table 5) while units for plastics, aluminum, magnesium, and some of the copper alloys must function at much greater speeds.

The power band saw possesses such diverse utility because: it may have a deep throat—up to 60 in.; it can accept stock stacked several inches deep; it will saw to a contour, external or internal; and it can be arranged for cutting off.

Types of Saw Bands. Proper choice of saw bands depends on work material, cutting speed, and finish desired. Six types of saw bands (Fig. 6) are:

1. Standard pitch, or "precision," as in Table 5, is used in raker and wave set for

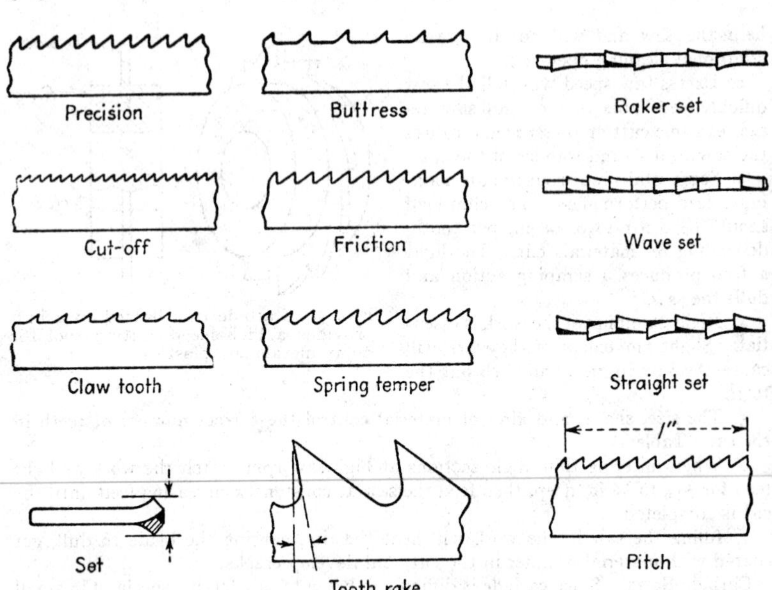

FIG. 6. Six types of band-saw bands are employed for cutting metals, plastics, wood, etc., and three types of set are encountered.

relatively low cutting speeds on ferrous and nonferrous metals, and in some instances on woods and plastics.

2. Buttress, or skip-tooth, coarse-pitch bands are used in raker set for heavy work thicknesses of ferrous metals, also fast cutting of wood, plastics, and nonferrous metals.

3. Claw-tooth bands are made in the same pitches as buttress blades but have a positive 10° rake angle for free-cutting action on light metals, steel, wood, and plastics.

4. Spring-temper metal-cutting bands are coarse pitch for trimming light-metal castings when abrasive elements are not present. Teeth have a 10° positive rake and straight set.

5. Cutoff bands have pitches from 6 to 24 teeth per inch, a raker set, and are used for production cutting of steel, aluminum, brass, copper, etc.

6. Friction bands have 10 or 12 teeth per inch, the teeth are specially heat-treated, and the band is operated at speeds up to 15,000 fpm to cut hard metals and alloys by a friction-softening and abrading process.

Standard bands are listed in Table 6.

Set Pattern. Set is the amount of bend given the teeth in order to create side clearance for the back of the band, when cutting through material. Three set patterns are, as shown in Fig. 6: (1) raker set—one unset tooth followed by two oppositely set teeth; (2) wave set—one group of teeth set to the right, the next group to the left, etc.; and (3) straight set—all teeth set symmetrically one to the left, the next to the right, etc.

Table 5. Guide for Band Sawing Various Materials*

Material	Thickness, In.	Saw Type† and Pitch	Speed, Fpm	Feed	Lubricant
Aluminum alloys:					
Sheet, plate, rod, press forgings, and tubing—2S, 3S, 4S, 11S, 17S	0–¼	P–18	6000	L	470
	¼–½	P–10	5000	L	470
Sheet, rod, tubing, shapes—24S, 52S	½–1	B–4	4000	L	470
Sand and permanent-mold castings—122, 214, A-214, 218, 220	1–3	C–3	3500	M	470
	3–6	C–3	3000	H	470
Die, sand, and permanent-mold castings—13, 43, 85, 108, A108, 113, C113, A-132, 138, 152, B-195, 212, 355, 356, 360, 380. Forgings—32S, A51S	0–¼	P–18	1800	L	120
	¼–½	P–10	1400	L	120
	½–1	B–4	800	L	120
	1–3	C–3	600	M	120
	3–6	C–3	300	H	120
Forgings and extruded and rolled shapes—14S, 18S, 25S, 53S	0–¼	P–18	3000	L	470
	¼–½	P–10	2800	L	470
Sheet, plate, tubing, bars, shapes—61S, 63S, 75S	½–1	B–4	2600	L	470
	1–3	C–3	2400	M	470
Sand and permanent-mold castings—142, 195, 750	3–6	C–3	2000	H	470
Magnesium alloys:					
Bars, rods, and shapes—SAE 52, 520, 522	0–¼	P–10	6000	L	M oil
	¼–½	P–6	5000	L	M oil
Sheets and plates—SAE 510, 511, 51	½–1	P–6	4000	L	M oil
	1–3	C–3	3000	M	M oil
	3–6	C–3	2000	M	M oil
Sand castings—SAE 500, 50	0–¼	P–14	4000	L	M oil
Die castings—SAE 501	¼–½	P–10	3500	L	M oil
Permanent-mold castings—SAE 502, 503	½–1	P–6	3000	L	M oil
	1–3	C–3	2500	M	M oil
Forgings—SAE 531, 532, 533	3–6	C–3	2000	M	M oil
Copper alloys:					
Copper, commercial brass sheet (SAE No. 70)—¼, ½, ¾, and hard tempers	0–¼	P–10	4000	L	None
	¼–½	P–6	3500	L	None
	½–1	B–3	3000	L	None
	1–3	B–3	2500	M	None
	3–6	B–3	2000	M	None
SAE No. 70 commercial brass sheet—extra hard, spring, and extra spring tempers	0–¼	P–10	3500	L	None
	¼–½	P–6	3000	L	None
	½–1	B–3	2500	L	None
Cartridge brass, naval brass, cast low brass, red brass, yellow brass, Muntz metal, SAE No. 64 phosphor bronze	1–3	B–3	2000	M	None
	3–6	B–3	1500	M	None

For footnotes see end of table.

TABLE 5. GUIDE FOR BAND SAWING VARIOUS MATERIALS* (*Continued*)

Material	Thickness, In.	Saw Type† and Pitch	Speed, Fpm	Feed	Lubricant
Aluminum bronze, cupronickel, copper-graphite, silicon bronze, manganese bronze	0–¼	P–18	1200	L	120
	¼–½	P–14	900	M	120
	½–1	P–10	600	M	120
	1–3	P–6	400	H	120
	3–6	B–3	200	H	120
Iron:					
Ductile irons—3.5% carbon	0–¼	P–18	150	M	None
	¼–½	P–14	125	M	None
	½–1	P–10	100	H	None
	1–3	P–6	80	H	None
High-test cast iron	0–¼	P–18	150	M	None
	¼–½	P–14	135	M	None
	½–1	P–10	125	H	None
	1–3	P–6	110	H	None
Kirksite	0–¼	P–18	200	L	120
	¼–½	P–10	175	L	120
	½–1	P–8	160	M	120
	1–3	P–6	150	H	120
	3–6	B–3	150	H	120
Pregwood	0–¼	P–14	6000	L	None
	¼–½	P–10	5000	L	None
	½–1	P–6	4000	L	None
	1–3	B–3	3500	M	None
	3–6	B–3	3000	H	None
Plastics:					
Phenol-formaldehyde cast compound (no filler), or with filler materials like wood flour or fabric sisal	0–¼	P–14	6000	L	None
	¼–½	P–10	5000	L	None
	½–1	P–6	4000	M	None
	1–3	B–3	3500	M	None
	3–6	B–3	3000	H	None
Phenol-formaldehyde cast compound—asbestos filler	0–¼	P–18	4000	L	None
	¼–½	P–14	3000	M	None
	½–1	P–10	2500	M	None
	1–3	C–4	2000	H	None
	3–6	C–3	1500	H	None
Phenol-formaldehyde cast compound—laminated with glass, mica, mineral	0–¼	P–14	150	M	None
	¼–½	P–10	75	M	None
	½–1	P–6	50	H	None
	1–3	B–3	50	H	None
	3–6	B–3	....	..	None

For footnotes see end of table.

TABLE 5. GUIDE FOR BAND SAWING VARIOUS MATERIALS* (*Continued*)

Material	Thickness, In.	Saw Type† and Pitch	Speed, Fpm	Feed	Lubricant
Cellulose acetate	0–¼	P–10	4500	L	470
	¼–½	P–6	3500	L	470
	½–1	P–4	2500	L	470
	1–3	B–3	1800	L	470
	3–6	B–3	1500	M	470
Polystyrene	0–¼	P–10	2500	L	470
	¼–½	P–6	2000	L	470
	½–1	P–4	1500	L	470
	1–3	B–3	1500	M	470
	3–6	B–3	1000	M	470
Nickel alloys:					
Nickel-chrome alloys—(Chromel A, Cimet, Cooper alloy, Inconel, Misco)	0–¼	P–18	70	L	470
	¼–½	P–14	60	M	470
	½–1	P–10	50	H	470
	1–3	P–6	50	H	470
	3–6	P–6	50	H	470
Nickel alloys—(Hastelloy, Illium, and Parr metal)	0–¼	P–18	70	M	240
	¼–½	P–14	60	H	240
	½–1	P–6	50	H	240
	1–3	B–4	50	H	240
	3–6	B–3	50	H	240
Nickel silver	0–¼	P–18	300	L	120
	¼–½	P–14	250	L	120
	½–1	P–10	200	M	120
	1–3	P–6	150	M	120
	3–6	B–3	150	M	120
Monel metal	0–¼	P–18	150	M	470
	¼–½	P–12	125	M	470
	½–1	P–10	75	H	470
	1–3	P–6	50	H	470
	3–6	B–4	50	H	470
SAE steels:					
Carbon steels—1006–1030	0–¼	P–24	250	M	240
	¼–½	P–14	200	M	240
	½–1	P–10	175	H	240
	1–3	P–6	150	H	240
	3–6	B–4	150	H	240
Medium and high carbon—1035–1095	0–¼	P–24	200	M	240
Free-cutting 1112, 1115, X1315	¼–½	P–14	150	M	240
Manganese—X1330–X1340; 1330–1335	½–1	P–10	125	H	240
Exceptions: high carbon—cuts at 80 fpm for 3–6 in. thick; free-cutting manganese at speeds 25 fpm higher for ½–6 in. thick.	1–3	P–6	100	H	240
	3–6	B–4	100	H	240

For footnotes see end of table.

TABLE 5. GUIDE FOR BAND SAWING VARIOUS MATERIALS* (*Continued*)

Material	Thickness, In.	Saw Type† and Pitch	Speed, Fpm	Feed	Lubricant
Manganese steels—1340, 1350	0–¼	P–18	150	M	240
Nickel steels—2315–2350	¼–½	P–14	125	M	240
	½–1	P–12	100	H	240
	1–3	P–8	75	H	240
	3–6	B–3	75	H	240
Nickel-chromium steels—3115, 3120	0–¼	P–18	175	M	240
	¼–½	P–14	150	M	240
	½–1	P–12	100	H	240
	1–3	P–8	75	H	240
	3–6	B–4	75	H	240
Nickel-chromium steels—3215–3250	0–¼	P–24	125	M	240
	¼–½	P–18	75	M	240
	½–1	P–14	65	H	240
	1–3	P–8	50	H	240
	3–6	P–6	50	H	240
Molybdenum steels—4140, 4145, 4150, 4320, 4340	0–¼	P–18	125	M	240
	¼–½	P–14	100	M	240
	½–1	P–10	75	H	240
	1–3	P–6	50	H	240
	3–6	B–3	50	H	240
Chrome-vanadium steels—6130–6150	0–¼	P–18	175	M	240
	¼–½	P–14	100	M	240
	½–1	P–12	75	H	240
	1–3	P–8	50	H	240
	3–6	B–4	50	H	240
Stainless steels:					
30–302, 30–304, 30–309, 30–316, 30–347, 51–410, 51–430	0–¼	P–18	100	M	470
	¼–½	P–14	80	H	470
	½–1	P–10	70	H	470
	1–3	P–6	60	H	470
	3–6	C–3	50	H	470
30–303, 51–416	0–¼	P–18	140	M	470
	¼–½	P–14	125	H	470
	½–1	P–10	110	H	470
	1–3	P–6	100	H	470
	3–6	C–3	75	H	470
High-speed steels					
	0–¼	P–18	125	L	240
	¼–½	P–14	90	M	240
	½–1	P–10	70	M	240
	1–3	P–6	50	H	240
	3–6	B–4	50	H	240

For footnotes see end of table.

TABLE 5. GUIDE FOR BAND SAWING VARIOUS MATERIALS* (*Continued*)

Material	Thickness, In.	Saw Type† and Pitch	Speed, Fpm	Feed	Lubricant
Tool and die steels:					
Oil-hardening, nondeforming—(DoAll, Ketos, Mansil, Paragon, Stentor)	0–¼	P–18	120	M	240
	¼–½	P–10	110	M	240
	½–1	P–6	100	H	240
	1–3	B–3	80	H	240
	3–6	B–3	70	H	240
High-carbon, high-chrome (Huron, Ontario, Superdie)	0–¼	P–24	125	M	240
	¼–½	P–18	100	M	240
	½–1	P–10	75	H	240
	1–3	P–6	50	H	240
	3–6	B–4	50	H	240

* Abstracted from "DoAll Band Tool Manual," by permission of The DoAll Co., Des Plaines, Ill.

† Saw types: P = precision. B = buttress. C = claw. Pitch = No. teeth per in.; for example, 18.

TABLE 6. STANDARD METAL-CUTTING BAND SAWS*

Width, In.	Thickness, In.	No. of Teeth per In.
Regular-type Band Saws		
1/16	0.025	24, 32
3/32	0.025	18, 32
⅛	0.025	14, 18, 24
3/16	0.025	10, 14, 18, 24, 32
¼	0.025	10, 12, 14, 18, 24, 32
⅜	0.025	8, 10, 14, 18, 24
½	0.025	6, 10, 14, 18, 24
⅝	0.032	8, 10, 14, 18, 24
¾	0.032	6, 8, 10, 12, 14, 18
1	0.035	6, 8, 10, 14
Skip-tooth Band Saws		
¼	0.025	4, 6
⅜	0.025	3, 4
½	0.025	3, 4
¾	0.032	3
1	0.035	2, 3

* U.S. Commodity Standard R214-48

Kerf Size. Width of kerf is determined by the set, pitch, band velocity, and feeding rate. A kerf approximately equal to the over-all set is produced by a narrow set, fine pitch, high speed, and slow feed. A narrower kerf is produced by coarse pitch and rapid feed.

Saw Width and Gage. For accuracy of cutting, saw-band stability must be maintained between the upper and lower guides. Use the thickest and widest blade possible, considering the radiuses to be cut and the diameter of the band wheels. Wider and thicker bands take heavier feeding force than narrower and thinner bands. Approximate thickness of band should be 0.001 to 0.00175 in. per in. of wheel diameter.

In radius cutting, the back of the saw band may be moved to either side of the kerf an amount equal to the set dimension on one side of the band. Width of saw band vs. minimum radius cut depends to some extent on the type of band:

Width of Saw Band	Min Radius Cut		
	Precision	Buttress	Spring Temper
1/16	Sq.		
3/32	1/16		
1/8	1/8		
3/16	5/16	5/16	1/2
1/4	5/8	5/8	3/4
3/8	1 7/16	1 7/16	1 1/2
1/2	2 1/2	2 1/2	2 1/4
5/8	3 3/4	....	3
3/4	5 7/16	5 7/16	4 1/2
1	7 1/4	7 1/4	7 1/2

Coolants and Lubricants. As with other metal-cutting tools, the performance of a band saw and the finish produced are aided by the use of coolants and lubricants. Materials such as 17 ST and 24 ST aluminum require only a coolant (for example, DoAll 470), whereas 14 ST and A 132 require a combination of coolant and lubricant, the latter added by spray at the point of cutting. Tool steels, mild steels, and conventional alloys are best cut with an agent having lubricating quality, like cutting oils.

Feeding Pressure. The fewer the teeth, the less the feeding pressure required. But there are three job requisites to be considered. For maximum results follow these suggestions:

MAXIMUM CUTTING RATE. Use fast speed, coarse pitch, and fast feed.

MAXIMUM TOOL LIFE. Use slower speed, finer pitch, and medium feed.

BEST FINISH. Use fast speed, fine pitch, and slow feed.

The recommendations in Table 5 represent a compromise between cutting rate, tool life, and finish, to give practical results in service. You can force a saw band to its ultimate to get high production for a short time, and get low tool life and a rough finish. You may neglect to use a coolant, and the finish suffers. The recommendations provided are a norm and should not be abandoned unless you have good reasons for so doing.

Conservative cutting rates are given in Table 7.

TABLE 7. BAND-SAW CUTTING RATES, LIN IN. PER MIN*

Standard-pitch, or "Precision," Saw Bands

Work Thickness, In.	Tool Steel	Cold-rolled	Cast Iron	High-carbon High-chrome Steel
1/4	4 1/2	9	16	2 1/4
1/2	2 1/8	4	7 1/2	1
1	1	1 3/4	3 1/4	1/2
1 1/2	5/8	1 1/8	2 1/8	1/4
3	5/16	1/2	1	1/8
6	1/8	1/4	7/16	1/32

Buttress Saw Bands

Work Thickness, In.	Wood (Soft)	Wood (Hard)	Plastic (Thermo-plastic) "Non-abrasive"	Plastic (Thermo-setting) "Non-abrasive"	Aluminum, Magnesium (Low Fe, Si, and Ni)	Carbon (Less Than 35 Scleroscope Hardness)
1	400–600	300–500	40–100	140–300	160–240	300–400
2	200–300	175–240	20–50	70–140	80–120	150–200
4	120–160	76–130	10–25	36–65	40–60	70–100
6	80–105	65–85	7–17	24–45	27–40	45–65
8	60–80	48–64	5–13	18–36	20–30	35–50

* The DoAll Co.

FRICTION SAWING

Two types of friction sawing are performed in industry: friction band sawing and friction circular sawing. In either case the principle is the same: the band or saw blade traveling at high speed while in frictional contact with the work develops heat faster than the material can absorb it, and the material is softened. The teeth in the tool act as small scoops to remove the softened metal in the form of "sparks." These are minute oxidized chips. The temperature reached in friction sawing is above red heat but below the melting point.

Friction Band Sawing. The blade used is shown in Fig. 6. Saw teeth need not be sharp. In fact, the saw cuts best after it has been dulled by use. Most cutting is done at speeds between 6000 and 14,000 fpm (see Table 8). Thicknesses above 1 in. can be cut, but special training and skill are considered necessary.

Friction Circular Sawing. This form of friction sawing has been known for over 125 years. A blade with a smooth rim can be used for cutting all the high-carbon steels, in hardened or annealed conditions, but for low-carbon steel a toothed rim as shown in Fig. 7 is necessary.

Efficient rim speeds are from 20,000 rpm for small blades to 28,000 rpm for largest ones. At such speeds, the rim will develop sidewise flutter unless rim tension is developed by peening both sides of the blade from center to within a few inches of the rim. Also cooling water is provided to reduce burr and cutting pressure. Usually, several streams at pressures up to 500 psi are required.

Smooth-rim blades which cut steels of 0.60 to 1.00% carbon are dressed after 8 to 10 hr of cutting, by removal of the rounded corners, about 1/32 in. of the rim.

TABLE 8. FRICTION BAND-SAW CUTTING RATES, LIN IN. PER MIN*

Materials	Saw Velocity Work Thickness			Saw Pitch Work Thickness			Approx Lin Cutting Rate, In. per Min Work Thickness				
	1/16–1/4 In.	1/4–1/2 In.	1/2–1 In.	1/16–1/4 In.	1/4–1/2 In.	1/2–1 In.	1/16 In.	1/4 In.	1/2 In.	3/4 In.	1 In.
Carbon steel—1010–1095	6,000	9,000	12,500	14	10	10	1,400	60	30	8	6
Manganese steel—T1330–1350	6,000	9,000	12,500	14	10	10	1,200	55	25	7	5
Free-machining steel—X1112–X1340	6,000	9,000	12,500	14	10	10	1,400	60	25	7	5
Nickel steel—2015–2515	8,000	12,000	14,000†	14	10	10†	800	45	20	5	3
Nickel chromium steel—3115–3415	8,000	12,000	14,000†	14	10	10†	800	45	18	5	3
Molybdenum steel—4023–4820	8,000	12,000	14,000†	14	10	10†	1,000	50	20	6	3
Chromium steel—5120–5150	8,000	12,000	14,000†	14	10	10†	800	45	18	5	3
Chromium steel—51210–52100	8,000	12,000	14,000†	14	10	10†	800	40	15	5	3
Chromium vanadium—6115–6195	8,000	12,000	14,000†	14	10	10†	700	40	15	5	3
Tungsten steel—7260–7360	8,000	12,000	14,000†	14	10	10†	600	38	12	4	3
Silicon steel—9255–9260	8,000	12,000	14,000†	14	10	10†	600	45	16	4	3
Armor plate	7,000	10,000	13,500	14	10	10	1,200	60	30	6	5
Stainless steel	7,000	10,000	13,500	14	10	10	1,200	60	30	8	6
Cast steel	7,000	10,000	13,500	14	10	10	1,000	50	16	7	5
Gray cast iron	7,000	10,000	13,500	14	10	10	1,000	55	20	7	5
Malleable cast iron	7,000	10,000	13,500	14	10	10	1,000	55	20	7	5
Meehanite castings	7,000	10,000	13,500	14	10	10	1,000	55	20	7	5

* The DoAll Co.
† Use rocking technique on maximum thickness.

TABLE 9. PERFORMANCE OF FRICTION SAWS WITH BLADE PRESSURE IN LINE WITH STROKE

Blade dia, in.	24	36	48
Rim thickness, in.	5/32	7/32	9/32
Blade speed, rpm	3500	2375	1750
Motor load, hp	25	50	75
Av cutting time, sec:			
1-in. solid round	2	1.3	1
2-in. solid round	9	6	5
3-in. solid round	22	14	11
4-in. solid round	....	27	22
5-in. solid round	....	46*	38
6-in. solid round	....	....	60
7-in. solid round	....	....	92*

Hollow-ground, hobbed, average-sharp blade. Constant-speed motor, providing uniform blade velocity.

* Max contact area for complete heat dissipation and no metal adherence to blade. Higher feed pressures permissible with shorter contact areas.

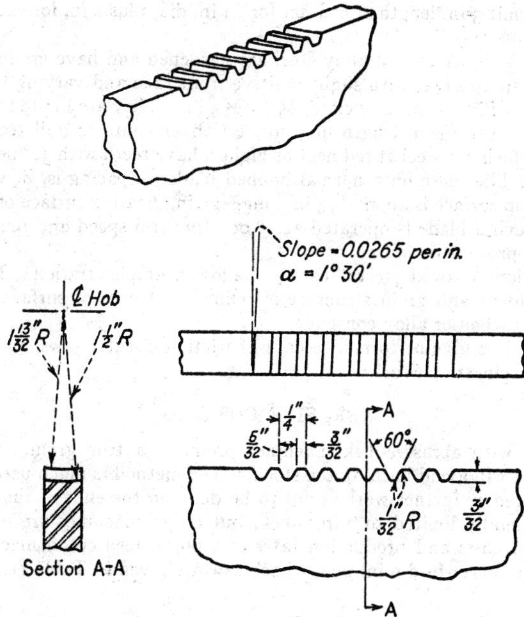

Fig. 7. "Hobbed" teeth are cut in the rim of a circular friction saw intended for low-carbon steels. These teeth scoop the friction-softened metal from the cut.

Table 10. Performance of Friction Saws with Adjustable Horizontal Feed Angle of Contact with Work below Center

Blade dia, in.	24	46	48	56	58
Blade thickness, in.	5/32	1/4	9/32	3/8	3/8
Blade speed, rpm	3500	1750	1750	1750	1750
Motor rated hp	10	30	40	75	125
Av motor cutting hp	13	39	52	97.5	162.5
Av cutting time, sec:					
8-in. × 25.5-lb I-beam	46	21.1	17.3	11.5	6.9
12-in. × 55-lb I-beam	....	47.2	38.7	25.8	15.3
18-in. × 70-lb I-beam	....	....	48.8	32.5	19.4
24-in. × 100-lb I-beam	....	....	....	42.0	28.1
8-in. × 13.75-lb channel	24.3	11.1	9.1	6.1	3.7
12-in. × 30-lb channel	....	24.8	20.5	13.6	8.2
15-in. × 50-lb channel	....	....	34.6	23.0	13.8

Av motor load not over 130% rating, peak not over 200%, duty cycle not over 50% of total cycle time, constant-speed motor. Feed rate varies directly with blade pressure, motor load with change of pressure angle; thus raising work to horizontal center line reduces length of contact, motor load, and blade pressure; raises efficiency.

On low-carbon steels, the blade rim tends to glaze smooth rather than to roughen. Therefore, teeth are "hobbed" in blanks with not over 0.4 C. These must be hollow-ground, but not edge-hardened. Such blades are operated at 24,000 fpm rim speed, and permit 3-in. length of contact for 24 in. dia, plus 2 in. for each foot additional saw diameter.

Tube saws of higher carbon alloy steel are hardened and have ground teeth like those on slow-speed saws, with slight positive front rake and varying top rake and length of land. Pitch is usually close, 1/4 to 5/16 in. They are limited to small sections with not over 3/4-in. length of contact (longer contacts dull teeth rapidly).

Hot saws, which cut steel at red heat or higher, have teeth with 30° negative front and rear angle like those on a normal hobbed blade. Spacing is, however, 3/4 in. or more and top surface is about 1/16 in. long—giving heating surface of only 8% of periphery. Such a blade is operated at 18,000 fpm rim speed and must be cooled adequately to prevent buildup.

Edges of friction-sawed pieces sustain shallow heat penetration. No hardness cracks are evident and grain structure is refined. Degree of surface hardness is dependent on carbon or alloy content.

Tables 9 and 10 give performance data of friction circular saws, as compiled by Kling Bros. Engineering Works.

ABRASIVE CUTTING

Cutting off with abrasive-disk machines produces a true ground surface, and when done wet will avoid burning the stock. The method is much used when good fit-up is required or further work is not to be done on the ends of the piece. The method is generally limited to 2-in. stock, but larger machines are made. Comparative cutoff costs and production rates of a power-feed cutoff machine of 6 in. capacity and a power-feed 6-in. power hack saw are given in Table 11.

Table 11. Comparative Production of Abrasive Cutoff and Power Hack-saw Machines

Rod Diameter and Material	Time, Sec per Cut	Approx No. Cuts per Wheel	Approx Cost per Cut
Abrasive Machine			
1 in. SAE 1035	10	610	0.006
2 in. SAE 1035	50	169	0.02
3 in. SAE 1035	150	52	0.07
4 in. SAE 1035	210	35	0.10
6 in. SAE 1035	360	10	0.36
Modern Hack Saw			
1 in. SAE 1035	15	3700	0.0004
2 in. SAE 1035	50	950	0.0014
3 in. SAE 1035	90	425	0.003
4 in. SAE 1035	140	240	0.006
6 in. SAE 1035	300	105	0.013

Blade—High-speed edge, composite-welded, hack-saw blade 14 in. long × 1 1/4 in. × 0.068 in., 6 teeth per in. Approximate cost—$1.40 each.

Wheel—20 in. dia × 1/8 in. thick. Approximate cost—$3.60 each.

Section 10

THREADING PROCESSES AND THREAD SYSTEMS

SECTION 10

THREADING PROCESSES AND THREAD SYSTEMS

Production of threads is a major machining problem. Threading is an intricate forming operation. It ties together a number of dimensions in such a manner that, if one is wrong, the whole result is wrong. Other complicating factors are the various thread forms (Fig. 1) used in industry, both standard and special (see further illustrations and dimensional data starting on page 10-29), the several classes of fit that may be required, the selection of the proper threading process and tools, and the choice of the inspection method.

THREADING PROCESSES

Threads are produced by six machining and forming processes:

1. Single-point cutting.
2. Solid adjustable, or button, dies, and acorn dies.
3. Self-opening die heads and chasers.
4. Thread rolling.
5. Thread milling and hobbing.
6. Thread grinding.

Single-point Cutting

Thread cutting with a single-point tool on a lathe is obtained by a combination of rotary motion of the workpiece and longitudinal motion of the carriage. Motion of the carriage is provided by the lead screw. Rotary speed of the lead screw relative to workpiece speed is determined by the change gears within the headstock. Most engine lathes are "geared even," so that, if equal gears are placed on the stud and the lead screw, the machine will cut a thread of the same pitch as the lead screw. If the lathe will not do this, find the pitch cut with even gears and consider this as the real pitch of the lead screw.

Change-gear Selection. The lead screw commonly has eight threads per inch. To cut a "faster" thread (fewer threads per inch), the lead screw must turn faster than the work. Put the larger gear on the stud and the smaller one on the lead screw. To cut a slower thread (finer pitch) the smaller gear goes on the stud and the larger gear on the lead screw.

Example: Given a lead screw with 8 threads per inch, and we must cut a thread with 16 threads per inch. What change gears should be used?

Solution: The ratio of lead screw to work is 8:16. If the change gears in the set vary by 4 teeth, multiply the ratio by 4:

$$\frac{8}{16} = \frac{8 \times 4}{16 \times 4} = \frac{32}{64}$$

Put the 32-tooth gear on the stud and the 64-tooth gear on the lead screw.

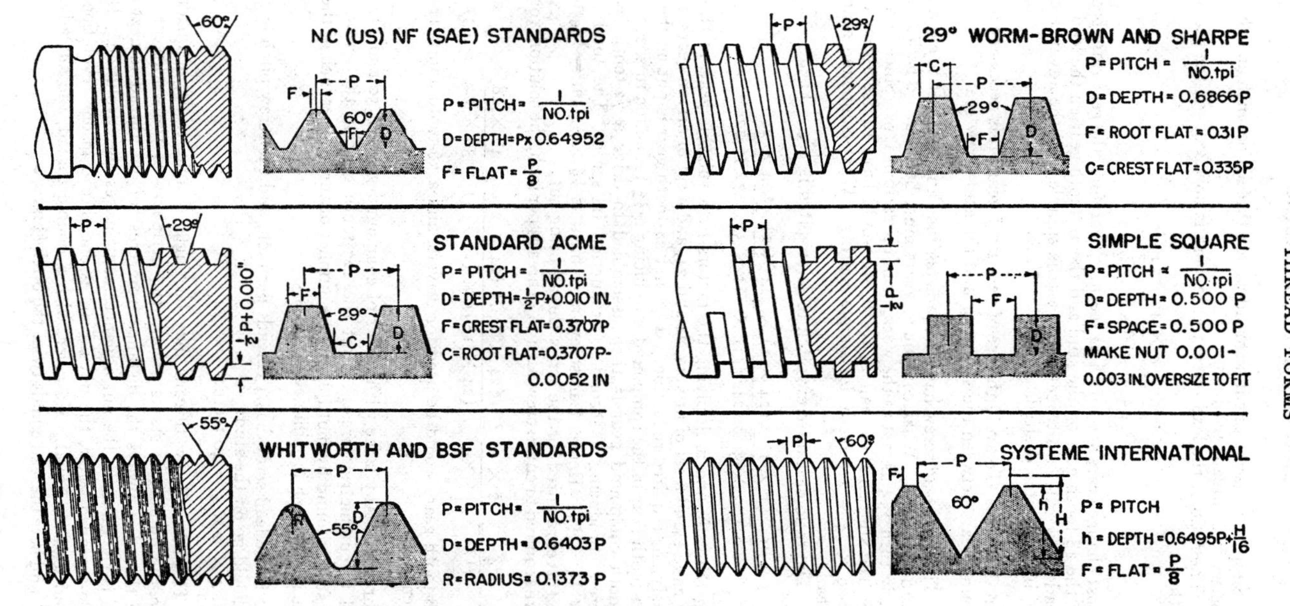

FIG. 1. Most of these thread forms can be cut with die heads, or ground, or rolled, but the unmodified square thread shown can be produced only with a lathe tool.

EXAMPLE: Suppose we want to cut a thread with 4 threads per inch and the pitch of the lead screw is 8 threads per inch.

SOLUTION: The ratio of the lead screw to the work is now 8:4. If the gears vary by 4 teeth, then

$$\frac{8}{4} = \frac{8 \times 4}{4 \times 4} = \frac{32}{16}$$

However, most change gears start with 24 teeth. Therefore, the gear ratio becomes 48:24. Put the 48-tooth gear on the stud and the 24-tooth gear on the lead screw. Thus, the lead screw will turn twice as fast as the spindle, and the advance equals $2 \times \frac{1}{8} = \frac{1}{4}$ in., the pitch desired.

Compound Gearing. To avoid the use of very large gears, it is necessary to use a set of compound gears between the stud gear and the lead-screw gear.

EXAMPLE: The lathe lead screw has 4 threads per inch and we want to cut a screw with 36 threads per inch. What compound gears must be used? The lead screw must obviously turn $\frac{4}{36}$ as fast as the work, or in the ratio of 1:9. Since we seldom use a gear on the stud smaller than 24 teeth, we would have to use a gear on the lead screw of $9 \times 24 = 216$ teeth, for simple gearing. What compound gears can be used instead?

SOLUTION: The ratio of lead-screw pitch to screw pitch is 1:9. Then

$$\frac{1}{9} = \frac{1 \times 1}{3 \times 3} = \frac{(24 \times 1) \times (24 \times 1)}{(24 \times 3) \times (24 \times 3)} = \frac{24 \times 24}{72 \times 72}$$

By factoring the ratio and multiplying the factors by a suitable gear, in this case 24 teeth, we have found a set of drivers, 24 and 24, and a set of driven gears, 72 and 72, that satisfy the required reduction. The first 24-tooth gear is put on the stud, to drive a 72-tooth gear in the compound, producing a $\frac{1}{3}$ reduction in speed. The second 24-tooth compound gear is then assembled to drive a 72-tooth gear on the lead screw, producing a second $\frac{1}{3}$ reduction in speed. The total reduction $= \frac{1}{3} \times \frac{1}{3} = \frac{1}{9}$. Hence, lead-screw movement is $\frac{1}{9} \times \frac{1}{4}$ in. $= \frac{1}{36}$ in., or the carriage travel required for the screw thread being cut.

If you were required to cut a screw with $1\frac{1}{2}$ threads per inch, the solution would be

$$\frac{4}{1\frac{1}{2}} = \frac{2 \times 2}{1\frac{1}{2} \times 1} = \frac{(24 \times 2) \times (24 \times 2)}{(24 \times 1\frac{1}{2}) \times (24 \times 1)} = \frac{48 \times 48}{36 \times 24}$$

Here two 48-tooth gears would be used for drivers, and a 36-tooth and a 24-tooth gear as driven gears.

Metric Threads. Any lathe with a pair of compound gears of 50 and 127 teeth can be used to cut metric threads. Put the 127-tooth gear on the screw. The gear for the stud is found by multiplying the number of threads per inch of the lead screw by the screw pitch in millimeters times 5.

EXAMPLE: If the lead screw has 4 threads per inch, and the screw pitch is 2 mm, what gear should go on the stud?

SOLUTION: $4 \times 2 \times 5 = 40$ teeth.

Multiple Threads. Screw threads may be single, double, triple, or quadruple, according to the number of starts around the periphery of the workpiece. The *pitch* of the thread is always the distance from crest of one thread to next crest, but the *lead* is the distance the nut advances as it makes one revolution. Hence, with a single-threaded screw, the nut advances one thread in one revolution, and the pitch and lead are equal. However, with a double thread, the nut advances two threads, and here the lead equals twice the pitch (Fig. 2).

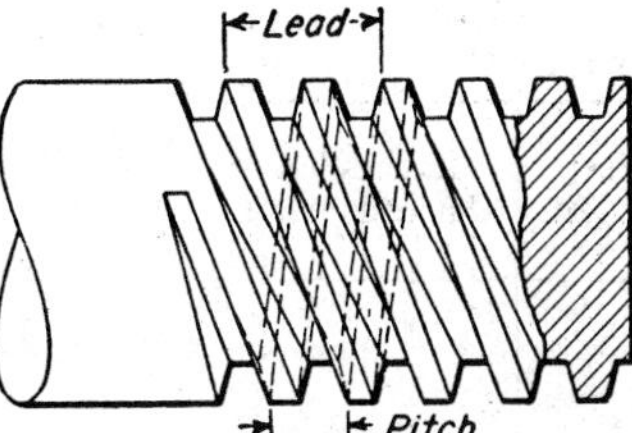

FIG. 2. Multiple-screw thread having two grooves (double thread). The lead is twice the pitch. Rule: multiple-thread lead = pitch × No. of starts.

Multiple threads are cut in two ways.

METHOD 1. BY DIVIDING THE FACEPLATE. In this method, slots are milled for the number of starts desired, or dowel pins inserted. The lathe dog is clamped to the work, set in the first slot, and a thread is cut. Then, without unclamping the dog from the work, the dog is set in the next slot and another thread is cut. This process is repeated until all desired threads are cut. Special precision faceplates are available for this operation.

Figure 3 shows a faceplate fixture used for various numbers of threads. On an ordinary driving plate is fitted a plate having, as shown, twelve holes enabling one to get two, three, four, or six leads if required. This ring carries the driving stud and is clamped at the back of the plate by two bolts as an extra safeguard. All that is necessary in operation is to slack off the bolts, withdraw the index pin, move the plate number of holes required, and retighten the bolts. It is used on different lathes as occasion requires, by making the driving plates alike and drilling a hole for the index pin. It is found that the index pin works best when made taper, and a light tap is sufficient to loosen or fix it. Multiple threads can all be cut at the same time using a properly located tool for each lead.

METHOD 2. BY RELEASING THE CHANGE GEARS OR LEAD SCREW FEED NUT FOR EACH START. The work is marked around the periphery for number of starts. The nut is engaged; the first thread is cut. The nut is released (the spindle being stopped) and the carriage is moved to the start for the next thread. The nut is reengaged, the next thread is cut, and so on. Table 1 gives the carriage movement

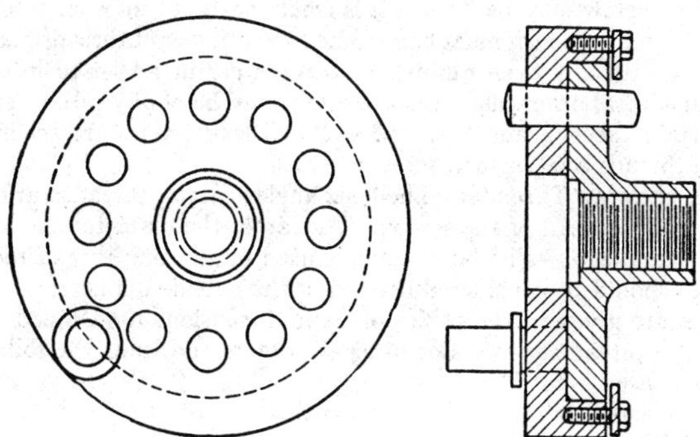

FIG. 3. Faceplate for multiple-thread cutting.

TABLE 1. CARRIAGE MOVEMENT BETWEEN STARTS WHEN MULTIPLE-THREAD CUTTING

Threads per In.	Thread Multiple	Carriage Movement, In.	Lathe Screw Threads per In.
$1\frac{1}{2} = \frac{3}{2}$	2	1	Any
$1\frac{1}{2} = \frac{3}{2}$	3	Cannot be split this way	
$1\frac{1}{2} = \frac{3}{2}$	4	$\frac{1}{2}$	Any even thread
$1\frac{3}{4} = \frac{7}{4}$	2	2	Any
$1\frac{3}{4} = \frac{7}{4}$	3	$1\frac{1}{3}$	6
$1\frac{3}{4} = \frac{7}{4}$	4	1	Any
$2\frac{1}{4} = \frac{9}{4}$	2	2	Any
$2\frac{1}{4} = \frac{9}{4}$	3	Cannot be split this way	
$2\frac{1}{4} = \frac{9}{4}$	4	1	Any
$2\frac{1}{2} = \frac{5}{2}$	2	1	Any
$2\frac{1}{2} = \frac{5}{2}$	3	$\frac{2}{3}$	3, 6, 9, 12
$2\frac{1}{2} = \frac{5}{2}$	4	$\frac{1}{2}$	Any even thread
$3 = \frac{3}{1}$	2	$\frac{1}{2}$	Any even thread
$3 = \frac{3}{1}$	3	Cannot be split this way	
$3 = \frac{3}{1}$	4	$\frac{1}{4}$	4

(by hand) between cuts. These data give a quick and sure method of starting the second, third, and fourth threads where the lead screw is of the pitch given in the table.

To determine whether multiple threads can be spaced by carriage movement, change the threads to be cut into a common fraction. Divide the numerator by the thread multiple. If it divides evenly, the thread cannot be spaced by moving the carriage. If it does not divide evenly, move the carriage the distance obtained by dividing the thread multiple into the denominator.

The carriage can, of course, be moved 1 in. and the nut closed, no matter what the pitch of the lead screw may be (unless it is fractional), but in order to close the nut after moving ½ in., the screw must have some even number of threads per inch.

A lead screw with any even number of threads per inch is used in a number of cases, while in several other instances the screw may be of any pitch—either odd or even. In certain cases, 4 and 8 thread per inch lead screws are specified; and in cutting triple threads a 6 per inch screw is usual.

Helix Angle of Screw Threads. The helix angle of screw threads varies with both the diameter and the lead of the screw. The larger the diameter the less the helix angle for the same lead. The helix angle is used in grinding thread tools, and the pitch diameter should be considered instead of the outside diameter.

Helix angles are given in the tables of basic dimensions for Unified coarse, fine, extra-fine, and 8-thread series. For other diameters and leads the following helix-angle formula can be used:

P = single threads per in. $1/P$ = lead = L.
D = pitch dia of work, in.
C = pitch circumference of work, in. $= \pi D$.

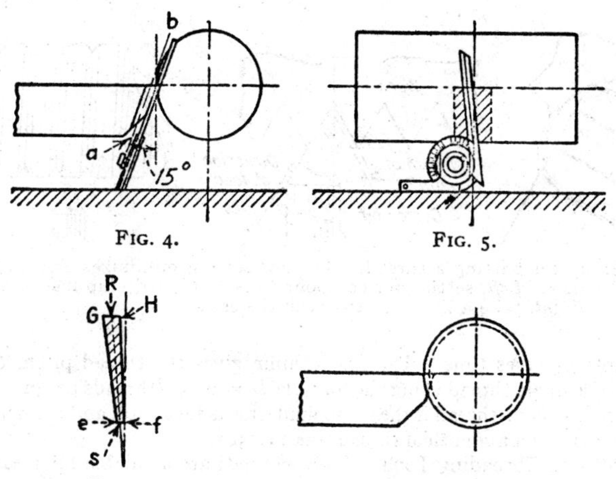

FIG. 4. FIG. 5.

FIG. 6. FIG. 7.

FIGS. 4 to 7. Use of protractor to establish proper side clearance on tool to cut right-hand thread.

$$\frac{L}{C} = \frac{\text{lead}}{\text{circumference of work}} = \frac{1/P}{D} = \text{tangent of helix angle}$$

Find angle in table of tangents.

The formula is given for single threads, but it can be used for double or triple threads by considering the lead equal to the advance of the work in one revolution instead of $1/P$.

Figures 4 and 5 show side and front elevations of the thread tool and of the protractor as applied to obtain the proper angle of side clearance to cut a right-hand screw thread. The front edge of the thread tool is used to determine the angle of side clearance. Figure 6 shows a section taken along the line *ab* (Fig. 4). It will be noticed that line *ef* is shorter than *GH* to give clearance to the cutting edges of the thread tool, and also that *GR* is equal to *HR*, and *es* is equal to *fs*. The thread tool must be set at the center as in Fig. 7 in order to cut a thread at the proper angle.

Setting the Single-point Tool. To cut 60° threads, set the V-shaped tool in the toolholder so that only a trifle more than the ground portion projects. Next, set the holder in the toolpost and adjust the toolbit point vertically to the exact center of the work. Place a center gage with its backedge in contact with the work and adjust the tool horizontally so that the tool point fits exactly into the proper angular notch in the front edge of the gage.

Cuts can be taken in two ways: (1) by infeeding the compound rest a few thousandths, and traversing the carriage by means of the lead screw (the compound rest being set at 0°); and (2) by setting over the compound rest 29 to 30°, as in Fig. 8, then setting the tool to the center gage, then feeding the carriage inward at the preset angle a few thousandths, and finally traversing the carriage. The second method, usually employing a tool with an included angle of 59° and top rake, is employed for ductile materials. Angular setting of the rest minimizes chatter and improves finish, because most of the chip load is toward the headstock.

When machining National Standard threads on an engine lathe, a thread-depth

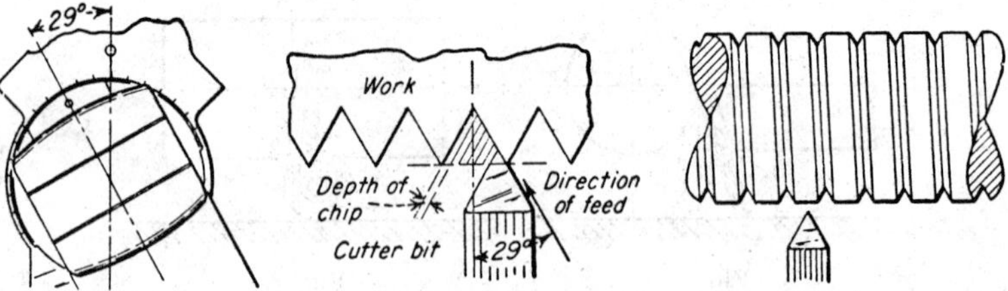

FIG. 8. Setup for cutting a thread. The 29° setting minimizes chatter, improves the thread finish. *Left*, setting for compound rest. *Center*, chip load is toward the headstock. *Right*, take a light cut to check the setup.

chart (Table 2) saves time. The first column gives the thread pitch, the second the depth of a single thread using the formula $D = 0.650/\text{threads per in.}$ The third gives depth of a single thread if the compound slide is set at 30°, and the depth setting on the feed dial is set at depth of thread times 1.1547.

Flat Width on Threading Tools. Two methods are available for measuring the width of flat, or point, on threading tools.

In the first method, a V-block is machined to the proper thread angle, and a hole is drilled through the back to receive a depth micrometer. The tool is placed in the V-block, and the drop A from the sharp V to the face of the tool is measured. The result (for a tool with 10° front clearance angle) is compared with the values given in Table 3.

EXAMPLE: A screw with 8 threads per inch is to be cut. What is the correct width of the flat? From the table, B = width of flat = 0.0156 in. when the drop $A = 0.0133$ in. The correct width of flat of any pitch = $p/8$, or $0.125 \div 8 = 0.0156$ in this example.

The second method employs a gear-tooth vernier caliper with depth vernier set to rest on the tool face, while the jaws contact the flanks at some distance h below the face of the tool. The general formula is

$$x = w - 2h\left(\tan\frac{C}{2}\right)$$

where x = width of flat, w = vernier measurement of jaw opening, h = setting of depth vernier, C = included angle of thread. If $h = 1/16$ in., then $2h\,(\tan C/2)$ becomes $\tan C/2 \div 8$. Values of this expression for various thread angles are:

Thread Angle, Deg	*Value of* $\tan C/2 \div 8$
29	0.0322
59	0.0707
60	0.0721

The general formula above is useful for measuring the crest width of any truncated cone or frustum.

Threading with Solid-adjustable Dies

Button Dies. Adjustable, round split (or button) dies (Fig. 9) are designed for use in hand stocks and screw machines. The button die is actually an adjustable split nut in which flutes have been machined and hardened, to provide cutting edges. Adjustment of pitch diameter is obtained by (1) three screws in the holder, one to

TABLE 2. DEPTH SETTINGS WHEN CUTTING THREADS ON ENGINE LATHE

Threads per In.	Compound Rest at		Threads per In.	Compound Rest at		Threads per In.	Compound Rest at		Threads per In.	Compound Rest at	
	0°	30°		0°	30°		0°	30°		0°	30°
80	0.0081	0.0094	24	0.0270	0.0312	8	0.0812	0.0935	2⅞	0.226	0.2610
72	0.0090	0.0104	20	0.0325	0.0375	7	0.093	0.1074	2¾	0.2363	0.2725
64	0.0101	0.0117	18	0.0361	0.0417	6	0.1083	0.1247	2⅝	0.2475	0.2858
56	0.0116	0.0134	16	0.0405	0.0468	5½	0.118	0.1363	2½	0.2595	0.2996
48	0.0135	0.0156	14	0.0465	0.0537	5	0.130	0.1501	2⅜	0.2735	0.3158
44	0.0147	0.0170	13	0.050	0.0577	4½	0.1445	0.1669	2¼	0.280	0.3233
40	0.0162	0.0187	12	0.0545	0.0629	4	0.1625	0.1876	2	0.325	0.3753
36	0.0180	0.0208	11	0.059	0.0681	3½	0.1855	0.2142			
32	0.0203	0.0234	10	0.065	0.0751	3¼	0.200	0.2309			
28	0.0234	0.0270	9	0.0722	0.0831	3	0.2116	0.2443			

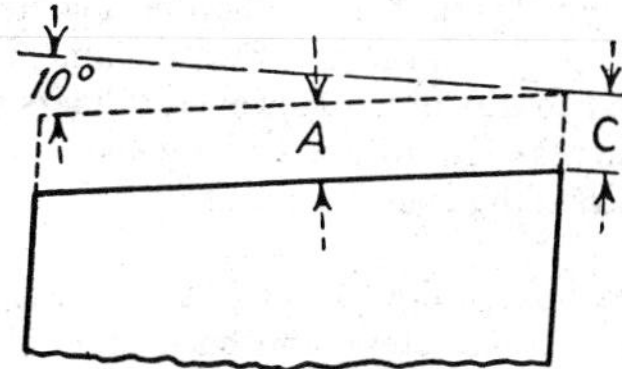

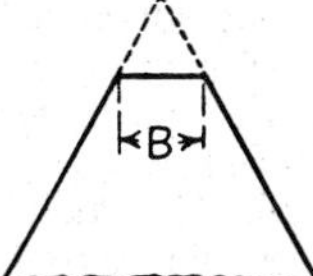

TABLE 3. WIDTH OF FLAT, OR POINT, ON THREADING TOOLS

Threads per In.	Pitch	*A*	*B*	*C*
4	0.2500	0.0266	0.0312	0.0270
5	0.2000	0.0213	0.0250	0.0216
6	0.1666	0.0177	0.0208	0.0180
7	0.1428	0.0152	0.0178	0.0154
8	0.1250	0.0133	0.0156	0.0135
9	0.1111	0.0118	0.0138	0.0120
10	0.1000	0.0106	0.0125	0.0108
11	0.0909	0.00963	0.0113	0.0098
12	0.0833	0.00886	0.0104	0.0090
13	0.0769	0.00818	0.0096	0.0083
14	0.0714	0.00758	0.0089	0.0077
16	0.0625	0.00673	0.0079	0.0068
18	0.0555	0.00588	0.0069	0.0059
20	0.0500	0.00530	0.0062	0.0054
24	0.0416	0.00433	0.0052	0.0045
32	0.0312	0.00332	0.0039	0.0034

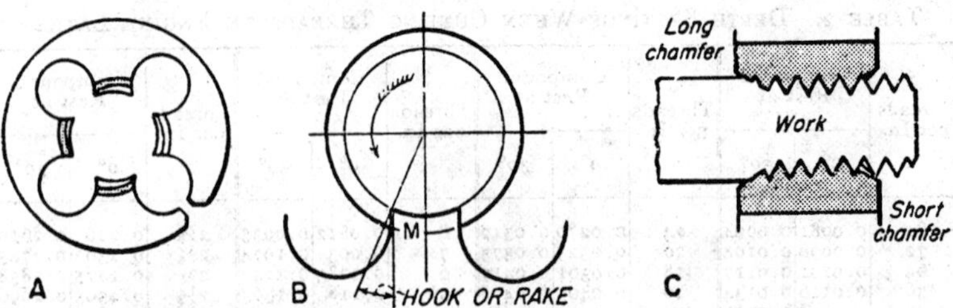

FIG. 9. Equal chamfering of a button die, as at *A*, provides balanced cutting. Usually cutting surface *M* at *B* is sharpened with a pencil wheel. Short chamfer at *C* is used for threading close to a shoulder; the long chamfer on all jobs where two incomplete threads can be tolerated.

enter the split for expanding and the two on either side for compressing; (2) a setscrew in the die to change its size. For the latter, a single-screw stock is used, although the die will fit in a three-screw stock as well.

As the die is solid it is nonreleasing; it must be backed off the newly cut thread or the workpiece must be passed completely through it. The die holder, for power threading, may be rotating or nonrotating. In lathe or screw machine, it is usually the latter, and the workpiece is reversed to unwind. However, holders are used in drillpresses or with threading attachments that rotate and reverse the die. The workpiece is either held stationary or threading speed is a differential between workpiece rpm and die rpm.

Lead-screw control will always help a button die to work better; it is practically essential for very fine threads, very small diameters, and hard or tough materials. Float should always be provided when threading the first piece, to allow the die to find its own center; then the die is clamped firmly. When threading coarser threads, using lead screw, a little float should be left in the holder. This will compensate for small differences between lead screw, work surface, and die lead.

Acorn or DuoCone Dies. The solid-adjustable spring die consists of a solid nut in which grooves have been machined to produce fingers (usually four) so pitch-diameter adjustments can be obtained by a collar that fits over, and draws in, the fingers. "Acorn" is the familiar shop term, a trade name of Greenfield Tap & Die. "DuoCone" is a comparatively recent development of Pratt & Whitney, different in design but similar in principle.

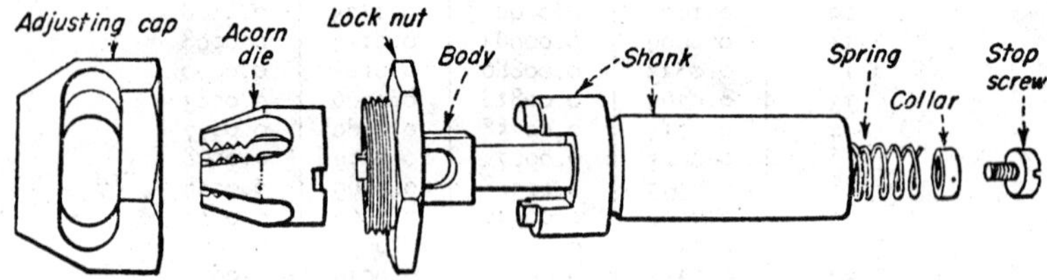

FIG. 10. Reversing, or regular, holders are made for both Acorn and DuoCone dies, but not for their interchangeable use.

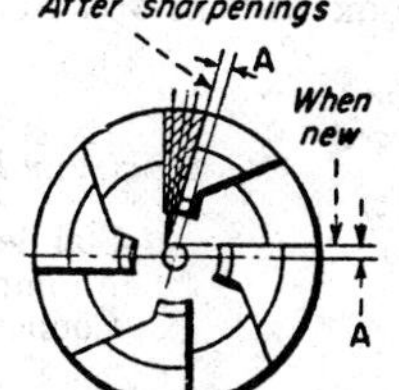

When sharpening acorn-style dies for soft steel, hold distance *A* ahead of center.

TABLE 4. TOP-RAKE MAINTENANCE ON ACORN DIES

Fractional

Size, In.	*A*	Size, In.	*A*	Size, In.	*A*
1/16	0.002	1/4	0.010	13/16	0.045
5/64	0.002	5/16	0.013	7/8	0.050
3/32	0.002	3/8	0.016	15/16	0.053
7/64	0.003	7/16	0.019	1	0.058
1/8	0.005	1/2	0.023	1 1/8	0.066
9/64	0.006	9/16	0.027	1 1/4	0.074
5/32	0.007	5/8	0.031	1 3/8	0.082
3/16	0.008	11/16	0.035	1 1/2	0.090
7/32	0.010	3/4	0.040		

Machine Screw				Pipe	
Size, No.	*A*	Size, No.	*A*	Size, In.	*A*
0	0.002	7	0.006	1/8	0.018
1	0.002	8	0.007	1/4	0.025
2	0.002	9	0.008	3/8	0.033
3	0.002	10	0.008	1/2	0.050
4	0.003	12	0.010	3/4	0.060
5	0.004	14	0.010	1	0.076
6	0.005				

Grinding cutting faces is necessary, occasionally, during the life of the die. When grinding the cutting faces on Acorn or DuoCone dies, refer to this table and sketch to insure maintenance of the proper top-rake angle for threading soft steels. For brass, grind cutting faces radial, or zero distance *A*, ahead of center. A thin resinoid or rubber wheel is used for this operation.

TABLE 5. HOOK ANGLES FOR ACORN DIES

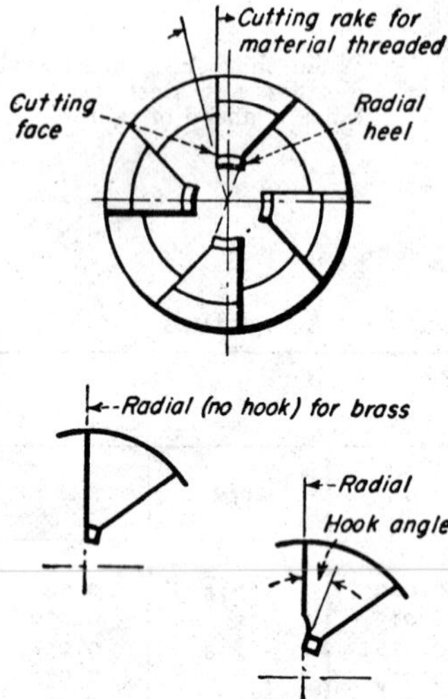

Maximum hook angle:

Stainless steel	6°
Aluminum	20°
Copper	30°

Hook limit by size of thread:

5° for 3⁄16-in. or No. 10 machine screw and smaller

10° for larger than No. 10 and including 3⁄8 in.

15° for larger than 3⁄8 in. and including 3⁄4 in.

20 to 30° for larger than 3⁄4 in.

Radial heel shown is Acorn feature patented by Greenfield Tool & Die, which, it is claimed, cleans chips from thread and prevents scoring on reversal.

Operating very much as do the button dies, these dies are not intended for hand use but are used in a holder for threading on lathes, screw machines, drillpresses, and power-threading attachments.

Holders for these dies are of two types, reversing (Fig. 10) and releasing. There is also, for the Acorn die, an adapter for use in round or spring die holders. The reversing regular holder has longitudinal float which allows the die to follow its own lead independent of any lag in the machine. The holder is used on practically all automatic screw machines and other machines which provide for automatic reversing of the die or rod at the instant when the desired length of thread has been cut. The releasing holder is designed for use on hand screw machines and other machines where work or tool reverse depends upon the operator. After a predetermined length is threaded, the head of the die releases and turns with the work. When the turret is backed, the die automatically drops back into the locked position and screws itself off.

One advantage of this die over the button die is that, when the adjusting cap is screwed up, its taper forces all the threaded lands "in" an equal amount, called concentric alignment. When the button die is adjusted the lands may be thrown slightly off center by the unequal squeeze at the split. Another advantage is the space increased between the lands or jaws to get coolant in and chips out.

The jaws do all their cutting on the chamfered threads at their front end. Usually the die is sharpened by regrinding the chamfers. Equal grinding of the lands is important (refer to Tables 4 and 5).

Threading with Die Heads

Die-head Types. There are two styles of die heads: (1) the stationary, or hand-operated, group and (2) the revolving or rotating group (see Fig. 11).

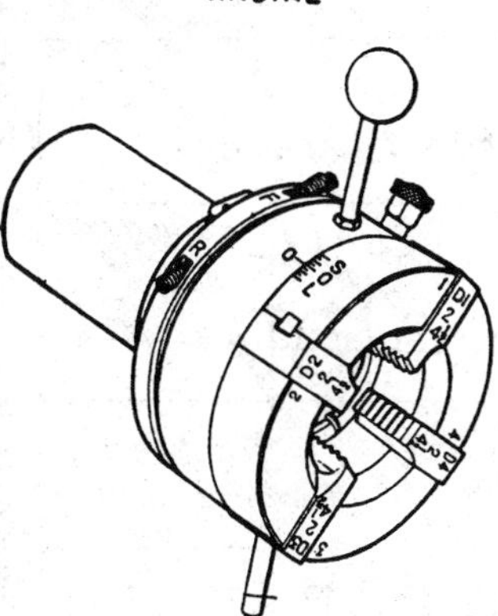

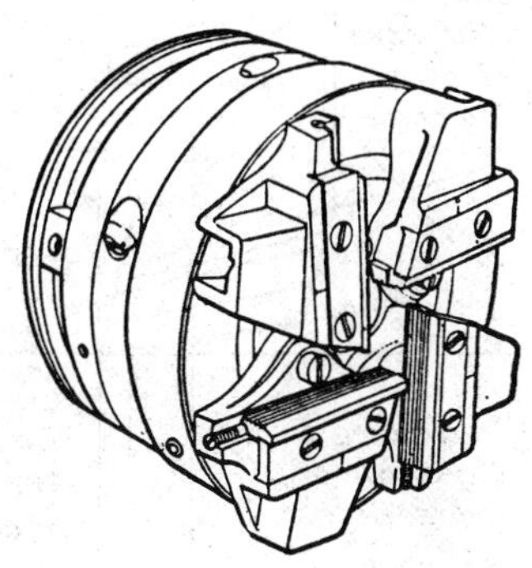

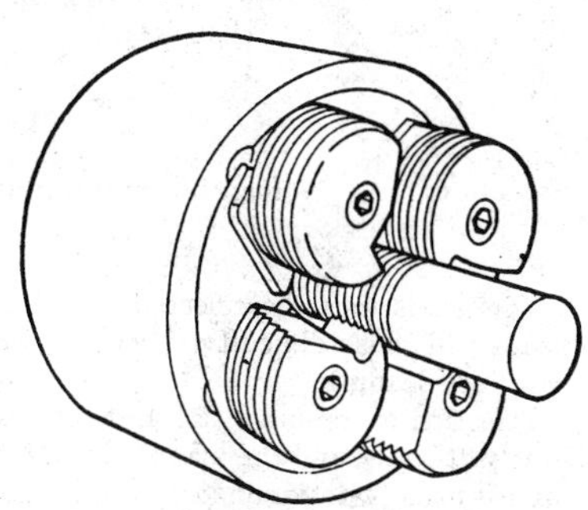

FIG. 11. Three types of die heads—radial, tangential, and circular, so-named for the style of chaser used—comprise the tools employed for die-head threading.

The hand or stationary die head has a handle to open or retract the chasers. It is never used where a die is required to revolve around its own axis. Chasers are locked by the handle either manually, as on the turret lathe, or by a finger or cam, as on automatics such as the Brown & Sharpe or Cleveland. This type of die head "pulls off" (opens up) at the end of the cut without help from an external mechanism, as is necessary with a revolving head.

Revolving die heads are used on machines where the die spindle rotates, as on automatic screw machines, drillpresses, and special threading machines. Actuation for opening and closing these dies is by a fork or yoke that fits into a groove around the outside of the head. This device is controlled by operating mechanisms which are an integral part of the various machines. All self-opening die heads must (1) position the particular chaser properly for correct cutting action, and (2) open automatically out of the thread, thus permitting the die to be retracted rapidly from the work.

Generally speaking, the rule for the number of chasers required in a die head is 4 chasers for diameters up to and including 4 in., then 1 chaser for each additional inch diameter. For 10-in. dia, 10 to 12 chasers are used.

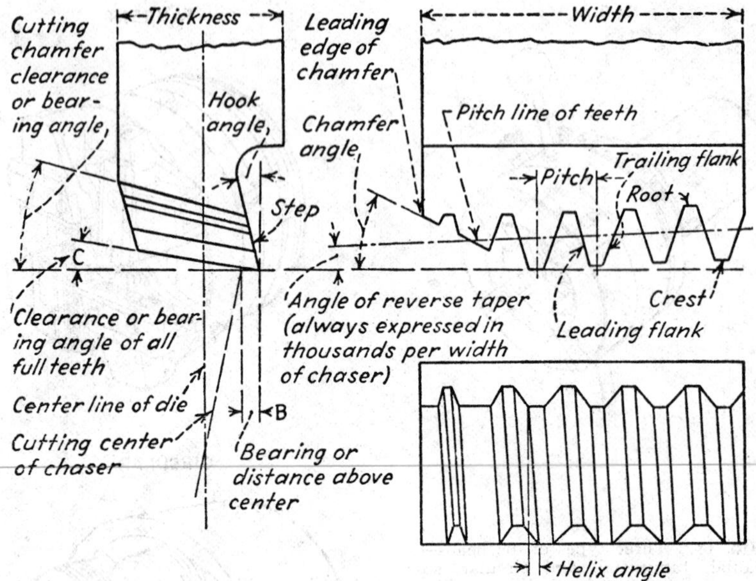

FIG. 12. Nomenclature commonly used in connection with a chaser.

Die heads must be selected in relation to size of thread to be chased. Heavy heads will cause difficulty if used for very fine (50 to 100 pitch) threads, especially when self-leading.

Selection of die-head size and style depends upon whether the die head revolves or is stationary and the manner in which it is opened or closed. Therefore, die-head style is usually determined by the type of machine on which it will be used.

Die-head size is selected by (1) size or clearances on machine the head is to be used on and (2) size of threads to be cut. If the die head is purchased for cutting relatively fine pitches on relatively short threads, then a smaller head is more suitable because it is lighter in weight and adjusts itself more readily to slight misalignment.

Coarse threads, tough material, and long threads call for a larger-capacity die head, if it will clear the machine. Standard-make die heads are not standard as to size. Size varies with style and purpose for each manufacturer. For example, circular-chaser die heads usually have a larger over-all diameter than radial or tangent chaser unit.

Types of Chaser. Three styles of chaser are used in die heads: radial, tangential, and circular.

RADIAL CHASERS. These are most suitable when large-diameter thread capacity is needed within a fixed OD of diehead. Also, many large-diameter coarse-pitch threads having a steep helix must be cut with radial chasers. Radial chasers are provided in two forms: milled or ground thread, and tapped or hobbed thread. Nomenclature is given in Fig. 12.

Advantages of the radial chaser are: Chasers can be supported more rigidly, can be wider to allow a longer chamfer and still have a number of full guiding teeth, more

TAPPED MILLED
STRAIGHT

TAPPED MILLED
ANGLE HOOK

Radial
TAPPED MILLED
RADIAL HOOK

Snub
TAPPED MILLED
SNUB

Radial Angle
(Tapped or milled)
RADIAL LIP HOOK
LIP HOOK
Lip hook grinds

TABLE 6. RADIAL CHASER GRINDS AND LUBRICANTS

Material	Grind No.	Lubricant	Material	Grind No.	Lubricant	Material	Grind No.	Lubricant
Aluminum:			Celluloid	5	B	Steel:		
Cast and die-cast	1	D	Copper	6	G	SAE 1040–1095	7	F
Rod	1	D	Everdur	2	A	SAE T1330–T1350	7	H
Stampings	1	D	Fiber	4	B	SAE 5120–52100	7	H
Bakelite	4	B	Iron:			SAE 6115–6195	7	H
Brass:			Cast	5	C	Forgings	7	E
Bar	3	G	Malleable	2	A	SAE 4130–4820	7	H
Cast	4	B	Wrought	2	A	SAE 2015–2515	7	H
Tubing	2	G	Magnesium	1	D	SAE 3115–3450	7	H
Naval	2	G	Monel	2	F	Nitralloy	7	H
Bronze:			Nickel	7	F	Stainless	7	H
Bar	2	G	Rubber	4	B	Stamping	8	F
Cast	5	B	Silver, German	2	G	Tool	7	F
Cast aluminum	2	G	Steel:			Tubing	8	H
Manganese	2	G	Bessemer screw			Semicasting	5	A
Naval	2	G	stock	2	F	Zinc die casting	1	A
Phosphor	2	G	Cast	2	A			
Tubing	2	G	SAE 1010–X1340	2	F			

Grind No.	Cutting-face Grinds for Die-head Chasers, Materials above: Milled, Straight Thread	Milled, Taper Thread	Tapped, Straight Thread	Tapped, Taper Thread	Lubricant: Symbol	Lubricant: Kind
1	15° radial hook	10° radial hook	15° radial hook	10° radial hook	A	Soluble oil
2	10° hook	5° hook	10° hook	5° hook	B	Dry*
3	5° hook	Straight	5° hook	Straight	C	Dry or soluble oil
4	5° snub	5° snub	5° snub	5° snub	D	Kerosene
5	Straight	Straight	Straight	Straight	E	Lard oil and white lead
6	15° radial lip	10° radial hook	15° radial lip	10° radial hook	F	Mineral lard
7	15° hook	10° hook	15° hook	10° hook	G	Paraffin oil
8	10° hook	5° hook	10° hook	5° hook	H	Mineral lard or sulfur base†

* Usually run dry but coolant is recommended.
† Mineral lard oil should contain about 20% lard oil.

chasers can be placed in a fixed OD of head, and the head is adaptable to rough and finish cuts. Right- or left-hand chasers can be used in the same head without changes or adapters.

Radial chasers, however, are short-lived. They are also inherently difficult to resharpen. Breakage is a problem in that a break in one chaser is likely to ruin the whole set. In close-to-shoulder threading the die head must be repositioned for correct thread length each time the chasers are resharpened in the chamfer.

Grinding instructions are given in Table 6. Only cutting-face grinding is described. All specifications are approximate and may have to be varied to obtain best results. Lip-hook grinds are to include first full tooth beyond the chamfer, except for shoulder work, in which case, use the same angle but omit the lip. Use this chart for all right-hand chasers and for all left-hand chasers when threads are National coarse and finer. For all other left-hand chasers follow original grind on chaser or request instructions from manufacturer. Taper die-head chasers are ground the same as straight thread chasers. Tapped-form die-head chasers can be sharpened on chamfer or cutting face, but milled-form die-head chasers should be sharpened on chamfer only. When it is necessary to touch up cutting face, do so very lightly.

Poor thread form may be caused by a number of conditions:

1. Misalignment.
2. Worn tools.
3. Worn chaser keys (in certain heads only).
4. An inexperienced operator may force the turret faster than the lead of the thread or allow the turret to drag.
5. Improper camming on automatic machines.
6. Wrong grind on chasers.
7. Insufficient bearing.
8. Shaving, resulting in imperfect threads, is often called poor thread form. Shaving means that not all the chasers in a set are tracking properly.

Conservative threading speeds with die heads are given in Tables 7 and 8.

Circular Chasers. Like tangent chasers, circular chasers are made with or without helix. The circular chaser with helix has the helix and spacing in the chaser independent of the holder. For this reason, the chaser requires only one set of holders, regardless of pitch, form, or diameter, for either right- or left-hand range, within the stated head capacity. Left-hand chasers are required for left-hand threads.

The circular chaser without helix takes its helix from the chaser holder and the chaser spacing. Theoretically, for accurate threading, a separate set of holders for each size of thread is required. Left-hand holders are required for left-hand threads.

Positive means of setting the chasers and measuring the bearing should be provided. It is not, however, possible to measure exactly the cutting bearing on a circular chaser as the cutting edge is the junction of two angles.

Circular chasers without lead can be used for cutting right- or left-hand threads. This style of chaser, like the others, should never be ground freehand.

Tangent Chasers. Two kinds of chaser are available for tangent cutting. One has the thread helix in the chaser and is mounted straight in the die head. The other has a straight thread on the chaser and hence must be mounted in the die head at the proper helix angle. The chaser without helix is more difficult to set for cutting bearing and spacing.

It is possible to replace individual chasers with the tangent chaser without helix, and they can be used for cutting left- or right-hand threads. Tangent chasers with helix must be reground and replaced as a set.

TABLE 7. THREADING SPEEDS FOR DIE HEADS*

Work Dia., In.	Materials											
	Tool Steel (Carbon)		Vanadium Steel, Stainless Steel, Forged Steel, Open Hearth, Monel Metal		Machine Steel, Drawn Steel, Tobin Bronze, Bessemer, Screw Stock		Malleable Iron, Cast Iron		Cast Brass, Bar Brass, Phosphor Bronze, Copper		Fiber, Bakelite, Aluminum, Lead	
	Rpm	Fpm	Rpm	Fpm	Rpm	Fpm	Rpm	Fpm	Rpm	Fpm	Rpm	Fpm
National Coarse Threads												
1/4	306	20	535	35	764	50	917	60	1070	70	1222	80
5/16	245		428		611		733		856		978	
3/8	183		306	30	458	45	560	55	662	65	764	75
7/16	157	18	262		393		481		568		656	
1/2	137		191		306		382		459		535	
9/16	102		170	25	272	40	340	50	407	60	475	70
5/8	92	15	153		245		306		367		428	
3/4	76		102		178		229		279		330	
7/8	65		87	20	153	35	196	45	241	55	285	65
1	46		76		134		172		210		258	
1 1/8	41	12	68		119		153		187		221	
1 1/4	37		55		92		123		153		183	
1 1/2	31		46	18	76	30	102	40	127	50	153	60
1 3/4	22		39		66		87		109		131	
2	19	10	29		48		67		86		105	
2 1/4	17		25	15	42	25	59	35	76	45	94	55
2 1/2	15		23		38		53		69		84	
2 3/4	11	8	21		35		49		62		77	
3	10		13	10	26	20	38	30	51	40	64	50
National Fine Threads												
1/4	382		611	40	764	50	917		1222		1222	
5/16	306	25	489		611		733		978	80	978	
3/8	255		357		458		611	60	815		815	
7/16	175		306		393		524		656	75	699	80
1/2	153	20	268	35	344	45	459		573		611	
9/16	136		238		306		373	55	475		543	
5/8	122		214		276		337		428	70	489	
3/4	92		153		203	40	254		357		381	
7/8	79	18	131	30	153		219	50	285	65	329	75
1	69		115		134		191		258		287	
1 1/8	61		85		119	35	153		204		255	
1 1/4	46		76	25	107		137	45	183	60	214	
1 1/2	38	15	64		89		115		153		178	70
1 3/4	33		44	20	66	30	87	40	120	55	153	
2	29		38		57		76		105		124	
2 1/4	20		25		42		59		85		111	65
2 1/2	18	12	23	15	38	25	54	35	76	50	99	
2 3/4	17		21		35		49		69		83	60
3	15		19		32		45		64		76	

Cutting speeds listed are for annealed stock only. When heat-treated material is threaded decrease cutting speed 20 to 30%.

* Landis Machine Co.

TABLE 8. DIE-HEAD THREADING SPEEDS FOR PIPE*

Pipe Dia, In.	Materials											
	Merchant Pipe, Stainless Steel		Cast Iron		Copper, Yellow Brass		Merchant Pipe, Stainless Steel		Cast Iron		Copper, Yellow Brass	
	Rpm	Fpm	Rpm	Fpm	Rpm	Fpm	Rpm	Fpm	Rpm	Fpm	Rpm	Fpm
1/8	377	40	471		566		17.2		24		27	
1/4	248	35	354	50	424	60	14.4	25	20	35	23	40
3/8	198		283		338		12.5		18		20	
1/2	150	33	204	45	250	55	11.0		11		15.4	
3/4	120		164		200		7.9		9.9		13.9	35
1	87		116		145		7.1		8.9		12.4	
1 1/4	69	30	92	40	115	50	6.5	20	8.1		11.4	
1 1/2	60		80		101		6.0		7.5		9.0	
2	48		64		81		5.5		6.8	25	8.2	30
2 1/2	33		47		60		5.1		6.4		7.6	
3	27		38		49	45	4.3		6.0		6.0	
3 1/2	24	25	33	35	43		4.0	18	5.6		5.6	25
4	21		30		38		3.8		5.3		5.3	
4 1/2	19		27		30	40	3.4		4.8		4.8	

* Landis Machine Co.

Tangent chasers are longer-lived than radial chasers, but they are not so well suited for the coarser of coarse threads. It is essential to trouble-free working of a die using tangent chasers with helix that a positive method of checking grinding and setting of the chasers be provided.

Threading Close to a Shoulder. Product designs that require threading close to a shoulder should be avoided. Designing short necks, without regard to the need for them or for the pitch of the thread, and the common practice of showing threads clear up to the shoulder, makes short chamfers necessary. A short chamfer increases the likelihood of chaser breakage, shortens chaser life, adversely affects thread quality, and necessitates operation at slower spindle speeds.

Chamfers. Long chamfers mean better threading performance. Costs, when threading against a shoulder, are in direct relationship to chamfer length, or clearance provided. Designers of threaded parts should keep in mind the advantage to the shop of maximum clearance at shoulders or in blind hole bottoms. Chamfers of 30° rather than 40 or 45° should be encouraged. Both form and width of a recess affect chaser performance. Recess width should be at least 2½ to 3 threads (see Fig. 13).

Thread Rolling

Thread rolling is a cold-forging process (Fig. 14). Screw threads, or other shapes, are formed by displacement of metal as opposed to cutting. The process is used for

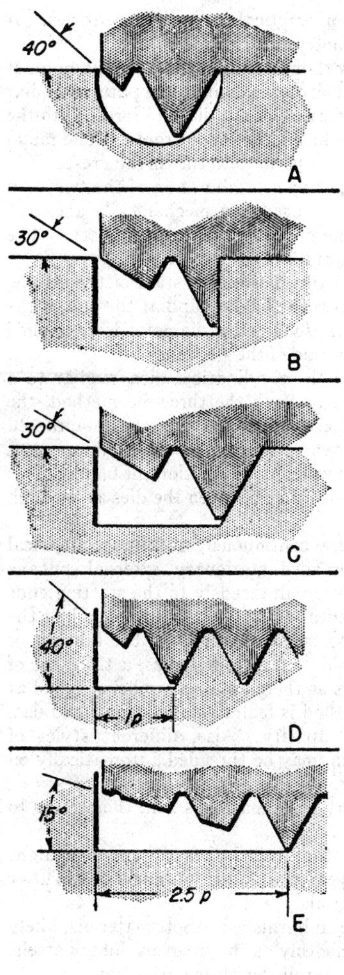

Form is altered to favor the tool. All three recesses are of equal width, but where a 40° chamfer is necessary for *A*, a 30° chamfer will thread into *B*. Still more desirable is *C*, as a slightly longer chamfer can be used, plus the advantage that the 30° side of the neck eliminates the sharp partial thread left in recess *B*. Recess width should be as wide as possible; at least 2½ to 3*p* (2½ to 3 threads) where practical. Using the preferred *C* recess, the effect of changing from 1 as at *D* to 2.5*p* in bottom view triples the number of cutting teeth so the 15° chamfered chaser will remove three times as many chips only one-third as thick. Lighter cuts improve thread quality and tool life.

FIG. 13. Both form and width of a recess affect chaser performance.

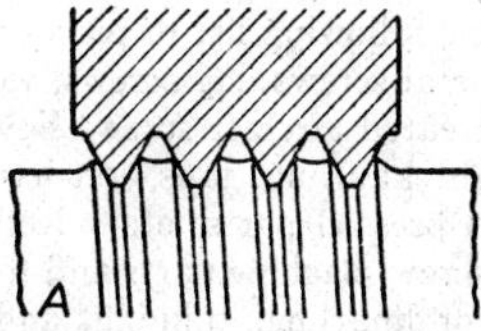

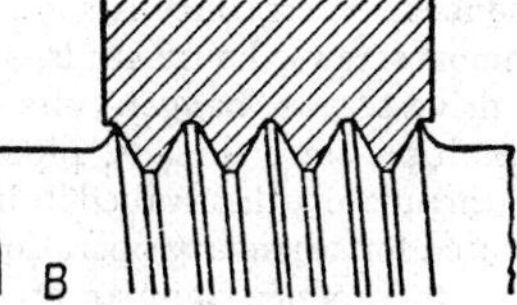

FIG. 14. Thread formation by rolling consists of a cold-working action. At *A*, the thread has just begun to roll up into the die root. At *B*, the thread has completely filled the die profile.

producing external threads only, and is seldom practical for small quantities. A separate set of dies must be used for each diameter and pitch.

A cylindrical blank of a diameter within a few thousandths of the pitch diameter of the desired thread is placed between hardened steel dies machined in a pattern similar to the desired finished thread. Ridges on the surface of the dies are forced into the blank to approximately half their depth. This forms the thread roots. The metal displaced flows up into the remaining spaces in the dies to form the thread crests.

Threads are rolled by either of two methods, or by a combination of the two. In the flat-die method, a set of two flat-faced dies is used in a mechanically actuated reciprocating machine. One of these dies is clamped and remains stationary. The other die is mounted on a ram facing the first die and reciprocates relative to it.

An unthreaded blank is wedged between the two dies at the start of the stroke, rotates about its own moving axis as it rolls between the dies, and at the end of the stroke, drops out completely threaded. The finished screw diameter is controlled by the distance between the die faces at the finish end of the stroke.

Cylindrical-die threading uses either two or three cylindrical dies, similar to a centerless-grinding setup, in a rotating machine. With the three-die method, the dies rotate synchronously at a predetermined speed on parallel shafts, cam-actuated alternately to come together and to withdraw. The two-die method operates similarly but is hydraulically actuated and has a workrest to position the blank.

Finished screw diameter is controlled by the distance between the dies at the high point of the feed.

Planetary or segmental thread rollers have a continuously rotating cylindrical inner die, or dies, and a series of internally threaded, stationary, external concave segment dies. Starter fingers introduce blanks simultaneously to the starting ends of the outer dies. When the screws have rolled in a planetary path about half the length of the outer dies, a second set of blanks is introduced. As the first set of screws is finished and reaches the leave-off points of the outer dies, a third set of blanks is started. Thus, twice as many screws as there are dies are being rolled at any given instant. It is claimed that this method is faster than conventional flat-die machines of comparable work-diameter capacity. Also, different styles of screws, all of the same pitch diameter and pitch, may be threaded automatically on one machine simultaneously.

Generally speaking, the flat-die and planetary-style machines lend themselves to automatic and semiautomatic feeding.

Laboratory and service tests have proved that rolled threads under straight tension loads, as a result of the cold working, average 10 to 25% stronger than cut or ground threads of corresponding size and material.

The surface metal of the thread is hardened and burnished. Soft materials, likely to tear if threads are cut, will roll-thread as smoothly as the freer-machining steels. Material with hardnesses to Rockwell 40 C may be rolled with smooth surfaces.

Parts Produced. Versatility is shown by the following list of pieces suitable for rolling: common screws, bolts, studs, sheet-metal screws, lag screws, wood screws, coat hooks, drive screws, hanger bolts, heat-treated aircraft screws having class 3 and 4 fits, multiple-start threads, pipe threads, knurling, taps, left-hand threads, oil grooves, threads on thin-walled tubes and pressed parts, etc. Roll threading may also be used for secondary operations on screw-machine parts and to straighten warped parts, to produce annular grooves, for knurling, splining, and crimping. Roll attachments make it possible to apply the process directly on screw machines and turret lathes within the regular work cycle.

Die Selection. Carbon and low-alloy steel dies give suitable production of brass and more ductile steel screws. However, alloy steel dies are usually more practical for threading any material. High-alloy die steel and high-speed steel are recommended for rolling screws from less ductile materials such as free-machining medium- and high-carbon, alloy, heat-treated, and stainless steels.

Chamfers are necessary in thread rolling. It is impossible to counterbalance entirely the overturning forces on die threads at the ends of blanks and along the top edges of the die, where the blanks press against one flank of the die thread only. To offset this, a chamfer of at least one thread is provided.

Blank Preparation. Blanks for rolling are usually prepared by cold heading (sometimes including extrusion), automatic screw machines, presses, and other such equipment. When accuracy is required, an additional sizing operation, usually centerless grinding, may be used. The rolling diameters, regardless of how prepared, must be kept as uniform as possible to assure uniform accuracy in the finished threads and maximum die life.

Bevel at blank ends should be 60° included angle rather than 90°. Necks adjacent to shoulders, commonly required for cut or ground threads, are unnecessary, and in some cases detrimental to good rolling. To obtain better results when using a neck, provide a 60° included angle instead of straight sides.

Approximate blank diameters may be found by subtracting the depth of a single thread from the completed thread OD, when thread form is balanced. Blank diameters, including taper and out of round, should be held at least as close as the pitch diameter required in the completed screws. Oversized blanks will result in oversized screws. Undersized blanks will result in undersized screws if the dies are squeezed tightly enough to complete a full thread. When undersized blanks are used together with less die pressure, the pitch diameter will be approximately correct but thread crests will not fill out and the characteristic rolled-thread smoothness of crest, flank, and root flank and root will be lacking. Though this may be permissible, the OD may vary more than the tolerances taken in the blanks.

Thread Milling

Conventional Thread Milling. Either a single-form or a multiple-form cutter may be used. With the single-form cutter, the work must traverse the cutter, or the cutter traverse the work for its full length. The cutter head is usually mounted on a compound slide. When the multiple-form cutter is used, the number of turns necessary to complete the thread is greatly reduced. The latter tool is sometimes called a hobbing cutter, or thread hob. However, unlike a gear hob, the teeth are annular and have no lead, which is obtained by either a lead screw or a cam.

Planetary Thread Milling. The part is held stationary, and the cutter both rotates and circumscribes the part. The planetary method is seldom suitable for feed and lead screws but is advantageous for odd-shaped parts such as forgings with angular projections.

Single-form or disk cutters are best suited to deep threads and coarse pitches but will produce a fine-pitch thread. This type is necessary if the thread length is greater than can be milled with multiple-form cutters. Usually, finer-pitch threads can be milled with single cutters than are possible with the multiple form. However, the cutter must run perfectly true sideways. The cutter spindle is set at the helix angle of the thread to be cut to prevent gashing or distorting the thread. Two-start threads can be produced in one pass, if the lead angle is not too steep, by spacing cutters one pitch apart and tilting the cutter head to the lead angle of the thread.

Thread milling is impractical, if not impossible, on materials harder than Rockwell

TABLE 9. SPEEDS AND FEEDS FOR THREAD MILLING

Cutter	High-speed Steel		Carbide			
			Cast Iron			
Material	Mild Steel	Tool Steel	Soft	Hard	Bronze	Aluminum
Cutter speed, sfpm	110–150	75–110	250–325	150–225	275–350	1200–1600
Feed, ipm (chip load), av finish	0.005–0.010	0.004–0.006	0.008–0.012	0.005–0.010	0.004–0.008	0.006–0.010
Feed, ipm (chip load), fine finish	0.003–0.004	0.002–0.003	0.003–0.005	0.002–0.004	0.002–0.004	0.002–0.004

37 to 40 C. Multiple thread-milling cutters have two distinct applications. First, they are used to produce an external or internal thread in one revolution of the workpiece, and second, they are used to mill the form on thread-rolling dies. They are especially valuable where a thread must be cut close to a shoulder, within an undercut or to the bottom of a recess. As the true thread form is held throughout the thread length, no undercut or relief is necessary at the thread bottom. An interrupted-tooth cutter is usually used for sharp V-threads on pipe. But this practice is not common for National or U.S. form.

Multiple cutters may be used for external threads which have lead angles under 3½° and for internal threads whose lead angles are less than 2½°. When thread forms have steeper sides or smaller included angles than American Standard, Whitworth, or Unified threads, it is usually necessary to mill them with a disk cutter. Although Acme threads are milled with hob cutters, it is done only when work specifications will tolerate the deformation caused by sidecutting.

Both single and multiple cutters are used at the speeds and feeds recommended in Table 9. An approximate feed of 0.020 in. per revolution of the cutter is usually satisfactory for commercial work.

As most threading cutters are form-relieved, only the face need be reground. Accurate spacing of the teeth and maintenance of the correct hook or rake are important. For general-purpose cutters a rake of 6 to 8° is usually satisfactory.

Profile cutters are sharpened by grinding the form rather than the flute and are usually made with radial teeth. The profile tooth should have a land of 4 to 6° clearance angle on top and on both sides.

Thread Grinding

Although "single-rib" thread grinding was the original method and still receives much attention, there is probably more productive potential in some of the more recent developments. Among the latter are the "multirib" plunge-cut, "multirib" traverse method, and the centerless method. The traverse method is used for headless parts and the plunge-cut method for cap screws, bolts, and other nonuniform-diameter parts.

The choice of rolling, milling, or thread grinding is primarily based on the production quantity, the material to be threaded, and the accuracy or precision demanded. Thread rolling is by far the fastest method but production must be high enough to pay for dies and machine.

When material hardness exceeds Rockwell 35 to 40 C, thread grinding is usually the preferred method. Likewise, if maximum accuracy and precision are required, grinding is the most likely choice. (Output of multirib, plunge, and feed-through centerless thread grinders is high.) Also, grinding can be done both externally and internally. However, the initial investment is high and, for automatic grinding, quantity must be sufficient to pay for the setup.

Single-rib Thread Grinding. Comparable with single-point chasing in versatility, the single-rib method is more valuable for precision than it is for high production. Periphery of the wheel is trued to the form of the thread to be ground, its shape being identical to that of a single-point cutting tool for cutting a thread. As the wheel is often 18 to 20 in. in diameter, the arc of contact at full depth is appreciable. Hence, for accuracy the wheel is inclined to the thread helix angle. This adjustment is imperative for helixes of 4° or more.

For threads such as large Acme, worm, or buttress, for precision work and for crush truing, the vitrified wheel is usually used. Resinoid wheels are preferred for commercial tolerances and for 40 threads per inch and finer; also for heavy feeds. They are not suitable for crush dressing. Single-rib wheels are usually dressed by means of diamonds in automatic compensating attachments.

Single-rib grinding must be used when:

1. Included angle is too sharp to admit multirib.
2. Extreme accuracy is desired.
3. Production demand is small (minimum setup time).
4. Threads to be ground are smaller than 32 threads per inch.

True the wheel carefully. Approximately 0.001- to 0.002-in. feed per pass is about right. Feeds not exceeding 0.0005 to 0.001 in. per pass must be used when dressing a very sharp edge on the wheel.

Wheel Speeds. Operating speeds between 8000 and 10,000 sfpm are commonly used for threading. On high-production commercial work, speeds up to 12,000 sfpm are sometimes used. When grinding 29° and Acme threads that have been previously rough ground or cut, for finish grinding, and on hardened steel, slower speeds may be advisable. Also, for precut threads it may be necessary to use a grade of wheel (vitrified) entirely different from that used for grinding threads from the solid piece. For this work an average surface speed of 7000 sfpm is recommended.

Work speed is about 6 fpm on work of medium pitch to finish "from the solid" in two or three passes. Some operators prefer 20 to 25 fpm with very light feed on hardened high-speed steel.

The approximate depth of an ordinary rough cut should be from 0.015 to 0.040 in. When two or more cuts are taken, it is generally advisable to take a minimum of 0.0015-in. depth of cut for the finish cut. During continuous grinding no finish cut of less than 0.0008 in. in diameter should be taken.

Generally speaking, it is practical to grind threads not exceeding 3 in. long productively from solid unthreaded blanks as coarse as:

10 pitch, up to 7/8 in. dia.
12 pitch, up to 1 in. dia.
13 pitch, up to 2 in. dia.
16 pitch, up to 3½ in. dia.

Threads coarser than 10 pitch are more economically produced when the blanks are rough threaded previous to hardening.

For long, thin, precision-threaded parts it may be necessary to reduce wheel and work surface speeds by 50%. Depths of cut may be taken which double the production otherwise obtainable on this grade of work using normal light cuts.

Root Width. It is possible to maintain the standard root width for most standard pitches. However, many classes of work require that a narrower width than this be maintained. The following factors may help to alleviate such problems:

Grit Size. On moderately coarse pitches (12 to 18) a coarser wheel will often hold the desired form better than one which is too fine.

Bond. Resinoid-bonded wheels tend to maintain a better edge than do vitrified wheels.

Wheel Speed. Narrow root width is often more easily maintained if wheel speed is increased above normal.

Coolant. Rapid breakdown of wheel edge will occur if coolant flow is inadequate. Beware of excessive sparking.

Distribution of Feed. If an excessively heavy roughing cut is taken, the point of the wheel is likely to crumble, leaving an excessive quantity of material for the point to remove on the finishing cut. On 12 pitch American National screws, a feed of 0.036 in. for roughing and 0.018 in. for finishing is usually satisfactory.

Wheel Truing. Poor thread form, other than root width, is usually a result of improper truing. Do not attempt heavy cuts when truing a resinoid wheel. The diamond, if dull, may cause sufficient wheel deflection to distort the thread form.

Lead error occurs occasionally when using resinoid wheels, particularly if the thread is precut and distorted in hardening. Vitrified wheels should be used when lead error occurs to a degree which cannot be tolerated. Thicker wheels and larger-diameter flanges will reduce the possibility of wheel flexing and thus also cure lead error.

Multirib Thread Grinding. The wheel used has two or more parallel annular grooves around its periphery. Each rib is trued to the thread form by diamond truing or by crush dressing.

When the wheel is made in a thickness as great as or greater than the required thread length, the thread can be completed in one revolution of the work plus one-half revolution for wheel infeed, or 1½ work revolutions total.

Another style of dressing is the alternate, or "skip-rib," profile. This is similar to the interrupted-tooth tap form, in that only every other tooth is formed, with a one tooth space between. The alternate style spaces the ribs to grind every other thread during the first revolution of the work, the thread being completed on the second revolution. Hence, about 2½ revolutions of the work are required for grinding a complete thread.

Centerless Grinding. This process is used on hardened headless setscrews and similar parts from ⅛ in.-40 to greater than 5 in.-16. The productive capacity of this method far exceeds any of the other methods discussed, except rolling.

The grinding wheel can be from 2 to 5 in. wide. Hardened blanks are hopper-fed to the starting position. The regulating wheel, set at the proper angle and rotating at a predetermined speed, then causes the workpiece to traverse the face of the grinding wheel. As the screw is completed, it emerges from the opposite side of the wheel face. Crush dressing is the best method so far for truing the wheel. Vitrified wheels are used. Helix-angle setting of the workrest blade depends upon the basic pitch diameter. If the basic pitch diameter is increased or decreased for any reason, the helix-angle setting must be recalculated.

Coolants. A good grade of grinding oil is recommended as the best coolant for thread grinding. Nonsoluble grinding oil is preferable to water-soluble compounds because it helps maintain better wheel form and promotes a higher surface finish. Keep the grinding oil clean. Use a filter. Remove settlings and flush out the entire coolant system frequently. Do not shut off the coolant entirely when truing or dressing.

Thread Generating

The thread-generating machine is used largely for the production of worm threads and similar forms: lead and feed screws, elevating screws, worms, screws used in woodworking machinery, and screws for taps and valves in the plumbing industry. Profiled shapes other than threads are produced with a cutter resembling a cam.

The thread-generating machine operates on a principle that is more readily visualized if the threads are considered as rack teeth that are wrapped around a cylinder in a helical path. In generating these threads, the work rotates upon its axis at right angles to the axis of a gear shaper cutter. The cutter and work are geared together in relation to the number of teeth on the cutter and the number of "starts" on the worm. Whereas in milling the planes of rotation are parallel, here they are perpendicular. There is a positive ratio between cutter speed and work speed. As the cutter is "rolled" in mesh with the work, it produces threads by the molding-generating process.

Both right- and left-hand threads can be cut. In cutting a right-hand thread the cutter travels from right to left under power feed, and when it reaches the end of the cut the machine stops automatically.

Depth of cut control is by a cam which determines the position of the head. This cam is made with different lengths of "dwell" to suit the lengths of thread to be cut and with different angles of "rise" to lower and raise the cutter at the required rate of feed.

SCREW-THREAD MEASUREMENT

Threaded products are checked with thread ring gages, optical gages and comparators, thread micrometers, and indicating thread gages.

In gaging of threads, the emphasis is always upon pitch diameter, because the thread fit largely depends upon the pitch-diameter tolerances and limits. Because direct measurement of pitch diameter of a thread ring gage is difficult, the usual practice is to fit the ring gage to a threaded setting plug.

When it becomes necessary to check the accuracy of the thread plug gage, or the setting plug, the three-wire method of measuring the pitch diameter is considered the most accurate procedure. If necessary, the procedure is also directly applicable to a threaded product.

Three-wire Method. By placing three wires of correct diameter in the threads, as shown in Fig. 15, and measuring over them, the pitch diameter can be found directly by subtracting the wire constant from the measured diameter. This simplified procedure results only when the "best wire" for the pitch of the screw is used. The best wire is the size of wire which touches the thread at the middle of the sloping sides, or at the pitch diameter.

The formula for the best wire is

$$G = \frac{\sec a}{2n}$$

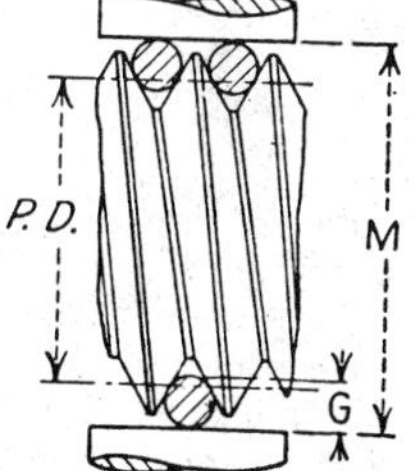

FIG. 15. How wires are used between micrometer anvils to measure the pitch diameter of threads.

The general formula for computing the measured pitch diameter of a screw is

$$E = M + \frac{\cot a}{2n} - G(1 + \csc a)$$

where M = measurement over wires, E (or PD) = pitch dia, G = dia "best wire," n = number of threads per in., a = one-half the included angle of thread.

For a 60° thread, which includes American coarse and National fine threads and also 60° metric threads, this formula reduces to

$$E = M + \frac{0.86602}{n} - 3G$$

To check pitch diameter E from outside measurement over the wires, transpose the formula to

$$M = E + 3G - \frac{0.86602}{n}$$

Now it so happens for a 60° thread that the depth of a sharp V-thread is $0.86602/N$ and that $3G$, or three times the diameter of the "best" wire, is just twice $0.86602/N$, so the formula is further simplified to

$$E = M - \text{depth of V-thread}$$
$$M = E + \text{depth of V-thread}$$

Best-size Wires. The diameter G of the "best" measuring wires for various thread systems is:

For 60° threads

$$G = \frac{0.57735}{n} \text{ or } 0.57735p \text{ (see Table 10)}$$

For 55° threads

$$G = \frac{0.563692}{n} \text{ or } 0.563692p$$

For 53°8′ threads

$$G = 0.55902p$$

For 47½° threads

$$G = 0.54625p$$

For 40° threads

$$G = \frac{0.5321}{n} \text{ or } 0.5321p$$

For 29° threads

$$G = \frac{0.51645}{n} \text{ or } 0.51645p$$

For 60° threads, the maximum wire size is $1.01036p$, and the minimum wire size is $0.50518p$.

For Acme 29° threads, the maximum wire size is $0.65001p$ and the minimum wire size is $0.48726p$.

Formulas for E, the pitch diameter in terms of G, M, and n for common thread systems, are given in Table 11.

TABLE 10. THREE-WIRE MEASUREMENT OF AMERICAN THREADS
Dimensions in Inches

Threads per In.	$G = \frac{0.57735}{n}$ Dia Best Wire	Best Wire Constant $H = \frac{0.86602}{n}$ Depth V-Thread	$h = \frac{0.64951}{n}$ Depth American Thread	Measuring Wires	
				Min Wire	Max Wire
80	0.00722	0.01083	0.00812	0.00722	0.01155
72	0.00802	0.01203	0.00902	0.00802	0.01312
64	0.00902	0.01353	0.01014	0.00902	0.01443
56	0.01031	0.01546	0.01160	0.01031	0.01604
48	0.01203	0.01804	0.01353	0.01155	0.02062
44	0.01312	0.01968	0.01476	0.01155	0.02221
40	0.01443	0.02165	0.01624	0.01312	0.02406
36	0.01604	0.02406	0.01804	0.01443	0.02624
32	0.01804	0.02706	0.02030	0.01604	0.02887
28	0.02062	0.03093	0.02319	0.01924	0.03207
26	0.02221	0.03331	0.02498	0.02062	0.03608
24	0.02406	0.03608	0.02706	0.02221	0.03608
22	0.02624	0.03936	0.02952	0.02406	0.04124
20	0.02887	0.04330	0.03248	0.02624	0.04441
19	0.03039	0.04558	0.03418	0.02887	0.04811
18	0.03207	0.04811	0.03608	0.02887	0.05020
16	0.03608	0.05413	0.04060	0.03207	0.05774
14	0.04124	0.06186	0.04640	0.04124	0.06415
13	0.04441	0.06662	0.04996	0.04124	0.06415
12	0.04811	0.07217	0.05413	0.04441	0.07217
11½	0.05020	0.07531	0.05648	0.04811	0.07217
11	0.05249	0.07873	0.05904	0.05020	0.08248
10	0.05774	0.08660	0.06495	0.05249	0.09622
9	0.06415	0.09622	0.07218	0.05774	0.09622
8	0.07217	0.10825	0.08119	0.06415	0.10497
7	0.08248	0.12372	0.09279	0.07217	0.12830
6	0.09622	0.14434	0.10825	0.09622	0.14434
5½	0.10497	0.15746	0.11805	0.09622	0.14434
5	0.11547	0.17320	0.12990	0.10497	0.14434
4½	0.12830	0.19245	0.14434	0.11547	0.14434
4	0.14434	0.21651	0.16238	0.12830	

TABLE 11. FORMULAS FOR THREE-WIRE MEASUREMENT OF VARIOUS THREADS

Designation	Included Angle Deg	Included Angle Min	Thread Depth	Formula for Finding Pitch Dia with 3-wire Method
National	60	..	$\frac{0.649519}{n}$	$E = M - \left(3G - \frac{0.86602}{n}\right)$
National pipe	60	..	$\frac{0.8}{n}$	$E = 1.00049M - \left(3G - \frac{0.86602}{n}\right)$
Sharp V	60	..	$\frac{0.86602}{n}$	$E = M - \left(3G - \frac{0.86602}{n}\right)$
International metric	60	..	$\frac{0.649519}{n}$	$E = M - \left(3G - \frac{0.86602}{n}\right)$
Whitworth	55	..	$\frac{0.64033}{n}$	$E = M - \left(3.16568G - \frac{0.96049}{n}\right)$
British Association	47½	..	$6p$	$E = M - (3.4829G - 1.13634p)$
Löwenherz	53	8	$0.75p$	$E = M - (3.23594G - p)$
Acme screws	29	..	$\frac{1}{2n} + 0.010$ in.	$E = M - \left(4.9939G - \frac{1.93334}{n}\right)$
Acme taps	29	..	$\frac{1}{2n} + 0.020$ in.	$E = M - \left(4.9939G - \frac{1.93334}{n}\right)$
29° worm	29	..	$\frac{0.6866}{n}$	Use wire $\frac{0.5149}{n}$ to come flush with top of thread
40° worm	40	..	$\frac{0.6866}{n}$	Use wire $\frac{0.51234}{n}$ to come flush with top of thread

NOTES: 1. On 29 and 40° worm threads the addendum above the pitch dia is $0.3183/n$. The thread depth of $0.6866/n$ provides a clearance of $0.05/n$ at the bottom of the thread.
2. In the above formula G can be any size wire that will fit in the thread.
3. Helix-angle corrections are not included in the foregoing formulas.

Measuring Included Angle of Thread. To measure the angle of a 60° thread, two measurements are made with two different-sized wires, usually as near as possible to the maximum-sized and minimum-sized wires that will fit into the thread. Table 10 gives the maximum and minimum wire size that can be used for any pitch of thread.

The general formula for determining the included angle of thread is

$$\csc a = \frac{M_1 - M_2}{G_1 - G_2} - 1 \qquad (1)$$

or

$$\frac{M_1 - M_2}{G_1 - G_2} = \csc a + 1 \qquad (2)$$

where M_1 = measurement over the three large-sized wires of dia G_1, M_2 = measurement over the three small-sized wires of dia. G_2.

Table 12 gives values of ($\csc a + 1$) for angles 2° greater and 2° less than the common angles used in screw threads.

For a thread exactly 60°, the half angle a being 30°, $\csc a = \csc 30° = 2.0000$, and Eq. (2) becomes

$$\frac{M_1 - M_2}{G_1 - G_2} = 2.0000 + 1 = 3.0000 \qquad \text{(result)}$$

The value of the included angle of the thread is found by interpolation in Table 12.

TABLE 12. VALUES OF ($\csc a + 1$)
For Measuring Included Angle of Threads

60° Threads			55° Threads			53°8′ Threads			47½° Threads			29° Threads		
Deg	Min	Result	Deg	Min	Result	Deg	Min	Result	Deg	Min	Result	Deg	Min	Result
58	..	3.0627	53	..	3.2411	51	8	3.3172	45	30	3.5859	27	..	5.2836
58	10	3.0573	53	10	3.2346	51	18	3.3101	45	40	3.5770	27	10	5.2579
58	20	3.0519	53	20	3.2282	51	28	3.3032	45	50	3.5681	27	20	5.2324
58	30	3.0466	53	30	3.2217	51	38	3.2962	46	..	3.5593	27	30	5.2072
58	40	3.0413	53	40	3.2153	51	48	3.2894	46	10	3.5506	27	40	5.1824
58	50	3.0360	53	50	3.2090	51	58	3.2825	46	20	3.5419	27	50	5.1578
59	..	3.0308	54	..	3.2027	52	8	3.2757	46	30	3.5333	28	..	5.1336
59	10	3.0256	54	10	3.1964	52	18	3.2690	46	40	3.5247	28	10	5.1096
59	20	3.0204	54	20	3.1902	52	28	3.2623	46	50	3.5163	28	20	5.0859
59	30	3.0152	54	30	3.1840	52	38	3.2556	47	..	3.5078	28	30	5.0625
59	40	3.0101	54	40	3.1778	52	48	3.2490	47	10	3.4998	28	40	5.0394
59	50	3.0050	54	50	3.1717	52	58	3.2425	47	20	3.4912	28	50	5.0165
60	..	3.0000	55	..	3.1657	53	8	3.2359	47	30	3.4829	29	..	4.9939
60	10	2.9950	55	10	3.1596	53	18	3.2294	47	40	3.4748	29	10	4.9716
60	20	2.9900	55	20	3.1536	53	28	3.2230	47	50	3.4666	29	20	4.9495
60	30	2.9850	55	30	3.1477	53	38	3.2166	48	..	3.4586	29	30	4.9277
60	40	2.9801	55	40	3.1418	53	48	3.2103	48	10	3.4506	29	40	4.9061
60	50	2.9752	55	50	3.1359	53	58	3.2039	48	20	3.4426	29	50	4.8848
61	..	2.9703	56	..	3.1300	54	8	3.1977	48	30	3.4347	30	..	4.8637
61	10	2.9654	56	10	3.1242	54	18	3.1914	48	40	3.4269	30	10	4.8428
61	20	2.9606	56	20	3.1185	54	28	3.1852	48	50	3.4191	30	20	4.8222
61	30	2.9558	56	30	3.1127	54	38	3.1791	49	..	3.4114	30	30	4.8018
61	40	2.9510	56	40	3.1070	54	48	3.1730	49	10	3.4037	30	40	4.7816
61	50	2.9463	56	50	3.1014	54	58	3.1669	49	20	3.3961	30	50	4.7617
62	..	2.9416	57	..	3.0957	55	..	3.1608	49	30	3.3886	31	..	4.7420

THREAD SYSTEMS

American Screw Threads

Machine Screws. Adequate standardization of the basic sizes, dimensions, and especially tolerances of machine-screw threads lagged many years after the development of production methods to make them. Machines for producing screw threads were introduced in 1846. But 103 years passed before the screw-thread systems of the English-speaking countries were consolidated in 1949 into a Unified system, and tolerances were applied by formula instead of as the outgrowth of practice. All threaded products for ordnance use in the United States, Great Britain, and Canada are now specified in the Unified system, and before long most commercial products will likewise be made to the same system. The net result will be interchangeability, rather than the confusion and expense of dual standards.

Inch-measure Systems. An understanding of existing inch-measure screw-thread systems is still required, despite the existence of the Unified system.

WHITWORTH. This, the first inch-measure system, was introduced in England in 1841 and gained general acceptance in about 20 years. The thread form is based on a 55° thread angle, and the crests and roots are rounded, as in British practice. Further data are given on page 10-68.

AMERICAN NATIONAL. In the United States, William Sellers introduced a screw thread with a 60° thread angle in 1864. Even as late as World War I, threaded products made throughout the United States were not interchangeable, and the American National screw-thread system was developed in 1933. This system is based on the 60° thread angle and flat crests and roots, and encompasses these groups: coarse-thread series in sizes 1 to 12 and ¼ to 4 in.; fine-thread series in sizes 0 to 12 and ¼ to 1½ in.; extra-fine thread series in sizes 12 and ¼ to 2 in.; an 8-pitch series in sizes from 1 to 6 in.; a 12-pitch series from ½ to 6 in.; and a 16-pitch series from ¾ to 6 in.

The American National thread system calls for four regular classes of fit:

Class I. The loosest fit, with no possibility for interference between screw and tapped hole.

Class II. Medium, or "free fit," but permitting slight interference in the worst combination of maximum screw and minimum nut.

Class III. Close tolerances on mating parts may require this fit, which applies to the highest grade of interchangeable work.

Class IV. A fine snug fit, where a screw driver or wrench may be necessary for assembly. To secure this fit, selective assembly of parts may be required.

An additional class V, or jam fit, is recognized for studs.

The Unified System. The Whitworth and National thread forms do not assemble because of the difference in thread angle. Military experiences in World Wars I and II with waste, confusion, and near disasters because of the lack of interchangeability of the two systems finally caused the governments of Great Britain, Canada, and the United States to agree to a Unified system. The 60° thread angle was adopted because American industry is too large to permit scrapping existing tools and gages. However, the British may use rounded crests and roots, and their product will assemble with that made in United States plants.

Threading costs will be reduced because of a realistic system of tolerance classes: three for screws—1A, 2A, and 3A, and three for nuts—1B, 2B, and 3B. The designer is free to use practically any combination of thread classes, new or old, that suits his needs. Formerly, "class" was misconstrued as "fit," the implication being that both components should have the same class tolerances, that class 2 external should be used with class 2 internal, and so on. In the Unified system, class signifies tolerance, or tolerance and allowance. Fit is determined by the selected combination of classes for mating external and internal threads. Disposition of tolerances for Unified and American National threads is shown in Figs. 16, 17, and 18.

The new classes are:

1A and 1B. A fit giving quick and easy assembly, even when threads are bruised or dirty. Applications: Ordnance and special uses.

2A and 2B. Suited for the vast majority of commercial fasteners. A moderate allowance on class 2A permits high-cycle wrenching with minimum galling and seizure.

3A and 3B. No allowance is provided, but tolerances on "go" gages are within limits of product size. Applications are those where close fit and accuracy of lead and thread angle are required. Consistent production of these threads demands accurate equipment and efficient inspection.

Obviously, in practice the combination of class 2A external with class 2B internal threads will satisfy the great majority of fastener requirements and these are interchangeable with the American National class II. Present gages can be used.

Basic dimensions, allowances, and tolerances are given in Tables 13 to 20 for coarse, fine, extra-fine, and 8-thread series, both American National and Unified threads. The 12- and 16-thread series are of limited application, likewise the special combinations of classes. Persons having need of thread data beyond the normal usage should make reference to a voluminous report published by the ASME and

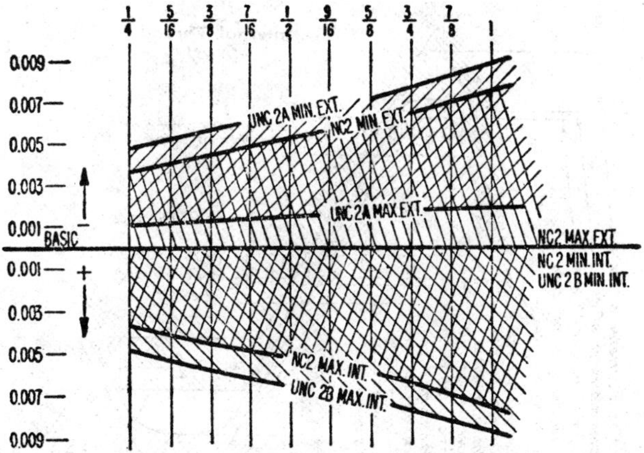

FIG. 16. Comparison of Unified and National pitch-diameter tolerances for coarse thread, class 2.

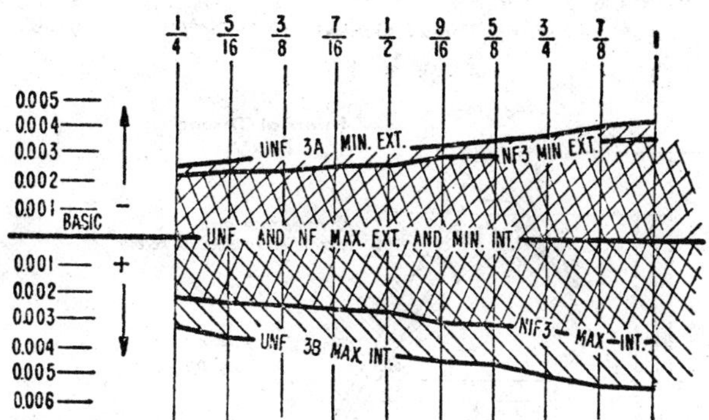

FIG. 17. Comparison of Unified and National pitch-diameter tolerances for fine thread, class 3.

entitled American Standard Unified and American Screw Threads: ASA B1.1-1949. See also Screw Thread Gages and Gaging: ASA B1.2-1951.

Limits for classes 2A and 2B in coarse, fine, and 8-thread series are given in Table 20 in sizes up to 1½ in. (the normal usage). Limits for larger sizes may be calculated from their basic dimensions and tolerances, using these rules:

For external threads:

Maximum diameters equal basic minus allowance (if any).
Minimum diameters equal maximum minus tolerances.

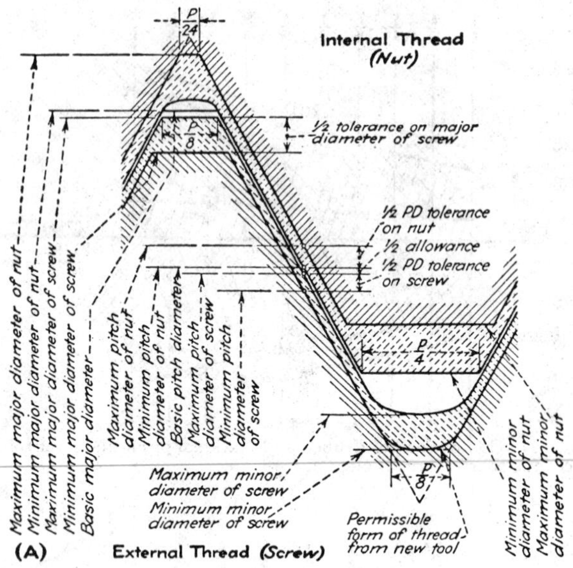

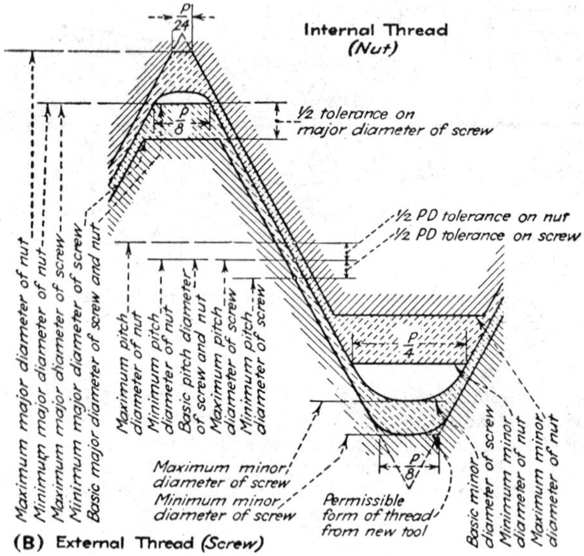

FIG. 18. Disposition of tolerances, allowances, and crest clearances for classes 1A, 1B, 2A, and 2B is shown in view *A*. Classes 3A and 3B, at view *B*, are for high-accuracy threads, because no allowance is provided, although in practice there is a slight clearance.

For internal threads:

Minimum diameters equal basic.
Maximum diameters equal minimum plus tolerances:

Thread formulas: The formulas for Unified and American threads are:

Number of threads = n
Pitch of thread = $p = 1/n$
Flat at crest and root of both internal and external threads = $f = 0.125p$
Truncation of external thread crest and internal thread root = $0.10825p$
Truncation of internal thread crest = $0.21651p$
Height of sharp V-thread = $H = 0.86603p$
Pitch dia $E = D - 0.64952/n$

TABLE 13. COARSE-THREAD SERIES—UNC AND NC
Basic Dimensions

Sizes	Basic Major Dia, D, In.	Threads per In., n	Basic Pitch Dia,* E, In.	Minor Dia External Threads, K_s, In.	Minor Dia Internal Threads, K_n, In.	Lead Angle at Pitch Dia, λ		Tensile Stress Area, Sq In.
						Deg	Min	
1 (0.073)	0.0730	64	0.0629	0.0538	0.0561	4	31	0.0026
2 (0.086)	0.0860	56	0.0744	0.0641	0.0667	4	22	0.0036
3 (0.099)	0.0990	48	0.0855	0.0734	0.0764	4	26	0.0048
4 (0.112)	**0.1120**	**40**	**0.0958**	**0.0813**	**0.0849**	**4**	**45**	**0.0060**
5 (0.125)	0.1250	40	0.1088	0.0943	0.0979	4	11	0.0079
6 (0.138)	**0.1380**	**32**	**0.1177**	**0.0997**	**0.1042**	**4**	**50**	**0.0090**
8 (0.164)	**0.1640**	**32**	**0.1437**	**0.1257**	**0.1302**	**3**	**58**	**0.0139**
10 (0.190)	**0.1900**	**24**	**0.1629**	**0.1389**	**0.1449**	**4**	**39**	**0.0174**
12 (0.216)	0.2160	24	0.1889	0.1649	0.1709	4	1	0.0240
1/4	**0.2500**	**20**	**0.2175**	**0.1887**	**0.1959**	**4**	**11**	**0.0317**
5/16	**0.3125**	**18**	**0.2764**	**0.2443**	**0.2524**	**3**	**40**	**0.0522**
3/8	**0.3750**	**16**	**0.3344**	**0.2983**	**0.3073**	**3**	**24**	**0.0773**
7/16	**0.4375**	**14**	**0.3911**	**0.3499**	**0.3602**	**3**	**20**	**0.1060**
1/2	**0.5000**	**13**	**0.4500**	**0.4056**	**0.4167**	**3**	**7**	**0.1416**
9/16	**0.5625**	**12**	**0.5084**	**0.4603**	**0.4723**	**2**	**59**	**0.1816**
5/8	**0.6250**	**11**	**0.5660**	**0.5135**	**0.5266**	**2**	**56**	**0.2256**
3/4	**0.7500**	**10**	**0.6850**	**0.6273**	**0.6417**	**2**	**40**	**0.3340**
7/8	**0.8750**	**9**	**0.8028**	**0.7387**	**0.7547**	**2**	**31**	**0.4612**
1	**1.0000**	**8**	**0.9188**	**0.8466**	**0.8647**	**2**	**29**	**0.6051**
1⅛	**1.1250**	**7**	**1.0322**	**0.9497**	**0.9704**	**2**	**31**	**0.7627**
1¼	**1.2500**	**7**	**1.1572**	**1.0747**	**1.0954**	**2**	**15**	**0.9684**
1⅜	**1.3750**	**6**	**1.2667**	**1.1705**	**1.1946**	**2**	**24**	**1.1538**
1½	**1.5000**	**6**	**1.3917**	**1.2955**	**1.3196**	**2**	**11**	**1.4041**
1¾	**1.7500**	**5**	**1.6201**	**1.5046**	**1.5335**	**2**	**15**	**1.8983**
2	**2.0000**	**4½**	**1.8557**	**1.7274**	**1.7594**	**2**	**11**	**2.4971**
2¼	**2.2500**	**4½**	**2.1057**	**1.9774**	**2.0094**	**1**	**55**	**3.2464**
2½	**2.5000**	**4**	**2.3376**	**2.1933**	**2.2294**	**1**	**57**	**3.9976**
2¾	**2.7500**	**4**	**2.5876**	**2.4433**	**2.4794**	**1**	**46**	**4.9326**
3	**3.0000**	**4**	**2.8376**	**2.6933**	**2.7294**	**1**	**36**	**5.9659**
3¼	**3.2500**	**4**	**3.0876**	**2.9433**	**2.9794**	**1**	**29**	**7.0992**
3½	**3.5000**	**4**	**3.3376**	**3.1933**	**3.2294**	**1**	**22**	**8.3268**
3¾	**3.7500**	**4**	**3.5876**	**3.4433**	**3.4794**	**1**	**16**	**9.6546**
4	**4.0000**	**4**	**3.8376**	**3.6933**	**3.7294**	**1**	**11**	**11.0805**

Bold type indicates Unified threads—UNC (the four numbered sizes will still be designated NC).

* British, effective dia.

TABLE 14. COARSE-THREAD SERIES—UNC AND NC—ALLOWANCES AND TOLERANCES

Sizes	Basic Major Dia, D, In.	Threads per In., n	Allowances,* In.	Tolerances, In.													
				Major Dia External Threads†			Minor Dia Internal Threads‡		Pitch Dia§								
			Classes 1A and 2A	Class 1A	Classes 2A and 3A	Classes 2 and 3	Classes 1B, 2B, and 3B	Classes 2 and 3	Class 1A	Class 1B	Class 2A	Class 2B	Class 3A	Class 3B	Class 2	Class 3
1 (0.073)	0.0730	64	0.0006	......	0.0038	0.0038	0.0062	0.0062	......	......	0.0020	0.0026	......	......	0.0019	0.0014
2 (0.086)	0.0860	56	0.0006	......	0.0041	0.0040	0.0070	0.0070	......	......	0.0021	0.0028	......	......	0.0020	0.0015
3 (0.099)	0.0990	48	0.0007	......	0.0045	0.0044	0.0081	0.0077	......	......	0.0023	0.0030	......	......	0.0022	0.0016
4 (0.112)	**0.1120**	**40**	**0.0008**	......	**0.0051**	0.0048	**0.0090**	0.0089	......	......	**0.0025**	**0.0033**	......	......	0.0024	0.0017
5 (0.125)	0.1250	40	0.0008	......	0.0051	0.0048	0.0083	0.0083	......	......	0.0026	0.0033	......	......	0.0024	0.0017
6 (0.138)	**0.1380**	**32**	**0.0008**	......	**0.0060**	0.0054	**0.0098**	0.0103	......	......	**0.0028**	**0.0037**	......	......	0.0027	0.0019
8 (0.164)	**0.1640**	**32**	**0.0009**	......	**0.0060**	0.0054	**0.0087**	0.0082	......	......	**0.0029**	**0.0038**	......	......	0.0027	0.0019
10 (0.190)	**0.1900**	**24**	**0.0010**	......	**0.0072**	0.0066	**0.0106**	0.0110	......	......	**0.0033**	**0.0043**	......	......	0.0033	0.0024
12 (0.216)	0.2160	24	0.0010	......	0.0072	0.0066	0.0098	0.0092	......	......	0.0034	0.0044	......	......	0.0033	0.0024
1/4	**0.2500**	**20**	**0.0011**	**0.0122**	**0.0081**	0.0072	**0.0108**	0.0101	**0.0056**	**0.0073**	**0.0037**	**0.0048**	**0.0028**	**0.0036**	0.0036	0.0026
5/16	**0.3125**	**18**	**0.0012**	**0.0131**	**0.0087**	0.0082	**0.0106**	0.0106	**0.0061**	**0.0079**	**0.0040**	**0.0053**	**0.0030**	**0.0039**	0.0041	0.0030
3/8	**0.3750**	**16**	**0.0013**	**0.0142**	**0.0094**	0.0090	**0.0109**	0.0111	**0.0065**	**0.0085**	**0.0044**	**0.0057**	**0.0033**	**0.0043**	0.0045	0.0032
7/16	**0.4375**	**14**	**0.0014**	**0.0155**	**0.0103**	0.0098	**0.0115**	0.0119	**0.0071**	**0.0092**	**0.0047**	**0.0061**	**0.0035**	**0.0046**	0.0049	0.0036
1/2	**0.5000**	**13**	**0.0015**	**0.0163**	**0.0109**	0.0104	**0.0117**	0.0123	**0.0074**	**0.0097**	**0.0050**	**0.0065**	**0.0037**	**0.0048**	0.0052	0.0037
9/16	**0.5625**	**12**	**0.0016**	**0.0172**	**0.0114**	0.0112	**0.0120**	0.0127	**0.0078**	**0.0102**	**0.0052**	**0.0068**	**0.0039**	**0.0051**	0.0056	0.0040
5/8	**0.6250**	**11**	**0.0016**	**0.0182**	**0.0121**	0.0118	**0.0125**	0.0131	**0.0083**	**0.0107**	**0.0055**	**0.0072**	**0.0041**	**0.0054**	0.0059	0.0042
3/4	**0.7500**	**10**	**0.0018**	**0.0194**	**0.0129**	0.0128	**0.0128**	0.0136	**0.0088**	**0.0115**	**0.0059**	**0.0077**	**0.0044**	**0.0057**	0.0064	0.0045
7/8	**0.8750**	**9**	**0.0019**	**0.0208**	**0.0139**	0.0140	**0.0134**	0.0142	**0.0095**	**0.0123**	**0.0063**	**0.0082**	**0.0047**	**0.0061**	0.0070	0.0049
1	**1.0000**	**8**	**0.0020**	**0.0225**	**0.0150**	0.0152	**0.0150**	0.0148	**0.0101**	**0.0132**	**0.0068**	**0.0088**	**0.0051**	**0.0066**	0.0076	0.0054
1⅛	**1.1250**	**7**	**0.0022**	**0.0246**	**0.0164**	0.0170	**0.0171**	0.0154	**0.0109**	**0.0141**	**0.0072**	**0.0094**	**0.0054**	**0.0071**	0.0085	0.0059
1¼	**1.2500**	**7**	**0.0022**	**0.0246**	**0.0164**	0.0170	**0.0171**	0.0154	**0.0111**	**0.0144**	**0.0074**	**0.0096**	**0.0055**	**0.0072**	0.0085	0.0059
1⅜	**1.3750**	**6**	**0.0024**	**0.0273**	**0.0182**	0.0202	**0.0200**	0.0180	**0.0120**	**0.0155**	**0.0080**	**0.0104**	**0.0060**	**0.0078**	0.0101	0.0071
1½	**1.5000**	**6**	**0.0024**	**0.0273**	**0.0182**	0.0202	**0.0200**	0.0180	**0.0121**	**0.0158**	**0.0081**	**0.0105**	**0.0061**	**0.0079**	0.0101	0.0071
1¾	**1.7500**	**5**	**0.0027**	**0.0308**	**0.0205**	0.0232	**0.0240**	0.0216	**0.0134**	**0.0174**	**0.0089**	**0.0116**	**0.0067**	**0.0087**	0.0116	0.0082
2	**2.0000**	**4½**	**0.0029**	**0.0330**	**0.0220**	0.0254	**0.0267**	0.0241	**0.0143**	**0.0186**	**0.0095**	**0.0124**	**0.0071**	**0.0093**	0.0127	0.0089
2¼	**2.2500**	**4½**	**0.0029**	**0.0330**	**0.0220**	0.0254	**0.0267**	0.0241	**0.0146**	**0.0190**	**0.0097**	**0.0126**	**0.0073**	**0.0095**	0.0127	0.0089
2½	**2.5000**	**4**	**0.0031**	**0.0357**	**0.0238**	0.0280	**0.0300**	0.0270	**0.0155**	**0.0202**	**0.0104**	**0.0135**	**0.0078**	**0.0101**	0.0140	0.0097
2¾	**2.7500**	**4**	**0.0032**	**0.0357**	**0.0238**	0.0280	**0.0300**	0.0270	**0.0158**	**0.0206**	**0.0105**	**0.0137**	**0.0079**	**0.0103**	0.0140	0.0097

Sizes	Basic Major Dia, *D*, In.	Threads per In., *n*	Allowances,* In.	Tolerances, In.												
				Major Dia External Threads†			Minor Dia Internal Threads‡		Pitch Dia§							
			Classes 1A and 2A	Class 1A	Classes 2A and 3A	Classes 2 and 3	Classes 1B, 2B, and 3B	Classes 2 and 3	Class 1A	Class 1B	Class 2A	Class 2B	Class 3A	Class 3B	Class 2	Class 3
3	**3.0000**	**4**	**0.0032**	**0.0357**	**0.0238**	0.0280	**0.0300**	0.0270	**0.0161**	**0.0209**	**0.0107**	**0.0139**	**0.0080**	**0.0104**	0.0140	0.0097
3¼	**3.2500**	**4**	**0.0033**	**0.0357**	**0.0238**	0.0280	**0.0300**	0.0270	**0.0163**	**0.0212**	**0.0109**	**0.0141**	**0.0082**	**0.0106**	0.0140	0.0097
3½	**3.5000**	**4**	**0.0033**	**0.0357**	**0.0238**	0.0280	**0.0300**	0.0270	**0.0166**	**0.0215**	**0.0110**	**0.0143**	**0.0083**	**0.0108**	0.0140	0.0097
3¾	**3.7500**	**4**	**0.0034**	**0.0357**	**0.0238**	0.0280	**0.0300**	0.0270	**0.0168**	**0.0218**	**0.0112**	**0.0145**	**0.0084**	**0.0109**	0.0140	0.0097
4	**4.0000**	**4**	**0.0034**	**0.0357**	**0.0238**	0.0280	**0.0300**	0.0270	**0.0170**	**0.0221**	**0.0113**	**0.0147**	**0.0085**	**0.0111**	0.0140	0.0097

Bold type indicates Unified threads—UNC (the four numbered sizes will still be designated NC).
The above values are based on a length of engagement equal to the nominal dia.

* Allowances apply to external threads classes 1A and 2A only.
† Major dia of internal threads may extend to a $p/24$ flat.
‡ Minor dia of external threads may extend to a $p/8$ flat.
§ British, effective dia.

TABLE 15. FINE-THREAD SERIES—UNF AND NF (Basic Dimensions)

Sizes	Basic Major Dia, D, In.	Threads per In., n	Basic Pitch Dia,* E, In.	Minor Dia External Threads K_s, In.	Minor Dia Internal Threads K_n, In.	Lead Angle at Pitch Dia, λ		Tensile Stress Area, Sq In.
						Deg	Min	
0 (0.060)	0.0600	80	0.0519	0.0447	0.0465	4	23	0.0018
1 (0.073)	0.0730	72	0.0640	0.0560	0.0580	3	57	0.0027
2 (0.086)	0.0860	64	0.0759	0.0668	0.0691	3	45	0.0039
3 (0.099)	0.0990	56	0.0874	0.0771	0.0797	3	43	0.0052
4 (0.112)	0.1120	48	0.0985	0.0864	0.0894	3	51	0.0065
5 (0.125)	0.1250	44	0.1102	0.0971	0.1004	3	45	0.0082
6 (0.138)	0.1380	40	0.1218	0.1073	0.1109	3	44	0.0101
8 (0.164)	0.1640	36	0.1460	0.1299	0.1339	3	28	0.0146
10 (0.190)	**0.1900**	**32**	**0.1697**	**0.1517**	**0.1562**	**3**	**21**	**0.0199**
12 (0.216)	0.2160	28	0.1928	0.1722	0.1773	3	22	0.0257
1/4	**0.2500**	**28**	**0.2268**	**0.2062**	**0.2113**	**2**	**52**	**0.0362**
5/16	**0.3125**	**24**	**0.2854**	**0.2614**	**0.2674**	**2**	**40**	**0.0579**
3/8	**0.3750**	**24**	**0.3479**	**0.3239**	**0.3299**	**2**	**11**	**0.0876**
7/16	**0.4375**	**20**	**0.4050**	**0.3762**	**0.3834**	**2**	**15**	**0.1185**
1/2	**0.5000**	**20**	**0.4675**	**0.4387**	**0.4459**	**1**	**57**	**0.1597**
9/16	**0.5625**	**18**	**0.5264**	**0.4943**	**0.5024**	**1**	**55**	**0.2026**
5/8	**0.6250**	**18**	**0.5889**	**0.5568**	**0.5649**	**1**	**43**	**0.2555**
3/4	**0.7500**	**16**	**0.7094**	**0.6733**	**0.6823**	**1**	**36**	**0.3724**
7/8	**0.8750**	**14**	**0.8286**	**0.7874**	**0.7977**	**1**	**34**	**0.5088**
1†	1.0000	14	0.9536	0.9124	0.9227	1	22	0.6791
1	**1.0000**	**12**	**0.9459**	**0.8978**	**0.9098**	**1**	**36**	**0.6624**
1⅛	**1.1250**	**12**	**1.0709**	**1.0228**	**1.0348**	**1**	**25**	**0.8549**
1¼	**1.2500**	**12**	**1.1959**	**1.1478**	**1.1598**	**1**	**16**	**1.0721**
1⅜	**1.3750**	**12**	**1.3209**	**1.2728**	**1.2848**	**1**	**9**	**1.3137**
1½	**1.5000**	**12**	**1.4459**	**1.3978**	**1.4098**	**1**	**3**	**1.5799**

Bold type indicates Unified threads—UNF (the 10–32 size will still be designated NF).

* British, effective dia.

† This thread should be designated NS when used and is not recommended for new design.

TABLE 16. EXTRA-FINE-THREAD SERIES—NEF (Basic Dimensions)

Sizes	Basic Major Dia, D, In.	Threads per In., n	Basic Pitch Dia, E, In.	Minor Dia External Threads K_s, In.	Minor Dia Internal Threads K_n, In.	Lead Angle Deg	Lead Angle Min	Tensile Stress Area, Sq In.
12 (0.216)	0.2160	32	0.1957	0.1777	0.1822	2	55	0.0269
¼	0.2500	32	0.2297	0.2117	0.2162	2	29	0.0377
5/16	0.3125	32	0.2922	0.2742	0.2787	1	57	0.0622
⅜	0.3750	32	0.3547	0.3367	0.3412	1	36	0.0929
7/16	0.4375	28	0.4143	0.3937	0.3988	1	34	0.1270
½	0.5000	28	0.4768	0.4562	0.4613	1	22	0.1695
9/16	0.5625	24	0.5354	0.5114	0.5174	1	25	0.2134
⅝	0.6250	24	0.5979	0.5739	0.5799	1	16	0.2676
11/16	0.6875	24	0.6604	0.6364	0.6424	1	9	0.3280
¾	0.7500	20	0.7175	0.6887	0.6959	1	16	0.3855
13/16	0.8125	20	0.7800	0.7512	0.7584	1	10	0.4573
⅞	0.8750	20	0.8425	0.8137	0.8209	1	5	0.5352
15/16	0.9375	20	0.9050	0.8762	0.8834	1	0	0.6194
1	1.0000	20	0.9675	0.9387	0.9459	0	57	0.7095
1 1/16	1.0625	18	1.0264	0.9943	1.0024	0	59	0.7973
1⅛	1.1250	18	1.0889	1.0568	1.0649	0	56	0.8993
1 3/16	1.1875	18	1.1514	1.1193	1.1274	0	53	1.0074
1¼	1.2500	18	1.2139	1.1818	1.1899	0	50	1.1216
1 5/16	1.3125	18	1.2764	1.2443	1.2524	0	48	1.2420
1⅜	1.3750	18	1.3389	1.3068	1.3149	0	45	1.3684
1 7/16	1.4375	18	1.4014	1.3693	1.3774	0	43	1.5010
1½	1.5000	18	1.4639	1.4318	1.4399	0	42	1.6397
1 9/16	1.5625	18	1.5264	1.4943	1.5024	0	40	1.7846
1⅝	1.6250	18	1.5889	1.5568	1.5649	0	38	1.9357
1 11/16	1.6875	18	1.6514	1.6193	1.6274	0	37	2.0929
1¾	1.7500	16	1.7094	1.6733	1.6823	0	40	2.2382
2	2.0000	16	1.9594	1.9233	1.9323	0	35	2.9501

TABLE 17. FINE-THREAD SERIES—UNF AND NF—ALLOWANCES AND TOLERANCES

Sizes	Basic Major Dia, D, In.	Threads per In., n	Allowances In.[a]	Tolerances, In.												
				Major Dia External Threads[b]			Minor Dia Internal Threads[c]		Pitch Dia[d]							
			Classes 1A and 2A	Class 1A	Classes 2A and 3A	Classes 2 and 3	Classes 1B, 2B, and 3B	Classes 2 and 3	Class 1A	Class 1B	Class 2A	Class 2B	Class 3A	Class 3B	Class 2	Class 3
0 (0.060)	0.0600	80	0.0005	......	0.0032	0.0034	0.0049	0.0049	......	......	0.0018	0.0023	......	......	0.0017	0.0013
1 (0.073)	0.0730	72	0.0006	......	0.0035	0.0036	0.0055	0.0054	......	......	0.0019	0.0025	......	......	0.0018	0.0013
2 (0.086)	0.0860	64	0.0006	......	0.0038	0.0038	0.0062	0.0055	......	......	0.0020	0.0027	......	......	0.0019	0.0014
3 (0.099)	0.0990	56	0.0007	......	0.0041	0.0040	0.0068	0.0059	......	......	0.0022	0.0028	......	......	0.0020	0.0015
4 (0.112)	0.1120	48	0.0007	......	0.0045	0.0044	0.0074	0.0066	......	......	0.0024	0.0031	......	......	0.0022	0.0016
5 (0.125)	0.1250	44	0.0007	......	0.0048	0.0046	0.0075	0.0064	......	......	0.0025	0.0032	......	......	0.0023	0.0016
6 (0.138)	0.1380	40	0.0008	......	0.0051	0.0048	0.0077	0.0070	......	......	0.0026	0.0034	......	......	0.0024	0.0017
8 (0.164)	0.1640	36	0.0008	......	0.0055	0.0050	0.0077	0.0063	......	......	0.0028	0.0036	......	......	0.0025	0.0018
10 (0.190)	**0.1900**	**32**	**0.0009**	......	**0.0060**	0.0054	**0.0079**	0.0062	......	......	**0.0030**	**0.0039**	......	......	0.0027	0.0019
12 (0.216)	0.2160	28	0.0010	......	0.0065	0.0062	0.0084	0.0062	......	......	0.0032	0.0042	......	......	0.0031	0.0022
1/4	**0.2500**	**28**	**0.0010**	**0.0098**	**0.0065**	0.0062	**0.0077**	0.0060	**0.0050**	**0.0065**	**0.0033**	**0.0043**	**0.0025**	**0.0032**	0.0031	0.0022
5/16	**0.3125**	**24**	**0.0011**	**0.0108**	**0.0072**	0.0066	**0.0080**	0.0065	**0.0055**	**0.0071**	**0.0037**	**0.0048**	**0.0027**	**0.0036**	0.0033	0.0024
3/8	**0.3750**	**24**	**0.0011**	**0.0108**	**0.0072**	0.0066	**0.0073**	0.0065	**0.0057**	**0.0074**	**0.0038**	**0.0049**	**0.0029**	**0.0037**	0.0033	0.0024
7/16	**0.4375**	**20**	**0.0013**	**0.0122**	**0.0081**	0.0072	**0.0082**	0.0072	**0.0062**	**0.0081**	**0.0042**	**0.0054**	**0.0031**	**0.0041**	0.0036	0.0026
1/2	**0.5000**	**20**	**0.0013**	**0.0122**	**0.0081**	0.0072	**0.0078**	0.0072	**0.0064**	**0.0084**	**0.0043**	**0.0056**	**0.0032**	**0.0042**	0.0036	0.0026
9/16	**0.5625**	**18**	**0.0014**	**0.0131**	**0.0087**	0.0082	**0.0082**	0.0076	**0.0068**	**0.0089**	**0.0045**	**0.0059**	**0.0034**	**0.0044**	0.0041	0.0030
5/8	**0.6250**	**18**	**0.0014**	**0.0131**	**0.0087**	0.0082	**0.0081**	0.0076	**0.0070**	**0.0091**	**0.0047**	**0.0060**	**0.0035**	**0.0045**	0.0041	0.0030
3/4	**0.7500**	**16**	**0.0015**	**0.0142**	**0.0094**	0.0090	**0.0085**	0.0080	**0.0075**	**0.0098**	**0.0050**	**0.0065**	**0.0038**	**0.0049**	0.0045	0.0032
7/8	**0.8750**	**14**	**0.0016**	**0.0155**	**0.0103**	0.0098	**0.0091**	0.0085	**0.0081**	**0.0106**	**0.0054**	**0.0070**	**0.0041**	**0.0053**	0.0049	0.0036
1[e]	1.0000	14	0.0017	0.0155	0.0103	0.0098	0.0088	0.0085	0.0084	0.0109	0.0056	0.0073	0.0042	0.0054	0.0049	0.0036
1	**1.0000**	**12**	**0.0018**	**0.0172**	**0.0114**	......	**0.0100**	......	**0.0088**	**0.0114**	**0.0059**	**0.0076**	**0.0044**	**0.0057**		
1⅛	**1.1250**	**12**	**0.0018**	**0.0172**	**0.0114**	0.0112	**0.0100**	0.0090	**0.0090**	**0.0117**	**0.0060**	**0.0078**	**0.0045**	**0.0059**	0.0056	0.0040
1¼	**1.2500**	**12**	**0.0018**	**0.0172**	**0.0114**	0.0112	**0.0100**	0.0090	**0.0092**	**0.0120**	**0.0062**	**0.0080**	**0.0046**	**0.0060**	0.0056	0.0040
1⅜	**1.3750**	**12**	**0.0019**	**0.0172**	**0.0114**	0.0112	**0.0100**	0.0090	**0.0094**	**0.0123**	**0.0063**	**0.0082**	**0.0047**	**0.0061**	0.0056	0.0040
1½	**1.5000**	**12**	**0.0019**	**0.0172**	**0.0114**	0.0112	**0.0100**	0.0090	**0.0096**	**0.0125**	**0.0064**	**0.0083**	**0.0048**	**0.0063**	0.0056	0.0040

Bold type indicates Unified threads—UNF (the 10-32 size will still be designated NF).
The above values are based on a length of engagement equal to the nominal dia.

[a] Allowances apply to external threads classes 1A and 2A only.
[b] Major dia of internal threads may extend to a $p/24$ flat.
[c] Minor dia of external threads may extend to a $p/8$ flat.
[d] British, effective dia.
[e] **This thread should be designated NS when used and is not recommended for new design.**

TABLE 18. EXTRA-FINE-THREAD SERIES—UNEF AND NEF
Allowances and Tolerances, In.

Sizes	Threads per In., n	Allowances* Class 2A	Tolerances							
			Major Dia External Threads†		Pitch Dia‡				Minor Dia Internal Threads§	
			Class 2A	Classes 2 and 3	Class 2A	Class 2B	Class 2	Class 3	Class 2B	Classes 2 and 3
12 (0.216)	32	0.0009	0.0060	0.0054	0.0031	0.0041	0.0031	0.0022	0.0073	0.0053
1/4	32	0.0010	0.0060	0.0054	0.0032	0.0042	0.0032	0.0022	0.0067	0.0048
5/16	32	0.0010	0.0060	0.0054	0.0032	0.0042	0.0033	0.0023	0.0060	0.0048
3/8	32	0.0010	0.0060	0.0054	0.0034	0.0044	0.0034	0.0024	0.0057	0.0048
7/16	**28**	**0.0011**	**0.0065**	0.0062	**0.0036**	**0.0046**	0.0036	0.0025	**0.0063**	0.0056
1/2	**28**	**0.0011**	**0.0065**	0.0062	**0.0037**	**0.0048**	0.0037	0.0026	**0.0063**	0.0056
9/16	24	0.0012	0.0072	0.0066	0.0039	0.0051	0.0040	0.0028	0.0070	0.0064
5/8	24	0.0012	0.0072	0.0066	0.0040	0.0052	0.0041	0.0029	0.0070	0.0064
11/16	24	0.0012	0.0072	0.0066	0.0040	0.0052	0.0041	0.0029	0.0070	0.0064
3/4	**20**	**0.0013**	**0.0081**	0.0072	**0.0044**	**0.0057**	0.0046	0.0032	**0.0078**	0.0072
13/16	**20**	**0.0013**	**0.0081**	0.0072	**0.0044**	**0.0057**	0.0046	0.0032	**0.0078**	0.0072
7/8	**20**	**0.0013**	**0.0081**	0.0072	**0.0044**	**0.0057**	0.0047	0.0033	**0.0078**	0.0072
15/16	**20**	**0.0014**	**0.0081**	0.0072	**0.0045**	**0.0059**	0.0047	0.0033	**0.0078**	0.0072
1	**20**	**0.0014**	**0.0081**	0.0072	**0.0045**	**0.0059**	0.0048	0.0034	**0.0078**	0.0072
1 1/16	18	0.0014	0.0087	0.0082	0.0047	0.0062	0.0051	0.0036	0.0081	0.0076
1 1/8	18	0.0014	0.0087	0.0082	0.0047	0.0062	0.0052	0.0036	0.0081	0.0076
1 3/16	18	0.0015	0.0087	0.0082	0.0049	0.0063	0.0052	0.0036	0.0081	0.0076
1 1/4	18	0.0015	0.0087	0.0082	0.0049	0.0063	0.0053	0.0037	0.0081	0.0076
1 5/16	18	0.0015	0.0087	0.0082	0.0049	0.0063	0.0053	0.0037	0.0081	0.0076
1 3/8	18	0.0015	0.0087	0.0082	0.0049	0.0063	0.0054	0.0038	0.0081	0.0076
1 7/16	18	0.0015	0.0087	0.0082	0.0050	0.0065	0.0054	0.0038	0.0081	0.0076
1 1/2	18	0.0015	0.0087	0.0082	0.0050	0.0065	0.0055	0.0038	0.0081	0.0076
1 9/16	18	0.0015	0.0087	0.0082	0.0050	0.0065	0.0055	0.0039	0.0081	0.0076
1 5/8	18	0.0015	0.0087	0.0082	0.0050	0.0065	0.0056	0.0039	0.0081	0.0076
1 11/16	18	0.0015	0.0087	0.0082	0.0051	0.0066	0.0056	0.0039	0.0081	0.0076
1 3/4	**16**	**0.0016**	**0.0094**	0.0090	**0.0053**	**0.0069**	0.0059	0.0041	**0.0085**	0.0080
2	**16**	**0.0016**	**0.0094**	0.0090	**0.0054**	**0.0070**	0.0061	0.0043	**0.0085**	0.0080

Bold type indicates Unified threads—UNEF.
The above values are based on a length of engagement of nine threads.

* Allowances apply to external threads class 2A only.
† Major dia of internal threads may extend to a $p/24$ flat.
‡ British, effective dia.
§ Minor dia of external threads may extend to a $p/8$ flat.

TABLE 19. 8-THREAD SERIES—8N

Sizes, In.	Basic Dimensions						Allowances and Tolerances, In.								
	Basic Pitch Dia,* E, In.	Minor Dia External Threads K_s, In.	Minor Dia Internal Threads K_n, In.	Lead Angle at Pitch Dia, λ		Tensile Stress Area, Sq In.	Allowances† Class 2A	Tolerances							
								Major Dia External Threads‡		Pitch Dia*				Minor Dia Internal Threads§	
				Deg	Min			Class 2A	Classes 2 and 3	Class 2A	Class 2B	Class 2	Class 3	Class 2B	Classes 2 and 3
1	0.9188	0.8466	0.8647	2	29	0.6051	0.0020	0.0150	0.0152	0.0068	0.0088	0.0076	0.0054	0.0150	0.0148
1 1/8	1.0438	0.9716	0.9897	2	11	0.7896	0.0021	0.0150	0.0152	0.0069	0.0090	0.0079	0.0055	0.0150	0.0148
1 1/4	1.1688	1.0966	1.1147	1	57	0.9985	0.0021	0.0150	0.0152	0.0070	0.0092	0.0083	0.0058	0.0150	0.0148
1 3/8	1.2938	1.2216	1.2397	1	46	1.2319	0.0022	0.0150	0.0152	0.0072	0.0093	0.0086	0.0061	0.0150	0.0148
1 1/2	1.4188	1.3466	1.3647	1	36	1.4899	0.0022	0.0150	0.0152	0.0073	0.0095	0.0090	0.0063	0.0150	0.0148
1 5/8	1.5438	1.4716	1.4897	1	29	1.7723	0.0022	0.0150	0.0152	0.0074	0.0097	0.0093	0.0065	0.0150	0.0148
1 3/4	1.6688	1.5966	1.6147	1	22	2.0792	0.0023	0.0150	0.0152	0.0075	0.0098	0.0097	0.0068	0.0150	0.0148
1 7/8	1.7938	1.7216	1.7397	1	16	2.4107	0.0023	0.0150	0.0152	0.0077	0.0100	0.0100	0.0070	0.0150	0.0148
2	1.9188	1.8466	1.8647	1	11	2.7665	0.0023	0.0150	0.0152	0.0078	0.0101	0.0104	0.0073	0.0150	0.0148
2 1/8	2.0438	1.9716	1.9897	1	7	3.1469	0.0024	0.0150	0.0152	0.0079	0.0102	0.0107	0.0075	0.0150	0.0148
2 1/4	2.1688	2.0966	2.1147	1	3	3.5519	0.0024	0.0150	0.0152	0.0080	0.0104	0.0110	0.0077	0.0150	0.0148
2 1/2	2.4188	2.3466	2.3647	0	57	4.4352	0.0024	0.0150	0.0152	0.0082	0.0106	0.0117	0.0082	0.0150	0.0148
2 3/4	2.6688	2.5966	2.6147	0	51	5.4164	0.0025	0.0150	0.0152	0.0083	0.0108	0.0124	0.0087	0.0150	0.0148
3	2.9188	2.8466	2.8647	0	47	6.4957	0.0026	0.0150	0.0152	0.0085	0.0111	0.0130	0.0092	0.0150	0.0148
3 1/4	3.1688	3.0966	3.1147	0	43	7.6738	0.0026	0.0150	0.0152	0.0087	0.0113	0.0132	0.0093	0.0150	0.0148
3 1/2	3.4188	3.3466	3.3647	0	40	8.9504	0.0026	0.0150	0.0152	0.0088	0.0115	0.0133	0.0093	0.0150	0.0148
3 3/4	3.6688	3.5966	3.6147	0	37	10.3249	0.0027	0.0150	0.0152	0.0090	0.0117	0.0134	0.0094	0.0150	0.0148
4	3.9188	3.8466	3.8647	0	35	11.7975	0.0027	0.0150	0.0152	0.0091	0.0119	0.0135	0.0095	0.0150	0.0148
4 1/4	4.1688	4.0966	4.1147	0	33	13.3683	0.0028	0.0150	0.0152	0.0093	0.0121	0.0137	0.0096	0.0150	0.0148
4 1/2	4.4188	4.3466	4.3647	0	31	15.0372	0.0028	0.0150	0.0152	0.0094	0.0122	0.0138	0.0097	0.0150	0.0148
4 3/4	4.6688	4.5966	4.6147	0	29	16.8042	0.0029	0.0150	0.0152	0.0095	0.0124	0.0139	0.0098	0.0150	0.0148
5	4.9188	4.8466	4.8647	0	28	18.6694	0.0029	0.0150	0.0152	0.0097	0.0126	0.0140	0.0099	0.0150	0.0148
5 1/4	5.1688	5.0966	5.1147	0	26	20.6330	0.0029	0.0150	0.0152	0.0098	0.0127	0.0141	0.0099	0.0150	0.0148
5 1/2	5.4188	5.3466	5.3647	0	25	22.6945	0.0030	0.0150	0.0152	0.0099	0.0129	0.0142	0.0100	0.0150	0.0148
5 3/4	5.6688	5.5966	5.6147	0	24	24.8541	0.0030	0.0150	0.0152	0.0100	0.0130	0.0143	0.0101	0.0150	0.0148
6	5.9188	5.8466	5.8647	0	23	27.1118	0.0030	0.0150	0.0152	0.0102	0.0132	0.0144	0.0102	0.0150	0.0148

The above values for classes 2A and 2B are based on a length of engagement equal to the nominal dia and for classes 2 and 3 on a length of engagement equal to the nominal dia for sizes up to and including 3 in. and on a length of engagement of 3 in. for larger sizes.

* British, effective dia.

† Allowances apply to external threads class 2A only.

‡ Major dia of internal threads may extend to a $p/24$ flat.

§ Minor dia of external threads may extend to a $p/8$ flat.

TABLE 20. LIMITS FOR UNC, UNF, NC, NF, AND 8N THREAD SERIES

Size	Threads per In.	Allowance, In.	Class 2A—External Threads: Major Dia, In.: Limits Max*	Limits Min†	Limits Min‡	Tolerance †	Tolerance ‡	Pitch Dia, In.§: Limits Max*	Limits Min	Tolerance	Minor Dia, In.	Class 2B—Internal Threads: Minor Dia, In.: Limits Min	Limits Max	Tolerance	Pitch Dia, In.§: Limits Min	Limits Max	Tolerance	Major Dia, In.: Min
0	80	0.0005	0.0595	0.0563	......	0.0032	......	0.0514	0.0496	0.0018	0.0042	0.0465	0.0514	0.0049	0.0519	0.0542	0.0023	0.0600
1	64	0.0006	0.0724	0.0686	......	0.0038	......	0.0623	0.0603	0.0020	0.0532	0.0561	0.0623	0.0062	0.0629	0.0655	0.0026	0.0730
1	72	0.0006	0.0724	0.0689	......	0.0035	......	0.0634	0.0615	0.0019	0.0554	0.0580	0.0635	0.0055	0.0640	0.0665	0.0025	0.0730
2	56	0.0006	0.0854	0.0813	......	0.0041	......	0.0738	0.0717	0.0021	0.0635	0.0667	0.0737	0.0070	0.0744	0.0772	0.0028	0.0860
2	64	0.0006	0.0854	0.0816	......	0.0038	......	0.0753	0.0733	0.0020	0.0662	0.0691	0.0753	0.0062	0.0759	0.0786	0.0027	0.0860
3	48	0.0007	0.0983	0.0938	......	0.0045	......	0.0848	0.0825	0.0023	0.0727	0.0764	0.0845	0.0081	0.0855	0.0885	0.0030	0.0990
3	56	0.0007	0.0983	0.0942	......	0.0041	......	0.0867	0.0845	0.0022	0.0764	0.0797	0.0865	0.0068	0.0874	0.0902	0.0028	0.0990
4	**40**	**0.0008**	**0.1112**	**0.1061**	......	**0.0051**	......	**0.0950**	**0.0925**	**0.0025**	**0.0805**	**0.0849**	**0.0939**	**0.0090**	**0.0958**	**0.0991**	**0.0033**	**0.1120**
4	48	0.0007	0.1113	0.1068	......	0.0045	......	0.0978	0.0954	0.0024	0.0857	0.0894	0.0968	0.0074	0.0985	0.1016	0.0031	0.1120
5	40	0.0008	0.1242	0.1191	......	0.0051	......	0.1080	0.1054	0.0026	0.0935	0.0979	0.1062	0.0083	0.1088	0.1121	0.0033	0.1250
5	44	0.0007	0.1243	0.1195	......	0.0048	......	0.1095	0.1070	0.0025	0.0964	0.1004	0.1079	0.0075	0.1102	0.1134	0.0032	0.1250
6	**32**	**0.0008**	**0.1372**	**0.1312**	......	**0.0060**	......	**0.1169**	**0.1141**	**0.0028**	**0.0989**	**0.1042**	**0.1140**	**0.0098**	**0.1177**	**0.1214**	**0.3307**	**0.1380**
6	40	0.0008	0.1372	0.1321	......	0.0051	......	0.1210	0.1184	0.0026	0.1065	0.1109	0.1186	0.0077	0.1218	0.1252	0.0034	0.1380
8	**32**	**0.0009**	**0.1631**	**0.1571**	......	**0.0060**	......	**0.1428**	**0.1399**	**0.0029**	**0.1248**	**0.1302**	**0.1389**	**0.0087**	**0.1437**	**0.1475**	**0.0038**	**0.1640**
8	36	0.0008	0.1632	0.1577	......	0.0055	......	0.1452	0.1424	0.0028	0.1291	0.1339	0.1416	0.0077	0.1460	0.1496	0.0036	0.1640
10	**24**	**0.0010**	**0.1890**	**0.1818**	......	**0.0072**	......	**0.1619**	**0.1586**	**0.0033**	**0.1379**	**0.1449**	**0.1555**	**0.0106**	**0.1629**	**0.1672**	**0.0043**	**0.1900**
10	**32**	**0.0009**	**0.1891**	**0.1831**	......	**0.0060**	......	**0.1688**	**0.1658**	**0.0030**	**0.1508**	**0.1562**	**0.1641**	**0.0079**	**0.1697**	**0.1736**	**0.0039**	**0.1900**
12	24	0.0010	0.2150	0.2078	......	0.0072	......	0.1879	0.1845	0.0034	0.1639	0.1709	0.1807	0.0098	0.1889	0.1933	0.0044	0.2160
12	28	0.0010	0.2150	0.2085	......	0.0065	......	0.1918	0.1886	0.0032	0.1712	0.1773	0.1857	0.0084	0.1928	0.1970	0.0042	0.2160
1/4	**20**	**0.0011**	**0.2489**	**0.2408**	0.2367	**0.0081**	0.0122	**0.2164**	**0.2127**	**0.0037**	**0.1876**	**0.1959**	**0.2067**	**0.0108**	**0.2175**	**0.2223**	**0.0048**	**0.2500**
1/4	**28**	**0.0010**	**0.2490**	**0.2425**	......	**0.0065**	......	**0.2258**	**0.2225**	**0.0033**	**0.2052**	**0.2113**	**0.2190**	**0.0077**	**0.2268**	**0.2311**	**0.0043**	**0.2500**
5/16	**18**	**0.0012**	**0.3113**	**0.3026**	0.2982	**0.0087**	0.0131	**0.2752**	**0.2712**	**0.0040**	**0.2431**	**0.2524**	**0.2630**	**0.0106**	**0.2764**	**0.2817**	**0.0053**	**0.3125**
5/16	**24**	**0.0011**	**0.3114**	**0.3042**	......	**0.0072**	......	**0.2843**	**0.2806**	**0.0037**	**0.2603**	**0.2674**	**0.2754**	**0.0080**	**0.2854**	**0.2902**	**0.0048**	**0.3125**
3/8	**16**	**0.0013**	**0.3737**	**0.3643**	0.3595	**0.0094**	0.0142	**0.3331**	**0.3287**	**0.0044**	**0.2970**	**0.3073**	**0.3182**	**0.0109**	**0.3344**	**0.3401**	**0.0057**	**0.3750**
3/8	**24**	**0.0011**	**0.3739**	**0.3667**	......	**0.0072**	......	**0.3468**	**0.3430**	**0.0038**	**0.3228**	**0.3299**	**0.3372**	**0.0073**	**0.3479**	**0.3528**	**0.0049**	**0.3750**

7/16	**14**	**0.0014**	**0.4361**	**0.4258**	0.4206	**0.0103**	0.0155	**0.3897**	**0.3850**	**0.0047**	**0.3485**	**0.3602**	**0.3717**	**0.0115**	**0.3911**	**0.3972**	**0.0061**	**0.4375**
7/16	**20**	**0.0013**	**0.4362**	**0.4281**	......	**0.0081**	......	**0.4037**	**0.3995**	**0.0042**	**0.3749**	**0.3834**	**0.3916**	**0.0082**	**0.4050**	**0.4104**	**0.0054**	**0.4375**
1/2	**13**	**0.0015**	**0.4985**	**0.4876**	0.4822	**0.0109**	0.0163	**0.4485**	**0.4435**	**0.0050**	**0.4041**	**0.4167**	**0.4284**	**0.0117**	**0.4500**	**0.4565**	**0.0065**	**0.5000**
1/2	**20**	**0.0013**	**0.4987**	**0.4906**	......	**0.0081**	......	**0.4662**	**0.4619**	**0.0043**	**0.4374**	**0.4459**	**0.4537**	**0.0078**	**0.4675**	**0.4731**	**0.0056**	**0.5000**
9/16	**12**	**0.0016**	**0.5609**	**0.5495**	0.5437	**0.0114**	0.0172	**0.5068**	**0.5016**	**0.0052**	**0.4587**	**0.4723**	**0.4843**	**0.0120**	**0.5084**	**0.5152**	**0.0068**	**0.5625**
9/16	**18**	**0.0014**	**0.5611**	**0.5524**	......	**0.0087**	......	**0.5250**	**0.5205**	**0.0045**	**0.4929**	**0.5024**	**0.5106**	**0.0082**	**0.5264**	**0.5323**	**0.0059**	**0.5625**
5/8	**11**	**0.0016**	**0.6234**	**0.6113**	0.6052	**0.0121**	0.0182	**0.5644**	**0.5589**	**0.0055**	**0.5119**	**0.5266**	**0.5391**	**0.0125**	**0.5660**	**0.5732**	**0.0072**	**0.6250**
5/8	**18**	**0.0014**	**0.6236**	**0.6149**	......	**0.0087**	......	**0.5875**	**0.5828**	**0.0047**	**0.5554**	**0.5649**	**0.5730**	**0.0081**	**0.5889**	**0.5949**	**0.0060**	**0.6250**
3/4	**10**	**0.0018**	**0.7482**	**0.7353**	0.7288	**0.0129**	0.0194	**0.6832**	**0.6773**	**0.0059**	**0.6255**	**0.6417**	**0.6545**	**0.0128**	**0.6850**	**0.6927**	**0.0077**	**0.7500**
3/4	**16**	**0.0015**	**0.7485**	**0.7391**	......	**0.0094**	......	**0.7079**	**0.7029**	**0.0050**	**0.6718**	**0.6823**	**0.6908**	**0.0085**	**0.7094**	**0.7159**	**0.0065**	**0.7500**
7/8	**9**	**0.0019**	**0.8731**	**0.8592**	0.8523	**0.0139**	0.0208	**0.8009**	**0.7946**	**0.0063**	**0.7368**	**0.7547**	**0.7681**	**0.0134**	**0.8028**	**0.8110**	**0.0082**	**0.8750**
7/8	**14**	**0.0016**	**0.8734**	**0.8631**	......	**0.0103**	......	**0.8270**	**0.8216**	**0.0054**	**0.7858**	**0.7977**	**0.8068**	**0.0091**	**0.8286**	**0.8356**	**0.0070**	**0.8750**
1	**8**	**0.0020**	**0.9980**	**0.9830**	0.9755	**0.0150**	0.0225	**0.9168**	**0.9100**	**0.0068**	**0.8446**	**0.8647**	**0.8797**	**0.0150**	**0.9188**	**0.9276**	**0.0088**	**1.0000**
1	**12**	**0.0018**	**0.9982**	**0.9868**	......	**0.0114**	......	**0.9441**	**0.9382**	**0.0059**	**0.8960**	**0.9098**	**0.9198**	**0.0100**	**0.9459**	**0.9535**	**0.0076**	**1.0000**
1⅛	**7**	**0.0022**	**1.1228**	**1.1064**	1.0982	**0.0164**	0.0246	**1.0300**	**1.0228**	**0.0072**	**0.9475**	**0.9704**	**0.9875**	**0.0171**	**1.0322**	**1.0416**	**0.0094**	**1.1250**
1⅛	8	0.0021	1.1229	1.1079	1.1004	0.0150	0.0225	1.0417	1.0348	0.0069	0.9695	0.9897	1.0047	0.0150	1.0438	1.0528	0.0090	1.1250
1⅛	**12**	**0.0018**	**1.1232**	**1.1118**	......	**0.0114**	......	**1.0691**	**1.0631**	**0.0060**	**1.0210**	**1.0348**	**1.0448**	**0.0100**	**1.0709**	**1.0787**	**0.0078**	**1.1250**
1¼	**7**	**0.0022**	**1.2478**	**1.2314**	1.2232	**0.0164**	0.0246	**1.1550**	**1.1476**	**0.0074**	**1.0725**	**1.0954**	**1.1125**	**0.0171**	**1.1572**	**1.1668**	**0.0096**	**1.2500**
1¼	8	0.0021	1.2479	1.2329	1.2254	0.0150	0.0225	1.1667	1.1597	0.0070	1.0945	1.1147	1.1297	0.0150	1.1688	1.1780	0.0092	1.2500
1¼	**12**	**0.0018**	**1.2482**	**1.2368**	......	**0.0114**	......	**1.1941**	**1.1879**	**0.0062**	**1.1460**	**1.1598**	**1.1698**	**0.0100**	**1.1959**	**1.2039**	**0.0080**	**1.2500**
1⅜	**6**	**0.0024**	**1.3726**	**1.3544**	1.3453	**0.0182**	0.0273	**1.2643**	**1.2563**	**0.0080**	**1.1681**	**1.1946**	**1.2146**	**0.0200**	**1.2667**	**1.2771**	**0.0104**	**1.3750**
1⅜	8	0.0022	1.3728	1.3578	1.3503	0.0150	0.0225	1.2916	1.2844	0.0072	1.2194	1.2397	1.2547	0.0150	1.2938	1.3031	0.0093	1.3750
1⅜	**12**	**0.0019**	**1.3731**	**1.3617**	......	**0.0114**	......	**1.3190**	**1.3127**	**0.0063**	**1.2709**	**1.2848**	**1.2948**	**0.0100**	**1.3209**	**1.3291**	**0.0082**	**1.3750**
1½	**6**	**0.0024**	**1.4976**	**1.4794**	1.4703	**0.0182**	0.0273	**1.3893**	**1.3812**	**0.0081**	**1.2931**	**1.3196**	**1.3396**	**0.0200**	**1.3917**	**1.4022**	**0.0105**	**1.5000**
1½	8	0.0022	1.4978	1.4828	1.4753	0.0150	0.0225	1.4166	1.4093	0.0073	1.3444	1.3647	1.3797	0.0150	1.4188	1.4283	0.0095	1.5000
1½	**12**	**0.0019**	**1.4981**	**1.4867**	......	**0.0144**	......	**1.4440**	**1.4376**	**0.0064**	**1.3959**	**1.4098**	**1.4198**	**0.0100**	**1.4459**	**1.4542**	**0.0083**	**1.5000**

For a given size, the coarse thread is given first, then the fine thread.

Bold type indicates Unified threads—UNC and UNF.

NOTE: Class 3B holes are frequently used with class 2A screws, to obtain a closer fit, especially in cast-iron and aluminum machine parts. Class 3B has no numbered sizes; otherwise all values in Table 14 apply to 3B except those for the maximum pitch dia and the PD tolerance. To obtain a 3B PD tolerance multiply its 2B value by 0.75 and add that to its minimum to get the 3B maximum PD limit.

Dimensions shown apply to the finished product whether it be coated, plated, or unplated.

* Maximum limits are increased by the amount of the allowance for threads which are electroplated or have coatings of similar thicknesses.

† For semifinished and finished screws and bolts, threaded portion only.

‡ For unfinished hot-rolled material, threaded portion only. These values for sizes above 1½ in. dia may be obtained by increasing the 2A major-dia tolerance from Table 14, 50%.

§ British, effective dia.

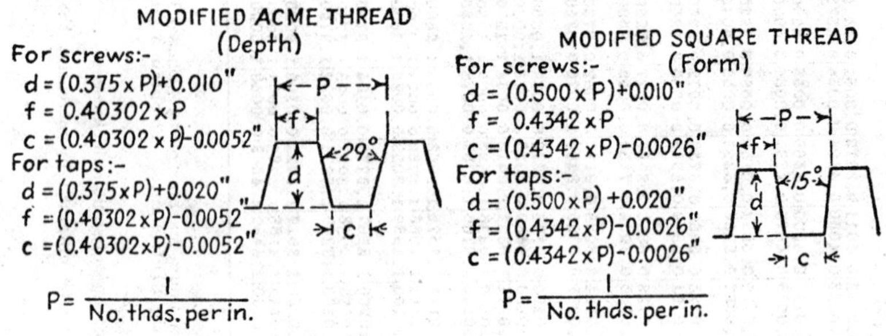

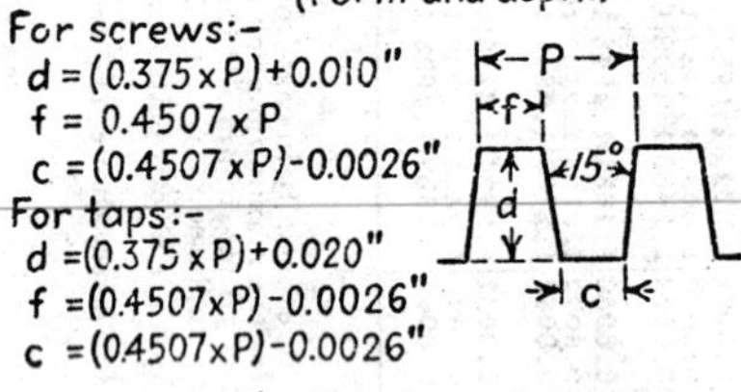

FIG. 19. Modified Acme and square threads are still used by industry despite introduction of a new Acme thread standard.

Acme Screw Threads. When formulated prior to 1895, Acme 29° threads were intended to replace square threads (in the unmodified form, square threads cannot be produced with dies, milling cutters, or grinding, and can be generated only with a single-point tool on a lathe). Acme threads were originally used to produce traversing motions on machine tools, valves, etc., but are now extensively employed for a variety of purposes.

OLD STANDARDS. The introduction of new Acme thread standards in 1952 caused confusion in industry, because the old and new standards, both of which use a 29° thread angle, do not produce an interchangeable product. Because the old standards have been in use for a long time, particularly by valve manufacturers and others, diagrams of modified Acme and square threads are reproduced (see Figs. 19 and 20), and basic dimensions and tolerances of old-standard general-purpose American National Acme threads given (see Tables 21 to 23).

NEW STANDARDS. The American Standard B1.5-1952 gives data on three classes of 29° general-purpose Acme threads and five classes of 29° centralizing threads.

In both these types, the height of thread, and likewise the thickness of thread, equals $p/2$, or one-half the pitch.

American Standard B1.8-1952 covers 29° stub Acme screw threads. Here the basic thickness of thread remains at $p/2$, but the height of thread has been reduced to $0.3p$, resulting in a coarse-pitch, shallow thread. Alternative stub threads call for a height of 0.375 pitch and 0.25 pitch, respectively.

General-purpose Acme threads (ASA B1.5-1952) encompass three classes with clearances on all diameters for free movement. These classes are 2G, 3G, and 4G. Class 2G is considered preferable, but the other two can be used if less backlash or end play is required.

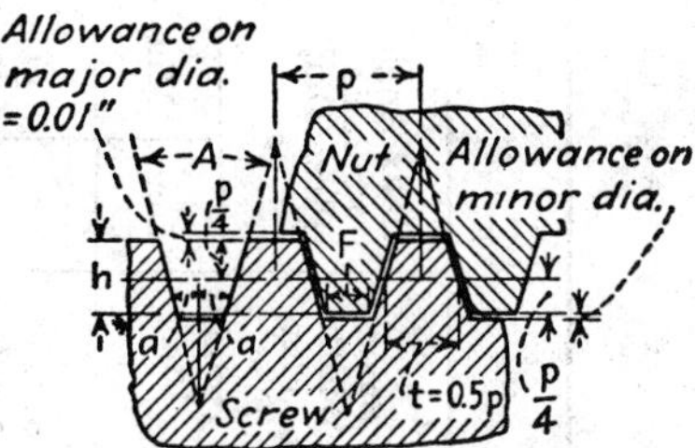

FIG. 20. Old standard American National Acme threads are likewise still required by industry.

The form of the general-purpose Acme thread is given in Fig. 21. Basic dimensions are given in Table 24, and limiting dimensions and tolerances in Table 25.

As will be noted from Fig. 21, the external and internal threads have a square root, but the crest corners of the external thread may be chamfered at 45° to a maximum depth of $p/15$.

Centralizing threads are standardized in five classes. They have limited clearance at the major diameters of the external and internal threads, so that a bearing at the major diameter maintains approximate alignment of the thread axis and prevents wedging on the thread flanks. Thread roots may be filleted and the crests of the external threads may be chamfered or rounded.

TABLE 21. OLD STANDARD ACME THREADS
Recommended Pitches

Threads per In., n	Recommended Range of Major Dia,* In.		Pitch, p, In.	Basic Depth of Thread, $h = 0.5p$, In.	Basic Width of Flat, $F = 0.37069p$, In.
	Least	Greatest			
1	4.5000	13.5000	1.00000	0.5000	0.3707
1⅓	3.5000	10.5000	0.75000	0.3750	0.2780
1½	3.0000	9.0000	0.66667	0.3333	0.2471
2	2.2500	6.7500	0.50000	0.2500	0.1853
2½	1.7500	5.2500	0.40000	0.2000	0.1483
3	1.5000	4.5000	0.33333	0.1667	0.1236
4	1.1250	3.3750	0.25000	0.1250	0.0927
5	0.8750	2.6250	0.20000	0.1000	0.0741
6	0.7500	2.2500	0.16667	0.0833	0.0618
8	0.5625	1.6875	0.12500	0.0625	0.0463
10	0.4375	1.3125	0.10000	0.0500	0.0371
12	0.3750	1.1250	0.08333	0.0417	0.0309
14	0.3125	0.9375	0.07143	0.0357	0.0265
16	0.2500	0.7500	0.06250	0.0312	0.0232

* These recommended least dia correspond to a maximum helix angle (at the minor dia) of approximately 5°. The recommended greatest dia is 3 times the least.

TABLE 22. OLD STANDARD ACME GENERAL-PURPOSE THREADS
Basic Dimensions

Identification		Basic Dia			Thread Data						
Sizes	Threads per In.	Major Dia, D, In.	Pitch Dia, E, In.	Minor Dia, K, In.	Pitch, p, In.	Thread Thickness at Pitch Line, In.	Basic Depth of Thread, $h = 0.5p$, In.	Depth of Thread with Clearance, In.	Basic Width of Flat, $F = 0.37069p$, In.	Helix Angle at Basic Pitch Dia, s	
										Deg	Min
1/4	16	0.2500	0.2187	0.1875	0.06250	0.03125	0.03125	0.03625	0.0232	5	12
5/16	14	0.3125	0.2768	0.2411	0.07143	0.03571	0.03571	0.04071	0.0265	4	42
3/8	12	0.3750	0.3333	0.2816	0.08333	0.04167	0.04167	0.04667	0.0309	4	33
7/16	12	0.4375	0.3958	0.3541	0.08333	0.04167	0.04167	0.04667	0.0309	3	50
1/2	10	0.5000	0.4500	0.4000	0.10000	0.05000	0.05000	0.06000	0.0371	4	3
5/8	8	0.6250	0.5625	0.5000	0.12500	0.06250	0.06250	0.07250	0.0463	4	3
3/4	8	0.7500	0.6875	0.6250	0.12500	0.06250	0.06250	0.07250	0.0463	3	19
7/8	8	0.8750	0.8125	0.7500	0.12500	0.06250	0.06250	0.07250	0.0463	2	48
1	5	1.0000	0.9000	0.8000	0.20000	0.10000	0.10000	0.11000	0.0741	4	3
1 1/8	5	1.1250	1.0250	0.9250	0.20000	0.10000	0.10000	0.11000	0.0741	3	33
1 1/4	5	1.2500	1.1500	1.0500	0.20000	0.10000	0.10000	0.11000	0.0741	3	10
1 3/8	5	1.3750	1.2750	1.1750	0.20000	0.10000	0.10000	0.11000	0.0741	2	52
1 1/2	4	1.5000	1.3750	1.2500	0.25000	0.12500	0.12500	0.13500	0.0927	3	19
1 3/4	4	1.750	1.6250	1.5000	0.25000	0.12500	0.12500	0.13500	0.0927	2	48
2	4	2.0000	1.8750	1.7500	0.25000	0.12500	0.12500	0.13500	0.0927	2	26
2 1/2	2	2.5000	2.2500	2.0000	0.50000	0.25000	0.25000	0.26000	0.1853	4	3
3	2	3.0000	2.7500	2.5000	0.50000	0.25000	0.25000	0.26000	0.1853	3	19
4	2	4.0000	3.7500	3.5000	0.50000	0.25000	0.25000	0.26000	0.1853	2	26
5	2	5.0000	4.7500	4.5000	0.50000	0.25000	0.25000	0.26000	0.1853	1	55

TABLE 23. OLD STANDARD ACME GENERAL-PURPOSE THREAD SERIES, LIMITING DIMENSIONS AND TOLERANCES

Sizes	Threads per In.	Screw Sizes, In.							Nut Sizes, In.					
		Major Dia		Pitch Dia		Tolerance Equivalent on Thickness of Threads	Minor Dia		Minor Dia		Pitch Dia		Tolerance Equivalent on Thickness of Threads	Major Dia, Min
		Max (Basic)	Min	Max (Basic)	Min		Max	Min	Min (Basic)	Max	Min	Max		
1/4	16	0.2500	0.2469	0.2187	0.2107	0.002	0.1775	0.1744	0.1875	0.1906	0.2267	0.2347	0.002	0.2600
5/16	14	0.3125	0.3089	0.2768	0.2688	0.002	0.2311	0.2275	0.2411	0.2447	0.2847	0.2928	0.002	0.3225
3/8	12	0.3750	0.3708	0.3333	0.3213	0.003	0.2816	0.2774	0.2816	0.2858	0.3453	0.3573	0.003	0.3850
7/16	12	0.4375	0.4333	0.3958	0.3838	0.003	0.3441	0.3399	0.3541	0.3583	0.4078	0.4198	0.003	0.4475
1/2	10	0.5000	0.4950	0.4500	0.4380	0.003	0.3800	0.3750	0.4000	0.4050	0.4620	0.4740	0.003	0.5200
5/8	8	0.6250	0.6187	0.5625	0.5465	0.004	0.4800	0.4737	0.5000	0.5063	0.5785	0.5945	0.004	0.6450
3/4	8	0.7500	0.7437	0.6875	0.6715	0.004	0.6050	0.5987	0.6250	0.6313	0.7035	0.7195	0.004	0.7700
7/8	8	0.8750	0.8687	0.8125	0.7965	0.004	0.7300	0.7237	0.7500	0.7563	0.8285	0.8445	0.004	0.8950
1	5	1.0000	0.9900	0.9000	0.8800	0.005	0.7800	0.7700	0.8000	0.8100	0.9200	0.9400	0.005	1.0200
1 1/8	5	1.1250	1.1150	1.0250	1.0050	0.005	0.9050	0.8950	0.9250	0.9350	1.0450	1.0650	0.005	1.1450
1 1/4	5	1.2500	1.2400	1.1500	1.1260	0.006	1.0300	1.0200	1.0500	1.0600	1.1740	1.1980	0.006	1.2700
1 3/8	5	1.3750	1.3650	1.2750	1.2510	0.006	1.1550	1.1450	1.1750	1.1850	1.2990	1.3230	0.006	1.3950
1 1/2	4	1.5000	1.4875	1.3750	1.3470	0.007	1.2300	1.2175	1.2500	1.2625	1.4030	1.4310	0.007	1.5200
1 3/4	4	1.7500	1.7375	1.6250	1.5980	0.007	1.4800	1.4675	1.5000	1.5125	1.6530	1.6810	0.007	1.7700
2	4	2.0000	1.9875	1.8750	1.8430	0.008	1.7300	1.7175	1.7500	1.7625	1.9070	1.9390	0.008	2.0200
2 1/2	2	2.5000	2.4800	2.2500	2.2060	0.011	1.9800	1.9600	2.0000	2.0200	2.2940	2.3380	0.011	2.5200
3	2	3.0000	2.9800	2.7500	2.7060	0.011	2.4800	2.4600	2.5000	2.5200	2.7940	2.8380	0.011	3.0200
4	2	4.0000	3.9800	3.7500	3.7060	0.011	3.4800	3.4600	3.5000	3.5200	3.7940	3.8380	0.011	4.0200
5	2	5.0000	4.9800	4.7500	4.7060	0.011	4.4800	4.4600	4.5000	4.5200	4.7940	4.8380	0.011	5.0200

TABLE 24. BASIC DIMENSIONS OF GENERAL-PURPOSE ACME THREADS*

Threads per In.	Pitch p	Height of Thread (Basic), $h = p/2$	Total Height of Thread, $h_s = h + \frac{1}{2}$ allowance	Thread Thickness (Basic), $t = p/2$	Width of Flat at Crest of Internal Thread, F_{cn}	Width of Flat at Root of Internal Thread, F_{rn}
16	0.06250	0.03125	0.0362	0.03125	0.0232	0.0206
14	0.07143	0.03571	0.0407	0.03571	0.0265	0.0239
12	0.08333	0.04167	0.0467	0.04167	0.0309	0.0283
10	0.10000	0.05000	0.0600	0.05000	0.0371	0.0319
8	0.12500	0.06250	0.0725	0.06250	0.0463	0.0411
6	0.16667	0.08333	0.0933	0.08333	0.0618	0.0566
5	0.20000	0.10000	0.1100	0.10000	0.0741	0.0689
4	0.25000	0.12500	0.1350	0.12500	0.0927	0.0875
3	0.33333	0.16667	0.1767	0.16667	0.1236	0.1184
2½	0.40000	0.20000	0.2100	0.20000	0.1483	0.1431
2	0.50000	0.25000	0.2600	0.25000	0.1853	0.1802
1½	0.66667	0.33333	0.3433	0.33333	0.2471	0.2419
1⅓	0.75000	0.37500	0.3850	0.37500	0.2780	0.2728
1	1.00000	0.50000	0.5100	0.50000	0.3707	0.3655

* ASA B1.5-1952.

Symbols:
$2\alpha = 29°$ $\alpha = 14°30'$ p = pitch
n = number of threads per in. N = number of turns per in. $h = p/2$ $t = p/2$
$F_{cn} = 0.3707p$ $F_{cs} = 0.3707p$
$F_{rn} = 0.3707p - 0.259 \times$ (major-dia allowance on internal thread)
$F_{rs} = 0.3707p - 0.259 \times$ (minor-dia allowance on external thread − pitch-dia allowance on external thread)

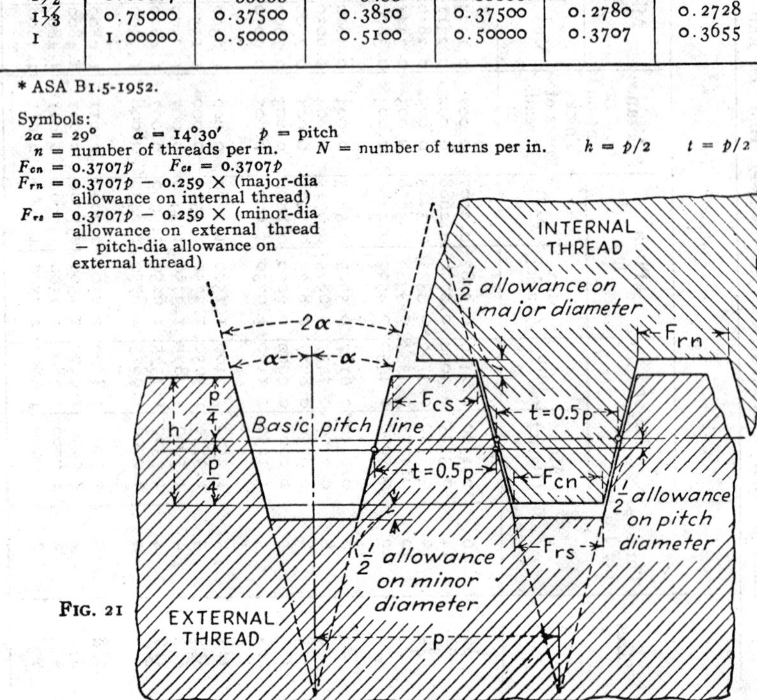

FIG. 21

TABLE 25. GENERAL-PURPOSE ACME SCREW THREADS, CLASSES 2G, 3G, AND 4G*

Limiting Dimension and Tolerances, In., of Preferred Diameter-pitch Combinations

Limiting Dia and Tolerances	Nominal Dia D, In.											
	1/4	5/16	3/8	7/16	1/2	5/8	3/4	7/8	1	1 1/8	1 1/4	1 3/8
	Threads per In.											
	16	14	12	12	10	8	6	6	5	5	5	4
External threads:												
Classes 2G, 3G, and 4G, major dia												
Max D	0.2500	0.3125	0.3750	0.4375	0.5000	0.6250	0.7500	0.8750	1.0000	1.1250	1.2500	1.3750
Min	0.2450	0.3075	0.3700	0.4325	0.4950	0.6188	0.7417	0.8667	0.9900	1.1150	1.2400	1.3625
Tolerance	0.0050	0.0050	0.0050	0.0050	0.0050	0.0062	0.0083	0.0083	0.0100	0.0100	0.0100	0.0125
Classes 2G, 3G, and 4G, minor dia max	0.1775	0.2311	0.2817	0.3442	0.3800	0.4800	0.5633	0.6883	0.7800	0.9050	1.0300	1.1050
Class 2G, minor dia min	0.1618	0.2140	0.2632	0.3253	0.3594	0.4570	0.5371	0.6615	0.7509	0.8753	0.9998	1.0719
Class 3G, minor dia min	0.1702	0.2231	0.2730	0.3354	0.3704	0.4693	0.5511	0.6758	0.7664	0.8912	1.0159	1.0896
Class 4G, minor dia min	0.1722	0.2254	0.2755	0.3379	0.3731	0.4723	0.5546	0.6794	0.7703	0.8951	1.0199	1.0940
Class 2G, pitch dia												
Max	0.2148	0.2728	0.3284	0.3909	0.4443	0.5562	0.6598	0.7842	0.8920	1.0165	1.1411	1.2406
Min	0.2043	0.2614	0.3161	0.3783	0.4306	0.5408	0.6424	0.7663	0.8726	0.9967	1.1210	1.2186
Tolerance	0.0105	0.0114	0.0123	0.0126	0.0137	0.0154	0.0174	0.0179	0.0194	0.0198	0.0201	0.0220
Class 3G, pitch dia												
Max	0.2158	0.2738	0.3296	0.3921	0.4458	0.5578	0.6615	0.7861	0.8940	1.0186	1.1433	1.2430
Min	0.2109	0.2685	0.3238	0.3862	0.4394	0.5506	0.6534	0.7778	0.8849	1.0094	1.1339	1.2327
Tolerance	0.0049	0.0053	0.0058	0.0059	0.0064	0.0072	0.0081	0.0083	0.0091	0.0092	0.0094	0.0103
Class 4G, pitch dia												
Max	0.2168	0.2748	0.3309	0.3934	0.4472	0.5593	0.6632	0.7880	0.8960	1.0208	1.1455	1.2453
Min	0.2133	0.2710	0.3268	0.3892	0.4426	0.5542	0.6574	0.7820	0.8895	1.0142	1.1388	1.2380
Tolerance	0.0035	0.0038	0.0041	0.0042	0.0046	0.0051	0.0058	0.0060	0.0065	0.0066	0.0067	0.0073

TABLE 25. GENERAL-PURPOSE ACME SCREW THREADS, CLASSES 2G, 3G, AND 4G* (*Continued*)

Limiting Dia and Tolerances	Nominal Dia *D*, In.											
	1/4	5/16	3/8	7/16	1/2	5/8	3/4	7/8	1	1 1/8	1 1/4	1 3/8
	Threads per In.											
	16	14	12	12	10	8	6	6	5	5	5	4
Internal threads:												
Classes 2G, 3G, and 4G, major dia												
Min.	0.2600	0.3225	0.3850	0.4475	0.5200	0.6450	0.7700	0.8950	1.0200	1.1450	1.2700	1.3950
Max.	0.2700	0.3325	0.3950	0.4575	0.5400	0.6650	0.7900	0.9150	1.0400	1.1650	1.2900	1.4150
Classes 2G, 3G, and 4G, minor dia												
Min.	0.1875	0.2411	0.2917	0.3542	0.4000	0.5000	0.5833	0.7083	0.8000	0.9250	1.0500	1.1250
Max.	0.1925	0.2461	0.2967	0.3592	0.4050	0.5062	0.5916	0.7166	0.8100	0.9350	1.0600	1.1375
Tolerance	0.0050	0.0050	0.0050	0.0050	0.0050	0.0062	0.0083	0.0083	0.0100	0.0100	0.0100	0.0125
Class 2G, pitch dia												
Min.	0.2188	0.2768	0.3333	0.3958	0.4500	0.5625	0.6667	0.7917	0.9000	1.0250	1.1500	1.2500
Max.	0.2293	0.2882	0.3456	0.4084	0.4637	0.5779	0.6841	0.8096	0.9194	1.0448	1.1701	1.2720
Tolerance	0.0105	0.0114	0.0123	0.0126	0.0137	0.0154	0.0174	0.0179	0.0194	0.0198	0.0201	0.0220
Class 3G, pitch dia												
Min.	0.2188	0.2768	0.3333	0.3958	0.4500	0.5625	0.6667	0.7917	0.9000	1.0250	1.1500	1.2500
Max.	0.2237	0.2821	0.3391	0.4017	0.4564	0.5697	0.6748	0.8000	0.9091	1.0342	1.1594	1.2603
Tolerance	0.0049	0.0053	0.0058	0.0059	0.0064	0.0072	0.0081	0.0083	0.0091	0.0092	0.0094	0.0103
Class 4G, pitch dia												
Min.	0.2188	0.2768	0.3333	0.3958	0.4500	0.5625	0.6667	0.7917	0.9000	1.0250	1.1500	1.2500
Max.	0.2223	0.2806	0.3374	0.4000	0.4546	0.5676	0.6725	0.7977	0.9065	1.0316	1.1567	1.2573
Tolerance	0.0035	0.0038	0.0041	0.0042	0.0046	0.0051	0.0058	0.0060	0.0065	0.0066	0.0067	0.0073

Limiting Dia and Tolerances	Nominal Dia D, In.										
	1½	1¾	2	2¼	2½	2¾	3	3½	4	4½	5
	Threads per In.										
	4	4	4	3	3	3	2	2	2	2	2
External threads:											
Classes 2G, 3G, and 4G, major dia											
Max D	1.5000	1.7500	2.0000	2.2500	2.5000	2.7500	3.0000	3.5000	4.0000	4.5000	5.0000
Min	1.4875	1.7375	1.9875	2.2333	2.4833	2.7333	2.9750	3.4750	3.9750	4.4750	4.9750
Tolerance	0.0125	0.0125	0.0125	0.0167	0.0167	0.0167	0.0250	0.0250	0.0250	0.0250	0.0250
Classes 2G, 3G, and 4G, minor dia max	1.2300	1.4800	1.7300	1.8967	2.1467	2.3967	2.4800	2.9800	3.4800	3.9800	4.4800
Class 2G, minor dia min	1.1965	1.4456	1.6948	1.8572	2.1065	2.3558	2.4326	2.9314	3.4302	3.9291	4.4281
Class 3G, minor dia min	1.2144	1.4640	1.7136	1.8783	2.1279	2.3776	2.4579	2.9574	3.4568	3.9563	4.4558
Class 4G, minor dia min	1.2188	1.4685	1.7183	1.8835	2.1333	2.3831	2.4642	2.9638	3.4634	3.9631	4.4627
Class 2G, pitch dia											
Max	1.3652	1.6145	1.8637	2.0713	2.3207	2.5700	2.7360	3.2350	3.7340	4.2330	4.7319
Min	1.3429	1.5916	1.8402	2.0450	2.2939	2.5427	2.7044	3.2026	3.7008	4.1991	4.6973
Tolerance	0.0223	0.0229	0.0235	0.0263	0.0268	0.0273	0.0316	0.0324	0.0332	0.0339	0.0346
Class 3G, pitch dia											
Max	1.3677	1.6171	1.8665	2.0743	2.3238	2.5734	2.7395	3.2388	3.7380	4.2373	4.7364
Min	1.3573	1.6064	1.8555	2.0620	2.3113	2.5607	2.7248	3.2237	3.7225	4.2215	4.7202
Tolerance	0.0104	0.0107	0.0110	0.0123	0.0125	0.0127	0.0147	0.0151	0.0155	0.0158	0.0162
Class 4G, pitch dia											
Max	1.3701	1.6198	1.8693	2.0773	2.3270	2.5767	2.7430	3.2425	3.7420	4.2415	4.7409
Min	1.3627	1.6122	1.8615	2.0685	2.3181	2.5676	2.7325	3.2317	3.7309	4.2302	4.7294
Tolerance	0.0074	0.0076	0.0078	0.0088	0.0089	0.0091	0.0105	0.0108	0.0111	0.0113	0.0115

TABLE 25. GENERAL-PURPOSE ACME SCREW THREADS, CLASSES 2G, 3G, AND 4G* (*Continued*)

Limiting Dia and Tolerances	Nominal Dia D, In.										
	1½	1¾	2	2¼	2½	2¾	3	3½	4	4½	5
	Threads per In.										
	4	4	4	3	3	3	2	2	2	2	2
Internal threads:											
Classes 2G, 3G, and 4G, major dia											
Min	1.5200	1.7700	2.0200	2.2700	2.5200	2.7700	3.0200	3.5200	4.0200	4.5200	5.0200
Max	1.5400	1.7900	2.0400	2.2900	2.5400	2.7900	3.0400	3.5400	4.0400	4.5400	5.0400
Classes 2G, 3G, and 4G, minor dia											
Min	1.2500	1.5000	1.7500	1.9167	2.1667	2.4167	2.5000	3.0000	3.5000	4.0000	4.5000
Max	1.2625	1.5125	1.7625	1.9334	2.1834	2.4334	2.5250	3.0250	3.5250	4.0250	4.5250
Tolerance	0.0125	0.0125	0.0125	0.0167	0.0167	0.0167	0.0250	0.0250	0.0250	0.0250	0.0250
Class 2G, pitch dia											
Min	1.3750	1.6250	1.8750	2.0833	2.3333	2.5833	2.7500	3.2500	3.7500	4.2500	4.7500
Max	1.3973	1.6479	1.8985	2.1096	2.3601	2.6106	2.7816	3.2824	3.7832	4.2839	4.7846
Tolerance	0.0223	0.0229	0.0235	0.0263	0.0268	0.0273	0.0316	0.0324	0.0332	0.0339	0.0346
Class 3G, pitch dia											
Min	1.3750	1.6250	1.8750	2.0833	2.3333	2.5833	2.7500	3.2500	3.7500	4.2500	4.7500
Max	1.3854	1.6357	1.8860	2.0956	2.3458	2.5960	2.7647	3.2651	3.7655	4.2658	4.7662
Tolerance	0.0104	0.0107	0.0110	0.0123	0.0125	0.0127	0.0147	0.0151	0.0155	0.0158	0.0162
Class 4G, pitch dia											
Min	1.3750	1.6250	1.8750	2.0833	2.3333	2.5833	2.7500	3.2500	3.7500	4.2500	4.7500
Max	1.3824	1.6326	1.8828	2.0921	2.3422	2.5924	2.7605	3.2608	3.7611	4.2613	4.7615
Tolerance	0.0074	0.0076	0.0078	0.0088	0.0089	0.0091	0.0105	0.0108	0.0111	0.0113	0.0115

* ASA B1.5-1952.

Buttress Threads.[1] The buttress form of thread is used for thrust purposes involving high stresses in one direction only. Radial component of the thrust is reduced to a minimum, because the thrust side of the thread is nearly perpendicular. Applications are military uses, and where tubular members are screwed together.

BASIC FORM OF THREAD. The basic form of the buttress thread (Fig. 22) has the following characteristics:

1. A pressure flank angle, measured in an axial plane, of 7° from the normal to the axis.
2. A trailing flank angle, measured in an axial plane, of 45°.
3. Equal truncations at the crests of the internal and external threads such that the basic depth of engagement (assuming no allowance) is equal to 0.6 of the pitch.
4. Equal radiuses at the roots of the internal and external threads tangential to the pressure flank and the trailing flank.
5. Crest corners of external thread broken to approximately a radius of 0.02 of the pitch.

Basic dimensions are given in Table 26.

Pitch Selection. Suitable associations of diameters and pitches are:

Dia Range, In.	*Associated Pitches, Threads per In.*
From ½ to 11/16	20, 16, 12
Over 11/16 to 1	16, 12, 10
Over 1 to 1½	16, 12, 10, 8, 6
Over 1½ to 2½	16, 12, 10, 8, 6, 5, 4
Over 2½ to 4	16, 12, 10, 8, 6, 5, 4
Over 4 to 6	12, 10, 8, 6, 5, 4, 3
Over 6 to 10	10, 8, 6, 5, 4, 3, 2½, 2
Over 10 to 16	10, 8, 6, 5, 4, 3, 2½, 2, 1½, 1¼
Over 16 to 24	8, 6, 5, 4, 3, 2½, 2, 1½, 1¼, 1

TOLERANCES. Tolerances on external threads shall be minus, and on internal threads shall be plus.

Pitch (Effective) Diameter. The following formula is used for determining pitch diameter tolerances:

Class 2 (Medium) Pitch (Effective) Diameter:

$$PD \text{ tolerance} = 0.002\sqrt[3]{D} + 0.00278\sqrt{L_e} + 0.00854\sqrt{p}$$

where D = major dia of thread, L_e = length engagement, p = pitch of thread.

Class 3 (close) tolerances are ⅔ of class 2 (medium) tolerances, and class 1 (free) tolerances are 1½ times class 2 (medium) tolerances.

If the length of engagement is taken as $10p$, the formula can be further simplified to

$$0.002\sqrt[3]{D} + 0.0173\sqrt{p}$$

Watch and Instrument Threads. Among the thread forms and sizes used in the manufacture of instruments and watches are the standard American National fine thread from size 00-96 and larger, AN coarse thread to a more limited extent, and a variety of special forms and sizes, each manufacturer of such equipment having developed his own system. Thus, leading American watch manufacturers have had series that resembled one another only in having 60° thread forms. Pitch diameters and pitches varied widely. Hamilton Watch Co., however, adopted a new series based on the Swiss NHS thread but transposed into inch units from the metric

[1] Proposed American Standard.

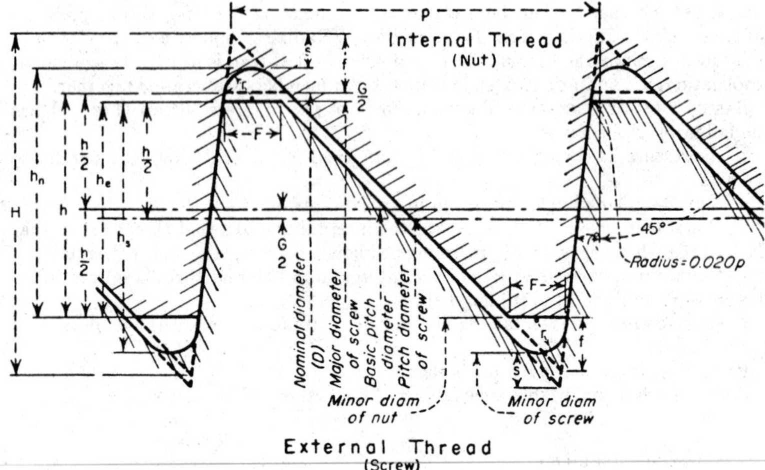

FIG. 22. Adoption of a 7° flank for the buttress thread makes this thread form more practical to make for its limited applications.

TABLE 26. BASIC DIMENSIONS, IN., FOR BUTTRESS SCREW THREADS

Threads per In.	Pitch p	Basic Depth of Thread, $h = 0.6p$	Depth of Sharp V-thread, $H = 0.89604p$	Crest Truncation, $f = 0.14532p$	Depth of Thread, h_s or $h_n = 0.66271p$	Root Truncation, $s = 0.08261p$
20	0.0500	0.0300	0.0445	0.0073	0.0331	0.0041
16	0.0625	0.0375	0.0557	0.0091	0.0414	0.0052
12	0.0833	0.0500	0.0742	0.0121	0.0552	0.0069
10	0.1000	0.0600	0.0891	0.0145	0.0663	0.0083
8	0.1250	0.0750	0.1113	0.0182	0.0828	0.0103
6	0.1667	0.1000	0.1484	0.0242	0.1105	0.0138
5	0.2000	0.1200	0.1781	0.0291	0.1325	0.0165
4	0.2500	0.1500	0.2227	0.0363	0.1657	0.0207
3	0.3333	0.2000	0.2969	0.0484	0.2209	0.0275
2½	0.4000	0.2400	0.3563	0.0581	0.2651	0.0330
2	0.5000	0.3000	0.4453	0.0727	0.3314	0.0413
1½	0.6667	0.4000	0.5938	0.0969	0.4418	0.0551
1¼	0.8000	0.4800	0.7125	0.1163	0.5302	0.0661
1	1.0000	0.6000	0.8906	0.1453	0.6627	0.0826

Root radius $r = 0.07141p$. Width of flat at crest $F = 0.16316p$.

TABLE 27. DIMENSIONS OF WATCH AND INSTRUMENT THREADS*

Nominal Size or Designation	Pitch Threads per In. $1/h$	Pitch Dia d_f, In.	Screw		Nut		Depth of Threads t_o, In.	Depth of Engagement t_l, In.	Clearance a	Root Radius r, In.
			Major Dia d, In.	Minor Dia d_k, In.	Major Dia D, In.	Minor Dia D_k, In.				
				50° Thread Angle						
30 NHS	$338\frac{2}{3}$	0.00990	0.0118	0.0077	0.0121	0.0080	0.00205	0.00190	0.00015	0.00035
35 NHS	$338\frac{2}{3}$	0.01185	0.0138	0.0097	0.0141	0.0100	0.00205	0.00190	0.00015	0.00035
40 NHS	254	0.01320	0.0157	0.0102	0.0161	0.0106	0.00275	0.00255	0.00020	0.00046
45 NHS	254	0.01515	0.0177	0.0122	0.0181	0.0126	0.00275	0.00255	0.00020	0.00046
50 NHS	$203\frac{1}{5}$	0.01650	0.0197	0.0128	0.0202	0.0133	0.00345	0.00320	0.00025	0.00058
55 NHS	$203\frac{1}{5}$	0.01845	0.0217	0.0148	0.0222	0.0153	0.00345	0.00320	0.00025	0.00058
60 NHS	$169\frac{1}{3}$	0.01980	0.0236	0.0153	0.0242	0.0159	0.00415	0.00385	0.00030	0.00070
70 NHS	$145\frac{1}{7}$	0.02310	0.0276	0.0180	0.0283	0.0187	0.00480	0.00445	0.00035	0.00081
80 NHS	127	0.02640	0.0315	0.0205	0.0323	0.0213	0.00550	0.00510	0.00040	0.00093
90 NHS	$112\frac{8}{9}$	0.02970	0.0354	0.0230	0.0363	0.0239	0.00620	0.00575	0.00045	0.00104
				60° Thread Angle						
100 NHS	$101\frac{3}{5}$	0.03300	0.0394	0.0256	0.0404	0.0266	0.00690	0.00640	0.00050	0.00057
110 NHS	$101\frac{3}{5}$	0.03690	0.0433	0.0295	0.0443	0.0305	0.00690	0.00640	0.00050	0.00057
120 NHS	$101\frac{3}{5}$	0.04085	0.0472	0.0334	0.0482	0.0344	0.00690	0.00640	0.00050	0.00057
130 NHS	$84\frac{2}{3}$	0.04350	0.0512	0.0347	0.0524	0.0359	0.00825	0.00765	0.00060	0.00068
140 NHS	$84\frac{2}{3}$	0.04745	0.0551	0.0386	0.0563	0.0398	0.00825	0.00765	0.00060	0.00068
150 NHS	$84\frac{2}{3}$	0.05140	0.0591	0.0426	0.0603	0.0438	0.00825	0.00765	0.00060	0.00068
160 NHS	$72\frac{4}{7}$	0.05405	0.0630	0.0437	0.0644	0.0451	0.00965	0.00895	0.00070	0.00080
170 NHS	$72\frac{4}{7}$	0.05800	0.0669	0.0476	0.0683	0.0490	0.00965	0.00895	0.00070	0.00080
180 NHS	$72\frac{4}{7}$	0.06190	0.0709	0.0516	0.0723	0.0530	0.00965	0.00895	0.00070	0.00080
200 NHS	$63\frac{1}{2}$	0.06850	0.0787	0.0567	0.0803	0.0583	0.01100	0.01020	0.00080	0.00091

* This screw thread series, based on the Swiss NHS thread series but transposed into inch units, has been adopted by Hamilton Watch Co. for all new watch designs.

TABLE 28. DIAMETER TOLERANCES, IN., FOR WATCH SCREW THREAD*

Nominal Size or Designation	Major Dia Screw		Pitch Dia Screw		Minor Dia Nut		3-wire Measurement		
	Max	Min	Max	Min	Min	Max	Wire Size	Max	Min
30 NHS	0.0118	0.0112	0.00990	0.00945	0.0080	0.0088	0.00180	0.01278	0.01233
35 NHS	0.0138	0.0132	0.01185	0.01140	0.0100	0.0108	0.00180	0.01478	0.01433
40 NHS	0.0157	0.0149	0.01320	0.01260	0.0106	0.0116	0.00240	0.01700	0.01640
45 NHS	0.0177	0.0169	0.01515	0.01455	0.0126	0.0136	0.00240	0.01900	0.01840
50 NHS	0.0197	0.0188	0.01650	0.01580	0.0133	0.0145	0.00300	0.02133	0.02063
55 NHS	0.0217	0.0208	0.01845	0.01775	0.0153	0.0165	0.00300	0.02333	0.02263
60 NHS	0.0236	0.0225	0.01980	0.01895	0.0159	0.0173	0.00360	0.02555	0.02470
70 NHS	0.0276	0.0263	0.02310	0.02215	0.0187	0.0203	0.00420	0.02988	0.02893
80 NHS	0.0315	0.0301	0.02640	0.02535	0.0213	0.0231	0.00480	0.03410	0.03305
90 NHS	0.0354	0.0338	0.02970	0.02850	0.0239	0.0259	0.00540	0.03833	0.03713

* These tolerances, established by Hamilton Watch Co., include sizes 30 to 90 NHS (50° series). The wire sizes and maximum and minimum measurements given in the table hold the required tolerance on the pitch dia.

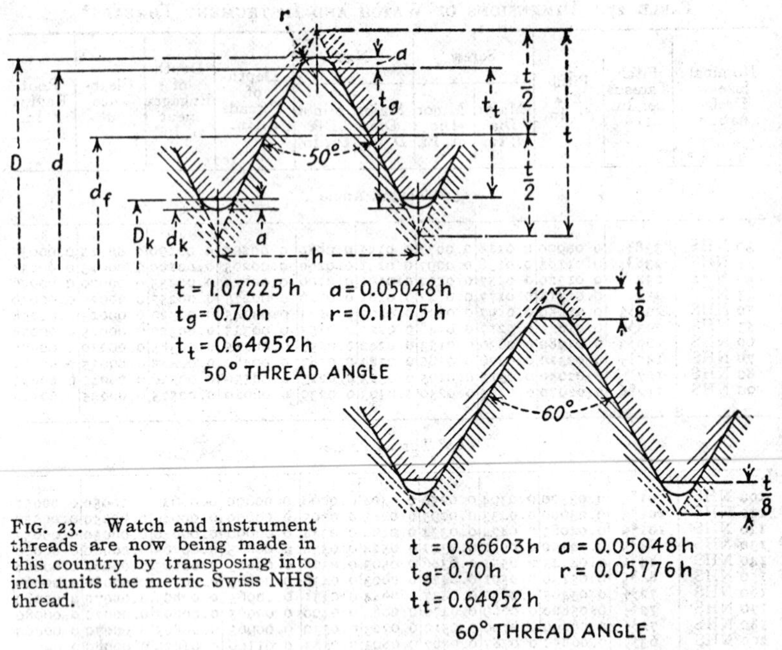

FIG. 23. Watch and instrument threads are now being made in this country by transposing into inch units the metric Swiss NHS thread.

system (Fig. 23). American Bosch, Hamilton, and the Kollsmann Instrument Co. have urged adoption of data in Tables 27 and 28 by the American Standards Association. The thread has a round bottom and a flat top, which is easily made, and has many advantages. Table 27 gives minor and major diameters for both screw and nut, as well as clearance.

American Pipe Threads

By ELMER ZOOK, *Chief Engineer, Detroit Tap & Tool Co.*

Three systems of American pipe threads are currently in use. These three systems are basically the same in that they have the 60° thread form with flattened crests and roots. Rather than to provide three complete thread specifications, it is less confusing to show the identical characteristics and then to elaborate upon the differences. Basic dimensions common to the three systems are therefore given in Table 29 for sizes to 6 in. Larger pipe sizes are 8, 10, and 12 in. ID and 14, 16, 18, 20, and 24 in. OD.

American Standard or "American National" pipe threads specifications (ASA B2.1-1945) cover tapered (¾ in. per ft) external and internal threads and several classes of straight external and internal threads. The taper-threaded joints and those with straight internal threads and tapered external threads will generally produce a pressuretight joint only when a sealing compound is used in the assembly. Several types of American Standard straight pipe threads are also used for assembling pipe when fluid pressure is not a factor.

Table 29. Basic Pipe-thread Dimensions Common to (A) American Standard, (B) American Standard Dryseal, and (C) Military Aeronautical Specifications

Applies to	Nominal Pipe Size and Threads per In.	OD of Pipe D, In.	Pitch Dia at Small End of External Thread E_0, In.	Hand Engagement			Effective Thread External			Wrench Make-up Length for Internal Threads		
				Pitch Dia E_1, In.	Length L-1		Pitch Dia E_2, In.	Length L-2		Pitch Dia E_3, In.	Length L-3	
					In.	Threads		In.	Threads		In.	Threads
A, B, C	1/16—27	0.3125	0.27118	0.28118	0.160	4.32	0.2875	0.2611	7.05	0.2642	0.1111	3
A and C	1/8—27	0.4050	0.36351	0.37476	0.180	4.86	0.3800	0.2639	7.12	0.3566	0.1111	3
B				0.37360	0.1615	4.36						
A and C	1/4—18	0.540	0.47739	0.48989	0.200	3.60	0.5025	0.4018	7.23	0.4670	0.1667	3
B				0.49163	0.2278	4.10						
All three systems	3/8—18	0.675	0.61201	0.62701	0.240	4.32	0.6375	0.4078	7.34	0.6016	0.1667	3
	1/2—14	0.840	0.75843	0.77843	0.320	4.48	0.7918	0.5337	7.47	0.7450	0.2143	3
	3/4—14	1.050	0.96768	0.98887	0.339	4.75	1.0018	0.5457	7.64	0.9543	0.2143	3
	1—11½	1.315	1.21363	1.23863	0.400	4.60	1.2563	0.6828	7.85	1.1973	0.2609	3
	1¼—11½	1.660	1.55713	1.58338	0.420	4.83	1.6013	0.7068	8.13	1.5408	0.2609	3
	1½—11½	1.900	1.79609	1.82234	0.420	4.83	1.8413	0.7235	8.32	1.7798	0.2609	3
	2—11½	2.375	2.26902	2.29627	0.436	5.01	2.3163	0.7565	8.70	2.2527	0.2609	3
A	2½—8	2.875	2.71953	2.76216	0.682	5.46	2.7906	1.1375	9.10	2.7039	0.2500	2
B and C										2.6961	0.3750	3
A	3—8	3.500	3.34062	3.38850	0.766	6.13	3.4156	1.2000	9.60	3.3250	0.2500	2
B and C										3.3172	0.3750	3
A	3½—8	4.000	3.83750	3.88881	0.821	6.57	3.9156	1.2500	10.00	3.8219	0.2500	2
A	4—8	4.500	4.33438	4.38712	0.844	6.75	4.4156	1.3000	10.40	4.3188	0.2500	2
A	5—8	5.563	5.39073	5.44929	0.937	7.50	5.4786	1.4063	11.25	5.3751	0.2500	2
A	6—8	6.625	6.44609	6.50597	0.958	7.66	6.5406	1.5125	12.10	6.4305	0.2500	2

Note: Dimensions for 3½ in. and larger nominal size are for American Standard pipe threads only.
A, American Standard.
B, Dryseal pipe threads.
C, Aeronautical pipe threads.

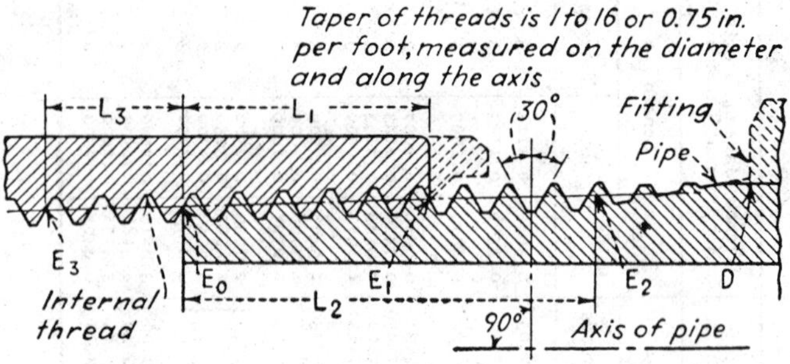

FIG. 24. American taper pipe threads have a 60° angle and a taper of ¾ in. per ft.

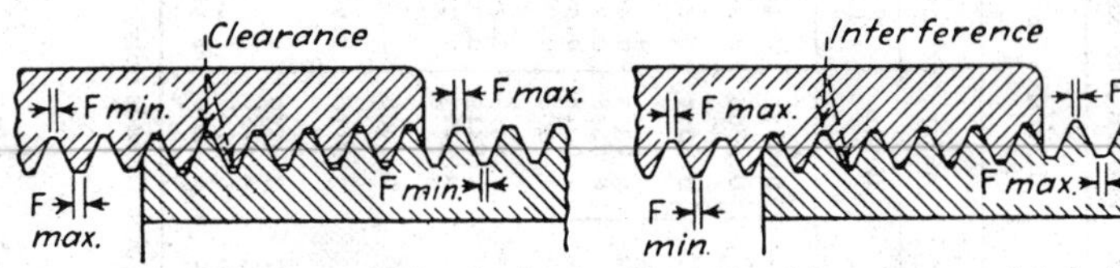

FIG. 25. Clearance is normally obtained between flats; so a sealer is required to avoid spiral leakage.

FIG. 26. Most commercial fittings are made so that interference does not occur at crest and root of mating threads.

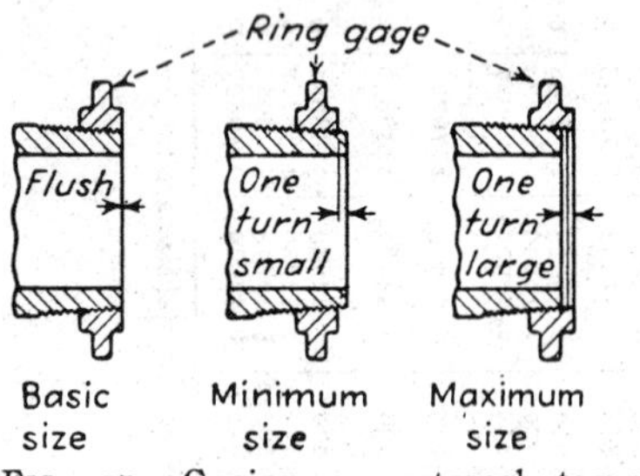

FIG. 27. Gaging an external taper thread with ring gage.

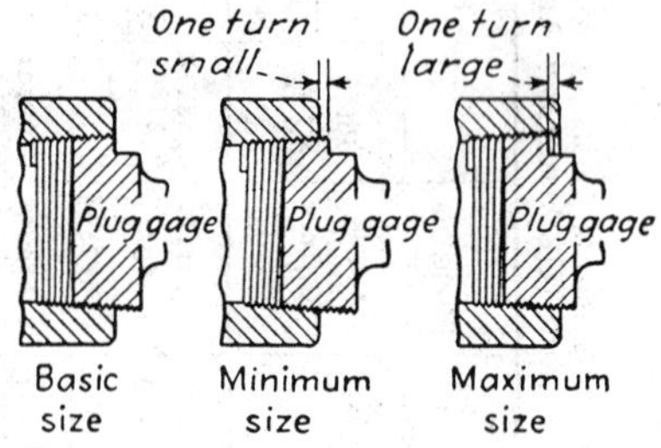

FIG. 28. Gaging an internal taper thread with a plug gage with gaging notch.

Taper Pipe Threads. American Standard taper pipe threads (Fig. 24) have maximum and minimum flats which are the same for both crest and root of both mating parts as shown in Table 30. If the crest flats are larger than the adjacent root flats, a clearance results between the flats as shown in Fig. 25. This clearance must be filled with a sealing compound to avoid spiral leakage.

If, on the other hand, the mating parts happen to have minimum-crest flats adjoining maximum-root flats at both the major and minor diameters of the thread, interference results (Fig. 26). This condition is usually not obtained on commercial fittings.

Taps. Internal-taper pipe threads are generally produced with standard ground-thread taps marked NPT.

Gaging. External-taper pipe (NPT) threads are gaged with Standard NPT L-1 (thin) ring gages. The ring is screwed hand-tight, and correct size is indicated if the gaging face is not more than one turn, large or small, from being flush with the end of the product thread as shown in Fig. 27. The part should have a minimum of approximately three threads with full roots past the large end of the ring gage to allow for wrench make-up (see L-2 length in Table 29).

Table 30. Limits on Crest and Root Width, In.—External and Internal Threads

Threads per In.	F	
	Max	Min
27	0.0041	0.0014
18	0.0057	0.0021
14	0.0064	0.0027
11½	0.0073	0.0033
8	0.0090	0.0048

Internal-taper pipe (NPT) threads are gaged with standard NPT L-1 plug gages. The plug gage is screwed up hand-tight into the fitting or coupling. Correct size is indicated when the gaging notch is not more than one turn, large or small, from being flush with the end of the part (see Fig. 28). On chamfered fittings, the gaging reference should be to the intersection of the pitch line with the chamfer cone, rather than to the end of the part. As the gaging notch is located at the pitch line of the gage, its position in relation to the chamfer cone is easily noted. Parts that do not have a clearance beyond the thread should have approximately three full threads below the small end of the L-1 plug gage to allow for wrench make-up (see L-3 length in Table 29).

Straight Pipe Threads. Several types of American Standard straight pipe threads are commonly used. Choice depends on the intended service. Most of them have the same basic thread form and dimensions as American Standard taper pipe threads, and they are usually designated by NPS, with an additional letter added to indicate the exact type of thread (NPSM, NPSC, etc.). *Straight pipe threads are truncated more than tapered threads.*

NPSC. Many low-pressure pipe couplings are made with American Standard straight pipe coupling internal threads, and assembled with tapered (standard NPT) external threads, using a suitable sealer. Such joints are not recommended if subject to vibration or if a rigid connection is required.

NPSC pipe couplings are usually tapped in high-production machines requiring special taps. Such tools must be made to the customer's specifications. For special applications, standard basic straight pipe taps will produce satisfactory holes for assembly with tapered pipe threads.

Gaging procedure for straight-pipe-thread couplings is the same as for tapered threads, using NPTL–1 plug gages, except that the tolerance is 1½ turns, large or small, rather than 1 turn as in checking tapered threads.

NPSM. The American Standard straight pipe thread for free-fitting mechanical joints (NPSM) is used for assembling pipe and fittings which are not required to withstand fluid pressures. Such joints consist of straight internal and external threads made to the limits shown in Table 31.

Standard basic straight pipe taps are used to produce threaded holes to the specified limits, while straight go and not-go thread plug gages and thread ring gages are used for gaging. These gages are special and are made to the product pitch diameters shown in Table 31.

NPSL. The American Standard straight pipe thread for lock-nut connections (NPSL) is used where a loose-fitting connection is desired. It was developed for the seal joint commonly used in tank nipple connections. The internal thread will assemble freely with the largest full thread which can be cut on the outside diameter of a standard pipe. Thread dimensions are shown in Table 32.

Oversized straight pipe taps for tapping pipe lock-nut threads and the straight go and not-go thread plug and ring gages are special and are made to the product pitch diameters shown in Table 32.

Dryseal Pipe Threads. American Standard Dryseal pipe threads are used for both external and internal threads where a pressuretight joint is required and the use of a sealer may be objectionable. The external threads are tapered while the internal threads may be tapered or straight.

Dryseal threads, in most cases, have the same basic dimensions as American Standard pipe thread (see basic dimensions Table 29), except the crest and root truncations are modified to produce a pressuretight joint.

American Standard Dryseal pipe threads have crest flats which are equal to or smaller than the root flats of the mating part, as shown in Table 33. When fittings producing the minimum-interference condition as shown in Fig. 29 are assembled, a complete mating of the form results so all spiral leakage is eliminated.

In the maximum-interference condition (Fig. 30) a clearance results at the flank of the thread when the joint is made up hand-tight. This clearance is eliminated by a crushing of the crests and roots when the joint is screwed up wrench-tight so complete mating of the form again results.

Pipe joints consisting of American Standard Dryseal taper internal and external threads (NPTF) are generally considered to be superior for strength and sealing as compared with joints using straight internal (NPSF or NPSI) and tapered external (NPTF) threads.

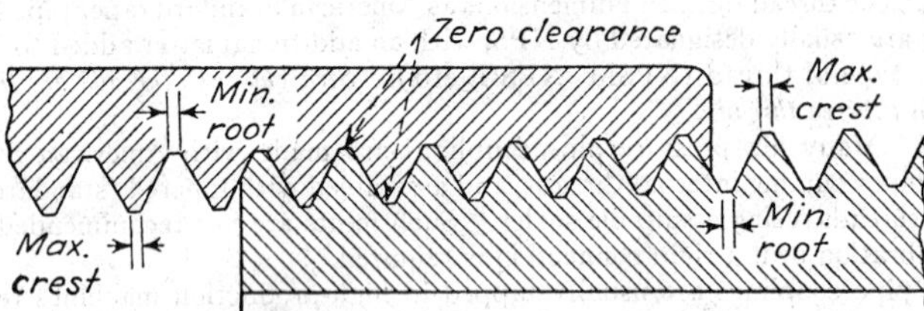

FIG. 29. Spiral leakage is avoided by Dryseal threads even with the minimum-interference condition.

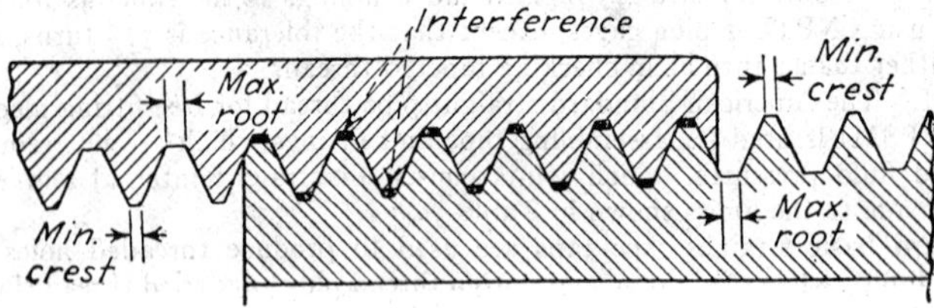

FIG. 30. In the maximum-interference condition, any clearance between Dryseal thread flanks in the hand-tight condition is nullified by crushing of crests and roots at wrenching.

Table 31. Dimensions, In., External and Internal Straight Pipe Threads for Mechanical Joints

Nominal Pipe Size and Threads per In.	External Thread					Internal Thread				
	Major Dia		Pitch Dia		Minor Dia Max	Major Dia Min	Pitch Dia		Minor Dia	
	Max	Min	Max	Min			Max	Min	Max	Min
⅛—27	0.3995	0.3926	0.3748	0.3713	0.3501	0.3995	0.3783	0.3748	0.3570	0.3501
¼—18	0.5269	0.5165	0.4899	0.4847	0.4529	0.5269	0.4951	0.4899	0.4633	0.4529
⅜—18	0.6640	0.6536	0.6270	0.6218	0.5900	0.6640	0.6322	0.6270	0.6004	0.5900
½—14	0.8260	0.8126	0.7784	0.7717	0.7308	0.8260	0.7851	0.7784	0.7442	0.7308
¾—14	1.0365	1.0231	0.9889	0.9822	0.9413	1.0365	0.9956	0.9889	0.9547	0.9413
1—11½	1.2965	1.2802	1.2386	1.2305	1.1807	1.2965	1.2468	1.2386	1.1970	1.1807
1¼—11½	1.6413	1.6250	1.5834	1.5753	1.5255	1.6413	1.5916	1.5834	1.5418	1.5255
1½—11½	1.8802	1.8639	1.8223	1.8142	1.7644	1.8802	1.8305	1.8223	1.7807	1.7644
2—11½	2.3542	2.3379	2.2963	2.2882	2.2384	2.3542	2.3044	2.2963	2.2547	2.2384
2½— 8	2.8455	2.8221	2.7622	2.7505	2.6789	2.8455	2.7739	2.7622	2.7023	2.6789
3— 8	3.4718	3.4484	3.3885	3.3768	3.3052	3.4718	3.4002	3.3885	3.3286	3.3052
3½— 8	3.9721	3.9487	3.8888	3.8771	3.8055	3.9721	3.9005	3.8888	3.8289	3.8055
4— 8	4.4704	4.4470	4.3871	4.3754	4.3038	4.4704	4.3988	4.3871	4.3272	4.3038
5— 8	5.5326	5.5092	5.4493	5.4376	5.3660	5.5326	5.4610	5.4493	5.3894	5.3660
6— 8	6.5893	6.5659	6.5060	6.4943	6.4227	6.5893	6.5177	6.5060	6.4461	6.4227

Taps. The internal threads for such joints are produced with standard Dryseal taper pipe taps (NPTF). These are furnished with ground threads and dimensions identical to American Standard taper pipe taps (NPT) except for the large crest flat and slight difference in the maximum root flat.

Gaging. Although there is very little difference between NPT and NPTF threads, it is necessary to control the various thread dimensions within specified Dryseal tolerances so leakproof joints are assured.

Thread plug and thread ring gages are used for checking size and taper. Thread form and truncation may be checked by projecting a percentage of the product or chasers on an optical comparator in the case of external threads, and by similar inspection of the taps themselves in the case of internal threads. Casts of internal threads can be projected, but this method is not practical for extensive use. Taps used under reasonable conditions reproduce their own form quite accurately, so inspection of the tools together with proper use of plug gages will insure properly threaded holes.

Gages. NPTF basic L-1 tapered plug and ring thread gages are used to check the portion of the threaded product normally engaged by hand. These gages are similar to standard NPT gages except the thread crests are truncated an additional amount to clear the wide roots in the Dryseal thread. The L-1 plug gages normally have a basic step but can be supplied with limit steps (maximum and minimum) if

TABLE 32. DIMENSIONS, IN., EXTERNAL AND INTERNAL STRAIGHT PIPE THREADS FOR LOCK-NUT CONNECTIONS

Nominal Pipe Size and Threads per In.	External Thread					Internal Thread				
	Major Dia		Pitch Dia		Minor Dia Max	Major Dia Min	Pitch Dia		Minor Dia	
	Max	Min	Max	Min			Max	Min	Max	Min
1/8—27	0.4087	0.4018	0.3840	0.3805	0.3593	0.4110	0.3898	0.3863	0.3685	0.3616
1/4—18	0.5408	0.5304	0.5038	0.4986	0.4668	0.5443	0.5125	0.5073	0.4807	0.4703
3/8—18	0.6779	0.6675	0.6409	0.6357	0.6039	0.6814	0.6496	0.6444	0.6178	0.6074
1/2—14	0.8439	0.8305	0.7963	0.7896	0.7487	0.8484	0.8075	0.8008	0.7666	0.7532
3/4—14	1.0543	1.0409	1.0067	1.0000	0.9591	1.0588	1.0179	1.0112	0.9770	0.9636
1—11½	1.3183	1.3020	1.2604	1.2523	1.2025	1.3237	1.2739	1.2658	1.2242	1.2079
1¼—11½	1.6630	1.6467	1.6051	1.5970	1.5472	1.6685	1.6187	1.6106	1.5689	1.5526
1½—11½	1.9020	1.8857	1.8441	1.8360	1.7862	1.9074	1.8576	1.8495	1.8079	1.7916
2—11½	2.3759	2.3596	2.3180	2.3099	2.2601	2.3813	2.3315	2.3234	2.2818	2.2655
2½— 8	2.8767	2.8533	2.7934	2.7817	2.7101	2.8845	2.8129	2.8012	2.7413	2.7179
3— 8	3.5031	3.4797	3.4198	3.4081	3.3365	3.5109	3.4393	3.4276	3.3677	3.3443
3½— 8	4.0034	3.9800	3.9201	3.9084	3.8368	4.0112	3.9396	3.9279	3.8680	3.8446
4— 8	4.5017	4.4783	4.4184	4.4067	4.3351	4.5095	4.4379	4.4262	4.3663	4.3429
5— 8	5.5638	5.5404	5.4805	5.4688	5.3973	5.5716	5.5001	5.4884	5.4285	5.4051
6— 8	6.6205	6.5971	6.5372	6.5255	6.4539	6.6283	6.5567	6.5450	6.4857	6.4617

TABLE 33. CREST AND ROOT FLATS, IN., OF AMERICAN STANDARD DRYSEAL THREADS

Threads per In.	Root Flat		Crest Flat	
	Max	Min	Max	Min
27	0.006	0.004	0.004	0.002
18	0.007	0.005	0.005	0.003
14	0.007	0.005	0.005	0.003
11½	0.009	0.006	0.006	0.004
8	0.011	0.008	0.008	0.006

the limit method of gaging is to be used. L-1 ring gages can also be supplied with limit steps where required.

The NPTF L-3 tapered plug thread gage is designed to check the make-up threads and taper of the tapped hole. This plug gage has only four threads at the small end, the balance of its length being relieved below the root so that any error in taper is indicated by a variation in the basic engagement difference between the L-1 and L-3 gages. The basic gaging step is located on the relieved portion and has the same relationship to the face of the product as the basic step on the L-1 gage. Limit steps similar to those on the L-1 plug gage can also be provided.

The NPTF L-2 tapered ring thread gage checks the make-up threads as well as the taper of the external Dryseal threads. It has a width equal to L-2 in Table 29 of basic dimensions but is counterbored at the small end so that any errors in taper will cause a variation in the basic engagement difference between the L-1 and L-2 gages. Limit steps similar to those on the L-1 ring gage can also be provided.

GAGING METHODS. Three methods of gaging with thread plug and thread ring gages are in use: the turns-engagement method, position method, and limit method. Regardless of the method used, care should always be taken so that gages are screwed up *only hand-tight.*

Turns-engagement Method. The turns-engagement method involves counting the number of turns necessary to unscrew the plug or ring gage from the product. It eliminates the need for estimating the standoff from a basic step and is accurate although other methods may be faster. The turns-engagement method is the only means of checking holes which are recessed or where a view of the gage cannot be obtained for other reasons. Maximum and minimum turn limits are given in Table 34.

External parts are of correct size when both the L-1 and L-2 ring gages engage the part within the maximum and minimum number of turns shown. To insure that the part has correct taper, the difference in turns engagement counted on the L-2 and L-1 rings must be within ½ turn of the difference in basic turns shown. Likewise, when gaging internal threads, the L-1 and L-3 plug gages must engage the part within the specified number of turns and the difference of their engagement must be within ½ turn of the difference in basic turns shown in Table 34. The decimal parts of a turn shown for plug and ring gages are theoretical, and in practice the nearest ¼ turn is usually employed.

The Position Method. The position method of gaging is the one most commonly used, although it is probably the least accurate. Standard gages having a basic gaging face or step are used. Location of the step in relation to the gaging point of the product must be estimated. In the case of internal threads, the gaging point is the intersection of the pitch line and the chamfer cone, while external threads should be gaged to the first scratch of the thread on the chamfer. The tolerances for the position of the step or face of the gage in relation to the gaging point are shown in Table 35. Also, to insure correct taper, the position of the L-3 plug-gage notch must not vary more than ½ thread from that of the L-1 plug, and the L-2 ring-gage position must be within ½ thread of the L-1 ring-gage position.

The Limit Method. The limit method of gaging American Standard Dryseal taper pipe threads is recommended as a practical and fast method. Standar[illegible] L-1 and L-3 plug gages and L-1 and L-2 ring gages are used, except that gaging s[illegible] are provided which represent the maximum and minimum sizes. When usin[illegible]t type L-1 plug gage, it is only necessary to ascertain that the gaging po[illegible]e internal thread falls between the gage steps and to note its approximate r[illegible] to those steps. The L-3 plug is then applied, and again the gaging poin[illegible]

Table 34. Tolerances for Turns-engagement Method

For Gaging American Standard Dryseal Taper Pipe Threads (NPTF) Using L-1 and L-3 Basic Step Plug Gages and L-1 and L-2 Basic Face Ring Gages

PTF SAE Short Size	External Threads					Internal Threads				
	Turns Engagement L-1 Ring		Turns Engagement L-2 Ring		Difference in Basic Turns Engagement L-2 − L-1	Turns Engagement L-1 Plug		Turns Engagement L-3 Plug		Difference in Basic Turns Engagement L-3 − L-1
	Max Size	Min Size	Max Size	Min Size		Max Size	Min Size	Max Size	Min Size	
1/16—27	1.82	3.32	4.30	5.80	2.48	3.82	2.32	6.57	5.07	2.75
1/8—27	2.36	3.86	4.37	5.87	2.01	3.86	2.36	6.61	5.11	2.75
1/4—18	1.10	2.60	4.48	5.98	3.38	3.60	2.10	6.35	4.85	2.75
3/8—18	1.82	3.32	4.59	6.09	2.77	3.82	2.32	6.57	5.07	2.75
1/2—14	1.98	3.48	4.72	6.22	2.74	3.98	2.48	6.73	5.23	2.75
3/4—14	2.25	3.75	4.89	6.39	2.64	4.25	2.75	7.00	5.50	2.75
1—11½	2.10	3.60	5.10	6.60	3.00	4.10	2.60	6.85	5.35	2.75
1¼—11½	2.33	3.83	5.38	6.88	3.05	4.33	2.83	7.08	5.58	2.75
1½—11½	2.33	3.83	5.57	7.07	3.24	4.33	2.83	7.08	5.58	2.75
2—11½	2.51	4.01	5.95	7.45	3.44	4.51	3.01	7.26	5.76	2.75

Table 35. Tolerances for Position Method

NPTF Size	External Threads		Internal Threads	
	Face of Gage Position, Threads		Notch Position, Threads	
	Off	On	In	Out
1/16—27	1	1	1	1
1/8—27	1	1	½	1½
1/4—18	1	1	1½	½
3/8—18	1	1	1	1
1/2—14	1	1	1	1
3/4—14	1	1	1	1
[illegible]11½	1	1	1	1
[illegible]½	1	1	1	1
[illegible]	1	1	1	1
[illegible]	1	1	1	1

must fall between the limit steps and in the same relationship as the L-1 gage within ½ thread. The same procedure is used in gaging external threads, in which case the gaging point of the external thread must gage between the limit steps on both the L-1 and L-2 ring gages and have the same relationship to the steps within ½ thread on both gages.

Dryseal Pipe Threads PTF (SAE Short). In some applications, economy of materials, space restrictions, or other factors make it advisable to use a shorter thread engagement than provided by the American Standard Dryseal thread. For such purposes the PTF (SAE short) thread is used. The form, taper, and general specifications are the same as NPTF threads except that one thread has been eliminated from the small end of the external thread while the internal thread is shortened by eliminating one thread at the large end.

PTF (SAE short) external threads are intended for assembly with Dryseal intermediate internal straight pipe threads (NPSI). They will also assemble with American Standard Dryseal taper pipe internal threads (NPTF). They are not designed for and may not assemble with PTF (SAE short) internal threads. The PTF (SAE short) internal thread is intended only for assembly with American Standard Dryseal taper pipe external thread (NPTF).

TAPS. PTF (SAE short) internal threads are generally used where the depth of hole is limited so that special taps are required. Such taps are made oversize on the pitch diameter so that the length from the gage line to the front of the tap is shorter than standard and are marked "PTF (SAE short)."

GAGING. The method of gaging PTF (SAE short) threads is identical with that used for NPTF threads except that the turns engagement or position is changed as shown in Tables 36 and 37 to compensate for the shortening. The product tolerance is also reduced to insure proper assembly of the shortened thread. NPTF basic step L-1 and L-3 plug gages are used for PTF (SAE short) internal threads and NPTF basic face L-1 and L-2 ring gages for the PTF (SAE short) external thread.

If the limit method of checking is used, plug and ring thread gages with limit steps adjusted for the PTF (SAE short) tolerances are supplied so that gaging then becomes identical to NPTF practice.

Dryseal Straight Pipe Threads. In the manufacture of commercial products where cost is an important factor, it is often preferable to use a straight internal pipe thread mating with a tapered external thread. Such internal threads are known as American Standard Dryseal fuel straight pipe threads (NPSF) and are intended only for assembly with Standard Dryseal taper pipe (NPTF) external threads. As the most common applications are on the smaller sizes of pipe and tubing, the specifications in Table 38 cover only sizes up to and including 1 in.

NPSI. American Standard Dryseal intermediate straight pipe threads are identical to NPSF threads except that the major and pitch diameters are larger by the amount of taper in one thread of the external part. They are intended primarily for assembly with PTF (SAE short) external threads. NPTF fittings are sometimes used in NPSI threads when the holes are in hard or brittle materials of heavy section so there there is little give for wrench make-up. In such cases, the larger NPSI internal thread provides a more satisfactory length of engagement in the wrench-tight position.

TAPS. Standard taps suitable for tapping NPSF threads are available. Special taps which are identical to NPSF taps except for major and pitch diameters are used for NPSI threads.

GAGING. As NPSF and NPSI threads are assembled with tapered external threads, the only accurate gaging method is by the use of tapered plug gages. Stand-

Table 36. Tolerances for Turns-Engagement Method

For Gaging PTF (SAE Short) Dryseal Pipe Threads Using L-1 and L-3 Basic Step Plug Gages and L-1 and L-2 Basic Face Ring Gages

NPTF Size	External Threads					Internal Threads				
	Turns Engagement L-1 Ring		Turns Engagement L-2 Ring		Difference in Basic Turns Engagement L-2 − L-1	Turns Engagement L-1 Plug		Turns Engagement L-3 Plug		Difference in Basic Turns Engagement L-3 − L-1
	Max Size	Min Size	Max Size	Min Size		Max Size	Min Size	Max Size	Min Size	
1/16—27	2.32	4.32	4.80	6.80	2.48	4.82	2.82	7.57	5.57	2.75
1/8—27	2.86	4.86	4.87	6.87	2.01	4.86	2.86	7.61	5.61	2.75
1/4—18	1.60	3.60	4.98	6.98	3.38	4.60	2.60	7.35	5.35	2.75
3/8—18	2.32	4.32	5.09	7.09	2.77	4.82	2.82	7.57	5.57	2.75
1/2—14	2.48	4.48	5.22	7.22	2.74	4.98	2.98	7.73	5.73	2.75
3/4—14	2.75	4.75	5.39	7.39	2.64	5.25	3.25	8.00	6.00	2.75
1—11 1/2	2.60	4.60	5.60	7.60	3.00	5.10	3.10	7.85	5.85	2.75
1 1/4—11 1/2	2.83	4.83	5.88	7.88	3.05	5.33	3.33	8.08	6.08	2.75
1 1/2—11 1/2	2.83	4.83	6.07	8.07	3.24	5.33	3.33	8.08	6.08	2.75
2—11 1/2	3.01	5.01	6.45	8.45	3.44	5.51	3.51	8.26	6.26	2.75

Table 37. Tolerances for Position

For Gaging PTF (SAE Short) Dryseal Pipe Threads Using L-1 and L-3 Basic Step Plug Gages and L-1 and L-2 Basic Face Ring Gages

PTF (SAE Short) Size	External Threads Face of Gage Position, Threads		Internal Threads Notch Position, Threads	
	Off	On	In	Out
1/16—27	1 1/2	0	0	1 1/2
1/8—27	1 1/2	0	..	1/2 to 2
1/4—18	1 1/2	0	1/2	1
3/8—18	1 1/2	0	0	1 1/2
1/2—14	1 1/2	0	0	1 1/2
3/4—14	1 1/2	0	0	1 1/2
1—11 1/2	1 1/2	0	0	1 1/2
1 1/4—11 1/2	1 1/2	0	0	1 1/2
1 1/2—11 1/2	1 1/2	0	0	1 1/2
2—11 1/2	1 1/2	0	0	1 1/2

ard NPTF basic L-1 plug gages can be used in conjunction with the turns-engagement method or the position method can be used as in the case of NPTF threads. Maximum and minimum turn limits as well as position limits are shown in Table 39. The L-3 gages cannot be used, of course. The decimal parts of turn engagement shown are theoretical, and in practice the nearest ¼ turn is usually used.

LIMIT-TYPE GAGES. Limit-type gages having maximum and minimum steps can also be supplied for checking NPSF and NPSI threads. When using this type of gage it is only necessary to ascertain that the gaging point of the work falls between the maximum and minimum steps of the gage.

STRAIGHT THREAD GAGES. Some manufacturers prefer to gage NPSF and NPSI tapped holes with go and not-go straight thread gages. While this method is not recommended and is not an accurate check, it can be used with some degree of accuracy, provided that it is used in conjunction with standard NPTF gages as a reference. Straight gages inspect the full length of thread while tapered gages check only the first few threads normally engaged so that a bellmouthed hole which is small on pitch diameter near the bottom may be rejected by a straight go thread plug, although it is entirely acceptable and will gage properly with an NPTF gage. For this reason, a percentage of parts passed by the straight gages and all parts rejected by straight gages should receive a final inspection with NPTF gages.

As straight gages are not recommended for NPSF or NPSI threads, there is no standard on pitch diameters for such gages. The actual size of a straight thread gage must be considerably smaller than the size indicated by a tapered gage. The amount of undersize depends upon the diameter and pitch of thread and also the amount of bellmouth, taper, etc., produced in the hole. Pitch-diameter limits have been set up which, under average conditions, will provide a reasonable degree of agreement with the proper NPTF gages.

It must be remembered, however, that the use of straight thread gages made to the pitch diameters shown in Table 38 does not necessarily indicate whether or not a hole is of correct size. As pointed out above, such gages should be used only in conjunction with NPTF gages.

Aeronautical Pipe Threads

Aeronautical Taper Pipe Threads. Aeronautical National form taper pipe threads as given in Military Specification MIL-P-7105 (formerly AN-P-363) are identical in all respects with American Standard taper pipe threads except for the E-3 and L-3 dimensions in the 2½- and 3-in. sizes as shown in Table 29. Only tapered external and internal threads are found in this system. Crest and root flats are the same as shown for American Standard pipe threads (Table 30). A sealer is necessary to insure a pressuretight joint. Standard NPT ground thread taps are used to produce the internal threads.

The importance of the Aeronautical pipe thread specification lies in the gaging system specified which provides a *closer* dimensional control of the various thread elements than the gages used for American Standard taper pipe threads or American Standard Dryseal pipe threads.

GAGING INTERNAL THREADS. A set of three plug gages is required to gage the internal thread. An L-1 plug gage having a crest flat equal to the maximum root flat of the internal thread is used to gage the pitch diameter of the thread normally engaged by hand, as well as checking the maximum root truncation. This gage must enter so that the basic notch is flush with the face of the fitting with a tolerance of plus or minus 1 turn.

An L-3 plug gage having a crest flat considerably larger than the L-1 is used to gage the make-up threads and taper and it must also check to the basic notch within plus or minus 1 turn. In addition, it must check in the same relationship as the L-1 gage within ½ turn.

The fact that the L-1 and L-3 gages have different amounts of truncation at the crest also allows them to detect errors in form. Errors in thread angle, or a root flat greater than the maximum specified, will cause a variation in the standoff of the

TABLE 38. SUGGESTED PITCH DIAMETERS
For Straight Thread Gages for NPSF

NPSF or NPSI Size	NPSF Gages Pitch Dia, In.		NPSI Gages Pitch Dia, In.	
	Go	Not Go	Go	Not Go
1/16—27	0.2756	0.2789	0.2779	0.2812
1/8—27	0.3680	0.3713	0.3703	0.3736
1/4—18	0.4842	0.4890	0.4877	0.4925
3/8—18	0.6196	0.6244	0.6231	0.6279
1/2—14	0.7692	0.7754	0.7737	0.7799
3/4—14	0.9797	0.9859	0.9842	0.9904
1—11½	1.2278	1.2352	1.2332	1.2406

TABLE 39. TOLERANCES FOR TURNS ENGAGEMENT AND POSITION
For Gaging NPSF and NPSI Threads Using L-1 Basic Step Plug Gage

NPSF or NPSI Size	NPSF				NPSI			
	Turns Engagement L-1 Plug		Notch Position L-1 Plug, Threads		Turns Engagement L-1 Plug		Notch Position L-1 Plug, Threads	
	Max Size	Min Size	In	Out	Max Size	Min Size	In	Out
1/16—27	3.82	2.32	0	1½	4.82	3.32	1	½
1/8—27	3.86	2.36	..	½ to 2	4.86	3.36	½	1
1/4—18	3.60	2.10	½	1	4.60	3.10	1½	0
3/8—18	3.82	2.32	0	1½	4.82	3.32	1	½
1/2—14	3.98	2.48	0	1½	4.98	3.48	1	½
3/4—14	4.25	2.75	0	1½	5.25	3.75	1	½
1—11½	4.10	2.60	0	1½	5.10	3.60	1	½

L-1 and L-3 gages. When this variation exceeds ½ turn the fittings are rejected. L-1 and L-3 thread plug gages for Aeronautical taper pipe threads are marked ANPT and are normally supplied with one gaging step representing the basic size. They can also be supplied with three gaging steps representing the minimum, basic, and maximum sizes.

A six-step plain tapered plug gage is used to check crest truncation of the internal thread. This gage has three sets of two steps, each set representing the limits of truncation for a thread of minimum, basic, or maximum pitch diameter as determined by the L-1 and L-3 gages. As an example, when the L-1 and L-3 gages enter the fitting so as to indicate maximum size, the face of the part must fall between the two steps of the plain plug gage representing maximum size. Taper errors in the minor diameter can be detected by a shaky fit of the plain gage.

GAGING EXTERNAL THREADS. A set of three ring gages is required to gage external ANPT threads. An ANPT L-2 ring gage having a crest flat equal to the maximum root flat of the external thread is first screwed on the part by hand. To be within proper limits the end of the product should be flush with the face of the ring with a tolerance of plus or minus 1 turn. The operation is then repeated using an ANPT L-1 ring gage having a crest flat considerably larger than that of the L-2 ring. The end of the product must again be flush with the face of the ring within 1 turn. Also, the relative positions of the L-1 and L-2 rings must not vary more than ½ turn to insure that the thread angle and taper are within proper tolerances.

A six-step plain tapered ring gage is used to check crest truncation of the external thread. This gage has three sets of two steps, each set representing the limits of truncation for a thread of minimum, basic, or maximum pitch diameter as determined by the L-2 and L-1 ring gages. As an example, when the L-1 and L-2 gages indicate that the fitting is minimum size, the end of the product must fall between the two steps of the plain ring gage representing the truncation for minimum size. Taper errors on the outside diameter of the external threads can be detected by a shaky fit of the plain ring gage.

FOREIGN SCREW THREADS

British Association (B.A.) Threads. In 1951 the British Standards Institution issued a revised specification on British Association (B.A.) screw threads for fine-pitch screws and nuts under ¼ in. Actually these threads are based on the metric system (see Table 45). Thread form (Fig. 31) is a symmetrical V of 47½° included angle, and the crests and roots are rounded with equal radiuses, such that the basic depth is 0.6 pitch. Provision is made in the new standard for one class of fits for nuts and two for bolts—close and normal. A summary of tolerance formulas is:

	Major dia	Effective* dia	Minor dia
Close class bolts 0 to 10 B.A., inc.	$0.15p$	$0.08p + 0.02$	$2(0.08p + 0.02)$
Normal class bolts 0 to 10 B.A., inc.	$0.20p$	$0.10p + 0.025$	$0.20p + 0.$[illegible]
Normal class bolts 11 to 16 B.A., inc.	$0.25p$	$0.10p + 0.025$	$0.20p + 0$[illegible]
B.A. nuts	None	$0.12p + 0.03$	$0.375p$

* Pitch diameter in American terminology.

Units given are in millimeters. Calculated tolerances can be converted to i[illegible] relationship 1 mm = 0.03937 in.

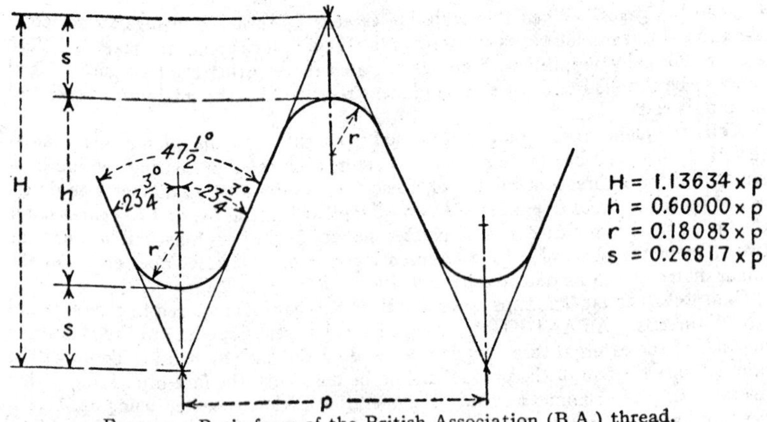

FIG. 31. Basic form of the British Association (B.A.) thread.

Whitworth Threads. The British Standard Whitworth form of thread is a symmetrical V-form with a thread angle of 55°. One-sixth of the sharp V is truncated top and bottom, the crests and roots being rounded. Theoretical depth of thread is 0.64033 times the nominal pitch. Various relationships of thread elements are given in Fig. 32.

The basic thread form applied to three groups of Whitworth threads: British Standard Whitworth, known as B.S.W., and being a coarse series from ⅛–40 to 6–2.5; a British Standard Fine (B.S.F.) in sizes from 3⁄16–32 to 4¼–4, and a British Standard Pipe (Parallel Threads), and known as B.S.P., in sizes from ⅛–28 to 1½–11 threads.

The screw threads are designated as close, medium, or free fit. These fits are described as:

CLOSE FIT. Applies to screw threads requiring a fine, snug fit, recommended only for special work where refined accuracy of pitch and thread form is required.

MEDIUM FIT. Applies to the better class of interchangeable screw threads.

FREE FIT. Used for the great bulk of screw threads of ordinary commercial quality.

Table 41 gives the basic sizes of B.S.W. threads, and Table 42 gives the limits and tolerances for bolts and nuts made to the free fit. Comparable data for B.S.F. threads are given in Tables 43 and 44. These data should prove ample for the majority of uses, considering the current application of Unified threads in Great Britain, Canada, and the United States. If need for data on other classes of fit should arise, consult "British Standards for Workshop Practice—Handbook No. 2," published by the British Standards Institution, 2 Park St., London, W 1.

Continental Screw Threads

a ... Système International 60° metric threads are used in Europe
t ... rdized internationally in pitches from 6 mm upward. In
... ic threads are used for spark plugs. Figure 33 gives the
... and Table 45 the basic dimensions.

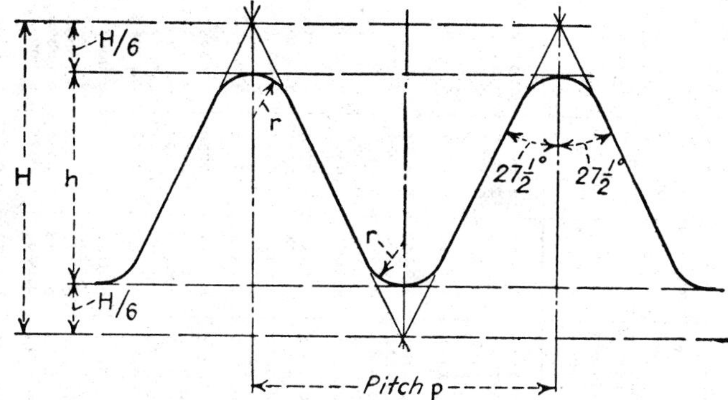

H = 0.960491 X p h = 0.640327 X p r = 0.137329 X p

FIG. 32. Basic form of the Whitworth thread.

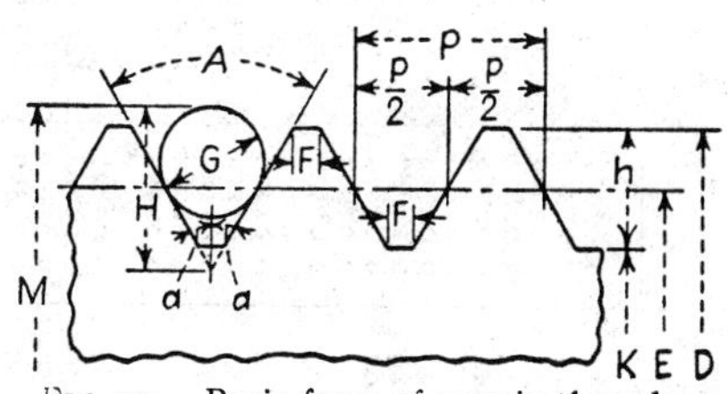

FIG. 33. Basic form of metric threads.

60° *International*

$A = 60°$
$a = 30°$
p = pitch in millimeters
$H = 0.866p$
$h = 0.6495p$
$F = 0.125p$
$E = D - h$ = pitch dia
$= M - (3G - 0.866/n)$
$G = 0.577p$
$K = D - 2h$

Löwenherz

$A = 53° 8'$
$a = 26° 34'$
p = pitch in millimeters
$H = p$
$h = 0.75p$
$F = 0.125p$
$E = D - h$ = pitch dia
$= M - (3.2359G - p)$
$G = 0.559p$
$K = D - 2h$

Löwenherz Thread. This thread form (Fig. 33) has an angle of 53°8′ and is based on the metric system, the diameters ranging from 1 to 40 mm and the pitches from 0.25 to 4.4 mm. Originally used in Germany for instrument threads, the Löwenherz is now largely superseded by the metric system, since both the German (DIN) and the French metric thread systems carry diameters down to 1 mm, other details varying slightly.

Table 40. Basic Sizes of British Association (B.A.) Screw Threads

No.	Metric Dimensions					Approx In. Equivalents						
	Pitch p, Mm	Depth of Thread $0.6p$, Mm	Major Dia, Mm	Effective* Dia, Mm	Minor Dia, Mm	Approx Pitch, In.	Approx Threads per In.	Depth of Thread, In.	Major Dia, In.	Effective* Dia, In.	Minor Dia, In.	Approx Area at Bottom of Thread, Sq In.
			Bolt and Nut D	Bolt and Nut $D - h$	Bolt and Nut $D - 2h$				Bolt and Nut	Bolt and Nut	Bolt and Nut	
0	1.0000	0.600	6.00	5.400	4.80	0.03937	25.4	0.0236	0.2362	0.2126	0.1890	0.0281
1	0.9000	0.540	5.30	4.760	4.22	0.03543	28.2	0.0213	0.2087	0.1874	0.1661	0.0217
2	0.8100	0.485	4.70	4.215	3.73	0.03189	31.3	0.0191	0.1850	0.1659	0.1468	0.0169
3	0.7300	0.440	4.10	3.660	3.22	0.02874	34.8	0.0173	0.1614	0.1441	0.1268	0.0126
4	0.6600	0.395	3.60	3.205	2.81	0.02598	38.5	0.0156	0.1417	0.1262	0.1106	0.0096
5	0.5900	0.355	3.20	2.845	2.49	0.02323	43.1	0.0140	0.1260	0.1120	0.0980	0.0075
6	0.5300	0.320	2.80	2.480	2.16	0.02087	47.9	0.0126	0.1102	0.0976	0.0850	0.0057
7	0.4800	0.290	2.50	2.210	1.92	0.01890	52.9	0.0114	0.0984	0.0870	0.0756	0.0045
8	0.4300	0.260	2.20	1.940	1.68	0.01693	59.1	0.0102	0.0866	0.0764	0.0661	0.0034
9	0.3900	0.235	1.90	1.665	1.43	0.01535	65.1	0.0093	0.0748	0.0656	0.0563	0.0025
10	0.3500	0.210	1.70	1.490	1.28	0.01378	72.6	0.0083	0.0669	0.0587	0.0504	0.0020
11	0.3100	0.185	1.50	1.315	1.13	0.01220	81.9	0.0073	0.0591	0.0518	0.0445	0.0016
12	0.2800	0.170	1.30	1.130	0.96	0.01102	90.7	0.0067	0.0512	0.0445	0.0378	0.0011
13	0.2500	0.150	1.20	1.050	0.90	0.00984	102.0	0.0059	0.0472	0.0413	0.0354	0.0010
14	0.2300	0.140	1.00	0.860	0.72	0.00905	110.0	0.0055	0.0394	0.0339	0.0283	0.0006
15	0.2100	0.125	0.90	0.775	0.65	0.00827	121.0	0.0049	0.0354	0.0305	0.0256	0.0005
16	0.1900	0.115	0.79	0.675	0.56	0.00748	134.0	0.0045	0.0311	0.0266	0.0220	0.0004
17	0.1700	0.100	0.70	0.600	0.50	0.00669	149.0	0.0039	0.0276	0.0236	0.0197	0.0003
18	0.1500	0.090	0.62	0.530	0.44	0.00590	169.0	0.0035	0.0244	0.0209	0.0173	0.0002
19	0.1400	0.085	0.54	0.455	0.37	0.00551	182.0	0.0033	0.0213	0.0179	0.0146	0.0002
20	0.1200	0.070	0.48	0.410	0.34	0.00472	213.0	0.0028	0.0189	0.0161	0.0134	0.0001
21	0.1100	0.065	0.42	0.355	0.29	0.00433	232.0	0.0026	0.0165	0.0140	0.0114	0.0001
22	0.1000	0.060	0.37	0.310	0.25	0.00394	256.0	0.0024	0.0146	0.0122	0.0098	0.00008
23	0.0900	0.055	0.33	0.275	0.22	0.00354	286.0	0.0022	0.0130	0.0108	0.0087	0.00006
24	0.0800	0.050	0.29	0.240	0.19	0.00315	322.0	0.0020	0.0114	0.0094	0.0075	0.00004
25	0.0700	0.040	0.25	0.210	0.17	0.00276	357.0	0.0016	0.0098	0.0083	0.0067	0.00004

* Pitch dia in American terminology.

TABLE 41. BRITISH STANDARD WHITWORTH SCREW THREADS—BASIC SIZES

Nominal Dia, In.	Threads per In.	Pitch, In.	Depth of Thread, In.	Major Dia, In.	Effective Dia, In.	Minor Dia, In.	Cross-sectional Area at Bottom of Thread, Sq In.
1/8*	40	0.025 00	0.0160	0.1250	0.1090	0.0930	0.0068
3/16	24	0.041 67	0.0267	0.1875	0.1608	0.1341	0.0141
1/4	20	0.050 00	0.0320	0.2500	0.2180	0.1860	0.0272
5/16	18	0.055 56	0.0356	0.3125	0.2769	0.2413	0.0457
3/8	16	0.062 50	0.0400	0.3750	0.3350	0.2950	0.0683
7/16	14	0.071 43	0.0457	0.4375	0.3918	0.3461	0.0941
1/2	12	0.083 33	0.0534	0.5000	0.4466	0.3932	0.1214
9/16	12	0.083 33	0.0534	0.5625	0.5091	0.4557	0.1631
5/8	11	0.090 91	0.0582	0.6250	0.5668	0.5086	0.2032
11/16†	11	0.090 91	0.0582	0.6875	0.6293	0.5711	0.2562
3/4	10	0.100 00	0.0640	0.7500	0.6860	0.6220	0.3039
7/8	9	0.111 11	0.0711	0.8750	0.8039	0.7328	0.4218
1	8	0.125 00	0.0800	1.0000	0.9200	0.8400	0.5542
1 1/8	7	0.142 86	0.0915	1.1250	1.0335	0.9420	0.6969
1 1/4	7	0.142 86	0.0915	1.2500	1.1585	1.0670	0.8942
1 1/2	6	0.166 67	0.1067	1.5000	1.3933	1.2866	1.300
1 3/4	5	0.200 00	0.1281	1.7500	1.6219	1.4938	1.753
2	4.5	0.222 22	0.1423	2.0000	1.8577	1.7154	2.311
2 1/4	4	0.250 00	0.1601	2.2500	2.0899	1.9298	2.925
2 1/2	4	0.250 00	0.1601	2.5000	2.3399	2.1798	3.732
2 3/4	3.5	0.285 71	0.1830	2.7500	2.5670	2.3840	4.464
3	3.5	0.285 71	0.1830	3.0000	2.8170	2.6340	5.449
3 1/4	3.25	0.307 69	0.1970	3.2500	3.0530	2.8560	6.406
3 1/2	3.25	0.307 69	0.1970	3.5000	3.3030	3.1060	7.577
3 3/4	3	0.333 33	0.2134	3.7500	3.5366	3.3232	8.674
4	3	0.333 33	0.2134	4.0000	3.7866	3.5732	10.03
4 1/2	2.875	0.347 83	0.2227	4.5000	4.2773	4.0546	12.91
5	2.75	0.363 64	0.2328	5.0000	4.7672	4.5344	16.15
5 1/2	2.625	0.380 95	0.2439	5.5000	5.2561	5.0122	19.73
6	2.5	0.400 00	0.2561	6.0000	5.7439	5.4878	23.65

* Dimensionally the 1/8 in. × 40 threads per in. thread belongs more appropriately to the British Standard fine series, but it has for so long been associated with the Whitworth series that it is now included herein.

† To be dispensed with wherever possible.

TABLE 42. BRITISH STANDARD WHITWORTH LIMITS AND TOLERANCES, FREE FIT, BOLTS AND NUTS

Nominal Dia, In.	Threads per In.	Bolts									Nuts						
		Major Dia, In.			Effective Dia, In.			Minor Dia, In.			Major Dia, In.	Effective Dia, In.			Minor Dia, In.		
		Max	Tol	Min	Max	Tol	Min	Max	Tol	Min	Min	Max	Tol	Min	Max	Tol	Min
1/8	40	0.1250	0.0059	0.1191	0.1090	0.0043	0.1047	0.0930	0.0075	0.0855	0.1250	0.1133	0.0043	0.1090	0.1020	0.0090	0.0930
3/16	24	0.1875	0.0072	0.1803	0.1608	0.0052	0.1556	0.1341	0.0093	0.1248	0.1875	0.1660	0.0052	0.1608	0.1474	0.0133	0.1341
1/4	20	0.2500	0.0080	0.2420	0.2180	0.0058	0.2122	0.1860	0.0103	0.1757	0.2500	0.2238	0.0058	0.2180	0.2030	0.0170	0.1860
5/16	18	0.3125	0.0087	0.3038	0.2769	0.0063	0.2706	0.2413	0.0110	0.2303	0.3125	0.2832	0.0063	0.2769	0.2594	0.0181	0.2413
3/8	16	0.3750	0.0093	0.3657	0.3350	0.0068	0.3282	0.2950	0.0118	0.2832	0.3750	0.3418	0.0068	0.3350	0.3145	0.0195	0.2950
7/16	14	0.4375	0.0100	0.4275	0.3918	0.0073	0.3845	0.3461	0.0126	0.3335	0.4375	0.3991	0.0073	0.3918	0.3674	0.0213	0.3461
1/2	12	0.5000	0.0106	0.4894	0.4466	0.0077	0.4389	0.3932	0.0135	0.3797	0.5000	0.4543	0.0077	0.4466	0.4169	0.0237	0.3932
9/16	12	0.5625	0.0109	0.5516	0.5091	0.0080	0.5011	0.4557	0.0138	0.4419	0.5625	0.5171	0.0080	0.5091	0.4794	0.0237	0.4557
5/8	11	0.6250	0.0114	0.6136	0.5668	0.0084	0.5584	0.5086	0.0144	0.4942	0.6250	0.5752	0.0084	0.5668	0.5338	0.0252	0.5086
11/16*	11	0.6875	0.0116	0.6759	0.6293	0.0086	0.6207	0.5711	0.0146	0.5565	0.6875	0.6379	0.0086	0.6293	0.5963	0.0252	0.5711
3/4	10	0.7500	0.0122	0.7378	0.6860	0.0090	0.6770	0.6220	0.0153	0.6067	0.7500	0.6950	0.0090	0.6860	0.6490	0.0270	0.6220
7/8	9	0.8750	0.0129	0.8621	0.8039	0.0096	0.7943	0.7328	0.0163	0.7165	0.8750	0.8135	0.0096	0.8039	0.7620	0.0292	0.7328
1	8	1.0000	0.0137	0.9863	0.9200	0.0102	0.9098	0.8400	0.0173	0.8227	1.0000	0.9302	0.0102	0.9200	0.8720	0.0320	0.8400
1 1/8	7	1.1250	0.0145	1.1105	1.0335	0.0107	1.0228	0.9420	0.0183	0.9237	1.1250	1.0442	0.0107	1.0335	0.9776	0.0356	0.9420
1 1/4	7	1.2500	0.0149	1.2351	1.1585	0.0111	1.1474	1.0670	0.0187	1.0483	1.2500	1.1696	0.0111	1.1585	1.1026	0.0356	1.0670
1 1/2	6	1.5000	0.0161	1.4839	1.3933	0.0120	1.3813	1.2866	0.0202	1.2664	1.5000	1.4053	0.0120	1.3933	1.3269	0.0403	1.2866
1 3/4	5	1.7500	0.0174	1.7326	1.6219	0.0129	1.6090	1.4938	0.0218	1.4720	1.7500	1.6348	0.0129	1.6219	1.5408	0.0470	1.4938
2	4.5	2.0000	0.0184	1.9816	1.8577	0.0137	1.8440	1.7154	0.0231	1.6923	2.0000	1.8714	0.0137	1.8577	1.7668	0.0514	1.7154
2 1/4	4	2.2500	0.0194	2.2306	2.0899	0.0144	2.0755	1.9298	0.0244	1.9054	2.2500	2.1043	0.0144	2.0899	1.9868	0.0570	1.9298
2 1/2	4	2.5000	0.0199	2.4801	2.3399	0.0149	2.3250	2.1798	0.0249	2.1549	2.5000	2.3548	0.0149	2.3399	2.2368	0.0570	2.1798
2 3/4	3.5	2.7500	0.0210	2.7290	2.5670	0.0157	2.5513	2.3840	0.0264	2.3576	2.7500	2.5827	0.0157	2.5670	2.4481	0.0641	2.3840
3	3.5	3.0000	0.0214	2.9786	2.8170	0.0161	2.8009	2.6340	0.0268	2.6072	3.0000	2.8331	0.0161	2.8170	2.6981	0.0641	2.6340

* To be dispensed with wherever possible.

TABLE 43. BRITISH STANDARD FINE SCREW THREADS—BASIC SIZES

Nominal Dia, In.	Threads per In.	Pitch, In.	Depth of Thread, In.	Major Dia, In.	Effective Dia, In.	Minor Dia, In.	Cross-sectional Area at Bottom of Thread, Sq In.
3/16	32	0.031 25	0.0200	0.1875	0.1675	0.1475	0.0171
7/32	28	0.035 71	0.0229	0.2188	0.1959	0.1730	0.0235
1/4	26	0.038 46	0.0246	0.2500	0.2254	0.2008	0.0317
9/32	26	0.038 46	0.0246	0.2812	0.2566	0.2320	0.0423
5/16	22	0.045 45	0.0291	0.3125	0.2834	0.2543	0.0508
3/8	20	0.050 00	0.0320	0.3750	0.3430	0.3110	0.0760
7/16	18	0.055 56	0.0356	0.4375	0.4019	0.3663	0.1054
1/2	16	0.062 50	0.0400	0.5000	0.4600	0.4200	0.1385
9/16	16	0.062 50	0.0400	0.5625	0.5225	0.4825	0.1828
5/8	14	0.071 43	0.0457	0.6250	0.5793	0.5336	0.2236
11/16	14	0.071 43	0.0457	0.6875	0.6418	0.5961	0.2791
3/4	12	0.083 33	0.0534	0.7500	0.6966	0.6432	0.3249
13/16	12	0.083 33	0.0534	0.8125	0.7591	0.7057	0.3911
7/8	11	0.090 91	0.0582	0.8750	0.8168	0.7586	0.4520
1	10	0.100 00	0.0640	1.0000	0.9360	0.8720	0.5972
1 1/8	9	0.111 11	0.0711	1.1250	1.0539	0.9828	0.7586
1 1/4	9	0.111 11	0.0711	1.2500	1.1789	1.1078	0.9639
1 3/8	8	0.125 00	0.0800	1.3750	1.2950	1.2150	1.159
1 1/2	8	0.125 00	0.0800	1.5000	1.4200	1.3400	1.410
1 5/8	8	0.125 00	0.0800	1.6250	1.5450	1.4650	1.686
1 3/4	7	0.142 86	0.0915	1.7500	1.6585	1.5670	1.928
2	7	0.142 86	0.0915	2.0000	1.9085	1.8170	2.593
2 1/4	6	0.166 67	0.1067	2.2500	2.1433	2.0366	3.258
2 1/2	6	0.166 67	0.1067	2.5000	2.3933	2.2866	4.106
2 3/4	6	0.166 67	0.1067	2.7500	2.6433	2.5366	5.054
3	5	0.200 00	0.1281	3.0000	2.8719	2.7438	5.913
3 1/4	5	0.200 00	0.1281	3.2500	3.1219	2.9938	7.039
3 1/2	4.5	0.222 22	0.1423	3.5000	3.3577	3.2154	8.120
3 3/4	4.5	0.222 22	0.1423	3.7500	3.6077	3.4654	9.432
4	4.5	0.222 22	0.1423	4.0000	3.8577	3.7154	10.84
4 1/4	4	0.250 00	0.1601	4.2500	4.0899	3.9298	12.13

NOTE: It is recommended that for larger diameters in this series four threads per inch be used.

TABLE 44. BRITISH STANDARD FINE LIMITS AND TOLERANCES, FREE FIT, BOLTS AND NUTS

Nominal Dia, In.	Threads per In.	Bolts									Nuts						
		Major Dia, In.			Effective Dia, In.			Minor Dia, In.			Major Dia, In.	Effective Dia, In.			Minor Dia, In.		
		Max	Tol	Min	Max	Tol	Min	Max	Tol	Min	Min	Max	Tol	Min	Max	Tol	Min
3/16	32	0.1875	0.0068	0.1807	0.1675	0.0050	0.1625	0.1475	0.0085	0.1390	0.1875	0.1725	0.0050	0.1675	0.1577	0.0102	0.1475
7/32	28	0.2188	0.0072	0.2116	0.1959	0.0053	0.1906	0.1730	0.0091	0.1639	0.2188	0.2012	0.0053	0.1959	0.1841	0.0111	0.1730
1/4	26	0.2500	0.0076	0.2424	0.2254	0.0056	0.2198	0.2008	0.0095	0.1913	0.2500	0.2310	0.0056	0.2254	0.2125	0.0117	0.2008
9/32	26	0.2812	0.0078	0.2734	0.2566	0.0058	0.2508	0.2320	0.0097	0.2223	0.2812	0.2624	0.0058	0.2566	0.2437	0.0117	0.2320
5/16	22	0.3125	0.0083	0.3042	0.2834	0.0062	0.2772	0.2543	0.0105	0.2438	0.3125	0.2896	0.0062	0.2834	0.2684	0.0141	0.2543
3/8	20	0.3750	0.0088	0.3662	0.3430	0.0066	0.3364	0.3110	0.0111	0.2999	0.3750	0.3496	0.0066	0.3430	0.3280	0.0170	0.3110
7/16	18	0.4375	0.0094	0.4281	0.4019	0.0070	0.3949	0.3663	0.0117	0.3546	0.4375	0.4089	0.0070	0.4019	0.3844	0.0181	0.3663
1/2	16	0.5000	0.0099	0.4901	0.4600	0.0074	0.4526	0.4200	0.0124	0.4076	0.5000	0.4674	0.0074	0.4600	0.4395	0.0195	0.4200
9/16	16	0.5625	0.0102	0.5523	0.5225	0.0077	0.5148	0.4825	0.0127	0.4698	0.5625	0.5302	0.0077	0.5225	0.5020	0.0195	0.4825
5/8	14	0.6250	0.0108	0.6142	0.5793	0.0081	0.5712	0.5336	0.0134	0.5202	0.6250	0.5874	0.0081	0.5793	0.5549	0.0213	0.5336
11/16	14	0.6875	0.0111	0.6764	0.6418	0.0084	0.6334	0.5961	0.0137	0.5824	0.6875	0.6502	0.0084	0.6418	0.6174	0.0213	0.5961
3/4	12	0.7500	0.0117	0.7383	0.6966	0.0088	0.6878	0.6432	0.0146	0.6286	0.7500	0.7054	0.0088	0.6966	0.6669	0.0237	.6432
13/16	12	0.8125	0.0119	0.8006	0.7591	0.0090	0.7501	0.7057	0.0148	0.6909	0.8125	0.7681	0.0090	0.7591	0.7294	0.0237	0.7057
7/8	11	0.8750	0.0123	0.8627	0.8168	0.0093	0.8075	0.7586	0.0153	0.7433	0.8750	0.8261	0.0093	0.8168	0.7838	0.0252	0.7586
1	10	1.0000	0.0131	0.9869	0.9360	0.0099	0.9261	0.8720	0.0162	0.8558	1.0000	0.9459	0.0099	0.9360	0.8990	0.0270	0.8720
1 1/8	9	1.1250	0.0137	1.1113	1.0539	0.0104	1.0435	0.9828	0.0171	0.9657	1.1250	1.0643	0.0104	1.0539	1.0120	0.0292	0.9828
1 1/4	9	1.2500	0.0141	1.2359	1.1789	0.0108	1.1681	1.1078	0.0175	1.0903	1.2500	1.1897	0.0108	1.1789	1.1370	0.0292	1.1078
1 3/8	8	1.3750	0.0148	1.3602	1.2950	0.0113	1.2837	1.2150	0.0184	1.1966	1.3750	1.3063	0.0113	1.2950	1.2470	0.0320	1.2150
1 1/2	8	1.5000	0.0151	1.4849	1.4200	0.0116	1.4084	1.3400	0.0187	1.3213	1.5000	1.4316	0.0116	1.4200	1.3720	0.0320	1.3400

TABLE 45. BASIC DIMENSIONS OF INTERNATIONAL 60° METRIC THREADS

Millimeter Sizes					Approx Threads per In. n	Equivalent Sizes, In.			
Dia of Screw D	Pitch p	Depth of Thread $0.6495p$ (h)	Pitch Dia $D - h$ (E)	Minor Dia $D - 2h$ (k)		Depth of Thread $0.6495p$ (h)	Major Dia D	Pitch Dia $D - h$ (E)	Minor Dia $D - 2h$ (K)
3	0.50	0.32475	2.67525	2.350	50.8	0.01279	0.11811	0.10532	0.09253
3.5	0.60	0.38970	3.11030	2.721	42.3	0.01534	0.13780	0.12246	0.10712
4	0.70	0.45465	3.54535	3.091	36.3	0.01790	0.15748	0.13958	0.12168
4.5	0.75	0.48712	4.01288	3.526	33.9	0.01918	0.17716	0.15798	0.13880
5	0.80	0.51960	4.48040	3.961	31.8	0.02046	0.19685	0.17639	0.15593
5.5	0.90	0.58455	4.91545	4.331	28.2	0.02301	0.21654	0.19353	0.17052
6	1.00	0.64950	5.35050	4.70	25.4	0.02557	0.23622	0.21065	0.18508
7	1.00	0.64950	6.35050	5.70	25.4	0.02557	0.27559	0.25002	0.22445
8	1.25	0.81188	7.18813	6.38	20.3	0.03196	0.31496	0.28300	0.25104
9	1.25	0.81188	8.18813	7.38	20.3	0.03196	0.35433	0.32237	0.29041
10	1.5	0.97425	9.026	8.05	16.9	0.03836	0.39370	0.35534	0.31698
11	1.5	0.97425	10.026	9.05	16.9	0.03836	0.43307	0.39471	0.35635
12	1.5	0.97425	11.026	10.05	16.9	0.03836	0.47244	0.43408	0.39572
12	1.75	1.13662	10.863	9.73	14.5	0.04475	0.47244	0.42769	0.38294
14	2.0	1.2990	12.701	11.40	12.7	0.05114	0.55118	0.50004	0.44890
16	2.0	1.2990	14.701	13.40	12.7	0.05114	0.62992	0.57878	0.52764
18	2.5	1.62375	16.376	14.75	10.2	0.06393	0.70866	0.64473	0.58080
20	2.5	1.62375	18.376	16.75	10.2	0.06393	0.78740	0.72347	0.65954
22	2.5	1.62375	20.376	18.75	10.2	0.06393	0.86614	0.80221	0.73828
22	3.0	1.9485	20.052	18.10	8.5	0.07671	0.86614	0.78943	0.71272

TABLE 45. BASIC DIMENSIONS OF INTERNATIONAL 60° METRIC THREADS (*Continued*)

Millimeter Sizes					Approx Threads per In. n	Equivalent Sizes, In.			
Dia of Screw D	Pitch p	Depth of Thread $0.6495p$ (h)	Pitch Dia $D - h$ (E)	Minor Dia $D - 2h$ (k)		Depth of Thread $0.6495p$ (h)	Major Dia D	Pitch Dia $D - h$ (E)	Minor Dia $D - 2h$ (K)
24	3.0	1.9485	22.052	20.10	8.5	0.07671	0.94488	0.86817	0.79146
26	3.0	1.9485	24.052	22.10	8.5	0.07671	1.02362	0.94691	0.87020
27	3.0	1.9485	25.052	23.10	8.5	0.07671	1.06299	0.98628	0.90957
28	3.0	1.9485	26.052	24.10	8.5	0.07671	1.10236	1.02565	0.94894
30	3.5	2.27325	27.727	25.45	7.3	0.08950	1.18110	1.09160	1.00210
32	3.5	2.27325	29.727	27.45	7.3	0.08950	1.25984	1.17034	1.08084
33	3.5	2.27325	30.727	28.45	7.3	0.08950	1.29921	1.20971	1.12021
34	3.5	2.27325	31.727	29.45	7.3	0.08950	1.33858	1.24908	1.15958
36	4.0	2.5980	33.402	30.80	6.4	0.10228	1.41732	1.31504	1.21276
38	4.0	2.5980	35.402	32.80	6.4	0.10228	1.49606	1.39378	1.29150
39	4.0	2.5980	36.402	33.80	6.4	0.10228	1.53543	1.43315	1.33087
40	4.0	2.5980	37.402	34.80	6.4	0.10228	1.57480	1.47252	1.37024
42	4.5	2.92275	39.077	36.15	5.6	0.11507	1.65354	1.53847	1.42340
44	4.5	2.92275	41.077	38.15	5.6	0.11507	1.73228	1.61721	1.50214
45	4.5	2.92275	42.077	39.15	5.6	0.11057	1.77165	1.65658	1.54151
46	4.5	2.92275	43.077	40.15	5.6	0.11057	1.81102	1.69595	1.58088
48	5.0	3.2475	44.752	41.50	5.1	0.12785	1.88976	1.76191	1.63406
50	5.0	3.2475	46.752	43.50	5.1	0.12785	1.96850	1.84065	1.71280
52	5.0	3.2475	48.752	45.50	5.1	0.12785	2.04724	1.91939	1.79154
56	5.5	3.57225	52.428	48.86	4.6	0.14064	2.20472	2.06408	1.92344
60	5.5	3.57225	56.428	52.86	4.6	0.14064	2.36220	2.22156	2.08092
64	6.0	3.8970	60.103	56.21	4.2	0.15342	2.51968	2.36626	2.21284
68	6.0	3.8970	64.103	60.21	4.2	0.15342	2.67716	2.52374	2.37032
72	6.5	4.22175	67.778	63.56	3.9	0.16621	2.83464	2.66843	2.50222
76	6.5	4.22175	71.778	67.56	3.9	0.16621	2.99212	2.82591	2.65970
80	7.0	4.5465	75.454	70.91	3.6	0.17900	3.14960	2.97060	2.79160

Section 11

TAPS AND TAPPING

SECTION 11

TAPS AND TAPPING

A tap is a single-point internal-threading tool with the number of points multiplied to reduce the number of passes necessary to produce a completed thread and to make threading possible without a lead screw. All straight-tap cutting is done by the chamfer and first full thread (Fig. 1). Following threads provide lead and support only, plus material for resharpening the tap. Tapped threads will always be better if the tap is lead screw controlled.

KINDS OF TAPS

Taps fall in two groups. These are the solid taps, either cut or ground from solid stock for use manually or with a machine; and collapsible taps, which depend for cutting on inserted chasers. They are used with automatic lathes, turret lathes, screw machines, and special threading machines. They may also be used with an engine lathe for quantity production of large threads. Collapsible taps are of two types: revolving and stationary. Chasers are either radial or circular.

TYPES OF SOLID TAPS

Hand Taps

Hand taps were originally designed for threading by hand. Now, most hand taps are suitable for machine tapping as well. However, it is usually better practice to select the tap specifically for a machine operation. Also, with most materials, the spiral-point, or "gun," grind is more efficient.

Probably the best-known hand-tap group is the popular set consisting of taper, plug, and bottoming taps (Fig. 2). These are identical in size, length, and vital measurements; differ only in the chamfered portion of the tip. The taper, or "starting," tap has a chamfer length of 8 to 10 threads and is the longest chamfer of all. The plug tap (by far the most popular) has about 3 to 5 threads chamfered, and the bottoming tap only 1 to 1½ threads chamfered.

Taper hand taps are intended for use when tapping coarse threads by hand in open or through holes, especially for harder metals. As tool life and breakage are largely determined by chamfer length, taper taps should always be considered when the job permits their use. Blind holes seldom provide enough room for taper taps to develop a full thread.

Plug taps are designed for use in through or open holes when tapping softer metals or fine-pitch threads. Blind holes may be tapped with this style, if sufficient clearance is provided below the thread to clear chamfer and chips. Plug hand taps are also used for machine tapping but should be purposely selected for a specific job If the material is such that cuttings crumble or break easily, a hand tap will perform effectively. Taps for machine use should be chosen to provide more chip clearance, deeper flutes, or spiral flutes, or to shoot the chips forward through the hole.

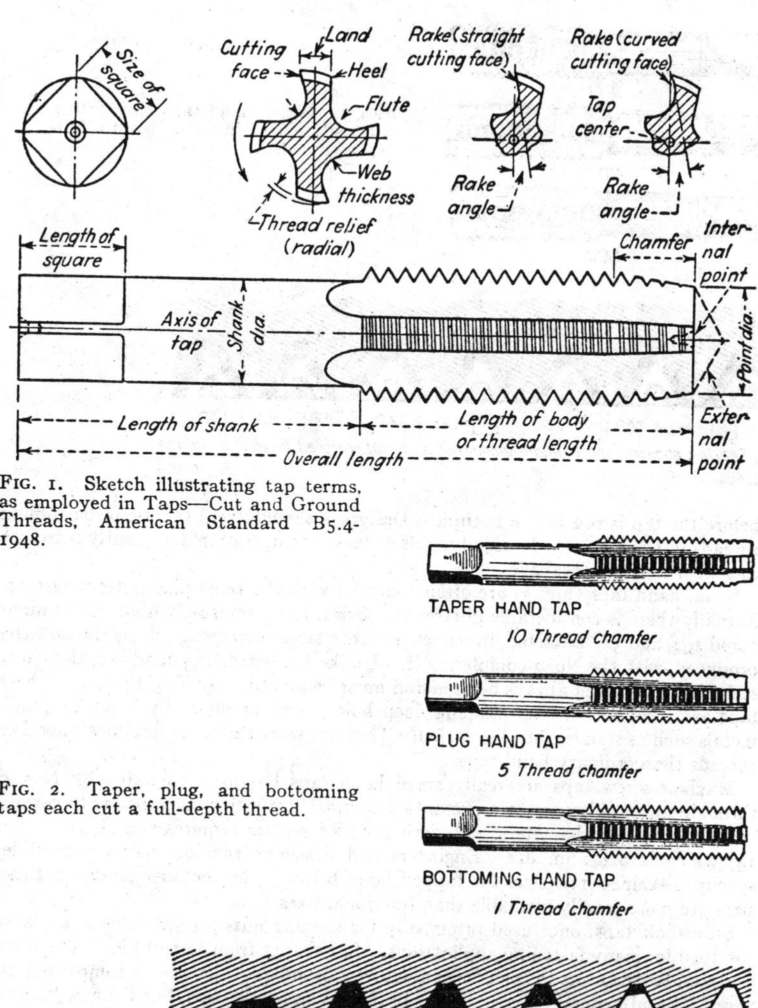

FIG. 1. Sketch illustrating tap terms, as employed in Taps—Cut and Ground Threads, American Standard B5.4-1948.

FIG. 2. Taper, plug, and bottoming taps each cut a full-depth thread.

Bottoming hand taps are seldom necessary. Their one use is for threading to the bottom of a blind hole. To do this they are permitted a 1- to 1½-thread chamfer only. Chip load per tooth is tremendous. The bottoming tap should always be preceded by a plug tap. For coarse threads or hard material, both taper and plug taps should precede the bottoming tap. Chips must be removed from the hole

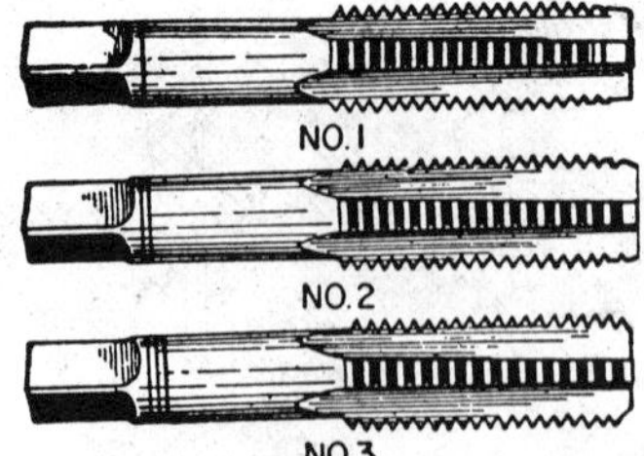

FIG. 3. Serial hand taps Nos. 1 to 3 must all be used in series to cut a full-depth thread.

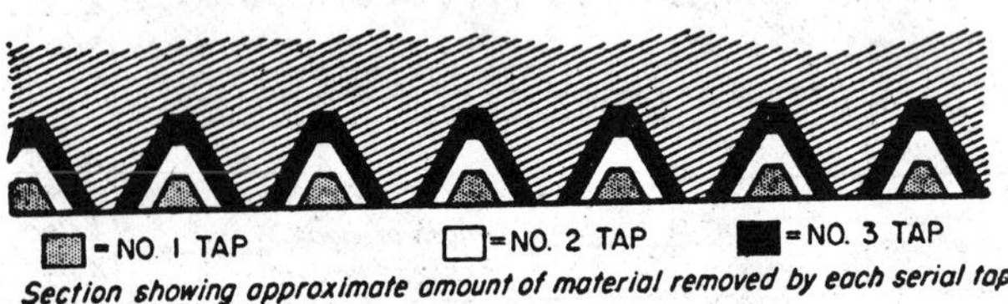

before the tap is run to the bottom. Designers should avoid the use of bottoming threads in blind holes. It usually adds at least one operation and greatly increases tool wear and tap breakage.

Serial hand taps (Fig. 3) are often confused with the taper-plug-bottoming taps. Actually there is considerable difference. Serial taps are usually used in sets numbered 1, 2, and 3. Each tap, in sequence, cuts a larger percentage of the thread to be produced, with the No. 3 completing it. Unlike the bottoming hand tap, the No. 3 serial hand tap must always be used and must be preceded by Nos. 1 and 2. These taps are recommended for tapping deep holes, open or blind, by hand, in tough metals such as stainless steel or nickel. They cut more easily and produce smoother threads than ordinary hand taps.

Machine-screw taps are really small hand taps but are designated by Nos. 0 through 30. Each number represents a decimal size which corresponds to certain "machine-screw" diameters. Modern practice seldom requires sizes beyond No. 14, which is 0.242 in. dia. Engineers and designers should, wherever possible, specify machine-screw sizes for tapped holes below ¼ in., because screws in these sizes are more readily available than fractional sizes.

Stove-bolt taps, once used principally for tapping nuts for stove bolts, are now specified for many fastening applications. Sizes range from ⅛ to ½ in. There are two standards: One is the old "Manufacturers Standard," the other a comparatively new ASA and Federal specification. Both are commonly used, but the new "Amer Std" permits the use of standard-stock machine-screw and hand taps in both carbon and HSS. The old "Manufacturers Std" taps are similar in physical dimensions to plug hand taps but have a different pitch for all diameters except 3/16-24, 5/16-18, and ⅜-16, plus a slightly larger pitch dia than hand taps of similar size and pitch. These differences must be considered, and the proper standard specified, when taps are ordered.

Taper pipe taps (Fig. 4) are used for pipe fittings and in other places where extremely tight fits are necessary. To accomplish this, the threaded portion is tapered. Tap diameter, from bottom to top of threaded portion, increases at ¾ in. per ft, but the angle formed by the sides of the thread is 60° when measured in an

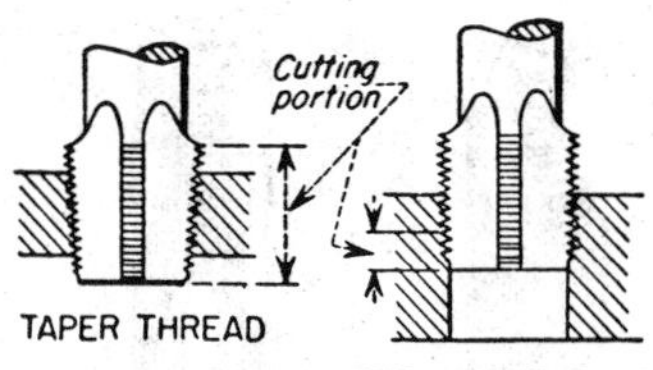

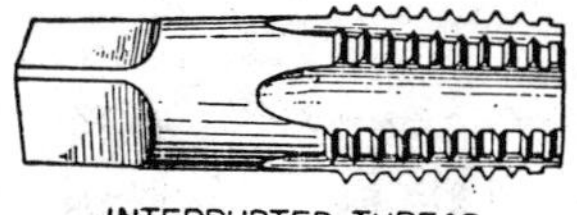

FIG. 4. Taper ream straight holes before taper tapping. Interrupted-thread taper pipe tap provides cleaner cutting.

axial plane, and a line bisecting this angle is perpendicular to the tap axis. These taps differ in action from straight taps as all the threads or teeth cut, rather than just the chamfered portion.

Taper pipe taps are also made with interrupted threads and may be applied wherever tough material must be tapped. By removing alternate teeth in successive lands, a groove is provided which permits chip escape, plus thorough lubrication of the material just ahead of the cutting teeth. In addition to better cutting, the interrupted thread on a taper pipe tap produces a sharper crest in the minor dia of the part, which improves the pressuretight fit.

Where necessary, the same principle may be applied to straight taps for threading tough materials, if a reasonably long chamfer is usable. However, chamfer must be of sufficient length to prevent overloading the reduced number of cutting teeth. Coarse threads will require lead-screw control.

Machine Taps

Tapping by machine differs from hand tapping in one important way. Cutting action, in machine tapping, is continuous rather than intermittent. Therefore, means must be provided to clear the continuous chip formed and to get lubricant to the tool interface. This is more important when chips are stringy and tough than it is when they are brittle and readily crumble. Tapping-torque tests have shown that chip interference has a tremendous influence on tap efficiency. The increased speed of machine tapping would appear also to have an important effect.

"Gun" Taps. Most efficient of the machine taps is the angular-fluted chamfer, spiral-pointed, or "gun" tap (Fig. 5). This is a straight-fluted tap, similar to a plug tap, which has had an angle ground into the flutes at the chamfer. As it cuts, the chips are deflected by this angle so they leave the hole ahead of the tap and hence do not travel up the flute to cause binding. This means that flutes are left clear for the passage of lubricant.

This tap design has very little improving effect in materials such as cast iron, but torque tests show it will increase the efficiency of tapping steel or nickel by as much as 70%. Blind holes should not be tapped with this tap unless there is plenty of clearance at the bottom. All chips will be packed there and will cause the tap to jam and break. Also, this design cannot be used for taper threads. On through holes, higher tapping speeds may be used with less fear of chip jamming.

Taps selected for machine use often have one less land than taps of the same size for hand use. This practice increases chip space but increases the cutting load.

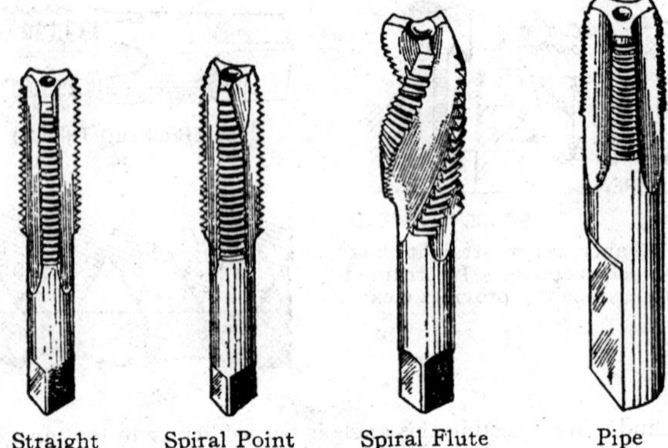

FIG. 5. Common styles of taps. The spiral-point and the spiral-flute types have the most efficient cutting action.

"Stub" Taps. Thin parts or shallow holes may be tapped with "stub," machine-screw, "hand," or "angular-fluted chamfer" taps to reduce the danger of breakage.

Automatic tapping machines employed for these jobs run at high speeds. Hence greater tap rigidity is desirable. "Stub" taps, in machine-screw sizes 3 to 12, are shorter than regular taps. Also, they have tapered flutes which provide for a gradual increase of the flute-core diameter toward the shank. To provide for maximum chip room, all sizes of "stub" machine-screw hand taps have three flutes; "gun" taps have two flutes.

These taps are recommended for use in electric or pneumatic hand drills and tappers.

Spiral-fluted Taps. These should not be confused with spiral-pointed taps. A spiral-fluted tap resembles a spiral-fluted reamer—with teeth cut in the lands. These may be used for hand or machine tapping and are another design for improving chip removal. They are recommended for tough, soft, and other "stringy" materials such as nickel, stainless, copper, and aluminum alloys; also for certain plastics. Spiral-fluted taps for cutting a right-hand thread should have a right-hand spiral of about the same helix angle as that used on an ordinary twist drill. Both bottoming threads and taper threads may be cut with this tap design.

Taps for fine threads in relatively large holes should have more than four lands. One land per ¼ in. dia is a good general rule—with land width one-half of flute width. Taps such as these are special, not stock items.

Tapper Taps. Two styles are straight shank and bent shank. They are generally used by nut manufacturers for production tapping. In primary function they are the same as the nut tap, but structurally they differ in several respects. Their over-all lengths are greater; their thread lengths and chamfers are shorter. Furthermore, they often have three flutes in the sizes in which nut taps have four flutes.

Bent-shank tapper taps are more widely used in production nut tapping. Nut-tapping machines are manufactured, at present, in seven sizes. As capacities of some of these overlap, bent-shank tap orders should specify machine size.

Special-purpose Taps

Literally thousands of designs have been produced to accomplish specific nonroutine threading problems. Some designs are for special tools used so widely that manufacturers produce them as standard stock items. Pulley taps, straight and taper boiler taps, and mud or washout taps are in this category.

Stay-bolt taps are used principally in boiler, locomotive, and railroad shops for tapping the stay-bolt holes in the inner and outer plates or shells of boilers. Distances between these shells vary, and the tap must have sufficient length to tap both holes in one operation so the threads in both holes will be in alignment and have the same lead.

Spindle stay-bolt taps find use largely in repair work for retapping stay-bolt holes. A sliding internal spindle having a tapered guide is provided to insure that both tapped holes are properly lined up. The first hole is tapped; then the spindle is removed, and a nonfluted thread section at the back of the tap guides threading of the second hole.

Straight boiler taps are a special class of hand taps. They all have 12 threads per in. and range in dia from ½ to 1½ in. A pilot is provided, ahead of the chamfered portion of the thread, to simplify starting.

Taper boiler taps have no pilot and the threaded portion is tapered ¾ in. per ft. Over-all and shank diameters are similar to those of straight boiler taps. Tap dia is measured ⅝ in. from the large dia of the thread.

Acme thread taps are usually made to special order as required.

Ordering Special Taps. Blueprints, sketches, or samples of the tools to be duplicated should always be furnished when special taps are ordered. Blueprints and sketches should contain all the following information, or, if the blueprint is not available, the details must be carefully listed:

1. Exact cutting size, threads per inch, and thread form. If multiple thread, number of starts wanted.
2. Right- or left-hand thread.
3. Style and design of tap—such as taper, plug, bottoming, serial, and hand, spiral point, nut, or tapper.
4. High-speed or carbon steel.
5. Cut or ground threads.
6. If ground thread, class of fit required.
7. Over-all length; thread length; style and dia of shank.
8. Number of flutes, if not standard, and whether straight, spiral, or spiral-pointed.
9. If spiral flutes, right- or left-hand spiral.
10. Length of thread to be cut.
11. Through or blind holes.
12. Hole size before tapping and depth of hole to be tapped.
13. Material to be threaded.
14. Taps to be used manually or by machine. If machine, what type.
15. Any special conditions of part design that the tap must meet.

COLLAPSIBLE TAPS

Collapsible taps are made for stock, but a large percentage are designed specifically for the job. They are designed with hob or radial, and circular chasers only. At present, the radial chaser (Fig. 6) has wider use. Circular chasers have greater tool life but are limited to a larger-diameter hole.

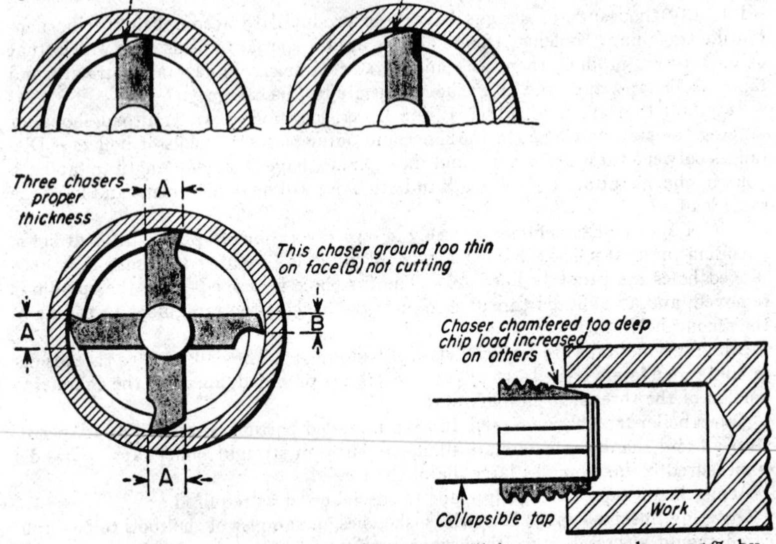

FIG. 6. Cost of chasers for collapsible taps will be cut as much as 25% by centralized grinding instead of operator sharpening.

Both rotating and stationary collapsible taps are manufactured for application as needed—automatic machine or turret lathe. Universal models are also available which can be converted to serve as rotating or stationary taps.

Radial chasers have the lead in the chaser; hence when one is ground, all must be ground to exactly the same length.

Circular chasers have annular grooves and the helix angle is provided by the mounting block. Therefore, individual chasers may be ground and reset without touching the others in the series.

Collapsible taps are used according to the same rules as solid taps. These rules must often be followed more strictly with collapsible than with solid taps. All taps work best where tap and workpiece are carefully aligned. If this is impossible, plenty of parallel float should be provided.

TAP CHASERS

Tap chasers are made to many different designs and in both straight and overhang styles. The latter style permits tapping to a shoulder or in a blind hole where the hole dia will not permit the tap nose to enter. Clearance is manufactured into a radial or hob tap chaser, according to the material to be cut. When this is changed, the thread form must be changed too. Taper and straight threading collapsible taps are made, and one special design provides a cam-actuated tap for cutting a reverse taper.

Sharpening of tap chasers is the same as for die-head chasers, except that a right-hand die-head chaser is a left-hand tap chaser.

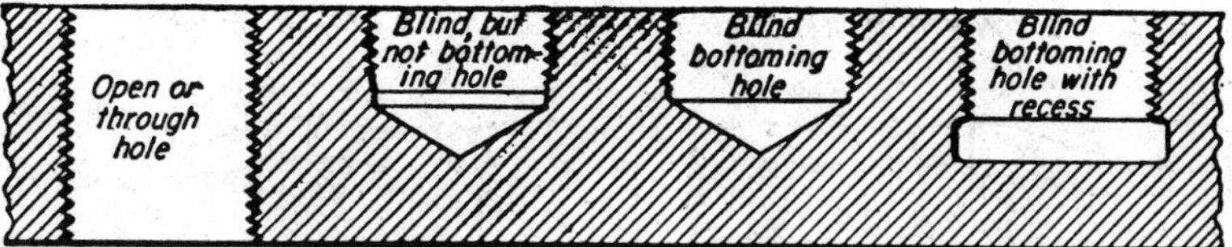

FIG. 7. Open or through holes are best for tapping, but if the hole must be blind avoid having the tap bottom.

TAP SELECTION

Selection of the right tap for the job involves three major points:

1. Correct design of tap.
2. Best size of tap.
3. Most efficient superficial surface finish for a production tap, considering the material being tapped.

Design Factors. It is recognized that a spiral-point or gun-point tap with the conventional plug chamfer of approximately 15° is recommended for through holes. Use of a bottoming tap on a through hole will dull the tap more rapidly than necessary, because most of the work is being done by one or two cutting teeth. Use of a spiral-point tap in a blind hole will force the chips down into the bottom of the hole ahead of the tap and usually cause the tap to break. See Fig. 7 for design suggestions for holes.

Two-flute taps are stronger than three-flute taps, particularly the gun-point tap, and are to be preferred for small sizes for that reason.

Three-flute taps, however, cut closer to their own size and are easier to line up in screw-machine use.

All taps smaller than ⅜ in. dia are concentrically ground; that is, without eccentric relief on the tooth form. For types of relief, see Fig. 8.

Eccentrically relieved taps in sizes larger than ⅜ in. dia cut much more over their own size than do the smaller concentric taps.

The correct amounts of hook and chamfer relief are very important on high-production jobs. Recommended hook and chamfer-relief angles are:

	Hook	Relief
Aluminum	18	12
Copper	18	12
Magnesium	18	12
Stainless steel	12	10
Mild steel	8	8
Plastics	8	12
Cast iron	5	6
Brass	5	10

Correct Pitch-diameter Tolerance. All taps cut over their own size because of unavoidable misalignment, uneven rate of feed, and other factors. The amount of oversize can be controlled, however, by precision-grinding the hook and chamfer-

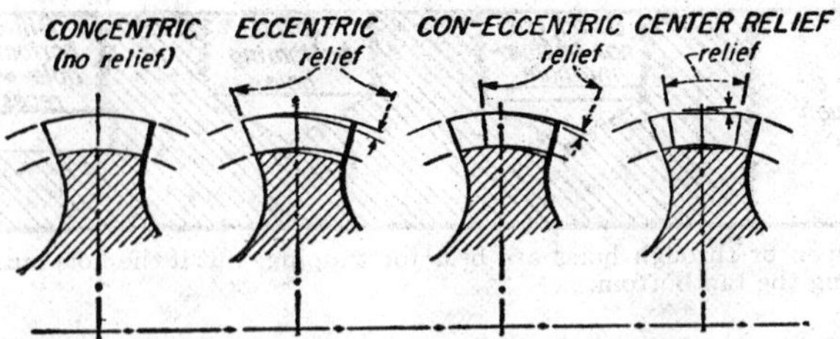

FIG. 8. Types of relief applied to taps: Concentric—no relief other than back taper; eccentric—continuous relief from cutting edge to heel; con-eccentric—usually first third concentric, remainder relieved; center relief—concentric at cutting edge and heel, relieved at center, used where tap teeth must pass over a keyway or slot.

relief angles. Machine-screw-size taps properly ground may be expected to cut no more than 0.0015 to 0.002 in. over their own size. As the tap starts to get dull, it gradually cuts closer to its own size because of the build-up of pressure between the stock and the cutting teeth as the chip is formed.

Many materials, particularly the soft ductile materials, also close in after the chip is formed, and this close-in increases as the tap becomes dull. This close-in effect is particularly troublesome when tapping plastics. To compensate for the loss in size, plastic taps are usually made 0.002 in. or more larger on pitch diameter (PD) than basic size.

Ground-thread taps are made in several ranges of PD tolerance (also see Tables 1 and 2):

Precision-ground No. 01 tap PD limits = basic − 0.0005 to basic.
Precision-ground No. 1 tap PD limits = basic to basic + 0.0005.
Precision-ground No. 2 tap PD limits = basic + 0.0005 to basic + 0.0010.
Precision-ground No. 3 tap PD limits = basic + 0.0010 to basic + 0.0015.
Commercial-ground tap PD limits = basic + 0.0005 to basic + 0.0015.
Bakelite (West Lynn Standard) = basic + 0.0002 to basic + 0.0025.

Considering the close-in factor inherent in various materials, it is possible to recommend the largest PD tolerance range which may be used in that material:

	Recommended Tap
Brass	No. 2 or commercial-ground
Aluminum	No. 2 or commercial-ground
Magnesium	No. 3 or commercial ground
Copper	No. 3 or commercial ground
Mild steel	No. 1
Stainless steel	No. 1
Cast iron	No. 1
Plastic	Bakelite

Because the commercial-ground range includes both precision-ground No. 1 and precision-ground No. 2, it may be assumed that half of a group of commercial-ground taps will fall in the precision-ground No. 1 range. Thus if commercial-ground taps are used where precision-ground No. 2 or precision-ground No. 3 is

TABLE 1. RECOMMENDED TAPS IN MACHINE-SCREW SIZES† FOR CLASSES 2, 3, AND 2B UNIFIED AND AMERICAN THREADS

Size	THREADS PER INCH		*RECOMMENDED TAP FOR CLASS OF THREAD			PITCH DIAMETER LIMITS FOR CLASS OF THREAD			
	NC and UNC	NF and UNF	Class 2	Class 3	Class 2B	Min. All Classes (Basic)	Max. Class 2	Max. Class 3	Max. Class 2B
0		80	PG1	PG1	PG1	.0519	.0536	.0532	.0542
1	64		PG1	PG1	PG1	.0629	.0648	.0643	.0655
1		72	PG1	PG1	PG1	.0640	.0658	.0653	.0665
2	56		PG1	PG1	PG1	.0744	.0764	.0759	.0772
2		64	PG1	PG1	PG1	.0759	.0778	.0773	.0786
3	48		PG1	PG1	CG	.0855	.0877	.0871	.0885
3		56	PG1	PG1	CG	.0874	.0894	.0889	.0902
4	40		CG	PG1	CG	.0958	.0982	.0975	.0991
4		48	PG1	PG1	CG	.0985	.1007	.1001	.1016
5	40		CG	PG1	CG	.1088	.1112	.1105	.1121
5		44	PG1	PG1	CG	.1102	.1125	.1118	.1134
6	32		CG	PG1	CG	.1177	.1204	.1196	.1214
6		40	CG	PG1	CG	.1218	.1242	.1235	.1252
8	32		CG	PG1	CG	.1437	.1464	.1456	.1475
8		36	CG	PG1	CG	.1460	.1485	.1478	.1496
10	24		CG	PG1	CG	1629	.1662	.1653	.1672
10		32	CG	PG1	CG	.1697	.1724	.1716	.1736
12	24		CG	PG1	CG	.1889	.1922	.1913	.1933
12		28	CG	PG1	CG	.1928	.1959	.1950	.1970

*The above recommended taps normally produce the class of thread indicated in average materials when used with reasonable care. However, if the tap specified does not give a satisfactory gage fit in the work, a choice of some other limit tap will be necessary.

† Tap & Die Division, Metal Cutting Tool Institute, 1952.

recommended, some of the taps may not produce so many good holes per grind as might be expected. In other words, use of commercial-ground taps cuts down the manufacturing tolerance. Whether it will be economical to purchase the more expensive precision-ground tap for the job depends on the accuracy of the setup and control of the operation.

Fundamentally, the ground-thread tap is intended for threading to close tolerances which cannot be held by conventional cut-thread taps. However, their clean accurate thread form is so much more efficient that it will often prove economical to use them for less demanding production operations.

Effect of Unified Threads. In development of the Unified thread system, the purpose was to minimize tooling changes in the United States, because of the size of its industry relative to those of England and Canada. Pitch-diameter tolerances of the nut and screw in the American National thread form are equal, with the nut

TABLE 2. RECOMMENDED TAPS IN FRACTIONAL SIZES* FOR CLASSES 2, 3, 1B,† 2B, AND 3B UNIFIED AND AMERICAN THREADS

Size, In.	Threads per Inch		Recommended Tap for Class of Thread‡				Pitch Diameter Limits for Class of Thread				
	NC, UNC	NF, UNF	Class 2	Class 3	Class 2B	Class 3B	Min. All Classes (Basic)	Max. Class 2	Max. Class 3	Max. Class 2B	Max. Class 3B
1/4	20	..	CG	PG1	CGH	CG	0.2175	0.2211	0.2201	0.2223	0.2211
1/4	..	28	CG	PG1	CGH	CG	0.2268	0.2299	0.2290	0.2311	0.2300
5/16	18	..	CG	CG	CGH§	CG	0.2764	0.2805	0.2794	0.2817	0.2803
5/16	..	24	CG	PG1	CGH§	CG	0.2854	0.2887	0.2878	0.2902	0.2890
3/8	16	..	CGH	CG	CGH§	CG	0.3344	0.3389	0.3376	0.3401	0.3387
3/8	..	24	CG	PG1	CGH§	CG	0.3479	0.3512	0.3503	0.3528	0.3516
7/16	14	..	CGH	CG	CGH§	CG	0.3911	0.3960	0.3947	0.3972	0.3957
7/16	..	20	CG	PG1	CGH§	CG	0.4050	0.4086	0.4076	0.4104	0.4091
1/2	13	..	CGH	CG	CGH§	CG	0.4500	0.4552	0.4537	0.4565	0.4548
1/2	..	20	CG	PG1	CGH§	CG	0.4675	0.4711	0.4701	0.4731	0.4717
9/16	12	..	CGH§	CG	CGH§	CG	0.5084	0.5140	0.5124	0.5152	0.5135
9/16	..	18	CG	PG1	CGH§	CG	0.5264	0.5305	0.5294	0.5323	0.5308
5/8	11	..	CGH§	CG	CGH§	CG	0.5660	0.5719	0.5702	0.5732	0.5714
5/8	..	18	CG	PG1	CGH§	CG	0.5889	0.5930	0.5919	0.5949	0.5934
3/4	10	..	CGH§	CG	CGH§	CGH	0.6850	0.6914	0.6895	0.6927	0.6907
3/4	..	16	CG	PG1	CGH§	CG	0.7094	0.7139	0.7126	0.7159	0.7143
7/8	9	..	CGH§	CG	CGH§	CGH	0.8028	0.8098	0.8077	0.8110	0.8089
7/8	..	14	CG	PG1	CGH§	CG	0.8286	0.8335	0.8322	0.8350	0.8339
1	8	..	CGH§	CG	CGH§	CGH§	0.9188	0.9264	0.9242	0.9276	0.9254
1	..	12	CG	CG	CGH§	CG	0.9459	0.9515	0.9499	0.9535	0.9516
1	14NS		CG	CG	CGH§	CG	0.9536	0.9585	0.9572	0.9609	0.9590
1 1/8	7	..	CGH§	CG	CGH§	CGH§	1.0322	1.0407	1.0381	1.0416	1.0393
1 1/8	..	12	CG	CG	CGH§	CG	1.0709	1.0765	1.0749	1.0787	1.0768
1 1/4	7	..	CGH§	CG	CGH§	CGH§	1.1572	1.1657	1.1631	1.1668	1.1644
1 1/4	..	12	CG	CG	CGH§	CG	1.1959	1.2015	1.1999	1.2039	1.2019
1 3/8	6	..	CGH§	CG	CGH§	CGH§	1.2667	1.2768	1.2738	1.2771	1.2745
1 3/8	..	12	CG	CG	CGH§	CG	1.3209	1.3265	1.3249	1.3291	1.3270
1 1/2	6	..	CGH§	CG	CGH§	CGH§	1.3917	1.4018	1.3988	1.4022	1.3996
1 1/2	..	12	CG	CG	CGH§	CG	1.4459	1.4515	1.4499	1.4542	1.4522

* Tap A Die Division, Metal Cutting Tool Institute, 1952.

† 1B tapped holes can be produced with CGH or cut thread taps.

‡ The above recommended taps normally produce the class of thread indicated in average materials when used with reasonable care. However, if the tap specified does not give a satisfactory gage fit in the work, a choice of some other limit tap will be necessary.

§ Cut thread taps may be used under normal conditions and in average materials for producing tapped holes to this classification in the sizes indicated.

above and the screw below basic. On the other hand, in the Unified series, to compensate for manufacturing difficulties, the tolerance of the tapped hole is always 1.3 times the tolerance of the screw for the corresponding class. Tapped holes passed as satisfactory for the National classes 2 and 3 are satisfactory for Unified classes 2B and 3B.

The Metal Cutting Tool Institute has recognized, however, that a ground-thread tap larger than the original commercial ground-thread tap may give improved performance on class 2B holes in sizes 1/4 to 1 1/2 in., if the tap size is properly selected and tap is kept sharp. Thus "commercial-ground high," or CGH, taps have been introduced (Tables 1 to 4). Four series of taps are therefore provided: (1) the usual cut thread; (2) a PG1 series, known as "precision ground No. 1 limit," with the minimum pitch diameter set at basic; (3) an altered CG, or commercial-ground series,

TABLE 3. AVAILABLE HAND TAPS IN MACHINE-SCREW SIZES† FOR TAPPING UNIFIED, NC, AND NF THREADS

Size	Threads per Inch		Tap Limits									
			Major Diameter				Pitch Diameter					
	NC UNC	NF UNF	Cut Thd		Ground Thd.		Ground Thd. PG1 Tol.		Ground Thd. CG Tol.		Cut Thd	
			Min.	Max.	Min.	Max.	*Min.	Max.	Min.	Max.	Min.	Max.
0		80	.0609	.0624	.0605	.0615	.0519	.0524	**	**	.0521	.0531
1	64		.0740	.0755	.0735	.0745	.0629	.0634	**	**	0631	.0641
1		72	0740	0755	0735	.0745	.0640	0645	**	**	.0642	.0652
2	56		.0872	0887	0865	,0875	.0744	.0749	**	**	.0746	.0756
2		64	0870	0885	.0865	.0875	.0759	.0764	**	**	.0761	.0771
3	48		.1003	1018	1000	1010	0855	.0860	0857	.0867	.0857	.0867
3		56	1002	1017	.0995	.1005	.0874	.0879	0876	0886	.0876	.0886
4	40		1136	1156	1135	1145	.0958	.0963	.0960	.0970	.0960	.0975
4		48	1133	1153	1130	1140	0985	.0990	.0987	.0997	.0987	1002
5	40		1266	1286	1265	.1275	1088	.1093	1090	.1100	1090	.1105
5		44	.1264	1284	1260	.1270	1102	.1107	.1104	.1114	1104	.1119
6	32		.1402	1422	1400	.1410	1177	1182	.1182	1192	.1182	.1197
6		40	.1396	1416	.1395	.1405	1218	.1223	.1220	1230	.1220	.1235
8	32		.1662	.1682	.1660	1670	.1437	.1442	.1442	.1452	.1442	.1457
8		36	1657	1677	.1655	.1665	.1460	.1465	.1462	1472	.1462	.1477
10	24		.1928	.1948	.1930	.1940	.1629	.1634	.1634	.1644	.1634	1649
10		32	.1922	1942	.1920	.1930	.1697	.1702	.1702	.1712	.1702	.1717
12	24		.2188	.2208	.2190	.2200	.1889	.1894	1894	.1904	.1894	.1909
12		28	.2184	.2204	.2185	.2195	.1928	1933	.1933	.1943	.1933	.1948

*Minimum is Basic.

**CG taps are not standard in sizes smaller than No. 3.

† Tap & Die Division, Metal Cutting Tool Institute, 1952.

where the tolerance spread has been shifted a few thousandths toward the basic pitch diameter; and (4) a CGH series, or commercial-ground high, where the tolerance spread has been shifted so its maximum is above the former maximum for the commercial-ground series.

These CGH and CG taps, in addition to the revised precision-ground taps, are adequate for tapping all class 2B and class 3 Unified holes, and they will also tap class 2 and class 3 National holes formerly tapped with the original commercial-ground and precision-ground taps. Tables 1 and 2 give recommendations for type of tap for the various classes of thread, thereby saving involved calculations and references.

Tap Materials and Surface Finish. Taps are manufactured in carbon steel, regular high-speed steel, and surface-treated high-speed steel. A possible fourth is

TABLE 4. AVAILABLE HAND TAPS IN FRACTIONAL SIZES* FOR TAPPING UNIFIED, NC, AND NF THREADS

Size, In.	Threads per Inch		Tap Limits											
			Major Diameter				Pitch Diameter							
			Cut Thread		Ground Thread		Ground Thread, PG1 Tol		Ground Thread, CG Tol		Ground Thread, CGH Tol		Cut Thread	
	NC, UNC	NF, UNF	Min	Max	Min	Max	Min†	Max	Min	Max	Min	Max	Min	Max
1/4	20	..	0.2532	0.2557	0.2540	0.2550	0.2175	0.2180	0.2177	0.2187	0.2186	0.2196	0.2180	0.2200
1/4	..	28	0.2524	0.2549	0.2525	0.2535	0.2268	0.2273	0.2269	0.2279	0.2277	0.2287	0.2273	0.2288
5/16	18	..	0.3160	0.3185	0.3170	0.3180	0.2764	0.2769	0.2767	0.2777	0.2777	0.2787	0.2769	0.2789
5/16	..	24	0.3153	0.3178	0.3155	0.3165	0.2854	0.2859	0.2856	0.2866	0.2865	0.2875	0.2859	0.2874
3/8	16	..	0.3789	0.3814	0.3800	0.3810	0.3344	0.3349	0.3347	0.3357	0.3358	0.3368	0.3349	0.3369
3/8	..	24	0.3778	0.3803	0.3780	0.3790	0.3479	0.3484	0.3481	0.3491	0.3491	0.3501	0.3484	0.3499
7/16	14	..	0.4419	0.4449	0.4435	0.4445	0.3911	0.3916	0.3915	0.3925	0.3927	0.3937	0.3916	0.3941
7/16	..	20	0.4407	0.4437	0.4415	0.4425	0.4050	0.4055	0.4052	0.4062	0.4063	0.4073	0.4055	0.4075
1/2	13	..	0.5047	0.5077	0.5065	0.5075	0.4500	0.4505	0.4505	0.4515	0.4518	0.4528	0.4505	0.4530
1/2	..	20	0.5032	0.5062	0.5040	0.5050	0.4675	0.4680	0.4678	0.4688	0.4689	0.4699	0.4680	0.4700
9/16	12	..	0.5675	0.5705	0.5690	0.5700	0.5084	0.5089	0.5089	0.5100	0.5102	0.5113	0.5089	0.5114
9/16	..	18	0.5660	0.5690	0.5670	0.5680	0.5264	0.5269	0.5268	0.5278	0.5279	0.5289	0.5269	0.5289
5/8	11	..	0.6304	0.6334	0.6320	0.6330	0.5660	0.5665	0.5666	0.5677	0.5679	0.5690	0.5665	0.5690
5/8	..	18	0.6285	0.6315	0.6295	0.6305	0.5889	0.5894	0.5893	0.5904	0.5905	0.5916	0.5894	0.5914
3/4	10	..	0.7559	0.7599	0.7575	0.7590	0.6850	0.6855	0.6857	0.6868	0.6871	0.6882	0.6855	0.6885
3/4	..	16	0.7539	0.7579	0.7550	0.7560	0.7094	0.7099	0.7098	0.7109	0.7111	0.7122	0.7099	0.7124
7/8	9	..	0.8820	0.8860	0.8835	0.8850	0.8028	0.8033	0.8035	0.8047	0.8051	0.8063	0.8038	0.8068
7/8	..	14	0.8799	0.8839	0.8810	0.8820	0.8286	0.8291	0.8291	0.8302	0.8305	0.8316	0.8296	0.8321
1	8	..	1.0078	1.0118	1.0095	1.0110	0.9188	0.9193	0.9196	0.9208	0.9213	0.9225	0.9198	0.9228
1	..	12	1.0055	1.0095	1.0065	1.0075	0.9459	0.9464	0.9465	0.9477	0.9479	0.9491	0.9469	0.9499
1	14 NS		1.0049	1.0089	1.0060	1.0070	0.9536	0.9541	0.9541	0.9553	0.9555	0.9567	0.9546	0.9571
1 1/8	7	..	1.1337	1.1382	1.1350	1.1370	1.0322	‡	1.0331	1.0344	1.0349	1.0362	1.0332	1.0367
1 1/8	..	12	1.1305	1.1350	1.1315	1.1325	1.0709	‡	1.0715	1.0727	1.0730	1.0742	1.0719	1.0749
1 1/4	7	..	1.2587	1.2632	1.2600	1.2620	1.1572	‡	1.1581	1.1594	1.1600	1.1613	1.1582	1.1617
1 1/4	..	12	1.2555	1.2600	1.2565	1.2575	1.1959	‡	1.1966	1.1978	1.1981	1.1993	1.1969	1.1999
1 3/8	6	..	1.3850	1.3895	1.3870	1.3890	1.2667	‡	1.2678	1.2691	1.2698	1.2711	1.2677	1.2712
1 3/8	..	12	1.3805	1.3850	1.3815	1.3825	1.3209	‡	1.3216	1.3229	1.3231	1.3244	1.3219	1.3249
1 1/2	6	..	1.5100	1.5145	1.5120	1.5140	1.3917	‡	1.3928	1.3941	1.3949	1.3962	1.3927	1.3962
1 1/2	..	12	1.5055	1.5100	1.5065	1.5075	1.4459	‡	1.4466	1.4479	1.4481	1.4494	1.4469	1.4499

* Tap & Die Division, Metal Cutting Tool Institute, 1952.
† Minimum is basic.
‡ PG1 taps are not standard in sizes larger than 1 in.

carbide, but little has been done in the way of quantity production of carbide taps. In the large tap or chaser sizes considerably higher than normal cutting speeds are made possible.

Surface-treated HSS taps have had a secondary hardening, or wear-resistance, treatment. This may consist of superhardening, a hard chromium plate of 0.0001 to 0.0003 in., nitriding, shot-peening, or some other special surface treatment to slow the effect of wear and abrasion. Treatments such as these make it possible to increase tapping speeds 10 to 20% over those for HSS.

Nitriding is good if done properly so it does not affect surface finish. Nitriding is used on steel and continuous-chip materials very successfully.

Steam-homo or steam-oxide finish is good on many materials including cast iron but is not so good on materials like copper and aluminum which have a tendency to "load" or weld to cutting-tooth faces.

Precision chrome plate is particularly good for plastics and soft materials such as copper and aluminum. Precision chrome produces a very shiny slippery surface which minimizes loading. Use of chrome plate to increase the size of a tap by any appreciable amount is not recommended, because when a thickness of from 0.001 to 0.005 in. of plate is applied, a noticeable change in tooth form takes place and sharp cutting edges become rounded by built-up chrome.

Carbon-steel taps are not recommended for production tapping hard, tough, or abrasive materials. When hand tapping a finished tool-steel die, a newly sharpened carbon tap should be used, then resharpened before reuse. If the tap breaks in the die, it is much more easily removed than a HSS tap. A similar suggestion is that, prior to die tapping, tap hardness always be checked. If the tap is drawn to Rockwell 55 to 58 C, some tool life is lost, but the danger of breakage is greatly reduced.

Ground and polished HSS taps, whether surface treated or not, may be used at speeds of 5 to 20% higher than those recommended before polishing. In addition, polished flutes and teeth will produce cleaner, smoother threads and longer tool life.

TAPPING SUGGESTIONS

The ground-thread tap is a precision tool. To get the most out of a tap, the spindle of the tapping machine and the tapping head should be free from runout and end play. A simple check of the runout of the spindle can be made by inserting a cylindrical plug in the chuck or collet and rotating the spindle slowly by hand. If the tap is held rigidly in the chuck, a runout of 0.001 in. may cause the tap to cut oversize by as much as 0.002 in., reducing materially the manufacturing tolerance for the tapped hole.

To compensate for such runout, it is common practice to use a floating tap holder which allows the tap to align itself with the drilled hole. This expedient should only be used when exact alignment is impossible or impractical. If a floating holder is used, it should allow the tap to float parallel to the spindle axis, because if a tap "angles" into a drilled hole, it will either break or cut bellmouthed. The chuck or collet in the tapping head should contact the tap shank for a considerable distance. The rubber-flex type of collet is practical and inexpensive.

When to Float the Part. It is sometimes customary to float the part, but this is practical only when the part is very light in comparison with the tap size. It is not reasonable, for instance, to expect a 4-48 tap to align a floating part which weighs ½ lb or more, or a workpiece held in a jig.

Never tap through a bushing because the crests of the tap threads are damaged by contact with the bushing. A countersink on the drilled hole will help start the tap,

TABLE 5. TAPPING SPEEDS AND LUBRICANTS

Material	Tap Speed, Fpm			Lubricant
	Carbon Steel	Regular HSS	Surface-treated	
Allegheny metal	*	15–25	20–30	Sulfur-base oil
Aluminum	45–50	90–100	100–110	Kerosene and lard oil
Bakelite	*	60–70	70–80	Dry
Brass	45–50	90–100	100–110	Compound or light-base oil
Bronze	20–30	40–60	50–70	Compound or light-base oil
Bronze-manganese	*	30–45	35–50	Light-base oil
Copper	45–50	90–100	100–110	Light-base oil
Die-cast aluminum	30–35	60–70	70–80	Kerosene and lard oil
Die-cast zinc	30–35	60–70	70–80	Compound
Duralumin	45–50	90–100	100–110	Compound or kerosene and lard oil
Fiber	*	80–90	90–100	Dry
Iron-cast	*	70–80	80–90	Dry or compound
Malleable	*	35–60	45–70	Compound or sulfur-base oil
Monel metal	*	20–25	25–30	Sulfur-base oil or kerosene and lard oil
Nickel silver	*	75–85	85–95	Sulfur-base oil or kerosene and lard oil
Rubber (hard)	*	80–90	90–100	Dry
Steel, cast	*	20–30	25–35	Sulfur-base oil
Chromium	*	20–30	25–35	Sulfur-base oil
Machinery	20–30	40–60	50–70	Compound, sulfur-base oil, or kerosene and paraffin
Manganese	*	10–15	15–20	Compound, sulfur-base oil, or kerosene and paraffin
Molybdenum	*	20–30	25–35	Sulfur-base oil
Nickel	*	25–35	30–40	Sulfur-base oil
Stainless	*	15–25	20–30	Sulfur-base oil
Tool	15–20	25–35	30–40	Sulfur-base oil or kerosene and lard oil
Tungsten	*	20–30	25–35	Sulfur-base oil
Vanadium	*	25–35	30–40	Sulfur-base oil

* Carbon-steel taps are not recommended for these materials.

particularly a bottoming tap. In screw-machine work, the drilled hole should always be countersunk.

The tap-drill size has an effect on the percentage of engagement and the torque required to tap the hole. Tap-drill size does not affect class of fit, however.

Too small a tap-drilled hole will cause the minor diameter of the tapped hole to be torn and ragged because of excessive crowding of the chips. This is a condition to be especially avoided when tapping stainless steel. Too large a tap-drilled hole decreases the percentage of engagement and, if carried below 50% of engagement, may materially decrease the resistance to stripping of the threads. See Tap Drilling below, for further details.

The length of full thread in a tapped hole, while affected by the kind of material being tapped, should be held within reasonable limits. On machine-screw sizes, five full threads of engagement are usually strong enough to break off the head of the screw before stripping. A maximum of ten full threads of engagement is suggested. Tapping to a greater depth is difficult and expensive. Where deep threads are necessary, a counterbored section at the top of the drilled hole may be used to reduce the number of full threads to practical limits.

Through holes are best tapped with spiral-point or gun-point taps, which are designed to force the chip ahead of the tap.

A taper chamfer on a tap is seldom used except on short holes through sheet stock where the length of the tapped holes is less than five threads. Taper chamfers in deeper holes usually set up excessive torque which has a tendency to break small-sized taps.

Tapping Speeds. Tapping speed parallels, and is usually 50 to 65% of, drilling speed. Speed recommendations for drilling and tapping are always expressed in surface feet per minute and may be converted to rpm by the formula

$$\text{rpm} = \text{sfpm} \times \frac{3.82}{\text{OD}}$$

where OD = outside diameter of tap, in.

As in drilling, the recommended sfpm (Table 5) should be maintained as closely as possible, particularly on small sizes. It is a mistake to run small-sized taps at slower speeds than recommended because the torsional strains introduced into the tap at slow speeds are not so readily absorbed by the body of the tap.

Obviously, alignment becomes more important as the size of the tap is decreased.

For tapping holes of finer than 48 threads per inch and in all pitches on sizes smaller than 6, a lead-screw tapper is recommended. For the hard-to-tap materials, such as stainless steel, a lead-screw tapper is almost essential.

Tap Drilling

The tap is not a corrective device for poorly drilled holes. Actually, there are only two important things to know about the holes you plan to thread, depth and diameter. Again, there are two varieties of holes: open or through, and blind or bottoming. Through holes are more readily tapped than blind, as they permit using the spiral-pointed or plug tap. Blind holes can be threaded efficiently only when plenty of clearance is provided at the bottom.

In all cases and in all types of holes, good clean drilling and a reasonable degree of size maintenance will ease considerably the tapping of good threads, will improve size control, minimize loading of tap threads, and reduce the strain on the tap. All these mean greater tap life and more threads per grind.

Holes for taper-thread tapping should be bored or reamed taper before tapping. Every tooth cuts on a taper-thread tap; hence the load gets progressively heavier as the tapping progresses. If the hole is reamed with a standard-taper pipe reamer, before threading, the tap will be required to cut only the full thread depth.

Go and not-go plug gages, or at least a set of standard drill blanks (hardened), are a valuable investment. Holes drilled to these controls bring you one big step further toward better threading.

Percentage of Thread. A drilled hole should be of sufficient diameter to produce a thread depth of approximately 75%. Adoption of this 75% depth was based on two reasons: lower power consumption and less tapping trouble. It has been proved that, in tapping a full, or 100%, thread depth, three times more power is necessary than for a 75% depth, but the full thread is only 5% stronger.

This 75% is only an average. Actually the percentage may vary from 50 to 53% for small or deep holes, to a maximum of 83⅓% on any size. A definite thread depth for all sizes, under all conditions, is not practical. The tap user must determine which is most suitable for his own requirements. By analyzing the following conditions, the problem is easily solved:

1. Diameter and pitch of tapped hole.
2. Nature of material being tapped.
3. Depth of tapped hole.

Fine or Coarse Pitch. The coarser the pitch, the smaller the thread depth should be. Try the 75% depth first. If tap breakage results, reduce the thread depth gradually until satisfactory performance is obtained. This is not likely to jeopardize thread strength. Remember that a nut with only 50% thread depth will break the bolt before it strips the thread.

Threads are "standardized" in two series, coarse and fine, but, as a tap size becomes smaller, the percentage of double-thread depth in relation to basic major diameter becomes larger.

Basic Major Dia	NC, %	NF, %
1.0 in.	16	9
0.750 in.	17	11
0.500 in.	20	13
0.375 in.	22	14
0.250 in.	26	19
No. 5 machine screw	32	26

Influence of Material. Soft stringy metals such as copper, drawn aluminum, monel metal, nickel silver, and some other low-melting-point alloys, because of their malleable nature, have a tendency to flow toward the crest of the minor diameter while being tapped. The minor diameter of tapped holes in such materials will be smaller after tapping. Take this into consideration when selecting tap-drill size.

Materials that are very tough or of high hardness can be tapped more successfully with less than the normal 75% thread depth. Use the smallest thread depth possible for these.

Effect of Hole Depth. Larger tap drills may be used when holes are deeper than 1½ times tap dia, particularly when using machine-screw sizes. Drills which will give a 50% depth are found to be entirely satisfactory.

Punched holes often cause binding or loading, and tap breakage, especially in thin sheet metal, which creates an "oil-can" effect. When the hole is punched, the metal is flared out. As the tap is reversed after threading it draws in this flare and binds. This should be considered when selecting punch size. Of course, the larger the punched hole, the less tendency for the tap to load.

Cored holes in castings and forged holes in forgings should be checked for this 75% depth, and reamed to size if necessary.

Lubricants for Tapping

Many tap users do not appreciate the importance of proper lubricants, or cutting fluids, in successful tapping practice. Summarized, these effects include: longer tap life, greater production, better size control, smoother and more accurate threads, and more efficient chip removal. However, there is no one lubricant that suits all requirements (see Table 5). Also, lubricants and "coolants" are often confused. Tapping almost always calls for a lubricant, very rarely for a coolant. A simple rule is: light lubricants for light materials and, progressively through the range, to heavy lubricants for hard or tough materials.

Lubricant should be forced into the hole under pressure, the amount of which is variable, determined by tapping method, hole depth, and tapping speed. An ideal condition is to have the stream enter the hole parallel to the tap axis, or at the smallest permissible angle. Get the lubricant to the cutting portion of the tap, in volume and at all times. Automatic lubrication should be timed so that it reaches the hole before the tap starts to cut. This is especially important if flow is shut off during reversal.

TAP SHARPENING

Grinding a tap properly for the job usually lengthens tap life several hundred per cent, with many regrinds possible before the tap is worn beyond salvage. Freehand grinding is not recommended (it must be used occasionally). Taps should be resharpened on a good tap grinder, even if the number of production taps is relatively small. Cost analysis of production tapping operations on the following basis will clearly bring out the value of effective tap control and sharpening:

First tap cost per 1000 holes.
Cost of regrinding per 1000 holes.
Machine down time per 1000 holes.
Machine setup time per 1000 holes.
Spoilage and breakage per 1000 holes.

A tap wears on the chamfer in direct relation to the work done by each tooth. The second cutting tooth on the chamfer produces the largest chip.

Close inspection of the first and second cutting teeth will determine whether these edges are still sharp or have been rolled over, worn, or broken. Sharpening the tap by rechamfering when such wear first becomes apparent will restore the tap to its original usefulness.

If rechamfering has shortened the spiral-point grind to the extent that it does not include the first full tooth beyond the chamfer, the spiral point should be reground.

When a tap is used without proper sharpening, the OD is sometimes worn until it is less than the OD of the plug thread go gage, preventing the part from passing as "good." Thus taps which have given considerable use should be checked on the OD. Such taps are impossible to sharpen as a finishing tap but might be used as a roughing tap in a hand-tapping operation.

The practice sometimes encountered of grinding down the tap OD to make it cut a hole which feels tight on the go gage or reducing the OD by any other means is not to be tolerated. Such practices contribute to excessive costs and poor quality.

As the tap-thread length is shortened by chamfering, the shank directly behind the thread should be reduced to the minor diameter of the thread, so that the tap may be used to the depth originally planned. This grind, called "necking," may be continued until the remaining length of full thread is no less than half the original length, or a minimum of five full threads back of the chamfer.

Broken teeth do not necessarily make a tap uneconomical to salvage. If the broken teeth can be removed by cylindrical grinding so that the remaining cutting edges on each land are equal in number, the tap may be salvaged. However, the remaining teeth should be so spaced that at least five full threads are in contact with the work at one time.

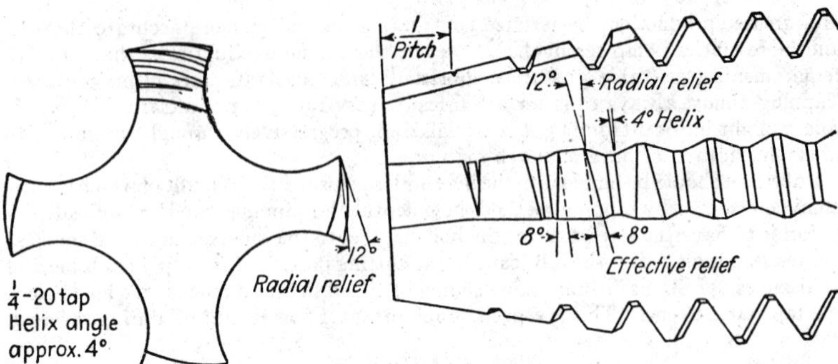

FIG. 9. Chamfer relief affects only the crests of the chamfered threads. The effective relief is important to good cutting action.

If broken teeth are not consecutive or on the same land, the tap may be salvaged by grinding away the broken teeth.

Excessive runout between the cutting teeth and shank will cause a tap to cut a bellmouth hole, even though the tap is held in a floating holder. The amount of runout that can be tolerated is certainly no more than the tolerance on pitch diameter.

Chamfering should be done either on centers or by holding the tap in a collet by the OD of the thread as close to the chamfer as possible, rather than by holding the shank. This will compensate for any runout between shank and threaded portion of the tap, and the chamfer will be concentric with the threads.

If the use of taps is under reasonably effective control, a tap may be sharpened repeatedly by removing only a few thousandths of stock and will produce as many acceptable tapped holes after each sharpening as when first used. After the setup is made, taps may be chamfered at the rate of 24 to 60 per hour with proper equipment. The cost of sharpening will be found to be considerably less than the cost of tap replacement.

Grinding Chamfer Relief

The accuracy of the chamfer relief affects the tapped-hole size, the finish, and the tap life more than is generally recognized.

In either case, relief is provided by an eccentric or cam that feeds the tap toward the wheel in a plane at 90° to the tap axis. Correspondingly, chamfer relief is usually measured by the amount of drop-off from the cutting edge to the heel in the same plane, at 90° to the tap axis.

Effective Relief. The above method does not give a true measurement of effective relief, which is correctly measured as the drop-off along the helix angle, and is something less than the radial relief.

On larger taps, the difference is negligible, but on small-sized machine-screw taps, the true effective relief equals the radial relief angle minus the helix angle. Thus a radial relief of 8° on a 1/4-20 tap represents a true effective relief of 8° minus the helix angle of 4°11″, or less than 4° of actual effective relief (Fig. 9).

The amount of effective cutting relief modifies cutting pressure and must be sufficient to prevent excessive rubbing. It varies inversely with toughness of the material. Softer materials require more relief than harder materials.

Cutting pressure must be balanced if the tap is to cut to size. Therefore, cutting edges and cutting relief must be evenly spaced. If flutes have been ground accurately with an indexing head, the chamfer relief may be ground in the same manner. If flute spacing is uneven, the chamfer relief must be ground on a finger to provide consistent cutting relief on all lands. The latter method is not so consistent as index grinding.

Finish of the chamfer relief should be good and without chatter marks. A180-KV or A220-JV wheels will grind more slowly than coarse wheels but will do a better job, particularly if coolant is used.

The resulting fine finish will retard loading and help retain a keen edge on the chamfered threads.

Figure 9 shows how a chamfered tap should look if correctly ground. Crests of chamfered threads are wider at the heel than at the front. One thread groove toward the point extends only partway across one land. This indicates drop-off, or relief, back of the cutting edge.

In Fig. 9, radial relief is approximately 12°, making effective cutting relief approximately 8°

Recommended chamfer relief is:

Hard materials................	4 to 6°
Average materials.............	6 to 8°
Soft materials................	8 to 12°

The amount of eccentricity necessary to produce the relief angle can be figured by

$$\text{Eccentricity} = \text{tap dia} \times \sin \text{ of half relief angle desired}$$

Values of eccentricity, in inches, for grinding radial chamfer relief for various sizes of taps are given in Table 6.

It must be remembered that this angle is not the true effective cutting angle. The latter is obtained by subtracting the helix angle of the tap thread from the relief angle.

Tap End. The end diameter must be less than the hole size. However, the tap point should be cut off no shorter than one full pitch or thread ahead of the first thread groove.

Chamfer Angle. The practice of expressing the chamfer as "so many threads" is inaccurate. If the chamfer grind is expressed in degrees, the meaning is perfectly clear. A "plug" chamfer is usually 10 to 15°, a "bottoming" chamfer 22 to 32°, and a "taper" chamfer may be as small as 5°.

The smaller the chamfer angle, the more cutting teeth share the work of removing stock, and, theoretically, the longer the tap will stay sharp. Thus a tap chamfered at 15° can be expected to last longer than a tap chamfered at 30°.

On the other hand, the smaller the chamfer angle, the farther the tap must be run into the hole to produce a full thread to the depth desired. This increases the time cycle and frequently increases the chip-disposal problem. From a practical standpoint, therefore, plug chamfer of approximately 15° is usually most satisfactory on through holes.

Flute Grinding

The majority of ground taps, particularly "general-purpose" taps, as purchased, are not consistent in respect to the flute grind. By correcting the flute grind it is possible to increase tap life and tapped-hole quality.

Table 6. Eccentricity, in Inches, for Grinding Radial Chamfer Relief

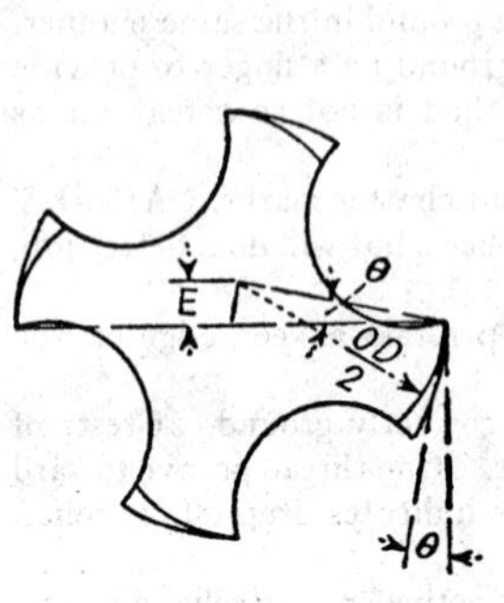

Tap Size	OD	Chamfer Relief Angle 4°	6°	8°	10°	12°
0	0.060	0.002	0.003	0.004	0.005	0.006
1	0.073	0.003	0.004	0.005	0.006	0.008
2	0.086	0.003	0.005	0.006	0.007	0.009
3	0.099	0.003	0.005	0.007	0.009	0.010
4	0.112	0.004	0.006	0.008	0.010	0.012
5	0.125	0.004	0.007	0.009	0.011	0.013
6	0.138	0.005	0.007	0.010	0.012	0.014
8	0.164	0.006	0.009	0.011	0.014	0.017
10	0.190	0.007	0.010	0.013	0.017	0.020
12	0.216	0.008	0.011	0.015	0.019	0.023
14	0.242	0.009	0.013	0.017	0.021	0.025
1/4	0.250	0.009	0.013	0.017	0.022	0.026
5/16	0.312	0.011	0.016	0.022	0.027	0.032
3/8	0.375	0.013	0.020	0.026	0.033	0.039
7/16	0.437	0.015	0.023	0.030	0.038	0.045
1/2	0.500	0.018	0.026	0.034	0.043	0.052
9/16	0.562	0.020	0.030	0.039	0.049	0.053
5/8	0.625	0.022	0.033	0.043	0.054	0.065

Recommended radial chamfer relief angles: hard materials, 4 to 6°; average materials, 6 to 8°; soft materials, 8 to 12°.

Formula:

$$\text{Eccentricity} = \frac{OD}{2} \times \sin\theta$$

θ = radial chamfer relief angle

To obtain *effective* chamfer relief, subtract thread helix angle from the *radial* chamfer relief angle.

Table 7. Recommended Tangential Hook Angles

	Deg
Brass	5
Cast iron	5
Mild steel	8
Stainless steel	12
Aluminum	18
Copper	18
Magnesium	18
Plastics	8

It is obvious that the value of the hook angle, which corresponds to the back-rake angle on a turning tool, should be varied for the kind of material being machined. Table 7 gives hook angles found effective in some of the more common materials.

The tap hook angle is often measured as the angle between the cutting center line and a line intersecting the crest and the root of the cutting tooth. This method of measurement (chordal hook angle, Fig. 10) produces different values of hook angle with variations in depth of thread or pitch of the tap. This method also does not take into consideration the radius of the flute curvature which materially modifies the effectiveness of the hook angle.

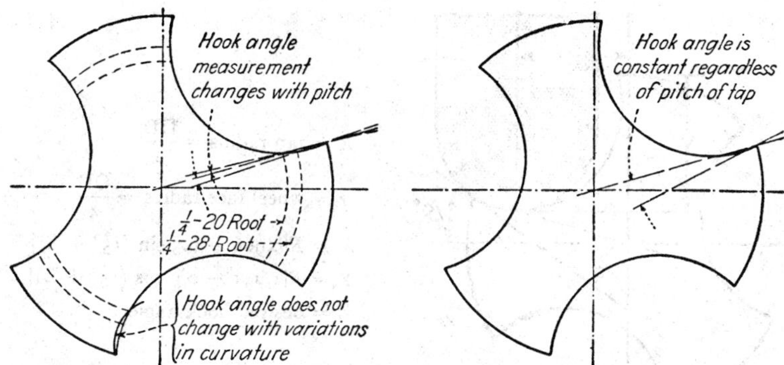

FIG. 10. Chordal hook-angle measurement.

FIG. 11. Tangential hook-angle measurement.

A more practical method of measuring the effective hook angle (Fig. 11) coincides with the accepted method of measuring hook on a single-point boring tool.

By this method (tangential hook angle) the effective hook angle is the angle between the cutting center line and a tangent to the flute curvature at the crest of the cutting tooth. Measurements taken by the tangential-hook method are independent of the pitch of the tap.

When the entire flute is ground to correct the hook angle, there is a tendency to burn the stock because of the large area of contact. It has also been found that grinding only the cutting face for a distance back from the crest, equal to twice or three times the depth of thread, is effective and can be done without burning the cutting edges.

A practical setup locates the cutting edge of the tap flute 15° above or ahead of a center line perpendicular to the plane of the wheel (see illustration with Table 8).

The wheel is offset from the center line of the tap by a definite amount X and a true radius r is dressed on the wheel. The wheel is then fed into the tap to a prescribed depth Y to produce a hook of the desired value.

Table 8 gives values of r, X, and Y to produce the more commonly used values of hook. Formulas by which these values may be calculated for any particular job not covered are also shown.

Flute grinding should be done on centers, or the tap may be back-rested on the OD of the threads.

In any event, the setup should be accurately positioned and closely aligned. Micrometer feed screws should be sufficiently accurate to position the tap and the wheel within ±0.001 in.

As in thread grinding, flute grinding should always be done under coolant.

Accurate control of the hook angle permits the tool planner to specify the correct design of tap for the job. It also permits modification of stock taps to suit the job. For instance, bottoming taps may be converted into spiral-point plug taps and the reverse.

Greatest value of controlled hook grinding lies in the fact that, once the optimum

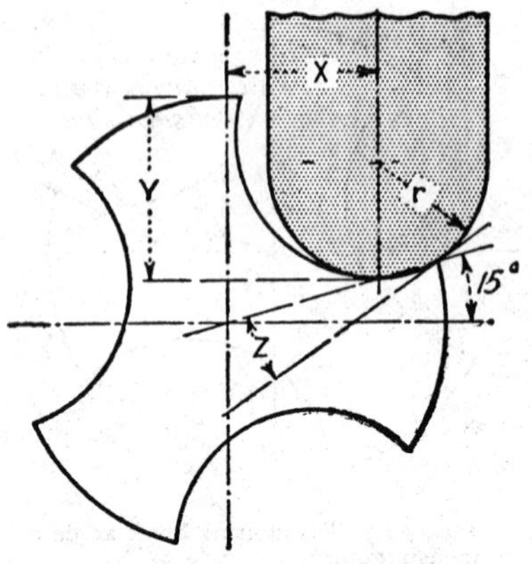

$$R = \text{tap radius} = \frac{\text{OD}}{2}$$

$$r = \text{wheel face radius} = \frac{\text{OD}}{4}$$

$$X = R[0.966 - 0.5 \sin (15° + Z)]$$

$$Y = R[1.241 - 0.5 \cos (15° + Z)]$$

$$Z = \text{desired hook angle}$$

TABLE 8. MACHINE SETTINGS FOR GRINDING HOOK ON TWO- AND THREE-FLUTE TAPS (CUTTING EDGE 15° AHEAD OF CENTER)

Tap Size	OD (2R)	Wheel Radius r	Offset X for Hook Angle Z of				Depth Y for Hook Angle Z of			
			5°	8°	12°	18°	5°	8°	12°	18°
0	0.060	0.015	0.024	0.023	0.022	0.021	0.023	0.023	0.024	0.025
1	0.073	0.018	0.029	0.028	0.027	0.025	0.028	0.029	0.029	0.030
2	0.086	0.022	0.034	0.033	0.032	0.030	0.033	0.034	0.034	0.035
3	0.099	0.025	0.040	0.039	0.037	0.035	0.039	0.039	0.040	0.041
4	0.112	0.028	0.045	0.043	0.041	0.039	0.043	0.044	0.045	0.046
5	0.125	0.031	0.050	0.048	0.046	0.043	0.048	0.049	0.050	0.051
6	0.138	0.035	0.055	0.053	0.051	0.048	0.053	0.054	0.055	0.057
8	0.164	0.041	0.065	0.063	0.061	0.057	0.063	0.064	0.065	0.067
10	0.190	0.047	0.076	0.073	0.070	0.066	0.073	0.074	0.076	0.078
12	0.216	0.054	0.086	0.083	0.080	0.075	0.083	0.084	0.086	0.089
1/4	0.250	0.063	0.099	0.096	0.092	0.087	0.096	0.098	0.099	0.103
5/16	0.313	0.078	0.124	0.120	0.115	0.108	0.120	0.122	0.124	0.128
3/8	0.375	0.094	0.149	0.145	0.139	0.130	0.145	0.146	0.149	0.154
7/16	0.438	0.109	0.174	0.169	0.162	0.152	0.169	0.171	0.174	0.180
1/2	0.500	0.125	0.199	0.193	0.185	0.174	0.193	0.195	0.199	0.205
9/16	0.563	0.141	0.223	0.217	0.208	0.195	0.217	0.220	0.223	0.231
5/8	0.625	0.156	0.248	0.240	0.231	0.217	0.240	0.244	0.248	0.256

grind has been developed on the job, any number of taps may be ground which will closely duplicate each other in performance. Barring accidental breakage, it is reasonable to expect the same life from all taps out of a group within 10%.

In summary, controlled hook grinding in both the flute and the spiral point of taps not only increases the number of good tapped holes per grind, but greatly increases the average of tapped holes per tap, and permits the obtaining of similar, if not equal, life from taps from any of the accepted manufacturers.

Grinding the Spiral Point

In a tap used for through holes, a spiral point is ground into the chamfered threads and the first full thread, to increase the effective hook, thus reducing the torque.

The spiral point (Fig. 12) is produced by passing the tap across the periphery of a wheel on which a radius has been formed somewhat smaller than the radius in the tap flute. The plane of the wheel is slightly offset from a parallel plane through the axis of the tap toward the outside of the tap. Direction of the spiral-point grind is also at an angle to the tap axis, and this angle is called the "spiral-point angle."

By setting the cutting edge of the tap approximately 15° ahead of center, a positive hook results on all the chamfered teeth. This setup also produces maximum hook not at the first full thread, but at some point on the chamfer between the root diameter and full diameter.

The spiral point is ground to include at least one full tooth back of the chamfer. This provides an opportunity to sharpen the tap by chamfering one or more times before it is necessary to regrind the spiral point. The effective hook angle in the spiral point will not be changed materially by chamfering as long as at least one full tooth is included in the spiral point.

Care should be taken to prevent weakening the web at the point of a tap. This can be avoided by keeping the spiral-point angle (pressure angle) approximately the same as the chamfer angle. Frequently the chamfer angle may exceed the spiral-

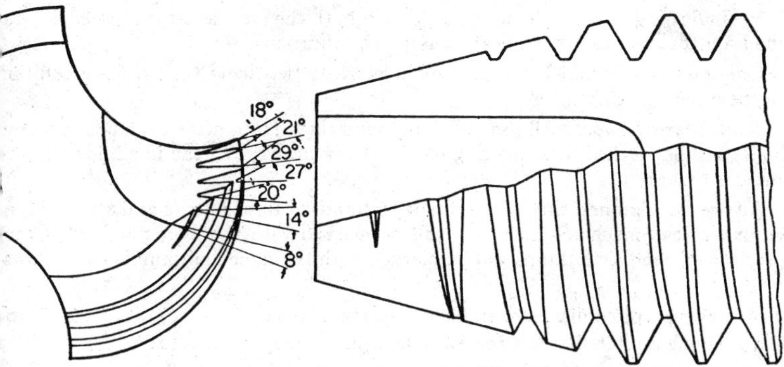

FIG. 12. In a spiral-pointed tap, the radial hook angle varies along the chamfered teeth. The drawing represents a ¼-20 tap with a 15° chamfer angle, 12° radial chamfer relief angle, 18° flute hook angle, and 15° spiral-point angle. The first full tooth is 15° ahead of center.

point angle by 5 to 10°, but, if the spiral-point angle is ground at a greater angle than the chamfer angle, poor design is the result.

The center points on spiral-point taps may be cut off as recommended for chamfering bottoming taps, that is, a minimum of one thread length in front of the point where the chamfer line crosses the root diameter of the thread.

Values of wheel radius r and offset X may be taken from Table 8 for flute grinding.

Good surface finish in the spiral-point grind is necessary to reduce chip welding. For this reason, 120- to 180-grit wheels are recommended. On difficult jobs in stringy materials, it may pay to lap the spiral-point grind.

STANDARD TAPS

The current American standard for cut and ground thread taps—ASA B5.4-1948—gives thread and general dimensions, plus working tolerances, for the following types of taps: hand taps (Table 9), machine-screw taps (Table 10), taper and straight pipe taps (Table 11), tapper taps, nut taps, pulley taps, boiler taps, mud or washout taps, and stay-bolt taps. For details of taps not covered in this handbook, refer to the above standard, which is published by the American Society of Mechanical Engineers, 29 West 39th St., New York 18.

All these taps are made to the 60° American National Thread Form.

Marking. All taps, dies, and other threading tools should be marked with the nominal size, number of threads per inch, and a symbol to indicate the thread form or series:

UNC—indicating Unified National Coarse.

UNF—indicating Unified National Fine.

NC—indicating American Standard Coarse Thread Series.

NF—indicating American Standard Fine Thread Series.

N—indicating American Standard 8-pitch, 12-pitch, and 16-pitch series.

NS—indicating special threads with the American National form.

NPT—indicating American Standard Taper Pipe Threads.

NPS—indicating American Standard Straight Pipe Threads.

V—indicating a 60° V-thread usually with both the crest and root flatted several thousandths from the theoretical to users' specifications.

Left-hand Taps. Left-hand taps will be marked "left hand" or "LH" in addition to the other marking.

Ground-thread Taps. All commercial-ground thread taps made to standard thread limits will be marked with one ring on the shank near the thread in addition to the standard marking.

All precision-ground thread taps made to standard thread limits and all precision-ground thread machine-screw taps will be marked with the limit number. Other precision-ground thread taps will be marked with the same limit number as follows:

Taps having a pitch dia between basic and minus 0.0005 01
Taps having a pitch dia between basic and plus 0.0005 1
Taps having a pitch dia between 0.0005 and 0.0010 over basic 2

Ground thread pipe taps will be marked "CG."

Other special ground thread taps will be marked "CG" if the pitch dia grinding

TABLE 9. DIMENSIONS OF FRACTIONAL-SIZE HAND TAPS

Tap Dia	Threads per In.						Flutes	Dimensions		
	Carbon Steel			HSS				Over-all Length	Length Full Thread	Shank Dia
	NC	NF	NS	NC	NF	NS				
Regular (Standard) Hand Taps										
1/16	....	..	64	..	..	.....	3	1 5/8	5/16	0.141
3/32	....	..	48	..	..	.....	3	1 3/4	7/16	0.141
1/8	....	..	40	..	..	.....	3	1 15/16	5/8	0.141
5/32	....	..	32, 36	..	..	.....	4	2 1/16	3/4	0.160
3/16	....	..	24, 32	..	..	.....	4	2 3/8	7/8	0.192
7/32	....	..	24, 32	..	..	.....	4	2 3/8	15/16	0.223
1/4	20	28	24, 32	20	28	.....	4	2 1/2	1	0.255
5/16	18	24	32	18	24	.....	4	2 23/32	1 1/8	0.318
3/8	16	24	.....	16	24	.....	4	2 15/16	1 1/4	0.381
7/16	14	20	.....	14	20	.....	4	3 5/32	1 7/16	0.323
1/2	13	20	.....	13	20	.....	4	3 3/8	1 21/32	0.367
9/16	12	18	.....	12	18	.....	4	3 19/32	1 21/32	0.429
5/8	11	18	.....	11	18	.....	4	3 13/16	1 13/16	0.480
11/16	....	..	11, 16	..	..	11, 16	4	4 1/32	1 13/16	0.542
3/4	10	16	.....	10	16	.....	4	4 1/4	2	0.590
7/8	9	14	.....	9	14	.....	4	4 11/16	2 7/32	0.697
1	8	14	.....	8	14	.....	4	5 1/8	2 1/2	0.800
1 1/8	7	12	.....	7	12	.....	4	5 7/16	2 9/16	0.896
1 1/4	7	12	.....	7	12	.....	4*	5 3/4	2 9/16	1.021
1 3/8	6	12	.....	6	12	.....	4*	6 1/16	3	1.108
1 1/2	6	12	.....	6	12	.....	4*	6 3/8	3	1.233
1 5/8	....	..	5 1/2	..	..	.....	6	6 11/16	3 3/16	1.305
1 3/4	5	..	.....	..	..	.....	6	7	3 3/16	1.430
1 7/8	....	..	5	..	..	.....	6	7 5/16	3 9/16	1.519
2	4 1/2	..	.....	..	..	.....	6	7 5/8	3 9/16	1.644
Three-flute Hand Taps										
1/4	20	..	.....	20	28	.....	3	2 1/2	1	0.255
5/16	18	..	.....	18	24	.....	3	2 23/32	1 1/8	0.318
3/8	16	..	.....	16	24	.....	3	2 15/16	1 1/4	0.381
7/16	14	..	.....	14	20	.....	3	3 5/32	1 7/16	0.323
1/2	13	..	.....	13	20	.....	3	3 3/8	1 21/32	0.367
Spiral-pointed Hand Taps										
1/8	....	..	40	..	..	.....	2	1 15/16	5/8	0.141
3/16	....	..	24, 32	..	..	.....	2	2 3/8	7/8	0.192
1/4	20	28	.....	20	28	.....	2	2 1/2	1	0.255
5/16	18	24	.....	18	24	.....	2	2 23/32	1 1/8	0.318
3/8	16	24	.....	16	24	.....	3	2 15/16	1 1/4	0.381
7/16	14	20	.....	14	20	.....	3	3 5/32	1 7/16	0.323
1/2	13	20	.....	13	20	.....	3	3 3/8	1 21/32	0.367
Serial Hand Taps										
1/4	20	..	.....	..	..	.....	4	2 1/2	1	0.255
5/16	18	..	.....	..	..	.....	4	2 23/32	1 1/8	0.318
3/8	16	..	.....	..	..	.....	4	2 15/16	1 1/4	0.381
7/16	14	..	.....	..	..	.....	4	3 5/32	1 7/16	0.323
1/2	13	..	.....	..	..	.....	4	3 3/8	1 21/32	0.367
9/16	12	..	.....	..	..	.....	4	3 19/32	1 21/32	0.429
5/8	11	..	.....	..	..	.....	4	3 13/16	1 13/16	0.480
3/4	10	..	.....	..	..	.....	4	4 1/4	2	0.590
7/8	9	..	.....	..	..	.....	4	4 11/16	2 7/32	0.697
1	8	..	.....	..	..	.....	4	5 1/8	2 1/2	0.800

* American Standard Fine Thread taps have six flutes in these sizes.

ASA B5.4-1948.

Table 10. Dimensions of Machine-Screw Taps

Screw Gage No.	Basic Major Dia	Threads per In.						Flutes			Dimensions		
		Carbon Steel			HSS			Standard	Optional		Over-all Length	Length Full Thread	Shank Dia
		NC	NF	NS	NC	NF	NS		Carbon Steel	HSS			
Regular (Standard) Machine-screw Taps													
0	0.060	..	80	.....	..	80	.....	2	..	...	1⅝	5/16	0.141
1	0.073	64	72	56	64	72	56	2	..	...	1 11/16	⅜	0.141
2	0.086	56	64	.....	56	64	.....	3	..	2	1¾	7/16	0.141
3	0.099	48	56	.....	48	56	.....	3	..	2	1 13/16	½	0.141
4	0.112	40	48	32, 36	40	48	36	3	..	2	1⅞	9/16	0.141
5	0.125	40	44	.....	40	44	.....	3	..	2	1 15/16	⅝	0.141
6	0.138	32	40	36	32	40	.....	3	..	2	2	11/16	0.141
8	0.164	32	36	40	32	36	.....	4	..	2, 3	2⅛	¾	0.168
8	0.164	32	..	.....	..	..	.....	..	3	...	2⅛	¾	0.168
10	0.190	24	32	30	24	32	.....	4	..	2, 3	2⅜	⅞	0.194
10	0.190	24	32	.....	..	..	.....	..	3	...	2⅜	⅞	0.194
12	0.216	24	28	32	24	28	.....	4	..	...	2⅜	15/16	0.220
14	0.242	..	..	20, 24	..	..	20, 24	4	..	...	2½	1	0.247
Spiral-pointed Machine-screw Taps													
3	0.099	48	56	.....	48	56	.....	2	..	...	1 13/16	½	0.141
4	0.112	40	48	36	40	48	36	2	..	...	1⅞	9/16	0.141
5	0.125	40	44	.....	40	44	.....	2	..	...	1 15/16	⅝	0.141
6	0.138	32	40	.....	32	40	.....	2	..	...	2	11/16	0.141
8	0.164	32	36	.....	32	36	.....	2	..	...	2⅛	¾	0.168
10	0.190	24	32	.....	24	32	.....	2	..	...	2⅜	⅞	0.194
12	0.216	24	28	.....	24	28	.....	2	..	...	2⅜	15/16	0.220
14	0.242	..	..	20, 24	..	..	20, 24	2	..	...	2½	1	0.247

ASA B5 4-1948.

TABLE 11. DIMENSIONS OF PIPE TAPS

Nominal Size	Threads per In.		Flutes		Dimensions		
	Carbon Steel	HSS	Regular	Interrupted	Over-all Length	Length Full Thread	Shank Dia
Taper Pipe Taps							
1/8	27	27	4	5	2 1/8	3/4	0.3125
1/8	27	27	4	5	2 1/8	3/4	0.4375
1/4	18	18	4	5	2 7/16	1 1/16	0.5625
3/8	18	18	4	5	2 9/16	1 1/16	0.7000
1/2	14	14	4	5	3 1/8	1 3/8	0.6875
3/4	14	14	5	5	3 1/4	1 3/8	0.9063
1	11 1/2	11 1/2	5	5	3 3/4	1 3/4	1.1250
1 1/4	11 1/2	11 1/2	5	5	4	1 3/4	1.3125
1 1/2	11 1/2	11 1/2	6	7	4 1/4	1 3/4	1.5000
2	11 1/2	11 1/2	6	7	4 1/2	1 3/4	1.8750
2 1/2	8	....	8	..	5 1/2	2 9/16	2.2500
3	8	....	8	..	6	2 5/8	2.6250
3 1/2	8	....	9	..	6 1/2	2 11/16	2.8125
4	8	....	9	..	6 3/4	2 3/4	3.0000
Straight Pipe Taps							
1/8	27	27	4	..	2 1/8	3/4	0.3125
1/8	27	27	4	..	2 1/8	3/4	0.4375
1/4	18	18	4	..	2 7/16	1 1/16	0.5625
3/8	18	18	4	..	2 9/16	1 1/16	0.7000
1/2	14	14	4	..	3 1/8	1 3/8	0.6875
3/4	14	14	5	..	3 1/4	1 3/8	0.9063
1	11 1/2	11 1/2	5	..	3 3/4	1 3/4	1.1250
1 1/4	11 1/2	....	5	..	4	1 3/4	1.3125
1 1/2	11 1/2	....	6	..	4 1/4	1 3/4	1.5000
2	11 1/2	....	6	..	4 1/2	1 3/4	1.8750

ASA B5.4-1948.

tolerance is equal to or greater than shown below, and will be marked "PG" if it is less.

	In.
4 to 5¼ threads per in., inclusive	0.0020
6 threads per in.	0.0018
7 threads per in.	0.0015
8 threads per in.	0.0014
9 threads per in.	0.0012
10 and 11½ threads per in.	0.0011
12 threads per in. and finer	0.0010

Availability of Standard Taps. By industry usage and custom, standard taps are available in materials and styles, as follows:

Regular (Standard) Hand Taps

Styles. Taper, plug, or bottoming.

Materials. Cut thread in carbon and HSS. Commercial-ground in HSS. Precision-ground in HSS.

Three-flute Hand Taps

Styles. Carbon steel in plug only. HSS taps in plug or bottoming.

Materials. Cut thread in carbon and HSS. Commercial-ground in HSS. Precision-ground in HSS.

Spiral-pointed Hand Taps

Styles. Plug, except ¼- and 5/16-in. sizes in plug or bottoming.

Materials. Same as regular hand taps.

Serial Hand Taps (Set of Three)

Material. Carbon steel.

Regular (Standard) Machine-screw Taps

Styles. Taper, plug or bottoming. Plug only for carbon-steel taps. Plug or bottoming for HSS taps with optional flutes.

Materials. Cut thread in carbon steel and in Nos. 3 to 14 for HSS. Commercial-ground in HSS, Nos. 3 to 14. Precision-ground in HSS, Nos. 0 to 12.

Spiral-pointed Machine-screw Taps

Styles. Plug in cut thread only. Plug and bottoming in ground thread.

Materials. Cut thread in carbon and HSS. Commercial-ground in HSS. Precision-ground in HSS.

Taper Pipe Taps

Standard Styles and Sizes

1. ⅛ to 4 in. carbon steel, regular thread, right hand.
2. ⅛ to 4 in. carbon steel, regular thread, left hand.
3. ⅛ to 2 in. HSS, regular cut thread, right hand.
4. ⅛ to 2 in. HSS, interrupted cut thread, right hand.
5. ⅛ to 2 in. HSS, regular ground thread, right hand.

Straight Pipe Taps

Styles. Plug only.

Materials. Cut thread in carbon and HSS. Ground thread in HSS.

Section 12

TURNING AND BORING

SECTION 12

TURNING AND BORING

In turning and boring, single-point tools are used to generate a surface of revolution—cylindrical, tapered, or contoured. How accurately and efficiently these operations are performed, and the quality of finish obtained, depend on a number of factors. Among these are the tool material, how well the tool is ground and set, the precision of its mounting and alignment, and use of suitable feeds, speeds, and depth of cut.

The purpose of this section is to correlate from reliable sources information on these topics: correct terminology and standards for single-point tools; data on kinds of tool materials and their selection; application of coolants; how to diagnose machining troubles; pertinent information on machinability of metals; and up-to-date information on tool angles, feeds, and speeds for machining various materials. Emphasis on selection of material has been on the practical rather than the theoretical aspects of matters related to turning and boring.

SINGLE-POINT TOOLS

The following material is taken from an American Standard sponsored by the Metal Cutting Tool Institute, Society of Automotive Engineers, National Machine Tool Builders Association, and the American Society of Mechanical Engineers. These groups define a single-point tool as "a cutting tool for use in a lathe, turret lathe, planer, shaper, boring mill, etc., having one face and one continuous cutting edge which produces the machined surface."

Terminology and Definitions—Single-point Cutting Tools

Tool Elements

The **size** of a tool of square or rectangular section is expressed by giving the width of shank W, the height of shank H, and the total tool length L, in inches, such as $\frac{3}{4} \times 1\frac{1}{2} \times 8$ in. The same method of designation is used for toolbit holders to which is added the size of the bits.

The **shank** is that part of the tool on one end of which the point is formed or the tip or bit is supported. The shank in turn is supported on the toolpost of the machine (Figs. 1 and 2).

The **base** is that surface of the shank which bears against the support and takes the tangential pressure of the cut (Figs. 1 and 2).

The **heel** consists of the areas adjacent to the intersection of the base and flank (Figs. 1 and 2).

The **face** is that surface on which the chip apparently impinges as it is separated from the work. The face may be provided with a *ridge* or narrow band ground along the cutting edge to support the built-up edge. This ridge is usually of less rake than that of the balance of the face. The face rake then is often greater than normal. For sintered-carbide tools the ridge may be ground to a negative rake.

The tool **point** is all that part of the tool which is shaped to produce the cutting edges and face.

The **cutting edge** is that portion of the face edge along which the chip is separated from the work. The cutting edge consists usually of the side-cutting edge, the nose, and the end-cutting edge (Figs. 1, 2, 4, and 5).

The **nose** is the corner, arc, or chamfer joining the side-cutting and end-cutting edges (Fig. 6).

The **shape** of the tool is the contour of the face when viewed in a direction at right angles to the base.

The **flank** of the tool is the surface or surfaces below and adjacent to the cutting edge (Fig. 1).

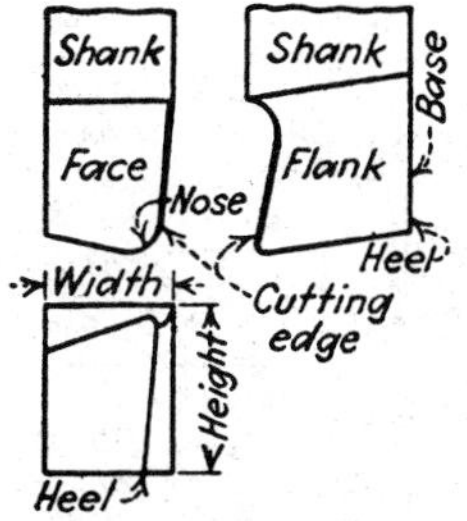

FIG. 1. Right-cut, straight-shank tool.

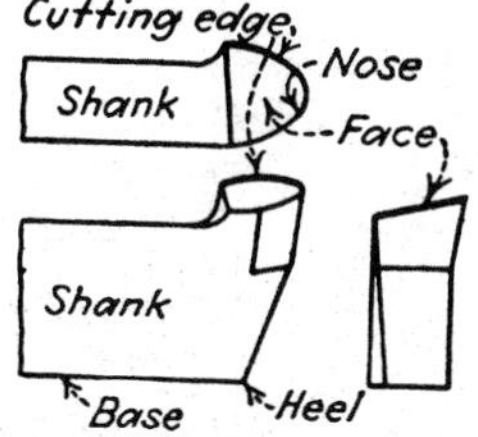

FIG. 2. Forged-type tool.

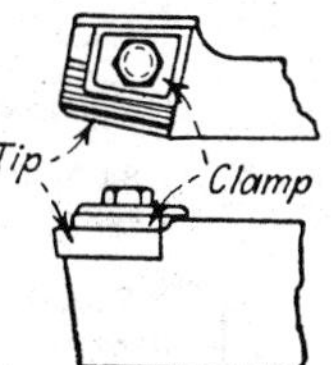
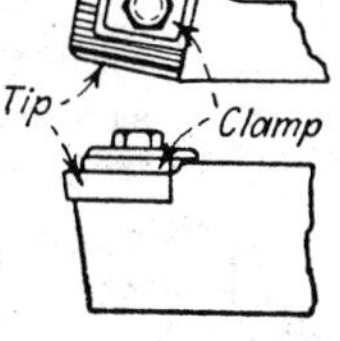

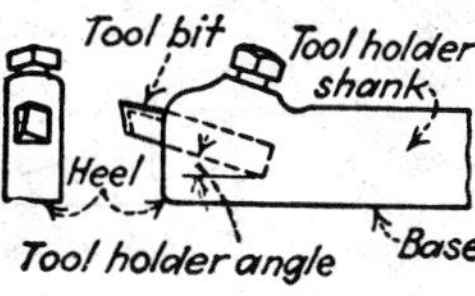

FIG. 3. Toolbit and holder.

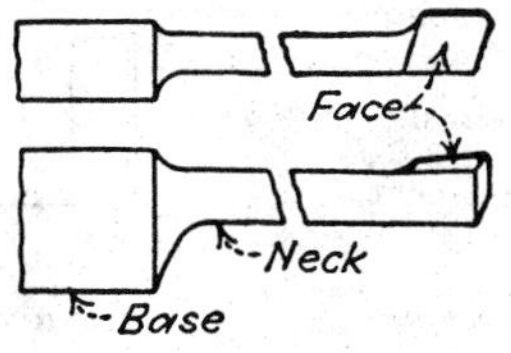
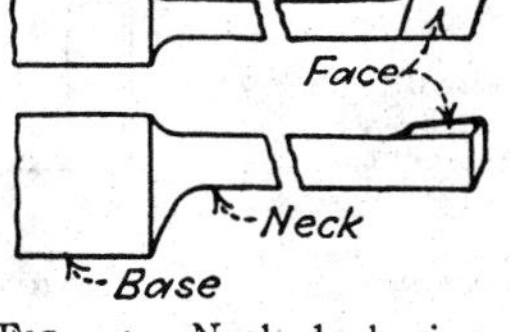

FIG. 4. Necked boring tool, forged.

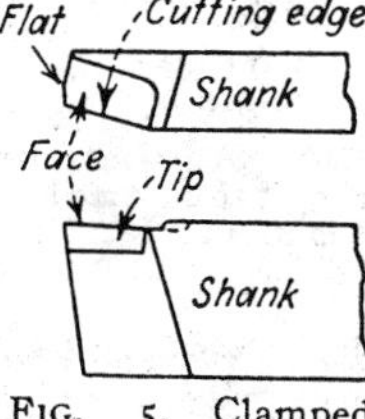

FIG. 5. Clamped and brazed tools.

The **neck** is an extension of the shank of reduced sectional area. A relatively small point, as required in boring, is sometimes attached to the shank by a neck (Fig. 4).

The **flat** (or drag) is the straight portion of the end-cutting edge at 0° angle, intended to eliminate feed marks and produce a smooth machined surface (Fig. 5).

The **chip breaker** is an irregularity in the face of a tool, or a separate piece fastened to the tool or toolholder, to cause the chip to break into short sections and prevent long curls.

TOOL ANGLES

The **toolholder angle** is that angle between the bottom of the bit slot and the base of the toolholder shank (Fig. 3).

The **shank angle** is the angle by which the center line of the point of a bent tool deviates from the straight portion of the shank (Figs. 8*A* and 9).

The **back-rake angle** is the angle between the face of a tool and a line parallel to the base of the shank or holder measured in a plane parallel to the center line of the point and at right angles to the base. The angle is positive if the face slopes downward from the point toward the shank, tending to reduce the lip angle, and is negative if the face slopes upward toward the shank (Figs. 7 and 10).

The **side-rake angle** is the angle between the face of a tool and a line parallel to the base. It is measured in a plane at right angles to the base, and at right angles to the center line of the point (Fig. 7). For convenience in grinding, the rake may be given as *normal rake,* that is, indicated perpendicular to the side- and end-cutting edges (Fig. 10).

The **relief angle** is the angle between a plane perpendicular to the base of a tool or toolholder and the ground flank immediately adjacent to the cutting edge. For convenience in grinding, the relief is often measured in a plane normal to the cutting edge so there would be *normal-side* and *normal-end* relief as shown in Fig. 10. Relief may also be expressed as side and end relief, as defined below.

The **side-relief angle** is the angle between the portion of the flank immediately below the cutting edge and a line drawn through this cutting edge perpendicular to the base. It is measured in a plane at right angles to the center line of the point (Fig. 7). *Normal side-relief angle* is measured in a plane perpendicular to the base of the shank and the cutting edge (Fig. 10).

The **end-relief angle** is the angle between the portion of the end flank immediately below the cutting edge and a line drawn through that cutting edge perpendicular to the base (Fig. 7). It is measured in a plane parallel to the center line of the point. *Normal end-relief angle* is measured in a plane perpendicular to the base of the shank and the end cutting edge (Fig. 10).

The **clearance angle** is greater than the relief angle. It is the angle between a plane perpendicular to the base of a tool and that portion of the flank immediately below the relieved flank. The side clearance angle is measured in the plane of the side-rake angle. The end clearance angle is measured in the plane of the back-rake angle.

The **side-cutting-edge angle** (SCEA) is the angle between the straight side-cutting edge and

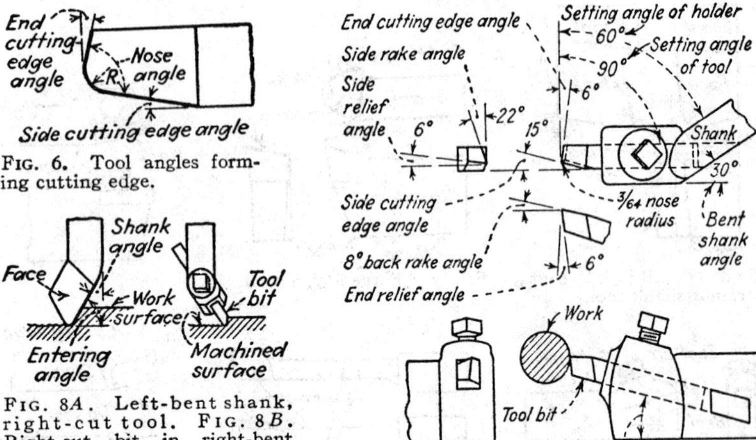

FIG. 6. Tool angles forming cutting edge.

FIG. 8A. Left-bent shank, right-cut tool. FIG. 8B. Right-cut bit in right-bent holder.

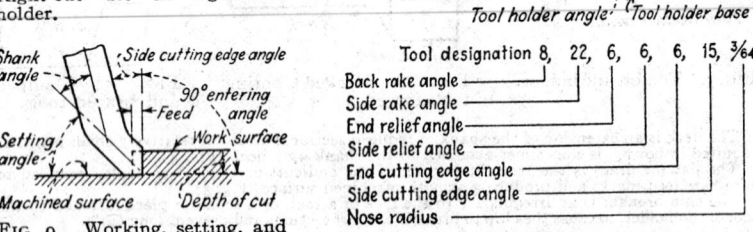

FIG. 9. Working, setting, and entering angles of a right-bent shank, right-cut tool.

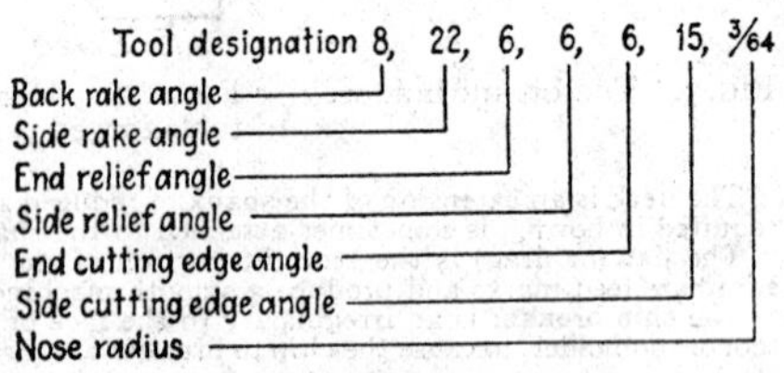

FIG. 7. Typical toolholder and ASA designation of tool elements.

the side of the tool shank (Fig. 6). In the case of a bent tool this angle is measured from the straight portion of the shank (Fig. 9).

The **end-cutting-edge angle** (ECEA) is the angle between the cutting edge on the end of the tool and a line at right angles to the side edge of the straight portion of the tool shank (Fig. 6).

As the setting angle is changed, the relation of the end-cutting edge to the work also is changed. See under Working Angles.

The **nose angle** is the angle included between the side-cutting edge and end-cutting edge (Fig. 6). The chamfer angle is the angle made by the chamfer from a perpendicular to the center line of the point.

Tool Character. A convenient abbreviated designation for specifying the angles of a single-point tool is shown in Fig. 7. Whenever a relief and clearance below the cutting edges are used, this tool designation may be written as 8, 22, 6, 6, 6, 15, $\frac{3}{64}$, as illustrated in Fig. 7. In case the relief and clearance angles are given as being normal to the cutting edges, then the tool designation may be written as indicated in Fig. 10.

Working Angles

The **working angles** are those angles between tool and work, which depend not only on the shape of the tool, but also on its position with respect to the work (Fig. 11).

The **setting angle** is the angle made by the straight portion of the shank of a tool with the machined surface of the work (Fig. 7).

The **entering angle** is the angle which the side-cutting edge of a tool makes with the machined surface of the work, measured on the cutting-edge side of the tool point, that is, right side of a right-cut tool (Figs. 8A and 9).

The **true-rake angle** (or "top rake"), under actual cutting conditions, is the slope of the tool face toward the tool base from the active cutting edge in the direction of chip flow. It is a combination of the back-rake and side-rake angles and varies with the setting of the tool and with the feed and depth of cut (Fig. 11). Rakes normal to the cutting edges are illustrated in Fig. 10.

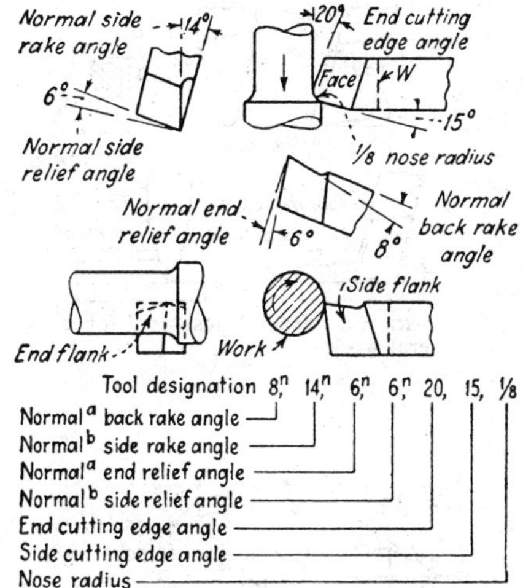

FIG. 10. Typical tool for turning steel of 250 to 300 brinell.

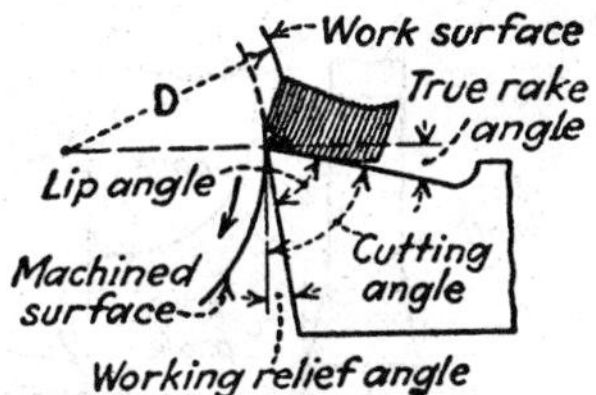

FIG. 11. Chip flow, illustrating working angles.

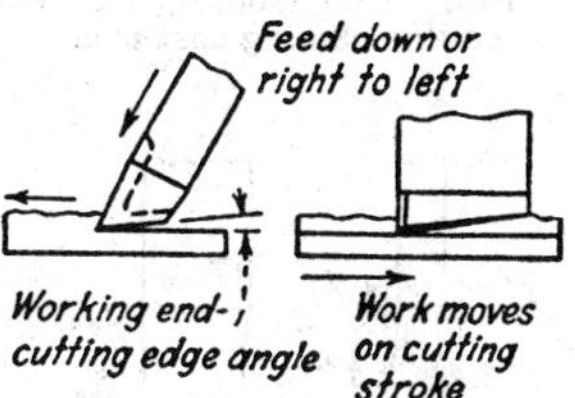

FIG. 12. Left-cut, corner-finishing tool for planer or shaper work.

The **cutting angle** is the angle between the face of the tool and a tangent to the machined surface at the point of action. It equals 90° minus the true-rake angle (Fig. 11).

The **lip angle** is the included angle of the tool material between the face and the relieved ground flank measured in a plane at right angles to the cutting edge. When measured in a plane perpendicular to the cutting edge at the end of the tool, it is called the end lip angle. When measured at the point of chip flow, it is called the true lip angle (Fig. 11).

The **working-relief angle** is the angle between the relieved ground flank of the tool and a line tangent to the machined surface passing through the active cutting edge (Fig. 11).

The **working-end–cutting-edge angle** is the angle between the straight-end-cutting edge and a plane tangent to the machined surface at the point of cutting (Fig. 12).

General Terms

The **cutting speed** is the peripheral or surface speed of the work with respect to the tool. In turning, it is usually measured on the uncut or work surface of the work ahead of the tool.

EXAMPLE: Cutting speed S, in turning in fpm $= 3.1416DN$, in which D = diameter of work, ft, N = rpm of work.

The **depth of cut** is the distance between the bottom of the cut and the uncut surface of the work, measured in a direction at right angles to the machined surface of the work. This is the difference in height between the machined and work surfaces.

The **feed** is the relative amount of motion of the tool into the work for each revolution, stroke, or unit of time.

The **machined surface** is the surface left by the cutting tool.

The **work surface** refers to the surface to be machined.

Classes of Single-point Tools

Solid tools are defined as those having the full section of the cutting end (point) consisting of the metal cutting material. The point and shank may be of the same uniform material (see Fig. 10), or the point may consist of the cutting-tool material welded to the shank as indicated by the dashed lines *W* in Fig. 10. The points of solid tools may be formed by grinding one end of the shank or forging the point roughly tc shape before hardening and grinding.

Tipped tools are those having a relatively small piece of metal-cutting material attached to the tool shank of noncutting material. The tip is ground to form the tool face and cutting edge (Fig. 5).

A **tip** is a piece of cutting-tool material of any shape for attachment by brazing, welding, or the like, to a supporting shank to form the cutting edge and working surfaces of a cutting tool.

A **raised-face tool** is one having its face above the top of the shank. Forged tools, as in Fig. 2, frequently have the face raised above the shank top so that less material need be removed in regrinding the point and so that the total tool life will be increased.

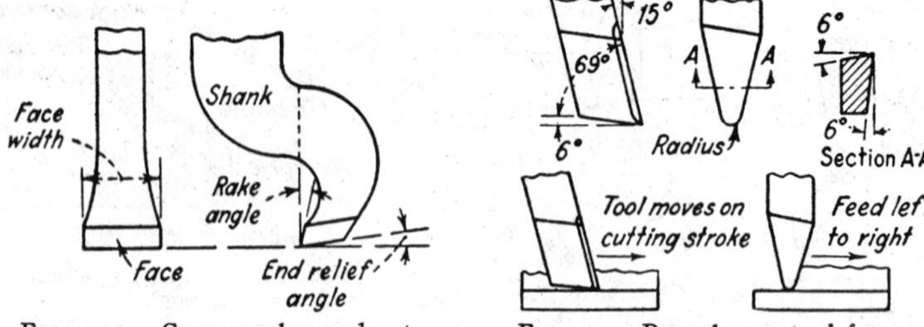

FIG. 13. Goose-neck end-cut tool for planing cast iron.

FIG. 14. Round-nose tool for general work.

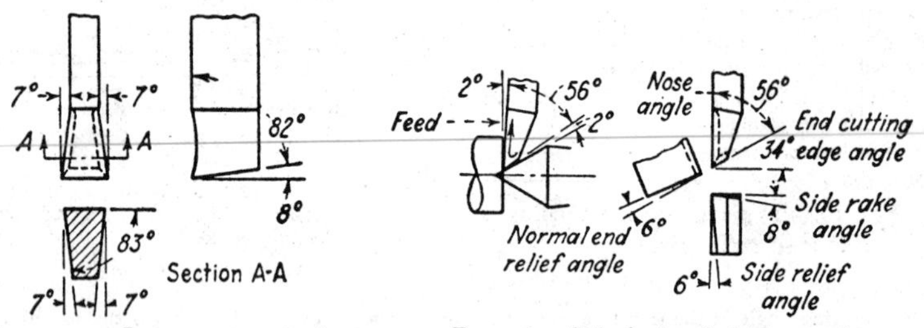

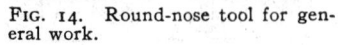

FIG. 15. Square-nose tool with two side-cutting edges for finish turning, chamfering, and slotting.

FIG. 16. Side-facing finishing tool.

A **toolholder** is any piece of material made to form a holder for a toolbit as shown in Figs. 3 and 8*B*, or to hold any special type of solid or tipped tool in a toolpost. They vary greatly in types and sizes. The holder angle is zero for cast nonferrous metals and sintered carbide bits.

A **toolbit** is a relatively small piece of cutting material, clamped in a tool shank or holder in such a way that it can readily be removed and replaced (Figs. 3, 8*B*). Toolbits may be of solid metal-cutting material or they may be tipped.

A **toolbit blank** is a piece of cutting-tool material of any shape from which a bit is made by grinding to final shape and size.

A **bit tool** is one in which toolbits of square, rectangular, or other section, or forged to special shapes, are held in the end of a holder or shank (Fig. 3). It comprises the bit and holder.

Radial Tools.—Most single-point tools are held in toolposts with the axis of the shank substantially in a radial position with respect to the work (Fig. 10) or at right angles to the machined surface.

Tangential tools are usually in the form of small tools or bits held in holders so that the bit is located substantially tangent to the machined surface, as in a bar-turning lathe or screw machine. The face of the tool point is on the end of the tool.

A **straight tool** has the point on one end of a straight shank (Figs. 1 and 5) so that there is no substantial deviation from the straight shank.

A **bent tool** has the point bent to the left or right (Fig. 8*A* and *B*) to make its operation more convenient. These tools are called left-bent tools if the point is bent to the left when looking at the tool from the point end with the face upward and the shank pointing away, and vice versa.

An **offset tool** has the point at either side of, but parallel to, the shank. It is known as a right-offset tool if the point is offset to the right of the shank when looking at the tool from the point end with the face upward and the shank pointing away.

Goosenecked Tools (End-cutting). A goosenecked finishing tool is used primarily for taking finishing cuts on cast iron in a planer. Very shallow cuts with very coarse feeds are most satisfactory. The neck is curved so as to locate the cutting edge nearly in the plane of the base of the shank to avoid digging in. The straight end-cutting edge should be honed to provide a keen straight edge, and it should be placed parallel to the machined surface of the work (see Fig. 13).

A **right-cut single-point tool** is one which, when viewed from the point end of the tool, with the face up, has the cutting edge on the right side (Figs. 1, 2, 4, 8*A*, 8*B*, and 10). When such a tool is used in an engine lathe, the cutting edge is on the left side. It is fed into the work from right to left, as when cutting a right-hand screw thread.

A **left-cut tool** has the cutting edge on the left when looking at the point end with the face upward.

An **end-cut single-point** tool is one having its principal cutting edge on the end (Fig. 15).

A **roughing tool** is any tool of an adequate size for relatively heavy cuts (see Figs. 1, 2, and 10). The tool may be ground to various shapes, depending upon the use to which it is put. In planing, for all tool materials, zero or negative back-rake angles are commonly used, whereas in turning, positive back-rake and side-rake angles are more generally employed. In turning cast iron with scale, a zero or small side-cutting edge angle is desirable so as to crack rather than cut the scale. In turning steel, a large side-cutting-edge angle is desirable, as greater tool life is obtained. The nose radius may be relatively small, with large side-cutting-edge angles.

A **finishing tool** is one keenly ground to some specific shape as required by a given job. It is ground usually with greater rake than roughing tools to take light cuts and produce good surface quality to final size and shape, such as sharp corners.

A **curved-cutting-edge tool** has variable side-cutting-edge angles (Fig. 2). They are used usually for rough-cutting steel and cast iron or for producing forms. Figure 14 shows a round-nose tool with back rake but with little side rake for general-purpose work on all metals. The narrow point makes it possible to machine in corners or in recesses.

Square-nosed Tools (End-cutting). A square-nosed tool is an end-cut tool usually ground with the end-cutting edge at approximately 90° to the axis of the point; that is, it has a zero end-cutting-edge angle. These tools usually have right and left side-cutting edges with small axial clearance angles so that the face at the cutting edge is wider than that adjacent to the shank (Fig. 15). This tool may be used for chamfering or for roughing cuts on flat surfaces where sharp corners are to be obtained, in which case the tool is fed axially parallel to the surface being machined so that end cutting maintains.

The square-nosed tool is also used for finishing flat surfaces on cast iron. After the scale is removed by a roughing tool, the finishing tool removes a shallow depth of metal from 0.002 to 0.006 in., employing a feed per stroke or per revolution slightly less than the width of the end-cutting edge. The end-cutting edge is parallel to the machined surface.

Cutoff (or Parting) Tools. The cutting-off tool is usually either forged or provided as bits in the form of thin blades (see page 12-8).

These tools are end-cutting and have a square-cutting edge at right angles to the center line of the tool point, or the cutting edge may be beveled on the end so as to have an end-cutting-edge angle of approximately 15° to prevent burrs on the cutoff work. These tools have right- and left-hand side-cutting edges which are provided with axial clearance of from 1 to 4°, as the cutting is supposed to be done principally by the end-cutting edge. They usually have no side rake but may or may not have a slight back rake or be provided with a chip breaker groove parallel to but slightly back of the end-cutting edge.

Side-facing tools are used for finishing the tailstock end of work in lathes or the face of work mounted in a chuck of a lathe, planer, or shaper (see Fig. 16). These tools are either of the forged type or furnished in formed bits to be held in special toolholders. If they are ground with a side-cutting-edge angle of 0° and an end-cutting-edge angle of 32°, they will have a nose angle of 56°, which will admit the point between the end of the work and the 60° tailstock center. They normally cut on the side-cutting edge which is placed parallel to or within 1 or 2° clearance from the surface being faced. They may be right- or left-cut and right- or left-bent shank to facilitate their application.

A **dovetailing tool** is similar to side-facing tool. It normally has a 30° end-cutting-edge angle and a 0° side-cutting-edge angle to give a 60° nose angle. It is used primarily to finish dovetails for machine construction. This tool may be made right- or left-cut, or for roughing or finishing.

An **intermittent cutting tool,** particularly when tipped with sintered carbide, is provided with the normal side rake but with excessive negative back rake. In this way the shoulder of the work engaging the cutting edge meets the tip back of the nose and prevents tool-tip fracture.

Carbide-tool Standards

STANDARD STRAIGHT CUTOFF BLADES FOR LATHES AND SCREW MACHINES*

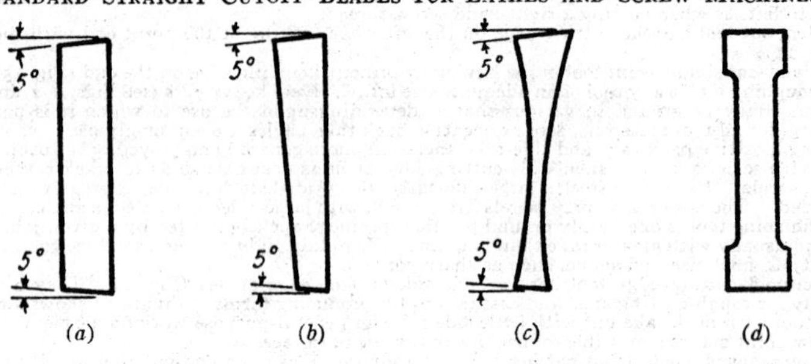

Optional shapes for cutoff blade stock. (*a*) Without side clearance. (*b*) With straight side clearance. (*c*) With concaved side clearance. (*d*) With channeled sides (heavy duty) only made 1 in. height or greater.

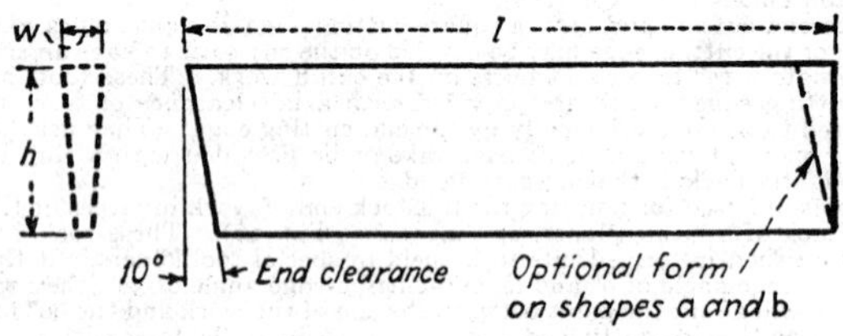

DIMENSIONS OF STRAIGHT CUTOFF BLADES*

Width (*w*)	Height (*h*) × Length (*l*)					
1/16	1/2 × 4 1/2†	11/16 × 5	13/16 × 6			
5/64	1/2 × 4 1/2					
3/32	1/2 × 4 1/2†	11/16 × 5	13/16 × 6	....	1 × 6 1/2	
1/8	1/2 × 4 1/2†	11/16 × 5†	13/16 × 6	....	1 × 6 1/2	
5/32	....	11/16 × 5†	13/16 × 6			
3/16	....	11/16 × 5†	13/16 × 6	7/8 × 6†	1 × 6 1/2	1 1/8 × 6 1/2†
1/4	....	....	11/16 × 6	7/8 × 6†	1 × 6 1/2	1 1/8 × 6 1/2†

All dimensions are given in inches.
Cross-sectional shape and dimensional limits shall be in accordance with the manufacturer's standards.
Height dimensions are nominal.

* American Standard ASA B5.21-1949.
† Preferred blade sizes for future designs and development. These should not be construed as indicating that other listed sizes are special, nonprocurable, or subject to special prices.

Terminology for Standard Brazed Carbide Tools

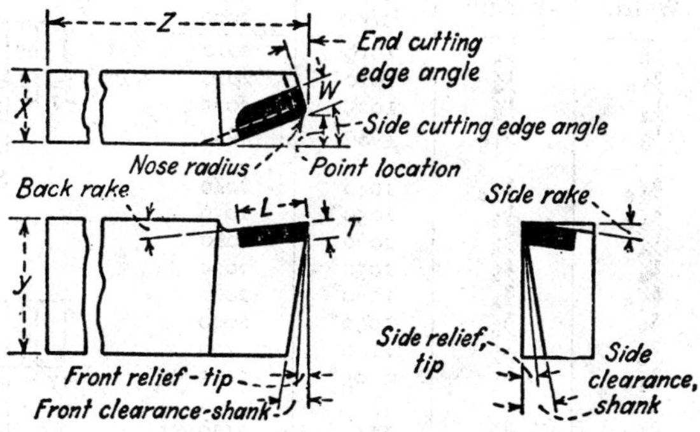

General Specifications

Unless otherwise specified by purchaser, the following tolerances apply to all standard and special carbide-tipped cutting tools.

Tip:	
End-cutting edge angle	±1°
Side-cutting edge angle	±1°
Side relief	±1°
Front relief	±1°
Back rake	±2°
Side rake	±2°
Nose radius	Maximum 1/32 in.
Point location	±1/64 in.
Tip thickness *T*	±1/64 in.
Tip width *W*	±1/64 in.
Tip length *L*	±1/64 in.
Shank:	
Front clearance	±1°
Side clearance	±1°
Shank thickness *X*	1 in. or less +0.000 −0.010
	Over 1 in. +0.000 −0.015
Shank width *Y*	1 in. or less +0.000 −0.010
	Over 1 in. +0.000 −0.015
Shank length *Z*	Up to 1 in. square and 1 × 1¼ in., ±1/8
	Over 1 in. square and over 1 × 1¼ in., ±1/4

Standard Inserts for Brazed Carbide Tools

Dimension, Inches			Tip Number	
Thickness	Width	Length	Style 1000	Style 2000
1/16	1/8	5/8	1010	2010
	3/16	1/4	1020	2020
	1/4	1/4	1025	2025
	1/4	5/16	1030*	2030
3/32	3/16	5/16	1040	2040
	3/16	1/2	1050*	2050
	1/4	3/8	1060	2060
	1/4	1/2	1070*	2070
	5/16	3/8	1080*	2080
	3/8	3/8	1090*	2090
	3/8	1/2	1100	2100
	7/16	1/2	1105*	2105
1/8	3/16	3/4	1110*	2110
	1/4	1/2	1120*	2120
	1/4	5/8	1130	2130
	1/4	3/4	1140*	2140
	5/16	7/16	1150	2150
	5/16	1/2	1160*	2160
	5/16	5/8	1170	2170
	3/8	1/2	1180*	2180
	3/8	3/4	1190	2190
	1/2	1/2	1200*	2200
	1/2	3/4	1210†	2210
	3/4	3/4	1215†	2215
5/32	3/8	9/16	1220	2220
	3/8	3/4	1230	2230
	5/8	5/8	1240*	2240
3/16	5/16	7/16	1250	2250
	5/16	5/8	1260*	2260
	3/8	1/2	1270*	2270
	3/8	5/8	1280	2280
	3/8	3/4	1290	2290
	7/16	5/8	1300	2300
	7/16	1 3/16	1310	2310
	1/2	1/2	1320*	2320
	1/2	3/4	1330†	2330
	3/4	3/4	1340*	2340

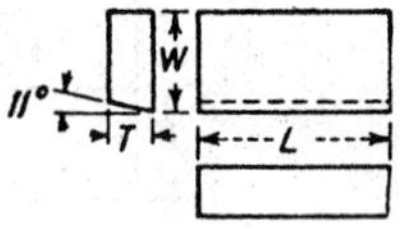

Thickness	Width	Length	Style 0000	Style 1000	Style 3000	Style 4000
1/4	3/8	9/16		1350	3350	4350
	3/8	3/4		1360	3360	4360
	7/16	5/8		1370	3370	4370
	1/2	3/4	0380	1380	3380	4380
	9/16	1	0390	1390	3390	4390
	5/8	5/8		1400*	3400	4400
	3/4	3/4		1405*	3405	4405
	3/4	1	0410	1410†	3410	4410

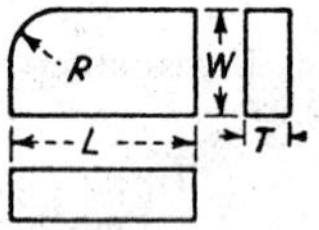

Style 0000

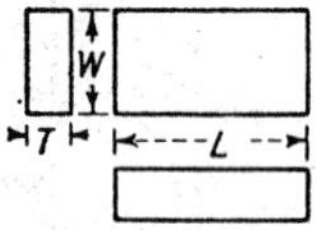

Style 1000

Style 2000

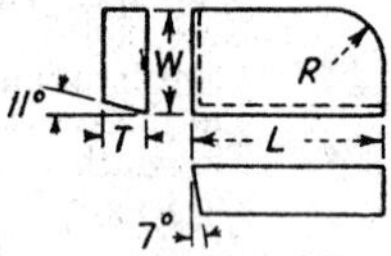

Style 3000

STANDARD INSERTS FOR BRAZED CARBIDE TOOLS (*Continued*)

Thickness	Width	Length	Style 0000	Style 1000	Style 3000	Style 4000
5/16	7/16	5/8		1420	3420	4420
	7/16	15/16		1430	3430	4430
	1/2	3/4		1440†	3440	4440
	1/2	1		1450	3450	4450
	5/8	1	0460	1460†	3460	4460
	3/4	3/4		1470*	3470	4470
	3/4	1		1475†	3475	4475
	3/4	1 1/4	0480	1480†	3480	4480
3/8	1/2	3/4		1490	3490	4490
	1/2	1	0500	1500	3500	4500
	5/8	1	0510	1510†	3510	4510
	3/4	1 1/4		1520†	3520	4520
	3/4	1 1/2	0525	1525	3525	4525
1/2	3/4	1		1530	3530	4530
	3/4	1 1/4	0540	1540†	3540	4540
	3/4	1 1/2	0550	1550	3550	4550

Tolerances:
+0.015—0.000 on all tip dimensions up to 3/8 in.
+0.020—0.000 on all tip dimensions over 3/8 in. through 1 in.
+0.040—0.000 on all tip dimensions over 1 in. through 2 in.

* Extra grinding stock on width to permit grinding on both sides.
† Extra grinding stock on length to permit grinding on both ends.

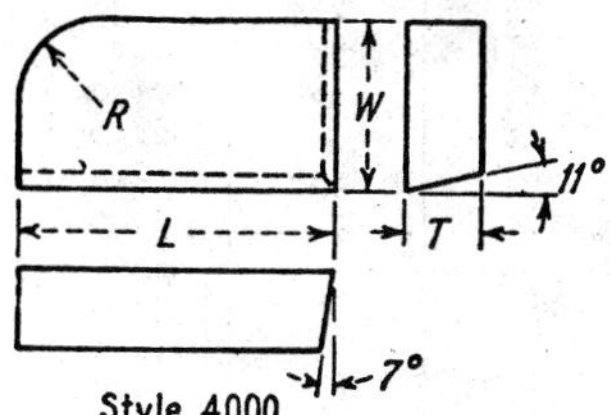

Style 4000

TIP RADIUSES
Styles 2000, 3000, and 4000

Width	*Radiuses*
1/8 through 1/4	1/8
9/32 through 3/8	3/16
Over 3/8	1/4

STYLES 5000 AND 6000 INSERTS

Tip Number	Tip Dimensions			Tip Number	Tip Dimensions		
	T	*W*	*L*		*T*	*W*	*L*
5030	1/16	1/4	5/16	6080	3/32	5/16	3/8
5080	3/32	5/16	3/8	6100	3/32	3/8	1/2
5100	3/32	3/8	1/2	6200	1/8	1/2	1/2
5105	3/32	7/16	1/2	6240	5/32	5/8	5/8
5200	1/8	1/2	1/2	6340	3/16	3/4	3/4
5240	5/32	5/8	5/8				
5340	3/16	3/4	3/4				
5410	1/4	3/4	1				

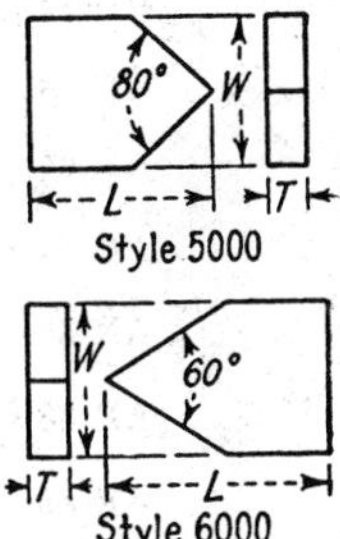

Style A Brazed Carbide Tools
Side Cutting Edge Angle = 0°

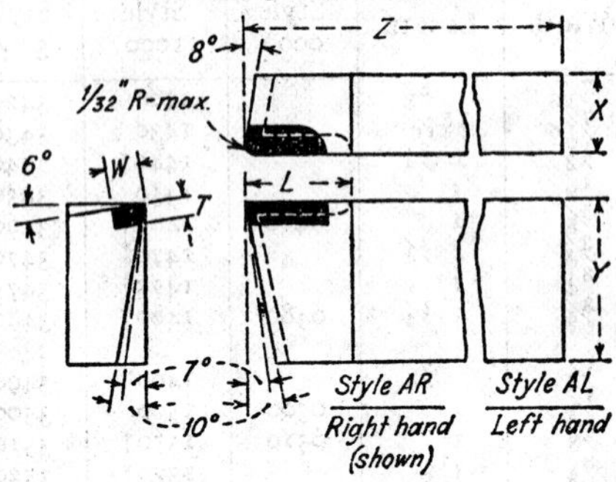

Dimensions of Style AR and AL Brazed Tools

Catalog Number		Shank Dimensions, In.			Tip Catalog Number
Style AR	Style AL	Width X	Height Y	Length Z	
AR 4	AL 4	1/4	1/4	1 1/2	2040
AR 5	AL 5	5/16	5/16	2 1/4	2070
AR 6	AL 6	3/8	3/8	2 1/2	2070
AR 7	AL 7	7/16	7/16	3	2070
AR 8	AL 8	1/2	1/2	3 1/2	2170
AR 10	AL 10	5/8	5/8	4	2230
AR 12	AL 12	3/4	3/4	4 1/2	2310
AR 16	AL 16	1	1	7	3390 4390
AR 20	AL 20	1 1/4	1 1/4	8	3460 4460
AR 24	AL 24	1 1/2	1 1/2	8	3510 4510
AR 44	AL 44	1/2	1	7	2260
AR 54	AL 54	5/8	1	6	3360 4360
AR 55	AL 55	5/8	1 1/4	8	3360 4360
AR 64	AL 64	3/4	1	6	3380 4380
AR 66	AL 66	3/4	1 1/2	8	3430 4430
AR 85	AL 85	1	1 1/4	8	3460 4460
AR 86	AL 86	1	1 1/2	8	3510 4510
AR 88	AL 88	1	2	10	3510 4510
AR 90	AL 90	1 1/2	2	8	3540 4540

Style B Brazed Carbide Tools

Side Cutting Edge Angle = 15°

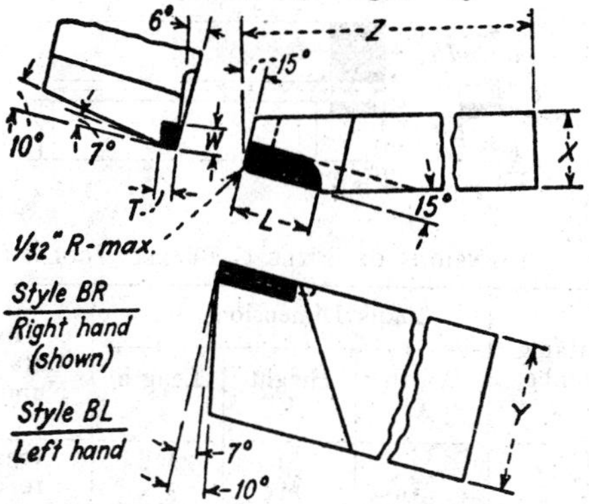

6° Z 15° 10° 7° W X T L 15° 1/32" R-max.

Style BR / Right hand (shown)

Style BL / Left hand

7° 10° Y

Dimensions of Style BR and BL Brazed Tools

Catalog Number		Shank Dimensions, In.			Tip Catalog Number
Style BR	Style BL	Width X	Height Y	Length Z	
BR 6	BL 6	3/8	3/8	2 1/2	1070
BR 8	BL 8	1/2	1/2	3 1/2	2170
BR 10	BL 10	5/8	5/8	4	2320
BR 12	BL 12	3/4	3/4	4 1/2	2310
BR 16	BL 16	1	1	7	3390 4390
BR 20	BL 20	1 1/4	1 1/4	8	3460 4460
BR 24	BL 24	1 1/2	1 1/2	8	3510 4510
BR 44	BL 44	1/2	1	7	2260
BR 54	BL 54	5/8	1	6	3360 4360
BR 55	BL 55	5/8	1 1/4	8	3360 4360
BR 64	BL 64	3/4	1	6	3380 4380
BR 66	BL 66	3/4	1 1/2	8	3430 4430
BR 85	BL 85	1	1 1/4	8	3460 4460
BR 86	BL 86	1	1 1/2	8	3510 4510
BR 88	BL 88	1	2	10	3510 4510
BR 90	BL 90	1 1/2	2	8	3540 4540

Style C Brazed Carbide Tools
Square End-cutting

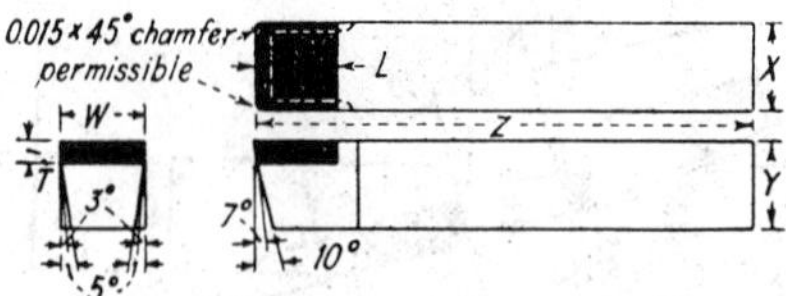

Dimensions of Style C Brazed Tools

Catalog Number	Shank Dimensions, In.			Tip Catalog Number
	Width X	Height Y	Length Z	
C 4	1/4	1/4	1 1/2	1030
C 5	5/16	5/16	2 1/4	1080
C 6	3/8	3/8	2 1/2	1090
C 7	7/16	7/16	3	1105
C 8	1/2	1/2	3 1/2	1200
C 10	5/8	5/8	4	1240
C 12	3/4	3/4	4 1/2	1340
C 16	1	1	7	1410
C 20	1 1/4	1 1/4	8	1480
C 44	1/2	1	7	1320
C 54	5/8	1	6	1400
C 55	5/8	1 1/4	8	1400
C 64	3/4	1	6	1405
C 66	3/4	1 1/2	8	1470
C 86	1	1 1/2	8	1475

Style D Brazed Tools
80° Nose Angle

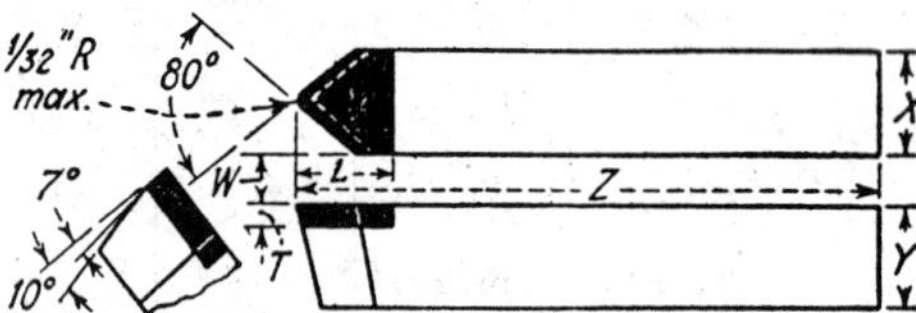

Dimensions of Style D Brazed Tools

Catalog Number	Shank Dimensions, In.			Tip Catalog Number
	Width X	Height Y	Length Z	
D 4	1/4	1/4	1 1/2	5030
D 5	5/16	5/16	2 1/4	5080
D 6	3/8	3/8	2 1/2	5100
D 7	7/16	7/16	3	5105
D 8	1/2	1/2	3 1/2	5200
D 10	5/8	5/8	4	5240
D 12	3/4	3/4	4 1/2	5340
D 16	1	1	7	5410

Style E Brazed Tools
60° Nose Angle

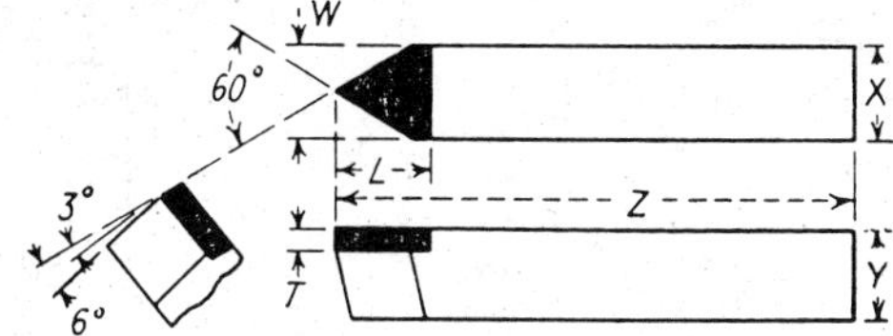

Catalog Number	Shank Dimensions, In.			Tip Catalog Number
	Width X	Height Y	Length Z	
E 5	5/16	5/16	2¼	6080
E 6	3/8	3/8	2½	6100
E 8	½	½	3½	6200
E 10	5/8	5/8	4	6240
E 12	¾	¾	4½	6340

Style F Brazed Tools
Offset, End-cutting

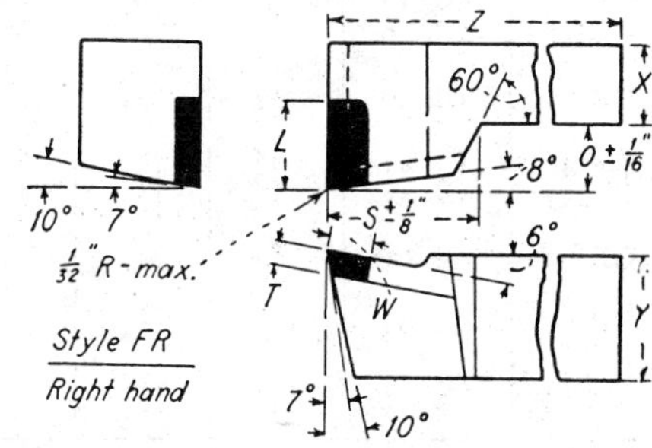

Catalog Number		Shank Dimensions, In.					Tip Catalog Number
Style FR	Style FL	Width X	Height Y	Length Z	Offset O	Length of Offset S	
FR 12	FL 12	¾	¾	4½	5/8	1⅛	2310
FR 16	FL 16	1	1	7	¾	1 7/16	3390 4390
FR 20	FL 20	1¼	1¼	8	¾	1⅜	3460 4460
FR 44	FL 44	½	1	6	½	1	2260
FR 55	FL 55	5/8	1¼	7	5/8	1¼	3360 4360
FR 64	FL 64	¾	1	6	5/8	1 5/16	3380 4380
FR 85	FL 85	1	1¼	8	¾	1½	3460 4460
FR 86	FL 86	1	1½	8	¾	1 9/16	3510 4510

Style G
Brazed Tools
Offset, Side-cutting

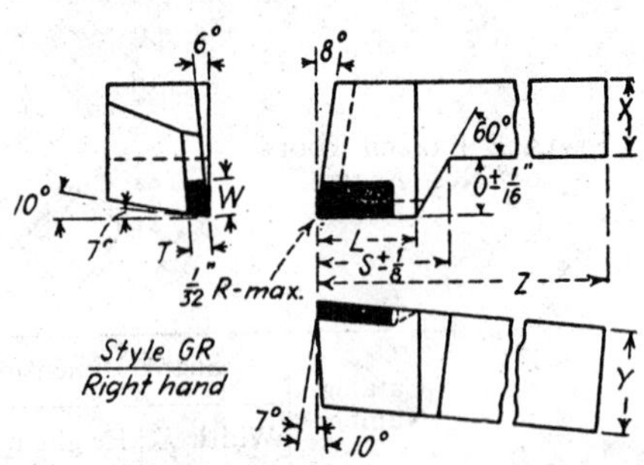

Catalog Number		Shank Dimensions, In.					Tip Catalog Number
Style GR	Style GL	Width *X*	Height *Y*	Length *Z*	Offset *O*	Length of Offset *S*	
GR 12	GL 12	¾	¾	4½	⅜	1⅜	2310
GR 16	GL 16	1	1	7	½	1⅝	3390, 4390
GR 20	GL 20	1¼	1¼	8	¾	1¾	3460, 4460
GR 44	GL 44	½	1	6	¼	1⅛	2260
GR 55	GL 55	⅝	1¼	7	⅜	1⅛	3360, 4360
GR 64	GL 64	¾	1	6	½	1⅜	3380, 4380
GR 85	GL 85	1	1¼	8	½	1⅝	3460, 4460
GR 86	GL 86	1	1½	8	½	1⅝	3510, 4510

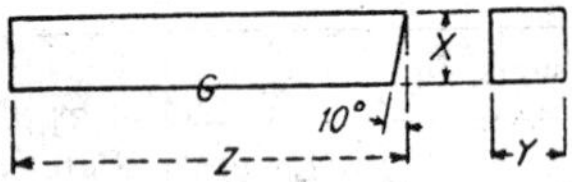

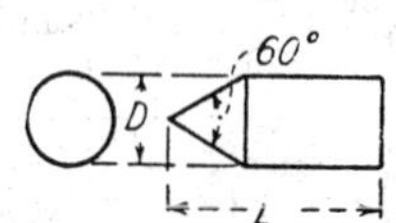

Square Boring Bits

Designation	Size, In.		
	X +0.000 −0.010	*Y* +0.000 −0.005	*Z*
SQ-310	3/16	3/16	1¼
SQ-410	¼	¼	1¼
SQ-512	5/16	5/16	1½
SQ-514	⅜	⅜	1¾

Standard Center Inserts

Designation	Insert Size, In.		For Use with Taper Number		
	D	*L*	Morse	B & S	Jarno
CT-40	¼	7/16	1	5, 6	4, 5
CT-50	5/16	9/16	2	7	
CT-60	⅜	11/16	...	8	6, 7
CT-80	½	⅞	3, 4	9, 10	8, 9, 10, 11
CT-100	⅝	1 1/16	5, 6	11, 12	12, 13, 14
CT-120	¾	1¼	...	13	
CT-140	⅞	1⅜	7		

Solid Triangular Inserts and Toolholders

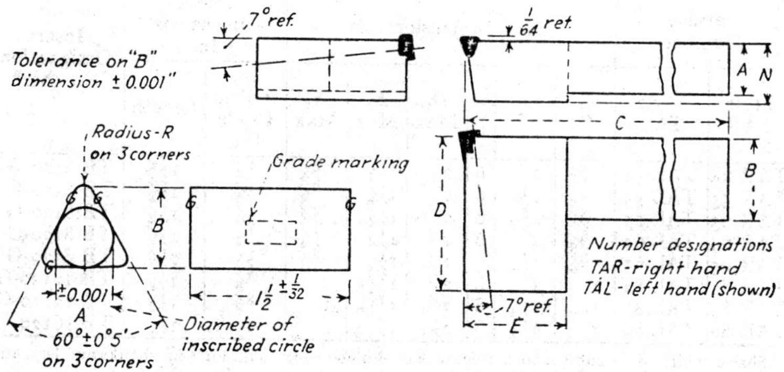

Dimensions of Triangular Inserts

Diameter of Inscribed Circle A	Height B	Standard Corner Radius* R	Nonstandard Corner Radiuses for Each Insert Size			
1/4	0.344	1/32	R = 0	1/64	3/64	1/16
			B = 0.375	0.359	0.328	0.313
3/8	0.516	3/64	R = 0	1/64	1/32	1/16
			B = 0.563	0.547	0.531	0.500
1/2	0.688	1/16	R = 0	1/32	1/8	
			B = 0.750	0.719	0.625	

All dimensions are given in inches.

* Standard corner radiuses specified are the standard dimensions and will be carried in stock by the carbide fabricators. Inserts with other specified corner radiuses (nonstandard) will not be carried in stock but will be furnished upon request.

Style Numbers for Triangular Inserts

Dia of Inscribed Circle	Corner Radiuses—Ground Inserts						Unground Inserts without Radius*
	0	1/64	1/32	3/64	1/16	1/8	
1/4	TB 8120-G	TB 8121-G	TB 8122-G	TB 8123-G	TB 8124-G	....	TB 812
3/8	TB 12120-G	TB 12121-G	TB 12122-G	TB 12123-G	TB 12124-G	....	TB 1212
1/2	TB 16120-G	....	TB 16122-G	TB 16123-G	TB 16124-G	TB 16128-G	TB 1612

All dimensions are given in inches.

* On all unground inserts grinding stock shall be provided.

Dimensions and Triangular Inserts for Style TAR and TAL Toolholders

Catalog Number		Shank Dimensions, In.						Insert Dimensions, In.		Insert Designation Number*
Style TAR	Style TAL	A	B	C	D, Max	E, Max	N, Max	Inscribed Circle Dia	Length	
TAR 54	TAL 54	5/8	1	6	2	1 1/4	3/4	1/4	1 1/2	TB 8120-G
TAR 55	TAL 55	5/8	1 1/4	6	2	1 1/4	3/4	1/4	1 1/2	TB 8120-G
TAR 63	TAL 63	3/4	3/4	6	2	1 1/4	3/4	1/4	1 1/2	TB 8120-G
TAR 64	TAL 64	3/4	1	6	2	1 1/4	3/4	1/4	1 1/2	TB 8120-G
TAR 65	TAL 65	3/4	1 1/4	6	2	1 1/4	3/4	1/4	1 1/2	TB 8120-G
TAR 84	TAL 84	1	1	8	2	1 7/16	1	3/8	1 1/2	TB 12120-G
TAR 85	TAL 85	1	1 1/4	8	2	1 7/16	1	3/8	1 1/2	TB 12120-G
TAR 86	TAL 86	1	1 1/2	8	2 1/4	1 7/16	1	1/2	1 1/2	TB 16120-G

Shank width *A* is subject to a tolerance of +0.000 —0.010 in. for 1-in. shanks and less, and +0.000 —0.015 in. for shanks over 1 in.

* Designation numbers cover inserts with zero corner radius. Designation numbers for inserts with other corner radiuses are shown above.

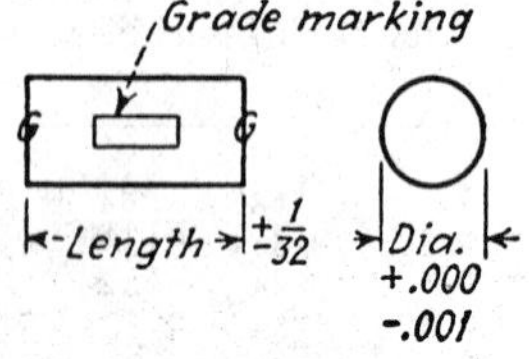

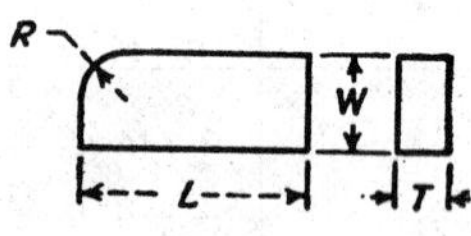

Carbide Cylinders for Turning and Boring

Designation	*Size, In.*
SC-33	3/32 dia × 3/8
SC-34	3/32 dia × 1/2
SC-44	1/8 dia × 1/2
SC-46	1/8 dia × 3/4
SC-48	1/8 dia × 1
SC-55	5/32 dia × 5/8
SC-64	3/16 dia × 1/2
SC-66	3/16 dia × 3/4
SC-69	3/16 dia × 1 1/8
SC-77	7/32 dia × 7/8
SC-88	1/4 dia × 1
SC-810	1/4 dia × 1 1/4
SC-1010	5/16 dia × 1 1/4
SC-1212	3/8 dia × 1 1/2
SC-1612	1/2 dia × 1 1/2

Reamer Tips

Tip Style No.	T	W	L	R
RT-1	1/32	1/16	1/2	1/4
RT-2	3/64	3/32	1/2	1/4
RT-3	3/64	3/32	11/16	1/4
RT-4	1/16	1/8	11/16	1/4
RT-5	5/64	1/8	7/8	1/4
RT-6	3/32	3/16	7/8	5/16
RT-7	1/16	13/64	3/4	5/16
RT-8	1/16	17/64	3/4	3/8
RT-9	1/16	5/16	3/4	3/8
RT-10	1/16	3/8	3/4	3/8
RT-11	3/64	3/32	1	1/4

Tolerances on tip dimensions:
+0.015 —0.000 up to 3/8 in.
+0.20 —0.000 over 3/8 through 1 in.
+0.040 —0.000 over 1 through 2 in.

Styles SBR and SBL Insert Toolholders

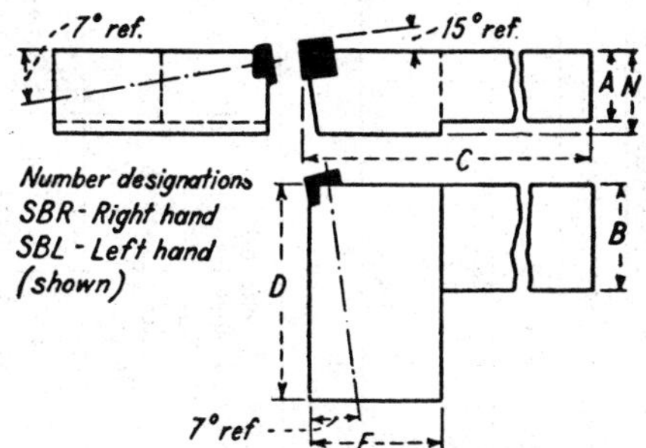

Dimensions of Square Inserts

Nominal Size A	Measurement B	Measurement C	Standard Corner Radius* R	Nonstandard Corner Radiuses for Each Insert Size		
3/8	0.692	0.598	1/32	R = 0	3/64	1/16
				B = 0.692	0.692	0.692
				C = 0.611	0.592	0.585
1/2	0.780	0.724	3/64	R = 0	1/32	1/16
				B = 0.780	0.780	0.780
				C = 0.743	0.730	0.717
3/4	0.957	0.983	1/16	R = 0	1/32	1/8
				B = 0.957	0.957	0.957
				C = 1.009	0.996	0.957

All dimensions are given in inches.

* Standard corner radiuses specified are the standard dimensions and will be carried in stock by the carbide fabricators. Inserts with other specified corner radiuses (nonstandard) will not be carried in stock but will be furnished upon request.

Solid Square Insert

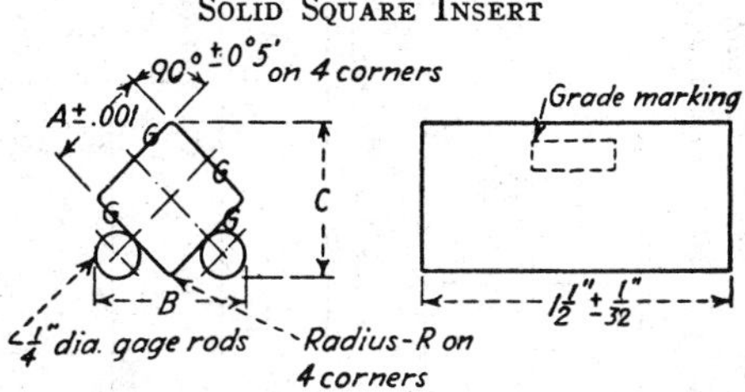

Style Numbers for Square Inserts

Size A	Corner Radiuses—Ground Inserts: 0	1/32	3/64	1/16	1/8	Unground Inserts without Radius*
3/8	SQ 12120-G	SQ 12122-G	SQ 12123-G	SQ 12124-G	....	SQ 1212
1/2	SQ 16120-G	SQ 16122-G	SQ 16123-G	SQ 16124-G	....	SQ 1612
3/4	SQ 24120-G	SQ 24122-G	....	SQ 24124-G	SQ 24128-G	SQ 2412

* On all unground inserts, grinding stock shall be provided.

Styles RAR and RAL Insert Toolholders

Inserts

Number designations:
RAR - right hand
RAL - left hand (shown)

Designation	D, In.	L, In.
SC-1212	3/8	1 1/2
SC-1612	1/2	1 1/2

Solid Carbide Squares

Designation	Size, In.
SQ-310	3/16 square × 1 1/4
SQ-410	1/4 square × 1 1/4
SQ-512	5/16 square × 1 1/2
SQ-614	3/8 square × 1 3/4

Standard Carbide Scraper Tips

Tip Style No.	T	W	L
ST-1	3/64	3/3	1
ST-2	1/16	1/4	1
ST-3	3/32	1/4	1
ST-4	1/8	1/4	1 1/4

Dimensions and Inserts for Toolholders Receiving Square or Round Inserts

Shank Dimensions, In.						Insert Dimension, In.		Square-insert Holders		Insert No.†	Round-insert Holders		Insert No.
A	B	C	D, Max	E, Max	N, Max	Dia or Sq	Length	Style SBR	Style SBL		Style RAR	Style RAL	
5/8	1	6	2	1 1/4	3/4	3/8	1 1/2	SBR 54	SBL 54	SQ 12120-G	RAR 54	RAL 54	SC-1212
5/8	1 1/4	6	2	1 1/4	3/4	3/8	1 1/2	SBR 55	SBL 55	SQ 12120-G	RAR 55	RAL 55	SC-1212
3/4	3/4	6	2	1 1/4	3/4	3/8	1 1/2	SBR 63	SBL 63	SQ 12120-G	RAR 63	RAL 63	SC-1212
3/4	1	6	2	1 1/4	3/4	3/8	1 1/2	SBR 64	SBL 64	SQ 12120-G	RAR 64	RAL 64	SC-1212
3/4	1 1/4	6	2	1 1/4	3/4	3/8	1 1/2	SBR 65	SBL 65	SQ 12120-G	RAR 65	RAL 65	SC-1212
1	1	8	2	1 7/16	1	1/2	1 1/2	SBR 84	SBL 84	SQ 16120-G	RAR 84	RAL 84	SC-1612
1	1 1/4	8	2	1 7/16	1	1/2	1 1/2	SBR 85	SBL 85	SQ 16120-G	RAR 85	RAL 85	SC-1612
1	1 1/2	8	2 1/4	1 7/16	1	*	1 1/2	SBR 86	SBL 86	SQ 24120-G	RAR 86	RAL 86	SC-1612

* 1/2-in. dia or 3/4-in. square, according to type of toolholder.

† Designation numbers cover inserts with zero corner radius. Designation numbers for inserts with other corner radiuses are shown above.

Shank width A is subject to a tolerance of 0.000 to 0.010 in. for 1-in. shanks or less and 0.000 to 0.015 in. for shanks over 1 in.

TOOLPOSTS

Nomenclature for Toolposts for Lathes, Turret Lathes, Boring Mills, Planers, and Shapers

A **toolholder** is a device for holding a cutting tool in a definite position with respect to the toolslide of a machine tool. Some holders are provided with a means for adjusting the tool point with respect to the center line of the work, while others require the use of shims or parallels.

The **single-screw toolpost** as used in most lathes (Fig. 17), shapers (Fig. 18), and some light planers consists of a screw, a round post, and a base.

The **toolpost screw** is fitted in the post. It allows for tool shanks of various sizes and furnishes the means for clamping.

The **lathe toolpost** carries the screw at one end and an integral circular flange at the other. This flange is recessed in the bottom of a T-nut fitted into the T-slot of the lathe tool rest (Fig. 17).

The **lathe toolpost collar** is usually a circular ring fitted about the base of the post. It is sometimes a ring with diametrally spaced steps on the upper face, on which the tool rests. The steps furnish a means for vertical adjustment of the tool point.

The **toolpost ring collar** also may consist of a ring provided with a circular seat on the upper face, into which is fitted a rocker seat on which the tool rests.

The **rocker base** (Fig. 17) is that part which fits into the circular seat of the toolpost collar. It supports the tool and furnishes a means of adjusting the height of the tool point. [Shims are also used for this purpose (Fig. 20).]

The **shaper toolpost circular flange** (Fig. 18) is recessed into the back face of the clapper and passes through the clapper and a serrated-tool baseplate mounted on the face of the clapper, against which the tool base bears.

TABLE 1. SHANK DIMENSIONS OF TOOLHOLDERS

Turning Tools $W \times H \times L$	Cutoff and Side-cutting Tools $W \times H \times L$	Boring Tools $W \times H$	Knurling Tools $W \times H \times L$	Threading Tools $W \times H \times L$
$\frac{5}{16} \times \frac{1}{2} \times 4$				
$\frac{5}{16} \times \frac{1}{2} \times 4\frac{1}{2}$	$\frac{5}{16} \times \frac{3}{4} \times 4\frac{1}{2}$	$\frac{5}{16} \times \frac{3}{4}$	$\frac{5}{16} \times \frac{3}{4} \times 5$	$\frac{5}{16} \times \frac{3}{4} \times 5$
$\frac{3}{8} \times \frac{7}{8} \times 5$	$\frac{3}{8} \times \frac{7}{8} \times 5$	$\frac{3}{8} \times \frac{7}{8}$	$\frac{3}{8} \times \frac{7}{8} \times 5$	$\frac{3}{8} \times \frac{7}{8} \times 5$
$\frac{1}{2} \times 1\frac{1}{8} \times 5$	$\frac{1}{2} \times 1\frac{1}{8} \times 6$	$\frac{1}{2} \times 1\frac{1}{8}$	$\frac{1}{2} \times 1\frac{1}{8} \times 6$	$\frac{1}{2} \times 1\frac{1}{8} \times 6$
$\frac{5}{8} \times 1\frac{3}{8} \times 7$	$\frac{5}{8} \times 1\frac{3}{8} \times 7$	$\frac{5}{8} \times 1\frac{3}{8}$	$\frac{5}{8} \times 1\frac{3}{8} \times 7$	$\frac{5}{8} \times 1\frac{3}{8} \times 7$
$\frac{3}{4} \times 1\frac{5}{8} \times 8$	$\frac{3}{4} \times 1\frac{5}{8} \times 8$	$\frac{3}{4} \times 1\frac{5}{8}$		
$\frac{7}{8} \times 1\frac{3}{4} \times 9$	$\frac{7}{8} \times 1\frac{3}{4} \times 9$	$\frac{7}{8} \times 1\frac{3}{4}$	$\frac{7}{8} \times 1\frac{3}{4} \times 9$	
$1 \times 2 \times 11$		1×2		
$1\frac{1}{4} \times 2\frac{1}{4} \times 13$				

Carbide-tipped Toolholder $W \times H \times L$	Planer Tools $W \times H \times L$
$\frac{3}{8} \times \frac{15}{16} \times 6$	
$\frac{1}{2} \times 1\frac{1}{4} \times 7$	$\frac{1}{2} \times 1 \times 6$
$\frac{5}{8} \times 1\frac{1}{2} \times 8$	$\frac{5}{8} \times 1\frac{1}{4} \times 8\frac{1}{2}$
$\frac{3}{4} \times 1\frac{3}{4} \times 9$	$\frac{3}{4} \times 1\frac{1}{2} \times 10$
$\frac{7}{8} \times 1\frac{7}{8} \times 10$	
$1 \times 2\frac{1}{8} \times 12$	$1\frac{1}{8} \times 1\frac{3}{4} \times 13$
	$1\frac{3}{8} \times 2 \times 16$
	$1\frac{7}{8} \times 2\frac{1}{4} \times 19$
	$2\frac{1}{8} \times 2\frac{3}{4} \times 22$

A **screw plate** is frequently provided for use between the screw and tool shank to distribute the pressure and to prevent damage to the shank (Fig. 18).

Serrated wedges are sometimes used with square toolposts in screw machines or in open-side toolposts for vertical adjustment of the tools.

An **adjustable collar and nut** fitted about the base of a round toolpost is sometimes used in screw machines to provide the vertical adjustment of the tools.

The **open-side toolpost for lathes** (Fig. 19) carries one or more binding screws. It may be provided with either a rocker base or flat-tool baseplate.

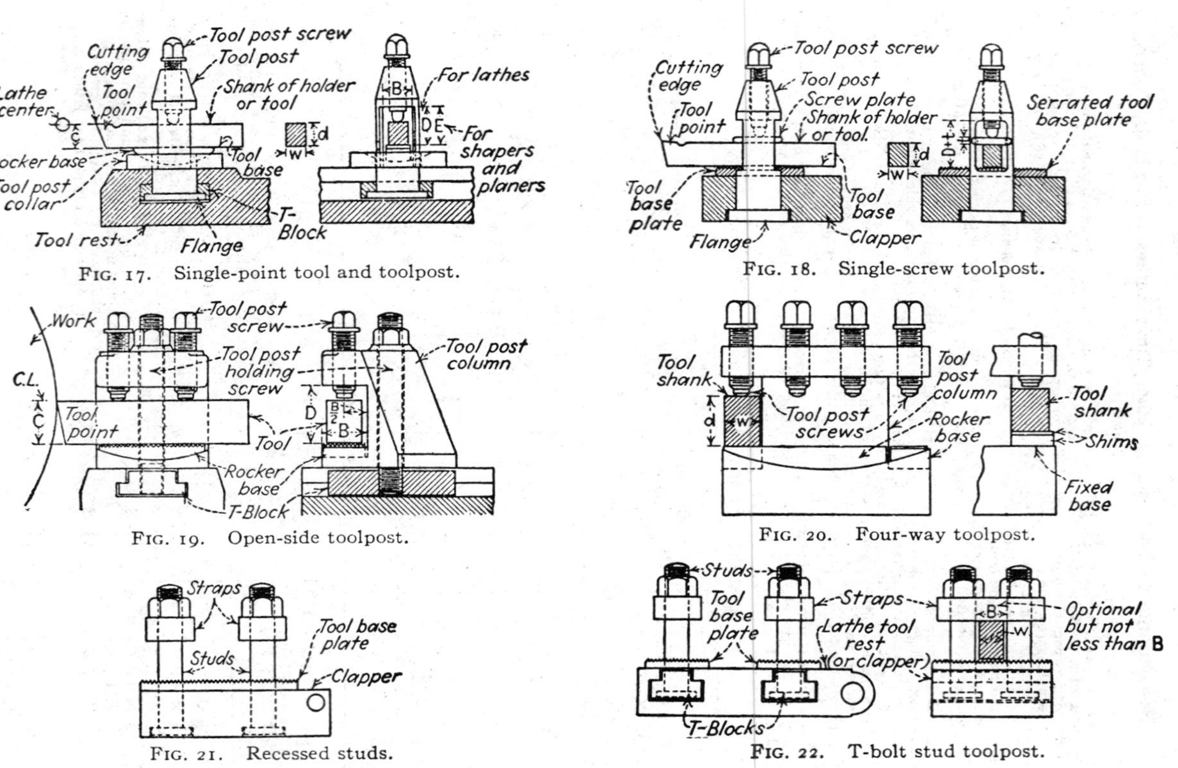

FIG. 17. Single-point tool and toolpost.

FIG. 18. Single-screw toolpost.

FIG. 19. Open-side toolpost.

FIG. 20. Four-way toolpost.

FIG. 21. Recessed studs.

FIG. 22. T-bolt stud toolpost.

TABLE 2. DIMENSIONS OF TOOL SHANKS, TOOLPOST OPENINGS, AND LATHE CENTER HEIGHT

Shank Section (Figs. 17, 18, 20, 21)			Lathe Opening (Figs. 17, 19, 20, 21)			Planer and Shaper Opening (Figs. 17, 18, 21)			Lathe Center Height (Figs. 17, 19)		
Max W	Nominal $W \times d$	Max d	Min B	Nominal $B \times D$	Min D	Min B	Nominal $B \times E$	Min E	Max C	Nominal C	Min C
0.40	$\frac{3}{8} \times \frac{3}{4}$*	0.85	0.49	$\frac{1}{2} \times 1\frac{3}{8}$	1.27	0.49	$\frac{1}{2} \times 1\frac{1}{16}$	1.04	0.93	$\frac{7}{8}$	0.85
0.48	$\frac{7}{16} \times \frac{7}{8}$	0.99	0.57	$\frac{9}{16} \times 1\frac{1}{2}$	1.48	0.57	$\frac{9}{16} \times 1\frac{1}{4}$	1.19	1.09	1	0.99
0.56	$\frac{1}{2} \times 1$*	1.15	0.68	$\frac{11}{16} \times 1\frac{3}{4}$	1.72	0.68	$\frac{11}{16} \times 1\frac{7}{16}$	1.38	1.26	$1\frac{3}{16}$	1.15
0.67	$\frac{5}{8} \times 1\frac{1}{4}$*	1.34	0.81	$\frac{13}{16} \times 2$	2.00	0.81	$\frac{13}{16} \times 1\frac{5}{8}$	1.61	1.47	$1\frac{3}{8}$	1.34
0.80	$\frac{3}{4} \times 1\frac{1}{2}$*	1.56	0.96	$1 \times 2\frac{3}{8}$	2.34	0.96	$1 \times 1\frac{7}{8}$	1.87	1.72	$1\frac{9}{16}$	1.56
0.95	$\frac{7}{8} \times 1\frac{3}{4}$†	1.81	1.14	$1\frac{3}{16} \times 2\frac{3}{4}$	2.71	1.14	$1\frac{3}{16} \times 2\frac{3}{16}$	2.17	1.99	$1\frac{13}{16}$	1.81
1.13	1×2*	2.11	1.35	$1\frac{3}{8} \times 3\frac{1}{8}$	3.16	1.35	$1\frac{3}{8} \times 2\frac{9}{16}$	2.53	2.32	$2\frac{1}{8}$	2.11
1.34	$1\frac{1}{4} \times 2\frac{1}{4}$†	2.43	1.61	$1\frac{5}{8} \times 3\frac{11}{16}$	3.65	1.61	$1\frac{5}{8} \times 3$	2.92	2.67	$2\frac{7}{16}$	2.43
1.60	$1\frac{1}{2} \times 2\frac{3}{4}$†	2.86	1.91	$1\frac{7}{8} \times 4\frac{1}{4}$	4.29	1.91	$1\frac{7}{8} \times 3\frac{7}{16}$	3.43	3.15	$2\frac{7}{8}$	2.86

* Size listed in Table 3.
† Size listed in Table 1.

TABLE 3. TOOL LENGTHS

Square	Length L	Rectangular	Length L
$\frac{3}{16}$	2	$\frac{1}{4} \times \frac{1}{2}$	4
$\frac{1}{4}$	$2\frac{1}{2}$	$\frac{5}{16} \times \frac{5}{8}$	$4\frac{1}{2}$
$\frac{5}{16}$	$2\frac{1}{2}$	$\frac{3}{8} \times \frac{3}{4}$*	5
$\frac{3}{8}$	3		
		$\frac{1}{2} \times \frac{3}{4}$	$5\frac{1}{4}$
$\frac{1}{2}$	4	$\frac{1}{2} \times 1$*	$5\frac{1}{2}$ or 7†
$\frac{5}{8}$	$4\frac{1}{2}$	$\frac{5}{8} \times 1\frac{1}{4}$*	$6\frac{1}{2}$ or 8†
$\frac{3}{4}$	6	$\frac{3}{4} \times 1\frac{1}{2}$*	$6\frac{1}{2}$ or 9†
1	7	$1 \times 1\frac{1}{2}$	10
$1\frac{1}{4}$	8	1×2*	12
$1\frac{1}{2}$	10	$1\frac{1}{2} \times 2$	

* These sizes are given in Table 2.
† Shorter lengths normally for turret lathes, boring mills, etc.

The **four-way turret toolpost** used on manufacturing lathes, turret lathes, and vertical boring mills is an indexing, multiple, open-side toolpost with fixed base or rocker (Fig. 20.)

The **strap and stud clamp type** of toolholder (Fig. 21) is generally fitted on large lathes, boring mills, shaper, slotters, and planers.

The **studs or headed bolts** are fixed to a T-nut fitted into the T-slot of the lathe tool rest or planer clapper, a distance apart indicated as B to accommodate the size of tool shanks to be used. Large lathes are provided with two sets of studs.

The **strap** is that member fitted over the studs bearing on the top of the tool shank. The tool rest is supported by the surface of the compound tool rest or on parallels used to provide the vertical adjustment of the tool.

Studs in a Planer. In a planer, the studs are attached (threaded or recessed) to the face of the clapper (Fig. 22) or in sliding T-nuts (Fig. 21) and pass through the serrated-tool baseplate against which the tool bears.

CUTTING-TOOL MATERIALS

Selection of Cutting-tool Materials

Six kinds of cutting materials are used for single-point tools. The choice depends on the operation, machinability of the material, speed and feed, condition of machine, finish, and dimensional accuracy requirements of the workpiece. The cutting materials are:

1. Carbon tool steels (0.90 to 1.30% carbon) can be sharpened to a keen cutting edge but lose hardness, and hence cutting ability, if the tool-tip temperature exceeds 400 to 500 F. For this reason, carbon tool steels are generally restricted to operations on brass.

2. Fast-finishing steels are similar to carbon tool steels but have small percentages of tungsten, chromium, or vanadium to increase wear resistance. However, they are also unable to maintain hardness at red heat, although used to machine chilled iron rolls at slow speeds.

3. Steels containing from 14 to 22% tungsten or 6 to 9% molybdenum plus 1.5 to 6% tungsten are called high-speed steels (HSS), because they do not lose their hardness when cutting at speeds sufficient to generate red heat. Cobalt is added to these steels to impart additional red hardness under heavy cuts where excessive heat is generated. High-speed steels are widely used throughout industry for machining all sorts of materials. For tool angles, speeds, and feeds, for a given material, refer to Machining Data for Various Materials, starting on page 12–80.

4. Cast nonferrous materials (Stellite, Rexalloy, and Tantung) are capable of withstanding cutting speeds 25 to 80% greater than the maximum for high-speed steels. These materials are often used for heavy and intermittent cuts on chilled iron castings but are, of course, used on all sorts of materials.

5. Cemented carbides enable attainment of the greatest metal removal per unit of time, because they can be used at extremely high speeds. Feeds are customarily lighter than with high-speed steels or cast-alloy tools. Straight tungsten carbides are used for machining cast iron, aluminum, nonferrous alloys, plastics, and fiber. The combined carbides, tungsten carbide plus titanium or tantulum carbide, or both, are suggested for machining all types of steels. But the manufacturer's recommendation is the safe guide to grade selection (see Tables 10 to 14).

6. Diamond tools compete to an extent with carbides where surface finish and dimensional accuracy requirements are high. But diamond tools are chipped by hard spots and poor handling. Metals, hard rubber, and plastics can be finish-turned and bored with diamond tools.

To summarize: cutting materials of widest usage are high-speed steels, cast alloys, and cemented carbides. Their proper utilization depends on the skill in application to the machine and job and the care they receive in the shop. Further details follow for the application and grinding of high-speed-steel, cast-alloy, and carbide tools.

Types of High-speed Steels

Twenty basic analyses of high-speed steels are recognized by industry. Eight steels in which tungsten is the major alloying element are designated by the letter T followed by a numeral.

Example: The most common high-speed steel of the tungsten series, commonly called 18-4-1, is identified by the symbol T-1. Twelve high-speed steels in which molybdenum is the major alloying element are designated by the letter M followed by a number.

Tungsten High-speed Steels, T

Chemical Composition, %				Analysis Symbol
Tungsten	Chromium	Vanadium	Cobalt	
18.00	4.00	1.00	.....	T-1
18.00	4.00	2.00	.....	T-2
18.00	4.00	3.25	.....	T-3
18.00	4.00	1.00	4.00	T-4
18.00	4.00	2.00	8.00	T-5
22.00	5.00	1.50	12.00	T-6
14.00	4.00	2.00	.....	T-7
14.00	4.00	2.00	5.00	T-8

Molybdenum High-speed Steels, M

Chemical Composition, %						Analysis Symbol
Molybdenum	Chromium	Vanadium	Tungsten	Cobalt	Boron	
8.00	4.00	1.00	1.50	.....	......	M-1
5.00	4.00	2.00	6.00	.....	......	M-2
5.00	4.00	3.00	6.00	.....	......	M-3
4.50	4.50	4.00	5.50	.....	......	M-4
5.00	4.50	1.50	4.00	12.00	......	M-6
8.00	4.00	2.00	....	.....	......	M-10
8.00	4.00	1.00	....	2.50	Added	M-20
8.00	4.00	1.00	1.50	4.00	......	M-30
8.00	4.00	1.00	2.00	5.00	......	M-32
8.50	4.00	2.00	2.00	8.00	......	M-35
6.00	4.00	2.00	6.00	8.00	......	M-36
8.00	4.00	1.50	....	8.00	Added	M-40

Grinding of High-speed-steel Tools

Proper sharpening of standard tools produces a surface condition that promotes a substantial increase in number of pieces per grind, better surface finish and accuracy of the work, and a reduction in setup and sharpening time. To secure the desired keenness of the cutting edge, grind first with an aluminum oxide wheel of 46 to 60 grit, and follow with a 320-grit resinoid- or shellac-bonded wheel. The second step should remove only 0.0005 to 0.001 in. of material. Or hand honing can be employed if the cutter grinder is not in the best of condition. However, an experienced operator is required for honing of tools, and he should use a hard Arkansas or medium India stone. Many more pieces per grind are obtained if honing is properly done.

Grinding is preferably done wet, provided that the coolant floods the work.

TABLE 4. CUTTING ANGLES FOR HIGH-SPEED-STEEL TOOLS*

Material	Side Relief Angle†	Front Relief Angle	Back-rake Angle‡	Side-rake Angle
High-speed steels (annealed)	10	8	8	12
Tool steels (annealed)—alloy	10	8	8	12
Tool steels (annealed)—high-carbon	10	8	8	12
Stainless steels	10	8	10	15–20
SAE steels:				
1020	12	8	16	14
X1020	12	8	16	14
1035	12	8	16	14
1040	12	8	16	14
1045	10	8	12	14
1095	10	8	8	12
1112	12	8	16	22
X1112	12	8	16	22
1120	12	8	16	18
X1314	12	8	16	22
X1315	12	8	16	22
T1335	10	8	12	14
X1335	12	8	16	18
2315	10	8	12	14
2320	10	8	12	14
2330	10	8	12	14
2335 (annealed)	10	8	12	14
2340 (annealed)	10	8	10	12
2345 (annealed)	10	8	10	12
2350 (annealed)	10	8	10	12
3115	10	8	12	14
3120	10	8	12	14
3130	10	8	10	12
3135 (annealed)	10	8	10	12
3140 (annealed)	10	8	10	12
3250 (annealed)	10	8	8	12
4140	10	8	12	14
4340	10	8	8	12
6140 (annealed)	10	8	10	12
6145 (annealed)	10	8	8	12
Aluminum	12	8	35	15
Bakelite and pressure-molded plastics	12	8	0	0
Brass:				
Free-cutting	10	8	0	1 to 5
Red and yellow	10	8	0	0 to −4
Bronze:				
Cast and commercial	10	8	0	0 to −4
Free-cutting	10	8	0	0 to 5
Hard	10	8	0	0 to −2
Phosphor	10	12	10	0 to −2
Cast iron—gray	10	8	5	12
Celluloid and cast plastics	14	10	0 to −5	0
Copper	14	12	16	20
Copper alloys:				
Harder	10	8	0	0 to −2
Softer	10	8	0	0 to −4
Fiber	15	12	0	0
Formica gear material	15	10	16½	10
Nickel iron	15	13	8	14
Micarta	15	10	16½	10
Monel metal	15	13	8	14
Nickel	15	13	8	14
Nickel silvers	10	12	10	0 to −2
Rubber—hard	20	15	0 to −5	0 to −7

* Allegheny Ludlum Steel Corporation.

† A front relief of 8° and a side relief of 10 to 12° are fairly standard for hand-ground tools. In some cases smaller relief angles can be employed for production jobs. Front relief of 4 to 5° is suggested for shaper and planer tools set vertical.

‡ In general, where negative side- and back-rake angles are mentioned, these are to be used when evidence of hogging is manifested.

TABLE 5. GRINDING PRACTICE FOR CAST-ALLOY TOOLS*

Operation	Abrasive	Grain	Grade	Structure	Bond	Wheel Speed, Sfpm
Offhand:						
Roughing	Aluminum oxide	60	L	6	Vit.	3800–4200
Finishing	Aluminum oxide	60	I	6	Vit.	3800–4200
Machine:						
Roughing	Aluminum oxide	60	I	8	Vit.	3800–4200
Finishing	Aluminum oxide	150	I	6	Vit.	3800–4200
Forming	Aluminum oxide	150	I	6	Vit.	3800–4200
Crush forming	Aluminum oxide	220	J	13	Vit.	3800–4200†
Cutting-off	....	30	A	5	Rubber or resinoid	11,000

* Data for Tantung produced by the Vascoloy-Ramet Corporation.
† Use liberal flow of No. 6 oil.

TABLE 6. CUTTING ANGLES FOR CAST-ALLOY TURNING TOOLS*

Material to Be Cut	Back-rake Angle	Side-rake Angle	Side-cutting-edge Angle	End-cutting-edge Angle	Relief Angles (Side and End)
Cast iron	0	5	0–15	10	5–6
Malleable iron	6–8	8–12	8–10	15	6
Steel:					
Cast	8	8	10	10	6
Soft	15	15	8–15	15	7
Medium	10	10	8–10	15	6
Hard	6–8	6	8	15	6
Stainless steel:					
Soft	15	15	10	15	7
Medium	10	10	10	15	7
Hard	8	8	10	15	7
Aluminum	10–20	12–15	10	10	7–8
Brass	4	4	10	10	5–6
Bronze	4	4	10	10	5–6

* Data for Tantung produced by Vascoloy-Ramet Corporation.

TABLE 7. CUTTING SPEEDS AND DEPTH OF CUT WITH CAST-ALLOY TOOLS*
Feed—0.025 to 0.035 ipr

Material Cut	Depth of Cut, In.					
	0.004–1/32	1/32–3/32	1/8–7/32	1/4–3/8	3/8–1/2	1/2–5/8
	Cutting Speed, Sfpm					
Cast iron:						
Soft	190	175	165	150	125	100
Medium	150	140	135	120	100	75
Malleable iron	250	225	200	175	140	100
Steel cast:						
Soft	190	175	160	150	125	100
Medium	150	135	120	110	100	70
Steel:						
SAE 1010-1040	250	220	180	150	125	100
SAE 1045-1070	175	160	140	120	100	75
SAE 1080-1095	155	140	120	100	75	60
Free-cutting	300	265	225	180	145	125
Manganese	130	120	100	80	65	50
SAE 2015-2515	150	140	125	100	80	60
SAE 3115-3450	135	125	115	100	75	50
SAE 4130-4820	150	135	125	100	80	60
SAE 6120-6145	120	115	100	85	75	60
SAE 52100	125	115	100	85	75	60
Tool steel	125	115	100	80	75	50
Stainless steel, free-cutting	225	200	180	150	130	100
Aluminum	1250	1100	950	800	650	550
Brass	500	450	400	350	275	200
Bronze:						
Soft	200	175	150	125	100	75
Hard	150	140	125	100	80	60

NOTE: When cuts are intermittent, reduce speed 25 to 50%. When coolant is used, speed may be increased 25% or more.

* Data for Tantung produced by Vascoloy-Ramet Corporation.

ALTERNATE GRINDING PROCEDURE FOR HIGH-SPEED-STEEL TOOLS

1. Rough-grind side and end faces.
2. Semifinish-grind nose radius, side, and end faces.
3. Grind chip breaker on surface grinder.
4. Superfinish chip breaker on surface grinder.
5. Superfinish nose radius, side, and end faces.

Grinding Cast-alloy Tools. The nonferrous cast alloys require no tempering after grinding and should not be quenched in water. Use a light pressure when moving the tool across the wheel face, to prevent localized overheating and cracking at the cutting edge. Steps in reconditioning tools are:

1. If the tool is tipped, grind back the steel shank to an angle of 10 to 15°.
2. Rough-grind top rake, side, and front clearance angles.
3. Finish-grind top rake, side, and front clearance angles.
4. Hand-hone the cutting edge. Note that frequent honing of the cutting edge during use, without changing the setup, tends to increase the life of the tool between grinds and thus saves grinding and setup time and tool material. Suggestions for suitable grinding wheels are given in Table 5.

Cast-alloy tools can be ground either wet or dry. If ground wet, a generous flow of coolant should be directed at low velocity to the point of grinding. It is safer to grind dry if an abundant supply of coolant is not available.

Carbide-tool Recommendations

Types of Carbide Tools. Cemented-carbide tools are available in several forms: brazed toolbits and tools, mechanically clamped tools, and solid-insert toolholders.

Brazed tools are used in the greatest quantity, because they conveniently replace tools and toolbits made of other cutting materials. They may be purchased completely fabricated in the standard sizes listed under Single-point Tools, page 12–12, or made up in the shop, by brazing a standard insert into a suitably recessed shank. The recess must be flat, but not too smooth. Satisfactory shank materials are SAE 2340, SAE 9250, any straight 0.50 to 0.95% carbon steel, or Meehanite cast to shape. It is good practice to mill the shank with required clearance, rake, and cutting-edge angles before applying the tip, in order to reduce grinding afterward.

Carbide inserts (flat) can be clamped to a recessed toolholder (Fig. 23). This type of construction is often used for hogging cuts on lathes and planers. It is said that less potential chance of tool damage is likely, because no brazing strains are present in the carbide.

Solid inserts—round, triangular, and square in cross-section—are coming into wide use for production tooling. The inserts are held in standard toolholders devised for the purpose. Advantages are: setup is undisturbed while new cutting positions of tool are selected (there are three to six positions on each end of insert), minimum grinding is required, and maximum amount of carbide is obtained per tool dollar.

Design of Carbide Tools. SIDE- AND END-CUTTING-EDGE ANGLES. Cemented carbides, as with the cast-alloy tool materials, are more likely to fail by chipping and breaking than the high-speed steels. Their edge strength is dependent upon the support given to the cutting edge. If possible, the side-cutting-edge angle should be

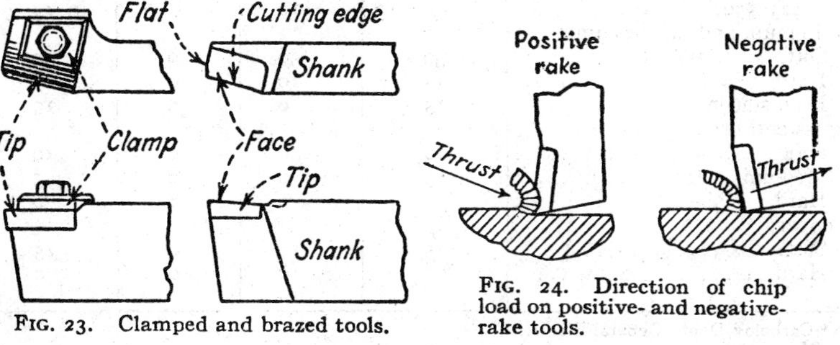

FIG. 23. Clamped and brazed tools.

FIG. 24. Direction of chip load on positive- and negative-rake tools.

great enough to allow the starting load to be taken at a point back of the nose, which is the weakest part of the tool. For most jobs a side-cutting-edge angle of 15° will be satisfactory, but irregular workpieces may require an angle of 20 to 45°.

The end-cutting-edge angle, on the other hand, should be just great enough to avoid dragging on the work. For most turning and facing work, an end-cutting-edge angle of 8 to 15° will be found suitable.

SIDE AND BACK RAKES. Positive side and back rakes are suggested for most materials (see Table 8). Interrupted cuts and scaly cuts on steels require negative rakes. Side-rake and back-rake angles depend on several factors: material being cut and its hardness and cutting characteristics, rigidity of tool mounting, and machine condition. On side-cutting tools, the side rake is the cutting rake and the back rake serves as the control angle. Conversely, on end-cutting tools, the back rake is the cutting rake and side rake may not be required.

NEGATIVE RAKES. Current usage of negative rakes requires explanation of the reasons for using them. A negative back rake protects the cutting edge. With positive-rake tools, chips are formed by shearing the metal ahead of the tool. Thrust is directed against the cutting edge (Fig. 24); but, with negative rake, the thrust is directed back into the body of the carbide tip.

TABLE 8. RAKE ANGLES FOR CARBIDE TOOLS*
(In Degrees)

Material	Side-cutting Tools		End-cutting Tools	
	Side Rake	Back Rake	Side Rake	Back Rake
Steel or cast iron:				
Continuous cut, clean metal, brinell:				
100–200	15	0	..	10
200–325	8	0	..	8
325–425	3 to 5	0	..	5
425–550	0	0	..	0
Interrupted cut and/or scale, brinell:				
100–200	0	0	..	0
200–325	−3	−3	..	−5
325–425	−5	−5	..	−8
425–550	−8 to −10	−10	..	−10
Aluminum and magnesium:				
Soft	30	30	0	30
Hard	15	10	0	15
High silicon	15	0	0	10
Brass and bronze:				
Soft	10	0	0	10
Medium	5	0	0	5
Hard	0	0	0	0
Copper:				
Soft	15	15	0	15
Hard	0	0	0	0

* Carboloy Dept., General Electric Co.

Use of negative rake permits harder, but more brittle, carbides to be used to resist abrasive chip flow over the tool face. And this gain in abrasion resistance allows faster cutting speeds to be used. Tool loading is automatically reduced with increase in speed, and a fast-flowing thin chip is produced. Eventually, of course, the tool face will be cratered by chip abrasion and must be reground. However, cratering action is retarded by the use of the correct grade of carbide.

The effect of back rake on various types of cuts can be seen in Fig. 25. At *A*, the side-cutting-edge angle protects the nose by having the first contact with the work about the center of the carbide tip. The other illustrations show different conditions depending on the shape of the tool and of the work as it makes contact. In plain turning, as at *A* and *B*, some consider the zero rake desirable, while others prefer negative rake in nearly all cases where the cut is very heavy.

Illustrations *C*, *D*, and *E* show different contacts with interrupted cuts in which the angle of contact varies with the work itself. Here an angle of 3° is shown. The main object is to secure a shear cut beginning behind the cutting point. This has been very successful in machining armor plate. One difficulty in turning armor plate is the tendency to work harden so that tools should be as free cutting as possible.

Nose radius on a negative-rake tool should be of the same proportions as on a positive-rake tool, and it should be set on or slightly above the center line of the work.

Production runs show that negative-rake tools used on rough turning will give longer life between grinds if the cutting edge and nose are honed to a 45° chamfer about 0.005 to 0.010 in. wide. Honing removes slight irregularities on the cutting edge invisible to the eye and avoids a potential source of breakdown. Dubbing off the cutting edges of finishing tools is likewise of advantage on interrupted cuts.

More heat is generated with negative-rake turning than with conventional turning

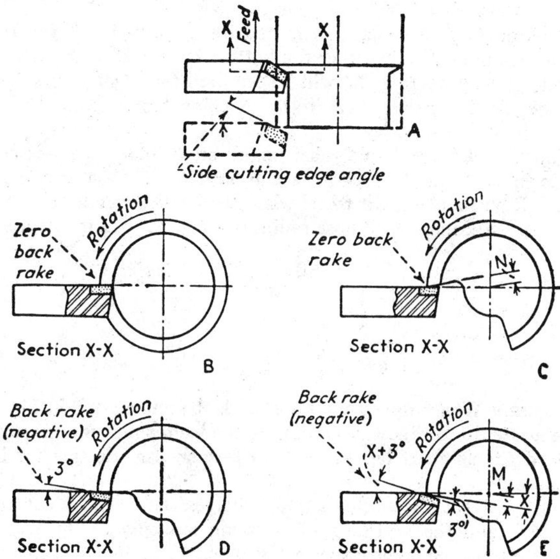

FIG. 25. Zero- and negative-rake turning tools.

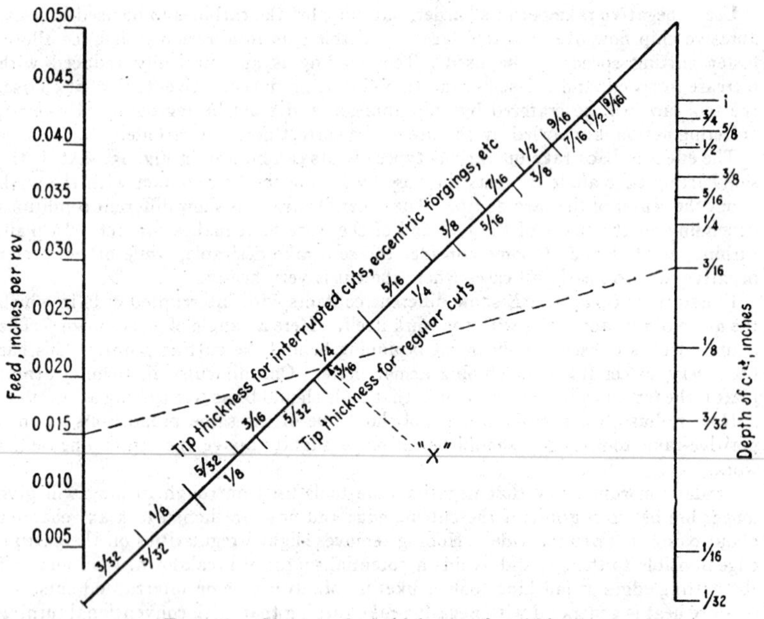

FIG. 26. Chart to find thickness of carbide tip.

practice. Consequently, a coolant is more often required. A generous flow of coolant should be directed from either the bottom or one side.

More power, perhaps 10 to 15%, will be required for negative-rake turning, but no difficulty should be encountered if the machine tool is suitable for use with carbides.

RELIEF ANGLES. End and side relief angles on carbide tools are generally 7°, except that down to 5° may be used in case of hard steel. A clearance of 10° keeps the steel shank from touching the wheel when the tip is ground.

NOSE RADIUS. Recommended nose radiuses are related to the depth of cut:

Depth of Cut, In.	*Nose Radius*
To $\frac{1}{8}$	$\frac{1}{32}$
$\frac{3}{16}$–$\frac{3}{8}$	$\frac{3}{64}$
$\frac{7}{16}$–$\frac{3}{4}$	$\frac{1}{16}$
$\frac{13}{16}$–$1\frac{1}{4}$	$\frac{3}{32}$

TIP THICKNESS. Where tipped tools are used, the nomograph in Fig. 26 enables one to choose a safe tip thickness, according to the load, which is a function of depth of cut and feed. This chart has scales for tip-thickness selection for interrupted and continuous cuts.

SHANK SIZE. Correct selection of a tool involves the depth of cut and feed (or load on the tool) and the overhang (or bending moment). The required shank size is obtainable from the nomograph in Fig. 27 by the following procedure.

Assume a depth of cut of $\frac{1}{8}$ in., a feed of 0.015 in., and an overhang of $1\frac{1}{2}$ in. Con-

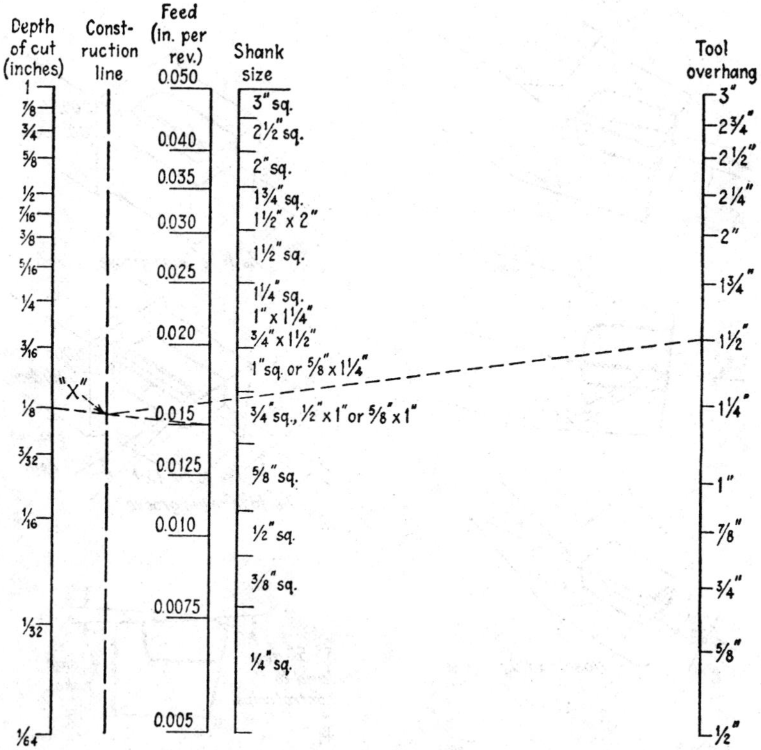

FIG. 27. Chart to find shank size of carbide tool.

nect these values on the appropriate scales. Then from the intersection with the construction line, run another line to $1\frac{1}{2}$-in. overhang. The shank-size scale is cut in the block for a $\frac{3}{4}$ in. sq, $\frac{1}{2}$- or $\frac{5}{8}$- by 1-in. shank.

Chip Breakers for Carbide Tools.[1] There are four basic methods of creating an obstruction to chip flow on single-point, carbide-tipped tools. One of these methods is essential to control the shape and the length of the chip, particularly when machining steels, and some of the bronzes and the aluminum alloys. A chip breaker is not required when machining the "short-chip" materials, such as cast iron, brass, and nonmetallics.

GROUND-IN STEP TYPE. This is the type of chip breaker most widely used.

Angular Type. Forces chips to curl away from tool point and clear the shoulder of the cut. Chip characteristics are controlled by angle, depth, radius, and width of step. Angle, as at *A* (Fig. 28), is usually 8°; depth between 0.015 and 0.020 in.; radius should not exceed depth; width is varied to control chip curl—narrow for tight curls; wider for loose curls and heavy feeds. For finishing cuts, a chip breaker, *B*, may be ground at 45° across the tool nose and $\frac{1}{16}$ in. wide. If nose radius exceeds

[1] Carboloy Dept., General Electric Co.

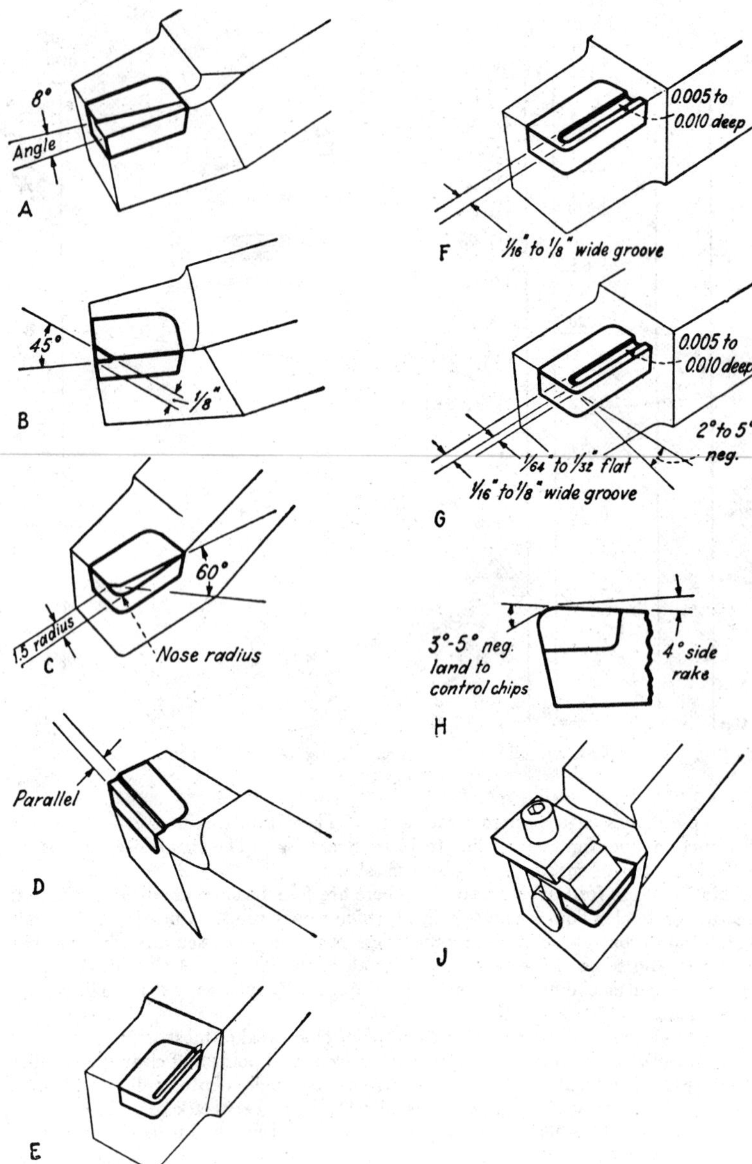

FIG. 28. Various types of chip breakers are ground into carbide-tool tips.

TABLE 9. CHIP-BREAKER DIMENSIONS

Depth of Cut (Inches)	Feed—in. per rev.				
	.008–.012	.013–.017	.018–.022	.023–.027	.028–.032
	Width of Chip Breaker				
1/64–3/64	1/16	5/64	3/32	7/64	1/8
1/16–1/4	3/32	1/8	5/32	11/64	3/16
5/16–1/2	1/8	5/32	3/16	13/64	7/32
9/16–3/4	5/32	3/16	7/32	15/64	1/4

the width, a second step, C, at 60° to the cutting edge is ground, and of a length equal to $1\frac{1}{2}$ times radius.

Parallel Type. When stock is eccentric, producing irregular depth of cut, a parallel ground-in step, as at D, is most effective.

GROUND-IN GROOVE TYPE. This type of chip breaker is extensively used for roller turner, or box tools, because the well-coiled chip is unlikely to fall between the roller and the finished work surface.

A shallow groove, as at E, is ground parallel to the cutting edge and often stops short of the end.

Ordinarily the groove, as at F, is 0.005 to 0.010 in. deep by $\frac{1}{16}$ to $\frac{1}{8}$ in. wide. The flat, as at G, between the cutting edge and the groove, is 0.015 to 0.030 in. wide, and a negative side rake of 2 to 5° is used.

NEGATIVE-RAKE-ANGLE TYPE. Chip control can be obtained with negative-rake angles ranging from 3 to 5°, as at H.

MECHANICAL CHIP BREAKERS. A clamped-on chip breaker J is feasible if the tool is large. Precision fit of the contact surfaces is required to prevent entrance of metal particles between the chip breaker and tool.

Cutting Speeds for Carbides. One set of recommendations, chosen from up-to-date reliable sources, is ordinarily given in this book. See Notes on Machining Various Materials, page 12–80. But the turning of materials with carbides has assumed such importance that the editors feel justified in supplementing these data with the information compiled by the Carboloy Dept., General Electric Co. (see Tables 10 to 13).

The Carboloy recommendations are more specific than is usually the case for data derived from industry. One valuable feature is the inclusion of recommendations for chip cross-sections (depth of cut and feed), instead of the less definite common expressions "roughing" and "finishing"; for the finishing cut of one industry may be the roughing cut of another industry.

To use Tables 10 to 13, find the material, then read the cutting speed for the selected depth of cut and feed. Note that lightface type indicates the maximum and minimum cutting speeds and that often a recommended grade of Carboloy is also given in lightface type. These entries should be used in connection with production runs where it is feasible to make some adjustments in order to obtain

TABLE 10. FEEDS AND SPEEDS FOR CUTTING FERROUS CASTINGS WITH SINTERED CARBIDES*

	MATERIAL TO BE CUT	POWER CONSTANT	7/16" to 3/4" Cut .020" to .030" Feed		3/16" to 3/8" Cut .015" to .025" Feed		5/64" to 5/32" Cut .010" to .020" Feed		1/64" to 1/16" Cut .008" to .012" Feed		.005" to 1/64" Cut .002" to .008" Feed	
			Speed F.P.M.	Carbide Grade	Speed F.P.M.	Carbide Grade	Speed F.P.M.	Carbide Grade	Speed F.P.M.	Carbide Grade	Speed F.P.M.	Carbide Grade
CAST IRON	Hard (No Alloy)	4	125-225 **175**	**44A**	150-250 **200**	**44A** 883	175-275 **225**	**883**	200-300 **250**	**883** 905	225-325 **275**	**999**
	Medium (No Alloy)	3	150-250 **200**	**44A**	175-275 **225**	**44A** 883	200-300 **250**	**883**	224-325 **275**	**883** 905	250-350 **300**	**999**
	Soft (No Alloy)	3	175-225 **225**	**44A**	200-300 **250**	**44A** 883	225-325 **275**	**883**	250-350 **300**	**883** 905	275-375 **325**	**999**
	Hard (Alloy)	4	100-200 **150**	**44A**	**175**	**44A** 883	**200**	**883** 907	**225**	**883** 905	**250**	**999**
	Medium (Alloy)	3	**175**	**44A**	**200**	**44A** 883	**225**	**883** 907	**250**	**883** 905	**275**	**999**
	Soft (Alloy)	3	**200**	**44A**	**225**	**44A** 883	**250**	**883** 907	**275**	**883** 905	**300**	**999**
	Brake Drums Heat Treated	4			**225**	**44A** 883	**250**	**883** 907	**275**	**883** 905	**300**	**999**
	Brake Drums Centrifugal	4			**200**	**44A** 883	**225**	**883** 907	**250**	**883** 905	**275**	**999**
	Chilled Rolls	5	10-45 **15**	**44A**	10-45 **15**	**44A** 883	15-50 **20**	**44A** 883	15-50 **20**	**883**	20-60 **25**	**883** 905
SEMI-STEEL	Up to 25% Semi-Steel	3	150-250 **200**	**44A**	**225**	**44A** 883	**250**	**883** 907	**275**	**883** 905	**300**	**999**
	Over 25% Semi-Steel	3	100-200 **150**	**44A**	**175**	**44A** 883	**200**	**883** 907	**225**	**883** 905	**250**	**999**
MALLEABLE IRON	Hard	5	**150**	**44A** 78B	**175**	**44A** 883	**200**	**883** 907	**225**	**883** 907	**250**	**883** 907
	Medium	4	**175**	**44A** 78B	**200**	**44A** 883	**225**	**883** 907	**250**	**883** 907	**275**	**883** 907
	Soft	3	**200**	**44A** 78B	**225**	**44A** 883	**250**	**883** 907	**275**	**883** 907	**300**	**883** 907

* Carboloy Dept., General Electric Co.

TABLE II. FEEDS AND SPEEDS FOR CUTTING NONFERROUS AND NONMETALLIC MATERIALS WITH SINTERED CARBIDES*

MATERIAL TO BE CUT		POWER CONSTANT	7/16" to 3/4" Cut .020" to .030" Feed		3/16" to 3/8" Cut .015" to .025" Feed		5/64" to 5/32" Cut .010" to .020" Feed		1/64" to 1/16" Cut .008" to .012" Feed		.005" to 1/64" Cut .002" to .008" Feed	
			Speed F.P.M.	Carbide Grade	Speed F.P.M.	Carbide Grade	Speed F.P.M.	Carbide Grade	Speed F.P.M.	Carbide Grade	Speed F.P.M.	Carbide Grade
BRASS AND BRONZE	Hard	10	100-250 **175**	**44A**	125-275 **200**	**44A** 883	150-350 **250**	**883**	200-400 **300**	**883** 905	300-600 **400**	**883** 905
	Soft	4	150-300 **225**	**44A**	175-325 **250**	**44A** 883	200-400 **300**	**883**	300-500 **400**	**883** 905	350-1000 **500**	**883** 905
ALUMINUM	Castings	3	175-325 **250**	**44A**	200-500 **300**	**44A** 883	250-600 **350**	**883** 905	300-1000 **450**	**883** 905	400-1500 **600**	**883** 905
	Bar Stock	4	150-300 **225**	**44A**	200-500 **275**	**44A** 883	250-600 **325**	**883** 905	300-1000 **400**	**883** 905	400-1500 **500**	**883** 905
ZINC ALLOY	Die Castings	3	....	....	200-500 **300**	**44A** 883	250-500 **350**	**883** 905	300-600 **450**	**883** 905	400-1000 **600**	**883** 905
RUBBER	Hard		....	....	250-450 **350**	**883**	300-500 **400**	**883** 905	350-600 **475**	**883** 905	400-800 **600**	**905** 999
	Soft		....	....	300-600 **450**	**883**	400-800 **550**	**883** 905	500-1000 **650**	**883** 905	600-1200 **800**	**905** 999
COPPER		4	100-300 **200**	**44A** 883	150-350 **250**	**44A** 883	200-500 **350**	**883** 905	300-600 **450**	**883** 905	400-1000 **600**	**883** 905
COMMUTATORS		4	....	....	....	....	....	....	100-300 **200**	**44A** 883	200-600 **300**	**883**
FIBRE			....	....	200-400 **300**	**44A** 883	300-500 **400**	**883** 905	300-800 **500**	**883** 905	350-1000 **600**	**883** 905
PLASTICS			....	....	250-500 **350**	**44A** 883	300-800 **450**	**883** 905	400-1000 **600**	**883** 905	500-1500 **800**	**883** 905
MONEL METAL		12	....	....	175-275 **225**	**78B** 44A	200-275 **250**	**78B** 907	200-300 **250**	**78B** 78	225-325 **275**	**78B** 78

* Carboloy Dept., General Electric Co

TABLE 12. FEEDS AND SPEEDS FOR CUTTING STEEL WITH SINTERED CARBIDES—AVERAGE WORK*

Material	To be cut	Power constant	7/16" to 3/4" Cut .020" to .030" Feed		3/16" to 3/8" Cut .015" to .025" Feed		5/64" to 5/32" Cut .010" to .020" Feed		1/64" to 1/16" Cut .008" to .012" Feed		.005" to 1/64" Cut .002" to .008" Feed	
			Speed F.P.M.	Carbide Grade	Speed F.P.M.	Carbide Grade	Speed F.P.M.	Carbide Grade	Speed F.P.M.	Carbide Grade	Speed F.P.M.	Carbide Grade
CARBON	S.A.E. 1010-1025	6	175-325 **250**	**78B**	200-400 **300**	**78B** 78	275-475 **375**	**78**	375-625 **500**	**78** 831	500-1500 **675**	**831**
CARBON	S.A.E. 1030-1095	8	125-275 **200**	**78B**	150-350 **250**	**78B** 78	200-400 **300**	**78**	300-500 **400**	**78** 831	400-1000 **550**	**831**
FREE CUTTING	S.A.E. 1112-1120	6	175-325 **250**	**78B**	200-400 **300**	**78B** 78	275-475 **375**	**78**	375-625 **500**	**78** 831	500-1500 **675**	**831**
FREE CUTTING	S.A.E. X1314-X1340	6	150-300 **225**	**78B**	175-375 **275**	**78B** 78	250-450 **350**	**78**	350-550 **450**	**78** 831	400-1200 **600**	**831**
Mn.	S.A.E. T1330-T1350	9	100-250 **175**	**78B**	125-300 **200**	**78B** 78	150-350 **250**	**78**	250-450 **350**	**78** 831	300-800 **475**	**831**
NICKEL	S.A.E. 2015-2320	7	150-300 **225**	**78B**	175-375 **275**	**78B** 78	250-450 **350**	**78**	350-550 **450**	**78** 831	400-1200 **600**	**831**
NICKEL	S.A.E. 2330-2515	9	125-275 **200**	**78B**	150-350 **250**	**78B** 78	200-400 **300**	**78**	300-500 **400**	**78** 831	400-1000 **550**	**831**
NICKEL-CHROME	S.A.E. 3115-3140	8	125-275 **200**	**78B**	150-350 **250**	**78B** 78	200-400 **300**	**78**	300-500 **400**	**78** 831	400-1000 **550**	**831**
NICKEL-CHROME	S.A.E. 3145-3450	9	100-250 **175**	**78B**	125-300 **200**	**78B** 78	150-350 **250**	**78**	250-450 **350**	**78** 831	300-800 **475**	**831**
Mo.	S.A.E. 4130-4820	9	100-250 **175**	**78B**	125-300 **200**	**78B** 78	150-350 **250**	**78**	250-450 **350**	**78** 831	300-800 **475**	**831**
Cr.	S.A.E. 5120-52100	10	100-250 **175**	**78B**	125-300 **200**	**78B** 78	150-350 **250**	**78**	250-450 **350**	**78** 831	300-800 **475**	**831**
Cr. V.	S.A.E. 6115-6195	10	100-250 **175**	**78B**	125-300 **200**	**78B** 78	150-350 **250**	**78**	250-450 **350**	**78** 831	300-800 **475**	**831**
	S.A.E. Cast Steel	9	125-275 **200**	**78B**	150-350 **250**	**78B** 78	200-400 **300**	**78**	300-500 **400**	**78**	300-800 **550**	**831**

* Carboloy Dept., General Electric Co.

TABLE 13. FEEDS AND SPEEDS FOR CUTTING STAINLESS STEELS WITH SINTERED CARBIDES*

MATERIAL TO BE CUT	Power Constant	7/16" to 3/4" Cut .020" to .090" Feed		1/8" to 3/8" Cut .015" to .040" Feed		1/16" to 1/8" Cut .010" to .030" Feed		1/64" to 1/16" Cut .007" to .020" Feed		.007" to 1/64" Cut .004" to .012" Feed	
		Speed F.P.M.	Carbide Grade	Speed F.P.M.	Carbide Grade	Speed F.P.M.	Carbide Grade	Speed F.P.M.	Carbide Grade	Speed F.P.M.	Carbide Grade
*A.I.S.I. Type 302	11	140-180 **160**	**78B**	140-210 **175**	**78B**	140-240 **190**	**78**	150-275 **215**	**78**	150-300 **225**	**905**
A.I.S.I. Type 303	10	150-250 **200**	**78B** 78	150-275 **215**	**78B** 78	175-325 **225**	**78B** 78	200-375 **285**	**78**	200-400 **300**	**831**
A.I.S.I. Type 304	11	140-180 **160**	**78B**	140-210 **175**	**78B** 78	140-240 **190**	**78**	150-275 **215**	**78**	150-300 **225**	**905**
A.I.S.I. Type 309	11	140-180 **160**	**78B**	140-210 **175**	**78B** 78	140-240 **190**	**78**	150-275 **215**	**78**	150-300 **225**	**905**
A.I.S.I. Type 316	11	140-180 **160**	**78B**	140-210 **175**	**78B** 78	140-240 **190**	**78**	150-275 **215**	**78**	150-300 **225**	**905**
A.I.S.I. Type 321	11	140-180 **160**	**78B**	140-210 **175**	**78B** 78	140-240 **190**	**78**	150-275 **215**	**78**	150-300 **225**	**905**
A.I.S.I. Type 347	11	140-180 **160**	**78B**	140-210 **175**	**78B** 78	140-240 **190**	**78**	150-275 **215**	**78**	150-300 **225**	**905**
A.I.S.I. Type 403	10	150-200 **175**	**78B** 78	150-250 **200**	**78B** 78	175-300 **235**	**78B** 78	200-350 **275**	**78**	200-400 **300**	**831**
A.I.S.I. Type 410	10	150-200 **175**	**78B** 78	150-250 **200**	**78B** 78	175-300 **235**	**78B** 78	200-350 **275**	**78**	200-400 **300**	**831**
A.I.S.I. Type 416	10	150-200 **175**	**78B** 78	150-250 **200**	**78B** 78	175-300 **235**	**78B** 78	200-350 **275**	**78**	200-400 **300**	**831**
A.I.S.I. Type 420	12	100-150 **125**	**78B**	100-175 **140**	**78B** 78	125-200 **165**	**78**	150-225 **185**	**78**	150-250 **200**	**905**
A.I.S.I. Type 420F	12	100-150 **125**	**78B** 78	100-175 **140**	**78B** 78	125-200 **165**	**78B** 78	150-225 **185**	**78**	150-250 **200**	**831**
A.I.S.I. Type 430	10	150-200 **175**	**78B** 78	150-250 **200**	**78B** 78	175-300 **235**	**78B** 78	200-350 **275**	**78**	200-400 **300**	**831**
A.I.S.I. Type 430F	10	150-200 **175**	**78B** 78	150-250 **200**	**78B** 78	175-300 **235**	**78B** 78	200-350 **275**	**78**	200-400 **300**	**831**
A.I.S.I. Type 431	11	140-180 **160**	**78B**	140-215 **175**	**78B** 78	140-250 **195**	**78**	150-350 **250**	**78**	150-350 **250**	**905**
A.I.S.I. Type 440A, B, C	12	100-150 **125**	**78B**	100-175 **140**	**78B** 78	125-200 **165**	**78**	150-225 **185**	**78**	150-250 **200**	**905**
A.I.S.I. Type 440F	10	100-150 **125**	**78B**	125-150 **135**	**78B** 78	125-175 **150**	**78**	150-250 **200**	**78**	150-200 **175**	**905**
A.I.S.I. Type 446	11	140-180 **160**	**78B**	140-215 **175**	**78B** 78	140-250 **195**	**78**	150-350 **185**	**78**	150-350 **250**	**905**

* Carboloy Dept., General Electric Co.

the greatest cutting efficiency. Entries in boldface type give the safe cutting speed under average conditions and the appropriate grade of Carboloy.

NOTE: If the shop also uses cemented carbides made by other manufacturers refer to Table 14, which compares grades from the several sources.

Horsepower Requirements. The horsepower needed for the cut is found by picking the constant for the material from Table 10 and substituting values in this formula:

$$\text{Hp} = D \times F \times S \times C$$

where D = depth of cut, in.; F = feed per revolution, in.; S = sfpm; and C = power constant.

Add 30% to the calculated horsepower to allow for machine friction and dull tools.

NOTE: There is some evidence that the power required when turning steel tends to drop off when speeds of 1000 sfpm or over are used.

Carbides on Older Machines. It is stated by authorities that there is no reason why older types of machines in good condition cannot be adapted readily to the use of carbide tooling. The main consideration in old equipment, as well as in the new types of machine tools, is that the machine must be able to run fast enough—and smoothly enough at that faster speed.

The main objective to keep in mind in cutting steels with carbides is that the cutting speed be high enough to prevent the forming of a "built-up" edge. This means an average cutting speed in the neighborhood of 200 sfpm (the lower the carbon, usually, the higher the speed).

CHECKING OLD MACHINES. The adaptability of an available tool to the use of carbides can be checked by the following considerations:

It takes more power to run at the higher speeds required—to remove metal at a faster rate. It takes more power, also, to cut steel than to cut nonferrous metals or cast iron. Horsepower requirements may be calculated by the above formula.

Then belts, clutches, etc., should be checked for ability to transmit the horsepower to the spindle. Clutch fingers should be adjusted to prevent slipping and stalling. When the machine is equipped with a flat belt, it may be changed to a V-belt drive, making sure that the number of belts is adequate.

NOTE: If the machine stalls in the cut, loosen the holding screws and remove the tool from the cut to prevent breakage. Do not attempt to move the work or try to back the tool out of the cut.

The high speeds make it advisable to use an antifriction tailstock center. Spindles should be checked for adequate lubrication at the higher speeds at which they will operate. Solid supports should be used under the tools in place of rocker tool plates, and shims should be used to keep the tool at proper cutting height.

To handle the increased volume of chips, where openings in machine bases and around tool blocks are small, sheet-metal chutes are useful to prevent chips from clogging slots and pockets; and chip breakers can be used where size of openings requires production of small chips.

Excessive tool clearances that might cause chatter at the higher cutting speeds and worn bearings, slides, and ways should be corrected. Where it is impractical to tighten the machine up sufficiently to eliminate all chatter, a certain amount can be corrected by the use of negative rakes in the tools.

Coolants for Carbide Tools. Usually, increased cutting fluid capacity and flow are required when machining with carbide-tipped tools, because of the very high speeds at which machines can and should be operated when cutting with such tools. Not only should the pump have sufficient capacity to supply large volumes of the fluid under sufficient pressure to cool the tool and work adequately, but the cutting

TABLE 14. CARBIDE MANUFACTURER'S GRADE RECOMMENDATIONS

Applications*	Adamas	Carboloy	Carmet	Firthite	Kennametal	Talide	Vascoloy-Ramet	Wesson	Willey
C-1	B	44A	CA3	H	K6	C89	2A68	GS	E8
C-2	A	883	CA4	HA	K6	C91	2A5-2A8	GI	E6
C-3	AA	905	CA7	HF	K8	C93	2A7	GA	E5
C-4	AA	999	CA8	HF	K8	C93	2A7	GF	E3
C-5	D	78C	CA5	T-04, T-89	K2S	S88	EE	WS	945
C-6	D	78B	CA1	TA, T-89	K2S	S90	EM	WM	710
C-7	C	78	CA2	T-16	K3H	S92	E	WH	606
C-8	CC	831	CA6	T-31	K5H	S92	EH	WH	509
C-9	A	883	CA4	HA	K8	C89	2A68 1WR	GI	E8
C-10	B	44A	CA3	H	K6	C88	2A3 2WR	GS	E12
C-11	HD-20	55B	CA10	HC	K1	C8515	2A16 3WR	M	E18
C-12	RDB	55A	CA10	DC-1, DC-2	K1	C8515	2A3 AW	GS	E12
C-13	HD-20	55B	CA11	DCX, DC-3	K18	C8020	2A16 AX	M	E18
C-14	HD-25	190	CA20	DC-4	K25	C7525	2A20 AXAY	M	E25

NOTE: This chart presents the manufacturer's recommendations for carbides for the uses indicated, and it is not intended as a grade comparison.

* Chip-removal applications:
- C-1 Roughing cuts—cast iron and nonferrous materials
- C-2 General-purpose—cast iron and nonferrous materials
- C-3 Light finishing—cast iron and nonferrous materials
- C-4 Precision boring—cast iron and nonferrous materials
- C-5 Roughing cuts—steel
- C-6 General-purpose—steel
- C-7 Finishing cuts—steel
- C-8 Precision boring—steel

Wear applications:
- C-9 Wear surface—no shock
- C-10 Wear surface—light shock
- C-11 Wear surface—heavy shock

Impact applications:
- C-12 Impact—light
- C-13 Impact—medium
- C-14 Impact—heavy

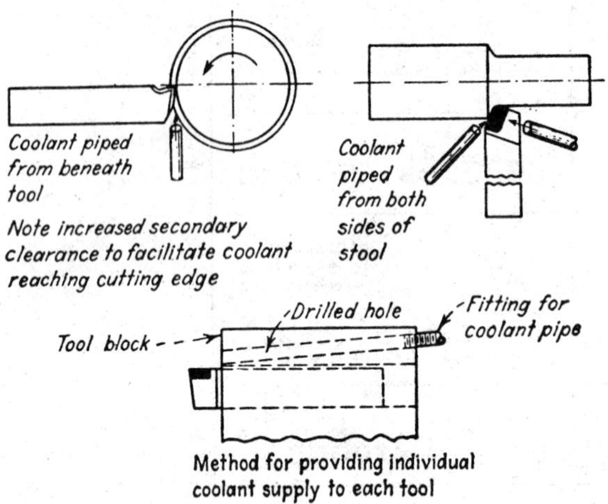

FIG. 29. Coolants are preferably directed at carbide tip by methods shown.

fluid must be so directed that it will not be carried away from the tool by the fast-moving chip. Several ways of directing the fluid are shown in Fig. 29.

The prime requisite of a cutting fluid for use with carbide-tipped tools is cooling quality. A good soluble oil works well. Cutting oils are used but in some cases are objectionable because of the smoke developed. It is reported that straight cutting oils are not good coolants at speeds in excess of 200 sfpm; however, where chip pressures are high and lubricating qualities of oils are desirable, large volumes can be used to enhance cooling.

From 3 to 5 gal of cutting liquid per minute should be supplied to each single-point tool. Some report that sulfurized oils are detrimental to carbide tools. On complicated machines such as automatics and gear cutters a rich emulsion, using only 10 parts of water instead of the usual 20 to 40 parts, is recommended.

Trouble-shooting with Carbides

Chipped Cutting Edge

When the cutting edge of a carbide-tipped tool chips while under cut, check each of the following:

1. If a heavy cut is being taken, particularly on a rough casting or forging, honing of the cutting edge helps prevent chipping. Scale and heavy chips are likely to flake off a sharp cutting edge. By slightly rounding or chamfering the edge with a silicon carbide or diamond hand hone, you give it sufficient strength to stand up under severe use.

2. Check the width of the chip breaker. If the chip is too tight, it may exert excessive pressure on the tool and cause chipping. See recommendations for width of chip breaker.

3. On heavy castings, an increase in feed may break up the scale ahead of the tool—and prevent chipping.

4. On steel-cutting jobs, slow speeds result in the formation of a built-up edge, which causes chipping. Try a higher speed.

5. Alternate heating and quenching of the tip, caused by interruptions in the flow of coolant, will cause chipping. Make sure flow is sufficient and that it is directed so the chip does not prevent the coolant from reaching the tip at all times.

6. If none of the above suggestions correct the conditon, try a different grade of carbide for the tip—a tougher grade may give better results.

Chatter

The following points should be checked if the tool chatters:

1. Excessive overhang will cause chatter. Make sure tool is well supported.
2. Next, check the work support. If the work vibrates, chatter will result.
3. If the nose radius is too large, the tool is likely to chatter.
4. Check the feed. Too light a feed may result in the tool rubbing instead of cutting. This will cause chatter.
5. Is the chip breaker too narrow? A chip that is too tight will cause chatter.
6. Is the tool sharp? On many materials, dull tools will chatter.
7. Check the end-cutting-edge angle. If this angle is too small, the tool will drag—causing chatter.
8. Vary the side-cutting-edge angle—this changes the direction of the load on the tool and may eliminate the condition.
9. If the cut is spread over a long length of cutting edge, chatter frequently results. This is particularly true with grooving tools. It often is better to use two tools—a narrow tool, followed by a wider one. If the bottom of the groove does not need to be flat, grind the cutting edge to break up the cut.

Worn Cutting Edge

When the cutting edge wears rapidily, try the following steps:

1. First, examine the tool with a magnifying glass to make sure the trouble is not minute chipping instead of wear. If chipping is indicated, correct as suggested above.
2. If the tool is actually wearing, check the feed—light feeds tend to wear the cutting edge more rapidly.
3. Excessive speeds cause rapid wear—try a slower speed.
4. Check the tool relief angles—if they are too small, the heel of the tool may be dragging.
5. Check the nose radius—too large a radius accelerates wear. See recommendations for size of nose radius.
6. If none of the above suggestions helps, check the grade of carbide used. Perhaps a harder grade should be employed.

Poor Breakup of Chips

If the chip does not break properly when cutting steel, the trouble probably lies in the chip breaker. Try the following:

1. Widen the chip breaker if the chip is too tight.
2. Narrow the chip breaker, by grinding along the cutting edge, if the chip is too loose.
3. Check the nose radius. Too large a radius distorts the chip and makes it hard to control.
4. Check the coolant supply. A heavy flow of coolant at high velocity makes it easier to break the chip.

5. Check the depth of the chip breaker—a shallow chip breaker results in a loose chip, while a chip breaker that is too deep will curl the chip tightly and may lead to chipped cutting edges or broken tips.

Poor Surface Finish on Work

If the finish produced on the work is not satisfactory, check the following:

1. If the finish is rough and torn—increase the cutting speed to minimize or eliminate the built-up edge.

2. If the finish is marred with small flakes of metal—try a smaller nose radius on the tool.

Tool Brazing. Before applying a cemented-carbide tip to a recessed shank, remove all grease and dirt from the shank with carbon tetrachloride. Tips should likewise be cleaned, but they need not be ground. Flux the recess, the Easy-Flo No. 3, Tobin bronze, copper, or sandwich braze material, both sides of tip, and assemble after fluxing. Flux top of tip liberally, and braze with a nonoxidizing green-tip oxyacetylene flame. Heat the shank first, then out to tool tip, and withdraw the flame when the tip and the shank end are a bright cherry red.

Torch brazing, as described, is the most common method, and after some experience an operator can produce satisfactory tools. Other brazing methods are: induction heating, furnace brazing, and use of special gas burners.

If carbide tips are $\frac{3}{4}$ in. long or more, some users advocate a sandwich braze, which consists of silver foil on both sides of a copper or constantan core. The core is said to have a cushioning effect to absorb brazing strains. Annealing of fabricated tools tipped with a steel-cutting grade of carbide containing titanium carbide is recommended by some manufacturers to remove brazing strains.

Grinding of Carbide Single-point Tools. Because of extreme hardness, cemented carbides cannot be ground with the same wheels as high-speed steel and cast-alloy tool materials. Two types of wheels are required: aluminum oxide for grinding clearance on the steel shank and silicon carbide for grinding the cemented-carbide tip.

Offhand Grinding. Sharpening of carbide tools by the offhand method (Fig. 30) is not recommended for production tools and should be done only by trained operators possessing adequate grinding and checking equipment.

Clearance on the steel shank is ground with an A24-L6-V straight wheel having the face crowned or straight as the work demands.

Roughing of the carbide tip is done on a silicon carbide C60-18-V straight wheel with the face crowned about $\frac{1}{16}$ in. The crown reduces the contact between wheel and carbide, provides faster and cooler cutting action, and promotes grinding a straight cutting edge.

Finish grinding of the carbide tip may be done with a silicon carbide cup wheel C (80 or 100)-H8-V, which has been dressed to a $\frac{1}{32}$-in. crown. This step is often disregarded in favor of finishing with diamond wheels.

To produce the high-class finish which lengthens the life of carbide tools, supplement grinding with silicon carbide wheels, as above, with a diamond wheel having a flat face (Fig. 31). An average finish is secured with a 100- to 120-grit wheel; a fine finish with a 220- to 240-grit wheel. Lapping of the carbide may be required and is done with a 320- to 400-grit wheel or a cast-iron lapping disk charged with No. 4 diamond dust. When diamond wheels become loaded, apply pumice stone or a silicon carbide stick to the revolving wheel.

Wheel speeds between 4500 and 5500 fpm are used for both abrasive and diamond wheels.

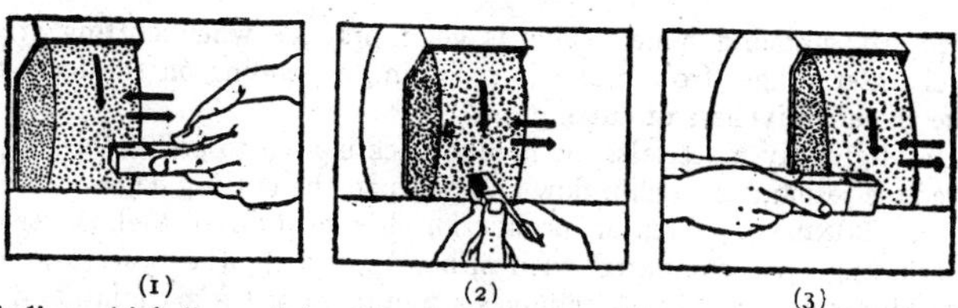

Rough grinding—(1) hollow grind top face, (2) front clearance, and (3) side clearance, all on periphery of a crowned straight wheel.

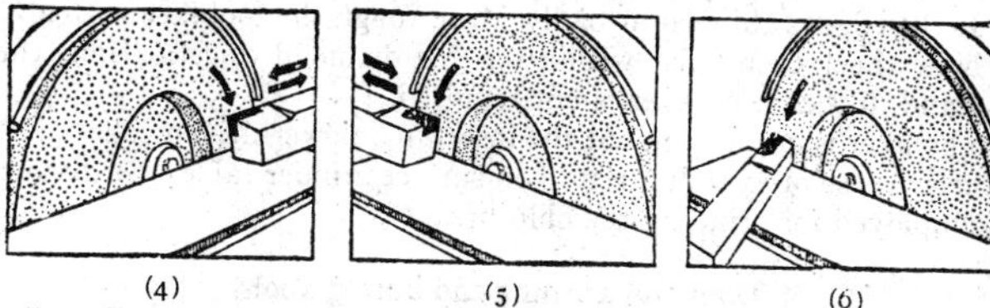

Finish grinding—Using crowned face of cup wheel, (4) grind top face of toolbit tip, (5) grind side clearance, and (6) grind front clearance.

FIG. 30. Offhand grinding of carbide-tipped tools requires six steps.

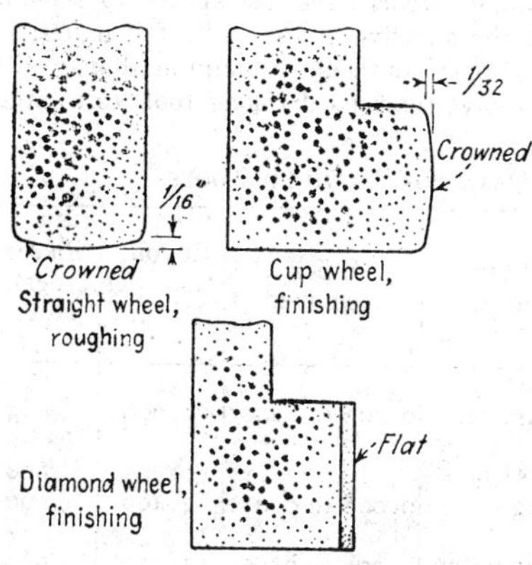

FIG. 31. Straight and cup wheels for roughing and finishing carbide-tipped tools should be crowned as shown in order to get straight cutting edges. The diamond finishing wheel must have a flat face.

General pointers in grinding carbide tools are:

1. Keep the tool constantly in motion during grinding.
2. Grind the top face first, then the front, then the side.
3. Grind against the cutting edges, that is, from tip to shank.
4. Dry grinding is preferred, because tool tip is visible and chances of unequal temperature rise are lessened. A good exhaust system is desirable. Wet grinding is permissible if a full flow of coolant is maintained at all times.

Chip breakers are ground in a surface grinder.

HONING CARBIDE TOOLS. If examined under a microscope, a sharp carbide edge will be seen to be irregular, because of the nature of the material. In service, minute particles of carbide will break out of the matrix because of cutting strain. To avoid this condition and lengthen the effective cutting life, the cutting edge is chamfered with a 320-grit diamond hone. This is good practice when cutting steel. The width of chamfer ranges from 0.002 to 0.005 in., depending on the feed, and does not reduce the effectiveness of cutting.

On soft and gummy materials, the honing stick is used to polish the top face and relief to reduce resistance to chip flow and sharpen the cutting edge.

MACHINE GRINDING. Greater production life of brazed tools is secured by machine grinding. As compared with offhand grinding, the process gives better finish, reproduces angles exactly, removes a minimum of tip material, and is faster.

Insert tools of round, triangular, or square cross-section are sharpened by dressing the ends square with the longitudinal axis. This operation is done on a carbide grinder. If the edge is chipped or badly worn, rough the tool face with a C60-T8-V silicon carbide wheel, then finish with a 100-grit diamond wheel. Hone the edge to a 0.005-in. land, if a chip breaker is not required.

If a chip breaker is needed, mount the insert in a power-driven collet of a machine built for the purpose or in collet fixture on surface grinder table. A diamond wheel is usually employed for grinding the chip breaker.

Diamond Turning and Boring Tools

By HARRY STRAUSS, JR., *National Diamond Hone & Wheel Co.*

Parts produced on precision turning and boring machines must be finished with a light cut at extremely high speeds. Diamond tools have the ability to remain sharp over long runs, because they resist the abrasive action of "soft" materials like bearing metals, plastics, commutator copper, and some aluminum alloys. On these applications, cemented-carbide tools have replaced diamond tools to a large

TABLE 15. PERFORMANCE DATA FOR DIAMOND TOOLS

Operation	Material	Work Size in Inches	Stock Removed	Revolutions per Minute	Pieces per Hour
Turning roll....	Bearing bronze	2½ diameter × 18	0.003 to 0.005	1,000	3
Recessing......	Catalin	3/64 × 31/32	....	6,500	600
Turning........	Lynite (aluminum alloy)	4½ long	0.008 to 0.010	2,400	96
Boring (finish)..	Bronze	15/16 diameter	0.008 to 0.010	3,500	160
Boring (finish)..	Aluminum	1 × 1	0.008 to 0.010	3,400	150
Turning........	Copper	2 diameter × 1	0.005 to 0.008	2,000	60
Turning........	Aluminum	3½ long	0.008 to 0.010	1,100	52
Facing.........	Hard rubber	3¼ diameter	0.020	3,600	110
Reaming.......	Hard rubber	3/16 diameter × 15/16	0.010	3,600	1,800
Reaming.......	Celluloid	3/16 diameter × 15/16	0.010	3,600	700
Turning........	Celluloid	7/16 × 5/8	0.008 to 0.010	3,600	1,000
Turning rod....	Bakelite	1¼ diameter × 8¾ long	0.008	3,500	30

Data from Arthur A. Crafts & Co.

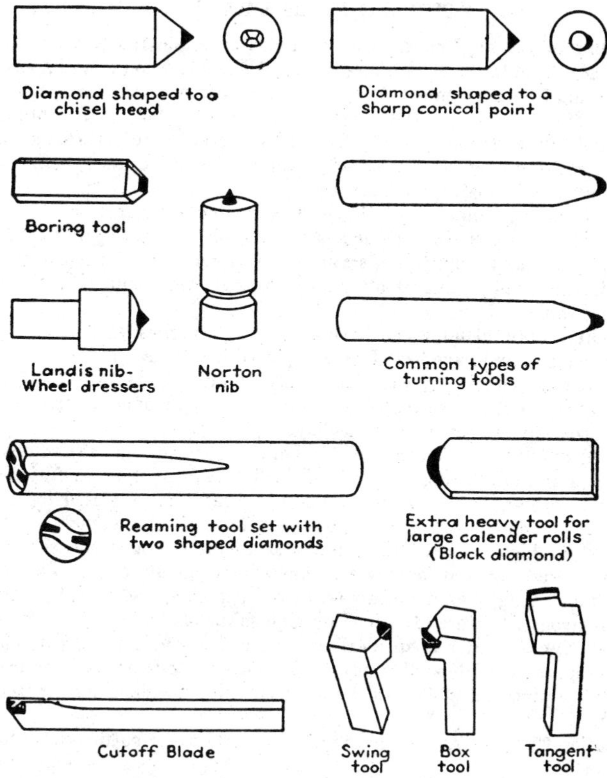

FIG. 32. Diamond-tipped tools are still used for a variety of purposes.

extent, especially where impact may be involved, but even so a good percentage of precision boring and turning jobs is still done with diamond tools.

Depth of cut in most diamond-tool applications is 0.010 in. max, and the general range is 0.002 to 0.004 in. Feed is normally 0.0015 to 0.003 in. per revolution. Cutting speed is high, being limited in many cases by the machine and the dynamic balance of the part.

Excellent finish plus maintenance of geometrical accuracy of the product can be secured with diamond tools. Finishes of 30 μin. have been obtained on aluminum pistons, and down to 2 μin. in special setups in machining copper printing shells.

Tool angles are customarily akin to those used with cemented carbides, except that positive top rake cannot exceed 10°, and front rake and side clearance are somewhat less.

Some actual case examples of diamond-tool results are shown in Table 15. They are not unusual cases but were taken at random from many shop studies, made recently under average shop conditions and with usual equipment. Figure 32 shows a number of tools with diamond tips.

CUTTING-FLUID SELECTION

Functions of Cutting Fluids. Cutting fluids have four functions: cooling, lubrication, rust prevention, and flushing chips. Chip formation is an important gage of cutting-fluid requirements. A segmental, or discontinuous, chip indicates that the principal requirement of the cutting fluid is cooling ability. A continuous chip without built-up edge indicates low friction on the tool face, whereas a continuous chip with built-up edge shows that high sliding friction occurs. Both lubrication and cooling are essential in the latter case.

Types of Cutting Fluids. General types of cutting fluids are:

STRAIGHT MINERAL OILS. These oils are suitable for light operations on certain steels and brass and for difficult operations on white metals. They combine cooling and lubricating qualities but are often mixed with a base cutting oil with still higher lubricating value.

LARD OIL. Straight fatty oils have excellent properties, except for a tendency toward rancidity, and have been largely replaced by mineral lard oils.

MINERAL LARD OILS. A blend of fatty oil and mineral oil is often used, especially on automatics, to obtain better finish than with straight mineral oils. This material is noncorrosive to copper and its alloys.

SULFURIZED OILS. Sulfur additions to mineral oils and mineral lard oils increase their cooling and lubricating qualities. Film strength is high. Sulfurized mineral oil and sulfurized-base oils which include some fatty oil are used for machining straight carbon and alloy steels, stainless steels, and high-nickel alloys. Soft stringy steels require a high sulfur content; harder and more brittle steels need less sulfur. Broaching, threading, and tapping are done with high-sulfur oils. Low-sulfur oils are used for drilling, reaming, shaping, turning, milling, and hobbing. Brass and other nonferrous alloys are blackened by high-sulfur oils.

SOLUBLE OILS. Water is an ideal coolant, but to avoid rusting it is mixed in various proportions with soluble oils. "All-purpose" oils now on the market are mixed with water and even used as a coolant and machine lubricant on screw machines.

Application Chart. Table 16 gives recommendations for cutting fluids to be used

NOTES FOR TABLE 16

METALS DESIGNATIONS. Group 1: (steel, 70–100 machinability). Group 2: (steel, 55–70 machinability). Group 3: (steel, 35–55 machinability). Group 4: (cast iron). Group 5: (copper-base alloys, 80–100 machinability). Group 6: (copper-base alloys, 50–70 machinability). Group 7: (copper-base alloys, machinability below 50). Group 8: (aluminum and magnesium).

KEY TO FLUIDS RECOMMENDATION:

A—Straight mineral oil
B—Mineral lard oil, 10% lard oil
C—Mineral lard oil, 40% lard oil
D-1 and D-2—Transparent active sulfur oil, in ascending viscosity and compounding
E-1, E-2, and E-3—Dark active sulfur oil viscosity range 110 to 250 sec Saybolt at 100 F, mild types
G-1 and G-2—Dark, active sulfur oils, heavy duty, compounded with lard oils
H—Dark active sulfur oil, heavily compounded, for severest operations
I—Noncorrosive transparent oil, equivalent to lard oil of 40 to 100% concentration
J-1, J-2, and J-3—Noncorrosive transparent oil, with petroleum-base substitute for lard oil, viscosity and compounding increasing
K-1 and K-2—Thread grinding oil, 160 and 300 sec Saybolt at 100 F viscosity, noncorrosive and transparent
L—Soluble oil, grinding type (ratios indicated)
M—Soluble oil, cutting type (ratios indicated)
N—For aluminum and magnesium, transparent and odorless soluble oil

NOTE: If in the machining of copper-base alloys, sulfur-base oils are required, stains on parts should be removed by soaking them in a 5 to 10 % solution of sodium cyanide. Alternate recommendations are given in the table, and selection will be based finally on severity of operation and performance tests.

TABLE 16. CUTTING-FLUID APPLICATIONS

Operation	Group 1	Group 2	Group 3	Group 4	Group 5	Group 6	Group 7	Group 8
Broaching (roughing) up to 1-in. broach	20:1 M, E-3, B, J-2	15:1 M, E-3, B, J-2	10:1 M, H, E-1, D	10:1 M	15:1 L, A, N, B	15:1 L, A, N, C	10:1 L, B, C, I	N
Broaching (roughing) over 1-in. broach	15:1 M, E-2, B, J-2	15:1 M, E-1, E-2, B, J-3	10:1 M, G-1, G-2, D-1, I	10:1 M	15:1 L, A, N, B, I	15:1 L, A, B	10:1 L, C, J-1, I	N
Broaching (finishing)	E-2, B, J-3	I, G-1, D-1	I, G-2, G-1, D-2	10:1 L, N	15:1 L, A, N, B	15:1 L, A, B, J-2, I	10:1 L, C, J-2, I D-1	N, J-1
Broaching, heavy-duty	G-2, H, D-2	G-2, H	G-2, H	10:1 M, N, J-2	15:1 L, A, B, D-2	15:1 L, A, B, D-2, I	10:1 L, C, J	N, J-1
Die threading—tapping	E-2, E-1, C, J	E-1, G-1, D, J-1	G-2, D-2, I	10:1 M, N	15:1 L, A, N, C, J	10:1 L, A, C, J-2	10:1 L, C, J-2, I	N, J-1
Pipe threading, automatic tapping machines	E-1, G-2, D-2	G-1, G-2, D-2	G-2, D-2	....	15:1 L, A, N, C, J	10:1 L, A, C, J-2	10:1 L, C, J-1, I	N, J-1
Gear shaving	G-2, G-1, C, I	G-2, G-1, C, N	G-2, G-1					
Gear cutting	E-3, C, J	E-2, B, C, I, J	G-1, E-2, D-1, C, B, I	....	A, N, B, J	A, N, B, J	J-2, I, C, D	N
Reaming	15:1 M, E-2, E-1, B	10:1 M, E-2, B, J-1, G-1	10:1, G-2, D-1, C, I	20:1 M	15:1 L, A, N, J-1	15:1 L, A, B, J-1	10:1 L, A, B, C, J-2	N
Drilling	20:1 M, E-3, B, J-1	15:1 M, E-2, B, J-1	10:1 M, G-1, D-1, J-2	20:1 M	15:1 L, A, N, J-1	15:1 L, A, B, J-1	10:1 L, J-2, I, J-1	N
Hobbing	....	....	....	20:1 M	....	....	....	N
Boring	20:1 M, E-3, B, J-1	15:1 M, E-2, B, J-1	10:1 M, G-1, D-1, J-2	20:1 M	15:1 L, A, N, J-1	15:1 L, A, B, J-1	10:1 L, J-2, I, J-1	N
Milling	20:1 M, E-3, B, J-2	15:1 M, E-2, B, J-2	10:1 M, C-2, D-2, J-2	20:1 M	15:1 L, A, N, J-2	15:1 L, A, B, J-2	10:1 L, J-2, I, J-1	N
Shaping	30:1 M, E-3, B, J-1	30:1 M, E-3, B, J-2	30:1 M, E-3, J-1, I	20:1 M	15:1 L, A, N, J-1	15:1 L, A, N, J-1	10:1 L, J-1, I, D-1	N
Turning	30:1 M, E-3, B, J-1	30:1 M, E-3, B, J-2	30:1 M, E-3, J-1, I	20:1 M	15:1 L, A, N, J-1	15:1 L, A, N, J-1	10:1 L, J-1, I, D-1	N
Cold sawing, high-speed	30:1 M, A, E-3	30:1 M, A, E-3	20:1 M, A, E-3	....	20:1 L, A, N	20:1 L, A, N	20:1 L, A, N, J-1	A
Cold sawing, low-speed	10:1 M, A, E-3, BJ-1	10:1 M, A, E-2, B, J-1	10:1 M, A, E-2, B, J-1	20:1 M	20:1 L, A, N	20:1 L, A, N	20:1 L, A, J-1, J-2	N
Grinding—plain	50:1 L, A	50:1 L, A	50:1 L, A	30:1 L, A	50:1 L, N	50:1 L, N	50:1 L, N	N
Grinding—form thread, etc.	K-1, K-2	K-1, K-2	K-1, K-2	N	K-1, J-2	K-1, J-2	K-1, J-2	N, J-2, K-2

with various machining operations on eight groups of materials, broken down according to machinability and nature of the metals. To use this chart, refer to Table 24 on machinability ratings of ferrous metals, page 12–71.

Notes on Practice. Cutting fluids that have worked well in practice are also given below.

Cast iron is usually worked dry, but when hard cast-iron gears are to be cut, as with three cutters, the first cut through will work better with strong soda water. It makes an objectionable mess, but the work will be done faster and the cutters keep sharper longer than with the dry process of cutting.

Brass and babbitt are usually cut dry, but to hand-ream brass and babbitt is sometimes difficult if the reamer is a little dull. Kerosene and turpentine are used with good results. Cast iron can be hand-reamed easily with tallow and graphite, mixed, and the hole will be kept just the size of the reamer. Copper can be worked well with lard oil and turpentine mixed.

In boring babbit bushings and rod boxes in a lathe or boring mill, it is very difficult to work the material dry as the chips have a great tendency to roll around the tool and into a hard ball, tearing the metal and making a rough ragged hole. In this case kerosene and lard oil mixed will work well.

Turpentine is good in some cases where fitting is done, such as scraping layout plates or faceplates. Oil will form a coating so that marks cannot be seen plainly, but turpentine will prove beneficial on this kind of work if used freely. The marks can be seen plainly, and the work is a great deal easier to scrape than with an oil surface, as the oil glazes over the surface and makes it hard to start the scraper cutting.

DIAGNOSIS OF MACHINING TROUBLES

The best approach to overcoming cutting-tool troubles, according to the Gorham Tool Co., consists of an on-the-job analysis of the reasons for tool failure or poor finish. When the type of trouble has been determined, steps can be taken to remedy the conditions responsible for the failure. Often more than one of the troubles (page 12-51) will appear at the same time, but careful study of the actual operation of the single-point flat or circular forming tool will show how to correct the various troubles as they are isolated.

TURRET-LATHE TOOLING

Boring-tool Design for Turret Lathes

For Cast Iron and Bronze. Tools *A* and *B* (Fig. 33) are alike, except for end-cutting-edge angles of 23 and 15°, and secondary front clearance angles of 18° and 10°. These tools, designed for use in holes of different diameter, prevent dragging, as tabulated. Also, tool *A* is used in a bar arranged to bore holes that do not go clear through the work.

For Steel. Tools *C* and *D* (Fig. 33) have zero back rake and a chip breaker 0.005 in. deep, and are otherwise the same as tools *A* and *B*. The combination of zero back rake and a chip breaker has been found to produce satisfactory boring of steel and proper control of chips. Further, these features reduce the grinding time and also extend tool life, because a minimum of carbide is removed.

Calculation of Boring Cuts on Turret Lathes

The chart (Fig. 36) developed by the methods department, Cleveland Automatic Machine Co., enables the rate setter to determine the number of boring cuts to

Diagnosis of Machining Troubles

Probable Cause	Suggestions for Correction
Breakage	
Excessive feed	Reduce feed to a point where the tool will stand up under the work pressure or redesign tool.
Tool deflection	Reduce overhang. Increase tool support rigidity.
Excessive vibration	Keep machine tool, toolholders, work fixtures in first class condition. Reduce relief angles to minimum required. Follow suggestions for correcting for tool deflection.
Clamping difficulty	Make sure that methods of holding the tool do not induce stresses causing breakage under cut.
Improper resharpening	Check resharpening methods to make sure that tool is not cracked by improper resharpening.
Dull tool	Do not run tool past its resharpening time.
Chipping of Cutting Edge	
Excessive feed	Reduce feed rate or redesign tool to strengthen cutting edge.
Improperly ground chip breakers or clearances	Avoid on-the-job additions of chip breakers or curlers. Maintain specified clearances.
Excessive vibration	Keep machine tool, toolholders, work fixtures etc., in first class condition. Reduce relief angles to minimum required.
Excessive relief	Do not exceed recommended relief when resharpening.
Burning of Cutting Edge	
Excessive speed	Reduce surface speed to a point at which tool will stand up; use tool material of higher red hardness.
Hard work material	Check hardness of work material to keep it within range of machinability under the operating conditions.
Inadequate cutting fluid	Increase flow of cutting fluid or improve direction of flow to cool cutting edge in the cut.
Cratering	
Excessive speed or feed eroding top of tool	Reduce speed or feed to eliminate excessive cratering. Possibly use a more abrasion-resistant tool material.
Insufficient rake angle	Increase rake angles to reduce friction of chip against tool.
Chip Clogging	
Chips too long	Use separate chip breakers or grind chip breaker or curler.
Insufficient cutting fluid	Increase pressure of fluid or direct flow to wash chips away.
Rubbing	
Too little relief or clearance	Increase relief or clearance until tool does not drag and rub in cut. Do not weaken tool by exceeding necessary relief.
Tool cocked in holder	Make sure that tool is held in correct relation to the work.

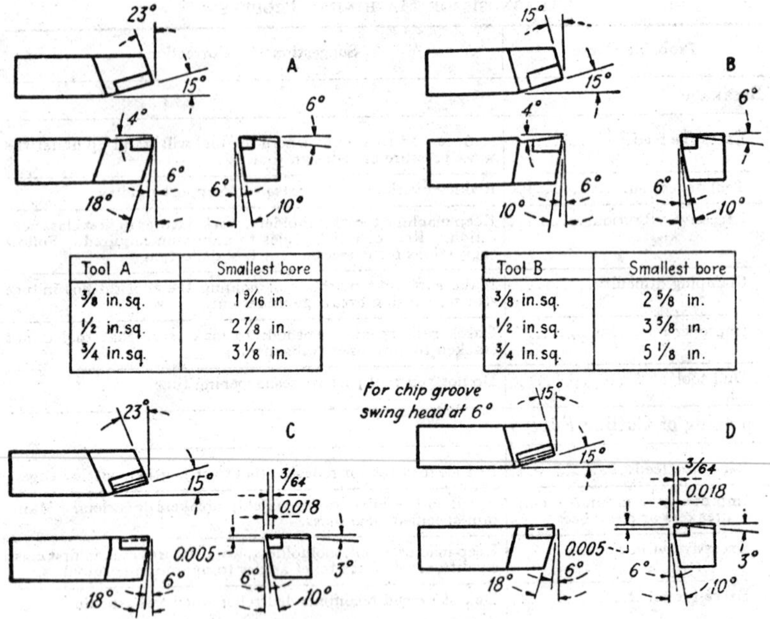

Tool A	Smallest bore
3/8 in. sq.	1 9/16 in.
1/2 in. sq.	2 7/8 in.
3/4 in. sq.	3 1/8 in.

Tool B	Smallest bore
3/8 in. sq.	2 5/8 in.
1/2 in. sq.	3 5/8 in.
3/4 in. sq.	5 1/8 in.

FIG. 33. Boring tools for cast iron, bronze, and steel.

allow with carbide tooling on the turret lathe for a given piece. Data that must be known are the total amount of stock to be removed from the hole, the length of the hole, and the size of the boring bar.

EXAMPLE: Hole size, $2\frac{1}{2}$ in. to be increased to $3\frac{1}{2}$ in., or total stock removal, 1 in. Length of hole, 8 in. Boring-bar size, 2 in. Estimated boring-bar overhang, 10 in.
SOLUTION: From the 2-in. bar size, project vertically to the curve for 10-in. overhang. Then from the point of intersection, project to the left. The permissible increase in diameter is $\frac{3}{8}$ in. (reading to nearest figure). Number of cuts equals total stock removal divided by increase in diameter per cut, or $1 \div \frac{3}{8} = 3$ cuts. The column "Approximate depth of cut" (Fig. 36) is useful in further calculations of machining time.

Horsepower Calculations for Turret-lathe Work

Table 18 was computed by the methods department, Cleveland Automatic Machine Co., to determine how many cuts can be combined and still remain within the capacity of a given turret lathe. Entries were computed from the Carboloy formula

$$\text{Hp} = D \times F \times S \times C \times 1.33$$

where D = depth of cut, F = feed per revolution, S = sfpm, C = material constant, and 1.33 = a safety factor. The rate setter picks out a tabulated figure for appropriate conditions of feed per revolution, depth of cut, and sfpm, and multiplies by the material constant, to obtain the horsepower for a given cut. If the result is

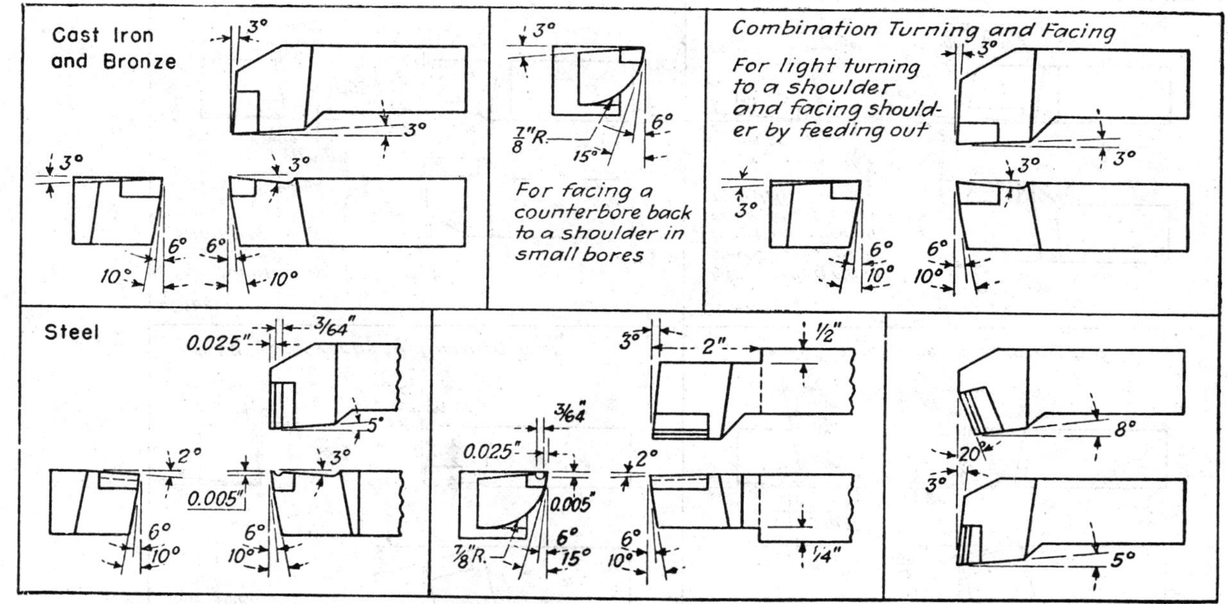

FIG. 34. Carbide-tipped tools used by Cleveland Automatic Machine Co. for facing operations.

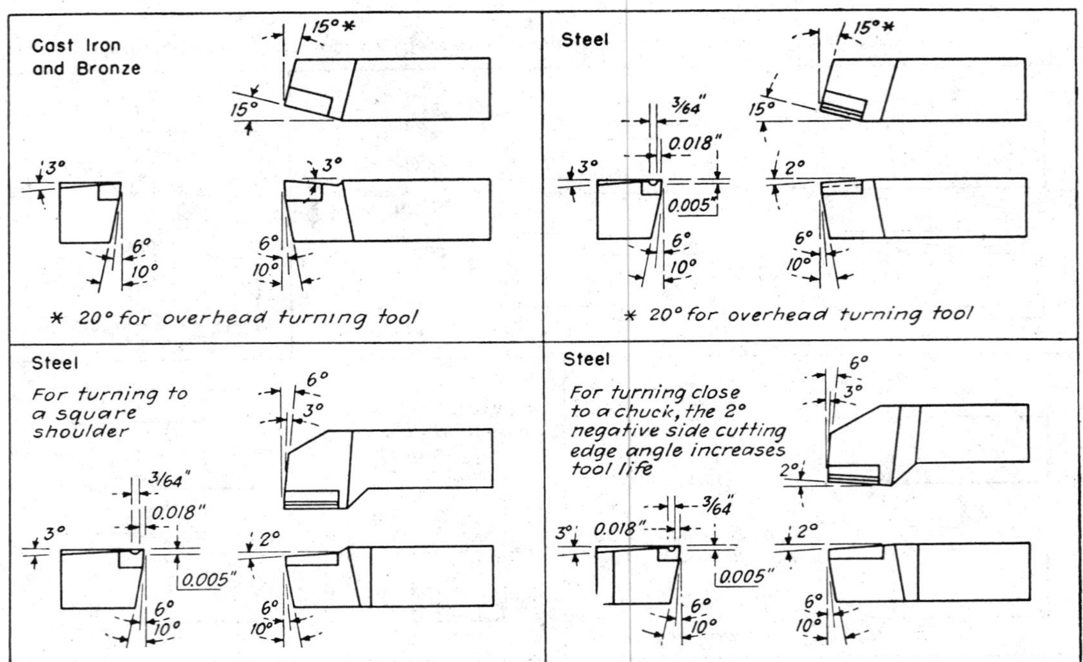

FIG. 35. Carbide turning tools used on turret lathes by the Cleveland Automatic Machine Co.

TABLE 17. SPEEDS AND FEEDS FOR TURRET-LATHE OPERATIONS

Material	Surface Feet per Minute						Feeds			
	H. S. S.		J Stellite		Carbide		Ram-Type Machine		Saddle-Type Machine	
	Rough	Finish	Rough	Finish	Rough	Finish	Rough	Finish	Rough	Finish
Cast iron	50 to 60	80 to 110	90 to 120	130 to 160	180 to 200	350 to 400	0.016 to 0.025	0.016 to 0.025	0.032 to 0.060	0.032 to 0.125
Semisteel, hard	40 to 50	65 to 90	75 to 100	100 to 130	140 to 160	250 to 300	0.016 to 0.025	0.016 to 0.025	0.032 to 0.060	0.032 to 0.125
Malleable iron	80 to 100	110 to 130	120 to 140	150 to 200	250 to 300	400 to 500	0.016 to 0.025	0.016 to 0.025	0.032 to 0.060	0.032 to 0.060
Steel casting (0.35 carbon)	45 to 60	70 to 90	70 to 80	90 to 130	150 to 180	200 to 250	0.010 to 0.020	0.012 to 0.020	0.018 to 0.030	0.018 to 0.045
Brass (commercial 85-5-5)	200 to 300	200 to 300			600 to 1000	600 to 1000	Maximum Feed of Machine			
Bronze (80-10-10)	110 to 150	150 to 180			600	1000	0.016 to 0.030	0.010 to 0.016	0.016 to 0.050	0.010 to 0.030
Aluminum	400	700			800	1000	Fine Feeds to Produce Good Finish			
S.A.E. 1020 (coarse feed)	60 to 80	60 to 80			300	300	0.020 to 0.030	0.020 to 0.030	0.024 to 0.044	0.024 to 0.044
(fine feed)	100 to 120	100 to 120			450	450	0.007 to 0.020	0.007 to 0.020	0.010 to 0.030	0.010 to 0.030
S.A.E. 1035	75 to 90	90 to 120			250	250	0.020 to 0.030	0.020 to 0.030	0.024 to 0.044	0.024 to 0.044
S.A.E.-X-1315	175 to 200	175 to 200			400 to 500	400 to 500	0.010 to 0.015	0.010 to 0.015	0.010 to 0.015	0.010 to 0.015
S.A.E. 1050	60 to 80	100			200	200	0.012 to 0.020	0.012 to 0.020	0.025 to 0.090	0.015 to 0.045
S.A.E. 2315	90 to 110	90 to 110			300	300	0.012 to 0.020	0.012 to 0.020	0.025 to 0.090	0.015 to 0.045
S.A.E. 3150	50 to 60	70 to 90			200	200	0.012 to 0.020	0.012 to 0.020	0.025 to 0.090	0.015 to 0.045
Stainless steel (selenium content)	100 to 120	100 to 120			240 to 300	240 to 300	0.010 to 0.015	0.010 to 0.015	0.010 to 0.015	0.010 to 0.015

This table is offered for general guidance only, and the cited values must be used with care, particularly feeds. Much depends on the type of machine used, different conditions for various jobs, the material being cut, method of holding tools, and other job factors.

TABLE 18. HORSEPOWER FOR TURNING AND BORING WITH CARBIDE*

DEPTH OF CUT	S.F.P.M.	FEED 0.009	0.010	0.011	0.012	0.013	0.014	0.015	0.016	0.017	0.018	0.019	0.020	0.021	0.022	0.023	0.024	0.025
1/8	125	0.19	0.21	0.23	0.25	0.27	0.29	0.31	0.33	0.35	0.37	0.40	0.42	0.44	0.46	0.48	0.50	0.52
	150	0.23	0.25	0.28	0.30	0.33	0.35	0.38	0.40	0.43	0.45	0.48	0.50	0.53	0.55	0.58	0.60	0.63
	175	0.27	0.29	0.32	0.35	0.38	0.41	0.44	0.47	0.50	0.53	0.56	0.59	0.62	0.64	0.67	0.70	0.73
	200	0.30	0.33	0.37	0.40	0.44	0.47	0.50	0.54	0.57	0.60	0.64	0.67	0.70	0.73	0.77	0.80	0.84
	225	0.34	0.38	0.41	0.45	0.49	0.53	0.56	0.60	0.64	0.68	0.71	0.75	0.79	0.83	0.86	0.90	0.94
	250	0.38	0.42	0.46	0.50	0.54	0.58	0.63	0.67	0.71	0.75	0.79	0.83	0.87	0.92	0.96	1.00	1.04
	275	0.42	0.46	0.51	0.55	0.60	0.64	0.69	0.73	0.78	0.83	0.87	0.92	0.96	1.01	1.06	1.10	1.15
	300	0.45	0.50	0.55	0.60	0.65	0.70	0.75	0.80	0.85	0.90	0.95	1.00	1.05	1.10	1.15	1.20	1.25
	325	0.49	0.54	0.60	0.65	0.71	0.76	0.81	0.87	0.92	0.98	1.03	1.08	1.14	1.19	1.25	1.30	1.36
	350	0.53	0.59	0.64	0.70	0.76	0.82	0.88	0.94	1.00	1.05	1.11	1.17	1.23	1.29	1.34	1.40	1.46
	375	0.56	0.63	0.69	0.75	0.81	0.88	0.94	1.00	1.06	1.12	1.19	1.25	1.31	1.37	1.44	1.50	1.56
	400	0.60	0.67	0.73	0.80	0.87	0.93	1.00	1.07	1.13	1.20	1.27	1.33	1.40	1.47	1.53	1.60	1.67
1/4	125	0.38	0.42	0.46	0.50	0.54	0.58	0.62	0.66	0.70	0.74	0.80	0.84	0.88	0.92	0.96	1.00	1.04
	150	0.46	0.50	0.56	0.60	0.66	0.70	0.76	0.80	0.86	0.90	0.96	1.00	1.06	1.10	1.16	1.20	1.26
	175	0.54	0.58	0.64	0.70	0.76	0.81	0.88	0.94	1.00	1.06	1.12	1.18	1.24	1.28	1.34	1.40	1.46
	200	0.60	0.66	0.74	0.80	0.88	0.94	1.00	1.08	1.14	1.20	1.28	1.34	1.40	1.46	1.54	1.60	1.68
	225	0.68	0.76	0.82	0.90	0.98	1.06	1.12	1.20	1.28	1.36	1.42	1.50	1.58	1.66	1.72	1.80	1.88
	250	0.76	0.84	0.92	1.00	1.08	1.16	1.26	1.34	1.42	1.50	1.58	1.66	1.74	1.84	1.92	2.00	2.08
	275	0.84	0.92	1.02	1.10	1.20	1.28	1.38	1.46	1.56	1.66	1.74	1.82	1.92	2.02	2.12	2.20	2.30
	300	0.90	1.00	1.10	1.20	1.30	1.40	1.50	1.60	1.70	1.80	1.90	2.00	2.10	2.20	2.30	2.40	2.50
	325	0.98	1.08	1.20	1.30	1.42	1.52	1.62	1.74	1.84	1.96	2.06	2.16	2.28	2.38	2.50	2.60	2.72
	350	1.06	1.18	1.28	1.40	1.52	1.64	1.76	1.88	2.00	2.10	2.22	2.34	2.46	2.58	2.68	2.80	2.92
	375	1.12	1.26	1.38	1.50	1.62	1.76	1.88	2.00	2.12	2.24	2.38	2.50	2.62	2.74	2.88	3.00	3.12
	400	1.20	1.34	1.46	1.60	1.74	1.86	2.00	2.14	2.26	2.40	2.54	2.66	2.80	2.94	3.06	3.20	3.34

DEPTH OF CUT	S.F. P.M.	FEED 0.009	0.010	0.011	0.012	0.013	0.014	0.015	0.016	0.017	0.018	0.019	0.020	0.021	0.022	0.023	0.024	0.025
3/8	125	0.57	0.63	0.69	0.75	0.81	0.87	0.93	0.99	1.05	1.11	1.20	1.26	1.32	1.38	1.44	1.50	1.56
	150	0.69	0.75	0.84	0.90	0.99	1.05	1.14	1.20	1.29	1.35	1.44	1.50	1.59	1.65	1.74	1.80	1.89
	175	0.81	0.87	0.96	1.05	1.14	1.23	1.32	1.41	1.50	1.59	1.68	1.77	1.86	1.92	2.01	2.10	2.18
	200	0.90	0.99	1.11	1.20	1.32	1.41	1.50	1.62	1.71	1.80	1.92	2.01	2.10	2.19	2.31	2.40	2.52
	225	1.02	1.14	1.23	1.35	1.47	1.59	1.68	1.80	1.92	2.04	2.13	2.25	2.37	2.49	2.58	2.70	2.82
	250	1.14	1.26	1.38	1.50	1.62	1.74	1.89	2.01	2.13	2.25	2.37	2.49	2.61	2.76	2.88	3.00	3.12
	275	1.26	1.38	1.53	1.65	1.80	1.92	2.07	2.19	2.34	2.48	2.61	2.76	2.88	3.03	3.18	3.30	3.45
	300	1.35	1.50	1.65	1.80	1.95	2.10	2.25	2.40	2.55	2.70	2.85	3.00	3.15	3.30	3.45	3.60	3.75
	325	1.47	1.62	1.80	1.95	2.13	2.28	2.43	2.61	2.76	2.94	3.09	3.24	3.42	3.57	3.75	3.90	4.08
	350	1.59	1.77	1.92	2.10	2.28	2.46	2.64	2.82	3.00	3.15	3.33						
1/2	125	0.76	0.84	0.92	1.00	1.08	1.16	1.24	1.32	1.40	1.48	1.60	1.68	1.76	1.84	1.92	2.00	2.08
	150	0.92	1.00	1.12	1.20	1.32	1.40	1.52	1.60	1.72	1.80	1.92	2.00	2.12	2.20	2.32	2.40	2.52
	175	1.08	1.16	1.28	1.40	1.52	1.64	1.76	1.88	2.00	2.12	2.24	2.36	2.48	2.56	2.68	2.80	2.92
	200	1.20	1.32	1.48	1.60	1.76	1.88	2.00	2.16	2.28	2.40	2.56	2.68	2.80	2.92	3.08	3.20	3.36
	225	1.36	1.52	1.64	1.80	1.96	2.12	2.24	2.40	2.56	2.72	2.84	3.00	3.16	3.32	3.44	3.60	3.76
	250	1.52	1.68	1.84	2.00	2.16	2.32	2.52	2.68	2.84	3.00	3.16						
	275	1.68	1.84	2.04	2.20	2.40	2.56	2.76	2.92	3.12	3.32	3.48						
	300	1.80	2.00	2.20	2.40	2.60	2.80	3.00	3.20									
5/8	125	0.94	1.05	1.15	1.25	1.35	1.45	1.55	1.65	1.75	1.85	2.00	2.10	2.20	2.30	2.40	2.50	2.60
	150	1.15	1.25	1.40	1.50	1.65	1.75	1.90	2.00	2.15	2.25	2.40	2.50	2.65	2.75	2.90	3.00	3.15
	175	1.35	1.45	1.60	1.75	1.90	2.05	2.20	2.35	2.50	2.65	2.80	2.95	3.10	3.20	3.35	3.50	3.65
	200	1.50	1.55	1.85	2.00	2.20	2.35	2.50	2.70	2.85	3.00	3.20	3.35	3.50	3.65	3.84	4.00	4.20
	225	1.70	1.90	2.05	2.25	2.45	2.65	2.80	3.00	3.20	3.40	3.55						
	250	1.90	2.10	2.30	2.50	2.70	2.90	3.14	3.35	3.55	3.75	3.95						
	275	2.10	2.30	2.55	2.75	3.00	3.20	3.45										

* Cleveland Automatic Machine Co.

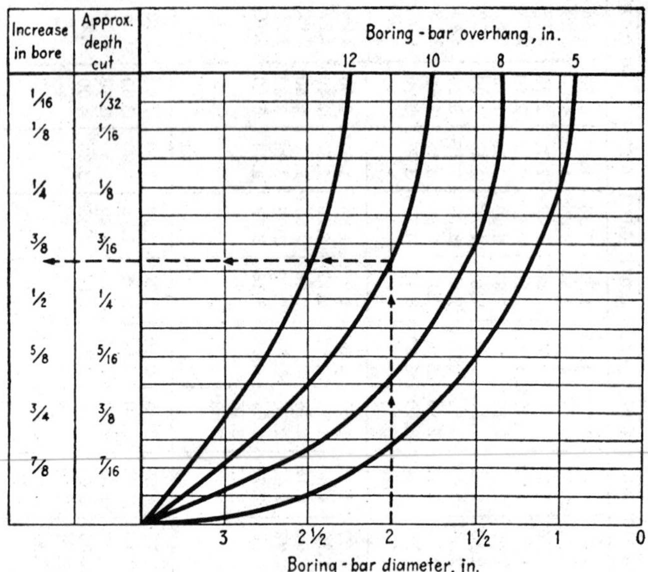

FIG. 36. Chart for estimating the number of boring cuts required on a turret lathe, when the overhang is known.

well within the capacity of the machine, he considers combining another cut.

Hp = chart value times material constant

Material Constants:

Material	Constant	Material	Constant
Cast iron, Bronze	3	4150, 3250, Steel castings, Steel forgings	9
X1315, 4615, 1020, Straight tubing	6	HT 4150, 52100	10
1045	8		

BORING AND BORING TOOLS

Boring is the machining process relied upon to generate a truly round straight hole. It is applied to either a drilled hole or a cored hole. Turning and boring are similar in that they both use single-point tools. Rigidity of tool mounting favors turning in some cases; hence in boring, some reduction in feed, speed, or depth of cut may be required. Furthermore, chip interference with the tool is often more pronounced with boring, thereby reducing tool life, unless slower machining rates are used.

A number of machine tools are used for boring purposes: lathes, horizontal boring machines, boring mills, precision boring machines, jig boring machines, special equip-

TABLE 19. SPEEDS AND FEEDS FOR BORING ON HORIZONTAL BORING MACHINES*

	Type of Hole	Tolerance (Inches)	Speed (fpm)	Feed (In. Per Rev.)	No. of Cuts
Flycutter in Stub Bar	Cored or Cut Cored or Cut Cored or Cut Drilled Hole to Be Reamed	.001-.003 .003-.010 .010-.030	30-40 30-40 30-40 30-40	.010-.020 .010-.020 .010-.020 .012	4 3 2 2
Flycutter in Boring Head Mounted on Line Bar	Cored or Cut Cored or Cut Cored or Cut	.001-.003 .003-.010 .010-.030	30-40 30-40 30-40	.015-.030 .015-.030 .015-.030	4 3 2
Block Type Double Cutters Stub Bar	Cored or Cut Cored or Cut Cored or Cut	.001-.003 .003-.010 .010-.030	30-40 30-40 30-40	.020-.040 .020-.040 .020-.040	3 2 1
Block Type Double Cutters Line Bar, Fixtures, etc.	Cored or Cut Cored or Cut Cored or Cut	.001-.003 .003-.010 .010-.030	30-40 30-40 30-40	.025-.050 .025-.050 .025-.050	3 2 1

Feeds for cemented carbides are about the same as for high-speed steels. Speeds may be increased to 200 to 400 fpm.

	Type of Hole	Depth of Cut, Radius (Inches)	Speed (fpm)	Feed (In. Per Rev.)	No. of Cuts
Flycutter in Stub Bar	Cored Hole to .005 in., Tolerance Cored Hole to .0005 in., Tolerance Drilled Hole to Be Reamed	3/8 max. 3/8 max. 1/16 max.	40-45 40-45 40-45	.012-.016 .012-.016 .012	3 4 2
Flycutter in Boring Head Mounted on Line Bar	Cored Rough Semi-finish	 1/4-3/8 1/16	 35-40 35-40	 .012-.025 .035-.050	
Block Type Double Cutters, Fixture Support	Cored Rough Semi-finish Finish	 1/4-3/8 .018 .007	 35-40 35-40 35-40	 .019-.025 .031-.050 .031-.050	

* Giddings & Lewis Machine Tool Co.

ment for gun boring. Boring tools are of many varieties: forged-shank types held in toolposts, toolbits held in boring bars, forged or tipped tools for heavy-lathe and boring-mill operations. Except for precision boring, tool design and tool angles for boring operations are generally similar to those for turning work.

Precision Boring with Carbides

By BRUNO HOLMSTROM, *Tool Supervisor, Heald Machine Company*

Carbide-tipped boring tools (Fig. 37) are widely used on all ferrous and most nonferrous and nonmetallic materials. Extremely fine finishes can be obtained on a production basis.

Tool Design. Chief variables in design of tools are radius, clearance angles, and rake angles. Radius may be selected to produce a specific surface finish (Fig. 38).

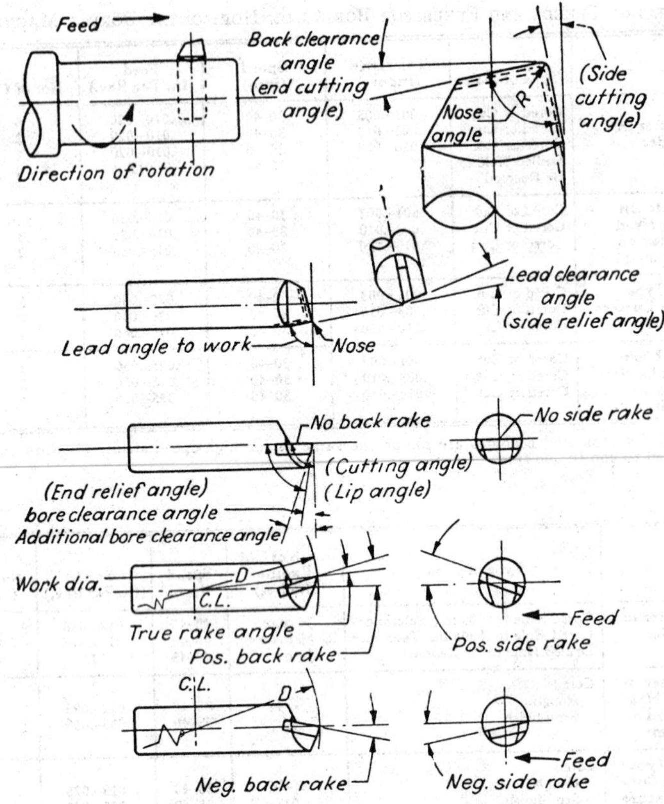

FIG. 37. Terminology for a single-point tool used on precision boring machines.

Clearance angles vary with the bore diameter, the material, and the height of tool point above center. Bore clearance angle *C* (Fig. 39) decreases as the tool is set at increasing distance above center. Proper clearance must be maintained. Too great an angle causes rapid tool wear and chatter; too small an angle results in poor finish and added friction. Table 20 shows the clearance angle *F* for the leading and trailing edge (angle *L*) of round-nose boring tools when lapped at a bore-clearance angle *C*.

In establishing a setup, one must consider the effects of (1) setting tool above center upon back-rake and bore-clearance angle, and (2) actual end relief to grind on tool to provide proper side clearance when tool is set at an angle in the quill.

In connection with the first problem, consult Table 20 of neutral angles and bore-clearance angles. The table of neutral angles gives the back rake that must be ground on the tool to obtain an effective rake of 0° when the tool is set at a specified height above center in a hole of a given size.

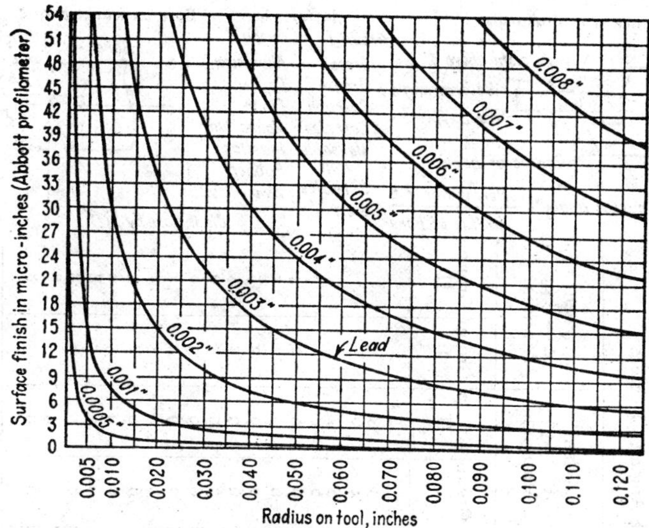

FIG. 38. Chart for predicting the surface finish secured by precision boring, when the tool-nose radius and lead are known.

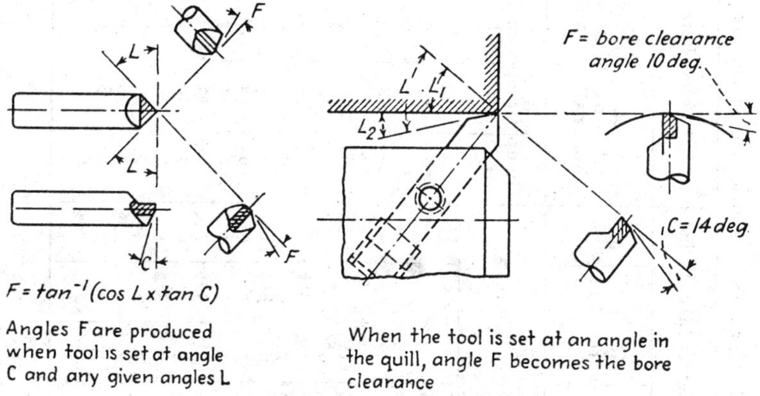

FIG. 39. Blending angles for precision-boring tools.

For example, to obtain a true rake of 0° when a tool is set $\frac{1}{32}$ in. above center in a 1-in. hole, it is necessary to grind a $3\frac{1}{2}$° back rake on the toolbit.

When a toolbit with side angles L is ground with bore clearance C, the side-clearance angle $F = \tan^{-1} (\cos L \times \tan C)$. Values of F are provided by Table 21. If a tool is set an angle on the quill, side clearance F becomes the bore clearance. Under these conditions, it is essential to find the angle C to which the tool must be tilted during lapping. Assume a tool with side angles L of 45° and set at an angle

Back rake or neutral angles

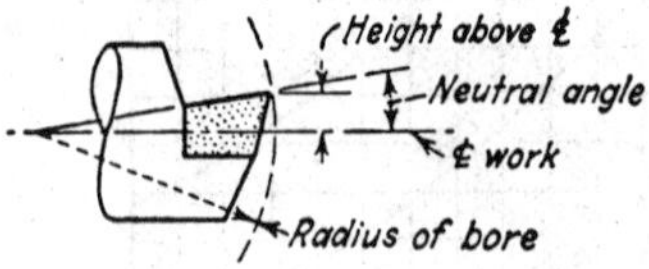

Height above ℄

Neutral angle

℄ work

Radius of bore

Sine "neutral angle" = $\frac{\text{Height above ℄}}{\text{Radius of bore}}$

Materials

Group I	Group II
Brass	Aluminum
Bronze	Copper
Cast iron	Fiber
Cast steel	Magnesium
Malleable iron	Plastics
Semi-steel	Rubber
Steel	Zinc alloy

Bore clearance angles

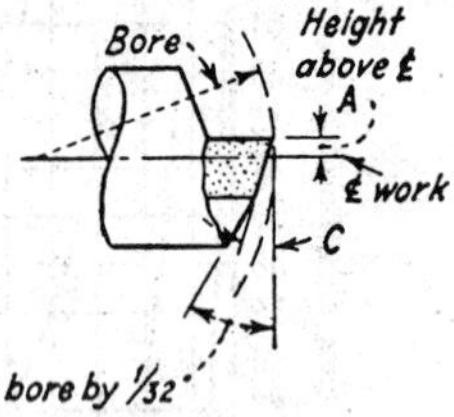

Bore

Height above ℄

A

℄ work

C

Clear bore by 1/32

TABLE 20. RAKE AND CLEARANCE ANGLES FOR PRECISION BORING TOOLS SET ABOVE CENTER

Bore Dia, In.	Height above Center Line, Inches						First Bore Clearance Angle C, Degrees							
	0.010	1/64	1/32	1/16	3/32	1/8	When A = 0 Material		When A = 0.010 Material		When A = 1/32 Material		When A = 1/16 Material	
	Neutral Angle, Degrees*						Group I	Group II	Group I	Group II	Group I	Group II	Group I	Group II
5/16	4	6	11½				10	13	9	12	3	5		
3/8	3½	5	9½				10	13	9	12	4	6		
7/16	3	4	8				10	13	9	12	5	7		
1/2	2½	3½	7				10	13	9	12	6	8		
5/8	2	3	5½	11½			10	13	9	12	6	9	2	4
3/4	1½	2½	5	9½			10	13	9	12	6	9	2	5
7/8	1	2	3½	7½	12½		10	13	9	12	6	9	2	5
1	1	2	3½	7	11		10	12	8	11	6	9	2	5
1¼		1½	3	6	8½	11	9	12	8	11	6	9	3	6
1½		1	2½	5	7	9½	9	12	7	10	6	9	3	6
1¾		1	2	4	6	8	8	11	7	10	6	9	4	7
2		1	2	3½	5½	7	8	11	7	10	6	9	4	7
2½		1	1½	3	4½	6	7	10	6	9	6	9	4	7
3			1	2½	3½	5	7	10	6	9	6	9	5	8
3½			1	2	3	4	7	10	6	9	6	9	5	8
4			1	2	3	3½	6	8	6	9	6	9	5	8
5			1	1½	2	3	6	8	6	9	6	9	5	8
6				1	2	2½	6	8	5	8	6	9	5	8

* To nearest ½°.

TABLE 21. NORMAL RELIEF OR "BLENDING" ANGLES FOR PRECISION BORING TOOLS

Bore Clearance Angle C, Deg	L Values, Deg								
	5	10	15	20	25	30	35	40	45
	F Values Deg and Min								
1	1°00′	0°59′	0°58′	0°56′	0°54′	0°52′	0°49′	0°46′	0°42′
2	2°00′	1°58′	1°56′	1°53′	1°49′	1°44′	1°38′	1°32′	1°25′
3	2°59′	2°57′	2°54′	2°49′	2°43′	2°36′	2°27′	2°18′	2°07′
4	3°59′	3°56′	3°52′	3°46′	3°38′	3°28′	3°17′	3°04′	2°50′
5	4°59′	4°55′	4°50′	4°42′	4°32′	4°20′	4°06′	3°50′	3°32′
6	5°59′	5°55′	5°48′	5°38′	5°26′	5°12′	4°55′	4°36′	4°15′
7	6°58′	6°54′	6°46′	6°35′	6°21′	6°04′	5°45′	5°22′	4°58′
8	7°58′	7°53′	7°44′	7°31′	7°16′	6°56′	6°34′	6°09′	5°41′
9	8°58′	8°52′	8°42′	8°28′	8°10′	7°49′	7°24′	6°55′	6°23′
10	9°58′	9°51′	9°40′	9°24′	9°05′	8°41′	8°13′	7°42′	7°06′
11	10°58′	10°50′	10°38′	10°21′	9°59′	9°33′	9°03′	8°28′	7°50′
12	11°57′	11°49′	11°36′	11°18′	10°54′	10°26′	9°53′	9°15′	8°33′
13	12°57′	12°49′	12°34′	12°14′	11°49′	11°18′	10°43′	10°02′	9°16′
14	13°57′	13°48′	13°32′	13°11′	12°44′	12°11′	11°33′	10°49′	10°00′
15	14°57′	14°47′	14°31′	14°08′	13°39′	13°04′	12°23′	11°36′	10°44′
16	15°57′	15°46′	15°29′	15°05′	14°34′	13°57′	13°13′	12°23′	11°28′
17	16°56′	16°45′	16°27′	16°02′	15°29′	14°50′	14°04′	13°11′	12°12′
18	17°56′	17°45′	17°25′	16°59′	16°25′	15°43′	14°54′	13°59′	12°56′
19	18°56′	18°44′	18°24′	17°56′	17°20′	16°36′	15°45′	14°47′	13°41′
20	19°56′	19°43′	19°22′	18°53′	18°15′	17°30′	16°36′	15°35′	14°26′
21	20°56′	20°42′	20°21′	19°50′	19°11′	18°23′	17°27′	16°23′	15°11′
22	21°56′	21°42′	21°19′	20°47′	20°07′	19°17′	18°19′	17°12′	15°57′

with the bore axis, as in Fig. 39. The chart of blending angles can be used to find angle C, or in this case the angle to which the tool is tilted during lapping; so read down under F values until 10° is reached. This is the actual bore-clearance angle; so move left to the corresponding value of C and find that 14° is the angle to which the tool must be tilted for lapping.

Surface Finish. To obtain a specified surface, refer to Fig. 38. Assuming that the desired surface finish is 12 μin., and a tool of 0.060 in. radius is available, what is the proper lead? From the chart, it will be seen that a lead of 0.003 in. per work revolution should be satisfactory.

Boring Speeds. When starting a new job in steel, refer to Fig. 40 for a safe boring speed. The chart is based on the best finish, not on tool life. Selection of the final speed will be affected by tool shape, chip thickness, and work material. To use the chart, find the key number of the steel in the table below.

EXAMPLE: SAE 3120 steel with a Rockwell 25 C hardness has a key number of 9. On the chart, from 9 draw a horizontal line until it intersects the Rockwell 25 C curve, and then drop line to the surface-feet scale and read that the best finishing speed is about 550 sfpm.

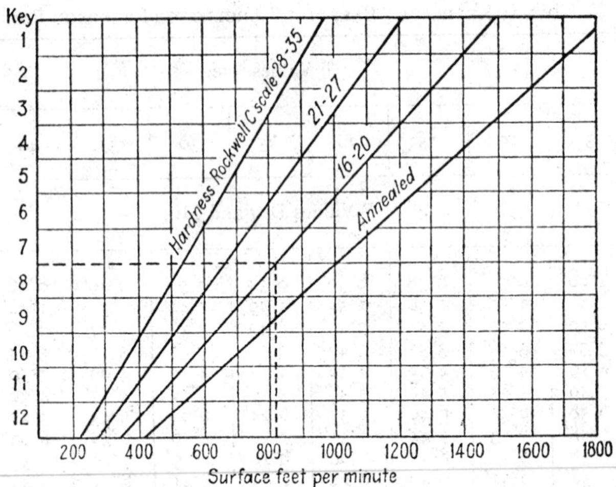

FIG. 40. Chart for finding boring speed for steels.

KEY NUMBERS FOR SAE STEELS

1010	4	1112	1	T1345	8	2350	11	3150	11	3415	10	4340	12	5140	11
1015	4	1115	2	T1350	9	2512	10	3230	11	3430	12	4345	12	5150	11
1020	4	1120	2	2105	7	2515	11	3240	12	3435	12	4615	11	6115	9
1030	6	X1314	3	2115	7	3115	8	3245	12	3450	12	4620	11	6120	10
1035	7	X1315	3	2315	7	3120	9	3250	12	4115	8	4640	12	6125	11
1040	8	X1330	5	2320	10	3125	10	3312	11	4130	10	4650	12	6130	11
1045	9	X1335	7	2330	11	3130	10	3325	11	X4130	10	4815	11	6135	11
1050	10	T1330	7	2335	11	3140	11	3330	11	4135	12	4820	11	6140	11
1060	11	T1335	8	2340	11	X3140	11	3335	12	4140	12	5120	10	6145	12
1095	12	T1340	8	2345	11	3145	11	3340	12	4150	12	5130	10	6150	12
														51210	9

Cutting fluids. Use of a cutting fluid in precision boring is helpful to maintain size control. By its use, the tendency for a temperature rise in the work from tool friction and pressure is greatly reduced. Good finish is promoted by reducing the possibility of work material welding to the tool face.

Cast iron may be bored dry, or use an air blast to blow away chips and cool the work and tool. If coolant is used on steels, brass, bronzes, aluminum, magnesium, and babbitt, the first choice is soluble oil. For a second choice on steels and bronzes, use mineral-oil compound. For zinc use kerosene and lard oil, this mixture being second choice for aluminum and magnesium. Silver is bored with an oil spray.

Carbide Boring Bars. **Since the stiffness of cemented carbides is approximately 2.8 times that of steel, it is possible to bore holes with depths up to eight times the diameter without runout from the boring-bar spring.**

TABLE 22. TOOL SHAPES FOR PRECISION BORING VARIOUS METALS WITH CARBIDES*

Clearance angle C varies with height of cutting point above tool center line; see Table 20.
The side clearance angle F is controlled by angle C; see Fig. 39 and Table 21.
D = back-rake angle; E = side-rake angle; H = distance above center.

Material	Lead per Rev	Depth of Cut, In.	Cutting Speed, Sfpm	D, Deg	E Deg	Nose Radius R		H, In.
						Small Bores	Larger Bores	
Aluminum	0.002–0.006	0.005–0.015	800–6000	0–15	5–15	0.015	0.015–0.060	$\frac{1}{32}$
Copper alloys:†								
Class 1			800–3000	0	5	0.015–0.030	0.015–0.030	$\frac{1}{32}$
Class 2			450–1500	0–5	5–10	0.015–0.030	0.015–0.030	
Class 3			200–800	5–10	15–20	0.015–0.030	0.015–0.030	
Cast iron	0.003–0.006	0.005–0.015	300–700	0	0	0.015–0.030	0.030–0.060	0–$\frac{1}{32}$
Steel:†			See Fig. 40					
Class 1	0.003–0.007	0.005–0.015		0 to −6	−3 to −8	Equal to depth of cut	Equal to depth of cut	See Table 20
Class 2	0.003–0.007	0.005–0.015		−5 to −35	0 to −8	Equal to depth of cut	Equal to depth of cut	See Table 20
Class 3	0.003–0.007	0.005–0.015		−3 to −10	0 to −8	Equal to depth of cut	Equal to depth of cut	See Table 20

* Heald Machine Co.

† Classes of materials: Copper-base alloys: class 1 (free-cutting)—leaded copper, common bronze, leaded red brass, free-cutting yellow brass, forging brass, leaded naval brass; class 2 (readily machinable)—red brass, yellow brass, Muntz metal, naval brass, tobin bronze, leaded phosphor bronze; class 3—copper, commercial bronze, nickel silver, beryllium copper, chrome copper. Steels: class 1—high-carbon and alloy steels; class 2—interrupted cuts and tough steels; class 3—low-carbon steels and free-cutting alloys.

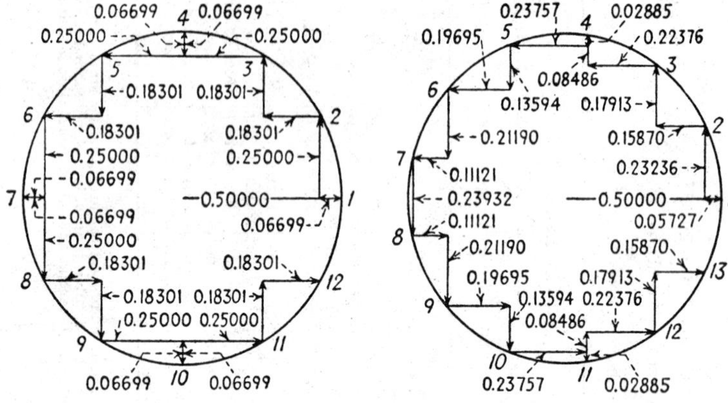

FIG. 41. Layout of 12 holes on a 1-in. circle can be done accurately by using the horizontal and vertical movements shown.

FIG. 42. Layout of 13 or any other number of holes on a 1-in. circle is done by selecting horizontal and vertical values from tables.

Diamond Boring Tools. Although cemented carbide tools have largely replaced diamond tools for precision boring operations, there are places where diamond tools are still used to advantage. Because of their hardness, diamonds do not wear in the accepted sense; failure is due to chipping. Shaped diamonds are fixed to a holder by brazing, cold-setting, or powder metallurgy. A common boring tool has an included angle of 90°, end relief of 8 to 10°, and nose radius of 0.010 to 0.020 in., depending on bore diameter. On aluminum, back rake may be 4 to 5° positive, on brass several degrees negative, and on copper no rake is needed. Speeds up to 3000 work rpm, tight-fitting bearings for spindles, and stock removal of 0.003 to 0.004 in. per side are considered good practice. Diamond boring tools are good on plastics, because low heat generation minimizes or avoids distortion of the parts.

Sapphire tools are much less hard than diamond tools but are harder than cemented carbide. The sapphire (synthetic) takes a high polish, has no grain boundaries, and works well on brass and similar metals.

Hole-circle Layouts for Jig Boring. Accurate layout of holes on a given circumference is made possible by the use of Table 23. The values given are constants for a 1-in. circle. For larger or smaller circles, multiply the values by the diameter of the given circle. All values are given in ten-thousandths of an inch.

The tables are useful on precision machines, particularly the jig borer. They are to be used as follows:

Have the work fastened in position and locate the true center. Then move to the right a distance equal to the radius of the circle to be divided. Bore hole No. 1. Then move horizontally to the left the required number of ten-thousandths for hole No. 2. Follow this by moving vertically the required number of ten-thousandths for hole No. 2, thus locating this hole. Repeat these horizontal and vertical movements, locating each hole in turn. At each outer intersection bore the hole, and repeat until the circle is complete.

For example, in laying out 12 holes on a 1-in. circle as shown in Fig. 41, hole No. 1

TABLE 23. CONSTANTS FOR HOLE-CIRCLE LAYOUTS

Hole No.	Horizontal	Vertical
Three holes:		
1	0.50000	
2	0.75000	0.43301
3	....	0.86602
Five holes:		
1	0.50000	
2	0.34549	0.47553
3, 5	0.55902	0.18164
4	....	0.58778
Six holes:		
1, 3, 6	0.50000	
2, 4, 5	0.25000	0.43301
Seven holes:		
1	0.50000	
2	0.18826	0.39091
3, 7	0.42300	0.09655
4, 6	0.33923	0.27052
5	....	0.43388
Eight holes:		
1	0.50000	
2, 5, 6	0.14645	0.35355
3, 4, 7, 8	0.35355	0.14645
Nine holes:		
1	0.50000	
2	0.11698	0.32139
3, 9	0.29620	0.17101
4, 8	0.33682	0.05939
5, 7	0.21984	0.26200
6	....	0.34202
Ten holes:		
1	0.50000	
2, 6, 7	0.09549	0.29389
3, 5, 8, 10	0.25000	0.18164
4, 9	0.30902	
Eleven holes:		
1	0.50000	
2	0.07937	0.27032
3, 11	0.21292	0.18450
4, 10	0.27887	0.04009
5, 9	0.25626	0.11704
6, 8	0.15233	0.23701
7	....	0.28172
Twelve holes:		
1	0.50000	
2, 7, 8	0.06699	0.25000
3, 6, 9, 12	0.18301	0.18301
4, 5, 10, 11	0.25000	0.06699
Thirteen holes:		
1	0.50000	
2	0.05727	0.23236
3, 13	0.15870	0.17913
4, 12	0.22376	0.08486
5, 11	0.23757	0.02885
6, 10	0.19695	0.13594
7, 9	0.11121	0.21190
8	....	0.23932
Fourteen holes:		
1	0.50000	
2, 8, 9	0.04951	0.21694
3, 7, 10, 14	0.13875	0.17397
4, 6, 11, 13	0.20048	0.09655
5, 12	0.22252	
Fifteen holes:		
1	0.50000	
2	0.04323	0.20337
3, 15	0.12221	0.16820
4, 14	0.18005	0.10396
5, 13	0.20677	0.02173
6, 12	0.19774	0.06425
7, 11	0.15451	0.13912
8, 10	0.08456	0.18994
9	....	0.20790
Sixteen holes:		
1	0.50000	
2, 9, 10	0.03806	0.19134
3, 8, 11, 16	0.10839	0.16221
4, 7, 12, 15	0.16221	0.10839
5, 6, 13, 14	0.19134	0.03806
Seventeen holes:		
1	0.50000	
2	0.03377	0.18062
3, 17	0.09672	0.15623
4, 16	0.14664	0.11073
5, 15	0.17674	0.05028
6, 14	0.18296	0.01695
7, 13	0.16449	0.08190
8, 12	0.12379	0.13580
9, 11	0.06637	0.17134
10	....	0.18374
Eighteen holes:		
1	0.50000	
2, 10, 11	0.03016	0.17101
3, 9, 12, 18	0.08682	0.15038
4, 8, 13, 17	0.13302	0.11162
5, 7, 14, 16	0.16318	0.05939
6, 15	0.17364	

TABLE 23. CONSTANTS FOR HOLE-CIRCLE LAYOUTS (*Continued*)

Hole No.	Horizontal	Vertical
Nineteen holes:		
1	0.50000	
2	0.02709	0.16235
3, 19	0.07834	0.14475
4, 18	0.12110	0.11148
5, 17	0.15073	0.06612
6, 16	0.16403	0.01358
7, 15	0.15956	0.04039
8, 14	0.13779	0.09003
9, 13	0.10110	0.12989
10, 12	0.05344	0.15567
11	....	0.16460
Twenty holes:		
1	0.50000	
2, 11, 12	0.02447	0.15451
3, 10, 13, 20	0.07102	0.13938
4, 9, 14, 19	0.11062	0.11062
5, 8, 15, 18	0.13938	0.07102
6, 7, 16, 17	0.15451	0.02447
Twenty-one holes:		
1	0.50000	
2	0.02221	0.14738
3, 21	0.06467	0.13428
4, 20	0.10138	0.10925
5, 19	0.12908	0.07452
6, 18	0.14530	0.03317
7, 17	0.14862	0.01114
8, 16	0.13874	0.05445
9, 15	0.11652	0.09293
10, 14	0.08397	0.12314
11, 13	0.04393	0.14242
12	....	0.14904
Twenty-two holes:		
1	0.50000	
2, 12, 13	0.02025	0.14086
3, 11, 14, 22	0.05912	0.12946
4, 10, 15, 21	0.09321	0.10755
5, 9, 16, 20	0.11971	0.07695
6, 8, 17, 19	0.13655	0.04009
7, 18	0.14232	
Twenty-three holes:		
1	0.50000	
2	0.01854	0.13490
3, 23	0.05425	0.12489
4, 22	0.08593	0.10562
5, 21	0.11125	0.07853
6, 20	0.12830	0.04560
7, 19	0.13585	0.00930
8, 18	0.13331	0.02771
9, 17	0.12091	0.06264
10, 16	0.09951	0.09295
11, 15	0.07076	0.11634
12, 14	0.03673	0.13112
13	....	0.13616
Twenty-four holes:		
1	0.50000	
2, 13, 14	0.01704	0.12941
3, 12, 15, 24	0.04995	0.12059
4, 11, 16, 23	0.07946	0.10355
5, 10, 17, 22	0.10355	0.07946
6, 9, 18, 21	0.12059	0.04995
7, 8, 19, 20	0.12941	0.01704
Twenty-five holes:		
1	0.50000	
2	0.01508	0.12434
3, 25	0.04677	0.11653
4, 24	0.07367	0.10140
5, 23	0.09657	0.07989
6, 22	0.11340	0.05337
7, 21	0.12312	0.02348
8, 20	0.12508	0.00787
9, 19	0.11920	0.03873
10, 15	0.10582	0.06716
11, 17	0.08580	0.09136
12, 16	0.06038	0.10983
13, 15	0.03116	0.12140
14	....	0.12532
Twenty-six holes:		
1	0.50000	
2, 14, 15	0.01454	0.11966
3, 13, 16, 26	0.04273	0.11270
4, 12, 17, 25	0.06848	0.09920
5, 11, 18, 24	0.09022	0.07993
6, 10, 19, 23	0.10673	0.05601
7, 9, 20, 22	0.11703	0.02885
8, 21	0.12054	
Twenty-seven holes:		
1	0.50000	
2	0.01348	0.11530
3, 27	0.03971	0.10910
4, 26	0.06379	0.09699
5, 25	0.08444	0.07967
6, 24	0.10054	0.05805
7, 23	0.11121	0.03329
8, 22	0.11589	0.00675
9, 21	0.11433	0.02016
10, 20	0.10660	0.04598
11, 19	0.09312	0.06933
12, 18	0.07462	0.08893
13, 17	0.05210	0.10374
14, 16	0.02678	0.11297
15	....	0.11608

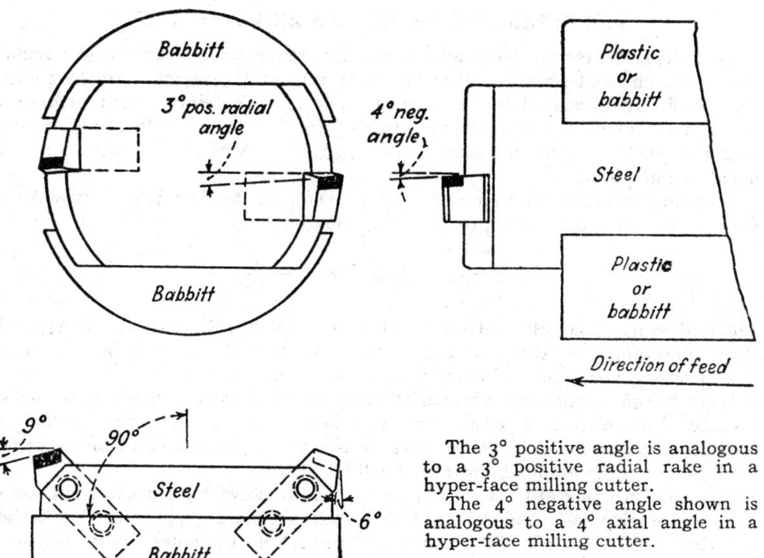

The 3° positive angle is analogous to a 3° positive radial rake in a hyper-face milling cutter.

The 4° negative angle shown is analogous to a 4° axial angle in a hyper-face milling cutter.

FIG. 43. Arrangement of modern gun-boring cutter head with plastic or babbitt wear strips.

is bored 0.5000 in. to the right of the center. Hole No. 2 is located by first moving 0.06699 in. to the left and then 0.2500 in. vertically. Hole No. 3 is located by moving 0.18301 in. horizontally to the left and 0.18301 in. vertically. This is repeated, using the proper constants, for locating hole No. 4. But, when locating hole No. 5, move vertically downward 0.06699 in. and then horizontally to the left 0.2500 in. This is repeated for holes Nos. 6 and 7, using the proper constants. Hole No. 8 is located by moving horizontally to the right 0.06699 in. and then vertically downward 0.2500 in. Thus in laying out the holes on the circumference, the mechanic must observe whether the next movement is vertically upward or downward, or horizontally left or right. He selects the proper movement by observing the development of the circle as the holes are bored.

Gun-boring Cutter Heads. Cutting heads for boring guns and long bores were formerly packed with hardwood to bear in the gun bore. However, to obtain longer wear, babbitt or plastic blocks are often used (Fig. 43). The plastic, called Insuroc, is made of cotton fabric impregnated with graphite and bonded with a synthetic resin.

For a gun-boring tool to run true through a long bore, it is essential that the tool be properly guided at the start of the cut. This has been done successfully with a long, tight-fitting bushing in a heavy support placed right next to the end of the piece. The tool is constrained to cut in line for several inches of depth, with the result that the backing strips then fit snugly in a true bore and can guide the tool properly for the rest of the cut. Of course, the piece should have been properly straightened and qualifying cuts taken for the chuck jaws and steadyrests.

MACHINABILITY OF METALS AND TOOL LIFE

Tool life, power consumption, and work finish are the practical aspects of machinability, which has been the subject of a vast amount of research. Machinability ratings of various materials are given in Table 24. Cold-drawn AISI 1112 steel is taken as having a machinability rating of 100%. A steel with a 50% rating would be machined at approximately half the speed for AISI 1112 in order to obtain equal tool life, if desired.

According to James Sorenson and Marlowe Peters of the Four Wheel Drive Auto Co., years of shop tests have produced this equation:

$$M = A + K_1C - K_2E - \frac{T}{K_3}$$

where M = machinability relative to AISI 1112, taken at 100%; A, K_1, K_2, and K_3 = constants (see Table 25); T = tensile strength of material in 1000 psi; C = percentage of carbon, as points; and E = elongation, %.

It is common practice to use brinell hardness as a guide to machinability, probably because of its relation to tensile strength. But brinell hardness may not be an accurate measure of machinability because two unlike steels may be of the same hardness but their machinability rating will be different.

The microstructure of the steel has a great deal to do with its machinability and the finish obtainable. Evaluations of alloy steels are more complicated than those of carbon steels, but G. P. Witteman of Bethlehem Steel Co. notes relative machinability according to three carbon ranges and four structures (Table 26).

Effect of Microstructure. Extensive investigation of carbon and alloy steels has proved that all steels with the same microstructure have about the same machinability. Importance of these studies lies in the economies in machining that can be realized by putting bar stock and forgings into the most desirable microstructure by heat-treatment. These data were compiled by a group of researchers acting under the direction of the Curtiss-Wright Corp., Wood Ridge, N.J. That company served as contractor to the U.S. Air Force and published the "U.S. Air Force Machinability Reports—1950 and 1951" in two volumes.* By taking advantage of the data and suggestions in those books, the progressive shop can attain far higher metal-removal rates than realized with the generally accepted data under "Machining Various Materials," page 12-80.

Microconstituents of Irons and Steels. Irons and steels have many common constituents, but cast iron contains free graphite, while steel does not.

FLAKY GRAPHITE. Black streaks indicate flaky graphite, which acts in iron as a free-machining agent.

NODULAR GRAPHITE. Free graphite in nodular or rounded form, instead of flakes, makes the metal ductile, increases its strength, and improves tool life.

FERRITE. White areas indicate practically carbon-free iron. Presence of free ferrite is extremely beneficial to tool life because it is relatively soft.

PEARLITE. Alternate white and black stripes represent white for ferrite and black for iron carbide. Pearlite is 50% harder than ferrite, thus shortens tool life.

FREE CARBIDE. Definite grain boundaries separate free carbide (white background) from other constituents. Presence of as little as 3 to 5% free carbide is detrimental to tool life.

* These books are out of print. For the gist of the information published in them, consult *American Machinist's* Special Reports: "Air Force Report Proves Machinability Depends on Microstructure," Nov. 27, 1950, page 109, and "U.S. Air Force Machinability Report 1951," Oct. 1, 1951, page 161.

TABLE 24. APPROXIMATE MACHINABILITY RATINGS OF VARIOUS METALS

(Based on Tool Life, Finish Produced, and Power Required When Cutting Is Done under Constant Conditions)

AISI No.	Approx. Brinell Hardness	Machinability Rating, %	AISI No.	Approx. Brinell Hardness	Machinability Rating, %	AISI No.	Approx. Brinell Hardness	Machinability Rating, %	Nonferrous Alloy	Machinability Rating, %
C1008*	113–126	45	B1112*	170	100	A4135†	180–195	56	Manganese alloys	500–2000
C1010*	113–130	45	B1113*	180	135	A4150†	215–225	54	Aluminum, 11-S	500–2000
C1015*	130–143	50	C1115*	180–190	90	A4320†	175–185	55	Aluminum, 2-S and 17-S	300–1500
C1020*	143–149	60	C1117*	160–165	94	E4340†	225–240	58		
C1025*	150–162	65	C1132*	195–205	57	E4615†	160–172	58		
C1030*	160–179	70	C1132†	175–182	75	E4640†	200–210	60	Brass, leaded	150–600
C1035*	180–200	62	C1137†	183–190	70	A4815†	205–215	55	Brass, yellow	200
C1040*	200–221	62	A2317†	160–170	50	A5120†	140–150	50	Brass, rod	200
C1045*	220–235	58	A2330†	195–210	45	A5150†	200–210	50	Bronze, leaded	200–500
C1050*	220–235	58	A2340†	220–230	40	E52100†	230–240	45	Zinc	200
C1070*	220–250	48	A2514†	170–190	40	A6120†	165–185	50	Mn bronze	40
C1085*	220–250	45	A3115†	145–155	50	A6140†	190–230	48	Copper, cast	70
C1095*	220–250	45	A3130†	200–220	45	A6152†	200–235	45	Copper, rolled	60
C1010†	100–115	40	A3135†	190–200	57	A9255†	300–310	30	Nickel	20
C1020†	130–140	52	A3140†	220–230	50	A9263†	325–350	25	Monel cast	35
C1025†	125–135	58	A3150†	220–230	50	Ni resist		30	Monel rolled	45
C1030†	130–140	60	A3240†	190–205	46	Stainless		25–45	K monel	50
C1035†	170–180	60	E3310†	170–229	40	Mn steel		30	Inconel	45
C1040†	185–195	60	A4023†	156–175	70	Tool steel		30	Everdur	60
C1045†	185–200	55	A4027†	150–175	70	High-speed steel		30		
C1055†	190–210	53	A4032†	165–185	65	High carbon—high chrome		25		
C1074†	208–215	45	A4037†	185–205	60	Cast iron	160–193	50		
C1080†	200–220	44	A4040†	190–235	55					
C1095†	190–210	45	A4119†	175–185	60					
			A4130†	175–185	58					

A = basic open-hearth alloy steel
B = acid bessemer carbon steel
E = electric furnace alloy steel

* = cold-drawn steel
† = as rolled or annealed

White Iron. A structure of 50% fine pearlite and 50% free carbide, that is, no free graphite, is very hard and is virtually unmachinable.

Free ferrite is present in most low-carbon steels, while pearlite is present in small quantities in low-carbon steels but the amount increases with carbon content. Thus SAE 1010 contains about 15% pearlite, slowly cooled SAE 1040 about 50%, and slowly cooled SAE 1080 steel up to 100%.

Other constituents of steels are:

Spheroidite. By heat-treatment iron carbide in pearlite can be altered from a platelike form to a spherical shape to improve tool life when machined with HSS tools.

Table 25. Constants for Calculating Relative Machinability

SAE Steels*	Condition of Steel	Constants				Microstructure of Annealed SAE Steels
		A	K_1	K_2	K_3	
Series 1000: carbon steels	As-rolled	108	2.00	1.00	1.00	SAE 1020, 1030, 1040: medium-sized pearlite and ferrite
	Annealed	173	1.00	2.00	1.00	SAE 1050 and 1060: ferrite and spheroidized pearlite
	Quenched and tempered to 300 brinell max	185	0.50	2.00	1.00	SAE 1070, 1080, 1095: spheroidal, carbide plus ferrite
	Cold-drawn	94	1.20	1.00	2.00	
Series 2300: nickel steels	Annealed	177	0.20	1.50	1.00	SAE 2320, 2330: medium-sized pearlite and ferrite
	Quenched and tempered to 300 Brinell max	175	0.50	1.00	1.00	SAE 2340, 2350: ferrite and spheroidized pearlite
	Cold-drawn	155	1.00	1.00	1.00	
	Annealed and cold-drawn	96	1.00	1.00	2.00	
Series 3100: low-chrome nickel steels	Annealed	124	2.00	1.00	1.00	SAE 3115, 3120, 3130, 3140, 3150; medium-sized pearlite and ferrite
	Quenched and tempered to 300 Brinell max	113	0.50	1.00	2.00	
Series 3200: medium-chrome nickel steels	Annealed	130	1.60	1.00	1.00	SAE 3200: medium-sized pearlite and ferrite
	Quenched and tempered to 300 brinell max	202	0.20	2.00	1.00	SAE 3230, 3240, 3250: ferrite and spheroidized pearlite
Series 3300: high-chrome nickel steels	Annealed	147	1.20	1.00	1.00	SAE 3312: medium-sized pearlite and ferrite SAE 3340; ferrite and spheroidized pearlite
Series 4100: chrome molybdenum steels	Annealed	147	1.00	1.00	1.00	SAE 4120, 4130; medium-sized pearlite and ferrite SAE 4140, 4150; ferrite and spheroidized pearlite
Series 4300: chrome nickel molybdenum steels	Annealed	105	1.00	1.00	2.00	SAE 4340; ferrite and spheroidized pearlite
	Quenched and tempered to 300 brinell max	160	0.50	3.00	2.00	
Series 4600; nickel molybdenum steels	Annealed	151	1.00	1.00	1.00	SAE 4620, 4630, 4640: medium-sized pearlite and ferrite
	Quenched and tempered to 300 brinell max	118	0.50	1.00	2.00	
Series 6100: chrome vanadium steels	Annealed	98	1.00	1.00	2.00	SAE 6120: medium-sized pearlite and ferrite
	Quenched and tempered to 300 brinell max	203	0.50	2.00	1.00	SAE 6140, 6150: ferrite and spheroidized pearlite
	Annealed and cold-drawn	175	1.00	1.00	2.00	

* Grain size: 5 to 8, according to classification E-19 of the ASTM.

TABLE 26. MACHINABILITY OF ALLOY STEELS—SUGGESTED STRUCTURES

Carbon Range	Process	Structure	Turning	Forming	Drilling	Broaching
Low (0.08 to 0.30).....	Normalize and anneal	Blocky ferrite	Good	Good	Good	Good
Medium (0.30 to 0.50)	Anneal	Spheroidized	Good	Poor	Fair	Poor
Medium (0.30 to 0.50)	Anneal	Lamellar	Fair	Good	Good	Good
Medium (0.30 to 0.50)	Heat-treat	Sorbitic	Fair	Fair	Fair	Fair
High (0.50 to 0.80)....	Anneal	Spheroidized	Good	Good	Good	Fair
High (0.50 to 0.80)....	Anneal	Lamellar	Fair	Poor	Poor	Poor
High (0.50 to 0.80)....	Heat-treat	Sorbitic	Good	Fair	Good	Good

WIDMANSTATTEN STRUCTURE. A ferrite-pearlite structure of cross-hatched appearance is obtained by cooling at a critical rate from high temperatures. When steel so treated is machined with KH 3 or 78 carbides, tool life up to three times that for machining annealed or normalized structures is obtained.

Alloy content and heat-treatment, also prior processing, affect the actual structures obtained in steels. Seldom will all the bars of any lot of steel, as received, be of identical structure, and furthermore the structure may vary within a single bar. The cost of heat-treatment versus machining rate must be balanced, and a profit shown, to make machining of controlled-microstructure material worthwhile.

Mechanics of Cutting. There is more to machinability than microstructure of the material. In forming a chip, the shear process produces three forces on the tool's cutting edge—cutting force, feed force, and radial force. The cutting, or vertical, force is the one that requires the most energy or horsepower, but as the tool wears, the feed and radial forces increase.

In general, cutting force is constant with speed but varies with respect to depth of cut and to feed.

Cutting speeds are restricted by the ability of the tool and coolant to remove heat from the chip-tool interface. Under extreme temperatures and heavy feeds, tool wear is accelerated by welding of the chip to tool and consequent cratering of the tool surface. However, microstructure does affect the frictional heat developed. Normalized and annealed pearlitic structures give longer tool life above 700 sfpm than do spheroidized steels. Cutting temperature does not rise in proportion to feed. Therefore, the trick of increasing metal-removal rate by increasing the feed has less effect on the cutting temperature than increasing the cutting speed.

An increase in depth of cut has little effect on chip-tool temperature, but is limited by the amount of stock on the work and finish tolerances.

Effective tooling for high-velocity turning requires rigid machines and toolholders, higher positive rake angles, a grade of carbide that resists cratering, and suitable chipbreaker design. Then, a balance between good tooling, proper work microstructure, and feed rate will reduce the frictional heat generated in cutting and permit an over-all increase in cutting speed and hence metal-removal rate with reasonable tool wear and frequency of tool changes.

When a seemingly satisfactory cutting speed has been determined, rapid tool-life tests can be made to determine the optimum feed rate. This practice is especially important for production operations to be carried out at cutting speeds above 500 fpm.

ESTIMATING MACHINING TIME

Formulas

CUTTING SPEED

$$S = 3.1416D \times R$$

where S = cutting speed, sfpm; D = diameter of work, ft; and R = work speed, rpm.

If diameter is expressed in inches,

$$S = 3.1416D' \times R \div 12 = 0.2618D' \times R$$

EXAMPLES: $D = 1$ ft or $D' = 12$ in. $R = 100$ rpm.
Then

$$S = 3.1416D \times R = 3.1416 \times 1 \times 100 = 314.16 \text{ sfpm}$$

or

$$S = 0.2618 \times 12 \times 100 = 314.16 \text{ sfpm}$$

CUTTING TIME

$$T = \frac{3.1416 \times D \times L}{12 \times F \times S}$$

where T = time, min; D = diameter, in.; L = length of cut, in.; F = feed, ipr; and S = cutting speed, sfpm.

EXAMPLE: $D = 12$ in., $L = 10$ in., $F = \frac{1}{8}$ in. per revolution, and $S = 200$ sfpm.
Then

$$T = \frac{12 \times 3.1416 \times 10}{12 \times \frac{1}{8} \times 200} = \frac{31.416}{25} = 1.26$$

Known factors in the above formula for cutting time can be resolved into a constant, and a table prepared for a wide range of speed and feed combinations. Table 27 has two parts. The upper section gives constants for two cuts, the pairs ranging from $\frac{1}{64}$- and $\frac{1}{32}$-in. feeds up to $\frac{1}{8}$- and $\frac{1}{4}$-in. feeds. The lower section gives con-

TABLE 27. CONSTANTS FOR CUTTING TIME IN MINUTES

Feed, Inches	Cutting Speed, Feet per Minute—Turning						
	20	25	30	35	45	55	65
Two cuts at:							
1/64 and 1/32	1.257	1.005	0.8378	0.7181	0.5581	0.4572	0.386
1/32 and 1/16	0.6283	0.5027	0.4189	0.3590	0.2793	0.2286	0.193
1/16 and 1/8	0.3142	0.2513	0.2094	0.1795	0.1396	0.1143	0.097
1/8 and 1/4	0.1571	0.1257	0.1047	0.0898	0.0698	0.0571	0.048
One cut at:							
1/64	0.8378	0.6702	0.5585	0.4787	0.3723	0.3046	0.257
1/32	0.4189	0.3351	0.2793	0.2394	0.1862	0.1523	0.128
1/16	0.2094	0.1676	0.1396	0.1197	0.0931	0.0761	0.064
1/8	0.1047	0.0838	0.0698	0.0598	0.0465	0.0381	0.032
1/4	0.0524	0.0419	0.0349	0.0299	0.0233	0.0191	0.016

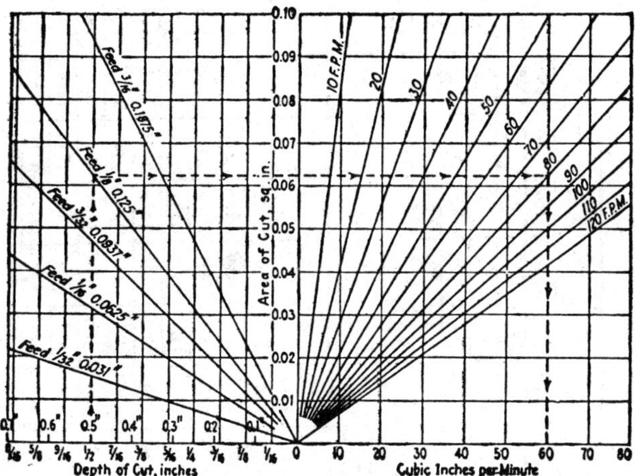

FIG. 44. Chart for estimating metal removal at various cuts, speeds, and feeds.

stants for single cuts ranging from $\frac{1}{64}$ to $\frac{1}{4}$ in. Cutting speeds range from 20 to 65 sfpm, and the selected values are such that constants for many others can easily be secured. Constants for any multiple of the cutting speeds stated are obtained by dividing by the multiple involved. For example, constants for 200 sfpm are found by dividing the constants under 20 sfpm by 10. Similarly, for 210 sfpm, divide constants under 35 sfpm by 6.

A typical computation is as follows:

A piece 4 in. dia, 10 in. long, is turned with two cuts at $\frac{1}{16}$- and $\frac{1}{32}$-in. feed, each with a cutting speed of 20 sfpm. Diameter × length in inches × constant = time in minutes. $4 \times 10 \times 0.6283 = 26$ min.

If, for the purpose of accuracy it is thought advisable, in connection with these two cuts, to run another cut over, the constant 0.6283 is added to a constant for the third feed used. If this feed is $\frac{1}{64}$ in., then we have $0.6283 + 0.8378 = 1.4661$, the constant for three cuts, one of $\frac{1}{16}$-, one of $\frac{1}{32}$-, and one of $\frac{1}{64}$-in. feed.

Cutting-speed Data. Information on cutting speeds for various work materials and tool materials can be looked up under Machining Various Materials, starting on page 12–80. Additional data will be found under Cutting-tool Materials, starting on page 12–24.

Conversion of cutting speed to work rpm, and vice versa, for work diameters of $\frac{1}{8}$ to 18 in. dia, is accomplished conveniently by use of Table 28. This table covers cutting speeds from 40 to 1000 sfpm, and other conversion values can be interpolated readily.

Volume of Metal Removed. The chart shown in Fig. 44 makes it possible to estimate rapidly the amount of metal removed at various cuts, feeds, and speeds. To use the chart, start with the depth of cut at bottom left.

EXAMPLE: Follow dash line upward from $\frac{1}{2}$-in. depth of cut to $\frac{1}{8}$-in. feed, go right to intersect the cutting-speed curve for 80 fpm, and drop down to find that 60 cu in. of metal are removed per minute. Figures in the center of the chart show the area of the cut in square inches.

TABLE 28. WORK OR TOOL RPM FOR CUTTING SPEEDS FROM 40 TO 1000 SFPM
Calculated by Gorham Tool Co.

Tool or Work Dia. (Inches)	SURFACE FEET PER MINUTE																		
	40	50	60	70	80	90	100	125	150	175	200	250	300	400	500	600	700	800	900	1000
1/16	2445	3056	3667	..	..	..	..	..	..	..	..	..	..	..	..	..	..	..	..	..
1/8	1222	1529	1833	2139	2445	2750	3056	..	..	..	..	..	..	..	..	..	..	..	..	..
3/16	815	1019	1222	1426	1630	1833	2037	2544	3056	3565	..	..	..	..	..	..	..	..	..	..
1/4	611	764	917	1070	1222	1375	1528	1908	2292	2674	3056	..	..	..	..	..	..	..	..	..
5/16	489	611	733	856	978	1100	1222	1526	1833	2139	2444	3056	..	..	..	..	..	..	..	..
3/8	407	509	611	713	815	917	1019	1272	1528	1783	2036	2547	3056	..	..	..	..	..	..	..
7/16	349	437	524	611	698	786	873	1090	1310	1528	1748	2183	2619	3492	..	..	..	..	..	..
1/2	306	382	458	535	611	688	764	954	1146	1337	1528	1910	2292	3056	..	..	..	..	..	..
5/8	244	306	367	428	489	550	611	763	917	1070	1224	1528	1834	2445	3056	..	..	..	..	..
3/4	204	255	306	357	407	458	509	636	764	891	1016	1273	1528	2037	2547	3053	3565	..	..	..
1	153	191	229	267	306	344	382	477	573	668	764	955	1146	1528	1910	2292	2674	3056	3438	..
1 1/4	122	153	183	214	244	275	306	382	458	535	612	764	917	1222	1528	1832	2139	2445	2750	3056
1 1/2	102	127	153	178	204	229	255	318	382	445	510	637	764	1019	1273	1527	1783	2037	2292	2547
1 3/4	87	109	131	153	175	196	218	273	327	382	436	546	655	873	1091	1308	1528	1746	1964	2183
2	76	96	115	134	153	172	191	239	287	334	382	477	573	764	955	1145	1337	1528	1719	1910
2 1/4	68	85	102	119	136	153	170	212	255	297	340	424	509	679	849	1018	1188	1358	1528	1698
2 1/2	61	76	92	107	122	138	153	191	229	268	306	382	458	612	764	916	1070	1222	1375	1528
2 3/4	56	70	83	97	111	125	139	174	208	243	278	347	417	556	694	833	972	1111	1250	1389
3	51	64	76	89	102	115	127	159	191	223	255	318	382	510	637	763	891	1019	1146	1273
3 1/4	47	59	71	62	94	106	118	147	176	206	234	294	353	470	588	705	823	940	1058	1175

Tool or Work Dia. (Inches)	SURFACE FEET PER MINUTE																		
	40	50	60	70	80	90	100	125	150	175	200	250	300	400	500	600	700	800	900	1000
3½	44	55	66	76	87	98	109	136	164	191	218	273	328	437	546	654	764	873	982	1091
3¾	41	51	61	71	82	92	102	127	153	178	204	255	306	407	509	611	713	815	917	1019
4	38	58	57	67	76	86	96	119	143	167	191	239	287	382	478	572	668	764	859	955
4½	34	42	51	59	68	76	85	106	127	149	170	212	255	340	425	509	594	679	764	849
5	31	38	46	54	61	69	76	96	115	134	153	191	229	306	382	458	535	611	688	764
5½	28	35	42	49	56	63	70	87	104	122	139	174	209	278	348	416	486	556	625	695
6	25	32	38	45	51	57	64	80	96	112	127	159	191	255	319	382	446	509	573	637
6½	23	29	35	41	47	53	59	74	88	103	118	147	176	235	294	352	411	470	529	588
7	22	27	33	38	44	49	55	68	82	96	109	136	164	218	273	327	382	436	491	546
7½	20	26	31	36	41	46	51	64	76	89	102	127	153	204	255	305	356	407	458	509
8	19	24	29	33	38	43	48	60	72	84	95	119	143	191	239	286	334	382	430	478
8½	18	23	27	32	36	40	45	56	67	78	90	112	135	180	225	269	315	359	404	449
9	17	21	26	30	34	38	42	53	64	74	85	106	127	170	212	254	297	339	382	424
9½	16	20	24	28	32	36	40	50	60	70	80	100	121	161	201	241	281	322	362	402
10	15	19	23	27	31	34	38	48	57	67	76	95	115	153	191	229	267	306	344	382
11	14	17	21	24	28	31	35	43	52	61	69	87	104	139	174	208	243	278	312	347
12	13	16	19	22	26	29	32	40	48	56	64	79	96	127	159	191	223	255	286	318
13	12	15	18	21	23	26	29	37	44	51	59	73	68	117	147	176	206	235	264	294
14	11	14	16	19	22	24	27	34	41	48	55	68	82	109	136	163	191	218	245	273
15	10	13	15	18	20	23	25	32	38	44	51	64	76	102	127	153	178	204	229	255
16	9	12	14	17	19	21	24	30	36	42	48	60	72	95	119	143	167	191	215	239
17	9	11	13	16	18	20	22	28	34	39	45	56	67	90	112	135	157	180	202	225
18	8	11	13	15	17	19	21	26	32	37	42	53	64	85	106	127	148	170	191	212

Table 29. Lathe and Boring-mill Time
In Minutes when Length of Cut = 12 In.

Dia of Work	25 Feeds, Ipr				30 Feeds, Ipr				35 Feeds, Ipr				40 Feeds, Ipr			
	1/32	1/8	1/4	1/2	1/32	1/8	1/4	1/2	1/32	1/8	1/4	1/2	1/32	1/8	1/4	1/2
2	8.042	2.011	1.005	0.503	6.702	1.675	0.838	0.419	5.745	1.436	0.718	0.359	5.027	1.257	0.628	0.314
3	12.064	3.016	1.508	0.754	10.053	2.513	1.257	0.628	8.617	2.154	1.077	0.539	7.540	1.885	0.942	0.471
4	16.085	4.021	2.011	1.005	13.404	3.351	1.676	0.838	11.489	2.872	1.436	0.718	10.053	2.513	1.257	0.628
5	20.106	5.027	2.513	1.257	16.755	4.189	2.094	1.047	14.362	3.590	1.795	0.898	12.566	3.142	1.571	0.785
6	24.127	6.032	3.016	1.508	20.106	5.027	2.513	1.257	17.234	4.308	2.154	1.077	15.080	3.770	1.885	0.942
7	28.149	7.037	3.519	1.759	23.457	5.864	2.932	1.466	20.106	5.027	2.513	1.257	17.593	4.398	2.199	1.100
8	32.170	8.042	4.021	2.011	26.808	6.702	3.351	1.676	22.979	5.745	2.872	1.436	20.106	5.027	2.513	1.257
9	36.191	9.048	4.524	2.262	30.159	7.540	3.770	1.885	25.851	6.463	3.231	1.616	22.620	5.655	2.827	1.414
10	40.212	10.053	5.027	2.513	33.510	8.378	4.189	2.094	28.723	7.181	3.590	1.795	25.133	6.283	3.142	1.571
11	44.234	11.058	5.529	2.765	36.861	9.215	4.608	2.304	31.596	7.890	3.949	1.975	27.646	6.912	3.456	1.728
12	48.255	12.064	6.032	3.016	40.122	10.053	5.027	2.513	34.468	8.617	4.308	2.154	30.159	7.540	3.770	1.885
13	52.276	13.069	6.535	3.267	43.564	10.891	5.445	2.723	37.340	9.335	4.668	2.334	32.673	8.168	4.084	2.042
14	56.297	14.074	7.037	3.519	46.915	11.729	5.864	2.932	40.212	10.053	5.027	2.513	35.186	8.796	4.398	2.199
15	60.319	15.080	7.540	3.770	50.266	12.566	6.283	3.142	43.085	10.771	5.386	2.693	37.699	9.425	4.712	2.356
16	64.340	16.085	8.042	4.021	53.617	13.404	6.702	3.351	45.957	11.489	5.745	2.872	40.212	10.053	5.027	2.513
17	68.361	17.090	8.545	4.273	56.968	14.242	7.121	3.560	48.829	12.207	6.104	3.052	42.726	10.681	5.341	2.670
18	72.382	18.096	9.048	4.524	60.319	15.080	7.540	3.770	51.702	12.925	6.463	3.231	45.239	11.310	5.655	2.828
19	76.404	19.101	9.550	4.775	63.670	15.917	7.959	3.979	54.574	13.644	6.822	3.411	47.752	11.938	5.969	2.985
20	80.425	20.106	10.053	5.027	67.021	16.755	8.378	4.189	57.446	14.362	7.181	3.590	50.266	12.566	6.283	3.142
21	84.446	21.112	10.556	5.278	70.372	17.593	8.796	4.398	60.319	15.080	7.540	3.770	52.779	13.195	6.597	3.299
22	88.467	22.117	11.058	5.529	73.723	18.431	9.215	4.608	63.191	15.798	7.899	3.949	55.292	13.823	6.912	3.456
23	92.489	23.122	11.561	5.781	77.074	19.268	9.634	4.817	66.063	16.516	8.258	4.129	57.805	14.451	7.226	3.613
24	96.510	24.127	12.064	6.032	80.425	20.106	10.053	5.027	68.936	17.234	8.617	4.308	60.319	15.080	7.540	3.770
25	100.53	25.133	12.566	6.283	83.776	20.944	10.472	5.236	71.808	17.952	8.976	4.488	62.832	15.708	7.854	3.927
26	104.55	26.138	13.069	6.535	87.127	21.782	10.891	5.445	74.680	18.670	9.335	4.668	65.345	16.336	8.168	4.084
27	108.57	27.143	13.572	6.786	90.478	22.620	11.310	5.655	77.553	19.388	9.694	4.847	67.859	16.956	8.482	4.241
28	112.59	28.149	14.074	7.037	93.829	23.457	11.729	5.864	80.425	20.106	10.053	5.027	70.372	17.593	8.796	4.398
29	116.62	29.154	14.577	7.289	97.180	24.295	12.148	6.074	83.297	20.824	10.412	5.206	72.885	18.221	9.111	4.555
30	120.64	30.159	15.080	7.540	100.53	25.133	12.566	6.283	86.170	21.542	10.771	5.386	75.398	18.850	9.425	4.712

Column group headings: Cutting Speed, Fpm*

* For speeds of 50, 60, 70, and 80 fpm, divide these time values by 2. For speeds of 75, 90, 105, and 120 fpm, divide by 3. Other speed values can be found in the same way.

Table 30. Time Required for Tool to Travel 1 In. When Feed Is 1/100 In. per Revolution

In Minutes and Seconds

Dia, Inches	Surface Speed per Minute							
	20	25	30	35	40	45	50	60
1/4	0 20	0 16	0 13	0 11	0 10	0 9	0 8	0 7
5/16	0 25	0 20	0 17	0 14	0 13	0 11	0 10	0 9
3/8	0 29	0 23	0 19	0 17	0 15	0 13	0 12	0 10
7/16	0 34	0 27	0 23	0 19	0 17	0 15	0 14	0 12
1/2	0 39	0 31	0 26	0 22	0 20	0 17	0 16	0 13
9/16	0 44	0 35	0 29	0 25	0 22	0 19	0 18	0 15
5/8	0 49	0 39	0 32	0 28	0 25	0 22	0 20	0 16
11/16	0 54	0 43	0 36	0 31	0 27	0 24	0 22	0 18
3/4	0 59	0 47	0 39	0 34	0 30	0 26	0 24	0 20
13/16	1 4	0 51	0 42	0 36	0 32	0 28	0 26	0 21
7/8	1 9	0 55	0 46	0 39	0 35	0 30	0 28	0 23
15/16	1 14	0 59	0 49	0 42	0 37	0 33	0 30	0 25
1	1 19	1 3	0 52	0 45	0 40	0 35	0 32	0 26
1 1/8	1 28	1 10	0 58	0 50	0 44	0 39	0 35	0 29
1 1/4	1 38	1 18	1 5	0 56	0 49	0 43	0 39	0 33
1 3/8	1 48	1 26	1 11	1 2	0 54	0 48	0 43	0 36
1 1/2	1 58	1 34	1 18	1 7	0 59	0 52	0 47	0 39
1 5/8	2 8	1 42	1 24	1 13	1 4	0 56	0 51	0 42
1 3/4	2 18	1 50	1 31	1 19	1 9	1 1	0 55	0 46
1 7/8	2 27	1 58	1 37	1 24	1 14	1 5	0 59	0 49
2	2 37	2 6	1 44	1 29	1 19	1 9	1 3	0 52
2 1/8	2 46	2 13	1 50	1 34	1 23	1 13	1 6	0 55
2 1/4	2 56	2 21	1 56	1 40	1 28	1 17	1 11	0 58
2 3/8	3 6	2 29	2 3	1 46	1 33	1 22	1 15	1 2
2 1/2	3 16	2 37	2 9	1 52	1 38	1 26	1 19	1 5
2 3/4	3 37	2 54	2 23	2 4	1 49	1 35	1 27	1 12
3	3 56	3 9	2 36	2 15	1 58	1 44	1 34	1 18
3 1/4	4 16	3 25	2 49	2 26	2 8	1 53	1 42	1 25
3 1/2	4 35	3 40	3 2	2 37	2 18	2 1	1 50	1 31
3 3/4	4 56	3 57	3 15	2 49	2 28	2 10	1 58	1 38
4	5 14	4 11	3 27	2 59	2 37	2 18	2 6	1 44
4 1/4	5 33	4 26	3 40	3 10	2 47	2 27	2 13	1 50
4 1/2	5 52	4 42	3 52	3 21	2 56	2 35	2 21	1 56
4 3/4	6 12	4 58	4 6	3 32	3 6	2 44	2 29	2 3
5	6 32	5 14	4 19	3 43	3 16	2 52	2 37	2 10
5 1/2	7 15	5 48	4 47	4 8	3 37	3 11	2 54	2 24
6	7 52	6 18	5 12	4 29	3 56	3 28	3 9	2 36
6 1/2	8 33	6 50	5 39	4 52	4 17	3 46	3 25	2 50
7	9 10	7 20	6 3	5 14	4 35	4 2	3 40	3 2
7 1/2	9 54	7 55	6 32	5 39	4 57	4 21	3 58	3 16
8	10 28	8 22	6 54	5 58	5 14	4 36	4 11	3 27
8 1/2	11 7	8 54	7 20	6 20	5 34	4 53	4 27	3 40
9	11 46	9 25	7 46	6 42	5 53	5 11	4 42	3 53
9 1/2	12 25	9 56	8 12	7 5	6 12	5 28	4 58	4 6
10	13 9	10 31	8 40	7 30	6 34	5 47	5 16	4 20
10 1/2	13 40	10 56	9 2	7 47	6 50	6 1	5 28	4 31
11	14 30	11 36	9 34	8 16	7 15	6 23	5 48	4 47
11 1/2	15 2	12 2	9 55	8 34	7 31	6 37	6 1	4 58
12	15 44	12 35	10 23	8 58	7 52	6 55	6 18	5 14
12 1/2	16 24	13 7	10 49	9 21	8 12	7 13	6 34	5 25
13	17 4	13 39	11 16	9 44	8 32	7 31	6 50	5 38
13 1/2	17 32	14 2	11 34	10 0	8 46	7 43	7 1	5 47
14	18 20	14 40	12 6	10 27	9 10	8 4	7 20	6 3
14 1/2	18 52	15 6	12 28	10 46	9 26	8 19	7 33	6 14
15	19 38	15 41	12 57	11 10	9 49	8 38	7 50	6 29

MACHINING VARIOUS MATERIALS

Aluminum Alloys

Machining Characteristics. Commercially pure aluminum and some alloys are soft and gummy; others produce long stringy chips that foul the tool or machine. But many alloys can be cut at high speed, produce small chips, can readily be machined to smooth surfaces, and give long tool life.

Wrought alloys that do not respond to heat-treatment give continuous chips and offer little resistance to cutting but require care to obtain good finish because they are inclined toward gumminess. In full-hard temper the alloys are easier to machine to a good finish than when in the annealed state.

Most heat-treatable wrought alloys contain a fairly high percentage of copper and can be machined to a good finish with or without lubrication. However, a good coolant-lubricant is recommended for most operations.

Alloys containing more than 10% silicon are the most difficult to machine, and none of those containing over 5% silicon will finish to a bright lustrous surface. Chips are likely to tear rather than shear from the work.

Thermal Expansion. The thermal expansion of aluminum is about twice that of steel—a good rule is 0.0001 in. of expansion for each 8 F temperature rise. When dimensional accuracy is important, overheating is kept at a minimum with sharp well-designed tools, a coolant, and feeds that are not too heavy. Even so, it may be necessary to make a thermal allowance in measuring. This will be required if parts have been put through a washer immediately after machining.

Aluminum has a high coefficient of friction with steel. Tool surfaces must therefore be highly polished and free from scratches.

Because thermal expansion is a major factor, internal expanding chucks should be checked often to see that the expansion does not loosen the grip. Expansion may put heavy end thrust on the tailstock, so that live centers are preferable to fixed centers.

The fullest economy in machining aluminum is realized by use of high speeds but moderate feeds and cuts. In consequence, the machine, fixture, and tool must be rigid to avoid vibration. Tools should be supported close to the cutting edge.

Springiness. Aluminum "springs" more under load than steel. This means care is necessary in clamping or chucking the work to avoid distortion. Both tool- and workholders should be of heavy construction. Soft liners between work and jaw faces prevent marring of the work.

Tool Design. In general, tools should have more rake than for steel, with additional space for chips. Tools should be designed so that chips are directed away from the finished work.

Most authorities agree on round-nose tools with large rake angles set at or slightly above the center of the stock. If the rough and finish cuts are made with the same tool, the cutting edge should be restored before the finishing cut.

Chip disposal may become a problem because of the continuous chips produced. If rake angles are decreased or feed increased, more curl may be caused, thus resulting in better chip breakage. These changes tend to reduce the quality of surface produced, and their extent varies with different alloys. When satisfactory chip control is not possible by these means, a chip breaker may be ground in the tool.

Cutting Oil. Whenever heavy cuts and feeds are employed, a coolant is desirable. For this purpose soda water or a lean soluble-oil solution, perhaps with a little lard oil or kerosene added, is satisfactory.

When lubricating characteristics are required, a straight mineral oil with viscosity about 60 sec Saybolt Universal at 100 F gives good results at little expense. Better results can be obtained by adding 5 to 10% of fatty oil such as lard oil. Heavy cuts may require higher viscosity lubricants.

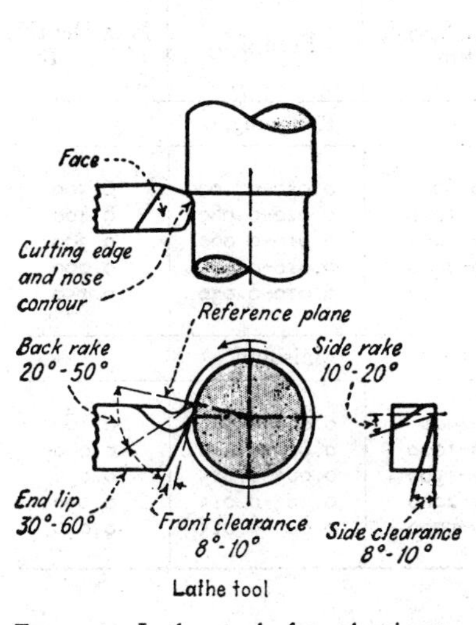

FIG. 45. Lathe tool for aluminum.

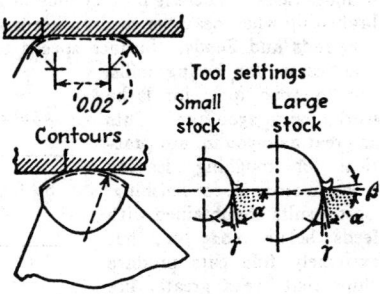

$\alpha = 74\text{-}90°;\ \beta = 10\text{-}0°;\ \gamma = 6\text{-}0°$

FIG. 46. These angles have been suggested for finishing aluminum with diamond tools.

ROUGHING

Diamond-tipped tools are not recommended.

FINISHING

	Soft	Hard	High sil.
Cut.......	0.010	0.006	0.006
Sfpm.....	Max without vibration		
Feed......	0.005	0.004	0.003

TABLE 31. SPEEDS AND FEEDS FOR ALUMINUM ALLOYS

	HIGH-CARBON OR HIGH-SPEED STEEL				CEMENTED-CARBIDE TOOLS		
	Alloy type	Max. cut, in.	Speed fpm.	Feed ipr.	Max. cut, in.	Speed fpm.	Feed ipr.
ROUGHING	Soft	0.250	700-1600	up to 0.050	0.250	4000-7000	up to 0.012
	Hard	0.200	up to 650	0.007-0.020	0.200	500-1300	up to 0.010
	High silicon	0.120	up to 400	0.007-0.020	0.120	500-1000	up to 0.008
FINISHING	Soft	0.040	1500-3500	0.004-0.015	0.020	6000-8000	up to 0.006
	Hard	0.020	600-2000	0.002-0.010	0.020	700-2500	up to 0.010
	High silicon	0.020	up to 600	0.002-0.004	0.020	500-1500	up to 0.004

Magnesium Alloys

Machining Characteristics. Magnesium alloys machine so readily that it is not often possible to obtain full advantage of this characteristic without specially designed machine tools. Excellent surface finish can be obtained, to avoid grinding in most cases. There is no tendency of the metal to tear or drag. Chips are well broken up when heavy or medium feeds are taken.

Speeds and Feeds. Surface speeds to 5000 fpm have been used, but general practice is: for boring mills, 150 to 1000 fpm; for lathe work, 400 to 2500 fpm. Cuts as great as 0.500 in. are practical for roughing, leaving 0.003 to 0.004 in. for finishing. Best results are obtained with feeds below 0.025 in., but extremely fine cuts produce chips that are a greater fire hazard.

Tool Design. Tools of the same design as used for brass and steel may be used, but larger relief angles and attention to chip flow produce better results. Back- and side-rake angles are usually 0 to 15° with high-speed tools and 0 to 8° with carbide tools. Relief angles of 8 to 12° are generally required. Side- and end-cutting-edge angles can be varied widely, but the latter

TABLE 32. RECOMMENDED TURNING SPEEDS, FEEDS, AND CUTS

Surface Speed, Fpm	Feed, Ipr	Max Depth of Cut, In.
Roughing		
300–600	0.030–0.100	0.500
600–1000	0.020–0.080	0.400
1000–1500	0.010–0.060	0.300
1500–2000	0.010–0.040	0.200
2000–5000	0.010–0.030	0.150
Finishing		
300–600	0.005–0.020	0.100
600–1000	0.005–0.020	0.080
1000–1500	0.003–0.015	0.050
1500–2000	0.003–0.015	0.050
2000–5000	0.003–0.015	0.050

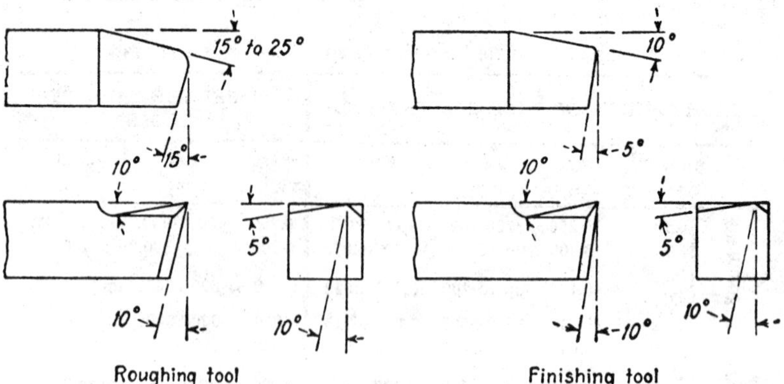

FIG. 47. Typical turning tools for magnesium.

should not exceed 40° to avoid chatter. Finishing tools should have a round nose to insure smooth finish, but keep the radius small. Parting tools should have at least 6° front and side relief. Back rake is often 15 to 20° but can be less.

Fire Control. Fine chips are easily ignited. Good housekeeping is the major preventive means. Some shops use vacuum cleaners at frequent intervals to remove small chips and dust from worker's clothing and machines, then deposit chips in closed containers for immediate removal to a safe storage area. A graphite powder seems to extinguish small fires as quickly as anything. Fires in large masses of ingots or castings have been brought under control with heavy streams of water. An efficient sprinkler system is indicated to protect the shop. Cutting fluids are used primarily to cool the work and reduce the fire hazard when taking fine cuts; otherwise machining can be done dry in most cases.

Copper and Its Alloys

Machining Characteristics. Although the machining of copper and its alloys has been considered less difficult than the machining of steels, the problems of the shop have mounted in proportion to the increasing number of compositions. For convenience, the 36 standard compositions recognized by the Copper & Brass Research Association are divided into three groups according to machinability rating (see Sec. 26 for compositions).

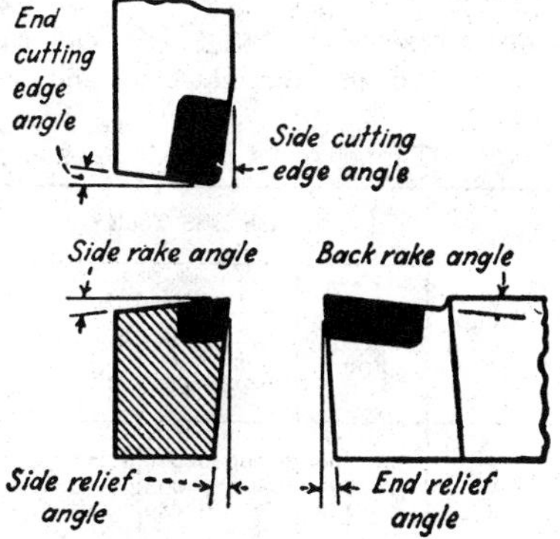

FIG. 48. Key to tool angles

GROUP 1. Free-machining alloys (usually brasses) with machinability ratings from 70 to 100. Free-cutting brass has an index of 100.

GROUP 2. Readily machinable alloys have a machinability index of over 30,

TABLE 33. TOOL ANGLES FOR COPPER ALLOYS

Group	Carbon and High-speed Turning Tools*				Carbide-tipped Turning Tools†			
	Relief Angles		Rake Angles		Relief Angles		Rake Angles	
	Side	End	Back	Side	Side	End	Back	Side
1	0–5	6	0	0–3	4–6	4–6	0	2–6
2	5–10	6–15	5–10	5–10	4–8	4–8	0–5	4–8
3	10–20	10–15	10–20	20–30	7–10	7–10	4–8	15–25

* An 8 to 15° end-cutting angle is proposed for the three groups with a side-cutting-edge angle of 10 to 15° for groups 1 and 2 and a 15° for group 3.

† End-cutting-edge angle to suit or from 8 to 15° and side-cutting-edge angle from 10 to 15° or to suit.

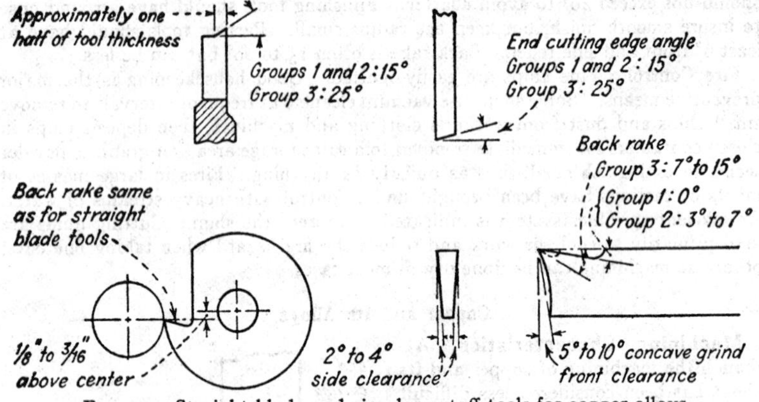

FIG. 49. Straight-blade and circular cutoff tools for copper alloys.

TABLE 34. CUTTING SPEEDS AND FEEDS FOR COPPER ALLOYS

Copper Group	With HSS Tools*			With Carbide-tipped Tools†			
	Cutting Speed, Fpm	Feed, Ipr		Cutting Speed, Fpm		Feed, Ipr	
		Roughing	Finishing	Roughing	Finishing	Roughing	Finishing
1	300–700	0.006–0.020	0.003–0.015	400–800	500–1000	0.015–0.025	0.005–0.015
2	150–300	0.015–0.035	0.005–0.015	300–500	400–600	0.015–0.030	0.005–0.015
3	75–150	0.015–0.040	0.005–0.020	250–600	300–800	0.015–0.030	0.008–0.015

* When using carbon-steel cutting tools, reduce suggested speeds by one-half. A roughing cut of approximately 0.045 to 0.125 and finishing cut of 0.015 to 0.030 were used as a basis for this table.

† A roughing cut of approximately 0.045 to 0.125 and finishing cut of 0.015 to 0.30 were used as a basis for this table.

less than 70, and include nonleaded brasses from Muntz metal or naval brass with 60% copper to red brass with 85% copper.

GROUP 3. Alloys with a machinability rating of 20 include nonleaded coppers, nickel, silvers, and phosphor bronzes. These alloys produce tough, long, stringy turnings.

Although demarcation between the three groups is not clear-cut, their machinability ratings are a reasonable guide to tool life and power required for cutting. But, as shown by Table 33, with a decrease in machinability index of the copper alloy, the cutting speed must be reduced and the rake and relief angles increased.

With the exception of sand castings, use the highest practical cutting speed, a light feed, and a moderate depth of cut. The castings usually retain hard scale after cleaning, and hence it is best to use a relatively coarse feed and a cut deep enough to cope with casting eccentricities.

Nose Radius. Use a nose radius of $\frac{1}{16}$ to $\frac{1}{8}$ in. on carbon or high-speed-steel tools. The radius on carbide-tipped tools should be held to a minimum and should not exceed $\frac{1}{16}$ in. High finishes on the tools will help to reduce heat and loading.

Cutting Fluids. It is generally advisable to use a coolant with carbon- or high-speed-steel tools. The carbide tools usually function best without coolant, but some jobs require it.

Suggestions for cutting compounds are: group 1 alloys—20:1 or more soluble oil; group 2—soluble oil or a coolant with a mineral-oil base fortified with 5 to 15% lard oil, especially on alloys producing long stringy chips; and group 3—mineral-oil base with 10 to 20% lard oil.

Common Problems. A *chattered surface* is caused by improperly distributed stock or an interrupted surface; by tool faults such as too much relief, too large a radius, or improper rake; and by lack of rigidity of tool or work. A *scratched surface* is caused by a tool not ground for chip control.

Meehanite Castings

Meehanite processed cast irons are produced under rigidly controlled conditions by a number of foundries licensed by the Meehanite Metal Corporation. All six grades are readily machined with high-speed steels, cast-alloy, and cemented-carbide tool materials.

TOOL ANGLES FOR MACHINING MEEHANITE IRONS

	High-speed Steel Tools			Tungsten Carbide Tools		
	Lathe	Planer	Boring Mill	Lathe	Planer	Boring Mill
Side-cutting-edge angle	6–10°	8–10°	6–10°	8–10°	5–10°	6–10°
End-cutting-edge angle	8–12°	8–12°	5–8°	8–10°	8–10°	10–12°
Front-clearance angle	2–4°	2–4°	4–6°	4–6°	4–6°	2–6°
Side-clearance angle	2–5°	2–5°	2–8°	4–6°	4–6°	4–6°
Back-rake angle	4–8°	3–5°	0–4°	0–4°	0–8°	0–2°
Side-rake angle	6–10°	6–10°	6–8°	2–6°	2–6°	2–10°
Nose radius	1/8–1/4 in.	1/4 in.	1/8–3/16 in.	1/8 in.	1/8 in.	3/32–1/4 in.

CUTTING SPEEDS FOR MEEHANITE IRONS

Type of Tool Material	Meehanite Types GE and GD Sfm	Meehanite Types GC and GB Sfm	Meehanite Types GA and Gm Sfm
High-speed steel	80	60	40
Cobalt high-speed steel	100	75	50
Stellite and Rexalloy	150	100	75
Tungsten carbide	300	250	200

Cast Irons

Although irons are used in tremendous quantities, lack of control in foundry practice and specification has made it difficult for industry to compile data that can be applied across the board. Besides Tables 35 and 36, refer to *American Machinist* Special Report "Machinability Depends on Microstructure," Nov. 27, 1950.

TABLE 35. TOOL ANGLES FOR MACHINING IRONS*

Material	High-speed Steel				Carbides			
	Back Rake	Side Rake	End Relief	Side Relief	Back Rake	Side Rake	End Relief	Side Relief
Standard malleable	3	3	6	6	0	0–2	4–6	4–6
Pearlitic malleable	6	6	6	6	0	0–2	4–6	4–6
Soft cast iron, 160–193 brinell	3	3	6	6	0	0–2	4–6	4–6
Medium cast iron, 200–220 brinell	2–3	2–3	6	6	0	0–2	4–6	4–6
Hard cast iron, 220–240 brinell	2	2	6	6	0	0–2	4–6	4–6

NOTE: For cast iron, Carboloy Dept., General Electric Co., advocates: side rake, 8° for depth of cut under $\frac{1}{4}$ in.; 4° for depth of cut over $\frac{1}{4}$ in.; back rake, 0°; end relief, 7°; side relief, 7°.

* Malleable Founders' Society.

TABLE 36. FEEDS AND SPEEDS FOR IRONS*

Material	Depth of Cut, In.	Feeds					
		Up to $\frac{1}{32}$	$\frac{1}{32}$–$\frac{3}{32}$	Up to $\frac{1}{32}$	$\frac{1}{32}$–$\frac{3}{32}$	Up to $\frac{1}{64}$	$\frac{1}{64}$–$\frac{1}{32}$
		High-speed Steel		Stellite J Metal		Carbides	
Malleable iron	$\frac{1}{32}$–$\frac{1}{8}$	120–160	90–120	170–250	140–170	220–500	175–350
	$\frac{1}{8}$–$\frac{1}{2}$	90–120	55–90	140–170	110–140	175–400	175–300
Pearlitic	$\frac{1}{32}$–$\frac{1}{8}$	110–140	80–110	150–220	130–160	200–400	150–300
Malleable	$\frac{1}{32}$–$\frac{1}{2}$	80–110	50–85	120–155	100–130	150–350	150–250
Soft cast iron	$\frac{1}{32}$–$\frac{1}{8}$	80–120	50–70	130–175	100–135	250–400	200–350
	$\frac{1}{8}$–$\frac{1}{2}$	50–65	40–55	90–130	80–105	200–350	150–350
Semisteel	$\frac{1}{32}$–$\frac{1}{8}$	75–115	45–60	120–150	90–125	200–400	200–350
	$\frac{1}{8}$–$\frac{1}{2}$	50–65	35–50	85–120	65–110	175–300	175–250
Hard cast iron	$\frac{1}{32}$–$\frac{1}{8}$	65–100	40–55	100–145	80–115	200–350	150–300
	$\frac{1}{32}$–$\frac{1}{2}$	40–55	30–45	75–100	55–95	150–250	150–200

* Malleable Founders' Society.

Nickel-alloy Steels

By International Nickel Co.

The following data apply to machining the nickel-bearing SAE steels in these series: 2300, 2500, 3100, 3300, 4600, 4800, 8600, also Nitralloy.

In a qualitative way, the machinability of steel is a function of hardness. Suggested speeds and feeds for rough turning (Table 38) are based on that convenient property. Nine hardness ranges are considered representative:

Group No.	Brinell hardness	Group No.	Brinell hardness
1	Up to 163	5	255–285
2	163–192	6	285–321
3	192–223	7	321–352
4	223–255	8	352–388
		9	388–415

It will be found that most as-rolled, normalized, or annealed nickel-alloy steels will fall in the first four groups. The other five groups cover chiefly steels that have been quenched and tempered. Machining of heat-treated parts has advantages that may offset any loss in machinability. Heat-treated parts may be finish-machined without warpage, distortion, scaling, or decarburization.

Table 38 represents good starting feeds and speeds for rough turning in everyday practice. For light finishing cuts, the maximum speeds given in the table for each group may be increased 50 to 100% or even more.

Suitable tool angles and nose radiuses for single-point lathe tools are given in Table 37. For interrupted cuts it is desirable to use a negative back rake of 2 to 8°; average 5°. When a negative back rake is used, the side-rake angle should be positive and be 2 to 5° greater than the negative back-rake angle.

Table 37. Tool Angles for Nickel-alloy Steels

	High-speed Steel		Cast Nonferrous Tool Materials		Cemented Carbides	
	Range	Av	Range	Av	Range	Av
Side-rake angle	5 to 15	10	5 to 15	12	0 to 15	6*
Back-rake angle	5 to 10	8	0 to 7	0	0 to 10	0
Side relief angle	6 to 10	8	7 to 10	7	4 to 12	7
End relief angle	6 to 10	8	7 to 10	7	4 to 12	7
End-cutting-edge angle	6 to 15	10	6 to 15	10	8 to 15	10
Side-cutting-edge angle	0 to 30	15	0 to 45	15	0 to 20	15
Nose radius	10% of cut depth		1/32 in.		1/32 in.	

* Average for depths of cut less than 1/2 in. is 8°; for heavier cuts, 4°.

TABLE 38. CUTTING SPEEDS FOR ROUGH-TURNING NICKEL STEELS
(In Sfpm)

Group No.	Depth of Cut	High-speed Steel					Cast Nonferrous Tool Materials					Sintered Carbides				
		Feed, Ipr														
		0.015	0.021	0.030	0.045	0.060	0.015	0.021	0.030	0.045	0.060	0.015	0.021	0.030	0.045	0.060
Group 1, brinell up to 163	1/16	180	150	130	110	95	250	220	195	175	155	580	500	435	365	325
	1/8	145	125	110	95	80	210	190	170	150	135	490	425	365	315	270
	1/4	125	110	95	80	70	190	170	150	135	120	420	365	310	260	230
	1/2	110	95	80	70	60	170	150	135	120	110	355	305	260	220	190
Group 2, brinell 163–192	1/16	130	110	95	85	75	190	165	145	130	120	470	410	350	295	260
	1/8	110	95	85	75	65	165	145	130	115	105	400	350	295	250	220
	1/4	95	85	75	65	55	145	130	115	105	95	335	290	250	215	185
	1/2	85	75	65	55	50	130	115	105	95	85	285	250	215	180	160
Group 3, brinell 192–223	1/16	100	90	80	70	60	150	135	120	105	95	395	340	290	245	220
	1/8	90	80	70	60	50	125	115	105	95	85	330	285	245	205	180
	1/4	80	70	60	50	40	115	105	95	85	75	280	240	205	175	155
	1/2	70	60	50	40	35	105	95	85	75	70	235	200	170	145	130
Group 4, brinell 223–255	1/16	85	75	65	55	45	120	105	95	85	75	335	290	245	210	180
	1/8	75	65	55	45	40	100	90	80	75	65	280	245	210	175	155
	1/4	65	55	45	40	35	90	80	70	65	60	240	205	175	150	130
	1/2	55	45	40	35	30	80	70	65	60	55	200	175	150	125	110
Group 5, brinell 255–285	1/16	75	65	55	45	40	105	95	85	75	65	295	255	215	180	160
	1/8	60	55	45	40	35	95	85	75	65	55	250	215	180	155	135
	1/4	55	45	40	35	30	85	75	65	55	50	210	180	155	130	115
	1/2	50	40	35	30	25	75	65	60	50	45	180	155	130	110	100
Group 6, brinell 285–321	1/16	65	55	50	40	35	95	85	75	65	60	255	220	185	160	135
	1/8	55	45	40	35	30	85	75	65	60	50	215	185	160	135	115
	1/4	45	40	35	30	25	75	65	60	50	45	180	155	135	110	100
	1/2	40	35	30	25	20	65	60	50	45	40	150	130	110	95	85
Group 7, brinell 321–352	1/16	60	50	45	40	35	85	75	70	60	55	225	195	165	140	120
	1/8	50	45	40	35	30	75	65	60	55	50	190	160	140	120	100
	1/4	45	40	35	30	25	65	60	55	50	45	160	140	120	100	90
	1/2	40	35	30	25	20	60	55	50	45	40	135	115	100	85	75
Group 8, brinell 352–388	1/16	55	45	40	35	30	75	70	60	55	50	200	170	145	125	110
	1/8	45	40	35	30	25	65	60	55	50	45	165	145	125	105	90
	1/4	40	35	30	25	20	60	55	50	45	40	140	120	105	90	80
	1/2	35	30	25	20	20	55	50	45	40	35	120	105	90	75	65
Group 9, brinell 388–415	1/16	50	40	35	30	25	70	65	55	50	45	180	160	135	115	100
	1/8	40	35	30	25	20	60	55	50	45	40	155	135	115	100	85
	1/4	35	30	25	20	20	55	50	45	40	40	130	115	100	80	75
	1/2	35	30	25	20	15	50	45	40	40	35	110	95	80	70	60

When employing cemented-carbide tools on continuous cuts at high speeds, use a chip breaker to break the chips or coil them in convenient form. Use a chip-breaker depth of 0.020 in. and a width of $\frac{3}{64}$ to $\frac{5}{32}$ in. for 0.015-in. feed and $\frac{1}{32}$- to $\frac{1}{2}$-in. depth of cut. For a 0.030-in. feed increase the width of the chip breaker from $\frac{1}{8}$ to $\frac{7}{32}$ in.

Heat-treated nickel-alloy steels, when machined with carbide tools at high speeds, often show such a smooth finish that no subsequent grinding is necessary. In other cases, grinding allowances can be reduced, thereby saving wheel wear and time.

Nitriding Steels

The nitriding steels (Nitralloy) are heat-treated to produce the sorbitic grain structure necessary for nitriding. In this condition the brinell hardness generally ranges from 225 to 260, but it may run over 300 for special purposes.

After rough machining, the parts must be subjected to a subcritical anneal, usually 100 F higher than the nitriding temperature, to relieve internal stress set up by heat-treating, rough machining, and grinding. Light machining and grinding are permissible after the anneal.

Machining Requirements. To minimize internal stress, machining cuts should not be deep and small feeds should be used. Feeds most commonly used are 0.015 to 0.030 for roughing and 0.005 to 0.007 for finishing. For precision boring, feeds of 0.001 in. or less are used.

Tool Design. With high-speed tools, surface cutting speeds usually run from 50 to 75 fpm for heavy cuts, and from 90 to 130 fpm for light cuts. Single-point tools in general use have side-rake angles of 16 to 20°, back-rake angles from 3 to 5°, side-cutting-edge angles of 10°, and end and side relief angles from 3 to 6°. Cobalt high-speed tools are generally recommended for machining forgings.

Cemented-carbide cutting tools are used for fine feeds and light cuts, with surface speeds running as high as 600 sfpm for very fine finishing cuts. Back rake on these tools varies from 6 to 15°, while side-rake angles are from 8 to 30°. The relief angle around the cutting-edge angle is usually 15°, while the side-cutting-edge angle varies from 5 to 20°.

For interrupted cuts on lathes and planers, it is desirable to use a tool with negative back rake as great as 8° in combination with 10° or more side rake. Some users recommend turning tools with 6° side rake and 2° negative back rake. Standard tools with 12° positive back rake are made in common boring tool sizes for bars holding the tool at 90, 45, and 30°. In larger bores, the cutting edge may be leveled off slightly to 8 or 6°, or whatever is necessary to give the desired negative back rake with the cut surface of the bore.

Stainless Steels

All the chromium-nickel grades (types 301 to 347) machine about the same, except for type 303. Troubles encountered are stringy chips and glazing of the work. On these grades, never let the tool ride. Get under the work-hardened skin and keep cutting. Machinability compared with mild steel averages 50%.

Type 303 steel has a machinability index of 80, the addition of sulfur giving a brittle chip and lubricating the tool.

The straight chromium steels show wide variations in machinability. Many machinists prefer type 430 F (free-machining) to type 416, on which high cutting rates are reported.

Types 403, 410, and 416 can be machined when hardened to approximately 300 brinell, by reducing the cutting speed about 20%. At this hardness range, light

TABLE 39. TURNING SPEEDS FOR STAINLESS STEELS

Type	Carbide Tooling		High-cobalt or Cast Alloy		High-speed Tooling	
	Rough	Finish	Rough	Finish	Rough	Finish
403, 405, 410	150/200	200/400	100/130	100/150	80/100	80/130
416	150/200	200/400	100/150	150/200	80/100	100/150
420	100/150	150/250	80/100	100/150	60/80	80/120
420 F	110/165	165/275	90/110	110/165	65/90	90/130
430	150/200	200/400	100/130	100/150	80/100	80/130
430 F	150/200	200/400	100/150	150/200	80/100	100/150
414, 431	140/180	150/350	90/120	90/140	60/80	80/100
440, ABC	100/150	150/200	60/80	80/100	40/60	60/80
440 F	110/165	165/220	70/90	90/110	45/65	65/90
446	140/180	150/350	100/130	100/150	60/90	90/120
301, 302, 304	130/180	150/300	100/130	100/150	60/90	100/120
303	150/250	200/400	100/150	150/200	70/90	100/140
309	130/180	150/300	100/130	100/150	60/90	100/120
316, 317	130/180	150/300	100/130	100/150	60/90	100/120
321, 347	130/180	150/300	100/130	100/150	60/90	100/120

cuts produce a fine finish. On all stainless types, cold-drawn bars appear to machine better than annealed stock.

Turning speeds are given in Table 39. Feeds vary greatly, according to the machine and operation. On rigid machines, roughing feeds may be as high as 0.015 ipr. On light work, finishing cuts of 0.005 in., or less, are required. For best finish, finishing cuts of 0.005 to 0.015 in. should be taken at 0.001- to 0.005-in. feed. Carbide tools can take such cuts at speeds from 150 to 500 sfpm.

The proper tool angles for turning stainless steel have been found to be:

Back rake	5 to 10°
Side rake	5 to 10°
End relief	7 to 10°
Side relief	5 to 15°
End-cutting-edge angle	8 to 15°
Side-cutting-edge angle	10 to 15°

Sulfur-chlorinated oils are recommended for cutting stainless steel. Generally they are combined with paraffin oil on a 1:1 basis for heavy work, and in a 1:5 mixture for light cuts. Soluble oil in 10:1 mixture is also used for light cuts at high speed, when carbide tooling is employed.

Nickel Alloys

Nickel-base alloys are used for a wide variety of products that must have excellent corrosion resistance as well as high mechanical properties. These materials (for analyses, see Sec. 26) are made in wrought and cast forms, and have high strength, ductility, and toughness.

Machining Characteristics. High-nickel materials are quite machinable. Good results can be obtained by careful selection of speeds, feeds, and coolants.

Two matters require careful attention: sharp tools plus plenty of coolant to carry away the high frictional heat developed, and good finish. Torn surfaces sometimes provide focal points for acceleration of corrosion.

High-nickel castings are nearly always machined in the as-cast condition, because only the S monel can be rehardened after annealing.

Tool Design. Slightly larger rake angles are required for high-nickel alloys,

TABLE 40. TURNING SPEEDS FOR NICKEL ALLOYS USING HIGH-SPEED STEEL*

Depth of cut, in.	1/32	1/32	1/32	1/16	1/16	1/16	1/8	1/8	1/8	1/8	1/4	1/4	1/4	3/8	3/8	3/8	1/2	1/2	1/2
Feed, in.	0.008	1/64	1/32	1/64	1/32	1/16	1/64	1/32	1/16	1/8	1/32	1/16	1/8	1/32	1/16	1/8	1/32	1/16	1/8
Wrought monel, R monel, KR monel (unhardened), and nickel	168	139	118	121	104	60	110	68	48	34	57	39	29	52	34	24	49	31	21
Wrought K monel (unhardened) and inconel	115	85	75	75	65	40	70	45	35	25	37	27	20	30	25	15	32	20	15
Wrought K monel (hardened)								25	15										
Cast nickel		110	95	95	80	55	85	60	55	35									
Cast monel		90	80	80	70	45	75	50	40	30									
Cast H monel and Cast S monel (annealed)		60	55	55	45	30	50	35	25	20									
Cast S monel (as cast)		30	27	27	22	15	25	17	13	10									

Above data apply to use of high-speed-steel tools. If cast-tool materials are used, increase speeds about 25 to 35%.

* International Nickel Co.

TABLE 41. MACHINING HIGH-NICKEL ALLOYS WITH CARBIDES

Material	Cast Alloys		Wrought Alloys, Roughing-Finishing Range
	Roughing,* Sfpm	Finishing,† Sfpm	
Nickel	250	325	200–350
Monel	225	275	175–350
H monel	160	235	
S monel	100	150	
K monel	...	...	150–325
K monel (hardened)	...	...	100–275
Inconel	...	...	100–250

NOTE: For roughing wrought materials, use feeds up to 0.020 ipr.

* Feed, 0.015 to 0.020 ipr.
† Feed, 0.005 to 0.010 ipr.

as compared with other tough materials. Tool angles recommended for turning with HSS tools are:

MEDIUM-DUTY

Side-rake angle	12 to 15°
Side relief angle	6°
End-cutting-edge angle	10°
Nose radius	$\frac{1}{16}$ to $\frac{3}{32}$ in.
Back-rake angle	8 to 10°
End relief angle	7°
Side-cutting-edge angle	15°

ROUGHING HEAVY WORK

Side-rake angle	10 to 12°
Side relief angle	6°
End-cutting-edge angle	15°
Nose radius	$\frac{1}{8}$ to $\frac{3}{16}$ in.
Back-rake angle	6 to 8°
End relief angle	8°
Side-cutting-edge angle	15°

End and side relief angles should be held to a minimum to afford free cutting with roughing tools. However, finishing tools may have a front clearance angle of 15°, together with large back rake to give sharp cutting angles. Back rake as high as 25° is therefore used on finishing tools.

Disposal of the tough chip is a problem. A heavy, continuous spiral chip cannot be handled economically when chip breakers are not used. Tool-design principles may be disregarded in part and rake angles reduced so that the heavy chip packs against the tool face, becomes embrittled, and breaks when it curls back against the work.

For cemented-carbide tools used on the nickel alloys, the Carboloy Dept., General Electric Co., recommends these tool angles:

Back rake	0°
Side rake	8°
Front and side relief on carbide	7°
Front and side relief on steel shank	10°
End-cutting-edge angle	15°
Side-cutting-edge angle	15°
Nose radius:	
For cuts $\frac{1}{8}$ to $\frac{1}{4}$ in. deep	$\frac{3}{64}$ to $\frac{1}{16}$ in.
For finish cuts $\frac{1}{16}$ in. and under	0.010 to 0.020 in.

Toughest alloys may require reduction of the relief angles on the carbide insert to 4°, to give more support to the cutting edge. The 15° side-cutting-edge angle should be maintained for interrupted cuts and for machining castings and forgings.

Chip breakers are essential for optimum chip control when machining most of the wrought materials. Step-type chip breakers have widths tabulated and depth of 0.010 to 0.015 in. (See Table 9.)

Titanium

Sharp turning tools are essential for turning titanium. A dull tool will work-harden or glaze the work surface, so that cutting becomes difficult or impossible.

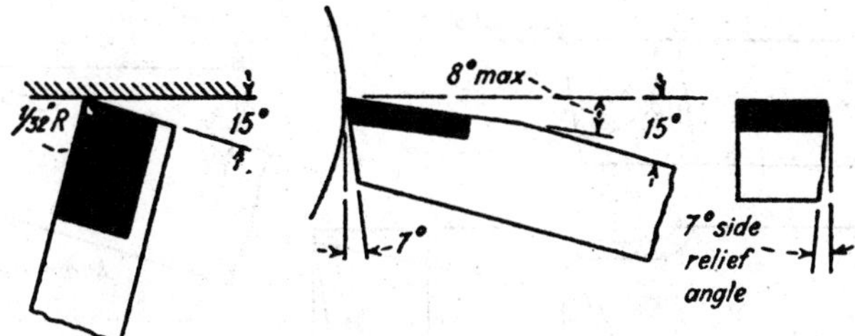

FIG. 50. Carbide tools for sprayed metals are set as shown. The side-cutting-edge angle is unimportant because cuts are light.

Recommended cutting speeds are from 85 to 95 sfpm and feeds of 0.003 to 0.008 ipr for heavy cuts. For finishing cuts, metal removal should be less than 0.020 in. In general, the machinability of titanium is similar to that of stainless steel.

Sprayed Metals

Sprayed metals differ from the original metals in many physical properties, including hardness, density, ductility, and tensile strength. Failure to recognize this situation is the major cause of difficulty in finishing operations. Different tool angles, tool settings, speeds, and feeds must be used. Recommendations for carbide tools are given in Fig. 50 and Table 42; for high-speed steel in Fig. 51 and

TABLE 42. SPEEDS AND FEEDS FOR CARBIDE TOOLS ON SPRAYED METAL

Material	Speed, Sfpm		Feed, In. per Rev	
	Rough	Finish	Rough	Finish
Yellow brass	250–300	300–350	0.006	0.002
Commercial bronze	250–300	300–350	0.006	0.002
Manganese bronze	250–300	300–350	0.006	0.003
Phosphor bronze	250–300	300–350	0.006	0.003
Tobin bronze	250–300	300–350	0.006	0.003
Copper	250–300	300–350	0.006	0.003
Iron	75–100	100–125	0.006	0.003
18-8 stainless	100–125	125–175	0.006	0.003
Hi-chrome hi-carbon stainless	30–40	30–40	0.004	0.003
Monel	200–250	250–300	0.004	0.002
Nickel	200–250	250–300	0.004	0.002
Steel, 10 carbon	75–100	75–100	0.006	0.003
Steel, 25 carbon	50–75	50–75	0.004	0.003
Steel, 80 carbon	30–40	30–40	0.004	0.003

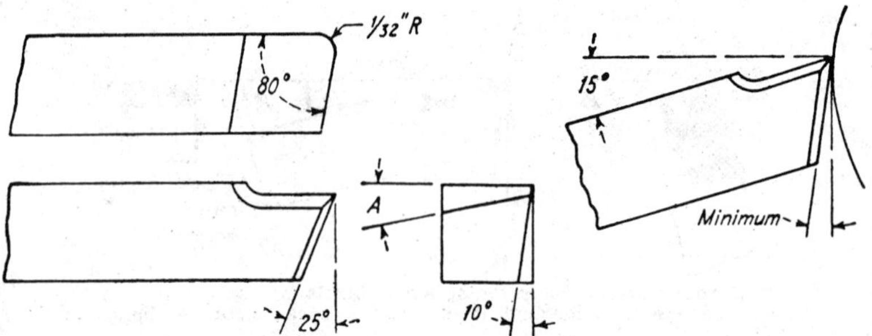

FIG. 51. Tool angles recommended for turning sprayed metals with high-speed-steel tools. Tool 1, $A = 12$ to $15°$; tool 2, $A = 8$ to $12°$; tool 3, $A = 0$ to $5°$.

TABLE 43. RECOMMENDED ANGLES, FEEDS, AND SPEEDS FOR HIGH-SPEED-STEEL TOOLS ON SPRAYED METAL

Metals	Tool Number	Speed, Sfpm	Feed, In. per Rev
Aluminum	1	125–200	0.003–0.005
Babbitt	1	150–250	0.005–0.008
Brass, yellow	2	100–125	0.003–0.005
Bronze, commercial	1	100–125	0.003–0.005
Bronze, phosphor	3	100–125	0.003–0.005
Bronze, tobin	3	100–125	0.003–0.005
Copper	1	100–125	0.003–0.005
Lead	1	150–250	0.005–0.012
Monel	1	100–125	0.003–0.005
Nickel	1	100–125	0.003–0.005
Iron, Swedes	1	50–100	0.003–0.005
Steel—10 carbon	1	75–100	0.003–0.005
Steel—25 carbon	2	50–100	0.003–0.005
Steel—40 carbon	2	50–100	0.003–0.005
Steel—80 carbon	0	Grind	0
Steel—120 carbon	0	Grind	0
Stainless—18-8	1	100–125	0.003–0.005
Stainless—Hi chrome	0	Grind	0
Tin	1	150–225	0.006–0.008
Zinc	1	150–225	0.006–0.008

Table 43. Carbide tools resist the abrasive action of inclusions in the sprayed metal.

In turning, tools are set slightly above center with front clearance at a minimum. Although sprayed bores are best finished by internal grinding, they can be bored at light cut and feed with a tool ground approximately the same as for turning.

Plastics, Hard Rubber, and Fiber

Machining Characteristics. Nonmetallic materials like plastics, hard rubber, and fiber are generally machined in a manner similar to brass. But these materials have low heat conductivity and are abrasive. Tool materials commonly used are: high-speed steel, which gives better service if nitrided; cast-alloy and cemented-carbide tools for production runs; and diamond tools for fine finishing cuts.

Plastics fall into two categories: *thermoplastic* materials which will soften under heat after molding, and *thermosetting* resins which undergo a chemical change at molding temperature and do not subsequently soften if exposed to heat. Materials in the second class, usually the phenolics, are commonly cast or molded with a filler content of up to 50% wood flour, asbestos, mica, powder metal, or chopped fabric. Phenolics are also made into laminates with paper, cotton cloth, glass fabric, or asbestos. Hence machining practices may vary widely and certain precautions should be observed. Among the latter are: in turning and facing, the heel of the cutting edge should cut ahead of the nose to avoid throwing up a fin or burr; and in turning the laminates, female centers should be used to avoid separation by pressure of male centers.

Necessity for Aging. Plastics should be aged if close tolerances are required. For tolerances of ±0.002 in., age the material for 24 hr; for very accurate work, set aside the material for periods up to 6 weeks.

Thermoplastic Materials. Recommendations for machining several groups are:

ACRYLIC GROUP. No rake, but plenty of clearance, is used turning for Plexiglas and other resins in this group. Set tools 60° to work axis; use about 65 fpm and 0.020-in. depth of cut for smooth finish.

CELLULOSE GROUP. These materials can be cut at speeds to 4000 fpm.

POLYSTYRENE GROUP. Select cutting speed to 600 fpm; use feeds to 0.005 in. Tools should have 0 to 3° negative back rake, up to 7° negative side rake, from 7 to 12° end and side relief angles, and up to $\frac{1}{32}$ in. nose radius.

Thermosetting Materials. Cast phenolics require tools with zero to slightly negative back rake and 15 to 18° end relief. With tools set at 60° to the work, cutting speeds range between 700 and 900 fpm. Light feeds are used. A nonalkaline coolant can be used.

Laminates require varied turning practices. Formica gear material (cotton duck impregnated with phenolic resin) is turned at 750 fpm, with depth of cut from $\frac{1}{16}$ to $\frac{1}{8}$ in., and a feed of 0.030 in., regardless of depth of cut. Tool angles and setting are shown in Fig. 52. The tool should overlap the feed.

Micarta, with a paper or fabric base, should be turned at 100 to 125 fpm with HSS tools, and about 0.010 in. is left for the finishing cut.

Textolite, a canvas laminate, is turned with tools having no back rake and 6° end and 6° side relief. Cutting speed is about 500 to 600 fpm, roughing cuts range between $\frac{1}{16}$ and $\frac{1}{8}$ in., and finishing cuts are not less than $\frac{1}{64}$ in.

When carbide tools are used to machine fabric-base laminates, the cutting speed may be as high as 800 fpm.

Glass-reinforced plastic is machined with carbide tools at 400 fpm and a feed of 0.010 in. Tool grind is: 0° back rake, 13° side rake, and 33° clearance. In chasing

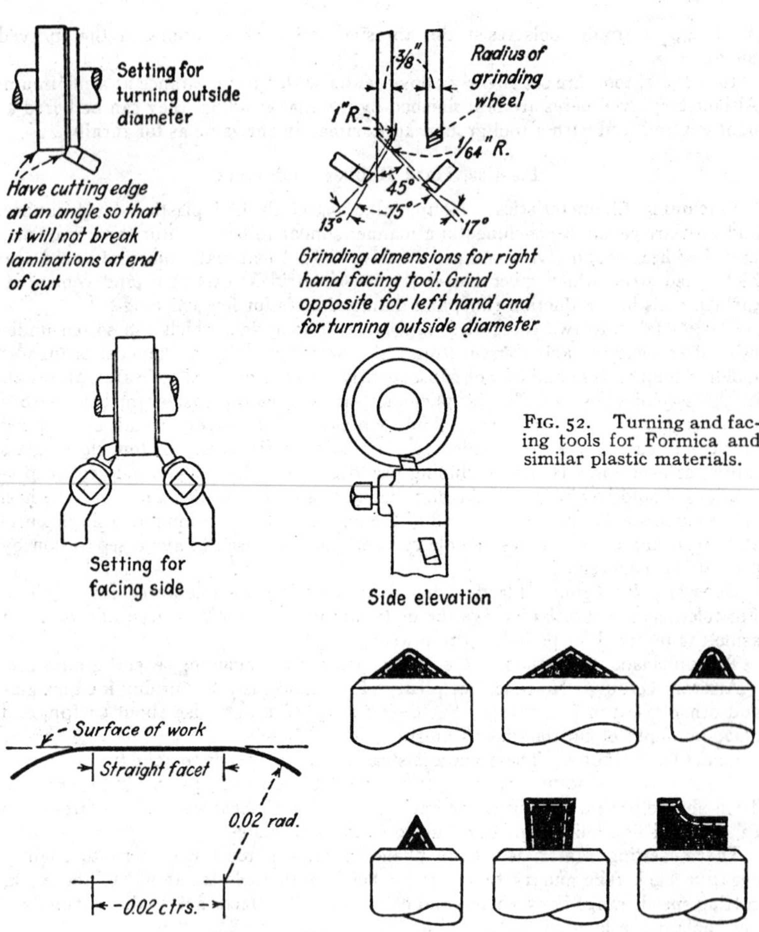

Fig. 52. Turning and facing tools for Formica and similar plastic materials.

Fig. 53. Diamond tools for plastics.

threads on the lathe, swing the slide rest 30° and cut on one side of the tool only. In facing, cut toward the center to keep the material under compression.

Diamond Tools for Plastics. Resistance to abrasion, low heat generation, and capacity to hold keen cutting edges on long runs make diamond tools effective and practical for machining plastics. Figure 53 shows a number of shapes for diamond tools and also an enlarged profile of the cutting edge of a typical tool point. Relief angles of 5 to 10° are used in turning tools. A point angle of 110 to 120° is satisfactory for turning, but shoulders on the work may make it necessary to reduce this to 80°. Angles as low as 55° have been used in thread cutting but should be avoided if possible. The rake angle varies from 0 to 3°. A zero rake is easier to check in setting the tool to the work.

TABLE 44. TOOL ANGLES, DEPTH OF CUT, AND FEED FOR TURNING HARD RUBBER*

Tool	Top Rake, Deg	Side Rake, Deg	Clearance, Deg	Notes	Depth of Cut, In.	Feed, Ipr
			Turning		Roughing	
Steel (lathe)	None	None	10 to 20	If consistency of stock causes tool to tear, tilt forward to increase clearance and provide negative rake of 5 to 10°	$\frac{1}{16}$	0.025
					$\frac{3}{32}$	0.023
					$\frac{1}{8}$	0.020
Form tools	None	None	15 to 20		$\frac{5}{32}$	0.018
					$\frac{3}{16}$	0.017
					$\frac{7}{32}$	0.016
Tungsten carbide tip	None	None	10		$\frac{1}{4}$	0.015
			Facing		Finishing	
Roughing diamond	None	None	10	Round nose $\frac{3}{32}$ in. radius	$\frac{1}{32}$	0.012
					$\frac{1}{16}$	0.012
Finishing diamond	None	None	10	Round nose $\frac{1}{64}$ in. radius	$\frac{3}{32}$	0.011
					$\frac{1}{8}$	0.010

* American Hard Rubber Co.

Plastics materials can be machined with diamonds at speeds from 4000 to 4200 fpm, but vibration must be avoided. Feeds may range between 0.0012 and 0.040 in., increasing with work size, and may be by either hand or power.

Hard Rubber. Tool angles for turning hard rubber are given in Table 44. High-speed steel is recommended for turning tools. Tungsten carbide tools are used when the savings justify the expense and because they eliminate frequent resharpening and improve the finish and accuracy. Where accuracy demands, diamond tools are used, but they cannot be employed on interrupted cuts. With coolant, turning, facing, and boring operations can be done at 300 fpm. If work is cut dry, the speed is reduced to 200 fpm. Use a large coolant screen to remove rubber particles.

Fiber. Since fiber is hard and tough, tools must be kept sharp to obtain best machining results. The material is slightly elastic and tends to impinge against the back of the tool and generate heat. Tools should be ground about the same as for brass; that is, no rake and large relief should be used. With a wide-nosed tool and coarse feed, the cutting speed should be about 30% higher than for cast iron.

TURNING TAPERS BETWEEN CENTERS

Tapers can be cut between lathe centers either by setting over the tail center or by using a taper attachment. In either case the angle must be one-half the difference in diameters between the two ends of the taper, if it extends the whole length of the piece. To turn a piece 20 in. long, 5 in. at one end, and 3 in. at the other, as in Fig. 54, the tailstock must be set over 1 in., this being one-half of the difference between 5 and 3.

If the taper extends only half the length of the piece, with the same difference in diameter, the tailstock must be set over twice as far, or 2 in., for the true taper is now from 3 in. at the small end and 7 in. at the large end. The difference is 4 in., making it necessary to set the tailstock over 2 in. (see Fig. 55).

A third case is shown in Fig. 56, where the taper is not at either end. With the same difference between large and small diameters, the taper is again twice as great as in Fig. 55, the dotted lines showing the taper extended the whole length of the piece. This would be 1 in. at the small end and 9 in. at the large end, a difference of 8 in., requiring a setover of 4 in.

Extreme tapers of this kind are not easy to turn by setting over the tailstock. The point of the cutting tool must be set at the same height as the lathe centers. Figures 57, 58, and 59 show work set over in lathe.

Accurate Tool Setting with Compound Rest

A compound rest fitted with a micrometer dial may be used to take very minute cuts by swiveling the rest to the proper point. Starting with the rest set to feed straight in and moving to the right, the infeed will be the cosine of the angle. If the rest is moved 30° from center, the infeed, or cut taken, will be 0.86603 of the feed as shown on the dial. Or, for every 0.001 in. on the dial the actual infeed is 0.000866 in. at 30°. With a 60° setover the actual advance is one-half the amount shown on dial. The formulas are:

A = actual advance of tool in radial direction.
B = amount removed on diameter of work = $2A$.
C = number of degrees of setover of compound rest, from center.
D = advance as read on dial in 0.001 in.
$A = D \times \cos C$. $B = 2A$.

$\cos C = \frac{A}{D}$, or C = angle whose cos is $\frac{A}{D}$.

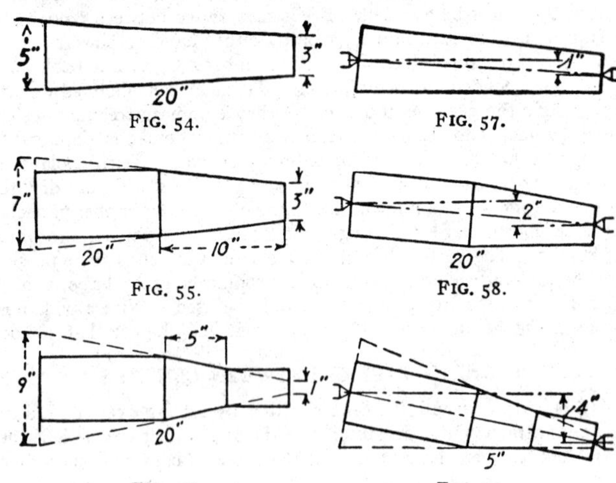

FIGS. 54–59. Problems in turning tapers on the lathe and amount of tailstock setover to use.

Section 13

SCREW-MACHINE WORK

SECTION 13

SCREW-MACHINE WORK

TYPES OF TOOLS

Proper selection and setting of tools insures producing accurate screw-machine work at high output rates. A considerable number of types of tools are used.

External Turning Tools. Balance turning tools, hollow mills, box tools, swing tools, and knee tools are used in the turret.

Internal Cutting Tools. Drills, counterbores, and reamers are held on nonfloating or floating holders.

Threading Tools. Taps are retained in tap holders, dies in die holders, chasers in opening die holders, and thread rolls in cross-slide knurl holders or in knurling swing tools.

Knurling Tools. Top and side knurl holders are used on cross-slides, adjustable knurl holders, and knurling swing tools in the turret.

Forming and Cutting-off Tools. Circular form tools and circular cutting-off tools are used on the regular cross-slide toolposts; square tools and cutoff blades are used on the cross-slides; angular cutting-off tools are held in the turret.

Balance Turning Tools

The balance turning tool (Fig. 1) is best for straight roughing. The blades cut tangentially, and adjustment is obtained by swinging them up or down. The cutting edge of each blade is ground square with the face of the tool. Usually, the blades are set on the body at some predetermined angle, such as 15°. Since the tools may be ground with a rake of 30°, the effective clearance is 15°. This clearance angle is ordinarily satisfactory for steel but is varied to suit conditions.

Hollow Mills

Plain hollow mills (Fig. 2) have been largely superseded for roughing work by the better designed balance turning tools. A round hole, extending through the tool, together with three milled spaces, form the teeth. Undercut teeth are used for steel, straight teeth for brass. Because of its construction, the plain hollow mill machines only a single size of work.

Adjustable hollow mills have three inserted blades with independent radial adjustment.

Box Tools

Box tools (Fig. 3) consist of one or more blades mounted in a boxlike frame and set tangent to the work. V-shaped or roller supports placed opposite the blades prevent the work from springing away from the blades. Although box tools will turn one, two, or three diameters simultaneously, they are preferred for finishing cuts. The cutting point a (Fig. 4) should be set on the diameter of the work, which is perpendicular to the face f of the blade. Edge k may be ground square with f, or tipped back slightly a small angle b. This edge should never lean forward of the cutting point a.

The blade is set back on the rear end to achieve a clearance angle e, usually 8°. Angle d ranges from 15 to 40° to obtain a good chip-cleaving action and may be less than 15° on brass.

Drills, Reamers, and Taps

For the general run of work, commercial two-fluted drills of the spiral type (and with straight shanks) are used, according to Brown & Sharpe. Commercial drills with two straight flutes are employed when drilling deep holes in brass, because chips work out of the hole better than with spiral-fluted tools. If drill breakage is still experienced, a half-round flat drill is usually satisfactory after the hole has been started to a depth of two or three diameters with a conventional drill.

Straight-fluted rose reamers are chosen for most work, but the spiral-fluted type is required for reaming deep holes. Spiral-fluted reamers have ability to remove chips left by a flat drill in a deep hole.

Two-fluted taps are customarily specified for holes up to $\frac{3}{8}$ in. in steel, to avoid breakage from seizure with chips clogged behind the cutting edge, particularly at

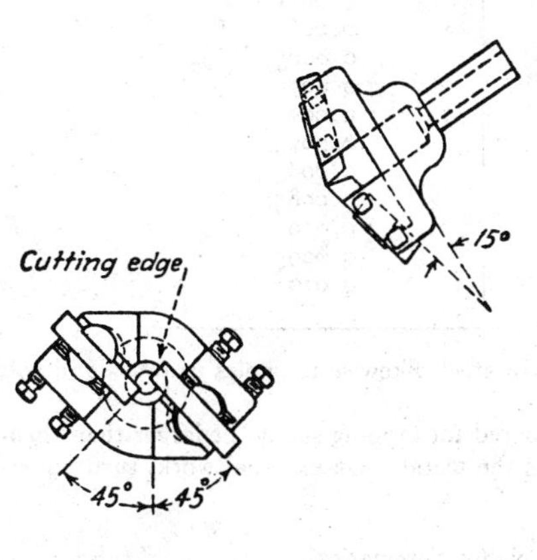

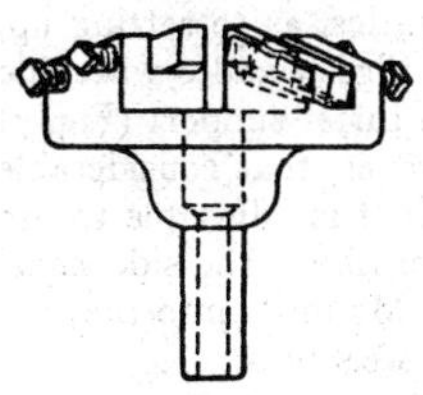

FIG. 1. Brown & Sharpe balance turning tool.

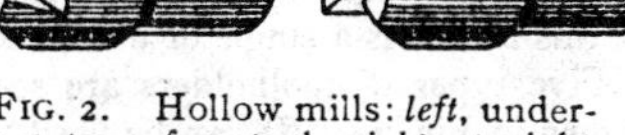

FIG. 2. Hollow mills: *left*, undercut type for steel; *right*, straight for brass.

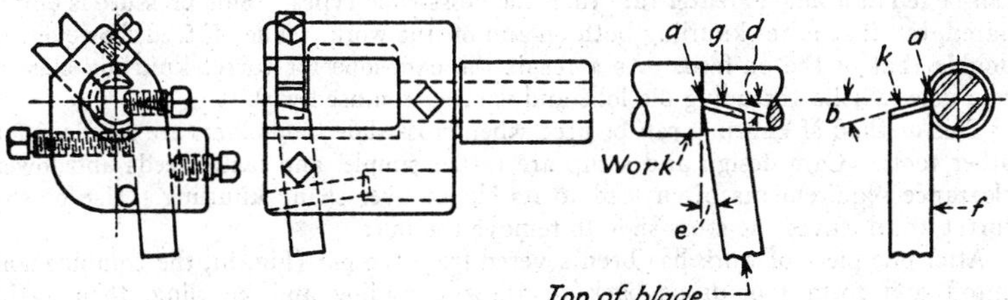

FIG. 3. Box tool with single blade.

FIG. 4. Angles of box tool blade with reference to the work.

TABLE 1. ALLOWANCES FOR THREADING IN SCREW MACHINES

Threads per Inch	External Work Turn Undersize	Internal Work Increase over Theoretical Bottom of Thread
28	0.002	0.004
24	0.002	0.0045
22	0.0025	0.005
20	0.0025	0.0055
16	0.003	0.006
14	0.003	0.0065
13	0.0035	0.007
12	0.0035	0.007
11	0.0035	0.0075
10	0.004	0.008
9	0.004	0.0085
8	0.0045	0.009
7	0.0045	0.0095
6	0.005	0.010

reverse. For tapping holes over $\frac{3}{8}$ in. in steel, likewise for holes in brass and soft metals, regular taps are used.

Boring Work or Threading. Holes bored for tapping should be larger than theoretical diameter because the tap crowds the metal. On external work, turn undersize. See Table 1 for allowances.

Knurling Tools for Automatics

A simple method of knurling on Brown & Sharpe automatics, as to setting up, adjustment, and cam design, for plain parts, is as shown in Figs. 5 and 6 with side knurl holders where there is no need of turret tools. With a turret support (Fig. 7) wide knurling can be done. A. Ainsworth points out, however, that considerable side pressure results from this method of knurling, and that it is advisable to use light feeds and to provide for a dwell at the end of the cam rise. The side knurl holder is a circular disk similar to cross-slide circular tool. Mounted into the face of this holder is a single or double knurl rotating on ground steel pins.

Five types of toolholders are seen in Figs. 16 to 20. A very versatile type is shown in Fig. 18. It can be used to produce straight, diamond, or helical knurled diameters with a pair of straight knurls, by adjusting the knurls in their blocks. It can be fed at a much greater rate than the cross-slide types. Side pressure is eliminated, but it requires knurling both on and off the work. The off-feed, however, is double that of the on-feed. As a result the cam lobe for turret knurling closely resembles a solid threading die lobe and is often confused with it.

This method of knurling can be used when cross-slide toolholders are required for other tools. Cam design and setup are rather simple, and faster feeds and lower clearance requirements often lead to its choice over swing knurling. Use of the turret knurl leaves the cross-slide to remove the burr.

After one piece of work has been severed from the bar (Fig. 8), the combination cutoff and form tool drops back to allow threading and knurling, then again approaches the work to remove the burr. Again dropping away, the cam allows the finished piece to be fed to length and the cutoff tool approaches to sever the part.

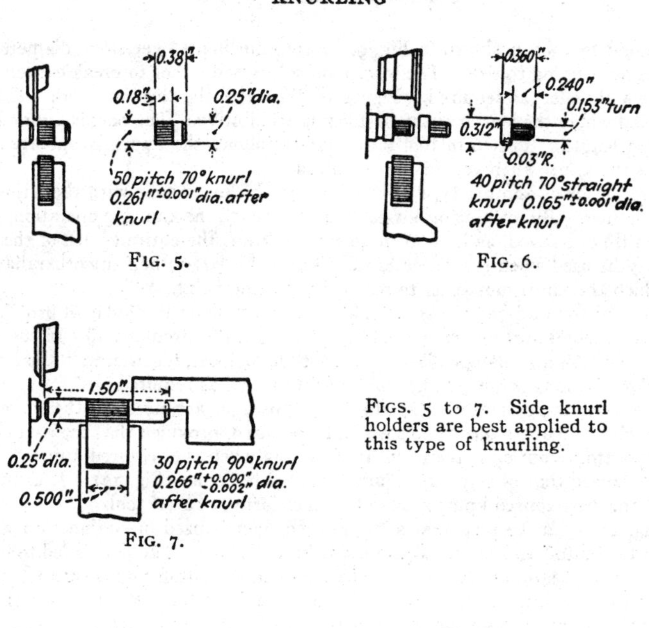

FIG. 5.

FIG. 6.

FIG. 7.

FIGS. 5 to 7. Side knurl holders are best applied to this type of knurling.

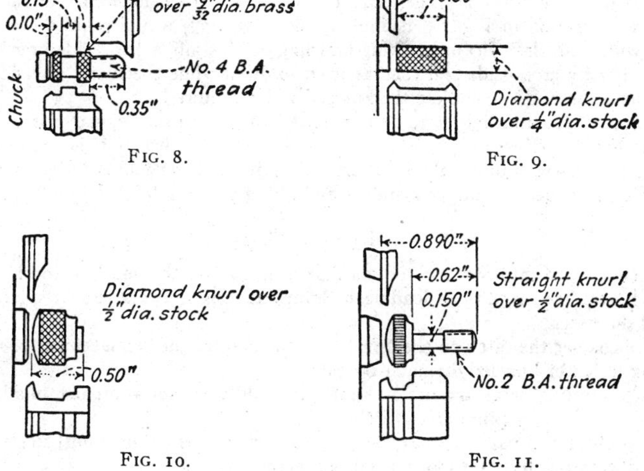

FIG. 8.

FIG. 9.

FIG. 10.

FIG. 11.

FIGS. 8 to 11. Knurls like these are preferably produced with adjustable-turret knurl holders.

A second method is shown in Fig. 9. Here knurling is over stock diameter and requires no forming to size. The form tool is designed either to break corners or to produce a chamfer, as required. Figure 10 illustrates the third method. This can be applied where the length of the part is not prohibitive. The part is knurled over its entire length. The form tool immediately follows the knurling operation and removes the knurl wherever it is not required.

Figure 11 shows another type of knurling which can be done with the adjustable turret method. By forming or box-tooling previous to the knurling operation, much machine time is saved, as it is not necessary to knurl the entire length of the part. This may be used when the diameter to be knurled is larger than any other diameter over which the knurl must pass to reach the starting point.

The use of top and bottom knurl holders for cross-slides is similar in principle to the use of a side knurl holder. A hub is ground on the circular tool and the knurl inserted over this and clamped into the cross-slide toolpost, together with the circular tool. An adjusting screw permits the knurl to be set as required.

This method is applied when all the turret positions are close to other necessary tools or where no other turret tools would be required, provided that the part can be top and bottom knurled. It can be applied only when the required knurled width does not exceed that of standard knurls mounted in holder (Fig. 12). It cannot be used where the required knurl is close to either form or cutoff tool.

In Fig. 13 it can be seen that a top knurl holder is used in conjunction with a combination cutoff and form tool and a side knurl holder. The part is fed to length into a combined stop and support. The combination cutoff and form tool, which contains top knurl, advances and knurls the large diameter. Passing over the top of the bar, the knurl loses contact and the cutoff tool severs one piece and finish forms the next. Drop-back on the cam allows this holder to clear the side knurl holder, advancing to knurl the small diameter. At the completion of this knurl the cutoff tool again returns to remove the burr.

Figures 14 and 15 show two types of work that can be done more easily, or only, by the swing tool method. The knurl in Fig. 14 could be done only in this way, while the operation in Fig. 15, if done by side knurling, would require the use of special width knurls. In turret swing knurling on this job, a $\frac{1}{4}$ in. wide knurl is fed into the bar by cross-slide and then allowed to dwell while the turret advances the knurl along the bar to complete the knurled width required.

In swing turret knurling, a guide is required to replace the regular raising block on right-hand cutting jobs. When a left-hand spindle direction is required, no block can be inserted under the front cross-slide if a cutting tool is to be applied in this position; a special guide is installed by bolting to the side of the toolholder.

Good Knurl-tool Usage

1. Feed a knurl into the work only a distance equal to the depth of tooth on the knurl. Extra feed results in loading of knurl, tearing of work, and reduction of knurled diameter.

2. The coarser the pitch on the knurl tool, the greater the displacement of metal, resulting in larger knurled diameter on part.

3. In computing necessary rise on knurl cam, add 0.010 for approach, in addition to rise as required by pitch of knurl chosen.

4. A dwell of 0.01 cam space at the end of knurl cam will allow knurl to clean up and produce a much better looking knurled diameter.

5. Steady even knurl feed is necessary to keep the knurl in track. A hesitating or uneven rise on the cam will result in torn, poor knurling.

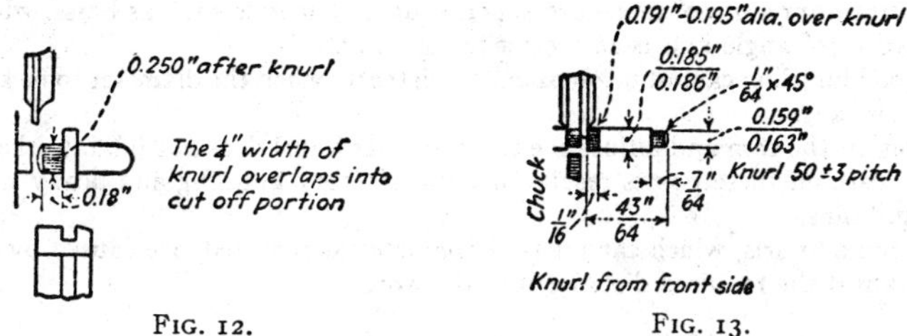

FIG. 12. FIG. 13.

FIGS. 12, 13. Top and bottom knurl holders for cross-slides handle these jobs.

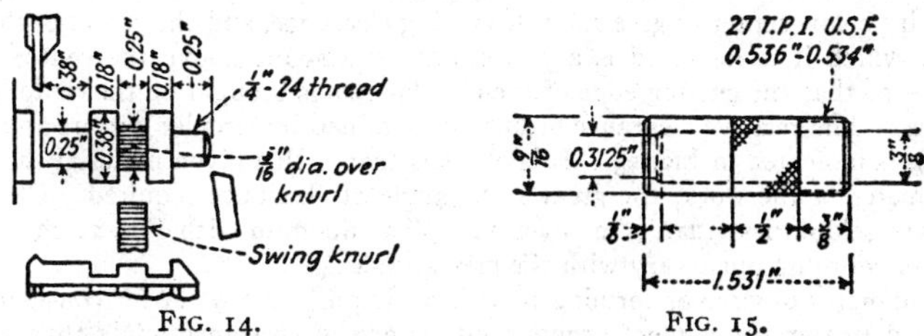

FIG. 14. FIG. 15.

FIGS. 14, 15. Turret-swing toolholders are needed to knurl these pieces.

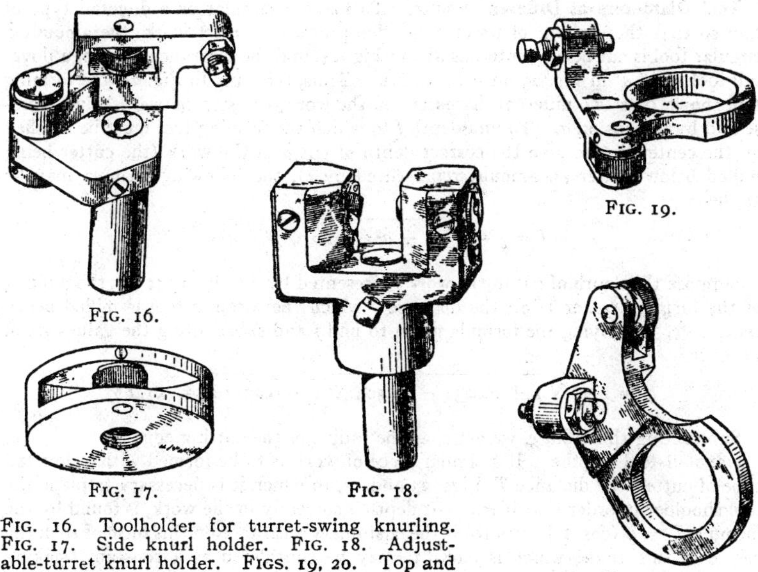

FIG. 16. FIG. 17. FIG. 18. FIG. 19. FIG. 20.

FIG. 16. Toolholder for turret-swing knurling. FIG. 17. Side knurl holder. FIG. 18. Adjustable-turret knurl holder. FIGS. 19, 20. Top and bottom knurl holders for the cross-slide.

6. Blunt knurls of 90° angle are superior on soft metals such as brass, while the more acute 70° angle knurls are better for use on steel.

7. Good knurling cannot be obtained consistently when the diameter to be knurled is not constant.

8. Design the cam and adjust the setup so that knurl does not jab into the work. Lack of care on these points results in unsatisfactory knurling and heavy wear on knurl and pins.

9. Double knurls, which cannot raise diameter as required, are caused by failure to feed knurl the required distance into the work.

Forming Tools

The two types of forming cutters commonly used in the screw machine are shown in Figs. 21 and 22. The circular forming cutter in Fig. 21 is usually cut away from $\frac{1}{8}$ to $\frac{3}{16}$ in. below center to give suitable cutting clearance, and the center of the tool-post on which it is mounted is a corresponding amount above the center of the machine, so that the cutting edge of the circular tool is brought on the center line of the work. The relative clearance ordinarily obtained by circular cutters and dovetail tools is indicated in Fig. 23. It is obvious that with a given material the larger the diameter of the work, the greater the angle of clearance required. Clearance angles are seldom less than 7° or over 12°. The diagram with Fig. 22 shows why form-tool contours must vary with the clearance angle.

The diameter of circular forming tools is an important matter for consideration. A small diameter has a more pronounced change of clearance angle than a large diameter. In fact, when of an exceedingly large diameter, the circular tool approaches in cutting action the dovetail type of tool which is usually provided with about 10° clearance.

Tool Diameters at Different Points. To make a circular or a dovetail type of tool so that the contour of its cutting edge produces correct work, the amount a circular tool is cut below center, as at c in Fig. 24, and the clearance angle of a dovetail tool as at A' in Fig. 23 must be known. Thus, referring to Fig. 24, the forming tool shown cuts two different diameters on the work, the step between being represented by dimension a. To find depth f to which the forming tool must be finished on the center line to give the correct depth of cut a in the work (the cutter being milled below center an amount represented by c), the following formula may be applied:

$$f = g - \sqrt{g^2 + a^2 - (2a\sqrt{g^2 - c^2})}$$

Suppose the depth of cut in the work represented by a to be 0.152 in.; the radius g of the forming cutter 1 in.; the distance c which the forming tool is milled below center, $\frac{3}{16}$. Applying the formula given to find f and substituting the values given we have

$$f = 1 - \sqrt{1 + 0.0231 - (0.304\sqrt{1 - 0.03516})} = 0.1488$$

Subtracting this from g, we get e, as the radius of the cutting edge.

Dovetail-tool Depths. If a similar piece of work is to be formed with a dovetail type of cutter, the distance T, Figs. 23 and 25, to which it is necessary to plane the tool shoulder in order that it may cut depth a correctly in the work, is found by the formula $T = a\ (\cos A')$. As 10° is the customary clearance on this form of tool, the cosine of this angle, which is 0.98481, may be considered as a constant, making reference to a table of cosines unnecessary as a rule. Assuming the same depth for a

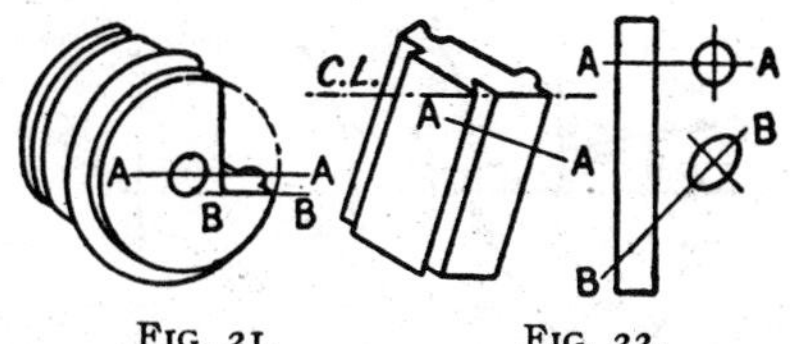

FIG. 21. FIG. 22.

FIGS. 21, 22. Circular and dovetail form tools.

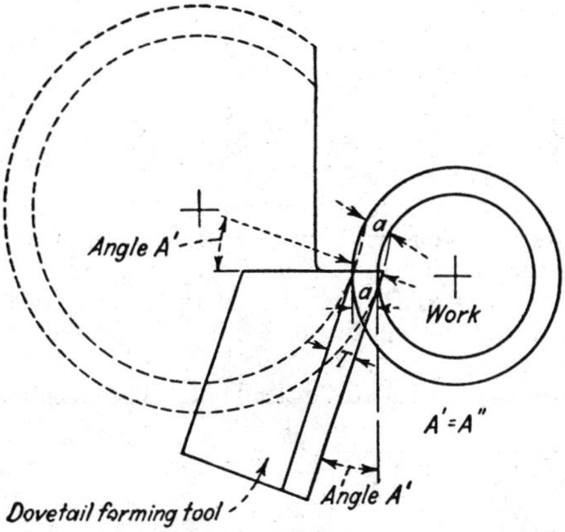

$$T = a\,(\text{cosine } A')$$

FIG. 23.

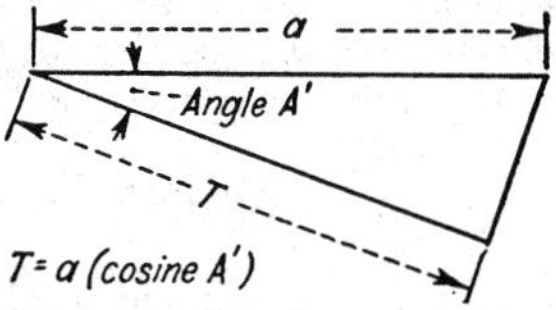

FIG. 25.

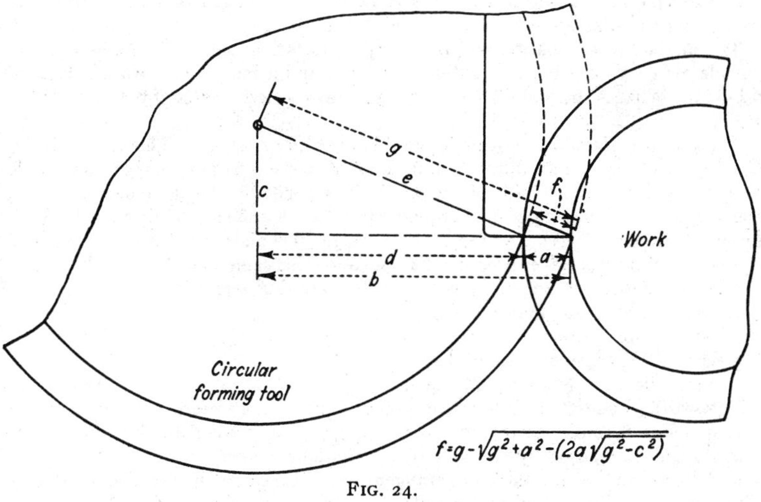

$$f = g - \sqrt{g^2 + a^2 - (2a\sqrt{g^2 - c^2})}$$

FIG. 24.

FIGS. 23–25. Forming-tool diameters and depths.

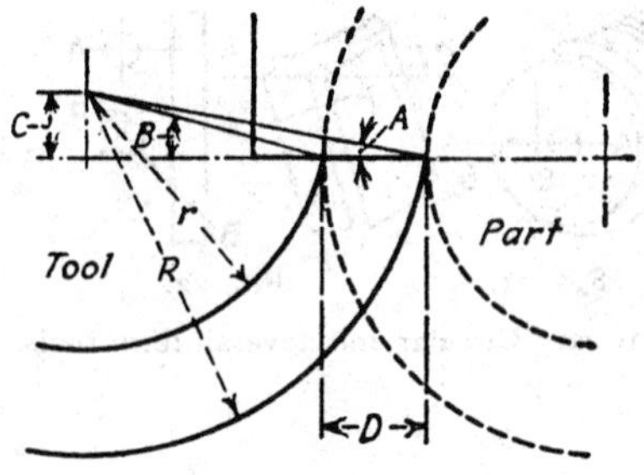

Fig. 26. Corrections for circular form tools without rake can be found by a simple trigometric solution.

as in the previous case, that is, 0.152 in., and multiplying by 0.98481, we have 0.1496 in. as the depth of T to which the tool must be planed.

Form Tools without Rake. As mentioned, a circular form tool is cut back from center by an amount that usually bears a relationship to the outside diameter of the tool blank, in order to provide clearance behind the cutting edge. Suppose the radius of the work is to be reduced by the step D (Fig. 26). If the forming-tool radius is cut back by this exact amount, the tool will cut too deep. This should be obvious from the equation for f (Fig. 24).

Ordinarily, the corrected radius r (Fig. 26) is calculated instead of f. The formula is

$$r = \sqrt{(\sqrt{R^2 - C^2} - D)^2 + C^2}$$

where r = corrected radius of a cutting edge, R = largest radius of tool, C = offset of cutting edge, and D = step on work.

A faster method of calculation devised by B. S. Sample is based upon the use of cotangents and cosecants (see Fig. 26 and Table 2).

The method of calculation will be described for an actual case. Assume a tool outside diameter of 3.375 in., an offset of $\frac{1}{4}$ in., and a minimum work diameter of 0.426 in. What is the tool diameter necessary to give a diameter of 0.817 in. on the work?

The step D is found by subtraction and halving to be 0.1955. This is divided by C, or $\frac{1}{4}$, obtaining 0.7820, which is subtracted from the constant 6.6755 (from Table 2), giving 5.8935 for the value of cot B. This is looked up in a table of cotangents, and the corresponding cosecant is found to be 5.9778, which is multiplied by $\frac{1}{2}$, or twice the value of C, giving 2.9889 as the corrected tool diameter. The "square and square-root" method gives a value of 2.98890 for this same figure.

The steps for the cotangent-cosecant method are thus seen to be:

1. Division by a simple fraction.
2. Subtraction.
3. Consulting trigonometric tables.
4. Multiplication by a simple fraction.

Noteworthy Points. 1. Where the tool offset is a simple fraction, the calculation is easy. For example, to divide 0.1955 by $\frac{1}{4}$, multiply by 4 and obtain 0.7820. Also, to multiply 5.9778 by $\frac{1}{2}$, divide by 2, obtaining 2.9889.

2. All constants in the table of cotangent A are based on the fractional value of the offset, to facilitate longhand calculation. If a calculating machine is used, the value of C should be carried to four decimal places in order to obtain four-place

TABLE 2. CALCULATIONS FOR CORRECTING CIRCULAR FORM TOOLS WITHOUT RAKE

Offset C	Tool OD	cot A	Offset C	Tool OD	cot A
$\frac{1}{32}$	1.250	19.97498	$\frac{5}{32}$	1.938	6.12044
$\frac{1}{16}$	1.688	13.46692	$\frac{5}{32}$	2.250	7.13022
$\frac{1}{16}$	2.312	18.46895	$\frac{5}{32}$	2.312	7.33051
$\frac{1}{16}$	2.375	18.97367	$\frac{5}{32}$	2.500	7.93725
$\frac{1}{16}$	2.750	21.97726	$\frac{5}{32}$	2.688	8.54327
$\frac{1}{16}$	3.000	23.97916	$\frac{5}{32}$	3.500	11.15527
			$\frac{5}{32}$	3.750	11.95826
$\frac{3}{32}$	1.125	5.91608	$\frac{3}{16}$	2.000	5.23875
$\frac{3}{32}$	1.250	6.59124	$\frac{3}{16}$	2.250	5.91608
$\frac{3}{32}$	1.375	7.26483	$\frac{3}{16}$	2.500	6.59124
$\frac{3}{32}$	1.500	7.93725	$\frac{3}{16}$	2.625	6.92820
$\frac{3}{32}$	1.750	9.27961	$\frac{3}{16}$	2.750	7.26483
			$\frac{3}{16}$	3.000	7.93725
$\frac{1}{8}$	1.500	5.91608	$\frac{1}{4}$	2.840	5.59128
$\frac{1}{8}$	1.750	6.92820	$\frac{1}{4}$	3.000	5.91608
$\frac{1}{8}$	1.875	7.43303	$\frac{1}{4}$	3.375	6.67552
$\frac{1}{8}$	2.000	7.93725	$\frac{1}{4}$	3.875	7.68521
$\frac{1}{8}$	2.312	9.19378	$\frac{1}{4}$	4.375	8.69267
$\frac{1}{8}$	2.375	9.44722	$\frac{1}{4}$	5.000	9.94988
$\frac{1}{8}$	2.500	9.94988	$\frac{1}{4}$	5.250	10.45227
$\frac{1}{8}$	2.750	10.95445			
$\frac{1}{8}$	3.000	11.95826			
$\frac{1}{8}$	3.250	12.96148	$\frac{5}{16}$	4.000	6.32139

accuracy in the answer. Although the values of cotangent A are given to five decimal places, only four places need be used in the calculations.

3. The tabular difference between cotangents and cosecants is practically the same, so exact interpolation of the tables is not necessary.

4. The answer obtained by this method is accurate within ± 1 in the fourth decimal place, which is sufficient for most form tools.

EXAMPLE: For a tool $\frac{1}{4} \times 3.375$, calculations follow:

Part dia	Step D	$\frac{D}{C}$	cot $A - \frac{D}{C}$ or cot B	cosec B	Corr. dia
0.426 min					
0.817	0.1955	0.7820	5.8935	5.9778	2.9889
0.932	0.253	1.0120	5.6635	5.7512	2.8756
1.000	0.287	1.1480	5.5275	5.6171	2.8085
1.250	0.412	1.6480	5.0275	5.1260	2.5630

Brown & Sharpe Method for Correcting Radiuses. The problem of finding corrected radiuses for circular form tools used on Brown & Sharpe screw machines is simplified by standardized dimensions of tool blanks (Table 3) and use of Table 4. The amount of cutback is always figured from the full diameter of the tool blank. Suppose the piece in Fig. 27 is to be produced on a No. 2 machine, which uses a 3-in. tool blank.

This is the procedure: Start with the smallest work diameter, which is 0.375 in. The difference between the first and second steps is 0.655 − 0.375 = 0.280 in. In Table 4, we find that 0.280 falls between entries of 0.250 and 0.3125 in the column headed Difference in Diameter of Piece. Interpolating in the related entries in the column headed Amount of Radial Difference, we can say that 0.0045 should be subtracted from 0.280, giving 0.2755. By subtracting this figure from the full diameter of 3 in., we find that the diameter of the first step on the tool is 2.725 in.

The second step is found by going back to the smallest work diameter and subtracting from 0.875. By going through the process outlined above, the second step diameter on the tool is found to be 2.508 in.

Circular Form Tools with Rake. Corrections for circular form tools with rake require more computations than tools without rake. Assume a workpiece (Fig. 28) with four diameters. The tool is set to the center line of the work. For every diameter of the part, it is necessary to find S, m, n, r_1, and d_1 for corresponding tool diameters. The formulas are

$$S = R \cos A - r \cos E$$

$$\cos A = \cos \sin^{-1} \frac{r \sin E}{R}$$

$$m = G + S \sin E$$

$$n = R_1 \cos F - S \cos E$$

$$r_1 = \sqrt{m^2 + n^2}$$

$$d_1 = 2 r_1$$

Circular and Dovetail Forming-tool Blanks and Holders. The American Standard ASA B5. 7-1948 includes the same data on circular forming-tool blanks and holders as found in the 1943 publication, but a larger range is now given for dovetail forming tools and holders in order to handle heavier work on certain sizes of machines. See Table 5 on classification of automatic and other screw machines.

To obtain the advantages of a minimum number of interchangeable blank sizes, the machines are classified into six groups of comparable stock capacities. Each group takes a definite size of tool. To facilitate procurement of commercial tool blanks, circular tools are designated by outside diameter and width, dovetailed tool blanks by group number and width of blank.

Interchangeability of holders among various makes of machines is not provided, except for mounting or clamping details of the circular or dovetailed tool member.

Definitions

A circular tool blank is a disk with a centrally located mounting hole, plain or threaded, and a number of adjusting holes on one side. Circular blanks are used for making forming and cutoff tools by machining the periphery to the form of the product to be made. A notch is ground in the periphery to provide a cutting edge.

A *circular forming toolholder* is a device for holding circular cutting tools in the proper relation to the work as it is actuated back and forth on the tool slide of an automatic screw machine. A center bolt and hook bolt are generally provided for clamping the tool in the holder.

TABLE 3. DIMENSIONS OF CIRCULAR FORM TOOLS FOR BROWN & SHARPE AUTOMATICS

Mach. No.	Approx. Diam.	Max. Radius	Max. Base Line	Distance above or below Center
2 & 2G	3	1.49998	1.479	0.250
0 & 0G	2.250	1.12474	1.114	0.15626
00 & 00G	1.750	0.87497	0.866	0.125

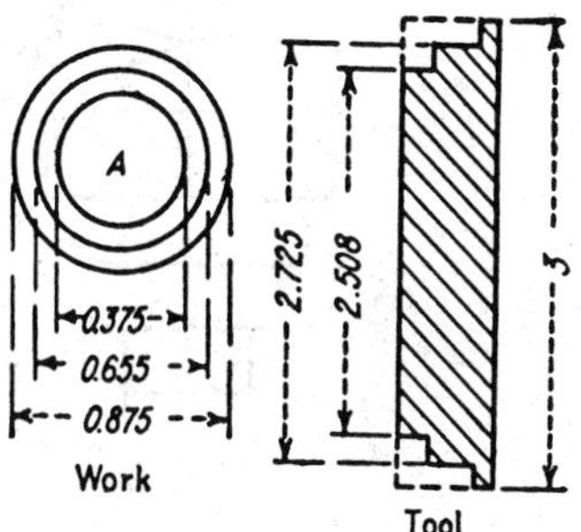

FIG. 27. Example for finding diameters of circular form tools without rake, using Brown & Sharpe data in Table 4.

TABLE 4. DATA FOR CALCULATING DIAMETERS OF CIRCULAR FORM TOOLS FOR BROWN & SHARPE AUTOMATICS

	Machine Where Used		
	00 and 00G	0 and 0G	2 and 2G
Diameter of Tool	1.750	2.250	3.000
Dimension A	1/8	5/32	1/4

Difference in Diameter of Piece	Difference in Diameter of Tool			Amount of Radial Difference		
	Machine					
	00 and 00G	0 and 0G	2 and 2G	00 and 00G	0 and 0G	2 and 2G
0.0625	0.062	0.062	0.0615	0.0005	0.0005	0.001
0.125	0.124	0.124	0.123	0.001	0.001	0.002
0.1875	0.1855	0.1855	0.1855	0.002	0.002	0.0025
0.250	0.247	0.2475	0.246	0.003	0.0025	0.004
0.3125	0.3085	0.309	0.3075	0.004	0.0035	0.005
0.375	0.370	0.371	0.369	0.005	0.004	0.006
0.4375		0.432	0.4305		0.0055	0.007
0.500		0.494	0.492		0.006	0.008
0.5625		0.555	0.553		0.0075	0.0095
0.625		0.617	0.614		0.008	0.011
0.6875			0.675			0.0125
0.750			0.736			0.014
0.8125			0.797			0.0155
0.875			0.858			0.017
0.9375			0.9185			0.019
1.000			0.979			0.021
1.0625			1.040			0.0225
1.125			1.100			0.025

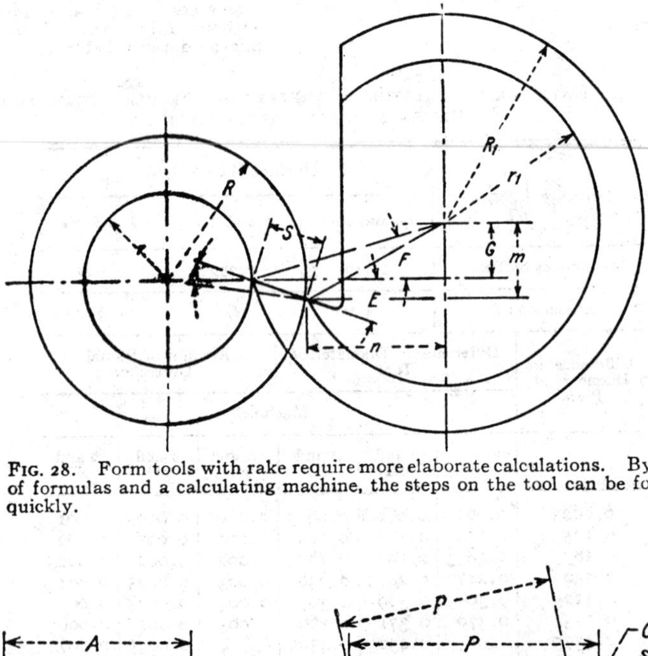

FIG. 28. Form tools with rake require more elaborate calculations. By use of formulas and a calculating machine, the steps on the tool can be found quickly.

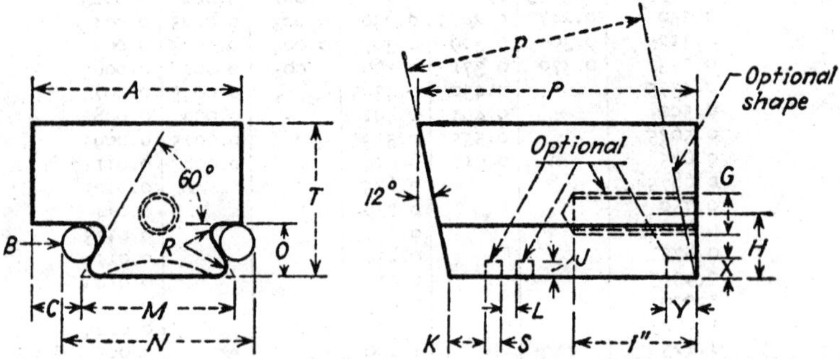

FIG. 29. Blanks for dovetail form tools, according to current American standards.

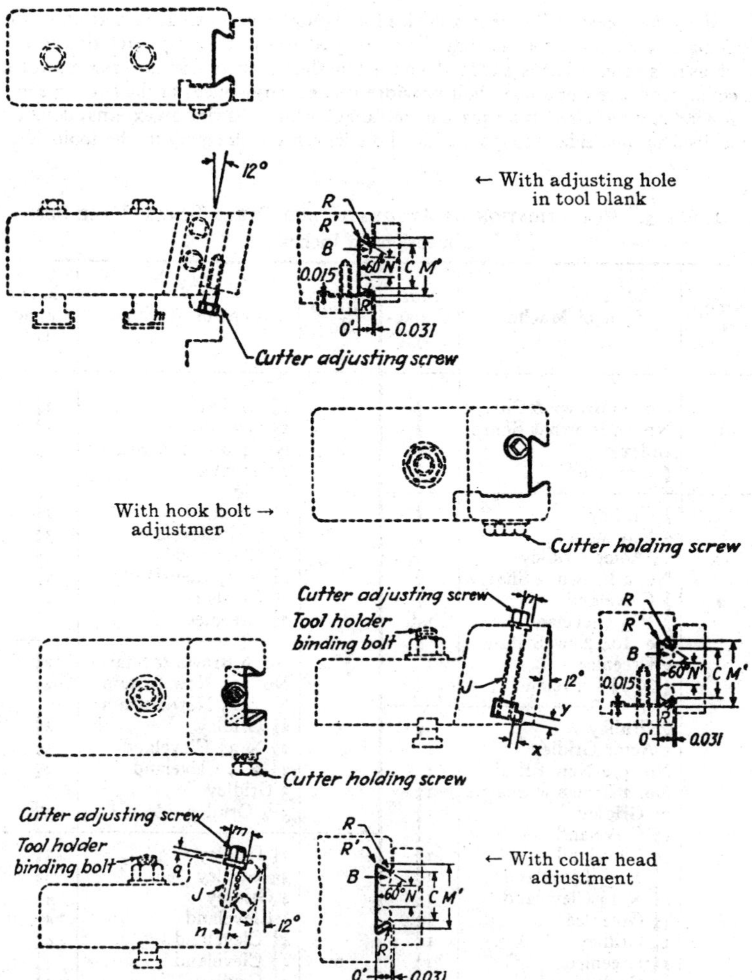

FIG. 30. ASA designs for holders for dovetail forming tools.

Circular tool blanks for machines in group 1 have no adjusting pinholes. A friction plate may be provided for adjusting the cutting edge of the tool to the center of the work. The holders for the larger machines have a positive means of adjusting the circular tool to the center of the work. A plate with a pin which engages in one of the six holes on the side of the circular tool provides for a circular motion to the cutter to raise and lower the cutting edge. This movement is transmitted to the mounting plate by a screw.

A *dovetailed form toolholder* is a device for clamping the dovetailed portion of the forming or cutting-off tool so that the cutting edge is in the proper relation to the work as it is actuated back and forth on the tool slide of an automatic screw machine. A collar-head screw or a hook bolt provides vertical adjustment of the cutting edge.

A *dovetailed tool blank* is a square or rectangular block having an external dovetail machined on one side throughout its whole length for clamping in the toolholder.

TABLE 5. CLASSIFICATION OF AUTOMATIC AND OTHER SCREW MACHINES
Dimensions in Inches

Group No.	Type of Machine	Max Capacity*
1	No. 00 Brown & Sharpe	3/8
	No. 19 Brown & Sharpe	3/8
	Index 0	7/16
	5/8 Cleveland	5/8
2	3/8 Gridley	3/8
	1/2 Davenport	7/8
	9/16 Acme Gridley	9/16
	No. 0 Brown & Sharpe	5/8
	5/8 Cleveland	3/4
	5/8 × 7/8 Cleveland	1 1/16
	No. 204 New Britain	5/8
	7/8 Greenlee	1
	7/8 × 1 1/4 Cleveland	7/8
3	7/8 Gridley	7/8
	1 Acme Gridley	1
	No. 172 New Britain	1
	No. 2 Brown & Sharpe	1 1/8
	1 1/4 Gridley	1 1/4
	1 1/4 Cleveland	1 1/4
	1 1/4 Cleveland	1 3/8
	1 1/4 × 1 1/2 Cleveland	1 3/4
	1 1/4 × 1 1/2 Cleveland	1 1/2
	1 1/4 Greenlee	1 1/4
	1 3/8 Gridley	1 3/8
	1 1/2 Greenlee	1 1/2
	1 5/8 Gridley	1 5/8
4	1 5/8 Gridley	1 5/8
	1 5/8 Acme Gridley	1 5/8
	No. 206 New Britain	1 5/8
	No. 415 New Britain	1 5/8
	No. 410 New Britain	...
4	1 3/4 Gridley	1 3/4
	1 3/4 Greenlee	1 3/4
	No. 4 Brown & Sharpe	1 7/8
	2 Greenlee	2
	2 Gridley	2
	2 Cleveland	2 1/2
	2 × 2 3/4 Cleveland	3 1/4
	2 1/4 Cleveland	2 1/2
	2 1/4 × 2 3/4 Cleveland	3 1/4
	2 1/4 Gridley	2 1/4
	2 1/4 Greenlee	2 1/4
5	No. 6 Brown & Sharpe	2 3/8
	No. 208 New Britain	2 1/2
	No. 425 New Britain	2 1/2
	2 5/8 Gridley	2 5/8
	2 3/4 × 3 3/4 Cleveland	3 1/4
	2 3/4 × 4 Cleveland	2 3/4
	3 Gridley	3
	3 5/16 Gridley	3 5/16
6	3 1/4 Gridley	3 1/4
	3 1/2 Gridley	3 1/2
	4 Gridley	4
	4 Cleveland	4
	4 1/4 Cleveland	4 1/4
	4 1/2 Cleveland	4 1/2
	4 1/4 Gridley	4 1/4
	4 1/2 Gridley	4 1/2
	5 Gridley	5
	5 1/2 Cleveland	5 1/2
	6 3/4 Cleveland	6 3/4
	7 3/4 Cleveland	7 3/4

NOTE: Technical Committee No. 10, ASME, in preparing these machine capacities and classifications used all the information available.

* The group classification numbers apply to all machine models of the respective makes listed having the maximum capacities indicated.

The dovetail is slotted to provide adjustment. Dovetailed blanks are used for making forming and cutting-off tools by machining the desired formed profile in the side opposite the dovetail and throughout its entire length. The top of the formed section is the cutting edge.

TABLE 6. TYPICAL CIRCULAR FORM TOOLHOLDER FOR MACHINES IN GROUP 1, TABLE 5
Dimensions in Inches

Machine Group Number	Width of Blank	Bolt		Tool Post
		Length	Thread Size	
	A	*f*	*d*	*e*
1	1/4	3/4	3/8 to 16	7/16
	3/8	7/8	3/8 to 16	7/16
	1/2	1	3/8 to 16	7/16
	3/4	1 1/4	3/8 to 16	7/16
	1	1 1/2	3/8 to 16	7/16

This holder is intended for cutters without adjusting pin holes.

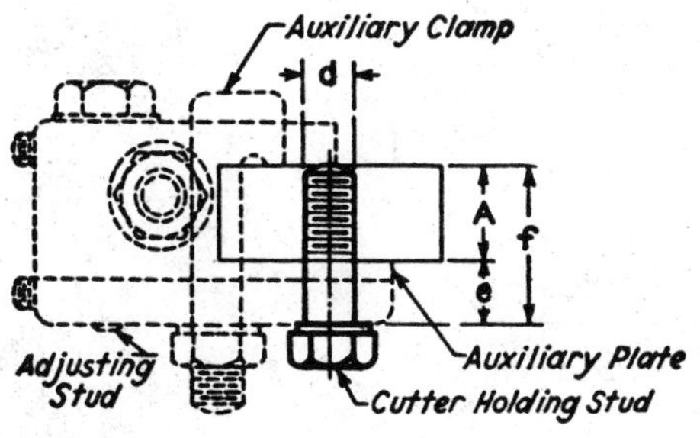

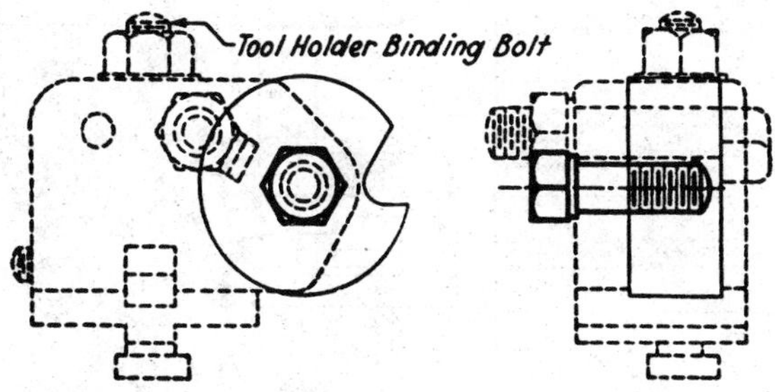

For details of other toolholders and blanks, see:
Page 13-18: Circular form toolholders for machines in groups 2 to 6.
Page 13-19: Dimensions of blanks for circular form tools.
Page 13-20: Dimensions of blanks for dovetail form tools.
Page 13-21: Dimensions of holders for dovetail form tools.

TABLES 7 AND 8

Machine Group Number	Width of Blank	Bolt		Adjusting Pin		Tool Post
		Length	Thread Size	Radius	Diameter	
	A	f	d	g	h	e
2	3/8	1 1/8	1/2–13	11/16	0.185	3/4
	1/2	1 1/4	1/2–13	11/16	0.185	3/4
	3/4	1 1/2	1/2–13	11/16	0.185	3/4
	1	1 3/4	1/2–13	11/16	0.185	3/4
	1 1/4	2	1/2–13	11/16	0.185	3/4
3	1/2	1 3/8	5/8–11	3/4	0.185	13/16
	3/4	1 5/8	5/8–11	3/4	0.185	13/16
	1	1 7/8	5/8–11	3/4	0.185	13/16
	1 1/4	2 1/8	5/8–11	3/4	0.185	13/16
	1 1/2	2 3/8	5/8–11	3/4	0.185	13/16

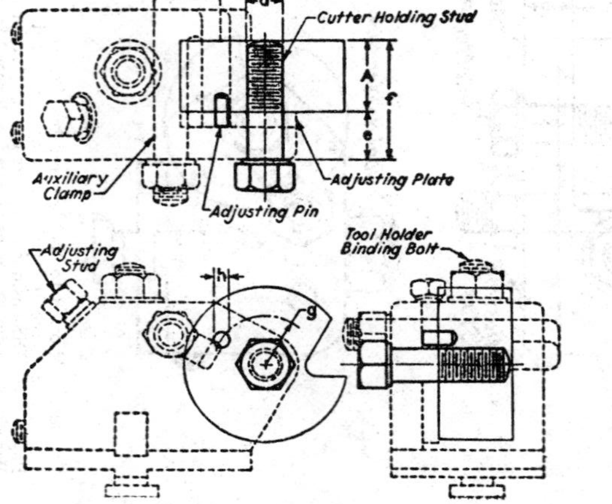

Machine Group Number	Width of Blank	Bolt					Adjusting Pin		Tool Post
		Length	Thread	Body	Head		Radius	Diameter	
	A	f	d	b	k	L	g	h	e
4	1/2	2 1/2	3/4–10	0.749	1	1/4	15/16	0.248	1 1/8
	3/4	2 3/4	3/4–10	0.749	1	1/4	15/16	0.248	1 1/8
	1	2 3/4	3/4–10	0.749	1	1/4	15/16	0.248	1 1/8
	1 1/2	3 1/2	3/4–10	0.749	1	1/4	15/16	0.248	1 1/8
	2	3 1/2	3/4–10	0.749	1	1/4	15/16	0.248	1 1/8
	2 1/2	3 1/2	3/4–10	0.749	1	1/4	15/16	0.248	1 1/8
5	5/8	3	1–8	0.999	1 5/16	5/16	1 1/16	0.310	1 3/8
	1	3 1/4	1–8	0.999	1 5/16	5/16	1 1/16	0.310	1 3/8
	1 1/2	3 1/4	1–8	0.999	1 5/16	5/16	1 1/16	0.310	1 3/8
	2	4 1/4	1–8	0.999	1 5/16	5/16	1 1/16	0.310	1 3/8
	2 1/2	4 1/2	1–8	0.999	1 5/16	5/16	1 1/16	0.310	1 3/8
	3	4 1/2	1–8	0.999	1 5/16	5/16	1 1/16	0.310	1 3/8
6	5/8	3 1/4	1–8	0.999	1 5/16	5/16	1 1/16	0.310	1 5/8
	1	3 3/4	1–8	0.999	1 5/16	5/16	1 1/16	0.310	1 5/8
	1 1/2	3 3/4	1–8	0.999	1 5/16	5/16	1 1/16	0.310	1 5/8
	2	4 3/4	1–8	0.999	1 5/16	5/16	1 1/16	0.310	1 5/8
	3	4 3/4	1–8	0.999	1 5/16	5/16	1 1/16	0.310	1 5/8
	4	4 3/4	1–8	0.999	1 5/16	5/16	1 1/16	0.310	1 5/8

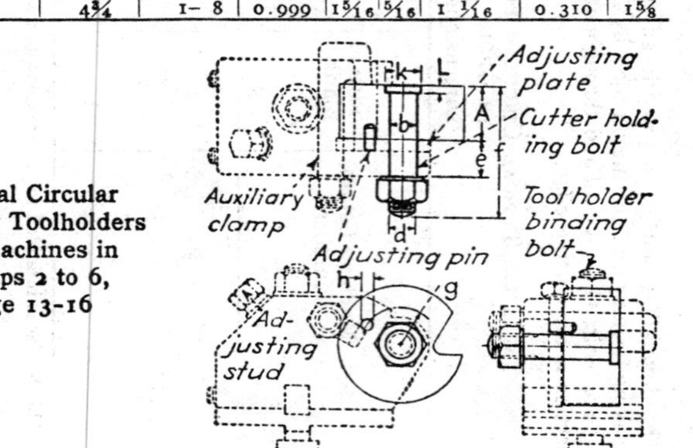

Typical Circular Forming Toolholders for Machines in Groups 2 to 6, Page 13-16

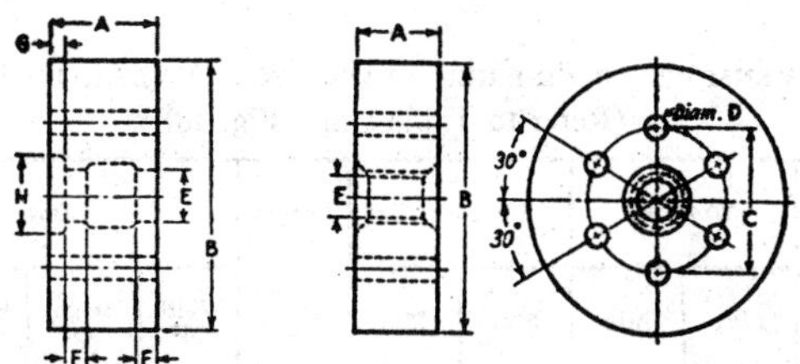

Table 9. Dimensions for Finished Blanks for Circular Forming Tools with Threaded Mounting Hole

Group Number	Basic Blank Size	Diameter B		Width A		Adjusting Holes		Dia of Threaded Hole E	Pitch of Threaded Hole E
		Max	Min	Max	Min	C	D		
1	1 3/4 × 1/4	1 25/32	1 49/64	9/32	17/64	No pin hole		3/8	16
	1 3/4 × 3/8	1 25/32	1 49/64	13/32	25/64			3/8	16
	1 3/4 × 1/2	1 25/32	1 49/64	17/32	33/64			3/8	16
	1 3/4 × 3/4	1 25/32	1 49/64	25/32	49/64			3/8	16
	1 3/4 × 1	1 25/32	1 49/64	1 1/32	1 1/64			3/8	16
2	2 1/4 × 3/8	2 9/32	2 17/64	13/32	25/64	1 3/8	3/16	1/2	13
	2 1/4 × 1/2	2 9/32	2 17/64	17/32	33/64	1 3/8	3/16	1/2	13
	2 1/4 × 3/4	2 9/32	2 17/64	25/32	49/64	1 3/8	3/16	1/2	13
	2 1/4 × 1	2 9/32	2 17/64	1 1/32	1 1/64	1 3/8	3/16	1/2	13
	2 1/4 × 1 1/4	2 9/32	2 17/64	1 9/32	1 17/64	1 3/8	3/16	1/2	13
3	3 × 1/2	3 1/32	3 1/64	17/32	33/64	1 1/2	3/16	5/8	11
	3 × 3/4	3 1/32	3 1/64	25/32	49/64	1 1/2	3/16	5/8	11
	3 × 1	3 1/32	3 1/64	1 1/32	1 1/64	1 1/2	3/16	5/8	11
	3 × 1 1/4	3 1/32	3 1/64	1 9/32	1 17/64	1 1/2	3/16	5/8	11
	3 × 1 1/2	3 1/32	3 1/64	1 17/32	1 33/64	1 1/2	3/16	5/8	11

Table 10. Dimensions for Finished Blanks for Circular Forming Tools with Counterbored Mounting Hole

Group Number	Basic Blank Size	Diameter B		Width A		Adjusting Holes		Mounting Hole			
								Diameter	Bearing Surface	Counterbore	
	B × A	Max	Min	Max	Min	C	D	E	F	Diameter H	Depth G
4	3 1/2 × 1/2	3 17/32	3 33/64	17/32	33/64	1 7/8	1/4	3/4	...	1 1/32	3/16
	3 1/2 × 3/4	3 17/32	3 33/64	25/32	49/64	1 7/8	1/4	3/4	...	1 1/32	3/16
	3 1/2 × 1	3 17/32	3 33/64	1 1/32	1 1/64	1 7/8	1/4	3/4	...	1 1/32	7/16
	3 1/2 × 1 1/2	3 17/32	3 33/64	1 17/32	1 33/64	1 7/8	1/4	3/4	5/16	1 1/32	1/4
	2 1/2 × 2	3 17/32	3 33/64	2 1/32	2 1/64	1 7/8	1/4	3/4	5/16	1 1/32	3/4
	3 1/2 × 2 1/2	3 17/32	3 33/64	2 17/32	2 33/64	1 7/8	1/4	3/4	5/16	1 1/32	1 1/4
5	4 × 5/8	4 1/32	4 1/64	21/32	41/64	2 1/8	5/16	1	...	1 11/32	1/4
	4 × 1	4 1/32	4 1/64	1 1/32	1 1/64	2 1/8	5/16	1	...	1 11/32	1/4
	4 × 1 1/2	4 1/32	4 1/64	1 17/32	1 33/64	2 1/8	5/16	1	...	1 11/32	3/4
	4 × 2	4 1/32	4 1/64	2 1/32	2 1/64	2 1/8	5/16	1	7/16	1 11/32	1/4
	4 × 2 1/2	4 1/32	4 1/64	2 17/32	2 33/64	2 1/8	5/16	1	7/16	1 11/32	3/4
	4 × 3	4 1/32	4 1/64	3 1/32	3 1/64	2 1/8	5/16	1	7/16	1 11/32	1 1/4
6	5 × 5/8	5 1/32	5 1/64	21/32	41/64	2 1/8	5/16	1	...	1 11/32	1/4
	5 × 1	5 1/32	5 1/64	1 1/32	1 1/64	2 1/8	5/16	1	...	1 11/32	1/4
	5 × 1 1/2	5 1/32	5 1/64	1 17/32	1 33/64	2 1/8	5/16	1	...	1 11/32	3/4
	5 × 2	5 1/32	5 1/64	2 1/32	2 1/64	2 1/8	5/16	1	7/16	1 11/32	1/4
	5 × 3	5 1/32	5 1/64	3 1/32	3 1/64	2 1/8	5/16	1	7/16	1 11/32	1 1/4
	5 × 4	5 1/32	5 1/64	4 1/32	4 1/64	2 1/8	5/16	1	7/16	1 11/32	2 1/4

Table 11. Dimensions of Finished Blanks for Dovetailed Forming Tools
(Refer to Table 5 and Fig. 29)

Group No.	Basic Blank Size[1]	Width A		Thickness T		Length	Dovetail			
		Max	Min	Max	Min	P	Width M[2]	Height O	Radius R	To Face C
1	1	1 1/32	1 1/64	29/32	57/64	1 1/2	0.732	9/32	1/32	1/4
	1 1/4	1 9/32	1 17/64	29/32	57/64	1 1/2	0 951	19/64	1/32	19/64
2	1 1/4	1 9/32	1 17/64	29/32	57/64	2	0.951	19/64	1/32	19/64
	1 1/2	1 17/32	1 33/64	29/32	57/64	2	0.951	19/64	1/32	19/64
	1 3/4	1 25/32	1 49/64	1 5/32	1 9/64	2	1.250	13/32	1/16	7/16
	2 3/4	2 25/32	2 49/64	1 5/32	1 9/64	2 7/16	2.000	33/64	1/16	1/2
3	1 3/4	1 25/32	1 49/64	1 5/32	1 9/64	2 7/16	1.250	13/32	1/16	7/16
	2 1/4	2 9/32	2 17/64	1 5/32	1 9/64	2 7/16	1.250	13/32	1/16	7/16
	2 3/4	2 25/32	2 49/64	1 5/32	1 9/64	2 7/16	1.250	13/32	1/16	7/16
4	2 3/4	2 25/32	2 49/64	2 17/32	2 33/64	2 5/8	1.614	35/64	1/16	1/2
	2 3/4	2 25/32	2 49/64	2 17/32	2 33/64	2 5/8	1.882	35/64	1/16	1/2
	3	3 1/32	3 1/64	2 17/32	2 33/64	2 5/8	1.882	35/64	1/16	1/2
5	2 3/4	2 25/32	2 49/64	1 9/32	1 17/64	2 7/16	2.000	33/64	1/16	1/2
	2 3/4	2 25/32	2 49/64	3 1/32	3 1/64	3	2.000	33/64	1/16	1/2
	3	3 1/32	3 1/64	3 1/32	3 1/64	3	2.000	33/64	1/16	1/2
	3 1/4	3 9/32	3 17/64	3 1/32	3 1/64	3	2.000	33/64	1/16	1/2
6	3 1/4	3 9/32	3 17/64	3 17/32	3 33/64	4	2.238	35/64	1/16	9/16
	3 1/2	3 17/32	3 33/64	3 17/32	3 33/64	4	2.883	43/64	1/16	5/8
	4	4 1/32	4 1/64	3 17/32	3 33/64	4	2.883	43/64	1/16	5/8
	4 1/2	4 17/32	4 33/64	3 17/32	3 33/64	4	2.883	43/64	1/16	5/8

Group No.	Basic Blank Size[1]	Across Plugs	Plug Diam	Adjusting Holes		Slots					Notch for Hook Bolt Adj. Nut	
		N[3]	B	G	H[4]	Number	J[4]	K	L	S[5]	Width Y	Depth X
1	1	0.834	5/32	5/16–18	3/8	2	1/8	5/16	1/8	1/8	1/8	1/8
	1 1/4	1.035	5/32	5/16–18	1/2	2	1/8	5/16	1/8	1/8	3/16	3/16
2	1 1/4	1.035	5/32	5/16–18	1/2	3	1/8	11/32	1/8	1/8	3/16	3/16
	1 1/2	1.035	5/32	5/16–18	1/2	...	..	....	...	...	3/16	3/16
	1 3/4	1.464	1/4	5/16–18	1/2	3	1/8	11/32	1/8	1/8	1/4	1/4
	2 3/4	2.771	1/2	7/16–14	3/8	3	1/8	7/16	5/32	5/32	5/16	5/16
3	1 3/4	1.464	1/4	5/16–18	1/2	3	1/8	7/16	5/32	5/32	1/4	1/4
	2 1/4	1.464	1/4	5/16–18	1/2	...	..	....	...	...	1/4	1/4
	2 3/4	1.464	1/4	5/16–18	1/4	...	..	....	...	...	1/4	1/4
4	2 3/4	2.349	1/2	5/16–18	1/2	...	..	....	...	...	1/4	1/4
	2 3/4	2.617	1/2	7/16–14	3/8	...	..	....	...	...	1/4	1/4
	3	2.617	1/2	7/16–14	3/8	...	..	....	...	...	1/4	1/4
5	2 3/4	2.771	1/2	7/16–14	3/8	...	..	....	...	...	5/16	5/16
	2 3/4	2.771	1/2	7/16–14	3/8	...	..	....	...	...	5/16	5/16
	3	2.771	1/2	7/16–14	3/8	...	..	....	...	...	5/16	5/16
	3 1/4	2.771	1/2	7/16–14	3/8	...	..	....	...	...	5/16	5/16
6	3 1/4	2.973	1/2	7/16–14	3/8	...	..	....	...	...	5/16	5/16
	3 1/2	3.815	5/8	7/16–14	3/8	...	..	....	...	...	5/16	5/16
	4	3.815	5/8	7/16–14	3/8	...	..	....	...	...	5/16	5/16
	4 1/2	3.815	5/8	7/16–14	3/8	...	..	....	...	...	5/16	5/16

Table 12. Dimensions of Holders for Dovetailed Forming Tools
(See Fig. 30)

Common Dimensions							Collar Head Screw					Hook Bolt			
								Adjusting Screw				Adjusting Screw and Nut			
Width	Width	Between Plugs	Depth	Plug Dia	Radius	Radius	Screw Location	Collar Dia	Collar Thickness	Thread Size	Screw Location	Length of Toe	Thickness of Toe	Thread Size
M'	C	N'	O'	B	R	R'	n	m	q	J	n	x	y	J
0.732	0.489	0.305	1/4	5/32	1/32	1/64	11/64	9/16	0.122	1/4–28	7/32	1/8	1/8	1/4–28
0.951	0.690	0.524	17/64	5/32	1/32	1/64	13/64	5/8	0.122	5/16–24	1/4	3/16	3/16	5/16–24
1.250	0.9085	0.567	3/8	1/4	1/16	1/32	15/64	11/16	0.152	5/16–24	1/4	1/4	1/4	5/16–24
1.614	1.110	0.248	33/64	1/2	1/16	1/32	17/64	13/16	0.184	3/8–24	9/32	1/4	1/4	3/8–24
1.882	1.378	0.516	33/64	1/2	1/16	1/32	17/64	13/16	0.184	3/8–24	9/32	5/16	5/16	3/8–24
2.000	1.532	0.634	31/64	1/2	1/16	1/32	21/64	1	0.245	7/16–20	5/16	5/16	5/16	7/16–20
2.238	1.734	0.872	33/64	1/2	1/16	1/32	21/64	1	0.245	7/16–20	5/16	5/16	5/16	7/16–20
2.883	2.235	1.175	41/64	5/8	1/16	1/32	21/64	1	0.245	7/16–20	5/16	5/16	5/16	7/16–20

SCREW-MACHINE CAM DESIGN

Laying Out Brown & Sharpe Cams

By C. W. HINMAN

1. Draw vertical line *AB* (Fig. 31) and through its center draw horizontal line *CD*. The latter represents the center line of the work and spindle. Against *AB* and to the right, draw the finished work, with its small end toward *D*. To the left of *AB* (¼ in. or more) draw parallel line *EF*. This is the face line of the spindle chuck. Extend the largest diameter of the work lines to the left, cutting *EF*. Draw form-tool groove *G* in work.

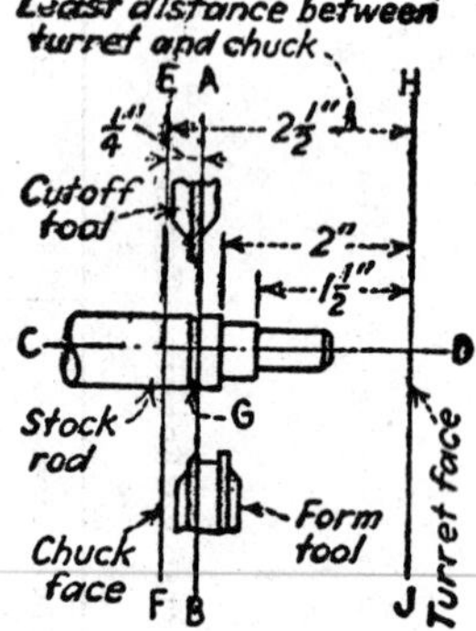

FIG. 31. Method for laying out cams.

2. Determine the method for machining; whether entirely formed by the front or back tools, and cut off, or if turned from the turret, and then formed and cut off. This depends upon ratio between width of form cut and work dia.

3. Determine the cutting travel, or throw, and feed per revolution for the front and back tools; whether these tools must work alone, or if either or both can start together with a turret tool.

4. Select the machine. This involves diameter of stock, length of finished piece, extreme length that can be turned, rigidity of machine for the work, lowest speed of machine (whether it is too fast for threading), greatest length of tools the turret will swing, and the diameter of special tools the turret will swing, if used.

5. Determine the working position of the form and cutoff tools and draw them on the sketch.

6. Draw another vertical line *HJ* to the right of the work, representing the turret face, at the nearest position between turret and chuck, and so dimension it. Add dimensions from the turret face to each shoulder.

7. Determine the rpm of the spindle, forward and reverse. Reverse is needed only when threading.

8. Determine the order of operations. Name each one that requires time to complete, and write them in orderly sequence, one below the other, on a trial layout sheet.

9. Determine the cutting travel or throw for each turret tool. Write these after their respective operations on the trial layout.

10. Select the feed per revolution for each turret tool. Write these after their respective operations.

11. Calculate the number of revolutions for each operation. Revolutions = travel ÷ feed.

12. Complete the trial layout. Estimate about 20 revolutions for each operation of indexing and to feed the stock, then total with all the machining operations. Compare the total revolutions with the table[1] of actual number of revolutions to make one piece on the machine selected. Choose the next higher number in the table, and substitute it on the layout. Also determine from the table the exact number of revolutions for feeding and indexing, and substitute. Spread the extra

[1] This refers to the tables in Brown & Sharpe screw-machine treatise.

revolutions, if any, on the heaviest cuts, and recalculate the feeds. The total number of revolutions must equal the actual number of revolutions chosen.

13. Determine from the table the time in seconds to make one piece and select the change gears.

14. Insert the number of hundredths of cam surface for each operation. The number of hundredths is found by dividing the number of revolutions for the operation by the actual number of revolutions to make one piece, with two places pointed off from the right.

15. Total divisions must be 100.

16. Finish the layout, allowing for dwells and tool clearances. When using the slotting arm, one turret hole, previous to feeding, must be vacant to clear the arm.

17. Determine where the form and cutoff tools are to begin and end their work. The cutting-off lobe usually ends at 99½.

18. There should be from 5 to 7 hundredths cam clearance between all large turret tools and the cross-slide front or back tools, if either of the latter follow next.

19. Draw the cams from the layout, cutting down the heights of the lead cam lobes, if necessary, to suit the tool body lengths (plus ¼ in.) as compared with the shoulder dimensions in the sketch. Lobes on front and back cams are usually the same height as the cam-blank radius.

20. There should be sufficient time to index the turret past its idle holes while cutting off.

Example: For Brown & Sharp No. 0 machine, it requires ⅝ sec to feed stock or to revolve the turret once. If the spindle speed is 1,207 rpm, it will require 13 revolutions to index once; 26, twice; and 39, three times.

21. Check the feeding distance for the extreme length of the feed for one throw of the feeding finger.

22. When right-handed threading is performed, the fast speed of the spindle will be backward, and the slow forward. For left-handed threading, conditions are reversed.

23. The number of revolutions for threading off is the same as the number of threads in the length to be threaded plus 4.

24. Number of revolutions for threading on in the layout is the number for threading off, multiplied by the ratio found by dividing the fast spindle speed by the slow.

25. If taps are coarser than 18 pitch and dies coarser than 16 pitch, for screw stock, provide two threading lobes, one for roughing and one for finishing. On tool steel, the same applies to taps coarser than 24 pitch and to dies coarser than 20 pitch.

26. Check the threading operation, whether a releasing or a nonreleasing holder is required.

Thread Lobes for Brown & Sharpe Cams

By A. Ainsworth

The following five methods show the difference in required die-lobe rises on Brown & Sharpe cams for a screw threaded for a distance of 0.875 in. at forty threads per inch, with a self-opening die head. Each method results in the same ultimate thread.

1. A die lobe representing the exact duplication of a leadscrew in an engine lathe; a 40-pitch thread advances 0.025 in. per revolution of spindle or part. (1.000/40 = 0.025.) Thread length of ⅞ in. equals 35 revolutions. (40 × 7/8 = 35

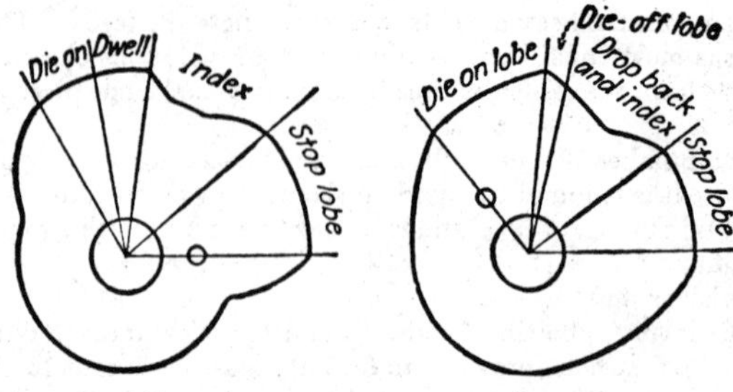

FIG. 32 FIG. 33

FIG. 32. Cam for self-opening die head. FIG. 33. Cam for solid die head. Both cams produce identical parts at an index cycle of 3 sec and same spindle speed.

revolutions.) To this add three revolutions for approach to the work, making a total of 38 revolutions to thread the part. Multiply the 38 revolutions by the 0.025-in. rise per revolution to determine that 0.950-in. die-lobe rise is required.

2. The general recommendation of self-opening die head manufacturers is the actual lead as shown by method 1, plus 5%. Required die-lobe rise is then 0.950 in. times 1.05, or 0.997-in. die-lobe rise.

3. Another principle holds that the die head must advance 12.5% ahead of lead on the cam as compiled by actual required lead on the part to be threaded. After this is determined, divide results of 12.5% less than actual lead into the length of thread required. This result equals actual revolutions required to complete the thread. To this add four revolutions for approach and the total number of revolutions is determined. To find the required die-lobe rise multiply the actual $\frac{1}{8}$ lead by the total revolutions and then subtract 10%. Converting this into figures we find that a lead of 0.025 minus 12.5% equals 0.022 in. lead. A threaded portion of 0.875 in. divided by 0.022-in. lead equals 39 revolutions. This plus four revolutions for approach equals 43 required revolutions of spindle to complete the thread; multiply 43 revolutions by 0.022 lead equals 0.946-in. rise; this less 10% equals 0.852 in.—the required die-lobe rise.

4. By a fourth method the following information must be known: (*a*) actual number of revolutions required to thread; (*b*) 10% of this figure to add for approach; (*c*) ratio between threading and cutting spindle speeds; (*d*) complete machine cycle time to produce one part; (*e*) spindle speeds at which job is to be run; (*f*) die-lobe revolutions converted into hundredths of cam space, and (*g*) revolutions of spindle per hundredths of cam space at threading spindle rpm.

Then $\frac{7}{8}$ in. of 40 threads per inch thread requires 35 revolutions; adding 10%, this equals 39 revolutions; multiplied by ratio of difference between spindle speeds (1,474 rpm divided by 663) or 39 × 2.22 equals 86 revolutions. With a known machine index time of 18$\frac{1}{3}$ sec we can determine the total revolutions at 1,474 rpm to complete the job and determine the hundredths of cam space required to thread. Converted, this is 18$\frac{1}{3}$ sec at 1,474 revolutions or 450 revolutions (4.5 revolutions per 0.01 of cam space). Thus 86 revolutions are required to complete the thread; therefore 86 total revolutions divided by 4.5 equals 0.19 of cam space required to thread. Converted into revolutions at threading spindle speed the 450 revolutions required equal 205 or 2.05 revolutions per cam space.

The formula then used is: Pitch in thousandths × revolutions per cam space at threading spindle speed × cam space to thread in hundredths = the required die-lobe rise.

Numerically this is: 6.000/40 × 2.05 × 19 = 0.973 in. required die-lobe rise.

5. Another widely used formula is

$$\text{Die-lobe rise} = \frac{\text{revolutions required to thread}}{\text{number of threads per inch}} \times \begin{cases}\text{a variable constant for groups of} \\ \text{threads known as } A \text{ and } B\end{cases}$$

A = 0.85 for 14 to 24 threads per inch.
B = 0.88 for 28 to 48 threads per inch.

Translated to our example this is 38/40 × 0.88 = 0.836 in. required die-lobe rise.

Methods 1, 2, and 4 Recommended. In checking the figures shown in Table 13 it will be noted that method 3 shows a higher cam lobe to thread than method 5. However, the former method allows more revolutions to thread and when proportioned out to equal advance per revolution it is found to be less than No. 5. This means that the thread in method 3 will cover more cam space but will have less lead on the lobe as compared with No. 5.

TABLE 13. RESULTS OF CALCULATIONS FOR THREAD LOBE RISE BY FIVE METHODS DESCRIBED

Figured for Self-opening Die Heads

Method	Req. Rev. to Thread	Req. Die Rise Lobe	Die Head Adv. per Spindle Rev.	Lgth. of Thd. Portion	Threads per Inch
[1]1	38	0.950	0.025	0.875	40
[1]2	38	0.997	0.026	0.875	40
3	43	0.852	0.019	0.875	40
[1]4	39	0.973	0.025	0.875	40
5	38	0.836	0.022	0.875	40

[1] Recommended.

VARIATION IN THREAD LOBE RISE AS CALCULATED BY THE FIVE METHODS

Applied to Cams for Use on Solid Die Heads

Method	Req. Rev. to Thread	Req. Die Rise Lobe	Die Head Adv. per Spindle Rev.	Lgth. of Thd. Portion	Threads per Inch
1	38	0.950	0.025	0.875	40
2	38	0.997	0.026	0.875	40
[1]3	43	0.852	0.019	0.875	40
4	39	0.876	0.0225	0.875	40
[1]5	38	0.753	0.0198	0.875	40

[1] Recommended.

A self-opening die head has a bumper spring which allows lead on the cam either to equal or to exceed the advance of the die head per work spindle revolution. Another die head feature compensates for worn index parts to eliminate variances in thread length. This is the outside trip. When a cam is designed to equal or exceed required lead, the outside trip will assure equal threaded lengths regardless of machine wear.

Experience recommends that method 1, 2, or 4 be used to determine self-opening die thread lobes.

With solid dies, where the die must be threaded on and off the work, different factors must be taken into consideration. The die head must lead on the thread-on and thread-off portion of the lobe, as faulty threads may be produced during either operation. Generally speaking there is no bumper or compensating spring in solid die heads, thus the lead on the die lobe must permit a slight lead-off from the start. However, too much lead would result in the die head pulling out to its extreme and causing stripped threads, as the die head pulls the turret forward, rather than the cam pushing the turret forward. On the thread-off operation the same difficulties are encountered. An incorrect thread-off lobe will cause the turret to pull the die head off. This is caused by the cam dropping away too quickly. Crowding of die head back into the holder before it is clear of work is caused by a cam designed with drop away too slow. Either condition results in poor threads.

A cam lobe for a solid die head requires more care and consideration than is required for the construction of one for a self-opening die.

Reviewing the requirements of the two types of die heads we find that the first two methods of determining the thread-lobe rise cannot be applied to calculations for the solid die heads.

Method 3. This is entirely satisfactory for solid die head lobe construction, but care must be taken to use the extra revolutions, which require more cam space and thereby decrease the actual lead on the cam.

In applying method 4, take this same formula and use only 90% of the die-lobe rise computation obtained. On the example used this would result in a die-lobe rise of 0.876 in.

Method 5. The use of this method will, of course, result in the same calculation as for the self-operating die lobe, or a required die-lobe rise of 0.836. Experience has shown that the use of this method, with a further reduction of from 8 to 10% on the results of its calculation, produces a desirable thread lobe for solid die head threading. On the example shown this would mean a required die-lobe rise of approximately 0.753 in. Figures 32 and 33 show cams for both types of dies.

Points on Threading in Screw Machines

By A. Ainsworth

The following are some practical points to keep in mind in camming automatics for threading:

1. A job so cammed that the die head must pull the turret along, instead of leading out in front, will cause defective threading.

2. Cams with excessive lead will result in distorted or stripped threads. A correct thread lobe will present the die head to the work with a positive start, then recede slightly and follow up the die head to completion of the thread length.

3. Self-opening die lobes require a dwell at the finish of the lobe to allow the die head to open correctly.

4. With the exception of one make, solid die heads have no allowance for the

addition of a bumper spring. Therefore, the die lobe must allow the die head to lead ahead from the start of the rise.

5. Self-opening die lobes are cammed more heavily than those required for solid die threading.

6. Misalignment between the spindle and the turret will usually result in tapered threads.

7. Worn die head parts or bent shanks will also result in a bad thread.

8. Outside trips should be used if incorporated in die head.

9. A solid die head incorporating both bumper spring and adjustments for misalignment of the spindle is highly recommended.

10. Grinds on chasers will greatly affect threads produced.

11. A die lobe designed for use with a solid die head usually will not allow the substitution of a self-opening die head. The cam will contain no dwell at top of lobe and clearances to approaching cross-slides are usually insufficient, because of larger diameter of the self-opening die head.

12. A solid die can never be used in conjunction with a die lobe designed for a self-opening head, as no provision is made for a thread-off portion on cam lobe.

13. Threads of coarser pitch require greater allowance when deducting from actual rise for computing rise on a thread lobe as they require more power and are apt to cause belt or clutch slippage.

SCREW-MACHINE LIMITS

Work made on screw machines cannot be made practically to a closer limit than 0.0005 in. either side of a fixed figure (0.001-in. limit). If closer work is required, it must be made large and ground to limit required. If no limit is given on the large part of piece and if not otherwise stated, it can be left stock size.

Tolerances for Piece Made in Screw Machines
(Outside Diameters)

Running Fits Not to Be Ground

To ½-in. dia inclusive	0.0005–0.0015 in. small
To 1-in. dia inclusive	0.001 –0.002 in. small
To 2-in dia inclusive	0.0015–0.0025 in. small
To 3½-in. dia inclusive	0.0015–0.003 in. small

Standard Fits Not to Be Ground—All Sizes

Standard to 0.001 in. small

Driving Fits Not to Be Ground—All Sizes

0.0005 to 0.0015 in. large

Pieces to Be Ground—Amount of Stock to Be Left for Grinding

To 1-in. dia inclusive	0.005–0.007 in. large
To 2-in. dia inclusive	0.007–0.010 in. large
To 3½-in. dia inclusive	0.010–0.014 in. large

ESTIMATING SCREW-MACHINE WORK

The National Screw Machine Products Association has issued net weight, no allowance data (Table 14) based on the actual densities of C1117 steel, free-cutting brass, and 11ST aluminum. Conversion factors (Table 15) are used when other grades of material are involved, or the shape is a hexagon or octagon.

The advantage of using a table based on actual densities is this: weight of materials needed in screw-machine products is a manufacturing-cost factor. Relation of material cost to total cost can vary all the way from 5 to over 80%.

TABLE 14. WEIGHT TABLES FOR SCREW-MACHINE PRODUCTS

Size, Inches	Steel				Brass				Aluminum			
	Round		Hex		Round		Hex		Round		Hex	
	Lb per 1000 In.	Lb per Ft	Lb per 1000 In.	Lb per Ft	Lb per 1000 In.	Lb per Ft	Lb per 1000 In.	Lb per Ft	Lb per 1000 In.	Lb per Ft	Lb per 1000 In.	Lb per Ft
1/32	0.2167	0.00260	0.2389	0.002866	0.257	0.00308	0.309	0.00371	0.08074	0.00097	0.09548	0.00114
3/64	0.4875	0.00585	0.5375	0.006450	0.563	0.00675	0.658	0.00790	0.1843	0.00221	0.2073	0.00249
1/16	0.8667	0.01040	0.9556	0.011468	0.982	0.01179	1.138	0.01366	0.3239	0.00389	0.3628	0.00461
5/64	1.3542	0.01625	1.4934	0.017920	1.524	0.01829	1.744	0.02093	0.5091	0.00611	0.5686	0.00682
3/32	1.950	0.02340	2.150	0.02580	2.185	0.02622	2.491	0.02989	0.7281	0.00874	0.8122	0.00975
7/64	2.654	0.03185	2.927	0.03512	2.959	0.03551	3.354	0.04025	0.9882	0.01186	1.099	0.01319
1/8	3.467	0.01460	3.822	0.04588	3.847	0.04620	4.342	0.05211	1.286	0.01543	1.429	0.01715
9/64	4.388	0.05264	4.838	0.05806	4.850	0.05821	5.462	0.06559	1.621	0.01946	1.801	0.02162
5/32	5.417	0.06500	5.972	0.07168	5.976	0.07176	6.717	0.08061	1.998	0.02398	2.219	0.02662
11/64	6.554	0.07865	7.228	0.08672	7.223	0.08667	8.100	0.09721	2.414	0.02897	2.678	0.03214
3/16	7.800	0.09360	8.602	0.10320	8.593	0.1031	9.619	0.1154	2.872	0.03448	3.182	0.03818
13/64	9.154	0.10985	10.094	0.12110	10.05	0.1206	11.26	0.1351	3.359	0.04032	3.728	0.04473
7/32	10.615	0.12740	11.700	0.14050	11.65	0.1398	13.03	0.1564	3.895	0.04675	4.317	0.05180
15/64	12.190	0.14620	13.440	0.16120	13.37	0.1605	14.93	0.1792	4.470	0.05365	4.949	0.05938
1/4	13.87	0.1664	15.29	0.1835	15.20	0.1824	16.96	0.2035	5.081	0.06100	5.623	0.06750
17/64	15.66	0.1878	17.26	0.2071	17.15	0.2058	19.12	0.2294	5.733	0.06880	6.348	0.07616
9/32	17.56	0.2106	19.35	0.2322	19.21	0.2306	21.41	0.2569	6.423	0.07709	7.108	0.08530
19/64	19.56	0.2346	21.56	0.2587	21.40	0.2568	23.83	0.2860	7.152	0.08584	7.913	0.09498
5/16	21.67	0.2600	23.89	0.2867	23.69	0.2843	26.38	0.3165	7.920	0.09506	8.762	0.1051
21/64	23.89	0.2866	26.34	0.3161	26.11	0.3133	29.05	0.3486	8.728	0.1047	9.652	0.1158
11/32	26.22	0.3146	28.91	0.3469	28.65	0.3438	31.86	0.3824	9.577	0.1149	10.59	0.1271
23/64	28.65	0.3438	31.60	0.3792	31.29	0.3756	34.80	0.4176	10.46	0.1255	11.58	0.1388
3/8	31.20	0.3744	34.41	0.4128	34.07	0.4088	37.86	0.4544	11.39	0.1367	12.59	0.1511
25/64	33.85	0.4063	37.33	0.4479	36.95	0.4432	41.06	0.4926	12.35	0.1482	13.66	0.1639
13/32	36.62	0.4394	40.38	0.4845	39.99	0.4796	44.38	0.5326	13.36	0.1603	14.76	0.1772
27/64	39.49	0.4738	43.55	0.5225	43.09	0.5168	47.84	0.5740	14.40	0.1728	15.92	0.1910
7/16	42.47	0.5097	46.85	0.5619	46.32	0.5555	51.43	0.6170	15.48	0.1858	17.12	0.2054

29/64	45.56	0.5467	50.25	0.6028	49.68	0.5958	55.13	0.6615	16.60	0.1993	18.35	0.2202
15/32	48.78	0.5850	53.75	0.6450	53.14	0.6375	58.97	0.7076	17.76	0.2132	19.63	0.2356
31/64	52.05	0.6246	57.40	0.6887	56.75	0.6807	62.97	0.7554	18.97	0.2277	20.96	0.2516
1/2	55.47	0.6656	61.16	0.7339	60.45	0.7251	67.05	0.8046	20.20	0.2425	22.32	0.2679
33/64	59.00	0.7080	65.05	0.7805	64.40	0.7727	71.55	0.8585	21.50	0.2583	23.78	0.2854
17/32	62.62	0.7514	69.06	0.8285	68.33	0.8198	75.92	0.9110	22.84	0.2741	25.24	0.3029
35/64	66.36	0.7963	73.17	0.8779	72.43	0.8692	80.42	0.9650	24.21	0.2906	26.73	0.3208
9/16	70.20	0.8424	77.41	0.9288	76.60	0.9191	85.04	1.0208	25.60	0.3073	28.28	0.3394
37/64	74.16	0.8900	81.77	0.9812	80.89	0.9703	89.81	1.078	27.04	0.3245	29.86	0.3588
19/32	78.23	0.9387	86.25	1.035	85.32	1.023	94.67	1.136	28.52	0.3423	31.50	0.3780
39/64	82.40	0.9887	90.85	1.090	89.82	1.077	99.69	1.196	30.02	0.3603	33.17	0.3981
5/8	86.67	1.040	95.58	1.147	94.50	1.133	104.8	1.258	31.59	0.3791	34.88	0.4186
41/64	91.05	1.093	100.40	1.205	99.24	1.190	110.1	1.321	33.18	0.3981	36.65	0.4398
21/32	95.54	1.147	105.37	1.264	104.1	1.249	115.5	1.386	34.81	0.4177	38.45	0.4614
43/64	100.15	1.202	110.44	1.325	109.1	1.309	121.0	1.452	36.43	0.4378	40.29	0.4835
11/16	104.9	1.258	115.65	1.387	114.3	1.370	126.7	1.520	38.20	0.4584	42.18	0.5061
45/64	109.7	1.316	120.95	1.451	119.5	1.433	132.5	1.589	39.95	0.4794	44.10	0.5291
23/32	114.6	1.375	126.40	1.517	124.8	1.498	138.4	1.661	41.74	0.5009	46.09	0.5530
47/64	119.7	1.436	131.95	1.583	130.3	1.564	144.5	1.734	43.56	0.5228	48.11	0.5771
3/4	124.8	1.498	137.60	1.651	135.9	1.631	150.7	1.809	45.44	0.5452	50.17	0.6020
49/64	129.9	1.560	143.4	1.721	141.6	1.699	157.0	1.883	47.34	0.5681	52.26	0.6271
25/32	135.2	1.625	149.4	1.792	147.4	1.769	163.4	1.960	49.28	0.5914	54.42	0.6530
51/64	141.0	1.691	155.4	1.864	153.3	1.840	170.0	2.039	51.28	0.6153	56.61	0.6793
13/16	146.5	1.758	161.5	1.938	159.4	1.913	176.7	2.120	53.30	0.6396	58.84	0.7061
53/64	152.0	1.826	167.8	2.014	165.6	1.987	183.5	2.202	55.34	0.6643	61.13	0.7335
27/32	157.9	1.896	174.2	2.090	171.9	2.063	190.4	2.285	57.48	0.6897	63.45	0.7614
55/64	163.9	1.966	180.7	2.168	178.3	2.140	197.5	2.370	59.61	0.7153	65.80	0.7896
7/8	169.8	2.038	187.3	2.247	184.8	2.218	204.7	2.456	61.80	0.7413	68.21	0.8186
57/64	176.0	2.112	194.1	2.329	191.5	2.298	212.0	2.545	64.02	0.7680	70.67	0.8480
29/32	182.2	2.186	200.9	2.411	198.3	2.379	219.5	2.634	66.28	0.7953	73.16	0.8780
59/64	188.6	2.263	207.9	2.495	205.2	2.462	227.1	2.725	68.58	0.8229	75.70	0.9085
15/16	194.9	2.340	215.0	2.580	212.2	2.546	234.8	2.818	70.92	0.8509	78.27	0.9393

TABLE 14. WEIGHT TABLES FOR SCREW-MACHINE PRODUCTS (*Continued*)

Size, Inches	Steel				Brass				Aluminum			
	Round		Hex		Round		Hex		Round		Hex	
	Lb per 1000 In.	Lb per Ft	Lb per 1000 In.	Lb per Ft	Lb per 1000 In.	Lb per Ft	Lb per 1000 In.	Lb per Ft	Lb per 1000 In.	Lb per Ft	Lb per 1000 In.	Lb per Ft
61/64	201.6	2.417	222.3	2.667	219.3	2.631	242.7	2.913	73.30	0.8794	80.90	0.9708
31/32	208.2	2.498	229.6	2.755	226.5	2.718	250.7	3.009	75.71	0.9085	83.56	1.003
63/64	215.0	2.580	237.0	2.844	233.8	2.806	258.9	3.106	78.16	0.9379	86.28	1.035
1	221.9	2.662	244.6	2.935	241.3	2.895	267.1	3.205	80.67	0.9679	89.03	1.068
1 1/64	228.8	2.746	252.4	3.028	249.1	2.989	276.1	3.316	83.28	0.9992	91.91	1.103
1 1/32	236.0	2.832	260.2	3.122	256.8	3.081	284.5	3.414	85.85	1.030	94.73	1.137
1 1/16	250.5	3.006	276.2	3.314	272.6	3.271	302.0	3.623	91.14	1.094	100.6	1.207
1 3/32	265.4	3.185	292.7	3.512	288.8	3.465	319.8	3.837	96.55	1.158	106.6	1.279
1 1/8	280.8	3.370	309.6	3.715	305.6	3.667	338.4	4.060	102.2	1.226	112.7	1.353
1 5/32	296.7	3.560	327.1	3.924	322.7	3.873	357.6	4.288	107.8	1.294	119.0	1.428
1 3/16	312.9	3.755	345.0	4.140	340.4	4.084	376.9	4.523	113.8	1.365	125.6	1.507
1 7/32	329.5	3.955	363.4	4.361	358.5	4.302	396.7	4.761	119.9	1.438	132.2	1.587
1 1/4	346.7	4.160	382.3	4.587	377.1	4.525	417.4	5.008	126.1	1.513	139.1	1.669
1 9/32	364.2	4.370	401.6	4.819	396.1	4.753	438.5	5.261	132.4	1.589	146.1	1.753
1 5/16	382.3	4.587	421.5	5.058	415.7	4.987	460.0	5.520	138.9	1.667	153.3	1.840
1 11/32	400.6	4.807	441.8	5.301	435.7	5.228	482.0	5.784	145.6	1.747	160.6	1.928
1 3/8	419.5	5.034	462.6	5.550	456.1	5.474	504.7	6.056	152.5	1.829	168.1	2.019
1 13/32	438.8	5.265	483.8	5.805	476.9	5.723	528.0	6.333	159.4	1.913	175.9	2.111
1 7/16	458.5	5.502	505.5	6.068	498.4	5.980	551.4	6.617	166.6	1.999	183.8	2.205
1 15/32	478.6	5.743	527.8	6.334	520.4	6.245	575.4	6.905	174.0	2.088	191.8	2.302
1 1/2	499.2	5.990	550.5	6.605	542.6	6.512	600.2	7.203	181.4	2.177	200.1	2.401
1 17/32	520.2	6.243	573.6	6.883	565.4	6.785	625.3	7.504	190.3	2.284	212.1	2.545
1 9/16	541.6	6.499	597.3	7.168	588.6	7.064	651.3	7.814	198.1	2.378	220.7	2.649
1 19/32	563.6	6.764	621.4	7.458	612.4	7.348	677.3	8.127	206.1	2.473	229.6	2.754
1 5/8	586.0	7.031	646.0	7.752	636.8	7.638	704.1	8.449	214.3	2.572	238.6	2.863
1 21/32	608.6	7.302	671.2	8.053	661.3	7.935	731.3	8.775	222.5	2.669	247.7	2.973
1 11/16	631.8	7.582	696.7	8.360	686.4	8.236	759.2	9.110	231.0	2.772	257.1	3.085
1 23/32	655.4	7.865	722.7	8.673	711.8	8.545	787.3	9.447	239.6	2.875	266.5	3.198
1 3/4	679.4	8.153	749.3	8.992	737.8	8.856	816.1	9.793	248.3	2.980	276.3	3.316

1 25/32	704.0	8.448	776.2	9.315	764.7	9.176	845.5	10.15	257.3	3.087	286.1	3.434
1 13/16	728.9	8.746	803.6	9.645	791.9	9.502	875.5	10.50	266.3	3.195	296.1	3.554
1 27/32	754.3	9.052	831.5	9.980	819.2	9.830	905.8	10.86	275.4	3.306	306.3	3.676
1 7/8	780.0	9.360	860.0	10.32	847.3	10.17	936.7	11.24	284.9	3.418	316.8	3.801
1 29/32	806.3	9.675	889.0	10.67	875.5	10.51	968.2	11.62	294.4	3.533	327.4	3.928
1 15/16	832.9	9.995	918.4	11.02	904.5	10.85	1000.0	12.00	304.1	3.650	338.1	4.057
1 31/32	860.0	10.32	948.3	11.38	933.9	11.20	1032.0	12.39	313.9	3.767	348.9	4.187
2	887.5	10.65	978.6	11.74	963.7	11.56	1065.0	12.78	324.0	3.888	360.1	4.320
2 1/16	943.8	11.33	1041.0	12.49	1025.0	12.31	1133.0	13.60	344.4	4.133	384.1	4.610
2 1/8	1002	12.02	1105	13.26	1088	13.05	1202	14.43	365.6	4.387	407.7	4.892
2 3/16	1062	12.74	1171	14.05	1152	13.83	1275	15.29	387.3	4.647	431.7	5.180
2 1/4	1124	13.48	1239	14.86	1220	14.64	1348	16.17	409.7	4.916	456.7	5.481
2 5/16	1187	14.24	1308	15.70	1288	15.46	1423	17.07	432.6	5.191	482.0	5.784
2 3/8	1252	15.02	1380	16.56	1359	16.31	1501	18.01	456.3	5.473	508.2	6.098
2 7/16	1318	15.82	1454	17.44	1431	17.17	1581	18.97	480.5	5.766	535.0	6.420
2 1/2	1387	16.64	1529	18.35	1505	18.07	1663	19.95	505.4	6.064	562.6	6.751
2 5/8	1529	18.35	1686	20.23	1661	19.94	1837	22.04	557.0	6.684	619.8	7.438
2 3/4	1678	20.14	1850	22.20	1823	21.88	2013	24.16	611.2	7.334	679.8	8.158
2 7/8	1834	22.02	2022	24.26	1993	23.91	2204	26.44	667.9	8.014	742.5	8.910
3	1997	23.96	2202	26.42	2170	26.04	2400	28.79	727.1	8.724	807.9	9.696
3 1/8	2167	26.00	2389	28.67	2354	28.25	2603	31.24	788.6	9.463		
3 1/4	2344	28.12	2584	31.01	2546	30.56	2816	33.79	852.9	10.23		
3 3/8	2527	30.33	2787	33.44	2746	32.95	3036	36.43	919.5	11.03		
3 1/2	2718	32.61	2997	35.96	2953	35.44	3265	39.18	1002	12.02		
3 5/8	2916	34.99	3215	38.58	3168	38.04	3503	42.03	1074	12.89		
3 3/4	3120	37.45	3441	41.28	3390	40.67	3749	44.99	1149	13.79		
3 7/8	3332	39.98	3674	44.08	3620	43.44	4003	48.04	1226	14.71		
4	3550	42.60	3914	46.97	3857	46.28	4265	51.18	1305	15.67		
4 1/8	3775	45.30	4163	49.95	4104	49.24	4536	54.43	1388	16.65		
4 1/4	4008	48.09	4419	53.02	4357	52.26	4816	57.79	1472	17.67		
4 3/8	4247	50.96	4683	56.19	4617	55.36	5103	61.23	1560	18.72		
4 1/2	4493	53.92	4954	59.45	4882	58.57	5398	64.79	1650	19.80		
4 5/8	4746	56.95	5234	62.80	5157	61.87	5703	68.43	1742	20.90		
4 3/4	5006	60.07	5520	66.24	5439	65.26	6015	72.18	1837	22.04		
4 7/8	5273	63.25	5814	69.77	5729	68.74	6335	76.04	1934	23.20		
5	5547	66.56	6116	73.39	6027	72.32	6665	79.97	2034	24.41		

TABLE 15. CONVERSION FACTORS FOR WEIGHT CALCULATIONS

MATERIAL	ROUNDS BASED ON ROUNDS	HEX BASED ON HEX	SQUARE BASED ON HEX	OCTAGON BASED ON HEX
Steel C-1117	1.000	1.000	1.155	.957
Steel C-1137	.998	.998	1.153	.955
Stainless Steel—No. 303C	1.002	1.002	1.157	.959
Stainless Steel No. 416	.978	.978	1.130	.936
Free-Cutting Brass	1.000	1.000	1.155	.957
Naval Brass	.990	.990	1.143	.947
Type KR Monel Cold Rolled	.993	.998	1.152	.955
Type R Cold Drawn Monel	1.039	1.044	1.205	.999
Grade "A" Cold Drawn Nickel	1.046	1.051	1.214	1.005
Aluminum 11-ST	1.000	1.000	1.155	.957
Aluminum 17-ST	.987	.987	1.140	.944
Aluminum 24-ST	.981	.981	1.133	.939
Magnesium (Dow Metal FS-1)	.628	.628	.725	.601
Magnesium (Dow Metal M)	.627	.627	.724	.600
Magnesium (Dow Metal J-1)	.635	.635	.733	.608

For Round multiply weight for Round in table by factor.
Magnesium based on Aluminum 11-ST
For other shapes multiply weights for Hex in table by factor.

For convenience the NSMPA tables are expressed in pounds per 1000 in. By knowing the length per piece and multiplying this by the tabulated weight for 1000 in. of material, one automatically gets the weight per 1000 pcs. But also weights in pounds per foot are tabulated. The loss in material for bar ends that must be added to the calculation is given in Table 16.

One example of weight-estimating procedure follows:

Material:
Brass—$\frac{11}{32}$-in. round
Bar end length—$2\frac{1}{2}$ in.
Length of bar—12 ft

Quantity:
125,000 pcs

Dimensions:

Part length	1.041
Facing	0.010
Cutoff	0.085
Total	1.136 in.

From tables:

Weight per 1000 pcs = 1.136 × 28.65 × 4069 lb

Bar ends (interpolated)	1.73%
Rejection and setup loss	5.00%
Total	6.73%

Total weight required = 4069 × 106.73 = 4343 lb

A calculation of this type can be performed in 2 min.

Hourly Output. In the past, many estimators have used a machine efficiency of 85%, to allow for downtime arising from tool changes, personal needs of the operator, lack of stock, etc. But in current practice, one may find it necessary to use machine efficiencies ranging from 60 to 80%. Table 17 makes it easy to calculate net hourly outputs for parts requiring $\frac{1}{2}$ to 49 sec at the mentioned efficiencies.

TABLE 16. BAR-END LOSS IN PER CENT

	6′	8′	10′	12′	14′	16′	18′	20′
1″	1.39	1.04	.83	.69	.59	.52	.46	.42
2″	2.78	2.08	1.66	1.39	1.19	1.04	.93	.83
3″	4.17	3.13	2.50	2.08	1.79	1.56	1.39	1.25
4″	5.56	4.17	3.33	2.78	2.38	2.08	1.85	1.67
5″	6.94	5.21	4.17	3.47	2.98	2.60	2.31	2.08
6″	8.35	6.25	5.00	4.17	3.57	3.13	2.78	2.50
7″	9.72	7.29	5.83	4.86	4.17	3.65	3.24	2.92
8″	11.11	8.33	6.66	5.56	4.76	4.16	3.70	3.33
9″	12.50	9.38	7.50	6.25	5.36	4.68	4.17	3.75
10″	13.88	10.41	8.33	6.95	5.95	5.20	4.62	4.16
11″	15.27	11.45	9.17	7.64	6.55	5.73	5.09	4.58
12″	16.66	12.50	10.00	8.33	7.14	6.25	5.56	5.00

COLOR CODE FOR STEEL SCREW-MACHINE STOCK

The following represents the 1948 code of the National Screw Machine Products Association for bar-stock identifying colors.

COLOR CODE FOR STEEL BARS

Steel	Color Marking	Steel	Color Marking
B1112	Green	C1117	Aluminum
B1113	Orange	C1118	Red
C1019	Yellow	C1137	Gold
C1020	White		

MACHINING VARIOUS MATERIALS ON SCREW MACHINES

Aluminum. Both high speed and carbide are used, and commonly on Alcoa 11S-T3 and 17S-T. Form tools are used without rake for the first material, but with 5° tap rake on 17S-T. Both form and cutoff tools are made with ½° side clearance. No chip curler is needed on box tools for 11S-T3, but it is frequently ground in tools for 17S-T alloy.

Magnesium. Feeds 50 to 75% higher than brass and maximum speed of machine are used. Form tools should have 0 to 10° rake, with better chip breakage resulting from the low side. Clearance should be at least 8°; side relief of 1° minimizes rubbing. Ratio of maximum width of cut to work diameter is 2:1.

Monel Metal. Cold-drawn R monel has been handled at 100 to 125 sfpm and feeds two-thirds those for mild steel. With unhardened KR monel, the range is 60 to 80 sfpm. Cutting edges must be kept sharp and cutting when against the work, or it will glaze.

Plastics. Screw-machine operations on laminated stock are performed dry, except that 60% paraffin oil, 40% kerosene may be used for deep-hole drilling and threading. Feeds of 0.0005 to 0.001 in. are used for forming and recessing. Form-tool width should not exceed 75% of the work diameter. Tools are ground the same as for brass, except for slightly greater clearance.

Stainless Steel. Form tools should have 7 to 10° top rake. A 3 to 5° taper on side of a tool helps to balance end thrust and to prevent galling.

Tabular Data. For purposes of cam layout and estimating, it is convenient to use tabulated data for screw-machine operations on various materials. Table 18 gives such data for brass, Table 19 gives speeds and feeds for steels (using HSS tools), and Table 20 covers similar information for working the stainless steels.

TABLE 17. HOURLY OUTPUT AT 60 TO 80% EFFICIENCY*

Sec per Piece	Gross Product per Hr	60 %	65 %	70 %	75 %	80 %	Sec per Piece	Gross Product per Hr	60 %	65 %	70 %	75 %	80 %
½	7200	4320	4680	5040	5400	5760	12½	288	173	187	202	216	230
⅝	5760	3456	3744	4032	4320	4608	13	276	166	179	193	207	221
¾	4800	2880	3120	3360	3600	3840	13½	267	160	174	187	200	214
⅞	4114	2468	2674	2880	3086	3291	14	257	154	167	180	193	206
1	3600	2160	2340	2520	2700	2880	14½	248	149	161	174	186	198
1¼	2880	1728	1872	2016	2160	2304	15	240	144	156	168	180	192
1½	2400	1440	1560	1680	1800	1920	15½	232	139	151	162	174	186
1¾	2057	1234	1337	1440	1543	1646	16	225	135	146	158	169	180
2	1800	1080	1170	1260	1350	1440	16½	218	131	142	153	164	174
2¼	1600	960	1040	1120	1200	1280	17	212	127	138	148	159	170
2½	1440	864	936	1008	1080	1152	17½	206	124	134	144	155	165
2¾	1309	785	851	916	982	1047	18	200	120	130	140	150	160
3	1200	720	780	840	900	960	18½	195	117	127	137	146	156
3¼	1107	664	720	775	830	886	19	189	113	123	132	142	151
3½	1028	617	668	720	771	822	19½	185	111	120	130	139	148
3¾	960	576	624	672	720	768	20	180	108	117	126	135	144
4	900	540	585	630	675	720	21	171	103	111	120	128	137
4¼	847	508	551	593	635	678	22	164	98	107	115	123	131
4½	800	480	520	560	600	640	23	156	94	101	109	117	125
4¾	757	454	492	530	568	606	24	150	90	98	105	113	120
5	720	432	468	504	540	576	25	144	86	94	101	108	115
5¼	686	412	446	480	515	549	26	138	83	90	97	104	110
5½	654	392	425	458	491	523	27	133	80	86	93	100	106
5¾	626	376	407	438	470	501	28	128	77	83	90	96	102
6	600	360	390	420	450	480	29	124	74	81	87	93	99
6¼	576	346	374	403	432	461	30	120	72	78	84	90	96
6½	553	332	359	387	415	442	31	116	70	75	81	87	93
6¾	533	320	346	373	400	426	32	112	67	73	78	84	90
7	514	308	334	360	386	411	33	109	65	71	76	82	87
7¼	497	298	323	348	373	398	34	106	64	69	74	80	85
7½	480	288	312	336	360	384	35	103	62	67	72	77	82
7¾	465	279	302	326	349	372	36	100	60	65	70	75	80
8	450	270	293	315	338	360	37	97	58	63	68	73	78
8¼	436	262	283	305	327	349	38	95	57	62	67	71	76
8½	423	254	275	296	317	338	39	92	55	60	64	69	74
8¾	411	247	267	288	308	329	40	90	54	59	63	68	72
9	400	240	260	280	300	320	41	88	53	57	62	66	70
9¼	389	233	253	272	292	311	42	86	52	56	60	65	69
9½	379	227	246	265	284	303	43	84	50	55	59	63	67
9¾	369	221	240	258	277	295	44	82	49	53	57	62	66
10	360	216	234	252	270	288	45	80	48	52	56	60	64
10½	342	205	222	239	257	274	46	78	47	51	55	59	62
11	327	196	213	229	245	262	47	77	46	50	54	58	62
11½	313	188	203	219	235	250	48	75	45	49	53	56	60
12	300	180	195	210	225	240	49	73	44	47	51	55	58

* Calculated by National Screw Machine Products Association.

TABLE 18. SCREW-MACHINE FEEDS FOR BRASS*

Tool	Cut: Width or Depth	Cut: Dia of Hole	Feed
Boring tools	0.005		
Box tools—roller rest	$\frac{1}{32}$	....	0.012
1 chip finishing	$\frac{1}{16}$	....	0.010
	$\frac{1}{8}$	....	0.008
	$\frac{3}{16}$	....	0.008
	$\frac{1}{4}$	....	0.006
Finishing	0.005	....	0.010
Center drills	......	Under $\frac{1}{8}$	0.003
		Over $\frac{1}{8}$	0.006
Cutoff tools:			
Angular	......	....	0.0015
Circular	$\frac{3}{64}$–$\frac{1}{8}$	....	0.0035
Straight	$\frac{1}{16}$–$\frac{1}{8}$	....	0.0035
Dia stock under $\frac{1}{8}$ in.	......	....	0.002
Drills	......	.02	0.0014
Twist cut	......	.04	0.002
		$\frac{1}{16}$	0.004
		$\frac{3}{32}$	0.006
		$\frac{1}{8}$	0.009
		$\frac{3}{16}$	0.012
		$\frac{1}{4}$	0.014
		$\frac{5}{16}$	0.016
		$\frac{3}{8}$	0.016
		$\frac{1}{2}$	0.016
		$\frac{5}{8}$	0.016
Form tools—circular	$\frac{1}{8}$–$\frac{1}{4}$	....	0.002
	$\frac{3}{8}$–$\frac{1}{2}$	....	0.0015
			0.0012
	$\frac{5}{8}$–$\frac{3}{4}$	....	0.001
	1	....	0.001

Tool	Cut: Width or Depth	Cut: Dia of Hole	Feed
Hollow mills, turned dia under 5/32 in.	$\frac{1}{32}$	....	0.012
	$\frac{1}{16}$	....	0.010
Balance turning tools, turned dia over 5/32 in.	$\frac{1}{32}$	....	0.017
	$\frac{1}{16}$	....	0.015
	$\frac{1}{8}$	....	0.012
	$\frac{3}{16}$	....	0.010
	$\frac{1}{4}$	....	0.009
Knurl tools turret	On	....	0.020
	Off	....	0.040
Side or swing	......	....	0.004
	......	....	0.006
Top	......	....	0.005
	......	....	0.008
Pointing and facing tools	......	....	0.001
			0.0025
Reamers and bits	0.003–0.004	$\frac{1}{8}$ or less	0.010–0.007
	0.004–0.008	$\frac{1}{8}$ or over	0.010
Recessing tools, end cut	......	....	0.001
			0.005
Inside cut	$\frac{1}{16}$–$\frac{1}{8}$	....	0.0025
			0.0008
Swing tools, forming	$\frac{1}{8}$–$\frac{1}{4}$	....	0.002
			0.0012
	$\frac{3}{8}$–$\frac{1}{2}$	....	0.001
			0.0008
Turning straight	$\frac{1}{32}$	....	0.008
	$\frac{1}{16}$	....	0.006
	$\frac{1}{8}$	....	0.005
	$\frac{3}{16}$	....	0.004

NOTE: The maximum speed of the machine is used for brass. Figures for feeds are averages; if the work has unusual features, alter them accordingly.

* Brown & Sharpe.

TABLE 19. SPEEDS AND FEEDS FOR STEELS*

Based on the use of high-speed tools and an average 8-hr tool life. Feeds are based on the use of multiple-spindle-type machines using both rough and finish forming tools on cold-drawn material. On lighter, single-spindle machines using only one forming tool, feeds are usually much lower

Groups:
A. Union Supercut; AISI B1113
B. Union Free Cut; AISI B1112; AISI C1113, C1119
C. AISI B1111; AISI C1116
D. AISI C1111, C1117, C1118

Tool Name	Size of Hole, In.	Width or Depth of Cut, In.	Group A		Group B		Group C		Group D	
			Sfpm	Feed, Ipr	Sfpm	Feed, Ipr	Sfpm	Feed, Ipr	Sfpm	Feed, Ipr
		Width								
Form: Circular or dovetail	....	0.500	225	0.003	165	0.0025	155	0.0023	150	0.0022
	....	1.000	210	0.0025	160	0.002	150	0.0019	145	0.0018
	....	1.500	210	0.0025	160	0.0018	150	0.0017	145	0.0016
	....	2.000	205	0.0018	155	0.0015	146	0.0014	141	0.0014
	....	2.500	200	0.0015	150	0.0012	141	0.0011	136	0.0011
Twist drills	0.250	.....	125	0.0054	105	0.0045	99	0.0042	95	0.0041
	0.500	.....	125	0.006	105	0.005	99	0.0047	95	0.0045
	0.750	.....	140	0.007	115	0.006	108	0.0056	105	0.0055
	1.000	.....	140	0.008	115	0.007	108	0.0066	105	0.0064
	1.250	.....	145	0.009	120	0.008	113	0.0076	119	0.0073
		Depth								
Box-tool blades	....	0.125	225	0.0085	165	0.007	155	0.0066	150	0.0064
	....	0.250	210	0.008	160	0.0065	150	0.0061	145	0.0059
	....	0.375	205	0.0065	155	0.0055	146	0.0052	141	0.0050
	....	0.500	200	0.0055	150	0.0045	141	0.0042	136	0.0041
Hollow mills	....	0.062	200	0.012	150	0.010	141	0.0094	136	0.0091
	....	0.125	186	0.0096	140	0.008	132	0.0075	127	0.0073
	....	0.187	179	0.0084	135	0.007	127	0.0066	123	0.0064
	....	0.250	172	0.0073	130	0.0065	122	0.0061	118	0.0059

Tool Name	Size of Hole, In.	Width or Depth of Cut, In.	Group A		Group B		Group C		Group D	
			Sfpm	Feed, Ipr	Sfpm	Feed, Ipr	Sfpm	Feed, Ipr	Sfpm	Feed, Ipr
Knurl tools in turret:										
On	....	.....	225	0.020	165	0.015	155	0.0014	150	0.0014
Off	....	.....	225	0.040	165	0.030	155	0.0028	150	0.0028
Knurl—cross-slide	....	.....	225	0.020	165	0.015	155	0.0014	150	0.0014
Chamfer and facing	....	.....	280	0.008–0.010	200	0.005–0.007	188	0.0047–0.0066	182	0.0045–0.0064
Reamers	Under ½ in.	.....	180	0.0085	145	0.007	136	0.0066	132	0.0064
	½ in. or over	.....	180	0.012	145	0.010	136	0.0094	132	0.0091
Cutoff		Width								
	....	0.062	225	0.003	165	0.002	155	0.0019	150	0.0018
	....	0.125	230	0.0035	175	0.0025	164	0.0023	160	0.0023
	....	0.187	235	0.0035	180	0.0025	169	0.0023	164	0.0023
	....	0.250	250	0.004	190	0.003	179	0.0028	173	0.0027

* Union Drawn Steel Division, Republic Steel Corp.

TABLE 19. SPEEDS AND FEEDS FOR STEELS* (*Continued*)

Groups:
E. AISI C1144* (C1144, annealed, can be machined at about 10% higher speeds than shown if the feeds are lowered accordingly)
F. AISI C1108, C1109, C1114, C1115, C1120, C1125, C1126, C1141,* C1151*
G. AISI C1016, C1019, C1022, C1145*; AISI B1010; AISI 4017, 4023, 4024
H. AISI C1023, C1138, C1144; AISI 4032,* 5120

Tool Name	Size of Hole, In.	Width or Depth of Cut, In.	Group E		Group F		Group G		Group H	
			Sfpm	Feed, Ipr	Sfpm	Feed, Ipr	Sfpm	Feed, Ipr	Sfpm	Feed, Ipr
		Width								
Form: circular or dove-tail	....	0.500	140	0.0021	135	0.0020	130	0.0019	125	0.0019
	....	1.000	136	0.0017	130	0.0017	125	0.0016	121	0.0015
	....	1.500	136	0.0015	130	0.0015	125	0.0014	121	0.0014
	....	2.000	132	0.0013	127	0.0012	121	0.0012	117	0.0011
	....	2.500	127	0.001	122	0.0010	117	0.0009	113	0.0009
Twist drills	0.250	.....	89	0.004	86	0.0040	82	0.0038	79	0.0037
	0.500	.....	89	0.0045	86	0.0045	82	0.0043	79	0.0042
	0.750	.....	98	0.0055	94	0.0054	90	0.0052	87	0.0050
	1.000	.....	98	0.0064	94	0.0063	90	0.006	87	0.0058
	1.250	.....	102	0.007	98	0.0072	94	0.0068	91	0.0066
		Depth								
Box-tool blades	....	0.125	140	0.0059	135	0.0057	130	0.0055	125	0.0052
	....	0.250	136	0.0055	130	0.0053	125	0.0051	121	0.0049
	....	0.375	132	0.0047	127	0.0045	121	0.0043	117	0.0041
	....	0.500	127	0.004	122	0.0037	117	0.0035	113	0.0034
Hollow mills	....	0.062	127	0.0085	122	0.0081	117	0.0078	113	0.0075
	....	0.125	119	0.0068	114	0.0065	109	0.0062	106	0.0060
	....	0.187	115	0.006	110	0.0057	105	0.0055	102	0.0053
	....	0.250	110	0.0055	106	0.0053	102	0.0051	98	0.0049

Tool Name	Size of Hole, In.	Width or Depth of Cut, In.	Group E		Group F		Group G		Group H	
			Sfpm	Feed, Ipr	Sfpm	Feed, Ipr	Sfpm	Feed, Ipr	Sfpm	Feed, Ipr
Knurl tools in turret:										
On	....	.....	140	0.012	135	0.0125	130	0.012	125	0.0110
Off	....	.....	140	0.024	135	0.0250	130	0.024	125	0.0220
Knurl—cross-slide	....	.....	140	0.012	135	0.0125	130	0.012	125	0.0110
Chamfer and facing	....	.....	170	0.005–0.0072	163	0.0045–0.0063	156	0.0047–0.0066	151	0.0042–0.0058
Reamers	Under ½ in.	.....	123	0.006	118	0.0057	113	0.0055	110	0.0053
	½ in. or over	.....	123	0.0085	118	0.0081	113	0.0078	110	0.0075
Cutoff		Width								
	....	0.062	140	0.0017	135	0.0016	130	0.0016	125	0.0015
	....	0.125	149	0.0021	142	0.0020	136	0.0019	132	0.0019
	....	0.187	153	0.0021	147	0.0020	140	0.0019	136	0.0019
	....	0.250	161	0.0025	155	0.0024	148	0.0023	144	0.0022

* Annealed.

TABLE 19. SPEEDS AND FEEDS FOR STEELS (*Continued*)

Groups:

J. AISI C1015, C1017, C1020, C1025, C1045,* C1137, C1140; AISI 3130,* 4028, 4037,* 4130,* 5132,* 5135,* 8630;* AISI E4132. These steels may be machined at 120 sfpm with feeds as given in Group J or at 100 to 110 sfpm with 10 to 20% heavier feeds.

K. AISI C1010, C1012

L. AISI C1030, C1035, C1050,* C1141, C1146, C1151; AISI 2330,* 2335,* 3135,* 4042,* 4137,* 5045,* 5140,* 8615, 8632,* 8635,* 8735;* AISI E4135*

M. AISI C1008, C1145; AISI 2317, 3115, 3120, 3140,* 4027, 4047,* 4140,* 4142,* 4608, 4615, 4620, 4621, 4640,* 5145,* 5147,* 6145,* 8617, 8620, 8622, 8637,* 8640,* 8642,* 8720, 8740,* 8742,* 9437,* 9440,* 9442;* AISI E4137*

Tool Name	Size of Hole, In.	Width or Depth of Cut, In.	Group J		Group K		Group L		Group M	
			Sfpm	Feed, Ipr	Sfpm	Feed, Ipr	Sfpm	Feed, Ipr	Sfpm	Feed, Ipr
		Width								
Form: circular or dovetail	....	0.500	120	0.0018	120	0.0015	115	0.0017	110	0.0016
	....	1.000	115	0.0014	115	0.0012	112	0.0014	106	0.0013
	....	1.500	115	0.0013	115	0.0011	112	0.0013	106	0.0012
	....	2.000	112	0.0011	112	0.0009	108	0.0011	102	0.0010
	....	2.500	108	0.0009	108	0.0007	105	0.0008	99	0.0008
Twist drills	0.250	.....	76	0.0035	76	0.0031	73	0.0034	69	0.0033
	0.500	.....	76	0.004	76	0.0034	73	0.0038	69	0.0036
	0.750	.....	83	0.0047	83	0.0041	80	0.0045	76	0.0044
	1.000	.....	83	0.0055	83	0.0048	80	0.0053	76	0.0051
	1.250	.....	86	0.0064	86	0.0055	84	0.0062	79	0.0058
		Depth								
Box-tool blades	....	0.125	120	0.005	120	0.0042	115	0.0049	110	0.0046
	....	0.250	115	0.0047	115	0.0039	112	0.0045	106	0.0043
	....	0.375	112	0.004	112	0.0033	108	0.0038	102	0.0036
	....	0.500	108	0.0032	108	0.0027	105	0.0031	99	0.0030

Tool Name	Size of Hole, In.	Width or Depth of Cut, In.	Group J		Group K		Group L		Group M	
			Sfpm	Feed, Ipr	Sfpm	Feed, Ipr	Sfpm	Feed, Ipr	Sfpm	Feed, Ipr
Hollow mills	....	0.062	108	0.0072	108	0.0060	105	0.007	99	0.0066
	....	0.125	101	0.0058	101	0.0048	98	0.0056	92	0.0053
	....	0.187	97	0.005	97	0.0042	95	0.0049	89	0.0046
	....	0.250	94	0.0047	94	0.0039	91	0.0045	86	0.0043
Knurl tools in turret:										
On..............	....	.....	120	0.011	120	0.009	115	0.011	110	0.0099
Off..............	....	.....	120	0.022	120	0.018	115	0.022	110	0.0198
Knurl—cross-slide...	....	.....	120	0.011	120	0.009	115	0.011	110	0.0099
Chamfer and facing	....	.....	144	0.004–0.006	144	0.0034–0.0048	140	0.004–0.006	132	0.0038–0.0053
Reamers	Under ½ in.	.....	104	0.005	104	0.0042	102	0.0049	96	0.0046
	½ in. or over	.....	104	0.0072	104	0.0060	102	0.007	96	0.0066
Cutoff		Width								
	....	0.062	120	0.0015	120	0.0012	115	0.0014	110	0.0013
	....	0.125	126	0.002	126	0.0015	122	0.0017	116	0.0016
	....	0.187	130	0.002	130	0.0015	126	0.0017	119	0.0016
	....	0.250	137	0.0024	137	0.0018	133	0.0022	126	0.0020

* Annealed.

TABLE 19. SPEEDS AND FEEDS FOR STEELS (Continued)

Groups:
N. AISI C1040, C1043; AISI 3141,* 3145,* 4145,* 4147,* 5150,* 5152,* 8625, 8627, 8645,* 8745,* 9445,* 9747;* AISI E4617, E4620
P. AISI C1045; AISI 1320, 1340,* 2340,* 3130, 4340,* 5130, 6120, 8655*
O. AISI 1330,* 1335,* 3150,* 4150,* 4317,* 4320,* 6150,* 6152,* 8647,* 8650,* 8750,* 8757*
Q. AISI C1050; AISI 2330, 3135, 8660,* 9255,* 9763; AISI E4337*

Tool Name	Size of Hole, In.	Width or Depth of Cut, In.	Group N		Group O		Group P		Group Q	
			Sfpm	Feed, Ipr	Sfpm	Feed, Ipr	Sfpm	Feed, Ipr	Sfpm	Feed, Ipr
		Width								
Form: circular or dovetail	....	0.500	105	0.0015	100	0.0015	95	0.0014	90	0.0014
	....	1.000	101	0.0012	96	0.0012	91	0.0012	87	0.0011
	....	1.500	101	0.0011	96	0.0011	91	0.0010	87	0.0010
	....	2.000	98	0.0009	93	0.0009	88	0.0009	84	0.0008
	....	2.500	95	0.0007	90	0.0007	85	0.0007	81	0.0007
Twist drills	0.250	.....	67	0.0032	63	0.0031	60	0.0028	57	0.0028
	0.500	.....	67	0.0035	63	0.0034	60	0.0031	57	0.0031
	0.750	.....	73	0.0042	69	0.0041	65	0.0037	62	0.0037
	1.000	.....	73	0.0049	69	0.0048	65	0.0044	62	0.0044
	1.250	.....	76	0.0056	72	0.0055	68	0.0050	65	0.0050
		Depth								
Box-tool blades	....	0.125	105	0.0044	100	0.0042	95	0.0040	90	0.0038
	....	0.250	101	0.0041	96	0.0039	91	0.0037	87	0.0035
	....	0.375	98	0.0034	93	0.0033	88	0.0031	84	0.0030
	....	0.500	95	0.0028	90	0.0027	85	0.0026	81	0.0024
Hollow mills	....	0.062	95	0.0063	90	0.0060	85	0.0057	81	0.0054
	....	0.125	89	0.0050	84	0.0048	80	0.0046	76	0.0043
	....	0.187	86	0.0044	81	0.0042	77	0.0040	73	0.0038
	....	0.250	83	0.0041	78	0.0039	74	0.0037	70	0.0035

Tool Name	Size of Hole, In.	Width or Depth of Cut, In.	Group N		Group O		Group P		Group Q	
			Sfpm	Feed, Ipr	Sfpm	Feed, Ipr	Sfpm	Feed, Ipr	Sfpm	Feed, Ipr
Knurl tools in turret:										
On	.....	.....	105	0.0094	100	0.0090	95	0.0085	90	0.0089
Off	.....	.....	105	0.0188	100	0.0180	95	0.0170	90	0.0178
Knurl—cross-slide	.....	.....	105	0.0094	100	0.0090	95	0.0085	90	0.0089
Chamfer and facing	.....	.....	127	0.0036–0.0050	120	0.0034–0.0048	114	0.0035–0.0048	108	0.0031–0.0043
Reamers	Under ½ in.	.....	92	0.0044	87	0.0042	83	0.0040	78	0.0038
	½ in. or over	.....	92	0.0063	87	0.0060	83	0.0057	78	0.0054
Cutoff		Width								
	.....	0.062	105	0.0012	100	0.0012	95	0.0011	90	0.0011
	.....	0.125	112	0.0015	105	0.0015	100	0.0014	95	0.0013
	.....	0.187	115	0.0015	108	0.0015	103	0.0014	98	0.0013
	.....	0.250	121	0.0019	114	0.0018	108	0.0017	103	0.0016

* Annealed.

TABLE 19. SPEEDS AND FEEDS FOR STEELS (*Continued*)

Groups:

R. AISI C1055,* C1060;* AISI 2335, 2345,* 2515,* 4063,* 4812,* 4815,* 9260,* 9840; AISI E2512,* E2517,* E3310,* E4640,* E9310*

S. AISI C1065,* C1070;* AISI 4068,* 4817,* 4820,* 9261,* 9262,* 9845;* AISI E3316, E9315,* E9317*

T. AISI C1080,* C1085,* C1095;* AISI 3140; AISI E50100

U. AISI 3141; AISI E51100, E52100*

* Annealed.

Tool Name	Size of Hole, In.	Width or Depth of Cut, In.	Group R		Group S		Group T		Group U	
			Sfpm	Feed, Ipr	Sfpm	Feed, Ipr	Sfpm	Feed, Ipr	Sfpm	Feed, Ipr
Form: circular or dovetail		Width								
	....	0.500	85	0.0013	80	0.0012	70	0.0011	65	0.0010
	....	1.000	83	0.0010	78	0.0009	68	0.0008	63	0.0008
	....	1.500	83	0.0009	78	0.0008	68	0.0007	63	0.0007
	....	2.000	80	0.0008	75	0.0007	66	0.0006	61	0.0006
	....	2.500	77	0.0006	73	0.0006	64	0.0005	59	0.0005
Twist drills	0.250	.....	54	0.0027	51	0.0025	45	0.0023	41	0.0022
	0.500	.....	54	0.0030	51	0.0028	45	0.0025	41	0.0024
	0.750	.....	59	0.0035	56	0.0033	49	0.0030	45	0.0029
	1.000	.....	59	0.0041	56	0.0038	49	0.0035	45	0.0034
	1.250	.....	62	0.0047	58	0.0044	51	0.0040	47	0.0038
Box-tool blades		Depth								
	....	0.125	85	0.0036	80	0.0034	70	0.0030	65	0.0028
	....	0.250	83	0.0033	77	0.0031	68	0.0027	63	0.0026
	....	0.375	80	0.0028	75	0.0026	66	0.0023	61	0.0022
	....	0.500	77	0.0023	73	0.0021	64	0.0019	59	0.0018
Hollow mills	....	0.062	77	0.0051	73	0.0047	64	0.0042	59	0.0040
	....	0.125	73	0.0041	68	0.0038	60	0.0034	55	0.0032
	....	0.187	70	0.0036	65	0.0034	57	0.0030	53	0.0028
	....	0.250	67	0.0033	63	0.0031	55	0.0027	51	0.0026

TABLE 20. SPEEDS AND FEEDS FOR STAINLESS STEELS*

Groups:
A-ST. Enduro FC (free-cutting) (AISI 416); Enduro AA-FM (free-machining) (AISI 430-F)
B-ST. Enduro 18-8 FM (free-machining stainless) (AISI 303)
C-ST. Enduro AA (AISI 430); Enduro S-1 (AISI 410)
D-ST. Enduro 18-8 stainless steel (AISI 302); Enduro 18-8-S stainless steel (AISI 304)

Based on the use of high-speed-steel tools and an average 8-hr tool life. Feeds are based on the use of multiple-spindle-type machines using both rough and finish forming tools on cold-drawn material. On lighter, single-spindle machines using only one forming tool, feeds are usually much lower

Tool Name	Size of Hole, In.	Width or Depth of Cut, In.	Group A-ST		Group B-ST		Group C-ST		Group D-ST	
			Sfpm	Feed, Ipr	Sfpm	Feed, Ipr	Sfpm	Feed, Ipr	Sfpm	Feed, Ipr
Form: circular or dovetail		Width								
	....	0.500	150	0.0018	100	0.0021	90	0.0014	75	0.0016
	....	1.000	145	0.0014	96	0.0017	87	0.0011	73	0.0013
	....	1.500	145	0.0013	96	0.0015	87	0.0010	73	0.0012
	....	2.000	140	0.0011	93	0.0013	84	0.0008	70	0.0010
	....	2.500	135	0.0009	90	0.001	81	0.0007	68	0.0008
Twist drills	0.250	.....	100	0.0035	63	0.004	57	0.0028	48	0.0033
	0.500	.....	100	0.0040	63	0.0045	57	0.0031	48	0.0036
	0.750	.....	105	0.0047	69	0.0055	62	0.0037	52	0.0044
	1.000	.....	105	0.0055	69	0.0064	62	0.0044	52	0.0051
	1.250	....	110	0.0064	72	0.007	65	0.0050	54	0.0058
Box-tool blades		Depth								
	....	0.125	150	0.0050	100	0.0059	90	0.0038	75	0.0046
	....	0.250	145	0.0047	96	0.0055	87	0.0035	72	0.0043
	....	0.375	140	0.0040	93	0.0047	84	0.0030	70	0.0036
	....	0.500	135	0.0032	90	0.004	81	0.0024	68	0.0030

TABLE 20. SPEEDS AND FEEDS FOR STAINLESS STEELS* (*Continued*)

Tool Name	Size of Hole, In.	Width or Depth of Cut, In.	Group A-ST		Group B-ST		Group C-ST		Group D-ST	
			Sfpm	Feed, Ipr	Sfpm	Feed, Ipr	Sfpm	Feed, Ipr	Sfpm	Feed, Ipr
Hollow mills	....	0.062	135	0.0072	90	0.0085	81	0.0054	68	0.0060
	....	0.125	127	0.0058	84	0.0068	76	0.0043	64	0.0053
	....	0.187	123	0.0050	81	0.006	73	0.0038	61	0.0046
	....	0.250	118	0.0047	78	0.0055	70	0.0035	59	0.0043
Knurl tools in turret:										
On...............	....	.....	150	0.011	100	0.012	90	0.0089	75	0.0099
Off...............	....	.....	150	0.022	100	0.024	90	0.0178	75	0.0198
Knurl—cross-slide...	....	.....	150	0.011	100	0.012	90	0.0089	75	0.0099
Chamfer and facing	....	.....	182	0.004–0.006	120	0.005–0.0072	108	0.0031–0.0043	91	0.0038–0.0053
Reamers	Under ½ in.	.....	132	0.0050	87	0.006	78	0.0038	66	0.0046
	½ in. or over	.....	132	0.0072	87	0.0085	78	0.0054	66	0.0066
Cutoff		Width								
	....	0.062	150	0.0015	100	0.0017	90	0.0011	75	0.0013
	....	0.125	159	0.0020	105	0.0021	95	0.0013	79	0.0016
	....	0.187	163	0.0020	108	0.0021	98	0.0013	82	0.0016
	....	0.250	172	0.0024	114	0.0025	103	0.0016	86	0.0020

* Union Drawn Steel Division, Republic Steel Corp.

Part 2

METAL-FORMING METHODS

Section 14

METAL SPINNING

SECTION 14

METAL SPINNING

Spinning is one of the oldest metalworking arts. The process is applicable to small-quantity production that does not justify the cost of deep-drawing dies, to experimental work, and to manufacture of parts that cannot be formed on a draw press because of design or size.

In the spinning process, a lathe is used to rotate a circular metal blank at high speed, while it is pressed against a form block that rotates with the spindle. The blank is forced against the block by suitably formed tools, which are manipulated by either hand or power.[1] Tallow, lard oil, wax, and yellow soap are used as lubricants.

Steps Required

Three steps are involved in the spinning process: (1) breakdown, wherein approximately 95% of the deformation from the flat blank or previous form block is achieved; (2) laying the metal down, or realizing the last 5% of deformation so that the metal is placed in contact with the form block; and (3) planishing, or the removal of tool marks with a flat, highly polished, hardened tool.

[1] See Spinning Takes Know-how, *Am. Machinist*, May 1, 1950.

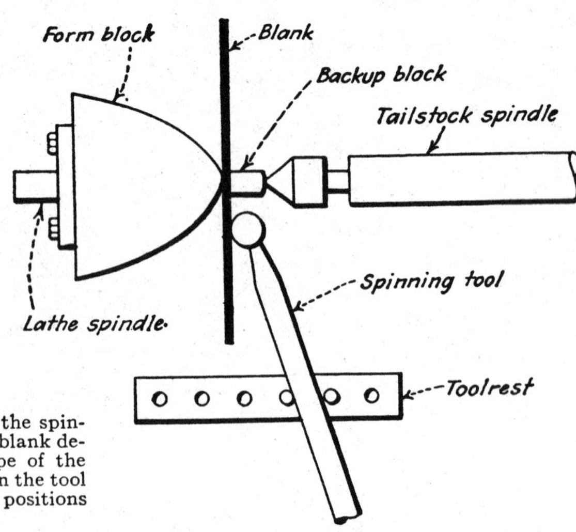

FIG. 1. Pressure of the spinning tool against the blank deforms it to the shape of the form block. Holes in the tool rest provide various positions for the fulcrum pin.

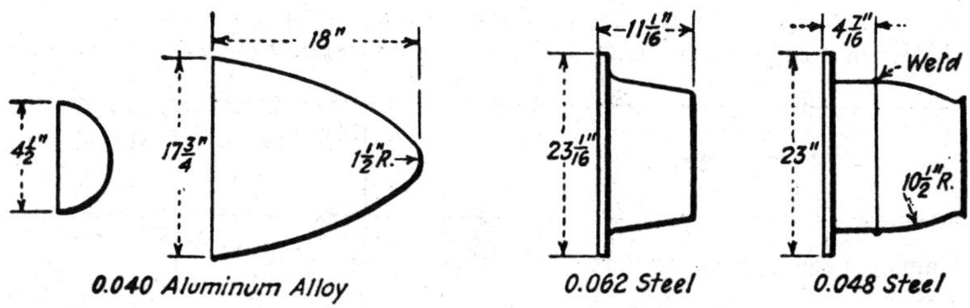

FIG. 2. Common spun parts.

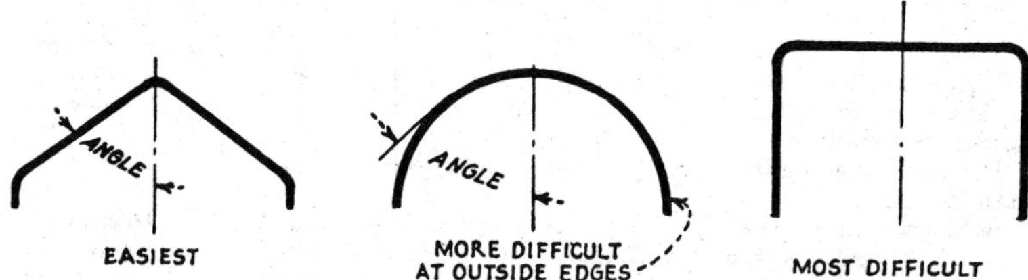

FIG. 3. Basic shapes. Spinning difficulty is increased as the angle between the metal and the axis of rotation is reduced.

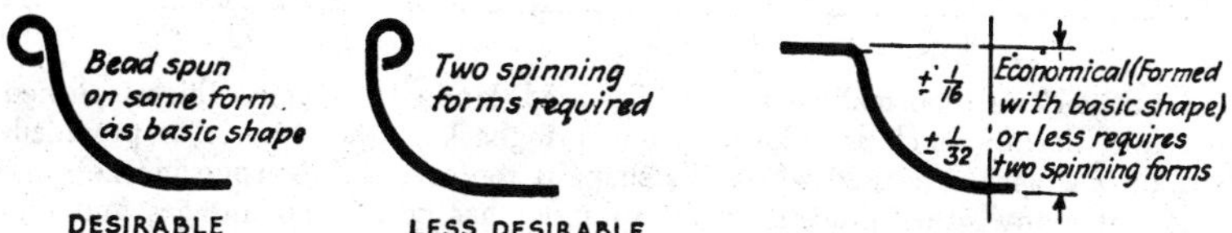

FIG. 4. Beads. External beads can be formed in one spinning; internal beads require two forms.

FIG. 5. Flanges. Try to specify a $\frac{1}{16}$-in. tolerance on flange location.

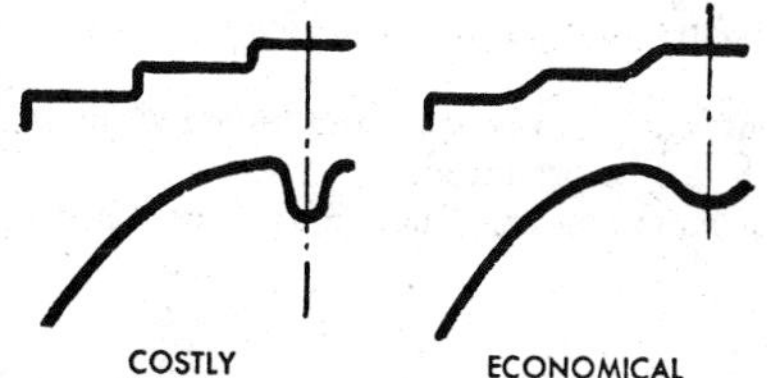

FIG. 6. Design economy. Sharp corners often require extra annealing operations. Avoid reentrant shapes, which require costly collapsible split forms.

SPINNING RATIOS

Depth—Dia

Shallow—less, than 1:4
Moderate—1:4 to 3:4
Deep—3:4 to 5:4

MATERIALS THAT MAY BE SPUN

Listings in each group are made in the order of formability for shallow, moderate, and deep draws

Material	Max Thickness, In.	Max Depth of Spinning
Aluminum and magnesium alloys:		
2SO	1/4	Deep
3SO	1/4	Deep
52SO	1/4	Deep
61SO	1/4	Deep
2S½H	3/16	Deep
52S¼H	3/16	Moderate
24SO	1/4	Moderate
75SO	1/4	Moderate
MAG AM3SO	1/4	Moderate
52S½H	1/8	Shallow
Low-carbon steels:		
Deep-drawing quality 40,000 psi tensile	3/16	Deep
Stainless steel:		
AISI 3002, deep-drawing quality	0.102	Deep
AMS-5515 AISI 347, annealed	0.102	Deep
Corrosion-resistant steel	0.102	Moderate
Nickel alloy, inconel (annealed)	0.102	Moderate
Copper and brass:		
Copper (soft temper)	1/4	Deep
Commercial brass (soft)	1/4	Deep

Form blocks are normally made of kiln-dried, laminated maple, which is turned to shape, during which an allowance for springback of the material is provided. Breakdown blocks are used where the shape is too severe to be spun in one pass.

As with any other process, metal spinning has certain advantages and disadvantages:

Advantages

1. Tool cost is low, because form blocks are not costly.
2. Spinning can be applied to a wide variety of forming jobs involving elements of rotation about a common axis.
3. Tolerances can be held closer on job lots than is economically feasible for other methods of manufacture.
4. Where good finish is a basic requirement, parts can be spun and buffed without scratches or draw marks, and a mirror finish can be produced.
5. Larger shapes than practical with other manufacturing methods can be shaped on spinning lathes of suitable swing.

Disadvantages

1. Spinning is limited to symmetrical articles that are circular in cross-section normal to the axis of rotation.
2. The process is normally considered limited to low-quantity production.

Design Information

1. Dimension parts to a surface adjacent to the form block, so that these dimensions can apply directly to the form block.

2. Minimum permissible thickness of the part should be specified, because some reduction of material occurs during spinning. Material about 30% thicker than the minimum permissible part thickness should be specified.

3. Commercial tolerances are: large work, $\pm\frac{1}{16}$ in.; average work, $\pm\frac{1}{32}$ in.; and small work, $\pm\frac{1}{64}$ in. Tolerances of $\pm$0.005 in. can be obtained but are expensive. Generally speaking, tolerances closer than $\pm$0.020 in. for aluminum alloy over 0.064 in. thick or $\pm$0.015 in. for other materials or thicknesses should not be specified.

Aluminum is spun at speeds up to 4000 sfpm. For work from 5 to 8 in. dia a work speed of 2800 to 2600 rpm will suffice, while for work down to 4 in. dia the speed will rise to 3200 rpm.

Monel is harder to spin than copper, brass, steel, or silver and requires more frequent annealing. Spinning tools should be brass, bronze, wood, or tool steel, but not soft steel. Lubricate with tallow.

Breakdown Procedure

To maintain control of metal flow in spinning, it is often necessary to break down the job into several operations. Factors that determine need for extra operations are: height to diameter ratio, h/d; the spinnability and thickness of the metal; contour of shell; and operator experience.

Parallel-walled shells are more difficult to spin than those with tapered walls and domed shapes, because better control of metal flow is possible in the latter. The number of operations cited in the adjacent table are average for hand-spun shells of 2SO, 3SO, and 57SO aluminum, but must be applied cautiously because of the factors cited.

Blank-size Calculations

Determining the exact size of a blank required for a spun shell is largely a matter of cut and try, mainly because an increase in area occurs when thickness is reduced during spinning, and this increase varies with the spinner and his technique.

When calculating the blank size for a drawn shell it can be assumed that no appreciable change occurs in the thickness during drawing, and that the area of the blank is approximately equal to the area of the shell.

For a spun shell, neither of these assumptions can be made, except in the case of the simplest spinnings. A wall thickness reduction does occur in spinning and, to produce a shell of specified thickness, a thicker blank must be used to start with. Because of this thickness reduction, an increase in area occurs. Hence, the blank for a spinning would be greater in thickness but smaller in diameter than the blank for a drawn shell of the same size.

NUMBER OF OPERATIONS FROM BLANK TO FINISHED SHELL

$\frac{h}{d}$	Number of Operations		
	Parallel Shells	Tapered Shells	Domed Shells
Up to 1.0	1	1	1
1 to 1.5	1 or 2	1	1
1.5 to 2.5	2 or 3	1 or 2	1 or 2
2.5 to 3.5	3 or 4	2 or 3	2 or 3
3.5 to 4.5	4 or 5	3 or 4	3
5.0 to 6.0	5 or 6	4	4

h = height of shell.
d = dia of parallel or domed shells or bottom dia of tapered shells.

No. of Operations

From table, $\frac{h}{d} = 3$. *Therefore 3 operations would be required. (2 breakdown, 1 final)*

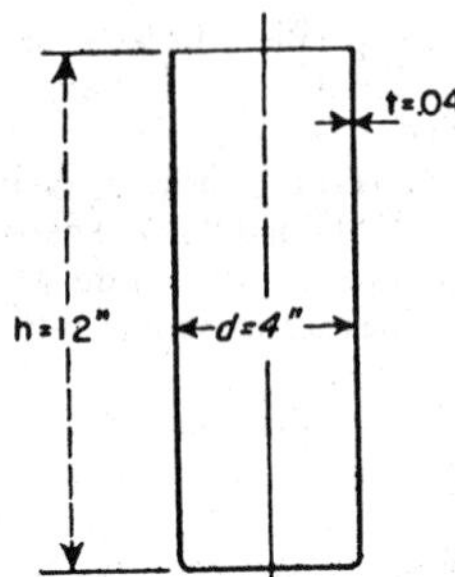

Gage Thickness for Blank

Gage required in shell, no 18 B&S. (0.040 in.) = t
(Add 1 gage for 2 breakdowns and 1 gage for final operation) Use No. 16 B&S ga. (0.064 in.) = t_1

Volume of Shell Metal

$$\text{Sides} = \pi dht = 3.14 \times 4 \times 12 \times 0.040 = 6.03 \text{ cu. in.}$$

$$\text{Bottom} = \frac{\pi d^2}{4} t_1 = \frac{3.14 \times 16 \times 0.064}{4} = 0.80 \text{ cu. in.}$$

$$\text{(total)} \quad 6.83 \text{ cu. in.}$$

Blank Area

$$\text{Volume} = 6.83, \text{Thickness} = .064 \therefore \text{Area} \frac{6.83}{.064} = 106.7 \text{ sq. in}$$

Blank Diameter

$$\text{Since } A = \frac{\pi}{4} D^2 \therefore D = \sqrt{\frac{A \times 4}{\pi}} = \sqrt{\frac{106.7 \times 4}{3.14}} = 11.63 \text{ dia.} \times 0.064$$

FIG. 7. Method required to calculate blank diameter required to produce a spun shell. Theoretical size would be 11⅝ by 0.064 in. To allow for variables a 12-in. blank would be used for trial. Blank for drawing would be 14⅜ by 0.040 in.

The amount of work on the metal necessary to spin the blank to shape may be used as a general guide to the allowance for thickness reduction. Simple spinnings, requiring one operation, will not thin appreciably and need no allowance. For deeper spinnings produced in one or two operations, allow one gage. For those requiring more than two operations, allow one gage for each two breakdown operations, and one gage for the final operation. The general procedure for blank calculations for spinnings is thus based on volume rather than on area, keeping in mind the fact that the bottom of the shell will not change in thickness, and that the sides of the shell will be reduced somewhat.

Delivery and Cost

The critical factor with many jobs is quick delivery. Spinning tools can be built much quicker than press tools for drawing. For example, spinning tools for a 6- by 6-in. aluminum shell might be made in 2½ weeks, whereas drawing dies might take 7 weeks. If the spinning work involves a first spin and a second spin and bead, output might approximate 100 shells per day. Drawing, on the other hand, could be handled at the rate of 1000 pieces per day. Unit costs in this particular case would become equal at 2300 pieces. And at 5000 pieces, the drawn shells would cost about one-half as much.

If future orders of large quantities of shells are a reasonable possibility, the economics are changed somewhat. Tooling costs are usually absorbed on the first order and are not a factor on repeat orders. This circumstance would give drawing a decided cost advantage. If future possibilities are considered, the break-even point between spinning and drawing costwise is in effect lowered.

Section 15

PRESSWORKING AND COLD-ROLL FORMING

Consultant: J. R. PAQUIN, Tool Engineering Consultant

SECTION 15

PRESSWORKING AND COLD-ROLL FORMING

PRESSWORKING OPERATIONS

Pressworking is the art of producing stampings by cutting and shaping sheet and strip between the members of a die.

Stampings may be divided into two classifications:

1. Those that are made by cutting or shearing the metal, either to change the outlines of the edges or to cut holes in the interior of the piece. The operations used are:

BLANKING. The process of cutting a flat piece to the outline and size necessary to produce the finished part or to serve as the workpiece upon which secondary operations are performed.

PUNCHING OR PIERCING. The cutting of openings, holes, or slots in the part. The slug produced by piercing or punching is scrap.

TRIMMING. A secondary cutting or shearing operation to cut off superfluous metal, as around the edges of drawn pieces or formed or forged parts to achieve the desired dimensions.

SHAVING. A trimming operation where a thin shaving of metal is removed to square up blanked edges or to achieve a smooth hole of uniform diameter.

2. The second classification of stampings includes those made by forming the material to the desired shape without cutting. Appropriate operations are:

BENDING AND FORMING. Shaping a portion of or the entire blank by bending, flanging, folding, twisting, or offsetting, and usually without materially changing the thickness of the material.

DRAWING. The metal in a flat blank or sheet is stretched or drawn over a form to produce the desired shape. Usually a slight change, at least, takes place in thickness of the metal at the drawn portion. A more applicable definition of drawing, for most work, would be "producing a cuplike form from a sheet-metal disk by holding it firmly between blank-holding surfaces while punch travel gives the required shape."

COINING. By application of extreme pressure to metal placed between relatively flat dies, the metal is forced to flow to fill the die impressions. Sometimes, the dies are truly flat and the squeezing action thins the metal between the dies to the required thickness tolerance.

DESIGN OF STAMPINGS[1]

In any small mechanism, the application of metal stampings is desirable from the standpoints of low cost, light weight, high strength, and volume production of identical parts. What follows applies particularly to the field of mechanisms but also warrants consideration in the design of many stampings.

[1] A. A. Laurich, *Product Eng.*, January, 1951, p. 120.

Design of Blank Edges and Notches

Edges

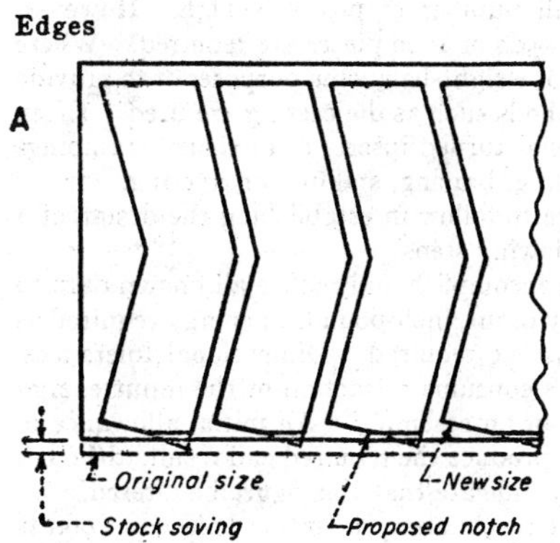

A. Parts should be studied for projecting corners or flanges that can be cut off without sacrificing strength or welding surface.

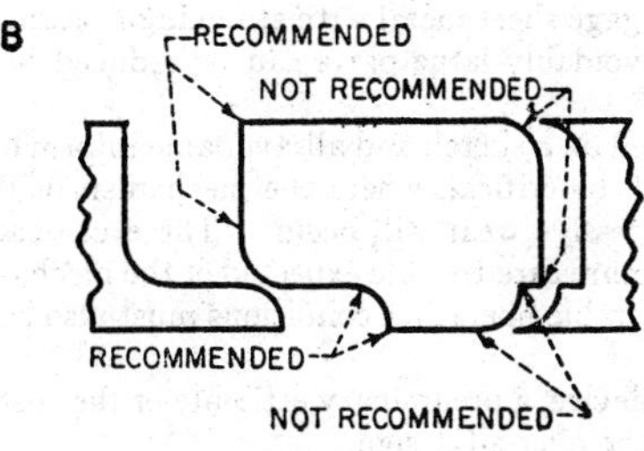

B. Parts cut from strip stock should have corners which are partly formed by the edge of the stock, sharp and approximately 90°; corners not partly formed by the edge should be round, wherever possible.

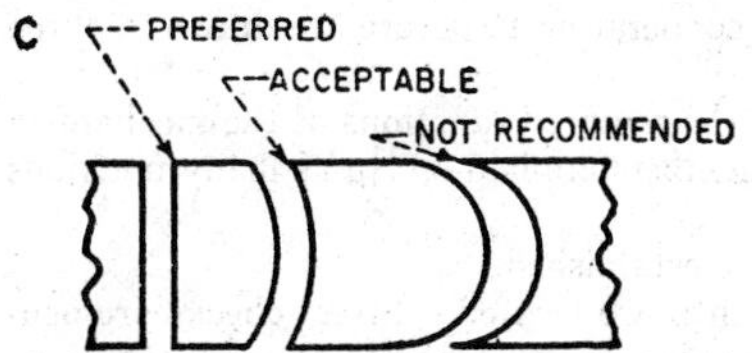

C. Feather edges should be avoided to minimize die maintenance. These edges are difficult to maintain as they tend to break off in the die and thus increase cost and maintenance.

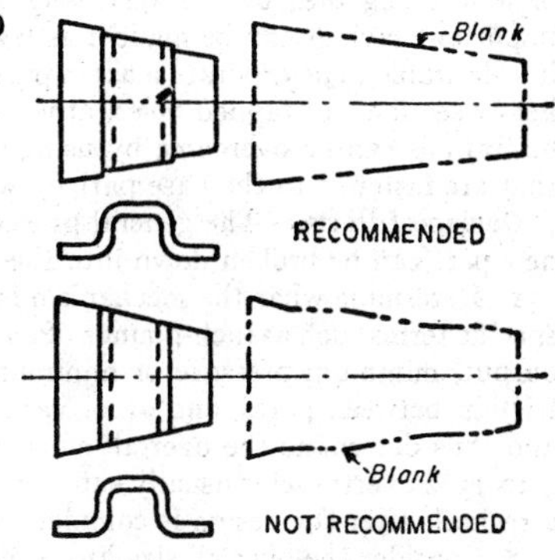

D. To reduce cost of blanking dies, parts should be designed so that straight edges can be maintained on the flat blanks.

Notches

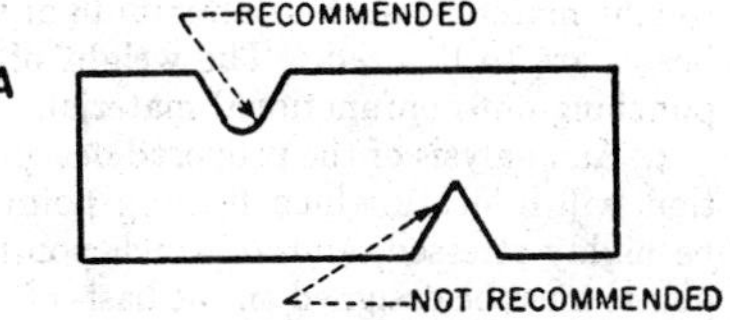

A. The radius of the vertex of a notch should be as large as possible to prevent tearing; a minimum of twice metal thickness and a 60° angle notch is recommended. Notches with a sharp V at the vertex are not recommended as this is where tearing usually starts.

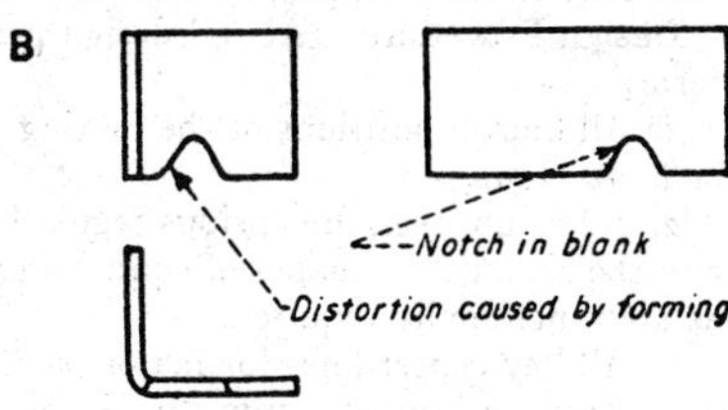

B. If a notch is to be put in a blank requiring subsequent forming in the vicinity of the notch, allowance should be made in the design for distortion that might be caused by the forming operation.

Number of Pieces. Stamping applications are sometimes limited because the cost of amortizing dies over a relatively small number of pieces is high. However, important savings can be made if as few as 500 or 1000 pieces are required. Where it is desirable to incorporate mass in parts for weight-balancing purposes or to provide heavy sections for tapped holes, other methods such as die casting are used. Often limitations can be overcome by using special turned inserts or auxiliary stampings that are fastened to the base part by welding, brazing, staking, or riveting.

Design of Parts. The general procedure to follow in establishing the design of a new part can be broken down into the following steps:

1. Determine what the mechanism is to accomplish and reduce all known data to specific terms such as inch-pounds of energy input, inch-pounds of energy required as output, minimum pressure or minimum stroke required. Dimensional tolerances, friction between parts, and other variables consume a fraction of the input energy and thus determine the over-all efficiency of an assembly. An initial allowance of 5 to 15% overtravel is usually sufficient to produce the required end result, although a recheck after the design is complete may indicate that this figure be altered.

2. Consider the over-all size limits of the mechanism, mounting holes, and points of attachment to controls in the light of probable dimensional variations of the mechanism itself and cooperating parts. This survey may indicate that one or more holes should be slotted, or that adjustable connections such as turnbuckles, serrated devices, spacers, or similar devices are needed.

3. If a weight limit is imposed, a preliminary estimate will indicate whether light-weight materials such as aluminum or thin-gage sheet metal with stiffening ribs and bosses are to be used. The weight of unavoidably large parts can be reduced by punching out nonfunctional material.

4. An analysis of the proposed design based on a sketch and all available information will indicate which bearing points will be critical, where the mechanism will be highly stressed, and at which points excessive wear will occur. These critical points must be designed on the basis of minimum care-free life expected of the mechanism. Corrosion, erosion, and other unfavorable operating conditions must also be considered.

5. If a definite cost limit is placed on the device, a preliminary estimate of the cost is made and used as a guide in developing the over-all design.

6. Any general limitation such as size and type of available equipment, size and type of material available, ease of fabrication and assembly, tool and die costs must be established and used as guides during development of a design.

7. Any finish or appearance conditions must also be borne in mind.

Design Procedure. Actual design procedure from this point takes the following form:

1. All known positions of the mating or cooperating structure are drawn in three or more views.

2. All established dimensions regulating the required functions of the mechanism and the established shapes of connecting members are laid out in as many positions as are functionally required.

3. All key center lines for major parts are established.

4. Any bearings or points considered critical are located. Stress checks are usually limited to simple shear tests of rivets used as pivot points or as connections between parts. Factors such as 10 to 15% variation in the thickness of steel, differences in hardness or temper of this steel, and tendency of the metal to check or crack at the edges during forming make the testing of handmade samples a more reliable criterion. Parts must also be designed to resist deformation during tumbling and plating operations, handling in tote boxes, and ejection from the dies.

Design Limitations for Holes

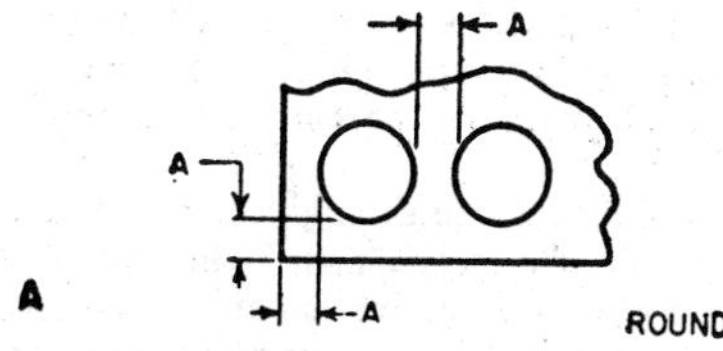

Metal Thickness, Inches	Distance "A" Min, Inches
Up to 0.062	0.12
Over 0.062	Twice Metal thickness

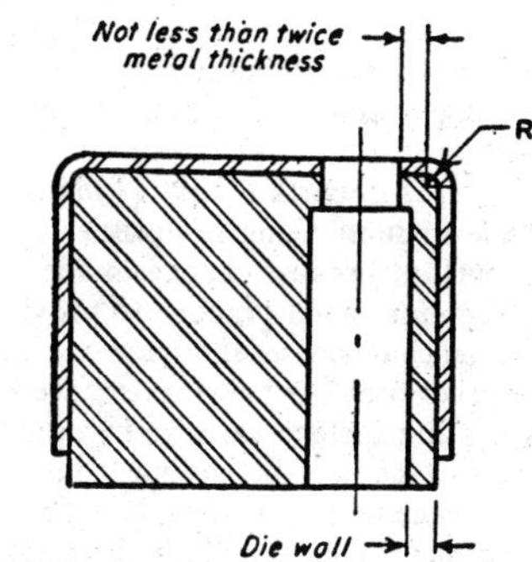

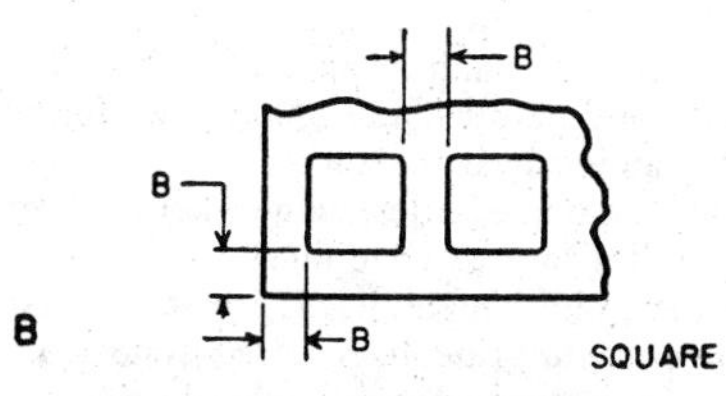

Metal Thickness, Inches	Distance "B" Min, Inches
Up to 0.090	0.18
Over 0.090	Twice Metal thickness

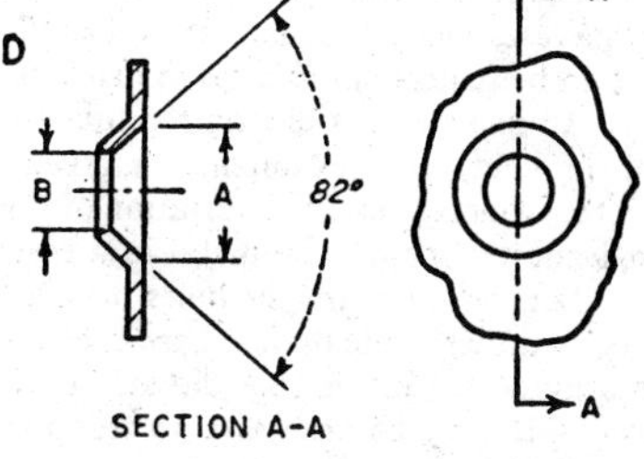

Screw Size	Diameter Counterpunch, In.	Diameter Hole, In.
No.4	0.24	0.14
No.5	0.26	0.15
No.6	0.29	0.17
No.7	0.32	0.18
No.8	0.34	0.19
No.9	0.37	0.21
No.10	0.40	0.22
No.12	0.45	0.25
No.14	0.50	0.27
1/4"	0.52	0.28
5/16"	0.65	0.34
3/8"	0.77	0.40
7/16"	0.90	0.46
1/2"	1.03	0.53

A and *B*. The distance between holes or between a hole and a trimmed edge is governed by stock thickness, hole size, punch shank size, type of hole, and die-steel strength. *C*. Hole edges should be separated from a line normal to the metal face and through the center of the radius at a bend by an amount equal to at least twice metal thickness. *D*. Standard sizes for counterpunch and hole diameters. The charted dimensions are for screw heads only; when a lock washer is used, allowance must be made for the washer.

5. The following special considerations apply to stamping designs:

a. The principal dimensional differences in stamping occur as a result of metal-thickness variation and the manner in which a die is set in the press. For a given batch of parts produced from stock that runs close to the extreme high or extreme low thickness limit and a single setting of the die, all parts will be uniformly near the high or low limit, as the case may be.

b. It is important that as many parts as possible operate through their complete cycle without being too precisely restricted within the mechanism. This factor is important because changes made in dies after a part is in production are costly. It is therefore good practice to provide enough "free travel" at one or more points in the mechanism to allow for accumulated tolerances if the mechanism is complex. Another way to provide for accumulated tolerances is to make a given hole or operating surface variable by adding an extra operation to pierce the hole or shave the surface in a die having adjustable gaging.

c. Whenever practical, it is desirable to keep the parts of a mechanism as flat as possible. This simplifies the dies, avoids variables caused by forming, and eases inspection.

d. If a design calls for several irregularly shaped parts that have large unit areas, contours should be modified so that one or more of these parts will "nest" well, thus reducing scrap.

e. From the standpoint of die construction and die maintenance, it is desirable to make as many "forms" or "offsets" straight and parallel, and to have two forms or two bent lugs opposite each other so that the die can be balanced.

f. An analysis of the required tolerances must be made to determine whether they can be held. For example, if a given part with a formed edge must be held to a tolerance closer than the variation of the thickness of the metal itself, provision must be made to "coin" the metal in a restrike operation, grind to required dimension, and "shave" or ream the holes in relation to the formed edge as required.

6. Samples made by hand are used for installation checks, destruction testing, and as a guide for estimating die and unit costs. If the results indicate questionable results, it may be necessary to test special samples having high- and low-tolerance gage thicknesses and to limit dimensions at critical points.

7. Any revisions are made in layout and detail drawings.

8. All drawings are given a final check, and in some cases, a complete new checking layout is made to include high- and low-tolerance dimensions at critical points. It is usually helpful to make a ten-times-size layout of the whole mechanism or of critical points to check geometrical complications that arise under certain tolerance conditions.

Design Tips. Five matters relating to die design are:

1. Dimensional tolerance factors. The basic factor in stamping design is that, whether a part is made in a series of single-operation dies or in elaborate automatic progressive dies, all dies are built to accommodate steel of the greatest thickness allowed by tolerance. When steel of the least thickness allowable is used, bends will not be square. If steel beyond the high limit of thickness is used the metal will squeeze or "iron" out, causing the part to be dimensionally distorted and possibly "galled" or cracked from the force of the die.

When metal is formed at right angles, a certain amount of stock is used as a bend allowance, depending on the design of the dies and the material used.

Considering these conditions, it is well to avoid the use of several right-angle bends parallel to one another because of the build-up of tolerance variations.

If a forming operation is to be gaged from a locating pilot through a working hole,

Design of Flanges

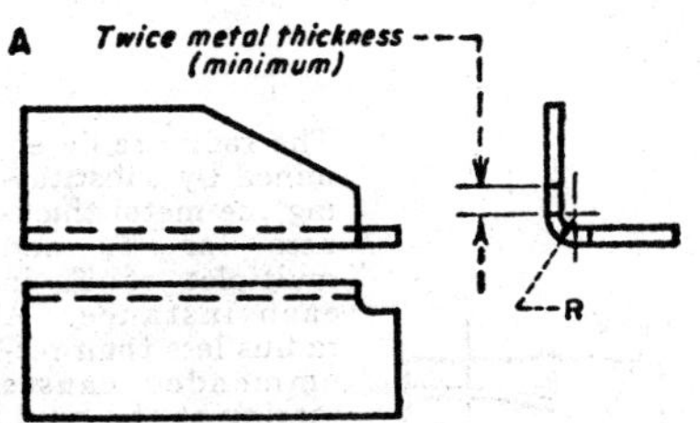

A. Tapered flanges should not taper to the end of the blank but should be cut off so that the narrowest section will be at least twice metal thickness measured from the radius center.

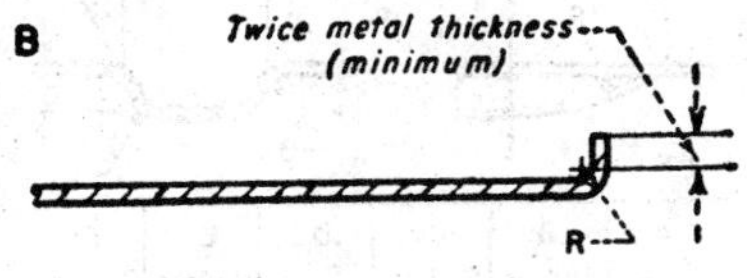

B. Outside flanges and flanges around openings should have a height of at least twice the metal thickness measured from the radius center. This serves two purposes: (1) eases the forming of the lip, and (2) provides a surface for attaching covers and other members.

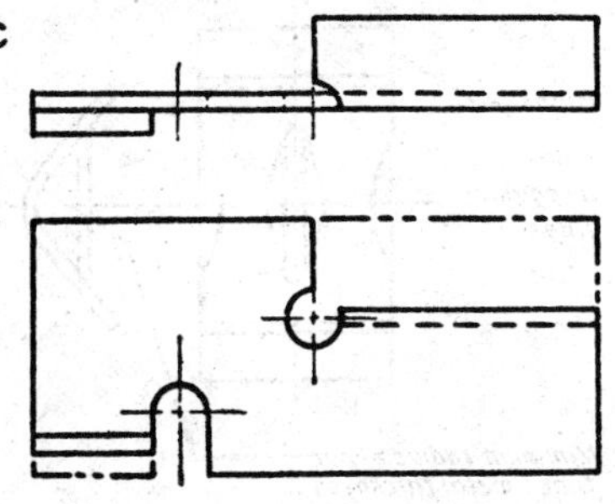

C. Where flanges extend over only a portion of a part, it is desirable to punch a notch or hole at the end of the flange to eliminate tearing of the metal. This is also a method of reducing stress-concentration factors at the intersection of the two surfaces.

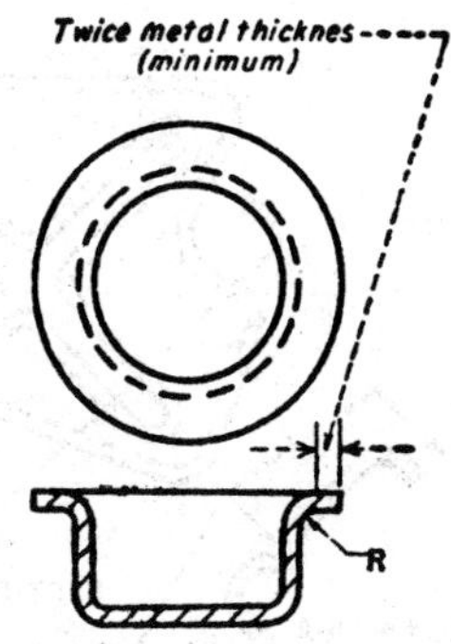

D. Flanges left by trimming should have a width of at least twice the metal thickness measured from the radius center.

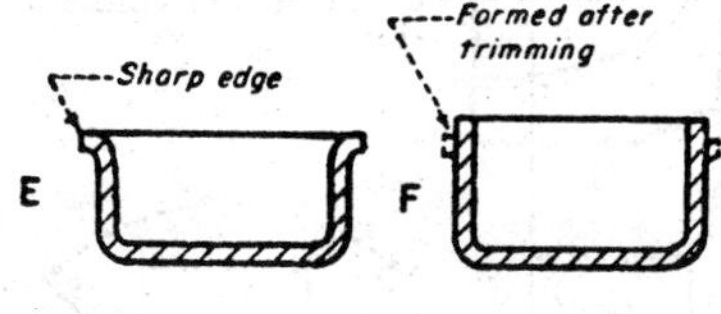

E. Where a sharp edge is not objectionable on a drawn part, the flange may be trimmed close. *F*. Flanges can be eliminated on some drawn parts by adding an operation to form them up after trimming.

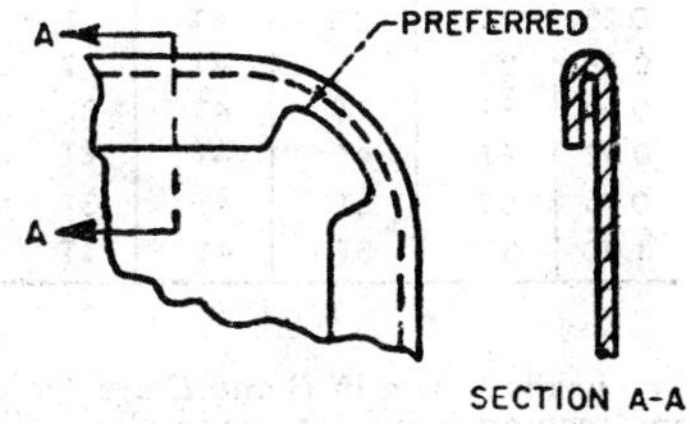

G

G. Notch the corner of a hemmed edge to eliminate gathering of the metal in the flanging operation.

Design of Strengthening Beads and Ribs

Beads

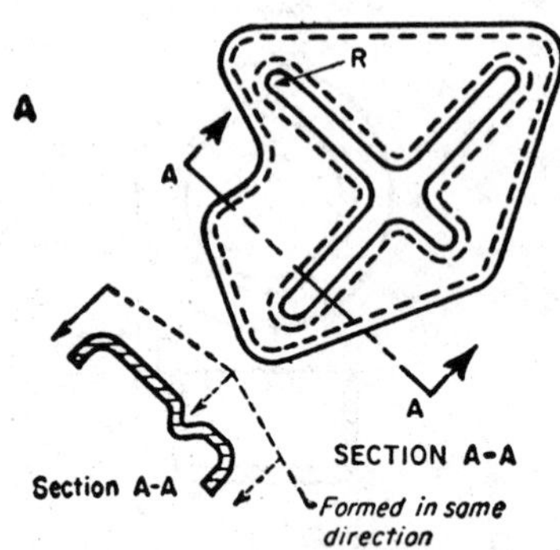

Beads of the type shown in *A* should be formed in the same direction as the outer flanges to facilitate die construction.

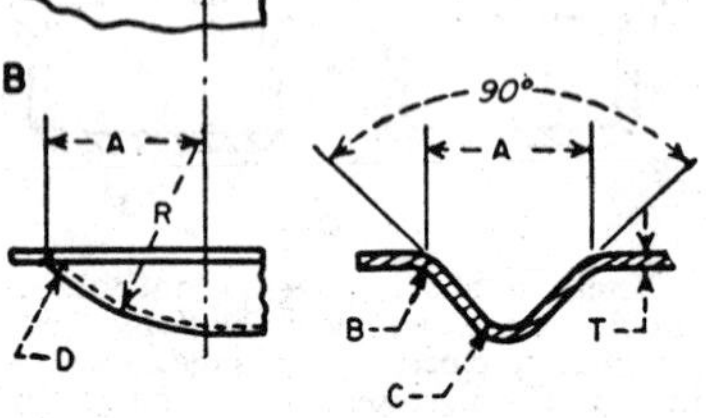

A	B Radius	C Radius	D Radius	E Radius
0.25	2T	2T	4T	T
0.38	2T	2T	4T	T
0.50	2T	2T	4T	2T
0.62	4T	4T	4T	2T
0.75	5T	5T	4T	3T
1.00	5T	5T	4T	3T

The beads shown in *B* and *C* are typical and recommended for use in all possible applications. These have been designed so that the angle of the sides will normally allow their use on flat, angular, and curved surfaces without producing backdraft. Chart information is given for depressed or internal beads.

The radii are determined by substituting the metal thickness for *T*, and multiples of *T*, in each instance. A radius less than recommended causes tearing at the bead.

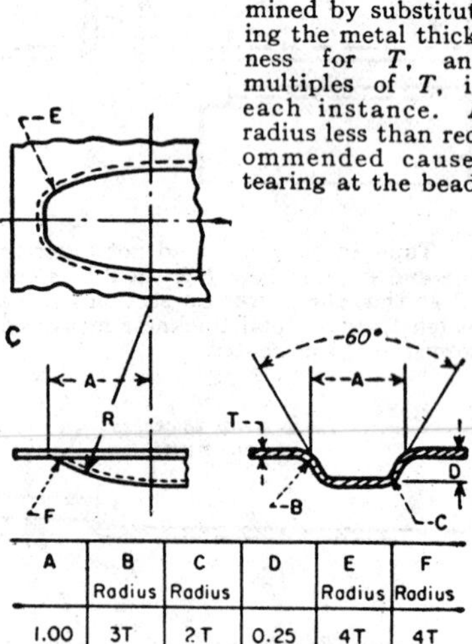

A	B Radius	C Radius	D	E Radius	F Radius
1.00	3T	2T	0.25	4T	4T
1.50	3T	2T	0.31	4T	4T

Ribs

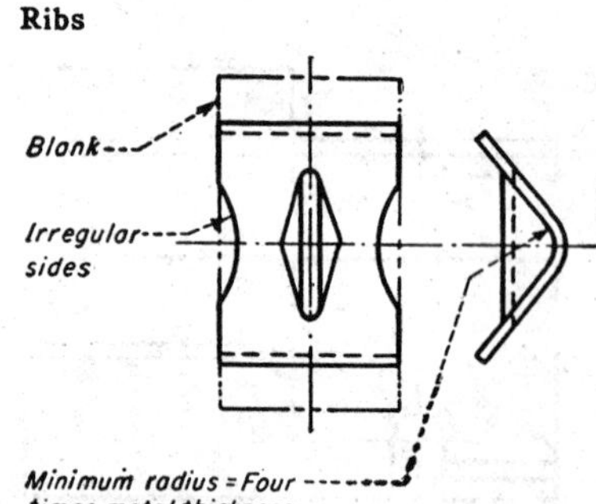

Properly designed ribs add rigidity to angles and flanges. Irregular sides allow for rib when part is made from strip stock. A radius of less than four times metal thickness will usually tear and cause part failure.

Design of Lanced Lugs

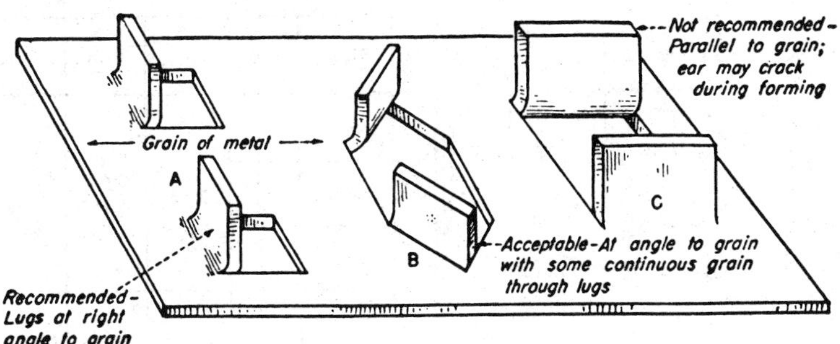

When lugs are used, they should be formed at right angles to the grain of the metal as in *A*. If this is not practical, then *B* is permissible. *C* is not recommended and may necessitate the use of premium-grade steel.

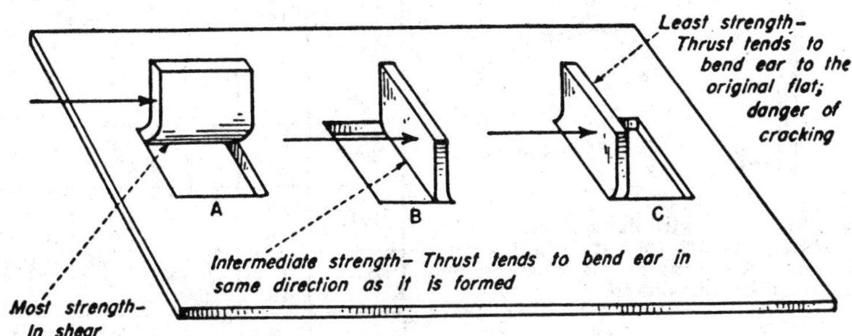

Direction of load on lug is a factor in positioning the lug. Loading as in *A* is recommended because the strength in shear exceeds that in bending. If *B* and *C* are only possibilities, ribs or gussets can be used as reinforcements.

Bearing Area for Shafts

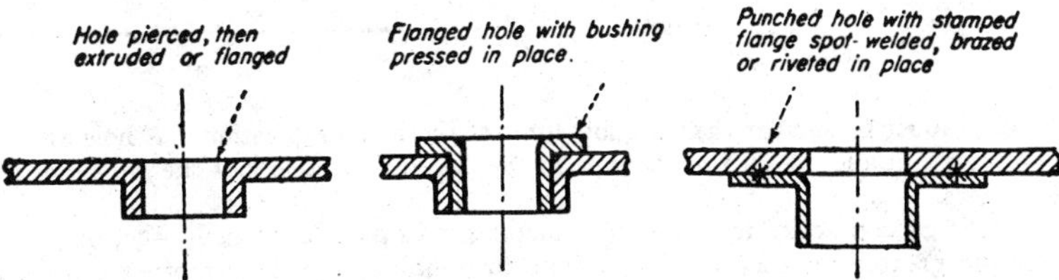

Increased bearing area for supporting shafts can be provided in various ways. The three methods shown can be used singly or in combination.

Clinch Nuts—Depression and Hole Dimensions

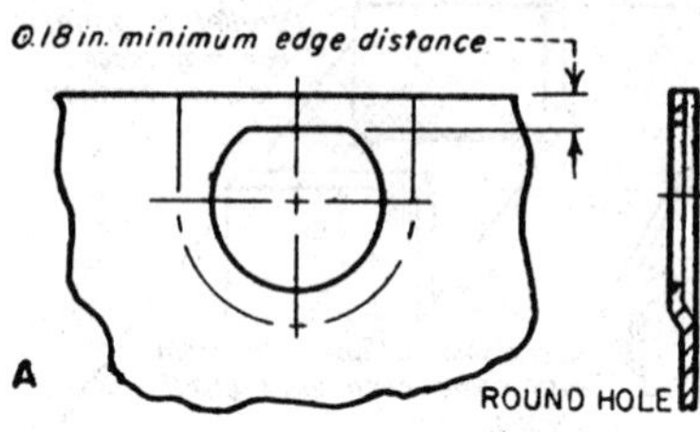

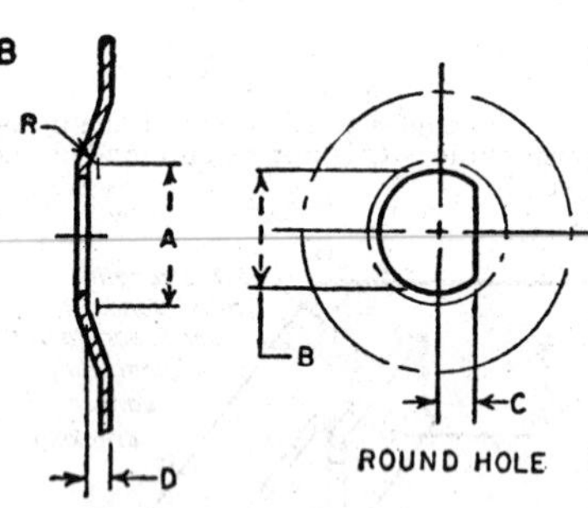

A. Clinch nuts with flat surfaces should be assembled with the flat surface facing the edge of the metal. *B*. Dimensions for round depressions for clinch nuts.

Size	Th'd	A, in.	B, in.	C, in.
No. 10	24 32	0.57	0.450 0.453	0.156 0.159
No. 12	24 28			
1/4"	20 28			
5/16"	18 24	0.76	0.635 0.645	0.255 0.260
3/8"	16 24			
7/16"	14 20	0.88	0.760 0.770	0.282 0.288
7/16"	14 20	0.95	0.822 0.832	0.350 0.357
1/2"	13 20			
1/2"	13 20	1.00	0.885 0.895	0.382 0.390
9/16"	12 18	1.12	1.016 1.026	0.445 0.453
5/8"	11 18			
9/16"	12 18	1.06	0.950 0.960	0.413 0.422
3/4"	10 16	1.32	1.203 1.213	0.450 0.550

D = {0.06" Minimum; 0.09" Preferred}

R = Metal Thickness

the pilot must be smaller than the low limit of the hole in question. A hole at the high limit will allow a shift of the part in the die equal to one-half the difference in diameters.

2. It is good practice to make all formed sections parallel to each other or at an angle no greater than 45°. This is done to minimize the danger of ears or lugs fracturing because of being formed parallel to the grain of the metal. If it is imperative that forms be made at right angles to one another, parts can be blanked diagonally in the strip to avoid forming parallel to the grain. If neither of these meas-

Square Clinch Nuts—Hole Dimensions

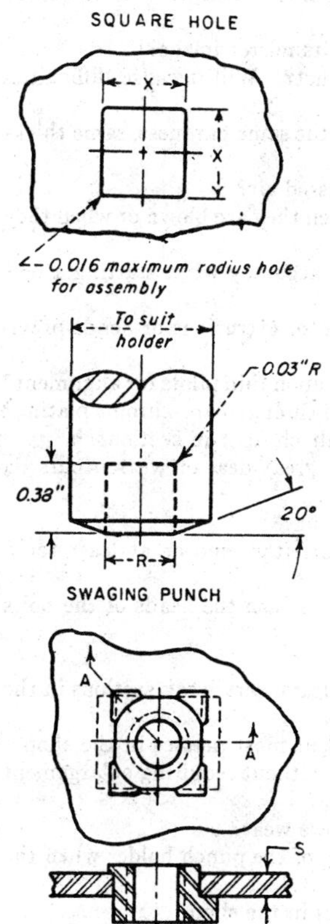

Nut Size	Threads per Inch	S Maximum Metal Thickness, Inches	X Hole Size, Inches	R Swaging Punch, Inches
No.6 No.8 No.10 No.12 1/4"	32 32 32 24 28	0.037	0.378 0.375	0.392 0.388
No.12 1/4"	24 28	0.046	0.378 0.375	0.392 0.388
No.12 1/4" 1/4"	24 20 28	0.078	0.378 0.375	0.392 0.388
1/4"	28	0.094	0.378 0.375	0.392 0.388
5/16" 5/16"	18 24	0.062	0.440 0.437	0.445 0.451
5/16" 5/16"	18 24	0.078	0.503 0.500	0.518 0.514
5/16" 5/16" 3/8" 3/8"	18 24 16 24	0.078	0.565 0.562	0.586 0.582

Dimensions for square holes for clinch nuts. The swaging punch sizes are provided for reference purposes only, but such information may be required to determine tool clearances.

ures can be taken, it may be necessary to purchase premium-grade steel and/or anneal the blanks before forming.

3. The number of formed or raised sections in any one part should be kept low. This results in economy during die construction and simplified inspection procedures. If each stamped part in an assembly is designed with this factor in mind, the completed mechanism will be compact and dimensional variations caused by forming operations will be less likely to occur.

4. All dimensioning should be from base lines established with regard to the functional requirements of the parts, thus avoiding accumulative tolerances between working points. Such dimensioning is also useful to the die designer and inspector.

5. If possible, the die side (rounded edges) of stampings should work together so that any burrs present will not cause sticking or rough action in the mechanism.

CHECK LIST FOR DIE DESIGNERS

In die design, it is easy to overlook some essential point. This list was developed as a final check list to avoid any oversights.

1. Will the die produce the piece or will it make its mirror image?[1]
2. Where is the sharp edge on the finished product? Will it cause difficulties there as small burrs develop?
3. Have the bends actually been tried on stock of the same hardness, same thickness, in the same relation to the grain?
4. Can the finished piece be blown off by compressed air?
5. Will the finished pieces clear the pillar posts when they are blown or when they drop to the rear?
6. Is there a pilot in the second stage for greater accuracy when starting a new strip?
7. Do succeeding stages allow adequate clearance for extrusions or bends previously formed?
8. Is the stock stop positive or does the die depend upon thin pilots for alignment?
9. Which sections would be considerably improved through hard-chrome plating?
10. Are there any large die parts that can be built cheaper in sections?[2]
11. Are those sections easily replaceable where a great deal of wear occurs on vertical walls used for forming?
12. Are the knockout holes in the center of the shank?[2]
13. Is the die shoe large enough for a new stage at either end or at least for a "hitch feed?[2,3]
14. Are the punches so designed that they will not smash the heads of the bolts holding the stripper plate?[2]
15. Are the stripper-plate spacers high enough?
16. Are any punches so constructed that they will leave very weak sections in the stripper plate?
17. Is the "shut height" of the die so that it will fit most presses in the shop?[4]
18. Will most presses in the shop take the new die without requiring enlargement of the hole in the bolster plate?
19. Will it be difficult to grind the die when it shows wear?
20. Will the top of the pillar posts be below the top of the punch holder when the die is fully closed?
21. Will the knockout arrangement fit most presses in the shop?
22. Do all screw or dowel holes clear the shank of the punch holder?
23. Is there room for springs long enough to withstand high-speed work?[5]
24. Are all pilot holes drilled through?
25. Can all dowel holes be drilled through or is there at least room for a smaller through hole as a knockout hole?
26. Will those scrap pieces which are not punched through the die drop to the side or rear in sufficient time?
27. Are there adequate provisions for separation of scrap from product?[6]

[1] A vellum strip can be run through the die design and the operations performed on it with a scissor. This is the best check possible.

[2] A full-sized drawing of the die on transparent paper in colored pencil and one drawing of the punches in another color will show this clearly.

[3] Space for an extra stage will help in the event of small product changes.

[4] It is often a good idea to use a screw and lock nut to prevent the die setters from closing the die beyond the optimum position.

[5] Springs should be used according to manufacturer's specifications; if those are not available, an empirical rule can be used. Springs should not be depressed more than two-thirds of total deflection.

[6] Holes should be provided for scrap-separator chutes; standard chutes should be available.

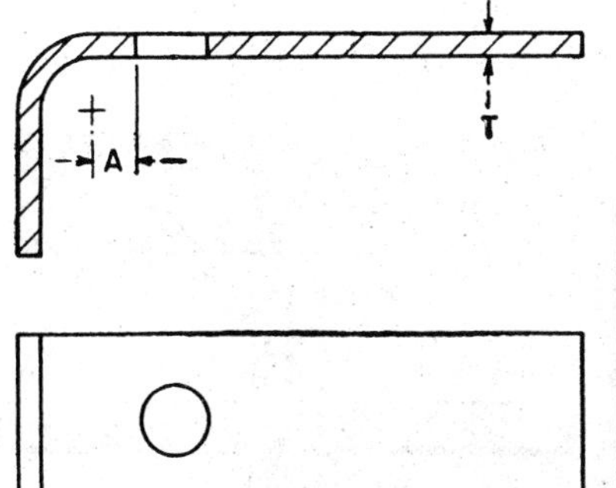

FIG. 1. Dimension A must not be less than $1\frac{1}{2}T$, when piercing stock that is to be bent to a radius.

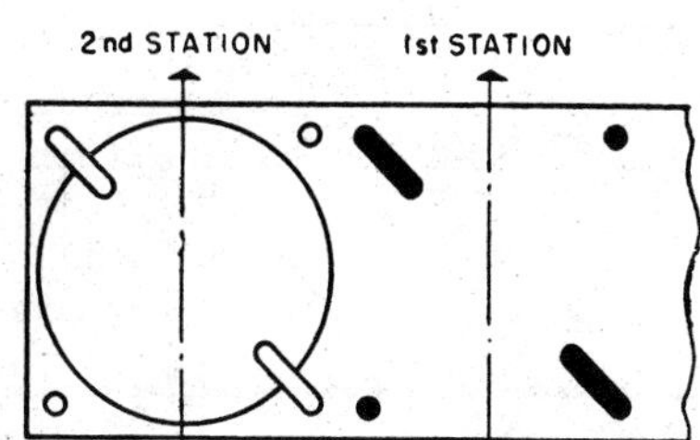

FIG. 2. Frail inserts in the blanking punch are avoided by punching the slots and pilot holes in the first station, and blanking the part at the second station.

58 POINTERS FOR DESIGN OF PROGRESSIVE DIES[1]

Use of design procedure outlined will insure that no important part of the finished die is omitted, and that design decisions are made in proper sequence. Many "false starts" will be eliminated.

In use, design decisions are made in chronological order given, starting with 1 and following through to 58. When the last step is reached, the progressive die will have been designed in an efficient and accurate way—no matter how intricate or elaborate it may be.

Design Procedure

1. Develop the blank with dimensions in decimals.

2. Establish a tentative sequence of operations, allowing idle stations for adequate strength of sections. Consider the required part accuracy. Progressive dies cannot hold distance between holes and outside of the blank closer than 0.005 to 0.010 in., depending on size of the blank.

3. When piercing stock "in the flat" (which is to be bent later) the edge of the hole must be at least $1\frac{1}{2}$ times the stock thickness from the center line of the blending radius (Fig. 1).

4. Can frail inserts be eliminated by punching out certain portions of the blank at one station, and blanking the rest in a later operation (Fig. 2)?

5. Lay out the scrap strip, using three cardboard templets.

Consider:

a. Relation of bend to grain. It must not be less than 30° across the grain.
b. Relation of burr side to bend. Burr must always be on the inside.
c. Pilot interference.
d. Half holes.
e. Partial blanks.
f. Location of pilot holes. Pierce these at first station.
g. Material economy.

6. Establish the final sequence of operations.

[1] J. R. Paquin.

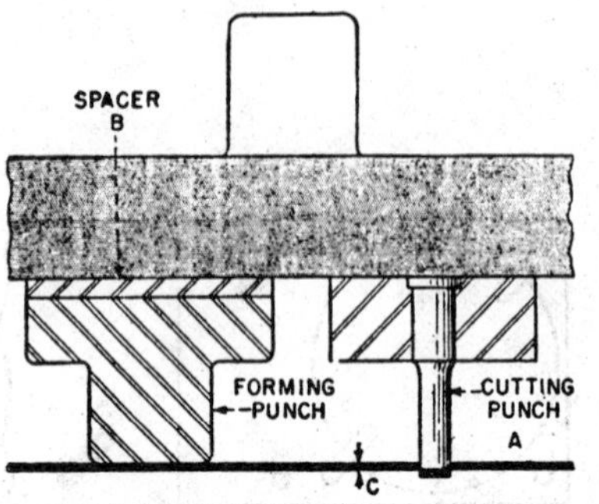

Fig. 3. When sharpening the cutting punch *A*, the same amount is ground off spacer *B*, to maintain the relative distance *C*.

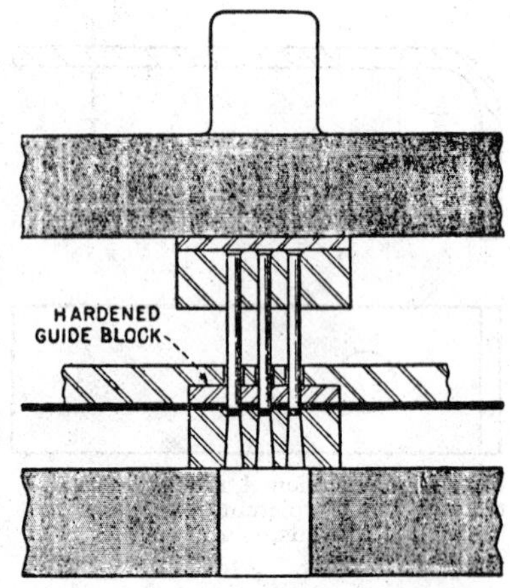

Fig. 4. If delicate punches must be placed close together, use a hardened guide block with the required number of holes for the punches.

7. Establish the blanking centers and the strip width.

8. Mark, in color, the bolster-plate opening for scrap disposal to insure scrap holes in the die are over this opening.

9. Establish the location of the punches for the pilot holes at the first station. On drawn work, drawing takes precedence and is done at the first station. Drawing depth must not exceed $3/16$ in., with the diameter of the cup twice the depth, or more.

10. Pilot in the pierced holes at the second station. Decide the succeeding pilot sequence.

11. Draw the notching punches. Incorporate backup heels. These are usually placed at the first or second station, or both.

12. Decide the method of bending. Two procedures are:

a. Bend the ears down.
b. Ramp the strip and form the ears up.

At the last station, when cutting off and bending simultaneously with a pressure pad, the difference in height between the cutoff punch and the bending punch must be one stock thickness. Provide a pressure pin in the bending punch to hold the stock while cutting takes place.

13. Design any forming punches. Forming is performed, usually, immediately before or after cutting off (for one form). Provide shedder pins as required.

14. Design the spanking punch, if one is necessary. It is usually placed at the station immediately preceding the last operation, for one spank. Combine with the bending or forming operation, if possible.

15. Decide if a scrap cutter is necessary, and where applied.

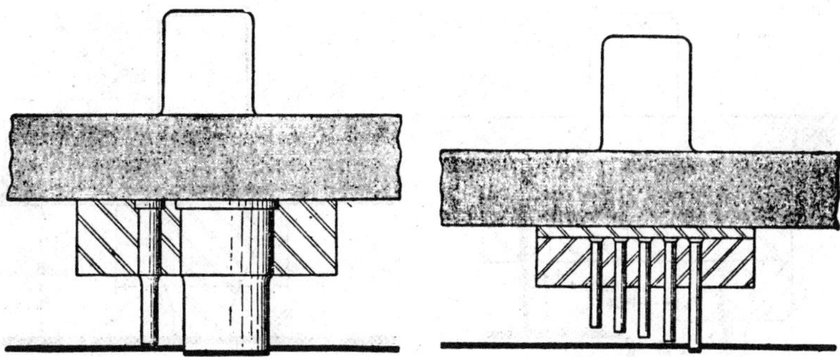

FIG. 5. A slender piercing punch (upper left) is made shorter than the adjacent large punch by the material thickness plus 0.010 in. in order to avoid deflection of the smaller punch. Closely grouped small punches (above right) are also stepped in the same manner.

16. Sketch suitable method of strip location:

a. Channel stripper.

b. Channel and bar pusher.

c. Channel and locating roller.

The length of actual guide before the first operation should be at least twice the stock width.

17. Sketch the die block. The minimum distance between the die-hole edges and the outside edge of the block should be from 1 to 1½ times the thickness of the block, or more, depending on the intricacy of the block opening. Establish the number of sections for easy machining, hardening, and replacement. Decide the method of keying and, whether the key should be "set into" the die holder, or "set on top." Select the screws (and dowels, if the die is to be set on top). If possible to assemble the die parts improperly, show one dowel off center for foolproof purposes.

18. Locate the finger stops.

19. Choose the automatic stop.

20. Incorporate inserts at weak places, if any are required.

21. Select a die set from the manufacturer's catalog. Use one longer than necessary to provide for possible future additions. Allow ⅝ in., minimum, between the die posts and die block for grinding clearance. Use a semisteel die set for average work. In severe applications, specify an all-steel die set.

SECTION VIEW

22. Decide the grinding allowance—¼ in. normally.

23. Consider how sharpening will affect relative heights of bend and form punches. Spacer pieces may be required (Fig. 3).

24. Compute the spring sizes, and list on drawing.

PUNCHES

25. Where a bent lug on the part would be injured by crowding, provide a punch long enough to push the blank through the die.

26. Must punches be contained in a quill? If possible, mount all quills in individual punch plates.

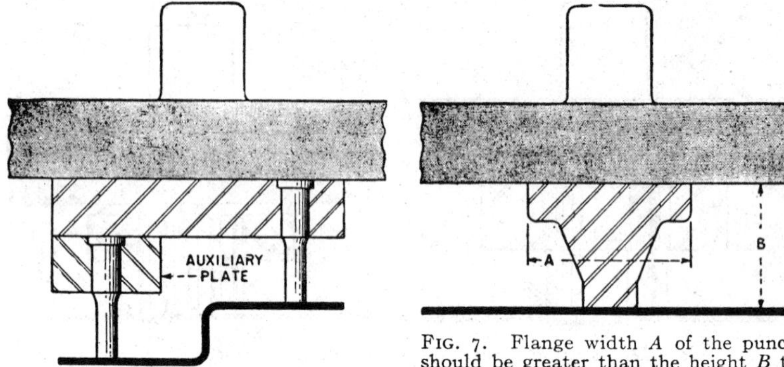

FIG. 6. When punches must protrude more than 4 in. beyond the punch holder, use an auxiliary plate to maintain stiffness.

FIG. 7. Flange width *A* of the punch should be greater than the height *B* to provide stability for unguided punches.

27. All delicate punches must be guided in the stripper. If they come too close together, apply a tool-steel block (Fig. 4).

28. Slender piercing punches, set close to a large punch, should be made shorter, by the material thickness plus 0.010 in. at least (Fig. 5). This practice avoids deflection by metal crowding to one side. Step small punches set close together.

29. For a small punch producing a hole 30 to 35% deeper than its own diameter, the shank should be at least twice the hole diameter. The cutting end should be hole size for only two stock thicknesses.

30. For angular-headed punches made of drill rod, use a 60° included angle. This type of punch may be used only when the diameter of the hole to be pierced is at least twice the stock thickness. It should always be guided in the stripper.

31. Punches should not be longer than 3 to 4 in. over-all. Use auxiliary plates, if necessary, as shown in Fig. 6.

32. All punches used in high-grade dies should be guided in the stripper. For low-production requirements, they need not be guided if the point diameter exceeds twice the material thickness.

33. For punches too small in diameter for push-off pins, but with a hole diameter from two to three times the material thickness, rounding the punch face will prevent slugs from pulling up. If smaller, make the die-hole wall straight for twice the material thickness, at least, and reduce clearance between the punch and die.

34. The flange width of unguided blanking punches should not be less than the punch height (Fig. 7).

35. In larger piercing punches, provide holes for push-off pins, to prevent slugs from pulling up (see Fig. 8).

36. Never press a hardened punch into a hardened punch plate. Use a soft plug in applications of this nature (Fig. 9).

37. In detailing make the punch body a *light* press fit in the punch plate. It should be a slip fit except where it actually bears in the punch plate.

38. For other than round punches, lock the punch by a milled flat in the punch head.

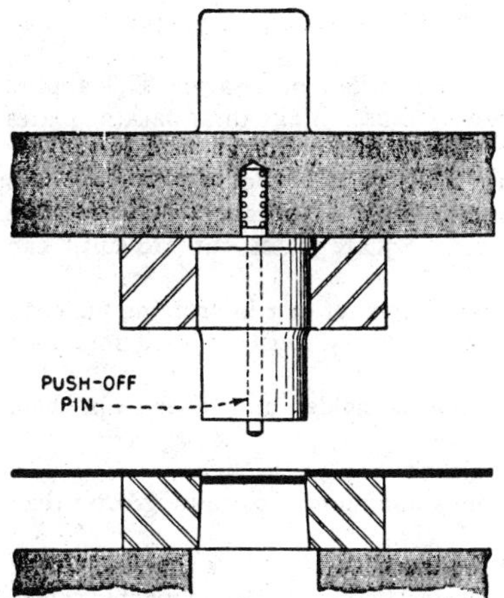

FIG. 8. Push-off pins in large punches prevent slugs from pulling up.

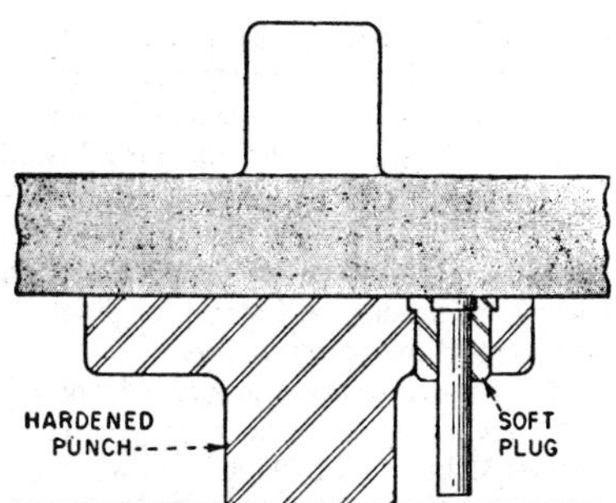

FIG. 9. To avoid cracking a large hardened punch, or punch plate, never press a small punch directly into either of these members, but use a soft plug or insert instead.

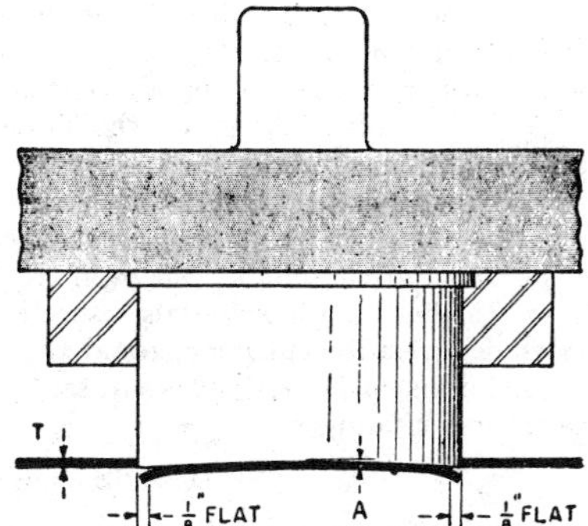

FIG. 10. Hollow-grind long slotting punches so that dimension *A* equals the metal thickness, to put shear on the punch. However, the ends should be flat for ⅛ in. to avoid bending the stock.

39. Long slotting punches should be made low in the center, so the ends will cut first. Leave the face flat for ⅛ in. at either end (Fig. 10).

40. Blanking punches with pilots must be set ahead of the piercing punches, so that the pilot can locate the strip before the piercing punches bind it. As in item 28, a height difference of 0.010 in. should be allowed.

General Details

41. When "shearing" punches and dies, the distance between high and low points should be from two-thirds to one thickness of stock. For piercing dies, apply shear to the punch. For blanking dies (where the blank is saved), apply it to the die.

42. Taper ream scrap holes ¼° on a side, or ½° included angle, leaving straight the first ⅛ in. from the cutting edge.

43. In piercing, when using tool-steel bushings set into the die block, make all holes in the block, stripper, and punch plate the same size, if possible, so they can be bored together.

44. When pressing die buttons into a hardened block, use a *light* press fit and "shoulder" them to prevent pulling out.

45. Use a hardened plate in back of all punches, where necessary. This applies to dies for long runs, and also for all small punch heads. Make these backing plates ¼ in. thick. On a semisteel punch holder, use them for No. 11 gage (0.120) or over. For steel punch holders, use them for No. 7 gage (0.180) or over. Always use them for very small punch heads, however. Round plates, counterbored into the punch holder, will serve for single punches. Square plates, cut "to suit," can be used for multiple punches.

46. If possible the pilots should be made removable for ease in grinding punches.

47. Avoid spring pilots, if you can. However, they must be used for stock heavier than 1/16 in. thick.

48. Carry the pilot holes through the die and die holder, and taper ream them ¼° per side to compensate for misfeeds.

49. Compute the required stripper-plate thickness.

50. Make the channel stripper width (inside) a minimum of 0.004 in. greater than the strip width.

51. The punch-plate thickness should not be less than 1½ times the diameter of the punch shank. The material should be tool steel, left soft, or good machine steel.

52. Punch plates and die plates must be from 1¼ to 1½ times wider than they are high, minimum, for stability.

53. Decide location of any air-vent holes.

54. Bolt heads in die plates should be set ¼ or ⅜ in. below the top surface to allow for sharpening the die.

55. Tapped holes for fastening gages should be tapped ¼ or ⅜ in. deeper than necessary to allow for sharpening the die. This lowers the gages.

56. Incorporate bumper blocks, if required.

57. In the "bill of materials," specify steel for cutting members so they can be machined with the cutting edges at the end of the grain.

58. Furnish a jig-boring layout, set to the correct blank angle, and incorporating the proper clearances.

PRESSWORK FORMULAS

Shearing Operations. Estimation of pressures is easiest with respect to the shearing operations, namely, shearing proper, blanking, punching, piercing, notching, trimming, and cutting off. Pressure requirements for any of these operations depend on the area to be severed and the shear strength of the material (see Table 1). The formulas are:

$$P_s = L \times t \times S = \pi \times D \times t \times S$$

where P_s = shear resistance, tons; L = length of cut, in.; t = thickness of stock, in.; S = shear strength of stock, tons per sq in.; and D = sum of the diameters of round openings.

Severance is obtained usually by penetration of the cutting edge less than halfway through the material, although it may range from 40 to 70%, the latter for very soft material.

Stripping load is 3½% of P_s.

Dull edges on the cutting tools may increase the shearing pressure by 25%.

If the punches are stepped, or shear is used on the die, the cutting pressure will be reduced.

Bending Operations. To bend metal, it must be subjected to stresses in the plastic range. Generally, it is assumed that compression stresses at the inside of the neu-

TABLE 1. STRENGTH OF MATERIALS, PLATES, SHEETS, AND STRIPS
In Tons per Sq In.

Material	Strength		
	Tensile		Shear
	Yield	Ultimate	
Aluminum:			
2SO annealed	2.5	6.5	4.7
¼ hard (H12)	6.5	7.5	5.0
½ hard (H14)	7.0	8.5	5.5
¾ hard (H16)	8.5	10.0	6.0
Hard (H18)	10.5	12.0	6.5
3SO annealed	3.0	8.0	5.5
¼ hard (H12)	7.5	9.0	6.0
½ hard (H14)	9.0	10.5	7.0
¾ hard (H16)	10.5	12.5	7.5
Hard (H18)	12.5	14.5	8.0
17SO annealed	5.0	13.0	9.0
T4 heat-treated	20.0	31.0	19.0
T Alclad, heat-treated	16.0	27.5	16.0
24SO annealed	5.5	12.5	9.0
T heat-treated	23.0	34.0	20.5
T Alclad, heat-treated	21.5	32.0	20.5
52SO annealed	7.0	14.5	9.0
¼ hard (H32)	13.0	17.0	10.0
½ hard (H34)	14.5	18.5	10.5
¾ hard (H36)	17.0	19.5	11.5
Hard (H38)	18.0	20.5	12.0
53SO annealed	3.5	8.0	5.5
T4 heat-treated	10.0	16.5	10.0
61SO annealed	4.0	9.0	6.2
T4 heat-treated	10.5	17.5	6.7
75S Alclad, annealed	7.0	16.0	
T Alclad, heat-treated	33.0	38.0	23.0
Brasses:			
Gilding metal, 95%, annealed	7.0	19.0	15.0
¼ hard	16.0	21.0	16.0
½ hard	20.0	24.0	17.0
Hard	25.0	28.0	18.5
Commercial bronze, 90%			
Annealed	7.5	20.5	16.0
¼ hard	17.5	22.5	16.5
½ hard	22.5	26.0	17.5
Hard	27.0	30.5	19.0
Commercial bronze, 90%			
Annealed	7.5	20.5	16.0
¼ hard	17.5	22.5	16.5
½ hard	22.5	26.0	17.5
Hard	27.0	30.5	19.0

TABLE I. STRENGTH OF MATERIALS, PLATES, SHEETS, AND STRIPS (*Continued*)

Material	Strength		
	Tensile		Shear
	Yield	Ultimate	
Brasses (continued):			
Red brass, 85%			
Annealed	9.0	22.5	16.5
¼ hard	19.5	25.0	17.5
½ hard	24.5	28.5	18.5
Hard	28.5	35.0	21.0
Low brass, 80%			
Annealed	10.0	25.0	16.5
¼ hard	20.0	26.5	18.0
½ hard	25.0	30.5	19.5
Hard	29.5	37.0	21.5
Cartridge brass, 70%			
Annealed	11.0	26.5	17.5
¼ hard	20.0	27.0	18.0
½ hard	26.0	31.0	20.0
Hard	31.5	38.0	22.0
Yellow brass, 65%			
Annealed	11.0	26.5	
¼ hard	20.0	27.0	18.0
½ hard	25.0	31.5	20.0
Hard	30.0	37.0	21.5
Bronzes:			
High silicon, A, 3%			
Annealed	15.0	31.5	22.5
¼ hard	17.0	34.0	23.5
½ hard	22.5	39.0	25.0
Hard	29.0	47.0	28.5
Cardboard:			
Soft	....	....	2.0
Hard	....	....	4.0
Copper:			
Electrolytic, tough pitch			
Annealed	5.0	16.0	11.0
¼ hard	15.0	19.0	12.5
½ hard	18.0	21.0	13.0
Hard	23.5	25.0	14.0
Fiber:			
Vulcanized: Soft	....	7.5	10.0
Hard	....	....	14.0
Inconel:			
Annealed, hot-rolled	22.5	47.5	32.5
As-rolled hot-rolled	30.0	60.0	37.0
Annealed, cold-rolled	18.5	45.0	31.5
Hard, cold-rolled	54.0	78.0	41.5

TABLE 1. STRENGTH OF MATERIALS, PLATES, SHEETS, AND STRIPS (*Continued*)

Material	Strength		
	Tensile		Shear
	Yield	Ultimate	
Leather:			
Soft	....	....	4.3
Hard	....	....	7.1
Lead, commercial pure	0.6	1.2	0.9
Magnesium:			
ASTM No. AZ31X annealed	11.0	18.0	10.5
ASTM No. AZ31X hard-rolled	17.5	22.5	10.5
ASTM No. AZ61X annealed	11.5	20.0	10.0
ASTM No. AZ61X hard-rolled	17.0	23.5	11.0
ASTM No. MI annealed	8.0	16.0	9.5
Hard-rolled	13.5	18.5	10.0
Monel:			
Soft, cold-rolled	17.5	37.5	25.0
¼ hard	27.5	40.0	
½ hard	....	45.0	
¾ hard	....	50.0	
Hard	50.0	55.0	32.5
K Monel:			
Annealed	30.0	60.0	37.5
Cold-worked	60.0	80.0	50.0
Nickel:			
A and L annealed, hot-rolled	10.0	31.5	26.0
A and L as-hot-rolled	24.0	42.5	
A and L annealed, cold-rolled	10.0	31.5	26.0
A and L as-cold-rolled	15.0	36.0	
D hot-rolled	17.0	43.0	
Z hot-rolled	17.0	45.0	
Nickel-silver, A, 18%:			
Annealed	15.0	30.0	28.5
¼ hard	25.0	32.5	
½ hard	31.0	37.0	
Hard	37.0	42.5	39.0
Paper:			
Soft	....	....	2.0
Hard	....	....	2.5
Steel:			
Hot-rolled, Nos. 4 to 18 gage	15.0	26.0	20.0
Hot-rolled, Nos. 19 to 24 gage	15.0	25.0	20.0
Cold-rolled strip:			
Grade 1: hard	....	40.0	30.5
Grade 2: ½ hard	....	32.0	25.0
Grade 3: ¼ hard	16.0	27.0	21.5
Grade 4: soft	....	24.0	21.0
Grade 5: dead soft	....	22.0	20.5
Stainless: No. 302 (18-8)	15.0	37.0	28.5

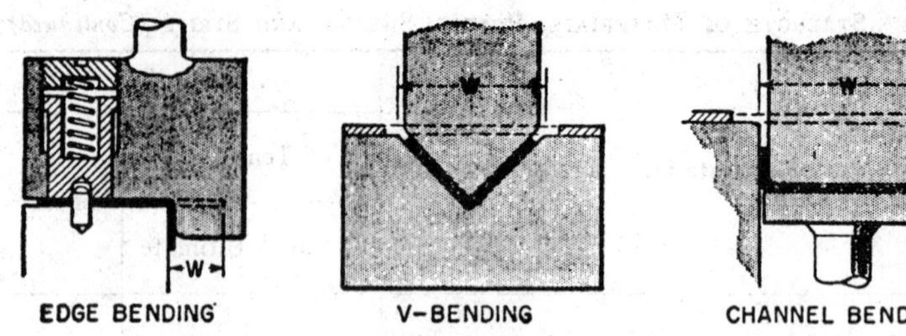

FIG. 11. Conditions to which bending formulas apply.

tral plane and the tension stresses at the outisde are equal to the ultimate tensile strength of the material. Formulas are (Fig. 11):

Edge bends,

$$P_b = \frac{1 \times t^2}{2W} \times S_t$$

Free V-bends with centrally located load,

$$P_b = \frac{1 \times t^2}{W} \times S_t$$

Channel and U-bends,

$$P_b = \frac{2l \times t^2}{W} \times S_t$$

where P_b = bending force, tons; l = length of bend, in.; t = stock thickness, in.; W = width of unsupported metal, in.; and S_t = ultimate tensile strength of the metal, tons per sq in.

For U-ing and channel bending, presses are often equipped with die cushions. If so, the resistance of the cushions must be added to the pressure found by the formulas.

Drawing. To draw metal into a cup or shell, the metal must be caused to flow. In producing round shells without flange, the factors that govern the possible height of single-operation drawing of a shell are:

1. Ratio of height to diameter of shell.
2. Ductility of the material.
3. Corner radius.

When drawing a round shell (without flange) in one operation, the maximum ratio of height divided by diameter may vary between 1:4 and a possible 3:4, depending on the corner radius and ductility of the material. A generous corner radius helps to secure greater height in one operation, while too small a corner radius may cause the shell to fracture at the radius. Soft, ductile material will permit drawing to a greater height in one operation.

In general, materials such as deep-drawing steel, annealed sheet steel (SAE 1005 to 1015), and dead-soft cold-rolled strip steel (SAE 1010 to 1020) will allow drawing to a maximum height in one operation. Other less ducile steels may require one or more precupping operations; that is, additional drawing dies for gradually reducing the blank to the diameter required, and possibly annealing between operations.

Table 2. Draw-reduction Ratios for Round Shells Without Flange
Shell height = inside height of shell
Shell dia = ID of shell plus one thickness of material

Height-dia Ratio	% Reduction of Dia	Blank-draw Ratio	Height-dia Ratio	% Reduction of Dia	Blank-draw Ratio	Height-dia Ratio	% Reduction of Dia	Blank-draw Ratio
0.01	2.0	1.02	0.48	41.5	1.70	0.95	54.4	2.194
0.02	3.8	1.03	0.49	41.9	1.72	0.96	54.5	2.201
0.03	5.5	1.05	0.50	42.3	1.73	0.97	54.7	2.208
0.04	7.1	1.07	0.51	42.7	1.745	0.98	54.9	2.216
0.05	8.7	1.09	0.52	43.1	1.758	0.99	55.1	2.224
0.06	10.2	1.11	0.53	43.5	1.770	1.00	55.3	2.231
0.07	11.5	1.13	0.54	43.8	1.780	1.05	56.2	2.28
0.08	13.0	1.15	0.55	44.2	1.790	1.10	57.0	2.32
0.09	14.3	1.16	0.56	44.5	1.800	1.15	57.8	2.37
0.10	15.4	1.18	0.57	44.8	1.810	1.20	58.5	2.41
0.11	16.6	1.20	0.58	45.1	1.820	1.25	59.2	2.45
0.12	17.8	1.21	0.59	45.4	1.830	1.30	59.9	2.49
0.13	18.8	1.23	0.60	45.8	1.840	1.35	60.5	2.53
0.14	19.9	1.24	0.61	46.0	1.850	1.40	61.0	2.56
0.15	21.0	1.26	0.62	46.4	1.862	1.45	61.6	2.60
0.16	21.9	1.28	0.63	46.7	1.873	1.50	62.2	2.64
0.17	22.8	1.29	0.64	47.0	1.885	1.55	62.7	2.68
0.18	23.7	1.31	0.65	47.3	1.898	1.60	63.2	2.72
0.19	24.6	1.32	0.66	47.6	1.910	1.65	63.7	2.76
0.20	25.5	1.34	0.67	47.8	1.920	1.70	64.2	2.79
0.21	26.3	1.35	0.68	48.1	1.930	1.75	64.6	2.83
0.22	27.1	1.37	0.69	48.4	1.940	1.80	65.1	2.86
0.23	27.8	1.38	0.70	48.7	1.950	1.85	65.5	2.90
0.24	28.6	1.40	0.71	49.0	1.960	1.90	65.9	2.93
0.25	29.3	1.41	0.72	49.2	1.970	1.95	66.3	2.97
0.26	30.0	1.43	0.73	49.5	1.980	2.00	66.7	3.00
0.27	30.7	1.44	0.74	49.8	1.990	2.05	67.0	3.03
0.28	31.3	1.45	0.75	50.0	2.000	2.10	67.4	3.06
0.29	31.9	1.47	0.76	50.2	2.010	2.15	67.8	3.10
0.30	32.6	1.48	0.77	50.5	2.020	2.20	68.1	3.13
0.31	33.2	1.49	0.78	50.7	2.030	2.25	68.4	3.16
0.32	33.8	1.51	0.79	51.0	2.040	2.30	68.7	3.20
0.33	34.4	1.52	0.80	51.2	2.050	2.35	69.0	3.23
0.34	35.0	1.53	0.81	51.4	2.060	2.40	69.3	3.26
0.35	35.5	1.55	0.82	51.6	2.070	2.45	69.6	3.29
0.36	36.0	1.56	0.83	51.8	2.080	2.50	69.9	3.32
0.37	36.5	1.57	0.84	52.1	2.090	2.55	70.2	3.35
0.38	37.0	1.58	0.85	52.4	2.100	2.60	70.4	3.38
0.39	37.5	1.60	0.86	52.6	2.110	2.65	70.7	3.41
0.40	38.0	1.61	0.87	52.8	2.120	2.70	70.9	3.44
0.41	38.5	1.62	0.88	53.0	2.130	2.75	71.2	3.47
0.42	39.0	1.63	0.89	53.2	2.140	2.80	71.5	3.50
0.43	39.4	1.65	0.90	53.4	2.150	2.85	71.7	3.53
0.44	39.8	1.66	0.91	53.6	2.160	2.90	71.9	3.56
0.45	40.2	1.67	0.92	53.8	2.170	2.95	72.1	3.58
0.46	40.7	1.68	0.93	54.0	2.178	3.00	72.2	3.60
0.47	41.1	1.69	0.94	54.2	2.185			

How to use table: Divide height of shell by dia; find corresponding ratio in col. 1; % reduction is given in col. 2 (use to determine number of reductions required); blank-draw ratio is given in col. 3 (blank-draw ratio times shell dia equals blank dia approximately).

Corner radius of the shell has a considerable effect on the possible height of single-operation drawing. Corner radius should, when possible, be specified at a minimum of four times the thickness of material when the height of the shell exceeds one-third of the diameter. When a smaller radius is specified, additional drawing or flattening operations may be required. In no case should the corner radius be less than the thickness of the material for a one-operation draw.

When the ratio of "height divided by diameter" exceeds 5:8, it will be necessary in most cases to reduce the flat blank to the finished shell by using two or more draw dies of proportionately decreasing diameters. In some cases, one or more annealing operations will be necessary between first and finish draw operations.

Determination of the number of reductions necessary to draw a shell with ratios (height divided by diameter) greater than 5:8 cannot be done by hard-and-fast rules. In general, for ductile materials, with generous corner radius in the shell, the requirements are:

1. Height equals ⅝ to 1⅛ times the diameter of the shell—two reductions will be required.
2. Height equals 1⅛ to 2 times the diameter of the shell—three reductions will be required.
3. Height equals 2 to 3 times the diameter of the shell—four reductions will be required.

Finish of edge depends on the "height divided by diameter" ratio and on the material being drawn. For relatively shallow shells where the "height divided by diameter" ratio is not over 1:3, it is possible to produce an edge within commercial tolerances without requiring finishing operations. That is, the height and uniformity of the edge depend on the size of blank used. For higher shells it is not possible to do this, and one of the following finishing operations will be required:

1. Flange trim and finish draw.
2. Pinch trim.
3. Machine trim.
4. Wedge or "shimmy"-die trim (not suitable for small quantities).

Table 2 gives a simple means of determining percentages of draw reduction and the flat-blank diameter. This table can be used only with round, straight-sided shells or cups.

Draw Beads. The pressure required to hold the steel blank flat while deep-drawing round work varies from zero for relatively thick blanks, up to a third or a half of the drawing pressure. When the blank-holding pressure exceeds a third of the drawing pressure, it is best to use draw beads to permit the blank-holding pressure to be decreased.

Mechanical grip on the edges of blanks, as obtained in toggle and cam presses, is, for practical purposes, positive and nonyielding. It is desirable for domed and tapered shallow shapes and for shallow stretching, especially where draw beads must be stamped to help grip the metal.

Rubber bumper pressures may be approximated at 10 psi of area per 1% of compression. More uniform pressures and more convenient control of pressures are obtained with die cushions. Pneumatic die cushions may be selected on a basis of 100 psi of piston area, maximum; hydropneumatic cushions for drawing purposes can be selected on a basis of about 500 psi.

The total percent reduction accomplished between annealing operations, whether in one or more operations, should be kept within the percentage of reduction in area established by tensile tests of the properly annealed steel.

MATERIAL LAYOUT

Because from 50 to 70% of piece-part cost is for material, the procedures employed in laying out the scrap strip directly influence the financial success or failure of press operations. The part must be laid out so a maximum area of the strip is utilized in production of the blank. This blank layout must be drawn before any work is done on the die itself. In fact, the material layout will govern the shapes and sizes of all die members.

Basic Blank Shapes. The shape of most blanks will fall into one of the classifications shown in Fig. 12. The accompanying scrap-strip designs for these typical part outlines provide a starting point in establishing the correct material layout for any part to be blanked. As shown, the scrap strip for a circular blank is laid out for a double row. This is far more economical than a single row and should be specified for all dies except those scheduled for very low production.

Many blanks have elaborate contours that cannot be readily classified. Upon study these will be found to be made up of two or more of the basic forms given.

Blank Positioning. Parts can be positioned eleven different ways in the strip. Choice of the correct method depends upon part shape, production requirements, and any bends that must be applied. Figure 13 shows the "single-row, one-pass" and "single-row, two-pass" methods. In the first case the blanks are arranged in one row only, and the strip is passed through the die once to cut all the blanks. At

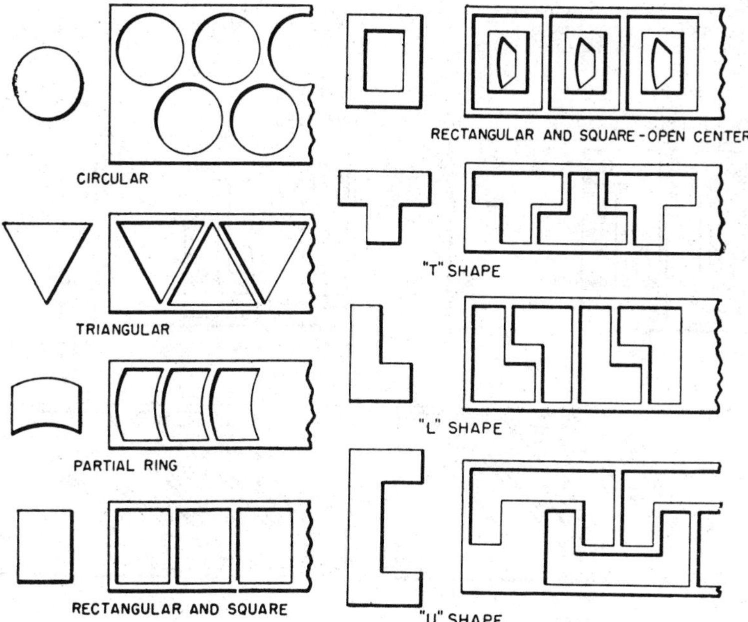

FIG. 12. Blank contours are made up of one or more basic forms, and scrap-strip layout should take into account the particular shape combinations.

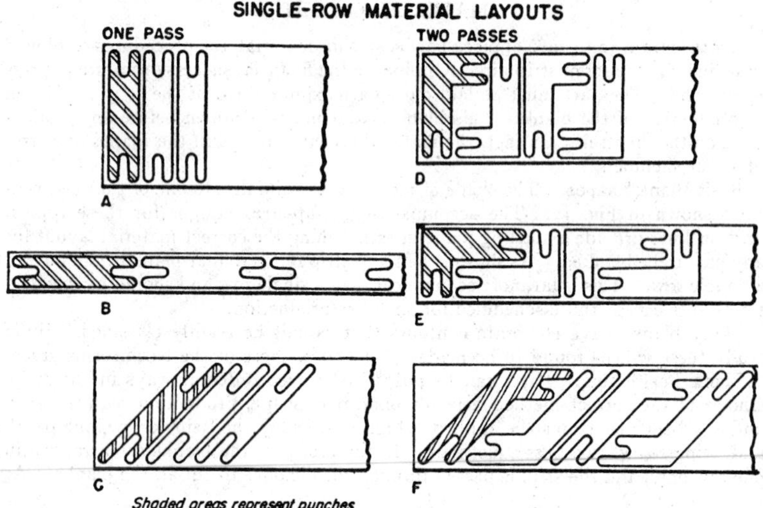

FIG. 13. Blanks may be arranged in a single row and cut out with either one or two passes of the strip through the die. Material savings obtained by the double-pass method must be balanced against extra labor cost.

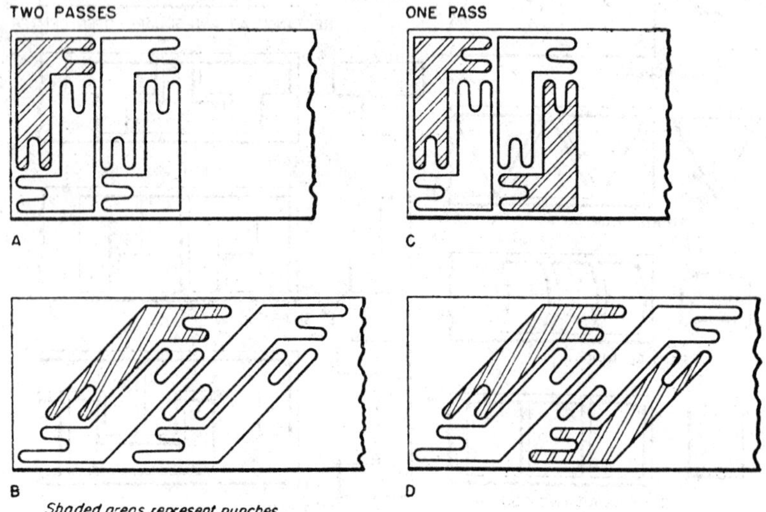

FIG. 14. Greater operating speed is secured by the double-row, two-pass method than by the single-row method. The double-row, one-pass method is even faster because the strip is passed once through gang dies.

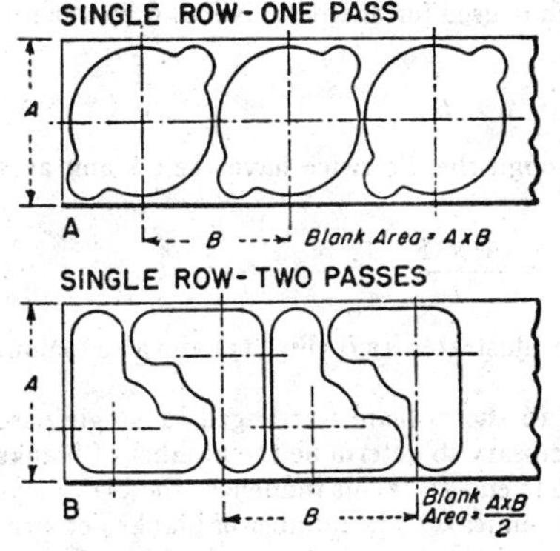

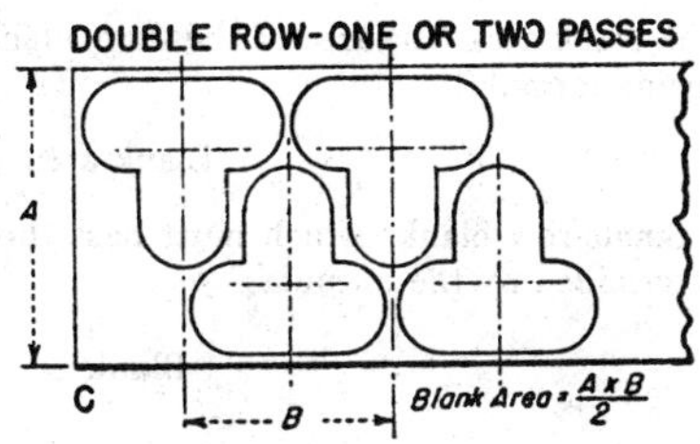

FIG. 15. Blank areas must be calculated for the various positioning methods in order to obtain maximum economy in use of material.

A, the parts are located in a vertical position in the strip. This is the preferred method where no bends are required, because the maximum number of blanks can be cut from one strip, and fewer strips must be handled.

Where severe bends are required in subsequent operations, as in flat springs made from strip, method *B* must be used. This procedure involves handling more strips to produce the same number of blanks.

Shape of the piece part will often lend itself to angular positioning, as at *C*. For some contours, this method is economical of material. Also, it has the further advantage of allowing bends to be made without fracture.

An excellent means of saving material is provided by the "single-row, two-pass" method (Fig. 13) when used in conjunction with certain part shapes. In this case, the parts are positioned in the strip in a single row. But alternate blanks are turned upside down as shown, and the strip must be passed through the die twice to remove all the blanks from it. As the strip goes through the first time, the blanks in the upper row are cut. The strip is then turned over and run through the die again, removing the rest of the blanks.

A 10 to 15% higher labor cost will occur in double-run layouts of the type described.

Vertical positioning is employed at *D* (Fig. 13) as the means of part location. In this case, it is obvious that horizontal positioning, as shown at *E*, would consume a greater percentage of the strip area. The contour and bends found in many piece parts will dictate use of angular positioning as shown at *F*.

Further economy in material can be achieved by use of the "double-row, two-pass" method shown in Fig. 14. Here, the strip is run through the die twice, but blanking centers are much shorter, giving greater operating speed.

The same positioning method may be used for a double-row gang die. An extra punch and die opening is applied to the die, and the strip is run through only once. Gang dies are high-speed tools and should be used where the added expense of the extra punch and die hole is warranted.

Blank-area Calculations. In material layout, four cardboard or celluloid templets are cut out to conform to the contour of the blank required. These templets are laid side by side in the various positions described. The blank area for each method is found, and the one most economical of material is chosen for the die.

At *A* in Fig. 15 is shown a blank laid out for "single-row, one-pass" positioning.

The blank area (the area of the strip which is used for one part) is found by the following formula:

$$\text{Blank area} = A \times B$$

Single-row blanks which must pass through the die twice have their blank area determined by the formula

$$\text{Blank area} = \frac{A \times B}{2}$$

This formula applies to the blank layout illustrated at B (Fig. 15) and also to double-row blanks at C.

Number of Blanks per Strip. Figure 16 shows blanks arranged in single-pass layout in the strip for blanking. It is necessary to determine the number of blanks in each strip to find out the extent of waste end D. This influences the blank layout, because too great a waste end is uneconomical. The number of blanks per strip is found by means of the formula

$$A = \frac{S - [X + Y + 2E]}{B} + 1$$

where A = number of blanks in each strip.

For the waste end,

$$D = S - [B(A - 1) + X + Y + 2E]$$

When strips must make two passes through the die (Fig. 16) the following formulas determine the number of blanks in the strip and the extent of the waste end:

For number of blanks,

$$A = \frac{S - [X + Y + 2E]}{0.5B} + 1$$

For the waste end,

$$D = S - [0.5B(A - 1) + X + Y + 2E]$$

SCRAP-STRIP ALLOWANCES FOR BLANKING DIES

From the economy standpoint, and that of successful die performance, it is important that correct scrap-strip allowances be applied to the material layout. Too great an allowance is wasteful of material on long runs. On the other hand, insufficient stock between blanks, and between the blank and strip edge, will make the scrap strip weak and subject to breakage, and hence slow down press operation. A weak scrap area around the blank can also bring about severe dishing of the part.

Curved-outline Blanks. Piece-part peripheries fall under several distinct outline shapes. The most common is the curved outline (Fig. 17). Where curves on adjacent blanks are opposite, or the curve on one part is adjacent to a straight line on the next one, the correct allowance is 70% of the strip thickness for "one-pass" layouts. In Fig. 17, all dimensions A fall in this classification, and they are made 70% of the material thickness, or as close to it as possible, consistent with the application of a commercial dimension to the center-to-center distance B and strip width C.

Straight-sided Blanks. The blanks in Fig. 18 have straight sides adjacent to each other and to the outside edges of the strip. For one-pass layouts, if the straight portion of the blank outline is shorter in length than 2½ in., apply one thickness of stock as the scrap-strip allowance A.

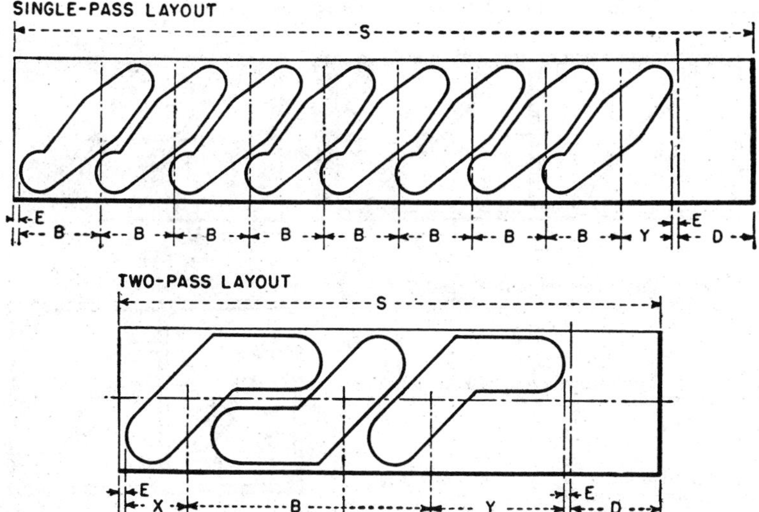

Fig. 16. Formulas are given to determine the number of blanks in a strip and the extent of the waste end *D*.

Scrap-strip Allowances for One-pass Layouts

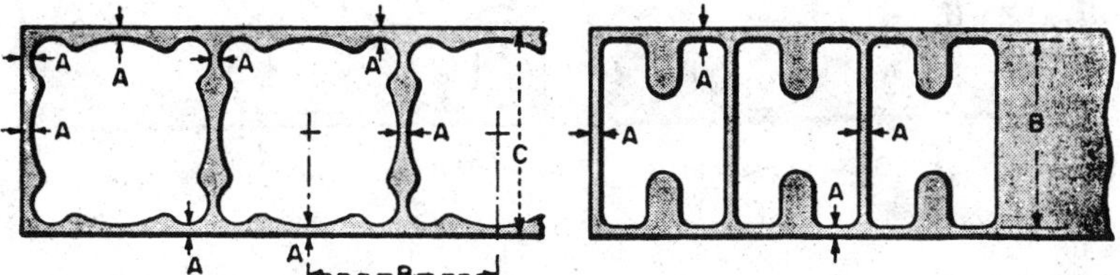

Fig. 17. For work with curved outlines, dimensions $A = 70\%$ of strip thickness T.

Fig. 18. For straight-edge blanks:
Where B is less than $2\frac{1}{2}$ in., $A = 1T$
Where B is $2\frac{1}{2}$ to 8 in., $A = 1\frac{1}{4}T$
Where B is over 8 in., $A = 1\frac{1}{2}T$

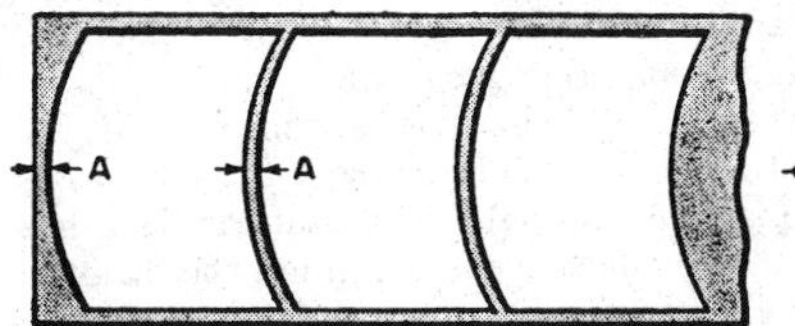

Fig. 19. For work with parallel curves, the same rules for *A* apply as in Fig. 18.

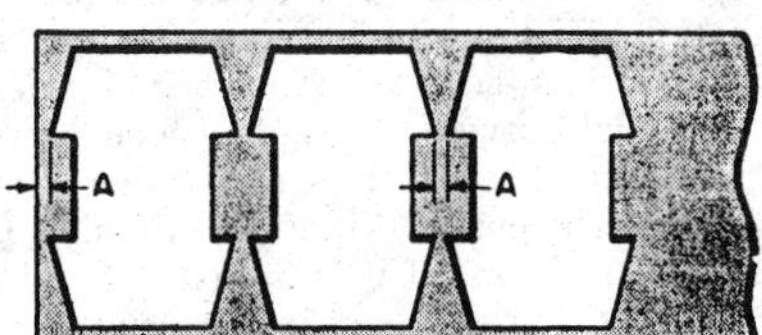

Fig. 20. For layouts with sharp corners of blanks adjacent, $A = 1\frac{1}{4}T$.

Scrap-strip Allowances for Two-pass Layouts

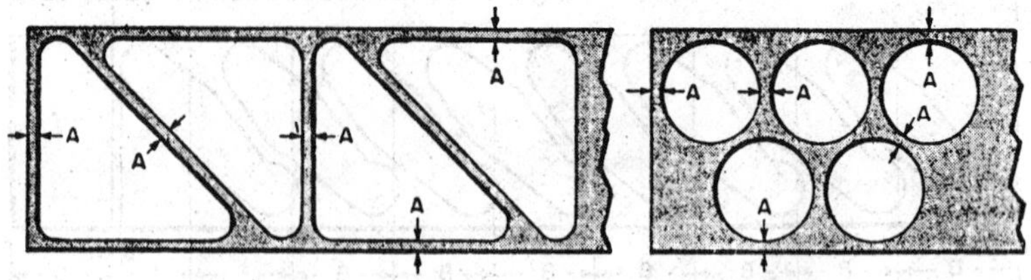

Fig. 21. Fig. 22.

Fig. 21. Single-row layout intended for two passes through die: $A = 1\frac{1}{2}T$.

Fig. 22. Double-row layout of blanks with curved outlines: $A = 1\frac{1}{4}T$.

Fig. 23. Double-row layout of parts with straight and curved outlines: $A = 1\frac{1}{4}T$.

Fig. 23.

Minimum Scrap-strip Allowances for Thin Materials

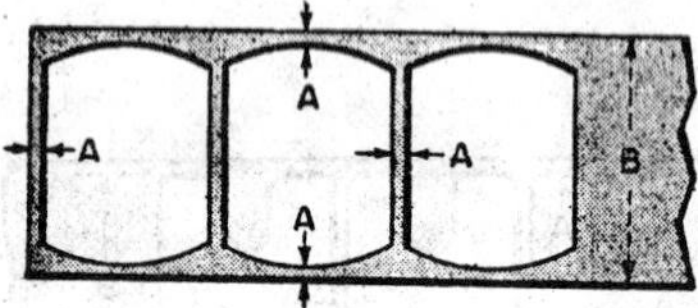

Fig. 24. One-pass layout.

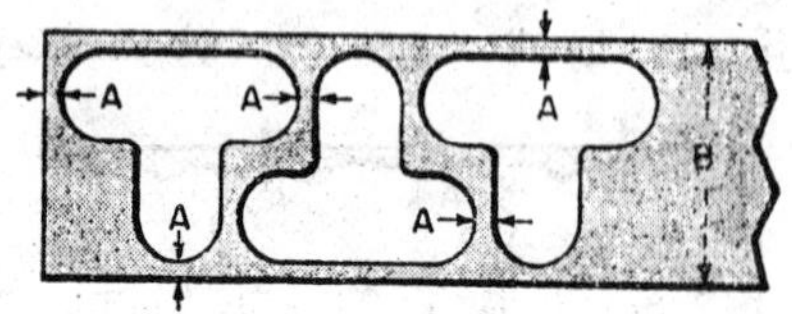

Fig. 25. Double-pass layout.

Strip Width B	*Space A*
0 to 3 in.	1/32 in.
3 to 6 in.	1/16 in.
6 to 12 in.	3/32 in.
Over 12 in.	1/8 in.

Strip Width B	*Space A*
0 to 3 in.	1/16 in.
3 to 6 in.	3/32 in.
6 to 12 in.	1/8 in.
Over 12 in.	5/32 in.

Figs. 24 and 25. For thin gages of stock, the preceding rules do not apply, and minimum scrap-strip allowances should be selected from the tables.

Where the extent of the straight edge is from 2½ to 8 in., apply 1¼ times the material thickness as the scrap-strip allowance A; if blank length is greater than 8 in., apply a scrap-strip allowance of 1½ times the material thickness.

Parallel Curves. If edges of some parts form parallel curves, as shown in Fig. 19, these are treated in the same manner as straight edges (Fig. 18) and the same rules apply. Dimension A in Fig. 19 should equal one material thickness if part length is less than 2½ in., if it is from 2½ to 8 in., the allowance is 1¼ times the material thickness; but if it is over 8 in. long, the allowance is 1½ times the strip thickness. Dimension B, on the other hand, follows the rule for parts with curved outlines as given in Fig. 17, and is 70% of the strip thickness.

Adjacent Sharp Corners. Sharp corners form a focal point for fractures and should be given considerable allowances. In Fig. 20, dimensions *A* should be 1¼ times the strip thickness, at least, for one-pass layouts.

Double-pass Layouts. Where the strip must be passed through the die twice in order to remove all the blanks from it, as shown in Fig. 21, and the blanks are arranged in a single row, the scrap-strip allowance should be 1½ times the stock thickness, at least. Blanking the first row of parts distorts the strip. Hence the specified allowance is necessary so that enough material will be left for cutting full blanks on the second pass.

Where piece parts are arranged in the strip in two rows, as shown in Figs. 22 and 23, allow 1¼ times the material thickness as the scrap-strip allowance *A* for blanks with both curved and straight outlines. The wider strip does not distort as much in cutting the first row, and slightly less scrap-strip allowance is required. Because of this minimum distortion of the strip, double-row layouts are preferred wherever they can be used.

Minimum Allowances. Tables in Figs. 24 and 25 give minimum scrap-strip allowances for both one-pass and double-pass layouts. These will apply to the thinner gages of stock, where use of the previous rules would give such a small allowance as to be impractical.

DIE CLEARANCES

Clearance between punch and die is essential for satisfactory action and life of cutting dies. Correct space must exist between the edge of the punch and the edge of the die for a blank to part cleanly from the material. Excessive clearance will dish the blank and produce long stringy burrs all around the edge. Application of the correct clearance will result in a blank with little burr and with the burnished portion extending about one-third of the blank thickness.

Proper clearance to apply depends upon the material, and its hardness and thickness. Table 3 gives the die clearance in terms of percentage of stock thickness and

TABLE 3. DIE CLEARANCE IN TERMS OF PERCENTAGE OF STOCK THICKNESS
Over-all Clearance, Not "per Side" Type

Material	Constant	Percentage
Aluminum:		
To ⅛ in.	10	10
Over ⅛ in.	8	12
Magnesium	25	4
Brass and copper	20	5
Phosphor bronze	16	6
Steel:		
Soft, to ¼ in.	17	5½
Soft, over ¼ in.		8
Medium hard, to ¼ in.	16	6
Medium hard, over ¼ in.	10	10
Hard	14	7
Stainless steel:		
To ¼ in.	29	3½
Over ¼ in.	20	5

Allowance = stock thickness ÷ constant.

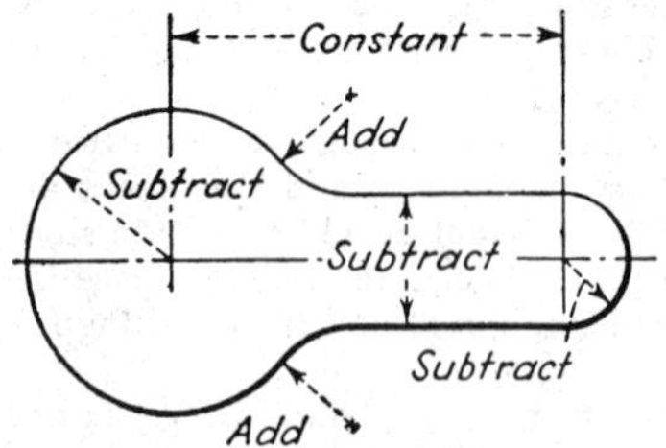

FIG. 26. When clearance is applied to punch: Subtract clearance from all radiuses with centers inside punch. Add clearance to all radiuses with centers outside. Subtract from all dimensions between parallel lines. Angles and dimensions between centers remain constant.

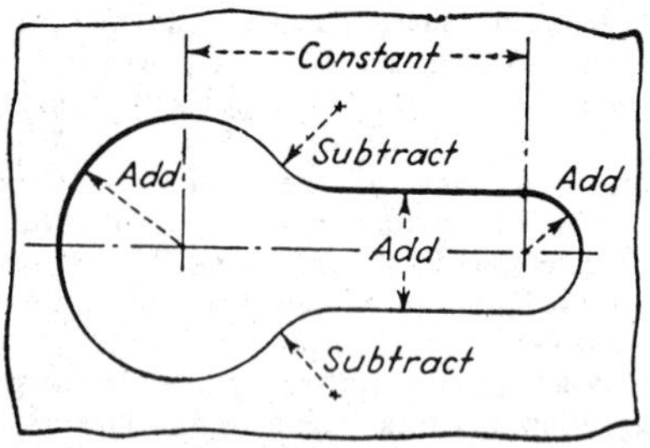

FIG. 27. When clearance is applied to die: Add clearance to all radiuses with centers inside die. Subtract clearance from all radiuses with centers outside. Add to all dimensions between parallel lines. Angles and dimensions between centers remain constant.

hardness. The values given have been used for thousands of successful high-quality dies to produce clean burn-free blanks. Note that the clearance given is the over-all clearance, not the clearance per side.

Clearance affects the condition of the blank edge and press tonnage. Broken piercing punches, chipped cutting edges, and short tool life are seldom associated with clearance as such, either inadequate or excessive, but are almost invariably the result of structure defects in the die, caused by faulty design or construction.

Table 4 gives standard die clearances based on decimal thicknesses of stock, to avoid the complications introduced by considering the different gage system used for various materials (see Sec. 28 for gage systems). If you know the decimal thickness of the metal, you can interpolate the clearance from Table 4 or calculate it from these relationships:

Clearance = decimal thickness of stock times the allowance in percent, Table 3

or

Clearance = stock thickness divided by the constant in Table 3

Application of Clearance. The required clearance is applied to the punch or the die, but never to both. The rules are:

1. A piercing punch is made to size, and clearance is applied to the die.
2. A blanking die is made to size and the clearance is applied to the blanking punch.

These rules apply when the slug or the blank is of a regular shape. When laying out dies of an irregular shape or when cutting out only portions of a blank, the clearance per side is obtained by dividing the over-all clearance by 2 and applying said clearance in the manner shown in Figs. 26 and 27.

Secondary Allowances. Many books state that holes close in a certain amount after being punched, and that blanks under 1 in. dia tend to swell. Therefore "secondary allowances" have been suggested—to be added to the punch or subtracted from the die-hole dimension to compensate for the alleged conditions and produce an accurate piece. Recent tests described below indicate that secondary allowance can be neglected.

TABLE 4. STANDARD PUNCH AND DIE CLEARANCES*

Stock Thickness	Soft Steel	Medium Steel	Hard Steel	Stainless	Phosphor Bronze	Brass	Copper	Aluminum
0.010	0.0006	0.0006	0.0007	0.0008	0.0006	0.0005	0.0005	0.001
0.020	0.0011	0.0012	0.0014	0.0016	0.0012	0.001	0.0009	0.002
0.030	0.0017	0.0018	0.0021	0.0024	0.0018	0.0015	0.0014	0.003
0.040	0.0023	0.0025	0.0028	0.0032	0.0025	0.002	0.0019	0.004
0.050	0.0029	0.0031	0.0035	0.004	0.0031	0.0025	0.0023	0.005
0.060	0.0035	0.0037	0.0043	0.0048	0.0037	0.003	0.0028	0.006
0.070	0.0041	0.0043	0.005	0.0056	0.0043	0.0035	0.0033	0.007
0.080	0.0047	0.005	0.0057	0.0064	0.005	0.004	0.0038	0.008
0.090	0.0052	0.0056	0.0064	0.0072	0.0056	0.0045	0.0042	0.009
0.100	0.0058	0.0062	0.0071	0.008	0.0062	0.005	0.0047	0.010
0.110	0.0064	0.0069	0.0078	0.0088	0.0069	0.0055	0.0052	0.011
0.120	0.007	0.0075	0.0085	0.0096	0.0075	0.006	0.0057	0.012
0.130	0.0076	0.0081	0.0093	0.0104	0.0081	0.0065	0.0062	0.0162
0.140	0.0082	0.0087	0.010	0.0112	0.0087	0.007	0.0066	0.0175
0.150	0.0088	0.0093	0.0107	0.012	0.0093	0.0075	0.0071	0.0187
0.160	0.0094	0.010	0.0114	0.0128	0.010	0.008	0.0076	0.020
0.170	0.010	0.0106	0.0121	0.0136	0.0106	0.0085	0.008	0.0212
0.180	0.0105	0.0112	0.0128	0.0144	0.0112	0.009	0.0085	0.0225
0.190	0.0111	0.0118	0.0135	0.0152	0.0118	0.0095	0.009	0.0237
0.200	0.0117	0.0125	0.0142	0.016	0.0125	0.010	0.0095	0.025
0.210	0.0123	0.0131	0.015	0.0168	0.0131	0.0105	0.010	0.0262
0.220	0.0129	0.0137	0.0157	0.0176	0.0137	0.011	0.0104	0.0275
0.230	0.0135	0.0143	0.0164	0.0184	0.0143	0.0115	0.0109	0.0287
0.240	0.0141	0.015	0.0171	0.0192	0.015	0.012	0.0114	0.030
0.250	0.0147	0.0156	0.0178	0.020	0.0156	0.0125	0.0119	0.0312

* J. R. Paquin.

The values given in this table apply to over-all clearances, or diameters. Stock thicknesses are used by industry are in terms of gages or decimal thicknesses. In either case, the clearance can be interpolated by reference to values tabulated above. For clearances "on a side," divide the given values by 2.

Tests on Die Clearance. Considerable disagreement exists in respect to the proper amount of clearance between a punch and die. Recommendations vary from 5% total (2½% per side) to 12% per side on the same type of material. In an attempt to determine the exact effects of clearance, an experimental die was constructed with an adjustable die opening, and several hundred stampings were made.

The actual effect of varying the clearance is shown in Fig. 28. The edge conditions illustrated are representative of either the pierced hole or the blanked part.

The type III edge is usually desired for ordinary work and is produced with what would normally be called "optimum" clearance. It shows a typical "die-cut" edge, comprising two main areas: (1) the sheared and burnished portion, and (2) the "break" or tensile fractured portion. The latter surface is tapered at angle A, which will be referred to as the "break angle." Any burr, as at x, will be referred to

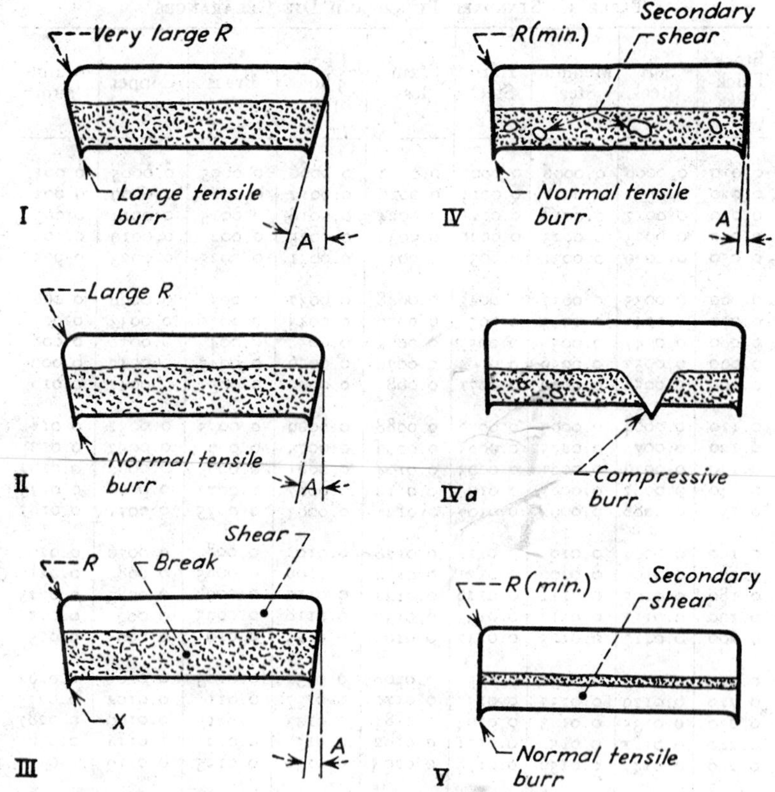

FIG. 28. Types of burr and edge condition produced with varying clearance.

as a "tensile burr," because it is obviously the tail end of the break in tension. Radius R is also common to all stamped parts.

With somewhat greater values of clearance, the type II edge is produced in addition to an increase in edge, except that radius R and angle A are considerably increased. Note that there is no appreciable increase in the tensile burr.

By increasing the clearance past a critical point, the type I edge is produced. In addition to an increase in angle A and radius R, there is now a marked increase in the tensile burr. This edge would be generally unacceptable.

Reductions in clearance below the optimum value also reach a critical point, at which slight secondary shearing takes place. This is evidenced by small burnished spots appearing in the break area, as shown in the type IV edge. The break angle is very small, and radius R is much reduced. There is also a slight increase in the depth of the shear area. The tensile burr is unchanged. This is the most desirable type of edge for high-quality work. Further reductions in clearance result in a type V edge. In this edge, increased secondary shearing produces almost com-

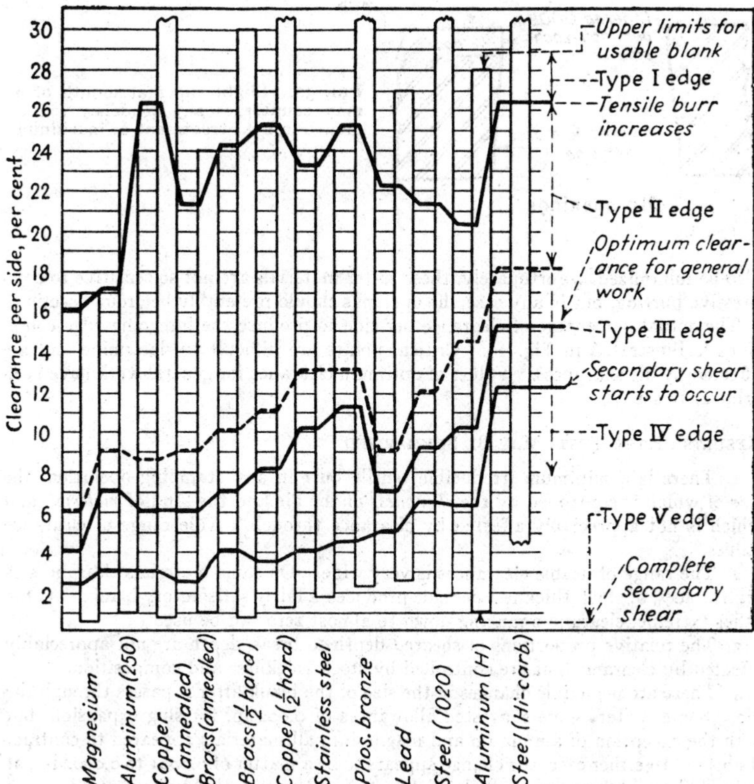

FIG. 29. Percentage of clearance *per side* to produce five types of blank edge.

plete burnishing of the break area, and the edge somewhat resembles that from a shaving operation. The break angle is practically zero, and radius R is at its smallest or minimum value for a given material.

A special case is the type IV*a* edge, which is produced under the same conditions as for type IV, that is, clearance below the critical secondary shear point. However, here a defect in the die opening, either bellmouth, reversal of taper, or rough surfaces, produced sufficient pressures to "smear" some of the high spots in the break area into another type of burr—the compressive burr. This type of burr is burnished and usually extends from the shear surface. The remedy is to increase the clearance to eliminate the secondary shear. Strictly speaking, this does not correct the real cause of the trouble.

While types IV and V edges are not required in ordinary work, there are some special applications, notably high-quality work, very small parts, and in notching prior to a shave operation to minimize the amount of shaving allowance required. On softer materials which are inclined to produce a large radius R, this condition will

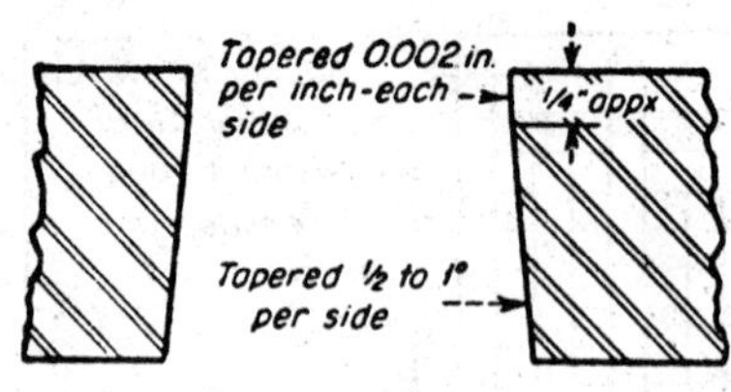

FIG. 30. Slight taper at mouth of die may counteract any tendency to bellmouth in heat-treatment or machining.

also be minimized. Fortunately, these softer materials are not so sensitive to compressive burring, but in any case, die openings should preferably be ground sections.

The actual percentage of clearance per side to produce the foregoing edge conditions is illustrated in Fig. 29. Critical points are difficult to determine and are affected by die sharpness. Values of optimum clearance for general work have been arbitrarily set.

RESULTS OF THE TESTS MAY BE SUMMARIZED

1. There is a minimum irreducible tensile burr in any stamping operation, the size of which is controlled by the *sharpness* of the die and the kind of *material*, and which is not appreciably affected by clearance through a wide range of clearance values.

2. The range of usable clearances is very wide. On some materials clearances as high as 40% of stock thickness per side produced a fairly satisfactory blank. At the other extreme, clearances ranging down to almost zero can be used.

3. The relative proportions ot sheared depth to break depth are not appreciably affected by clearance but are controlled by stock thickness and composition.

4. There are negligible changes in the size of the blank after it passes through the die. Some writers have suggested allowances of 0.002 in. for slug expansion, but with the exception of aluminum and magnesium all materials appeared to contract slightly. In either case, the change appears to be a matter of 0.0001 to 0.0002 in., at least in the middle range of stock thicknesses, and may probably be ignored.

5. Radius R varies with the type of material but is at a minimum for a given material with the smallest possible clearance.

6. Compressive-type burrs not *primarily* caused by too little clearance are due to defects in the die opening. The condition is aggravated by small clearances and may be corrected by increasing clearance.

The condition of bellmouth, or accidental reversals of taper in die openings, appears to be of such importance that a modification in the standard specifications for filed die openings has been suggested. As shown in Fig. 30, instead of the customary straight portion or "die life," the sides of the die would have a very small angle specified for approximately ¼ in. of depth. This angle would have a negligible effect on resharpening yet would counteract any tendency to bellmouth either in machining or heat-treatment. This specification should, of course, not be applied to ground die sections.

Effect of Die Grinding on Clearance. It is of value to know how much increase in blanking-die size occurs as the face is ground down in resharpening. Table 5 has been worked out to cover various angles of taper besides the conventional slope of ½° on a side.

Ordinarily, the angle of ½° on a side will allow free discharge of slugs from per-

TABLE 5. DIE CLEARANCES OR RELIEF

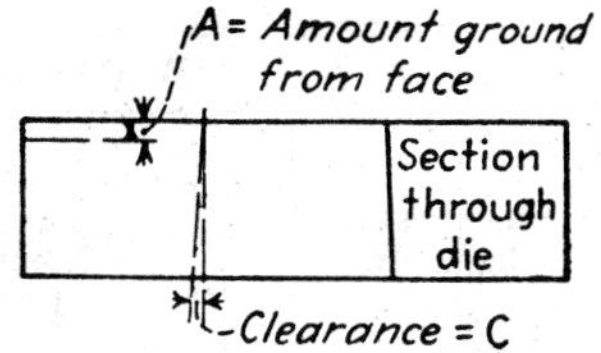

A Amount Ground from Face of Die in Thousandths In.	*C* Clearance in Thousandths Corresponding to Degrees on Each Side				
	½°	1°	1½°	2°	2½°
0.010	0.000087	0.000175	0.000262	0.000349	0.000437
0.020	0.000175	0.000350	0.000524	0.000698	0.000873
0.030	0.000241	0.000525	0.000786	0.001047	0.001310
0.040	0.000378	0.000700	0.001048	0.001396	0.001746
0.050	0.000437	0.000875	0.001310	0.001745	0.002183
0.060	0.000524	0.001050	0.001572	0.002094	0.002620
0.070	0.000611	0.001225	0.001834	0.002443	0.003016
0.080	0.000698	0.001400	0.002096	0.003092	0.003493
0.090	0.000776	0.001575	0.002358	0.003141	0.003929
0.100	0.000870	0.001750	0.002620	0.003493	0.004370
0.110	0.000960	0.001925	0.002882	0.003839	0.004803
0.120	0.001047	0.002100	0.003144	0.004188	0.005239
0.130	0.001135	0.002275	0.003406	0.004537	0.005677
0.140	0.001222	0.002450	0.003668	0.004886	0.006112
0.150	0.001309	0.002625	0.003930	0.005235	0.006549
0.160	0.001417	0.002800	0.004192	0.005584	0.006986
0.170	0.001484	0.002975	0.004554	0.005933	0.007422
0.180	0.001571	0.003150	0.004712	0.006282	0.007859
0.190	0.001658	0.003325	0.004978	0.006631	0.008295
0.200	0.001746	0.003500	0.005240	0.006980	0.008732
0.210	0.001827	0.003675	0.005502	0.007329	0.009169
0.220	0.001914	0.003850	0.005764	0.007678	0.009605
0.230	0.002001	0.004025	0.006026	0.008027	0.010042
0.240	0.002088	0.004200	0.006288	0.008376	0.010478
0.250	0.002175	0.004375	0.006550	0.008725	0.010915

forated blanks, but materials like aluminum and similar soft metals sometimes produce slugs that, in the small sizes particularly, tend to swage into the walls of the piercing die when forced down by the punch. These slugs may then stick and stack up and, becoming welded together, cause the punch to break or the die to become chipped around the edges of the small hole.

TABLE 6. AMOUNT TO ALLOW ON A SIDE FOR SHAVING CONTOUR WHERE ONLY ONE SHAVE IS TAKEN

Thickness of Blank, In.	Soft Steel, In.	Half-hard Steel, In.	Hard Steel, In.	German Silver, In.	Brass, In.	Thickness of Blank, In.
3/64 (0.0468)	0.0025	0.003	0.004	0.005	0.005	3/64 (0.0468)
1/16 (0.0625)	0.003	0.004	0.005	0.006	0.006	1/16 (0.0625)
5/64 (0.0782)	0.0035	0.005	0.006–0.007	0.007	0.007	5/64 (0.0782)
3/32 (0.0938)	0.004	0.006	0.007–0.008	0.008	0.008	3/32 (0.0938)
7/64 (0.1094)	0.005	0.007	0.009–0.011	0.010	0.010	7/64 (0.1094)
1/8 (0.125)	0.007	0.009	0.012–0.014	0.014	0.014	1/8 (0.125)

Table 5 is arranged to show the additional amount of clearance at each side of a die for every ten-thousandths ground off its top face.

This is much more than the average amount removed in one grinding, but it is a convenient increment to work from.

Consider the ½° column: The increase at each side of the die is entirely negligible when the first 0.010 in. has been ground from the face. Only after the die has been ground down 0.120 in., or almost ⅛ in., does the diameter of the die opening become 0.001 in. greater on each side, or 0.002 in. more over all.

ALLOWANCES FOR SHAVING

As for any given material, or grade of stock, the condition of the contour of the blank will vary with the thickness of the metal, the amount left for shaving should likewise vary and with a fair degree of uniformity from the thinner gages to the thick material. Similarly the allowance for any given thickness should vary for soft, half-hard, and hard material. In order to cover these allowances for steel blanks of the three grades noted, Table 6 has been developed and has been given thorough tests in connection with numerous shaving dies operating on different classes of work.

This table covers thicknesses of metal from 3/64 to ⅛ in., inclusive, and also includes allowances for german silver and brass. For the latter two materials, it will be noted, the shaving allowances are double those for steel of the same thickness

BEND ALLOWANCES FOR SHEET METAL

When a blank or sheet is to be bent, it is necessary to consider the effect of stretching the metal at the outside of the bend, when developing the size of the piece. Various formulas are in existence for bend allowance. Three methods are given herewith:

1. The assumption that the neutral plane of a formed metal part lies 20% of the material thickness from the inside face (Figs. 31 to 34 and Tables 7 and 8). Since the neutral plane is one where no stretching takes place, it is obvious that the length of a formed part along the neutral plane will be the correct length, or cutting size, of the blank or sheet. These tables are worked out for 90° bends, and the allowance is considered to be 0.4 thickness.

2. Where the radius of a 90° bend is less than 1/64 in., the custom is to apply an

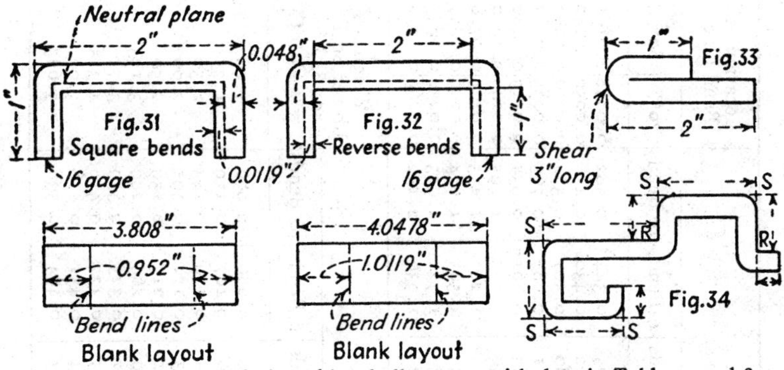

FIGS. 31 to 34. Calculation of bend allowances with data in Tables 7 and 8.

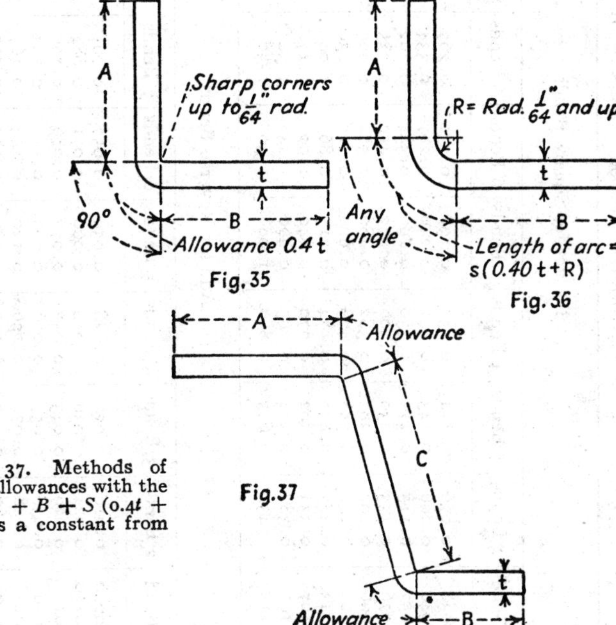

FIGS. 35 to 37. Methods of figuring bend allowances with the formula $L = A + B + S\,(0.4t + R)$, where S is a constant from Table 9.

allowance of 0.4 t in a well-known business-machine company's plant (Figs. 35 to 37). This is added to the inside dimensions. For bends having a radius of greater than 1/64 in., and of any angle, the formula is

$$L = A + B + S(0.4t + R)$$

where S is a constant taken from Table 9. The developed length of a blank is computed by considering each bend separately.

TABLE 7. ALLOWANCES FOR 90° BENDS IN STEEL*

Mfgrs Std Gage		Deduct for Square Bends										Location of Bend Line	
No.	Equivalent	1 Bend	2 Bends	3 Bends	4 Bends	5 Bends	6 Bends	7 Bends	8 Bends	9 Bends	10 Bends	20%	80%
24	0.0239	0.038	0.076	0.114	0.152	0.191	0.230	0.267	0.305	0.344	0.382	0.0048	0.0191
22	0.0299	0.048	0.096	0.143	0.191	0.240	0.287	0.334	0.382	0.430	0.478	0.0059	0.0240
20	0.0359	0.057	0.115	0.172	0.230	0.290	0.344	0.402	0.460	0.507	0.574	0.0072	0.0287
18	0.0478	0.076	0.153	0.230	0.306	0.382	0.459	0.535	0.611	0.688	0.765	0.0095	0.0382
16	0.0598	0.096	0.192	0.287	0.382	0.480	0.574	0.669	0.765	0.861	0.957	0.0119	0.0480
14	0.0747	0.120	0.240	0.358	0.478	0.600	0.717	0.836	0.956	1.075	1.195	0.0149	0.0600
13	0.0897	0.143	0.287	0.430	0.574	0.717	0.861	1.00	1.14	1.292	1.435	0.0179	0.0718
12	0.1046	0.167	0.334	0.502	0.670	0.840	1.00	1.17	1.33	1.505	1.674	0.0209	0.0837
11	0.1196	0.191	0.382	0.574	0.765	0.960	1.14	1.34	1.53	1.721	1.913	0.0239	0.0957
10	0.1345	0.215	0.430	0.645	0.860	1.08	1.30	1.50	1.72	1.936	2.152	0.0269	0.1076

Mfgrs Std Gage		Add for Reverse Bends										Weight, Lb per Sq Ft 501.84 Lb = 1 Cu Ft
No.	Equivalent	1 Bend	2 Bends	3 Bends	4 Bends	5 Bends	6 Bends	7 Bends	8 Bends	9 Bends	10 Bends	
24	0.0239	0.0096	0.019	0.029	0.038	0.050	0.057	0.067	0.076	0.086	0.096	1.00
22	0.0299	0.0120	0.024	0.036	0.048	0.060	0.071	0.084	0.096	0.107	0.119	1.25
20	0.0359	0.014	0.029	0.043	0.057	0.072	0.086	0.100	0.115	0.129	0.143	1.50
18	0.0478	0.019	0.038	0.057	0.076	0.095	0.114	0.134	0.153	0.172	0.191	2.00
16	0.0598	0.024	0.048	0.072	0.096	0.120	0.143	0.167	0.191	0.215	0.239	2.50
14	0.0747	0.030	0.060	0.090	0.120	0.150	0.180	0.209	0.239	0.269	0.299	3.123
13	0.0897	0.036	0.072	0.108	0.143	0.180	0.215	0.251	0.287	0.323	0.359	3.75
12	0.1046	0.042	0.084	0.126	0.167	0.210	0.251	0.293	0.335	0.377	0.418	4.375
11	0.1196	0.048	0.096	0.144	0.191	0.240	0.287	0.335	0.383	0.431	0.478	5.00
10	0.1345	0.054	0.108	0.161	0.215	0.270	0.322	0.377	0.403	0.484	0.538	5.625

* Alf J. Abrahamsen, Virginia Metal Products Corp.

TABLE 8. ALLOWANCES FOR COMBINATIONS OF SQUARE AND REVERSE BENDS

Square		0	1							0	2						
Reverse		1	0	1	2	3	4	5	6	2	0	1	2	3	4	5	6
Constant		+0.40	1.60	1.20	0.80	0.40	0.00	+0.40	+0.80	+0.80	3.20	2.80	2.40	2.00	1.60	1.20	0.80
Gage	Decimal																
24	0.0239	+0.0096	0.038	0.029	0.019	0.0096	.....	+0.0096	+0.019	+0.019	0.076	0.067	0.057	0.050	0.038	0.029	0.019
22	0.0299	+0.0120	0.048	0.036	0.024	0.0121	.....	+0.0121	+0.024	+0.024	0.096	0.083	0.071	0.060	0.048	0.036	0.024
20	0.0359	+0.014	0.057	0.043	0.029	0.014	.....	+0.014	+0.029	+0.029	0.115	0.100	0.086	0.071	0.057	0.043	0.029
18	0.0478	+0.019	0.076	0.057	0.038	0.019	.....	+0.019	+0.038	+0.038	0.153	0.133	0.114	0.095	0.076	0.057	0.038
16	0.0598	+0.024	0.096	0.072	0.048	0.024	.....	+0.024	+0.048	+0.048	0.192	0.167	0.143	0.120	0.096	0.072	0.048
14	0.0747	+0.030	0.120	0.090	0.060	0.030	.....	+0.030	+0.060	+0.060	0.240	0.210	0.180	0.150	0.120	0.090	0.060
13	0.0897	+0.036	0.143	0.108	0.072	0.036	.....	+0.036	+0.072	+0.072	0.287	0.251	0.215	0.180	0.143	0.108	0.072
12	0.1046	+0.042	0.167	0.126	0.084	0.042	.....	+0.042	+0.084	+0.084	0.334	0.292	0.251	0.210	0.167	0.126	0.084
11	0.1196	+0.048	0.191	0.144	0.096	0.048	.....	+0.048	+0.096	+0.096	0.382	0.334	0.287	0.240	0.191	0.144	0.096
10	0.1345	+0.054	0.215	0.161	0.108	0.054	.....	+0.054	+0.108	+0.108	0.430	0.377	0.322	0.270	0.215	0.161	0.108
Square		0	3							0	4						
Reverse		3	0	1	2	3	4	5	6	4	0	1	2	3	4	5	6
Constant		+1.20	4.80	4.40	4.00	3.60	3.20	2.80	2.40	+1.60	6.40	6.00	5.60	5.20	4.80	4.40	4.00
Gage	Decimal																
24	0.0239	+0.029	0.114	0.105	0.095	0.086	0.076	0.070	0.057	+0.038	0.152	0.143	0.133	0.124	0.114	0.105	0.095
22	0.0299	+0.036	0.143	0.131	0.119	0.107	0.096	0.083	0.071	+0.048	0.191	0.180	0.167	0.155	0.143	0.131	0.119
20	0.0359	+0.043	0.172	0.157	0.143	0.129	0.114	0.100	0.086	+0.057	0.230	0.215	0.201	0.187	0.172	0.157	0.143
18	0.0478	+0.057	0.230	0.210	0.191	0.172	0.152	0.133	0.114	+0.076	0.306	0.287	0.268	0.249	0.229	0.210	0.191
16	0.0598	+0.072	0.287	0.263	0.239	0.215	0.191	0.167	0.143	+0.096	0.382	0.360	0.334	0.310	0.287	0.263	0.239
14	0.0747	+0.090	0.358	0.328	0.298	0.268	0.240	0.210	0.180	+0.120	0.478	0.448	0.418	0.388	0.360	0.328	0.298
13	0.0897	+0.108	0.430	0.394	0.358	0.322	0.287	0.251	0.215	+0.143	0.574	0.538	0.502	0.466	0.430	0.394	0.358
12	0.1046	+0.126	0.502	0.460	0.418	0.377	0.334	0.292	0.251	+0.167	0.670	0.628	0.585	0.544	0.502	0.460	0.418
11	0.1196	+0.144	0.574	0.526	0.478	0.430	0.382	0.334	0.287	+0.191	0.765	0.718	0.670	0.621	0.574	0.526	0.478
10	0.1345	+0.161	0.645	0.591	0.538	0.484	0.430	0.377	0.322	+0.215	0.860	0.807	0.753	0.699	0.645	0.591	0.538

TABLE 8. ALLOWANCES FOR COMBINATIONS OF SQUARE AND REVERSE BENDS (*Continued*)

Square		0	5							0	6						
Reverse		5	0	1	2	3	4	5	6	6	0	1	2	3	4	5	6
Constant		+2.00	8.00	7.60	7.20	6.80	6.40	6.00	5.60	+2.40	9.60	9.20	8.80	8.40	8.00	7.60	7.20
Gage	Decimal																
24	0.0239	+0.050	0.191	0.181	0.172	0.162	0.152	0.143	0.133	+0.057	0.230	0.220	0.210	0.200	0.191	0.181	0.172
22	0.0299	+0.060	0.240	0.227	0.215	0.203	0.191	0.180	0.167	+0.071	0.287	0.275	0.263	0.251	0.240	0.227	0.215
20	0.0359	+0.072	0.290	0.272	0.258	0.244	0.230	0.215	0.201	+0.086	0.344	0.330	0.315	0.301	0.287	0.272	0.258
18	0.0478	+0.095	0.382	0.363	0.344	0.325	0.306	0.287	0.268	+0.114	0.459	0.440	0.420	0.401	0.382	0.363	0.344
16	0.0598	+0.120	0.480	0.454	0.430	0.407	0.382	0.360	0.334	+0.143	0.574	0.550	0.526	0.502	0.478	0.454	0.430
14	0.0747	+0.150	0.600	0.568	0.538	0.508	0.478	0.448	0.418	+0.180	0.717	0.687	0.657	0.627	0.600	0.568	0.538
13	0.0897	+0.180	0.717	0.681	0.646	0.610	0.574	0.538	0.502	+0.215	0.861	0.825	0.790	0.753	0.717	0.681	0.646
12	0.1046	+0.210	0.840	0.794	0.753	0.711	0.670	0.630	0.585	+0.251	1.00	0.962	0.920	0.879	0.837	0.794	0.753
11	0.1196	+0.240	0.960	0.910	0.861	0.813	0.765	0.718	0.670	+0.287	1.14	1.10	1.05	1.00	0.957	0.910	0.861
10	0.1345	+0.270	1.08	1.02	0.968	0.914	0.860	0.807	0.753	+0.322	1.30	1.23	1.18	1.12	1.08	1.02	0.968

Square		0	7							0	8						
Reverse		7	0	1	2	3	4	5	6	8	0	1	2	3	4	5	6
Constant		+2.60	11.20	10.80	10.40	10.00	9.60	9.20	8.80	+3.00	12.80	12.40	12.00	11.60	11.20	10.80	10.40
Gage	Decimal																
24	0.0239	+0.067	0.267	0.258	0.248	0.240	0.230	0.220	0.210	+0.076	0.305	0.296	0.286	0.277	0.267	0.258	0.248
22	0.0299	+0.084	0.334	0.322	0.310	0.300	0.287	0.275	0.263	+0.096	0.382	0.370	0.358	0.346	0.334	0.322	0.310
20	0.0359	+0.100	0.402	0.387	0.373	0.359	0.344	0.330	0.315	+0.115	0.460	0.445	0.430	0.416	0.402	0.387	0.373
18	0.0478	+0.134	0.535	0.516	0.497	0.478	0.460	0.440	0.420	+0.153	0.611	0.592	0.573	0.554	0.535	0.516	0.497
16	0.0598	+0.167	0.669	0.645	0.621	0.598	0.574	0.550	0.526	+0.191	0.765	0.741	0.717	0.693	0.669	0.645	0.621
14	0.0747	+0.209	0.836	0.806	0.776	0.747	0.717	0.687	0.657	+0.239	0.956	0.926	0.896	0.866	0.836	0.806	0.776
13	0.0897	+0.251	1.00	0.968	0.932	0.897	0.861	0.825	0.790	+0.287	1.14	1.11	1.07	1.04	1.00	0.968	0.932
12	0.1046	+0.293	1.17	1.12	1.08	1.04	1.00	0.962	0.920	+0.335	1.33	1.29	1.25	1.21	1.17	1.12	1.08
11	0.1196	+0.335	1.34	1.29	1.24	1.19	1.14	1.10	1.05	+0.383	1.53	1.48	1.43	1.38	1.34	1.29	1.24
10	0.1345	+0.377	1.50	1.45	1.40	1.34	1.30	1.23	1.18	+0.403	1.72	1.66	1.61	1.56	1.50	1.45	1.40

Add all the dimensions which must be on one side of sheet. Then deduct the dimension in the table which corresponds to the allowance for the combination of the square bends and reverse bends under consideration (see Fig. 34).

TABLE 9. CONSTANTS FOR ANGLE BENDS

Angle, Deg	Constant = S	Angle, Deg	Constant = S
0°10′	0.0029	35	0.6109
0°30″	0.0087	40	0.6981
1	0.0175	45	0.7854
2	0.0349	50	0.8728
3	0.0524	55	0.9599
4	0.0698	60	1.0472
5	0.0873	65	1.1345
10	0.1745	70	1.2217
15	0.2618	75	1.3090
20	0.3491	80	1.3963
25	0.4363	85	1.4835
30	0.5236	90	1.5708

3. In the automotive industry, the bend allowance for a 90° bend in steel is taken as $\pi/2(R + \frac{1}{3}t)$. Table 10 gives allowances for various angles of bend and considers the radius.

The first method cited has not only proved satisfactory for general use in connection with 90° bends, but Table 8 gives effect to deductions or allowances for combinations of square and reverse bends in material from Nos. 10 to 24 gage.

SQUARE BENDS. When the dimensions are given to the *outside* of a 90° bend, it is considered square, and the bend allowance is deducted (Fig. 31).

REVERSE BENDS. When the dimensions are given to the *inside* of a 90° form, the bend is considered reverse and the allowance is added to the part-print dimensions (Fig. 32).

FOLDS. A fold (Fig. 33) requires no deduction or allowance.

Sample Calculations. To show how Tables 7 and 8 are computed, consider two square bends (Fig. 31) in 16-gage sheet. The lengths of the three sides along the neutral axes are:

$$\begin{aligned} 2 - 2 \times 0.048 &= 1.904 \\ 1 - 0.048 &= 0.952 \\ 1 - 0.048 &= \underline{0.952} \\ & 3.808 \text{ in.} \end{aligned}$$

Correct cutting length of the sheet is 3.808 in., not 4 in. The deduction to be made from the sum of the dimensions (which must all be on one side of the sheet) is 0.192 in. This amount checks with that given in Table 8 for two square bends in 16-gage material.

Calculation of the cutting length for a part with reverse bends will show that material must be added to compensate for the bends. From Fig. 32 the lengths of the sides along the neutral plane are:

$$\begin{aligned} 2 + 2 \times 0.0119 &= 2.0238 \\ 1 + 0.119 &= 1.0119 \\ 1 + 0.119 &= \underline{1.0119} \\ & 4.0476 \text{ in.} \end{aligned}$$

The amount to be added to the dimensions of the piece in order to get the correct cutting length is 0.0476 in. This value checks with the table for the allowance to be added when considering two reverse bends in 16-gage material. A blank layout, not to scale, is given for Figs. 31 and 32 to show the location of the bend lines.

TABLE 10. BEND ALLOWANCES FOR SHEET STEEL AS USED IN THE AUTOMOTIVE INDUSTRY

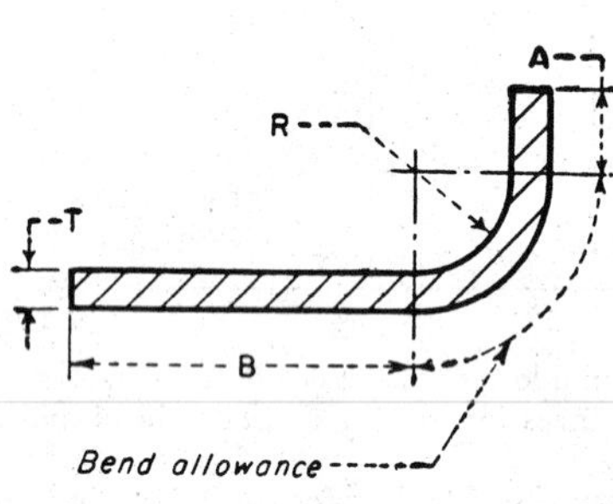

Blank Development. It is not general practice to show blank developments on production drawings. Therefore, the following formulas and illustration are presented for reference purposes only.

The formula below may be used for blank development:

Total length = $A + B$ + bend allowance

Bend Allowance. The bend allowances shown in the table were computed by the formula

$$\text{Length for } 90° \text{ bend} = \frac{\pi}{2}\left(R + \frac{1}{3}T\right)$$

See illustration. The chart also indicates the bend allowance for each degree of bend. For instance, when a 30° bend is specified, it is only necessary to multiply 30 by the length given for 1° in order to determine the bend allowance.

METAL THICKNESS		.024	.027	.030	.033	.035	.047	.059	.067	.075	.089	.105	.120	.135
RAD.	DEG.	BEND ALLOWANCE												
.016	1	.0004	.0004	.0004	.0004	.0005	.0005	.0006	.0007	.0007	.0008	.0009	.0010	.0011
	90	.038	.039	.041	.042	.044	.050	.055	.060	.064	.070	.080	.088	.096
.031	1	.0007	.0007	.0007	.0007	.0008	.0008	.0008	.0008	.0009	.0010	.0011	.0012	.0013
	90	.061	.063	.064	.066	.068	.075	.077	.080	.088	.090	.104	.111	.119
.047	1	.0009	.0010	.0010	.0010	.0010	.0010	.0011	.0012	.0012	.0013	.0014	.0015	.0016
	90	.086	.088	.090	.091	.093	.097	.105	.108	.113	.121	.129	.137	.144
.062	1	.0012	.0012	.0012	.0012	.0012	.0013	.0014	.0014	.0015	.0016	.0017	.0017	.0018
	90	.110	.112	.113	.115	.116	.123	.129	.132	.137	.145	.152	.160	.168
.078	1	.0015	.0015	.0015	.0015	.0015	.0016	.0017	.0017	.0018	.0018	.0019	.0020	.0021
	90	.135	.137	.138	.140	.141	.148	.154	.157	.162	.170	.178	.185	.193
.094	1	.0018	.0018	.0018	.0018	.0018	.0019	.0019	.0020	.0020	.0021	.0022	.0023	.0024
	90	.160	.162	.163	.165	.167	.173	.179	.182	.187	.195	.203	.211	.218
.109	1	.0020	.0020	.0020	.0021	.0021	.0021	.0022	.0022	.0023	.0024	.0025	.0026	.0027
	90	.184	.185	.187	.189	.190	.196	.203	.206	.211	.218	.226	.234	.242
.125	1	.0023	.0023	.0023	.0023	.0023	.0024	.0025	.0025	.0026	.0027	.0028	.0028	.0028
	90	.209	.211	.212	.214	.215	.222	.228	.231	.236	.244	.251	.259	.267
.141	1	.0026	.0026	.0026	.0026	.0026	.0027	.0028	.0028	.0029	.0029	.0030	.0031	.0032
	90	.234	.236	.237	.239	.240	.247	.253	.256	.261	.269	.276	.284	.292
.156	1	.0028	.0028	.0029	.0029	.0029	.0030	.0030	.0031	.0031	.0032	.0033	.0034	.0035
	90	.258	.259	.261	.262	.264	.270	.276	.280	.284	.292	.300	.308	.315
.172	1	.0031	.0031	.0031	.0031	.0032	.0032	.0033	.0033	.0034	.0035	.0036	.0037	.0038
	90	.283	.284	.286	.287	.289	.295	.302	.305	.309	.317	.325	.333	.341
.188	1	.0034	.0034	.0034	.0034	.0034	.0035	.0036	.0036	.0037	.0038	.0039	.0039	.0040
	90	.308	.309	.311	.313	.314	.320	.327	.330	.335	.342	.350	.358	.366
.203	1	.0037	.0037	.0037	.0037	.0037	.0038	.0038	.0039	.0039	.0040	.0041	.0042	.0043
	90	.331	.333	.335	.336	.338	.344	.350	.353	.358	.366	.374	.382	.390
.219	1	.0039	.0039	.0040	.0040	.0040	.0041	.0041	.0042	.0042	.0043	.0044	.0045	.0046
	90	.357	.358	.360	.361	.363	.369	.375	.379	.383	.391	.399	.407	.415
.234	1	.0041	.0041	.0043	.0043	.0043	.0044	.0044	.0045	.0045	.0046	.0047	.0048	.0049
	90	.380	.382	.383	.385	.386	.393	.399	.402	.407	.415	.423	.430	.438
.250	1	.0045	.0045	.0045	.0046	.0046	.0046	.0047	.0047	.0048	.0049	.0049	.0051	.0051
	90	.405	.407	.408	.410	.412	.418	.424	.427	.432	.440	.448	.456	.463

Square and Reverse Bends. Tables 7 and 8 of deductions or allowances for combinations of square and reverse bends (Fig. 34) are also based on gages Nos. 10 to 24, inclusive. They are set up to give the answer directly for combinations of from one square bend and one reverse bend to eight square bends and six reverse bends, thus taking in practically every case that will come up. In addition, values are given for the condition of no square bend and one reverse bend, and vice versa. The zeros in the headings indicate the absence of one or the other kind of bend.

The figures given in the tables are actually the algebraic sum of the allowances or deductions for the individual bends. In some cases the result is positive and is indicated by a plus sign, meaning that this amount must be added to the dimensions on the piece. Where no plus sign is given, the figure must be deducted from the sum of the dimensions.

SPRINGBACK CONTROL

Allowances. In forming ears and other right-angle parts on stampings, the harder the material the greater amount of "springback" will be found in the work. This has to be offset by allowances in the press tools. Quarter-hard cold-rolled steel has approximately 1 to 2° springback; half-hard has approximately 3 to 4°; hard steels have, in most instances, more than 5° springback; and annealed spring steel has as high as 15 to 20° springback.

The Pad Form Die. In general with dies where simple V tools cannot be used, the pad form die, as shown in Fig. 38, is best known and is very satisfactory. As illustrated, the part to be formed is located by nests or pilot pins on the pad of the form die. At the downstroke of the press, the part is held between the punch and pad of the die and is formed up, as shown in Fig. 39. The forms thus obtained are deformations of the metal which are exceeding the elastic limit, thus remaining in the formed-up condition after the punch is releasing the part at the upstroke of the press.

Figure 40 shows at *A* a part formed up to exactly 90° in the form die. The part shows a decided springback, the amount of which depends on the characteristics of the metal from which the part is made and on the radius *R* in the corner of the bend.

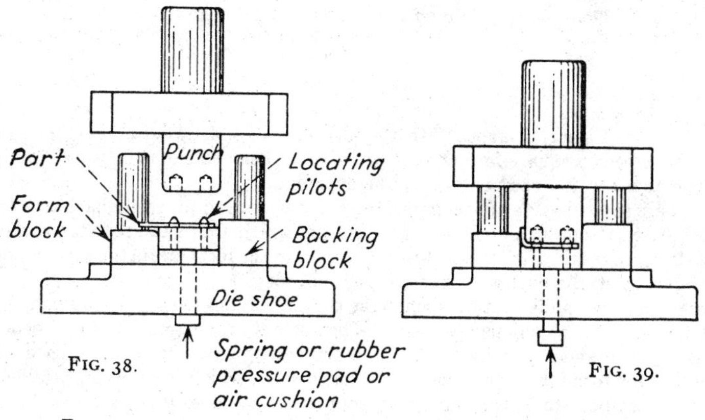

FIGS. 38 and 39. Pad form die aids in controlling springback.

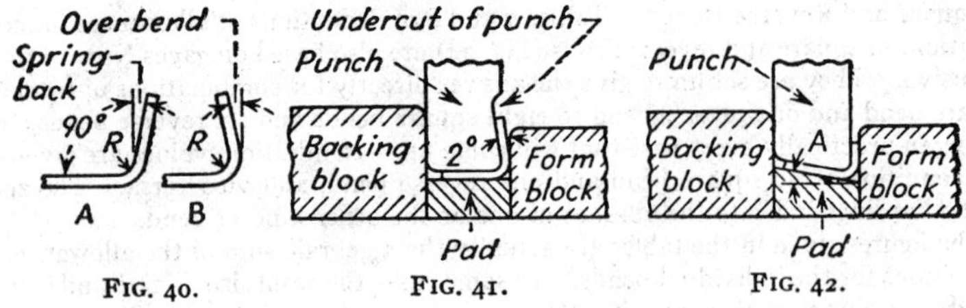

FIG. 40. FIG. 41. FIG. 42.

FIGS. 40 to 42. Overforming helps to overcome springback.

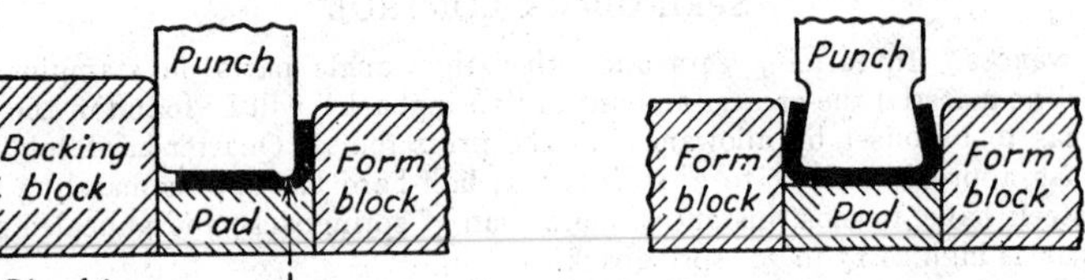

FIG. 43. Pinching sets metal.

FIG. 44. U-ing punch is undercut.

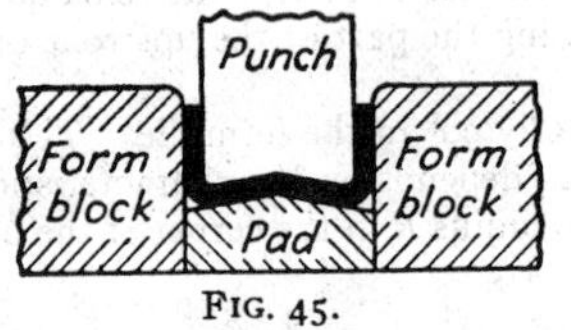

FIG. 45.

FIG. 45. Double-angle punch sets the form.

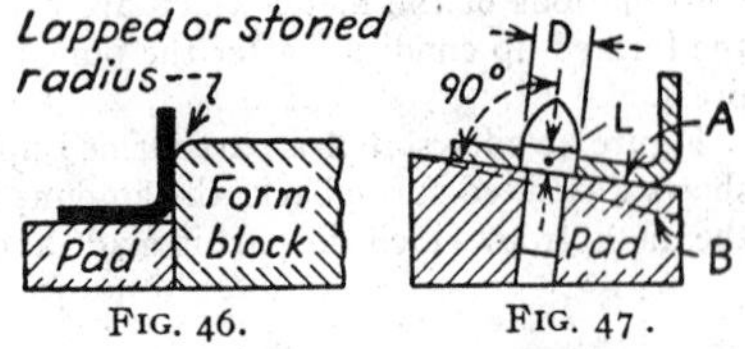

FIG. 46. FIG. 47.

FIGS. 46 and 47. Highly polished corners on die and a pilot help to control forming.

If the part had been "overbent" in the form die, as illustrated at *B* (Fig. 40), it would have been a square-formed part after being removed from the form die, since the springback would compensate for the overbend.

There are several ways of accomplishing a compensation for springback in a form die. Figure 41 shows an undercut of the punch, which allows the part to overbend. Since the springback, after the part is removed from the form die, is the same amount as the overbend, the part will have an exact 90° form.

Figure 42 shows a die that overbends the part by having the punch and pad of the form die ground with angular surfaces. The angle *A* is the amount of overbending and is to be found again by trial. Two-degree angular surface is again a good dimension for forming quarter-hard cold-rolled steel, which is sometimes called C, or No. 3, temper stock, and possesses approximately Rockwell 70 to 75 B scale hardness.

Figure 43 shows another method of getting square-formed parts. The bottom of the punch is ground and possesses a slightly higher surface in the corner, which gives the metal a greater "setting" during the forming, therefore destroying elasticity of the metal and consequent springback. Since with this method the structure of the metal loses certain desired strength qualities, this method should be eliminated as much as possible.

U-bends. The method of forming a square-corner U-bend is shown in Fig. 44 and is practically the same as in Fig. 41, using an undercut punch. Another U-bend method is shown in Fig. 45 and is largely the same as Fig. 42, the double angle being found by trial.

It is very important in the making of a good working form die that the corner over which the part is formed be highly polished. Figure 46 shows the radius, which should be lapped or stoned to a very highly polished surface, thus allowing the part to be formed without galling up or scratching. The polishing of the radius should be done in the direction of the bend and not in the lengthwise direction of the form block. In case of a U-form, it is important that the radii on both form blocks be the same. A small radius on one form block and a large radius on the other form block would cause a side strain on the part, as well as on the punch, allowing variations and inaccuracies in the form.

Parts are sometimes located on a form pad of a form die by pilots, and if the pad has an angular surface for a compensating forming operation, the pilot should be located at a vertical (90°) position to this surface, as shown in Fig. 47. Any appreciable tilting of the pilot caused by regrinding of the angular surface should be corrected by regrinding or reestablishing the pilot hole in the pad. This will put the pilot again in a vertical position to the angular surface.

The best forms, without breakage or cracks in the corner, are obtained with quarter-hard, or No. 3 (sometimes also called C temper), cold-rolled strip steel, which bends 150° "across" the grain, or at a sharp angle of 90° "with" the grain, without showing any marks of failure. Half-hard, or No. 2 temper, stock will allow a sharp bend "across" the grain, but "with" the grain only a bend with a large radius can be made without showing signs of failure. Hard, or No. 1 temper, stock is mostly used for flat work without forms. Any forming of this stock should always be done "across" the grain, and with a very large radius in the corner.

SHEAR ON DIES

Several factors make shear desirable in cutting dies. The principal factor is the permissible reduction in power required to operate the press.

Where shear is correctly applied, the pressure required to blank or cut off is reduced by one-quarter for metals thicker than ¼ in. When thinner stock is blanked, the pressure reduction is as large as one-third.

A common method of applying shear to a die producing a rectangular part is shown in Fig. 48. The face of the die is ground to a taper either side of center, to angle A, but the center is rounded to radius C to avoid a focal point for stock fracture. Angular height B should be twice the stock thickness for thin, light materials and the same as the stock thickness when 20-gage or over is run.

A better method of providing angular shear is shown in Fig. 49. Here, the die face is concave. With this condition, the punch cuts from the outside toward center. The corners are flat for a distance of ¼ to ½ in. in area D, depending on the size of the die. View X-X, Fig. 50, shows this flat area.

Rule for Shear. The prime rule to follow in application of shear is: If the blank which the punch pushes through is required, and the stock around the punch is scrap,

Shear Applied to Die

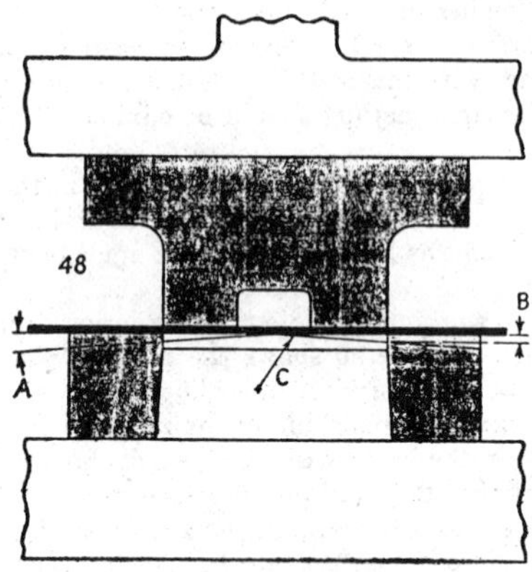

FIG. 48. Dimension *B* should equal twice the stock thickness for thin, light materials, and be same as the stock thickness for No. 20 gage or over.

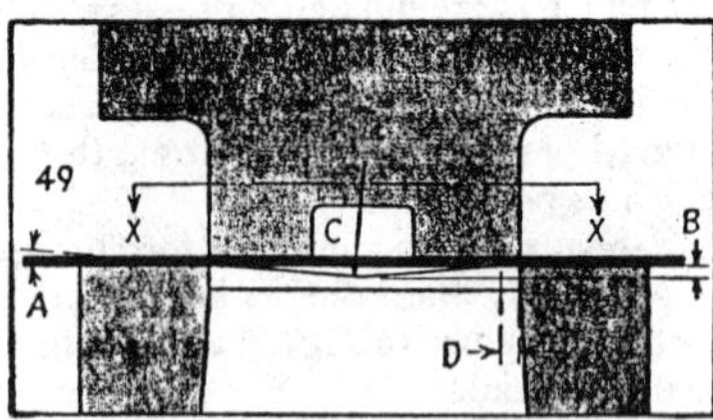

FIG. 49. When shear is applied so that the die face is concave, cutting progresses from outside to center.

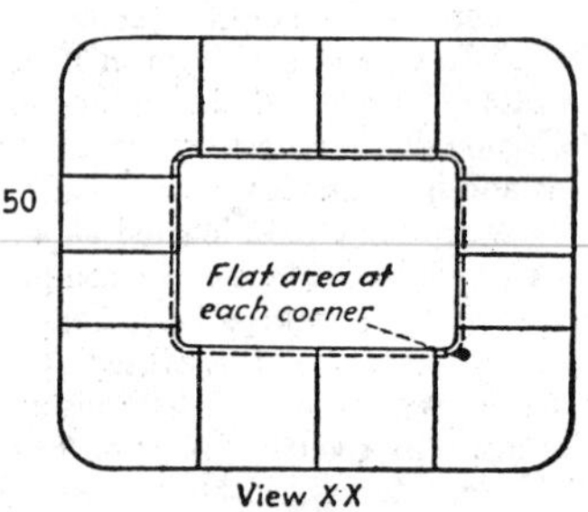

FIG. 50. Flat corners at *D* aid in starting the cutting at the corners.

Shear Applied to Punch

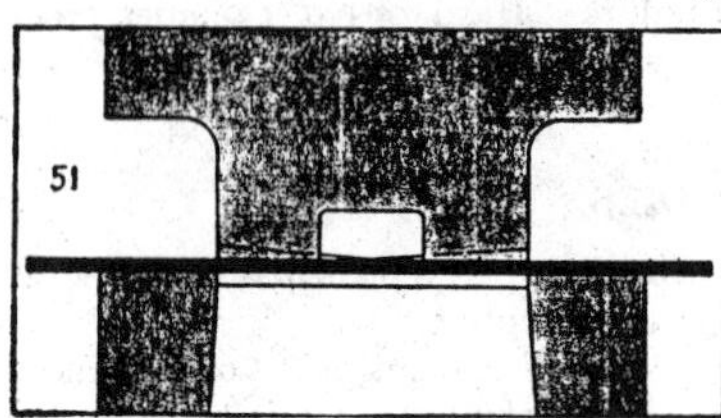

FIG. 51. By applying shear to the punch as shown the cutting action starts at the center.

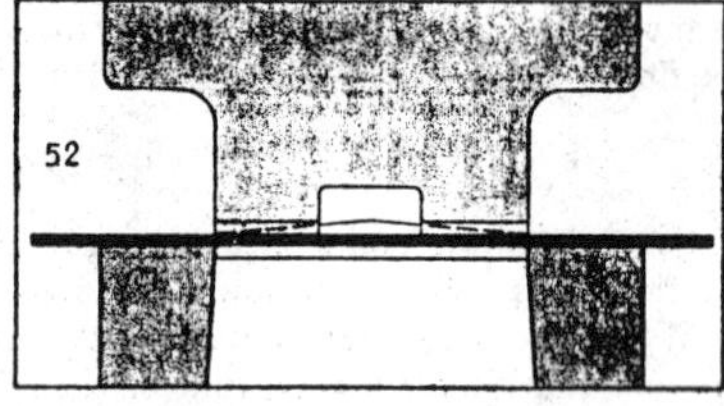

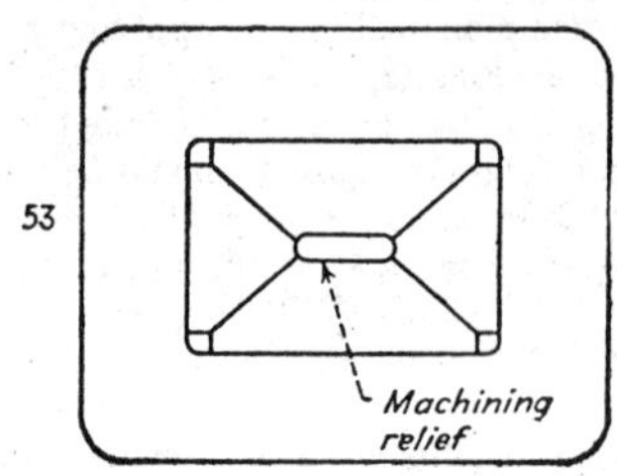

FIGS. 52 and 53. Outer corners of punch are made flat to cut first into strip.

apply shear to the die. If the blank under the punch is scrap, and the material around the punch is kept, apply shear to the punch.

When shear has been applied to either the punch or die member, the portion of the strip that bears on the sheared member will be distorted. Most blanks must be produced to a fair degree of flatness. It is therefore important to apply shear to the correct member, so that distortion will apply to the scrap portion of the strip.

Shear is applied to the punch in the same manner as to the die. Figure 51 illustrates the usual method.

A preferable method is shown in Fig. 52. Here, the outer corners cut into the strip first. The design is similar to the one outlined in Fig. 49, and the same rules apply. Figure 53 is a plan view of this punch.

Round punches and dies are given a shearing action by machining a series of waves around the periphery.

Stepped Punches. Where a number of holes must be punched out in a piece, stepping of the punches achieves the same effect as applying shear. The punches are stepped slightly more than the thickness of the strip.

PRESS-TOOL COMPONENTS

Punch and Die Sets

This American Standard was developed by the SAE, the Metal Cutting Tool Institute, the National Machine Tool Builders Association, and the ASME.

Types. Two series of punch and die sets are covered in this standard consisting of back-post sets and diagonal-post sets. The punch and die sets shown are of the conventional type with pressed-in guide posts and guide-post bushings; however, the standardized die areas may also apply to die sets having patented mountings for the guide posts and guide-post bushings.

Punch and Die Sets. A punch and die set is an assembly consisting of an upper member called the punch holder and a lower member called the die holder. The lower surface of the punch holder and the upper surface of the die holder are those on which the punch and die details of a finished punch-press tool are mounted. In use, the die holder is clamped to the bed or bolster plate of the punch press, while the punch-holder shank is clamped in the clamping hole in the punch-press ram. The mating guide posts and bushings of an assembled die set assist in maintaining tool alignment during die setting and the operation of the tool in the punch press. The size of punch and die set is generally designated by its available die area.

Back-post Sets. Back-post punch and die sets have two rear guide posts and have a square or rectangular die area, and with the exception of the sets having an extra narrow die area, the maximum available round die area is also shown. The respective die holders have slotted clamping flanges on two sides with bossed surfaces machined to the desired height to accommodate the clamping screws.

Diagonal-post Sets. Diagonal-post sets differ from back-post sets only in that they have one rear guide post and one front guide post located on a diagonal center line in relation to the die area.

ROUND SERIES. These punch and die sets are a combination of a round die holder and a rectangular punch holder. This design permits their use either as a diagonal-post or center-post set. The available die area may be either round or rectangular, provided the maximum diameter or width of the die member clears the guide-post bushings. The die holder is furnished with an unslotted flange all around machined to the proper height. This continuous flange facilitates clamping in any position with the clamp and cap screws fitting tapped holes in the bolster plate.

RECTANGULAR SERIES. These punch and die sets have a rectangular die holder and rectangular punch holder. The front guide post is always in the left-hand

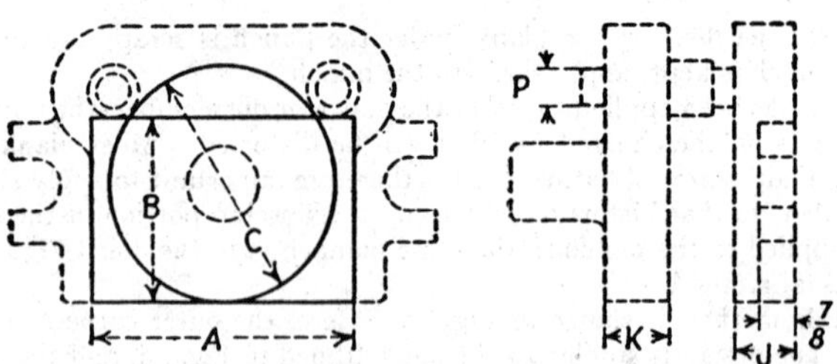

TABLE 11. DIMENSIONS OF BACK-POST DIE SETS

Die Area			Thickness				Min Guide-post Dia *P*
Right to Left *A*	Front to Back *B*	Dia *C*	Die Holder *J* From	Die Holder *J* To	Punch Holder *K* From	Punch Holder *K* To	
3	3	3	1	1¼	1	...	¾
4	4	4	1⅜	1¾	1¼	...	1
4	6	....	1½	2¾	1¼	2¼	1
5	4	....	1⅜	1¾	1¼	...	1
5	5	5	1½	2	1¼	1¾	1
5	8	....	1½	3	1¼	2¼	1
6	3	....	1½	2	1¼	1¾	1
6	4	5	1½	2¾	1¼	2¼	1
6	6	6½	1½	2½	1¼	2¼	1
6	9	....	1½	3¼	1¼	2¼	1¼
7	5	5¾	1½	3	1¼	2¼	1
7	7	7½	1½	2½	1¼	2¼	1
7	10	....	1⅝	3¼	1⅜	2¼	1¼
8	4	....	1½	2½	1¼	2¼	1
8	6	7	1½	3	1¼	2¼	1
8	8	8½	1½	2½	1¼	2¼	1
9	12	....	1¾	3½	1½	2¼	1½
10	5	....	1½	2½	1¼	1¾	1¼
10	7	....	1⅜	2¾	1⅜	2¼	1¼
10	10	10	1⅝	2¾	1⅜	2¼	1¼
10	14	....	1⅞	3¾	1⅝	2¾	1½
11	9	10	1¾	3½	1½	2¼	1¼
12	4	....	1¾	2¼	1½	2	1¼
12	6	....	1½	2½	1½	2	1¼
12	12	12½	1¾	3½	1¾	2¼	1½
12	16	....	2	3¾	1¾	2¾	1½
14	8	....	1¾	3¼	1⅝	2¾	1½
14	10	11¼	1¾	3¼	1⅝	2¾	1½
14	14	14	1¾	3¼	1⅝	2¼	1½
15	5	....	1½	2½	1½	2	1½
15	9	....	1½	2½	1½	2	1½
18	8	....	1½	2½	1½	2	1½
18	10	....	1½	2½	1½	2¼	1½
18	14	15	2	3	1¾	2¼	1½
18	16	17	2	3	1¾	2¼	1½
20	5	....	1¾	2½	1½	2	1½
22	6	....	1¾	2½	1½	2¼	1½
22	12	....	2	3	1½	2	1½
25	7	....	1¾	3	1½	2¼	1½
25	14	....	1¾	3	1½	2¼	1½

TABLE 12. SHANK DIAMETERS AND LENGTHS
Back-post Die Sets

Dia	1½	1 9/16*	2	2½	3
Length	2⅛	2⅛	2⅞	2⅞	2⅞

All dimensions are given in inches.
Allowable tolerance on diameter shall be +0.000, −0.002 in.
Allowable tolerance on length shall be ±0.010 in.
Lengths shown are standard but shorter shanks will be furnished when specified.
Steel shanks screwed in place (to be locked by user) will be furnished instead of integrally cast shanks when specified.
Back-post die sets may be obtained without the shank.
* Not a preferred shank diameter.

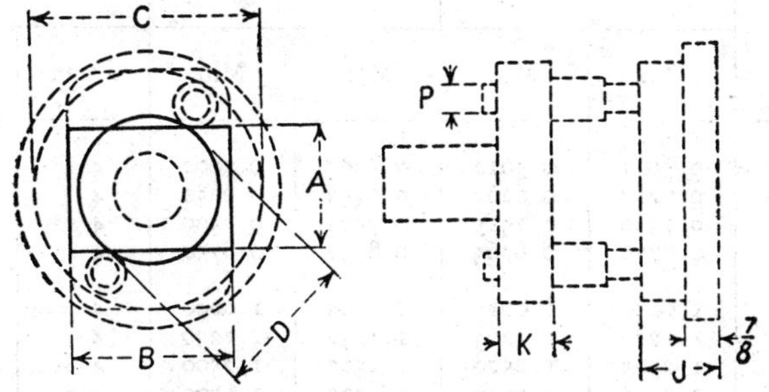

TABLE 13. DIMENSIONS OF ROUND DIAGONAL-POST DIE SETS

Die Area			Die-holder Dia	Thickness		Min Guide-post Dia
		Dia				
A	*B*	*D*	*C*	*J*	*K*	*P*
1¾	3½	2¾	5	1¾	1¼	½
2¼	4½	3½	6	1¾	1¼	⅝
2¾	5½	4	7	2	1½	¾
3½	7	5¼	9	2	1½	1
4½	9	7	11	2	1½	1⅛

Back-post die set tolerances apply.

TABLE 14. SHANK DIAMETERS AND LENGTHS
Round Diagonal-post Die Sets

Dia	1½	1 9/16*	2
Length	2⅛	2⅛	2⅞

Lengths shown are standard but shorter shanks will be furnished when specified.
* Not a preferred shank diameter.

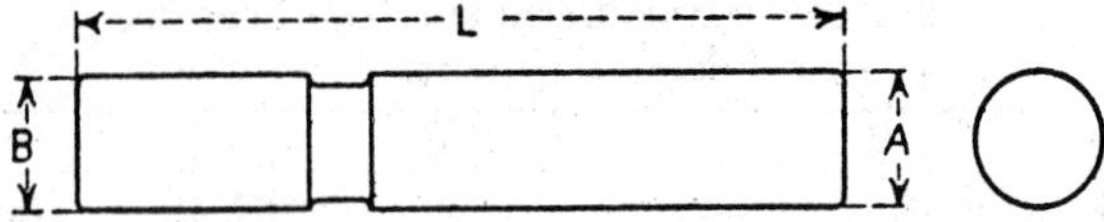

TABLE 15. DIMENSIONS OF GUIDE POSTS—REGULAR SERIES

Nominal Dia	Dia				Length L	
	Ground B		Lapped A			
	Max	Min	Max	Min	From	To
½	0.5017	0.5012	0.5002	0.5000	4	4½
⅝	0.6267	0.6262	0.6252	0.6250	4	4½
¾	0.7520	0.7515	0.7502	0.7500	4	5
⅞	0.8770	0.8765	0.8752	0.8750	4	8
1	1.0020	1.0015	1.0002	1.0000	4	9
1⅛	1.1270	1.1265	1.1252	1.1250	4	9
1¼	1.2525	1.2520	1.2502	1.2500	4½	12
1½	1.5025	1.5020	1.5002	1.5000	4½	12
1¾	1.7525	1.7520	1.7502	1.7500	6	14
2	2.0025	2.0020	2.0003	2.0000	6	20
2½	2.5030	2.5025	2.5003	2.5000	8	20
3	3.0030	3.0025	3.0003	3.0000	8	20

NOTE: Longer lengths of guide posts than those tabulated in this standard are available. Oil groove in the guide post may be optional with the supplier.

All dimensions are given in inches.

Material: This standard does not specify the material for guide posts. They are generally made from X1314 steel or equivalent, heat-treated Rockwell 62 to 65 C.

corner of the die set. The die holder is generally furnished with flanges machined on opposite sides to the proper height to provide means of clamping to the bolster plate.

Punch-holder Shank. The punch-holder shank is that portion of the punch holder that is clamped in the ram of the punch press. Punch holders for small sets generally have a cylindrical shank which is an integral part of the holder. In some of the larger sets removable punch shanks are screwed in a threaded hole in the punch holder and, after being assembled, are generally keyed in place to prevent turning and thus become an integral part of the punch holder.

Guide Posts. Guide posts are cylindrical in shape, ground and lapped to size, and furnished without oil grooves. They are made to be a press fit in the die holder and a working fit in the guide-post bushings. Guide posts generally range in diameter from ½ to 3 in. as required by the several types and sizes of punch and die sets, and for each diameter a selected series of lengths is provided.

Guide-post Bushings. Guide-post bushings are cylindrical in shape with a longitudinal hole furnished with circular oil grooves for retaining lubricating oil. The outside diameter is a press fit in the punch holder and the inside diameter a working fit on the guide post. The internal diameter may be relieved for the length of the portion which is pressed into the punch holder in order to compensate for any closure that may occur when the bushing is pressed into the hole. Three types of guide-post bushings, "regular," "long," and "shoulder," are provided as optional designs.

Flanges. The flange portion of a punch and die set is an extended ledge of metal at the die-holder base extending beyond the die area and machined to a convenient thickness to facilitate clamping the die holder to the punch-press bolster plate.

Punch-press Tool Height. The punch and die set height or "shut height" of the tool, when assembled, is measured from the lower surface of the die holder to the upper surface of the punch-holder top when the tool is closed as in the operating position. The length of the guide posts is determined to suit this "shut height" so as conveniently to enter the guide bushings in the die holder, so that the upper ends of the guide posts are a reasonable distance below the upper face of the punch holder, in order to compensate for subsequent tool sharpenings. The maximum "shut height" of the tool to suit the die space of the punch press is determined by measuring from the upper surface of the bolster plate on the bed to the ram with the stroke down and the adjustment up.

Die Area. Die area is the surface on the die holder that may be utilized for the mounting of tool members in a punch and die set.

Tolerances for Die Sets

	6 × 6 and Under	6 × 6 to 12 × 12	12 × 12 to 20 × 20	Over 20 × 20
Die Holders				
Parallelism—top and bottom, in. per ft	0.001	0.001	0.001	0.0015
Thickness	$+\frac{1}{64}$ $-\frac{1}{32}$	$+\frac{1}{64}$ $-\frac{1}{32}$	$+\frac{1}{32}$ $-\frac{1}{16}$	$+\frac{1}{32}$ $-\frac{1}{16}$
Punch Holders with Solid Shanks				
Parallelism, in. per ft	0.0015	0.0015	0.0015	
Thickness	$\pm\frac{1}{32}$ in.	$\pm\frac{1}{32}$ in.	$+\frac{3}{16}$ $-\frac{1}{16}$	
Punch Holders without Shanks				
Parallelism, in. per ft	0.001	0.0015	0.0015	0.002
Thickness	$\pm\frac{1}{32}$	$\pm\frac{1}{32}$	$\pm\frac{1}{16}$	$\pm\frac{1}{16}$

Tolerance for parallelism applies to die space or contact surfaces.

Punch or Die Holders. Flange thickness: not machined on top, $\pm\frac{1}{16}$ in.; machined, $+\frac{1}{64}$ or $-\frac{1}{32}$ in.

Partial Assembly. Guide posts, with punch holder removed, must be parallel and at right angles to bottom surface of die holder within 0.001 in. in 6 in.

Guide bushings must be parallel and at right angles to top surface of punch holder within 0.001 in. in 6 in.

Differences in center distances of guide posts and guide bushings shall not exceed 0.002 in. on any set.

Shank diameter tolerance, +0.000 or −0.002 in.

TABLE 16. PIERCING PUNCHES

Dia Punch D, In.	A	B	C	E	F
0–1/4	5/16	0.251	3/32	5/8	1 3/4
5/16	7/16	0.3135	3/32	5/8	1 3/4
3/8	1/2	0.376	3/32	5/8	1 3/4
7/16	9/16	0.4385	3/32	5/8	1 3/4
1/2	5/8	0.501	3/32	5/8	1 3/4
9/16	11/16	0.5635	3/32	5/8	1 3/4
5/8	3/4	0.626	3/32	5/8	1 3/4
11/16	13/16	0.6885	3/32	5/8	1 3/4
3/4	7/8	0.751	3/32	5/8	1 3/4

TABLE 17. PUNCHES HELD BY SETSCREWS

D = Dia, In.	A	B	C	E	F	G	H
1/4	1/8	5/8	1/2	1	1/2	9/16	3/8
3/8	1/8	5/8	1/2	1	1/2	9/16	3/8
1/2	1/8	3/4	1/2	1	1/2	9/16	3/8
9/16	1/8	3/4	5/8	1 1/4	5/8	9/16	3/8
5/8	1/8	7/8	5/8	1 1/4	5/8	9/16	3/8
3/4	1/8	7/8	5/8	1 1/4	5/8	9/16	3/8
1/8	1/8	1	3/4	1 1/2	3/4	5/8	7/16
1	1/8	1	3/4	1 1/2	3/4	5/8	7/16
1 1/8	1/8	1 1/8	3/4	1 1/2	3/4	5/8	7/16
1 1/4	1/8	1 1/4	1	1 3/4	7/8	11/16	1/2
1 1/2	1/8	1 1/2	1	1 3/4	7/8	11/16	1/2
1 3/4	1/8	1 3/4	1 1/4	1 7/8	7/8	13/16	5/8
2	1/8	2	1 3/8	1 7/8	7/8	13/16	5/8

Assembled Sets. Parallelism of top surface of punch holder and bottom surface of die holder shall be 6 by 6 and under, 0.002 in. per ft; 6 by 12 by 12, 0.0025 in. per ft; 12 by 2 to 20 by 20—0.0025 in. per ft, and over 20 by 20—0.003 in. per ft.

Types of Punches

Table 16 covers a series of punches for various sizes of pierced holes. The punch bodies at a point immediately under the head are 0.001 in. large for fitting snugly in the punch plate as indicated. Such punches are readily ground to size and fitted to their places.

The punches in Table 17 are useful where they are close to the edge of the punch holder and therefore reached by setscrews. The sizes given range from 1/4 to 2 in., and they are suited for blanking as well as piercing operations.

A substantial type of punch for either piercing or blanking is covered in various

Table 18. Punches Held by Screw from Top

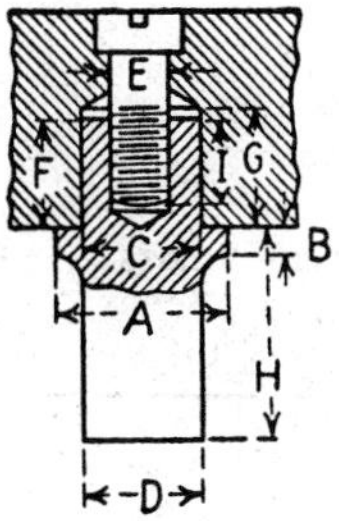

D = Dia of Punch	A	B	C	E	F	G	H	I
1/4	5/8	1/8	1/2	5/16	1	1 1/8	1 1/8	5/8
3/8	5/8	1/8	1/2	5/16	1	1 1/8	1 1/8	5/8
1/2	3/4	1/8	1/2	5/16	1	1 1/8	1 1/8	5/8
9/16	3/4	1/8	5/8	3/8	1	1 1/8	1 1/8	5/8
5/8	7/8	1/8	5/8	3/8	1	1 1/8	1 1/8	5/8
3/4	7/8	1/8	5/8	3/8	1	1 1/8	1 1/8	5/8
7/8	1	1/8	3/4	3/8	1	1 1/8	1 1/8	5/8
1	1 1/8	1/8	3/4	3/8	1	1 1/8	1 1/8	5/8
1 1/8	1 1/4	1/8	3/4	3/8	1	1 1/8	1 1/8	5/8
1 1/4	1 3/8	5/32	1	3/8	1	1 1/8	1 1/8	5/8
1 1/2	1 5/8	5/32	1	3/8	1	1 1/8	1 1/8	5/8
1 3/4	1 7/8	5/32	1 1/4	3/8	1	1 1/8	1 1/8	5/8
2	2 3/16	5/32	1 3/8	3/8	1	1 1/8	1 1/8	5/8

Table 19. Pilots for Progressive Blanking Punches

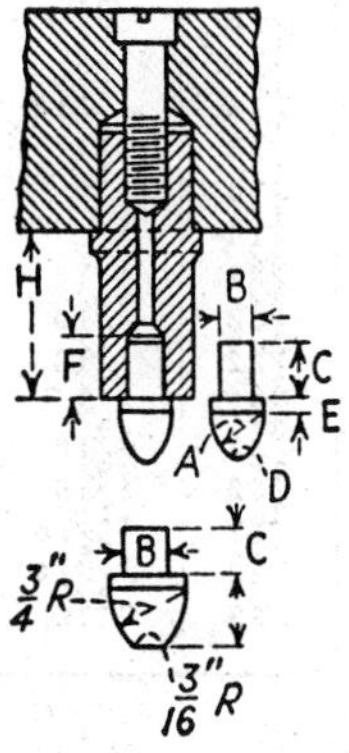

A, In.	B	C	D	E	F	H
3/16	1/8	5/16	1/32	1/16	7/16	1 3/16
1/4	3/16	7/16	1/32	1/16	9/16	1 3/16
5/16	1/4	1/2	1/16	1/16	5/8	1 3/16
3/8	1/4	1/2	1/16	1/16	5/8	1 3/16
7/16	1/4	1/2	3/32	1/16	5/8	1 3/16
1/2	1/4	1/2	1/8	1/16	5/8	1 3/16
5/8	5/16	9/16	5/32	1/16	11/16	1 3/16
11/16	5/16	9/16	Flat end	1/16	11/16	1 3/16
3/4	3/8	5/8	Flat end	1/16	3/4	1 3/16
7/8	3/8	5/8	Flat end	1/16	3/4	1 3/16
1	7/16	11/16	Flat end	1/16	13/16	1 3/16
1 1/4	1/2	3/4	Flat end	1/16	7/8	1 3/16

sizes by Table 18. This construction provides for the securing of the punch in its holder by means of a fillister-head screw.

Pilots. The pilot is inserted in the blanking punch and is made to a long radius, equal at least to the diameter of the pilot body. This question of pilots is of interest as they are so extensively used in connection with progressive dies. Table 19 has been laid out to cover pilot dimensions for the type of punch which is held by a fillister-head screw tapped in from the top of the punch holder. The same dimensions apply also to pilots for the other classes of punches whose proportions are given

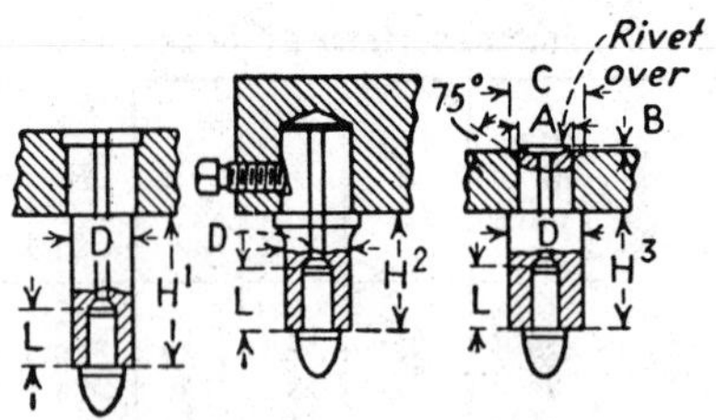

TABLE 20. PROGRESSIVE BLANKING PUNCHES WITH PILOTS
For Pilot Dimensions Refer to Table 19
Make L for All Punches Equal to Pilot Shank + 1/8 In.

D = Dia, In.	H^1	H^2	H^3	A	B	C
1/4	1 13/16	1 3/16				
5/16	1 13/16	1 3/16				
3/8	1 13/16	1 3/16				
7/16	1 13/16	1 3/16				
1/2	1 13/16	1 3/16				
9/16	1 13/16	1 3/16				
5/8	1 13/16	1 3/16	1 3/16	1/2	1/16	5/8
11/16	1 13/16	1 3/16	1 3/16	1/2	1/16	5/8
3/4	1 13/16	1 3/16	1 3/16	1/2	1/16	5/8
7/8	.....	1 3/16	1 3/16	9/16	1/16	11/16
1	.....	1 3/16	1 3/16	5/8	1/16	3/4
1 1/8	.....	1 3/16	1 3/16	3/4	1/16	7/8
1 1/4	.....	1 3/16	1 3/16	7/8	1/16	1
1 1/2	.....	1 3/16	1 3/16	1 1/8	3/32	1 5/16
1 5/8	.....	1 3/16	1 3/16	1 1/4	3/32	1 7/16
1 3/4	.....	1 3/16	1 3/16	1 5/16	3/32	1 7/16
1 7/8	.....	1 3/16	1 3/16	1 5/16	3/32	1 1/2
2	.....	1 3/16	1 3/16	1 3/8	3/32	1 9/16
2 1/8	.....	....	1 3/16	1 3/8	3/32	1 9/16
2 3/8	.....	....	1 3/16	1 1/2	3/32	1 11/16
2 5/8	.....	....	1 3/16	1 3/4	3/32	1 15/16
2 7/8	.....	....	1 3/16	2	3/32	2 3/16

in Tables 16, 17, and 20, and in all cases the pilots should be ground or otherwise finished on the stem or shank to a press fit in the ends of the blanking punches.

Ordinarily the pilots can be made of drill rod. In the smaller sizes, as indicated in the tables, the point of the pilot is rounded to a definite radius given in column D, Table 19. For larger sizes of pilots, the bottom end is flat, reducing the length accordingly, and the corners are rounded to a radius of 3/16 in. These proportions are varied, where necessary, but for a wide range of work, they have proved satisfactory.

Piloting in Progressive Dies

There are two methods of piloting in progressive dies. Direct piloting consists of piloting in holes punched in the part at a previous station. This is the ideal method

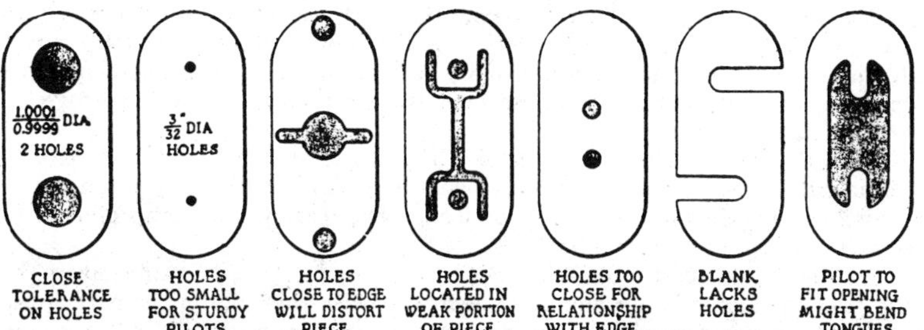

FIG. 54. Direct piloting in holes punched at a previous station assists location in subsequent operations, but is impractical for thin stock or the cases illustrated.

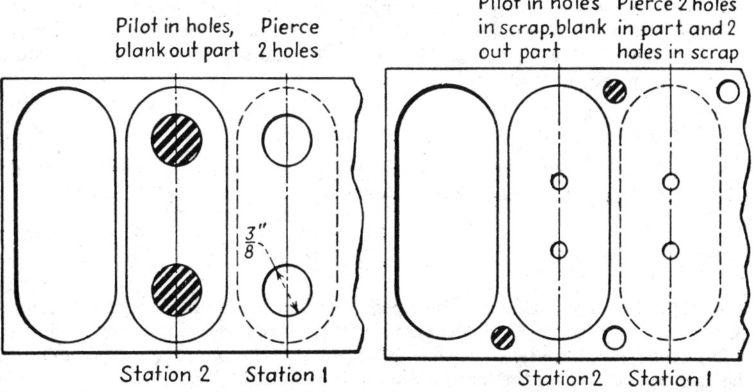

FIG. 55. Large, properly located pierced holes are ideal for piloting while blanking.

FIG. 56. Indirect piloting consists of piercing holes in the scrap strip and locating from these holes with pilots (cross-hatched) in a subsequent operation.

for locating the part in subsequent die operations. Unfortunately, ideal conditions may not exist, and in such cases indirect piloting must be used to achieve the desired results of part accuracy and high production speed. Indirect piloting consists of piercing holes in the scrap strip and locating these holes with pilots at later operations.

There are several reasons for avoiding direct piloting (Fig. 54). The chief one is: The stock is too thin (especially so with holes of small diameter). In fact, stock can be so thin that use of a progressive die is impractical, and it is necessary to resort to the slower compound die, despite its cost.

In the indirect piloting method, it is possible to use pilots of considerably greater diameter than might be true if holes in the part are used for piloting. The greater the diameter of the pilot, the less chance there is of distortion of either the strip or the pilot.

The second objection to the use of direct piloting arises when holes in the part must be held to great accuracy. It is possible for the pilots to destroy the hole size in their effort to pull heavy material strip for part location.

When holes in the part are too close to the edges, the weak outer portions of the part are likely to distort upon contact with the pilots, instead of the strip moving to the correct position. This possibility is often overlooked in planning a progressive die and gives rise to subsequent runs of scrap parts and expensive die alterations.

A similar problem exists when the part holes are located in a weak portion of the inside area of the part. Here, there is great possibility of the part buckling before the pilots can pull the material strip along. The remedy again is to pilot in the scrap strip with two sturdy pilots.

Put Pilots Far Apart. To achieve accurate part location, it is imperative that the pilots be placed as far apart as possible. When the holes in the part are set too close together they will not be very accurately located with respect to the outer edges of the part. Indirect piloting allows the pilot holes to be placed as close to the edges of the strip as is consistent with strength and proper strip location.

When no holes exist in the part, it is necessary to pilot in the strip. If irregular slots are punched out at a previous station, the strip must be located subsequently by pilots that enter holes pierced in the strip at the same time the slots were cut. The part is then blanked out of the strip. The pilots insure accurate location of the slots with the outer edges of the part.

Stripper Springs

A satisfactory formula for calculating the pressure required to strip the work from the punch, assuming normal punch clearance, is

$$P = \frac{L \times T}{0.00117}$$

where L = length of cut (in piercing operations, this is the sum of the perimeters of all the perforator faces); T = thickness of stock; P = pressure lb, to effect stripping.

In shaving operations, or on small subpresses (where no clearance is allowed), the figure calculated by the formula should be doubled at least.

This formula is for "full stripping." In progressive dies, where only strips are to be run, and bridges, or weak spots, exist in the scrap strip, as at A, Fig. 57, the calculated figure may be materially reduced, depending on the number and length of the weak spots. But the value of P should never be reduced in applications where there is any possibility of scrap pieces from other operations being run through the die.

Three classes of deflection must be taken into consideration (Fig. 58). These are: (1) initial deflection to store up stripping energy; (2) working deflection, which is a fixed value, depending on the stock thickness; and (3) grinding deflection, which is normally ¼ in. but may be reduced to as little as ⅛ in. where space is limited.

Standard springs are rated in pounds per ⅛-in. deflection. Therefore, if a spring 1 in. dia by 12 in. long is selected, and initial deflection is ¼ in., the total stored energy per spring will be twice the listed amount for that size of spring.

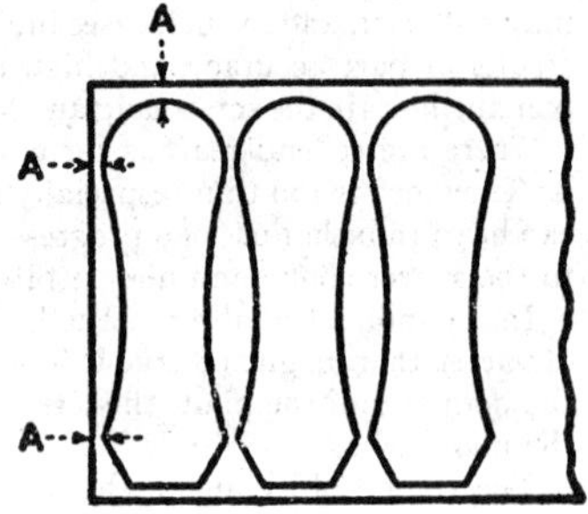

FIG. 57. Weak sections affect stripping.

To this sum is added the pressure for the first ⅛ in. of working deflection. Should the working deflection be less than ⅛ in., say 1/16 in., only one-quarter of the rated pressure could be used in the circulations.

EXAMPLE:

Pressure per ⅛-in. deflection = 20 lb
¼-in. initial deflection, (2 × 20) = 40 lb
1/16-in. working deflection, (¼ × 20) = 5 lb
¼-in. grinding deflection, (2 × 20) = 40 lb
Total = 85 lb

Only the first ⅛ in. of working deflection can be used in the computation of stripping pressure.

The number of springs required is found by dividing the total pressure required to effect stripping by the "pressure per spring." If the number arrived at is out of proportion to the die, use fewer larger springs, or perhaps, more springs of a smaller size. Or pressure may be increased by applying more initial deflection.

Rules for Selecting Springs

1. For high-speed press work, both "medium" and "high-pressure" springs should never be deflected more than one-fourth their free length.

2. For heavy, slow-moving presses, use "medium-pressure" springs with a total deflection not to exceed three-eighths of the free length.

Spring strippers have many advantages over solid strippers for many classes of die work:

1. In blanking, the strip is flattened and all waves are eliminated. Therefore, a more accurate part is produced.

2. The stock is held firmly between the die and pressure pad, thus insuring a cleaner fracture in piercing and blanking and reducing to a minimum the rounded edges caused by a slight drawing of the metal—typical of thin-stock work.

3. Reduced setup time. Since there are no knockouts or extra attachments to connect and adjust, setup time is held to a minimum. An incidental advantage is the fact that there are no loose parts to get lost.

4. Since stripping is accomplished simultaneously with punch retraction, the danger of breaking slender piercing punches and frail die inserts is minimized. Also, the uneven wear on small punches, caused by use of solid strippers, is eliminated.

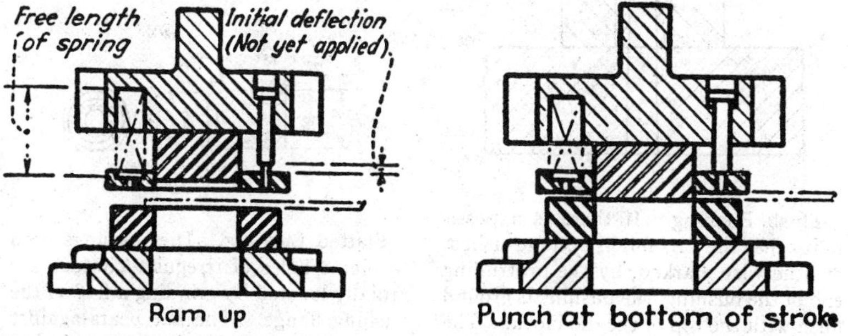

FIG. 58. Three deflections—initial, working, and grinding—must be considered in applying stripper springs.

Applications of Die Bushings

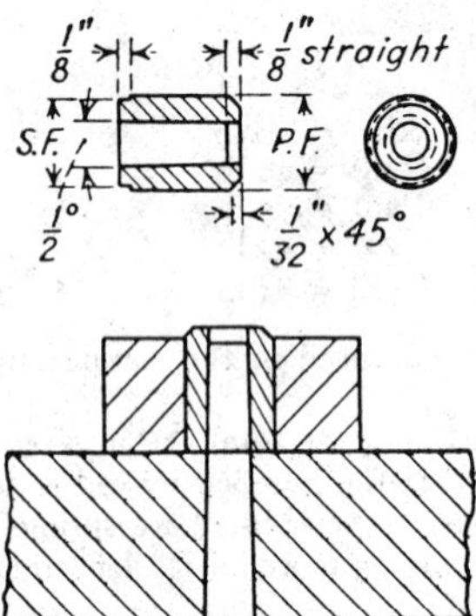

Straight Bushing. The straight, press-fit type of bushing is commonly used, particularly for large dies, to conserve tool steel. Should the die hole become chipped, replacement of the bushing is simple and economical.

The bushing is made slip fit for approximately ⅛ in. at its lower end. This assures correct alignment while pressing, an important feature for accurate center-to-center distances between holes. The inside of the hole is ground straight for ⅛ in.; the remainder is given ½° included angle for correct slug disposal.

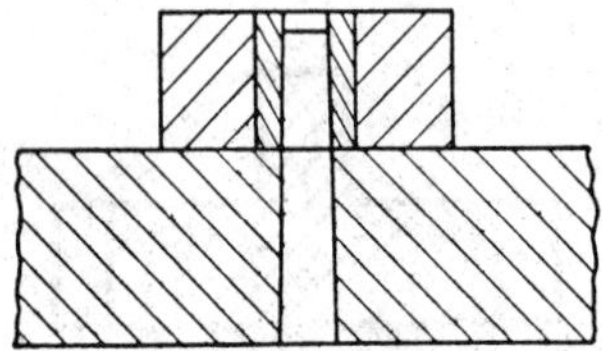

Flush Bushing. If there is a possibility that the part might become dished, or otherwise marked, by the protruding end of the bushing, the bushing is ground flush with the top of the die block. The flush-type bushing is commonly used in secondary-operation dies for formed and drawn parts; also, where heavy spring strippers are used to hold down the blank or strip. Otherwise it is made in the same manner as the straight type.

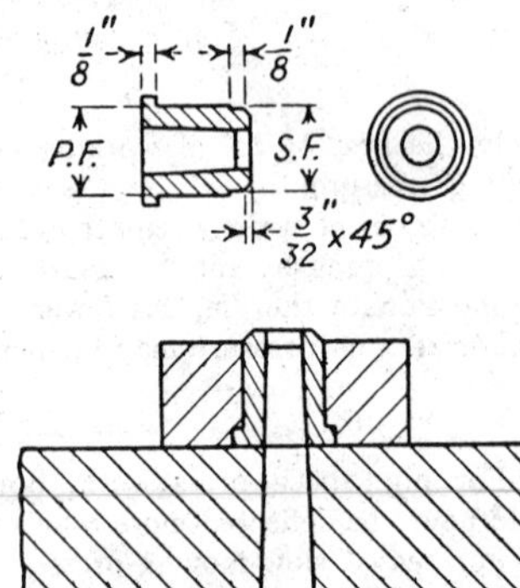

Flange Type. For heavy cuts through thick stock, the flange-type bushing provides a greater seat area, with less possibility of the bushing sinking into the die holder under the cutting pressure. It is also used where upward pressure exists. Such pressure might eventually raise the bushing if no positive means, such as a shoulder, were provided to retain it. Since this type of bushing is pressed in from the bottom, the ⅛-in. slip-fit portion is applied to the top of the bushing for accurate starting.

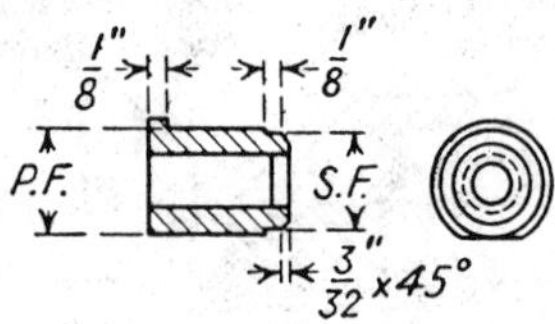

Flatted Bushing. Die bushings used to pierce holes of irregular contour are radially located by grinding a flat on the bushing flange. The flat bears against one wall of a milled slot in the die block. This positively keeps the bushing from

Applications of Die Bushings (*Continued*)

turning. Because there are no frail parts or dowels to become loose and drop out, this construction is recommended for first-class dies.

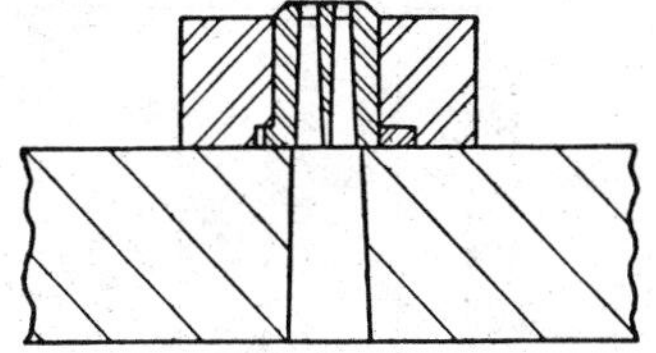

Duplex-hole Bushing. Small holes close to each other may be incorporated in a single bushing. In this case, a key located in an end-milled slot in the die block locates the bushing radially. This method is often used for multiple-station progressive dies where space is limited. Where more than one hole is incorporated in a bushing, it should be made of a nondeforming, oil-hardening tool steel.

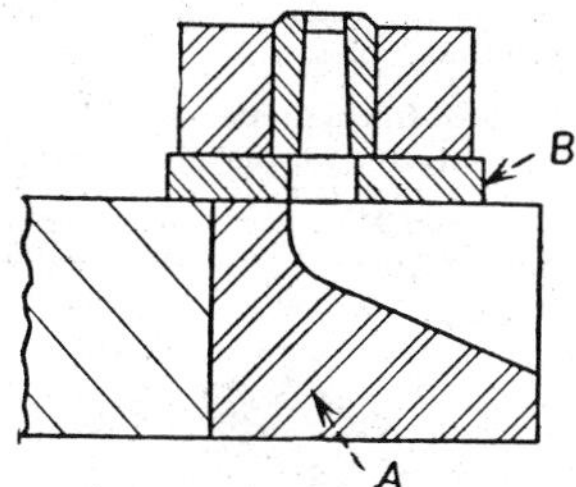

Slug Channels. When laying out large dies, slugs must often be channeled out through the side of the die holder. This happens when the hole in the bolster plate is small for the size of the die set, and the solid portion of the bolster is directly under the die bushing. A slot is milled in the side of the die holder and a spacer *A*, fastened in this slot, deflects slugs to the side of the die. A hardened plate *B* supports the bushing.

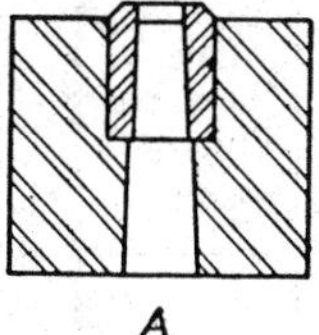

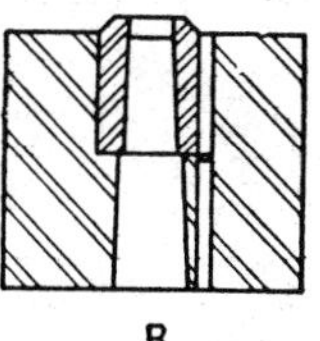

Bushings in Die Block. When the die has a thick section, the bushings are pressed in counterbored holes in the die block, as shown at *A*. The through hole in the die block is relieved to a ½° angle, as is the hole in the bushing. This prevents possible jamming of slugs. At *B*, the hole in the bushing is of irregular contour. Hence, the bushing is kept from turning by a "dutchman," which is pressed half in the bushing and half in the die block.

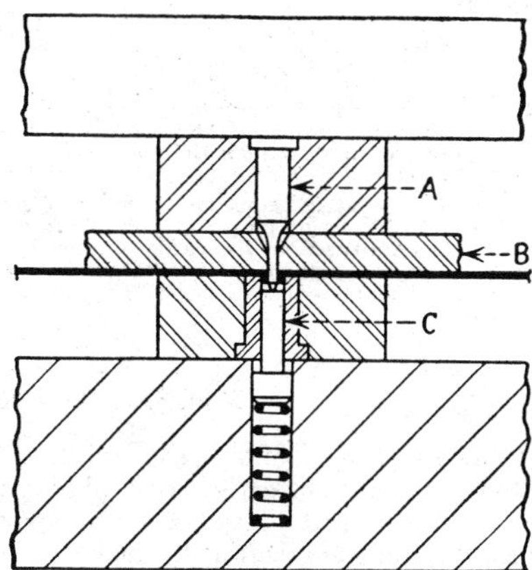

Extruded Holes. When holes in sheet-metal parts are extruded by a cone-pointed extruding punch *A*, spring stripper *B* first descends and securely holds the strip. Knockout pin *C* raises the extruded portion for further movement of the strip.

TABLE 21. RATED CAPACITIES, OR BOTTOM-SLIDE PRESSURE, OF CRANKSHAFT PRESSES

Rated Capacity, Tons	Crank-shaft Dia,* In.	Over-hanging Pin Dia,† In.	Rated Capacity, Tons	Crank-shaft Dia,* In.	Tie-rod Dia,‡ In.	Over-hanging Pin Dia,† In.
3	1	1⅛	43	3½	2	4
4	1⅛	...	50	3¾	2	4½
4½	1¼	1½	56	4	2¼	
6	1⅜	...	65	4¼	2¼	
7½	1½	1¾	71	4½	2½	5
9	1⅝	...	80	4¾	2½	
10½	1¾	2	86	5	2¾	5½
12	1⅞	...	106	5½	3	6
14	2	2¼	126	6	3¼	7
16	2⅛	...	150	6½	3½	
18	2¼	2½	180	7	3¾	
20	2⅜	...	215	7½	4	
22	2½	2¾	255	8	4½	
24	2⅝	...	300	8½	4¾	
26½	2¾	3	345	9	5	
28	2⅞	...	400	9½	5½	
31½	3	...	450	10	5¾	
37	3¼	3½				

* Diameter of crankshaft in single- or double-crank, open-frame presses.
† Diameter of overhanging pin in end-wheel presses.
‡ Diameter of tie rods in one- or two-point suspension, closed-frame presses.

Pressure capacity figures are given for all single-action presses and for the inner, or drawing, slide of double-action presses. Crankshaft diameters are measured at main bearings. These capacities are based on standards adopted by the Committee on Metalworking Machinery of the Technical Committee on Standard Commodity Classification, Division of Statistical Standards, Bureau of the Budget, Washington, D.C.

MECHANICAL-PRESS SELECTION

Rated capacity of a mechanical press is the pressure that the slide is expected to exert at the bottom (lower dead point) of the stroke.

For crankshaft presses, bottom capacities of the slide are determined by the diameter of crankshaft at the main bearings (see Table 21)

In gear-eccentric presses and in all inclosed presses rated bottom capacities are usually indicated on the name plates. If not so identified, tie-rod diameters may be used.

End-wheel presses are rated by the diameters of the overhanging pins.

Presses intended for work near the bottom of the slide stroke are selected according to their rated pressure capacities.

Slide-pressure Fluctuation. Presses starting their work at some distance above the bottom of their slide stroke are selected according to their midstroke pressure capacity. This is done because in mechanical power presses the slide pressure fluctuates during its working stroke. These fluctuations are due to variations in the angular position of the crankpin or the gear eccentric at various distances of the slide from the bed, and upon the ratio of the slide stroke to the length of the oscillating press pitman or connection.

These conditions can be expressed mathematically:

$$M_t = P\frac{s}{2}\left(\sin \alpha + \frac{k}{2}\sin 2\alpha\right)$$

where M_t = torsional moment of crankshaft or eccentric created by slide pressure, in.-tons; P = slide pressure, tons; s = stroke of the slide, in.; α = angle determining the position of crankpin or gear eccentric from lower dead point; k = ratio of the stroke of the slide to length of pitman.

The above is the basic formula for determining the slide pressure in an ideal press (without frictional losses or bending stresses) at any position of the slide stroke.

Angular position of the crankpin or gear eccentric at the midstroke of the slide is about 90° from the lower dead point. Assume that it is at 90°. Then

$$\sin \alpha = \sin 90^\circ = 1$$

$$\sin 2\alpha = \sin 180^\circ = 0$$

$$M_t = P\frac{s}{2}\left(\sin \alpha + \frac{k}{2}\sin 2\alpha\right) = P\frac{s}{2}$$

Crankshaft presses are built on the basis of a selected torsional stress which their crankshafts must withstand under severe conditions of presswork.

The torsional moment that a round shaft (crankshaft included) can withstand is

$$M_{max} = \frac{\pi d^3}{16} S_{sh} = 0.2d^3 S_{sh}$$

where M_{max} = torsional moment, in.-tons; d = dia of shaft (at main bearings), in.; S_{sh} = selected shearing resistance of shaft material, tons per sq in.

Midstroke pressure capacity P_m of the press slide may then be expressed by the formula

$$M_{max} = P_m \frac{S}{2}$$

$$P_m = \frac{2M_{max}}{S} = \frac{2 \times 0.2d^3}{S} S_{sh}$$

$$= \frac{0.4d^3}{S} S_{sh}$$

where S = stroke of the slide, in.

In most presses, crankshafts are forged from 0.35 to 0.45% carbon steel and heat-treated. Accepted shearing strength is usually:

6 tons per sq in. in all single-driven presses with a flywheel or a main gear on one end of crankshaft.

9 tons per sq in. in twin-driven presses with main (bull) gears on each end of the crankshaft. This drive is used on presses with crankshafts over 7 in. in diameter with slide stroke longer than twice the diameter at main bearings. Same rule is applied. in addition, to all wide (right to left) double-crank presses.

Thus simplified formulas for maximum permissible midstroke capacity of the slide of crankshaft presses are

Single-driven presses,

$$P_m = \frac{2.4d^3}{S}$$

Twin-driven presses,

$$P_m = \frac{3.6d^3}{S}$$

In gear-eccentric presses, calculation of the midstroke pressure capacity of the press slide presumes a knowledge of the maximum torsional moment on the gear eccentric used in designing the press. This maximum torsional moment depends on the rated pressure capacity of the press, on the initial position of the press slide where the rated pressure must be exerted, and on the ratio k. Therefore, the midstroke pressure capacity of the slide in the gear-eccentric presses must be ascertained from the maker.

However, since gear-eccentric presses are built in competition with crankshaft presses, it may be assumed that their midstroke pressure capacity is within the same limits as in crankshaft presses. This assumption is supported by the fact that the diameters of the fixed pins on which the gear eccentrics revolve are usually made to the same diameter as the crankshafts in crankshaft presses of the same rated capacity.

The selected press should always be checked as to actual working dimensions by consulting a "die-space" plan (Fig. 59). The usual die-space plan for an inclinable, open-back power press is shown. On this plan are marked the points in the construction of *any* selected press which should be checked when either the dies are available or a decision on their general shapes is to be made.

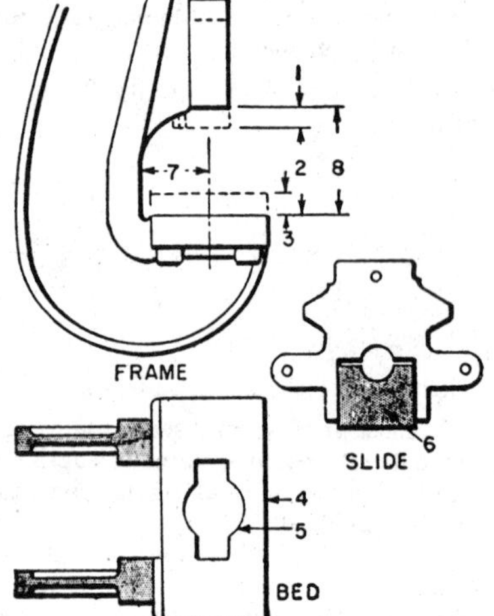

FIG. 59. Check die-space plan for actual working dimensions, after the press is selected. Points to watch are:

1. Stroke of slide (or slides in double-action presses).
2. Die space: distance bed to slide, stroke down, adjustment up (shut height).
3. Bolster-plate thickness.
4. Bed area.
5. Bed opening.
6. In gap-frame presses: area of slide or slides.
7. Gap: distance back from center of slide (stem hole).
8. Distance bed to gibs.
9. Width between frames.

PRESSWORKING OF VARIOUS MATERIALS

Aluminum Alloys

Blanking and Piercing. Blanking of aluminum sheet is comparable with blanking of other metals, with the exception of the punch and die clearances. These can be related to shear strength of a particular alloy. See blanking-tool clearances listed below and the table of mechanical properties of various alloys (Table 22).

Blanking punches should be made of annealed tool steel and dies of hardened tool steel. Surfaces of both members should be ground. Press loading may be reduced by grinding shear on the punch or die. Put shear on the punch if the material being blanked must be kept flat.

With flat dies and punches, the maximum blanking pressure in tons equals $L \times t \times S/2000$, where L is the length of the cutting edge, in.; t is the material thickness, in.; and S is the ultimate shear strength, psi. If shear is provided, the blanking pressure in tons is

$$\text{Pressure} = \frac{L \times t \times S}{2000} \times t \div (t + Sh)$$

where Sh = shear, in.

Clearances for piercing or perforating should be kept to an absolute minimum. Suggested die size is the punch size plus 5% of the metal thickness. Die relief should be at least $\frac{3}{4}°$ for quick "get-away" of slugs. Smooth sheared edges of small holes, similar to a reamed hole, can be obtained by prepiercing slightly undersize and then shaving to the required size, using shear on the punch.

Shearing edges of blanking tools should be kept well lubricated, as well as the metal itself. The metal may be fed through felt rollers or pads well saturated with the lubricant.

Required stripping pressure is generally considered to be approximately one-eighth of the maximum blanking pressure.

BLANKING-TOOL CLEARANCES

Shear Strength, Psi	*Die Size*
Up to 10,000	Punch size + 0.12t
10,000 to 12,000	Punch size + 0.13t
12,000 to 15,000	Punch size + 0.14t
15,000 to 18,000	Punch size + 0.15t
18,000 to 24,000	Punch size + 0.16t
24,000 to 30,000	Punch size + 0.17t
30,000 and up	Punch size + 0.18t

Drawing. The non-heat-treatable aluminum alloys, such as 2S, 3S, 4S, and 52S, are most commonly used for drawn shells. The strong heat-treatable alloys should be used only where their higher ultimate strength is required in service.

Alloys 2S and 3S can usually be drawn and finished completely when starting from an annealed blank, without an intermediate annealing. However, severe draws in 4S and 52S may sometimes require annealing. Intermediate annealing should always be avoided wherever possible because of the added expense and the chance that the reduced tensile strength may result in tensile fractures later.

Simple and shallow draws in 2S, 3S, 4S, and 52S may be accomplished in H12 (H32) or H14 (H34) tempers in cases where one operation only is employed, provided that the draw size (diameter, width, length, etc.) is not less than 75% of the blank size.

Drawing Pressure. When drawing aluminum, the metal gradually hardens as it is worked; so allowance for this must be made in estimating proper press size. Also proper clearance between punch and die is necessary to prevent ironing, which increases drawing pressure. Best results are obtained with a press having a tonnage rating only slightly above the actual requirements, as determined by this formula:

$$\text{Drawing pressure} = \left(\frac{D}{d} - 0.7\right) 3.1416 \times d \times t \times st$$

where D = initial dia of blank or shell, in.; d = drawn shell dia, in.; st = tensile strength of material; t = thickness of material, in.

Drawing speed is a factor of great importance in drawing aluminum successfully. The metal must be given sufficient time to flow; otherwise failures will occur. Consider depth and shape of the part when determining approximate drawing speed. As a guide, relatively deep-drawn parts should not be attempted at a speed over 100 fpm.

Reductions for first draw can be determined by use of the factors listed below, which are to be multiplied by the blank size:

For annealed alloys:

Material ⅛ in. and up, 0.55 to 0.58
Material under ⅛ in, 0.58 to 0.60

For H12 (H32) common alloys:

Material ⅛ in. and up, 0.64 to 0.66
Material under ⅛ in, 0.66 to 0.68

For H14 (H34) common alloys:

Material ⅛ in. and up, 0.68 to 0.72
Material under ⅛ in., 0.72 to 0.75

For H16 (H36) and H18 (H38) common alloys and 61ST6, 24ST3, 75ST6, R301T6 strong alloys:

Material ⅛ in. and up, 0.75 to 0.78
Material under ⅛ in., 0.78 to 0.80

Reductions for subsequent draws are given by the following factors multiplied by the previous draw size:

For annealed alloys:

Material ⅛ in. and up, 0.72 to 0.80
Material under ⅛ in., 0.78 to 0.85

For H12 (H32) and H14 (H34) common alloys; 61ST4, R301T3 strong alloys:

Material any thickness; 0.85 and up

It should be noted that F temper material should not be used for drawn parts. Multiple drawing is not recommended for common alloys in the H16 (H36) and H18 (H38) temper or for strong alloys 61ST6, 24ST3, 75ST6, R301T6.

TABLE 22. COMPARATIVE RATINGS OF ALUMINUM ALLOYS FOR DRAWING

Alloy and Temper	Forming Classification	Shear Strength	Yield Strength	Ultimate Tensile Strength	% Elongation	Cost Classification
2SO	Class I	9,500	5,000	13,000	35	Class I
3SO	Class I	11,000	6,000	16,000	30	Class I
4SO	Class III	16,000	10,000	26,000	20	Class III
C50SO	Class I	15,000	9,000	22,000	25	Class II
52SO	Class II	18,000	13,000	28,000	25	Class V
61SO	Class II	12,500	8,000	18,000	22	Class III
24SO	Class II	18,000	11,000	27,000	20	Class VI
R301-O	Class II	18,000	10,000	25,000	21	Class IV
75SO	Class IV	22,000	15,000	33,000	17	Class VII
2S-H12	Class I	10,000	14,000	16,000	12	Class I
3S-H12	Class I	12,000	17,000	19,000	10	Class I
4S-H32	Class IV	17,000	22,000	31,000	10	Class III
C50S-H32	Class II	16,000	21,000	25,000	10	Class II
52S-H32	Class III	20,000	27,000	34,000	11	Class IV
2S-H14	Class I	11,000	16,000	18,000	9	Class I
3S-H14	Class II	14,000	19,000	22,000	8	Class I
4S-H34	Class IV	18,000	27,000	34,000	9	Class III
C50S-H34	Class II	17,000	24,000	28,000	9	Class II
52S-H34	Class III	21,000	31,000	37,000	9	Class IV
2S-H16	Class I	12,000	18,000	21,000	6	Class I
3S-H16	Class II	15,000	22,000	26,000	5	Class I
4S-H36	Class IV	20,000	31,000	37,000	5	Class III
C50S-H36	Class II	18,000	27,000	30,500	8	Class II
52S-H36	Class III	20,000	34,000	39,000	8	Class IV
61S-T4	Class III	24,000	21,000	35,000	22	Class V
2S-H18	Class I	13,000	22,000	24,000	5	Class I
3S-H18	Class II	16,000	26,000	29,000	4	Class I
4S-H38	Class IV	21,000	34,000	40,000	5	Class III
C50S-H38	Class II	19,000	29,000	32,500	7	Class II
52S-H38	Class III	24,000	36,000	41,000	7	Class IV
61S-T6	Class II	30,000	41,000	45,000	12	Class V
R301-T3	Class III	37,000	40,000	62,000	20	Class VI
24S-T3	Class I	41,000	48,000	69,000	18	Class I
75S-T6	Class II	49,000	72,000	82,000	11	Class II
R301-T6	Class I	41,000	60,000	68,000	12	Class I

All mechanical properties and elongation values are typical. Shear-strength values for C50S tempers are unofficial.

Mechanical properties of F tempers are not guaranteed. Therefore these are omitted.

Class I rated material has best formability in its respective group; class II is next.

Class I rated material under "Cost Classification" is the least expensive; class II is the next cheapest.

Clad alloys are not listed. If considered, use the same "Forming Classification" as shown for the bare alloys having the same alloy symbol. Mechanical properties of clad alloys are slightly less than those of bare alloy. Elongation values are approximately the same.

The minimum and maximum values shown in both the first draw and subsequent draw reduction formulas should be used in relationship to the drawability classification appearing in the table, using the lower figure as the maximum for class I rating, an intermediate figure for class II, etc.

It should be remembered that, the thinner the material is, the more difficult it is to draw it successfully without rupture. Allowances should be made for this.

Blank-holding Pressure. Each specific draw job requires careful adjustment of the blank-holder pressure. It must be merely enough to prevent the formation of excessive wrinkles.

Formulas

Drawing-tool clearances, cylindrical shells:

First-draw die bore = punch dia + 2.2 × blank thickness
Second-draw die bore = punch dia + 2.3 × blank thickness
Third and subsequent draw die bore = punch dia + 2.4 × blank thickness
For ironing a shell = punch dia + 2.0 × blank thickness

Drawing-tool clearances, square and rectangular shells:

First-draw die size = punch size + 2.2 × blank thickness
Second and subsequent = same as above
Final-draw die size = punch size + 2.0 × blank thickness

Punch and die draw radiuses:

Minimum draw radius = 4 × blank thickness

Note that, when shear strength of metal is lower than yield strength, the punch and die radiuses should be increased somewhat.

Example

Problem: Assume that you want to draw a straight-sided shell in a single draw from 0.064-in.-thick aluminum. There are no restrictions, from the standpoint of finishing or corrosion, as to choice of alloy. The diameter is 10 in., the depth 3 in. There is to be a 1-in.-wide flange at the top edge. You want the material in the finished shell to have as high a yield strength as possible.

What alloy and temper should you select?

Solution: First calculate the blank diameter required. In this problem it will be approximately 16 in. Divide 10 by 16 and get a reduction factor of 0.625. You could use any annealed alloy—but you want something harder if possible. You find, however, that the H12 (H32) tempers should be drawn on a 0.66 or higher reduction factor. Therefore, you will have to use an annealed alloy.

Refer to Table 22 and scan the "forming classification" ratings. Also compare the mechanical properties of the various alloys, and the "cost classification." You immediately eliminate 75SO because of its high cost and also because it has the worst forming classification. The next strongest alloy, as far as yield strength is concerned, is 52SO. You note that it is in class II in formability and class V in cost.

If you decide to sacrifice yield strength in favor of cost and formability, you will probably end up by choosing C50SO. If you compromise between strength, workability, and cost, you will probably choose R301O or 4SO. Your best choice, disregarding the cost factor, would be 52SO.

Causes for Defective Work

Causes of wrinkling:

Lubrications too heavy.
Insufficient blank-holder pressure.
Reduction too great.

Causes of puckers:

Insufficient blank-holder pressure.
Excessive metal clearance.
Material too thin.
Shape of shell conducive to puckers.

Causes of fractures:

Poor lubrication.
Excessive blank-holder pressure or unequal pressure.
Tool radiuses too small.
Reductions too great.
Insufficient metal clearance.
Material too thin.
Wrong material or of poor drawing quality.
Punch off center with die.
Drawing surfaces of tools bad.
Blank larger than die and blank holder.
Poorly sheared edge on blank.

Necking. Tops, or open ends, of drawn shells can be reduced by necking. The metal is stressed entirely in compression, and only a limited amount of reduction in diameter can be accomplished in one operation. The percentage of reduction is determined by metal characteristics before necking. Case records indicate that these data will serve as a guide:

Temper before Necking	*Reduction in Dia*
H12 or H32	12–15%
O	9–12%
H14, H34, T4	6–9%

Necking angle should not exceed 35° (Fig. 60). After necking, the wall thickness is determined by $t\sqrt{D/d}$, where t is wall thickness before necking, D is mean dia of shell after necking, and d is mean dia of shell after necking.

Working pressure in tons to neck a shell is

$$T = \frac{t \times S(D - d)}{\text{cos necking angle}} \div 2000$$

where t is thickness of metal after necking; S is ultimate compressive strength, psi; D is mean dia before necking; and d is mean dia after necking.

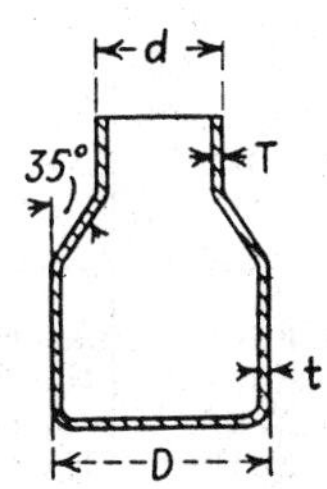

FIG. 60. Necking angle should not exceed 35°.

Magnesium Alloys

Magnesium-base alloys harden rapidly when cold-worked. Cold forming around small radiuses may lead to trouble, but most of the alloys may be formed easily into intricate shapes if heated between 500 and 700 F. Temperatures as low as 400 F may be employed for moderate forming. Forming tools should be clean, smooth, and well lubricated. Lard oil frequently is used, since it withstands heat well and can be removed easily with alkaline cleaners.

With ordinary tools, blanked or sheared edges more than 0.065 in. thick tend to have a flaky structure. Special designs in punches and shear often will permit smooth cuts in stock up to several times this thickness.

When blanking or punching sheet, clearance between punch and die should be held to the minimum permitted by the accuracy of the press. Smooth edges on sheet thicker than 0.065 in. can be obtained by warming sheet before blanking. A slight rake or shear (about 3°) on the punch also will improve the sheared surface. If heating is necessary in this and subsequent operations, dimensional changes due to expansion during heating, and subsequent contraction, must be considered.

Bending and Forming. Sharp corners and burrs must be removed from the edges of the sheet near the bend line if short-radius bends are to be made. Bend lines should not be scribed or prick-punched, because such marks may lead to fatigue cracks. Minimum cold-bending radius for sheet stock varies with the alloy, temper, thickness, and rate of deformation. Type of equipment used and relation of axis of bend to direction of sheet rolling also influence results.

Annealed sheet stock has the best cold-forming properties, a bend radius of four times stock thickness being permissible. Hard-rolled sheet can usually be bent around a radius of five to ten times stock thickness.

Bends with radiuses of 2 to 2½ times sheet thickness can be made by hot bending regardless of alloy or temper. In either hot or cold bending a slow rate of deformation will give best results. Heating for hot forming changes properties of hard-rolled sheet but does not change room-temperature properties of annealed sheet. Time of heating should be just sufficient to heat the sheet thoroughly.

Pressing and Drawing. Although each case must be considered individually, draws and forms of medium depth can be made successfully. Best results are obtained by heating the dies. Design of the dies follows closely that used for other metals. Where the forming operation is not severe and the work can be done cold, it sometimes is possible to adapt dies developed for other metals.

TABLE 23. RADIUSES FOR 90° BENDS IN MAGNESIUM ALLOY SHEET

Dowmetal Alloy	Condition	Min Radiuses at 90 F	Typical Radiuses at Temp, F			
			70	300	400	600
FSI-0	Annealed.........	5t	4.5t	.....	1–2t	
FSI-H24	Hard-rolled......	10t	8t	4.5–6t		
M-0	Annealed.........	7t	5.5t	.....	3–4t	1–2t
M-H24	Hard-rolled.......	12t	9t	.....	6–9t	

Dow Chemical Co.
Springback 8 to 10° for annealed sheet, 12 to 15° for hard-rolled sheet.

Nickel Alloys

Drawing. Monel, nickel, and inconel deep-drawing-quality sheets and strip may be drawn into any shape feasible with deep-drawing steel. These high-nickel materials have physical characteristics different from those of deep-drawing steel, but not sufficiently so as to require different manipulation of dies for the average pressing jobs encountered. A great majority of the deep-drawn shapes that have been produced in the high-nickel materials have been drawn over dies and formed with tools that had been laid out to handle steel or the softer copper-base alloys. On some jobs, however, where the shapes to be drawn have been particularly intricate, with accurate finished dimensions required, it has been necessary to make minor die alterations, such as increasing the clearance or enlarging the radius of the die ring or punch, in order to adjust the setup to handle the high-nickel materials successfully. These materials have good deep-drawing qualities, but they have their own characteristic working and physical properties which must be considered in order to obtain the full value of their ductility in deep drawing.

The contributing factors essential for satisfactory large production runs and difficult draws are:

1. Special lubricant and die material to keep friction at a minimum.
2. Dies constructed so as to permit freest possible flow of material without wrinkles forming.
3. Uniform and constant blank hold-down pressure.
4. Slow and smooth-action press with ample power.
5. Satisfactory annealing equipment.

Types, Limits, and Tolerances. *Monel, nickel, or inconel sheet or strip that is to be deep drawn should be so specified in ordering.*

Two types of sheet in each material are furnished for deep drawing and stamping. In addition, "35" monel sheet, which is specially rolled and finished for counter and cabinet-top construction, is suitable for press brake forming and for perforating, but is *not* suitable for stamping or pressing operations.

TABLE 24. COMPARATIVE DRAWING CHARACTERISTICS OF NICKEL ALLOYS AND OTHER MATERIALS (0.062 IN. THICK)

Material	Shore Hardness	Tensile Properties		Olsen Ductility Test		Pressure Factor at 0.375 In. Cup Depth, Taking Aluminum at 1
		Yield Strength 0.2% Offset, Psi	Elongation in 2 In., %	Depth of Cup, In.	Beam Load at Fracture, Lb	
Monel	16	27,450	52	0.495	11,800	5.15
Nickel	10	21,350	47	0.545	11,650	4.70
Inconel	21	39,965	43.5	0.480	13,750	6.30
Ingot iron	18	26,150	49	0.425	6,500	3.40
Deep-drawing steel	20	31,550	48.5	0.430	7,510	3.90
Copper	7	11,740	68.5	0.490	5,110	2.15
Aluminum	4	8,710	41	0.430	1,880	1.00

Standard cold-rolled sheets are rolled hot, then lightly cold rolled to gage, bright annealed, leveled, and resquared. This type of sheet is produced in thicknesses from 0.018 to 0.250 in. (0.025 in. is the minimum thickness for inconel), widths from 24 to 60 in., and lengths from 60 to 172 in., depending on the width and gage. Standard cold-rolled sheets are supplied with two finishes, namely, regular finish which is the "as-rolled" finish, and No. 9 finish, which is a ground, satin finish. Standard sheets are furnished in one temper only, which is soft temper. They are not so ductile as special cold-rolled sheet or strip, soft temper, but they can be used for moderate drawing and stamping operations.

Special cold-rolled sheets and strip are cold-rolled heavily and carry a highly finished, rolled surface. These have higher ductility than the standard cold-rolled sheets. Cold-rolled sheets and strip can be obtained in a range of tempers from dead soft to full hard. Soft-temper sheets and strip with fine to medium grain are the most suitable for deep drawing. Dead-soft sheets have the greatest ductility, but material of this temper is likely to have large grain size, which may result in a pebble surface on the finished drawn article. Special cold-rolled sheets are available in the same gages and sizes as standard cold-rolled sheets.

Lubricants. Beef tallow and castor oil are satisfactory, likewise water-soluble oils, but lead- or sulfur-bearing lubricants should not be used if the pieces are to be annealed. Nickel alloys require better lubricants than light oils or thin soapy solutions frequently used on steel and soft metals. The high-nickel materials do not form thin oxide films, such as are found on steel sheets, and which prevent actual metal-to-metal contact, and allow the steel to flow between the die and pressure plate with relatively low frictional resistance. To avoid wrinkling, more hold-down pressure is required on the pressure plate for the high-nickel materials than is needed usually for steel and the softer metals. To reduce to a minimum the factors conducive to galling, lubricants with high film strength and low coefficient of friction are required. Most of the satisfactory lubricants contain a filler of some inert material.

Circular Shells

Reduction. Wherever the term "reduction" is used hereafter it applies to reduction of shell or blank diameter, and not to area reduction. For example, a shell drawn from 10 to 6 in. has received 40% reduction.

Double-action Operation. On double-acting presses, a well-balanced series of reductions for light gage (1⁄16 in. thick and under) cylindrical shells, with no "ironing," would be 35 to 40% on the first, or cupping, operation, and 15 to 25% on redraws. If the walls are held to gage, the first and second operation may be the same as suggested above, but on further redrawing, the amount of reduction should be diminished about 5% on each successive redraw.

While reductions up to 50% have been made in one operation on a production run, it is necessary, in order to keep shell breakage low, to employ (for materials of high nickel content) all the known factors contributing to an ideal drawing setup.

Single-action Operation. Depth to which the high-nickel materials can be drawn in single-action presses, without hold-down mechanism, is controlled by thickness of the material. With properly designed dies and sufficiently thick material the reduction with a single-action setup on the first (cupping) operation may be made equal to those recommended for double-action dies, that is, 35 to 40%. Redraws should not exceed 20% reduction. If the walls are to be "ironed," it is necessary, because of the pressure exerted on the bottom, to decrease the reductions to avoid breaking the shells. With very slight reductions in diameter, 5% or less, reductions as high as 30% in wall thickness may be obtained in one draw. With medium

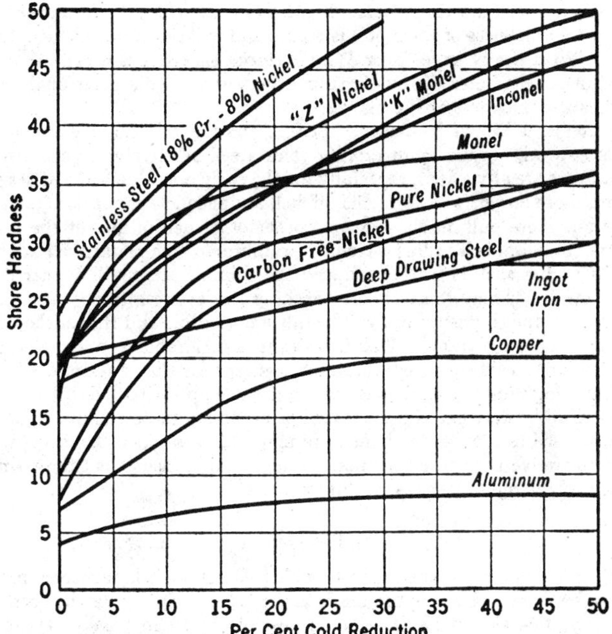

FIG. 61. Hardness of metals increases with cold reduction, especially the stainless steels and high-nickel alloys.

reductions of around 12% in diameter, the wall thickness may be reduced in the neighborhood of 15%. If there is a considerable amount of wall reduction to be made, it is better practice to draw the shape first to approximately the required inside diameter with little or no wall thinning, and to do the "ironing" operation last. If a good surface finish is desired, the final reduction in wall thickness should be very slight, merely a burnishing operation.

Clearances and Draw-ring Radiuses. Because monel, nickel, and inconel possess higher physical properties than deep-drawing-quality steel, they give greater resistance to the wall thinning caused by the pressure of the punch on the bottom of the shell as it is pushed through the die. Consequently, greater die clearance is required than for steel under similar conditions, if the natural flow of the metal is not to be resisted. The clearance required for the high-nickel materials differs only slightly from that for steel and, since dies designed for the ordinary run of cylindrical steel shapes usually are given much more than the minimum requirement, such tools are suitable for monel, nickel, and inconel also.

For the purpose of maintaining a uniform wall thickness or for burnishing, dies developed for accurate deep drawing of steel are designed frequently with little or no clearance. With this type of die it may be necessary to increase the clearance for drawing high-nickel materials of the same gage, unless the reduction is within the range that will permit the final "ironing" without causing rupture.

For ordinary deep drawing of cylindrical shells, an over-all clearance of about 40 to 50% of the thickness of the metal is ample and will not result in the formation of wrinkles. With heavy gage (over 1/16 in. thick) sheets it is a general practice to have the inside diameter of the die ring larger than the diameter of the punch by three times the thickness of the blanks.

Draw-ring and Punch Radiuses. Because of the tendency of the high-nickel materials to work-harden more rapidly than steel, relatively large draw-ring and punch radiuses are advocated, especially for the early operations of a series of draws. It requires more power to draw monel, nickel, and inconel than it does to draw steel, which means more pull on the bottom corner of a shell of any of the high-nickel materials. Small punch radiuses result in thinning of the metal shell at the line of contact; if such a shell is further reduced the thinned areas will appear farther up in the shell wall and may cause visible "necking," or even rupture. Also, if the shell is to be buffed, the thinned areas will be quite noticeable and the finished shell wall will have a wavy appearance. For redraws it is preferable to draw over a beveled edge. This will avoid round edged punches except for the final draw.

The draw-ring radius for a circular die is governed principally by the thickness of the material to be drawn and the amount of reduction to be made. A general rule that can be used as a guide for light gage material is to make the draw-ring radius from five to twelve times the thickness of the metal. Insufficient draw-ring radius may result in galling and excessive thinning of the wall.

Square and Rectangular Shells

As with other materials, the depth to which monel, nickel, or inconel rectangular shapes can be drawn in one operation is governed principally by the corner radius; consequently, this should be as large as possible to avoid possible trouble. The depth of draw should be limited to from two to five times the corner radius for monel and nickel, and to not greater than four times for inconel. The depth of draw for thin stock (0.025 in. and under) should not be over about three times the corner radius for monel and nickel and less for inconel.

The corner draw-edge radius should be from four to ten times the sheet thickness.

Stainless Steel

Blanking and Punching. Power requirements for blanking and punching stainless will vary in proportion to the shear strength of the materials being worked. The chromium-nickel types of lower tensile strength (approximately 100,000 psi) will require at least 40% more power than similar thicknesses of mild steel. Speeds should be lowered to about two-thirds that required for mild steel. Because straight chromium steels break out about the same amount as mild steels, dies and punches may be set conventionally.

In blanking and punching, as in shearing, the clearances for chromium-nickel steels should be set closer than for mild steels.

Shear of the punch should put the metal in a slight state of tension just before cutting begins. Present shop practice indicates that the shear is no greater than the thickness of the metal.

In one-level punching of holes, suggested pressures may be multiplied by the number of holes to be punched. The resultant tonnage should not exceed about two-thirds the rated capacity of the equipment. If punches are stepped one-half of the metal thickness, or more, the total pressures can be divided by the number of steps attained.

TABLE 25. DRAWING CLEARANCES AND RADIUSES FOR STAINLESS STEELS

Gage	Metal Thickness, In.	Chromium-nickel			Straight Chromium		
		Die Clearance, In.	Die Radius, In.*	Punch Radius, In.†	Die Clearance, In.	Die Radius, In.*	Punch Radius, In.†
30	0.0120	0.0144	0.0600	0.0360	0.0138	0.0480	0.0264
28	0.0149	0.0178	0.0745	0.0447	0.0171	0.0596	0.0328
26	0.0179	0.0215	0.0895	0.0537	0.0205	0.0646	0.0471
24	0.0239	0.0287	0.1195	0.0717	0.0280	0.0956	0.0526
22	0.0299	0.0359	0.1495	0.0897	0.0344	0.1196	0.0658
20	0.0418	0.0431	0.1795	0.1077	0.0413	0.1436	0.0790
18	0.0500	0.0600	0.2390	0.1434	0.0550	0.1913	0.1052
16	0.0598	0.0738	0.2990	0.1794	0.0688	0.2392	0.1316
14	0.0747	0.0896	0.3735	0.2241	0.0859	0.2988	0.1643
12	0.1046	0.1255	c.5230	0.3138	0.1202	0.4184	0.2301
10	0.1345	0.1614	0.6725	0.4035	0.1547	0.5420	0.2959
8	0.1644	0.1973	0.8220	0.4932	0.1891	0.6576	0.3617
6	0.1943	0.2332	0.9750	0.5829	0.2234	0.7772	0.4275
4	0.2242	0.2690	1.1210	0.6726	0.2578	0.8968	0.4932

* For first drawing operation.
† This dimension dependent on design requirements.

By reducing press speed to about one-half to one-third of mild-steel practice and increasing the power at least 75%, stainless steel can be satisfactorily perforated. It is inadvisable to attempt to perforate straight chromium steels where the holes are smaller in diameter than stock thickness, or chromium-nickel steels where diameter of the holes is less than 1½ to 2 times metal thickness. After perforating, it may be necessary to flatten or anneal the work for certain applications.

Bending and Forming. Bending and forming equipment, rated at a definite capacity for mild steel, is limited to stainless steel several gages lighter. Aside from the added power and reduced speed for chromium-nickel steel, increased springback must be compensated for in design of the tools. Springback is overcome by bending the work to an angle greater than that desired. Angular springback is sometimes avoided in thick metals by rounding the V in the die to conform with the outside bend in the work. The sharp angle on the punch is made about 1 to 2° less than the die angle for conventional 90° bends. The punch in coming down then scores a line in the finished bend and sets the metal over a positive support.

Drawing. High ductility of chromium-nickel stainless provides good deep-drawing qualities, but the high strength and tendency to work-harden require modified drawing procedures and greater power. Straight high-chromium steels have lower ductility and a tendency to become brittle at high temperatures. They can be drawn but may require more intermediate annealing.

Because chromium-nickel stainless work-hardens, it is necessary to use a lower press speed, say 25 to 30 fpm. A number of fabricators have increased speeds to as high as 50 fpm by careful attention to die design, lubrication, and close annealing control.

Reductions as high as 50 to 55% have been made under certain conditions, but the normal drawing reduction in the initial draw is about 40%. This varies with gage and diameter of blank. On 14 gage, 18-8 stainless with a blank diameter of over 36 in., 30 to 35% is a safe limit. Subsequent draws will be between 20 and 25%, depending on size of blank and material being handled. Straight-chromium steels can be reduced about 20 to 25% in the first operation. For extra deep-drawing qualities, a hardness not exceeding Rockwell 80 B is desirable.

Solid dies of high-carbon, high-chrome steel are suitable for drawing chromium-nickel stainless. Chromium-nickel cast-iron dies stand up well for severe reductions on fairly long runs. Tungsten carbide dies for small parts have long tool life. On larger parts, tungsten carbide inserts are used on draw rings, but cost is high and breakage possible. Cast-iron tools are suitable for short-run work on the lower-tensile stainless steels.

Cast bronze of 300 to 325 brinell hardness is almost a standard draw-ring material with many companies. The centrifugally cast grade has longer life. Chromium-nickel cast-iron draw rings are recommended by certain press manufacturers and fabricators for long-run drawing. Polishing of both the draw ring and punch in the direction of drawing is recommended, and a growing practice is to stone the draw ring on the radius.

Preheating the dies to about 200 F increases the ductility of straight-chromium steels and makes them more adaptable to deep drawing.

Draw-ring radiuses should be four to six times the thickness of the steel. This avoids stretching the metal and allows it to flow more freely into the die, at the same time decreasing the rate of work hardening.

For a first operation, with a moderate draw, the punch diameter may be 60% of the blank diameter. Following draws are made, after intermediate cleaning and annealing, with punches approximately 20% smaller in diameter for each successive operation.

SELECTION OF DRAWING COMPOUNDS

As a basis of selection of lubricants, a thumbnail classification on the basis of severity is useful. Thus shaping operations, in which wall thickness is reduced only slightly or not at all, are much less severe than where the wall thickness is greatly reduced by ironing or forcing the metal into a restricted area between two tool surfaces. Operations will be classified by severity as:

Mild. Shaping under good conditions.

Medium. Severe shaping or sinking as in wire drawing.

Severe. Ironing of metal as in seamless-tube drawing or cartridge-case manufacture.

Drawing Compounds for Steel. Good drawing practice on steel requires that the scale be removed and the surface be slightly roughened by pickling, while avoiding hydrogen embrittlement. In selection of the drawing compound, consider both the severity of the operation and need for lubrication. A good lubricant permits wide variation in other job considerations. Compounds for mild, medium, and severe drawing are:

MILD

Most automobile-body forming and shallow draws on low-carbon steel.

1. Mineral oil of medium heavy to heavy viscosity.
2. Soap solutions (0.03 to 2.0%, high-titer soap).
3. Fat, fatty oil, or fatty and mineral-oil emulsions in soap-base emulsions.
4. Lard oil or other fatty oil blends (10 to 30% fatty oil).

MEDIUM

Deep draws on low-carbon sheet steel and drawing wire of low-carbon steel.

1. Fat or oil in soap-base emulsions containing finely divided fillers such as whiting or lithopone.
2. Fat or oil in soap-base emulsions containing sulfurized oils.
3. Fat or oil in soap-base emulsions with fillers and sulfurized oils.
4. Dissimilar metals deposited on steel plus emulsion lubricant or soap solution.
5. Rust or phosphate deposits plus emulsion lubricants or soap solution.
6. Dried soap film.

SEVERE

Wire-drawing medium- to high-carbon steel, seamless-tube drawing and cartridge-case drawing.

1. Dried soap or wax film, with light rust, phosphate, or dissimilar metal coatings.
2. Sulfide or phosphate coatings plus emulsions with finely divided fillers and sometimes sulfurized oils.
3. Emulsions, lubricants containing flowers of sulfur as combination filler and sulfide former.
4. Oil-base sulfurized blends containing finely divided fillers.

Stainless Steel

MILD

1. Corn oil or castor oil.
2. Castor oil plus emulsified soap.
3. Waxed or oiled paper.

MEDIUM

1. Powdered graphite, suspension dried on work before operation (to be removed before annealing).
2. Filler bearing emulsion lubricant at heavy concentration.
3. Solid wax films.

SEVERE

1. Lithopone and boiled linseed oil.
2. White lead and linseed oil to a consistency of 600 W oil

Brass

MILD

1. Soap solution (0.03 to 2% high-titer soap).
2. Fat or oil emulsions with soap emulsifier.
3. Lard-oil blends (10 to 20% lard oil in mineral oil).

MEDIUM

1. Soap solution. Soap should be high titer (39 to 42%); fatty acids and free alkali should be less than 0.07%. Solution should be low concentration (0.3 to 1.0%), and lubricant should contact work at least 30 sec.

2. Fairly rich fat or fatty-oil emulsions with soap emulsifiers. Free fatty acid in the paste base, if diluted with 8 parts water, should be at least 2%.

3. Lard-oil blends (25 to 100% lard oil in mineral oil). Free fatty-acid content should be 1.5 to 5%. Addition of about 1% melted tallow, 0.25% stearic acid to 1 to 2% soap solution is desirable for medium to severe draws.

Severe

1. Soap solution of 0.3 to 1% or one of 1 to 2%, containing 1 to 2% tallow or 0.25% stearic acid. Lubricant and work should be in contact longer than with less severe draws.

2. Rich lard-oil blends (75 to 100%).

3. Dried soap properly added.

In most instances, brass recommendations apply to copper.

Magnesium

Mild

1. Graphite (usually colloidal) in mineral oil, high flash point.
2. Beeswax or paraffin and tallow.

Medium

Flake or colloidal graphite in a volatile solvent such as carbon tetrachloride, naphtha, or alcohol spread on the work and solvent-evaporated. Add 20% graphite in tallow on the die.

Aluminum

Mild

1. Mineral oil, increasing in viscosity as severity of operation increases.
2. Fatty-oil blends in mineral oil (10 to 20% fatty oil) or petroleum jelly.

Medium

1. Tallow and paraffin.
2. Sulfurized fatty-oil blends (10 to 15%), preferably enriched with 10% fatty oil.

Severe

1. Dried soap film or wax films.

2. Mineral-oil or fatty-oil blends or sulfurized-oil blends, plus finely divided fillers.

3. Fat emulsions in soap water plus finely divided fillers.

COLD-ROLL FORMING

By Elmer J. Vanderploeg, *Chief Engineer, The Yoder Co.*

Definition. Cold-roll forming has been defined as a process whereby a flat strip of metal, by passing through a series of rolls arranged in tandem, is progressively formed into the ultimate desired shape. Ordinarily the metal requires no heating or heat-treatment before, during, or after forming. Standard roll-forming machines are usually equipped for a constant speed of 100 fpm but may be designed for a higher or lower speed to meet special requirements

Where to Use Cold-roll Forming. Inherently, cold-roll forming is applicable mainly in the mass production of shapes or uniform profile. A limited amount of transverse forming may be done, but the width must be uniform, unless blanks are first cut and fed into the machine.

The minimum practical radius of bends is equal to the thickness of the stock, but it should preferably be double the stock thickness or better. With some sacrifice of strength, sharp inside and outside corners are, however, obtainable. In forming wide shapes from thin stock, some bending at or near the outside edges is necessary to stiffen them and avoid waviness.

Conversely, maximum thickness of stock suitable for forming depends on the material used and on the angle or radius of bends or curves. Plain angles, channels, and other structurals up to ½ in. thick have been and are being produced, and also pipe up to ¾ in. wall thickness.

As a general rule, it may safely be assumed that, if a thing can be roll formed as is, or redesigned to make it suitable for roll forming, this method will result in reduced unit cost, because of saving in either weight and raw material alone, or in conversion, cost, or both. A roll-forming machine will also provide a much greater production per hour than any other machine available.

As in most applications, the material cost is much higher than the conversion cost, and any saving in weight made possible by roll forming is frequently a most important factor in unit cost.

The number of rolls required is determined by the character and intricacy of the shape, also the thickness and kind of material to be formed; hence it is customary where a number of different shapes are to be made on the same machine to use one having a base long enough to accommodate the maximum number of rolls ever likely to be required for any shape likely to be made on that machine.

Machines. The typical roll-forming machine consists of a welded steel base on which are mounted identical roll stands, each designed for holding one pair of rolls. For forming light narrow shapes, the so-called outboard-type machine is used. In this type the roll spindles are supported only at one end, affording better visibility and access to the machine in changing and adjusting the tooling. However, the pressures required in forming most commercial shapes are such that the spindles are best supported at both ends in adjustable antifriction bearings. The latter also provide the means for keeping the spindles in accurate alignment.

The inboard type, being the machine most extensively used, has been standardized by one or two manufacturers in four basic sizes, for forming stock from 0.010 up to 0.156 in. thick and widths up to 16 in. By providing a wider base, these machines can also be used for widths up to 30 in. or more, although the capacity ratings in respect to stock thickness are thereby reduced by deflection of the spindles.

Each of the four basic sizes can be made as long or as short as desired, with anywhere from 2 or 3 up to 20 roll stands, although more than 12 to 14 roll passes are seldom required.

When sections to be formed have deep profiles, such as deep channel, box and tubular shapes, the upper, or male, rolls have a greater pitch or driving diameter than the lower ones. The pitch diameter is defined as the widest and most nearly horizontal area of the roll profile. This area is important in providing most effective, balanced traction for the stock. In deep forming, the male rolls, being of greater diameter, and therefore having a greater peripheral speed than the lower, will exert a greater pull on the stock. This will result in excessive friction and heat, scoring of the stock, and excessive wear of the rolls. Leading makes of machines are available with different gear ratios for the upper and lower spindles, so as to equalize differences in peripheral roll speeds.

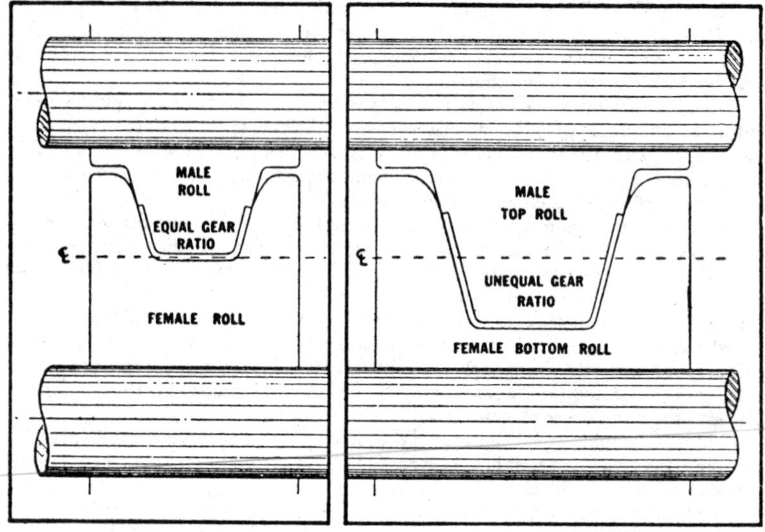

FIG. 62. Shifting the pitch diameters of rolls by unequal gear ratios will increase the depth of flange that can be cold-roll formed in one pass. Surface speeds of the two rolls are identical.

Figure 62 shows the relative capabilities of equal and unequal gear-ratio machines. On the left is a cross-section of a pair of rolls of equal pitch diameter, mounted in a machine with equal gear ratio for forming a shallow channel. The pitch diameters of the upper and lower rolls here are identical; that is, it lies midway between the upper and lower spindle.

To form a channel with flanges twice as deep would be impossible because of lack of headroom. The flanges would strike the spindles after they had been bent up a little better than halfway toward the perpendicular.

Compare this with the drawing on the right (Fig. 62), which illustrates the same pass in an unequal-gear machine, forming an angle of the same width but with flanges twice as deep. The pitch diameters of the rolls here are placed way below the center line between the top and bottom spindles, and the gear ratio is so fixed that the surface speed of the large upper roll is identical with that of the smaller, lower roll. By this arrangement enough space has been gained to make it possible to form channels and other profiles with flanges or legs much deeper than with the equal-gear arrangement.

In designing and making rolls for all profiles, whether deep or shallow, to be made on a machine with unequal gearing, it is necessary only to make sure that the diameter ratios between the upper and lower rolls will be the same as in the original set of rolls for which the gearing was first designed.

Forming Bright and Precoated Stock. When roll forming is not to be followed by other operations, such as trimming and forming the ends, welding, and bending, it is common practice to use cut lengths or blanks and even coiled stock which have already been given the finish ultimately desired. As the change of profile between

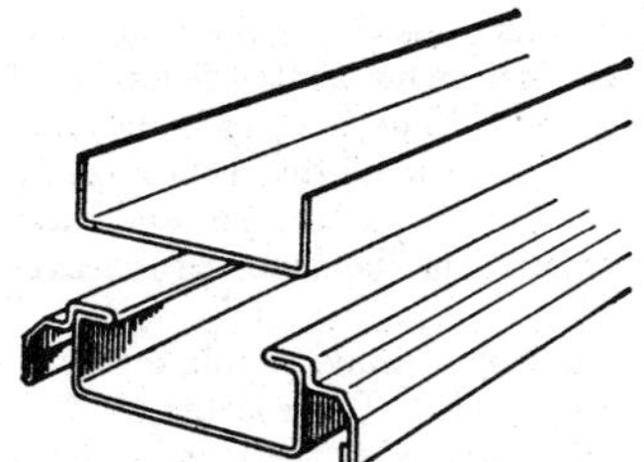

FIGS. 63 and 64. Open side or central section of the shape should face up.

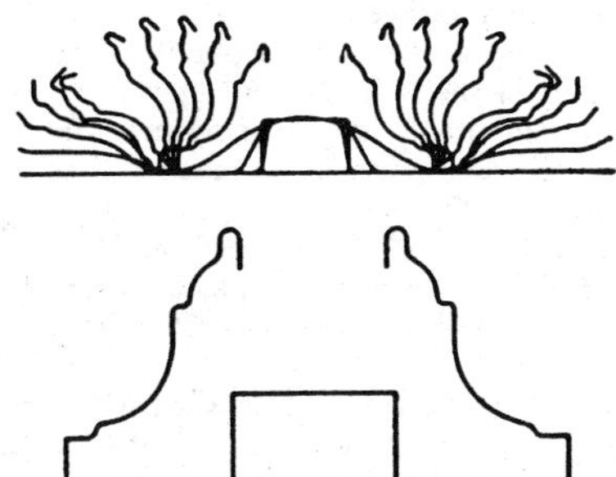

FIG. 65. A "flower," or layout of successive passes, is developed to assist roll design.

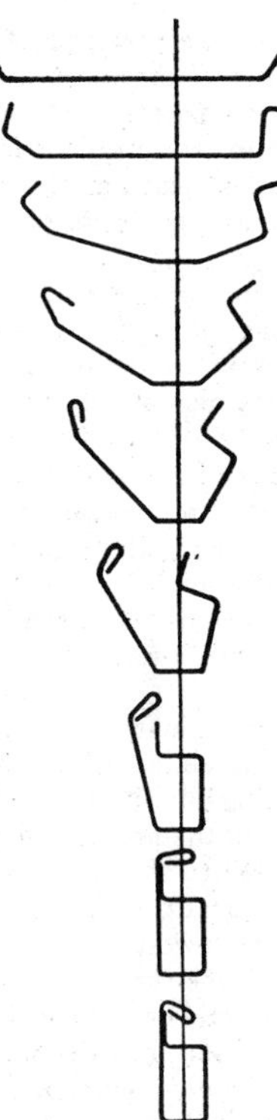

FIG. 66. Development of window-screen section.

roll passes is limited, the roll pressures, in cold forming light gages of metal, are low enough to permit successful forming on a large scale of galvanized, electrogalvanized, and electroplated stock. Metals which have been subjected to mechanical and chemical surface treatments such as buffing, polishing, and burnishing may also be cold-roll-formed successfully without objectionable marring of the surface.

Hot-dipped galvanized stock is being extensively used in cold-roll forming but is subject to greater limitations than are the other finishes just mentioned. Baked enamels and paints may be used to precoat metal used for cold-roll forming of venetian blinds, awnings, and other light shapes.

Ordinary paint, enamel, and other organic finishes are not well adapted to cold-roll forming.

Tooling for Roll Forming

First and most essential in tooling for cold-roll forming are, of course, the main, or driven, rolls, with the spacers for holding them in proper lateral alignment on the spindles.

Depending on the character of the shape to be formed, there may also be a pair of entry rolls or a guide to keep the stock, as it is fed into the machine, in proper alignment with the first pair of forming rolls; also idler rolls, bar guides, or shoes for pinching in of the edges and also to prevent vertical or horizontal deflection of the stock in its passage between successive roll stands. The idler rolls may also be designed to exert pressure from the sides, as an aid in forming the vertical surfaces of deep sections, such as channels and box shapes. Finally, at the exit end there is a straightening guide, to prevent curving and twisting of the shape as it leaves the machine.

Small rolls or dies may also be provided for curving, coiling, or ring forming; dies for cutting to length, perforating, and notching the ends. These latter are mounted in an automatic flying cutoff machine installed in line with the forming machine. There may also be rolls for making lock-seam tubing; cutters for trimming stock to the exact width required; embossing rolls; in fact, a number of attachments with tooling may be provided for performing operations other than strictly roll forming, with little or no increase in labor cost or reduction in speed.

Roll Design. In designing the main rolls, the open side of the section to be formed is generally made to face up; that is, the edges of the strip are bent upward rather than downward, as shown in Fig. 63. When the profile is to have openings facing both up and down, the larger or most central opening is usually made to face up (Fig. 64).

As a preliminary to designing a set of rolls by the most advanced method, one makes a rough layout or "flower" (Fig. 65) of the successive roll passes, after first establishing a horizontal and a vertical guide line for developing the profiles from the flat strip to the finished shape.

In regard to the guide lines, one is known as the horizontal pitch line, the other as the vertical pass line. The former is placed at the most advantageous level, at or near the lowest point of the shape to be made, and extends in a straight, almost horizontal line through from the first to the last roll pass. This line establishes the pitch diameters of the male top rolls and the female bottom rolls, respectively. In the case of forming a simple angle, the pitch line would be at the bottom; in other words, it would correspond with the outside corner of the angle which, in this case, would be the lowest point in forming.

The vertical pass line intersects the horizontal pitch line, or vice versa, depending on which is made first.

As forming in the first roll pass generally starts near the middle of the strip and in the following passes progresses toward the outside edges, the vertical pitch line for any given section is established as the central starting point or dividing line with reference to the number and severity of all the bends to be made on either side. The object is to balance or distribute the amount of forming work to be done by the rolls as equally as possible to the left and right of this center line.

In the case of the angle, it would be at the exact center of the strip, while in the window-screen section in Fig. 66 it would be off center, namely, slightly to the right of the actual or geometrical center line of the flat strip width. You will observe that the number of bends on either side of this line is the same, namely, three; also that the amount of work to be done by the rolls on either side is about the same. For, while the top bend on the left virtually requires doubling the metal back on itself, the radius of bend is fairly easy, involving about the same amount of work as each of the other four bends to the left and right of the vertical pass line. These, as will be seen, form angles of only about 90°, but are made sharper, that is, with smaller inside radiuses.

In respect to sequence, the forming, as already mentioned, usually starts at the center, or as near to it as the profile permits, proceeding in the successive roll

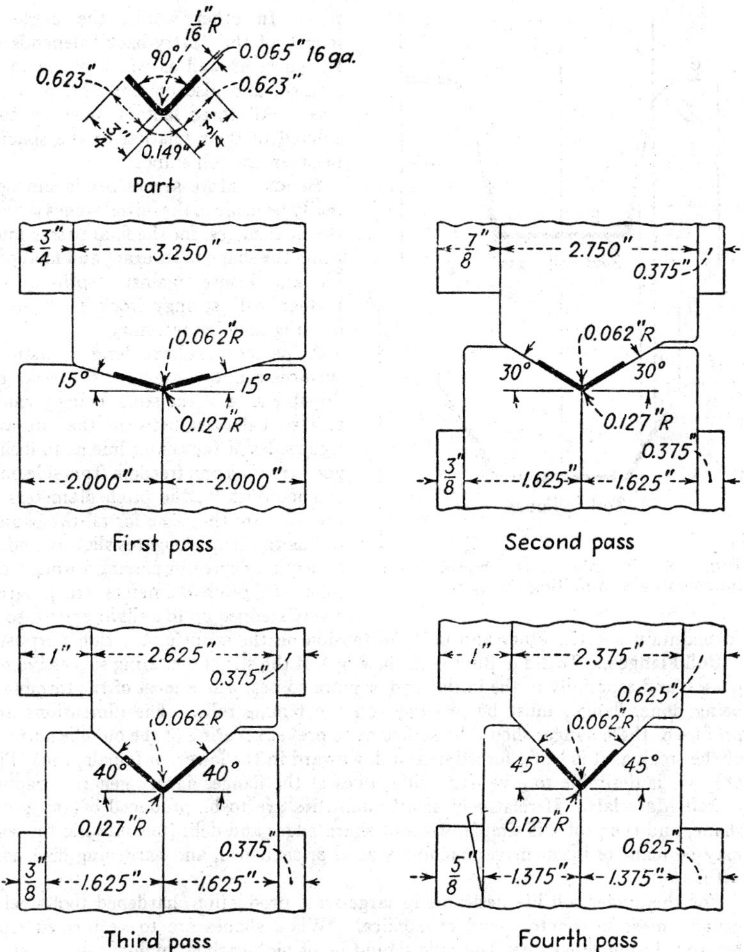

FIG. 67. In rolling a section, forming action must not "carry back" in the strip through the preceding roll.

passes toward the edges. The reason is that many bends or beads, if first formed at or near the edges, would lock the metal in place between the rolls, causing it to tear if subsequent forming is to be done nearer to the middle. Such forming obviously cannot be done without pulling the stock from both edges in toward the middle.

Number of Roll Passes. The amount of forming that can be done in any roll pass is limited (Fig. 67). A basic rule is that under no circumstances should a bend be great enough to "carry back" through the strip into and beyond the preceding roll

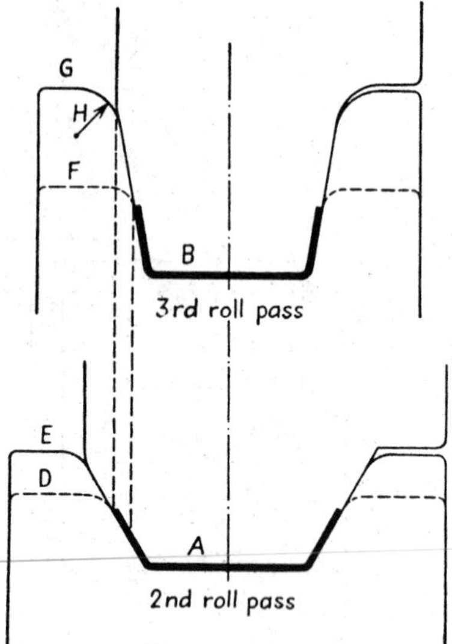

FIG. 68. Female rolls should have flanges to aid in pulling the strip.

pass. In other words, the angle or length of the "carry-back" depends on thickness as well as width of leg to be formed, and also on the kind of stock used. All three factors must be considered in their relation to the spacing between the roll stands.

Bends. More severe bends can normally be made in the initial stages than in the final stages, for the final passes must bring the shape to its exact and final profile and insure against "springback." In fact, with springy stock, slight overforming may be necessary.

All passes must be so designed that roll surfaces are in contact with the largest possible area of the stock being formed, and so balanced between the left and right sides of the center line as to insure good traction and freedom from side pull.

Theoretically, the pitch diameters of the rolls are the same for all the passes, but as the strip elongates slightly and to a varying degree in passing through the rolls, the pitch diameters are progressively stepped up to a slight extent, so as to maintain a satisfactory and uniform tension on the strip for traction purposes.

Roll Flanges. To aid in pulling in the edges of the strip in entering successive roll passes, and especially in the initial two or more passes, where most of the forming is being done, flanges must be provided on the female rolls. The dimensions and profiles of these flanges should be sufficient to prevent scoring of the outside surfaces of the stock as it is being forced in and downward in the lower, or female, rolls (Fig. 68). It is desirable to give the inside edges of the flanges a very generous radius.

Roll Materials. If relatively small quantities are to be produced of any given shape, and the profile is simple, without sharp edges and delicate contours, the rolls may be made of inexpensive machinery steel or semisteel, and hardening dispensed with.

For the longer roll life desirable in large-scale production, hardened tool steel is usually most satisfactory and economical. When shapes are to be formed from hot-rolled, unpickled steel, the rolls should be of high-carbon, high-chrome steel, so as to resist the abrasive action of the scaly surfaces of this stock.

Wide rolls are often made in sections, to the end that different parts thereof may be made of different materials, according to service requirements.

Section 16

FORGING, UPSETTING, AND COLD HEADING

SECTION 16

FORGING, UPSETTING, AND COLD HEADING

Forging may be defined as the shaping of metal by plastic deformation, usually at an elevated temperature. Basically, metal can be forged by impact or pressure. When forged by impact in hammers, the part is formed by dies that are flat or slightly shaped, as in smith forging, or in closed-impression dies (drop forging or impacter forging). Forgings made by pressure included those produced in mechanical or hydraulic presses, upsetters, and forging rolls.

By the various processes available, forging offers an excellent method of obtaining structurally sound parts particularly suited to mechanical construction. Forgings possess the quality of high strength for a given weight, plus superior resistance to loads and impacts. They may be produced in a wide variety of sizes and shapes but are limited in respect to undercuts and cored sections, although some excellent cored forging has been done in recent years. By careful techniques many forgings can be made of less raw material than a machined part, and the amount of machining left to be done is greatly reduced.

TYPES OF FORGINGS

The kinds of forgings produced and the types of equipment on which they are produced may be listed as:

Smith Forgings. These (sometimes known as hand, hammered, flat-die, or blacksmith forgings) are commonly produced on steam hammers. Small forgings may be made with motor-driven pneumatic hammers or helve hammers.

The rate of production is relatively low in most cases, and smith forging is principally used where the quantity is too low to justify dies or the work is needed too quickly to permit die sinking.

Drop Forgings. Drop forgings (also known as impact-die forgings) account for most of the tonnage produced commercially. They are formed by impact pressure on board drop, steam drop, or gravity drop hammers. The hammers are built to maintain alignment between top and bottom dies. The part is processed through a series of die impressions, generally combined in a single set of dies but sometimes requiring two or more sets.

Machine Forgings. These (also known as upset forgings) are produced on horizontal double-acting presses known as forging machines or upsetters. Operations such as bolts, cap screws, and rivets are made in cold headers, but most work is done hot. The heated bar stock is placed between a pair of gripper dies which close on the work before the header slide pushes a punch against the plastic metal to force it into the die impression. Simple parts can be produced in one pass. Several passes can be incorporated in one set of dies and punches when required.

Press Forgings. Mechanical or hydraulic presses are used for press forgings. The action is similar to that of the forging machine except that ram movement is vertical. Split lower dies as in the forging machine are possible but are rarely

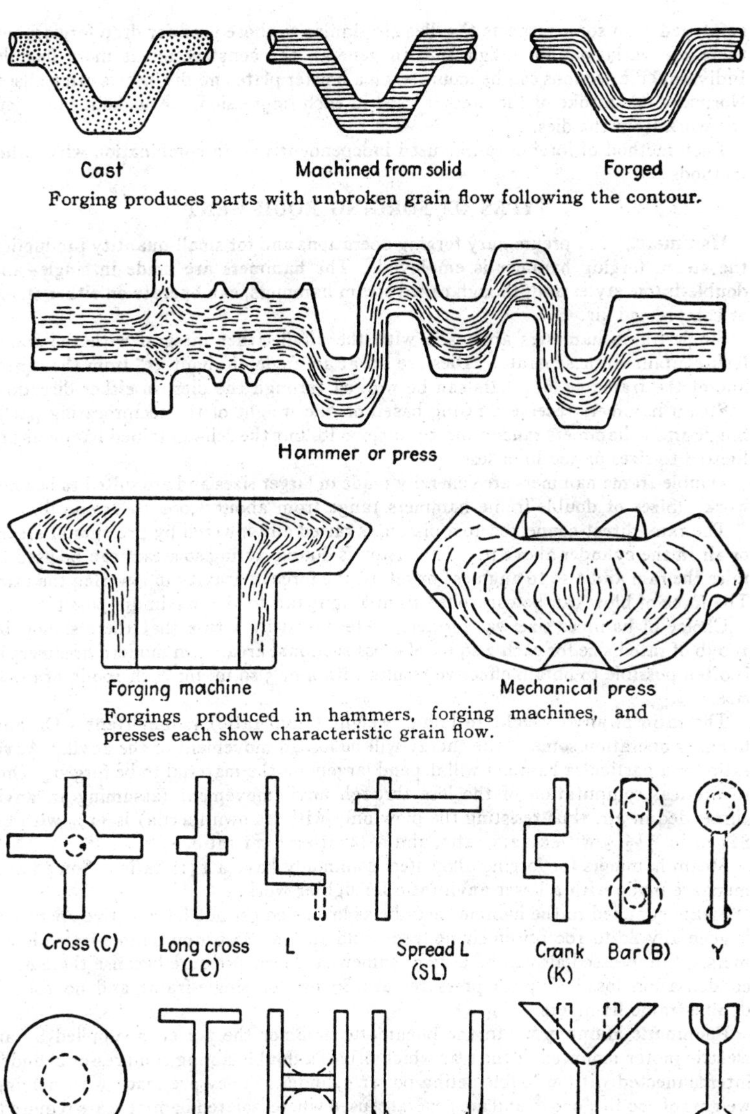

Forging produces parts with unbroken grain flow following the contour.

Forgings produced in hammers, forging machines, and presses each show characteristic grain flow.

Forging shapes frequently encountered and their designations.

employed. In some respects the dies are similar to those used for drop forging, but there are variations in design and, in general, the construction is more flexible. Individual die sections can be mounted on a bolster plate and die inserts can be used. Normally one stroke of the press is used in each impression. Knockout pins eject the work from the dies.

Each method of forging can be used independently or in combination with other methods.

TYPES OF FORGING EQUIPMENT

Hammers. For preliminary forging operations and for small-quantity production the steam forging hammer is employed. The hammers are made in single- and double-frame styles and, though called steam hammers, can operate on either steam or compressed air.

Single-frame hammers are made with the anvil either integral with the frame (self-contained) or separate. Dies are set at an angle of about 35° from the center line of the frame so long bars can be worked through the dies, in either direction.

Steam hammers receive a rating based on the weight of the reciprocating parts. Single-frame hammers range from 50 to 5,000 lb, but the self-contained hammers are limited to sizes of 300 lb or less.

Double-frame hammers are generally made in larger sizes and are suited to heavier work. Sizes of double-frame hammers range from about 1,000 to 25,000 lb.

The ram, direct-connected to a piston, is raised and lowered by pressure of steam or air in the cylinder above it. The piston is double-acting so steam can be used to raise the ram and also to augment or retard the force of gravity in lowering the ram. The force of blow can be controlled from a light tap to the maximum possible.

Choice of hammer size was formerly determined by a rule that there should be 100 lb of rated size for each 2 sq in. of cross-sectional area. On modern hammers it is often possible to obtain effective results with 4 or 5 sq in. for each 100 lb of rated size.

The ratio of anvil weight to falling weight (anvil ratio) is important. On any hammer operation some of the energy will be lost in movement of the anvil. Anvil ratio for a particular hammer will depend largely on the material to be forged. One theoretical computation of the loss through anvil movement (assuming an anvil suspended in air, thus resisting the blow only with its own inertia) is 11% with an 8:1 ratio, 7½% with a 12:1 ratio, and 6% with a 15:1 ratio.

Steam hammers for forging alloy steel commonly have a 15:1 ratio. Some hammers are made with a lower anvil ratio for lighter work.

Steam supplied to the hammer may be as low as 60 psi but for effective operation is generally 80 to 100 psi on single-frame and 100 to 120 psi on double-frame hammers. Compressed air can be used at somewhat lower pressure because there is no condensation loss. Typical pressures are 80 psi for single-frame and 90 psi for double-frame hammers.

Pneumatic Hammers. In the pneumatic hammer the power is supplied by an electric motor mounted at the rear which drives a double-acting compressor cylinder interconnected with a double-acting power cylinder. They are made with moving weights of 200 to 5,000 lb and are generally used where isolated hammers are required.

In the pneumatic hammer, a uniform number of blows per minute is struck, and this rate is not affected by varying the force of the blow.

Drop Hammers. Impact die forgings, or drop forgings, are produced in closed-impression dies in drop hammers. Drop hammers have the guides and other parts designed to maintain alignment between the dies. The principal types are the board drop hammer and the steam drop hammer.

Board Drop Hammers. Gravity is the force employed in the board drop hammer. The ram is raised to striking position by means of boards, the lower ends of which are wedged into the ram. The upper ends of the boards pass between one or two pairs of rolls. With the two rolls in a pair rotating in opposite directions, the boards and ram are raised when the rolls move together. As the ram reaches the top of the stroke the rolls separate and the boards are released.

When the treadle is depressed the ram falls at once, but when the treadle is not depressed the boards are held by a set of eccentrics or board clamps so the ram is in position to deliver the next blow when the treadle is depressed. As long as the treadle is held down the board drop hammer will automatically deliver blows at a uniform rate.

The only way the force of the blow can be varied in a board drop hammer is by changing the distance of fall.

The hammers are rated by the weight of the ram, or falling weight. This weight may range from 100 to 10,000 lb, but in most commercial forging plants is between 600 and 6,500 lb.

The ratio of anvil weight to falling weight is about 20:1 in most cases.

Steam Drop Hammers. The steam drop hammer is similar in operation to the steam forging hammer except that the frame is mounted on the anvil so top and bottom dies can be held in alignment. Anvil ratios are higher, usually 20:1 as with board drop hammers, though the ratio may be 25 or more to 1.

Steam drop hammers are rated by the weight of reciprocating parts, not including the top die. They are made in sizes from 500 to 50,000 lb.

The throttle valve is usually of the rotary type and can be opened to any point to control the force of the blow.

Gravity Drop Hammers. This hammer uses compressed air or steam to lift the ram and lets it fall by gravity as in the board drop hammer. The ram is lifted by a steel rod connecting it to the piston. A separate air circuit operates a clamp which holds the ram in the raised position between strokes. The rate of the gravity hammer is faster than with the board drop hammer.

Forging Presses. Machines for forging steel are of massive rigid construction with steel frames and heavy crankshafts with eccentrics to operate the pitman and ram.

On small and medium presses, positive clutches of the rolling-key or jaw type, or friction clutches, operate the press. Heavy presses use air-operated clutches.

Mechanical ejectors or knockouts are provided in the top and bottom dies which are operated by cams and levers. When a longer, more powerful, or delayed knockout is needed on the lower die, it can be operated by an air cylinder.

On a mechanical press the speed of the ram is greatest at the center of the stroke. The pressure is greatest, however, at the bottom of the stroke as the pitman reaches bottom dead center. The rating of forging presses, based on the estimated pressure at this point, varies from about 200 to 6,000 tons. In some instances the designation is limited to a size number. Such ratings are from No. 1½ to No. 10 (producing in one blow in each die impression the size of work regularly handled on 500- to 12,000-lb steam drop hammers).

One method of comparing the capacity of mechanical presses with hammers is to multiply the tonnage rating by 2 for steam-drop and by 2½ for board-drop ratings in pounds. By this method a 1,000-ton press would have approximately the capacity of a 2,000-lb steam or a 2,500-lb board drop hammer.

Hydraulic Presses. Hydraulic forging presses move the ram by the pressure of a fluid. Steam and water are used, but oil is more common in the newer presses. The hydraulic circuit is designed so the ram has a fast speed for traversing and a slow speed for working. The control is extremely flexible, permitting adjustment of the length of stroke and pressure attained. Frequently the system can be adjusted to reverse automatically when a preset pressure is reached.

On the very large hydraulic presses, accumulator bottles are used to build up the pressure as in heavy presses for forging aluminum parts for aircraft. Water is the operating fluid.

Forging Rolls. Stock can be drawn out to long slender sections in forging rolls. Straight or tapered work can be forged completely at high speed. Parts such as axle shafts, leaf springs, and automotive shift levers are suited to production in forging rolls. More often the operation is used as a preliminary to finish-forging in a press or hammer. Blanks can be produced rapidly which will reduce the number of subsequent operations.

Circular products are rolled between dies in a mill. The blank is pierced at one end to position it on the vertical die. When the blank is loaded, the vertical die advances under forging pressure and both dies revolve. Some upsetting from the initial pressure plus rolling action shapes the work to the die contours. Gear blanks, wheels, brake drums, and turbine rotors are typical of the work produced by this method of forging.

Forging Machines or Upsetters. This is a horizontal double-acting press originally developed to gather or upset metal to form boltheads. This original purpose has been broadened to include a variety of forging work including not only upset but displacement operations. A stationary and a moving die plate grip the work while a heading tool moves against the stock, forming it into the cavities in the dies. As the heading slide recedes, the dies open. When work requires more than one pass, several cavities are included in one set of dies and the work is shifted to the position for the next pass.

The size of a forging machine is indicated in inches. Sizes range from ½ to 10 in.

An air clutch connects the flywheel to the crankshaft from which an eccentric operates the header slide through a pitman. A cam or cams on the crankshaft operate the movable die through a cam slide which by toggle action closes the grip slide. A grip relief on the cam slide releases if stock is misplaced in the dies or for other reasons the dies cannot close. This is designed so that when it relieves, the movable die opens at once. The grip relief automatically resets on the next stroke.

When production is sufficient, hot-forging machines can be equipped with transfer mechanisms for automatic operation. In most cases, the variety of work handled makes it necessary to feed and control the machine by hand.

COLD HEADERS

Cold heading is the process whereby products like screws, bolts, rivets, and similar round-cross-section heads are made from wire or rod on a horizontal press arranged for continuous fast operation and automatic feed and cutoff. The cold header merely upsets the end of the stock. If a square or hexagon head is required, a subsequent trimming operation is needed.

Header dies may, or may not, include cavities for shaping the part (Fig. 1). Most simple head forming is done on the face of the die, with the punch cavity shaping the head. For countersunk-head parts, however, the punch fact is flat and the cavity is entirely in the die. If two or more diameters are required under a head, and the differential is greater than obtainable by extrusion, the head is formed by the punch, and the underhead diameters are upset in the die cavity.

Standard headers are designated as single, double, triple, or multiple stroke.

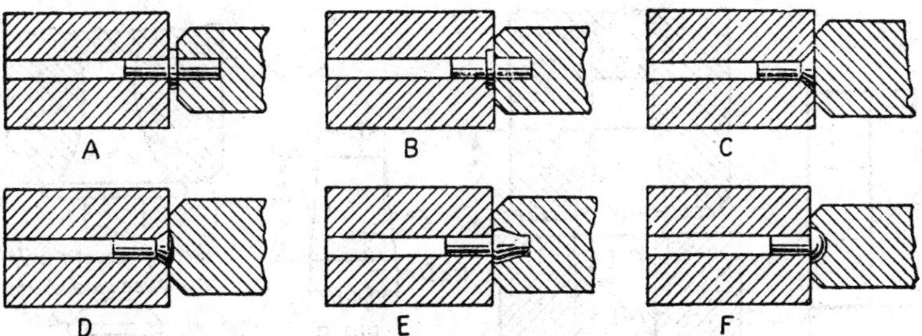

FIG. 1. Header dies may or may not have cavities:
A. Forming a collar without a die cavity is satisfactory if a radius on the OD is permissible.
B. Square edges on the collar necessitate a cavity in either the die or punch.
C. Countersunk heads are formed entirely in a die cavity.
D. Oval heads are produced partly in die and partly in punch.
E. Coning punches start the upset of the material.
F. Round-head parts are formed with a flat-faced die, the entire cavity being in the punch.

When two or more strokes are required to make the piece, the first and intermediate punches are known as containing punches and serve to gather the metal into approximately a conical shape, allowing it partially to head next to the die. The final shape is obtained in the final stroke. Formulas for coning punches are given in Fig. 2.

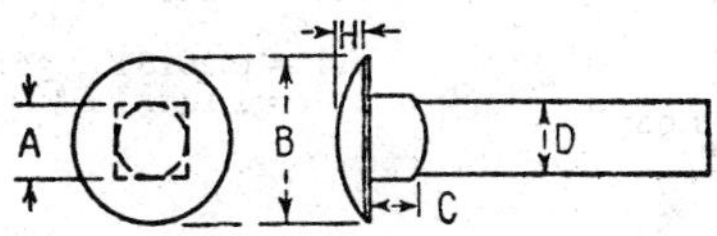

TABLE 1. ALLOWABLE HEAD DIAMETERS IN RELATION TO WIRE DIAMETERS

Nominal Dia *D*	Dia Head, In. *B*	Thickness of Head, In. *H*	Width of Square, In. *A*	Depth of Square, In. *C*	Pressure Required in Tons to Upset
3/16	0.530	0.120	0.205	0.125	40
1/4	0.610	0.142	0.260	0.125	50
5/16	0.730	0.168	0.320	0.187	65
3/8	0.844	0.194	0.382	0.187	100
7/16	0.970	0.220	0.450	0.250	115
1/2	1.094	0.250	0.510	0.250	125
5/8	1.343	0.343	0.630	0.250	200
3/4	1.593	0.375	0.735	0.250	300

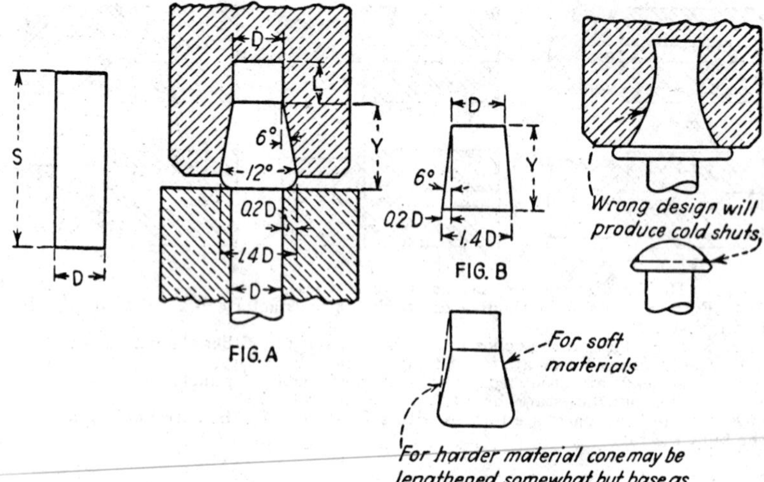

FIG. 2. Formulas for coning punches used on cold headers.

Formula 1

$$\tan 6° = \frac{0.2D}{Y}$$

$$0.1051 = \frac{0.2D}{Y}$$

$$Y = \frac{0.2D}{0.1051} = 1.903D$$

Formula 2

Volume of cone Fig. B

$$\text{Vol} = 0.2618Y\left(\frac{2}{D} + D \times 1.4D + \frac{2}{1.4D}\right)$$

$$\text{Vol} = 0.2618Y(4.36D^2)$$
$$= 1.1414YD^2$$

Formula 3

$$\frac{\text{Vol of cone}}{\text{Area } D} = \text{length of } D \text{ wire to make cone}$$

$$\frac{1.1414YD^2}{\pi D^2/4} = 1.453Y$$

Formula 4

$$\frac{\text{Length}}{\text{Wire dia}} = \text{No. of diameters in cone portion}$$

EXAMPLE: Let $D = 0.375$ in.
From formula 1, $Y = 1.903 \times 0.375 = 0.71358$ in.

From formula 2, $\text{Vol} = 1.1414 \times 0.7136 \times \frac{2}{0.375} = 0.0114489.$

From formula 3, $\text{Length} = 1.453 \times 0.71358 = 1.0368$ in.

From formula 4, $\text{No. of diameters} = \frac{1.0368}{0.375} = 2.76$ diameters of wire in Fig. 3.

Where base of cone $= 1.3D$ then (4) $= 1.9$ diameters; where base of cone $= 1.4D$ then (4) $= 2.76$ diameters; where base of cone $= 1.5D$ then (4) $= 3.76$ diameters; all irrespective of D.

To length of wire from (3) add straight portion L to obtain S, or total length of wire in cone.

Selection of Headers. The type of header required for a particular job depends upon the amount of material needed to form the upset, the length of the shank, and the location and shape of the head. Allowable head diameters in relation to wire diameters are given in Table 1. Normally in a single-blow header it is not advisable to attempt a head diameter exceeding 2¼ times the wire diameter. The selection can also be expressed as:

Single-stroke header: Max $L = 2\frac{1}{4}D$
Double-stroke header: Max $L = 4\frac{1}{2}D$
Triple-stroke header: Max $L = 6$ to $8D$

where L = length of wire to form the head and D = wire dia.

In a solid-die header, the shank length S of the part should not exceed 8 to $10D$, for a machine operated at normal speed. Special lengths up to $12D$ are possible, but the machine speed must be reduced to avoid trouble in clearing the work from the dies.

In an open-die header, length S should not be less that than $4D$. Solid dies are merely steel or carbide cylinders with a hole through the center; open dies are two rectangular blocks with matching half holes on each of the four sides.

Heading Wire. Open-hearth bright-drawn basic wire gives the best results in cold heading. The usual tolerance is ±0.002 in. A good low-carbon heading wire has this analysis: 0.14 to 0.20 carbon, 0.30 to 45 manganese, 0.10 silicon, and phosphorus and sulfur, 0.04 max. High-carbon wire for stock to be heat-treated has this analysis: 0.27 to 0.37 carbon, 0.79 to 0.90 manganese, 0.10 silicon, and phosphorus and sulfur, 0.04 max. If there is only a small upset, and some machining must be done after heading, a low-sulfur bessemer wire may be used.

Many materials can be cold-headed successfully, including aluminum and its alloys in 2S, 3S, 17S, A-17S, 24S, and 53S, and yellow screw brass and cartridge brass.

Carbide Cold-heading Dies. Successful application of cold-header dies made of carbide extends from No. 4 to ¾-in. screws, also rivets of corresponding sizes. Average life of carbide dies is 15 to 20 times that of steel dies (10 times is the break-even point). An advantage of carbide is ability to hold closer tolerances on extruded shanks for longer runs, which increases life of thread-rolling dies. Carboloy Co. recommends Grade 190 carbide for cold-heading die nibs and hammer inserts, because this grade is easily drilled and turned, minimizing costly lapping when making new dies or recutting inserts to larger size. This tough, impact-resisting grade of carbide is also suitable for inserts in coining and swaging dies for mild-steel parts, and inserts in cold-swaging dies. Proportions of die inserts for header tools are shown in Fig. 3.

FORGING DIES

There are three principal types of forging dies (Fig. 4):

Single impression—one half of the die contains the complete impression; the other half is flat.

Double impression—part of the impression is sunk in each half of the die, but no part of one die extends past the parting line into the other die.

Interlocked—a projection from either die extends past the parting line into the mating die.

Drop-forging Dies. The elements for a die block for a drop forging are shown in Fig. 5. The forging drawing should show the parting lines, draft angles, fillet and corner radiuses, forging tolerances, and machining allowances. If these have been approved by a forging engineer, mistakes will be avoided and time saved. The

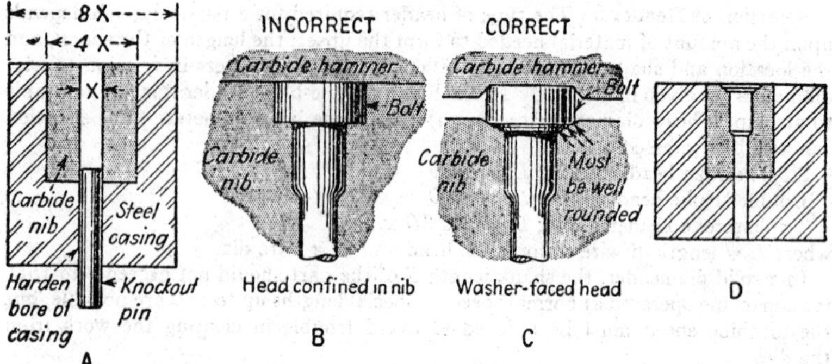

FIG. 3. Carbide cold-header dies should have the proportions at *A*. The bolthead is not completely confined in the carbide, *B*. Sketch *C* shows correct setup for washer-faced bolt. Pointing can be combined with heading, as at *D*.

Best ratios between nib ID and OD and casing OD for carbide cold-heading dies are indicated above. OD of casing should be at least twice nib OD to insure adequate support for the carbide insert. Work shank should preferably be confined within the nib, with the knockout pin entered part way into the nib bore. But when the bolt to be made is long, wear can be equalized by hardening the casing bore.

Damage can be avoided only if the bolthead is not completely confined in carbide. Sketches *A* and *C* show correct setups for plain and washer-faced bolts. Never inset cavity for head in nib as at *B*. Eliminate all sharp corners in nib or hammer to avoid stress concentrations likely to cause die failure. Corners should be rounded off or filleted as indicated above for best results.

In many operations, pointing can be combined with heading by restricting bore of carbide die nib as indicated above. Allowance for this practice has been made when developing standard nibs.

actual layout of the dies can then proceed, such as planning the number of operations and designing the impressions. For drop forgings, it is customary to place the finish impression at the center if possible, because this operation requires the greatest force.

Die blocks are drilled with holes into which handling pins can be inserted. Blocks are then paired and the shank planed to suit the hammer on which they are to be used. Blocks are turned over and machined parallel to the face of the shank, removing about ¼ in. to provide a clean sound surface for the impressions.

Two edges (designated the match edges) are then machined at right angles to the face and to each other. This may remove metal about ¼ in. inward and 1 in. downward. All measurements are made from the match edges. When dies are mounted in the hammer, these match edges are aligned.

The finishing impression is machined first. It is laid out on the die face from the templet and the impression machined. Bench work follows.

Templets or a model of the part (called a "type") are used to check the work. To use the type, the impression is coated with prussian blue and the type is inserted. The diemaker strikes the type with a hammer, then removes it. Where high spots exist, the dye has been removed. Additional bench work to remove the high spots is followed by another typing operation. The process is continued until even bearing is obtained.

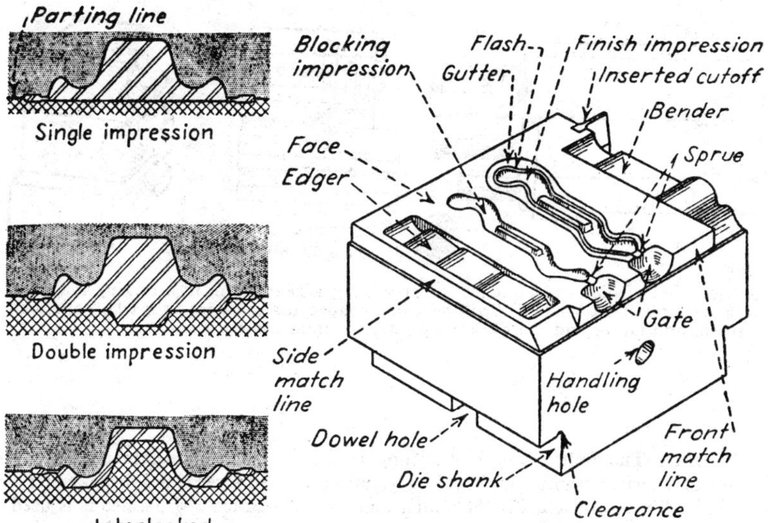

FIG. 4. Three common types of forging dies.

FIG. 5. Elements of a die block for drop forging.

When the finish impression is machined, the die blocks are clamped together and a lead cast is made by pouring molten metal through the sprue. This lead cast is checked for dimensions.

A gutter for flash is machined around the finished impression. Impressions for preliminary operations are machined in the blocks. A cutoff can be milled in one corner of the blocks or inserted in a slot machined in the side of the block.

The final step is to mill dowel holes into the shanks.

Press Dies. When designed for a forging press, dies are frequently of the die-set type with guide pins, to maintain alignment. Separate die blocks are customarily used for each impression. Press forgings are produced from slugs or roll-forged billets, the latter reducing the amount of work to be done in the press. Lifter pins or knockouts help to raise the work out of an impression so that it can be manipulated into the next impression or pushed out of the press.

Forging-machine Dies. Parts upset in a forging machine may require from one to five operations. Tooling consists of a pair of rectangular die blocks, one stationary and the other moving, and a punch slide mounting the required number of heading punches. Figure 6 shows a set of dies and punches for producing a gear blank in a forging machine. Stock is gathered in the bottom impression, upset to final dimensions at the center, and punched from the bar at the top. The wad punched from the center stays on the bar and is used in making the next piece.

Definite rules for design of forging machine dies to work solid stock have been developed by National Machine Co.:

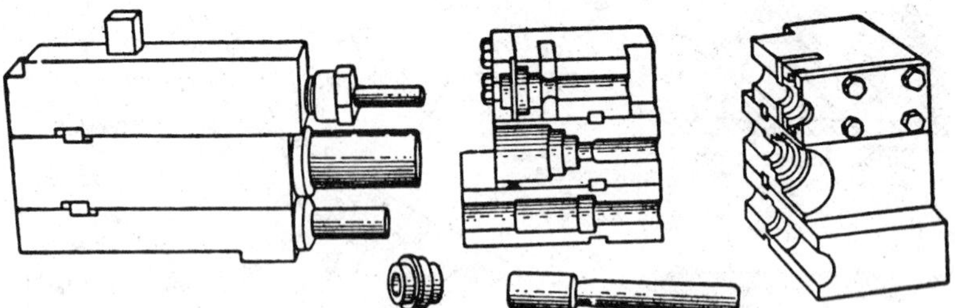

FIG. 6. Set of dies and punches for producing a gear blank in a forging machine, or upsetter. This kind of machine can produce many types of parts, solid and pierced, and larger and more complex than the item shown.

RULE 1. The limit of length of unsupported stock that can be gathered or upset in one blow, without injurious buckling, is not more than three times the diameter of the bar. Three diameters is the actual limit, so in practice 2½ diameters is used. This rule holds true whether the upset is in the open or in a cavity in the punch or grip dies, or both.

RULE 2. Lengths of stock more than three times the diameter of the bar can be successfully upset in one blow, provided the diameter of the upset made is not more than 1½ times the diameter of the bar.

Again the rule indicates the limit. In practice 1.3 times the diameter is used as the maximum. Since stock will buckle near the center, more than half the stock should be inside the die cavity at the start of the upset. The amount upset by this method can be increased by making the first impression square or by increasing the cavity near the grip die beyond the limit.

RULE 3. In an upset requiring more than three diameters of stock in length, and in which the diameter of the upset is 1½ times the diameter of the bar, the amount of unsupported stock beyond the face of the die must not exceed one diameter of the stock.

If the diameter of the hole is reduced, the length of unsupported stock can be correspondingly increased. For a 1¼-dia hole, unsupported stock may be 1½ diameters. Since this provides limited travel of the heading tool, taper upsets are frequently used. A good angle of taper is 4°. Using a taper, Rule 2 applies to the midpoint of the taper. Rules advocated by Ajax Mfg. Co., and illustrated in Fig. 7, are similar in general to those published by National Machinery Co. Rules for upsetting tubing are available.

When designing for sliding dies, these three rules apply. In this case, unsupported stock refers to the stock between sliding and grip dies.

Once the necessary sequence of passes is established, the body blocks (or grip dies) and punches can be designed. Perhaps the most common practice is to locate the passes in sequence from top to bottom, where work is to be handled by hand, although some designers prefer to work from the bottom up. In some cases where unusual pressure is required for the final pass, it is located at the center to receive full pressure of the header slide without tending to distort the punch. Other operations will then be in sequence above and below.

I-Unsupported Working Stock-One Blow

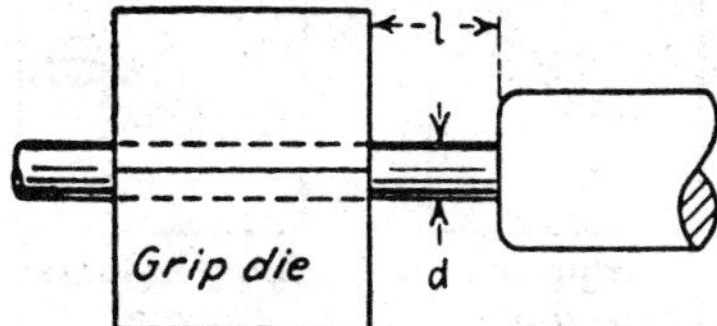

Working recommendations:

l should not exceed $2\frac{1}{2}d$

Exposed working stock up length of $3d$ can be upset without support. Stock is "exposed" even when within die impression or heading-tool recess too large to lend support. Stock projecting into tight recess in tool is not considered working stock or exposed.

II-Stock Supported in Die Impression-One Blow

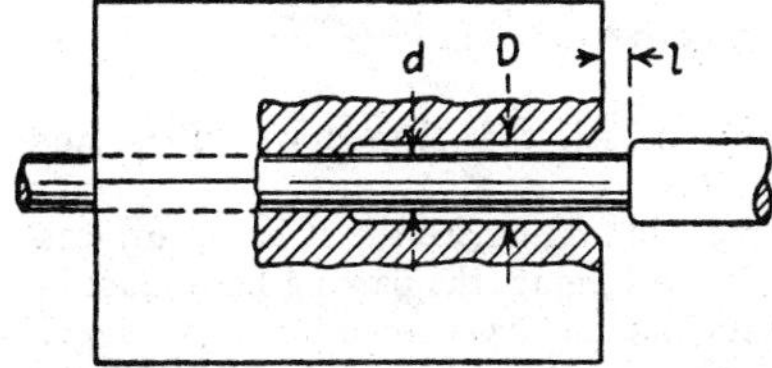

Working recommendations:

D should not exceed $1.4d$

l should not exceed $D-d$

Working stock longer than $3d$ will be supported against injurious bending by a die diameter not exceeding $1\frac{1}{2}d$. Length of stock projecting beyond the impression should not exceed $\frac{1}{2}d$.

III-Stock Supported in Tool Recess-One Blow

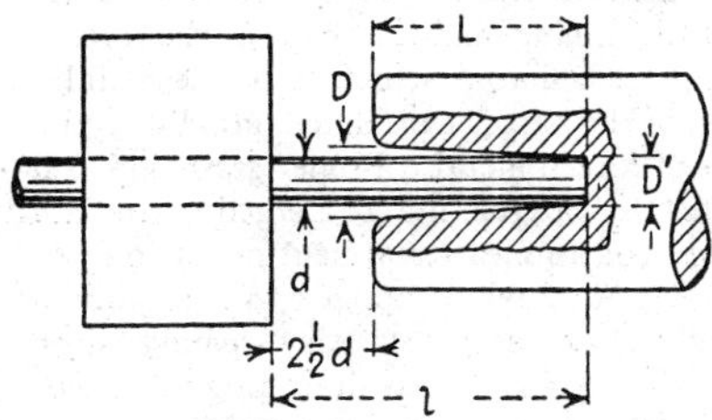

Working recommendations:

D' should not exceed $1\frac{1}{8}d$

D should not exceed $1.4d$

L should not be less than $\frac{2}{3}l$

Working stock longer than $3d$ will be supported against injurious bending by a heading-tool recess not exceeding $1\frac{1}{2}d$ at the mouth and $1\frac{1}{8}d$ at the bottom, and a recess length not less than $\frac{2}{3}l$, or not less than $l - 2\frac{1}{2}d$.

FIG. 7. Rules for upsetting bar stock as developed by Ajax Mfg. Co. The working rules given are not maximums as in the text.

Frequently, individual die blocks will be used for each step in machine forging. These are keyed to maintain alignment when clamped in the forging machine. This arrangement simplifies replacement if dies wear faster on some steps than on others.

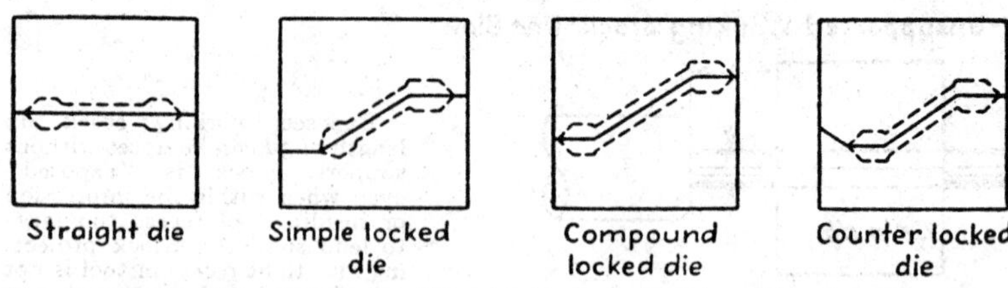

FIG. 8. Types of parting lines. Straight die is preferable. Side thrust is present in locked dies and may be in compound locked dies. Counterlock prevents side thrust.

FORGING DESIGN

In planning a forging, a parting line must be selected, then draft, fillets, and tolerances applied.

Parting Line. The surfaces of dies that meet in forging are the striking surfaces, and the line of meeting is the parting line. On a forging, the parting lines must be determined in order to establish draft and its location. On some forgings, several possible parting lines exist. Parting lines are classified as straight, single lock, compound lock, and counterlock (Fig. 8). Although the straight parting line is most economical, most dies have locked parting lines.

Parting lines are located so that little or no side thrust is imparted to the dies. This is done by placing them at right angles to the direction of die thrust or by adding a balancing angle or counterlock to the die, in excess of part requirements, in order to obtain the required thrust balance. The best location and type of parting line are determined by die cost, the need for smooth trimming, and the shape of the forging.

Grain Flow. Hot working of metal imparts a fibrous structure to the metal in the direction of working. When the design of the part is considered, grain direction offers one of the most important properties of forged metal. Proper grain flow contributes to fatigue life of forged parts and helps to control heat-treatment distortion.

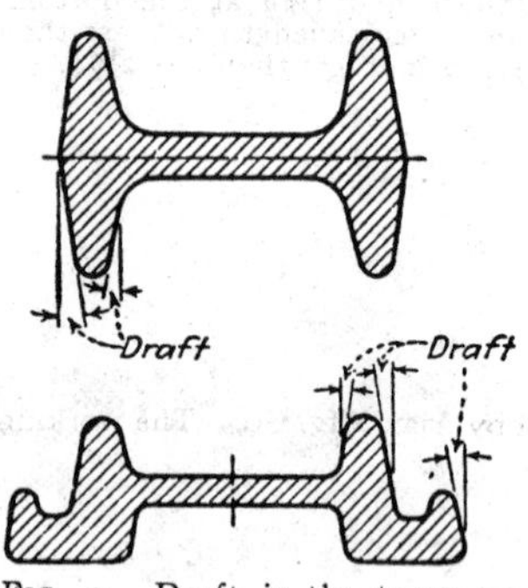

FIG. 9. Draft is the taper applied to side walls of a forging, inside and outside, so that the part can be removed from the dies.

Draft. An angle is applied to side walls of a forging, to permit its easy removal from the dies, and is known as draft (Fig. 9). Normal draft or taper on exterior surfaces of drop forgings is 7°; for interior contours, 10°, because the latter may grip the dies during cooling. In special cases, and for thin sections, where shrinkage will not wedge the forging in the dies, a 5° draft may be used. On machine and press forging, draft may not be required, but commercial practice is to allow 3°.

Die-draft Equivalent. The amount of offset that results from applying draft to a forging is known as the die-draft equivalent. Table 2 gives draft equivalents for various draft angles and depth of draft.

Fillets and Corner Radiuses. Adequate

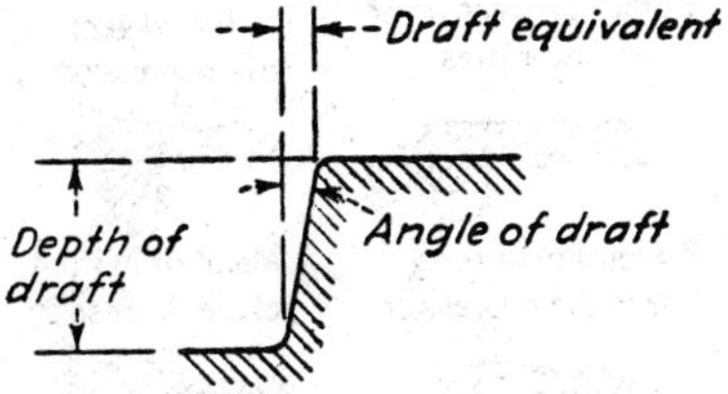

Draft equivalent
Depth of draft
Angle of draft

TABLE 2. DIE-DRAFT EQUIVALENT

Depth of Draft	Draft Equivalent for Angle of		
	5°	7°	10°
0.03	0.0026	0.0037	0.0053
0.06	0.0052	0.0074	0.0106
0.09	0.0079	0.0110	0.0159
0.12	0.0105	0.0147	0.0212
0.16	0.0140	0.0196	0.0282
0.19	0.0166	0.0233	0.0335
0.22	0.0192	0.0270	0.0388
0.25	0.0219	0.0307	0.0441
0.28	0.0245	0.0344	0.0494
0.31	0.0271	0.0381	0.0547
0.34	0.0298	0.0418	0.0599
0.38	0.0332	0.0467	0.0670
0.41	0.0359	0.0503	0.0723
0.44	0.0385	0.0540	0.0776
0.47	0.0411	0.0577	0.0829
0.50	0.0438	0.0614	0.0882
0.56	0.0490	0.0689	0.0987
0.62	0.0542	0.0761	0.1093
0.69	0.0604	0.0847	0.1216
0.75	0.0656	0.0921	0.1322
0.81	0.0709	0.0995	0.1428
0.88	0.0770	0.1081	0.1551
1.00	0.0875	0.1228	0.1763

TABLE 3. THICKNESS TOLERANCES

Net Weight up to, Lb	Commercial, In.		Close, In.	
	−	+	−	+
0.2	0.008	0.024	0.004	0.012
0.4	0.009	0.027	0.005	0.015
0.6	0.010	0.030	0.005	0.015
0.8	0.011	0.033	0.006	0.018
1	0.012	0.036	0.006	0.018
2	0.015	0.045	0.008	0.025
3	0.017	0.051	0.009	0.027
4	0.018	0.054	0.009	0.027
5	0.019	0.057	0.010	0.030
10	0.022	0.066	0.011	0.033
20	0.026	0.078	0.013	0.039
30	0.030	0.090	0.015	0.045
40	0.034	0.102	0.017	0.051
50	0.038	0.114	0.019	0.057
60	0.042	0.126	0.021	0.063
70	0.046	0.138	0.023	0.069
80	0.050	0.150	0.025	0.075
90	0.054	0.162	0.027	0.081
100	0.058	0.174	0.029	0.087

Fillet and corner tolerances apply to all meeting surfaces unless larger radiuses are specified.

TABLE 4. FILLET AND CORNER TOLERANCES

Net Weight up to, Lb	Commercial Radius, In.	Close Radius, In.
0.3	3/32	3/64
1	1/8	1/16
5	5/32	5/64
10	3/16	3/32
30	7/32	7/64
100	1/4	1/8

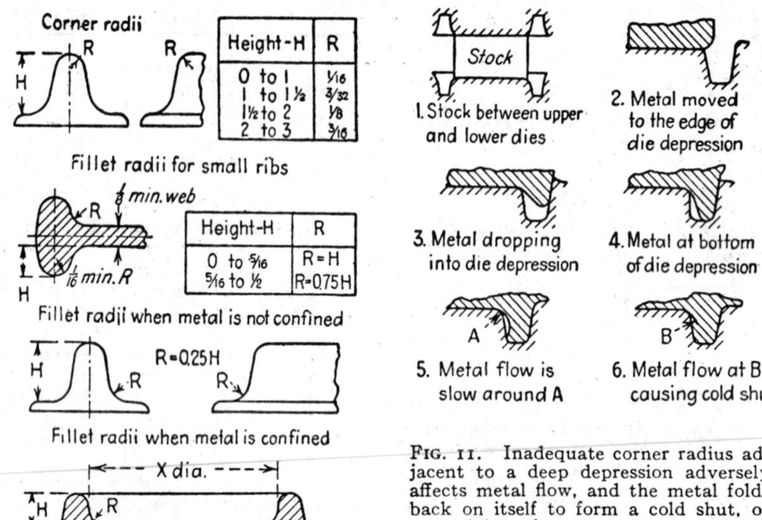

FIG. 11. Inadequate corner radius adjacent to a deep depression adversely affects metal flow, and the metal folds back on itself to form a cold shut, or potential crack.

FIG. 10. Corner radiuses and fillet radiuses on forgings are important matters for the forging designer.

radiuses are required where two surfaces of a forging meet. Corner and fillet radiuses should be as large as possible to assist flow of metal, to promote economy, and to avoid cold shuts, die cracking, and heat-quench cracking. Figure 10 illustrates minimum corner radiuses and fillets. Larger values increase die life.

Sharp fillets cause formation of cold shuts (Fig. 11) which form weak spots, because two surfaces of metal have folded against each other and may open into a crack at heat-treatment. Cold shuts are likely to form at fillets in deep depressions or in deep sections, especially where the metal is confined. In such cases, larger radiuses are required.

Forging Tolerances. Sufficient allowance must be made in the design of the forged part for variation in outline dimensions and weight of the finished part. These variations are caused by die wear and mismatch, or lateral misalignment of the two halves of the die. Die wear causes the greatest variation in contours and planes of and parallel to the parting line. In many parts, variations in contours measured perpendicular to the parting line will cause large variations in weight. These variations can be controlled only by resinking the die. Mismatch is controlled by setup and machine conditions.

Tolerances for impression-die forgings up to 100 lb weight have been established by the Drop Forging Association. Thickness tolerances are given in Table 3, fillet and corner tolerances in Table 4, and draft-angle tolerances in Table 5.

Width and length tolerances are alike and are divided into shrinkage and die wear, mismatching, and trimmed-size tolerances.

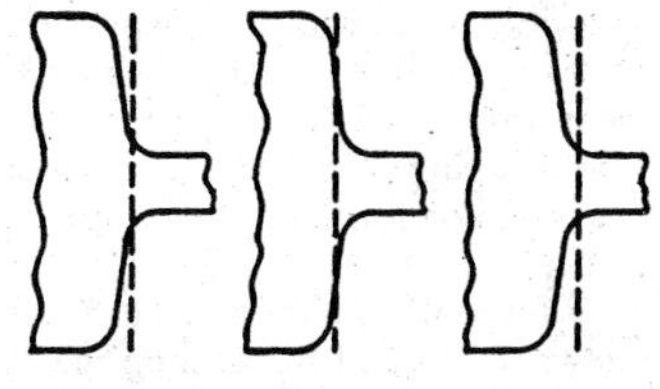

FIG. 12. Normal trimming is customarily specified on forgings.

TABLE 5. DRAFT-ANGLE TOLERANCES

	Nominal Angle	Commercial Limit Max	Close Limit Max
Drop forgings:			
Outside	7	10	8
Inside holes and depressions:			
Commercial limits	10	13	..
Close limits	7	..	8
Upset forgings:			
Outside	3	5	4
Inside holes and depressions	5	8	7

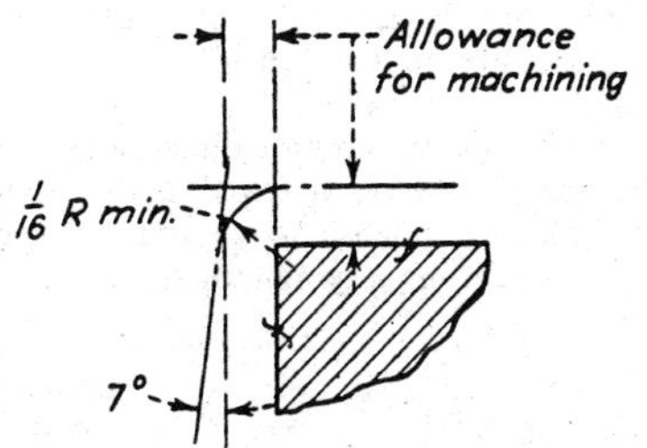

FIG. 13. Machining allowance and draft must be added to forgings.

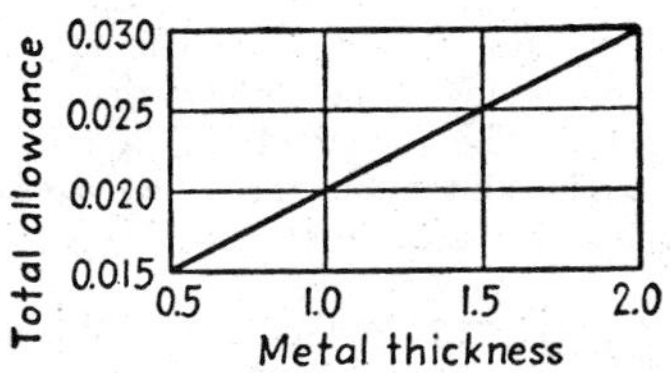

FIG. 14. Coining allowance normally follows this chart.

Shrinkage and die-wear tolerances apply to dimensions formed within a single die block. Neither of them is applied separately. The sum of the two is applied but not to include draft or draft variation. Shrinkage tolerances for the commercial standard are ±0.003 in. per in. and for the close standard are ±0.0015 in. per in. Commercial die-wear tolerances are ±0.032 in. for the first pound of net weight and ±0.003 in. for each additional pound or fraction of a pound. Close die-wear tolerances are ±0.016 in. for the first pound of net weight and ±0.0015 in. for each additional pound or fraction of a pound.

Mismatching tolerance is displacement of a point formed in one die block from its desired position as located from the part of the forging formed in the other die block. It is measured parallel to the main parting plane. For forgings up to 1 lb the tolerance is 0.015 in. for commercial standard and 0.010 in. for close standard. For each additional 6 lb or fraction thereof 0.003 in. is added to the commercial tolerance and 0.002 in. is added to the close tolerance.

Trimmed size must fall within the limits imposed at the parting plane by the sum of the draft angle, shrinkage, and die-wear tolerances.

Trimming Allowance. Drop and press forgings usually have flash on the parting line that must be trimmed off. Types of trim are: normal, close, and loose (Fig. 12). In some cases a close trim is relatively easy to make, but in others it may seriously complicate the forging operation. Trim loss in weight of material per forging varies from a few per cent to a considerable amount, depending on the type of material and the forging design. Advice from the forging engineer should be obtained on this matter.

Machining Allowance. When a forging is to be machined, allowance must be made for additional metal. If surfaces carry draft, the draft is additional to the machining allowance (Fig. 13). Forgings of normal size ordinarily require machining allowance of 1/16 in. while large and more intricate forgings need say 1/8 and more. Die wear must also be considered in applying machining allowances. The greatest amount of die wear occurs on the slant of the draft.

Coining Allowance. Coining may be applied to forgings to gain a smooth finish and close tolerance. The amount of coining allowance to add for coining may be taken from Fig. 14.

FORGING ALUMINUM

Shrinkage Allowance. For ordinary parts, allow 1/10 in. per ft; for large parts, up to 1/8 in. per ft.

Coining Tolerances. Boss thickness can be held to ±0.005 in. Total area to be coined should not exceed 8 sq in., if this tolerance is to be maintained.

Refer to Tables 6 to 11 for other tolerances.

Computing Forging Dimensions. An example of how to compute dimensions of a forging, giving effect to the various tolerances and allowances required, is shown in Fig. 15. The first step toward establishing the forging dimensions W_F and H_F is to determine all the tolerances involved as given in the accompanying tables. For this forging they are:

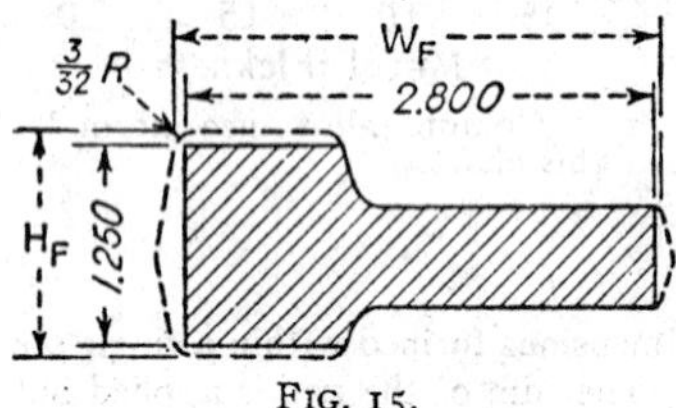

Fig. 15.

Die closure	+0.032	−0.015
Length or width	+0.032	−0.016
Straightness	0.016	
Mismatch	0.015	

If the minimum cut is set at 0.015 in. per side, the machining allowance for both sides of the forging would be:

Minimum cut (2 sides)	0.030
Length or width tolerance	0.016
Straightness tolerance (2 sides)	0.032
Mismatch tolerance (2 sides)	0.030
Total machining allowance	0.108

The width of the forging then is $W_F = 2.800 + 0.108 = 2.908$ or $2^{29}/_{32}$ in.

The machining allowance for the top and bottom of the heavy lug would be:

Minimum cut (2 sides)	0.030
Die-closure tolerance	0.015
Straightness tolerance (2 sides)	0.032
Total machining allowance	0.077

The height of the forging then is $H_F = 1.250 + 0.077 = 1.327$ or $1^{21}/_{64}$ in.

The 3/32 in. radiuses, however, might not clean up with sharp machined corners, when these forging dimensions are used. If sharp corners are imperative, machining allowances equal to the full edge radius would be ample. Adding twice the radius, or 3/16 in., to the 1.250-in. machined height gives for H_F a practical forging height of $1^7/_{16}$ in. Similarly, adding twice the radius, or 3/16 in., to the 2.800-in. finished width gives 2.988 in. for W_F, or about 3 in., for the width of the forging.

TABLE 6. STANDARD DIE CLOSURE TOLERANCES FOR ALUMINUM

Magnesium Net Weight of Forging, Lb		Aluminum Net Weight of Forging, Lb		Die Closure Tolerance, In.	
From	To	From	To	+	−
0	1/4	0	1/4	0.032	0.010
1/4	1	1/4	1	0.032	0.015
1	3	1	4	0.045	0.032
3	11	4	17	0.062	0.032
11	16	17	24	0.078	0.032
16	33	24	50	0.093	0.032
33	67	50	100	0.125	0.045
67	170	100	250	0.187	0.062
170	...	250	...	0.250	0.062

TABLE 7. STANDARD MISMATCH TOLERANCES

Magnesium Net Weight of Forging, Lb		Aluminum Net Weight of Forging, Lb		Tolerance, In.
From	To	From	To	
0	1	0	1	0.015
1	5	1	7	0.018
5	9	7	13	0.021
9	13	13	19	0.024
13	17	19	25	0.027
17	21	25	31	0.030
21	25	31	37	0.033
25	29	37	43	0.036
29	33	43	49	0.039
33	37	49	55	0.042
For each 1 1/3 lb exceeding 37		For each 2 lb exceeding 55		Add 0.001

TABLE 9. DRAFT ANGLE TOLERANCES

Draft Angle	*Tolerance*
Up to 3°	±1/2°
3° and over	± 1°

TABLE 8. STANDARD LENGTH OR WIDTH TOLERANCES

Dimension, In.	Exterior Dimensions *A*	Interior Dimensions *B*	Step and Center Dimensions *C*
Up to 8	+0.032 −0.016	+0.016 −0.032	±0.016
Over 8 add per in.	+0.004 −0.002	+0.002 −0.004	±0.002

TABLE 10. STANDARD STRAIGHTNESS TOLERANCES FOR ALUMINUM

Length, In.		Tolerance, In.
From	To	
0	9	1/64
9	18	1/32
18	30	3/64
30	45	1/16
45	60	3/32
60	80	1/8

TABLE 11. STANDARD FLASH EXTENSION TOLERANCES

Magnesium Net Weight of Forging, Lb		Aluminum Net Weight of Forging, Lb		Max Flash Extension, In.
From	To	From	To	
0	1/4	0	1/4	1/64
1/4	3	1/4	4	1/32
3	11	4	17	1/16
11	16	17	24	3/32
16	..	24	..	1/8

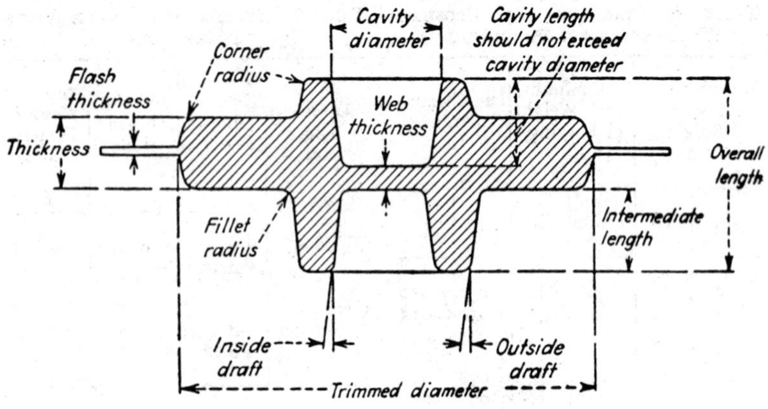

TABLE 12. MINIMUM TOLERANCES FOR NONFERROUS FORGINGS**

	Copper	Forging Brass Naval Brass	Aluminum Bronze Silicon Bronze	Aluminum Alloys	Magnesium Alloys
Outside draft,* deg:					
Hammer	3 to 7	3 to 7	3 to 7	3 to 7	3 to 7
Press	1 to 5	1 to 5	1 to 5	1 to 5	1 to 5
Fillet and corner radiuses, in.†	1/8	1/8	1/8	1/8	1/8
Dimensions, in.‡					
Up to 1 in.	0.007	0.005	0.007	0.007	0.007
Up to 2 in.	0.010	0.008	0.010	0.010	0.010
Up to 4 in.	0.015	0.010	0.015	0.015	0.015
Up to 6 in.	0.020	0.015	0.020	0.020	0.020
Over 6 in.	0.031	0.031	0.031	0.031	0.031
Flash thickness, in.	0.080	0.035§ 0.045‖	0.080	0.080	0.100
Web thickness, in.	1/4	3/32§ 1/8‖	1/4	1/4	3/8
Flatness per in.	0.005	0.005	0.005	0.005	0.005
Machining allowance, in.	1/32	1/32	1/32	1/32	1/32

** Recommended by The Brass Forging Association.
* Inside draft will be somewhat greater than outside draft.
† These radiuses are minimum desirable for maximum die life; fillets with much smaller radiuses can sometimes be made.
‡ Minimum of ±0.007 should be allowed for all dimensions crossing the parting line.
§ Forging brass.
‖ Naval brass.

TABLE 13. FORGING TEMPERATURES FOR SAE STEELS

SAE No.	Forge,* °F	Normalize, °F	Die-life Ratio
1010	2400	1650–1800	105
1020	2350	1650–1750	100
1030	2320	1600–1675	98
1035	2320	1575–1650	
1040	2280	1575–1650	95
1045	2280	1550–1650	
1050	2220	1550–1625	90
1060	2160	1525–1600	85
1070	2100	1500–1575	
1080	2050	1500–1575	
1095	2020	1500–1575	
2317	2220	Carburizing	87
2330	2200	1600–1700	84
2340	2200	1600–1700	
2345	2200	1600–1700	
3115	2220	Carburizing	92
3120	2220	Carburizing	
3125	2180	1600–1700	88
3130	2180	1600–1700	
3135	2180	1600–1700	85

SAE No.	Forge,* °F	Normalize, °F	Die-life Ratio
3140	2180	1600–1700	
3141	2180	1600–1700	
3145	2180	1600–1700	82
3150	2180	1600–1700	81
3240	2180	1600–1700	
4120	2200	Carburizing	90
4130	2200	1600–1700	87
4140	2200	1600–1700	84
4150	2200	1600–1700	80
4340	2200	1600–1700	65
4615	2200	Carburizing	
4640	2200	1600–1700	
4815	2200	Carburizing	84
5120	2200	Carburizing	90
5140	2200	1600–1700	84
5150	2200	1600–1700	86
52100	2200		
6150	2220	1650–1750	78
9260	2150		

* Maximum temperatures for reducing atmospheres.

TABLE 14. FORGING TEMPERATURES FOR LIGHT METALS

Aluminum Alloys					Magnesium Alloys		
Alloy	Blocking, °F		Finishing, °F		Alloy	Blocking, °F	Finishing, °F
	Min	Max	Min	Max			
14S........	650	830	650	830	AT35......	820	800
17S........	780	810	800	830	AZ61X....	720	700
25S........	..	860	...	860	AZ80X....	720	650
A51S......	..	870	...	870			
61S........	..	880	...	880			
75S (R303).	675	730	700	750			

TABLE 15. FORGING AND HEAT-TREATING TEMPERATURES FOR STAINLESS STEELS

Type	Group*	Preheating, °F	Begin Forging, °F	Finish Forging, °F	Annealing Temp, °F	Stress-relieve
302....	C	1500–1600	2100–2200	Above 1700	1850–2050	400–750
302B...	C	1500–1600	2100–2150	Above 1700	1850–2050	400–750
303....	C	1500–1600	2100–2200	Above 1700	1850–2050	400–750
304....	C	1500–1600	2100–2200	Above 1700	1850–2050	400–750
309....	C	1500–1600	2200–2300	Above 1800	1850–2050	400–750
310....	C	1500–1600	2100–2200	Above 1800	1900–1950	400–750
321....	C	1500–1600	2150–2200	Above 1700	1700–1950	450–750
347....	C	1500–1600	2150–2200	Above 1800	1700–2000	400–750
403....	A	1400–1500	2000–2100	1450–1500	1550–1650†	450–700
410....	A	1400–1500	2000–2100	1600–1700	1550–1650†	450–700
416....	A	1400–1500	2100–2200	1500–1600	1550–1650†	450–700
420....	A	1400–1500	1950–2050	1650–1700	1600–1650†	300–700
430....	B	1400–1500	1900–2000	1350–1400	1450–1550	
430F...	B	1400–1500	2000–2100	1350–1400	1250–1450	
440....	A	1400–1500	1850–2000	1700–1750	1625–1675†	300–700
442....	B	1400–1500	1900–2000	1300–1400	1450–1550	
446....	B	1400–1500	1900–2000	1300–1400	1450–1550	
501....	A	1400–1500	1950–2000	1650–1700	1550–1575	

* Group: A, hardenable chromium; B, nonhardenable chromium; C, austenitic chrome-nickel.
† Process annealing temperatures are about 200 F lower.

TABLE 16. FORGING AND HEAT-TREATING TEMPERATURES FOR COPPER ALLOYS

Alloy	Forging Range, °F	Annealing, °F	Die-life Ratio	Machinability	
				SAE 1020 = 100	Free-cutting Brass = 100
Copper..............	1400–1600	490–500 up	110	125	20
Forging brass..........	1250–1550	800–1100	115	200	80
Naval brass...........	1200–1350	800–1100	110	160	30
Leaded naval brass.....	1200–1350	800–1100	...	...	70
Manganese bronze......	1200–1300	800–1100	85	80	
600 bearing metal......	1250–1350				
Muntz metal..........	1150–1300	800–1100	...	75	40
Silicon bronze (type A).	1200–1500	900–1300	83	75	30
Silicon bronze (type B).	1200–1500	900–1250	83	75	30
Aluminum bronze......	1450–1700	950–1500	...	110	20
Nickel silver..........	1300–1500	....	...	...	80

Section 17

COLD WORKING OF METALS

SECTION 17

COLD WORKING OF METALS[1]

ECONOMICS OF PROCESSES

The ultimate aim in manufacturing is to proceed from raw material to finished product with the least number of operations. Cold-working processes for metals constitute one of the most direct means of accomplishing this objective. The first operation is to cut, shear, or blank the slug to be worked. Next, the slug is cold-worked by one of the processes, and finally it may be trimmed, sized, or machined. Fewer operations mean less equipment, less invested capital, less labor, and less total handling of parts. Besides, most cold-working operations are faster than customary production setups.

The greatest advantage in cold-working metal comes from the saving of raw material. There is virtually no scrap. Since in present-day manufacturing, 50% of the finished-product cost is for material, cost reduction is best realized by elimination of material waste. Equipment used for cold working consists of cold headers, rotary swagers, toggle presses, hydraulic presses, and crank and eccentric presses. Waste on cold headers ranges from 4 to 5% as compared with 60 to 70% for most automatic screw-machine jobs.

The problem of how to make a part by cold working is usually more difficult than to determine the cost savings or other advantages. Some parts, like screw-machine products, present only minor difficulty in switching to cold heading. But on other parts that are to be flat-swaged or extruded, where the flow of metal is quite large, tooling and equipment problems may be difficult. The decision with respect to a switch to cold working must be made for each job rather than for a class of work. Many parts have similar characteristics, but material, quantity, tolerances, and finish must be considered. There must be high production on the part to justify the tooling expense. Customarily, the savings should be great enough to pay for initial tooling costs in 1 or 2 years.

NATURE OF COLD WORKING

The cold working of metals may be defined as those operations which produce changes in the shape of a piece of metal by pressing, pulling, or bending it under pressure at a temperature below the recrystallization point. The cold working of metals may be classified in three main groups: compression, drawing, and bending. The last two groups are covered in Sec. 15.

The cold working of metals by compression consists of four main processes. These are, in order of increasing severity of metal flow: (1) sizing, (2) flat and rotary swaging, (3) coining, and (4) plain and impact extrusion. All these processes place metal under pressure to cause plastic flow. The other cold-working processes sub-

[1] W. A. Fletcher, Chief Process Engineer, and W. P. Bowman, Process Engineer, Delco-Remy Div., General Motors Corp.

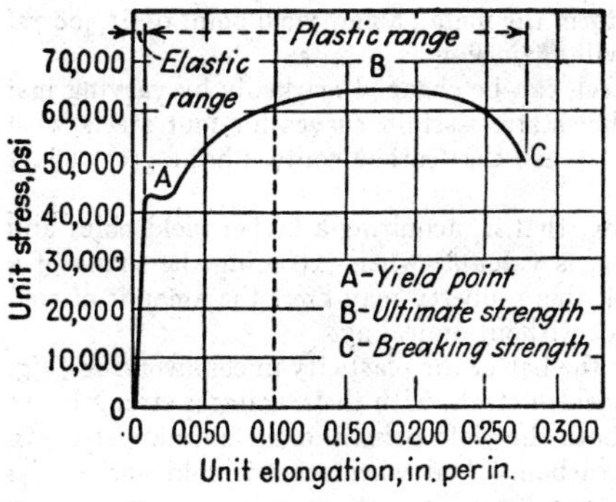

FIG. 1. To cause plastic flow for any of the cold-working processes, the metal must be stressed past the yield point A.

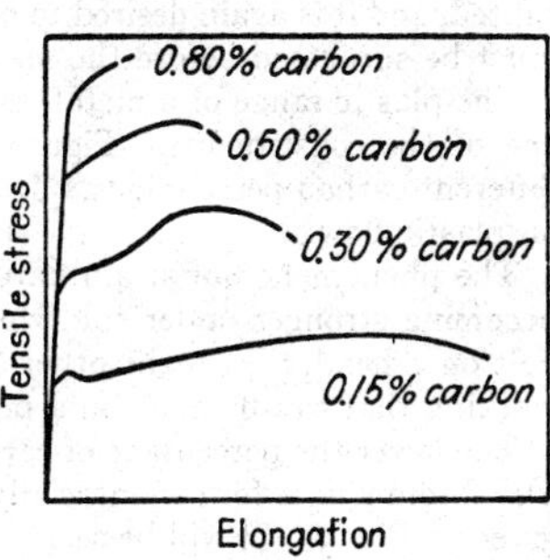

FIG. 2. Carbon content of the steel has a marked effect on working pressure.

ject the metal being worked to tensile or a combination of tensile and compressive forces. The stresses set up in the metal by the four compressive cold-working processes are not necessarily compressive, but the forces causing them are.

THEORY OF COLD WORKING

When pressure is applied to metal, it must deform a slight amount in order to exert a counteracting force. Deformation occurs at the weakest, or "slip," planes. After all the slip-plane movement has been used by cold working, any additional pressure causes rupture.

Study of the stress-strain diagram of a metal helps to understand cold-working theory. In the elastic range between zero pressure and point A in Fig. 1, the metal deforms uniformly with increasing pressure, according to Hooke's law. But at point A a great increase in strain (deformation) occurs for only a slight increase in pressure, or stress. This is the yield point. Metal that has been stressed past this point will retain the shape it had at time of pressure release. Between points A and C the metal is in the plastic range but takes progressively more pressure to keep it moving. For example, if a piece of steel is cold-worked until its length increases 0.100 in., the pressure required will be 61,300 psi, or well within the plastic range. If the process is inter-

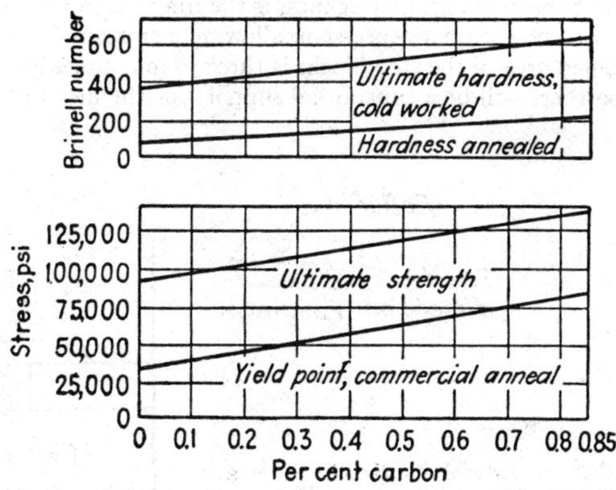

FIG. 3. Yield points of commercially annealed steels rise with carbon content.

rupted, and it is again desired to cold-form the metal, a new yield point of 61,300 psi must be surpassed before the metal will again flow.

The plastic range of a metal, say steel, can be changed markedly by varying just one of the constituents. Figure 2 shows stress-strain curves for four steels with different carbon percentages. It will be seen that carbon content has a great effect on plastic flow.

The phenomenon of strain hardening, that is, acquiring a higher yield point and becoming stronger under cold working, is valuable when extruding the shank of a bolt on a header. On the other hand, this property may limit the amount of cold working that can be done on a product without annealing.

The lower the percentage of carbon, the better the plasticity in cold working (Fig. 2). Yield points for commercially annealed steels, with their ultimate strength, are given in Fig. 3. It will be seen that both the yield point and the ultimate strength increase directly with percentage of carbon. And when steel is cold-worked, its strength and hardness increase.

SIZING

Sizing is the process of finishing forged, rolled, stamped, and cold-worked parts to the proper thickness by working the metal to a fixed dimension. It is done between flat, curved, or angular die surfaces on toggle or hydraulic presses. Flow of metal is on the surface of the piece, and movement is in the horizontal direction unrestricted by the dies. Sizing not only gives a smooth wear-resistant surface, but it is also much more economical to do than milling or shaping.

The stroke on a sizing operation is usually only ½ in. or slightly more. Therefore, the press can be run at a fairly high speed, usually 30 to 35 strokes per min. On most sizing jobs the parts are fed from the front of the press by an automatic horizontal feeder and are ejected at the rear.

Sizing Allowance. Small parts may have as much as 1/32 in. to size, while large parts should not have more than 1/64 in. In addition to the part being sized, there should also be at least two sizing stops between the dies to maintain the same distance between them on each stroke of the press. The stops should be located as close as possible to the part, to minimize spring in the tools.

A simple sizing operation is illustrated in Fig. 4. Because the only surface that must be held to close thickness is the rim of each boss, the rest of each boss is relieved by a cone-shaped depression, allowing a smaller size of press to be used. The combined area of the stop blocks is three to four times greater than the area being sized; so there will be a substantial support for the dies to keep them from bending.

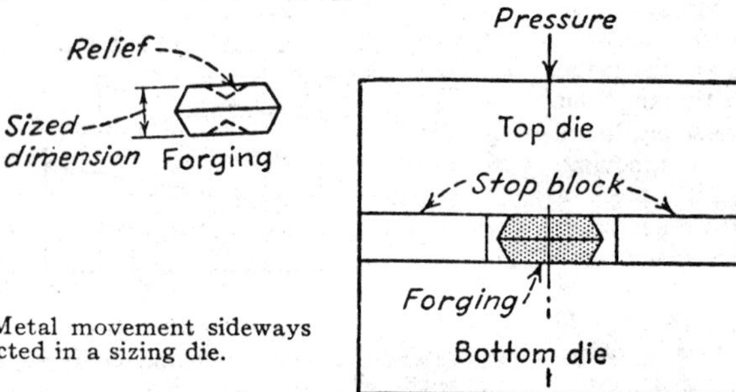

FIG. 4. Metal movement sideways is unrestricted in a sizing die.

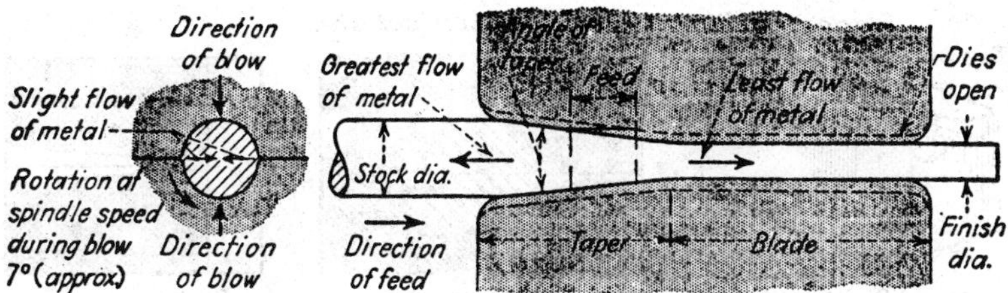

FIG. 5. Action of metal when rotary-swaging solid work. In swaging tubing, part of the metal flow is inward, increasing the wall thickness.

Relieved Sections. In order to relieve some of the pressure that might build up in sizing, it is sometimes possible to relieve the part, or the dies, so that only the area on the part that actually has to be sized is worked. When a solid surface is sized, the metal in the middle of the piece is surrounded by other metal and is trapped to some extent. The metal on the edge of the part, however, has an unrestricted space into which it can flow. This condition, known as pyramiding of pressure, causes the middle of most sized parts to be thicker than the edges.

As the amount of flow in the metal during sizing is very small, there is no need to lubricate the part or the sizing surfaces. As a matter of fact, having oil films between the part being sized and the dies will result in more variation of the finished thickness than without them.

Sizing dies or tools are usually made from high-carbon high-chromium steels of the air-hardening type. The dies are heat-treated to give fairly hard sizing surfaces, usually in the Rockwell 56 to 60 C range. This type of steel gives a combination of toughness and hardness which results in dies that are about the best compromise between wearing qualities and strength. Plain steels of 1.25% carbon have also been used successfully on sizing tools.

ROTARY SWAGING

Rotary swaging is a process of shaping work with many blows applied by rotating dies. The dies (two dies in most cases, but four dies on some large machines) reciprocate rapidly as the spindle in which they are mounted rotates. This means that the finished work must be round. Swaging is applied to pointing, tapering, reducing, sizing, and assembly operations.

Work can be swaged either hot or cold, but it is done cold whenever possible because of greater ease of handling and superior finish.

Minimum practical finished diameter of swaged work is 0.020 in.

The maximum finished diameter is limited only by the size of the machine, the material, amount of reduction, whether solid or tube, and whether the operation is done hot or cold. Machines available are rated up to 6 in. dia for tubing; a few have larger capacity. Solid-work capacity is about half that for tubing.

Because swaging is a hammering operation (Fig. 5), it has the same beneficial effect on work as forging. It produces a desirable grain structure and results in increased tensile strength and elasticity. Cold swaging work-hardens most materials. There is no waste except on jobs that require trimming on one or both ends.

The degree of accuracy depends on condition of dies and machine and on the size. Solid work of ¼ in. dia could probably be swaged to ±0 001 in.

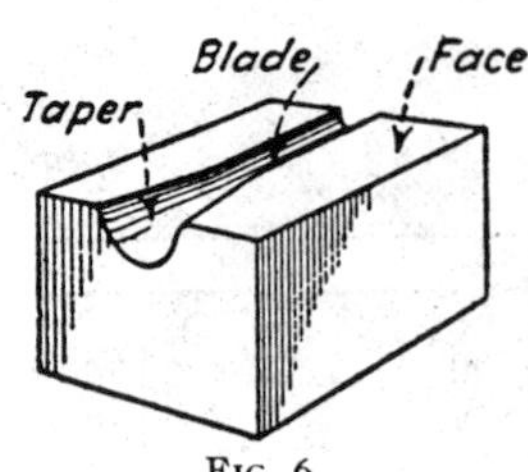

FIG. 6.

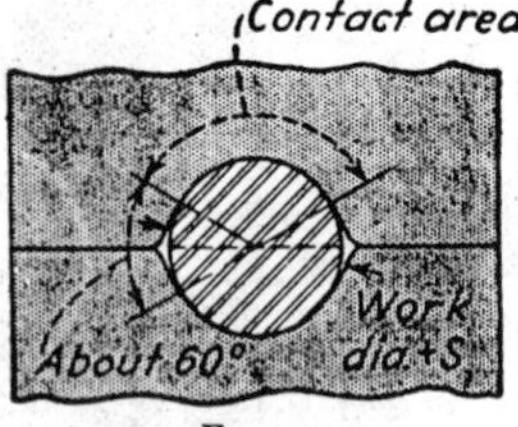

FIG. 7.

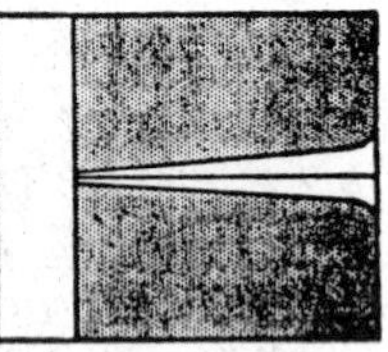

FIG. 8.

FIG. 6. A rotary-swaging die has a tapered entrance and a straight section. FIG. 7. Die oval is required to reduce contact area. When the contact area becomes excessive, the dies are likely to split. FIG. 8. For pointing operations, stop blocks can be placed behind the rotary-swaging dies or fitted into a slot in the dies.

Sharp tapers or swaging close to a shoulder of the work are difficult. The length of a taper that can be swaged is limited by the die length the machine will take.

When tubing is swaged, wall thickness is increased unless the tubing is worked over a mandrel.

Rotary-swaging Dies. Swaging dies are of simple rectangular shape. Typical dies will have a straight groove or blade and a tapered approach (Fig. 6). Other dies may have a taper all the way through or a cone-shaped cavity only part way into the die. For most work the dies are made with a slight oval in the cavity (Fig. 7) rather than with a round cross-section.

Die materials vary widely. For most cold swaging, high-carbon steel will give excellent results. One recommendation is for 1.40 C on small dies and 0.90 to 1.00 C on the medium sizes. For larger dies another firm recommends 0.55 to 0.60 C manganese steel.

Dies are hardened to a depth of about 9⁄64 in. with a surface hardness of Rockwell 62 to 64 C for small dies, 58 to 61 for larger dies. The steel should give a gradual change of hardness between case and core.

Light alloy or high-speed steels may be used for dies which are subject to heavy wear, especially in small sizes. For hot swaging, alloys or low-carbon steel with a hard-facing material applied are common. Carbide gives excellent results in hot-swaging tungsten and molybdenum. Carbide inserts are used on large dies, but the small sizes are usually better with solid carbide. Carbides should be of relatively tough grade with a high percentage of cobalt. Carbide has also been used with success in cold swaging.

Dies with cavities from 1⁄16 to 1⁄4 in. dia are clamped together and the groove is drilled and reamed, or bored, leaving 0.005 to 0.010 in. for finishing. After the dies are hardened, they are lapped to size.

Oval. An oval opening (called simply "oval") is required for most work so dies will not split under the high pressures encountered. Table 1 gives "oval" for dies of various diameters. Oval is obtained by placing shims between the dies during processing and increasing the diameter to which the dies are finished by the thickness of the shims. When the shims are removed, the groove is of the correct diameter in the direction of die travel but is large (by the amount of shimming) at right angles. This amount of oval is increased, in practice, by rounding the die corners. On very small dies, rounding the corners sometimes gives all the oval required.

TABLE 1. OVAL IN ROTARY-SWAGER DIES*

Amount of Blocking for Solid Work

<table>
<tr><th>Dia</th><th>Alloy Steel, Drill Rod</th><th>Low-carbon Steel</th><th>Hard Brass</th><th>Copper</th><th>Lead</th></tr>
<tr><td>3/4</td><td>—</td><td>—</td><td colspan="2" rowspan="3"></td><td rowspan="13">No blocking</td></tr>
<tr><td>11/16</td><td>—</td><td>—</td></tr>
<tr><td>5/8</td><td>—</td><td></td></tr>
<tr><td>9/16
1/2
7/16
3/8
5/16
1/4</td><td>1.25 times value at right</td><td colspan="3">Taper
0.001 in. per deg of taper plus 0.5% of max dia
Blade
Value for taper less 0.003 to 0.004 in.</td></tr>
<tr><td>3/16</td><td>2%</td><td colspan="3" rowspan="3">of average dia</td></tr>
<tr><td>1/8</td><td>4%</td></tr>
<tr><td>1/16</td><td>3%</td></tr>
<tr><td>1/32</td><td colspan="4">Stone edges of groove</td></tr>
</table>

Amount of Blocking for Tubing

Ratio of OD to Wall Thickness	*Blocking Required*
25 or over	None
10 to 24	60% of solid
9 or less	100% of solid

NOTE: Corner radius on the groove should be 1/16 of blade diameter to nearest 0.005 in. or equal to thickness of tube.

* The Torrington Co.

More oval is required in the taper than in the blade; so the dies must be shimmed different amounts for these operations. The radius between taper and blade must then be skillfully blended into the straight sections.

All the reduction takes place in the tapered section of the dies, with final sizing taking place in the blade section. This means that most of the wear takes place in the taper.

When dies are new they do not contact the work near the edges of the die because of the oval. As dies wear the area of contact increases. It should not approach nearer the face of the die than about one-eighth of the groove diameter. If it does (with too little oval in the dies as a result), the dies are likely to split.

Die Design for Rotary Swaging. When the finished work is to have a taper, the angle will be determined by the work. When the taper is only a means to bring about reduction in the work, considerable latitude is possible.

For reducing tungsten, the preliminary dies may have a blade length 1½ times the work diameter. Finishing dies have blade length about three times diameter and have a carefully rounded die entrance.

For cold swaging on solid work, when hand feeding is employed, it is desirable not to exceed 8° of included angle in the taper. This amount can be increased to 12 or

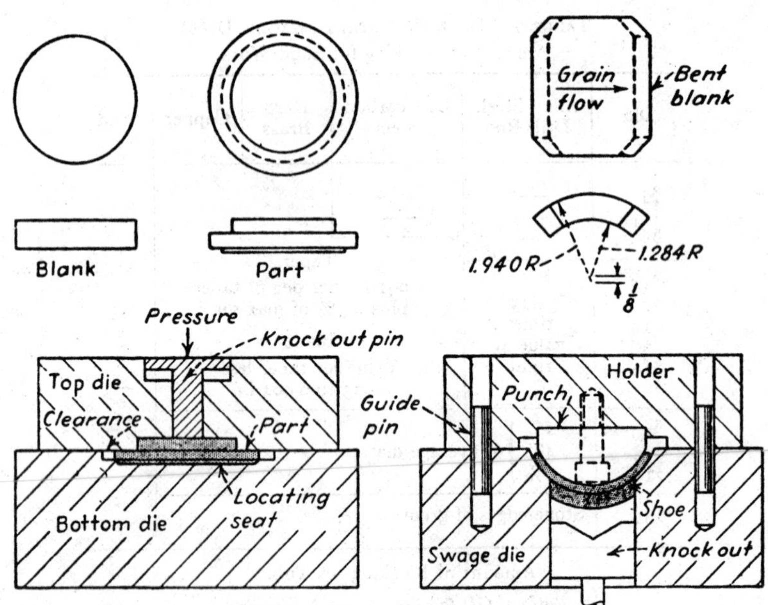

FIG. 9. In flat-swaging a gear blank to create a tooth rim, clearance must be allowed in the dies.

FIG. 10. A pole piece had to be prebent before flat-swaging the wings at the sides.

15° in some cases, but it is more important in such cases that the work be free of oil. On large work, feeding devices are customary and tapers as high as 20° are possible.

Somewhat steeper tapers are possible with tubing. Copper tubing has been swaged with tapers as high as 34°.

Whenever long thin tapers are to be produced, care must be exercised not to exceed the capacity of the machine. The maximum length of taper on solid work is limited not only by the die length of the machine but by the requirement that the length of taper must not exceed ten times the average diameter.

Multiple tapers are difficult to swage with one set of dies.

FLAT SWAGING

Flat swaging is the process of reducing the thickness of metal between dies in a press or header. The process starts with a blank as thick as the thickest section of the finished part, and consists of reducing the other portions to proper thickness. A simple swaging operation is shown in Fig. 9. The gear blank is placed in the locating seat in the bottom die and then compressed to reduce the thickness of the outer ring of metal. The center portion retains the original thickness to form the gear hub. Thus, the diameter of the blank is greater than the hub and less than that of the swaged portion. On the upstroke, the gear part is retained in the upper die and must be ejected by a knockout.

Stock used in flat swaging should be quarter or half hard to keep the edges of the thickest portion of the blank from mushrooming downward.

A more difficult swaging operation is that of making the pole shoe (Fig. 10). This part could not be hot-forged or drawn or flat-swaged directly from a flat blank. First, the blank had to be bent. To bend a piece of metal and make it stay put, it is necessary to cause metal flow of at least 3% of the dimension being worked. In this case cold flow was induced at the center and bending accomplished in tools having a punch with a radius smaller than the radius in the die and not concentric with it. Therefore, the metal was pinched more at the center. Plastic flow toward the sides of the blank was induced during the bending operation. With the present iblank, metal flow occurred only in the downward direction during flat swaging.

Pyramiding of Pressure. In flat swaging and sizing operations, pyramding of pressure occurs with an increasing ratio of width to thickness. Pyramiding is due to tool friction and trapping of metal at the center of the part by surrounding metal. If the ratio of width to thickness exceeds 10:1, only a feathering of the work edges will be achieved.

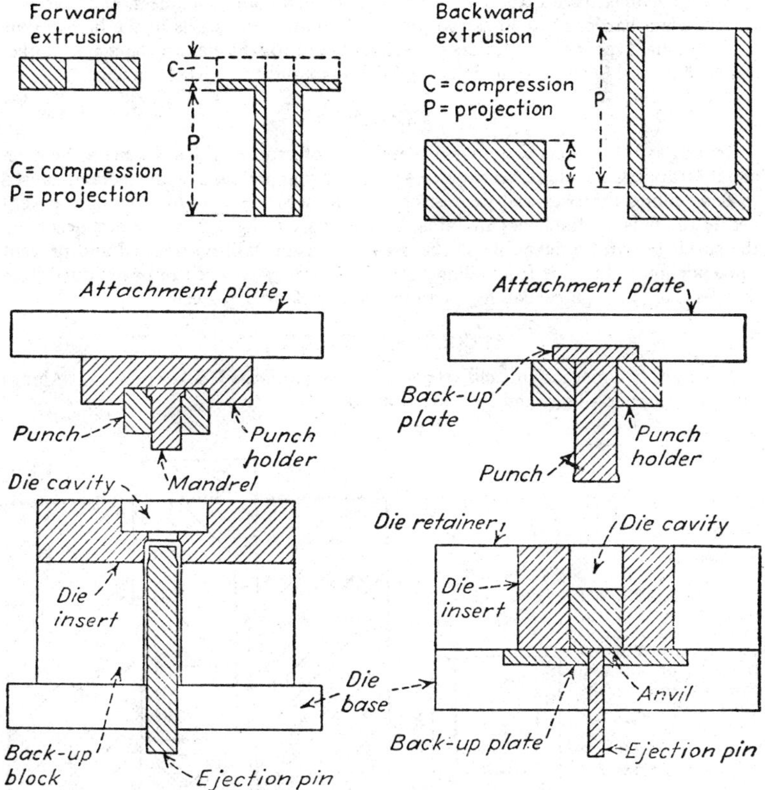

FIG. 11. Forward extrusion creates metal flow in direction of punch travel. Projection ratio = P/C.

FIG. 12. Backward extrusion causes metal to flow back over the punch.

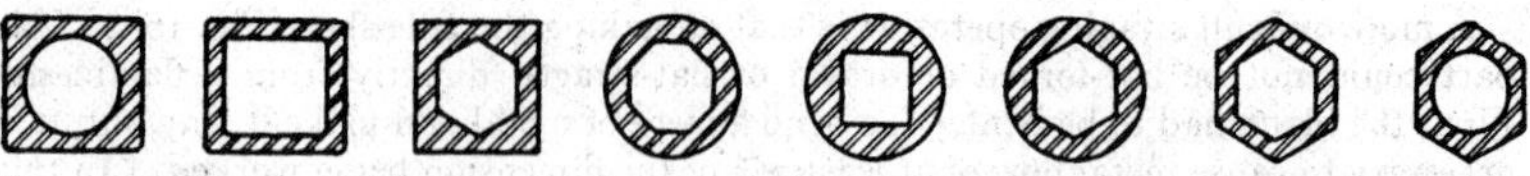

FIG. 13. It is possible to extrude a part with any combination of these inside and outside shapes, plus splines, bottoms with bosses, etc.

Blank Quality. The blanks used in swaging operations should be fairly uniform in volume in order to obtain uniform finished parts. Any irregularity on the edges of sheared blanks will be exaggerated by the swaging operation.

Lubrication. Tools and blanks must be well lubricated. For low-pressure jobs a coating of trisodium phosphate with light machine oil is satisfactory. On severe work zinc-phosphate coatings with a soap film as lubricant work extremely well.

Steels for Swaging Dies. The tools used for swaging must be tough to withstand shock, and they must be hard to resist wear. Chromium-tungsten steels are easy to machine, have little distortion, possess maximum toughness, and give best results when hardened to Rockwell 58 C. Air-hardening steels of the high-carbon high-chrome type are not so shock-resistant and are harder to machine. Other steels that work well on some jobs are SAE 3250 and "Air Kool" steel.

COINING

Coining is the process of embossing designs, patterns, or ridges on metal by completely trapping the blank between dies and applying pressure. The metal is caused to flow, filling the rises and depressions in the dies. Blanks for coining are very nearly the finished thickness and area, because only a slight flow of metal occurs on the surface. No lubricant should be used, because it will be trapped and prevent some portion of the part from filling out. Satisfactory service can be obtained from die steels of the high-carbon high-chrome or 1.00% carbon steels.

COLD EXTRUSION

There are two methods of cold extrusion: impact and the Hooker process. Aluminum, tin, copper alloys, and steel are worked.

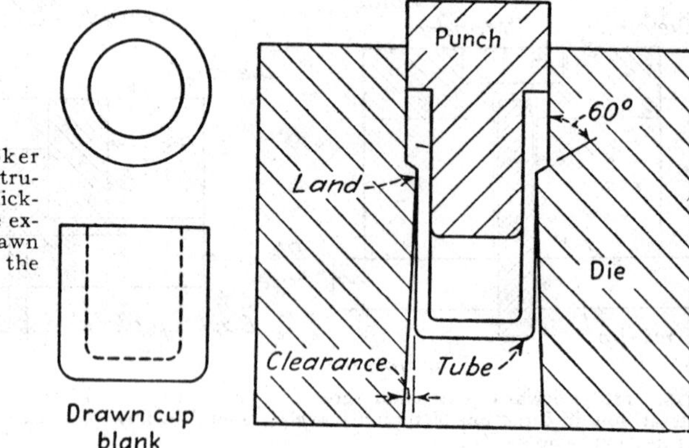

FIG. 14. Hooker process for extrusion causes a thick-walled cup to be extruded or drawn over a land in the die.

Impact Extrusion. There are three variants of the impact extrusion process: *forward*, or *direct*, wherein the metal flows (Fig. 11) in the same direction as the applied force; (2) *backward*, or *indirect*, wherein the metal flow (Fig. 12) is opposite to the applied force, or back over the punch; and (3) *opposed*, wherein the metal is forced to flow simultaneously with and opposite to the direction of applied force. It is possible to form a part with any combination of the inside and outside shapes shown in Fig. 13.

Hooker Process. A cupped blank is used in this process (Fig. 14). The blank is placed in the die, the punch enters, and pressure causes the metal to flow between the nose of the punch and the land of the die. Both large and small tubing can be extruded to considerable length, and very thin, if required. Copper radiator tubing is an example of products made by the Hooker process.

Punch Design. The slenderness ratio (punch length divided by punch diameter) is the controlling factor in design of punches for backward extrusion (Fig. 15).

Percentage Reduction. The percentage reduction in extrusion is related to the malleability of the material. Given the projected blank area A and the minimum cross-sectional area a of the finished part, the percentage reduction can be expressed as $100\left(1 - \frac{a}{A}\right)$. For aluminum alloys, the practical top limit is around 80%; for steel about 50%.

Lubrication. Success in cold extrusion of steel was achieved with application of zinc phosphate coating to overcome tool and work friction. The coating is applied to the blank. Oil may be sprayed onto the work during the extruding cycle.

A practical example of forward extrusion is shown in Fig. 18. The automobile generator frame is made from $1\frac{1}{32} \times 6$-in. stock. Essential parts of the dies are: the carbide extrusion ring with a 45° throat angle and a land for sizing the final OD, the arbor for sizing the ID, and the collar. The collar applies pressure to the end of the work. Plastic flow causes the metal to thicken 9% above the extrusion ring, to achieve the desired ID. Flow between the arbor and extrusion ring reduces the thickness 50%. Since the metal can flow faster than ram movement, the arbor must have float or it can be ruptured by the clamping action of the moving metal.

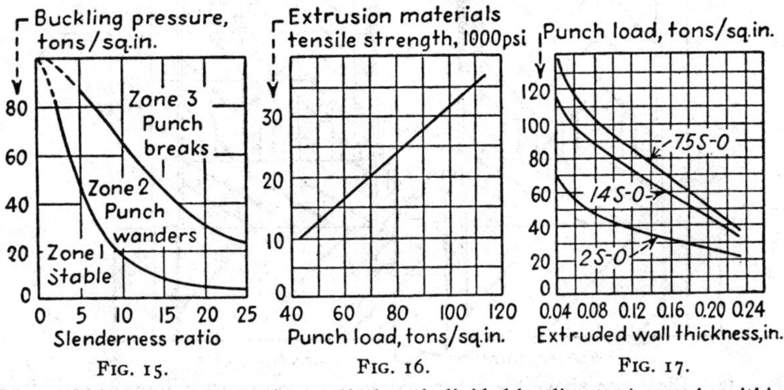

FIG. 15. FIG. 16. FIG. 17.

FIG. 15. Slenderness ratio of a punch (length divided by diameter) must be within correct limits to avoid wander. FIG. 16. Tensile strength of the alloy extruded (aluminum) influences the pressure required on the punch. FIG. 17. Extruding pressure rises as wall thickness is reduced, or the percentage of reduction is increased.

Slug Development for Aluminum. Slugs or blanks are generally developed by computing the volume of metal in the finished extrusion and translating this volume into a solid piece of metal to fit within the die cavity.

In actual practice, the above procedure may be safely followed for reverse extrusion, provided that a small amount of extra metal is added to the slug mass to allow for irregularities in extrusion flow and material and die tolerances.

Tolerances on slug-length dimensions for reverse extrusions should not be greater than +0.030 − 0.000 in. on the final developed dimensions. Punch-length and part-length tolerances are the determining factors for slug-length tolerances and the die-cavity diameter governs slug-diameter tolerances.

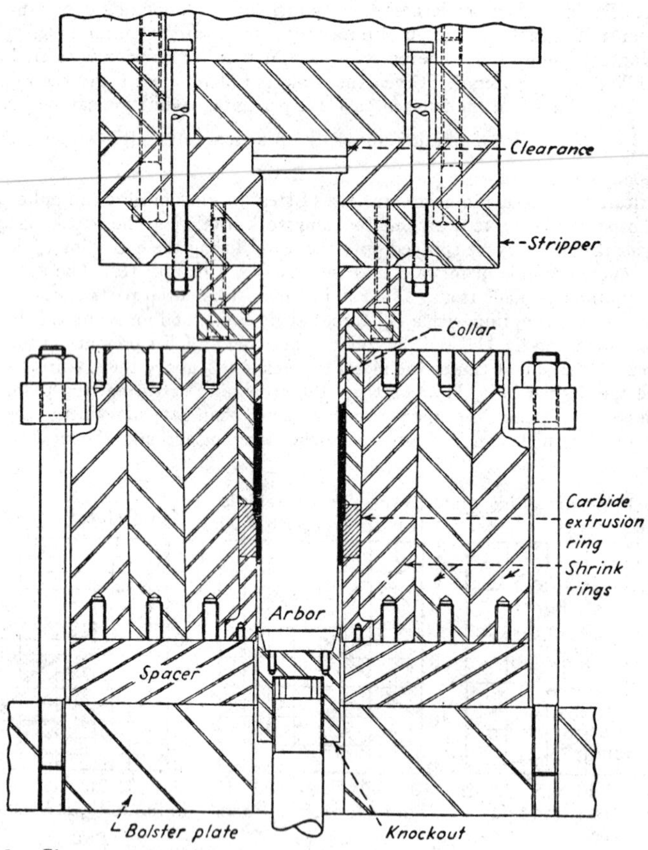

FIG. 18. Clearance above the arbor may be vital to success in forward extruding of steel. Here a generator frame is being sized.

Section 18

DIE CASTING

By: J. C. Fox, Chief Metallurgist, Doehler-Jarvis Corporation

SECTION 18

DIE CASTING

Die casting is the metalworking process by which molten metal is forced into steel molds under pressure, in order to form simple or complex parts with a rapid cycle and little labor. Metals that are die-cast are: alloys of zinc, aluminum, and magnesium principally; and tin, lead, and copper. Production rates up to 500 casting cycles per hour are possible with suitable automatic machines.

Process Requirements. The die-casting process involves chiefly (1) the use of a suitable casting mechanism capable of holding and operating a die under pressure, (2) a properly designed die, and (3) a suitable alloy.

TYPES OF MACHINES

Essentially, a die-casting machine is a horizontal press with two die halves accurately registered and fastened to the stationary and movable platens. The function of the machine is to hold the die halves tightly together while molten metal is injected into the die cavity and then to open the die halves to permit removal of the finished "gate." The gate includes the single or multiple parts made in the cast, plus the sprues, runners, and flash. The latter account for about 50% of the gate and are broken off and remelted.

Modern die-casting machines are operated by a hydraulic cylinder and piston or a hydraulic cylinder and toggle. They are differentiated by the type of metal-injection system used.

Submerged-plunger Machines. With the submerged-plunger method of hot-metal injection, sometimes called the gooseneck method, a plunger and cylinder are permanently submerged in a pot of molten alloy. The ports in the cylinder admit alloy, when the plunger is retracted by an air or hydraulic cylinder. On the die-filling stroke, the plunger first closes the ports and then pushes the molten metal through a passage to the die. This type of machine is used for casting the lower melting materials: lead, tin, and zinc.

Cold-chamber Machines. In these machines, the gooseneck is eliminated and the molten alloy is ladled by hand or automatically into an injection cylinder prior to each shot. This type of machine is used for casting the higher melting point alloys like aluminum, magnesium, and copper, because a steel plunger or cylinder submerged in these alloys in the molten state is not practical.

Injection pressures up to 30,000 psi are used, so that the metal will be forced into small and remote sections of the die.

Die-locking Methods. Three types of die closing and locking methods are in use: hydraulic with toggles, straight hydraulic, and mechanical means. The closing pressure must be sufficient to stress the tiebars connecting the stationary and movable platens. Provision is made for adjusting the locked toggle position to

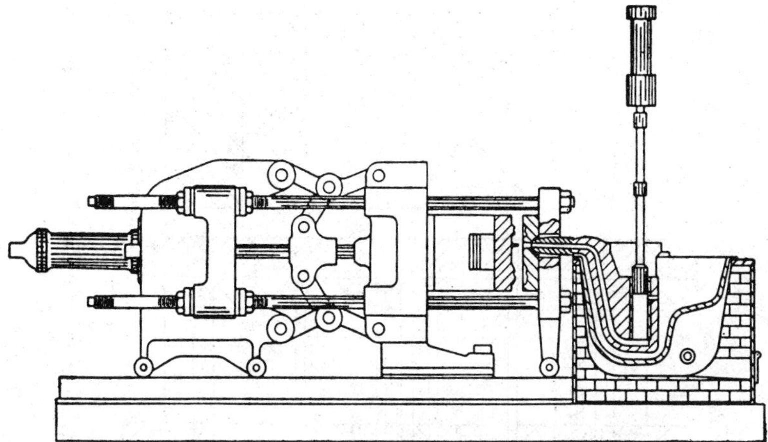

FIG. 1. In the submerged-plunger die-casting machine, the furnace for melting metal is attached to the front end and the plunger and cylinder are submerged in the molten alloy. This type of machine die-casts tin, lead, and zinc.

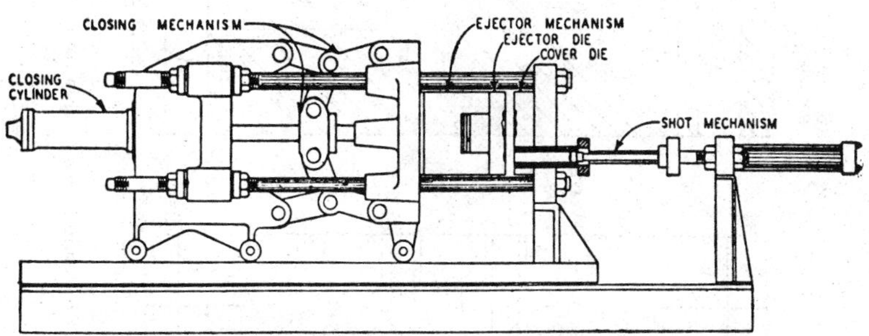

FIG. 2. Aluminum, magnesium, and copper-base alloys are cast by the cold-chamber machine, wherein the furnace or metal pot is a separate entity. Molten metal is in contact with plunger and cylinder for only short intervals to avoid iron pickup.

suit different die heights. Some machines are adjusted by turning a nut on each of the four tension bars, but sometimes movement of these nuts is mechanically synchronized. In smaller machines, a centered screw eliminates platen misalignment.

Cycle Control. With all automatic and semiautomatic die-casting machines, the timing controls can be adjusted to obtain the desired closing and opening intervals, to suit conditions in the dies. Hydraulically operated core pulling and ejector mechanisms can likewise be tied in with the opening and closing cycles.

Automatic cycle control, aside from manual operation when dies are being tried

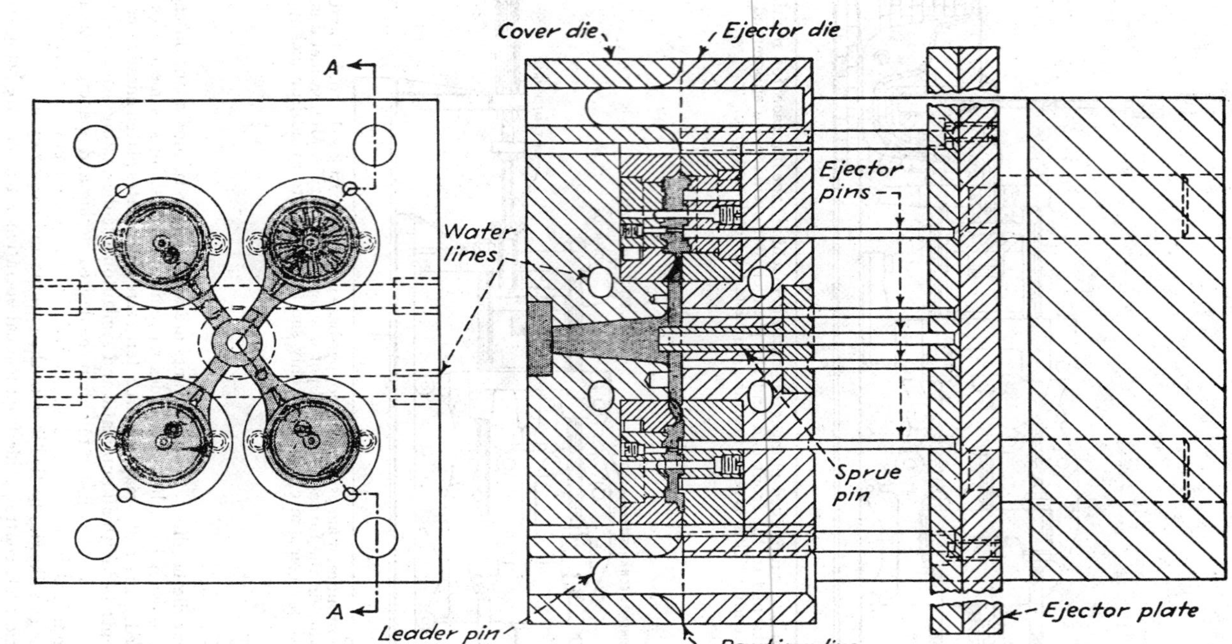

FIG. 3. Principal components of die-casting dies. This four-cavity die is built up of inserted duplicate cavities for high production.

out, has definite advantages. First, the die cavity must be properly filled. Second, the dies must remain closed long enough for the metal to solidify properly to avoid injury to the operator and distortion of the casting. Third, the casting must not remain too long in the die or shrinkage will occur to stress the part severely. Fourth, the correct cycle maintains an even die temperature and good finish on the casting and assists scheduling of refilling of the furnace.

Safety Devices. To avoid injury of personnel and damage to equipment, it is good practice to place an electrical interlock in a circuit with the hydraulic core pull or ejection valve. Then if limit switches are placed on the cores and ejector plates, these cannot be operated before the ejector pins are retracted.

DIE CONSTRUCTION

Major die components are: the *cover half* block which is bolted to the stationary platen and which contains the sprue hole to fit the machine's injection nozzle; and the *ejector half* block which travels with the movable platen and which usually contains the ejector pins and the ejector-actuating mechanism, cores, and other movable parts.

Parting Line. The die cavity is seldom machined half in the cover plate and half in the ejector plate, because the parting plane, or mating surface of the die, must occur at the greatest cross-sectional area of the part. Items to be die-cast are seldom symmetrical. While it is desirable to have the parting line in one plane, design considerations may require that it be curved, slanted, or irregular. This is often necessary to avoid an undercut on an irregularly shaped part.

Die Register. Dowel pins are placed in the cover half of the die to maintain the two die halves in exact register. Sometimes, holding locks are used to keep the parting surfaces from shifting.

Ejection Means. Headed ejector pins are the most common means for stripping the casting from the die. Sleeve-type ejectors are used for thin-wall castings, in order to avoid product distortion. Such devices are placed to bear against surplus metal that will be trimmed off, or around a hole in the finished part. Manual or automatic operation of the ejector pins or sleeves may be provided.

Cores. Holes of almost any shape, as well as splines, can be cored. The cores may be stationary or movable. Fixed cores are used primarily at right angles to the parting line. Movable cores, which must be withdrawn before casting removal, are generally located at an angle to or parallel to the parting lines. Rotary cores are pulled on an arc and provide the only method of coring a casting with a curved hole. Collapsible cores are practical only when the casting cannot be produced more cheaply by using loose pieces in the die or machining the hole. Hinged cores are employed when an insert must be incorporated in the casting and the die face on which the insert is to be positioned is inaccessible to the operator.

Core Actuating and Locking Devices. The means selected for core actuation and holding the core in position determines the success of the operation. If the core is difficult to pull, the speed of operation is reduced; if the core is forced out of position by molten metal, the part must be rejected. Common core pulls are the *rack-and-pinion type* core pull, which consists of a rack attached to a hydraulic cylinder and a pinion that actuates two racks to advance and retract the core plate; and the *pin core pull* for small dies, which involves a long dowel pin fastened to the cover half of the die. This dowel pin engages an angled hole in the core backup block. When the die is closed, no backward movement of the core backup can take place.

Cores are frequently operated by hydraulic means and infrequently by cams.

Gating. The sprue is the metal that fills the tapered sprue hole in the cover half

TABLE 1. FREQUENTLY USED DIE-CASTING ALLOYS

	Zinc		Aluminum				Magnesium	Brass		
Designation										
ASTM	B86-46		B85-46T				B94-44T		B176-42T	
N.J. Zinc Co.	XXIII Zamak 3	XXV Zamak 5	S-9	SC-6	SG2	G2	A291	A	B	C
Alcoa	....	....	13	380	360	218	AM263			
Dow	....	....	....	....	....	....	R			
Doehler-Jarvis Corp.	Doler zink 3	Doler zink 5	Alsiloy 1	Alsiloy 9	Alsiloy 10	Alum. 8	Doler mag 10	Doler brass 2	Doler brass 1	Doler brass 4
Composition										
Aluminum	3.5–4.3	3.5–4.3	Remainder	Remainder	Remainder	Remainder	8.3–9.7	0.25 max	0.15 max	0.15 max
Magnesium	0.03–0.08	0.03–0.08	0.10 max	0.10 max	0.4–0.6	7.5–8.5	Remainder			
Copper	0.10 max	0.75–1.25	0.60 max	3.0–4.0	0.60 max	0.20 max	0.25 max	57.0 min	63.0–67.0	80.0–83.0
Zinc	Remainder	Remainder	0.50 max	0.60 max	0.50 max	0.10 max	0.4–1.0	30.0 min	Remainder	Remainder
Tin	0.005 max	0.005 max	0.10 max	0.30 max	0.10 max	0.10 max	....	1.50 max	0.25 max	0.25 max
Lead	0.007 max	0.007 max	....	....	....	....		1.50 max	0.25 max	0.15 max
Silicon	....	....	11.0–13.0	7.5–9.5	9.0–10.	0.50 max	0.50 max	0.25 max	0.75–1.25	3.75–4.25
Nickel	....	....	0.50 max	0.50 max	0.50 max	0.30 max	0.3 max			
Manganese	....	....	0.30 max	0.50 max	0.30 max	0.3 max	0.3 min	0.25 max	0.15 max	0.15 max
Cadmium	0.005 max	0.005 max								
Iron	0.100 max	0.100 max	1.3 max	1.3 max	1.3 max	1.8 max	....	0.25 max	0.15 max	0.15 max
Properties										
Tensile strength	41,000	47,000	39,000	46,000	43,000	45,000	33,000	50,000	75,000	85,000
Yield strength	....	....	21,000	25,000	23,000	25,000	21,000	25,000	35,000	50,000
Impact strength, Charpy	43	48	2.0	3.0	5.0	5.0	2	30	40	40
Elongation %	10	7	2.0	3.0	5.0	5.0	3	10	20	20
Brinell hardness	82	91	80	80	75	80	60	120	130	170
Sp. gr.	6.6	6.7	2.66	2.76	2.68	2.55	1.81	8.5	8.6	8.45
Melt pt., F	717.6	717.7	1060	1100	1110	1160	1120	1650	1575	1550

of the die. Runners are die passages connecting the sprue hole to the gates, and gates are passages connecting a runner with a die cavity. Sprue, runner, and gate arrangement must be as short as possible and have few bends to reduce turbulence and defects in the casting.

No formulas exist for the best type of gating and its relationship to runners, feeders, and vents. The gate should not be larger than the maximum wall section of the casting. A heavy gate is usually over 0.050 in., a thin gate about 0.025 in. A heavy gate allows metal to flow into the die cavity as a stream instead of a spray and helps to promote soundness of the part, but a heavy gate is difficult to trim. A thin gate is useful for producing good surface finish if soundness is of secondary importance, but soldering or welding of metal to the die surface is more likely to occur than with a heavy gate.

Metal feeders are small openings at the edge of an impression and are served by the gate runner to fill the cavities.

Venting. Air-escape channels about 0.005 to 0.006 in. dia are placed at the parting line to avoid porous or blistered castings from entrapped air. Draw pockets or overflows may be arranged around the impression for additional escape of air.

Water Cooling. Heavy-wall die sections, gating areas, slides, cores, and other movable elements of die-casting dies may be water-cooled to avoid seizure from localized overheating. The water is carried through these sections by passages drilled in the die block. Water passages should not generally be closer than $\frac{3}{4}$ in. to any part of the die impression. Soft water is preferred for cooling purposes to avoid deposit of heat-insulating scale.

Die Classification. Single-impression dies are the most common and are cheapest to make. Multiple dies have two or more similar impressions and are used for quantity production of small parts. Combination dies produce two or more parts or unlike shape, such as sets of castings for an assembly. Unit dies are replaceable impression blocks assembled to form a multiple die.

DIE-CASTING ALLOYS[1]

Zinc alloys are used for approximately 60% by weight of all die-cast parts, largely because of economies from ease and speed of casting. Speeds up to 500 cycles can be maintained in many instances. Zinc alloys have good mechanical properties and can be machined readily and finished economically.

Aluminum die castings account for about 30% of total volume, because they are light in weight, have good electrical and thermal conductivity, are tarnish-resistant, and are competitive in cost with cast iron and steel. Aluminum die castings are used in the as-cast condition or simply polished. They may also be electroplated.

Magnesium alloys have a growing field, because of the relative lightness of the material coupled with good strength. Cost has been a factor in limiting wider adoption for many uses of the metal.

Copper-alloy die castings, particularly brass, have the highest strength, toughness, corrosion resistance, and wear resistance of all die-casting alloys.

Consumption of tin-base alloy die castings is very limited, chiefly because of the strategic value of the metal for more important usage, the high cost of tin, and the relatively low strength of the alloys. They are used, however, for some parts where corrosion resistance is of importance.

Lead alloys, like tin, represent only a small part of die castings made and are employed where a low-cost noncorrodible metal must be used and where strength and hardness are not of importance.

[1] Doehler, H. H., "Die Casting," McGraw-Hill Book Company, Inc., New York, 1951.

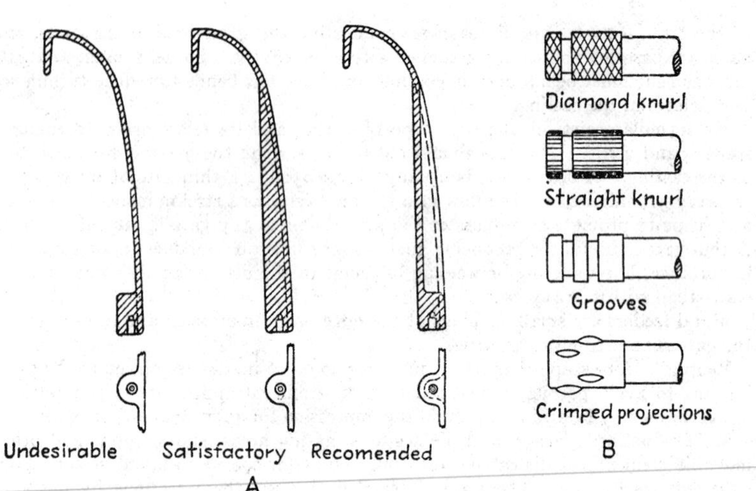

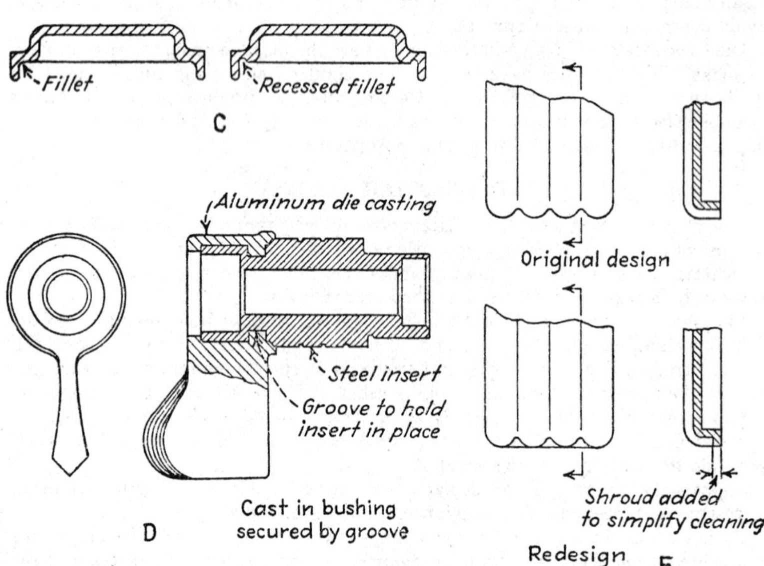

FIG. 4. Design pointers: *A*. *Undesirable*, undercut requires movable or loose die members; *satisfactory*, no undercut but requires more metal; and *recommended*, neither disadvantage. *B*. Methods of locking inserts. *C*. When assembly requires sharp corners, a fillet minimizes shrinkage stresses. *D*. A groove secures insert. *E*. Shroud at parting line of complex part reduces trimming-die intricacy.

DIE-CASTING DIE STEELS

Uniformity and Soundness. The surface of a die casting faithfully reflects the condition of a die surface, and any surface defects in the steel such as minute holes, cracks, or tool marks left in the die will readily show on the casting. It is obvious that the die steel must be clean and free from segregation, cracks, hammer bursts, seams, scabs, and other mechanical defects and impurities.

TABLE 2. LIMITS FOR DESIGN OF DIE CASTINGS*

Casting detail	Die-casting Alloy					
	Zinc	Aluminum	Magnesium	Tin	Lead	Copper
Max weight, lb	35	20	10	10	15	5
Min wall thickness, in.:						
Large castings	0.05	0.08	0.08	$\frac{1}{16}$	$\frac{1}{16}$	$\frac{1}{8}$
Small castings	0.015	0.035	0.035	$\frac{1}{32}$	$\frac{1}{32}$	0.050
Shrinkage per in. dia or length, in.	0.001†	0.0015†	0.0015†	0.001†	0.001†	0.002†
Max number of cast threads per in.:						
External	24	20	16–20	32	32	10
Internal	24	None	None	32	32	None
Min draft, in. per in., of length or dia:						
On cores	0.003	0.010	0.010	None	None	0.020
On side walls	0.005	0.010	0.010	0.005	0.005	0.020
Min dia cast holes, in.‡	0.031	$\frac{3}{32}$	$\frac{3}{32}$	0.031	0.031	$\frac{1}{8}$

* Data are for average conditions. A considerable range for all data can be obtained by special design techniques. Especially is this true for casting weight and wall thickness.
† Depends to a large extent on design details and casting conditions.
‡ Depends somewhat on length of core.

TABLE 3. DEPTH-DIAMETER RELATIONSHIP OF CORED HOLES

Alloys	Diameter, In.	Max Depth in Relation to Diameter
Magnesium	Smaller than 0.093	Not cored
	0.093 to 0.250	3 times diameter
	0.250 to 1.000	3 to 5 times diameter
Aluminum	Smaller than 0.093	Not cored
	0.093 to 0.250	3 times diameter
	0.250 to 1.000	3 to 5 times diameter
Zinc	Smaller than 0.093	Not generally cored
	0.093 to 0.250	3 to 8 times diameter
	0.250 to 1.000	6 to 8 times diameter
Copper (brass)	Smaller than 0.186	Not cored
	0.186 to 0.500	1½ to 3 times diameter
	0.500 to 1.000	3 to 5 times diameter

Machinability. Die-casting dies are usually fabricated from steel in the fully annealed condition, chiefly for ease in machining.

Heat Checking. Heat checking is the greatest single cause of die failure. Heat checking results from forcing molten metal under pressure into intimate contact with the die, thus causing the die surface to become very hot in relation to adjacent parts of the die. Thermal conductivity and thermal expansion have considerable influence in the rate of heat checking. As would be expected, the most serious heat-checking problems occur in the casting of alloys of high melting points, such as aluminum, magnesium, and copper-base alloys.

Cleavage Cracking. Cleavage cracking or fracture of die-casting dies occurs when the die material is stressed beyond its ultimate strength by the formation of high stresses at sharp corners. These stress concentrations cause minute cracks that later progress to serious fractures. The ability of a die steel to absorb stress is a function of its toughness and notch sensitivity.

Deformation. Die steels must have sufficient hardness and strength to resist deformation and defacement caused by high closing and casting pressures. The hardness of dies usually ranges from 375 to 460 brinell with the most desirable hardness of 440 brinell for all-round use.

Stability in Heat-treatment. Heat-treating characteristics of die steels are critical, and extreme care must be exercised to prevent undue dimensional changes, distortion, warpage, and the formation of scale, pitting, and other similar surface defects. Also, any decarburization or carburization must be guarded against.

Cores and Ejector Pins. Movable cores and ejector pins must have a very hard wearing surface. Steel that can be nitrided—Nitralloy G—is commonly used. Stainless steel with about 0.65% carbon, 17% chromium, 0.30% manganese, and 0.04% silicon is also specified.

TABLE 4. COMPOSITION OF STEELS FOR DIE-CASTING DIES*

Steel No.	Composition, %								Uses
	C	Mn	Si	Cr	Mo	V	W	Others	
1	0.30	0.30	0.45	3.25	....	0.50	9.00	....	Brass impressions
2	0.30	0.30	0.50	1.40	0.40	....	4.0	5.0 Co	Brass impressions
3	0.40	0.30	1.00	5.00	1.00	0.50	....	....	Aluminum impressions
4	0.40	0.40	1.00	5.25	1.00	1.00	....	....	Aluminum impressions
5	0.35	0.40	1.00	5.0	1.5	0.30	1.25	....	Aluminum impressions
6	0.40	0.30	1.00	5.0	....	....	5.0	....	Aluminum impressions
7	0.30	0.75	0.50	0.80	0.30	....	....	....	Zinc impressions
8	0.05	0.15	....	....	....	....	....	....	Hubbed impressions
9	0.10	0.20	....	0.60	....	....	....	1.25 Ni	Hubbed impressions
10	0.65	0.30	0.40	17.00	....	....	....	....	Small cores and ejector pins
11	0.35	0.50	0.75	1.25	0.20	....	....	1.25 Al	Small cores and ejector pins
12	0.40	0.70	0.25	0.90	0.25	....	....	....	Holding blocks
13	0.45	0.70	0.25	....	....	....	....	....	Holding blocks
14	0.40	0.80	0.50	0.80	0.30	....	....	1.00 Ni	Cast holding blocks

* The compositions given are type compositions only. The actual compositions vary with manufacturers; also other elements than indicated may be added.

Section 19

BABBITTING OF BEARINGS

By: B. J. Esarey, Chief Engineer, National Bearing Division, American Brake Shoe Co., St. Louis, Mo.

SECTION 19

BABBITTING OF BEARINGS

The object of bonding babbitt to a bronze or steel shell is to obtain a composite bearing, which as a whole will have the strength and rigidity of a steel or bronze back and a bearing surface with these properties: (1) antifriction characteristics, (2) imbeddability, (3) conformability, (4) corrosion resistance, (5) fatigue strength, (6) compressive strength, and (7) bondability.

In the ideal case, perfect metallic continuity would exist, and there would be gradual transition of mechanical and thermal properties on passing from the lining to the backing material.

Bonding Problem. Bearings for modern machinery should have the babbitt chemically bonded to the backing material. The problem is to prepare a chemically clean surface for babbitting. The importance of cleanliness cannot be overemphasized for all steps of the babbitting process. The best soft solder available or economically feasible should be used to tin the surface, depending on conditions under which the work must be done. Proper flux must be used to insure the solder's "wetting" the surfaces to be bonded. The shell surface, solder, and babbitt must be at proper temperature. If pyrometic equipment is not available, "Tempil" marking crayons afford a fairly close approximation of temperature of the parts.

After the babbitt is selected, keep it clean and at proper pouring temperature. If babbitt should be overheated, reduce its temperature, clean its surface with sal ammoniac, and skim off dross before pouring.

Selection of Babbitt. End use of the bearing and conditions under which the babbitt must be applied are of chief importance in selecting babbitt. Table 1 gives three compositions that cover most applications. Tin-base alloys are used extensively where high speeds and moderately high bearing pressures are involved. They have good resistance to impact. The principal hardening elements are 4.5 to 12% antimony and 1.5 to 8.3% copper. Trend is toward standardization on, say, 10% antimony, 5% copper, balance tin, for most applications. Alloys containing less antimony and copper can be used mainly for thin linings in bearing shells. Lead lowers melting temperature of tin-base alloys and is kept below 0.35%.

Most lead-base babbitts contain: antimony 10 to 15%, tin 2 to 20%, balance lead. Small amounts of copper (1 to 2%) are added to alloys containing around 20% tin for the purpose of limiting segregation. Above 10% tin content, little is gained in strength. Lead-base babbitts have good imbeddability and conformability. Where thickness of linings can be cut down, these babbitts can carry loads comparable with heavier tin-base linings.

Silver-lead babbitt known as No. 397 is a high-grade lead-base babbitt that will handle practically all applications where tin-base material has been used. The tin and antimony content inhibit corrosion of this alloy.

Babbitt compositions given in Table 2 cover ASTM specifications for bearing materials that can be used where a low pouring temperature is necessary. This may

TABLE 1. BABBITTS THAT COVER MOST APPLICATIONS

Metal	Tin	Antimony	Copper	Lead	Silver	Physical Properties	
						Compressive Strength at 1% Deformation	Brinell Hardness at Temp
1. Tin-base babbitt...	85%	10%	5%	...	...	10,600 psi Liquidus—691 F Solidus—460 F Pouring temp—850 F	70 F—23.8 100 F—22.3 150 F—17.8 212 F—12.6 250 F—10.1 300 F— 8.9 350 F— 6.6
2. No. 397 silver-lead-base babbitt	3%	10%	...	Bal.	2.6	8,000 psi Liquidus—497 F Solidus—470 F Pouring temp—650 F	70 F—15.3 100 F—13.7 150 F—11.1 212 F— 9.5 250 F— 9.1 300 F— 7.6 350 F— 6.5
3. Lead-base (Rex) babbitt	10%	15%	...	Bal.	...	10,000 psi Liquidus—514 F Solidus—464 F Pouring temp—640 F	70 F—24.5 100 F—18.3 150 F—15.3 212 F—11.1 250 F—10.9 300 F— 9.9 350 F— 6.4

NOTE: Solidus temperature is equivalent to melting point; liquidus temperature is equivalent to complete liquefaction.

Applications:

Metal No. 1—Good for all babbitt applications. Expensive and not needed in most applications.

Metal No. 2—This babbitt will handle nearly all applications where high tin-base babbitt is used, except where room-temperature hardness is the most important property.

Metal No. 3—A cheaper babbitt—good enough to handle many applications.

be required because very little equipment is available to babbitt the bearing. An alloy of low melting point is an advantage under such conditions.

Hardness of Babbitt. In many cases, hardness at specific temperature is the most important property of a babbitt. Strength figures for babbitt may vary according to how the specimen was prepared. However, if hardness is plotted against temperature, the babbitt giving the nearest approach to a straight line will have the best properties. In other words, the babbitt having the greatest remaining strength at running temperatures will give the best results.

Most babbitt-lined bearings fail in fatigue, by shelling out. The weakness in babbitt that causes such failures is the property that provides necessary conformability and imbeddability. With this exception, babbitt is still the best bearing material known.

Preparation of Shell. The first step is preparation of the surface to be tinned. Depending on the material, it can be handled in various ways. The ideal condition for preparation of a chemically clean surface is:

A cast-iron or mild-steel shell should be degreased and then anodically etched in dilute solution of sulfuric acid (30 to 50% sulfuric); then rinsed in running water; and then either copper, iron, or tin is plated on the surface to be babbitted (0.0002 to 0.001 in. thick), to provide a chemically clean surface. The shell should finally be rinsed in hot water and dried by air blast.

Successful preparation of cast iron depends upon the quality of etching. This treatment must remove carbon and oxides from the surface, leaving pure iron

TABLE 2. ASTM SPECIFICATIONS FOR WHITE-METAL BEARING ALLOYS

Alloy Grade No.	Nominal Composition, %				Sp Gr	Yield Point, Psi		Ultimate Compressive Strength, Psi		Brinell Hardness		Melting Point, F	Complete Liquefaction, F	Pouring Temp, F
	Tin	Antimony	Lead	Copper		At 20 C	At 100 C	At 20 C	At 100 C	At 20 C	At 100 C			
1	91.0	4.5	....	4.5	7.34	4000	2650	12,850	6950	17.0	8.0	433	700	825
2	89.0	7.5	....	3.5	7.39	6100	3000	14,900	8700	24.5	12.0	466	669	795
3	83.33	8.33	....	8.33	7.46	6600	3150	17,600	9900	27.0	14.5	464	792	915
4	75.0	12.0	10.0	3.0	7.52	5550	2150	16,150	6900	24.5	12.0	363	583	710
5	65.0	15.0	18.0	2.0	7.75	5050	2150	15,050	6750	22.5	10.0	358	565	690
6	20.0	15.0	63.5	1.5	9.33	3800	2050	14,550	8050	21.0	10.5	358	531	655
7	10.0	15.0	75.0	....	9.73	3550	1600	15,650	6150	22.5	10.5	464	514	640
8	5.0	15.0	80.0	....	10.05	3400	1750	15,600	6150	20.0	9.5	459	522	645
10	2.0	15.0	83.0	....	10.07	3350	1850	15,450	5750	17.5	9.0	468	507	630
11	.....	15.0	85.0	....	10.28	3050	1400	12,800	5100	15.0	7.0	471	504	630
12	.....	10.0	90.0	....	10.67	2800	1250	12,900	5100	14.5	6.5	473	498	625
15	1.0	15.0	82.5	0.5	10.05	....	....	......	....	21.0	13.0	479	538	662
16	10.0	12.5	77.0	0.5	9.88	....	....	......	....	27.5	13.6	471	495	620
19	5.0	9.0	86.0	....	10.50	....	....	15,600	6100	17.7	8.0	462	495	620

NOTE: Yield-point values were taken from the stress-strain curves at a deformation corresponding to 0.125 % reduction of gage length. Ultimate-strength values were taken at a unit load necessary to produce a deformation of 25 % of the length of the specimen.

(ferrite) for bonding. In the patented Kolene process, cast iron is immersed in molten salts instead of acid. The Tin Research Institute has developed baths of fused salts for the same purpose. The choice of acid or molten salts depends on the job to be done and the equipment available.

To clean medium-carbon steel (0.35% carbon and less than 0.5% nickel), pickle in 10% hydrochloric acid at 212 F for 1 min., rinse, dip a few seconds in 60% cold nitric acid and then anodically etch in 30% sulfuric acid.

In the case of nickel steel (2.75 to 3.75 Ni, 0.25 to 0.35 C, with sulfur and phosphorus below 0.05%), it is necessary to degrease, then dip 2 min in boiling 10% HCl, rinse; dip 2 min. in boiling 10% H_2SO_4, rinse; dip 20 min. in cold 10% H_2SO_4, rinse; dip 20 sec in 60% nitric acid; anodic etch 3 min at high current density in 30% H_2SO_4; and finally plate with iron, copper, or tin to about 0.001- to 0.002-in. thickness.

After cleaning, the surface to be babbitted should be fluxed with eutectic mixture of zinc chloride and ammonium chloride (71% $ZnCl_2$, 29% NH_4Cl) and then tinned at 750 F. The shell is then ready for babbitting.

Often it is not possible to plate, and then the procedure should be as follows:

Very greasy or oily shells may be swabbed with a suitable solvent to remove the bulk of the oil. The remainder can be removed by degreasing solvents or by boiling in a strong caustic solution for 10 to 15 min, then rinsed.

Where possible, the shell should have been freshly turned, using a fine cut. Sand or shot blasting should be avoided because these processes cause nontinnable particles of grit to become imbedded in the surface, or they force oil into the surface.

Pickle the shells to provide a reactive surface. Steel shells should be given 5 min in 10% hydrochloric acid solution. Cast iron and steel containing nickel and chromium should be given an anodic pickling in 30% sulfuric acid. Alternate procedure for cast iron and nickel-bearing steel is to use a 10% hydrofluoric acid pickling solution to remove the silicon to produce a surface in which the pure ferrite is available for reaction.

Flux the shell in zinc chloride solution (3 oz zinc chloride, 7 oz water), and transfer to the tin bath. For mild-steel shells the tin bath should be kept at 750 F and parts immersed up to 2 min. For all other steels containing nickel or manganese, the tinning is best done in two steps: Soak the shell in a tin bath at 750 to 850 F for ½ to 1 hr. Remove the shell, wipe with asbestos, and plunge into a second tin bath at 570 F for a few seconds.

In the case of bonding to bronze, the treatment is much simpler. The surface to be tinned should be degreased, fluxed with the chemically active eutectic mixture of zinc chloride and ammonium chloride, and then tinned at 650 to 700 F.

To secure a good bond on pure copper, it is sometimes necessary to degrease, followed by cathodic treatment in hot 5% sodium hydroxide solution, rinse, dip in 20% nitric acid, rinse, and flux.

Aluminum bronze is not readily bonded, probably because of a superficial coating of aluminum oxide. Hence, clean the surface with dilute hydrofluoric acid. Then scrub the work surface with a wire brush or scraper while the shell is submerged in the tin bath. Electrodeposition of about 0.001 in. copper avoids further difficulty in bonding babbitt to aluminum bronze.

Brass can be handled like bronze, but the tin bath tends to pick up zinc, which may cause difficulties in the bonding of other metals.

Tinning of aluminum is difficult. Mechanical scrubbing under molten tin or use of fluxes given in "Hot Tinning," published by the Tin Research Institute, assists bonding.

TABLE 3. TIN-LEAD SOLDERS FOR BONDING BABBITT

Composition,		Melting Range, F		Tin-bath Temp,* F
Tin	Lead	Solidus	Liquidus	
65	35	361	374	650
50	50	361	421	610
40	60	361	441	630
30	70	361	460	650

* Tin-bath temperatures are higher in some cases in order to heat shells more quickly.

Alloy steels can be tinned by expedients involving pickling solutions, scrubbing to remove pickling smut, strong fluxes, and long immersion periods.

Tinning Alloys. Choice of a tinning alloy (Table 3) depends on conditions of use. For babbitting purposes a soft solder means a tin solder, which is essentially an alloy of tin and lead, with or without a small amount of antimony.

The 65-35 tin-lead solder alloy has good mechanical strength, fair ductility, and maximum fluidity. It represents the practical eutectic composition with the lowest melting point of the series and a single solidification point. This general-purpose solder is particularly useful where the temperature requirements are critical.

The 50-50 solder alloy is the most popular for general-purpose use. It is good for hand soldering and has excellent mechanical strength and fair ductility and fluidity. It has a freezing range of 60 F.

The 40-60 solder is slightly more sluggish than 50-50 solder when used with a bit because of its freezing range of 92 F. However, it is an excellent alloy for stick soldering where parts to be tinned are too large to put into a tin pot. It has good ductility and good strength.

The 30-70 alloy has fair strength and ductility but requires higher heat input into the work. It is usually employed where cost is the deciding factor. It requires torch soldering to get the part to be soldered up to the temperature of about 486 to 500 F. This alloy has the wide freezing range of 125 F.

Tinning Procedure. After the bearing back has been thoroughly cleaned and the tinning alloy is ready, a flux (a eutetic mixture of zinc chloride and ammonium chloride) is applied to the surface to be babbitted and is dried.

The bearing back is immersed in the tinning bath to produce a coating consisting of a chemically bonded layer and the tinning compound. The chemically bonded layer consists of an intermetallic compound $FeSn_2$ in the case of iron or steel, while that on copper is Cu_6Sn_5 and Cu_3Sn.

When tinning has been successfully carried out, the coating comprises a continuous layer of alloy upon the base metal and a covering layer of smooth bright solder. Inadequate preparation of the surface before tinning leads to a lack of continuity or uniformity in one or both layers.

In exceptional cases there may be uncoated spots of appreciable size where the tin has not wetted or adhered to the backing. More often, an apparently continuous compound layer is formed, but the solder does not form a uniform layer over it. This is a principal defect in tinning, and it is first observed when the part is removed from the tinning bath. At the moment the work is withdrawn, the coating is continuous and uniform, but with draining the coating breaks and gathers in drops and streaks. Between drops of solder there is no visible exposure of the base metal, but the coating is matte and consists of little more than a very thin layer of compound. Sprinkle sal ammoniac on these spots, and brush the surfaces heavily until the drops and streaks are removed and the surface is wetted uniformly. Correct the surface preparation of the balance of the parts to prevent recurrence of the trouble.

Babbitting Procedure. Assume for initial purposes that a bronze bushing (about 32 lb, 8 in. dia by 8.5 in. long) is to be lined with babbitt. The part is bored with a smooth finish (free of casting oxides) on the surface to be tinned. The back and surfaces not to be tinned can be protected against the action of tin by various coatings. Ordinary whitewash may be used. The following coating has been used extensively: 1 gal dry talc, 2 qt Spanish Brown, ½ pt silicate soda, 5 gal water. This mixture must be used within 2 days after mixing; otherwise it becomes too alkaline and causes a gummy scum formation on the solder pot.

Another masking coating is a slurry of: 1 lb magnesia; 2 pt water glass, 1 pt water. One or more coats of these masking solutions may be brushed on and allowed to dry. The coating may be removed by hot water and brushing.

A bath of 65-35 tin-lead solder is held at a temperature of 650 F. The lining mandrel (Fig. 1) has been masked with a coat of whitewash and warmed to about 150 to 200 F.

The bushing is cleaned, fluxed with a eutectic mixture of ammonium chloride–zinc chloride, placed on the side of the solder bath to dry, and then immersed in the bath. After about 25 to 30 sec the bushing is removed and the tinned surface is inspected. Streaks or spots, as described above, are brushed with sal ammoniac until the surface is uniformly wetted. Then the bushing is lowered into the solder pot to regain heat. It is then taken out and put in place on the mandrel, and babbitt at the proper pouring temperature is poured into the bearing before the tinned surface can solidify.

In case bearings are too large for available tinning pots, clean the surface by pickling, using the bearing shell as its own container. Then apply heat uniformly to the

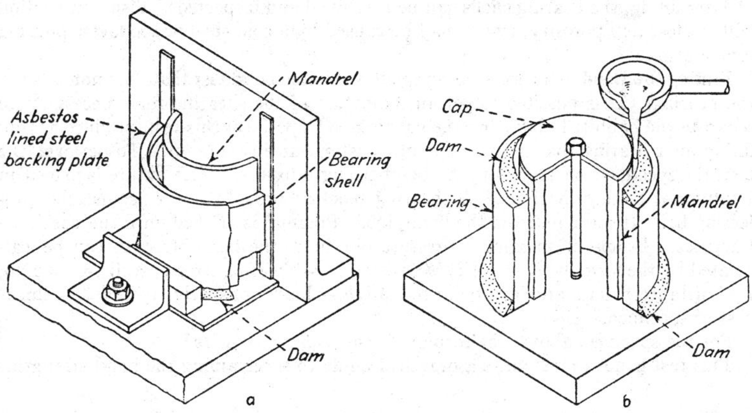

FIG. 1. Setups for pouring babbitt into a bearing shell and a solid bearing.

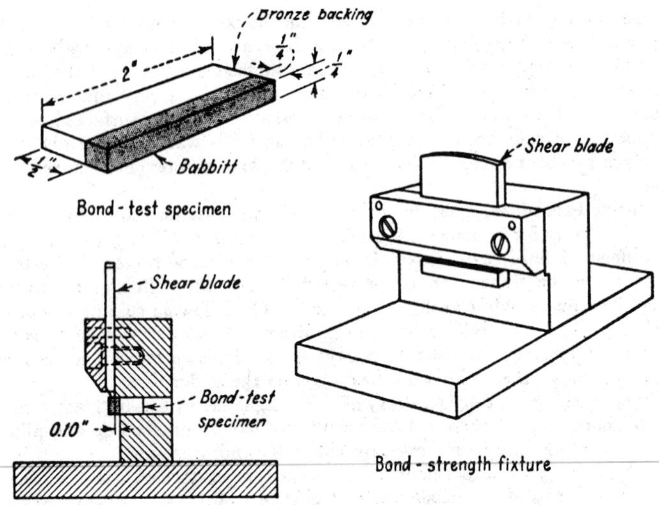

FIG. 2. Bond strength of a babbitt bearing can be determined with reasonable accuracy by cutting specimens from a shell and testing them in a bond-strength fixture set up in a universal testing machine.

back to avoid hot spots, until the shell reaches 370 to 450 F. Then tin, using 40-60 stick solder and plenty of flux and sal ammoniac. The bearing shell is kept heated, the mandrel is warmed, and the babbitt is poured.

In centrifugal lining of shells with babbitt the same tinning procedure is followed. The speed of rotation should be 500 to 600 fpm. The bearing should be close to speed of rotation when babbitt is put in the bearing. Cooling is desirable as soon as possible with an air blast or water, depending on the rate of cooling needed.

After lining, the bearing shells can be machined and inspected. Visual inspection will disclose any porosity, which may be caused by a cold shell or too low a pouring temperature.

Testing the Bond. There is no completely reliable bond test that does not destroy the bearing. It is possible to get an indication of the shearing force necessary to separate the babbitt lining from the backing. To perform this test, produce a $\frac{1}{4}$-in. lining on a bearing that is to be cut into test specimens (Fig. 2). To perform the test, the specimen is clamped in the bond-strength fixture. This fixture is placed in a universal testing machine. A shearing block is brought down against the projecting babbitt 0.100 in. from the bond line. Pressure is applied until the specimen fractures. From the maximum pressure obtained, the bond strength can be calculated by the formula $S_s = 6WI/bd^2$, where S_s = shearing stress, psi; W = load, lb; I = distance from stationary point to load; b = length of specimen; and d = thickness of specimen.

For the specimen above, the formula becomes $S_s = 0.3W/d^2$.

This test is not exact but is a practical guide to ascertaining the bond strength.

Part 3

ASSEMBLY METHODS

Section 20

ARC WELDING AND SURFACING

By: Robert Wilson, Engineer in Charge of Welding Applications, Lincoln Electric Company

SECTION 20

ARC WELDING AND SURFACING

PROCESSES AVAILABLE

Electric arc welding employs the heat of an electric arc to bring the metals to a molten state. In the majority of arc-welding methods, the work is made part of the welding circuit, which has its source in a welding generator or transformer. One cable from the power source is attached to the work and another cable to an electrode holder. An arc is established between the electrode and the work and is moved along the work to melt and fuse the metal. Since the arc is the hottest known commercial source of heat, melting action takes place almost instantaneously as the arc is applied to the metal.

All arc-welding methods have one common problem—shielding the arc.

Molten steel has a strong affinity for oxygen and nitrogen. If the arc and molten-metal pool are exposed to atmosphere during welding, the metal picks up oxygen and nitrogen to form oxides and nitrides in the weld as it solidifies. These impurities embrittle a weld and weaken it. All generally used welding methods shield the arc and molten pool from the atmosphere, obtaining welds (when correctly made) that are as strong or stronger than the metal being welded.

Carbon-arc welding employs a carbon rod as an electrode. The arc is formed between the carbon and the work, creating a molten pool in the work. This pool is kept molten by playing the arc across it. If extra metal is needed, it is supplied by introducing a filler rod into the arc. This is a puddling process and is not applicable to vertical or overhead welding. Shielding may be obtained by introducing a paste, powder, or fibrous flux into the arc.

Carbon-arc welding is used only for specialized applications and with automatic equipment. The carbon arc is also used for cutting where a precision cut is not necessary or on alloys that cannot be cut by the gas process.

Inert-gas arc welding forms the arc directly between the work and a tungsten electrode. Inert gas, helium or argon, is introduced through a small orifice in the end of the electrode holder to shield the arc. The tungsten electrode is not consumed, and a filler rod is used if extra metal is needed in the joint. This welding method is used to good advantage in welding aluminum and magnesium.

Atomic-hydrogen welding differs from other arc-welding methods in that the arc is formed between two tungsten electrodes and the work is not part of the welding circuit. A stream of hydrogen passing through the arc is changed from molecular to atomic form, giving off intense heat close to the work. A filler rod supplies additional metal to the joint. The process has some advantages in welding thin sheet where high finish is needed.

Metal arc welding is the most widely used method. The electrode is a metal rod between which, and the work, an electric arc is formed. Intense heat of the arc, as high as 13,000 F, melts both the work and metal electrode. Tiny globules of metal form on the tip of the electrode and are forced across with the arc stream into the molten pool. This arc force makes overhead welding possible.

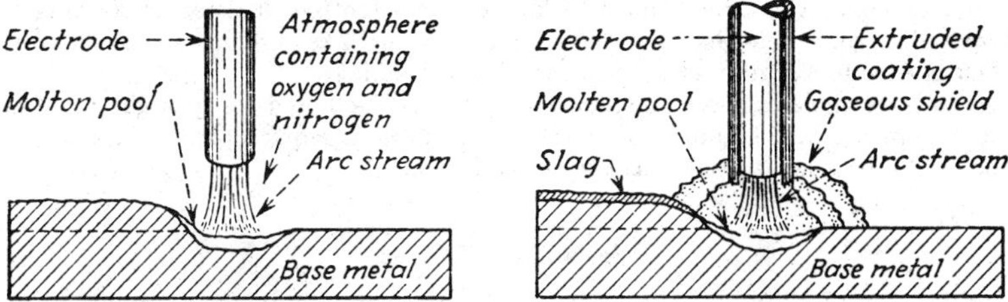

FIG. 1. Comparison of arc with bare electrode (left) and shielded arc with coated electrode (right).

The metal electrode is covered with a dried chemical coating which also burns to form a gaseous cloud that shields molten metal from contact with the atmosphere. The coating also forms a protective slag, which floats on the molten pool and solidifies there as the weld cools (Fig. 1).

The electrode coating, the gas, and the slag have several functions which improve weld quality. The coating controls the chemical analysis of the weld by preventing the vaporization or oxidation of essential elements in the electrode core wire. In some electrodes the coating serves to supply alloying elements not present in the core wire.

Primary purpose of the coating, however, is to form protective atmosphere and slag. The slag also acts as a fluxing agent in some cases. The coating is also used to improve the ease in striking and maintaining the arc, the welding speed, and the penetration.

Automatic arc welding maintains constant arc voltage as the electrode is fed mechanically toward the work and along the joint. Depending on the setup, the work may be held stationary while the automatic head is moved, or the work may be moved under the arc while the head remains stationary. Shielding is obtained by introducing fibrous flux into the arc or putting a paste or powder along the seam being welded. As these added elements burn they form a protective inert gas.

For a given cross-section of electrode wire, much higher welding amperage can be applied with automatic welding than with hand welding. This is true because the current travels only a very short distance along the bare metal electrode wire, contact being made with the current-carrying jaws close to the arc. The higher current densities used with automatic welding result in a deeply penetrating arc, which permits faster welding speeds. Distortion is minimized and uniformly good weld quality is assured.

Equipment is also manufactured for semiautomatic use of hidden arc welding. With this equipment the electrode wire is fed to the work automatically, but movement of the arc along the joint can be controlled manually. Flexibility of this method permits its use on irregular shapes and contours.

WHAT IS A GOOD WELD?

A weld joining two pieces of metal may be called upon to serve any one or a combination of the three generally recognized weld functions:

1. To be adequately strong.
2. To be liquid- or gastight.
3. To meet an acceptable standard of appearance.

A good weld adequately performs the required function or functions, but it must be made at minimum cost. It is not good welding to overweld.

For example, where welds must be liquid- or gastight but little or no strength is required, a good weld does not exceed actual strength requirements. Extra cost, such as edge preparation or extra passes, is not justified.

A slightly undercut weld might be satisfactory for liquid- or gastight requirements, but if a high standard of external appearance is required, it would not be a good weld, because it would have to be made over.

Weldability of Metals

Some metals can be welded more readily than others. Behavior of the metal under the heat cycle of welding may or may not be critical.

Adverse Factors. Economy and quality of welding on various metals may be affected by one or more of the following factors:

1. Oxidation products: (*a*) gaseous oxide of some one of the elements, causing gas holes in the weld metal; (*b*) solid oxides which have a melting temperature higher than the metal, thus causing slag inclusions; (*c*) oxides which are soluble or which are heavier and sink in the molten metal, and which render the weld metal brittle or of low strength.
2. Vaporization of some element at a temperature lower than the melting point of the metal.
3. Nonmetallic inclusions which have a melting point higher than the metal and did not coalesce when the metal was refined, but melt and coalesce under the high temperature of the arc to form visible slag inclusions.
4. Change of structure or arrangement of elements may take place within the metal during arc welding, to cause change of physical properties or resistance to corrosion, etc.
5. Gas solubility of metal depends on content of different elements, and porosity may result in weld metal. Fluxing out or elimination of an element during welding may reduce a metal's absorption power, cause the gas to be given up, and produce porosity in weld metal. Absorption of gases which form stable compounds with metallic elements will alter composition and physical properties of the weld metal.
6. High coefficient of thermal expansion—or high contraction of weld metal upon cooling.
7. "Hot shortness"—or low strength of the metal at high temperatures.
8. Thermal conductivity—or rate of transfer of heat from fusion zone.
9. Hardenability—tendency of metal to become hard and brittle in the weld or fusion zone during heat cycle of welding.

Corrective Measures. Study of these factors indicates that most of the potential undesirable characteristics can be corrected by one or more of these methods:

1. Selection of metal within the permissible class most suitable for arc welding.
2. Use of a proper shielded arc.
3. Use of proper fluxing material.
4. Use of proper electrode or filler metal.
5. Proper welding procedure.
6. In some cases, subsequent heat-treatment may be required.

A weld depends largely upon the characteristics of the weld metal which may come from two sources: the base metal and electrode of filler metal.

If little or no electrode or filler metal is used, the proper selection of the base metal becomes of prime importance. If the weld metal comes mostly from the electrode or filler metal, then selection of the proper electrode or filler metal becomes of prime importance. But both electrode and base metal are subjected to similar require-

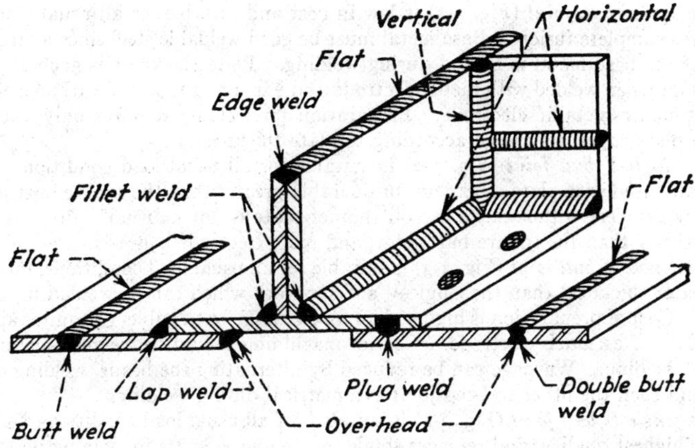

FIG. 2. Examples of arc welds and locations.

ments during arc welding and both should be of best arc-welding quality, although in many cases the electrode or filler metal serves as a corrective for the base metal.

Carbon Content of Metal. Carbon content of steel is the most important factor in determining heat-treatment procedures. When the carbon content is low the weldment will require no special consideration. When it is high, preheating is generally necessary at 450 to 650 F to control the rate of cooling.

Failure to preheat high-carbon steel before welding usually results in hardening of the steel adjacent to the weld. Since this reduces ductility, cracking next to the fusion zone is likely to result. This steel should be welded while it is hot and then allowed to cool slowly, increasing the cooling time proportionately as the carbon content of the steel is greater. For steel above 0.40% carbon, overnight cooling in a furnace is sometimes necessary to avoid a hardened zone. Types E6015, E6016, E7015, and E7016 low-hydrogen coating-type electrodes are recommended for welding high-carbon steels, although the normal mild-steel electrodes can be used.

Definite limitations and specifications for preheating should be established only after experience. Research into use of low-hydrogen coatings shows that preheating is unnecessary in many applications generally thought required.

TYPES OF JOINTS

Various types of joints are shown in Fig. 2. The use of each type of joint is governed by three factors.

1. Whether the load is in tension or compression, and whether bending, fatigue, or impact stresses are present in any combination.
2. Whether load application is steady, variable, or sudden.
3. Cost of joint preparation and welding.

Aid in selecting the best joint for given service conditions and cost is provided in the following detailed discussion of the principal types of joints.

Types of Butt Joints. Butt joints are of several types, each having a number of variations. However, the general classification lists butt joints as square, V-bevel, U, and J.

The *square butt joint* (Fig. 3-1) is low in cost and suitable for all usual loads but requires complete fusion. Base metal must be good weldable steel since a large portion of the base metal is melted during welding. Plate thickness is generally $\frac{3}{8}$ in. or lighter when welded with metal electrode and $\frac{3}{4}$ in. or lighter with carbon electrode or automatic metallic electrode. Preparation for welding requires only matching of the plate edges, separated according to plate thickness.

The *single-V butt joint* (Fig. 3-2) is suitable for all usual load conditions. It is generally used with plate thickness considerably greater than the square butt joint—$\frac{3}{8}$ in. or heavier—although its use on thinner plate is not unusual. Preparation is more costly than the square butt joint, and more electrode is used in welding.

The *double-V butt joint* (Fig. 3-3) is suitable for all usual load conditions, for plates of greater thickness than the single-V and for work which can be welded from both sides. Cost of preparation is higher than for single-V, but double-V requires approximately half as much electrode. Cost of machining should be weighed against the cost of welding. Warpage can be reduced by alternating the beads, welding on one side and then the other to keep joint symmetrical during welding.

The *single-U butt joint* (Fig. 3-4) is suitable for all usual load conditions, for work of the highest quality, and replaces single- or double-V joints for joining plates $\frac{1}{2}$ to $\frac{3}{4}$ in. thick, although it is also used on heavier plate. Machining plates to a single-U reduces weld metal needed but increases machining costs. The joint is welded from one side, except for a single bead which is put in last on opposite side.

The *double-U butt joint* (Fig. 3-5) is suitable for all load conditions and is used for welding heavy plates—$\frac{3}{4}$ in. and thicker, where the welding can be done from both sides. Joint requires less weld metal than single U but costs more to machine.

Types of T-Joints. The *square-T joint* (Fig. 3-6) like the square-butt joint requires no machining of plates; it is used for all ordinary plate thicknesses, principally for loads which place the welds in longitudinal shear. For severe impact or heavy transverse loads, keep in mind nonuniform stress distribution of the joint. The square T requires considerable weld metal, and therefore the welding cost may be higher than for other types of T-joints.

The *single-bevel T-joint* (Fig. 3-7) is suitable for much more severe loads than the square T, because of its better distribution of stress. It is employed, in most instances, for welding plates $\frac{1}{2}$ in. or thinner in work which can be welded from one side only. While more costly to machine than the plain T, the single-bevel T-joint is lower in electrode costs.

The *double-bevel T-joint* (Fig. 3-8) is suitable for heavy loads in longitudinal or transverse shear and in joining heavy plate where welding can be done from both sides. Machining cost may exceed the single-bevel T-joint, but electrode cost is lower than some other types such as plain T.

The *single-J T-joint* (Fig. 3-9) is suitable for severe loads, and while it may be used for smaller sizes, it is generally applied to plates 1 in. and heavier. Welding is done from one side, but put in a finish bead on the side opposite the J. Although somewhat more costly to machine than the single V, the single-J T-joint is lower in electrode cost.

The *double-J T-joint* (Fig. 3-10) is suitable for exceedingly severe loads of all types in heavy plate—$1\frac{1}{2}$ in. and heavier—where welding can be done from both sides. Machining costs are higher than for other T-joints; less electrode is required.

Types of Lap Joints. The *single-fillet lap joint* (Fig. 3-11), frequently used, requires practically no machining to fit the plate edges. Study stress distribution when fatigue or impact loads are encountered. Where loading is not too severe, the joint is suitable for welding plate of all thicknesses

The *double-fillet lap joint* (Fig. 3-12) is suitable for more severe load conditions. In general, the two fillets should be full size, although one fillet may be smaller than the other in some instances. Because of its low cost, the double-fillet lap joint is widely used.

Types of Corner Joints. The *flush corner joint* (Fig. 3-13) is suitable where loads

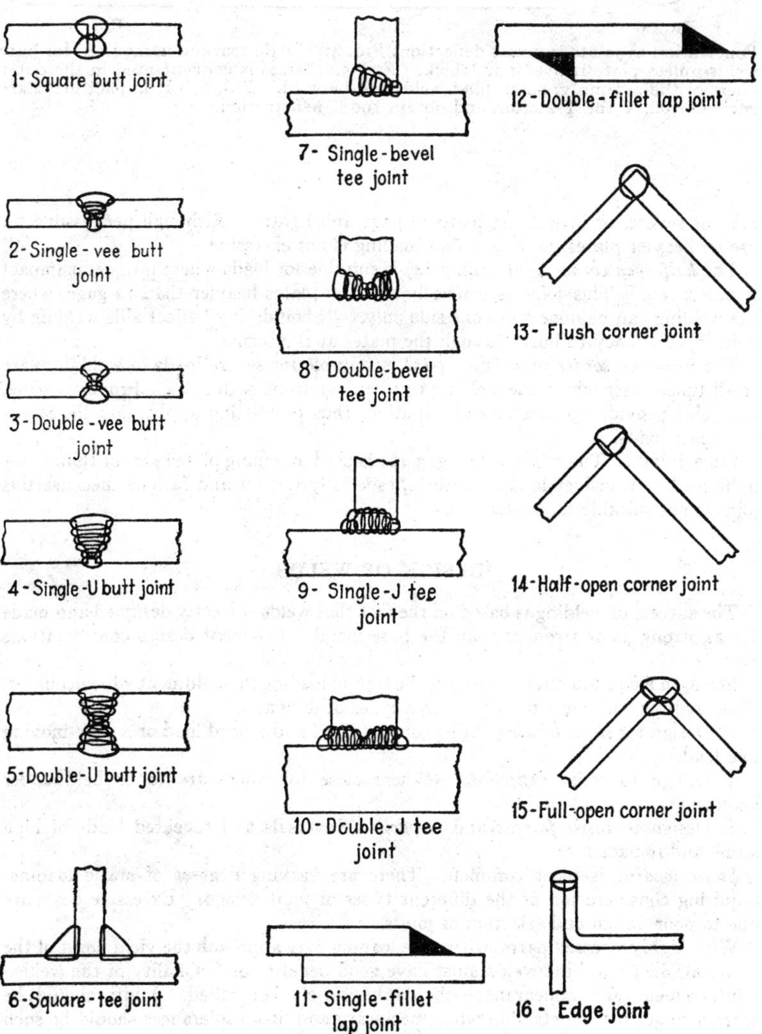

FIG. 3. Types of arc-welded joints.

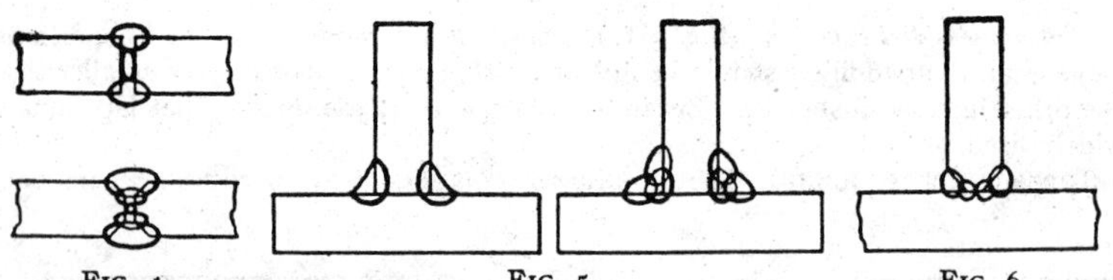

FIG. 4. FIG. 5. FIG. 6.

Penetration in joints to resist deflection: FIG. 4. Little scarfing is required for butt welds unless plate is over 1 in. thick. FIG. 5. Stress is concentrated in the outer fibers of the assembly; small fillet welds can be used. FIG. 6. If impact loads are encountered, V the T-section and obtain 100% penetration.

are not severe, or in welding plate 12 gage and lighter. Although permissible for use on heavier plates, take care that loading is not excessive.

The *half-open corner joint* (Fig. 3-14) is suitable for loads where fatigue or impact is not severe. This joint is generally used on plates heavier than 12 gage, where the welding can be done from one side only. "Shouldering" effect aids welding by reducing tendency to burn through the plates at the corner.

The *full-open corner joint* (Fig. 3-15) is suitable for severe loads in welding plate of all thicknesses where the welding can be done from both sides. Properly made, this joint provides good stress distribution, thus permitting application to fatigue or impact loads.

Edge Joints. The *edge joint* (Fig. 3-16) is used in joining plates $\frac{1}{4}$ in. or thinner for light loads. Consider load conditions, especially impact and fatigue, because this joint is not suitable for severe loads.

DESIGN OF WELDS

The success of welding is based on the fact that welds correctly designed and made are as strong as or stronger than the base metal. Pertinent design considerations are:

Strength Requirements. Analyze the type of loading the weldment will encounter. This analysis will point to one of three types of design.

1. Design for *static loading*. (To support itself and a dead load or a slow-moving live load.)

2. Design to resist *deflection*. (Where close tolerances are to be held under loading.)

3. Design to resist *fatigue* and *impact*. (Reversals and repeated loads of high stress and impact.)

Static loading is most common. There are varying degrees of static loading, requiring consideration of the different types of joint design. Excessive costs are due to poor design and selection of joints.

With highly stressed parts, where the loading may approach the yield point of the material, fillet and butt welds must have good penetration. Quality of the weld—reinforcement and appearance—should be closely controlled. Joints should be scarfed to get full penetration where necessary and fit-up tolerances should be such as to obtain top-quality welds.

Parts of a weldment not subject to high stresses and merely supporting the structure should be welded speedily and at low cost. For example, ¾-in. plate can be butt-welded, using a V'd joint, at 9.8 ft of joint per hr. Using a square-butt joint increases the rate to 45 ft per hr. The economical joint would not have 100% penetration. If 100% penetration is not necessary, it is wasteful to design for it.

To Resist Deflection. Here the structure is made heavy enough to withstand high loading without deformation. Since there is very little stress except in the extreme fibers, the savings gained from proper design are tremendous. In most cases plates up to 1 in. need not be scarfed; plates over 1 in. need very little scarfing (see Fig. 4).

Since deflection is mainly a function of section modulus, welding done in the center of the plate near the neutral axis has little effect upon rigidity. In most cases much less than 100% fusion is required. Fillet welds can be much smaller. The reduced amount of welding minimizes warpage from excessive heat, making it much easier to hold close tolerances (see Fig. 5).

To Resist Fatigue and Impact. Highest quality weldments, with 100% penetration, are necessary. Avoid fillet welds if possible. Instead of fillet welds on a T-weld section, V the leg of the T and weld to the flanges with 100% penetration (see Fig. 6).

Structures should have gradual change of section, to avoid stress concentration. Change the direction of stress gradually, by rounding out corners; sharp corners tend to concentrate stress and fatigue cracks result. Such errors as undercut, too much reinforcement, and insufficient penetration are to be avoided.

TABLE 1. SAFE ALLOWABLE LOADS FOR FILLET WELDS (IN SHEAR)*

For dynamic or vibrational loads it may be desirable to reduce unit stress, depending on severity of load

Size of Fillet Weld, In.	*Lb per Lin In.*
$\frac{1}{8}$	1200
$\frac{3}{16}$	1800
$\frac{1}{4}$	2400
$\frac{5}{16}$	3000
$\frac{3}{8}$	3600
$\frac{1}{2}$	4800
$\frac{5}{8}$	6000
$\frac{3}{4}$	7200

* "Fusion Code" (structural)—American Welding Society.

TABLE 2. LENGTH OF FILLET WELD TO REPLACE RIVETS

Rivet Dia, In.	Rivet Shear Value at 15,000 Psi	Length of Fillet Welds (to nearest $\frac{1}{8}$ in.)—"Fusion Code" (Structural) Shielded Arc Welding				
		$\frac{1}{4}$-in. Fillet	$\frac{5}{16}$-in. Fillet	$\frac{3}{8}$-in. Fillet	$\frac{1}{2}$-in. Fillet	$\frac{5}{8}$-in. Fillet
$\frac{1}{2}$	2,950	$1\frac{1}{2}$	$1\frac{1}{4}$	$1\frac{1}{8}$	$\frac{7}{8}$	$\frac{3}{4}$
$\frac{5}{8}$	4,600	$2\frac{1}{4}$	$1\frac{3}{4}$	$1\frac{1}{2}$	$1\frac{1}{4}$	1
$\frac{3}{4}$	6,630	3	$2\frac{1}{2}$	$2\frac{1}{8}$	$1\frac{5}{8}$	$1\frac{3}{8}$
$\frac{7}{8}$	9,020	$4\frac{1}{8}$	$3\frac{3}{8}$	$2\frac{7}{8}$	$2\frac{1}{8}$	$1\frac{3}{4}$
1	11,780	$5\frac{1}{4}$	$4\frac{1}{4}$	$3\frac{5}{8}$	$2\frac{3}{4}$	$2\frac{1}{4}$

NOTE: $\frac{1}{4}$ in. is added to calculated length of bead for starting and stopping the arc.

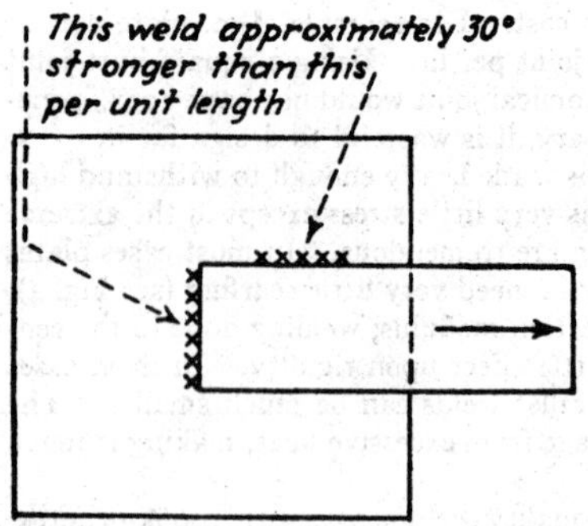

FIG. 7. Welds are stronger in transverse shear than in longitudinal shear.

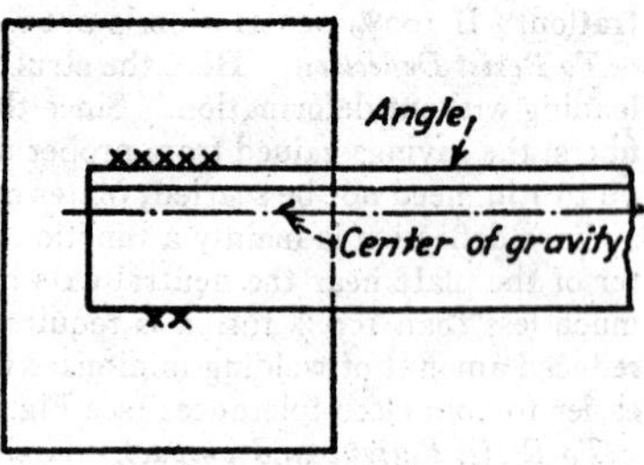

FIG. 8. Better load distribution results when length of weld is proportioned to loading.

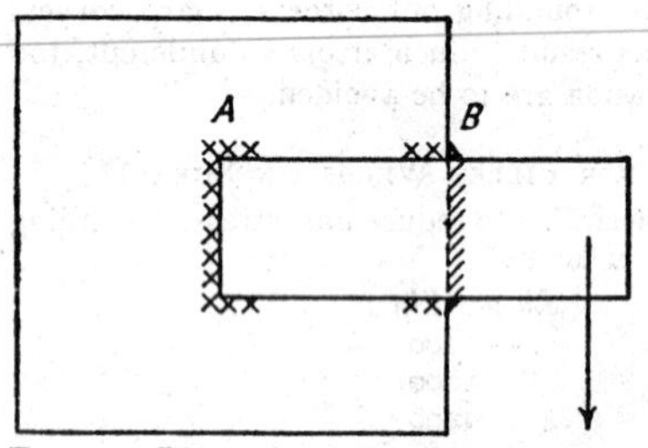

FIG. 9. Placement of welds to resist turning.

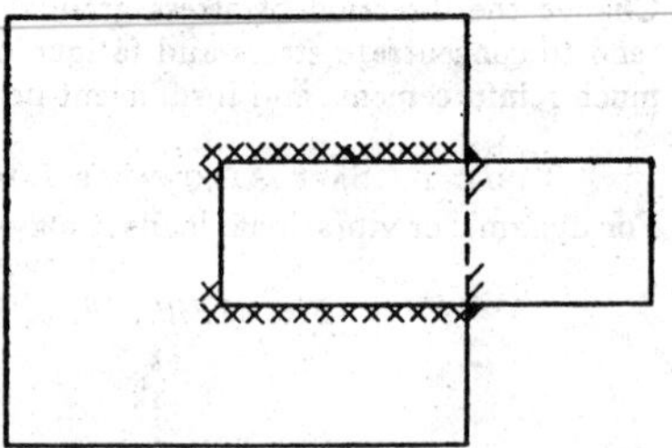

FIG. 10. Resistance to tearing by eccentric loads is secured by hooking welds around corners.

Location of fillet welds affects joint strength.

Make load analyses and stress calculations where nature of the job warrants it In many cases it is unnecessary. Tables 1 and 2 will aid in figuring allowable loads Figures 7 to 10 indicate the effect of the location of a weld with relation to its strength.

Design for Economy. The most economical weld requires: (1) the least edge preparation, (2) the minimum amount of deposited weld metal, and (3) maximum deposition rates.

Avoid overwelding. Build-up should be kept at a minimum. Use short intermittent welds rather than long continuous welds where possible.

In some cases, however, small continuous welds produce strength equal to large intermittent welds, require only half as much deposited metal, and are made at twice the speed.

Unnecessary welding can be eliminated by forming corners rather than welding them. Special steel shapes, formed heads, castings, and forgings can also be incorporated in a welded assembly to eliminate unnecessary welding.

The added expense required to secure good fit-up will be repaid many times over by economies in welding. Figure 11 indicates the savings that can be realized.

ELECTRODE SELECTION

There are two aspects to the problem of selecting the correct electrode for making a good weld under given conditions. These are selection of an electrode according to (1) coating and core-wire analysis and (2) diameter.

Electrode Requirements. It is necessary to know:

1. Position in which the work is to be welded.
2. Type and thickness of metal used.
3. Work preparation with regard to fit-up.
4. Type of welding current.
5. Class of work; that is, whether chief essential is deep penetration, surface quality, required physical properties, or code requirements.

The American Welding Society and the American Society for Testing Metals have jointly established specifications for the manufacture of welding electrodes:

Mild Steel Arc Welding Electrodes (A233-48T).

Low-alloy Steel Arc Welding Electrodes (A316-48T).

Corrosion Resisting Steel Welding Electrodes (A298-48T).

Copper Arc Welding Electrodes (B225-48T).

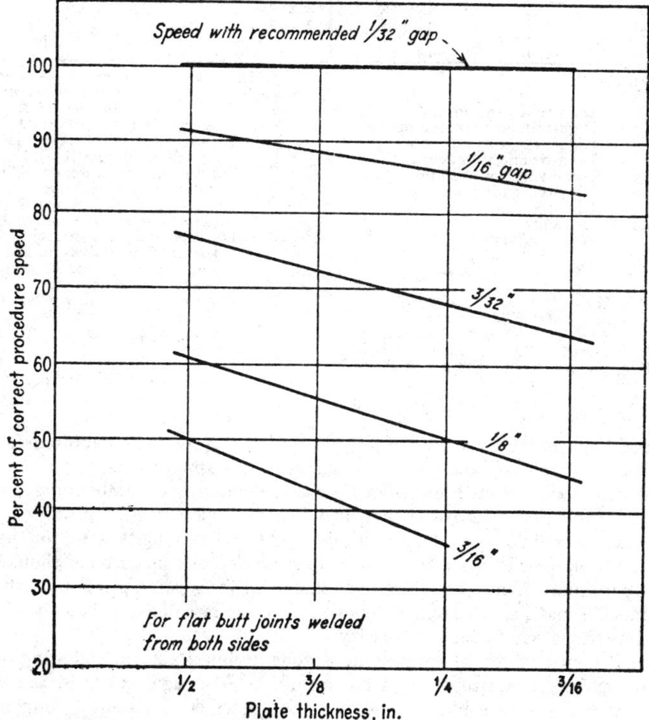

FIG. 11. Effect of fit-up on welding speed for a downhand butt joint (square or grooved) in various plate thicknesses.

TABLE 3. ELECTRODE CLASSIFICATIONS

No.	Covering*	Welding Positions†	Current Type‡
E45 series—min tensile 45,000 psi			
E4510	Light	1	G
E4520	Light	2, 3	G
E60 series—min tensile 62,000 psi			
E6010	1	1	A
E6011	2	1	B
E6012	3	1	C
E6013	4	1	C
E6015	5	1	D
E6016	6	1	B
E6020	7	2, 3	E
E6030	7	3	F

No.	Covering*	Welding Positions†	Current Type‡
E70 series—min tensile, 70,000 psi E80 series—min tensile, 80,000 psi E90 series—min tensile, 90,000 psi E100 series—min tensile, 100,000 psi			
Exx10	1	1	A
Exx11	2	1	B
Exx13	4	1	C
Exx15	5	1	D
Exx16	6	1	B
Exx20	7	2, 3	E
Exx25	5	2, 3	D
Exx26	6	2, 3	B
Exx30	7	3	F

* Covering:

No.	*Type*
1	High-cellulose sodium
2	High-cellulose potassium
3	High-titania sodium
4	High-titania potassium
5	Low-hydrogen sodium
6	Low-hydrogen potassium
7	High-iron oxide

† Welding positions:

No.	*Symbol*
1	*F, V, OH, H*
2	*C*-fillets
3	*F*

where *F* = flat, *H* = horizontal, *H*-fillets = horizontal fillets

V = vertical, *OH* = overhead } for electrodes 3/16 in. and under, except in Exx15 and Exx16, 5/32 in. and under

‡ Current type:

Symbol	*Current and Polarity*
A	For use with dc reversed polarity (electrode positive) only.
B	For use with ac or dc, reversed polarity (electrode positive).
C	For use with ac or dc, straight polarity (electrode negative).
D	For use with dc, reversed polarity (electrode positive) only.
E	For use with dc, straight polarity (electrode negative), or ac for horizontal fillet welds; or ac for flat-position welding.
F	For use with dc, either polarity, or ac
G	Not specified, but generally dc, straight polarity (electrode negative).

In addition to these classifications, electrodes are also manufactured for hard surfacing, welding cast iron, and other miscellaneous applications.

Mild and low-alloy steel electrodes are classified with a numbering system for simple identification. E6010 is a typical four-digit classification number. Prefix E designates a metal arc-welding electrode. The first two digits stand for the minimum allowable tensile strength of stress-relieved deposits in terms of thousands of pounds per square inch. The third digit stands for the welding position in which the electrode will make a satisfactory deposit, while the last digit indicates various arc characteristics, among which is polarity.

Since at least 90% of all arc welding is done in mild steel, the following descriptions of mild-steel electrode types are included. The significance of the various classification digits as explained for these electrodes is consistent throughout the E70, 80, 90, and 100 series of steel electrodes. Tables 3 and 4 give classification characteristics for steel electrodes, and Table 5 provides properties of base metals.

TABLE 4. YIELD POINT AND DUCTILITY OF WELD METAL

Electrode Classification No.	Tensile Strength, Min, Psi	Yield Point, Min, Psi	Elongation in 2 In., Min, %	Electrode Classification No.	Tensile Strength, Min, Psi	Yield Point, Min, Psi	Elongation in 2 In., Min, %
E4510	45,000	Not specified	5	E8016	80,000	67,000	19
E4520	45,000	Not specified	5	E8020	80,000	67,000	22
E6010	62,000	52,000	22	E8025	80,000	67,000	22
E6011	62,000	52,000	22	E8026	80,000	67,000	22
E6012	68,000	55,000	17	E8030	80,000	67,000	22
E6013	68,000	55,000	17	E9010	90,000	77,000	17
E6015	68,000	55,000	22	E9011	90,000	77,000	17
E6016	68,000	55,000	22	E9013	90,000	77,000	14
E6020	62,000	52,000	25	E9015	90,000	77,000	17
E6030	62,000	52,000	25	E9016	90,000	77,000	17
E7010	70,000	57,000	22	E9020	90,000	77,000	20
E7011	70,000	57,000	22	E9025	90,000	77,000	20
E7013	70,000	57,000	18	E9026	90,000	77,000	20
E7015	70,000	57,000	22	E9030	90,000	77,000	20
E7016	70,000	57,000	22	E10010	100,000	87,000	16
E7020	70,000	57,000	25	E10011	100,000	87,000	16
E7025	70,000	57,000	25	E10013	100,000	87,000	13
E7026	70,000	57,000	25	E10015	100,000	87,000	16
E7030	70,000	57,000	25	E10016	100,000	87,000	16
E8010	80,000	67,000	19	E10020	100,000	87,000	18
E8011	80,000	67,000	19	E10025	100,000	87,000	18
E8013	80,000	67,000	16	E10026	100,000	87,000	18
E8015	80,000	67,000	19	E10030	100,000	87,000	18

Electrode Characteristics. Types E6010–E6011—these general-purpose electrodes possess high average mechanical characteristics.

E6010 is best suited for dc, electrode positive. In sizes $\frac{3}{16}$ in. and smaller, it is suitable for use in all positions. It has deep penetration qualities and is used satisfactorily on square-groove butt joints where the electrodes actually scarf or melt the plates.

The E6010 electrode has a heavy covering low in minerals. The arc is very penetrating; the metal quickly solidifies, tending to give a flat bead surface (Fig. 12). There is small slag; the weld metal has excellent quality, including corrosion resistance. Some applications are welding piping, ships, machinery, structures (especially field or erection), and jigs and fixtures.

E6011 electrode is similar to E6010 but is designed for ac or dc operation in all positions. Excellent results are obtained with ac or dc, either polarity, depending on type of work. Characteristics are: heavy coating, low in minerals; penetrating arc; the metal solidifies quickly; slag action similar to E6010; protection of molten metal obtained principally by gases; stable arc with ac, therefore less sensitive than E6010 to polarity changes.

The E6011 electrode is well suited for making vertical and overhead fillet and butt welds, and it is generally recommended if welding must be done with ac when the structure cannot be placed so that the metal can be deposited in flat or downhand position. Deposited metal has good strength, high elongation, and excellent corrosion resistance. Applications are the same as for E6010.

E6012 electrode has a heavy coating, is used with dc (electrode negative) or with ac. Sizes $\frac{3}{16}$ in. and smaller are suitable for all positions, and in larger sizes for

TABLE 5. PROPERTIES OF BASE METALS

Metal	Tensile Strength, Psi	Yield Point, Psi	% Elongation in 2 In.	Endurance Limit, Psi	Impact Value, Ft Lb (Izod)
Mild steel	55,000–65,000	27,500–37,500	20–40	28,000–32,000	40–80
Cast iron	15,000–30,000	None	None	7,500–15,000	5
High-tensile steel (typical)	90,000	60,000	20	45,000	40
18-8 stainless steel	80,000–95,000*	30,000–45,000	55–60	†	70–100
18-8 MO stainless steel	80,000–100,000*	35,000–50,000	50–60	....	70–110
25-12 stainless steel	90,000–110,000*	40,000–60,000	35–50	†	
4–6% chromium steel	65,000–70,000*	30,000–35,000	30–40	†	

* Annealed.
† Fatigue value is usually 40 to 50% of ultimate strength.

welding in flat positions. Electrode may be used for fillet welding, single or multiple pass, and for butt welds of the V-groove or U-groove type. Because of deposition characteristics and ability to build up, it is used to fill gaps in cases of poor fit-up.

Arc of E6012 is less penetrating than E6010 and E6011 but adequate for relatively poor fit-up. Larger amount of slag gives a better coverage, produces a finer ripple with more pleasing bead surface. Melting rate is higher, spatter lower than E6010 or E6011. It is ideally suited for horizontal and flat fillet welds, can be used on steel having poor welding action with electrodes producing greater penetration. Because it does not penetrate deeply, it is used in cases where an inwash of a base metal is not desirable or required. It produces a somewhat convex bead (Fig. 13). Some applications are welding sheet-metal ducts, tanks, and machine guards.

E6013 has penetration similar to E6012, works well for poor fit-up, but is more suitable for light-gage metals. Bead has a tendency to be convex in horizontal fillets. Applications are similar to E6012.

E6015 electrode consists of rimmed-steel core wire with covering of carbonate of soda and lime type plus other compounds low in hydrogen. This electrode is slightly more difficult to use than previous types because a shorter arc must be maintained.

This dc reversed-polarity electrode was developed for welding higher strength, high-carbon, alloy steels in which the ordinary coverings produce "underbead cracking" in the unfused parent metal, usually just below center of the weld metal. These cracks are caused by hydrogen present in the conventional electrode covering. Elimination of the hydrogen permits the welding of "difficult to weld steels" with little or no preheat, thus making for better welding conditions. Although these cracks do not occur in ordinary steels, they may occur whenever an ordinary electrode is used on high-tensile steels.

Many of the newer high-tensile steels contemplated for future production will require the use of E6015 electrodes.

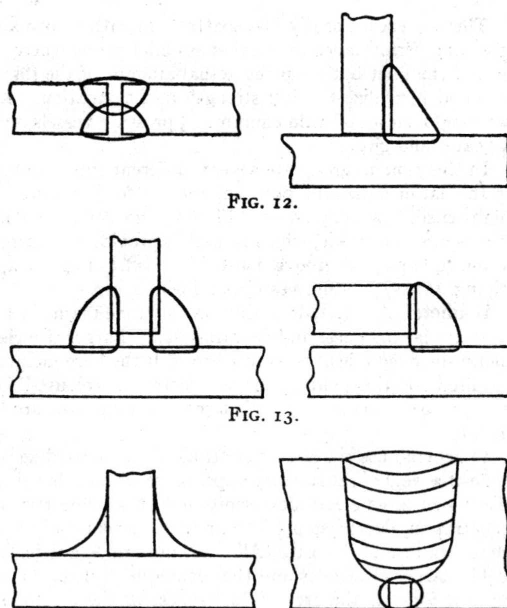

Bead characteristics of various electrodes: FIG. 12. Flat beads produced by E6010–E6011 electrodes. FIG. 13. Convex bead secured with E6012 electrode. FIG. 14. Concave bead obtained with E6020 electrode. FIG. 15. "Finish-pass" type of E6020 electrode produces flat bead.

FIG. 12.

FIG. 13.

FIG. 14. FIG. 15.

Another use for the E6015 electrode is the welding of high-sulfur (0.10 to 0.25%) steels without difficulty. Ordinary electrode deposit on these steels is badly honeycombed. Typical analysis of the deposit is 0.08% carbon, 0.56% manganese, and 0.25% silicon. The arc is moderately penetrating, the slag heavy, friable, and easily removed, and the deposited metal lies in a flat or slightly convex bead.

E6015 electrodes up to and including $\frac{5}{32}$ in. are used in all positions; larger diameters for fillet welds in horizontal and flat positions.

As-welded mechanical and impact properties of deposits made with low-hydrogen electrodes are superior to those of E6010 and E6011 electrodes. Necessity for preheat and postheat of weldments is reduced, thus making for better welding conditions and reducing thermal-treatment costs.

Other uses for E6015 electrode are welding of armor plate, malleable iron, spring steels, and the mild-steel side of clad plates. Another extensive use is welding of steels that are subsequently enameled and in all steels that contain selenium.

E6016 electrode has all the characteristics of the E6015 classification. Coating includes a certain amount of potassium silicate or other potassium salts to facilitate use on ac.

E6020 and E6030 electrodes have heavy coating, can be used with dc with either positive or negative polarity, and can also be used with ac. They are used in the flat position, but under special conditions as to setup, such as 30° from vertical, they may be used for fillet welding downward. They are used for fillet or butt joints of the V-groove or U-groove types, flow readily with a heavy slag covering the weld, produce a very smooth bead, slightly concave, and in some cases very concave (Fig. 14).

There is exceptionally low spatter loss with E6020 and E6030 electrodes. They give fine performance for single-pass fillet welds where strength is not based on the size of the fillet but upon the actual amount of the fused base metal. Fit-up must be good to realize the full strength and elongation. Applications are: production welding of ships, oil-field equipment, pressure vessels, machine bases, and fabricated I-beams and girders.

In this general group are several different types. One is generally used for fillets in the flat or horizontal position, another for U-groove work. One type is made for high-tensile low-alloy steels. Finally, another type is known as a "finish-pass" or "last-pass" electrode, which is used to provide an exceptionally smooth last bead on a multiple-pass U-groove joint (Fig. 15). Typical applications: the welding of piping, tanks, pressure vessels, and conveyors.

Diameter of Electrode. Correct electrode diameter for a given job is determined by material thickness and operator skill. Larger size electrodes are used for thicker metal since more heat is required to melt the base metal and more deposited metal is required for the joint. Smaller electrodes are used for thin material, because if enough current is used to melt large electrodes, the arc burns through the thin base metal.

In general, use the largest electrode that the welding operator can handle and still deposit a weld of satisfactory appearance without burning through. Results will be: the use of large electrode permits higher welding currents, faster travel, increased penetration, decreased arc time, reduced amount of metal deposited in a joint, and increased rate of deposit. All these are important to low-cost production welding.

The larger electrodes and the technique of using them take full advantage of the natural force of the arc. Effectiveness of the arc force is the product of welding current and arc travel speed. Increased arc force is most effective when arc speed is fast enough to prevent a pool of molten metal from getting under the arc and absorbing its force. Direct the arc against the base metal by keeping the tip of the electrode ahead of the molten pool at all times. This gives the arc force full opportunity to dig deep into the root of the joint. The increased current also increases the melt-off rate of the electrode so that sufficient metal is deposited at the increased speed of travel. Less metal will be required, since penetration is increased and the joint consists of more fused plate than is the case at lower speeds and currents.

Table 6 gives a typical range of currents for the mild-steel electrodes. Currents that can be used vary with brand. Lower limits of the range are used in vertical and overhead welding.

CONTROL OF DISTORTION

Distortion caused by the heat of welding may be a problem with sheet metal or unrestrained large sections. Three simple precautions which may be applied singly or together are:

1. Reduce the effective shrinkage force:

a. Avoid overwelding. Use as little weld metal as possible by taking full advantage of penetrating effect of arc force.

b. Use correct edge preparation and fit-up to obtain required fusion at root of weld.

c. Use few passes.

d. Place welds near neutral axis.

e. Use intermittent welds.

f. Use backstep welding method.

2. Make shrinkage forces work to minimize distortion:

a. Locate parts out of position so that when weld shrinks they will be in correct position.

b. Space parts to allow for shrinkage.

c. Prebend parts so that contraction will pull parts into alignment.

3. Balance shrinkage forces with other forces (where natural rigidity of parts is insufficient to resist contraction):

a. Balance one force with another by correct welding sequence so that contraction caused by weld counteracts forces of welds previously made.

b. Peen beads to stretch weld metal. Care must be used not to damage weld metal.

c. Use jigs and fixtures to hold work in a rigid position with sufficient strength to prevent parts from distorting. Fixtures actually cause weld metal to stretch, thus preventing distortion.

TABLE 6. CURRENT AND VOLTAGE RANGES FOR VARIOUS ELECTRODES

E6010-7010-8010-9010-10010			E6015-7015-8015-9015-10015		
Electrode, Dia, In.	Current, Amp	Voltage, Arc V	Electrode Dia, In.	Current, Amp	Voltage, Arc V
$\frac{1}{16}$	20 to 40	20 to 22	$\frac{3}{32}$	70 to 110	20 to 22
$\frac{5}{64}$	25 to 60	20 to 22	$\frac{1}{8}$	100 to 150	20 to 22
$\frac{3}{32}$	30 to 80	22 to 24	$\frac{5}{32}$	135 to 200	21 to 23
$\frac{1}{8}$	80 to 120	24 to 26	$\frac{3}{16}$	160 to 240	22 to 24
$\frac{5}{32}$	120 to 160	24 to 26	$\frac{7}{32}$	260 to 320	23 to 25
$\frac{3}{16}$	140 to 220	26 to 30	$\frac{1}{4}$	300 to 375	24 to 27
$\frac{7}{32}$	170 to 250	26 to 30	$\frac{5}{16}$	350 to 450	24 to 28
$\frac{1}{4}$	200 to 300	28 to 32			
$\frac{5}{16}$	250 to 450	28 to 32			

E6012-7012-8012-9012-10012			E6020-7020-8020-9020-10020		
Electrode Dia, In.	Current, Amp	Voltage, Arc V	Electrode Dia, In.	Current, Amp	Voltage Arc V
$\frac{1}{16}$	20 to 40	17 to 20	$\frac{1}{8}$	100 to 140	24 to 28
$\frac{5}{64}$	20 to 60	17 to 21	$\frac{5}{32}$	120 to 180	26 to 30
$\frac{3}{32}$	30 to 80	17 to 21	$\frac{3}{16}$	175 to 250	30 to 36
$\frac{1}{8}$	80 to 130	18 to 22	$\frac{7}{32}$	200 to 325	30 to 36
$\frac{5}{32}$	120 to 180	18 to 22	$\frac{1}{4}$	250 to 400	30 to 36
$\frac{3}{16}$	140 to 250	20 to 24	$\frac{5}{16}$	350 to 450	32 to 38
$\frac{7}{32}$	170 to 300	20 to 24			
$\frac{1}{4}$	200 to 400	20 to 24			
$\frac{5}{16}$	250 to 500	22 to 26			

PROCEDURE VARIABLES

In addition to joint selection, design, electrode selection, and control of distortion, other variables must be controlled to produce good welds at low cost.

Choice of Current. Welding can be accomplished efficiently with either ac or dc, but certain factors may necessitate using one to the exclusion of the other, or in some cases advantages may be gained through the use of one in preference to the other.

Many stainless steels, low-alloy high-tensile steels, and nonferrous metals cannot readily be welded with ac. For certain types of welding an electrode for use with ac will cost more per pound than a comparable electrode for use with dc. In sheet-metal work or when smaller diameter electrodes are used, ac will generally result in slower welding speeds.

When larger electrodes, over $\frac{3}{16}$ in., are to be used with higher currents, ac will sometimes increase the speed of welding. Arc blow is reduced to a minimum by ac when welding into corners in heavy plate. Power cost and maintenance costs are generally lower with an ac machine, although these two factors are a relatively insignificant portion of the total cost of welding.

Polarity. The polarity of welding circuit is significant only when welding with dc. Direction of current flow in the circuit establishes the relative polarity of the terminals, the work, and the electrode. The polarity to use with any particular electrode is determined by the manufacturer. Polarity controls the heat balance of the arc. In most cases, with "electrode negative" polarity (or "straight" polarity as it is sometimes called) more heat is concentrated at the tip of the electrode than on the work. With "electrode positive" polarity, or "reverse" polarity, the opposite is true, and more heat is concentrated on the work side of the arc. With electrode negative, more electrode is melted per minute than with electrode positive.

An ac electrode is suitable for welding with either polarity. Many electrodes are designed and manufactured for operation with either ac or dc and are so classified. An electrode classified for dc operation only cannot be used with ac.

Arc Length. Wherever possible, the tip of the electrode *coating* should be kept in contact lightly with the work, dragging the electrode along the joint. This technique prevents the heat and force of the arc from being dissipated into the air. A short arc also reduces the arc voltage, hence increases the welding amperage, and it also reduces tendency to spatter.

Edge Preparation. The method used must be a balance between cost of preparation and cost of welding. Shearing and blanking are the most economical methods, but flame cutting is widely used where presses and press brakes are not available or the nature of the work makes press work impossible. Modern flame-cutting equipment can make a clean cut to very close tolerances. Machining may be necessary for precision work.

It is good practice to perform as many machining operations as possible on components before welding, to avoid machining larger and sometimes awkward weldments.

Cleaning. Rust, scale, grease, and other foreign matter should be removed from joint edges before welding, especially automatic welding. Greater loss of time in repairing rejected work will be avoided.

Positioning. The position of the joint affects speed and ease of welding. With the exception of sheet-metal welding, welds should be made in the downhand position wherever practical. Change from vertical to downhand position can increase welding speed as much as 400%. Sheet-metal welding is done fastest with the work tilted at 45°. Positioners for moving joints into the downhand position are an essential part of every welding setup.

TABLE 7. PRINCIPAL WELDING CODES AND REGULATIONS

PRESSURE VESSELS:

ASME Power Boiler Code for welding drums and shells of power boilers.

ASME Unfired Pressure Vessel Code for welding all types of pressure vessels.

API-ASME Unfired Pressure Code for welding tanks, etc., for petroleum, liquids, and gases.

Amended Rules I and II of the general rules and regulations of the U.S. Department of Commerce, Bureau of Navigation and Steamboat Inspection—51st supplement to general rules and regulations containing a section on fusion welding, AWS Rules and Fusion Welding of Drums and Shells of Marine Boilers and Pressure Vessels.

Codes used by United States Government, including specifications of Bureau of Engineering, U.S. Navy; many codes confidential.

Requirements for Repairs by Fusion Welding of Boilers or Other Pressure Vessels —a brief set of rules formulated by the National Bureau of Casualty and Surety Underwriters in cooperation with the National Board of Boiler and Pressure Vessel Inspectors, to indicate the extent to which fusion welding will be acceptable to the authorities for repairing steam boilers.

TANKS:

AWS Tentative Rules for the Fusion Welding of Gravity Tanks, Tank Risers and Towers.

MACHINERY:

AWS Code for Fusion Welding and Flame Cutting in Machinery Construction.

Operating Conditions. Maintenance of a high operating factor is important to good welding. Working conditions can be subjected to standard method studies with savings. Work flow through the welding station must be maintained at a high level. A helper setting up the job in one fixture while the welder works at another increases output. Operator fatigue can be reduced by providing work tables of correct height. Welding smoke should be removed by efficient exhaust.

WELDING INSPECTION

Inspection for weld quality varies from plant to plant depending on nature of the work. Inspection methods range from visual inspection through physical measurement, X-ray, gamma ray, magnetic flux, and sound-wave inspection. The type of inspection for some classes of work is indicated in the codes and regulations recommending procedures for obtaining specified results that have been established by societies, institutes, associations, and the government. The principal codes and regulations are listed in Table 7. One of these listed is the Standard Methods for Mechanical Testing of Welds, prepared by the American Welding Society.

For the majority of welding work, visual inspection is adequate. The inspector can be assured that the strength requirements will be met when the following factors are in accordance with recommended procedures:

1. Plate preparation.
2. Electrode size and type.
3. Current.
4. Surface appearance of weld.

All factors important for making a good weld can be set up by the inspector before the weld is made and occasional checks can be made.

It is also the duty of the supervisor or foremen to see that all welding specifications are followed. The instructions used by the welding department should specify:

1. Plate preparation.
2. Electrode size and type.
3. Current.
4. Arc travel speed.

Arc travel speed must be maintained at specified rates to insure maximum penetration and minimum cost. This can be measured in arc speed in inches per minute or in feet of joint welded per hour. Current specified is the current at the cable terminal, not necessarily that indicated on the welding-machine dials. This current can be measured using tong-type meters or by checking the burn-off rate of an electrode and comparing it with manufacturers' recommendations.

Appearance Factors. The following factors define good appearance:

No cracks, serious undercut, surface holes, or slag inclusion.

Ripples and width of bead should be uniform.

Butt welds should be flush or slightly above the plate surface, without excessive build-up.

Fillet welds should have equal legs on each plate and should not have overlap.

Checking Trouble. Slight variations in the above can be expected, but major difficulties should be investigated:

See that proper plate preparation is used.

See that gap is within limits—minimum and maximum.

Check polarity.

Make sure recommended current is used.

Make sure that travel speeds are those recommended.

See that proper angle of electrode is used with proper technique (short arc, proper striking, etc.).

Try the same weld on steel of known good welding analysis, to see if difficulty is due to admixture of plate metal.

If the steel is not suitable for high-speed welding, follow the procedure recommended in the paragraph on Steel Outside Recommended Analysis, below.

In addition, there also is the possibility of arc blow as a factor in weld appearance.

TROUBLE-SHOOTING SUGGESTIONS

Moisture Pickup. Electrodes exposed to damp atmosphere may pick up enough moisture to cause undercut, rough welds, porosity, or cracking. This condition is usually corrected by storing the electrodes in a cabinet or room heated to about 10 F above the surrounding atmosphere. If electrodes become wet, spread out to dry at 200 F for 1 hr.

Arc Blow. This can be reduced greatly by the following steps:

Weld toward a heavy tack or toward a weld already made.

Use backstepping on long welds.

Place ground connection as far from joint to be welded as possible.

On small pieces place ground connection at starting end and weld toward a heavy tack if possible.

Hold short arc, so that the electrode coating touches the plates, directing the tip opposite to the arc blow so that arc force counteracts arc blow.

Tack small plate across seam at finish end.

Steel Outside Recommended Analysis. If for some reason it becomes necessary to weld steel other than that covered by recommendations in Table 8, difficulties may be encountered, requiring the following corrective measures:

TABLE 8. LIMITS OF STEEL ANALYSIS FOR HIGH-SPEED WELDING

	Low	Preferred	High
Carbon	0.10%	0.13 to 0.20%	0.25%
Manganese	0.30%	0.40 to 0.60%	0.90%
Silicon	......	0.10% or under	0.15% max
Sulfur	......	0.035% or under	0.05% max
Phosphorus	......	0.04% or under	0.04% max

1. CRACKING. High content of sulfur, silicon, or carbon is likely to cause weld cracks. This usually can be corrected by:

a. Leave $\frac{1}{32}$-in. gap between plates to allow free movement.

b. Weld toward unrestrained end of joint.

c. Change to less penetrating type of electrode.

d. Position and weld uphill 4° to increase weld section on first pass.

e. Decrease welding current, using additional passes if necessary. (Fillet welds made with less current than that recommended in high-speed welding procedures must be measured by the conventional fillet gage.)

f. Decrease welding speed.

g. Use low-hydrogen-type electrode.

If cracks appear in the crater only, use the backstepping method so that the weld is finished on top of the bead put in previously and then the crater will penetrate into weld metal instead of plate metal, and concentration of undesirable elements will be low enough to prevent cracking.

2. SURFACE HOLES. These can usually be eliminated by increasing arc length, using a less penetrating electrode, or decreasing the current.

3. POROSITY. Low carbon and manganese content increases the tendency for internal porosity. Change to an electrode less likely to give internal holes, use a short arc, and decrease the current.

WELDING ALLOY STEELS AND NONFERROUS METALS

Practically all metals are weldable. Some, however, require special welding procedures to preserve the properties and characteristics of the metal.

If a metal cannot be welded with mild-steel E6010–E6012 electrodes, some degree of preheat is the next step. Following this, the next alternative is to use a low-hydrogen electrode, and finally a stainless-type electrode.

High-tensile Low-alloy Steels. This group of steels can readily be welded with electrodes designed for them. Excellent joints are obtained, and it is not necessary to have a coated electrode of composition similar to each of the alloys.

Stainless Steels. Commonly used types of stainless for welded structures are 18-8, 25-12, and 25-20 groups. The 18-8 group with 0.08% carbon maximum is commonly specified for welding because carbon content is so low that carbides cannot be precipitated, to cause corrosion. Electrodes should contain 0.05 to 0.07% carbon.

Procedures as in welding mild steel are used, but taking into account the higher electrical resistance, lower thermal conductivity, and higher thermal expansion of stainless steels. Fit work carefully, clean all edges. Clamp light-gage work to prevent distortion and buckling. Use small-diameter short electrodes to prevent chromium loss and electrode overheating. Core-wire deposit should be approximately the same analysis as the plate.

Stainless-clad Steel. Significant precautions in welding this material are in joint design (including edge preparation), procedure, and choice of electrode. Choose an electrode of the correct analysis for the cladding. Prepare joint and weld it to prevent dilution of clad surface by the steel backing material. The backing material is welded with a mild-steel electrode in multiple passes to prevent penetration into the cladding. The clad side is also welded in small passes to prevent penetration into the backing material. Where in thin-gage material it is necessary to weld in one pass, 25-12 stainless electrode should be used for the steel and stainless sides.

Chromium Steels. The intense air-hardening property of these steels (proportional to carbon and chromium content) is the chief consideration in establishing welding procedures. Considerable care must be taken to keep work warm during welding and annealed afterward; otherwise welds and adjacent area will be brittle. Consult steel supplier for specific heat-treatment, temperatures, and treatment.

High-manganese Steel. For building up parts of high-manganese steel, an electrode should be used of a type such that physical characteristics of deposited metal will be approximately the same as the base metal.

High-carbon Steel. Welding raises steel above the critical temperature and cold metal surrounding the weld area creates a quenching affect. Hardness and lack of ductility result in cracking as the weld cools and contracts. Preheating to 400 to 600 F and slow cooling will prevent cracking.

For steels over 0.30% carbon, special electrodes are recommended. The lime-ferritic low-hydrogen electrodes (E6015–16 and E7015–16) can be used to good advantage in overcoming the cracking difficulties in high-carbon steels. These electrodes deposit a more ductile weld because of the absence of hydrogen in the deposit. An 18-8 stainless electrode can also be used.

Cast Iron. The brittleness and the uneven contraction and expansion of cast iron are of principal concern in welding it.

Each job must be analyzed to predetermine the effect of welding heat and procedures devised to suit the case. A curved bead reduces the cumulative effect of the strain of contraction produced by a straight weld. Weld can also be deposited in short lengths, each being allowed to cool. Peening of the weld metal will stretch the deposit.

Steel or cast-iron electrodes may be used, as well as carbon electrodes and nonferrous rods. All oil, dirt, and foreign matter must be removed from the joint before welding. With steel electrodes, intermittent welds no longer than 3 in. and light peening should be used. To reduce contraction the work should never be allowed to get too hot in one spot. Preheating will help to soften the deposit to make it more machinable.

For welds of easy machinability, a nonferrous alloy rod should be used. A two-layer deposit has a softer fusion zone than a single-layer deposit. When it is practical, heating of the entire casting (often to a dull red heat) is recommended further to soften the fusion zone and burn out dirt and foreign matter. When the weld to be made is in a deep groove, use a steel electrode to fill the joint to within $\frac{1}{8}$ in. of the surface. Then finish the weld with a more machinable nonferrous deposit.

Aluminum. Pure aluminum and various aluminum alloys in sheet, forged, extruded, and cast forms can be welded with either metal or carbon arc and inert-gas arc welding. With metal arc welding, the high melting rate of the aluminum electrode necessitates rapid welding and presents the problem of getting enough heat into the work. Rapid freezing rate of the metal may also trap gases in the weld to cause porosity. For metal arc welding, the minimum thickness of metal is generally recommended as ⅛ in., although thinner sections are successfully welded. Material ¼ in. thick is generally regarded as the minimum thickness for gastight metal arc welds in aluminum.

On welding applications involving thick pieces or complicated welds, it is desirable to preheat to 250 to 400 F, to prevent porosity arising from too rapid cooling of the weld metal. Welding techniques differ little from those of welding steel, except that there is no necessity for weaving. Striking the arc must be accomplished by a match-striking motion rather than simply by tapping the work. Direct current with electrode positive is used.

Inert-gas arc welding generally employs ac and argon gas for shielding. The method has proved valuable in welding thin sheet.

Copper and Copper Alloys. Copper and its alloys can be welded with either metal-arc or carbon-arc welding, but these materials absorb gas when molten, causing porosity. Rapid oxidation may also affect mechanical properties of the metal. Decrease in tensile strength with temperature rise, and high coefficient of contraction may also complicate welding of copper.

Preheat is usually necessary on thicker sections because of the high heat conductivity of the metal. Keep the work hot, pointing the electrode at an angle so the flame is directed back over the work, and put down as much metal per bead as is practical.

SURFACING BY ARC WELDING

Arc welding has many applications in building up worn parts and depositing wear-resisting surfaces. Success depends on choice of surfacing material, which is affected by:

1. Material to be surfaced.
   a. Composition.
   b. Physical condition such as hardness and previous heat-treatment.
2. Dimensions.
   a. Size and shape of parts.
   b. Size and location of the area to be surfaced.
   c. Thickness of weld deposit.
3. Finish required.
   a. As welded.
   b. Machined.
   c. Rough grind.
   d. Fine grind.
4. Service conditions.
   a. Hardness required.
   b. Corrosion and the corrosive medium.
   c. Abrasion—its extent and nature.
   d. Impact—its extent and nature.

Material to Be Surfaced. With respect to their suitability for weld surfacing, alloys fall into two groups:

Group A metals or alloys have physical characteristics that are not greatly changed by heating and cooling, and they withstand sudden localized temperature changes without cracking. This group includes plain carbon steel of 0.30% carbon maximum, low-carbon alloy steels, austenitic steels such as stainless chrome-nickel, and high-manganese steels. Copper and most of its alloys are included.

Group B materials are changed considerably (particularly in hardness) by application of welding heat and subsequent cooling, or crack with sudden application of localized heat. This group includes medium-to-high carbon steels; tool steels; medium-to-high carbon, low-alloy ferritic steels; cast irons; and, in general, all hard metals and alloys.

To avoid cracking due to thermal shock, either the hardness of group B metals must be reduced through annealing or the thermal shock reduced by preheating. In some cases both treatments may be necessary. To minimize weld-hardening adjacent to the weld, preheat, slow cool, and postheat the article as required.

Dimensional Changes. Size and shape determine the heat capacity of parts to be surfaced and to some extent the thermal cycle. Articles of large mass and large heat capacity heat up slowly, reach a relatively low temperature, and draw the heat away from the weld area rapidly. If welding heat is applied to a large mass of metal, capable of quench hardening, the hardening effect will be drastic and cracking is likely to result. If the mass is small, cooling will be more uniform, resulting in less severe thermal stresses. If the weld-metal hardness is affected by the thermal cycle, the hardness of the deposit will be greater on a large mass because of the quench effect.

Heat input is determined by the size and location of the area to be surfaced.

Electrodes for hard surfacing are not well adapted for vertical or overhead welding. Work should be positioned as nearly flat as possible. Thick deposits should generally be avoided. If thick deposits are desired, electrodes designed for such applications must be used. If machining is to be done on the deposit, it may be necessary to anneal the weld; otherwise grinding may be the only way to work the deposit.

Service conditions to which a weld-faced deposit may be subjected are abrasion, impact, and corrosion. Before proper choice of weld-surfacing material can be made for a given application, the service conditions must be analyzed.

Table 9 is used for determining the economy of weld surfacing. Surfacing metal is expensive. By depositing only a limited quantity on a low-cost base metal great savings and greater life can be secured than by making a part of a single relatively costly material.

TABLE 9. ESTIMATING ECONOMY OF ARC SURFACING VS. USING A SINGLE COSTLY MATERIAL

Cost of Single Metal, ¢ per Lb	Percentage of Total Weight Which May Be Surface Material without Exceeding Cost of Using One Metal Throughout				
	Cost of Surfacing Materials				
	$1 per Lb	$2 per Lb	$3 per Lb	$4 per Lb	$5 per Lb
20	15.8	7.7	5.1	3.8	3.0
40	37.0	18.0	11.9	8.9	7.1
60	58.0	28.2	18.6	13.9	11.1
80	79.0	38.5	25.4	19.0	15.1
100	100.0	48.7	32.2	24.0	19.2
120		59.0	39.0	29.1	23.2
140		69.2	45.7	34.2	27.3

Section 21

GAS WELDING AND OXYGEN CUTTING

By: Arthur N. Kugler, Mechanical Engineer, Air Reduction Sales Company

SECTION 21

GAS WELDING AND OXYGEN CUTTING

PROCESSES AVAILABLE

The gas-welding and oxygen-cutting processes, while seemingly opposite in results, are actually closely related. The same gases are used in both groups of processes; much of the equipment is common.

The American Welding Society defines gas welding[1] as "a group of welding processes wherein coalescence is produced by heating with a gas flame or flames, with or without the application of pressure, and with or without the use of filler metal." From this definition, gas welding is seen to be a general term covering a number of specific processes: oxyacetylene welding, oxyhydrogen welding, air-acetylene welding, etc., and in fact any combination of gases to produce a flame for melting and fusing metals.

Also, the AWS defines oxygen cutting as "A group of cutting processes wherein the severing of metals is effected by means of chemical reaction of oxygen with the base metal at elevated temperatures. In the case of oxidation-resistant metals the reaction is facilitated by use of a flux." Again, the general term covers specific processes: oxyacetylene cutting, oxyhydrogen cutting, etc., and in fact, any combination of a source of preheating with oxygen for the purpose of severing metals. Even the electric arc as a source of preheating is covered by this definition.

OXYACETYLENE FLAMES

The oxyacetylene flame is most generally employed for both welding and cutting, because the required gases are almost universally available in transportable cylinders.

Commercial oxygen, O_2, in usual purity of 99.5% minimum, is compressed in steel cylinders. Cylinders come in two sizes.

Acetylene, C_2H_2, cannot be simply compressed in cylinders because dissociation occurs at certain temperatures and pressures. In consequence, acetylene is dissolved in acetone under pressure, and the cylinders contain a porous filler. In view of this method of "packaging" acetylene, simple pressure and volumetric relations do not apply. It is necessary to measure cylinder contents by weight, 1 lb being approximately equal to 14.5 cu ft of gas at normal pressure and temperature. Thus cylinders which apparently contain equal volumes may show variations in pressures and gas contents. It must be remembered, however, that the actual gas volume is indicated in both pounds and cubic feet.

Combination of oxygen and acetylene in correct proportions in a properly designed torch results (when ignited) in a flame with a temperature of about 6300 F. This combustion reaction releases 1433 Btu per cu ft. Combination of high temperature and heat content enables the flame to melt or fuse, locally, many metals and alloys.

[1] "The Welding Handbook," 1949.

FIG. 1. Neutral flame.

FIG. 2. Excess acetylene flame.

FIG. 3. Excess oxygen flame.

Flame Adjustments. Three flame adjustments have significance: the neutral flame, the excess acetylene flame, and the oxidizing flame.

The neutral flame (Fig. 1) is attained when approximately equal volumes of oxygen and acetylene are burned at the torch tip. It is characterized by a sharply defined inner luminous cone surrounded by a pale bluish envelope. The highest temperature occurs at the end of the inner cone where primary combustion takes place. Oxygen for primary combustion comes from the cylinder. The envelope provides a reducing atmosphere and is of lower temperature since it is the secondary combustion zone. Oxygen for secondary combustion is derived from the atmosphere. The neutral flame is usable on most metals and is preferred on many. When in doubt, it is wise to start with a neutral flame for either welding or cutting.

The excess acetylene flame (Fig. 2) is secured by admitting more acetylene than is employed for a neutral flame. It is characterized by a brilliant incandescent streamer burning at the end of the inner cone. The degree of excess acetylene adjustment is determined by the size of the streamer with relation to the inner cone; that is, with a $2X$ excess acetylene flame, the total length of the streamer from torch tip to end of the streamer is twice the length of the normal inner cone. Excess acetylene flames, in adjustments up to $1\frac{1}{2}X$, are used extensively to avoid oxidation when welding steel with the low-alloy welding rods. Greater values of excess acetylene adjustment are frequently used in hardfacing or hard surfacing.

The oxidizing flame (Fig. 3) is characterized by a sharp inner cone somewhat shorter than the normal neutral-flame inner cone. It is obtained by using more oxygen than is needed for a neutral flame. However, it is generally adjusted by first securing a neutral flame and then reducing the amount of acetylene until the desired condition exists. In welding, this flame adjustment should be avoided in all cases except when welding copper and copper-base alloys.

OXYHYDROGEN FLAMES

The oxyhydrogen flame is limited to welding light-gage aluminum and magnesium alloys, and to underwater cutting. Its flame temperature is lower than that for the oxyacetylene flame.

Hydrogen, at 99.5% purity, is supplied in steel cylinders similar to oxygen cylinders but distinguishable by labels. Two pressures are employed: 1800 and 2000 psi, and approximate volumetric contents are 176 and 194 cu ft, respectively.

Torches, tips, and other equipment used for oxyacetylene welding are employed for oxyhydrogen welding, but a hydrogen regulator replaces the acetylene regulator. Flame adjustments are more difficult to distinguish because the oxyhydrogen flame is not normally incandescent. In actual welding, the theoretical proportions of hydrogen to oxygen (2:1) are increased to three or four to one in order to secure a reducing atmosphere.

OXYGEN CUTTING

If iron is heated to about 1500 F, and brought in contact with high-purity oxygen, the metal will burn rapidly, with evolution of much heat. Theoretically, the combustion process is self-supporting; in actual practice, continuous preheating is necessary. This preheating is secured from oxyacetylene, oxyhydrogen flames, or other fuel gases burned with oxygen; or from an electric arc drawn between a hollow shielded-arc electrode or a hollow carbon electrode.

The purity of the oxygen used for cutting is of the greatest importance. A reduction of oxygen purity from 99.5 to 99.0%—only 0.50%—can result in a reduction in cutting efficiency of over 10%.[1]

EQUIPMENT FOR WELDING

The equipment necessary for oxyacetylene welding and oxygen cutting is designed and constructed especially for safe use under service conditions encountered. Information on correct installation and safe operation of oxyacetylene equipment is available from the IAA.[2] For manual operation, the essential equipment consists of a source of oxygen (usually from cylinders or pipe lines), a source of acetylene (usually from cylinders, generators, or pipe lines); regulators, hoses, and the torches.

Oxygen Supply. When operations are of limited extent or widely scattered, or where it is necessary to move operations frequently, cylinders provide the most economical and practical source of oxygen supply. When work involves moderate volumes of oxygen concentrated at a few spots, it is desirable to manifold the oxygen cylinders and supply the using points by pipe line. For larger consumptions of oxygen, pipe lines are supplied from bulk delivery stations.

Acetylene Supply. Limited or scattered operations are supplied from individual cylinders; larger and more centralized operations are fed from pipe lines supplied from either acetylene manifolds or, in the case of very large requirements, from acetylene generators.

Cylinders. The cylinders are manufactured, tested, and maintained in conformance with ICC regulations. Users are prohibited from tampering with, altering, or otherwise changing these cylinders. If a cylinder in use or storage develops any unusual condition, make certain the valve is closed, remove it to outdoors (away from any source of ignition), and notify the supplier immediately.

Regulators. Cylinder pressures are far too high for direct use in torches. Gas pressures are reduced to a safe level by pressure regulators: two-stage and single-stage. The former type is generally used since it provides the best regulation. Single-stage regulators are most frequently used in pipe lines where pressure reduction is relatively small.

Regulators must be used only with the gas and for the pressure conditions for which they are designed. Regulators must never be interchanged or used with other gases. Regulators for welding carry working-pressure gages reading lower maximum pressures than those employed for cutting. Flows obtainable with cutting regulators are higher. Combination welding and cutting regulators are available.

Hose. Rubber and fabric hose specially constructed for oxyacetylene service is necessary for conducting the gases from the regulators to the torches. Hoses are generally identified by colors: red for acetylene, and green or black for oxygen. Twin hose is acceptable for oxyacetylene operations, but use only those types which have two distinct and complete gas passages.

Torches (Blowpipes), Mixers, Tips. The torch will here be considered as one unit consisting of the handle proper (including needle valves), mixer, and tip, although

[1] Crowe, J. J., and G. L. Walker, Economics through the Use of High Purity Oxygen in Cutting, *Welding J.*, March, 1925.

[2] "Safe Practices for Installation and Operation of Oxyacetylene Welding and Cutting Equipment," International Acetylene Association, New York.

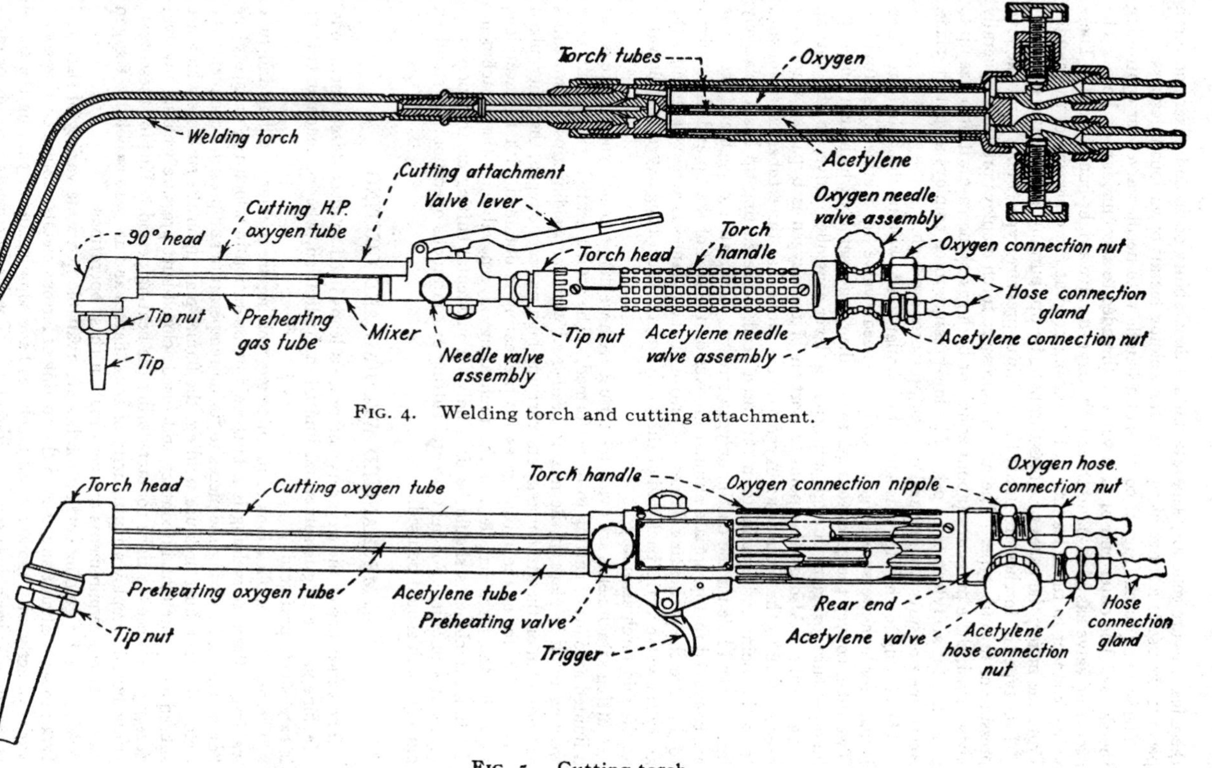

Fig. 4. Welding torch and cutting attachment.

Fig. 5. Cutting torch.

these features are generally separate units. Separability of tip and mixer and torch permits the use of a range of tip sizes for welding metals of various thickness.

The welding torch (Fig. 4) consists of a handle with needle valves for controlling the gases. The outer end of each valve is threaded to accommodate its hose. The acetylene valve has a left-hand thread, the oxygen valve a right-hand thread. These valves connect to tubes leading to the mixing head at the other end of the handle. From the mixer, the gases progress through the tip, burning at the outer end.

Two types of torches are used: the medium-pressure type (Fig. 4) and the low-pressure or injector type. The medium-pressure torch employs pressures generally above 1 psi; these pressures (about equal for both gases) increase with tip size up to safe limit for acetylene of 15 psi. The low-pressure torch employs a venturi mixer. Acetylene pressures are low and oxygen pressures are high.

The cutting torch (Fig. 5) is somewhat similar to the welding torch, but provision is made for high-pressure oxygen. Only one needle valve (for acetylene) is provided. Two tubes convey the gases through the handle to an assembly which distributes the high-pressure oxygen, through a trigger or lever-controlled valve, to the cutting oxygen tube. Also at this point, a portion of the oxygen is bypassed through a needle valve to the preheating oxygen tube. Note that three tubes carry forward from this point and terminate in the torch head. The cutting tip has a relatively large central hole for the cutting-oxygen stream. Preheating oxygen and acetylene are conveyed to appropriate annular grooves on the seat of the tip and there mixed to produce the preheating flames. Control of these flames is achieved through the acetylene valve at the back end of the torch and the preheating oxygen needle valve at the mid-torch assembly.

Cutting attachments are available for use with welding torches, where there is but a limited amount of cutting to be performed. The attachment (Fig. 4) fits into the welding torch in place of the mixing head used with the welding tips. Two tubes are provided on the attachment, one carrying high-pressure oxygen, the other mixed gases. Control of the mixed gases for preheating is obtained with the acetylene valve on the torch handle and the preheating oxygen valve on the attachment Cutting attachments should be used only for incidental cutting on thicknesses up to about 4 to 6 in. For heavy-metal cutting or continuous operation, the standard cutting torch should be used.

MECHANIZED WELDING AND CUTTING

The versatility of the oxyacetylene process lends itself to mechanized operations. Machine gas welding is somewhat limited in application, whereas machine cutting is extensively practiced.

The principal applications of machine gas welding are pressure welding and tube welding. Pressure welding is a butt-welding operation in which the oxyacetylene flames are used to bring the pieces to fusion temperature and pressure completes the weld. In open-butt pressure welding, the flames play upon the surfaces which are subsequently brought in contact and fused; closed-butt pressure welding involves keeping the pieces in tight contact during the heating period. Tips for these operations usually employ a multiplicity of small flames and are water-cooled. Oxygen and acetylene requirements are very high in these operations.

Machines for guiding cutting torches are available in a wide variety of sizes and types to serve almost any need in the cutting of ferrous metals. Since cutting speed varies directly with thickness, all machines have some means of varying the travel speed, usually variable-speed electric motors. Torch adjustments provide for racking up and down. In addition, angular adjustments permit of positioning the torch

for bevel cutting. For straight-line cutting, tracks are commonly employed to guide the machine, while for circles and arcs a radius rod and center point are used.

Irregular shape cutting imposes more complicated requirements which are met by the following systems:

1. A manual tracer consisting of a pivoted, motor-driven wheel which follows an outline of the cut and by direct connection to the torch causes it to traverse the same path.

2. An aluminum strip bent to the contour of the cut and fastened to a base; motor-driven rollers, contacting the strip, guide the torch in the same outline as the pattern.

3. A steel or iron pattern or "cam" shaped to the outline of the cut and a magnetized, motor-driven roller which contacts and follows the cam, thus causing the torch to follow the same path.

4. An electronic device equipped with a photoelectric cell capable of following a black and white outline and, through direct connection to the torch, causing it to follow the same path.

System 1 is the best suited to a small production, and the accuracy is entirely dependent upon the skill of the operator. Systems 2 and 3 are practical where a large number of identical pieces are to be cut so that the cost of the patterns may be spread over a number of pieces; accuracy may be controlled in the making of the patterns. System 4 is suitable for either small or large production, since a paper drawing will suffice for one or two pieces and more durable and expensive patterns may be used for large-quantity production.

The mechanisms for translating tracer movement into torch movement are required to move freely along axes at right angles to each other and in the area between. These machines generally employ one of the following two methods:

1. A pantograph arrangement.
2. An arrangement of two sliding boxes on shafts at right angles to each other.

Arc-oxygen Cutting. This method of cutting employs an arc between an electrode and the work to raise the metal to the ignition point, after which oxygen performs the actual cutting. Special hollow electrodes are necessary to permit passage of oxygen. Two types of electrodes are available: (1) a hollow steel electrode provided with a heavy, extruded coating, and (2) a hollow carbon electrode.

JOINTS FOR GAS WELDING

The general range of joint designs suitable for gas welding is illustrated in Fig. 6. Note the absence of metal backing strips at the weld root. Such devices are not considered good practice, because gas welding against such backing is difficult and results are uncertain.

Joint designs *A*, *B*, and *C* (Fig. 6) are used principally for sheet metal up to about $\frac{5}{32}$ in. and, except for *C*, require no special preparation other than edges must be straight and true. Filler metal may be omitted for joint *A* but the weld will lack reinforcement. Filler metal is necessary with joint *B*. Edges of joint *C* are flanged by an amount about equal to the thickness and held tightly together by clamps, jigs, or tack welding. The upstanding edges are melted and fused to form the weld. The joint may be so proportioned (size of flange) as to have the weld throughout the thickness and thus avoid the notch illustrated. The dotted, vertical lines indicate the form of the joint when used for an edge weld. On joints *A* and *B* welding is generally performed from one side, but if necessary welding may be applied from both sides.

The single V-groove butt joint *D* (Fig. 6) is probably the most commonly used joint design in gas welding. It is suitable for metal thicknesses from about $\frac{5}{32}$ to 1

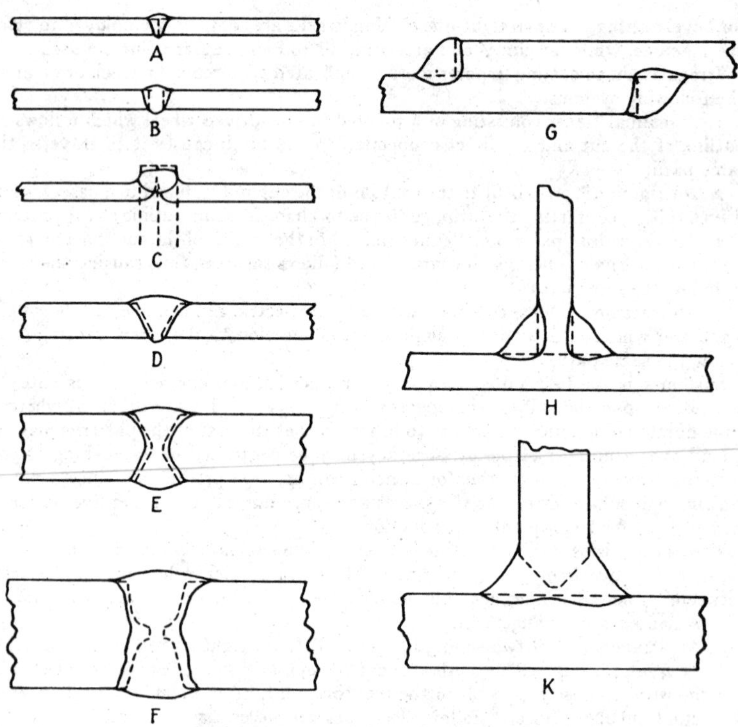

FIG. 6. Joint designs for gas welding.

in. If the joint must be welded from one side, greater thicknesses can be handled. Normally, this joint is welded from one side. If necessary, the root may be flame-gouged or chipped and backing bead applied. When the thickness exceeds $\frac{3}{8}$ in., multilayer welding should provide one layer for each $\frac{1}{4}$ in. of thickness. Bevel angles for this joint may range from 45 to 30°, making 90 to 60° V's. In pipe welding, a bevel angle of $37\frac{1}{2}$°, making a 70° V, has been established as standard. These bevels may be prepared by oxyacetylene cutting.

For heavy metal thicknesses, above about $\frac{3}{4}$ in. and under conditions that permit of welding from two sides, either the double V-groove butt joint *E* or the double U-groove butt joint *F* should be used. These joint designs provide appreciable savings in filler metal and welding time. Multilayer welding will be found advantageous. To assure sound welds, flame-gouge or chip the root of the deposit from the first side before depositing metal from the second side.

Fillet welds as applied to lap and T-joints are illustrated in *G* and *H* (Fig. 6). These joint designs require simple preparation—merely cutting the plates to size. Against this saving must be balanced the increased weld metal and welding time necessary when compared with the butt joint at *D*. Neither of these joints is considered satisfactory under shock loading, as butt joints (*D*, *E*, and *F*), particularly those welded from both sides. The double-bevel groove weld on the T-joint *K* (Fig. 6) provides the most reliable welding for this type of connection.

In all these joint designs, there are only two fundamental weld types: the groove weld and the fillet weld.

GAS-WELDING TECHNIQUES

The performance of gas welding involves two techniques: *forehand* welding and *backhand* welding.

Forehand Welding. In forehand welding (Fig. 7) the torch points ahead in the direction of welding and the rod precedes the torch. To distribute the heat and molten metal, oscillate the torch and rod in opposite, semicircular paths. Thus it is possible alternately to expose one side and then the other side of the molten puddle to the air, thereby permitting the metal to oxidize. This technique may be necessary when welding metals in which it is important to control oxides and slag, as in cast-iron welding. It is also preferred for braze welding.

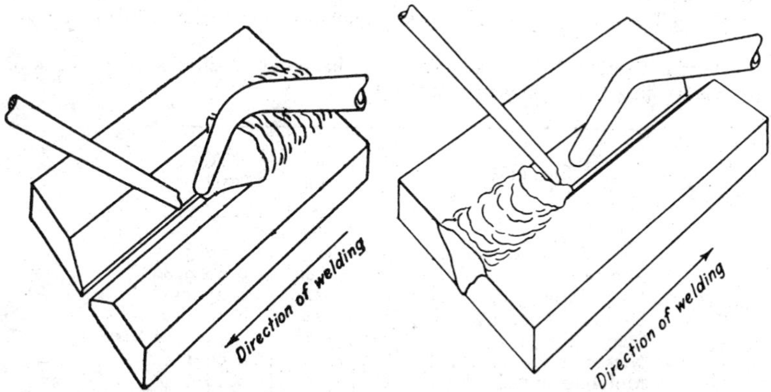

FIG. 7. Forehand welding.

FIG. 8. Backhand welding.

Backhand Welding. In backhand welding (Fig. 8) the torch points back at the completed weld and the rod is interposed between the torch and weld. Distribution of heat and molten metal is simpler in this technique. Relatively small movements of the torch suffice to distribute the heat on the edges of the joint. Distribution of the welding rod may be secured by a slight "rolling" of the rod or vertical "in and out" movement in the line of the rod. Since there is less manipulation in this method, narrower V's are preferred, normally 60°.

Most steel welding and pipe welding can be best accomplished with the backhand technique. Faster speeds and higher quality are achieved.

Braze Welding. Braze welding is defined by the AWS as "A method of welding using a nonferrous filler metal having a melting point below that of the base metals but above 800 F. The filler metal is not distributed in the joint by capillary attraction." Whereas capillary brazing requires closely fitted parts, braze welding merely employs the conventional groove and fillet welds.

In braze welding, brass or so-called bronze rods are generally employed. Braze welds on mild steel with brass filler rods will consistently show tensile strengths of 60,000 to 70,000 psi.

Hardfacing. Hardfacing is the technique of welding hard wear-resistant alloys to softer metals to increase service life. It is important to avoid deep fusion of the filler metal with the base metal. Instead, "sweat" the alloys on the parts. Hard-facing alloys are available in compositions to provide varying degrees of hardness, red hardness, toughness, wear resistance, and shock resistance.

WELDING RODS

Steel Welding Rods. Oxyacetylene welding rods are covered in ASTM-AWS Tentative Specifications for Iron and Steel Gas Welding Rods, ASTM A251-46T, AWS A5.2-46T. Six classifications are GA65, GA60, GA50, GB65, GB60, and GB45. The significance of the symbols is as follows: G—gas welding rods; A or B—the index of relative ductility of weld deposit, where A is higher than B; numbers (65, etc.) are the first two digits of the value of the tensile requirements as measured on all-weld-metal specimens, stress relieved. Rods of the GA60 classification are considered best for producing quality welds and are generally of low-alloy composition. The common low-carbon (0.06% C max) steel welding rods are classified as GA50. The above is the only general industry specification available for gas welding rods.

Cast-iron Welding Rods. Common usage has established two compositions of cast-iron welding rods as industry standards:

Elements	Gray cast iron	Alloy cast iron
Carbon	3.00–3.75	3.00–3.75
Manganese	0.40–0.80	0.40–0.80
Phosphorus	0.75 max	0.45 max
Sulfur	0.10 max	0.10 max
Silicon	2.75–3.50	1.75–2.75
Nickel	....	1.00–1.50
Molybdenum	....	0.10–0.50

The relatively high silicon contents are employed to assure (when proper welding techniques are used) soft machinable welds. A flux designed for cast-iron welding must also be used.

Brass (Bronze) Welding Rods. Brass rods are used principally for the welding of brasses and the braze welding of ferrous and nonferrous metals and dissimilar combinations thereof. While no standard industry specification exists for this group, several compositions, all developed from the basic 60% Cu-40% Zn analysis, have been evolved. The first of these variations contains about 0.75% tin, which acts principally as a hardener. To this same analysis may be added iron in about 1.00% and manganese in about 0.30% to secure the so-called "manganese bronze" rods. Additions of silicon to the extent of 0.10% produces the low-fuming types of rods. Another variety of brass rod used for high compressive strength at steam temperatures contains additions of about 2.5% tin and 0.2% silicon. This rod is used extensively for building up worn steam pistons. Nickel in amounts up to 0.50% is added to the Cu-Sn-Fe-Mn composition to assure uniform distribution of the iron and secure uniform hardness.

A flux specially designed for braze welding is necessary for this operation. It serves to clean base metal surfaces and control the oxidation of the elements of the rod, particularly zinc. This same flux is used for welding coppei.

Copper Welding Rods. Only deoxidized, or oxygen-free, copper welding rods should be used in welding of copper. Copper rods may be deoxidized with either phosphorus or silicon. Flux of the type used in braze welding is frequently used. An oxidizing flame permits welding without a flux.

Aluminum Welding Rods. Two analyses of aluminum welding rods are available. Commercially pure aluminum rod contains 99.0% Al, min, is designated as 2S, and is used for welding 2S and 3S base metals. The alloy rod contains an average of 5% silicon, carries the designation 43S, and is used for welding 52S, 53S, and 61S and the weldable casting alloys. The 43S wire may also be used for brazing 2S and 3S base metals. Another brazing wire, No. 718, contains 13% silicon and is used for the brazing of alloy base metals.

A flux specially compounded for aluminum welding is essential. For aluminum brazing a special low-melting aluminum brazing flux is necessary.

Welding Rods for Other Metals. In selecting welding rods for a given application, the first criterion is to match the chemical analysis of the base metal. Next, the mechanical properties should be duplicated or exceeded. Thus welding rods are available for nickel and nickel alloys such as monel and inconel. Copper alloys (other than those discussed under brasses) should employ rods of matching analyses. For stainless-steel welding, the rod should at least match the base-metal analysis, and in some cases it may be necessary to resort to a higher chrome-nickel composition to offset dilution, particularly when welding clad steels. For corrosive conditions, rods with an inhibitor such as columbium are preferred.

Hardfacing Rods. At present, there is no industry standard for hardfacing rods, although an excellent summary and classification[1] may be abstracted thus:

Group 1. Iron-base alloys with less than 20% of alloying ingredients; Cr, W, Mn, Si, and C. Hardness is lower than for other groups but toughness and shock resistance are better; wear resistance better than for machine steels. Generally low in cost.

Group 2. Iron-base alloys with more than 20% of alloying ingredients; these include Ni, Co, and others in addition to those in group 1. Compared with group 1, these rods show better abrasion resistance and longer service but reduced resistance to shock. Some alloys in this group exhibit "red hardness." Cost is higher; hence rods are used principally for final layer.

Group 3. Nonferrous alloys of Co, Cr, W, and other elements are characterized by red hardness. Available in several grades which provide a range in strength and toughness.

Group 4. So-called diamond substitutes; virtually pure tungsten carbide or mixtures of 90 to 95% tungsten carbide with Co, Ni, or similar elements.

Group 5. Crushed tungsten carbides of various screen sizes held with fabricated steel tubes or fused to strips of mild or low-alloy steel. In welding, the tube or strip is melted and fused to the base metal, where it serves to anchor the carbide particles.

CUTTING PROCEDURES

Oxygen Cutting. Manual oxygen cutting of iron and steel is a simple operation. The first step is lighting the torch and adjusting the cutting-oxygen flow. Correct pressures for oxygen and acetylene for any given job may be found in tables provided by manufacturers; also in the section on Operating Data, page 21-15. After the preheating flames have been adjusted to the neutral condition, the cutting-oxygen stream should be turned on and the flames checked to make certain that the

[1] "Hardfacing by the Oxyactylene Process," International Acetylene Association, New York, 1947.

neutral condition is maintained. If the flame changes, it should be readjusted with the cutting-oxygen stream flowing.

Oxygen cuts are preferably started at the edge of the metal, with the torch and tip positioned at 90° to the surface. Preheating flames are held about $\frac{1}{8}$ in. above the edge surface. When the metal becomes bright red, the cutting-oxygen stream is turned on. Immediately, a kerf is created under the heated spot by oxygen burning the iron. If the torch is moved along the desired line, the piece will be severed. Uniform steady movement is essential to the production of good cuts.

When it is necessary to make internal cuts away from an edge, it can be accomplished by starting at a drilled hole, or a hole may be pierced with the torch.

Bevel cutting proceeds in much the same manner as straight cutting, except the effective depth of cut is increased for a given metal thickness. This requires that greater pressures and slower travel speeds be employed.

Machine gas cutting is fundamentally the same as the manual operation. However, the fact that the torch is held steady and traverses at a uniform speed results in cuts of higher quality. With special, heavy-duty, water-cooled equipment, machine cuts have been made in thicknesses up to 72 in.

Stack Cutting. Cutting through an assembly of many thin plates, held substantially in a compact bundle, is possible because machine gas cutting produces smooth accurate cuts. If the stack is made up of sheet-metal thicknesses, it is good practice to apply heavier "waster plates" top and bottom to preserve sharp edges on the thin sheets.

Cast-iron Cutting. Cast iron is cut only by using a special technique, and the difficulty increases as the uncombined or graphitic carbon content increases. Preheating flames must be adjusted to a strongly carburizing condition; the length of the excess acetylene streamer should be roughly equal to the thickness to be cut. Oscillate the tip in a rather wide semicircular path. Severance of the metal is attained by the normal oxygen cutting reaction plus a washing action created by the molten products of the combustion. The resulting cut is rough when compared with a cut in steel. Consumptions of oxygen and acetylene are much higher.

Flux-injection and Powder Cutting. Stainless and straight chromium steels cannot be cut by the techniques described earlier, because the presence of chromium oxide in the kerf stops cutting action. To overcome this, two methods have been developed: (1) flux-injection cutting and (2) powder (iron) cutting.

In the flux-injection method, a powdered chemical flux is injected into the cutting oxygen stream by means of an electrically operated vibrator distributor. Three-hose cutting torches (both hand and machine) are necessary to keep the solid flux particles out of the relatively restricted preheating passages. This flux, reacting in the kerf, permits fluid slagging of chromium and other oxides resulting from the cutting reaction. Rather heavy preheats are necessary. A six-preheat flame tip should be used.

In the powder cutting method, powdered iron is introduced into the kerf through an auxiliary gas jet—usually air or nitrogen. A special device is necessary to introduce the powder in the kerf since it is not carried in the oxygen stream. Combustion of the iron powder in the cut maintains the oxides fluid for removal with the cutting slag.

WELDING PROCEDURES

Similar metals may be welded, braze welded, or brazed. Dissimilar metals, except in a few special cases, cannot be welded but instead require braze welding or brazing.

Steel. Low-carbon (0.30% C max) and low-carbon, low-alloy steels are readily oxyacetylene welded by forehand or backhand techniques. Preheating is not normally necessary unless base-metal analysis requires such treatment, or ambient temperature is below 32 F, or mass of parts indicates the necessity therefor.

Oxyacetylene welding of steels may be performed in all positions—flat, horizontal, vertical, and overhead. It is generally necessary to reduce heat (over that used for flat position) for vertical and overhead welding by selection of smaller tip size or reduction of operating pressures on tip (reduce pressures only within the operating range for tip as specified by manufacturer).

It is standard practice to select filler metal on the basis of mechanical properties. An exception is welding of stainless steels intended for corrosion resistance. The GA60 rods provide most reliable results for welding most carbon and alloy steels. The GA50 and GB45 rods have lower properties and cost less but are used extensively. Rods of the GA65 or the GB65 classifications are used for welding higher strength carbon and alloy steels. Genuine wrought iron should be welded with the GA60 rods.

The braze-welding technique is used on steel assemblies when it is necessary to control distortion. However, braze welding finds extensive application in the joining of sheet steels, including galvanized steels. Preferred rods are manganese bronze and low-fuming types with suitable braze-welding flux. In the case of galvanized iron, liquid fluxes and flux-coated bronze rods are used.

Cast Iron. Oxyacetylene welding of cast iron provides the best welded connection, although it is somewhat more costly than other methods. Preheating and slow cooling after welding are necessary to provide soft machinable welds and avoid cracking. Rods should approximate the casting analysis, gray-iron rods for gray-iron castings and alloy rods for alloy castings. Preheating temperatures should range from 900 to 1200 F. Welding should be followed by reheating the casting to 1100 F and cooling slowly. Multilayer welding is recommended for thicknesses above $\frac{1}{4}$ to $\frac{3}{8}$ in., using a layer for about each $\frac{1}{4}$ in. of thickness. Flux designed for cast iron must be used.

Braze welding is less costly than welding and the strength and reliability are completely satisfactory, but difference in color must be acceptable. Brass rods of the Cu-Zn-Sn composition are satisfactory for strength joints although the manganese and low-fuming varieties are preferred by some. Flux is necessary to secure "tinning" of the brass on the cast iron and to assist in controlling the brass.

For either welding or braze welding, it is necessary to prepare a V along the joint. This may be done by chipping or machining. Flame-cut surfaces require additional cleaning by chipping or sandblasting.

Aluminum. Oxyacetylene and oxyhydrogen welding are used in the welding of aluminum and its alloys. The choice is based upon metal thickness. Lighter gages are handled more readily with the oxyhydrogen flame. Always use a neutral or reducing flame; with oxyhydrogen flames in the ratio of H_2 to O_2 of 3 or 4 to one. Weldable compositions of aluminum and its alloys are 2S, 3S, 52S, and 61S.

Joint designs *A*, *B*, *C*, *D*, and *E* (Fig. 6) are the most commonly used in aluminum welding. Design *B* may be employed in thicknesses up to $\frac{3}{16}$ in. without beveling. On square and V-groove butt joints in thicknesses heavier than $\frac{1}{16}$ in., notch the edges about $\frac{1}{16}$ in. deep and $\frac{3}{16}$ in. apart with a cold chisel.

Cleaning prior to welding to remove grease, dirt, and other foreign substances is essential; use mechanical or chemical cleaners. Also the use of suitable welding fluxes is necessary to remove the aluminum oxide from the weld zone. The dry powdered flux is mixed with water or alcohol and applied to the top, bottom, and edges of the joint and the filler metal. After welding, flux residue must be removed to avoid attack on the metal, particularly in the presence of moisture.

Aluminum in the compositions noted above may also be brazed. Compositions 2S and 3X may be brazed, using 43S filler metal as the brazing wire. The alloys require No. 718 wire.

Copper Alloys. Copper for welded construction should be of either the deoxidized or oxygen-free varieties, in order to achieve optimum joint strengths. Electrolytic or tough-pitch copper contains oxides which under welding heat tend to separate out, creating a zone of weakness alongside the weld. Welds in deoxidized or oxygen-free copper will show strengths of 28,000 to 30,000 psi, the annealed strength of the base metal, whereas electrolytic copper will show only about 12,000 psi because of the weakening effect of the copper oxide.

Oxyacetylene flames should be about one tip size larger than for equal steel thicknesses. Heavy sections may require supplementary heating as from additional torches or continuous preheating. Flux of the type used for braze welding will be helpful, although flux may be omitted if slightly oxidizing flames are used. Welding rods must also be of the deoxidized variety. Copper may be welded in all positions by the oxyacetylene process.

Copper alloys as a group are weldable by the oxyacetylene process. Copper alloys, for welding, should contain less than 0.05% lead for good weldability and reliable results. Filler metal should match the analysis of the base metal. With proper filler metals, weld strengths will match the annealed strength of the base metal. Flux, in general, will be necessary although the oxidizing flame is usable on some compositions.

Nickel Alloys. Nickel, monel, and inconel may all be welded by the oxyacetylene process. The joint designs in Fig. 6 are, in general, applicable. The welding rods should be of the same composition as the base metal. Flux is unnecessary in the welding of nickel. However, flux is necessary for monel and inconel and should be specifically designed for this work.

Stainless Steels. The oxyacetylene process is employed principally on the lighter gages of stainless steel. Joint designs *A*, *B*, and *C* in Fig. 6 are the types that will be encountered. Welding rods, where used, should match the chemistry of the base metal. A flux for stainless-steel welding is essential.

Other Metals. Magnesium alloys are welded with either oxyhydrogen or oxyacetylene flames. Welding rods that match base-metal compositions and flux are essential. After welding, all flux residue must be removed to avoid attack on the metal. Lead, in its various compositions, is welded with oxyacetylene, oxyhydrogen, oxypropane, oxynatural gas, or oxycity gas flames. The oxyacetylene flame is preferred for the heavier thicknesses, while the lower temperature flames are used for the lighter sections. Flux is unnecessary. Filler metal is in rod form of matching analysis or is secured by cutting strips of the metal to be welded.

Dissimilar Metals. As a general rule, dissimilar metals cannot be welded unless the melting points are within about 50 F, a rare situation. Brazing and braze welding must ordinarily be used. Aluminum, magnesium, and zinc alloys cannot, in general, be joined to other metals except by special soldering techniques, not always satisfactory.

Silver-alloy brazing provides an excellent method of joining dissimilar metals where the lap- or shear-type joint is acceptable. One alloy of silver, copper, and phosphorus is used for joining copper and the brasses. Another alloy of silver, copper, cadmium, and zinc is usable on any metals except aluminum and magnesium. Braze welding with the standard brass rods is another means of solving these problems. The brass deposit will not match the color of steel and similar metals and may prove unsuitable under some corrosion conditions. Braze welding may also be accomplished with other rods, for example, stainless-steel rods used for joining high-carbon and high-alloy steels or stainless steels to other steels.

Fluxes are generally necessary for these operations. For silver-alloy brazing, the flux should be designed specifically for this type of brazing. For braze welding, the filler metal will usually determine the flux type, although mixing of standard fluxes may be necessary on some combinations. Using either of the techniques described above, the following metals may be joined to themselves or one another in dissimilar combinations: iron, steel, alloy steels, stainless steels, copper, copper alloys, nickel, nickel alloys.

CUTTING OF METALS

Steel and Wrought Iron. Steels containing up to about 0.25% C and normal amounts of Mn, P, S, and Si (e.g., SAE 1025 steel, boiler plate, structural steel, etc.) may be cut very readily. Wrought iron may also be cut without difficulty. Coated steels (galvanized, etc.) may in general be cut, but it is important to provide ventilation and protection for the operators, since the metals used for coating produce objectionable or poisonous fumes. Coatings will be burned off adjacent to the cut, but this may be offset in the case of galvanizing by painting liquid braze-welding flux along the line of cut (top and bottom) before cutting.

The presence of carbon in excess of about 0.25% and alloying ingredients above the normal values commonly employed will, in general, call for special treatments. Table 1 provides a listing of the more commonly encountered elements and their influence on cutting. This table must be used only as a guide, since it is impossible to provide all the data in such a condensed summary. Detailed discussions are available in "The Welding Handbook" (3d ed., American Welding Society, New York, 1950).

OPERATING DATA

Jigs and Fixtures. In the performance of any welding work, the alignment of the parts is of the utmost importance.

Welded seams in flat sheets or plates are best held by clamping arrangements. For adequate clamping, the jig should have a substantial backup member and two clamping elements, one on each side of the seam. A simple way of assembling such a unit would be to use either an I-beam or a solid steel bar for the backup and two angles for the clamps. If the jig were needed for but two or three jobs, the elements could be held together by means of C-clamps. If water cooling is needed, the various jig elements may be constructed hollow and circulating water supplied.

Alignment of cylindrical elements—solid or tubular—is best accomplished by means of V-blocks or modifications thereof. The location of pads, bosses, and similar features is obtained by using toggle clamps equipped with the necessary features to hold the part. Hydraulic and pneumatic clamps are also employed, particularly where it is necessary to exert strong pressure over large areas.

Welding Costs. In determining the amount of oxygen, acetylene, and welding rods and the time necessary to make a given weld, the most accurate method is by actual welding of samples. Consumptions of oxygen and acetylene are obtained by weighing the cylinders before and after each weld and noting the differences in weights. These weights are converted to cubic feet of gases, using 12.1 cu ft per lb for oxygen and 14.7 cu ft per lb for acetylene; these volumes will be at 70 F and normal pressure. Quantity of welding rods is determined by weighing the rods before and

TABLE 1. INFLUENCE OF ALLOYING ELEMENTS ON THE CUTTING OF STEELS*

Element	*Results*
Carbon	Steels containing C above 0.25% require preheating in range 300 to 600° F to avoid hardening and cracking of cut edges.
Manganese	Manganese steels (14% Mn and 1.5% C) are difficult to cut and require preheating.
Silicon	Silicon creates no special problems in cutting. If combined with high C and Mn, careful preheating and postheating are essential for optimum results.
Chromium	Straight chrome steels up to about 5% Cr may be cut without much difficulty; clean surfaces essential. Above 5% Cr, use flux injection or powder cutting.
Nickel	Nickel contents up to 7% readily cut; 20 to 30% Ni with moderate C may also be cut.
Stainless Steels	The Cr-Ni stainless steels cannot be cut with any degree of reliability with conventional techniques. Use of flux injection cutting or powder cutting will give results roughly comparable to mild-steel cutting. Arc-oxygen cutting also usable on these alloys with special shielded arc cutting electrodes.
Molybdenum	Molybdenum acts similarly to Cr but is usually present in small quantities; hence causes little trouble. SAE 4130, etc., readily cut but may harden at edges; careful preheating and postheating needed to avoid this. High Mo-W steels cut with difficulty and require special techniques.
Copper	Steels containing up to 2% Cu may be cut readily.
Aluminum	Aluminum in the amount usually present in steel does not cause any trouble. If up around 10%, will cause trouble.
Phosphorus	Phosphorus in the amount generally permissible in steels does not cause trouble.
Sulfur	Sulfur in small amounts permissible in steels not troublesome; higher amounts will slow cutting speeds.
Vanadium	Vanadium in the amounts usually encountered does not cause any trouble.

* Adapted from "The Welding Handbook," 3d ed., p. 528, American Welding Society, New York, 1950.

after welding, the difference being considered the amount used since there is virtually no spatter loss. Welding time is, of course, taken by stop watch.

Tabular data such as those provided in Table 2 are helpful for preliminary and rough estimates. However, certain assumptions must be made in recording data of this type, and these must be fully understood if estimating errors are to be avoided. The data here recorded represent the work of average skilled operators; those of lesser skill will not equal the speeds listed, and similarly those of greater skill, particularly on specialized operations, will exceed these figures.

The welding speeds recorded are for 100% duty factor, that is, no time out. To avoid proprietary references to equipment, tip sizes are specified by means of the range of acetylene flows. Manufacturers have available, on request, charts showing the rated acetylene flows for their various sizes of tips. The data are for single-layer welds as indicated. Multilayer welds in thicknesses $\frac{3}{8}$ in. and greater will show reductions in gas consumptions in the order of 5 to 10% and increases in welding speeds of about 5 to 7%.

TABLE 2. OPERATING DATA EOR OXYACETYLENE WELDING

Metal thickness, In.	Joint	Rod Dia, In.	Tip Size,* Cu Ft per Hr	Oxygen, Cu Ft per Ft of Weld†	Acetylene, Cu Ft per Ft of Weld†	Welding Speed, Ipm	Welding Rod Consumption, Lb per Ft of Weld
1/64	Square groove	None	0.2–0.7	0.03	0.03	5.2–6.0	
1/32	Square groove	None	0.6– 1.6	0.05– 0.04	0.05– 0.04	4.4–5.0	
1/16	Square groove	1/16	1.0– 3.0	0.13– 0.11	0.13– 0.11	3.6–4.2	0.013
3/32	Square groove	3/32	2.0– 6.0	0.36– 0.30	0.36– 0.29	2.8–3.4	0.030
1/8	Square groove	1/8	3.0– 10.0	0.80– 0.68	0.77– 0.65	2.2–2.6	0.053
3/16	90° V	3/16	6.0– 18.0	2.36– 2.08	2.27– 2.00	1.5–1.7	0.150
1/4	90° V	3/16	11.0– 30.0	4.50– 3.86	4.33– 3.72	1.2–1.4	0.265
5/16	90° V	1/4	20.— 36.0	7.40– 6.05	7.11– 5.82	0.9–1.1	0.414
3/8	90° V	1/4	30.0– 50.0	11.42– 9.13	11.00– 8.80	0.8–1.0	0.597
1/2	60° V	1/4	37.0– 60.0	11.65– 9.70	11.20– 9.33	1.0–1.2	0.580
5/8	60° V	5/16	49.0– 75.0	21.10–16.42	20.30–15.79	0.7–0.9	0.872
3/4	60° V	5/16	62.0–103.0	36.60–26.16	35.20–25.17	0.5–0.7	1.307

* Select tip normally rated for the flow shown; consult manufacturer's data for tip-flow ratings.

† Low flow for fast welding speeds; high flows for slow welding speeds.

Fundamental deposition data for oxyacetylene welding are dependent upon the type of welding and the flame adjustment. For neutral-flame forehand welding, it will require approximately 18 to 20 cu ft of acetylene to deposit 1 lb of steel weld metal; oxygen is calculated at 10% greater (1.1 times) consumption. For backhand welding with the slightly reducing flame, the value is 16 to 18 cu ft of acetylene per lb of weld metal deposited; with this flame adjustment oxygen is taken at 5% less (0.95 times) than the acetylene consumption.

Cutting Costs. Many factors influence the speed, economy, and quality of oxyacetylene cutting. In consequence, it is difficult to provide a single set of data to cover all conditions. Table 3 records data for both manual and machine cutting of clean steel by reasonably competent operators. The presence of dirty, rusty, or painted steel will slow down the cutting speed and raise the gas consumptions. The higher cutting speeds correspond to the lower oxygen and acetylene consumptions, and conversely the lower speeds result in the higher consumptions. The speeds recorded are for 100% duty factor, that is, no time out.

Bibliography

"Hard-facing by the Oxyacetylene Process," International Acetylene Association, New York.

"Oxyacetylene Cutting," International Acetylene Association, New York.

"The Oxyacetylene Handbook," Linde Air Products Co., New York.

"Oxyacetylene Welding and Cutting Instruction Course—Lectures and Exercises," Air Reduction Company, Inc., New York.

"Safe Practices for Installation and Operation of Oxyacetylene Welding and Cutting Equipment," International Acetylene Association, New York.

"The Welding Encyclopaedia," McGraw-Hill Book Company, Inc., New York.

"The Welding Handbook," 3d ed., American Welding Society, New York, 1950.

TABLE 3. OPERATING DATA FOR OXYACETYLENE CUTTING OF CLEAN STEEL

Thickness, In.	Manual Cutting							Machine Cutting						
	Cutting Tip*		Oxygen, Total (Cutting and Preheating)		Acetylene		Cutting Speed,† Ipm	Cutting Tip*		Oxygen, Total (Cutting and Preheating)		Acetylene		Cutting, Speed,† Ipm
	Oxygen Orifice, In.	4 Preheat Holes, In.	Psi	Cu Ft per Ft of Cut†	Psi	Cu Ft per Ft of Cut†		Oxygen Orifice, In.	4 Preheat Holes, In.	Psi	Cu Ft per Ft of Cut†	Psi	Cu Ft per Ft of Cut†	
1/4	0.0465	0.031	20	0.65–0.58	2.5	0.13–0.11	16.0–18.0	0.038	0.028	35	0.46–0.38	4.5	0.10–0.08	22.0–27.0
3/8	0.0465	0.031	25	0.83–0.73	3.0	0.15–0.13	14.5–16.5	0.0465	0.031	40	0.78–0.63	4.0	0.12–0.10	21.0–26.0
1/2	0.0465	0.031	30	1.10–0.91	3.0	0.18–0.15	12.0–14.5	0.0465	0.031	55	1.05–0.87	4.5	0.15–0.13	20.0–24.0
3/4	0.0595	0.0465	35	2.27–1.87	4.0	0.48–0.40	12.0–14.5	0.0595	0.0465	50	1.77–1.44	2.0	0.22–0.18	18.0–22.0
1	0.0595	0.0465	35	3.20–2.37	4.0	0.68–0.50	8.5–11.5	0.0595	0.0465	55	2.49–1.93	2.25	0.30–0.23	14.0–18.0
1 1/2	0.0595	0.0465	40	4.93–3.95	4.0	0.97–0.77	6.0–7.5	0.067	0.0465	55	4.00–3.20	3.0	0.42–0.33	12.0–15.0
2	0.067	0.0465	50	8.40–6.60	4.5	1.13–0.89	5.5–7.0	0.067	0.0465	60	5.47–4.33	3.5	0.57–0.45	9.5–12.0
3	0.067	0.0465	55	12.40–9.88	4.5	1.55–1.24	4.0–5.0	0.086	0.0550	50	8.30–6.64	3.5	0.80–0.64	8.0–10.0
4	0.086	0.0550	60	19.40–15.50	4.5	1.90–1.52	4.0–5.0	0.086	0.0550	60	11.82–9.03	4.0	1.08–0.82	6.5–8.5
5	0.086	0.0550	70	25.00–19.50	4.5	2.17–1.69	3.5–4.5	0.086	0.0550	65	14.94–11.76	4.5	1.35–1.06	5.5–7.0
6	0.098	0.0550	80	37.80–28.40	5.0	2.67–2.00	3.0–4.0	0.098	0.0550	65	21.30–17.43	4.5	1.60–1.31	4.5–5.5

* Select tips with these dimensions.
† Low consumptions for fast cutting speeds; high consumptions for low speeds.

Section 22

RESISTANCE WELDING

SECTION 22

RESISTANCE WELDING

In resistance welding the welding heat is generated by resistance of the parts to passage of electric current, and mechanical pressure is employed to forge the heated parts together. No fillers or fluxes are used.

Process Classifications. The two major classifications of resistance-welding processes (Fig. 1) are:

Lap welding comprising three methods:

1. Spot welding. Current is passed through the overlapped parts by electrodes that contact opposite sides of the work and apply pressure.

2. Seam welding. Roller electrodes or wheels pass along the overlapped joint, transmitting current and pressure to make a series of overlapping spot welds.

3. Projection welding. Projections or embossings on one or both workpieces act as localized current paths, and when these projections reach welding temperature the flat-faced electrodes force the parts together.

Butt welding consists of two methods:

1. Upset butt welding. Current-carrying dies grip the butted pieces under end pressure, and when the ends are sufficiently heated the weld is made by push-up pressure.

2. Flash-butt welding. The gripper dies hold the pieces together lightly at first, or strike an arc, and when the ends are molten the application of push-up pressure forces out the molten metal and makes a weld in the material behind.

SPOT WELDING

Advantages. High production speeds are possible with speed dependent on the skill of the operator, when equipment is either foot- or hand-operated. Completely automatically controlled apparatus is available. Where items to be welded are bulky, portable equipment is available. Good appearance is possible.

Applications. Joining thin to medium-thick parts of either similar or dissimilar metals; tacking assemblies prior to furnace brazing or arc welding. Used for products like cabinet, sheet-metal assemblies, equipment frames.

Three general types of joints can be made: (1) coach joint—joining flanges formed at a 90° angle; (2) lap joint—weld placed in center of overlapping edges; (3) butt strap joint.

Pulsation welding, a variant of spot welding, permits increased electrode life from water cooling in off-time cycles; thicker sections can be welded in production using the same equipment; scaly steels and oxide coatings can be penetrated by pulsation; high-carbon and other brittle steels can be welded and then annealed.

Disadvantages. Thickness limitations vary from ½- to 1-in. thickness (two plates or sheets) depending upon the alloy type. Weldability is also limited by such factors as low electrical resistance and high heat conductivity. Flanges and

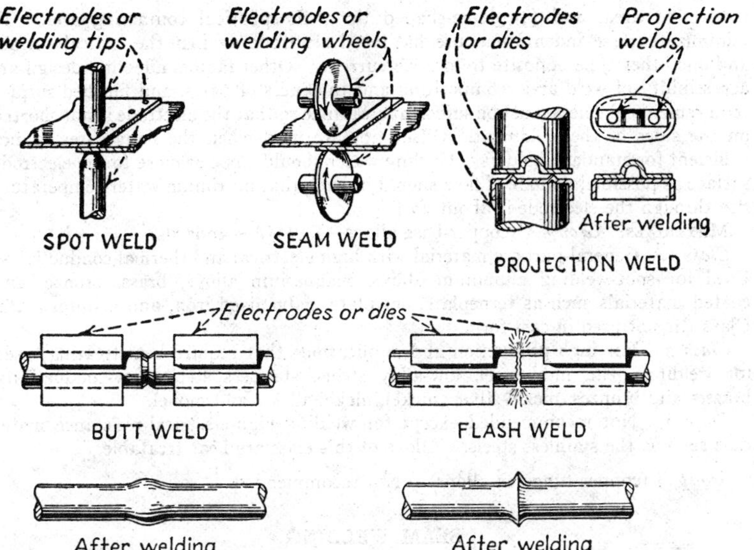

FIG. 1. Methods of making the five principal types of resistance welds. (*Resistance Welding Manual of the RWMA.*)

overlaps of proper design are required for accessibility and maximum weld strength. Equipment needs high power supply.

Design Considerations: LOADING. Applied loads should be shear loads; spot welds inherently develop stress concentrations at the edge of the weld and the inner faces of the sheets when subjected to tension or angle loading.

SPACING. Edge distance should be sufficient to retain all fused metal in one place and prevent expulsion at the edge. Spot spacing should be at least five times weld diameter.

NUMBER OF SPOTS. Increasing the number of spots in a structure will increase shear strength. To get shear strength as nearly equal to tensile strength as possible, use additional rows of spot welds.

PATTERN. Arrangements similar to rivet spacing supply high joint efficiency. Three types are used: (1) single row—spot aligned on center dimension, requires the least amount of overlap; (2) double row—requires greater overlap but more evenly distributes the loading to increase service life; (3) double row staggered—requires less overlap than the straight double row.

NUGGET SIZE. Electrode-tip size controls weld-nugget size. As a rule of thumb, tip diameter = (0.1 + 2 × sheet thickness).

Up to 0.060-in. material, weld nugget is 98% of tip diameter.

Over 0.060-in. material, weld nugget is 90% of tip diameter.

Over 0.125-in. material, weld nugget is 110% of tip diameter.

PRESSURE. 6000 × (two thicknesses).

Electrode Selection. Tip shapes fall into four classifications: (1) flat face, (2) truncated-cone form with flat contacting area, (3) dome-shaped tip with flat contacting area, and (4) dome-shaped tip with spherical contacting area. To minimize surface indentation, use flat-faced electrode against the critical surface and one other type opposite to localize current. Other factors affecting design are accessibility of weld area, composition and thickness of parts, and finished-surface requirements. There must be sufficient clearance so that the electrode is not shorted on the sides of the weldment. Offset tips are used when the flange area is not sufficient for standard holders. Cooling water should come as close to the electrode surface as possible; volume of flow should be such that maximum water-temperature rise through the electrode is about 20 F.

MATERIALS. *Group A* (copper-base alloys—RWMA standards).

Class 1. General-purpose material with high electrical and thermal conductivity. Used for spot-welding aluminum alloys, magnesium alloys, brass, bronze, and coated materials such as terneplate, tin plate, galvanized iron, and cadium plate. Class 1 is not used in cast form.

Class 2. For high production and applications that require heat-treating; used for welding clean mild steel, low-alloy steels, stainless steels, low-conductivity brasses and bronzes, nickel-silver, nickel, nickel alloys, and monel.

Class 3. Not recommended except for welding high-electrical-resistance materials such as the stainless steels. Alloys in this class are heat-treatable.

Group B (copper-tungsten alloys). Not recommended.

SEAM WELDING

Advantages. Seam welding is used where long linear welds or spaced multiple shots are desired; for large sheet welding, tube welding, and similar products. Parts manufactured in large volumes are especially adapted to automatic feed. Some small parts requiring continuous seams and parts that require liquid- or gastight seams are made by this process.

There are three types: (1) *continuous*—gas- and watertight weld made by continuous wheel rotation and uninterrupted current; (2) *overlapping*—more stable than continuous, made by interrupting current so individual overlapping spot welds form; (3) *roll spot welds*—appear like carefully spaced spot welds, made by interrupting current for longer intervals between welds.

Design Considerations. As a rule of thumb, to produce a gastight weld use spacing ranging from 12 spots per inch with a welding speed of 12 fpm for light gage to 8 spots per inch with a welding speed of 2 fpm for heavy gage.

The width of the welding wheel depends on the thickness of the stock and the type of current (see Spot Welding, page 22-2, for edge spacing and nugget size).

Design of joints is important for economical use of seam welding. Work clearances should permit use of standard arms and wheels. Abundant cooling water is required to improve weld quality and retard wheel wear. In addition, water will retard warpage and minimize surface burning and pickup on wheels.

Mash seam welding requires a lap 1½ times the thickness of one sheet. The weld wheel is about 2½ times the width used for standard seam welding. Pressures are about three times that required in ordinary seam welding. Final mash thickness is about 25% greater than the original thickness of one sheet.

Electrode Selection. Many factors involved in spot welding apply here with these additional considerations: wheel diameters vary from 7 to 10 in.; the edge or face in contact with the work has a slight crown and rounded corners; the width of the wheel usually varies between ⅜ to ¾ in., with a minimum value of ⅛ to 3⁄16 in. When welding coated stock use knurled wheels with frequent dressing.

MATERIALS. *Group A* (copper-base alloys).

Class 1. Same as spot welding.

Class 2. Same as spot welding.

Class 3. Not recommended.

Group B (copper-tungsten alloys).

Class 10. Not recommended.

Class 11. Recommended for seam welder bearing inserts.

Class 12. Principally for welding nonferrous metals having relatively high electrical conductivity.

PROJECTION WELDING

Advantages. Projection welding is usually performed in multiples. High production speeds and good joint strengths are possible, provided that the design of projections is within recommended practices. Projection welding is used in joining sheet to sheet, sheet to tube, and sheet to solid bar or wire and previously formed sections made from either sheet or bar.

Applications. When several welds must be made within a concentrated area; to weld fasteners and to join parts having natural projections.

Limitations. Tooling is usually required to form projections unless projection is natural (line or point contact between parts). Metal must have sufficient strength to support the projection—brass is not usually satisfactory; aluminum parts are limited to extrusions; and copper is unsatisfactory because of low electrical resistance, high heat conductibility, and low strength.

Design Considerations. For a given thickness of stock there is an optimum height and diameter of projection. They should be round or as close to it as possible. There are three common types:

Button. Used in flat sheets with thickness from 24 to 13 gage, or 0.025 to 0.0937 in. thick.

Cone. More rigid than the button; used in flat sheets from 12 to 5 gage, or 0.01093 to 0.218 in. thick.

Spherical. Used in heavier sections and forgings, as well as sheet metal and plate from 0.012 to 0.500 in. (see Tables 1 and 2).

Three rules are fundamental: (1) projection must be sufficiently rigid to support pressure applied by the electrodes or dies; (2) it must have sufficient mass to heat the opposing surface to proper welding temperature; and (3) it must not be sheared or distorted when initially formed.

Basic rules of heat transfer control the location of a projection. When welding dissimilar metals, place projection on part having the highest thermal and electrical conductivity. If welding light to heavy gage, make projection diameter and height as per light gage specification. If welding heavy to light gage, make projection diameter as per heavy gage specifications, and projection height as to light gage specifications.

Electrode Selection. DESIGN. Flat surfaces of dies have larger areas than spot-welding electrodes. Opposing faces of the dies must maintain a parallel relationship. Design depends primarily on the nature of the job. If dies have gages, clips, or locators attached, design should prevent the accessories from carrying current.

MATERIALS. *Group A* (copper-base alloys).

Classes 1 and 2. Not recommended but can be used for dies.

Class 3. Specifically recommended, particularly when cast for highly stressed structural and current-carrying parts of the electrode dies.

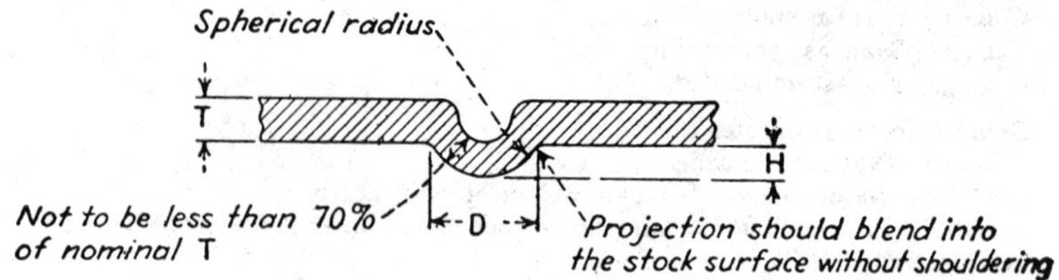

D- Represents the total diameter raised above the metal surface

TABLE 1. EMBOSSED PROJECTIONS FOR RESISTANCE WELDING

Material Thickness T		Projection Dia D		Projection Height H	
Inch	U.S.S.G.	Min	Max	Min	Max
0.012	30	0.075	0.080	0.015	0.020
0.014	29	0.075	0.080	0.015	0.020
0.015	28	0.075	0.080	0.015	0.020
0.016	27	0.075	0.080	0.015	0.020
0.018	26	0.090	0.095	0.020	0.025
0.021	25	0.090	0.095	0.020	0.025
0.024	24	0.090	0.095	0.020	0.025
0.027	23	0.105	0.110	0.025	0.030
0.030	22	0.105	0.110	0.025	0.030
0.033	21	0.105	0.110	0.025	0.030
0.036	20	0.120	0.125	0.030	0.035
0.042	19	0.120	0.125	0.030	0.035
0.048	18	0.135	0.140	0.035	0.040
0.054	17	0.135	0.140	0.035	0.040
0.060	16	0.150	0.160	0.040	0.045
0.067	15	0.150	0.160	0.040	0.045
0.076	14	0.170	0.185	0.045	0.050
0.090	13	0.170	0.185	0.045	0.050
0.105	12	0.200	0.215	0.050	0.055
0.120	11	0.220	0.235	0.050	0.055
0.135	10	0.230	0.250	0.052	0.060
0.150	9	0.250	0.270	0.052	0.060
0.165	8	0.270	0.290	0.057	0.065
0.180	7	0.290	0.310	0.057	0.065
0.195	6	0.310	0.330	0.062	0.070
0.210	5	0.330	0.350	0.062	0.070
0.225	4	0.350	0.370	0.067	0.075
0.245	3	0.370	0.390	0.067	0.075
0.265	2	0.390	0.410	0.072	0.080
0.281	1	0.410	0.430	0.077	0.085
0.312	0	0.450	0.480	0.080	0.090
0.343	2-0	0.490	0.520	0.085	0.095
0.375	3-0	0.530	0.560	0.090	0.100
0.406	4-0	0.560	0.590	0.095	0.105
0.437	5-0	0.600	0.630	0.100	0.110
0.468	6-0	0.640	0.670	0.105	0.115
0.500	7-0	0.680	0.710	0.110	0.120

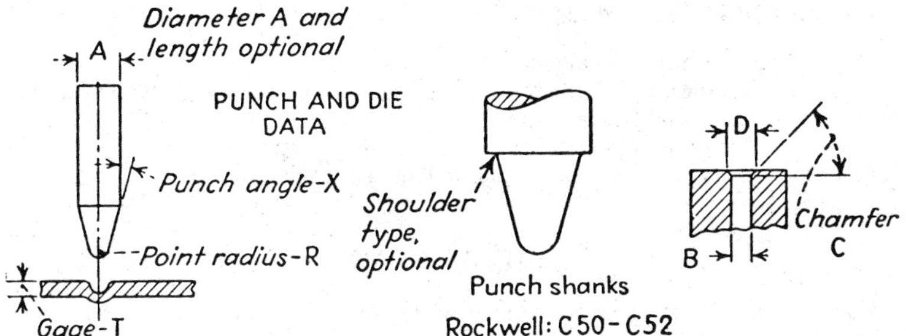

TABLE 2. DIES FOR WELDING PROJECTIONS

Gage T	Punch			Die	
	Radius R ±0.002	Angle X, deg	Shank A	Dia B +0.005 −0.000	Chamfer C, deg
0.012–0.016	0.031	15	3/8	0.062	45
0.018–0.024	0.031	15	3/8	0.079	45
0.027–0.033	0.047	15	3/8	0.095	45
0.036–0.042	0.047	15	3/8	0.110	45
0.048–0.054	0.047	15	3/8	0.125	45
0.060–0.067	0.062	15	3/8	0.140	45
0.076–0.090	0.062	15	1/2	0.140	45
0.105	0.078	15	1/2	0.172	45
0.120	0.094	15	1/2	0.191	45
0.135	0.094	15	1/2	0.171	45
0.150	0.109	15	1/2	0.188	45
0.165	0.109	15	1/2	0.207	45
0.180	0.109	15	1/2	0.227	45
0.195	0.125	15	1/2	0.250	45
0.210	0.141	15	1/2	0.272	45
0.225	0.141	15	1/2	0.290	45
0.245	0.156	15	1/2	0.312	45
0.265	0.156	15	1/2	0.328	45
0.281	0.187	15	1/2	0.348	45
0.312	0.187	15	1/2	0.328	40
0.343	0.187	15	1/2	0.368	40
0.375	0.203	15	1/2	0.406	40
0.406	0.219	15	1/2	0.437	40
0.437	0.219	15	1/2	0.484	40
0.468	0.234	15	1/2	0.515	40
0.500	0.234	15	1/2	0.562	40

Electrode Materials (*Continued*)

Group B (copper-tungsten alloys).

Class 10. Recommended for die facings and inserts when high electrical conductivity and malleability are desired.

Class 11. For general-purpose projection-welding electrodes.

Class 12. For heavy-duty projection-welding electrodes.

BUTT WELDING

Advantages. UPSET METHOD. Limited to relatively condensed sections; however, can be used in joining tubes, bars, etc. This process, because of its limitations, has been largely superseded by flash butt welding.

FLASH METHOD. More widely used than the upset-butt method, because it can also weld sheets and other extended sections. Other advantages include greater weld strength, smaller power demand, less power consumption, greater speed, less heat developed in the body of the work, no special preparation of weld surfaces required, and dissimilar metals varying widely in individual fusing temperatures can be joined.

Limitations. UPSET METHOD. Abutting sections must be identical in size and shape (maximum variation of dimension allowable, 15%). Tight-fitting clean surfaces are required. Design of part welded is critical. Tooling costs are high. Equipment-maintenance costs are relatively high. Power requirements limit size of section that can be welded. Clean, smooth, and uniform surfaces are required for die contact.

FLASH METHOD. Abutting edges also should be identical in size and shape with same maximum variation in dimension as upset. Beveled edges on one or both parts are required to obtain good heat balance. Design of parts is critical. Tooling costs are high and maintenance costs relatively high. There is considerable spattering during the process. The upset area is larger than in upset-butt welding.

Design Considerations. FLASH METHOD. Sections must be shaped so that both pieces will attain the same degree of plasticity and depth of plastic zone during flashing action. Sections of weldments should be nearly identical in shape and size. Also parts must be shaped so they can be held to alignment by the clamping dies during the forging action. They require enough area to allow application of the clamping force. Also in the shape of the parts, reactive resistance forces should not tend to destroy alignment during forging. Forging force should be resisted in the workpiece by forces that are parallel to the axis of the workpiece and to the direction of the welder forging force.

Large sections require bevels. In flash welding, joint design is important to obtain proper heat balance when dissimilar metals are welded.

Electrode Selection. Ability to resist compression usually is more important than conductivity, except in welding nonferrous metals that require good current distribution in the dies. Electrode dies always fit the work exactly to insure intimate contact and proper alignment.

MATERIALS. *Group A* (copper-base alloys).

Class 1. Can be used for flash- and upset-welding dies.

Class 2. Can be used for flash- and upset-welding dies and current-carrying members.

Class 3. Specifically recommended for butt-welding dies, current-carrying shafts, and bushings.

Group B (copper-tungsten alloys).

Class 10. Recommended for facings and inserts in dies where high electrical conductivity and malleability are required.

Class 11. Harder material than class 10 for use in facings and inserts of butt-welding dies.

Class 12. For heavy-duty upsetting of rivets and studs (not usually used).

Class 13. Not usually used for upset or flash welding.

STUD WELDING

This is a modified arc-welding technique for joining a stud to a flat or curved sheet or plate (Fig. 2). High amperage and low voltage are discharged through the stud as the tip touches the sheet, causing an arc. The heat of the arc melts the adjoining surfaces of the stud and workpiece, which are then firmly joined by spring pressure supplied by the welding gun. The stud, usually a machined screw, hook, or eyebolt, has a flux-filled tip.

Advantages. A weld is made in about ½ sec. Inexperienced operators can do stud welding, producing from 500 to 1000 welds per 8-hr day. Good fusion is possible between sheet and stud. Studs can be placed even in difficult locations.

Applications. Widely used in shipbuilding, transportation, and oil industries. Can be used on mass-produced products like appliances, automobiles, and name-plates as well as shorter-run industrial products. As in any arc-welding process, low carbon content is required to prevent weld brittleness.

Limitations. There is no limitation on maximum thickness of sheet, but minimum thickness for satisfactory welding is 0.02 in. Strength of the weld is normally measured only by the strength of the material or the stud.

Design Considerations. Studs can be spaced as close as ¼ in. With automatic equipment stud locations are maintained within a tolerance of 0.003 in. Size and design depend on application. Diameters range from 0.060 to ½ in. Studs may be threaded or unthreaded, produced as screw-machine or cold-headed products. Length should be at least twice the diameter to permit automatic selection.

Electrode. None required; proper stud-holding collet depends on stud diameter.

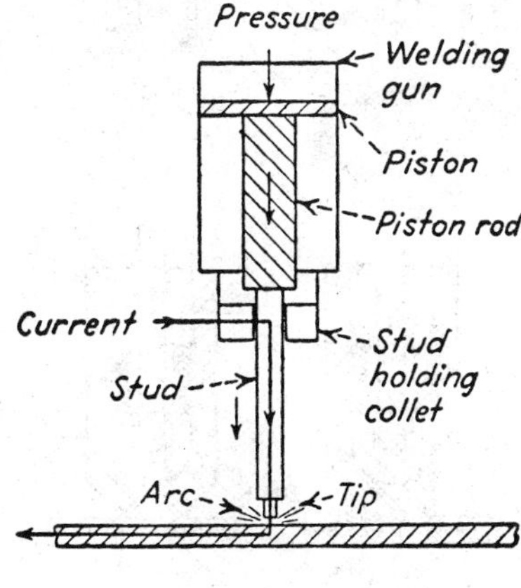

FIG. 2. Stud welding is assuming more importance in the manufacture of mass-produced goods.

RESISTANCE-WELDING APPLICATIONS

(With Examples of Good and Bad Practice)

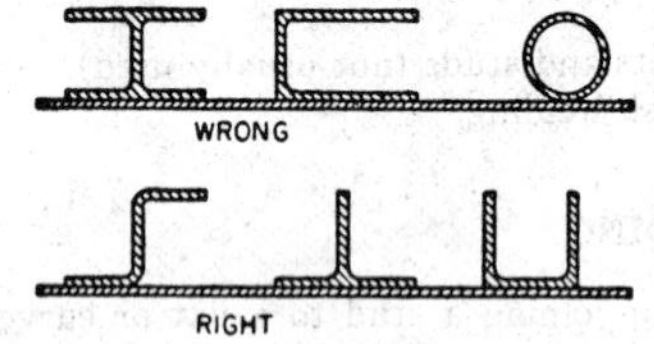

Spot Welding. SECTION SHAPE. Channels, tubes, and I-beam sections with wide flanges are difficult to weld to a flat sheet. They require special electrodes and equipment. Welding costs and maintenance are increased, and the results are erratic. Such troubles can be avoided by choosing simple sections that are easily reached with standard electrodes.

FIT-UP. Poor fit-up of flanged members cannot be corrected by the welding machine, because copper electrodes lack sufficient hardness. Use of poorly formed components results in a product that is out of square or may be deformed at the welds, or a hole is burned through the parts. Variation from the optimum 90° flange may amount to 3 to 5°.

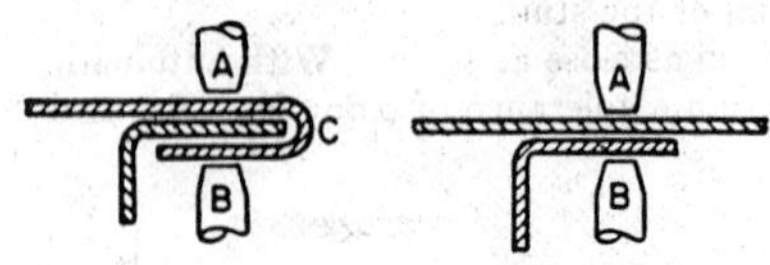

CURRENT PATH. In a lock seam, current does not pass directly from electrode *A* to electrode *B*, but flows along the path of least resistance *C*. For this reason, a straight lap or flange joint is preferred for spot- or seam-welded products. The members can be forged without excessive pressure.

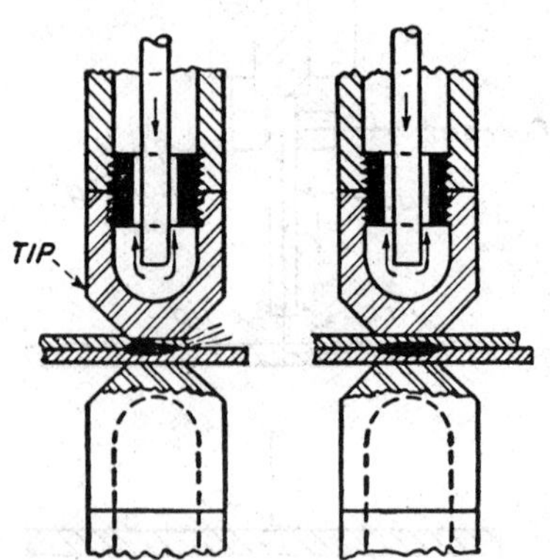

AMOUNT OF OVERLAP. Insufficient overlap of a spot-welded joint results in a weak, porous joint, and perhaps even a void. Not enough cold metal surrounds the spot to confine the molten metal properly, if the correct size of electrode is used.[1]

[1] For the proper amount of overlap and electrode size, see Recommended Practices for Resistance Welding, Tables 1.1 to 1.8, American Welding Society; "Welding Handbook," 3d ed., pp. 403–418, American Welding Society.

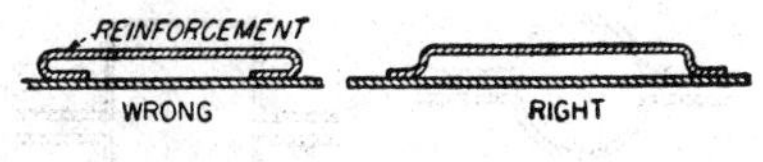

Blind Joints. Reinforcing members with turned-under flanges are difficult to weld. In the first place a copper arbor, which is difficult to remove and replace, is required to prevent crushing the part and to provide an electrical path. And secondly, erratic welding results are produced, because of varying contact resistance along the seam. Preferred method is to keep the reinforcement simple and its flanges accessible to the welder.

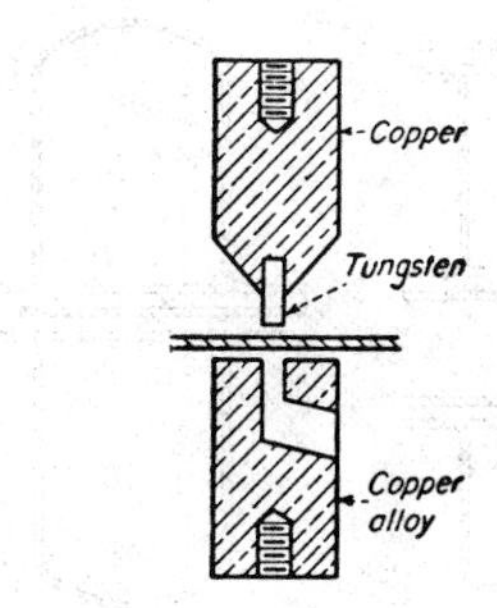

Upsetting and Burning Holes in Hard Materials. Holes may be burned through hard metals up to 0.090 in. thick by spot-welding electrodes consisting of a tungsten punch and a copper alloy die (RWMA alloy, group A, class 3). When the punch gently contacts the material, current is applied and the hole is burned out. Current is turned off as soon as the punch penetrates the material to avoid burning away the edges of the hole. This method is used for materials impossible or difficult to machine, and where they would be cracked or the edges of the hole shattered by punching.

Seam Welding. **Finned Tubing.** Heat-transfer tubing is economically produced by seam welding. Any reasonable length of single-fin tubing can be produced. Make the weld close to the tube or trim the fin to the edge of the weld, in order to minimize fin width. For proper flange width refer to the "Welding Handbook."

To make spiral-wound finned tubing, the fin may be slit or formed with a solid flange. Wall thickness of the tubing must be great enough to withstand welding pressure. In most cases, one welding wheel contacts the fin and the other the tubing directly opposite. Accurate fixtures are required to form the fin and guide it and the tube through the machine.

Flanged washers, forced over the tube,

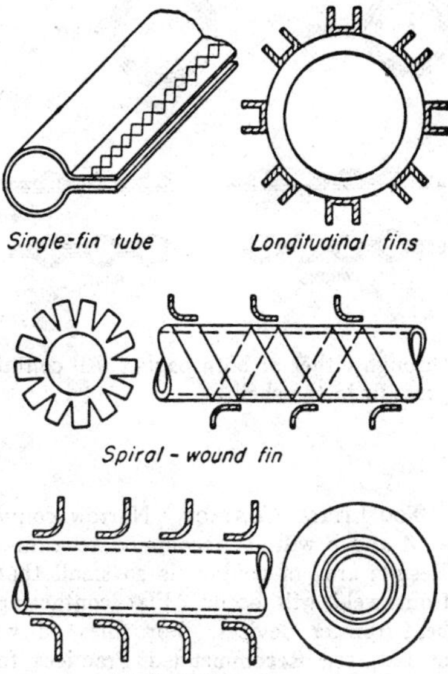

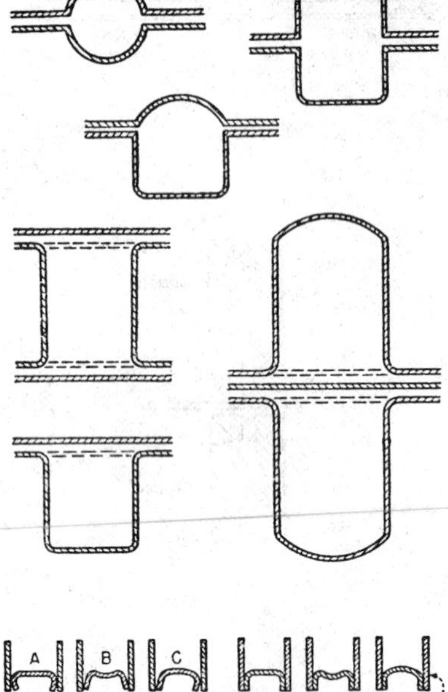

are seam-welded by contact wheels on opposite sides. Two welds are made in series so that one-half revolution of the tube completes the weld.

If longitudinal fins are required, an even number must be specified. Two fins on opposite sides of the tube are welded in series at one time.

Containers. Many sizes and shapes of containers have been seam-welded pressuretight to gases or liquids, using drawn or spun parts of most ferrous and nonferrous alloys. A two-compartment vessel is produced by placing a plate between the flanges of two drawn shapes and seam-welding. Baffles may also be seam- or spot-welded into one or both halves of a tank prior to joining the two main parts. Container size is limited only by the equipment available and the material thickness. Gages may range from 0.010 to 0.187 in.

Container Heads. Poor forming or incorrect size of container heads and bottoms will result in defective seam welding of tanks. Common faults are: flange toes in, *A*, toes out, *B*, or is skewed, *C*, preventing good metal-to-metal contact. If the head is too small, as at *D*, the surplus metal in the tank wall is gathered into a wrinkle, the joint cannot be pushed together by the machine, and current burns a hole through either or both parts. All container ends should be properly formed and a press fit in the shell.

Too Little Contact. Narrow convolutions like the curved contact surfaces at *A* and *B* will not accommodate seam-welding wheels and spot-welding electrodes. Result: area of contact is so small that excessive distortion and metal expulsion from weld will occur. Flat contacting surfaces *C* and *D* produce satisfactory heat-transfer devices. For rules on wheel widths, electrode sizes, and widths of flat, see Recommended Practices for Resistance Welding or the "Welding Handbook."

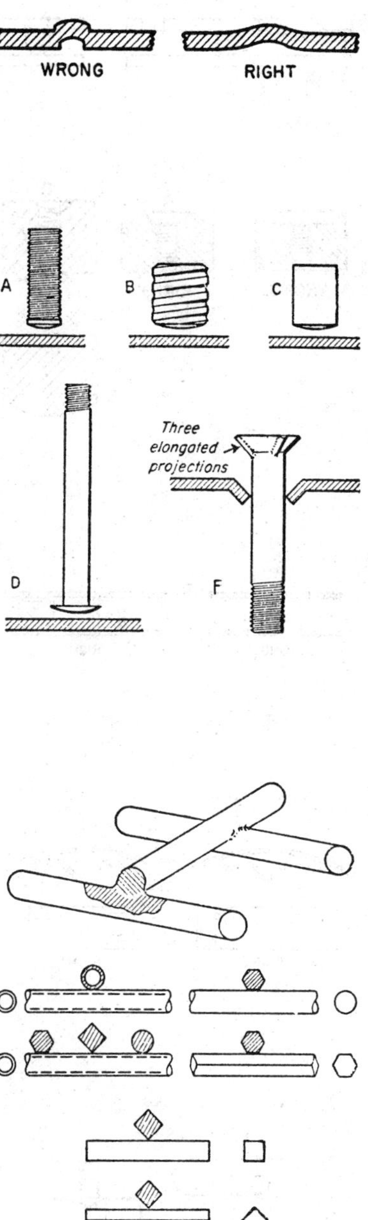

Projection Welding. PROJECTION CONTOUR. If a projection is practically punched out of the sheet, it may drop out during welding, or burn off just above the straight section because of local high resistance, or flatten and act to separate the parts. A good projection weld starts with point contact and has a continuously increasing cross-section to the full area of the weld.

STUDS AND SCREWS. Small, short studs *A* are projection-welded to sheet-metal structures for attachment of brackets, instruments, knobs, hinges, nameplates, etc. Heavy stud *B* with coarse-pitch round-bottomed thread serves as a spring seat or retainer. Heavy, plain stud *C* is employed for parts that must turn or be adjustable for several positions. Thin-head screws *D* are projection-welded to cabinets for fastening racks or thermal insulation. Flat-head screws *F*, without a screw slot, and having three elongated projections under the head, can be projection-welded flush with thin panels having pierced and dimpled holes. This type of welding screw works well on 0.010- to 0.1251-in. sheet and is easily produced on a header.

CROSS-WIRE WELDING. Cross-wire welding is the most simple form of projection welding, because the crossed wires form natural projections. Wire, rod, and tubing of almost any cross-section can be welded to itself or to nearly any other shape. By using proper equipment and control, including slope control, nearly all the nonferrous metals may be joined in like or dissimilar combinations, and many combinations of ferrous and nonferrous materials are weldable by the method.

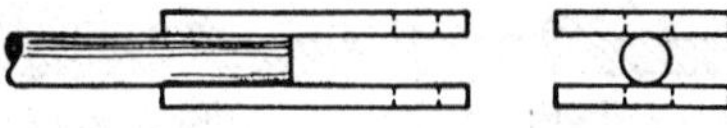

CLEVIS. Two pieces of strip stock can be projection-welded to opposite sides of a round rod in one operation.

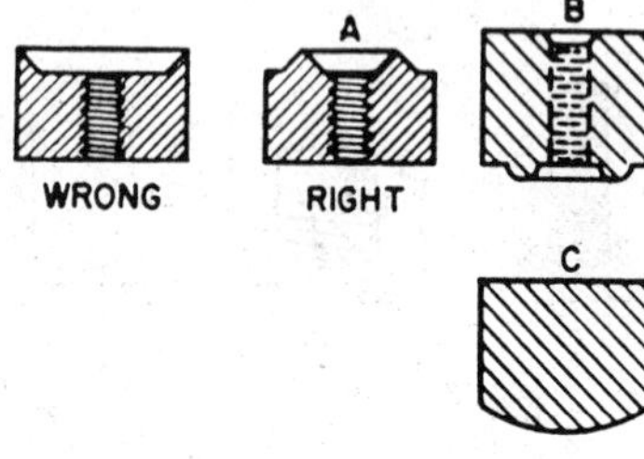

STUDS AND BOSSES. An unsupported edge on a projection-welding stud or boss tends to spread outward under heat and pressure, preventing a good weld. Symmetrical design *A* will produce good welds in sheet up to 0.091 in. thick but may collapse on thicker material. Design *B* is then preferred, because its larger cross-section permits generation of adequate heat in the sheet or plate. Solid studs *C* are suitable for a wide range of stock thicknesses.

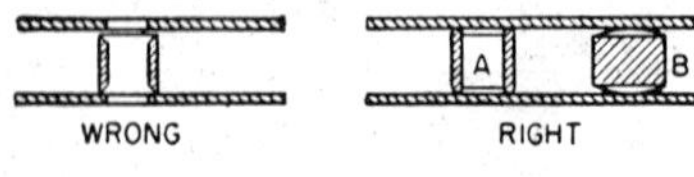

PROJECTION-WELDED SPACERS. Unsymmetrical projections on spacers cause the same difficulties as with studs. If the spacer is to be a tubular part (to carry a rod or bolt as in conveyor chain), round the ends of the tube, as at *A*. If a solid member is desired, round the ends of the spacer as at *B*.

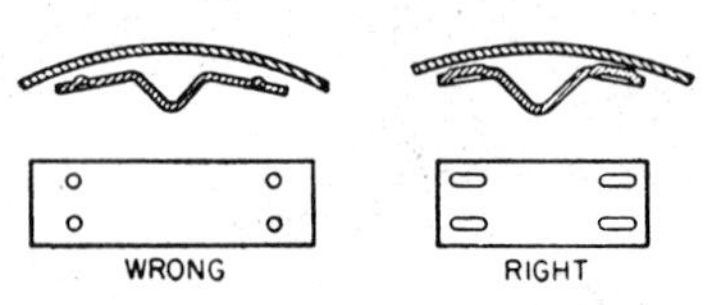

PROJECTIONS FOR CURVED WORK. No weld, or a cold weld, results when the usual round projection is employed for a part that is to be attached to a curved surface. As pressure forces the parts together, hot metal in the bracket contacts cold metal on the curved surface. This change in contact relationship is overcome by using projections elongated in the direction of movement.

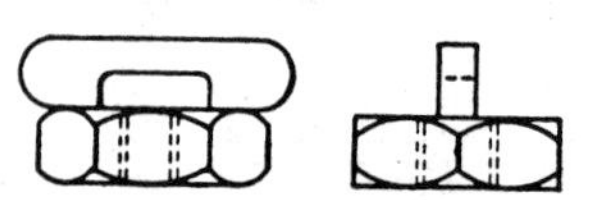

WING NUTS. Sheet stock from 0.035 to 0.125 in. may be punched to form wings of the desired size and shape and then edge-welded to common nuts, using a slotted or clamping welding fixture or electrode.

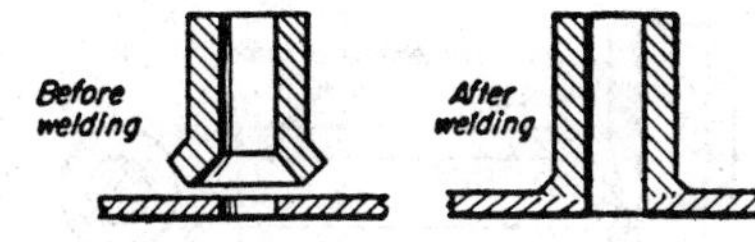

NIPPLES. No crevices or openings are permissible in the joint, when nipples or tubes are welded to a structure that is later vitreous enameled. Such defects trap air bubbles to cause holes or defective spots in the finish. By flaring the nipple as shown, a smooth joint and a sound weld are produced.

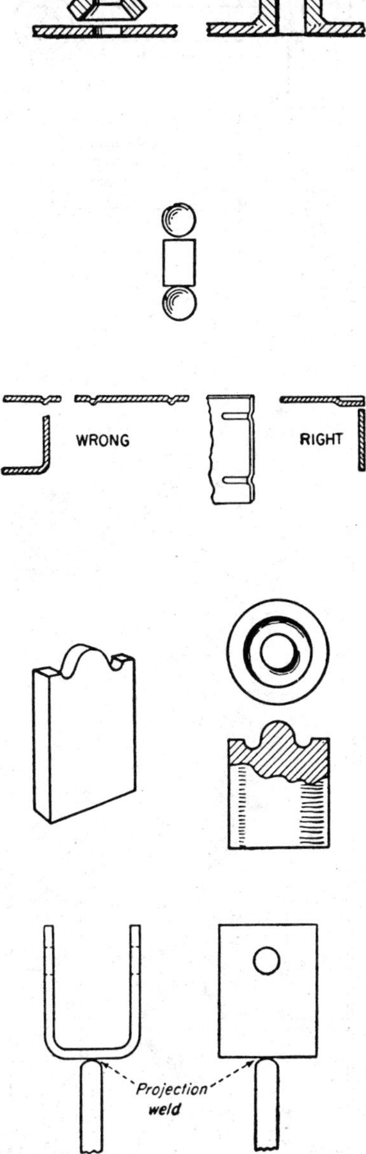

BALLS. A ball produces ideal projection-welding conditions and may be welded readily to one or both ends of a stud or bar, likewise to sheet and plate. Typical application: a ball-valve assembly.

SKIDPROOF PROJECTIONS. Typical round projections on a box cover will very likely skid off the mating edge of the container, when current and welding pressure are applied. Preferred practice in such work is to form elongated projections in the cover and to use an adequate fixture for locating the parts on the welding machine.

FLASH-FREE ASSEMBLIES. It is often necessary to projection-weld a rod or bar to sheet or plate in such fashion that close contact is obtained between the two, but no upset or flash is permissible. This is accomplished by the center projection which forms the weld and a surrounding undercut of equal or slightly greater volume to accommodate metal displaced by the forging or upsetting action.

TRUNNION. Plate and rod in thicknesses and diameters up to ½ in. can be assembled to form trunnions and universal joints. Projection welding is a cheap, fast process of manufacture in such cases.

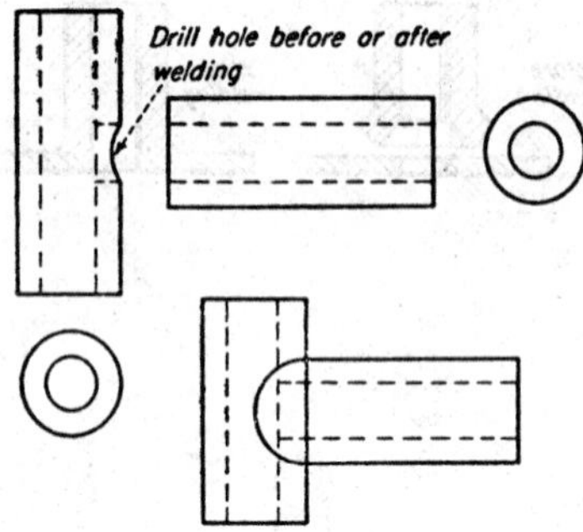

Flash Welding. T-WELDS. Tubing, solid rod, and plates may be flash-welded for T-shaped connections. The end of long arm of the T is cut square. During flashing, the parts will flash to a fit. If plate stock is properly welded, a fillet will be formed at the joint. In the case of tubing, the hole in one part may be drilled before welding, but a cleaner hole will result if drilled after welding, assuming the long arm of the T is short enough.

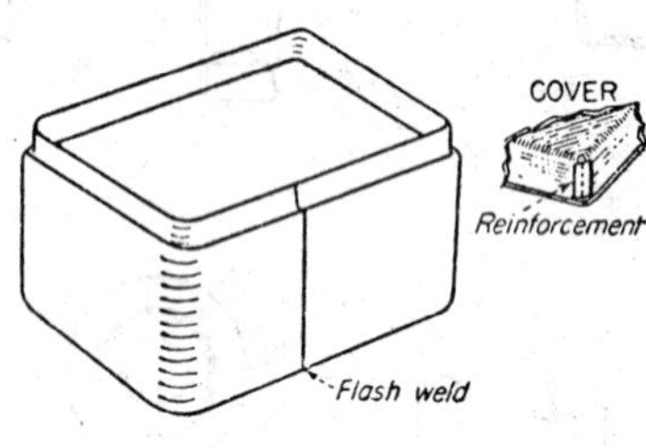

BOXES. A single set of tools will suffice to make a complete line of boxes. First, blank the side of the box from strip. Form the cover ledge and a bottom flange. Bend the corners in a brake. Flash-weld the ends; spot-weld a bottom in place.

To produce the cover, blank with four corner notches, form the sides on a brake. Then in a die common to all sizes of corners, bring the edges together. The metal is caused to flow or be drawn together, and is thinned. So weld in a reinforcement at the back of the seam and finish the outside surface on a disk grinder. This practice produces a cover that is adequate and cheaper for many purposes.

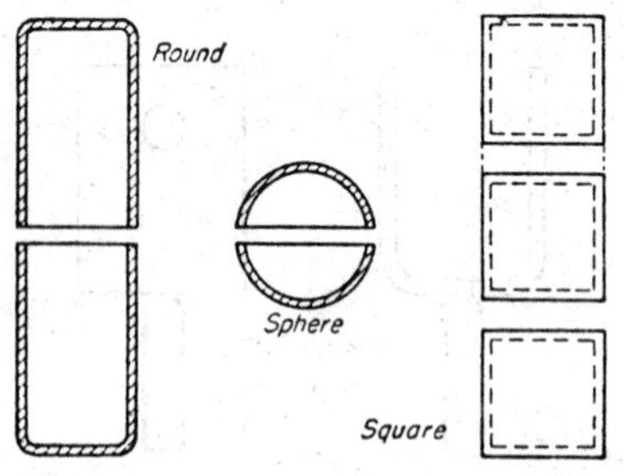

FLOATS. Containers or floats of any size up to 3½ in. dia (depending on material thickness) may be made by drawing two halves, trimming the edges square, and flash-welding them together. Mounting or operating brackets may be spot- or projection-welded to either or both halves before they are flash-welded.

Butt Welding. COLLARS OR FLANGES. A head, collar, or flange can be formed on a long rod by means of a butt welder. The bar of any standard shape is held in the machine with sufficient material between the clamps to form the upset. Current and pressure are applied and the upset is produced.

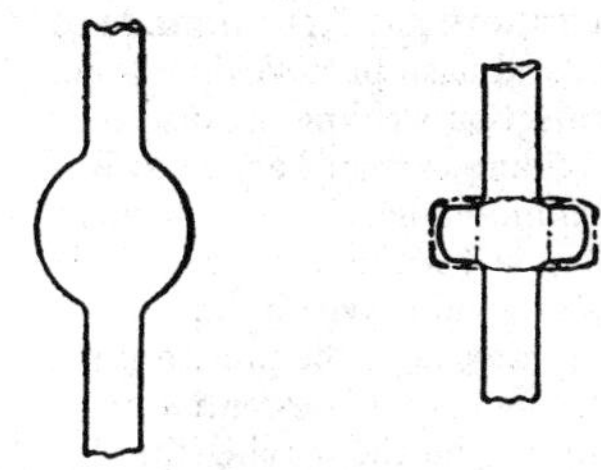

FUSING AND POINTING. Stranded cable may be burned off and the ends fused to prevent untwisting, using a butt welder. The cable is clamped with sufficient material length between dies; current is applied without pressure or motion. Any insulation must be stripped first. Ferrous and nonferrous cables can be burned off, including those with copper or aluminum cores.

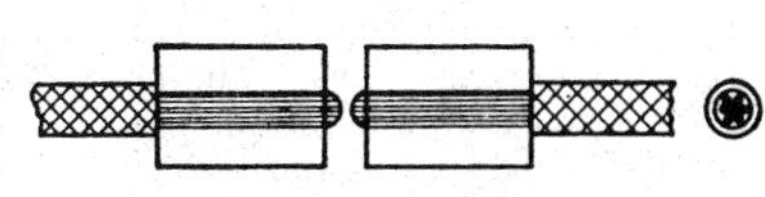

Solid rods are burned off and the ends pointed by a modification of this process. The material is clamped, current brings the stock to the forging temperature, the clamps are slowly moved apart, and points are drawn on each end.

WELDABILITY OF METALS

The response of various metals to resistance welding and their limitations may be understood from Tables 3 and 4—and the following notes.

LOW-CARBON STEEL. The metals most commonly resistance-welded are 0.10 and 0.15% low-carbon steels in plate, sheet, rod, wire, or extruded form. Plates and sheets may be spot-welded in thickness from 0.001 to 1.0 in., or seam-welded in thicknesses from 0.001 to about 3/8 in.

Projection welding is satisfactory for parts ranging from 0.010 to 1/2 in. in thickness, and for wire and rod in sizes from 0.002 to 1.0 in. dia. Also, low-carbon-steel products from 0.015 to 0.40 in. thick may be flash-welded with proper equipment.

STAINLESS STEELS. These are readily welded. Greater electrode pressure is required in spot and projection welding than when welding low-carbon steels; a welding interval is recommended to avoid carbide precipitation; and precise control must be exercised. Synchronous control is recommended.

LOW-ALLOY AND TRADE-NAME STEELS. Because of hardening of constituent alloys many well-known special steel alloys require special welding techniques, and sequencing controls are required in the welder, so that post heat-treatment can be performed.

HEAT-TREATABLE STEELS. Alloy steels and high-carbon steels can be easily welded, provided that a special type of heat-treatment is performed immediately after welding. Parts having a high-speed-steel section flash-welded to a low-alloy section are usually removed to a furnace for tempering and avoidance of cracking.

STAINLESS STEEL WELDED TO CARBON STEELS. To compensate for difference in rate of rise to welding temperature, special techniques must be employed.

Aluminum and Aluminum Alloys. These have been extensively welded by spot, seam, and flash-butt welding processes. Little success has been achieved, however, in projection welding because the projection usually collapses under pressure before the welding current has raised it to welding temperature.

Aluminum has a narrow plastic range and requires an accurate and precise welding control to avoid complete melting. Aluminum also has high electrical and thermal conductivities, and very low contact resistance. Because of the narrow plastic range and the precise timing required, rapid "follow-up" of the moving tip on the spot weld is essential and can be accomplished only by minimizing friction and inertia on the moving tip.

The inherent film of oxide must be removed by either mechanical or chemical means within 3 to 5 hr before spot or seam welding. Surface preparation is not so critical for flash welding.

Brass. Resistance-welding processes can be used on nearly all brasses except those high in lead content. Brass should be cleaned before welding because the oxides create varying contact resistance between the electrodes. Welding current must be accurately controlled, because excessive welding time often vaporizes zinc from zinc-bearing alloys. The current requirement is greater than that needed for steel, but a lower electrode pressure and a shorter welding time are required.

Brass may be projection-welded in thicknesses ranging from 0.010 to 3⁄16 in. Seam welding is practical in thicknesses ranging from 0.001 to about 3⁄16 in. Flash-butt or upset-butt welding is practical for brass rods from 0.005 to 2 in. dia. The range of thicknesses for spot-welding brass is between 0.005 and 1⁄4 in.; thicknesses in excess of 1⁄4 in. require large machines and excessive currents.

Bronze. Most bronzes can be resistance-welded except those containing lead and some containing high amounts of copper. Phosphor bronze and silicon are relatively as easy to weld as low-carbon steel.

Sheet can be spot-, projection-, seam-, flash-, and upset-welded, while rod and wire can be cross-wire projection-welded or flash-butt- and upset-butt-welded.

Magnesium. Practically the same types of welding machines, precise controls, and working techniques are required in the welding of magnesium as for aluminum. Magnesium also has a very high electrical and thermal conductivity, a relatively low melting point, and a very sensitive plastic range. It is particularly sensitive to overburning.

Zinc. Zinc and zinc-base die-cast materials have been welded satisfactorily by spot, projection, seam, and upset-butt welding. Although welds are satisfactory, the strength is not equal to the parent metal. Zinc and zinc-base die-cast materials should be cleaned before welding. Zinc requires a low electrode force to avoid extreme indentation or complete penetration. Lower currents and a longer weld time are used than for steel. Precise synchronous controls are necessary, and pickup on electrodes slows production. Molybdenum or Elkonite tips may help.

Lead. Procedures are similar to those used in welding of zinc and zinc-base die-casting materials. Tip pressure and welding current must be very low. In the welding of both zinc and lead, the welding machine should have a low-inertia moving tip and minimum-friction-type welding-head mechanism.

Copper. Copper has been successfully upset-butt or flash-butt welded by a special technique involving a short flashing period (to provide a good clean surface) followed by a regular upset-butt weld. Copper chain, rods, bars, and wire ranging from 0.010 to 1.0 in. dia have been successfully upset-butt-welded in this manner. Copper has seldom been successfully spot-, projection-, or seam-welded.

BERYLLIUM COPPER. This alloy is easily weldable by all processes. Special controls and welding machines like those used in the welding of brass and the bronzes are required.

CAST IRON. Only by performing special operations and preheating to between 1200 and 1400 F have spot welds been made on untreated cast iron as thick as ¼ in. Ordinarily, the material cracks at the welding area because of carbon solution and quenching. However, when annealed to give low combined carbon, machined, and then surface-treated in molten alkalis to remove free carbon, it may be spot-welded to sheet steel.

COATED METALS. Aluminum-coated steel, galvanized iron, tin plate, terneplate, cadmium-plated steel, and others can be joined, although the coatings cause higher contact resistance and necessitate frequent cleaning of the welding tips.

NICKEL. Nickel requires a slightly higher current density and electrode pressure than steel. Precise synchronous welding controls should be used in conjunction with antifriction low-inertia types of rams in the welders.

ALNICO. Alnico is not satisfactory for resistance welding.

TANTALUM. Special techniques are required to avoid absorption of various gases into the weld. These include immersion in water or carbon tetrachloride during welding.

TUNGSTEN AND MOLYBDENUM. A true resistance weld of tungsten is, for the present, not practical. Molybdenum can be resistance-welded by spot, projection, seam, and upset-butt welding. A short welding time is required at a high power level.

PRECIOUS METALS. Gold and many of the gold alloys can be spot-, projection-, seam-, and upset-butt-welded, but the techniques and special skills of achieving such welds are not generally used outside the jewelry-manufacturing industry.

Silver has been extensively projection-welded to base metals of copper, phosphor bronze, beryllium copper, and brass in the manufacture of electrical contacts; a fine line is drawn between the resistance-welding processes and joining by silver soldering. Silver has a low electrical resistance and therefore requires relatively larger amounts of welding current to raise the temperature to the welding point.

Platinum and its alloys containing iridium, osmium, palladium, and other rare metals are nearly all equally easy to weld.

ANTIMONY, BERYLLIUM, BISMUTH, CADMIUM, TELLURIUM, AND TIN. Each of these metals has been spot-, projection-, seam-, and upset-butt-welded but only with special equipment and in experimental laboratories.

DISSIMILAR METALS. Charts on the weldability of various metals disclose that many metals are readily weldable to similar metals but are not so easily weldable to metals of dissimilar characteristics (see Table 5).

The spot welding of dissimilar metals of greatly different electrical-resistance characteristics or welding temperatures is accomplished by the use of dissimilar tip or electrode materials; electrode materials are available in varying degrees of resistances from pure copper to tungsten. Another method of achieving heat balance is by the application of different sizes of contacting tips, electrodes, or dies. The smallest contacting area is used against the material having the lowest resistance or the highest melting temperature.

Table 3. General Weldability of Metals for Resistance Welding

Metals	Aluminum	Stainless Steel	Brass	Copper	Galvanized Iron	Steel	Monel	Nickel	Nichrome	Tin Plate	Zinc	Phosphor Bronze	Nickel Silver	Terneplate
Aluminum	B	E	D	E	C	D	D	D	D	C	C	C	F	C
Stainless steel	F	A	E	E	B	A	C	C	C	B	F	D	D	B
Brass	D	E	C	D	D	D	C	C	C	D	E	C	C	D
Copper	E	E	D	F	E	E	D	D	D	E	E	C	C	E
Galvanized iron	C	B	D	E	B	B	C	C	C	B	C	D	E	B
Steel	D	A	D	E	B	A	C	C	C	B	F	C	D	A
Lead	E	F	F	E	D	E	E	E	E		C	E	E	D
Monel	D	C	C	D	C	C	A	B	B	C	F	C	B	C
Nickel	D	C	C	D	C	C	B	A	B	C	F	C	B	C
Nichrome	D	C	C	D	C	C	B	B	A	C	F	D	B	C
Tin plate	C	B	D	E	B	B	C	C	C	C	C	D	D	C
Zinc	C	F	E	E	C	F	F	F	F	C	C	D	F	C
Phosphor bronze	C	D	C	C	D	C	C	C	D	D	D	B	B	D
Nickel silver	F	D	C	C	E	D	B	B	B	D	F	B	A	D
Terneplate	C	B	D	E	B	A	C	C	C	C	C	C	D	B

A, excellent; B, good; C, fair; D, poor; E, very poor; and F, impractical.

TABLE 4. ELECTRODES FOR SPOT WELDING SIMILAR MATERIALS
Read Block under Metal to Be Welded

Tin Plate		Terneplate		Galvanized Iron		Cadmium-plated Steel		Chrome-plated Steel		Stainless Steel 18-8 Type		Scaly H.R. Steel		C.R. Steel H.R. Steel Clean	
B	I	A	I	A	I or II	B	I	A	II	A	III or II	B	I or II	A	II
I	3	I	3	I or II	3	I	3	II	3	III or II		I or II	2	II	

Aluminum		Aluminum Alloys		Cupro-nickel		Nickel Silver		Nickel		Nickel Alloys		Brass Yellow 25–40% Zinc		Phosphor Bronze Grades A, C, and D		Silicon Bronze	
B	I or II	B	I or II	A	II	B	II	A	II	A	II	B	II	B	II	B	II
I or II	2	I or II	2	II		II		II		II		II		II		II	

Block Interpretation:

Weldability	Electrode Against
Electrode Against	Special Information

Weldability: A, excellent; B, good.
Special Information:
1. Special conditions required.
2. Good practice recommends cleaning before welding.
3. If plating is heavy, weld strength is questionable.

Electrodes: RWMA specifications, I, group A, class 1; II, group A, class 2; III, group A, class 3.
NOTE: The second-listed electrode material is definitely second choice.

Table 5. Electrodes for Spot Welding Dissimilar Materials

Ferrous Alloys	Stainless Steel 18-8 Type		Chrome-plated Steel		Cadmium-plated Steel		Galvanized Iron		Terneplate		Tin Plate	
Cold-rolled, hot-rolled, clean	A	II or III	A	II	B	II	B	I	A	I or II	B	I
	II		II	3	II	3	II	3	II	3	II	3
Tin plate	B	II or III	B	II	B	I or II	B	I	B	I or II		
	I		I	3	I	3	I	3	I	3		
Terneplate	B	II	B	II	B	I or II	B	I				
	I	3	I	3	I	3	I	3				
Galvanized iron	B	II	B	II	B	I						
	I	3	I	3	I	3						
Cadmium plate	B	II	B	II								
	I	3	I	3								
Chrome plate	A	III or II										
	II	3										

Nonferrous Alloys	Nickel Alloys		Nickel		Phosphor Bronze		Silicon Bronze		Yellow Brass		Nickel Silver	
Cupronickel	B	II	B	II	B	II	B	II	B	II	B	II
	II		II		II		II		II		II	
Silicon bronze	B	II	B	II	B	II	B	II	B	II		
	II		II		II		II		II			
Nickel silver	B	II	B	II	B	II	B	II				
	II		II		II		II					
Nickel alloys	A	II	B	II								
	II		II									
Stainless steel 18-8 type	B	II	B	II								
	III or II		II or III									

	Aluminum	
Aluminum alloys, duraluminum	B	I or II
	I or II	2

Note: See Block Interpretation under Table 4.

Section 23

BRAZING

SECTION 23

BRAZING

Definition. According to the American Welding Society, brazing is a group of welding processes, whereby similar or dissimilar metals are joined by heating above 800 F, but below their fusion temperatures, and using a nonferrous filler metal that has a melting point below that of the materials being joined. The filler metal is distributed in the joint by capillary attraction and coalesces or "grows together" with the joined metals. Some people consider that an alloying action has taken place.

Brazed joints can be made to develop high shear strength, assuming proper fit-up and selection of filler metal. Sound application of brazing involves proper design of the joint, and making the joint or assembly with a good brazing procedure. Joints can be neat, as well as strong, they may be gas- or pressure-tight, possess electrical and thermal conductivity and good corrosion resistance, as requirements demand.

Brazing Processes. Common brazing processes are copper brazing, bronze brazing, or "bronze welding," and silver brazing. Actually, seven groups of filler metals are included in the tentative specifications of the ASTM and AWS for filler materials (see Table 1). The brazing-temperature range varies widely, according to the nature of the brazing alloy (see Table 2).

Brazing can be performed with several sources of heat: gas-air torches or machines, the electric arc, resistance welders, induction coils, dipping in molten brazing material or flux, and in furnaces with or without protective atmospheres.

Base Materials. One or more filler metals can be used for brazing common base materials (see Table 3, which applies only to general applications). If dissimilar metals are joined, be sure that unequal rates of expansion and contraction will not introduce a stress that will cause the joint to fail, especially in parts of larger size.

Table 4 will serve to cross-index the filler-metal recommendations in Table 3 with the ASTM-AWS specifications in Table 1.

LIMITATIONS. In respect to the joint materials to be brazed, certain limitations and considerations should be observed:

Carbon Steels. In general, low-carbon steel up to 0.40% carbon max and free of scale, heavy oxide, dirt, grease, or foreign material can be brazed.

Low-alloy Steels. This refers to steels low in carbon (0.40 C max) which may have low percentages of one or more of the following elements: nickel, chromium, molybdenum, tungsten, vanadium, and cobalt.

Stainless Steels. If corrosion resistance is required after brazing then specify 347 or 321. Carbide precipitation will be eliminated if either of these two types are used, and this type of stainless will retain corrosion resistance after brazing. Note: Brazing alloys high in silver may attack high nickel-chromium steel intergranularly if the stainless parts are under stress during brazing. Parts should be annealed before brazing to relieve stresses.

Nichrome. **When using silver brazing alloys, the nichrome should be free of**

TABLE 1. PROPOSED TENTATIVE SPECIFICATIONS FOR ASTM-AWS BRAZING ALLOYS

Aluminum-silicon

AWS-ASTM Classification	Si	Cu	Fe	Zn	Mg	Mn	Cr	Ti	Al	Other Elements, Each	Other Elements, Total
BAlSi-1	4.0– 6.0	0.30	0.80	0.10	0.05	0.05		0.20	Remainder	0.05	0.15
BAlSi-2	6.8– 8.2	0.25	0.80	0.20					Remainder	0.05	0.15
BAlSi-3	9.3–10.7	3.3–4.7	0.80	0.20	0.15	0.15	0.15		Remainder	0.05	0.15
BAlSi-4	11.0–13.0	0.30	0.80	0.20	0.10	0.15			Remainder	0.05	0.15

Copper-phosphorus

AWS-ASTM Classification	P	Ag	Cu	Other Elements, Total
BCuP-1	4.75–5.25		Remainder	0.15
BCuP-2	6.75–7.50		Remainder	0.15
BCuP-3	6.00–6.50	4.75– 5.25	Remainder	0.15
BCuP-4	6.75–7.80	5.75– 6.25	Remainder	0.15
BCuP-5	4.75–5.25	14.50–15.50	Remainder	0.15

Silver

AWS-ASTM Classification	Ag	Cu	Zn	Cd	Ni	Sn	Other Elements, Total
BAg-1	44–46	14–16	14–18	23–25			0.15
BAg-2	34–36	25–27	19–23	17–19			0.15
BAg-3	49–51	14.5–16.5	13.5–17.5	15–17	2.5–3.5		0.15
BAg-4	39–41	29–31	26–30		1.5–2.5		0.15
BAg-5	44–46	29–31	23–27				0.15
BAg-6	49–51	33–35	14–18				0.15
BAg-7	55–57	21–23	15–19			4.5–5.5	0.15
BAg-8	71–73	27–29					0.15
BAg-9	64–66	19–21	13–17				0.15
BAg-10	69–71	19–21	8–12				0.15
BAg-11	74–76	21–23	2.5–3.5				0.15

TABLE 1. PROPOSED TENTATIVE SPECIFICATIONS FOR ASTM-AWS BRAZING ALLOYS (*Continued*)

Copper and Copper-zinc

AWS-ASTM Classification	Cu	Sn	Fe	Mn	Ni	P	Pb	Al	Si	Ag	Zn	Other Elements, Total
BCu	99.90 min*					0.075	0.02	0.01				0.10
BCuZn-1	58.0–62.0						0.05	0.01			Remainder	0.50
BCuZn-2	57.0 min	0.25–1.0					0.05	0.01			Remainder	0.50
BCuZn-3†	56.0 min	1.1	0.25–1.25	1.0	1.0		0.05	0.01	0.25		Remainder	0.50
BCuZn-4	50.0–55.0		0.10				0.50				Remainder	0.50
BCuZn-5	50.0–53.0	3.0–4.5	0.10				0.50				Remainder	0.50
BCuZn-6	46.0–50.0		0.25		9.0–11.0		0.05	0.005	0.15		Remainder	0.50
BCuZn-7	46.0–48.0				10.0–11.0	0.20–0.50			0.15	0.30–1.00	Remainder	0.10

Magnesium

AWS-ASTM Classification	Al	Mn	Zn	Si	Cu	Ni	Mg	Other Elements, Total
BMg	8.3–9.7	0.10 min	1.7–2.3	0.3	0.05	0.01	Remainder	0.03

Heat-resisting Materials

AWS-ASTM Classification	Ni	Cr	B	Fe	Si	C	Ag	Mn	Other Elements, Total
BNiCr	65–75	13–20	2.75–4.75	‡	‡	‡			0.50
BAgMn							84–86	14–16	0.15

* Copper plus silver.

† RCuZn-B and RCuZn-C welding rod (see AWS Specifications for Copper and Copper Alloy Welding Rod) meet these requirements. BCuZn-3 brazing filler metal can be used as RCuZn-B or RCuZn-C welding rod only if it meets the specification requirements of these rods.

‡ Total Fe + Si + C = 10.0 max.

stresses. The silver in the brazing alloy B1 penetrates the grain boundaries causing the nichrome to break or crack when stressed. It is therefore necessary that the nichrome be annealed before brazing. If, because of the application, the nichrome is stressed slightly, then use a lower silver content brazing alloy (B5).

Silver. This metal will anneal at brazing temperatures. A minimum of brazing alloy should be used as the filler metal alloys rapidly with the silver.

TABLE 2. BRAZING TEMPERATURES FOR ASTM-AWS BRAZING ALLOYS

AWS-ASTM Classification	Solidus, F	Liquidus, F	Brazing Range, F
Aluminum-silicon:			
BAlSi-1	1070	1165	1150–1185
BAlSi-2	1070	1135	1120–1140
BAlSi-3	970	1085	1060–1185
BAlSi-4	1070	1080	1090–1185
Copper-phosphorus:			
BCuP-1	1305	1650	1450–1700
BCuP-2	1305	1485	1350–1550
BCuP-3	1195	1500	1300–1550
BCuP-4	1185	1380	1300–1500
BCuP-5	1185	1500	1300–1500
Silver:			
BAg-1	1125	1145	1145–1400
BAg-2	1125	1295	1295–1550
BAg-3	1195	1270	1270–1500
BAg-4	1240	1435	1435–1650
BAg-5	1250	1370	1370–1550
BAg-6	1270	1425	1425–1600
BAg-7	1145	1205	1205–1400
BAg-8	1435	1435	1435–1650
BAg-9	1280	1325	1325–1550
BAg-10	1335	1390	1390–1600
BAg-11	1365	1450	1450–1650
Copper-gold:			
BCuAu-1	1755	1815	1815–2000
BCuAu-2	1620	1630	1630–1850
Copper and copper-zinc:			
BCu	1980	1980	2000–2100
BCuZn-1	1650	1660	1670–1750
BCuZn-2	1630	1650	1670–1750
BCuZn-3	1590	1630	1670–1750
BCuZn-4	1570	1595	1600–1700
BCuZn-5	1585	1610	1620–1700
BCuZn-6	1690	1715	1720–1800
BCuZn-7	1685	1710	1690–1800
Magnesium:			
BMg	770	1110	1120–1160
Heat-resistant:			
BNiCr	1850	1950	2000–2150
BAgMn	1760	1780	1780–2100

TABLE 3. SELECTION CHART FOR BRAZING ALLOYS TO JOIN VARIOUS COMBINATIONS OF BASE MATERIALS

Base Material*	Carbon Steels	Low-alloy Steels	Stainless Steels	Nichrome	Silver	Copper	Yellow Brass	Red Brass	Phosphor Bronze 5 %	Beryllium Copper	Nickel Silver	Cupro-nickel	Nickel	Monel	Inconel	Aluminum	Aluminum Alloys
Carbon steels	A, B1, B4	A, B1, B4	A, B1, B4	B1, B5	B1	B1, B4	B1, B4	B1, B4	B1	B1	B1	B1	B1	B1	B1		
Low-alloy steels	A, B1, B4	B, B1, B4, B3	A, B1, B4, B3	B1, B5	B1	B1, B4	B1, B4	B1, B4	B1	B1	B1	B1	B1	B1	B1		
Stainless steels	A, B1, B4	A, B1, B4, B3	A, B1, B4, B3	B1, B5	B1	B1, B4	B1, B4	B1, B4	B1	B1	B1	B1	B1	B1	B1		
Nichrome	B1, B5	B1, B5	B1, B5	B1, B5	B1	B1	B1	B1	B1	B1	B1	B1	B1	B1	B1		
Silver	B1	B1	B1	B1	B1, B2	B1, B2	B1, B2	B1, B2	B1, B2	B1, B2	B1	B1	B1	B1	B1		
Copper	B1, B4	B1, B4	B1, B4	B1	B1, B2	B1, B2	B1, B2	B1, B2	B1, B2	B1, B2	B1	B1	B1	B1	B1		
Yellow brass	B1, B4	B1, B4	B1, B4	B1	B1, B2	B1, B2	B1, B2	B1, B2	B1, B2	B1, B2	B1	B1	B1	B1	B1		
Red brass	B1, B4	B1, B4	B1, B4	B1	B1, B2	B1, B2	B1, B2	B1, B2	B1, B2	B1, B2	B1	B1	B1	B1	B1		
Phosphor bronze 5 %	B1	B1	B1	B1	B1, B2	B1, B2	B1, B2	B1, B2	B1, B2	B1, B2	B1	B1	B1	B1	B1		
Beryllium copper	B1	B1	B1	B1	B1, B2	B1, B2	B1, B2	B1, B2	B1, B2	B1, B2	B1	B1	B1	B1	B1		
Copper nickel, zinc alloy (nickel-silver)	B1	B1	B1	B1	B1	B1	B1	B1	B1	B1	B1	B1	B1	B1	B1		
Cupronickel	B1	B1	B1	B1	B1	B1	B1	B1	B1	B1	B1	B1	B1	B1	B1		
Nickel	B1	B1	B1	B1	B1	B1	B1	B1	B1	B1	B1	B1	B1	B1	B1		
Monel	B1	B1	B1	B1	B1	B1	B1	B1	B1	B1	B1	B1	B1	B1	B1		
Inconel	B1	B1	B1	B1	B1	B1	B1	B1	B1	B1	B1	B1	B1	B1	B1		
Aluminum	.....	.....	.....	.....	.....	.....	.....	.....	.....	.....	..	..	..	..	..	C	C
Aluminum alloys	.....	.....	.....	.....	.....	.....	.....	.....	.....	.....	..	..	..	..	..	C	C

* Refer to Limitations in text.

Copper. Commercial tough pitch copper should not be used where hydrogen gas is given off, such as in an oxygen-hydrogen torch flame, reducing atmosphere in a furnace, etc. In cases where a slight amount of hydrogen is evolved during the brazing process use deoxidized copper. Where a hydrogen atmosphere is needed, use oxygen-free copper.

Yellow Brass. This material should not be used where the sections are so heavy that a long time is required to bring the parts to brazing temperature. Zinc vaporizes from yellow brass when it is held at brazing temperatures for long periods of time. This condition is aggravated even more in the free-machining brasses containing small percentages of lead. Since this material is hot short, it should not be stressed during the brazing cycle.

Red Brass. The same as for yellow brass. Not quite so critical as there is less zinc present.

Phosphor Bronze 5%. Phosphor bronze will anneal or soften at brazing temperatures. The more rapidly the joint can be brazed the smaller the annealed area.

Beryllium Copper. Beryllium copper fully heat-treated will soften under ordinary brazing conditions. By silver plating beryllium copper the base metal can be brazed without a flux when using alloy B2.

Nickel. The softening of nickel can be kept to a minimum by brazing as rapidly as possible and by keeping brazing temperatures below approximately 1382 F.

Monel. This material is subject to intergranular attack by silver brazing alloys even if under low stress at temperatures above approximately 1382 F. The softening of monel can be kept to a minimum by brazing as rapidly as possible and by keeping brazing temperatures below approximately 1382 F.

Inconel. The softening of inconel can be kept to a minimum by brazing as rapidly as possible and by keeping brazing temperatures below approximately 1382 F.

Electrical Conductivity Grade Aluminum. Any method used to bring this grade of aluminum to brazing temperature will cause annealing of the base metal.

Aluminum Alloys 2S, 3S, 53S, 61S, *and* 63S. The 2S or 3S alloys will be annealed when heated to brazing temperature. 53S, 61S, or 63S will be annealed when heated to brazing temperature but can be hardened and aged to temper. Brazing sheets having a core of 3S or J51S may be used for special application.

TABLE 4. ASTM-AWS BRAZING ALLOYS CORRESPONDING TO FILLER METALS IN TABLE 3

Filler Metal (Table 3)	Type	ASTM-AWS Nos. (Table 1)
A	Copper	BCu
B1	Silver alloy	BAg-1
B2	Silver alloy	*
B3	Silver alloy	BAg-3
B4	Silver alloy	BAg-6 or 8
B5	Silver alloy	BAg-2
C	Aluminum alloys	BAlsi-2 or 4

* 15% silver, 80% copper, 5% phosphorus. Not an ASTM-AWS specification.

Applications of Brazing Alloys (ASTM-AWS)

BAlSi (Aluminum-silicon). Brazing filler metals of the BAlSi classifications (ASTM-AWS specifications) are used for joining aluminum and aluminum alloys: 2S, 3S, 4S, B50S, J51S, 53S, 61S, 62S, 63S, and cast alloys A612 and C612. The BAlSi-1, -2, and -3 filler metals are most suitable for brazing with the furnace and dip processes; the BAlSi-4 is most suitable for the torch brazing process but can be used with the furnace and dip processes. These filler metals are generally used with lap-type rather than butt-type joints.

Clearances of 0.006 to 0.010 in. are common for laps less than ¼ in. long; clearances up to 0.025 in. are used for longer laps. Fluxing is essential for all processes.

BAlSi-1 brazing filler metal is a general-purpose metal.

BAlSi-2 brazing filler metal is available only as coating on 3S and J51S aluminum-alloy core sheet, coated on one or both sides.

BAlSi-4 brazing filler metal is a general-purpose metal used in some cases for its relatively high corrosion-resistant properties.

BCuP (Copper-phosphorus). BCuP classifications are used primarily for joining copper and copper alloys with some limited use on silver, tungsten, and molybdenum. Because of the phosphorus content, these filler metals should not be used on ferrous metals. Their use on alloys containing more than 10% nickel should also be avoided. They are suitable for all brazing processes. Lap- and tee-type joints are generally used. These filler metals have self-fluxing properties when used on copper; however, a flux is recommended when used on all other metals, including copper alloys. Joints made with these filler metals on copper have excellent electrical and thermal conductivities. Color after brazing is light gray. Immersion in 10% sulfuric acid will restore the copper color.

BCuP-1 brazing filler metal is used primarily for preplacing in the joint and is suited particularly for resistance brazing and some furnace brazing applications.

BCu-2 filler metal is extremely fluid at brazing temperatures and will penetrate joints with very little clearance. Best results are obtained with clearances of 0.001 to 0.003 in.

BCuP-3 filler metal is also extremely fluid at brazing temperatures. Joint clearances of 0.002 to 0.005 in. are recommended.

BCuP-4 filler metal is best used with joint clearances of 0.001 to 0.003 in.

BCuP-5 filler metal is particularly suitable for use where very close fits cannot be held. Joint clearances of 0.003 to 0.005 in. are recommended.

BAg (Silver). BAg classifications are used for joining all ferrous and nonferrous metals except aluminum, magnesium, titanium, and some other metals melting below 1500 F. They are used with all brazing processes. They are generally rapid melting and free flowing. Lap-type joints are recommended, but butt-type joints may be used. Clearances should be between 0.002 to 0.005 in. Flux is generally required.

BAg-1 brazing filler metal is free-flowing and also has other qualities making it well suited for general-purpose work. The color after brazing is light yellow.

BAg-2 filler metal is, like BAg-1, free-flowing and suited for general-purpose work. Its broader melting range is helpful where clearances are not uniform, but, unless heating is rapid, care must be taken that the lower melting constituents do not separate out by liquation. Color after brazing is a light yellow.

BAg-3 filler metal is used primarily for joining carbide tool tips to tool shanks, because it wets the carbide well. It has a wide melting range in which solid and liquid do not tend to separate excessively. This makes it a good metal for bridging gaps or forming fillets. It has generally good corrosion-resistant properties but is not free-flowing to the degree BAg-1 and -2 are. Color after brazing is whitish yellow.

BAg-4 filler metal is, like BAg-3, used extensively for carbide tip brazing. It is freer flowing than BAg-3. The color after brazing is light yellow.

BAb-5 and -6 filler metals are used particularly for brazing in the electrical industry.

BAg-7 filler metal is a general-purpose low-melting metal used particularly for furnace brazing. Its whitish color after brazing makes it suitable for joining whitish metals like stainless steel. It is much used by the dairy and food industries.

BAg-8 filler metal is used primarily in the assembling of electronic and vacuum tubes.

BAg-9, -10, and -11 filler metals are used particularly for joining sterling silver. These three metals have different brazing temperatures and so can be used for step brazing of consecutive joints. Color after brazing is whitish.

BCu and BCuZn (Copper and Copper-zinc). Brazing filler metals of the BCu and BCuZn classifications are used for joining various ferrous and nonferrous metals by various brazing processes. Lap- and butt-type joints are commonly used.

BCu brazing filler metal is used for joining ferrous metals, nickel, and copper-nickel alloys. It is very free-flowing. It is much employed in furnace brazing, using a hydrogen or dissociated ammonia atmosphere and generally no flux. On metals that have constituents with difficult-to-reduce oxides (chromium, manganese, silicon, titanium, vanadium, aluminum, and zinc), a flux may be required.

BCuZn-1, -2, -3, -4, and -5 brazing filler metals are used on steels, copper, copper alloys, nickel, nickel alloys, and stainless steel. They are used with the torch, furnace, and induction processes. Fluxing is required. A borax–boric acid flux is common. Joint clearances from 0.002 to 0.005 in. are suitable. With the BCuZn filler metals, overheating must be guarded against since voids may be formed in the joint by entrapped zinc vapors. Corrosion resistance of BCuZn filler metals is generally inadequate for joining copper, silicon-bronze, copper-nickel, or stainless steel.

BCuZn-6 and -7 brazing filler metals (called white brasses, formerly called nickel silvers) are used with steels, nickel, and nickel alloys. They can be used with all brazing processes.

BMg (Magnesium). Brazing filler metal of the BMg classification is used for joining Ml magnesium-base metal. It is used most with the torch, furnace, and dip brazing processes and also somewhat with the other brazing processes, except resistance. Heating must be closely controlled to prevent melting of the base metal. Lap- and tee-type joints are most common. A flux is required with all processes. Clearances from 0.004 to 0.010 in. are best for most applications. Corrosion resistance is good, if the flux is removed after brazing. For furnace brazing, this filler metal is supplied with a small amount of beryllium, which prevents possible ignition.

BNiCr and BAgMn (Heat-resistant). Brazing filler metal of the BNiCr heat-resistant classification is used chiefly for joining stainless steels and high-nickel alloys to be used at elevated temperatures, as in jet engines. It also can be used on carbon- and low-alloy steels. BNiCr filler metal retains its heat-resistant properties at temperatures up to 2000 F. It has excellent corrosion-resistant properties. It is used primarily for brazing in a controlled dry-hydrogen-atmosphere furnace, where no flux is required. It can be used in a controlled standard-atmosphere furnace but then flux is required.

Brazing filler metal of the BAgMn heat-resistant classification is used chiefly to join stainless steel and high-nickel alloys. It does not have the high-temperature strength of the BNiCr but has good strength in the 500 to 900 F range. Furnace brazing with a reducing atmosphere is most commonly used. Other brazing processes can be used. However, these processes require a flux, and results are usually not so satisfactory.

Applications of Brazing Processes

Resistance Brazing. A resistance welder can be used, employing the heat generated by (1) the resistance of the parts or (2) the resistance of carbon blocks transmitted to the joint. The first method provides pressure follow-up; the second is less costly.

ADVANTAGES

1. Rapid localized heating is possible.
2. Work may be set up for a fairly high rate of production.
3. This method of brazing is ideally suited for use with low-melting silver-alloy filler metals. It is especially applicable for use with self-fluxing filler metals on nonferrous base metals such as copper and phosphor bronze.

LIMITATIONS

1. Accurate temperature control is sometimes difficult.
2. Localized heating may cause some distortion of the work.
3. The size of an assembly that can be brazed is limited.

Induction Brazing. Heat is obtained by the resistance of the work to the flow of induced electric current.

ADVANTAGES

1. Good control of heat, with automatic timing, is possible.
2. Increased output is available through extremely fast heating.
3. Heating of the base material can be localized, preventing overheating of parts of the assembly other than those adjacent to the joint.
4. Versatility of induction heating permits low production, using single heating coils, or high production, using multiple sets of coils.

LIMITATIONS

1. Localized heating may cause some distortion of the work.
2. Brazing with high-melting filler metals is not practical at present for high production work.

Furnace Brazing. Parts that can be assembled with preplaced alloy can be run in a furnace with or without controlled atmosphere. Use controlled atmosphere for parts that must be free of oxide or scale. Those parts requiring pickling and plating can be run without special atmosphere at expense of flux and a poorer joint.

ADVANTAGES

1. Uniform heating of parts, with minimum distortion.
2. No residual stresses left in work.
3. Good temperature control is possible.

4. Good control of the surface condition (carburized, decarburized) of the parts is possible, depending upon the temperature and type of atmosphere used.
5. A wide range of filler metals is possible with the furnace brazing process.
6. Furnace brazing is ideal for mass production.

LIMITATIONS

1. Localized heating is not possible.
2. Parts joined by furnace brazing must be self-positioning, or jigs or fixtures must be used.

Torch Brazing. Four flames are available: neutral oxyacetylene, natural gas and oxygen, premixed gas, and natural gas and air.

ADVANTAGES

1. Cost of equipment is considerably lower than for other brazing processes.
2. Heating can easily be localized.
3. Torch setups can be made which will allow use of turntable and conveyor production systems.

LIMITATIONS

1. This brazing process usually requires skilled operators.
2. Torch brazing can be used only with low- or medium-temperature brazing filler metal.
3. Localized heating may cause distortion.
4. The work will oxidize quite rapidly.

Dip Brazing (Metal Bath). Metal-bath dip brazing is less widely used than torch, furnace, induction, or resistance brazing, primarily because of the necessity of special joint design to limit "drag-out" of the filler metal.

ADVANTAGES

1. Rapid localized heating of the joint area is possible.
2. Accurate temperature control is provided.

LIMITATIONS

1. Parts must be so designed that the "drag-out" of filler metal is held to a minimum.
2. Only certain filler metals can be used.
3. The length of time of parts in the metal bath is critical.
4. The danger of explosions exists if wet parts are submerged in the metal bath.
5. Fixtures are usually required.

Dip Brazing (Chemical Bath). Dip brazing in neutral salts is possible if the brazing alloy can be replaced.

ADVANTAGES

1. Flux is not required for most applications.
2. Most of the available filler metals may be used in this process. Both high- and low-temperature alloys are included.
3. Single or multiple joints may be brazed on one assembly and will depend upon the design.

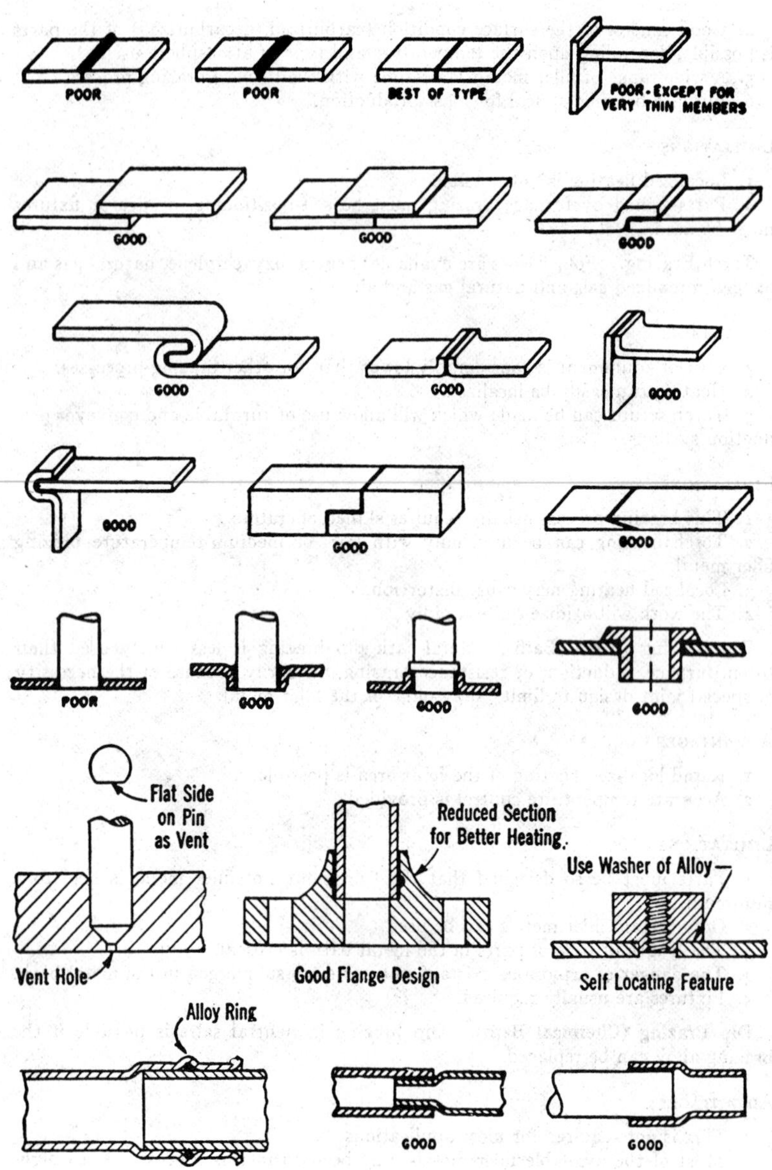

FIG. 1. Lap joints are always preferred to butt joints because of the greater strength obtained in brazed assemblies. In some joints, the alloy may be preplaced, thereby giving a better control of cost.

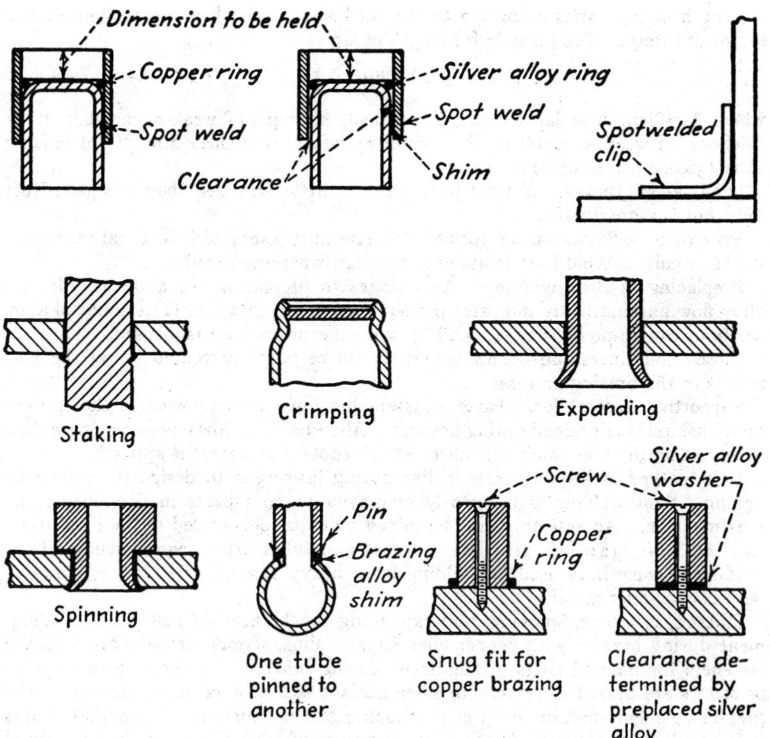

FIG. 2. Support of the parts during brazing is important to making a good joint. A spot weld, pin, screw, or staking is used.

LIMITATIONS

1. After brazing, the parts usually must be washed in order to remove the salts.
2. Excessive oxidation may occur upon cooling.
3. The danger of an explosion may exist if wet work is submerged in the bath.
4. Jigs or fixtures are usually required.
5. The chemical bath may etch or pit the base metal unless proper control of time and temperature is maintained.

Chemical-bath dip brazing has been used successfully with the brazing flux as the chemical bath. It has been used to braze steel parts with copper-filler metal, etc.

Design of Joints for Brazing

Three types of joints are used in brazing: lap or shear, scarf, and butt. Each of these may be used with flat, round, or tubular members.

LAP JOINTS. Lap or shear joints are the most reliable, simple, and easy to braze. Joints should be designed for shear wherever possible. Moreover, lap joints give a better opportunity to support the members properly and maintain correct clearances.

Length of lap varies according to the thickness, strength of metals joined, and factor of safety. The formula for length of lap is

$$X = YTW \div L$$

where X = length of lap (shear), T = tensile strength of weakest member, W = thickness of weakest member, Y = factor of safety, L = shear strength of brazing alloy (30,000 psi is common value).

SCARF-BUTT JOINTS. A scarf-butt joint is little stronger than a square-butt joint but is more ductile.

STRAIGHT OR SQUARE-BUTT JOINTS. Square-butt joints add electrical resistance in the circuit. Avoid butt joints of any design wherever possible.

Preplacing of Brazing Alloy. Advantages to preplacing the alloy include: (1) alloy flow automatically indicates proper brazing temperature, (2) amount of alloy per joint can be predetermined, and (3) a homogeneous joint is insured.

Closed containers and hollow bodies should be properly vented to get the best results in the brazing process.

Supporting Assemblies. Parts or assemblies must be supported so that proper relationship is maintained during brazing. Alloy tends to flow downward more than in any other direction and will collect at low spots if an excess is applied.

Best method of keeping parts in line during brazing is to design them for self-jigging. Spot welding is frequently employed to hold parts in line during the brazing cycle. An assembly may be joined by clips spot-welded to the main parts and the joint brazed to give leaktightness and fulfill other requirements. Tack welding is sometimes used for holding large heavy assemblies in line for brazing with copper-filler metal.

Staking, crimping, spinning, and expanding can be used to hold parts in alignment during brazing with copper-filler metal. Pins, screws, or bolts are versatile fasteners for holding parts in alignment during brazing. In most cases the pins or screws are brazed in place. In some cases it is not practical to use any of the previously mentioned methods of positioning and supporting. When that occurs it is usually necessary to devise a jig or fixture which will do the job. If metal fixtures or jigs are used, their expansion and contraction during the brazing cycle must be considered.

Brazing Fluxes

With brass or bronze rod:

Boric acid, 83%; borax, 17%.

For aluminum:

Lithium, not less than 1.75%.
Fluorine, not less than 5%.
Potassium to sodium: ratio of 1:2.
Mixed with distilled water to a thin paste.

For cast iron:

Boric acid, 85%.
Borax, 10%.
Sodium fluoride, 5%.

Low-temperature flux:

Potassium tetraborate, 30%.
Boric acid, 30%.
Potassium acid fluoride, 40%.

Dilute with 20% water for torch or induction brazing; make thin consistency for resistance welding.

Flux remaining after brazing must be removed either with water or acid.

Bibliography

Proposed Tentative Specifications for Brazing Filler Metals, *Welding J.*, August, 1952.
VanNatten, W. J., Design Data for Brazing, *Welding J.*, 1951.

Section 24

SOLDERING AND SOFT SOLDERS

SECTION 24

SOLDERING AND SOFT SOLDERS

Definition. Soft solders are alloys of tin-lead, tin-lead-antimony, tin-lead-silver-antimony, tin-lead-cadmium, and silver-lead that are fusible at temperatures below 700 F. The hard solders (brazing alloys and silver solders) are fusible at temperatures above 1200 F.

Materials Soldered. Metals to be joined by soldering are not heated to their melting points. Hence the melting point of solder must be lower than that of the metals being joined. Relative solderability of base metals in descending order is: silver, copper, brass or bronze, tin, lead, german silver, zinc, cadmium, iron, steel, and stainless steel.

Solder Compositions. Tables 1 to 4 give the compositions of common alloys, according to ASTM, SAE, and federal and military identification systems. Manufacturers may supply other compositions as well to suit particular needs, notably to conserve tin. For military radio and radar assemblies, the 50 tin–50 lead solder is undoubtedly desirable in order to make highly reliable joints. Commercial users have leaned toward 40 tin–60 lead solders, but current solders containing antimony and silver and only 20% tin are found to have excellent to superior wetting power, which controls joint-forming properties.

Melting Characteristics. Tin-lead solders usually have three states: (1) solid, (2) semiliquid, and (3) liquid. The alloys have solidus and liquidus points that are lower than the melting points of the metals being joined, and these points vary with the composition of the solder (see Table 5).

It is important to distinguish between the solidus and liquidus temperatures of a specific solder. For any solder except the eutectic alloy of 62 tin–38 lead, there is a temperature range in which the solder is mushy. To wet the surfaces of the joint and produce a strong bond, it is essential to raise the temperature of both the work and the solder to above the liquidus temperature. Otherwise, alloying with the joint metals will not occur. Once this temperature has been reached, the joint must not be disturbed while it is cooling down through the semiliquid range.

PHASE DIAGRAM. Melting ranges of tin-lead alloys are given in Fig. 1. Lead melts at 621 F (left boundary line). Tin melts at 450 F (right boundary). By adding tin to lead, its melting point is progressively lowered, until the eutectic point is reached with 63% lead content. This alloy of pure lead and pure tin is the only one that has a definite melting point; that is, the temperature remains constant during melting or freezing. All other compositions of these two metals melt or freeze through a range of temperatures. *Solidus* is the temperature at which melting begins or freezing ends, and *liquidus* is the temperature at which melting ends or freezing begins.

Antimony can be added to solder to replace tin and lower the melting point or, at least, to contract the freezing range. Thus low-tin solders can be made to approximate high-tin solders in freezing range. Antimony raises the tensile strength, shear strength, hardness, and density but impairs the electrical conductivity, elongation, and impact strength of the solder composition.

Low-tin Solders. The popular 20 tin–80 lead solder has a solidus point of 361 F and a liquidus point of 531 F. Hence it is essential to heat the joint higher than for a tin-rich solder. However, the heat input is not directly in proportion to the higher temperature involved, and it is unlikely that it will be necessary to use heating equipment different from that employed for tin-rich solders, except possibly when the joint is very large. Nevertheless, good soldering technique is necessary with low-tin solders.

Effect of Coatings. Coatings found on work to be soldered are: (1) electroplate; (2) grease; (3) insulation; (4) enamel, paint, varnish; (5) oxides, sulfides, carbonates; (6) dirt and smuts.

Electroplated tin coatings are difficult to solder and very often produce poor joint characteristics even if the solder joint holds. On the other hand, a tinned coating is excellent and promotes good soldering. Nickel and chromium plate are detrimental to good soldering. Cadmium plate may dissolve in the solder and cause various kinds of trouble.

Insulation must be stripped and finishes removed from wire before soldering, because such substances cannot be removed by fluxing or burning off. In stripping insulation, make sure that it is removed to a point where heat will not cause sulfurous fumes, as from rubber, because such fumes attack the metal and interfere with good joints.

Oxides, sulfides, etc., can be removed with suitable fluxes. Grease, dirt, and smuts should be wiped off carefully, using materials that will not tend to spread them.

Fluxes. Functions of a flux are: (1) to clean the work surfaces by dissolving oxides, etc., (2) to prevent further oxidation when the joint is brought to the solder-

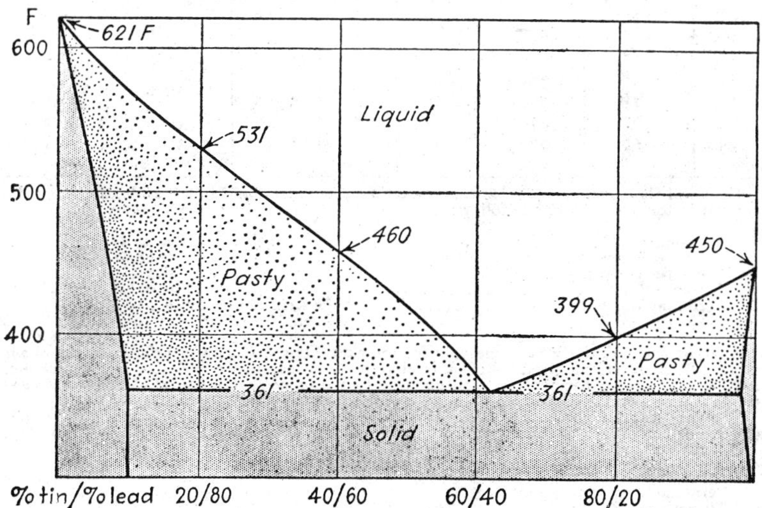

FIG. 1. Constitution diagram for tin-lead solders.

TABLE 1. ASTM SPECIFICATIONS FOR SOFT-SOLDER METAL (B32-1949)

Alloy Grade	Tin, Desired, %	Lead, Nominal, %	Antimony, %			Silver, %		
			Min	Desired	Max	Min	Desired	Max
70A	70	30			0.12			
70B	70	30			0.50			
60A	60	40			0.12			
60B	60	40			0.50			
50A	50	50			0.12			
50B	50	50			0.50			
45A	45	55			0.12			
45B	45	55			0.50			
40A	40	60			0.12			
40B	40	60			0.50			
40C	40	58	1.8	2.0	2.4			
35A	35	65			0.25			
35B	35	65			0.50			
35C	35	63.2	1.6	1.8	2.0			
30A	30	70			0.25			
30B	30	70			0.50			
30C	30	68.4	1.4	1.6	1.8			
25A	25	75			0.25			
25B	25	75			0.50			
25C	25	73.7	1.1	1.3	1.5			
20B	20	80			0.50			
20C	20	79	0.8	1.0	1.2			
15B	15	85			0.50			
10B	10	90			0.50			
5A	5*	95			0.12			
5B	5*	95			0.50			
2A	2†	98			0.12			
2B	2†	98			0.50			
2.5S	0‡	97.5			0.40	2.3	2.5	2.7
1.5S	1§	97.5			0.40	1.3	1.5	1.7

Chemical Composition: For elements other than those mentioned in the table, the maximum content in the alloy shall be as follows:

Bismuth	0.25 %
Copper alloy grades 70A to 2B, incl	0.08 %
Copper alloy grades 2.5S and 1.5S	0.3 %
Iron	0.02 %
Aluminum, zinc	Each shall not exceed 0.005 % when determined on a 10-g sample

Analysis shall regularly be made only for the elements specifically mentioned in the above table. If, however, the presence of other elements is suspected, or indicated in the course of routine analysis, further analysis shall be made to determine that the total of these other elements is not in excess of 0.08 %.

The chemical requirements of SAE Specifications 1A, 2A, 2B, 3A, 3B, 4A, 4B, 5A, 5B, 6A, and E—07 conform substantially to the requirements for alloy grade 45B, 40B, 40C, 30B, 30C, 25B, 25C, 20B, 20C, 15B, and 2.5S, respectively.

* Permissible tin range, 4.5 to 5.5 %.
† Permissible tin range, 1.5 to 2.5 %.
‡ Tin maximum, 0.25 %.
§ Permissible tin range, 0.75 to 1.25 %.

TABLE 2. SAE STANDARD SOLDERS

% Composition*

Composition†	Tin‡ (Range)	Antimony, Max	Silver, Max	Copper, Max	Iron, Max	Bismuth, Max	Zinc, Max	Aluminum, Max	Others, Max	Lead, Max
Sn70	69.5 to 71.5	0.50		0.08	0.02	0.25	0.005	0.005	0.08 total	Remainder
Sn60	59.5 to 61.5	0.50		0.08	0.02	0.25	0.005	0.005	0.08 total	Remainder
Sn50	49.5 to 51.5	0.50		0.08	0.02	0.25	0.005	0.005	0.08 total	Remainder
Sn40	39.5 to 41.5	0.50		0.08	0.02	0.25	0.005	0.005	0.08 total	Remainder
Sn35	34.5 to 36.5	1.6 to 2.0		0.08	0.02	0.25	0.005	0.005	0.08 total	Remainder
Sn30	29.5 to 31.5	1.4 to 1.8		0.08	0.02	0.25	0.005	0.005	0.08 total	Remainder
Sn20	19.5 to 21.5	0.8 to 1.2		0.08	0.02	0.25	0.005	0.005	0.08 total	Remainder
Ag2.5		0.40	2.3 to 2.7	0.30	0.02	0.25	0.005	0.005	0.30 total	Remainder
Ag5.5		0.40	5.0 to 6.0	0.30	0.02	0.25	0.005	0.005	0.30 total	Remainder
Sb5	94.0 min	4.0 to 6.0		0.08	0.08	Cadmium 0.03	0.03	0.03	0.30 total	0.2

* Specified percentage is for solder metal only. In flux-cored wire solder, the weight of the flux shall be subtracted from the total weight to obtain the weight of the solder metal.

† Tin-lead solders (prefixed by "Sn") may be furnished as flux-cored wire as well as plain wire and other forms. The weight of the flux in rosin-flux-cored wire shall not exceed 4 % of the total weight. The weight of the flux in chloride-flux-cored wire shall not exceed 6 % of the total weight.

‡ When tin-lead solders (prefixed by "Sn") are furnished as flux-cored wire, the minimum permissible tin content shall be 0.5 % less than the minimum values specified in the table.

TABLE 3. FEDERAL SPECIFICATIONS QQ-S-571b FOR SOFT SOLDERS

SAE No.	Tin	Antimony	Lead	Similar Specifications ASTM B32
1A	45.0, −1.0	0.4 max	Remainder	Alloy 45B
1B	43.0, +0.5	1.5–2.00	Remainder	
2A	40.0, −1.0	0.4 max	Remainder	Alloy 40B
2B	38.0, +0.5	1.5–2.00	Remainder	
3A	30.0, −1.0	0.4 max	Remainder	Alloy 30B
3B	28.0, +0.5	1.5–2.00	Remainder	
4A	25.0, −1.0	0.4 max	Remainder	Alloy 25B
4B	25.0, −1.0	1.25–1.75	Remainder	
4A	20.0, −1.0	0.4 max	Remainder	Alloy 20B
5B	20.0, −1.0	1.25–1.75	Remainder	
6A	15.0, −1.0	0.4 max	Remainder	Alloy 15B
6B	15.0, −1.0	As specified*	Remainder	

Impurities, %, Max

Bismuth	0.25	Zinc	0.005
Copper†	0.08	Aluminum	0.005
Iron	0.02	Other elements, total	0.08

Class B solders are wiping type.

The choice of the type and grade of solder for any specified purpose depends on the material in connection with which it is to be used and the method of applying. For galvanized iron and zinc, only class A should be used. Class A solder should be furnished under this specification unless otherwise specified.

It is recommended that the grade of solder metal be selected that contains the least amount of tin required to give suitable flowing and adhesive qualities for the work in hand.

* Max, 2.75 %.

† In dipping solders 0.5 % max copper is permissible because of pickup in bath.

TABLE 4. MILITARY SPECIFICATION JAN-S-627

Composition	Tin, %	Lead, %	Bismuth, %	Cadmium, %	Other Elements, Total %	Freezing-point range, F
A	42–44	41–45	13–15			330–285
B	16–18	15–17	66–68			305–200
C	10.8–13.2	25.2–30.8	45–55	9–11	1.00	165–155
D	36–44	18–22	36–44		1.00	220–210

An analysis of each lot shall be furnished by the contractor, showing the percentage of each element specified.

Solder covered by this specification is intended for application where a melting-point range of 155 to 330 F is required.

TABLE 5. PROPERTIES OF SOFT-SOLDER ALLOYS

Nominal Composition, %			Sp. Gr.*	Melting Ranges		Uses
Tin	Lead	Antimony		Solidus, F	Liquidus, F	
Tin-Lead Alloys						
70	30		8.32	361	378	For coating metals
60	40		8.65	361	374	"Fine solder," for general purposes, but particularly where the temperature requirements are critical
50	50		8.85	361	421	For general purposes. Most popular of all
45	55		8.97	361	441	For automobile radiator cores and roofing seams
40	60		9.30	361	460	Wiping solder for joining lead pipes and cable sheaths. For automobile radiator cores and heating units
35	65		9.50	361	477	General-purpose and wiping solder
30	70		9.70	361	491	For machine and torch soldering
25	75		10.00	361	511	For machine and torch soldering
20	80		10.20	361	531	For coating and joining metals. For filling dents or seams in automobile bodies
15	85		10.50	440†	550	For coating and joining metals
10	90		10.80	514†	570	For coating and joining metals
5	95		11.30	518	594	For coating and joining metals
Tin-Lead-Antimony Alloys						
40	58	2	9.23	365	448	Same uses as (50-50) tin-lead but not recommended for use on galvanized iron
35	63.2	1.8	9.44	365	470	For wiping and all uses except on galvanized iron
30	68.4	1.6	9.65	364	482	For torch soldering or machine soldering except on galvanized iron
25	73.7	1.3	9.96	364	504	For torch and machine soldering, except on galvanized iron
20	79	1	10.17	363	517	For machine soldering and coating of metals, dipping, and like uses but not recommended for use on galvanized iron
Silver-Lead Alloys						
Tin	Lead	Silver				
0	97.5	2.5	11.35	579	579	For use on copper, brass, and similar metals with torch heating. Not recommended in humid environments, because of its known susceptibility to corrosion
1	97.5	1.5	11.28	588	588	For use on copper, brass, and similar metals with torch heating

* The specific gravity multiplied by 0.0361 equals the density in pounds per cubic inch.

† For some engineering-design purposes it is well to consider these alloys as having practically no mechanical strength at 361 F.

ing temperature, (3) to reduce surface tension so that the molten solder will wet the work surfaces, and (4) to assist alloying action of the solder with the work surfaces. Further requirements of a flux are: (1) it should not decompose or volatilize at soldering temperature; (2) any residue must not be corrosive, electrically conductive, or prone to pick up moisture.

Design of Joints. Good fit-up is required to make sound soldered joints. For best strength, the solder thickness should be between 0.003 and 0.005 in. Nothing is gained by using more solder than is required for the joint. It is true that a heavy fillet may add some shear resistance, but too often a fillet covers up a partially filled joint that has no real strength.

How a Solder Joint Is Made. In making a joint, the solder goes through four stages: (1) it melts, (2) it wets the joint and flows, (3) it is drawn into the joint by capillary attraction, and (4) it cools and solidifies. Precautions are:

If a soldering iron is used, the bit should be considerably hotter than the minimum working temperature for proper melting. Thus a number of joints can be made before reheating. The degree of heating depends on the work. When working in confined spaces, as in radio or electrical assemblies, it is undesirable to employ excessive soldering-iron temperature because of the chance of spoiling insulation. For such work a 50-50 solder is best.

Capillarity usually decreases when the solder has less than 40% tin. If a tight seam must be soldered to some depth, use a 50-50 alloy.

Cooling the joint to a temperature well below the solidus temperature is essential for a good joint. There must be no movement of the joint members during the freezing process, from either mechanical disturbance or unequal cooling contraction of the joint itself or the fixture.

Soldering Temperature. If heat input can be controlled accurately, it is desirable to solder at around 535 F, regardless of solder alloy or the material. Preliminary tinning assists in making a sound joint.

Equipment for Soldering. Several kinds of tools and equipment are used for soldering: soldering irons, carbon-electrode soldering tools, torches, baths, and induction-heating coils.

Soldering irons are copper bits, heated over a plumber's torch or furnace. A stubby iron with a short point is preferred. Heat the bit behind the point. It is too hot if dull red when shielded from the light. Heat to possibly 100 F higher than temperature required to melt stick solder (not flux cored) of the type used.

Electrical soldering irons are made in a variety of wattages for light to heavy work, and with replaceable tips. Heat is drawn at irregular intervals from the iron by the work, while the iron itself receives current at a constant rate. If it overheats, it may be necessary to incorporate a suitable means to control current flow.

Gas-air irons have a tube to feed gas to a chamber and orifices for air to support combustion. Such irons have been used satisfactorily for heavy work like soldering electrical connections and commutators.

Acetylene torches are employed for making heavy copper connections, and small blowpipes have been used for delicate work.

Carbon-electrode units use a low-voltage high-current amperage. The carbon electrode is touched to the work surface and it quickly raises the temperature.

Solder baths are used for assembly of many small parts at once, or assembly of radiators and the like.

Induction heating is also applied to soldering operations, and normally on a mass-production basis. Coil design is important, for it is essential to raise the temperature of the parts evenly.

Part 4

MATERIALS

Section 25

STEELS AND IRONS

SECTION 25

STEELS AND IRONS

Steels may be classified into five categories: (1) carbon, (2) alloy, (3) high-strength low-alloy, (4) stainless, and (5) tool and die. Prior to 1941, thousands of compositions of steel were in use, but most of them had little significance. A simplification program undertaken by the American Iron & Steel Institute and the Society of Automotive Engineers reduced the list, except tool and die steels, to around 800 items. Standard and tentative standard steels in the first four categories above are now recognized under AISI-SAE numbers. The listing given, starting on page 25-6, closely corresponds to these numbers but is taken from Federal Standard 66-1954. The federal standard has the advantage that availability of carbon steel is recognized. Similar information is not obtainable for alloy steels, because the availability and usage requirements are subject to greater change. However, it should be said that the list of alloy steels presumably supplies far greater choice than is needed, and that the user will be advised to ascertain the state of warehouse and distributor stocks before he writes steel numbers into product specifications.

STANDARD STEELS

The simplified list of carbon, alloy, and stainless and heat-resisting steel compositions is recognized for bars, semifinished products up to and including 200 sq in. in cross-sectional area, seamless tubular products, wire, and alloy steel plates.

Alloy steels ordered to hardenability requirements, identified as H-steels, are also standard steels.

A system of symbols is used to identify the grades of standard steels. Numbers are used to indicate grades of steel.

Numerical Designations of Grades

Carbon Steels. A four-numeral series is used to designate graduations of chemical composition of carbon steel, the last two numbers of which indicate the *approximate middle* of the carbon range. For example, in the grade designation 1035, 35 represents a carbon range of 0.32 to 0.38% (see Table 1).

It is necessary, however, to deviate from this rule and to interpolate numbers in the case of some carbon ranges; and for variations in manganese, phosphorus, or sulfur with the same carbon range.

The first two digits of the four-numeral series of the various grades of carbon steel and their meanings are as follows:

10xx or MT10xx..... Nonresulfurized, basic open hearth, and acid bessemer carbon steel grades; MT refers to mechanical tubing grades

11xx............... Resulfurized basic open hearth and acid bessemer carbon steel grades

12xx............... Rephosphorized and resulfurized basic open hearth carbon steel grades

Alloy Steels (Excluding Plates). A four-numeral series designates most of the alloy steels specified to chemical composition ranges (Tables 2 and 3). Five numerals are used to designate certain types of alloy steels.

As with carbon steels, the last two digits of the four-numeral series are intended to indicate the *approximate middle* of the carbon range.

The first two digits of the four-numeral series for the various grades of alloy steel and their meanings are as follows:

Series Designation	*Types (Av Content)*
13xx	Manganese 1.75%
14Bxx	Carbon steel, boron treated
23xx	Nickel 3.50%
25xx	Nickel 5.00%
31xx	Nickel 1.25% or 0.80%
33xx	Nickel 3.50% chromium 1.55%
40xx	Molybdenum 0.25%
41xx	Chromium 0.95% molybdenum 0.20%
43xx	Nickel 1.80% chromium 0.50 or 0.80% molybdenum 0.25%
46xx	Nickel 1.80% molybdenum 0.25%
47xx	Nickel 1.05% chromium 0.45% molybdenum 0.20%
48xx	Nickel 3.50% molybdenum 0.25%
50xx	Chromium 0.30 or 0.60%
51xx	Chromium 0.80, 0.95, or 1.05%
50xxx, 51xxx, 52xxx	Carbon 1.00% chromium 0.50, 1.00, or 1.45%
61xx	Chromium 0.80 or 0.95% vanadium 0.10% or 0.15% min
81xx	Nickel 0.30% chromium 0.40% molybdenum 0.11%
86xx	Nickel 0.55% chromium 0.50% molybdenum 0.20%
87xx	Nickel 0.55% chromium 0.50% molybdenum 0.25%
92xx	Manganese 0.85% silicon 2.00%
93xx	Nickel 3.25% chromium 1.20% molybdenum 0.12%
94xx	Manganese 1.00% nickel 0.45% chromium 0.40% molybdenum 0.12%
97xx	Nickel 0.55% chromium 0.17% molybdenum 0.20%
98xx	Nickel 1.00% chromium 0.80% molybdenum 0.25%

H-steels. As a means of identifying steels specified to hardenability band limits (Table 4), the suffix letter H has been added to the conventional series number. It is quite important that purchasers use the suffix letter in specification requirements as there is no other means of determining when hardenability band limits apply.

The letter B between the second and third digits of the grade number indicates a boron steel (example, 94B17). See Tables 4 and 5.

The letters BV between the second and third digits of the grade number indicate a boron-vanadium steel (43BV14).

Alloy-steel Plates. Same as indicated above for 13xx, 23xx, 25xx, 31xx, 40xx, 41xx, 43xx, 46xx, 48xx, 51xx, 61xx, 86xx, 87xx, and 94xx. In addition, the following series designations apply only to alloy plates:

Series Designation	*Types (Av Composition)*
28xx	Nickel 9.00%
99xx	Nickel 1.15% chromium 0.50% molybdenum 0.25%

Stainless and Heat-resisting Steels. A system of numbers is used to identify stainless and heat-resisting steels by type and according to three general groups (see Table 19). In a three-numeral number, the first numeral indicates the group and the last two numerals indicate type. Modifications of types are indicated by suffix letters.

The meaning of these numbers is as follows:

Series Designation	
3xx or TP-3xx.......	Chromium-nickel steels; nonhardenable by heat-treatment, austenitic and nonmagnetic
4xx or TP-4xx.......	Chromium steels; hardenable by heat-treatment, martensitic and magnetic
4xxx or TP-4xxx.....	Chromium steels; nonhardenable by heat-treatment, ferritic and magnetic
5xx................	Chromium steels; low-chromium, heat-resisting.

NOTE: TP refers to tubular products grades.

Steels Other Than Standard. It is recognized that compositions other than those listed as standard steels may infrequently be needed for carbon and alloy bars, semifinished products, seamless tubular products, alloy plates, and wire. Wider ranges of chemistry are in most cases necessary because of infrequent production and limited outlets for off-analysis heats.

Types of Steel

Control of the amount of gas evolved during solidification in the mold determines the type of steel. Silicon, aluminum, etc., are used to deoxidize molten steel and stop the evolution of gas formed by combination of carbon and oxygen. If practically no gas is evolved, the steel is termed "killed" because it lies quietly. Increasing degrees of gas evolution result in semikilled, capped, or rimmed steels.

Some carbon steels and high-strength low-alloy steels can be supplied in all four types; alloy steels and stainless and heat-resisting steels are normally manufactured only as killed steel.

Killed steels, because of greater uniformity in chemical composition and soundness, are used for forging, carburizing, heat-treating, and other applications.

Semikilled steels have characteristics intermediate between those of killed and rimmed steels, and they are used principally for general structural applications.

Capped steels have characteristics similar to those of rimmed steels but to a degree intermediate between those of rimmed and semikilled steels.

Rimmed steels normally have a carbon content less than 0.25% and manganese content less than 0.60%. Satisfactorily rimmed steels do not retain any significant percentages of highly oxidizable elements such as aluminum, silicon, or titanium.

Rimmed steel, because of surface and other characteristics, is used advantageously for a large number of applications and is particularly adapted to cold-bending, cold-forming, and cold-heading applications.

Austenitic Grain Size

The grain size of carbon and alloy steels is understood to mean austenitic or inherent grain size. Austenitic grain size should be distinguished from ferritic grain size, which is the size of the grains in the as-rolled or as-forged condition with the exception of those steels which are austenitic at room temperature.

When steel is heated through the critical range (approximately 1350 to 1600 F for most steels, depending on the composition) transformation to austenite takes place. The austenite grains are extremely small when first formed but grow in size as the temperature above the critical range and, to a limited extent, the time are increased.

When temperatures are raised materially above the critical range, different steels show wide variations in grain size, depending on the chemical composition and the deoxidation practice used. Steels which are made using aluminum or other deoxidizers in carefully controlled amounts maintain a slow rate of grain growth at 1700 F, while heats finished with certain other elements, usually ferrosilicon, develop relatively large austenitic grain size at temperatures somewhat below 1700 F. Vanadium or titanium is not normally used for grain control unless specifically designated in the composition.

The McQuaid-Ehn test as described in ASTM Specification E19 is the usual criterion for the determination of austenitic grain size. An arbitrary set of eight grain-size charts is set up, numbered 1 to 8. Number 1 grain size is the coarsest and contains up to 1½ grains per sq in. of area examined at 100 diameters magnification. The finest is number 8, which shows 96 and over grains in the same area and the same magnification.

Because it is impractical to control austenitic grain size within narrow limits, killed steels may be ordered as either fine grain (number 5 or finer) or coarse grain (number 5 or coarser). Intermediate grain sizes such as 4 to 6 cannot be consistently obtained.

Limitations. Because of certain manufacturing conditions, the above grain-size distinctions cannot always be maintained (1) for sizes larger than 100 sq in. in cross-sectional area and (2) when magnetic-particle inspection requirements apply. In the latter case, negotiation of magnetic-particle inspection requirements between the purchaser and producer may be necessary, because grain-size control involves deoxidation practices which must be taken into consideration in establishing standards for magnetic-particle inspection.

Austenitic-grain-size control generally is not applicable to stainless and heat-resisting steels.

The following ultimate properties of carbon and alloy steels should be considered when grain size is to be specified:

Fine-grain steels do not harden so deeply and have less tendency to crack when quenched in heat-treating operations as compared with coarse-grain steels of identical analysis. Fine-grain steels exhibit greater toughness and shock resistance and they are usually to be preferred in all applications involving moving loads and high impact resistance. Practically all alloy steels and high-carbon carbon steels are made using fine-grain practice.

Coarse-grain steels exhibit definite machining superiority. For this reason, most resulfurized carbon steels are coarse-grain.

CARBON STEELS

Steel is considered to be carbon steel when no minimum content is specified or required for aluminum, boron, chromium, cobalt, columbium, molybdenum, nickel, titanium, tungsten, vanadium, or zirconium, or any other element added to obtain a desired alloying effect; when the specified minimum for copper does not exceed 0.40%; or when the maximum content specified for any of the following elements does not exceed the percentage noted: manganese 1.65, silicon 0.60, copper 0.60.

Ordinary carbon steels of the AISI-SAE grades (see Table 1) are intended for structural parts, not for use as tool and die steels. For the latter see Table 21.

Effects of Alloying Elements. The effect of a single alloying element depends upon the effects of other elements on the steel, and to evaluate a given composition,

TABLE 1. CARBON STEELS—BASIC OPEN HEARTH AND ACID BESSEMER*

Steel No.	Chemical Composition Ranges and Limits, %				Hot-rolled Bars	Cold-finished Bars	Semifinished Steel for Forging	Wire
	Carbon	Manganese	Phosphorus	Sulfur				
1005	0.06 max	0.35 max	0.040 max	0.050 max	x	..	..	y
1006	0.08 max	0.25–0.40	0.040 max	0.050 max	x	x	x	y
1008	0.10 max	0.25–0.50	0.040 max	0.050 max	X	x	X	y
1010	0.08–0.13	0.30–0.60	0.040 max	0.050 max	X	X	X	y
1011	0.08–0.13	0.60–0.90	0.040 max	0.050 max	x	..	x	
1012	0.10–0.15	0.30–0.60	0.040 max	0.050 max	X	x	X	y
1013	0.11–0.16	0.50–0.80	0.040 max	0.050 max	x	x	x	y
1015	0.13–0.18	0.30–0.60	0.040 max	0.050 max	X	X	X	y
1016	0.13–0.18	0.60–0.90	0.040 max	0.050 max	X	X	X	y
1017	0.15–0.20	0.30–0.60	0.040 max	0.050 max	X	x	X	y
1018	0.15–0.20	0.60–0.90	0.040 max	0.050 max	X	X	X	y
1019	0.15–0.20	0.70–1.00	0.040 max	0.050 max	X	X	x	y
1020	0.18–0.23	0.30–0.60	0.040 max	0.050 max	X	X	X	y
1021	0.18–0.23	0.60–0.90	0.040 max	0.050 max	X	x	X	y
1022	0.18–0.23	0.70–1.00	0.040 max	0.050 max	X	X	X	y
1023	0.20–0.25	0.30–0.60	0.040 max	0.050 max	X	x	X	y
1024	0.19–0.25	1.35–1.65	0.040 max	0.050 max	x	X	x	y
1025	0.22–0.28	0.30–0.60	0.040 max	0.050 max	X	X	X	y
1026	0.22–0.28	0.60–0.90	0.040 max	0.050 max	X	X	X	y
1027	0.22–0.29	1.20–1.50	0.040 max	0.050 max	x	x	x	y
1029	0.25–0.31	0.60–0.90	0.040 max	0.050 max	X	x	X	y
1030	0.28–0.34	0.60–0.90	0.040 max	0.050 max	X	X	X	y
1031	0.28–0.34	0.30–0.60	0.040 max	0.050 max	x	..	X	
1032	0.30–0.36	0.60–0.90	0.040 max	0.050 max	x	..	x	
1033	0.30–0.36	0.70–1.00	0.040 max	0.050 max	x	..	x	y

1034	0.32–0.38	0.50–0.80	0.040 max	0.050 max	x	..	..	y
1035	0.32–0.38	0.60–0.90	0.040 max	0.050 max	X	X	X	y
1036	0.30–0.37	1.20–1.50	0.040 max	0.050 max	x	x	x	y
1037	0.32–0.38	0.70–1.00	0.040 max	0.050 max	X	..	x	
1038	0.35–0.42	0.60–0.90	0.040 max	0.050 max	X	X	X	y
1039	0.37–0.44	0.70–1.00	0.040 max	0.050 max	X	..	..	y
1040	0.37–0.44	0.60–0.90	0.040 max	0.050 max	X	X	X	y
1041	0.36–0.44	1.35–1.65	0.040 max	0.050 max	x	x	x	y
1042	0.40–0.47	0.60–0.90	0.040 max	0.050 max	X	X	X	
1043	0.40–0.47	0.70–1.00	0.040 max	0.050 max	X	x	X	
1045	0.43–0.50	0.60–0.90	0.040 max	0.050 max	X	X	X	
1046	0.43–0.50	0.70–1.00	0.040 max	0.050 max	X	X	X	
1049	0.46–0.53	0.60–0.90	0.040 max	0.050 max	X	X	X	
1050	0.48–0.55	0.60–0.90	0.040 max	0.050 max	X	X	X	
1051	0.45–0.56	0.85–1.15	0.040 max	0.050 max	x	..	X	
1052	0.47–0.55	1.20–1.50	0.040 max	0.050 max	x	x	x	
1053	0.48–0.55	0.70–1.00	0.040 max	0.050 max	x	..	X	
1054	0.50–0.60	0.50–0.80	0.040 max	0.050 max	x			
1055	0.50–0.60	0.60–0.90	0.040 max	0.050 max	X	x	X	
1057	0.50–0.61	0.85–1.15	0.040 max	0.050 max	x	..	x	
1059	0.55–0.65	0.50–0.80	0.040 max	0.050 max	x			
1060	0.55–0.65	0.60–0.90	0.040 max	0.050 max	X	x	x	
1061	0.54–0.65	0.75–1.05	0.040 max	0.050 max	x	..	x	
1062	0.54–0.65	0.85–1.15	0.040 max	0.050 max	x	x	x	
1064	0.60–0.70	0.50–0.80	0.040 max	0.050 max	x	..	x	
1065	0.60–0.70	0.60–0.90	0.040 max	0.050 max	x	x	x	
1066	0.60–0.71	0.85–1.15	0.040 max	0.050 max	x	..	x	
1069	0.65–0.75	0.40–0.70	0.040 max	0.050 max	x	..	x	
1070	0.65–0.75	0.60–0.90	0.040 max	0.050 max	X	x	x	
1071	0.65–0.76	0.75–1.05	0.040 max	0.050 max	x			

TABLE 1. CARBON STEELS—BASIC OPEN HEARTH AND ACID BESSEMER* (*Continued*)

Steel No.	Chemical Composition Ranges and Limits, %				Hot-rolled Bars	Cold-finished Bars	Semifinished Steel for Forging	Wire
	Carbon	Manganese	Phosphorus	Sulfur				
1072	0.65–0.76	1.00–1.30	0.040 max	0.050 max	x			
1074	0.70–0.80	0.50–0.80	0.040 max	0.050 max	x			
1075	0.70–0.80	0.40–0.70	0.040 max	0.050 max	x	..	x	
1078	0.72–0.85	0.30–0.60	0.040 max	0.050 max	X			
1080	0.75–0.88	0.60–0.90	0.040 max	0.050 max	X	x	X	
1084	0.80–0.93	0.60–0.90	0.040 max	0.050 max	x			
1085	0.80–0.93	0.70–1.00	0.040 max	0.050 max	X	x	x	
1086	0.80–0.93	0.30–0.50	0.040 max	0.050 max	x	..	x	
1090	0.85–0.98	0.60–0.90	0.040 max	0.050 max	X			
1095	0.90–1.03	0.30–0.50	0.040 max	0.050 max	X	X	x	
B 1006†	0.08 max	0.45 max	0.07–0.12	0.060 max	x	..	..	y
B 1010†	0.13 max	0.30–0.60	0.07–0.12	0.060 max	x	x	..	y
Basic Open-hearth Resulfurized Steels								
1106	0.08 max	0.30–0.60	0.040 max	0.08–0.13	x	..	..	y
1108	0.08–0.13	0.50–0.80	0.040 max	0.08–0.13	x	x	..	y
1109	0.08–0.13	0.60–0.90	0.040 max	0.08–0.13	X	x	..	y
1110	0.08–0.13	0.30–0.60	0.040 max	0.08–0.13	x	x	..	y
1111	0.08–0.13	0.60–0.90	0.040 max	0.16–0.23	x	x		
1113	0.10–0.16	1.00–1.30	0.040 max	0.24–0.33	x	x		
1114	0.10–0.16	1.00–1.30	0.040 max	0.08–0.13	x	x		
1115	0.13–0.18	0.60–0.90	0.040 max	0.08–0.13	X	X		
1116	0.14–0.20	1.10–1.40	0.040 max	0.16–0.23	x	x		
1117	0.14–0.20	1.00–1.30	0.040 max	0.08–0.13	X	X	x	

1118	0.14–0.20	1.30–1.60	0.040 max	0.08–0.13	X	X	x	
1119	0.14–0.20	1.00–1.30	0.040 max	0.24–0.33	x	X		
1120	0.18–0.23	0.70–1.00	0.040 max	0.08–0.13	X	X	x	
1125	0.22–0.28	0.60–0.90	0.040 max	0.08–0.13	X	..	X	
1126	0.23–0.29	0.70–1.00	0.040 max	0.08–0.13	x	x	X	
1132	0.27–0.34	1.35–1.65	0.040 max	0.08–0.13	x	x		
1137	0.32–0.39	1.35–1.65	0.040 max	0.08–0.13	X	X	x	
1138	0.34–0.40	0.70–1.00	0.040 max	0.08–0.13	x	x	x	
1140	0.37–0.44	0.70–1.00	0.040 max	0.08–0.13	X	x	x	
1141	0.37–0.45	1.35–1.65	0.040 max	0.08–0.13	X	X	x	
1144	0.40–0.48	1.35–1.65	0.040 max	0.24–0.33	X	X	x	
1145	0.42–0.49	0.70–1.00	0.040 max	0.04–0.07	x	..	X	
1146	0.42–0.49	0.70–1.00	0.040 max	0.08–0.13	x	X	x	
1148	0.45–0.52	0.70–1.00	0.040 max	0.04–0.07	x	..	x	
1151	0.48–0.55	0.70–1.00	0.040 max	0.08–0.13	x	x		
Basic Open-hearth Rephosphorized and Resulfurized Steels								
1211	0.13 max	0.60–0.90	0.07–0.12	0.08–0.15	x	x		
1212	0.13 max	0.70–1.00	0.07–0.12	0.16–0.23	X	X		
1213	0.13 max	0.70–1.00	0.07–0.12	0.24–0.33	X	X		
Acid Bessemer Resulfurized Steels								
B 1111	0.13 max	0.60–0.90	0.07–0.12	0.08–0.15	X	X		
B 1112	0.13 max	0.70–1.00	0.07–0.12	0.16–0.23	X	X		
B 1113	0.13 max	0.70–1.00	0.07–0.12	0.24–0.33	X	X		

* Availability: X designates steels more frequently used for hot-rolled and cold-finished bars and semifinished steel products for forging. They are more readily available than items identified by x. For wire, steel numbers are identified by y without distinction as to use or availability.

† These are bessemer steels.

this interrelationship of elements must be considered. For simplicity, though, the properties introduced by the various elements are:

Carbon is the principal hardening element in steel. It increases hardness and tensile strength of as-rolled or normalized steels proportionately to carbon content, up to 0.85%, and less so above this figure. Maximum hardness of quenched steel is proportional to carbon content up to 0.60%, and less so thereafter. Ductility decreases with increase in carbon content, and weldability is impaired above certain carbon levels.

Manganese is not so effective as carbon in respect to increasing hardness and strength, but it is beneficial to surface quality, particularly for resulfurized steels, and it increases the rate of carbon penetration in carburizing.

Phosphorus increases strength and hardness of hot-rolled steels but impairs ductility and toughness, particularly in higher-carbon steels. However, phosphorus promotes machinability in lower carbon steels.

Sulfur aids machinability but is detrimental to surface quality, particularly in low-carbon and low-manganese steels, and to transverse ductility, impact resistance, and weldability.

Silicon content is related to the type of steel. No significant amounts are found in rimmed and capped steels, but semikilled steels contain a moderate amount, fully killed steels up to 0.30%. Silicon is less effective than manganese in increasing strength and hardness, promotes adherences of zinc coatings to hot-dipped galvanized wire, and is detrimental to surface quality of low-carbon steels, particularly the resulfurized steels.

Aluminum deoxidizes steel, promotes fine grain size, and is sometimes used to prevent recurrence of stretcher strains in sheet and strip.

Copper in amounts greater than 0.20% increases resistance to atmospheric corrosion, and in small amounts has a minor effect on mechanical properties.

ALLOY STEELS

Steel is considered to be alloy steel when the maximum of the range given for the content of alloying elements exceeds one or more of the following limits: manganese 1.65%, silicon 0.60%, copper 0.60%, or in which a definite range or a definite minimum quantity of any of the following elements is specified or required within the limits of the recognized field of constructional alloy steels: aluminum, boron, chromium up to 3.99%, cobalt, columbium, molybdenum, nickel, titanium, tungsten, vanadium, zirconium, or any other alloying element added to obtain a desired alloying effect.

Effects of Alloying Elements. Common alloying elements have these effects upon steel:

Manganese contributes to strength and is of major importance in increasing hardenability (the depth of hardness penetration after quenching).

Silicon increases the resiliency of steel for spring applications and raises the critical temperature for heat-treatment. Increasing the silicon content promotes the susceptibility of steel to decarburization. When the maximum silicon content is specified within the limits of 0.60 to 2.20%, the product is classed as alloy steel.

Aluminum, when in amounts approximating 1%, promotes nitriding properties.

Nickel lowers the critical temperature of steel and widens the temperature range for successful heat-treatment. Nickel is used to promote resistance to corrosion.

When nickel is present in appreciable amounts it results in higher-strength steels with improved shock resistance. It counteracts the brittleness which develops in most pearlitic steels at subnormal temperatures.

Chromium is used in constructional steels primarily to increase hardness, improve hardenability, and promote the formation of carbides. Chromium enhances corrosion resistance and heat resistance, and it is the essential element in stainless and heat-resisting steels. Chromium steels are relatively stable at elevated temperatures and have exceptional wear resistance.

Molybdenum, in common with manganese and chromium, has a major effect on increasing hardenability, and a strong effect in increasing the high-temperature tensile and creep strength. Steels containing molybdenum are considered to be less susceptible to temper brittleness.

Vanadium promotes a fine austenitic grain size. The amount used in constructional steels ranges from about 0.03 to 0.25%, although larger quantities are used in tool steels and other special steels.

Hardenability of medium-carbon steels is increased with a minimum effect upon grain size with vanadium additions of about 0.04 to 0.05%; above this content, the hardenability decreases with normal quenching temperatures. However, the hardenability can be increased with the higher vanadium contents by increasing the austenitizing temperatures.

Titanium acts as a deoxidizer in pearlitic steels. When present in amounts of 0.02 to 0.05% it increases the yield point of plain carbon steels. Weldability is promoted without the necessity for normalizing. In austenitic stainless steels the element is utilized to retard intergranular corrosion.

Boron is added to steel for one purpose: to increase hardenability. It is effective only when added to fully killed steels. Boron steels are evaluated by increased hardenability rather than chemical content.

Boron intensifies the hardenability characteristics of elements which are already present in the steel. It makes possible a large degree of alloy conservation when employed with steels containing small amounts of alloying elements. Boron is very effective when used with low-carbon alloy steels, but the effect diminishes as the carbon increases. Its use is not suggested above the 0.60% carbon level.

Boron is employed most frequently in the 0.30 to 0.50% carbon range. It is used less extensively in the 0.20% carbon carburizing grades, because it improves only the core properties and does not increase the depth hardness of the case.

Multiply Alloying Elements. A combination of two or more of the above alloying elements usually imparts some of the characteristic properties of each. Constructional chromium-nickel-alloy steels, for example, develop good hardening properties, with excellent ductility, while chromium-molybdenum combinations develop excellent hardenability with satisfactory ductility and a certain amount of heat resistance.

The combined effect of two or more alloying elements on the hardenability of a steel is considerably greater than the sum of the effects of the same alloying elements used separately. The general effectiveness of the nickel-chromium-molybdenum steels, both with and without boron, is accounted for in this way.

HIGH-STRENGTH LOW-ALLOY STEELS

High-strength low-alloy steel comprises a group of trade-name steels with chemical compositions specially developed to impart improved mechanical properties and greater resistance to atmospheric corrosion than are obtainable from conventional carbon structural steels containing copper. High-strength low-alloy steels are produced to mechanical property requirements rather than to chemical composition limits. The steels are readily adaptable to fabrication by shearing, gas cutting, punching, hot and cold forming, riveting, and welding.

Continued on page 25-17

TABLE 2. STANDARD ALLOY STEELS

Steel No.	Chemical Composition Ranges and Limits, %							
	Carbon	Manganese	Phosphorus Max	Sulfur Max	Silicon	Nickel	Chromium	Molybdenum
1330	0.28–0.33	1.60–1.90	0.040	0.040	0.20–0.35			
1335	0.33–0.38	1.60–1.90	0.040	0.040	0.20–0.35			
1340	0.38–0.43	1.60–1.90	0.040	0.040	0.20–0.35			
1345	0.43–0.48	1.60–1.90	0.040	0.040	0.20–0.35			
2515	0.12–0.17	0.40–0.60	0.040	0.040	0.20–0.35	4.75–5.25		
E 2517	0.15–0.20	0.45–0.60	0.025	0.025	0.20–0.35	4.75–5.25		
3140	0.38–0.43	0.70–0.90	0.040	0.040	0.20–0.35	1.10–1.40	0.55–0.75	
E 3310	0.08–0.13	0.45–0.60	0.025	0.025	0.20–0.35	3.25–3.75	1.40–1.75	
E 3316	0.14–0.19	0.45–0.60	0.025	0.025	0.20–0.35	3.25–3.75	1.40–1.75	
4023	0.20–0.25	0.70–0.90	0.040	0.040	0.20–0.35	....	....	0.20–0.30
4024	0.20–0.25	0.70–0.90	0.040	0.035–0.050	0.20–0.35	....	....	0.20–0.30
4027	0.25–0.30	0.70–0.90	0.040	0.040	0.20–0.35	....	....	0.20–0.30
4028	0.25–0.30	0.70–0.90	0.040	0.035–0.050	0.20–0.35	....	....	0.20–0.30
4032	0.30–0.35	0.70–0.90	0.040	0.040	0.20–0.35	....	....	0.20–0.30
4037	0.35–0.40	0.70–0.90	0.040	0.040	0.20–0.35	....	....	0.20–0.30
4042	0.40–0.45	0.70–0.90	0.040	0.040	0.20–0.35	....	....	0.20–0.30
4047	0.45–0.50	0.70–0.90	0.040	0.040	0.20–0.35	....	....	0.20–0.30
4053	0.50–0.56	0.75–1.00	0.040	0.040	0.20–0.35	....	....	0.20–0.30
4063	0.60–0.67	0.75–1.00	0.040	0.040	0.20–0.35	....	....	0.20–0.30
4068	0.63–0.70	0.75–1.00	0.040	0.040	0.20–0.35	....	....	0.20–0.30
4118	0.18–0.23	0.70–0.90	0.040	0.040	0.20–0.35	....	0.40–0.60	0.08–0.15
4130	0.28–0.33	0.40–0.60	0.040	0.040	0.20–0.35	....	0.80–1.10	0.15–0.25

4135	0.33–0.38	0.70–0.90	0.040	0.040	0.20–0.35	....	0.80–1.10	0.15–0.25
4137	0.35–0.40	0.70–0.90	0.040	0.040	0.20–0.35	....	0.80–1.10	0.15–0.25
4140	0.38–0.43	0.75–1.00	0.040	0.040	0.20–0.35	....	0.80–1.10	0.15–0.25
4142	0.40–0.45	0.75–1.00	0.040	0.040	0.20–0.35	....	0.80–1.10	0.15–0.25
4145	0.43–0.48	0.75–1.00	0.040	0.040	0.20–0.35	....	0.80–1.10	0.15–0.25
4147	0.45–0.50	0.75–1.00	0.040	0.040	0.20–0.35	....	0.80–1.10	0.15–0.25
4150	0.48–0.53	0.75–1.00	0.040	0.040	0.20–0.35	....	0.80–1.10	0.15–0.25
4320	0.17–0.22	0.45–0.65	0.040	0.040	0.20–0.35	1.65–2.00	0.40–0.60	0.20–0.30
4337	0.35–0.40	0.60–0.80	0.040	0.040	0.20–0.35	1.65–2.00	0.70–0.90	0.20–0.30
4340	0.38–0.43	0.60–0.80	0.040	0.040	0.20–0.35	1.65–2.00	0.70–0.90	0.20–0.30
E 4340	0.38–0.43	0.65–0.85	0.025	0.025	0.20–0.35	1.65–2.00	0.70–0.90	0.20–0.30
4608	0.06–0.11	0.25–0.45	0.040	0.040	0.25 max	1.40–1.75	....	0.15–0.25
4615	0.13–0.18	0.45–0.65	0.040	0.040	0.20–0.35	1.65–2.00	....	0.20–0.30
4617	0.15–0.20	0.45–0.65	0.040	0.040	0.20–0.35	1.65–2.00	....	0.20–0.30
4620	0.17–0.22	0.45–0.65	0.040	0.040	0.20–0.35	1.65–2.00	....	0.20–0.30
4621	0.18–0.23	0.70–0.90	0.040	0.040	0.20–0.35	1.65–2.00	....	0.20–0.30
4640	0.38–0.43	0.60–0.80	0.040	0.040	0.20–0.35	1.65–2.00	....	0.20–0.30
4812	0.10–0.15	0.40–0.60	0.040	0.040	0.20–0.35	3.25–3.75	....	0.20–0.30
4815	0.13–0.18	0.40–0.60	0.040	0.040	0.20–0.35	3.25–3.75	....	0.20–0.30
4817	0.15–0.20	0.40–0.60	0.040	0.040	0.20–0.35	3.25–3.75	....	0.20–0.30
4820	0.18–0.23	0.50–0.70	0.040	0.040	0.20–0.35	3.25–3.75	....	0.20–0.30
5015	0.12–0.17	0.30–0.50	0.040	0.040	0.20–0.35	....	0.30–0.50	
5046	0.43–0.50	0.75–1.00	0.040	0.040	0.20–0.35	....	0.20–0.35	
5117	0.15–0.20	0.70–0.90	0.040	0.040	0.20–0.35	....	0.70–0.90	
5120	0.17–0.22	0.70–0.90	0.040	0.040	0.20–0.35	....	0.70–0.90	
5130	0.28–0.33	0.70–0.90	0.040	0.040	0.20–0.35	....	0.80–1.10	
5132	0.30–0.35	0.60–0.80	0.040	0.040	0.20–0.35	....	0.75–1.00	
5135	0.33–0.38	0.60–0.80	0.040	0.040	0.20–0.35	....	0.80–1.05	
5140	0.38–0.43	0.70–0.90	0.040	0.040	0.20–0.35	....	0.70–0.90	
5145	0.43–0.48	0.70–0.90	0.040	0.040	0.20–0.35	....	0.70–0.90	
5147	0.45–0.52	0.70–0.95	0.040	0.040	0.20–0.35	....	0.85–1.15	

TABLE 2. STANDARD ALLOY STEELS (*Continued*)

Steel No.	Chemical Composition Ranges and Limits, %							
	Carbon	Manganese	Phosphorus Max	Sulfur Max	Silicon	Nickel	Chromium	Molybdenum
5150	0.48–0.53	0.70–0.90	0.040	0.040	0.20–0.35	....	0.70–0.90	
5152	0.48–0.55	0.70–0.90	0.040	0.040	0.20–0.35	....	0.90–1.20	
5155	0.50–0.60	0.70–0.90	0.040	0.040	0.20–0.35	....	0.70–0.90	
5160	0.55–0.65	0.75–1.00	0.040	0.040	0.20–0.35	....	0.70–0.90	
E 50100	0.95–1.10	0.25–0.45	0.025	0.025	0.20–0.35	....	0.40–0.60	
E 51100	0.95–1.10	0.25–0.45	0.025	0.025	0.20–0.35	....	0.90–1.15	
E 52100	0.95–1.10	0.25–0.45	0.025	0.025	0.20–0.35	....	1.30–1.60	
								V
6117	0.15–0.20	0.70–0.90	0.040	0.040	0.20–0.35	....	0.70–0.90	*
6120	0.17–0.22	0.70–0.90	0.040	0.040	0.20–0.35	....	0.70–0.90	*
6145	0.43–0.48	0.70–0.90	0.040	0.040	0.20–0.35	....	0.80–1.10	†
6150	0.48–0.53	0.70–0.90	0.040	0.040	0.20–0.35	....	0.80–1.10	†
								Mo
8615	0.13–0.18	0.70–0.90	0.040	0.040	0.20–0.35	0.40–0.70	0.40–0.60	0.15–0.25
8617	0.15–0.20	0.70–0.90	0.040	0.040	0.20–0.35	0.40–0.70	0.40–0.60	0.15–0.25
8620	0.18–0.23	0.70–0.90	0.040	0.040	0.20–0.35	0.40–0.70	0.40–0.60	0.15–0.25
8622	0.20–0.25	0.70–0.90	0.040	0.040	0.20–0.35	0.40–0.70	0.40–0.60	0.15–0.25
8625	0.23–0.28	0.70–0.90	0.040	0.040	0.20–0.35	0.40–0.70	0.40–0.60	0.15–0.25
8627	0.25–0.30	0.70–0.90	0.040	0.040	0.20–0.35	0.40–0.70	0.40–0.60	0.15–0.25
8630	0.28–0.33	0.70–0.90	0.040	0.040	0.20–0.35	0.40–0.70	0.40–0.60	0.15–0.25
8635	0.33–0.38	0.75–1.00	0.040	0.040	0.20–0.35	0.40–0.70	0.40–0.60	0.15–0.25
8637	0.35–0.40	0.75–1.00	0.040	0.040	0.20–0.35	0.40–0.70	0.40–0.60	0.15–0.25
8640	0.38–0.43	0.75–1.00	0.040	0.040	0.20–0.35	0.40–0.70	0.40–0.60	0.15–0.25
8641	0.38–0.43	0.75–1.00	0.040	0.040–0.060	0.20–0.35	0.40–0.70	0.40–0.60	0.15–0.25
8642	0.40–0.45	0.75–1.00	0.040	0.040	0.20–0.35	0.40–0.70	0.40–0.60	0.15–0.25
8645	0.43–0.48	0.75–1.00	0.040	0.040	0.20–0.35	0.40–0.70	0.40–0.60	0.15–0.25

8650	0.48–0.53	0.75–1.00	0.040	0.040	0.20–0.35	0.40–0.70	0.40–0.60	0.15–0.25
8653	0.50–0.56	0.75–1.00	0.040	0.040	0.20–0.35	0.40–0.70	0.50–0.80	0.15–0.25
8655	0.50–0.60	0.75–1.00	0.040	0.040	0.20–0.35	0.40–0.70	0.40–0.60	0.15–0.25
8660	0.55–0.65	0.75–1.00	0.040	0.040	0.20–0.35	0.40–0.70	0.40–0.60	0.15–0.25
86B45‡	0.43–0.48	0.75–1.00	0.040	0.040	0.20–0.35	0.40–0.70	0.40–0.60	0.15–0.25
8715	0.13–0.18	0.70–0.90	0.040	0.040	0.20–0.35	0.40–0.70	0.40–0.60	0.20–0.30
8717	0.15–0.20	0.70–0.90	0.040	0.040	0.20–0.35	0.40–0.70	0.40–0.60	0.20–0.30
8720	0.18–0.23	0.70–0.90	0.040	0.040	0.20–0.35	0.40–0.70	0.40–0.60	0.20–0.30
8735	0.33–0.38	0.75–1.00	0.040	0.040	0.20–0.35	0.40–0.70	0.40–0.60	0.20–0.30
8740	0.38–0.43	0.75–1.00	0.040	0.040	0.20–0.35	0.40–0.70	0.40–0.60	0.20–0.30
8742	0.40–0.45	0.75–1.00	0.040	0.040	0.20–0.35	0.40–0.70	0.40–0.60	0.20–0.30
8750	0.48–0.53	0.75–1.00	0.040	0.040	0.20–0.35	0.40–0.70	0.40–0.60	0.20–0.30
9255	0.50–0.60	0.70–0.95	0.040	0.040	1.80–2.20			
9260	0.55–0.65	0.70–1.00	0.040	0.040	1.80–2.20			
9261	0.55–0.65	0.75–1.00	0.040	0.040	1.80–2.20	....	0.10–0.25	
9262	0.55–0.65	0.75–1.00	0.040	0.040	1.80–2.20	....	0.25–0.40	
E 9310	0.08–0.13	0.45–0.65	0.025	0.025	0.20–0.35	3.00–3.50	1.00–1.40	0.08–0.15
E 9314	0.11–0.17	0.40–0.70	0.025	0.025	0.20–0.35	3.00–3.50	1.00–1.40	0.08–0.15
9840	0.38–0.43	0.70–0.90	0.040	0.040	0.20–0.35	0.85–1.15	0.70–0.90	0.20–0.30
9845	0.43–0.48	0.70–0.90	0.040	0.040	0.20–0.35	0.85–1.15	0.70–0.90	0.20–0.30
9850	0.48–0.53	0.70–0.90	0.040	0.040	0.20–0.35	0.85–1.15	0.70–0.90	0.20–0.30

NOTE: The frequency of use and availability of these steels are not indicated.

* 0.10 % min vanadium.
† 0.15 % min vanadium.
‡ This steel can be expected to contain a minimum of 0.0005 % boron. Boron is not normally reported in a ladle analysis.

TABLE 3. TENTATIVE STANDARD ALLOY STEELS

Steel No.	Chemical Composition Ranges and Limits, %							
	Carbon	Manganese	Phosphorus Max	Sulfur Max	Silicon	Nickel	Chromium	Molybdenum
TS 4012	0.09–0.14	0.75–1.00	0.040	0.040	0.20–0.35	....	....	0.15–0.25
TS 4130	0.28–0.33	0.45–0.65	0.040	0.040	0.20–0.35	....	0.90–1.20	0.08–0.15
TS 4132	0.30–0.35	0.45–0.65	0.040	0.040	0.20–0.35	....	0.90–1.20	0.08–0.15
TS 4137	0.35–0.40	0.75–1.00	0.040	0.040	0.20–0.35	....	0.90–1.20	0.08–0.15
TS 4140	0.38–0.43	0.80–1.05	0.040	0.040	0.20–0.35	....	0.90–1.20	0.08–0.15
TS 4142	0.40–0.45	0.80–1.05	0.040	0.040	0.20–0.35	....	0.90–1.20	0.08–0.15
TS 4145	0.43–0.48	0.80–1.05	0.040	0.040	0.20–0.35	....	0.90–1.20	0.08–0.15
TS 4150	0.48–0.53	0.80–1.05	0.040	0.040	0.20–0.35	....	0.90–1.20	0.08–0.15
TS 4720	0.17–0.22	0.50–0.70	0.040	0.040	0.20–0.35	0.90–1.20	0.35–0.55	0.15–0.25
TS 8115	0.13–0.18	0.70–0.90	0.040	0.040	0.20–0.35	0.20–0.40	0.30–0.50	0.08–0.15
TS 8117	0.15–0.20	0.70–0.90	0.040	0.040	0.20–0.35	0.20–0.40	0.30–0.50	0.08–0.15
TS 8120	0.18–0.23	0.70–0.90	0.040	0.040	0.20–0.35	0.20–0.40	0.30–0.50	0.08–0.15
TS 8122	0.20–0.25	0.70–0.90	0.040	0.040	0.20–0.35	0.20–0.40	0.30–0.50	0.08–0.15
TS 8123	0.20–0.25	0.70–0.90	0.040	0.035–0.050	0.20–0.35	0.20–0.40	0.30–0.50	0.08–0.15
TS 8125	0.23–0.28	0.70–0.90	0.040	0.040	0.20–0.35	0.20–0.40	0.30–0.50	0.08–0.15
TS 8126	0.23–0.28	0.70–0.90	0.040	0.035–0.050	0.20–0.35	0.20–0.40	0.30–0.50	0.08–0.15
TS 8127	0.25–0.30	0.70–0.90	0.040	0.040	0.20–0.35	0.20–0.40	0.30–0.50	0.08–0.15
TS 8128	0.25–0.30	0.70–0.90	0.040	0.035–0.050	0.20–0.35	0.20–0.40	0.30–0.50	0.08–0.15
TS 8615	0.13–0.18	0.70–0.90	0.040	0.040	0.20–0.35	0.30–0.60	0.55–0.75	0.08–0.15
TS 8617	0.15–0.20	0.70–0.90	0.040	0.040	0.20–0.35	0.30–0.60	0.55–0.75	0.08–0.15
TS 8620	0.18–0.23	0.70–0.90	0.040	0.040	0.20–0.35	0.30–0.60	0.55–0.75	0.08–0.15
TS 8622	0.20–0.25	0.70–0.90	0.040	0.040	0.20–0.35	0.30–0.60	0.55–0.75	0.08–0.15
TS 8625	0.23–0.28	0.70–0.90	0.040	0.040	0.20–0.35	0.30–0.60	0.55–0.75	0.08–0.15
TS 8627	0.25–0.30	0.70–0.90	0.040	0.040	0.20–0.35	0.30–0.60	0.55–0.75	0.08–0.15

Various combinations of alloying elements are employed:

Carbon is generally maintained at a level to insure freedom from excessive hardening after welding and to retain ductility.

Manganese is used principally as a strengthening element.

Phosphorus is sometimes employed as a strengthening element and to enhance resistance to atmospheric corrosion.

Copper is used to enhance resistance to atmospheric corrosion and as a strengthening element.

Silicon, nickel, chromium, molybdenum, vanadium, aluminum, titanium, zirconium, and other elements sometimes are used, singly or in combination, for their beneficial effects on strength, toughness, corrosion resistance, and other desirable properties.

SELECTION OF H-STEELS

Until the war years, the alloy-steel consumer bought material on chemical specifications. Scarcity of alloying elements led to development of National Emergency, or lean-alloy, steels. The purpose of these steels was to provide adequate hardening power while saving elements in short supply. Some of these NE steels have been continued under standard numbers. But the value of making steels to guaranteed hardenability ranges—in other words, H-steels—has prompted the development of them in standard and tentative numbers paralleling most steels made to chemistry limits. Presumably the H-steels are easier for the mill to make, because it is concerned primarily with producing a material within a guaranteed range of hardness at a given depth, and can change the chemistry as needed in the melting process. In contrast, steels made to chemistry limits are more difficult to produce, because of residual alloys carried over from melting scrap of diversified analyses. Thus, the user of H-steels gets what he actually needs—proper hardness at desired depths, regardless of chemistry, and presumably at lower cost.

Carbon steels are not available as H-steels. Except in small sizes, they are not through hardening and therefore should not be used in large sections where heat-treated properties must be essentially the same throughout the cross-section.

The surface hardness of a steel after quenching is largely a function of carbon content, whereas the depth of hardness depends on the carbon content, total content of alloying elements, and the grain size. To make any steel through hardening, it is necessary only to add the sufficient quantity of the proper alloying elements, taking into consideration the quenching medium that will be used.

Steels are considered to have optimum properties in the as-quenched condition when they exhibit 90% martensite, the hardest constituent. In respect to hardening properties, there are three types of steels: shallow-hardening—steels like 8630 quench out at the center of 1-in. sections; medium hardening—grades like 4140 quench out in oil sections up to 1½ in. dia; and deep-hardening—steels like 4340 will harden throughout in oil up to about 4 in. dia. Note, however, that through hardening is not desirable for many applications. A strong, hard outer section and a softer, tough core, that is, a hardness gradient from surface to center, is generally preferred for shock applications, for example.

Jominy End-quench Test. The hardness gradient, or curve, for a given steel is determined by the Jominy end-quench test. In conducting this test, a 1-in. round specimen about 4 in. long is heated to the proper quenching temperature and is then quenched by a water jet (Fig. 1) that is confined to the bottom face. Decarburization is removed by grinding flats on the specimen, and Rockwell C hardness readings are made at 1/16-in. intervals from the end. Depth of hardening is the distance along the specimen from the quenched end to a given hardness.

Hardenability Bands. The minimum and maximum values of hardness at progressive depths are plotted as a hardenability band (Fig. 2). Tables 6 to 18 give the band limits for J-distances from 1/16-in. depth to 2-in. depth. Experiments have been conducted which indicate the various points on the hardenability curve which approximate the cooling rates at the center, midradius, and surface of rounds of various sizes quenched *in a variety of cooling mediums.* End-quench curves normally are based on two of these mediums—agitated oil and agitated water, which are the ones most frequently used.

Selection of H-steel. Suppose that the product to be heat-treated is 1½ in. in dia. Evidently an oil-hardening steel is required. For maximum life of the part, it has been determined that the hardnesses at surface, midradius, and center of the part must be Rockwell 55 C, 50 C, and 48 C as minimums. Since oil will be used as the quench, and hardenability-band limits are reported for water quenching, it is necessary to use the cooling-rate curves in Fig. 3 to establish the locations in the round where the cooling rate with oil is the same as at various positions along the end-quenched hardenability test bar. From inspection of the cooling-rate curves for an agitated oil quench, it is seen that equivalent positions on the Jominy test bar are: surface—3/16 in.; midradius—7/16 in.; and center—8/16 in. Now by inspection of Tables 6 to 18 and their band limits, we find that three typical compositions will produce adequate minimum hardenability:

	3/16 in.	7/16 in.	8/16 in.
8750H	58	55	53
4147H	56	55	54
8650H	57	53	50

Final choice would depend on cost and processing factors, but any one of the steels would possess adequate hardenability.

For H-steel specification purposes, two points should be used to designate hardenability limits. Four methods are illustrated in connection with Fig. 2. The manner of designating them is shown below.

Four Methods of Designating Hardenability Limits of H-steels

Method	*Example*
A. Min and max hardness values at a desired distance	*A-A.* J39/52 = 4/16 in.
B. A desired hardness value at min and max distances	*B-B.* J42 = 3/16 to 8/16 in. (min distance to nearest 1/16 in. at left and max to nearest 1/16 in. at right)
C. Two max hardness values at two desired distances	*C-C.* J50 = 5/16 in. max J34 = 12/16 in. max
D. Two min hardness values at two desired distances	*D-D.* J35 = 5/16 in. min J21 = 16/16 in. min

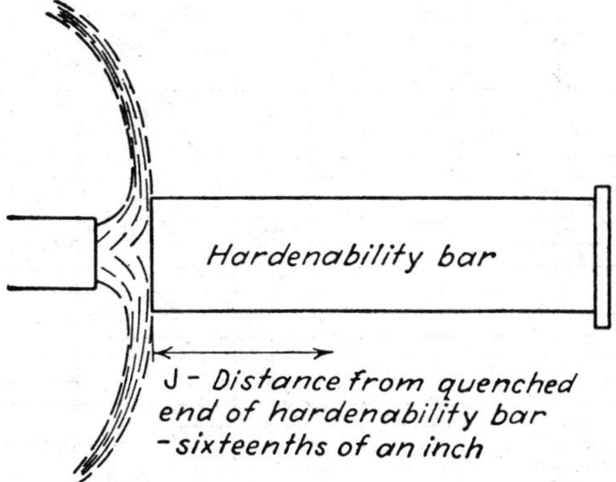

FIG. 1. J-distance from quenched end of hardenability bar, sixteenths of an inch.

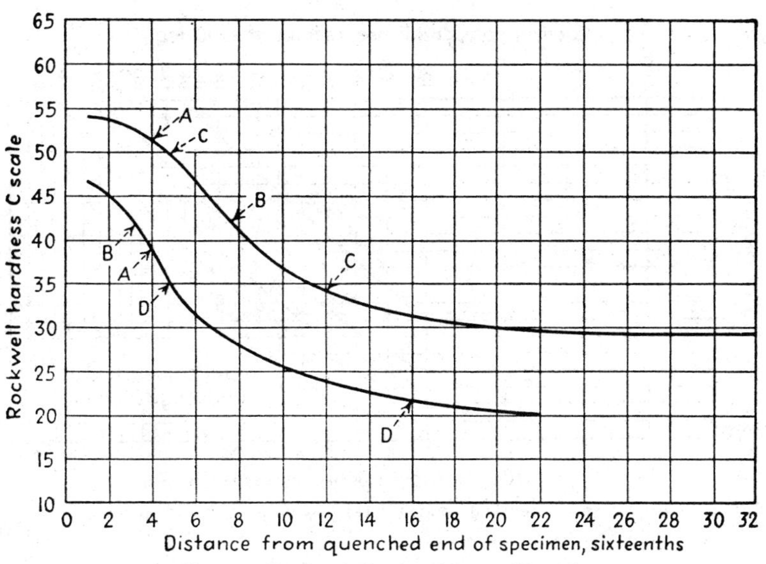

FIG. 2. Hardenability band for an H-steel.

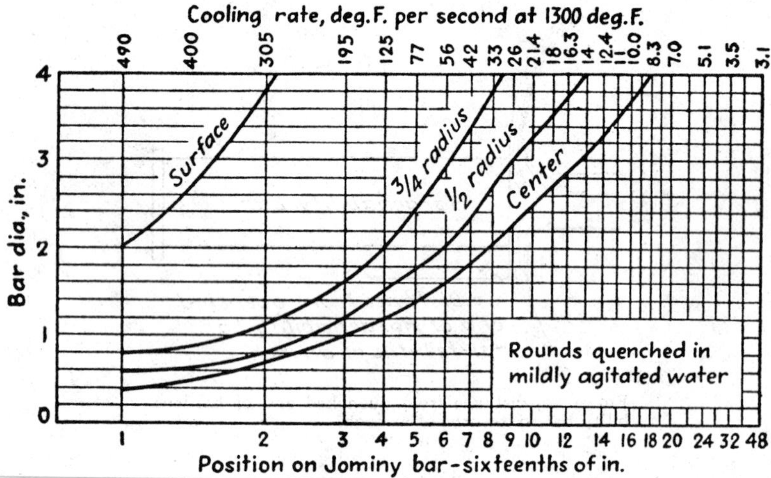

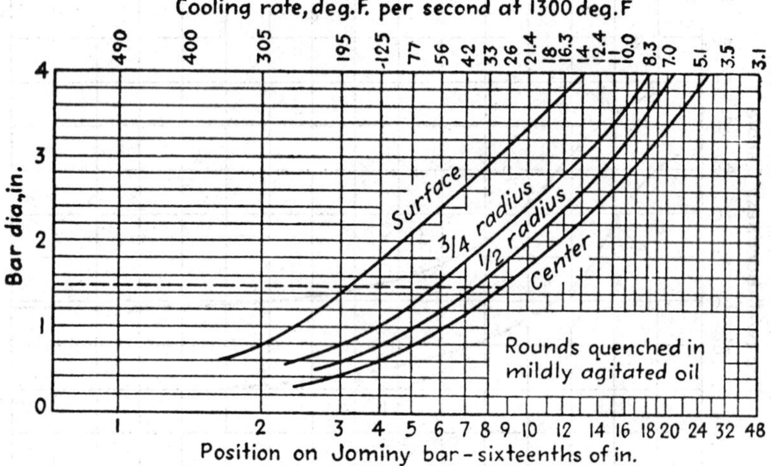

FIG. 3. Correlation of identical cooling rates in Jominy bar and quenched round bars.

TABLE 4. STANDARD AND TENTATIVE STANDARD H-STEELS[a]

Steel No.	Chemical Composition Ranges and Limits, %					
	Carbon	Manganese	Silicon	Nickel	Chromium	Molybdenum
1330H	0.27–0.33	1.45–2.05	0.20–0.35			
1335H	0.32–0.38	1.45–2.05	0.20–0.35			
1340H	0.37–0.44	1.45–2.05	0.20–0.35			
2515H	0.12–0.18	0.30–0.70	0.20–0.35	4.70–5.30		
2517H	0.14–0.20	0.30–0.70	0.20–0.35	4.70–5.30		
3140H	0.37–0.44	0.60–1.00	0.20–0.35	1.00–1.45	0.45–0.85	
3310H	0.07–0.13	0.30–0.70	0.20–0.35	3.20–3.80	1.30–1.80	
3316H	0.13–0.19	0.30–0.70	0.20–0.35	3.20–3.80	1.30–1.80	
4032H	0.29–0.35	0.60–1.00	0.20–0.35	....	....	0.20–0.30
4037H	0.34–0.41	0.60–1.00	0.20–0.35	....	....	0.20–0.30
4042H	0.39–0.46	0.60–1.00	0.20–0.35	....	....	0.20–0.30
4047H	0.44–0.51	0.60–1.00	0.20–0.35	....	....	0.20–0.30
4053H	0.49–0.56	0.65–1.10	0.20–0.35	....	....	0.20–0.30
4063H	0.59–0.69	0.65–1.10	0.20–0.35	....	....	0.20–0.30
4068H	0.62–0.72	0.65–1.10	0.20–0.35	....	....	0.20–0.30
4118H	0.17–0.23	0.60–1.00	0.20–0.35	....	0.30–0.70	0.08–0.15
4130H	0.27–0.33	0.30–0.70	0.20–0.35	....	0.75–1.20	0.15–0.25
4135H	0.32–0.38	0.60–1.00	0.20–0.35	....	0.75–1.20	0.15–0.25
4137H	0.34–0.41	0.60–1.00	0.20–0.35	....	0.75–1.20	0.15–0.25
4140H	0.37–0.44	0.65–1.10	0.20–0.35	....	0.75–1.20	0.15–0.25
4142H	0.39–0.46	0.65–1.10	0.20–0.35	....	0.75–1.20	0.15–0.25
4145H	0.42–0.49	0.65–1.10	0.20–0.35	....	0.75–1.20	0.15–0.25
4147H	0.44–0.51	0.65–1.10	0.20–0.35	....	0.75–1.20	0.15–0.25
4150H	0.47–0.54	0.65–1.10	0.20–0.35	....	0.75–1.20	0.15–0.25
TS 4130H	0.27–0.33	0.35–0.75	0.20–0.35	....	0.85–1.30	0.08–0.15
TS 4132H	0.29–0.35	0.35–0.75	0.20–0.35	....	0.85–1.30	0.08–0.15
TS 4137H	0.34–0.41	0.65–1.10	0.20–0.35	....	0.85–1.30	0.08–0.15
TS 4140H	0.37–0.44	0.70–1.20	0.20–0.35	....	0.85–1.30	0.08–0.15
TS 4142H	0.39–0.46	0.70–1.20	0.20–0.35	....	0.85–1.30	0.08–0.15
TS 4145H	0.42–0.49	0.70–1.20	0.20–0.35	....	0.85–1.30	0.08–0.15
TS 4150H	0.47–0.54	0.70–1.20	0.20–0.35	....	0.85–1.30	0.08–0.15
4320H	0.17–0.23	0.40–0.70	0.20–0.35	1.55–2.00	0.35–0.65	0.20–0.30
4337H	0.34–0.41	0.55–0.90	0.20–0.35	1.55–2.00	0.65–0.95	0.20–0.30
4340H	0.37–0.44	0.55–0.90	0.20–0.35	1.55–2.00	0.65–0.95	0.20–0.30
E 4340H	0.37–0.44	0.60–0.95	0.20–0.35	1.55–2.00	0.65–0.95	0.20–0.30
4620H	0.17–0.23	0.35–0.75	0.20–0.35	1.55–2.00	....	0.20–0.30
4621H	0.17–0.23	0.60–1.00	0.20–0.35	1.55–2.00	....	0.20–0.30
4640H	0.37–0.44	0.50–0.90	0.20–0.35	1.55–2.00	....	0.20–0.30

TABLE 4. STANDARD AND TENTATIVE STANDARD H-STEELS[a] (*Continued*)

Steel No.	Chemical Composition Ranges and Limits, %					
	Carbon	Manganese	Silicon	Nickel	Chromium	Molybdenum
TS 4720H	0.17–0.23	0.45–0.75	0.20–0.35	0.85–1.25	0.30–0.60	0.15–0.25
4812H	0.09–0.15	0.30–0.70	0.20–0.35	3.20–3.80	....	0.20–0.30
4815H	0.12–0.18	0.30–0.70	0.20–0.35	3.20–3.80	....	0.20–0.30
4817H	0.14–0.20	0.30–0.70	0.20–0.35	3.20–3.80	....	0.20–0.30
4820H	0.17–0.23	0.40–0.80	0.20–0.35	3.20–3.80	....	0.20–0.30
5120H	0.17–0.23	0.60–1.00	0.20–0.35	....	0.60–1.00	
5130H	0.27–0.33	0.60–1.00	0.20–0.35	....	0.75–1.20	
5132H	0.29–0.35	0.50–0.90	0.20–0.35	....	0.65–1.10	
5135H	0.32–0.38	0.50–0.90	0.20–0.35	....	0.70–1.15	
5140H	0.37–0.44	0.60–1.00	0.20–0.35	....	0.60–1.00	
5145H	0.42–0.49	0.60–1.00	0.20–0.35	....	0.60–1.00	
5147H	0.45–0.52	0.60–1.05	0.20–0.35	....	0.80–1.25	
5150H	0.47–0.54	0.60–1.00	0.20–0.35	....	0.60–1.00	
5152H	0.48–0.55	0.60–1.00	0.20–0.35	....	0.85–1.30	
5160H	0.55–0.65	0.65–1.10	0.20–0.35	....	0.60–1.00	
6120H	0.17–0.23	0.60–1.00	0.20–0.35	....	0.60–1.00	[b]
6145H	0.42–0.49	0.60–1.00	0.20–0.35	....	0.75–1.20	[c]
6150H	0.47–0.54	0.60–1.00	0.20–0.35	....	0.75–1.20	[c]
TS 8122H	0.19–0.25	0.60–0.95	0.20–0.35	0.15–0.45	0.25–0.55	0.08–0.15
8617H	0.14–0.20	0.60–0.95	0.20–0.35	0.35–0.75	0.35–0.65	0.15–0.25
8620H	0.17–0.23	0.60–0.95	0.20–0.35	0.35–0.75	0.35–0.65	0.15–0.25
8622H	0.19–0.25	0.60–0.95	0.20–0.35	0.35–0.75	0.35–0.65	0.15–0.25
8625H	0.22–0.28	0.60–0.95	0.20–0.35	0.35–0.75	0.35–0.65	0.15–0.25
8627H	0.24–0.30	0.60–0.95	0.20–0.35	0.35–0.75	0.35–0.65	0.15–0.25
8630H	0.27–0.33	0.60–0.95	0.20–0.35	0.35–0.75	0.35–0.65	0.15–0.25
8635H	0.32–0.38	0.70–1.05	0.20–0.35	0.35–0.75	0.35–0.65	0.15–0.25
8637H	0.34–0.41	0.70–1.05	0.20–0.35	0.35–0.75	0.35–0.65	0.15–0.25
8640H	0.37–0.44	0.70–1.05	0.20–0.35	0.35–0.75	0.35–0.65	0.15–0.25
8641H[d]	0.37–0.44	0.70–1.05	0.20–0.35	0.35–0.75	0.35–0.65	0.15–0.25
8642H	0.39–0.46	0.70–1.05	0.20–0.35	0.35–0.75	0.35–0.65	0.15–0.25
8645H	0.42–0.49	0.70–1.05	0.20–0.35	0.35–0.75	0.35–0.65	0.15–0.25
8650H	0.47–0.54	0.70–1.05	0.20–0.35	0.35–0.75	0.35–0.65	0.15–0.25
8653H	0.49–0.56	0.70–1.05	0.20–0.35	0.35–0.75	0.50–0.85	0.15–0.25
8655H	0.50–0.60	0.70–1.05	0.20–0.35	0.35–0.75	0.35–0.65	0.15–0.25
8660H	0.55–0.65	0.70–1.05	0.20–0.35	0.35–0.75	0.35–0.65	0.15–0.25
86B45H[e]	0.42–0.49	0.70–1.05	0.20–0.35	0.35–0.75	0.35–0.65	0.15–0.25
TS 8617H	0.14–0.20	0.60–0.95	0.20–0.35	0.25–0.65	0.50–0.80	0.08–0.15
TS 8620H	0.17–0.23	0.60–0.95	0.20–0.35	0.25–0.65	0.50–0.80	0.08–0.15
TS 8622H	0.19–0.25	0.60–0.95	0.20–0.35	0.25–0.65	0.50–0.80	0.08–0.15
TS 8625H	0.22–0.28	0.60–0.95	0.20–0.35	0.25–0.65	0.50–0.80	0.08–0.15
TS 8627H	0.24–0.30	0.60–0.95	0.20–0.35	0.25–0.65	0.50–0.80	0.08–0.15

TABLE 4. STANDARD AND TENTATIVE STANDARD H-STEELS[a] (*Continued*)

Steel No.	Chemical Composition Ranges and Limits, %					
	Carbon	Manganese	Silicon	Nickel	Chromium	Molybdenum
8720H	0.17–0.23	0.60–0.95	0.20–0.35	0.35–0.75	0.35–0.65	0.20–0.30
8740H	0.37–0.44	0.70–1.05	0.20–0.35	0.35–0.75	0.35–0.65	0.20–0.30
8742H	0.39–0.46	0.70–1.05	0.20–0.35	0.35–0.75	0.35–0.65	0.20–0.30
8750H	0.47–0.54	0.70–1.05	0.20–0.35	0.35–0.75	0.35–0.65	0.20–0.30
9260H	0.55–0.65	0.65–1.10	1.70–2.20			
9261H	0.55–0.65	0.65–1.10	1.70–2.20	....	0.05–0.35	
9262H	0.55–0.65	0.65–1.10	1.70–2.20	....	0.20–0.50	
9310H	0.07–0.13	0.40–0.70	0.20–0.35	2.95–3.55	1.00–1.45	0.08–0.15
9840H	0.37–0.44	0.60–0.95	0.20–0.35	0.80–1.20	0.65–0.95	0.20–0.30
9850H	0.47–0.54	0.60–0.95	0.20–0.35	0.80–1.20	0.65–0.95	0.20–0.30
Boron H-steels						
TS 14B35H	0.32–0.38	0.65–1.10	0.20–0.35			
TS 14B50H	0.47–0.54	0.65–1.10	0.20–0.35			
TS 40B37H	0.34–0.41	0.60–1.00	0.20–0.35	....	....	0.08–0.15
TS 46B12H	0.09–0.15	0.35–0.75	0.20–0.35	1.55–2.00	....	0.20–0.30
TS 50B46H	0.43–0.50	0.65–1.10	0.20–0.35	....	0.13–0.43	
TS 50B60H	0.55–0.65	0.65–1.10	0.20–0.35	....	0.30–0.70	
TS 51B60H	0.55–0.65	0.65–1.10	0.20–0.35	....	0.60–1.00	
TS 81B40H	0.37–0.44	0.70–1.05	0.20–0.35	0.15–0.45	0.30–0.60	0.08–0.15
TS 81B45H	0.42–0.49	0.70–1.05	0.20–0.35	0.15–0.45	0.30–0.60	0.08–0.15
86B45H	0.42–0.49	0.70–1.05	0.20–0.35	0.35–0.75	0.35–0.65	0.15–0.25
TS 94B17H	0.14–0.20	0.70–1.05	0.20–0.35	0.25–0.65	0.25–0.55	0.08–0.15

[a] These steels are specified to end-quench hardenability requirements, not to chemical composition.
[b] 0.10 % min vanadium.
[c] 0.15 % min vanadium.
[d] Sulfur content, 0.040–0.060 %.
[e] This steel can be expected to contain a minimum of 0.0005 % boron. Boron is not normally reported in a ladle analysis.

TABLE 5. TENTATIVE STANDARD AND ALTERNATE BORON STEELS

Steel No.	Chemical Composition Ranges and Limits, %						
	Carbon	Manganese	Silicon	Nickel	Chromium	Molybdenum	Vanadium
TS 14B35	0.33–0.38	0.75–1.00	0.20–0.35				
TS 14B50	0.48–0.53	0.75–1.00	0.20–0.35				
TS 40B37	0.35–0.40	0.70–0.90	0.20–0.35	....	....	0.08–0.15	
42B35	0.32–0.39	0.70–1.00	0.20–0.35	....	0.40–0.65	0.08–0.15	
42B40	0.37–0.45	0.70–1.00	0.20–0.35	....	0.40–0.65	0.08–0.15	
42B45	0.42–0.50	0.70–1.00	0.20–0.35	....	0.40–0.65	0.08–0.15	
42B50	0.47–0.55	0.70–1.00	0.20–0.35	....	0.40–0.65	0.08–0.15	
TS 43BV12	0.08–0.13	0.75–1.00	0.20–0.40	1.65–2.00	0.40–0.60	0.20–0.30	0.03 min
TS 43BV14	0.10–0.15	0.45–0.65	0.20–0.35	1.65–2.00	0.40–0.60	0.08–0.15	0.03 min
TS 46B12	0.10–0.15	0.45–0.65	0.20–0.35	1.65–2.00	....	0.20–0.30	
50B15	0.12–0.18	0.70–1.00	0.20–0.35	....	0.35–0.60		
50B20	0.17–0.23	0.70–1.00	0.20–0.35	....	0.35–0.60		
50B30	0.27–0.34	0.70–1.00	0.20–0.35	....	0.35–0.60		
50B35	0.32–0.39	0.70–1.00	0.20–0.35	....	0.35–0.60		
50B40	0.37–0.45	0.70–1.00	0.20–0.35	....	0.35–0.60		
50B44	0.42–0.50	0.70–1.00	0.20–0.35	....	0.35–0.60		
TS 50B46	0.43–0.50	0.75–1.00	0.20–0.35	....	0.20–0.35		
TS 50B50	0.48–0.53	0.75–1.00	0.20–0.35	....	0.40–0.60		
TS 50B60	0.55–0.65	0.75–1.00	0.20–0.35	....	0.40–0.60		
TS 51B60	0.55–0.65	0.75–1.00	0.20–0.35	....	0.70–0.90		
80B20	0.17–0.23	0.60–0.90	0.20–0.35	0.20–0.40	0.15–0.35	0.08–0.15	
80B30	0.27–0.34	0.55–0.80	0.20–0.35	0.20–0.40	0.15–0.35	0.08–0.15	
80B35	0.32–0.39	0.65–0.95	0.20–0.35	0.20–0.40	0.15–0.35	0.08–0.15	
TS 80B37	0.35–0.40	0.75–1.00	0.20–0.35	0.20–0.40	0.20–0.35	0.08–0.15	
TS 80B40	0.38–0.43	0.75–1.00	0.20–0.35	0.20–0.40	0.20–0.35	0.08–0.15	
TS 80B45	0.43–0.48	0.75–1.00	0.20–0.35	0.20–0.40	0.20–0.35	0.08–0.15	
80B50	0.47–0.55	0.70–1.00	0.20–0.35	0.20–0.40	0.25–0.50	0.08–0.15	
80B55	0.50–0.60	0.70–1.00	0.20–0.35	0.20–0.40	0.30–0.55	0.08–0.15	
80B60	0.55–0.65	0.70–1.00	0.20–0.35	0.20–0.40	0.30–0.55	0.08–0.15	
81B35	0.32–0.39	0.70–1.00	0.20–0.35	0.20–0.40	0.30–0.55	0.08–0.15	
TS 81B40	0.38–0.43	0.75–1.00	0.20–0.35	0.20–0.40	0.35–0.55	0.08–0.15	
TS 81B45	0.43–0.48	0.75–1.00	0.20–0.35	0.20–0.40	0.35–0.55	0.08–0.15	
81B50	0.47–0.55	0.75–1.05	0.20–0.35	0.20–0.40	0.35–0.60	0.08–0.15	
86B45*	0.43–0.48	0.75–1.00	0.20–0.35	0.40–0.70	0.40–0.60	0.15–0.25	
TS 94B15	0.13–0.18	0.75–1.00	0.20–0.35	0.30–0.60	0.30–0.50	0.08–0.15	
TS 94B17	0.15–0.20	0.75–1.00	0.20–0.35	0.30–0.60	0.30–0.50	0.08–0.15	
TS 94B20	0.17–0.22	0.75–1.00	0.20–0.35	0.30–0.60	0.30–0.50	0.08–0.15	
TS 94B30	0.28–0.33	0.75–1.00	0.20–0.35	0.30–0.60	0.30–0.50	0.08–0.15	
TS 94B40	0.38–0.43	0.75–1.00	0.20–0.35	0.30–0.60	0.30–0.50	0.08–0.15	

NOTE: These steels contain a minimum of 0.0005 % boron.

* This is a standard steel.

TABLE 6. HARDENABILITY BAND LIMITS FOR 1300H AND 2500H STEELS

J-distance, In.	1330H		1335H		1340H		2515H		2517H	
	Max	Min	Max	Min	Max	Min	Max	Min	Max	Min
1/16	56	49	58	51	60	53	45	38	46	39
1/8	56	47	57	49	60	52	44	37	46	38
3/16	55	44	56	47	59	51	43	33	46	35
1/4	53	40	55	44	58	49	42	30	45	31
5/16	52	35	54	38	57	46	41	27	44	28
3/8	50	31	52	34	56	40	40	24	43	25
7/16	48	28	50	31	55	35	38	22	42	23
1/2	45	26	48	29	54	33	37	20	41	21
9/16	43	25	46	27	52	31	35	..	39	20
5/8	42	23	44	26	51	29	34	..	37	
11/16	40	22	42	25	50	28	32	..	35	
3/4	39	21	41	24	48	27	31	..	34	
13/16	38	20	40	23	46	26	30	..	33	
7/8	37	..	39	22	44	25	29	..	32	
15/16	36	..	38	22	42	25	29	..	32	
1	35	..	37	21	41	24	28	..	31	
1 1/8	34	..	35	20	39	23	27	..	30	
1 1/4	33	..	34	..	38	23	26	..	29	
1 3/8	32	..	33	..	37	22	25	..	28	
1 1/2	31	..	32	..	36	22	24	..	27	
1 5/8	31	..	31	..	35	21	24	..	27	
1 3/4	31	..	31	..	35	21	23	..	26	
1 7/8	30	..	30	..	34	20	23	..	26	
2	30	..	30	..	34	20	22	..	25	

These values were adjusted to the nearest Rockwell C point and are used when points are selected and specified.

TABLE 7. HARDENABILITY BAND LIMITS FOR 3100H AND 3300H STEELS

J-distance, In.	Steel No.					
	3140H		3310H		3316H	
	Max	Min	Max	Min	Max	Min
1/16	60	53	43	36	47	40
1/8	60	52	43	36	47	39
3/16	59	50	43	35	47	38
1/4	59	49	42	35	46	38
5/16	58	47	42	34	46	37
3/8	57	45	42	33	46	37
7/16	57	43	41	32	45	36
1/2	56	41	41	31	45	35
9/16	55	38	41	30	45	34
5/8	54	36	40	30	45	33
11/16	53	34	40	29	45	33
3/4	52	33	40	29	45	32
13/16	51	32	39	28	45	32
7/8	50	31	39	28	44	32
15/16	49	30	38	27	44	31
1	48	30	38	27	44	31
1 1/8	46	29	37	26	44	31
1 1/4	44	28	37	26	43	31
1 3/8	43	28	37	26	43	31
1 1/2	42	27	36	26	43	31
1 5/8	41	27	36	25	42	31
1 3/4	40	26	36	25	42	30
1 7/8	40	26	35	25	42	30
2	39	25	35	25	41	30

These values were adjusted to the nearest Rockwell C point and are used when points are selected and specified.

TABLE 8. HARDENABILITY BAND LIMITS FOR 4000H STEELS

J-distance, In.	Steel No.													
	4032H		4037H		4042H		4047H		4053H		4063H		4068H	
	Max	Min	Max	Min	Max	Min	Max	Min	Max	Min	Max	Min	Max	Min
1/16	57	50	59	52	62	55	64	57	65	59	..	60	..	60
1/8	54	45	57	49	60	52	62	55	65	59	..	60	..	60
3/16	51	36	54	42	58	48	60	50	64	57	65	58	..	60
1/4	46	29	51	35	55	40	58	42	62	53	65	56	..	59
5/16	39	25	45	30	50	33	55	35	61	45	65	50	65	56
3/8	34	23	38	26	45	29	52	32	59	38	64	39	64	45
7/16	31	22	34	23	39	27	47	30	57	34	62	36	63	39
1/2	29	21	32	22	36	26	43	28	55	32	61	35	62	36
9/16	28	20	30	21	34	25	40	28	51	31	59	34	60	35
5/8	26	..	29	20	33	24	38	27	47	30	57	33	58	34
11/16	26	..	28	..	32	24	37	26	44	29	54	32	55	33
3/4	25	..	27	..	31	23	35	26	42	29	51	32	52	33
13/16	24	..	26	..	30	23	34	25	40	28	48	32	49	33
7/8	24	..	26	..	30	23	33	25	38	28	46	31	47	32
15/16	23	..	26	..	29	22	33	25	37	28	44	31	45	32
1	23	..	25	..	29	22	32	25	36	28	43	31	44	32
1 1/8	23	..	25	..	28	22	31	24	34	27	41	30	42	31
1 1/4	22	..	25	..	28	21	30	24	33	27	40	30	41	31
1 3/8	22	..	25	..	28	20	30	23	33	27	39	29	40	30
1 1/2	21	..	24	..	27	20	30	23	32	26	38	29	39	30
1 5/8	21	..	24	..	27	..	30	22	32	26	38	28	38	29
1 3/4	20	..	24	..	27	..	29	22	32	26	37	28	38	29
1 7/8	..	..	23	..	26	..	29	21	31	25	37	27	38	28
2	..	..	23	..	26	..	29	21	31	25	36	27	37	28

These values were adjusted to the nearest Rockwell C point and are used when points are selected and specified.

TABLE 9. HARDENABILITY BAND LIMITS FOR 4100H STEELS

J-distance, In.	Steel No.																	
	4118H		4130H		4135H		4137H		4140H		4142H		4145H		4147H		4150H	
	Max	Min	Max	Min	Max	Min	Max	Min	Max	Min	Max	Min	Max	Min	Max	Min	Max	Min
1/16	48	41	56	49	58	51	59	52	60	53	62	55	63	56	64	57	65	59
1/8	46	36	55	46	58	50	59	51	60	53	62	55	63	55	64	57	65	59
3/16	41	27	53	42	57	49	58	50	60	52	62	54	62	55	64	56	65	59
1/4	35	23	51	38	56	48	58	49	59	51	61	53	62	54	64	56	65	58
5/16	31	20	49	34	56	47	57	49	59	51	61	53	62	53	63	55	65	58
3/8	28	..	47	31	55	45	57	48	58	50	61	52	61	53	63	55	65	57
7/16	27	..	44	29	54	42	56	45	58	48	60	51	61	52	63	55	65	57
1/2	25	..	42	27	53	40	55	43	57	47	60	50	61	52	63	54	64	56
9/16	24	..	40	26	52	38	55	40	57	44	60	49	60	51	63	54	64	56
5/8	23	..	38	26	51	36	54	39	56	42	59	47	60	50	62	53	64	55
11/16	22	..	36	25	50	34	53	37	56	40	59	46	60	49	62	52	64	54
3/4	21	..	35	25	49	33	52	36	55	39	58	44	59	48	62	51	63	53
13/16	21	..	34	24	48	32	51	35	55	38	58	42	59	46	61	49	63	51
7/8	20	..	34	24	47	31	50	34	54	37	57	41	59	45	61	48	62	50
15/16	..	..	33	23	46	30	49	33	54	36	57	40	58	43	60	46	62	48
1	..	..	33	23	45	30	48	33	53	35	56	39	58	42	60	45	62	47
1 1/8	..	..	32	22	44	29	46	32	52	34	55	37	57	40	59	42	61	45
1 1/4	..	..	32	21	42	28	45	31	51	33	54	36	57	38	59	40	60	43
1 3/8	..	..	32	20	41	27	44	30	49	33	53	35	56	37	58	39	59	41
1 1/2	..	..	31	..	40	27	43	30	48	32	53	34	55	36	57	38	59	40
1 5/8	..	..	31	..	39	27	42	30	47	32	52	34	55	35	57	37	58	39
1 3/4	..	..	30	..	38	26	42	29	46	31	51	34	55	35	57	37	58	38
1 7/8	..	..	30	..	38	26	41	29	45	31	51	33	55	34	56	37	58	38
2	..	..	29	..	37	26	41	29	44	30	50	33	54	34	56	36	58	38

These values were adjusted to the nearest Rockwell C point and are used when points are selected and specified.

TABLE 10. HARDENABILITY BAND LIMITS FOR TS 4100H STEELS

J-distance, In.	Steel No.													
	TS 4130H		TS 4132H		TS 4137H		TS 4140H		TS 4142H		TS 4145H		TS 4150H	
	Max	Min	Max	Min	Max	Min	Max	Min	Max	Min	Max	Min	Max	Min
1/16	56	49	57	50	59	52	60	53	62	55	63	56	65	59
1/8	55	46	56	48	59	51	60	53	62	55	63	55	65	59
3/16	53	42	55	45	58	50	60	52	62	54	62	55	65	59
1/4	51	38	54	42	58	49	59	51	61	53	62	54	65	58
5/16	49	34	52	37	57	49	59	51	61	53	62	53	65	58
3/8	47	31	50	34	56	48	58	50	61	52	61	53	65	57
7/16	44	29	47	32	55	45	58	48	60	51	61	52	65	57
1/2	42	27	45	31	55	43	57	47	60	50	61	52	64	56
9/16	40	26	43	30	55	40	57	44	60	49	60	51	64	56
5/8	38	26	41	29	54	39	56	42	59	47	60	50	64	55
11/16	36	25	39	28	53	37	56	40	59	46	60	49	64	54
3/4	35	25	38	28	52	36	55	39	58	44	59	48	63	53
13/16	34	24	37	27	51	35	55	38	58	42	59	46	63	51
7/8	34	24	37	27	50	34	54	37	57	41	59	45	62	50
15/16	33	23	36	26	49	33	54	36	57	40	58	43	62	48
1	33	23	36	26	48	33	53	35	56	39	58	42	62	47
1 1/8	32	22	35	25	46	32	52	34	55	37	57	40	61	45
1 1/4	32	21	34	24	45	31	51	33	54	36	57	38	60	43
1 3/8	32	20	33	23	44	30	49	33	53	35	56	37	59	41
1 1/2	31	..	33	22	43	30	48	32	53	34	55	36	59	40
1 5/8	31	..	33	21	42	30	47	32	52	34	55	35	58	39
1 3/4	30	..	32	21	42	29	46	31	51	34	55	35	58	38
1 7/8	30	..	32	20	41	29	45	31	51	33	55	34	58	38
2	29	..	32	20	41	29	44	30	50	33	54	34	58	38

These values were adjusted to the nearest Rockwell C point and are used when points are selected and specified.

TABLE 11. HARDENABILITY BAND LIMITS FOR 4300H AND 4600H STEELS

J-distance, In.	Steel No.													
	4320H		4337H		4340H		E4340H		4620H		4621H		4640H	
	Max	Min	Max	Min	Max	Min	Max	Min	Max	Min	Max	Min	Max	Min
1/16	48	41	59	52	60	53	60	53	48	41	48	41	60	53
1/8	47	38	59	52	60	53	60	53	45	35	47	38	60	52
3/16	45	35	59	52	60	53	60	53	42	27	46	34	59	51
1/4	43	32	59	52	60	53	60	53	39	24	44	30	58	50
5/16	41	29	59	52	60	53	60	53	34	21	41	27	57	48
3/8	38	27	58	51	60	53	60	53	31	..	37	25	56	44
7/16	36	25	58	51	60	53	60	53	29	..	34	23	55	41
1/2	34	23	58	51	60	52	60	53	27	..	32	22	53	37
9/16	33	22	58	50	60	52	60	53	26	..	30	20	51	34
5/8	31	21	57	50	60	52	60	53	25	..	28	..	49	32
11/16	30	20	57	50	59	51	60	53	24	..	27	..	46	30
3/4	29	20	57	49	59	51	60	53	23	..	26	..	44	29
13/16	28	..	57	48	59	50	60	52	22	..	26	..	42	28
7/8	27	..	57	47	58	49	59	52	22	..	25	..	41	27
15/16	27	..	57	46	58	49	59	52	22	..	25	..	40	27
1	26	..	57	46	58	48	59	51	21	..	24	..	39	27
1 1/8	25	..	56	44	58	47	58	51	21	..	24	..	38	26
1 1/4	25	..	56	42	57	46	58	50	20	..	23	..	37	26
1 3/8	24	..	55	41	57	45	58	49	..	..	23	..	36	26
1 1/2	24	..	55	40	57	44	57	48	..	..	22	..	35	25
1 5/8	24	..	55	39	57	43	57	47	..	..	22	..	35	25
1 3/4	24	..	54	39	56	42	57	46	..	..	22	..	34	25
1 7/8	24	..	54	39	56	41	57	45	..	..	21	..	34	24
2	24	..	53	39	56	40	57	44	..	..	21	..	33	24

These values were adjusted to the nearest Rockwell C point and are used when points are selected and specified.

TABLE 12. HARDENABILITY BAND LIMITS FOR TS 4720H AND 4800H STEELS

J-distance, In.	Steel No.									
	TS 4720H		4812H		4815H		4817H		4820H	
	Max	Min	Max	Min	Max	Min	Max	Min	Max	Min
1/16	48	41	44	37	45	38	46	39	48	41
1/8	47	39	43	34	44	37	46	38	48	40
3/16	43	31	42	30	44	34	45	35	47	39
1/4	39	27	41	26	42	30	44	32	46	38
5/16	35	23	39	23	41	27	42	29	45	34
3/8	32	21	37	21	39	24	41	27	43	31
7/16	29	..	35	..	37	22	39	25	42	29
1/2	28	..	33	..	35	21	37	23	40	27
9/16	27	..	31	..	33	20	35	22	39	26
5/8	26	..	29	..	31	..	33	21	37	25
11/16	25	..	28	..	30	..	32	20	36	24
3/4	24	..	27	..	29	..	31	20	35	23
13/16	24	..	26	..	28	..	30	..	34	22
7/8	23	..	25	..	28	..	29	..	33	22
15/16	23	..	25	..	27	..	28	..	32	21
1	22	..	24	..	27	..	28	..	31	21
1 1/8	21	..	24	..	26	..	27	..	29	20
1 1/4	21	..	23	..	25	..	26	..	28	20
1 3/8	21	..	23	..	24	..	25	..	28	
1 1/2	20	..	22	..	24	..	25	..	27	
1 5/8	..	..	22	..	24	..	25	..	27	
1 3/4	..	..	21	..	23	..	25	..	26	
1 7/8	..	..	21	..	23	..	24	..	26	
2	..	..	21	..	23	..	24	..	25	

These values were adjusted to the nearest Rockwell C point and are used when points are selected and specified.

TABLE 13. HARDENABILITY BAND LIMITS FOR 5100H STEELS

J-distance, In.	Steel No.																			
	5120H		5130H		5132H		5135H		5140H		5145H		5147H		5150H		5152H		5160H	
	Max	Min	Max	Min	Max	Min	Max	Min	Max	Min	Max	Min	Max	Min	Max	Min	Max	Min	Max	Min
1/16	48	40	56	49	57	50	58	51	60	53	63	56	64	57	65	59	65	59	..	60
1/8	46	34	55	46	56	47	57	49	59	52	62	55	64	56	65	58	65	58	..	60
3/16	41	28	53	42	54	43	56	47	58	50	61	53	63	55	64	57	65	58	..	60
1/4	36	23	51	39	52	40	55	43	57	48	60	51	62	54	63	56	64	57	65	59
5/16	33	20	49	35	50	35	54	38	56	43	59	48	62	53	62	53	63	56	65	58
3/8	30	..	47	32	48	32	52	35	54	38	58	42	61	52	61	49	63	55	64	56
7/16	28	..	45	30	45	29	50	32	52	35	57	38	61	49	60	42	62	53	64	52
1/2	27	..	42	28	42	27	47	30	50	33	56	35	60	45	59	38	62	51	63	47
9/16	25	..	40	26	40	25	45	28	48	31	55	33	60	40	58	36	61	48	62	42
5/8	24	..	38	25	38	24	43	27	46	30	53	32	59	37	56	34	60	45	61	39
11/16	23	..	37	23	37	23	41	25	45	29	52	31	59	35	55	33	60	42	60	37
3/4	22	..	36	22	36	22	40	24	43	28	50	30	58	34	53	32	59	39	59	36
13/16	21	..	35	21	35	21	39	23	42	27	48	30	58	33	51	31	59	38	58	35
7/8	21	..	34	20	34	20	38	22	40	27	47	29	57	32	50	31	58	37	56	35
15/16	20	..	34	..	34	..	37	21	39	26	45	28	57	32	48	30	58	36	54	34
1	..	..	33	..	33	..	37	21	38	25	44	28	56	31	47	30	57	35	52	34
1 1/8	..	..	32	..	32	..	36	20	37	24	42	26	55	30	45	29	56	34	48	33
1 1/4	..	..	31	..	31	..	35	..	36	23	41	25	54	29	43	28	55	32	47	32
1 3/8	..	..	30	..	30	..	34	..	35	21	39	24	53	27	42	27	53	31	46	31
1 1/2	..	..	29	..	29	..	33	..	34	20	38	23	52	26	41	26	51	30	45	30
1 5/8	..	..	27	..	28	..	32	..	34	..	37	22	51	25	40	25	50	29	44	29
1 3/4	..	..	26	..	27	..	32	..	33	..	37	21	50	24	39	24	48	27	43	28
1 7/8	..	..	25	..	26	..	31	..	33	..	36	..	49	22	39	23	47	26	43	28
2	..	..	24	..	25	..	30	..	32	..	35	..	48	21	38	22	45	25	42	27

These values were adjusted to the nearest Rockwell C point and are used when points are selected and specified.

TABLE 14. HARDENABILITY BAND LIMITS FOR 6100H AND TS 8122H STEELS

J-distance, In.	Steel No.							
	6120H		6145H		6150H		TS 8122H	
	Max	Min	Max	Min	Max	Min	Max	Min
1/16	48	40	63	56	65	59	50	43
1/8	47	38	63	55	65	58	49	37
3/16	45	33	62	54	64	57	45	30
1/4	42	29	62	54	64	56	39	25
5/16	39	26	61	52	63	55	34	22
3/8	36	24	61	49	63	53	31	20
7/16	34	23	60	45	62	50	28	
1/2	33	22	59	42	61	47	27	
9/16	32	21	58	39	61	43	26	
5/8	31	21	57	38	60	41	25	
11/16	31	20	56	37	59	39	24	
3/4	31	20	55	36	58	38	23	
13/16	30	..	54	35	57	37	22	
7/8	30	..	52	35	55	36	22	
15/16	29	..	51	34	54	35	21	
1	29	..	50	33	52	35	20	
1 1/8	28	..	49	32	50	34		
1 1/4	28	..	48	31	48	32		
1 3/8	27	..	47	30	47	31		
1 1/2	26	..	46	29	46	30		
1 5/8	25	..	45	27	45	29		
1 3/4	25	..	44	26	44	27		
1 7/8	24	..	43	25	43	26		
2	23	..	42	24	42	25		

These values were adjusted to the nearest Rockwell C point and are used when points are selected and specified.

Table 15. Hardenability Band Limits for 8600H Steels

J-distance, In.	8617H		8620H		8622H		8625H		8627H		8630H		8635H		8637H		8640H 8641H		8642H		8645H		8650H		8653H		8655H		8660H	
	Max	Min	Max	Min	Max	Min	Max	Min	Max	Min	Max	Min	Max	Min	Max	Min	Max	Min	Max	Min	Max	Min	Max	Min	Max	Min	Max	Min	Max	Min
1/16	46	39	48	41	50	43	52	45	54	47	56	49	58	51	59	52	60	53	62	55	63	56	65	59	65	59	..	60	..	60
1/8	44	33	47	37	49	39	51	41	52	43	55	46	57	49	58	51	60	53	62	54	63	56	65	58	65	59	..	59	..	60
3/16	41	27	44	32	47	34	48	36	50	38	54	43	56	47	58	50	60	52	62	53	63	55	65	57	65	59	..	59	..	60
1/4	38	24	41	27	44	30	46	32	48	35	52	39	55	45	57	48	59	51	61	52	63	54	64	57	65	58	..	58	..	60
5/16	34	20	37	23	40	26	43	29	45	32	50	35	54	42	56	45	59	49	61	50	62	52	64	56	65	58	..	57	..	60
3/8	31	..	34	21	37	24	40	27	43	29	47	32	53	39	55	42	58	46	60	48	61	50	63	54	64	57	..	56	..	59
7/16	28	..	32	..	34	22	37	25	40	27	44	29	51	35	54	39	57	42	59	45	61	48	63	53	64	57	..	55	..	58
1/2	27	..	30	..	32	20	35	23	38	26	41	28	50	33	53	36	55	39	58	42	60	45	62	50	64	56	..	54	..	57
9/16	26	..	29	..	31	..	33	22	36	24	39	27	48	31	51	34	54	36	57	39	59	41	61	47	63	55	..	52	..	55
5/8	25	..	28	..	30	..	32	21	34	24	37	26	46	30	49	32	52	34	55	37	58	39	60	44	63	53	65	49	..	53
11/16	24	..	27	..	29	..	31	20	33	23	35	25	45	29	47	31	50	32	54	34	56	37	60	41	63	49	65	46	..	50
3/4	23	..	26	..	28	..	30	..	32	22	34	24	43	28	46	30	49	31	52	33	55	35	59	39	62	47	64	43	..	47
13/16	23	..	25	..	27	..	29	..	31	21	33	23	41	27	44	29	47	30	50	32	54	34	58	37	62	45	64	41	..	45
7/8	22	..	25	..	26	..	28	..	30	21	33	22	40	26	43	28	45	29	49	31	52	33	58	36	62	44	63	40	..	44
15/16	22	..	24	..	26	..	28	..	30	20	32	22	38	25	41	27	44	28	48	30	51	32	57	35	62	43	63	39	..	43
1	21	..	24	..	25	..	27	..	29	20	31	21	37	25	40	26	42	28	46	29	49	31	56	34	61	42	62	38	65	42
1 1/8	21	..	23	..	25	..	27	..	28	..	30	21	36	24	39	25	41	26	44	28	47	30	55	33	61	40	61	37	64	40
1 1/4	20	..	23	..	24	..	26	..	28	..	30	20	35	23	37	25	39	26	42	28	45	29	53	32	60	39	60	35	64	39
1 3/8	..	..	23	..	24	..	26	..	28	..	29	20	34	23	36	24	38	25	41	27	43	28	52	31	59	38	59	34	63	38
1 1/2	..	..	23	..	24	..	26	..	27	..	29	..	33	23	36	24	38	25	40	27	42	28	50	31	59	37	58	34	62	37
1 5/8	..	..	23	..	24	..	26	..	27	..	29	..	33	23	35	24	37	24	40	26	42	27	49	30	59	36	57	33	62	36
1 3/4	..	..	22	..	24	..	25	..	27	..	29	..	33	22	35	24	37	24	39	26	41	27	47	30	59	35	56	33	61	36
1 7/8	..	..	22	..	24	..	25	..	27	..	29	..	32	22	35	23	37	24	39	26	41	27	46	29	58	35	55	32	60	35
2	..	..	22	..	24	..	25	..	27	..	29	..	32	22	35	23	37	24	39	26	41	27	45	29	58	34	53	32	60	35

These values were adjusted to the nearest Rockwell C point and are used when points are selected and specified.

TABLE 16. HARDENABILITY BAND LIMITS FOR TS 8600H AND 8700H STEELS

J-distance, In.	Steel No.																	
	TS 8617H		TS 8620H		TS 8622H		TS 8625H		TS 8627H		8720H		8740H		8742H		8750H	
	Max	Min	Max	Min	Max	Min	Max	Min	Max	Min	Max	Min	Max	Min	Max	Min	Max	Min
1/16	46	39	48	41	50	43	52	45	54	47	48	41	60	53	62	55	65	59
1/8	44	33	47	37	49	39	51	41	52	43	47	38	60	53	62	55	65	59
3/16	41	27	44	32	47	34	48	36	50	38	45	35	60	52	62	54	65	58
1/4	38	24	41	27	44	30	46	32	48	35	42	30	60	51	61	53	64	57
5/16	34	20	37	23	40	26	43	29	45	32	38	26	59	49	61	52	64	57
3/8	31	..	34	21	37	24	40	27	43	29	35	24	58	46	60	49	63	56
7/16	28	..	32	..	34	22	37	25	40	27	33	22	57	43	59	47	63	55
1/2	27	..	30	..	32	20	35	23	38	26	31	21	56	40	58	44	62	53
9/16	26	..	29	..	31	..	33	22	36	24	30	20	55	37	57	41	62	51
5/8	25	..	28	..	30	..	32	21	34	24	29	..	53	35	56	39	61	49
11/16	24	..	27	..	29	..	31	20	33	23	28	..	52	34	54	37	60	47
3/4	23	..	26	..	28	..	30	..	32	22	27	..	50	32	53	35	60	45
13/16	23	..	25	..	27	..	29	..	31	21	26	..	49	31	52	34	59	43
7/8	22	..	25	..	26	..	28	..	30	21	26	..	48	31	51	33	59	42
15/16	22	..	24	..	26	..	28	..	30	20	25	..	46	30	49	32	58	40
1	21	..	24	..	25	..	27	..	29	20	25	..	45	29	48	31	58	39
1 1/8	21	..	23	..	25	..	27	..	28	..	24	..	43	28	46	30	57	37
1 1/4	20	..	23	..	24	..	26	..	28	..	24	..	42	28	45	29	55	35
1 3/8	..	..	23	..	24	..	26	..	28	..	23	..	41	27	43	29	53	34
1 1/2	..	..	23	..	24	..	26	..	27	..	23	..	40	27	42	28	52	33
1 5/8	..	..	23	..	24	..	26	..	27	..	23	..	39	27	42	28	51	33
1 3/4	..	..	22	..	24	..	25	..	27	..	23	..	39	27	41	28	50	32
1 7/8	..	..	22	..	24	..	25	..	27	..	22	..	38	26	41	28	49	32
2	..	..	22	..	24	..	25	..	27	..	22	..	38	26	40	27	48	32

These values were adjusted to the nearest Rockwell C point and are used when points are selected and specified.

TABLE 17. HARDENABILITY BAND LIMITS FOR 9200H, 9300H, AND 9800H STEELS

J-distance, In.	Steel No.											
	9260H		9261H		9262H		9310H		9840H		9850H	
	Max	Min	Max	Min	Max	Min	Max	Min	Max	Min	Max	Min
1/16	..	60	..	60	..	60	43	36	60	53	65	59
1/8	..	60	..	60	..	60	43	35	60	53	65	59
3/16	65	57	..	60	..	60	43	35	60	53	65	59
1/4	64	53	65	59	..	60	42	34	60	53	65	59
5/16	63	46	65	56	..	58	42	32	60	53	65	59
3/8	62	41	64	52	65	56	42	31	60	53	65	59
7/16	60	38	64	47	65	53	42	30	60	53	65	58
1/2	58	36	63	42	64	48	41	29	60	52	65	58
9/16	55	36	62	39	63	42	40	28	59	51	65	58
5/8	52	35	60	37	62	39	40	27	59	51	65	58
11/16	49	34	57	36	61	37	39	27	59	49	65	58
3/4	47	34	54	36	59	37	38	26	58	48	65	58
13/16	45	33	49	35	57	36	37	26	58	46	65	57
7/8	43	33	45	35	55	36	36	26	58	45	64	57
15/16	42	32	43	34	51	35	36	26	57	44	64	57
1	40	32	42	34	48	35	35	26	57	43	64	56
1 1/8	38	31	39	33	45	34	35	26	56	41	63	54
1 1/4	37	31	38	32	43	33	35	25	55	39	62	52
1 3/8	36	30	37	31	41	33	34	25	55	38	62	50
1 1/2	36	30	37	31	39	32	34	25	55	36	61	49
1 5/8	35	29	36	30	38	31	34	25	54	36	61	48
1 3/4	35	29	36	30	37	31	34	25	54	35	61	47
1 7/8	35	28	35	29	36	30	33	24	53	34	60	47
2	34	28	35	29	36	30	33	24	53	34	60	47

These values were adjusted to the nearest Rockwell C value and are used when points are selected and specified.

TABLE 18. HARDENABILITY BAND LIMITS FOR BORON H-STEELS

J-distance, In.	Steel No. TS 14B35H		TS 14B50H		TS 40B37H		TS 46B12H		TS 50B46H		TS 50B60H		TS 51B60H		TS 81B40H		TS 81B45H		86B45H		TS 94B17H	
	Max	Min	Max	Min	Max	Min	Max	Min	Max	Min	Max	Min	Max	Min	Max	Min	Max	Min	Max	Min	Max	Min
1/16	57	50	63	56	59	52	44	37	63	56	..	60	..	60	60	53	63	56	63	56	46	39
1/8	56	49	62	55	58	50	44	37	62	54	..	60	..	60	60	53	63	56	63	56	46	38
3/16	55	31	61	51	57	46	43	35	61	52	..	60	..	60	60	53	63	56	62	55	46	36
1/4	53	24	60	46	56	38	42	33	60	50	..	60	..	60	60	52	63	56	62	54	45	33
5/16	50	22	59	35	55	31	42	28	59	41	..	60	..	60	60	52	63	55	62	54	45	29
3/8	45	20	57	30	53	28	41	24	58	32	..	59	..	59	59	51	63	54	61	53	44	26
7/16	36	..	54	29	52	26	40	22	57	31	..	57	..	58	59	49	62	53	61	52	43	24
1/2	30	..	51	28	50	25	39	21	56	30	65	53	..	57	58	48	62	51	60	52	41	22
9/16	27	..	46	27	48	24	38	20	54	29	65	47	..	54	58	45	61	48	60	51	40	20
5/8	26	..	41	27	45	23	37	..	51	28	64	42	..	50	57	42	60	44	60	51	38	
11/16	25	..	37	26	42	23	36	..	47	27	64	39	..	44	57	39	60	41	59	50	37	
3/4	24	..	35	26	38	22	35	..	43	26	64	37	65	41	56	37	59	39	59	50	35	
13/16	24	..	34	25	35	22	34	..	40	26	63	36	65	40	56	35	58	38	59	49	34	
7/8	23	..	33	25	32	22	33	..	38	25	63	35	64	39	55	34	57	37	59	48	32	
15/16	23	..	32	24	30	21	32	..	37	25	63	34	64	38	54	33	57	36	58	46	31	
1	22	..	31	23	29	21	31	..	36	24	62	34	63	37	54	32	56	35	58	45	30	
1 1/8	21	..	30	22	28	20	29	..	35	23	60	33	61	36	52	30	55	34	58	42	28	
1 1/4	20	..	29	21	27	..	27	..	34	22	58	31	59	34	50	29	53	32	58	39	27	
1 3/8	..	..	28	20	27	..	26	..	33	21	55	30	57	33	48	28	52	31	57	37	26	
1 1/2	..	..	27	..	26	..	25	..	32	20	53	29	55	31	46	27	50	30	57	35	25	
1 5/8	..	..	26	..	26	..	24	..	31	..	51	28	53	30	45	26	49	29	57	34	25	
1 3/4	..	..	25	..	25	..	23	..	30	..	49	27	51	28	43	26	47	28	57	32	24	
1 7/8	..	..	24	..	24	..	23	..	29	..	47	26	49	27	42	25	45	28	56	32	23	
2	..	..	23	..	24	..	22	..	28	..	44	25	47	25	40	25	43	27	56	31	23	

These values were adjusted to the nearest Rockwell C point and are used when points are selected and specified.

STAINLESS STEELS

Standard and tentative stainless steels fall into three categories: (1) ferritic, which are nonhardenable; (2) martensitic, which can be heat-treated; and (3) austenitic, which are work-hardenable. Typical compositions and applications are given in Table 19. Comparative fabricating qualities are given in Table 20. Further notes on the three types of stainless are:

Martensitic Types. 403, 410, 414, 416, 420, 431, 440A, 440B, 440C. The steels in this group are hardenable, because they have a high carbon-to-chromium ratio to undergo the necessary transformations. Their best mechanical as well as corrosion-resisting properties are found in the hardened condition. Tensile strengths run from 70,000 to 105,000 psi when annealed, to 125,000 to 200,000 psi when hardened.

Thermal conductivity of the martensitic stainless steels is low, but still the best of the stainless family. They are especially suitable for hot working or forging. These types are air-hardening and must be slowly cooled (or annealed) after forging to prevent cracking. Their cold-forming characteristics are fair.

The martensitic types are well suited for most moderately corrosive applications requiring high strength, hardness, and resistance to abrasion and wet and dry erosion.

Ferritic Types. 405, 430, 430F, 446. The ferritic types are not hardenable, because they have a lower carbon-to-chromium ratio which reduces the effect of the transformation responsible for hardening. But while they cannot be hardened by heat-treating, hardness can be increased slightly by cold working. In practice, however, these steels are only rarely used in the cold-worked condition, because the small increase in hardness obtainable is accompanied by a relatively large decrease in ductility.

The ferritics have a lower coefficient of expansion than the austenitic stainless types—about the same or slightly less than carbon steel. They also offer good resistance to oxidation and corrosion, and are often selected for these reasons.

While their ductility is less than that of the chromium-nickel types, they can be fabricated without difficulty by such methods as forming, bending, spinning, and shallow drawing. And they can be buffed to a high finish. Welding is possible, but welds may tend to be brittle—which usually can be overcome with proper heat-treatment. A special welding grade, type 430T, contains titanium and has excellent weldability.

These types have higher resistance to corrosion and oxidation than the martensitic types. Like the martensitic types, the ferritics also are magnetic in all conditions.

Austenitic Types. 301, 302, 302B, 303, 304, 304L, 305, 308, 309, 310, 310S, 314, 321, 347, 316, 316L, 317. Steels in this group possess the highest corrosion resistance in the family of stainless steels. Austenitic stainless steels are inherently tough and well adapted for fabrication by deep drawing and other similar means. They can be welded easily and also can be soldered by proper technique. Type 303, a special free-machining grade, contains such additives as sulfur or selenium for excellent machinability.

The austenitic stainless steels cannot be hardened by heat-treatment; they also are nonmagnetic. These steels have unusually high ductility. Tensile strength in the annealed condition is considerably above that of mild steel—approximately 90,000 psi as compared with approximately 50,000 psi. Austenitic stainless has a yield point of approximately 35,000 psi—comparable with that of mild steel.

These steels can be work-hardened to extremely high tensile and yield strengths

TABLE 19. STANDARD AND TENTATIVE STANDARD TYPES OF STAINLESS STEELS

Steel No.	Carbon	Manganese Max	Silicon Max	Chromium	Nickel	Other	Remarks*
301	Over 0.08–0.20	2.00	1.00	16–18	6–8	....	Rapid work-hardening
302	Over 0.08–0.20	2.00	1.00	17–19	8–10	....	General-purpose chromium-nickel type
302B	Over 0.08–0.20	2.00	2–3	17–19	8–10	....	Higher scaling resistance
303	0.15 max	2.00	1.00	17–19	8–10	†	Free-machining
304	0.08 max	2.00	1.00	18–20	8–11	....	General purpose—welding
304L	0.03 max	2.00	1.00	18–20	8–11	....	Weldments for severely corrosive conditions
305	0.12 max	2.00	1.00	17–19	10–13	....	Low work-hardening rate
308	0.08 max	2.00	1.00	19–21	10–12	....	Welding rod and electrodes
309	0.20 max	2.00	1.00	22–24	12–15	....	High scale resistance and good strength at high temperatures
309S	0.08 max	2.00	1.00	22–24	12–15	....	Weldments with high scale resistance, good strength
310	0.25 max	2.00	1.50	24–26	19–22	....	Like 309 but even higher heat resistance
310S	0.08 max	2.00	1.50	24–26	19–22	....	Like 309S but even higher heat resistance
314	0.25 max	2.00	1.50–3	23–26	19–22	....	Highest heat resistance
316	0.10 max	2.00	1.00	16–18	10–14	Mo 2–3	Higher resistance to certain corrosives which can affect other stainless steels, such as halide salts. Also high creep resistance at high temperatures. TS 316 is alloy conservation version. 316L is version for weldments. 317 is higher in corrosion and creep resistance
TS 316‡	0.10 max	2.00	1.00	16–18	10–14	Mo 1.75–2.5	
316L	0.03 max	2.00	1.00	16–18	10–14	Mo 1.75–2.5	
317	0.10 max	2.00	1.00	18–20	11–14	Mo 3–4	

TABLE 19. STANDARD AND TENTATIVE STANDARD TYPES OF STAINLESS STEELS (*Continued*)

Steel No.	Carbon	Manganese Max	Silicon Max	Chromium	Nickel	Other	Remarks*
321	0.08 max	2.00	1.00	17–19	8–11	Ti 5 × C min	For weldments subject to severely corrosive conditions and/or service in 800–1650 F temperature range
347	0.08 max	2.00	1.00	17–19	9–12	Cb 10 × C min	
TS 347‡	0.08 max	2.00	1.00	17–19	9–12	Cb 8 × C min	
TS 347A‡	0.08 max	2.00	1.00	17–19	9–12	Cb-Ta 8 × C min	
403	0.15 max	1.00	0.50	11.5–13	....	....	Steam-turbine blades
405	0.08 max	1.00	1.00	11.5–13.5	....	Al 0.10–0.30	Low hardenability
410	0.15 max	1.00	1.00	11.5–13.5	....	....	General-purpose. Hardenable by heat-treatment
414	0.15 max	1.00	1.00	11.5–13.5	1.25–2.5	....	High-strength version of 410
416	0.15 max	1.25	1.00	12–14	....	†	Free-machining
420	Over 0.15	1.00	1.00	12–14	....	....	High hardness from heat-treatment. Cutlery
430	0.12 max	1.00	1.00	14–18	....	....	General-purpose chromium type
430F	0.12 max	1.25	1.00	14–18	....	†	Free-machining
431	0.20 max	1.00	1.00	15–17	1.25–2.5	....	Special-purpose hardenable
440A	0.6–0.75	1.00	1.00	16–18	....	Mo 0.75 max	High strength and corrosion resistance. Hardenable by heat-treatment
440B	0.75–0.95	1.00	1.00	16–18	....	Mo 0.75 max	Cutlery grade
440C	0.95–1.20	1.00	1.00	16–18	....	Mo 0.75 max	Ball-bearing grade
446	0.35 max	1.50	1.00	23–27	....	Mo 0.25 max	Resistance to high-temperature scaling and hot sulfur-bearing gases

* Remarks are indicative of primary characteristics or uses, but cannot be definitive because of space limitations.
† Phosphorus or sulfur or selenium, 0.07 min; zirconium or molybdenum, 0.60 max.
‡ TS grades were developed to meet limitations set by the National Production Authority in 1951.

TABLE 20. COMPARATIVE FABRICATING PROPERTIES OF ANNEALED STAINLESS STEELS

	Ferritic Types Nonhardenable				Martensitic Types Hardenable by Heat-treating								
	405	430	430F	446	403	410	414	416	420	431	440A	440B	440C
Cutting:													
Shearing	2	2	2	2	2	2	2	2	2	2	2	2	2
Blanking	2	2	2	2	2	2	2	2	2	2	2	2	2
Perforating	2	2	2	2	2	2	2	2	2	2	2	2	2
Notching	2	2	2	2	2	2	2	2	2	2	2	2	2
Slitting	2	2	2	2	2	2	2	2	2	2	2	2	2
Expanding	2	2	2	3*a*	2	2	2	2	2	2	2	2	2
Punching	2	2	2	2	2	2	2	2	2	2	2	2	2
Machining:													
Turning	2	2	1	2	2	2	2	1	3*f*	2	3*f*	3*f*	3*f*
Milling	2	2	1	2	2	2	2	1	3*f*	2	3*f*	3*f*	3*f*
Drilling	2	2	1	2	2	2	2	1	2	2	3*f*	3*f*	3*f*
Tapping	2	2	1	2	2	2	2	1	2	2	3*f*	3*f*	3*f*
Boring	2	2	1	2	2	2	2	1	3*f*	2	3*f*	3*f*	3*f*
Reaming	2	2	1	2	2	2	2	1	3*f*	2	3*f*	3*f*	3*f*
Planing	2	2	1	2	2	2	2	1	3*f*	2	3*f*	3*f*	3*f*
Broaching	2	2	1	2	2	2	2	1	3*f*	2	3*f*	3*f*	3*f*
Working:													
Forging	2	2	3*d*	2	2	2	2	3*d*	2	2	2	2	2
Bending	2	2	3*g*	3*a*	2	2	2	3*g*	3*n*	2	3*n*	3*n*	3*n*
Roll forming	2	2	3*g*	3*a*	2	2	2	3*g*	3*n*	2	3*n*	3*n*	3*n*
Drawing	2	2	3*g*	3*a*	2	2	3*n*	3*g*	3*n*	3*n*	3*n*	3*n*	3*n*
Stamping	2	2	3*g*	2	2	2	3*n*	3*g*	3*n*	3*n*	3*n*	3*n*	3*n*
Hydroforming	2	2	3*g*	2	2	2	2	3*g*	3*n*	2	3*n*	3*n*	3*n*
Spinning	2	1	3*g*	3*a*	2	2	3*n*	3*g*	3*n*	3*n*	4	4	4
Joining:													
Fusion welding	1	2	4	3*a*	3*l*	3*l*	3*l*	4	3*l*	3*l*	3*l*	3*l*	3*l*
Resistance welding	2	2	4	3*a*	2	2	2	4	3*l*	2	2	2	2
Silver brazing	2	2	4	3*a*	2	2	2	4	2	2	2	2	2
High-temperature soldering	2	2	4	3*a*	2	2	2	4	2	2	2	2	2
Soldering	2	2	2	2	2	2	2	2	2	2	2	2	2
Riveting	1	1	4	2	1	1	2	4	4	2	4	4	4
Finishing:													
Polishing	2	2	3*m*	2	2	2	2	3*m*	2	2	2	2	2
Plating	2	2	3*m*	2	2	2	2	3*m*	2	2	2	2	2

CODE: 1. Exceptionally good.
2. Well suited.
3. Precautions necessary. See notes on p. 25-43.
4. Try another material.

TABLE 20. COMPARATIVE FABRICATING PROPERTIES OF ANNEALED STAINLESS STEELS (*Continued*)

	Austenitic Types Work-hardenable																	
	18-8's							Higher Cr-Ni						18-8 Stabilized		18-8 with Molybdenum		
	301	302	302B	303	304	304L	305	308	309	309S	310	310S	314	321	347	316	316L	317
Cutting:																		
Shearing	2	2	2	2	2	2	2	2	2	2	2	2	2	2	2	2	2	2
Blanking	2	2	2	2	2	2	2	2	2	2	2	2	2	2	2	2	2	2
Perforating	2	2	2	2	2	2	2	2	2	2	2	2	2	2	2	2	2	2
Notching	2	2	2	2	2	2	2	2	2	2	2	2	2	2	2	2	2	2
Slitting	2	2	2	2	2	2	2	2	2	2	2	2	2	2	2	2	2	2
Expanding	2	2	2	2	2	2	2	2	2	2	2	2	2	2	2	2	2	2
Punching	2	2	2	2	2	2	2	2	2	2	2	2	2	2	2	2	2	2
Machining:																		
Turning	2	2	2	1	2	2	3*b*	3*b*	3*b*	3*b*	3*b*	3*b*	3*b*	3*b*	3*b*	3*b*	3*b*	3*b*
Milling	2	2	2	1	2	2	3*b*	3*b*	3*b*	3*b*	3*b*	3*b*	3*b*	3*b*	3*b*	3*b*	3*b*	3*b*
Drilling	2	2	2	1	2	2	3*b*	3*b*	3*b*	3*b*	3*b*	3*b*	3*b*	3*b*	3*b*	3*b*	3*b*	3*b*
Tapping	2	2	2	1	2	2	3*b*	3*b*	3*b*	3*b*	3*b*	3*b*	3*b*	3*b*	3*b*	3*b*	3*b*	3*b*
Boring	2	2	2	1	2	2	3*b*	3*b*	3*b*	3*b*	3*b*	3*b*	3*b*	2	2	3*b*	3*b*	3*b*
Reaming	2	2	2	1	2	2	3*b*	2	2	2	2	2	2	2	2	2	2	2
Planing	2	2	2	1	2	2	3*b*	3*b*	3*b*	3*b*	3*b*	3*b*	3*b*	2	2	2	2	2
Broaching	2	2	2	1	2	2	3*b*	3*b*	3*b*	3*b*	3*b*	3*b*	3*b*	2	2	3*b*	3*b*	3*b*
Working:																		
Forging	2	2	2	3*d*	2	2	2	2	3*e*	3*e*	3*e*	3*e*	3*e*	2	2	3*e*	3*e*	3*e*
Bending	2	2	2	3*g*	2	2	2	2	2	2	2	2	2	2	2	2	2	2
Roll forming	2	2	2	3*g*	2	2	2	2	2	2	2	2	2	2	2	2	2	2
Drawing	3*h*	2		3*g*	2	2	1	3*i*	3*i*	3*i*	3*i*	3*i*	3*i*	2	2	2	2	2
Stamping	3*h*	2		3*g*	2	2	2	2	2	2	2	2	2	2	2	2	2	2
Hydroforming	3*h*	2		3*g*	2	2	2	2	2	2	2	2	2	2	2	2	2	2
Spinning	3*h*	2		3*g*	2	2	1	2	2	2	2	2	2	2	2	2	2	2
Joining:																		
Fusion welding	2	2	3*j*	4	2	1*k*	2	2	2	2	2	2	3*j*	1*k*	1*k*	2	1*k*	2
Resistance welding	2	2	3*j*	4	2	2	2	2	2	2	2	2	3*j*	2	2	2	2	2
Silver brazing	2	2	2	4	2	1*k*	2	2	2	2	2	2	2	1*k*	1*k*	2	1*k*	2
High-temperature soldering	2	2	2	4	2	1*k*	2	2	2	2	2	2	2	1*k*	1*k*	2	1*k*	2
Soldering	2	2	2	2	2	2	2	2	2	2	2	2	2	2	2	2	2	2
Riveting	3*h*	2	2	3*g*	2	2	1	2	2	2	1	1	1	2	2	2	2	2
Finishing:																		
Polishing	2	2	2	3*m*	2	2	2	2	2	2	2	2	2	3*m*	2	2	2	2
Plating	2	2	2	3*m*	2	2	2	2	2	2	2	2	2	3*m*	2	2	2	2

and still maintain good ductility. Tensile strengths up to approximately 200,000 psi can be developed in flat-rolled products, and in excess of 300,000 psi in fine-drawn wire. Because of this work-hardening characteristic, it is sometimes necessary to reanneal the material between operations in certain fabricating processes such as drawing or spinning. Some grades, especially types 301 and 302, are frequently used in harder tempers for structural and other high-strength purposes in aircraft, railway cars, and trucks and trailers.

Notes on Fabricating Properties (Table 20). Relative workability ratings for the different stainless steels in the austenitic, martensitic, and ferritic groups are indicated. In all cases, suitability is based on the realization that these are stainless steels of three separate families.

In the martensitic steels, type 410 is considered as the mean; variations are indicated as deviating from it. Similarly, in the ferritic steels, type 430 is the mean, and in the austenitic steels, type 302 is the mean.

There is no attempt to correlate among groups. The basic performance of one group may be better or worse than another.

NOTES:

a. Slightly greater care required because of the high chromium content.

b. Because of the "gumminess" of these austenitic steels, slower speeds and low feeds are necessary.

c. For reference, comparative machinability figures (based on cold-finished stock):

Bessemer screw stock	100
416	85
430F	90
303	75
302	45
430	54

d. Free-machining steels are somewhat susceptible to hot shortness. Temperatures must be controlled within close limits.

e. These steels, because of the elevated temperature strength, require higher pressures and special die considerations.

f. More power, smaller cuts necessary because of the greater hardness of annealed steel.

g. Additions to improve machining characteristics lessen formability.

h. Higher work-hardening characteristics of this grade make it less suitable for any but the more shallow draws.

i. Somewhat more difficult to draw than the basic type; generally drawn only when ultimate use necessitates these alloys.

j. Higher silicon content may necessitate special care to avoid cracking.

k. These steels show freedom from harmful carbide precipitation caused by welding heat. They are not necessarily easier to weld so far as the mechanical operations of welding are concerned.

l. Can be welded but preheating, postheating, or both usually will be required to prevent cracking.

m. Discontinuities in high polishes and finishes.

n. Formability restricted because of the greater hardness of annealed steel.

TOOL AND DIE STEELS

Selection of tool and die steels is complicated by their many special properties. The five principal ones are (1) heat resistance, (2) abrasion resistance, (3) shock resistance, (4) resistance to movement or distortion in hardening, and (5) cutting ability. No one steel can develop these properties to the maximum extent. Therefore, hundreds of steels have been developed. Besides variations in chemical composition, proprietary variations in methods of production complicate the selection of tool and die steels.

Classification System. To clarify the situation, several groups have worked for years to classify the tool and die steels, and they have issued various reports. These sources are the Gorham Tool Co., the American Iron & Steel Institute, the SAE, the Joint Industry Committee, and the ASM. Finally in 1954, the ASM adopted the designations that appear in Table 21. The six categories of tool and die steels are now listed as water-hardening, cold-work (these include oil- and water-hardening types, plus high-carbon high-chromium steels), shock-resisting, hot-work, high-speed, and special-purpose steels.

TABLE 21. CLASSIFICATION AND APPROXIMATE COMPOSITIONS OF TOOL AND DIE STEELS

Type	AISI-SAE	JIC	ASM	Carbon	Manganese	Silicon	Chromium	Vanadium	Tungsten	Molybdenum	Cobalt	Nickel
Water-hardening tool steels:												
0.80 carbon	W1-0.80C	W8	IA	0.70–0.85	*	*	*					
0.90 carbon	W1-0.90C	W9	IA	0.85–0.95	*	*	*					
1.00 carbon	W1-1.00C	W10	IA	0.95–1.10	*	*	*					
1.20 carbon	W1-1.20C	W12	IA	1.10–1.30	*	*	*					
0.90 carbon-V	W2-0.90C	W9V	IC	0.85–0.95	*	*	*	0.15–0.35				
1.00 carbon-V	W2-1.00C	W10V	IC	0.95–1.10	*	*	*	0.15–0.35				
1.00 carbon-VV	W3-1.00C	W10VV	IC	0.95–1.10	*	*	*	0.35–0.50				
Carbon-chromium	W4	......	IB	0.70–1.30	*	*	0.25					
Carbon-chromium	W5	......	IB	0.70–1.30	*	*	0.50					
Carbon-chrome-vanadium	W6	......	....	0.70–1.30	*	*	0.25	0.25				
Carbon-chrome-vanadium	W7	......	....	0.70–1.30	*	*	0.50	0.20				
Shock-resisting tool steels:												
Chromium-tungsten	S1	S1	IIID and E	0.50	0.25	0.35†	1.40	0.20	2.25	0.40‡		
Silicon-molybdenum	S2	S2	IIIB	0.50	0.40	1.00	.....	0.25‡	.....	0.50		
Low chromium-tungsten	S3	......	....	0.50	.....	.....	0.75	....	1.00			
Silicon-manganese	S4	......	IIIC	0.50	0.80	2.00						
Silicon-manganese-moly	S5	S5	IIIC	0.55	0.80	2.00	0.30‡	0.25‡	.....	0.40‡		
Cold-work tool steels:												
Oil-hardening:												
Low manganese	O1	O1	IIA2	0.90	1.20	0.25	0.50	0.20‡	0.50			
High manganese	O2	O2	IIA2	0.90	1.60	0.25	0.35‡	0.20‡	.....	0.30‡		
Molybdenum graphitic	O6	O6	....	1.45	0.75	1.00	.....	....	.....	0.25		
Tungsten	O7	......	IIA3	1.20	.....	.....	0.75	....	1.75	0.25‡		
Air-hardening:												
5 % chrome air-hardening	A2	A2	IIB2	1.00	0.60	0.25	5.25	0.40‡	.....	1.10		
2 % manganese	A4	......	IIB1	1.00	2.00	.....	1.00					
3 % manganese	A5	......	IIB1	1.00	3.00	.....	1.00					
Manganese	A6	......	IIB1	0.70	2.00	.....	1.00					
High carbon high chromium:												
High carbon high chromium	D1	......	IID1	1.00	.....	.....	12.00	....	.....	1.00		
High carbon high chromium (air)	D2	A1	IID2	1.50	0.40	0.40	12.00	0.80‡	.....	0.90	0.60‡	
High carbon high chromium (oil)	D3	O5	IIC1	2.15	0.35†	0.35†	12.00	0.80‡	0.75‡	0.80‡	.....	0.50‡
High carbon high chromium-moly	D4	......	IID3	2.25	.....	.....	12.00	....	.....	1.00		
High carbon high chromium-cobalt	D5	A3	IID2	1.50	0.40	0.40	12.00	0.80‡	.....	0.90	3.10	
High carbon high chromium-tungsten	D6	......	IIC1	2.25	.....	1.00	12.00	....	1.00			

Hot-work tool steels:												
Chromium base:												
Chrome-moly-V........	H11	H5	IVB	0.35	0.30	1.00	5.00	0.40	......	1.50		
Chrome-moly-tungsten......	H12	H1	IVB	0.35	0.30	1.00	5.00	0.25‡	1.25	1.50		
Chrome-moly-VV...........	H13	H6	IVB	0.35	0.30	1.00	5.00	0.90	......	1.50		
5 chrome 5 tungsten..	H14	......	IVC	0.40	.....	.....	5.00	....	5.00			
5 chrome 5 moly...........	H15	......	IVE1	0.40	.....	.....	5.00	....	......	5.00		
8 chrome 8 tungsten........	H16	......	IVD	0.55	.....	.....	7.00	....	7.00			
Tungsten base:												
9 tungsten 2 chromium......	H20	......	IVF1	0.35	.....	.....	2.00	....	9.00			
9 tungsten 4 chromium......	H21	H2	IVF1	0.32	0.30	0.20	3.25	0.40	9.00			
12 tungsten...............	H22	......	IVF2	0.35	.....	.....	2.00	....	11.00			
12 tungsten 12 chromium...	H23	......	....	0.30	.....	.....	12.00	....	12.00			
15 tungsten 3 chromium....	H24	......	IVF3	0.45	.....	.....	3.00	....	15.00			
15 tungsten 4 chromium....	H25	......	IVF3	0.25	.....	.....	4.00	....	15.00			
18 tungsten..............	H26	......	IVF4	0.50	.....	.....	4.00	1.00	18.00			
Molybdenum base:												
8 molybdenum.............	H41	......	....	0.65	.....	.....	4.00	1.00	1.50	8.00		
5 molybdenum.............	H42	......	....	0.60	.....	.....	4.00	2.00	6.00	5.00		
8 molybdenum 2 vanadium..	H43	......	....	0.55	.....	.....	4.00	2.00	......	8.00		
High-speed tool steels:												
Tungsten base:												
Tungsten 18-4-1............	T1	T1	VC1	0.70	0.30	0.30	4.10	1.10	18.00			
Tungsten 18-4-2............	T2	T2	VC2	0.80	0.30	0.30	4.10	2.10	18.50	0.80		
Tungsten 18-4-3............	T3	T3	VC3	1.05	0.30	0.30	4.10	3.25	18.50	0.70		
Cobalt-tungsten 18-4-1-5....	T4	T4	VD2	0.75	0.30	0.30	4.10	1.00	18.00	0.80	5.00	
Cobalt-tungsten 18-4-2-8....	T5	T5	VD3	0.80	0.30	0.30	4.10	1.75	18.50	0.80	8.00	
Cobalt-tungsten 18-4-2-12...	T6	T6	VD4	0.80	0.30	0.30	4.10	1.75	20.00	0.80	12.00	
Tungsten 14-4-2............	T7	......	....	0.80	.....	.....	4.00	2.00	14.00	0.80		
Cobalt-tungsten 13-4-5-5....	T8	T8	VD1	0.80	0.30	0.30	4.10	2.00	14.00	0.80	5.00	
Tungsten 18-4-4............	T9	......	....	1.20	.....	.....	4.00	4.00	18.00			
Cobalt-tungsten 13-4-5-5....	T15	......	....	1.55	.....	.....	4.00	5.00	12.00	.....	5.00	
Molybdenum base:												
Molybdenum 8-2-1.........	M1	M1	VA2	0.80	0.30	0.30	4.00	1.15	1.50	8.50		
Molybdenum-tungsten 6-6-2	M2	M2	VA3	0.83	0.30	0.30	4.10	1.90	6.25	5.00		
Molybdenum-tungsten 6-6-3	M3	M3	....	1.15	0.30	0.30	4.10	3.25	5.75	5.25		
Molybdenum-tungsten 6-6-4	M4	M4	VA4	1.30	0.30	0.30	4.25	4.25	5.75	5.25		
Cobalt-moly-tungsten 5-4-1-12.............	M6	......	....	0.80	.....	.....	4.00	1.50	4.00	5.00	12.00	
Moly-tungsten-columbium 4-5-1.............	M8	......	....	0.80	.....	.....	4.00	1.50	5.00	5.00	......	1.25C†
Molybdenum 8-0-2.........	M10	......	VA1	0.85	.....	.....	4.00	2.00	......	8.00		
Cobalt-moly-tungsten 8-2-1-5	M30	......	VB1	0.85	.....	.....	4.00	1.25	2.00	8.00	5.00	
Cobalt-moly-tungsten 8-2-2-8	M34	......	VB2	0.85	.....	.....	4.00	2.00	2.00	8.00	8.00	
Cobalt-moly-tungsten 5-6-2-5	M35	......	VB3	0.85	.....	.....	4.00	2.00	6.00	5.00	5.00	
Cobalt-moly-tungsten 6-6-2-8	M36	M36	VB4	0.85	.....	.....	4.00	2.00	6.00	5.00	8.00	

TABLE 21. CLASSIFICATION AND APPROXIMATE COMPOSITIONS OF TOOL AND DIE STEELS (*Continued*)

Type	AISI-SAE	JIC	ASM	Carbon	Manganese	Silicon	Chromium	Vanadium	Tungsten	Molybdenum	Cobalt	Nickel
Special-purpose tool steels:												
Low-alloy types:												
1% chromium	L1	...	...	1.00	...	...	1.25					
Chrome-vanadium	L2	...	IIIA	0.50–1.10	...	...	1.00	0.20				
Chrome-vanadium	L3	...	VIJ	1.00	...	...	1.50	0.20				
Chrome-vanadium	L4	...	...	1.00	0.60	...	1.50	0.20				
Chrome-manganese	L5	...	...	1.00	1.00	...	1.00	...	...	0.25		
Nickel-chromium	L6	O4	VIF1 and E	0.75	0.70†	0.25	0.85†	0.25‡	...	0.50‡	...	1.50†
Chromium	L7	O3	...	1.00	0.35	0.25	1.40	...	...	0.40		
Carbon-tungsten types:												
Carbon-W	F1	...	...	1.00	...	...	...	...	1.25			
Carbon-WW	F2	...	VIK	1.25	...	...	...	...	3.50			
Carbon-chrome-WW	F3	...	VIK	1.25	...	...	0.75	...	3.50			
Mold steels:												
Straight iron	P1	...	VIA	0.10 max								
Chrome-nickel-moly	P2	...	...	0.07 max	...	...	1.25	...	...	0.20	...	0.50
Nickel alloy	P3	...	VIB	0.10 max	...	...	0.60	...	...	...	...	1.25
5% chrome air-hard	P4	...	...	0.07 max	...	...	5.00					
2% chromium	P5	...	...	0.10 max	...	...	2.25					
Chrome-moly	P20	...	...	0.30	...	...	0.75	...	...	0.25		

* Limits on manganese, silicon, and chromium are not normally required on special and extra grades in lieu of Shepherd hardenability limits (see p. 25-52). On standard and commercial grades, limits are generally 0.35 Mn max, 0.35 Si max, 0.15 Cr max in standard grade, and 0.20 Cr max in commercial grade. Total of the three elements not to exceed 0.75%.

† May be present in percentages other than shown.

‡ Optional element.

To supplement the basic information that follows, a classified list of over 900 tool and die steels is given, starting on page 25-52. This listing provides the names of suppliers, the composition, typical properties, applications, and basic heat-treating data. When a new steel not yet listed by the ASM is involved, the suffix A has been added to the identification to avoid confusion with the existing or future standards covering tool and die steels.

Types of Tool Steels. By accepted practice, tool and die steels are considered to fall into these categories:

WATER-HARDENING. Plain carbon and the carbon-vanadium tool steels still have widest applications, because of their low cost and abrasion-resisting and shock-resisting qualities combined with ease of machinability and ability to take a keen cutting edge.

The 0.80 carbon steel (W8) is best for tools subject to shock, the 1.00 carbon steel (W10) is the general-purpose tool and die steel in widest use, and the 1.20 carbon steel (W12) provides maximum abrasion resistance and keenest cutting edge on tools.

The carbon-vanadium steels are shallower hardening and allow a wider range of hardening temperature without increase in grain size (which would cut toughness).

Water-hardening tool steels should not be used:

1. Where maximum safety in hardening is desired.
2. In applications requiring maximum shock resistance.
3. Where maximum abrasion resistance is required.
4. Where tools or dies operate substantially above room temperature.
5. Where tool or die must have high hardness all the way through.

OIL-HARDENING. This group of steels was developed for maximum safety in hardening and minimum dimensional change after heat-treatment. They are preferred, therefore, for tools or dies with adjacent thin and thick sections, sharp corners, or numerous holes. They have better wear resistance than the water-hardening grades but are not quite so good in shock resistance. Machining properties are good and material cost is relatively low.

AIR-HARDENING. Even safer hardening is provided by this group. In addition, these steels provide more wear resistance in general, but at a loss of shock resistance. More costly than oil-hardening steels, they are sometimes employed for hot work (such as bending dies) where wear resistance is of greater importance than heat resistance.

SHOCK-RESISTING. This group was developed for resistance to shock in cold-battering operations and where abrasion resistance is of secondary importance. The chromium-tungsten steel (S1) has some hot-work properties and can be used for hot-shearing and drop-hammer dies with operating temperatures from 500 to 800 F.

HOT-WORK. This group must combine red hardness with wear resistance and shock resistance, using less carbon and alloys than high-speed steels. Applications include header dies, gripper dies, extrusion dies, permanent-mold and die-casting dies, hot punches, shear blades, and trim dies. The low-alloy medium-carbon forging die blocks are included in this group by JIC (but not by SAE).

HIGH-SPEED. High-speed steels were developed primarily to provide red hardness and high abrasion resistance with some shock resistance for use in cutting tools. The familiar tungsten 18-4-1 (T1) is still the most widely used high-speed steel.

Molybdenum steels have equivalent qualities and are cheaper, but require more care in heat-treatment and have a greater tendency to decarburize. These are not serious objections if modern equipment is available.

Cobalt-bearing steels have higher red hardness with the usual loss of other properties.

Effect of Alloying Elements. CARBON. As carbon is added, up to about 0.85%, capacity of the steel to harden is increased. More carbon does not materially increase the hardness, but it does increase wear resistance. Carbide segregation occurs in varying degree in the more highly alloyed steels and may cause tool failure or difficulty in machining.

MANGANESE. When present in amounts above 0.60 to 0.80%, manganese counteracts the brittleness caused by sulfur and makes the steel easier to forge. As an alloying addition it provides an inexpensive method of increasing hardenability. Steels with alloying percentage of manganese should be oil-quenched.

SILICON. Small percentages of silicon, usually from 0.15 to 0.35%, deoxidize steels and make them easier to hot-forge and roll. For alloying purposes, silicon may range up to 2% or higher. In combination with molybdenum, manganese, or chromium it increases the strength and toughness of shock tools. Care must be taken in hardening and forging to avoid decarburization of high-silicon steels.

CHROMIUM. Chromium is added in amounts up to about 12% to increase hardening characteristics. In combination with high carbon, it contributes wear resistance and toughness. Its tendency to form carbides is greater than manganese, but less than tungsten. Low- and medium-chromium steels will frequently change size in hardening even more than plain carbon steels. Chromium raises the hardening temperature.

VANADIUM. Small amounts of vanadium refine the carbide structure of tool steel, which otherwise is difficult to break up during hot working. In larger amounts, up to about 4%, vanadium increases the red hardness of the steel. Vanadium has a very strong tendency to form carbides.

TUNGSTEN. Tungsten alloys with iron in all proportions, enabling it to resist the effects of high temperature, and forms hard, abrasion-resisting carbides. Tungsten enables steels to resist the softening that takes place during tempering. For this reason it is sometimes included in hot-work steels.

MOLYBDENUM. Molybdenum contributes to deep hardening, being second only to carbon in this respect. It raises the resistance to elevated temperatures, and in amounts between 0.25 and 1.50% it increases toughness. In some high-speed steels, molybdenum replaces part of the tungsten.

COBALT. Cobalt is added, usually to high-speed steels, to increase red hardness. This permits cutting tools to operate at higher speeds. Cobalt raises the temperature necessary for hardening, increases surface decarburization, and decreases toughness.

NICKEL. Nickel has little effect on the hardenability of tool steel, but it does add to the toughness and possibly to the wear resistance when used with a hardening alloy such as chromium.

Selecting a Tool Steel. The basic principles involved in selecting a tool or die steel can be easily learned. But the job of applying these principles to a particular operation requires skill and experience. Table 22 gives an approximate rating to each of the basic characteristics involved in selection except cost. Cost varies from about 25 cents to more than $1.50 a pound.

The basic characteristics are:

NONDEFORMING PROPERTIES. This column in Table 22 refers to the distortion normally obtained in quenching from the hardening temperature. Steels rated

TABLE 22. APPROXIMATE METALLURGICAL CHARACTERISTICS FOR SELECTING TOOL STEELS

AISI-SAE No.	Non-deforming Properties	Safety in Hardening	Toughness	Red Hardness	Wear Resistance	Cutting Ability	Machinability, Annealed	Resistance to Decarburization	Depth of Hardness	Hardness, Rockwell C
Water-hardening Tool Steels										
W1-0.80C	20	D	60	20	20	20	A	A	Shallow	56–65
W1-0.90C	20	D	58	20	22	24	A	A	Shallow	56–65
W1-1.00C	20	D	57	20	25	25	A	A	Shallow	56–65
W1-1.20C	20	D	54	20	27	28	A	A	Shallow	56–65
W2-0.90C	20	D	60	20	25	23	A	A	Shallow	56–65
W2-1.00C	20	D	59	20	27	25	A	A	Shallow	56–65
W3-1.00C	20	D	65	20	30	25	A	A	Shallow	56–65
Shock-resisting Tool Steels										
S1	42	C	82	35	21	22	D	D+	Medium	45–57
S2	42	O-C W-E	90	25	21	22	D	E	Medium	O-54–58 W-54–60
S5	42	O-C W-E	88	28	21	22	D	E	Medium	O-54–58 W-54–60
Cold-work Tool Steels										
O1	86	B	30	25	30	33	C	C	Medium	50–62
O2	90	B	37	28	32	33	C	C	Medium	57–62
O6	81	B	35	28	60	27	B	C	Medium	50–63
A2	78	A	20	45	68	34	D	D+	Deep	57–60
D2	78	A	10	50	85	40	E	D	Deep	58–60
D3	81	B	10	52	90	52	E	D	Deep	58–62
D5	78	A	13	50	85	40	E	D	Deep	57–59
Hot-work Tool Steels										
H11	30	C	50	60	60	10	D+	D	Deep	43–51
H12	30	C	50	60	60	10	D	D	Deep	43–51
H13	30	C	50	60	62	10	D+	D	Deep	43–51
H21	30	C	40	72	50	10	D	D	Deep	47–50
High-speed Tool Steels										
T1	30	C	25	72	75	63	D	C	Deep	63–65
T2	30	C	15	75	87	78	D	C	Deep	63–65
T3	30	C	10	78	102	95	D	C	Deep	63–65
T4	30	D	20	88	77	76	D	E	Deep	63–65
T5	30	D	8	92	84	96	D	E	Deep	63–65
T6	30	D	4	103	86	98	D	E	Deep	63–65
T8	30	D	14	85	90	85	D	E	Deep	63–65
M1	30	D	35	72	78	68	D	E	Deep	63–65
M2	30	D	35	76	78	68	D	E	Deep	63–65
M3	30	D	29	79	90	79	D	E	Deep	63–65
M36	30	D	17	91	88	96	D	E	Deep	63–65
Special-purpose Tool Steels										
L6	30	C	75	25	28	28	C	C	Medium	48–62
L7	87	C	32	30	32	33	C	C	Medium	60–62

Letter code: A, best; B, very good; C, good; D, fair; E, poor; O, oil-quenched; W, water-quenched.
Number code: Higher numbers represent better characteristics.

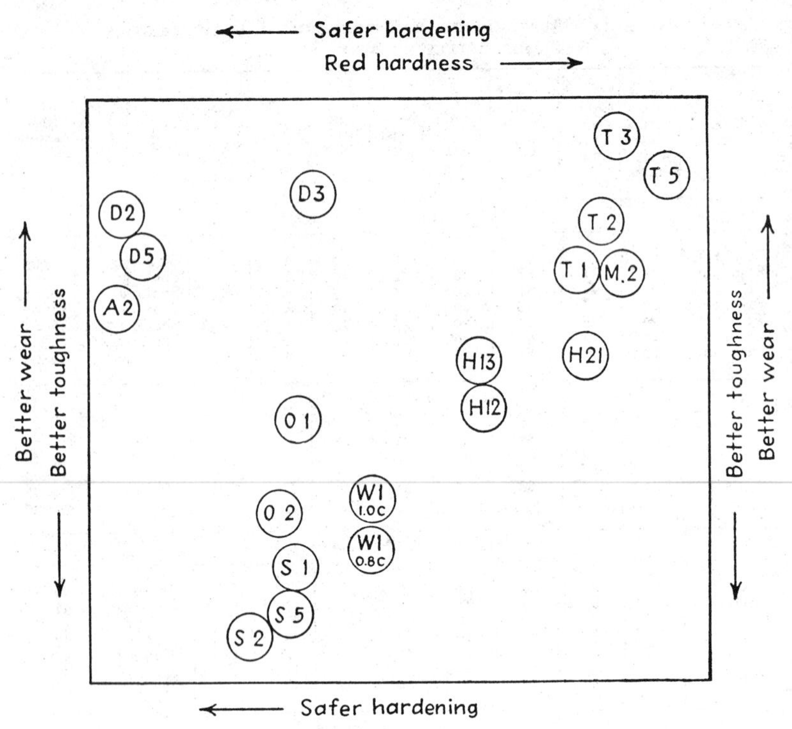

FIG. 4. Initial selection of tool steels can be made from this chart. More exact data are given in Table 22.

above 75 can be machined close to size before hardening. Water-hardening steels distort most, except when the depth of case is slight compared with core diameter.

SAFETY IN HARDENING. The major point here is freedom from cracking in hardening intricate sections.

TOUGHNESS. The ability to resist softening is important in hot-work tools. Toughness varies directly with the content of alloying elements that form hard, stable carbides.

WEAR RESISTANCE. This varies with alloy content, and markedly with hardness of a given steel. Thus wear resistance can be increased, in general, by operating at the upper end of the hardness range when other factors will permit.

CUTTING ABILITY. This factor is a combination of several of the other characteristics. A direct rating of the ability of the steel to hold a keen cutting edge is useful, however, in selecting steels for cutting tools.

MACHINABILITY. Many ratings of machinability exist in percentage terms. The available data differ so widely for a given tool steel, and vary so much with changes in hardness and variations in composition from the nominal, that no direct percentages are given in Table 22. The letter code gives a truer indication of the relative machinability of these steels.

RESISTANCE TO DECARBURIZATION. This factor influences the type of heat-treating equipment required and the amount of material that has to be removed after hardening. Steels rated E (poor) should be protected from decarburization during the heating cycle.

DEPTH OF HARDNESS. Steels quenched in water generally have a shallow case over a soft core. When high strength is required all the way through a large section, a steel with deep-hardening characteristics should be selected.

HARDNESS. The Rockwell C range indicates the normal hardness that can be obtained through the tempering range of the steel.

Combined Characteristics. To some extent the desirable characteristics of tool steels are mutually opposed. Of the four principal ones, toughness goes up as wear resistance goes down, red hardness goes up as a safe hardness goes down, and vice versa. This is not completely true because by alloy combinations it is possible to increase both toughness and wear resistance, for example, though usually at a loss in some other characteristic.

Figure 4 shows roughly the relationship of a number of widely used tool steels in relation to these four factors. It is impossible to position each steel accurately in relation to every other, because the factors do not always vary directly. Even so, this chart should help in learning the general relationships of the steels and in simplifying the first steps in selection.

Steps in Selection. In selecting a steel, the starting point is always a plain carbon steel such as W-10. This group is the cheapest and easiest to machine; so it should be selected when it will meet requirements. If slightly deeper hardness penetration is required, go to a lower-carbon steel such as W-9. If shallower hardness penetration or finer grain structure is required, go to a carbon-vanadium steel such as W-10V.

Check the metallurgical characteristics of the carbon steels in the table. If they are inadequate for the job, return to the chart and move in the desired direction. After selecting a steel in the required area, check the table to see if it will meet all the requirements. The metallurgical characteristics must be examined in the light of the requirements of the job, the quantity to be produced, and the relative cost of the steels.

If none of these steels seems quite right, it is advisable to consult with vendors.

The average plant can simplify its tool-steel problem by standardizing on as few steels as possible.

Surface Allowances. Because tool steels are a quality material, they are usually governed by rigid specifications. Many companies specify physical, chemical, and metallurgical tolerances when ordering these steels and test the steels upon receipt.

Surface allowances must be included for decarburization and surface defects in hot-rolled or cold-drawn steels. A typical allowance is:

Dimension, In.	*Allowance per Side, In.*
Up to ½	0.016
Over ½ to 1	0.031
Over 1 to 2	0.047
Over 2 to 4	0.063
Over 4 to 5	0.078

Centerless ground stock is normally specified to be free of surface defects, but an allowance must be made for decarburization. Typical allowance is 0.005 in.

It is also necessary to establish dimensional tolerances when steel is ordered to specified sizes.

DIRECTORY OF TOOL AND DIE STEELS

A Classified List of More Than 900 Tool and Die Steels Based on SAE Classification Methods with Names of Suppliers, Composition, Typical Properties, Applications, and Basic Heat-treat Data

Designed to aid in the selection of tool and die steels, this list supplements the basic information on pages 25-44 to 25-46. The AISI-SAE classification method is applied, supplemented by ASM classifications in the same pattern and the Gorham Tool Co. numbers for high-speed steels. Similar designations have been applied to all other groups, followed by the suffix A to avoid confusion with future standards.

Each group includes a brief description of properties and applications. The applications are not intended as a complete list or to indicate that the steels referred to are the best for those purposes. Temperature data for forging and heat-treating are included for groups not covered elsewhere. These indicate the general range and hence the equipment required but will not apply exactly to every steel in the group.

Specialties such as drill rod, ground flat stock, and forging die blocks are not included. The companies listed have cooperated by supplying the data on their steels. Their names and addresses appear in full on page 25-79.

Water-hardening Tool Steels

W1—Carbon

SAE specifications on page 25-44 and below.

Plain carbon tool steel is made in four grades of quality:* Special, Extra, Standard, and Commercial. Special and Extra conform to rigid macroscopic, microscopic, or hardenability specifications, Special being the highest quality. They are suitable for tools and dies requiring steel of uniform high quality. Standard and Commercial grades are not always made in electric furnaces and meet less rigid processing requirements. They are suitable for many general-purpose applications or for short-run jobs. Steels are listed below according to grade, with most of them available in any desired carbon content. The standard carbon range is usually 0.95 to 1.10.

* On Special and Extra grades limits on manganese, silicon, and chromium are not generally required in lieu of the following Shepherd hardenability limits:

0.70–0.85 C and 0.85–0.95 C			0.95–1.10 C and 1.10–1.30 C		
	Pene-tration	Fracture Grain Size		Pene-tration	Fracture Grain Size
Shallow	10 max	8 min	Shallow	8 max	9 min
Regular	9 to 13	8 min	Regular	7 to 11	9 min
Deep	12 min	8 min	Deep	10 to 16	8 min

On Standard and Commercial grades, the following limits on composition are generally required:

	Mn	Si	Cr
Standard	0.35 max	0.35 max	0.15 max
Commercial	0.35 max	0.35 max	0.20 max

Total of manganese, silicon, and chromium not to exceed 0.75%.

W1—Carbon Tool Steels (See Page 25-52)

Company	Grade 1 Special	Grade 2 Extra	Grade 3 Standard	Commercial
Allegheny Ludlum	Pompton Spe cial	Pompton Extra	Pompton	Corinth
Amalgamated	Creston No. 8	Creston No. 7	Creston No. 6	Creston No. 5
Atlantic	Atsco Special	Atsco Extra	Atsco	Standard
Atlas	XX-95, X-12, X-10	Refined 8, Refined 10	Alpha-8	Maple Leaf
Bethlehem	XXX	XX	XCL	X
Boyd-Wagner	Best		Standard	
Braeburn	Coldie*	Extra	Standard	
Carpenter	No. 11 Special	No. 11 Extra, H-9 Extra	No. 11 Comet	No. 11 Titan
Columbia	Special	Extra	Standard	Electrex
Craine-Schrage		Crasco Green Label		
Crucible	Crescent Special	Sanderson Extra	Black Diamond	Granada
	La Belle Cold Striking Die*	La Belle Extra		
Darwin & Milner		SSC Brand	Standard	
Delaware		Extra	Standard	
Diehl	Grade "A"			
Disston	Best	Extra	Standard	
Faitoute	Special	Extra	Carbon	Duplex
Firth Sterling	Special ASV, FS Special	FS Extra	Sterling	Silver Star
Great Western	Special, Cold Header*	Extra	Regular	
Hawkridge	Hawk Special	Hawk Brand	Standard	
Houghton & Richards	Special Carbon	Carbon		
	Special Heading Die*			
Hoyland	Special, Cold Header*	SS Extra	Regular	
Jessop	Washington	Lion Extra		
	New Process Cold Header*		Lion	
Kloster	Pure-Ore Special	Pure-Ore Extra	Pure-Ore Standard	
Latrobe	Special, Carbon Cold Header*	Extra	Standard	
McDonald	Macco Special			
McInnes	Special	Superior	Misco A	Misco B
Milne	Orange Label	Red Label	White Label	Green Label
Peninsular		Penco—W. H.		
Republic	Special	Extra	Standard	Tool Grade
Simonds	Blue Label	Diamond S	Green Label	
Tennessee	Special			
Uddeholm		UHB Extra	UHB	
Universal-Cyclops	Special	Extra	Standard	
Vanadium-Alloys	Colonial No. 14	Extra L	Red Star	
Vulcan Crucible	Special, Striking Die†	Extra	Fort Pitt	

* Specially selected heats or sections of the billet for cold-heading dies.
† Special control of hardenability.

Water-hardening Tool Steels (*Continued*)

W2—Carbon-vanadium

SAE specifications on page 25-44. Vanadium provides a finer grain, giving better toughness than the plain carbon steels. These steels have a wider hardening range than plain carbon steels.

Allegheny Ludlum—Python—0.25 V
Amalgamated—Car-Van Special, Car-Van—0.20 V
Atlas—Special Alloy 10, Special Alloy 8—0.20 V
Bethlehem—Best, Superior—0.20 V
Braeburn—Special—0.20 V
Carpenter—No. 11 Special Vanadium, No. 11 Extra Vanadium, Nitro
Columbia—Vanadium Special, Vanadium Extra, Vanadium Standard—0.20 V
Craine-Schrage—Crasco Special Vanadium—0.20 V
Crucible—Alva Extra, Granada Vanadium—0.20 V
Darwin & Milner—Darwin CV—0.20 V
Disston—Vatool—0.25 V
Firth Sterling—Sterling V—0.20 V
Hawkridge—Hawk Vanadium—0.20 V
Houghton & Richards—Vanadium—0.18 V
Jessop—Washington Special—0.20 V
Latrobe—Special Carbon with V, Carbon Cold Header with V, Renown, Standard Carbon with V—0.20 V
McDonald—Macco B-29—0.20 V
McInnes—Vanadium Crucible—0.20 V
Milne—Special Vanadium—0.20 V
Peninsular—Blue Label—0.20 V
Republic—Dumost—0.30 V
Ryerson—VD—0.18 V
Simonds—Blue Label V
Uddeholm—UHB-VA—0.20 V, UHB-19VA Cold Header—0.15 V
Universal-Cyclops—Draco—0.20 V
Vanadium-Alloys—Colonial No. 7, Elvandie, Red Star Vanadium—0.20 V
Vulcan Crucible—Special Vanadium, Vanadium Striking Die—0.30 V

W3—Carbon-vanadium

SAE specifications on page 25-44. Higher vanadium content than the V group provides better toughness.

Allegheny Ludlum—Double Vanadium—1.00 C, 0.45 V
Universal-Cyclops—Draco DV—as desired C, 0.45 V
Vanadium-Alloys—Colhead—1.00 C, 0.45 V

W4—Carbon-chromium

ASM composition: 0.60–1.40 C, 0.25 Cr. Chromium content increases the depth of hardness and reduces the danger of soft spots. A number of the steels are available in several carbon ranges.

Firth Sterling—Diamond M—1.30 C, 0.28 Mn, 0.20 Si, 0.28 Cr

W5—Carbon-chromium

ASM composition: 0.60–1.40 C, 0.50 Cr. Higher chromium content than W4 for increased depth of hardness.

Allegheny Ludlum—Crow—1.20 C, 0.50 Cr
Atlas—Q—1.20 C, 0.50 Cr
Columbia—Waterdie Extra—1.00 C, 0.50 Cr
Columbia—Waterdie Standard—1.00 C, 0.50 Cr
Disston—844-A—1.00 C, 0.32 Mn, 1.12 Cr
Disston—844-B—0.85 C, 0.32 Mn, 0.72 Cr
Kloster—Pure-Ore No. 14—1.00 C, 0.50 Cr
Latrobe—CFS—1.00 C, 0.50 Cr
Republic—C-C—1.10 C, 0.60 Cr
Vulcan Crucible—KR—1.10 C, 0.30 Mn, 0.60 Cr

W7—Carbon-chrome-vanadium

ASM composition: 0.60–1.40 C, 0.50 Cr, 0.20 V. Addition of vanadium to W5 provides more toughness because of the finer grain structure.

Bethlehem—Piston—1.15 C, 0.55 Cr, 0.20 V
Boyd-Wagner—Very Best—1.05 C, 0.30 Mn, 0.20 Si, 0.50 Cr, 0.10 V
Peninsular—Black Label—1.10 C, 0.55 Cr, 0.20 V
Universal-Cyclops—Hercules—1.00 C, 0.50 Cr, 0.20 V

W8A—Carbon-molybdenum

Molybdenum content provides deeper hardening, increased toughness, and red hardness.

Delaware—Molel—0.75 C, 0.35 Mn, 0.20 Si, 0.10 Mo
Ryerson—VD Chisel—0.77 C, 0.30 Mn, 0.23 Si, 0.18 Mo
Vanadium-Alloys—Croman—1.20 C, 0.80 Mn, 0.50 Cr, 0.60 Mo

Shock-resisting Tool Steels

S1—Chromium-tungsten

SAE composition: 0.50 C, 0.25 Mn, 0.35 Si, 1.40 Cr, 0.20 V, 2.25 W, 0.40 Mo opt. Excellent shock resistance for both hot and cold operation. Though red hardness is not as good as in the H-steels, it is better than other S-steels. Suitable for punches, chisels, pneumatic tools.

Allegheny Ludlum—Seminole Hard—0.52 C, 0.25 Mn, 0.70 Si, 1.50 Cr, 0.25 V, 2.00 W

Allegheny Ludlum—Seminole Medium—0.42 C, 0.25 Mn, 0.30 Si, 1.50 Cr, 0.25 V, 2.00 W

Amalgamated—Super-Shock—0.50 C, 0.75 Si, 1.15 Cr, 0.20 V, 2.50 W

Atlantic—Tuncro—0.50 C, 0.25 Mn, 0.25 Si, 1.40 Cr, 0.25 V, 2.00 W

Atlas—Falcon 4—0.45 C, 0.25 Mn, 0.30 Si, 1.50 Cr, 0.25 V, 2.00 W

Atlas—Falcon 6—0.55 C, 0.25 Mn, 0.30 Si, 1.50 Cr, 0.25 V, 2.00 W

Bethlehem—67 Chisel—0.50 C, 0.75 Si, 1.15 Cr, 0.20 V, 2.50 W

Braeburn—Vibro—0.50 C, 0.25 Mn, 0.25 Si, 1.40 Cr, 0.25 V, 1.90 W

Columbia—Buster—0.50 C or 0.60 C, 0.30 Mn, 0.70 Si, 1.20 Cr, 0.25 V, 2.15 W

Crucible—Atha Pneu—0.50 C, 1.25 Cr, 0.20 V, 2.75 W

Darwin & Milner—Ideor—0.50 C, 0.25 Mn, 0.30 Si, 1.50 Cr, 0.25 V, 2.00 W

Disston—Keystone—0.50 C, 0.30 Mn, 0.20 Si, 1.10 Cr, 0.20 V, 2.00 W

Firth Sterling—JS Punch—0.50 C, 0.30 Mn, 1.40 Cr, 0.25 V, 2.25 W

Houghton & Richards—No. 225—0.50 C, 0.25 Mn, 1.50 Cr, 0.25 V, 2.50 W, 0.50 max Mo

Jessop—Top Notch—0.50 C, 0.30 Mn, 0.70 Si, 1.15 Cr, 0.20 V, 2.40 W

Kloster—Chiz Alloy, Super Alloy—0.50 C, 0.35 Mn, 0.95 Si, 1.25 Cr, 0.22 V, 2.37 W

Latrobe—XL Chisel—0.55 C, 0.25 Mn, 0.25 Si, 1.50 Cr, 0.20 V, 2.00 W

McDonald—Macco Foolproof—0.55 C, 0.25 Mn, 0.25 Si, 1.40 Cr, 0.30 V, 2.40 W

McInnes—Special Punch and Chisel—0.50 C, 0.25 Mn, 0.70 Si, 1.50 Cr, 0.25 V, 2.15 W

Milne—A 020—0.50 C, 1.50 Cr, 0.25 V, 2.25 W

Peninsular—Brown Label—0.50 C, 0.20 Mn, 0.75 Si, 1.20 Cr, 0.20 V, 2.50 W

Republic—C-V—0.50 C, 1.20 Cr, 0.25 V, 2.20 W

Simonds—Commando 47—0.47 C, 0.25 Mn, 0.80 Si, 1.40 Cr, 0.25 V, 2.00 W, 0.25 Mo

Uddeholm—UHB 711—0.50 C, 0.20 Mn, 0.75 Si, 1.15 Cr, 0.20 V, 2.50 W

Universal-Cyclops—Alco M—0.45 C, 0.30 Mn, 0.25 Si, 1.50 Cr, 0.25 V, 2.25 W, 0.50 Mo

Vanadium-Alloys—Par Exc—0.53 C, 0.20 Mn, 0.25 Si, 1.60 Cr, 0.25 V, 2.00 W

Vulcan Crucible—QA—0.45 C, 1.20 Cr, 0.25 V, 2.20 W

S2—Silicon-moly

SAE composition: 0.50 C, 0.40 Mn, 1.00 Si, 0.25 V opt, 0.50 Mo. This steel has the highest toughness of any tool steel. Red hardness is not as good as S1. Other characteristics are about the same as S1. Suitable for punches, hammer dies, and swaging dies.

Braeburn—Triton—0.50 C, 0.35 Mn, 1.00 Si, 0.60 Mo

Carpenter—Solar—0.50 C, 0.40 Mn, 1.00 Si, 0.50 Mo

Delaware—Delsteel—0.50 C, 0.75 Mn, 2.00 Si, 0.20 V, 0.20 Mo

Houghton & Richards—Silico—0.50 C, 0.45 Mn, 1.10 Si, 0.20 V, 0.50 Mo

Jessop—RTS—0.55 C, 0.50 Mn, 0.80 Si, 0.45 Mo

Latrobe—Trident—0.55 C, 0.30 Mn, 0.70 Si, 0.60 Cr, 0.20 V, 0.40 Mo

Peninsular—Penco No. 71—0.50 C, 0.40 Mn, 1.05 Si, 0.20 V, 0.45 Mo

Simonds—Havoc—0.50 C, 0.40 Mn, 1.00 Si, 0.20 V, 0.50 Mo

Universal-Cyclops—Venango—0.50 C, 0.45 Mn, 1.10 Si, 0.20 V, 0.50 Mo

Universal-Cyclops—Venango Special—0.68 C, 0.55 Mn, 1.10 Si, 0.20 V, 0.55 Mo

S3—Low Chromium-tungsten

ASM composition: 0.50 C, 0.75 Cr, 1.00 W. Toughness is combined with fair cutting ability. Suitable for chisels, rivet sets, punches, screwdrivers, and shear blades. Forging temperature 1900 to 2100 F, annealing 1450 F, hardening 1500 to 1550 F, tempering 400 to 600 F.

Shock-resisting Tool Steels (*Continued*)

Darwin & Milner—W-Brand—0.45 C, 0.35 Mn, 0.30 Si, 0.80 Cr, 0.85 W
Great Western—422-Mirycal—0.50 C, 0.25 Mn, 0.20 Si, 0.95 Cr, 1.00 W, 0.20 Mo
Hoyland—422—0.50 C, 0.25 Mn, 0.20 Si, 0.95 Cr, 1.00 W, 0.20 Mo
Republic—M Tungsten—0.50 C, 0.35 Mn, 1.00 Cr, 1.00 W
Vulcan Crucible—Blue Edge—0.50 C, 1.00 Cr, 1.00 W

S4—Silicon-manganese

ASM composition: 0.55 C, 0.80 Mn, 2.00 Si, 0.30 Cr opt, 0.25 V opt, 0.30 Mo opt (see also S5). This is the basic silicon-manganese steel. Good toughness, suitable for shear blades, chisels, punches, and pneumatic tools. Note: S4 and S5 are similar. SAE does not at present list a composition for S4. For this list, steels with less than 0.30 Mo are S4, 0.30 Mo and higher are S5.

Allegheny Ludlum—609—0.60 C, 0.80 Mn, 2.00 Si, 0.25 Cr, 0.20 V, 0.25 Mo
Atlantic—Atsil—0.50 C, 0.60 Mn, 1.30 Si, 0.50 W, 0.30 Mo
Atlas—Monark 2—0.60 C, 0.75 Mn, 2.00 Si, 0.30 Cr, 0.20 Mo
Bethlehem—71 Alloy—0.60 C, 0.90 Mn, 2.00 Si
Braeburn—Alloy No. 10—0.55 C, 0.80 Mn, 2.00 Si
Crucible—La Belle 2-70—0.60 C, 0.75 Mn, 1.90 Si, 0.25 Cr
Delaware—Silicon-manganese—0.50 C, 0.80 Mn, 2.10 Si
Houghton & Richards—No. 8—0.55 C, 0.85 Mn, 2.10 Si, 0.25 Cr, 0.25 V
Houghton & Richards—No. 8M—0.55 C, 0.75 Mn, 2.00 Si, 0.20 Mo
McDonald—Macco Silicon Manganese—0.63 C, 0.78 Mn, 2.00 Si
Republic—Special Punch—0.55 C, 0.85 Mn, 2.00 Si, 0.25 Cr, 0.30 V
Universal-Cyclops—No. 67—0.55 C, 0.85 Mn, 2.00 Si, 0.20 Mo
Vanadium-Alloys—Silman—0.55 C, 0.85 Mn, 2.00 Si, 0.25 Cr, 0.20 V
Vulcan Crucible—4870—0.55 C, 0.85 Mn, 2.00 Si, 0.25 Cr, 0.20 V

S5—High-silicon Manganese

SAE composition: 0.55 C, 0.80 Mn, 2.00 Si, 0.30 Cr opt, 0.25 V opt, 0.40 Mo opt (see also S4). Toughness almost as good as S2 and better red hardness than S4 because of the molybdenum content. Suitable for chisels, heavy-duty punches, rivet busters, drift pins, spring collets, and nail sets.

Allegheny Ludlum—602—0.50 C, 0.70 Mn, 1.70 Si, 0.12 V, 0.40 Mo
Amalgamated—Duro-Chip—0.60 C, 0.70 Mn, 1.85 Si, 0.25 V, 0.50 Mo
Bethlehem—Omega—0.60 C, 0.70 Mn, 1.85 Si, 0.25 V, 0.50 Mo
Carpenter—No. 481—0.60 C, 0.80 Mn, 1.90 Si, 0.30 Cr, 0.30 Mo
Columbia—CEC Impact—0.55 C, 0.80 Mn, 2.00 Si, 0.30 Cr, 0.25 V
Crucible—La Belle Silicon No. 2—0.60 C, 0.75 Mn, 1.90 Si, 0.25 Cr, 0.30 Mo
Darwin & Milner—Extra Tough—0.60 C, 0.75 Mn, 1.90 Si, 0.25 Cr, 0.30 Mo
Delaware—Special Alloy—0.50 C, 0.80 Mn, 2.00 Si, 0.35 V, 0.40 Mo
Disston—D 29—0.55 C, 0.80 Mn, 1.90 Si, 0.10 V, 0.50 Mo
Faitoute—Omega—0.60 C, 0.70 Mn, 1.85 Si, 0.20 V, 0.45 Mo
Firth Sterling—Chimo—0.55 C, 0.80 Mn, 2.00 Si, 0.18 V, 0.50 Mo
Great Western—280-Tufkut—0.50 C, 0.70 Mn, 1.50 Si, 0.35 Mo
Houghton & Richards—Silco—0.55 C, 0.70 Mn, 2.15 Si, 0.20 V, 0.45 Mo
Hoyland—280—0.50 C, 0.70 Mn, 1.50 Si, 0.35 Mo
Jessop—Magic—0.52 C, 0.90 Mn, 2.00 Si, 1.00 Mo
Kloster—Pure-Oil V-76—0.55 C, 0.75 Mn, 1.90 Si, 0.25 V, 0.50 Mo
Milne—MSM—0.55 C, 0.80 Mn, 1.90 Si, 0.25 V, 0.50 Mo
Peninsular—Silver Label—0.57 C, 0.75 Mn, 1.85 Si, 0.20 V, 0.45 Mo
Uddeholm—UHB Resisto—0.60 C, 0.70 Mn, 1.85 Si, 0.25 V, 0.50 Mo
Vanadium-Alloys—Mosil—0.55 C, 0.90 Mn, 2.00 Si, 0.30 Cr, 0.30 Mo

S6A—Nontempering

A general-purpose tool steel for shock resistance where accurate temperature control is not available. Can be hardened over a wide range with satisfactory results. No tempering required. Forging temperature 2100 to 2150 F, annealing 1400 F, hardening 1500 to 1800 F.

Amalgamated—Malga Non-Tempering Special—0.35 C, 0.75 Mn, 0.35 Si, 0.80 Cr, 0.25 W, 0.50 Mo

Atlantic—33—0.33 C, 0.40 Mn, 0.65 Si, 0.75 Cr, 0.75 Mo, 0.75 Cu

Bethlehem—Non-Tempering—0.35 C, 0.70 Mn, 0.45 Si, 0.80 Cr, 0.30 Mo, 0.30 Cu

Houghton & Richards—Non Tempering—0.35 C, 0.70 Mn, 0.45 Si, 0.80 Cr, 0.25 Mo, 0.25 Cu

Milne—FPC—0.35 C, 0.70 Mn, 0.80 Cr, 0.45 Mo

Peninsular—Penco-Non-Tempering—0.35 C, 0.70 Mn, 0.45 Si, 0.80 Cr, 0.30 Mo, 0.30 Cu

Ryerson—Non Tempering—0.38 C, 0.75 Mn, 0.80 Cr, 0.25 Mo

S8A—Chrome-molybdenum

Better shock resistance than carbon steel for cutting dies for leather, rubber, paper, and fabrics. Also suitable for woodworking tools, hand chisels, and boring bars. Forging temperature 1750 to 1950 F, annealing 1425 F, hardening 1550 to 1600 F, tempering 400 to 900 F.

Allegheny Ludlum—Atlas 93—0.55 C, 0.55 Mn, 0.20 Si, 0.65 Cr, 0.35 Mo

Darwin & Milner—DSS—0.52 C, 0.60 Mn, 0.25 Si, 0.75 Cr, 0.35 Mo

S9A—Chrome-nickel

Shock resistance for hubs, forming rolls, short-run and experimental dies, cold-forming and bending dies.

Boyd-Wagner—Formaloy—0.75 C, 0.30 Mn, 0.25 Si, 1.00 Cr, 0.20 Mo, 1.75 Ni

Jessop—ET No. 4—0.70 C, 0.65 Mn, 0.20 Si, 0.55 Cr, 0.25 Mo, 1.30 Ni

Latrobe—NDS—0.75 C, 1.00 Cr, 1.70 Ni

Universal-Cyclops—N9—0.75 C, 0.45 Mn, 1.00 Cr, 1.50 Ni

S10A—Nickel

Nondeforming properties and wear resistance are combined with shock resistance. Suitable for intricate hobs, forming dies, and punches. Forging 1875 to 1975 F, annealing 1550 F, hardening 1600 F, tempering 375 or 850 F.

Darwin & Milner—Firex Special—0.50 C, 0.65 Mn, 0.30 Si, 0.85 Cr, 0.20 V, 0.60 Mo, 4.00 Ni

Great Western—L-97—0.55 C, 1.00 Cr, 0.35 Mo, 3.00 Ni

Hoyland—L-97—0.55 C, 1.00 Cr, 0.35 Mo, 3.00 Ni

Latrobe—Staminal—0.55 C, 1.00 Mn, 1.00 Si, 0.40 Cr, 0.40 Mo, 2.70 Ni

S11A—Chrome-manganese

Toughness combined with ease of heat-treatment and machining. Suitable for hard chisels and battering tools. Forging temperature 1850 to 1950 F, hardening 1650 F (oil quench) or 1850 F (water quench), tempering 300 to 500 F.

Amalgamated—Rexor—0.35 C, 0.70 Mn, 0.15 Si, 0.60 Cr, 0.35 Mo, 0.20 Ni

S12A—Chrome-manganese-tungsten

Toughness combined with moderate wear resistance. Suitable for forging dies and blacksmith tools. Higher-carbon version is harder and has better cutting ability suitable for shear blades, trim dies, and collets. Forging temperature 1800 to 2100 F, annealing 1450 F, hardening 1500 F (water quench), or 1600 F (oil), tempering 400 to 1000 F.

Republic—UA-4—0.50 C, 0.60 Mn, 1.00 Cr, 0.25 W

Republic—UA-6—0.65 C, 0.60 Mn, 1.00 Cr, 0.25 W

Cold-work Tool Steels

Oil-hardening Types

O1—Low Manganese

SAE composition: 0.90 C, 1.20 Mn, 0.25 Si, 0.50 Cr, 0.20 V opt, 0.50 W. The most popular oil-hardening nondeforming tool steel. Not so tough as plain carbon steel but with better cutting ability, safety in hardening, and medium depth of hardness. Suitable for blanking, forming, and drawing dies, taps, hobs, milling cutters, and gages.

Allegheny Ludlum—Saratoga—0.90 C, 1.20 Mn, 0.35 Si, 0.50 Cr, 0.50 W

Amalgamated—Dymal—0.90 C, 1.15 Mn, 0.50 Cr, 0.50 W

Atlantic—Atlan—0.90 C, 1.15 Mn, 0.50 Cr, 0.50 W

Cold-work Tool Steels

Oil-hardening Types (*Continued*)

Atlas—Keewatin—0.90 C, 1.15 Mn, 0.30 Si, 0.50 Cr, 0.50 W
Bethlehem—BTR—0.90 C, 1.20 Mn, 0.50 Cr, 0.20 V, 0.50 W
Boyd-Wagner—Arrow Swedish—0.95 C, 1.20 Mn, 0.20 Si, 0.50 Cr, 0.10 V, 0.50 W
Boyd-Wagner—Bowco—0.90 C, 1.20 Mn, 0.35 Si, 0.50 Cr, 0.50 W
Braeburn—Kiski—0.95 C, 1.10 Mn, 0.50 Cr, 0.20 V, 0.60 W
Columbia—Exl-Die—0.90 C, 1.15 Mn, 0.50 Cr, 0.50 W
Columbia—Oildie—1.00 C, 0.65 Mn, 0.40 Si, 1.60 Cr, 0.50 W
Craine-Schrage—Crasco Red Label—0.90 C, 1.15 Mn, 0.25 Si, 0.50 Cr, 0.50 W
Crucible—Ketos—0.90 C, 1.30 Mn, 0.50 Cr, 0.50 W
Darwin & Milner—OHT—0.90 C, 1.20 Mn, 0.35 Si, 0.50 Cr, 0.50 W
Diehl—Utex—0.90 C, 1.10 Mn, 0.25 Si, 0.50 Cr, 0.25 V, 0.70 W
Disston—Mansil—0.90 C, 1.15 Mn, 0.50 Cr, 0.50 W
Faitoute—Oilhard—0.90 C, 1.15 Mn, 0.30 Si, 0.50 Cr, 0.20 V, 0.50 W
Firth Sterling—Invaro—0.90 C, 1.15 Mn, 0.50 Cr, 0.20 V, 0.50 W
Great Western—CW Oil—0.90 C, 1.10 Mn, 0.35 Si, 0.50 Cr, 0.50 W
Houghton & Richards—Tungsten Oil Hardening—0.90 C, 1.10 Mn, 0.25 Si, 0.50 Cr, 0.20 V, 0.50 W
Houghton & Richards—Oil Hardening—0.95 C, 0.95 Mn, 0.20 Si, 0.55 Cr
Hoyland—CW Oil—0.90 C, 1.10 Mn, 0.35 Si, 0.50 Cr, 0.50 W
Jessop—Truform—0.90 C, 1.17 Mn, 0.30 Si, 0.50 Cr, 0.50 W
Kloster—Swed-Oil—0.97 C, 1.15 Mn, 0.30 Si, 0.47 Cr, 0.20 V, 0.47 W
Latrobe—Badger—0.95 C, 1.20 Mn, 0.30 Si, 0.50 Cr, 0.50 W
McDonald—Macco Royal Crown—0.90 C, 1.20 Mn, 0.30 Si, 0.50 Cr, 0.25 V, 0.50 W
McInnis—Cello Vanadium—0.90 C, 1.20 Mn, 0.35 Si, 0.50 Cr, 0.20 V, 0.50 W
Milne—Amcoh—0.95 C, 1.25 Mn, 0.50 Cr, 0.20 V, 0.50 W
Peninsular—Yellow Label—0.90 C, 1.20 Mn, 0.30 Si, 0.50 Cr, 0.20 V, 0.50 W
Republic—Special Oil Hardening—0.90 C, 1.20 Mn, 0.50 Cr, 0.50 W
Simonds—Teenax No. 46—0.90 C, 1.25 Mn, 0.20 Si, 0.50 Cr, 0.20 V, 0.50 W
Uddeholm—UHB-46—0.90 C, 1.20 Mn, 0.30 Si, 0.50 Cr, 0.10 V, 0.50 W
Universal-Cyclops—Wando—0.90 C, 1.20 Mn, 0.25 Si, 0.50 Cr, 0.20 V, 0.50 W
Vanadium-Alloys—Non-Shrinkable, Colonial, No. 6—0.95 C, 1.20 Mn, 0.30 Si, 0.50 Cr, 0.20 V, 0.50 W
Vulcan Crucible—Oil Hardening—0.90 C, 1.20 Mn, 0.50 Cr, 0.50 W

O2—High Manganese

SAE composition: 0.90 C, 1.60 Mn, 0.25 Si, 0.55 Cr opt, 0.20 V opt, 0.30 Mo opt. With higher manganese content than O1, this steel has slightly better cutting ability and nondeformation properties. Toughness is considerably better and is the best of any of the standard oil-hardening group. Suitable for blanking, forming, trimming, and molding dies, taps and threading dies, broaches, and reamers.

Allegheny Ludlum—Deward—0.90 C, 1.55 Mn, 0.25 Si, 0.30 Mo
Braeburn—SOD—0.90 C, 1.65 Mn
Carpenter—Stentor—0.90 C, 1.60 Mn, 0.25 Si
Crucible—Paragon—0.90 C, 1.60 Mn
Darwin & Milner—H Brand—0.90 C, 1.50 Mn, 0.25 Si, 0.30 Mo
Firth Sterling—Invaro No. 2—0.90 C, 1.65 Mn, 0.25 Si, 0.35 Mo, 0.15 V
Houghton & Richards—No. 19—0.92 C, 1.55 Mn, 0.32 Si, 0.18 Cr
Jessop—Special Oil Hardening—0.95 C, 1.75 Mn, 0.35 Si
Latrobe—Mangano—0.95 C, 1.60 Mn, 0.25 Si
Republic—Arrestite—0.90 C, 1.50 Mn, 0.20 Cr
Ryerson—Ry-Alloy—0.92 C, 1.55 Mn, 0.29 Si, 0.23 Mo
Simonds—No. 864—0.90 C, 1.40 Mn, 0.25 Mo
Vulcan Crucible—Non-Shrinkable—0.90 C, 1.50 Mn, 0.20 Cr

O6—Molybdenum Graphite

SAE composition: 1.45 C, 0.75 Mn, 1.00 Si, 0.25 Mo. Graphite particles amounting to about one-third of the carbon content are distributed through this steel. They make for easy machining and good wear resistance. Suitable for forming, blanking, and trimming dies, punches, and gages.

Allegheny Ludlum—Oilgraph—1.45 C, 0.65 Mn, 1.00 Si, 0.25 Mo
Peninsular—Graphmo—1.45 C, 0.65 Mn, 1.00 Si, 0.25 Mo
Timken—Graph-Mo—1.45 C, 1.00 Mn, 1.25 Si, 0.25 Mo

O7—Tungsten

ASM composition: 1.20 C, 0.75 Cr, 1.75 W, 0.25 Mo opt. Better cutting ability than the other oil-hardening steels except D3 with high hardness and fairly deep hardening. Suitable for taps, threading tools, drills, reamers, cutting tools for brass, and punches and dies for light stock. Forging temperature 1750 to 1900 F, annealing 1500 to 1550 F, hardening 1550 to 1650 F, tempering 325 to 500 F.

Allegheny Ludlum—Utica—1.25 C, 0.30 Mn, 0.30 Si, 0.40 Cr, 0.20 V, 1.40 W
Atlas—Badger—1.25 C, 0.25 Mn, 0.25 Si, 0.40 Cr, 0.20 V, 1.50 W
Bethlehem—67 Tap—1.25 C, 0.65 Cr, 0.20 V, 1.40 W
Firth Sterling—Meteor—1.20 C, 0.30 Mn, 0.25 Cr, 0.15 V, 1.50 W
Houghton & Richards—No. 60—1.20 C, 0.25 Mn, 0.30 Si, 0.70 Cr, 0.20 V, 1.60 W, 0.25 Mo
Latrobe—WT—1.20 C, 0.25 Mn, 0.30 Si, 0.40 Cr, 0.40 V, 1.50 W, 0.30 Mo
Simonds—OHD—1.20 C, 0.35 Mn, 0.25 Si, 0.60 Cr, 0.20 V, 1.50 W
Universal-Cyclops—Para—1.20 C, 0.30 Mn, 0.20 Si, 0.40 Cr, 0.20 V, 1.60 W
Vanadium-Alloys—Red Star Tungsten—1.20 C, 0.70 Cr, 0.20 V, 1.60 W, 0.25 Mo
Vulcan Crucible—Hardrite—1.05 C, 0.60 Cr, 0.25 V, 1.70 W

Air-hardening Types

A2—5% Chrome Air Hardening

SAE composition: 1.00 C, 0.60 Mn, 0.25 Si, 5.25 Cr, 0.40 V opt, 1.10 Mo. This steel is intermediate between D3 or D2 and O1 or O2. It is tougher than the former, more wear-resistant than the latter. Deep hardening. Suitable for punches, blanking and forming dies.

Allegheny Ludlum—Sagamore—1.00 C, 0.50 Mn ,0.30 Si, 5.00 Cr, 0.25 V, 1.00 Mo
Amalgamated—Mal-Die—1.00 C, 0.60 Mn, 5.25 Cr, 0.25 V, 1.10 Mo
Atlantic—Hardnair—1.30 C, 0.52 Mn, 0.95 Si, 5.00 Cr, 0.33 V, 1.25 Mo
Atlas—Cro-Mo-Loy—1.00 C, 1.00 Mn, 0.30 Si, 5.00 Cr, 0.25 V, 1.00 Mo
Bethlehem—A-H5—1.00 C, 0.60 Mn, 5.25 Cr, 0.25 V, 1.10 Mo
Braeburn—Airque—1.00 C, 0.70 Mn, 5.25 Cr, 0.25 V, 1.15 Mo
Carpenter—484—1.00 C, 0.70 Mn, 0.20 Si, 5.00 Cr, 0.20 V, 1.00 Mo
Columbia—E-Z-Die—1.00 C, 0.60 Mn, 5.25 Cr, 1.15 Mo, 0.25 V
Crucible—Airkool—1.00 C, 0.40 Mn, 5.25 Cr, 0.40 V, 1.15 Mo
Darwin & Milner—Mineor—1.00 C, 0.55 Mn, 0.40 Si, 5.00 Cr, 0.25 V, 1.00 Mo
Firth Sterling—Airvan Die—1.00 C, 0.65 Mn, 0.35 Si, 5.25 Cr, 0.25 V, 1.15 Mo
Great Western—CVM—1.00 C, 0.70 Mn, 0.20 Si, 5.00 Cr, 0.25 V, 1.10 Mo
Houghton & Richards—No. 80—1.00 C, 0.65 Mn, 0.20 Si, 5.25 Cr, 0.25 V, 1.10 Mo
Hoyland—CVM—1.00 C, 0.70 Mn, 0.20 Si, 5.00 Cr, 0.25 V, 1.10 Mo
Jessop—Windsor—1.00 C, 0.50 Mn, 0.45 Si, 5.00 Cr, 0.30 V, 1.25 Mo
Kloster—Air-Chrom—1.00 C, 0.57 Mn, 0.23 Si, 5.30 Cr, 0.26 V, 1.00 Mo
Latrobe—Select B FM—1.00 C, 5.00 Cr, 0.30 V, 1.00 Mo
McDonald—Macco 35 Air Hard—1.00 C, 0.60 Mn, 0.25 Si, 5.25 Cr, 0.25 V, 1.15 Mo
McInnes—M-10 Air Hardening—1.00 C, 0.70 Mn, 5.00 Cr, 1.05 Mo, 0.25 V
Milne—Milnair 5—1.00 C, 5.25 Cr, 0.25 V, 1.10 Mo
Peninsular—Penair No. 5—1.00 C, 0.60 Mn, 0.25 Si, 5.00 Cr, 0.25 V, 1.10 Mo

Cold-work Tool Steels

Air-hardening Types (*Continued*)

Ryerson—Ry-Die—1.00 C, 0.60 Mn, 5.25 Cr, 0.25 V, 1.00 Mo

Simonds—Airtrue 51—1.00 C, 0.65 Mn, 0.25 Si, 5.25 Cr, 0.30 V, 1.00 Mo

Uddeholm—UHB-151—1.00 C, 0.60 Mn, 0.25 Si, 5.25 Cr, 0.20 V, 1.10 Mo

Universal-Cyclops—Sparta—1.00 C, 0.65 Mn, 0.30 Si, 5.25 Cr, 0.25 V, 1.10 Mo

Vanadium-Alloys—Air Hard—1.00 C, 1.75 Mn, 0.30 Si, 5.00 Cr, 0.25 V, 1.10 Mo

Vulcan Crucible—Vuldie—1.00 C, 5.25 Cr, 0.25 V, 1.15 Mo

A4—Manganese

ASM composition: 1.00 C, 2.00 Mn, 1.00 Cr, 1.00 Mo. Low hardening temperature of this steel is combined with good toughness, though wear resistance is lower than the other nondeforming steels. Suitable for punches, blanking and forming dies, and gages. Forging temperature 2000 F, annealing 1350 to 1400 F, hardening 1500 to 1550 F, tempering 300 to 400 F. ASM classifies the steels with 3% Mn as A5 and those with 0.70% C as A6.

Allegheny Ludlum—Airloy—1.00 C, 3.00 Mn, 1.00 Cr, 1.00 Mo

Bethlehem—BA-H—0.95 C, 2.00 Mn, 2.20 Cr, 1.10 Mo

Carpenter—Vega—0.70 C, 2.00 Mn, 0.30 Si, 1.00 Cr, 1.35 Mo

Firth Sterling—Airmo—1.00 C, 0.30 Si, 2.00 Mn, 1.00 Cr, 1.00 Mo

Republic—Airaloy—1.00 C, 2.00 Mn, 0.90 Cr, 0.90 Mo

Timken—Graph-MNS C Graphitic Tool Steel—1.50 C, 1.25 Mn, 1.25 Si, 0.50 Cr, 0.50 Mo, 1.75 Ni

Vulcan Crucible—Vairloy—1.00 C, 2.00 Mn, 0.90 Cr, 0.90 Mo

High-carbon High-chromium Types

D2—High Carbon High Chromium

SAE composition: 1.50 C, 0.40 Mn, 0.40 Si, 12.00 Cr, 0.80 V opt, 0.90 Mo, 0.60 Co opt. Properties are similar to D3. This steel has better safety in hardening and high abrasive resistance. Deep hardening. Some corrosion resistance. Suitable for punches, blanking, forming, and drawing dies, and gages.

Allegheny Ludlum—Ontario—1.50 C, 0.40 Mn, 0.30 Si, 12.00 Cr, 0.20 V, 0.80 Mo

Amalgamated—Dykrome—1.50 C, 11.50 Cr, 0.25 V, 0.75 Mo

Atlantic—Atlan HCC—1.50 C, 12.00 Cr, 1.00 V, 0.80 Mo

Atlas—FNS—1.50 C, 0.30 Mn, 0.30 Si, 12.00 Cr, 0.25 V, 0.80 Mo

Bethlehem—Lehigh H—1.55 C, 11.50 Cr, 0.40 V, 0.80 Mo

Boyd-Wagner—Hypro 61—1.50 C, 0.25 Mn, 0.40 Si, 11.50 Cr, 0.85 V, 0.80 Mo, 0.40 Co

Braeburn—Superior No. 3—1.50 C, 12.00 Cr, 0.80 V, 0.80 Mo

Carpenter—610—1.50 C, 0.30 Mn, 0.30 Si, 12.00 Cr, 0.90 V, 0.80 Mo

Columbia—Atmodie—1.50 C, 12.00 Cr, 0.85 V, 0.85 Mo

Craine-Schrage—Crasco Black Label—1.50 C, 0.35 Mn, 0.25 Si, 12.00 Cr, 0.20 V, 0.80 Mo

Crucible—Airdie 150—1.50 C, 11.50 Cr, 0.20 V, 0.80 Mo

Darwin & Milner—Darwin No. 1—1.50 C, 0.35 Mn, 0.40 Si, 12.00 Cr, 0.25 V, 0.75 Mo, 0.40 Ni

Disston—Croloy—1.50 C, 12.00 Cr, 0.90 V, 0.80 Mo

Faitoute—Chromdie—1.55 C, 11.50 Cr, 0.40 V, 0.85 Mo

Firth Sterling—Cromovan Die—1.60 C, 12.50 Cr, 1.00 V, 0.80 Mo

Great Western—265 High Production—1.60 C, 0.30 Mn, 0.30 Si, 12.00 Cr, 0.20 V, 0.80 Mo

Hawkridge—Haldi No. 2—1.50 C, 11.50 Cr, 0.20 V, 0.80 Mo

Houghton & Richards—K-2—1.50 C, 0.25 Mn, 0.30 Si, 11.50 Cr, 0.25 V, 0.75 Mo

Hoyland—No. 265—1.60 C, 0.30 Mn, 0.30 Si, 12.00 Cr, 0.20 V, 0.80 Mo
Jessop—CNS No. 1—1.50 C, 0.30 Mn, 0.25 Si, 12.00 Cr, 0.25 V, 0.75 Mo
Kloster—Pure-Oil Hi-Run—1.55 C, 0.35 Mn, 0.45 Si, 12.00 Cr, 0.30 V, 0.83 Mo
Latrobe—Olympic FM—1.50 C, 0.30 Mn, 0.30 Si, 12.00 Cr, 1.00 V, 0.80 Mo
McDonald—Macco Kromax 1—1.50 C, 0.34 Mn, 0.23 Si, 11.90 Cr, 0.28 V, 0.90 Mo
Milne—High Production—1.55 C, 11.50 Cr, 1.00 V, 0.75 Mo
Peninsular—White Label—1.50 C, 12.00 Cr, 0.35 V, 0.80 Mo
Republic—404—1.55 C, 12.00 Cr, 0.25 V, 0.75 Mo
Simonds—CCM—1.55 C, 0.25 Mn, 0.35 Si, 12.00 Cr, 0.35 V, 0.80 Mo
Uddeholm—Tri-Mo—1.50 C, 0.45 Mn, 0.30 Si, 12.00 Cr, 0.20 V, 0.80 Mo
Universal-Cyclops—Ultradie No. 2—1.50 C, 0.25 Mn, 0.30 Si, 12.00 Cr, 0.25 V, 0.80 Mo
Universal-Cyclops—Ultradie No. 3—1.50 C, 0.25 Mn, 0.30 Si, 12.00 Cr, 0.90 V, 0.80 Mo
Vanadium-Alloys—Ohio Die—1.50 C, 0.25 Mn, 0.38 Si, 12.00 Cr, 0.85 V, 0.80 Mo, 0.40 Co
Vulcan Crucible—Alidie—1.55 C, 12.00 Cr, 0.75 Mo, 0.25 V

D3—High Carbon High Chromium

SAE composition: 2.15 C, 0.35 Mn, 0.35 Si, 12.00 Cr, 0.80 V opt, 0.75 W opt, 0.80 Mo opt, 0.50 Ni opt. Extremely high abrasion resistance makes this steel suitable for blanking dies and punches for high production. Hardens more deeply than other oil-hardening steels. Poor shock resistance. Suitable for wire- and deep-drawing dies, swaging and blanking dies, gages, forming tools, and rolls. ASM classifies those steels containing 1% Mo as D4, and those with 1% Si as D6.

Allegheny Ludlum—Huron—2.15 C, 0.30 Mn, 0.25 Si, 12.50 Cr, 1.00 V
Amalgamated—Dy-Krome Special—2.15 C, 12.50 Cr, 1.00 V
Atlas—NN—2.25 C, 0.30 Mn, 0.25 Si, 12.50 Cr, 0.25 V
Bethlehem—Lehigh S—2.05 C, 11.75 Cr, 0.60 V
Bethlehem—Lehigh L—0.85 C, 11.50 Cr, 0.30 V, 0.45 Mo, 1.00 Ni
Boyd-Wagner—Hypro 62—2.00 C, 0.60 Mn, 0.40 Si, 12.00 Cr
Boyd-Wagner—Hypro 63—1.70 C, 0.20 Mn, 18.00 Cr
Braeburn—Superior No. 1—2.15 C, 12.50 Cr, 0.60 V
Carpenter—Hampden—2.10 C, 0.25 Mn, 0.25 Si, 12.50 Cr, 0.50 Ni
Columbia—Super Die—2.10 C, 0.30 Mn, 0.90 Si, 11.50 Cr, 1.00 W
Crucible—HYCC—2.25 C, 11.50 Cr, 0.20 V, 0.80 Mo
Darwin & Milner—Neor—2.15 C, 0.50 Mn, 0.45 Si, 13.25 Cr, 0.50 Ni
Disston—Mix 812—2.10 C, 0.30 Mn, 0.25 Si, 12.00 Cr
Firth Sterling—Triple Die—2.20 C, 12.00 Cr, 0.30 V
Great Western—265-H—2.05 C, 0.40 Mn, 11.50 Cr, 0.60 V
Hawkridge—Marathon—2.25 C, 11.50 Cr, 0.20 V, 0.80 Mo
Houghton & Richards—K—2.32 C, 0.32 Mn, 0.22 Si, 13.00 Cr, 0.22 V
Hoyland—265—1.60 C, 0.30 Mn, 0.30 Si, 12.00 Cr, 0.20 V, 0.80 Mo
Hoyland—265-H—2.05 C, 0.40 Mn, 11.50 Cr, 0.60 V
Jessop—CNS No. 2—2.25 C, 0.30 Mn, 0.25 Si, 12.00 Cr, 0.20 V, 0.80 Mo
Kloster—Pure-Ore Kapo—2.00 C, 12.00 Cr, 2.00 W
Latrobe—GSN FM—2.20 C, 0.50 Mn, 0.50 Si, 13.00 Cr
McInnes—High Carbon High Chrome—2.15 C, 13.00 Cr, 1.00 V, 0.50 Ni
Milne—Double Six—2.25 C, 13.00 Cr
Peninsular—White Label-S—2.15 C, 0.40 Mn, 0.60 Si, 12.00 Cr
Simonds—No. 12225—2.10 C, 0.40 Mn, 0.60 Si, 13.00 Cr
Uddeholm—Tri-Van—2.05 C, 11.75 Cr, 0.60 V
Universal-Cyclops—Ultradie No. 1—2.25 C, 0.25 Mn, 0.20 Si, 12.00 Cr, 0.20 V
Universal-Cyclops—Ultradie No. 1M—2.25 C, 0.25 Mn, 0.20 Si, 12.00 Cr, 0.20 V, 0.80 Mo
Vanadium-Alloys—Crocar—2.20 C, 0.20 Mn, 0.30 Si, 12.00 Cr, 0.80 V, 0.50 Co

D5—High Carbon High Chrome Cobalt

SAE composition: 1.50 C, 0.40 Mn, 0.40 Si, 12.00 Cr, 0.80 V opt, 0.90 Mo, 3.10 Co. Good abrasion resistance and the ability to hold a keen cutting edge are characteristics of this steel. Safe hardening and good nondeforming properties. Suitable for blanking and forming dies, burnishing tools, and some hot-work applications.

Allegheny Ludlum—FCC-66 (forgings and castings only)—1.50 C, 12.00 Cr, 0.50 V, 0.80 Mo, 3.25 Co
Amalgamated—Cobaltloy—1.50 C, 12.00 Cr, 0.50 V, 0.90 Mo, 3.25 Co
Braeburn—Superior No. 2—1.50 C, 0.40 Mn, 0.40 Si, 12.00 Cr, 0.90 Mo, 3.10 Co
Darwin & Milner—PRK-33—1.40 C, 0.30 Mn, 0.60 Si, 13.00 Cr, 0.60 Mo, 3.30 Co, 0.50 Ni
Houghton & Richards—No. 61—1.35 C, 0.25 Mn, 0.50 Si, 12.12 Cr, 0.64 Mo, 3.04 Co
Jessop—3C Special—1.45 C, 0.30 Mn, 0.40 Si, 12.50 Cr, 0.80 Mo, 3.20 Co
Latrobe—Cobalt Chrome FM—1.30 C, 12.00 Cr, 0.80 Mo, 3.00 Co
McDonald—Macco Kromax 2—1.40 C, 0.32 Mn, 0.42 Si, 12.50 Cr, 0.85 Mo, 3.25 Co
Milne—Double Seven—1.35 C, 12.50 Cr, 0.60 Mo, 3.00 Co
Universal-Cyclops—EK 81—1.35 C, 0.30 Mn, 0.65 max Si, 12.75 Cr, 0.80 Mo, 3.00 Co

D7A—High Chrome Vanadium

A cold-work die steel with high abrasion resistance. Suitable for lamination dies, cold-work dies and punches, and wear-resisting tools and dies.

Allegheny Ludlum—Huron V—2.40 C, 12.00 Cr, 0.80 Mo, 4.00 V
Crucible—HYCV—2.45 C, 12.25 Cr, 3.75 V, 1.00 Mo
Houghton & Richards—K-3—2.40 C, 12.75 Cr, 4.00 V, 1.10 Mo
Latrobe—BR-4 FM—2.40 C, 12.75 Cr, 4.00 V, 1.10 Mo

D8A—High Chrome Vanadium Moly

An air-hardening steel that may also be hardened by oil quenching for extreme abrasive wear. Suitable for brick mold liners, sand-slinger liners, and blanking dies.

Crucible—Airkool V—2.25 C, 0.75 Si, 5.25 Cr, 3.75 V, 3.00 Mo

Hot-work Tool Steels

Chromium-base Types

H3A—Chrome-nickel-moly

Hammer dies, heading and gripper dies, piercing punches, die-casting dies.

Atlas—Ultimo 6—0.55 C, 0.55 Mn, 0.25 Si, 0.75 Cr, 0.75 Mo, 1.60 Ni
Kloster—KLS-44—0.57 C, 0.51 Mn, 0.28 Si, 1.01 Cr, 0.92 Mo, 1.43 Ni

H5A—2 Chromium, 2 Tungsten

Suitable for die-casting and plastic-molding dies. Forging temperature 1900 F, annealing 1450 F, hardening 1750 F, tempering 1200 F. Oil quench.

Bethlehem—67 Chisel—0.50 C, 0.30 Mn, 0.75 Si, 1.15 Cr, 0.20 V, 2.50 W
McDonald—Hardtuf—0.50 C, 0.30 Mn, 0.60 Si, 1.60 Cr, 0.25 V, 2.60 W
Uddeholm—UHB 711—0.50 C, 0.20 Mn, 0.75 Si, 1.15 Cr, 0.20 V, 2.50 W
Universal-Cyclops—Alco S—0.50 C, 0.30 Mn, 1.00 Si, 1.75 Cr, 0.25 V, 2.25 W, 0.50 Mo

H6A—2½% Chromium

Suitable for short-service die-casting dies.

Latrobe—Aldivan—0.45 C, 0.65 Mn, 0.25 Si, 2.50 Cr, 0.30 V

H7A—4% Chromium

Designed for hot forming by hydraulic or compressive methods with heavy pressure while in contact with the hot metal. Will stand water cooling in operation. Forging temperature 1800 F, annealing 1500 F, hardening 1700 to 1800 F, tempering 600 to 1000 F. Air or oil quench.

Allegheny Ludlum—EB—0.70 C, 0.30 Mn, 3.75 Cr, 0.55 V, 0.70 Mo
Crucible—La Belle 89—0.55 C, 3.90 Cr, 0.90 V, 0.45 Mo
Firth Sterling—CYW Choice—0.95 C, 0.50 Mn, 3.60 Cr

Houghton & Richards—Hot Work No. 4—0.97 C, 0.35 Mn, 0.35 Si, 3.90 Cr
Jessop—J—0.62 C, 0.30 Mn, 0.35 Si, 3.80 Cr, 0.55 V, 0.55 Mo
Jessop—JJ—0.62 C, 0.30 Mn, 0.35 Si, 3.80 Cr, 0.55 V, 0.55 Mo
Latrobe—Select M—0.90 C, 3.50 Cr, 0.20 V, 0.60 Mo
Republic—6-H-W—0.60 C, 4.00 Cr, 0.75 V, 0.45 Mo
Republic—XX—0.90 C, 4.00 Cr, 0.75 V, 0.45 Mo
Simonds—Chrome Hot Die—1.00 C, 4.00 Cr
Universal-Cyclops—Ajax No. 2—1.00 C, 4.25 Cr, 0.50 V, 0.45 Mo
Vanadium-Alloys—Choice—1.00 C, 4.00 Cr
Vulcan Crucible—4-HW—0.90 C, 4.00 Cr, 0.75 V, 0.45 Mo
Vulcan Crucible—6-HW—0.60 C, 4.00 Cr, 0.75 V, 0.45 Mo

H8A—Chrome-silicon

Suitable for hot-work dies and plastic-mold dies. Forging temperature 1600 to 1950 F, annealing 1500 F, hardening 1650 to 1750 F, tempering 350 to 600 F. Oil quench.

Houghton & Richards—MYA—0.42 C, 0.30 M, 1.45 Si, 1.45 Cr, 0.25 V

H9A—Chrome-nickel-aluminum

Precipitation-hardening alloy steel for special applications in die casting and plastic injection molding.

Crucible—PHV Die Steel—0.27 C, 0.60 Mn, 0.80 Si, 1.15 Cr, 0.40 V, 0.25 Mo, 2.80 Ni, 1.15 Al

H11—Chrome-moly-V

SAE composition: 0.35 C, 0.30 Mn, 1.00 Si, 5.00 Cr, 0.40 V, 1.50 Mo. Better toughness at elevated temperatures than H21. It is resistant to heat checking when water-cooled in operation. Suitable for aluminum and magnesium extrusion and forging dies, punches, die inserts, and shell-piercing dies.

Bethlehem—Cr-Mo-V—0.38 C, 1.00 Si, 5.00 Cr, 0.45 V, 1.25 Mo
Braeburn—Pressuredie No. 3-L—0.39 C, 1.00 Si, 5.50 Cr, 0.50 V, 1.10 Mo
Columbia—Firedie—0.37 C, 5.25 Cr, 0.50 V, 1.35 Mo
Crucible—Halcomb 218—0.40 C, 1.00 Si, 5.00 Cr, 0.35 V, 1.35 Mo
Firth Sterling—HWD No. 2—0.38 C, 0.40 Mn, 1.00 Si, 5.25 Cr, 0.30 V, 1.35 Mo
Hawkridge—Nu-Die Mix 218—0.40 C, 1.00 Si, 5.00 Cr, 0.35 V, 1.35 Mo
Houghton & Richards—Hot Work No. 5—0.35 C, 0.35 Mn, 1.00 Si, 5.00 Cr, 0.40 V, 1.00 Mo
Jessop—Dica B Modified—0.36 C, 0.95 Si, 5.00 Cr, 0.20 V, 1.50 Mo
Latrobe—Dycast No. 1—0.40 C, 1.00 Si, 5.00 Cr, 0.50 V, 1.00 Mo
McDonald—Macco NLS—0.50 C, 0.60 Mn, 0.25 Si, 5.20 Cr, 0.30 V, 1.20 Mo
Peninsular—Penco-Cr-Mo-V—0.40 C, 1.10 Si, 5.25 Cr, 0.50 V, 1.20 Mo
Universal-Cyclops—Thermold A—0.35 C, 0.45 Mn, 1.00 Si, 5.00 Cr, 0.40 V, 1.35 Mo
Vanadium-Alloys—Hotform No. 2—0.35 C, 0.30 Mn, 0.90 Si, 5.00 Cr, 0.50 V, 1.40 Mo
Vulcan Crucible—Magal—0.35 C, 1.00 Si, 5.00 Cr, 0.50 V, 1.25 Mo

H12—Chrome-moly-tungsten

SAE composition: 0.35 C, 0.30 Mn, 1.00 Si, 5.00 Cr, 0.25 V opt, 1.25 W, 1.50 Mo. General-purpose hot-work steel with good toughness, red hardness, and wear resistance. Resists softening up to 1000 F. Suitable for dies, dummy blocks, and other parts for extrusion of aluminum, magnesium, and brass; gripper and header dies; punches, hot billet shears, and die-casting dies.

Allegheny Ludlum—Potomac—0.33 C, 0.30 Mn, 0.85 Si, 5.00 Cr, 0.23 V, 1.25 W, 1.45 Mo
Amalgamated—Kro-tung—0.33 C, 0.90 Si, 4.75 Cr, 0.25 V, 1.30 W, 1.50 Mo
Atlas—Crodi—0.35 C, 0.40 Mn, 1.00 Si, 5.00 Cr, 0.40 V, 1.50 Mo
Bethlehem—Cr-Mo-W—0.35 C, 1.05 Si, 5.15 Cr, 1.55 W, 1.65 Mo
Braeburn—Pressuredie No. 2—0.35 C, 0.35 Mn, 1.00 Si, 5.00 Cr, 0.35 V, 1.20 W, 1.45 Mo
Carpenter—345—0.35 C, 1.00 Si, 5.00 Cr, 1.25 W, 1.50 Mo
Crucible—Chro-Mow—0.30 C, 1.00 Si, 5.00 Cr, 0.25 V, 1.25 W, 1.35 Mo
Darwin & Milner—HWS-Hot Work—0.35 C, 1.00 Si, 5.00 Cr, 1.00 V, 1.00 Mo

Hot-work Tool Steels

Chromium-base Types (*Continued*)

Delaware—Exelent—0.50 C, 0.70 Mn, 0.20 Si, 3.50 Cr, 1.50 Mo

Disston—Mix 873—0.35 C, 0.32 Mn, 1.05 Si, 5.00 Cr, 0.40 V, 1.20 W, 1.35 Mo

Firth Sterling—HWD No. 1—0.35 C, 0.60 Mn, 1.00 Si, 5.00 Cr, 0.25 V, 1.40 W, 1.55 Mo

Great Western—99-Hot Work—0.35 C, 0.35 Mn, 0.90 Si, 4.75 Cr, 0.25 V, 1.10 W, 1.50 Mo

Houghton & Richards—Hot Work No. 6—0.35 C, 0.35 Mn, 1.00 Si, 5.00 Cr, 1.35 W, 1.75 Mo

Hoyland—33—0.35 C, 0.35 Mn, 0.90 Si, 4.75 Cr, 0.25 V, 1.10 W, 1.50 Mo

Jessop—Dica B—0.34 C, 0.35 Mn. 0.95 Si, 4.85 Cr, 0.20 V, 1.15 W, 1.50 Mo

Kloster—D-C-33—0.33 C, 0.30 Mn, 1.05 Si, 5.00 Cr, 1.55 W, 1.65 Mo

Latrobe—LPD—0.35 C, 0.30 Mn, 1.00 Si, 5.00 Cr, 1.50 W, 1.50 Mo

Latrobe—MGR—0.55 C, 0.30 Mn, 1.00 Si, 5.00 Cr, 1.00 W, 1.00 Mo

McDonald—Macco ML—0.35 C, 0.30 Mn, 1.00 Si, 5.00 Cr, 1.50 W, 1.65 Mo

Milne—CMW—0.35 C, 1.00 Si, 5.00 Cr, 1.30 W, 1.65 Mo

Peninsular—Penco-Cr-Mo-W—0.33 C, 1.05 Si, 5.00 Cr, 1.55 W, 1.65 Mo

Republic—10-H-W—0.35 C, 1.00 Si, 5.00 Cr, 1.25 W, 1.50 Mo

Uddeholm—UHB Special Hot Work—0.35 C, 1.05 Si, 5.00 Cr, 1.55 W, 1.65 Mo

Universal-Cyclops—Thermold B—0.35 C, 0.30 Mn, 1.00 Si, 5.00 Cr, 0.25 V, 1.35 W, 1.60 Mo

Vanadium-Alloys—Hotform No. 1—0.35 C, 0.30 Mn, 0.90 Si, 5.00 Cr, 0.50 V, 1.40 W, 1.40 Mo

Vulcan-Crucible—TCM—0.35 C, 1.00 Si, 5.00 Cr, 0.25 V, 1.25 W, 1.50 Mo

H13—Chrome-moly-VV

SAE composition: 0.35 C, 0.30 Mn, 1.00 Si, 5.00 Cr, 0.90 V, 1.50 Mo. Higher vanadium content than H11 gives slightly improved wear resistance. Developed primarily for die casting and plastic molding. Also suitable for forging dies and inserts, punches, piercers, and extrusion tools.

Allegheny Ludlum—Potomac M—0.40 C, 0.35 Mn, 1.00 Si, 5.25 Cr, 1.00 V, 1.15 Mo

Bethlehem—Cr-Mo-V (High Vanadium)—0.40 C, 1.10 Si, 5.25 Cr, 0.90 V, 1.20 Mo

Braeburn—Pressuredie No. 3—0.39 C, 1.00 Si, 5.50 Cr, 1.00 V, 1.10 Mo

Carpenter—883—0.40 C, 0.35 Mn, 1.10 Si, 5.00 Cr, 0.90 V, 1.35 Mo

Columbia—Vanadium Firedie—0.37 C, 5.25 Cr, 1.00 V, 1.35 Mo

Crucible—Nu Die V—0.40 C, 1.00 Si, 5.00 Cr, 1.10 V, 1.35 Mo

Darwin & Milner—HWA Hot Work—0.35 C, 1.00 Si, 5.00 Cr, 1.00 V, 1.00 Mo

Firth Sterling—HWD No. 3—0.40 C, 0.40 Mn, 1.00 Si, 5.25 Cr, 1.05 V, 1.25 Mo

Great Western—99-HV—0.40 C, 1.00 Si, 5.25 Cr, 1.25 Mo, 1.05 V

Houghton & Richards—Hot Work No. 5 V—0.40 C, 0.30 Mn, 1.00 Si, 5.00 Cr, 1.00 V, 1.00 Mo

Hoyland—33-HV—0.40 C, 1.00 Si, 5.25 Cr, 1.05 V, 1.25 Mo

Jessop—Dica B. Vanadium—0.36 C, 0.95 Si, 5.00 Cr, 1.00 V, 1.50 Mo

Kloster—Pure-Ore D-C-33-VA—0.38 C, 1.00 Si, 5.25 Cr, 1.05 V, 1.25 Mo

Latrobe—VDC—0.40 C, 0.30 Mn, 1.00 Si, 5.00 Cr, 1.00 V, 1.00 Mo

McDonald—Macco 33—0.40 C, 0.30 Mn, 1.00 Si, 5.50 Cr, 1.00 V, 1.40 Mo

Peninsular—Hi Van—0.40 C, 1.10 Si, 5.25 Cr, 0.90 V, 1.20 Mo

Uddeholm—UHB Orvar—0.38 C, 1.00 Si, 5.00 Cr, 1.00 V, 1.50 Mo

Universal-Cyclops—Thermold AV—0.35 C, 0.45 Mn, 1.00 Si, 5.00 Cr, 1.00 V, 1.35 Mo

Vulcan Crucible—Vulcast—0.40 C, 1.00 Si, 5.00 Cr, 1.00 V, 1.25 Mo

H14—5 Chromium, 5 Tungsten

Composition: 0.40 C, 5.00 Cr, 5.00 W. Primarily developed for long-run die casting. Also suitable for brass forging, extrusion, and heating. Forging 1950 to 2050 F, annealing 1650 F, hardening 1850 F, tempering 1050 to 1100 F, air or oil quench.

Allegheny Ludlum—Grade 593—0.40 C, 0.90 Si, 5.00 Cr, 5.00 W, 0.40 Mo, 0.25 V

Atlas—Red Indian—0.35 C, 0.30 Mn, 1.00 Si, 5.00 Cr, 0.30 V, 4.50 W, 0.30 Mo, 0.50 Co

Braeburn—Pressuredie No. 1—0.38 C, 0.30 Mn, 0.90 Si, 5.00 Cr, 0.20 V, 5.00 W, 0.25 Mo, 0.50 Co

Crucible—CCS—0.40 C, 1.15 Si, 5.25 Cr, 4.25 W

Great Western—515-Hot Work—0.35 C, 0.25 Mn, 1.00 Si, 5.00 Cr, 5.00 W, 0.20 Mo

Houghton & Richards—No. 55—0.35 C, 0.25 Mn, 0.90 Si, 5.25 Cr, 0.20 V, 5.25 W, 0.20 Mo, 0.50 Co

Hoyland—515—0.35 C, 0.25 Mn, 1.00 Si, 5.00 Cr, 5.00 W, 0.20 Mo

Jessop—Dica D—0.36 C, 0.30 Mn, 1.00 Si, 5.00 Cr, 5.00 W, 0.25 Mo

Latrobe—Lumdie—0.40 C, 1.00 Si, 5.00 Cr, 4.50 W

McDonald—Macco Lensmold—0.40 C, 0.25 Mn, 1.00 Si, 5.25 Cr, 4.65 W

Universal-Cyclops—KS—0.40 C, 0.90 Si, 5.25 Cr, 5.25 W

H16—7 Chromium, 7 Tungsten

Composition: 0.55 C, 7.00 Cr, 7.00 W. With its high chromium content, this steel is tougher than the other tungsten steel. Suitable for shear blades, die inserts, forming dies, heading dies, and punches. Forging 1950 to 2000 F, annealing 1650 F, hardening 2000 to 2100 F, tempering 600 to 1300 F. Air or oil quench.

Firth Sterling—WCR—0.60 C, 0.60 Mn, 1.00 Si, 8.50 Cr, 7.50 W

Universal-Cyclops—KL—0.35 C, 0.60 Mn, 1.50 Si, 7.50 Cr, 7.50 W

Universal-Cyclops—KM—0.45 C, 0.60 Mn, 1.50 Si, 7.50 Cr, 7.50 W

Universal-Cyclops—KR—0.55 C, 0.60 Mn, 0.60 Si, 7.50 Cr, 7.50 W

Tungsten-base Types

H21—9% Tungsten

SAE composition: 0.32 C, 0.30 Mn, 0.20 Si, 3.25 Cr, 0.40 V, 9.00 W. Excellent red hardness permits this steel to hold its tempered hardness at temperatures to 1100 F. Suitable for brass forging dies and inserts, extrusion dies, nut punches, permanent molds, extrusion mandrels, and hot trim dies. ASM classifies the 2% Cr steels as H20.

Allegheny Ludlum—Atlas A—0.30 C, 0.30 Mn, 0.30 Si, 3.00 Cr, 0.45 V, 9.00 W

Amalgamated—Thermal B—0.35 C, 2.75 Cr, 0.30 V, 9.00 W

Atlas—Hodi—0.28 C, 0.30 Mn, 0.30 Si, 3.25 Cr, 0.40 V, 10.00 W

Atlas—Seneca—0.35 C, 0.30 Mn, 0.30 Si, 3.25 Cr, 0.40 V, 10.00 W

Bethlehem—57 HW—0.30 C, 3.25 Cr, 0.30 V, 9.35 W

Boyd-Wagner—EZ 9 W—0.40 C, 0.30 Mn, 0.20 Si, 3.00 Cr, 0.25 V, 9.00 W

Braeburn—T-Alloy A—0.33 C, 0.25 Mn, 0.25 Si, 3.50 Cr, 0.50 V, 9.60 W

Carpenter—T-K—0.35 C, 0.30 Mn, 0.30 Si, 3.50 Cr, 0.40 V, 9.00 W

Columbia—Formite No. 2—0.32 C, 3.25 Cr, 9.40 W, 0.30 Mo, 0.50 V

Crucible—Peerless A—0.28 C, 3.25 Cr, 0.25 V, 9.00 W

Darwin & Milner—Darwin 93—0.30 C, 3.00 Cr, 0.45 V, 9.00 W

Firth Sterling—LT Forging Die—0.33 C, 0.20 Mn, 0.30 Si, 3.50 Cr, 0.45 V, 9.50 W

Great Western—310-Hot Work—0.30 C, 0.25 Mn, 0.30 Si, 3.25 Cr, 0.35 V, 10.00 W

Houghton & Richards—Hot Work No. 2—0.33 C, 0.20 Mn, 0.30 Si, 3.50 Cr, 0.50 V, 9.25 W

Hoyland—310—0.30 C, 0.25 Mn, 0.30 Si, 3.25 Cr, 0.35 V, 10.00 W

Jessop—2B-LC—0.30 C, 0.30 Mn, 0.30 Si, 3.10 Cr, 0.30 V, 10.00 W

Kloster—D-C-66—0.30 C, 0.35 Mn, 0.33 Si, 3.50 Cr, 0.27 V, 8.75 W

McDonald—Macco P-175—0.31 C, 0.29 Mn, 0.43 Si, 3.30 Cr, 0.50 V, 9.50 W

Milne—3074—0.35 C, 4.00 Cr, 0.50 V, 9.00 W

Peninsular—Penco 57 HW—0.35 C, 2.75 Cr, 0.30 W, 9.00 W

Republic—Air Hardening No. 30—0.30 C, 3.50 Cr, 0.25 V, 9.00 W

Simonds—D. V. V. Hot Work—0.32 C, 0.30 Mn, 0.35 Si, 3.25 Cr, 0.40 V, 9.50 W

Hot-work Tool Steels

TUNGSTEN-BASE TYPES (*Continued*)

Universal-Cyclops—B44J—0.33 C, 0.30 Mn, 0.35 Si, 3.35 Cr, 0.50 V, 9.50 W, 1.00 max Mo

Vanadium-Alloys—Marvel—0.33 C, 3.50 Cr, 0.50 V, 9.00 W

Vanadium-Alloys—Hotpress—0.35 C, 2.00 Cr, 0.50 V, 9.00 W

Vulcan Crucible—Calo Ferro 0.30—0.30 C, 3.50 Cr, 0.50 V, 9.50 W

H22—12% Tungsten

Composition: 0.35 C, 2.00 Cr, 11.00 W. Better wear resistance and less toughness than H21. Good hardness at elevated temperatures. Principal applications in brass industry for extrusions, forging, and die casting.

Allegheny Ludlum—Atlas B—0.40 C, 0.30 Mn, 0.25 Si, 3.00 Cr, 0.45 V, 12.00 W

Braeburn—T-Alloy—0.35 C, 0.25 Mn, 0.25 Si, 3.50 Cr, 0.40 V, 10.50 W

Crucible—Peerless LCT 2—0.40 C, 2.00 Cr, 0.35 V, 11.50 W

Houghton & Richards—Hot Work—0.32 C, 0.28 Si, 3.25 Cr, 0.40 V, 10.25 W

Jessop—2-B-HC—0.49 C, 0.30 Mn, 0.30 Si, 2.90 Cr, 0.35 V, 11.25 W

Jessop—2-B-MC—0.39 C, 3.10 Cr, 0.35 V, 13.30 W

Republic—Air Hard No. 40—0.45 C, 3.00 Cr, 0.25 V, 13.00 W

Universal-Cyclops—B 44—0.35 C, 3.50 Cr, 0.45 V, 13.00 W

H23—12 Chromium, 12 Tungsten

Composition: 0.30 C, 12.00 Cr, 12.00 W. Suitable for brass extrusion and forging dies. Forging temperature 2150 F, annealing 1550 F, hardening 2300 F, tempering 1300 F. Oil quench.

Braeburn—HCA—0.30 C, 0.30 Mn, 0.50 Si, 12.00 Cr, 0.90 V, 12.00 W

Crucible—Halcomb 236—0.30 C, 0.35 Mn, 0.50 Si, 12.00 Cr, 0.90 V, 12.00 W

Firth Sterling—BDC—0.33 C, 0.35 Mn, 0.35 Si, 12.00 Cr, 1.05 V, 12.00 W

Houghton & Richards—Hot Work No. 12—0.30 C, 12.00 Cr, 1.00 V, 12.00 W

Latrobe—Kalkos—0.30 C, 12.00 Cr, 1.00 V, 12.00 W

Universal-Cyclops—Thor—0.30 C, 0.30 Mn, 0.50 Si, 12.00 Cr, 1.00 V, 12.00 W

H24—15% Tungsten

Composition: 0.45 C, 3.00 Cr, 15.00 W. More wear resistance and less toughness than H22. Not suitable for battering applications. Similar to H25. Suitable for hot shear blades, drawing, and gripper dies, and extrusion dies.

Allegheny Ludlum—Mohawk—0.45 C, 0.30 Mn, 0.30 Si, 3.50 Cr, 0.70 V, 14.00 W

Bethlehem—57 Spec—0.40 C, 3.50 Cr, 0.30 V, 14.00 W

Boyd-Wagner—EZ 14 W—0.45 C, 0.30 Mn, 0.20 Si, 3.00 Cr, 0.30 V, 14.00 W

Braeburn—T-Alloy B—0.50 C, 0.25 Mn, 0.25 Si, 3.00 Cr, 0.50 V, 15.00 W

Darwin & Milner—IWI—0.45 C, 0.25 Mn, 0.30 Si, 4.00 Cr, 0.70 V, 15.00 W

Latrobe—CHW—0.50 C, 3.00 Cr, 0.50 V, 15.00 W

McDonald—Macco P-150—0.51 C, 2.90 Cr, 0.60 V, 15.30 W

McInnes—Record A—0.50 C, 3.75 Cr, 0.75 V, 14.00 W

Universal-Cyclops—B 4—C as desired, 2.75 Cr, 0.50 V, 15.00 W

Vulcan Crucible—0.50 Calo Ferro—0.50 C, 3.00 Cr, 0.50 V, 14.50 W

H25—15% Tungsten

Composition: 0.25 C, 4.00 Cr, 15.00 W. More wear resistance and less toughness than H22. Not suitable for battering applications. Similar to H24. Suitable for hot shear blades, drawing, and gripper dies, and extrusion dies.

Braeburn—T-Alloy C—0.25 C, 0.25 Mn, 0.25 Si, 4.00 Cr, 0.50 V, 15.00 W

Columbia—Formite No. 1—0.35 C, 4.00 Cr, 0.35 V, 13.50 W, 0.40 Mo

Crucible—LLCT—0.25 C, 4.00 Cr, 0.50 V, 15.00 W

Firth Sterling—XDL—0.38 C, 0.25 Mn, 0.30 Si, 4.00 Cr, 0.45 V, 15.50 W

Great Western—313-Hot Work—0.35 C, 0.30 Mn, 0.30 Si, 3.00 Cr, 13.50 W

Houghton & Richards—Hot Work No. 15—0.25 C, 0.29 Mn, 0.26 Si, 4.03 Cr, 0.51 V, 15.10 W

Hoyland—313—0.35 C, 0.30 Mn, 0.30 Si, 3.00 Cr, 13.50 W
Latrobe—EHW No. 1—0.25 C, 0.25 Mn, 0.25 Si, 4.00 Cr, 0.50 V, 15.00 W
McDonald—Macco P-125—0.25 C, 4.20 Cr, 0.50 V, 5.50 W
Vanadium Alloys—Forge Die—0.25 C, 3.50 Cr, 0.50 V, 14.00 W

H26—18% Tungsten

Composition: 0.50 C, 4.00 Cr, 1.00 V, 18.00 W. A lower carbon version of T1 with better shock resistance but other properties essentially the same. Shock resistance is still quite low for a hot-work steel. Suitable for shear blades, swaging dies, gripper dies, and glass molds.

Allegheny Ludlum—LXX 6 Temper—0.60 C, 4.00 Cr, 18.00 W, 1.00 V
Atlas—Spartan 5—0.50 C, 0.25 Mn, 0.30 Si, 4.00 Cr, 1.20 V, 18.00 W
Braeburn—Vinco—0.62 C, 0.25 Mn, 0.25 Si, 4.00 Cr, 1.00 V, 18.00 W
Carpenter—Star Zenith Low Carbon—Low C, 4.00 Cr, 1.15 V, 18.25 W
Columbia—Clarite HW—0.57 C, 4.00 Cr, 0.70 V, 18.00 W
Crucible—Rex AA "OX" temper—0.63 C, 4.00 Cr, 1.10 V, 18.00 W
Crucible—Rex AA "PX" temper—0.53 C, 4.00 Cr, 1.10 V, 18.00 W
Disston—Kutkwik—0.63 or 0.68 C, 4.00 Cr, 1.00 V, 18.00 W
Firth Sterling—XDH—0.58 C, 0.25 Mn, 0.35 Si, 4.00 Cr, 1.00 V, 18.00 W
Firth Sterling—XDM—0.45 C, 0.25 Mn, 0.35 Si, 4.00 Cr, 0.45 V, 17.50 W
Houghton & Richards—No. 50—0.60 C, 0.25 Mn, 0.25 Si, 4.00 Cr, 1.00 V, 12.00 W
Latrobe—Electrite No. 5—0.60 C, 0.25 Mn, 0.25 Si, 4.00 Cr, 1.00 V, 18.00 W
Universal-Cyclops—B-6 X—0.50 C, 0.25 Mn, 0.30 Si, 4.00 Cr, 1.10 V, 18.00 W

Vanadium-Alloys—Red Cut Superior J Temper—0.50 C, 0.20 Mn, 0.32 Si, 4.00 Cr, 1.00 V, 18.00 W
Vulcan Crucible—Wolfram—0.55 C, 4.00 Cr, 1.00 V, 18.00 W

H31A—Nickel-tungsten

Modified form of H2 with low carbon and an appreciable nickel content. Suitable for piercing heads, die-casting dies for copper alloys, dummy blocks, and forming dies. Forging 2000 to 2200 F, annealing 1600 F, hardening 2000 to 2300 F, tempering 1000 to 1300 F. Oil or air quench.

Firth Sterling—LTL—0.25 C, 0.25 Mn, 0.20 Si, 2.75 Cr, 0.25 V, 10.00 W, 1.50 Ni
Republic—Resco—0.28 C, 2.75 Cr, 9.50 W, 0.25 Mo, 1.75 Ni
Uddeholm—Valand 1—0.30 C, 0.30 Mn, 0.30 Si, 3.00 Cr, 0.30 V, 9.50 W, 1.75 Ni
Universal-Cyclops—K 390—0.24 C, 0.40 Mn, 0.40 Si, 2.75 Cr, 10.00 W, 2.00 Ni
Vulcan Crucible—A-42—0.30 C, 2.75 Cr, 9.75 W, 0.25 Mo, 1.75 Ni

H32A—Cobalt-tungsten

Maintains shock and wear resistance at elevated temperatures. Suitable for brass extrusion tools, press-forging dies, and hot punches. Forging 1900 to 2100 F, annealing 1600 to 1650 F, hardening 2100 to 2250 F, tempering 900 to 1300 F. Air or oil quench.

Allegheny Ludlum—B-47 Hot Die—0.40 C, 0.35 Mn, 0.25 Si, 4.25 Cr, 2.25 V, 4.25 W, 0.40 Mo, 4.25 Co
Braeburn—Pressuredie "C"—0.40 C, 0.30 Mn, 0.30 Si, 4.25 Cr, 2.00 V, 4.25 W, 0.40 Mo, 4.25 Co
Uddeholm—Valand 2—0.32 C, 0.30 Mn, 0.30 Si, 2.50 Cr, 0.35 V, 8.00 W, 2.00 Co

Molybdenum-base Types

H41—8% Molybdenum

ASM composition: 0.65 C, 4.00 Cr, 1.00 V, 1.50 W, 8.00 Mo. Low-carbon version of M1. High temperature use.

Houghton & Richards—Molyhi—C per temper, 4.00 Cr, 1.15 V, 1.50 W, 8.50 Mo
Latrobe—MCH—0.50 C, 3.75 Cr, 0.75 V, 1.00 W, 6.25 Mo
Latrobe—MCL—0.30 C, 3.75 Cr, 0.75 V, 1.00 W, 6.25 Mo
Universal-Cyclops—Mo Tung—C per temper, 0.25 Mn, 0.30 Si, 4.00 Cr, 1.15 V, 1.50 W, 8.50 Mo

Hot-work Tool Steels

Molybdenum-base Types (*Continued*)

H42—5% Molybdenum

ASM composition: 0.60 C, 4.00 Cr, 2.00 V, 6.00 W, 5.00 Mo. Low-carbon version of M2. Suitable for extrusion dies, header inserts, and bolt trimmers.

Allegheny Ludlum—DBL-Low Carbon —0.62 C, 4.00 Cr, 6.25 W, 5.00 Mo, 1.90 V

Houghton & Richards—No. 45—0.65 C, 4.00 Cr, 2.00 V, 6.50 W, 5.00 Mo

Latrobe—Electrite No. 7—0.65 C, 4.00 Cr, 2.00 V, 6.00 W, 5.00 Mo

H43—8 Molybdenum 2 Vanadium

ASM composition: 0.55 C, 4.00 Cr, 2.00 V, 8.00 Mo. Similar to H41, but with higher vanadium content for higher red hardness.

Allegheny Ludlum—VLM-Low Carbon —0.65 C, 4.00 Cr, 1.80 V, 8.00 Mo

Bethlehem—HW 8—0.60 C, 3.60 Cr, 1.75 V, 8.50 Mo

Columbia—Molite HW10—0.63 C, 4.00 Cr, 2.00 V, 0.25 W max, 8.25 Mo

H51A—1% Molybdenum

A tough steel with resistance to heat checking. Suitable for hot-header dies, upset punches, extrusion dies, and press-forging dies. Forging temperature 1900 to 2100 F, annealing 1450 F, hardening 1650 to 1750 F, tempering 900 to 1300 F. Oil quench.

Republic—550—0.45 C, 0.65 Mn, 1.60 Cr, 0.25 V, 1.10 Mo

Vulcan Crucible—A-41—0.45 C, 1.60 Cr, 0.25 V, 1.10 Mo

H52A—2% Molybdenum

Combines red hardness and toughness. Suitable for punches, mandrels, forming and forging dies, extrusion and swaging dies, shear knives, and die-casting dies and plastic molds. Forging 1950 to 2150 F, annealing 1450 to 1550 F, hardening 1800 to 1900 F, tempering 950 to 1200 F.

Amalgamated—Kropunch—0.40 C, 1.10 Si, 5.00 Cr, 0.30 V, 2.50 W, 2.00 Mo

High-speed Tool Steels

Tungsten-base Types

T1—Tungsten, 18-4-1

SAE composition: 0.70 C, 0.30 Mn, 0.30 Si, 4.10 Cr, 1.10 V, 18.00 W. A general-purpose high-speed steel with a balanced combination of shock resistance and abrasion resistance. It is the easiest high-speed grade to machine. Has high red hardness. Principal application is cutting tools. In a lower carbon range it is listed under H26.

Allegheny Ludlum—LXX—0.70 C, 0.25 Mn, 0.25 Si, 4.00 Cr, 1.00 V, 18.00 W

Amalgamated—Malax AA—0.73 C, 4.00 Cr, 1.10 V, 18.00 W

Atlantic—High Speed—0.50 C, 0.20 Mn, 0.20 Si, 4.00 Cr, 1.00 V, 18.00 W, 0.15 Mo

Atlas—Spartan 7—0.72 C, 0.25 Mn, 0.30 Si, 4.00 Cr, 1.20 V, 18.00 W

Bethlehem—Special HS—0.73 C, 4.00 Cr, 1.10 V, 18.00 W

Boyd-Wagner—Record Superior—0.70 C, 0.30 Mn, 0.20 Si, 4.00 Cr, 1.00 V, 18.00 W, 0.30 Co

Braeburn—Vinco—0.70 C, 0.25 Mn, 0.25 Si, 4.00 Cr, 1.00 V, 18.00 W

Carpenter—Star Zenith—0.72 C, 0.25 Mn, 0.20 Si, 4.00 Cr, 1.15 V, 18.25 W

Columbia—Clarite—0.72 C, 0.25 Mn, 0.30 Si, 4.00 Cr, 1.15 V, 18.00 W

Crucible—Rex AA—0.73 C, 4.00 Cr, 1.15 V, 18.00 W

Darwin & Milner—Cannon—0.70 C, 0.25 Mn, 0.35 Si, 4.00 Cr, 1.00 V, 18.00 W

Delaware—18-4-1—0.70 C, 0.30 Mn, 0.20 Si, 4.00 Cr, 1.00 V, 18.00 W

Diehl—Blue Streak—0.72 C, 4.00 Cr, 1.00 V, 18.00 W

Disston—Kutkwik—0.73 or 0.78 C, 4.00 Cr, 1.00 V, 18.00 W

Firth Sterling—Blue Chip—0.72 C, 0.52 Mn, 0.20 Si, 4.10 Cr, 1.15 V, 18.20 W

Great Western—Silver Stripe—0.70 C, 0.20 Mn, 0.30 Si, 4.00 Cr, 1.00 V, 18.00 W

Houghton & Richards—No. 1—0.70 C, 0.20 Mn, 0.30 Si, 4.00 Cr, 1.00 V, 18.00 W

Hoyland—18-4-1—0.70 C, 0.20 Mn, 0.30 Si, 4.00 Cr, 1.00 V, 18.00 W

Jessop—Supremus—0.73 C, 0.28 Mn, 0.28 Si, 4.00 Cr, 1.10 V, 18.00 W

Kloster—Clipper—0.65 C, 0.27 Si, 4.02 Cr, 1.07 V, 18.12 W

Latrobe—Electrite No. 1—0.72 C, 0.25 Mn, 0.25 Si, 4.00 Cr, 1.00 V, 18.00 W

McDonald—Macco Superior—0.75 C, 0.28 Mn, 0.22 Si, 4.00 Cr, 1.00 V, 18.00 W

McInnes—"V" High Speed—0.75 C, 4.00 Cr, 1.10 V, 18.25 W

Milne—AMC—4.00 Cr, 1.00 V, 18.00 W

Peninsular—Penco XX—0.72 C, 0.25 Mn, 0.30 Si, 4.00 Cr, 1.10 V, 18.00 W

Republic—B-F High Speed—0.72 C, 0.30 Mn, 4.00 Cr, 1.00 V, 18.00 W

Simonds—Red Streak—0.72 C, 4.00 Cr, 1.00 V, 18.00 W

Universal-Cyclops—B 6—0.75 C, 0.25 Mn, 0.30 Si, 4.00 Cr, 1.10 V, 18.00 W

Vanadium-Alloys—Red Cut Superior—0.72 C, 0.20 Mn, 0.30 Si, 4.00 Cr, 1.00 V, 18.00 W

Vulcan Crucible—Wolfram—0.71 C, 4.00 Cr, 1.00 V, 18.00 W

T2—Tungsten, 18-4-2

SAE composition: 0.80 C, 0.30 Mn, 0.30 Si, 4.10 Cr, 2.10 V, 18.50 W, 0.80 Mo. With higher carbon and vanadium content than T1 and a small molybdenum addition, this steel provides a harder and more durable tool edge. Often more economical than cobalt steels, it hardens without a soft skin. Not as tough as T1. Suitable for fine-edge tools such as hobs and threading dies, form tools, twist drills, reamers, broaches, and milling cutters.

Allegheny Ludlum—ML—0.82 C, 0.30 Mn, 0.30 Si, 4.10 Cr, 2.10 V, 18.50 W, 0.75 Mo

Atlas—Trojan—0.80 C, 0.25 Mn, 0.30 Si, 4.00 Cr, 2.00 V, 18.50 W, 0.50 Mo

Braeburn—Twinvan—0.82 C, 0.25 Mn, 0.25 Si, 4.25 Cr, 2.10 V, 18.50 W, 0.65 Mo

Columbia—Vanite—0.83 C, 4.00 Cr, 2.15 V, 18.50 W, 1.00 Mo max

Crucible—Rex Supervan—0.85 C, 4.00 Cr, 2.10 V, 18.50 W, 0.75 Mo

Darwin & Milner—Cannon Special—0.80 C, 0.25 Mn, 0.30 Si, 4.00 Cr, 2.00 V, 18.00 W, 0.60 Mo

Firth Sterling—HV Blue Chip—0.82 C, 0.25 Mn, 0.30 Si, 4.25 Cr, 2.10 V, 18.50 W, 0.65 Mo

Houghton & Richards—No. 2—0.80 C, 0.35 Si, 4.25 Cr, 2.15 V, 18.50 W, 0.65 Mo

Jessop—Supremus Extra—0.85 C, 0.28 Mn, 0.28 Si, 4.00 Cr, 2.00 V, 18.50 W

Kloster—Prior—0.81 C, 0.30 Mn, 0.35 Si, 4.25 Cr, 2.12 V, 18.50 W, 0.65 Mo

Latrobe—Electrite No. 19—0.83 C, 0.25 Mn, 0.25 Si, 4.10 Cr, 2.20 V, 18.50 W, 0.60 Mo

Milne—Milvan—4.10 Cr, 2.25 V, 19.00 W

Republic—IXL High Speed—0.80 C, 0.30 Mn, 4.00 Cr, 2.00 V, 19.00 W, 0.75 Mo

Simonds—Lockport Special—0.80 C, 4.25 Cr, 2.25 V, 18.50 W, 0.70 Mo

Universal-Cyclops—B 9—0.84 C, 0.25 Mn, 0.30 Si, 4.25 Cr, 2.25 V, 18.50 W, 0.50 Mo

Vanadium-Alloys—EVM—0.85 C, 0.25 Mn, 0.35 Si, 4.25 Cr, 2.00 V, 18.10 W

Vulcan Crucible—Super—0.80 C, 4.25 Cr, 2.25 V, 18.50 W, 0.70 Mo

T3—Tungsten, 18-4-3

SAE composition: 1.05 C, 0.30 Mn, 0.30 Si, 4.10 Cr, 3.25 V, 18.50 W, 0.70 Mo. The triple vanadium and high carbon content of this steel provide the highest wear resistance of any tool steel. It is suitable for cutting hard wrought metals or castings, material that work-hardens, and soft, gummy materials where wear resistance is a major factor.

Crucible—Rex 939—1.05 C, 4.00 Cr, 3.00 V, 18.00 W

Houghton & Richards—No. 3—1.04 C, 0.26 Mn, 0.27 Si, 4.18 Cr, 3.41 V, 18.46 W, 0.84 Mo

Latrobe—Electrite Vanadium—1.10 C, 0.25 Mn, 0.25 Si, 4.10 Cr, 3.40 V, 18.50 W, 0.85 Mo

Peninsular—Penco XXX—1.05 C, 0.25 Mn, 0.30 Si, 4.35 Cr, 3.00 V, 18.50 W

T4—Cobalt-tungsten 18-4-1-5

SAE composition: 0.75 C, 0.30 Mn,

High-speed Tool Steels

Tungsten-base Types (*Continued*)

0.30 Si, 4.10 Cr, 1.00 V, 18.00 W, 0.80 Mo, 5.00 Co. The addition of 5% cobalt to T1 increases cutting ability at high temperatures, making this steel suitable for machining abrasive materials and for hogging cuts where high heats develop. Requires higher hardening temperature and so tends to develop a soft skin. Should be used where tools are well supported, not subject to shock, and ground all over after hardening.

Allegheny Ludlum—Panther Special—0.75 C, 0.25 Mn, 0.25 Si, 4.50 Cr, 1.00 V, 19.00 W, 5.00 Co

Amalgamated—Malax AAA—0.74 C, 4.50 Cr, 1.25 V, 18.50 W, 0.75 Mo, 5.00 Co

Atlas—Powhatan—0.75 C, 0.25 Mn, 0.30 Si, 4.00 Cr, 1.25 V, 19.00 W, 5.00 Co

Bethlehem—Comokut—0.74 C, 4.50 Cr, 1.25 V, 18.50 W, 0.75 Mo, 5.00 Co

Braeburn—Cobalt—0.74 C, 0.25 Mn, 0.25 Si, 4.00 Cr, 1.00 V, 18.00 W, 0.50 Mo, 5.00 Co

Crucible—Rex AAA—0.75 C, 4.00 Cr, 1.15 V, 18.00 W, 0.75 Mo, 5.00 Co

Darwin & Milner—505—0.70 C, 0.30 Mn, 0.25 Si, 4.00 Cr, 1.15 V, 18.00 W, 0.60 Mo, 5.00 Co

Firth Sterling—Red Chip—0.75 C, 0.25 Mn, 0.30 Si, 4.00 Cr, 1.00 V, 18.00 W, 0.75 Mo, 5.00 Co

Great Western—Super-Kut—0.70 C, 0.25 Mn, 0.30 Si, 4.00 Cr, 1.00 V, 18.00 W, 0.50 Mo, 5.00 Co

Houghton & Richards—Cobalt—0.78 C, 4.00 Cr, 1.00 V, 18.00 W, 5.00 Co

Hoyland—Cobalt 5—0.70 C, 0.25 Mn, 0.30 Si, 4.00 Cr, 1.00 V, 18.00 W, 0.50 Mo, 5.00 Co

Jessop—Purple Label—0.74 C, 0.28 Mn, 0.28 Si, 4.20 Cr, 1.10 V, 18.50 W, 0.50 Mo, 5.00 Co

Latrobe—Electrite Cobalt—0.72 C, 0.25 Mn, 0.25 Si, 4.00 Cr, 1.00 V, 18.00 W, 0.65 Mo, 5.00 Co

Milne—Hyco—4.25 Cr, 1.30 V, 18.50 W, 5.00 Co

Republic—Cobalt High Speed—0.72 C, 4.00 Cr, 1.00 V, 18.00 W, 0.50 Mo, 5.00 Co

Simonds—Tunco—0.73 C, 4.50 Cr, 1.10 V, 19.00 W, 5.00 Co

Universal-Cyclops—B 7—0.75 C, 0.25 Mn, 0.30 Si, 4.25 Cr, 1.10 V, 18.00 W, 0.50 Mo, 5.00 Co

Vanadium-Alloys—Red Cut Cobalt—0.73 C, 0.20 Mn, 0.30 Si, 4.25 Cr, 1.08 V, 18.25 W, 0.65 Mo, 4.75 Co

Vulcan Crucible—Wolfram Cobalt—0.72 C, 4.00 Cr, 1.00 V, 18.00 W, 0.50 Mo, 5.00 Co

T5—Cobalt-tungsten 18-4-2-8

SAE composition: 0.80 C, 0.30 Mn, 0.30 Si, 4.10 Cr, 1.75 V, 18.50 W, 0.80 Mo, 8.00 Co. An intermediate cobalt steel with high red hardness and good wear resistance. Cutting speeds can be about 25% faster than with T1 with longer tool life. Suited for heavy, dry cuts on heat-treated forgings.

Allegheny Ludlum—Super Panther—0.80 C, 0.25 Mn, 0.25 Si, 4.25 Cr, 2.00 V, 19.00 W, 1.00 Mo, 7.25 Co

Amalgamated—Malax AAAA—0.80 C, 4.25 Cr, 2.00 V, 19.00 W, 1.00 Mo, 7.25 Co

Atlas—Nipigon—0.78 C, 0.25 Mn, 0.30 Si, 4.00 Cr, 2.00 V, 19.00 W, 0.80 Mo, 8.00 Co

Boyd-Wagner—Record Eminent—0.75 C, 0.30 Mn, 0.20 Si, 5.00 Cr, 1.50 V, 18.50 W, 8.50 Co

Braeburn—Bonded Carbide Jr.—0.77 C, 0.25 Mn, 0.25 Si, 4.25 Cr, 1.95 V, 18.50 W, 0.75 Mo, 7.60 Co

Crucible—Rex Supercut—0.80 C, 4.00 Cr, 2.00 V, 18.50 W, 0.60 Mo, 8.00 Co

Darwin & Milner—505 Special—0.80 C, 0.30 Mn, 0.25 Si, 4.00 Cr, 2.00 V, 18.00 W, 0.80 Mo, 8.00 Co

Firth Sterling—Circle C—0.77 C, 0.25 Mn, 0.25 Si, 4.50 Cr, 1.90 V, 18.50 W, 1.00 Mo, 9.00 Co

Hawkridge—Dreadnought 18-8—0.80 C, 4.00 Cr, 2.00 V, 18.50 W, 0.60 Mo, 8.00 Co

Houghton & Richards—No. 4—0.80 C, 4.50 Cr, 1.75 V, 18.50 W, 0.80 Mo, 7.50 Co

Jessop—Purple Label Extra—0.78 C, 0.28 Mn, 0.28 Si, 4.20 Cr, 1.95 V, 18.50 W, 0.75 Mo, 7.90 Co

Latrobe—Electrite Super Cobalt—0.83 C, 0.25 Mn, 0.30 Si, 4.00 Cr, 2.00 V, 18.00 W, 9.00 Co

McInnes—Duro—0.80 C, 4.00 Cr, 18.50 W, 4.00 Mo, 7.00 Co

Milne—Milco 9—0.80 C, 0.30 Mn, 0.30 Si, 4.25 Cr, 1.75 V, 18.50 W, 0.75 Mo, 9.00 Co

Simonds—Super Cobalt—0.75 C, 4.25 Cr, 2.00 V, 18.50 W, 0.60 Mo, 7.75 Co

Universal-Cyclops—B 10—0.80 C, 0.25 Mn, 0.30 Si, 4.50 Cr, 2.00 V, 18.50 W, 1.00 Mo, 9.00 Co

Vanadium-Alloys—Red Cut Cobalt B—0.78 C, 0.25 Mn, 0.30 Si, 4.25 Cr, 1.85 V, 18.50 W, 0.75 Mo, 8.75 Co

T6—Cobalt-tungsten, 18-4-2-12

SAE composition: 0.80 C, 0.30 Mn, 0.30 Si, 4.10 Cr, 1.75 V, 20.00 W, 0.80 Mo, 12.00 Co. A high-cobalt steel having the highest red hardness of any tool steel. Wear resistance is better than the lower cobalt steels. Suitable for heavy-duty lathe and planer tools. Careful heat-treating practice is essential and salt baths or controlled atmospheres desirable.

Atlas—ACX—0.83 C, 0.25 Mn, 0.30 Si, 18.50 W, 4.00 Cr, 1.75 V, 1.00 Mo, 10.50 Co

Braeburn—Bonded Carbide—0.70 C, 0.25 Mn, 0.25 Si, 4.50 Cr, 1.50 V, 18.00 W, 0.70 Mo, 12.00 Co

Crucible—Rex 440—0.80 C, 4.00 Cr, 2.00 V, 19.50 W, 0.60 Mo, 12.00 Co

Darwin & Milner—Darwin 1366—0.80 C, 0.30 Mn, 0.25 Si, 4.00 Cr, 2.00 V, 18.00 W, 0.80 Mo, 13.00 Co

Hawkridge—Halcomb No. 440—0.80 C, 4.00 Cr, 2.00 V, 19.50 W, 0.60 Mo, 12.00 Co

Houghton & Richards—Super Cobalt—0.80 C, 0.20 Mn, 0.32 Si, 4.25 Cr, 1.35 V, 20.50 W, 0.60 Mo, 12.25 Co

Jessop—King Cobalt—0.78 C, 0.30 Mn, 4.20 Cr, 1.75 V, 19.50 W, 0.75 Mo, 11.50 Co

Latrobe—Electrite Ultra Cobalt—0.83 C, 0.25 Mn, 0.25 Si, 4.00 Cr, 2.00 V, 18.00 W, 0.80 Mo, 12.00 Co

McDonald—Macco Enormous—0.83 C, 0.25 Mn, 0.27 Si, 4.50 Cr, 2.00 V, 22.00 W, 1.00 Mo, 10.00 Co

Milne—Major—4.50 Cr, 1.80 V, 21.00 W, 0.75 Mo, 13.00 Co

Vanadium-Alloys—Gray Cut Cobalt—0.80 C, 0.25 Mn, 0.30 Si, 4.25 Cr, 1.60 V, 20.50 W, 0.60 Mo, 12.25 Co

T7—Tungsten, 14-4-2

Lowered tungsten content gives increased toughness with less wear resistance. Suitable for intermittent cutting and for sand castings, hard alloys, or gritty materials. Forging 1900 to 2000 F, annealing 1650 F, hardening 2325 to 2350 F. Double temper 2 hr at 1050 F.

Boyd-Wagner—Bowco One Star—0.70 C, 0.35 Mn, 0.20 Si, 4.00 Cr, 2.00 V, 14.00 W

Crucible—Rex Champion—0.73 C, 4.00 Cr, 2.00 V, 14.00 W

Firth Sterling—Star Blue Chip—0.73 C, 4.00 Cr, 2.00 V, 14.00 W

Hawkridge—Super Dreadnought—0.73 C, 4.00 Cr, 2.00 V, 14.00 W

Latrobe—Electrite U—0.70 C, 4.00 Cr, 2.00 V, 14.00 W

T8—Cobalt-tungsten, 14-4-2-5

SAE composition: 0.80 C, 0.30 Mn, 0.30 Si, 4.10 Cr, 2.00 V, 14.00 W, 0.80 Mo, 5.00 Co. Wear resistance exceeded only by T3 combined with good red hardness make this steel suitable for severe cutting operations, especially stainless steels. It has also given good results on hard die blocks, manganese steel castings, and chilled cast iron.

Carpenter—Gold Star—0.77 C, 0.25 Mn, 0.30 Si, 3.75 Cr, 2.00 V, 13.75 W, 5.00 Co

Columbia—Maxite—0.73 C, 0.25 Mn, 0.30 Si, 4.00 Cr, 2.15 V, 14.00 W, 0.60 Mo, 4.75 Co

Crucible—Rex 95—0.80 C, 4.00 Cr, 2.00 V, 14.00 W, 0.75 Mo, 5.25 Co

Firth Sterling—FS 2-5—0.80 C, 4.00 Cr, 2.00 V, 14.00 W, 0.75 Mo, 5.25 Co

Hawkridge—Halcomb No. 999—0.80 C, 4.00 Cr, 2.00 V, 14.00 W, 0.75 Mo, 5.25 Co

Jessop—T-8—0.79 C, 4.00 Cr, 2.00 V, 14.00 W, 0.75 Mo, 5.00 Co

Kloster—Cobalt—0.76 C, 0.27 Mn, 0.25 Si, 4.25 Cr, 2.25 V, 14.00 W, 0.50 Mo, 5.00 Co

Latrobe—Electrite UB—0.75 C, 0.30 Mn, 0.25 Si, 4.00 Cr, 2.00 V, 14.00 W, 0.50 Mo, 5.00 Co

High-speed Tool Steels

Tungsten-base Types (*Continued*)

Universal-Cyclops—B 8—0.78 C, 0.25 Mn, 0.30 Si, 4.25 Cr, 2.25 V, 14.00 W, 0.50 Mo, 5.00 Co

T9—Tungsten, 18-4-4

A high-vanadium steel for extremely abrasive conditions. Runs best at high speeds with light cuts. Forging temperature 1950 to 2050 F, annealing 1650 F, hardening 2275 to 2325 F, double temper 2 hr at 1000 to 1050 F.

Crucible—Rex 4-V—1.25 C, 4.00 Cr, 4.00 V, 0.75 Mo, 18.50 W
Hawkridge—Dreadnought 4-V—1.25 C, 4.00 Cr, 4.00 V, 0.75 Mo, 18.50 W

T12—Tungsten, 14-4-3

A tough high-speed steel designed for high resistance to impact. Suitable for variable cutting, such as turning through scale and broaching. Forging temperature 1900 to 2000 F, annealing 1650 F, hardening 2275 to 2325 F, double temper 2 hr at 1000 to 1050 F.

Crucible—Rex 3-V—1.00 C, 4.00 Cr, 3.00 V, 14.00 W, 0.75 Mo

T15—Cobalt-tungsten, 13-4-5-5

High-vanadium steel for maximum abrasive resistance, principal application is for cutting tools. Annealing temperature 1625 F, hardening 2270 F, tempering 1000 to 1200 F.

Allegheny Ludlum—Panther 5—1.55 C, 4.75 Cr, 5.00 V, 12.50 W, 5.00 Co
Houghton & Richards—No. 445—1.50 C, 4.00 Cr, 5.00 V, 14.00 W, 5.00 Co
Latrobe—Electrite Dyna-Van XL—1.50 C, 4.00 Cr, 5.00 V, 14.00 W, 5.00 Co
Vanadium-Alloys—Vasco Supreme—1.50 C, 0.25 Mn, 0.25 Si, 4.75 Cr, 5.00 V, 12.50 W, 5.00 Co

Molybdenum-base Types

M1—Molybdenum, 8-2-1

SAE composition: 0.80 C, 0.30 Mn, 0.30 Si, 4.00 Cr, 1.15 V, 1.50 W, 8.50 Mo. A general-purpose high-speed steel that can be substituted for T1 for many applications. It is cheaper and has a lower hardening temperature, but requires more care in heating to avoid decarburization. Both toughness and wear resistance are slightly better than T1.

Allegheny Ludlum—LMW—0.80 C, 0.25 Mn, 0.35 Si, 3.75 Cr, 1.20 V, 1.65 W, 8.75 Mo
Atlantic—Amotun—0.85 C, 4.00 Cr, 1.75 V, 1.50 W, 8.00 Mo
Atlas—Mohican—0.80 C, 0.25 Mn, 0.30 Si, 4.00 Cr, 1.20 V, 1.50 W, 9.00 Mo
Bethlehem—HM—0.78 C, 3.85 Cr, 1.05 V, 1.60 W, 8.50 Mo
Braeburn—Mocut—0.80 C, 0.25 Mn, 0.25 Si, 4.00 Cr, 1.10 V, 1.55 W, 8.00 Mo
Carpenter—Star Max—0.80 C, 0.30 Mn, 0.25 Si, 3.75 Cr, 1.10 V, 1.50 W, 8.50 Mo
Crucible—Rex TMO—0.80 C, 3.75 Cr, 1.15 V, 1.55 W, 8.70 Mo
Disston—Di-Mol—0.85 C, 3.75 Cr, 1.00 V, 1.50 W, 9.00 Mo
Firth Sterling—Hi Mo—0.81 C, 3.80 Cr, 1.15 V, 1.65 W, 8.80 Mo
Houghton & Richards—Molyhi—0.80 C, 4.00 Cr, 1.15 V, 1.50 W, 8.50 Mo
Jessop—Mogul—0.80 C, 0.25 Mn, 0.30 Si, 3.80 Cr, 1.15 V, 1.50 W, 8.70 Mo
Latrobe—Electrite Tatmo XL—0.80 C, 0.25 Mn, 0.25 Si, 4.00 Cr, 1.00 V, 1.50 W, 8.00 Mo
McDonald—Macco Super Moly—0.80 C, 0.25 Mn, 0.24 Si, 3.90 Cr, 1.10 V, 1.70 W, 8.80 Mo
Peninsular—Penco Hi Moly—0.81 C, 0.25 Mn, 0.35 Si, 3.75 Cr, 1.15 V, 1.65 W, 8.60 Mo
Simonds—STM—0.80 C, 3.75 Cr, 1.25 V, 1.50 W, 8.75 Mo
Universal-Cyclops—Mo Tung—0.80 C, 0.25 Mn, 0.30 Si, 4.00 Cr, 1.15 V, 1.50 W, 8.50 Mo
Vanadium-Alloys—8-N-2—0.79 C, 0.20 Mn, 0.25 Si, 3.75 Cr, 1.10 V, 1.50 W, 8.60 Mo

M2—Molybdenum-tungsten, 6-6-2

SAE composition: 0.83 C, 0.30 Mn, 0.30 Si, 4.10 Cr, 1.90 V, 6.25 W, 5.00

Mo. A general-purpose high-speed steel with higher red hardness than M1. There is less difficulty with decarburization than in the higher molybdenum steels, but controlled atmospheres or salt baths are desirable. Especially suitable for taps, reamers, twist drills, broaches, milling cutters, and lathe and planer tools.

Allegheny Ludlum—DBL-2—0.82 C, 0.25 Mn, 0.30 Si, 4.25 Cr, 1.90 V, 6.40 W, 6.00 Mo

Amalgamated—Malax A—0.83 C, 4.15 Cr, 1.90 V, 6.35 W, 5.00 Mo

Atlas—Sixix—0.82 C, 0.25 Mn, 0.30 Si, 4.00 Cr, 2.00 V, 6.50 W, 5.00 Mo

Bethlehem—66HS—0.83 C, 4.15 Cr, 1.90 V, 6.35 W, 5.00 Mo

Boyd-Wagner—Record 66—0.80 C, 0.30 Mn, 0.20 Si, 4.50 Cr, 1.50 V, 5.50 W, 4.50 Mo

Braeburn—Braemow M-2—0.82 C, 0.25 Mn, 0.25 Si, 4.20 Cr, 1.90 V, 6.50 W, 5.00 Mo

Carpenter—Speed Star—0.82 C, 0.25 Mn, 0.25 Si, 4.25 Cr, 1.90 V, 6.25 W, 5.00 Mo

Columbia—Molite—0.82 C, 0.25 Mn, 0.30 Si, 4.15 Cr, 1.90 V, 6.50 W, 5.00 Mo

Crucible—Rex M-2—0.83 C, 4.15 Cr, 1.90 V, 6.40 W, 5.00 Mo

Darwin & Milner—MT-6—0.85 C, 0.25 Mn, 0.25 Si, 4.00 Cr, 1.50 V, 6.00 W, 6.00 Mo

Delaware—M2-HS—0.60 C, 0.30 Mn, 0.20 Si, 4.00 Cr, 2.00 V, 6.00 W, 6.00 Mo

Disston—6N6—0.80 C, 4.00 Cr, 2.00 V, 6.00 W, 5.00 Mo, 0.50 Cu

Firth Sterling—Star-Mo M2—0.82 C, 4.20 Cr, 1.90 V, 6.50 W, 5.00 Mo

Great Western—6-6-2—0.80 C, 0.25 Mn, 0.30 Si, 4.00 Cr, 1.50 V, 5.75 W, 5.00 Mo

Hawkridge—M-2 Dreadnought—0.83 C, 4.15 Cr, 1.90 V, 6.40 W, 5.00 Mo

Houghton & Richards—No 57—0.80 C, 0.25 Mn, 0.25 Si, 4.00 Cr, 1.90 V, 6.00 W, 5.00 Mo

Hoyland—6-6-2—0.80 C, 0.25 Mn, 0.30 Si, 4.00 Cr, 1.50 V, 5.75 W, 5.00 Mo

Jessop—Mustang—0.84 C, 4.20 Cr, 1.95 V, 6.35 W, 5.00 Mo

Latrobe—Electrite Double-Six M-2 XL—0.83 C, 0.25 Mn, 0.30 Si, 4.00 Cr, 2.00 V, 6.50 W, 5.00 Mo

McDonald—Macco Radio—0.84 C, 0.24 Mn, 0.26 Si, 4.00 Cr, 1.90 V, 6.50 W, 5.00 Mo

McInnis—Moly High Speed—0.80 C, 4.00 Cr, 1.50 V, 5.50 W, 4.25 Mo

Milne—MM 6&6—0.85 C, 4.15 Cr, 1.90 V, 6.40 W, 5.00 Mo

Peninsular—Penco 6-6—0.83 C, 4.15 Cr, 1.90 V, 6.35 W, 5.00 Mo

Republic—Special M-O—0.83 C, 0.30 Mn, 4.00 Cr, 1.90 V, 6.40 W, 5.00 Mo

Simonds—Molva-T—0.80 C, 4.00 Cr, 1.60 V, 5.75 W, 4.50 Mo

Universal-Cyclops—Mo Tung 652—0.82 C, 0.25 Mn, 0.30 Si, 4.00 Cr, 1.90 V, 6.50 W, 5.00 Mo

Vanadium-Alloys—Vasco M-2—0.83 C, 0.25 Mn, 0.30 Si, 4.20 Cr, 1.90 V, 6.35 W, 5.00 Mo

Vulcan Crucible—TM-6—0.83 C, 4.15 Cr, 1.90 V, 6.40 W, 5.00 Mo

M3—Molybdenum-tungsten, 6-6-3

SAE composition: 1.15 C, 0.30 Mn, 0.30 Si, 4.10 Cr, 3.25 V, 5.75 W, 5.25 Mo. Higher carbon and vanadium content combine for better wear resistance than M2. Suitable for difficult cutting and shearing jobs. Requires special care in heating.

Allegheny Ludlum—DBL-3—1.15 C, 4.00 Cr, 6.00 W, 5.50 Mo, 3.10 V

Braeburn—Braevan—1.02 C, 0.20 Mn, 0.20 Si, 4.00 Cr, 2.50 V, 6.25 W, 5.75 Mo

Carpenter—Super Speed Star—1.05 C, 0.25 Mn, 0.30 Si, 4.00 Cr, 2.50 V, 6.25 W, 5.50 Mo

Crucible—Rex M3—1.00 C, 4.00 Cr, 2.75 V, 6.00 W, 5.00 Mo

Darwin & Milner—Darwin M-3—1.10 C, 4.00 Cr, 3.00 V, 5.75 W, 5.00 Mo

Firth Sterling—Van-Chip M3—1.15 C, 4.10 Cr, 3.00 V, 6.00 W, 5.75 Mo

Hawkridge—M-3 Dreadnought—1.00 C 4.00 Cr, 2.75 V, 6.00 W, 5.00 Mo

Houghton & Richards—No. 7—1.15 C, 0.25 Mn, 0.25 Si, 4.00 Cr, 3.00 V, 6.00 W, 6.00 Mo

Jessop—M-3—1.00 C, 4.00 Cr, 2.50 V, 6.00 W, 5.50 Mo

Latrobe—Electrite Crusader XL—1.15 C, 0.25 Mn, 0.25 Si, 4.00 Cr, 3.00 V, 6.00 W, 6.00 Mo

Universal-Cyclops—Unicut—1.00 C, 0.25 Mn, 0.30 Si, 4.00 Cr, 2.40 V, 6.25 W, 6.25 Mo

High-speed Tool Steels

MOLYBDENUM-BASE TYPES (*Continued*)

M4—Molybdenum-tungsten, 4-5-4

SAE composition: 1.30 C, 0.30 Mn, 0.30 Si, 4.25 Cr, 4.25 V, 5.75 W, 5.25 Mo. Higher carbon and vanadium content than M3 for higher wear resistance. Annealing temperature 1600 C, hardening 2225 F, double temper 2 hr at 1000 to 1200 F.

Vanadium-Alloys—Neatro—1.27 C, 0.25 Mn, 0.30 Si, 4.50 Cr, 4.00 V, 5.50 W, 4.50 Mo

M6—Cobalt-moly-tungsten, 5-4-1-12

ASM composition: 0.80 C, 4.00 Cr, 1.50 V, 4.00 W, 5.00 Mo, 12.00 Co. A steel with high red hardness, having properties similar to T6. Suitable for cutting hard materials and heat-treated forgings. Operates at higher speeds and feeds than regular high-speed steels. Forging temperature 1800 to 1900 F, annealing 1550 to 1600 F, hardening 2150 to 2200 F, double temper 1050 to 1100 F.

Braeburn—Congon—0.78 C, 0.25 Mn, 0.25 Si, 4.00 Cr, 1.40 V, 4.00 W, 5.00 Mo, 12.00 Co

M8—Moly-tungsten-columbium, 4-5-1

ASM composition: 0.80 C, 4.00 Cr, 1.50 V, 5.00 W, 5.00 Mo, 1.25 Cb. A columbium-bearing high-speed steel with unusually high wear resistance. For general-purpose cutting. Resists decarburization in hardening. Forging temperature 1950 to 2050 F, annealing 1550 to 1575 F, hardening 2225 to 2325 F, temper 1000 to 1025 F.

Carpenter—Star Columbium—0.80 C, 4.25 Cr, 1.50 V, 5.50 W, 4.50 Mo, 1.25 Columbium

M10—Molybdenum, 8-0-2

ASM composition: 0.85 C, 4.00 Cr, 2.00 V, 8.00 Mo. A tungsten-free high-speed steel offering marked economy for small tools. Requires care in heating to avoid decarburization. Forging temperature 1900 to 2000 F, annealing 1550 to 1600 F, hardening 2150 to 2250 F, tempering 950 to 1050 F.

Allegheny Ludlum—VLM—0.85 C, 4.00 Cr, 2.00 V, 8.00 Mo

Bethlehem—M-10—0.87 C, 4.00 Cr, 2.05 V, 8.35 Mo

Crucible—Rex VM—0.85 C, 4.00 Cr, 1.90 V, 8.00 Mo

Firth Sterling—FSM-10—0.85 C, 4.25 Cr, 2.00 V, 8.25 Mo

Houghton & Richards—Moly Van—0.82 C, 4.00 Cr, 2.20 V, 9.00 Mo

Jessop—M-10—0.88 C, 4.00 Cr, 2.00 V, 8.00 Mo

Latrobe—Electrite TNW XL—0.85 C, 4.00 Cr, 2.00 V, 8.00 Mo

Universal-Cyclops—Mo-Van—0.85 C, 0.25 Mn, 0.30 Si, 4.00 Cr, 2.10 V, 8.00 Mo

Vanadium-Alloys—Van-Lom—0.87 C, 0.20 Mn, 0.30 Si, 4.00 Cr, 1.95 V, 8.25 Mo

M20—Cobalt-moly-boron, 8-5-2

Low hardening temperature and toughness are combined in this economical high-speed steel. Suitable for taps, threading dies, form tools, and broaches. Forging temperature 1750 to 1950 F, annealing 1550 to 1600 F, hardening 2025 to 2100 F, tempering 1000 to 1050 F.

Firth Sterling—Mo-Chip—0.59 C, 0.25 Mn, 0.25 Si, 5.00 Cr, 1.25 V, 8.00 Mo, 2.50 Co, 0.25 boron

M30—Cobalt-moly-tungsten, 8-2-1-5

ASM composition: 0.85 C, 4.00 Cr, 1.25 V, 2.00 W, 8.00 Mo, 5.00 Co. High red hardness and wear resistance with loss of toughness. Recommended for turning chilled iron, locomotive tires, and heat-treated forgings and castings. Subject to decarburization. Forging temperature 1850 to 2000 F, annealing 1550 F, hardening 2225 to 2250 F, tempering 1050 to 1100 F.

Allegheny Ludlum—Super LMW—0.80 C, 4.00 Cr, 1.50 W, 8.00 Mo, 5.00 Co

Braeburn—Como—0.77 C, 0.25 Mn, 0.25 Si, 4.00 Cr, 1.20 V, 1.55 W, 8.50 Mo, 5.00 Co

Crucible—TMO-5—0.88 C, 4.00 Cr, 1.25 V, 2.00 W, 8.00 Mo, 5.00 Co

Firth Sterling—Super Hi-Mo—0.84 C, 0.25 Mn, 0.25 Si, 4.00 Cr, 1.25 V, 1.80 W, 8.50 Mo, 5.00 Co
Latrobe—Electrite Lacomo—0.80 C, 0.30 Mn, 0.30 Si, 4.00 Cr, 1.20 V, 1.50 W, 8.50 Mo, 5.00 Co
Houghton & Richards—Super Molyhi—0.82 C, 0.25 Mn, 0.32 Si, 4.00 Cr, 1.25 V, 1.50 W, 8.50 Mo, 5.00 Co
Universal-Cyclops—Super Mo-Tung—0.82 C, 0.25 Mn, 0.30 Si, 4.00 Cr, 1.25 V, 1.50 W, 9.50 Mo, 5.00 Co

M34—Cobalt-moly-tungsten, 8-2-2-8

ASM composition: 0.85 C, 4.00 Cr, 2.00 V, 2.00 W, 8.00 Mo, 8.00 Co

Crucible—TMO-8—0.88 C, 4.00 Cr, 2.00 V, 2.00 W, 8.00 Mo, 8.00 Co

M36—Cobalt-moly-tungsten, 6-6-2-8

SAE composition: 0.85 C, 0.30 Mn, 0.30 Si, 4.10 Cr, 2.00 V, 6.00 W, 5.00 Mo, 8.00 Co. Analysis of this steel is essentially the same as M2 with the addition of 8% cobalt. This provides higher red hardness and considerably better wear resistance. Toughness is low, though better than the tungsten equivalent, T5. Suitable for cutting gritty materials, heat-treated alloy steels, and stainless steels.

Allegheny Ludlum—Super DBL—0.82 C, 0.40 Mn, 0.40 Si, 4.25 Cr, 1.65 V, 5.50 W, 4.25 Mo, 7.75 Co
Crucible—Rex M2-5—0.83 C, 4.15 Cr, 1.90 V, 6.40 W, 5.00 Mo, 5.00 Co
Firth Sterling—Circle "M"—0.84 C, 0.25 Mn, 0.25 Si, 4.10 Cr, 1.80 V, 5.50 W, 4.50 Mo, 9.00 Co
Houghton & Richards—Cobalt Moly—0.88 C, 0.25 Mn, 0.25 Si, 4.10 Cr, 1.90 V, 6.00 W, 6.00 Mo, 9.00 Co
Latrobe—Electrite CO-6—0.90 C, 0.25 Mn, 0.30 Si, 4.00 Cr, 2.00 V, 6.00 W, 5.00 Mo, 9.00 Co

M38A—Cobalt-moly-tungsten, 6-6-5-5

Similar to M36 but with only 5% cobalt and increased vanadium for better wear resistance.

Latrobe—Electrite UB-4M—1.50 C, 4.50 Cr, 4.75 V, 6.50 W, 5.00 Mo, 5.00 Co

M40—Cobalt-moly-boron, 8-4-1-8

More highly alloyed than M20, this steel has wear resistance said to be several times that of other high-speed steels. Suitable for heat-treated steel, cast iron, brass, plastics, and other abrasive materials. Hardening temperature 2125 to 2175 F, double temper 2 hr at 1025 to 1050.

Firth Sterling—Super Mo-Chip—0.60 C, 0.25 Mn, 0.25 Si, 4.20 Cr, 1.80 V, 1.70 W, 8.20 Co, 0.50 boron

M52—Moly-chrome-vanadium, 4-4-2

Low alloy content of this group of steels suits them for applications not requiring full high-speed properties. Applications are body stock for carbide-tipped drills and reamers, wood cutters, pipe taps, thread chasers, and small drills.

Houghton & Richards—No. 59—0.88 C, 0.30 Mn, 0.25 Si, 4.00 Cr, 2.00 V, 4.30 Mo

Special-purpose Tool Steels

Low-alloy Types

L2—Chrome-vanadium

ASM composition: 0.50–1.10 C, 0.60 Mn, 0.25 Si, 1.00 Cr, 0.20 V. Toughness is combined with ease of heat-treatment and some hot-work properties. Suitable for die-casting, dies (for zinc), chisels, shear blades, and punches. Forging temperature 2000 to 2050 F, annealing 1425 F, hardening 1675 F, tempering 350 or 950 F.

Allegheny Ludlum—Caroga—0.50 C, 0.70 Mn, 0.95 Cr, 0.20 V
Bethlehem—Tough M—0.45 C, 0.55 Mn, 0.20 Si, 0.95 Cr, 0.20 V
Craine-Schrage—Crasco Yellow Label—0.50 C, 0.80 Mn, 1.00 Cr, 0.20 V
Crucible—Halvan—0.50 C, 0.80 Mn, 1.00 Cr, 0.20 V
Disston—Mix 874—0.50 C, 0.70 Mn, 0.85 Cr, 0.18 V
Firth Sterling—Demmler D—0.50 C, 0.85 Mn, 0.25 Si, 1.00 Cr, 0.20 V
Houghton & Richards—No. 85—0.50 C, 0.80 Mn, 1.00 Cr, 0.20 V

Special-purpose Tool Steels

Low-alloy Types (*Continued*)

Jessop—ET No. 6—0.50 C, 0.80 Mn, 0.20 Si, 0.95 Cr, 0.17 V

Latrobe—Crown—0.50 C, 0.80 Mn, 1.00 Cr, 0.20 V

Latrobe—Superb—0.75 C, 1.00 Cr, 0.20 V

Peninsular—Green Label—0.50 C, 0.80 Mn, 1.00 Cr, 0.20 V

Universal-Cyclops—Orion—0.50 C, 0.60 Mn, 0.25 Si, 1.00 Cr, 0.20 V

Vanadium-Alloys—Types D, G, H, K, N—0.80 Cr, 0.20 V; D, 0.50 C; G, 0.60 C; H, 0.70 C; K, 0.80 C; N, 0.90 C

Vulcan Crucible—Auto—0.35 or 0.50 C, 1.00 Cr, 0.25 V

Vulcan Crucible—Hecla—0.50 C, 0.65 Mn, 1.00 Cr, 0.25 V, W

L3—Chromium

ASM composition: 1.00 C, 1.50 Cr, 0.20 V

Firth Sterling—AW Special—1.00 C, 0.20 Mn, 0.25 Si, 1.40 Cr, 0.20 V opt

Vanadium-Alloys—Type BB—1.00 C, 1.40 Cr, 0.20 V

L6—Chrome-nickel

SAE composition: 0.75 C, 0.70 Mn, 0.25 Si, 0.85 Cr, 0.25 V opt, 0.50 Mo opt, 1.50 Ni. Toughness is the principal characteristic of this steel, being nearly as good as in the S group. Nondeforming properties are not so good as the other steels in this group, but are adequate for many applications. Suitable for punches, blanking, and forming dies, shear blades and forming rolls.

Allegheny Ludlum—Tioga—0.68 C, 0.60 Mn, 0.25 Si, 0.65 Cr, 0.20 Mo, 1.40 Ni

Atlantic—Die Steel—0.70 C, 0.40 Mn, 0.25 Si, 1.00 Cr, 1.60 Ni

Bethlehem—Bethalloy—0.75 C, 0.75 Mn, 0.90 Cr, 0.35 Mo, 1.75 Ni

Carpenter—RDS—0.75 C, 0.35 Mn, 0.25 Si, 1.00 Cr, 1.75 Ni

Crucible—Champaloy—0.75 C, 0.70 Mn, 0.75 Cr, 0.30 Mo, 1.50 Ni

Darwin & Milner—Temper Tough—0.75 C, 0.70 Mn, 0.75 Cr, 0.30 Mo, 1.50 Ni

Disston—**Nicroman**—**0.70 C, 0.45 Mn, 0.20 Si, 1.00 Cr, 1.65 Ni**

Houghton & Richards—N 150—C per temper, 0.55 Mn, 0.30 Si, 0.85 Cr, 0.42 Mo, 1.40 Ni

Jessop—Extra Tough No. 4—0.70 C, 0.65 Mn, 0.55 Cr, 1.30 Ni

McInnis—Special Chrome-Nickel—0.70 C, 0.65 Mn, 0.55 Cr, 1.50 Ni

Peninsular—Pen-O-Four—0.70 C, 0.55 Mn, 0.30 Si, 0.65 Cr, 0.30 Mo, 1.35 Ni

Universal-Cyclops—N 9—0.75 C, 0.45 Mn, 0.20 Si, 1.00 Cr, 1.50 Ni

Vanadium-Alloys—Nikro M—0.70 C, 0.55 Mn, 0.30 Si, 0.85 Cr, 0.25 Mo, 1.40 Ni

L7—Chromium

SAE composition: 1.00 C, 0.35 Mn, 0.25 Si, 1.40 Cr, 0.40 Mo. Better red hardness than O1 and O2. Toughness is slightly better than O1, not so good as O2. Safety in hardening slightly less than other oil-hardening grades. Suitable for some hot-press work, ring dies, swaging tools and dies, taps, broaches, and gages.

Allegheny Ludlum—Teton—1.00 C, 1.35 Cr

Allegheny Ludlum—Ludlum XCM—1.20 C, 1.40 Cr, 0.45 Mo

Atlas—KK—1.10 C, 0.40 Mn, 0.20 Si, 1.40 Cr, 0.40 Mo

Boyd-Wagner—Bowco 7720—1.20 C, 0.55 Mn, 0.30 Si, 1.65 Cr, 0.45 Mo

Braeburn—Viking—1.00 C, 0.35 Mn, 0.25 Si, 1.40 Cr, 0.40 Mo

Crucible—Halcomb SS Tool—1.00 C, 0.35 Mn, 1.20 Cr, 0.30 Mo

Darwin & Milner—Chrome Roll—1.00 C, 0.30 Mn, 1.20 Cr, 0.30 Mo

Firth Sterling—AW Special—1.00 C, 0.25 Si, 0.35 Mn, 1.45 Cr, 0.20 V

Hawkridge—SS Extra—1.00 C, 0.35 Mn, 1.20 Cr, 0.30 Mo

Houghton & Richards—No. 135—1.00 C, 0.35 Mn, 0.25 Si, 1.50 Cr

Latrobe—M Chrome—1.00 C, 0.35 Mn, 0.25 Si, 1.50 Cr

Peninsular—Hollobar—1.00 C, 0.40 Mn, 1.10 Cr, 0.30 Mo

Republic—UA-8—1.00 C, 0.35 Mn, 1.35 Cr, 0.35 Mo

Universal-Cyclops—Alloy B—1.05 C, 0.35 Mn, 0.30 Si, 1.35 Cr

Carbon-tungsten Types

F1—Carbon-W (Finishing)

ASM composition: 1.00 C, 1.25 W. These steels become intensely hard after quenching in water or brine. They maintain a sharp cutting edge under abrasive conditions with light loads and low speeds. Primarily they are for light finishing cuts on hard materials when an exceptionally smooth surface is required.

Atlas—Denine—1.25 C, 0.25 Mn, 0.25 Si, 1.40 W
Carpenter—Berkshire—1.20 C, 1.35 W

F2—Carbon-WW (Finishing)

ASM composition: 1.25 C, 3.50 W. This steel has better abrasion resistance than F1 because of the higher tungsten, but is more difficult to grind.

Bethlehem—BFS—1.30 C, 0.28 Mn, 3.50 W
Carpenter—K-W—1.30 C, 0.30 Mn, 0.30 Si, 3.50 W
Crucible—Double Special—1.30 C, 3.50 W
Darwin & Milner—E-E—1.42 C, 0.25 Mn, 0.22 Si, 0.15 Cr, 0.30 V, 4.00 W
Great Western—350-Fast Finishing—1.30 C, 0.30 Mn, 0.45 Si, 3.50 W
Houghton & Richards—Gold Label—1.33 C, 0.35 Mn, 0.49 Si, 4.25 W, 0.35 Mo
Hoyland—350—1.30 C, 0.30 Mn, 0.45 Si, 3.50 W
Republic—Fast Finishing—1.35 C, 3.50 W
Timken—Graph-Tung (graphitic tool steel)—1.50 C, 0.50 Mn, 0.65 Si, 2.80 W, 0.50 Mo
Universal-Cyclops—Saturn—1.25 C, 0.25 Mn, 0.30 Si, 3.50 W
Vanadium-Alloys—Colonial No. 4—1.30 C, 3.50 W
Vulcan Crucible—Regal—1.35 C, 3.50 W

F3—Carbon-chrome-WW (Finishing)

ASM composition: 1.25 C, 0.75 Cr, 3.50 W. Addition to F2 of chromium provides deeper hardening, making this steel more suitable for large tools.

Atlas—XXX—1.35 C, 0.30 Mn, 0.35 Cr, 3.75 W
Columbia—Double Special—1.30 C, 0.30 Mn, 0.25 Si, 0.30 Cr, 3.50 W
Firth Sterling—RT Steel—1.30 C, 0.35 Mn, 0.25 Cr, 3.50 W
Jessop—Rapid Finishing—1.35 C, 0.75 Cr, 3.75 W
Latrobe—ESA—1.40 C, 0.50 Cr, 0.30 V, 4.00 W

Mold Steels

P1—Straight Iron

ASM composition: 0.10 max C. A case-hardening mold steel for cold-hubbed cavities. Annealing temperature 1550 to 1600 F, carburize at 1600 to 1650 F, reheat to 1425 to 1450 F and quench in water or brine. Temper at 300 F.

Atlas—Hobbing Iron—0.06 C, 0.15 Mn, 0.15 Si
Bethlehem—Duramold C—0.10 C, 0.15 Mn, 0.10 Si
Carpenter—Mirromold—0.10 C, 0.20 Mn, 0.10 V
Darwin & Milner—Hobalite—0.05 C, 0.15 Mn, 0.10 Si
Great Western—Rema—0.05 C, 0.20 Mn, 0.10 Si
Hoyland—Rema—0.05 C, 0.20 Mn, 0.10 Si
Latrobe—LCX—0.05 max C, 0.03 max Cr
McDonald—Macco Hobomold—0.04 C, 0.20 Mn, 0.16 Si
Peninsular—Hob-A-Form—0.06 max C, 0.15 Mn, 0.10 max Si
Republic—Plastic Die—0.08 max C, 0.10 max Mn
Uddeholm—UHB Forma—0.05 C, 0.10 Mn, 0.10 Si
Vulcan Crucible—Plastic Die—0.07 C.

P2—Chrome-nickel-moly

ASM composition: 0.07 max C, 1.25 Cr, 0.20 Mo, 0.50 Ni. Higher core strength than P1, but not quite so good hubability. Forging 1900 to 2000 F, annealing 1450 F, carburizing 1650 F, hardening 1525 to 1550 F, tempering 350 to 400 F. Oil quench.

Special-purpose Tool Steels

MOLD STEELS (*Continued*)

Crucible—Formold—0.07 max C, 2.00 Cr, 0.20 Mo, 0.55 Ni
Vulcan Crucible—Vulmold—0.10 max C, 0.30 Mn, 1.40 Cr, 0.25 Mo, 0.50 Ni

P3—Nickel Alloy

ASM composition: 0.10 max C, 0.60 Cr, 1.25 Ni. Properties similar to P2. Carburize and oil quench.

Peninsular—Pen Hob—0.10 max C, 0.50 Mn, 0.60 Cr, 1.25 Ni

P4—5% Chrome Air Hard

ASM composition: 0.07 max C, 5.00 Cr. Better core strength and hubability than P2 or P3. Best safety in hardening and good wear resistance. Suitable not only for plastic molds, but for zinc and aluminum die casting. Forging 2000 to 2050 F, annealing 1600 F, carburizing 1725 to 1750 F, tempering 800 F. Air quench.

Bethlehem—Duramold A—0.07 C, 0.40 Mn, 0.25 Si, 4.50 Cr, 0.45 Mo
Carpenter—Super Samson—0.10 C, 0.30 Mn, 0.20 Si, 5.00 Cr, 0.25 V, 0.90 Mo
Peninsular—Penco ACS—0.07 C, 0.40 Mn, 0.25 Si, 4.50 Cr
Uddeholm—UHB Premo—0.04 C, 0.10 Mn, 0.10 Si, 3.90 Cr, 0.50 Mo

P5—2% Chromium

ASM composition: 0.10 max C, 2.25 Cr. Good hubability and high strength. Annealing temperature 1600 F, process annealing 1325 F, carburizing 1600 F, tempering 200 to 300 F. Water or brine quench.

Bethlehem—Duramold B—0.06 C, 0.30 Mn, 0.15 Si, 1.00 Cr, 0.25 Mo, boron added
Carpenter—Samson Extra—0.10 C, 0.30 Mn, 0.20 Si, 2.30 Cr

P6A—Nickel-vanadium

An age-hardening die-casting steel for zinc die casting and plastic molds.

Latrobe—Cascade—0.14 C, 0.12 Cr, 0.50 V, 4.00 Ni

P7A—Chrome-nickel

High core strength. Can be hubbed slightly but not suitable for deep hubbing. Good machinability. Forging 1900 to 2000 F, carburizing 1625 F, hardening 1450 F, tempering 300 F. Oil quench.

Bethlehem—Duramold N—0.10 C, 0.50 Mn, 0.25 Si, 1.50 Cr, 3.50 Ni
Carpenter—No. 158—0.10 C, 0.50 Mn, 1.50 Cr, 3.50 Ni
Crucible—Crusco 12-B—0.10 C, 0.50 Mn, 1.60 Cr, 3.50 Ni

P8A—1% Chromium

Suitable for cold-hobbed plastic mold dies. Oil hardening.

Great Western—Rema B—0.06 C, 0.30 Mn, 0.15 Si, 1.00 Cr, 0.25 Mo
Hoyland—Rema B—0.06 C, 0.30 Mn, 0.15 Si, 1.00 Cr, 0.25 Mo

P20—1% Chromium

Suitable for short-service die-casting dies or plastic mold dies.

Allegheny Ludlum—FCC-6 (Forgings and castings only)—0.38 C, 0.80 Mn, 0.95 Cr, 0.18 V
Bethlehem—Multimold—0.35 C, 0.70 Mn, 0.45 Si, 0.80 Cr, 0.30 Mo
Crucible—CSM-2—0.30 C, 0.75 Mn, 0.50 Si, 0.80 Cr, 0.25 Mo
Jessop—Dica A—0.42 C, 0.75 Mn, 0.25 Si, 1.00 Cr, 0.20 Mo
McDonald—99—0.35 C, 0.80 Mn, 0.60 Si, 0.85 Cr, 0.35 Mo
Vanadium-Alloys—Speed-Cut—0.40 C, 0.90 Mn, 1.10 Cr, 0.50 Mo

P21A—13% Chromium

Shallow hubbing is possible after a special anneal. For maximum corrosion resistance harden at 1800 to 1900 F. Temper at 300 to 400 F.

Carpenter—Stainless No. 2 Mold steel—0.30 C, 13.00 Cr
Uddeholm—UHB Stainless 31—0.18 C, 13.50 Cr, 0.70 Ni

Basic Suppliers of Tool and Die Steels Listed on Pages 25-53 to 25-78.

Allegheny Ludlum Steel Corp.
Pittsburgh 22, Pa.
Amalgamated Steel Corp.
Broadway and Wire Ave., Newburgh Mill Dist., Cleveland 5, Ohio
Anchor Drawn Steel Co.; see Vanadium-Alloys
Atlantic Steel Corp.
1775 Broadway, New York 19, N.Y.
Atlas Steels Ltd.
Welland, Ontario, Canada
Bethlehem Steel Co.
Bethlehem, Pa.
Boyd-Wagner Co.
1440 W. Lake St., Chicago 7, Ill.
Braeburn Alloy Steel Corp.
Braeburn, Pa.
The Carpenter Steel Co.
Reading, Pa.
Colonial Steel Div.; see Vanadium-Alloys
Columbia Tool Steel Co.
Lincoln Highway and State St., Chicago Heights, Ill.
Craine-Schrage Steel Div.
Detroit Steel Corp., 13770 Joy Rd., Detroit 28, Mich.
Crucible Steel Co. of America
Oliver Bldg., Pittsburgh 30, Pa.
Darwin & Milner, Inc.
2345 St. Clair Ave., Cleveland 14, Ohio
Delaware Tool Steel Corp.
Wilmington, Del.
Diehl Steel Co.
236 Broadway, Cincinnati 2, Ohio
Henry Disston & Sons, Inc.
Philadelphia 35, Pa.
Faitoute Iron Steel Co. Inc.
182–188 Frelinghuysen Ave., Newark 5, N.J.
Firth Sterling Inc.
3113 Forbes St., Pittsburgh 30, Pa.
Great Western Steel Co., Inc.
1011 East 61st St., Los Angeles 1, Calif.
Hawkridge Brothers Co.
303 Congress St., Boston, Mass.
Houghton & Richards Inc.
19 Jersey St., Boston 15, Mass.
Hoyland Steel Co.
405 Lexington Ave., New York 17
Jessop Steel Co.
Washington, Pa.
Kloster Steel Corp.
224–228 N. Justine St., Chicago 7, Ill.
Latrobe Electric Steel Co.
Latrobe, Pa.
P. F. McDonald & Co.
17 King Terminal, Boston 27, Mass.
McInnes Steel Co.
441 E. Main St., Corry, Pa.
A. Milne & Co.
741 Washington St., New York 14, N.Y.
The Peninsular Steel Co.
1030–40 McDougall Ave., Detroit 7
Republic Steel Corp.
1970 Carter Rd., Cleveland, Ohio
Joseph T. Ryerson & Son, Inc.
16th & Rockwell Sts., Chicago, Ill.
Simonds Saw & Steel Co.
Lockport, N.Y.
Tennessee Coal & Iron Div.
U.S. Steel, Fairfield, Ala.
The Timken Roller Bearing Co.
Steel & Tube Div., Canton 6, Ohio
Uddeholm Co. of America Inc.
155 East 44th St., New York 17, N.Y.
Universal-Cyclops Steel Corp.
Universal Div., Bridgeville, Pa.
Vanadium-Alloys Steel Co.
Latrobe, Pa.
Vulcan Crucible Steel Co.
Aliquippa, Pa.

QUALITY DESCRIPTIONS OF STEEL

The word "quality" relates to the performance of a product to a broader extent than is indicated by chemical composition and mechanical properties of carbon or alloy steels. Where wording such as "should be negotiated between the purchaser and producer" is used, the intent is to warn that such additional requirements are not regularly supplied in normal mill operations and that it then becomes necessary to negotiate for such additional requirements.

Additional Requirements. In most product categories, qualities are described which are regularly manufactured in normal mill operations. There are further "additional requirements" which describe quality requirements adapted to conditions encountered in fabrication or use of the steel. These additional requirements normally entail one or more of the following special practices:

1. Selection of raw materials for melting.
2. Steelmaking practices.
3. Selection of heats or portions of heats.
4. Heating and rolling practices.
5. Percentage of discard.
6. Conditioning and processing for surface.
7. Extensive testing.

The special practices used to obtain these requirements are left to the discretion of the producer because of variations in practice.

Many additional requirements are common to more than one product. The specification of more than one requirement for one product may be conflicting or impracticable. Indicated below are descriptions which will be merely referred to in connection with various products:

GRAIN SIZE. When required, austenitic grain size may be specified in killed steels as either coarse (grain size 5 or coarser) or fine (grain size 5 or finer).

COLD HEADING. This term applies to steel required for applications involving severe cold plastic deformation by upsetting, heading, or forging. Cold-headed products are obtained from steel produced by closely controlled steelmaking practices and are subject to mill testing and inspection designed to assure internal soundness, uniformity of chemical composition, and freedom from injurious surface imperfections. Grades of steel with a maximum specified carbon of 0.30% or over are generally specified annealed or spheroidize-annealed to effect proper hardness or microstructure for cold heading.

SPECIAL SURFACE. When hot-rolled steel products, or parts made therefrom, are subject to surface-inspection procedures more restrictive than visual inspection, special surface should be specified.

A special surface necessitates exacting control in manufacturing operations and surface preparation to minimize frequency and severity of seams and other surface imperfections. Different end uses and manufacturing practices in the purchaser's plant, which should be made known to the producer, require various methods and degrees of inspection.

HEAT-TREATING REQUIREMENTS. Certain end uses may require particular mechanical properties after purchaser's heat-treatment. The details of purchaser's heat-treatment procedure and mechanical-property requirements should be made known to the producer, and the mechanical properties should be obtainable throughout the full range of the specified chemical limits.

RESTRICTIVE CHEMICAL LIMITS OR RANGES. This requirement applies when chemical composition limits or ranges more restrictive than those in the fundamental product specifications are required. Such limits or ranges should be negotiated.

SEGREGATION LIMITS AFFECTED BY METHODS OF SAMPLING refers to any method of sampling for check analysis more restrictive than common.

Additional testing requirements which are common to more than one product are:

DECARBURIZATION TEST. When a limiting depth of decarburization is specified, the location and number of tests, method of measurement, acceptable limits, and interpretation should be negotiated.

MACROETCH TEST. This test consists of etching a representative cross-section in suitable hot acid solution to evaluate soundness and homogeneity of the steel. The location, number of tests, area of the specimens to be examined, details of testing technique, and interpretation of test results should be negotiated.

FRACTURE TEST. This test consists of fracturing a sample or representative section to evaluate the soundness and homogeneity of the product.

NONMETALLIC INCLUSION TEST. When this requirement is specified, the nature of the test requires that samples be taken on a longitudinal plane midway between center and surface of the piece. Generally the test specimen is heated and quenched before being polished to avoid polishing pits. The area of the sample to be examined and the rating of the inclusion count at a magnification of 100 diameters should be negotiated.

By their nature, free-machining steels are not subject to inclusion ratings.

EXTENSOMETER TEST. The measurement of elastic properties such as proportional limit, proof stress, or yield strength by the offset method requires the use of special testing equipment and testing procedures, such as the use of an extensometer or the plotting of a stress-strain diagram.

QUALITY DESCRIPTIONS OF PRODUCTS

Bars, Hot-rolled

Carbon Steel. Two qualities are merchant quality and special quality.

MERCHANT QUALITY. Hot-rolled carbon-steel bars of this quality are supplied to standard nonresulfurized chemical grades or ranges to ladle analysis only, or to standard mechanical properties. Merchant quality is not recommended when forging, heat-treating, or other operations requiring special steel-producing practices are involved.

When required, and if so specified by the purchaser, ladle analysis or mechanical tests are furnished.

This quality is usually supplied rimmed, capped, semikilled, or killed at the producer's option and is not restricted by requiring silicon content, grain size, or any other restriction.

Bars of this quality should be free from visible pipe; however, they may contain pronounced chemical segregation. Seams and other surface irregularities are generally to be expected in this quality of steel, which may make the material unsuitable for certain applications.

This quality is supplied only in round, square, or hexagon sizes under 3 in. and in other standard bar sizes under 40.8 lb per ft.

Merchant-quality bars are specified for a wide range of uses involving simple longitudinal cold bending, mild hot forming, punching and welding as used in the production of noncritical parts of bridges, buildings, ships, road-building equipment railway equipment, and general machinery.

SPECIAL QUALITY. Special-quality hot-rolled carbon-steel bars are produced for applications involving forging, heat-treating, cold drawing, machining, etc., in standard or other than standard chemical grades or to mechanical property specifications.

Rimmed, capped, semikilled, and killed steels are produced in this quality, dependent upon chemical composition, manufacturing facilities, purchaser's methods of fabrication, and end use.

Special-quality hot-rolled bars may contain some surface imperfections which may make the steel unsuitable for certain applications. Lower-carbon or lower-manganese grades, especially in killed steels or resulfurized steels, are most susceptible to surface imperfections. In general, the severity of surface imperfections tends to increase with bar size.

ADDITIONAL REQUIREMENTS:

Cold Heading.
Special Surface.
Heat-treating Requirements.

ADDITIONAL TESTING REQUIREMENTS:

Decarburization Test.
Macroetch Test.
Fracture Test.
Nonmetallic-inclusion Test.
Extensometer Test.

Alloy Steel. Standard alloy steels may be specified to chemistry, grain size (coarse or fine), and limited end-quench hardenability requirements.

Standard H-steels are specified to wider chemistry ranges, grain size (coarse or fine), and hardenability bands.

Billets and bars may contain some surface imperfections which may make the steel unsuitable for certain applications. The frequency and severity of the surface imperfections are influenced by the chemical composition and size. Steel is inspected for center soundness.

AIRCRAFT QUALITY OR STEEL SUBJECT TO MAGNETIC-PARTICLE INSPECTION. This quality designation applies to steels requiring closely controlled restrictive and special practices in their manufacture, in which several of the following requisites are involved:

Careful selection of raw materials for melting.
Exacting steelmaking practices.
Additional discard.
Selection of heats or portions of heats.
Exacting heating and rolling practices.
Magnetic-particle inspection.
Macroetch tests.
Grain-size tests.
Mechanical and hardenability tests.

AXLE-SHAFT QUALITY. Applies to bars intended for the manufacture of power-driven axle shafts of the automotive or truck type, which by their design or method of manufacture are not machined all over or have insufficient stock removed for proper clean-up of normal surface defects, or where inspection standards impose restrictive surface requirements. To meet these requirements special rolling practices, special billet and bar conditioning, and selective inspection are involved.

BEARING QUALITY. Bearing-quality steels are usually limited to the standard alloy carburizing grades, the high-carbon-chromium series, or to ASTM specification A295 or its equivalent. Steels of this quality are subjected to restricted melting and special teeming, heating, rolling, cooling, and conditioning practices to meet the quality requirements.

These requirements necessitate thorough examinations for imperfections by one or more of the following methods:

1. Microscopic examination.
2. Macroetch testing.
3. Fracture testing.

It is not practicable to furnish bearing-quality steel in sizes exceeding 100 sq in. in cross-sectional dimension to the same rigid quality requirements as for small sizes.

ADDITIONAL REQUIREMENTS:

Cold Heading. The regular description applies, except alloy-steel bars are usually specified annealed or spheroidize-annealed to effect proper hardness or microstructure for cold heading.

Special Hardenability Requirements. Hardenability is sometimes desired for chemical compositions other than H-steels. In such a case a minimum or maximum hardenability value can be specified, but the hardenability requirement should be compatible with the chemical composition ordered.

Restricted Hardenability Requirements. Restricted hardenability limits are sometimes specified for H-steels. A restricted hardness range of not less than three-fourths of that shown by the H-band can be produced, if necessary, at any one specified distance 4/16 in. or greater on the standard end-quench specimen.

Restrictive Chemical Limits or Ranges.

Segregation Limits Affected by Methods of Sampling.

ADDITIONAL TESTING REQUIREMENTS:

Magnetic-particle Test. This test consists of suitably magnetizing the steel and applying a prepared magnetic powder which adheres to it along lines of flux leakage. On properly magnetized steel, flux leakage develops along surface or subsurface nonuniformities.

When the test is applied to bar stock, short-length samples should be completely machined or ground as nearly as feasible to the cross-section and surface condition corresponding to the machined finished part they are to represent.

Special Hardenability Test. This test consists of using any hardenability or equivalent mechanical-property test other than the standard 1-in. end-quench test, and the details of testing are negotiated.

Extensometer Test.

Fracture Test.

Macroetch Test.

Nonmetallic-inclusion Test.

High-strength Low-alloy Steel. The characteristics of high-strength low-alloy steel described on page 25-11 represent basic requirements. Sometimes bars are specified for special applications in which certain additional requirements or additional testing requirements may be involved.

Stainless and Heat-resisting Steels. AIRCRAFT QUALITY OR STEEL SUBJECT TO MAGNETIC-PARTICLE INSPECTION. If magnetic-particle testing is specified, it may be applied to all type numbers of the 400 or 500 series except the free-machining ones. Steel of the 300 series does not respond to the magnetic-particle inspection.

ADDITIONAL TESTING REQUIREMENTS:

Special Tests. Mechanical, physical, or corrosion tests, other than tensile, bend, or hardness tests, are sometimes specified, such as the fatigue test, impact test, determination of electrical resistance, nitric acid test, and similar tests.

Macroetch Tests.

Nonmetallic-inclusion Tests.

Bars, Cold-finished

When close tolerances are desired or improved surface finish or mechanical properties developed by cold working are required, bars may be cold-drawn, cold-rolled, centerless-ground or polished, or smooth-turned, singly or in combinations that are appropriate. Such bars are all termed "cold-finished."

Cold-finishing operations which do not affect the mechanical properties are turning or grinding, for example, turning or grinding of hot-rolled steel bars. The term "cold finishing" includes cold working as well as turning and grinding.

The term "cold-worked" refers to steel that is cold-drawn or cold-rolled, and the mechanical properties are significantly affected by the cold working. Marked changes in the mechanical properties of hot-rolled steel are developed by cold drawing or cold rolling. These changes include an increase in tensile strength, yield strength, torsional strength, hardness, and wear resistance, and are accompanied by some decrease in elongation and reduction of area. Cold-worked steel does not give the stress-strain characteristic that is generally considered to be the yield point.

Cold-worked carbon steel is sometimes specified to meet mechanical-property requirements in addition to chemical composition. The effects of cold work depend upon such factors as chemical composition, cross-section, method of steel manufacture, and type of thermal treatment. By a suitable combination of those factors, cold-worked steel is produced having mechanical properties which in many respects are comparable with those of heat-treated bars.

Carbon Steel. *Standard-quality* cold-finished carbon-steel bars are produced from special-quality hot-rolled carbon-steel bars. These may contain surface imperfections which may make the steel unsuitable for certain applications.

The characteristic surface finish of *cold-drawn* bars precludes close visual inspection for surface imperfections. For applications requiring freedom from surface imperfections on the finished-part surface, the purchaser should consult the producer regarding bar-size requirements, so that sufficient machining allowance is provided.

ADDITIONAL REQUIREMENTS:

Special-surface bars are produced from steel which has been processed under exacting control to minimize the frequency and severity of seams and other surface imperfections. For bars or parts subject to surface inspection, or when the machining allowance will clean up imperfections in standard-quality bars, specify the special-surface quality. Special-surface bars are usually inspected by magnetic testing.

Grain Size.

Cold Heading.

Heat-treating Requirements.

ADDITIONAL TESTING REQUIREMENTS:

Magnetic testing should not be confused with magnetic-particle inspection. In magnetic testing, each bar is passed through a magnetic field and imperfections at the surface, such as seams, cracks, laps, or other discontinuities, are detected and the indications are obtained by suitable magnetic measurements.

Decarburization Test.
Macroetch Test.
Fracture Test.
Nonmetallic-inclusion Test.
Extensometer Test.

Alloy Steel. Materials of aircraft, axle-shaft, or bearing quality are subject to the same tests as hot-rolled carbon-steel bars of the same quality, except that the several references to large sizes having cross-sectional dimensions over 100 sq in. should be disregarded.

ADDITIONAL REQUIREMENTS:

Grain Size.
Cold Heading.
Also
Special Hardenability Requirements.
Restricted Hardenability Requirements.
Restrictive Chemical Limits or Ranges.
Segregation Limits Affected by Methods of Sampling.

ADDITIONAL TESTING REQUIREMENTS:

Magnetic-particle Test.
Special Hardenability Test.
Extensometer Test.
Fracture Test.
Macroetch Test.
Nonmetallic-inclusion Test.

High-strength low-alloy steel in cold-finished grades can be produced only subject to negotiation between the purchaser and producer.

Stainless and Heat-resisting Steels. Aircraft quality or steel subject to magnetic-particle inspection (see notes under Hot-rolled Bars).

ADDITIONAL TESTING REQUIREMENTS:

Special Tests.
Macroetch Test.
Nonmetallic-inclusion Test.

Plates, Hot-rolled

Carbon Steel. REGULAR QUALITY is specified to chemical composition ranges and limits, not mechanical property, cold bend, or ductility requirements. When stock steel plates are specified, or when no chemical composition limits are specified, plates having a maximum of 0.33% carbon, based on ladle analysis, are produced.

Plates of this quality do not have the high degree of uniformity in chemical composition, internal soundness, or freedom from surface imperfections that are associated with other plate-steel qualities.

STRUCTURAL-QUALITY plates are intended for application in bridges, buildings, structural steel for locomotives, railroad cars, and other mobile equipment. A typical specification is: Structural Steel for Locomotives and Cars (Structural Grades), ASTM A113.

HOT-PRESSING-QUALITY plates are intended for ordinary hot pressing, flanging, or bending work. Plates of this quality are specified to mechanical property requirements and standard ladle phosphorus and sulfur requirements.

Cold-pressing-quality plates are made of soft steel, equivalent to the cold-pressing grade of ASTM A113, which can be bent or formed either longitudinally or transversely at ordinary temperature by good shop practice. Cold-bending-quality plates as defined in ASTM A131 are of higher tensile strength and are used where greater design stresses with less severe forming are contemplated.

Forging-quality plates are intended for forging, heat-treating, or similar purposes in which uniformity of composition and freedom from injurious imperfections are essential. Plates of this quality are produced by a killed-steel practice and rolled from slabs or ingots conditioned to eliminate injurious surface imperfections. It is customary to specify chemical ranges and limits.

Flange-quality plates are intended for pressure vessels and similar purposes except when exposed to fire or radiant heat.

It is industry practice to take one longitudinal tension and one transverse bend test representing each plate as rolled.

Firebox-quality plates are intended for application in pressure vessels when exposed to fire or radiant heat where they are subject to thermal and mechanical stresses; for unfired pressure vessels in lieu of flange quality, and for similar purposes.

Marine-quality plates are intended for application in pressure vessels and combustion chambers of marine boilers. This quality necessitates a very close control during manufacture because of the exceptionally rigid requirements.

Additional Requirements:

Killed steel, when specified or required, should be negotiated between the purchaser and producer.

Incidental Elements. Maximum limitations for incidental elements, or request for reporting incidental elements, require special control of raw materials, and additional analytical determinations by the producer. Such requirements should be negotiated between purchaser and producer.

Specified Discard. A sufficient amount of discard is made from ingots for all qualities of plates to secure freedom from injurious piping and undue segregation. The amount of discard from the top of the ingot varies for the different plate qualities. Degrees of limitations on chemical requirements, and additional segregation tests (chemical and mechanical), also affect the amount of discard.

Surface Imperfections. Purchasers sometimes pickle or blast clean plates prior to surface inspection. Plates for such applications necessitate special surface conditioning and closer inspection than ordinarily employed, and those requirements should be negotiated between purchaser and producer.

Grain Size.

Additional Testing Requirements:

Homogeneity (or nick and break) test for firebox-quality plates is normally made on one of the broken tension-test specimens. The specimen is nicked and broken transversely to disclose harmful laminations by visual examination.

Segregation Test. Specifications for plates of particular qualities impose checks on segregation in several ways, for example, by restricting the specified limits for chemical composition, or by designating the number or location of the samples to be taken for chemical or mechanical tests. The checks on the control of segregation contained in ASTM A285 are the customary standards for special-quality plates.

Magnetic-particle inspection, for effectiveness, depends largely upon the condition of plate edges. Therefore, all edges of plates to be so tested must be machined.

Macroetch Test. This test is not used in evaluating the quality of carbon-steel plates.

Alloy Steel. *Aircraft Quality* or *Steel Subject to Magnetic Particle Inspection.* Such steel is of a quality requiring closely controlled and selected practices in manufacture. To meet requirements, unusual manufacturing risks are encountered.

Marine Quality.

Flange Quality.

Firebox Quality.

ADDITIONAL REQUIREMENTS:

Special Discard.

Surface Imperfections Disclosed by Pickling or Blast Cleaning.

ADDITIONAL TESTING REQUIREMENTS:

Segregation Test.

Impact Tests, longitudinal and transverse directions. Impact tests are sometimes specified for special-purpose applications.

High-strength Low-alloy Steel. In addition to basic requirements, high-strength low-alloy steel plates are sometimes specified for special applications in which certain additional requirements or additional testing requirements may be involved.

Stainless and Heat-resisting Steels. There are no quality designations for this product.

Semifinished Steels

Carbon Steel. Hot-rolled carbon-steel semifinished products are ordinarily produced in one quality, namely, *forging quality*. Semifinished material cannot be supplied to mechanical-property requirements. Applications require manufacturing control for chemical composition, deoxidation, mold practice, pouring, rolling discard, cooling, surface preparation, testing, and inspection. Products are produced in sizes up to and including 200 sq in. in cross-section. They are produced to ladle chemical ranges or limits, subject to check-analysis limits. For larger sizes the chemical ranges or limits for ladle analysis and provision for check analysis are developed by the purchaser and producer.

The type of steel, such as killed or semikilled steel, produced in this quality is dependent upon chemical composition, producer's manufacturing facilities, purchaser's methods of fabrication, and end uses. Sufficient discard is made to obtain freedom from injurious pipe and harmful segregation.

ADDITIONAL REQUIREMENTS:

Special Surface.

Heat-treating Requirements.

ADDITIONAL TESTING REQUIREMENTS:

Macroetch Test.

Fracture Test.

Nonmetallic-inclusion Test.

Alloy Steel

Aircraft Quality or Steel Subject to Magnetic-particle Inspection.

Axle-shaft Quality.

Bearing Quality.

ADDITIONAL REQUIREMENTS FOR ALLOY STEELS:

Grain Size.
Restricted Hardenability Requirements.
Restrictive Chemical Limits or Ranges.
Segregation Limits Affected by Methods of Sampling.

ADDITIONAL TESTING REQUIREMENTS:

Magnetic-particle Test.
Special Hardenability Test.
Fracture Test.
Macroetch Test.
Nonmetallic-inclusion Test.

High-strength Low-alloy Steel. This material is produced subject to negotiation between the purchaser and producer.

Stainless and Heat-resisting Steels. These steels are produced subject to negotiation between the purchaser and producer.

Sheets, Hot-rolled

Carbon Steel. Hot-rolled and hot-rolled annealed sheets are produced in three principal qualities:

COMMERCIAL QUALITY. These low-carbon-steel sheets are suitable for purposes where presence of oxide and normal surface imperfections is not objectionable. When a carbon content is not specified, it is assumed that carbon not exceeding 0.15% by ladle analysis is desired.

Sheets should not contain abnormal imperfections visible on the surface as produced. The oxide or scale may have varying degrees of adherence, color, or type. When oxide is removed or the sheet deformed, surface imperfections may be disclosed which were not visible, and these conditions are normal.

This quality is subject to stretcher strains, fluting, and coil breaks but can be furnished with nonfluting properties when specified. Minimum-bend-test requirements should be met.

DRAWING QUALITY. Such sheets are produced for use in fabricating parts where surface finish is not of primary importance. To prevent excessive die scoring, the oxide should be removed by pickling.

Sheets of this grade should produce identified parts too difficult for the fabricating properties of commercial-quality material. The material is not ordinarily specified to chemical composition, but instead for ability to draw a specified article.

The production of drawing quality, with uniform drawing properties, usually requires the special selection of raw materials, the use of specially produced or selected steel, and exacting control of the processing operations.

Sheets should not contain visible abnormal imperfections on the surface as produced or after fabrication. Oxide or scale may have varying degrees of adherence, color, or type. When the oxide or scale is removed, normal surface imperfections may be disclosed.

Heat-treatment by the purchaser before or between draws may detrimentally affect drawing characteristics of the steel.

Requirements to meet both mechanical test values and drawing performances are sometimes necessary, but must be compatible.

PHYSICAL QUALITY. These sheets are produced when uniformity of temper is required or when mechanical properties are specified or required other than the bend tests of commercial quality. Such properties or values include those indicated by tension tests, hardness tests, or other generally accepted mechanical tests. It is customary to specify only one kind of test requirement on any one item.

Physical-quality sheets are sometimes specified to structural specifications or to tensile ranges. The composition of steel is related to the required tensile properties; hence, a range for carbon is not specified.

The surface characteristics of this quality are the same as those of commercial-quality steel.

ADDITIONAL REQUIREMENTS:

Hot-rolled special surface has a better surface than do hot-rolled and hot-rolled annealed sheets and should be specified when sheets having one smooth clean surface and adherent oxide are required, or when the purchaser requires a surface equivalent to that of commercial quality after pickling or blast cleaning.

When required, special surface is specified together with one of the three principal qualities.

Special killed steel is usually associated with cold-rolled sheets that are to be essentially free from significant changes in mechanical properties over a period of time or from stretcher strains without roller leveling, but can be specified for hot-rolled sheets when such sheets are to be cold-reduced and are to have the characteristics described above.

Special-soundness steel has a high degree of internal soundness, homogeneity, and uniformity of chemical composition and grain size. These sheets should produce parts, within properly established limits and allowances, that can be fabricated or treated by the following typical methods: pickling or metallic coating with freedom from blisters; high-speed or critical welding methods; heat-treatment (quenching and tempering); and controlled carburizing.

Low-metalloid steel is an open-hearth iron in which the total percentage of the carbon, manganese, phosphorus, sulfur, and silicon does not exceed an extraordinarily low maximum limit, say 0.10 to 0.20%.

Grain Size.

ADDITIONAL TESTING REQUIREMENTS:

Specification Requirements. Carbon-steel sheets are sometimes specified to the following specifications:

Heavy-gage Structural-quality Flat-rolled: ASTM A245 for sheets 0.299 to 0.0478 in. in thickness.

Light-gage Structural-quality Flat-rolled: ASTM A246 for sheets 0.0477 to 0.0225 in. in thickness.

Extensometer Test.

Alloy Steel. Such sheets are produced only in certain compositions and qualities subject to negotiation between purchaser and producer.

High-strength Low-alloy Steel. Hot-rolled sheets carry hot mill oxide. When the oxide is removed or the sheet deformed, surface defects or surface disturbances may be disclosed which were not visible. These conditions are normal.

When specified or required, hot-rolled sheets can be processed to minimize surface disturbances known as fluting, stretcher strains, or coil breaks.

When required, sheets may be produced in the normalized or annealed condition, or as drawing quality. In those cases, mechanical-property requirements should be negotiated between the purchaser and producer with the yield point and ultimate strength modified to levels compatible with the fabricating requirements involved.

Stainless and Heat-resisting Steels. ADDITIONAL REQUIREMENTS:

Mechanical, physical, or corrosion tests, other than tensile, bend, or hardness tests, such as the fatigue test, impact test, determination of electrical resistance, nitric acid corrosion test, and similar tests, may be specified.

Minimum tensile strength or yield strength, or hardness higher than that normally obtained on sheets in the annealed condition, or any combination of those properties, may be specified.

Strip, Hot-rolled

Carbon Steel. Like hot-rolled sheets, hot-rolled strip is produced in three principal qualities:

COMMERCIAL-QUALITY STRIP is similar to hot-rolled sheets of similar quality.

DRAWING-QUALITY STRIP is designed for the fabrication of identified parts that involves forming operations too severe for commercial-quality strip to withstand within acceptable limits of breakage. Where surface finish of the fabricated part is of importance, it may be advisable to consider the use of cold-rolled steel.

This quality of strip is normally produced from steel with a carbon content of 0.10% maximum by ladle analysis, and does not usually exceed 0.25% carbon by ladle analysis.

When chemical composition is specified it should be compatible with the fabricating requirements involved.

Drawing-quality strip, as hot-rolled, is subject to stretcher strains, fluting, and coil breaks, and in certain types of drawing, the parts may develop scalloping (earing). Strip of this quality can also be produced with nonscalloping properties when specified.

PHYSICAL-QUALITY STRIP. The notes for comparable sheet generally apply. Physical-quality strip is sometimes specified to structural specifications, such as ASTM A303.

ADDITIONAL REQUIREMENTS:

These are similar to hot-rolled sheets.

ADDITIONAL TESTING REQUIREMENTS:

Extensometer Test.

Alloy Steel. Produced in certain compositions by some manufacturers subject to negotiation between the purchaser and producer.

High-strength Low-alloy Steel. Same as for sheet.

Stainless and Heat-resisting Steels. These are produced mainly for cold-rolled conversion.

Sheets, Cold-rolled

Carbon Steel. Cold-rolled sheets have a surface finish superior to hot-rolled pickled finish. The hot-rolled coils are pickled, cold-reduced in thickness and usually annealed, temper- or skin-rolled and roller-leveled. For more lustrous finish than normally furnished, specify cold-rolled luster finish.

GRADING. In cut lengths, sheets are graded as either cold-rolled sheets or cold-rolled primes.

Cold-rolled sheets may contain surface imperfections of such a character that the sheets can be used for identified parts with a reasonable amount of metal finishing by the purchaser.

Cold-rolled primes are inspected to meet specific surface requirements on one side without metal finishing by the purchaser, except surface imperfections caused by the purchaser's handling and fabrication. Cold-rolled primes are not supplied in coils because it is not practical to cut the coils to remove surface imperfections.

Cold-rolled sheets are produced in three principal qualities: commercial quality, drawing quality, and physical quality.

COMMERCIAL QUALITY is low-carbon steel suitable for exposed parts requiring a good surface finish. The dull surface texture is intended for application of paints, enamels, or lacquers. Sheets of this quality are not suitable for electroplating where surface uniformity is essential.

If not specified, carbon content does not exceed 0.15%, and the hardness does not ordinarily exceed Rockwell 60 B. However, sheets with up to 0.25% carbon maximum, and sometimes higher, are produced.

When specified, commercial-quality sheets are processed to be free from fluting or stretcher straining during fabrication, provided that the sheets are properly roller-leveled immediately before fabrication.

Commercial-quality sheets do not have a high degree of uniformity in chemical composition and mechanical properties but should not contain any abnormal imperfections which are clearly visible on the surface. It is normal experience that some imperfections will appear after deformation.

Cold-rolled commercial-quality sheets should meet minimum-bend-test requirements. If greater ductility than the minimum indicated by that bend test is required, or if uniformity of properties is necessary, commercial quality is not recommended.

DRAWING QUALITY. This grade should produce identified parts too difficult for the drawing properties of sheets of any other quality, within the breakage allowance as commonly negotiated between the purchaser and producer.

This quality of sheet is not ordinarily specified to chemical composition. Requirements to meet both mechanical-test values and drawing performance are sometimes necessary but should be compatible.

Abnormal imperfections should not be clearly visible on the surface of the sheet as produced or after drawing. When required, sheets of drawing quality should be free from surface disturbances in fabrication when properly roller-leveled immediately beforehand. Special killed steel should be specified if roller leveling will not be used.

Processing or heat-treatment by the purchaser before or during fabrication or delays between draws may detrimentally affect the drawing characteristics of the steel.

PHYSICAL QUALITY. This quality is produced when mechanical test values are specified or required, other than those described under other qualities of cold-rolled sheets. Such test values include those indicated by tension tests, hardness tests, or other accepted mechanical tests. It is customary to specify only one kind of test requirement. Surface characteristics are the same as for commercial quality.

Requirements to meet both mechanical tests and drawing performance are sometimes necessary but should be compatible.

ADDITIONAL REQUIREMENTS:

Cold-rolled luster finish is particularly applicable for parts requiring a lustrous finish combined with simple forming. The luster is not ordinarily retained after fabrication, and the formed parts should be polished and buffed to make them suitable for bright plating.

Special-soundness Steel.

Special Killed Steel.

ADDITIONAL TESTING REQUIREMENTS:

Extensometer Test.

Alloy Steel. Cold-rolled sheets are produced by some manufacturers only in certain compositions and qualities subject to negotiation.

High-strength Low-alloy Steel. Modifications of the minimum yield point of 50,000 psi and minimum tensile strength of 70,000 psi may be necessary for high-strength low-alloy steel cold-rolled sheets, which normally are thermally treated after cold reduction.

For applications involving severe deformation, cold-rolled sheets may be specified to drawing quality. In those cases the yield point and ultimate strength may have to be further modified to levels compatible with fabricating requirements.

When specified or required, cold-rolled sheets are processed to minimize fluting or stretcher straining, provided that the sheets are properly roller-leveled immediately before fabrication.

Stainless and Heat-resisting Steels. ADDITIONAL REQUIREMENTS: Mechanical and physical properties and tests may be required.

Strip, Cold-rolled

Carbon Steel. Cold-rolled carbon-steel strip is commonly produced from rimmed steel. Capped or semikilled steels are also made, their use being generally left to the discretion of the producer.

ADDITIONAL REQUIREMENTS:

Killed steel is specified: (1) when the strip is required to have a more uniform composition than that obtainable from rimmed, capped, or semikilled steels; (2) when fine or coarse grain size is specified. Killed steel is not recommended for drawing, forming, and difficult bending operations.

Aluminum killed steel is sometimes specified when drawing and forming properties are required. This type of steel combines uniformity of chemical composition with mechanical properties suitable for these operations. The application of the appropriate chemical composition should be left to the discretion of the producer, in order to produce strip of suitable characteristics to make the identified part.

Nonscalloping strip is intended to have no marked degree of unevenness or "ears" on the edges of drawn articles such as cups or shells. This type of strip requires additional production and metallurgical controls and increased inspection.

ADDITIONAL TESTING REQUIREMENTS:

Extensometer Test.

Alloy Steel. Cold-rolled strip is made in certain compositions by some producers subject to negotiations between the purchaser and producer.

High-strength Low-alloy Steel. See notes for sheet. It is common practice to inspect the surface on one side of sheet or strip. Strip or sheet in cut lengths should not contain visible defects on the inspected surface, or laminations resulting from pipe. Production of coils does not afford the same opportunity for the producer to reject portions containing defects. It is probable that some imperfections will appear after deformation.

Stainless and Heat-resisting Steels. There are no quality designations for this product in strip form.

WIRE

Carbon Steel

Coarse Round Wire

Merchant wire, sometimes called "fence wire," is a soft all-purpose wire produced as soft galvanized and black annealed wire and is specified in 100-, 50-, 25-, and 10-lb even-weight coils and may contain one or more pieces. The size is designated by steel wire gage numbers (Sec. 28).

Manufacturers' wire is used in the manufacture of a wide variety of products and is ordinarily specified according to the product applications, such as "mechanical spring wire," "cold heading wire," and "tying wire." The size is designated either by gage number or by diameter expressed in decimals of an inch.

Grades and qualities of manufacturers' wire are divided into four classes based on chemical composition or mechanical properties, as well as special purposes for which wire is used, as follows:

1. Low-carbon-steel wires such as standard steels 1006, 1008, 1010, 1012, B1006,* and B1010.*
2. Medium-low-carbon-steel wires such as standard steels 1013 to 1022, inclusive.
3. Medium-high-carbon-steel wires such as standard steels 1025 to 1041, inclusive.
4. Special-purpose wires including both high- and low-carbon steels.

A manufacturers' wire can be produced either to chemical composition or to physical and mechanical test specifications.

Bright Basic Wire. This name designates a coarse round, dry-drawn, low-carbon, basic-open-hearth steel wire, produced without any intermediate heat-treatment or special processing for the specific purpose of altering the normal properties imparted by the cold drawing of low-carbon steel. This type of wire is confined to applications that do not require a finish other than that obtained by regular dry-drawing practice. Those applications also do not require specific tempers, tensile-strength requirements, or softness limitations.

Bright Bessemer Wire. Bright bessemer wire is, in general, harder and stiffer than bright basic wire because of the inherent differences in the character of steel produced by the two steelmaking processes. Bright bessemer wire is used for applications which do not require special finishes or specific physical or mechanical properties.

Medium-low-carbon Bright Wire. This wire is stiffer and harder than bright basic wire because of its higher carbon content, but its use is confined to applications where special finishes or specific physical or mechanical properties are not required or necessary for proper performance.

Specification Wire. Some applications for low-carbon- and medium-low-carbon-steel manufacturers' wire require specific mechanical properties or finishes not normally possessed by bright wire as described.

The production of drawn wire to definite tensile-strength values or to other specified ranges of mechanical properties in general involves some type of process heat-treatment or other special wire-mill procedure.

Specification wire is sometimes furnished with special finishes for spot welding, electroplating, tinning, etc.

High-carbon-steel specification wire is generally produced as specific commodities.

* Bessemer grades.

Special Finishes

Clean bright wire is dry-drawn wire which requires a finish sufficiently clean and bright to assure satisfactory performance for such end uses as spot welding and japanning.

Extra-clean smooth bright wire is wire dry-drawn from rods selected with respect to surface characteristics, and drawn by varying the lubrication and drafting practice in order to obtain an especially clean, smooth, bright surface for electroplating, tinning, or similar processes. Wire with this surface quality is also produced with coppered and liquor finish.

Coppered wire and liquor-finished wire are produced by wet-drawing wire which has been immersed in a copper sulfate or copper-tin sulfate solution. The solution used is dependent upon whether copper finish, brass finish, or white finish is desired. The temper of the finished wire is controlled by intermediate heat-treatment and wire-mill processing prior to wet drawing.

Heat-treatments. Coarse round manufacturers' wire can be given a variety of heat-treatments: annealed in process; annealed at finish size; spheroidize-annealed in process; spheroidize-annealed at finish size; patented in process; patented at finish size; oil tempered at finish size; and lead annealed in process of galvanizing. Rods from which coarse round manufacturers' wire is drawn may be normalized, annealed, spheroidize-annealed, or patented.

Commodity Descriptions. There are many wires for specific applications that have been developed for definite components of machines and equipment or for some particular end use.

Metal stitching wire is used in the automobile, sheet metal, and allied industries for joining or assembling low-stressed light metal parts or fabrics. For uniform and satisfactory performance, selected high-carbon steel of suitable chemical analysis, patentings, drafting, and finishing techniques is employed.

Sizes 18 (0.0475 in.) and 20 (0.0348 in.) gage are practically the only ones used. Normally, tinned finish is required. However, drawn galvanized, galvanized at finish size, galvanized and black enameled, and dry bright finish may be produced.

Mechanical Spring Wires. WMB, WHB, and Extra WHB designate acid open-hearth steels.

Chemical composition and mechanical properties for wire for the manufacture of mechanical springs are the same for both basic and acid steel.

These types are not intended to meet surface-inspection requirements more rigid than visual examination. Only aircraft steel-spring wire and valve-spring wire are subject to critical surface inspection involving magnetic-particle or etch tests.

Oil-tempered spring-steel wire of MB, WMB, HB, WHB, Extra HB, and Extra WHB types has mechanical properties developed by the heat-treatment of the wire at its finished size. Oil-tempered wire is more suitable to precision forming and coiling operations than hard-drawn wire.

Tensile strengths are shown in Table 23.

Hard-drawn spring-steel wire of MB, WMB, HB, and WHB types is intended for use in springs subjected to static loads or infrequent stress repetitions. The fibrous structure is intended to withstand more severe deformations than oil-tempered wire and is more adaptable to certain wire-spring formations.

Spring-steel wires for heat-treated components are used for springs or other components that are to be subsequently heat-treated. Mechanical properties are developed by heat-treatment. The severity of deformation encountered in the fabrication of the spring guides the selection of the particular wire quality.

Untempered spring-steel wire of MB, WMB, HB, and WHB types is intended for use in springs the fabrication of which involves moderate deformation.

Spheroidize-annealed and lightly drawn spring-steel wire of MB, WMB, HB, WHB, Extra HB, and Extra WHB types is intended for use in springs the fabrication of which involves severe deformation.

Spheroidize-annealed spring-steel wire of MB, WMB, HB, WHB, Extra HB, and Extra WHB types is intended for use in springs the fabrication of which involves the most severe deformation. The wires are spheroidize-annealed at finished size.

Aircraft steel-spring wire designates wire that involves unusual restrictions in manufacture to make the wire suitable for aircraft applications.

TABLE 23. TENSILE STRENGTH OF CARBON-STEEL WIRE FOR MECHANICAL SPRINGS, 1000 PSI

Dia, In.	MB and WMB Hard-drawn	HB and WHB Hard-drawn	MB and WMB Oil-tempered	HB and WHB Oil-tempered	Extra HB and Extra WHB Oil-tempered
0.020	283–320	310–350	293–323	308–338	318–348
0.028	271–311	293–333	286–316	301–331	311–341
0.032	265–305	287–327	280–310	295–325	305–335
0.035	261–301	282–322	273–303	288–318	298–328
0.041	255–293	276–314	266–296	281–311	291–321
0.047	248–286	268–306	259–289	274–304	284–314
0.054	243–279	264–300	253–283	268–298	278–308
0.062	237–272	258–293	247–277	262–292	272–302
0.072	232–266	253–287	241–271	256–286	266–296
0.080	227–261	248–282	235–265	250–280	260–290
0.091	221–254	244–277	230–260	245–275	255–285
0.105	216–248	238–270	225–255	240–270	250–280
0.120	210–241	232–263	220–250	235–265	245–275
0.135	206–237	227–258	215–240	230–255	240–265
0.148	203–234	222–253	210–235	225–250	235–260
0.162	200–230	219–249	205–230	220–245	230–255
0.177	195–225	215–245	200–225	215–240	225–250
0.192	192–221	212–242	195–220	210–235	220–245
0.207	190–218	210–238	190–215	205–230	215–240
0.225	186–214	206–234	188–213	203–228	213–238
0.250	182–210	202–230	185–210	200–225	210–235
0.312	174–200	194–220	183–208	198–223	208–233
0.375	167–193	184–210	180–205	195–220	205–230
0.437	162–187	175–200	175–200	190–215	200–225
0.500	156–180	168–192	170–195	185–210	195–220
0.562	152–176	......	165–190	180–205	190–215
0.625	147–170	....	165–190	180–205	190–215

As a guide to the coiling properties of mechanical spring wires, except hard drawn, HB, and WHB (which are not designed for close coiling characteristics), a wrap test is conventionally performed by winding the wire as a close-wound helix on a mandrel the diameter of which is related to the wire diameter. For sizes to 0.162 in., inclusive, the wire withstands wrapping around a mandrel equivalent to the diameter of the wire. For sizes larger than 0.162 to 0.312 in., inclusive, the wire can be expected to wind on a mandrel of twice the wire diameter. When the wire size exceeds 0.312 in. in diameter the resistance to winding on conventional test equipment is so great that the wrap test is generally not practical.

Music spring-steel wire is one of the highest-quality wires of the several types used for mechanical springs. High mechanical properties are dependent upon a greater amount of cold reduction after the final patenting operation than is ordinarily employed. Bright smooth surface luster is characteristic and essential for music spring-steel wire.

The tensile-strength ranges are shown in Table 24 for sizes up to 0.207 in.; however, sizes in common use are 0.148 in. and smaller.

Wire for Heading, Forging, and Roll Threading. Wire for heading, forging, and roll threading is produced by specially controlled manufacturing practices. Steel grades included in all the first three composition groups named under Manufacturers' Wire are employed.

For severe cold-upsetting or cold-forging operations, best results are obtained from wire that has been heat-treated in process.

When the steel is harder than 1018 or 1025, inclusive, wire heat-treated in process or, in the case of moderate upsetting, wire drawn from heat-treated rods should be used. When the grade of steel is harder than the equivalent of 1025 and also for severe upsetting of steel harder than the equivalent of 1018, the most satisfactory results are provided by spheroidize-annealing in the wire-mill processing.

Decarburization tests, when required on cold-heading wire of 0.28% C and over, are made by the microscopic method. Given below are the lowest practicable average limits for free ferrite and total affected depth, respectively.

Wire Size, In.	Av Depth Free Ferrite, In.	Av Total Affected Depth, In. (Free Ferrite Plus Partial Decarburization)
Up to 3/8, incl.	0.004	0.010
Over 3/8 to 1/2, incl.	0.005	0.012
Over 1/2 to 45/64, incl.	0.006	0.014

Finishes or Coatings on Wire for Cold Heading and Cold Forging. Cold-heading wire is produced with special finishes to provide proper lubrication in the header dies. Solid-die heading machines, especially those used for extrusion heading, require a coating of special consistency; whereas with open- or split-die heading machines a light coating will perform satisfactorily. Cold-heading finishes often vary considerably even for the same type of heading, in order to meet individual cold-heading requirements.

Aircraft-quality bolt and screw wire is produced for use in the manufacture of vital or highly stressed parts of airplanes and for similar or corresponding purposes.

Fine Round Wire

The various uses of fine round wire require different chemical compositions. Specifications requiring both chemical composition and physical or mechanical properties are sometimes impracticable. When it is necessary to specify both features, the limits or ranges should be negotiated between the purchaser and producer.

When mechanical properties are included in the following commodity decriptions, it is generally understood that those values refer to fine wire in coils. It is to be noted that when wire is straightened and cut the mechanical properties of the wire are altered.

Finishes. Fine wire is produced in the finishes available for coarse wire as follows:

Bright: to 25 gage or finer, depending upon equipment of manufacturer.
Liquor finish: all sizes.
Tinned: all sizes.
Galvanized: all sizes.
Coppered: all sizes.
Bright annealed: all sizes.
Annealed: all sizes.

Gages. Fine wire is designated by the steel wire gage, as shown in Sec. 28, or by decimals of an inch.

TABLE 24. TENSILE STRENGTH OF CARBON-STEEL MUSIC SPRING-STEEL WIRE

Dia, In.	Min Tensile Strength, Psi	Max Tensile Strength, Psi	Dia, In.	Min Tensile Strength, Psi	Max Tensile Strength, Psi
0.004	439,000	485,000	0.051	303,000	335,000
0.005	426,000	471,000	0.055	300,000	331,000
0.006	415,000	459,000	0.059	296,000	327,000
0.007	407,000	449,000	0.063	293,000	324,000
0.008	399,000	441,000	0.067	290,000	321,000
0.009	393,000	434,000	0.072	287,000	317,000
0.010	387,000	428,000	0.076	284,000	314,000
0.011	382,000	422,000	0.080	282,000	312,000
0.012	377,000	417,000	0.085	279,000	308,000
0.013	373,000	412,000	0.090	276,000	305,000
0.014	369,000	408,000	0.095	274,000	303,000
0.015	365,000	404,000	0.100	271,000	300,000
0.016	362,000	400,000	0.102	270,000	299,000
0.018	356,000	393,000	0.107	268,000	296,000
0.020	350,000	387,000	0.110	267,000	295,000
0.022	345,000	382,000	0.112	266,000	294,000
0.024	341,000	377,000	0.121	263,000	290,000
0.026	337,000	373,000	0.125	261,000	288,000
0.028	333,000	368,000	0.130	259,000	286,000
0.030	330,000	365,000	0.135	258,000	285,000
0.032	327,000	361,000	0.140	256,000	283,000
0.034	324,000	358,000	0.145	254,000	281,000
0.036	321,000	355,000	0.150	253,000	279,000
0.038	318,000	352,000	0.156	251,000	277,000
0.040	315,000	349,000	0.162	249,000	275,000
0.042	313,000	346,000	0.177	245,000	270,000
0.045	309,000	342,000	0.192	241,000	267,000
0.048	306,000	339,000	0.207	238,000	264,000

Tempers. The tempers produced in fine round wire are designated as hard, soft, annealed, or oil-tempered, which refer to characteristics developed for given end uses.

Aircraft cord wire is a hard-drawn, high-carbon, high-tensile-strength wire designed for use in the manufacture of flexible cords and nonflexible multiple-wire strands for aircraft controls and other products. This wire is produced in tinned or zinc-coated finishes.

Alloy Steel

This wire is supplied in coils only and should not be confused with the straight-length products of cold-finished alloy-steel bars which incorporate features that distinctly differ from wire products.

Alloy-steel spring wire is used for the manufacture of springs operating at moderately elevated temperatures, not over 450 F. Depending on requirements, the steel is of either open-hearth or electric-furnace manufacture, of compositions shown in Table 25. This wire is produced in the following conditions and sizes.

SPHEROIDIZE-ANNEALED AT FINISHED SIZE is used in sizes up to ⅝ in., inclusive, for very severe cold forming or coiling before hardening and tempering. Tensile strength is not specified.

SPHEROIDIZE ANNEALED IN PROCESS AND LIGHTLY DRAWN TO FINISHED SIZE (bright soft) is used in sizes up to ⅝ in., inclusive, for severe cold forming or uniform coiling before hardening and tempering. Tensile strength is not specified.

OIL-TEMPERED WIRE is intended for very light forming and is generally used for coiling into springs of the usual designs. The wire is produced in sizes up to ½ in., inclusive, to tensile-strength or hardness requirements (Table 26 shows the tensile strengths of various sizes).

Wire in the above three conditions is also produced as *aircraft alloy-steel spring wire.*

Alloy-steel Cold-heading Wire. For severe upsetting or cold-forging operations, alloy-steel wire is produced in one of the following types, depending upon the grade of steel selected and the severity of the cold work:

Wire drawn from spheroidize-annealed hot-rolled coils.

Wire spheroidize-annealed in process.

Wire spheroidize-annealed at finish size.

TABLE 25. CHEMICAL COMPOSITION OF ALLOY-STEEL SPRING WIRE

	Chromium-Vanadium Spring Wire	Chromium-Vanadium Aircraft Spring Wire	Chromium-Silicon Spring Wire	Chromium-Silicon Aircraft Spring Wire	Silicon-Manganese Spring Wire	Silicon-Manganese Aircraft Spring Wire
Carbon	0.45–0.55	0.48–0.53	0.50–0.60	0.50–0.60	0.55–0.65	0.55–0.65
Manganese	0.70–0.90	0.70–0.90	0.50–0.80	0.50–0.80	0.70–1.00	0.70–1.00
Phosphorus	0.040 max	0.030 max	0.040 max	0.030 max	0.040 max	0.030 max
Sulfur	0.040 max	0.030 max	0.040 max	0.030 max	0.040 max	0.030 max
Silicon	0.20–0.35	0.20–0.35	1.20–1.60	1.20–1.60	1.80–2.20	1.80–2.20
Chromium	0.80–1.10	0.80–1.10	0.50–0.80	0.50–0.80		
Vanadium	0.15 min	0.15–0.30				

TABLE 26. TENSILE STRENGTH OF OIL-TEMPERED ALLOY-STEEL SPRING WIRE, 1000 PSI

Wire Dia, In.	Chromium-Vanadium	Chromium-Silicon	Silicon-Manganese
0.500	190-210		
0.437	195-215	235-260	220-245
0.375	200-220	240-265	225-250
0.312	203-223	245-270	230-255
0.283	205-225	248-273	233-258
0.250	210-230	250-275	235-260
0.192	220-240	260-285	238-263
0.162	225-245	265-290	242-267
0.135	235-255	270-295	245-270
0.105	245-265	275-300	248-273
0.093	250-270	280-305	250-275
0.080	255-275	285-310	255-280
0.072	260-280	288-313	260-285
0.062	265-290	290-315	265-290
0.048	280-305	295-320	270-295
0.035	290-315	300-325	275-300
0.020	300-325		

Experience has shown that thermal treatment is required in the higher-carbon grades. For very severe cold heading, the lower-carbon grades are spheroidized in process, and in some cases spheroidized at the finish size.

When alloy steel harder than the equivalent of 0.20% carbon is required for cold heading or cold forging, the wire should be spheroidized.

Cold-heading requirements in steel grades higher than 0.20% carbon and also in the case of extremely severe heading operations in even the lower-carbon grades require a high degree of surface perfection and uniformity of composition. To meet such requirements necessitates special manufacturing practices. Cold-heading wire is produced with special finishes to provide proper lubrication in the header dies.

Decarburization. Data below give the lowest practicable average limits for free ferrite and total affected depth, respectively.

Wire Size, In.	Av Depth Free Ferrite, In.	Av Total Affected Depth, In. (Free Ferrite Plus Partial Decarburization)
Up to 3/8, incl.	0.004	0.010
Over 3/8 to 1/2, incl.	0.005	0.012
Over 1/2 to 3/4, incl.	0.006	0.014

Stainless and Heat-resisting Steels

Stainless and heat-resisting steel wire is a coiled product derived by cold-finishing hot-rolled and annealed coils. Wire is sometimes straightened and cut subsequent to cold finishing, and is known as cut wire.

Free-machining stock is produced for machining in automatic screw machines. Principal types used are 303, 416, and 430 F. The wire is produced in straightened and cut lengths with a cold drawn or centerless ground finish and with variable hardness depending upon the machining operations involved.

Cold-heading wire can be produced in all the standard types, but 302, 303, 304, 305, 410, and 430 are most generally used for cold-headed bolts, rivets, screws, and nails. It is customary to produce cold-heading wire annealed, pickled, coated, and lightly drawn. The application of a lubricating coating is made to facilitate cold heading.

Spring wire is produced in types 302 and 304 to the tensile-strength ranges shown in Table 27. When specified in straightened and cut lengths, the tensile strength may be as much as 10% less. Torsional modulus for wire produced in these types may vary from 8,500,000 to 11,000,000 psi. The magnetic permeability is low as compared with carbon-steel wire. Springs made from this corrosion-resisting wire retain their physical and mechanical properties at elevated temperatures up to approximately 600 F.

MANUFACTURING PRACTICES FOR STEEL PRODUCTS

Surface condition, type of finish, temper, and edge detail are of major importance with certain steel products. Details for several products are:

Plates, Hot-rolled

Carbon Steel. EDGES. Plates are termed sheared plates or sheared mill plates when rolled between horizontal rolls and trimmed on all edges. They are termed universal plates or universal mill plates (abbreviation U.M. plates) when rolled between horizontal and vertical rolls and trimmed on the ends. Universal plates are sometimes rolled in grooved rolls. They are termed mill-edge plates when rolled between horizontal rolls and trimmed only on the ends.

CUTTING PRACTICE. The difficulty in cutting steel plates by shearing, the hazard of breakage, and the development of strains or of incipient cracks increase with the thickness and hardness of the steel. Production experience has shown that the mill shearing of plates can be performed within the chemical and mechanical property limits and the maximum thickness limits set forth in Table 28.

Sheets, Hot-rolled

Thickness and Weight Data. The U.S. Standard gage for sheet and plate (1893) is based upon the weight per square foot in wrought iron for each gage number and is inaccurate for steel. The Manufacturers' Standard Gage for Steel Sheet takes into account the true weight of steel, shearing allowances, and the greater thickness of sheet at the center than at the edges. Sheets specified to the manufacturers' standard gage number are produced to the inch equivalent of that gage number. On the other hand, sheets specified to unit weight are produced to the corresponding thickness. Sheets should be ordered in decimals of an inch wherever possible (see Sec. 28).

EDGES. The normal edge produced in hot rolling does not conform to any definite contour and is designated as *mill edge.* Hot-rolled and hot-rolled pickled sheets are made with mill edge or cut edge. Mill-edge sheets may contain some edge defects, the more common types being cracked edges, thin edges (feather), and damage due to handling or processing. When the consumer intends to shear or blank sheets, a sufficient allowance should be made in the specified dimensions of

TABLE 27. TENSILE STRENGTH OF TYPES 302 AND 304 STAINLESS-STEEL SPRING WIRE

Dia, In.	Range of Tensile Strength, Psi		Dia, In.	Range of Tensile Strength, Psi	
	Min	Max		Min	Max
0.009 and smaller	325,000	355,000	0.041	269,000	299,000
0.010	320,000	350,000	0.047	262,000	292,000
0.011	318,000	348,000	0.054	260,000	290,000
0.012	316,000	346,000	0.062	255,000	285,000
0.013	314,000	344,000	0.072	250,000	280,000
0.014	312,000	342,000	0.080	245,000	275,000
0.015	310,000	340,000	0.092	240,000	270,000
0.016	308,000	338,000	0.105	232,000	262,000
0.017	306,000	336,000	0.120	225,000	255,000
0.018	304,000	334,000	0.148	210,000	240,000
0.020	300,000	330,000	0.162	205,000	235,000
0.022	296,000	326,000	0.177	195,000	225,000
0.024	292,000	322,000	0.207	185,000	215,000
0.026	289,000	319,000	0.225	180,000	210,000
0.028	286,000	316,000	0.250	175,000	205,000
0.032	277,000	307,000	0.312	160,000	190,000
0.036	273,000	303,000	0.375	140,000	170,000

TABLE 28. MILL-SHEARING LIMITS FOR CARBON-STEEL PLATES

Specified Max Carbon or Min Tensile Strength	Max Thickness to Be Sheared, In.			
	Rectangular Sheared Plates	U.M. Plates Sheared to Length		
		Circular Plates	To 12 In. Wide, Incl.	Over 12 In. Wide
0.35 and under, %	1½	1	2½	2
0.36 to 0.50%	1¼	¾	1¾	1½
0.51 to 0.60%	1	½	1½	1¼
Over 0.60%*				
Min tensile strength less than 65,000 psi	1½	1	2½	2
Min tensile strength 65,000 psi or equivalent hardness	1¼	¾	1¾	1½

* Plates over 0.60% carbon are generally gas-cut to size; but rectangular plates less than 1¼ in. thick and over 0.60% carbon can be hot-sheared in certain combinations of composition, thickness, and size. When plates specified to a maximum carbon limit exceeding 0.39% or of an equivalent hardness are to be gas-cut, edge treating is generally necessary.

the sheets to assure obtaining the desired shape and size of the pattern sheet, and the producer should be consulted.

SURFACE. It is common practice to inspect sheets on the premise that only one side need meet the surface requirements indicated by the appropriate quality description. In certain cases, where both surfaces of the finished part are exposed, two-side inspection may be necessary, but it can be performed only on cut lengths. Sheets may contain surface imperfections of such a character that the product can be used with a reasonable amount of metal finishing by the purchaser.

Surface inspection of coils does not afford the same opportunity for the producer to reject portions containing defects. Coils contain some defects, such as welds and holes, which render a portion unusable. In addition, coils contain more minor defects than cut lengths, and require more metal finishing.

COILS. Coils are produced to decimal thickness and are subject to thickness tolerances. However, thickness and width tolerances do not apply to the uncropped ends of hot-rolled mill-edge coils.

Coil diameters vary and are limited by individual mill equipment.

COIL WEIGHTS. Sheets in coils are specified to maximum coil-weight limits. The ability of producers to meet the maximum coil weights depends upon individual mill equipment.

High-strength Low-alloy Steel. SURFACE. Same as for carbon steel.

Stainless and Heat-resisting Steels. FINISH. Sheets are produced in the following finish: No. 1 hot rolled, annealed, and pickled.

Sheets, Cold-rolled

Carbon Steel. SURFACE. Same notes as for hot-rolled sheet.

HARDNESS RANGES. Limits are Rockwell 60 B maximum for soft low-carbon cold-rolled sheets and Rockwell 84 B minimum for full-hard sheets. For some applications, both minimum and maximum Rockwell B limits are required. In such cases, it is necessary to employ restricted chemical-composition limits or special processing practices, or both. The following two tempers are in common use:

Quarter-hard temper, Rockwell 60 B minimum to approximately Rockwell 75 B maximum.

Half-hard temper, Rockwell 70 B minimum to approximately Rockwell 85 B maximum.

When cold-rolled sheets are specified to hardness ranges, a variation of two Rockwell points, over the approximate maximum or under the minimum, is allowed. The sheets should not be subject to any definite forming requirements or to flatness tolerances.

High-strength Low-alloy Steel. FINISH. Sheets are produced with a dull surface texture intended for the application of paints, enamels, or lacquers but are not intended for electroplating without grinding and polishing.

Stainless and Heat-resisting Steels. FINISHES. Sheets are ordinarily produced in the following finishes:

No. 2D full finish (dull cold-rolled).

No. 2B full finish (bright cold-rolled).

Chromium-nickel steels may not have the same appearance as to color as straight chromium steels. Polished sheets are:

No. 4 polished.

No. 6 polish tampico brushed.

No. 7 high luster polished.

No. 8 mirror-finished.

Conditions and Tempers. High-tensile sheets are produced in type 301 in quarter-hard temper (125,000 tensile strength, 75,000 yield strength minimum) and half-hard temper (150,000 tensile strength, 110,000 yield strength minimum). Tempers are based on specified minimum values for tensile strength or yield strength or both.

Strip, Hot-rolled

Carbon Steel. Edge. Three types of edges are mill edge, square edge, and slit edge.

Mill Edge. This is the normal edge and it does not conform to any definite contour.

Square Edge. This edge is produced by hot edge rolling and normally is available only in thicker sections. The corners are not so square as those of square-edge bars. The width tolerances applicable to square-edge strip are the same as those for mill-edge strip.

Slit Edge. Minimum variation from specified width is obtained by slitting, in either single or multiple widths, usually by means of rotary knives. Slit-edge strip is desirable for machine welding and for feeding into machines having limited guide clearance.

Surface. It is industry practice to inspect one side of the strip.

Strip, Cold-rolled

Carbon Steel. Temper. Cold rolling changes the mechanical properties of the strip and produces certain useful combinations of hardness, strength, stiffness, ductility, and other characteristics. Through control it is possible to produce mechanical characteristics that are known as tempers.

Temper numbers indicate degrees of strength, hardness, and ductility produced in cold-rolled strip, and are associated with the ability of each temper to withstand certain degrees of cold forming. Typical illustrations of the forming characteristics of each temper are shown in Fig. 5. Usefulness of the various tempers is not completely indicated by any one mechanical test; however, the Rockwell hardness is most widely used.

The general hardness characteristics (also see Table 29) are:

Tempers 1, 2, and 3 provide maximum hardness, or rigidity (stiffness).

Tempers 4 and 5 are used in production of parts involving difficult forming or drawing. Hardness-value limitations are less significant than control of steelmaking and processing operations. For that reason, minimum hardness values are not shown in the temper descriptions for cold-rolled strip below, and the maximum hardness values shown are approximate for a generalized comparison with other tempers.

In producing the foregoing tempers it is customary not to exceed 0.60% manganese by ladle analysis.

No. 1 (hard temper) is a very stiff, springy, cold-rolled strip suitable for flat work not requiring ability to withstand cold forming. This temper is produced in chemical compositions of less than 0.25% carbon (ladle analysis) and Rockwell 84 B minimum for thicknesses 0.070 in. and greater, or Rockwell 90 B minimum for thicknesses less than 0.070 in.

No. 2 (half-hard temper) is a moderately stiff cold-rolled strip intended for limited bending. Strip of this temper can be bent 90° across the direction of rolling around a radius equal to the thickness. This temper is produced in chemical compositions of less than 0.25% carbon (ladle analysis) and Rockwell 70 B minimum and approximately Rockwell 85 B maximum.

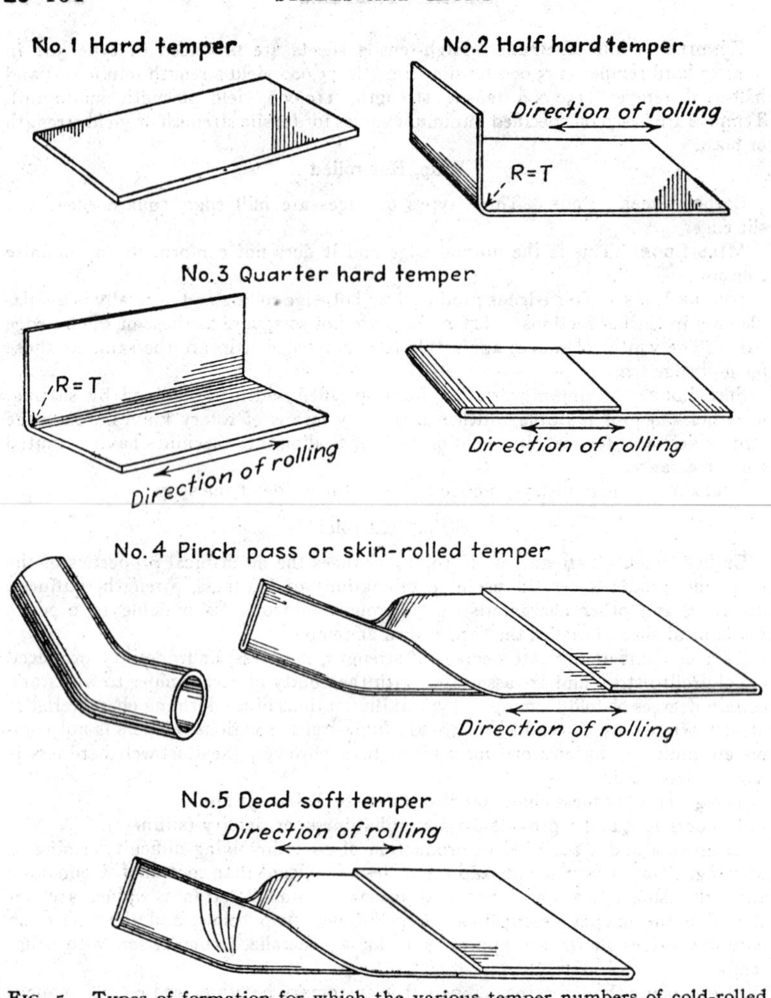

FIG. 5. Types of formation for which the various temper numbers of cold-rolled carbon-steel strip are suited.

No. 3 (quarter-hard temper) is a medium-soft cold-rolled strip intended for limited bending and forming and drawing. Strip of this temper can be bent 180° across the direction of rolling and 90° in the direction of rolling around a radius equal to the thickness. This temper is produced in chemical compositions of less than 0.25% carbon (ladle analysis) and Rockwell 60 B minimum and approximately Rockwell 75 B maximum.

No. 4 (skin-rolled temper) is a soft, ductile, cold-rolled strip intended for fairly deep drawing where surface disturbances such as stretcher strains are objectionable. It is capable of being bent flat upon itself in any direction. Skin-rolled, planish-

TABLE 29. TEMPER DESCRIPTIONS FOR COLD-ROLLED STRIP

Temper	Ladle Carbon, %	Rockwell Hardness	
		Min	Max
No. 1 (hard temper).........	0.25 max	84 B for thicknesses 0.070 in. and greater 90 B for thicknesses less than 0.070 in.	
No. 2 (half-hard temper).....	0.25 max	70 B	85 B approx
No. 3 (quarter-hard temper)..	0.25 max	60 B	75 B approx
No. 4 (skin-rolled temper)....	0.15 max	....	65 B approx
No. 5 (dead-soft temper).....	0.15 max	....	55 B approx

rolled, and pinch-passed are equivalent terms with respect to temper. This temper is produced in chemical compositions of less than 0.15% carbon (ladle analysis) and approximately Rockwell 65 B maximum (refer to Aging Phenomenon below).

No. 5 (dead-soft temper) is a soft, ductile, cold-rolled strip produced without definite control of stretcher straining and fluting (refer to Aging Phenomenon). It is intended for difficult drawing applications where such surface disturbances are not objectionable. It is suitable for bending flat upon itself in any direction. This temper is produced in chemical compositions of less than 0.15% carbon (ladle analysis) and approximately Rockwell 55 B maximum.

AGING PHENOMENON. Although the maximum ductility is obtained in steel strip in its dead-soft or annealed condition, it is unsuited to many forming operations because of its tendency to stretcher strain. A small amount of cold rolling will prevent this tendency, but the effect is only temporary because of a phenomenon called aging. The phenomenon of aging is accompanied by a loss of ductility with an increase in hardness, yield point, and tensile strength. For those uses in which stretcher straining or breakage due to aging of the steel is likely to occur, the steel should be fabricated as promptly as possible after temper rolling.

TYPES OF EDGES

No. 1 edge is a prepared edge of a specified contour (round, square, or beveled), which is produced when a very accurate width is required, or when an edge finish suitable for electroplating is required, or both.

No. 2 edge is a natural-mill edge.

No. 3 edge is an approximately square edge produced by slitting.

No. 4 edge is a rounded edge produced by edge rolling either the natural edge of hot-rolled strip or slit-edge strip. This edge is produced when the finish of the edge is not important.

No. 5 edge is an approximately square edge produced by rolling or filing of a slit edge to remove burr.

No. 6 edge is a square edge produced by edge rolling the natural edge of hot-rolled strip or slit-edge strip, when the width tolerance and finish required are not so exacting as for No. 1 edge.

FINISHES. For the majority of end uses the surface of one side only is of prime importance. When the finish of both surfaces is of equal importance, the producer should be informed accordingly. It is usual practice to place the inspected surface on the outside of the coil or top side of cut lengths.

No. 1 (dull finish) is a finish without luster, produced by rolling on rolls roughened by mechanical or chemical means. This finish is especially suitable for lacquer or paint adhesion and is beneficial in drawing operations by reducing contact friction between die and strip.

No. 2 (regular bright finish) is produced by rolls having a moderately high finish. It is suitable for many requirements, but not generally applicable to plating.

No. 3 (best bright finish) is generally of high luster produced by selective rolling practices, is the highest-quality finish produced, and is particularly suited for electroplating.

High-strength Low-alloy Steel. FINISHES. The two finishes customarily produced are: No. 1 (dull finish) and No. 2 (regular bright finish).

Stainless and Heat-resisting Steels

No. 1 edge is a rolled edge, either round or square as specified.

No. 3 edge is an edge produced by slitting.

No. 5 edge is a square edge as produced by rolling or filing after slitting.

FINISHES

No. 1 finish is cold-rolled, annealed, and pickled.

No. 2 finish is cold-rolled, annealed, pickled, and rerolled.

CONDITIONS AND TEMPERS. High-tensile cold-rolled strip is produced in type 301 in quarter-hard temper (125,000 tensile strength, 75,000 yield strength minimum), half-hard temper (150,000 tensile strength, 110,000 yield strength minimum), three-fourths-hard temper (175,000 tensile strength, 135,000 yield strength minimum), and full-hard temper (185,000 tensile strength, 140,000 yield strength minimum). Tempers are based on specified minimum values for tensile strength or yield strength or both.

Cast Iron

Cast iron forms the structural base of most heavy machinery and innumerable smaller components or complete items. The bulk of the production machining in many plants is on cast iron, for more than 15 million tons of ferrous castings (not including steel castings) are processed each year in the United States.

All cast irons are alloys of iron, carbon, and silicon that have a carbon content in excess of the amount that can be retained in solid solution in austenite at the eutectic temperature. Thus all contain some carbon in the form of free graphite, with the total carbon usually ranging between 2.4 and 3.8%. The silicon content will generally range from 0.5 to 3.0%, and there will be appreciable quantities of manganese, phosphorus, and sulfur. In addition, alloying elements may be added to cast iron.

It has been the general practice to classify cast irons by their mechanical properties rather than their chemical composition. This is not only because widely varying compositions can be made to produce similar properties, but because the properties for a given composition may vary so widely depending on the casting procedure and the size and shape of the casting itself.

Thus experienced foundries are able to provide the specified mechanical properties for a given casting by their choice of composition and casting procedure. It follows

TABLE 30. TYPICAL PROPERTIES OF GRAY CAST IRON*

ASTM Class	Carbon Content, %			Hardness, Brinell	Tensile, M Psi	Shear, M Psi	Compression, M Psi	Modulus of Elasticity, M Psi	Endurance Limit, M Psi	Impact, Ft-lb
	Graphitic	Combined	Total							
20	2.90	0.60	3.50	180	20	32	95	12,000	10	55
25	....	....	3.40	190	27.5	37	100	13,000	12.5	55
30	2.65	0.70	3.35	200	32.5	44	115	15,000	14.5	60
35	2.55	0.70	3.25	210	37.5	43	125	16,000	17.5	60
40	2.25	0.85	3.10	220	45	57	143	17,000	21	70
50	2.00	0.90	2.90	240	55	59	150	19,000	25	80
60	2.00	0.85	2.85	290	65	..	170	20,000	....	115

* Data are for medium section castings; light sections will generally have higher properties; heavy sections will have lower properties.

TABLE 31. AUTOMOTIVE GRAY IRONS

SAE No.	Typical Composition (Not a Specification), %					Brinell Hardness	Min Tensile, M Psi	Min Transverse Load, Lb	Min Deflection, In.
	Carbon	Silicon	Manganese	Sulfur	Phosphorus				
110	3.40–3.70	2.30–2.80	0.50–0.80	0.15	0.25	187 max	20	1800	0.15
111	3.25–3.50	2.00–2.30	0.60–0.90	0.15	0.20	170–223	30	2200	0.20
113*	3.40 min	1.00–1.70	0.60–0.90	0.14	0.20	179–229	30	2200	0.20
114*	3.40 min	1.10–1.70	0.60–0.90	0.14	0.20	207–269	40	2600	0.27
120	3.20–3.40	1.90–2.20	0.60–0.90	0.15	0.15	187–241	35	2400	0.24
121	3.10–3.30	1.80–2.10	0.60–0.90	0.15	0.12	202–255	40	2600	0.27
122	3.00–3.20	1.80–2.10	0.70–1.00	0.15	0.10	217–269	45	2800	0.30

* Nos. 113 and 114 are for brake drums and clutch plates. Minimum carbon content is a specification for these types. Alloying elements may be added as required. Microstructure is specified: type 113, graphite ASTM type A, sizes 2 to 4, lamellar pearlite with not over 15% ferrite. Type 114, graphite ASTM type A, sizes 3 to 5, fine lamellar pearlite with not over 5% free cementite or free ferrite.

that two castings of the same hardness or even the same tensile strength may have quite different compositions, microstructures, and machinability.

Microstructure, then, is more of a key to the character of cast iron than is brinell hardness. More use is made of the information it provides, and the practice of specifying by microstructure is growing.

The basic groups of cast irons are white iron, gray iron, and malleable iron. These basic groups include such important subdivisions as nodular or ductile iron, inoculated iron, and alloy iron.

White Iron. White or chilled iron may occur either accidentally or intentionally in portions of a casting or as a complete casting. Most of the carbon is in combined form chiefly as cementite and pearlite. The free cementite (iron carbide) makes white iron hard and brittle. White iron is a step in the manufacture of malleable iron; beyond this it has limited applications where extremely hard surfaces are required. Such surfaces usually must be ground. White iron is obtained intentionally by adjusting the composition and by accelerating the cooling.

Gray Iron. The composition is largely pearlite (alternate layers of ferrite and iron carbide) and flake graphite. Coarse graphite and coarse pearlite are usually

TABLE 32. TYPICAL PROPERTIES OF INOCULATED CAST IRON

Meehanite Grade	Hardness, Brinell, Min	Tensile, M Psi, Min	Shear, M Psi	Compression, M Psi	Modulus of Elasticity, M Psi	Endurance Limit, M Psi	Impact, Ft-lb
GM	217	55	55	200	22,000	25	8.0
GA	207	50	48	175	20,000	22	7.2
GB	196	45	44	160	18,000	19	5.8
GC	192	40	40	150	17,000	17.5	4.5
GD	183	35	35	130	14,500	15	3.2
GE	174	30	30	120	12,000	13.7	2.1
HE	223	30	31	...	10,000		
HD	223	33	34	145	15,000		
HA	223	50	48	...	20,000		
HB	300	38	40	160	18,000		
HR	300	40	42.5	162	21,000		
SC	300	27	28	130	17,000		

TABLE 33. TYPICAL PROPERTIES OF STANDARD MALLEABLE IRONS

ASTM Grade	Total Carbon	Hardness, Brinell	Tensile, M Psi	Shear, M Psi	Modulus of Elasticity, M Psi	Endurance Limit, M Psi	Elongation, % in 2 In.	Impact, Ft-lb
32510	2.25 to 2.70	110 to 135	52	48	25,000	25 to 26.5	12.5	16.5
35018	1.75 to 2.30	110 to 145	55	48	25,000	22 to 30.5	20	16.5

TABLE 34. MINIMUM PROPERTIES OF STANDARD PEARLITIC MALLEABLE IRONS

ASTM Grade	Hardness, Brinell	Tensile, M Psi	Yield, M Psi	Elongation, % in 2 In.
43010	163–207	60	43	10
48005	179–228	70	48	5
53004	197–241	80	53	4
60003	197–241	90	60	3
70002	241–285	90	70	2

associated with slow cooling of an iron that is not highly alloyed. Fine graphite and pearlite are usually the result of rapid cooling, low carbon, low silicon, and the presence of such alloying elements as chromium, vanadium, and silicon.

Gray iron may also contain free ferrite, steadite, free carbide, manganese sulfide, and various inclusions.

It is the graphite flakes that impart the characteristic gray color to a fracture and give gray iron its name. A wide range of strengths may be obtained. Seven classes, from 20,000 to 60,000 psi tensile, have been established by ASTM (see Table 30). Seven classes of automotive gray irons have been similarly standardized (see Table 31).

It is often thought that gray iron must be used as cast, but this is not true. Heat-treatment is possible and is being increasingly employed. Stress relieving and annealing are the most frequent treatments, but hardening, tempering, and various specialized heat-treatments are possible.

Inoculated Iron. Inoculants are materials that, when added to molten cast iron, change the structure and hence the properties without materially changing the composition.

An early inoculant was ferrosilicon. Melting the iron with a lower silicon content than desired, then adding the missing silicon in the ladle, improves the structure by providing better mechanical properties and a reduced tendency toward chilled edges. Other inoculants are calcium-silicon, ferromanganese-silicon, and zirconium-silicon.

Proprietary inoculants are available that are more effective than ferrosilicon when used in combination with careful foundry control. A familiar product is Meehanite, produced by licensed foundries, in which a specially made white-cast-iron composition is graphitized in the ladle with a calcium silicide inoculant. Meehanite (Table 32) is produced in a number of grades including a group for high-temperature applications.

Alloy Cast Iron. In addition to silicon, manganese, sulfur, and phosphorus, which are normally present in all cast irons and not generally considered alloy additions, alloying elements are added to modify the physical or mechanical properties and to accelerate or retard the formation of graphite. The most frequent alloying elements are nickel, chromium, molybdenum, and copper.

Because the properties of cast iron are influenced by so many factors besides the composition, there has been little attempt to standardize compositions of the alloy cast irons.

Malleable Iron. Standard malleable iron consists almost entirely of ferrite and nodular graphite; that is, the graphite is in the form of nodules, rather than the flakes found in gray and alloy iron. This is accomplished by casting white iron and graphitizing it in a special annealing process, perhaps requiring as long as 10 days. The exact time and the procedure will vary with the grade desired, the size and nature of the original castings, and the type of equipment available. The shorter times (as low as 15 hr) require small batches and mechanized equipment, so that longer times in larger furnaces may be more economical.

The resulting product is basically different from cast iron. It provides toughness, impact resistance, ductility, high resistance to corrosion, and easy machinability. Standard malleable iron is produced in two grades (Table 33). The ASTM designations for these irons are in five digits, with the first three indicating the minimum yield point in hundreds of pounds and the last two indicating the percentage elongation in 2 in.

Grade 32510 is extremely fluid and easily cast into light sections. **Malleable**

TABLE 35. PHYSICAL PROPERTIES OF NODULAR IRONS

Grade	Tensile, Min, M Psi	Yield, Min, M Psi	Elongation, % in 2 In., Min	Hardness, Brinell
ASTM 80-60-03	80	60	3	
ASTM 60-45-10	60	45	10	
AMS 5315	60	45	10†	190 max
AMS 5316	80	60	3†	202 to 269
Mil-I-17166 (Ships)	60	45	15	190 max

foundries produce it in sections as light as 1/32 in. in limited areas, but it is not generally used if there are large areas more than 1½ in. thick.

Grade 35018 has a lower carbon content providing higher strength and ductility. It is regularly cast in sections from 3/32 to 2½ in. thick.

Pearlitic malleable irons are produced from the same or similar compositions required for standard malleable irons, but a different annealing process leaves some of the carbon content in the form of combined carbides. (The pearlitic designation is a convenient label and does not mean that the microstructure is always such that the carbides are in the form of pearlite.)

Minimum properties of the ASTM grades of pearlitic malleable (Table 34) are typical of the range available. In addition to these standard grades, a number of pearlitic malleable irons are available under various trade names. In all cases they provide higher strength and wear resistance than the standard malleable irons, but with some loss of ductility, shock resistance, and machinability.

Nodular Iron. Within the past few years a practical method has been developed of producing cast iron in which the graphite has a nodular or spheroid form, rather than a flake form, in the original casting. Variously called ductile, nodular, or spheroidal iron, it is obtained by the retention of about 1 lb magnesium in 1 ton iron. Because magnesium vaporizes below the melting point of iron, difficulties were encountered. Several methods have been devised for introducing the magnesium into the melt just before casting, either alone or in combination with other materials.

There has been a rapid growth in the number of licensed foundries producing this new material, and nodular iron has already become an important and fairly common material.

Specifications for several standard and tentative standard types are shown in Table 35.

References

SAE Handbook Supplement.

Supply and Logistics Handbook H8, "Steel and Iron Wrought Products," developed by the Office of the Assistant Secretary of Defense with association of the American Iron & Steel Institute.

Federal Standard 66-1954, Steel—Chemical Compositions and Hardenability.

American Machinist Special Reports 267 and 271, How to Work Tool and Die Steels.

Section 26

NONFERROUS METALS

SECTION 26

NONFERROUS METALS

ALUMINUM ALLOYS

Pure aluminum is silvery in luster; its crystal structure is face-centered cubic. Hence, the abundance of slip planes makes it very malleable and ductile. In addition, this material is a good conductor of electricity and heat, provides a good reflective surface, is corrosion-resistant, and is light in weight compared with most other structural materials. Aluminum has a density of only 0.0975 lb per cu in.

Low tensile strength of *pure* aluminum is a drawback. The initial strength of 9000 psi can be doubled by working the metal, but it will not respond to heat-treatment.

The strength of aluminum can be greatly increased by alloying, without too severely reducing ductility and low weight. Wrought alloys are of two types: One group can be hardened by cold-working and the second group responds to both working and heat-treatment or either one individually. Other aluminum alloys are primarily used in the form of castings, some of which are heat-treatable while some are not heat-treatable.

Numerous alloys in the above categories can be obtained in many conditions of temper, the condition produced by heat or mechanical treatment. The aluminum alloys are identified by numbers rather than by names. A particular alloy is seldom used for both casting and wrought (cold-worked) products. The primary constituent in an alloy is indicated by the following system:

No.	*Constituent* (*Alloying Element*)
0 to 10	Manganese, Mn
10 to 30	Copper, Cu (the hardening constituent)
30 to 50	Silicon, Si
50 to 70	Mg_2Si
70	Zinc, Zn

When the letter S follows the above numbers, the alloy is used for wrought products such as sheet, rivets, and rolled rod. If the letter S is not present, the alloy is used to make castings. When an existing alloy's analysis has been changed to a degree that produces different properties, the material may be designated by a letter preceding the original number. Thus, a modified 17S alloy would be identified as A17S.

Temper Designations. The alloy numerical designation, as described above, has its temper indicated by a letter following a hyphen after the number. This letter can, in turn, be followed by one or more digits to define the temper specifically. Here is the nomenclature:

-O annealed, recrystallized (wrought products only).

-F as fabricated.

-H strain-hardened.

-H1, plus one or more digits—strain-hardened only.
-H2, plus one or more digits—strain-hardened and then partially annealed.
-H3, plus one or more digits—strain-hardened and then stabilized.
-W solution heat-treated—unstable temper.
-T treated to produce stable tempers other than -O, -F, or -H.
-T2 annealed (cast products only).
-T3 solution heat-treated and then cold-worked.
-T4 solution heat-treated.
-T5 artificially aged only.
-T6 solution heat-treated, then artificially aged.
-T7 solution heat-treated, then stabilized.
-T8 solution heat-treated, cold-worked, and then artificially aged.
-T9 solution heat-treated, artificially aged, and then cold-worked.
-T10 artificially aged and then cold-worked.

Tables 1 and 2 show in more detail both the old and new temper designations for non-heat-treatable and heat-treatable wrought-1 alloys, respectively.

Wrought Alloys. NON-HEAT-TREATABLE ALLOYS. These wrought alloys are non-heat-treatable: 2S, 3S, 4S, and 52S (see Table 3 for compositions), and can be hardened only by cold-working. Actual strength depends upon the amount of deformation performed on them after the last annealing, or softening, operation.

Supplied in six tempers, these alloys vary from annealed or soft temper, O, to the extra-hard temper, as designated for the 2S and 3S alloys by H19, and for the 52S alloy by H39. Besides these tempers the F designation is also used. It indi-

TABLE 1. TEMPER DESIGNATIONS FOR STRAIN-HARDENED WROUGHT ALUMINUM ALLOYS THAT ARE NOT HEAT-TREATABLE*

Old Temper Designation	New Temper Designation†		
	Strain-hardened Only	Strain-hardened and Then Partially Annealed	Strain-hardened and Then Stabilized
Alloys 2S and 3S:			
¼H	-H12	-H22	
½H	-H14	-H24	
¾H	-H16	-H26	
H	-H18	-H28	
Extra hard (not standard)	-H19		
Alloys 4S, 52S, and 56S:‡			
¼H	.....	.....	-H32
½H	.....	.....	-H34
¾H	.....	.....	-H36
H	.....	.....	-H38
Extra hard (not standard)	.....	.....	-H39

* Alloys listed are also available in -O and -F tempers.
† The tempers shown are standard for the alloys listed.
‡ The alloys 4S, 52S, and 56S may also be obtained in the -H1 and -H2 type conditions.

TABLE 2. TEMPER DESIGNATIONS FOR HEAT-TREATED WROUGHT ALUMINUM ALLOYS

Alloy	Old Temper Desig-nation	New Temper Designation											
		Sheet and Plate[a]				Extrusions[a]		Wire Rod and Bar[a]	Tubing[a,b]		Forgings[c]	Rivets	
		Flat Sheet	Coiled Sheet	Plate									
		A	A	A	C	A	C	A	A	C	A	A	D
11S....................	-W	....	...	....	...	....	....	-T4					
	-T3	....	...	....	...	....	....	-T3					
	-58	....	...	....	...	....	....	-T8					
14S and Clad 14S[d]..	-W	-T3	-T4	-T4	-T4	-T4	-T42	-T4	...	...	-T4		
	-T	-T6	-T6	-T6	-T6	-T6	-T62	-T6	...	...	-T6		
	-T	....	...	....	...	....	....	....	...	...	-T61[e] F.B.		
17S..................	-T	....	...	....	...	....	....	....	...	...	....	....	-T3[f] F.V.
	-T	....	...	....	...	....	....	-T4	...	...	....	-T4	-T3[g] F.S.
	Any temper	....	...	....	...	....	....	....	...	...	....	....	-T41[h]
A17S..............	-T	....	...	....	...	....	....	....	...	...	....	-T4	-T3[f] F.V.
18S................	-T	....	...	....	...	....	....	....	...	...	-T61[e] F.B.		
24S and Clad 24S[d]..	-T	-T3	-T4	-T4	-T4	-T4	-T42	-T4	-T3	-T4	....	-T4	-T31[g] F.S.
	-RT	-T36	...	-T36	...	....	....	-T36					
	-T80	....	-T6	....	-T6								
	-T81	-T81											
	-T86	-T86	...	-T86									
25S and 32S........	-T	....	...	....	...	....	....	....	...	...	-T6		
A 51S............	-W	....	...	....	...	....	....	....	...	...	-T4		
	-T	....	...	....	...	....	....	....	...	...	-T6		

Alloy	Old Temper Designation	New Temper Designation											
		Sheet and Plate[a]				Extrusions[a]		Wire Rod and Bar[a]	Tubing[a,b]		Forgings[c]	Rivets	
		Flat Sheet	Coiled Sheet	Plate									
		A	A	A	C	A	C	A	A	C	A	A	D
53S	-W	....	...	....	...	-T4	-T4	....	...	...	-T4	-T4	-T4
	Any temper	....	...	....	...	....	....	....	...	...	....	....	-T41[h]
	-T	....	...	....	...	-T6	-T6	....	...	...	-T6	-T6	-T6
	-T5	....	...	....	...	-T5							
	-T61	....	...	....	...	....	....	....	...	...	....	-T61	-T61
61S	-W	-T4	-T4	-T4	-T4	-T4	-T4	-T4	-T4	-T4			
	-T	-T6	-T6	-T6	-T6	-T6	-T6	-T6	-T6	-T6	....	-T6	-T6
	-T5	....	...	....	...	-T5							
	-T62	....	...	....	...	-T62							
	-T81	....	...	....	...	....	....	-T81					
63S	-T	....	...	....	...	-T6	-T6						
	-T5	....	...	....	...	-T5	-T5						
75S and Alclad 75S[d]	-W	-W[i]	-W[i]	-W[i]	-W[i]	-W[i]	-W[i]						
	-T	-T6	-T6	-T6	-T6	-T6	-T6	-T6					

A, heat-treated by the producer. C, heat-treated by the manufacturer of the finished product. D, driven rivets.

[a] Products listed are available also in -O and -F tempers.
[b] For extruded tubing see columns for extrusions.
[c] Forgings are available also in -F temper.
[d] Available only in sheet and plate.
[e] F.B., boiling-water quench.
[f] F.V., driven cold after full natural aging.
[g] F.S., driven cold immediately after solution heat-treatment, or when refrigerated to defer natural aging.
[h] Driven hot, at the solution-heat-treating temperature.
[i] To be specific, the time of natural aging must be stated: for example, 75 S-W (2 hr), 75 S-W (2 months).

cates the as-fabricated condition, the condition for which no specific effort has been made to control the mechanical properties of the alloy.

For similar tempers, the strengths of the alloys 2S, 3S, 4S, and 52S increase in that order; ease of deforming decreases in the same order. The 3S alloy, though it has greater strength than the 2S alloy, costs the same. Hence it is more commonly used than 2S. Typical mechanical properties of all these alloys are shown in Table 4.

Typical uses of non-heat-treatable alloys are:

2S. Known for good forming properties. It has a low yield strength and good corrosion resistance; is used for cooking utensils, sheet, and tubing.

3S. Similar to the 2S alloy, but slightly stronger and less ductile; is used to make cooking utensils and sheet-metal work.

4S. This alloy has fairly high strength and its forming properties are fair. Uses include forms of sheet and tubing.

52S. It has forming properties which are fair plus high strength. 52S is used

TABLE 3. NOMINAL COMPOSITION OF WROUGHT ALUMINUM ALLOYS*

Alloy	% of Alloying Elements—Aluminum and Normal Impurities Constitute Remainder								
	Copper	Silicon	Manganese	Magnesium	Zinc	Nickel	Chromium	Lead	Bismuth
EC	99.45% minimum aluminum								
2S	99% minimum aluminum								
3S†	....	....	1.2						
4S†	....	....	1.2	1.0					
11S†	5.5	....	...	...	...	...	....	0.5	0.5
14S†	4.4	0.8	0.8	0.4					
17S	4.0	....	0.5	0.5					
A17S	2.5	....	...	0.3					
8S	4.0	....	...	0.6	...	2.0			
B18S	4.0	....	...	1.5	...	2.0			
24S†	4.5	....	0.6	1.5					
25S	4.5	0.8	0.8						
32S	0.9	12.2	...	1.1	...	0.9			
43S	....	5.0							
B50S	....	....	...	1.2					
A51S	...	1.0	...	0.6	...	...	0.25		
52S	....	....	...	2.5	...	...	0.25		
53S	....	0.7	...	1.3	...	...	0.25		
56S	....	....	0.1	5.2	...	...	0.1		
61S	0.25	0.6	...	1.0	...	...	0.25		
62S	0.25	0.6	...	1.0					
63S	....	0.4	...	0.7					
72S	....	....	...	...	1.0				
75S†	1.6	....	...	2.5	5.6	...	0.3		

* Heat-treatment symbols have been omitted because composition does not vary for different heat-treatment practices.

† The clad form of these alloys consists of a "core" of the basic alloy coated with pure aluminum or a suitable alloy.

TABLE 4. MECHANICAL PROPERTIES OF WROUGHT ALUMINUM ALLOYS*

Alloy and Temper	Tensile Strength, Psi	Yield Strength (Offset = 0.2%), Psi	Elongation, % in 2 In.		Brinell Hardness, 500-kg Load 10-mm Ball	Shearing Strength, Psi	Endurance Limit,† Psi
			Sheet Specimen (1/16 In. Thick)	Round Specimen (1/2 In. Dia)			
			Non-heat-treatable				
2S-O	13,000	5,000	35	45	23	9,500	5,000
2S-H12	15,500	14,000	12	25	28	10,000	6,000
2S-H14	17,500	16,000	9	20	32	11,000	7,000
2S-H16	20,000	18,000	6	17	38	12,000	8,500
2S-H18	24,000	22,000	5	15	44	13,000	8,500
3S-O	16,000	6,000	30	40	28	11,000	7,000
3S-H12	19,000	17,000	10	20	35	12,000	8,000
3S-H14	21,500	19,000	8	16	40	14,000	9,000
3S-H16	25,000	22,000	5	14	47	15,000	9,500
3S-H18	29,000	26,000	4	10	55	16,000	10,000
Clad 3S	Properties substantially same as for 3S						
4S-O	26,000	10,000	20	25	45	16,000	14,000
4S-H32	31,000	22,000	10	17	52	17,000	14,500
4S-H34	34,000	27,000	9	12	63	18,000	15,000
4S-H36	37,000	31,000	5	9	70	20,000	15,500
4S-H38	40,000	34,000	5	6	77	21,000	16,000
Clad 4S	Properties substantially same as for 4S						
11S-T3‡	55, 00	48,000	..	15	95	32,000	18,000
11S-T6	57,000	39,000	..	17	97	34,000	18,000
11S-T8	59,000	45,000	..	12	100	35,000	18,000
A51S-T6	48,000	43,000	..	17	100	32,000	11,000
52S-O	27,000	12,000	25	30	45	18,000	17,000
52S-H32	34,000	27,000	12	18	62	20,000	17,500
52S-H34	37,000	31,000	10	14	67	21,000	18,000
52S-H36	39,000	34,000	8	10	74	23,000	18,500
52S-H38	41,000	36,000	7	8	85	24,000	19,000
			Heat-treatable				
14S-O	27,000	14,000	..	18	45	18,000	13,000
14S-T4	62,000	40,000	..	20	105	38,000	20,000
14S-T6	70,000	60,000	..	13	135	42,000	18,000
Clad 14S-O	25,000	10,000	21	..	...	18,000	
Clad 14S-T3	63,000	40,000	20	..	...	37,000	
Clad 14S-T4	61,000	37,000	22	..	...	37,000	
Clad 14S-T6	68,000	60,000	11	..	...	41,000	
17S-O	26,000	10,000	..	22	45	18,000	13,000
17S-T4	62,000	40,000	..	22	105	38,000	18,000
A17S-T4	43,000	24,000	..	27	70	28,000	13,500
18S-T61	61,000	46,000	..	12	120	39,000	17,000
B18S-T72	48,000	37,000	..	11	95	30,000	
24S-O	27,000	11,000	19	22	47	18,000	13,000
24S-T3	70,000	50,000	18	..	120	41,000	20,000
24S-T4	68,000§	48,000	20	19	120	41,000	20,000
24S-T36	72,000	57,000	14	..	130	42,000	18,000
Clad 24S-O	26,000	11,000	19	..	...	18,000	
Clad 24S-T3	64,000	44,000	18	..	...	40,000	
Clad 24S-T4	64,000	42,000	19	..	...	40,000	
Clad 24S-T36	67,000	53,000	11	..	...	41,000	

TABLE 4. MECHANICAL PROPERTIES OF WROUGHT ALUMINUM ALLOYS* (*Continued*)

Alloy and Temper	Tensile Strength, Psi	Yield Strength (Offset = 0.2%), Psi	Elongation, % in 2 In.		Brinell Hardness, 500-kg Load 10-mm Ball	Shearing Strength, Psi	Endurance Limit,† Psi
			Sheet Specimen (1/16 In. Thick)	Round Specimen (1/2 In. Dia)			
Clad 24S-T81	65,000	60,000	6				
Clad 24S-T86	70,000	66,000	6				
53S-O	16,000	8,000	..	35	26	11,000	8,000
53S-T4	30,000	20,000	..	21	62	18,000	13,000
53S-T5	27,000	21,000	..	15	60	17,000	
53S-T6	37,000	32,000	..	13	80	23,000	13,000
61S-O	18,000	8,000	22	30	30	12,500	9,000
61S-T4	35,000	21,000	22	25	65	24,000	13,500
61S-T6	45,000	40,000	12	17	95	30,000	13,500

* The values given in this table are averages which take into account the variations introduced by size, shape, or method of manufacture. For guaranteed minimum values, see other sources.

† Based on 500,000,000 cycles of completely reversed stress using the R. R. Moore type machine and specimen.

‡ For sizes up to 1½ in. For larger sizes, the strengths will be somewhat lower.

§ The strengths of extrusions more than about ¾ in. thick will be 15 to 20% higher.

in a variety of forms including plate, sheet, rod, wire, pipe, tubing, and bar. A new alloy designated as A54S has the highest strength of any non-heat-treatable alloy and is used in the form of sheet for welded structures.

HEAT-TREATABLE ALLOYS. There are a number of alloys which can be heat-treated to various tempers. Among these are A17S, 17S, 14S, 24S, 53S, and 61S. Mechanical properties of these alloys are given in Table 4. Some characteristics and uses are:

24S. Known for its high strength; obtainable in practically all wrought forms and can be made into strong extrusions of good ductility. Obtained in clad form, the alloy designation is Clad 24S.

14S. An alloy which can be made into high-strength extrusions with fair ductility, as well as into high-strength forgings.

17S. Used to make free-cutting screw-machine products.

53S. Produces extrusions of highest strength in this group of alloys, with fair ductility.

61S. Forming properties are very good; strength is fair. Resistance to corrosion is intermediate. With the exception of forgings, 61S is available in all wrought forms.

Aluminum Alloy Castings.[1] To obtain lowest over-all costs, several factors must be considered when choosing between aluminum alloy castings of the die, sand, and permanent-mold types. By recognizing these factors and the data in Tables 5 to 9, the engineer should obtain the proper alloy and casting to meet service conditions. The principal factors are:

1. Estimate the number and rate of production so that the most economical casting method can be chosen.

[1] Floyd A. Lewis, Aluminum Association, as reported in *Product Engineering*.

TABLE 5. COMPARISON OF SAND, PERMANENT-MOLD, AND DIE CASTINGS

	Sand Castings	Permanent-mold Castings	High-pressure Die Castings
Strength	2	1	2*
Structural density	2	1	3
Reproducibility of successive castings	3	2	1
Pressuretightness (after machining)	2	1	2
Cost per piece†	3	2	1
Speed of production†	3	2	1
Flexibility as to alloys	1	2	3
Tolerances	3	2	1
Flexibility of design	1	2	3
Size limitation	1	2	3
Surface	3	2	1
Speed of getting into production	1	2	2
Pattern or mold cost	1	2‡	3§
Thickness of section	3	2	1

Ratings 1, 2, and 3 indicate the relative advantages of the three methods for each factor, number 1 being the most advantageous.

* Thin die castings of uniform section may have first classification in the "as-cast" condition

† Although this rating covers the majority of castings, in some cases either sand or permanent-mold castings may take first place.

‡ The cost of permanent-mold equipment may be equal to or less than production sand-cast equipment for large or complicated castings adaptable only to these two processes.

§ The cost of dies plus the cost of tooling for machining or finishing is frequently less than the same over-all cost for sand or permanent-mold castings.

Table 5 shows the relative advantages of three casting methods on the basis of production cost and casting characteristics.

Sand casting is the most flexible method and is generally employed for large or intricate castings, and where the quantity will not justify construction of a metal mold or die. Sand castings weighing up to 7000 lb have been made.

Permanent-mold and die-casting processes are inherently mass-production methods, but as few as 500 pieces have been produced economically.

In general, permanent-mold castings have higher mechanical properties, greater uniformity, smoother surfaces, closer dimensional tolerances, and better pressure-tightness than sand castings. The smoother surfaces and closer dimensional tolerances mean lower machining and finishing costs.

Designs too intricate for the full permanent-mold process can be cast using iron molds and sand cores. This combination results in the "semipermanent" mold method. Permanent- and semipermanent-mold castings from a few ounces to 300 lb in weight are in regular production.

Ability to cast thin sections, accuracy, uniformity of reproduction, and low unit cost characterize the die-casting process. In reproduction of surface detail and surface finish, die casting is superior to other methods. Thinner sections possible with this method permit substantial savings in metal. Dimensional accuracy, smooth surfaces, and ability to cast threads and core holes contribute to minimum finishing cost. The weight range is from a fraction of an ounce up to about 20 lb. Die castings having over-all dimensions up to 36 by 12 by 9 in. are being produced, but the dimensions in one direction may exceed these, castings up to 84 in. in length having been produced.

(Continued on page 26-15)

TABLE 6. CLASSIFICATION OF COMMERCIAL ALUMINUM CASTING ALLOYS BY TYPE OF APPLICATION

Casting Method		
Die	Sand	Permanent Mold
General-purpose Alloys		
S12A SC54A SC84B SG100A	C4A CS43A SC64B SC64C ZG61A	CS42A SC64B SC64C
General-purpose Alloys Where Pressure-tightness Is Required		
SC84B SG100A	CS72A S5A SC51A SG70A	CS72A S5A SC64A SC51A SG70A
Alloys for Architectural and Decorative Purposes		
G8A S12A SG100A	S5A G4A GS42A SG70A ZC81A ZG32A GZ42A ZG61A ZG61B	S5A GS42A GZ42A SG70A ZC60A ZC81B ZG32A ZG42A
Alloys with High Resistance to Corrosion		
G8A S5C S12A SG100A	S5A S5B G4A G10A GS42A SG70A ZC81A ZG32A ZG42A ZG61A ZG61B	S5A S5B GZ42A SG70A ZC81B ZG32A ZG32A ZG42A
Alloys Retaining Strength at Elevated Temperatures		
SC84A SC84B SG100A	CG100A CN42A SC51A SC82A	CG100A CN42A SN122A SC51A CS104B SC122A
Piston Alloys		
SC84B	CG100A CN42A	CG100A CN42A CS66A SN122A SC122A

NOTE: When service temperatures are above 600 F, die castings may develop surface blisters.

TABLE 7. RELATIVE PROPERTIES AND CHARACTERISTICS OF COMMERCIAL ALUMINUM CASTING ALLOYS BY CASTING METHOD

Alloy (ASTM)	Corrosion Resistance	Casting Properties	Normally Heat-treated	Machin-ability	Suitability for	
					Welding	Brazing
Die-casting Alloys						
G8A	1	3	....	2		
S5C	2	3	....	3		
S12A, B	2	1	....	1		
SC54A, B	2	3	....	3		
SC84A, B	3	1	....	1		
SG100A, B	1	1	....	1		
Sand-casting Alloys						
C4A	2	2	Yes	1	2	No
CG100A	3	2	Yes	1	3	No
CN42A	2	3	Yes	1	3	No
CS43A	3	1	No	1	1	No
CS72A	3	1	No	1	2	No
G4A	1	3	No	1	3	No
G10A	1	3	Yes	1	3	No
GS42A	1	3	No	1	3	No
S5A	1	1	No	3	1	Limited
S5B	2	1	No	3	1	Limited
SC51A	2	1	Yes	2	1	No
SC64B	3	2	Yes	2	1	No
SC64C	2	1	Yes	2	1	No
SC82A	2	1	Yes	2	1	No
SG70A	1	1	Yes	2	1	No
ZC81A	1	3	Aged only	1	3	Yes
ZG32A	1	3	Aged only	1	3	Yes
ZG42A	1	3	Yes	1	3	Yes
ZG61A	1	3	Aged only	1	3	Yes
ZG61B	1	3	Aged only	1	3	Yes
Permanent-mold Alloys						
CG100A	3	2	Yes	1	3	No
CN42A	2	3	Yes	1	3	No
CS42A	2	2	Yes	2	3	No
CS104B	3	2	No	1	2	No
CS66A	3	2	Yes	2	3	No
CS72A	3	1	No	1	3	No
GS42A	1	3	No	1	3	No
GZ42A	1	3	No	1	3	No
S5A	1	1	No	3	1	Limited
S5B	2	1	No	3	1	Limited
SC51A	2	1	No	2	1	No
SC64A	3	1	No	2	1	No
SC64B	3	2	Yes	2	1	No
SC64C	2	1	Yes	2	1	No
SC122A	2	3	Yes	3	1	No
SG70A	1	1	Yes	2	1	No
SN122A	2	3	Yes	3	1	No
ZC60A	1	3	Aged only	1	3	Yes
ZC81B	1	3	Aged only	1	3	Yes
ZG32A	1	3	Aged only	1	3	Yes
ZG42A	1	3	Yes	1	3	Yes

NOTE: Alloy properties and characteristics are ranked within casting class on the basis of excellent 1, good 2, and fair 3.

TABLE 8. MECHANICAL PROPERTIES OF COMMERCIAL ALUMINUM CASTING ALLOYS (TYPICAL VALUES)

Alloy (ASTM)	Condition	Yield Strength, Psi	Ultimate Strength, Psi	Elongation % in 2 In.	Compression Yield Strength	Brinell 500-kg Load, 10-mm Ball	Shear Strength, Psi	Endurance Limit, Psi
Die-casting Alloys								
G8A	As cast	27,000	45,000	8.0	27,000	...	27,000	23,000
S5C	As cast	16,000	30,000	9.0	16,000	...	19,000	17,000
S12A	As cast	21,000	39,000	2.0	21,000	...	25,000	19,000
S12B	As cast	......	37,000	1.8				
SC54A	As cast	24,000	40,000	5.0	24,000	...	26,000	22,000
SC54B	As cast	22,000	38,000	5.5	22,000	...	25,000	22,000
SC84A	As cast	25,000	46,000	3.0	25,000	...	29,000	19,000
SC84B	As cast	26,000	45,000	2.0	26,000	...	29,000	20,000
SG100A	As cast	27,000	44,000	3.0	27,000	...	28,000	19,000
SG100B	As cast	......	43,000	3.0				
Sand-casting Alloys								
C4A	Solution heat-treated	16,000	32,000	8.5	16,000	60	24,000	6,000
C4A	Solution heat-treated and aged	30,000	40,000	2.0	38,000	95	31,000	7,000
CG100A	Aged	20,000	27,000	1.0	20,000	80	21,000	9,500
CG100A	Solution heat-treated and aged	30,000	40,000	0.5	43,000	115	29,000	8,500
CN42A	Aged	18,000	27,000	1.0	18,000	70	21,000	6,500
CN42A	Solution heat-treated and aged	25,000	28,000	2.0	......	75	24,000	9,500
CS43A	As cast	14,000	19,000	1.5				
CS72A	As cast	15,000	24,000	1.5	17,000	70	20,000	9,000
G4A	As cast	12,000	25,000	9.0	12,000	50	20,000	5,500
G10A	Solution heat-treated	25,000	46,000	14.0	26,000	75	33,000	7,000
GS42A	As cast	13,000	20,000	2.0	15,000	50	17,000	
S5A	As cast	9,000	19,000	6.0	10,000	40	14,000	6,500
S5B	As cast	9,000	19,000	6.0	10,000	40	14,000	6,500
SC51A	Artificially aged	23,000	28,000	1.5	24,000	65	22,000	7,000
SC51A	Solution heat-treated and aged	25,000	35,000	2.5	29,000	80	30,000	8,500
SC51A	Solution heat-treated and aged	29,000	35,000	1.5	......	75	......	10,000
SC64B	As cast	18,000	27,000	2.0	18,000	70	......	10,000
SC64B	Solution heat-treated and aged	24,000	36,000	2.0	24,000	80	......	10,000
SC64C	As cast	18,000	23,000					
SC64C	Solution heat-treated and aged	24,000	31,000	1.5				
SC82A	Solution heat-treated	20,000	33,000	2.9	20,000	82		
SC82A	Solution heat-treated and aged	28,000	36,000	2.0	28,000	81		
SG70A	Artificially aged	20,000	25,000	2.0	22,000	60	18,000	7,500
SG70A	Solution heat-treated and aged	24,000	33,000	4.0	24,000	70	27,000	8,000
ZC81A	Artificially aged	......	30,000	3.0				
ZG32A	As cast	13,000	29,000	12.0	......	50		

ZG32A..........	Artificially aged	19,000	35,000	9.0	......	65		
ZG42A..........	As cast	19,000	30,000	5.0	......	65		
ZG42A..........	Artificially aged	27,000	37,000	3.0	......	85		
ZG42A..........	Solution heat-treated	40,000	44,000	1.5	......	80		
ZG42A..........	Solution heat-treated and aged	38,000	43,000	1.5	......	80		
ZG61A..........	Artificially aged	25,000	32,000	3.0	25,000	75	26,000	9,000
ZG61B..........	As cast	......	32,000	2.0				
Permanent-mold Casting Alloys								
CG100A..........	Artificially aged	35,000	37,000	0.5	40,000	115	27,000	8,500
CG100A..........	Solution heat-treated and aged	36,000	48,000	0.5	36,000	140	30,000	9,000
CN42A..........	Artificially aged	34,000	40,000	1.0	34,000	105	26,000	10,500
CN42A..........	Solution heat-treated and aged	42,000	47,000	0.5	46,000	110	31,000	9,500
CS42A..........	Solution heat-treated	22,000	37,000	9.0	22,000	75	30,000	9,500
CS42A..........	Solution heat-treated and aged	26,000	40,000	5.0	26,000	90	32,000	
CS42A..........	Solution heat-treated and aged	20,000	39,000	4.5	20,000	80		
CS66A..........	Artificially aged	16,000	29,000	1.0	16,000	95	22,000	
CS72A..........	As cast	19,000	28,000	2.0	19,000	70	23,000	
CS104B..........	As cast	24,000	32,000	1.5	32,000	100	22,000	
GZ42A..........	As cast	16,000	27,000	7.0	17,000	60	22,000	
SC51A..........	Solution heat-treated and aged	27,000	43,000	4.0	27,000	90	30,000	9,000
SC64A..........	As cast	16,000	28,000	2.0	16,000	70	25,000	
SC64B..........	As cast	19,000	34,000	2.5	19,000	85	24,000	
SC64B..........	Solution heat-treated and aged	27,000	40,000	3.0	27,000	95		
SC64C..........	As cast	19,000	26,000					
SC82A..........	Solution heat-treated and aged	40,000	49,000	3.0	......	97		
SC122A..........	Solution heat-treated	27,000	41,000	2.5	27,000	90		
SC122A..........	Artificially aged	26,000	40,000	0.3	26,000	97		
SC122A..........	Solution heat-treated and aged	38,000	43,000	1.0	38,000	92		
SG70A..........	Solution heat-treated and aged	27,000	40,000	5.0	24,000	90		
SN122A..........	Artificially aged	28,000	36,000	0.5	28,000	105	24,000	
SN122A..........	Solution heat-treated and aged	43,000	47,000	0.5	43,000	125	27,000	
ZC60A..........	Aged	......	28,000	7.0				
ZC81B..........	As cast	......	33,000	3.0				
ZG32A..........	As cast	15,000	33,000	22.0	......	55		
ZG32A..........	Artificially aged	21,000	42,000	18.0	......	70		
ZG92A..........	As cast	24,000	42,000	14.0	......	75		
ZG42A..........	Artificially aged	29,000	47,000	8.0	......	95		
ZG42A..........	Solution heat-treated	36,000	49,000	9.5	......	95		
ZG42A..........	Solution heat-treated and aged	43,000	53,000	6.5	......	95		

1. Tension and hardness values determined from standard ½-in. dia specimens individually cast in sand and permanent molds, and tested without machining off the surface. Tension values for die-casting alloys determined from standard ¼-in. die-cast specimens; hardness values are not shown for die castings because they cannot be reliably determined.
2. Yield strength is the stress which produces a permanent set of 0.2% of the initial gage length (ASTM Standard Methods of Tension Testing).
3. Endurance limits are based on 500,000,000 cycles of completely reversed stress using the R. R. Moore type of machine and specimens.
4. Compression tests made on specimens having an l/r ratio of 12.

TABLE 9. COMPOSITION OF COMMERCIAL ALUMINUM CASTING ALLOYS

Alloy Designations			Nominal Composition, %, Balance Aluminum							
ASTM†	ALCOA	SAE	Copper	Silicon	Manganese	Magnesium	Zinc	Nickel	Titanium	Others
Die-casting Alloys										
G8A	218	...	0.2	0.3	0.3	3.0	0.1	0.1	....	0.1Sn
S5C	43	304	0.6	5.0	0.3	0.1	0.5	0.5	....	0.1Sn
S12A,*B	13	305	0.6	12.0	0.3	0.1	0.3	0.5	....	0.1Sn
SC54A, B*	85	...	3.5	5.0	0.5	0.1	0.9	0.5	....	0.3Sn
SC84A,B	380	...	3.5	8.5	0.5	0.1	0.9	0.5	....	0.3Sn
SG100A,B*	360	...	0.6	9.5	0.3	0.5	0.5	0.5	....	0.1Sn
Sand-casting Alloys										
C4A	195	38	4.5	1.5	0.3	0.03	0.3	...	0.2	
CG100A	122	34	10.0	2.0	0.5	0.25	0.4	0.3	0.2	
CN42A	142	39	4.0	0.7	0.3	1.50	0.3	2.0	0.2	0.02Cr
CS43A	108	380	4.0	3.0	0.5	0.05	1.0	0.3	0.2	
CS72A	C113	33	7.0	2.5	0.5	0.07	2.5‡	0.3	0.2	
G4A	214	320	0.1	0.3	0.3	4.0	0.1	...	0.2	
G10A	220	324	0.2	0.2	0.1	10.0	0.1	...	0.2	
GS42A	A214	...	0.3	1.8	0.8	4.0	0.3	...	0.2	0.2Cr
S5A§	43	35	0.1	5.0	0.3	0.05	0.3	...	0.2	
S5B§	43	35	0.3	5.0	0.3	0.05	0.3	...	0.2	0.2Cr
SC64B	319	326	3.7	6.3	0.5	0.1	1.0‡	0.5‡	0.2	
SC64C	....	...	3.7	6.3	0.8	0.5	1.0	0.5	0.2	
SC51A	355	322	1.3	5.0	0.5	0.5	0.3	...	0.2	0.2Cr
SC82A	....	327	1.5	7.8	0.4	0.4	1.0	0.2	0.2	0.3Cr
SG70A	356	323	0.2	7.0	0.3	0.3	0.3	...	0.2	
ZG61A	605	310	0.3	0.25	0.3	0.6	5.5	...	0.2	0.5Cr
SG61B	....	...	0.5	0.15	0.5	0.7	6.5	...	0.2	
ZC81A	....	...	0.7	0.25	0.6	0.35	7.5	0.1	0.2	0.3Cr
ZG32A	....	...	0.2	0.2	0.5	1.6	3.0	...	0.2	0.3Cr
ZG42A	....	...	0.2	0.2	0.5	2.1	4.25	...	0.2	0.3Cr
Permanent-mold Alloys										
CG100A‖	122	34	10.0	2.0						
CN42A‖	142	39	4.0	.....	....	1.50				
CS42A	B195	380	4.0	0.7	0.3	0.05	0.3	0.3	0.02	0.02Cr
CS104B	138	...	10.0	4.0	0.5	1.0	0.5	1.0	0.02	
CS72A‖	C113	33	7.0	2.5	....	.....	2.50			
CS66A	152	300	6.5	5.5	0.8	0.3	0.8	...	0.2	
GS42A‖	A214	...	....	1.8	....	4.0				
GZ42A	B214	...	0.1	0.3	0.3	4.0	1.8	...	0.2	
S5A‖	43	35	....	5.0						
S5B‖	43	35	....	5.0						
SC64A	A108	...	4.5	5.5	0.5	0.1	1.0	...	0.2	
SC64B‖	319	326	3.7	6.3	....	.....	1.0			
SC64C	....	...	3.7	6.3	....	.....	1.0			
SC51A‖	355	322	1.3	5.0						
SC122A	....	328	1.5	12.0	0.7	0.7	0.4	0.5	0.2	
SG70A‖	356	323	....	7.0						
SN122A	A132	321	1.0	12.0	0.1	1.1	0.01	2.5	0.02	
ZC60A	....	...	0.5	0.3	0.05	0.3	6.5	0.1	0.2	
ZC81B	....	...	0.7	0.25	0.6	0.35	7.5	0.1	0.2	
ZG32A	....	...	....	.....	....	1.6	3.0			
ZG42A‖	....	...	....	.....	....	2.1	4.25			

* Impurities in these die-casting alloys are more closely controlled than in the other die-casting alloys of otherwise identical composition. These alloys can be cast only in cold-chamber machines.

† From specifications of ASTM.

‡ Maximum values.

§ Alloys S5A and S5B are identical except that impurities in S5A are held to lower limits giving it greater resistance to corrosion. S5C is a similar alloy for die casting.

‖ Also used for sand casting.

2. Determine the mechanical and physical requirements of the part so that the proper alloy can be selected.

First decision is whether a heat-treated alloy is required. Heat-treatment increases the strength but often lowers ductility. Die-casting alloys are not usually heat-treated. In some cases, however, aging treatments are given at relatively low temperature.

Castings of lowest cost are obtained by keeping specifications as simple as possible. In other words, a heat-treated alloy should not be specified if a non-heat-treated one will do. An alloy having difficult casting characteristics should not be selected if one easier for the foundryman to handle will serve.

Tables 6 to 8 list data to aid in the selection of the proper alloy to meet physical requirements and service conditions. Table 9 shows approximate chemical composition of the aluminum casting alloys which are cross-referenced with respect to common industry designations.

3. Obtain bids from several foundries of good reputation. Both the total quantity and required rate of delivery must be specified, for these are important factors in determining costs.

4. Submit detailed toleranced drawings and a model or test pattern when available to each bidder. If a reorder, submit a sample casting as previously made. Each bidder should be given full information on service conditions.

5. Specify service conditions in detail, and indicate machining locating points. Points of high stress should be specified, as well as pressuretightness, if that is required. If the castings are to be used where corrosive influences are present, these conditions should be clearly and fully described. If temperatures at which the castings will function differ materially from ordinary room temperature, these must be clearly indicated.

6. Determine if prospective suppliers have the necessary equipment and facilities to assure the attainment of specified physical and mechanical properties consistently. One of the important matters is whether the prospective supplier has an adequate quality-control setup.

7. If the casting requires a high-strength alloy, make sure prospective suppliers have adequate heat-treating equipment with accurate temperature control.

Other Production Methods. A few parts can best be produced by other methods, such as the plaster-mold process, investment or "lost-wax" method, and centrifugal casting. Parts requiring smoother surfaces and greater accuracy than can be obtained with any of the three principal processes can be cast in plaster molds. The investment process is also capable of great accuracy and is sometimes used for small complicated parts that can be produced in groups or clusters. Aluminum parts of special character are cast by the centrifugal method.

MAGNESIUM ALLOYS

Light weight is probably the best known characteristic of magnesium. Aluminum is 1½ times heavier, iron and steel are four times heavier, and copper and nickel alloys are five times heavier. Pure metal has a specific gravity of only 1.74, and alloying does not increase the specific gravity to more than 1.83.

Magnesium can be cast and fabricated by practically every known method, and the alloys have excellent machinability. They possess relatively high thermal and electrical conductivities and are nonmagnetic. Coefficient of thermal expansion of magnesium alloys is approximately 0.0000145 in. per deg F in the temperature range of 70 to 212 F. Good thermal stability is obtained at temperatures up to 200 F. Modulus of elasticity is 6,500,000 psi, and modulus of rigidity or modulus of shear is

TABLE 10. PROPERTIES AND FORMS OF MAGNESIUM ALLOYS

Form	Dow-metal	Condition[a]	Nominal Composition, % (Mg the Remainder)					Physical and Mechanical Properties[b]										
			Aluminum	Zinc	Manganese	Rare Earths	Zirconium	Melting Point, F	Tensile Strength, 1000 Psi, Typical	Tensile Yield Strength, 1000 Psi, Typical	Elongation in 2 In., % Typical	Compressive Yield Strength, 1000 Psi, Typical	Shear, 1000 Psi	Bearing Strength, 1000 Psi: Ultimate	Bearing Strength, 1000 Psi: Yield	Hardness: Brinell	Hardness: Rockwell E
Sand and permanent-mold casting alloys	C	-F	9.0	2.0	0.10 min	...	....	1110	24	14	2	14	19	50	46	65	77
		-T4							40	14	10	14	19	68	46	63	75
		-T5							24	14	2	..	20	50	46		
		-T6							40	23	2	23	22	80	65	84	90
	G[f]	-F	10.0	...	0.10 min	...	....	1100	22	12	2	12	17	..	..	54	65
		-T4							40	13	10	13	19	..	..	52	62
		-T6							40	16	4	16	21	..	..	69	80
		-T61							40	22	1	22					
	H	-F	6.0	3.0	0.15 min	...	....	1135	29	14	6	14	18	60	40	50	59
		-T4							40	14	12	14	18	60	44	55	66
		-T5							29	14	5	..	19	60	40		
		-T6							40	19	5	19	21	75	52	73	83
	AZ91C	-F	8.7	0.7	0.13 min	...	....	1120	24	14	2	14	18	50	30	52	62
		-T4							40	14	11	14	18	50	36	55	66
		-T6							40	19	4	19	21	65	45	73	83
	EK30A	-T6	....	0.3 max	....	3.0	0.35	....	23	16	3	16	..	..	..	45	49
	EK41A	-T5	....	0.3 max	....	4.0	0.7	....	23	16	1	16	..	..	..	50	59
		-T6							25	18	2	18	..	..	..	50	59
	EZ33A	-T5	....	2.7	....	3.0	0.7	....	23	16	..	16	..	..	..	50	59
Die-casting alloys	R and RC	-F	9.0	0.6	0.13 min	...	....	1120	33	22	3	22	20	..	..	60	72
Sheet[d]	FS1 and FS1W	-F	3.0	1.0	....	...	....	1160	37	22	21	13	21				
		-H24							42	32	15	26	23	72	47	73	83
		-O							37	22	21	16	21	70	42	56	67
	M	-F	....	...	1.2	...	....	1200	33								
		-H24							36	27	7	18	17	..	..	56	67
		-O							34	17	16	11	18	56	28	48	55

Form	Dow-metal	Condition[a]	Nominal Composition, % (Mg the Remainder)					Physical and Mechanical Properties[b]									
			Aluminum	Zinc	Manganese	Rare Earths	Zirconium	Melting Point, F	Tensile Strength, 1000 Psi, Typical	Tensile Yield Strength,[c] 1000 Psi, Typical	Elongation in 2 In., % Typical	Compressive Yield Strength, 1000 Psi, Typical	Shear, 1000 Psi	Bearing Strength, 1000 Psi		Hardness	
														Ultimate	Yield	Brinell	Rockwell E
Extruded bars, rods, and solid shapes[e]	FS and FS1	-F	3.0	1.0	....	...	....	1160	38	29	15	14	19	56	33	49	57
	J1	-F	6.5	1.0	....	...	....	1145	45	33	16	19	20	68	41	60	72
	M	-F	....	...	1.2	...	....	1200	37	26	11	12	18	51	28	44	45
	O1	-F -T5	8.5	0.5	....	...	....	1130	49 55	36 40	11 7	.. 35	22 24	68 ..	48 ..	60 82	77 88
	ZK60A	-F -T5	....	5.7	....	...	0.55	1175	49 53	38 44	14 11	33 36	27 26	80 83	57 60	75 82	84 88
Extruded hollow shapes and tubing	FS and FS1	-F	3.0	1.0	....	...	....	1160	36	24	16	12	..	..	..	46	51
	J1	-F	6.5	1.0	....	...	....	1145	41	24	14	16	..	..	..	50	60
	M	-F	....	...	1.2	...	....	1200	35	21	10	9	..	..	..	42	41
	ZK60A	-F -T5	....	5.7	....	...	0.55	1175	46 50	34 40	12 11	25 29	.. ..	.. ..	.. ..	75 82	84 88

[a] Condition -T4 = solution heat-treated. -T6, -T61 = solution heat-treated and artificially aged. -O = annealed. -H24 = hard rolled. -F = as fabricated. -T5 = artificially aged.

[b] Unless otherwise indicated, values are typical.

[c] Yield strength is defined as the stress at which the stress-strain curve deviates 0.2% from the modulus line.

[d] Properties apply to following thickness ranges: FSI-F and M-F = 0.080 to 0.500 in.; FSI-H24 = 0.065 to 0.250 in.; FSI-O = 0.064 to 0.250 in.; M-H24 and M-O = 0.020 to 0.250 in.

[e] The properties for bars, rods, and solid shapes apply to extrusions with a minimum dimension of 1/4 in. to a 1½-in. maximum. ZK60A properties are based on sections less than 2 sq in. in cross-sectional area.

[f] Permanent-mold casting alloy only.

TABLE 11. PRODUCT SPECIFICATIONS FOR MAGNESIUM ALLOYS

Form	Dow-metal	ASTM		Form	Dow-metal	ASTM	
		Spec.	Alloy			Spec.	Alloy
Sand castings	A	B80-51T	AM80A	Extruded tubing	FS1	B217-49T	AZ31B
	C	B80-51T	AZ92A		J1	B217-49T	AZ61A
	G	B80-51T	AM100A		M	B217-49T	M1A
	H	B80-51T	AZ63A				
	M	B80-51T	M1B	Sheet	FS1	B90-51T	AZ31A
	AZ91C	B80-51T	AZ91C		M	B90-51T	M1A
Permanent-mold castings	C	B199-51T	AZ92A	Forgings*	FS1	B91-49T	TA54A
	G	B199-51T	AM100A		J1	B91-49T	AZ31B
					M	B91-49T	AZ61A
Die castings	R	B94-52	AZ91A		O1	B91-49T	AZ80A
	RC	B94-52	AZ91B				
				Welding rod	C		
Extruded round rods, bars, and solid shapes	FS1	B107-49T	AZ31B		J1		
	J1	B107-49T	AZ61A		M		
	M	B107-49T	M1A				
	O1	B107-49T	AZ80A				
	ZK60A						

* Forgings should be made the subject of special inquiry.

2,400,000 psi. Value of Poisson's ratio for magnesium alloys is 0.35. Product specifications are given in Table 11.

Bearing strengths for magnesium alloys are given in Table 10. The bearing yield strength is defined as the stress at which the stress-hole-elongation curve deviates 2% of the hole diameter from the initial straight-line portion of the curve.

WROUGHT COPPER ALLOYS

Copper-base alloys constitute one of the most useful groups of nonferrous materials—primarily for corrosion resistance, but also for strength, ductility, spring properties, etc. They are called brasses, bronzes, aluminum bronzes, nickel silvers, and a legion of other names, proprietary and those bestowed by custom or association over a long period, that is, gilding metal, admiralty bronze, etc. Thirty-nine standard compositions are recognized by the Copper and Brass Research Association (see Tables 12 to 14).

Brasses. When copper is alloyed with zinc in various proportions the resulting material is known as "brass," which is stronger than either constituent. As the copper content decreases, the color changes from a rich bronze (90-10) to yellow (70-30) and finally to yellow-red (60-40). Alloys with more than 63% copper are the "alpha" type and are suitable for cold-working, such as cupping, drawing, stamping, and cold forging. Alloys in the low copper range, 60 ± 2% copper, are best suited for hot extrusion and hot forging. Brasses of the "beta" type with less than 58% copper have limited applications.

Leaded Brasses. The addition of lead to brass aids machinability but impairs the cold-working, hot-working, and welding properties in most instances (see Tables 13 and 14).

Bronzes. When alloying elements, with or without zinc, are added to copper, the product may be known as a "bronze," because of color, but the characteristics are so changed that exact specifications, not the word "bronze," should be used.

Table 12. Compositions of Wrought Copper Alloys

Name	Composition, %, Nominal	Form Most Commonly Ordered				
		Rolled Flat	Rod and Drawn Flat	Shapes	Wire (Except Flat)	Tube
Coppers						
Electrolytic tough pitch	Cu 99.92, O 0.04	X	X	X	X	X
Deoxidized	Cu 99.94, P 0.02	X	X	X		X
Plain Brasses, Nonleaded						
Gilding, 95 %	Cu 95, Zn 5	X	X		X	
Commercial bronze, 90 %	Cu 90, Zn 10	X	X		X	X
Red brass, 85 %	Cu 85, Zn 15	X	X		X	X
Low brass, 80 %	Cu 80, Zn 20	X	X		X	
Cartridge brass, 70 %	Cu 70, Zn 30	X	X		X	X
Yellow brass (flat products)	Cu 65, Zn 35	X	X			
Yellow brass (wire)	Cu 65.3, Zn 34.7	X	X		X	
Muntz metal	Cu 60, Zn 40	X	X		X	X
Leaded Brasses						
Leaded commercial bronze	Cu 89, Pb 1.75, Zn 9.25		X	X		
Low-leaded brass (tube)	Cu 67, Pb 0.5, Zn 32.5					X
Low-leaded brass	Cu 64.5, Pb 0.5, Zn 35	X				
Medium-leaded brass	Cu 64.5, Pb 1, Zn 34.5	X	X	X	X	
High-leaded brass (tube)	Cu 67, Pb 1.6, Zn 31.4					X
High-leaded brass	Cu 62.5, Pb 1.75, Zn 35.75	X	X			
Extra-high-leaded brass	Cu 62.5, Pb 2.5, Zn 35	X				
Free-cutting brass	Cu 61.5, Pb 3, Zn 35.5		X	X		
Leaded Muntz metal	Cu 60, Pb 0.5, Zn 39.5	X				
Free-cutting Muntz metal	Cu 60.5, Pb 1.1, Zn 38.4					X
Forging brass	Cu 60, Pb 2, Zn 38	X	X	X		
Architectural bronze	Cu 57, Pb 3, Zn 40		X	X		
Tin, Aluminum Brasses						
Admiralty	Cu 71, Sn 1, Zn 28	X			X	X
Naval brass	Cu 60, Zn 0.75, Zn 39.25	X	X	X	X	X
Leaded naval brass	Cu 60, Zn 0.25, Pb 1.75, Zn 37.5					
Manganese bronze (A)	Cu 58.5, Zn 1, Fe 1 Mn 0.3, Zn 39.2	X	X X	X X	X	
Aluminum brass	Cu 76, Al 2, Zn 22					X
Phosphor Bronzes (Tin Bronzes)						
Phosphor bronze, 5 % (A)	Cu 95, Sn 5, P 0.03–0.35	X	X		X	X
Phosphor bronze, 8 % (C)	Cu 92, Zn 8, P 0.03–0.35	X	X		X	
Phosphor bronze, 10 % (D)	Cu 90, Sn 10, P 0.03–0.25	X	X		X	
Phosphor bronze, 1.25 % (E)	Cu 98.75, Sn 1.25, P trace	X				

TABLE 12. COMPOSITIONS OF WROUGHT COPPER ALLOYS (*Continued*)

Name	Composition, %, Nominal	Form Most Commonly Ordered				
		Rolled Flat	Rod and Drawn Flat	Shapes	Wire (Except Flat)	Tube
Cupronickel and Nickel Silvers						
Cupronickel, 30%	Cu 70, Ni 30	X	X			X
Nickel silver, 18% (A)	Cu 65, Ni 18, Zn 17	X	X		X	
Nickel silver, 18% (B)	Cu 55, Ni 18, Zn 27	X	X		X	
Nickel silver 65-15	Cu 65, Ni 15, Zn 20	X				
Nickel silver 65-12	Cu 65, Ni 12, Zn 23	X	X		X	
Nickel silver 65-10	Cu 65, Ni 10, Zn 25	X	X		X	
Silicon Bronzes (Copper-silicon Alloys)						
High-silicon bronze (A)	Traces of six metals Cu 94.8 min, Si 3	X	X	X	X	X
Low-silicon bronze (B)	Traces of six metals Cu 96 min, Si 1.5	X	X	X	X	X

Effects of alloying elements are:

TIN. Improves corrosion resistance, increases strength, modifies the color of copper-zinc alloys.

ALUMINUM. Improves resistance to impingement corrosion from turbulent water containing entrapped air (used for condenser tubes), and improves strength.

SILICON. Lowers thermal and electrical conductivity, makes brass suitable for spot welding, and reduces fuming of bronze welding rod. When added to copper, silicon increases strength and toughness; when added to aluminum bronze, it increases strength, corrosion resistance, and machinability.

IRON. Increases strength and hardness, retards grain growth during annealing, but must be limited in brass strip intended for drawing, to avoid adverse effects on ductility.

NICKEL. Whitens brass to a silvery color, increases strength, decreases electrical conductivity for high-resistance purposes.

BERYLLIUM. Produces alloys that have nonaging characteristics in springs and can be precipitation-hardened.

Machinability Ratings. Free-cutting brass (copper 60.0 to 63.0, lead 2.50 to 3.75, zinc remainder) is given the top rating of 100, and each alloy is compared with this standard. For ease of handling copper and copper-base alloys, it is possible to group them into three main divisions with regard to machinability.

GROUP 1: THE FREE-CUTTING ALLOYS. Usually brasses to which a small percentage of lead has been added. This group includes those alloys with a machinability rating of 70 to 100 as given in Table 13. These alloys are most frequently used in rod form for the production of screw-machine parts and produce short brittle chips.

GROUP 2: READILY MACHINABLE ALLOYS. With a machinability rating of 30 or more, but less than 70, include the group of nonleaded brasses from Muntz metal or naval brass with 60% copper, to red brass with 85% copper in the alloy.

GROUP 3: ALLOYS WITH MACHINABILITY RATING OF 20. This group includes non-

Table 13. Fabricating Properties of Wrought Copper Alloys

Name	Machinability Rating	Fabricating Properties			
		Cold-working Properties	Hot-working Properties	Hot-working Temp, F	Annealing Temp, F
Coppers					
Electrolytic tough pitch	20	Excellent	Excellent	1400–1600	700–1200
Deoxidized	20	Excellent	Excellent	1400–1600	700–1200
Plain Brasses, Nonleaded					
Gilding, 95%	20	Excellent	Good	1400–1600	800–1450
Commercial bronze, 90%	20	Excellent	Good	1400–1600	800–1450
Red brass, 85%	30	Excellent	Good	1450–1650	800–1350
Low brass, 80%	30	Excellent	Fair	1500–1650	800–1300
Cartridge brass, 70%	30	Excellent	Fair	1350–1550	800–1400
Yellow brass (flat products)	30	Excellent	Poor	....	800–1300
Yellow brass (wire)	30	Excellent	Poor	....	800–1300
Muntz metal	40	Fair	Excellent	1150–1450	800–1100
Leaded Brasses					
Leaded commercial bronze	80	Good	Poor	....	800–1200
Low-leaded brass (tube)	60	Excellent	Poor	....	800–1200
Low-leaded brass	60	Good	Poor	....	800–1300
Medium-leaded brass	70	Good	Poor	....	800–1200
High-leaded brass (tube)	80	Fair	Poor	....	800–1200
High-leaded brass	90	Fair	Poor	....	800–1100
Extra-high-leaded brass	100	Poor	Fair	1300–1450	800–1100
Free-cutting brass	100	Poor	Fair	1300–1450	800–1100
Leaded Muntz metal	60	Fair	Excellent	1150–1450	800–1100
Free-cutting Muntz metal	70	Fair	Excellent	1150–1450	800–1100
Forging brass	80	Fair	Excellent	1200–1500	800–1100
Architectural bronze	90	Poor	Excellent	1150–1350	800–1100
Tin, Aluminum Brasses					
Admiralty	30	Excellent	Fair	1200–1450	800–1100
Naval brass	30	Fair	Excellent	1200–1500	800–1100
Leaded naval brass		Poor	Good	1200–1400	800–1100
Manganese bronze (A)	70	Poor	Excellent	1150–1450	800–1100
	30				
Aluminum brass	30	Excellent	Fair	1400–1600	800–1100
Phosphor Bronzes (Tin Bronzes)					
Phosphor bronze, 5% (A)	20	Excellent	Poor	....	900–1250
Phosphor bronze, 8% (C)	20	Good	Poor	....	900–1250
Phosphor bronze, 10% (D)	20	Good	Poor	....	900–1250
Phosphor bronze, 1.25% (E)	20	Excellent	Good	1450–1600	900–1200
Cupronickel and Nickel Silvers					
Cupronickel, 30%	20	Good	Good	1700–1900	1200–1500
Nickel silver, 18% (A)	20	Excellent	Poor	....	1100–1500
Nickel silver, 18% (B)	30	Good	Poor	....	1100–1500
Nickel silver 65-15	20	Excellent	Poor	....	1100–1500
Nickel silver 65-12	20	Excellent	Poor	....	1100–1500
Nickel silver 65-10	20	Excellent	Poor	....	1100–1400
Silicon Bronzes (Copper-silicon Alloys)					
High-silicon bronze (A)	30	Excellent	Excellent	1300–1600	900–1300
Low-silicon bronze (B)	30	Excellent	Excellent	1300–1600	900–1250

TABLE 14. WELDING CHARACTERISTICS OF WROUGHT COPPER ALLOYS

Name	Suitability for Being Joined by				
	Soft Soldering	Silver Alloy Brazing	Oxyacetylene Welding	Carbon-arc Welding	Resistance Welding
Coppers					
Electrolytic tough pitch	Excellent	Good	Poor	Fair	Poor
Deoxidized	Excellent	Excellent	Fair	Good	Poor
Plain Brasses, Nonleaded					
Gilding, 95%	Excellent	Excellent	Fair	Good	Poor
Commercial bronze, 90%	Excellent	Excellent	Good	Good	Poor
Red brass, 85%	Excellent	Excellent	Good	Good	Poor
Low brass, 80%	Excellent	Good	Good	Fair	Poor
Cartridge brass, 70%	Excellent	Good	Good	Fair	Fair
Yellow brass (flat products)	Excellent	Good	Good	Fair	Fair
Yellow brass (wire)	Excellent	Good	Good	Fair	Fair
Muntz metal	Excellent	Good	Good	Fair	Fair
Leaded Brasses					
Leaded commercial bronze	Excellent	Fair	Fair	Fair	Poor
Low-leaded brass (tube)	Excellent	Good	Fair	Fair	Fair
Low-leaded brass	Excellent	Good	Fair	Fair	Fair
Medium-leaded brass	Excellent	Good	Fair	Fair	Poor
High-leaded brass (tube)	Excellent	Good	Fair	Fair	Poor
High-leaded brass	Excellent	Good	Fair	Fair	Poor
Extra-high-leaded brass	Excellent	Good	Fair	Poor	Poor
Free-cutting brass	Excellent	Good	Fair	Poor	Poor
Leaded Muntz metal	Excellent	Good	Fair	Poor	Poor
Free-cutting Muntz metal	Excellent	Good	Fair	Poor	Poor
Forging brass	Excellent	Good	Fair	Poor	Poor
Architectural bronze	Excellent	Good	Fair	Poor	Poor
Tin, Aluminum Brasses					
Admiralty	Excellent	Good	Good	Fair	Fair
Naval brass	Excellent	Good	Good	Fair	Fair
Leaded naval brass	Excellent	Good	Fair	Poor	Poor
Manganese bronze (A)	Excellent	Good	Good	Fair	Good
Aluminum brass	Good	Good	Good	Good	Good
Phosphor Bronzes (Tin Bronzes)					
Phosphor bronze, 5% (A)	Excellent	Good	Good	Good	Good
Phosphor bronze, 8% (C)	Excellent	Good	Good	Good	Excellent
Phosphor bronze, 10% (D)	Excellent	Good	Good	Good	Excellent
Phosphor bronze, 1.25% (E)	Excellent	Excellent	Good	Good	Fair
Cupronickel and Nickel Silvers					
Cupronickel, 30%	Excellent	Excellent	Good	Fair	Excellent
Nickel silver, 18% (A)	Excellent	Excellent	Good	Fair	Excellent
Nickel silver, 18% (B)	Excellent	Excellent	Good	Poor	Excellent
Nickel silver 65-15	Excellent	Excellent	Good	Fair	Excellent
Nickel silver 65-12	Excellent	Excellent	Good	Fair	Excellent
Nickel silver 65-10	Excellent	Excellent	Good	Fair	Excellent
Silicon Bronzes (Copper-silicon Alloys)					
High-silicon bronze (A)	Excellent	Excellent	Excellent	Excellent	Excellent
Low-silicon bronze (B)	Excellent	Excellent	Excellent	Excellent	Excellent

leaded coppers, nickel silvers, and phosphor bronzes. These usually produce long, tough, stringy turnings.

These machinability ratings are arbitrary values to be considered only as an approximate indication. They are a reasonable guide to relative tool life and the amount of power required for cutting. The demarcation line between the groups is not clear-cut.

As the machinability of the copper alloys decreases, rake and clearance angles on the cutting tools increase, as well as surface speed. In group 1, where the chip is brittle, it is necessary to reduce the rake angles to prevent "hogging."

Temper of Flat-drawn Products. Stiffness, hardness, and strength of wrought metals correspond to the amount of cold-working after the last anneal. In standard mill practice, the reduction by rolling is calculated on the basis of thickness only, because the metal does not widen very much during the rolling operation. The degree of rolling is expressed in Brown & Sharpe gage numbers, which have a definite relationship. For example, in the B&S system, the gages are reduced 50% for every six numbers. The following standards have been adopted for temper by the brass industry for sheet and strip:

Temper	Rolled B&S	% Reduction in Thickness
Quarter hard	1 B&S No. hard	10.95
Half hard	2 B&S No. hard	20.7
Three-quarter hard	3 B&S No. hard	29.4
Hard	4 B&S No. hard	37.1
Extra hard	6 B&S No. hard	50.15
Spring	8 B&S No. hard	60.5
Extra spring	10 B&S No. hard	68.65

Tensile strengths and Rockwell hardness of various kinds of brass sheets are given in Table 15.

Temper of Drawn Wire. When wire is drawn through a die, the percent reduction is based on the difference in cross-sectional areas. Here B&S gage numbers are used to represent reduction to indicate the temper of drawn material.

Temper	Nominal Reduction B&S Gage Nos.	% Reduction Cross-sectional Area
Eighth hard	½	10.9
Quarter hard	1	20.7
Half hard	2	37.1
Three-quarter hard	3	50.0
Hard	4	60.5
Extra hard	6	75.0
Spring	8	84.4

TABLE 15. TENSILE STRENGTH AND HARDNESS OF BRASS SHEET BY ANALYSIS AND TEMPER*

Rolled Temper	Tensile Strength Psi		Approximate Rockwell Hardness†					
			B Scale		F Scale		Superficial 30 T	
	Min	Max	Min	Max	Min	Max	Min	Max
ALLOY 1 (95 Cu 5 Zn)								
Quarter hard	37,000	47,000	20	52	69‡	88‡	29	51
Half hard	42,000	52,000	40	60	82‡	95‡	43	56
Three-quarter hard	46,000	56,000	50	64	...	....	50	60
Hard	50,000	59,000	57	67	...	....	54	62
Extra hard	56,000	64,000	64	72	...	....	60	65
Spring	60,000	68,000	68	75	...	....	62	67
Extra spring	61,000	69,000	69	76	...	....	63	68
ALLOY 2 (90 Cu 10 Zn)								
Quarter hard	40,000	50,000	27	56	70‡	90‡	34	54
Half hard	47,000	57,000	50	66	88‡	98‡	50	61
Three-quarter hard	52,000	62,000	59	71	...	....	56	64
Hard	57,000	66,000	65	75	...	....	60	67
Extra hard	64,000	72,000	72	79	...	....	65	70
Spring	69,000	77,000	76	81	...	....	68	71
Extra spring	72,000	80,000	78	83	...	....	69	72
ALLOY 3 (85 Cu 15 Zn)								
Quarter hard	44,000	54,000	33	62	78‡	94‡	38	58
Half hard	51,000	61,000	56	71	92‡	102‡	54	64
Three-quarter hard	57,000	67,000	66	76	...	....	60	68
Hard	63,000	72,000	72	80	...	....	65	70
Extra hard	72,000	80,000	78	85	...	....	69	74
Spring	78,000	86,000	82	87	...	....	72	75
Extra spring	82,000	90,000	84	89	...	....	73	76
ALLOY 4 (80 Cu 20 Zn)								
Quarter hard	48,000	58,000	38	65	80‡	95‡	42	60
Half hard	55,000	65,000	59	73	93‡	102‡	56	66
Three-quarter hard	61,000	71,000	69	79	...	....	63	70
Hard	68,000	77,000	76	84	...	....	68	73
Extra hard	78,000	87,000	83	89	...	....	72	76
Spring	85,000	93,000	87	92	...	....	75	78
Extra spring	89,000	97,000	88	93	...	....	76	79
ALLOY 6 (70 Cu 30 Zn)								
Quarter hard	49,000	59,000	40	65	...	....	43	60
Half hard	57,000	67,000	60	77	...	....	56	68
Three-quarter hard	64,000	74,000	72	82	...	....	65	72
Hard	71,000	81,000	79	86	...	....	70	74
Extra hard	83,000	92,000	85	91	...	....	74	77
Spring	91,000	100,000	89	93	...	....	76	78
Extra spring	95,000	104,000	91	95	...	....	77	79
ALLOY 8 (65 Cu 35 Zn)								
Quarter hard	49,000	59,000	40	65	...	....	43	60
Half hard	55,000	65,000	57	74	...	....	54	66
Three-quarter hard	62,000	72,000	70	80	...	....	65	71
Hard	68,000	78,000	76	84	...	....	68	73
Extra hard	79,000	89,000	83	89	...	....	73	76
Spring	86,000	95,000	87	92	...	....	75	78
Extra spring	90,000	99,000	88	93	...	....	76	79

* ASTM B 36-1952.

† Rockwell hardness values apply as follows: The B and F scales apply to metal 0.020 in. in thickness and over. The 30-T scale applies to metal 0.012 in. in thickness and over.

‡ The Rockwell B scale is preferred for testing material in these tempers.

NOTE: Plate is generally available in only the soft, quarter-hard, and half-hard tempers. Required properties for other tempers shall be agreed upon between the manufacturer and the purchaser at the time of placing the order.

Typical Uses of Copper Alloys

ELECTROLYTIC TOUGH-PITCH COPPER. Gutters, roofing, screening, bus bars, conductivity wire, electrical contacts, radio parts, switches, terminals, anodes, nails, rivets, soldering copper, and tacks.

DEOXIDIZED COPPER. Gas lines, heater lines and units, oil-burner tubes, plumbing pipe and tube, condenser evaporator and heat-exchanger tubes, dairy tubes, distiller tubes, steam and water lines, air, gasoline, hydraulic, and oil lines.

GILDING, 95%. Coins, medals, tokens, bullet jackets, firing-pin support shells, fuse caps, primers, emblems, jewelry, plaques, base for gold plate and vitreous enamel.

COMMERCIAL BRONZE, 90%. Etching bronze, grillwork, screen cloth, weather stripping, compacts, marine hardware, rivets, screws, screw shells, costume jewelry, ornamental strip, screen wire.

RED BRASS, 85%. Weather strip, conduit, screw shells, electrical sockets, eyelets, fasteners, fire extinguishers, condenser and heat-exchange tubes, flexible hose, pickling crates, plumbing pipe, service lines, traps, badges, costume jewelry, dials, etched articles.

LOW BRASS, 80%. Ornamental metalwork, medallions, battery caps, bellows, musical instruments, clock dials, flexible hose, pump lines, tokens.

CARTRIDGE BRASS, 70%. Radiator cores and tanks, reflectors, bead chain, flashlight shells, lamp fixtures, socket shells, screw shells, eyelets, fasteners, pins, rivets, spring tubes, ammunition components.

YELLOW BRASS. Grillwork, radiator cores, reflectors, electrical socket shells, stencils, plumbing accessories, sink strainers, pins, rivets, screws, springs.

MUNTZ METAL. Architectural trimming, large nuts and bolts, brazing rod, condenser plates, condenser, evaporator and heat-exchanger tubes, hot forgings, valve stems.

LEADED COMMERCIAL BRONZE. Screws, screw-machine parts, pickling crates.

LOW-LEADED BRASS (TUBE). Used where some degree of machinability is required, together with modern cold-working properties, plumbing J-bends, pipe, pump lines, trap tubes.

LOW-LEADED BRASS. Butts, hinges, watch backs.

MEDIUM-LEADED BRASS. Butts, gears, nuts, rivets, screws, dials, engravings, instrument plates.

HIGH-LEADED BRASS (TUBE). General-purpose screw-machine products.

HIGH-LEADED BRASS. Clock plates and nuts, clock and watch backs, clock gears, wheels, channel plate.

EXTRA-HIGH-LEADED BRASS. Clock plates and nuts, clock and watch backs, clock gears and wheels, channel plate.

FREE-CUTTING BRASS. Gears, pinions, automatic high-speed screw-machine parts.

LEADED MUNTZ METAL. Condenser-tube plates.

FREE-CUTTING MUNTZ METAL. Automatic screw-machine products.

FORGING BRASS. Forgings and pressings of all kinds.

ARCHITECTURAL BRONZE. Architectural trim, butts, hinges, lock bodies, forgings.

ADMIRALTY. Condenser, evaporator and heat-exchanger tubes, condenser-tube plates, distiller tubes, ferrules.

NAVAL BRASS. Aircraft turnbuckle barrels, balls, bolts, marine hardware, nuts, propeller shafts, rivets, structural uses, valve stems, condenser plates, welding rod.

LEADED NAVAL BRASS. Marine hardware, screw-machine products, valve stems.

MANGANESE BRONZE (A). Clutch disks, pump rods, shafting balls, valve stems, bodies, welding rods.

ALUMINUM BRASS. Condenser, evaporator and heat-exchanger tubes, distiller tubes, ferrules.

PHOSPHOR BRONZE, 5% (A). Bridge bearing plates, bellows, beater bars, clutch disks, cotter pins, diaphragms, fuse clips, fasteners, lock washers, sleeve bushings, springs, switch parts, truss wire, wire brushes, chemical hardware, perforated sheets, textile machinery, welding rods.

PHOSPHOR BRONZE, 8% (C). For identical use as 5% (A), but for more severe service conditions.

PHOSPHOR BRONZE, 10% (D). Heavy bars and plates for severe compression, good wear and corrosion resistance; bridge expansion plates and fittings, and articles requiring extra spring qualities, greatest resiliency, particularly in fatigue.

PHOSPHOR BRONZE, 1.25% (E). Electrical contacts, flexible hose, pole-line hardware.

CUPRONICKEL, 30%. Condensers, condenser plates, distiller tubes, evaporator and heat-exchanger tubes, ferrules.

NICKEL SILVER, 18% (A). Rivets, screws, table flatware, truss wire, zippers, optical bows, camera parts, core bars, base for silver plate, costume jewelry, etching stock, hollow ware, nameplates, radio dials.

NICKEL SILVER, 18% (B). Optical goods, springs, resistance wires.

HIGH-SILICON BRONZE (A). Hydraulic pressure lines, bolts, butts, clamps, cotter pins, hinges, marine hardware, nails, nuts, pole-line hardware, screws, hot-water tanks, bearing plates, bushings, cable, channels, chemical equipment, heat-exchanger tubes, piston rings, screen cloth and wire, propeller shafts (marine).

LOW-SILICON BRONZE (B). Hydraulic pressure lines, anchor screws, bolts, cable clamps, cap and machine screws, marine hardware, nuts and rivets.

BERYLLIUM COPPER

Beryllium copper has the highest strength and the greatest wear resistance of any nonmagnetic material. In addition, it offers good electrical and thermal conductivity, plus the corrosion resistance of deoxidized copper. Moreover, the alloy can be severely formed and heat-treated to give good spring properties. Since the material is nonsparking, it is used for tools used in working about gas equipment.

As compared with a steel spring, for example, the beryllium spring retains its desired characteristics at normal temperatures for a much greater length of time—it does not lose its recovery characteristics after being stretched the designed amount. However, when service temperatures exceed 300 F, relaxation may occur in spring members, so that some allowance must be made in design stresses. On the other hand, beryllium copper has increased elastic properties at subzero temperatures with no loss of shock resistance. These properties make the alloy particularly useful in aircraft instruments, etc., as well as in commercial applications where reliability of spring members is more important than increase in cost over more common spring metals.

Current ASM specifications cover an alloy with 1.80 to 2.05% beryllium; nickel or cobalt, or both, 0.20% minimum; nickel plus cobalt plus iron, 0.60% maximum; and the balance copper. The nominal beryllium content is 1.95%. Forms available are plate, sheet, strip, bar, rod and wire, forgings, and castings.

Tempers. Beryllium copper is supplied solution-treated, or solution-treated and cold-worked, ready for fabrication and precipitation-hardening by the user. The four tempers of commercially supplied material are: solution annealed—A; quarter hard—¼ H; half hard—½ H; and hard—H. When these materials have been precipitation-hardened, the suffix T is added to the temper designation. See Tables 16 and 17 for mechanical properties of commercially supplied and of heat-treated beryllium copper. Table 18 will aid you in selection of the temper for various applications while Table 19 gives physical properties of heat-treated strip.

Design Considerations. The engineering properties of beryllium copper shown in the tables result from tests conducted under inspection or laboratory procedures. Some care should be exercised in attempting to incorporate these values in design work.

For static loads, the stress will be limited by the amount of offset or permanent

TABLE 16. TYPICAL MECHANICAL PROPERTIES OF BERYLLIUM COPPER STRIP AS SUPPLIED

	Solution Annealed A	Quarter Hard ¼ H	Half Hard ½ H	Hard H
Tensile strength, psi	60,000–78,000	75,000–88,000	85,000–100,000	100,000–120,000
Proportional limit (0.002% offset), psi	15,000–20,000	40,000–60,000	55,000– 70,000	70,000– 85,000
Yield strength (0.01% offset), psi	19,000–30,000	47,000–67,000	64,000– 76,000	82,000– 99,000
Yield strength (0.2% offset), psi	28,000–36,000	60,000–80,000	75,000– 90,000	100,000–112,000
Yield strength (0.5% extension), psi	30,000–37,000	63,000–79,000	75,000– 90,000	90,000–102,000
Tension modulus, psi	17,000,000	17,000,000	17,000,000	17,000,000
Elongation, %	35–60	10–30	5–15	2–7
Rockwell hardness:				
Strip 0.032 in, thick and over (B scale)	45–78	68–90	88–96	96–102
Strip 0.015 in. thick and over (30 T scale)	46–67	62–75	74–79	79–83

TABLE 17. TYPICAL MECHANICAL PROPERTIES OF HEAT-TREATED BERYLLIUM COPPER STRIP

	Solution Annealed AT	Quarter Hard ¼ HT	Half Hard ½ HT	Hard HT
Tensile strength, psi	165,000–195,000	175,000–200,000	185,000–210,000	190,000–215,000
Proportional limit (0.002% offset), psi	100,000–125,000	110,000–135,000	120,000–145,000	125,000–155,000
Yield strength (0.01% offset), psi	110,000–135,000	120,000–145,000	130,000–155,000	135,000–165,000
Yield strength (0.2% offset), psi	140,000–170,000	150,000–180,000	160,000–190,000	165,000–195,000
Yield strength (0.5% extension), psi	90,000–100,000	90,000–100,000	90,000–100,000	90,000–100,000
Tension modulus, psi	19,000,000	19,000,000	19,000,000	19,000,000
Elongation, %	5–10	3–7	2–5	1–4
Rockwell hardness:				
Strip 0.032 in. thick and over (B scale)	36–42	38–43	39–44	40–45
Strip 0.015 in. thick and over (30 T scale)	56–62	58–63	59–46	60–54

TABLE 18. TEMPER SELECTION FOR BERYLLIUM COPPER STRIP

Temper	Cold-rolled Reduction in B&S Numbers	ASTM Designation	% Reduction in Thickness	Relative Formability
Solution annealed	0	A	0	Best for deep-drawn, cupped, or severely formed parts
¼ hard	1 hard	¼ H	11	Almost as formable as solution-annealed stock and gives better spring properties
½ hard	2 hard	½ H	21	Moderate formability with good strength for parts incorporating liberal radiuses or light drawing
Hard	4 hard	H	37	Max mechanical properties with reduced cold-forming characteristics for parts essentially flat

TABLE 19. TYPICAL PHYSICAL PROPERTIES OF HEAT-TREATED BERYLLIUM COPPER STRIP

Specific gravity	8.30
Density, lb per cu in	0.300
Magnetic properties	Nonmagnetic
Melting range, F	1600 to 1800
Increase in density on heat-treatment, %	0.6
Decrease in length on heat-treatment, %	0.2
Specific heat, cal per g per deg C	0.10
Thermal expansion coefficient, per deg F:	
(68 to 212 F)	0.0000093
(68 to 393 F)	0.0000094
(68 to 572 F)	0.0000099
Poisson's ratio	0.35
Elastic modulus, psi (68 F)	19,000,000
Elastic modulus temperature coefficient, % change per deg F	0.02
Electrical resistivity, microhm cm (20 C)	7.8 to 5.7
Temperature coefficient of resistivity (20 to 200 C)	0.0009
Thermal conductivity, Btu per sq ft per in. per hr per deg F (68 F)	750 to 900

set permissible in the application—never by ultimate strength. When no regularly measurable deviation from the elastic condition is allowable, working stresses must be kept below the proportional limit. When space permits, a maximum design stress of at least 10,000 psi below the proportional limit is recommended to insure positive action in statically loaded parts.

Though frequently employed for copper alloys, the yield strength at 0.5% extension under load has limited value as an index of plastic damage. The proportional limit always exceeds this yield value for heat-treated strip; so the specified extension can be produced elastically, and the yield value is dependent solely on the tension modulus. Even in the case of unhardened strip, the hard temper has a proportional limit greater than the yield strength at 0.5% extension. Consequently, yield strengths qualified by 0.01 or 0.2% offset may prove more useful for design purposes.

In the case of endurance loads, however, it is necessary to reduce design stresses to prevent failure from fatigue. Relatively high strength in the fully heat-treated condition does not indicate brittleness or loss of fatigue resistance. Failure on the part of strip of a given temper to precipitation-harden to a certain level may be an indication of inferior structure. Such material frequently offers decreased resistance to shock and fatigue.

For applications requiring added ductility in the hardened condition, to permit forming during assembly or in the field, it is frequently advisable for the user to select a suitable under- or overaging treatment.

HIGH-NICKEL ALLOYS

Nickel, a white, malleable, noncorrodible metal, has mechanical properties similar to those of mild steel. In addition, a number of important nickel-base alloys have unusual resistance to corrosive agents of varying types and concentrations. Therefore, these alloys are used in manufacture of a wide variety of parts, which must have high mechanical properties as well as corrosion resistance. For mechanical properties see Table 20.

Resistance of Nickel Alloys to Corrosive Media

Wrought Alloys

Nickel. Has outstanding resistance to corrosion by caustic soda. Does not readily discharge hydrogen from any of common acids—a supply of some oxidizing agent, such as dissolved air, is necessary for corrosion to proceed. Oxidizing conditions generally favor corrosion of nickel while reducing conditions retard it.

Low-carbon Nickel. Has essentially the same immunity to rust and corrosion as wrought nickel.

D-Nickel. Superior to nickel under oxidizing sulfurous conditions at elevated temperatures. Oxidizes more rapidly at first, but resists attack by sulfur. Not generally used under high-temperature reducing conditions, in either the presence or absence of sulfur. This alloy is highly resistant to conditions of electrical sparking.

Monel. More resistant than nickel under reducing conditions and more resistant than copper under oxidizing conditions. Monel is resistant to most acids, alkalis, salts, waters, food products, organic substances, and atmospheric conditions, both at normal and at elevated temperatures. It is not used with highly oxidizing acids.

R Monel. Has essentially the same immunity to rust and corrosion as monel.

K Monel. Generally corrosion resistance is the same as for monel in mineral and organic acids, alkalis, salts, potable and industrial waters, food products, organic compounds, and atmospheric oxidation at normal and high temperatures.

KR Monel. Has essentially the same immunity to rust and corrosion as K monel.

Inconel. Chromium content makes inconel superior to pure nickel under oxidizing conditions while its high nickel content enables it to retain good resistance under reducing conditions. It is highly resistant to attack in many strongly oxidizing acid solutions, and has excellent resistance to corrosion by alkaline solutions.

Hastelloy A. Withstands action of hydrochloric acid in all concentrations at temperatures up to 158 F. Resistant to sulfuric acid of any concentration up to 158 F and to all concentrations below 50% up to boiling point. In addition, is resistant to acetic, formic, and other organic acids, but not to oxidizing agents.

Hastelloy B. Particularly well suited for equipment handling boiling hydrochloric acid and wet hydrochloric acid gas. Resistance to sulfuric acid better than that of alloy A. Has good resistance to phosphoric acid and is practically unaffected by alkalis. Not recommended for use with nitric acid or other acid oxidizers.

Table 20. Characteristics of Wrought and Cast Nickel-base Alloys

Alloys Available (Trade Names)	Average Mechanical Properties					Relative Suitability for Hot Working[a]	Relative Suitability for Cold Working[a]	Relative Suitability for Being Welded[a]			Relative Machinability[a]	Av Physical Constants				
	Form and Condition	Tensile Strength 1000 Psi	Yield Strength in Tension (0.2 % Offset) 1000 Psi	Elongation in 2 In., %	Hardness, Brinell 10-mm Ball (3000 Kg)			Oxyacetylene	Metallic Arc	Resistance		Density, Lb per Cu In.	Specific Heat (32 to 212 F), Btu per Lb per F	Coeff. of Thermal Expansion (32 to 212 F) $(+10^{-6})$ In. per In. per F	Thermal Conductivity (32 to 212 F), Btu (Sq Ft per Hr per F per In.)	Electrical Resistivity (32 F), Ohms per Cir Mil Ft
Wrought Alloys																
Nickel	Annealed	70	20	40	100	A	A	A	A	A	C	0.321	0.130	7.2	420	57
	Hot-rolled	75	25	40	110											
	Cold-drawn	95	70	25	170											
	Cold-rolled[b]	105	95	5	210											
Low-carbon nickel	Annealed	65	20	45	......	A	A	A	A	A	C	0.321	0.130	7.2	420	57
D nickel	Annealed	75	35	40	140	A	A	A	A	A	C	0.317	0.130	7.4	335	110
	Hot-rolled	90	50	35	150											
	Cold-drawn	100	80	25	190											
Duranickel	Annealed	105	50	35	180	A	B	X	D	C	C	0.298	0.104	7.1	128	280
	Hot-rolled[c]	170	130	15	320	..	..	..	..	..	..	.....	.....	7.3	137	260
	Cold-drawn	120	90	25	220											
	Cold-drawn[c]	175	135	15	340											
Duranickel R	Hot-rolled	105	50	35	180	X	B	X	X	X	B	0.298	0.104	7.1	128	280
	Hot-rolled[c]	170	130	15	320	..	..	..	..	..	..	.....	.....	7.3	137	260
Monel	Annealed	75	35	40	125	A	A	A	A	A	B	0.319	0.130	7.8	180	290[d]
	Hot-rolled	90	50	35	150											
	Cold-drawn	100	80	25	190											
	Cold-rolled[b]	110	100	5	240											
R monel	Hot-rolled	85	45	35	145	A	A	A	A	A	A	0.319	0.130	7.8	180	290[d]
	Cold-drawn	90	75	25	180											
	Cold-drawn[d]	75	30	40	115											
K monel	Hot-rolled	100	45	40	160	A	A	A	A	A	C[e]	0.306	0.130	7.8	130	350
	Hot-rolled[c]	150	110	25	280											
	Cold-drawn	115	85	25	210											
	Cold-drawn[e]	155	115	20	290											

KR monel	Hot-rolled	105	65	40	185	X	B	X	X	X	B[e]	0.306	0.13	7.8	130	350
	Hot-rolled[c]	150	105	25	280											
	Cold-drawn	115	85	25	215											
	Cold-drawn[c]	155	115	25	290											
326 monel	Annealed	75	35	40	125	A	A	A	A	A	B	0.320	0.13	7.8	180	290
	Hot-rolled	85	45	35	150											
Inconel	Annealed	85	35	45	150	A	A	A	A	A	C	0.307	0.110	6.4	104	590
	Hot-rolled	100	60	35	180											
	Cold-drawn	115	90	20	200											
	Cold-rolled[b]	135	110	2	260											
Inconel X	Annealed	115	50	50	200	A	B	X	B	A	D	0.298	0.105	7.6	102	750
	Hot-rolled[c]	180	120	25	360											
Inconel W	Annealed	115	50	50	200	A	B	X	B	A	D	0.298	0.105	7.6	102	750
	Hot-rolled	180	120	25	360											
Incoloy	Annealed	80	40	40	150	A	B	A	A	A	C	0.290	0.120	8.0	104	584
	Hot-rolled	100	60	30	175											
	Cold-drawn	110	90	20	200											
Hastelloy A	Hot-rolled[d]	110–120	47–52	40–48	200–215	C	C	A	A	A	B	0.318	0.094	6.1	116	760
Hastelloy B	Hot-rolled[d]	130–140	60–65	40–45	210–235	C	C	A	A	A	C	0.334	0.091	5.6	78.5	813
Hastelloy C	Hot-rolled[d]	115–128	55–65	25–50	160–210	C	C	A	A	A	C	0.323	0.092	6.3	87	800
Ilium R	Hot-rolled[d]	100	50	32	190	C	C	B	A	B	B	0.310	0.110	5.1	.....	740
	Cold-rolled[b]	150	100	12	365											
Casting Alloys																
Nickel	Sand-cast[e]	50	25	25	100	X	X	B	B	X	C	0.303	0.130	7.4	410	125
Monel	Sand-cast[e]	80	35	35	135	X	X	B	B	X	B	0.312	0.130	6.8	180	320
H monel	Sand-cast[e]	100	60	15	210	X	X	X	X	X	C	0.307	0.130	6.8	180	370
S monel	Sand-cast[d]	90	70	3	275	X	X	X	X	X	C	0.303	0.130	6.8	180	380
	Sand-cast[e]	130	100	3	320											
	Sand-cast[e]	130	100	3	350											
Inconel	Sand-cast[e]	85	40	20	175	X	X	[illegible]	B	X	C	0.300	0.110			
Hastelloy A	Sand-cast[e]	69–78	43–45	8–12	155–200	D	D	A	A	A	B	0.318	0.094	6.1	116	760
Hastelloy B	Sand-cast[e]	75–82	55–57	6–9	190–230	D	D	A	A	A	C	0.334	0.091	5.6	78	813
Hastelloy C	Sand-cast[e]	72–80	45–48	10–15	175–215	D	D	A	A	A	C	0.323	0.092	6.3	87	800
Hastelloy D	Sand-cast[e]	36–41	36–41	0	....	D	D	A	D	X	D	0.282	0.109	6.1	145	680
Illium	Sand-cast[e]	68	50	9.5	180	X	X	B	A	B	C	0.310	0.105	7.5	.....	735
Illium R	Sand-cast[e]	70	50	9.5	180	X	X	B	A	B	C	0.310	0.110	5.1	.....	740

[a] Relative hot and cold workability, weldability, and machinability are indicated as follows: A = excellent, B = good, C = fair, D = poor, and X = not recommended. Ratings are based on the high-nickel alloys shown, as a group, and are not to be used in comparison with other metals, even though they may have a high nickel content.

[b] Hard temper.

[c] Age-hardened.

[d] Annealed.

[e] As-cast.

HASTELLOY C. Is unique in its resistance to hypochlorite solutions and to moist chlorine at temperatures up to 105 F. Resistant to hydrochloric acid up to 122 F, to dilute nitric acid up to 150 F, and to oxidizing solutions.

ILLIUM R. Resistant to all concentrations of nitric and phosphoric acids at all temperatures, also resistant to oxidizing acid mixtures. Usefully resistant to sulfuric acid under all conditions of concentration, aeration, agitation, and temperature. Resistant to hydrochloric acid to about 15% concentration at room temperature, not above.

CAST ALLOYS

In general, the corrosion resistance of cast alloys of the above composition have corrosion resistance comparable with wrought material.

HASTELLOY D. Exceptionally resistant to sulfuric acid of all concentrations up to boiling point. Resistance to hydrochloric acid fair at moderate temperatures. Practically unaffected by acetic or formic acid, and has low corrosion rate in phosphoric acid. It is not resistant to strong oxidizing agents such as nitric acid.

ILLIUM G. Resistant to all concentrations of nitric and phosphoric acid at all temperatures, also resistant to oxidizing acid mixtures. Usefully resistant to sulfuric acid under all conditions of concentration, aeration, agitation, and temperature. Resistant to hydrochloric acid to about 15% concentration at room temperature, not above.

TITANIUM ALLOYS

Properties. Titanium may be classed as a light metal; although it is 60% heavier than aluminum, it is only 56% as heavy as alloy steel. Absolutely pure titanium is extremely difficult to produce, and is soft and ductile with a yield strength of 15,000 psi, ultimate strength of 35,000 psi, and elongation of 55%.

Properties of commercially pure titanium vary somewhat, but all are considerably stronger and less ductile than absolutely pure titanium. Typical properties of several grades are shown in Table 21. The metal does not respond to heat-treatment but may be cold-worked to above 100,000 psi tensile and 85,000 psi yield strength, with a drop in elongation to 12%.

Titanium Alloys. Alloys are available with tensile strengths as high as 175,000 psi, and have been developed up to 200,000 psi with fair ductility. Tables 22 and 23 give nominal compositions and approximate properties of the alloys now offered commercially.

Effects of alloying elements are:

Iron, in the amounts normally occurring in sponge metal, has little or no effect.

Carbon up to 0.25% increases tensile strength without materially affecting ductility, weldability, or formability. From 0.25 to 1.0% carbon, the tensile strength is not materially increased, but ductility, formability, weldability, and surface characteristics are seriously impaired.

Nitrogen lowers ductility and formability but raises tensile strength rapidly.

Oxygen appears to have more effect on tensile strength and ductility than carbon, less than nitrogen.

Copper causes a slight increase in hardness.

Aluminum increases hardness and oxidation resistance at elevated temperatures. It also strengthens the alloy.

Chromium increases strength and hardness, does not cause brittleness, and decreases oxidation resistance at elevated temperatures in usual alloying amounts. Above 15% chromium, oxidation resistance again equals that of unalloyed titanium.

Manganese hardens and strengthens the alloy without harming ductility.

Corrosion Resistance. Titanium is immune to most corrosives, but when it is attacked, the rate is likely to be unusually severe.

TABLE 21. PROPERTIES OF ANNEALED COMMERCIALLY PURE TITANIUM

	Ti-75A	RC-70	RC-55
Ultimate tensile strength, psi	70,000–80,000	90,000	75,000
Tensile yield strength, psi	55,000–60,000	80,000	65,000
Elongation, % in 2 in.	20–30	20	25
Hardness, Vhn	175	225–275	
Reduction in area, %	45–70	55	
Proportional limit, psi	....	52,000	50,000
Modulus of elasticity, psi, $\times 10^6$	....	15–5	15–5

TABLE 22. COMPOSITIONS OF TITANIUM ALLOYS

Ti-100A	0.1 max Fe, 0.02 max C, trace O, trace N (O + N = 0.16), bal. Ti
Ti-125A	1.8 Cr, 0.9 Fe, 0.15 O, 0.02 C, 0.04 N, bal. Ti
Ti-150A	2.7 Cr, 1.3 Fe, 0.25 O, 0.02 N, 0.02 C, bal. Ti
Ti-175A	3.0 Cr, 1.5 Fe, 0.5 O, 0.04 N, 0.02 C, bal. Ti
L-2748	3.0 Al, 5.0 Cr, 0.5 C, bal. Ti
L-2749	0.5 C, bal. Ti
RC-130-A	7.0 Mn, bal. Ti
RC-130-B	4.0 Mn, 4.0 Al, bal. Ti

In atmosphere, tough oxide and nitride coatings form at all service temperatures. Up to 1050 F, prolonged exposure to air does not seriously affect properties.

Galvanically, titanium is near the noble end of the scale and galvanic couples behave like austenitic stainless steel.

Among chemicals that titanium resists are chloride salts, nitric acid (at room temperature), sulfur compounds, strong alkalis, and chlorinated solvents. It is best to test titanium under service conditions.

High-temperature Properties. Despite its high melting point (3150 F), the ultimate and yield strengths drop rapidly with increasing temperatures; so the material becomes too weak for structural uses at temperatures at which the hot-air corrosion resistance is entirely adequate. It does have possibilities in short-time applications, such as aircraft firewalls, up to 2000 F.

Fire Precautions. Fine turnings can be ignited in air with an oxyacetylene torch, and will continue to burn with a hot glow after the flame has been removed. Titanium dust in air suspension is, like most dusts, explosive. Therefore, some precautions in the disposal of dust and fine turnings are necessary.

Grinding Problems. Some of the more serious problems in machining titanium occur in grinding. The metal may be ground in much the same manner as stainless steel, but wheel wear is exceptionally fast. One theory about this high wear is that there is a chemical reaction between the wheel and work. Titanium is a relatively poor conductor of heat, so temperatures in the grinding area are high.

Some improvement in grinding has been obtained by running the wheel at a much slower speed than customary, say, around 3000 fpm. Because titanium dust is highly explosive, it is desirable to grind wet. Dust-collection methods as used with magnesium are desirable. Aside from the dust hazard, the rapid wheel breakdown causes smoke and fumes. Workers are likely to experience discomfort unless greater ventilation than customary is supplied.

TABLE 23. PROPERTIES OF TITANIUM ALLOYS

	L-2748 (Forged)	L-2749 (Forged)	RC-130-A	RC-130-B (Forged)	Ti-100A	Ti-125A	Ti-150A	Ti-175A
Ultimate tensile strength, psi	165,000	90,000	150,000	145,000	100,000	125,000	150,000	175,000
Tensile yield strength psi (0.2% offset)	153,000	75,000	140,000	135,000	75,000	80,000	130,000	160,000
Elastic modulus ($\times 10^6$), psi	18.0	16.3	15.5	15.5	....	....	14	
Elongation, % in 2 in.	10	18	15	16*	20–25	18–20	12–20	10
Endurance limit, fatigue	95,000	63,000	70,000 bending, 110,000 tension	80,000	....	....	75,000	
Coefficient of linear thermal exp (C) $\times 10^{-6}$	9.0	10.4						
Electrical resistivity, microhm cm	140	60						
Reduction in area, %	....	....	32	40	45	45	40	30
Compressive yield strength, psi (0.2% offset)	....	....	140,000					
Proportional limit, tension, psi	....	....	105,000	105,000				
Hardness, Rockwell A	71	62						
Hardness, Vickers	....	....	330–350	330–350	245	300	345	380
Hardness, Rockwell C	....	....	33–36	33–36	24	33	39	44
Density, lb per cu in	0.16	0.16	0.17	0.17				
Melting range, C	....	....	1400–1500†	1600–1700†				
Melting range, F	....	....	2550–2730†	2910–3090†				

* 1-in. gage length.
† Estimated.

Section 27

HEAT-TREATMENT OF METALS

SECTION 27

HEAT-TREATMENT OF METALS

DEFINITIONS OF HEAT-TREATING TERMS[1]

These definitions were prepared by the Joint Committee on Definitions of Terms Relating to Heat-treatment appointed by the American Society for Testing Materials, the American Society for Metals, the American Foundrymen's Association, and the Society of Automotive Engineers.

Ac_{cm}, Ac_1, Ac_3, Ac_4. Defined under Transformation Temperature.

Ae_{cm}, Ae_1, Ae_3, Ae_4. Defined under Transformation Temperature.

Age Hardening. Hardening by aging, usually after rapid cooling or cold working. See Aging.

Aging. In a metal or alloy, a change in properties that generally occurs slowly at room temperature and more rapidly at higher temperatures. See also Age Hardening, Artificial Aging, Interrupted Aging, Natural Aging, Overaging, Precipitation Hardening, Precipitation Heat-treatment, Progressive Aging, Quench Aging, and Strain Aging.

Annealing. Heating to and holding at a suitable temperature and then cooling at a suitable rate, for such purposes as reducing hardness, improving machinability, facilitating cold working, producing a desired microstructure, or obtaining desired mechanical, physical, or other properties. When applicable, the following more specific terms should be used: black annealing, blue annealing, box annealing, bright annealing, flame annealing, full annealing, graphitizing, intermediate annealing, isothermal annealing, malleablizing, process annealing, quench annealing, recrystallization annealing, and spheroidizing.

Definitions of the above terms are given in their alphabetic positions in this glossary.

When applied to ferrous alloys, the term "annealing," without qualification, implies full annealing.

When applied to nonferrous alloys, the term "annealing" implies a heat-treatment designed to soften a cold-worked structure by recrystallization or subsequent grain growth or to soften an age-hardened alloy by causing a nearly complete precipitation of the second phase in relatively coarse form.

Any process of annealing will usually reduce stresses, but if the treatment is applied for the sole purpose of such relief it should be designated *stress relieving*.

Ar_{cm}, Ar_1, Ar_3, Ar_4, Ar''. Defined under Transformation Temperature.

Artificial Aging. Aging above room temperature. See Aging and Precipitation Heat-treatment. Compare with Natural Aging.

Austempering. Quenching a ferrous alloy from a temperature above the transformation range, in a medium having a rate of heat abstraction high enough to prevent the formation of high-temperature transformation products, and then holding the alloy, until transformation is complete, at a temperature below that of pearlite formation and above that of martensite formation.

Austenitizing. Forming austenite by heating a ferrous alloy into the transformation range (partial austenitizing) or above the transformation range (complete austenitizing).

[1] SAE Iron and Steel Standards and Specifications.

Black Annealing. Box annealing or pot annealing ferrous alloy sheet, strip, or wire. See Box Annealing.

Blue Annealing. Heating hot-rolled ferrous sheet in an open furnace to a temperature within the transformation range and then cooling in air, in order to soften the metal. The formation of a bluish oxide on the surface is incidental.

Bluing. Subjecting the scale-free surface of a ferrous alloy to the action of air, steam, or other agents at a suitable temperature, thus forming a thin blue film of oxide and improving the appearance and resistance to corrosion.

Note: This term is ordinarily applied to sheet, strip, or finished parts. It is used also to denote the heating of springs after fabrication, in order to improve their properties.

Box Annealing. Annealing a metal or alloy in a sealed container under conditions that minimize oxidation. In box annealing a ferrous alloy, the charge is usually heated slowly to a temperature below the transformation range, but sometimes above or within it, and is then cooled slowly; this process is also called "close annealing" or "pot annealing." See Black Annealing.

Bright Annealing. Annealing in a protective medium to prevent discoloration of the bright surface.

Burning. Permanently damaging a metal or alloy by heating to cause either incipient melting or intergranular oxidation. See Overheating.

Carbonitriding. Introducing carbon and nitrogen into a solid ferrous alloy by holding above Ac_1 in an atmosphere that contains suitable gases such as hydrocarbons, carbon monoxide, and ammonia. The carbonitrided alloy is usually quench hardened.

Carburizing. Introducing carbon into a solid ferrous alloy by holding above Ac_1 in contact with a suitable carbonaceous material. The carburized alloy is usually quench hardened.

Case. In a ferrous alloy, the outer portion that has been made harder than the inner portion or *core* by *casehardening*.

Casehardening. Hardening a ferrous alloy so that the outer portion, or case, is made substantially harder than the inner portion, or core. Typical processes used for case hardening are *carburizing*, *cyaniding*, *carbonitriding*, *nitriding*, *induction hardening*, and *flame hardening*.

Cementation. The introduction of one or more elements into the outer portion of a metal object by means of diffusion at high temperature.

Cold-treatment. Cooling to a low temperature for the purpose of obtaining desired conditions or properties, such as dimensional or structural stability.

Controlled Cooling. Cooling from an elevated temperature in a predetermined manner, to avoid hardening, cracking, or internal damage, or to produce a desired microstructure. This cooling usually follows a hot-forming operation.

Core. In a ferrous alloy, the inner portion that is softer than the outer portion, or *case*.

Critical Range or **Critical Temperature Range.** Synonymous with *transformation range*, which is preferred.

Cyaniding. Introducing carbon and nitrogen into a solid ferrous alloy by holding above Ac_1 in contact with molten cyanide of suitable composition. The cyanided alloy is usually quench hardened.

Decarburization. The loss of carbon from the surface of a ferrous alloy as a result of heating in a medium that reacts with the carbon.

Differential Heating. Heating that produces a temperature distribution within an object in such a way that, after cooling, various parts have different properties as desired.

Direct Quenching. Quenching carburized parts directly from the carburizing operation.

Drawing. A misnomer for *tempering*.

Flame Annealing. Annealing in which the heat is applied directly by a flame.

Flame Hardening. Quench hardening in which the heat is applied directly by a flame.

Fog Quenching. Quenching in a fine vapor or mist.

Full Annealing. Annealing a ferrous alloy by austenitizing and then cooling slowly through the transformation range. The austenitizing temperature for hypoeutectoid steel is usually above Ac_3 and for hypereutectoid steel usually between Ac_1 and Ac_{cm}.

Gas Cyaniding. A misnomer for *carbonitriding.*

Graphitizing. Annealing a ferrous alloy in such a way that some or all of the carbon is precipitated as graphite.

Hardenability. In a ferrous alloy, the property that determines the depth and distribution of hardness induced by quenching.

Hardening. Increasing the hardness by suitable treatment, usually involving heating and cooling. When applicable, the following more specific terms should be used: ***age hardening, casehardening, flame hardening, induction hardening, precipitation hardening,*** and ***quench hardening.***

Heat-treatment. Heating and cooling a solid metal or alloy in such a way as to obtain desired conditions or properties. Heating for the sole purpose of hot working is excluded from the meaning of this definition.

Homogenizing. Holding at high temperature to eliminate or decrease chemical segregation by diffusion.

Hot Quenching. Quenching in a medium at an elevated temperature.

Induction Hardening. Quench hardening in which the heat is generated by electrical induction.

Induction Heating. Heating by electrical induction.

Intermediate Annealing. Annealing wrought metals at one or more stages during manufacture and before final thermal treatment.

Interrupted Aging. Aging at two or more temperatures, by steps, and cooling to room temperature after each step. See Aging and compare with Progressive Aging.

Interrupted Quenching. Quenching in which the metal object being quenched is removed from the quenching medium while the object is at a temperature substantially higher than that of the quenching medium. See also Time Quenching.

Isothermal Annealing. Austenitizing a ferrous alloy and then cooling to and holding at a temperature at which austenite transforms to a relatively soft ferrite-carbide aggregate.

Isothermal Transformation. A change in phase at any constant temperature.

Malleablizing. Annealing white cast iron in such a way that some or all of the combined carbon is transformed to graphite or, in some instances, part of the carbon is removed completely.

Martempering. Quenching an austenitized ferrous alloy in a medium at a temperature in the upper part of the martensite range, or slightly above that range, and holding in the medium until the temperature throughout the alloy is substantially uniform. The alloy is then allowed to cool in air through the martensite range.

Martensite Range. The temperature interval between M_s and M_f.

M_f. Defined under Transformation Temperature.

M_s (or **Ar″**). Defined under Transformation Temperature.

Natural Aging. Spontaneous aging of a supersaturated solid solution at room temperature. See Aging and compare with Artificial Aging.

Nitriding. Introducing nitrogen into a solid ferrous alloy by holding at a suitable temperature (below Ac_1 for ferritic steels) in contact with a nitrogenous material, usually ammonia or molten cyanide of appropriate composition. Quenching is not required to produce a hard case.

Normalizing. Heating a ferrous alloy to a suitable temperature above the transformation range and then cooling in still air to room temperature.

Overaging. Aging under conditions of time and temperature greater than those required to obtain maximum change in a certain property, so that the property is altered in the direction of the initial value. See Aging.

Postheating. Heating weldments immediately after welding, for tempering, for stress relieving, or for providing a controlled rate of cooling to prevent formation of a hard or brittle structure.

Precipitation Hardening. Hardening caused by the precipitation of a constituent from a supersaturated solid solution. See also Age Hardening and Aging.

Precipitation Heat-treatment. Artificial aging in which a constituent precipitates from a supersaturated solid solution. See Artificial Aging, Interrupted Aging, and Progressive Aging.

Preheating. Heating before some further thermal or mechanical treatment. For tool steel, heating to an intermediate temperature immediately before final austenitizing. For some nonferrous alloys, heating to a high temperature for a long time in order to homogenize the structure before working.

Process Annealing. In the sheet and wire industries, heating a ferrous alloy to a temperature close to, but below, the lower limit of the transformation range and then cooling, in order to soften the alloy for further cold working.

Progressive Aging. Aging by increasing the temperature in steps or continuously during the aging cycle. See Aging and compare with Interrupted Aging.

Quench Aging. Aging induced by rapid cooling after *solution heat-treatment.*

Quench Annealing. Annealing an austenitic ferrous alloy by *solution heat-treatment.*

Quench Hardening. Hardening a ferrous alloy by austenitizing and then cooling rapidly enough so that some or all of the austenite transforms to martensite. The austenitizing temperature for hypoeutectoid steels is usually above Ac_3 and for hypereutectoid steels usually between Ac_1 and Ac_{cm}.

Quenching. Rapid cooling. When applicable, the following more specific terms should be used: *direct quenching, fog quenching, hot quenching, interrupted quenching, selective quenching, spray quenching,* and *time quenching.*

Recrystallization Annealing. Annealing cold-worked metal to produce a new grain structure without phase change.

Secondary Hardening. Tempering certain alloy steels at certain temperatures so that the resulting hardness is greater than that obtained by tempering the same steel at some lower temperature for the same time.

Selective Heating. Heating only certain portions of an object so that they have the desired properties after cooling.

Selective Quenching. Quenching only certain portions of an object.

Soaking. Prolonged holding at a selected temperature.

Solution Heat-treatment. Heating an alloy to a suitable temperature, holding at that temperature long enough to allow one or more constituents to enter into solid solution, and then cooling rapidly enough to hold the constituents in solution. The alloy is left in a supersaturated, unstable state and may subsequently exhibit *quench aging.*

Spheroidizing. Heating and cooling to produce a spheroidal or globular form of carbide in steel. Spheroidizing methods frequently used are:

1. Prolonged holding at a temperature just below Ae_1.

2. Heating and cooling alternately between temperatures that are just above and just below Ae_1.

3. Heating to a temperature above Ae_1 or Ae_3 and then cooling very slowly in the furnace or holding at a temperature just below Ae_1.

4. Cooling at a suitable rate from the minimum temperature at which all carbide is dissolved, to prevent the re-formation of a carbide network, and then reheating in accordance with method 1 or 2 above (applicable to hypereutectoid steel containing a carbide network).

Spray Quenching. Quenching in a spray of liquid.

Stabilizing Treatment. Any treatment intended to stabilize the structure of an alloy or the dimensions of a part. (1) Heating austenitic stainless steels that contain titanium, columbium, or tantalum to a suitable temperature below that of a full anneal in order to inactivate the maximum amount of carbon by precipitation as a carbide of titanium, columbium, or tantalum. (2) Transforming retained austenite in parts made from tool steel. (3) Precipitating a constituent from a nonferrous solid solution to improve the workability, to decrease the tendency of certain alloys to age harden at room temperature, or to obtain dimensional stability.

Strain Aging. Aging induced by cold working. See Aging.

Stress Relieving. Heating to a suitable temperature, holding long enough to reduce residual stresses, and then cooling slowly enough to minimize the development of new residual stresses.

Temper Brittleness. Brittleness that results when certain steels are held within, or are cooled slowly through, a certain range of temperature below the transformation range. The brittleness is revealed by notched-bar impact tests at or below room temperature.

Tempering. Reheating a quench-hardened or normalized ferrous alloy to a temperature below the transformation range and then cooling at any rate desired.

Time Quenching. Interrupted quenching in which the duration of holding in the quenching medium is controlled.

Transformation Ranges or Transformation Temperature Ranges. Those ranges of temperature within which austenite forms during heating and transforms during cooling. The two ranges are distinct, sometimes overlapping but never coinciding. The limiting temperatures of the ranges depend on the composition of the alloy and on the rate of change of temperature, particularly during cooling. See Transformation Temperature.

Transformation Temperature. The temperature at which a change in phase occurs. The term is sometimes used to denote the limiting temperature of a transformation range. The following symbols are used for iron and steels:

Ac_{cm}. In hypereutectoid steel, the temperature at which the solution of cementite in austenite is completed during heating.

Ac_1. The temperature at which austenite begins to form during heating.

Ac_3. The temperature at which transformation of ferrite to austenite is completed during heating.

Ac_4. The temperature at which austenite transforms to delta ferrite during heating.

Ae_1, Ae_3, Ae_{cm}, Ae_4. The temperatures of phase changes at equilibrium.

Ar_{cm}. In hypereutectoid steel, the temperature at which precipitation of cementite starts during cooling.

Ar_1. The temperature at which transformation of austenite to ferrite or to ferrite plus cementite is completed during cooling.

Ar_3. The temperature at which austenite begins to transform to ferrite during cooling.

Ar_4. The temperature at which delta ferrite transforms to austenite during cooling.

M_s (or **Ar″**). The temperature at which transformation of austenite to martensite starts during cooling.

M_f. The temperature at which martensite formation finishes during cooling.

Note: All these changes except the formation of martensite occur at lower temperatures during cooling than during heating, and depend on the rate of change of temperature.

HEAT-TREATMENT OF CONSTRUCTIONAL STEELS

Steel is heat-treated for a variety of purposes, such as:

1. To improve machinability.
2. To restore ductility for cold forming and extrusion.
3. To provide the hardness, tensile strength, and other mechanical properties required by service usage of the part.
4. To produce a wear-resistant hard case.

Of necessity, heat-treating processes will vary with purpose. Typical treatments for "heat-treating" grades of carbon and alloy steels are given in Tables 1 and 2.

TABLE 1. TREATMENTS FOR HEAT-TREATING GRADES OF CARBON STEELS

SAE Steels	Normalizing Temperature, F	Annealing Temperature, F	Hardening Temperature, F	Quenching Medium
1025, 1030	—	—	1575-1650	Water or brine
1033*, 1035, 1037	—	—	1525-1575	Oil or water
1036	1600-1700	—	1525-1575	Oil or water
	—	—	1525-1575	Oil or water
1038, 1039*, 1040	1600-1700	—	1525-1575	Oil or water
	—	—	1525-1575	Oil or water
1041	1600-1700 and/or	1400-1500	1475-1550	Oil
1042, 1043*, 1045*, 1046*, 1050*	1600-1700	—	1475-1550	Oil or water
1052, 1055	1550-1650 and/or	1400-1500	1475-1550	Oil
1060, 1064, 1065, 1070, 1074	1550-1650 and/or	1400-1500	1475-1550	Oil
1078	—	1400-1500†	1450-1500	Water or brine
1080, 1084, 1085, 1086, 1090	1550-1650 and/or	1400-1500†	1450-1500	Oil‡
1095	—	1400-1500†	1450-1500	Oil, water, or brine
	—	1400-1500†	1500-1600	Oil
1132, 1137	1600-1700 and/or	1400-1500	1525-1575	Oil or water
1138, 1140	—	—	1500-1550	Oil or water
	1600-1700	—	1500-1550	Oil or water
1141, 1144	—	1400-1500	1475-1550	Oil
	1600-1700	1400-1500	1475-1550	Oil
1145, 1146, 1151	—	—	1475-1550	Oil or water
	1600-1700	—	1475-1550	Oil or water

Tempering temperatures are not given.

Even where recommended draw temperatures are shown, the draw is not mandatory on many applications. Tempering is generally employed for a partial stress relief and improves resistance to cracking from grinding operations. Higher temperatures than those shown may be employed where the hardness specification on the finished parts permits.

* Commonly used on parts where induction hardening is employed. However, all steels from SAE 1030 up may have induction-hardening applications.

† Spheroidal structures are often required on these high-carbon steels for machining purposes and should be cooled very slowly or be isothermally transformed to produce the desired structure.

‡ May be water- or brine-quenched by special techniques such as partial immersion, or time-quenched; otherwise they are subject to quench cracking.

TABLE 2. HEAT-TREATMENTS FOR DIRECTLY HARDENABLE GRADES OF ALLOY STEELS*

SAE Steels	Normalizing Temperature, F	Annealing Temperature, F	Hardening Temperature, F	Quenching Medium
1330	—	—	1525–1575	Water or oil
	1600–1700 and/or	1500–1600	1525–1575	Water or oil
1335, 1340	—	—	1500–1550	Oil
	1600–1700 and/or	1500–1600	1525–1575	Oil
3130	1600–1700	—	1500–1550	Water or oil
3135, 3140	—	—	1500–1550	Oil
	1600–1700 and/or	1450–1550	1500–1550	Oil
4037, 4042	—	1525–1575	1500–1575	Oil
	—	1525–1575	1500–1575	Oil
4047, 4053	—	1450–1550	1500–1575	Oil
4063, 4068	—	1450–1550	1475–1550	Oil
4130	1600–1700 and/or	1450–1550	1600–1650	Water or oil
4137, 4140	1600–1700 and/or	1450–1550	1550–1600	Oil
4145, 4150	1600–1700 and/or	1450–1550	1500–1600	Oil
4340	1600–1700 and temper	1100–1225	1475–1525	Oil
4640	1600–1700 and/or	1450–1550	1450–1500	Oil
	1600–1700 and/or	1450–1500	1450–1500	Oil
5046	1600–1700 and/or	1450–1550	1475–1500	Oil
5130, 5132	1650–1750 and/or	1450–1550	1500–1550	Water or caustic solution
	1650–1750 and/or	1450–1550	1500–1550	Water, caustic solution, or oil
5135, 5140, 5145	1650–1750 and/or	1450–1550	1500–1550	Oil
	1650–1750 and/or	1450–1550	1500–1550	Oil
5147, 5150, 5152, 5155	1650–1750 and/or	1450–1550	1475–1550	Oil
	1650–1750 and/or	1450–1550	1475–1550	Oil
50100, 51100, 52100	—	1350–1450	1425–1475	Water
	—	1350–1450	1500–1600	Oil
6150	1650–1750 and/or	1550–1650	1600–1650	Oil
9254, 9255, 9260, 9261, 9262	—	—	1500–1650	Oil

TABLE 2. HEAT-TREATMENTS FOR DIRECTLY HARDENABLE GRADES OF ALLOY STEELS* (*Continued*)

SAE Steels	Normalizing Temperature, F	Annealing Tempera-ture, F	Hardening Tempera-ture, F	Quenching Medium
8627, 8630	1600–1700 and/or	1450–1550	1550–1650	Water or oil
8635, 8637, 8640, 8641	1600–1700 and/or	1450–1550	1525–1575	Oil
8642, 8645, 8650, 8653	1600–1700 and/or	1450–1550	1500–1550	Oil
8655, 8660	1650–1750 and/or	1450–1550	1475–1550	Oil
8740	1600–1700 and/or	1450–1550	1525–1575	Oil
8750	1600–1700 and/or	1450–1550	1500–1550	Oil
9840	1600–1700 and/or	1450–1550	1500–1550	Oil

* Except as noted, the steel is to be tempered to the desired hardness. The exceptions are gears made from steels 4037, 4640, 5147, 5150, 5152, and 5155; temper at 350 to 400 F. Steel 5046 may be tempered at 250 to 300 F.

The effect of these processes is better understood by realizing that:

1. Steel is an alloy of iron with up to 1.7% carbon.

2. Microstructures, or grain formations, present in the steel are determined by the percentage of carbon and the heat-treatment.

3. Hardness of steel is primarily a function of the carbon content, the other alloys being present for desired effects on heat-treatment or mechanical properties.

Hardening

Phases of Steel. The wide range of hardness possible in steels is a consequence of the carbon present, the manner in which it is associated with iron and the aggregation, or amounts of the resultant phases. A phase relates to any homogenous constituent of an alloy. Important phases of steel are:

FERRITE. A solid solution of about 0.025% carbon in (body-centered cubic) iron.

CEMENTITE. An intermetallic compound consisting of a definite lattice arrangement of iron and carbon atoms, the relative numbers of each of the atoms present in a given sample being in accordance with the formula Fe_3C.

AUSTENITE. A solid solution of up to 1.7% carbon in (face-centered cubic) iron.

Iron-carbon Diagram. A phase diagram is a chart which shows the number and nature of phases that are present in a given alloy at any temperature and composition under equilibrium conditions. The iron-carbon diagram in Fig. 1 is such a phase diagram for steel. This chart merely tells which of the three steel phases (ferrite, cementite, or austenite) is present at a given temperature and carbon concentration when the alloy is cooled or heated slowly enough so that it remains in a state of equilibrium.

A phase diagram reveals nothing about the state of aggregation of the phases present. In the temperature region in which both ferrite and cementite are present, the iron-carbon diagram does not tell us the relative sizes of particles.

Use of Phase Diagram. If a steel with 0.5% carbon is cooled slowly from a high temperature, the phase transformations that occur may be followed on the iron-carbon diagram. At (1) in Fig. 1 the metal is entirely austenite and remains in this state until temperature (2) is reached at line A_3. Now, ferrite starts to form along the grain boundaries of the austenite and increases as the temperature falls from (2) to (3). At temperature A_1 the remaining austenite has 0.80% carbon content and will transform at constant temperature into alternate plates of ferrite and cementite. This platelike, or lamellar, structure is known as pearlite. At all temperatures below the A_1 line (1330 F) we still have a mixture of ferrite and pearlite.

The changes that occur when a eutectoid steel (steel of 0.8% carbon) is cooled are also shown in Fig. 1. Nothing happens until a temperature of 1330 F is reached. Then all the metal transforms at constant temperature into pearlite. When a steel having a carbon content of 1.2% carbon is cooled, cementite first precipitates out along the grain boundaries at point (4) (A_{cm}). This continues until temperature A_1 is reached, when all the remaining material transforms into pearlite. There is no further change in structure below temperature A_1.

The ability of steel to harden depends upon the difference in carbon solubility of austenite and ferrite, and the tendency for excess carbon to precipitate in the form of cementite when austenite transforms to ferrite. If the steel is quickly cooled as it crosses the 1330 F transformation line (A_1), the precipitated particles of cementite will be very small and closely spaced. On the other hand, if the steel is slowly cooled as it crosses the A_1 line, the particles of Fe_3C will be larger and more widely spaced.

Types of Cementite. As revealed by microscopic examination, shapes of three general types of cementite particles are:

LAMELLAR TYPE. The layered structure already described, consisting of alternate plates of cementite and ferrite, and called pearlite. There are coarse and fine pearlites, depending upon the relative spacing of the cementite plates.

SPHEROIDAL TYPE. Roughly spherical globules of cementite in a matrix of ferrite. When the globules are relatively large, this structure is spheroidite.

ACICULAR TYPE. A crosshatched, needlelike structure of ferrite needles in very fine pearlite. This structure is known as the Widmanstaetten structure.

If 0.5% carbon steel is very slowly cooled from above the A_3 line, the resulting structure will be spheroidite, while if it is cooled somewhat less slowly pearlite will be formed. Coarse pearlite is obtained at a slower cooling rate than fine pearlite. The Widmanstaetten structure may be obtained if the steel is cooled at a critical rate from a high temperature.

Isothermal Transformation. When steel is heated to 1500 F or so, the carbide is dissolved in iron to form austenite, as shown in the iron-carbon diagram. But this diagram does not tell us what happens to the structure by use of various rates of cooling. That is the function of the S-curve, or time-temperature-transformation diagram (Fig. 2). If the austenite in a eutectoid steel (0.80% carbon) is cooled to 1200 F and held there long enough, it will transform entirely to pearlite, which is relatively soft. However, if the cooling rate reduces the austenite quickly to 600 F, pearlite will be avoided, and bainite, a much harder microstructure, is produced. And if the cooling rate is sufficiently rapid, formation of both pearlite and bainite will be bypassed, and the microstructure will be martensite, the hardest constituent that can be produced. When the austenite reaches 400 F or so, martensite begins to form, and reaches 90% or more at 200 F.

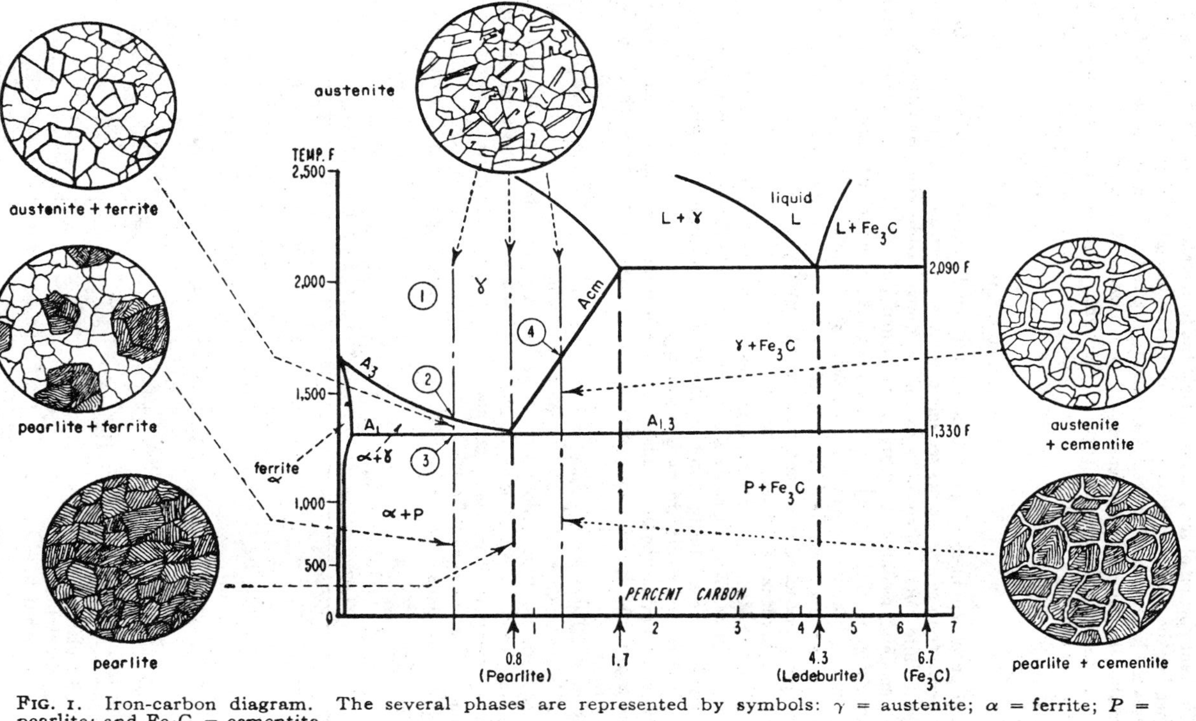

FIG. 1. Iron-carbon diagram. The several phases are represented by symbols: γ = austenite; α = ferrite; P = pearlite; and Fe_3C = cementite.

The bent portion of the S-curve nearest the temperature ordinate is known as the "knee" or the "nose." If the steel is cooled fast enough so that the rate of cooling passes the knee without intersecting it, no ferrite and carbide will occur, and transformation to bainite or martensite will be realized. Alloys are added to carbon steel to push the knee of the curve to the right. This means that a slower cool will miss the knee, give a fully martensitic structure, and allow one to obtain satisfactory depth of hardness on thicker sections.

Quenching

The purpose of quenching is to cool the steel rapidly enough from the austenitizing temperature to secure the desired microstructure—normally martensite, but sometimes bainite. Rate of cooling a workpiece is governed by the quenching power of the quenching medium, the degree of agitation, and the size of the article. Heat is abstracted from the piece by cooling the surface. Quenching severity is a measure of how fast heat is removed from the surface; the time is a function of how fast the heat can flow from the interior to the surface. Cooling rates for various sizes of sections, quenched in oil and in water, were given in connection with the discussion of hardenability bands (Sec. 25).

Quenching Media. The quenchants most commonly used in heat-treating plants are water, brine solutions, oils, air, and dilute sodium hydroxide solutions. Increasing applications are being found for molten salts and molten metals. Quenching powers of several media are given in Table 3.

WATER. Ordinary water finds widest use for quenching carbon steels ranging from 0.50 to as high as 1.20% carbon. Temperature of the water should not be over 100 F.

BRINES. There is a growing tendency to substitute a 9% (by weight) sodium chloride (common salt) brine for water as a general quenchant for steels of the water-hardening type. More uniform hardening can be obtained with brine, and the presence of sodium chloride or calcium chloride prevents water from dissolving atmospheric gas. The brine takes hold and wets the part all over immediately so quenching can proceed uniformly.

Brine also seems to be able to crack away the insulating furnace scale cleanly and

TABLE 3. QUENCHING POWER OF VARIOUS MEDIA

	Air	Severity of Quench*		
		Oil	Water	Brine
No circulation of liquid or agitation of piece	0.02	0.25 to 0.30	0.9 to 1.0	2
Mild circulation (or agitation)	....	0.30 to 0.35	1.0 to 1.1	2 to 2.2
Moderate circulation	....	0.35 to 0.40	1.2 to 1.3	
Good circulation	....	0.4 to 0.5	1.4 to 1.5	
Strong circulation	....	0.5 to 0.8	1.6 to 2.	
Violent circulation	....	0.8 to 1.1	4	5

* Nominally, the severity of quench H is based on a value of 1.0 for still water.

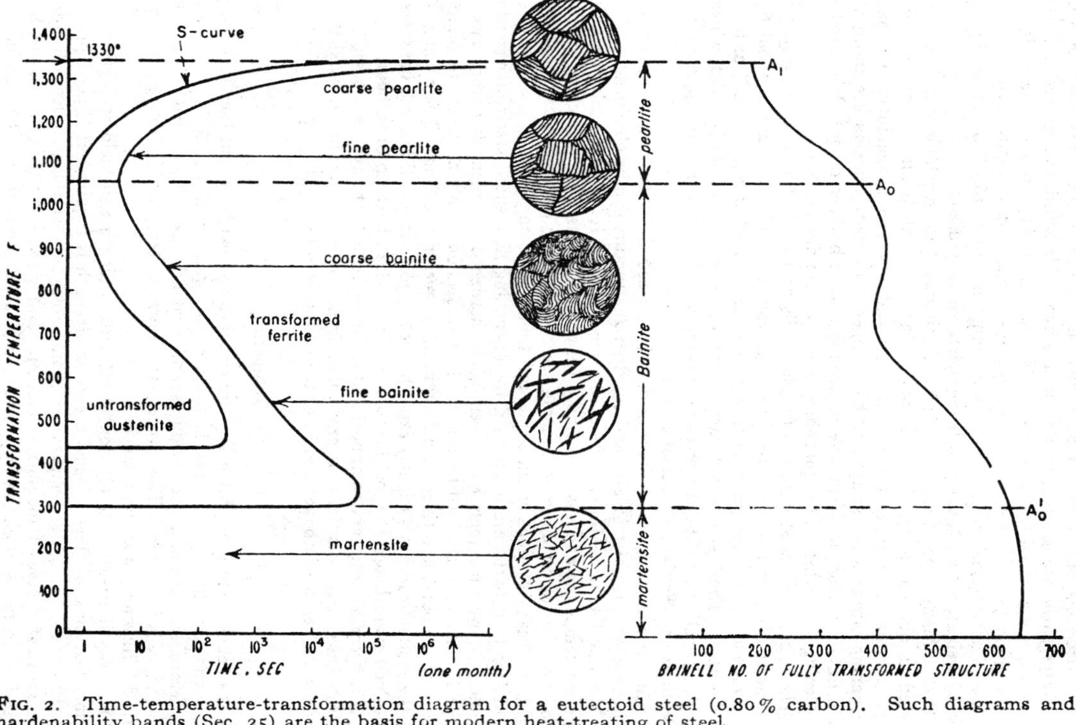

FIG. 2. Time-temperature-transformation diagram for a eutectoid steel (0.80% carbon). Such diagrams and hardenability bands (Sec. 25) are the basis for modern heat-treating of steel.

thus get right down to the metal surface immediately. The temperature of brine baths should be maintained between 70 and 100 F.

CAUSTIC-SODA SOLUTION. Where quenching speeds higher than those provided by water baths are desired, dilute sodium hydroxide solutions are used. Cooling rates are less affected by temperature rise than those of water, and their vigorous action upon heated steels effectively removes scale and so tends to prevent soft spots. Caustic-soda solutions usually contain from 2.5 to 16.5% of sodium hydroxide by weight, with 5% solution being most generally used.

OILS. Many combinations of animal, mineral, and vegetable oils have been tested as quenching oils, but mineral oils and mineral-oil blends have generally replaced the other types because of their stability and availability. Oil has a quenching rate that is less affected by changes in the bath temperature and is able to maintain a more uniform rate of quench over a broader range. It is used for hardening steel when water quenching would be too drastic.

The usual slow-speed mineral quenching oils cool steel at 1500 F about 30% as fast as cold water, and high-speed mineral quenching oils reach a rate about 65% that of cold water. At 400 F, each type of oil cools steel only about 10% as fast as cold water. The fast oils will harden types and sections of steels which will not harden effectively in the slower oils. Viscosity of the oil should be approximately 100 to 150 sec Saybolt Universal at 210 F.

AIR. Air hardening, in still air or dry compressed air, is used where distortion of other methods is objectionable and the desired structure can be obtained.

MOLTEN BATHS. Molten salts or hot oils are used for hot-quenching processes like martempering and austempering. Oil at 325 F is used for quenching intricate tools and dies of certain oil-hardening steels, where the part has a combination of light and heavy sections or close tolerances must be maintained.

Molten-lead baths are often used for quenching high-speed steels of hazardous shape. Many such tools have been safely hardened by quenching from 2300 to 2350 F, holding in the lead bath until the part has uniformly assumed the temperature of the bath, and then quenching in oil or even air cooling. The cooling rate of the molten lead is sufficiently rapid between 1300 and 900 F and full hardness is obtained even though the tool may be held in the lead bath up to an hour.

If it becomes necessary to anneal and reharden tools that have been quenched in a lead bath, it is essential that any adhering particles of metal be removed before the steel is reheated above 1400 F, or the surface will be pitted.

Tempering

As-quenched steels exhibit maximum strength and hardness, but they lack toughness and are therefore unsuitable for use. By reheating the steels to a low or moderate temperature, the hardness and strength are "drawn back," and the toughness is increased. This process is known as tempering, although it is also called "drawing" in shop parlance.

Figure 3 illustrates the relation of ultimate strength and yield strength to the brinell hardness of steel. When a certain strength is required, it is good practice to temper at the highest temperature permissible, because this practice will induce the greatest degree of toughness (Fig. 4). But what is toughness? It has been described in terms of elongation, reduction in area, the energy to break a notched test piece (Charpy or Izod impact tests), torsion strength, bending strength, etc. Actually, a part has toughness if it has some plasticity in the actual shape used and under the actual stresses imposed, which leads to the reasoning that tests should be made on pieces, not test specimens or bars, when possible.

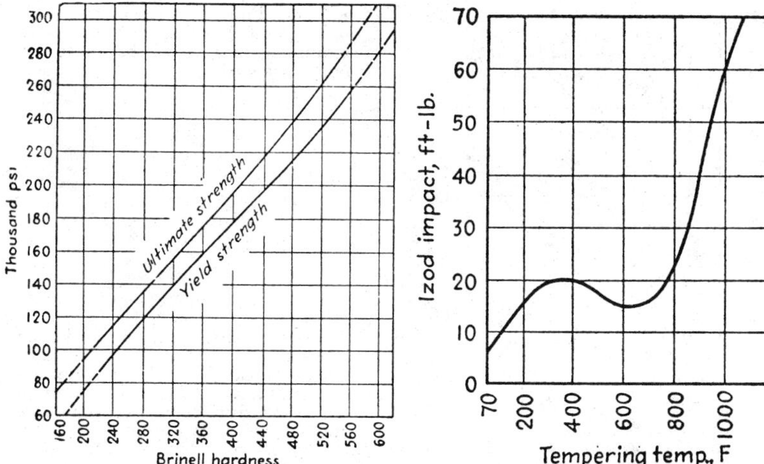

FIG. 3. Strength of steel is proportional to the hardness.

FIG. 4. Impact strength, which is a laboratory measure of toughness, is a function of tempering temperature. Dip in curve is region not customarily used for tempering.

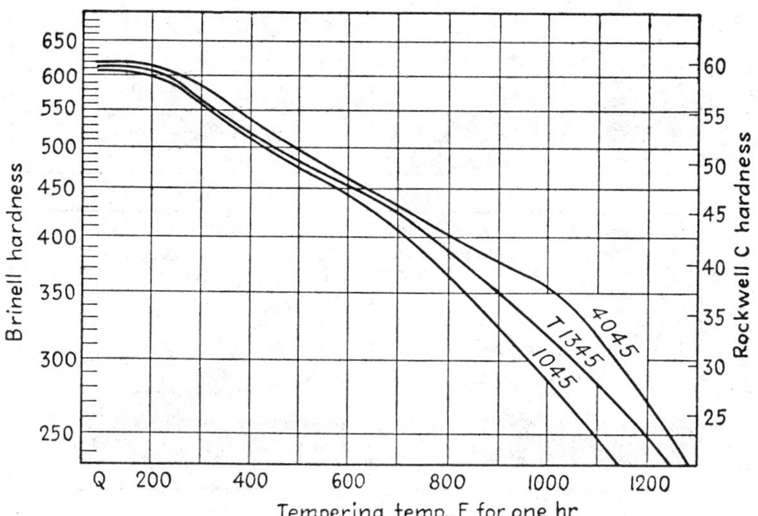

FIG. 5 Alloy content affects the hardness developed by tempering temperature.

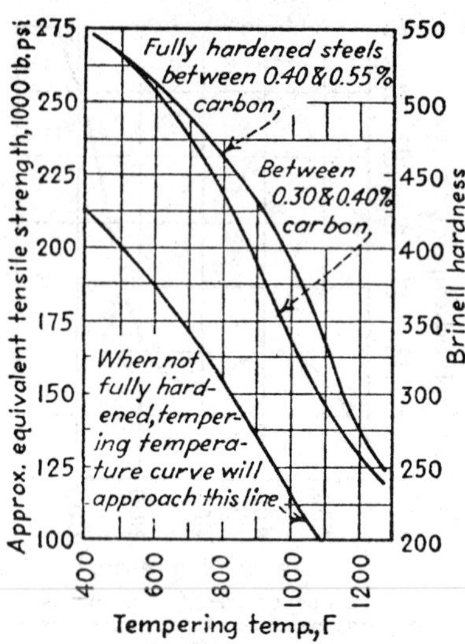

Fig. 6. Fully hardened steels will have greater residual hardness after tempering than steels that were not fully hardened.

Fig. 7. Comparison of hardening processes in relation to the S-curve (shaded area). In the conventional process, the steel is quenched through the martensitic-transformation range and then reheated for tempering. In the other three processes, the steel is caused to transform to bainite (austempering) in hot salt or to martensite, and tempered in salt.

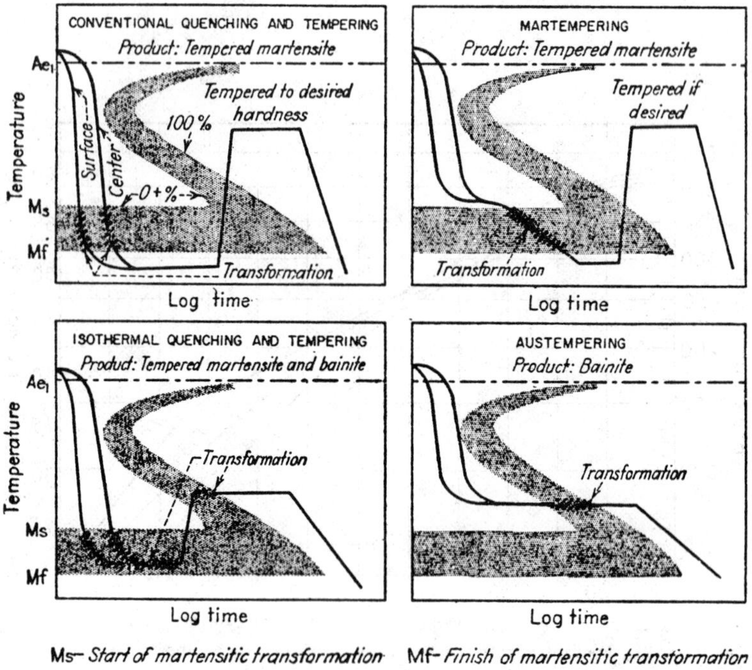

CONVENTIONAL QUENCHING AND TEMPERING
Product: Tempered martensite

MARTEMPERING
Product: Tempered martensite

ISOTHERMAL QUENCHING AND TEMPERING
Product: Tempered martensite and bainite

AUSTEMPERING
Product: Bainite

Ms— *Start of martensitic transformation* Mf— *Finish of martensitic transformation*

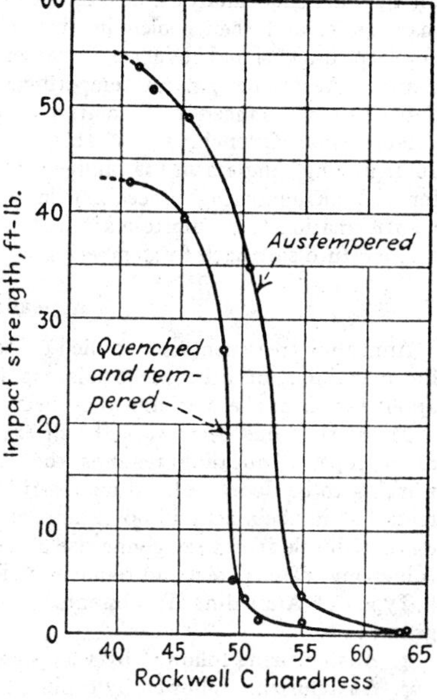

FIG. 8. Austempering provides greater toughness than conventional quenching and tempering. Test performed on 0.180-in. rod having 0.74% carbon, 0.37% manganese, 0.145% silicon. (*U.S. Steel Corp.*)

All steels do not exhibit the same hardness for a given tempering temperature. In fact, steels with the same carbon content will temper at different rates if they have different alloy contents. This statement is borne out by Fig. 5. Here SAE 1045 has a brinell hardness of 280 when tempered at 1000 F, whereas SAE 4045 has a brinell hardness of 350, when tempered at the same temperature.

Response to tempering is also a function of the degree of hardening of the steel (Fig. 6).

If you depend upon the notch-bar test as a measure of toughness, it is well to know that Izod values dip for tempering temperatures in the 425 to 750 F region, approximately. Alloy steels are seldom tempered in this "blue-brittle" range. The tempering range is 200 to 1200 F, and normally the heat-treater avoids the temperatures between 450 and 650 F. As a rule, articles subject to wear, such as cutting tools, ball and roller bearings, and gears, are tempered between 300 and 400 F, and items that must possess great toughness, such as structural parts of machinery, are tempered in the 800 to 1100 F range.

Interrupted Quenching

In normal quenching practice, the work is quenched down nearly to room temperature and then tempered while still warm to avoid cracking. But with isothermal heat-treatments, or interrupted quenching procedures, residual thermal contraction of the steel and expansion of martensite are minimized in the following three processes (see Fig. 7).

Austempering. Steel heated to the austenitizing temperature is rapidly cooled by plunging into molten salt held at a temperature between 400 and 800 F, perhaps about 650 F, where it is held long enough to transform the austenite to bainite. Austempered steel is extraordinarily tough as compared with quenched and tempered steel of the *same* analysis at the *same* hardness (see Fig. 8).

The process, which is patented by U.S. Steel Corp., is not limited with respect to section thickness, but products are normally ½ in. thick or less, especially with higher-hardenability steels, which would take too long to convert. So far austempering is used mainly for plain carbon steels and alloy steels of low alloy content.

Martempering. This process seeks to develop martensite instead of bainite, as in austempering. The steel is quenched into salt maintained at about the Ms point for the particular analysis, held long enough to equalize the temperature and create martensite, and then cooled in air. High hardness is obtained without quench cracks or dimensional changes. Heavier sections can be processed by martempering than by austempering. A tempering operation may follow if lower hardness is desired than produced in the martempered structure.

Isothermal Quenching. In this process the steel after being heated to the austenitizing temperature is plunged into a salt bath at a lower temperature than for the austempering process, usually 450 F or thereabouts. Upon isothermal transformation to a bainite-martensite structure, the steel is transferred to a higher-temperature salt bath for tempering. Minimum dimension change occurs.

Annealing of Steel

Annealing treatments are applied to steel to put it into the most suitable condition for machining or other cold-finishing operations. The treatments may also be applied to alter the mechanical or electrical properties of the material.

Metal that has been worked will exhibit a distorted grain structure under the microscope. Annealing restores the original structure and allows further cold working to be done. Mixed microstructures, arising from previous working of the metal, as in the steel-mill operations or hot forging, are changed to a uniform and more desirable structure, giving rise to more predictable results in further processing. Annealing softens steel and removes residual stresses.

Types of Annealing Treatments. Three types of annealing treatments are in general use:[1]

1. Austenitizing followed by very slow cooling.
2. Austenitizing followed by cooling to a predetermined temperature and *holding* for sufficient time for the steel to transform to the desired structure.
3. Heating to a temperature below the austenitizing temperature and cooling at any desired rate.

The first two processes involve heating the steel to, say, 100 F above its critical temperature, or Ae_1 point (Fig. 9), which is usually near or well over 1300 F, depending on composition, to put the steel in the austenitic condition. When the steel is cooled below the critical point, it will transform to ferrite and carbide. In the isothermal diagram, the top curve represents the time required for free ferrite to start to separate from the austenite, and it is called the "ferrite" line. It occurs on diagrams for steels with less than 0.90% carbon. The middle curve is sometimes called the "pearlite line" and represents the time required for the beginning of transformation to a ferrite-carbide aggregate. The third, or right-hand curve, represents the time for transformation at any temperature. Numbers at the right of the chart show the hardness of the steel after it has been completely isothermally transformed at the corresponding temperature and then cooled to room temperature. These numbers show that a soft steel is obtained by transforming at a high temperature. On the other hand, the upper portion of the curve is relatively flat, so that a much shorter time can be used if a slight increase in hardness can be tolerated.

Transformation to ferrite carbide does not take place between the austenitizing temperature and the critical temperature. Thus, there is no advantage in slow cooling between these temperatures. But the proper cooling rate to use below the critical temperature, for satisfactory annealing by continuous cooling, depends on the transformation characteristics of the *particular* steel. Four cooling-rate curves are superimposed on the *T-T-T*-diagrams for 2340 and 4340 steels (Fig. 9). For practical purposes, the end of transformation occurs 50° below the intersection of the cooling rate-curve and the isothermal curve.

[1] International Nickel Co.

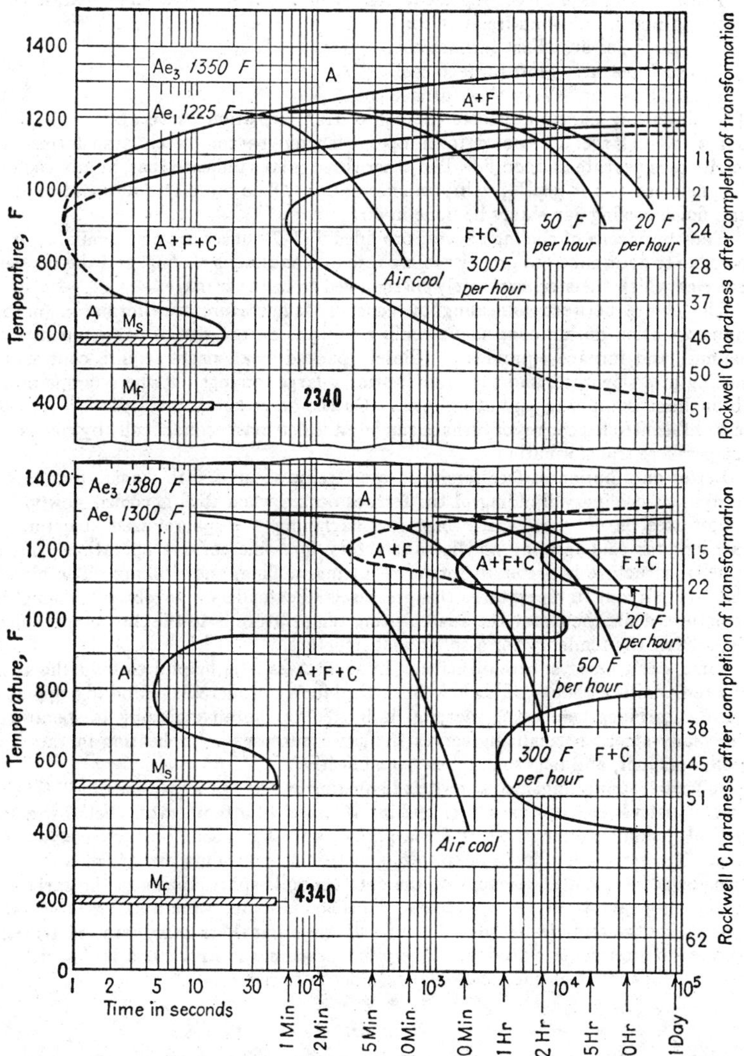

FIG. 9. Cooling-rate curves superimposed on the time-temperature-transformation diagram enable one to determine the most effective annealing cycle. Note that higher alloy content of 4340 steel pushes the S-curve to the right of the diagram, as compared with 2340 steel.

These cooling-rate curves represent, for a 2-in. round, the following conditions:

Air cool. Cooling freely in air.
300 F per hr. Fast-furnace cool.
50 F per hr. Slow-furnace cool.
20 F per hr. Very-slow-furnace cool.

The air-cooling curve intersects the end-of-transformation curve at 1050 F; so a bar 2 in. or larger air-cooled from 1450 F should be completely transformed to ferrite plus pearlite at 1000 F. The other three curves indicate that, at any cooling rate between 20 and 300 F per hr, transformation will be completed at about 1050 F, and finish cooling can safely be done in air.

Thus by means of cooling curves, provided *T-T-T*-curves are also available, it is possible to determine the proper cooling rate for continuous cooling, and the temperature at which the steel can safely be removed from the furnace.

The choice between annealing by constant-temperature transformation or by transformation during continuous cooling depends on the size of parts, manner of loading, and furnace equipment. Closely packed small parts require continuous cooling at a slow rate, and the same is true of large forgings, or where temperature throughout the furnace is not uniform. On the other hand, small parts processed in a well-controlled conveyor furnace can be annealed most economically by constant-temperature transformation.

Better distribution of free ferrite in the steel is secured with transformation by continuous cooling. Banding of the ferrites occurs when steel is cooled slowly.

Soft steel is not necessarily best for machining. A spheroidized structure is excellent for cold forming and is satisfactory for some turning operations, but a lamellar structure is preferred for form cutting, milling, and drilling. The higher the austenitizing temperature, the greater is the tendency to obtain a lamellar structure, while the closer the austenitizing temperature is to the Ae_1 temperature, the greater the tendency for a spheroidal structure.

Some steels resist spheroidization. In such cases it is helpful to hold the steel at a temperature about 50 F below the critical for a considerable period to agglomerate the carbides before the material is heated to the austenitizing temperature. For many steels, normalizing serves the same purposes as preheating in the suggested manner, and it is sometimes more effective.

Subcritical annealing, also known as tempering and as "process annealing" (in the wire industry), consists merely of heating the steel to a temperature below the Ae_1 critical, holding at temperature for a period, and then cooling at any convenient rate. The structure will not be so soft as realized by annealing procedures involving transformed austenite. Generally, the only action is spheroidizing of the carbides. If the original structure was coarse lamellar pearlite, subcritical heating will spheroidize the carbides to some extent. If the original structure was martensite, bainite, an intermediate product, or very fine pearlite, the structure after subcritical heating will consist of fine spheroidal carbides in ferrite.

Normalizing

In this process, the steel is heated to a temperature range approximately the same as used for hardening and quenching, but the material is usually cooled in still air. The process recrystallizes the metal and refines the grain structure. The softening effect of annealing and its adverse effect on machinability can be avoided in the case of forgings by normalizing. Sometimes, the work is cooled rapidly in the furnace to the pearlite transformation zone of the *T-T-T*-curve, held for transformation, and then cooled as desired.

Stress Relieving

Plastically deformed parts and welded structures frequently contain internal residual stresses that would be harmful to service life or further working. Such stresses are reduced by a stress-relieving treatment at a temperature that is normally just below the transformation range for the steel in question. For example, carbon steels are usually soaked at 1100 to 1250 F for 1 hr per in. of thickness. Higher-alloy steels, like chrome-moly types would be soaked up to 3 hr per in. of thickness at a temperature of 1350 to 1400 F. Obviously, stress relief cannot be applied to hardened and tempered parts, even though they contain residual stresses from quenching, because to do so would destroy the hardness created.

Carburizing[1]

Carburizing is a casehardening procedure whereby carbon is added to steel from the surface inward to a specified depth. The carbon content in the affected zone must be sufficiently high to impart the desired wear resistance, hardness, and strength after a subsequent hardening heat-treatment. The underlying metal retains its original toughness after treatment. Progressive diminution of carbon from surface to core is termed the "carbon gradient."

Development of the desired physical properties is obtained only by quenching the carburized case from prescribed temperature above the transformation range of the steel. When this is done the carbon-enriched case becomes hard and wear-resistant.

The process is applicable to carbon and alloy steels of low carbon content, and to a lesser extent to steels of medium carbon content (see Tables 4 and 5).

The layer of carbon enrichment is termed the "case," the original or underlying material the "core," and the blend zone in between is known as the gradation zone. At customary carburizing temperatures, iron in the gamma phase is capable of dissolving carbon and alloying elements to form a solid solution known as austenite. As the operation proceeds, carbon continues to be added to the surface and migrates inward at a rate dependent upon such factors as diffusion coefficient, temperature, and the carbon potential of the carburizing medium.

Up to 1.7% carbon can be dissolved in the austenite but this is seldom done. Maximum hardness is obtained when carbon content is about 0.80% and decreases progressively somewhat beyond 0.90%. To be on the safe side, the aim is generally to obtain 0.85 to 1.20% carbon. Carburizing temperatures in excess of 1750 F are seldom used. The common range is 1675 to 1725 F, and 1700 F is the predominant temperature.

Influence of Hardenability. The hardenability of steel is an important factor in influencing case characteristics. In general, as the hardenability of steel increases, the hardness of the case decreases, whereas the core hardness increases.

Several factors are influential in increasing the hardenability of steel. Important factors that affect hardenability are carbon content, alloying elements, grain size, and homogeneity, or likeness throughout, of the austenite prior to quenching. As hardenability of the steel increases, the start of the transformation on cooling is correspondingly delayed.

Depending upon chemical composition and rate of cooling, austenite may be directly transformed to martensite. Except when low-hardenability plain carbon steel is used for large sections, the case usually quenches out with a basic martensitic structure interspersed with residual austenite. The core transforms to softer constituents.

Where maximum resistance to indentation and abrasion is desired, high-harden-

[1] T. A. Frischman, Chief Metallurgist, Axle Division. Eaton Manufacturing Co.

TABLE 4. HEAT-TREATMENTS FOR CARBURIZING GRADES OF CARBON STEELS

SAE Steels	Normalizing Temperature, F	Carburizing Temperature, F	Cooling Method	Reheat Temperature, F	Cooling Medium	Second Reheat, F	Cooling Medium	Temper,* F
1010, 1012, 1015, 1016, 1017, 1018, 1019, 1020, 1021, 1022, 1023	—	1650–1700	Water or brine	—	—	—	—	250–400
	—	1650–1700	Oil or water	1400–1450	Water or brine	—	—	250–400
	—	1650–1700	Cool slowly	1400–1450	Water or brine	—	—	250–400
	—	1650–1700	Cool slowly	1650–1700	Oil or water	1400–1450	Water or brine	250–400
	—	1500–1650†	Oil or water	—	—	—	—	Optional
	—	1350–1575‡	Air or oil	—	—	—	—	Optional
1024	1650–1750§	1650–1700	Oil	—	—	—	—	250–400
	—	1350–1575‡	Air or oil	—	—	—	—	Optional
1025, 1026	—	1650–1700	Water or brine	—	—	—	—	250–400
	—	1500–1650†	Oil or water	—	—	—	—	Optional
1027	—	1350–1575‡	Air or oil	—	—	—	—	Optional
1030	—	1500–1650†	Oil or water	—	—	—	—	Optional
	—	1350–1575‡	Air or oil	—	—	—	—	Optional
1111, 1112, 1113	—	1500–1650†	Oil or water	—	—	—	—	Optional
	—	1350–1575‡	Air or oil	—	—	—	—	Optional
1108, 1109, 1115, 1117, 1118, 1119, 1120	—	1650–1700	Water or brine	—	—	—	—	250–400
	—	1650–1700	Oil or water	1400–1450	Water or brine	—	—	250–400
	—	1650–1700	Cool slowly	1400–1450	Water or brine	—	—	250–400
	—	1650–1700†	Cool slowly	1650–1700	Oil or water	1400–1450	Water or brine	250–400
	—	1500–1650†	Oil or water	—	—	—	—	Optional
	—	1350–1575‡	Air or oil	—	—	—	—	Optional
1126	—	1500–1650†	Oil or water	—	—	—	—	Optional
	—	1350–1575‡	Air or oil	—	—	—	—	Optional

* Even where recommended draw temperatures are shown, the draw is not mandatory on many applications. Tempering is generally employed for a partial stress relief and improves resistance to cracking from grinding operations. Higher temperatures than those shown may be employed where the hardness specification on the finished parts permits.

† This treatment is for activated or cyanide baths. Parts may be given refining heat as indicated for other heat-treating processes.

‡ This treatment is for carbonitriding atmospheres. Parts may be given refining treatments as indicated for other heat-treating processes.

§ Normalizing temperatures at least 50 F above the carbonizing temperature are sometimes recommended where heat-treat distortion is of vital importance.

ability carburized steels should not be used in the direct-quenched condition. Instead, they should be rehardened if at all possible, freedom from distortion permitting.

Carburizing Materials and Methods. General industrial practices are: (1) pack method, (2) gas method (including oil vapors), and (3) liquid-salt method. Sometimes, the impression is given that one or another method is faster. Based on total elapsed time this is true, but the essential difference in time is the interval necessary in bringing the work to the carburizing temperature. When the carburizing temperature is once attained the speed thereafter to obtain a given case depth is much the same for any of the three methods, provided that the temperature and time are the same and that the carburizing media are of the same carbon potential.

Pack Carburizing

Also known as box carburizing, pack carburizing is the oldest and most widely used. It still remains the most foolproof and is perhaps the most versatile method in accommodating wide ranges in geometry of parts.

Case Depth. Use of pack carburizing is generally restricted to case depths greater than 0.025 in., because it is difficult to heat uniformly and quickly through carburizing compound. Also, case depth is usually specified to wider ranges than with other carburizing processes. Experience has shown that the difference in carbon content of work in large containers adjacent to the furnace wall and at the center of the charge will vary by 0.10% carbon at the outermost layer of the case and 0.13% carbon at a depth of 0.050 in.

Pack carburizing consists of heating the work in containers or pots. The work is surrounded by a carbonaceous compound. When the compound is heated to carburizing temperatures, carbon monoxide is formed. It reacts as nascent or atomic carbon at the surface of the steel and migrates inward by a solid-diffusion process.

Most carburizing compounds fall into two classes: (1) charcoal-coke and (2) coke type. Both kinds contain 15 to 20% alkali carbonates or "energizers," whose action is similar to a catalyst.

Working mixtures of the charcoal-coke type generally run from 1 part new to 3 parts old compound to as high as 1 part new to 5 parts old. While charcoal-coke compound is still used in volume, the high-percentage-coke type known as the "nonburning" is rapidly becoming popular. The essential difference is the replacement of part or all of the charcoal ingredient with coke of petroleum or bituminous origin. Coke burns slowly in comparison with charcoal. Hence, working ratios as high as 8 parts old to 1 part new compound are employed, and there is greater freedom from dust. The coke compound is cleaner to handle and makes for better shop conditions. Where direct quenching from the carburizing pots is employed, the nonburning type is generally used because of its slow-burning characteristics.

Particle size is governed by the product carburized and should be in proportion to the product size. Maximum particle size should be sought so that free circulation of gases and rapid heat transfer are realized.

Packing procedures are somewhat the same for either type of compound and are governed considerably by the shrinkage factor. Shrinkage may run from 15 down to 5%. A good rule is a spacing of ½ to 1 in. between container and part and between individual parts, after having gone through the carburizing cycle. This distance should guarantee sufficient gas generation to carburize the part.

Current practice in up-to-date shops is to employ direct quenching wherever possible. If pot cooling is used, means should be provided to cool the compound

TABLE 5. HEAT-TREATMENTS FOR CARBURIZING GRADES OF ALLOY STEELS

SAE Steels	Alternate Pretreatments		Cycle Anneal[c]	Carburizing Temperature, F	Cooling Method	Reheat Temperature, F	Cooling Medium	Tempering[d] Temperature, F
	Normalize[a]	Normalize and Temper[b]						
2317	Yes	—	Yes	1650–1700	Quench in oil	1375–1425[e]	Oil	250–350
	Yes	—	Yes	1650–1700	Quench in oil	1450–1500[f]	Oil	250–350
	Yes	—	Yes	1650–1700	Cool slowly	1375–1425[e]	Oil	250–350
	Yes	—	Yes	1650–1700	Cool slowly	1475–1525[f]	Oil	250–350
	Yes	—	Yes	1650–1700	Quench in oil[g]	—	—	250–350
	Yes	—	Yes	1450–1650	Quench in oil	—	—	250–350
2515, 2517	—	Yes	—	1650–1700	Cool slowly	1325–1375[e]	Oil	250–350
	—	Yes	—	1650–1700	Cool slowly	1425–1475[f]	Oil	250–350
3120	Yes	—	—	1650–1700	Quench in oil	1400–1450[e]	Oil	250–350
	Yes	—	—	1650–1700	Quench in oil	1475–1525[f]	Oil	250–350
	Yes	—	—	1650–1700	Cool slowly	1400–1450[e]	Oil	250–350
	Yes	—	—	1650–1700	Cool slowly	1500–1550[f]	Oil	250–350
	Yes	—	—	1650–1700	Quench in oil[g]	—	—	250–350
	Yes	—	—	1500–1650	Quench in oil[h]	—	—	250–350
3310, 3316	—	Yes	—	1650–1700	Quench in oil	1400–1450[e]	Oil	250–350
	—	Yes	—	1650–1700	Cool slowly	1475–1500[f]	Oil	250–350
4023, 4024, 4027, 4028, 4032	Yes	—	Yes	1650–1700	Quench in oil[g]	—	—	250–350
4118, 4119	Yes	—	—	1650–1700	Quench in oil[g]	—	—	250–350
4320, 4608, 4615, 4617, 4620, X4620, 4621	Yes	—	Yes	1650–1700	Quench in oil	1425–1475[e]	Oil	250–350
	Yes	—	Yes	1650–1700	Quench in oil	1475–1525[f]	Oil	250–350
	Yes	—	Yes	1650–1700	Cool slowly	1425–1475[e]	Oil	250–350
	Yes	—	Yes	1650–1700	Cool slowly	1475–1525[f]	Oil	250–350
	Yes	—	Yes	1650–1700	Quench in oil[g]	—	—	250–350
	Yes	—	—	1500–1650[h]	Quench in oil	—	—	250–350
4812, 4815, 4817, 4820	—	Yes	Yes	1650–1700	Quench in oil	1375–1425[e]	Oil	250–350
	—	Yes	Yes	1650–1700	Quench in oil	1450–1500[f]	Oil	250–350
	—	Yes	Yes	1650–1700	Cool slowly	1375–1425[e]	Oil	250–350
	—	Yes	Yes	1650–1700	Cool slowly	1450–1500[f]	Oil	250–350
	—	—	—	1650–1700	Quench in oil[g]	—	—	250–350

SAE Steels	Alternate Pretreatments		Cycle Anneal[c]	Carburizing Temperature, F	Cooling Method	Reheat Temperature, F	Cooling Medium	Tempering[d] Temperature, F
	Normalize[a]	Normalize and Temper[b]						
5115, 5117, 5120, 6117	Yes	—	—	1650–1700	Quench in oil	1425–1475[e]	Oil	250–350
	Yes	—	—	1650–1700	Quench in oil	1500–1550[f]	Oil	250–350
	Yes	—	—	1650–1700	Cool slowly	1425–1475[e]	Oil	250–350
	Yes	—	—	1650–1700	Cool slowly	1500–1550[f]	Oil	250–350
	Yes	—	—	1500–1650[h]	Quench in oil	—	—	250–350
8615, 8617, 8620, 8622, 8625, 8715, 8720	Yes	—	Yes	1650–1700	Quench in oil	1475–1525[e]	Oil	250–350
	Yes	—	Yes	1650–1700	Quench in oil	1525–1575[f]	Oil	250–350
	Yes	—	Yes	1650–1700	Cool slowly	1475–1525[e]	Oil	250–350
	Yes	—	Yes	1650–1700	Cool slowly	1525–1575[f]	Oil	250–350
	Yes	—	Yes	1650–1700	Quench in oil[g]	—	—	250–350
	Yes	—	Yes	1500–1650[h]	Quench in oil	—	—	250–350
9310, 9315, 9317	—	Yes	—	1650–1700	Quench in oil	1400–1450[e]	Oil	250–350
	—	Yes	—	1650–1700	Cool slowly	1500–1525	Oil	250–350

[a] Normalizing temperatures should be not less than 50 F higher than the carburizing temperature followed by air cooling.

[b] After normalizing, reheat to temperature of 1000 to 1200 F and hold approximately 4 hr.

[c] Where cycle annealing is desired, heat to normalizing temperature, hold for uniformity, cool rapidly to 1000 to 1250 F. Hold 1 to 3 hr, then air cool or furnace cool to obtain a structure suitable for machining and finish.

[d] Tempering treatment is optional. Tempering is generally employed for partial stress relief and improved resistance to cracking from grinding operations. Temperatures higher than those shown are used in some instances where application requires.

[e] This treatment is applicable when casehardness only is paramount.

[f] This treatment is applicable when higher core hardness is desired.

[g] This treatment is applicable to fine-grained steels only. When fine-grained steels are employed, a second reheat is often unnecessary.

[h] This treatment is for activated or cyanide baths, and parts may be given refining heats as indicated for other heat-treating processes.

and contents to black heat as quickly as possible to overcome decarburization. In some instances, work can be dumped and air-cooled.

Carburizing Pots. Containers must be strong enough to hold the work and withstand handling at elevated temperatures, particularly when removing from the furnace for direct quenching. They should not be too bulky or massive for handling. The lightest possible wall thickness commensurate with good design should be used to permit rapid heat penetration to the center of the charge to eliminate differences in case depth throughout the pot.

Two basic pot designs seem most popular: (1) complete casting, and (2) fabricated rolled sheets. For the same cubical content, sheet boxes are lighter than cast boxes.

Carburizing pots or boxes are made of heat-resisting alloy such as 25% chromium 12% nickel, 25% chromium 20% nickel, 25% nickel 20% chromium, 35% nickel 15% chromium, 60% nickel 12% chromium, and 80% nickel 13% chromium. The higher the nickel content the less the carbon absorption by the alloy, and hence the lower the tendency for the carburizing compound to carburize through the container walls. Such action causes brittleness and lowered resistance to thermal shock. The 80% nickel 13% chromium alloy has been known to give from two to three times the life of 35% nickel 15% chromium alloy, especially in fabricated rolled-sheet containers.

Gas Carburizing

Gas carburizing involves the use of fuel gases like natural gas, propane, butane, manufactured gas, and other compositions as the carburizing medium. Hydrocarbon liquids which break down into oil vapors are also employed. Workpieces are exposed to these gases in retorts, muffles, or tight furnace chambers under slight pressure and maintained at the same temperature ranges as used for the pack method.

Types of furnace equipment in general use are: (1) continuous, (2) stationary batch, and (3) retort, either vertical pit or horizontal rotary. Most installations are gas-fired, especially the continuous type. Since the carburizing-gas reactions are dependent upon definite percentages of various constituents, the products of combustion arising from firing of fuels must not be allowed to contaminate the carburizing gases.

Carburizing with straight natural gas is the simplest and most economical of the various gas-carburizing methods. If rigid avoidance of fairly high carbon concentration in the case is not necessary, a straight carburizing cycle, omitting a diffusion period, can be used for the batch and retort-type furnaces. The natural gas is simply admitted at a safe furnace temperature, above 1300 F, at a high rate of flow, then cut back until the contents of the furnace approach the carburizing temperature. As the furnace temperature and workpieces reach equilibrium, the gas is admitted at a constant rate until the desired case depth is obtained.

If diffusion is employed (the practice of restraining the formation of high-carbon-content cases), gas is cut off at about two-thirds the duration of the carburizing cycle, the discharge pipe is capped, and the contents and furnace are allowed to reach equilibrium with the carbon potential of the gas during the last third of the cycle. Thus, the carbon diffuses inward without building up to an excess much above the eutectoid or 0.85% carbon range in the outer layers of the case.

Carburizing with oil vapors is much the same as with natural gas, and both were widely used until the introduction of carrier gases.

Carrier Gases. As a rule, carrier gases are not intended to be very reactive with the material being protected or treated in the furnace. One of their important functions is to act as a vehicle for the active carburizing gas such as methane, propane, or other gas and thus make up most of the volume of the mixture introduced into the furnace. Carrier gases have different analyses, depending on whether they are generated exothermically or endothermically.

Natural gas, when available, is almost invariably used to supply carbon in the gas-carburizing process. It is generally added in measured amounts to the carrier gas, and the mixture admitted into the furnace chamber. However, either can be added unmixed. Where natural gas is not available, propane or butane can be used, as can manufactured or artificial gas. The latter present more of a problem because the percentages of some of their important constituents change from time to time even though the Btu content remains constant.

Dew Point. Dew point can be defined as that temperature on cooling at which moisture or water vapor will condense out of a gas. Dew points of gases vary considerably, depending upon their chemical composition. Control of moisture content is achieved by correct balancing of air-gas ratios to prevent either the formation of carbon dioxide, or excessive sooting of the reaction chamber, or in other words, cracking either too lean or too rich a mixture. In some instances, low percentages of carbon dioxide are permissible, and they are actually desired in some plants but rarely exceed 0.50% for carburizing operations. Natural gas usually has a low dew point, in the neighborhood of −30 F, but can vary as high as +30 F. By keeping the dew point within a range of +10 to +25 F, the process can be controlled within the desired limit for case depth and carbon concentration.

Furnace-chamber Pressure. If the pressure is not positive, air will find its way into the furnace to form water vapor or carbon dioxide, both of which are decarburizing agents. General practice is to aim for a furnace pressure of 0.10- to 0.50-in. water column to prevent ingress of contaminants with opening and closing of doors.

Decarburization. Decarburization may be defined as loss of, or lowering of, surface carbon concentration below that of the immediate interior portion of the case. Means of combating decarburizing tendencies in gas carburizing are many, but the main methods are proper control of the carburizing gases and the prevention of the infiltration of contaminating mediums, such as products of combustion, air, and/or water vapor.

Rate of Case Formation. Time at heat to produce a given effective case depth is a function of the carburizing temperature and the carbon potential of the carburizing medium. A figure in common use is approximately 0.007 in. per hr up to 0.040-in. case depth, and diminishing to approximately 0.005 in. per hr upon reaching 0.060 in. These figures are estimates and are to be used for a temperature of 1700 F assuming the steel has reached the carburizing temperature.

Carbon Stop-off. Effective carbon stop-off, or selective carburizing as the term is also known, is difficult with gas carburizing. Copper-plating procedures employed for carbon stop-off in pack carburizing do not ordinarily suffice for gas carburizing, unless the quality of the copper plate and its depth are carefully watched. Plating time will average at least four times longer for effective stop-off when gas carburizing than for the pack method.

Dew-point Determinations. With gas carburizing, it is a good plan to control the chemical composition and dew point of both the generator gas and the mixture of gas in the furnace chamber. Daily dew-point determinations and weekly gas analyses are recommended. In analyzing the gas, frequently only the carbon dioxide, oxygen, and carbon monoxide are determined.

Advantages and Disadvantages. The favorable factors for gas carburizing are: (1) shorter time than pack method for equivalent case depths; (2) carburizing boxes,

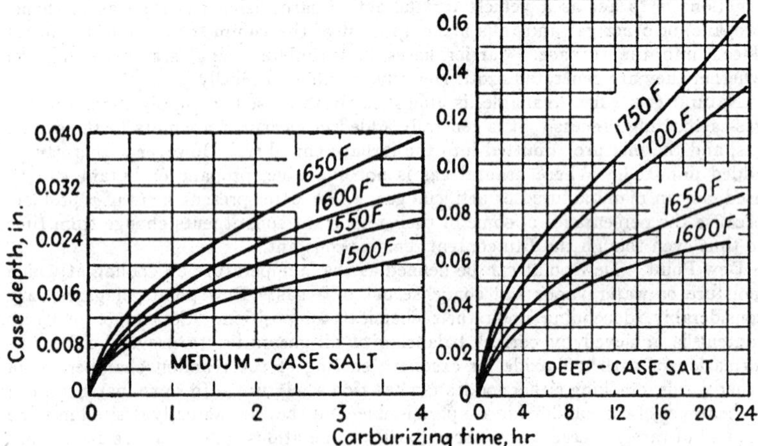

FIGS. 10 and 11. Time-penetration curves for two types of carburizing salts. Case depth is affected by both the bath characteristics and the operating temperature and time. (*American Cyanamid.*)

covers, and carburizing compound are not needed; (3) shop conditions are cleaner; (4) quenching of tray loads of materials is permitted instead of individual handling of the pieces; (5) low labor cost; and (6) more positive control of type of case (carbon concentration). In continuous carburizing, the process becomes a more or less production-line operation.

Disadvantages are: (1) more technical control is required when using mixtures of carrier or diluent gases and hydrocarbon fuels as the carburizing medium; (2) protective atmosphere becomes lost in case of gas-supply failure and to a lesser degree, power failure; (3) furnace must be kept tight and in good state of operation; (4) wide variety of work-supporting fixtures are needed unless parts can be bulk-loaded; (5) copper plate for carbon stop-off is less effective than for the pack method, unless greater depth of plate is provided; (6) heat-resisting alloy items with tendencies to be porous deteriorate faster because carbon enters the voids.

CARBURIZING SALT BATHS

Salt-bath carburizing is an outgrowth of the well-known cyaniding bath but produces much deeper cases higher in carbon content and lower in nitrogen content. With liquid-salt carburizing baths, heavy cases ranging from 0.040- to 0.250-in. depth can be produced in comparison with the straight cyanide bath where 0.010-in. depth is generally regarded as the maximum.

By varying the chemical composition of the baths, they can be made to produce light, medium, or heavy case depths. Light case depth is generally considered to range from 0.003 to 0.015 in., medium from 0.015 to 0.040 in., and heavy from 0.040 up to as high as 0.250 in.

Carbon content at various depths of case is affected by the carburizing temperature and the characteristics of the salt bath. For example, SAE 1020 steel held for 2 hr in medium-case salt at 1550 F has a carbon content of 0.48% and a nitrogen content of 0.09% at a case depth of 0.010 in. For the same exposure in deep-case salt held at 1750 F, the carbon content is 0.90% and the nitrogen content about 0.12% at a case depth of 0.012 in. (see Figs. 10 and 11).

Furnace Equipment. Prevailing furnaces for liquid-salt bath carburizing are: (1) externally heated with either gas, oil, or electricity; and (2) internally heated, either immersed electrode or immersed fuel-fired heater. The externally heated salt-bath furnaces are mostly of the round-pot type. Rectangular furnaces incorporating mechanization for continuous advancement of workpieces through the salt bath are also built. Either of these types of furnaces can be heated by gas, oil, or electricity.

Immersed-electrode furnaces use resistance of the liquid salt to create the heat necessary to maintain the bath temperature and to heat the charge. This type of furnace is efficient from the standpoint of heat input and requires rather small floor space.

Rapid heating rate of liquid carburizing baths is the prime reason for producing equivalent case depths so much more quickly than with the other methods of carburizing. Light and medium case depths are obtained through the range from 1550 to 1650 F; deep cases are generally developed with carburizing temperatures from 1700 to 1750 F.

Versatility of Process. There appears to be no limit to the size and shape, either large or small, of the article to be liquid-carburized, so long as the container or pot will accommodate it. Brazing can be combined with the carburizing operation. Where selective carburizing is required, liquid carburizing provides an easy solution. If, for instance, only one extremity of the part requires casehardening, that portion only need be immersed in the bath.

Control of Bath. Deep-case carburizing baths are as a rule controlled to 10% cyanide, and medium to light case at from 15 to 20%. If liquid carburizing baths must be used intermittently, or not over week ends, best practice is to idle the bath at reduced temperature, even though the fuel expense might seem high.

Advantages of Salt Baths. Because of its even heating characteristics, the method is unexcelled for producing uniform carburized cases, light or medium in depth, all over intricately shaped parts, where even small differences in case depth cannot be permitted. Work comes from the operation scale-free and in many instances relatively bright, depending upon the bath composition. Distortion of parts is found to be generally less than from other methods of carburizing. Martempering, a process that reduces crackage and distortion on hardening, is especially adaptable to the liquid-carburizing process.

Other advantages are: (1) parts can be direct-quenched into either oil or water; (2) in batch installations particularly, different case depths can be produced simultaneously by the simple expedient of lengthening the immersion periods; (3) no soot problem is encountered; and (4) the process can be made entirely independent of fuel-gas shortages by heating with electricity.

Disadvantages. Hardly anyone who has used the process needs any precautionary instructions concerning the dangers of charging workpieces containing moisture. Laxity can result in serious splashing and even explosion of molten salt, causing painful burning of the skin. Goggles should be worn at all times by furnace operators to prevent eye injury. Fumes should be vented.

Cyanides are deadly poisons. Every care should be taken, so that even minute portions are not consumed accidentally. In storage, always keep cyanide-containing salts separate from acids in case of fire, explosion, or other unforeseen occurrences, because mixing will form deadly hydrocyanic gas. Do not add nitrate or nitrite salts to those containing cyanides because of the danger of explosion.

Postcarburizing Treatments

After carbon enrichment of steel to specified concentration and depth of penetration, a choice of subsequent treatments is available. These postcarburizing treatments may be: (1) cool the work in air, in carburizing compound, or in gaseous medium for either intermediate machining operations or for one or more reheating treatments; or (2) quench from the carburizer, either directly or after an intentional delay.

Slow Cooling. Work cooled in carburizing compound, especially where large containers are used, has a tendency to become decarburized. Where sufficient grinding stock is provided or where controlled-atmosphere furnaces are employed for rehardening operations, carbon loss is no obstacle.

If air cooling of pieces individually or within containers or within gas-filled chambers does not leave the material soft enough for intermediate machining operations, a tempering treatment must be employed. Although this might seem expensive, it is often cheaper than incurring high tool costs from work with hardness above the economical machining range.

Slow-cooling Applications. Most slow-cooling procedures are employed for subsequent rehardening heat-treatments, either single or double. On slowly cooled material, a good combination of case and core properties can be obtained by reheating to a temperature at or slightly above the upper transformation temperature of the core. This is referred to as a single quenching treatment. Thus both the case and core become hardened, giving the part increased strength and resistance to shock. Treatments such as the foregoing are employed where parts are to be fixture-quenched, and where it is not always feasible to quench parts individually from the carburizer because of the large quantity being discharged at once.

Direct Quenching. The term "direct quenching" implies quenching directly from the carburizing medium into brine, water, oil, or liquid salt. The effect of direct quenching is often modified by a deliberate time delay, before quenching, or by reducing the temperature somewhat at the end of the carburizing cycle.

Extensive use is made of direct quenching, in order to maintain minimum distortion, as compared with single or double reheating treatments. Martempering, a method that involves quenching into hot liquid baths (oil or salt), is outstanding in hardening applications where distortion must be kept to an absolute minimum. However, hot-salt or hot-oil quenching tends to increase the amount of retained austenite. For this reason the carbon content of the outer surface layers of the case must be substantially lowered to avoid this condition.

Tempering Procedure for Carburized Work

Tempering, as used in conjunction with carburizing and hardening, is a low-temperature heat-treatment employed primarily for the benefit of the case. It increases the resistance to impact, helps to prevent grinding cracks or checks, aids in reducing residual stress, minimizes chipping of the ends of clash gear teeth, and also facilitates lapping of tooth contacts in gear manufacture. General range of temperature for tempering carburized work is from 275 to 375 F. Customary time at temperature is 2 to 3 hr.

For maximum assurance that cracking or checking will not develop during grinding of carburized and hardened work, it is essential that the tempering temperature be at least 325 F, or better still 350 to 375 F. As an added precaution, grinding is sometimes carried out with the piece completely submerged in oil.

Final hardness after tempering is related to the type of steel and its microstructure and level of hardness in the as-quenched condition. If the hardness is high enough originally and the microstructure is martensitic, relatively high tempering temperatures are required to lower the Rockwell hardness below 59 C.

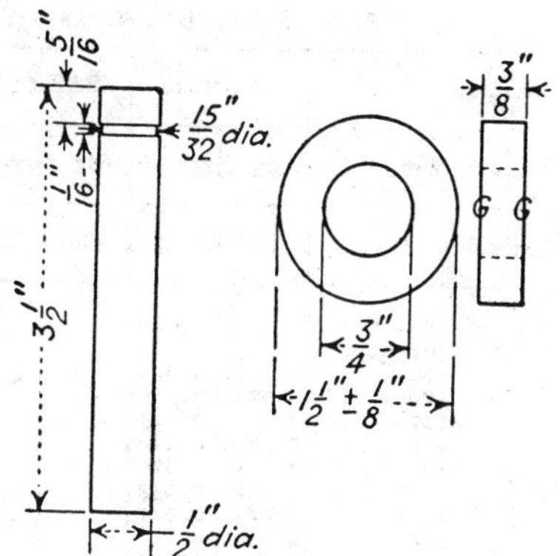

FIG. 12. Machined pins and rings are frequently used with production carburizing setups to check case depth.

CASE-DEPTH MEASUREMENT

For production measurement of case depth, the most common methods used are: (1) reading of fractured hardened case with a brinell microscope; (2) the same as 1 but etching to reveal the case more clearly; (3) step grinding and determining Rockwell hardness at predetermined depths; (4) bluing of the fractured surface; (5) polishing and etching of a hardened fractured or cut surface and reading the depth with a brinell microscope; (6) carbon-cut analyses; (7) microscopical examination; and (8) the martensite-start method.

In the foregoing methods of measurement, the difficulty is to find the fine point of demarcation between case and core.

Production Method. A production method to bring out sharp distinctions between case and core is to air-cool the test pin (Fig. 12) from the carburizing temperature, reheat to 1475 F, and quench in oil. This treatment refines the so-called effective case and leaves the core somewhat coarse-grained. Etching the fractured surface in a solution of 7% concentrated nitric acid in water produces good contrast between case and core.

Other Methods. More specialized methods of case-depth testing are also in use (Fig. 13). They may require more technical skill than the customary fracture and etch method, and often require grinding and measuring equipment. Among these methods are:

1. By step grinding, hardness is measured at predetermined levels from the surface of the case.
2. The taper-grinding method achieves the same result, because the angle is selected to permit readings at spaced increments corresponding to depth increments.
3. The traverse method is akin to the taper-grinding method but is done at an angle across a section through case and core.

SURFACE HARDNESS VS. CASE DEPTH[1]

It is not always practical to prepare test specimens for depth of carburized case by polishing and etching samples in a 5% nital solution. Table 6 for several different steels shows the brinell hardness at various depths of carburized case. For example, if the hardness reading for a SAE 1020 carburized and hardened part is 321 brinell, the depth of the effective case is 0.074 in.

Some variation in results may be encountered because of the size and shape of the work. Separate data may be required for special cases. With this reservation, the table can be used for establishing values for routine checking of carburized case depths on individual gears and similar parts after carburizing and hardening.

[1] James Sorenson, Chief Metallurgist, and Alfred Hendrickson, Four Wheel Drive Auto Co.

TABLE 6. CASE DEPTH AS DETERMINED BY BRINELL HARDNESS

Steel	SAE 1020		SAE 6120		SAE 43BV17		SAE 1020		SAE 43B17	
Quench temp.	1675 F	1500 F	1675 F	1500 F	1675 F	1500 F	1675 F	1500 F	1675 F	1400 F
Quench med.	Oil	Oil	Oil	Oil	Oil	Oil	Water	Water	Oil	Oil
Brinell										
156	......	0.000								
167	......	0.010								
170	......	0.014								
179	0.000	0.019								
187	0.007	0.024								
192	0.009	0.026	.....	....	.....	....	.....	0.000		
197	0.011	0.029	.....	....	.....	....	.....	0.003		
207	0.016	0.035	.....	....	.....	....	.....	0.005		
217	0.023	0.041	.....	....	.....	....	.....	0.009		
229	0.030	0.049	.....	....	.....	....	0.000	0.013		
241	0.035	0.069	.....	0.000	.....	....	0.005	0.018		
255	0.040	0.077	.....	0.003	.....	....	0.008	0.023		
269	0.046	.....	.....	0.007	.....	....	0.015	0.027		
285	0.053	.....	.....	0.013	.....	0.000	0.022	0.034		
302	0.060	.....	.....	0.018	.....	0.004	0.027	0.039		
321	0.074	.....	0.000	0.023	0.000	0.008	0.034	0.046	.....	0.000
341	......	.....	0.004	0.029	0.004	0.012	0.041	0.054	.....	0.010
363	......	.....	0.009	0.036	0.009	0.016	0.048	0.062	.....	0.015
387	......	.....	0.014	0.043	0.017	0.020	0.056	.....	.....	0.021
412	......	.....	0.019	0.048	0.024	0.025	0.062	.....	0.000	0.027
444	......	.....	0.026	0.060	0.034	0.031	.....	.....	0.019	0.034
460	......	.....	0.030	.....	0.039	0.034	.....	.....	0.024	0.038
477	......	.....	0.034	.....	0.044	0.037	.....	.....	0.028	0.042
495	......	.....	0.038	.....	0.050	0.040	.....	.....	0.033	0.047
512	......	.....	0.042	.....	0.055	0.043	.....	.....	0.038	0.051
532	......	.....	0.046	.....	0.062	0.047	.....	.....	0.043	0.055
555	......	.....	0.051	.....	.....	0.051	.....	.....	0.049	0.061
578	......	.....	0.057	.....	.....	0.054	.....	.....	0.055	
600	......	.....	0.062	.....	.....	0.060	.....	.....	0.061	
627										

Steel	SAE 4615		SAE 8620		SAE 3310		SAE 4820		FWD Krupp	
Quench temp.	1675 F	1500 F	1675 F	1500 F	1675 F	1500 F	1675 F	1500 F	1675 F	1500 F
Quench med.	Oil	Oil	Oil	Oil	Oil	Oil	Oil	Oil	Oil	Oil
Brinell										
217	......	0.000								
229	......	0.004								
241	......	0.009								
255	......	0.015								
269	0.0000	0.021								
285	0.0070	0.024	.....	0.000						
302	0.0100	0.028	.....	0.005						
321	0.0130	0.032	0.000	0.009						
341	0.0170	0.035	0.003	0.012	.....	.....	.....	0.000		
363	0.0210	0.039	0.005	0.016	0.000	0.000	0.000	0.004	0.000	0.000
387	0.0240	0.041	0.010	0.022	0.012	0.005	0.008	0.010	0.016	0.009
412	0.0300	0.045	0.016	0.027	0.018	0.009	0.012	0.015	0.030	0.021
444	0.0330	0.048	0.022	0.035	0.033	0.018	0.024	0.022	0.046	0.030
460	0.0350	0.051	0.026	0.039	0.039	0.022	0.028	0.026	0.051	0.035
477	0.0370	0.054	0.030	0.043	0.045	0.026	0.033	0.030	0.062	0.039
495	0.0390	0.056	0.034	0.047	0.052	0.031	0.038	0.035	0.076	0.045
512	0.0430	0.058	0.037	0.050	0.058	0.035	0.043	0.039	.....	0.048
532	0.0510	0.060	0.042	0.055	0.065	0.040	0.049	0.043	.....	0.055
555	0.0640	.....	0.047	0.061	.....	0.045	0.055	0.049	.....	0.060
578	......	.....	0.051	0.066	.....	0.051	0.063	0.054	.....	0.068
600	......	.....	0.055	.....	.....	0.060	.....	0.060	.....	0.075
627	......	.....	0.061	.....	.....	0.066	.....		.....	

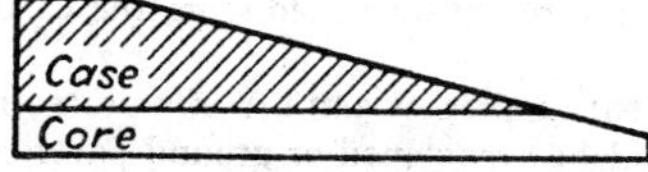

Taper sectioning of specimens provides more accurate readings on light and medium cases. Angle is selected so readings at spaced distances represent desired depth increments.

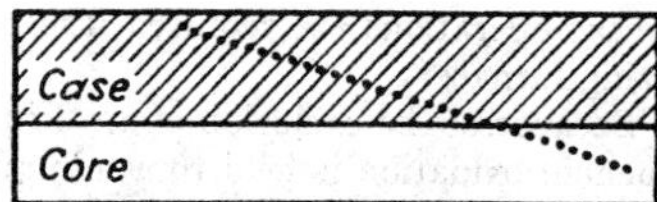

Traverse procedure for reading hardness at various depths on light, medium, and heavy cases, the specimen not being taper ground. Distance from center of indentation to surface is measured with microscope.

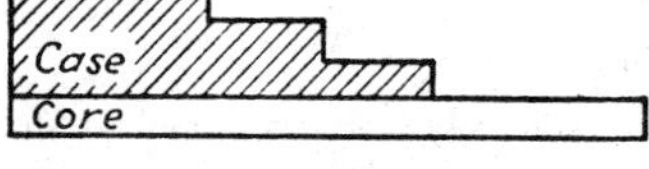

Step-ground section is recommended for medium and heavy cases. Hardness readings are taken on the steps, which are at known distances from the surface.

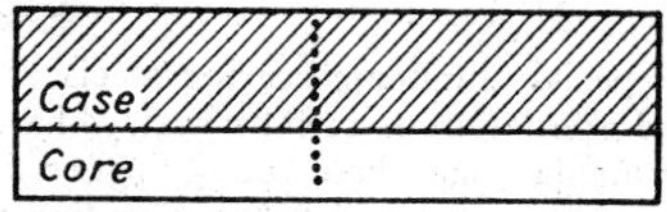

Alternate method for traversing heavy cases.

FIG. 13. Three methods of exploring case depth on specimens require good grinding and measuring equipment.

Carbonitriding

Carbonitriding is a gas casehardening process that produces a thin duplex case of nitrides and carbides on wrought and cast steels, cast iron, and malleable iron, when exposed to a furnace atmosphere consisting of ammonia and a carbon-rich gas. The case may range from 0.001 to 0.035 in. thick, but the normal range is 0.005 to 0.020 in. The process is cleaner than cyaniding and requires somewhat less time and a lower temperature than gas carburizing. It is said that parts which distort when oil-quenched from salt will show virtually no distortion when carbonitrided.

In carbonitriding the ammonia content of the furnace atmosphere varies from say 1% for work to be quenched to 15% for furnace-cooled work.

Wear resistance of carbonitrided cases is said to excel cyanided or carburized cases. Case hardness of Rockwell 62 to 64 C is possible, and a core hardness of Rockwell 32 C.

Steels Used. Straight carbon steels: SAE-AISI, 1008, 1010, 1015, 1016, 1020, and manganese compositions; 1320 and the alloy steels usually used in carburizing.

Carbonitrided Products. These presently include thin-wall tubing, ratchet wrenches, sheet-metal screws, bolts, washers, studs, water pump shafts, and gears.

Advantages. Economical production of thin (up to 0.010 in.) and, in some cases, light (0.010 to 0.020 in.) case depths which have exceptional high hardness.

With adjustment of ammonia to the carburizing medium, file hardness can be obtained with quenching. File hardness up to temperatures of 600 to 800 F can be expected of cases that have been produced with a relatively high ammonia content within the carburizing medium.

Operating costs can be as low as one-fourth the cost of cyanide. Usually no reheat operations are required, the parts being directly quenched from the carburizing furnace.

Because of the exclusion of air during the heating process and use of a protective quench, oxidation is held to a minimum. Finished machined or ground parts can be surface-hardened.

Disadvantages. The process is limited to light case depths (up to 0.020 in.). Bright work can be obtained only on light pieces. Process is limited to small sections.

A case obtained by high percentage of ammonia for an air quench is not economically advisable unless other types are ruled out because of excessive distortion.

Maintenance and replacement costs of equipment are high. Close chemical control of the carburizing medium is essential.

Gas Nitriding

Special aluminum-bearing steels (Nitralloy), some AISI steels, stainless steels, and special irons can be surface-hardened without quenching by heating in an ammonia atmosphere at 925 to 1000 F for a long period, upwards of 20 hr. Workpieces are machined before nitriding to eliminate any decarburized layer, and the nitriding operation is therefore done before final grinding. Tin plating only a few "tenths" thick will serve as an effective stop-off for selective nitriding. The case produced is in the order of 0.02 in. when nitrided for 20 hr, and 0.0275 for 60 hr. Exceptional wear resistance and hardness (800 to 1000 brinell) make gas nitriding suitable for aircraft cylinder barrels, paper-machining rolls, reduction-drive gears, stainless exhaust valves, roller chains, and sealing rings, and cast-iron cylinder sleeves.

Materials Nitrided. Specialized nitriding compositions:

Nitralloy 135 (type G).

Nitralloy 135 modified.

Nitralloy N (3.0% nickel).

Nitralloy EZ.

Nitralloy GR.

Alloy steels containing chromium, molybdenum, nickel, vanadium, and tungsten: SAE-AISI 4130, 4140, 4340, 8630, and 9440.

Aluminum-free cast steels, one composition being 0.18 C, 0.40 Si, 0.40 Mo, 0.50 Mn, 2.50 Cr, and 0.20 Va.

AISI stainless steels: austenitic types 302, 304, 321, and 347; martensitic types 410, 416, and 420; ferritic type 430.

Typical cast irons containing alloys, in percentages:

	No. 1	No. 2
Total carbon	2.70	2.75
Silicon	1.55	2.58
Manganese	0.60	0.73
Aluminum	1.10	1.01
Chromium	0.30	1.22
Molybdenum	0.70	0.24
Vanadium	....	0.16

Case Depth. In a total case of 0.015 in. on Nitralloy, about 0.005 to 0.007 in. in depth will exceed 900 Vickers hardness. With total case depths of 0.030 in. (100-hr cycle) approximately 0.010 to 0.012 will be in excess of 900 Vickers hardness.

Nitriding of standard SAE compositions results in a hardness considerably lower than with the Nitralloy compositions. On AISI 410 and 302 stainless a surface hardness of Rockwell 15 92 to 94 N is possible.

Cast steels will develop a surface hardness of 850 to 900 Vickers after 45 hr, while the cast irons will have a similar hardness after 60 hr nitriding.

Post-heat-treatment. No heat-treatment is required after nitriding, because full hardness is realized. However, heat-treatment prior to nitriding is necessary to obtain a structure amenable to the nitriding cycle and obtain a constant growth factor. A typical procedure on Nitralloy 135 and Nitralloy 135 modified, annealed bar stock is:

1. Rough machine annealed bar; leave 1/32 to 1/16 in. for final machining after heat-treatment.
2. Heat to 1700 to 1750 F, followed by quench in oil or water. Water is used when the section exceeds 1 in. in thickness or diameter. Temper at 1000 to 1300 F, depending upon the desired core hardness. Time at temperature at least 1 hr per in. thickness.
3. Finish machine or grind.
4. Nitride.

A similar heat-treatment is used on forgings; however, where heat-treated bar stock is obtained the operations would be:

1. Rough machine, 1/32 to 1/16 in. for finish machining or grinding.
2. Stress-relieve by reheating to 1100 to 1200 F. Time at temperature approximately 1 hr per in. thickness.
3. Finish machine or grind.
4. Nitride.

The SAE-AISI steels obtain full hardness upon nitriding, and prior treatment is necessary similar to Nitralloy. This consists of a high-temperature quench (1750 to 1800 F) followed by a draw at 1000 to 1150 F.

Advantages. Practically no distortion or warpage, as a result of the low temperature in nitriding and the elimination of a quench.

Good corrosion resistance to the action of certain types of corrosion, such as alkalis, atmosphere, crude oil, natural gas, combustion products, tap water, and still salt water.

The SAE-AISI alloy and medium-carbon steels have properties superior to those not nitrided.

The stainless steels have good abrasive resistance, and retention of hardness at elevated temperatures.

Disadvantages. Method is expensive because of (1) length of time required; (2) necessity of pretreatment operations to obtain a suitable structure for nitriding with sufficient core strength; and (3) need of specialized furnace equipment containing high-alloy muffles which mean high maintenance cost and close chemical control of gases.

All decarburization must be removed prior to nitriding to prevent a brittle case and excessive and uncontrolled growth. Parts should be made undersized to counteract the normal growth which occurs in nitriding; this factor is constant but must be established by trial.

In nitriding stainless steels, the protective layer of chromium oxide should be reduced or dissolved since its presence tends to inhibit or prevent nitriding. To improve corrosion resistance further a black passivate coating can be used after nitriding.

Care must be taken in selection of the stainless to be nitrided—some compositions are susceptible to intergrown bar carbides which may result in cracking upon exposure to the nitriding influence for long periods of time.

Liquid Nitriding

High-speed-steel tools, preferably those that have previously been tempered, may be given an extremely hard case from 0.0005 to 0.003 in. thick by the "liquid-nitriding" process. A cyanide bath is operated at 1040 to 1050 F, and the time at temperature ranges from 5 min to 1 hr, although the usual cycle is 20 to 30 min to give a case depth of say 0.0012 in. After the proper nitriding period, the tools are removed from the bath, air-cooled, washed, and dried.

The process is particularly applicable to taps, chasers, form tools, broaches, and reamers, and examples are cited of a tenfold increase in tool life. Nitriding may be beneficially applied to drawing dies or dies subject to galling. Tools should be nitrided in an aged bath to avoid brittleness of the case. The process should not be used on impact tools, because it lowers the impact and bending strengths.

Flame Hardening

Selective surface hardening is done by localized application of the high-temperature flames of natural gas, acetylene, or butane-propane gases burned with air or oxygen—usually in commercial or specially built flame-hardening machines. With this process, it is necessary to control the temperature of the heated area within very narrow limits and to control the area to which heat energy is applied. Further, the heat cycle must be a matter of seconds ordinarily. These requirements are being met, and it is now possible, with proper selection of steel or iron, to heat-treat parts with less distortion and at lower cost.

Combinations of gases used are:

Acetylene-oxygen. Flame temperatures are up to 6000 F, but ordinarily the range of 5200 to 5800 F is used. For gas consumption see Table 7.

Propane-air. This mixture produces a lower flame temperature.

Propane-oxygen. Flame temperature is normally 4500 to 5000 F.

TABLE 7. GAS CONSUMPTION FOR OXYACETYLENE FLAME HARDENING

Hardness Penetration, In.*	Torch Speed, Ipm	4-in. Tip Coverage, Sq In. per Min	Cu Ft Acetylene per Sq In. of Surface†	Cu Ft Oxygen per Sq In. of Surface†
5/16	2	8	0.55+	0.61−
1/4	3	12	0.37−	0.40+
3/16	4	16	0.28−	0.30+
1/8	5	20	0.22+	0.24+
1/16	6	24	0.18+	0.20+

* Depths subject to variation with the thickness of the object heated.

† Based on using 4-in.-wide tip. If a 3-in.-wide tip is used, add 10% to gas consumption. If a 2-in.-wide tip is used, add 15% to gas consumption. If a 1-in.-wide tip is used, add 20% to gas consumption. Other sizes proportionately.

Natural gas and air. Flame-temperature range is 2800 to 3500 F because of the high ratio of air to gas that must be used.

Any type of hardenable carbon or alloy steel, including some of the stainless steels and irons, can be flame-hardened. Fine-grain steel is preferred. Steels from 1035 to 1050 are most commonly selected, next the pearlitic and medium-alloy steels. With high-alloy steels, when the alloy content is over 1.75%, the steel should be preheated slightly. In fact, preheating any steel will cut flame-hardening time by 20%.

Quenching. For a maximum quench of some items, water at 60 to 100 psi is used, but in other instances a fine air-water spray is employed.

Costs. Many parts that are furnace-hardened require alloy steels because their shape or design does not permit a vigorous or full quench for full surface hardness, if made in plain carbon steels. Selective hardening of carbon steels will yield surface hardnesses 100 to 150 brinell points higher than with fully quenched work. If core properties must be higher than those of annealed carbon steels, alloy steels are employed. They are heat-treated for core properties, machined, and then flame-hardened.

Cast Iron. Wear resistance of pearlitic irons can be increased by raising the surface hardness to 600 brinell. These irons should have 0.60 to 0.80% combined carbon, total carbon of not more than 3.25%, and low silicon.

Induction Hardening

Induction heating is a speedy clean method of selective casehardening. Electromagnetic energy from a suitable work coil raises the temperature of the surface layers of the steel to the hardening range in a matter of seconds. The time is closely controlled to obtain the desired heat penetration and pattern, and then the work is quenched with water or oil or is allowed to cool in air.

Heat Pattern. Although 60-cycle equipment can be used, most induction-heating equipment is of the following types:

MOTOR-GENERATOR SETS. 1000 to 10,000 cycles per min, normally 9600 cycles for deeper cases, and where a difficult contour need not be hardened to uniform thickness.

VACUUM-TUBE OSCILLATORS. 200,000 to 500,000 cycles, and higher, for thin-case applications, and those where uniform depth is needed. The higher the frequency, the greater is the tendency for the depth of penetration to follow the work contour uniformly. Thus, in some applications of induction hardening, such as fine-pitch gears, a vacuum-tube oscillator might be preferable, while for a cam surface or a coarse-pitch gear, the motor-generator set will do nicely. Motor-generator sets undoubtedly fill most needs for induction heating for hardening.

Steels Used. Medium-carbon steels are preferred with carbon above 0.30%. Alloy steels containing strong carbide formers (such as chromium and molybdenum) respond with difficulty because of possible overheating. Other alloy steels are used where either rapid quenching cannot be used or deep hardening is essential.

Gray-iron castings may be used and treated as in the flame-hardening process.

Pearlitic malleable or ductile iron is recommended for induction hardening.

Case Depth. The case depth can be controlled by the length of time cycle. Minimum case depths are 0.015 in. Common hardening depths are from 0.030 up to 0.125 in. In iron castings, the case depths are usually 0.070 in.

Heat-treatment. Tempering operations may follow induction hardening.

Advantages. Both external and internal surfaces can be selectively hardened. Automatic controls assure uniformity of heat-treated parts. Efficient heat utilization is high because of localizing the heat. Equipment is available for immediate

TABLE 8. COMPOSITION AND NATURE OF TYPICAL CONTROLLED ATMOSPHERES*

Atmosphere	Description	Air-to-gas Ratio	N_2	H_2	CO	CO_2	CH_4	O_2	Dew Point	Nature of Atmosphere
1	Completely reacted fuel	2.75:1†	41.7	38.0	19.0	0.0	1.3	0.0	−10 F	Most reducing; combustible; toxic
2	Completely burned fuel	10:1†	89.0	0.5	0.5	10.0	0.0	0.0	‡	Slightly reducing; noncombustible
3	Partially cracked fuel	6:1†	69.0	15.0	10.0	5.0	1.0	0.0	‡	Medium reducing; combustible; toxic
4	Scrubbed atmosphere No. 2	10:1†	94.0	3.0	3.0	0.0	0.0	0.0	−50 F	Inert; noncombustible
5	Scrubbed atmosphere No. 3	6:1†	72.0	16.0	11.0	0.0	1.0	0.0	−50 F	Reducing; combustible; toxic
6	Dissociated ammonia	No air	25.0	75.0	0.0	0.0	0.0	0.0	−60 F	Reducing; combustible
7	No. 6 completely burned	1.88:1§	99.0	1.0	0.0	0.0	0.0	0.0	‡	Inert; noncombustible
8	No. 6 partially burned	1.25:1§	80.0	20.0	0.0	0.0	0.0	0.0	‡	Slightly reducing; combustible

* Westinghouse Electric Corp.
† Air-to-gas ratios are representative for natural gas containing practically nothing but methane.
‡ Dew points correspond to room temperatures unless auxiliary drying equipment is added.
§ Dissociated ammonia.

use and can be applied to other operations such as brazing or annealing. It is possible to substitute less expensive steels than on other methods. A short heating cycle is good for high production; this also minimizes both distortion and oxidation.

Disadvantages. Initial equipment investment for complex parts may be high. Extra-large sections may present difficulties because of the limitations in the equipment available.

As a rule, for thin sections, higher frequencies are used; for extremely heavy sections, the low frequencies are advantageous.

Heating and quenching cycles may be critical for consistent results. The limitations would be in the automatic timing equipment and not in the induction heater. Experimentation is required to establish time cycles. Close material control is usually necessary.

Controlled Atmospheres for Heat-treatment

In heat-treating, brazing, and sintering operations, where quality and uniformity of product are required, the furnace atmosphere is usually separately generated to controlled chemical characteristics. This is true, because a furnace can be considered a chemical retort, wherein the reaction depends on both the temperature and the nature of the gaseous atmospheres surrounding the work. With due care in selecting the atmosphere, one can control or prevent oxidation, carburization, or decarburization. And as a consequence of establishing proper atmosphere control, expensive cleaning and pickling and also grinding after hardening can be eliminated.

Atmosphere furnaces are fabricated to be gastight. When fuel-fired furnaces are used, radiant tubes or muffles are necessary to keep the products of combustion from mixing with the prepared atmosphere. Other factors permitting, resistor-type electric furnaces may be used without muffles.

Separately produced atmospheres (Tables 8 and 9) are now being widely applied

TABLE 9. ATMOSPHERES SUITABLE FOR HEAT-TREATMENT OF DIFFERENT METALS*

Material	Process	Temperature Range, F	Cycle Time (Long If Over 2 Hr)	Required Surface	Atmosphere No. (See Table 8)
		Bright or Clean Annealing			
Low-carbon steels	Anneal	1200 to 1350	Long	Bright	3
Medium-carbon steels	Anneal (no decarburization)	1200 to 1450	Long	Bright	4
High-carbon steels	Anneal (no decarburization)	1200 to 1450	Long	Bright	4
Alloy steel, medium- and high-carbon	Anneal (no decarburization)	1300 to 1600	Long	Bright or clean	4
High-speed tool steels, including molybdenum high speeds	Anneal (no decarburization)	1400 to 1600	Long	Bright or clean	4
Stainless steels, chromium and nickel chromium	Anneal	1800 to 2100	Short and long	Bright	6
High-silicon steel, electrical sheet	Anneal	1900 to 2100	Long	Clean	4 or 6
Copper	Anneal	400 to 1200	Long or short	Bright	2
Various brasses	Anneal	800 to 1350	Long or short	Clean	2
Copper-nickel alloys	Anneal	800 to 1400	Long or short	Bright	2 or 4
Silicon copper alloys	Anneal	1200 to 1400	Long or short	Bright	4
Nickel	Anneal	1600 to 2000	Long or short	Bright	2 or 6
		Bright Hardening and Tempering			
Medium-carbon steels	Hardening	1400 to 1600	Short	Bright or clean	1
High-carbon steels	Hardening	1400 to 1800	Short	Bright or clean	1
Alloy steels, medium- and high-carbon	Hardening	1400 to 1800	Short	Bright or clean	1
High-speed tool steels, including molybdenum	Hardening	1800 to 2400	Short	Bright or clean	1
All classes of ferrous metals	Tempering or drawing	400 to 1200	Short	Bright or clean	2

* Westinghouse Electric Corp.

to pusher, roller-hearth, box, bell, belt-conveyor, and elevator furnaces. The principal heat-treating processes to which controlled atmospheres can be applied are bright annealing, bright hardening, gas carburizing, furnace brazing, and sintering of powder metals.

In general, no single atmosphere is universally applicable to all processes. The closest approach to a universal heat-treating atmosphere is completely burned fuel gas, which has high nitrogen content, is oxygen-free, and contains sufficient reducing properties to overcome the effect of impurities from the metal and from furnace brickwork. The equipment necessary to produce this atmosphere is more expensive than that for alternate choices of atmospheres suitable for a given process.

General types of controlled atmospheres may be broadly classified as reacted fuel-gas atmospheres, neutral or inert atmospheres, and dissociated ammonia atmospheres.

The general problem in applying controlled atmospheres to metals usually resolves itself into partial or total prevention of oxidation of the metal surfaces and prevention of metallurgical changes in metal, such as loss of carbon from steel surfaces. In some processes the atmosphere is applied purposely to affect the metal structure and composition, such as gas carburizing, whereby controlled mounts of carbon are added to the steel surface.

Check for Carbon Content of Steel

Conventional chemical analysis of steels is too costly and time-consuming for the heat-treater to segregate steels from mixed stock of known grades.

It has been established that the degree of hardness attainable upon quenching of steels increases with carbon content. The maximum hardness attainable is largely independent of other alloying constituents.

The procedure for determining carbon content involves (1) quenching a sample from above the critical temperature to known maximum obtainable hardness, (2) measurement of hardness, and (3) direct conversion of hardness to percentage carbon content using Fig. 14.

A suitable and convenient specimen is generally 3/8 × 3/8 × 1/4 in. or less, and with opposite sides ground to approximate flatness. Heating to effect complete solution may be carried on in any furnace capable of temperatures from 200 to 300 F higher than normal quenching temperatures without incurring undue scaling, decarburization, carburization, or absorption of gaseous elements.

Test samples are quenched with violent agitation into water, brine, or iced brine, depending upon the steel in question. For most medium-alloy and low- to medium-carbon steels the water quench will produce maximum hardness, but for many water-hardening grades of low hardenability the brine quenches will be necessary.

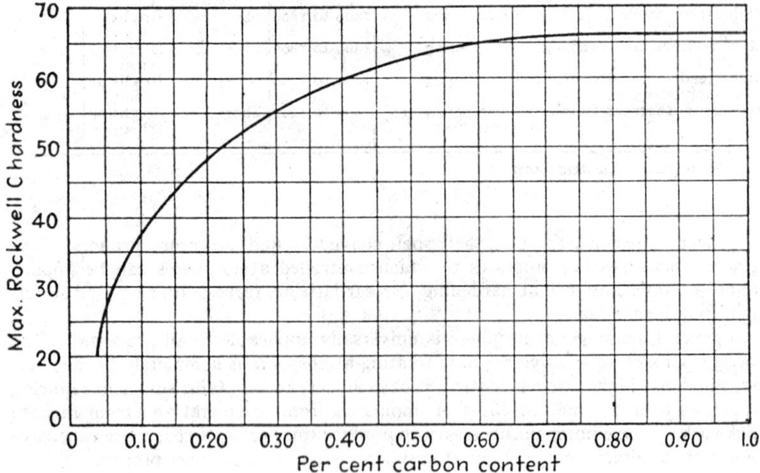

FIG. 14. Carbon content is a function of the Rockwell hardness of as-quenched steel and can be used to check mixed lots of steel so that appropriate heat-treating cycles are used.

Following quenching to room temperature (or preferably below, in some instances) specimens are reground to parallelism of the flat sides. Testing for hardness of quenched specimens is generally done with a standard Rockwell machine with conical diamond indenter (A or C scale).

Indentations should be examined at magnifications of 5 to 20 to verify that accuracy of hardness values has not been adversely affected by superficial cracking not visible to the eye.

In the conversion of maximum hardness to terms of carbon content, the curve shown in Fig. 14 is employed. Use of this curve at extremes of carbon content and hardness is not advocated, but its accuracy with common heat-treating grades (0.15 to 0.45% carbon) is satisfactory.

Value of the test lies in predicting the response of a particular steel to heat-treatment.

Hardness-testing Methods[1]

Values for hardness of metals are usually reported in terms of Rockwell or brinell numbers, but Shore and Vickers numbers are also used. Table 10 gives a means for converting Rockwell numbers into approximately equivalent numbers in the other systems. These hardness numbers have been found reliable on practically all constructional steels and tool steels in the as-forged, annealed, normalized, and quenched and tempered conditions. High-manganese steels, 18-8 stainless, and other austenitic steels, as well as constructional alloy steels and tool steels in the cold-worked condition may not conform so accurately to the relationships given.

Boldface numbers in Table 10 were prepared jointly by the ASTM, the ASM, and the SAE, while the values in regular type were taken from the Army-Navy Approximate Hardness–Tensile Strength Relationship of Carbon and Low-alloy Steels (AN-QQ-H-201).

Use of Table. The conversions given in Table 10 should be applied only to tests on flat surfaces, and the thickness should be roughly ten times the depth of indentation. Conversions from brinell hardness to shallow-impression-type tests, such as the Rockwell superficial and diamond pyramid hardness, should be made only on materials of uniform hardness to a depth of at least ten times the depth of indentation. Such conversions should not be made on surface-hardened, coated, or decarburized surfaces. Details of the several testing procedures follow.

Rockwell Hardness. The tester measures hardness by depth of penetration of a steel ball or a sphero-conical diamond. A minor load (10 kg) is first applied to cause initial penetration and set the penetrator in position. The dial is then set at zero on the black-figure scale and the major load is applied. After the major load is removed, the depth reading is taken while the minor load is still in position.

Several Rockwell scales are used. The black figures are used only for the diamond penetrator (Brale) with various loads. Scale A applies with a major load of 60 kg, scale C with a major load of 150 kg, and scale D with a major load of 100 kg. The red figures are used for all readings taken with steel-ball penetrators, regardless of size or the magnitude of the major load. Scale B applies when a 1/16-in. steel ball is used with a major load of 100 kg. Note that all readings should be followed by a letter showing whether the hardness values are in terms of the A, B, C, or D scales.

The Rockwell superficial hardness tester employs a minor load of 3 kg and major loads of 15, 30, or 45 kg. The more sensitive measuring system is needed for testing thin strip or sheet, nitrided or lightly carburized pieces, finished parts on which large test marks would be objectionable, areas near edges, extremely small

[1] SAE Iron and Steel Standards and Specifications.

TABLE 10. EQUIVALENT HARDNESS NUMBERS FOR ROCKWELL C HARDNESS NUMBERS*

Rockwell C-scale Hardness No.†	Diamond Pyramid Hardness No., Vickers	Brinell Hardness No. 10-mm Ball, 3000-kg Load			Rockwell Hardness No.†			Rockwell, Superficial Hardness No., Superficial Brale Penetrator			Shore Scleroscope Hardness No.	Tensile Strength (Approx) in 1000 Psi
		Standard Ball	Hultgren Ball	Tungsten Carbide Ball	A-scale, 60-kg Load, Brale Penetrator	B-scale, 100-kg Load, 1/16-in. Dia Ball	D-scale, 100-kg Load, Brale Penetrator	15-N Scale, 15-kg Load	30-N Scale, 30-kg Load	45-N Scale, 45-kg Load		
68	940	...	...	...	85.6	....	76.9	93.2	84.4	75.4	97	
67	900	...	...	...	85.0	....	76.1	92.9	83.6	74.2	95	
66	865	...	...	...	84.5	....	75.4	92.5	82.8	73.3	92	
65	832	...	...	739	83.9	....	74.5	92.2	81.9	72.0	91	
64	800	...	...	722	83.4	....	73.8	91.8	81.1	71.0	88	
63	772	...	...	705	82.8	....	73.0	91.4	80.1	69.9	87	
62	746	...	...	688	82.3	....	72.2	91.1	79.3	68.8	85	
61	720	...	...	670	81.8	....	71.5	90.7	78.4	67.7	83	
60	697	...	613	654	81.2	....	70.7	90.2	77.5	66.6	81	
59	674	...	599	634	80.7	....	69.9	89.8	76.6	65.5	80	326
58	653	...	587	615	80.1	....	69.2	89.3	75.7	64.3	78	315
57	633	...	575	595	79.6	....	68.5	88.9	74.8	63.2	76	305
56	613	...	561	577	79.0	....	67.7	88.3	73.9	62.0	75	295
55	595	...	546	560	78.5	....	66.9	87.9	73.0	60.9	74	287
54	577	...	534	543	78.0	....	66.1	87.4	72.0	59.8	72	278
53	560	...	519	525	77.4	....	65.4	86.9	71.2	58.6	71	269
52	544	500	508	512	76.8	....	64.6	86.4	70.2	57.4	69	262
51	528	487	494	496	76.3	....	63.8	85.9	69.4	56.1	68	253
50	513	475	481	481	75.9	....	63.1	85.5	68.5	55.0	67	245
49	498	464	469	469	75.2	....	62.1	85.0	67.6	53.8	66	239
48	484	451	455	455	74.7	....	61.4	84.5	66.7	52.5	64	232
47	471	442	443	443	74.1	....	60.8	83.9	65.8	51.4	63	225
46	458	432	432	432	73.6	....	60.0	83.5	64.8	50.3	62	219
45	446	421	421	421	73.1	....	59.2	83.0	64.0	49.0	60	212
44	434	409	409	409	72.5	....	58.5	82.5	63.1	47.8	58	206
43	423	400	400	400	72.0	....	57.7	82.0	62.2	46.7	57	201
42	412	390	390	390	71.5	....	56.9	81.5	61.3	45.5	56	196
41	402	381	381	381	70.9	....	56.2	80.9	60.4	44.3	55	191

40	392	371	371	371	70.4	....	55.4	80.4	59.5	43.1	54	186
39	382	362	362	362	69.9	....	54.6	79.9	58.6	41.9	52	181
38	372	353	353	353	69.4	....	53.8	79.4	57.7	40.8	51	176
37	363	344	344	344	68.9	....	53.1	78.8	56.8	39.6	50	172
36	354	336	336	336	68.4	(109.0)	52.3	78.3	55.9	38.4	49	168
35	345	327	327	327	67.9	(108.5)	51.5	77.7	55.0	37.2	48	163
34	336	319	319	319	67.4	(108.0)	50.8	77.2	54.2	36.1	47	159
33	327	311	311	311	66.8	(107.5)	50.0	76.6	53.3	34.9	46	154
32	318	301	301	301	66.3	(107.0)	49.2	76.1	52.1	33.7	44	150
31	310	294	294	294	65.8	(106.0)	48.4	75.6	51.3	32.5	43	146
30	302	286	286	286	65.3	(105.5)	47.7	75.0	50.4	31.3	42	142
29	294	279	279	279	64.7	(104.5)	47.0	74.5	49.5	30.1	41	138
28	286	271	271	271	64.3	(104.0)	46.1	73.9	48.6	28.9	41	134
27	279	264	264	264	63.8	(103.0)	45.2	73.3	47.7	27.8	40	131
26	272	258	258	258	63.3	(102.5)	44.6	72.8	46.8	26.7	38	127
25	266	253	253	253	62.8	(101.5)	43.8	72.2	45.9	25.5	38	124
24	260	247	247	247	62.4	(101.0)	43.1	71.6	45.0	24.3	37	121
23	254	243	243	243	62.0	100.0	42.1	71.0	44.0	23.1	36	118
22	248	237	237	237	61.5	99.0	41.6	70.5	43.2	22.0	35	115
21	243	231	231	231	61.0	98.5	40.9	69.9	42.3	20.7	35	113
20	238	226	226	226	60.5	97.8	40.1	69.4	41.5	19.6	34	110
(18)	230	219	219	219	....	96.7	....	....	....	....	33	106
(16)	222	212	212	212	....	95.5	....	....	....	....	32	102
(14)	213	203	203	203	....	93.9	....	....	....	....	31	98
(12)	204	194	194	194	....	92.3	....	....	....	....	29	94
(10)	196	187	187	187	....	90.7	....	....	....	....	28	90
(8)	188	179	179	179	....	89.5	....	....	....	....	27	87
(6)	180	171	171	171	....	87.1	....	....	....	....	26	84
(4)	173	165	165	165	....	85.5	....	....	....	....	25	80
(2)	166	158	158	158	....	83.5	....	....	....	....	24	77
(0)	160	152	152	152	....	81.7	....	....	....	....	24	75

* The values in this table shown in **boldface type** correspond to the values shown in the corresponding joint SAE-ASM-ASTM Committee Report on Hardness Conversions.

† Values in parentheses are beyond normal range and are given for information only.

TABLE 11. TEMPERATURE-CONVERSION TABLE

General formula: $F = (C \times 9/5) + 32$; $C = (F - 32) \times 5/9$

The numbers in boldface type refer to the temperature (in either centigrate or Fahrenheit degrees) which it is desired to convert into the other scale. If converting from Fahrenheit degrees to centigrade degrees, the equivalent temperature is in the left column, while if converting from degrees centigrade to degrees Fahrenheit, the equivalent temperature is in the column on the right.

C		F	C		F	C		F	C		F	C		F	C		F
−273.1	**−459.4**		−17.8	**0**	32	10.0	**50**	122.0	38	**100**	212	260	**500**	932	538	**1000**	1832
−268	**−450**		−17.2	**1**	33.8	10.6	**51**	123.8	43	**109**	230	266	**510**	950	543	**1010**	1850
−262	**−440**		−16.7	**2**	35.6	11.1	**52**	125.6	49	**120**	248	271	**520**	968	549	**1020**	1868
−257	**−430**		−16.1	**3**	37.4	11.7	**53**	127.4	54	**130**	266	277	**530**	986	554	**1030**	1886
−251	**−420**		−15.6	**4**	39.2	12.2	**54**	129.2	60	**140**	284	282	**540**	1004	560	**1040**	1904
−246	**−410**		−15.0	**5**	41.0	12.8	**55**	131.0	66	**150**	302	288	**550**	1022	566	**1050**	1922
−240	**−400**		−14.4	**6**	42.8	13.3	**56**	132.8	71	**160**	320	293	**560**	1040	571	**1060**	1940
−234	**−390**		−13.9	**7**	44.6	13.9	**57**	134.6	77	**170**	338	299	**570**	1058	577	**1070**	1958
−229	**−380**		−13.3	**8**	46.4	14.4	**58**	136.4	82	**180**	356	304	**580**	1076	582	**1080**	1976
−223	**−370**		−12.8	**9**	48.2	15.0	**59**	138.2	88	**190**	374	310	**590**	1094	588	**1090**	1994
−218	**−360**		−12.2	**10**	50.0	15.6	**60**	140.0	93	**200**	392	316	**600**	1112	593	**1100**	2012
−212	**−350**		−11.7	**11**	51.8	16.1	**61**	141.8	99	**210**	410	321	**610**	1130	599	**1110**	2030
−207	**−340**		−11.1	**12**	53.6	16.7	**62**	143.6	100	**212**	413	327	**620**	1184	604	**1120**	2048
−201	**−330**		−10.6	**13**	55.4	17.2	**63**	145.4	104	**220**	428	332	**630**	1166	610	**1130**	2066
−196	**−320**		−10.0	**14**	57.2	17.8	**64**	147.2	110	**230**	446	338	**640**	1184	616	**1140**	2084
−190	**−310**		− 9.44	**15**	59.0	18.3	**65**	149.0	116	**240**	464	343	**650**	1202	621	**1150**	2102
−184	**−300**		− 8.89	**16**	60.8	18.9	**66**	150.8	121	**250**	482	349	**660**	1220	627	**1160**	2120
−179	**−290**		− 8.33	**17**	62.6	19.4	**67**	152.6	127	**260**	500	354	**670**	1238	632	**1170**	2138
−173	**−280**		− 7.78	**18**	64.4	20.0	**68**	154.4	132	**270**	518	360	**680**	1256	638	**1180**	2156
−169	**−273**	−459.4	− 7.22	**19**	66.2	20.6	**69**	156.2	138	**280**	536	366	**690**	1274	643	**1190**	2174
−168	**−270**	−454	− 6.67	**20**	68.0	21.1	**70**	158.0	143	**290**	554	371	**700**	1292	649	**1200**	2192
−162	**−260**	−436	− 6.11	**21**	69.8	21.7	**71**	159.8	149	**300**	572	377	**710**	1310	654	**1210**	2210
−157	**−250**	−418	− 5.56	**22**	71.6	22.2	**72**	161.6	154	**310**	590	382	**720**	1328	660	**1220**	2228

−151	**−240**	−400	− 5.00	**23**	73.4	22.8	**73**	163.4	160	**320**	608	388	**730**	1346	666	**1230**	2246
−146	**−230**	−382	− 4.44	**24**	75.2	23.3	**74**	165.2	166	**330**	626	393	**740**	1364	671	**1240**	2264
−140	**−220**	−364	− 3.89	**25**	77.0	23.9	**75**	167.0	171	**340**	644	399	**750**	1382	677	**1250**	2282
−134	**−210**	−346	− 3.33	**26**	78.8	24.4	**76**	168.8	177	**350**	662	404	**760**	1400	682	**1260**	2300
−129	**−200**	−328	− 2.78	**27**	80.6	25.0	**77**	170.6	182	**360**	680	410	**770**	1418	688	**1270**	2318
−123	**−190**	−310	− 2.22	**28**	82.4	25.6	**78**	172.4	188	**370**	698	416	**780**	1436	693	**1280**	2336
−118	**−180**	−292	− 1.67	**29**	84.2	26.1	**79**	174.2	193	**380**	716	421	**790**	1454	699	**1290**	2354
−112	**−170**	−274	− 1.11	**30**	86.0	26.7	**80**	176.0	199	**390**	734	427	**800**	1472	704	**1300**	2372
−107	**−160**	−256	− 0.56	**31**	87.8	27.2	**81**	177.8	204	**400**	752	432	**810**	1490	710	**1310**	2390
−101	**−150**	−238	0	**32**	89.6	27.8	**82**	179.6	210	**410**	770	438	**820**	1508	716	**1320**	2408
− 95.6	**−140**	−220	0.56	**33**	91.4	28.3	**83**	181.4	216	**420**	788	443	**830**	1526	721	**1330**	2426
− 90.0	**−130**	−202	1.11	**34**	93.2	28.9	**84**	183.2	221	**430**	806	449	**840**	1544	727	**1340**	2444
− 84.4	**−120**	−184	1.67	**35**	95.0	29.4	**85**	185.0	227	**440**	824	454	**850**	1562	732	**1350**	2462
− 78.9	**−110**	−166	2.22	**36**	96.8	30.0	**86**	186.6	232	**450**	842	460	**860**	1580	738	**1360**	2480
− 73.3	**−100**	−148	2.78	**37**	98.6	30.6	**87**	188.6	238	**460**	860	466	**870**	1598	743	**1370**	2498
− 67.8	**− 90**	−130	3.33	**38**	100.4	31.1	**88**	190.4	243	**470**	878	471	**880**	1616	749	**1380**	2516
− 62.2	**− 80**	−112	3.89	**39**	102.2	31.7	**89**	192.2	249	**480**	896	477	**890**	1634	754	**1390**	2534
− 56.7	**− 70**	− 94	4.44	**40**	104.0	32.2	**90**	194.0	254	**490**	914	482	**900**	1652	760	**1400**	2552
− 51.1	**− 60**	− 76	5.00	**41**	105.8	32.8	**91**	195.8				488	**910**	1670	766	**1410**	2570
− 45.6	**− 50**	− 58	5.56	**42**	107.6	33.3	**92**	197.6				493	**920**	1688	771	**1420**	2588
− 40.0	**− 40**	− 40	6.11	**43**	109.4	33.9	**93**	199.4				499	**930**	1706	777	**1430**	2606
− 34.4	**− 30**	− 22	6.67	**44**	111.2	34.4	**94**	201.2				504	**940**	1724	782	**1440**	2624
− 28.9	**− 20**	− 4	7.22	**45**	113.0	35.0	**95**	203.0				510	**950**	1742	788	**1450**	2642
− 23.3	**− 10**	14	7.78	**46**	114.8	35.6	**96**	204.8				516	**960**	1760	793	**1460**	2660
			8.33	**47**	116.6	36.1	**97**	206.6				521	**970**	1778	799	**1470**	2678
			8.89	**48**	118.4	36.7	**98**	208.4				527	**980**	1796	804	**1480**	2696
			9.44	**49**	120.2	37.2	**99**	210.2				532	**990**	1814	810	**1490**	2714
						37.8	**100**	212.0									

parts, and shapes that would collapse under the loads of the regular Rockwell tester. If a diamond penetrator is used, the hardness value is followed by the letter N, and if a 1/16-in. steel ball is used, the values are followed by the letter T.

Brinell Hardness. A 10-mm steel ball is pressed into a flat surface under a load of 3000 kg for at least 10 sec (on iron and steel). Hardness reading is the average diameter of the impression as obtained from two measurements at right angles. This test should not be used on soft steels less than 1/2 in. thick or on areas small enough to permit deflection of the edges, owing to the flow from the ball impression.

Diamond Pyramid Hardness (Vickers, DPH). A square-base diamond pyramid with an apex angle of 136° is pressed into the specimen under loads of 5 to 50 kg. Diagonals of the impressions are measured and the DP hardness = $2L \sin a/2 \div d^2$, where L = the load, kg; d = length of average diagonal, mm; and a = apex angle = 136°.

Shore Hardness. The Shore hardness number is the reading of rebound of a small diamond-pointed hammer dropped onto the specimen from a fixed height. The arbitrary scale reads from 0 to 120 and may be read directly on some instruments or from a recording dial on others.

Temperature-conversion Table (Table 11)

Formulas:

$$\text{Degrees F} = \text{degrees C} \times 1.8 + 32$$
$$\text{Degrees C} = (\text{degrees F} - 32) \div 1.8$$

How to Use Table 11. Boldface numbers refer to temperatures that are to be converted from Fahrenheit to centigrade, or vice versa.

EXAMPLE: What is the centigrade equivalent of 70 F? At boldface 70 look left to centigrade column and read 21.1.

Colors of Steel at Various Temperatures

F	Color	F	Color
420	Very faint yellow	600	Very dark blue
430	Very pale yellow	752	Red—visible in dark
440	Light yellow	885	Red—visible in twilight
450	Pale straw yellow	975	Red—visible in daylight
460	Deep straw yellow	1077	Red—visible in sunlight
470	Dark yellow—straw yellow	1292	Dark red
480	Deep straw	1472	Dull cherry red
490	Yellow-brown	1652	Cherry red
500	Brown yellow	1832	Bright cherry red
510	Spotted red-brown	2012	Orange-red
520	Brown purple	2192	Orange-yellow
530	Light purple	2372	Yellow-white
540	Full purple	2552	White—welding
550	Dark purple	2732	Brilliant white
560	Full blue	2912	Bluish white
570	Dark blue		

HEAT-TREATMENT OF TOOL AND DIE STEELS

Normalizing. It is advisable to normalize tool steels after forging, if they do not have strong air-hardening tendencies. The process results in more uniform structure and grain size. The work is heated slowly (see heating precautions) to about 100 F above the critical temperature and held only long enough to reach a uniform temperature through the piece. Prolonged soaking is undesirable.

After heating, the work is removed from the furnace and cooled in still air. Normalizing should always be followed by annealing.

Oil Quench before Anneal. Carbon and carbon-vanadium tool steels with more than 1.10% carbon and size over 2 in. are generally oil-quenched rather than normalized before annealing. The work is heated as for normalizing, then quenched in oil rather than cooled in air. This promotes the formation of a spheroidized structure upon annealing.

Annealing. Tool steels are normally purchased already annealed. But if they have been forged or hardened they must be annealed before further heat-treatment. They should be annealed in controlled atmospheres or packed in sealed containers filled with cast-iron chips, lime, mica, or other neutral material. The work must not be in contact with the container.

Heat slowly and uniformly to the temperature (see Table 12). Hold at this temperature long enough for complete penetration and transformation, generally 1 to 4 hr.

Cool slowly, preferably in the furnace, down to 1000 F. Proper cooling rates range from 10 to 40 F per hr, with the slower rates being required for higher alloy content. Below 1000 F the cooling rate can be a little faster.

Stress Relieving. After heavy machining or any cold working of annealed tool steels, they should be stress-relieved before hardening. This is done by heating to about 1200 F and cooling slowly. Any change in dimensions should be corrected by further machining before hardening. To hold warpage in hardening to a minimum, stress-relieve between rough and finish machining.

Hardening. The rate of heating for hardening should be slower for alloy steels than for plain carbon steels. The higher the alloy content, the slower the heating rate should be. Much difficulty with warping or size change can be reduced or eliminated by slow uniform heating.

Molten baths provide the fastest method for heating. Open or semimuffle fuel-fired furnaces are slower. Electric or complete muffle furnaces are slowest of all.

Large or intricate shapes should always be charged into a furnace below 1000 F.

Preheating is not always necessary for the water-hardening or oil-hardening groups.

Preheating will reduce the time in a hardening furnace without atmosphere control and reduce scaling and decarburization.

In general, large pieces are heated to the high side of the hardening range and small pieces to the low side. Steel must be held at temperature long enough to insure uniform heating. Quenching when the center is cooler than the surface is likely to cause spalling of the corners. Longer time at temperature is required for the high-alloy steels.

The manganese oil-hardening steels (O1 and O2) tend to quench with a soft skin on the surface. A slightly oxidizing atmosphere will prevent this condition. The high-speed and high-chromium steels should be heated in a reducing atmosphere.

When pack hardening, it is advisable to insert a thermocouple in the pack, near or in contact with the tools. Otherwise it is difficult to estimate the time required.

TABLE 12. HEAT-TREATING AND FORGING OF TOOL AND DIE STEELS

AISI-SAE No.	Forging				Normalize, °F	Annealing		Hardening		Quenching Medium	Tempering Temperature, °F	Hardness, Rockwell C
	Heat Slowly to, °F	Start Forging, °F	Stop Forging, °F	Hardness as Rolled or Forged, Brinell		Annealing Temperature, °F	Annealed Hardness, Brinell	Preheat Temperature, °F	Hardening Temperature, °F			
Water-hardening Tool Steels												
W1-0.80C	1450	1875–1925	1500	275	1500–1600	1400–1450	187	....	1450–1500	W, B	300–650	56–65
W1-0.90C	1450	1850–1900	1475	275	1500–1600	1375–1425	187	....	1425–1475	W, B	300–650	56–65
W1-1.00C	1450	1850–1875	1475	275	1550–1650	1375–1425	187	....	1375–1450	W, B	300–650	56–65
W1-1.20C	1450	1825–1850	1450	275	1550–1650	1375–1425	187	....	1375–1450	W, B	300–650	56–65
W2-0.90C	1450	1850–1900	1475	275	1500–1600	1375–1425	187	....	1425–1475	W, B	300–650	56–65
W2-1.00C	1450	1850–1875	1475	275	1550–1650	1375–1425	187	....	1400–1450	W, B	300–650	56–65
W3-1.00C	1450	1850–1875	1475	275	1550–1650	1375–1425	187	....	1400–1450	W, B	300–650	56–65
Shock-resisting Tool Steels												
S1	1500	1900–2100	1600	400	Don't	1450–1500	212	1200	1700–1800	O	{ 300–500	55–58
											1000–1200	45–52
S2	1500	1900–2100	1600	...	1500–1600	1400–1450	...	....	1550–1650	B, W	300–800	50–60
S5	1500	1850–1950	1650	350	1500–1600	1400–1450	229	....	1600–1700	O	400–650	55–59
Cold-work Tool Steels												
O1	1500	1800–1950	1500	325	Don't	1400–1450	202	1200	1450–1500	O	350–450	60–63
O2	1500	1800–1900	1550	...	1500–1550	1375–1425	202	1200	1400–1440	O	325–475	58–62

O6	1500	1950	1500	...	1650	1450	217	....	1450–1550	O	300–800	50–64
A2	1650	1900–2050	1700	500	Don't	1600–1650	212	1200	1725–1775	A	350–600	58–62
D2	1250	1900–2000	1650	550	Don't	1600–1650	241	1225	1800–1850	A	400–900	57–62
D3	1500	1900–1950	1700	400	Don't	1600–1650	241	1500	1750–1825	O	{ 350–500 800–1000	55–63
D5	1650	1800–2000	1650	...	Don't	1625–1650	241	1225	1825–1850	A	450–980	60–62
Hot-work Tool Steels												
H11	1600	2000–2100	1650	500	Don't	1550–1600	217	1350	1825–1875	A	1050–1150	45–55
H12	1650	2050–2150	1650	500	Don't	1550–1600	217	1350	1800–1875	O, A	1000–1150	45–55
H13	1650	2050–2150	1650	500	Don't	1550–1600	217	1350	1825–1875	A	1050–1100	48–52
H21	1600	1950–2000	1700	500	Don't	1550–1600	229	1500	2050–2150	O, A	1100–1250	40–53
High-speed Tool Steels												
T1	1600	2000–2150	1700	575	Don't	1600–1650	241	1600	2325–2375	O	1025–1100	62–65
T2	1600	2000–2150	1700	575	Don't	1600–1650	241	1600	2325–2375	O	1025–1100	63–65
T3	1600	2000–2050	1750	575	Don't	1600–1650	250	1550	2300–2350	O	1020–1050	64–66
T4	1600	2000–2100	1750	575	Don't	1600–1650	255	1550	2350–2400	O	1000–1100	64–66
T5	1600	1950–2050	1700	575	Don't	1600–1650	255	1550	2350–2400	O	1050–1100	64–66
T6	1600	1950–2050	1700	575	Don't	1600–1650	302	1550	2350–2425	O	1050–1100	64–66
T8	1600	1900–2100	1700	575	Don't	1600–1650	250	1550	2300–2350	O	1050	64–66
M1	1500	1900–1950	1700	575	Don't	1550–1600	241	1550	2175–2240	O	1000–1050	63–65
M2	1500	1900–2100	1700	575	Don't	1600–1625	241	1550	2250–2300	O	1000–1075	63–67
M3	1500	1950–2050	1800	575	Don't	1550–1600	241	1500	2200–2250	O	1020–1080	65–66
M4	1500	1950–2050	1700	575	Don't	1550–1600	241	1500	2150–2225	O	1025–1075	63–65
M36	1500	1950–2050	1750	575	Don't	1550–1600	241	1500	2200–2275	O	1020–1080	64–65
Special-purpose Tool Steels												
L6	1500	1800–2000	1600	...	1550–1650	1400–1450	...	....	1500–1600	O	400–800	48–62
L7	1500	1800–2000	1550	...	1550–1650	1450–1500	...	....	1525–1550	O	350–500	60–62

Quenching medium: W, water; B, brine; O, oil; A, air.

Higher temperatures than necessary or overlong soaking at hardening temperatures will lead to grief. They cause decarburization and coarse grain structure and are likely to result in cracking in the quench.

Quenching Media. WATER. Fresh water is undesirable because of air content. Water that has been boiled or frequently used for quenching will give better results. Temperature should be 60 to 80 F.

BRINE. A more uniform and drastic quench is provided by brine (not over 10% salt by weight) than by water, especially in still or mildly agitated baths. It produces a cleaner, more uniform surface on the tools. Temperature should be 60 to 80 F. Rinse parts after quenching to avoid corrosion.

OIL. Best results with oil baths are at temperatures from 100 to 150 F. Mineral oils are preferable to animal or vegetable oils. Check the bath frequently for presence of water.

AIR. Air hardening can be done in still air, fan air, or compressed-air blast applied evenly over the work.

SODIUM HYDROXIDE. Carbon-tool steels may be quenched in 5 to 10% sodium hydroxide. It has a tendency to overcome soft spots, gives a bright finish, and does not corrode the steel. It will irritate the skin and must be handled with care.

Quenching Methods. Formation of gas pockets must be avoided when heated work is placed in the quench tank. This can be done by vigorous agitation or by sprays or geysers. The quenching medium must come in contact with all parts of the work (this means the work must not rest on the bottom of the tank).

The work should not be removed from the quench until it is below 200 F (see exceptions below). If a large mass is involved, place the work in an oil bath until the temperature is equalized throughout the work.

After quenching, the steel is highly strained and cracking is always imminent. The work should never be allowed to get completely cold but should be placed immediately in the tempering furnace.

INTERRUPTED QUENCHING. High-speed and other highly alloyed steels frequently give better results with an interrupted quench. The work is quenched into a salt or lead bath at 1050 to 1150 F. When the steel has cooled to bath temperatures it is removed and cooled in still air to 100 to 125 F ready for tempering.

MARTEMPERING. Oil- and air-hardening tool steels can be quenched into a salt bath slightly above the temperature where martensite begins to form, then cooled in still air. Bath temperature is 400 to 425 F. The bath should be of sufficient size and circulation for the work to reach this temperature in not more than 10 min. This method avoids warpage but must be performed with accurate temperature control if cracking is to be avoided. Martempering is followed by tempering in the usual way.

AUSTEMPERING. Austempering is a patented method that can be applied to some tools. They are quenched into the low-temperature bath which has a rate of heat abstraction high enough to prevent the formation of high-temperature transformation products. It is limited to section sizes smaller than 1 in. and tools which do not require a hardness higher than Rockwell 60.

Tempering. Satisfactory tempering depends on slow even heating. Liquid baths are generally best—oil at low temperatures and salt or lead for high temperatures. In a liquid, the work should be placed in wire baskets to prevent contact with sides or bottom of the tempering pot.

Bring the temperature *slowly* to the desired level. Time at temperature is of special importance. At 300 F, time should be about 4 hr per in. of section; at 400 F, only about 2 hr is required. In the higher range, above 1000 F, 45 to 60 min per in. is adequate. An extra hour in the tempering bath may avoid breakage of the tool or die. After tempering, the work is removed and cooled in still air.

High-speed and hot-work steels require double tempering; that is, the tempering cycle is repeated twice. Failure to perform this second tempering may cause early failure. Hot-work steels should be tempered about 50 F above their expected operating temperature.

Ring dies are generally flush-quenched on the center only and are not tempered because the stresses are in a favorable direction.

HEAT-TREATMENT OF NONFERROUS METALS

Aluminum

The heat-treatments for aluminum alloys are fundamentally the same. The alloy is heated to a high temperature, rapidly cooled, and then hardened by being reheated again but to a lower temperature than that initially used. The metallurgist refers to these operations as a "solution anneal," followed by a "quench," and then by "artificial aging," respectively.

Solution Annealing. When alloys are added to aluminum, they can form new phases such as solid solutions with increase in temperature. Hence, a solution anneal consists of heating a heat-treatable alloy to about 950 F (see Table 13) so that all the new phase material is in solution with the matrix. As the temperature of solution anneal is increased, this process becomes more rapid. The primary limitation to continued increase of temperature is avoidance of melting within the alloy, or "burning," which results in loss of ductility. In severe cases blisters are produced which reduce the alloy's strength.

Too low a solution-anneal temperature either does not dissolve all the new phase or else takes too long.

Quenching. To let the solution-heat-treated alloy cool slowly to room temperature would allow the new phase just placed in solution to come out as relatively

TABLE 13. HEAT-TREATING CYCLES FOR WROUGHT ALUMINUM ALLOYS

Solution Heat-treatment				Precipitation Treatment (Aging)		
Alloy	Soaking Temperature, F	Quench	Temper Designation	Aging Temperature, F	Aging Time, Hr	Temper Designation
14S	930–945	Cold water	-T4	360 ± 5 350 ± 5 340 ± 5 320 ± 5	5 8 10 18	-T6
17S	930–950	Cold water	-T4			
24S	910–930	Cold water	-T4	375 ± 5	12 or 9	-T81 or -T86
53S	960–980	Cold water	-T4	320 ± 5 350 ± 5	18 8	-T6
61S	960–980	Cold water	-T4	320 ± 5 350 ± 5	18 8	-T6

TABLE 14. ANNEALING CYCLES FOR ALUMINUM ALLOYS

Alloy	To Soften after Heat-treatment*			To Remove Cold Work		
	Soaking Temperature, F	Soaking Time, Hr	Cooling Rate†	Soaking Temperature, F	Soaking Time, Hr	Cooling Rate†
2S	....	Not heat-treated	..	650 ± 15	½–2	A or B
3S	....	Not heat-treated	..	750 ± 15	½–2	A or B
14S	775 ± 25	2	B	650 ± 10	½–2	A
17S	775 ± 25	2	B	650 ± 10	2	A
24S clad	775 ± 25	2	B	650 ± 10	2	A
24S	775 ± 25	2	B	650 ± 10	2	A
52S	....	Not heat-treated	..	650 ± 10	2	A or B
53S	775 ± 25	2	B	650 ± 10	2	A or B
61S	775 ± 25	2	B	650 ± 10	2	A or B

* Maximum drawability cannot be obtained without mechanical working and subsequent reannealing.

† Annealing cooling rates: A, air cool; B, furnace cool 50 F per hr to 500 F; C, air cool to 450 F; soak 4 hr at 450 F.

large particles. Large particles in the matrix would bring the alloy back to where it was prior to the solution anneal, and nothing would be gained. To "freeze" the atoms of the new material in solution with the aluminum, the material is cooled rapidly by quenching in cold water. The condition that now exists is one of supersaturation, and the alloy is striving to get out of solution.

Aging or Precipitation Hardening. A low-temperature heat-treatment, or aging treatment, following the quench of the solution-annealed alloy, causes the new phase to come out of solution and precipitate, but at a controlled size in a controlled dispersion. Thus, the particles are no longer distributed randomly in large particle sizes but rather as very tiny platelets, evenly distributed throughout the matrix. This condition causes the alloy's hardness and strength to increase perceptibly. Ductility usually declines while resistance to corrosion also decreases.

Some heat-treatable aluminum alloys do not require a low-temperature precipitation treatment. They start precipitating at room temperature. Thus, alloys that are aged at a slightly elevated temperature, say 350 F for about 8 hr, are termed "artificially aged," while those which precipitate at room temperature are referred to as "naturally aged."

Frozen Alloys. To retard aging of the naturally aging alloys, freezing is employed, using mechanical refrigeration, or dry ice. At 32 F aging is retarded several hours. Dry ice (−50 to −100 F) delays aging for a longer period. If an artificially aged alloy is aged at too high a temperature or for too long a period, the precipitated particles grow. This condition is referred to as "overaging," and advantages gained by a proper precipitation treatment are lost.

One of the best-known precipitation-hardened aluminum alloys is 24S. One very serious disadvantage of the precipitated 24S alloy is its poor resistance to corrosion. During aging, $CuAl_2$ particles form at grain boundaries, depleting the surrounding metal in copper. The $CuAl_2$ particles act cathodic; the surrounding aluminum particles behave as anodes. Hence, when moisture is present, galvanic corrosion sets in. To avoid intergranular corrosion, cladding is done with a thin layer of pure aluminum, and the product is known as Alclad or Duraluminum.

Annealing cycles to soften aluminum alloys after heat-treatment or to remove the effects of cold work are given in Table 14.

Magnesium

The heat-treatments to which magnesium alloys are subject include annealing, quenching, solution heat-treatment, aging, and stabilizing.

Sheet and plate are annealed at the rolling mill. Annealing gives a lower yield strength but greater ductility than is obtained in hard-rolled sheet. It is common when the sheet is to be formed.

The solution heat-treatment is given to put as much of the alloying ingredients as possible into solid solution. It results in a high tensile strength and the maximum ductility. In typical treatments, the temperature is raised from 500 F to the soaking temperature in about 2 hr. A10 is soaked at 780 F for 18 hr, AZ63 is soaked at 730 F for 10 hr, and AZ92 at 770 F for 18 hr.

Research has been conducted with quenching following the solution heat-treatment. It has been found that, if the work is removed from the furnace and allowed to cool slightly in air, then quenched in water at 180 to 200 F, about 20% increase in mechanical properties is obtained.

Aging is applied to castings following heat-treatment where maximum hardness and yield strength are desired. Forgings made of AZ80X may be aged directly after forging or following heat-treatment. Typical aging consists of holding at 350 to 425 F for 8 to 16 hr. With the higher temperatures less time is required and elongation is higher, but some yield strength is sacrificed.

Stabilizing is a treatment given castings and may be applied to the metal as cast or following heat-treatment. When applied to as-cast work it provides higher creep strength and less growth at elevated temperatures; after heat-treatment it also gives some increase in yield strength. Stabilizing is accomplished by holding the castings (either as cast or after solution heat-treatment) at 450 to 550 F for 2 to 6 hr.

Copper Alloys

Work-hardened copper alloys can be softened by annealing. Two objectives are to be accomplished by annealing: (1) relief of internal stresses and (2) recrystallization or softening of the metal. At relatively low temperatures, the metal is relieved of residual stresses imparted by cold working but no change occurs in microstructure. However, when the temperature is increased to about 570 F, the deformed grains start to reform (recrystallize) into new small grains of fairly uniform shape. At 750 F, the entire deformed structure is transformed. However, heating the cold-worked structure to 1200 F creates annealed alpha brass, but causes the recrystallized grains to grow in size. A fine grain structure results in a smoother finished product with lower finishing costs.

Beryllium Copper

Beryllium copper is commercially supplied as solution-annealed, or solution-annealed and cold-worked. Therefore, the fabricator will ordinarily be concerned only with the low-temperature hardening or aging operation.

For most applications, the standard aging treatment at 600 F, which requires 3 hr for solution-annealed material, or 2 hr for cold-rolled tempers, is perhaps the most useful. This treatment is not critical with respect to time, and produces the highest degree of strength and hardness. Following hardening, parts can be cooled at any convenient rate.

Fixture Hardening. For springs and small structural parts requiring close dimensional control, a heat-treating fixture may offer economies through the elimination of hand adjustment during assembly. Heat-treating in a fixture at a time and temperature selected to give a high degree of stress relief causes parts to set to the shape induced by the fixture.

Fixtures are made of cold-rolled steel. Because the forming-face dimensions determine size and shape of the finished part, they must be closely held. Properly designed fixtures will produce small flat springs to a flatness tolerance of ±0.0001 in., a straightness tolerance of ±0.003 in., while in blanked and formed parts, angles can be held to ±½ to 2°. As no allowance for spring-back is necessary in forming dies, the wide tolerances possible permit considerable die wear and long production runs.

In the two types of fixtures in general use, parts are either stacked or nested side by side. The fixture is held together by bolts or clamps, and dowel pins can be used to provide alignment where necessary.

For diaphragms, a stacking fixture may be used, and the parts stacked in series, with separating washers or forms to hold the desired shape. Although in certain critical operations, contour-fitting washers may be needed, it is usually sufficient to hold only the center and rim of the diaphragm. A clamp holds the stack securely during heat-treating.

To obtain maximum return on fixture investment, capacity of the fixture and production rate are important. The higher the heat-treating temperature, the shorter the hardening time, and the greater the degree of stress relief and fixture conformity. Because aging temperatures over 700 F usually do not develop full strength and hardness, the treatment selected is usually a compromise. The more rapid heat transfer offered by the salt bath, as compared with an atmosphere furnace, may prove to be an advantage.

When they are not in use, fixtures should be coated with oil to prevent rusting, but must be degreased before reuse. When used in a salt bath, the fixtures should be carefully rinsed, then the parts removed and rinsed, and finally the parts and fixtures should be thoroughly dried.

Heat-treating Equipment. Formed parts may be precipitation-hardened in a circulating air furnace, a muffle furnace, or a salt bath. Because the heating rate is slower in a muffle furnace, hardening time will be longer than in a circulating air furnace. A salt bath, by giving faster heat transfer, requires from 25 to 50% less time than either of the other types of furnaces.

Any furnace that provides temperature control to ±10 F is generally suitable for aging beryllium copper. If wide temperature variations exist in the furnace, variations in hardness response can be expected. Controlled or gas atmospheres are generally not necessary, because the discoloration produced by heating in air does not affect the mechanical properties.

When salt baths are used, a commercial mixture of sodium and potassium nitrates is generally used. The chocolate-colored film produced on the work is not detrimental to desired properties of the material, but can be readily removed if appearance reasons demand this step.

If solution annealing is required, any furnace with control up to 1475 F is satisfactory, but most salt baths will attack the metal rapidly at this temperature.

Solution Annealing. Certain severe forming operations, such as deep drawing, may require an intermediate anneal. In most cases, good results can be obtained by heating in the range 1450 to 1475 F for 15 to 60 min, with time dependent upon stock thickness and furnace charge. Parts should be quenched in water immediately after removal from the furnace.

Parts should be free from dirt, oil, or grease when placed in the furnace, and cadmium plate should be stripped prior to annealing. When additional processing follows, the oxide film formed during heating can be removed by pickling, or can be prevented by bright annealing in moisture-free atmospheres.

High-nickel Alloys

Three annealing treatments may be applied to nickel, low-carbon nickel, monel, K monel, R monel, inconel, and nickel-, monel- and inconel-clad steels. These are: (1) soft annealing so that cold working can be continued; (2) temper or partial annealing, limited to light sections and usually applied only to strip and wire where fully annealed material is not required; and (3) stress-equalizing annealing of cold-worked and hot-worked alloys that require a low-temperature thermal treatment to develop the optimum combination of strength and ductility, and to insure against distortion and warping upon subsequent machining.

If a bright, unoxidized surface is desired, care must be taken to provide a sulfur-free reducing atmosphere. Box annealing is not suitable for softening K monel as this alloy must be cooled rapidly to prevent age hardening.

The amount of previous cold work has a critical influence upon the ductility of nickel and nickel alloys after annealing; this is true regardless of the type of cold work. A minimum of 20% of cold working between anneals is required to insure maximum ductility and softness following annealing.

The following general recommendations for annealing the individual alloys should be supplemented by shop trials.

Nickel. Box annealing is done most satisfactorily at 1350 to 1450 F, for 2 to 6 hr at temperature; the total time in the furnace will depend on the rate of heating. The range for open annealing is 1600 to 1750 F for 2 to 5 min at temperature, when mechanical work is to follow. If manual operations, such as spinning, are to follow, the annealing time should be about 50% longer in order to soften the material fully. Open-annealed material should be quenched in a solution of 2 to 3% denatured alcohol in water to reduce the oxide flash.

D Nickel. The temperature for annealing D nickel is 1300 to 1400 F.

Monel. Open-annealing range is 1650 to 1800 F for 2 to 5 min at temperature, when mechanical work is to follow. Spinning and other manual operations may require up to 7 min at temperature. Box annealing is done at 1350 to 1450 F for 2 to 6 hr at temperature. An alcohol quenching bath, in the proportion of 1 gal methyl or denatured alcohol to 50 gal water, will reduce the oxide flash that results when the work is brought out into the air, and give the part a silvery-white surface. A pink color after the alcohol quenching indicates oxidation in the furnace and improper heating conditions, or an undue delay in quenching. R monel requires approximately the same treatment as monel for annealing.

Inconel. Internal stresses are almost completely relieved by heating for 1½ hr at 1400 F, with only slight softening. Softening by annealing begins at about 1600 F and is reasonably complete in 10 to 15 min by heating at 1800 F. The rate of cooling is unimportant. Welded joints in inconel do not require annealing or other heat-treatment to insure best corrosion resistance.

Illium R. Annealing is accomplished by bringing illium R to a temperature of

1900 F and allowing it to soak at this temperature for 3 to 5 min, depending on the piece. In the temperature range just above the annealing temperature illium is "hot short." Unless it is annealed in an inert or reducing atmosphere, oxide forms.

K Monel. The alloy may be annealed by the following procedure: Heat to 1600 F for 2 to 5 min at temperature, or to 1800 F for ½ to 2 min at temperature, and cool rapidly in a water quench containing 2% by volume of alcohol. If the material is held at temperature for a sufficient period of time it may be partially or completely softened by quenching from 1100 F or higher. Rapid air cooling effects a mild quench and results in partial softening.

The correct temperature and time to be used in hardening will vary according to the initial temper of the material. Practically complete hardening will result if a moderately slow cooling rate, such as 25 to 50 F per hr, is used.

1. Soft K monel (140 to 180 brinell) is hardened by holding it for 16 hr at 1080 to 1100 F, followed by furnace-cooling to 900 F at a rate not exceeding 15 F per hr. Cooling from 900 F to room temperature may be carried out by furnace cooling or air cooling, or by quenching, without regard to the cooling rate.

2. Moderately cold-worked K monel (175 to 250 brinell) is hardened by holding it for 8 hr or longer at 1080 to 1100 F, followed by furnace cooling to 900 F at a rate not exceeding 15 F per hr. Higher hardness may be obtained by holding it for as long as 16 hr at temperature, particularly if the material has been only slightly cold-worked.

3. Fully cold-worked K monel (260 to 315 brinell or Rockwell 25 to 32 C) is hardened by holding it at 980 to 1000 F for 6 hr, or longer, followed by furnace cooling to 900 F at a rate not exceeding 15 F per hr.

S Monel. For extensive machining, this 4% silicon casting alloy may be softened by heating at 1600 F for 1 hr, air cooling to 1200 F, and then quenching in oil. This lowers the hardness to about 200 brinell. After machining, the parts can be hardened by aging at 1100 F for 4 to 6 hr, followed by furnace cooling. The aging treatment produces hardness as high as or higher than that of the alloy in the as-cast condition, and mechanical properties of the same order as those of the as-cast material.

Titanium

Annealing. Stress-relief anneal can be performed by heating at not over 570 F for 1 hr for bar or forgings, 15 min for sheet. A stress-relief anneal of cold-worked sheet will improve the minimum bend radius.

A full anneal of commercially pure titanium requires 1 hr for each inch of thickness at 1300 to 1350 F. The alloys should be soaked for a slightly longer time at a lower temperature—1250 to 1300 F. Tight scale results when annealing in air.

After heavy grinding, especially with the alloys, an anneal for 1 hr at 1200 F is desirable to avoid heat cracking.

Hardening. Titanium and a number of its alloys are not noticeably hardened or strengthened by heat-treatment, although their properties can be affected by working.

Three treatments for Ti-150A have been developed:

1. Air-cool from 1400 F for the best combination of yield and tensile strength.

2. Quench from 1600 F and temper at 1300 F to retain the 22% elongation of annealed material with slightly increased strength.

3. Air-cool from 1800 F for the highest yield strength and the highest yield tensile ratio.

Section 28

GAGES, TOLERANCES, AND WEIGHTS OF METAL PRODUCTS

SECTION 28

GAGES, TOLERANCES, AND WEIGHTS OF METAL PRODUCTS

GAGING SYSTEMS

Commodities like sheet, strip, wire, and tubing are produced with thicknesses, diameters, or wall thicknesses according to several gaging systems, depending on the article and metal. This situation is the result of natural development and preferences of the several producing industries.

Since 1926, efforts have been made by standardizing groups to arrive at a single standard for at least a commodity, for example, the ASA preferred thicknesses for uncoated metals and alloys up to 0.236 in. thickness. The theoretical reasons favoring preferred thicknesses are: (1) fewer items have to be handled; (2) only one table of raw material sizes has to be consulted, instead of several; (3) it should be easier to mate different materials, because they would all be made to the same thickness standards; and (4) one material can be substituted for another in times of scarcity without redesign of product or tooling.

The practical considerations that stand in the way of adoption of preferred thicknesses, for example, are: (1) in the case of steel, large users are thoroughly familiar with the behavior of existing gages in tooling, especially dies, and do not intend that their shop personnel shall be burdened with learning how preferred thicknesses behave; and (2) warehouses catalog and stock existing gages, and the sum total of their orders is an important factor in the metals business. The mills can roll any thickness desired, but for the reasons cited it is unlikely that much steel, except a limited amount of stainless, will be rolled to ASA numbers. In the brass industry, the ASA numbers are said to be preferred for simplicity of stocking, but actually most of the metal is still made to B&S gage numbers.

The engineer and purchasing agent should keep abreast of any change in availability of metals in common gaging systems vs. simplified systems. In the meantime, his problems in selecting the right system and the proper decimal equivalent of the gage number are simplified by two tables, Index of Gaging Systems Used for Various Metals and Commodities, and Wire and Sheet-metal Gages in Inch Equivalents.

The second table has gage numbers arranged from 7/0 to 50, and corresponding inch equivalents for the several gaging systems in common use. In specifying material on drawings or purchase orders, *always* use the inch equivalent to at least three significant figures, and *not* the gage number.

Sheet-metal Gaging Systems

Several gaging systems are employed for sheet and strip:

Manufacturers' Standard Gage for steel sheets is currently used for carbon and alloy sheets. This system is based on steel weighing 41.82 psf 1 in. thick.

U.S. Standard Gage for sheet, plate iron, and steel is obsolete so far as carbon and alloy sheets are concerned, because it is not based on the true weight of steel in

INDEX OF GAGING SYSTEMS USED FOR VARIOUS METALS AND COMMODITIES

Metal	Commodity			
	Sheet	Strip	Wire	Tubing
Steel:				
Carbon (hot-rolled)	Mfrs Std	BWG	SWG	BWG
Carbon (cold-rolled)	Mfrs Std	U.S. Std		
Alloy (hot-rolled)	Mfrs Std	BWG	SWG	BWG
Alloy (cold-rolled)	Mfrs Std	U.S. Std		
Stainless	*	*	SWG	U.S. Std
Copper	Oz per sq ft	B&S	B&S	BWG†
Copper alloys‡ (brass, bronze)	B&S	B&S	B&S	BWG
Aluminum	B&S	No strip	B&S	Stubs
Magnesium	B&S	....	....	Stubs
Nickel alloys	U.S. Std	U.S. Std	B&S, U.S. Std	Stubs

* Eastern warehouses stock cold-rolled stainless in U.S. Standard Gage, whereas Pacific Coast and some Middle Western warehouses follow ASA preferred thicknesses.

† For most sizes.

‡ The copper and brass industry prefers ASA preferred thicknesses, but most material is still fabricated to customer orders in terms of the Brown & Sharpe gaging system.

sheet form. Nevertheless, the U.S. Standard Gage is still used somewhat for stainless sheets and is continued for cold-rolled strip, both carbon and alloy, for stainless tubing, and for nickel-alloy sheet and strip.

Birmingham Wire Gage, also known as the *Stubs Iron Wire Gage*, is used for hot-rolled carbon and alloy strip and steel tubing.

The *Brown & Sharpe*, or *American Wire Gage*, is used for copper strip, brass and bronze sheet and strip, and aluminum and magnesium sheet.

ASA preferred thicknesses, page 28-6, have received some acceptance for stainless sheet, and for brass-mill sheet and strip.

Wire Gaging Systems

The *Steel Wire Gage*, or *Washburn & Moen*, is used for manufacturers' wire, carbon-steel mechanical spring wire, alloy spring wire, stainless-steel wire, etc.

Music wire is nominally specified to the sizes in the *American Steel & Wire Co.* music-wire sizes (page 28-8), although there are a number of other names to be found in steel catalogs.

Brown & Sharpe, or *American Wire Gage*, is used for copper, copper alloy, aluminum, magnesium, nickel alloy, and other nonferrous commercial wires.

Gaging Systems for Rods

Brown & Sharpe gage is used for copper, brass, and aluminum rods. Steel rods are nominally listed in fractional sizes, but drill rod may be listed in *Stubs Steel Wire Gage* or the *Twist Drill and Steel Wire Gage*. It is considered preferable to refer to twist-drill sizes in inch equivalents (Sec. 2) than to either the Stubs or Twist Drill gages.

Wire and Sheet-metal Gages in Inch Equivalents

Gage No.	U.S. Std	Mfrs Std for Sheet Steel		Birmingham or Stubs Wire Gage BWG	Steel Wire Gage SWG (Washburn & Moen)		AWG or B&S
		Nominal	Limits		Nominal	Limits	
7/0	0.500	......	....	.....	0.4900	0.490 –0.469	
6/0	0.469	......	....	.....	0.4615	0.4615–0.438	0.5800
5/0	0.438	......	....	.....	0.4305	0.4305–0.403	0.5165
4/0	0.406	......	....	0.454	0.3938	0.3938–0.370	0.4600
3/0	0.375	......	....	0.425	0.3625	0.3625–0.339	0.4096
2/0	0.344	......	....	0.380	0.3310	0.331 –0.313	0.3648
1/0	0.312	......	....	0.340	0.3065	0.3065–0.289	0.3249
1	0.281	......	....	0.300	0.2830	0.283 –0.268	0.2893
2	0.266	......	....	0.284	0.2625	0.2625–0.248	0.2576
3	0.250	0.2391	0.2465–0.2317	0.259	0.2437	0.2437–0.230	0.2294
4	0.234	0.2242	0.2316–0.2168	0.238	0.2253	0.2253–0.212	0.2043
5	0.219	0.2092	0.2167–0.2018	0.220	0.2070	0.207 –0.196	0.1819
6	0.203	0.1943	0.2017–0.1869	0.203	0.1920	0.192 –0.181	0.1620
7	0.188	0.1793	0.1868–0.1719	0.180	0.1770	0.177 –0.166	0.1443
8	0.172	0.1644	0.1718–0.1570	0.165	0.1620	0.162 –0.152	0.1285
9	0.156	0.1495	0.1569–0.1420	0.148	0.1483	0.1483–0.138	0.1144
10	0.141	0.1345	0.1419–0.1271	0.134	0.1350	0.135 –0.124	0.1019
11	0.125	0.1196	0.1270–0.1121	0.120	0.1205	0.1205–0.109	0.0907
12	0.109	0.1046	0.1120–0.0972	0.109	0.1055	0.1055–0.095	0.0808
13	0.0938	0.0897	0.0971–0.0822	0.095	0.0915	0.0915–0.083	0.0720
14	0.0781	0.0747	0.0821–0.0710	0.083	0.0800	0.080 –0.074	0.0641
15	0.0703	0.0673	0.0709–0.0636	0.072	0.0720	0.072 –0.065	0.0571
16	0.0625	0.0598	0.0635–0.0568	0.065	0.0625	0.0625–0.056	0.0508
17	0.0562	0.0538	0.0567–0.0509	0.058	0.0540	0.054 –0.0491	0.0453
18	0.0500	0.0478	0.0508–0.0449	0.049	0.0475	0.0475–0.0426	0.0403
19	0.0438	0.0418	0.0448–0.0389	0.042	0.0410	0.0410–0.0363	0.0359
20	0.0375	0.0359	0.0388–0.0344	0.035	0.0348	0.0348–0.0325	0.0320
21	0.0344	0.0329	0.0343–0.0314	0.032	0.0317	0.0317–0.0294	0.0285
22	0.0312	0.0299	0.0313–0.0284	0.028	0.0286	0.0286–0.0265	0.0253
23	0.0281	0.0269	0.0283–0.0255	0.025	0.0258	0.0258–0.0237	0.0226
24	0.0250	0.0239	0.0254–0.0225	0.022	0.0230	0.0230–0.0211	0.0201
25	0.0219	0.0209	0.0224–0.0195	0.020	0.0204	0.0204–0.0187	0.0179
26	0.0188	0.0179	0.0194–0.0172	0.018	0.0181	0.0181–0.0175	0.0159
27	0.0172	0.0164	0.0171–0.0157	0.016	0.0173	0.0173–0.0165	0.0142
28	0.0156	0.0149	0.0156–0.0142	0.014	0.0162	0.0162–0.0153	0.0126
29	0.0141	0.0135	0.0141–0.0128	0.013	0.0150	0.0150–0.0143	0.0113
30	0.0125	0.0120	0.0127–0.0113	0.012	0.0140	0.0140–0.0134	0.0100
31	0.0109	0.0105	0.0112–0.0101	0.010	0.0132	0.0132–0.0129	0.0089
32	0.0102	0.0097	0.0100–0.0094	0.009	0.0128	0.0128–0.0121	0.0080
33	0.00938	0.0090	0.0093–0.0086	0.008	0.0118	0.0118–0.0108	0.0071
34	0.00859	0.0082	0.0085–0.0079	0.007	0.0104	0.0104–0.0097	0.0063
35	0.00781	0.0075	0.0078–0.0071	0.005	0.0095	0.0095–0.0091	0.0056
36	0.00703	0.0067	0.0070–0.0066	0.004	0.0090	0.0090–0.0086	0.0050
37	0.00664	0.0064	0.0065–0.0062	.....	0.0085	0.0085–0.0081	0.0045
38	0.00625	0.0060	0.0061–0.0058	.....	0.0080	0.0080–0.0076	0.0040
39	....	......	....	.....	0.0075	0.0075–0.00712	0.0035
40	....	......	....	.....	0.0070	0.007 –0.0067	0.0031
41	....	......	....	.....	0.0066	0.0066–0.0063	0.0028
42	....	......	....	.....	0.0062	....	0.0025
43	....	......	....	.....	0.0060	....	0.0022
44	....	......	....	.....	0.0058	....	0.0020
45	....	......	....	.....	0.0055	....	0.0018
46	....	......	....	.....	0.0052	....	0.0016
47	....	......	....	.....	0.0050	....	0.0014
48	....	......	....	.....	0.0048	....	0.0012
49	....	......	....	.....	0.0046	....	0.0011
50	....	......	....	.....	0.0044	....	0.0010

Manufacturers' Standard Gage for Sheet Steel

Gage-thickness equivalents are based on 0.0014945 in. per oz per sq ft; 0.023912 in. per lb per sq ft (reciprocal of 41.82 lb per sq ft per in. thick); 3.443329 in. per lb per sq in.

Mfrs Std Gage No.	Oz per Sq Ft	Psi	Psf	In. Equivalent for Steel Sheet Thickness
3	160	0.069444	10.0000	0.2391
4	150	0.065104	9.3750	0.2242
5	140	0.060764	8.7500	0.2092
6	130	0.056424	8.1250	0.1943
7	120	0.052083	7.5000	0.1793
8	110	0.047743	6.8750	0.1644
9	100	0.043403	6.2500	0.1495
10	90	0.039062	5.6250	0.1345
11	80	0.034722	5.0000	0.1196
12	70	0.030382	4.3750	0.1046
13	60	0.026042	3.7500	0.0897
14	50	0.021701	3.1250	0.0747
15	45	0.019531	2.8125	0.0673
16	40	0.017361	2.5000	0.0598
17	36	0.015625	2.2500	0.0538
18	32	0.013889	2.0000	0.0478
19	28	0.012153	1.7500	0.0418
20	24	0.010417	1.5000	0.0359
21	22	0.0095486	1.3750	0.0329
22	20	0.0086806	1.2500	0.0299
23	18	0.0078125	1.1250	0.0269
24	16	0.0069444	1.0000	0.0239
25	14	0.0060764	0.8750	0.0209
26	12	0.0052083	0.7500	0.0179
27	11	0.0047743	0.6875	0.0164
28	10	0.0043403	0.6250	0.0149
29	9	0.0039062	0.5625	0.0135
30	8	0.0034722	0.5000	0.0120
31	7	0.0030382	0.43750	0.0105
32	6.5	0.0028212	0.40625	0.0097
33	6	0.0026042	0.37500	0.0090
34	5.5	0.0023872	0.34375	0.0082
35	5	0.0021701	0.31250	0.0075
36	4.5	0.0019531	0.28125	0.0067
37	4.25	0.0018446	0.26562	0.0064
38	4	0.0017361	0.25000	0.0060

ASA Preferred Thicknesses for Uncoated Metals and Alloys

	0.125	**0.063**	**0.032**	**0.016**	**0.008**	**0.004**
0.236	0.118	0.060	0.030	0.015		
0.224	**0.112**	**0.056**	**0.028**	**0.014**	**0.007**	
0.212	0.106	0.053	0.026	0.013		
0.200	**0.100**	**0.050**	**0.025**	**0.012**	**0.006**	
0.190	0.095	0.048	0.024			
0.180	**0.090**	**0.045**	**0.022**	**0.011**		
0.170	0.085	0.042	0.021			
0.160	**0.080**	**0.040**	**0.020**	**0.010**	**0.005**	
0.150	0.075	0.038	0.019			
0.140	**0.071**	**0.036**	**0.018**	**0.009**		
0.132	0.067	0.034	0.017			

All dimensions are given in inches.
20-series numbers are in boldface type.

Nearest Gages to ASA Preferred Thicknesses

ASA 20 Series	Substitutes*			
	Mfrs Std Gage	U.S. Std Gage	AWG or B&S	BWG
0.004	....	....	0.004 (No. 38)	0.004 (No. 36)
0.005	....	....	0.005 (No. 36)	0.005 (No. 35)
0.006†	0.006 (No. 38)	0.0062 (No. 38)	0.0063 (No. 34)	
0.007	0.0067 (No. 36)	0.007 (No. 36)	0.0071 (No. 33)	0.007 (No. 34)
0.008†	0.0082 (No. 34)	0.0078 (No. 35)	0.008 (No. 32)	0.008 (No. 33)
0.009	0.009 (No. 33)	0.0086 (No. 34)	0.0089 (No. 31)	0.009 (No. 32)
0.010†	0.0097 (No. 32)	0.0102 (No. 32)	0.010 (No. 30)	0.010 (No. 31)
0.011	0.0105 (No. 31)	0.0109 (No. 31)	0.0113 (No. 29)	
0.012†	0.012 (No. 30)	0.0125 (No. 30)	0.0126 (No. 28)	0.012 (No. 30)
0.014	0.0135 (No. 29)	0.0141 (No. 29)	0.0142 (No. 27)	0.014 (No. 28)
0.016	0.0164 (No. 27)	0.0156 (No. 28)	0.0159 (No. 26)	0.016 (No. 27)
0.018	0.0179 (No. 26)	0.0172 (No. 27)	0.0179 (No. 25)	0.018 (No. 26)
0.020†	0.0209 (No. 25)	0.0188 (No. 26)	0.0201 (No. 24)	0.020 (No. 25)
0.022	....	0.0219 (No. 25)	0.0226 (No. 23)	0.022 (No. 24)
0.025†	0.0239 (No. 24)	0.025 (No. 24)	0.0253 (No. 22)	0.025 (No. 23)
0.028	0.0269 (No. 23)	0.0281 (No. 23)	0.0285 (No. 21)	0.028 (No. 22)
0.032†	0.0329 (No. 21)	0.0312 (No. 22)	0.032 (No. 20)	0.032 (No. 21)
0.036	0.0359 (No. 20)	0.0344 (No. 21)	0.0359 (No. 19)	0.035 (No. 20)
0.040†	0.0418 (No. 19)	0.0375 (No. 20)	0.0403 (No. 18)	0.042 (No. 19)
0.045	....	0.0438 (No. 19)	0.0453 (No. 17)	
0.050†	0.0478 (No. 18)	0.050 (No. 18)	0.0508 (No. 16)	0.049 (No. 18)
0.056	0.0538 (No. 17)	0.0562 (No. 17)	0.0571 (No. 15)	0.058 (No. 17)
0.063†	0.0598 (No. 16)	0.0625 (No. 16)	0.0641 (No. 14)	0.065 (No. 16)
0.071	0.0673 (No. 15)	0.0703 (No. 15)	0.072 (No. 13)	0.072 (No. 15)
0.080	0.0747 (No. 14)	0.0781 (No. 14)	0.0808 (No. 12)	0.083 (No. 14)
0.090†	0.0897 (No. 13)	0.0938 (No. 13)	0.0907 (No. 11)	0.095 (No. 13)
0.100	0.1046 (No. 12)	....	0.1019 (No. 10)	
0.112	....	0.1094 (No. 12)	0.1144 (No. 9)	0.109 (No. 12)
0.125†	0.1196 (No. 11)	0.125 (No. 11)	0.1285 (No. 8)	0.120 (No. 11)
0.140	0.1345 (No. 10)	0.1406 (No. 10)	0.1443 (No. 7)	0.134 (No. 10)
0.160	0.1644 (No. 8)	0.1562 (No. 9)	0.162 (No. 6)	0.165 (No. 8)
0.180†	0.1793 (No. 7)	0.1875 (No. 7)	0.1819 (No. 5)	0.180 (No. 7)
0.200	0.1943 (No. 6)	0.2031 (No. 6)	0.2043 (No. 4)	0.203 (No. 6)
0.224	0.2242 (No. 4)	0.2188 (No. 5)	0.2294 (No. 3)	0.220 (No. 5)

* Some corrosion-resistant-steel suppliers use the U.S. Standard Gage.
† Sizes preferred by large electrical manufacturers.

Steel Wire Gage (Split Gage Numbers and Decimal Equivalents)

7/0	0.490	6	0.192	18	0.0475	30	0.0140
7/0 ¼	0.483	6¼	0.188	18¼	0.0459	30¼	0.0138
7/0 ½	0.476	6½	0.185	18½	0.0443	30½	0.0136
7/0 ¾	0.469	6¾	0.181	18¾	0.0426	30¾	0.0134
6/0	0.4615	7	0.177	19	0.0410	31	0.0132
6/0 ¼	0.454	7¼	0.173	10¼	0.0394	31¼	0.0131
6/0 ½	0.446	7½	0.170	19½	0.0379	31½	0.0130
6/0 ¾	0.438	7¾	0.166	19¾	0.0363	31¾	0.0129
5/0	0.4305	8	0.162	20	0.0348	32	0.0128
5/0 ¼	0.421	8¼	0.159	20¼	0.0340	32¼	0.0126
5/0 ½	0.412	8½	0.155	20½	0.0332	32½	0.0123
5/0 ¾	0.403	8¾	0.152	20¾	0.0325	32¾	0.0121
4/0	0.3938	9	0.1483	21	0.0317	33	0.0118
4/0 ¼	0.386	9¼	0.145	21¼	0.0309	33¼	0.0115
4/0 ½	0.378	9½	0.142	21½	0.0301	33½	0.0111
4/0 ¾	0.370	9¾	0.138	21¾	0.0294	33¾	0.0108
3/0	0.3625	10	0.135	22	0.0286	34	0.0104
3/0 ¼	0.355	10¼	0.131	22¼	0.0279	34¼	0.0102
3/0 ½	0.347	10½	0.128	22½	0.0272	34½	0.0100
3/0 ¾	0.339	10¾	0.124	22¾	0.0265	34¾	0.0097
2/0	0.331	11	0.1205	23	0.0258	35	0.0095
2/0 ¼	0.325	11¼	0.117	23¼	0.0251	35¼	0.0094
2/0 ½	0.319	11½	0.113	23½	0.0244	35½	0.0093
2/0 ¾	0.313	11¾	0.109	23¾	0.0237	35¾	0.0091
1/0	0.3065	12	0.1055	24	0.0230	36	0.0090
1/0 ¼	0.301	12¼	0.102	24¼	0.0224	36¼	0.0089
1/0 ½	0.295	12½	0.099	24½	0.0217	36½	0.0087
1/0 ¾	0.289	12¾	0.095	24¾	0.0211	36¾	0.0086
1	0.283	13	0.0915	25	0.0204	37	0.0085
1¼	0.278	13¼	0.089	25¼	0.0198	37¼	0.0084
1½	0.273	13½	0.086	25½	0.0193	37½	0.0083
1¾	0.268	13¾	0.083	25¾	0.0187	37¾	0.0081
2	0.2625	14	0.080	26	0.0181	38	0.0080
2¼	0.258	14¼	0.078	26¼	0.0179	38¼	0.0079
2½	0.253	14½	0.076	26½	0.0177	38½	0.0078
2¾	0.248	14¾	0.074	26¾	0.0175	38¾	0.0076
3	0.2437	15	0.072	27	0.0173	39	0.0075
3¼	0.239	15¼	0.070	27¼	0.0170	39¼	0.00737
3½	0.235	15½	0.067	27½	0.0168	39½	0.00725
3¾	0.230	15¾	0.065	27¾	0.0165	39¾	0.00712
4	0.2253	16	0.0625	28	0.0162	40	0.007
4¼	0.221	16¼	0.060	28¼	0.0159	40¼	0.0069
4½	0.216	16½	0.058	28½	0.0156	40½	0.0068
4¾	0.212	16¾	0.056	28¾	0.0153	40¾	0.0067
5	0.207	17	0.054	29	0.0150	41	0.0066
5¼	0.203	17¼	0.052	29¼	0.0148	41¼	0.0065
5½	0.200	17½	0.051	29½	0.0145	41½	0.0064
5¾	0.196	17¾	0.0491	29¾	0.0143	41¾	0.0063

MUSIC-WIRE SIZES

Gage No.	Dia	Gage No.	Dia	Gage No.	Dia	Gage No.	Dia
6/0	0.004	8	0.020	21	0.047	34	0.100
5/0	0.005	9	0.022	22	0.049	35	0.106
4/0	0.006	10	0.024	23	0.051	36	0.112
3/0	0.007	11	0.026	24	0.055	37	0.118
2/0	0.008	12	0.029	25	0.059	38	0.124
0	0.009	13	0.031	26	0.063	39	0.130
1	0.010	14	0.033	27	0.067	40	0.138
2	0.011	15	0.035	28	0.071	41	0.146
3	0.012	16	0.037	29	0.075	42	0.154
4	0.013	17	0.039	30	0.080	43	0.162
5	0.014	18	0.041	31	0.085	44	0.170
6	0.016	19	0.043	32	0.090	45	0.180
7	0.018	20	0.045	33	0.095		

GALVANIZED-SHEET GAGE NUMBERS, UNIT WEIGHTS, AND THICKNESSES

Galvanized Sheet Gage No.	Oz per Sq Ft	Psf	Psi	Thickness Equivalent for Galvanized Sheet Gage No.
8	112.5	7.03125	0.048828	0.1681
9	102.5	6.40625	0.044488	0.1532
10	92.5	5.78125	0.040148	0.1382
11	82.5	5.15625	0.035807	0.1233
12	72.5	4.53125	0.031467	0.1084
13	62.5	3.90625	0.027127	0.0934
14	52.5	3.28125	0.022786	0.0785
15	47.5	2.96875	0.020616	0.0710
16	42.5	2.65625	0.018446	0.0635
17	38.5	2.40625	0.016710	0.0575
18	34.5	2.15625	0.014974	0.0516
19	30.5	1.90625	0.013238	0.0456
20	26.5	1.65625	0.011502	0.0396
21	24.5	1.53125	0.010634	0.0366
22	22.5	1.40625	0.0097656	0.0336
23	20.5	1.28125	0.0088976	0.0306
24	18.5	1.15625	0.0080295	0.0276
25	16.5	1.03125	0.0071615	0.0247
26	14.5	0.90625	0.0062934	0.0217
27	13.5	0.84375	0.0058594	0.0202
28	12.5	0.78125	0.0054253	0.0187
29	11.5	0.71875	0.0049913	0.0172
30	10.5	0.65625	0.0045573	0.0157
31	9.5	0.59375	0.0041233	0.0142
32	9.0	0.56250	0.0039062	0.0134

Decimal Thickness Table for Sheet Copper

Oz per Sq Ft	Decimal Thickness, In.	Oz per Sq Ft	Decimal Thickness, In.
3	0.004	64	0.083
4	0.005	70	0.095
6	0.008	81	0.109
8	0.010	89	0.120
10	0.013	100	0.134
12	0.016	110	0.148
14	0.018	123	0.165
16	0.022	134	0.180
18	0.025	151	0.203
20	0.028	164	0.220
24	0.032	177	0.238
32	0.042	193	0.259
40	0.049	211	0.284
48	0.065	223	0.300
56	0.072	253	0.340

Sheet copper is frequently designated by the weight in ounces of a square foot instead of by its thickness.

Illinois Zinc Company's Zinc Gage

No.	Psf	Thickness, In.	No.	Psf	Thickness, In.
3	0.22	0.006	16	1.68	0.045
4	0.30	0.008	17	1.87	0.050
5	0.37	0.010	18	2.06	0.055
6	0.45	0.012	19	2.25	0.060
7	0.52	0.014	20	2.62	0.070
8	0.60	0.016	21	3.00	0.080
9	0.67	0.018	22	3.37	0.090
10	0.75	0.020	23	3.75	0.100
11	0.90	0.024	24	4.70	0.125
12	1.05	0.028	25	9.40	0.250
13	1.20	0.032	26	14.00	0.375
14	1.35	0.036	27	18.75	0.500
15	1.50	0.040	28	37.50	1.000

DIMENSIONAL TOLERANCES

Carbon-steel Bars, Hot-rolled

Rounds, Squares, and Round-cornered Squares

Specified Sizes, In.	Size Tolerances, In.		Out-of-round or Out-of-square Section, In.
	Over	Under	
To 5/16 incl	0.005	0.005	0.008
Over 5/16 to 7/16 incl	0.006	0.006	0.009
Over 7/16 to 5/8 incl	0.007	0.007	0.010
Over 5/8 to 7/8 incl	0.008	0.008	0.012
Over 7/8 to 1 incl	0.009	0.009	0.013
Over 1 to 1 1/8 incl	0.010	0.010	0.015
Over 1 1/8 to 1 1/4 incl	0.011	0.011	0.016
Over 1 1/4 to 1 3/8 incl	0.012	0.012	0.018
Over 1 3/8 to 1 1/2 incl	0.014	0.014	0.021
Over 1 1/2 to 2 incl	1/64	1/64	0.023
Over 2 to 2 1/2 incl	1/32	0	0.023
Over 2 1/2 to 3 1/2 incl	3/64	0	0.035
Over 3 1/2 to 4 1/2 incl	1/16	0	0.046
Over 4 1/2 to 5 1/2 incl	5/64	0	0.058
Over 5 1/2 to 6 1/2 incl	1/8	0	0.070
Over 6 1/2 to 8 1/4 incl	5/32	0	0.085

Hexagons

Specified Sizes between Opposite Sides, In.	Size Tolerances, In.		Out-of-hexagon Section, In.
	Over	Under	
To 1/2 incl	0.007	0.007	0.011
Over 1/2 to 1 incl	0.010	0.010	0.015
Over 1 to 1 1/2 incl	0.021	0.013	0.025
Over 1 1/2 to 2 incl	1/32	1/64	1/32
Over 2 to 2 1/2 incl	3/64	1/64	3/64
Over 2 1/2 to 3 1/2 incl	1/16	1/64	1/16

Square- and Round-edge Flats

Specified Widths, In.	Thickness Tolerances, for Thicknesses Given, Over and Under, In.					Width Tolerances, In.	
	Under 1/4	1/4 to 1/2, Incl.	Over 1/2 to 1, Incl.	Over 1 to 2, Incl.	Over 2	Over	Under
To 1 incl	0.007	0.008	0.010	...	...	1/64	1/64
Over 1 to 2 incl	0.007	0.012	0.015	1/32	...	1/32	1/32
Over 2 to 4 incl	0.008	0.015	0.020	1/32	3/64	1/16	1/32
Over 4 to 6 incl	0.009	0.015	0.020	1/32	1/16	3/32	1/16

Straightness: 1/4 in. in any 5 ft, or 1/4 in. times number of feet divided by 5.

Carbon-steel Bars, Cold-finished

Dimensional Tolerances (Under Size Only)

Size, In.	Max of Carbon Range 0.28 % or Less	Max of Carbon Range over 0.28 % to 0.55 % Incl.	All Grades, Stress-relieved after Cold-finishing	Max of Carbon Range over 0.55 % or All Grades, Heat-treated* before Cold-finishing
	Size, Tolerances Under, In.			
Rounds, Cold-drawn or Turned and Polished†				
To 1 incl.	0.002	0.003	0.004	0.006
Over 1 to 2 incl.	0.003	0.004	0.006	0.008
Over 2 to 2 15/16 excl.	0.004	0.005	0.008	0.010
Rounds, Turned and Polished				
From 2 15/16 to 4 incl.	0.004	0.005	0.008	0.010
Over 4 to 6 incl.	0.005	0.006	0.010	0.012
Over 6 to 7 3/4 incl.	0.006	0.008	0.012	0.016
Hexagons, Cold-drawn				
To 5/16 incl.	0.002	0.003	0.004	0.006
Over 5/16 to 1 incl.	0.003	0.004	0.006	0.008
Over 1 to 2 1/2 incl.	0.004	0.005	0.008	0.010
Over 2 1/2 to 3 1/8 incl.	0.005	0.006	0.010	0.012
Squares, Cold-drawn				
To 5/16 incl.	0.003	0.004	0.006	0.008
Over 5/16 to 1 incl.	0.004	0.005	0.008	0.010
Over 1 to 2 1/2 incl.	0.005	0.006	0.010	0.012
Over 2 1/2 to 4 incl.	0.006	0.008	0.012	0.016
Flats, Cold-finished, Thickness and Width, In.				
To 3/4 incl.	0.003	0.004	0.006	0.008
Over 3/4 to 1 1/2 incl.	0.004	0.005	0.008	0.010
Over 1 1/2 to 3 incl.	0.005	0.006	0.010	0.012
Over 3 to 4 incl.	0.006	0.008	0.011	0.016
Over 4 to 6 incl.	0.008	0.010	0.012	0.020
Over 6	0.013			

Rounds, Turned, Ground, and Polished or Cold-drawn, Ground, and Polished

Diameter Tolerances, In.	Over	Under
Less than 2 1/2 in.	0.000	0.002
2 1/2 in. and larger	0.000	0.003

Size tolerances for round bars apply to individual measurements.

* In this column, "Heat-treated" does not include stress relieving. This column does not include tolerances for bars that are heat-treated after cold finishing.

† Some producers have facilities for cold-drawing rounds in sizes larger than 2 7/8 in. For sizes larger than 2 7/8 in. the tolerances for cold-drawn rounds should be negotiated between the purchaser and producer.

Alloy-steel Bars, Cold-finished

Size Tolerances. All Tolerances Are in Inches and Are Minus

Size, In.	Max of Carbon Range 0.28% or Less	Max of Carbon Range over 0.28 to 0.55%*	Max of Carbon Range over 0.55%
Rounds, Cold-drawn or Turned and Polished			
To 1 incl.	0.003	0.005	0.007
Over 1 to 2 incl.	0.004	0.006	0.009
Over 2 to 4 incl.	0.005	0.007	0.011
Over 4 to 6 incl.	0.006	0.008	0.013
Over 6 to 7¾	0.007	0.010	0.017
Hexagons, Cold-drawn			
To 5/16 incl.	0.003	0.005	0.007
Over 5/16 to 1 incl.	0.004	0.006	0.009
Over 1 to 2½ incl.	0.005	0.007	0.011
Over 2½ to 3⅛ incl.	0.006	0.008	0.013
Squares, Cold-drawn			
To 5/16 incl.	0.004	0.006	0.009
Over 5/16 to 1 incl.	0.005	0.007	0.011
Over 1 to 2½ incl.	0.006	0.008	0.013
Over 2½ to 4	0.007	0.010	0.017
Flats, Cold-finished, Width, In.			
To ¾ incl.	0.004	0.006	0.009
Over ¾ to 1½ incl.	0.005	0.007	0.011
Over 1½ to 3 incl.	0.006	0.008	0.013
Over 3 to 4 incl.	0.007	0.010	0.017
Over 4 to 6 incl.	0.009	0.012	0.021
Over 6	0.014		

The above tolerances provide for undersize tolerances only.
The tolerances for flats apply to thickness as well as width.
Size tolerances for round bars apply to individual measurements.

* Tolerances in this column also apply to all carbons up to 0.55 % maximum annealed before drawing.

Carbon-steel Sheets

Thickness Tolerances, Hot-rolled and Hot-rolled Annealed, Coils and Cut Lengths, Including Pickled Sheets

Specified Widths, In.	Thickness Tolerances for Specified Widths and Thicknesses, Over or Under, In.														
	0.2299 0.1875	0.1874 0.1800	0.1799 0.1420	0.1419 0.0972	0.0971 0.0822	0.0821 0.0710	0.0709 0.0568	0.0567 0.0509	0.0508 0.0389	0.0388 0.0344	0.0343 0.0314	0.0313 0.0255	0.0254 0.0195	0.0194 0.0142	0.0141 and Thinner
To 3½ incl	.....	.....	.....	.....	.....	.....	.....	.....	.....	.....	.....	.....	0.003	0.002	0.002
Over 3½ to 6 incl	.....	.....	.....	.....	.....	.....	.....	.....	.....	.....	0.004	0.003	0.003	0.002	0.002
Over 6 to 12 incl	.....	.....	.....	.....	.....	.....	.....	0.005	0.005	0.004	0.004	0.003	0.003	0.002	0.002
Over 12 to 15 incl	0.008	0.007	0.007	0.007	0.006	0.006	0.006	0.005	0.005	0.004	0.004	0.003	0.003	0.002	
Over 15 to 20 incl	0.008	0.008	0.008	0.008	0.007	0.007	0.006	0.006	0.005	0.004	0.004	0.003	0.003	0.002	
Over 20 to 32 incl	0.009	0.009	0.009	0.008	0.007	0.007	0.006	0.006	0.005	0.004	0.004	0.003	0.003	0.002	
Over 32 to 40 incl	0.009	0.009	0.009	0.009	0.008	0.007	0.006	0.006	0.005	0.004	0.004	0.003	0.003	0.002	0.002
Over 40 to 48 incl	0.010	0.010	0.010	0.010	0.008	0.007	0.006	0.006	0.005	0.004	0.004	0.003	0.003	0.002	0.002
Over 48 to 60 incl	.....	.....	0.010	0.010	0.008	0.007	0.007	0.006	0.005	0.004	0.004				
Over 60 to 70 incl	.....	.....	0.011	0.011	0.009	0.008	0.007	0.007	0.006	0.005	0.005				
Over 70 to 80 incl	.....	.....	0.012	0.012	0.009	0.008									
Over 80 to 90 incl	.....	.....	0.012	0.012	0.010										
Over 90	.....	.....	0.012	0.012											

Thickness is measured at any point on the sheet not less than ⅜ in. from a cut edge and not less than ¾ in. from a mill edge. The above table does not apply to the uncropped ends of mill-edge coils.

Carbon-steel Sheets

Width Tolerances, Coils and Cut Lengths, Including Pickled Sheets

Mill Edge Hot-rolled

Specified Widths, In.	*Tolerance Over Specified Width, In., No Tolerance Under*
Over 12 to 20 incl	7/16
Over 20 to 30 incl	1¼
Over 30 to 40 incl	1⅜
Over 40 to 50 incl	1⅝
Over 50 to 72 incl	1¾
Over 72 to 78 incl	2
Over 78	2½

Cut Edge Hot-rolled and Hot-rolled Annealed

Specified Widths, In.	*Tolerance Over Specified Width, In. No Tolerance Under*
To 20 incl	⅛
Over 20 to 30 incl	3/16
Over 30 to 50 incl	¼
Over 50 to 80 incl	5/16
Over 80	⅜

Carbon-steel Sheets

Length Tolerances, Hot-rolled and Hot-rolled Annealed
(Sheets Not Resquared, Including Pickled Sheets)

Specified Length, In.	*Tol Over Specified Length, In., No Tol Under*	*Specified Length, In.*	*Tol Over Specified Length, In., No Tol Under*
To 15 incl.	1/8	Over 120 to 156 incl.	1 1/4
Over 15 to 30 incl.	1/4	Over 156 to 192 incl.	1 1/2
Over 30 to 60 incl.	1/2	Over 192 to 240 incl.	1 3/4
Over 60 to 96 incl.	3/4	Over 240	2
Over 96 to 120 incl.	1		

Camber Tolerances for Cut Lengths
(Coils: Camber tolerance for coils = 1 in. per 20 ft)

Sheet Length, Ft	Camber	Sheet Length, Ft	Camber
To 4 incl.	1/8	Over 14 to 16 incl.	5/8
Over 4 to 6 incl.	3/16	Over 16 to 18 incl.	3/4
Over 6 to 8 incl.	1/4	Over 18 to 20 incl.	7/8
Over 8 to 10 incl.	5/16	Ovef 20 to 30 incl.	1 1/4
Over 10 to 12 incl.	3/8	Over 30 to 40 incl.	1 1/2
Over 12 to 14 incl.	1/2		

Flatness Tolerances, Hot-rolled and Hot-rolled Annealed, Not Stretcher-leveled*

Specified Weight, Psf	Specified Thickness, In.	Specified Width, In.	Flatness Tolerance (Max Deviation from a Horizontal Flat Surface), In.
2.375 (16 gage) and heavier	0.0568 and thicker	To 60 incl.	1/2
		Over 60 to 72 incl.	3/4
		Over 72	1
2.374 (17 gage) and lighter	0.0567 and thinner	To 36 incl.	1/2
		Over 36 to 60 incl.	3/4
		Over 60	1

Hot-rolled and Hot-rolled Annealed, Stretcher-leveled

Specified Weight, Psf	Specified Thickness, In.	Specified Width, In.	Specified Length, In.	Flatness Tolerance (Max Deviation from a Horizontal Flat Surface), In.
1.188 (22 gage) and heavier	0.0284 and thicker	To 48 incl.	To 96 incl.	1/8
		Wider or longer sheets		1/4

* Does not apply to coils.

Carbon-steel Sheets, Cold-rolled

Thickness Tolerances, Coils and Cut Lengths

Specified Widths, In.	Thickness Tolerances for Specified Widths and Thicknesses Given, Over or Under, In.												
	0.1875 and Thicker	0.1874 0.1420	0.1419 0.0972	0.0971 0.0822	0.0821 0.0710	0.0709 0.0568	0.0567 0.0509	0.0508 0.0389	0.0388 0.0314	0.0313 0.0255	0.0254 0.0195	0.0194 0.0142	0.0141 0.0113	0.0112 and Thinner
Up to 15 incl	0.007	0.006	0.006	0.006	0.005	0.005	0.005	0.004	0.003	0.003	0.003	0.002		
Over 15 to 20 incl	0.007	0.007	0.007	0.006	0.005	0.005	0.005	0.004	0.003	0.003	0.003	0.002		
Over 20 to 24 incl	0.007	0.007	0.007	0.006	0.005	0.005	0.005	0.004	0.003	0.003	0.003	0.002		
Over 24 to 32 incl	0.008	0.008	0.008	0.006	0.006	0.005	0.005	0.004	0.003	0.003	0.003	0.002		
Over 32 to 40 incl	0.009	0.009	0.009	0.007	0.006	0.005	0.005	0.004	0.0035	0.003	0.003	0.002	0.002	0.0015
Over 40 to 48 incl	0.010	0.010	0.009	0.007	0.006	0.005	0.005	0.004	0.0035	0.003	0.003	0.002	0.002	
Over 48 to 60 incl	0.011	0.010	0.010	0.008	0.007	0.006	0.005	0.004	0.0035	0.0035	0.003	0.002		
Over 60 to 70 incl	0.012	0.011	0.010	0.009	0.007	0.006	0.006	0.005	0.004	0.004				
Over 70 to 80 incl	0.013	0.012	0.011	0.009	0.007	0.006	0.006	0.005	0.004	0.004				
Over 80 to 90 incl	0.014	0.012	0.012											
Over 90	0.015	0.012	0.012											

Thickness is measured at any point on the sheet not less than ⅜ in. from an edge.

Width Tolerances, Sheets Not Resquared

Specified Width, In.	*Tolerance Over Specified Width, In. No Tolerance Under*
Up to 20 incl	⅛
Over 20 to 32 incl	3/16
Over 32 to 48 incl	¼
Over 48 to 80 incl	5/16
Over 80	⅜

Other tolerances:
Length, same as hot-rolled.
Camber, same as hot-rolled.
Flatness, same as hot-rolled.

Alloy-steel Sheets, Hot-rolled

Width Tolerances, Sheets Not Resquared, Coils and Cut Lengths

All Tolerances Are Plus

Specified Widths, In.	0.2299 0.1800	0.1799 0.1091	0.1090 0.0621	0.0620 0.0568	0.0567 0.0410	0.0409 0.0344	0.0343 0.0255	0.0254 and Under
Mill Edge								
To 2 incl.	...	...	...	...	...	...	...	1/16
Over 2 to 3½ incl.	...	...	...	...	...	...	...	1/8
Over 3½ to 5 incl.	...	...	...	...	...	...	1/8	1/8
Over 5 to 6 incl.	...	...	...	...	...	...	1/8	1/8
Over 6 to 10 incl.	...	...	...	...	3/16	3/16	3/16	3/16
Over 10 to 15 incl.	...	...	...	...	3/16	3/16	3/16	3/16
Over 15 to 20 incl.	...	...	...	...	1/4	1/4	1/4	1/4
Over 20 to 24 excl.	...	...	...	...	3/8	3/8	3/8	3/8
24 to 30 incl.	3/8	3/8	3/8	3/8	3/8	3/8	3/8	3/8
Over 30 to 48 incl.	1/2	1/2	1/2	1/2	1/2	1/2	1/2	1/2
Over 48 to 50 incl.	...	1/2	1/2	1/2	1/2	1/2	1/2	1/2
Over 50 to 80 incl.	...	5/8	5/8	5/8	5/8	5/8	5/8	5/8
Over 80	...	3/4	3/4	3/4	3/4	3/4	3/4	3/4
Sheared Edge								
To 2 incl.	...	...	...	...	...	...	...	1/16
Over 2 to 3½ incl.	...	...	...	...	...	...	...	1/8
Over 3½ to 5 incl.	...	...	...	...	...	...	1/8	1/8
Over 5 to 6 incl.	...	...	...	...	...	...	1/8	1/8
Over 6 to 10 incl.	...	...	...	...	1/8	1/8	1/8	1/8
Over 10 to 15 incl.	...	...	...	...	1/8	1/8	1/8	1/8
Over 15 to 20 incl.	...	...	...	...	1/8	1/8	1/8	1/8
Over 20 to 24 excl.	...	...	...	...	3/16	3/16	3/16	3/16
24 to 30 incl.	3/16	3/16	3/16	3/16	3/16	3/16	3/16	3/16
Over 30 to 38 incl.	1/4	1/4	1/4	1/4	1/4	1/4	1/4	1/4
Over 48 to 50 incl.	...	1/4	1/4	1/4	1/4	1/4	1/4	1/4
Over 50 to 80 incl.	...	5/16	5/16	5/16	5/16	5/16	5/16	5/16
Over 80	...	3/8	3/8	3/8	3/8	3/8	3/8	3/8

No tolerance under.

Flatness Tolerances*

Specified Thickness, In.	Specified Width, In.	Flatness Tolerances, In.
From 0.0195 to 0.0567	To 36 incl.	½
	Over 36 to 60 incl.	¾
	Over 60	1
From 0.0568 to 0.2299 incl.	To 60 incl.	½
	Over 60 to 72 incl.	¾
	Over 72	1

* Tolerances apply to annealed sheets only.

ALLOY-STEEL SHEETS, HOT- AND COLD-ROLLED
Thickness Tolerances

Specified Widths, In.	Thickness, In.																		
	.2299 .1910	.1909 .1800	.1799 .1420	.1419 .1180	.1179 .1090	.1089 .0972	.0971 .0822	.0821 .0710	.0709 .0610	.0609 .0568	.0567 .0509	.0508 .0410	.0409 .0389	.0388 .0344	.0343 .0314	.0313 .0255	.0254 .0195	.0194 .0142	.0141 and Under
	Thickness Tolerances for Specified Widths and Thicknesses, Over or Under, In.																		
To 3½ incl																	0.003	0.002	0.002
Over 3½ to 5 incl															0.004	0.003	0.003	0.002	0.002
Over 5 to 6 incl															0.004	0.003	0.003	0.002	0.002
Over 6 to 10 incl											0.005	0.005	0.005	0.004	0.004	0.003	0.003	0.002	0.002
Over 10 to 15 incl											0.005	0.005	0.005	0.004	0.004	0.003	0.003	0.002	0.002
Over 15 to 24 excl											0.006	0.005	0.005	0.004	0.004	0.003	0.003	0.002	
24 to 32 incl	0.009	0.009	0.009	0.008	0.008	0.008	0.007	0.007	0.006	0.006	0.006	0.005	0.005	0.004	0.004	0.003	0.003	0.002	
Over 32 to 40 incl	0.009	0.009	0.009	0.009	0.009	0.009	0.008	0.007	0.006	0.006	0.006	0.005	0.005	0.004	0.004	0.003	0.003	0.002	0.002
Over 40 to 48 incl	0.010	0.010	0.010	0.010	0.010	0.010	0.008	0.007	0.006	0.006	0.006	0.005	0.005	0.004	0.004	0.003	0.003	0.002	0.002
Over 48 to 60 incl			0.010	0.010	0.010	0.010	0.008	0.007	0.007	0.006	0.006	0.005	0.005	0.004	0.004				
Over 60 to 70 incl			0.011	0.011	0.011	0.011	0.009	0.008	0.007	0.007	0.007	0.006	0.006	0.005	0.005				
Over 70 to 80 incl			0.012	0.012	0.012	0.012	0.009	0.008											
Over 80 to 90 incl			0.012	0.012	0.012	0.012	0.010												
Over 90			0.012	0.012	0.012	0.012													

NOTE: Thickness is measured at any point on the sheet not less than ⅜ in. from an edge.

Stainless and Heat-resisting Sheets, Cold-rolled

Thickness Tolerances

Specified Thickness, In.	*Thickness Tolerances, In. Over and Under*
0.146 to less than 3/16	0.014
0.131 to 0.145	0.012
0.115 to 0.130	0.010
0.099 to 0.114	0.009
0.084 to 0.098	0.008
0.073 to 0.083	0.007
0.059 to 0.072	0.006
0.041 to 0.058	0.005
0.027 to 0.040	0.004
0.017 to 0.026	0.003
0.008 to 0.016	0.002
0.006 to 0.007	0.0015
0.005	0.001

NOTE: Thickness measurements are taken at least 3/8 in. from the edge of the sheet.

Width Tolerances*

Specified Thickness, In.	Specified Width, In.	
	24 to 48, Excl.	48 and Over
	Width Tolerances	Width Tolerances
Less than 3/16	1/16 over, 0 under	1/8 over, 0 under

* Not resquared.

Length Tolerances

Length, Ft	*Tolerance, In.*
Up to 10	1/4 over, 0 under
Over 10 to 20	1/2 over, 0 under

Camber Tolerances

Specified Width, In.	*Tolerance, In. per Unit Length of 8 Ft*
24 to 36, excl.	1/8
Over 36	3/32

Flatness Tolerances

Sheets Not Stretcher-leveled and Exclusive of Dead-soft and Deep-drawing Sheets

Specified Thickness, In.	Width, In.	Flatness Tolerance (Max Deviation from a Horizontal Flat Surface), In.
0.062 and over	To 60 incl.	1/2
	Over 60 to 72 incl.	3/4
	Over 72	1
Under 0.062	To 36 incl.	1/2
	Over 36 to 60 incl.	3/4
	Over 60	1

NOTE: The above tolerances do not apply to 1/4 and 1/2 hard tempers of 300 series.

Carbon-steel Strip, Hot-rolled

Thickness Tolerances

Specified Widths, In.	Tolerances for Specified Thickness for Widths Given, Over or Under, In.						
	0.2299 to 0.2031 Incl.	0.2030 to 0.1875 Incl.	0.1874 to 0.1180 Incl.	0.1179 to 0.0568 Incl.	0.0567 to 0.0449 Incl.	0.0448 to 0.0344 Incl.	0.0343 to 0.0255 Incl.
Up to 3½ incl.	.....	0.006	0.005	0.004	0.003	0.003	0.003
Over 3½ to 6 incl.	.....	0.006	0.005	0.005	0.003	0.003	
Over 6 to 12 incl.	0.006	0.006	0.005	0.005	0.004		

Measurements are taken ⅜ in. from edge of strip on 1 in. or wider; at any place when narrower.

Crown Tolerances for Above Thicknesses

Strip may be thicker at the center than at a point ⅜ in. in from the edge by the following maximum amounts.

Over 1 to 3½ incl.	.....	0.001	0.002	0.002	0.002	0.002	0.002
Over 3½ to 6 incl.	.....	0.002	0.002	0.003	0.003	0.003	
Over 6 to 12 incl.	0.002	0.003	0.003	0.004	0.004		

Camber Tolerances

For strip wider than 1½ in. . . . . ¼ in. in any 8 ft
For strip 1½ in. and narrower . . . . ½ in. in any 8 ft

Length Tolerances

Specified Widths, In.	Length Tolerances for Given Widths and Lengths, In.					
	To 5 Ft Incl.	Over 5 to 10 Ft Incl.	Over 10 to 20 Ft Incl.	Over 20 to 30 Ft Incl.	Over 30 to 40 Ft Incl.	Over 40 Ft
To 3 incl.	¼	⅜	½	¾	1	1½
Over 3 to 6 incl.	⅜	½	⅝	¾	1	1½
Over 6 to 12 incl.	½	¾	1	1¼	1½	1¾

No tolerance under.

Width Tolerances

Specified Widths, In.	Tolerances for Specified Width for Thicknesses Given, Over or Under, In.		
	Mill Edge and Square Edge, All Thicknesses	Slit Edge	
		To 0.109 Incl.	Over 0.109
To 2 incl.	1/32	0.008	0.016
Over 2 to 5 incl.	3/64	0.008	0.016
Over 5 to 10 incl.	1/16	0.010	0.016
Over 10 to 12 incl.	3/32	0.016	0.016

Carbon-steel Strip, Cold-rolled

Thickness Tolerances
Measured ⅜ in. from edge on 1 in. or wider; any place otherwise

Specified Thickness, In.		Tolerances for Specified Thickness, Plus or Minus, In. — Widths, In.							
Over	To and Incl.	Over ½ to Less Than 1	1 to Less Than 3	3 to 6 Incl.	Over 6 to 9 Incl.	Over 9 to 12 Incl.	Over 12 to 16 Incl.	Over 16 to 20 Incl.	Over 20 to 23^15^⁄16 Incl.
0.160	0.2499	0.002	0.003	0.0035	0.0035	0.0035	0.0045	0.005	0.005
0.099	0.160	0.002	0.002	0.003	0.003	0.003	0.0035	0.0045	0.005
0.068	0.099	0.002	0.002	0.0025	0.003	0.003	0.0035	0.0035	0.0035
0.049	0.068	0.002	0.002	0.0025	0.0025	0.0025	0.003	0.0035	0.0035
0.039	0.049	0.002	0.002	0.0025	0.0025	0.0025	0.003	0.003	0.003
0.034	0.039	0.002	0.002	0.002	0.002	0.002	0.002	0.002	0.002
0.031	0.034	0.0015	0.0015	0.002	0.002	0.002	0.002	0.002	0.002
0.028	0.031	0.0015	0.0015	0.0015	0.002	0.002	0.002	0.002	0.002
0.025	0.028	0.001	0.0015	0.0015	0.002	0.002	0.002	0.002	0.002
0.019	0.025	0.001	0.001	0.0015	0.0015	0.0015	0.002	0.002	0.002
0.012	0.019	0.001	0.001	0.001	0.0015	0.0015	0.0015	0.0015	0.0015
0.011	0.012	0.001	0.001	0.001	0.001	0.0015	0.0015	0.0015	0.0015
0.009	0.011	0.001	0.001	0.001	0.001	0.001	0.001	0.001	0.001
0.005	0.009	0.00075	0.00075	0.00075	0.001	0.001	0.001	0.001	0.001
	0.005	0.0005	0.0005	0.0005					

Crown Tolerances
Maximum thickness at center equals edge measurement plus:

Thickness, In.	Width, inches — 1 to 5 Incl.	Over 5 to 12 Incl.	Over 12 to 23^15^⁄16 Incl.
	Additional Thickness at Center, In.		
0.005 to 0.010 incl.	0.00075	0.001	0.0015
Over 0.010 to 0.025 incl.	0.001	0.0015	0.002
Over 0.025 to 0.065 incl.	0.0015	0.002	0.0025
Over 0.065 to 0.187 incl.	0.002	0.0025	0.003
Over 0.187 to 0.2499 incl.	0.002	0.0025	0.003

Camber Tolerances

Width, In.	*Camber Tolerances*
Over 2½	¼ in. in any 8 ft
Over 1½ to 2½	⅜ in. in any 8 ft
Over ½ to 1½	½ in. in any 8 ft

Carbon-steel Strip, Cold-rolled

Width Tolerances for No. 2 Edge (Mill Edge)

Specified Width, In.		Tolerances for Specified Width, Plus or Minus, In.
Over	Up to and Including	
1/2	2	1/32
2	5	3/64
5	10	5/64
10	15	3/32
15	20	1/8
20	23 15/16	5/32

Width Tolerances for No. 3 Edge (Slit Edge)

Specified Thickness, In.		Width, In.				
Over	To and Incl.	Over 1/2 to 6 Incl.	Over 6 to 9 Incl.	Over 9 to 12 Incl.	Over 12 to 20 Incl.	Over 20 to 23 15/16 Incl.
		Tolerances for Specified Width, Plus or Minus, In.				
0.160	0.2499	0.016	0.020	0.020	0.031	0.031
0.099	0.160	0.010	0.016	0.016	0.020	0.020
0.068	0.099	0.008	0.010	0.010	0.016	0.020
0.016	0.068	0.005	0.005	0.010	0.016	0.020
Up to	0.016	0.005	0.005	0.010	0.016	0.020

Width Tolerances for Special Edges

Edge No.	Width, In.	Thickness, In.	Tolerances for Specified Width, Plus or Minus, In.
1	Over 1/2 to 3/4 incl.	3/32 and thinner	0.005
1	Over 3/4 to 5 incl.	1/8 and thinner	0.005
4 and 6	Over 1/2 to 1 incl.	3/16 to 0.025 incl.	0.015
4 and 6	Over 1 to 2 incl.	0.2499 to 0.025 incl.	0.025
4 and 6	Over 2 to 4 incl.	0.2499 to 0.035 incl.	0.047
4 and 6	Over 4 to 6 incl.	0.2499 to 0.047 incl.	0.047
5	Over 1/2 to 3/4 incl.	3/32 and thinner	0.005
5	Over 3/4 to 5 incl.	1/8 and thinner	0.005
5	Over 5 to 9 incl.	1/8 to 0.008 incl.	0.010
5	Over 9 to 20 incl.	0.105 to 0.015	0.010
5	Over 20 to 23 15/16 incl.	0.080 to 0.023	0.015

Alloy-steel Strip, Hot-rolled

Thickness Tolerances, over or under Specified Thickness

Specified Widths, In.	Thickness, In.																	
	.2299 .2031	.2030 .1719	.1718 .1570	.1569 .1420	.1419 .1271	.1270 .1121	.1120 .0972	.0971 .0822	.0821 .0710	.0709 .0636	.0635 .0568	.0567 .0509	.0508 .0449	.0448 .0389	.0388 .0344	.0343 .0314	.0313 .0284	.0283 .0255
To 3½ incl	.....	0.006	0.005	0.005	0.005	0.005	0.004	0.004	0.004	0.004	0.004	0.004	0.003	0.003	0.003	0.003	0.003	0.003
Over 3½ to 6 incl	.....	0.006	0.005	0.005	0.005	0.005	0.005	0.005	0.005	0.005	0.005	0.003	0.003	0.003	0.003			
Over 6 to 12 incl	0.007	0.006	0.006	0.006	0.006	0.006	0.005	0.005	0.005	0.005	0.005							
Over 12 to 15 incl	0.008	0.007	0.007	0.007	0.007	0.007	0.006	0.006	0.006	0.006	0.006							
Over 15 to 24 excl	0.009	0.009	0.008	0.008	0.008	0.008	0.007	0.007	0.007	0.006	0.006							

Note: Thickness measurements are taken ⅜ in. from edge of strip. The above tolerances do not include crown tolerances.

Length Tolerances

Specified Length, In.	Tolerance Over, None Under, In.		
	3 In. Wide or Under	3 to 6 In. Wide Incl.	Over 6 to 24 In. Wide Excl.
Over 15 to 60 incl	¼	⅜	½
Over 60 to 120 incl	⅜	½	¾
Over 120 to 240 incl	½	⅝	1
Over 240 to 360 incl	¾	¾	1¼
Over 360 to 480 incl	1	1	1½
Over 480	1½	1½	1¾

Crown Tolerances

Hot-rolled alloy strip may be thicker at the center than at the edges by the following amounts:

Width, In.	*Crown Tolerance, In.*
Up to 2	0.002
Over 2 to 5 incl	0.003
Over 5 to 10 incl	0.004
Over 10 to 15 incl	0.005
Over 15 to 24 excl	0.006

Alloy-steel Strip, Hot-rolled

Width Tolerances

Specified Widths, In.	Mill Edge					
	0.2299 0.2031	0.2030 0.0621	0.0620 0.0568	0.0567 0.0410	0.0409 0.0344	0.0343 0.0255
To 2 incl.	......	±1/32	±1/32	±1/32	±1/32	±1/32
Over 2 to 3½ incl.	......	±3/64	±3/64	±3/64	±3/64	±3/64
Over 3½ to 6 incl.	......	±3/64	±3/64	±3/64	±3/64	
Over 6 to 12 incl.	±1/16	±1/16	±1/16			
Over 12 to 15 incl.	±3/32	±3/32	±3/32			
Over 15 to 20 incl.	±1/8	±1/8	±1/8			
Over 20 to 24 excl.	±5/32	±5/32	±5/32			
	Sheared Edge					
To 2 incl.	......	±0.008	±0.008	±0.008	±0.008	±0.008
Over 2 to 3½ incl.	......	±0.008	±0.008	±0.008	±0.008	±0.008
Over 3½ to 6 incl.	......	±0.008	±0.008	±0.008	±0.008	
Over 6 to 12 incl.	±0.016	±0.010	±0.010			
Over 12 to 15 incl.	±0.016	±0.016	±0.016			
Over 15 to 20 incl.	±0.016	±0.016	±0.016			
Over 20 to 24 excl.	±1/32	±1/32	±1/32			

Other tolerances:
Camber, ¼ in. max in 8 ft.
Flatness, to be negotiated.

Alloy-steel Strip, Cold-rolled

Thickness Tolerances

Specified Thickness, In.	Thickness Tolerances for the Thicknesses and Widths Given, Over and Under, In.							
	Under 1 to 3/16 Incl.	Under 3 to 1 Incl.	3 to 6 Incl.	Over 6 to 9 Incl.	Over 9 to 12 Incl.	Over 12 to 16 Incl.	Over 16 to 20 Incl.	Over 20 to 23 15/16 Incl.
0.187 to 0.161 incl.	0.002	0.003	0.004	0.004	0.004	0.005	0.006	0.006
0.160 to 0.100 incl.	0.002	0.002	0.003	0.004	0.004	0.004	0.005	0.005
0.099 to 0.069 incl.	0.002	0.002	0.003	0.003	0.003	0.004	0.004	0.004
0.068 to 0.050 incl.	0.002	0.002	0.003	0.003	0.003	0.003	0.004	0.004
0.049 to 0.040 incl.	0.002	0.002	0.0025	0.003	0.003	0.003	0.004	0.004
0.039 to 0.035 incl.	0.002	0.002	0.0025	0.003	0.003	0.003	0.003	0.003
0.034 to 0.029 incl.	0.0015	0.0015	0.002	0.0025	0.0025	0.0025	0.003	0.003
0.028 to 0.026 incl.	0.001	0.0015	0.0015	0.002	0.002	0.002	0.0025	0.003
0.025 to 0.020 incl.	0.001	0.001	0.0015	0.002	0.002	0.002	0.0025	0.0025
0.019 to 0.017 incl.	0.001	0.001	0.001	0.0015	0.0015	0.002	0.002	0.002
0.016 to 0.013 incl.	0.001	0.001	0.001	0.0015	0.0015	0.0015	0.002	0.002
0.012	0.001	0.001	0.001	0.001	0.0015	0.0015	0.0015	0.0015
0.011	0.001	0.001	0.001	0.001	0.001	0.0015	0.0015	0.0015
0.010	0.001	0.001	0.001	0.001	0.001	0.001	0.0015	0.0015

Alloy-steel Strip, Cold-rolled (*Continued*)

Width Tolerances for No. 3 Edge (Slit Edge)

Specified Thickness, In.	Tolerances for Specified Width, Over and Under, In.					
	Under ½ to 3/16 Incl.	½ to 6 Incl.	Over 6 to 9 Incl.	Over 9 to 12 Incl.	Over 12 to 20 Incl.	Over 20 to 23 15/16 Incl.
0.187 to 0.161 incl.	.....	0.016	0.020	0.020	0.031	0.031
0.160 to 0.100 incl.	0.010	0.010	0.016	0.016	0.020	0.020
0.099 to 0.069 incl.	0.008	0.008	0.010	0.010	0.016	0.020
0.068 and under	0.005	0.005	0.005	0.010	0.016	0.020

Camber Tolerances

Specified Width, In.	*Tolerance per Unit Length of 8 Ft*
To 1½ incl.	½ in. in any 8 ft length
Over 1½ to 24 excl.	¼ in. in any 8 ft length

Stainless Strip, Cold-rolled

Thickness Tolerances

Specified Thickness, In.	Tolerances for Thicknesses and Widths Given, Plus and Minus, In.							
	Under 1 to 3/16 Incl.	Under 3 to 1 Incl.	3 to 6 Incl.	Over 6 to 9 Incl.	Over 9 to 12 Incl.	Over 12 to 16 Incl.	Over 16 to 20 Incl.	Over 20 to 23 15/16 Incl.
Under 3/16 to 0.161 incl.	0.002	0.003	0.004	0.004	0.004	0.005	0.006	0.006
0.160 to 0.100 incl.	0.002	0.002	0.003	0.004	0.004	0.004	0.005	0.005
0.099 to 0.069 incl.	0.002	0.002	0.003	0.003	0.003	0.004	0.004	0.004
0.068 to 0.050 incl.	0.002	0.002	0.003	0.003	0.003	0.003	0.004	0.004
0.049 to 0.040 incl.	0.002	0.002	0.0025	0.003	0.003	0.003	0.004	0.004
0.039 to 0.035 incl.	0.002	0.002	0.0025	0.003	0.003	0.003	0.003	0.003
0.034 to 0.029 incl.	0.0015	0.0015	0.002	0.0025	0.0025	0.0025	0.003	0.003
0.028 to 0.026 incl.	0.001	0.0015	0.0015	0.002	0.002	0.002	0.0025	0.003
0.025 to 0.020 incl.	0.001	0.001	0.0015	0.002	0.002	0.002	0.0025	0.0025
0.019 to 0.017 incl.	0.001	0.001	0.001	0.0015	0.0015	0.002	0.002	0.002
0.016 to 0.013 incl.	0.001	0.001	0.001	0.0015	0.0015	0.0015	0.002	0.002
0.012	0.001	0.001	0.001	0.001	0.0015	0.0015	0.0015	0.0015
0.011	0.001	0.001	0.001	0.001	0.001	0.0015	0.0015	0.0015
0.010	0.001	0.001	0.001	0.001	0.001	0.001	0.0015	0.0015

Note: For thicknesses under 0.010 in. in widths up to and including 16 in., a tolerance of plus or minus 10% of the thickness is to apply. For thicknesses under 0.010 in. in widths over 16 in. to 23 15/16 in. incl., a tolerance of plus or minus 15% of the thickness is to apply.

Width Tolerances for No. 3 Edge

Specified Thickness, In.	Tolerances for Specified Width, Over and Under, In.					
	Under ½ to 3/16 Incl.	½ to 6 Incl.	Over 6 to 9 Incl.	Over 9 to 12 Incl.	Over 12 to 20 Incl.	Over 20 to 23 15/16 Incl.
Under 3/16 in. to 0.161 incl.	.....	0.016	0.020	0.020	0.031	0.031
0.160 to 0.100 incl.	0.010	0.010	0.016	0.016	0.020	0.020
0.099 to 0.069 incl.	0.008	0.008	0.010	0.010	0.016	0.020
0.068 and under	0.005	0.005	0.005	0.010	0.016	0.020

Stainless-steel Strip, Cold-rolled
Width Tolerances for No. 1 or 5 Edges

Specified Edge No.	Width, In.	Thickness, In.	Tolerance, In.	
			Over	Under
1 and 5	9/32 and under	1/16 and under	0.005	0.005
1 and 5	Over 9/32 to 3/4 incl.	3/32 and under	0.005	0.005
1 and 5	Over 3/4 to 5 incl.	1/8 and under	0.005	0.005
5	Over 5 to 9 incl.	1/8 to 0.008 incl.	0.010	0.010
5	Over 9 to 20 incl.	0.105 to 0.015	0.010	0.010
5	Over 20 to 23 15/16 incl.	0.080 to 0.023	0.015	0.015

Crown Tolerances
Cold-rolled strip may be thicker at the middle than at the edges by the following amounts:

Specified Thickness, In.	To 5 Incl.	Over 5 to 12 Incl.	Over 12 to 24 Excl.
0.005 to 0.010 incl.	0.00075	0.001	0.0015
Over 0.010 to 0.025 incl.	0.001	0.0015	0.002
Over 0.025 to 0.065 incl.	0.0015	0.002	0.0025
Over 0.065 to 3/16 excl.	0.002	0.0025	0.003

Camber Tolerances

Specified Width, In.	*Tolerance per Unit Length of 8 Ft*
To 1½ incl.	½ in. in any 8 ft length
Over 1½ to 24 excl.	¼ in. in any 8 ft length

Wire Tolerances

Uncoated Fine Round Wire in Coils*
Low-carbon, Medium-low-carbon, and Medium-high-carbon Grades†

Size, In.	*Tolerances, Plus and Minus, In.*
0.0625/0.0348	0.001
0.0347/0.0271	0.0008
0.0270/0.0200	0.0006
0.0199/0.0151	0.0005
0.0150/0.0101	0.0004
0.0100/0.0060	0.0003
0.0059/0.0044	0.0002

* These tolerances do not apply to special wires which have been annealed as a separate operation following cold drawing or immediately prior to coating.
† Low-carbon grades such as 1006, 1008, 1010, 1012, B1006 and B1010; medium-low-carbon grades such as 1013 to 1022, inclusive; medium-high-carbon grades such as 1025 to 1041, inclusive.

Alloy-steel Wire

Size, In.	*Tolerances, Plus and Minus, In.*
Coarser than 0.499	0.003
0.499 to 0.076	0.002
0.0759 to 0.0348	0.001

Out of round is one-half the total size tolerance.

Carbon-steel Spring Wire

Dia, In.	MB and WMB, HB, WHB, Extra HB, and Extra WHB	Music	Valve Spring
	Tolerances, In., Plus and Minus		
0.026 and smaller	....	0.0003	
0.020 to 0.027	0.00075		
0.027 to 0.063	....	0.0005	
0.028 to 0.075	0.001		
0.064 to 0.075	....	0.001	
0.076 to 0.092	0.002	0.001	
0.093 to 0.148	0.002	0.001	0.001
0.149 to 0.177	0.002	0.001	0.0015
0.178 to 0.250	0.002	0.001	0.002
0.251 to 0.375	0.002		
0.376 and larger	0.003		

Out of round is one-half the total size tolerance.
Note: Tolerances indicate the range of sizes furnished in various qualities.

Alloy-steel Spring Wire

Wire Dia, In.	Alloy-steel Spring Wire Chromium-Vanadium, Chromium-Silicon, Silicon-Manganese, Plus and Minus, In.	Aircraft Alloy-steel Spring Wire Plus and Minus, In.
0.625 to 0.501	0.004	0.003
0.500 to 0.376	0.003	0.002
0.375 to 0.149	0.002	0.0015
0.148 to 0.076	0.002	0.001
0.075 to 0.020	0.001	0.00075

WEIGHTS

Carbon Bar Steel
Weight per Lin Ft, Lb

Size, In.	Round	Square	Octagon	Hexagon	Size, In.	Round	Square	Octagon	Hexagon
1/16	0.010	0.013	0.011	......	2 1/16	11.36	14.46	12.00	12.51
1/8	0.042	0.053	0.044	0.046	2 1/8	12.06	15.35	12.74	13.30
3/16	0.094	0.119	0.099	0.10	2 3/16	12.78	16.27	13.50	14.08
1/4	0.167	0.212	0.177	0.18	2 1/4	13.52	17.22	14.29	14.91
5/16	0.261	0.333	0.276	0.29	2 5/16	14.28	18.19	15.10	15.74
3/8	0.375	0.478	0.397	0.41	2 3/8	15.07	19.18	15.92	16.62
7/16	0.511	0.651	0.540	0.56	2 7/16	15.86	20.20	16.77	17.50
1/2	0.667	0.850	0.706	0.74	2 1/2	16.69	21.25	17.64	18.41
9/16	0.845	1.076	0.893	0.93	2 9/16	17.53	22.33	18.53	19.35
5/8	1.043	1.328	1.102	1.15	2 5/8	18.40	23.43	19.45	20.30
11/16	1.262	1.608	1.325	1.40	2 11/16	19.29	24.56	20.38	21.28
3/4	1.502	1.913	1.588	1.66	2 3/4	20.20	25.00	20.75	22.28
13/16	1.763	2.245	1.863	1.94	2 13/16	21.12	26.90	22.33	23.30
7/8	2.044	2.603	2.161	2.25	2 7/8	22.07	28.10	23.32	24.34
15/16	2.347	2.989	2.481	2.59	2 15/16	23.04	29.34	24.35	25.40
1	2.670	3.400	2.822	2.94	3	24.03	30.60	25.40	26.51
1 1/16	3.014	3.838	3.186	3.32	3 1/16	25.04	31.89	26.47	27.89
1 1/8	3.379	4.303	3.572	3.73	3 1/8	26.08	33.20	27.56	28.77
1 3/16	3.766	4.795	3.980	4.15	3 3/16	27.13	34.55	28.68	29.90
1 1/4	4.173	5.312	4.409	4.60	3 1/4	23.20	35.92	29.81	31.10
1 5/16	4.600	5.857	4.861	5.06	3 5/16	29.30	37.31	30.97	32.29
1 3/8	5.049	6.428	5.335	5.54	3 3/8	30.42	38.73	32.15	33.75
1 7/16	5.518	7.026	5.832	6.06	3 7/16	31.56	40.18	33.35	34.75
1 1/2	6.008	7.650	6.350	6.63	3 1/2	32.71	41.65	34.57	36.08
1 9/16	6.520	8.301	6.890	7.17	3 9/16	33.90	43.14	35.81	37.34
1 5/8	7.051	8.978	7.452	7.78	3 5/8	35.09	44.68	37.08	38.70
1 11/16	7.604	9.682	8.036	8.37	3 11/16	36.31	46.24	38.38	40.00
1 3/4	8.178	10.41	8.640	9.02	3 3/4	37.56	47.82	39.69	41.43
1 13/16	8.773	11.17	9.271	9.67	3 13/16	38.81	49.42	41.02	42.75
1 7/8	9.388	11.95	9.919	10.36	3 7/8	40.10	51.05	42.37	44.20
1 15/16	10.02	12.76	10.59	11.05	3 15/16	41.40	52.71	43.75	45.65
2	10.68	13.60	11.29	11.78	4	42.73	54.40	45.15	47.13

Flat-rolled Steel
Weights per Lin Ft

Thickness, In.	Width, In.										
	1/2	5/8	3/4	7/8	1	1 1/8	1 1/4	1 3/8	1 1/2	1 5/8	1 3/4
1/16	0.1060	0.1381	0.1594	0.1859	0.212	0.2391	0.2656	0.292	0.319	0.346	0.372
1/8	0.2125	0.2656	0.3188	0.3720	0.4250	0.4782	0.5312	0.585	0.638	0.692	0.744
3/16	0.319	0.399	0.478	0.558	0.638	0.717	0.797	0.875	0.957	1.04	1.15
1/4	0.425	0.531	0.636	0.743	0.850	0.957	1.06	1.17	1.28	1.38	1.49
5/16	0.531	0.664	0.797	0.929	1.06	1.20	1.33	1.46	1.59	1.73	1.86
3/8	0.638	0.797	0.957	1.116	1.28	1.43	1.59	1.76	1.92	2.08	2.23
7/16	0.744	0.929	1.116	1.302	1.49	1.68	1.86	2.05	2.23	2.42	2.60
1/2	0.850	1.06	1.275	1.487	1.70	1.92	2.12	2.34	2.55	2.72	2.98
9/16	0.957	1.20	1.434	1.674	1.92	2.15	2.39	2.63	2.87	3.11	3.35
5/8	1.06	1.33	1.594	1.859	2.12	2.39	2.65	2.92	3.19	3.46	3.72
11/16	1.17	1.46	1.753	2.045	2.34	2.63	2.92	3.22	3.51	3.80	4.09
3/4	1.28	1.60	1.913	2.232	2.55	2.87	3.19	3.51	3.83	4.15	4.47
13/16	1.38	1.73	2.072	2.417	2.76	3.11	3.45	3.80	4.14	4.49	4.84
7/8	1.49	1.86	2.232	2.604	2.98	3.35	3.72	4.09	4.47	4.84	5.20
15/16	1.60	1.99	2.391	2.789	3.19	3.59	3.99	4.39	4.78	5.18	5.58
1	1.70	2.13	2.55	2.98	3.40	3.83	4.25	4.68	5.10	5.53	5.95
1 1/8	1.91	2.39	2.868	3.347	3.83	4.304	4.78	5.26	5.74	6.22	6.70
1 1/4	2.12	2.66	3.19	3.72	4.25	4.79	5.31	5.85	6.38	6.91	7.44
1 3/8	2.34	2.92	3.51	4.09	4.67	5.26	5.84	6.43	7.02	7.60	8.18
1 1/2	2.55	3.19	3.83	4.47	5.10	5.74	6.38	7.02	7.65	8.29	8.93
1 5/8	2.76	3.45	4.15	4.84	5.52	6.22	6.90	7.60	8.29	8.98	9.67
1 3/4	2.98	3.72	4.45	5.21	5.95	6.70	7.44	8.19	8.92	9.67	10.42
1 7/8	3.19	3.99	4.79	5.58	6.38	7.17	7.97	8.77	9.57	10.36	11.15
2	3.40	4.25	5.10	5.95	6.80	7.65	8.50	9.35	10.20	11.05	11.90

Thickness, In.	Width, In.										
	2	2 1/4	2 1/2	2 3/4	3	3 1/4	3 1/2	3 3/4	4	4 1/4	4 1/2
1/16	0.425	0.478	0.531	0.584	0.638	0.691	0.744	0.80	0.85	0.90	0.96
1/8	0.850	0.96	1.06	1.17	1.28	1.38	1.49	1.59	1.70	1.81	1.91
3/16	1.28	1.44	1.59	1.75	1.91	2.07	2.23	2.39	2.55	2.71	2.87
1/4	1.70	1.92	2.12	2.34	2.55	2.76	2.98	3.19	3.40	3.61	3.83
5/16	2.12	2.39	2.65	2.92	3.19	3.45	3.72	3.99	4.25	4.52	4.78
3/8	2.55	2.87	3.19	3.51	3.83	4.15	4.47	4.78	5.10	5.42	5.74
7/16	2.98	3.35	3.72	4.09	4.46	4.83	5.20	5.58	5.95	6.32	6.70
1/2	3.40	3.83	4.25	4.67	5.10	5.53	5.95	6.38	6.80	7.22	7.65
9/16	3.83	4.30	4.78	5.26	5.74	6.22	6.70	7.17	7.65	8.13	8.61
5/8	4.25	4.78	5.31	5.84	6.38	6.91	7.44	7.97	8.50	9.03	9.57
11/16	4.67	5.26	5.84	6.43	7.02	7.60	8.18	8.76	9.35	9.93	10.52
3/4	5.10	5.75	6.38	7.02	7.65	8.29	8.93	9.57	10.20	10.84	11.48
13/16	5.50	6.21	6.90	7.60	8.29	8.98	9.67	10.36	11.05	11.74	12.43
7/8	5.95	6.69	7.44	8.18	8.93	9.67	10.41	11.16	11.90	12.65	13.39
15/16	6.38	7.18	7.97	8.77	9.57	10.36	11.16	11.95	12.75	13.55	14.34
1	6.80	7.65	8.50	9.35	10.20	11.05	11.90	12.75	13.60	14.45	15.30
1 1/8	7.65	8.61	9.57	10.52	11.48	12.43	13.39	14.34	15.30	16.26	17.22
1 1/4	8.50	9.57	10.63	11.69	12.75	13.81	14.87	15.94	17.00	18.06	19.13
1 3/8	9.35	10.52	11.69	12.85	14.03	15.20	16.36	17.53	18.70	19.87	21.04
1 1/2	10.20	11.48	12.75	14.03	15.30	16.58	17.85	19.13	20.40	21.68	22.95
1 5/8	11.05	12.43	13.81	15.19	16.58	17.96	19.34	20.72	22.10	23.48	24.87
1 3/4	11.90	13.40	14.88	16.37	17.85	19.34	20.83	22.32	23.80	25.29	26.78
1 7/8	12.75	14.34	15.94	17.53	19.13	20.72	22.31	23.91	25.50	27.10	28.69
2	13.60	15.30	17.00	18.70	20.40	22.10	23.80	25.50	27.20	28.90	30.60

Weights are based on 489.6 lb per cu ft of steel.

Weights of Brass, Copper, and Aluminum Bars per Foot

Dia or Distance Across Flats, In.	Brass			Copper		Aluminum	
	Weight per Ft			Weight per Ft		Weight per Ft	
	Round	Square	Hexagon	Round	Square	Round	Square
1/16	0.011	0.014	0.013	0.012	0.015	0.003	0.004
1/8	0.045	0.055	0.048	0.047	0.060	0.014	0.018
3/16	0.100	0.125	0.108	0.106	0.135	0.032	0.041
1/4	0.175	0.225	0.194	0.189	0.241	0.057	0.072
5/16	0.275	0.350	0.301	0.296	0.377	0.089	0.114
3/8	0.395	0.510	0.436	0.426	0.542	0.128	0.163
7/16	0.540	0.690	0.592	0.579	0.737	0.174	0.222
1/2	0.710	0.905	0.773	0.757	0.964	0.227	0.290
9/16	0.900	1.15	0.978	0.958	1.22	0.288	0.367
5/8	1.10	1.40	1.24	1.18	1.51	0.356	0.453
11/16	1.35	1.72	1.45	1.43	1.82	0.430	0.548
3/4	1.66	2.05	1.73	1.70	2.17	0.516	0.652
13/16	1.85	2.40	2.03	2.00	2.54	0.601	0.766
7/8	2.15	2.75	2.36	2.32	2.95	0.697	0.888
15/16	2.48	3.15	2.71	2.66	3.39	0.800	1.02
1	2.85	3.65	3.10	3.03	3.86	0.911	1.16
1 1/16	3.20	4.08	3.49	3.42	4.35	1.03	1.31
1 1/8	3.57	4.55	3.91	3.81	4.88	1.15	1.47
1 3/16	3.97	5.08	4.38	4.27	5.44	1.28	1.64
1 1/4	4.41	5.65	4.82	4.72	6.01	1.42	1.81
1 5/16	4.86	6.22	5.33	5.21	6.63	1.57	2.00
1 3/8	5.35	6.81	5.76	5.72	7.24	1.72	2.19
1 7/16	5.86	7.45	6.38	6.26	7.97	1.88	2.40
1 1/2	6.37	8.13	6.92	6.81	8.67	2.05	2.61
1 9/16	6.92	8.83	7.54	7.39	9.41	2.22	2.83
1 5/8	7.48	9.55	8.15	7.99	10.18	2.41	3.06
1 11/16	8.05	10.27	8.80	8.45	10.73	2.59	3.30
1 3/4	8.65	11.00	9.47	9.27	11.80	2.79	3.55
1 13/16	9.29	11.82	10.15	9.76	12.43	2.99	3.81
1 7/8	9.95	12.68	10.86	10.64	13.55	3.20	4.08
1 15/16	10.58	13.50	11.68	11.11	14.15	3.41	4.35
2	11.25	14.35	12.36	12.11	15.42	3.64	4.64
2 1/8	12.78	16.27	13.92	13.67	17.42	4.11	5.24
2 1/4	14.32	18.24	15.72	15.33	19.51	4.61	5.87
2 3/8	15.96	20.32	17.52	17.08	21.74	5.14	6.54
2 1/2	17.68	22.53	19.44	18.92	24.09	5.69	7.25
2 5/8	19.50	24.83	21.24	20.86	26.56	6.27	7.99
2 3/4	21.40	27.25	23.40	22.89	29.05	6.89	8.53
2 7/8	23.39	29.78	25.82	25.02	31.86	7.52	9.58
3	25.47	32.43	27.84	27.24	34.69	8.20	10.44
3 1/4	30.45	38.77	32.76	31.97	40.71	9.62	12.25
3 1/2	35.31	44.96	37.80	37.08	47.22	11.16	14.21
3 3/4	40.07	51.01	43.56	42.11	53.61	12.81	16.31
4	46.12	58.73	49.44	48.43	61.67	14.56	18.56

Weights of Nonferrous Sheets and Wire

B&S Gage	Sheets, Lb per Sq Ft			Wire, Lb per 1000 Ft	
	Aluminum	Brass	Copper	Brass	Copper
0000	6.48	19.688	20.838	605.18	640.51
000	5.77	17.533	18.557	479.91	507.95
00	5.14	15.613	16.525	380.67	402.83
0	4.58	13.904	14.716	301.82	319.45
1	4.08	12.382	13.105	239.35	253.34
2	3.63	11.027	11.670	189.82	200.91
3	3.23	9.819	10.392	150.52	159.32
4	2.88	8.745	9.255	119.38	126.35
5	2.56	7.788	8.242	94.666	100.20
6	2.28	6.935	7.340	75.075	79.462
7	2.03	6.175	6.536	59.545	63.013
8	1.81	5.499	5.821	47.219	49.976
9	1.61	4.898	5.183	37.437	39.636
10	1.44	4.631	4.616	29.687	31.426
11	1.28	3.884	4.110	23.549	24.924
12	1.14	3.458	3.660	18.676	19.766
13	1.01	3.080	3.260	14.809	15.674
14	0.903	2.743	2.903	11.746	12.435
15	0.804	2.442	2.585	9.315	9.859
16	0.716	2.175	2.302	7.587	7.819
17	0.638	1.937	2.050	5.857	6.199
18	0.568	1.725	1.825	4.645	4.916
19	0.506	1.536	1.626	3.684	3.899
20	0.450	1.367	1.448	2.920	3.094
21	0.401	1.218	1.289	2.317	2.452
22	0.357	1.085	1.148	1.838	1.945
23	0.318	0.966	1.023	1.457	1.542
24	0.283	0.860	0.910	1.155	1.223
25	0.252	0.766	0.811	0.9163	0.9699
26	0.225	0.682	0.722	0.7267	0.7692
27	0.200	0.608	0.643	0.5763	0.6099
28	0.178	0.541	0.573	0.4570	0.4837
29	0.159	0.482	0.510	0.3624	0.3835
30	0.141	0.429	0.454	0.2874	0.3042
31	0.126	0.382	0.404	0.2280	0.2413
32	0.113	0.340	0.360	0.1808	0.1913
33	0.100	0.303	0.321	0.1434	0.1517
34	0.0888	0.269	0.286	0.1137	0.1204
35	0.0790	0.240	0.254	0.0901	0.0956
36	0.0704	0.214	0.226	0.0715	0.0757
37	0.0627	0.191	0.202	0.0567	0.0600
38	0.0558	0.170	0.180	0.0449	0.0475
39	0.0497	0.151	0.160	0.0356	0.0375
40	0.0442	0.135	0.142	0.0282	0.0299

Part 5

FINISHING OF METALS

Section 29

METAL-CLEANING PROCESSES

SECTION 29

METAL-CLEANING PROCESSES

Metal surfaces must be cleaned before subsequent processing: some machining operations, heat-treating, galvanizing, phosphate coating, enameling, plating, etc. The cleaning process depends on what must be removed from the surface such as scale and rust, and on the nature and adherence of dirt, oils, greases, drawing compounds, polishing compounds, and the like. The degree of required cleanliness will depend on subsequent operations. Visible cleanliness may suffice for painting, but a chemically clean surface (one devoid of all traces of oils, greases, solid particles, oxides, and corrosion products) may be necessary for satisfactory plated work.

Control of "dirt" on work should properly start with study of raw material and all processing operations. Drawing compounds, polishing compounds, oils, etc., should all be studied to reduce the difficulty and number of cleanings.

CLEANING FERROUS MATERIALS

No one cleaner can be applied to all cleaning jobs on irons and steels. Available cleaning materials fall into four groups: (1) organic or hydrocarbon solvents, (2) self-emulsifying or emulsifiable organic solvents, (3) alkalis, and (4) acids. More than one of these types of cleaners may be needed. Oil might be removed by a solvent or an alkaline cleaner, but removal of oxides would also necessitate use of an acid.

Solvent Cleaners. Cleaners of this type are kerosene, gasoline, naphthas, other petroleum spirits, Stoddard's solvent, and even coal tar or aromatic solvents.

The purpose of solvent cleaning is to remove light coats of oil, grease, and processing dirts, but wiping is considered essential in removing solid dirts, as the solvent has no effect on them. Surface condition produced is not physically clean, but the method is applicable to parts cleaned prior to inspection or assembly. A particular advantage lies in the cleaning of delicate parts that should retain a slight oil film to resist rusting.

Disadvantages are many. Parts that should be entirely freed from oil and grease seldom are, because the solvent, in tank dipping, retains all oil and grease removed from preceding parts.

Most of these solvents are considered toxic, especially if noninflammable. Others are inflammable, with low flash points that make it advisable to have adequate fire-preventive and protective devices installed on the equipment. Not least of the disadvantages is possible dermatitis.

Vapor Degreasing. Degreasing of steel is usually done with trichlorethylene, but sometimes perchlorethylene may be preferred because of its higher boiling point. These chlorinated hydrocarbons have a high and rapid solvent action on oils, fats, and waxes.

The primary method consists in suspending the part in hot vapor over solvent that has been heated to its boiling point of 188 F. Solvent condenses on the metal surface, dissolves soluble oils and greases, and even removes many solid particles as it drips back into the tank.

Rate of flow over the metal surfaces is not great enough to flush off all solid particles. Insoluble compounds, such as limes, pigments, and water-soluble soaps, are not removed. Thus physically clean surfaces are not obtained with vapor degreasing.

A second method of degreasing employs a liquid immersion in warm liquid. Parts are transferred to the vapor rinse to remove remaining oils and greases.

A more effective cycle is to immerse parts in boiling solvent in one tank, transfer them to a second compartment where cooler solvent cools them 10 to 15°, and then raise them into vapor for a final rinse. Liquid-vapor degreasing is particularly adapted to the cleaning of lightweight parts.

Most popular degreasing method is the vapor spray, especially for large parts for which liquid immersion is impractical or for parts that have recesses that would trap air if immersed in liquid solvent. Parts are first vapor-cleaned and then sprayed with warm solvent that is cool enough to allow for a final condensation to effect a pure rinse before the metal reaches vapor temperature.

Degreasing must be followed by some other means of cleaning (usually alkaline) before electroplating or other finishing operations.

Vapor degreasing is considered to be more expensive than other cleaning methods, because of vapor loss, quantity of cooling water needed, and the relatively high cost of solvent. Moreover, its very removal of practically all oils and greases may, in some cases, be a disadvantage, because steel cleaned by degreasing will rust rapidly unless promptly treated with a protective coating.

Emulsifiable Cleaners. Self-emulsifying, or emulsifiable, organic solvents combine the action of a solvent and the emulsifying properties of a solvent-soluble dispersing agent or a soap in the presence of a blending agent. The hydrocarbon solvent may be kerosene, naphtha, or safety solvent, the emulsifying agent (a soap such as potassium oleate) and the blending agent an organic compound such as butyl cellosolve, cresylic acid, a cyclohexanol derivative, or a synthetic surface-active material.

These cleaners have been particularly useful in precleaning operation where surfaces are covered with smut or such adherent dirts as buffing compounds. On the latter, however, an emulsion dissolved in water with agitation is sometimes more effective than soaking in full-strength emulsifiable cleaner.

Applied either by tank dipping or washing-machine spraying, emulsifiable cleaners combine with oil dirts and the combination can be rinsed off readily, particularly with hot water.

If the work is only oily, a short immersion of 10 sec to 1 min may be sufficient, while buffing compound may require 5 to 15 min and agitation of the bath for good cleaning.

After a brief drain, parts are rinsed, preferably in a hot-water spray, to remove all residues. The parts leave the rinse with only a slight film of solvent on their surfaces. This film must be removed by alkaline cleaning before electroplating but does not interfere with painting, lacquering, blackening, or phosphate treatments.

Principal advantage of emulsifiable cleaners is their removal of dirt that is not affected by alkaline cleaning or vapor degreasing. They are effective for removing stubborn pigmented drawing compounds, as well as buffing and polishing compounds and slushing oils. Nesting of parts should be avoided.

Alkaline Cleaners. Various combinations include the older types, such as caustic soda, trisodium phosphate, and soda ash, and the newer salts, such as the soluble sodium silicates. Present-day cleaners obtain their required alkalinity by salts that are less dangerous to handle than the caustics. In case of hard water, water-softening agents are added, among them being sodium carbonates and phosphates. Phosphates, such as trisodium phosphate, are more expensive than carbonates but form flocculent calcium or magnesium phosphates that are free-rinsing.

An alkaline salt alone cannot serve as a cleaner for the widely diversified dirts encountered in process cleaning, however. A wetting agent is needed that will bring the solution into intimate contact with the dirty surface and assist in displacing the dirt from the metal surface. Soap is difficult to rinse and is precipitated by the calcium and magnesium salts present in hard water. Synthetic wetting agents are complex organic compounds of many types. The pH for steel cleaning should be held above 11. See page 30-6 for pH data.

Soak-tank Cleaning. Simplest of the alkaline cleaning methods is soak-tank cleaning in which the solution is kept at its boiling point. Solution agitation, which may be controlled by proper location of the heating element, is particularly important in cleaning recessed or nested parts. Common methods of agitating soak-tank solutions are by use of agitation shields, steam-jet heaters, circulator heaters, propellers, centrifugal pumps, and work movement by hand or mechanically.

Of all heating methods, steam is usually considered the most efficient. Most commonly used are closed coils placed on the *side* of the tank, preferably nearest the operator so rolling of the solution away from him provides a clean area for removal of the work.

When work is quite dirty, a two-tank system is of value. Most of the dirt is removed in the first tank, while final cleaning is done in the second tank. This insures that work is given the second soaking in comparatively clean solution, reduces oil and dirt carry-over into rinse, and lengthens solution life.

Electrolytic Cleaning. In electrolytic cleaning, also called electrocleaning, the alkaline solution is the electrolyte, the work one pole, and the tank or a number of steel plates the other pole. When current passes through the electrolyte, water is decomposed to form oxygen bubbles at the anode (+) and hydrogen bubbles at the cathode (−). This continuous evolution of gas at the surface of the work breaks up the oil film and, in effect, pushes off the dirt.

More agitation on the work surface is provided when the work is made the cathode.

Alkaline cleaners used in electrolytic cleaning generally differ in composition from soak cleaners. Electrocleaners must have a high electrical conductivity. Moreover, cleaners should not contain any ingredients such as chlorides that would dissociate and attack either electrode, nor should any soaps and colloidal or other materials be present that might plate out and deposit as smut on electrodes or work. Also, organic materials that might decompose from electrolytic action must be avoided.

Another important requirement of the cleaner is light foaming and proper gassing action. If a heavy foam blanket is formed, there is danger of explosion as the hydrogen, in the presence of oxygen, may be detonated by a spark when parts are removed from bus bars.

Electrolytic cleaning is carried on at 6 to 12 v and 10 to 100 amp. Small parts in batches or recessed parts, for example, may require raising the voltage to 12 v. Current densities generally range from 15 to 45 amp per sq ft. In calculating the metal area, it is important to include both sides of the part. A density too high may result in pitting, especially of cast iron and high-carbon steel.

Temperature of an electrolytic cleaning solution should be kept at 212 F for cleaning steel, as detergency increases with temperature.

Proper control of cleaner concentration, current, voltage, and temperature will result in a clean surface; but electrocleaning, in most cases, should be preceded by soak cleaning or mechanical washing to remove heavy accumulations of dirt and so lengthen the life of the electrocleaning solution.

Acid Pickling. Pickling is the chemical removal of iron oxides from the surfaces of iron and steel parts by dipping in an acid solution.

The surface required by electroplating and similar finishing operations must be free of all oxides. Preparation of such a surface requires an acid dip, after alkaline cleaning and prior to plating. Whenever steel is heated to temperatures required for hot working or heat-treating, its surface is rapidly oxidized to form a blue-black "mill scale." Even annealing temperatures cause formation of scale. Moreover, even when parts are not heated, oxides may still exist on the surface from common rusting.

Physical characteristics of these oxides are so different from those of the steel that they must be removed prior to electroplating, enameling, lacquering, japanning, galvanizing, tinning, black-oxide finishing, and similar finishing operations. Removal of these oxides also facilitates operations. After heat-treating, cutting-tool life can be appreciably increased if parts are descaled before machining.

Descaling also saves time and reduces cost before grinding or lapping operations.

PICKLING SOLUTIONS. The two most common types of pickling solutions are based on sulfuric and hydrochloric acids, although nitric and hydrofluoric acids are used for some applications.

Sulfuric acid, which removes scale by attacking the base metal under the scale and so loosening it, is used in concentrations from about 6 to 15%. This solution is used hot, from 140 to 180 F, to maintain a rapid pickling rate. The foregoing values are common for mild steel. Alloy and high-carbon steels, however, are more subject to attack during pickling from an electrolytic action between steel and scale, and are usually pickled at lower temperatures.

Concentrations of hydrochloric acid solutions run from 5 to 50% of 18 to 20 Bé acid. These are generally used at room temperature to avoid fumes and loss of the volatile hydrogen chloride. While hydrochloric acid is more effective in removing scale, because scale is more soluble in it than in sulfuric acid, hydrochloric acid costs and fumes more. Its fast action makes it especially preferable for removal of light oxide films formed on polished steels.

Advantages of both acids may be obtained by adding sodium chloride to a sulfuric acid solution. The salt combines with the acid to form a low percentage of hydrochloric acid.

Nitric acid solutions are far less common than sulfuric or hydrochloric acid baths, being adapted particularly for pickling stainless steel and other corrosion-resistant steels. Concentrations generally vary from 10 to 25%, although up to 50% concentrations are used. Typical solutions are 15 to 25% nitric acid plus 2 to 10% sodium chloride for stainless steel, and 25% nitric acid plus 2% hydrofluoric acid for corrosion-resistant steel. Hydrofluoric acid is added to accelerate the pickling action in the latter bath, which operates at room temperature.

Organic materials, known as inhibitors, are added to the bath. They retard acid action on the descaled metal but not on the scaly area. One disadvantage in the use of an inhibitor is that the film deposited on the work should be removed before electroplating, vitreous enameling, phosphate coating, and similar operations. To remove it effectively requires an alkaline or strong uninhibited acid dip or a high-pressure hot-water rinse.

When iron content increases, action of a sulfuric acid bath decreases to such an extent that an 8% iron content reduces pickling action by about 50%. Therefore, when the iron content reaches from 5 to 6%, further acid additions have slight effect so the bath is usually discarded.

Hydrochloric acid solutions, however, become more active as the iron chloride content increases. But as iron salts are hard to rinse from the work, these baths too are generally discarded when the iron content rises above 5%.

NEUTRALIZING PICKLED PARTS. When steel parts are removed from a pickling bath, all traces of acid and iron salts must be removed from their surfaces to prevent rusting. Removal of all traces of acid is difficult by water rinsing alone; so an alkaline rise is recommended to neutralize any remaining acid.

Neutralizing baths are of various compositions. A successful one consists of 75% soda ash and 25% borax with an active sodium oxide content of 0.25 to 0.3% and an operating temperature of about 160 F. This is followed by a borax dip at 170 F with a sodium oxide content of about 0.08%. Another bath giving good results consists of sodium cyanide (0.2 oz per gal) with enough caustic soda to give an active sodium oxide content of 0.3%. Operating temperature is from 120 to 130 F. Ventilation with a cyanide bath must be good.

Electrolytic Pickling. Pickling action can be accelerated and descaling time reduced by electric current. Electrolytic pickling is especially adapted to rapid removal of black magnetic iron oxide, Fe_3O_4, which in a still pickling solution is difficult to remove. Average electrolytic-pickling time is about one-fifth to one-sixth that required for still pickling.

The process is carried on, usually in a sulfuric acid solution, by making the work the cathode. Procedure and solutions vary according to the process employed.

The Bullard-Dunn process uses a 10% acid solution containing a small quantity of tin (or zinc) and an additive to reduce surface tension and improve the throwing power of the bath. The work is made the cathode, while anodes are of high-silicon iron except one or two that are made of tin to furnish the tin content to the solution. The bath is held at 140 to 150 F with a current density of about 60 amp per sq ft. Current density of this value is generally obtainable at 4 to 6 v for racked work and 6 to 9 v for barrel work.

When the scale and rust are removed from work surfaces by combined action of acid and gas evolved from electrolysis, a thin continuous film of tin is plated on the cleaned base metal. This layer of tin protects the steel surfaces from pitting and etching.

When steel parts are to be electroplated, phosphate coated, blackened, or similarly finished, the tin film must be removed before any of these coatings is applied. This is done in an alkaline electrolytic cleaning bath in a ½- to 1-min treatment, with the work made the anode.

Hot dipping, painting, soldering, and such operations do not require defilming. Painting and lacquering over the tin film, for example, gives better corrosion resistance than over the base metal.

Another electrolytic-pickling process is the Hanson–Van Winkle–Munning bright dip. This is carried out in two steps—the first dip with the work the cathode and the second dip with the work the anode. The result is quicker pickling and much less attack on the metal than in still pickling.

The first solution is 10 to 20% sulfuric acid in which the scale is removed, but parts are coated with a smut—"acid black." This is removed in the second bath, 40 to 50% sulfuric acid. Current densities in both baths are kept high, from 100 to 150 amp per sq ft., and solution temperatures low, up to about 90 F. If temperatures are too high and current densities too low, workpieces may come from the bath clean but not bright.

Sodium Hydride Descaling. The du Pont sodium hydride descaling process is based on the fact that oxygen-hungry sodium hydride, dissolved in fused caustic, reduces oxides in scale and tears it into a loose and flaky powdery mass. This is blasted off by the steam formed when the metal part heated to 700 F is quenched in cold water. The process is suitable for all metals not affected by molten caustic at elevated temperatures. The descaling operation is uniform and results in clean bright surfaces, even in recesses. As only the scale is attacked, original dimensions are maintained, and pitting or embrittlement of the metals being treated is avoided. Alloys of chromium, copper, nickel, tungsten, and cobalt, as well as plain carbon steels, have been processed satisfactorily.

Sand and Shot Blasting. Scale, rust, and dirt are often removed from castings and forgings by blasting with sand, steel grit, and steel shot. Sand introduces a silicosis hazard. Hence steel grit or shot have come into wide use in both the foundry and machine shop. In SAE standards for shot and grit, it is stated that round shot stand up better than irregularly shaped particles. The shot number (Table 1) is roughly the size of the shot pellets in ten thousandths of an inch.

Vapor blasting, or "liquid honing," is a process of spraying a fine abrasive suspended in water onto workpieces at high pressure. This process is used to remove fine burrs from precision parts and to remove grinding lines, leaving a matte surface.

Soft-grit Blasting. Efficient removal of paint, dirt, grease, carbonaceous deposits, and scale can be achieved with soft grits—crushed nut shells, fruit pits, oat hulls, rice hulls, corncobs, and hardwood sawdust.

Tumble Finishing. Improvement of surface quality and removal of burrs and sharp edges can be accomplished by tumble or barrel finishing, in addition to removal of scale and rust. Barrel treatment of bulk parts, or parts loaded into fixtures within the barrel if their mass would create damage by striking one another, is an economical method of cleaning and finishing. Depending on the nature of the operation—cutting down, deburring, rolling, or burnishing—the abrasive may be metal slugs, jacks, stars or punchings, steel balls, granite chips, loose abrasive, sawdust, etc. Most tumbling is done wet, with water, acids, or cyanides.

CLEANING COPPER ALLOYS

Cleaning and Deburring. In a horizontal octagonal cast-iron barrel, or an obliquely tilting steel barrel:

1. Add hot water to cover 50 to 250 lb of work; add approximately 5 lb caustic soda and 4 oz sodium cyanide.
2. To remove oil, roll for 10 min at approximately 30 rpm.
3. With water piped to mouth of barrel, wash parts until free from alkalis.
4. Add 3 lb of No. 2 pumice and 6 lb of sea sand (screened through 40 mesh) with a cover of water.
5. Roll until burrs are removed. The time may vary from 2 to 24 hr.
6. Dump work into a perforated centrifuge basket and wash free from abrasive.
7. Dry work in a centrifuge equipped with a hot-air blast. Work will have a smooth dull matte finish.

The abrasive used may be different grades of pumice, rottenstone, sea sand, silex, emery, aluminum oxides, silicon carbides, or similar materials.

Bright or Water Rolling. In obliquely tilting steel barrel with narrowed mouth and centrifugal drier:

Operations 1, 2, and 3 for bright rolling are similar to those for deburring, or the workpieces may have been deburred already.

4. Add about 6 oz cream of tartar to 20 gal water and roll at 30 to 40 rpm until

TABLE 1. SAE CLEANING SHOT NUMBERS AND SCREENING TOLERANCES

% of Total Sample by Weight Retained by On Screen and Passed by Through Screen
4 (.187) Denotes Screen No. 4 with .187-in. Aperture

Shot No.	On Screen	% Max	Through Screen	On Screen	% Max	Through Screen	On Screen	% Min	Through Screen	On Screen	% Max	Through Screen	% Max
1320	4 (.187)	0				4 (.187)	6 (.132)	90	6 (.132)	7 (.111)	7	7 (.111)	3
1110	5 (.157)	0				5 (.157)	7 (.111)	90	7 (.111)	8 (.0937)	7	8 (.0937)	3
930	6 (.132)	0				6 (.132)	8 (.0937)	90	8 (.0937)	10 (.0787)	7	10 (.0787)	3
780	7 (.111)	0				7 (.111)	10 (.0787)	85	10 (.0787)	12 (.0661)	12	12 (.0661)	3
660	8 (.0937)	0				8 (.0937)	12 (.0661)	85	12 (.0661)	14 (.0555)	12	14 (.0555)	3
550	10 (.0787)	0				10 (.0787)	14 (.0555)	85	14 (.0555)	16 (.0469)	12	16 (.0469)	3
460	10 (.0787)	0	10 (.0787)	12 (.0661)	5	12 (.0661)	16 (.0469)	80	16 (.0469)	18 (.0394)	11	18 (.0394)	4
390	12 (.0661)	0	12 (.0661)	14 (.0555)	5	14 (.0555)	18 (.0394)	80	18 (.0394)	20 (.0331)	11	20 (.0331)	4
330	14 (.0555)	0	14 (.0555)	16 (.0469)	5	16 (.0469)	20 (.0331)	80	20 (.0331)	25 (.028)	11	25 (.028)	4
230	18 (.0394)	0	18 (.0394)	20 (.0331)	10	20 (.0331)	30 (.0232)	75	30 (.0232)	35 (.0197)	12	35 (.0197)	3
170	20 (.0331)	0	20 (.0331)	25 (.028)	10	25 (.028)	40 (.0165)	75	40 (.0165)	45 (.0138)	12	45 (.0138)	3
110	30 (.0232)	0	30 (.0232)	35 (.0197)	10	35 (.0197)	50 (.0117)	70	50 (.0117)	80 (.007)	10	80 (.007)	10
70	40 (.0165)	0	40 (.0165)	45 (.0138)	10	45 (.0138)	80 (.007)	70	80 (.007)	120 (.0049)	10	120 (.0049)	10

the desired bright finish is obtained. It is usually necessary to replace the dirty solution with clean water and more cream of tartar.

5. Rinse with running water for 5 to 10 min while barrel is rotating.

6. Dump work into a perforated centrifuge basket, place in heated centrifuge, and dry.

Ball Burnishing. Use a horizontal wood barrel, steel barrel with wood lining, steel barrel with brass lining, or an oblique tilting stainless-steel barrel. Procedure:

1. Work should have been deburred or bright rolled, then treated by "bright acid dipping."

2. For each volume of work, add two volumes of steel burnishing balls, $\frac{1}{16}$ to $\frac{1}{2}$ in. dia.

3. If using the preferred horizontal barrel, fill with water to within a few inches of cover and add approximately 2 oz soap per gal. If the water is hard, add a water softener.

4. Roll at approximately 30 rpm for 6 to 48 hr, depending on luster required and the kind of workpieces.

5. Dump contents and riddle out balls while washing the parts.

6. Dry work by rolling for about 20 min in a horizontal wooden barrel that contains sawdust slightly moistened with Stoddard's solvent.

7. Riddle sawdust from the work.

Ball burnishing produces the highest luster that can be obtained in bulk finishing. Such burnishing may be applied to copper alloys before plating or lacquering, or may be used after plating to improve the luster. When balls are not in use, they should be kept in a soap solution or other alkaline solution to prevent rusting.

Wheel Polishing. Wheel materials are 18-in. cotton cloth sections, felt or sheepskin leather. Abrasive is 80-220 grain emery or aluminum oxide. Wheels are set up with hide glue, then "cracked" by striking lightly with a round bar to produce flexibility. Polishing-stick tallow is applied to wheel to lubricate cutting. Wheel's speed is around 2500 rpm.

Wheel Buffing. Highest luster is obtained with wheel made of 12-in. sections of cotton buffs, and run at 1200 to 3600 rpm. For cutting down to remove polishing lines, a tripoli composition is used on the wheel; for coloring, a white lime or red oxide rouge to develop luster.

Solvent Cleaning. Immersion in Stoddard's solvent, or similar, removes the greater part of oil adhering to workpieces. Centrifuging removes excess solvent.

Degreasing. See degreasing of steels, page 29–2.

Alkaline Scouring. Formulas for satisfactory cleaners for removal of oils, greases, solid particles of dirt, and metal particles are:

FORMULA 1

85% sodium orthosilicate
10% sodium carbonate
5% sodium resinate

FORMULA 2

46% sodium carbonate (anhydrous)
32% trisodium phosphate
16% sodium hydroxide
6% rosin

Concentrations of cleaner may range from 3 to 8 oz per gal; bath temperature from 120 to 200 F.

Emulsion Scrubbing. Emulsifiable cleaners, miscible with oil, can be washed off with water. A slight oil film remaining on the parts may have to be removed by alkaline cleaning.

Removal of Tarnish. During standing or in alkaline cleaning, parts may become tarnished. Immerse them in a water solution containing 4 to 8 oz of sodium cyanide per gal and rinse thoroughly. Or use 5 to 10% solution of hydrochloric or sulfuric acid.

Pickling and Bright Dipping. Surface oxides are removed by pickling in 5 to 10% by volume of sulfuric acid at a temperature of 125 to 150 F. Red stains can be removed by dipping in 4 to 10% sulfuric acid solution plus 4 to 8 oz of sodium bichromate per gal, at 80 to 120 F. Rinse thoroughly, several times and both hot and cold.

To obtain a bright lustrous finish after pickling, a "scale dip" and a "bright dip" are required. The scale dips are:

Dip A: 40% nitric acid, 30% sulfuric acid, 0.5% hydrochloric acid, and 29.5% water.

Dip B: 50% nitric acid; 50% water.

The bright dip is composed of 25% nitric acid, 60% sulfuric acid, 0.2% hydrochloric acid; water, the remainder.

These solutions are used at room temperature. Immerse parts in scale dip, rinse, immerse in bright dip, rinse, and dip in 2 to 4 oz of sodium cyanide per gal to remove acid stains, rinse again, and dry.

A matte surface can be obtained after the above procedure by using this dip at 180 F: 65% nitric acid, 35% sulfuric acid, 1 lb zinc sulfate per gal. Rinse and dry.

If aluminum bronzes are sulfuric acid pickled, they will be clean but dull. A solution containing 3% sulfuric acid and 3 to 4 oz of sodium bichromate may be tried. A bright finish is obtained only by mechanical cleaning.

For silicon bronzes requiring a clean bright surface, first sulfuric acid pickle, then dip in 5 to 10% sulfuric, 10 to 20% nitric, and 3 to 5% commercial hydrofluoric acid, using a wax- or carbon-lined tank and exhaust hood. Then rinse.

CLEANING ALUMINUM ALLOYS

One formula for acid cleaning is 10% of 85% phosphoric acid, 40% butyl alcohol, and a wetting agent. Light oxide films will be removed, leaving a phosphate film to aid paint adhesion.

Alkaline cleaning is more common than acid cleaning. Proprietary cleaners are used. An inhibited mild solution at 160 to 180 F will remove light grease and oil films, dirt, chips, and other foreign material in 3 to 5 min. Many of the cleaners now available consist of tetrasodium pyrophosphate with an inhibitor such as sodium metasilicate. Others are sodium carbonate or trisodium phosphate inhibited with sodium disilicate.

Sometimes aluminum is cleaned by applying petroleum solvent that boils at about 350 F, with the object of spreading grease into a thin film. Then the work is passed through a tunnel oven at 425 F for about 5 min to oxidize the grease film. This method may soften the temper.

CLEANING MAGNESIUM ALLOYS

Like aluminum alloys, the magnesium alloys may be cleaned with organic solvents, emulsion cleaners, or vapor degreasers. Alkaline cleaning with proprietary cleaners of the type used for steel are employed at 180 F. Acid cleaning is done with bath consisting of 1.5% chromic acid and water to make 1 gal used at 190 to 212 F for 1 to 15 min immersion. After sandblasting, parts are cleaned with 3% sulfuric acid.

Section 30

PLATING AND METAL COATING

SECTION 30

PLATING AND METAL COATING

CLEANING BEFORE PLATING

Unless the work is scrupulously cleaned from the outset, no amount of care and skill in the plating operation can produce satisfactory results. Apparent cleanliness is not enough—the work must be chemically clean.

If there is considerable oil or grease present, it is advisable to remove this with a solvent degreaser to avoid excessive contamination of the chemical bath and destruction of its cleaning properties.

Alkali cleaners are most common and may be used either in simple soaking baths or in an electric cleaning cycle. Cleaners available operate on the same principle, namely, the reaction between sodium salts and oils to form soaps. In many cases soaps are added to the alkali cleaner to improve its performance by increasing its emulsifying and wetting ability. Where very hard water is encountered, however, the soap may form insoluble compounds that are difficult to rinse off. In such cases, wetting agents are added in place of soap.

Alkali-soap cleaners are used for cleaning iron, steel, and brass, where some risk of slight surface attack may be tolerated. A broad field of usefulness is as a precleaner before electric-alkali cleaning. Bulk cleaning in baskets, and as a step in barrel plating, is commonly done with cleaners of this type.

When a direct current flows through the work in an alkali cleaner bath, either oxygen or hydrogen is released on the work surface. This speeds up cleaning by removing more of the inert dirt with the oils. If the work is made the cathode, large quantities of hydrogen are released, but the inert-dirt removal may not be complete, foreign matter may plate out onto the work, and there may be some tarnishing.

If the work is easily attacked by alkalis, it must be made cathodic, but the above-mentioned troubles can be eliminated by making it anodic for a few seconds in a separate bath of the same cleaner. This is commonly called reverse-current cleaning. Steel, which is not attacked by alkali, can be cleaned by making the work the anode.

Selecting a Cleaner. As a rule, the same alkali cleaners may be used for soak and electric cleaning, but it is advisable to avoid the use of alkali-soap mixtures in electric cleaning because of the froth produced by the soap. In some cases, also, the soap or wetting agents may retard the operation and also may carry over on the work to contaminate subsequent baths.

The same cleaners cannot always be used for all metals; some are suitable only for ferrous materials, others for nonferrous, and still others on several kinds of material when used at different concentrations. In any event, all alkali cleaners are used very hot and should be followed by a flowing-water rinse to remove all traces of the cleaner from the surface of the work. Electric cleaners require a minimum of 6 v for efficient operation.

Other common cleaners include special soaps for removing heavy accumulations of dirt and grease, and for cleaning highly polished surfaces by removing buffing compounds. They are particularly recommended for brass, aluminum, and die castings, as they will not cause surface attack.

Pickling. Removal of scale and rust presents a different cleaning problem and is usually handled by a pickling process. This consists of immersing the work in a solution of acid, usually sulfuric, at a fairly high temperature. The base metal reacts with the acid, and the resultant evolution of hydrogen mechanically lifts the flakes of scale from the surface.

Inhibitors are not normally used with acid pickle in the plating cycle, because they make it necessary to follow the pickling operation with thorough electrocleaning to remove all traces of the inhibitor from the work.

Where heavy scale or rust is present, a pickling operation is often performed prior to the regular cleaning and plating cycle. Depending on the condition of the work, it may be necessary to degrease it before pickling, as the presence of oil will appreciably slow the pickling process. If an alkali degrease is used, the work must be well rinsed to avoid carrying over any alkali into the acid bath and thus tending to neutralize it in a short time.

With continued use, acid pickling baths gradually become exhausted. The bath is usually discarded when crystallization of iron sulfate occurs on the bottom and sides of the tank as an indication of excessive concentration.

The problem of acid etching can be overcome by the use of the patented Bullard-Dunn process in which a film of tin is deposited on the work surface as soon as it is descaled, rendering it immune from attack by the acid.

Another electric process is the Hanson-Munning bright dip, in which the work is made cathodic in an acid bath to remove scale, and then anodic in a second acid bath.

PLATING PRACTICE

Electrodeposition of metals may be carried on in a variety of ways. Whatever the equipment or bath, the fundamental principle remains the same, namely, the passage of an electric current from an anode, usually, but not necessarily, of the plating metal, to the workpiece, through an electrolyte containing salts of the plating metal in solution.

The rate of deposition varies directly as the current density; that is, the higher the amperage per square foot of the work, the faster the plate will be applied. There are definite limits to current density, depending on the plating metal and the work material (see Table 1). Too high a density produces burning of the work and excessively large crystalline structure. If there are flaws and blemishes in the work, they will appear on the plated surface. If a highly lustrous surface is desired, the surface must be polished or buffed before plating. Most plate is dead matte as it comes from the bath but can be polished to the same degree as the base metal.

A relatively new plating process known as periodic-reverse-current plating is rapidly being adopted and has proved most effective in producing a polishing effect on relatively rough base metal. The process consists of plating in the normal way for a period of from 2 to 40 sec, then deplating by current reversal for from ½ to 5 sec. Each layer of plate is exceedingly thin but combines to create a denser and more homogeneous deposit than can be produced by continuous plating.

Copper Plating. Copper plating is used for decorative or protective purposes on the base metal, for preventing carbon penetration during carburizing operations, and as an undercoat preparatory to chromium, nickel, and some other platings.

TABLE 1. CURRENT REQUIRED FOR VARIOUS PLATING SOLUTIONS

Type of Plating Solution	Volts	Amp per Sq Ft
Brass	2–3	3–5
Bronze	2–3	2–4
Cadmium	1–4	15–45
Decorative chrome	3.5–5	100
Hard chrome	6	250
Cobalt-sulfate	....	30–165
Copper-cyanide	1.5–2	5–15
High-speed copper cyanide	2–4	35–60
Rochelle copper cyanide	2–3	20–60
Acid copper	0.75–2	10–40
High-speed acid copper	3.5–4.5	50–70
Gold, 24 karat	1–3	1–6
Gold, flash	6–12	10–20
White gold	6–10	20–40
Pink gold	2–3	3–10
Iron chloride	....	100–180
Iron, double sulfate	....	20–30
Lead, fluoborate	....	5–20
Lead, perchlorate	....	20–30
Dull white nickel	2–3	5–20
Bright nickel	3–5	30–65
Platinum	3–4	9.3
Rhodium	2.5–5	10–80
Silver	1–2	5–15
High-speed silver	3–4	50–60
Tin	2.6–4	10–25
Zinc	3–6	40–50

Two general types of copper plating are in use: the acid sulfate and the alkali cyanide process. Of these, the acid process is the simpler but cannot be used for direct deposition on iron, zinc, or several other metals. If this is attempted, copper will be deposited by mere immersion in the solution, and a spongy nonadherent coating will be produced. If it is desired to use acid copper on steel, the work must first be given a thin coat, or flash, in an alkali plating solution, then rinsed and transferred to the acid bath.

The acid process can be used either hot or cold, although the cold process is generally preferred for its simplicity, as the heating of acid solutions always poses corrosion problems.

Hot-alkali copper plating is standard in most shops. Deposits obtained are smooth and homogeneous, but the surface is dull and requires polishing if it is to be used as an undercoat for nickel or other plates. Brighter finishes can be obtained by adding lead carbonate or sodium thiosulfate, but such baths are somewhat difficult to control. More commonly the rochelle salt bath is used and this bath is operated at 140 to 160 F, and approximately 3.6 min will be required to deposit 0.0001 in. of copper at 20 amp per sq ft and 2.2 min at 60 amp.

If full brightness is required it is preferable to use one of the proprietary bright copper baths. For heavy deposits at high-speed application, the copper fluoborate process is frequently employed. This is particularly useful for electroforming and electrotyping and may be adjusted to produce either soft deposits that are easily buffed to a high luster, or hard deposits of high strength.

Nickel Plating. Standard nickel plating is performed in a Watts-type solution. The bath may be either hot or cold, and current densities may vary between 15 and 50 amp. Wetting agents may be added to control pitting, but brittleness of the deposit may result either from excess wetting agents or their decomposition products. Filtration of the bath through activated carbon will usually remove the organic materials which cause this. Agitation will permit the use of higher current densities, as will increased temperature, but may produce surface roughness unless the solution is regularly or continuously filtered and the anodes are bagged.

This bath will produce a dull plate that will require buffing if a luster is desired. For a bright plate, organic or inorganic brighteners may be added to the standard bath, but more satisfactory results are usually obtained by using one of the proprietary solutions.

Black nickel is a purely decorative finish having poor ductility and corrosion resistance and is usually applied in very thin deposits. It may be obtained by first plating with zinc or cadmium and then depositing the black nickel by simple immersion, or it may be applied by electroplating either directly over copper or brass, or over an undercoating of white nickel.

Zinc and Cadmium. For most purposes, zinc and cadmium are virtually interchangeable, but there is a strong preference for zinc on the grounds of its substantially lower cost and greater availability. Both metals afford excellent rust protection to steel because, under corrosive conditions, the plate itself will corrode rather than the steel, and, even if the surface is scratched, rust will not spread beyond the immediate area of the scratch itself.

Zinc may be deposited from an acid bath if a dead-white full-matte surface is desired, but the throwing power of an acid solution is very poor and the process cannot be used for workpieces which have irregular surfaces. In general, acid zinc plating is used only for wire, sheet, and woven screen material.

The cyanide process is most commonly used in plating shops and gives excellent results. Standard solutions produce a dull finish, but brightness may be obtained either by adding brighteners to the bath or by using a bright dip after plating. Several proprietary bright-plating baths are available and also several proprietary bright dips. To preserve the luster, the deposits may be lacquered.

Zinc plate can also be colored, usually black or olive drab, by simply dipping in solutions that both protect the zinc from corrosion and can act as bonding surfaces to improve paint adhesion.

Zinc fluoborate solutions may also be used and are particularly suitable for plating directly on cast and malleable iron. The process is much faster than with other solutions and is well adapted to high-speed continuous plating of wire and strip.

Cadmium plating solutions are similar to zinc, except that acid solutions are not used. Bright plates can be produced by the use of additives in the standard bath or by one of the proprietary bright-plating baths. Bright dips and coloring dips may be used in the same way as for zinc.

Chromium Plating. Chromium plating has achieved popularity as a decorative finish because of its blue-white color and high resistance to corrosion and tarnish, but its extreme hardness and excellent wear resistance make it of even greater importance in the industrial field.

It has become customary to differentiate between chromium plating for decorative and industrial uses by designating the latter as hard chromium. Actually, the hardness of both is approximately the same; the difference lies in the hardness of the base metal and the deposit thickness. Decorative plate is usually applied over undercoatings of copper and nickel or of nickel alone, and has a thickness of only 0.00001 to 0.00002 in. Industrial chromium is usually applied directly to hardened steel and may reach thicknesses as high as 0.010 in.

Tin Plating. Tin may be electrodeposited from acid, alkali, or fluorborate solutions. The acid process will not produce as adherent a coating as the alkali, nor is the throwing power as good. Rate of deposition, however, is much higher, and acid baths can be operated at room temperature, while alkali must be heated. Fluorborate baths are extremely fast and simple to operate at room temperature.

Deposits are always silvery white in color, and it is not possible to obtain a bright finish direct from the bath. This finish, however, may be produced by heating the plated parts at a sufficient temperature to cause the tin coating to flow, resulting in a finish equivalent in appearance to hot-dipped tin but with only a fraction of the thickness.

Other Plating Metals. Lead, lead-tin alloy, brass, bronze, and iron are among the common metals most frequently plated; silver, gold, rhodium, indium, arsenic, palladium, and platinum are also used to some extent industrially, although their greatest field is in decoration. Silver, however, is being used to a considerable extent in electrical work and for engine bearings, and gold in various types of electronic work. Indium is frequently used over lead-plated bearings to protect the lead from attack by acids in lubricating oil, and sometimes on the surface of stamping dies to increase their useful life. In both cases the indium is diffused into the base metal by heating in a low-temperature furnace.

Plate can be applied to almost every known metal, although some are more difficult than others and may require a preliminary coat of one type of metal before the desired coat can be applied. Aluminum and magnesium, for example, must be given a zinc-immersion plate, followed by a copper strike, before other plates can be applied.

CONTROL OF pH

Success of plating processes depends upon the close control of pH, that is, hydrogen-ion concentration. Roughly speaking, pH is a measure of the effective acidity or alkalinity of a solution and, therefore, is intimately tied in with the chemistry of processing. The speed and completeness of chemical reactions are affected by the following principal conditions: (1) temperature, (2) pressure, (3) pH, (4) concentration of reactants, (5) nature of reactants, and (6) presence of catalysts.

In the electrodeposition of base metals such as copper, zinc, and nickel, small variations in pH make or mar the efficiency of the process. Unless the pH is closely held, the metallic hydroxides will precipitate rather than the desired metal.

Means for Measuring pH. There are two basic ways to measure pH, namely: (1) the colorimetric method and (2) the electrometric method. Litmus, phenolphthalein, and methyl orange are common examples of colorimetric indicators. In the electrometric method two electrodes are immersed in the unknown solution. One electrode—the measuring electrode—is sensitive to the presence of hydrogen ions; the other electrode—the reference electrode—is required to complete an electric circuit through the solution and to supply a constant potential.

Electrodes for Measuring pH. Four electrodes have been developed which are sensitive to hydrogen-ion concentration, namely: (1) the hydrogen electrode, (2) the quinhydrone electrode, (3) the antimony electrode, and (4) the glass electrode. The glass electrode is the modern and most practical means of pH measurement and is rapidly replacing the other electrodes for industrial applications.

Limits of the pH scale are determined by the maximum and minimum concentrations of hydrogen ions. But in addition to the neutral point, a pH of 7, two other points on the scale are of particular interest, namely, a pH of 0, and a pH of 14. A pH of 0 corresponds to an acid solution of unit strength. Likewise, a pH of 14 corresponds to a basic solution of unit strength. Since the pH scale is a logarithmic function of concentration, a change of one pH unit represents a tenfold change in concentration. Thus, a solution with a pH of 4 is ten times more acid than a solution with a pH of 5.

PROCESS SHEETS FOR PLATING OPERATIONS

Process Sheet for Cadmium Plating

Because of its relatively high price, cadmium is usually used indoors or in protected areas where thin coatings will suffice. However, aircraft, marine, and military outdoor uses are common because of the high degree of corrosion resistance offered. Cadmium coatings are attractive but relatively soft; they are resistant to alkalis and are therefore used to protect washing equipment and steel imbedded in concrete. The electrical industry has made wide use of cadmium because it is easily soldered and has good conductivity. Simple dip aftertreatments assure good paint adhesion.

Steps in the process are:

1. DEGREASE. Various types of cleaners are available for the removal of loose dirt, grease, and oils. Petroleum solvents, chlorinated solvents, and solvent emulsion cleaners have all been used. Present trend for production-line work seems to favor vapor degreasing, using trichlorethylene and perchlorethylene vapors.
2. RINSE.
3. CLEAN. In widest use today are the various alkali soak cleaners which are used at elevated temperatures. Higher temperatures increase cleaning speed, and present trend is toward high-conductivity cleaners which permit cathodic and anodic cleaning.
4. RINSE.
5. PICKLE. Pickling operations are used for the removal of scale and rust. Industry seems divided between the use of sulfuric and hydrochloric acid. The action of HCl at room temperature is more rapid, but sulfuric acid is lower in cost. To minimize attack on the base metal, many proprietary inhibitors are widely used.
6. RINSE.
7. PLATING. Cadmium plating is conducted almost exclusively through the medium of cyanide baths. Addition agents are added to produce bright or semi-bright deposits. The solution is operated at room temperature with cadmium bars or balls as anodes. Steel tanks may be used, but rubber-lined tanks are recommended.
8. RINSE.
9. AFTERDIP. To secure good paint adhesion, cadmium coatings are usually dipped in dilute chromic acid or in various proprietary chromate or phosphate solutions. Clear dips are available for coatings which are not to be painted; it is claimed that increased corrosion resistance can be obtained.
10. RINSE.

Process Sheet for Copper Plating

Copper plating is used most widely as an undercoating on steel parts and zinc-base die castings prior to the deposition of nickel coatings. The excellent throwing power, high deposition rate, and good adhesion achieved from commercial baths make copper ideal for this application. In addition, the ease with which copper lends itself to polishing and buffing has been instrumental in cutting down the time required to secure a lustrous finish on steel surfaces. Steps are:

1. Polish. The degree of polishing and buffing depends upon the material being plated and the luster of the finish desired. Nonferrous parts are usually completely buffed at this stage, but steel parts may often be buffed more economically after plating.
2. Degrease.
3. Rinse.
4. Alkaline Cleaner. Hot alkaline cleaners are in common use. Anodic cleaning is generally sufficient, but for heavy-soil removal cathodic cleaning usually precedes it.
5. Rinse.
6. Acid Dip. Both hydrochloric and sulfuric acid dips are common for steel parts; sulfuric acid is recommended for zinc-base die castings.
7. Rinse.
8. Cyanide Dip or Copper Strike. An ordinary cyanide dip is often sufficient prior to plating steel parts with cyanide copper. The copper strike is required before plating from acid sulfate or fluoborate baths.
9. Copper Plate. Copper plating has been successfully performed from acid sulfate, cyanide, fluoborate, and pyrophosphate solutions. Acid copper baths are used primarily for electroforming, because they make possible rapid build-up of thick deposits. Cyanide baths are most common for the thin deposits required prior to nickel plating. Proprietary plating baths have been developed which permit mirror-bright deposits, resulting in an extremely attractive and decorative finish.
10. Rinse.

Process Sheet for Decorative Chromium Plating

Seldom is it realized that the chromium portion of "chrome plate" is no more than a flash deposit, approximately 0.00001 in. thick, applied over a nickel or copper, plus nickel, undercoat. Although the nickel supplies the corrosion protection, it is the thin layer of chromium which is responsible for the permanent luster and beauty. The following steps are required for applying chromium to a buffed nickel surface.

1. Alkaline Cleaner. Wet cleaning is recommended whenever the size and shape of the part permit immersion treatment. Alkaline cleaning with the work as the cathode gives excellent results.
2. Warm-water Rinse.
3. Cold-water Rinse.
4. Acid Dip. Dilute sulfuric or hydrochloric acid dips at room temperatures are widely used. Good results have been obtained with sulfuric acid electrolytic baths in which the work is made the cathode at 4 to 6 v.
5. Rinse.
6. Chromium Plating. Commercial decorative chromium plating is done exclusively from a bath containing chromic acid plus a sulfate or fluoride radical. The baths have poor throwing power, making the use of auxiliary anodes mandatory. Bright deposits can be obtained by plating at 105 F with a current density of 115 amp per sq ft. The brilliance of the plate is determined by the condition of the nickel surface upon which it is applied.
7. Rinse.

NOTE: The use of modern bright-nickel solutions, in some instances, permits the transfer of nickel-plated items, after rinsing, directly into the chromium-plating bath without buffing, in which case these cleaning steps may be omitted.

Process Sheet for Hard Chromium Plating

General practice in the deposition of heavy coatings is to plate from 0.001 to 0.003 in. more than is desired, and then to grind or polish to the required dimensions. Very thin coatings, 0.0005 in. and thinner, have been employed successfully on metal cutting tools; the heavier deposits are used for rolls, drums, drawing dies, molds, shafts, and similar parts. Industrial, or hard chromium, deposits are used for the building of worn surfaces, the creation of bearing surfaces, and the facing of parts to resist erosion, corrosion, or wear. Steps are:

1. DEGREASE. Removal of grease and oil may be accomplished by organic solvents, emulsion cleaners, or vapor degreasers.
2. RINSE.
3. ALKALINE CLEANER. Best results have been obtained with commercial electrolytic cleaners. The work is made the anode at 6 v.
4. RINSE.
5. ETCH. To assure adhesion, electrolytic etching is recommended in a chromic or sulfuric acid solution. When using the CrO_2 etch, a separate solution, or the plating solution itself, may be used. The latter method is more convenient but may cause the bath to become contaminated. Treatment time in electrolytic etches is approximately ½ to 1 min at a current density of 100 to 400 amp per sq ft. As an alternate, for highly finished steel, when maximum adhesion is not required, room-temperature immersion in 10 to 50% by volume of concentrated HCl or 5 to 15% by volume of concentrated H_2SO_4 may be substituted for anodic etching.
6. RINSE.
7. CHROMIUM PLATE. Composition of the plating bath is the same as specified for decorative chromium plating. It consists of chromic acid plus a sulfate radical, with the weight ratio maintained, respectively, at 100:1. Insoluble lead anodes are used. Operating temperatures of 104 to 140 F within a current-density range from 70 to 450 amp pei sq ft should result in bright hard deposits (approximately 1000 brinell).
8. RINSE.
9. BAKE. Baking at 300 to 500 F for 1 to 5 hr will prevent embrittlement of the steel, removing the hydrogen which may have been absorbed during processing.

Hard Chrome on Gages. Despite the plating of new gages by leading manufacturers, the larger part of gage plating comes under the heading of salvage. The procedure is to grind 0.003 in. or so undersize, plate 0.003 to 0.004 in. oversize, then grind and lap to size. This will provide a good surface with low frictional characteristics and usually extends the life from three to seven times.

On new gages, there are two schools of thought. One calls for a light plate of approximately 0.0001 in., the other a heavier plate as in salvaging gages.

The light-plate method is to lap 0.00005 to 0.0001 in. undersize, plate 0.00015 in. oversize, and lap to size. This method is particularly desirable on close-tolerance gages, as the part can be continuously replated economically with the light plate over a long period of time.

A practice of stress relieving after grinding and after plating has greatly enhanced gage life from a cracking and breaking standpoint.

Hard Chrome on Cutting Tools. Hard-chrome deposits have been tried on all types of cutting tools. After much experimentation, it was found that a light film, around 0.00002 to 0.00005 in., provides the necessary reduction in friction without any dulling of the cutting edge.

The only function of chrome on a cutting tool is to provide freer chip removal. This means a cooler running tool, reduced breakage, etc. Usually the best results are obtained on the softer steels and nonferrous materials.

Plated taps have produced gratifying results. A light plate, 0.00002 to 0.00005 in., works well on 1010 or 1020 low-carbon cold-rolled steel. Life is increased 10 to 20 times, and taps can be reground and replated. Less life is reported when plating is done for the second time. Plated taps have been tried on die-cast aluminum with spotty results.

Plated taps, when used in brass, have proved favorable. Plating thicknesses have been used up to 0.0015 in., increasing tool life and improving work finish.

Hard Chrome on Molds. White and colored materials are not subjected to dirt and iron oxide pickup when preformed in chrome-plated setups. The die is usually plated on both faces and through the ID, and on the faces of the punches only. This is to eliminate galling by running chrome against chrome.

On injection molds, chrome is confined to eliminating drag marks on deep cavities and for corrosion protection. Compression and transfer cavities and forces are chrome-plated. A freer flow, higher finish on the molded product, corrosion protection of the mold surfaces from gases of reaction, corrosion protection during down time, easier flash removal, and no down time for polishing are direct benefits of chrome surfaces. Transfer pots, plungers, pressure plates, and sprues are plated.

A procedure now becoming standard is to put 0.004 in. of chrome into these gates in new transfer molds and to salvage worn molds by building up with chrome.

The thickness of chrome on molds depends on size and also on types of compounds being worked. Abrasive compounds will require heavier deposits for wear protection. A good rule to follow is to deposit at least 0.0005 in. on all small molds, up to 0.005 in. on large housings.

Process Sheet for Decorative Nickel Plating

Nickel plating is used primarily to protect metallic objects from corrosion. General practice is to nickel-plate directly on the base metal or to use a composite coating of copper and nickel. In both processes, the nickel plate is followed by the deposition of a thin coating of chromium, which supplies the abrasion- and tarnish-resistant qualities.

Investigations have shown that the protective value of the nickel deposit is directly proportional to the thickness applied, and the modern trend is toward the use of heavier, more protective coatings (supplies of nickel permitting). The following chart traces the steps recommended for the application of bright-nickel plate to steel, zinc, and brass surfaces.

1. Polish. The degree of polishing and buffing depends upon the material being plated and the luster desired. Generally, nonferrous parts are completely buffed at this stage. Steel items, and others requiring a very high luster, may be buffed after deposition of the copper or the nickel plate.

2. Degrease.

3. Rinse.

4. Electroclean. Hot alkaline cleaners, in which the work is made the cathode and then the anode, are in general use.

5. Rinse.

6. Acid Dip. Dilute solutions of hydrochloric or sulfuric acid are recommended at room temperature. Overetching must be avoided.

7. Rinse.

8. Nickel Plate. Nickel may be deposited directly on the base metal or on previous deposits of copper (on zinc). The composition and operating conditions of the bath will vary with the metal being plated. Items which are buffed after copper plating must be returned to the cleaning cycle before being reentered in the plating sequence. For the treatment of buffed nickel plate prior to chromium plating, see process sheet pertaining to decorative chromium plating.

Process Sheet for Zinc Electroplating

Electrodeposited zinc is one of the most widely used protective finishes for small articles, because it is relatively simple to apply, economical in initial cost and application, decorative and highly corrosion-resistant. Its high corrosion resistance has resulted in wide use for metal screening, steel strip, pipe couplings, and wire. Insufficient tarnish resistance formerly provided a major drawback to the use of zinc plate for decorative consumer items. Now, recently developed bright dips and aftertreatments, which maintain the brilliance of the plated coating, have given new impetus to its use in the decorative field. Steps are:

1. Degrease. The removal of loose dirt, grease, and oils may be accomplished by petroleum solvents, chlorinated solvents, and solvent-emulsion cleaners. Modern trend is toward the use of trichlorethylene and perchlorethylene degreasers.

2. Rinse.

3. Alkaline Cleaner. Alkali-soak cleaners, operated at elevated temperatures, are in widest use. For higher cleaning speed, electrolytic cleaning is practiced.

4. Rinse.

5. Acid Pickle. The use of dilute sulfuric acid at room temperature is most common, although there has been a decided increase in the trend toward the use of the more expensive hydrochloric acid pickle. The action of the latter is more rapid and results in a smut-free surface, which is particularly advantageous on the production line.

6. Rinse.

7. Plating. Zinc electroplate is applied through the medium of acid and cyanide baths. Acid-zinc deposits are widely used for plating wire, strip, pipe, and other objects which do not require a bright deposit, or wherever rust resistance is more important than the decorative effect. The throwing power of acid-zinc baths is poor; they are not recommended for articles of irregular shape. Cyanide baths with suitable addition agents result in bright lustrous deposits, and the throwing power is greatly superior to the other method.

8. Rinse.

9. Afterdip. In order to secure satisfactory paint adhesion, and to retard the formation of white corrosion products, zinc coatings are dipped in various proprietary chromate or phosphate solutions. Clear dips are available for coatings which are not to be painted.

10. Rinse.

Process Sheet for Zinc-immersion Treatment

The zinc-immersion process is a method for applying an extremely thin coating of zinc on the surface of aluminum and its alloys, in order to facilitate subsequent electroplating. Usually, a copper strike is applied over the zinc coating before plating with other metals, but procedures have been worked out for direct deposition of silver, brass, chromium, nickel, and other metals. The main features of the process are: adaptability to bulk plating; simplicity of operation; low cost; excellent adhesion of the deposit; and effectiveness on wrought and cast materials, with variations in the pickling bath. Generally speaking, zinc-immersion treatment is essential for good adhesion of subsequent plating.

1. Degrease. Vapor-degreaser or solvent-immersion treatments may be used.

2. Rinse.

3. Alkali Cleaner. Soaking for 1 to 3 min in a sodium carbonate–sodium phosphate solution maintained at 180 F is recommended.

4. Rinse.

5. Acid Treatment. For magnesium-aluminum and magnesium-silicon-aluminum alloys, a 5-min immersion in sulfuric acid solution at 175 F is suggested. Room-temperature dip in 1:1 nitric acid is suitable for relatively pure aluminum such as 2S and 3S. Cast alloys are usually dipped in a solution of nitric and hydrofluoric acid.

6. Rinse.

7. Zinc Immersion. Immerse for 1 to 2 min in a mildly agitated sodium zincate solution at room temperature. The reaction is self-arresting, thus preventing the formation of objectionable rough deposits.

8. Rinse. Double rinsing is suggested to assure complete removal of the viscous solution.

PLATING OF STAINLESS STEEL

Operations. The following operations are presented for plating any type of stainless steel. After operation 5, choice of treatment is optional (see Table 2).

1. Degrease with trichlorethylene degreaser or any other satisfactory method.
2. Electroclean: 180 to 190 F, 6 v; (*a*) direct, 30 sec; (*b*) reverse, 30 sec; (*c*) direct, 30 sec.
3. Water rinse.
4. Acid dip: fluoboric acid 25%—30 sec, or hydrochloric acid 25%—30 sec.
5. Water rinse.
6. Electroactivator solution: 70 F, 6 v, carbon anodes; (*a*) hydrochloric acid 50%—2 min, or (*b*) sulfuric acid 10%—2 min.
7. Nickel strike solution: 70 F, 3 to 4 v, 2 min; nickel anodes.
8. Water rinse.
9. Nickel-sulfate rinse: 32 oz per gal, pH 5.6.
10. Nickel plate: 70 to 110 F, 1.75 to 2 v, 10 min, nickel sulfate 24 oz per gal, nickel chloride 9 oz per gal; boric acid 2.5 oz per gal; pH 5.6 to 5.8.
11. Water rinse.
12. Acid copper: 70 F, 0.5 to 0.75 v, 30 min, copper sulfate 32 oz per gal, sulfuric acid 8 fluid oz per gal.
13. Electrocyanide: 6 oz per gal, 110 F, 15 sec.
14. Cyanide brass: standard brass, pH 12.2, 10 to 20 min.
15. Cyanide (du Pont H.S.) copper: pH 11.2 to 11.8.
16. Chromium solution: standard solution; standard operating conditions; 2 to 3 min.
17. Water rinse.
18. Hot water dry.

Notes

1. Sulfuric acid, as an activating agent, is not recommended in strength over 10%, because in greater concentrations it makes subsequent plating questionable.

TABLE 2. RECOMMENDED PRACTICES FOR PLATING STAINLESS STEEL

Method	Operation No.												
	6	7	8	9	10	11	12	13	14	15	16	17	18
1	X	X	X	X	X							X	X
2	X	X	X	X	X	X	X					X	X
3	X	X	X	X	X	X		X	X			X	X
4	X	X	X	X	X	X		X		X		X	X
5	X	X	X	X	X	X					X	X	X
6	X		X	X	X							X	X
7	X		X	X	X	X	X					X	X
8	X		X	X	X	X		X	X			X	X
9	X		X	X	X	X		X		X		X	X
10	X		X	X	X	X					X	X	X
11		X	X	X	X							X	X
12		X	X	X	X	X	X					X	X
13		X	X	X	X	X		X	X			X	X
14		X	X	X	X	X		X		X		X	X
15		X	X	X	X	X					X	X	X
16		X	X				X					X	X
17		X	X					X	X			X	X
18		X	X					X		X		X	X
19	X		X				X					X	X
20	X		X						X			X	X
21	X		X							X		X	X
22	X		X								X	X	X
23	X										X	X	X
24											X	X	X

2. When the final coating is nickel, plating time should be sufficient to give a thickness that can be color-buffed.

3. Before chromium plating on nickel, color-buff nickel after a 15-min plate.

4. In method 23 of the table, no electric current is used; merely 2 min agitated immersion.

5. Chromium plating directly on stainless steel presents no problem, because the chromium solution is an activating agent.

6. By considerable experiment on plating of stainless steel, and testing the various deposits for adhesion, after operation 5, the favored operations are 6, 7, 8, 9, 10 in above list of operations (Table 2), after which deposits of copper, du Pont H.S. copper, brass, or chromium can be applied.

DESIGN CONSIDERATIONS IN PLATING

Finish. There are several types of plating finishes to be considered. Among those most frequently used are the following:

Lustrous. Luster may be described as the sheen, brilliance, or reflective quality of a plated surface. This type of plating is generally applied for pleasing appearance. Lustrous finish is obtained by polishing and buffing.

Partially Lustrous. Many plated parts require that only specified areas need be bright or lustrous.

Nonlustrous. A flat or dull plate added without any brightening, prepolishing, or subsequent buffing produces a nonlustrous finish. This type of plating is usually applied for corrosion-resistant purposes.

Coloring of Metallic Surfaces. By the electrical or chemical treatment of plated surfaces, it is possible to obtain a desired appearance such as black nickel or zinc. For example, parts might be zinc-plated for corrosion resistance and the plating then oxidized black if the brightness is considered objectionable.

When designing articles to be electroplated, consideration should be given to the following:

Base Material. The surface of the base metal to be plated should be smooth and of suitable finish to receive the metal coating, since coatings do not level out surface roughness or hide defects.

Steel stampings, which are to be chromium-plated, should be made from cold-rolled steel rather than hot-rolled steel. The hot-rolled steel may be a more economical material, but the added cost for subsequent surface preparation will more than offset the cost of using cold-rolled steel.

Zinc-base die castings are frequently plated. Other types, such as sand castings of brass, bronze, and malleable iron, are also plated but the cost of surface conditioning before plating is a factor of prime importance.

Flexing. Hard and brittle plating materials should be avoided on surfaces that will flex in service because of the danger of the plating chipping or cracking. Chromium is brittle; nickel has some flexibility; cadmium, zinc, copper, and tin are ductile.

All hardened steel parts, such as springs, are embrittled in the electroplating process and should be reheated to approximately 400 F to remove brittleness. Cadmium has less tendency to cause brittleness than other coatings.

Sharp Edges and Recesses. Except under closely controlled conditions, the plating metal is not deposited in a uniform thickness but follows the distribution of the electric current. Points, sharp edges, and protrusions are concentration points for the current and will receive an excessive metal coating while recessed areas will receive light deposits. Ample use of large radiuses and fillets assists in producing a uniform thickness of plate.

Holes and Cavities. Deep holes and cavities should be avoided because they collect cleaning and finishing compounds which are difficult to remove prior to plating and consequently result in poor plating deposits. Also such designs result in thin uneven deposits in recesses because the electric current concentrates on the more exposed surfaces.

Racking. In production, where conveyers are used for transporting large parts through the plating solution, design consideration must be given to racking on conveyors. The addition of notches or holes may facilitate this operation.

Flat and Concave Surfaces. Large, flat, and concave surfaces should be avoided because they often result in unattractive appearance relative to high lights and are difficult to plate to the required thicknesses.

Designation of Significant Surfaces. The drawings should specifically show the areas that must be bright in order to avoid unnecessary costs of finishing operations on plated areas which are not visible, as shown in Fig. 1. Also, those areas on which

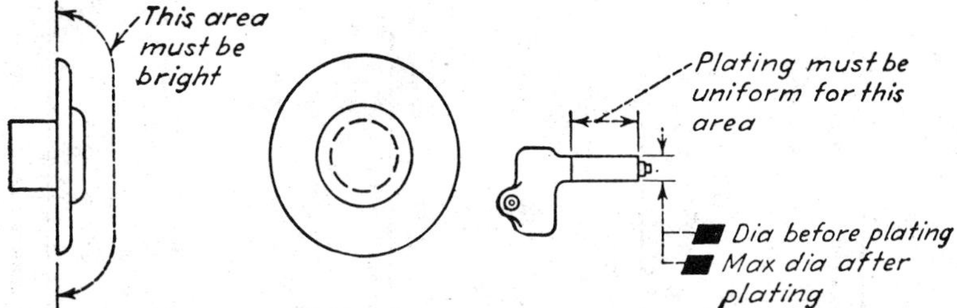

FIG. 1. Identify areas that should be bright to avoid unnecessary plating costs.

FIG. 2. Components of close-fitting assemblies should have allowance for plating.

the specified thickness of plating must be maintained are indicated on the drawing as Significant Surfaces.

Welding, Soldering, and Brazing. The surfaces of steel parts in those areas which are to be welded, soldered, or brazed should be free from plating when possible. Assemblies can often be plated after these processes are performed.

Cadmium- and zinc-coated surfaces can both be soldered; however, cadmium is easier to solder.

While cadmium- and zinc-coated surfaces can be welded, all welding operations involving cadmium-coated surfaces are prohibited in most plants because the cadmium volatilizes, producing a deadly poisonous gas.

Polishing. Parts which require a bright and decorative finish usually are polished and buffed prior to plating. Designs of complex and unusual shapes are difficult to polish, buff, and plate and therefore result in high costs. Consideration should be given to the use of existing automatic polishing equipment wherever possible.

Allowance for Plating. Plating is a build-up of metal which causes a measurable increase to the material thickness. Therefore, allowances for plating must be made on components of close-fitting assemblies. They should be specified as shown in Fig. 2. Proper allowances should be established in order to meet drawing requirements after plating. In the case of threaded parts, which are plated, proper allowances must be made in the pitch diameters.

Type, Code, and Drawing Notes. The drawing of a part requiring a metallic coating of any kind must indicate the coating or plating material and any other related pertinent fact such as type and code. The type and class of plating are usually specified by the engineer or are determined by mutual agreement between the customer and the supplier.

ESTIMATING PLATING TIME

The accompanying chart (Fig. 3) permits quick estimates of plating time and also can be used to find the finish thickness obtained during a given interval. Only factors that need be known are current density and desired finish thickness, or the time available for finishing.

EXAMPLE 1: What time is required to deposit 0.001 in. thick zinc at a current density of 15 amp per sq ft?

SOLUTION: Connect Zinc on the finishing-metal scale with 0.001 in. on the thickness scale and prolong line *A* to intersect the reference line. Through this point

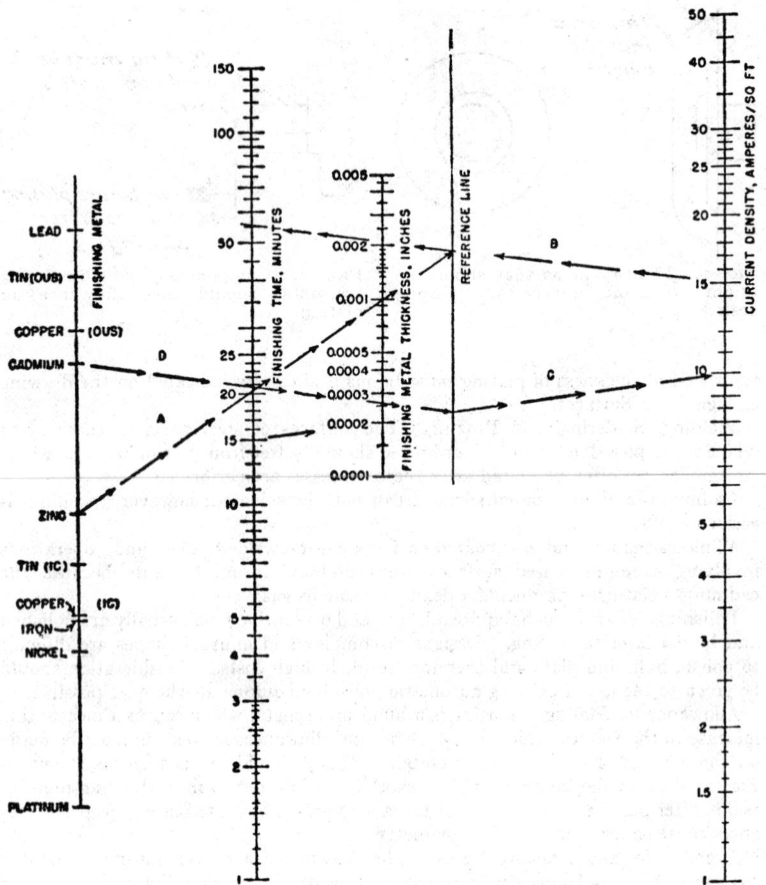

FIG. 3. Chart for estimating plating time.

draw line *B* from a current density of 15 amp per sq ft to the finishing-time scale and read about 57 min.

EXAMPLE 2: Production time allows 15 min for cadmium plating of metal sheets. How thick a finish can be produced by a current density of 10 amp per sq ft?

SOLUTION: Connect 15 min with 10 amp per sq ft, line *C*. Through the intersection on the reference line, draw line *D* from cadmium on the finishing-metal scale. On the thickness scale read 0.00026 in. finish.

This chart is based on cathode efficiency of 100%. Many finishing cathodes approach this value closely. Where actual efficiency is considerably less, divide the time found by efficiency expressed as a decimal, or multiply the thickness found

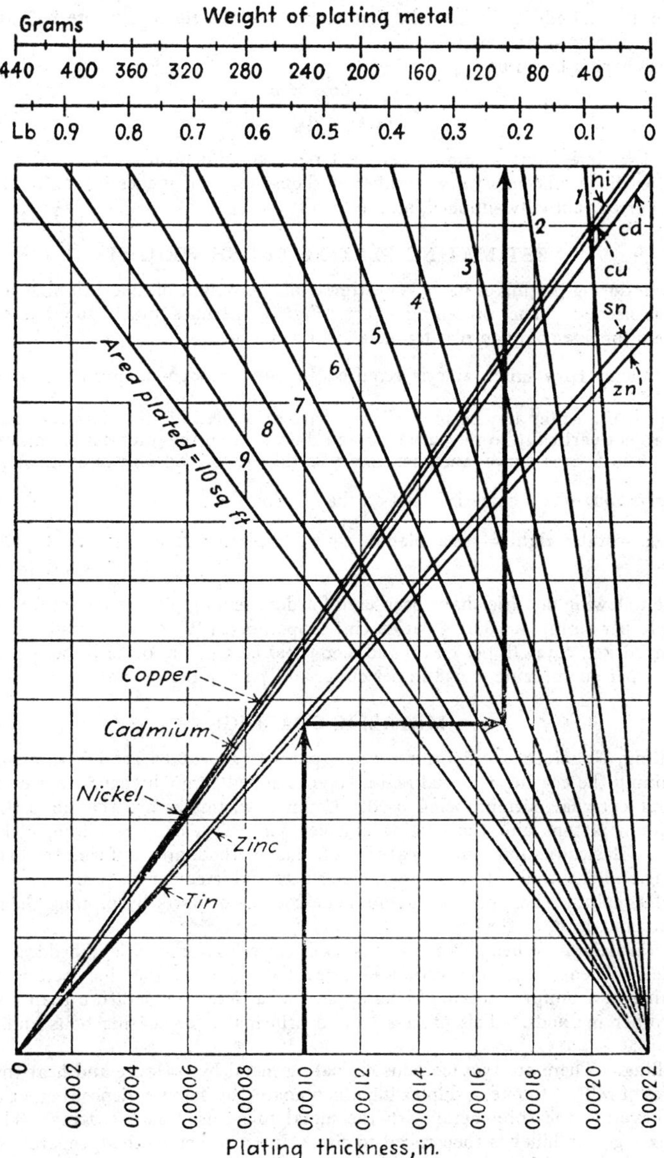

FIG. 4. Chart for estimating weight of metal for plating.

by the decimal efficiency. In the first example, if the efficiency were 80%, actual time = 57/0.80 = 71.4 min.

This chart is based on

$$t = \frac{60A \times n}{A_s}$$

where t = time, min; A = amp per sq ft required to deposit 0.001 in. of metal at 100% cathode efficiency; n = number of thousandths of inches in actual plating; A_s = current actually supplied, amp per sq ft.

ESTIMATING PLATING FINISH WEIGHT

Often during plating of parts, an estimate of the weight of metal to plate a given area is needed. The preceding chart (Fig. 4) permits quick solution of this problem for five common plating metals. Use it thus:

EXAMPLE: How much zinc is required to put a 0.0010-in. finish on an area of 6 sq ft?

SOLUTION: Enter the bottom of the chart at a plating thickness of 0.0010 in. and project vertically to the curve marked Zinc. From the intersection, move right to the 6 sq ft area curve, and then move vertically, to read 0.23 lb or about 104 g.

Derivation. This chart is based on the relation

Plating metal weight = area plated, sq in. × plating thickness, in. ÷ plating metal density, lb per cu in.

The following densities have been used for determining plating weight: cadmium, 0.313 lb per cu in. and 8.65 g per cu cm; copper, 0.324 lb per cu in. and 8.96 g per cu cm; nickel, 0.322 lb per cu in. and 8.90 g per cu cm; tin, 0.264 lb per cu in. and 7.298 g per cu cm; zinc, 0.258 lb per cu in. and 7.133 g per cu cm.

METALLIC COATINGS

Plating, Nonelectric. By this process, parts to be coated are left in a solution containing the coating material and a layer is added through chemical electrolysis, without external current being used. Common examples are the tin plating of aluminum pistons and the coating of steel parts with copper sulfate, nickel, or cobalt. Basically, important features of this method are uniform thickness of coating and the ability to coat internal surfaces and irregular shapes.

Hot Dipping. This process involves the coating of parts by dipping them in a molten zinc, lead, or tin.

Sherardizing. When this method is used, clean iron or steel is imbedded in zinc powder. Then the temperature is elevated to just below the zinc melting point, resulting in a uniform coating being applied at a lower temperature than possible with other methods. This process is used principally for coating bolts and small castings.

Rolling. There are two methods of coating metal by pressure and heat through a series of rolls. In one, a duplex ingot is prepared by casting, depositing, or otherwise covering the core metal with the metal to be used as a coating. The hot duplex ingot or billet is then rolled to size. In the other method, separate sheets of the two materials are heated in intimate mechanical contact to the proper temperature and then hot-rolled and reduced as in conventional practice to produce a "clad" sheet.

Sputtering. In this method of plating, the material to be deposited as a film is used as a cathode in a vacuum chamber. The parts to be coated are placed adjacent to the cathode. Passing of high voltage disintegrates the cathode metal. The disintegrated coating material diffuses and deposits on the adjacent parts and forms an extremely thin uniform coating.

Thermal Evaporation. This method uses a vacuum chamber and is similar in results to cathode sputtering. In this process, the coating material is heated by a coil and evaporates, condensing on the parts to be coated. Metal coating by this process is more rapid than by cathode sputtering, and the thickness of the coating can be better controlled.

Spraying. In this method, a wire of the coating material is melted in an electric arc or gas torch and sprayed under pressure on the object to be plated.

Calorizing Steel. Calorizing covers a variety of methods for applying and diffusing aluminum to steel (and other metal) surfaces. It is intended primarily for the protection of iron from oxidation at elevated temperatures. Calorized metals will withstand severe deformation; they are resistant to sulfurous gases and to fused salts at high temperatures. Calorizing is recommended for oil-refining equipment such as tube stills for cracking and topping, as well as for valves, retorts, and condenser parts. It is also widely used for furnace parts, including conveyors, stokers, roasters, carburizing and heat-treatment boxes, and kiln parts, and is enjoying increased use in diesel-engine construction and in miscellaneous fields where resistance to oxidation and sulfur attack at high temperatures is involved.

Process steps are:

1. DEGREASE. Cleaning procedure will vary with the condition of the metal being treated. As a general rule, the precautions set forth for cleaning prior to plating should be followed.
2. RINSE.
3. ALKALINE CLEANER.
4. RINSE.
5. ACID PICKLE.
6. RINSE.
7. CALORIZE. Four methods are used commercially:

a. Heat in mixture of powdered aluminum, aluminum oxide, and from 1 to 5% ammonium chloride. Hydrogen atmosphere is used; temperature maintained between 850 and 950 C. Thick protective films are produced.

b. Heat in inert or reducing atmosphere containing dry aluminum chloride gas. Keep temperature between 750 and 1100 C. Coating essentially the same as obtained through method *a*.

c. Coat steel by hot dipping and then heat so as to diffuse the aluminum coating with the underlying surface. Coating produced is thinner and less protective than that formed by methods *a* and *b*. Finish is bright.

d. Spray part with aluminum coating and then heat in air to form tough adherent oxide layer. This process produces an inferior coating but is economical to apply, involving low initial and low renewal costs.

Chromising Steel. Chromising is a general term covering various methods for producing on the surface of iron and steel parts an integral layer high in chromium. The coating is formed by means of diffusion at relatively high temperatures, the most common method consisting of exposing articles to be treated to a chromous chloride atmosphere under controlled temperature. Work that has been treated includes exhaust manifolds, bolts, milk cans, and tubing. Advantages claimed for chromised surfaces are excellent corrosion and wear resistance (greater than plating) minimum of surface preparation; complete coverage of the surface; extended service life; low consumption of chromium; and freedom from flaking or pitting because the coating is integral with the surface. The treated material can be rolled, bent, or flanged without damage.

Preparation of chromising agent:

1. Prepare Mixture. A steel retort is filled with a mixture of broken unglazed porcelain crockery, which acts like a sponge, and ferrochromium in approximately equal parts.

a. Displace Air. After inclining the retort at 45°, hydrogen is passed through the contents until all air is displaced.

b. Heat in Furnace. Hold for 5 hr in gas-fired furnace at 1000 C, and pump hydrogen chloride into retort. At the end of this time, hydrogen is again passed through.

c. Cool in Air. The porcelain mass is removed after cooling and stored until needed for chromising. The chromous chloride content is easily measured by specific-gravity tests.

2. Chromising

a. Degrease. Conventional equipment may be used.

b. Chromise. Parts to be chromised are packed into retort along with the chromous chloride charged porcelain mass. The retort is then placed in the furnace and heated for 5 hr at 1000 to 1050 C. At this temperature, the chromous chloride reacts with iron in the surface of the material, forming ferrous chloride, which passes off as a gas, leaving chromium to deposit on and diffuse into the surface of the material.

FILTRATION AND AGITATION

Good plating results cannot be obtained with dirty solutions caused by dirt carried into the bath or by sludges formed from chemical reactions within the plating bath. Filtration of the plating solution is essential for maintenance of top-quality work in all solutions, with the exception of chrome plate. Continuous automatic filters are desirable on large installations, but for smaller tanks filtration can be handled with a small portable unit used at controlled intervals.

Care must be taken in selecting a filter to see that it is suitable for the various fluids to be handled, particularly in respect to the housing and pump. Stainless steel is commonly used for parts exposed to plating acids.

Agitation of a plating bath is desirable in most cases. More uniform plating and faster deposition are obtained. Power-driven agitators of the propeller type with corrosion-resistant shafts and propellers are widely used. These are suitable for most plating operations. Air agitation is also employed by permitting compressed air to escape through a perforated pipe at the bottom of the tank. This may be the less expensive method of agitation, but great care must be taken to see that the air is free from oil or other contaminants.

The work may be agitated too. This method is desirable for the production of a uniform coating, because it tends to average the distance between the work and the anodes and prevents local exhaustion of the solution. Agitation of the work is readily performed by reciprocating the cathode bars on which the work is suspended. Small gear-motor reciprocators handling one or more cathode bars are usually employed for the purpose. Double-acting air cylinders with automatic valves are sometimes applied. In small-tank work, manual agitation may also be used.

Types of Anodes. Brass tubing may be used for anode and cathode rods, but solid copper bars are essential if heavy currents, as in chromium plating, are to be carried. The anodes, however, in most cases are bars of the metal being plated and are hung from the anode rods by hooks.

Section 31

PAINT AND PAINTING

SECTION 31

PAINT AND PAINTING

Paint is cheaper than a plated coating and can be compounded to suit virtually any set of specifications. Care must be exercised to select the correct type of paint or finish, to prepare or clean the metal surface, and to choose the most efficient method of application.

Cleaning. The surface should be clean and dry. Rusted steel surfaces may be wire brushed. Other contamination may be removed by proprietary alkaline cleaners, rinsed with hot water (twice preferably) and then neutralized in a solution of 1 to 2 fluid oz of phosphoric acid to 100 gal of water, or $7\frac{1}{2}$ oz by weight of chromic acid to 100 gal of water to passivate the surface and retard rusting. Solvent cleaning is practical for small lots. Oxides can be removed in proprietary cleaners. Passivating in the phosphoric acid dip or, better, the conversion of the surface to a crystalline phosphate coating (like bonderizing) should follow.

Spangled zinc, matte galvanized stock, and zinc-base die castings require special phosphate treatments in most instances. Cadmium-plated surfaces should receive similar treatment.

Aluminum must be cleaned and passivated to retain a good paint bonding surface. Immersion in a boiling solution of 5% sodium or potassium dichromate will produce a passive oxide coating. But this is less effective than an anodic treatment, in which the work is immersed in chromic or sulfuric acid solution and made the anode in an electric circuit. Treatment with phosphoric acid and alcohol forms an insoluble and adherent layer of aluminum phosphate.

Tin may be vapor degreased or solvent washed, and best adherence is probably obtained with synthetic finishes.

Brass should be cleaned, then passivated by dipping in a hot solution of 5% sodium dichromate, $2\frac{1}{2}$% sulfuric acid, balance water, for 30 sec, rinsing and drying.

Nickel, after cleaning and rinsing, should be passivated as for brass, or with the phosphoric or chromic acid rinses, as for steel.

Adhesion of paint to chromium plating can be improved by sandblasting to a dull surface or immersing in hot 0.1% solution of phosphoric acid for a few seconds.

Magnesium must be cleaned by vapor degreasing, solvent wipe, or alkali cleaner, then protected with a chromate treatment like Dow No. 7.

PAINT SELECTION

"Paint" is an all-inclusive term. There are marked differences between oil paints, enamels, lacquers, and japans.

Primers. An undercoat, or primer, is usually required on high-quality work. Red lead is widely used to prime structural steel. For better work, zinc chromate primer is preferred, in either the air-drying or baking types. Zinc-dust primer serves to give good adhesion on galvanized iron and is useful on copper and bronze to prevent staining of the finish coat. For black enamels and japans, the primer can be the same material as the finish coat but thinned down and applied lightly.

Synthetic enamels are usually self-priming, but a baked primer is desirable with lacquer enamels. If metal surfaces are not smooth, a primer surfacer is applied and sanded.

Bituminous Coatings

This classification includes: (1) japans—natural asphalt with drying oils dissolved in petroleum solvents; (2) asphalt varnish—similar to japan but made with petroleum asphalt; (3) bituminous enamels—high-melting-point asphalt or coal tar containing mineral fillers such as clay or asbestos.

Method of Application. Spray, brush, or dip. Japans can be air-dried, but when baked at high temperatures (300 F or higher) hard semitransparent finish is obtained. Bituminous enamels are usually brushed on hot, giving extremely heavy films.

Properties. Japans are hard finishes. Applied varnishes have excellent water resistance, especially when aluminum powder is added to them. Bituminous enamels give excellent corrosion and water resistance.

Appearance. Japans are transparent amberlike films. Asphalt varnishes and bituminous enamels are black. The latter are very dense; hence they are applied after melting.

Uses. Japans are used where inexpensive but hard finishes are required; one of the oldest types of protective coatings for metals. Asphalt varnish and bituminous enamels are used to protect iron pipes, water mains, and interiors of ships.

Surface Preparation. A clean surface is obtained by sanding or rubbing with steel wool. It must be oil-free. Japan should never be applied over primer.

Lacquers

Cellulose Types

Composed of nitrocellulose or other cellulosic resins—ethyl cellulose, cellulose acetate, and acetate butyrate—plasticizers, and pigments dissolved in volatile organic solvents. Coating film is formed on evaporation of solvents. However, many so-called "lacquers" are combinations of such thermoplastic resins modified with alkyd or other oxidizing resins which provide special properties.

Method of Application. Usually sprayed, but can be dipped, roller coated, or brushed. Heavy coatings, such as tank linings or coatings for protection against shipping and handling (vinyls and cellulose acetate butyrate), are usually applied by hot dipping. Lacquers are usually air-dried, but force drying—applying heat at relatively low temperature—can also be used to speed up drying. In spray application, extreme care must be exercised to avoid "orange peeling," which can be removed from dry film only by rubbing. In avoiding this condition, viscosity and solids control are important, along with a well-developed spray technique.

Properties. Suitable for indoor use and when properly formulated also for outdoor conditions. Speed of drying is outstanding. Toughness and flexibility in a wide range of temperature conditions are excellent. Nitrocellulose is inflammable, but ethyl cellulose, cellulose acetate, and cellulose acetate butyrate have low inflammability and good heat resistance. Best adhesion to metals and best water resistance are obtained with nitrocellulose-type lacquer. For a given film thickness, more coats of lacquer are required than varnish or synthetic-base paints, but this is being partly overcome by hot lacquering technique.

Appearance. Supplied in clear and pigmented coatings that range in gloss from high to dead flat. With proper choice of solvents, extremely smooth finishes can be obtained. Novelty finishes such as crackle, webbing, and crystal can be produced by carefully controlled application techniques.

Uses. Unpigmented (transparent) nitrocellulose lacquers are most widely used. Pigmented metallic lacquers used as automotive finish are also widely used on aircraft because of their light weight. The new trend in application is toward hot lacquering, by means of which the spray viscosity of lacquer is reduced by heat instead of thinner, thus permitting application of heavier films per spray coat. A normal two-coat operation can be completed with one coat of "hot lacquer." Cellulose acetate butyrate can be used as a hot dip where extremely heavy coatings are required—such as for flashlight batteries. Although expensive, lacquers heavily loaded with metallic powders, such as stainless-steel powder, are effective one-coat protection for small ferrous and nonferrous parts.

Surface Preparation. Solvent or vapor degreasing is usually necessary. For resistance to moisture and corrosive conditions, preliminary chromate or phosphate treatments are essential. Oil or grease must be completely removed. Appropriate zinc yellow primer should be used prior to application to magnesium or aluminum.

Other Types of Lacquers

Same as cellulosic, but instead of cellulosic resins, other thermoplastics such as vinyls and acrylics are used.

Method of Application. Spray, dip, or roller coat. Usually air-dried but can be force-dried for better adhesion.

Properties. Vinyl coatings have excellent resistance to outdoor conditions, especially to continuous water immersion. Can be sprayed on as a "strippable" coating which is used to protect metals in storage for indefinite periods of time. Acrylic coatings are light- and heat-stable and are resistant to water and some chemicals.

Appearance. Usually used as clear finishes; acrylics can be colored.

Uses. Hot-dip and cold-sprayed vinyl lacquers are widely used for protection of metallic parts during fabrication, shipment, and storage. Vinyl coatings are used in protection of tank-car interiors, fuel and chemicals storage tanks, and the like. Acrylic coatings are available as a proprietary "one-step" finish for small metallic parts such as die castings; they are applied by brush, dip, or roller coat, with short bake; outstanding moisture and chemical resistance are claimed for them.

Surface Preparation. A cleaning or degreasing operation is essential before applying lacquers. Solvent wipe or vapor degreasing is generally used.

Synthetic-resin-base Coatings

Phenolic—Pure and Modified Types

Method of Application. Brush, spray, or dip. High baking temperature (300 to 350 F) gives fast production.

Properties. Phenolic finishes are relatively fast drying, and have excellent durability and chemical and moisture resistance. Modified phenolics have excellent abrasion resistance.

Appearance. Usually pigmented only in dark shades because of dark color of vehicle.

Uses. Marine finishes, electrical insulation, food- and chemical-processing equipment, tank-car linings, food-handling equipment, and automotive primers and surfaces. Modified furfuryl-alcohol-resin base, air-drying coatings are commercially available. They are similar to phenolics in many properties but are claimed to be tougher and harder than unmodified phenolics. Electrical properties are outstanding. Cured films are black and opaque. Cost is higher than phenolics.

Alkyd

Phenolic anydride–polyhydric alcohol–oil combinations. A recent development is the addition of styrene as a modifier, thereby reducing the drying time of alkyds to almost lacquer speed.

Method of Application. Brush, spray, dip, or roller coat. Addition of metallic driers is required for air drying. Baking is required for special effects, such as crackle or wrinkle.

Properties. Superior to phenolics and oleoresinous-type finishes in gloss retention. Possess excellent durability, hardness, heat resistance; good adhesion, flexibility, and oil and grease resistance. Alkali and water resistance fair.

Appearance. Gloss enamels are possible in a wide range of colors. Almost every type of special-effect finish (except for lacquer novelty finishes mentioned above) are obtained with alkyd. These include metallic effects (automotive) and various types of hammer, wrinkle, and spatter finishes. All of these can be applied to smooth metal. Wrinkle finishes hide imperfections in metallic surfaces to a very great extent.

Uses. Enamels for wide variety of kitchen and outdoor products, including refrigerators, traffic signals, porch furniture. Widely used as automotive and truck finish. Typical vehicle for an automotive finish contains the following resins (percent):

Dying alkyd (short oil)	65
Melamine	15
Urea formaldehyde	20

Urea and Melamine

Never used alone—usually in conjunction with alkyd resins. Urea and melamine thermosetting resins have high color stability at elevated temperatures and are usually available in solutions of 50% solids in mixtures of coal tar and petroleum derivatives.

Method of Application. Spray or dip application usually recommended. Drying by baking only ranges between 225 and 325 F. Some forms containing catalysts may be cured by force drying from 125 to 140 F.

Properties. Good hardness, chemical and abrasion resistance, excellent oil and grease resistance. Brittle when used alone, they must be combined with alkyds for flexibility and adhesion. Melamines provide for wider compatibility with pigments and other resins, rendering them somewhat more versatile. Melamines have outstanding resistance to electrical arcs and are stable to heat and light.

Appearance. Wide range of colors, including white. Good gloss retention in wide variety of conditions, even after prolonged exposure to excessive temperatures and light. This makes them especially adaptable to light-colored baking finishes.

Uses. Finishes for refrigerators, automobiles, hospital equipment, stove parts, and kitchen appliances. Urea resins, combined with alkyds, and titanium dioxide are used as a baking refrigerator enamel.

Styrene

"Styrenated" oils and alkyds.

Method of Application. Spray, brush, dip, or roller coat. Air drying and baking. Some types of oil-modified styrenated polyesters dry "tack-free" in 10 to 15 min.

Properties. High gloss, good stability and color retention. Limited compatibility with other film-forming resins.

Appearance. Wide range of colors. Excellent decorative effects.

Uses. Decorative and protective coating for smooth metallic parts, where appearance is the main factor, such as metal cabinets. Enamel finishes that air-dry as quickly as lacquers can be made from oil-modified styrene copolymer solutions. Baking-type enamels are also produced from such solutions.

Silicones

Method of Application. Spray or dip—always baked. Baking schedule varies from 15 to 30 min, at temperatures from 350 to 425 F.

Properties. Outstanding heat resistance—far superior to conventional finish in this respect. Also has good hardness and resistance to weathering. However, silicone enamels not containing metals have poor abrasion resistance. Good wet insulation properties.

Appearance. Wide range of colors, including white, which are stable at high temperatures. Compatible with pigments providing finishes of high gloss, with purity of color and tint.

Uses. Finishes for refrigerators, gas heaters, and stoves. Modified silicone aluminum paint used on a wide variety of heating equipment. Widely applied to electrical equipment.

Surface Preparation for Alkyd, Urea, Styrene, and Silicone Finishes

Surface treatments for metals to be coated with phenolic or alkyd-type finishes usually include the following: solvent vapor degreasing, rust removal, flush with water, dry, spray with zinc chromate-type primer, air-dry or bake. Although rough castings can be coated with almost any type of finish, phenolic or alkyd types are most often used.

A typical finishing system for cast-iron machine parts is:

1. The casting is first sanded or shot blasted.
2. If oil or grease is present standard alkaline cleaners can be used. One commercial cleaning preparation contains both solvents and phosphoric acid.
3. Immediately after cleaning, a primer should be applied. Corrosion-resistant primer containing red lead or zinc chromate should be applied.
4. If casting is porous, a filler is sprayed on, then sanded, to smooth out surface roughness.
5. A synthetic-resin lacquer is then applied as a base, to provide a smooth surface which will have minimum paint absorption.
6. The final coat of synthetic-resin-base paint is then applied by dipping, spraying, or brushing. Enamel can be air-dried or baked (if it contains urea or melamine).

VINYLS

Polyvinyl butyral, vinyl chloride, vinyl chloride acetate, and vinyl chloride–vinylidene chloride, dissolved in ketones and alcohols.

Method of Application. Spray, dip and brush. Dries by solvent evaporation but often force-dried at 200 F for better adhesion.

Properties. Flexible and tough. Resistant to oils, foodstuffs, and many chemicals. Adhesion only fair; improved by incorporating an alkyd resin.

Appearance. Supplied as clear, pigmented, and dyed coatings. One type of vinyl-dispersion coating ("organosols") is used as decorative and protective finish.

Uses. Coating based on vinyl resins, and containing lead chromate and basic zinc chromate, is used as a wash primer on aluminum, galvanized iron, and steel. Baked alkyd finishes are applied over the vinyl primers. Chemical-resistant coatings containing modified vinyl resins developed for finishing such products as refrigerators and washing machines. Vinyl "organosols," in which the resin is suspended in a nonsolvent, have limited use as metallic coatings. A variation of the technique, in which a polishing operation is included, produces a coating suitable for outdoor metallic furniture.

Surface Preparation. Precleaning and degreasing absolutely necessary.

SPRAY PAINTING

Spray painting is the most widely used method of applying paint in industry. Spray guns vary in size from a pencil-type air brush to a heavy-duty gun for continuous operation. All of them consist of the same essential elements, and all require a source of compressed air—hoses for the air and paint, a container for the paint, and a means of adjusting the flow of paint and air. For relatively small parts or a limited amount of work, a simple pressure-feed or suction-feed gun with a paint container of 1-qt size will suffice. Pressure-feed tanks or paint-recirculating systems are employed for high-production painting installations.

Spraying Technique. As much as 60% of paint can be wasted by improper painting methods. Temperature of the paint is important. There is a correct, economical air and fluid pressure for each type of material. The regulator should be set so that there is sufficient fluid pressure to obtain the desired film thickness at the proper production rate. Never atomize the paint more than necessary. When pressures are specified, remember that they apply at the gun, not at the tank. Depending upon the length and diameter of the hose, there is a substantial pressure drop between tank and gun.

Viscosity is important. Paint that is too thin will produce an excessive overspray and sagging films. Paint that is too thick requires heavy pressures and may produce a heavier film than is desired.

The gun must be clean and provided with the proper spray head and nozzle for the material being used. Adjust the spreader valve to obtain a spray pattern most suitable for the job. The pattern should usually be narrower than the work surface.

Hold the gun perpendicular to the work and 6 to 10 in. away, as a rule. However the distance should be checked by trial, because it will vary with different paint materials. Once the distance from the work is established, keep it constant during the entire stroke. This means that you should move the gun in a straight line, not in an arc or tilted. If the gun is moved in an arc, the center of the sprayed area will be coated too heavily and the ends too lightly. Also, if the gun is tilted, the spray pattern will not be uniform, and in extreme cases the sagging will occur at one portion and the remainder will be too light. For best results, make strokes with an arm and wrist motion, not with the wrist alone.

Overspray can be reduced by triggering the gun only when on the target. The trick of "banding in" will save much paint. This consists of spraying both ends of each work panel with vertical strokes and the face, or center, with horizontal strokes. The bands act as signals for triggering the gun.

Compressed Air for Spray Painting

For successful spray painting, a steady supply of compressed air is essential. A variation of even a few pounds in pressure will make a perceptible difference in the rate of paint flow. The accompanying table gives the minimum pipe sizes for connecting compressors, and also the pressure drop in air hoses.

MINIMUM PIPE-SIZE RECOMMENDATIONS FOR AIR COMPRESSORS

Compressor		Main-air-line Pipe Size		Compressor		Main-air-line Pipe Size	
Size	Capacity, cfm	Length, ft	Size	Size	Capacity, cfm	Length, ft	Size
1½ and 2 hp	6 to 9	Over 50	¾	5 to 10 hp	20 to 40	Up to 100 100 to 200 Over 200	¾ 1 1¼
3 and 5 hp	12 to 20	Up to 200 Over 200	¾ 1	10 to 15 hp	40 to 60	Up to 100 100 to 200 Over 200	1 1¼ 1½

PRESSURE DROP IN AIR HOSES

ID of Air Hose, ¼ in.	Air-pressure Drop at Spray Gun, Lb					
	5-ft length	10-ft length	15-ft length	20-ft length	25-ft length	50-ft length
At 40 psi	6	8	9½	11	12¾	24
At 50 psi	7½	10	12	14	16	28
At 60 psi	9	12½	14½	16¾	19	31
At 70 psi	10¾	14½	17	19½	22½	34
At 80 psi	12¼	16½	19½	22½	25½	37
At 90 psi	14	18¾	22	25¼	29	39½
ID of Air Hose, 5/16 in.	**Air-pressure Drop at Spray Gun**					
	5-ft length	10-ft length	15-ft length	20-ft length	25-ft length	50-ft length
At 40 psi	2¼	2¾	3¼	3½	4	8½
At 50 psi	3	3½	4	4½	5	10
At 60 psi	3¾	4½	5	5½	6	11½
At 70 psi	4½	5¼	6	6¾	7¼	13
At 80 psi	5½	6¼	7	8	8¾	14½
At 90 psi	6½	7½	8½	9½	10½	16

NOTE: These pressure drops apply to hoses used with guns equipped with air cap consuming approximately 12 cfm at 60 psi.

Section 32

ELECTROLYTIC AND CHEMICAL FINISHES

SECTION 32

ELECTROLYTIC AND CHEMICAL FINISHES FOR METALS

Functional and decorative purposes are achieved by applying electrolytic and chemical finishes to metals: (1) protection against atmospheric and chemical corrosion, (2) provision of a satisfactory base for other finishes, (3) improved appearance, and (4) protection and improved appearance. These finishes fall into these categories: electrolytic oxide coatings, chemically etched finishes, chemically colored finishes, phosphate coatings, and bright-dip finishes. The base metal must always be considered along with the finish.

ALUMINUM

There are many types of chemical and electrolytic finishes for aluminum. Many of these are proprietary, but some are not. These finishes are used for: (1) surface-preparation treatments, and (2) final finishes.

Finishes Produced by Chemical Treatment. Important objectives of chemical treatments are:

1. To improve appearance of a product or to obtain either a bright reflecting surface, a smooth diffusing surface, or a rough diffusing surface.
2. To remove dirt, grease, annealing and heat-treating films, buffing compounds, and welding flux by removing a surface layer of metal.
3. To clean without roughening or otherwise changing the surface.
4. To produce a surface which has increased resistance to corrosion, and to provide good adhesion for paints and enamels.

Chemical finishing methods include: bright dip, frosted, diffused, reflector, deep etched, and Alrok treatments. The bright-dip finish is produced by immersing the part in an acid mixture at elevated temperatures (Alcoa R5 bright-dip process). It can be used instead of buffing or polishing to obtain bright finishes on a variety of aluminum-alloy products, including most wrought-aluminum alloys. In general, it is not recommended for casting alloys having a high silicon content. The bright-dip treatment is particularly effective over embossed or patterned surfaces, because such surfaces are ordinarily difficult to buff.

A frosted finish for aluminum alloys is used on small and intricate items or large flat surfaces not adapted to machine methods of finishing. The frosted or matte effect results from an etching action in hot caustic soda. Because a surface of this type will fingerprint or stain easily, a coat of transparent lacquer should be applied over it.

Other alkaline solutions may be used to obtain less attack and, consequently, a smoother finish. These solutions are generally mixtures of soda ash and trisodium phosphate, and are used at 150 to 180 F. Their effect is to remove a limited amount of grease or oily film and to produce a slightly etched surface such as desired for aluminum reflectors of the diffusing type.

Alkaline solutions can also be used for cleaning aluminum surfaces; a typical solution consists of 1 oz of sodium disilicate to 1 gal of 3% soda ash–sodium triphosphate solution. Another satisfactory method of cleaning is by cathodic action in dilute solutions of caustic soda or sodium silicate. However, cleaners containing silicates should not be used prior to anodizing or Alrok treatments, unless they are followed by a dip in strong nitric acid to remove the film formed by the silicate.

Acid treatments may be used to produce a variety of frosted or matte finishes on aluminum alloys. They are also used to remove undesirable annealing or heat-treating film, buffing composition, and welding flux from the surface. Generally, these acid solutions are mixtures of nitric acid and hydrofluoric acid. If the solutions are heated, the etching action is increased, but the problem of finding suitable containers becomes serious as the temperature is increased.

Another use of nitric and hydrofluoric acid is to prepare certain aluminum alloys, such as Alclad 24S, for spot welding. An effective solution for removing annealing and heat-treating stains from aluminum surfaces consists of 10% by volume of sulfuric acid and 3% by weight of chromic acid. With this solution, which is used at from 160 to 180 F, little etching attack takes place.

One series of chemical finishes for aluminum alloys is identified by the trademark Alrok. Surfaces so treated possess appreciable resistance to corrosion and serve as an excellent base for paint. This type of finish is especially adapted to the bulk treatment of small parts.

Coatings differ in color and appearance, depending upon the process used and aluminum alloy coated. The color of the finish, which is applied in an alkaline solution, ranges from gray to greenish gray, depending on the concentration and length of exposure. A sealing treatment may change this color to a yellowish green. Generally, the coatings have a glossy appearance, but they will be dull if the surface has been previously etched.

For greatest resistance to corrosion and as a base for paint, Alrok 12 and 14 are used. Unpainted A17S-T4 rivets treated by the Alrok 14 process will pass the salt-spray test of 250 hr required by the federal specification for anodic coatings. Type 24S-T4 rivets similarly treated will also meet this specification.

Anodic Finishes. One patented process is the Alumilite finish. By making aluminum or its alloys the anode in a suitable electrolyte under controlled conditions of voltage, current density, and temperature, a surface coating is formed which is essentially aluminum oxide.

The anodic process is applicable to all aluminum alloys, but the structure and color of the resulting oxide film will vary with alloy and temper. Anodic coatings on high-purity aluminum are continuous and transparent. Silicon is particularly detrimental, causing the coating to change from brown to a gray gun-metal color.

This type of coating as formed has a cellular structure that can be used to advantage in coloring the coating by means of water-soluble dyestuffs or pigments.

The finish reproduces the texture of the surface on which it is formed. If it is desired to maintain the silvery metallic appearance, a sealing treatment is used to close the pores and make the coating stain-resistant. When protection is the main consideration, special sealing techniques are used to absorb corrosion inhibitors into the coating. Such a system is employed when the coating is used as a base for protective paints or enamels.

By variations in the procedure, coatings with a variety of hardness may be produced. These hard coatings will give excellent resistance to rubbing wear.

Any anodic finish gives substantial protection against the weather, and for this reason they are used extensively in the architectural field. Films having a thickness of 0.4 mil are generally used for surfaces having routine maintenance. However, for maximum resistance to corrosion and weathering, films having a thickness of 0.8 mil are recommended.

Several anodic treatments for aluminum and its alloys, known as "Electrobright" processes, are used to obtain the maximum in surface brightness. These processes remove a thin surface film of metal together with any contaminating substances such as oxide, buffing compound, or alloy constituent. There is also a tendency for electrolytic brightening treatments to smooth the surface. Some Electrobright treatments will remove burrs and other minor irregularities from the surface.

The best known commercial application of Electrobrightening is the Alzak process, which consists of an anodic treatment in a fluoboric acid electrolyte and is used for finishing aluminum lighting reflectors. An anodic finish is then applied to protect the surface. Reflectors made by this process are characterized by a combination of permanence and high reflectivity.

MAGNESIUM

Chemical treatments for magnesium are used mainly as a base for an organic finish or for protection.

Dichromate Process. Purpose is to produce a corrosion-resistant coating, especially to salt water. Coatings vary in color from dark brown to black.

TREATMENT PROCEDURE

1. Clean work by alkaline cleaning, preceded by vapor degreasing if required.
2. Rinse in cold water.
3. Immerse work for 5 min in solution containing 15 to 20% by weight of hydrofluoric acid at room temperature.
4. Rinse in cold water.
5. Immerse in a boiling-water solution of sodium dichromate (10 to 15% by weight) for 45 min.
6. Rinse in cold and hot water.

NOTES AND COMMENTS

1. The dichromate treatment is known by proprietary designations: Dow No. 7 Treatment (Dow Chemical Co.), and Treatment AMCG (American Magnesium Co.).
2. No dimensional changes are caused by this treatment.
3. The treatment is applied after machining and before painting.
4. Brass, bronze, and steel inserts are unaffected by this treatment, but aluminum and cadmium are rapidly attacked during the hydrofluoric acid dip.
5. The sodium dichromate solution may be controlled by maintaining its pH between 4.2 and 5.5 by additions of chromic acid.

Alkali-chromate Process. Gives better abrasion resistance than the dichromate process. Coatings vary from dark gray to black.

TREATMENT PROCEDURE. Same as for dichromate process through the acid dip, then:

1. Immerse for at least 45 min in a boiling-water solution containing:

Ammonium sulfate	4 oz per gal
Sodium dichromate	4 oz per gal
Ammonia (sp. gr. 0.880)	$\frac{1}{3}$ fluid oz per gal

2. Rinse thoroughly in cold running water.

3. Immerse for at least 5 min in a boiling aqueous solution containing 1% by weight of arsenious oxide.

4. Rinse in cold running water followed by hot water.

NOTES AND COMMENTS

1. This treatment is also known as the acid-alkaline dichromate treatment, Dow Treatment No. 8, and Treatment AMCH.

2. Where heavy oxide films appear on the parts, they should be removed by sandblasting, pickling, wire brushing, etc.

Anodic Process. Provides a corrosion-resistant paint base. Coating may be black on sand and permanent-mold castings, and somewhat grayer on die castings.

TREATMENT PROCEDURE

1. Clean by vapor degreasing, followed by alkaline cleaning.
2. Rinse in cold water.
3. Pickle in a chromic, sulfuric, or nitric sulfuric acid.
4. Rinse in cold water.
5. Immerse for 5 min in a 15% hydrofluoric acid solution at room temperature.
6. Rinse in cold water.
7. Anodically treat in the following bath and under the following conditions:

Sodium dichromate	13.4 oz
Monosodium phosphate	2.67 oz
Water to make 1 gal	
Current density	5 to 10 amp per sq ft
Tank voltage	3 to 6 v
Temperature	122 F
Time	45 min
pH	4.2 to 4.8

8. Rinse in cold water, then hot water and dry.

NOTES AND COMMENTS

1. Anodic throwing power is only fair.

2. The process finds its widest application on products made of Dowmetal H or AMC 265 alloys.

Sealed Chrome-pickle Process. Purpose: to produce a corrosion-resistant coating or a paint base. Coatings are iridescent, range from matte gray through yellow-red to brassy or silvery white.

TREATMENT PROCEDURE

1. Clean by degreasing or alkaline cleaning and rinsing.
2. Immerse for ½ to 2 min at 70 to 90 F in:

Sodium dichromate	1.5 lb
Nitric acid	0.9 pt
Water	To make 1 gal

3. Rinse in cold water.

4. Immerse for 30 min in a boiling 10 to 15% sealing solution of potassium or sodium dichromate.

5. Rinse in cold water, then hot water, and dry.

NOTES AND COMMENTS

1. The treatment is also known as Dow Treatment No. 10 (Dow Chemical Co.), AMC-L (American Magnesium Co.), and An-M-12 Type II (Army-Navy Joint Specification).

2. Etching occurs during treatment. As much as 0.0006 to 0.001 in. of metal per surface may be removed. To minimize this, particularly on finish machined surfaces, the following dip for 15 sec can be used in place of the solutions under Treatment Procedure (2):

Sodium dichromate	0.75 lb
Nitric acid	0.50 pt
Water	To make 1 gal

3. The coating has low electrical resistance and is therefore useful where passage of current is required through mating parts.

4. Sealing does not affect a very great increase in corrosion resistance of sand castings, but the increase for die castings is considerable. By incorporating a preliminary 5-min dip in a 15% hydrofluoric acid solution before chrome pickling, the corrosion resistance of sand castings is materially increased.

5. The abrasion resistance of the coating is poor.

COLORING OF COPPER ALLOYS

Articles Containing 85% Cu or More

Statuary Bronze (Dark Brown to Reddish Bronze). Yellow brass articles can be copper-plated and then given the above finish, depth of shade being controlled by solution strength and immersion time, using either solution

Solution 1		Solution 2	
Liquid sulfur	1 oz	Potassium sulfuret	2 oz
Potassium sulfuret	2 oz	Caustic soda	3 oz
Aqua ammonia (sp. gr., 0.89)	$\frac{1}{4}$ oz	Water	1 gal
Water	1 gal	Temperature	170 F
Room temperature			

Rinse in cold, then hot water; dry in hot sawdust or air-blast. Desired finish is secured by scratch brushing, using a fire wire wheel dry, and lacquering.

Brown

Solution 1		Solution 2	
Copper sulfate	4 oz	"Liquid sulfur"	1 oz
		or	
Potassium chlorate	8 oz	Potassium sulfuret	2 oz
Water	1 gal	Water	1 gal

Immerse in solution 1 for 1 min; transfer to Solution 2. Then rinse in cold water. Repeat process until desired color is obtained. Finally rinse, dry, scratch-brush, and lacquer.

Steel Black

Arsenious oxide	4 oz
Hydrochloric acid	8 fluid oz
Water	1 gal

Immerse in solution at 180 F until uniform color is obtained, scratch-brush while wet, dry, and lacquer.

Coloring Brass Articles

Old English Finish (Light Brown)

Solution 1		Solution 2	
"Liquid sulfur"	½ oz	Copper sulfate	2 oz
or			
Potassium sulfuret	1 oz	Water	1 gal
Water	1 gal		

Immerse in solution 1, and without rinsing, immerse in solution 2. Rinse in cold water. Repeat dipping in solutions until light color is produced. Scratch-brush. Dip in solution 1, then 2, until desired color is obtained. Rinse in cold and hot water, dry in sawdust, scratch-brush, and lacquer.

Blue Black

Copper carbonate	1 lb
Ammonium hydroxide	1 qt
Water	3 qt

Immerse work in thoroughly mixed solution, maintained at 175 F, until desired color is obtained, about 1 min. Rinse in warm water. Immerse in hot cleaning solution comprised of 2 to 4 oz of alkaline cleaner per gal of water. Rinse in cold, then hot, water; dry and lacquer.

Black. Tumble for 15 to 30 min in oblique stainless-steel barrel containing 3 to 5 gal of warm water, to which has been added 3 oz of copper sulfate and 6 oz of sodium thiosulfate. Drain barrel, wash contents. Remove, dry, and lacquer if desired.

Black Anodizing. Place in steel tank containing 16 oz caustic soda per gal of water, maintained at 180 to 210 F, and with steel, carbon, or graphite anodes. Apply current of 2 to 10 amp per sq ft at 6 v for a cycle of 45 to 225 sec. Wash in hot and cold water, rinse in hot water, dry, buff lightly with soft wheel, and lacquer if desired.

STEEL

Oxide Coating. Decorative and inhibitive finish on steel, gray iron, and stainless steel. Subsequent treatment with oil or wax usually employed. Little dimension change occurs. Decorative coats are very thin (0.00003 to 0.00007 in.).

APPEARANCE. Attractive, dull, dense black. Lighter gray on nickel and chromium alloy steels. Forms good base for paint. Clean steel, heated in air, can be temper colored to yellow bronze or blue, depending on temperatures.

CORROSION PROTECTION. Attacked by acids. Bare coat is too porous to be satisfactory. Performance helped greatly by oil, wax, or paint overlay. Corners are not covered as well as with electroplate. Not so protective as phosphate coatings in marine conditions.

SURFACE PREPARATION. Surface is degreased and acid-pickled to remove rust and scale. Mechanical polishing not used. Gray iron, stainless and alloy steels are not treated. Preliminary dip in 5 to 10% hydrofluoric acid allows coating to be applied in some cases.

APPLICATION METHODS. Decorative blacks applied on steel by immersion in molten sodium nitrate for several minutes at 700 to 800 F, or in nitrate-alkali solutions at 285 to 300 F. Stainless steel is dipped in molten sodium dichromate for 20 to 30 min at 730 to 750 F. Steel is heated in air or molten lead or salt to 410 F (yellow) up to 610 F (dark blue) for temper colors. Thick coats applied to gray iron by steam at 1100 F.

REMARKS. Steel or iron surface is tranformed to Fe_3O_4, to inhibit formation of rust Fe_2O_3. Coating is primarily used for decoration rather than corrosion or abrasion resistance. Blackening process for stainless steel is patented. Reproduces profile of base metal.

MILITARY SPECIFICATIONS FOR PHOSPHATE COATINGS[1]

Protecting Ferrous Metal Surfaces from Rust

Specification Number. U.S.A. No. 57-0-2C.
Specification Title. Finishes, protective, for iron and steel parts.
Type of Finish. Phosphate coatings finished with a rust-preventive suitably reduced for application and containing corrosion inhibitors. Described in this specification as type II, class B.
Metal Products Treated. Nuts, bolts, screws, hardware, tools, guns, cartridge clips, and other nonmoving parts.
Summary of Process. Class B Finish. The properly cleaned articles should be immersed for not less than 30 min in a hot balanced phosphate rustproofing solution containing nitrate as an accelerating agent, until a uniform crystalline phosphate coating is formed, after which the parts should be immediately rinsed in clean water, followed by a second rinse in a controlled dilute chromic acid solution. After drying a suitable corrosion-resistant finish should be applied. (Concentration such that 10 ml of the phosphating solution should require not less than 30 ml of $N/10$ sodium hydroxide for neutralization.)
Precleaning Notes. Parts to be processed should be free from all oil, grease, burrs, rust, and loose scale.
Supplementary Finish. A thorough coating of rust-inhibiting oil or compound which after application is centrifuged or drained.

Specification Number. U.S.A. No. 51-70-1A.
Specification Title. Painting and finishing of fire-control instruments; general specification for.
Type of Finish. Similar to above, but described as "finish 22.02, class C. Phosphate, steel."
Metal Products Treated. Fire-control instruments and accessories.
Summary of Process. Chemical surface treatment in accordance with type II of U.S. Army Specification No. 57-0-2.
Precleaning Notes. Parts to be processed should be free from all oil, grease, burrs, rust, and loose scale. Same methods as those mentioned for JAN-C-490.
Supplementary Finish. Various finishes; depending on individual requirements.

Specification Number. U.S.A. No. 51-70-1A.
Specification Title. Painting and finishing of fire-control instruments; general specification for.
Type of Finish. Similar to above, but described as "Finish 22.02, Class B Phosphate, steel."
Metal Products Treated. Fire-control instruments and accessories.
Summary of Process. Chemical surface treatment in accordance with type II of U.S. Army Specification No. 57-0-2.
Precleaning Notes. Parts to be processed should be free from all oil, grease, burrs, rust, and loose scale.
Supplementary Finish. Rust-preventive oil.

Specification Number. U.S.A. No. 50-60-1.
Specification Title. Containers, metal, for artillery and rocket ammunition.
Type of Finish. Phosphate coating meeting the requirements of Specification JAN-C-490 (grade 1 finish).
Metal Products Treated. All types and sizes of complete round containers and cartridge storage cases used for artillery ammunition, rockets, and components.
Summary of Process. Upon completion of the machining operations, the cartridge storage cases or complete round containers shall be cleaned to free them of all grease, dirt, chips, and other foreign material and shall be given a phosphate coating preparatory to painting.
Precleaning Notes. The preliminary cleaning to remove contamination may be any method which is adequate and which does not interfere with the phosphate coating process.
Supplementary Finish. Both interior and exterior surfaces of the case or container body and cover assemblies as indicated on the drawings shall be coated with one coat of olive drab enamel, U.S.A. Specification No. 3-181.

Specification Number. MIL-E-917A (Ships).
Specification Title. Equipment, electric power, basic requirements for (naval shipboard use).

[1] Electrolytic and Chemical Finishes. *Product Eng.*, February, 1951, pp. 141–148 and chart.

Type of Finish. Phosphate coating meeting the requirements of specification JAN-C-490 (grade 1 finish).
Metal Products Treated. Electric (but not electronic or interior communication) equipment used for naval shipboard applications.
Summary of Process. A hot-dip-tank phosphate or chromate-phosphate treatment conforming to Specification JAN-C-490, grade 1, followed by one coat of primer.
Precleaning Notes. Remove all rust, and other visible corrosion products. Remove all grease, oil, and dirt by solvent wiping, vapor degreasing, or caustic washing and rinsing.
Supplementary Finish. One coat of primer in accordance with any of the following specifications, JAN-P-72, JAN-P-735, or HIL-P-6889 applied as a continuous film 0.0002 to 0.001 in. thick, followed by two coats of gray enamel, conforming to type II or III, Class 2 of Specification MIL-E-15090, applied as continuous films, each approximately 0.001 in. thick.

Specification Number. MIL-C-16232.
Specification Title. Coatings: phosphate; oiled, slushed, or waxed (for ferrous metal surfaces) and phosphate treating compounds.
Type of Finish. Phosphate coatings on surfaces of nonmoving parts or where bearing friction is not likely to be encountered. Described in this specification as type II.
Metal Products Treated. Nuts, bolts, screws, hardware, tools, guns, cartridge clips, and other nonmoving parts.
Summary of Process. After sufficient time of immersion in the phosphate treating bath, the treated parts shall be thoroughly rinsed in a clean overflowing water bath for not less than 60 sec. Flow of water through the rinse shall be so regulated with the rate of production that at no time will the main body of the rinse become contaminated. Immediately following the water rinse, the parts shall be treated in an acidified final rinse, according to the instructions of the manufacturer of the phosphate coating compounds.
Precleaning Notes. The base metal shall be subjected to such finishing and cleaning procedures as are necessary to insure phosphate coatings having the required appearance and quality. The preparation shall be controlled so as not to affect adversely the surface condition, form, or shape of the part as required by the applicable drawing.

Specification Number. MIL-S-5002.
Specification Title. Surface treatments (except priming and painting) for metal and meta. parts in aircraft.
Type of Finish. Phosphate coatings shall be applied in such a manner as to produce coatings in accordance with Specification JAN-C-490, grade 1.
Metal Products Treated. Land airplanes (all aircraft operated entirely with wheel-type landing gear) whether shore-based or carrier-based.
Summary of Process. The properly cleaned articles should be subjected to a balanced phosphate solution, containing nitrate as an accelerating agent (or other phosphate solution which is known to accomplish equivalent results), until a uniform crystalline phosphate coating is obtained. The article should then be rinsed in clean water, followed by a second rinse in a dilute chromic acid solution, and dried.
Precleaning Notes. Cleaning shall be accomplished with materials and processes which have no deterious effect on the metal and which produce surfaces satisfactory for the application of the surface treatment required. All parts after cleaning shall be free from rust, scale, paint, grease, oil, flux, and other foreign matter.

Specification Number. M-364.
Specification Title. Navy aeronautical process specification for compound phosphate rust-proofing process.
Metal Products Treated. Metal aircraft parts after machining.
Summary of Process. The parts are inserted in the bath and coating is continued until the evolution of hydrogen bubbles practically ceases.
Precleaning Notes. Iron and steel parts shall be cleaned and, if necessary, pickled in accordance with specification PT-4 or by an approved electrolytic descaling process. Sandblasting may be used where the size and tolerances of the parts will permit and where the carbon content is not over 0.6 %. Heavy oxide scale shall be removed by pickling in a solution of sulfuric acid containing approximately 1 lb of acid per gal. All parts shall be thoroughly rinsed in clean water after each cleaning and pickling operation.
Supplementary Finish. Paralketone, type B of Navy Aeronautical Specification RM-61, or equivalent.

Specification Number. JAN-C-490.
Type of Finish. A treatment which produces an adherent crystalline phosphate deposit on a previously clean, ferrous, metal surface. Described in the specification as grade 1.
Metal Products Treated. Projectiles, rockets, bombs, rifles, small arms, belt links, cartridge tanks, vehicular sheet metal, tank bolts and links, recoil-less guns.
Summary of Process. The properly cleaned articles should be subjected to a balanced phosphate solution, containing nitrate as an accelerating agent (or other phosphate solution which is known to accomplish equivalent results), until a uniform crystalline phosphate coating is obtained. The article should then be rinsed in clean water, followed by a second rinse in a dilute chromic acid solution and dried.
Precleaning Notes. Any common degreasing method can be used; alkali cleaning; phosphoric acid cleaning, emulsion-alkali cleaning; vapor degreasing; solvent wiping; etc. Acid cleaning may be required after other cleaning methods if rust or scale is present.

Protecting Friction Surfaces and Providing Rust Resistance

Specification Number. Mil-C-16232, Type I.

Specification Title. Coatings: phosphate; oiled, slushed, or waxed (for ferrous metal surfaces) and phosphate treating compounds.

Type of Finish. Coatings on bearing surfaces where a finish with a moderate degree of corrosion resistance is required to prevent wear and assist in "break-in" of bearing surfaces. Described in this specification as type I.

Metal Products Treated. Pistons, piston rings, cylinders, cylinder liners, gears, camshafts, crankshafts, and other friction surfaces.

Summary of Process. After sufficient time of immersion in the phosphate treating bath, the treated parts shall be thoroughly rinsed in a clean overflowing water bath for not less than 60 sec. The flow of water through the water rinse shall be so regulated in conjunction with the rate of production that at no time will the main body of the rinse become contaminated. Immediately following the water rinse, the parts shall be treated in an acidified final rinse, according to the instructions of the manufacturer of the phosphate-coating compounds.

Precleaning Notes. The base metal shall be subjected to such finishing and cleaning procedures as are necessary to insure phosphate coatings having the required appearance and quality. The preparation shall be controlled so as not to adversely affect the surface condition, form, or shape of the part as required by the applicable drawing.

Supplementary Finish. Unless otherwise approved, type I phosphate coatings shall be impregnated with oil conforming to MIL-L-3150.

Specification Number. U.S.A. No. 57-0-2C.

Specification Title. Finishes, protective, for iron and steel parts.

Type of Finish. Phosphate coatings finished with nondrying petroleum oils containing corrosion inhibitors, suitable for use on sliding and bearing surfaces and with the build-up limited to a maximum of 0.0003 in. per surface. Described in this specification as type II, class A.

Metal Products Treated. Friction surfaces such as pistons, piston rings, gears, cylinder liners, camshafts, tappets, crankshafts, rocker arms. Also small arms, weapon components.

Summary of Process. Class A finish. Immerse the properly cleaned parts for not less than 15 min in a hot (205 to 210 F) balanced manganese acid-phosphate rustproofing solution containing nitrate as an accelerating agent, until a uniform and finely crystalline coating is obtained; after which the articles should be rinsed in clean hot water, followed by a second rinse in a controlled dilute chromic acid solution. After drying, a suitable corrosion-resistant finish should be applied. This coating should be used where a nongalling finish is desired.

Precleaning Notes. Parts to be processed should be free from all oil, grease, burrs, rust, and loose scale.

Supplementary Finish. A thorough coating of rust-inhibiting oil or compound, which after each application is centrifuged or drained.

ZINC AND GALVANIZED METAL

Chromate Process. The big use is in process industries and on consumers' goods, not structural applications. Produces yellow, golden, iridescent olive green or clear finish depending on process, usually proprietary. Coating decreases atmospheric and solution attack and inhibits formation of corrosion products, which on machine parts might hamper functioning of mechanisms. Therefore used on zinc-coated steel-stamped parts.

APPLICATION METHODS. Dip in either chromic acid or dichromate salt solutions at from 70 to 110 F, with or without acid or alkalis added. Dimensional changes up to 0.0006 in. are possible with some processes. Solution concentration and temperature are not critical.

ZINC DIE CASTINGS

Chromate Treatment. To improve resistance of zinc die castings to salt spray or stagnant moisture, a chromate treatment may be applied. Film is bronze or golden in color and should always be iridescent.

Parts freed of oil or grease are soaked for 1 to 2 in. in an alkaline cleaner, rinsed twice in cold running water, and then immersed for 10 to 20 sec in the dichromate solution. This consists of 16 oz of sodium dichromate and 0.6 fluid oz of concentrated sulfuric acid to 1 gal of water. Rinse thoroughly, and air-dry at a temperature under 150 F.

Part 6

INSPECTION

Section 33

LIMITS, FITS, AND TOLERANCES

SECTION 33

LIMITS, FITS, AND TOLERANCES

INTERCHANGEABLE MANUFACTURE

Reason for Interchangeable Parts. Strict interchangeability consists in making the different parts of a mechanism so uniform in size and contour that each part will fit and properly function in any one of the whole number of mechanisms, no matter when or where it is made. If the quantities being manufactured are large, interchangeable manufacture is economical because it entails a correct system of gaging as well as manufacture.

How Fits Are Established. To achieve interchangeability, mating parts must be made to certain design specifications for fit; that is, part dimensions that will create the desired clearance or interference. The basic, or nominally standard, size governs how the fit is established. Either of two systems can be selected: the *basic hole* or the *basic shaft* (for definitions see page 33-4). The basic hole system is most commonly selected when both parts are machined, because standard drills and reamers can often be employed. However, where finished shafting is employed in machinery to minimize machining, the basic shaft system may be preferred.

Why Tolerances Are Applied. It is impractical to expect that parts can all be manufactured to an exact dimension, because of machine inaccuracies, tool wear, etc. Thus, after prescribing the allowance, or difference in dimensions, of two mating parts, the designer must apply tolerances to the dimensions for the mating parts. Tolerances are of two types: *unilateral*, when the dimensional spread for a part is given in one direction from the basic, or design, size, and *bilateral*, when the dimensional spread is given in both directions.

Explanation of Tolerances. In mating parts that fit and move freely, there is clearance between the surfaces. If this space is reduced gradually, a condition will ultimately be reached in which the mating parts are metal to metal. Any movement of the parts one within the other will require some force. If this condition of tightness is represented by a horizontal line, which we might call "zero," then this zero line marks the low limit in the dimension of the external member and the high limit of the dimension of the internal member.

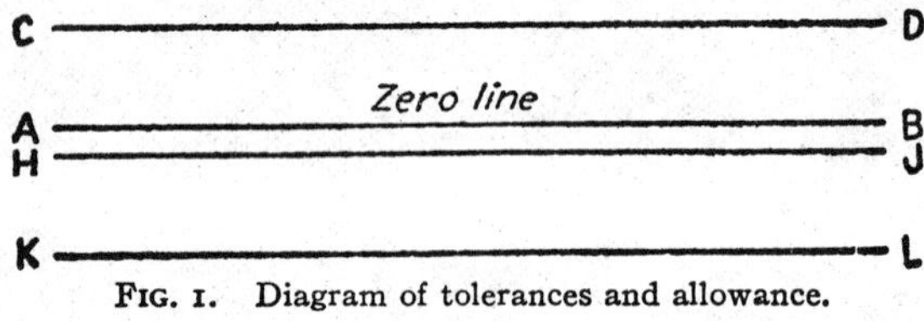

FIG. 1. Diagram of tolerances and allowance.

Any overlapping of this line by the dimensions of either member of a mating pair tends to produce tightness, and any deviation away from this line without crossing it will tend to produce looseness. This zero line, then, is the logical place to coincide with the standard. The best experience in interchangeable manufacture has verified this, and the best shops use it.

In Fig. 1, the zero line *AB* corresponds to the exact standard size and the line *CD* represents the largest allowable dimension on the tools for external members. The space between the lines *AB* and *CD* is the manufacturing tolerance and also the wear limit on these tools. As soon as the tools have worn so that the standard gage will not go in the hole, their use is stopped. It might be asked, Why not continue their use to, say, 0.001 in. undersize before stopping and so get more wear out of the tool? The obvious answer is that the standard gage will not check it, and hence the gage is of no use. This is called "unilateral" tolerance.

The tolerance on the tool above the minimum size determines the amount of wear on the tool and not the relation of the minimum size to the standard. If 0.002-in. wear is desired on a tool, then there must be 0.002-in. more looseness when the tool begins to work and this should be above standard, as nothing is gained by starting it at 0.001 in. above standard and stopping at 0.001 in. under standard. This would be "bilateral" tolerance.

The line *HJ* represents the largest allowable dimension on the shaft or internal member, and the space between this line and the zero line is the minimum allowance, which must never be intentionally encroached upon. It is intended for a predetermined looseness for oil and freedom of fits to prevent seizing.

This space between the zero line and line *HJ* is the difference allowed between the dimensions of the smallest external member and the largest internal member by the designer for the express purpose of establishing the tightest permissible fit.

It is necessary to have a manufacturing tolerance when a large number of these are made. The line *KL* represents the smallest allowable dimension on the shaft or internal member, and the space between the lines *HJ* and *KL* is the tolerance allowed for manufacturing the internal members.

The maximum looseness is represented by the distance between the lines *CD* and *KL*. The greatest tightness is represented by the distance between the zero line *AB* and *HJ*. In this case, the limits for the external member are from standard to *CD* oversize, and the limits for internal member are from *HJ* to *KL*, both undersize.

The space between the lines *AB* and *HJ* is commonly called neutral space, indicating that it is not to be encroached upon by the intrusion of the dimensions of either member of a pair of mating parts. In force or shrink fits, there is no neutral space.

LIMITS AND FITS FOR CYLINDRICAL PARTS

Fits between parts control the efficiency of assembly and operation of machine elements. They can be of several kinds: (1) running and sliding fits, (2) locational fits, and (3) force or shrink fits. Interchangeable manufacture is created by applying tolerances and allowances to establish the desired fit, or in other words defining the limiting sizes of mating parts.

Development of a national standard on preferred limits and fits for cylindrical parts has taken many years to effect. A tentative standard issued in 1925 was used to some extent in industry but was not complete enough to cover the large variety of fits required by different branches. A 1947 standard merely covered preferred basic sizes but not fits. The current proposal is reproduced in the following pages because little opposition has arisen, and it appears that industrial units will adopt portions of the proposed standard that cover their specific needs.

The tables in the proposed standard include in boldface type the recommendations of American-British-Canadian (ABC) conferences on data to serve as the basis for development of national standards, which would have the further military and commercial values of a unified system.

PROPOSED STANDARD FOR CYLINDRICAL FITS

Scope and Application. This standard presents definitions of terms applying to fits between plain (nonthreaded) cylindrical parts and makes recommendations on preferred sizes, allowances, tolerances, and fits for use wherever they are applicable.

Many factors, such as length of engagement, bearing load, speed, lubrication, temperature, humidity, and materials, must be taken into consideration in the selection of fits for a particular application, and modifications in these recommendations might be required to satisfy extreme conditions. Subsequent adjustments might also be desired as the result of experience in a particular application to suit critical functional requirements or to permit optimum manufacturing economy. Selection or departure from these recommendations will depend upon consideration of the engineering and economic factors that might be involved.

Definitions. *Dimension.* A dimension is a geometrical characteristic such as a diameter, length, angle, circumference, or center distance, of which the size is specified.

Size. Size is a designation of magnitude.

Nominal Size. The nominal size is the size which is used for purposes of general identification.

Basic Size. The basic size of a dimension is the theoretical size from which the limits for that dimension are derived by the application of the allowance and tolerance.

Design Size. The design size of a dimension is the size in relation to which the limits of tolerance for that dimension are assigned.

Actual Size. The actual size of a dimension is the measured size of that dimension on an individual part.

Limits of Size. These limits are the maximum and minimum sizes permissible for a specific dimension.

Tolerance. The tolerance on a dimension is the total permissible variation in its size. The tolerance is the difference between the limits of size.

Unilateral Tolerance System. The unilateral tolerance system is a system in which the tolerance is given in one direction only; plus for the hole and minus for the shaft.

Fit. The fit between two mating parts is the relationship existing between them with respect to the amount of clearance or interference which is present when they are assembled.

Allowance. Allowance is a prescribed difference in dimensions of mating parts. It is a minimum clearance (positive allowance) or maximum interference (negative allowance) between mating parts.

Clearance Fit. A fit between mating parts having limits of size so prescribed that a clearance always results in assembly.

Interference Fit. A fit between mating parts having limits of size so prescribed that an interference always results in assembly.

Transition Fit. A fit between mating parts having limits of size so prescribed as to partially or wholly overlap, so that either a clearance or interference may result in assembly.

Basic Hole System. The basic hole system of fits is a system in which the minimum limit of each hole size is basic. The fit desired is obtained by varying the allowance of the shaft and the tolerances of the mating parts.

Basic Shaft System. The basic shaft system of fits is a system in which the maximum limit of each shaft size is basic. The fit desired is obtained by varying the allowance of the hole and the tolerances of the mating parts.

Preferred Basic Sizes. In specifying fits, the basic size of mating parts shall be chosen from Table 1.

Preferred Series for Tolerances and Allowances. All fundamental tolerances and allowances of all shafts and holes have been taken from the series given in Table 2. All dimensions are given in inches.

Acceptance of Parts. *Acceptability.* A part shall be acceptable if its actual size does not exceed the limits of size specified in numerical values on the drawing or in writing. It does not meet dimensional specification if its actual size exceeds those limits.

Reference Temperature. The actual size of a part shall be considered to be that size which the part has when it is measured at the International standard reference temperature of 68 F (20 C).

Selection of Fits. In selecting limits of size for any application, the type of fit is determined first, based on the use of service required from the equipment being designed; then the limits of size of the mating parts are established, to insure that

TABLE 1. PREFERRED BASIC SIZES

...	0.0100	5/8	0.6250	3 1/4	3.2500	11	11.0000
...	0.0125	11/16	0.6875	3 1/2	3.5000	11 1/2	11.5000
1/64	0.015625	3/4	0.7500	3 3/4	3.7500	12	12.0000
...	0.0200	7/8	0.8750	4	4.0000	12 1/2	12.5000
...	0.0250	1	1.0000	4 1/4	4.2500	13	13.0000
1/32	0.03125	1 1/8	1.1250	4 1/2	4.5000	13 1/2	13.5000
...	0.0400	1 1/4	1.2500	4 3/4	4.7500	14	14.0000
...	0.0500	1 3/8	1.3750	5	5.0000	14 1/2	14.5000
1/16	0.0625	1 1/2	1.5000	5 1/4	5.2500	15	15.0000
...	0.0800	1 5/8	1.6250	5 1/2	5.5000	15 1/2	15.5000
3/32	0.09375	1 3/4	1.7500	5 3/4	5.7500	16	16.0000
...	0.1000	1 7/8	1.8750	6	6.0000	16 1/2	16.5000
1/8	0.1250	2	2.0000	6 1/2	6.5000	17	17.0000
5/32	0.15625	2 1/8	2.1250	7	7.0000	17 1/2	17.5000
3/16	0.1875	2 1/4	2.2500	7 1/2	7.5000	18	18.0000
1/4	0.2500	2 3/8	2.3750	8	8.0000	18 1/2	18.5000
5/16	0.3125	2 1/2	2.5000	8 1/2	8.5000	19	19.0000
3/8	0.3750	2 5/8	2.6250	9	9.0000	19 1/2	19.5000
7/16	0.4375	2 3/4	2.7500	9 1/2	9.5000	20	20.0000
1/2	0.5000	2 7/8	2.8750	10	10.0000	20 1/2	20.5000
9/16	0.5625	3	3.0000	10 1/2	10.5000	21	21.0000

TABLE 2. PREFERRED SERIES FOR TOLERANCES AND ALLOWANCES

0.0001	0.0003	0.001	0.003	0.010	0.030
....	......	0.0012	0.0035	0.012	0.035
0.00015	0.0004	0.0014	0.004	0.014	0.040
....	......	0.0016	0.0045	0.016	0.045
0.0002	0.0005	0.0018	0.005	0.018	0.050
....	0.0006	0.002	0.006	0.020	0.060
0.00025	0.0007	0.0022	0.007	0.022	.....
....	0.0008	0.0025	0.008	0.025	....
....	0.0009	0.0028	0.009	0.028	.....

the desired fit will be produced. The small number of standard fits shown here should cover most applications.

Standard Fits. Tables 3 to 7 have been developed to give a series of standard types and classes of fits on a unilateral hole basis, such that the fit produced by mating parts in any one class will produce approximately similar performance throughout the range of sizes. These tables prescribe the fit for any given size, or type of fit; they also prescribe the standard limits for the mating parts which will produce the fit.

In developing Tables 3 to 7 it has been recognized that any fit will usually be required to perform one of three functions, as indicated by the three general types of fits: running fits, locational fits, and force fits.

The fits listed in Tables 3 to 7 contain all those in the approved ABC proposal but have been extended to include a wider range of sizes. Fits in exact agreement with the proposal are shown in boldface type.

Designation of Standard Fits. Standard fits are designated by means of the symbols given below to facilitate reference to classes of fit for educational purposes. These symbols are not intended to be shown on manufacturing drawings; instead, sizes should be specified on drawings.

The letter symbols used are as follows:

RC. Running or sliding fit.
LC. Locational clearance fit.
LT. Transition fit.
LN. Locational interference fit.
FN. Force or shrink fit.

These letter symbols are used in conjunction with numbers representing the class of fit; thus FN 4 represents a class 4 force fit.

Each of these symbols (two letters and a number) represent a complete fit, for which the minimum and maximum clearance or interference, and the limits of size for the mating parts, are given directly in the tables.

Description of Fits. RUNNING AND SLIDING FITS. Running and sliding fits, for which limits of clearance are given in Table 3, represent a special type of clearance fit. These are intended to provide a similar running performance, with suitable lubrication allowance, throughout the range of sizes. The clearances for the first two classes, used chiefly as slide fits, increase more slowly with diameter than the other classes, so that accurate location is maintained even at the expense of free relative motion.

These fits may be described briefly as follows:

RC 1. *Close sliding fits* are intended for the accurate location of parts which must assemble without perceptible play.

RC 2. *Sliding fits* are intended for accurate location, but with greater maximum clearance than class RC 1. Parts made to this fit move and turn easily but are not intended to run freely, and in the larger sizes may seize with small temperature changes.

RC 3. *Precision running fits* are about the closest fits which can be expected to run freely, and are intended for precision work at slow speeds and light journal pressures, but are not suitable where appreciable temperature differences are likely to be encountered.

RC 4. *Close running fits* are intended chiefly for running fits on accurate machinery with moderate surface speeds and journal pressures where accurate location and minimum play are desired.

RC 5 and RC 6. *Medium running fits* are intended for higher running speeds, or heavy journal pressures, or both.

RC 7. *Free running fits* are intended for use where accuracy is not essential, or where large temperature variations are likely to be encountered, or both conditions.

RC 8 and RC 9. *Loose running fits* are intended for use where materials such as cold-rolled shafting and tubing, made to commercial tolerances, are involved.

LOCATIONAL FITS. Locational fits are fits intended to determine only the location of the mating parts; they may provide rigid or accurate location, as with interference fits, or provide some freedom of location, as with clearance fits. Accordingly they are divided into three groups: clearance fits, transition fits, and interference fits.

These are more fully described as follows:

LC. *Locational clearance fits* are intended for parts which are normally stationary but which can be freely assembled or disassembled. They run from snug fits for parts requiring accuracy of location, through the medium clearance fits for parts such as spigots, to the looser fastener fits where freedom of assembly is of prime importance (see Table 4).

LT. *Transition fits* are a compromise between clearance and interference fits, for application where accuracy of location is important, but either a small amount of clearance or interference is permissible (see Table 5).

LN. *Location interference fits* are used where accuracy of location is of prime importance, and for parts requiring rigidity and alignment with no special requirements for bore pressure. Such fits are not intended for parts designed to transmit frictional loads from one part to another by virtue of the tightness of fit, as these conditions are covered by force fits (see Table 6).

FORCE FITS. Force or shrink fits constitute a special type of interference fit, normally characterized by maintenance of constant bore pressures throughout the range of sizes. The interference therefore varies almost directly with diameter, and the difference between its minimum and maximum value is small, to maintain the resulting pressures within reasonable limits (see Table 7).

These fits may be described briefly as follows:

FN 1. *Light drive fits* are those requiring light assembly pressures, and produce more or less permanent assemblies. They are suitable for thin sections or long fits, or in cast-iron external members.

FN 2. *Medium drive fits* are suitable for ordinary steel parts, or for shrink fits on light sections. They are about the tightest fits that can be used with high-grade cast-iron external members.

FN 3. *Heavy drive fits* are suitable for heavier steel parts or for shrink fits in medium sections.

FN 4 and FN 5. *Force fits* are suitable for parts which can be highly stressed, or for shrink fits where the heavy pressing forces required are impractical.

Standard Tolerances. The series of standard tolerances shown in Table 8 are so arranged that for any one grade they represent approximately similar production difficulties throughout the range of sizes. The table provides a suitable range from which appropriate tolerances for holes and shafts can be selected. This enables the use of standard gages. These tolerances have been used in arranging the fits given in Tables 3 to 7.

Modified Standard Fits. Fits having the same limits of clearance or interference as those shown in Tables 3 to 7 may sometimes have to be produced by using holes or shafts other than those shown in the tables. This may be accomplished by using one of the following:

a. BILATERAL HOLES (SYMBOL B). This will result in nonstandard holes and shafts.

b. A BASIC SHAFT SYSTEM (SYMBOL S). This will result in nonstandard holes and shafts.

RC1 — RC2 — RC3 — RC4 — RC5 — RC6 — RC7 — RC8 — RC9

Basic size

Holes — Shafts

Scale: 0.001 in. for diameter of 1 in.

Table 3. Running and Sliding Fits

Limits are in 0.001 in.

Limits for hole and shaft are applied algebraically to the basic size to obtain the limits of size for the parts.

Data in boldface accord with ABC agreements.

Nominal Size Range, In.		Class RC 1			Class RC 2			Class RC 3			Class RC 4		
		Limits of Clearance	Standard Limits		Limits of Clearance	Standard Limits		Limits of Clearance	Standard Limits		Limits of Clearance	Standard Limits	
Over	To		Hole	Shaft		Hole	Shaft		Hole	Shaft		Hole	Shaft
0.04–	0.12	0.1	+0.2	− 0.1	0.1	+ 0.25	− 0.1	0.3	+ 0.25	− 0.3	0.3	+ 0.4	− 0.3
		0.45	0	− 0.25	0.55	0	− 0.3	0.8	0	− 0.55	1.1	0	− 0.7
0.12–	0.24	0.15	+0.2	− 0.15	0.15	+ 0.3	− 0.15	0.4	+ 0.3	− 0.4	0.4	+ 0.5	− 0.4
		0.5	0	− 0.3	0.65	0	− 0.35	1.0	0	− 0.7	1.4	0	− 0.9
0.24–	0.40	0.2	+0.25	− 0.2	0.2	+ 0.4	− 0.2	0.5	+ 0.4	− 0.5	0.5	+ 0.6	− 0.5
		0.6	0	− 0.35	0.85	0	− 0.45	1.3	0	− 0.9	1.7	0	− 1.1
0.40–	0.71	0.25	+0.3	− 0.25	0.25	+ 0.4	− 0.25	0.6	+ 0.4	− 0.6	0.6	+ 0.7	− 0.6
		0.75	0	− 0.45	0.95	0	− 0.55	1.4	0	− 1.0	2.0	0	− 1.3
0.71–	1.19	0.3	+0.4	− 0.3	0.3	+ 0.5	− 0.3	0.8	+ 0.5	− 0.8	0.8	+ 0.8	− 0.8
		0.95	0	− 0.55	1.2	0	− 0.7	1.8	0	− 1.3	2.4	0	− 1.6
1.19–	1.97	0.4	+0.4	− 0.4	0.4	+ 0.6	− 0.4	1.0	+ 0.6	− 1.0	1.0	+ 1.0	− 1.0
		1.1	0	− 0.7	1.4	0	− 0.8	2.2	0	− 1.6	3.0	0	− 2.0
1.97–	3.15	0.4	+0.5	− 0.4	0.4	+ 0.7	− 0.4	1.2	+ 0.7	− 1.2	1.2	+ 1.2	− 1.2
		1.2	0	− 0.7	1.6	0	− 0.9	2.6	0	− 1.9	3.6	0	− 2.4
3.15–	4.73	0.5	+0.6	− 0.5	0.5	+ 0.9	− 0.5	1.4	+ 0.9	− 1.4	1.4	+ 1.4	− 1.4
		1.5	0	− 0.9	2.0	0	− 1.1	3.2	0	− 2.3	4.2	0	− 2.8

Nominal Size Range, In. Over – To	Class RC 1 Limits of Clearance	Class RC 1 Standard Limits Hole	Class RC 1 Standard Limits Shaft	Class RC 2 Limits of Clearance	Class RC 2 Standard Limits Hole	Class RC 2 Standard Limits Shaft	Class RC 3 Limits of Clearance	Class RC 3 Standard Limits Hole	Class RC 3 Standard Limits Shaft	Class RC 4 Limits of Clearance	Class RC 4 Standard Limits Hole	Class RC 4 Standard Limits Shaft
4.73– 7.09	0.6	+0.7	− 0.6	0.6	+ 1.0	− 0.6	1.6	+ 1.0	− 1.6	1.6	+ 1.6	− 1.6
	1.8	0	− 1.1	2.3	0	− 1.3	3.6	0	− 2.6	4.8	0	− 3.2
7.09– 9.85	0.6	+0.8	− 0.6	0.6	+ 1.2	− 0.6	2.0	+ 1.2	− 2.0	2.0	+ 1.8	− 2.0
	2.0	0	− 1.2	2.6	0	− 1.4	4.4	0	− 3.2	5.6	0	− 3.8
9.85– 12.41	0.8	+0.9	− 0.8	0.8	+ 1.2	− 0.8	2.5	+ 1.2	− 2.5	2.5	+ 2.0	− 2.5
	2.3	0	− 1.4	2.9	0	− 1.7	4.9	0	− 3.7	6.5	0	− 4.5
12.41– 15.75	1.0	+1.0	− 1.0	1.0	+ 1.4	− 1.0	3.0	+ 1.4	− 3.0	3.0	+ 2.2	− 3.0
	2.7	0	− 1.7	3.4	0	− 2.0	5.8	0	− 4.4	7.4	0	− 5.2
15.75– 19.69	1.2	+1.0	− 1.2	1.2	+ 1.6	− 1.2	4.0	+ 1.6	− 4.0	4.0	+ 2.5	− 4.0
	3.0	0	− 2.0	3.8	0	− 2.2	7.2	0	− 5.6	9.0	0	− 6.5
19.69– 30.09	1.6	+1.2	− 1.6	1.6	+ 2.0	− 1.6	5.0	+ 2.0	− 5.0	5.0	+ 3.0	− 5.0
	3.7	0	− 2.5	4.8	0	− 2.8	9.0	0	− 7.0	11.0	0	− 8.0
30.09– 41.49	2.0	+1.6	− 2.0	2.0	+ 2.5	− 2.0	6.0	+ 2.5	− 6.0	6.0	+ 4.0	− 6.0
	4.6	0	− 3.0	6.1	0	− 3.6	11.0	0	− 8.5	14.0	0	−10.0
41.49– 56.19	2.5	+2.0	− 2.5	2.5	+ 3.0	− 2.5	8.0	+ 3.0	− 8.0	8.0	+ 5.0	− 8.0
	5.7	0	− 3.7	7.5	0	− 4.5	14.0	0	−11.0	18.0	0	−13.0
56.19– 76.39	3.0	+2.5	− 3.0	3.0	+ 4.0	− 3.0	10.0	+ 4.0	−10.0	10.0	+ 6.0	−10.0
	7.1	0	− 4.6	9.5	0	− 5.5	18.0	0	−14.0	22.0	0	−16.0
76.39–100.9	4.0	+3.0	− 4.0	4.0	+ 5.0	− 4.0	12.0	+ 5.0	−12.0	12.0	+ 8.0	−12.0
	9.0	0	− 6.0	12.0	0	− 7.0	22.0	0	−17.0	28.0	0	−20.0
100.9 –131.9	5.0	+4.0	− 5.0	5.0	+ 6.0	− 5.0	16.0	+ 6.0	−16.0	16.0	+10.0	−16.0
	11.5	0	− 7.5	15.0	0	− 9.0	28.0	0	−22.0	36.0	0	−26.0
131.9 –171.9	6.0	+5.0	− 6.0	6.0	+ 8.0	− 6.0	18.0	+ 8.0	−18.0	18.0	+12.0	−18.0
	14.0	0	− 9.0	19.0	0	−11.0	34.0	0	−26.0	42.0	0	−30.0
171.9 –200	8.0	+6.0	− 8.0	8.0	+10.0	− 8.0	22.0	+10.0	−22.0	22.0	16.0	−22.0
	18.0	0	−12.0	24.0	0	−12.0	42.0	....	−32.0	54.0	0	−38.0

TABLE 3. RUNNING AND SLIDING FITS (*Continued*)

Nominal Size Range, In.		Class RC 5			Class RC 6			Class RC 7			Class RC 8			Class RC 9		
		Limits of Clearance	Standard Limits		Limits of Clearance	Standard Limits		Limits of Clearance	Standard Limits		Limits of Clearance	Standard Limits		Limits of Clearance	Standard Limits	
Over	To		Hole	Shaft		Hole	Shaft		Hole	Shaft		Hole	Shaft		Hole	Shaft
0.04–	0.12	0.6	+ 0.4	− 0.6	0.6	+ 0.6	− 0.6	1.0	+ 1.0	− 1.0	2.5	+ 1.6	− 2.5	4.0	+ 2.5	− 4.0
		1.4	0	− 1.0	1.8	0	− 1.2	2.6	0	− 1.6	5.1	0	− 3.5	8.1	0	− 5.6
0.12–	0.24	0.8	+ 0.5	− 0.8	0.8	+ 0.7	− 0.8	1.2	+ 1.2	− 1.2	2.8	+ 1.8	− 2.8	4.5	+ 3.0	− 4.5
		1.8	0	− 1.3	2.2	0	− 1.5	3.1	0	− 1.9	5.8	0	− 4.0	9.3	0	− 6.0
0.24–	0.40	1.0	+ 0.6	− 1.0	1.0	+ 0.9	− 1.0	1.6	+ 1.4	− 1.6	3.0	+ 2.2	− 3.0	5.0	+ 3.5	− 5.0
		2.2	0	− 1.6	2.8	0	− 1.9	3.9	0	− 2.5	6.6	0	− 4.4	10.7	0	− 7.2
0.40–	0.71	1.2	+ 0.7	− 1.2	1.2	+ 1.0	− 1.2	2.0	+ 1.6	− 2.0	3.5	+ 2.8	− 3.5	6.0	+ 4.0	− 6.0
		2.6	0	− 1.9	3.2	0	− 2.2	4.6	0	− 3.0	7.9	0	− 5.1	12.8	0	− 8.8
0.71–	1.19	1.6	+ 0.8	− 1.6	1.6	+ 1.2	− 1.6	2.5	+ 2.0	− 2.5	4.5	+ 3.5	− 4.5	7.0	+ 5.0	− 7.0
		3.2	0	− 2.4	4.0	0	− 2.8	5.7	0	− 3.7	10.0	0	− 6.5	15.5	0	− 10.5
1.19–	1.97	2.0	+ 1.0	− 2.0	2.0	+ 1.6	− 2.0	3.0	+ 2.5	− 3.0	5.0	+ 4.0	− 5.0	8.0	+ 6.0	− 8.0
		4.0	0	− 3.0	5.2	0	− 3.6	7.1	0	− 4.6	11.5	0	− 7.5	18.0	0	− 12.0
1.97–	3.15	2.5	+ 1.2	− 2.5	2.5	+ 1.8	− 2.5	4.0	+ 3.0	− 4.0	6.0	+ 4.5	− 6.0	9.0	+ 7.0	− 9.0
		4.9	0	− 3.7	6.1	0	− 4.3	8.8	0	− 5.8	13.5	0	− 9.0	20.5	0	− 13.5
3.15–	4.73	3.0	+ 1.4	− 3.0	3.0	+ 2.2	− 3.0	5.0	+ 3.5	− 5.0	7.0	+ 5.0	− 7.0	10.0	+ 9.0	− 10.0
		5.8	0	− 4.4	7.4	0	− 5.2	10.7	0	− 7.2	15.5	0	− 10.5	24.0	0	− 15.0
4.73–	7.09	3.5	+ 1.6	− 3.5	3.5	+ 2.5	− 3.5	6.0	+ 4.0	− 6.0	8.0	+ 6.0	− 8.0	12.0	+ 10.0	− 12.0
		6.7	0	− 5.1	8.5	0	− 6.0	12.5	0	− 8.5	18.0	0	− 12.0	28.0	0	− 18.0
7.09–	9.85	4.0	+ 1.8	− 4.0	4.0	+ 2.8	− 4.0	7.0	+ 4.5	− 7.0	10.0	+ 7.0	− 10.0	15.0	+ 12.0	− 15.0
		7.6	0	− 5.8	9.6	0	− 6.8	14.3	0	− 9.8	21.5	0	− 14.5	34.0	0	− 22.0
9.85–	12.41	5.0	+ 2.0	− 5.0	5.0	+ 3.0	− 5.0	8.0	+ 5.0	− 8.0	12.0	+ 8.0	− 12.0	18.0	+ 12.0	− 18.0
		9.0	0	− 7.0	11.0	0	− 8.0	16.0	0	− 11.0	25.0	0	− 17.0	38.0	0	− 26.0
12.41–	15.75	6.0	+ 2.2	− 6.0	6.0	+ 3.5	− 6.0	10.0	+ 6.0	− 10.0	14.0	+ 9.0	− 14.0	22.0	+ 14.0	− 22.0
		10.4	0	− 8.2	13.0	0	− 9.5	19.5	0	− 13.5	29.0	0	− 20.0	45.0	0	− 31.0

Nominal Size Range, In. Over – To	Class RC 5 Limits of Clearance	Class RC 5 Standard Limits Hole	Class RC 5 Standard Limits Shaft	Class RC 6 Limits of Clearance	Class RC 6 Standard Limits Hole	Class RC 6 Standard Limits Shaft	Class RC 7 Limits of Clearance	Class RC 7 Standard Limits Hole	Class RC 7 Standard Limits Shaft	Class RC 8 Limits of Clearance	Class RC 8 Standard Limits Hole	Class RC 8 Standard Limits Shaft	Class RC 9 Limits of Clearance	Class RC 9 Standard Limits Hole	Class RC 9 Standard Limits Shaft
15.75– 19.69	8.0	+ 2.5	− 8.0	8.0	+ 4.0	− 8.0	12.0	+ 6.0	− 12.0	16.0	+10.0	− 16.0	25.0	+ 16.0	− 25.0
	13.0	0	−10.5	16.0	0	−12.0	22.0	0	− 16.0	32.0	0	− 22.0	51.0	0	− 35.0
19.69– 30.09	10.0	+ 3.0	−10.0	10.0	+ 5.0	−10.0	16.0	+ 8.0	− 16.0	20.0	+12.0	− 20.0	30.0	+ 20.0	− 30.0
	16.0	0	−13.0	20.0	0	−15.0	29.0	0	− 21.0	40.0	0	− 28.0	62.0	0	− 42.0
30.09– 41.49	12.0	+ 4.0	−12.0	12.0	+ 6.0	−12.0	20.0	+10.0	− 20.0	25.0	+16.0	− 25.0	40.0	+ 25.0	− 40.0
	20.0	0	−16.0	24.0	0	−18.0	36.0	0	− 26.0	51.0	0	− 35.0	81.0	0	− 56.0
41.49– 56.19	16.0	+ 5.0	−16.0	16.0	+ 8.0	−16.0	25.0	+12.0	− 25.0	30.0	+20.0	− 30.0	50.0	+ 30.0	− 50.0
	26.0	0	−21.0	32.0	0	−24.0	45.0	0	− 33.0	62.0	0	− 42.0	100	0	− 70.0
56.19– 76.39	20.0	+ 6.0	−20.0	20.0	+10.0	−20.0	30.0	+16.0	− 30.0	40.0	+25.0	− 40.0	60.0	+ 40.0	− 60.0
	32.0	0	−26.0	40.0	0	−30.0	56.0	0	− 40.0	81.0	0	− 56.0	125	0	− 85.0
76.39–100.9	25.0	+ 8.0	−25.0	25.0	+12.0	−25.0	40.0	+20.0	− 40.0	50.0	+30.0	− 50.0	80.0	+ 50.0	− 80.0
	41.0	0	−33.0	49.0	0	−37.0	72.0	0	− 52.0	100	0	− 70.0	160	0	−110
100.9 –131.9	30.0	+10.0	−30.0	30.0	+16.0	−30.0	50.0	+25.0	− 50.0	60.0	+40.0	− 60.0	100	+ 60.0	−100
	50.0	0	−40.0	62.0	0	−46.0	91.0	0	− 66.0	125	0	− 85.0	200	0	−140
131.9 –171.9	35.0	+12.0	−35.0	35.0	+20.0	−35.0	60.0	+30.0	− 60.0	80.0	+50.0	− 80.0	130	+ 80.0	−130
	59.0	0	−47.0	75.0	......	−55.0	110.0	0	− 80.0	160	0	−110	260	0	−180
171.9 –200	45.0	+16.0	−45.0	45.0	+25.0	−45.0	80.0	+40.0	− 80.0	100	+60.0	−100	150	+100	−150
	77.0	0	−61.0	95.0	0	−70.0	145.0	0	−105.0	200	0	−140	310	0	−210

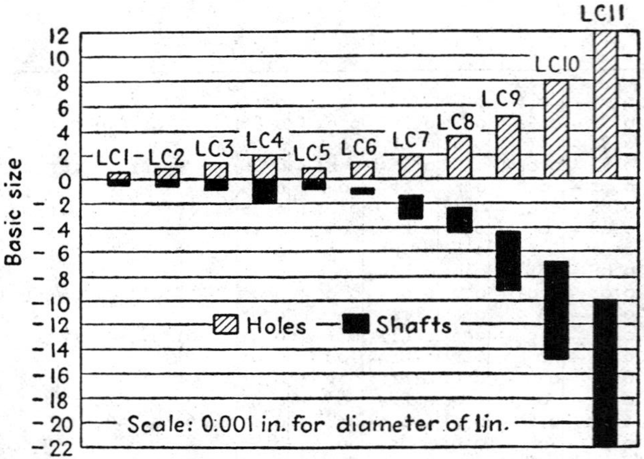

Basic size
LC1 LC2 LC3 LC4 LC5 LC6 LC7 LC8 LC9 LC10 LC11
12 10 8 6 4 2 0 −2 −4 −6 −8 −10 −12 −14 −16 −18 −20 −22
Holes — Shafts
Scale: 0.001 in. for diameter of 1 in.

Table 4. Clearance Locational Fits

Limits are in 0.001 in.
Limits for hole and shaft are applied algebraically to the basic size to obtain the limits of size for the parts.

Nominal Size Range, In.		Class LC 1			Class LC 2			Class LC 3			Class LC 4			Class LC 5		
		Limits of Clearance	Standard Limits		Limits of Clearance	Standard Limits		Limits of Clearance	Standard Limits		Limits of Clearance	Standard Limits		Limits of Clearance	Standard Limits	
Over	To		Hole	Shaft		Hole	Shaft		Hole	Shaft		Hole	Shaft		Hole	Shaft
0.04–	0.12	0 0.45	+ 0.25 − 0	+ 0 − 0.2	0 0.65	+ 0.4 − 0	+ 0 − 0.25	0 1	+ 0.6 − 0	+ 0 − 0.4	0 2.0	+ 1.0 − 0	+ 0 − 1.0	0.1 0.75	+ 0.4 − 0	− 0.1 − 0.35
0.12–	0.24	0 0.5	+ 0.3 − 0	+ 0 − 0.2	0 0.8	+ 0.5 − 0	+ 0 − 0.3	0 1.2	+ 0.7 − 0	+ 0 − 0.5	0 2.4	+ 1.2 − 0	+ 0 − 1.2	0.15 0.95	+ 0.5 − 0	− 0.15 − 0.45
0.24–	0.40	0 0.65	+ 0.4 − 0	+ 0 − 0.25	0 1.0	+ 0.6 − 0	+ 0 − 0.4	0 1.5	+ 0.9 − 0	+ 0 − 0.6	0 2.8	+ 1.4 − 0	+ 0 − 1.4	0.2 1.2	+ 0.6 − 0	− 0.2 − 0.6
0.40–	0.71	0 0.7	+ 0.4 − 0	+ 0 − 0.3	0 1.1	+ 0.7 − 0	+ 0 − 0.4	0 1.7	+ 1.0 − 0	+ 0 − 0.7	0 3.2	+ 1.6 − 0	+ 0 − 1.6	0.25 1.35	+ 0.7 − 0	− 0.25 − 0.65

Nominal Size Range, In.		Class LC 1			Class LC 2			Class LC 3			Class LC 4			Class LC 5		
		Limits of Clearance	Standard Limits		Limits of Clearance	Standard Limits		Limits of Clearance	Standard Limits		Limits of Clearance	Standard Limits		Limits of Clearance	Standard Limits	
Over	To		Hole	Shaft		Hole	Shaft		Hole	Shaft		Hole	Shaft		Hole	Shaft
0.71-	1.19	0 0.9	+ 0.5 - 0	+ 0 -0.4	0 1.3	+ 0.8 - 0	+ 0 - 0.5	0 2	+ 1.2 - 0	+ 0 - 0.8	0 4	+ 2.0 - 0	+ 0 - 2.0	0.3 1.6	+ 0.8 - 0	- 0.3 - 0.8
1.19-	1.97	0 1.0	+ 0.6 - 0	+ 0 -0.4	0 1.6	+ 1.0 - 0	+ 0 - 0.6	0 2.6	+ 1.6 - 0	+ 0 - 1	0 5	+ 2.5 - 0	+ 0 - 2.5	0.4 2.0	+ 1.0 - 0	- 0.4 - 1.0
1.97-	3.15	0 1.2	+ 0.7 - 0	+ 0 -0.5	0 1.9	+ 1.2 - 0	+ 0 - 0.7	0 3	+ 1.8 - 0	+ 0 - 1.2	0 6	+ 3 - 0	+ 0 - 3	0.4 2.3	+ 1.2 - 0	- 0.4 - 1.1
3.15-	4.73	0 1.5	+ 0.9 - 0	+ 0 -0.6	0 2.3	+ 1.4 - 0	+ 0 - 0.9	0 3.6	+ 2.2 - 0	+ 0 - 1.4	0 7	+ 3.5 - 0	+ 0 - 3.5	0.5 2.8	+ 1.4 - 0	- 0.5 - 1.4
4.73-	7.09	0 1.7	+ 1.0 - 0	+ 0 -0.7	0 2.6	+ 1.6 - 0	+ 0 - 1.0	0 4.1	+ 2.5 - 0	+ 0 - 1.6	0 8	+ 4 - 0	+ 0 - 4	0.6 3.2	+ 1.6 - 0	- 0.6 - 1.6
7.09-	9.85	0 2.0	+ 1.2 - 0	+ 0 -0.8	0 3.0	+ 1.8 - 0	+ 0 - 1.2	0 4.6	+ 2.8 - 0	+ 0 - 1.8	0 9	+ 4.5 - 0	+ 0 - 4.5	0.6 3.6	+ 1.8 - 0	- 0.6 - 1.8
9.85-	12.41	0 2.1	+ 1.2 - 0	+ 0 -0.9	0 3.2	+ 2.0 - 0	+ 0 - 1.2	0 5	+ 3.0 - 0	+ 0 - 2.0	0 10	+ 5 - 0	+ 0 - 5	0.7 3.9	+ 2.0 - 0	- 0.7 - 1.9
12.41-	15.75	0 2.4	+ 1.4 - 0	+ 0 -1.0	0 3.6	+ 2.2 - 0	+ 0 - 1.4	0 5.7	+ 3.5 - 0	+ 0 - 2.2	0 12	+ 6 - 0	+ 0 - 6	0.7 4.3	+ 2.2 - 0	- 0.7 - 2.1
15.75-	19.69	0 2.6	+ 1.6 - 0	+ 0 -1.0	0 4.1	+ 2.5 - 0	+ 0 - 1.6	0 6.5	+ 4 - 0	+ 0 - 2.5	0 12	+ 6 - 0	+ 0 - 6	0.8 4.9	+ 2.5 - 0	- 0.8 - 2.4
19.69-	30.09	0 3.2	+ 2.0 - 0	+ 0 -1.2	0 5.0	+ 3 - 0	+ 0 - 2	0 8	+ 5 - 0	+ 0 - 3	0 16	+ 8 - 0	+ 0 - 8	0.9 5.9	+ 3.0 - 0	- 0.9 - 2.9
30.09-	41.49	0 4.1	+ 2.5 - 0	+ 0 -1.6	0 6.5	+ 4 - 0	+ 0 - 2.5	0 10	+ 6 - 0	+ 0 - 4	0 20	+10 - 0	+ 0 -10	1.0 7.5	+ 4.0 - 0	- 1.0 - 3.5
41.49-	56.19	0 5.0	+ 3.0 - 0	+ 0 -2.0	0 8.0	+ 5 - 0	+ 0 - 3	0 13	+ 8 - 0	+ 0 - 5	0 24	+12 - 0	+ 0 -12	1.2 9.2	+ 5.0 - 0	- 1.2 - 4.2
56.19-	76.39	0 6.5	+ 4.0 - 0	+ 0 -2.5	0 10	+ 6 - 0	+ 0 - 4	0 16	+10 - 0	+ 0 - 6	0 32	+16 - 0	+ 0 -16	1.2 11.2	+ 6.0 - 0	- 1.2 - 5.2
76.39-	100.9	0 8.0	+ 5.0 - 0	+ 0 -3.0	0 13	+ 8 - 0	+ 0 - 5	0 20	+12 - 0	+ 0 - 8	0 40	+20 - 0	+ 0 -20	1.4 14.4	+ 8.0 - 0	- 1.4 - 6.4
100.9 -	131.9	0 10.0	+ 6.0 - 0	+ 0 -4.0	0 16	+10 - 0	+ 0 - 6	0 26	+16 - 0	+ 0 -10	0 50	+25 - 0	+ 0 -25	1.6 17.6	+10.0 - 0	- 1.6 - 7.6
131.9 -	171.9	0 13.0	+ 8.0 - 0	+ 0 -5.0	0 20	+12 - 0	+ 0 - 8	0 32	+20 - 0	+ 0 -12	0 60	+30 - 0	+ 0 -30	1.8 21.8	+12.0 - 0	- 1.8 - 9.8
171.9 -	200	0 16.0	+10.0 - 0	+ 0 -6.0	0 26	+16 - 0	+ 0 -10	0 41	+25 - 0	+ 0 -16	0 80	+40 - 0	+ 0 -40	1.8 27.8	+16.0 - 0	- 1.8 -11.8

TABLE 4. CLEARANCE LOCATIONAL FITS (*Continued*)

Nominal Size Range, In.		Class LC 6			Class LC 7			Class LC 8			Class LC 9			Class LC 10			Class LC 11		
		Limits of Clearance	Standard Limits		Limits of Clearance	Standard Limits		Limits of Clearance	Standard Limits		Limits of Clearance	Standard Limits		Limits of Clearance	Standard Limits		Limits of Clearance	Standard Limits	
Over	To		Hole	Shaft		Hole	Shaft		Hole	Shaft		Hole	Shaft		Hole	Shaft		Hole	Shaft
0.04–	0.12	0.3	+ 0.6	− 0.3	0.6	+ 1.0	− 0.6	1.0	+ 1.6	− 1.0	2.5	+ 2.5	− 2.5	4	+ 4	− 4	5	+ 6	− 5
		1.5	− 0	− 0.9	2.6	− 0	− 1.6	2.0	− 0	− 2.0	7.5	− 0	− 5.0	12	− 0	− 8	17	− 0	− 11
0.12–	0.24	0.4	+ 0.7	− 0.4	0.8	+ 1.2	− 0.8	1.2	+ 1.8	− 1.2	2.8	+ 3.0	− 2.8	4.5	+ 5	− 4.5	6	+ 7	− 6
		1.8	− 0	− 1.1	3.2	− 0	− 2.0	4.2	− 0	− 2.4	8.8	− 0	− 5.8	14.5	− 0	− 9.5	20	− 0	− 13
0.24–	0.40	0.5	+ 0.9	− 0.5	1.0	+ 1.4	− 1.0	1.6	+ 2.2	− 1.6	3.0	+ 3.5	− 3.0	5	+ 6	− 5	7	+ 9	− 7
		2.3	− 0	− 1.4	3.8	− 0	− 2.4	5.2	− 0	− 3.0	10.0	− 0	− 6.5	17	− 0	− 11	25	− 0	− 16
0.40–	0.71	0.6	+ 1.0	− 0.6	1.2	+ 1.6	− 1.2	2.0	+ 2.8	− 2.0	3.5	+ 4.0	− 3.5	6	+ 7	− 6	8	+ 10	− 8
		2.6	− 0	− 1.6	4.4	− 0	− 2.8	6.4	− 0	− 3.6	11.5	− 0	− 7.5	20	− 0	− 13	28	− 0	− 18
0.71–	1.19	0.8	+ 1.2	− 0.8	1.6	+ 2.0	− 1.6	2.5	+ 3.5	− 2.5	4.5	+ 5.0	− 4.5	7	+ 8	− 7	10	+ 12	− 10
		3.2	− 0	− 2.0	5.6	− 0	− 3.6	8.0	− 0	− 4.5	14.5	− 0	− 9.5	23	− 0	− 15	34	− 0	− 22
1.19–	1.97	1.0	+ 1.6	− 1.0	2.0	+ 2.5	− 2.0	3.6	+ 4.0	− 3.0	5	+ 6	− 5	8	+ 10	− 8	12	+ 16	− 12
		4.2	− 0	− 2.6	7.0	− 0	− 4.5	9.5	− 0	− 5.5	17	− 0	− 11	28	− 0	− 18	44	− 0	− 28
1.97–	3.15	1.2	+ 1.8	− 1.0	2.5	+ 3.0	− 2.5	4.0	+ 4.5	− 4.0	6	+ 7	− 6	10	+ 12	− 10	14	+ 18	− 14
		4.8	− 0	− 3.0	8.5	− 0	− 5.5	11.5	− 0	− 7.0	20	− 0	− 13	34	− 0	− 22	50	− 0	− 32
3.15–	4.73	1.4	+ 2.2	− 1.4	3.0	+ 3.5	− 3.0	5.0	+ 5.0	− 5.0	7	+ 9	− 7	11	+ 14	− 11	16	+ 22	− 16
		5.8	− 0	− 3.6	10.0	− 0	− 6.5	13.5	− 0	− 8.5	25	− 0	− 16	39	− 0	− 25	60	− 0	− 38
4.73–	7.09	1.6	+ 2.5	− 1.6	3.5	+ 4.0	− 3.5	6	+ 6	− 6	8	+ 10	− 8	12	+ 16	− 12	18	+ 25	− 18
		6.6	− 0	− 4.1	11.5	− 0	− 7.5	16	− 0	− 10	28	− 0	− 18	44	− 0	− 28	68	− 0	− 43
7.09–	9.85	2.0	+ 2.8	− 2.0	4.0	+ 4.5	− 4.0	7	+ 7	− 7	10	+ 12	− 10	16	+ 18	− 16	22	+ 28	− 22
		7.6	− 0	− 4.8	13.0	− 0	− 8.5	18.5	− 0	− 11.5	34	− 0	− 22	52	− 0	− 34	78	− 0	− 50
9.85–	12.41	2.2	+ 3.0	− 2.2	4.5	+ 5.0	− 4.5	7	+ 8	− 7	12	+ 12	− 12	20	+ 20	− 20	28	+ 30	− 28
		8.2	− 0	− 5.2	14.5	− 0	− 9.5	20	− 0	− 12	36	− 0	− 24	60	− 0	− 40	88	− 0	− 58
12.41–	15.75	2.5	+ 3.5	− 2.5	5	+ 6	− 5	8	+ 9	− 8	14	+ 14	− 14	22	+ 22	− 22	30	+ 35	− 30
		9.5	− 0	− 6.0	17	− 0	− 11	23	− 0	− 14	42	− 0	− 28	66	− 0	− 44	100	− 0	− 65
15.75–	19.69	2.8	+ 4.0	− 2.8	5	+ 6	− 5	9	+ 10	− 9	16	+ 16	− 16	25	+ 25	− 25	35	+ 40	− 35
		10.8	− 0	− 6.8	17	− 0	− 11	25	− 0	− 15	48	− 0	− 32	75	− 0	− 50	115	− 0	− 75
19.69–	30.09	3.0	+ 5.0	− 3.0	6	+ 8	− 6	10	+ 12	− 10	18	+ 20	− 18	28	+ 30	− 28	40	+ 50	− 40
		13.0	− 0	− 8.0	22	− 0	− 14	20	− 0	− 18	58	− 0	− 38	88	− 0	− 58	140	− 0	− 90
30.09–	41.49	3.5	+ 6.0	− 3.5	7	+ 10	− 7	12	+ 16	− 12	20	+ 25	− 20	30	+ 40	− 30	45	+ 60	− 45
		15.5	− 0	− 9.5	27	− 0	− 17	38	− 0	− 22	70	− 0	− 45	110	− 0	− 70	165	− 0	− 105

Nominal Size Range, In.		Class LC 6			Class LC 7			Class LC 8			Class LC 9			Class LC 10			Class LC 11		
		Limits of Clearance	Standard Limits		Limits of Clearance	Standard Limits		Limits of Clearance	Standard Limits		Limits of Clearance	Standard Limits		Limits of Clearance	Standard Limits		Limits of Clearance	Standard Limits	
Over	To		Hole	Shaft		Hole	Shaft		Hole	Shaft		Hole	Shaft		Hole	Shaft		Hole	Shaft
41.49	56.19	4.0	+ 8.0	− 4.0	8	+12	− 8	14	+20	−14	25	+ 30	− 25	40	+ 50	− 40	60	+ 80	− 60
		20.0	− 0	−12.0	32	− 0	−20	46	− 0	−26	85	− 0	− 55	140	− 0	− 90	220	− 0	−140
56.19	76.39	4.5	+10.0	− 4.5	9	+16	− 9	16	+25	−16	30	+ 40	− 30	50	+ 60	− 50	70	+100	− 70
		24.5	− 0	−14.5	41	− 0	−25	57	− 0	−32	110	− 0	− 70	170	− 0	−110	270	− 0	−170
76.39	100.9	5	+12	− 5	10	+20	−10	18	+30	−18	35	+ 50	− 35	50	+ 80	− 50	80	+125	− 80
		29	− 0	−17	50	− 0	−30	68	− 0	−38	135	− 0	− 85	210	− 0	−130	330	− 0	−205
100.9	131.9	6	+16	− 6	12	+25	−12	20	+40	−20	40	+ 60	− 40	60	+100	− 60	90	+160	− 90
		38	− 0	−22	62	− 0	−27	85	− 0	−45	160	− 0	−100	260	− 0	−160	410	− 0	−250
131.9	171.9	7	+20	− 7	14	+30	−14	25	+50	−25	50	+ 80	− 50	80	+125	− 80	100	+200	−100
		47	− 0	−27	74	− 0	−44	105	− 0	−55	210	− 0	−130	330	− 0	−205	500	− 0	−300
171.9	200	7	+25	− 7	14	+40	−14	25	+60	−25	50	+100	− 50	90	+160	− 90	125	+250	−125
		57	− 0	−32	94	− 0	−54	125	− 0	−65	250	− 0	−150	410	− 0	−250	625	− 0	−375

Limits are in 0.001 in.
Limits for hole and shaft are applied algebraically to the basic size to obtain the limits of size for the parts.

Basic size
2
1
0
−1
−2
LT1
LT2
LT3
LT4
LT6
LT7
Holes
Shafts
Scale: 0.001 in. for diameter of 1 in.

Table 5. Transition Locational Fits

Limits are in 0.001 in.

Limits for hole and shaft are applied algebraically to the basic size to obtain the limits of size for mating parts.

Data in boldface accord with ABC agreements.

"Fit" represents the maximum interference (minus values) and the maximum clearance (plus values).

Nominal Size Range, In.	Class LT 1			Class LT 2			Class LT 3			Class LT 4			Class LT 6			Class LT 7		
	Fit	Standard Limits		Fit	Standard Limits		Fit	Standard Limits		Fit	Standard Limits		Fit	Standard Limits		Fit	Standard Limits	
Over To		Hole	Shaft		Hole	Shaft		Hole	Shaft		Hole	Shaft		Hole	Shaft		Hole	Shaft
0.04– 0.12	−0.15 +0.5	+0.4 −0	+0.15 −0.1	−0.3 +0.7	+0.6 −0	+0.3 −0.1							−0.55 +0.45	+0.6 −0	+0.55 +0.15	−0.5 +0.15	+0.4 −0	+0.5 +0.25
0.12– 0.24	−0.2 +0.6	+0.5 −0	+0.2 −0.1	−0.4 +0.8	+0.7 −0	+0.4 −0.1							−0.7 +0.5	+0.7 −0	+0.7 +0.2	−0.6 +0.2	+0.5 −0	+0.6 +0.3
0.24– 0.40	−0.3 +0.7	+0.6 −0	+0.3 −0.1	−0.4 +1.1	+0.9 −0	+0.4 −0.2	−0.5 +0.5	+0.6 −0	+0.5 +0.1	−0.7 +0.8	+0.9 −0	+0.7 +0.1	−0.8 +0.7	+0.9 −0	+0.8 +0.2	−0.8 +0.2	+0.6 −0	+0.8 +0.4
0.40– 0.71	−0.3 +0.8	+0.7 −0	+0.3 −0.1	−0.5 +1.2	+1.0 −0	+0.5 −0.2	−0.5 +0.6	+0.7 −0	+0.5 +0.1	−0.8 +0.9	+1.0 −0	+0.8 +0.1	−1.0 +0.7	+1.0 −0	+1.0 +0.3	−0.9 +0.2	+0.7 −0	+0.9 +0.5
0.71– 1.19	−0.3 +1.0	+0.8 −0	+0.3 −0.2	−0.5 +1.5	+1.2 −0	+0.5 −0.3	−0.6 +0.7	+0.8 −0	+0.6 +0.1	−0.9 +1.1	+1.2 −0	+0.9 +0.1	−1.1 +0.9	+1.2 −0	+1.1 +0.3	−1.1 +0.2	+0.8 −0	+1.1 +0.6
1.19– 1.97	−0.4 +1.2	+1.0 −0	+0.4 −0.2	−0.6 +2.0	+1.6 −0	+0.6 −0.4	−0.7 +0.9	+1.0 −0	+0.7 +0.1	−1.1 +1.5	+1.6 −0	+1.1 +0.1	−1.4 +1.2	+1.6 −0	+1.4 +0.4	−1.3 +0.3	+1.0 −0	+1.3 +0.7
1.97– 3.15	−0.4 +1.5	+1.2 −0	+0.4 −0.3	−0.7 +2.3	+1.8 −0	+0.7 −0.5	−0.8 +1.1	+1.2 −0	+0.8 +0.1	−1.3 +1.7	+1.8 −0	+1.3 +0.1	−1.7 +1.3	+1.8 −0	+1.7 +0.5	−1.5 +0.4	+1.2 −0	+1.5 +0.8
3.15– 4.73	−0.5 +1.8	+1.4 −0	+0.5 −0.4	−0.8 +2.8	+2.2 −0	+0.8 −0.6	−1.0 +1.3	+1.4 −0	+1.0 +0.1	−1.5 +2.1	+2.2 −0	+1.5 +0.1	−1.9 +1.7	+2.2 −0	+1.9 +0.5	−1.9 +0.4	+1.4 −0	+1.9 +1.0
4.73– 7.09	−0.6 +2.0	+1.6 −0	+0.6 −0.4	−0.9 +3.2	+2.5 −0	+0.9 −0.7	−1.1 +1.5	+1.6 −0	+1.1 +0.1	−1.7 +2.4	+2.5 −0	+1.7 +0.1	+2.2 +1.9	+2.5 −0	+2.2 +0.6	−2.2 +0.4	+1.6 −0	+2.2 +1.2
7.09– 9.85	−0.7 +2.3	+1.8 −0	+0.7 −0.5	−1.0 +3.6	+2.8 −0	+1.0 −0.8	−1.4 +1.6	+1.8 −0	+1.4 +0.2	−2.0 +2.6	+2.8 −0	+2.0 +0.2	−2.4 +2.2	+2.8 −0	+2.4 +0.6	−2.6 +0.4	+1.8 −0	+2.6 +1.4
9.85–12.41	−0.7 +2.6	+2.0 −0	+0.7 −0.6	−1.0 +4.0	+3.0 −0	+1.0 −1.0	−1.4 +1.8	+2.0 −0	+1.4 +0.2	−2.2 +2.8	+3.0 −0	+2.2 +0.2	−2.8 +2.2	+3.0 −0	+2.8 +0.8	−2.6 +0.6	+2.0 −0	+2.6 +1.4
12.41–15.75	−0.7 +2.9	+2.2 −0	+0.7 −0.7	−1.0 +4.7	+3.5 −0	+1.2 −1.0	−1.6 +2.0	+2.2 −0	+1.6 +0.2	−2.4 +3.3	+3.5 −0	+2.4 +0.2	−3.0 +2.7	+3.5 −0	+3.0 +0.8	−3.0 +0.6	+2.2 −0	+3.0 +1.6
15.75–19.69	−0.8 +3.2	+2.5 −0	+0.8 −0.7	−1.3 +5.2	+4.0 −0	+1.3 −1.2	−1.8 +2.3	+2.5 −0	+1.8 +0.2	−2.7 +3.8	+4.0 −0	+2.7 +0.2	−3.4 +3.1	+4.0 −0	+3.4 +0.9	−3.4 +0.7	+2.5 −0	+3.4 +1.8

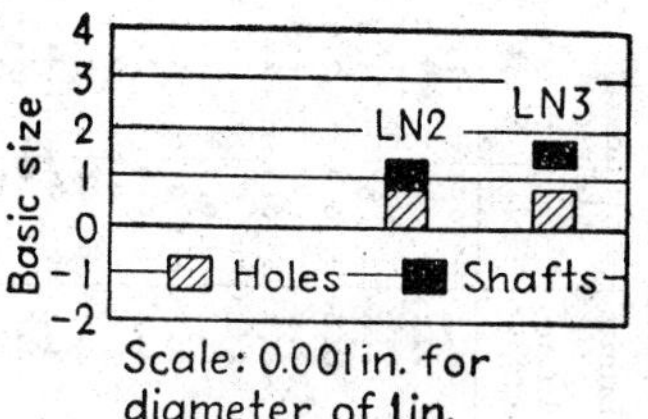

TABLE 6. INTERFERENCE LOCATIONAL FITS

Nominal Size Range, In.		Class LN 2			Class LN 3		
		Limits of Interference	Standard Limits		Limits of Interference	Standard Limits	
Over	To		Hole	Shaft		Hole	Shaft
0.04–	0.12	0 0.65	+ 0.4 − 0	+ 0.65 + 0.4	0.1 0.75	+ 0.4 0	+ 0.75 + 0.5
0.12–	0.24	0 0.8	+ 0.5 − 0	+ 0.8 + 0.5	0.1 0.9	+ 0.5 0	+ 0.9 + 0.6
0.24–	0.40	0 1.0	+ 0.6 − 0	+ 1.0 + 0.6	0.2 1.2	+ 0.6 − 0	+ 1.2 + 0.8
0.40–	0.71	0 1.1	+ 0.7 − 0	+ 1.1 + 0.7	0.3 1.4	+ 0.7 − 0	+ 1.4 + 1.0
0.71–	1.19	0 1.3	+ 0.8 − 0	+ 1.3 + 0.8	0.4 1.7	+ 0.8 − 0	+ 1.7 + 1.2
1.19–	1.97	0 1.6	+ 1.0 − 0	+ 1.6 + 1.0	0.4 2.0	+ 1.0 − 0	+ 2.0 + 1.4
1.97–	3.15	0.2 2.1	+ 1.2 − 0	+ 2.1 + 1.4	0.4 2.3	+ 1.2 − 0	+ 2.3 + 1.6
3.15–	4.73	0.2 2.5	+ 1.4 − 0	+ 2.5 + 1.6	0.6 2.9	+ 1.4 − 0	+ 2.9 + 2.0
4.73–	7.09	0.2 2.8	+ 1.6 − 0	+ 2.8 + 1.8	0.9 3.5	+ 1.6 − 0	+ 3.5 + 2.5
7.09–	9.85	0.2 3.2	+ 1.8 − 0	+ 3.2 + 2.0	1.2 4.2	+ 1.8 − 0	+ 4.2 + 3.0
9.85–	12.41	0.2 3.4	+ 2.0 − 0	+ 3.4 + 2.2	1.5 4.7	+ 2.0 − 0	+ 4.7 + 3.5
12.41–	15.75	0.3 3.9	+ 2.2 − 0	+ 3.9 + 2.5	2.3 5.9	+ 2.2 − 0	+ 5.9 + 4.5
15.75–	19.69	0.3 4.4	+ 2.5 − 0	+ 4.4 + 2.8	2.5 6.6	+ 2.5 − 0	+ 6.6 + 5.0
19.69–	30.09	0.5 5.5	+ 3 − 0	+ 5.5 + 3.5	4 9	+ 3 − 0	+ 9 + 7
30.09–	41.49	0.5 7.0	+ 4 − 0	+ 7.0 + 4.5	5 11.5	+ 4 − 0	+11.5 + 9
41.49–	56.19	1 9	+ 5 − 0	+ 9 + 6	7 15	+ 5 − 0	+15 +12
56.19–	76.39	1 11	+ 6 − 0	+11 + 7	10 20	+ 6 − 0	+20 +16
76.39–	100.9	1 14	+ 8 − 0	+14 + 9	12 25	+ 8 − 0	+25 +20
100.9 –	131.9	2 18	+10 − 0	+18 +12	15 31	+10 − 0	+31 +25
131.9 –	171.9	4 24	+12 − 0	+24 +16	18 38	+12 − 0	+38 +30
171.9 –	200	4 30	+16 − 0	+30 +20	24 50	+16 − 0	+50 +40

Limits are in 0.001 in.

Limits for hole and shaft are applied algebraically to the basic size to obtain the limits of size for the parts.

Data in boldface are in accordance with ABC agreements.

TABLE 7. FORCE AND SHRINK FITS

Limits are in 0.001 in.

Limits for hole and shaft are applied algebraically to the basic size to obtain limits of size for the parts.

Data in boldface accord with ABC agreements.

FN1 FN2 FN3 FN4 FN5

Basic size 3 2 1 0 −1 −2

Holes Shafts

Scale: 0.001 in. for diameter of 1in.

Nominal Size Range, In.		Class FN 1			Class FN 2			Class FN 3			Class FN 4			Class FN 5		
Over	To	Limits of Interference	Standard Limits Hole	Standard Limits Shaft	Limits of Interference	Standard Limits Hole	Standard Limits Shaft	Limits of Interference	Standard Limits Hole	Standard Limits Shaft	Limits of Interference	Standard Limits Hole	Standard Limits Shaft	Limits of Interference	Standard Limits Hole	Standard Limits Shaft
0.04–	0.12	0.05	+ 0.25	+ 0.5	0.2	+ 0.4	+ 0.85				0.3	+ 0.4	+ 0.95	0.5	+ 0.4	+ 1.3
		0.5	− 0	+ 0.3	0.85	− 0	+ 0.6				0.95	− 0	+ 0.7	1.3	− 0	+ 0.9
0.12–	0.24	0.1	+ 0.3	+ 0.6	0.2	+ 0.5	+ 1.0				0.4	+ 0.5	+ 1.2	0.7	+ 0.5	+ 1.7
		0.6	− 0	+ 0.4	1.0	− 0	+ 0.7				1.2	− 0	+ 0.9	1.7	− 0	+ 1.2
0.24–	0.40	0.1	+ 0.4	+ 0.75	0.4	+ 0.6	+ 1.4				0.6	+ 0.6	+ 1.6	0.8	+ 0.6	+ 2.0
		0.75	− 0	+ 0.5	1.4	− 0	+ 1.0				1.6	− 0	+ 1.2	2.0	− 0	+ 1.4
0.40–	0.56	0.1	+ 0.4	+ 0.8	0.5	+ 0.7	+ 1.6				0.7	+ 0.7	+ 1.8	0.9	+ 0.7	+ 2.3
		0.8	− 0	+ 0.5	1.6	− 0	+ 1.2				1.8	− 0	+ 1.4	2.3	− 0	+ 1.6
0.56–	0.71	0.2	+ 0.4	+ 0.9	0.5	+ 0.7	+ 1.6				0.7	+ 0.7	+ 1.8	1.1	+ 0.7	+ 2.5
		0.9	− 0	+ 0.6	1.6	− 0	+ 1.2				1.8	− 0	+ 1.4	2.5	− 0	+ 1.8
0.71–	0.95	0.2	+ 0.5	+ 1.1	0.6	+ 0.8	+ 1.9				0.8	+ 0.8	+ 2.1	1.4	+ 0.8	+ 3.0
		1.1	− 0	+ 0.7	1.9	− 0	+ 1.4				2.1	− 0	+ 1.6	3.0	− 0	+ 2.2
0.95–	1.19	0.3	+ 0.5	+ 1.2	0.6	+ 0.8	+ 1.9	0.8	+ 0.8	+ 2.1	1.0	+ 0.8	+ 2.3	1.7	+ 0.8	+ 3.3
		1.2	− 0	+ 0.8	1.9	− 0	+ 1.4	2.1	− 0	+ 1.6	2.3	− 0	+ 1.8	3.3	− 0	+ 2.5
1.19–	1.58	0.3	+ 0.6	+ 1.3	0.8	+ 1.0	+ 2.4	0.8	+ 1.0	+ 2.4	1.5	+ 1.0	+ 3.1	2.0	+ 1.0	+ 4.0
		1.3	− 0	+ 0.9	2.4	− 0	+ 1.8	2.4	− 0	+ 1.8	3.1	− 0	+ 2.5	4.0	− 0	+ 3.0
1.58–	1.97	0.4	+ 0.6	+ 1.4	0.8	+ 1.0	+ 2.4	1.2	+ 1.0	+ 2.8	1.8	+ 1.0	+ 3.4	3.0	+ 1.0	+ 5.0
		1.4	− 0	+ 1.0	2.4	− 0	+ 1.8	2.8	− 0	+ 2.2	3.4	− 0	+ 2.8	5.0	− 0	+ 4.0
1.97–	2.56	0.6	+ 0.7	+ 1.8	0.8	+ 1.2	+ 2.7	1.3	+ 1.2	+ 3.2	2.3	+ 1.2	+ 4.2	3.8	+ 1.2	+ 6.2
		1.8	− 0	+ 1.3	2.7	− 0	+ 2.0	3.2	− 0	+ 2.5	4.2	− 0	+ 3.5	6.2	− 0	+ 5.0
2.56–	3.15	0.7	+ 0.7	+ 1.9	1.0	+ 1.2	+ 2.9	1.8	+ 1.2	+ 3.7	2.8	+ 1.2	+ 4.7	4.8	+ 1.2	+ 7.2
		1.9	− 0	+ 1.4	2.9	− 0	+ 2.2	3.7	− 0	+ 3.0	4.7	− 0	+ 4.0	7.2	− 0	+ 6.0
3.15–	3.94	0.9	+ 0.9	+ 2.4	1.4	+ 1.4	+ 3.7	2.1	+ 1.4	+ 4.4	3.6	+ 1.4	+ 5.9	5.6	+ 1.4	+ 8.4
		2.4	− 0	+ 1.8	3.7	− 0	+ 2.8	4.4	− 0	+ 3.5	5.9	− 0	+ 5.0	8.4	− 0	+ 7.0
3.94–	4.73	1.1	+ 0.9	+ 2.6	1.6	+ 1.4	+ 3.9	2.6	+ 1.4	+ 4.9	4.6	+ 1.4	+ 6.9	6.6	+ 1.4	+ 9.4
		2.6	− 0	+ 2.0	3.9	− 0	+ 3.0	4.9	− 0	+ 4.0	6.9	− 0	+ 6.0	9.4	− 0	+ 8.0
4.73–	5.52	1.2	+ 1.0	+ 2.9	1.9	+ 1.6	+ 4.5	3.4	+ 1.6	+ 6.0	5.4	+ 1.6	+ 8.0	8.4	+ 1.6	+ 11.6
		2.9	− 0	+ 2.2	4.5	− 0	+ 3.5	6.0	− 0	+ 5.0	8.0	− 0	+ 7.0	11.6	− 0	+ 10.0
5.52–	6.30	1.5	+ 1.0	+ 3.2	2.4	+ 1.6	+ 5.0	3.4	+ 1.6	+ 6.0	5.4	+ 1.6	+ 8.0	10.4	+ 1.6	+ 13.6
		3.2	− 0	+ 2.5	5.0	− 0	+ 4.0	6.0	− 0	+ 5.0	8.0	− 0	+ 7.0	13.6	− 0	+ 12.0

6.30– 7.09	1.8	+1.0	+3.5	2.9	+1.6	+5.5	4.4	+1.6	+7.0	6.4	+1.6	+9.0	10.4	+1.6	+13.6
	3.5	−0	+2.8	5.5	−0	+4.5	7.0	−0	+6.0	9.0	−0	+8.0	13.6	−0	+12.0
7.09– 7.88	1.8	+1.2	+3.8	3.2	+1.8	+6.2	5.2	+1.8	+8.2	7.2	+1.8	+10.2	12.2	+1.8	+15.8
	3.8	−0	+3.0	6.2	−0	+5.0	8.2	−0	+7.0	10.2	−0	+9.0	15.8	−0	+14.0
7.88– 8.86	2.3	+1.2	+4.3	3.2	+1.8	+6.2	5.2	+1.8	+8.2	8.2	+1.8	+11.2	14.2	+1.8	+17.8
	4.3	−0	+3.5	6.2	−0	+5.0	8.2	−0	+7.0	11.2	−0	+10.0	17.8	−0	+16.0
8.86– 9.85	2.3	+1.2	+4.3	4.2	+1.8	+7.2	6.2	+1.8	+9.2	10.2	+1.8	+13.2	14.2	+1.8	+17.8
	4.3	−0	+3.5	7.2	−0	+6.0	9.2	−0	+8.0	13.2	−0	+12.0	17.8	−0	+16.0
9.85– 11.03	2.8	+1.2	+4.9	4.0	+2.0	+7.2	7.0	+2.0	+10.2	10.0	+2.0	+13.2	16.0	+2.0	+20.0
	4.9	−0	+4.0	7.2	−0	+6.0	10.2	−0	+9.0	13.2	−0	+12.0	20.0	−0	+18.0
11.03– 12.41	2.8	+1.2	+4.9	5.0	+2.0	+8.2	7.0	+2.0	+10.2	12.0	+2.0	+15.2	18.0	+2.0	+22.0
	4.9	−0	+4.0	8.2	−0	+7.0	10.2	−0	+9.0	15.2	−0	+14.0	22.0	−0	+20.0
12.41– 13.98	3.1	+1.4	+5.5	5.8	+2.2	+9.4	7.8	+2.2	+11.4	13.8	+2.2	+17.4	19.8	+2.2	+24.2
	5.5	−0	+4.5	9.4	−0	+8.0	11.4	−0	+10.0	17.4	−0	+16.0	24.2	+0	+22.0
13.98– 15.75	3.6	+1.4	+6.1	5.8	+2.2	+9.4	9.8	+2.2	+13.4	15.8	+2.2	+19.4	22.8	+2.2	+27.2
	6.1	−0	+5.0	9.4	−0	+8.0	13.4	−0	+12.0	19.4	−0	+18.0	27.2	−0	+25.0
15.75– 17.72	4.4	+1.6	+7.0	6.5	+2.5	+10.6	9.5	+2.5	+13.6	17.5	+2.5	+21.6	25.5	+2.5	+30.5
	7.0	−0	+6.0	10.6	−0	+9.0	13.6	−0	+12.0	21.6	−0	+20.0	30.5	−0	+28.0
17.72– 19.69	4.4	+1.6	+7.0	7.5	+2.5	+11.6	11.5	+2.5	+15.6	19.5	+2.5	+23.6	27.5	+2.5	+32.5
	7.0	−0	+6.0	11.6	−0	+10.0	15.6	−0	+14.0	23.6	−0	+22.0	32.5	−0	+30.0
19.69– 24.34	6.0	+2.0	+9.2	9.0	+3.0	+14.0	15.0	+3.0	+20.0	22.0	+3.0	+27.0	32.0	+3.0	+38.0
	9.2	−0	+8.0	14.0	−0	+12.0	20.0	−0	+18.0	27.0	−0	+25.0	38.0	−0	+35.0
24.34– 30.09	7.0	+2.0	+10.2	11.0	+3.0	+16.0	17.0	+3.0	+22.0	27.0	+3.0	+32.0	37.0	+3.0	+43.0
	10.2	−0	+9.0	16.0	−0	+14.0	22.0	−0	+20.0	32.0	−0	+30.0	43.0	−0	+40.0
30.09– 35.47	7.5	+2.5	+11.6	14.0	+4.0	+20.5	21.0	+4.0	+27.5	31.0	+4.0	+37.5	46.0	+4.0	+54.0
	11.6	−0	+10.0	20.5	−0	+18.0	27.5	−0	+25.0	37.5	−0	+35.0	54.0	−0	+50.0
35.47– 41.49	9.5	+2.5	+13.6	16.0	+4.0	+22.5	24.0	+4.0	+30.5	36.0	+4.0	+43.5	56.0	+4.0	+64.0
	13.6	−0	+12.0	22.5	−0	+20.0	30.5	−0	+28.0	46.5	−0	+40.0	64.0	−0	+60.0
41.49– 48.28	11.0	+3.0	+16.0	17.0	+5.0	+25.0	30.0	+5.0	+38.0	45.0	+5.0	+53.0	65.0	+5.0	+75.0
	16.0	−0	+14.0	25.0	−0	+22.0	38.0	−0	+35.0	53.0	−0	+50.0	75.0	−0	+70.0
48.28– 56.19	13.0	+3.0	+18.0	20.0	+5.0	+28.0	35.0	+5.0	+43.0	55.0	+5.0	+63.0	75.0	+5.0	+85.0
	18.0	−0	+16.0	28.0	−0	+25.0	43.0	−0	+40.0	63.0	−0	+60.0	85.0	−0	+80.0
56.19– 65.54	14.0	+4.0	+20.5	24.0	+6.0	+34.0	39.0	+6.0	+49.0	64.0	+6.0	+74.0	94.0	+6.0	+106
	20.5	−0	+18.0	34.0	−0	+30.0	49.0	−0	+45.0	74.0	−0	+70.0	106	−0	+100
65.54– 76.39	18.0	+4.0	+24.5	29.0	+6.0	+39.0	44.0	+6.0	+54.0	74.0	+6.0	+84.0	114	+6.0	+126
	24.5	−0	+22.0	39.0	−0	+35.0	54.0	−0	+50.0	84.0	−0	+80.0	126	−0	+120
76.39– 87.79	20.0	+5.0	+28.0	32.0	+8.0	+45.0	52.0	+8.0	+65.0	82.0	+8.0	+95.0	132	+8.0	+148
	28.0	−0	+25.0	45.0	−0	+40.0	65.0	−0	+60.0	95.0	−0	+90.0	148	−0	+140
87.79–100.9	23.0	+5.0	+31.0	37.0	+8.0	+50.0	62.0	+8.0	+75.0	92.0	+8.0	+105	152	+8.0	+168
	31.0	−0	+28.0	50.0	−0	+45.0	75.0	−0	+70.0	105	−0	+100	168	−0	+160
100.9 –115.3	24.0	+6.0	+34.0	40.0	+10.0	+56.0	70.0	+10.0	+86.0	110	+10.0	+126	170	+10.0	+190
	34.0	−0	+30.0	56.0	−0	+50.0	86.0	−0	+80.0	126	−0	+120	190	−0	+180
115.3 –131.9	29.0	+6.0	+39.0	50.0	+10.0	+66.0	80.0	+10.0	+96.0	130	+10.0	+146	190	+10.0	+210
	39.0	−0	+35.0	66.0	−0	+60.0	96.0	−0	+90.0	146	−0	+140	210	−0	+200
131.9 –152.2	37.0	+8.0	+50.0	58.0	+12.0	+78.0	88.0	+12.0	+108	148	+12.0	+168	208	+12.0	+232
	50.0	−0	+45.0	78.0	−0	+70.0	108	−0	+100	168	−0	+160	232	−0	+220
152.2 –171.9	42.0	+8.0	+55.0	68.0	+12.0	+88.0	108	+12.0	+128	168	+12.0	+188	238	+12.0	+262
	55.0	−0	+50.0	88.0	−0	+80.0	128	−0	+120	188	−0	+170	262	−0	+250
171.9 –200	50.0	+10.0	+66.0	74.0	+16.0	+100	124	+16.0	+150	184	+16.0	+210	284	+16.0	+316
	66.0	−0	+60.0	100	−0	+90	150	−0	+140	210	−0	+200	316	−0	+300

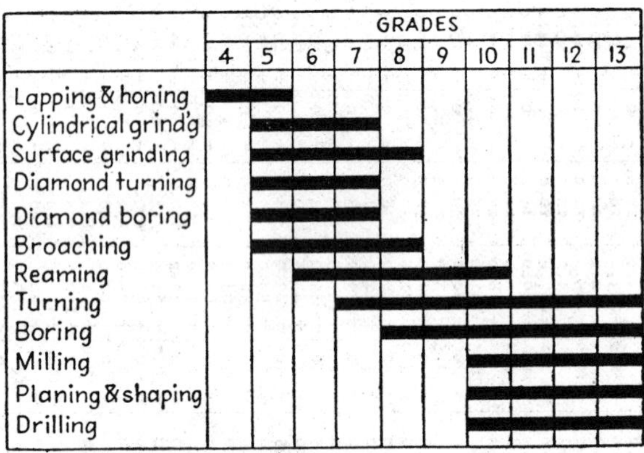

TABLE 8. STANDARD TOLERANCES

Nominal Size Range, In. Over – To	Grade 4	Grade 5	Grade 6	Grade 7	Grade 8	Grade 9	Grade 10	Grade 11	Grade 12	Grade 13
0.04– 0.12	0.00015	0.00020	0.00025	0.0004	0.0006	0.0010	0.0016	0.0025	0.004	0.006
0.12– 0.24	0.00015	0.00020	0.0003	0.0005	0.0007	0.0012	0.0018	0.0030	0.005	0.007
0.24– 0.40	0.00015	0.00025	0.0004	0.0006	0.0009	0.0014	0.0022	0.0035	0.006	0.009
0.40– 0.71	0.0002	0.0003	0.0004	0.0007	0.0010	0.0016	0.0028	0.0040	0.007	0.010
0.71– 1.19	0.00025	0.0004	0.0005	0.0008	0.0012	0.0020	0.0035	0.0050	0.008	0.012
1.19– 1.97	0.0003	0.0004	0.0006	0.0010	0.0016	0.0025	0.0040	0.006	0.010	0.016
1.97– 3.15	0.0003	0.0005	0.0007	0.0012	0.0018	0.0030	0.0045	0.007	0.012	0.018
3.15– 4.73	0.0004	0.0006	0.0009	0.0014	0.0022	0.0035	0.005	0.009	0.014	0.022
4.73– 7.09	0.0005	0.0007	0.0010	0.0016	0.0025	0.0040	0.006	0.010	0.016	0.025
7.09– 9.85	0.0006	0.0008	0.0012	0.0018	0.0028	0.0045	0.007	0.012	0.018	0.028
9.85– 12.41	0.0006	0.0009	0.0012	0.0020	0.0030	0.0050	0.008	0.012	0.020	0.030
12.41– 15.75	0.0007	0.0010	0.0014	0.0022	0.0035	0.006	0.009	0.014	0.022	0.035
15.75– 19.69	0.0008	0.0010	0.0016	0.0025	0.004	0.006	0.010	0.016	0.025	0.040
19.69– 30.09	0.0009	0.0012	0.0020	0.003	0.005	0.008	0.012	0.020	0.030	0.050
30.09– 41.49	0.0010	0.0016	0.0025	0.004	0.006	0.010	0.016	0.025	0.040	0.060
41.49– 56.19	0.0012	0.0020	0.003	0.005	0.008	0.012	0.020	0.030	0.050	0.080
56.19– 76.39	0.0016	0.0025	0.004	0.006	0.010	0.016	0.025	0.040	0.060	0.100
76.39–100.9	0.0020	0.003	0.005	0.008	0.012	0.020	0.030	0.050	0.080	0.125
100.9 –131.9	0.0025	0.004	0.006	0.010	0.016	0.025	0.040	0.060	0.100	0.160
131.9 –171.9	0.003	0.005	0.008	0.012	0.020	0.030	0.050	0.080	0.125	0.200
171.9 –200	0.004	0.006	0.010	0.016	0.025	0.040	0.060	0.100	0.160	0.250

Dimensions in inches.
Data in boldface accord with ABC agreement.

Modified Standard Fits (*Continued from page 33-7*)

Bilateral Hole Fits (*Symbol B*). The common case is where holes are produced with fixed tools, such as drills or reamers; to provide a longer wear life for such tools a bilateral tolerance is desired.

The symbols used for these fits are identical with standard fits except that they are followed by the letter B. Thus LC 4B is a clearance locational fit, class 4, except that it is produced with a bilateral hole.

The limits of clearance or interference are identical with those shown in Tables 3 to 7 for the corresponding fits.

The hole tolerance is changed so that the plus limit is that for one grade finer than the value shown in the tables, the minus limit equals the amount by which the plus limit was lowered, and the shaft limits are both lowered by the same amount as the lower limit of size of the hole. The finer grade of tolerance can be found in Table 8.

Basic Shaft Fits (*Symbol S*). For these fits the maximum size of the shaft is basic. The limits of clearance or interference are identical with those shown in Tables 3 to 7 for the corresponding fits. The symbols used for these fits are identical with those used for standard fits except that they are followed by the letter S. Thus LC 4S is a clearance locational fit, class 4, except that it is produced on a basic shaft basis.

The limits for hole and shaft as given in Tables 3 to 7 are increased for clearance fits, or decreased for transition or interference fits, by the value of the upper shaft limit, that is, by the amount required to change the maximum shaft to the basic size.

Machining Processes. To indicate the machining processes which may normally be expected to produce work within the tolerances indicated by the grades given in this standard, see the illustration with Table 8. This information is intended merely as a guide.

HOW TO MAKE INTERFERENCE FITS

The proposed American Standard for fits gives five classes of interference fits (force and shrink), Table 7, and they are explained on page 33-7. Interference fits are created in three ways: (1) a force or press fit produced by pressing a shaft into a smaller hole; (2) a shrink fit, where the outer member is heated, slipped onto the second member, and the grip is secured by contraction of the outer member on cooling; and (3) the expansion fit, wherein the inner member is cooled, slipped into the outer member, and the grip is secured by expansion of the inner member when the assembly attains room temperature. Further details of these three methods of making interference fits follow.

Force or Press Fits

In making a force or press fit, a number of considerations must be taken into account: (1) the actual allowance, (2) the finish on the two members, and (3) the kind of lubrication provided. The values for coefficient of friction vary with the lubricant, the condition of the surfaces, and the physical properties of the materials being assembled. The variation in pressure due to lubricant may be as great as 1:6, unless care is exercised to hold this condition constant by using a definite lubricant. The effect of lubricant upon pressure to force a 1.001-in. pin into a 1.000 bushing is shown in Fig. 2.

Allowance for Force Fits. The usual allowance is 0.001 to 0.0015 in. interference per inch of diameter. When the hole is surrounded by a heavy hub, and both the hole and shaft are round and smooth, an allowance of 0.0005 in. per in. dia has been used satisfactorily. If the surfaces are ground to a high-quality finish, the parts can be assembled and forced apart several times without appreciable change in the pressing pressure.

It is a mistake to assume that a tighter fit will be secured by increasing the amount of interference. This is true only so long as the elastic limit of the material of the part containing the hole is not exceeded. An allowance of 0.002 in., for example, creates a tensile stress of say 60,000 psi in a steel hub and from 15,000 to 29,000 psi in a cast-iron hub and is probably unsafe in both instances, except when the highest class of material is used.

Formulas for Press Fits. Many investigators have devised formulas for pressures to make press fits, the radial stress in the hub, etc. The formulas developed by

A. Lewis Jenkins (*American Machinist*, Mar. 4, 1915, p. 377) from analyses of many tests and the work of others are:

CAST-IRON HUB ON STEEL SHAFT

$$P' = \frac{100{,}000(D/d + 0.3)JL}{33D/d + 209}$$

Tangential hoop stress in ring, psi:

$$f = \frac{15{,}000{,}000z}{1 + [(10D/d + 3) \div (D/d + 66.6)]}$$

STEEL HUB ON STEEL SHAFT

$$P' = \frac{2060JL[(D/d)^2 - 1]}{(D/d)^2}$$

$$f = \frac{15{,}000{,}000z[(D/d)^2 + 1]}{(D/d)^2}$$

where P' = pressure, tons; D = OD of ring; d = shaft or plug dia; L = length of fit; J = total interference; z = interference per in. dia, in.; w = coefficient of friction = 0.085 for cast iron on steel, 0.0875 for steel on steel.

Torsional resistance of a forced or shrink fit is

$$T = \frac{2000P'rw'}{w} \quad \text{lb-in.}$$

where w' is the coefficient of friction that resists relative rotation and w is the coefficient of friction parallel to the axis. These can be assumed to be in the ratio of 3:1. Hence if $w'/w = 2.5$, the torsional resistance becomes

$$T = \frac{5000P'd}{2} \quad \text{lb-in.}$$

Unit shear stress in the shaft, when the ring is about to slip, is

$$q = \frac{12{,}750P'}{d^2} \quad \text{psi}$$

The charts in Figs. 3 and 4 can be used to solve the Jenkins formulas.

Additional charts (Figs. 5 and 6) have been devised from the formulas on pages 442 and 448 of "Design of Machine Members" by Vallance and Doughtie (McGraw-Hill Book Company, Inc., New York, 1951). These formulas are based on a straight mathematical analysis of the problem, whereas the Jenkins formulas take into account various empirical constants derived from many tests.

The chart for steel-to-steel assemblies (Fig. 5) is based on the equations:

$$p_c = \frac{AE(d_c^2 - d_1^2)(d_o^2 - d_c^2)}{2d_c^3(d_o^2 - d_i^2)}$$

and

$$F = \pi\, dLf\, p_c$$

where A = total allowance, in.; E = modulus of elasticity; d_o = OD of external member, in.; d_c = nominal dia of contact surfaces, in.; d_i = ID of internal member,

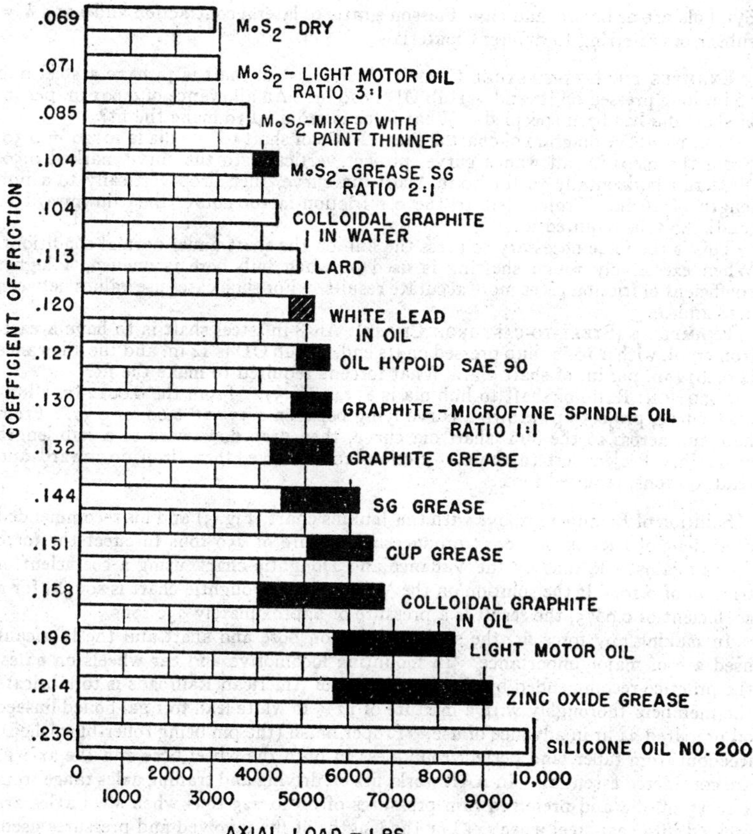

FIG. 2. Effect of lubricant on load required to force a 1.001-in. pin into a 1.000-in. bushing. Black areas represent load range where chatter occurred. MoS_2 is molybdenum sulfide.

in.; p_c = radial pressure between members, psi; F = press-fit force, lb; π = 3.14; d = nominal shaft dia, in.; L = hub length, in.; f = coefficient of friction.

Graphical solution eliminates the necessity of the radial pressure calculation, which is of little interest to the man in the shop. Instead he obtains directly the press-fit force which he must know in order to choose the proper press.

The chart (Fig. 6) for press-fitting a steel shaft to a cast-iron hub is based on these equations:

$$p_c = \frac{A}{d_c}\left[\frac{d_c^2 + d_i^2}{E_s(d_c^2 - d_i^2)} + \frac{d_o^2 + d_c^2}{E_h(d_o^2 - d_c^2)} - \frac{m_s}{E_s} + \frac{m_h}{E_h}\right]$$

and

$$F = \pi\, dLf\, p_c$$

Symbols are as before, and m = Poisson's ratio of lateral contraction and s and h = subscripts referring to different materials.

EXAMPLE 1 (STEEL-TO-STEEL CHART): A 10-in. steel shaft is to have a steel hub 16 in. long pressed on its end. Hub OD is 20 in. An allowance of 0.001 in. per in. of shaft dia has been specified. What force is required to make the fit?

SOLUTION: See diagram of chart usage. Ratio of shaft to hub dia is 10:20 = 0.50. From the 0.001-in. allowance curve, project vertically to the curve marked 0.50. Next run horizontally to the 10-in. shaft dia curve, then drop vertically to a hub length of 16 in. Project left to the 0.1 friction-factor curve, then downward to read: 280 tons required force.

This is the force necessary to press the hub on the shaft under normal conditions. When excessively rough shafting is used or when hub bore is uneven, a higher coefficient of friction gives more accurate results. For such cases use values between 0.18 and 0.25.

EXAMPLE 2 (STEEL-TO-CAST-IRON CHART): An 8-in. steel shaft is to have a cast-iron crank with a 10-in. hub pressed on its end. Hub OD is 14 in. and the allowance is 0.0012 in. per in. of shaft dia. What force is required to make the fit?

SOLUTION: Ratio of shaft to hub dia is 8:14 = 0.57. From the 0.0012-in. allowance curve, project upward to a curve lying between 0.55 and 0.60, or 0.57. From here run across to the 8-in. shaft dia curve, then drop downward to a hub length of 14 in. Project left to the 0.1 friction-factor curve, then drop downward and read: 90 tons, required force.

Solution of Example 1 above with the Jenkins chart (Fig. 3) and his recommended coefficient of friction of 0.0875 produces a pressure of 250 tons to effect the force fit, as against 280 tons by the Vallance and Doughtie chart, using a coefficient of friction of 0.10. If the solution on the Vallance and Doughtie chart is sought for a coefficient of 0.0875, the result is a pressure of approximately 230 tons.

In making any force fit, the surface finish on bore and shaft and the lubricant used are of major importance. In mounting locomotive and car wheels on axles, the practice recommended by the Association of American Railroads is to lubricate the members thoroughly with a mixture of 12½ lb white lead to 1 gal boiled linseed oil prepared 24 hr in advance of use. Proper finish (the pin being roller-burnished), freedom from taper, and perfect roundness in both the wheel bore and the axle fit are considered essential. In AAR work, fits on driving and trailing axles range from 4 to 15 in. dia and preferred ram pressures of 60 to 225 tons when steel axles are pressed into cast-steel wheels. For the lengths of fits involved and pressures used, it appears that the calculated pressures obtained with charts in Figs. 3 to 6 are reasonably consistent with railroad experience.

Shrink Fits

The temperature needed to expand a metal ring a given amount before making a shrink fit is found by the formula

$$T = \frac{N}{KD}$$

where T = temperature rise F, *above* room temperature, needed to expand piece N thousandths; N = number of thousandths of an inch expansion desired, expressed as a decimal; K = linear coefficient of expansion of the metal; D = hole dia, in.

Coefficients of expansion are not uniform for temperatures of say −200 to 1000 F. For accurate results the reader is referred for data to the "International Critical Tables" (McGraw-Hill Book Company, Inc., New York) or equivalent works. These data are far too extensive for reproduction in a shopman's handbook.

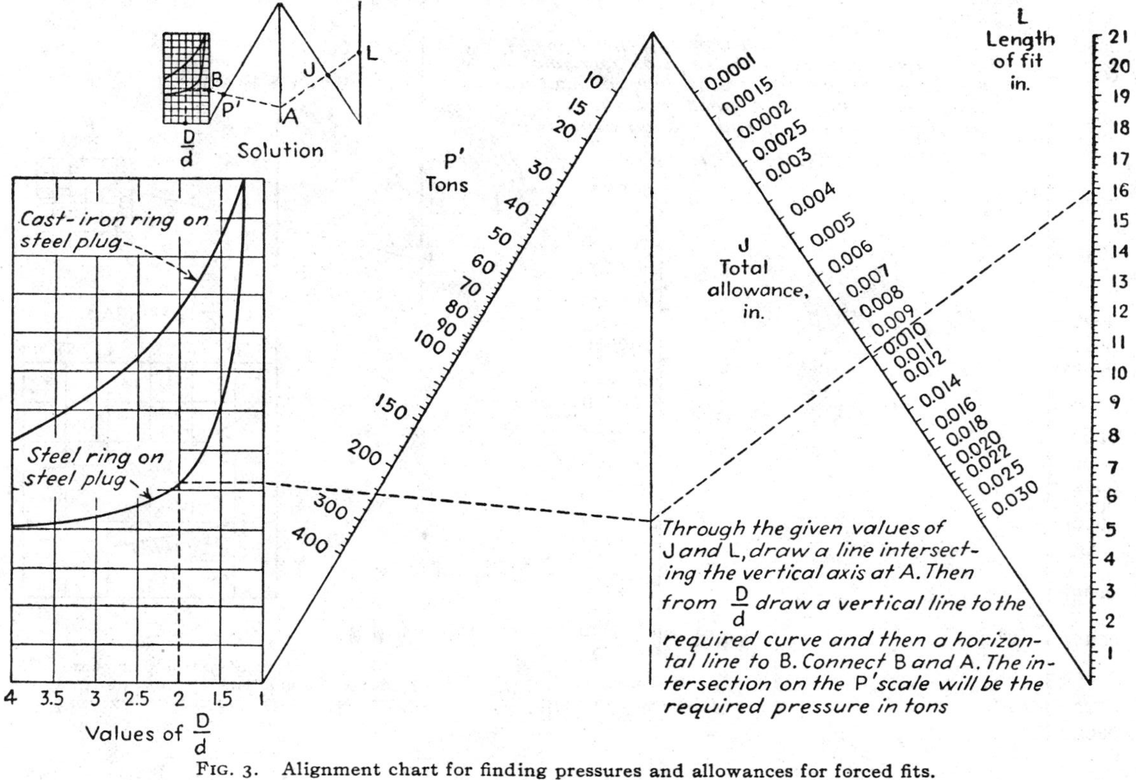

FIG. 3. Alignment chart for finding pressures and allowances for forced fits.

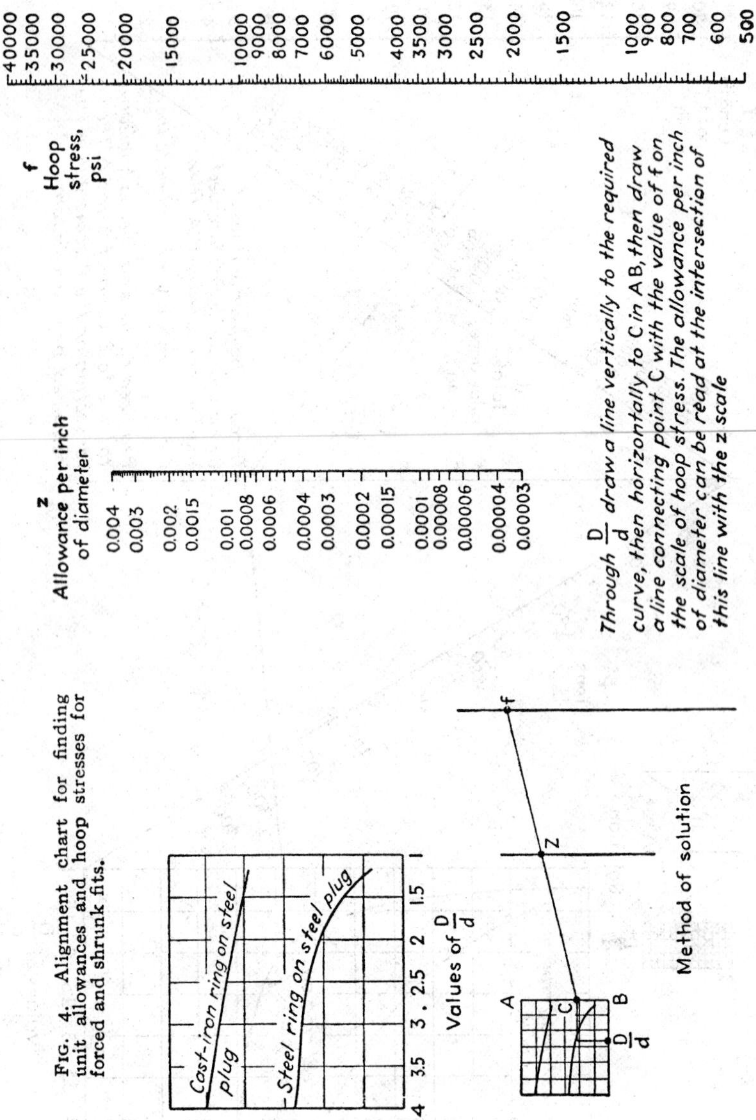
f
Hoop stress, psi
40000 35000 30000 25000 20000 15000 10000 9000 8000 7000 6000 5000 4000 3500 3000 2500 2000 1500 1000 900 800 700 600 500
z
Allowance per inch of diameter
0.004 0.003 0.002 0.0015 0.001 0.0008 0.0006 0.0004 0.0003 0.0002 0.00015 0.0001 0.00008 0.00006 0.00004 0.00003
Cast-iron ring on steel plug
Steel ring on steel plug
4 3.5 3 2.5 2 1.5 1
Values of $\frac{D}{d}$
A
B
C
$\frac{D}{d}$
Z
f
Method of solution
Through $\frac{D}{d}$ draw a line vertically to the required curve, then horizontally to C in AB, then draw a line connecting point C with the value of f on the scale of hoop stress. The allowance per inch of diameter can be read at the intersection of this line with the z scale

FIG. 4. Alignment chart for finding unit allowances and hoop stresses for forced and shrunk fits.

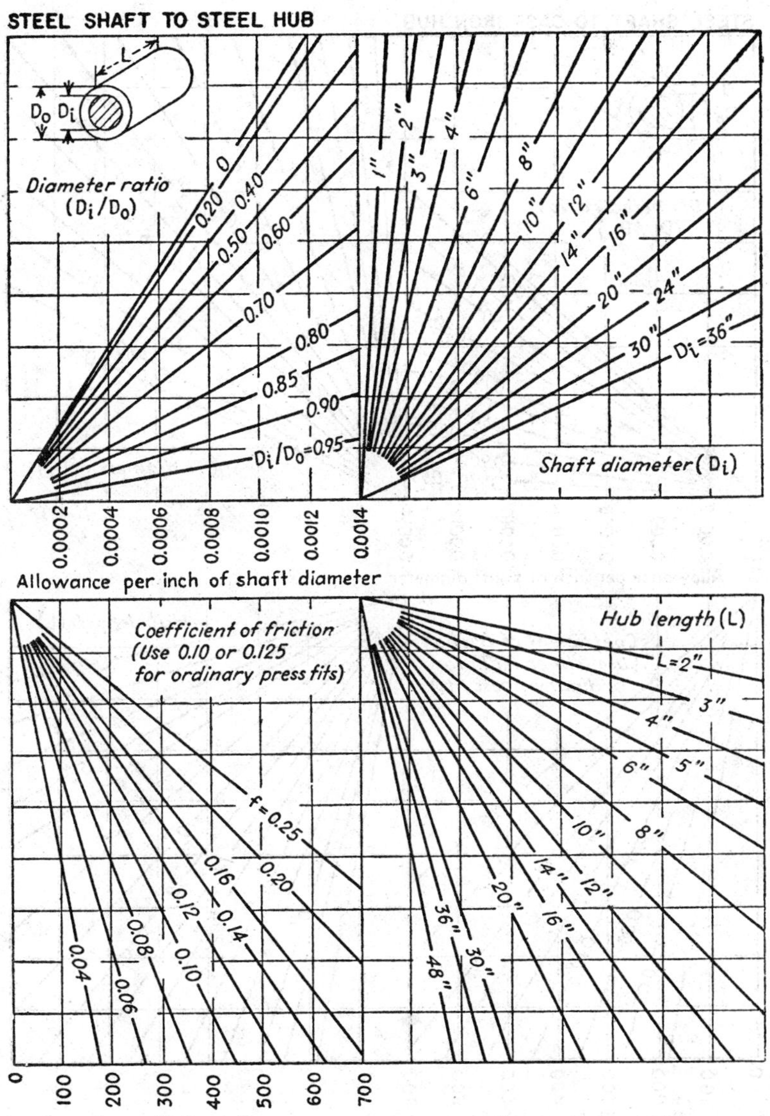

FIG. 5. Chart for determining pressure to press-fit steel shaft to steel hub.

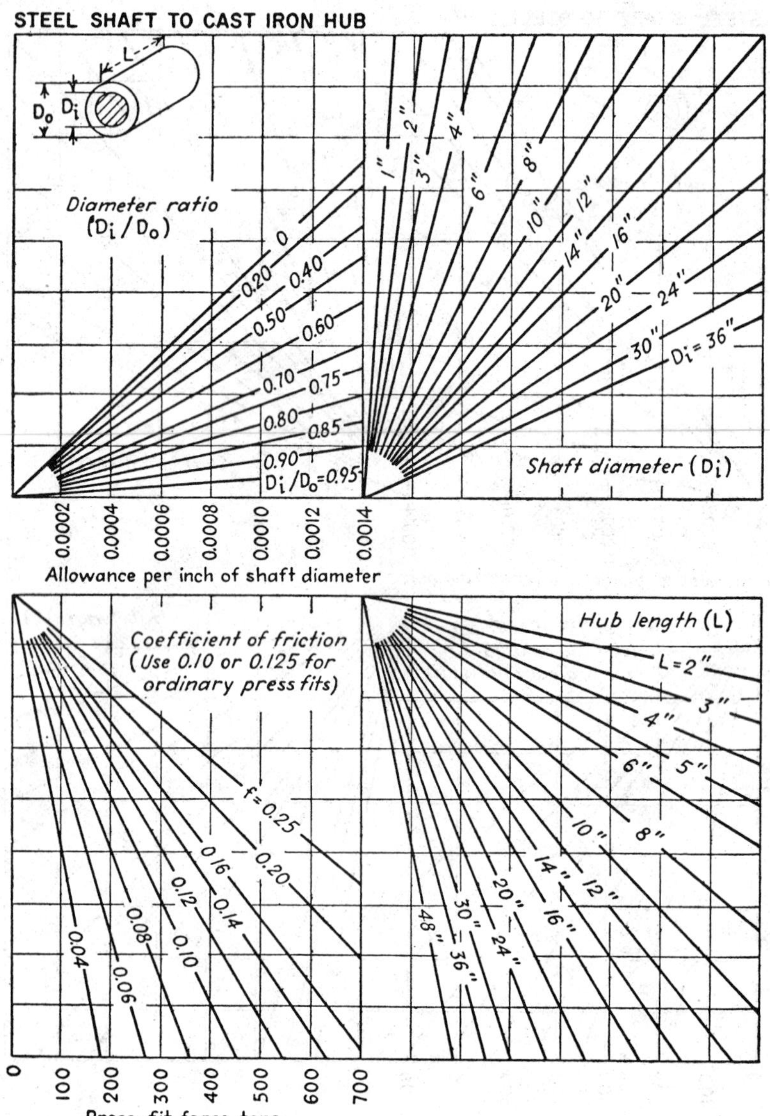

Fig. 6. Chart for determining pressure to press-fit steel shaft to cast-iron hub.

Coefficients of linear expansion for various metals, over the range 68 to 212 F. are:

	Coefficient of Expansion In. per In. per Deg F
Aluminum alloys:	
2 S	0.0000130
24 S	0.0000127
Copper alloys:	
Pure copper	0.0000091
Gilding metal	0.0000101
Red brass	0.0000104
Cartridge brass	0.0000110
Free-cutting brass	0.0000114
Silicon bronze	0.0000100
Silver	0.0000109
Nickel alloys:	
Monel	0.0000077
Inconel	0.0000064
Irons:	
Gray cast iron	0.0000058
Malleable	0.0000066
Magnesium	0.0000143
Steels:	
SAE 1010	0.0000068
SAE 1040	0.0000062
SAE 3140	0.0000065
SAE 3240	0.0000066

Shrinking Steel Tires on Locomotive Wheels. Tires, when heated for application to or removal from wheel centers, should be heated slowly and uniformly around their entire circumference, and the heat applied indirectly so as to avoid the impingement of flames against the surface of the tires. A temperature of 500 to 600 F is sufficient for tires of any size, and in no case should tires be heated hotter than 800 F, which is somewhat below the lowest visible red heat. Before placing tires on wheel centers, the bore must be clean and free from soot, rust, fins, and other obstructions.

Standard shrinkage allowances for steel tires, as revised in 1933 by the Association of American Railroads, are given in Table 9.

Expansion Fits

Expansion fits are made by cooling the inner member, which is larger than the hole, so that it contracts sufficiently to enter the hole. When the assembly warms to room temperature, an interference fit is created. Dry ice (solid carbon dioxide), which gasifies at −110 F, liquid air (−310 F), and liquid nitrogen (−320 F) can be used to cool the work. Alcohol can be used in conjunction with dry ice to speed the rate of heat transfer in the part being shrunk.

Expansion Fits with Liquid Nitrogen. High-purity liquid nitrogen avoids fire danger inherent in liquid air or oxygen, yet gives lower temperatures. It offers top possibilities for expansion fits when the external member is massive or likely to be damaged by heating an equivalent amount. Liquid nitrogen offers 390 F differential from room temperature, which, if obtained by heating (though cheaper), brings temperatures of the external element to 460 F, close to tempering temperature.

Physical Characteristics. Colorless, odorless, nontoxic, relatively inert. As a

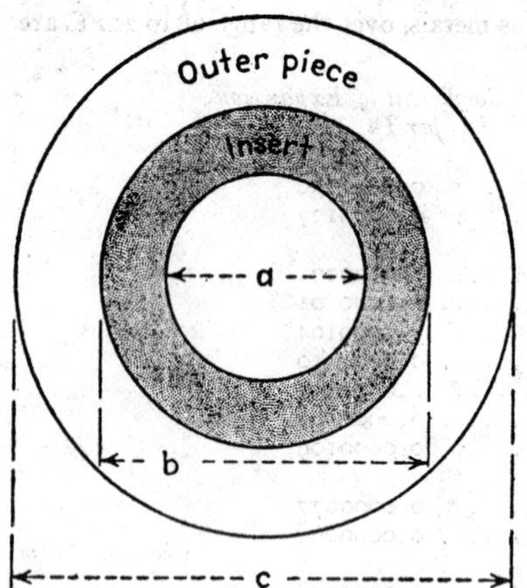

FIG. 7. Insert is expansion-fitted into outer piece.

liquid, resembles clear water. It boils at −320.4 F (carbon dioxide −109.3 F). Liquid density 50.45 lb per cu ft. Latent heat, 85.6 Btu per lb (CO_2 = 265.3); total heat (through 60 F) 179.8 Btu per lb (CO_2 = 330 Btu per lb).

EQUIVALENTS. 1 lb liquid nitrogen gives 13.6 cu ft of vapor, 1 gal gives 93.0 cu ft, weighs 6.73 lb.

TYPICAL TOLERANCES. 1-in.-dia (nominal) steel shaft to enter a 1-in. core: Total shrinkage (see Table 10) 0.0022 in. Shaft dia of 1.0018 plus 0.000 minus 0.0005, and bore dia of 1.000 plus 0.0005 minus 0.000 will give maximum interference of 0.0018 in., minimum interference of 0.0008 in.

2-in.-dia (nominal) shaft: Total shrinkage 0.0044 in. Shaft dia of 2.0035 plus 0.000 minus 0.001, and bore dia of 2.000 plus 0.001 minus 0.000 will give maximum interference of 0.0035 in., minimum interference of 0.0015.

ILLUSTRATIVE PROBLEMS. Expansion-fit an aluminum ring 1.505 in. OD × 1 in. ID × 5/16 in. thick into a brass ring 1.500 in. ID × 2.25 in. OD × 5/16 in. thick. Aluminum rings weigh 0.1 lb each, and have average specific heat of 0.20.

1. High-purity liquid to cool 100 Al rings from +75 F to −320 F (total temperature differential 395 F). For those who wish to make the calculations, the procedure is given below. It is normally easier, however, to utilize the consumption figure given in Table 10.

Heat content of 100 parts = specific heat × No. parts × weight of parts × temp change = 0.20 × 100 × 0.1 × 395 = 780 Btu

$$\text{Liquid nitrogen used} = \frac{\text{heat absorbed}}{\text{latent heat of nitrogen}} = \frac{780}{85} = 9.16 \text{ lb}$$

Conversion to gas = 9.16 lb × 13.6 = 125 cu ft (see equivalents)

Or from Table 10:

100 parts × 0.1 × 12.5 = 125 cu ft

2. Force required to separate members

$$\text{Radial stress } q = \frac{d}{b} \times \frac{1}{\frac{1}{E_o}\left(\frac{c^2+b^2}{c^2-b^2} + M_o\right) + \frac{1}{E_1}\left(\frac{b^2+a^2}{b^2-a^2} - M_1\right)}$$

where d = diametral interference; E_o, E_1 = Young's modulus for outer and inner members; M_o, M_1 = Poisson's ratio for outer and inner members. Other symbols are shown in Fig. 7.

SOLUTION:

$$q = \frac{0.005}{1.5} \times \frac{1}{\frac{1}{15{,}000{,}000}\left(\frac{2.25^2+1.5^2}{2.25^2-1.5^2} \times 0.35\right) + \frac{1}{10{,}000{,}000}\left(\frac{1.5^2+1}{1.5^2-1} - 0.33\right)}$$

$$= 0.0033 \times 235{,}000{,}000 = 7850 \text{ psi}$$

TABLE 9. STANDARD SHRINKAGE ALLOWANCES FOR STEEL TIRES APPLIED TO LOCOMOTIVE WHEELS

Shrinkage 1/80 in. per ft for 38-in. centers, 1/60 in. per ft for 90-in. centers; increasing uniformly between these limits

Center Exact Dia, In.	Tire		Center Exact Dia, In.	Tire	
	Shrinkage, In.	Exact Bore, In.		Shrinkage, In.	Exact Bore, In.
20	0.021	19.979	50	0.056	49.944
22	0.023	21.977	52	0.059	51.941
24	0.025	23.975	54	0.062	53.938
26	0.027	25.973	56	0.065	55.935
28	0.029	27.971	58	0.068	57.932
30	0.031	29.969	60	0.071	59.929
32	0.033	31.967	62	0.075	61.925
34	0.035	33.965	64	0.078	63.922
36	0.038	35.962	66	0.081	65.919
38	0.040	37.960	68	0.084	67.916
40	0.042	39.958	70	0.088	69.912
42	0.045	41.955	72	0.091	71.909
44	0.048	43.952	74	0.095	73.905
46	0.050	45.950	76	0.098	75.902
48	0.053	47.947	78	0.102	77.898

TABLE 10. CONSUMPTION OF LIQUID NITROGEN AND SHRINKAGE FOR EXPANSION FITS

Metal	Consumption, Cu Ft per Lb		Metal Shrinkage,† In. per In. of Dia	Young's Modulus	Poisson's Ratio
	Immersion in Liquid	Precool with Gas, Then Immerse*			
Magnesium alloys..........	14.3	8.6	0.0046	6,000,000	0.28
Aluminum alloys..........	12.5	7.5	0.0042	10,000,000	0.33
Copper alloys............	5.7	3.4	0.0033	15,000,000	0.35
Cr-Ni alloys (18-8 to 18-12)	6.0	3.6	0.0029	30,000,000	0.30
Monel metals............	6.0	3.6	0.0023	26,000,000	0.32
SAE steels...............	6.0	3.6	0.0022	30,000,000	0.29
Cr steels (5 to 27% Cr)....	6.0	3.6	0.0019	30,000,000	0.33
Cast iron (not alloyed).....	7.0	4.2	0.0017	15,000,000	0.25

* Multiply values in first column by 0.60.

† From 75 to −321 F.

TABLE 11. COEFFICIENTS OF FRICTION

Metal	Mg Alloys	Al Alloys	Cu Alloys	Cr-Ni Alloys	SAE Steels	Cr Steels (5-27%)	Cast Iron
Magnesium alloys	0.8	0.8	0.6	0.5	0.5	0.5	0.2
Aluminum alloys		0.9	0.7	0.6	0.6	0.6	0.2
Copper alloys			0.6	0.5	0.5	0.5	0.3
Cr-Ni alloys (18-8 to 18-12)				0.4	0.4	0.4	0.2
SAE steels					0.4	0.4	0.2
Cr steels (5 to 27% Cr)						0.4	0.2
Cast iron (not alloyed)							0.3

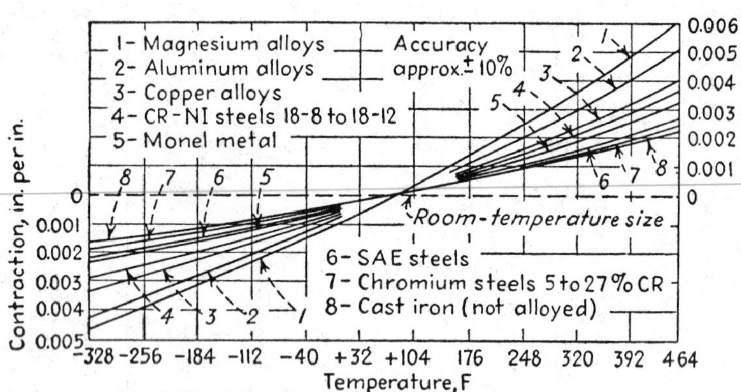

FIG. 8. Unit linear variation of materials with temperature change.

Force required to separate is equal to the strength in tension T, so:

$$T = \text{radial stress} \times \text{contact area} \times \text{coefficient of friction (Table 11)}$$
$$= 7850 \times \pi \times 1.5 \times 5/16 \times 0.6 = 6940 \text{ lb}$$

Strength in torsion S:

$$S = \text{radial stress} \times \text{contact area} \times \text{coefficient of friction} \times \text{radius}$$
$$= 7850 \times 1.5 \times 5/16 \times 0.6 \times 3/4 = 5220 \text{ in. per lb}$$

NOTE: Young's moduli, Poisson's ratios, and coefficients of friction are given in Table 10.

MOUNTING FITS FOR BEARINGS

Successful application of antifriction bearings depends as much upon their mounting as upon proper selection. Each bearing manufacturer has long had his own tables of recommended fits, and these are in close agreement. As a result the Anti-friction Bearing Manufacturers Association, New York, has been able to publish extensive tables on shaft and housing fitting recommendations for annular ball bearings, cylindrical roller bearings, and tapered roller bearings, plus the proper sizes of shaft and housing shoulders. Likewise, this material is found in ASA B3.8-1951, available from the American Standards Association, New York.

Table 12. Shaft Fits for ABEC-1 Ball Bearings, in Inches

Bearing Bore			Shaft Revolving						Shaft Stationary			Bore, Mm	Min Dia of Shaft Shoulder, Single-row Bearings, Bearing Series*			
			Light Loads, or Load Direction Indeterminate			Shaft Rotating with Relation to Load Direction										
Mm	Max	Min	Max	Min	Mean Fit	Max	Min	Mean Fit	Max	Min	Mean Fit		10 Series	02 Series	03 Series	04 Series
10	0.3937	0.3934	0.3937	0.3935	......	0.3939	0.3936	......	0.3935	0.3931	......	10	0.47	0.50	0.50	
12	0.4724	0.4721	0.4724	0.4721	......	0.4726	0.4723	......	0.4722	0.4717	......	12	0.55	0.58	0.63	
15	0.5906	0.5903	0.5906	0.5903	......	0.5908	0.5905	......	0.5904	0.5899	......	15	0.67	0.69	0.75	
17	0.6693	0.6690	0.6693	0.6690	......	0.6695	0.6692	......	0.6691	0.6686	......	17	0.75	0.77	0.83	0.95
20	0.7874	0.7870	0.7878	0.7872	0.0003	0.7878	0.7875	0.0005	0.7871	0.7866	0.0004	20	0.89	0.94	0.94	1.06
25	0.9843	0.9839	0.9847	0.9841	......	0.9847	0.9844	......	0.9840	0.9835	......	25	1.08	1.14	1.14	1.34
30	1.1811	1.1807	1.1815	1.1809	......	1.1815	1.1812	......	1.1808	1.1803	......	30	1.34	1.34	1.34	1.54
35	1.3780	1.3775	1.3784	1.3778	0.0004	1.3785	1.3781	0.0006	1.3776	1.3770	0.0005	35	1.53	1.53	1.69	1.73
40	1.5748	1.5743	1.5752	1.5746	......	1.5753	1.5749	......	1.5744	1.5738	......	40	1.73	1.73	1.93	1.97
45	1.7717	1.7712	1.7721	1.7715	......	1.7722	1.7718	......	1.7713	1.7707	......	45	1.94	1.94	2.13	2.17
50	1.9685	1.9680	1.9689	1.9683	......	1.9690	1.9686	......	1.9681	1.9675	......	50	2.13	2.13	2.36	2.44
55	2.1654	2.1648	2.1659	2.1651	......	2.1660	2.1655	0.0007	2.1650	2.1643	......	55	2.33	2.47	2.56	2.64
60	2.3622	2.3616	2.3627	2.3619	......	2.3628	2.3623	......	2.3618	2.3611	......	60	2.53	2.67	2.84	2.84
65	2.5591	2.5585	2.5596	2.5588	......	2.5597	2.5592	......	2.5587	2.5580	......	65	2.72	2.86	3.03	3.03
70	2.7559	2.7553	2.7564	2.7556	......	2.7565	2.7560	......	2.7555	2.7548	......	70	2.91	3.06	3.23	3.31
75	2.9528	2.9522	2.9533	2.9525	......	2.9534	2.9529	......	2.9524	2.9517	......	75	3.11	3.25	3.43	3.50
80	3.1496	3.1490	3.1501	3.1493	......	3.1502	3.1497	......	3.1492	3.1485	......	80	3.31	3.55	3.62	3.70
85	3.3465	3.3457	3.3470	3.3461	0.0005	3.3472	3.3466	0.0008	3.3460	3.3452	......	85	3.50	3.75	3.90	4.06
90	3.5433	3.5425	3.5438	3.5429	......	3.5440	3.5434	......	3.5428	3.5420	......	90	3.84	3.94	4.09	4.25
95	3.7402	3.7394	3.7407	3.7398	......	3.7409	3.7403	......	3.7397	3.7389	......	95	4.05	4.21	4.29	
100	3.9370	3.9362	3.9375	3.9366	......	3.9377	3.9371	......	3.9365	3.9357	......	100	4.23	4.41	4.49	
105	4.1339	4.1331	4.1344	4.1335	......	4.1346	4.1340	......	4.1334	4.1326	......	105	4.53	4.61	4.69	
110	4.3307	4.3299	4.3312	4.3303	......	4.3314	4.3308	......	4.3302	4.3294	......	110	4.72	4.80	4.88	

* Bearing series: 10 series = extra light; 02 series = light; 03 series = medium; and 04 series = heavy-duty bearings.

TABLE 13. HOUSING FITS FOR BALL BEARINGS (ABEC-1 TOLERANCE)

Bearing Bore, Mm for Bearings in			Bearing OD			Housing Stationary						Housing Rotates		
				In.		Solid Housing			Split Housing					
10 Series	02 Series	03 Series	Mm	Max	Min	Min	Max	Loose*	Min	Max	Loose	Min	Max	Tight
10	...	...	26	1.0236	1.0232	1.0236	1.0241	0.0004	1.0236	1.0244	0.0006	1.0228	1.0236	
	10	...	30	1.1811	1.1807	1.1811	1.1816	0.0004	1.1811	1.1819	0.0006	1.1803	1.1811	
15	12	...	32	1.2598	1.2593	1.2598	1.2604	0.0005	1.2598	1.2608	0.0007	1.2588	1.2598	0.0003
17	15	10	35	1.3780	1.3775	1.3780	1.3786	0.0005	1.3780	1.3790	0.0007	1.3770	1.3780	
		12	37	1.4567	1.4562	1.4567	1.4573	0.0005	1.4567	1.4577	0.0007	1.4557	1.4567	
	17	...	40	1.5748	1.5743	1.5748	1.5754	0.0005	1.5748	1.5758	0.0007	1.5738	1.5748	
20	...	15	42	1.6535	1.6530	1.6535	1.6541	0.0005	1.6535	1.6545	0.0007	1.6525	1.6535	
25	20	17	47	1.8504	1.8499	1.8504	1.8510	0.0005	1.8504	1.8514	0.0007	1.8494	1.8504	
	25	20	52	2.0472	2.0467	2.0472	2.0479	0.0006	2.0472	2.0484	0.0009	2.0460	2.0472	
30	...	...	55	2.1654	2.1649	2.1654	2.1661	0.0006	2.1654	2.1666	0.0009	2.1642	2.1654	
35	30	25	62	2.4409	2.4404	2.4409	2.4416	0.0006	2.4409	2.4421	0.0009	2.4397	2.4409	
40	...	...	68	2.6772	2.6767	2.6772	2.6779	0.0006	2.6772	2.6784	0.0009	2.6760	2.6772	
	35	30	72	2.8346	2.8341	2.8346	2.8353	0.0006	2.8346	2.8358	0.0009	2.8334	2.8346	
45	...	...	75	2.9528	2.9523	2.9528	2.9535	0.0006	2.9528	2.9540	0.0009	2.9516	2.9528	
50	40	35	80	3.1496	3.1491	3.1496	3.1503	0.0006	3.1496	3.1508	0.0009	3.1484	3.1496	
	45	...	85	3.3465	3.3459	3.3465	3.3474	0.0008	3.3465	3.3479	0.0010	3.3451	3.3465	0.0004
55	50	40	90	3.5433	3.5427	3.5433	3.5442	0.0008	3.5433	3.5447	0.0010	3.5419	3.5433	
60	...	...	95	3.7402	3.7396	3.7402	3.7411	0.0008	3.7402	3.7416	0.0010	3.7388	3.7402	
65	55	45	100	3.9370	3.9364	3.9370	3.9379	0.0008	3.9370	3.9384	0.0010	3.9356	3.9370	
70	60	50	110	4.3307	4.3301	4.3307	4.3316	0.0008	4.3307	4.3321	0.0010	4.3293	4.3307	
75	...	...	115	4.5276	4.5270	4.5276	4.5285	0.0008	4.5276	4.5290	0.0010	4.5262	4.5276	
	65	55	120	4.7244	4.7238	4.7244	4.7253	0.0008	4.7244	4.7258	0.0010	4.7230	4.7244	
80	70	...	125	4.9213	4.9205	4.9213	4.9223	0.0009	4.9213	4.9229	0.0012	4.9197	4.9213	
85	75	60	130	5.1181	5.1173	5.1181	5.1191	0.0009	5.1181	5.1197	0.0012	5.1165	5.1181	
90	80	65	140	5.5118	5.5110	5.5118	5.5128	0.0009	5.5118	5.5134	0.0012	5.5102	5.5118	
95	...	...	145	5.7087	5.7079	5.7087	5.7097	0.0009	5.7087	5.7103	0.0012	5.7071	5.7087	
100	85	70	150	5.9055	5.9047	5.9055	5.9065	0.0009	5.9055	5.9071	0.0012	5.9039	5.9055	
105	90	75	160	6.2992	6.2982	6.2992	6.3002	0.0009	6.2992	6.3008	0.0013	6.2976	6.2992	0.0003
110	95	80	170	6.6929	6.6919	6.6929	6.6939	0.0009	6.6929	6.6945	0.0013	6.6913	6.6929	
120	100	85	180	7.0866	7.0856	7.0866	7.0876	0.0009	7.0866	7.0882	0.0013	7.0850	7.0866	
	105	90	190	7.4803	7.4791	7.4803	7.4814	0.0012	7.4803	7.4821	0.0015	7.4785	7.4803	
130	110	95	200	7.8740	7.8728	7.8740	7.8751	0.0012	7.8740	7.8758	0.0015	7.8722	7.8740	
140	...	...	210	8.2677	8.2665	8.2677	8.2688	0.0012	8.2677	8.2695	0.0015	8.2659	8.2677	
	120	100	215	8.4646	8.4634	8.4646	8.4657	0.0012	8.4646	8.4664	0.0015	8.4628	8.4646	

* Mean fit (applies also to "tight").

TABLE 13. HOUSING FITS FOR BALL BEARINGS (ABEC-1 TOLERANCE) (*Continued*)
Max Housing Shoulder Dia

Bore, Mm	10 Series	02 Series	03 Series
10	0.95	0.98	1.18
12	1.02	1.06	1.22
15	1.18	1.18	1.42
17	1.30	1.34	1.61
20	1.46	1.61	1.77
25	1.65	1.81	2.17
30	1.93	2.21	2.56
35	2.21	2.56	2.80
40	2.44	2.87	3.19
45	2.72	3.07	3.58
50	2.91	3.27	3.94
55	3.27	3.58	4.33
60	3.47	3.98	4.65
65	3.66	4.37	5.04
70	4.06	4.57	5.43
75	4.25	4.76	5.83
80	4.65	5.12	6.22
85	4.84	5.51	6.54
90	5.16	5.91	6.93
95	5.35	6.22	7.32
100	5.55	6.61	7.91
105	5.91	7.01	8.31
110	6.30	7.40	8.90

It is impractical in a handbook to cover all sizes and all types of bearings. Therefore, the information here is restricted to:

For ball bearings—mounting fits for ABEC-1 tolerance bearings in metric sizes from 10- to 110-mm bore (see Tables 12 and 13).

For cylindrical roller bearings—minimum shaft and maximum shoulder diameters (see Table 14).

These restrictions are based upon these reasons. First, in ball bearings, ABEC-3, -5, and -7 tolerance bearings, also inch-series bearings, account for only a very small percentage of bearings used in machinery building. For example, the No. 5 and the No. 7 tolerance bearings are mostly used for spindles and special precision jobs, where consultation with the manufacturer is important. So far as bore sizes go, the range from 10 to 110 mm, inclusive, represents the spread of common usage.

Shaft and Housing Fits. In the majority of ball-bearing applications, for example, the shaft rotates and the housing is stationary. Sometimes, as in certain pulley and wheel mountings, the shaft is the stationary member. The following rule covers the fits to be used for both cases: *Ball bearings should be applied with the rotating ring a firm press or interference fit, and the stationary ring a close push fit.* The

TABLE 14. MINIMUM SHAFT SHOULDER AND MAXIMUM HOUSING SHOULDER DIAMETERS FOR CYLINDRICAL ROLLER BEARINGS

	02 Dimension Series				03 Dimension Series			
Bore, Mm	Min*	Max†	Max‡	Min§	Min*	Max†	Max‡	Min§
10	0.55	0.60	1.08	1.02	0.59	0.65	1.24	1.12
12	0.62	0.66	1.15	1.08	0.66	0.72	1.31	1.27
15	0.74	0.75	1.26	1.20	0.80	0.82	1.49	1.41
17	0.84	0.86	1.44	1.41	0.90	0.97	1.67	1.58
20	1.02	1.04	1.67	1.59	1.02	1.10	1.82	1.77
25	1.22	1.23	1.86	1.85	1.25	1.31	2.20	2.13
30	1.42	1.48	2.24	2.18	1.50	1.51	2.55	2.50
35	1.66	1.71	2.61	2.50	1.73	1.77	2.83	2.76
40	1.87	1.96	2.88	2.83	1.94	1.99	3.19	3.13
45	2.08	2.15	3.06	3.04	2.20	2.24	3.54	3.50
50	2.28	2.31	3.24	3.20	2.42	2.52	3.90	3.86
55	2.52	2.60	3.60	3.55	2.65	2.74	4.25	4.13
60	2.73	2.85	3.96	3.88	2.88	2.99	4.61	4.46
65	3.00	3.02	4.32	4.20	3.11	3.21	4.96	4.88
70	3.19	3.23	4.50	4.41	3.32	3.46	5.31	5.25
75	3.37	3.44	4.68	4.62	3.57	3.74	5.67	5.63
80	3.60	3.73	5.05	4.94	3.79	3.95	6.02	6.00
85	3.85	3.96	5.40	5.29	4.04	4.22	6.38	6.25
90	4.05	4.11	5.76	5.64	4.25	4.44	6.73	6.63
95	4.29	4.46	6.12	6.06	4.54	4.73	7.09	7.00
100	4.56	4.71	6.48	6.35	4.83	4.97	7.62	7.50
105	4.77	4.96	6.84	6.66	5.05	5.24	7.97	7.75
110	5.01	5.21	7.20	7.07	5.37	5.48	8.50	8.25

* Minimum shaft shoulder dia that will satisfy the maximum inner race contour of any manufacturer.
† Maximum shaft shoulder dia that will clear the dia under the rollers of any manufacturer.
‡ Maximum housing shoulder dia that will satisfy the maximum outer-race corner contour of any manufacturer.
§ Minimum housing shoulder dia that will clear the dia over the rollers of any manufacturer.

reasons for this rule, according to the New Departure Manufacturing Co., are:

1. Under normal load conditions, a press-fitted ring will not slip or turn on or in a rotating shaft or housing, and wear on the latter is avoided.

2. A bearing having one ring push fitted and not clamped can move axially so as to avoid the imposition of excessive thrust loads, such as those caused by a change in shaft length due to expansion.

3. Machine assembly is easier if one of the rings is a push fit.

Of course, such rules do not apply in all cases—very heavy vibratory loads, precision applications, duplex mountings of single-row, angular contact bearings, etc. For average conditions, bearings are supplied with sufficient internal looseness, so that the correct bearing operating fit-up is obtained by use of the recommended press fit.

Working tolerances in the tables for shaft and housing fits have been established in relation to bearing tolerances in a manner such that the desired average fit should be obtained.

Shaft and Housing Shoulders. A shaft shoulder is provided to square up a bearing. An off-square shoulder will distort the inner ring of a ball bearing. Size of the shoulder is important, and the proper height is necessary to provide sufficient thrust support for the inner ring. But the height cannot be too great or the bearing ring will be difficult or impossible to remove. The inner ring must project enough for application of dismantling tools.

The maximum housing shoulder diameter is determined in the same way as the minimum shaft shoulder height. Shoulder squareness and good surface are also important to the life of the bearing.

TOLERANCES IN DESIGN[1]

Few designers realize that assigning tolerances to dimensions is as important as originating the design. Final cost of the product is closely related to the degree of accuracy established for the dimensions by the engineer when assigning tolerances.

The extent to which components of a product are made interchangeable is dependent upon many factors. Accurate detailed knowledge of the product must be available before attempting to tolerance the dimensions. The engineer should study the end use of the product, analyze competitive products, and then determine why and where dimensions must be held to close limits.

Probable wear life of the product is one of the first considerations. Wear life depends upon the strength of the weakest part or group of parts. Excessive wear may occur when maximum-metal assemblies do not have sufficient allowance between the mating parts. Conversely, too much freedom between mating parts may cause misalignment, eccentricity, and hence excessive wear in other parts of the assembly. Parts that must assemble under force or shrink fits must be provided with proper negative allowances to permit expected deformation or flow of material.

Tolerances set closer than necessary should always be broadened. The closer the tolerance, the higher the cost is the rule. Available equipment should be checked to determine whether it will produce machined parts within the desired accuracy. When production has been started and an operation cannot be held within tolerance, production time is lost, and the effect upon the assembly line is expensive. However, it is bad practice merely to change the tolerance to suit the accuracy of the machine. An inferior product will result. To avoid such situations, large and experienced manufacturers make tests of the accuracy and suitability of available equipment before it is assigned to a specific operation. However, machine-capability studies must rest upon a sound basis and procedure for gaging, which is understood and followed by all concerned—product designer, tool engineer, production, inspection, and quality control. Statistical quality-control methods are helpful in making machine-capability studies and in maintaining parts manufacture within established tolerances.

It is also important that the designer should consult with the tool engineer before assigning sizes and tolerances to dimensions that must be produced with purchased tools. Cutting-tool manufacturers supply stock items that can be applied to a variety of jobs. Ultimate cost of the product is increased when the product designer needlessly calls for special tooling. When the size will permit, the tolerance on drilled and reamed holes should permit use of standard tools.

[1] G. H. Simpson, Chief Engineer, Gage Div., Greenfield Tap & Die Corp.

APPLICATION OF LIMITS AND TOLERANCES

General. Proper application of limits and tolerances is important because it is impractical to produce parts exactly to dimensions without tolerances.

A nominal dimension is the theoretical size from which all variations are specified. It is also the desired size to which the part would be made if there were no variations in production.

Limits are specified maximum and minimum sizes.

Tolerances are maximum permissible variations from the specified nominal sizes.

Unilateral tolerances allow variations in only one direction from the nominal dimension.

Bilateral tolerances allow variations in either direction from the nominal dimension.

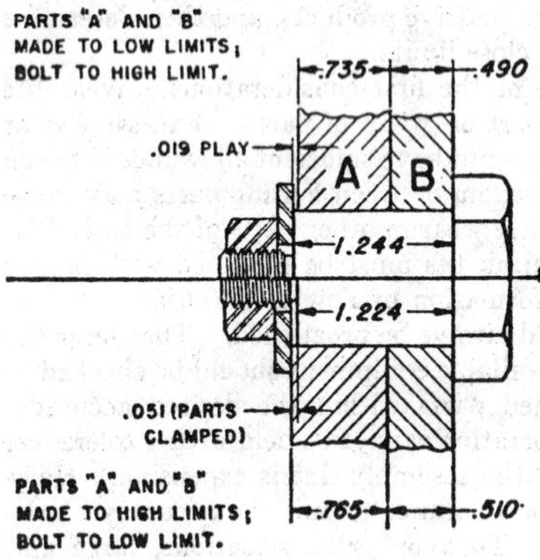

Responsibility. It is the responsibility of the draftsman, in collaboration with the checker, supervisor, and engineer, to decide intelligently what limits or tolerances are allowable, because they are usually most familiar with the surrounding or related parts and the intended functioning of the mechanism.

The experienced draftsman should be familiar with the various classes of fits used in the construction of mechanisms, and the degree of accuracy that can be maintained commercially by the various methods of manufacture. The proper fitting of mating and adjoining parts contributes largely to the successful operation of any mechanism. Because these fits are controlled by the limits and tolerances specified on the drawings, this phase of drafting is important.

Judgment. There are no rules or formulas for establishing limits and tolerances for all phases of engineering; hence the draftsman must rely on the judgment, experience, and knowledge of the persons responsible for the design. The design must therefore be analyzed carefully to determine the degree of accuracy which is essential to meet the functional requirements of each detail part.

The greater the permissible limits or tolerances, the less costly the part is to produce, because of reduced material scrap, lower labor costs, and less expensive tools. Limits or tolerances should never be specified closer than necessary, either by definite specification or by the inference of any of the general notes. In certain cases it may be advisable to use closer tolerances to facilitate assembly. The necessity for this should be determined by ascertaining where the greatest saving lies—by close tolerances and shorter assembly time, or by greater tolerances and more assembly time.

Investigation of current production practices is recommended as a guide in setting future limits and tolerances because in most cases they represent practical and successful manufacturing practices.

Accumulation of Limits. The possible accumulation of limits and tolerances should be carefully considered. It is possible that dimensions may vary to either the extreme high or low limits on related parts, causing an interference or undesirable condition.

To avoid such conditions it may sometimes be necessary to choose closer tolerances for individual dimensions than would normally be required where a small number of dimensions are involved. Frequently these conditions can be reduced by more intelligent dimensioning of the involved parts.

It must not be assumed that the "law

of averages" will prevent these extremes from occurring. See illustration.

Limits and tolerances are expressed as shown in the following illustrations.

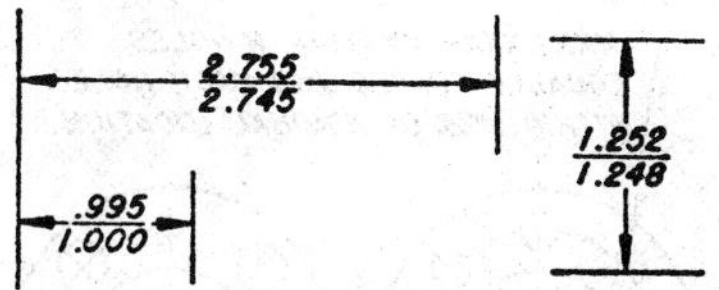

Expression of Limits. Limits are expressed in decimals showing maximum and minimum sizes. One limit is placed above the line and the other limit below the line. The decimal portion of both

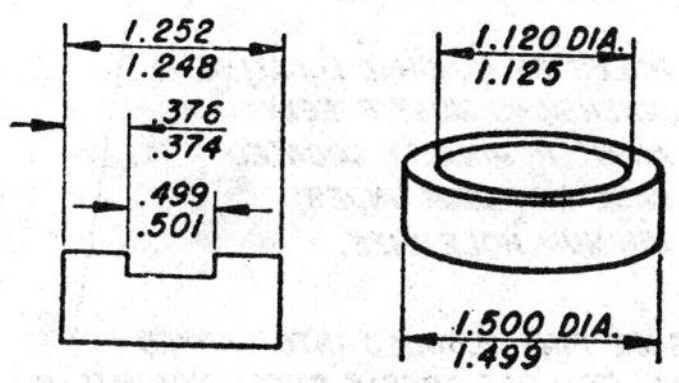

limits of any dimension contains the same number of digits. Limits on vertical dimensions are written horizontally and are separated by a line.

The minimum limit is placed above the line for all internal dimensions, and the maximum limit is placed above the line for all external dimensions.

$\frac{1.000}{.999}$ 1.000 - .999 $3\frac{1}{2}$ TO $4\frac{1}{2}$

Where limits are specified in notes, they may be written in either of the forms shown.

Expression of Tolerances. Tolerances are properly expressed by specifying the nominal dimension followed by either a bilateral or unilateral tolerance. A bilateral tolerance usually shows the plus or minus values equal, with the tolerance in line with the dimension.

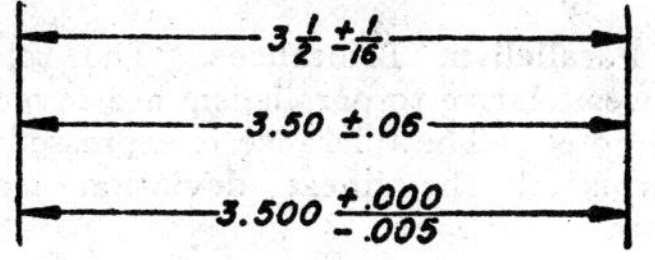

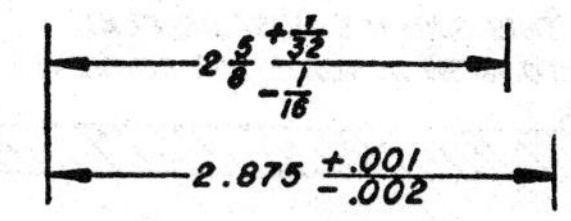

When the plus and minus values of a bilateral tolerance are not equal, the plus value is placed above the minus value in relation to the dimension line as shown.

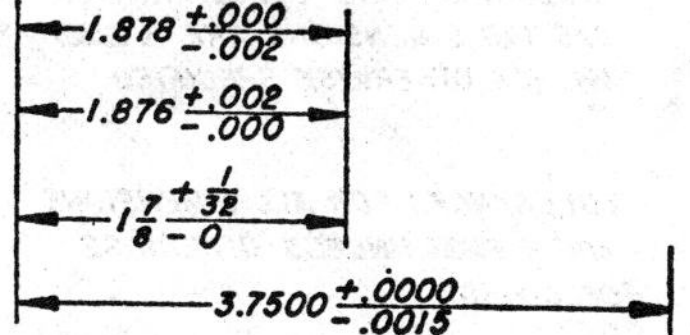

Unilateral tolerances are expressed with two figures. Where the dimension and one term of the tolerance are expressed in fractions, the zero tolerance is expressed as a single o without a decimal point. Wherever the nominal dimension is a fraction, the tolerance is in fractions; and where the nominal dimension is in decimals, the tolerance is in decimals. The decimal portion of both the nominal dimensions and the tolerance contains the same number of digits.

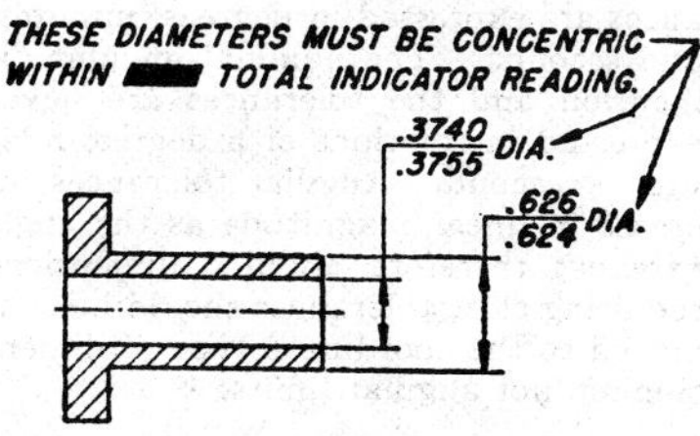

Concentricity Tolerances. The tolerances governing concentricity are expressed in notes. The note must always give the total indicator reading, which is the sum of positive and negative variations indicated. It is generally assumed that all circular surfaces not covered by a note are concentric within any general drawing tolerance which is specified on the drawing.

THIS SURFACE MUST BE FLAT
TO ▬ CONCAVE (OR CONVEX)

Flatness Tolerances. The tolerance relative to flatness is covered with a note, which specifies the direction and the amount of deviation from flat.

TOLERANCES FOR MACHINING DIMENSIONS ARE ± ▬ AND FOR CASTING DIMENSIONS ARE ± ▬ UNLESS OTHERWISE SPECIFIED.

TOLERANCES FOR ALL DIMENSIONS ARE ± ▬ UNLESS OTHERWISE SPECIFIED.

General Tolerances. Wherever possible, for purposes of simplicity, the predominant tolerance on any drawing is expressed in a general note rather than by individual tolerances applied to each dimension. The use of a standard general tolerance is acceptable, but care must be exercised that individual tolerances are specified on those dimensions which can have a larger tolerance than the general standard. This usually avoids unnecessary increases in manufacturing cost.

Angular Tolerances. Angular tolerances are expressed in degrees, minutes, and seconds. The nominal angular dimension and the tolerances are never expressed in fractions of a degree, minute, or second. Angular tolerances increase in linear magnitude as the angle extends; therefore angular dimensions requiring close tolerances should be converted to the coordinate type of dimensioning, not angular limits.

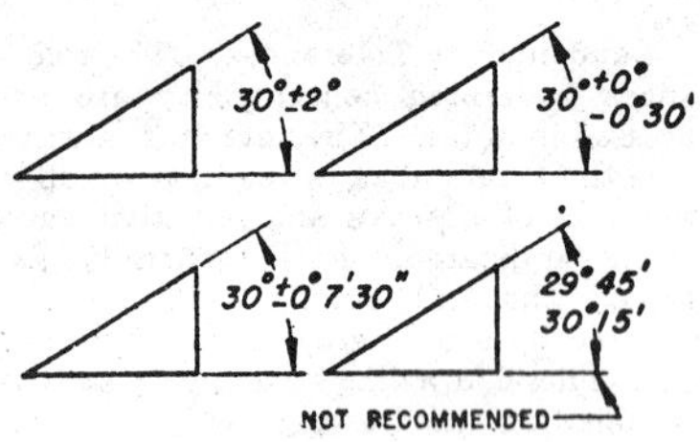
30°±2°
30° +0° -0°30'
30°±0°7'30"
29°45'
30°15'
NOT RECOMMENDED

Hole-spacing Tolerances. Tolerances on the distances between holes are usually expressed in bilateral terms. When it is desirable to control the location of a group of holes, this may be accomplished by the use of notes as shown. The notes may be applied either locally or as general drawing notes.

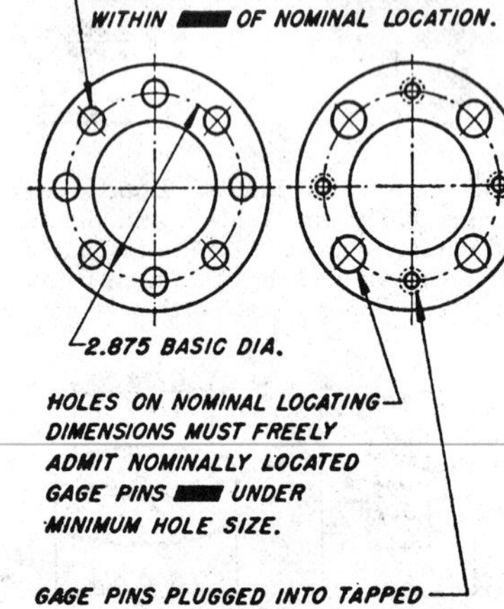
.XXX (.XXX-.XXX) DIA. 8 HOLES EQUALLY SPACED AND EACH HOLE WITHIN ▬ OF NOMINAL LOCATION.
2.875 BASIC DIA.
HOLES ON NOMINAL LOCATING DIMENSIONS MUST FREELY ADMIT NOMINALLY LOCATED GAGE PINS ▬ UNDER MINIMUM HOLE SIZE.
GAGE PINS PLUGGED INTO TAPPED HOLES MUST FREELY ENTER NOMINALLY LOCATED GAGE HOLES ▬ LARGER.

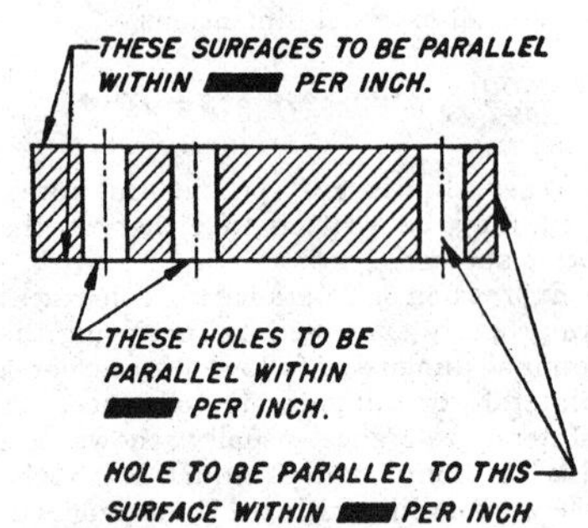
THESE SURFACES TO BE PARALLEL WITHIN ▬ PER INCH.
THESE HOLES TO BE PARALLEL WITHIN ▬ PER INCH.
HOLE TO BE PARALLEL TO THIS SURFACE WITHIN ▬ PER INCH

Parallelism Tolerances. The tolerances relative to parallelism are covered in notes. The tolerance is expressed in terms of the linear deviation from parallel per inch.

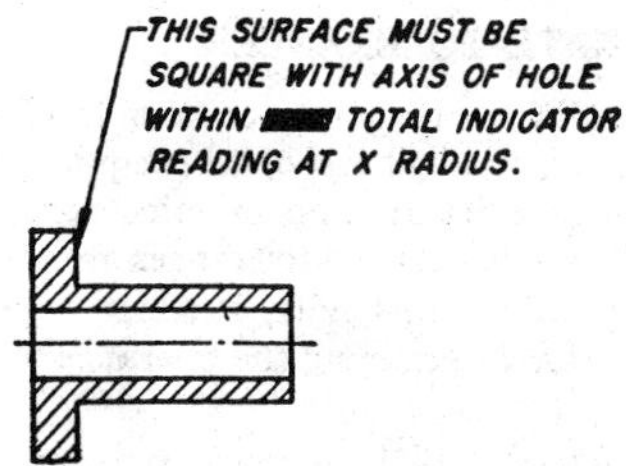
THIS SURFACE MUST BE
SQUARE WITH AXIS OF HOLE
WITHIN ▬ TOTAL INDICATOR
READING AT X RADIUS.

Squareness Tolerances. The tolerances governing the squareness of one surface to another are expressed in notes. The note gives the total indicator reading. It is generally assumed that all surfaces not covered by a note are square with the axis within any standard tolerance which is specified on the drawing, and that this tolerance applies at the outermost edge of the surface.

Drilled-hole Tolerances. Tolerances for drilled holes should be as wide as possible. This will often permit a variety of drill sizes to be used and will cover oversize holes made by incorrectly ground drills. Express the hole size in limits or tolerances wherever possible and avoid specifying the drill number or letter on the drawing.

Tolerances on Drilled Holes

Drill-size Range		Tolerance	
Smallest	Largest	Plus	Minus
0.0135 (No. 80)	0.042 (No. 58)	0.003	0.002
0.043 (No. 57)	0.093	0.004	0.002
0.0935 (No. 42)	0.156	0.005	0.002
0.1562	0.2656	0.006	0.002
0.266 (H)	0.4219	0.007	0.002
0.4375	0.6094	0.008	0.002
0.625	0.750	0.009	0.002
0.7656	0.8437	0.009	0.003
0.8594	2.000	0.010	0.003

Tap-drill sizes and tolerances must conform to the internal thread minor diameter.

The minimum commercial tolerances for drilled holes are shown in the table.

Reamed-hole Tolerances. The maximum limit of reamed holes is the nominal size plus 0.0005 in., wherever possible. This is the size hole which a new reamer will produce. The minimum size should be determined by considering the fit desired and proportioning the tolerances between the hole and the mating part so that maximum economy will result. This frequently means that the hole will carry a larger tolerance than the mating part, because it is more economical to hold an external diameter to closer tolerances than an internal one. See example.

Hole..........	0.5005	0.4975
Shaft..........	0.4955	0.4965
Clearance......	0.0050	0.0010

The 0.001-in. tolerance on the shaft usually does not penalize the cost of the grinding operation and permits the use of a new reamer until it has worn 0.003 undersize.

If the shaft can be made from standard purchased material without further machining, it may prove more economical to use a special reamer size.

Form Contour Tolerances. Where it is considered inadvisable to specify a tolerance on each dimension of a contour or shape due to possible objectionable accumulation of these tolerances, a note should be used to express them, as illustrated at "Preferred" below.

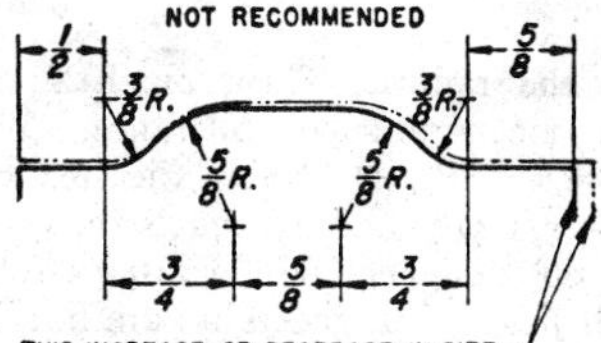
NOT RECOMMENDED
1/2
5/8
3/8 R.
3/8 R.
5/8 R.
5/8 R.
3/4
5/8
3/4
THIS INCREASE OR DECREASE IN SIZE RESULTS FROM THE ACCUMULATION OF TOLERANCES.

ALLOWABLE VARIATION ON ALL FRACTIONAL DIMENSIONS ± 1/32 UNLESS OTHERWISE SPECIFIED

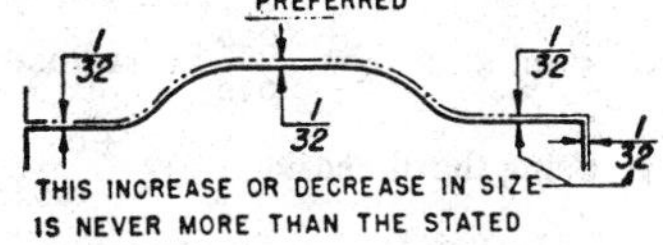
PREFERRED
1/32
1/32
1/32
1/32
THIS INCREASE OR DECREASE IN SIZE IS NEVER MORE THAN THE STATED TOLERANCE.

TOLERANCES OF FORM CONTOURS FROM NOMINAL SIZES ARE ± 1/32

HOW TO CONTROL COORDINATE TOLERANCES

Several tables exist for establishing the coordinates of points on a hole circle (see page 12-66 for an example) to permit machining on a jig borer or other equipment with lead-screw-controlled movements. Although efficient shop practice requires that the holes be located on a coordinate grid, it is true that the tolerances are often guessed at. Here is the method developed by Albert L. Levitt, Product Design Engineer, American District Telegraph Co., Inc., for determining the tolerance to be applied to coordinate dimensions.

Coordinate tolerance is controlled by two factors: radial tolerance and angular tolerance. When radial tolerance is greater than the angular tolerance, the formula $r_a = R \sin \frac{1}{2}A$, with R the mean radial tolerance, is used to determine the radius of the largest possible circle that can be scribed within the borders of angle A (Fig. 9). If, then, a square is circumscribed about the circle, this "tolerance" goes beyond the angular lines. However slight this distance may be, it should not be neglected. The phantom square represents the two lead screws of the machine's locating device; hence locating holes at other points on the circle will result in the corners of the square changing its relative position. Thus, the corners will protrude beyond the angular-tolerance lines and sometimes beyond the radial-tolerance lines. The only safe "tolerance square" for a hole located at any point on a circle and with any relation to the lead screws is the square *inscribed* in the circle.

The same relationship exists when the radial tolerance is less than the angular tolerance (Fig. 10)—where the prime concern is the minimum and maximum radius as determined by the workpiece tolerance. If one-half of the radial tolerance is used for the coordinate-dimension tolerance, the corners of the "tolerance square" will protrude beyond the outer circumference.

To solve the problem correctly, the mean of the workpiece tolerance *must* be obtained by

$$R = \frac{R_{max} - R_{min}}{2} + R_{min}$$

for the radius. Then, one-half of the full angular tolerance is taken for $\frac{1}{2}A$, the mean angular tolerance. With these established, $r_a = R \sin \frac{1}{2}A$ and $r_1 = R - R_{min}$. Taking the smaller value between r_a and r_1 for r, the tolerance $T = r/1.4142$ and $2T$ equals the coordinate tolerance (Fig. 11).

Tool designers should always follow this procedure, especially when designing drill jigs. Jigs, generally, are not designed so that they may be rotated to locate the bushing holes; hence coordinate dimensioning is necessary. Also, jigs are made to closer tolerances than the parts to be drilled in them; thus the tolerances are even more critical. The same conditions prevail in laying out templets for optical-comparator inspection.

EXAMPLE 1. Consider a part (Fig. 12) with three holes equally spaced within $\pm 0°15'$ on a $1.635 \begin{smallmatrix} +0.000 \\ -0.020 \end{smallmatrix}$ in. radius bolt circle.

1. Using the dimension $1.635 \begin{smallmatrix} +0.000 \\ -0.020 \end{smallmatrix}$ in., we obtain the mean dimension $R = 1.625$:

$$R = \frac{R_{max} - R_{min}}{2} + R_{min} = \frac{0.020}{2} + 1.615 = 1.625 \text{ in.}$$

(Continued on page 33-44)

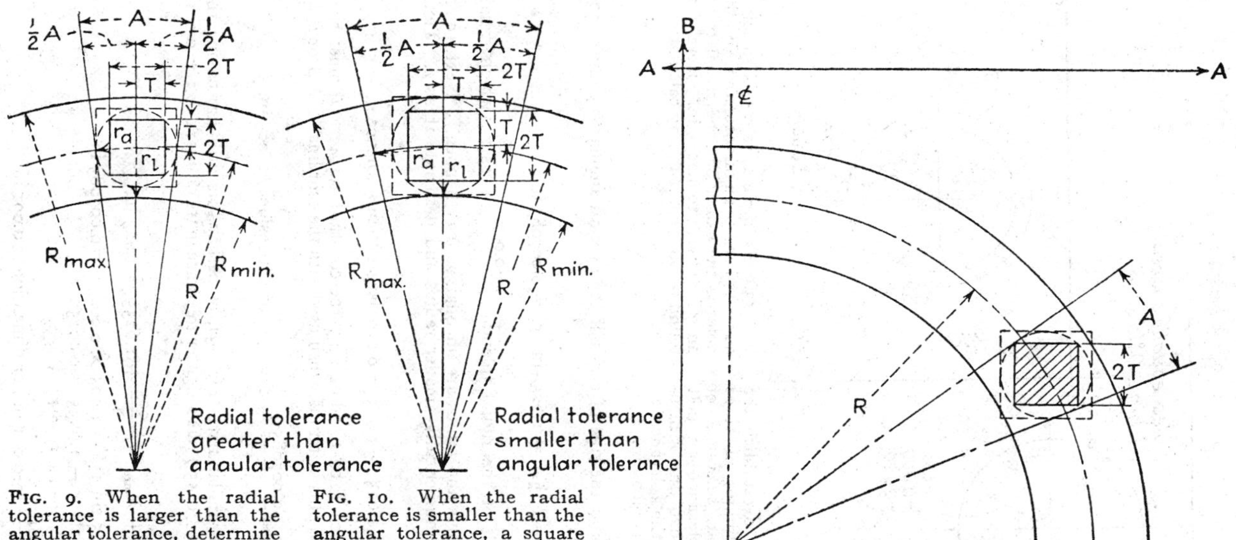

FIG. 9. When the radial tolerance is larger than the angular tolerance, determine the largest possible circle that can be scribed within the borders of angle *A*. This circle has a radius r_a. Then, if this circle is circumscribed by a square, corners of the square will fall outside the angular tolerance.

FIG. 10. When the radial tolerance is smaller than the angular tolerance, a square drawn to the equivalent radial tolerance will have corners protruding beyond the outer circumference. This circle has a radius r_1. Thus, neither tolerance is suitable in itself. To solve a problem, the mean of the workpiece tolerance must be obtained, relative to the *A* and *B* coordinates.

FIG. 11. Locating points for other holes on the circle will cause the corners of the oversize tolerance square to change its relative position. The only safe tolerance square for a hole located at any point on a circle and relative to the lead screws *A-A* and *B-B* is the one inscribed in the circle, the square whose side is $2T$.

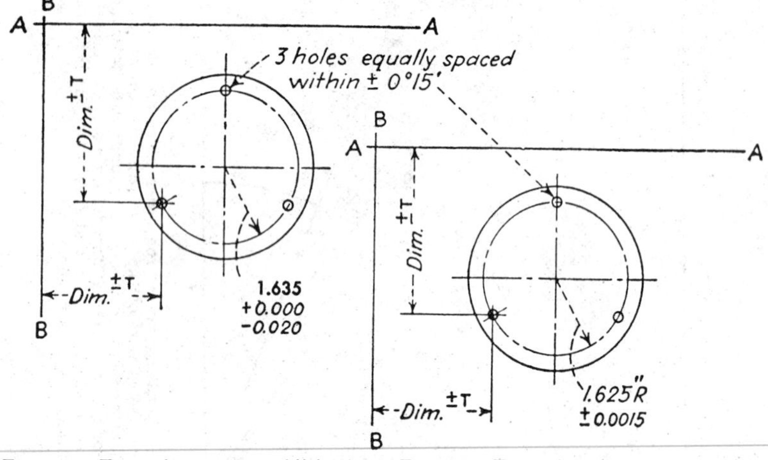

FIG. 12. Example 1: to establish the tolerance T for coordinates of three holes with an angular tolerance of ±15′ and a radial tolerance on bolt circle of +0.000 − 0.020 in. Calculations reveal that T, the tolerance on coordinate dimensions, is 0.005 in.

FIG. 13. Example 2: to establish tolerance T when the angular tolerance on the holes is the same, but the radial tolerance is reduced to 0.0015 in. Here, T is found to be ±0.001 in., by calculation.

2. Using $R = 1.625$ and the sine of one-half the over-all angular tolerance, 30′, (0.00436), $r_a = 0.007085$:

$$r_a = R\tfrac{1}{2}A = 1.625 \sin 15' = 0.007085 \text{ in.}$$

3. Designate one-half of the radial tolerance $\begin{smallmatrix}+0.000\\-0.020\end{smallmatrix}$ in. as r_1, which equals 0.010 in. (Note that plus and minus are unequal, as may happen.) Next, select between r_a and r_1 the numerically smaller radius and designate it as r. Thus, $r_a = r = 0.007085$ in.:

$$r_1 = R - R_{\min} = 1.625 - 1.615 = 0.010, \text{ larger than } r_a\text{; hence } r_a = r$$

4. Obtain the tolerance T by dividing r by the constant 1.4142. T must then be given the proper sign, plus or minus, when used with the coordinates A-A or B-B:

$$T = \frac{r}{1.4142} = \frac{0.007085}{1.4142} = 0.005 \text{ in. or dimension } \pm 0.005 \text{ in.}$$

EXAMPLE 2. Take a similar part (Fig. 13), or perhaps its drill jig, having the same angular tolerance (±0°15′) but a tighter tolerance in its radius: 1.625 ± 0.0015 in. Then,

1. $R = \dfrac{R_{\max} - R_{\min}}{2} + R_{\min} = \dfrac{0.003}{2} + 1.6235 = 1.625$ in.
2. $r_a = R\tfrac{1}{2}A = 1.625 \sin 15' = 1.625 \times 0.00436 = 0.007085$ in.
3. $r_1 = R - R_{\min} = 1.625 - 1.6235 = 0.0015$ in.
4. r_1 is smaller than r_a; hence $r_1 = r$.
5. $T = \dfrac{r}{1.4142} = \dfrac{0.0015}{1.4142} = 0.001$ in., or dimension ±0.001 in.

Section 34

MEASURING AND GAGING EQUIPMENT

SECTION 34

MEASURING AND GAGING EQUIPMENT

All metalworking operations involving a change in dimensions of a workpiece of necessity require that certain measurements be made. These may be approximate or exact to a tenth of a thousandth or less. The following material covers measuring equipment such as found in the worker's toolbox to the costly measuring and gaging devices used at the bench or machine by the skilled worker and in inspection departments.

MEASURING TOOLS

Steel Scales (Fig. 1). These are usually 6 or 12 in. long, although 4 in., 18 in., and longer are available.

As a rule, there are four sets of graduations, one on each edge of each side. The longest lines represent inch marks. On one edge, each inch is divided into eight equal spaces to represent $\frac{1}{8}$ in. The other edge of this side is divided into sixteenths. The $\frac{1}{4}$- and $\frac{1}{2}$- in. marks are commonly made longer than the smaller division marks to facilitate counting. The opposite side is similarly divided into 32 and 64 spaces per inch, and it is common practice to number every fourth division for easier reading.

Special graduations are available for patternmakers. If a part is to be made of cast iron, for example, which has a known shrinkage of $\frac{1}{8}$ in. per ft, a $\frac{1}{2}$-in. scale will be actually $12\frac{1}{8}$ in. long but will be graduated into 12 parts, each representing 1 in.

Variations in scales include those with one end tapered for about 2 in. This makes it easier to measure in holes, slots, or from shoulders. Some types are equipped with a removable hook at one end to form a fixed end support when setting dividers, or to assist in taking measurements when one cannot see whether the end is flush with the end of the work. This type is particularly useful for measuring from rounded corners or through the hubs of pulleys. Another type has a steel slider which may be pushed back and forth with the thumb and is useful for measuring against a shoulder.

For places where it is difficult to use an ordinary scale, a set of small sections of scale, from $\frac{1}{4}$ to 1 in. long, will be found convenient. The sections are designed to be held in a handle and may be adjusted to any desired position.

Numerous gadgets have been devised to increase the usefulness of an ordinary scale. Among these is a clamp for attaching two scales end to end for long measurements. The scales may be of the same or different widths. Clamps can also be had to connect two scales edge to edge to form a box square for laying out lines parallel to the axis of cylindrical work, as in scribing a keyway. Or a pair of small clamps can be attached to a single scale for the same purpose.

The edges of regular scales are made straight and true within reasonable limits of accuracy, but if an absolutely straight edge is required, a steel straightedge should be used. These usually have one edge beveled. Common use is for checking the

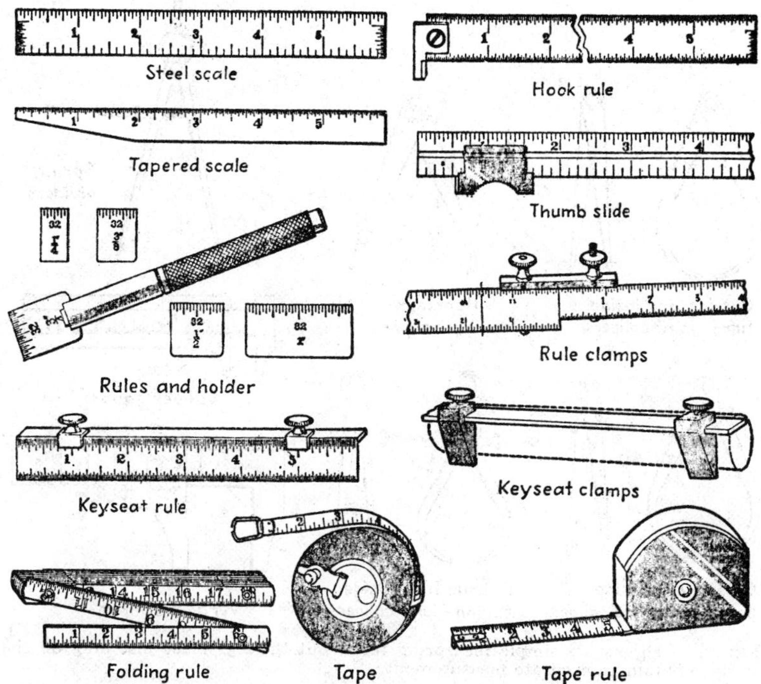

FIG. 1. Steel scales and tapes are made in many styles for a variety of purposes. The items shown are most of those commonly involved in machine-shop work.

straightness of a surface by placing the beveled edge on the surface and seeing if any light is visible between the two.

For lengths greater than about 18 in., folding steel, wood, and aluminum scales, sometimes called folding rules, are usually 2 to 6 ft long. However, they cannot be relied on for extremely accurate measurements because a certain amount of play develops at the joints after they have been used for a time.

Steel Tapes. Steel tapes run from 6 to about 100 ft. In the shorter lengths, these are frequently made with a curved cross-section so they are flexible enough to roll up but remain rigid when extended. Long flat tapes require support over their full length when measuring, or the natural sag will cause an error in reading.

Some tapes are designed to measure from the end of a ring attached to the start of the tape, while others have a blank space at the end, and readings are taken from a marked starting point.

Tapes should be handled carefully and kept lightly oiled to prevent rust.

Simple Calipers (Fig. 2). To measure diameters, calipers are used in conjunction with a steel scale. Outside calipers for measuring outside diameters are bowlegged; those used for inside diameters have straight legs with the feet turned out. If the

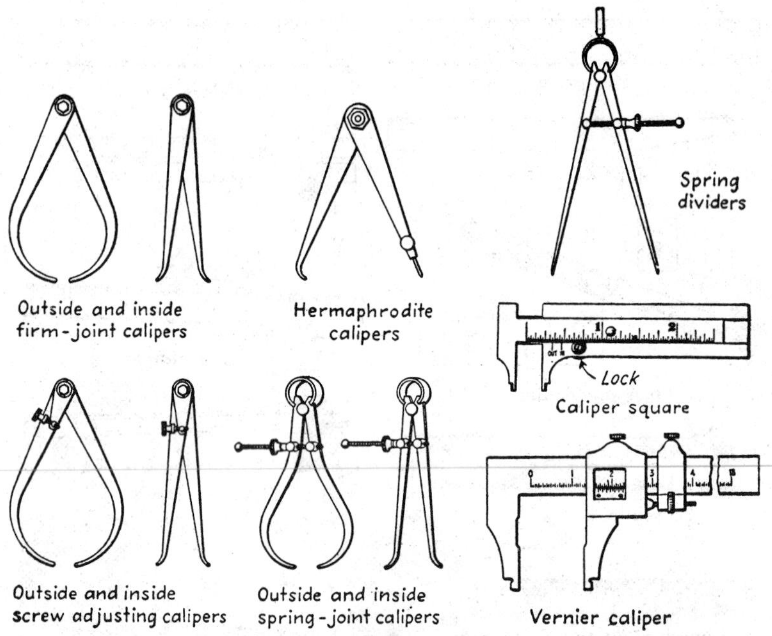

FIG. 2. Calipers are simple measuring tools, but the legs must just drag on the work to obtain an accurate measurement.

legs are joined at the top by a nut, they are called firm-joint calipers. Such types are adjusted by pulling or pushing the legs to open or close them.

Spring-joint calipers have the legs joined by a strong spring hinge and linked together by a screw and adjusting nut. Newer types are equipped with a special slip nut which grips the threads tightly when under pressure but slides freely over them when spring pressure is relieved.

A different type of caliper is the hermaphrodite, sometimes called "odd-leg." This has one straight leg ending in a sharp point, sometimes removable, and one bowleg. It thus combines the function of divider and outside caliper, and is used chiefly for locating the center of a shaft, or for locating a shoulder. To find the center of a shaft, the bowleg is held against the OD, and the legs opened to approximately half the diameter of the work. An arc is then scribed with the sharp point. This is repeated at two other positions, approximately 120° apart, so a small triangle is formed at the center. The exact center can be located by eye at the center of this triangle.

Dividers are somewhat similar to calipers but have both legs straight and terminating in sharp points. They are made in spring-joint and firm-joint types and may be used for measuring distances between two lines or points. Their principal use, however, is for scribing circles in layout work.

How to Use Calipers. In measuring a piece of work, the legs should pass over it with just a slight drag. If the caliper will just barely hang from the work without falling off, the pressure is about right.

When measuring with outside calipers, the axis of the calipers should be held perpendicular to the axis of the work. With inside calipers, the axes of work and caliper should coincide.

Caliper Squares. The chief drawback to ordinary calipers is that they do not give a direct reading; it is necessary to compare them with a standard scale. To overcome this, the caliper square is available. It consists of an L-shaped member with a scale. The toe of the L is flat on top and forms one of the measuring surfaces for outside measurements. The lower side is rounded to form a measuring surface for inside diameters. The shank of the L slides in a stock which carries a mating jaw for inside and outside measurements. Two witness lines are scribed on the face of the stock for reading the scale on the shank, one for inside and the other for outside dimensions.

These tools are commonly made in 3- and 5-in. sizes and are graduated to read in thirty-seconds and sixty-fourths. Such tools are valuable when extreme accuracy is not required and are frequently used for duplicating work when the expense of fixed gages is not warranted.

Vernier Calipers. The vernier caliper permits direct reading of the distance between two surfaces to within 0.001 in. It can be used for both internal and external surfaces and has a wide measuring range. Pocket models usually measure from zero to 3 in., but sizes are available all the way to 4 ft. It consists of an L-shaped member with a scale engraved on the long shank. A sliding member is free to move on the bar and carries a jaw which matches the arm of the L. A second, or vernier, scale is engraved on a small plate attached to the slider. External measurements are taken between the inner faces of the two jaws; internal measurements between the nibs at their outer ends. On some types it is necessary to know the width over the nibs, and add this to the reading obtained. On others, two separate scales are provided.

How to Read a Vernier. This method of dividing known distances into very small parts is said to be the invention of Pierre Vernier in 1631. The vernier consists of a small auxiliary scale attached to the main scale. A certain number of graduations on the vernier are equal in combined length to a different number of

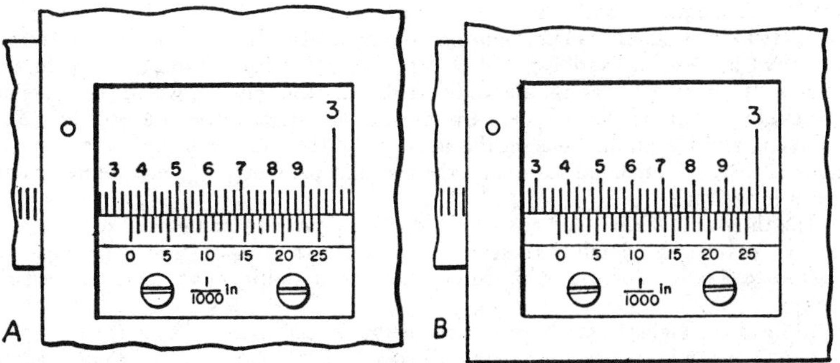

FIG. 3. Vernier scale readings show at *A* a reading in even hundredths, and at *B* a reading in thousandths.

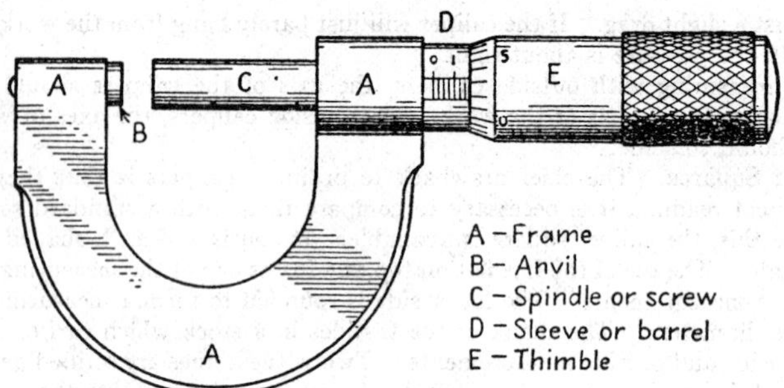

FIG. 4. Simple micrometer caliper for direct reading to thousandths.

graduations, one more or one less, on the main scale. With the vernier it is possible to read exactly a measurement between scale graduations.

Figure 3 shows a vernier used with a scale graduated into fortieths, or 0.025 in. The 25 divisions on the vernier equal 24 divisions on the scale, or $24 \times 0.025 = 0.600$ in. Thus, one vernier division $= 0.600 \div 25 = 0.024$ in. Therefore, the difference between vernier and scale divisions $= 0.025 - 0.024 = 0.001$ in.

If the zero on the vernier does not coincide with a scale division the dimension being measured is not exact to 0.025 in. It is then necessary to find which vernier division coincides with a scale division and to add that number of thousandths to the scale reading.

For example, illustration *A* shows the zero graduation on the vernier coinciding with a fortieth graduation on the scale (the second fortieth beyond an even tenth graduation). This indicates that the reading is exact with respect to fortieths of an inch. The reading therefore equals $2.000 + 0.300 + 0.050$ in. $= 2.350$ in.

In illustration *B*, however, the zero on the vernier scale lies between the second and third fortieths, and the eighteenth vernier graduation coincides with a line on the scale. This indicates that 0.018 in. should be added to the scale reading. The reading then equals $2.000 + 0.300 + 0.050 + 0.018$ in., or 2.368 in.

Micrometer Calipers. The commercial micrometer (Fig. 4) consists of a frame, the anvil or fixed measuring point, the spindle, which has a thread cut 40 to the inch on the portion inside the sleeve or barrel, and the thimble, which goes outside the sleeve and turns the spindle. One turn of the screw moves the spindle $\frac{1}{40}$ or 0.025 in., and the graduations on the sleeve show the number of turns the screw is moved. Every fourth graduation on the barrel is marked 1, 2, 3, etc., representing tenths of an inch.

The thimble has a beveled edge divided into 25 parts and numbered 0, 5, 10, 15, 20, and to 0 again. Each of these means $\frac{1}{25}$ of a turn or $\frac{1}{25}$ of $\frac{1}{40} = 1/1000$ in.

How to Read. To read a measurement to thousandths of an inch, proceed as follows:

1. Read the highest figure exposed on the barrel and multiply by 0.100 in.
2. Add 0.025 in. for each graduation between that figure and the thimble edge.
3. Add the number of 0.001 in. represented by the thimble graduation that coincides with or has just passed the axial line on the barrel scale.

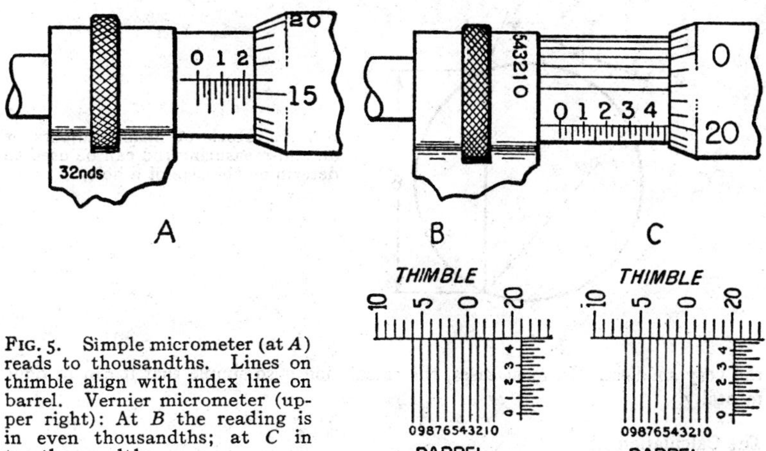

FIG. 5. Simple micrometer (at *A*) reads to thousandths. Lines on thimble align with index line on barrel. Vernier micrometer (upper right): At *B* the reading is in even thousandths; at *C* in ten-thousandths.

EXAMPLE (Fig. 5, left):

		MEASUREMENT
Highest figure on barrel....................	2	2 × 0.100 = 0.200 in.
Graduations visible after 2..................	1	1 × 0.025 = 0.025 in.
Thimble graduation........................	16	16 × 0.001 = 0.016 in.
		Measurement = 0.241 in.

The Ten-thousandth Micrometer. This adds a vernier to the micrometer sleeve or barrel, as shown in Fig. 5 (right), which is read the same as any vernier, as has been explained. First note the thousandths, as in the ordinary micrometer, and then look at the line on the sleeve which just matches a line on the thimble. If both zero lines on barrel match with lines on thimble, the measurement is in even thousandths as at *B*, which reads 0.469 in. At *C* the seventh line matches a line on the thimble so the reading is 0.4697 in.

On micrometers having an auxiliary thimble for direct readings to ten-thousandths, the procedure is somewhat different. The thousandths are read in the conventional manner, and to this reading is added the reading in ten-thousandths taken direct from the auxiliary thimble.

Micrometer calipers are made in standard sizes to read from 0 to ½ in., 0 to 1 in., 1 to 2 in., and by 1-in. steps to 60 in.

Inside Micrometers. The inside micrometer consists of a thimble and barrel similar to an outside micrometer, but a hardened anvil is provided on top of the thimble. Extension rods, furnished in a range of sizes, are attached to the end of the barrel. To use, the tool is inserted in the hole with the rod end on one wall and the thimble is turned until its anvil contacts the opposite wall. The tool must be at right angles to the hole axis.

Checking Holes with End-measuring Rods. The following is an approximate rule for obtaining the variation in the size of a hole corresponding to a given amount of side play of an end-measuring rod introduced into the hole. The rule has the merit of extreme simplicity and can be applied equally well to all diameters, except

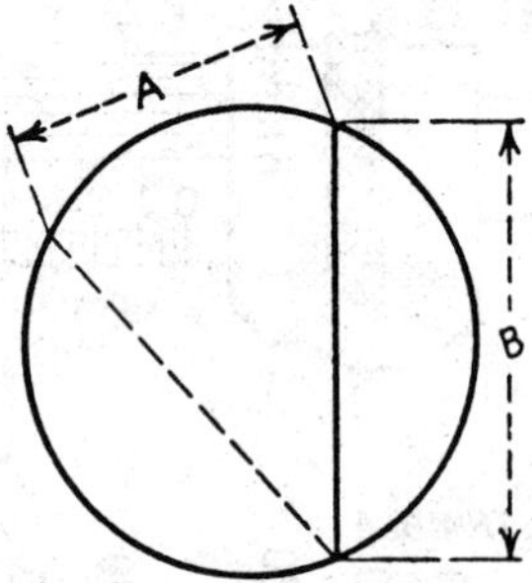

FIG. 6. Rock of inside calipers or an end-measuring rod can be used to determine the size of a hole.

the very smallest. In most cases, the calculation is so simple that it can be done mentally.

The Calculation

Let A in Fig. 6 = side play of calipers or end measuring rod, *in sixteenths of an inch.*
B = dimensions to which calipers are set, or length of measuring rod, *in inches.*
C = difference between diameter of hole and length of B, *in thousandths of an inch.*

Then $C = A^2/2B$, within a very small limit.

EXAMPLE: A standard-end measuring rod, $5\frac{1}{2}$ in. long, has $\frac{3}{8}$ in. of side play in a hole. What is the size of the hole? In this case $A = 6$, and $B = 5\frac{1}{2}$. Apply the above formula:

$$C = \frac{6 \times 6}{11} = \frac{36}{11} = 3.27 \text{ thousandths of an inch, or } 0.00327 \text{ in.}$$

The diameter of the hole, therefore, is $5\frac{1}{2} + 0.00327$ or 5.50327.

The method will be found to be correct within a limit of about 0.0002 in. if the amount of side play is not more than one-eighth of the diameter of the hole for holes up to 6 in. diameter; within 0.0005 in. for holes from 6 to 12 in.; and within 0.001 for holes from 12 to 24 in.

Bore Gaging by Pin Rock. In order to determine the bore of a bearing which is oversize so as to permit a running clearance, the length of the pin gage is made the same as the shaft diameter. After the rock of the pin gage is measured, the clearance may be found in Table 1 (expressed in thousandths of an inch).

EXAMPLE: The measurement of a nominal 6-in. bearing bore shows 0.410-in. rock. The table specifies 0.003-in. clearance for 0.379-in. rock and 0.004-in. clearance for 0.438-in. rock. By interpolation 0.410-in. rock results in a clearance of 0.0035 in.

Table 2 is used where the bore is undersize, such as may be the case when a press fit is required. The pin gage for such bores is made 0.01 in. less than the shaft.

EXAMPLE: For a 5-in. shaft the pin is made 4.99 in. long. Assuming a rock of 0.44 in., Table 2 shows 0.42-in. rock as indicating 0.0055-in. press fit; 0.45-in. rock as 0.005-in. press fit. Therefore, 0.44-in. rock would by interpolation be equivalent to a 0.0052-in. press fit.

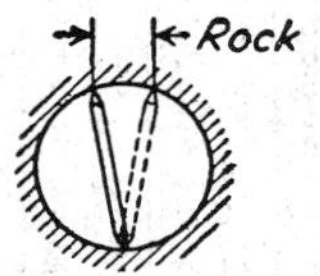

TABLE 1. BEARING CLEARANCE BY THE ROCK OF THE PIN GAGE

Shaft Dia or Length of Pin Gage, In.	Allowances, In.																				
	0.0005	0.001	0.002	0.003	0.004	0.005	0.006	0.007	0.008	0.009	0.010	0.011	0.012	0.013	0.014	0.015	0.016	0.017	0.018	0.019	0.020
	Rock of Pin Gage in Bore																				
3	0.110	0.155	0.219	0.268	0.309	0.346	0.379														
4	0.126	0.179	0.253	0.310	0.357	0.399	0.437	0.472													
5	0.141	0.200	0.283	0.346	0.400	0.447	0.489	0.528													
6	0.155	0.219	0.310	0.379	0.438	0.489	0.536	0.579	0.610												
7	0.167	0.237	0.334	0.410	0.473	0.529	0.579	0.625	0.668	0.709											
8	0.179	0.253	0.358	0.438	0.506	0.505	0.619	0.669	0.715	0.758	0.799										
9	0.190	0.268	0.379	0.465	0.536	0.600	0.657	0.709	0.758	0.804	0.847										
10	0.200	0.283	0.400	0.490	0.565	0.632	0.692	0.748	0.799	0.848	0.893	0.937									
11	0.210	0.297	0.419	0.514	0.593	0.663	0.726	0.784	0.838	0.889	0.937	0.983	1.026								
12	0.220	0.310	0.438	0.536	0.619	0.693	0.759	0.819	0.876	0.929	0.979	1.028	1.072	1.116							
13	0.229	0.322	0.456	0.558	0.645	0.721	0.789	0.853	0.911	0.967	1.019	1.069	1.116	1.161	1.205						
14	0.236	0.335	0.473	0.579	0.669	0.748	0.819	0.885	0.946	1.003	1.057	1.109	1.158	1.205	1.251	1.294					
15	0.245	0.347	0.490	0.600	0.693	0.774	0.848	0.916	0.979	1.039	1.095	1.148	1.199	1.248	1.295	1.340					
16	0.253	0.358	0.506	0.620	0.715	0.800	0.876	0.946	1.011	1.073	1.131	1.186	1.238	1.289	1.337	1.384	1.429				
17	0.261	0.368	0.522	0.639	0.737	0.824	0.903	0.975	1.042	1.106	1.165	1.222	1.276	1.328	1.379	1.427	1.473	1.519			
18	0.269	0.379	0.536	0.657	0.758	0.848	0.929	1.008	1.073	1.138	1.199	1.258	1.313	1.367	1.418	1.468	1.516	1.563	1.608		
19	0.276	0.390	0.551	0.675	0.779	0.871	0.955	1.031	1.102	1.169	1.232	1.292	1.350	1.404	1.457	1.509	1.558	1.606	1.652	1.697	
20	0.283	0.400	0.565	0.692	0.800	0.894	0.979	1.057	1.131	1.199	1.264	1.326	1.385	1.441	1.495	1.548	1.598	1.648	1.695	1.742	1.787

TABLE 2. PRESS FIT BY THE ROCK OF THE PIN GAGE

Dia of Shaft, In.	Length of Pin Gage, In.	Press-fit Allowances, 0.0001 In.															
		Less 9	Less 8	Less 7	Less 6½	Less 6	Less 5½	Less 5	Less 4½	Less 4	Less 3½	Less 3	Less 2½	Less 2	Less 1½	Less 1	Less ½
3	2.99	0.16	0.22	0.27	0.29	0.31	0.33	0.35	0.36	0.38	0.39	0.41	0.42	0.44	0.45	0.46	0.48
3½	3.49	0.17	0.24	0.29	0.31	0.33	0.35	0.37	0.39	0.41	0.42	0.44	0.46	0.47	0.49	0.50	0.51
4	3.99	0.18	0.25	0.31	0.33	0.36	0.38	0.40	0.42	0.44	0.46	0.47	0.49	0.51	0.52	0.54	0.55
4½	4.49	0.19	0.27	0.33	0.35	0.38	0.40	0.42	0.44	0.46	0.48	0.50	0.52	0.53	0.55	0.57	0.58
5	4.99	0.20	0.28	0.35	0.37	0.40	0.42	0.45	0.47	0.49	0.51	0.53	0.55	0.56	0.58	0.60	0.62
5½	5.49	0.21	0.30	0.36	0.39	0.42	0.44	0.47	0.49	0.51	0.53	0.55	0.57	0.59	0.61	0.63	0.64
6	5.99	0.22	0.31	0.38	0.41	0.44	0.46	0.49	0.51	0.54	0.56	0.58	0.60	0.62	0.64	0.66	0.67
6½	6.49	0.23	0.32	0.39	0.43	0.46	0.48	0.51	0.53	0.56	0.58	0.60	0.62	0.64	0.66	0.68	0.70
7	6.99	0.24	0.33	0.41	0.44	0.47	0.50	0.53	0.55	0.58	0.60	0.63	0.65	0.67	0.69	0.71	0.73
7½	7.49	0.24	0.35	0.42	0.46	0.49	0.52	0.55	0.57	0.60	0.62	0.65	0.67	0.69	0.71	0.73	0.75
8	7.99	0.25	0.36	0.44	0.47	0.51	0.54	0.57	0.59	0.62	0.65	0.67	0.69	0.72	0.74	0.76	0.78
8½	8.49	0.26	0.37	0.45	0.49	0.52	0.55	0.58	0.61	0.64	0.66	0.69	0.71	0.74	0.76	0.78	0.80
9	8.99	0.27	0.38	0.47	0.50	0.54	0.57	0.60	0.63	0.66	0.68	0.71	0.73	0.76	0.78	0.80	0.83
9½	9.49	0.28	0.39	0.48	0.52	0.55	0.58	0.62	0.65	0.67	0.70	0.73	0.75	0.78	0.80	0.83	0.85
10	9.99	0.28	0.40	0.49	0.53	0.57	0.60	0.63	0.66	0.69	0.72	0.75	0.77	0.80	0.82	0.85	0.87
10½	10.49	0.29	0.41	0.50	0.54	0.58	0.61	0.65	0.68	0.71	0.74	0.77	0.79	0.82	0.84	0.87	0.89
11	10.99	0.30	0.42	0.51	0.55	0.59	0.63	0.66	0.70	0.73	0.76	0.78	0.81	0.84	0.86	0.89	0.91
11½	11.49	0.30	0.43	0.52	0.57	0.60	0.64	0.68	0.71	0.74	0.77	0.80	0.83	0.86	0.88	0.91	0.93
12	11.99	0.31	0.44	0.54	0.58	0.62	0.66	0.69	0.73	0.76	0.79	0.82	0.85	0.88	0.90	0.93	0.95
12½	12.49	0.32	0.45	0.55	0.59	0.63	0.67	0.71	0.74	0.77	0.81	0.84	0.87	0.89	0.92	0.95	0.97
13	12.99	0.32	0.46	0.56	0.60	0.65	0.68	0.72	0.76	0.79	0.82	0.85	0.88	0.91	0.94	0.97	0.99
13½	13.49	0.33	0.46	0.57	0.61	0.66	0.70	0.73	0.77	0.80	0.84	0.87	0.90	0.93	0.96	0.99	1.01
14	13.99	0.34	0.47	0.58	0.63	0.67	0.71	0.75	0.78	0.82	0.85	0.89	0.92	0.95	0.98	1.00	1.03
14½	14.49	0.34	0.48	0.59	0.64	0.68	0.72	0.76	0.80	0.83	0.87	0.90	0.93	0.96	0.99	1.02	1.05
15	14.99	0.35	0.49	0.60	0.65	0.69	0.74	0.77	0.81	0.85	0.88	0.92	0.95	0.98	1.01	1.04	1.07
15½	15.49	0.35	0.50	0.61	0.66	0.70	0.75	0.79	0.83	0.86	0.90	0.93	0.96	1.00	1.03	1.06	1.08
16	15.99	0.36	0.51	0.62	0.67	0.72	0.76	0.80	0.84	0.88	0.91	0.95	0.98	1.01	1.04	1.07	1.10
16½	16.49	0.36	0.51	0.63	0.68	0.73	0.77	0.81	0.85	0.89	0.93	0.96	0.99	1.03	1.06	1.09	1.12
17	16.99	0.37	0.52	0.64	0.69	0.74	0.78	0.82	0.86	0.90	0.94	0.98	1.01	1.04	1.07	1.11	1.14
17½	17.49	0.37	0.53	0.65	0.70	0.75	0.79	0.84	0.88	0.92	0.95	0.99	1.02	1.06	1.09	1.12	1.15
18	17.99	0.38	0.54	0.66	0.71	0.76	0.80	0.85	0.89	0.93	0.97	1.00	1.04	1.07	1.11	1.14	1.17
18½	18.49	0.38	0.54	0.67	0.72	0.77	0.82	0.86	0.90	0.94	0.98	1.02	1.05	1.09	1.12	1.15	1.19
19	18.99	0.39	0.55	0.68	9.73	0.78	0.83	0.87	0.91	0.96	0.99	1.03	1.07	1.10	1.14	1.17	1.20
19½	19.49	0.39	0.56	0.68	0.74	0.79	0.84	0.88	0.93	0.97	1.01	1.04	1.08	1.12	1.16	1.18	1.22
20	19.99	0.40	0.57	0.69	0.75	0.80	0.85	0.89	0.94	0.98	1.02	1.06	1.10	1.13	1.17	1.20	1.23

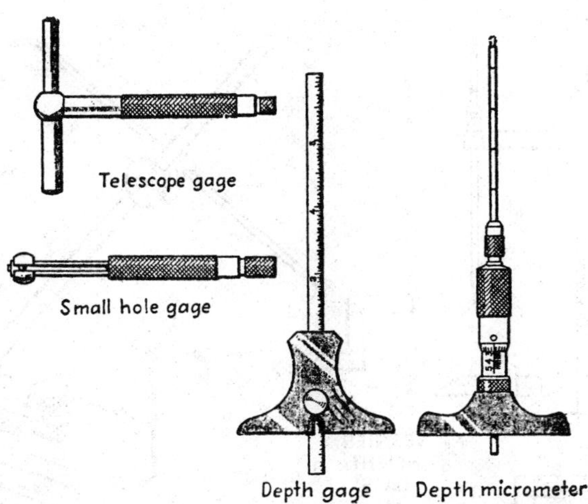

FIG. 7. Telescoping gages are usually furnished in sets for measuring hole diameters. Depth gages can be of the plain or micrometer pattern for checking depths of holes.

Other Inside Tools. The telescoping gage (Fig. 7) is a T-shaped tool in which the shaft of the T acts as a handle, and the crossarm is used for measuring. The crossarms telescope into each other and are held out by a light spring. To use, the arms are compressed, placed in the hole, and allowed to expand. A twist of the locknut on the handle locks the arms, and the tool may then be withdrawn and the distance across the arms measured with an outside micrometer.

For smaller work, a small hole gage may be employed. This consists of a small split ball-shaped member mounted on the end of a handle. The ball is expanded by turning a knurled knob until the proper feel is obtained, and then the size is measured with a micrometer. The smallest unit will measure from about $\frac{1}{8}$ to $\frac{3}{16}$ in.

Depth Gages. The depth of holes or slots is usually measured with a depth gage. This consists of a small flat or round scale attached to a flat head by a slotted screw.

To use this tool, the binding screw is loosened, the head placed on a flat surface, and the scale raised or lowered as needed. The screw is then tightened, the tool removed from the hole, and the depth read directly from the scale. If more exact measurements are required, a depth micrometer may be employed.

Height Gages. For measuring and marking off vertical distances from a plane surface, the vernier height gage (Fig. 8) is indispensable. The member which slides on the scale carries a jaw to which a hardened steel scriber may be clamped.

To use the tool for layout, the scriber is adjusted to the required distance above the foot by means of the vernier. The gage and the work are then placed on a surface plate, and the necessary layout lines scribed with the scriber. If the required lines are below the height of the gage foot, a special offset scriber will be required.

In taking a measurement, the lower face of the scriber is used, and the slider is brought down until the scriber just contacts the work surface.

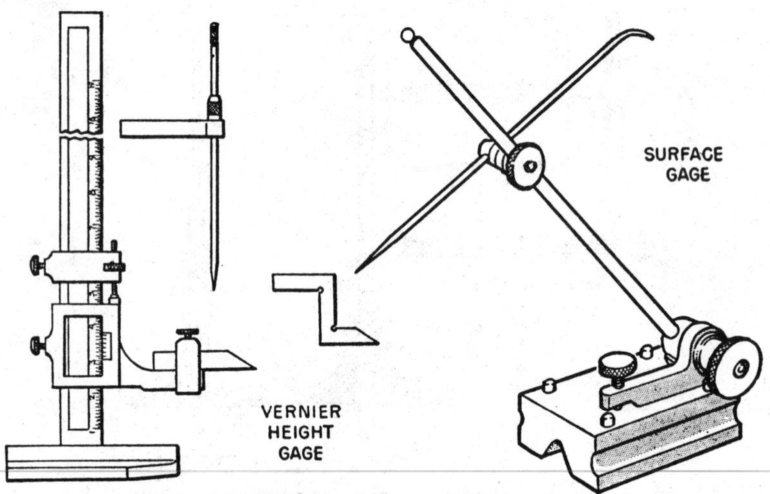

FIG. 8. For scribing lines at desired vertical distances from a plane surface (the surface plate), height gages and surface gages are employed.

For measuring inside recesses, a depth-gage attachment may be attached to the jaw by a clamp. It does not give a direct reading. The rod must first be adjusted to the upper face of the work and a reading taken on the vernier. Then, when the rod is adjusted to the bottom of the hole, the depth from the top face will be the difference between the two readings.

Surface Gages. If less precision is required, the conventional surface gage (Fig 8) is satisfactory for layout work. This consists of a heavy base and a pivoted upright to which a scriber is attached by a clamp. The scriber can be clamped at any point along its own length, or along the length of the upright, and may be rotated through a full circle. One end is bent to permit scribing on horizontal surfaces. The height of the scriber is set to a scale or other reference.

Miscellaneous Gages. Thickness gages, commonly called feeler gages, are thin strips of flexible steel of known thickness, usually made up in groups of from 10 to 25 different pieces, and hinged to fold into a handle like a pocket knife. Each leaf may be used separately or in combination with others. They are used for checking clearances between surfaces, and for similar work where no other tool is practicable.

Radius gages consist of groups of steel blades, made up in pocket-knife style, with accurately formed radiuses on their ends. Usually, the blades at one end of the holder are for convex radiuses, and those at the other end for concave radiuses. Individual blades are also available in sets with a detachable handle.

Angle gages are similar but have the ends of the blades ground to the more commonly used angles. They are particularly useful for checking angles on tools and dies, and in locations where a protractor cannot be applied conveniently.

Screw-pitch gages are used to determine the pitch of threads. They consist of thin leaves whose edges are toothed to correspond to standard thread sections.

Successive leaves may be placed on a thread of unknown pitch until some one leaf is found to coincide exactly, and the pitch can be read from the number stamped on it. These gages are available in all standard and metric thread forms.

For rapid checking of wire and sheet stock, a wire gage will be found convenient. This is a circular metal plate, having radial keyhole notches cut into its edge. Each notch is numbered with its corresponding wire size, and the readings are taken between the flat sides of the notch. Models are available for each of the different gage systems for wire and sheet and for different materials.

Tap and drill gages are hardened steel plates, having a number of holes corresponding to the standard drill diameters. They furnish an easy means of checking drill sizes, and usually have marked on them the correct size of drill for tapping and for clearance holes for standard machine screws.

When threads are cut with a single-point tool, it is imperative that the tool be exactly at right angles to the axis of the work. This can be checked with a center gage, a small steel blade ground at one end to a sharp 60° point, and having a deep 60° notch in the other end. Smaller 60° notches are formed in one or both edges. In use, the gage is held against the work, and the tool is adjusted until it fits exactly into one of the notches in the edge.

Gage Blocks. Gage blocks of hardened steel or carbide provide industry with basic references—in the plant's gage laboratory, in the inspection department, in the toolroom, and in the manufacturing areas. By assembling a stack of blocks with thicknesses adding up to the desired dimension, that dimension is obtained with utmost precision—any error will amount to only a very few millionths of an inch. The same dimension will also be obtained for any practical purpose if other blocks from a set are selected, provided that their nominal thicknesses add up to the desired dimension. In fact, gage blocks provide the common denominator for quality control within a plant, and from plant to plant.

Manufacturers of gage blocks produce their product in relationship to masters that are checked periodically by the National Bureau of Standards. The resulting

TABLE 3. REJECTION LIMITS FOR GAGE BLOCKS*

	Quality Classifications		
	A	B	C
Length:			
Tolerance, in.	±0.000004	±0.000008	±0.0000010
Rejection limit, in.	±0.000007	±0.000011	±0.000015
Parallelism (across width):			
Rejection limit, in.†	±0.000008	±0.000010	±0.000012
Flatness error (rejection limit):			
Across width, in.‡	0.000006	0.000008	0.000010
Along length	0.000008	0.000010	0.000012
Microflatness (rejection limit):			
Profilometer, rms	0.0000012	0.0000020	0.0000025

* National Bureau of Standards Letter Circular LC 725.

† Blocks are also rejected if sum of length and parallelism errors exceeds rejection limit for length.

‡ For square blocks the rejection limit on flatness error across the width is 0.000008 in.

blocks offered for sale are graded into four classifications. The AA blocks have a tolerance of ±0.000002 in. (two millionths) and are intended for reference purposes in temperature-controlled gage laboratories. These "masters" should be sent periodically to the National Bureau of Standards, Washington, D.C., for checking.

The A blocks are used in inspection departments and also may serve as masters. Blocks of B and C quality are used throughout the shop for accurate measurements and tool setting. Rejection limits for A, B, and C grade blocks are given in Table 3.

USE OF GAGE BLOCKS. Sets of gage blocks are available with various numbers and combinations of blocks. For example, a "standard" 83-block set will make more than 200,000 combinations in steps of 0.0001 in. from 0.200 to 12 in. Such a set would consist of 9 blocks in the 0.0001-in. series, ranging from 0.1001 to 0.1009 in.; 49 blocks in the 0.001-in. series, and ranging from 0.101 to 0.149 in.; 19 blocks in the 0.050-in. series, and ranging from 0.050 to 0.950 in.; 4 blocks ranging from 1.000 to 4.000 in. by 1-in. steps, and two 0.050-in. wear blocks.

EXAMPLE: Suppose with this set you are to make a combination for 4.6513 in. Employ as few blocks as possible. So work from right to left:

		PROOF
Dimension sought	4.6513	
Eliminate the 3 with	0.1003-in. block	0.1003
Result	4.5510	
Eliminate the 1 with	0.101-in. block	0.1010
Result	4.4500	
Eliminate the 45 with	0.450-in. block	0.450
Result	4.0000	
Last block	4.0000	4.000
	Total	4.6513

Optical Flats. Every working day, gages, tools, and highly accurate parts are measured for flatness, length, diameter, and taper to a few millionths of an inch. These operations are no more difficult than reading a micrometer.

No secret is involved. The inspector or toolmaker uses the optical flat, an inspection device in growing use since World War I, whereby the single unalterable standard of measurement—the light wave—is put to convenient practical application.

Like many important discoveries, the optical flat is inherently simple. It is nothing more than a transparent disk, preferably of quartz. The two sides of the disk are parallel. One side is polished for clear vision of the surface of the gage, part, or tool upon which the flat is placed. The other side of the disk is ground *optically flat.*

With proper conditions, the phenomenon of interference bands is created with an optical flat. As seen through the flat, these bands appear as a pattern of dark stripes on the illuminated work surface (Fig. 9). Interference bands can be used with great accuracy for two distinct purposes:

1. To determine the *flatness* of a surface, that is, the location and amount of concavity or convexity.

2. To measure the *linear difference* between a reference gage and an inspection gage, or between a gage block and a highly accurate part. Any linear difference can be detected between approximately 0.002 in. as a maximum and one to two millionths as a minimum.

The optical flat is thus a powerful inspection aid to gagemakers, toolmakers, and inspectors. With *one* flat, the user can see at a glance whether the surface of a

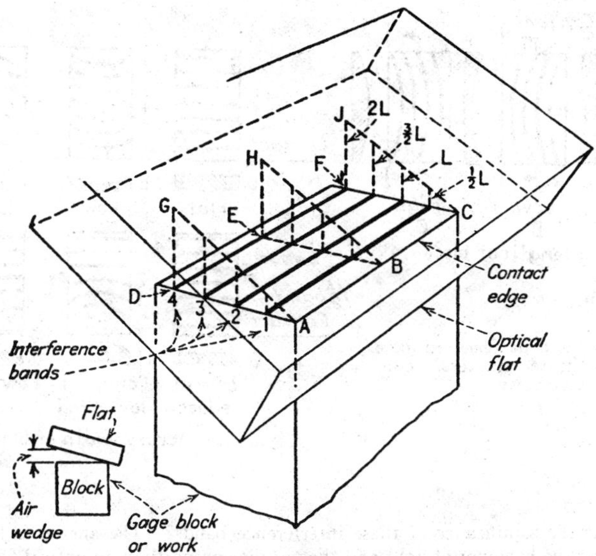

FIG. 9. When an optical flat is manipulated so that a wedge of air is created, dark bands, or interference bands, will be created at locations where the separation of flat and work equals a multiple of $L/2$, or one-half wavelength of the monochromatic light used.

gage block, micrometer anvil, or precision-lapped product is flat and, if not, the location and amount of high and low spots. With *two* flats he can make simple setups to check: gage blocks for height, flatness, and parallelism; plug gages for diameter and taper; thread plug gages for pitch diameter, and thread-measuring wires and balls for diameter. And optical flats are of use for inspecting other precision inspection and toolmaking accessories; namely, master squares, flats, and parallels.

Optical flats can also be used for quality control of ultra-precision products. At the present time, flat surfaces on various mechanical parts (for example, sealing members) are precision-lapped to extremely close tolerances for flatness and surface finish. In fact, these parts are produced to "gage block" accuracy, and the speed and accuracy of checking them by the optical-flat method present definite advantages.

HOW THE OPTICAL FLAT WORKS. Suppose two objects with nearly plane surfaces (an optical flat and gage block) are pressed together. It is practically impossible to squeeze out the air film between them. If uniform pressure is applied downward on the flat, so that the air film is not distorted, nothing will be seen on the gage block if it is illuminated with monochromatic, or single-color, light. But if the flat is prodded into contact with one edge of the gage block, two things happen: (1) the air film is distorted to a wedge shape (Fig. 9), and (2) dark bands will be seen on the surface of the block.

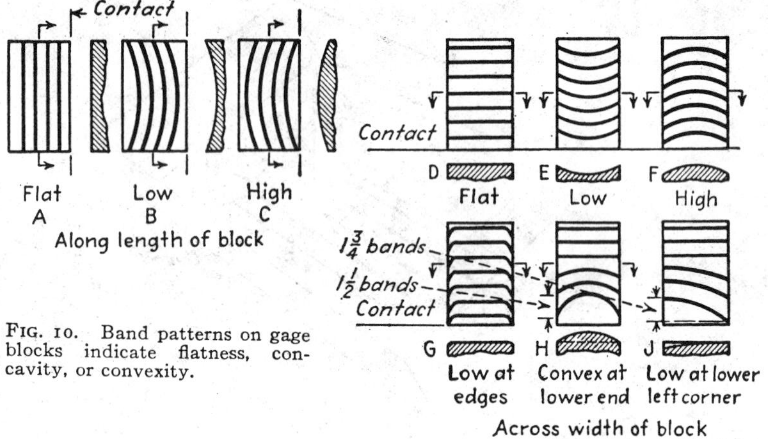

FIG. 10. Band patterns on gage blocks indicate flatness, concavity, or convexity.

What is the significance of these interference bands? The answer is found by considering the behavior of light and the peculiar property of an optical flat.

Light travels in a wave motion. For each kind of monochromatic light, the crest-to-crest distance of the wave motion, or wavelength L, is constant; that is, red light has one wavelength, yellow light another wavelength.

When monochromatic light falls on a flat, and there is a wedge of air between flat and work, interference bands are seen. These dark bands are caused by interference of light reflected from the work surface with light reflected from the undersurface of the flat. The thickness of the air gap at any band equals $T = N \times L/2$, where N is the number of bands counted from the contact edge.

As shown in Fig. 9, the first band from the contact edge occurs where the thickness of the air wedge equals $L/2$. Band 2 falls at an air gap of L, band 3 at $\frac{3}{2}L$, etc. The spacing of the bands varies according to the steepness of the air wedge and can be varied with finger pressure on the flat.

An interference-band pattern is analogous to a contour map. In the map, configuration of land is shown by lines connecting points of similar altitude above sea level. In the band pattern, a given band connects points where interference is a specific multiple of $L/2$. By observing a band pattern it is possible to tell whether a work surface is flat, convex, concave, or irregular (see Fig. 10).

Straight bands indicate a flat surface. Curved or irregular bands show lack of flatness. To tell whether a surface is concave or convex, use these rules:

Rule 1. Band ends that curve toward the line of contact indicate a convex surface.

Rule 2. Band ends that curve away from the line of contact indicate a concave surface.

Rule 3. The flatness error (amount of concavity or convexity) equals the fractional curvature times one band.

Connect the ends of a band with an imaginary dotted line. Then the flatness error $= H/B \times L/2$, where L is the wavelength of the monochromatic light used. Table 4 gives values of bands for three kinds of light—helium, sodium, and mercury.

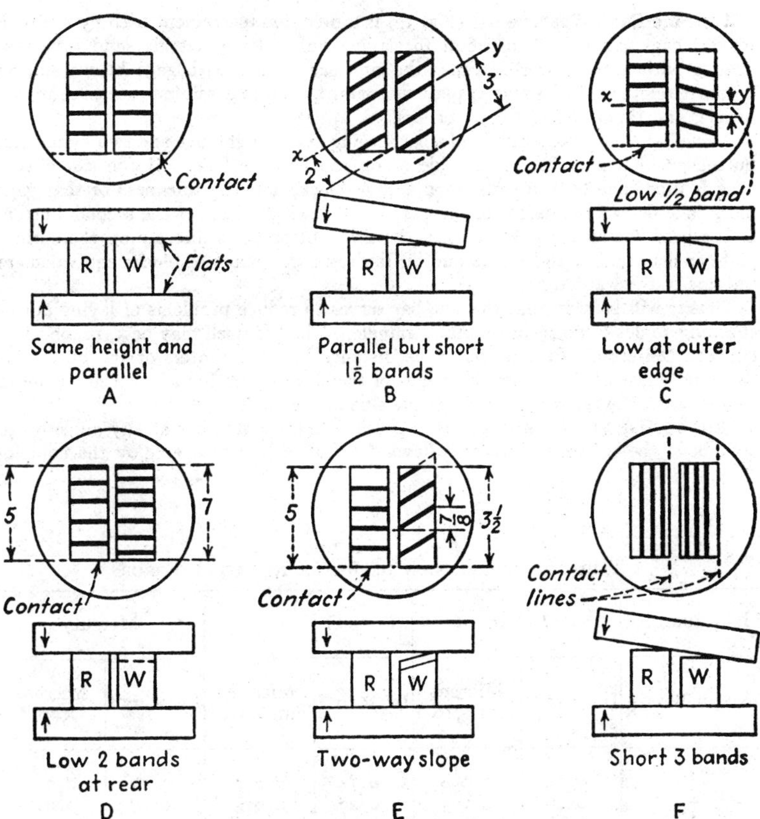

FIG. 11. Length and parallelism of gage blocks can be checked by this sequence of operations, or a workpiece can be checked in relation to a gage block.

With helium light, the value of one band $L/2 = 0.0000116$, or 11.6 millionths of an inch. By prodding the flat with the finger at the contact edge to space the bands properly, one can estimate the lack of flatness to tenths of a band with fair accuracy.

HOW TO CHECK LENGTH AND PARALLELISM. For convenience, the discussion will be confined to checking a working gage block against a reference block. Rejection limits for parallelism and length of gage blocks are given in Table 3. Select two optical flats with a diameter at least $1\frac{1}{2}$ times the greatest over-all dimension of the two gage blocks placed side by side. Clean the flats and blocks with tissue or cloth moistened with petroleum ether or alcohol. Rub the blocks in the palm and wring them to one flat. The reference block R is to the left at B, Fig. 11. Lay the second flat on top of the blocks; prod gently. Similar band patterns for the two blocks, as at A, are unlikely if the working block has seen much service. Compare the patterns obtained with those at B to F. Cross-check your results. Then calculate the errors in length and parallelism according to the methods illustrated.

The Sine Bar. The sine bar (Fig. 12) is a precision instrument used by the toolmaker, gagemaker, and inspector for laying out, setting, testing, and otherwise dealing with angular work. Since the sine bar is used with gage blocks and an accurate reference like a master plate flat or surface, it is possible to measure angles, bevels, and tapers down to 10″ or less.

A sine bar consists essentially of a bar serving as a straightedge and two cylindrical buttons, which may be on the side or on the undersurface. If the side-button type is used, one button rests upon a gage block, and the thickness of that gage block is added to the height of the gage-block stack used to set the second button. If the base-button sine bar is used, the first button rests directly on the surface plate or master flat, and the second button rests on a stack of blocks equivalent to the sine of the wanted angle.

Thus it will be seen that the sine bar serves to reduce problems of laying out or checking angles to terms of the right triangle. The bar itself may be 5, 10, or 20 in. between centers of the buttons. The distance between centers of the buttons is the hypotenuse of the angle; the height of the elevated button above the button in contact with the reference surface is the sine of the angle.

How to Use the Sine Bar. It is possible to set a sine bar at any angle up to 60° above the reference surface. Beyond 60° it is better to employ the comple-

Table 4. Conversion of Bands to Lineal Error

Light Source..	Helium		Sodium		Mercury	
No. of Bands	In. $\times 10^{-6}$	Microns mm $\times 10^{-3}$	In. $\times 10^{-6}$	Microns mm $\times 10^{-3}$	In. $\times 10^{-6}$	Microns $\times 10^{-3}$
1/8	0.14	0.038	0.14	0.037	0.13	0.034
1/4	0.29	0.073	0.29	0.074	0.27	0.068
1/2	0.58	0.147	0.58	0.147	0.54	0.136
1	1.16	0.294	1.16	0.295	1.07	0.273
2	2.31	0.588	2.32	0.589	2.15	0.546
3	3.47	0.881	3.48	0.884	3.22	0.819
4	4.63	1.175	4.64	1.179	4.30	1.092
5	5.78	1.469	5.80	1.473	5.37	1.365
6	6.94	1.736	6.96	1.768	6.45	1.638
7	8.10	2.057	8.12	2.063	7.52	1.911
8	9.27	2.351	9.29	2.357	8.60	2.184
9	10.41	2.645	10.44	2.652	9.67	2.457
10	11.57	2.938	11.60	2.946	10.75	2.730

Note: Inch measurements are expressed in hundred-thousandths of an inch.

The helium source is particularly recommended for work with optical flats for several reasons. First, a helium tube can be shaped in such a way and be made of sufficient length so that it is able to provide good illumination over a relatively large area. Second, these tubes (operated from an ordinary sign transformer) have a long useful life and can be easily replaced by a local sign manufacturer in case of breakage.

The character of the spectrum from the helium tube is such that sharp bands are obtained without the use of filters. At the level of brightness after reflection from the optical flat, the field is predominately yellow.

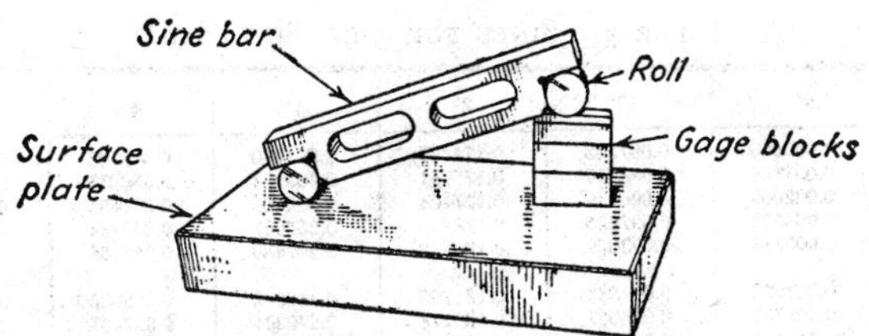

FIG. 12. With the sine bar, work can be located in accurate relationship to the reference plane, or surface plate.

mentary angle. With Table 5 giving the sines for the 5-in. bar to 60° and Table 6 giving constants per second, it is possible to use the 5-in. bar in two ways: (1) to set the bar at the required angle, or (2) to find the angle, if the height of the button above the reference plane is known.

EXAMPLE 1: To set the 5-in. bar at an angle of 36°19′42″.

$$\begin{aligned} \text{From Table 5, sine of } 36°19' &= 2.961238 \\ \text{From Table 6, } 42 \times 0.0000192 &= \underline{0.000806} \\ &\ 2.962044 \end{aligned}$$

The last figure is the height of a stack of gage blocks to be used in setting the 5-in. sine bar to the desired angle.

EXAMPLE 2: To find the angle corresponding to a height of 2.962044. Look down the table of sines for the 5-in. bar to find that this value lies between the sines for 36°19′ and 36°20′. The difference between the height setting and the sine for 36°19′ = 2.962044 − 2.961238 = 0.000806. Since the constant per second for an angle between 35 and 40° = 0.0000192, the number of seconds involved = 0.000806 ÷ 0.0000192 = 42″ approx. Therefore, the angle involved = 36°19′42″.

10- AND 20-IN. BARS. Although the 5-in. bar is most commonly used, the 10- and 20-in. bars are employed where greater accuracy is needed. The table of sines for a 5-in. bar can still be used: for a 10-in. bar multiply the tabulated value by 2; for a 20-in. bar multiply by 4.

GAGING EQUIPMENT

Purpose of Gaging. The purpose of gaging is to determine compliance of the component parts of a mechanical product with the dimensional requirements of contracts, drawings, and specifications. Gaging thereby assures proper functioning and interchangeability of parts.

Definition of a Gage. The term "gage" refers to those devices or instruments actually used in the acceptance or rejection of parts and assemblies insofar as dimensional limits or functional features are concerned. The term "gage" is not intended to cover test equipment or precision measuring equipment.

Why Gages Are Used. Normally any component that can be gaged can also be measured by general-purpose measuring equipment. However, the use of measuring equipment requires a skilled operator and is a slow process where the human element of error cannot be eliminated. The prime purpose of a gage is to provide the inspector with a positive single-purpose device which minimizes the human element of error and quickly indicates all unacceptable parts.

TABLE 5. SINES FOR 5-IN. SINE BAR*

Min.	0°	1°	2°	3°	4°	5°
0′	0.000000	0.087262	0.174497	0.261680	0.348782	0.435779
1′	0.001454	0.088716	0.175951	0.263132	0.350233	0.437228
2′	0.002909	0.090170	0.177404	0.264585	0.351684	0.438676
3′	0.004363	0.091625	0.178858	0.266037	0.353134	0.440125
4′	0.005818	0.093079	0.180312	0.267489	0.354586	0.441574
5′	0.007272	0.094533	0.181765	0.268942	0.356036	0.443023
6′	0.008727	0.095987	0.183218	0.270394	0.357487	0.444471
7′	0.010181	0.097441	0.184672	0.271846	0.358938	0.445920
8′	0.011635	0.098895	0.186125	0.273299	0.360389	0.447369
9′	0.013090	0.100350	0.187579	0.274751	0.361839	0.448817
10′	0.014544	0.101804	0.189032	0.276203	0.363290	0.450266
11′	0.015999	0.103258	0.190486	0.277655	0.364740	0.451714
12′	0.017453	0.104712	0.191939	0.279107	0.366191	0.453163
13′	0.018908	0.106166	0.193392	0.280560	0.367641	0.454611
14′	0.020362	0.107620	0.194846	0.282012	0.369092	0.456060
15′	0.021816	0.109074	0.196299	0.283464	0.370542	0.457508
16′	0.023271	0.110528	0.197752	0.284916	0.371993	0.458956
17′	0.024725	0.111983	0.199206	0.286368	0.373443	0.460405
18′	0.026180	0.113437	0.200659	0.287820	0.374894	0.461853
19′	0.027634	0.114891	0.202112	0.289272	0.376344	0.463301
20′	0.029089	0.116345	0.203565	0.290724	0.377794	0.464749
21′	0.030543	0.117799	0.205019	0.292176	0.379244	0.466197
22′	0.031997	0.119253	0.206472	0.293628	0.380695	0.467645
23′	0.033452	0.120707	0.207925	0.295080	0.382145	0.469093
24′	0.034906	0.122169	0.209378	0.296532	0.383595	0.470541
25′	0.036361	0.123615	0.210831	0.297984	0.385045	0.471989
26′	0.037815	0.125069	0.212284	0.299436	0.386495	0.473437
27′	0.039269	0.126523	0.213738	0.300887	0.387945	0.474885
28′	0.040724	0.127977	0.215191	0.302339	0.389395	0.476333
29′	0.042178	0.129431	0.216644	0.303791	0.390845	0.477781
30′	0.043633	0.130885	0.218097	0.305243	0.392295	0.479229
31′	0.045087	0.132339	0.219550	0.306694	0.393745	0.480676
32′	0.046541	0.133793	0.221003	0.308146	0.395195	0.482124
33′	0.047996	0.135246	0.222456	0.309598	0.396645	0.483572
34′	0.049450	0.136700	0.223909	0.311049	0.398095	0.485019
35′	0.050904	0.138154	0.225362	0.312501	0.399545	0.486467
36′	0.052359	0.139608	0.226815	0.313956	0.400995	0.487914
37′	0.053813	0.141062	0.228268	0.315404	0.402444	0.489362
38′	0.055268	0.142516	0.229721	0.316856	0.403894	0.490809
39′	0.056722	0.143970	0.231174	0.318307	0.405344	0.492257
40′	0.058176	0.145424	0.232626	0.319759	0.406793	0.493704
41′	0.059631	0.146877	0.234079	0.321210	0.408243	0.495151
42′	0.061085	0.148331	0.235532	0.322661	0.409692	0.496599
43′	0.062539	0.149785	0.236985	0.324113	0.411142	0.498046
44′	0.063994	0.151239	0.238438	0.325564	0.412591	0.499493
45′	0.065448	0.152692	0.239891	0.327016	0.414041	0.500940
46′	0.066902	0.154146	0.241343	0.328467	0.415490	0.502387
47′	0.068357	0.155600	0.242796	0.329918	0.416940	0.503834
48′	0.069811	0.157054	0.244249	0.331369	0.418389	0.505281
49′	0.071265	0.158507	0.245701	0.332821	0.419838	0.506728
50′	0.072719	0.159961	0.247154	0.334272	0.421288	0.508175
51′	0.074174	0.161415	0.248607	0.335723	0.422737	0.509622
52′	0.075628	0.162869	0.250059	0.337174	0.424186	0.511069
53′	0.077082	0.164322	0.251512	0.338625	0.425635	0.512516
54′	0.078537	0.165776	0.252965	0.340076	0.427085	0.513963
55′	0.079991	0.167229	0.254417	0.341527	0.428534	0.515409
56′	0.081445	0.168683	0.255870	0.342978	0.429983	0.516856
57′	0.082899	0.170137	0.257322	0.344429	0.431432	0.518303
58′	0.084354	0.171590	0.258775	0.345880	0.432881	0.519749
59′	0.085808	0.173044	0.260227	0.347331	0.434330	0.521196
60′	0.087262	0.174497	0.261680	0.348782	0.435779	0.522642

* Courtesy of The DoAll Co.

TABLE 5. SINES FOR 5-IN. SINE BAR (*Continued*)

Min.	6°	7°	8°	9°	10°	11°
0′	0.522642	0.609347	0.695865	0.782172	0.868241	0.954045
1′	0.524089	0.610790	0.697306	0.783609	0.869673	0.955473
2′	0.525535	0.612234	0.698746	0.785045	0.871105	0.956900
3′	0.526981	0.613677	0.700186	0.786482	0.872538	0.958328
4′	0.528428	0.615121	0.701626	0.787918	0.873970	0.959755
5′	0.529874	0.616564	0.703066	0.789354	0.875402	0.961182
6′	0.531320	0.618007	0.704506	0.790790	0.876834	0.962610
7′	0.532766	0.619451	0.705946	0.792226	0.878265	0.964037
8′	0.534213	0.620894	0.707386	0.793662	0.879697	0.965464
9′	0.535659	0.622337	0.708826	0.795098	0.881129	0.966891
10′	0.537105	0.623780	0.710265	0.796534	0.882561	0.968318
11′	0.538551	0.625223	0.711705	0.797970	0.883992	0.969745
12′	0.539997	0.626666	0.713145	0.799406	0.885424	0.971172
13′	0.541443	0.628109	0.714584	0.800842	0.886855	0.972598
14′	0.542888	0.629552	0.716024	0.802277	0.888286	0.974025
15′	0.544334	0.630995	0.717463	0.803713	0.889718	0.975452
16′	0.545780	0.632438	0.718902	0.805148	0.891149	0.976878
17′	0.547226	0.633880	0.720342	0.806583	0.892580	0.978304
18′	0.548671	0.635323	0.721781	0.808019	0.894011	0.979731
19′	0.550117	0.636766	0.723220	0.809454	0.895442	0.981157
20′	0.551563	0.638208	0.724659	0.810890	0.896873	0.982583
21′	0.553008	0.639651	0.726098	0.812325	0.898304	0.984009
22′	0.554454	0.641093	0.727537	0.813760	0.899734	0.985435
23′	0.555899	0.642536	0.728976	0.815195	0.901165	0.986861
24′	0.557345	0.643978	0.730415	0.816630	0.902596	0.988287
25′	0.558790	0.645420	0.731854	0.818064	0.904026	0.989712
26′	0.560235	0.646862	0.733293	0.819499	0.905457	0.991138
27′	0.561681	0.648305	0.734731	0.820934	0.906887	0.992563
28′	0.563126	0.649747	0.736170	0.822369	0.908317	0.993989
29′	0.564571	0.651189	0.737608	0.823803	0.909747	0.995414
30′	0.566016	0.652631	0.739047	0.825238	0.911178	0.996840
31′	0.567461	0.654073	0.740485	0.826672	0.912608	0.998265
32′	0.568906	0.655515	0.741924	0.828107	0.914038	0.999690
33′	0.570351	0.656957	0.743362	0.829541	0.915467	1.001115
34′	0.571796	0.658398	0.744800	0.830975	0.916897	1.002540
35′	0.573241	0.659840	0.746239	0.832410	0.918327	1.003965
36′	0.574686	0.661282	0.747677	0.833844	0.919757	1.005390
37′	0.576130	0.662724	0.749115	0.835278	0.921186	1.006814
38′	0.577575	0.664165	0.750553	0.836712	0.922616	1.008239
39′	0.579020	0.665607	0.751991	0.838146	0.924045	1.009663
40′	0.580464	0.667048	0.753428	0.839579	0.925475	1.011088
41′	0.581909	0.668489	0.754866	0.841013	0.926904	1.012512
42′	0.583354	0.669931	0.756304	0.842447	0.928333	1.013936
43′	0.584798	0.671372	0.757742	0.843880	0.929762	1.015361
44′	0.586243	0.672813	0.759179	0.845314	0.931191	1.016785
45′	0.587687	0.674255	0.760617	0.846747	0.932620	1.018209
46′	0.589131	0.675696	0.762054	0.848181	0.934049	1.019633
47′	0.590576	0.677137	0.763492	0.849614	0.935478	1.021056
48′	0.592020	0.678578	0.764929	0.851047	0.936906	1.022480
49′	0.593464	0.680019	0.766366	0.852481	0.938335	1.023904
50′	0.594908	0.681460	0.767804	0.853914	0.939764	1.025327
51′	0.596352	0.682900	0.769241	0.855347	0.941192	1.026751
52′	0.597796	0.684341	0.770678	0.856780	0.942621	1.028174
53′	0.599240	0.685782	0.772115	0.858213	0.944049	1.029598
54′	0.600684	0.687223	0.773552	0.859645	0.945477	1.031021
55′	0.602128	0.688663	0.774989	0.861078	0.946905	1.032444
56′	0.603572	0.690104	0.776426	0.862511	0.948333	1.033867
57′	0.605016	0.691544	0.777862	0.863943	0.949761	1.035290
58′	0.606459	0.692985	0.779299	0.865376	0.951189	1.036713
59′	0.607903	0.694425	0.780736	0.866808	0.952617	1.038136
60′	0.609347	0.695865	0.782172	0.868241	0.954045	1.039558

TABLE 5. SINES FOR 5-IN. SINE BAR (*Continued*)

Min.	12°	13°	14°	15°	16°	17°
0′	1.039558	1.124755	1.209609	1.294095	1.378187	1.461858
1′	1.040981	1.126172	1.211021	1.295500	1.379585	1.463249
2′	1.042404	1.127589	1.212432	1.296905	1.380983	1.464640
3′	1.043826	1.129006	1.213843	1.298309	1.382380	1.466031
4′	1.045248	1.130423	1.215254	1.299714	1.383778	1.467421
5′	1.046671	1.131840	1.216664	1.301118	1.385176	1.468811
6′	1.048093	1.133256	1.218075	1.302522	1.386573	1.470202
7′	1.049515	1.134673	1.219486	1.303927	1.387971	1.471592
8′	1.050937	1.136089	1.220896	1.305331	1.389368	1.472982
9′	1.052359	1.137506	1.222306	1.306735	1.390765	1.474371
10′	1.053781	1.138922	1.223717	1.308138	1.392162	1.475761
11′	1.055202	1.140338	1.225127	1.309542	1.393559	1.477151
12′	1.056624	1.141754	1.226537	1.310946	1.394955	1.478540
13′	1.058045	1.143170	1.227947	1.312349	1.396352	1.479930
14′	1.059467	1.144586	1.229357	1.313753	1.397749	1.481319
15′	1.060888	1.146002	1.230766	1.315156	1.399145	1.482708
16′	1.062310	1.147418	1.232176	1.316559	1.400541	1.484097
17′	1.063731	1.148833	1.233586	1.317962	1.401937	1.485486
18′	1.065152	1.150249	1.234995	1.319365	1.403333	1.486874
19′	1.066573	1.151664	1.236404	1.320768	1.404729	1.488263
20′	1.067994	1.153079	1.237814	1.322171	1.406125	1.489651
21′	1.069415	1.154494	1.239223	1.323573	1.407521	1.491040
22′	1.070835	1.155910	1.240632	1.324976	1.408916	1.492428
23′	1.072256	1.157325	1.242041	1.326378	1.410312	1.493816
24′	1.073677	1.158739	1.243449	1.327781	1.411707	1.495204
25′	1.075097	1.160154	1.244858	1.329183	1.413102	1.496592
26′	1.076517	1.161569	1.246267	1.330585	1.414498	1.497979
27′	1.077938	1.162984	1.247675	1.331987	1.415892	1.499367
28′	1.079358	1.164398	1.249084	1.333388	1.417287	1.500754
29′	1.080778	1.165812	1.250492	1.334790	1.418682	1.502142
30′	1.082198	1.167227	1.251900	1.336192	1.420077	1.503529
31′	1.083618	1.168641	1.253308	1.337593	1.421471	1.504916
32′	1.085038	1.170055	1.254716	1.338995	1.422866	1.506303
33′	1.086457	1.171469	1.256124	1.340396	1.424260	1.507690
34′	1.087877	1.172883	1.257532	1.341797	1.425654	1.509076
35′	1.089297	1.174297	1.258939	1.343198	1.427048	1.510463
36′	1.090716	1.175710	1.260347	1.344599	1.428442	1.511849
37′	1.092135	1.177124	1.261754	1.346000	1.429836	1.513236
38′	1.093555	1.178538	1.263161	1.347401	1.431229	1.514622
39′	1.094974	1.179951	1.264569	1.348801	1.432623	1.516008
40′	1.096393	1.181364	1.265976	1.350202	1.434016	1.517394
41′	1.097812	1.182778	1.267383	1.351602	1.435409	1.518780
42′	1.099231	1.184191	1.268790	1.353002	1.436803	1.520165
43′	1.100650	1.185604	1.270196	1.354402	1.438196	1.521551
44′	1.102068	1.187017	1.271603	1.355802	1.439588	1.522936
45′	1.103487	1.188429	1.273010	1.357202	1.440981	1.524321
46′	1.104906	1.189842	1.274416	1.358602	1.442374	1.525707
47′	1.106324	1.191255	1.275822	1.360002	1.443766	1.527092
48′	1.107742	1.192667	1.277229	1.361401	1.445159	1.528476
49′	1.109161	1.194080	1.278635	1.362801	1.446551	1.529861
50′	1.110579	1.195492	1.280041	1.364200	1.447943	1.531246
51′	1.111997	1.196904	1.281447	1.365599	1.449335	1.532630
52′	1.113415	1.198316	1.282853	1.366998	1.450727	1.534015
53′	1.114833	1.199728	1.284258	1.368397	1.452119	1.535399
54′	1.116251	1.201140	1.285664	1.369796	1.453511	1.536783
55′	1.117668	1.202552	1.287069	1.371195	1.454902	1.538167
56′	1.119086	1.203964	1.288475	1.372593	1.456294	1.539551
57′	1.120503	1.205375	1.289880	1.373992	1.457685	1.540935
58′	1.121921	1.206787	1.291285	1.375390	1.459076	1.542318
59′	1.123338	1.208198	1.292690	1.376789	1.460467	1.543702
60′	1.124755	1.209609	1.294095	1.378187	1.461858	1.545085

TABLE 5. SINES FOR 5-IN. SINE BAR (*Continued*)

Min.	18°	19°	20°	21°	22°	23°
0′	1.545085	1.627841	1.710101	1.791840	1.873033	1.953656
1′	1.546468	1.629216	1.711467	1.793197	1.874381	1.954994
2′	1.547851	1.630591	1.712834	1.794555	1.875730	1.956333
3′	1.549234	1.631966	1.714200	1.795913	1.877078	1.957671
4′	1.550617	1.633340	1.715566	1.797270	1.878426	1.959010
5′	1.552000	1.634715	1.716932	1.798627	1.879774	1.960348
6′	1.553382	1.636089	1.718298	1.799984	1.881121	1.961686
7′	1.554764	1.637464	1.719664	1.801341	1.882469	1.963023
8′	1.556147	1.638838	1.721030	1.802698	1.883816	1.964361
9′	1.557529	1.640212	1.722395	1.804054	1.885163	1.965698
10′	1.558911	1.641586	1.723761	1.805410	1.886510	1.967035
11′	1.560293	1.642960	1.725126	1.806767	1.887857	1.968373
12′	1.561675	1.644333	1.726491	1.808123	1.889204	1.969709
13′	1.563056	1.645707	1.727856	1.809479	1.890550	1 971046
14′	1.564438	1.647080	1.729221	1.810834	1.891897	1.972383
15′	1.565819	1.648453	1.730585	1.812190	1.893243	1.973719
16′	1.567200	1.649826	1.731950	1.813546	1.894589	1.975055
17′	1.568581	1.651199	1.733314	1.814901	1.895935	1.976394
18′	1.569962	1.652572	1.734678	1.816256	1.897281	1.977727
19′	1.571343	1.653945	1.736042	1.817611	1.898626	1.979063
20′	1.572724	1.655317	1.737406	1.818966	1.899972	1.980399
21′	1.574104	1.656689	1.738770	1.820321	1.901317	1.981734
22′	1.575485	1.658062	1.740133	1.821675	1.902662	1.983069
23′	1.576865	1.659434	1.741497	1.823030	1.904007	1.984404
24′	1.578245	1.660806	1.742860	1.824384	1.905352	1.985739
25′	1.579625	1.662177	1.744223	1.825738	1.906696	1.987074
26′	1.581005	1.663549	1.745586	1.827092	1.908041	1.988409
27′	1.582385	1.664921	1.746949	1.828446	1.909385	1.989743
28′	1.583764	1.666292	1.748312	1.829799	1.910729	1.991077
29′	1.585144	1.667663	1.749674	1.831153	1.912073	1.992411
30′	1.586523	1.669034	1.751037	1.832506	1.913417	1.993745
31′	1.587902	1.670405	1.752399	1.833859	1.914761	1.995079
32′	1.589282	1.671776	1.753761	1.835212	1.916104	1.996413
33′	1.590660	1.673147	1.755123	1.836565	1.917448	1.997746
34′	1.592039	1.674517	1.756485	1.837918	1.918791	1.999079
35′	1.593418	1.675888	1.757847	1.839270	1.920134	2.000412
36′	1.594796	1.677258	1.759208	1.840623	1.921477	2.001745
37′	1.596175	1.678628	1.760570	1.841975	1.922819	2.003078
38′	1.597553	1.679998	1.761931	1.843327	1.924162	2.004410
39′	1.598931	1.681368	1.763292	1.844679	1.925504	2.005743
40′	1.600309	1.682737	1.764653	1.846031	1.926846	2.007075
41′	1.601687	1.684107	1.766014	1.847382	1.928188	2.008407
42′	1.603065	1.685476	1.767374	1.848734	1.929530	2.009739
43′	1.604442	1.686845	1.768735	1.850085	1.930872	2.011071
44′	1.605820	1.688245	1.770095	1.851436	1.932213	2.012402
45′	1.607197	1.689584	1.771455	1.852787	1.933555	2.013733
46′	1.608574	1.690952	1.772815	1.854138	1.934896	2.015065
47′	1.609951	1.692321	1.774175	1.855489	1.936237	2.016396
48′	1.611328	1.693690	1.775535	1.856839	1.937578	2.017726
49′	1.612705	1.695058	1.776894	1.858189	1.938919	2.019057
50′	1.614082	1.696426	1.778254	1.859539	1.940259	2.020388
51′	1.615458	1.697794	1.779613	1.860890	1.941600	2.021718
52′	1.616835	1.699168	1.780972	1.862240	1.942940	2.023048
53′	1.618211	1.700530	1.782331	1.863589	1.944280	2.024378
54′	1.619587	1.701898	1.783690	1.864939	1.945620	2.025708
55′	1.620963	1.703265	1.785049	1.866288	1.946959	2.027038
56′	1.622339	1.704633	1.786407	1.867657	1.948299	2.028367
57′	1.623714	1.706000	1.787765	1.868987	1.949638	2.029696
58′	1.625090	1.707367	1.789124	1.870336	1.950978	2.031025
59′	1.626465	1.708734	1.790482	1.871684	1.952317	2.032354
60′	1.627841	1.710101	1.791840	1.873033	1.953656	2.033683

TABLE 5. SINES FOR 5-IN. SINE BAR (*Continued*)

Min.	24°	25°	26°	27°	28°	29°
0′	2.033683	2.113091	2.191856	2.269952	2.347358	2.424048
1′	2.035012	2.114409	2.193163	2.271248	2.348642	2.425320
2′	2.036340	2.115727	2.194470	2.272544	2.349926	2.426592
3′	2.037668	2.117045	2.195777	2.273839	2.351209	2.427863
4′	2.038997	2.118362	2.197083	2.275135	2.352493	2.429135
5′	2.040324	2.119680	2.198390	2.276430	2.353776	2.430406
6′	2.041652	2.120997	2.199696	2.277724	2.355059	2.431677
7′	2.042980	2.122314	2.201002	2.279019	2.356342	2.432948
8′	2.044307	2.123631	2.202308	2.280314	2.357625	2.434218
9′	2.045634	2.124948	2.203613	2.281608	2.358907	2.435488
10′	2.046961	2.126264	2.204919	2.282902	2.360190	2.436759
11′	2.048288	2.127580	2.206224	2.284196	2.361472	2.438029
12′	2.049615	2.128896	2.207529	2.285490	2.362754	2.439298
13′	2.050942	2.130212	2.208834	2.286783	2.364035	2.440568
14′	2.052268	2.131528	2.210139	2.288076	2.365317	2.441837
15′	2.053594	2.132844	2.211443	2.289370	2.366598	2.443106
16′	2.054920	2.134159	2.212748	2.290662	2.367879	2.444375
17′	2.056246	2.135474	2.214052	2.291955	2.369160	2.445644
18′	2.057572	2.136789	2.215356	2.293248	2.370441	2.446912
19′	2.058897	2.138104	2.216660	2.294540	2.371721	2.448180
20′	2.060223	2.139419	2.217963	2.295832	2.373002	2.449449
21′	2.061548	2.140733	2.219267	2.297124	2.374282	2.450716
22′	2.062873	2.142048	2.220570	2.298416	2.375562	2.451984
23′	2.064197	2.143362	2.221873	2.299707	2.376841	2.453251
24′	2.065522	2.144676	2.223176	2.300999	2.378121	2.454519
25′	2.066847	2.145989	2.224478	2.302290	2.379400	2.455786
26′	2.068171	2.147303	2.225781	2.303581	2.380679	2.457053
27′	2.069495	2.148616	2.227083	2.304872	2.381958	2.458319
28′	2.070819	2.149930	2.228385	2.306162	2.383237	2.459586
29′	2.072143	2.151243	2.229687	2.307453	2.384515	2.460852
30′	2.073466	2.152555	2.230989	2.308743	2.385794	2.462118
31′	2.074790	2.153868	2.232291	2.310033	2.387072	2.463384
32′	2.076113	2.155181	2.233592	2.311323	2.388350	2.464649
33′	2.077436	2.156493	2.234893	2.312612	2.389627	2.465914
34′	2.078759	2.157805	2.236194	2.313902	2.390905	2.467180
35′	2.080081	2.159117	2.237495	2.315191	2.392182	2.468445
36′	2.081404	2.160429	2.238795	2.316480	2.393459	2.469709
37′	2.082726	2.161740	2.240096	2.317769	2.394736	2.470974
38′	2.084048	2.163052	2.241396	2.319058	2.396013	2.472238
39′	2.085370	2.164363	2.242696	2.320346	2.397289	2.473502
40′	2.086692	2.165674	2.243996	2.321634	2.398565	2.474766
41′	2.088014	2.166985	2.245295	2.322922	2.399842	2.476030
42′	2.089335	2.168295	2.246595	2.324210	2.401117	2.477293
43′	2.090657	2.169606	2.247894	2.325498	2.402393	2.478557
44′	2.091978	2.170916	2.249193	2.326785	2.403669	2.479820
45′	2.093299	2.172226	2.250492	2.328073	2.404944	2.481082
46′	2.094619	2.173536	2.251791	2.329360	2.406219	2.482345
47′	2.095940	2.174846	2.253089	2.330646	2.407494	2.483608
48′	2.097260	2.176155	2.254388	2.331933	2.408768	2.484870
49′	2.098581	2.177465	2.255686	2.333220	2.410043	2.486132
50′	2.099901	2.178774	2.256984	2.334506	2.411317	2.487394
51′	2.101220	2.180083	2.258281	2.335792	2.412591	2.488655
52′	2.102540	2.181392	2.259579	2.337078	2.413865	2.489917
53′	2.103860	2.182700	2.260876	2.338364	2.415138	2.491178
54′	2.105179	2.184009	2.262173	2.339649	2.416412	2.492439
55′	2.106498	2.185317	2.263470	2.340934	2.417685	2.493699
56′	2.107817	2.186625	2.264767	2.342219	2.418958	2.494960
57′	2.109136	2.187933	2.266064	2.343504	2.420231	2.496220
58′	2.110455	2.189241	2.267360	2.344789	2.421503	2.497480
59′	2.111773	2.190548	2.268656	2.346073	2.422776	2.498740
60′	2.113091	2.191856	2.269952	2.347358	2.424048	2.500000

TABLE 5. SINES FOR 5-IN. SINE BAR (*Continued*)

Min.	30°	31°	32°	33°	34°	35°
0′	2.500000	2.575190	2.649596	2.723195	2.795964	2.867882
1′	2.501259	2.576437	2.650830	2.724415	2.797170	2.869073
2′	2.502519	2.577683	2.652063	2.725634	2.798376	2.870264
3′	2.503778	2.578929	2.653296	2.726853	2.799581	2.871455
4′	2.505037	2.580175	2.654528	2.728072	2.800786	2.872646
5′	2.506295	2.581421	2.655761	2.729291	2.801990	2.873836
6′	2.507554	2.582667	2.656993	2.730510	2.803195	2.875026
7′	2.508812	2.583912	2.658225	2.731728	2.804399	2.876216
8′	2.510070	2.585157	2.659457	2.732946	2.805603	2.877406
9′	2.511328	2.586402	2.660688	2.734164	2.806807	2.878595
10′	2.512585	2.587646	2.661919	2.735382	2.808010	2.879784
11′	2.513843	2.588891	2.663150	2.736599	2.809214	2.880973
12′	2.515100	2.590135	2.664381	2.737816	2.810417	2.882162
13′	2.516357	2.591379	2.665612	2.739033	2.811620	2.883350
14′	2.517613	2.592623	2.666842	2.740250	2.812822	2.884538
15′	2.518870	2.593866	2.668073	2.741466	2.814025	2.885726
16′	2.520126	2.595110	2.669302	2.742682	2.815227	2.886916
17′	2.521382	2.596353	2.670532	2.743898	2.816429	2.888101
18′	2.522638	2.597595	2.671762	2.745114	2.817630	2.889288
19′	2.523894	2.598838	2.672991	2.746330	2.818832	2.890475
20′	2.525149	2.600080	2.674220	2.747545	2.820033	2.891662
21′	2.526404	2.601323	2.675449	2.748760	2.821234	2.892848
22′	2.527659	2.602565	2.676677	2.749975	2.822434	2.894034
23′	2.528914	2.603807	2.677906	2.751189	2.823635	2.895220
24′	2.530169	2.605048	2.679134	2.752404	2.824835	2.896406
25′	2.531423	2.606289	2.680362	2.753618	2.826035	2.897591
26′	2.532677	2.607531	2.681589	2.754832	2.827235	2.898776
27′	2.533931	2.608771	2.682817	2.756045	2.828434	2.899961
28′	2.535185	2.610012	2.684044	2.757259	2.829633	2.901146
29′	2.536438	2.611253	2.685271	2.758472	2.830832	2.902330
30′	2.537692	2.612493	2.686498	2.759685	2.832031	2.903515
31′	2.538945	2.613733	2.687725	2.760898	2.833230	2.904699
32′	2.540198	2.614973	2.688951	2.762110	2.834428	2.905882
33′	2.541450	2.616212	2.690177	2.763322	2.835626	2.907066
34′	2.542703	2.617451	2.691403	2.764534	2.836824	2.908249
35′	2.543955	2.618691	2.692628	2.765746	2.838021	2.909432
36′	2.545207	2.619929	2.693854	2.766958	2.839219	2.910615
37′	2.546459	2.621168	2.695079	2.768169	2.840416	2.911797
38′	2.547710	2.622407	2.696304	2.769380	2.841613	2.912979
39′	2.548962	2.623645	2.697529	2.770591	2.842809	2.914161
40′	2.550213	2.624883	2.698753	2.771802	2.844006	2.915343
41′	2.551464	2.626121	2.699977	2.773012	2.845202	2.916525
42′	2.552715	2.627358	2.701202	2.774222	2.846398	2.917706
43′	2.553965	2.628596	2.702425	2.775432	2.847593	2.918887
44′	2.555215	2.629833	2.703649	2.776642	2.848789	2.920068
45′	2.556465	2.631070	2.704872	2.777851	2.849984	2.921248
46′	2.557715	2.632306	2.706095	2.779060	2.851179	2.922429
47′	2.558965	2.633543	2.707318	2.780269	2.852373	2.923609
48′	2.560214	2.634779	2.708541	2.781478	2.853568	2.924788
49′	2.561463	2.636015	2.709763	2.782687	2.854762	2.925968
50′	2.562712	2.637251	2.710986	2.783895	2.855956	2.927147
51′	2.563961	2.638486	2.712208	2.785103	2.857150	2.928326
52′	2.565210	2.639722	2.713429	2.786311	2.858343	2.929505
53′	2.566458	2.640957	2.714651	2.787518	2.859536	2.930683
54′	2.567706	2.642192	2.715872	2.788725	2.860729	2.931862
55′	2.568954	2.643426	2.717093	2.789933	2.861922	2.933040
56′	2.570202	2.644661	2.718314	2.791139	2.863115	2.934218
57′	2.571449	2.645895	2.719535	2.792346	2.864307	2.935395
58′	2.572696	2.647129	2.720755	2.793552	2.865499	2.936572
59′	2.573943	2.648363	2.721975	2.794759	2.866691	2.937749
60′	2.575190	2.649596	2.723195	2.795964	2.867882	2.938913

TABLE 5. SINES FOR 5-IN. SINE BAR (*Continued*)

Min.	36°	37°	38°	39°	40°	41°
0′	2.938913	3.009075	3.078307	3.146602	3.213938	3.280295
1′	2.940103	3.010236	3.079453	3.147732	3.215052	3.281393
2′	2.941279	3.011398	3.080599	3.148862	3.216166	3.282490
3′	2.942455	3.012559	3.081744	3.149992	3.217279	3.283587
4′	2.943631	3.013719	3.082890	3.151121	3.218392	3.284684
5′	2.944806	3.014880	3.084035	3.152250	3.219505	3.285780
6′	2.945982	3.016040	3.085179	3.153379	3.220618	3.286876
7′	2.947157	3.017200	3.086324	3.154508	3.221730	3.287972
8′	2.948332	3.018359	3.087468	3.155636	3.222843	3.289068
9′	2.949506	3.019519	3.088612	3.156764	3.223954	3.290163
10′	2.950680	3.020678	3.089755	3.157892	3.225066	3.291258
11′	2.951854	3.021837	3.090899	3.159019	3.226177	3.292353
12′	2.953028	3.022995	3.092042	3.160146	3.227288	3.293447
13′	2.954202	3.024154	3.093185	3.161273	3.228399	3.294541
14′	2.955375	3.025312	3.094327	3.162400	3.229510	3.295635
15′	2.956548	3.026470	3.095470	3.163527	3.230620	3.296729
16′	2.957721	3.027627	3.096612	3.164653	3.231730	3.297822
17′	2.958894	3.028785	3.097754	3.165779	3.232839	3.298915
18′	2.960066	3.029942	3.098895	3.166904	3.233949	3.300008
19′	2.961238	3.031099	3.100036	3.168030	3.235058	3.301101
20′	2.962410	3.032255	3.101177	3.169155	3.236167	3.302193
21′	2.963581	3.033412	3.102318	3.170280	3.237275	3.303285
22′	2.964753	3.034568	3.103459	3.171404	3.238384	3.304377
23′	2.965924	3.035724	3.104599	3.172528	3.239492	3.305468
24′	2.967094	3.036879	3.105739	3.173652	3.240599	3.306559
25′	2.968265	3.038034	3.106879	3.174776	3.241707	3.307650
26′	2.969435	3.039189	3.108018	3.175900	3.242814	3.308741
27′	2.970605	3.040344	3.109157	3.177023	3.243921	3.309831
28′	2.971775	3.041499	3.110296	3.178146	3.245028	3.310921
29′	2.972945	3.042653	3.111435	3.179269	3.246134	3.312011
30′	2.974114	3.043807	3.112573	3.180391	3.247240	3.313100
31′	2.975283	3.044961	3.113711	3.181513	3,248346	3.314189
32′	2.976452	3.046114	3.114849	3.182635	3.249452	3.315278
33′	2.977620	3.047268	3.115987	3.183757	3.250557	3.316367
34′	2.978788	3.048421	3.117124	3.184878	3.251662	3.317455
35′	2.979957	3.049573	3.118261	3.185999	3.252767	3.318543
36′	2.981124	3.050726	3.119398	3.187120	3.253871	3.319631
37′	2.982292	3.051878	3.120534	3.188240	3.254975	3.320718
38′	2.983459	3.053030	3.121671	3.189361	3.256079	3.321806
39′	2.984626	3.054182	3.122807	3.190481	3.257183	3.322893
40′	2.985793	3.055333	3.123942	3.191600	3.258286	3.323979
41′	2.986959	3.056484	3.125078	3.192720	3.259389	3.325066
42′	2.988126	3.057635	3.126213	3.193839	3.260492	3.326152
43′	2.989292	3.058786	3.127348	3.194958	3.261594	3.327237
44′	2.990457	3.059936	3.128483	3.196077	3.262697	3.328323
45′	2.991623	3.061086	3.129617	3.197195	3.263799	3.329408
46′	2.992788	3.062236	3.130751	3.198313	3.264900	3.330493
47′	2.993953	3.063386	3.131885	3.199431	3.266002	3.331578
48′	2.995118	3.064535	3.133019	3.200548	3.267103	3.332662
49′	2.996282	3.065685	3.134152	3.201666	3.268204	3.333746
50′	2.997447	3.066833	3.135285	3.202783	3,269304	3.334830
51′	2.998611	3.067982	3.136418	3.203899	3.270405	3.335914
52′	2.999774	3.069130	3.137551	3.205016	3.271505	3.336997
53′	3.000938	3.070278	3.138683	3.206132	3.272605	3.338080
54′	3.002101	3.071426	3.139815	3.207248	3.273704	3.339163
55′	3.003264	3.072574	3.140947	3.208364	3.274803	3.340245
56′	3.004427	3.073721	3.142078	3.209479	3.275902	3.341327
57′	3.005589	3.074868	3.143210	3.210594	3.277001	3.342409
58′	3.006751	3.076015	3.144341	3.211709	3.278099	3.343491
59′	3.007913	3.077161	3.145471	3.212824	3.279197	3.344572
60′	3.009075	3.078307	3.146602	3.213938	3.280295	3.345653

TABLE 5. SINES FOR 5-IN. SINE BAR (*Continued*)

Min.	42°	43°	44°	45°	46°	47°
0′	3.345653	3.409992	3.473292	3.535534	3.596699	3.656768
1′	3.346734	3.411055	3.474338	3.536562	3.597709	3.657760
2′	3.347814	3.412119	3.475384	3.537590	3.598719	3.658752
3′	3.348894	3.413182	3.476429	3.538618	3.599729	3.659743
4′	3.349974	3.414244	3.477474	3.539645	3.600738	3.660734
5′	3.351054	3.415307	3.478519	3.540672	3.601747	3.661724
6′	3.352133	3.416369	3.479564	3.541699	3.602755	3.662714
7′	3.353212	3.417431	3.480608	3.542726	3.603764	3.663704
8′	3.354291	3.418492	3.481652	3.543752	3.604772	3.664694
9′	3.355369	3.419553	3.482696	3.544778	3.605780	3.665683
10′	3.356447	3.420614	3.483739	3.545803	3.606787	3.666672
11′	3.357525	3.421675	3.484783	3.546829	3.607794	3.667661
12′	3.358603	3.422735	3.485825	3.547854	3.608801	3.668649
13′	3,359680	3.423796	3.486868	3.548878	3.609808	3.669637
14′	3.360757	3.424855	3.487910	3.549903	3.610814	3.670625
15′	3.361834	3.425915	3.488952	3.550927	3.611820	3.671612
16′	3.362910	3.426974	3.489994	3.551951	3.612825	3.672600
17′	3.363987	3.428033	3.491035	3.552974	3.613831	3.673586
18′	3.365062	3.429092	3.492076	3.553997	3.614836	3.674573
19′	3.366138	3.430150	3.493117	3.555020	3.615840	3.675559
20′	3.367213	3.431208	3.494158	3.556043	3.616845	3.676545
21′	3.368288	3.432266	3.495198	3.557065	3.617849	3.677531
22′	3.369363	3.433323	3.496238	3.558087	3.618853	3.678516
23′	3.370438	3.434381	3.497277	3.559109	3.619856	3.679501
24′	3.371512	3.435437	3.498317	3.560130	3.620859	3.680485
25′	3.372586	3.436494	3.499356	3.561151	3.621862	3.681470
26′	3.373659	3.437550	3.500394	3.562172	3.622865	3.682454
27′	3.374733	3.438606	3.501433	3.563193	3.623867	3.683437
28′	3.875806	3.439662	3.502471	3.564213	3.624869	3.684421
29′	3.376878	3.440718	3.503509	3.565233	3.625870	3.685404
30′	3.377951	3.441773	3.504546	3.566252	3.626872	3.686387
31′	3.379023	3.442828	3.505583	3.567271	3.627873	3.687369
32′	3.380095	3.443882	3.506620	3.568290	3.628874	3.688351
33′	3.381167	3.444937	3.507657	3.569309	3.629874	3.689333
34′	3.382238	3.445991	3.508693	3.570327	3.630874	3.690315
35′	3.383309	3.447044	3.509729	3.571346	3.631874	3.691296
36′	3.384380	3.448098	3.510765	3.572363	3.632873	3.692277
37′	3.385450	3.449151	3.511801	3.573381	3.633872	3.693257
38′	3.386520	3.450204	3.512836	3.574398	3.634871	3.694237
39′	3.387590	3.451256	3.513871	3.575415	3.635870	3.695217
40′	3.388660	3.452308	3.514905	3.576431	3.636868	3.696197
41′	3.389729	3.453360	3.515939	3.577448	3.637866	3.697176
42′	3.390798	3.454412	3.516973	3.578464	3.638864	3.698155
43′	3.391867	3.455463	3 518057	3.579479	3.639861	3.699134
44′	3.392935	3.456514	3.519040	3.580495	3.640858	3.700112
45′	3.394004	3.457565	3.520074	3.581510	3.641855	3.701091
46′	3.395072	3.458616	3.521106	3.582524	3.642851	3.702068
47′	3.396139	3.459666	3.522139	3.583539	3.643847	3.703046
48′	3.397206	3.460716	3.523171	3.584553	3.644843	3.704023
49′	3.398273	3.461765	3.524203	3.585567	3.645839	3.705000
50′	3.399340	3.462815	3.525234	3.586580	3.646834	3.705976
51′	3.400407	3.463864	3.526266	3.587594	3.647829	3.706952
52′	3.401473	3.464912	3.527297	3.588606	3.648823	3.707928
53′	3.402539	3.465961	3.528327	3.589619	3.649817	3.708904
54′	3.403604	3.467009	3.529358	3.590631	3.650811	3.709879
55′	3.404670	3.468057	3.530388	3.591643	3.651805	3.710854
56′	3.405735	3.469104	3.531418	3.592655	3.652798	3.711829
57′	3.406799	3.470152	3.532447	3.593667	3.653791	3.712803
58′	3.407864	3.471199	3.533476	3.594678	3.654785	3.713777
59′	3.408928	3.472245	3.534505	3.595688	3.655776	3.714751
60′	3.409992	3.473292	3.535534	3.596699	3.656768	3.715724

TABLE 5. SINES FOR 5-IN. SINE BAR (*Continued*)

Min.	48°	49°	50°	51°	52°	53°
0′	3.715724	3.773548	3.830222	3.885730	3.940054	3.993177
1′	3.716697	3.774502	3.831157	3.886645	3.940949	3.994053
2′	3.717670	3.775456	3.832091	3.887560	3.941844	3.994927
3′	3.718642	3.776409	3.833025	3.888474	3.942739	3.995802
4′	3.719614	3.777362	3.833959	3.889388	3.943633	3.996676
5′	3.720586	3.778315	3.834893	3.890302	3.944527	3.997550
6′	3.721558	3.779267	3.835826	3.891216	3.945420	3.998423
7′	3.722529	3.780219	3.836758	3.892129	3.946314	3.999296
8′	3.723500	3.781171	3.837691	3.893042	3.947207	4.000169
9′	3.724470	3.782123	3.838623	3.893954	3.948099	4.001042
10′	3.725440	3.783074	3.839555	3.894866	3.948991	4.001914
11′	3.726410	3.784025	3.840486	3.895778	3.949883	4.002785
12′	3.727380	3.784975	3.841418	3.896690	3.950775	4.003657
13′	3.728349	3.785925	3.842348	3.897601	3.951666	4.004528
14′	3.729318	3.786875	3.843279	3.898512	3.952557	4.005399
15′	3.730287	3.787825	3.844209	3.899422	3.953448	4.006269
16′	3.731255	3.788774	3.845139	3.900333	3.954338	4.007139
17′	3.732223	3.789723	3.846068	3.901242	3.955228	4.008009
18′	3.733191	3.790672	3.846998	3.902152	3.956118	4.008878
19′	3.734158	3.791620	3.847927	3.903061	3.957007	4.009747
20′	3.735125	3.792568	3.848855	3.903970	3.957896	4.010616
21′	3.736092	3.793515	3.849783	3.904879	3.958784	4.011484
22′	3.737058	3.794463	3.850711	3.905787	3.959673	4.012352
23′	3.738025	3.795410	3.851639	3.906695	3.960561	4.013220
24′	3.738990	3.796356	3.852566	3.907602	3.961448	4.014087
25′	3.739956	3.797303	3.853493	3.908510	3.962335	4.014954
26′	3.740921	3.798249	3.854420	3.909416	3.963222	4.015821
27′	3.741886	3.799195	3.855346	3.910323	3.964109	4.016687
28′	3.742850	3.800140	3.856272	3.911229	3.964995	4.017553
29′	3.743815	3.801085	3.857198	3.912135	3.965881	4.018419
30′	3.744779	3.802030	3.858123	3.913041	3.966767	4.019284
31′	3.745742	3.802974	3.859048	3.913946	3.967652	4.020149
32′	3.746705	3.803918	3.859972	3.914851	3.968537	4.021014
33′	3.747668	3.804862	3.860897	3.915755	3.969421	4.021878
34′	3.748631	3.805806	3.861821	3.916660	3.970306	4.022742
35′	3.749593	3.806749	3.862744	3.917564	3.971189	4.023606
36′	3.750555	3.807691	3.863668	3.918467	3.972073	4.024469
37′	3.751517	3.808634	3.864591	3.919370	3.972956	4.025332
38′	3.752478	3.809576	3.865513	3.920273	3.973839	4.026194
39′	3.753439	3.810519	3.866436	3.921176	3.974722	4.027057
40′	3.754400	3.811459	3.867358	3.922078	3.975604	4.027919
41′	3.755361	3.812401	3.868280	3.922980	3.976486	4.028780
42′	3.756321	3.813342	3.869201	3.923882	3.977367	4.029641
43′	3.757280	3.814282	3.870122	3.924783	3.978249	4.030502
44′	3.758240	3.815222	3.871043	3.925684	3.979129	4.031363
45′	3.759199	3.816162	3.871963	3.926585	3.980010	4.032223
46′	3.760158	3.817102	3.872883	3.927485	3.980890	4.033083
47′	3.761116	3.818041	3.873803	3.928385	3.981770	4.033942
48′	3.762074	3.818980	3.874722	3.929284	3.982650	4.034801
49′	3.763032	3.819919	3.875641	3.930184	3.983529	4.035660
50′	3.763990	3.820857	3.876560	3.931083	3.984408	4.036519
51′	3.764947	3.821795	3.877479	3.931981	3.985286	4.037377
52′	3.765904	3.822733	3.878397	3.932879	3.986164	4.038235
53′	3.766861	3.823670	3.879315	3.933777	3.987042	4.039092
54′	3.767817	3.824607	3.880232	3.934675	3.987920	4.039949
55′	3.768773	3.825544	3.881149	3.935572	3.988797	4.040806
56′	3.769728	3.826480	3.882066	3.936469	3.989674	4.041663
57′	3.770684	3.827416	3.882982	3.937366	3.990550	4.042519
58′	3.771639	3.828352	3.883898	3.938262	3.991426	4.043374
59′	3.772593	3.829287	3.884814	3.939158	3.992302	4.044230
60′	3.773548	3.830222	3.885730	3.940054	3.993177	4.045085

TABLE 5. SINES FOR 5-IN. SINE BAR (*Continued*)

Min.	54°	55°	56°	57°	58°	59°
0′	4.045085	4.095760	4.145188	4.193353	4.240240	4.285836
1′	4.045940	4.096594	4.146001	4.194145	4.241011	4.286585
2′	4.046794	4.097428	4.146814	4.194936	4.241781	4.287334
3′	4.047648	4.098261	4.147626	4.195728	4.242551	4.288082
4′	4.048502	4.099094	4.148438	4.196519	4.243320	4.288830
5′	4.049355	4.099927	4.149250	4.197309	4.244090	4.289577
6′	4.050208	4.100759	4.150061	4.198099	4.244858	4.290324
7′	4.051061	4.101591	4.150872	4.198889	4.245627	4.291071
8′	4.051913	4.102423	4.151683	4.199679	4.246395	4.291818
9′	4.052765	4.103254	4.152493	4.200468	4.247162	4.292564
10′	4.053617	4.104085	4.153303	4.201256	4.247930	4.293309
11′	4.054468	4.104916	4.154113	4.202045	4.248697	4.294054
12′	4.055319	4.105746	4.154922	4.202833	4.249463	4.294799
13′	4.056170	4.106576	4.155731	4.203621	4.250230	4.295544
14′	4.057020	4.107405	4.156540	4.204408	4.250996	4.296288
15′	4.057870	4.108235	4.157348	4.205195	4.251761	4.297032
16′	4.058719	4.109063	4.158156	4.205982	4.252526	4.297775
17′	4.059569	4.109892	4.158963	4.206768	4.253291	4.298519
18′	4.060418	4.110720	4.159771	4.207554	4.254055	4.299261
19′	4.061266	4.111548	4.160577	4.208340	4.254820	4.300004
20′	4.062114	4.112375	4.161384	4.209125	4.255583	4.300746
21′	4.062962	4.113203	4.162190	4.209905	4.256347	4.301487
22′	4.063810	4.114029	4.162996	4.210694	4.257110	4.302229
23′	4.064657	4.114856	4.163801	4.211478	4.257872	4.302970
24′	4.065504	4.115682	4.164606	4.212262	4.258635	4.303710
25′	4.066350	4.116507	4.165411	4.213045	4.259397	4.304450
26′	4.067196	4.117333	4.166215	4.213828	4.260158	4.305190
27′	4.068042	4.118158	4.167019	4.214611	4.260919	4.305930
28′	4.068888	4.118983	4.167823	4.215393	4.261680	4.306669
29′	4.069733	4.119807	4.168626	4.216176	4.262441	4.307407
30′	4.070578	4.120631	4.169429	4.216957	4.263201	4.308146
31′	4.071422	4.121454	4.170232	4.217738	4.263961	4.308884
32′	4.072266	4.122278	4.171034	4.218519	4.264720	4.309621
33′	4.073110	4.123101	4.171836	4.219300	4.265479	4.310359
34′	4.073953	4.123923	4.172637	4.220080	4.266238	4.311096
35′	4.074796	4.124746	4.173438	4.220860	4.266996	4.311832
36′	4.075639	4.125567	4.174239	4.221640	4.267754	4.312568
37′	4.076481	4.126389	4.175040	4.222419	4.268512	4.313304
38′	4.077323	4.127210	4.175840	4.223197	4.269269	4.314040
39′	4.078165	4.128031	4.176640	4.223976	4.270026	4.314775
40′	4.079006	4.128851	4.177439	4.224754	4.270782	4.315509
41′	4.079847	4.129672	4.178238	4.225532	4.271538	4.316244
42′	4.080688	4.130491	4.179037	4.226309	4.272294	4.316978
43′	4.081528	4.131311	4.179835	4.227086	4.273050	4.317711
44′	4.082368	4.132130	4.180633	4.227863	4.273805	4.318445
45′	4.083208	4.132949	4.181431	4.228639	4.274559	4.319177
46′	4.084047	4.133767	4.182228	4.229415	4.275314	4.319910
47′	4.084886	4.134585	4.183025	4.230191	4.276068	4.320642
48′	4.085724	4.135403	4.183821	4.230966	4.276821	4.321374
49′	4.086563	4.136220	4.184618	4.231741	4.277574	4.322105
50′	4.087400	4.137037	4.185414	4.232515	4.278327	4.322836
51′	4.088238	4.137854	4.186209	4.233289	4.279080	4.323567
52′	4.089075	4.138670	4.187004	4.234063	4.279832	4.324297
53′	4.089912	4.139486	4.187799	4.234836	4.280584	4.325027
54′	4.090749	4.140302	4.188594	4.235610	4.281335	4.325757
55′	4.091585	4.141117	4.189388	4.236382	4.282086	4.326486
56′	4.092420	4.141932	4.190181	4.237155	4.282837	4.327215
57′	4.093256	4.142746	4.190975	4.237927	4.283588	4.327944
58′	4.094091	4.143560	4.191768	4.238698	4.284338	4.328672
59′	4.094926	4.144374	4.192560	4.239469	4.285087	4.329400
60′	4.095760	4.145188	4.193353	4.240240	4.285836	4.330127

TABLE 6. CONSTANTS FOR SECONDS OF ARC

Degree	Constant	Degree	Constant
0 to 5	0.0000242	30 to 35	0.0000204
5 to 10	0.0000240	35 to 40	0.0000192
10 to 15	0.0000237	40 to 45	0.0000179
15 to 20	0.0000231	45 to 50	0.0000164
20 to 25	0.0000224	50 to 55	0.0000147
25 to 30	0.0000215	55 to 60	0.0000130

Gage Design Standards

American Gage Design Standard. The caption "American Gage Design Standard" has been adopted to designate gages made to the design specifications promulgated by the American Gage Design Committee and published in Commercial Standard CS-8.

Federal Specification GGG-G-61. This specification covers the minimum essential requirements for all types of gages to be manufactured for use by the Department of Defense.

Commercial Standard CS-8. This standard covers standard blanks for various standard plain plug, plain ring, thread plug, thread ring, adjustable snap, adjustable length, adjustable plug, and spline gages.

Military Standard (Mil-Std) Gage. A gage that can be used for inspecting a dimensional feature wherever that dimensional feature occurs. The stock number and complete specification for the gage are given in a Military Standard for that type of gage.

Military Standards have been prepared for several types of gages which employ American Gage Design Standard blanks.

Gage Terminology

Go Gage. A gage which represents maximum metal conditions.

Not-go Gage. A gage which represents the minimum metal condition. (Metal conditions = that of the component.)

Limit Gage. A gage that represents a limiting (maximum or minimum) size within which the work will be acceptable.

Wear-limit Gage. A gage used for determining when a limit gage or functional gage has worn to the maximum permitted.

Functional Gage. A gage which checks the piece not necessarily to the dimensions and tolerances shown on its drawing, but by simulating a mating piece and so designed as to insure successful operation or satisfactory installation of the piece under the particular service condition involved.

Plug Gage. A gage whose outside measuring surfaces are so designed as to test the size and/or contour of a hole or cavity.

Ring Gage. A receiver gage whose inside gaging surfaces are cylindrical.

Receiver Gage. One whose inside gaging surfaces are arranged to verify the specified uniformity of size and/or contour of the part.

Indicating Gage. One that exhibits visually the amount of variation in size from the basic dimensions.

Adjustable Gage. A type of gage which can be adjusted to any limiting dimension within a given general size range.

Fixed Gage. A gage which is finished to an exact size and cannot be adjusted in any manner.

Classification of Gages. Military Parlance

Government Inspection Gages. These are used by the Department of Defense for final inspection of the finished product. These gages are used to ensure that the product has been manufactured within the component limits specified on the drawing and/or that the product is functionally acceptable.

Manufacturing Gages. Manufacturing gages are those used by the contractor in production. Manufacturing gages cannot be used as a final means of acceptance without special approval from the responsible government agency. Gages of this type are also known as process or work gages.

The government inspection gage is the final gage upon which the acceptance or rejection of a part depends. The manufacturing gage is used to verify the dimension before the government inspection gage is used. To avoid delay and controversy, the manufacturing gage is generally dimensioned so that every part accepted by the manufacturing gage will surely be passed by the government inspection gage.

Master Gage. A gage made to one of the specified (max or min) product limits within a high degree of accuracy as related to the component tolerance. A master gage is used as a "referee" gage to accept or reject components which have been previously gaged and found to be borderline cases.

Master Check Gages. A type of check gage simulating the component dimensions that are to be gaged. The check gage is made accurately to within approximately 5% of the component tolerance and usually is made to the max and/or min conditions. Master check gages are for setting, acceptance, or surveillance.

Commercial Parlance

Master Gages. Master gages are made to their basic dimensions as accurately as possible and are used for reference, such as for checking or setting inspection or working gages.

Inspection Gages. These gages are used by the manufacturer or the purchaser in acceptance of the product. A gagemaker's tolerance will always be applied, and a wear allowance, where applicable, may be included in the design of these gages.

Working Gages. Working, or manufacturer's, gages are used for inspection of parts during production. So that the product will be within the limits of inspection gages, working gages should have dimensional limits, resulting from gage tolerances and wear allowances, slightly farther from the specified limits than inspection gages.

Gage Tolerances and Allowances. Military Policy[1]

Gage Tolerancing Policy. Gage tolerances and allowances shall be applied within the product limits; that is, the extreme limits of the gage must in all cases fall within the acceptable product limits.

Gage Tolerancing System. The unilateral system shall be used in applying tolerances to gaging dimensions on gages which control the extreme product limits. The bilateral system shall be used in applying tolerances to gaging dimensions which are based on mean or intermediate product dimensions, such as for location of holes.

[1] "Gage Design Manual" prepared by the Ordnance Corps, Department of the Army.

Tolerances for Gaging Dimensions. The tolerances applied to the functioning dimensions of gage designs prepared for the Department of Defense shall be in accordance with tolerances in Tables 7 and 8.

Tolerances for Maximum or Minimum Limit Gages. Where component dimensions are prescribed as maximum or minimum values without a given tolerance, the gage tolerance will be based on a component tolerance of 0.010 in. or the sum of the tolerances on the dimensions making up the over-all dimension to be gaged, whichever is the lesser value.

Wear Allowance. Wear allowance is applied on gage contact surfaces which are subject to a wearing action. The wear allowance provides a small amount of extra metal which lengthens the useful life of the gage. Excepted from this rule are the adjustable snap or length gage which may be reset; flush pin gages; height or depth gages on which wear occurs on both surfaces in the same direction; and certain classes of thread gages.

Specific Geometric Requirements. The tolerances directly specified on gage drawings for requirements such as concentricity, parallelism, perpendicularity, centrality, and flatness shall be to the maximum that will still insure an accurate gage, but shall in general not exceed 10% of the component-part tolerance on that requirement.

Implied Geometric Requirements. The general nature of gages requires that concentricity, parallelism, perpendicularity, centrality and flatness, and other requirements be maintained within general close limits. A section covering this is included in Federal Specification GGG-G-61 for the express purpose of controlling implied geometric requirements.

Commercial Policy

There is no universally accepted policy with respect to gaging in private industry. A statement of principles issued by the American Standards Association follows:

> The extreme sizes for all plain limit gages shall not exceed the extreme limits of the part to be gaged. All variations in the gages, whatever their cause or purpose, shall bring these gages within these extreme limits. Thus, a gage which represents a minimum limit may be larger, but never smaller, than the minimum size specified on the part to be gaged; likewise the gage which represents a maximum limit may be smaller, but never larger, than the maximum size specified for the part to be gaged.

The fact is, industry uses both the unilateral tolerance system (Ordnance policy) and the bilateral tolerance system, or modifications of these (for definitions see Sec. 33). It is also true that gagemakers must have a tolerance and that a wear allowance is desirable in many instances. As the gagemaker's tolerance is reduced, the cost of a fixed gage rises materially. Therefore, for plug and ring gages, the gagemaker's tolerance and the wear allowance are usually related to the component tolerance (Table 7).

Gagemaker's Tolerances. For working gages, the gage tolerance is usually limited to 10% of the component tolerance. If no gage tolerance is specified for inspection gages, the amount is usually 5%, except when the desired precision is not attained. The wear allowance is applied in addition.

Industrial gaging systems do not agree with respect to disposition of the gagemaker's tolerance. The tolerance may be taken plus or minus of the basic dimension, or split. For commercial plain plug gages, the tolerance on the go gage is placed inside the minimum hole limit, and for the not-go gage the tolerance is split

Table 7. Tolerances for Plug and Ring Gages*

Go Gages

Size Range: Above	Size Range: To and Incl.	Component Tolerance	Wear Allowance	Gagemaker Tolerance
0.031	0.825	Master	0	0.00002
0.825	1.510		0	0.00003
1.510	2.510		0	0.00004
2.510	4.510		0	0.00005
4.510	6.510		0	0.00006
6.510	9.010		0	0.00008
9.010	12.010		0	0.00010
0.031	0.825	0.0005	0	0.00004
0.825	1.510		0	0.00006
1.510	2.510		0	0.00008
2.510	12.010	0.0005	Use approved commercial measuring device	
0.031	0.825	0.001	0.00004	0.00004
0.825	1.510		0.00004	0.00006
1.510	2.510		0.00002	0.00008
2.510	4.510		0.00002	0.00010
4.510	12.010	0.001	Use approved commercial measuring device	
0.031	.825	0.002	0.00010	0.00007
0.825	1.510		0.00010	0.00009
1.510	2.510		0.00008	0.00012
2.510	4.510		0.00008	0.00015
4.510	6.510		0.00005	0.00019
6.510	12.510	0.002	Use approved commercial measuring device	
0.031	0.825	0.004	0.00020	0.00010
0.825	1.510		0.00020	0.00012
1.510	2.510		0.00020	0.00016
2.510	4.510		0.00020	0.00020
4.510	6.510	0.004	0.00020	0.00025
6.510	9.010		0.00010	0.00032
9.010	12.010		0.00010	0.00040
0.031	0.825	0.006	0.00030	0.00020
0.825	1.510		0.00030	0.00030
1.510	2.510		0.00020	0.00040
2.510	4.510		0.00020	0.00040
4.510	6.510		0.00020	0.00040
6.510	9.010		0.00010	0.00050
9.010	12.010		0.00010	0.00050
0.031	0.825	0.008	0.00040	0.00030
0.825	1.510		0.00040	0.00030
1.510	2.510		0.00040	0.00040
2.510	4.510		0.00030	0.00050
4.510	6.510		0.00030	0.00050
6.510	9.010		0.00020	0.00060
9.010	12.010		0.00020	0.00060
0.031	0.825	0.015 and up	0.00050	0.00050
0.825	1.510		0.00050	0.00050
1.510	2.510		0.00050	0.00050
2.510	4.510		0.00040	0.00060
4.510	6.510		0.00040	0.00060
6.510	9.010		0.00040	0.00070
9.010	12.010		0.00040	0.00080

Not-go Gages

Size Range: Above	Size Range: To and Incl.	Component Tolerance	Gagemaker's Tolerance
0.031	0.825	0.0005	0.00004
0.825	2.510		Snug fit on go
0.031	0.825	0.001	0.00004
0.825	1.510		0.00006
1.510	2.510		0.00008
2.510	4.510		0.00010
0.031	0.825	0.004	0.00007
0.825	1.510		0.00009
1.510	2.510		0.00012
2.510	4.510		0.00015
4.510	6.510		0.00019
6.510	9.010		0.00024
9.010	12.010		0.00030
0.031	0.825	0.010	0.00010
0.825	1.510		0.00012
1.510	2.510		0.00016
2.510	4.510		0.00020
4.510	6.510		0.00025
6.510	9.010		0.00032
9.010	12.010		0.00040
0.031	0.825	0.020 and up	0.00020
0.825	1.510		0.00024
1.510	2.510		0.00032
2.510	4.510		0.00040
4.510	6.510		0.00050
6.510	9.010		0.00060
9.010	12.010		0.00080

General:

1. This table is the basis for the wear allowances and tolerances in the military standards for plain plug and ring gages.
2. This table shall be used for selecting the wear allowances and tolerances for the following types of gages:
   a. Plain plug and ring gages that are prototypes of military standards.
   b. Special plug and ring gages having a simple modification such as a depth step or special chamfer.
   c. Minor-diameter plug gages.
   d. Flat cylindrical plug gages.

* "Military Gage Design Manual."

on the maximum hole limit. In the case of commercial ring gages, the tolerance on the go gage is placed outside the minimum work limit and for the not-go gage the tolerance is placed inside the maximum work limit.

Plug and Ring Gages (Figs. 13 and 14). Practical specification and use of these gages depend upon the degree of accuracy required in the gaging operation. To the basic dimension, there must be added gagemaker's and wear allowances. The American Gage Design Committee has established five standard classifications for gage tolerances: XX, X, Y, Z, and ZZ (see Table 9). The tolerance increases from classification XX to ZZ and with diameter of the gage.

TABLE 8. TOLERANCES FOR ADJUSTABLE SNAP AND LENGTH GAGES*

Range		Component Tolerance										
Above	To and Incl.	.001	.003	.005	.006	.007	.008	.009	.010	.012	.015	.020
0	2.500	.0001	.0002	.0003	.0003	.0003	.0003	.0003	.0004	.0004	.0004	.0004
2.500	5.687		.0002	.0004	.0004	.0004	.0004	.0004	.0005	.0005	.0005	.0005
5.687	12.000		.0003	.0005	.0005	.0005	.0005	.0005	.0006	.0006	.0006	.0006
12.000	18.750			.0005	.0005	.0005	.0005	.0005	.0006	.0006	.0007	.0007
18.750	25.500					.0005	.0005	.0005	.0006	.0006	.0007	.0008
25.500	30.125							.0005	.0006	.0006	.0007	.0008

* "Military Gage Design Manual."

TABLE 9. TOLERANCES FOR PLUG AND RING GAGES*

Size Range, In.		Class XX (Plug Gages Only), In.	Class X, In.	Class Y, In.	Class Z, In.	Class ZZ (Ring Gages Only), In.
Above	To and Incl.					
0.029	0.825	0.00002	0.00004	0.00007	0.00010	0.00020
0.825	1.510	0.00003	0.00006	0.00009	0.00012	0.00024
1.510	2.510	0.00004	0.00008	0.00012	0.00016	0.00032
2.510	4.510	0.00005	0.00010	0.00015	0.00020	0.00040
4.510	6.510	0.000065	0.00013	0.00019	0.00025	0.00050
6.510	9.010	0.00008	0.00016	0.00024	0.00032	0.00064
9.010	12.010	0.00010	0.00020	0.00030	0.00040	0.00080

XX—Precision-lapped masters, plugs only.
X—Precision-lapped plugs or rings.
Y—Lapped plugs or rings for inspection or working gages.
Z—Commercially ground and polished plugs and rings used for working gages.
ZZ—Ground rings only; applied to small-lot production where tolerances are liberal.
NOTE: Do not confuse the X and Y classifications of plain plug and ring gages with the X and Y classifications of thread gages.

* American Gage Design Standard.

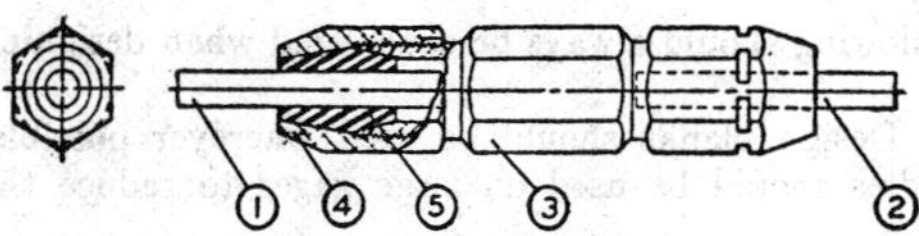

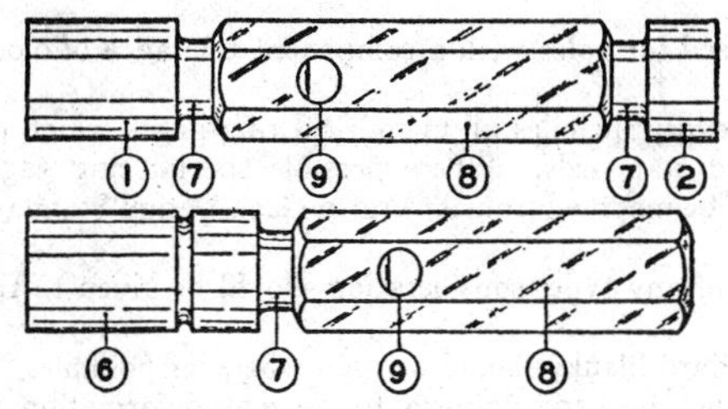

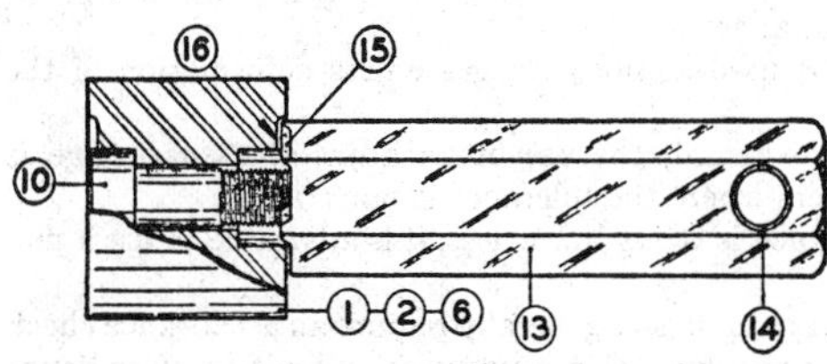

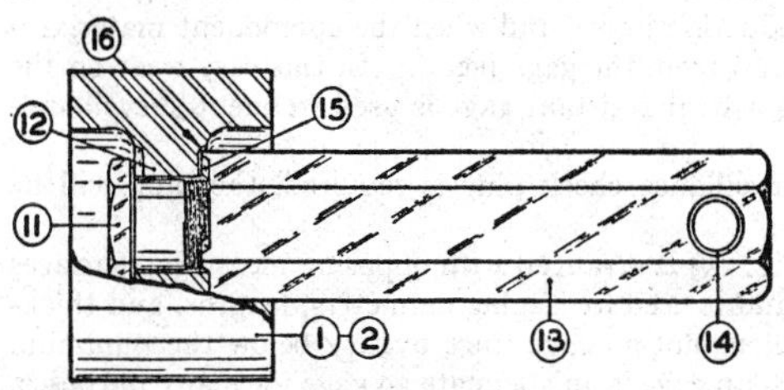

FIG. 13. Plain cylindrical plug gages, details of construction—ranges above 0.030 to and including 8.010 in.

1. Go gaging member.
2. Not-go gaging member.
3. Wire type handle.
4. Clamping nut.
5. Collet bushing.
6. Progressive gaging member.
7. Shank.
8. Taper lock handle.
9. Drift hole or slot.
10. Socket head screw.
11. Hexagon head screw.
12. Web.
13. Handle for reversible gage.
14. Cross-pin hole.
15. Locking prong.
16. Locking groove.

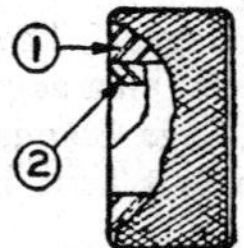

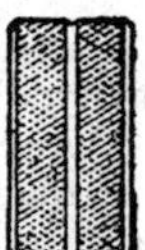

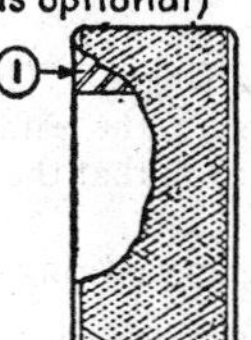

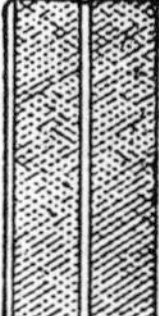

Range: Above 0.510 to and including 1.510 inches (optional above 0.059 to and including 0.510 inch)

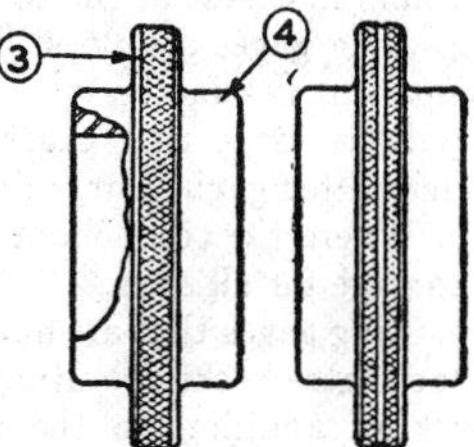

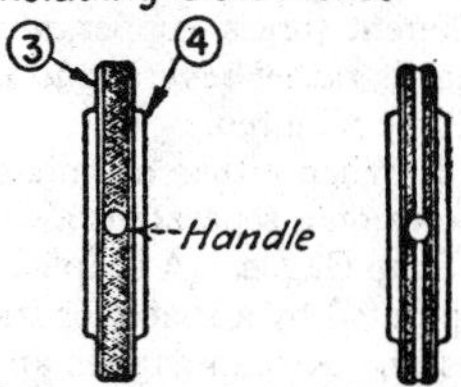

FIG. 14. Plain ring gages, details of construction.

1. Body.
2. Bushing.
3. Flange.
4. Hub.
5. Handle.

Special Plug Gages. The following should always be considered when designing any type of special plug gage:

1. American Gage Standard Design blanks should be used wherever possible.
2. Aluminum or plastic handles should be used on large gages to reduce the weight.
3. The use of carbide plugs or inserts will usually prolong the life of the gage.
4. Air grooves should be used on plug gages for blind holes on both go and not-go members.
5. Dial bore gages should be used for holes with a component tolerance of 0.001 or less.

Special Ring Gages. This category includes all cylindrical ring gages which do not conform with the established standards. Where possible special ring gages shall be designed so that they may be modified from American Gage Design Standard blanks.

In designing special ring gages of any type, consideration should be given to the following:

1. American Gage Design Standard blanks should be used wherever possible.
2. The thickness of section must be kept uniform to prevent deformation in heat-treating.
3. The depth of cross-section must be sufficient to prevent gage from going out of round in excess of the tolerance allowed.
4. Ring gages should seldom be used as not-go gages unless deformation of the component is a factor.
5. Since large ring gages are awkward, the use of an adjustable snap gage is preferred for gaging large diameters where the tolerance is not critical.
6. Where the component tolerance is 0.001 in. or less, it is advisable to use a dial snap gage for all sizes.
7. Ring gages that are under 0.250 in. in size generally require an acceptance check to facilitate checking. Ring gages of a small size may require a limit or surveillance check in addition to the acceptance check. A surveillance check is generally required when the component tolerance is less than 0.004 in., when the gage is used for 100% inspection on high-production items, and when the component material is different (brass, copper, aluminum) from the gage metal. In this case wear on the gage is rather severe, and unless a wear-resistant gage is used frequent surveillance will be required.
8. When either a limit or surveillance check plug is required, the plug will be shown on a separate drawing.

Snap Gages. A snap gage (Fig. 15) is arranged with opposing measuring surfaces separated by a spacer or frame and is used for gaging diameters, lengths, and thicknesses. A snap gage is an excellent not-go gage since ovality below the minimum limit can easily be detected. A snap gage is an adequate go gage for many purposes. However, a ring gage should be used if assembly is the prime consideration. There are three general classes of snap gages, that is, adjustable, fixed, and indicating.

Flush-pin Gages. Flush-pin gages (Fig. 16) are used for checking dimensional features such as the following:

1. Depth of holes, either straight or tapered.
2. Height of bosses, either straight or tapered.
3. Location and position of holes, bosses, etc.
4. Perpendicularity.

The application of flush-pin gages is generally confined to the gaging of component dimensions having tolerance of 0.005 in. or greater. When the component tolerance

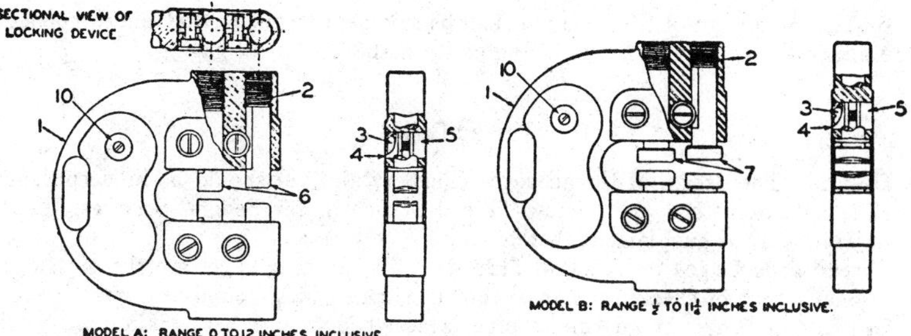

FIG. 15. Adjustable snap gages, details of construction.

1. Frame.
2. Adjusting screw.
3. Locking screw.
4. Locking bushing.
5. Locking nut.
6. Gaging pin.
7. Gaging button.
8. Anvil.
9. Anvil screw.
10. Marking disk.

is under 0.005 in., it is preferable to use a dial indicator or indicator flush pins rather than to rely on the accuracy of the inspector's sense of "feel."

The common type of flush pin consists of a body, movable gaging pin, and a pin-retaining device. The body is usually of a standard round, hexagonal, or rectangular bar stock. One surface of the body is ground to provide a gage bearing surface. The opposite face is ground to form two steps having a difference equal to the component tolerance. One end of the slide-fit pin bears on the part and the other end must fall between the steps on the body. A setscrew on a retaining pin is provided to limit overtravel and to prevent the flush pin from rotating.

In summary the following precautions should be considered:

1. The pin and/or bottom of the hole in the body shall be chamfered wherever necessary to clear the component and insure an accurate gage.
2. Diameter of the pin or diameter of the hole in the body must always clear the worst condition relative to the diameter of the component.
3. The component tolerance should be 0.004 in. or greater if a flush pin is to be used.
4. All "feel" steps must be easily accessible to the inspector when using the gage.
5. All feeler edges (both steps on body and top edge of pin) should be sharp to insure gaging accuracy. A note to this effect should appear on the drawing.

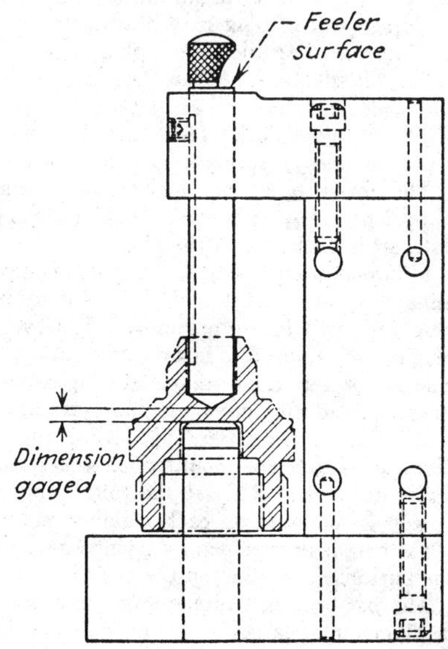

FIG. 16. This flush-pin gage has the feeler area at the top and is used to gage the minimum thickness of a part.

6. The movement of the flush pin shall be sufficient to insure that the component can always be applied to the gage and easily withdrawn after gaging.

Thread Gages

Design. The design of thread gages (Figs. 17 and 18) should be in accordance with specifications shown in Screw Thread Standards for Federal Services, Handbook H28 and the supplement thereto.

Thread Plug Gages for Internal Threads. Thread plug gages are the most conventional means of checking internal threads such as those found in a nut.

Go Thread Plug. A go thread plug is designed to check the minimum pitch diameter, the clearance at the major diameter, the lead, and the flank angle simultaneously. When a go gage enters an internal thread completely, assembly with the mating external thread is assured.

Not-go Thread Plug. A not-go thread plug is designed to check only one element, that is, the pitch diameter, to insure that it is not above the maximum limit. For this reason the thread form on the gage is truncated at the crest and cleared at the root so that the length of flank is reduced and will bear only on the central portion of the component thread flanks.

Depth Requirement. A large percentage of component tapped-hole specifications include a requirement for minimum length or depth of perfect thread. In rare instances a maximum depth of thread is specified. The depth of thread should not be confused with the specification for depth of tap drill or bore diameter. The depth shall be considered to be to the center line of a full-form thread space on the component internal thread unless otherwise specified.

The go thread plug may have a flat added for gaging a depth of thread. The leading thread on the thread plug must be removed to a full thread form.

Gage Blanks. American Gage Design Standard thread blanks should be specified wherever possible.

The not-go trilock blank should be specified for go thread gages in sizes above 1.510 in. unless a special requirement necessitates the use of the longer go blank.

The not-go taper lock blank should be specified for go thread plug gages which have depth steps, wherever the blank has sufficient length to allow for grinding a flat on the back face of the plug.

Minor-diameter Plugs. The conventional method of inspecting the limits of minor diameter of internal threads is by use of plain plug gages. The go thread plug gage will insure assembly. However, the use of the not-go minor-diameter plug is also required to insure sufficient depth of engagement and strength of thread. The use of a go minor-diameter plug is not required for most applications, because the go thread plug gives adequate insurance of assembly.

Roll Thread Plug, Bar Type. This type of gage utilizes two thread rolls of small diameter which are mounted on the spring-loaded telescoping center bar, which serves to expand and set the rolls firmly in the part being inspected. A gaging button is mounted on each member of the telescoping center bar. Acceptability of the part is determined by trying either a go or not-go feeler plug or both between the buttons depending on the design of the rolls.

The bar-type roll thread gage may be specified for gaging threaded holes 7 in. in diameter or larger. Gaging rolls can be made to check 60° form, Whitworth, Acme, Buttress, and other special forms. The thread form on the thread roll is ground as a cylinder rather than on the true helix and therefore cannot be successfully applied to threads having large helix angle (7° or greater or to multiple threads).

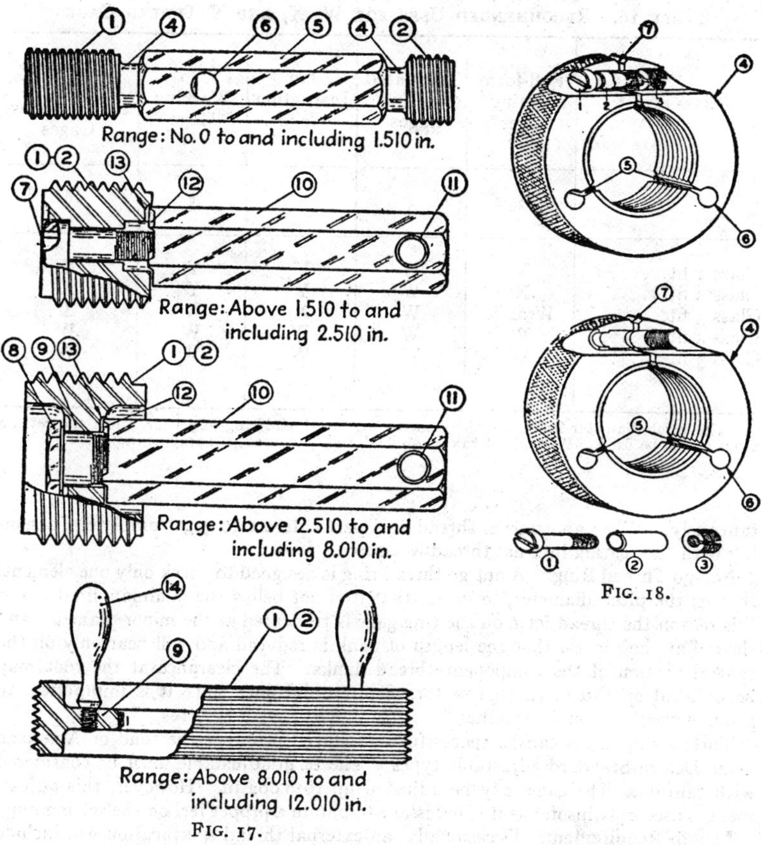

FIG. 17.

FIG. 18.

FIG. 17. Thread plug gages, details of construction.

1. Go gaging member.
2. Not-go gaging member.
4. Shank.
5. Taper lock handle.
6. Drift hole or slot.
7. Socket head screw.
8. Hexagon head screw.
9. Web.
10. Handle for reversible or trilock gage.
11. Cross-pin hole.
12. Locking prong.
13. Locking groove.
14. Ball handle.

FIG. 18. Thread ring gage locking device, details of construction, range No. 0 to $5\frac{1}{2}$ in., inclusive.

1. Locking screw.
2. Sleeve.
3. Adjusting screw.
4. Body.
5. Adjusting slots.
6. Adjusting slot terminal hole.
7. Locking slot.

Thread Ring Gages. Thread ring gages are the most conventional means of checking external threads.

Go Thread Ring. A go thread ring is designed to check the maximum pitch diameter, the clearance at the minor diameter, the lead, and the flank angle simul-

Table 10. Recommended Uses for W, X, and Y Thread Gages

Class of Fit	Full-form Setting Gages*	Truncated Setting Gages*	Go Inspection Gage	Go Working Gage	All Not-go Inspection and Working Gages
1	2	3	4	5	6
Class 1 fit.........	X	W	Y	Y	X
Class 2 fit.........	X	W	Y	Y	X
Class 3 fit.........	W or X	W	X	X	X
Class 4 fit.........	W	W	W	W	W
Class 5 fit.........	W	W	W	W	W

* The pitch-diameter limits are the same as on the thread ring gages for which the setting plugs are to be used. W or X tolerances on lead and thread angle apply as indicated.

taneously. When an external thread on a part enters a go gage completely, assembly with the mating internal thread is assured.

Not-go Thread Ring. A not-go thread ring is designed to check only one element, that is, the pitch diameter, to insure that it is not below the minimum limit. For this reason the thread form on the ring gage is truncated at the minor diameter and cleared at the root so that the length of flank is reduced and will bear only on the central portion of the component thread flanks. The clearance at the root may be omitted on thread ring gages for 28 pitch and finer since it is impractical to provide clearance on fine pitches.

Thread ring gages can be successfully produced in either the solid or American Gage Design Standard adjustable types. The term adjustable must be considered with caution. The gage may be adjusted up to 0.002 in. However, this adjustment is used only insofar as it is necessary to obtain a proper feel on the setting plug.

Length Requirement. Occasionally, an external thread specification will include a requirement for minimum length of perfect thread. The length shall be considered to be to the center line of a full-form thread space on the component external thread unless otherwise specified.

Major-diameter Adjustable Snaps. The most conventional method of inspecting the limits of major diameter of external threads is by use of adjustable snap gages. The go thread ring gage will insure assembly. However, the use of the not-go major-diameter snap is also required to insure sufficient depth of engagement and strength of thread. The use of a go major-diameter snap is not required for most applications, because the go thread ring gives adequate insurance of assembly.

Setting Plug Gages. Setting plug gages are required for the manufacture, acceptance inspection, and subsequent surveillance of solid- and adjustable-type thread rings, roll thread snaps, segment-type snaps, and other indicating-type gages.

Thread Ring Minor-diameter Check Plugs. Thread rings having minor diameters under 0.375 in. cannot be measured in all laboratories because of equipment limitations. Therefore, two plain plug gages must be provided to inspect the limits of the ring- or segment-type gage minor diameter on the go and not-go principle.

Taper Pipe Thread Gages. Taper pipe threads differ from the standard threads in that they are cut on a conical instead of a cylindrical surface. Consequently, the pipe thread gage is both go and not-go. The depth which the plug gage enters the hole or the distance which the ring gage goes on the external thread is determined by steps. The distance between steps is the tolerance on the pitch diameter expressed as a function of the taper and is the limit within which the thread will properly interchange with its mating thread. Obviously, this gaging can be functional only and checks none of the characteristics of the threads except that they will assemble.

Thread-gage Tolerances. Screw-thread tolerances for classes 1, 2, 3, 4, and 5 are classified according to accuracy as W, X, and Y, the W gages being the most accurate. Recommended uses for these classes are given in Table 10. The tolerance limits on W and X gages coincide with the extreme product limits. The tolerance limits on Y go gages are placed inside the extreme product limits to allow for wear of the gages. Tolerances on all minimum-metal limit, or not-go, gages, however, are applied from the extreme product limit.

Indicating-type Gages

Application. Most types of fixed gages such as built-up snaps, flush pins, length gages, and concentricity gages can be designed using ordinary dial indicators instead of fixed anvils or profiles. This conversion in design is preferred under the following conditions:

1. When the component tolerance is relatively small (under 0.005 in.) and the specific type of gage utilizes feel or sight an indicator can often be substituted to improve accuracy.
2. When it is desirable to know the exact dimensional limits of each component being inspected, as would be necessary under a quality-control setup.
3. When mating parts are fitted together by selective assembly.

Bore Gages

Gages for Straightness of Bore. The straightness of gun and other bores is checked by passing a plug gage of prescribed length through the bore. Hardened-tool-steel elements are used for small gages, while cast iron must be used for larger sizes to prevent the gage from seizing.

When designing bore-straightness gages, the designer should follow these rules:

1. Diameter of the gage is to be computed by subtracting 0.0005 in. per in. dia from the basic bore size.
2. The gage tolerance will be 10% of this value applied in a minus direction. Tolerances of less than 0.0002 in. or greater than 0.0006 in. shall not ordinarily be used.
3. Wear allowance is not used on this type of gage.
4. The length of the bore-straightness gage shall be equal to four times its diameter to the nearest $\frac{1}{16}$ in.

Star Gages. The star gage is an internal micrometer-type gage used to measure the diameters of lands and grooves at varying depths along the rifling of gun barrels. Each is inspected separately by employing two interchangeable sets of elements. In appearance the gage resembles a long plug gage with a center hole into which is threaded a graduated spindle. The front end of this spindle is tapered to actuate the diameter measuring elements into gaging position. An adjustable collar attachment is used as a stop and is locked at various points on the gage to control the depth at which each reading is taken. The diameter is read from a graduated scale at the rear end of the plug.

Optical-projection Methods

In gaging small- and medium-sized components there are many forms which are difficult or even impossible to check using fixed manual gages. Optical-projection methods verify that the components are within dimensional limits and, in a smaller percentage of cases, give the exact size of specific dimensions.

Optical projection consists of the projection of a sharply outlined and enlarged shadow silhouette of the component being inspected upon a translucent screen. This is accomplished by placing the component within an optical system consisting of a light source, condensing lens, objective lens, screen, and a first-surface mirror placed between the objective lens and the translucent screen. The component is placed in a staging fixture at a fixed distance from the objective lens.

The Autocollimator. The autocollimating telescope, or autocollimator, provides an extremely accurate method of detecting angular variation. By proper adaption it will test any surface for squareness, straightness, parallelism, angularity, and flatness.

Borescopes. The borescope employs varying types of industrial telescopes equipped with the necessary adapters to inspect internal surfaces, recesses, and hidden contours which could not otherwise be conveniently inspected.

Air Gages

An air gage is considered a comparison gaging device designed to measure the inside characteristics of holes by indicating on a scale the amount of air escaping between the side of the hole and a standard gaging spindle which has been inserted in the hole.

The spindle, a type of plug gage with air jets, is carried at the end of a flexible shaft connected to the indicating instrument. The inspector passes the gaging spindle through the bore in one continuous pass, watching the indicator float or dial as the case may be. Any variation in the bore diameter causes the float or indicator to change its position. If it remains between the markers previously set by masters for the maximum and minimum limits, the component is acceptable.

Electrical Gaging Units

The use of gages employing electrical gaging units is desirable for high-production items, because such devices are fast in operation, exact when properly adjusted, and positive in their acceptance of a component. The services of a skilled adjuster must be available.

In recent years several types of electrical gaging heads have been developed which have a wide range of application. The gaging heads consist of an electrical mechanism suitably contained, a gaging contact point or finger, and provision for mounting. The heads may be connected with lights which serve as visual aids to the inspector in segregating acceptable and nonacceptable parts. The heads may be connected to actuate sorting devices on automatic gaging machines. These units may also be used in multiple groups for checking a number of functions simultaneously.

A setting master or a gage-block build-up is required for setting the gage for each component limit. These same masters are also used by the inspector to check periodically the adjustment of the gage.

Section 35

TAPERS AND DOVETAILS

SECTION 35

TAPERS AND DOVETAILS

COMPUTING TAPERS

Taper is the gradual decrease of width or diameter of a piece from one end to the other. For example, a cone is tapered from the base to apex. The amount of taper depends on the diameters at the two ends of a part and its length. Some shops measure taper on each side, but the usual way is to take the total taper (Fig. 1).

The formula for taper per foot is

$$\text{TPF} = 12(D_1 - D_2) \div L$$

In Fig. 1, $\text{TPF} = 12(2 - 1\tfrac{1}{8}) \div 3\tfrac{1}{2} = 3$ in. per ft.

Taper per inch is simply

$$\text{TPI} = D_1 - D_2 \div L$$

On blueprints, the draftsman may indicate the taper in taper per foot, taper per inch, or as the angle measured from the center line, that is, one-half the included angle. For example, in Fig. 1, the tangent of one-half the included angle α is

$$\begin{aligned} \tan \tfrac{1}{2}\alpha &= \tfrac{1}{2} \times (2 - 1\tfrac{1}{8}) \div 3\tfrac{1}{2} \\ &= \tfrac{1}{2} \times \tfrac{7}{8} \div 3\tfrac{1}{2} = \tfrac{1}{2} \times 0.250 = 0.125 \end{aligned}$$

Then

$$\tfrac{1}{2}\alpha = 7°8'$$

Table for Use in Computing Tapers

In Table 1, the quantities when expressed in inches represent the taper per inch corresponding to various angles advancing by 10′ from 10′ to 90″. If an angle is given as, say, 27½° and it is desired to find the corresponding taper in inches, the amount, 0.4894, may be taken directly from the table. This is the taper per inch of length measured as in Fig. 2, along the axis. The taper in inches per foot of length is found by multiplying the tabulated quantity by 12, and in this particular case it would be 0.4894 in. × 12 = 5.8728 in. Where the included angle is not found directly in the table, the taper per inch is found as follows: Assume that the angle in question is 12¼°, then the nearest angles in the table are 12°10′ and 12°20′, the respective quantities tabulated under these angles being 0.21314 and 0.21610. The difference between the two is 0.00296, and as 12¼° is halfway between, one-half of 0.00296, or 0.00148, is added to 0.21314, giving 0.21462 in. as the taper of a piece 1 in. in length and of an included angle of 12¼°. The taper per foot equals 0.21462 in. × 12 = 2.5754 in.

In Fig. 1 the value of $\tfrac{1}{2}\alpha$ was 7°8′. Therefore, the included angle equals 14°16′.

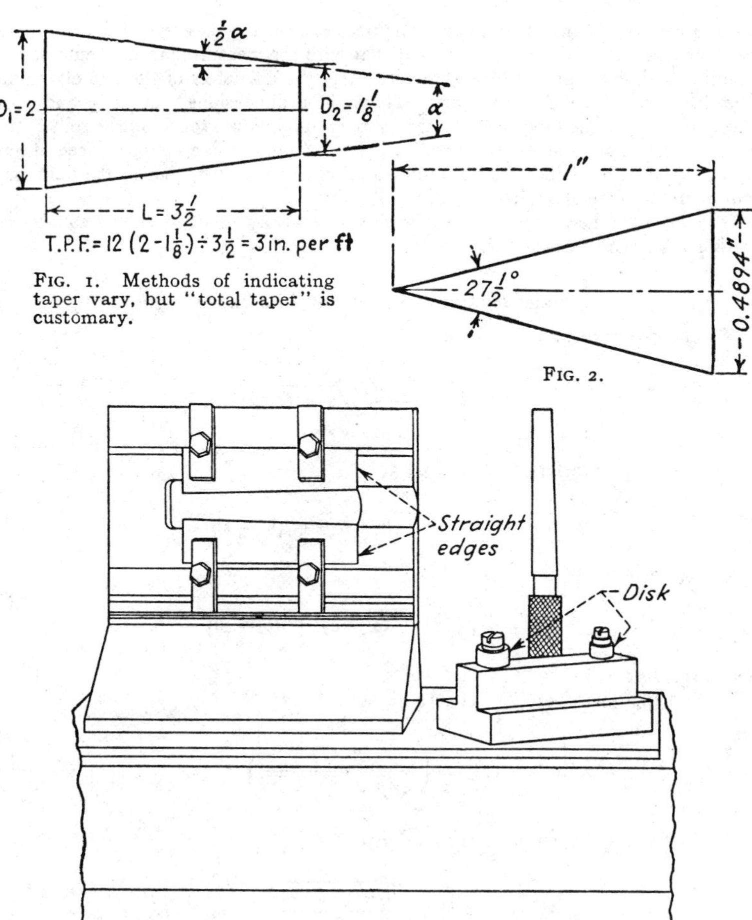

FIG. 1. Methods of indicating taper vary, but "total taper" is customary.

FIG. 2.

FIG. 3. An accurate taper-checking fixture consists of two straight edges that can be set to disks or buttons a known distance apart.

From Table 1, by inspection we see that the taper per inch for 14°16′ is approximately 0.25. Then the taper per foot = 12 × 0.25 = 3 in., as shown.

Table 2 gives the angles corresponding to tapers per foot of $\frac{1}{64}$ to 8 in.

MEASURING TAPERS

An Accurate Taper Gage

The gage illustrated in Fig. 3 is an exceedingly accurate device for the gaging of tapers.

It is evident that if two round disks of unequal diameter are placed on a surface plate a certain distance apart, two straightedges touching these two disks will represent a certain taper. It is also evident that with the measuring instruments now in use it is a simple matter to measure accurately the diameters of the two disks, and the distance these disks are apart. These three dimensions accurately and positively determine the taper represented by the straightedges touching the rolls. If a record is made of these three dimensions, these conditions can be reproduced at any time, thus making it possible to duplicate a taper piece even though the part may not at the time be accessible.

The formulas here and on the following pages may be of service in connection with a gage of this character.

Formulas for Use in Connection with Taper Gage

To find center distance (l), refer to Fig. 4.

$$l = \frac{R - r}{t} \sqrt{1 + t^2}$$

To find disk diameters, refer to Fig. 5.

$$r = \frac{a}{L}[\sqrt{L^2 + (b - a)^2} + (b - a)]$$

Dia small disk $= 2r$.

$$R = \frac{b}{L}[\sqrt{L^2 + (b - a)^2} - (b - a)]$$

Dia large disk $= 2R$.

To find taper per foot (T), refer to Fig. 6.

$$T = 24\left(\frac{R - r}{\sqrt{l^2 - (R - r)^2}}\right)$$

To find width of opening at ends, refer to Fig. 7.

$$a = r\sqrt{\frac{l - (R - r)}{l + (R - r)}}$$

Width of opening at small end $= 2a$.

$$b = R\sqrt{\frac{l + (R - r)}{l - (R - r)}}$$

Width of opening at large end $= 2b$.

Applications of Formulas

To Find Center Distance between Disks

Suppose there are two disks, as shown in Fig. 4, whose diameters are respectively 1¼ and 1 in. It is desired to construct a taper of ¾ in. to the foot, and the center distance l between disks must be determined in order that the gage jaws when touching both disks shall give that taper.

Let R = radius of large disk, or 0.625 in.

r = radius of small disk, or 0.500 in.

t = taper per in. on side, or

$$\frac{0.750}{24} = 0.03125 \text{ in.}$$

Then

$$l = \frac{R - r}{t}\sqrt{1 + t^2}$$

$$= \frac{0.125}{0.03125}\sqrt{1.000976}$$

$$= 4 \times 1.0005 = 4.002 \text{ in.}$$

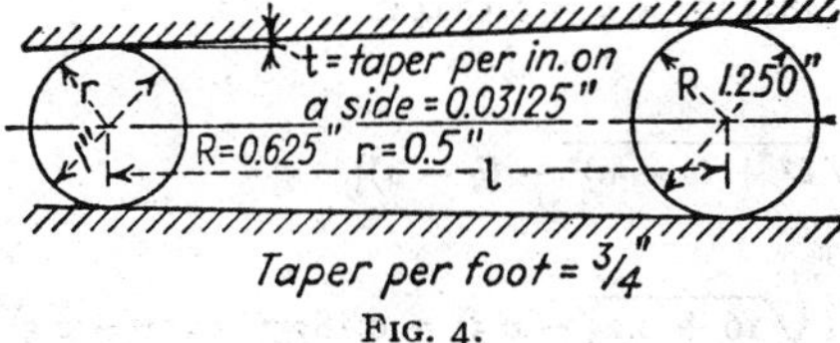

FIG. 4.

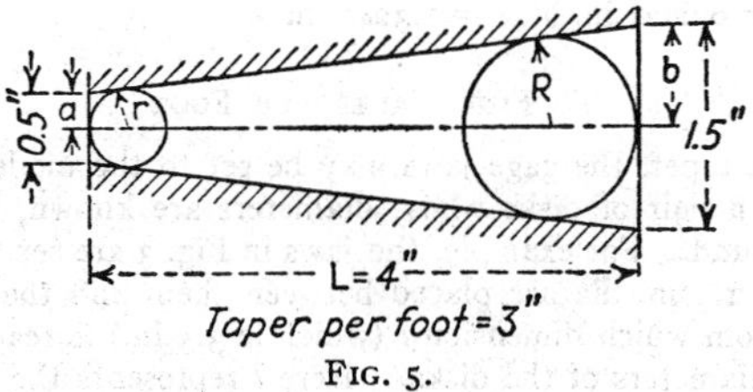

FIG. 5.

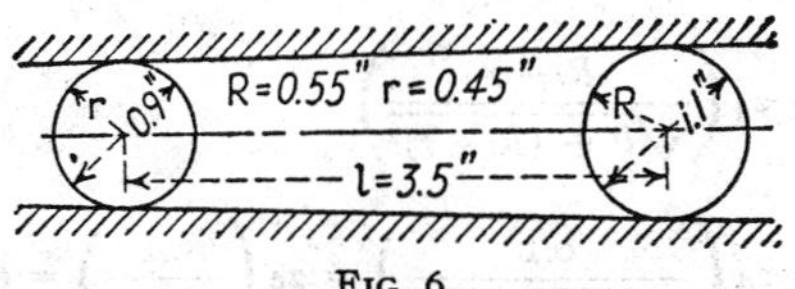

FIG. 6.

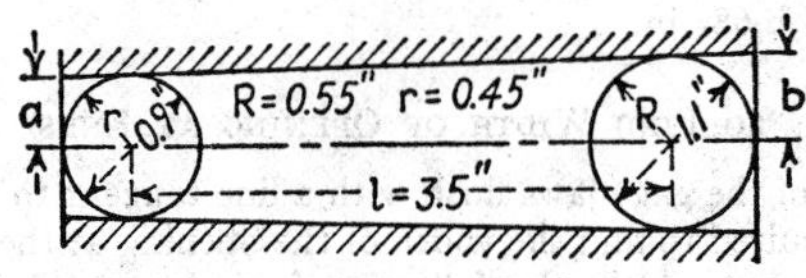

FIG. 7.

To Find Disk Diameters

Suppose the straightedges are to be set as in Fig. 5 for a 3 in. per ft taper whose length is to be 4 in. The small end is to be exactly ½ in. and the large end for this taper will, therefore, be 1½ in. What diameter must the disks be made so that when the jaws are in contact with them and the distance L over the disks measures 4 in., the taper will be exactly 3 in. per ft? Here a represents one-half the width of opening at the small end, and b one-half the width of opening at the large end. The radius of the small disk may be found by the formula

$$r = \frac{a}{L}[\sqrt{L^2 + (b - a)^2} + (b - a)]$$

Then

$$r = \frac{0.250}{4}(\sqrt{16 + 0.25} + 0.5)$$
$$= 0.0625(4.0311 + 0.5) = 0.2832$$

Dia small disk = 0.2832 in. × 2 = 0.5664 in.
For the large disk:

$$R = \frac{b}{L}[\sqrt{L^2 + (b - a)^2} - (b - a)]$$

Then

$$R = \frac{0.75}{4}(\sqrt{16 + 0.25} - 0.5) = 0.1875(4.0311 - 0.5) = 0.6621$$

Dia large disk = 0.6621 in. × 2 = 1.3242 in.

To Find Taper per Foot

In duplicating a taper, the gage jaws may be set to the model, and, by placing between the jaws a pair of disks whose diameters are known, the taper per foot may be readily found. For example, the jaws in Fig. 3 are set to a certain model, two disks 0.9 and 1.1 in. dia are placed between them and the distance over the disks measured, from which dimension l (which is 3.5 in.) is readily found by subtracting half the diameters of the disks. Here l represents the center distance as in Fig. 6. To determine the taper per foot which may be represented by T, the formula is

$$T = 24\left(\frac{R - r}{\sqrt{l^2 - (R - r)^2}}\right)$$

Then

$$T = 24\left(\frac{0.1}{\sqrt{12.25 - 0.01}}\right) = 24\left(\frac{0.1}{3.4985}\right) = 0.684$$

Taper per ft = 0.684 in.

To Find Width of Opening at Ends

If, with the ends of the gage jaws flush with a line tangent to the disk peripheries as in Fig. 7, it is required to find the width of the opening at the small end, where a represents one-half that width, the following formula may be applied, the disks being as in the last example 0.9 and 1.1 in. dia, respectively, and the center distance 3.5 in.:

$$a = r\sqrt{\frac{l - (R - r)}{l + (R - r)}}$$

Then

$$a = 0.45\sqrt{\frac{3.5 - (0.55 - 0.45)}{3.5 + (0.55 - 0.45)}} = 0.45\sqrt{\frac{3.4}{3.6}} = 0.45\sqrt{0.94444}$$
$$= 0.4373$$

0.4373 in. × 2 = 0.8746 in., width of opening at small end of gage.

Similarly the width of opening at the large end of the gage may be found as follows, where b = half the width of the large end.

$$b = R\sqrt{\frac{l + (R - r)}{l - (R - r)}}$$

Then

$$b = 0.55\sqrt{\frac{3.5 + (0.55 - 0.45)}{3.5 - (0.55 - 0.45)}} = 0.55\sqrt{\frac{3.6}{3.4}}$$
$$= 0.55\sqrt{1.02899} = 0.56595$$

0.56595 in. × 2 = 1.1319 in. = width of opening at large end.

DOVETAILS

Dimensioning Dovetail Slides and Gibs

Table 4 is calculated for machine-tool work so as to enable one to tell at a glance the amount to be added or subtracted in dimensioning dovetail slides and their gibs for the usual angles up to 60°. The column for 45° dovetails is omitted, as A and B would, of course, be alike for this angle.

In the application of the table, assuming a base with even dimensions, as in the sketch in Fig. 9, to obtain the dimensions x and y of the slide (Fig. 10), allowing for the gib which may be assumed to be ¼ in. thick, the perpendicular depth of the dovetail being ⅝ in., and the angle 60°, look under column A for ⅝ in., and it will be found opposite this that B is 0.360 in., which subtracted from 2 in. gives 1.640 in., the dimension x. To find y first get the dimension 1.640 in., then under the column for 60° gibs (where C is ¼ in.), D is found to be 0.289 in., which is added to 1.640, giving 1.929 in.

In practice, this dimension is usually made a little larger, say to the nearest sixty-fourth, to allow for fitting the gib.

Measuring External and Internal Dovetails

Table 5 gives constants for use with the plug method of sizing dovetail gages, etc. The constants are calculated for the plugs and angles most in use; and to use them a knowledge of arithmetic is all that is required. The formulas by which they were obtained are added for the convenience of those who may have an unusual angle to make.

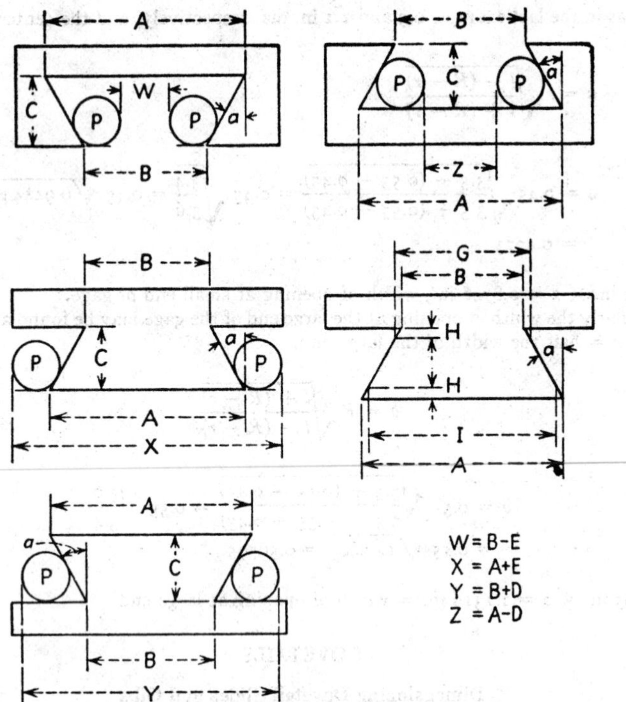

FIG. 8. Internal and external dovetails. See Table 5 for constants and formulas.

As an example of the use of the table, suppose that Z, Fig. 8, is the dimension wanted, and that the dimension A and the angle a are known. A glance at the formulas below shows that $Z = A - D$. Then the constant D corresponding to the size of plug and the angle used is subtracted from A and the remainder equals Z. For example, if $A = 4$ in., the plug used $= \frac{3}{8}$ in., and the angle $= 30°$, then $Z = A - D = 4$ in. $-$ 1.0245 in. $=$ 2.9755 in.

If A is not known, but B and C are given, as in the formula below Table 5, $A = B + CF$. Then if $B = 3.134$ in., $C = \frac{3}{4}$ in., and the angle is 30°, as before; $A = B + CF = 3.134$ in. + (0.75 in. × 1.1547) = 4 in., whence Z can be found, as already shown.

If the corners of the dovetail are flat, as shown in Fig. 8 at I and G, and the dimensions I and H and the angles are known, it will be found from the formulas that A also $= I + HF$; so that, if $I = 3.8557$ in., $H = \frac{1}{8}$ in., and the angle $= 30°$; then $A = I + HF = 3.8557$ in. + (0.125 in. × 1.1547) = 4 in., from which Z is found as before.

TABLE 1. TAPER PER INCH FOR VARIOUS INCLUDED ANGLES
The Tabulated Quantities = Twice the Tangent of Half the Angle

Deg	0′	10′	20′	30′	40′	50′	60′
0	0.00000	0.00290	0.00582	0.00872	0.01164	0.01454	0.01746
1	0.01746	0.02036	0.02326	0.02618	0.02910	0.03200	0.03492
2	0.03492	0.03782	0.04072	0.04364	0.04656	0.04946	0.05238
3	0.05238	0.05528	0.05820	0.06110	0.06402	0.06692	0.06984
4	0.06984	0.07276	0.07566	0.07858	0.08150	0.08440	0.08732
5	0.08732	0.09024	0.09316	0.09606	0.09898	0.10190	0.10482
6	0.10482	0.10774	0.11066	0.11356	0.11648	0.11940	0.12232
7	0.12232	0.12524	0.12816	0.13108	0.13400	0.13694	0.13986
8	0.13986	0.14278	0.14570	0.14862	0.15156	0.15448	0.15740
9	0.15740	0.16034	0.16326	0.16618	0.16912	0.17204	0.17498
10	0.17498	0.17790	0.18084	0.18378	0.18670	0.18964	0.19258
11	0.19258	0.19552	0.19846	0.20138	0.20432	0.20726	0.21020
12	0.21020	0.21314	0.21610	0.21904	0.22198	0.22492	0.22788
13	0.22788	0.23082	0.23376	0.23672	0.23966	0.24262	0.24556
14	0.24556	0.24852	0.25148	0.25444	0.25738	0.26034	0.26330
15	0.26330	0.26626	0.26922	0.27218	0.27516	0.27812	0.28108
16	0.28108	0.28404	0.28702	0.28998	0.29296	0.29502	0.29890
17	0.29890	0.30188	0.30486	0.30782	0.31080	0.31378	0.31676
18	0.31676	0.31976	0.32274	0.32572	0.32870	0.33170	0.33468
19	0.33468	0.33768	0.34066	0.34366	0.34666	0.34966	0.35266
20	0.35266	0.35566	0.35866	0.36166	0.36466	0.36768	0.37068
21	0.37068	0.37368	0.37670	0.37972	0.38272	0.38574	0.38876
22	0.38876	0.39178	0.39480	0.39782	0.40084	0.40388	0.40690
23	0.40690	0.40994	0.41296	0.41600	0.41904	0.42208	0.42512
24	0.42512	0.42816	0.43120	0.43424	0.43728	0.44034	0.44338
25	0.44338	0.44644	0.44950	0.45256	0.45562	0.45868	0.46174
26	0.46174	0.46480	0.46786	0.47094	0.47400	0.47708	0.48016
27	0.48016	0.48324	0.48632	0.48940	0.49248	0.49556	0.49866
28	0.49866	0.50174	0.50484	0.50794	0.51004	0.51414	0.51724
29	0.51724	0.52034	0.52344	0.52656	0.52966	0.53278	0.53590
30	0.53590	0.53902	0.54214	0.54526	0.54838	0.55152	0.55464
31	0.55464	0.55778	0.56092	0.56406	0.56720	0.57034	0.57350
32	0.57350	0.57664	0.57980	0.58294	0.58610	0.58926	0.59242
33	0.59242	0.59560	0.59876	0.60194	0.60510	0.60828	0.61146
34	0.61146	0.61464	0.61782	0.62102	0.62420	0.62740	0.63060
35	0.63060	0.63380	0.63700	0.64020	0.64342	0.64662	0.64984
36	0.64984	0.65306	0.65628	0.65950	0.66272	0.66596	0.66920
37	0.66920	0.67242	0.67566	0.67890	0.68216	0.68540	0.68866
38	0.68866	0.69192	0.69516	0.69844	0.70170	0.70496	0.70824
39	0.70824	0.71152	0.71480	0.71808	0.72136	0.72464	0.72794
40	0.72794	0.73124	0.73454	0.73784	0.74114	0.74446	0.74776
41	0.74776	0.75108	0.75440	0.75774	0.76106	0.76440	0.76772
42	0.76772	0.77106	0.77442	0.77776	0.78110	0.78446	0.78782
43	0.78782	0.79118	0.79454	0.79792	0.80130	0.80468	0.80806
44	0.80806	0.81144	0.81482	0.81822	0.82162	0.82502	0.82842
45	0.82842	0.83184	0.83526	0.83866	0.84210	0.84552	0.84894

TABLE 1. TAPER PER INCH FOR VARIOUS INCLUDED ANGLES (*Continued*)

Deg	0′	10′	20′	30′	40′	50′	60′
46	0.84894	0.85238	0.85582	0.85926	0.86272	0.86616	0.86962
47	0.86962	0.87308	0.87656	0.88002	0.88350	0.88698	0.89046
48	0.89046	0.89394	0.89744	0.90094	0.90444	0.90794	0.91146
49	0.91146	0.91496	0.91848	0.92202	0.92554	0.92908	0.93262
50	0.93262	0.93616	0.93970	0.94326	0.94682	0.95038	0.95396
51	0.95396	0.95752	0.96110	0.96468	0.96828	0.97186	0.97546
52	0.97546	0.97906	0.98268	0.98630	0.98990	0.99354	0.99716
53	0.99716	1.00080	1.00444	1.00808	1.01174	1.01538	1.01906
54	1.01906	1.02272	1.02638	1.03006	1.03376	1.03744	1.04114
55	1.04114	1.04484	1.04854	1.05226	1.05596	1.05970	1.06342
56	1.06342	1.06716	1.07090	1.07464	1.07840	1.08214	1.08592
57	1.08592	1.08968	1.09346	1.09724	1.10102	1.10482	1.10862
58	1.10862	1.11242	1.11624	1.12006	1.12388	1.12770	1.13154
59	1.13154	1.13538	1.13924	1.14310	1.14696	1.15082	1.15470
60	1.15470	1.15858	1.16248	1.16636	1.17026	1.17418	1.17810
61	1.17810	1.18202	1.18594	1.18988	1.19382	1.19776	1.20172
62	1.20172	1.20568	1.20966	1.21362	1.21762	1.22160	1.22560
63	1.22560	1.22960	1.23362	1.23764	1.24166	1.24570	1.24974
64	1.24974	1.25378	1.25784	1.26190	1.26598	1.27006	1.27414
65	1.27414	1.27824	1.28234	1.28644	1.29056	1.29468	1.29882
66	1.29882	1.30296	1.30710	1.31126	1.31542	1.31960	1.32378
67	1.32378	1.32796	1.33216	1.33636	1.34056	1.34478	1.34902
68	1.34902	1.35326	1.35750	1.36176	1.36602	1.37028	1.37456
69	1.37456	1.37984	1.38314	1.38744	1.39176	1.39608	1.40042
70	1.40042	1.40476	1.40910	1.41346	1.41782	1.42220	1.42658
71	1.42658	1.43098	1.43538	1.43980	1.44422	1.44864	1.45308
72	1.45308	1.45754	1.46200	1.46646	1.47094	1.47542	1.47992
73	1.47992	1.48442	1.48894	1.49348	1.49800	1.50256	1.50710
74	1.50710	1.51168	1.51624	1.52084	1.52544	1.53004	1.53466
75	1.53466	1.53928	1.54392	1.54856	1.55322	1.55790	1.56258
76	1.56258	1.56726	1.57196	1.57668	1.58140	1.58612	1.59088
77	1.59088	1.59562	1.60040	1.60516	1.60996	1.61476	1.61966
78	1.61956	1.62440	1.62922	1.63406	1.63892	1.64380	1.64868
79	1.64868	1.65356	1.65846	1.66338	1.66830	1.67324	1.67820
80	1.67820	1.68316	1.68814	1.69312	1.69812	1.70314	1.70816
81	1.70816	1.71320	1.71824	1.72332	1.72836	1.73348	1.73858
82	1.73858	1.74368	1.74882	1.75396	1.75910	1.76428	1.76946
83	1.76946	1.77464	1.77984	1.78506	1.79030	1.79554	1.80080
84	1.80080	1.80608	1.81138	1.81668	1.82198	1.82732	1.83266
85	1.83266	1.83802	1.84340	1.84878	1.85418	1.85960	1.86504
86	1.86504	1.87048	1.87594	1.88142	1.88690	1.89240	1.89792
87	1.89792	1.90346	1.90902	1.91458	1.92016	1.92576	1.93138
88	1.93138	1.93700	1.94266	1.94832	1.95400	1.95968	1.96540
89	1.96540	1.97112	1.97686	1.98262	1.98840	1.99420	2.00000
90	2						

TABLE 2. TAPERS PER FOOT IN INCHES AND CORRESPONDING ANGLES

Taper per Ft, In.	Included Angle			Angle with Center Line			Taper per Ft, In.	Included Angle			Angle with Center Line		
	Deg	Min	Sec	Deg	Min	Sec		Deg	Min	Sec	Deg	Min	Sec
1/64	0	4	28	0	2	14	1	4	46	18	2	23	9
1/32	0	8	58	0	4	29	1 1/8	5	22	40	2	41	50
1/16	0	17	53	0	8	57	1 1/4	5	57	48	2	58	54
3/32	0	26	52	0	13	26	1 3/8	6	33	26	3	16	43
1/8	0	35	48	0	17	54	1 1/2	7	9	10	3	34	35
5/32	0	44	44	0	22	22	1 5/8	7	44	48	3	52	24
3/16	0	53	44	0	26	52	1 3/4	8	20	26	4	10	13
7/32	1	2	36	0	31	18	1 7/8	8	56	2	4	28	1
1/4	1	11	36	0	35	48	2	9	31	36	4	45	48
9/32	1	20	30	0	40	15	2 1/4	10	42	42	5	21	21
5/16	1	29	30	0	44	45	2 1/2	11	53	36	5	56	48
11/32	1	38	26	0	49	13	2 3/4	13	4	24	6	32	12
3/8	1	47	24	0	53	42	3	14	15	0	7	7	30
13/32	1	56	24	0	58	12	3 1/4	15	25	24	7	42	42
7/16	2	5	18	1	2	39	3 1/2	16	35	40	8	17	50
15/32	2	14	16	1	7	8	3 3/4	17	45	40	8	52	50
1/2	2	23	10	1	11	35	4	18	55	24	9	27	42
17/32	2	32	4	1	16	2	4 1/4	20	5	2	10	2	31
9/16	2	41	4	1	20	32	4 1/2	21	14	20	10	37	10
19/32	2	50	2	1	25	1	4 3/4	22	23	22	11	11	41
5/8	2	59	0	1	29	30	5	23	32	12	11	46	6
21/32	3	7	56	1	33	58	5 1/4	24	40	42	12	20	21
11/16	3	16	54	1	38	27	5 1/2	25	48	48	12	54	24
23/32	3	25	50	1	42	55	5 3/4	26	56	46	13	28	23
3/4	3	34	44	1	47	21	6	28	4	20	14	2	10
25/32	3	43	44	1	51	52	6 1/4	29	11	34	14	35	47
13/16	3	52	38	1	56	19	6 1/2	30	18	26	15	9	13
27/32	4	1	32	2	0	46	6 3/4	31	25	2	15	42	31
7/8	4	10	32	2	5	16	7	32	31	12	16	15	36
29/32	4	19	26	2	9	43	7 1/4	33	37	44	16	48	32
15/16	4	28	24	2	14	12	7 1/2	34	42	30	17	21	15
31/32	4	37	20	2	18	40	7 3/4	35	47	32	17	53	46
							8	36	52	12	18	26	6

TABLE 3. DIFFERENCES IN DIAMETERS FOR TAPERS FROM 1/16 TO 1 1/4 IN. PER FT
Amount of Taper for Lengths up to 24 In.

Length of Tapered Portion	Taper per Ft									
	1/16	3/32	1/8	1/4	3/8	1/2	5/8	3/4	1	1 1/4
1/32	0.0002	0.0002	0.0003	0.0007	0.0010	0.0013	0.0016	0.0020	0.0026	0.0033
1/16	0.0003	0.0005	0.0007	0.0013	0.0020	0.0026	0.0033	0.0039	0.0052	0.0065
1/8	0.0007	0.0010	0.0013	0.0026	0.0039	0.0052	0.0065	0.0078	0.0104	0.0130
3/16	0.0010	0.0015	0.0020	0.0039	0.0059	0.0078	0.0098	0.0117	0.0156	0.0195
1/4	0.0013	0.0020	0.0026	0.0052	0.0078	0.0104	0.0130	0.0156	0.0208	0.0260
5/16	0.0016	0.0024	0.0033	0.0065	0.0098	0.0130	0.0163	0.0195	0.0260	0.0326
3/8	0.0020	0.0029	0.0039	0.0078	0.0117	0.0156	0.0195	0.0234	0.0312	0.0391
7/16	0.0023	0.0034	0.0046	0.0091	0.0137	0.0182	0.0228	0.0273	0.0365	0.0456
1/2	0.0026	0.0039	0.0052	0.0104	0.0156	0.0208	0.0260	0.0312	0.0417	0.0521
9/16	0.0029	0.0044	0.0059	0.0117	0.0176	0.0234	0.0293	0.0352	0.0469	0.0586
5/8	0.0033	0.0049	0.0065	0.0130	0.0195	0.0260	0.0326	0.0391	0.0521	0.0651
11/16	0.0036	0.0054	0.0072	0.0143	0.0215	0.0286	0.0358	0.0430	0.0573	0.0716
3/4	0.0039	0.0059	0.0078	0.0156	0.0234	0.0312	0.0391	0.0469	0.0625	0.0781
13/16	0.0042	0.0063	0.0085	0.0169	0.0254	0.0339	0.0423	0.0508	0.0677	0.0846
7/8	0.0046	0.0068	0.0091	0.0182	0.0273	0.0365	0.0456	0.0547	0.0729	0.0911
15/16	0.0049	0.0073	0.0098	0.0195	0.0293	0.0391	0.0488	0.0586	0.0781	0.0977
1	0.0052	0.0078	0.0104	0.0208	0.0312	0.0417	0.0521	0.0625	0.0833	0.1042
2	0.0104	0.0156	0.0208	0.0417	0.0625	0.0833	0.1042	0.125	0.1667	0.2083
3	0.0156	0.0234	0.0312	0.0625	0.0937	0.1250	0.1562	0.1875	0.250	0.3125
4	0.0208	0.0312	0.0417	0.0833	0.125	0.1667	0.2083	0.250	0.3333	0.4167
5	0.0260	0.0391	0.0521	0.1042	0.1562	0.2083	0.2604	0.3125	0.4167	0.5208
6	0.0312	0.0469	0.0625	0.125	0.1875	0.250	0.3125	0.375	0.500	0.625
7	0.0365	0.0547	0.0729	0.1458	0.2187	0.2917	0.3646	0.4375	0.5833	0.7292
8	0.0417	0.0625	0.0833	0.1667	0.250	0.3333	0.4167	0.500	0.6667	0.8333
9	0.0469	0.0703	0.0937	0.1875	0.2812	0.375	0.4687	0.5625	0.750	0.9375
10	0.0521	0.0781	0.1042	0.2083	0.3125	0.4167	0.5208	0.625	0.8333	1.0417
11	0.0573	0.0859	0.1146	0.2292	0.3437	0.4583	0.5729	0.6875	0.9167	1.1458
12	0.0625	0.0937	0.125	0.250	0.375	0.500	0.625	0.750	1.000	1.250
13	0.0677	0.1016	0.1354	0.2708	0.4062	0.5417	0.6771	0.8125	1.0833	1.3542
14	0.0729	0.1094	0.1458	0.2917	0.4375	0.5833	0.7292	0.875	1.1667	1.4583
15	0.0781	0.1172	0.1562	0.3125	0.4687	0.625	0.7812	0.9375	1.250	1.5625
16	0.0833	0.125	0.1667	0.3333	0.500	0.6667	0.8333	1.000	1.3333	1.6667
17	0.0885	0.1328	0.1771	0.3542	0.5312	0.7083	0.8854	1.0625	1.4167	1.7708
18	0.0937	0.1406	0.1875	0.3750	0.5625	0.750	0.9375	1.125	1.500	1.875
19	0.0990	0.1484	0.1979	0.3958	0.5937	0.7917	0.9896	1.1875	1.5833	1.9792
20	0.1042	0.1562	0.2083	0.4167	0.625	0.8333	1.0417	1.250	1.6667	2.0833
21	0.1094	0.1641	0.2187	0.4375	0.6562	0.875	1.0937	1.3125	1.750	2.1875
22	0.1146	0.1719	0.2292	0.4583	0.6875	0.9167	1.1458	1.375	1.8333	2.2917
23	0.1198	0.1797	0.2396	0.4792	0.7187	0.9583	1.1979	1.4375	1.9167	2.3958
24	0.125	0.1875	0.250	0.500	0.750	1.000	1.250	1.500	2.000	2.500

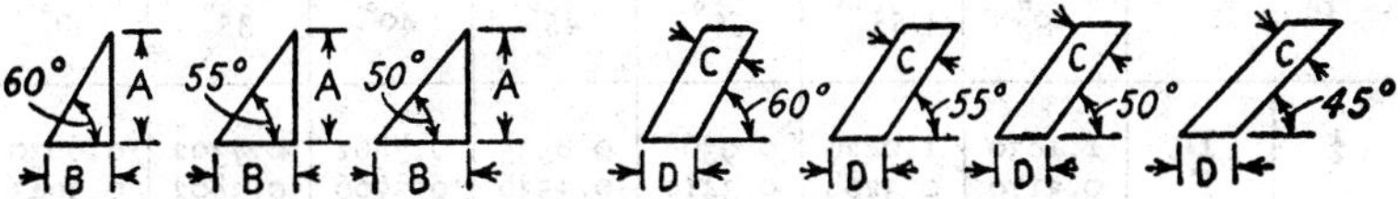

Table 4. Dimensioning Dovetail Slides and Gibs
All Dimensions in Inches

A	B 60°	B 55°	B 50°	C	D 60°	D 55°	D 50°	D 45°
1/32	0.018	0.022	0.027	1/8	0.144	0.152	0.163	0.176
1/16	0.036	0.044	0.053	3/16	0.216	0.228	0.244	0.264
1/8	0.072	0.087	0.105	1/4	0.289	0.305	0.326	0.353
1/4	0.144	0.175	0.210	5/16	0.361	0.381	0.407	0.442
3/8	0.216	0.262	0.314	3/8	0.433	0.457	0.489	0.530
1/2	0.288	0.350	0.420	1/2	0.577	0.610	0.652	0.707
5/8	0.360	0.437	0.525	5/8	0.721	0.762	0.815	0.883
3/4	0.433	0.525	0.629	3/4	0.866	0.915	0.979	1.060
7/8	0.505	0.612	0.734	7/8	1.010	1.067	1.142	1.237
1	0.577	0.700	0.839	1	1.154	1.220	1.305	1.414
1 1/8	0.649	0.787	0.944					
1 1/4	0.721	0.875	1.049					
1 3/8	0.794	0.962	1.153					
1 1/2	0.866	1.050	1.259					
1 3/4	1.010	1.225	1.469					
2	1.154	1.400	1.677					
2 1/4	1.298	1.575	1.888					
2 1/2	1.442	1.750	2.097					
2 3/4	1.588	1.925	2.307					
3	1.732	2.100	2.517					
3 1/2	2.020	2.450	2.937					
4	2.308	2.800	3.356					
4 1/2	2.598	3.150	3.776					
5	2.885	3.501	4.195					

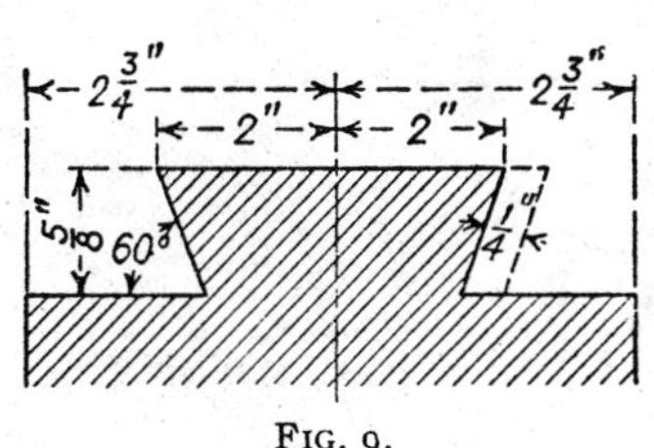

Fig. 9.

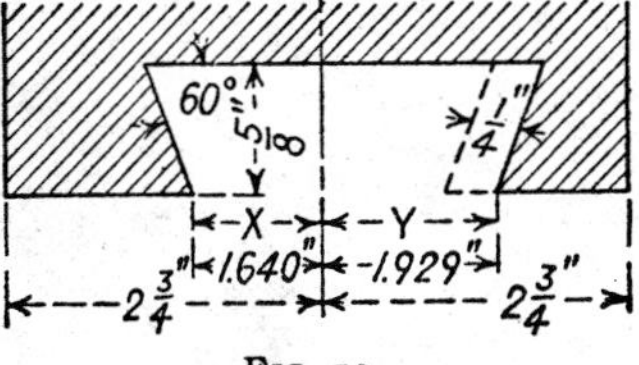

Fig. 10.

TABLE 5. CONSTANTS FOR DOVETAILS

Plug, In.		60°	55°	50°	45°	40°	35°	30°
$\frac{1}{4}$	D	1.1830	1.0429	0.9368	0.8535	0.7861	0.7302	0.6830
	E	0.3170	0.3288	0.3410	0.3536	0.3666	0.3802	0.3943
$\frac{3}{8}$	D	1.7745	1.5643	1.4053	1.2803	1.1792	1.0954	1.0245
	E	0.4755	0.4932	0.5115	0.5303	0.5499	0.5702	0.5915
$\frac{1}{2}$	D	2.3660	2.0858	1.8730	1.7070	1.5722	1.4604	1.3660
	E	0.6340	0.6576	0.6820	0.7072	0.7332	0.7603	0.7886
$\frac{3}{4}$	D	3.5490	3.1286	2.8106	2.5606	2.3584	2.1903	2.0490
	E	0.9510	0.9864	1.0230	1.0606	1.0998	1.1404	1.1830
	F	3.4641	2.8563	2.3836	2	1.6782	1.4004	1.1547

Formulas for dovetails (see Fig. 10):

$$A = B + CF = I + HF$$
$$B = A - CF = G - HF$$
$$E = P\left(\cot \frac{90 + a}{2}\right) + P$$
$$D = P\left(\cot \frac{90 - a}{2}\right) + P$$
$$F = 2 \tan a$$

Part 7

FASTENING DEVICES

Section 36

BOLTS, SCREWS, RIVETS, AND WASHERS

SECTION 36

BOLTS, SCREWS, RIVETS, AND WASHERS

DEFINITIONS FOR BOLTS AND SCREWS[1]

Bolt, Aircraft. See Bolt, Machine, also Screw, Machine.

Bolt, Assembled Washer. A machine bolt on which has been assembled a non-removable washer. See also Screw, Assembled Washer.

Bolt, Battery. See Bolt, Machine.

Bolt, Bent. See Bolt, Eye, also Bolt, Hook.

Bolt, Boiler Patch. An externally threaded fastener whose threaded portion is of one nominal diameter, without machined (or equivalent) surfaces, having a countersunk head surmounted by a square or hexagonal head to facilitate installation, after which the square of hexagonal head is removed.

Bolt, Cad. See Bolt, Tee-head.

Bolt, Cant Dog. See Bolt, Machine, also Screw, Machine.

Bolt, Captive. See Bolt, Externally Relieved Body, also Screw, Externally Relieved Body.

Bolt, Carriage. An externally threaded fastener whose threaded portion is of one nominal diameter, having a flat countersunk head with an included angle of 100° or more, a rounded head, or a head designed for driving by impact, and having a square, ribbed, or finned neck under the head to prevent turning. Excludes Bolt, Step. See also Bolt, Deck and Bolt, Ribbed-head.

Bolt, Clevis. An externally threaded fastener whose threaded portion is of one nominal diameter, but with a narrow circumferential groove between the threaded and unthreaded portions of the shank, and having a head designed to be held or driven by either an inserted driver or a wrench. See also Bolt, Externally Relieved Body.

Bolt, Close Tolerance. See Bolt (as modified).

Bolt, Coach. See Bolt, Lag.

Bolt, Coupling. See Bolt, Machine.

Bolt, Crossing. See Bolt, Machine.

Bolt, Cultivator. See Bolt, Carriage, also Bolt, Plow.

Bolt, Dardelet. See Bolt, Rivet.

Bolt, Dardelet Rivet. See Bolt, Rivet.

Bolt, Deck. An externally threaded fastener whose threaded portion is of one nominal diameter, having a flat head and a square neck under the head to prevent turning. See also Bolt, Carriage and Bolt, Step.

Bolt, Double-ended. See Stud, Threaded.

Bolt, Drive-grip. See Bolt, Carriage.

Bolt, Drop. See Bolt, Eye.

Bolt, Elevator. See Bolt, Carriage, also Bolt, Ribbed-head, also Bolt, Step.

Bolt, Externally Relieved Body. An externally threaded fastener whose threaded portion is of one nominal diameter, and whose unthreaded portion has been machined or ground in one or more places or over its entire length to a diameter approximating the root diameter of the thread. It is No. 10 (0.190 in.) or larger in size and has a head which is not designed to be held or driven with an inserted driver. Excludes Bolt, Internal Wrenching. See also Screw, Externally Relieved Body.

[1] Source: Munitions Board Cataloging Agency.

Bolt, Eye. An externally threaded device whose threaded portion is of one nominal diameter, without a head, but with the unthreaded end either bent more than 225° or cast, forged, drilled, or punched to resemble an eye. See also Bolt, Hook.

Bolt, Fin. See Bolt, Carriage, also Bolt, Plow, also Bolt, Ribbed-head.

Bolt, Firebrick Anchor. A fastener designed for securing firebrick to the outer wall or casing of furnaces or boilers. Normally consisting of a straight or bent shank with a square or rectangular head and designed for high-temperature application. The shank is sometimes threaded and furnished with a nut.

Bolt, Fitting Up. See Bolt, Machine.

Bolt, Flanged Coupling. See Bolt, Machine.

Bolt, Flush-head, Freight-car. See Bolt, Deck.

Bolt, Forcing. See Bolt (as modified).

Bolt, Frog. See Bolt, Machine.

Bolt, Guard. See Bolt (as modified).

Bolt, Hanger. A headless cylindrical threaded fastener which is threaded on both ends, one end being threaded to accommodate a nut and the other end having continuous wood-screw-type threads extending from a gimlet or cone point to the unthreaded portion. The unthreaded portion may have a ribbed shoulder.

Bolt, Heel. A carriage bolt complete with a nut having an extension on one side, usually in the shape of an "eye," to permit loosening by impact.

Bolt, Hex-head. See Bolt (as modified).

Bolt, Hook. An externally threaded device whose threaded portion is of one nominal diameter, without a head, but with the unthreaded end bent not over 225°. See also Bolt, Eye and Bolt, U.

Bolt, Hub. See Bolt, Carriage.

Bolt, Integral Lockwasher. See Bolt, Assembled Washer, also Screw, Assembled Washer.

Bolt, Internally Relieved Body. An externally threaded fastener whose threaded portion is of one nominal diameter, with a portion of its shank unthreaded, and a concentric hole through the head and unthreaded portion along the longitudinal center line. The head is designed to be held or turned by a wrench.

Bolt, Internal Wrenching. An externally threaded fastener whose threaded portion is of one nominal diameter. No. 10 (0.190 in.) or larger, whose head is designed with an internal socket or internal multiple spline for use with an inserted wrench. Excludes Screw, Machine and Setscrew.

Bolt, J. See Bolt, Hook.

Bolt, Key. See Bolt, Carriage, also Bolt, Plow.

Bolt, Key-head. See Bolt, Carriage, also Bolt, Plow.

Bolt, Key-neck. See Bolt, Carriage.

Bolt, Lag. An externally threaded fastener having a square or hexagon head and with a continuous thread (wood-screw type) extending from a gimlet or cone point for a distance slightly more than one-half the length of the bolt.

Bolt, Machine. An externally threaded fastener whose threaded and unthreaded portions are of one nominal diameter, No. 10 (0.190 in.) or larger, and having a square, hexagon, rounded, or flat circular countersunk type of head. The head is not designed to be held or driven with an inserted driver. Excludes Bolt, Carriage and Bolt, Plow. See also Bolt, Internal Wrenching.

Bolt, Oval-neck. An externally threaded fastener whose threaded portion is of one nominal diameter, with a round head and an unthreaded neck which is elliptical in cross-section.

Bolt, Peavey. See Bolt, Carriage, also Bolt, Machine.

Bolt, Plow. An externally threaded fastener whose threaded and unthreaded portions are of one nominal diameter, having a flat circular countersunk head with an included angle of less than 100° and with a square, ribbed, or finned neck under the head, or one rectangular or triangular projection attached to the underside of the head, or having a flat square countersunk head with an included angle of less than 100° to prevent turning. See also Bolt, Carriage.

Bolt, Rail. See Bolt, Oval-neck.

Bolt, Reduced Shank. See Bolt, Externally Relieved Body, also Screw, Externally Relieved Body.

Bolt, Ribbed-head. An externally threaded fastener whose threaded portion is of one nominal diameter, having a flat countersunk head with radial ridges or projections on the underside of the head to prevent turning. Excludes Bolt, Self-locking. See also Bolt, Carriage and Bolt, Plow.

Bolt, Rim. See Bolt, Carriage.

Bolt, Ring. An eyebolt having a nonremovable ring assembled in the eye.

Bolt, Rivet. An externally threaded fastener whose threaded portion is of one nominal diameter, with a rounded head and a tapered ribbed shank. The "Dardelet" type of Bolt, Rivet has a patented locking type thread but does not have a tapered shank.

Bolt, Self-locking. An externally threaded fastener whose threaded portion is of one nominal diameter, and with a locking feature incorporated in the design of the head. It is No. 10 (0.190 in.) or larger in size and has a head which is not designed to be held or driven with an inserted driver. It does not have a lockwasher attached to the head. Excludes Bolt, Ribbed-head. See also Screw, Self-locking.

Bolt, Set. See Setscrew.

Bolt, Shackle. See Bolt (as modified).

Bolt, Shaft. See Bolt, Carriage.

Bolt, Shoulder. An externally threaded fastener whose threaded portion is of one nominal diameter, and with a round unthreaded neck or shank, all or part of which is of greater diameter than the threaded portion. Excludes Bolt, Internal Wrenching.

Bolt, Sink. See Bolt, Machine, also Screw, Machine.

Bolt, Sleigh-shoe. See Bolt, Carriage.

Bolt, Spade. See Bolt, Eye.

Bolt, Spline-neck. See Bolt, Carriage, also Bolt, Rivet.

Bolt, Square-neck. See Bolt, Carriage, also Bolt, Plow, also Bolt, Step.

Bolt, Step. An externally threaded fastener, whose threaded portion is of one nominal diameter, similar to a carriage bolt, but having a rounded head of a diameter slightly more than three times the nominal diameter of the bolt. It has a square neck under the head to prevent turning. See also Bolt, Deck.

Bolt, Stove. See Screw, Machine.

Bolt, Stripper. See Bolt, Shoulder.

Bolt, Stud. See Stud, Threaded.

Bolt, T-slot. See Bolt, Machine, also Bolt, Tee-head.

Bolt, Tank. See Bolt, Machine, also Screw, Machine.

Bolt, Tap. See Bolt, Machine, also Screw, Machine.

Bolt, Tee-head. An externally threaded fastener whose threaded portion is of one nominal diameter and with a rectangular head designed to fit into a slot and hold against turning.

Bolt, Thimble-eye. See Bolt, Eye.

Bolt, Tire. See Bolt, Machine, also Bolt, Plow.

Bolt, Toggle. An internally threaded fastener whose threaded portion is of one nominal diameter having a wing or wings mounted on the bolt head or on a trunnion nut. The wing or wings upset when the toggle bolt is inserted in a constricted passage in a hollow wall; and, when the bolt is tightened, the wing or wings assume a transverse or athwart position. It is used for securely holding fixtures or devices to hollow walls.

Bolt, Track. See Bolt, Oval-neck.

Bolt, U. An externally threaded fastener bent approximately 180° in the shape of the letter U and with both ends threaded.

Bolt, Washer-head. See Bolt, Machine, also Screw, Machine.

Bolt, Wedge. See Bolt, Plow.

Bolt, Wedge-head. See Bolt, Plow.

Bolt, Welding. An externally threaded fastener whose threaded portion is of one nominal diameter having small raised projections on the top side or underside of the head to facilitate welding the head. May or may not be designed to be held or turned with a wrench or inserted driver.

Bolt, Wheel. See Bolt, Carriage.

Bolt, Wrenching. See Bolt, Internal Wrenching.

Bolt Blank. A cylindrical device having a head, eye, or hook; similar to a Bolt, Machine; Bolt, Eye; or Bolt, Hook, etc., but without threads, and normally intended for a subsequent threading operation.

Screw, Assembled Washer. A Screw, Machine; Screw, Tapping, Thread Cutting; or Screw, Tapping, Thread Forming to which has been assembled a nonremovable washer.

Screw, Cap. See Bolt (as modified), Screw (as modified).

The screw industry does not agree with the government cataloging agency on the definition for a capscrew but is otherwise in general agreement with the terms, as defined by that body. The screw industry's definition is: "**Screw, Cap (or Preferably Capscrew)**. An externally threaded fastener, whose threaded and unthreaded portions are of one nominal diameter, ¼ through 1½ in., with machined (or equivalent) surfaces, and having a hexagon, flat, fillister, or round type of head designed to be driven by a wrench or inserted driver. A capscrew has a chamfered point." In general, a capscrew is considered a more finished product than a bolt. Also, a capscrew is tightened by torquing the head, while a bolt is secured by tightening a nut.

Screw, Coach. See Bolt, Lag.

Screw, Drive. A headed, hardened cylindrical fastener with multiple spiral flutes on its shank. It also has an end smaller in diameter than the outside diameter of the spiral flutes, which acts as a pilot when driven into a drilled hole.

Screw, Externally Relieved Body. An externally threaded fastener whose threaded portion is of one nominal diameter, and whose unthreaded portion has been machined or ground in one or more places or over its entire length to a diameter approximating the root diameter of the thread. It is No. 0 (0.060 in.) or larger, and the head is designed to be held or driven with an inserted driver or wrench or both in sizes below No. 10 (0.190 in.). On No. 10 or larger sizes the head has either a slot, cross-recess, one-way slot, or clutch recess for use with an inserted driver. See also Bolt, Externally Relieved Body.

Screw, Hold-down. See Bolt, Hook.

Screw, Instrument. An externally threaded fastener similar to a Screw, Machine but smaller than No. 0 in size.

Screw, Integral Lockwasher. See Bolt, Assembled Washer, also Screw, Assembled Washer.

Screw, Lag. See Bolt, Lag.

Screw, Lockwasher. See Bolt, Assembled Washer, also Screw, Assembled Washer.

Screw, Machine. An externally threaded fastener whose threaded portion is of one nominal diameter, No. 0 (0.060 in.) or larger, designed to be held or driven with either a wrench or an inserted driver or both in sizes below No. 10 (0.190 in.) nominal diameter. No. 10 and larger sizes must have a head designed for any type of inserted driver (excluding internal socket or internal multiple spline types) but may also be designed for external wrenching. Excludes Bolt, Clevis; Bolt, Externally Relieved Body; Screw, Externally Relieved Body; and Screw, Assembled Washer. See also Screw, Instrument; Bolt, Machine; and Bolt, Internal Wrenching.

Screw, Self-locking. An externally threaded fastener whose threaded portion is of one nominal diameter, and with a locking feature incorporated in the design of the head. It is No. 0 (0.060 in.) or larger, and the head is designed to be held or driven with an inserted driver or wrench or both in sizes below No. 10 (0.190 in.). On No. 10 or larger sizes the head has either a slot, cross-recess, one-way slot, or clutch recess for use with an inserted driver.

Screw, Self-tapping. See Screw, Tapping, Thread Cutting; Screw, Tapping, Thread Forming.

Screw, Set. See Setscrew.

Screw, Sheet-metal. See Screw, Tapping, Thread Forming.

Screw, Shouldered. See Bolt, Shoulder.

Screw, Spade. See Bolt, Eye.

Screw, Tapping, Thread Cutting. A hardened externally threaded fastener whose thread extends from a tapered end to the bearing surface of the head and is interrupted by flutes or slots to permit cutting its own mating thread. Excludes Screw, Assembled Washer.

Screw, Tapping, Thread Forming. A hardened externally threaded fastener whose thread extends from a gimlet or dog-type point to the bearing surface of the head and designed to form its own mating thread. Excludes Screw, Assembled Washer.

Screw, Thread Cutting. See Screw, Tapping, Thread Cutting.

Screw, Thumb. See Thumbscrew.

Screw, Wood. An unhardened externally threaded fastener whose continuous thread extends from a gimlet point for a distance of approximately two-thirds of the length of the screw and which is designed to be driven with an inserted driver.

Setscrew. An externally threaded device whose threaded portion is of one nominal diameter with or without a head and having a cup, cone, or other type of machined point designed to prevent or restrict relative movement of parts and designed to be driven with either a wrench or inserted driver.

Stud, Threaded. A headless fastener which may be threaded externally or internally on one end, threaded continuously, or on both ends, with each threaded portion of one but not necessarily the same nominal diameter. Its unthreaded portion may be of any shape or size. The maximum length of the Stud, Threaded may not exceed 12 in. except that where the largest cross-sectional dimension is $\frac{3}{8}$ in. or more the maximum length will be limited to twenty times that maximum cross-sectional dimension.

Thumbscrew. An externally threaded fastener whose threaded portion is of one nominal diameter and with either a wing type, vertically flattened, circular knurled, or similar head, designed for rotation by the thumb and fingers.

Fastening devices have undergone considerable development and standardization in recent years. Consider these projects.

SLOTTED AND RECESSED-HEAD SCREWS

The original standard, approved in 1930, covered sizes from No. 2 to $\frac{3}{8}$ in., whereas the present standard ASA B18.6-1947 lists sizes from No. 0 to $\frac{3}{4}$ in. Scope of the work is: machine, cap, wood and tapping screws, and headless setscrews.

Slots. In addition to the plain slot, these screws are made with recessed heads of two types. One type of recessed head has two intersecting slots with parallel sides converging to a sharp apex at the bottom. The other type of recess has a large center opening, tapered wings, and a blunt bottom, with all edges relieved or rounded.

The recessed head has come into being because of the need for a design of slot that can be driven tighter and with less chance of marring the screw head, slot, and adjacent work area by slippage of the driver, especially in production work. It has been found that the recessed head can be driven up to one-third faster with power tools and that the chances of slot damage are slight. For these reasons, most of the slotted-head screws used in the automotive and other mass-production industries are of the recessed-head type.

Head Types. The 1947 standard includes a number of additional head types, as compared with the 1930 document. But head types are not a static matter. Already experts in the screw industry believe that three types of head are on the way out—the round head, the pan or stove head, and even the fillister head so fondly used by tool and diemakers. Screwmakers would prefer to offer the truss or binding heads in place of the above three types. The binding-head type is more difficult to make. Its features are: good appearance, greater covering area than the round head, and 30% more slot effectiveness in driving; it is stronger than the stove head, and it is as strong as the fillister head. If the binding-head screw is used to replace the fillister head, less depth will be required for counterboring, but the counterbore dia will be larger.

Details of the various types of head are:

ROUND HEAD. The round head has a semielliptical top surface and flat bearing surface. This style of head is standard for machine screws, capscrews, wood screws, and tapping screws.

FLAT HEAD. The flat head has a flat top surface and a conical bearing surface with an included angle of approximately 82°. This style of head is for machine screws, capscrews, wood screws, and tapping screws. Flat-head machine screws are also made with an included angle of approximately 100°.

OVAL HEAD. The oval head has a rounded top surface and a conical bearing surface with an included angle of approximately 82°. This style of head is standard for machine screws, wood screws, and tapping screws.

FILLISTER HEAD. The fillister head has a rounded surface, cylindrical sides, and a flat bearing surface. This style of head is for machine screws, capscrews, and tapping screws. Fillister-head machine screws are also made with a hole drilled through the head at right angles to the slot. Purpose of hole: to lock as with a wire.

TRUSS HEAD. The truss head has a low rounded top surface with a flat bearing surface having a larger diameter than the corresponding round head. This style of head is for machine screws and tapping screws.

BINDING HEAD. The binding head has a rounded top surface and slightly tapered sides. The bearing surface of the head is flat and usually undercut, but the use of the undercut is optional. This style of head is standard for machine screws.

PAN HEAD. The slotted pan head has a flat top surface with rounded corners, cylindrical sides, and a flat bearing surface. The recessed pan head has a rounded top surface, cylindrical sides, and a flat bearing surface. This style of head is for machine screws and tapping screws.

HEXAGON HEAD. The hexagon head has three sets of parallel sides with flat bearing surface. This style of head is for machine screws and tapping screws.

HEADLESS SETSCREW. The headless setscrew is a screw without a head but with a slot or recess in one end.

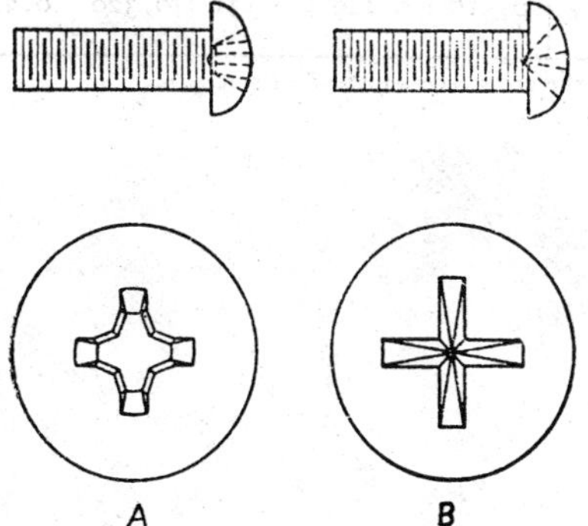

Recessed heads for machine screws: *A*—large center opening, tapered wings, and blunt bottom; *B*—intersecting slots with parallel sides.

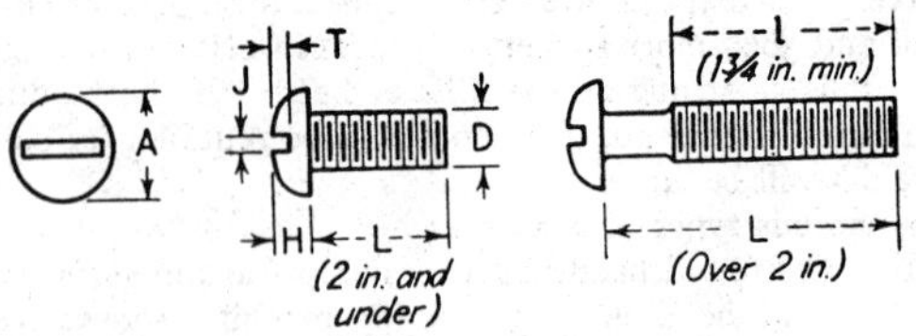

TABLE 1. ROUND-HEAD MACHINE SCREWS

Nominal Size	*D*	*A*		*H*		*J*		*T*	
	Max Dia	Max	Min	Max	Min	Max	Min	Max	Min
0	0.060	0.113	0.099	0.053	0.043	0.023	0.016	0.039	0.029
1	0.073	0.138	0.122	0.061	0.051	0.026	0.019	0.044	0.033
2	0.086	0.162	0.146	0.069	0.059	0.031	0.023	0.048	0.037
3	0.099	0.187	0.169	0.078	0.067	0.035	0.027	0.053	0.040
4	0.112	0.211	0.193	0.086	0.075	0.039	0.031	0.058	0.044
5	0.125	0.236	0.217	0.095	0.083	0.043	0.035	0.063	0.047
6	0.138	0.260	0.240	0.103	0.091	0.048	0.039	0.068	0.051
8	0.164	0.309	0.287	0.120	0.107	0.054	0.045	0.077	0.058
10	0.190	0.359	0.334	0.137	0.123	0.060	0.050	0.087	0.065
12	0.216	0.408	0.382	0.153	0.139	0.067	0.056	0.096	0.072
$\frac{1}{4}$	0.250	0.472	0.443	0.175	0.160	0.075	0.064	0.109	0.082
$\frac{5}{16}$	0.3125	0.590	0.557	0.216	0.198	0.084	0.072	0.132	0.099
$\frac{3}{8}$	0.375	0.708	0.670	0.256	0.237	0.094	0.081	0.155	0.117
$\frac{7}{16}$	0.4375	0.750	0.707	0.328	0.307	0.094	0.081	0.196	0.148
$\frac{1}{2}$	0.500	0.813	0.766	0.355	0.332	0.106	0.091	0.211	0.159
$\frac{9}{16}$	0.5625	0.938	0.887	0.410	0.385	0.118	0.102	0.242	0.183
$\frac{5}{8}$	0.625	1.000	0.944	0.438	0.411	0.133	0.116	0.258	0.195
$\frac{3}{4}$	0.750	1.250	1.185	0.547	0.516	0.149	0.131	0.320	0.242

NOTE: These screws are also available in the cross-recess instead of slot.

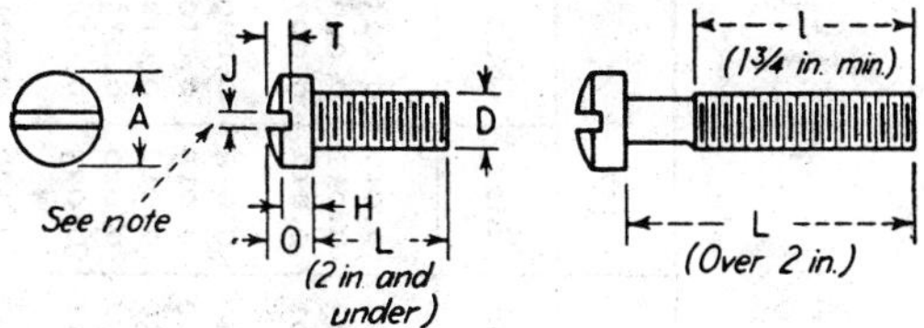

TABLE 2. FILLISTER-HEAD MACHINE SCREWS

Nominal Size	*D*	*A*		*H*		*O*		*T*	
	Max Dia	Max	Min	Max	Min	Max	Min	Max	Min
0	0.060	0.096	0.083	0.045	0.037	0.059	0.043	0.025	0.015
1	0.073	0.118	0.104	0.053	0.045	0.071	0.055	0.031	0.020
2	0.086	0.140	0.124	0.062	0.053	0.083	0.066	0.037	0.025
3	0.099	0.161	0.145	0.070	0.061	0.095	0.077	0.043	0.030
4	0.112	0.183	0.166	0.079	0.069	0.107	0.088	0.048	0.035
5	0.125	0.205	0.187	0.088	0.078	0.120	0.100	0.054	0.040
6	0.138	0.226	0.208	0.096	0.086	0.132	0.111	0.060	0.045
8	0.164	0.270	0.250	0.113	0.102	0.156	0.133	0.071	0.054
10	0.190	0.313	0.292	0.130	0.118	0.180	0.156	0.083	0.064
12	0.216	0.357	0.334	0.148	0.134	0.205	0.178	0.094	0.074
$\frac{1}{4}$	0.250	0.414	0.389	0.170	0.155	0.237	0.207	0.109	0.087
$\frac{5}{16}$	0.3125	0.518	0.490	0.211	0.194	0.295	0.262	0.137	0.110
$\frac{3}{8}$	0.375	0.622	0.590	0.253	0.233	0.355	0.315	0.164	0.133
$\frac{7}{16}$	0.4375	0.625	0.589	0.265	0.242	0.368	0.321	0.170	0.135
$\frac{1}{2}$	0.500	0.750	0.710	0.297	0.273	0.412	0.362	0.190	0.151
$\frac{9}{16}$	0.5625	0.812	0.768	0.336	0.308	0.466	0.410	0.214	0.172
$\frac{5}{8}$	0.625	0.875	0.827	0.375	0.345	0.521	0.461	0.240	0.193
$\frac{3}{4}$	0.750	1.000	0.945	0.441	0.406	0.612	0.542	0.281	0.226

Maximum and minimum widths of slot *J* are the same as for round-head machine screws. These screws are also available with cross-recess instead of slot.

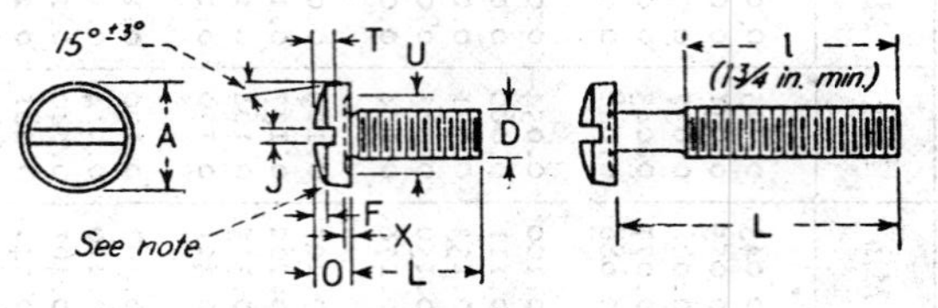

TABLE 3. BINDING-HEAD MACHINE SCREWS

Nominal Size	D	A		O		J		T		F		U		X	
		Max	Min	Max	Min	Max	Min	Max	Min	Max	Min	Max	Min	Max	Min
2	0.086	0.181	0.171	0.046	0.041	0.031	0.023	0.030	0.024	0.018	0.013	0.141	0.124	0.010	0.005
3	0.099	0.208	0.197	0.054	0.048	0.035	0.027	0.036	0.029	0.022	0.016	0.162	0.143	0.011	0.006
4	0.112	0.235	0.223	0.063	0.056	0.039	0.031	0.042	0.034	0.025	0.018	0.184	0.161	0.012	0.007
5	0.125	0.263	0.249	0.071	0.064	0.043	0.035	0.048	0.039	0.029	0.021	0.205	0.180	0.014	0.009
6	0.138	0.290	0.275	0.080	0.071	0.048	0.039	0.053	0.044	0.032	0.024	0.226	0.199	0.015	0.010
8	0.164	0.344	0.326	0.097	0.087	0.054	0.045	0.065	0.054	0.039	0.029	0.269	0.236	0.017	0.012
10	0.190	0.399	0.378	0.114	0.102	0.060	0.050	0.077	0.064	0.045	0.034	0.312	0.274	0.020	0.015
12	0.216	0.454	0.430	0.130	0.117	0.067	0.056	0.089	0.074	0.052	0.039	0.354	0.311	0.023	0.018
$\frac{1}{4}$	0.250	0.513	0.488	0.153	0.138	0.075	0.064	0.105	0.088	0.061	0.046	0.410	0.360	0.026	0.021
$\frac{5}{16}$	0.3125	0.641	0.609	0.193	0.174	0.084	0.072	0.134	0.112	0.077	0.059	0.513	0.450	0.032	0.027
$\frac{3}{8}$	0.375	0.769	0.731	0.234	0.211	0.094	0.081	0.163	0.136	0.094	0.071	0.615	0.540	0.039	0.034

These screws are available with cross-recess instead of slot.

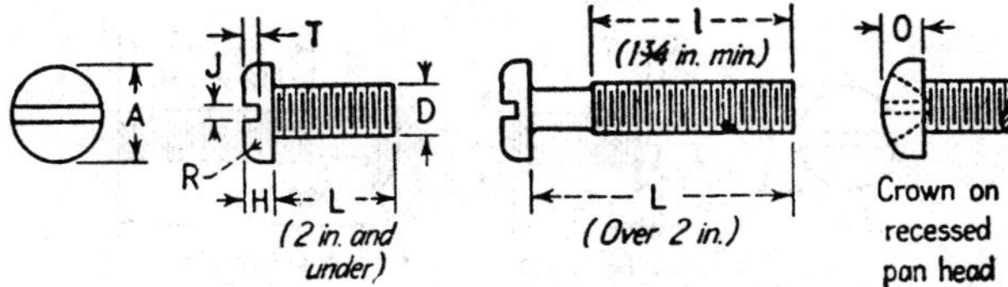

TABLE 4. PAN-HEAD MACHINE SCREWS

Nominal Size	D	A		H		J		T		R	O	
		Max	Min	Max	Min	Max	Min	Max	Min	Nominal	Max	Min
2	0.086	0.167	0.155	0.053	0.045	0.031	0.023	0.033	0.023	0.035	0.062	0.053
3	0.099	0.193	0.180	0.060	0.051	0.035	0.027	0.037	0.027	0.037	0.071	0.062
4	0.112	0.219	0.205	0.068	0.058	0.039	0.031	0.041	0.030	0.042	0.080	0.070
5	0.125	0.245	0.231	0.075	0.065	0.043	0.035	0.045	0.032	0.044	0.089	0.079
6	0.138	0.270	0.256	0.082	0.072	0.048	0.039	0.050	0.038	0.046	0.097	0.087
7*	0.151	0.296	0.281	0.089	0.079	0.048	0.039	0.055	0.040	0.049	0.106	0.096
8	0.164	0.322	0.306	0.096	0.085	0.054	0.045	0.058	0.043	0.052	0.115	0.105
10	0.190	0.373	0.357	0.110	0.099	0.060	0.050	0.067	0.050	0.061	0.133	0.122
12	0.216	0.425	0.407	0.125	0.112	0.067	0.056	0.077	0.060	0.078	0.151	0.139
$\frac{1}{4}$	0.250	0.492	0.473	0.144	0.130	0.075	0.064	0.087	0.070	0.087	0.175	0.162
$\frac{5}{16}$	0.3125	0.615	0.594	0.178	0.162	0.084	0.072	0.109	0.092	0.099	0.218	0.203
$\frac{3}{8}$	0.375	0.740	0.716	0.212	0.195	0.094	0.081	0.130	0.113	0.143	0.261	0.244

These screws are available with cross-recess instead of slot.

* Dimensions of No. 7 are included only for tapping screws.

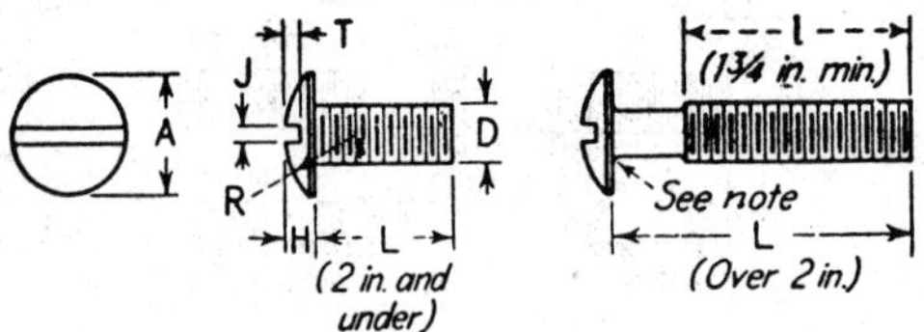

TABLE 5. TRUSS-HEAD MACHINE SCREWS

Nominal Size	A		H		J		T		R
	Max	Min	Max	Min	Max	Min	Max	Min	Max
2	0.194	0.180	0.053	0.044	0.031	0.023	0.031	0.022	0.129
3	0.226	0.211	0.061	0.051	0.035	0.027	0.036	0.026	0.151
4	0.257	0.241	0.069	0.059	0.039	0.031	0.040	0.030	0.169
5	0.289	0.272	0.078	0.066	0.043	0.035	0.045	0.034	0.191
6	0.321	0.303	0.086	0.074	0.048	0.039	0.050	0.037	0.211
7	0.352	0.333	0.094	0.081	0.048	0.039	0.054	0.041	0.231
8	0.384	0.364	0.102	0.088	0.054	0.045	0.058	0.045	0.254
10	0.448	0.425	0.118	0.103	0.060	0.050	0.068	0.053	0.283
12	0.511	0.487	0.134	0.118	0.067	0.056	0.077	0.061	0.336
$\frac{1}{4}$	0.573	0.546	0.150	0.133	0.075	0.064	0.087	0.070	0.375
$\frac{5}{16}$	0.698	0.666	0.183	0.162	0.084	0.072	0.106	0.085	0.457
$\frac{3}{8}$	0.823	0.787	0.215	0.191	0.094	0.081	0.124	0.100	0.538
$\frac{7}{16}$	0.948	0.907	0.248	0.221	0.094	0.081	0.142	0.116	0.619
$\frac{1}{2}$	1.073	1.028	0.280	0.250	0.106	0.091	0.161	0.131	0.701
$\frac{9}{16}$	1.198	1.149	0.312	0.279	0.118	0.102	0.179	0.146	0.783
$\frac{5}{8}$	1.323	1.269	0.345	0.309	0.133	0.116	0.196	0.162	0.863
$\frac{3}{4}$	1.573	1.511	0.410	0.368	0.149	0.131	0.234	0.182	1.024

These screws are available with cross-recess instead of slot.

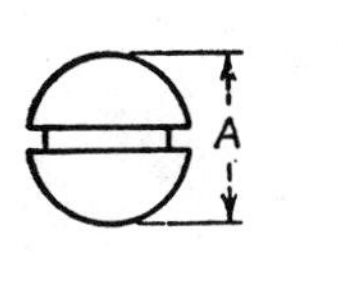

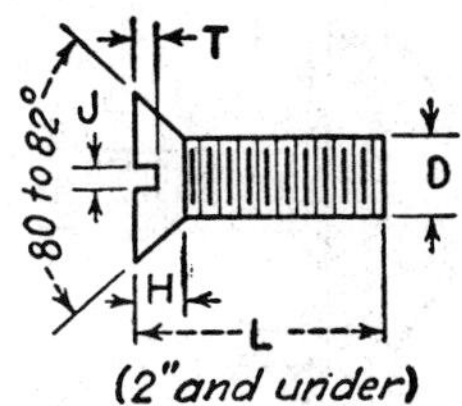

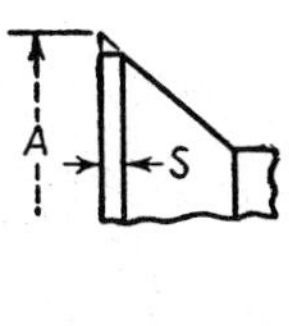

TABLE 6. FLAT-HEAD MACHINE SCREWS, 82°

Nominal Size	*D*	*A*			*S*	*H*		*J*		*T*	
	Max	Max Sharp	Min Sharp	Min with Max S	Flat on Min Screw	Max	Min	Max	Min	Max	Min
0	0.060	0.119	0.105	0.101	0.002	0.035	0.026	0.023	0.016	0.015	0.010
1	0.073	0.146	0.130	0.126	0.003	0.043	0.033	0.026	0.019	0.019	0.012
2	0.086	0.172	0.156	0.150	0.003	0.051	0.040	0.031	0.023	0.023	0.015
3	0.099	0.199	0.181	0.175	0.004	0.059	0.048	0.035	0.027	0.027	0.017
4	0.112	0.225	0.207	0.200	0.004	0.067	0.055	0.039	0.031	0.030	0.020
5	0.125	0.252	0.232	0.225	0.005	0.075	0.062	0.043	0.035	0.034	0.022
6	0.138	0.279	0.257	0.249	0.005	0.083	0.069	0.048	0.039	0.038	0.024
8	0.164	0.332	0.308	0.300	0.006	0.100	0.084	0.054	0.045	0.045	0.029
10	0.190	0.385	0.359	0.348	0.007	0.116	0.098	0.060	0.050	0.053	0.034
12	0.216	0.438	0.410	0.397	0.008	0.132	0.112	0.067	0.056	0.060	0.039
$\frac{1}{4}$	0.250	0.507	0.477	0.462	0.009	0.153	0.131	0.075	0.064	0.070	0.046
$\frac{5}{16}$	0.3125	0.635	0.600	0.581	0.011	0.191	0.165	0.084	0.072	0.088	0.058
$\frac{3}{8}$	0.375	0.762	0.722	0.700	0.013	0.230	0.200	0.094	0.081	0.106	0.070
$\frac{7}{16}$	0.4375	0.812	0.771	0.743	0.016	0.223	0.190	0.094	0.081	0.103	0.066
$\frac{1}{2}$	0.500	0.875	0.831	0.802	0.018	0.223	0.186	0.106	0.091	0.103	0.065
$\frac{9}{16}$	0.5625	1.000	0.950	0.919	0.020	0.260	0.220	0.118	0.102	0.120	0.077
$\frac{5}{8}$	0.625	1.125	1.069	1.035	0.023	0.298	0.253	0.133	0.116	0.137	0.088
$\frac{3}{4}$	0.750	1.375	1.306	1.267	0.027	0.372	0.319	0.149	0.131	0.171	0.111

These screws are available with cross-recess instead of slot.

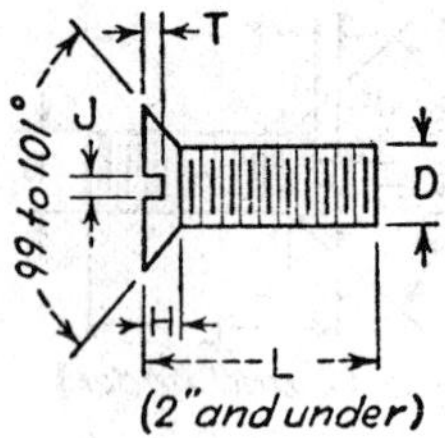

TABLE 7. FLAT-HEAD MACHINE SCREWS, 100°

Nominal Size	S	H		J		T	
	Flat on Min Screw	Max	Min	Max	Min	Max	Min
4	0.003	0.048	0.040	0.039	0.031	0.024	0.017
6	0.004	0.060	0.051	0.048	0.039	0.030	0.022
8	0.004	0.072	0.062	0.054	0.045	0.036	0.027
10	0.005	0.083	0.072	0.060	0.050	0.042	0.031
1/4	0.006	0.110	0.097	0.075	0.064	0.055	0.042
5/16	0.008	0.138	0.123	0.084	0.072	0.069	0.053
3/8	0.009	0.165	0.148	0.094	0.081	0.083	0.064

TABLE 8. HEXAGON-HEAD MACHINE SCREWS, PLAIN OR SLOTTED

Nominal Size	D Basic Diameter	Standard Trimmed or Upset Head			Optional Upset Type Head for Special Requirements			H		J		T	
		A		W	A		W						
		Max	Min	Min	Max	Min	Min	Max	Min	Max	Min	Max	Min
2	0.0860	0.125	0.120	0.134				0.050	0.040				
3	0.0990	0.187	0.181	0.202				0.055	0.044				
4	0.1120	0.187	0.181	0.202	0.219	0.213	0.238	0.060	0.049	0.039	0.031	0.036	0.025
5	0.1250	0.187	0.181	0.202	0.250	0.244	0.272	0.070	0.058	0.043	0.035	0.042	0.030
6	0.1380	0.250	0.244	0.272				0.080	0.067	0.048	0.039	0.046	0.033
8	0.1640	0.250	0.244	0.272	0.312	0.305	0.340	0.110	0.096	0.054	0.045	0.066	0.052
10	0.1900	0.312	0.305	0.340				0.120	0.105	0.060	0.050	0.072	0.057
12	0.2160	0.312	0.305	0.340	0.375	0.367	0.409	0.155	0.139	0.067	0.056	0.093	0.077
1/4	0.2500	0.375	0.367	0.409	0.437	0.428	0.477	0.190	0.172	0.075	0.064	0.101	0.083
5/16	0.3125	0.500	0.491	0.548				0.230	0.208	0.084	0.072	0.122	0.100
3/8	0.3750	0.562	0.552	0.616				0.295	0.270	0.094	0.081	0.156	0.131

Hexagon-head machine screws are usually not slotted. The slot is optional.

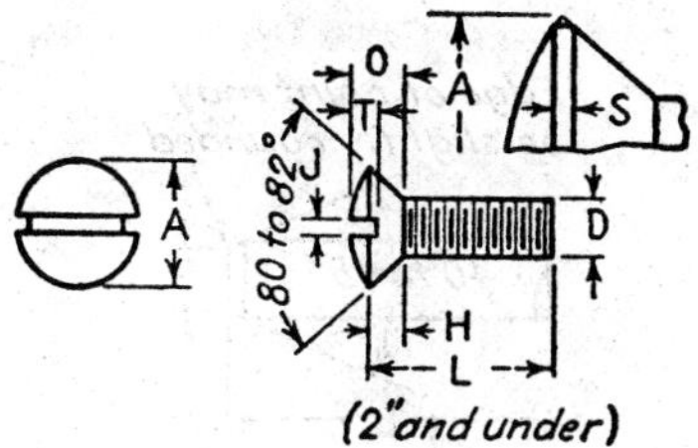

TABLE 9. OVAL-HEAD MACHINE SCREWS*

Nominal Size	*O*		*T*	
	Max	Min	Max	Min
0	0.056	0.041	0.030	0.025
1	0.068	0.052	0.038	0.031
2	0.080	0.063	0.045	0.037
3	0.092	0.073	0.052	0.043
4	0.104	0.084	0.059	0.049
5	0.116	0.095	0.067	0.055
6	0.128	0.105	0.074	0.060
8	0.152	0.126	0.088	0.072
10	0.176	0.148	0.103	0.084
12	0.200	0.169	0.117	0.096
1/4	0.232	0.197	0.136	0.112
5/16	0.290	0.249	0.171	0.141
3/8	0.347	0.300	0.206	0.170
7/16	0.345	0.295	0.210	0.174
1/2	0.354	0.299	0.216	0.176
9/16	0.410	0.350	0.250	0.207
5/8	0.467	0.399	0.285	0.235
3/4	0.578	0.497	0.353	0.293

* Dimensions *D*, *A*, *S*, *H*, and *J* are the same as for flat-head machine screws. These screws are available with cross-recess instead of slot.

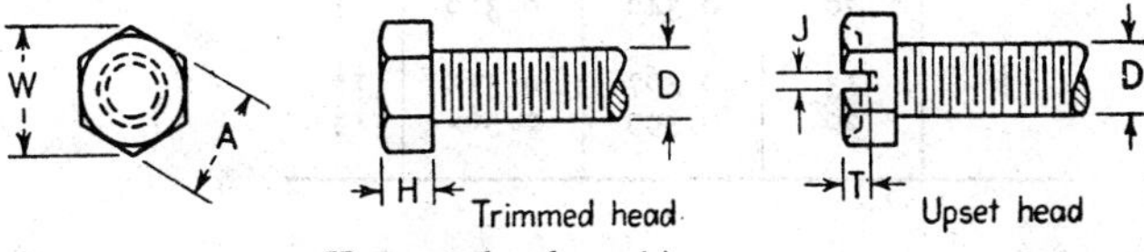

Hexagon-head machine screw.

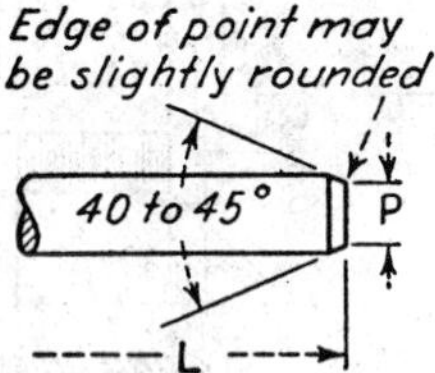

TABLE 10. HEADER POINTS FOR MACHINE SCREWS

Nominal Size	Threads per In.	P		L
		Max	Min	Max
4	40	0.074	0.065	½
	48	0.079	0.070	½
5	40	0.086	0.076	½
	44	0.088	0.079	½
6	32	0.090	0.080	¾
	40	0.098	0.087	¾
8	32	0.114	0.102	1
	36	0.118	0.106	1
10	24	0.125	0.112	1¼
	32	0.138	0.124	1¼
12	24	0.149	0.134	1⅜
	28	0.156	0.141	1⅜
¼	20	0.170	0.153	1½
	28	0.187	0.169	1½
5/16	18	0.221	0.200	1½
	24	0.237	0.215	1½
⅜	16	0.270	0.244	1½
	24	0.295	0.267	1½
7/16	14	0.316	0.287	1½
	20	0.342	0.310	1½
½	13	0.367	0.333	1½
	20	0.399	0.362	1½

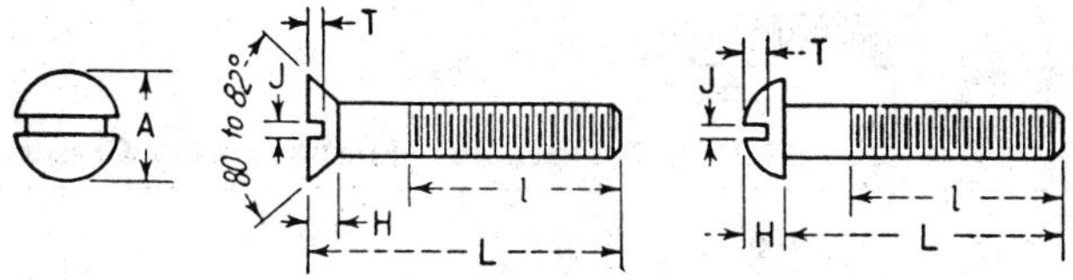

TABLE 11. FLAT-HEAD AND ROUND-HEAD CAPSCREWS

Nominal Size	*A*		*H*		*J*		*T*	
	Max	Min	Max	Min	Max	Min	Max	Min
Flat-head Capscrews								
1/4	0.500	0.477	0.140 av		0.075	0.064	0.069	0.046
5/16	0.625	0.598	0.176		0.084	0.072	0.086	0.057
3/8	0.750	0.720	0.210		0.094	0.081	0.103	0.069
7/16	0.8125	0.780	0.210		0.094	0.081	0.103	0.069
1/2	0.875	0.841	0.210		0.106	0.091	0.103	0.069
9/16	1.000	0.962	0.245		0.118	0.102	0.120	0.080
5/8	1.125	1.083	0.281		0.133	0.116	0.137	0.092
3/4	1.375	1.326	0.352		0.149	0.131	0.171	0.115
7/8	1.625	1.568	0.423		0.167	0.147	0.206	0.139
1	1.875	1.811	0.494		0.188	0.166	0.240	0.162
Round-head Capscrews								
1/4	0.438	0.418	0.191	0.175	0.075	0.064	0.117	0.097
5/16	0.562	0.541	0.246	0.226	0.084	0.072	0.151	0.126
3/8	0.625	0.602	0.273	0.252	0.094	0.081	0.168	0.138
7/16	0.750	0.725	0.328	0.302	0.094	0.081	0.202	0.167
1/2	0.812	0.786	0.355	0.328	0.106	0.091	0.219	0.179
9/16	0.937	0.908	0.410	0.379	0.118	0.102	0.253	0.208
5/8	1.000	0.970	0.438	0.405	0.133	0.116	0.270	0.220
3/4	1.250	1.215	0.547	0.506	0.149	0.131	0.337	0.277

NOTE: For body-diameter tolerance see Fillister-head Capscrews. Threads are either NC-3 or NF-3.

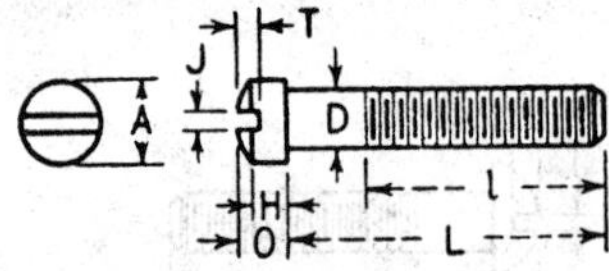

TABLE 12. FILLISTER-HEAD CAPSCREWS

Nominal Size	*D*		*A*		*H*		*O*		*J*		*T*	
	Max	Min	Max	Min	Max	Min	Max	Min	Max	Min	Max	Min
1/4	0.250	0.245	0.375	0.363	0.172	0.157	0.216	0.195	0.075	0.064	0.097	0.077
5/16	0.3125	0.307	0.437	0.424	0.203	0.186	0.253	0.230	0.084	0.072	0.115	0.090
3/8	0.375	0.369	0.562	0.547	0.250	0.229	0.314	0.285	0.094	0.081	0.143	0.113
7/16	0.4375	0.431	0.625	0.608	0.297	0.274	0.368	0.337	0.094	0.081	0.168	0.133
1/2	0.500	0.493	0.750	0.731	0.328	0.301	0.412	0.376	0.106	0.091	0.188	0.148
9/16	0.5625	0.555	0.812	0.792	0.375	0.347	0.466	0.428	0.118	0.102	0.214	0.169
5/8	0.625	0.617	0.875	0.853	0.422	0.392	0.521	0.480	0.133	0.116	0.240	0.190
3/4	0.750	0.742	1.000	0.976	0.500	0.466	0.612	0.566	0.149	0.131	0.283	0.233
7/8	0.875	0.866	1.125	1.098	0.594	0.556	0.720	0.669	0.167	0.147	0.334	0.264
1	1.000	0.990	1.312	1.282	0.656	0.613	0.802	0.744	0.188	0.166	0.372	0.292

TABLE 13. HEXAGON-HEAD CAPSCREWS

Nominal Size or Basic Major Dia of Thread	Body Dia, Min	Width across Flats *F*		Width across Corners *G*		Height *H*			Radius *R*	
		Max (Basic)	Min	Max	Min	Nominal	Max	Min	Max	Min
1/4	**0.2450**	**7/16**	**0.428**	**0.505**	**0.488**	**5/32**	**0.163**	**0.150**	**0.023**	**0.009**
5/16	**0.3065**	**1/2**	**0.489**	**0.577**	**0.557**	**13/64**	**0.211**	**0.195**	**0.023**	**0.009**
3/8	**0.3690**	**9/16**	**0.551**	**0.650**	**0.628**	**15/64**	**0.243**	**0.226**	**0.023**	**0.009**
7/16	**0.4305**	**5/8**	**0.612**	**0.722**	**0.698**	**9/32**	**0.291**	**0.272**	**0.023**	**0.009**
1/2	**0.4930**	**3/4**	**0.736**	**0.866**	**0.840**	**5/16**	**0.323**	**0.302**	**0.023**	**0.009**
9/16	**0.5545**	**13/16**	**0.798**	**0.938**	**0.910**	**23/64**	**0.371**	**0.348**	**0.041**	**0.021**
5/8	**0.6170**	**15/16**	**0.922**	**1.083**	**1.051**	**25/64**	**0.403**	**0.378**	**0.041**	**0.021**
3/4	**0.7410**	**1 1/8**	**1.100**	**1.299**	**1.254**	**15/32**	**0.483**	**0.455**	**0.041**	**0.021**
7/8	**0.8660**	**1 5/16**	**1.285**	**1.516**	**1.465**	**35/64**	**0.563**	**0.531**	**0.062**	**0.047**
1	**0.9900**	**1 1/2**	**1.469**	**1.732**	**1.675**	**39/64**	**0.627**	**0.591**	**0.062**	**0.047**
1 1/8	1.1140	1 11/16	1.631	1.949	1.859	11/16	0.718	0.658	0.125	0.110
1 1/4	1.2390	1 7/8	1.812	2.165	2.066	25/32	0.813	0.749	0.125	0.110
1 3/8	1.3630	2 1/16	1.994	2.382	2.273	27/32	0.878	0.810	0.125	0.110
1 1/2	1.4880	2 1/4	2.175	2.598	2.480	15/16	0.974	0.902	0.125	0.110

Bold type indicates products unified dimensionally with British and Canadian standards.

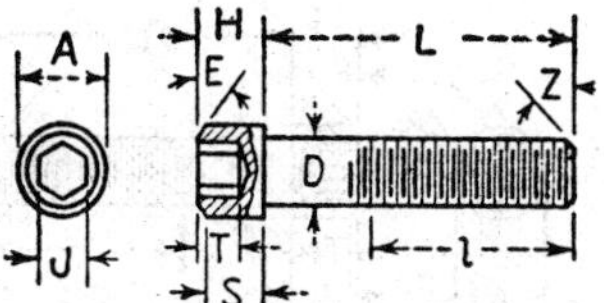

TABLE 14. HEXAGONAL SOCKET-HEAD CAPSCREWS

D			*A*		*H*		*S*		*J*	
Nominal	Max	Min	Max	Min	Max	Min	Max	Min	Max	Min
2	0.0860	0.0840	0.140	0.136	0.086	0.083	0.0803	0.0773	0.0635	$\frac{1}{16}$
3	0.0990	0.0968	0.161	0.157	0.099	0.096	0.0923	0.0891	0.0791	$\frac{5}{64}$
4	0.1120	0.1096	0.183	0.178	0.112	0.109	0.1043	0.1009	0.0791	$\frac{5}{64}$
5	0.1250	0.1226	0.205	0.200	0.125	0.122	0.1163	0.1129	0.0947	$\frac{3}{32}$
6	0.1380	0.1353	0.226	0.221	0.138	0.134	0.1284	0.1246	0.0947	$\frac{3}{32}$
8	0.1640	0.1613	0.270	0.265	0.164	0.160	0.1522	0.1484	0.1270	$\frac{1}{8}$
10	0.1900	0.1867	$\frac{5}{16}$	0.306	0.190	0.185	0.1765	0.1717	0.1582	$\frac{5}{32}$
12	0.2160	0.2127	$\frac{11}{32}$	0.337	0.216	0.211	0.2005	0.1957	0.1582	$\frac{5}{32}$
$\frac{1}{4}$	0.2500	0.2464	$\frac{3}{8}$	0.367	$\frac{1}{4}$	0.244	0.2317	0.2265	0.1895	$\frac{3}{16}$
$\frac{5}{16}$	0.3125	0.3084	$\frac{7}{16}$	0.429	$\frac{5}{16}$	0.306	0.2894	0.2834	0.2207	$\frac{7}{32}$
$\frac{3}{8}$	0.3750	0.3705	$\frac{9}{16}$	0.553	$\frac{3}{8}$	0.368	0.3469	0.3405	0.3155	$\frac{5}{16}$
$\frac{7}{16}$	0.4375	0.4326	$\frac{5}{8}$	0.615	$\frac{7}{16}$	0.430	0.4046	0.3974	0.3155	$\frac{5}{16}$
$\frac{1}{2}$	0.5000	0.4948	$\frac{3}{4}$	0.739	$\frac{1}{2}$	0.492	0.4620	0.4546	0.3780	$\frac{3}{8}$
$\frac{9}{16}$	0.5625	0.5569	$\frac{13}{16}$	0.801	$\frac{9}{16}$	0.554	0.5196	0.5116	0.3780	$\frac{3}{8}$
$\frac{5}{8}$	0.6250	0.6191	$\frac{7}{8}$	0.863	$\frac{5}{8}$	0.616	0.5771	0.5687	0.5030	$\frac{1}{2}$
$\frac{3}{4}$	0.7500	0.7436	1	0.987	$\frac{3}{4}$	0.741	0.6920	0.6830	0.5655	$\frac{9}{16}$
$\frac{7}{8}$	0.8750	0.8680	$1\frac{1}{8}$	1.111	$\frac{7}{8}$	0.865	0.8069	0.7971	0.5655	$\frac{9}{16}$
1	1.0000	0.9924	$1\frac{5}{16}$	1.297	1	0.989	0.9220	0.9112	0.6290	$\frac{5}{8}$
$1\frac{1}{8}$	1.1250	1.1165	$1\frac{1}{2}$	1.483	$1\frac{1}{8}$	1.113	1.0372	1.0254	0.7540	$\frac{3}{4}$
$1\frac{1}{4}$	1.2500	1.2415	$1\frac{3}{4}$	1.733	$1\frac{1}{4}$	1.238	1.1516	1.1398	0.7540	$\frac{3}{4}$
$1\frac{3}{8}$	1.3750	1.3649	$1\frac{7}{8}$	1.855	$1\frac{3}{8}$	1.361	1.2675	1.2533	0.7540	$\frac{3}{4}$
$1\frac{1}{2}$	1.5000	1.4899	2	1.979	$1\frac{1}{2}$	1.485	1.3821	1.3679	1.0040	1

ASA B18.3-1947. Published by the ASME.
NOTES: T must not exceed $\frac{3}{4}H$. $Z = 35° + 5°, -0°$. $E = 35° \pm 2°$.

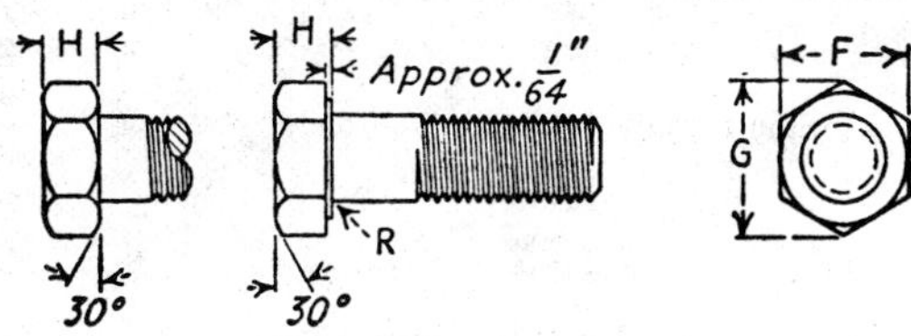

Hexagon-head capscrew.

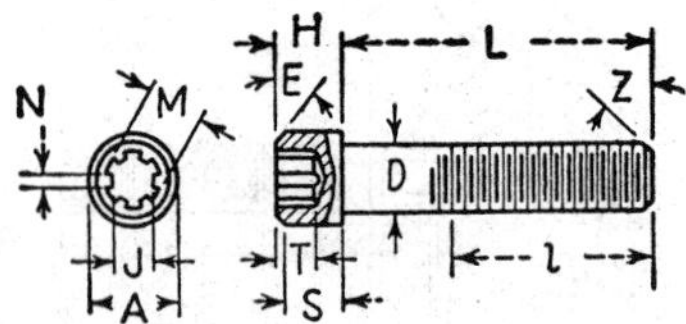

TABLE 15. FLUTED SOCKET-HEAD CAPSCREWS

D Nominal	No. of Flutes	J Max	J Min	M Max	M Min	N Max	N Min
2	6	0.064	0.063	0.074	0.073	0.016	0.015
3	6	0.082	0.080	0.098	0.097	0.022	0.021
4	6	0.082	0.080	0.098	0.097	0.022	0.021
5	6	0.098	0.096	0.115	0.113	0.025	0.023
6	6	0.098	0.096	0.115	0.113	0.025	0.023
8	6	0.128	0.126	0.149	0.147	0.032	0.030
10	6	0.163	0.161	0.188	0.186	0.039	0.037
12	6	0.163	0.161	0.188	0.186	0.039	0.037
$\frac{1}{4}$	6	0.190	0.188	0.221	0.219	0.050	0.048
$\frac{5}{16}$	6	0.221	0.219	0.256	0.254	0.060	0.058
$\frac{3}{8}$	6	0.319	0.316	0.380	0.377	0.092	0.089
$\frac{7}{16}$	6	0.319	0.316	0.380	0.377	0.092	0.089
$\frac{1}{2}$	6	0.386	0.383	0.463	0.460	0.112	0.109
$\frac{9}{16}$	6	0.386	0.383	0.463	0.460	0.112	0.109
$\frac{5}{8}$	6	0.509	0.506	0.604	0.601	0.138	0.134
$\frac{3}{4}$	6	0.535	0.531	0.631	0.627	0.149	0.145
$\frac{7}{8}$	6	0.604	0.600	0.709	0.705	0.168	0.164
1	6	0.685	0.681	0.801	0.797	0.189	0.185
$1\frac{1}{8}$	6	0.828	0.824	0.970	0.966	0.231	0.227
$1\frac{1}{4}$	6	0.828	0.824	0.970	0.966	0.231	0.227
$1\frac{3}{8}$	6	0.828	0.824	0.970	0.966	0.231	0.227
$1\frac{1}{2}$	6	1.007	1.003	1.275	1.271	0.298	0.294

ASA B18.3-1947. Published by the ASME.
NOTE: Dimensions *D*, *A*, *H*, and *S* are the same as shown in Table 14.

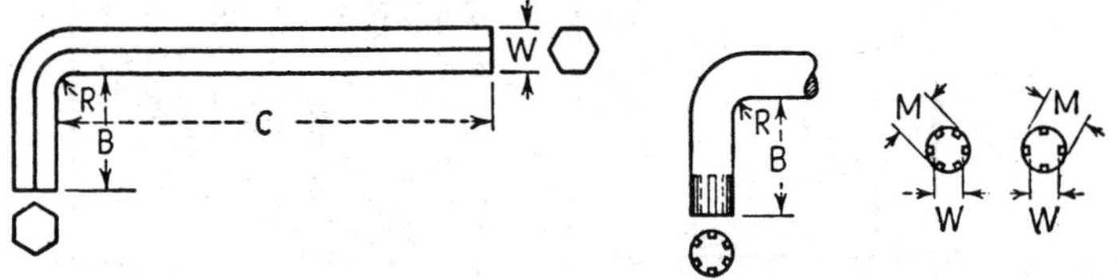

TABLE 16. KEYS FOR SOCKET-HEAD CAPSCREWS AND SETSCREWS

Hexagon Socket				Fluted Socket					Common Dimensions		
D		*W*		*D*		*M*	*W*		*B*	*C*	
Screw Size		Hexagon Width across Flats		Screw Size		Major Dia, Max	Minor Dia, Max	No. of Flutes	Length Short Arm, Max	Length Long Arm	
Cap	Set	Max	Min	Cap	Set					Short Series, Max	Long Series, Max
					5	0.0690	0.0510	4			
					6	0.0760	0.0530	4			
2	5	1/16	0.0615	2	5, 6	0.0720	0.0600	6	21/32	1 27/32	
2	6	1/16	0.0615	3, 4	8	0.0940	0.0780	6	21/32	1 27/32	
3, 4	8	5/64	0.0771	5, 6	10	0.1100	0.0940	6	45/64	1 31/32	
5, 6	10	3/32	0.0927	5, 6	12	0.1100	0.0940	6	3/4	2 3/32	
5, 6	12	3/32	0.0927						3/4	2 3/32	
				8	1/4	0.1440	0.1230	6			
8	1/4	1/8	0.1235	10, 12	5/16	0.1830	0.1580	6	27/32	2 11/32	3 27/32
10, 12	5/16	5/32	0.1547	1/4	3/8	0.2 60	0.1720	6	15/16	2 19/32	4 7/32
1/4	3/8	3/16	0.1860	5/16	7/16	0.2510	0.2010	6	1 1/32	2 27/32	4 19/32
5/16	7/16	7/32	0.2172		1/2	0.2910	0.2390	6	1 1/8	3 3/32	4 31/32
	1/2	1/4	0.2485						1 7/32	3 11/32	5 11/32
					9/16	0.2910	0.2390	6			
	9/16	1/4	0.2485	3/8, 7/16	5/8	0.3720	0.3100	6	1 7/32	3 11/32	5 11/32
3/8, 7/16	5/8	5/16	0.3110	1/2, 9/16	3/4	0.4540	0.3770	6	1 11/32	3 27/32	6 3/32
1/2, 9/16	3/4	3/8	0.3735	5/8	7/8	0.5950	0.5000	6	1 15/32	4 11/32	6 27/32
5/8	7/8	1/2	0.4985	3/4	1	0.6200	0.5240	6	1 23/32	5 11/32	8 11/32
	1	9/16	0.5600						1 27/32	5 27/32	9 3/32
				7/8	1 1/8	0.6980	0.5930	6			
3/4, 7/8	1 1/8	9/16	0.5600	1	1 1/4	0.7900	0.6740	6	1 27/32	5 27/32	9 3/32
1	1 1/4, 1 3/8	5/8	0.6225		1 3/8	0.8580	0.7330	6	1 31/32	6 11/32	9 27/32
1 1/8, 1 1/4, 1 3/8	1 1/2	3/4	0.7475	1 1/8, 1 1/4, 1 3/8	1 1/2	0.9590	0.8170	6	2 7/32	7 11/32	11 11/32
	1 3/4	1	0.9975		1 3/4	1.2640	0.9960	6	2 23/32	9 11/32	14 11/32
1 1/2	2	1	0.9975	1 1/2	2	1.2640	0.9960	6	2 23/32	9 11/32	14 11/32

ASA B18.3-1947. Published by the ASME.

TABLE 17. HEXAGONAL AND FLUTED SOCKET-HEAD STRIPPER BOLTS
Dimensions Given in Inches

D			A		H		S			D_1	E	L
Nominal	Max	Min	Max	Min	Max	Min	Nominal	Max	Min			
1/4	0.2480	0.2460	3/8	0.367	3/16	0.182	0.1615	0.1641	0.1589	10-24 NC-3	3/8 ± 0.010	3/4 to 2 1/2 ± 0.005
5/16	0.3105	0.3085	7/16	0.429	7/32	0.213	0.1869	0.1899	0.1839	1/4-20 NC-3	7/16 ± 0.010	1 to 3 ± 0.005
3/8	0.3730	0.3710	9/16	0.553	1/4	0.244	0.2123	0.2155	0.2091	5/16-18 NC-3	1/2 ± 0.010	1 to 4 ± 0.005
1/2	0.4980	0.4960	3/4	0.739	5/16	0.306	0.2631	0.2668	0.2594	3/8-16 NC-3	5/8 ± 0.015	1 1/4 to 5 ± 0.005
5/8	0.6230	0.6210	7/8	0.863	3/8	0.368	0.3139	0.3181	0.3097	1/2-13 NC-3	3/4 ± 0.015	1 1/2 to 6 ± 0.005
3/4	0.7480	0.7460	1	0.987	1/2	0.492	0.4296	0.4341	0.4251	5/8-11 NC-3	7/8 ± 0.015	1 1/2 to 8 ± 0.005
1	0.9980	0.9960	1 5/16	1.297	5/8	0.616	0.5312	0.5366	0.5258	3/4-10 NC-3	1 ± 0.015	1 1/2 to 8 ± 0.005
1 1/4	1.2480	1.2460	1 3/4	1.733	3/4	0.741	0.6327	0.6386	0.6268	7/8-9 NC-3	1 1/8 ± 0.015	1 1/2 to 8 ± 0.005
												Below 1- to 1/8 in. intervals 1 to 5 in. to 1/4-in. intervals 5 to 7 in. to 1/2-in. intervals Over 7 in. to 1-in. intervals

G		I	J_1		J_2		M		N		K*		F*	R*		Screw Dia
Max	Min		Max	Min	Max	Min	Max	Min	Max	Min	Max	Min		Max	Min	
0.145	0.136	1/16	0.1270	1/8	0.128	0.126	0.149	0.147	0.032	0.030	0.243	0.240	3/32	0.020	0.015	1/4
0.196	0.185	5/64	0.1582	5/32	0.163	0.161	0.188	0.186	0.039	0.037	0.305	0.302	3/32	0.020	0.015	5/16
0.252	0.240	5/64	0.1895	3/16	0.190	0.188	0.221	0.219	0.050	0.048	0.368	0.365	3/32	0.025	0.020	3/8
0.307	0.294	3/32	0.2520	1/4	0.254	0.252	0.298	0.296	0.068	0.066	0.493	0.490	3/32	0.025	0.020	1/2
0.417	0.400	1/8	0.3155	5/16	0.319	0.316	0.380	0.377	0.092	0.089	0.618	0.615	3/32	0.030	0.025	5/8
0.526	0.507	9/64	0.3780	3/8	0.386	0.383	0.463	0.460	0.112	0.109	0.743	0.740	3/32	0.030	0.025	3/4
0.642	0.620	5/32	0.5030	1/2	0.535	0.531	0.631	0.627	0.149	0.145	0.993	0.990	1/8	0.035	0.030	1
0.755	0.731	11/64	0.6290	5/8	0.685	0.681	0.801	0.797	0.189	0.185	1.243	1.240	1/8	0.035	0.030	1 1/4

ASA B18.3-1947. Published by the ASME.

* Neck under head (optional with manufacturers)
Neck diameter K:
Max No formula (see table for values)
Min = K (max) − 0.003
Neck width F No formula (see table for values)
Fillet R No formula (see table for values)

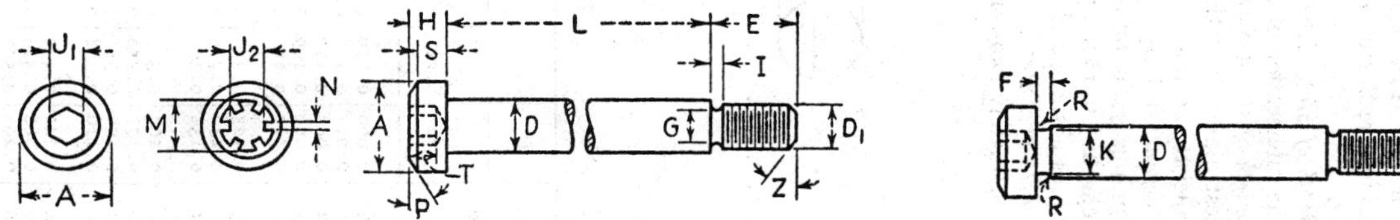

WOOD SCREWS

General Data

Radius of Fillet under Head. The fillet under the head for wood screws shall have a nominal radius of $\frac{1}{64}$ in. for sizes through No. 9, and $\frac{1}{32}$ in. for sizes larger than No. 9.

Depth of Slot. The depth of slots is measured from the highest part of the head to the intersection of the bottom of the slot with the head surface.

Length of Wood Screws. The length of wood screws shall be measured from the largest diameter of the bearing surface of the head to the extreme point in a line parallel to the axis.

Tolerance in Length. The tolerance in length of wood screws shall be plus nothing and minus $\frac{1}{32}$ in. on lengths of $\frac{1}{4}$ to $\frac{5}{8}$ in., inclusive; plus nothing and minus $\frac{3}{64}$ in. on lengths $\frac{3}{4}$ to $1\frac{1}{2}$ in.; plus nothing and minus $\frac{1}{16}$ in. on lengths $1\frac{1}{2}$ to $2\frac{3}{4}$ in., inclusive; and plus nothing and minus $\frac{3}{32}$ in. on lengths 3 to 5 in., inclusive.

Length of Thread. The length of the thread of wood screws shall be equal to approximately two-thirds of the length of the screw.

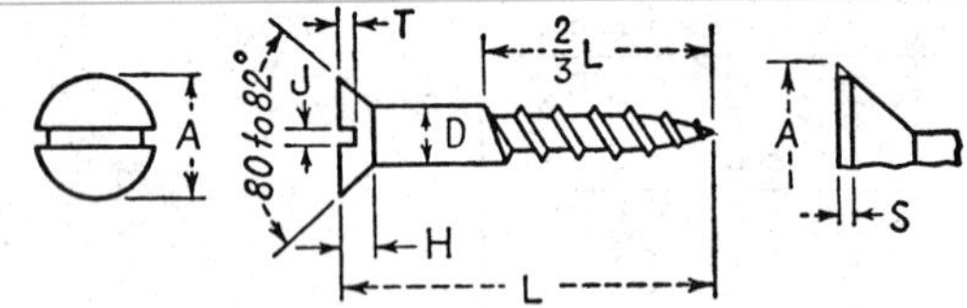

S = *Flat on minimum screw*

TABLE 18. FLAT-HEAD WOOD SCREWS

Size	D	A			S	H		J		T		Threads per In.
	Basic Dia	Max Sharp	Min Sharp	Min with Max S		Max	Min	Max	Min	Max	Min	
0	0.060	0.119	0.105	0.101	0.002	0.035	0.026	0.023	0.016	0.015	0.010	32
1	0.073	0.146	0.130	0.126	0.003	0.043	0.033	0.026	0.019	0.019	0.012	28
2	0.086	0.172	0.156	0.150	0.003	0.051	0.040	0.031	0.023	0.023	0.015	26
3	0.099	0.199	0.181	0.175	0.004	0.059	0.048	0.035	0.027	0.027	0.017	24
4	0.112	0.225	0.207	0.200	0.004	0.067	0.055	0.039	0.031	0.030	0.020	22
5	0.125	0.252	0.232	0.225	0.005	0.075	0.062	0.043	0.035	0.034	0.022	20
6	0.138	0.279	0.257	0.249	0.005	0.083	0.069	0.048	0.039	0.038	0.024	18
7	0.151	0.305	0.283	0.274	0.005	0.091	0.076	0.048	0.039	0.041	0.027	16
8	0.164	0.332	0.308	0.300	0.006	0.100	0.084	0.054	0.045	0.045	0.029	15
9	0.177	0.358	0.334	0.323	0.006	0.108	0.091	0.054	0.045	0.049	0.032	14
10	0.190	0.385	0.359	0.348	0.007	0.116	0.098	0.060	0.050	0.053	0.034	13
12	0.216	0.438	0.410	0.397	0.008	0.132	0.112	0.067	0.056	0.060	0.039	11
14	0.242	0.491	0.461	0.447	0.009	0.148	0.127	0.075	0.064	0.068	0.044	10
16	0.268	0.544	0.512	0.496	0.010	0.164	0.141	0.075	0.064	0.075	0.049	9
18	0.294	0.597	0.563	0.545	0.011	0.180	0.155	0.084	0.072	0.083	0.054	8
20	0.320	0.650	0.614	0.595	0.012	0.196	0.170	0.084	0.072	0.090	0.059	8
24	0.372	0.756	0.716	0.693	0.013	0.228	0.198	0.094	0.081	0.105	0.069	7

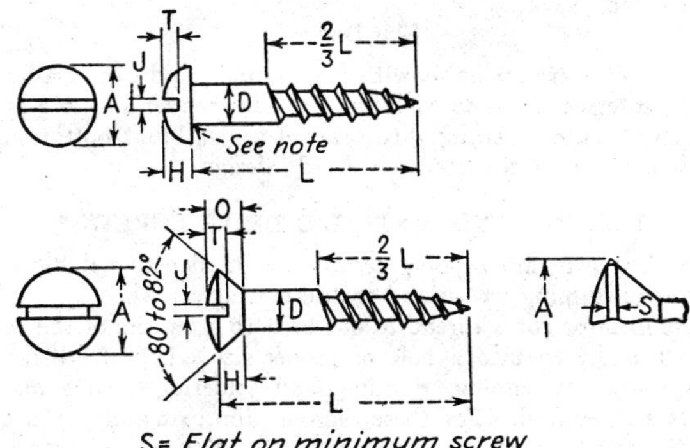

Table 19. Head Dimensions for Round- and Oval-head Wood Screws*

Size	Round Head						Oval Head			
	A		*H*		*T*		*O*		*T*	
	Max	Min	Max	Min	Max	Min	Max	Min	Max	Min
0	0.113	0.099	0.053	0.043	0.039	0.029	0.056	0.041	0.030	0.025
1	0.138	0.122	0.061	0.051	0.044	0.033	0.068	0.052	0.038	0.031
2	0.162	0.146	0.069	0.059	0.048	0.037	0.080	0.063	0.045	0.037
3	0.187	0.169	0.078	0.067	0.053	0.040	0.092	0.073	0.052	0.043
4	0.211	0.193	0.086	0.075	0.058	0.044	0.104	0.084	0.059	0.049
5	0.236	0.217	0.095	0.083	0.063	0.047	0.116	0.095	0.067	0.055
6	0.260	0.240	0.103	0.091	0.068	0.051	0.128	0.105	0.074	0.060
7	0.285	0.264	0.111	0.099	0.072	0.055	0.140	0.116	0.081	0.066
8	0.309	0.287	0.120	0.107	0.077	0.058	0.152	0.126	0.088	0.072
9	0.334	0.311	0.128	0.115	0.082	0.062	0.164	0.137	0.095	0.078
10	0.359	0.334	0.137	0.123	0.087	0.065	0.176	0.148	0.103	0.084
12	0.408	0.382	0.153	0.139	0.096	0.073	0.200	0.169	0.117	0.096
14	0.457	0.429	0.170	0.155	0.106	0.080	0.224	0.190	0.132	0.108
16	0.506	0.476	0.187	0.171	0.115	0.087	0.248	0.212	0.146	0.120
18	0.555	0.523	0.204	0.187	0.125	0.094	0.272	0.233	0.160	0.132
20	0.604	0.570	0.220	0.203	0.134	0.101	0.296	0.254	0.175	0.144
24	0.702	0.664	0.254	0.235	0.154	0.116	0.344	0.297	0.204	0.168

ASA B18.6-1947. Published by ASME.

* For dimensions not given, refer to Table 18.

Points. Wood screws shall be furnished with gimlet points.

Finish. Wood screws are regularly furnished with naturally bright finish and are not heat-treated.

Slot Design

Three types of slots are available with flat-, round-, and oval-head wood screws: single slot; two intersecting slots with parallel sides converging to a sharp apex at bottom; and large center opening with tapered wings, blunt bottom, and relieved edges. The latter two are the "recessed-head" design.

TAPPING AND METALLIC DRIVE SCREWS*

Screws described here are tapping screws and hardened metallic drive screws, classified as thread-forming screws and thread-cutting screws.

These screws form or cut a thread in one or both members of the assembly (see page 36-28) when driven into a hole of proper size. Use of self-tapping screws eliminates tapping, nut-running, riveting, and soldering. Thus machining and assembly costs are reduced. For these reasons, approximately 25% of all screws used, other than wood screws, are of the self-tapping type. They are made in all conventional and several special styles of head.

Thread-forming Screws

For applications in materials where large internal stresses are permissible or desirable to increase resistance to loosening, tapping screws are of the following types:

Type A. Spaced-thread screw with gimlet point primarily for use in light sheet metal (0.015 to 0.050 in.), resin-impregnated plywood, asbestos compositions, etc. Because of the gimlet point, the type A does not require exact alignment in workpieces, but the gimlet point causes the screw to "start all at once." Hence this type is not suitable for plastics or other brittle materials.

Type Z. Spaced-thread screw with pitches generally somewhat finer than type A with a blunt point for light and heavy sheet metal (0.015 to 0.200 in.), nonferrous castings, plastics, resin-impregnated plywood, asbestos compositions, etc. The blunt point requires good alignment of holes. Finer pitch and shallower threads as compared with type A make type Z easier to drive. Hence, type Z screws are better adapted to heavier metals and for fastening to plastics. Hex-head screws can be driven with more positive control and more torque and more rapidly than the slotted type. Hence, the hex-head type is much used for line production assembly.

Type Z Pointed. Spaced-thread screw, the same as type Z but having a cone point, used primarily in assemblies where holes are misaligned, etc.

Type C. Screws with threads approximating the machine-screw thread and with blunt point. Used where a machine-screw thread is preferable to the spaced-thread types of thread-forming screws. Also useful when chips from machine-screw thread-cutting screws are objectionable. In specific applications this type of screw may require extreme driving torques.

Thread-cutting Screws

For applications in materials where disruptive internal stresses are not desirable, or where excessive driving torques are encountered, thread-cutting screws are of the following types:

* Courtesy Parker-Kalon Corp.

Type F. Thread-cutting screws with threads approximating machine-screw threads, with blunt point, and with tapered entering threads having one or more cutting edges and chip-clearance cavities. These screws can be used in materials such as aluminum, zinc, and lead die castings, steel sheets and shapes, cast iron, brass, and plastics. The threads on these screws are approximately the same as a class 2 machine screw, and hence the self-tapping screw can be used as a direct replacement. Flutes provide cutting action in brittle material like plastics and in cast iron, which will not "give" or "move."

TABLE 20. TAPPING-SCREW DESIGNATIONS

Type	Auto-motive	ASA	Fed.	Screw Mfgr.	Type	Auto-motive	ASA	Fed.	Screw Mfgr.
A	A	A	A	A	1	D	1	CS Alt. 1	1
B*	B	B	B	B*	23		23	CG	23
B† Ptd.	BP	BP	BP	BP†	FZ		FZ	BF	FZ
C	C	C	C	C	25		25	BG	25
F	F	F	CF	F	H		H	BG	H
G	G	G	GS Pt. 2	G	U	U	U	U	U

* Also known as type Z.
† Also known as type Z pointed.

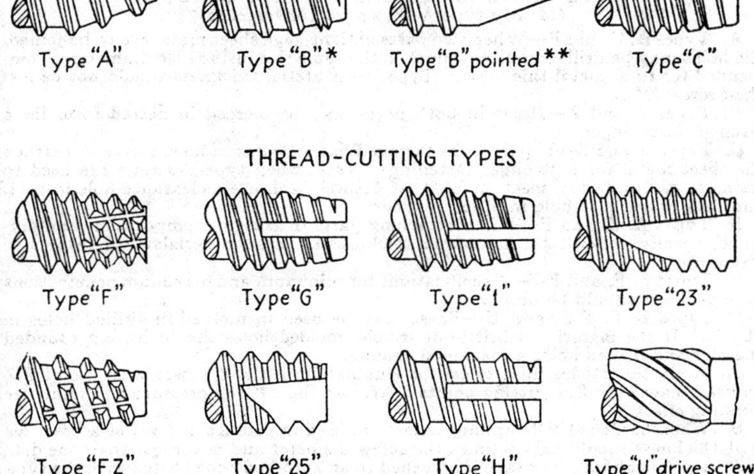

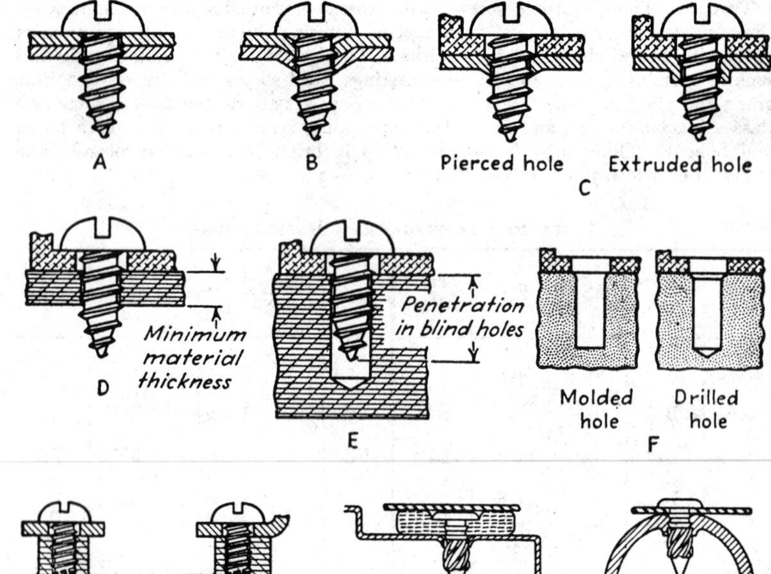

APPLICATIONS OF SELF-TAPPING SCREWS
(Mostly type A shown for convenience)

A. Types A, Z, and F—When two parts of light-gage sheet metal are to be joined, the holes may be drilled or clean-punched the same size, using hole diameter recommended for total metal thickness. Type U—Material thickness should not be less than screw OD.

B. Types A and Z—Holes in both parts may be pierced in nested form for a stronger fastening.

C. Types A and Z—If part to be fastened has a clearance hole, pierce or extrude the sheet metal for a stronger fastening. Note: when type-F screws are used to fasten a part to heavy sheet metal, or structural steel, use a clearance hole in part, and drill proper size hole in other member.

D. Types A, Z, and F—When attaching parts to asbestos compositions and plywood, specify at least the minimum thickness of these materials with respect to screw size.

E. Types Z, F, and F-Z—Specifications for minimum and maximum penetrations in blind holes should be observed.

F. Types Z, F. F-Z, and U—These may be used in molded or drilled holes in plastic. If the material is brittle or friable, molded holes should have a rounded chamfer and drilled holes a machined chamfer.

G. In plastic articles, inserts can be eliminated in island bosses with use of F-Z screws because the five cutting points distribute the cutting pressure and minimize bursting effect.

H. When hardened self-tapping screws are used in plastics in form of a boss, the wall thickness should be 1.5 times the screw diameter and never less than one dia.

J. Antisqueak lacing or tape is attached to an auto-body cowl ledge with the type 21 screw.

K. Type 21 screws can be driven by automatic hopper-fed machines.

Types FZ. Spaced-thread screws as in type Z with blunt points, with one or more cutting grooves, for use in plastics, die castings, metal-clad and resin-impregnated plywoods, asbestos, and other compositions. They have the thread-cutting flutes of the type F screw and the coarse pitch and shallow threads of the type Z to avoid cracking brittle materials like plastics. These screws are also used for making fastenings to comparatively thin sections.

Type U Metallic Drive Screws. Multiple-threaded screw with large helix angle, having a pilot, for use in metal and plastics. The screws are forced into the work by pressure. They are intended for making permanent fastenings.

Type 21 metallic drive screws are used to fasten upholstery, antisqueak, and windlace to metal structures, like automobile bodies. They are driven into a drilled or pierced hole and hold against severe vibration.

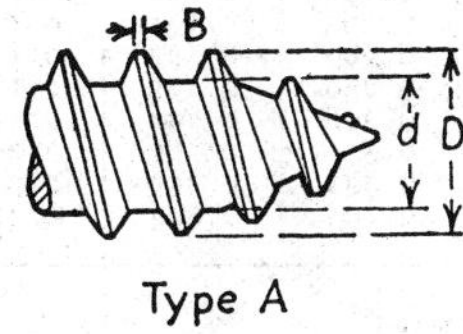

Type A

TABLE 21. TYPE A TAPPING SCREWS

Nominal size	Threads per In.	D Major Dia Max	D Major Dia Min	d Minor Dia Max	d Minor Dia Min
0	40	0.060	0.057	0.042	0.039
1	32	0.075	0.072	0.051	0.048
2	32	0.088	0.084	0.061	0.056
3	28	0.101	0.097	0.076	0.071
4	24	0.114	0.110	0.083	0.078
5	20	0.130	0.126	0.095	0.090
6	18	0.141	0.136	0.102	0.096
7	16	0.158	0.152	0.114	0.108
8	15	0.168	0.162	0.123	0.116
10	12	0.194	0.188	0.133	0.126
12	11	0.221	0.215	0.162	0.155
14	10	0.254	0.248	0.185	0.178
16	10	0.280	0.274	0.197	0.189
18	9	0.306	0.300	0.217	0.209
20	9	0.333	0.327	0.234	0.226
24	9	0.390	0.383	0.291	0.282

Maximum crest *B* up to size No. 8 equals 0.004 in.; over size No. 8, equals 0.006 in.

Table 22. Types Z* and Z Pointed† Tapping Screws

Nominal size	Threads per In.	D Major Dia Max	D Major Dia Min	d Minor Dia Max	d Minor Dia Min	P Point Dia Max	P Point Dia Min
0	48	0.060	0.057	0.036	0.033	0.031	0.027
1	42	0.075	0.072	0.049	0.046	0.044	0.040
2	32	0.088	0.084	0.064	0.060	0.058	0.054
3	28	0.101	0.097	0.075	0.071	0.068	0.063
4	24	0.114	0.110	0.086	0.082	0.079	0.074
5	20	0.130	0.126	0.094	0.090	0.087	0.082
6	20	0.139	0.135	0.104	0.099	0.095	0.089
7	19	0.154	0.149	0.115	0.109	0.105	0.099
8	18	0.166	0.161	0.122	0.116	0.112	0.106
8	18	0.189	0.183	0.141	0.135	0.130	0.123
12	14	0.215	0.209	0.164	0.157	0.152	0.145
1/4	14	0.246	0.240	0.192	0.185	0.179	0.171
5/16	12	0.315	0.308	0.244	0.236	0.230	0.222
3/8	12	0.380	0.371	0.309	0.299	0.293	0.285
7/16	10	0.440	0.431	0.359	0.349	0.343	0.335
1/2	10	0.504	0.495	0.423	0.413	0.407	0.399

* Also known as type B.
† Also known as type B pointed.

Table 23. Type C Tapping Screws

Nominal Size	Threads per In.	D Major Dia Max	D Major Dia Min	P Point Dia Max	P Point Dia Min
4	40	0.1120	0.1072	0.086	0.077
4	48	0.1120	0.1076	0.090	0.083
5	40	0.1250	0.1202	0.099	0.090
5	44	0.1250	0.1204	0.101	0.093
6	32	0.1380	0.1326	0.106	0.095
6	40	0.1380	0.1332	0.112	0.103
8	32	0.1640	0.1586	0.132	0.121
8	36	0.1640	0.1590	0.135	0.125
10	24	0.1900	0.1834	0.147	0.133
10	32	0.1900	0.1846	0.158	0.147
12	24	0.2160	0.2094	0.173	0.159
12	28	0.2160	0.2098	0.179	0.167
1/4	20	0.2500	0.2428	0.198	0.181
1/4	28	0.2500	0.2438	0.213	0.201
5/16	18	0.3125	0.3043	0.255	0.236
5/16	24	0.3125	0.3059	0.269	0.255
3/8	16	0.3750	0.3660	0.310	0.289
3/8	24	0.3750	0.3684	0.332	0.318

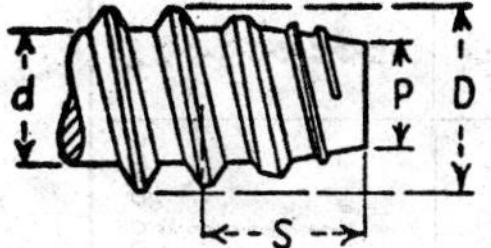

Types Z and Z pointed tapping screws.

TABLE 24. TYPES F, G, 1, AND 23 SCREWS

Screw size	No. of threads		OD	
	NC	NF	Max	Min
2	56		0.0860	0.0820
2		64	0.0860	0.0822
3	48		0.0990	0.0946
3		56	0.0990	0.0950
4	40		0.1120	0.1072
4		48	0.1120	0.1076
5	40		0.1250	0.1202
5		44	0.1250	0.1204
6	32		0.1380	0.1326
6		40	0.1380	0.1332
8	32		0.1640	0.1586
8		36	0.1640	0.1590
10	24		0.1900	0.1834
10		32	0.1900	0.1846
12	24		0.2160	0.2094
12		28	0.2160	0.2098
$\frac{1}{4}$	20		0.2500	0.2428
$\frac{1}{4}$		28	0.2500	0.2438
$\frac{5}{16}$	18		0.3125	0.3043
$\frac{5}{16}$		24	0.3125	0.3059
$\frac{3}{8}$	16		0.3750	0.3660
$\frac{3}{8}$		24	0.3750	0.3684

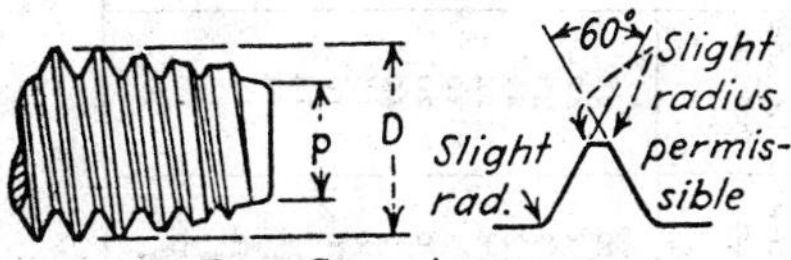

Type C tapping screw

Type F-Z Type 25 Type H

TABLE 25. TYPES FZ, 25, AND H TAPPING SCREWS

Screw Size	Threads per In.	Major Dia D		Root Dia d	
		Max	Min	Max	Min
0	48	0.060	0.057	0.036	0.033
1	42	0.075	0.072	0.049	0.046
2	32	0.088	0.084	0.064	0.060
3	28	0.101	0.097	0.075	0.071
4	24	0.114	0.110	0.086	0.082
5	20	0.130	0.126	0.094	0.090
6	20	0.139	0.135	0.104	0.099
7	19	0.154	0.149	0.115	0.109
8	18	0.166	0.161	0.122	0.116
10	16	0.189	0.183	0.141	0.135
12	14	0.215	0.209	0.164	0.157
1/4	14	0.246	0.240	0.192	0.185
5/16	12	0.315	0.308	0.244	0.236
3/8	12	0.380	0.371	0.309	0.299
7/16	10	0.440	0.431	0.359	0.349
1/2	10	0.504	0.495	0.423	0.413

TABLE 26. TYPE U DRIVE SCREWS

Screw Size	A	E	C	D	Use Drill Size
00	0.060 0.057	0.049 0.0465	0.099 0.090	0.034 0.026	55
0	0.075 0.072	0.063 0.060	0.127 0.118	0.049 0.041	51
2	0.100 0.097	0.083 0.080	0.162 0.146	0.069 0.059	44
4	0.116 0.112	0.096 0.092	0.211 0.193	0.086 0.075	37
6	0.140 0.136	0.116 0.112	0.260 0.240	0.103 0.091	31
7	0.154 0.150	0.126 0.122	0.285 0.264	0.111 0.099	29
8	0.167 0.162	0.136 0.132	0.309 0.287	0.120 0.107	27
10	0.182 0.177	0.150 0.146	0.359 0.334	0.137 0.123	20
12	0.212 0.206	0.177 0.173	0.408 0.382	0.153 0.139	11
14	0.242 0.236	0.202 0.198	0.457 0.429	0.170 0.155	2
5/16	0.315 0.309	0.272 0.267	0.590 0.557	0.216 0.198	M
3/8	0.378 0.371	0.334 0.329	0.708 0.670	0.256 0.237	T

B	F
1/8	3/64
3/16	3/64
1/4	3/64
5/16	3/64
3/8	1/16
1/2	1/16
5/8	5/64
3/4	5/64
1 and over	1/8

Table 27. Hole Sizes for Type A Self-tapping Sheet-metal Screws

In Sheet Metal

Screw No.	Steel, Stainless Steel, Monel Metal, Brass, Aluminum Alloy			
	Metal Thickness, In.	Pierced or Extruded Hole	Drilled or Clean-punched Hole	
		Hole Required	Hole Required	Drill No.
4	0.015		0.086	44
	0.018		0.086	44
	0.024	0.098	0.093	42
	0.030	0.098	0.093	42
	0.036	0.098	0.098	40
6	0.015		0.099	39
	0.018		0.099	39
	0.024	0.111	0.099	39
	0.030	0.111	0.101	38
	0.036	0.111	0.106	36
7	0.015		0.104	37
	0.018		0.104	37
	0.024	0.120	0.110	35
	0.030	0.120	0.113	33
	0.036	0.120	0.116	32
	0.048	0.120	0.120	31
8	0.018		0.113	33
	0.024	0.136	0.113	33
	0.030	0.136	0.116	32
	0.036	0.136	0.120	31
	0.048	0.136	0.128	30
10	0.018		0.128	30
	0.024	0.157	0.128	30
	0.030	0.157	0.128	30
	0.036	0.157	0.136	29
	0.048	0.157	0.149	25
12	0.024		0.147	26
	0.030	0.185	0.149	25
	0.036	0.185	0.152	24
	0.048	0.185	0.157	22
14	0.024		0.180	15
	0.030	0.209	0.189	12
	0.036	0.209	0.191	11
	0.048	0.209	0.196	9

In Plywood (Resin Impregnated)

Screw No.	Compreg, Pregwood, Etc.				
	Hole Required	Drill No.	Min Material Thickness	Penetration in Blind Holes	
				Min	Max
4	0.098	40	3/16	1/4	3/4
6	0.110	35	3/16	1/4	3/4
7	0.128	30	1/4	5/16	3/4
8	0.140	28	1/4	5/16	3/4
10	0.169	18	5/16	3/8	1
12	0.189	12	5/16	3/8	1
14	0.228	1	7/16	1/2	1

In Asbestos Compositions

Screw No.	Transite, Ebony, Asbestos, Etc.				
	Hole Required	Drill No.	Min Material Thickness	Penetration in Blind Holes	
				Min	Max
4	0.093	42	3/16	1/4	3/4
6	0.106	36	3/16	1/4	3/4
7	0.125	1/8	1/4	5/16	3/4
8	0.136	29	1/4	5/16	3/4
10	0.161	20	5/16	3/8	1
12	0.185	13	5/16	3/8	1
14	0.213	3	7/16	1/2	1

Because of varying conditions, it may be necessary to vary hole size to suit a particular application.

TABLE 28. HOLE SIZES FOR TYPE Z TAPPING SCREWS
In Sheet Metal

Screw No.	Steel, Stainless Steel, Monel Metal, Brass				Aluminum Alloy		
	Metal Thickness	Pierced or Extruded Hole	Drilled or Clean-punched Hole		Pierced or Extruded Hole	Drilled or Clean-punched Hole	
		Hole Required	Hole Required	Drill No.	Hole Required	Hole Required	Drill No.
2	0.015		0.063	52			
	0.018		0.063	52			
	0.024		0.067	51		0.063	52
	0.030		0.070	50		0.063	52
	0.036		0.073	49		0.063	52
	0.048		0.073	49		0.067	51
	0.060		0.076	48		0.070	50
4	0.015	0.086	0.086	44			
	0.018	0.086	0.086	44			
	0.024	0.098	0.089	43	0.086		
	0.030	0.098	0.093	42	0.086	0.086	44
	0.036	0.098	0.093	42	0.086	0.086	44
	0.048		0.096	41	0.086	0.086	44
	0.060		0.099	39		0.089	43
	0.075		0.101	38		0.089	43
	0.105					0.093	42
6	0.015	0.111	0.104	37			
	0.018	0.111	0.104	37			
	0.024	0.111	0.106	36	0.111		
	0.030	0.111	0.106	36	0.111	0.104	37
	0.036	0.111	0.110	35	0.111	0.104	37
	0.048		0.111	34	0.111	0.104	37
	0.060		0.116	32		0.106	36
	0.075		0.120	31		0.110	35
	0.105		0.128	30		0.111	34
	0.128 to ¼					0.120	31
7	0.018	0.120	0.113	33			
	0.024	0.120	0.113	33	0.120		
	0.030	0.120	0.116	32	0.120	0.113	33
	0.036	0.120	0.116	32	0.120	0.113	33
	0.048	0.120	0.120	31	0.120	0.116	32
	0.060		0.128	30		0.120	31
	0.075		0.136	29		0.128	30
	0.105		0.140	28		0.136	29
	0.128 to ¼					0.136	29
8	0.018	0.136					
	0.024	0.136	0.116	32	0.136		
	0.030	0.136	0.120	31	0.136	0.116	32
	0.036	0.136	0.120	31	0.136	0.120	31

TABLE 28. HOLE SIZES FOR TYPE Z TAPPING SCREWS (*Continued*)

Screw No.	Metal Thickness	Steel, Stainless Steel, Monel Metal, Brass			Aluminum Alloy		
		Pierced or Extruded Hole	Drilled or Clean-punched Hole		Pierced or Extruded Hole	Drilled or Clean-punched Hole	
		Hole Required	Hole Required	Drill No.	Hole Required	Hole Required	Drill No.
8	0.048	0.136	0.128	30	0.136	0.128	30
	0.060		0.136	29		0.136	29
	0.075		0.140	28		0.140	28
	0.105		0.149	25		0.147	26
	0.125		0.149	25		0.147	26
	0.135		0.152	24		0.149	25
	0.162 to $\frac{3}{8}$					0.152	24
10	0.018	0.157					
	0.024	0.157	0.144	27	0.157		
	0.030	0.157	0.144	27	0.157		
	0.036	0.157	0.147	26	0.157	0.144	27
	0.048	0.157	0.152	24	0.157	0.144	27
	0.060		0.152	24		0.144	27
	0.075		0.157	22		0.147	26
	0.105		0.161	20		0.147	26
	0.125		0.169	18		0.154	23
	0.135		0.169	18		0.154	23
	0.164		0.173	17		0.159	21
	0.200 to $\frac{3}{8}$					0.166	19
12	0.024	0.185	0.166	19			
	0.030	0.185	0.166	19			
	0.036	0.185	0.166	19			
	0.048	0.185	0.169	18		0.161	20
	0.060		0.177	16		0.166	19
	0.075		0.182	14		0.173	17
	0.105		0.185	13		0.180	15
	0.125		0.196	9		0.182	14
	0.135		0.196	9		0.182	14
	0.164		0.201	7		0.189	12
	0.200 to $\frac{3}{8}$					0.196	9
14	0.030	0.209	0.185	13			
	0.036	0.209	0.185	13			
	0.048	0.209	0.191	11			
	0.060		0.199	8		0.199	8
	0.075		0.204	6		0.201	7
	0.105		0.209	4		0.204	6
	0.125		0.228	1		0.209	4
	0.135		0.228	1		0.209	4
	0.164		0.234	$\frac{15}{64}$		0.213	3
	0.187		0.234	$\frac{15}{64}$		0.213	3
	0.194		0.234	$\frac{15}{64}$		0.221	2
	0.200 to $\frac{3}{8}$					0.228	

TABLE 28. HOLE SIZES FOR TYPE Z TAPPING SCREWS (*Continued*)

Screw No.	Compreg, Pregwood, Etc.					Phenol Formaldehyde		Cellulose Acetate, Cellulose Nitrate, Acrylic Resin, Styrene Resin		
	Hole Required	Drill No.	Min Material Thickness	Penetration in Blind Holes		Hole Required	Drill No.	Hole Required	Drill No.	Min Penetration in Blind Holes
				Min	Max					
2	0.073	49	$\frac{1}{8}$	$\frac{3}{16}$	$\frac{1}{2}$	0.078	47	0.078	47	$\frac{3}{16}$
4	0.099	39	$\frac{3}{16}$	$\frac{1}{4}$	$\frac{5}{8}$	0.099	39	0.093	42	$\frac{1}{4}$
6	0.125	$\frac{1}{8}$	$\frac{3}{16}$	$\frac{1}{4}$	$\frac{5}{8}$	0.128	30	0.120	31	$\frac{1}{4}$
7	0.136	29	$\frac{3}{16}$	$\frac{1}{4}$	$\frac{3}{4}$	0.136	29	0.128	30	$\frac{1}{4}$
8	0.144	27	$\frac{3}{16}$	$\frac{1}{4}$	$\frac{3}{4}$	0.149	25	0.144	27	$\frac{5}{16}$
10	0.173	17	$\frac{1}{4}$	$\frac{5}{16}$	1	0.177	16	0.169	18	$\frac{5}{16}$
12	0.193	10	$\frac{5}{16}$	$\frac{3}{8}$	1	0.199	8	0.191	11	$\frac{3}{8}$
14	0.228	1	$\frac{5}{16}$	$\frac{3}{8}$	1	0.234	$\frac{15}{64}$	0.221	2	$\frac{3}{8}$
	Transite, Ebony Asbestos, Etc.							Cast Aluminum, Magnesium, Zinc, Brass, Bronze		
2	0.076	48	$\frac{1}{8}$	$\frac{3}{16}$	$\frac{1}{2}$	....		0.078	47	$\frac{1}{8}$
4	0.101	38	$\frac{3}{16}$	$\frac{1}{4}$	$\frac{5}{8}$	....		0.104	37	$\frac{3}{16}$
6	0.120	31	$\frac{3}{16}$	$\frac{1}{4}$	$\frac{5}{8}$	....		0.128	30	$\frac{1}{4}$
7	0.136	29	$\frac{1}{4}$	$\frac{5}{16}$	$\frac{3}{4}$	....		0.144	27	$\frac{1}{4}$
8	0.147	26	$\frac{5}{16}$	$\frac{3}{8}$	$\frac{3}{4}$	....		0.152	24	$\frac{1}{4}$
10	0.166	19	$\frac{5}{16}$	$\frac{3}{8}$	1	....		0.177	16	$\frac{1}{4}$
12	0.196	9	$\frac{5}{16}$	$\frac{3}{8}$	1	....		0.199	8	$\frac{9}{32}$
14	0.228	1	$\frac{7}{16}$	$\frac{1}{2}$	1	....		0.234	$\frac{15}{64}$	$\frac{5}{16}$

Table 29. Hole Sizes for Type F Self-tapping Screws
In Sheet Metal and Structural Steel*

Screw No.	Sheet Steel, Structural Steel, Stainless Steel, Monel Metal, Brass, Aluminum Alloy		
	Metal Thickness	Hole Required	Drill No.
2-56	0.048	0.073	49
	0.060	0.073	49
	0.075	0.076	48
	0.105 to 0.156	0.078	47
4-40	0.048	0.093	42
	0.060	0.096	41
	0.075	0.096	41
	0.105 to 0.156	0.099	39
6-32	0.048	0.111	34
	0.060	0.113	33
	0.075	0.116	32
	0.105 to 0.375	0.120	31
8-32	0.048	0.140	28
	0.060	0.144	27
	0.075	0.144	27
	0.105	0.147	26
	0.125	0.147	26
	0.135	0.147	26
	0.164 to 0.375	0.149	25
10-32	0.048	0.159	21
	0.060	0.161	20
	0.075	0.166	19
	0.105	0.169	18
	0.125	0.169	18
	0.135	0.169	18
	0.164	0.173	17
	0.187 to 0.500	0.177	16
10-24	0.060	0.166	19
	0.075	0.169	18
	0.105	0.169	18
	0.125	0.173	17
	0.135	0.173	17
	0.164	0.173	17
	0.187 to 0.500	0.177	16
$\frac{1}{4}$-20	0.060	0.213	3
	0.075	0.221	2
	0.105	0.221	2
	0.125	0.228	1
	0.135	0.228	1
	0.164	0.228	1
	0.187	0.234	$\frac{15}{64}$
	0.250 to 0.625	0.238	B

* Drilled or clean-punched hole.

TABLE 29. HOLE SIZES FOR TYPE F SELF-TAPPING SCREWS (*Continued*)

In Castings

Screw No.	Aluminum, Magnesium, Zinc, Brass, Bronze, Gray Iron, Malleable Iron, Steel			
	Hole Required	Drill No.	Depth of Penetration	
			Min	Max
2-56	0.078	47	3/16	1/4
4-40	0.099	39	3/16	1/4
6-32	0.120	31	3/16	1/4
8-32	0.147	26	7/32	5/16
10-32	0.169	18	1/4	3/8
10-24	0.169	18	1/4	3/8
1/4-20	0.228	1	5/16	3/8

In Plywood (Resin Impregnated)

Screw No.	Compreg, Pregwood, Etc.				
	Hole Required	Drill No.	Min Material Thickness	Depth of Penetration	
				Min	Max
2-56	0.073	49	3/16	7/32	3/8
4-40	0.101	38	3/16	1/4	1/2
6-32	0.120	31	1/4	5/16	5/8
8-32	0.149	25	1/4	5/16	5/8
10-32	0.173	17	1/4	5/16	5/8
10-24	0.169	18	1/4	5/16	3/4
1/4-20	0.234	15/64	5/16	3/8	1

In Plastics

Screw No.	Phenol Formaldehyde				Cellulose Acetate, Cellulose Nitrate, Acrylic Resins, Styrene Resins			
	Hole Required	Drill No.	Depth of Penetration		Hole Required	Drill No.	Depth of Penetration	
			Min	Max			Min	Max
2-56	0.073	49	7/32	3/8	Not recommended (see type Z)			
4-40	0.098	40	1/4	5/16	0.093	42	1/4	5/16
6-32	0.116	32	1/4	5/16	0.116	32	1/4	5/16
8-32	0.144	27	5/16	1/2	0.144	27	5/16	1/2
10-32	0.166	19	3/8	1/2	0.166	19	3/8	1/2
10-24	0.161	20	3/8	1/2	0.161	20	3/8	1/2
1/4-20	0.228	1	3/8	5/8	0.228	1	3/8	1

TABLE 30. SLOTTED HEADLESS SETSCREWS*

Nominal Size	D	I	J	T
5	0.125	0.125	0.023	0.031
6	0.138	0.138	0.025	0.035
8	0.164	0.164	0.029	0.041
10	0.190	0.190	0.032	0.048
12	0.216	0.216	0.036	0.054
1/4	0.250	0.250	0.045	0.063
5/16	0.3125	0.313	0.051	0.078
3/8	0.375	0.375	0.064	0.094
7/16	0.4375	0.438	0.072	0.109
1/2	0.500	0.500	0.081	0.125
9/16	0.5625	0.563	0.091	0.141
5/8	0.625	0.625	0.102	0.156
3/4	0.750	0.750	0.129	0.188

* For other dimensions, see Table 31, except that W = 45 to 50°. Y = 116 to 120° when L is less than the nominal diameter, and 88 to 92° when L is greater than the nominal diameter. V and Z = 35 to 40°.

Flat point, Cone point, Oval point, Cup point, Dog point, Half point

Slotted headless setscrews.

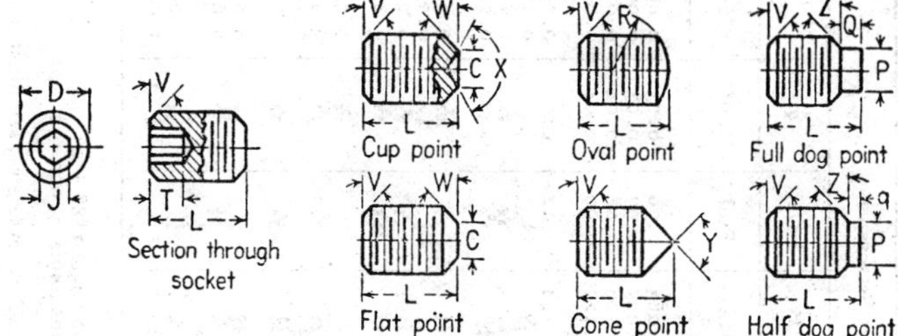

Hexagonal-socket headless setscrews.

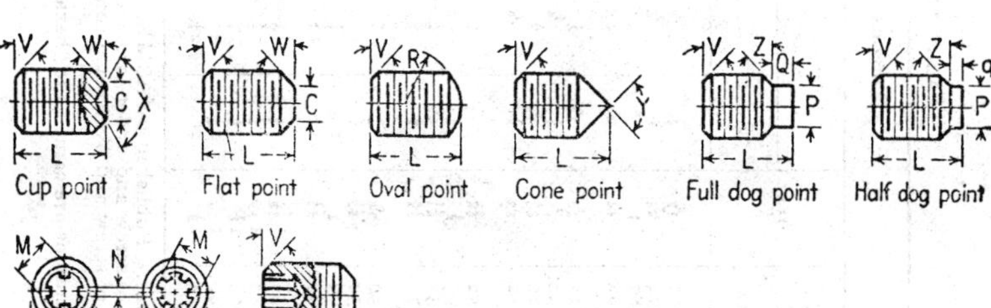

Fluted-socket setscrews.

Table 31. Hexagonal- and Fluted-socket Headless Setscrews

Hexagonal-socket Setscrews											Fluted-socket Setscrews*						
D	C	R	Y		P	Q	q	J		Flutes	J		M	N	
Nominal Dia	Max		L_1	L_2	Max	Full	Half	Max	Min		Max	Min	Max	Max	Min
5	0.067	3/32	1/8	3/16	0.083	0.06	0.03	0.0635	1/16	4	0.053	0.052	0.071	0.022	0.021
6	0.074	7/64	1/8	3/16	0.092	0.07	0.03	0.0635	1/16	4	0.056	0.055	0.079	0.023	0.022
8	0.087	1/8	3/16	1/4	0.109	0.08	0.04	0.0791	5/64	6	0.082	0.080	0.098	0.022	0.021
10	0.102	9/64	3/16	1/4	0.127	0.09	0.04	0.0947	3/32	6	0.098	0.096	0.115	0.025	0.023
12	0.115	5/32	3/16	1/4	0.144	0.11	0.06	0.0947	3/32	6	0.098	0.096	0.115	0.025	0.023
1/4	0.132	3/16	1/4	5/16	5/32	1/8	1/16	0.1270	1/8	6	0.128	0.126	0.149	0.032	0.030
5/16	0.172	15/64	5/16	3/8	13/64	5/32	5/64	0.1582	5/32	6	0.163	0.161	0.188	0.039	0.037
3/8	0.212	9/32	3/8	7/16	1/4	3/16	3/32	0.1895	3/16	6	0.190	0.188	0.221	0.050	0.048
7/16	0.252	21/64	7/16	1/2	19/64	7/32	7/64	0.2207	7/32	6	0.221	0.219	0.256	0.060	0.058
1/2	0.291	3/8	1/2	9/16	11/32	1/4	1/8	0.2520	1/4	6	0.254	0.252	0.298	0.068	0.066
9/16	0.332	27/64	9/16	5/8	25/64	9/32	9/64	0.2520	1/4	6	0.254	0.252	0.298	0.068	0.066
5/8	0.371	15/32	5/8	3/4	15/32	5/16	5/32	0.3155	5/16	6	0.319	0.316	0.380	0.092	0.089
3/4	0.450	9/16	3/4	7/8	9/16	3/8	3/16	0.3780	3/8	6	0.386	0.383	0.463	0.112	0.109
7/8	0.530	21/32	7/8	1	21/32	7/16	7/32	0.5030	1/2	6	0.509	0.506	0.604	0.138	0.134
1	0.609	3/4	1	1 1/8	3/4	1/2	1/4	0.5655	9/16	6	0.535	0.531	0.631	0.149	0.145
1 1/8	0.689	27/32	1 1/8	1 1/4	27/32	9/16	9/32	0.5655	9/16	6	0.604	0.600	0.709	0.168	0.164
1 1/4	0.767	15/16	1 1/4	1 1/2	15/16	5/8	5/16	0.6290	5/8	6	0.685	0.681	0.801	0.189	0.185
1 3/8	0.848	1 1/32	1 3/8	1 5/8	1 1/32	11/16	11/32	0.6290	5/8	6	0.744	0.740	0.869	0.207	0.203
1 1/2	0.926	1 1/8	1 1/2	1 3/4	1 1/8	3/4	3/8	0.7540	3/4	6	0.828	0.824	0.970	0.231	0.227
1 3/4	1.086	1 5/16	1 3/4	2	1 5/16	7/8	7/16	1.0040	1	6	1.007	1.003	1.275	0.298	0.294
2	1.244	1 1/2	2	2 1/4	1 1/2	1	1/2	1.0040	1	6	1.007	1.003	1.275	0.298	0.294

ASA 18.3-1947. Published by ASME.
L_1 = length of screw and under for cone point angle 118° ± 2°.
L_2 = length of screw and over when cone point angle = 90° ± 2.

* Other dimensions are same as for hexagon-socket setscrews; see illustrations on page 36-39.

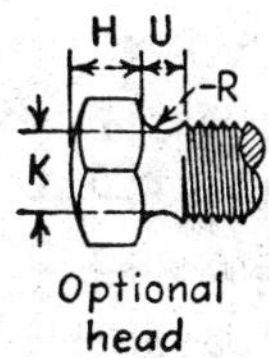

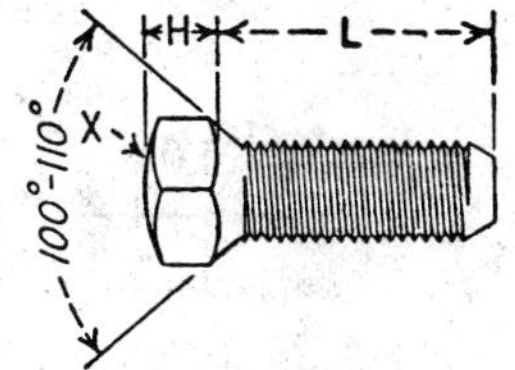

TABLE 32. SQUARE-HEAD SETSCREWS

Nominal Size	*F* Max	*F* Min	*G*, Min	*H*, Nominal	*K* Max	*K* Min	*X*, Nominal	*R*, Max	*U*, Max
No. 10	0.1875	0.180	0.247	9/64	0.145	0.140	15/32	0.027	0.083
No. 12	0.216	0.208	0.292	5/32	0.162	0.156	35/64	0.029	0.091
1/4	0.250	0.241	0.331	3/16	0.185	0.170	5/8	0.032	0.100
5/16	0.3125	0.302	0.415	15/64	0.240	0.225	25/32	0.036	0.111
3/8	0.375	0.362	0.497	9/32	0.294	0.279	15/16	0.041	0.125
7/16	0.4375	0.423	0.581	21/64	0.345	0.330	1 3/32	0.046	0.143
1/2	0.500	0.484	0.665	3/8	0.400	0.385	1 1/4	0.050	0.154
9/16	0.5625	0.545	0.748	27/64	0.454	0.439	1 13/32	0.054	0.167
5/8	0.625	0.606	0.833	15/32	0.507	0.492	1 9/16	0.059	0.182
3/4	0.750	0.729	1.001	9/16	0.620	0.605	1 7/8	0.065	0.200
7/8	0.875	0.852	1.170	21/32	0.731	0.716	2 3/16	0.072	0.222
1	1.000	0.974	1.337	3/4	0.838	0.823	2 1/2	0.081	0.250
1 1/8	1.125	1.096	1.505	27/32	0.939	0.914	2 13/16	0.092	0.283
1 1/4	1.250	1.219	1.674	15/16	1.064	1.039	3 1/8	0.092	0.283
1 3/8	1.375	1.342	1.843	1 1/32	1.159	1.134	3 7/16	0.109	0.333
1 1/2	1.500	1.464	2.010	1 1/8	1.284	1.259	3 3/4	0.109	0.333

ASA B18.2-1952. Published by ASME.

Threads shall be coarse-, fine-, or 8-thread series, class 2A. Square-head setscrews 1/4-in. size and larger are normally stocked in coarse-thread series only.

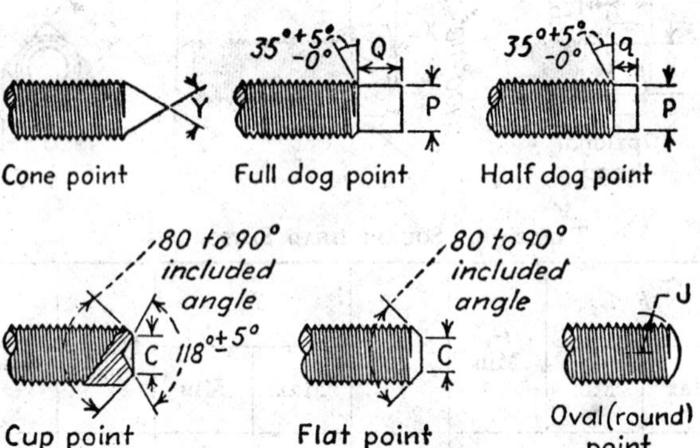

TABLE 33. SQUARE-HEAD SETSCREW POINTS

Nominal Size	Cup and Flat Point Dia C			Oval (Round) Point Radius J	Full-dog, Half-dog Pivot Point			
					Dia P		Full-dog Pivot Q	Half-dog Pivot q
	Nom	Max	Min	Nom	Max	Min		
No. 10	3/32	0.102	0.088	0.141	0.127	0.120	0.090	0.045
No. 12	7/64	0.115	0.101	0.156	0.144	0.137	0.110	0.055
1/4	1/8	0.132	0.118	0.188	0.156	0.149	0.125	0.063
5/16	11/64	0.172	0.156	0.234	0.203	0.195	0.156	0.078
3/8	13/64	0.212	0.194	0.281	0.250	0.241	0.188	0.094
7/16	15/64	0.252	0.232	0.328	0.297	0.287	0.219	0.109
1/2	9/32	0.291	0.270	0.375	0.344	0.334	0.250	0.125
9/16	5/16	0.332	0.309	0.422	0.391	0.379	0.281	0.140
5/8	23/64	0.371	0.347	0.469	0.469	0.456	0.313	0.156
3/4	7/16	0.450	0.425	0.563	0.563	0.549	0.375	0.188
7/8	33/64	0.530	0.502	0.656	0.656	0.642	0.438	0.219
1	19/32	0.609	0.579	0.750	0.750	0.734	0.500	0.250
1 1/8	43/64	0.689	0.655	0.844	0.844	0.826	0.562	0.281
1 1/4	3/4	0.767	0.733	0.938	0.938	0.920	0.625	0.312
1 3/8	53/64	0.848	0.808	1.031	1.031	1.011	0.688	0.344
1 1/2	29/32	0.926	0.886	1.125	1.125	1.105	0.750	0.375

ASA B18.2-1952. Published by ASME.
Where usable length of thread is less than the nominal diameter, half-dog point shall be used.
When length equals nominal diameter or less, $Y = 118° \pm 2°$; when length exceeds nominal diameter, $Y = 90° \pm 2°$.

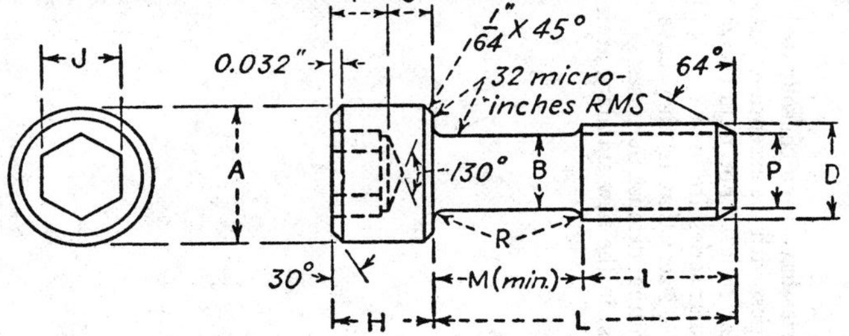

TABLE 34. HIGH-STRENGTH, HIGH-TEMPERATURE INTERNAL WRENCHING BOLTS

D	B		A		H		J		T	G	l, Max	P, Max	M, Min	R, Max
	Max	Min	Max	Min	Max	Min	Max	Min	Min					
3/8-16	0.325	0.320	0.750	0.738	0.375	0.355	0.378	0.375	0.227	0.125	0.750	0.278	0.188	0.032
1/2-13	0.440	0.435	0.875	0.883	0.500	0.480	0.503	0.500	0.290	0.188	0.875	0.376	0.250	0.032
5/8-11	0.496	0.491	1.000	0.988	0.625	0.605	0.565	0.562	0.352	0.250	1.000	0.483	0.312	0.063
3/4-10	0.608	0.603	1.125	1.113	0.750	0.730	0.629	0.625	0.414	0.312	1.125	0.562	0.375	0.063
7/8-9	0.718	0.713	1.312	1.300	0.875	0.855	0.754	0.750	0.477	0.375	1.250	0.668	0.438	0.063
1-8	0.825	0.820	1.500	1.488	1.000	0.980	0.880	0.875	0.539	0.438	1.375	0.756	0.500	0.094
1 1/4-8	1.074	1.069	1.875	1.859	1.250	1.230	1.130	1.125	0.665	0.562	1.625	1.006	0.625	0.094
1 1/2-8	1.323	1.318	2.250	2.234	1.500	1.480	1.380	1.375	0.790	0.688	1.875	1.256	0.750	0.094
1 3/4-8	1.572	1.567	2.625	2.609	1.750	1.730	1.630	1.625	0.915	0.812	2.125	1.506	0.875	0.094
2-8	1.822	1.817	3.000	2.980	2.000	1.980	1.880	1.875	1.040	0.938	2.375	1.756	1.000	0.094

ASA B18.8-1950. Published by the ASME.

These bolts are used in steam turbines and similar high-temperature applications in the order of 800 to 900 F. Head proportions are larger than for standard socket-head capscrews to provide greater bearing surface of the head.

SQUARE AND HEXAGON BOLTS[1]

The current standard ASA B18.2-1952 contains complete dimensional specifications for bolts and nuts, whereas earlier publications covered only the principal head proportions. Needless variety has been eliminated by consolidating types of bolts and nuts that have similar proportions. With the exception of heavy bolts, head dimensions of all series of hexagon bolts and capscrews have been consolidated from the former automotive and regular hexagon bolts in the following manner:

Selected as basic:

For sizes up to and including $\frac{9}{16}$ in., the across-the-flats dimensions of the former automotive hexagon-head bolts or capscrews.

For sizes $\frac{5}{8}$ in. and larger, the across-the-flats dimensions of the regular hexagon-head bolts.

Head height is predicated on a ratio of $\frac{2}{3}$ of the size dia.

For nuts up to $\frac{5}{8}$ in., the former light and regular series have been consolidated by selection of the dimensional proportions of the light series, with the exception that the $\frac{7}{16}$-in. nut is a modification of the light and regular series. For nuts above $\frac{5}{8}$-in. size, the dimensional proportions of the former regular series are used.

Items shown in bold type in the standard are Unified sizes; that is, they have been standardized with the British and Canadians for mutual defense purposes.

In short, the across-flats dimensions (wrench openings) in the new standard are:

Size	New Standards	
	Bolt	Nut
$\frac{1}{4}$	$\frac{7}{16}$	$\frac{7}{16}$
$\frac{5}{16}$	$\frac{1}{2}$	$\frac{1}{2}$
$\frac{3}{8}$	$\frac{9}{16}$	$\frac{9}{16}$
$\frac{7}{16}$	$\frac{5}{8}$	$\frac{11}{16}$
$\frac{1}{2}$	$\frac{3}{4}$	$\frac{3}{4}$
$\frac{5}{8}$	$\frac{15}{16}$	$\frac{15}{16}$
$\frac{3}{4}$	$1\frac{1}{8}$	$1\frac{1}{8}$
$\frac{7}{8}$	$1\frac{5}{16}$	$1\frac{5}{16}$
1	$1\frac{1}{2}$	$1\frac{1}{2}$
Larger sizes	Same as regular series	

The term "finished" hexagon bolt has been used to designate the consolidation of the automotive hexagon-head bolt and the close body-toleranced regular semi-finished bolt. The term "capscrew" is retained in the range of sizes from $\frac{1}{4}$ to $1\frac{1}{2}$ in. for products having the same proportions and characteristics as the "finished" bolts. The term "finished" also is used to designate the consolidation of light and regular series of washer-faced or double-chamfered nuts. In both instances, the term "finished" refers to the quality of manufacture and closeness of tolerance but does not indicate that surfaces are necessarily machined.

[1] For dimensions of nuts, see Tables 51 to 63

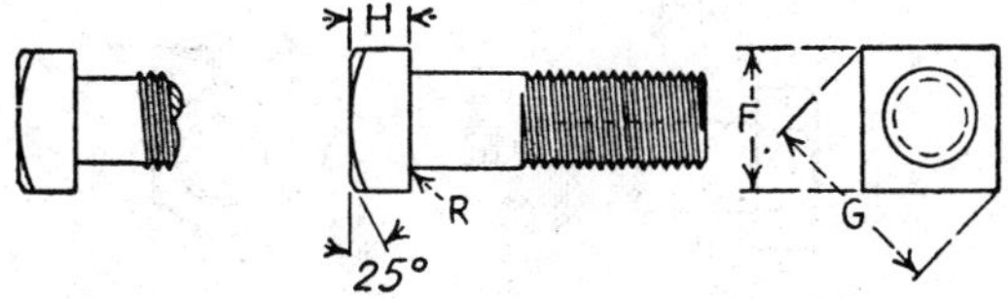

TABLE 35. REGULAR SQUARE BOLTS

Nominal Size	Body Dia, Max	Width across Flats F		Width across Corners G		Height H			Radius of Fillet R
		Max (Basic)	Min	Max	Min	Nominal	Max	Min	Max
$\frac{1}{4}$	0.280	$\frac{3}{8}$	0.362	0.530	0.498	$\frac{11}{64}$	0.188	0.156	0.031
$\frac{5}{16}$	0.342	$\frac{1}{2}$	0.484	0.707	0.665	$\frac{13}{64}$	0.220	0.186	0.031
$\frac{3}{8}$	0.405	$\frac{9}{16}$	0.544	0.795	0.747	$\frac{1}{4}$	0.268	0.232	0.031
$\frac{7}{16}$	0.468	$\frac{5}{8}$	0.603	0.884	0.828	$\frac{19}{64}$	0.316	0.278	0.031
$\frac{1}{2}$	0.530	$\frac{3}{4}$	0.725	1.061	0.995	$\frac{21}{64}$	0.348	0.308	0.031
$\frac{5}{8}$	0.675	$\frac{15}{16}$	0.906	1.326	1.244	$\frac{27}{64}$	0.444	0.400	0.062
$\frac{3}{4}$	0.800	$1\frac{1}{8}$	1.088	1.591	1.494	$\frac{1}{2}$	0.524	0.476	0.062
$\frac{7}{8}$	0.938	$1\frac{5}{16}$	1.269	1.856	1.742	$\frac{19}{32}$	0.620	0.568	0.062
1	1.063	$1\frac{1}{2}$	1.450	2.121	1.991	$\frac{21}{32}$	0.684	0.628	0.062
$1\frac{1}{8}$	1.188	$1\frac{11}{16}$	1.631	2.386	2.239	$\frac{3}{4}$	0.780	0.720	0.125
$1\frac{1}{4}$	1.313	$1\frac{7}{8}$	1.812	2.652	2.489	$\frac{27}{32}$	0.876	0.812	0.125
$1\frac{3}{8}$	1.469	$2\frac{1}{16}$	1.994	2.917	2.738	$\frac{29}{32}$	0.940	0.872	0.125
$1\frac{1}{2}$	1.594	$2\frac{1}{4}$	2.175	3.182	2.986	1	1.036	0.964	0.125
$1\frac{5}{8}$	1.719	$2\frac{7}{16}$	2.356	3.447	3.235	$1\frac{3}{32}$	1.132	1.056	0.125

ASA B18.2-1952. Published by ASME.
Bolt is not finished on any surface.

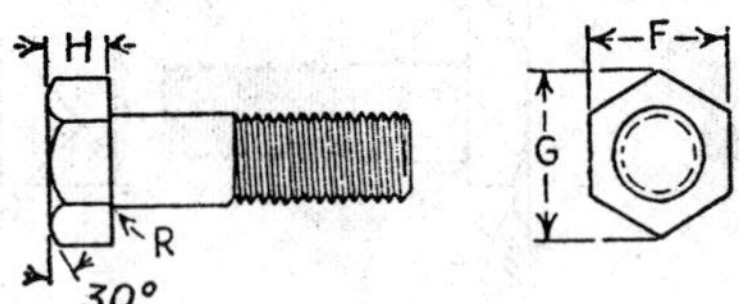

TABLE 36. REGULAR HEXAGON BOLTS

Nominal Size	Body Dia, Max	Width across Flats F		Width across Corners G		Height H			Radius of Fillet R
		Max (Basic)	Min	Max	Min	Nominal	Max	Min	Max
$\frac{1}{4}$	0.280	$\frac{7}{16}$	0.425	0.505	0.484	$\frac{11}{64}$	0.188	0.150	0.031
$\frac{5}{16}$	0.342	$\frac{1}{2}$	0.484	0.577	0.552	$\frac{7}{32}$	0.235	0.195	0.031
$\frac{3}{8}$	0.405	$\frac{9}{16}$	0.544	0.650	0.620	$\frac{1}{4}$	0.268	0.226	0.031
$\frac{7}{16}$	0.468	$\frac{5}{8}$	0.603	0.722	0.687	$\frac{19}{64}$	0.316	0.272	0.031
$\frac{1}{2}$	0.530	$\frac{3}{4}$	0.725	0.866	0.826	$\frac{11}{32}$	0.364	0.302	0.031
$\frac{5}{8}$	0.675	$\frac{15}{16}$	0.906	1.083	1.033	$\frac{27}{64}$	0.444	0.378	0.062
$\frac{3}{4}$	0.800	$1\frac{1}{8}$	1.088	1.299	1.240	$\frac{1}{2}$	0.524	0.455	0.062
$\frac{7}{8}$	0.938	$1\frac{5}{16}$	1.269	1.516	1.447	$\frac{37}{64}$	0.604	0.531	0.062
1	1.063	$1\frac{1}{2}$	1.450	1.732	1.653	$\frac{43}{64}$	0.700	0.591	0.062
$1\frac{1}{8}$	1.188	$1\frac{11}{16}$	1.631	1.949	1.859	$\frac{3}{4}$	0.780	0.658	0.125
$1\frac{1}{4}$	1.313	$1\frac{7}{8}$	1.812	2.165	2.066	$\frac{27}{32}$	0.876	0.749	0.125
$1\frac{3}{8}$	1.469	$2\frac{1}{16}$	1.994	2.382	2.273	$\frac{29}{32}$	0.940	0.810	0.125
$1\frac{1}{2}$	1.594	$2\frac{1}{4}$	2.175	2.598	2.480	1	1.036	0.902	0.125
$1\frac{5}{8}$	1.719	$2\frac{7}{16}$	2.356	2.815	2.686	$1\frac{1}{16}$	1.100	0.962	0.125
$1\frac{3}{4}$	1.844	$2\frac{5}{8}$	2.538	3.031	2.893	$1\frac{5}{32}$	1.196	1.054	0.125
$1\frac{7}{8}$	1.969	$2\frac{13}{16}$	2.719	3.248	3.100	$1\frac{7}{32}$	1.260	1.114	0.125
2	2.094	3	2.900	3.464	3.306	$1\frac{11}{32}$	1.388	1.175	0.125
$2\frac{1}{4}$	2.375	$3\frac{3}{8}$	3.262	3.897	3.719	$1\frac{1}{2}$	1.548	1.327	0.188
$2\frac{1}{2}$	2.625	$3\frac{3}{4}$	3.625	4.330	4.133	$1\frac{21}{32}$	1.708	1.479	0.188
$2\frac{3}{4}$	2.875	$4\frac{1}{8}$	3.988	4.763	4.546	$1\frac{13}{16}$	1.869	1.632	0.188
3	3.125	$4\frac{1}{2}$	4.350	5.196	4.959	2	2.060	1.815	0.188
$3\frac{1}{4}$	3.438	$4\frac{7}{8}$	4.712	5.629	5.372	$2\frac{3}{16}$	2.251	1.936	0.188
$3\frac{1}{2}$	3.688	$5\frac{1}{4}$	5.075	6.062	5.786	$2\frac{5}{16}$	2.380	2.057	0.188
$3\frac{3}{4}$	3.938	$5\frac{5}{8}$	5.437	6.495	6.198	$2\frac{1}{2}$	2.572	2.241	0.188
4	4.188	6	5.800	6.928	6.612	$2\frac{11}{16}$	2.764	2.424	0.188

ASA B18.2-1952. Published by ASME.
Bolt is not finished on any surface.

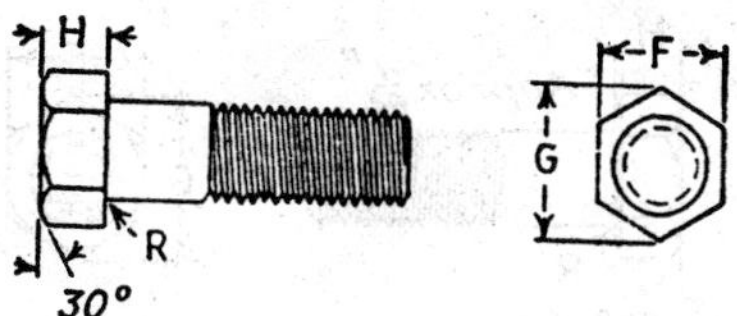

TABLE 37. HEAVY HEXAGON BOLTS

Nominal Size	Body Dia, Max	Width across Flats F		Width across Corners G		Height H			Fillet Radius R
		Max (Basic)	Min	Max	Min	Nom	Max	Min	Max
1/2	**0.530**	**7/8**	**0.850**	**1.010**	**0.969**	**7/16**	**0.458**	**0.386**	**0.031**
5/8	**0.675**	**1 1/16**	**1.031**	**1.227**	**1.175**	**17/32**	**0.553**	**0.478**	**0.062**
3/4	**0.800**	**1 1/4**	**1.212**	**1.443**	**1.383**	**5/8**	**0.649**	**0.570**	**0.062**
5/8	**0.938**	**1 7/16**	**1.394**	**1.660**	**1.589**	**23/32**	**0.745**	**0.662**	**0.062**
1	**1.063**	**1 5/8**	**1.575**	**1.876**	**1.796**	**13/16**	**0.840**	**0.722**	**0.062**
1 1/8	**1.188**	**1 13/16**	**1.756**	**2.093**	**2.002**	**29/32**	**0.936**	**0.814**	**0.125**
1 1/4	**1.313**	**2**	**1.938**	**2.309**	**2.209**	**1**	**1.032**	**0.906**	**0.125**
1 3/8	**1.469**	**2 3/16**	**2.119**	**2.526**	**2.416**	**1 3/32**	**1.128**	**0.997**	**0.125**
1 1/2	**1.594**	**2 3/8**	**2.300**	**2.742**	**2.622**	**1 3/16**	**1.224**	**1.089**	**0.125**
1 5/8	1.719	2 9/16	2.481	2.959	2.828	1 9/32	1.319	1.181	0.125
1 3/4	**1.844**	**2 3/4**	**2.662**	**3.175**	**3.035**	**1 3/8**	**1.415**	**1.272**	**0.125**
1 7/8	1.969	2 15/16	2.844	3.392	3.242	1 15/32	1.511	1.364	0.125
2	**2.094**	**3 1/8**	**3.025**	**3.608**	**3.449**	**1 9/16**	**1.606**	**1.394**	**0.125**
2 1/4	2.375	3 1/2	3.388	4.041	3.862	1 3/4	1.798	1.577	0.188
2 1/2	2.625	3 7/8	3.750	4.474	4.275	1 15/16	1.990	1.760	0.188
2 3/4	2.875	4 1/4	4.112	4.907	4.688	2 1/8	2.181	1.944	0.188
3	3.125	4 5/8	4.475	5.340	5.102	2 5/16	2.373	2.128	0.188

ASA B18.2-1952. Published by ASME.
Bold type indicates products unified dimensionally with British and Canadian standards.
Bolt is not finished on any surface.

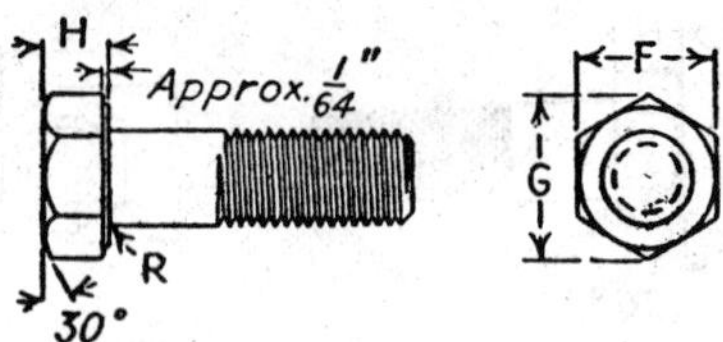

Table 38. Regular Semifinished Hexagon Bolts

Nominal Size	Body Dia, Max	Width across Flats F		Width across Corners G		Height H			Radius of Fillet R	
		Max (Basic)	Min	Max	Min	Nominal	Max	Min	Max	Min
1/4	0.280	7/16	0.425	0.505	0.484	5/32	0.163	0.150	0.031	0.016
5/16	0.342	1/2	0.484	0.577	0.552	13/64	0.211	0.195	0.031	0.016
3/8	0.405	9/16	0.544	0.650	0.620	15/64	0.243	0.226	0.031	0.016
7/16	0.468	5/8	0.603	0.722	0.687	9/32	0.291	0.272	0.031	0.016
1/2	0.530	3/4	0.725	0.866	0.826	5/16	0.323	0.302	0.031	0.016
5/8	0.675	15/16	0.906	1.083	1.033	25/64	0.403	0.378	0.031	0.016
3/4	0.800	1 1/8	1.088	1.299	1.240	15/32	0.483	0.455	0.047	0.031
7/8	0.938	1 5/16	1.269	1.516	1.447	35/64	0.563	0.531	0.047	0.031
1	1.063	1 1/2	1.450	1.732	1.653	39/64	0.627	0.591	0.047	0.031
1 1/8	1.188	1 11/16	1.631	1.949	1.859	11/16	0.718	0.658	0.062	0.047
1 1/4	1.313	1 7/8	1.812	2.165	2.066	25/32	0.813	0.749	0.062	0.047
1 3/8	1.469	2 1/16	1.994	2.382	2.273	27/32	0.878	0.810	0.062	0.047
1 1/2	1.594	2 1/4	2.175	2.598	2.480	15/16	0.974	0.902	0.062	0.047
1 5/8	1.719	2 7/16	2.356	2.815	2.686	1	1.038	0.962	0.062	0.047
1 3/4	1.844	2 5/8	2.538	3.031	2.893	1 3/32	1.134	1.054	0.062	0.047
1 7/8	1.969	2 13/16	2.719	3.248	3.100	1 5/32	1.198	1.114	0.062	0.047
2	2.094	3	2.900	3.464	3.306	1 7/32	1.263	1.175	0.062	0.047
2 1/4	2.375	3 3/8	3.262	3.897	3.719	1 3/8	1.423	1.327	0.062	0.047
2 1/2	2.625	3 3/4	3.625	4.330	4.133	1 17/32	1.583	1.479	0.062	0.047
2 3/4	2.875	4 1/8	3.988	4.763	4.546	1 11/16	1.744	1.632	0.062	0.047
3	3.125	4 1/2	4.350	5.196	4.959	1 7/8	1.935	1.815	0.062	0.047
3 1/4	3.438	4 7/8	4.712	5.629	5.372	2	2.064	1.936	0.062	0.047
3 1/2	3.688	5 1/4	5.075	6.062	5.786	2 1/8	2.193	2.057	0.062	0.047
3 3/4	3.938	5 5/8	5.437	6.495	6.198	2 5/16	2.385	2.241	0.062	0.047
4	4.188	6	5.800	6.928	6.612	1 1/2	2.576	2.424	0.062	0.047

ASA B18.2-1952. Published by ASME.
Semifinished bolt is processed to produce a flat bearing surface under head only.

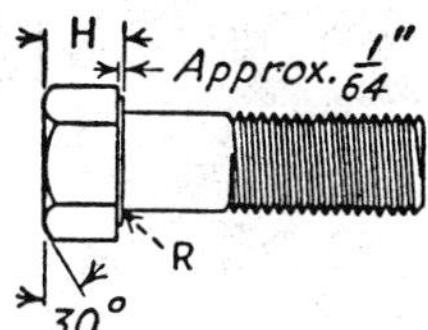

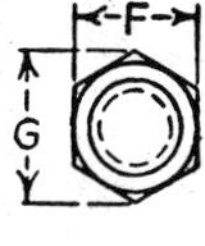

TABLE 39. HEAVY SEMIFINISHED HEXAGON BOLTS

Nominal Size	Body Dia, Max	Width across Flats F		Width across Corners G		Height H			Radius of Fillet R	
		Max (Basic)	Min	Max	Min	Nominal	Max	Min	Max	Min
1/2	**0.530**	**7/8**	**0.850**	**1.010**	**0.969**	**13/32**	**0.426**	**0.386**	**0.031**	**0.016**
5/8	**0.675**	**1 1/16**	**1.031**	**1.227**	**1.175**	**1/2**	**0.522**	**0.478**	**0.031**	**0.016**
3/4	**0.800**	**1 1/4**	**1.212**	**1.443**	**1.383**	**19/32**	**0.618**	**0.570**	**0.047**	**0.031**
7/8	**0.938**	**1 7/16**	**1.394**	**1.660**	**1.589**	**11/16**	**0.714**	**0.662**	**0.047**	**0.031**
1	**1.063**	**1 5/8**	**1.575**	**1.876**	**1.796**	**3/4**	**0.778**	**0.722**	**0.047**	**0.031**
1 1/8	**1.188**	**1 13/16**	**1.756**	**2.093**	**2.002**	**27/32**	**0.874**	**0.814**	**0.062**	**0.047**
1 1/4	**1.313**	**2**	**1.938**	**2.309**	**2.209**	**15/16**	**0.970**	**0.906**	**0.062**	**0.047**
1 3/8	**1.469**	**2 3/16**	**2.119**	**2.526**	**2.416**	**1 1/32**	**1.065**	**0.997**	**0.062**	**0.047**
1 1/2	**1.594**	**2 3/8**	**2.300**	**2.742**	**2.622**	**1 1/8**	**1.161**	**1.089**	**0.062**	**0.047**
$1\frac{5}{8}$	1.719	$2\frac{9}{16}$	2.481	2.959	2.828	$1\frac{7}{32}$	1.257	1.181	0.062	0.047
1 3/4	**1.844**	**2 3/4**	**2.662**	**3.175**	**3.035**	**1 5/16**	**1.352**	**1.272**	**0.062**	**0.047**
$1\frac{7}{8}$	1.969	$2\frac{15}{16}$	2.844	3.392	3.242	$1\frac{13}{32}$	1.448	1.364	0.062	0.047
2	**2.094**	**3 1/8**	**3.025**	**3.608**	**3.449**	**1 7/16**	**1.482**	**1.394**	**0.062**	**0.047**
$2\frac{1}{4}$	2.375	$3\frac{1}{2}$	3.388	4.041	3.862	$1\frac{5}{8}$	1.673	1.577	0.062	0.047
$2\frac{1}{2}$	2.625	$3\frac{7}{8}$	3.750	4.474	4.275	$1\frac{13}{16}$	1.864	1.760	0.062	0.047
$2\frac{3}{4}$	2.875	$4\frac{1}{4}$	4.112	4.907	4.688	2	2.056	1.944	0.062	0.047
3	3.125	$4\frac{5}{8}$	4.475	5.340	5.102	$2\frac{3}{16}$	2.248	2.128	0.062	0.047

ASA B18.2-1952. Published by ASME.
Bold type indicates products unified dimensionally with British and Canadian standards.
Semifinished bolt is processed to produce a flat bearing surface under head only.

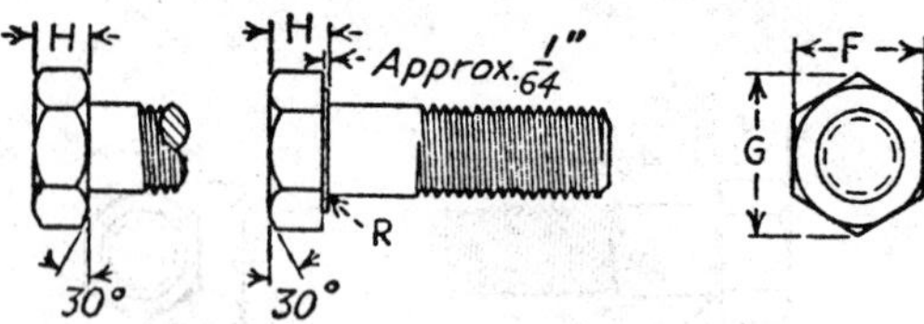

TABLE 40. FINISHED HEXAGON BOLTS

Nominal Size	Body Dia, Min	Width across Flats F		Width across Corners G		Height H			Radius R	
		Max (Basic)	Min	Max	Min	Nominal	Max	Min	Max	Min
1/4	**0.2450**	**7/16**	**0.428**	**0.505**	**0.488**	**5/32**	**0.163**	**0.150**	**0.023**	**0.009**
5/16	**0.3065**	**1/2**	**0.489**	**0.577**	**0.557**	**13/64**	**0.211**	**0.195**	**0.023**	**0.009**
3/8	**0.3690**	**9/16**	**0.551**	**0.650**	**0.628**	**15/64**	**0.243**	**0.226**	**0.023**	**0.009**
7/16	**0.4305**	**5/8**	**0.612**	**0.722**	**0.698**	**9/32**	**0.291**	**0.272**	**0.023**	**0.009**
1/2	**0.4930**	**3/4**	**0.736**	**0.866**	**0.840**	**5/16**	**0.323**	**0.302**	**0.023**	**0.009**
9/16	**0.5545**	**13/16**	**0.798**	**0.938**	**0.910**	**23/64**	**0.371**	**0.348**	**0.041**	**0.021**
5/8	**0.6170**	**15/16**	**0.922**	**1.083**	**1.051**	**25/64**	**0.403**	**0.378**	**0.041**	**0.021**
3/4	**0.7410**	**1 1/8**	**1.100**	**1.299**	**1.254**	**15/32**	**0.483**	**0.455**	**0.041**	**0.021**
7/8	**0.8660**	**1 5/16**	**1.285**	**1.516**	**1.465**	**35/64**	**0.563**	**0.531**	**0.062**	**0.047**
1	**0.9900**	**1 1/2**	**1.469**	**1.732**	**1.675**	**39/64**	**0.627**	**0.591**	**0.062**	**0.047**
1 1/8	1.1140	1 11/16	1.631	1.949	1.859	11/16	0.718	0.658	0.125	0.110
1 1/4	1.2390	1 7/8	1.812	2.165	2.066	25/32	0.813	0.749	0.125	0.110
1 3/8	1.3630	2 1/16	1.994	2.382	2.273	27/32	0.878	0.810	0.125	0.110
1 1/2	1.4880	2 1/4	2.175	2.598	2.480	15/16	0.974	0.902	0.125	0.110
1 5/8	1.6130	2 7/16	2.356	2.815	2.686	1	1.038	0.962	0.125	0.110
1 3/4	1.7380	2 5/8	2.538	3.031	2.893	1 3/32	1.134	1.054	0.125	0.110
1 7/8	1.8630	2 13/16	2.719	3.248	3.100	1 5/32	1.198	1.114	0.125	0.110
2	1.9880	3	2.900	3.464	3.306	1 7/32	1.263	1.175	0.125	0.110
2 1/4	2.2380	3 3/8	3.262	3.897	3.719	1 3/8	1.423	1.327	0.188	0.173
2 1/2	2.4880	3 3/4	3.625	4.330	4.133	1 17/32	1.583	1.479	0.188	0.173
2 3/4	2.7380	4 1/8	3.988	4.763	4.546	1 11/16	1.744	1.632	0.188	0.173
3	2.9880	4 1/2	4.350	5.196	4.959	1 7/8	1.935	1.815	0.188	0.173

ASA B18.2-1952. Published by ASME.
Maximum body dia equals nominal size.
Bold type indicates products unified dimensionally with British and Canadian standards.
"Finished" in the title refers to the quality of manufacture and the closeness of tolerance and does not indicate that surfaces are completely machined.

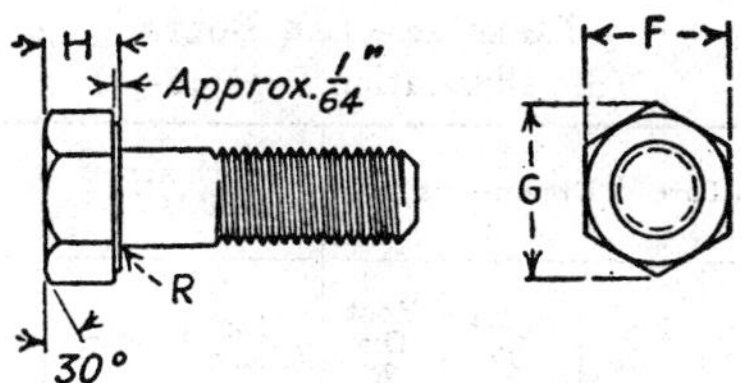

TABLE 41. HEAVY FINISHED HEXAGON BOLTS

Nominal Size	Body Dia, Min	Width across Flats F		Width across Corners G		Height H			Radius R	
		Max (Basic)	Min	Max	Min	Nominal	Max	Min	Max	Min
1/2	**0.4940**	**7/8**	**0.850**	**0.010**	**0.969**	**13/32**	**0.426**	**0.386**	**0.031**	**0.016**
5/8	**0.6190**	**1 1/16**	**1.031**	**1.227**	**1.175**	**1/2**	**0.522**	**0.478**	**0.031**	**0.016**
3/4	**0.7440**	**1 1/4**	**1.212**	**1.443**	**1.383**	**19/32**	**0.618**	**0.570**	**0.047**	**0.031**
7/8	**0.8690**	**1 7/16**	**1.394**	**1.660**	**1.589**	**11/16**	**0.714**	**0.662**	**0.047**	**0.031**
1	**0.9940**	**1 5/8**	**1.575**	**1.876**	**1.796**	**3/4**	**0.778**	**0.722**	**0.047**	**0.031**
1 1/8	**1.1170**	**1 13/16**	**1.756**	**2.093**	**2.002**	**27/32**	**0.874**	**0.814**	**0.062**	**0.047**
1 1/4	**1.2420**	**2**	**1.938**	**2.309**	**2.209**	**15/16**	**0.970**	**0.906**	**0.062**	**0.047**
1 3/8	**1.3670**	**2 3/16**	**2.119**	**2.526**	**2.416**	**1 1/32**	**1.065**	**0.997**	**0.062**	**0.047**
1 1/2	**1.4920**	**2 3/8**	**2.300**	**2.742**	**2.622**	**1 1/8**	**1.161**	**1.089**	**0.062**	**0.047**
1 5/8	1.6170	2 9/16	2.481	2.959	2.828	1 7/32	1.257	1.181	0.062	0.047
1 3/4	**1.7420**	**2 3/4**	**2.662**	**3.175**	**3.035**	**1 5/16**	**1.352**	**1.272**	**0.062**	**0.047**
1 7/8	1.8670	2 15/16	2.844	3.392	3.242	1 13/32	1.448	1.364	0.062	0.047
2	**1.9900**	**3 1/8**	**3.025**	**3.608**	**3.449**	**1 7/16**	**1.482**	**1.394**	**0.062**	**0.047**
2 1/4	2.2400	3 1/2	3.388	4.041	3.862	1 5/8	1.673	1.577	0.062	0.047
2 1/2	2.4900	3 7/8	3.750	4.474	4.275	1 13/16	1.864	1.760	0.062	0.047
2 3/4	2.7400	4 1/4	4.112	4.907	4.688	2	2.056	1.944	0.062	0.047
3	2.9900	4 5/8	4.475	5.340	5.102	2 3/16	2.248	2.128	0.062	0.047

ASA B18.2-1952. Published by ASME.
Bold type indicates products unified dimensionally with British and Canadian standards.
"Finished" in the title refers to the quality of manufacture and the closeness of tolerance and does not indicate that surfaces are completely machined.

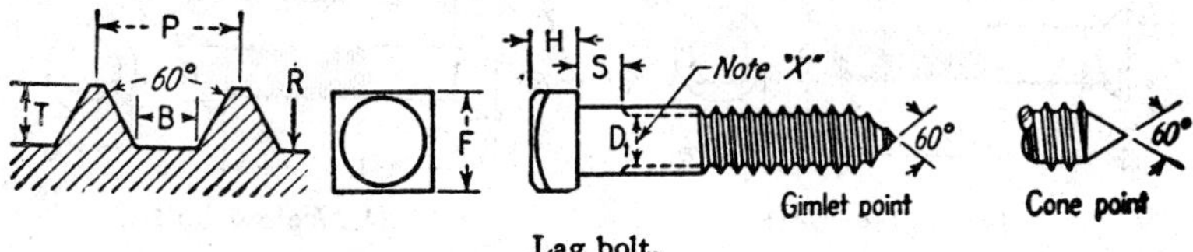

Lag bolt.

TABLE 42. LAG BOLTS

(See Illustration Page 36-51)

Bolt Dia	Threads per In.	Thread Dimensions				Width across Flats F		Height H		Shoulder S*
		Pitch P	Flat B	Depth T	Root Dia R	Max (Basic)	Min	Nominal	Max	Min
No. 10	11	0.091	0.039	0.035	0.120	9/32	0.271	1/8	0.140	0.094
1/4	10	0.100	0.043	0.039	0.173	3/8	0.362	11/64	0.188	0.094
5/16	9	0.111	0.048	0.043	0.227	1/2	0.484	13/64	0.220	0.125
3/8	7	0.143	0.062	0.055	0.265	9/16	0.544	1/4	0.268	0.125
7/16	7	0.143	0.062	0.055	0.328	5/8	0.603	19/64	0.316	0.156
1/2	6	0.167	0.072	0.064	0.371	3/4	0.725	21/64	0.348	0.156
5/8	5	0.200	0.086	0.077	0.471	15/16	0.906	27/64	0.444	0.312
3/4	4 1/2	0.222	0.096	0.085	0.579	1 1/8	1.088	1/2	0.524	0.375
7/8	4	0.250	0.108	0.096	0.683	1 5/16	1.269	19/32	0.620	0.375
1	3 1/2	0.286	0.123	0.110	0.780	1 1/2	1.450	21/32	0.684	0.625
1 1/8	3 1/4	0.308	0.133	0.119	0.887	1 11/16	1.631	3/4	0.780	0.625
1 1/4	3 1/4	0.308	0.133	0.119	1.012	1 7/8	1.812	27/32	0.876	0.625

ASA B18.2-1952. Published by ASME.
Thread formulas: Pitch = 1 ÷ No. threads per in. Flat at root = pitch × 0.4305.
Depth of single thread = pitch × 0.385.
X. Threads shall be rolled or cut. Diameter of rolled thread shank is not dimensioned but is indicated by D_1.

* Length of shoulder for rolled thread.

PLOW BOLTS

By a lengthy process of simplification, plow bolts have been standardized in the No. 3 and No 7 types (Table 43). These two types have, respectively, a round head with a square neck and a round head with reverse key. A No. 4 square head and a No. 6 round head with heavy key (not shown) are not standard plow bolts but will be found in the appendix to ASA B19.9-1950.

The No. 3 round countersunk head (80°) with square neck is used in steel parts where the holes are either dry sand or green sand cored. Since the height of the square section is added to the carrying height of the conical section, the No. 3 plow bolt is not well suited for very thin materials, and the square weakens the bolted part more than a round hole.

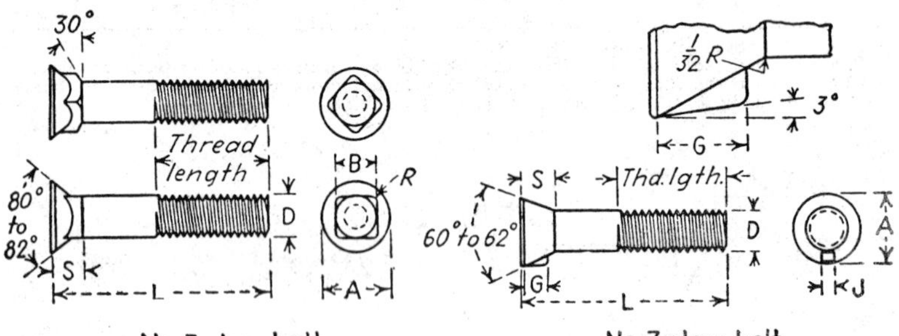

Plow bolt.

The No. 7 countersunk head (60°) with reverse key can be used in thinner parts than the No. 3. The key insures replacement of the bolt in its original position, and the key may be used opposite a narrow edge and avoids strength reduction at that point.

TABLE 43. PLOW BOLTS

No. 3 Regular Head—Round, Countersunk, Square Neck

D	*A*			*F*	*S*		*B*		
Nominal Dia	Dia of Head			Feed Thickness Max	Depth of Square and Head		Width of Square		*R*
	Max	Min Sharp	Min with Flat		Max	Min	Max	Min (Basic)	
5/16	0.605	0.578	0.563	0.025	0.269	0.243	0.325	0.313	1/32
3/8	0.708	0.671	0.656	0.031	0.312	0.281	0.387	0.375	3/64
7/16	0.826	0.781	0.766	0.036	0.364	0.328	0.450	0.438	3/64
1/2	0.945	0.890	0.875	0.042	0.417	0.375	0.515	0.500	3/64
9/16*	1.045	1.000	0.969	0.045	0.461	0.416	0.578	0.563	5/64
5/8	1.147	1.094	1.063	0.050	0.506	0.456	0.640	0.625	5/64
3/4	1.303	1.250	1.219	0.050	0.541	0.491	0.765	0.750	5/64
7/8	1.512	1.469	1.406	0.063	0.626	0.563	0.906	0.875	3/32
1	1.700	1.656	1.594	0.063	0.690	0.627	1.031	1.000	3/32

No. 7 Regular Head—Round, Countersunk, Reverse Key

D	*A*			*F*	*S*		*J*		*G*	
Nominal Dia	Dia of Head			Feed Thickness Max	Head Height		Width of Key		Key Length	
	Max	Min Sharp	Min with Flat		Max	Min	Max	Min	Min	Max
5/16	0.592	0.578	0.563	0.025	0.233	0.208	0.156	0.151	0.185	0.200
3/8	0.661	0.640	0.625	0.031	0.239	0.208	0.156	0.151	0.187	0.202
7/16	0.776	0.749	0.734	0.036	0.282	0.246	0.156	0.151	0.227	0.242
1/2	0.892	0.859	0.844	0.042	0.328	0.286	0.156	0.151	0.267	0.282
9/16*	1.021	0.984	0.969	0.045	0.383	0.338	0.156	0.151	0.321	0.336
5/8	1.121	1.078	1.063	0.050	0.414	0.364	0.156	0.151	0.348	0.363
3/4	1.277	1.234	1.219	0.050	0.440	0.390	0.156	0.151	0.375	0.390

ASA B18.9-1950. Published by ASME.

* This size is not recommended.

TABLE 44. SQUARE-NECK CARRIAGE BOLTS

Nominal Dia *D*	*A*, Max	*H*, Max	*P*: For Bolt Lengths	*P*: Min	*P*: Max	*B*, Max
No. 10	0.469	0.114	1⅛ and shorter	0.094	0.125	0.199
			1¼ and longer	0.188	0.219	
¼	0.594	0.145	1¼ and shorter	0.125	0.156	0.260
			1⅜ and longer	0.219	0.250	
5/16	0.719	0.176	1¼ and shorter	0.156	0.187	0.324
			1⅜ and longer	0.250	0.281	
⅜	0.844	0.208	1½ and shorter	0.188	0.219	0.388
			1⅝ and longer	0.281	0.312	
7/16	0.969	0.239	1½ and shorter	0.219	0.250	0.452
			1⅝ and longer	0.313	0.344	
½	1.094	0.270	1⅞ and shorter	0.250	0.281	0.515
			2 and longer	0.344	0.375	
9/16	1.219	0.312	1⅞ and shorter	0.281	0.312	0.579
			2 and longer	0.375	0.406	
⅝	1.344	0.344	1⅞ and shorter	0.313	0.344	0.642
			2 and longer	0.406	0.437	
¾	1.594	0.406	1⅞ and shorter	0.375	0.406	0.768
			2 and longer	0.469	0.500	
⅞	1.844	0.469	1⅞ and shorter	0.438	0.469	0.895
			2 and longer	0.531	0.562	
1	2.094	0.531	1⅞ and shorter	0.500	0.531	1.022
			2 and longer	0.594	0.625	

ASA B18.5-1952.

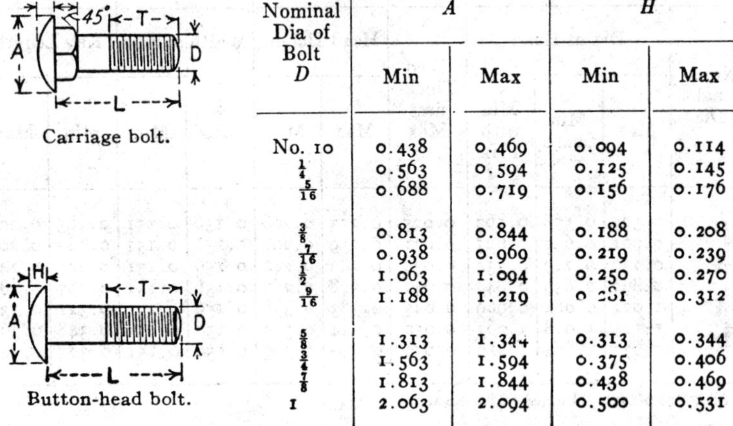

Carriage bolt.

Button-head bolt.

TABLE 45. BUTTON-HEAD BOLTS

Nominal Dia of Bolt *D*	*A*: Min	*A*: Max	*H*: Min	*H*: Max
No. 10	0.438	0.469	0.094	0.114
¼	0.563	0.594	0.125	0.145
5/16	0.688	0.719	0.156	0.176
⅜	0.813	0.844	0.188	0.208
7/16	0.938	0.969	0.219	0.239
½	1.063	1.094	0.250	0.270
9/16	1.188	1.219	0.281	0.312
⅝	1.313	1.344	0.313	0.344
¾	1.563	1.594	0.375	0.406
⅞	1.813	1.844	0.438	0.469
1	2.063	2.094	0.500	0.531

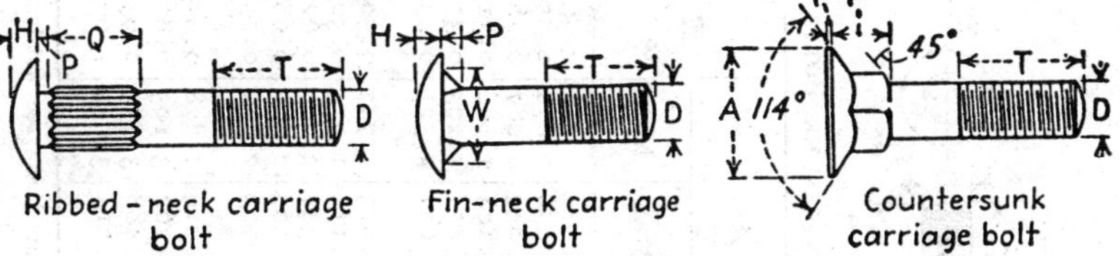

Ribbed-neck carriage bolt

Fin-neck carriage bolt

Countersunk carriage bolt

TABLE 46. DIMENSIONS OF RIBBED-NECK CARRIAGE BOLT

Nominal Dia D	A, Max	H, Max	P		Q			No. of Ribs
			For $L = \frac{7}{8}$ or Less	For $L = 1$ or More	$L = \frac{7}{8}$ or Less	$L = 1$ and $1\frac{1}{8}$	$L = 1\frac{1}{4}$ or More	
No. 10	0.469	0.114	0.031	0.063	0.188	0.313	0.500	9
$\frac{1}{4}$	0.594	0.145	0.031	0.063	0.188	0.313	0.500	10
$\frac{5}{16}$	0.719	0.176	0.031	0.063	0.188	0.313	0.500	12
$\frac{3}{8}$	0.844	0.208	0.031	0.063	0.188	0.313	0.500	12
$\frac{7}{16}$	0.969	0.239	0.031	0.063	0.188	0.313	0.500	14
$\frac{1}{2}$	1.094	0.270	0.031	0.063	0.188	0.313	0.500	16
$\frac{9}{16}$	1.219	0.312	0.094	0.094	0.188	0.313	0.500	18
$\frac{5}{8}$	1.344	0.344	0.094	0.094	0.188	0.313	0.500	19
$\frac{3}{4}$	1.594	0.406	0.094	0.094	0.188	0.313	0.500	22

ASA B18.5-1952.

TABLE 47. DIMENSIONS OF FIN-NECK CARRIAGE BOLT

Nominal Dia D	A, Max	H, Max	P, Max	W		M	
				Min	Max	Min	Max
No. 10	0.469	0.114	0.088	0.375	0.395	0.078	0.098
$\frac{1}{4}$	0.594	0.145	0.104	0.438	0.458	0.094	0.114
$\frac{5}{16}$	0.719	0.176	0.135	0.531	0.551	0.125	0.145
$\frac{3}{8}$	0.844	0.208	0.151	0.625	0.645	0.141	0.161
$\frac{7}{16}$	0.969	0.239	0.182	0.719	0.739	0.172	0.192
$\frac{1}{2}$	1.094	0.270	0.198	0.813	0.833	0.188	0.208

TABLE 48. DIMENSIONS OF COUNTERSUNK CARRIAGE BOLT

Nominal Dia of Bolt D	Dia of Head A		Feed Thickness F	Depth of Square and Countersink P		Width of Square B	
	Min	Max		Min	Max	Min	Max
No. 10	$\frac{1}{2}$ 0.500	0.520	0.016	$\frac{7}{32}$ 0.219	0.250	0.185	0.199
$\frac{1}{4}$	$\frac{5}{8}$ 0.625	0.645	0.016	$\frac{9}{32}$ 0.281	0.312	0.245	0.260
$\frac{5}{16}$	$\frac{3}{4}$ 0.750	0.770	0.031	$\frac{11}{32}$ 0.344	0.375	0.307	0.324
$\frac{3}{8}$	$\frac{7}{8}$ 0.875	0.895	0.031	$\frac{13}{32}$ 0.406	0.437	0.368	0.388
$\frac{7}{16}$	1 1.000	1.020	0.031	$\frac{15}{32}$ 0.469	0.500	0.431	0.452
$\frac{1}{2}$	$1\frac{1}{8}$ 1.125	1.145	0.031	$\frac{17}{32}$ 0.531	0.562	0.492	0.515
$\frac{9}{16}$	$1\frac{1}{4}$ 1.250	1.275	0.031	$\frac{19}{32}$ 0.594	0.625	0.554	0.579
$\frac{5}{8}$	$1\frac{3}{8}$ 1.375	1.400	0.031	$\frac{21}{32}$ 0.656	0.687	0.616	0.642
$\frac{3}{4}$	$1\frac{5}{8}$ 1.625	1.650	0.047	$\frac{25}{32}$ 0.781	0.812	0.741	0.768

Table 49. Step Bolts (Also Known as Oval-head Elevator Bolts)

Nominal Dia D	A		H		P			B	
	Min	Max	Min	Max	For Bolt Lengths	Min	Max	Min	Max
No. 10	0.625	0.656	0.094	0.114	$1\frac{1}{8}$ and shorter	0.094	0.125	0.185	0.199
					$1\frac{1}{4}$ and longer	0.188	0.219		
$\frac{1}{4}$	0.813	0.844	0.125	0.145	$1\frac{1}{4}$ and shorter	0.125	0.156	0.245	0.260
					$1\frac{3}{8}$ and longer	0.219	0.250		
$\frac{5}{16}$	1.000	1.031	0.156	0.176	$1\frac{1}{4}$ and shorter	0.156	0.187	0.307	0.324
					$1\frac{3}{8}$ and longer	0.250	0.281		
$\frac{3}{8}$	1.188	1.219	0.188	0.208	$1\frac{1}{2}$ and shorter	0.188	0.219	0.368	0.388
					$1\frac{5}{8}$ and longer	0.281	0.312		
$\frac{7}{16}$	1.375	1.406	0.219	0.239	$1\frac{1}{2}$ and shorter	0.219	0.250	0.431	0.452
					$1\frac{5}{8}$ and longer	0.313	0.344		
$\frac{1}{2}$	1.563	1.594	0.250	0.270	$1\frac{7}{8}$ and shorter	0.250	0.281	0.492	0.515
					2 and longer	0.344	0.375		

ASA B18.5-1952.

Table 50. Countersunk Bolts

Nominal Dia D	A			H
	Basic	Max	Min	
$\frac{1}{2}$	0.905	0.936	0.874	0.250
$\frac{5}{16}$	1.018	1.049	0.987	0.281
$\frac{5}{8}$	1.131	1.194	1.068	0.313
$\frac{3}{4}$	1.358	1.421	1.295	0.375
$\frac{7}{8}$	1.584	1.647	1.521	0.438
1	1.810	1.873	1.747	0.500
$1\frac{1}{8}$	2.036	2.114	1.973	0.563
$1\frac{1}{4}$	2.263	2.341	2.200	0.625
$1\frac{3}{8}$	2.489	2.567	2.426	0.688
$1\frac{1}{2}$	2.715	2.793	2.652	0.750
$1\frac{5}{8}$	2.941	3.019	2.878	0.813
$1\frac{3}{4}$	3.168	3.262	3.105	0.875
$1\frac{7}{8}$	3.394	3.488	3.425	0.938
2	3.620	3.714	3.651	1.000

For sizes smaller than $\frac{1}{2}$ in., see flat-head capscrews.

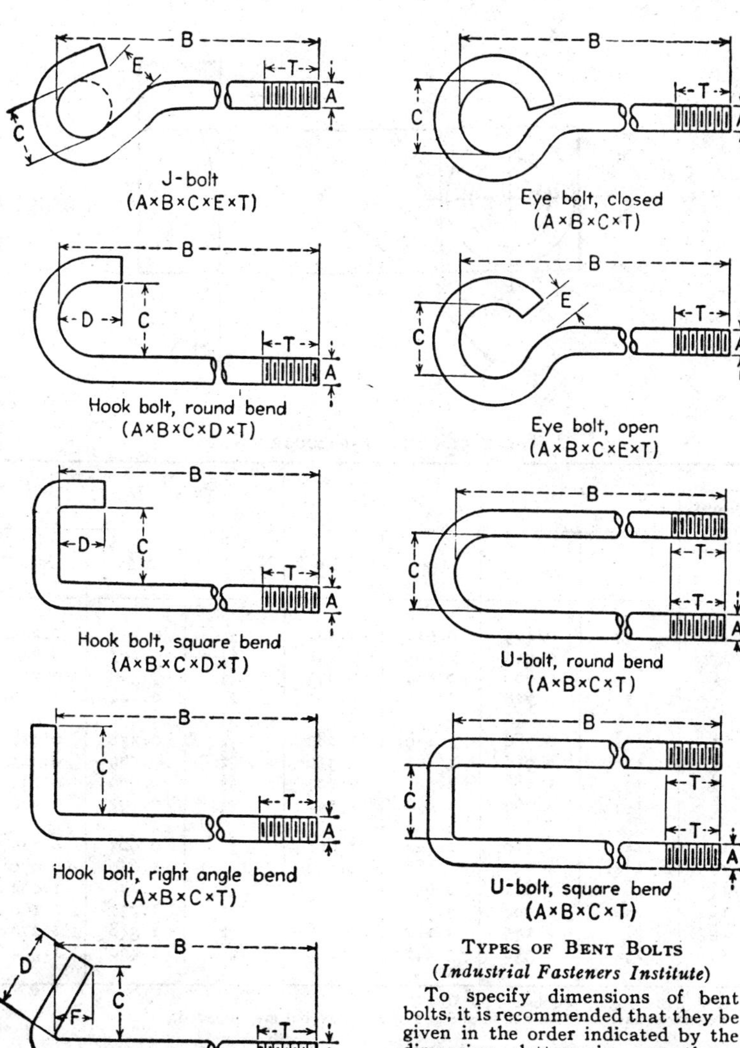

Types of Bent Bolts

(*Industrial Fasteners Institute*)

To specify dimensions of bent bolts, it is recommended that they be given in the order indicated by the dimension letters in parentheses below the legend for each bolt.

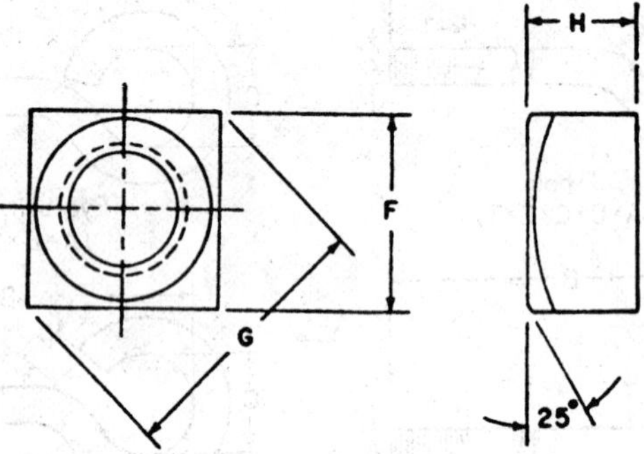

TABLE 51. REGULAR SQUARE NUTS

Nominal Size	F		G		H		
	Max	Min	Max	Min	Nominal	Max	Min
1/4	7/16	0.425	0.619	0.584	7/32	0.235	0.203
5/16	9/16	0.547	0.795	0.751	17/64	0.283	0.249
3/8	5/8	0.606	0.884	0.832	21/64	0.346	0.310
7/16	3/4	0.728	1.061	1.000	3/8	0.394	0.356
1/2	13/16	0.788	1.149	1.082	7/16	0.458	0.418
5/8	1	0.969	1.414	1.330	35/64	0.569	0.525
3/4	1 1/8	1.088	1.591	1.494	21/32	0.680	0.632
7/8	1 5/16	1.269	1.856	1.742	49/64	0.792	0.740
1	1 1/2	1.450	2.121	1.991	7/8	0.903	0.847
1 1/8	1 11/16	1.631	2.386	2.239	1	1.030	0.970
1 1/4	1 7/8	1.812	2.652	2.489	1 3/32	1.126	1.062
1 3/8	2 1/16	1.994	2.917	2.738	1 13/64	1.237	1.169
1 1/2	2 1/4	2.175	3.182	2.986	1 5/16	1.348	1.276
1 5/8	2 7/16	2.356	3.447	3.235	1 27/64	1.460	1.384

ASA B18.2-1952. Published by ASME.
Regular square nuts are not finished on any surface but are threaded.

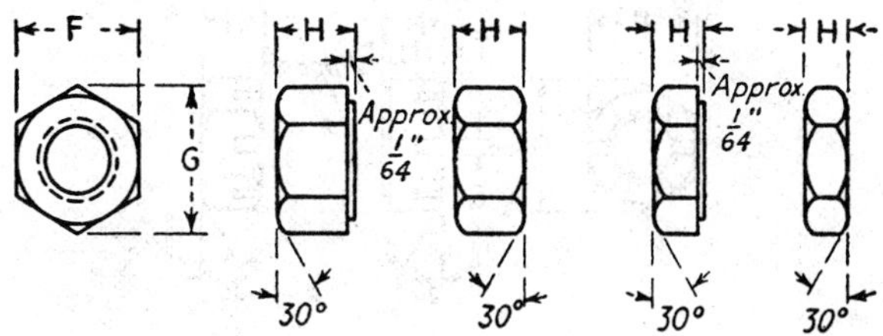

TABLE 52. FINISHED HEXAGON AND HEXAGON-JAM NUTS

Nominal Size	*F*		*G*	Thickness Nuts, *H*		Thickness Jam Nuts, *H*	
	Max	Min	Max	Max	Min	Max	Min
1/4	**7/16**	**0.428**	**0.505**	**0.226**	**0.212**	**0.163**	**0.150**
5/16	**1/2**	**0.489**	**0.577**	**0.273**	**0.258**	**0.195**	**0.180**
3/8	**9/16**	**0.551**	**0.650**	**0.337**	**0.320**	**0.227**	**0.210**
7/16	**11/16**	**0.675**	**0.794**	**0.385**	**0.365**	**0.260**	**0.240**
1/2	**3/4**	**0.736**	**0.866**	**0.448**	**0.427**	**0.323**	**0.302**
9/16	**7/8**	**0.861**	**1.010**	**0.496**	**0.473**	**0.324**	**0.301**
5/8	**15/16**	**0.922**	**1.083**	**0.559**	**0.535**	**0.387**	**0.363**
3/4	**1 1/8**	**1.088**	**1.299**	**0.665**	**0.617**	**0.446**	**0.398**
7/8	**1 5/16**	**1.269**	**1.516**	**0.776**	**0.724**	**0.510**	**0.458**
1	**1 1/2**	**1.450**	**1.732**	**0.887**	**0.831**	**0.575**	**0.519**
$1\frac{1}{8}$	$1\frac{11}{16}$	1.631	1.949	0.999	0.939	0.639	0.579
$1\frac{1}{4}$	$1\frac{7}{8}$	1.812	2.165	1.094	1.030	0.751	0.687
$1\frac{3}{8}$	$2\frac{1}{16}$	1.994	2.382	1.206	1.138	0.815	0.747
$1\frac{1}{2}$	$2\frac{1}{4}$	2.175	2.598	1.317	1.245	0.880	0.808
$1\frac{5}{8}$	$2\frac{7}{16}$	2.356	2.815	1.429	1.353	0.944	0.868
$1\frac{3}{4}$	$2\frac{5}{8}$	2.538	3.031	1.540	1.460	1.009	0.929
$1\frac{7}{8}$	$2\frac{13}{16}$	2.719	3.248	1.651	1.567	1.073	0.989
2	3	2.900	3.464	1.763	1.675	1.138	1.050
$2\frac{1}{4}$	$3\frac{3}{8}$	3.262	3.897	1.970	1.874	1.251	1.155
$2\frac{1}{2}$	$3\frac{3}{4}$	3.625	4.330	2.193	2.089	1.505	1.401
$2\frac{3}{4}$	$4\frac{1}{8}$	3.988	4.763	2.415	2.303	1.634	1.522
3	$4\frac{1}{2}$	4.350	5.196	2.638	2.518	1.763	1.643

ASA B18.2-1952. Published by ASME.
Bold type indicates products unified dimensionally with British and Canadian standards.
"Finished" refers to the quality of manufacture and the closeness of tolerance, and does not indicate that surfaces are completely machined.

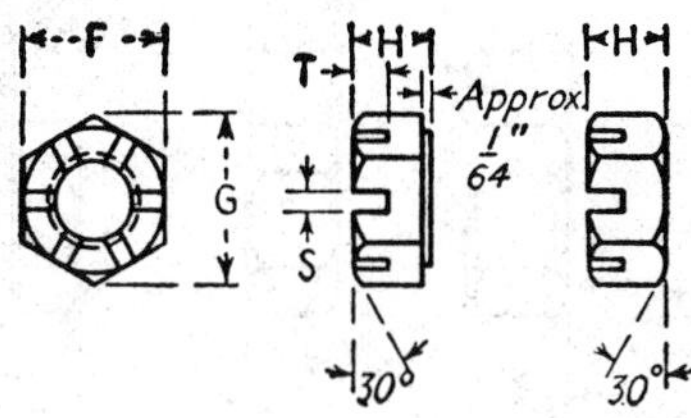

TABLE 53. FINISHED HEXAGON SLOTTED NUTS

Nominal Size	F		G, Max	H		Slot	
	Max (Basic)	Min		Max	Min	S	T
1/4	7/16	0.428	0.505	0.226	0.212	0.078	0.094
5/16	1/2	0.489	0.577	0.273	0.258	0.094	0.094
3/8	9/16	0.551	0.650	0.337	0.320	0.125	0.125
7/16	11/16	0.675	0.794	0.385	0.365	0.125	0.156
1/2	3/4	0.736	0.866	0.448	0.427	0.156	0.156
9/16	7/8	0.861	1.010	0.496	0.473	0.156	0.188
5/8	15/16	0.922	1.083	0.559	0.535	0.188	0.219
3/4	1 1/8	1.088	1.299	0.665	0.617	0.188	0.250
7/8	1 5/16	1.269	1.516	0.776	0.724	0.188	0.250
1	1 1/2	1.450	1.732	0.887	0.831	0.250	0.281
$1\frac{1}{8}$	$1\frac{11}{16}$	1.631	1.949	0.999	0.939	0.250	0.344
$1\frac{1}{4}$	$1\frac{7}{8}$	1.812	2.165	1.094	1.030	0.312	0.375
$1\frac{3}{8}$	$2\frac{1}{16}$	1.994	2.382	1.206	1.138	0.312	0.375
$1\frac{1}{2}$	$2\frac{1}{4}$	2.175	2.598	1.317	1.245	0.375	0.438
$1\frac{5}{8}$	$2\frac{7}{16}$	2.356	2.815	1.429	1.353	0.375	0.438
$1\frac{3}{4}$	$2\frac{5}{8}$	2.538	3.031	1.540	1.460	0.438	0.500
$1\frac{7}{8}$	$2\frac{13}{16}$	2.719	3.248	1.651	1.567	0.438	0.562
2	3	2.900	3.464	1.763	1.675	0.438	0.562
$2\frac{1}{4}$	$3\frac{3}{8}$	3.262	3.897	1.970	1.874	0.438	0.562
$2\frac{1}{2}$	$3\frac{3}{4}$	3.625	4.330	2.193	2.089	0.562	0.688
$2\frac{3}{4}$	$4\frac{1}{8}$	3.988	4.763	2.415	2.303	0.562	0.688
3	$4\frac{1}{2}$	4.350	5.196	2.638	2.518	0.625	0.750

ASA B18.2-1952. Published by ASME.
Bold type indicates products unified dimensionally with British and Canadian standards.

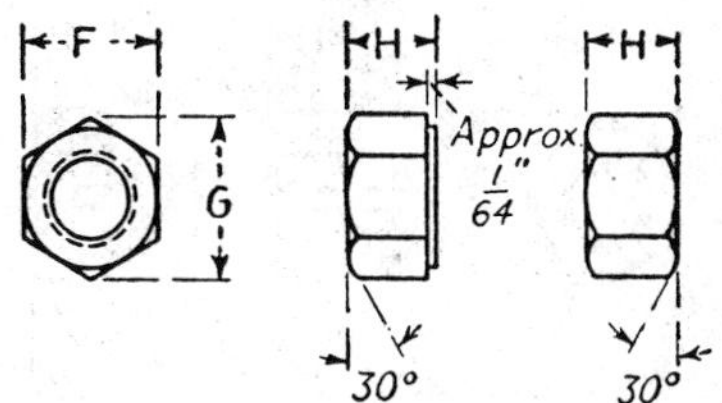

TABLE 54. FINISHED HEXAGON THICK NUTS

Nominal Size	Width across Flats F		Width across Corners G		Thickness H		
	Max (Basic)	Min	Max	Min	Nominal	Max	Min
$\frac{1}{4}$	$\frac{7}{16}$	0.428	0.505	0.488	$\frac{9}{32}$	0.288	0.274
$\frac{5}{16}$	$\frac{1}{2}$	0.489	0.577	0.557	$\frac{21}{64}$	0.336	0.320
$\frac{3}{8}$	$\frac{9}{16}$	0.551	0.650	0.628	$\frac{13}{32}$	0.415	0.398
$\frac{7}{16}$	$\frac{11}{16}$	0.675	0.794	0.768	$\frac{29}{64}$	0.463	0.444
$\frac{1}{2}$	$\frac{3}{4}$	0.736	0.866	0.840	$\frac{9}{16}$	0.573	0.552
$\frac{9}{16}$	$\frac{7}{8}$	0.861	1.010	0.982	$\frac{39}{64}$	0.621	0.598
$\frac{5}{8}$	$\frac{15}{16}$	0.922	1.083	1.051	$\frac{23}{32}$	0.731	0.706
$\frac{3}{4}$	$1\frac{1}{8}$	1.088	1.299	1.240	$\frac{13}{16}$	0.827	0.798
$\frac{7}{8}$	$1\frac{5}{16}$	1.269	1.516	1.447	$\frac{29}{32}$	0.922	0.890
1	$1\frac{1}{2}$	1.450	1.732	1.653	1	1.018	0.982
$1\frac{1}{8}$	$1\frac{11}{16}$	1.631	1.949	1.859	$1\frac{5}{32}$	1.176	1.136
$1\frac{1}{4}$	$1\frac{7}{8}$	1.812	2.165	2.066	$1\frac{1}{4}$	1.272	1.228
$1\frac{3}{8}$	$2\frac{1}{16}$	1.994	2.382	2.273	$1\frac{3}{8}$	1.399	1.351
$1\frac{1}{2}$	$2\frac{1}{4}$	2.175	2.598	2.480	$1\frac{1}{2}$	1.526	1.474

ASA B18.2-1952. Published by ASME.

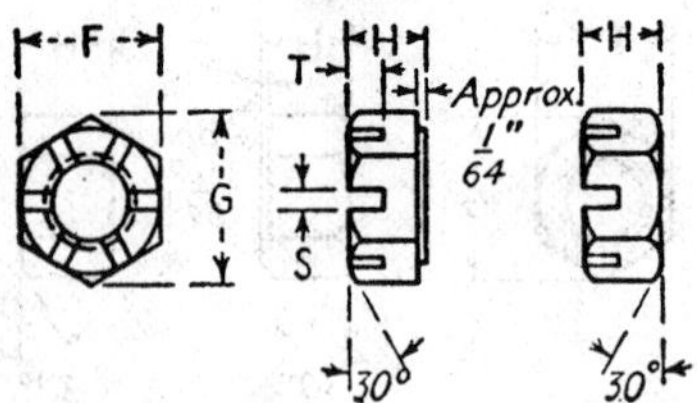

TABLE 55. FINISHED HEXAGON THICK SLOTTED NUTS

Nominal Size	F		G		H			Slot	
	Max (Basic)	Min	Max	Min	Nominal	Max	Min	Width S	Depth T
1/4	**7/16**	**0.428**	**0.505**	**0.488**	**9/32**	**0.288**	**0.274**	0.078	0.094
5/16	**1/2**	**0.489**	**0.577**	**0.557**	**21/64**	**0.336**	**0.320**	0.094	0.094
3/8	**9/16**	**0.551**	**0.650**	**0.628**	**13/32**	**0.415**	**0.398**	0.125	0.125
7/16	**11/16**	**0.675**	**0.794**	**0.768**	**29/64**	**0.463**	**0.444**	0.125	0.156
1/2	**3/4**	**0.736**	**0.866**	**0.840**	**9/16**	**0.573**	**0.552**	0.156	0.156
9/16	**7/8**	**0.861**	**1.010**	**0.982**	**39/64**	**0.621**	**0.598**	0.156	0.188
5/8	**15/16**	**0.922**	**1.083**	**1.051**	**23/32**	**0.731**	**0.706**	0.188	0.219
3/4	**1 1/8**	**1.088**	**1.299**	**1.240**	**13/16**	**0.827**	**0.798**	0.188	0.250
7/8	**1 5/16**	**1.269**	**1.516**	**1.447**	**29/32**	**0.922**	**0.890**	0.188	0.250
1	**1 1/2**	**1.450**	**1.732**	**1.653**	**1**	**1.018**	**0.982**	0.250	0.281
1 1/8	1 11/16	1.631	1.949	1.859	1 5/32	1.176	1.136	0.250	0.344
1 1/4	1 7/8	1.812	2.165	2.066	1 1/4	1.272	1.228	0.312	0.375
1 3/8	2 1/16	1.994	2.382	2.273	1 3/8	1.399	1.351	0.312	0.375
1 1/2	2 1/4	2.175	2.598	2.480	1 1/2	1.526	1.474	0.375	0.438

ASA B18.2-1952. Published by ASME.
Bold type indicates products unified dimensionally with British and Canadian standards.

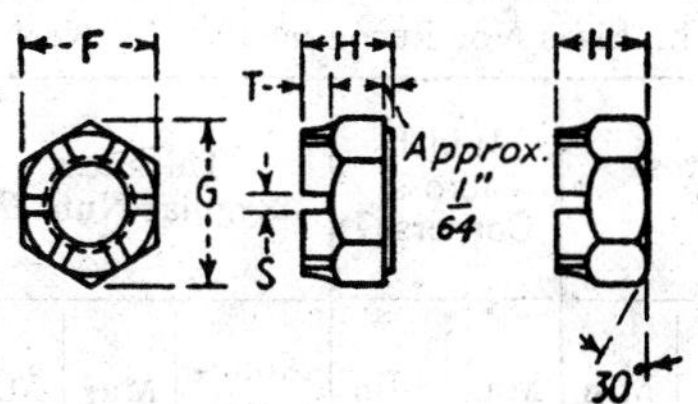

TABLE 56. FINISHED HEXAGON CASTLE NUTS

Nominal Size	F		G		H		Slot		Dia Cylindrical Part, Min
	Max (Basic)	Min	Max	Min	Max	Min	Width S	Depth T	
1/4	7/16	0.428	0.505	0.488	0.288	0.274	0.078	0.094	0.371
5/16	1/2	0.489	0.577	0.557	0.336	0.320	0.094	0.094	0.425
3/8	9/16	0.551	0.650	0.628	0.415	0.398	0.125	0.125	0.478
7/16	11/16	0.675	0.794	0.768	0.463	0.444	0.125	0.156	0.582
1/2	3/4	0.736	0.866	0.840	0.573	0.552	0.156	0.156	0.637
9/16	7/8	0.861	1.010	0.982	0.621	0.598	0.156	0.188	0.744
5/8	15/16	0.922	1.083	1.051	0.731	0.706	0.188	0.219	0.797
3/4	1 1/8	1.088	1.299	1.240	0.827	0.798	0.188	0.250	0.941
7/8	1 5/16	1.269	1.516	1.447	0.922	0.890	0.188	0.250	1.097
1	1 1/2	1.450	1.732	1.653	1.018	0.982	0.250	0.281	1.254
1 1/8	1 11/16	1.631	1.949	1.859	1.176	1.136	0.250	0.344	1.411
1 1/4	1 7/8	1.812	2.165	2.066	1.272	1.228	0.312	0.375	1.570
1 3/8	2 1/16	1.994	2.382	2.273	1.399	1.351	0.312	0.375	1.726
1 1/2	2 1/4	2.175	2.598	2.480	1.526	1.474	0.375	0.438	1.881

ASA B18.2-1952. Published by ASME.
"Finished" in the title refers to the quality of manufacture and the closeness of tolerance and does not indicate that surfaces are completely machined.

Table 57. Regular Hexagon and Hexagon-jam Nuts
$\frac{1}{4}$- to $\frac{5}{8}$-in. Sizes Not Recommended for New Designs*

Nominal Size or Basic Major Dia of Thread	Width across Flats F		Width across Corners G		Thickness Regular Nuts H			Thickness Regular Jam Nuts H		
	Max (Basic)	Min	Max	Min	Nominal	Max	Min	Nominal	Max	Min
$\frac{1}{4}$	$\frac{7}{16}$ 0.4375	0.425	0.505	0.484	$\frac{7}{32}$	0.235	0.203	$\frac{5}{32}$	0.172	0.140
$\frac{5}{16}$	$\frac{9}{16}$ 0.5625	0.547	0.650	0.624	$\frac{17}{64}$	0.283	0.149	$\frac{3}{16}$	0.204	0.170
$\frac{3}{8}$	$\frac{5}{8}$ 0.6250	0.606	0.722	0.691	$\frac{21}{64}$	0.346	0.310	$\frac{7}{32}$	0.237	0.201
$\frac{7}{16}$	$\frac{3}{4}$ 0.7500	0.728	0.866	0.830	$\frac{3}{8}$	0.394	0.356	$\frac{1}{4}$	0.269	0.231
$\frac{1}{2}$	$\frac{13}{16}$ 0.8125	0.788	0.938	0.898	$\frac{7}{16}$	0.458	0.418	$\frac{5}{16}$	0.332	0.292
$\frac{9}{16}$	$\frac{7}{8}$ 0.8750	0.847	1.010	0.966	$\frac{1}{2}$	0.521	0.479	$\frac{11}{32}$	0.365	0.323
$\frac{5}{8}$	1 1.0000	0.969	1.155	1.104	$\frac{35}{64}$	0.569	0.525	$\frac{3}{8}$	0.397	0.353
$\frac{3}{4}$	$1\frac{1}{8}$ 1.1250	1.088	1.299	1.240	$\frac{21}{32}$	0.680	0.632	$\frac{7}{16}$	0.462	0.414
$\frac{7}{8}$	$1\frac{5}{16}$ 1.3125	1.269	1.516	1.447	$\frac{49}{64}$	0.792	0.740	$\frac{1}{2}$	0.526	0.474
1	$1\frac{1}{2}$ 1.5000	1.450	1.732	1.653	$\frac{7}{8}$	0.903	0.847	$\frac{9}{16}$	0.590	0.534
$1\frac{1}{8}$	$1\frac{11}{16}$ 1.6875	1.631	1.949	1.859	1	1.030	0.970	$\frac{5}{8}$	0.655	0.595
$1\frac{1}{4}$	$1\frac{7}{8}$ 1.8750	1.812	2.165	2.066	$1\frac{3}{32}$	1.126	1.062	$\frac{3}{4}$	0.782	0.718
$1\frac{3}{8}$	$2\frac{1}{16}$ 2.0625	1.994	2.382	2.273	$1\frac{13}{64}$	1.237	1.169	$\frac{13}{16}$	0.846	0.778
$1\frac{1}{2}$	$2\frac{1}{4}$ 2.2500	2.175	2.598	2.480	$1\frac{5}{16}$	1.348	1.276	$\frac{7}{8}$	0.911	0.839

ASA B18.2-1952.
Nuts are not finished on any surface but are threaded.
* Dimensions furnished until change-over to Unified Series.

Table 58. Regular Semifinished Hexagon- and Hexagon-jam Nuts
$\frac{1}{4}$- to $\frac{5}{8}$-in. Sizes Not Recommended for New Designs*

Nominal Size	F		G		Regular Nuts H		Regular Jam Nuts H	
	Max (Basic)	Min	Max	Min	Max	Min	Max	Min
$\frac{1}{4}$	$\frac{7}{16}$	0.425	0.505	0.485	0.219	0.187	0.157	0.125
$\frac{5}{16}$	$\frac{9}{16}$	0.547	0.650	0.624	0.267	0.233	0.189	0.155
$\frac{3}{8}$	$\frac{5}{8}$	0.606	0.722	0.691	0.330	0.294	0.221	0.185
$\frac{7}{16}$	$\frac{3}{4}$	0.728	0.866	0.830	0.378	0.340	0.253	0.215
$\frac{1}{2}$	$\frac{13}{16}$	0.788	0.938	0.898	0.442	0.402	0.317	0.277
$\frac{9}{16}$	$\frac{7}{8}$	0.847	1.010	0.966	0.505	0.463	0.349	0.307
$\frac{5}{8}$	1	0.969	1.155	1.104	0.553	0.509	0.381	0.337

Semifinished nuts are finished on bearing surface and threaded.
* Dimensions furnished until change-over to Unified Series.

Right: Regular hexagon and hexagon-jam nuts.

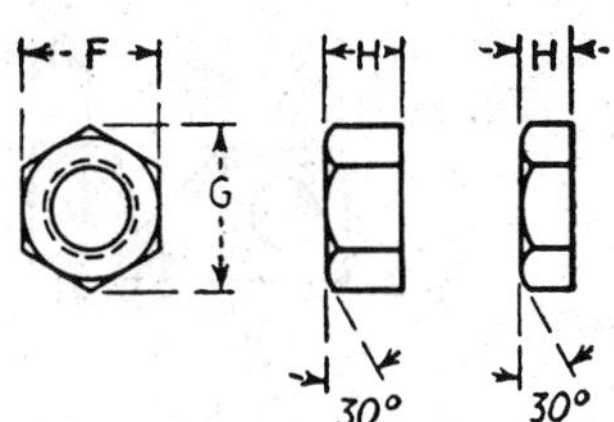

Below: Regular semifinished hexagon and hexagon-jam nuts.

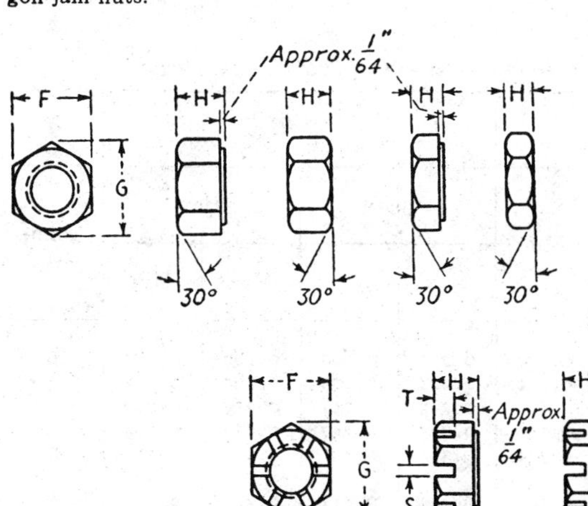

TABLE 59. REGULAR SEMIFINISHED HEXAGON SLOTTED NUTS

$\frac{1}{4}$- to $\frac{5}{8}$-in. Sizes Not Recommended for New Designs*

Nominal Size	*F*		*G*		*H*		Slot	
	Max (Basic)	Min	Max	Min	Max	Min	Width *S*	Depth *T*
$\frac{1}{4}$	$\frac{7}{16}$	0.425	0.505	0.485	0.219	0.187	0.078	0.094
$\frac{5}{16}$	$\frac{9}{16}$	0.547	0.650	0.624	0.267	0.233	0.094	0.094
$\frac{3}{8}$	$\frac{5}{8}$	0.606	0.722	0.691	0.330	0.294	0.125	0.125
$\frac{7}{16}$	$\frac{3}{4}$	0.728	0.866	0.830	0.378	0.340	0.125	0.156
$\frac{1}{2}$	$\frac{13}{16}$	0.788	0.938	0.898	0.442	0.402	0.156	0.156
$\frac{9}{16}$	$\frac{7}{8}$	0.847	1.010	0.966	0.505	0.463	0.156	0.188
$\frac{5}{8}$	1	0.969	1.155	1.104	0.553	0.509	0.188	0.219

Semifinished nuts are finished on bearing surface and threaded.

* Dimensions furnished until change-over to Unified Series.

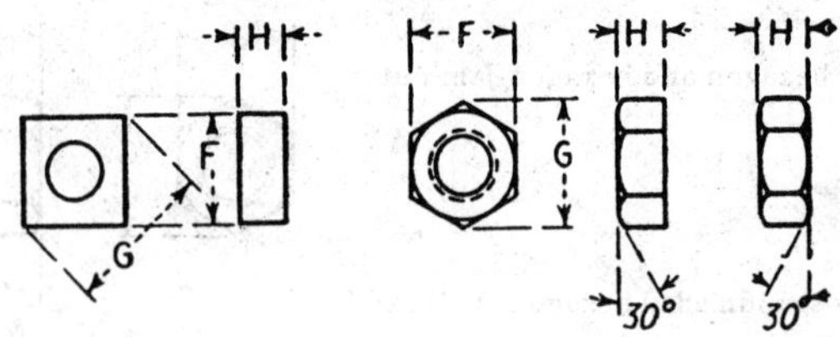

TABLE 60. MACHINE SCREW AND STOVE BOLT NUTS

Nominal Size		F		G				H
				Square		Hex		
		Max (Basic)	Min	Max	Min	Max	Min	Nominal
No. 0	0.0600	5/32	0.150	0.221	0.206	0.180	0.171	3/64
No. 1	0.0730	5/32	0.150	0.221	0.206	0.180	0.171	3/64
No. 2	0.0860	3/16	0.180	0.265	0.247	0.217	0.205	1/16
No. 3	0.0990	3/16	0.180	0.265	0.247	0.217	0.205	1/16
No. 4	0.1120	1/4	0.241	0.354	0.331	0.289	0.275	3/32
No. 5	0.1250	5/16	0.302	0.442	0.415	0.361	0.344	7/64
No. 6	0.1380	5/16	0.302	0.442	0.415	0.361	0.344	7/64
No. 8	0.1640	11/32	0.332	0.486	0.456	0.397	0.378	1/8
No. 10	0.1900	3/8	0.362	0.530	0.497	0.433	0.413	1/8
No. 12	0.2160	7/16	0.423	0.619	0.581	0.505	0.482	5/32
1/4	0.2500	7/16	0.423	0.619	0.581	0.505	0.482	3/16
5/16	0.3125	9/16	0.545	0.795	0.748	0.650	0.621	7/32
3/8	0.3750	5/8	0.607	0.884	0.833	0.722	0.692	1/4

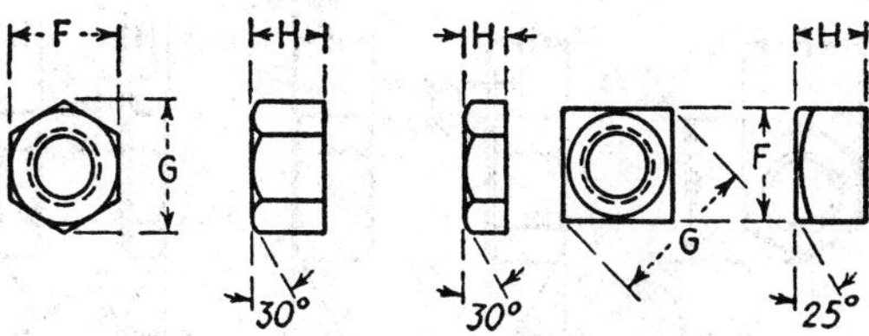

TABLE 61. HEAVY SQUARE, HEXAGON, AND HEXAGON-JAM NUTS

Nominal Size	Width across Flats F		Width across Corners G				Heavy Nuts H		Heavy Jam Nuts H	
			Square		Hex					
	Max (Basic)	Min	Max	Min	Max	Min	Max	Min	Max	Min
1/4	1/2	0.488	0.707	0.670	0.577	0.556	0.266	0.218	0.204	0.156
5/16	9/16	0.546	0.795	0.750	0.650	0.622	0.330	0.280	0.236	0.186
3/8	11/16	0.669	0.973	0.919	0.794	0.763	0.393	0.341	0.268	0.216
7/16	3/4	0.728	1.060	1.000	0.866	0.830	0.456	0.403	0.300	0.247
1/2	**7/8**	**0.850**	1.237	1.167	**1.010**	**0.969**	**0.520**	**0.464**	**0.332**	**0.277**
5/8	**1 1/16**	**1.031**	1.503	1.416	**1.227**	**1.175**	**0.647**	**0.587**	**0.397**	**0.337**
3/4	**1 1/4**	**1.212**	1.768	1.665	**1.443**	**1.382**	**0.774**	**0.710**	**0.462**	**0.398**
5/8	**1 7/16**	**1.394**	2.033	1.914	**1.660**	**1.589**	**0.901**	**0.833**	**0.526**	**0.458**
1	**1 5/8**	**1.575**	2.298	2.162	**1.876**	**1.796**	**1.028**	**0.956**	**0.590**	**0.519**
1 1/8	**1 13/16**	**1.756**	2.563	2.411	**2.093**	**2.002**	**1.155**	**1.079**	**0.655**	**0.579**
1 1/4	**2**	**1.938**	2.828	2.661	**2.309**	**2.209**	**1.282**	**1.187**	**0.782**	**0.687**
1 3/8	**2 3/16**	**2.119**	3.094	2.909	**2.526**	**2.416**	**1.409**	**1.310**	**0.846**	**0.747**
1 1/2	**2 3/8**	**2.300**	3.359	3.158	**2.742**	**2.622**	**1.536**	**1.433**	**0.911**	**0.808**
1 5/8	2 9/16	2.481			2.959	2.828	1.663	1.556	0.976	0.868
1 3/4	**2 3/4**	**2.662**			**3.175**	**3.035**	**1.790**	**1.679**	**1.040**	**0.929**
1 7/8	2 15/16	2.844			3.392	3.242	1.917	1.802	1.104	0.989
2	**3 1/8**	**3.025**			**3.608**	**3.449**	**2.044**	**1.925**	**1.169**	**1.050**
2 1/4	3 1/2	3.388	Sizes Not Standard	Sizes Not Standard	4.041	3.862	2.298	2.155	1.298	1.155
2 1/2	3 7/8	3.750			4.474	4.275	2.552	2.401	1.552	1.401
2 3/4	4 1/4	4.112			4.907	4.688	2.806	2.647	1.681	1.522
3	4 5/8	4.475			5.340	5.102	3.060	2.893	1.810	1.643
3 1/4	5	4.838			5.774	5.515	3.314	3.124	1.939	1.748
3 1/2	5 3/8	5.200			6.207	5.928	3.568	3.370	2.068	1.870
3 3/4	5 3/4	5.562			6.640	6.341	3.822	3.616	2.197	1.990
4	6 1/8	5.925			7.073	6.755	4.076	3.862	2.326	2.112

ASA B18.2-1952. Published by ASME.
Bold type indicates products unified dimensionally with British and Canadian standards.
Nuts are not finished on any surface but are threaded.

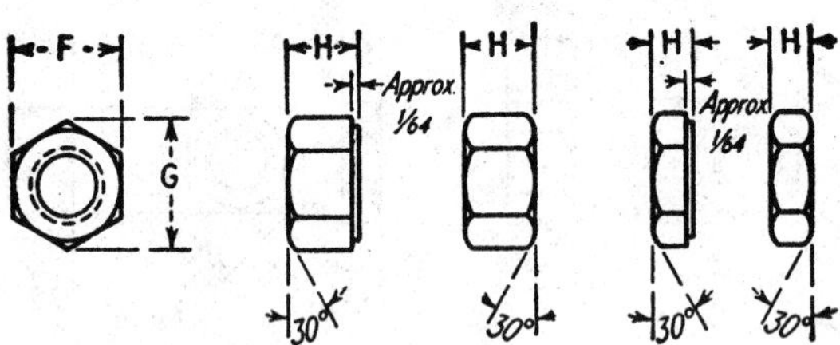

TABLE 62. HEAVY SEMIFINISHED HEXAGON AND HEXAGON-JAM NUTS

Nominal Size	F Max (Basic)	F Min	G, Max	Heavy Nuts H Max	Heavy Nuts H Min	Heavy Jam Nuts H Max	Heavy Jam Nuts H Min
1/4	1/2	0.488	0.577	0.250	0.218	0.188	0.156
5/16	9/16	0.546	0.650	0.314	0.280	0.220	0.186
3/8	11/16	0.669	0.794	0.377	0.341	0.252	0.216
7/16	3/4	0.728	0.866	0.441	0.403	0.285	0.247
1/2	**7/8**	**0.850**	**1.010**	**0.504**	**0.464**	**0.317**	**0.277**
9/16	5/16	0.909	1.083	0.568	0.526	0.349	0.307
5/8	**1 1/16**	**1.031**	**1.227**	**0.631**	**0.587**	**0.381**	**0.337**
3/4	**1 1/4**	**1.212**	**1.443**	**0.758**	**0.710**	**0.446**	**0.398**
7/8	**1 7/16**	**1.394**	**1.660**	**0.885**	**0.833**	**0.510**	**0.458**
1	**1 5/8**	**1.575**	**1.876**	**1.012**	**0.956**	**0.575**	**0.519**
1 1/8	**1 13/16**	**1.756**	**2.093**	**1.139**	**1.079**	**0.639**	**0.579**
1 1/4	**2**	**1.938**	**2.309**	**1.251**	**1.187**	**0.751**	**0.687**
1 3/8	**2 3/16**	**2.119**	**2.526**	**1.378**	**1.310**	**0.815**	**0.747**
1 1/2	**2 3/8**	**2.300**	**2.742**	**1.505**	**1.433**	**0.880**	**0.808**
1 5/8	2 9/16	2.481	2.959	1.632	1.556	0.944	0.868
1 3/4	**2 3/4**	**2.662**	**3.175**	**1.759**	**1.679**	**1.009**	**0.929**
1 7/8	2 15/16	2.844	3.392	1.886	1.802	1.073	0.989
2	**3 1/8**	**3.025**	**3.608**	**2.013**	**1.925**	**1.138**	**1.050**
2 1/4	3 1/2	3.388	4.041	2.251	2.155	1.251	1.155
2 1/2	3 7/8	3.750	4.474	2.505	2.401	1.505	1.401
2 3/4	4 1/4	4.112	4.907	2.759	2.647	1.634	1.522
3	4 5/8	4.475	5.340	3.013	2.893	1.763	1.643
3 1/4	5	4.838	5.774	3.252	3.124	1.876	1.748
3 1/2	5 3/8	5.200	6.207	3.506	3.370	2.006	1.870
3 3/4	5 3/4	5.562	6.640	3.760	3.616	2.134	1.990
4	6 1/8	5.925	7.073	4.014	3.862	2.264	2.112

ASA B18.2-1952. Published by ASME.
Bold type indicates products unified dimensionally with British and Canadian standards.
Semifinished nuts are finished on bearing surface and threaded.

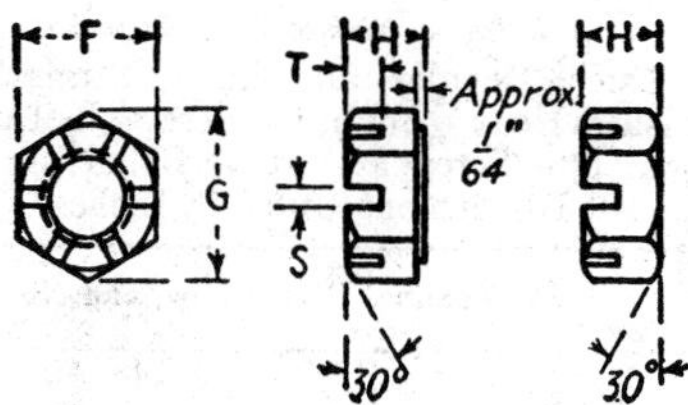

Table 63. Heavy Semifinished Hexagon Slotted Nuts

Nominal Size	F		G, Max	H		Slot	
	Max (Basic)	Min		Max	Min	Width *S*	Depth *T*
1/4	1/2	0.488	0.577	0.250	0.218	0.078	0.094
5/16	9/16	0.546	0.650	0.314	0.280	0.094	0.094
3/8	11/16	0.669	0.794	0.377	0.341	0.125	0.125
7/16	3/4	0.728	0.866	0.441	0.403	0.125	0.156
1/2	**7/8**	**0.850**	**1.010**	**0.504**	**0.464**	0.156	0.156
9/16	15/16	0.909	1.083	0.568	0.526	0.156	0.188
5/8	**1 1/16**	**1.031**	**1.227**	**0.631**	**0.587**	0.188	0.219
3/4	**1 1/4**	**1.212**	**1.443**	**0.758**	**0.710**	0.188	0.250
7/8	**1 7/16**	**1.394**	**1.660**	**0.885**	**0.833**	0.188	0.250
1	**1 5/8**	**1.575**	**1.876**	**1.012**	**0.956**	0.250	0.281
1 1/8	**1 13/16**	**1.756**	**2.093**	**1.139**	**1.079**	0.250	0.344
1 1/4	**2**	**1.938**	**2.309**	**1.251**	**1.187**	0.312	0.375
1 3/8	**2 3/16**	**2.119**	**2.526**	**1.378**	**1.310**	0.312	0.375
1 1/2	**2 3/8**	**2.300**	**2.742**	**1.505**	**1.433**	0.375	0.438
1 5/8	2 9/16	2.481	2.959	1.632	1.556	0.375	0.438
1 3/4	**2 3/4**	**2.662**	**3.175**	**1.759**	**1.679**	0.438	0.500
1 7/8	2 15/16	2.844	3.392	1.886	1.802	0.438	0.562
2	**3 1/8**	**3.025**	**3.608**	**2.013**	**1.925**	0.438	0.562
2 1/4	3 1/2	3.388	4.041	2.251	2.155	0.438	0.562
2 1/2	3 7/8	3.750	4.474	2.505	2.401	0.562	0.688
2 3/4	4 1/4	4.112	4.907	2.759	2.647	0.562	0.688
3	4 5/8	4.475	5.340	3.013	2.893	0.625	0.750
3 1/4	5	4.838	5.774	3.252	3.124	0.625	0.750
3 1/2	5 3/8	5.200	6.207	3.506	3.370	0.625	0.750
3 3/4	5 3/4	5.562	6.640	3.760	3.616	0.625	0.750
4	6 1/8	5.925	7.073	4.014	3.862	0.625	0.750

ASA B18.2-1952. Published by ASME.
Bold type indicates products unified dimensionally with British and Canadian standards.
Semifinished nuts are finished on bearing surface and threaded.

TABLE 64. WRENCH OPENINGS

For Finished, Finished Thick, Regular, and Heavy Series Nuts; Regular, Heavy, Finished, and Finished Heavy Bolts; Capscrews; Setscrews; Lag Bolts; Machine Screw and Stove Bolt Nuts
All Dimensions Given in Inches

Nominal* Size	Wrench Openings		Nominal Bolt, Screw, and Nut Diameters							
			Nuts			Regular Series Bolts, Finished Bolt, Hexagon Head Cap Screw	Heavy Series Bolts, Finished Heavy Bolt	Lag Bolts	Setscrews	Machine Screw Nuts and Stove Bolt Nuts
	Min	Max	Finished and Finished Thick Series Nuts	Regular Series Nuts	Heavy Series Nuts					
5/32	0.158	0.163								No. 0 and No. 1
3/16	0.190	0.195							No. 10	No. 2 and No. 3
1/4	0.252	0.257							1/4	No. 4
5/16	0.316	0.322						No. 10	5/16	No. 5 and No. 6
11/32	0.347	0.353								No. 8
3/8	0.378	0.384				1/4†		1/4	3/8	No. 10
7/16	0.440	0.446	1/4	1/4		1/4			7/16	No. 12 and 1/4
1/2	0.504	0.510	5/16		1/4	5/16		5/16	1/2	
9/16	0.566	0.573	3/8	5/16	5/16	3/8		3/8	9/16	5/16
19/32	0.598	0.605								
5/8	0.629	0.636		3/8		7/16		7/16	5/8	3/8
11/16	0.692	0.699	7/16		3/8					
3/4	0.755	0.763	1/2	7/16	7/16	1/2		1/2	3/4	
25/32	0.786	0.794								
13/16	0.818	0.826		1/2		9/16				
7/8	0.880	0.888	9/16	9/16	1/2		1/2		7/8	
15/16	0.944	0.953	5/8		9/16	5/8		5/8		
1	1.006	1.015		5/8					1	
1 1/16	1.068	1.077			5/8		5/8			
1 1/8	1.132	1.142	3/4	3/4		3/4		3/4	1 1/8	
1 1/4	1.257	1.267			3/4		3/4		1 1/4	
1 5/16	1.320	1.331	7/8	7/8		7/8		7/8		
1 3/8	1.383	1.394							1 3/8	
1 7/16	1.446	1.457			7/8		7/8			
1 1/2	1.508	1.520	1	1		1		1	1 1/2	
1 5/8	1.634	1.646			1		1			
1 11/16	1.696	1.708	1 1/8	1 1/8		1 1/8		1 1/8		
1 13/16	1.822	1.835			1 1/8		1 1/8			
1 7/8	1.885	1.898	1 1/4	1 1/4		1 1/4		1 1/4		
2	2.011	2.025			1 1/4		1 1/4			
2 1/16	2.074	2.088	1 3/8	1 3/8		1 3/8				
2 3/16	2.200	2.215			1 3/8		1 3/8			
2 1/4	2.262	2.277	1 1/2	1 1/2		1 1/2				
2 3/8	2.388	2.404			1 1/2		1 1/2			
2 7/16	2.450	2.466	1 5/8	1 5/8		1 5/8				

* Nominal size of wrench, also basic or maximum width across flats of bolt and screw heads and nuts.
† Regular square only.

TABLE 64. WRENCH OPENINGS (*Continued*)

Nominal* Size	Wrench Openings		Nominal Bolt, Screw, and Nut Diameters							
			Nuts			Regular Series Bolts, Finished Bolt, Hexagon Head Cap Screw	Heavy Series Bolts, Finished Heavy Bolt	Lag Bolts	Setscrews	Machine Screw Nuts and Stove Bolt Nuts
	Min	Max	Finished and Finished Thick Series Nuts	Regular Series Nuts	Heavy Series Nuts					
2 9/16	2.576	2.593			1 5/8		1 5/8			
2 5/8	2.639	2.656	1 3/4			1 3/4				
2 3/4	2.766	2.783			1 3/4		1 3/4			
2 13/16	2.827	2.845	1 7/8			1 7/8				
2 15/16	2.954	2.973			1 7/8		1 7/8			
3	3.016	3.035	2			2				
3 1/8	3.142	3.162			2		2			
3 3/8	3.393	3.414	2 1/4			2 1/4				
3 1/2	3.518	3.540			2 1/4		2 1/4			
3 3/4	3.770	3.793	2 1/2			2 1/2				
3 7/8	3.895	3.918			2 1/2		2 1/2			
4 1/8	4.147	4.172	2 3/4			2 3/4				
4 1/4	4.272	4.297			2 3/4		2 3/4			
4 1/2	4.524	4.550	3			3				
4 5/8	4.649	4.676			3		3			
5	5.026	5.055			3 1/4	3 1/4				
5 3/8	5.403	5.434			3 1/2	3 1/2				
5 3/4	5.780	5.813			3 3/4	3 3/4				
6 1/8	6.157	6.192			4	4				

Wrench Clearance[1]

Wrench clearance as shown in Tables 65 to 67 is the space required for the free movement of the wrench when applying torque to a fastener.

Bolt heads and nuts in general are manufactured to the dimensions established by the American Standard Association's standard B18.2-1952 under the categories of either light, regular, or heavy, or are manufactured to the dimensions of the various types cataloged as Air Force, Navy, and Military standards. The dimensions across the flats as indicated in these standards are not constant with the thread size but may vary according to the category or type of fastener selected.

Wrench-outline dimensions differ in varying amounts according to the design practices and manufacturing procedures inherent with the individual manufacturers. There are no established formulas for wrench configurations that are common to the industry.

Clearance dimensions shown in the tables are based on a wrench opening corresponding to the dimension across the flats of the fastener. The listed values were obtained from a composite study of the alloy-steel wrenches that are commercially available and Military Specifications MIL-W-15751 (Ships) and MIL-W-15838 (Ships). They are suitable for general use as minimum requirements.

[1] SAE Aeronautical Drafting Standard.

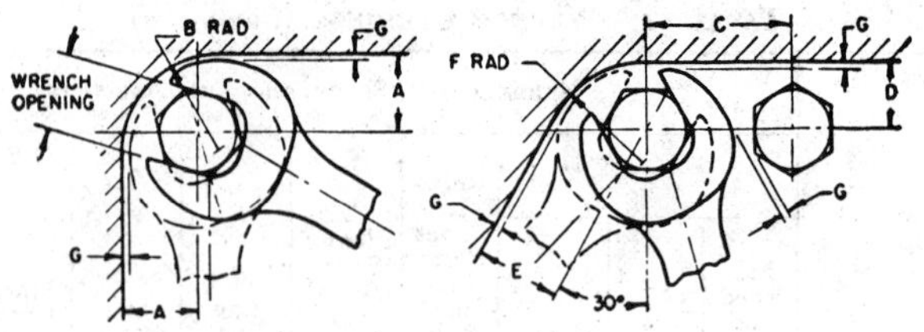

TABLE 65. WRENCH CLEARANCE REQUIRED FOR 15° OPEN-END WRENCHES

H = Thickness of Wrench Head

Wrench Opening	*A* Min	*B* Max	*C* Min	*D* Min	*E* Min	*F* Max	*G* Ref	*H* Max
0.156	0.220	0.250	0.390	0.160	0.250	0.200	0.030	0.094
0.188	0.250	0.280	0.430	0.190	0.270	0.230	0.030	0.172
0.250	0.280	0.340	0.530	0.270	0.310	0.310	0.030	0.172
0.312	0.380	0.470	0.660	0.280	0.390	0.390	0.050	0.203
0.344	0.420	0.500	0.750	0.340	0.450	0.450	0.050	0.203
0.375	0.420	0.500	0.780	0.360	0.450	0.520	0.050	0.219
0.438	0.470	0.590	0.890	0.420	0.520	0.640	0.050	0.250
0.500	0.520	0.640	1.000	0.470	0.580	0.660	0.050	0.266
0.562	0.590	0.770	1.130	0.520	0.660	0.700	0.050	0.297
0.594	0.640	0.830	1.210	0.530	0.700	0.700	0.050	0.344
0.625	0.640	0.830	1.230	0.550	0.700	0.700	0.050	0.344
0.688	0.770	0.920	1.470	0.660	0.880	0.800	0.060	0.375
0.750	0.770	0.920	1.510	0.670	0.880	0.800	0.060	0.375
0.781	0.830	0.950	1.550	0.690	0.890	0.840	0.060	0.375
0.812	0.910	1.120	1.660	0.720	0.970	0.860	0.060	0.406
0.875	0.970	1.150	1.810	0.800	1.060	0.910	0.060	0.438
0.938	0.970	1.150	1.850	0.810	1.060	0.950	0.060	0.438
1.000	1.050	1.230	2.000	0.880	1.160	1.060	0.060	0.500
1.062	1.090	1.250	2.100	0.970	1.200	1.200	0.080	0.500
1.125	1.140	1.370	2.210	1.000	1.270	1.230	0.080	0.500
1.250	1.270	1.420	2.440	1.080	1.390	1.310	0.080	0.562
1.312	1.390	1.690	2.630	1.170	1.520	1.340	0.080	0.562
1.438	1.470	1.720	2.800	1.250	1.590	1.340	0.090	0.641
1.500	1.470	1.720	2.840	1.270	1.590	1.450	0.090	0.641
1.625	1.560	1.880	3.100	1.380	1.750	1.560	0.090	0.641

Source: SAE Aeronautical Drafting Manual.
Ref. MIL-W-15751 (Ships) Table VI.

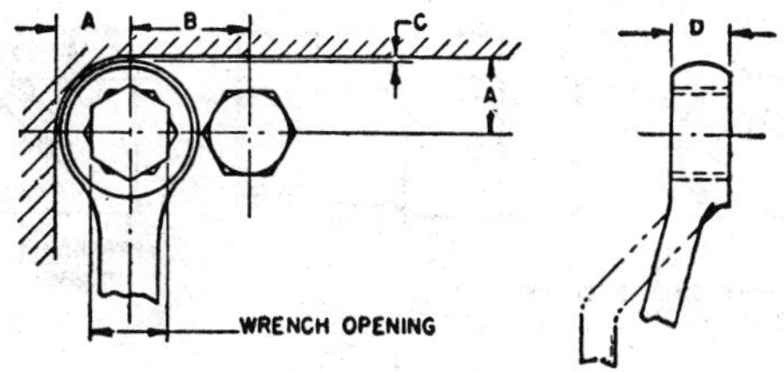

TABLE 66. WRENCH CLEARANCE REQUIRED FOR 12-POINT BOX WRENCHES

Wrench Opening	*A* Min	*B* Min	*C* Ref	*D* Max
0.156	0.190	0.280	0.030	0.156
0.188	0.200	0.309	0.030	0.172
0.250	0.270	0.410	0.030	0.250
0.312	0.300	0.480	0.030	0.281
0.344	0.300	0.500	0.030	0.281
0.375	0.340	0.560	0.030	0.344
0.438	0.400	0.650	0.030	0.359
0.500	0.450	0.740	0.030	0.375
0.562	0.500	0.830	0.030	0.406
0.594	0.530	0.870	0.030	0.469
0.625	0.560	0.920	0.030	0.469
0.688	0.590	0.990	0.030	0.531
0.750	0.660	1.090	0.030	0.594
0.781	0.690	1.140	0.030	0.594
0.812	0.720	1.190	0.030	0.594
0.875	0.750	1.260	0.030	0.594
0.938	0.780	1.320	0.030	0.656
1.000	0.810	1.390	0.030	0.718
1.062	0.840	1.450	0.030	0.781
1.125	0.950	1.600	0.030	0.844
1.250	0.980	1.700	0.030	0.875
1.312	1.090	1.850	0.030	0.906
1.438	1.220	2.050	0.030	1.000
1.500	1.270	2.140	0.030	1.062
1.625	1.340	2.280	0.030	1.156

Source: SAE Aeronautical Drafting Manual.
Ref. MIL-W-15751 (Ships) Tables I, II, and IV.

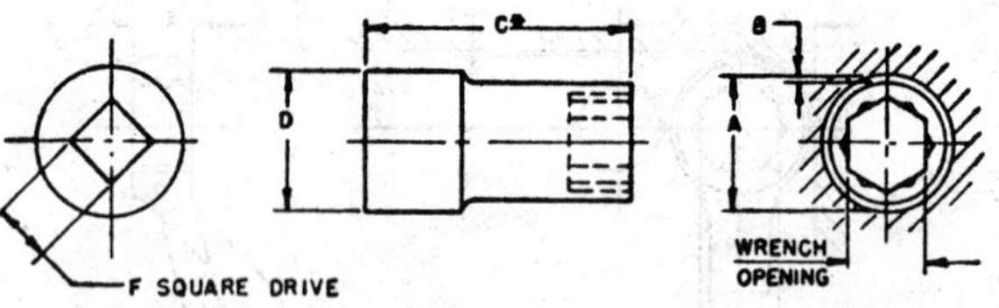

Table 67. Wrench Clearance for Regular-length Socket Wrenches

F Drive			0.250		0.375		0.500		0.750	
Wrench Opening	*A* Min	*B* Ref	*C* Max	*D* Max	*C* Max	*D* Max	*C* Max	*D* Max	*C* Max	*D* Max
0.188	0.370	0.030	1.000	0.510						
0.250	0.470	0.030	1.000	0.510	1.250	0.690				
0.312	0.550	0.030	1.000	0.510	1.250	0.690				
0.344	0.580	0.030	1.000	0.519	1.250	0.690				
0.375	0.620	0.030	1.000	0.580	1.250	0.690	1.500	0.880		
0.438	0.750	0.030	1.000	0.683	1.250	0.880	1.500	0.940		
0.500	0.810	0.030	1.000	0.692	1.250	0.880	1.500	0.940		
0.562	0.870	0.030			1.250	0.932	1.500	0.940		
0.594	0.920	0.030			1.250	0.963	1.562	0.970		
0.625	0.950	0.030			1.250	0.995	1.562	1.000		
0.688	1.030	0.030			1.250	1.058	1.562	1.065		
0.750	1.120	0.030			1.250	1.120	1.562	1.130		
0.781	1.150	0.030			1.250	1.126	1.625	1.130		
0.812	1.200	0.030			1.250	1.213	1.625	1.222		
0.875	1.280	0.030					1.750	1.285		
0.938	1.370	0.030					1.750	1.410		
1.000	1.470	0.030					1.750	1.410		
1.062	1.550	0.030					1.844	1.505		
1.125	1.610	0.030					1.938	1.567		
1.250	1.890	0.030					2.000	1.723	2.375	1.855
1.312	1.980	0.030							2.500	1.920
1.438	2.140	0.030							2.625	2.075
1.500	2.200	0.030							2.625	2.170
1.625	2.390	0.030							2.750	2.325

Source: SAE Aeronautical Drafting Manual.
Ref. MIL-W-15838 (Ships) Tables III, V, VII, and IX.

TABLE 68. FLAT-HEAD RIVETS

D			A		H	
Nominal	Max	Min	Max	Min	Max	Min
$\frac{1}{16}$	0.065	0.060	0.140	0.120	0.027	0.017
$\frac{3}{32}$	0.096	0.090	0.200	0.180	0.028	0.026
$\frac{1}{8}$	0.127	0.121	0.260	0.240	0.048	0.036
$\frac{5}{32}$	0.158	0.152	0.323	0.301	0.059	0.045
$\frac{3}{16}$	0.191	0.182	0.387	0.361	0.069	0.055
$\frac{7}{32}$	0.222	0.213	0.453	0.427	0.080	0.067
$\frac{1}{4}$	0.253	c.244	0.515	0.485	0.091	0.075
$\frac{9}{32}$	0.285	0.273	0.579	0.545	0.103	0.085
$\frac{5}{16}$	0.317	0.305	0.641	0.607	0.113	0.095
$\frac{11}{32}$	0.348	0.336	0.705	0.667	0.124	0.104
$\frac{3}{8}$	0.380	0.365	0.769	0.731	0.135	0.115
$\frac{13}{32}$	0.411	0.396	0.834	0.790	0.146	0.124
$\frac{7}{16}$	0.443	0.428	0.896	0.852	0.157	0.135

Approximate proportions. $A = 2.00 \times D$. $H = 0.33 \times D$.

TABLE 69. COUNTERSUNK-HEAD RIVETS

D			A			H
			Sharp	Sharp	Absolute Min of Rounded or Flat Edged	
Nominal	Max	Min	Max	Min		
$\frac{1}{16}$	0.065	0.060	0.118	0.115	0.110	0.027
$\frac{3}{32}$	0.096	0.090	0.176	0.171	0.163	0.040
$\frac{1}{8}$	0.127	0.121	0.235	0.227	0.217	0.053
$\frac{5}{32}$	0.158	0.152	0.293	0.284	0.272	0.066
$\frac{3}{16}$	0.191	0.182	0.351	0.340	0.326	0.079
$\frac{7}{32}$	0.222	0.213	0.413	0.400	0.384	0.094
$\frac{1}{4}$	0.253	0.244	0.469	0.455	0.437	0.106
$\frac{9}{32}$	0.285	0.273	0.528	0.511	0.491	0.119
$\frac{5}{16}$	0.317	0.305	0.588	0.569	0.547	0.133
$\frac{11}{32}$	0.348	0.336	0.646	0.626	0.602	0.146
$\frac{3}{8}$	0.380	0.365	0.704	0.682	0.656	0.159
$\frac{13}{32}$	0.411	0.396	0.763	0.738	0.710	0.172
$\frac{7}{16}$	0.443	0.428	0.823	0.797	0.765	0.186

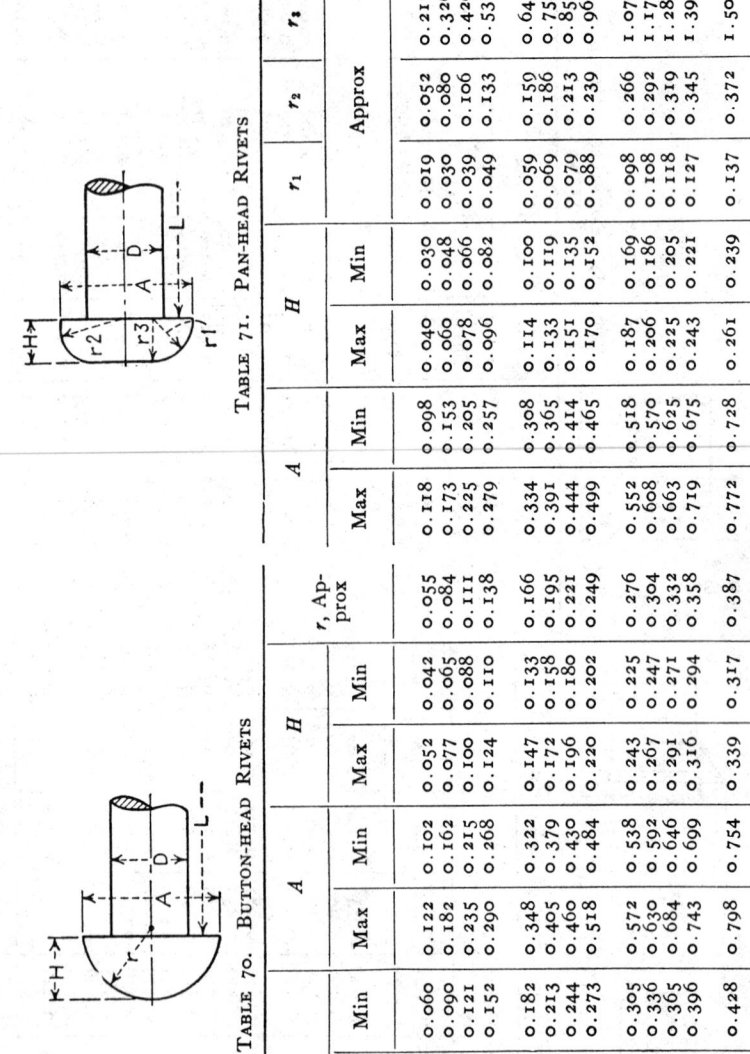

TABLE 70. BUTTON-HEAD RIVETS

Nominal	*D* Max	*D* Min	*A* Max	*A* Min	*H* Max	*H* Min	*r*, Approx
$\frac{1}{16}$	0.065	0.060	0.122	0.102	0.052	0.042	0.055
$\frac{3}{32}$	0.096	0.090	0.182	0.162	0.077	0.065	0.084
$\frac{1}{8}$	0.127	0.121	0.235	0.215	0.100	0.088	0.111
$\frac{5}{32}$	0.158	0.152	0.290	0.268	0.124	0.110	0.138
$\frac{3}{16}$	0.191	0.182	0.348	0.322	0.147	0.133	0.166
$\frac{7}{32}$	0.222	0.213	0.405	0.379	0.172	0.158	0.195
$\frac{1}{4}$	0.253	0.244	0.460	0.430	0.196	0.180	0.221
$\frac{9}{32}$	0.285	0.273	0.518	0.484	0.220	0.202	0.249
$\frac{5}{16}$	0.317	0.305	0.572	0.538	0.243	0.225	0.276
$\frac{11}{32}$	0.348	0.336	0.630	0.592	0.267	0.247	0.304
$\frac{3}{8}$	0.380	0.365	0.684	0.646	0.291	0.271	0.332
$\frac{13}{32}$	0.411	0.396	0.743	0.699	0.316	0.294	0.358
$\frac{7}{16}$	0.443	0.428	0.798	0.754	0.339	0.317	0.387

TABLE 71. PAN-HEAD RIVETS

Nominal	*A* Max	*A* Min	*H* Max	*H* Min	r_1 Approx	r_2 Approx	r_3 Approx
$\frac{1}{16}$	0.118	0.098	0.040	0.030	0.019	0.052	0.217
$\frac{3}{32}$	0.173	0.153	0.060	0.048	0.030	0.080	0.326
$\frac{1}{8}$	0.225	0.205	0.078	0.066	0.039	0.106	0.429
$\frac{5}{32}$	0.279	0.257	0.096	0.082	0.049	0.133	0.535
$\frac{3}{16}$	0.334	0.308	0.114	0.100	0.059	0.159	0.641
$\frac{7}{32}$	0.391	0.365	0.133	0.119	0.069	0.186	0.754
$\frac{1}{4}$	0.444	0.414	0.151	0.135	0.079	0.213	0.858
$\frac{9}{32}$	0.499	0.465	0.170	0.152	0.088	0.239	0.963
$\frac{5}{16}$	0.552	0.518	0.187	0.169	0.098	0.266	1.070
$\frac{11}{32}$	0.608	0.570	0.206	0.186	0.108	0.292	1.176
$\frac{3}{8}$	0.663	0.625	0.225	0.205	0.118	0.319	1.286
$\frac{13}{32}$	0.719	0.675	0.243	0.221	0.127	0.345	1.392
$\frac{7}{16}$	0.772	0.728	0.261	0.239	0.137	0.372	1.500

TABLE 72. TRUSS-HEAD RIVETS

	D		A		H		r, Approx
Nominal	Max	Min	Max	Min	Max	Min	
3/32	0.096	0.090	0.226	0.206	0.038	0.026	0.239
1/8	0.127	0.121	0.297	0.277	0.048	0.036	0.314
5/32	0.158	0.152	0.368	0.348	0.059	0.045	0.392
3/16	0.191	0.182	0.442	0.422	0.069	0.055	0.470
7/32	0.222	0.213	0.515	0.495	0.080	0.066	0.555
1/4	0.253	0.244	0.590	0.560	0.091	0.075	0.628
9/32	0.285	0.273	0.661	0.631	0.103	0.085	0.706
5/16	0.317	0.305	0.732	0.702	0.113	0.095	0.784
11/32	0.348	0.336	0.806	0.776	0.124	0.104	0.862
3/8	0.380	0.365	0.878	0.848	0.135	0.115	0.942
13/32	0.411	0.396	0.949	0.919	0.145	0.123	1.028
7/16	0.443	0.428	1.020	0.990	0.157	0.135	1.098

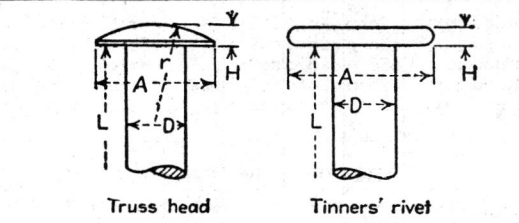

Truss head

Tinners' rivet

TABLE 73. TINNERS' RIVETS

	D		A		H		L, Nominal
Nominal*	Max	Min	Max	Min	Max	Min	
6 oz	0.085	0.075	0.213	0.193	0.028	0.016	1/8
8 oz	0.091	0.085	0.225	0.205	0.036	0.024	5/32
10 oz	0.097	0.091	0.250	0.230	0.037	0.025	11/64
12 oz	0.107	0.101	0.265	0.245	0.037	0.025	3/16
14 oz	0.111	0.105	0.275	0.255	0.038	0.026	3/16
1 lb	0.113	0.107	0.285	0.265	0.040	0.028	13/64
1¼ lb	0.122	0.116	0.295	0.275	0.045	0.033	7/32
1½ lb	0.132	0.126	0.316	0.294	0.046	0.034	15/64
1¾ lb	0.136	0.130	0.331	0.309	0.049	0.035	1/4
2 lb	0.146	0.140	0.341	0.319	0.050	0.036	17/64
2½ lb	0.150	0.144	0.311	0.289	0.069	0.055	9/32
3 lb	0.163	0.154	0.329	0.303	0.073	0.059	5/16
3½ lb	0.168	0.159	0.348	0.322	0.074	0.060	21/64
4 lb	0.179	0.170	0.368	0.342	0.076	0.062	11/32
5 lb	0.190	0.182	0.388	0.362	0.084	0.070	3/8
6 lb	0.206	0.197	0.419	0.393	0.090	0.076	25/64
7 lb	0.223	0.214	0.431	0.405	0.094	0.080	13/32
8 lb	0.227	0.218	0.475	0.445	0.101	0.085	7/16
9 lb	0.241	0.232	0.490	0.460	0.103	0.087	29/64
10 lb	0.241	0.232	0.505	0.475	0.104	0.088	15/32
12 lb	0.263	0.251	0.532	0.498	0.108	0.090	1/2
14 lb	0.288	0.276	0.577	0.543	0.113	0.095	33/64
16 lb	0.304	0.292	0.597	0.563	0.128	0.110	17/32
18 lb	0.347	0.335	0.706	0.668	0.156	0.136	19/32

* Size numbers refer to the approximate weight of 1000 rivets.

TABLE 74. COOPERS' RIVETS

Size No.*	D Max	D Min	A Max	A Min	H Max	H Min	E, Nominal	d, Nominal	L, Nominal
1 lb	0.111	0.105	0.291	0.271	0.045	0.031	Not pointed		0.226
$1\frac{1}{4}$	0.122	0.116	0.324	0.302	0.050	0.036	Not pointed		0.262
$1\frac{1}{2}$	0.132	0.126	0.324	0.302	0.050	0.036	Not pointed		0.262
$1\frac{3}{4}$	0.136	0.130	0.324	0.302	0.052	0.034	Not pointed		0.293
2	0.142	0.136	0.355	0.333	0.056	0.038	Not pointed		0.297
3	0.158	0.152	0.386	0.364	0.058	0.040	$\frac{1}{16}$	0.123	0.362
4	0.168	0.159	0.388	0.362	0.058	0.040	$\frac{1}{16}$	0.130	0.393
5	0.183	0.174	0.419	0.393	0.063	0.045	$\frac{1}{16}$	0.144	0.429
6	0.206	0.197	0.482	0.456	0.073	0.051	$\frac{3}{32}$	0.160	0.468
7	0.223	0.214	0.513	0.487	0.076	0.054	$\frac{3}{32}$	0.175	0.534
8	0.241	0.232	0.546	0.516	0.081	0.059	$\frac{3}{32}$	0.182	0.570
9	0.248	0.239	0.578	0.548	0.085	0.063	$\frac{3}{32}$	0.197	0.574
10	0.253	0.244	0.578	0.548	0.085	0.063	$\frac{3}{32}$	0.197	0.605
12	0.263	0.251	0.580	0.546	0.086	0.060	$\frac{3}{32}$	0.214	0.604
14	0.275	0.263	0.611	0.577	0.091	0.065	$\frac{3}{32}$	0.223	0.641
16	0.285	0.273	0.611	0.577	0.089	0.063	$\frac{3}{32}$	0.223	0.670
18	0.285	0.273	0.642	0.608	0.108	0.082	$\frac{1}{8}$	0.230	0.720
20	0.316	0.304	0.705	0.671	0.119	0.093	$\frac{1}{8}$	0.250	0.730
$\frac{3}{8}$	0.380	0.365	0.800	0.762	0.126	0.096	$\frac{1}{8}$	0.312	0.809

* Size numbers refer to the approximate weight of 1000 rivets.

TABLE 75. BELT RIVETS

Size No.*	D Max	A Max	A Min	H Max	H Min	d, Nominal	E, Nominal
14	0.085	0.260	0.240	0.042	0.030	0.065	$\frac{5}{64}$
13	0.097	0.322	0.302	0.051	0.039	0.073	$\frac{5}{64}$
12	0.111	0.353	0.333	0.054	0.040	0.083	$\frac{5}{64}$
11	0.122	0.383	0.363	0.059	0.045	0.097	$\frac{5}{64}$
10	0.136	0.417	0.395	0.065	0.047	0.109	$\frac{3}{32}$
9	0.150	0.448	0.426	0.069	0.051	0.122	$\frac{3}{32}$
8	0.167	0.481	0.455	0.072	0.054	0.135	$\frac{3}{32}$
7	0.183	0.513	0.487	0.075	0.056	0.151	$\frac{1}{8}$
6	0.206	0.606	0.580	0.090	0.068	0.165	$\frac{1}{8}$
5	0.223	0.700	0.674	0.105	0.083	0.185	$\frac{1}{8}$
4	0.241	0.921	0.893	0.138	0.116	0.204	$\frac{9}{64}$

* Size number refers to the Stubs iron wire gage number of the stock used in the body of the rivet.

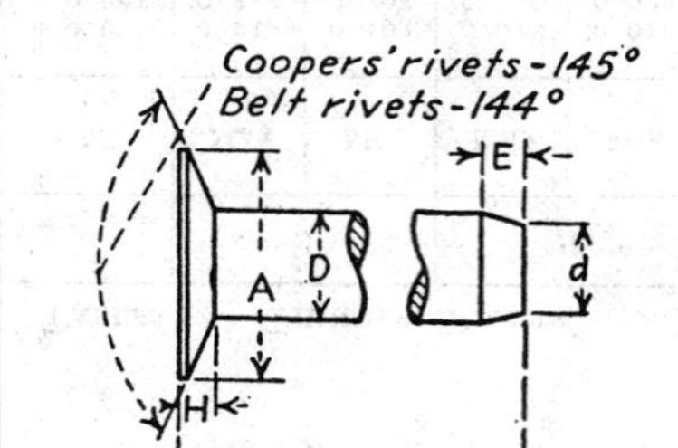

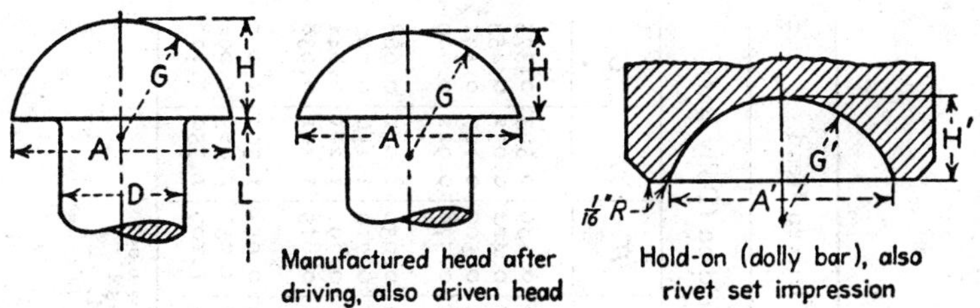

Manufactured head after driving, also driven head

Hold-on (dolly bar), also rivet set impression

TABLE 76. BUTTON HEAD
All Dimensions Given in Inches

Manufactured Shape								Head after Driving and Driven Head					
D			*A*		*H*		*G*	After Driving			Hold-on (Dolly Bar)		
Nominal	Max	Min	Max	Min	Max	Min (Basic)		*A*	*H*	*G*	*A'*	*H'*	*G'*
½	0.520	0.478	0.938	0.844	0.406	0.375	0.443	0.922	0.344	0.484	0.906	0.313	0.484
⅝	0.655	0.600	1.157	1.063	0.500	0.469	0.553	1.141	0.438	0.594	1.125	0.406	0.594
¾	0.780	0.725	1.391	1.282	0.594	0.563	0.664	1.375	0.516	0.719	1.344	0.484	0.719
⅞	0.905	0.850	1.609	1.500	0.687	0.656	0.775	1.594	0.609	0.844	1.578	0.563	0.844
1	1.030	0.975	1.828	1.719	0.781	0.750	0.885	1.828	0.688	0.953	1.813	0.641	0.953
1⅛	1.160	1.098	2.063	1.938	0.891	0.844	0.996	2.063	0.781	1.078	2.031	0.719	1.078
1¼	1.285	1.223	2.283	2.158	0.985	0.938	1.107	2.281	0.859	1.188	2.250	0.797	1.188
1⅜	1.415	1.345	2.500	2.375	1.078	1.031	1.217	2.516	0.953	1.313	2.469	0.875	1.313
1½	1.540	1.470	2.719	2.594	1.188	1.125	1.328	2.734	1.031	1.438	2.703	0.953	1.438
1⅝	1.665	1.588	2.938	2.813	1.282	1.219	1.439	2.969	1.125	1.547	2.922	1.047	1.547
1¾	1.790	1.713	3.172	3.032	1.376	1.313	1.549	3.203	1.203	1.672	3.156	1.125	1.672

Proportions (basic): $A = 1.75D$; $H = 0.75D$; $G = 0.885D$.
$A = 1.827D$; $H = 0.688D$; $G = 0.954D$.
$A' = 1.800D$; $H' = 0.641D$; $G' = 0.954D$.

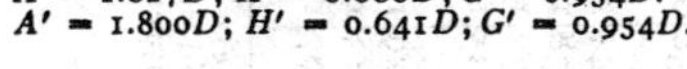

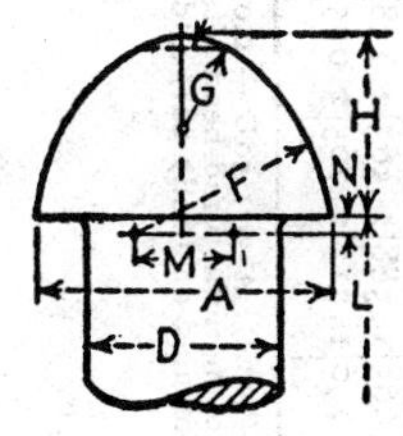

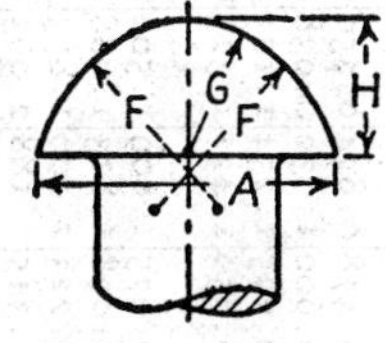

Manufactured head after driving, also driven head

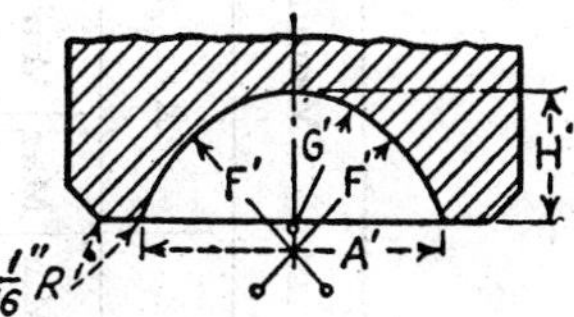

Hold-on (dolly bar), also rivet set impression

High-button-head rivet.

TABLE 77. HIGH BUTTON-HEAD RIVETS (ACORN)

Manufactured Shape											Head after Driving and Driven Head							
	D		*A*		*H*		*F*	*G*	*M*	*N*	Head after Driving				Hold-on (Dolly Bar)			
Nominal	Max	Min	Max	Min	Max	Min					*A*	*H*	*F*	*G*	*A'*	*H'*	*F'*	*G'*
1/2	0.520	0.478	0.844	0.750	0.531	0.484	0.656	0.094	0.500	0.094	0.875	0.375	0.563	0.375	0.859	0.359	0.563	0.375
5/8	0.655	0.600	1.032	0.938	0.625	0.578	0.750	0.188	0.500	0.094	1.063	0.453	0.672	0.453	1.047	0.422	0.672	0.453
3/4	0.780	0.725	1.234	1.125	0.719	0.657	0.844	0.282	0.500	0.094	1.250	0.531	0.797	0.531	1.234	0.500	0.797	0.531
7/8	0.905	0.850	1.422	1.313	0.812	0.750	0.937	0.375	0.500	0.094	1.438	0.609	0.922	0.609	1.422	0.578	0.922	0.609
1	1.030	0.975	1.609	1.500	0.906	0.844	1.031	0.469	0.500	0.094	1.625	0.688	1.031	0.688	1.609	0.656	1.031	0.688
1 1/8	1.160	1.098	1.813	1.688	1.016	0.938	1.125	0.563	0.500	0.094	1.813	0.766	1.156	0.766	1.797	0.719	1.156	0.766
1 1/4	1.285	1.223	2.000	1.875	1.110	1.032	1.219	0.657	0.500	0.094	2.000	0.844	1.281	0.844	1.984	0.797	1.281	0.844
1 3/8	1.415	1.345	2.188	2.063	1.203	1.125	1.312	0.750	0.500	0.094	2.188	0.938	1.391	0.938	2.172	0.875	1.391	0.938
1 1/2	1.540	1.470	2.375	2.250	1.313	1.219	1.406	0.844	0.500	0.094	2.375	1.000	1.500	1.000	2.344	0.953	1.500	1.000
1 5/8	1.665	1.588	2.563	2.438	1.407	1.313	1.500	0.938	0.500	0.094	2.563	1.094	1.625	1.094	2.531	1.031	1.625	1.094
1 3/4	1.790	1.713	2.765	2.625	1.501	1.407	1.594	1.032	0.500	0.094	2.750	1.172	1.750	1.172	2.719	1.109	1.750	1.172

Proportions (basic): $A = 1.50D + 0.031$; $H = 0.75D + 0.125$; $F = 0.75D + 0.281$; $G = 0.75D - 0.281$; $M = 0.50$; $N = 0.094$.
Proportions of manufactured head after driving and driven head: $A = 1.5D + 0.125$; $H = 0.425A$; $F = 1.5H$; $G = 0.425A$.
Proportions of hold-on (dolly bar) and rivet set impressions: $A' = 1.485D + 0.124$; $H' = 0.4A$; $F' = 1.5H$; $G' = 0.425A$.

TABLE 78. CONE-HEAD RIVETS

	Manufactured Shape								Head after Driving and Driven Head					
	D		*A*		*B*		*H*		After Driving			Hold-on (Dolly Bar)		
Nominal	Max	Min	Max	Min	Max	Min	Max	Min (Basic)	*A*	*B*	*H*	*A′*	*B′*	*H′*
½	0.520	0.478	0.938	0.844	0.532	0.438	0.469	0.438	0.922	0.469	0.406	0.891	0.469	0.391
⅝	0.655	0.600	1.157	1.063	0.649	0.555	0.578	0.547	1.141	0.594	0.516	1.109	0.594	0.484
¾	0.780	0.725	1.391	1.282	0.781	0.672	0.687	0.656	1.375	0.703	0.625	1.328	0.703	0.578
⅞	0.905	0.850	1.609	1.500	0.898	0.789	0.797	0.766	1.594	0.828	0.719	1.547	0.828	0.688
1	1.030	0.975	1.828	1.719	1.016	0.907	0.906	0.875	1.828	0.938	0.828	1.781	0.938	0.781
1⅛	1.160	1.098	2.063	1.938	1.149	1.024	1.031	0.984	2.063	1.063	0.938	2.016	1.063	0.875
1¼	1.285	1.223	2.283	2.158	1.266	1.141	1.141	1.094	2.281	1.172	1.031	2.219	1.172	0.969
1⅜	1.415	1.345	2.500	2.375	1.383	1.258	1.250	1.203	2.516	1.297	1.141	2.438	1.297	1.078
1½	1.540	1.470	2.719	2.594	1.500	1.375	1.376	1.313	2.734	1.406	1.250	2.672	1.406	1.172
1⅝	1.665	1.588	2.938	2.813	1.617	1.492	1.485	1.422	2.969	1.531	1.344	2.891	1.531	1.266
1¾	1.790	1.713	3.172	3.032	1.750	1.610	1.594	1.531	3.203	1.641	1.453	3.109	1.641	1.375

Proportions (basic): $A = 1.75D$; $H = 0.875D$; $B = 0.938D$.

Proportions of manufactured head after driving and driven head: $A = 1.827D$; $B = 0.938D$; $H = 0.828D$.

Proportions of hold-on (dolly bar) and rivet set impressions: $A' = 1.781D$; $B' = 0.938D$; $H' = 0.781D$.

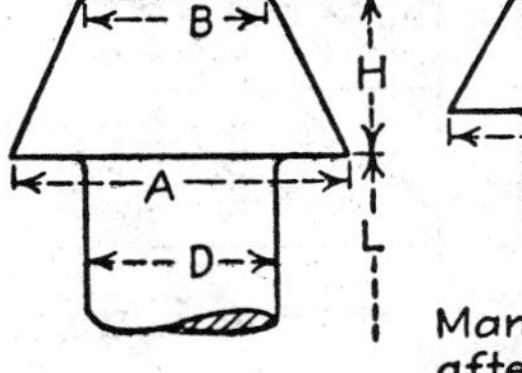

Manufactured head after driving, also driven head

Hold-on (dolly bar), also rivet set impression

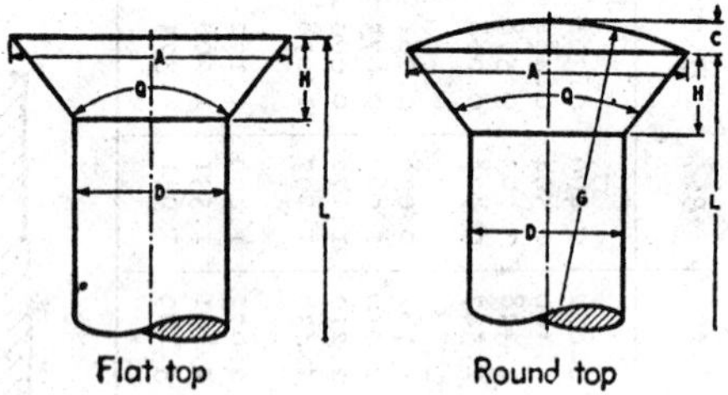

TABLE 79. FLAT-TOP AND ROUND-TOP COUNTERSUNK HEADS

All Dimensions in Inches

Dia of Body D			Dia of Head A		Depth of Head H		Height of Round Top C	Radius of Round Top G
Nominal	Max	Min	Max	Min	Max	Min (Basic)		
½	0.520	0.478	0.936	0.874	0.281	0.250	0.095	1.125
⅝	0.655	0.600	1.194	1.068	0.344	0.313	0.119	1.406
¾	0.780	0.725	1.421	1.295	0.406	0.375	0.143	1.688
⅞	0.905	0.850	1.647	1.521	0.469	0.438	0.166	1.969
1	1.030	0.975	1.873	1.747	0.531	0.500	0.190	2.250
1⅛	1.160	1.098	2.114	1.973	0.610	0.563	0.214	2.531
1¼	1.285	1.223	2.341	2.200	0.672	0.625	0.238	2.813
1⅜	1.415	1.345	2.567	2.426	0.751	0.688	0.261	3.094
1½	1.540	1.470	2.793	2.652	0.813	0.750	0.285	3.375
1⅝	1.665	1.588	3.019	2.878	0.876	0.813	0.309	3.656
1¾	1.790	1.713	3.262	3.105	0.938	0.875	0.333	3.938

Proportions (basic): $A = 1.81D$; $H = 0.50D$; $C = 0.19D$; $G = 2.25D$; $Q = 78°$.

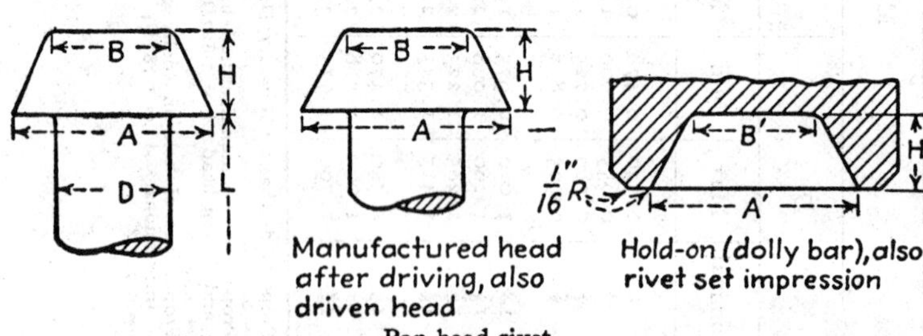

Pan-head rivet.

TABLE 80. PAN-HEAD RIVETS

	Manufactured Shape								Head after Driving and Driven Head					
	D		*A*		*B*		*H*		After Driving			Hold-on (Dolly Bar)		
Nominal	Max	Min	Max	Min	Max	Min	Max	Min (Basic)	*A*	*B*	*H*	*A′*	*B′*	*H′*
1/2	0.520	0.478	0.863	0.769	0.563	0.469	0.381	0.350	0.844	0.500	0.328	0.812	0.500	0.297
5/8	0.655	0.600	1.063	0.969	0.688	0.594	0.469	0.438	1.047	0.625	0.406	1.031	0.625	0.375
3/4	0.780	0.725	1.278	1.169	0.828	0.719	0.556	0.525	1.266	0.750	0.484	1.234	0.750	0.453
7/8	0.905	0.850	1.478	1.369	0.953	0.844	0.644	0.613	1.469	0.875	0.578	1.437	0.875	0.531
1	1.030	0.975	1.678	1.569	1.078	0.969	0.731	0.700	1.687	1.000	0.656	1.641	1.000	0.609
1 1/8	1.160	1.098	1.894	1.769	1.219	1.094	0.835	0.788	1.891	1.125	0.734	1.844	1.125	0.688
1 1/4	1.285	1.223	2.094	1.969	1.344	1.219	0.922	0.875	2.094	1.250	0.812	2.047	1.250	0.766
1 3/8	1.415	1.345	2.294	2.169	1.474	1.344	1.010	0.963	2.313	1.375	0.906	2.250	1.375	0.844
1 1/2	1.540	1.470	2.494	2.369	1.594	1.469	1.113	1.050	2.516	1.500	0.984	2.453	1.500	0.906
1 5/8	1.665	1.588	2.694	2.569	1.719	1.594	1.201	1.138	2.734	1.625	1.063	2.656	1.625	0.984
1 3/4	1.790	1.713	2.909	2.769	1.859	1.719	1.288	1.225	2.938	1.750	1.141	2.875	1.750	1.063

Proportions (basic): $A = 1.60D$; $B = D$; $H = 0.70D$.
Proportions of manufactured head after driving and driven head: $A = 1.681D$; $B = D$; $H = 0.656D$.
Proportions of hold-on (dolly bar) and rivet set impressions: $A' = 1.639D$; $B' = D$; $H' = 0.609D$.

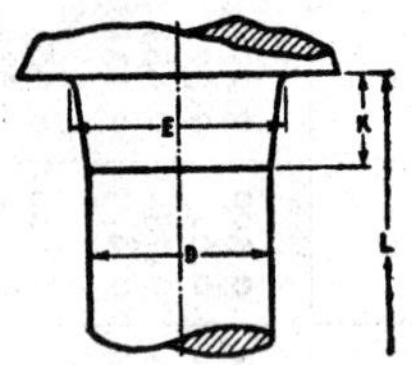

TABLE 81. SWELL-NECK RIVETS

D, Nominal	E Max (Basic)	E Min	K
½	0.563	0.543	0.250
⅝	0.688	0.658	0.313
¾	0.813	0.783	0.375
⅞	0.938	0.908	0.438
1	1.063	1.033	0.500
1⅛	1.188	1.153	0.563
1¼	1.313	1.278	0.625
1⅜	1.438	1.398	0.688
1½	1.563	1.523	0.750
1⅝	1.688	1.648	0.813
1¾	1.813	1.773	0.875

Proportions (basic): $E = D + 0.063$; $K = 0.50D$.

The swell neck is applicable to all standard large rivets except the flat-top countersunk and round-top countersunk types.

Table 82. Plain Washers

Bolt Size		Light				Medium				Heavy				Extra Heavy			
		ID	OD	Thickness	Gage	ID	OD	Thickness	Gage	ID	OD	Thickness	Gage	ID	OD	Thickness	Gage
0.060	0	5/64	3/16	0.020	25												
0.073	1	3/32	7/32	0.020	25												
0.086	2	3/32	1/4	0.020	25												
0.099	3	1/8	1/4	0.022	24												
0.112	4	1/8	1/4	0.022	24	1/8	5/16	0.032	21								
0.125	5	5/32	5/16	0.035	20	5/32	3/8	0.049	18								
0.138	6	5/32	5/16	0.035	20	5/32	3/8	0.049	18								
0.151	7	11/64	13/32	0.049	18	3/16	3/8	0.049	18								
0.164	8	3/16	3/8	0.049	18	3/16	7/16	0.049	18								
0.177	9	13/64	15/32	0.049	18	7/32	1/2	0.049	18								
0.187	3/16	7/32	7/16	0.049	18	7/32	1/2	0.049	18	1/4	9/16	0.049	18				
0.190	10	7/32	7/16	0.049	18	1/4	9/16	0.049	18	1/4	9/16	0.065	16				
0.203	11	15/64	17/32	0.049	18	1/4	9/16	0.049	18	1/4	9/16	0.065	16				
0.216	12	1/4	1/2	0.049	18	1/4	9/16	0.049	18	1/4	9/16	0.065	16				
0.242	14	17/64	5/8	0.049	18	5/16	3/4	0.065	16	5/16	7/8	0.065	16				
0.250	1/4	9/32	5/8	0.065	16	5/16	3/4	0.065	16	5/16	3/4	0.065	16	5/16	7/8	0.065	16
0.268	16	9/32	5/8	0.065	16	5/16	3/4	0.065	16	5/16	7/8	0.065	16	5/16	7/8	0.065	16
0.294	18	5/16	3/4	0.065	16	3/8	3/4	0.065	16	3/8	7/8	0.083	14	3/8	1 1/8	0.065	16
0.3125	5/16	11/32	11/16	0.065	16	3/8	3/4	0.065	16	3/8	7/8	0.083	14	3/8	1 1/8	0.065	16
0.320	20	11/32	11/16	0.065	16	3/8	3/4	0.065	16	3/8	7/8	0.083	14	3/8	1 1/8	0.065	16
0.372	24	13/32	13/16	0.065	16	7/16	7/8	0.083	14	7/16	1	0.083	14	7/16	1 3/8	0.083	14
0.375	3/8	13/32	13/16	0.065	16	7/16	7/8	0.083	14	7/16	1	0.083	14	7/16	1 3/8	0.083	14
0.4375	7/16	15/32	59/64	0.065	16	1/2	1 1/8	0.083	14	1/2	1 1/4	0.083	14	1/2	1 5/8	0.083	14
0.5000	1/2	17/32	1 1/16	0.095	13	9/16	1 1/4	0.109	12	9/16	1 3/8	0.109	12	9/16	1 7/8	0.109	12
0.5625	9/16	19/32	1 3/16	0.095	13	5/8	1 3/8	0.109	12	5/8	1 1/2	0.109	12	5/8	2 1/8	0.134	10

TABLE 82. PLAIN WASHERS (*Continued*)

Bolt Size		Light				Medium				Heavy				Extra Heavy			
		ID	OD	Thickness	Gage	ID	OD	Thickness	Gage	ID	OD	Thickness	Gage	ID	OD	Thickness	Gage
0.625	$\frac{5}{8}$	$\frac{21}{32}$	$1\frac{5}{16}$	0.095	13	$\frac{11}{16}$	$1\frac{1}{2}$	0.134	10	$\frac{11}{16}$	$1\frac{3}{4}$	0.134	10	$\frac{11}{16}$	$2\frac{3}{8}$	0.165	8
0.750	$\frac{3}{4}$	$\frac{13}{16}$	$1\frac{1}{2}$	0.134	10	$\frac{13}{16}$	$1\frac{3}{4}$	0.148	9	$\frac{13}{16}$	2	0.148	9	$\frac{13}{16}$	$2\frac{7}{8}$	0.165	8
0.875	$\frac{7}{8}$	$\frac{15}{16}$	$1\frac{3}{4}$	0.134	10	$\frac{15}{16}$	2	0.165	8	$\frac{15}{16}$	$2\frac{1}{4}$	0.165	8	$\frac{15}{16}$	$3\frac{3}{8}$	0.180	7
1.00	1	$1\frac{1}{16}$	2	0.134	10	$1\frac{1}{16}$	$2\frac{1}{4}$	0.165	8	$1\frac{1}{16}$	$2\frac{1}{2}$	0.165	8	$1\frac{1}{16}$	$3\frac{7}{8}$	0.239	4
1.125	$1\frac{1}{8}$					$1\frac{3}{16}$	$2\frac{1}{2}$	0.165	8	$1\frac{1}{4}$	$2\frac{3}{4}$	0.165	8				
1.250	$1\frac{1}{4}$					$1\frac{5}{16}$	$2\frac{3}{4}$	0.165	8	$1\frac{3}{8}$	3	0.165	8				
1.375	$1\frac{3}{8}$					$1\frac{7}{16}$	3	0.180	7	$1\frac{1}{2}$	$3\frac{1}{4}$	0.180	7				
1.500	$1\frac{1}{2}$					$1\frac{9}{16}$	$3\frac{1}{4}$	0.180	7	$1\frac{5}{8}$	$3\frac{1}{2}$	0.180	7				
1.625	$1\frac{5}{8}$					$1\frac{11}{16}$	$3\frac{1}{2}$	0.180	7	$1\frac{3}{4}$	$3\frac{3}{4}$	0.180	7				
1.750	$1\frac{3}{4}$					$1\frac{13}{16}$	$3\frac{3}{4}$	0.180	7	$1\frac{7}{8}$	4	0.180	7				
1.875	$1\frac{7}{8}$					$1\frac{15}{16}$	4	0.180	7	2	$4\frac{1}{4}$	0.180	7				
2.000	2					$2\frac{1}{16}$	$4\frac{1}{4}$	0.180	7	$2\frac{1}{8}$	$4\frac{1}{2}$	0.180	7				
2.250	$2\frac{1}{4}$									$2\frac{3}{8}$	$4\frac{3}{4}$	0.220	5				
2.500	$2\frac{1}{2}$									$2\frac{5}{8}$	5	0.238	4				
2.750	$2\frac{3}{4}$									$2\frac{7}{8}$	$5\frac{1}{4}$	0.259	3				
3.000	3									$3\frac{1}{8}$	$5\frac{1}{2}$	0.284	2				

ASA B27.1-1949. Published by the ASME.
Tolerances: Above division line—ID ± 0.005; OD ± 0.010
Below division line—ID ± 0.10; OD ± 0.010

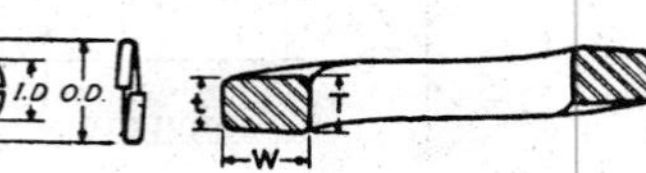

Spring lockwasher.

TABLE 83. SPRING LOCKWASHERS

Nominal Size	ID	Clearance of Nominal Bolt Size		Light		Medium		Heavy		Extra Heavy		OD, Max			
	Min	Min	Max	Width W	Thickness $\frac{T+t}{2}$	Width W	Thickness $\frac{T+t}{2}$	Width W	Thickness $\frac{T+t}{2}$	Width W	Thickness $\frac{T+t}{2}$	Light	Medium	Heavy	Extra Heavy
No. 2	0.088	0.002	0.011	0.030	0.015	0.035	0.020	0.040	0.025	0.053	0.027	0.165	0.175	0.185	0.211
No. 3	0.102	0.002	0.011	0.035	0.020	0.040	0.025	0.047	0.031	0.062	0.034	0.188	0.198	0.212	0.242
No. 4	0.115	0.003	0.012	0.035	0.020	0.040	0.025	0.047	0.031	0.062	0.034	0.202	0.212	0.226	0.256
No. 5	0.128	0.003	0.012	0.040	0.025	0.047	0.031	0.055	0.040	0.079	0.045	0.225	0.239	0.255	0.303
No. 6	0.141	0.003	0.013	0.040	0.025	0.047	0.031	0.055	0.040	0.079	0.045	0.237	0.251	0.267	0.315
No. 8	0.168	0.004	0.014	0.047	0.031	0.055	0.040	0.062	0.047	0.096	0.057	0.280	0.296	0.310	0.378
No. 10	0.194	0.004	0.015	0.055	0.040	0.062	0.047	0.070	0.056	0.112	0.068	0.323	0.337	0.353	0.437
No. 12	0.221	0.005	0.016	0.062	0.047	0.070	0.056	0.077	0.063	0.130	0.080	0.364	0.380	0.394	0.500
1/4	0.255	0.005	0.017	0.107	0.047	0.109	0.062	0.110	0.077	0.132	0.084	0.489	0.493	0.495	0.539
5/16	0.319	0.006	0.020	0.117	0.056	0.125	0.078	0.130	0.097	0.143	0.108	0.575	0.591	0.601	0.627
3/8	0.382	0.007	0.023	0.136	0.070	0.141	0.094	0.145	0.115	0.170	0.123	0.678	0.688	0.696	0.746
7/16	0.446	0.008	0.026	0.154	0.085	0.156	0.109	0.160	0.133	0.186	0.143	0.780	0.784	0.792	0.844
1/2	0.509	0.009	0.029	0.170	0.099	0.171	0.125	0.176	0.151	0.204	0.162	0.877	0.879	0.889	0.945
9/16	0.573	0.010	0.032	0.186	0.113	0.188	0.141	0.193	0.170	0.223	0.182	0.975	0.979	0.989	1.049
5/8	0.636	0.011	0.035	0.201	0.126	0.203	0.156	0.210	0.189	0.242	0.202	1.082	1.086	1.100	1.164
11/16	0.700	0.012	0.038	0.216	0.138	0.219	0.172	0.227	0.207	0.260	0.221	1.178	1.184	1.200	1.266
3/4	0.763	0.013	0.041	0.233	0.153	0.234	0.188	0.244	0.226	0.279	0.241	1.277	1.279	1.299	1.369
13/16	0.827	0.014	0.044	0.249	0.168	0.250	0.203	0.262	0.246	0.298	0.261	1.375	1.377	1.401	1.473
7/8	0.890	0.015	0.047	0.264	0.179	0.266	0.219	0.281	0.266	0.322	0.285	1.470	1.474	1.504	1.586
15/16	0.954	0.016	0.050	0.277	0.191	0.281	0.234	0.298	0.284	0.345	0.308	1.562	1.570	1.604	1.698
1	1.017	0.017	0.053	0.289	0.202	0.297	0.250	0.319	0.306	0.366	0.330	1.656	1.672	1.716	1.810
1 1/16	1.081	0.018	0.056	0.301	0.213	0.312	0.266	0.338	0.326	0.389	0.352	1.746	1.768	1.820	1.922
1 1/8	1.144	0.019	0.059	0.314	0.224	0.328	0.281	0.356	0.345	0.411	0.375	1.837	1.865	1.921	2.031
1 3/16	1.208	0.020	0.062	0.324	0.234	0.344	0.297	0.373	0.364	0.431	0.396	1.923	1.963	2.021	2.137
1 1/4	1.271	0.021	0.065	0.336	0.244	0.359	0.312	0.393	0.384	0.452	0.417	2.012	2.058	2.126	2.244
1 5/16	1.335	0.022	0.068	0.346	0.254	0.375	0.328	0.410	0.403	0.472	0.438	2.098	2.156	2.226	2.350
1 3/8	1.398	0.023	0.071	0.356	0.264	0.391	0.344	0.427	0.422	0.491	0.458	2.183	2.253	2.325	2.453
1 7/16	1.462	0.024	0.074	0.366	0.273	0.406	0.359	0.442	0.440	0.509	0.478	2.269	2.349	2.421	2.555
1 1/2	1.525	0.025	0.077	0.375	0.282	0.422	0.375	0.458	0.458	0.526	0.496	2.352	2.446	2.518	2.654

ASA B27.1-1950. Published by ASME.

SPRING LOCKWASHERS

Washer Section. Finished washers have a section that is slightly trapezoidal, with thickness at inner periphery greater than thickness at outer periphery.

Designation. Washers are designated by nominal size and series; that is, ½ in. light, ½ in. medium, ½ in. heavy, and ½ in. extra heavy. Washers in certain materials are not available in all four series.

Carbon-steel Washers. These are available in all four series. They are heat-treated to Rockwell 45 to 53 C. If made for hot-dip galvanizing, they are coiled to limits 0.020 in. in excess of those tabulated.

Stainless-steel Washers. These are available in two types of stainless steel, AISI 302 or 420, and light and medium series only. Rockwell hardness, 35 to 43 C.

Aluminum-zinc Alloy. Lockwashers are made in medium series only to ASTM B211-47T Specification, Alloy ZG42. They are heat-treated to Rockwell 75 to 67 B scale, and should have minimum tensile strength of 77,000 psi.

Phosphor-bronze Washers. Light and medium series lockwashers are available in phosphor bronze conforming to ASTM Specification B159-47T, Grade A. Minimum hardness, Rockwell 90 B.

Silicon-bronze Washers. Light and medium series are made from ASTM Specification B 99-47, Type B alloy. Hardness, Rockwell 90 B.

K-monel Washers. These are available in light and medium series in K-monel corresponding to Federal Specification QQ-N-286 or Navy Specification 46-N-5. Rockwell hardness 33 to 40 C; minimum tensile strength, 155,000 psi.

Details of finish, temper testing, and twist testing are given in the standard.

TOOTHED LOCKWASHERS

Manufacturing Detail. The number of teeth, the length of the teeth, the width of the rim, and the thickness of the washer over the teeth (free height) shall be optional with the manufacturer, with the provision, however, that the projection of the teeth on both sides of the washer shall be uniform within a tolerance equal to one-half of the projection on one side.

Material and Hardness. Washers shall be made from carbon steel, fabricated and heat-treated to a Rockwell hardness of 40 to 50 C scale or equivalent. It is recommended, however, that the lighter, more sensitive depth-reading Rockwell A scale be used when testing washers of thin section. Decarburization shall be removed before testing hardness.

Finish. Plated washers shall be baked 4 hr at 400 F to relieve hydrogen or acid embrittlement. If washers so treated fail to meet the prescribed tests, the baking time and/or the temperature shall be increased.

Designation. Washers shall be specified or designated by the nominal size, description, and type, for example, ½-in. internal tooth, type A; 9/16-in. external tooth, type B; No. 12 internal-external tooth, type A.

Testing. Specifications for testing these lockwashers are given in the complete standard.

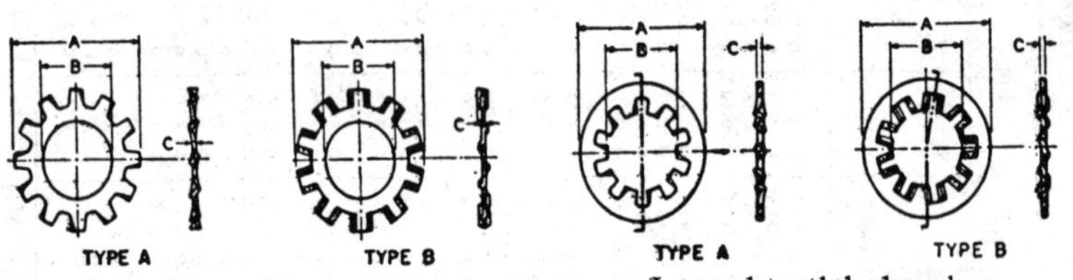

External-tooth lockwasher.

Internal-tooth lockwasher.

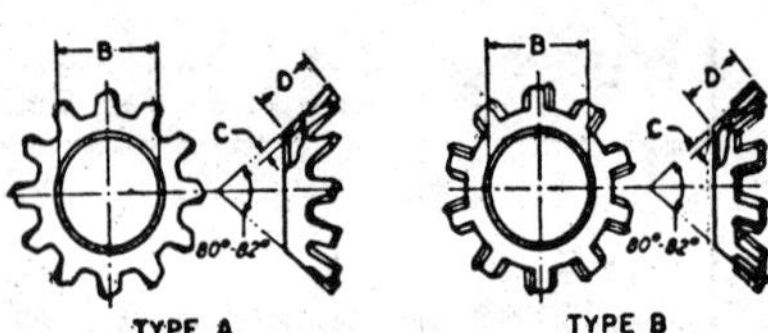

Countersunk-tooth lockwasher.

TABLE 84. TOOTHED LOCKWASHERS

Internal-tooth Lockwashers

	Size	No. 2	No. 3	No. 4	No. 5	No. 6	No. 8	No. 10	No. 12	1/4	5/16	3/8	7/16	1/2	9/16	5/8	11/16	3/4	13/16	7/8	1	1 1/8	1 1/4
A	Max	0.200	0.232	0.270	0.280	0.295	0.340	0.381	0.410	0.478	0.610	0.692	0.789	0.900	0.985	1.071	1.166	1.245	1.315	1.410	1.637	1.830	1.975
	Min	0.175	0.215	0.255	0.245	0.275	0.325	0.365	0.394	0.460	0.594	0.670	0.740	0.867	0.957	1.045	1.130	1.220	1.290	1.364	1.590	1.799	1.921
B	Max	0.095	0.109	0.123	0.136	0.150	0.176	0.204	0.231	0.267	0.332	0.398	0.464	0.530	0.596	0.663	0.728	0.795	0.861	0.927	1.060	1.192	1.325
	Min	0.089	0.102	0.115	0.129	0.141	0.168	0.195	0.221	0.256	0.320	0.384	0.448	0.512	0.576	0.640	0.704	0.769	0.832	0.894	1.019	1.144	1.275
C	Max	0.015	0.019	0.019	0.021	0.021	0.023	0.025	0.025	0.028	0.034	0.040	0.040	0.045	0.045	0.050	0.050	0.055	0.055	0.060	0.067	0.067	0.067
	Min	0.010	0.012	0.015	0.017	0.017	0.018	0.020	0.020	0.023	0.028	0.032	0.032	0.037	0.037	0.042	0.042	0.047	0.047	0.052	0.059	0.059	0.059

External-tooth Lockwashers

	Size	No. 2	No. 3	No. 4	No. 5	No. 6	No. 8	No. 10	No. 12	1/4	5/16	3/8	7/16	1/2	9/16	5/8	11/16	3/4	13/16	7/8	1	1 1/8	1 1/4
A	Max			0.290		0.320	0.381	0.410	0.475	0.510	0.610	0.694	0.760	0.900	0.985	1.070	1.155	1.260	1.315	1.410	1.620		
	Min			0.245		0.305	0.365	0.395	0.460	0.494	0.588	0.670	0.740	0.880	0.960	1.045	1.130	1.220	1.290	1.380	1.590		
B	Max			0.123		0.150	0.176	0.204	0.231	0.267	0.332	0.398	0.464	0.530	0.596	0.663	0.728	0.795	0.861	0.927	1.060		
	Min			0.115		0.141	0.168	0.195	0.221	0.256	0.320	0.384	0.448	0.513	0.576	0.641	0.704	0.768	0.833	0.897	1.025		
C	Max			0.019		0.022	0.023	0.025	0.028	0.028	0.034	0.040	0.040	0.045	0.045	0.050	0.050	0.055	0.055	0.060	0.067		
	Min			0.015		0.016	0.018	0.020	0.023	0.023	0.028	0.032	0.032	0.037	0.037	0.042	0.042	0.047	0.047	0.052	0.059		

Heavy Internal-tooth Lockwashers

	Size	1/4	5/16	3/8	7/16	1/2	9/16	5/8	3/4	7/8
A	Max	0.536	0.607	0.748	0.858	0.924	1.034	1.135	1.265	1.447
	Min	0.500	0.590	0.700	0.800	0.880	0.990	1.100	1.240	1.400
B	Max	0.267	0.332	0.398	0.464	0.530	0.596	0.663	0.795	0.927
	Min	0.256	0.320	0.384	0.448	0.512	0.576	0.640	0.768	0.894
C	Max	0.045	0.050	0.050	0.067	0.067	0.067	0.067	0.048	0.084
	Min	0.035	0.040	0.042	0.050	0.055	0.055	0.059	0.070	0.075

Countersunk External-tooth Lockwashers

	Size	No. 4	No. 6	No. 8	No. 10	No. 12	1/4	No. 16	5/16	3/8	7/16	1/2
B	Max	0.123	0.150	0.177	0.205	0.231	0.267	0.287	0.333	0.398	0.463	0.529
	Min	0.113	0.140	0.167	0.195	0.220	0.255	0.273	0.318	0.383	0.448	0.512
C	Max	0.019	0.021	0.021	0.025	0.025	0.025	0.028	0.028	0.034	0.045	0.045
	Min	0.015	0.017	0.017	0.020	0.020	0.020	0.023	0.023	0.028	0.037	0.037
D	Max	0.065	0.092	0.105	0.099	0.128	0.128	0.147	0.192	0.255	0.270	0.304
	Min	0.050	0.082	0.088	0.083	0.118	0.113	0.137	0.165	0.242	0.260	0.294

ASA B27.1-1950. Published by ASME.

STUDS[1]

Tap-end Studs

N = body diameter:

Type 1, unfinished—no standard body tolerance.

Type 2, finished, full or undersize body—tolerance from minimum pitch diameter of class 2A tolerance to basic major diameter of thread.

Type 3, finished, full body—tolerance equal to that on major diameter of class 2A thread tolerances.

Type 4, finished, close body—tolerance as specified by user (milled or ground body).

B = body diameter, no standard body tolerances.

D = nominal size or diameter of stud.

L = over-all length of stud. Tolerance on stud length for studs 6 in. and under in length shall be plus or minus $\frac{1}{32}$ in. for sizes $\frac{1}{4}$ to $\frac{3}{8}$ in., plus or minus $\frac{1}{16}$ in. for

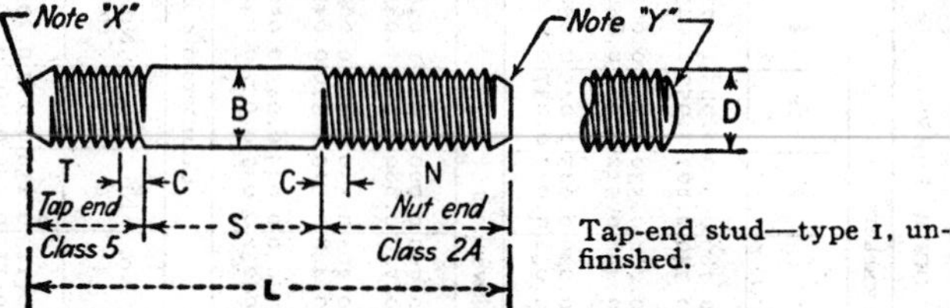

Tap-end stud—type 1, unfinished.

sizes $\frac{7}{16}$ and $\frac{1}{2}$ in., plus or minus $\frac{1}{8}$ in. for sizes $\frac{9}{16}$ to $1\frac{1}{4}$ in., and plus or minus $\frac{1}{4}$ in. for sizes $1\frac{3}{8}$ in. and over. Length tolerance for studs over 6 in. in length shall be plus or minus $\frac{1}{16}$ in. for sizes $\frac{1}{4}$ to $\frac{3}{8}$ in., plus or minus $\frac{3}{32}$ in. for sizes $\frac{7}{16}$ and $\frac{1}{2}$ in., plus or minus $\frac{3}{16}$ in. for sizes $\frac{9}{16}$ to $1\frac{1}{4}$ in., and plus or minus $\frac{1}{4}$ in. for sizes $1\frac{3}{8}$ in. and over.

T = length of thread on tap end shall be $1\frac{1}{2}D$ with plus tolerance of $2\frac{1}{2}$ threads, measured from extreme point to last scratch of thread.

C = thread runout or length of imperfect thread and shall not exceed $2\frac{1}{2}$ threads in length.

S = length of unthreaded body or shoulder.

N = length of thread on nut end and shall be $2D$ plus $\frac{1}{4}$ in. for studs up to and including 6 in. in length, and $2D$ plus $\frac{1}{2}$ in. for studs over 6 in. in length, measured from extreme point to last scratch of thread with plus tolerance of $2\frac{1}{2}$ threads. When the stud length is less than $4\frac{3}{4}D$, the thread length shall be $1\frac{1}{2}D$ minimum with a plus tolerance of $2\frac{1}{2}$ threads. For studs with a length of $3D$ or less, the thread length shall be as agreed upon by user and producer.

X—Point on tap end shall be flat and chamfered. The length of point shall be a minimum of 2 complete threads and a maximum of 3 complete threads.

Y—Point on the nut end shall be flat and chamfered or rounded (oval) at the option of the manufacturer.

THREADS. Tap-end threads for sizes $\frac{1}{4}$ to $1\frac{1}{2}$ in., inclusive, shall be in accordance with the thread limits for tentative class 5 fit for NC and NF thread series. Thread limits for sizes over $1\frac{1}{2}$ in. and for the 8N thread series shall be as agreed upon by user and producer. Class 5 fit for sizes smaller than $\frac{1}{4}$ in. is not recommended. Nut end threads shall be American Standard class 2A for NC and NF thread series.

[1] Source: "Bolt, Nut and Rivet Standards," 2d ed., Industrial Fasteners Institute, 1952.

Section 37

KEYS AND PINS

SECTION 37

KEYS AND PINS

KEYS

Plain Parallel Stock Keys. The dimensions and tolerances for plain parallel stock keys, both square and flat, and applicable to shafts up to 6 in. dia, are given in Table 1. These keys shall be cut from cold-finished stock and used without further machining, the dimensions and tolerances given in the table having been fixed with this consideration in mind.

Besides the key dimensions, Table 1 gives the distance from the bottom of the keyseat to the opposite side of the shaft, in order to facilitate gaging for the correct depth of keyseat when machining. This dimension is calculated on the basis of the keyseat being cut to a depth equal to one-half of the height of the key H, and it should be noted that this depth is measured at the side of the keyseat, not in the central plane of the key corresponding to the shaft diameter.

Table 2 gives the dimensions and tolerances of flat keys only, for shafts larger than 6 in. in dia, the square type having been omitted as it is not generally applicable in these larger sizes. Table 3 gives total keyway depth.

Plain and Gib-head Taper Stock Keys. The dimensions, tolerances, and stock lengths of plain and gib-head taper keys, both square and flat, for shaft diameters from $\frac{1}{2}$ to 6 in. are included. Tables 4 and 5 cover the plain taper and Table 6 the gib-head type. The taper for both types shall be $\frac{1}{8}$ in. in 12 in. and shall run the full length of the key (L).

The keyseat in the shaft, for a taper key, is the same as that for the corresponding size of parallel key, the depth of the keyseat being taken as one-half the height of the key, measured at a point, a distance equal to the key width (W), from the large end. At this point the cross-sections of corresponding sizes of parallel and taper keys are identical.

The distance from the bottom of the keyseat to the opposite side of the shaft for taper keys may, therefore, be obtained by referring to Table 1 and using the value of this dimension as given for the corresponding parallel key and shaft size.

Woodruff Keys, Keyslots, and Cutters. Keys. Standardization of Woodruff keys (ASA B17f-1930, and reaffirmed in 1949) produced a simplified series numbered 204 to 1212, inclusive, and ranging in size from $\frac{1}{16}$ by $\frac{1}{2}$ to $\frac{3}{8}$ by $1\frac{1}{2}$ in. Actually, Woodruff keys have been used for many years in sizes ranging from $\frac{1}{16}$ by $\frac{1}{4}$ to $\frac{3}{4}$ by $3\frac{1}{2}$ in. This situation is recognized by the SAE in its current standards. The numbers range from 201 down to 36. See column Old Std. No., Table 7. In ASA B5.3-1950, the sponsors of the milling-cutter standard, which includes keyslot cutters, recognized the factual situation and assigned American standard numbers to cutters—the range being Nos. 202 to 2428 (see Tables 11 and 12). The cutter number indicates the nominal key dimension or size of cutter; that is, the last two digits give the nominal diameter in eighths of an inch, and the digits preceding the last two give the nominal width in thirty-seconds of an inch. Thus, cutter No. 204 (or key No. 204) indicates a size $\frac{2}{32}$ by $\frac{4}{8}$ in. or $\frac{1}{16}$ in. thick by $\frac{1}{2}$ in. dia.

Keys may be round-bottomed or have the optional flat bottom. Maximum and minimum limits for key width (Table 7) are such that, when assembled with the appropriate keyslot (Table 8), the assembly can be made without distortion of the shaft or the necessity of fitting, when the maximum key is inserted in the minimum slot. Also, no shake will occur when the minimum key is inserted in the maximum slot. Such interchangeability facilitates mass production. For more particular work, the specified fit can be secured by gaging the finished keyslot and selecting the key by measurement.

To facilitate key layout in the drafting room, Table 7 gives dimension E, which is the distance from the top of the key to the center of the circle forming the key profile.

KEYSLOTS. Note from Table 8 that the depth B of the keyslot is measured from the sharp edge of the slot, not from the shaft circumference on the center line of the key. Therefore, B is not the distance, or depth, to which the cutter must be fed into the shaft. This amount varies with the shaft diameter and equals $B + G$, where G is the versed sine (see Table 9).

Various relationships are:

Shaft Diameter L. Decimal equivalents are given to four places and all figures are calculated from these basic dimensions. Any change in the shaft diameter from basic will necessarily change all other figures and, in this case, should accurate dimensions be required, the following formulas should be used.

Versed Sine G. The versed sines specified are determined from the following formula:

$$G = \frac{L}{2} \sqrt{\frac{L^2 - A^2}{4}}$$

where A = minimum width of the key.

Bottom of Keyslot to Opposite Side of Shaft H. Obtain by subtracting the versed sine G and depth of keyslot B, from the shaft diameter L.

Top of Key to Opposite Side of Shaft J. Obtain by subtracting the versed sine G, from the shaft diameter L, and then adding to this figure the height of key above shaft G.

Bottom of Keyway to Opposite Side of Bore K. Obtain by subtracting the versed sine G, from the shaft diameter L, and then adding to this figure the depth of keyway E.

Normally, for inspection purposes, dimension J (see illustration with Table 9) is required. According to the above relationships $J = L - G + C$. Optimum key width normally equals one-quarter of the shaft diameter. Assume $L = 1\frac{1}{2}$ in. Then A (width of key) $= \frac{3}{8}$ in. Probably a No. 1210 key ($\frac{3}{8}$ by $1\frac{1}{4}$ in.) will have sufficient shear area (see Table 7). From Table 8, the value of C for a No. 1210 key equals 0.1875, and the value of G (see Table 9) when $L = 1\frac{1}{2}$ in. and $A = \frac{3}{8}$ in. is 0.0238 in. Therefore,

$$J = 1.5000 - 0.0238 + 0.1875 = 1.6637 \text{ in.}$$

To save calculations, Table 10 can be used to find values of J for keys from $\frac{1}{16}$ to $\frac{3}{4}$ in. wide and shafts from $\frac{1}{2}$ to 3 in. dia.

Dimensional data on dowel, straight, clevis, taper, and cotter pins are given in Tables 13 to 21.

TABLE 1. SQUARE AND FLAT PLAIN PARALLEL STOCK KEYS

Shaft Dia	Square Key $W \times H$	Flat Key $W \times H$	Tolerance on W and H (−)	Bottom of Keyseat to Opposite Side of Shaft	
				Square Key S	Flat Key T
1/2	1/8 × 1/8	1/8 × 3/32	0.0020	0.430	0.445
9/16	1/8 × 1/8	1/8 × 3/32	0.0020	0.493	0.509
5/8	3/16 × 3/16	3/16 × 1/8	0.0020	0.517	0.548
11/16	3/16 × 3/16	3/16 × 1/8	0.0020	0.581	0.612
3/4	3/16 × 3/16	3/16 × 1/8	0.0020	0.644	0.676
13/16	3/16 × 3/16	3/16 × 1/8	0.0020	0.708	0.739
7/8	3/16 × 3/16	3/16 × 1/8	0.0020	0.771	0.802
15/16	1/4 × 1/4	1/4 × 3/16	0.0020	0.796	0.827
1	1/4 × 1/4	1/4 × 3/16	0.0020	0.859	0.890
1 1/16	1/4 × 1/4	1/4 × 3/16	0.0020	0.923	0.954
1 1/8	1/4 × 1/4	1/4 × 3/16	0.0020	0.986	1.017
1 3/16	1/4 × 1/4	1/4 × 3/16	0.0020	1.049	1.081
1 1/4	1/4 × 1/4	1/4 × 3/16	0.0020	1.112	1.144
1 5/16	5/16 × 5/16	5/16 × 1/4	0.0020	1.137	1.169
1 3/8	5/16 × 5/16	5/16 × 1/4	0.0020	1.201	1.232
1 7/16	3/8 × 3/8	3/8 × 1/4	0.0020	1.225	1.288
1 1/2	3/8 × 3/8	3/8 × 1/4	0.0020	1.289	1.351
1 9/16	3/8 × 3/8	3/8 × 1/4	0.0020	1.352	1.415
1 5/8	3/8 × 3/8	3/8 × 1/4	0.0020	1.416	1.478
1 11/16	3/8 × 3/8	3/8 × 1/4	0.0020	1.479	1.542
1 3/4	3/8 × 3/8	3/8 × 1/4	0.0020	1.542	1.605
1 13/16	1/2 × 1/2	1/2 × 3/8	0.0025	1.527	1.590
1 7/8	1/2 × 1/2	1/2 × 3/8	0.0025	1.591	1.654
1 15/16	1/2 × 1/2	1/2 × 3/8	0.0025	1.655	1.717
2	1/2 × 1/2	1/2 × 3/8	0.0025	1.718	1.781
2 1/16	1/2 × 1/2	1/2 × 3/8	0.0025	1.782	1.843
2 1/8	1/2 × 1/2	1/2 × 3/8	0.0025	1.845	1.908
2 3/16	1/2 × 1/2	1/2 × 3/8	0.0025	1.909	1.971
2 1/4	1/2 × 1/2	1/2 × 3/8	0.0025	1.972	2.034
2 5/16	5/8 × 5/8	5/8 × 7/16	0.0025	1.957	2.051
2 3/8	5/8 × 5/8	5/8 × 7/16	0.0025	2.021	2.114
2 7/16	5/8 × 5/8	5/8 × 7/16	0.0025	2.084	2.178
2 1/2	5/8 × 5/8	5/8 × 7/16	0.0025	2.148	2.242
2 5/8	5/8 × 5/8	5/8 × 7/16	0.0025	2.275	2.368
2 3/4	5/8 × 5/8	5/8 × 7/16	0.0025	2.402	2.495
2 7/8	3/4 × 3/4	3/4 × 1/2	0.0025	2.450	2.575
2 15/16	3/4 × 3/4	3/4 × 1/2	0.0025	2.514	2.639
3	3/4 × 3/4	3/4 × 1/2	0.0025	2.577	2.702
3 1/8	3/4 × 3/4	3/4 × 1/2	0.0025	2.704	2.829
3 1/4	3/4 × 3/4	3/4 × 1/2	0.0025	2.831	2.956
3 3/8	7/8 × 7/8	7/8 × 5/8	0.0030	2.880	3.005
3 7/16	7/8 × 7/8	7/8 × 5/8	0.0030	2.944	3.069
3 1/2	7/8 × 7/8	7/8 × 5/8	0.0030	3.007	3.132

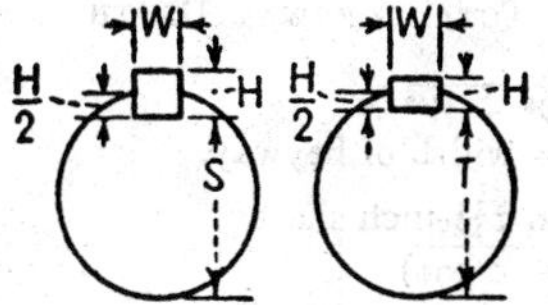

TABLE 1. SQUARE AND FLAT PLAIN PARALLEL STOCK KEYS (*Continued*)

Shaft Dia	Square Key $W \times H$	Flat Key $W \times H$	Tolerance on W and H (−)	Bottom of Keyseat to Opposite Side of Shaft: Square Key S	Bottom of Keyseat to Opposite Side of Shaft: Flat Key T
$3\frac{5}{8}$	$\frac{7}{8} \times \frac{7}{8}$	$\frac{7}{8} \times \frac{5}{8}$	0.0030	3.140	3.259
$3\frac{3}{4}$	$\frac{7}{8} \times \frac{7}{8}$	$\frac{7}{8} \times \frac{5}{8}$	0.0030	3.261	3.386
$3\frac{7}{8}$	1×1	$1 \times \frac{3}{4}$	0.0030	3.309	3.434
$3\frac{15}{16}$	1×1	$1 \times \frac{3}{4}$	0.0030	3.373	3.498
4	1×1	$1 \times \frac{3}{4}$	0.0030	3.437	3.562
$4\frac{1}{4}$	1×1	$1 \times \frac{3}{4}$	0.0030	3.690	3.815
$4\frac{7}{16}$	1×1	$1 \times \frac{3}{4}$	0.0030	3.881	4.006
$4\frac{1}{2}$	1×1	$1 \times \frac{3}{4}$	0.0030	3.944	4.069
$4\frac{3}{4}$	$1\frac{1}{4} \times 1\frac{1}{4}$	$1\frac{1}{4} \times \frac{7}{8}$	0.0030	4.042	4.229
$4\frac{15}{16}$	$1\frac{1}{4} \times 1\frac{1}{4}$	$1\frac{1}{4} \times \frac{7}{8}$	0.0030	4.232	4.420
5	$1\frac{1}{4} \times 1\frac{1}{4}$	$1\frac{1}{4} \times \frac{7}{8}$	0.0030	4.296	4.483
$5\frac{1}{4}$	$1\frac{1}{4} \times 1\frac{1}{4}$	$1\frac{1}{4} \times \frac{7}{8}$	0.0030	4.550	4.733
$5\frac{7}{16}$	$1\frac{1}{4} \times 1\frac{1}{4}$	$1\frac{1}{4} \times \frac{7}{8}$	0.0030	4.740	4.927
$5\frac{1}{2}$	$1\frac{1}{4} \times 1\frac{1}{4}$	$1\frac{1}{4} \times \frac{7}{8}$	0.0030	4.803	4.991
$5\frac{3}{4}$	$1\frac{1}{2} \times 1\frac{1}{2}$	$1\frac{1}{2} \times 1$	0.0030	4.900	5.150
$5\frac{15}{16}$	$1\frac{1}{2} \times 1\frac{1}{2}$	$1\frac{1}{2} \times 1$	0.0030	5.091	5.341
6	$1\frac{1}{2} \times 1\frac{1}{2}$	$1\frac{1}{2} \times 1$	0.0030	5.155	5.405

ASA, October, 1943.

TABLE 2. LARGE PLAIN PARALLEL STOCK KEYS*

$W \times H$	Tolerance*,† on W and H (−)	$W \times H$	Tolerance*,† on W and H (−)
$1\frac{3}{4} \times 1\frac{1}{4}$	0.0040	$3\frac{1}{2} \times 2\frac{1}{2}$	0.0050
$2 \times 1\frac{1}{2}$	0.0040	4×3	0.0050
$2\frac{1}{2} \times 1\frac{3}{4}$	0.0040	$5 \times 3\frac{1}{2}$	0.0050
3×2	0.0040	6×4	0.0050

* Stock keys are applicable to the general run of work and the tolerances have been set accordingly. They are not intended to cover the finer applications where a closer fit may be required.

† These tolerances are *negative* and represent the maximum allowable variation *below* the exact nominal size. For example, for the standard 3- by 2-in. key the maximum size is 3.000 × 2.000 in., and the minimum size is 2.996 × 1.996 in.

TABLE 3. FINDING TOTAL KEYWAY DEPTH

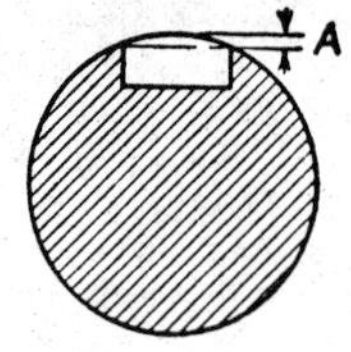

Formula: $A = \frac{1}{2}(D - \sqrt{D^2 - W^2})$.
D = shaft diameter. W = width of keyway.

EXAMPLE: $\frac{5}{8}$-inch keyway in $1\frac{11}{16}$-inch shaft

$$A = \frac{1}{2}(1.6875 - \sqrt{2.845 - 0.390})$$
$$= \frac{1}{2}(1.687 - 1.566)$$
$$= 0.060 \text{ in.}$$

In the column marked Size of Shaft find the number representing the size; then to the right find the column representing the keyway to be cut and the decimal there is the distance A, which added to the depth of the keyway will give the total depth from the point where the cutter first begins to cut.

Size of Shaft	$\frac{1}{4}$ Keyway	$\frac{5}{16}$ Keyway	$\frac{3}{8}$ Keyway	$\frac{7}{16}$ Keyway	$\frac{1}{2}$ Keyway
$\frac{1}{2}$	0.0325				
$\frac{9}{16}$	0.0289				
$\frac{5}{8}$	0.0254	0.0413			
$\frac{11}{16}$	0.0236	0.0379			
$\frac{3}{4}$	0.022	0.0346	0.0511		
$\frac{13}{16}$	0.0198	0.0314	0.0465		
$\frac{7}{8}$	0.0177	0.0283	0.042	0.0583	
$\frac{15}{16}$	0.0164	0.0264	0.0392	0.0544	
1	0.0152	0.0246	0.0365	0.0506	0.067
$1\frac{1}{16}$	0.0143	0.0228	0.0342	0.0476	0.0625
$1\frac{1}{8}$	0.0136	0.021	0.0319	0.0446	0.0581
$1\frac{3}{16}$	0.0131	0.0204	0.0304	0.0421	0.0551
$1\frac{1}{4}$	0.0127	0.0198	0.029	0.0397	0.0522
$1\frac{5}{16}$	0.0123	0.0191	0.0279	0.038	0.0499
$1\frac{3}{8}$	0.012	0.0185	0.0268	0.0364	0.0477
$1\frac{7}{16}$	0.0114	0.0174	0.0254	0.0346	0.0453
$1\frac{1}{2}$	0.011	0.0164	0.024	0.0328	0.0429
$1\frac{9}{16}$	0.0107	0.0158	0.0231	0.0309	0.0412
$1\frac{5}{8}$	0.0105	0.0153	0.0221	0.0291	0.0395
$1\frac{11}{16}$	0.0102	0.0147	0.0214	0.0282	0.0383
$1\frac{3}{4}$	0.0099	0.0142	0.0207	0.0274	0.0371
$1\frac{13}{16}$	0.0095	0.0136	0.0198	0.0265	0.0355
$1\frac{7}{8}$	0.0093	0.013	0.019	0.0257	0.0339
$1\frac{15}{16}$	0.009	0.0127	0.0184	0.025	0.0328
2	0.0088	0.0124	0.0179	0.0243	0.0317
$2\frac{1}{16}$	0.0083	0.0117	0.0173	0.0236	0.0308
$2\frac{1}{8}$	0.0078	0.0111	0.0168	0.0229	0.0299
$2\frac{3}{16}$	0.0073	0.0109	0.0163	0.0222	0.0291
$2\frac{1}{4}$	0.007	0.0107	0.0159	0.0216	0.0282
$2\frac{5}{16}$	0.0068	0.0104	0.0155	0.0209	0.0274
$2\frac{3}{8}$	0.0066	0.0102	0.0152	0.0202	0.0267
$2\frac{7}{16}$	0.0064	0.01	0.0149	0.0198	0.026
$2\frac{1}{2}$	0.0063	0.0098	0.0146	0.0194	0.0253

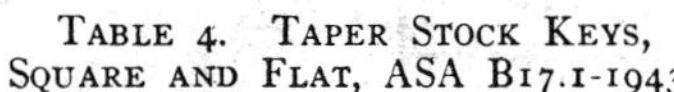

TABLE 4. TAPER STOCK KEYS, SQUARE AND FLAT, ASA B17.1-1943

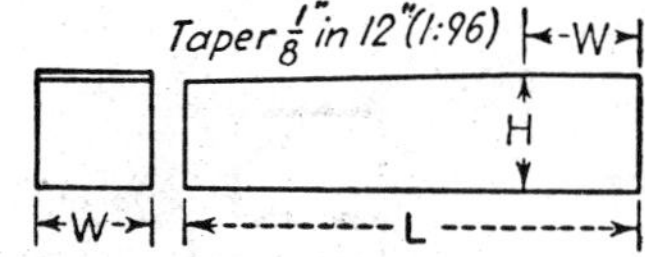

Diameters of Shafts (Inclusive)	Square Type		Flat Type		Tolerances on Keys	
	Max Width W	Height at Large End H	Max Width W	Height at Large End H	Width (−)	Height (+)
1/2 to 9/16	1/8	1/8	1/8	3/32	0.0020	0.0020
5/8 to 7/8	3/16	3/16	3/16	1/8	0.0020	0.0020
15/16 to 1 1/4	1/4	1/4	1/4	3/16	0.0020	0.0020
1 5/16 to 1 3/4	3/8	3/8	3/8	1/4	0.0020	0.0020
1 13/16 to 2 1/4	1/2	1/2	1/2	3/8	0.0025	0.0025
2 5/16 to 2 3/4	5/8	5/8	5/8	7/16	0.0025	0.0025
2 7/8 to 3 1/4	3/4	3/4	3/4	1/2	0.0025	0.0025
3 3/8 to 3 3/4	7/8	7/8	7/8	5/8	0.0030	0.0030
3 7/8 to 4 1/2	1	1	1	3/4	0.0030	0.0030
4 3/4 to 5 1/2	1 1/4	1 1/4	1 1/4	7/8	0.0030	0.0030
5 3/4 to 6	1 1/2	1 1/2	1 1/2	1	0.0030	0.0030

TABLE 5. STOCK LENGTHS OF PLAIN TAPER STOCK KEYS

Shaft Dia (Inclusive)	Length of Key,* L						
1/2 – 9/16	1/2	3/4	1	1 1/4	1 1/2	1 3/4	2
5/8 – 7/8	3/4	1 1/8	1 1/2	1 7/8	2 1/4	2 5/8	3
15/16–1 1/4	1	1 1/2	2	2 1/2	3	3 1/2	4
1 5/16–1 3/8	1 1/4	1 7/8	2 1/2	3 1/8	3 3/4	4 1/2	5 1/4
1 7/16–1 3/4	1 1/2	2 1/4	3	3 3/4	4 1/2	5 1/4	6
1 13/16–2 1/4	2	3	4	5	6	7	8
2 5/16–2 3/4	2 1/2	3 3/4	5	6 1/4	7 1/2	8 3/4	10
2 7/8 –3 1/4	3	4 1/2	6	7 1/2	9	10 1/2	12
3 3/8 –3 3/4	3 1/2	5 1/4	7	8 3/4	10 1/2	12 1/4	14
3 7/8 –4 1/2	4	6	8	10	12	14	16
4 3/4 –5 1/2	5	7 1/2	10	12 1/2	15	17 1/2	20
5 3/4 –6	6	9	12	15	18	21	24

* The minimum stock length of keys is equal to four times the key width, and the maximum stock length is equal to sixteen times the key width. The increments of increase in length are equal to twice the width

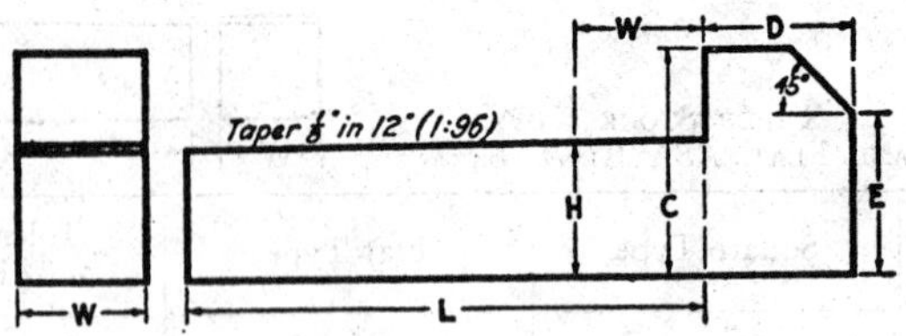

Table 6. Gib-head Taper Stock Keys

Shaft Dia	Square Type					Flat Type					Tolerance	
	Key		Gib Head			Key		Gib Head			Width (−)	Height (+)
(Incl)	*W*	*H*	*C*	*D*	*E*	*W*	*H*	*C*	*D*	*E*		
1/2 – 9/16	1/8	1/8	1/4	7/32	5/32	1/8	3/32	3/16	1/8	1/8	0.0020	0.0020
5/8 – 7/8	3/16	3/16	5/16	9/32	7/32	3/16	1/8	1/4	3/16	5/32	0.0020	0.0020
15/16 – 1 1/4	1/4	1/4	7/16	11/32	11/32	1/4	3/16	5/16	1/4	3/16	0.0020	0.0020
1 5/16 – 1 3/8	5/16	5/16	9/16	13/32	13/32	5/16	1/4	3/8	5/16	1/4	0.0020	0.0020
1 7/16 – 1 3/4	3/8	3/8	11/16	15/32	15/32	3/8	1/4	7/16	3/8	5/16	0.0020	0.0020
1 13/16 – 2 1/4	1/2	1/2	7/8	19/32	5/8	1/2	3/8	5/8	1/2	7/16	0.0025	0.0025
2 5/16 – 2 3/4	5/8	5/8	1 1/16	23/32	3/4	5/8	7/16	3/4	5/8	1/2	0.0025	0.0025
2 7/8 – 3 1/4	3/4	3/4	1 1/4	7/8	7/8	3/4	1/2	7/8	3/4	5/8	0.0025	0.0025
3 3/8 – 3 3/4	7/8	7/8	1 1/2	1	1	7/8	5/8	1 1/16	7/8	3/4	0.0030	0.0030
3 7/8 – 4 1/2	1	1	1 3/4	1 3/16	1 3/16	1	3/4	1 1/4	1	13/16	0.0030	0.0030
4 3/4 – 5 1/2	1 1/4	1 1/4	2	1 7/16	1 7/16	1 1/4	7/8	1 1/2	1 1/4	1	0.0030	0.0030
3/4 – 6	1 1/2	1 1/2	2 1/2	1 3/4	1 3/4	1 1/2	1	1 3/4	1 1/2	1 1/4	0.0030	0.0030

Note: Stock keys are applicable to the general run of work and the tolerances have been set accordingly. They are not intended to cover the finer applications where a closer fit may be required.

Note: Height of the flat key is measured at the distance *W*, equal to the width of the key, from the gib head.

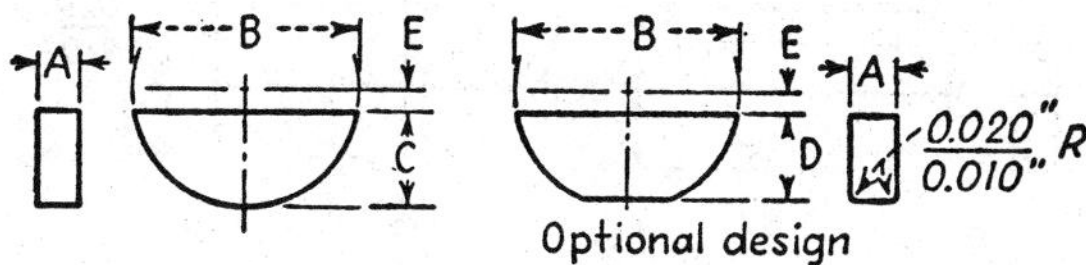

TABLE 7. WOODRUFF KEYS

Amer. Std. No.*	Old Std.† No.	Size	A +0.001 −0.000	B +0.000 −0.010	C +0.000 −0.005	D +0.000 −0.006	E Nominal	Key Area at Shear Line
202	201	1/16 × 1/4	0.0625	0.250	0.109	.....	1/64	0.0145
202½	206	1/16 × 5/16	0.0625	0.312	0.140	.....	1/64	0.0184
302½	207	3/32 × 5/16	0.0938	0.312	0.140	.....	1/64	0.0264
203	211	1/16 × 3/8	0.0625	0.375	0.172	.....	1/64	0.0225
303	212	3/32 × 3/8	0.0938	0.375	0.172	.....	1/64	0.0328
403	213	1/8 × 3/8	0.1250	0.375	0.172	.....	3/64	0.0420
204	1	1/16 × 1/2	0.0625	0.500	0.203	0.194	3/64	0.0296
304	2	3/32 × 1/2	0.0938	0.500	0.203	0.194	3/64	0.0434
404	3	1/8 × 1/2	0.1250	0.500	0.203	0.194	3/64	0.0512
305	4	3/32 × 5/8	0.0938	0.625	0.250	0.240	1/16	0.0523
405	5	1/8 × 5/8	0.1250	0.625	0.250	0.240	1/16	0.0716
505	6	5/32 × 5/8	0.1563	0.625	0.250	0.240	1/16	0.0871
605	61	3/16 × 5/8	0.1875	0.625	0.250	0.240	1/16	0.0105
406	7	1/8 × 3/4	0.1250	0.750	0.313	0.303	1/16	0.0884
506	8	5/32 × 3/4	0.1563	0.750	0.313	0.303	1/16	0.1086
606	9	3/16 × 3/4	0.1875	0.750	0.313	0.303	1/16	0.1279
806	91	1/4 × 3/4	0.2500	0.750	0.313	0.303	1/16	0.1623
507	10	5/32 × 7/8	0.1563	0.875	0.375	0.365	1/16	0.1294
607	11	3/16 × 7/8	0.1875	0.875	0.375	0.365	1/16	0.1531
707	12	7/32 × 7/8	0.2188	0.875	0.375	0.365	1/16	0.1813
807	A	1/4 × 7/8	0.2500	0.875	0.375	0.365	1/16	0.1976
608	13	3/16 × 1	0.1875	1.000	0.438	0.428	1/16	0.1781
708	14	7/32 × 1	0.2188	1.000	0.438	0.428	1/16	0.2100
808	15	1/4 × 1	0.2500	1.000	0.438	0.428	1/16	0.2320
1008	B	5/16 × 1	0.3125	1.000	0.438	0.428	1/16	0.2811
609	16	3/16 × 1 1/8	0.1875	1.125	0.484	0.475	5/64	0.2007
709	17	7/32 × 1 1/8	0.2188	1.125	0.484	0.475	5/64	0.2320
809	18	1/4 × 1 1/8	0.2500	1.125	0.484	0.475	5/64	0.2622
1009	C	5/16 × 1 1/8	0.3125	1.125	0.484	0.475	5/64	0.3193
610	19	3/16 × 1 1/4	0.1875	1.250	0.547	0.537	5/64	0.2284
710	20	7/32 × 1 1/4	0.2188	1.250	0.547	0.537	5/64	0.2608
810	21	1/4 × 1 1/4	0.2500	1.250	0.547	0.537	5/64	0.2955
1010	D	5/16 × 1 1/4	0.3125	1.250	0.547	0.537	5/64	0.3621
1210	E	3/8 × 1 1/4	0.3750	1.250	0.547	0.537	5/64	0.4243
811	22	1/4 × 1 3/8	0.2500	1.375	0.594	0.584	3/32	0.3259
1011	23	5/16 × 1 3/8	0.3125	1.375	0.594	0.584	3/32	0.4003
1211	F	3/8 × 1 3/8	0.3750	1.375	0.594	0.584	3/32	0.4705
812	24	1/4 × 1 1/2	0.2500	1.500	0.641	0.631	7/64	0.3562
1012	25	5/16 × 1 1/2	0.3125	1.500	0.641	0.631	7/64	0.4384
1212	G	3/8 × 1 1/2	0.3750	1.500	0.641	0.631	7/64	0.5166

* Numbers of standard shank-type cutters, and expected numbers in future revision of Woodruff-Key standard.
† SAE listing of manufacturers' part numbers.

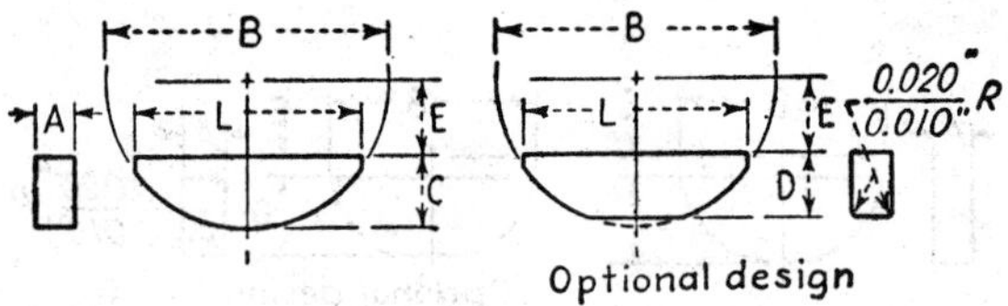

TABLE 7. WOODRUFF KEYS (*Continued*)

Amer. Std.‡ No.	Old Std.† No.	Size	*A* +0.001 −0.000	*B* +0.000 −0.010	*C* +0.000 −0.005	*D* +0.000 −0.006	*E* Nominal	*L* +0.000 −0.010	Key Area at Shear Line
	126	$\frac{3}{16} \times 2\frac{1}{8}$	0.1875	2.125	0.406	0.396	$\frac{21}{32}$	1.380	0.2578
	127	$\frac{1}{4} \times 2\frac{1}{8}$	0.2500	2.125	0.406	0.396	$\frac{21}{32}$	1.380	0.3437
	128	$\frac{5}{16} \times 2\frac{1}{8}$	0.3125	2.125	0.406	0.396	$\frac{21}{32}$	1.380	0.4296
	129	$\frac{3}{8} \times 2\frac{1}{8}$	0.3750	2.125	0.406	0.396	$\frac{21}{32}$	1.380	0.4833
617	26	$\frac{3}{16} \times 2\frac{1}{8}$	0.1875	2.125	0.531	0.521	$\frac{17}{32}$	1.723	0.3222
817	27	$\frac{1}{4} \times 2\frac{1}{8}$	0.2500	2.125	0.531	0.521	$\frac{17}{32}$	1.723	0.4178
1017	28	$\frac{5}{16} \times 2\frac{1}{8}$	0.3125	2.125	0.531	0.521	$\frac{17}{32}$	1.723	0.5062
1217	29	$\frac{3}{8} \times 2\frac{1}{8}$	0.3750	2.125	0.531	0.521	$\frac{17}{32}$	1.723	0.5868
	Rx	$\frac{1}{4} \times 2\frac{3}{4}$	0.2500	2.750	0.594	0.584	$\frac{25}{32}$	2.000	0.5000
	Sx	$\frac{5}{16} \times 2\frac{3}{4}$	0.3125	2.750	0.594	0.584	$\frac{25}{32}$	2.000	0.6286
	Tx	$\frac{3}{8} \times 2\frac{3}{4}$	0.3750	2.750	0.594	0.584	$\frac{25}{32}$	2.000	0.6943
	Ux	$\frac{7}{16} \times 2\frac{3}{4}$	0.4375	2.750	0.594	0.584	$\frac{25}{32}$	2.000	0.8253
	Vx	$\frac{1}{2} \times 2\frac{3}{4}$	0.5000	2.750	0.594	0.584	$\frac{25}{32}$	2.000	0.9094
822	R	$\frac{1}{4} \times 2\frac{3}{4}$	0.2500	2.750	0.750	0.740	$\frac{5}{8}$	2.317	0.5718
1022	S	$\frac{5}{16} \times 2\frac{3}{4}$	0.3125	2.750	0.750	0.740	$\frac{5}{8}$	2.317	0.7071
1222	T	$\frac{3}{8} \times 2\frac{3}{4}$	0.3750	2.750	0.750	0.740	$\frac{5}{8}$	2.317	0.8319
1422	U	$\frac{7}{16} \times 2\frac{3}{4}$	0.4375	2.750	0.750	0.740	$\frac{5}{8}$	2.317	0.9499
1622	V	$\frac{1}{2} \times 2\frac{3}{4}$	0.5000	2.750	0.750	0.740	$\frac{5}{8}$	2.317	1.0606
1228	30	$\frac{3}{8} \times 3\frac{1}{2}$	0.3750	3.500	0.938	0.927	$\frac{13}{16}$	2.880	1.0781
1428	31	$\frac{7}{16} \times 3\frac{1}{2}$	0.4375	3.500	0.938	0.927	$\frac{13}{16}$	2.880	1.2371
1628	32	$\frac{1}{2} \times 3\frac{1}{2}$	0.5000	3.500	0.938	0.927	$\frac{13}{16}$	2.880	1.3905
1828	33	$\frac{9}{16} \times 3\frac{1}{2}$	0.5625	3.500	0.938	0.927	$\frac{13}{16}$	2.880	1.5368
2028	34	$\frac{5}{8} \times 3\frac{1}{2}$	0.6250	3.500	0.938	0.927	$\frac{13}{16}$	2.880	1.6755
2228	35	$\frac{11}{16} \times 3\frac{1}{2}$	0.6875	3.500	0.938	0.927	$\frac{13}{16}$	2.880	1.8062
2428	36	$\frac{3}{4} \times 3\frac{1}{2}$	0.7500	3.500	0.938	0.927	$\frac{13}{16}$	2.880	1.9281

Material: Carbon steel or alloy heat-treated steel as specified.
Carbon steel keys to be 0.30 carbon min, with hardness of Rockwell 10 C min.
Alloy-steel keys to be SAE 2330 or 8630 steel, heat-treated to a hardness of Rockwell 40 to 50 C; or other alloy steels having equal physical properties at the same hardness.

Alloy heat-treated keys are marked with depressions on the top to distinguish them from carbon-steel keys.

‡ Numbers of standard arbor-type cutters, and expected numbers in future revision of Woodruff-Key standard.

† SAE listing of manufacturers' part numbers.

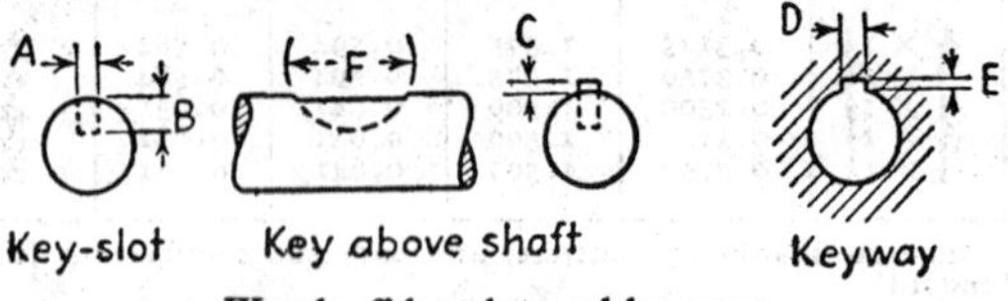

Woodruff keyslot and keyway.

TABLE 8. WOODRUFF KEYSLOT AND KEYWAY DIMENSIONS

Amer. Std. No.	Old Std.* No.	Size	Keyslot					Key above Shaft	Keyway	
			A		*B*	*F*		*C*	*D*	*E*
			Min	Max	+.005 −.000	Min	Max	±.005	+.002 −.000	+.005 −.000
202†	201	1/16 × 1/4	0.0615	0.0630	0.0728	0.255	0.260	0.0312	0.0635	0.0372
202½	206	1/16 × 5/16	0.0615	0.0630	0.1038	0.317	0.322	0.0312	0.0635	0.0372
302½	207	3/32 × 5/16	0.0928	0.0943	0.0882	0.317	0.322	0.0469	0.0948	0.0529
203	211	1/16 × 3/8	0.0615	0.0630	0.1358	0.380	0.385	0.0312	0.0635	0.0372
303	212	3/32 × 3/8	0.0928	0.0943	0.1202	0.380	0.385	0.0469	0.0984	0.0529
403	213	1/8 × 3/8	0.1240	0.1255	0.1045	0.380	0.385	0.0625	0.1260	0.0685
204	1	1/16 × 1/2	0.0615	0.0630	0.1668	0.510	0.515	0.0312	0.0635	0.0372
304	2	3/32 × 1/2	0.0928	0.0943	0.1511	0.510	0.515	0.0469	0.0948	0.0529
404	3	1/8 × 1/2	0.1240	0.1255	0.1355	0.510	0.515	0.0625	0.1260	0.0685
305	4	3/32 × 5/8	0.0928	0.0943	0.1981	0.635	0.640	0.0469	0.0948	0.0529
405	5	1/8 × 5/8	0.1240	0.1255	0.1825	0.635	0.640	0.0625	0.1260	0.0685
505	6	5/32 × 5/8	0.1553	0.1568	0.1669	0.635	0.640	0.0781	0.1573	0.0841
605	61	3/16 × 5/8	0.1863	0.1880	0.1513	0.635	0.640	0.0937	0.1885	0.0997
406	7	1/8 × 3/4	0.1240	0.1255	0.2455	0.760	0.765	0.0625	0.1260	0.0685
506	8	5/32 × 3/4	0.1553	0.1568	0.2299	0.760	0.765	0.0781	0.1573	0.0841
606	9	3/16 × 3/4	0.1863	0.1880	0.2143	0.760	0.765	0.0937	0.1885	0.0997
806	91	1/4 × 3/4	0.2487	0.2505	0.1830	0.760	0.765	0.1250	0.2510	0.1310
507	10	5/32 × 7/8	0.1553	0.1568	0.2919	0.887	0.892	0.0781	0.1573	0.0841
607	11	3/16 × 7/8	0.1863	0.1880	0.2763	0.887	0.892	0.0937	0.1885	0.0997
707	12	7/32 × 7/8	0.2175	0.2193	0.2607	0.887	0.892	0.1093	0.2198	0.1153
807	A	1/4 × 7/8	0.2487	0.2505	0.2450	0.887	0.892	0.1250	0.2510	0.1310
608	13	3/16 × 1	0.1863	0.1880	0.3393	1.012	1.017	0.0937	0.1885	0.0997
708	14	7/32 × 1	0.2175	0.2193	0.3237	1.012	1.017	0.1093	0.2198	0.1153
808	15	1/4 × 1	0.2487	0.2505	0.3080	1.012	1.017	0.1250	0.2510	0.1310
1008	B	5/16 × 1	0.3111	0.3130	0.2768	1.012	1.017	0.1562	0.3135	0.1622
609	16	3/16 × 1⅛	0.1863	0.1880	0.3853	1.137	1.142	0.0937	0.1885	0.0997
709	17	7/32 × 1⅛	0.2175	0.2193	0.3697	1.137	1.142	0.1093	0.2198	0.1153
809	18	1/4 × 1⅛	0.2487	0.2505	0.3540	1.137	1.142	0.1250	0.2510	0.1310
1009	C	5/16 × 1⅛	0.3111	0.3130	0.3228	1.137	1.142	0.1562	0.3135	0.1622
610	19	3/16 × 1¼	0.1863	0.1880	0.4483	1.265	1.270	0.0937	0.1885	0.0997
710	20	7/32 × 1¼	0.2175	0.2193	0.4327	1.265	1.270	0.1093	0.2198	0.1153
810	21	1/4 × 1¼	0.2487	0.2505	0.4170	1.265	1.270	0.1250	0.2510	0.1310
1010	D	5/16 × 1¼	0.3111	0.3130	0.3858	1.265	1.270	0.1562	0.3135	0.1622
1210	E	3/8 × 1¼	0.3735	0.3755	0.3545	1.265	1.270	0.1875	0.3760	0.1935
811	22	1/4 × 1⅜	0.2487	0.2505	0.4640	1.390	1.395	0.1250	0.2510	0.1310
1011	23	5/16 × 1⅜	0.3111	0.3130	0.4328	1.390	1.395	0.1562	0.3135	0.1622
1211	F	3/8 × 1⅜	0.3735	0.3755	0.4015	1.390	1.395	0.1875	0.3760	0.1935
812	24	1/4 × 1½	0.2487	0.2505	0.5110	1.515	1.520	0.1250	0.2510	0.1310
1012	25	5/16 × 1½	0.3111	0.3130	0.4798	1.515	1.520	0.1562	0.3135	0.1622
1212	G	3/8 × 1½	0.3735	0.3755	0.4485	1.515	1.520	0.0875	0.3760	0.1935
	126	3/16 × 2⅛	0.1863	0.1880	0.3073	2.125	2.135	0.0937	0.1885	0.0997
	127	1/4 × 2⅛	0.2487	0.2505	0.2760	2.125	2.135	0.1250	0.2510	0.1310
	128	5/16 × 2⅛	0.3111	0.3130	0.2448	2.125	2.135	0.1562	0.3135	0.1622
	129	3/8 × 2⅛	0.3735	0.3755	0.2135	2.125	2.135	0.1875	0.3760	0.1935
617‡	26	3/16 × 2⅛	0.1863	0.1880	0.4323	2.125	2.135	0.0937	0.1885	0.0997
817	27	1/4 × 2⅛	0.2487	0.2505	0.4010	2.125	2.135	0.1250	0.2510	0.1310
1017	28	5/16 × 2⅛	0.3111	0.3130	0.3698	2.125	2.135	0.1562	0.3135	0.1622
1217	29	3/8 × 2⅛	0.3735	0.3755	0.3385	2.125	2.135	0.1875	0.3760	0.1935
	Rx	1/4 × 2¾	0.2487	0.2505	0.4640	2.750	2.760	0.1250	0.2510	0.1310
	Sx	5/16 × 2¾	0.3111	0.3130	0.4328	2.750	2.760	0.1562	0.3135	0.1622

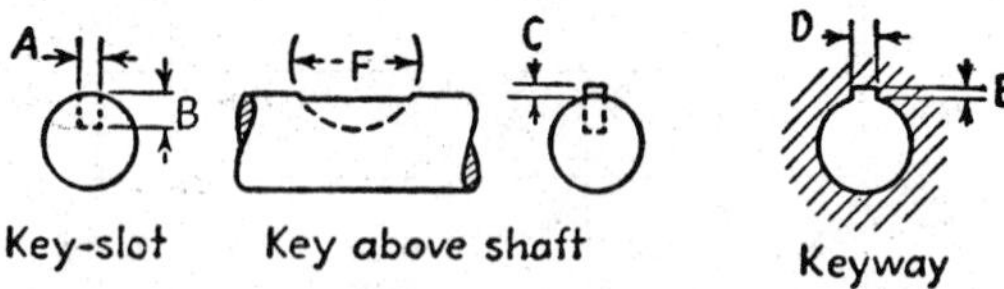

Table 8. Woodruff Keyslot and Keyway Dimensions (*Continued*)

Amer. Std. No.	Old Std.* No.	Size	Keyslot					Key above Shaft	Keyway	
			A		*B*	*F*		*C*	*D*	*E*
			Min	Max	+.005 −.000	Min	Max	±.005	+.002 −.000	+.005 −.000
	Tx	$\frac{3}{8} \times 2\frac{3}{4}$	0.3735	0.3755	0.4015	2.750	2.760	0.1875	0.3760	0.1935
	Ux	$\frac{7}{16} \times 2\frac{3}{4}$	0.4360	0.4380	0.3703	2.750	2.760	0.2187	0.4385	0.2247
	Vx	$\frac{1}{2} \times 2\frac{3}{4}$	0.4985	0.5005	0.3390	2.750	2.760	0.2500	0.5010	0.2560
822	R	$\frac{1}{4} \times 2\frac{3}{4}$	0.2487	0.2505	0.6200	2.750	2.760	0.1250	0.2510	0.1310
1022	S	$\frac{5}{16} \times 2\frac{3}{4}$	0.3111	0.3130	0.5888	2.750	2.760	0.1562	0.3135	0.1622
1222	T	$\frac{3}{8} \times 2\frac{3}{4}$	0.3735	0.3755	0.5575	2.750	2.760	0.1875	0.3760	0.1935
1422	U	$\frac{7}{16} \times 2\frac{3}{4}$	0.4360	0.4380	0.5263	2.750	2.760	0.2187	0.4385	0.2247
1622	V	$\frac{1}{2} \times 2\frac{3}{4}$	0.4985	0.5005	0.4950	2.750	2.760	0.2500	0.5010	0.2560
1228	30	$\frac{3}{8} \times 3\frac{1}{2}$	0.3735	0.3755	0.7455	3.500	3.510	0.1875	0.3760	0.1935
1428	31	$\frac{7}{16} \times 3\frac{1}{2}$	0.4360	0.4380	0.7143	3.500	3.510	0.2187	0.4385	0.2247
1628	32	$\frac{1}{2} \times 3\frac{1}{2}$	0.4985	0.5005	0.6830	3.500	3.510	0.2500	0.5010	0.2560
1828	‡33	$\frac{9}{16} \times 3\frac{1}{2}$	0.5610	0.5630	0.6518	3.500	3.510	0.2812	0.5635	0.2872
2028	34	$\frac{5}{8} \times 3\frac{1}{2}$	0.6235	0.6255	0.6205	3.500	3.510	0.3125	0.6260	0.3185
2228	35	$\frac{11}{16} \times 3\frac{1}{2}$	0.6860	0.6880	0.5893	3.500	3.510	0.3437	0.6885	0.3497
2428	36	$\frac{3}{4} \times 3\frac{1}{2}$	0.7485	0.7505	0.5580	3.500	3.510	0.3750	0.7510	0.3810

Width *A*. Dimensions shown are set with the maximum keyslot width as that figure which will receive a key with the greatest amount of looseness permissible to assure the key sticking in the slot.

Minimum keyslot width is that figure permitting the largest shaft distortion acceptable when assembling maximum key in minimum keyslot.

***B*, *C*, and *E*.** Dimensions to be taken at side intersection.

* SAE listing of manufacturers' part numbers.

† Numbers of standard shank-type cutters, and expected numbers in future revision of Woodruff-Key standard.

‡ Numbers of standard arbor-type cutters, and expected numbers in future revision of Woodruff Keyslot standard.

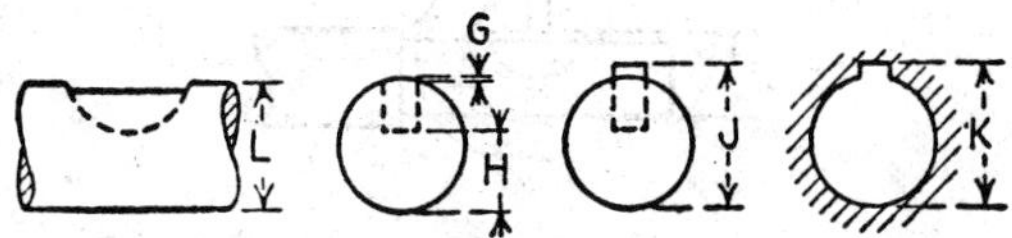

TABLE 9. WOODRUFF KEYSLOTS—VERSED SINE DIMENSION G*

Shaft Dia L	Keyway Width													
	1/16	3/32	1/8	5/32	3/16	1/4	5/16	3/8	7/16	1/2	9/16	5/8	11/16	3/4
0.3125	0.0032													
0.3437	0.0029	0.0065												
0.3750	0.0026	0.0060	0.0107											
0.4060	0.0024	0.0055	0.0099											
0.4375	0.0022	0.0051	0.0091											
0.4687	0.0021	0.0047	0.0085	0.0134										
0.5000	0.0020	0.0044	0.0079	0.0125										
0.5625		0.0039	0.0070	0.0111	0.0161									
0.6250		0.0035	0.0063	0.0099	0.0144									
0.6875		0.0032	0.0057	0.0090	0.0130	0.0235								
0.7500		0.0029	0.0052	0.0082	0.0119	0.0214	0.0341							
0.8125		0.0027	0.0048	0.0076	0.0110	0.0197	0.0312							
0.8750		0.0025	0.0045	0.0070	0.0102	0.0182	0.0288							
0.9375			0.0042	0.0066	0.0095	0.0170	0.0268	0.0391						
1.0000			0.0039	0.0061	0.0089	0.0159	0.0250	0.0365						
1.0625			0.0037	0.0058	0.0083	0.0149	0.0235	0.0342						
1.1250			0.0035	0.0055	0.0079	0.0141	0.0221	0.0322	0.0443					
1.1875			0.0033	0.0052	0.0074	0.0133	0.0209	0.0304	0.0418					
1.2500			0.0031	0.0049	0.0071	0.0126	0.0198	0.0288	0.0395					
1.3750				0.0045	0.0064	0.0115	0.0180	0.0261	0.0357	0.0471				
1.5000				0.0041	0.0059	0.0105	0.0165	0.0238	0.0326	0.0429				
1.6250				0.0038	0.0054	0.0097	0.0152	0.0219	0.0300	0.0394	0.0502			
1.7500					0.0050	0.0090	0.0141	0.0203	0.0278	0.0365	0.0464			
1.8750					0.0047	0.0084	0.0131	0.0189	0.0259	0.0340	0.0432	0.0536		
2.0000					0.0044	0.0078	0.0123	0.0177	0.0242	0.0318	0.0404	0.0501		
2.1250						0.0074	0.0116	0.0167	0.0228	0.0298	0.0379	0.0470	0.0572	0.0684
2.2500						0.0070	0.0109	0.0157	0.0215	0.0281	0.0357	0.0443	0.0538	0.0643
2.3750							0.0103	0.0149	0.0203	0.0266	0.0338	0.0419	0.0509	0.0608
2.5000								0.0141	0.0193	0.0253	0.0321	0.0397	0.0482	0.0576
2.6250								0.0135	0.0184	0.0240	0.0305	0.0377	0.0457	0.0547
2.7500									0.0175	0.0229	0.0291	0.0360	0.0437	0.0521
2.8750									0.0168	0.0219	0.0278	0.0344	0.0417	0.0498
3.0000										0.0210	0.0266	0.0329	0.0399	0.0476

* Listed for the different shaft sizes and keyway widths for reference in checking dimensions H, J, and K.

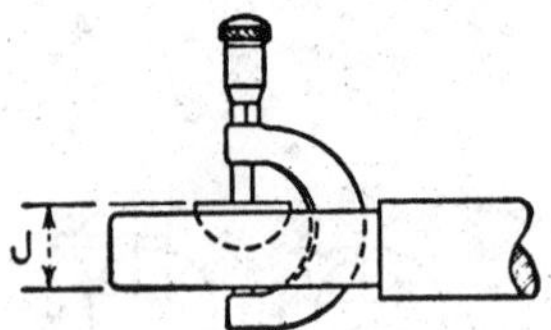

TABLE 10. VALUES OF *J* FOR ASSEMBLED-KEY INSPECTION

Keyway Width	1/16	3/32	1/8	5/32	3/16	1/4	5/16	3/8	7/16	1/2	5/8	3/4
Key Numbers	204	304 305	404 405 406	505 506 507	606 607 608 609	807 808 809 810 811 812	1008 1009 1010 1011 1012	1210 1211 1212				
Shaft Dia	**Values of *J***											
1/2	0.530	0.544	0.554									
9/16		0.605	0.617									
5/8		0.668	0.681	0.693								
11/16		0.731	0.744	0.756								
3/4		0.793	0.807	0.819	0.831							
13/16			0.870	0.883	0.895							
7/8			0.933	0.946	0.958							
15/16			0.995	1.009	1.021							
1			1.058	1.071	1.084	1.110						
1 1/16				1.134	1.147	1.173						
1 1/8				1.197	1.210	1.237						
1 3/16				1.260	1.273	1.299						
1 1/4				1.323	1.336	1.363	1.387					
1 5/16					1.339	1.425	1.450					
1 3/8					1.462	1.488	1.513					
1 7/16					1.525	1.551	1.576					
1 1/2					1.587	1.614	1.640	1.663				
1 9/16						1.677	1.704	1.727				
1 5/8						1.740	1.766	1.790				
1 11/16						1.802	1.830	1.854				
1 3/4						1.866	1.892	1.917	1.941			
1 13/16						1.928	1.955	1.981	2.005			
1 7/8						1.991	2.018	2.043	2.068			
1 15/16						2.053	2.081	2.107	2.131			
2						2.117	2.144	2.170	2.194	2.219		
2 1/16						2.180	2.207	2.223	2.258	2.282		
2 1/8						2.243	2.270	2.296	2.321	2.346		
2 3/16						2.305	2.333	2.359	2.384	2.408		
2 1/4						2.368	2.396	2.422	2.447	2.472		
2 5/16						2.431	2.458	2.485	2.511	2.535		
2 3/8						2.494	2.521	2.547	2.573	2.599		
2 7/16						2.556	2.583	2.611	2.637	2.661		
2 1/2						2.619	2.647	2.678	2.700	2.725	2.772	
2 5/8						2.744	2.772	2.800	2.825	2.851	2.899	
2 3/4						2.870	2.898	2.924	2.951	2.977	3.026	
2 7/8						2.995	3.023	3.050	3.077	3.103	3.153	
3						3.120	3.148	3.176	3.202	3.229	3.279	3.327
3 1/8						3.245	3.274	3.301	3.328	3.355	3.405	3.454
3 1/4						3.371	3.399	3.426	3.454	3.481	3.532	3.581
3 3/8						3.496	3.524	3.552	3.580	3.607	3.660	3.707

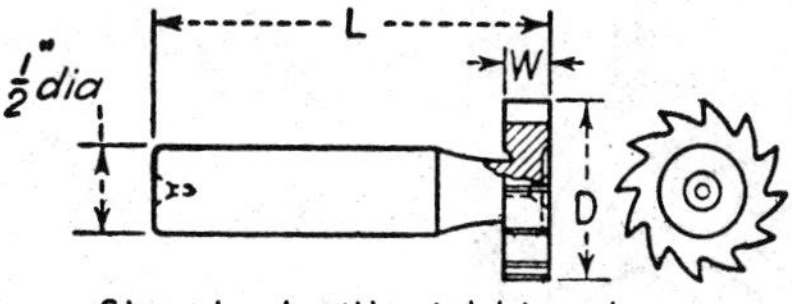

TABLE 11. SHANK-TYPE WOODRUFF KEYSLOT MILLING CUTTERS (HSS)

Numbers of Cutters		Dimensions, In.		
American* Standard	Old Standard	*D*	*W*	*L*
202	201	1/4	1/16	2 1/16
202 1/2	206	5/16	1/16	2 1/16
302 1/2	207	5/16	3/32	2 3/32
203	211	3/8	1/16	2 1/16
303	212	3/8	3/32	2 3/32
403	213	3/8	1/8	2 1/8
204	1	1/2	1/16	2 1/16
304	2	1/2	3/32	2 3/32
305	4	5/8	3/32	2 3/32
404	3	1/2	1/8	2 1/8
405	5	5/8	1/8	2 1/8
406	7	3/4	1/8	2 1/8
505	6	5/8	5/32	2 5/32
605	61	5/8	3/16	2 3/16
506	8	3/4	5/32	2 5/32
806	91	3/4	1/4	2 1/4
507	10	7/8	5/32	2 5/32
606	9	3/4	3/16	2 3/16
607	11	7/8	3/16	2 3/16
707	12	7/8	7/32	2 7/32
608	13	1	3/16	2 3/16
708	14	1	7/32	2 7/32
1208	152	1	3/8	2 3/8
609	16	1 1/8	3/16	2 3/16
807	A	7/8	1/4	2 1/4
808	15	1	1/4	2 1/4
709	17	1 1/8	7/32	2 7/32
809	18	1 1/8	1/4	2 1/4
610	19	1 1/4	3/16	2 3/16
710	20	1 1/4	7/32	2 7/32
810	21	1 1/4	1/4	2 1/4
811	22	1 3/8	1/4	2 1/4
812	24	1 1/2	1/4	2 1/4
1008	B	1	5/16	2 5/16
1009	C	1 1/8	5/16	2 5/16
1010	D	1 1/4	5/16	2 5/16
1011	23	1 3/8	5/16	2 5/16
1012	25	1 1/2	5/16	2 5/16
1210	E	1 1/4	3/8	2 3/8
1211	F	1 3/8	3/8	2 3/8
1212	G	1 1/2	3/8	2 3/8

For key No. 121, use cutter No. 807. For key No. 131, use cutter No. 1008.
For key No. 141, use cutter No. 808. For key No. 161, use cutter No. 1009.

* The cutter numbers shown under the column headed American Standard indicate the nominal key dimension or size cutter; that is, the last two digits give the nominal diameter in eighths of an inch and the digits preceding the last two give the nominal width in thirty-seconds of an inch. Thus, cutter No. 204 indicates a size 2/32 by 4/8 in. or 1/16 in. thick by 1/2 in. dia.

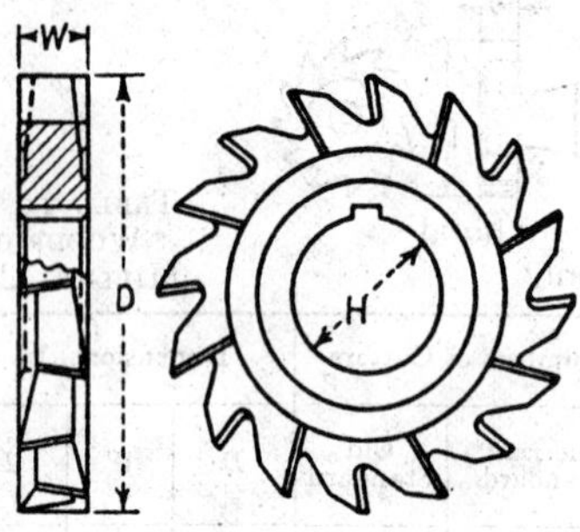

TABLE 12. ARBOR-TYPE WOODRUFF KEYSLOT MILLING CUTTERS (HSS)

Numbers of Cutters		Dimensions, In.		
American Standard	Old Standard	*D*	*W*	*H*
617	26	$2\frac{1}{8}$	$\frac{3}{16}$	$\frac{3}{4}$
817	27	$2\frac{1}{8}$	$\frac{1}{4}$	$\frac{3}{4}$
1017	28	$2\frac{1}{8}$	$\frac{5}{16}$	$\frac{3}{4}$
1217	29	$2\frac{1}{8}$	$\frac{3}{8}$	$\frac{3}{4}$
822	R	$2\frac{3}{4}$	$\frac{1}{4}$	1
1022	S	$2\frac{3}{4}$	$\frac{5}{16}$	1
1222	T	$2\frac{3}{4}$	$\frac{3}{8}$	1
1422	U	$2\frac{3}{4}$	$\frac{7}{16}$	1
1622	V	$2\frac{3}{4}$	$\frac{1}{2}$	1
1228	30	$3\frac{1}{2}$	$\frac{3}{8}$	1
1428	31	$3\frac{1}{2}$	$\frac{7}{16}$	1
1628	32	$3\frac{1}{2}$	$\frac{1}{2}$	1
1828	33	$3\frac{1}{2}$	$\frac{9}{16}$	1
2028	34	$3\frac{1}{2}$	$\frac{5}{8}$	1
2228	35	$3\frac{1}{2}$	$\frac{11}{16}$	1
2428	36	$3\frac{1}{2}$	$\frac{3}{4}$	1

For key No. 126 use cutter No. 617.
For key No. 127 use cutter No. 817.
For key No. 128 use cutter No. 1017.
For key No. 129 use cutter No. 1217.
For key No. RX use cutter No. 822.

For key No. SX use cutter No. 1022.
For key No. TX use cutter No. 1222.
For key No. UX use cutter No. 1422.
For key No. VX use cutter No. 1622.

TABLE 13. DOWEL PINS—HARDENED AND GROUND

Nominal dia	1/8	3/16	1/4	5/16	3/8	7/16	1/2	5/8	3/4	7/8
Std. ±0.0001	0.1252	0.1877	0.2502	0.3127	0.3752	0.4377	0.5002	0.6252	0.7502	0.8752
Oversize ±0.0001	0.1260	0.1885	0.2510	0.3135	0.3760	0.4385	0.5010	0.6260	0.7510	0.8760
Length L										
1/2	X	X	X	X						
5/8	X	X	X	X						
3/4	X	X	X	X	X					
7/8	X	X	X	X	X	X				
1	X	X	X	X	X	X				
1 1/4		X	X	X	X	X	X	X		
1 1/2		X	X	X	X	X	X	X	X	
1 3/4		X	X	X	X	X	X	X	X	
2		X	X	X	X	X	X	X	X	X
2 1/4				X	X	X	X			
2 1/2				X	X	X	X	X	X	X
3							X	X	X	X
3 1/2							X	X		
4							X	X	X	X
4 1/2								X	X	X
5									X	X
5 1/2									X	X

ASA B5.20-1947. Published by the ASME.

These pins are extensively used in the tool and machine industry and a machine reamer of nominal size may be used to produce the holes into which these pins tap or press fit. They must be straight and free from any defects that will affect their serviceability.

Standard pins are for original installation, and oversize pins are for replacement or repair. Heat-treatment: carburized, hardened, and tempered to Rockwell 60 to 64 C.

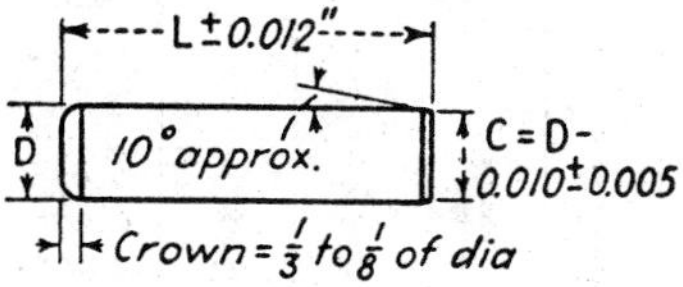

L ±0.012"

10° approx.

D

C = D-0.010±0.005

Crown = 1/3 to 1/8 of dia

Hardened dowel pin

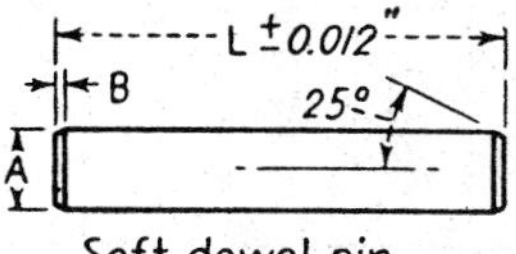

L ±0.012"

B

25°

A

Soft dowel pin

TABLE 14. DOWEL PINS—SOFT

Nominal Dia	Dia A Max	Dia A Min	B
0.062	0.0600	0.0595	0.010
0.094	0.0912	0.0907	0.010
0.109	0.1068	0.1063	0.010
0.125	0.1223	0.1218	0.010
0.156	0.1535	0.1530	1/64
0.188	0.1847	0.1842	1/64
0.219	0.2159	0.2154	1/64
0.250	0.2470	0.2465	1/64
0.312	0.3094	0.3089	1/32
0.375	0.3717	0.3712	1/32
0.438	0.4341	0.4336	1/32
0.500	0.4964	0.4959	1/32
0.625	0.6211	0.6206	3/64
0.750	0.7458	0.7453	3/64
0.875	0.8705	0.8700	1/16
1.000	0.9952	0.9947	1/16

Maximum diameters are graduated from 0.0005 on 1/16-in. pins to 0.0028 on 1-in. pins under the minimum commercial bar-stock sizes.

Table 15. Straight Pins

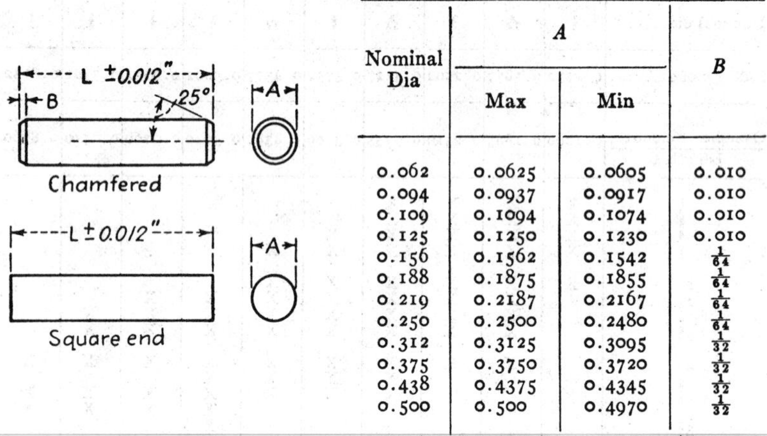

Nominal Dia	A Max	A Min	B
0.062	0.0625	0.0605	0.010
0.094	0.0937	0.0917	0.010
0.109	0.1094	0.1074	0.010
0.125	0.1250	0.1230	0.010
0.156	0.1562	0.1542	$\frac{1}{64}$
0.188	0.1875	0.1855	$\frac{1}{64}$
0.219	0.2187	0.2167	$\frac{1}{64}$
0.250	0.2500	0.2480	$\frac{1}{64}$
0.312	0.3125	0.3095	$\frac{1}{32}$
0.375	0.3750	0.3720	$\frac{1}{32}$
0.438	0.4375	0.4345	$\frac{1}{32}$
0.500	0.500	0.4970	$\frac{1}{32}$

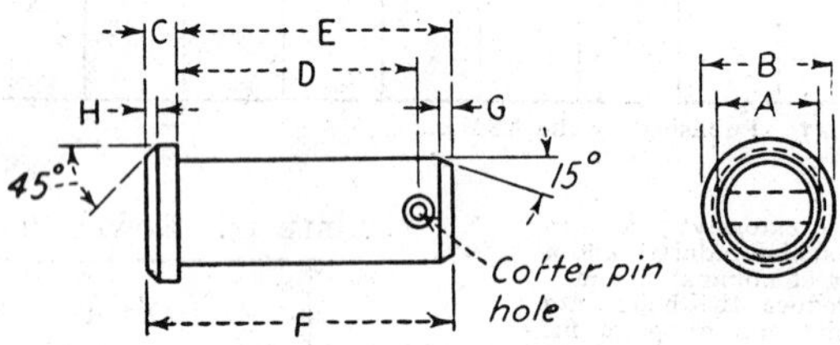

Table 16. Clevis Pins

A			B	C	D	E	F	G	H	Hole Sizes
Nominal	Max	Min								
0.188	0.186	0.181	$\frac{5}{16}$	$\frac{1}{16}$	$\frac{31}{64}$	$\frac{19}{32}$	$\frac{21}{32}$	$\frac{3}{64}$	$\frac{1}{64}$	0.0781
0.250	0.248	0.243	$\frac{3}{8}$	$\frac{3}{32}$	$\frac{43}{64}$	$\frac{51}{64}$	$\frac{57}{64}$	$\frac{1}{16}$	$\frac{1}{32}$	0.0781
0.312	0.311	0.306	$\frac{7}{16}$	$\frac{3}{32}$	$\frac{13}{16}$	$\frac{31}{32}$	$1\frac{1}{16}$	$\frac{5}{64}$	$\frac{1}{32}$	0.1094
0.375	0.373	0.368	$\frac{1}{2}$	$\frac{1}{8}$	$\frac{15}{16}$	$1\frac{3}{32}$	$1\frac{7}{32}$	$\frac{5}{64}$	$\frac{1}{32}$	0.1094
0.438	0.436	0.431	$\frac{9}{16}$	$\frac{5}{32}$	$1\frac{1}{16}$	$1\frac{15}{64}$	$1\frac{25}{64}$	$\frac{3}{32}$	$\frac{3}{64}$	0.1094
0.500	0.496	0.491	$\frac{5}{8}$	$\frac{5}{32}$	$1\frac{13}{64}$	$1\frac{27}{64}$	$1\frac{37}{64}$	$\frac{7}{64}$	$\frac{3}{64}$	0.1406
0.625	0.621	0.616	$\frac{13}{16}$	$\frac{13}{64}$	$1\frac{15}{32}$	$1\frac{23}{32}$	$1\frac{59}{64}$	$\frac{9}{64}$	$\frac{1}{16}$	0.1406
0.750	0.746	0.741	$\frac{15}{16}$	$\frac{1}{4}$	$1\frac{3}{4}$	$2\frac{3}{64}$	$2\frac{19}{64}$	$\frac{5}{32}$	$\frac{5}{64}$	0.1719
0.875	0.871	0.866	$1\frac{1}{16}$	$\frac{5}{16}$	$2\frac{1}{64}$	$2\frac{11}{32}$	$2\frac{21}{32}$	$\frac{3}{16}$	$\frac{3}{32}$	0.1719
1.000	0.996	0.991	$1\frac{3}{16}$	$\frac{11}{32}$	$2\frac{17}{64}$	$2\frac{5}{8}$	$2\frac{31}{32}$	$\frac{7}{32}$	$\frac{7}{64}$	0.1719

ASA B5.20-1947. Published by the ASME.
Cotter-pin holes are in accordance with accepted practice. These pins may be supplied either soft or cyanide-hardened as specified.

TABLE 17. STANDARD TAPER PINS

No.	Length of Taper Pin, In.																									
	3/8	1/2	5/8	3/4	7/8	1	1 1/4	1 1/2	1 3/4	2	2 1/4	2 1/2	2 3/4	3	3 1/4	3 1/2	3 3/4	4	4 1/4	4 1/2	4 3/4	5	5 1/4	5 1/2	5 3/4	6
7/0	X	X	X																							
6/0	X	X	X	X																						
5/0		X	X	X		X																				
4/0		X	X	X		X																				
3/0		X	X	X	X	X																				
2/0		X	X	X	X	X	X																			
0		X	X	X	X	X	X	X																		
1				X	X	X	X	X	X	X																
2				X	X	X	X	X	X	X	X	X														
3				X	X	X	X	X	X	X	X	X	X	X												
4						X	X	X	X	X	X	X	X	X												
5						X	X	X	X	X	X	X	X	X												
6							X	X	X	X	X	X	X	X	X	X	X	X								
7										X	X	X	X	X	X	X	X	X	X	X	X					
8										X	X	X	X	X	X	X	X	X	X	X	X	X				
9													X	X	X	X	X	X	X	X	X	X	X	X	X	X
10																X	X	X	X	X	X	X	X	X	X	X

The X entries indicate the standard lengths of pins made in the sizes indicated.

TABLE 18. TAPER-PIN DIAMETERS AT LARGE END*

No.	Size	No.	Size	No.	Size
7/0	0.0625	1	0.1720	8	0.4920
6/0	0.0780	2	0.1930	9	0.5910
5/0	0.0940	3	0.2190	10	0.7060
4/0	0.1090	4	0.2500	11	0.8600
3/0	0.1250	5	0.2890	12	1.032
2/0	0.1410	6	0.3410	13	1.241
0	0.1560	7	0.4090	14	1.523

* To find small diameter, multiply length by 0.02083 and subtract result from large diameter.

TABLE 19. TOLERANCES ON TAPER PINS

	Commercial Type	Precision Type
Sizes	7/0 to 14	7/0 to 10
Diameter	Plus 0.0013, minus 0.0007	Plus 0.0013, minus 0.0007
Taper	1/4 in. per ft	1/4 in. per ft
Length tolerance	Plus or minus 0.030	Plus or minus 0.030
Concavity tolerance	None	0.0005 up to 1 in. long 0.001, 1 1/16 to 2 in. long 0.002, 2 1/16 in. and longer

TABLE 20. SELECTION OF TAPER PIN FOR SHAFT DIAMETER

Pin Dia at Large End for Various Shaft Diameters

Pin No.	7/0	6/0	5/0	4/0	3/0	2/0	1/0	1	2	3	4	5	6	7	8	9	10
Shaft dia.....	$\frac{3}{16}$	$\frac{7}{32}$	$\frac{1}{4}$	$\frac{5}{16}$	$\frac{3}{8}$	$\frac{7}{16}$	$\frac{1}{2}$	$\frac{9}{16}$	$\frac{5}{8}$	$\frac{11}{16}$ and $\frac{3}{4}$	$\frac{13}{16}$	$\frac{7}{8}$	1	$1\frac{1}{4}$	$1\frac{1}{2}$	2	$2\frac{1}{2}$

The table gives shaft diameters on which the various sizes of pins may be used. The size of pin shown for a given shaft diameter will prove satisfactory for all ordinary applications.

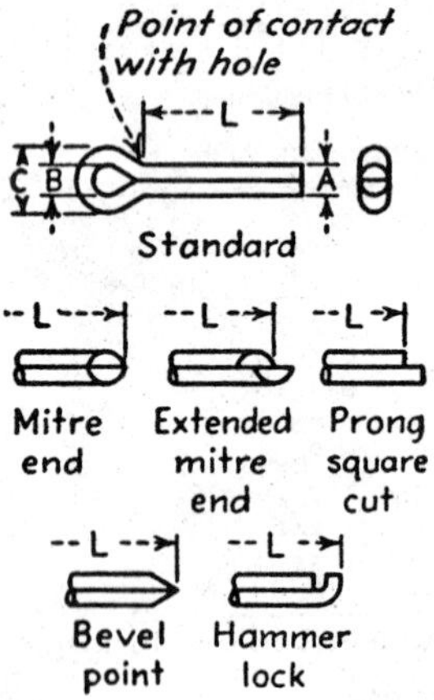

TABLE 21. COTTER PINS

Nominal Dia	Dia A Max	Dia A Min	B Min	C Min	For Holes
0.031	0.032	0.028	$\frac{1}{32}$	$\frac{1}{16}$	$\frac{3}{64}$
0.047	0.048	0.044	$\frac{3}{64}$	$\frac{3}{32}$	$\frac{1}{16}$
0.062	0.060	0.056	$\frac{1}{16}$	$\frac{1}{8}$	$\frac{5}{64}$
0.078	0.076	0.072	$\frac{5}{64}$	$\frac{5}{32}$	$\frac{3}{32}$
0.094	0.090	0.086	$\frac{3}{32}$	$\frac{3}{16}$	$\frac{7}{64}$
0.109	0.104	0.100	$\frac{7}{64}$	$\frac{7}{32}$	$\frac{1}{8}$
0.125	0.120	0.116	$\frac{1}{8}$	$\frac{1}{4}$	$\frac{9}{64}$
0.141	0.134	0.130	$\frac{9}{64}$	$\frac{9}{32}$	$\frac{5}{32}$
0.156	0.150	0.146	$\frac{5}{32}$	$\frac{5}{16}$	$\frac{11}{64}$
0.188	0.176	0.172	$\frac{3}{16}$	$\frac{3}{8}$	$\frac{13}{64}$
0.219	0.207	0.202	$\frac{7}{32}$	$\frac{7}{16}$	$\frac{15}{64}$
0.250	0.225	0.220	$\frac{1}{4}$	$\frac{1}{2}$	$\frac{17}{64}$
0.312	0.280	0.275	$\frac{5}{16}$	$\frac{5}{8}$	$\frac{5}{16}$
0.375	0.335	0.329	$\frac{3}{8}$	$\frac{3}{4}$	$\frac{3}{8}$
0.438	0.406	0.400	$\frac{7}{16}$	$\frac{7}{8}$	$\frac{7}{16}$
0.500	0.473	0.467	$\frac{1}{2}$	1	$\frac{1}{2}$
0.625	0.598	0.590	$\frac{5}{8}$	$1\frac{1}{4}$	$\frac{5}{8}$
0.750	0.723	0.715	$\frac{3}{4}$	$1\frac{1}{2}$	$\frac{3}{4}$

Part 8

TOOL ENGINEERING AND DRAFTING PRACTICE

Section 38

JIG AND FIXTURE DETAILS

SECTION 38

JIG AND FIXTURE DETAILS

STANDARD JIG BUSHINGS

Previous standards for jig bushings were revised in 1941 and approved by the American Standards Association as American Standard B5.6-1941.

Nomenclature

Press-fit Bushings. Press-fit wearing bushings to guide the tool are for installation directly in the jig without the use of a liner and are employed principally where the bushings are used for short production runs and will not require replacement. They are intended also for use where the closeness of the center distance of holes will not permit the installation of liners and renewable bushings. Press-fit bushings are made in two types, with heads and without.

Renewable Bushings. Renewable wearing bushings to guide the tool are for use in liners which in turn are installed in the jig. They are used where the bushing will wear out or become obsolete before the jig, or where several bushings are to be interchangeable in one hole. Renewable wearing bushings are divided into two classes, "fixed" and "slip."

Fixed renewable bushings are installed in the liner with the intention of leaving them in place until worn out.

Slip renewable bushings are interchangeable in a given size of liner and, to facilitate removal, they are usually made with a knurled head. They are most frequently used where two or more operations requiring different inside diameters are performed in a single jig, such as where drilling is followed by reaming, tapping, spotfacing, counterboring, or some other secondary operation.

Liner Bushings. Liner bushings are provided with and without heads and are permanently installed in a jig to receive the renewable wearing bushings. They are sometimes called "master bushings."

Bushing Specifications. The dimensions and tolerances of jig bushings shall conform to the specifications given in the following tables and notes.

Jig-plate Thickness. The standard lengths of the press-fit portion of jig bushings as established are based on standardized or uniform jig-plate thicknesses of $\frac{5}{16}$, $\frac{1}{2}$, $\frac{3}{4}$, 1, $1\frac{3}{8}$, and $1\frac{3}{4}$ in.

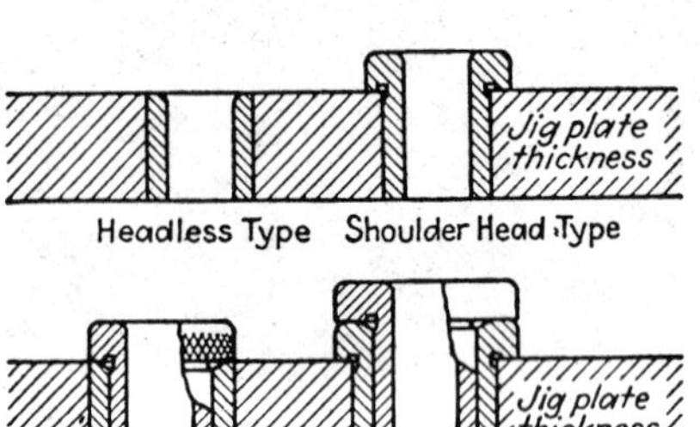

FIG. 1. Types of jig bushings.

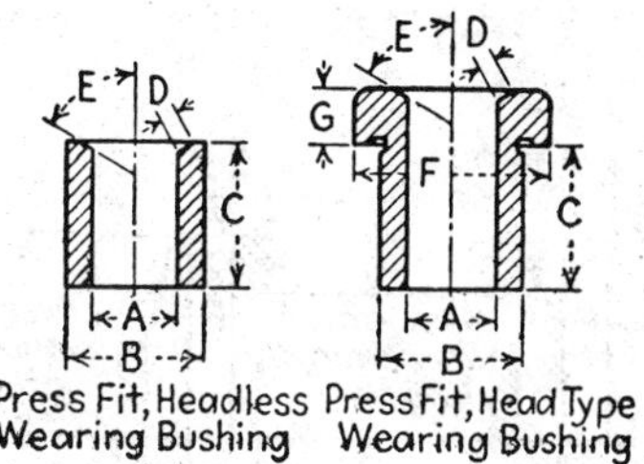

Press Fit, Headless Wearing Bushing Press Fit, Head Type Wearing Bushing

TABLE 1. PRESS-FIT WEARING BUSHINGS—HEADLESS AND HEAD TYPES

Range of Hole Size*† A		Body Dia B					Body Length§ C			Width of Chamfer¶ D	Head Dia‖ F max	Head Height G max
		Unfinished‡			Finished							
From	Up to and Including	Nominal	Max	Min	Max	Min	Short	Medium	Long			
0.0156	0.0625	5/32	0.166	0.161	0.1578	0.1575	5/16		1/2	1/32	1/4	3/32
0.0630	0.0995	13/64	0.213	0.208	0.2046	0.2043	5/16		1/2	1/32	5/16	3/32
0.1024	0.1378	1/4	0.260	0.255	0.2516	0.2513	5/16		1/2	1/32	3/8	3/32
0.1406	0.1875	5/16	0.327	0.322	0.3141	0.3138	5/16	1/2	3/4	1/32	7/16	1/8
0.1910	0.2500	13/32	0.421	0.416	0.4078	0.4075	5/16	1/2	3/4	1/16	17/32	5/32
0.2520	0.3125	1/2	0.520	0.515	0.5017	0.5014	5/16	1/2	3/4	5/64	5/8	7/32
0.3160	0.4219	5/8	0.645	0.640	0.6267	0.6264	1/2	3/4	1	3/32	13/16	7/32
0.4375	0.5000	3/4	0.770	0.765	0.7518	0.7515	1/2	3/4	1	7/64	15/16	7/32
0.5156	0.6250	7/8	0.895	0.890	0.8768	0.8765	5/8	1	1 3/8	7/64	1 1/8	1/4
0.6406	0.7500	1	1.020	1.015	1.0018	1.0015	3/4	1	1 3/8	7/64	1 1/4	5/16
0.7656	1.0000	1 3/8	1.395	1.390	1.3772	1.3768	3/4	1	1 3/8	9/64	1 5/8	3/8
1.0156	1.3750	1 3/4	1.770	1.765	1.7523	1.7519	1	1 3/8	1 3/4	9/64	2	3/8
1.3906	1.7500	2 1/4	2.270	2.265	2.2525	2.2521	1	1 3/8	1 3/4	7/32	2 1/2	3/8

All dimensions given in inches.
Tolerance on fractional dimensions where not otherwise specified shall be ±0.010 in.

* Hole sizes are in accordance with the American Standard for Twist Drill Sizes (ASA B5.12-1940).
† The maximum and minimum values of the hole size A shall be as follows:

Nominal Size of Hole	Max	Min
Above 0.0000 to 1/4 in. incl.	Nominal +0.0004 in.	Nominal +0.0001 in.
Above 1/4 to 3/4 in. incl.	Nominal +0.0005 in.	Nominal +0.0001 in.
Above 3/4 to 1 1/2 in. incl.	Nominal +0.0006 in.	Nominal +0.0002 in.
Above 1 1/2	Nominal +0.0007 in.	Nominal +0.0003 in.

‡ The body diameter B for unfinished bushings is larger than the nominal diameter in order to provide grinding stock for fitting to jig-plate holes. The grinding allowance is 0.005 to 0.010 in. for sizes 5/32, 13/64, and 1/4 in.; 0.010 to 0.015 in. for sizes 5/16 and 13/32 in.; and 0.015 to 0.020 in. for sizes 1/2 in. and up.
§ The length C is the over-all length for the headless type and the length underhead for the head type.
¶ The angle of chamfer E shall be 59° ± 1°, and a slight radius shall be provided at the intersection of this chamfer with the hole A.
‖ The head design shall be in accordance with the manufacturer's practice.

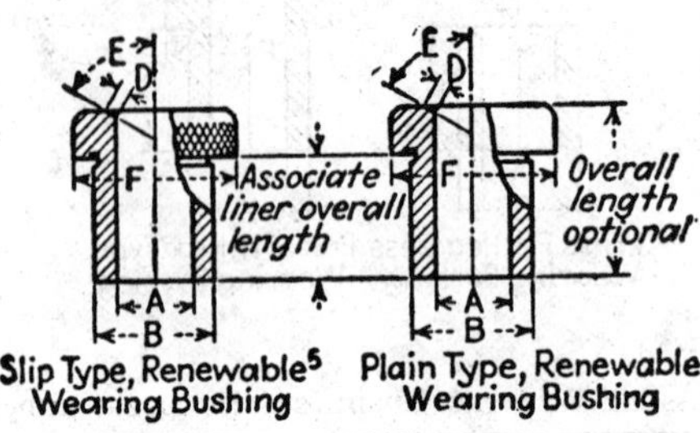

Table 2. Renewable Wearing Bushings—Slip and Plain Shoulder-Head Types

(Tolerance on fractional dimensions where not otherwise specified shall be ±0.010 in.)

Range of Hole*† Size *A*		Body Dia *B*			Width of Chamfer‡ *D*	Max head§ Dia *F*
From	Up to and Including	Nominal	Max	Min		
0.0000	0.1562	5/16	0.3125	0.3123	1/32	5/8
0.1610	0.3125	1/2	0.5000	0.4998	5/64	15/16
0.3160	0.5000	3/4	0.7500	0.7498	7/64	1 1/4
0.5156	0.7500	1	1.0000	0.9998	7/64	1 5/8
0.7656	1.0000	1 3/8	1.3750	1.3747	9/64	2
1.0156	1.3750	1 3/4	1.7500	1.7497	9/64	2 1/2
1.3906	1.7500	2 1/4	2.2500	2.2406	7/32	3

* Hole sizes are in accordance with the proposed American Standard for twist-drill sizes.

† The maximum and minimum values of hole size *A* shall be as follows:

Nominal Size of Hole, In.	Max, In.	Min, In.
Above 0.0000 to 1/4 incl.	Nominal +0.0004	Nominal +0.0001
Above 1/4 to 3/4 incl.	Nominal +0.0005	Nominal +0.0001
Above 3/4 to 1 1/2 incl.	Nominal +0.0006	Nominal +0.0002
Above 1 1/2	Nominal +0.0007	Nominal +0.0003

‡ The angle of chamfer *E* shall be 59° ± 1°, and a slight radius shall be provided at the intersection of this chamfer with the hole *A*.

§ The head design shall be in accordance with the manufacturer's practice.

Head of slip type is usually knurled.

When renewable wearing bushings are used with liner bushings of the head type, the length under the head should be increased over the jig-plate thickness by the thickness of the liner bushing head.

Headless Liner Bushing — Head Type Liner Bushing (Overall length optional)

TABLE 3. LINER BUSHINGS—HEADLESS AND HEAD TYPE

Range of Hole Size in Renewable Wearing Bushings*		ID A'			Body Dia B'					Jig-plate thickness‡ C'			Head dia § F'
						Unfinished†		Finished					
From	Up to and Including	Nominal	Max	Min	Nominal	Max	Min	Max	Min	Short	Medium	Long	Max
0.0000	0.1562	5/16	0.3129	0.3126	1/2	0.520	0.515	0.5017	0.5014	5/16	1/2	3/4	5/8
0.1610	0.3125	1/2	0.5005	0.5002	3/4	0.770	0.765	0.7518	0.7515	5/16	1/2	3/4	15/16
0.3160	0.5000	3/4	0.7506	0.7503	1	1.020	1.015	1.0018	1.0015	1/2	3/4	1	1 1/4
0.5156	0.7500	1	1.0007	1.0004	1 3/8	1.395	1.390	1.3772	1.3768	3/4	1	1 3/8	1 5/8
0.7656	1.0000	1 3/8	1.3760	1.3756	1 3/4	1.770	1.765	1.7523	1.7519	3/4	1	1 3/8	2
1.0156	1.3750	1 3/4	1.7512	1.7508	2 1/4	2.270	2.265	2.2525	2.2521	1	1 3/8	1 3/4	2 1/2
1.3906	1.7500	2 1/4	2.2515	2.2510	2 3/4	2.770	2.765	2.7526	2.7522	1	1 3/8	1 3/4	3

All dimensions given in inches.
Tolerance on fractional dimensions where not otherwise specified shall be ±0.010 in.

* For detail dimensions of renewable wearing bushings see Table 2.
† The body diameter B' for unfinished bushings is 0.015 to 0.020 in. larger than the nominal diameter in order to provide grinding stock for fitting to jig-plate holes.
‡ The length C' is the over-all length for the headless type and the length under head for the head type.
§ The head design shall be in accordance with the manufacturer's practice.

CLAMPS FOR JIGS

Strap Clamps. Strap clamps and the studs which hold them should be of approximately the same strength. Make the width of the clamp equal to the United States standard washer used with the stud, and select the strap thickness suitable to the span between the clamping point and the heel pin, or back rest. Use the chart in Fig. 3 where the stud is centrally located. The slot is made with a standard end mill one size larger than the stud.

Select the stud for the clamping pressure from Table 4, which is based on a fiber stress of 8000 psi.

EXAMPLE: Clamping pressure 400 lb; stud size ½-13 NC; span 8 in. Read up from 8 in. at bottom to ½-in. curve, then to left to find that strap thickness of 1⅜ in. is adequate.

If the stud is offset from center of clamp use the formula $P_0 = PL \div 2d$, where P_0 = equivalent pressure for use with chart to determine stud diameter, P = actual clamping pressure, L = actual span, and d = distance from stud to heel pin.

EXAMPLE: $P = 400$ lb, $L = 8$, and $d = 2$. Then $P_0 = 800$. From Table 4, it will be seen that a ⅝-11 NC stud will suffice for this situation.

But T is affected by the offset position of the stud and is found by taking L as twice the distance from stud to heel pin, or stud to clamping point, whichever is larger. In this case $L = 8 - 2 \times 2 = 12$ in. To find T, read up from span $L = 12$ to the curve (extended) for a ⅝-in. stud and across to left, where it is seen that T = 1¾ in., approximately.

Guide to Clamp Selection. Three considerations govern selection of a clamping device for a jig or fixture. These considerations are: (1) the device must be strong enough to hold the work, (2) the clamp must be rapid in operation, and (3) it must work properly every time.

Clamps shown in Figs. 4 and 5 are designed to hold the work down against the jig plate or table. A number of these devices can be made from standard parts. Wearing surfaces should be hardened.

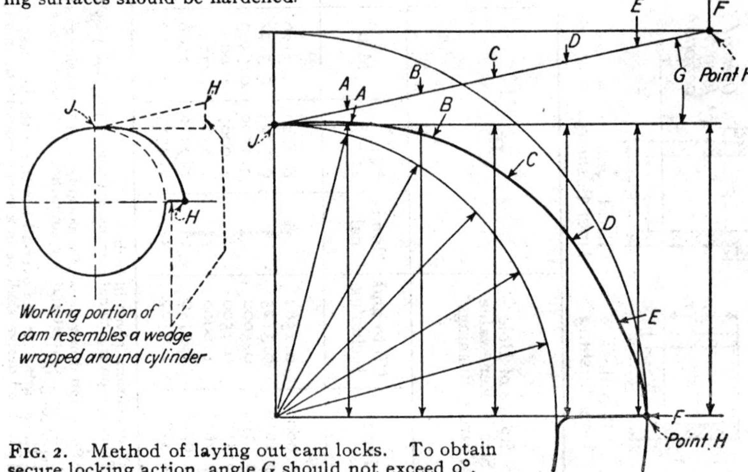

FIG. 2. Method of laying out cam locks. To obtain secure locking action, angle G should not exceed 9°.

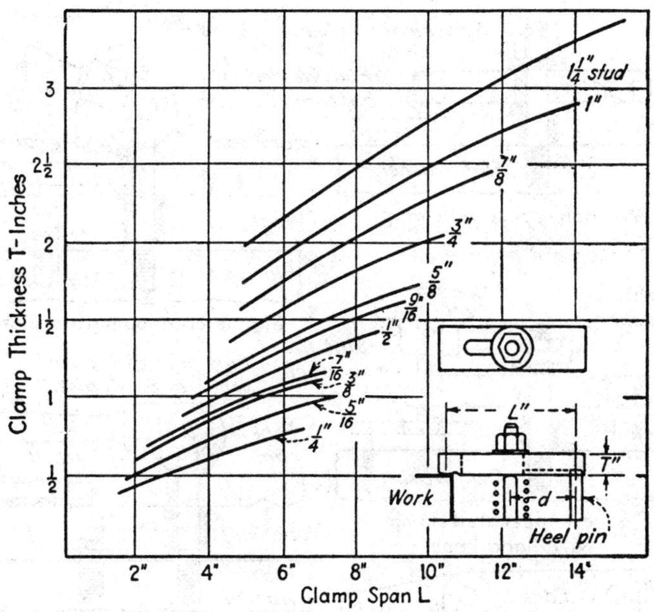

CLAMP DIMENSIONS

Stud	$\frac{1}{4}$	$\frac{5}{16}$	$\frac{3}{8}$	$\frac{7}{16}$	$\frac{1}{2}$	$\frac{9}{16}$	$\frac{5}{8}$	$\frac{3}{4}$	$\frac{7}{8}$	1	$1\frac{1}{4}$
Slot	$\frac{5}{16}$	$\frac{3}{8}$	$\frac{7}{16}$	$\frac{1}{2}$	$\frac{9}{16}$	$\frac{5}{8}$	$\frac{3}{4}$	$\frac{7}{8}$	1	$1\frac{1}{8}$	$1\frac{1}{2}$
Width	$\frac{3}{4}$	$\frac{7}{8}$	1	$1\frac{1}{4}$	$1\frac{3}{8}$	$1\frac{1}{2}$	$1\frac{3}{4}$	2	$2\frac{1}{4}$	$2\frac{1}{2}$	3

FIG. 3. Chart for finding dimensions of strap clamps for various spans *L*.

TABLE 4. STUD-SIZE SELECTION FOR STRAP CLAMPS

Clamping Pressure Required, Lb	*Stud Size NC Thread*
Under 100	$\frac{1}{4}$–20
110–185	$\frac{5}{16}$–18
185–275	$\frac{3}{8}$–16
275–375	$\frac{7}{16}$–14
375–500	$\frac{1}{2}$–13
500–650	$\frac{9}{16}$–12
650–800	$\frac{5}{8}$–11
800–1200	$\frac{3}{4}$–10
1200–1675	$\frac{7}{8}$– 9
1675–2250	1– 8
2250–3550	$1\frac{1}{4}$– 7

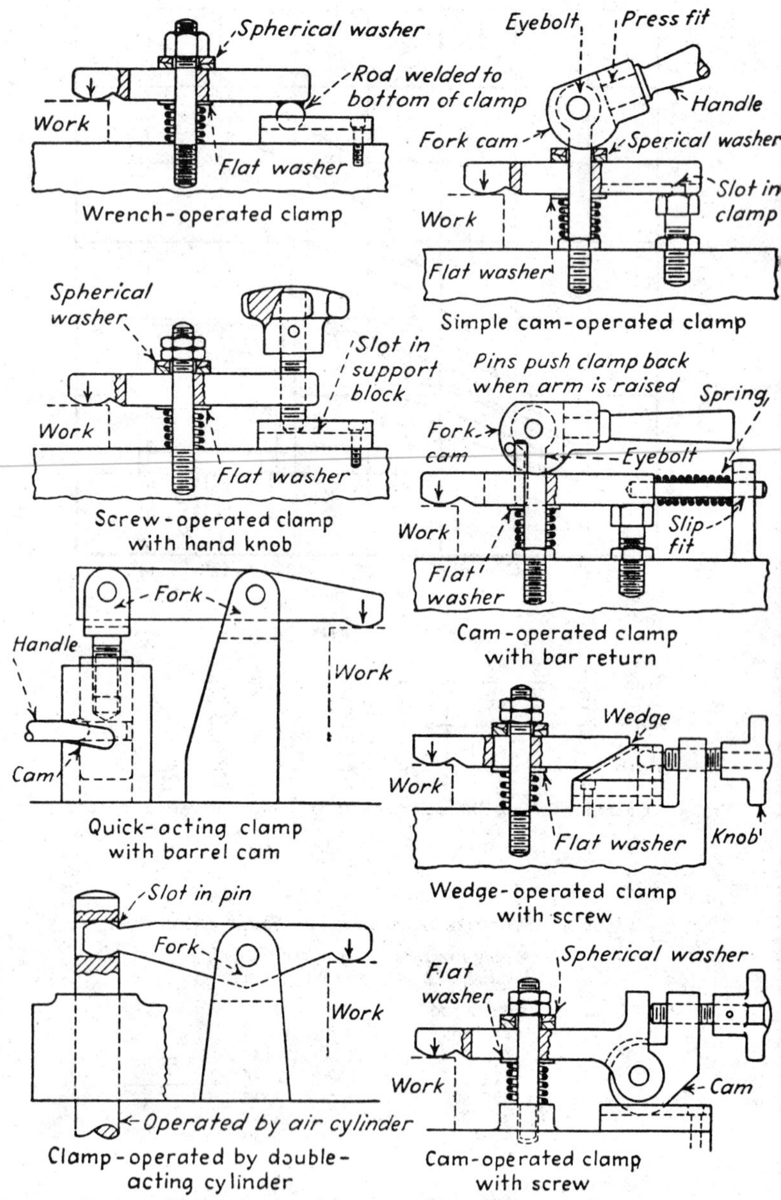

FIG. 4. These clamps will satisfy most jig and fixture requirements.

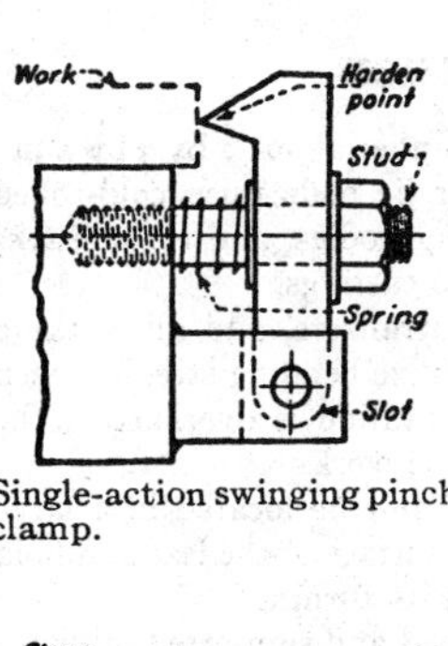

Single-action swinging pinch clamp.

Work
Harden points
45°
Bolt with special head
Flat on bolt
Body

Equalizing-pinch clamps hold work down.

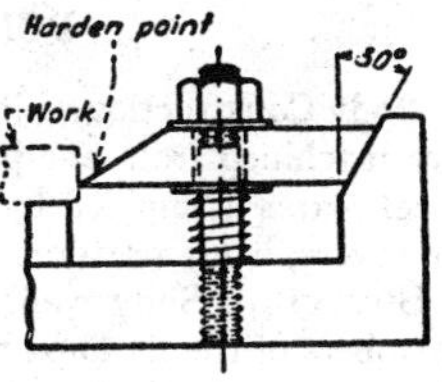

Wedge-operated pinch clamp.

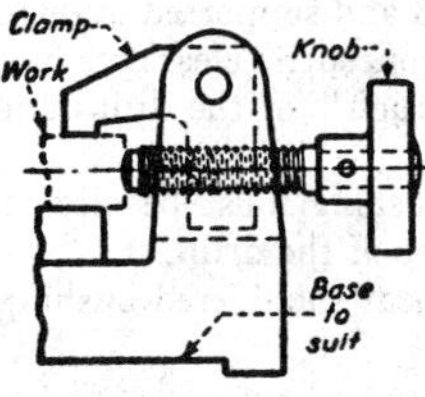

Double-acting screw clamp.

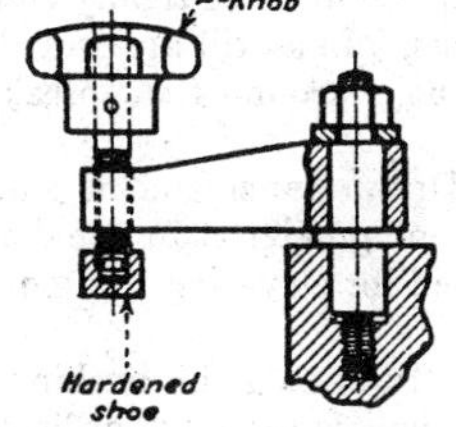

Swing clamp assembly has many uses.

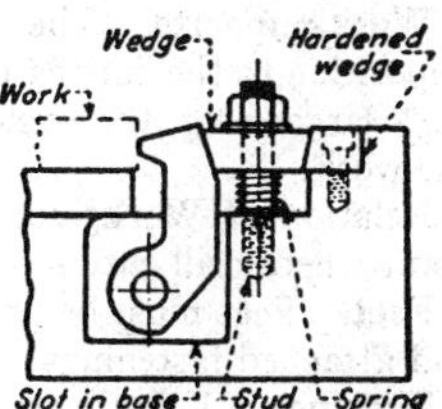

Frequently used wedge-type clamp.

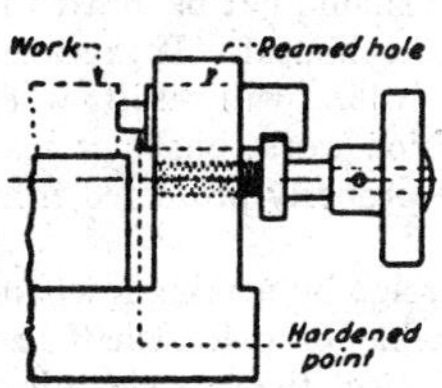

This clamp can be used as a locator.

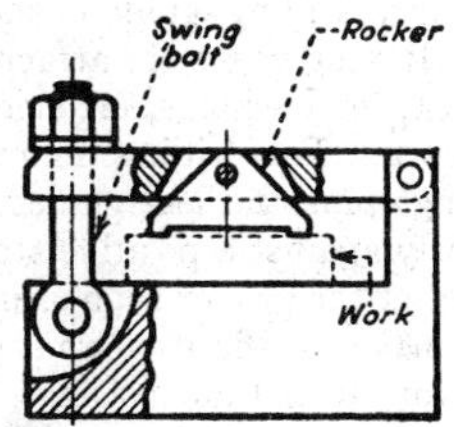

Simplified leaf-type clamp with rocker.

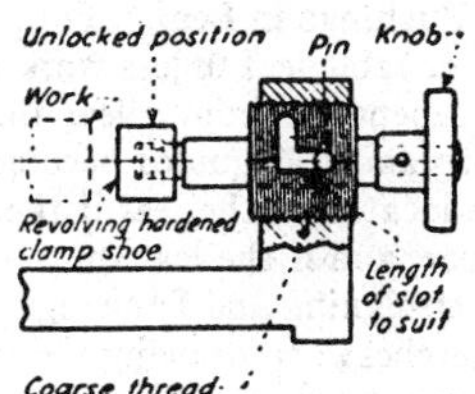

Quick-acting clamp with bayonet lock.

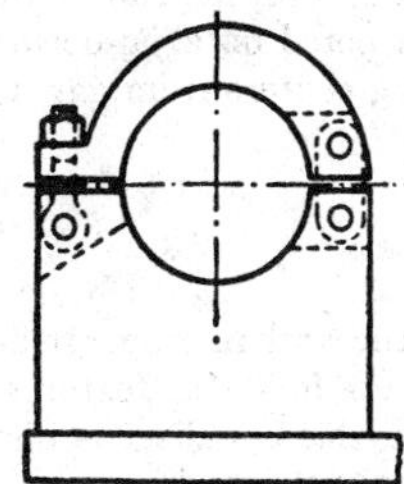

Cradle clamps are used for round work.

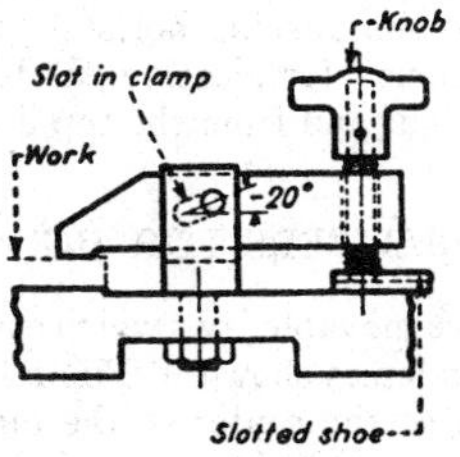

Quick-rising clamp adjusted by screw.

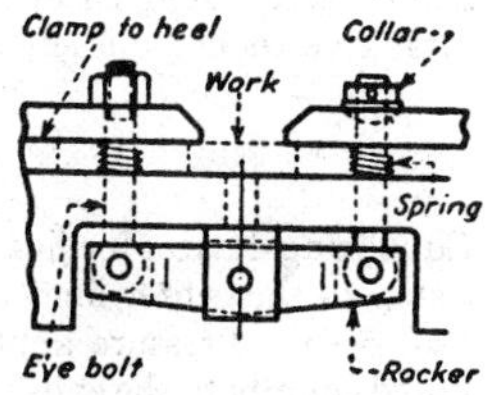

Rocker insures equal clamping.

FIG. 5. Additional clamps are shown for selection purposes.

POINTERS FOR DRILL-JIG DESIGN

Body Construction. Jig bodies of small channel type, under about 2 by 3 by 8 in., are machined from one piece of machinery steel. Larger jig bodies use cold-rolled steel plates assembled by arc welding. For still larger jig bodies, and their leaves or covers, wood patterns are made for obtaining semisteel castings.

Strength. Body walls, leaves, clamps, equalizers, centralizers, and all working members of the jig must have sufficient strength to neutralize bending stresses when the clamps are tightened against the work. If this precaution is overlooked, the positions of guide bushings may distort when clamping the work.

Equalizers. Equalizers for clamping the work should not be located in a large slot cut through the jig leaf. Mount the equalizer on the surface of the leaf and hold it by a centrally positioned screw and one pin to prevent its turning.

Work Supports. The workpiece must be rigidly clamped and supported in direct opposition to the cutting thrusts. In a drill jig, provide a concentric clearance hole, 10% larger than the bushing hole, around the "break-through" of the drill under the work.

Relation of Workpiece to Drill. Inner end of guide bushing must be close to point where drill enters work—no greater than one diameter of the drill.

Feet. Feet must be provided for elevating the tool to clear shouldered bushings and attached fastenings.

Bushings. Lengths of holes through guide bushings must be kept under 2½ to 3 times drill diameter. If guide hole is too long, drills will heat, clog with chips, distort the hole, and sometimes break. Where long bushings are necessary, the entrance hole diameter must be 30 to 50% larger than the drill, and the larger hole must continue from the mouth of the bushing down to where the drill guide begins.

Bushings in Leaf. For accurate drilling, guide bushings should not be located in a movable leaf unless work-positioning gages are attached on the leaf. If gages are independent from a leaf that carries bushings, erroneous drilling will result, when pins and leaf guides become worn. For drilling nonprecision work, bushings may be located in the leaf if the hinge pins and leaf guides are properly protected from wear, and if the leaf shuts tightly against a positive stop.

Designing and Drawing. Select the proper type of jig design by making freehand sketches. In drawing the jig, make full-sized views if possible, or at least half size. Lay out the plan view of the work in red dotted lines. Project the side elevations from the plan if drilling occurs there, or where work fastenings must be shown. Allow spaces between the views for drawing jig walls, thumbscrews or clamps, and for adding dimensions. Draw jig walls around views of the work and draw in gages and stops. Add a leaf, key, cams, clamps, thumbscrews, or whatever fastenings are necessary to hold the work. If the bushing holes are to be bored on a jig-boring machine, give "stepped dimensions" for locating the bushing centers from gaging points; give them from left to right and from the top down.

EQUALIZERS FOR JIGS

Equalizing devices for jigs have movable jaws which clamp the work in proper relationship to the center line. Examples shown in Fig. 6 have the following features:

FIG. 6—1. Pressure applied at the center of the plate in this equalizer is distributed equally to the ends of the plate and transmitted by pins to the work.

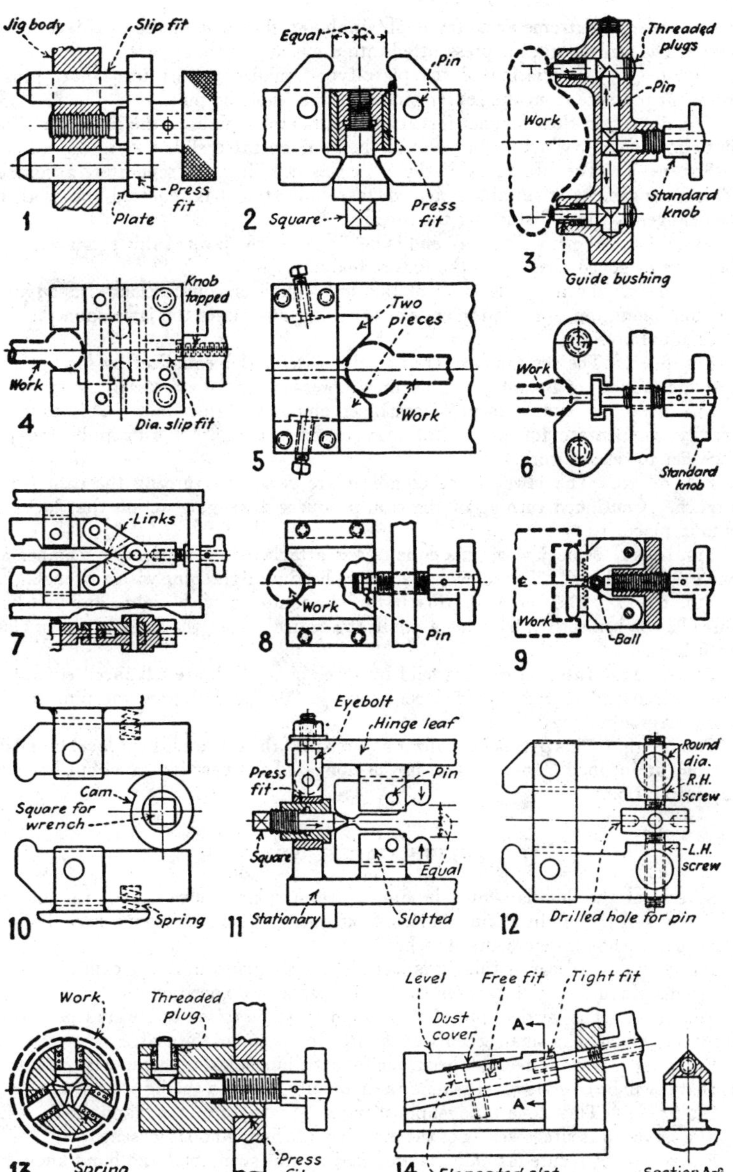

FIG. 6. Equalizers for jigs use various mechanisms.

FIG. 6—2. Extreme accuracy is obtainable in this form of the dual-lever equalizer. The screw socket is press-fitted into a bushing in the fixture.

FIG. 6—3. A variation of the plate-type equalizer uses intermediate pins to transmit pressure from the center pin to the workholding pins.

FIG. 6—4. In this compact form of the ball-crank equalizer, the crank arms have flat circular ends, which fit in a slot in the knob-operated slide.

FIG. 6—5. The two faces of the V in this equalizer are separately adjustable. This arrangement is suitable where delicate adjustment is required and work can be changed without disturbing the setting.

FIG. 6—6. Work with a ball end is held against the jaws of this equalizer. The jaws are attached directly to the fixture body.

FIG. 6—7. Turning the knob of this link equalizer counterclockwise advances the link mounting, spreads the ends of the links, and closes the jaws upon the work with equal pressure.

FIG. 6—8. The screw is attached to the slide of this equalizer by forcing a pin through the slide of assembly. This pin is a press fit against a recess in the screw.

FIG. 6—9. A simple form of equalizer employs a steel ball. This type will readily compensate for work that may be off center. A spring pulls the jaws together to release work.

FIG. 6—10. The jaws of the equalizer are closed by turning the cam with a wrench. Continued turning of the cam produces a wedging action that locks the cam in place.

FIG. 6—11. One jaw of this equalizer is attached to a stationary member and one to a hinged leaf. The screw has a square head for tightening with a wrench.

FIG. 6—12. The screw in this fixture is equipped with right- and left-hand threads. When turned by using a pin in the central collar, the screw opens or closes both jaws.

FIG. 6—13. Internal work is held by three pins which are advanced equally by the tapered end of the threaded center pin. Springs withdraw the pins so that work may be removed.

FIG. 6—14. This parallel equalizer assures that the surface is kept level by raising it along an inclined plane. A bolt in the elongated slot permits travel but holds the cap in position.

JIG LEAVES, JACKS, AND VISES

Several of the details shown in Fig. 7 are standard items. Constructions are varied to suit the job by taking an item from one, another idea from a second design. Purposes of the arrangements sketched are:

FIG. 7—1. The leaf of this jig is locked by a simple form of the cam lock which fits over a pin in the jig body when the leaf is dropped into position.

FIG. 7—2. The vertical pin of this support can be raised or lowered by turning the lever, which has a small eccentric mounted in the end of the shaft.

FIG. 7—3. Several methods are given for attaching a leaf or clamp with clearance, so that it will be free to swivel around the attaching point on the jig or fixture.

FIG. 7—4. This cone-type support jack is raised by the hand knob. The spring serves to assure lowering of the pin when the hand knob is loosened.

FIG. 7—5. A secure fastening is provided by this swing bolt and hand knob. A few turns of the knob let the bolt swing off to the side, freeing the leaf.

FIG. 7—6. In this arrangement, the lock is attached to the end of the leaf, and catches over a welded lug. A flat spring provides automatic locking.

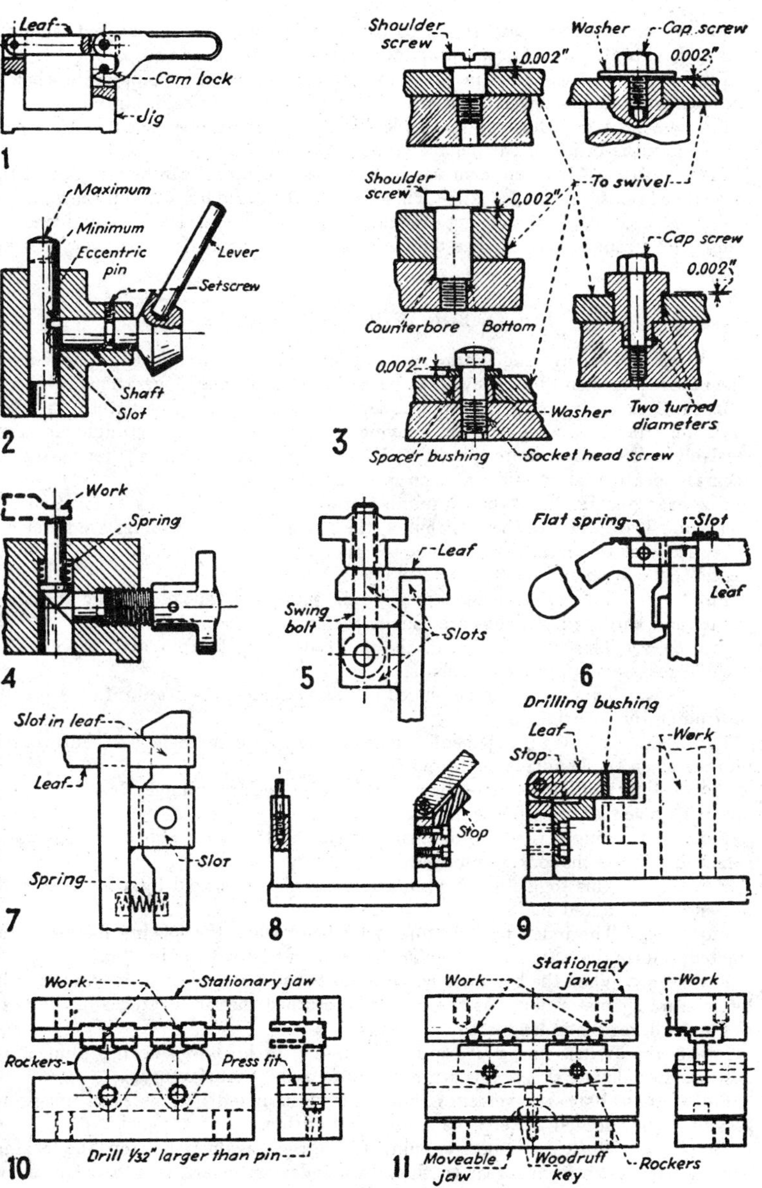

FIG. 7. Details of jig leaves, jacks, and vises.

FIG. 7—7. Pressure against the bottom of this lever releases the leaf from the lock. In closing, the leaf drops down into a slot which serves as a stop.

FIG. 7—8. This leaf stop, intended to keep the leaf from dropping down when in the open position, is screwed to the side of the jig to which the leaf is hinged.

FIG. 7—9. A drill bushing in this leaf is held in the proper position by a stop fastened on the inside of the member which holds the leaf.

FIG. 7—10. Parts of uniform size held to close tolerances can be clamped at the same time in this type of vise jaw. These jaws have tapped holes to suit the vise.

FIG. 7—11. Parts of varying size can be held in these vise jaws designed for milling four round pieces. The double-rocker action makes the vise very flexible in use.

LOCKS FOR INDEX MECHANISMS

Multiple-station jigs and fixtures are applied where lot sizes justify, particularly when the machine on which they are to be used has an automatic operation cycle.

In some cases, indexing is accomplished by rotating the jig or fixture; in other cases, indexing is accomplished by moving the jig or fixture in a straight line successively under two or more spindles. In either case, it is essential that means for accurate location be provided at each work station.

The examples in Fig. 8 show a number of means by which accurate location can be accomplished. In all these examples, the plunger and its bushings are hardened to reduce wear to a minimum. While the plunger should slide easily into the indexing hole, the fit should be snug if accuracy is desired.

FIG. 8—1. Pull-type index lockpins are the simplest means of locating an indexing fixture; hardened bushings are usually preferred to reamed holes.

FIG. 8—2. This modification of the pull-type index lockpin may be removed without disassembling other elements.

FIG. 8—3. The index lock employed in this mechanism is fitted into the handle controlling the indexing shaft.

FIG. 8—4. In this arrangement, a cam on the end of the handle withdraws the plunger when the handle is depressed.

FIG. 8—5. Hand-lever actuation of this index locking plunger insures speedy operation, even when spring is heavy.

FIG. 8—6. Finger pressure against the end of the lever above the spring is sufficient to raise the locking plunger.

FIG. 8—7. This toggle locking-pin arrangement is operated by a lever fixed to one end of the round shaft.

FIG. 8—8. The index plate is turned by hand when the locking handle is retracted; slots in the plate are ground for a snug fit on both sides of handle.

FIG. 8—9. When the handle is raised to index the plate, the cam moves the slide by pressure against a pin. As the handle is returned the latch catches in the next slot and rotates the index plate.

FIG. 8–10. When spring pressure is necessary to hold the locking plunger in position, in indexing slides, this arrangement is usually satisfactory.

FIG. 8—11. This arrangement is suitable for locking indexing slides where spring pressure is not needed to keep the plunger in position.

FIG. 8–12. When the handle is pulled to the right against the stop, the locking point is withdrawn from a gear tooth and the indexing finger turns the gear one or more tooth positions.

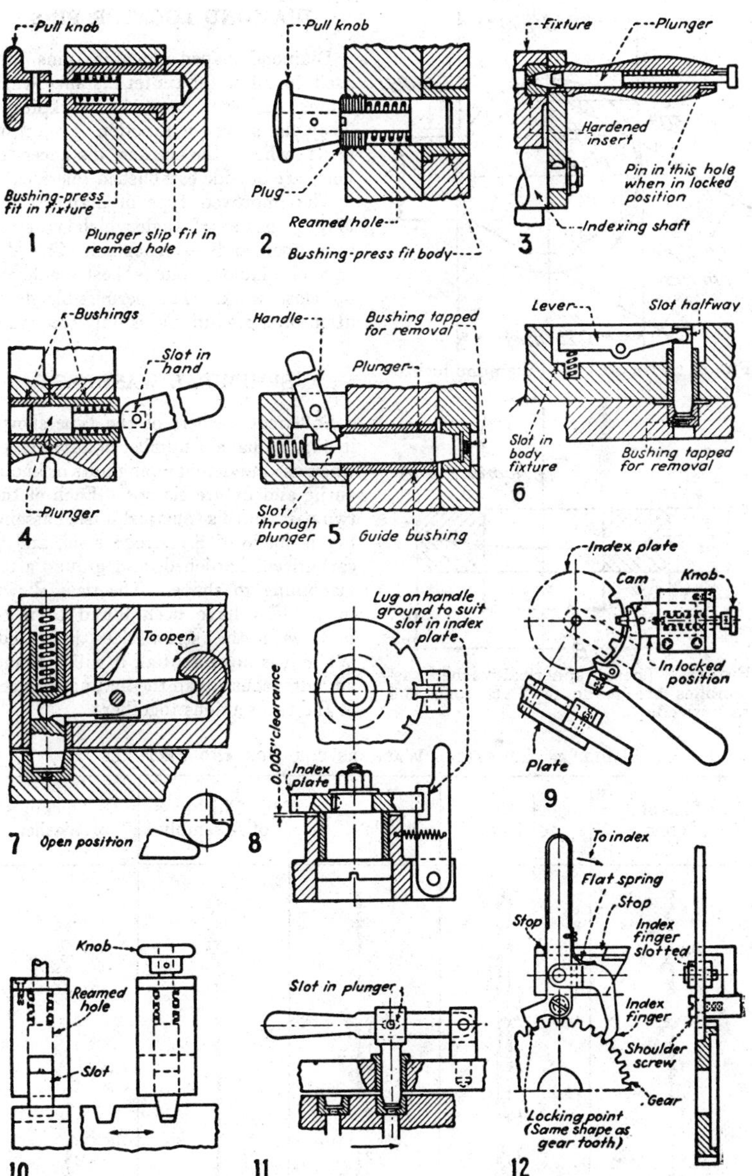

FIG. 8. Locking devices for jig-indexing mechanisms.

DIAMOND LOCATOR PINS

Diamond-shaped locator pins are used to allow for center-distance variations between holes in workpieces. They also serve the function of rejecting parts where the center distances of holes are outside established tolerances.

The improved type of diamond pin (Fig. 9) has a 30° angle which is easy to remember and to machine. This design of diamond pin is best employed on close work. The permissible deviation on a 0.25-in. pin is only 0.0025 in.

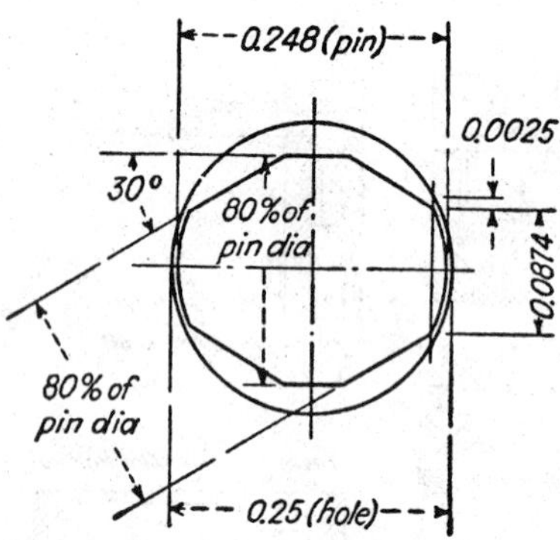

FIG. 9. Proportions of diamond locator pin.

SPHERICAL WASHERS

Spherical washers of the type shown in Fig. 10 have a number of advantages when used under clamping nuts or knobs on jig and fixture clamps. Each of the two parts of this spherical washer assembly is made of SAE 1020 steel, and is carburized, hardened, and ground after machining to shape. The sizes shown in Table 5 have been found to work best with the screw sizes indicated. While it is not essential, it will be found advantageous to lap the mating surfaces of the two sections together.

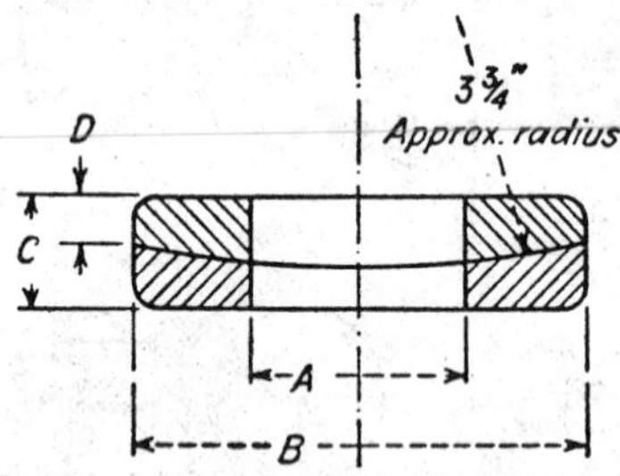

FIG. 10. These spherical-washer assemblies are made of SAE 1020 and carburized.

TABLE 5. SPHERICAL WASHERS FOR JIGS AND FIXTURES

Size of Screw	*A*, Dia of Hole	*B*, OD	*C*, Thickness of Assembly	*D*, Thickness of Washer
$\frac{1}{4}$	$\frac{9}{32}$	$\frac{5}{8}$	$\frac{9}{32}$	$\frac{9}{64}$
$\frac{5}{16}$	$\frac{11}{32}$	$\frac{3}{4}$	$\frac{9}{32}$	$\frac{9}{64}$
$\frac{3}{8}$	$\frac{13}{32}$	$\frac{7}{8}$	$\frac{5}{16}$	$\frac{5}{32}$
$\frac{7}{16}$	$\frac{15}{32}$	1	$\frac{5}{16}$	$\frac{5}{32}$
$\frac{1}{2}$	$\frac{17}{32}$	$1\frac{1}{8}$	$\frac{3}{8}$	$\frac{3}{16}$
$\frac{9}{16}$	$\frac{19}{32}$	$1\frac{3}{16}$	$\frac{3}{8}$	$\frac{3}{16}$
$\frac{5}{8}$	$\frac{21}{32}$	$1\frac{1}{4}$	$\frac{3}{8}$	$\frac{3}{16}$
$\frac{3}{4}$	$\frac{25}{32}$	$1\frac{1}{2}$	$\frac{3}{8}$	$\frac{3}{16}$
$\frac{7}{8}$	$\frac{15}{16}$	$1\frac{3}{4}$	$\frac{1}{2}$	$\frac{1}{4}$
1	$1\frac{1}{16}$	2	$\frac{1}{2}$	$\frac{1}{4}$
$1\frac{1}{8}$	$1\frac{3}{16}$	$2\frac{1}{4}$	$\frac{1}{2}$	$\frac{1}{4}$
$1\frac{1}{4}$	$1\frac{5}{16}$	$2\frac{1}{2}$	$\frac{1}{2}$	$\frac{1}{4}$
$1\frac{1}{2}$	$1\frac{9}{16}$	$2\frac{3}{4}$	$\frac{1}{2}$	$\frac{1}{4}$

Section 39

DRAFTING PRACTICE AND ENGINEERING

SECTION 39

DRAFTING PRACTICE AND ENGINEERING

DRAFTING-ROOM PRACTICE

Standard Drawing Practice. American Standard Z14.1-1946 covers accepted practice for arrangement of views, lines and line work, representation of screw threads, general rules for dimensioning, preparation of notes on drawings, trimmed sizes of drawing paper and cloth, and lettering. Some of this material has been briefed, but data are likewise taken from industrial sources.

Arrangement of Views. For many years, American drawing-room practice has followed the system of third-angle orthographic projection (Fig. 1). The British system is the reverse, in that first-angle projection is favored. Thus, when large manufacturing projects are transferred from the United States to England, or vice versa, it is customary to redraw the blueprints to avoid confusion from the arrangement of views and differences in writing tolerances and notes. Figure 2 compares the American and British arrangements of views for two simple objects.

Number of Views. It is possible to draw six principal views of an object (Fig. 1), but as a rule views are limited to those necessary to portray clearly the shape of the part. Views should be selected to give as few hidden lines as possible. Many cylindrical parts are adequately illustrated by one view, if necessary dimensions are indicated as diameters. Otherwise, two-view drawings or three-view drawings may be arranged as any two, or any three, adjacent views in the relation shown in Fig. 1

In a punch-and-die drawing the theoretical arrangement of views would be as in Fig. 3. To conserve space, however, the view of the bottom of the punch is placed to the right of and in line with the top view of the die. In cases like this, where rearrangement of views is made for convenience, the views should be identified by proper titles.

Lines and Line Work. Figure 4 gives the standard types of lines employed for the purposes indicated. Three widths of line, thick, medium, and thin, are considered desirable on finished drawings in ink. On penciled tracings dense-black medium and thin lines can be used. A large proportion of drawings are inexcusably bad in legibility and line weight, and the prints made from them waste much time in the shop and are productive of errors. Selection of the proper grade of paper, and particularly pencils, will do much to overcome this trouble.

Break lines (Fig. 5) may be used on detail and assembly drawings in order to shorten the view. Alternate positions, as, for example, limits of travel of a link or mechanism, are shown by an outline made up of long dashes of equal weight. Adjacent parts added on a drawing, to indicate their position or use, are likewise indicated with long dash lines.

Sectional Views. When interior construction cannot be shown clearly by outside views, sectional views are drawn. A sectional view (Fig. 6) should be made as if on that view the nearest part of the object had been broken or cut away. The exposed "cut surface" is indicated by section lining or crosshatching. In assemblies, when it is desirable to distinguish between different classes of materials without specifying their exact composition, assemblies may be crosshatched to indicate specific materials. Standard symbols are shown in Fig. 7.

Sectioning Detail Drawings. All parts of a detail drawing are generally sectioned as for cast iron, using fine parallel lines drawn normally at an angle of 45°, but at

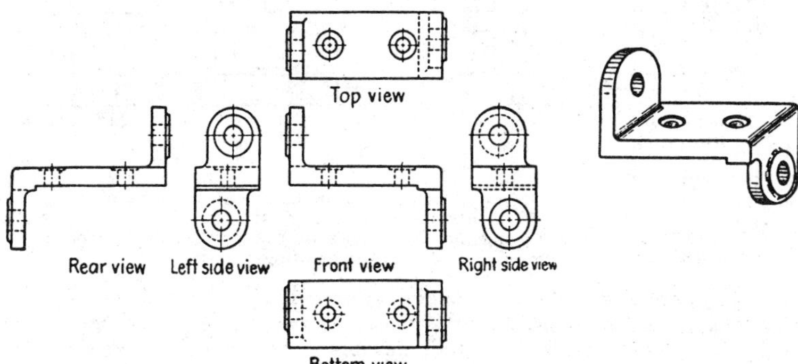

FIG. 1. Six views of a simple object, arranged according to the third-angle orthographic, or "American," system.

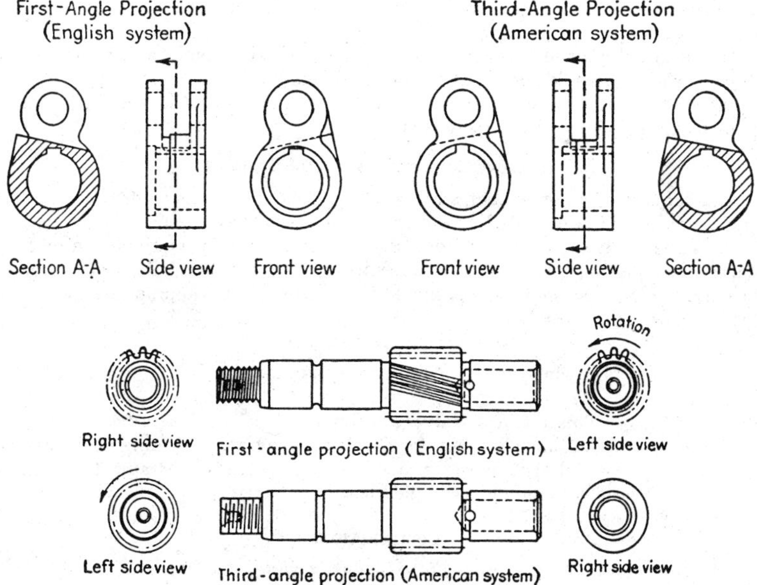

FIG. 2. Two simple objects are drawn with views according to the first-angle and third-angle systems, so that the differences in the English and American systems will be understood.

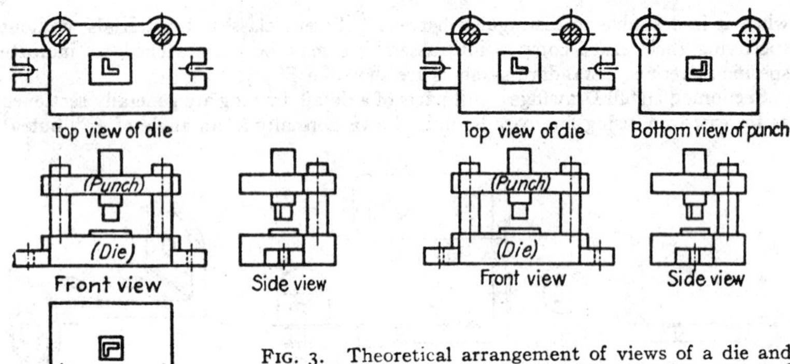

FIG. 3. Theoretical arrangement of views of a die and its components is shown at left, and the approved arrangement of views at the right. The latter makes maximum use of the drawing p per.

angles of 30 or 60° if confusion would otherwise result. Spacing of the lines may vary from $\frac{1}{32}$ to $\frac{1}{8}$ in. or wider, proportionate to the mass or size of the section. Section lining as for cast iron on detail drawings saves time and avoids confusion, because usually the exact composition of the material must be indicated by reference letters, symbols, or notes.

Sectioning Assembly Drawings. On assembly drawings it is desirable to distinguish between materials of the different parts. Hence the need for distinctive sectioning lining, as in Fig. 7. Section lines of adjacent parts are sloped in opposite directions.

Items Not Sectioned. Shafts, nuts, rods, rivets, keys, pins, and similar solid objects, whoses axes lie in the cutting plane, should not be sectioned.

When the section passes through a rib, web, or similar parallel element, section lines should be omitted from these parts.

Abbreviations and Graphical Symbols

Standards for the use of abbreviations and graphical symbols on drawings in the structural, mechanical, and electrical engineering fields have been approved by the ASA. The material in these standards is too voluminous for reproduction or condensation. Persons encountering matter incorporating abbreviations or symbols, or requiring their use on drawings, should obtain copies of the appropriate standards listed below.

GRAPHICAL SYMBOLS

Z14.1-1946. Drawings and Drafting-room Practice.
Z32.2.1-1949. Welding.
Z32.2.3-1949. Pipe Fittings, Valves and Piping.
Y32.1-1954. Graphical Symbols for Electrical Diagrams (covering such matters as electric power and control; telephone, telegraph, and radio; electronic devices; and electric apparatus).

ABBREVIATIONS

Y1 (Tentative). Abbreviations for Use on Drawings.

NOTE: These standards are under constant review, and if revised new designations will be applied.

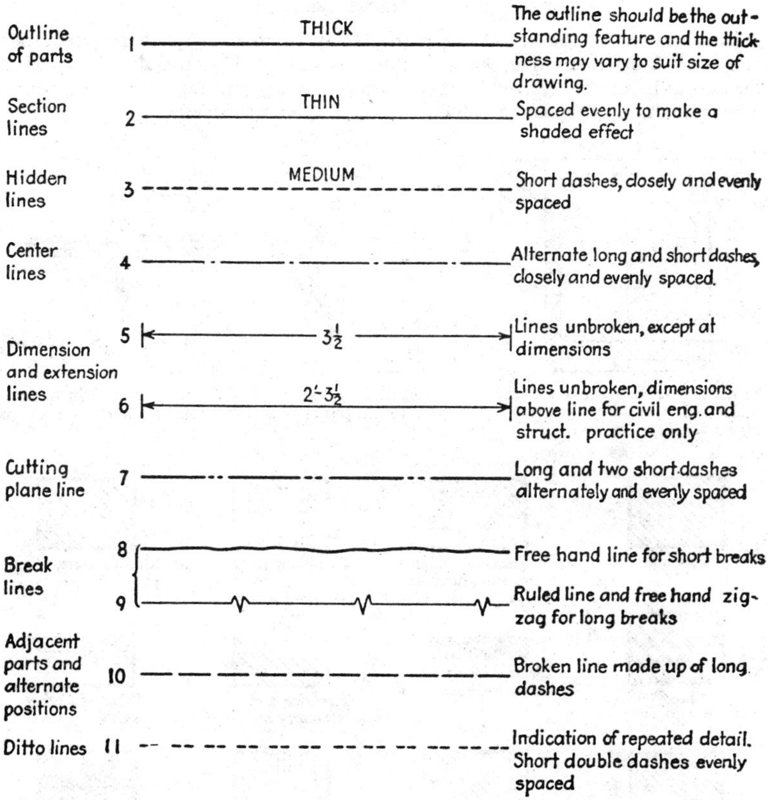

Line	No.	Weight / example	Description
Outline of parts	1	THICK	The outline should be the outstanding feature and the thickness may vary to suit size of drawing.
Section lines	2	THIN	Spaced evenly to make a shaded effect
Hidden lines	3	MEDIUM	Short dashes, closely and evenly spaced
Center lines	4		Alternate long and short dashes, closely and evenly spaced.
Dimension and extension lines	5	$3\frac{1}{2}$	Lines unbroken, except at dimensions
	6	$2'-3\frac{1}{2}$	Lines unbroken, dimensions above line for civil eng. and struct. practice only
Cutting plane line	7		Long and two short dashes alternately and evenly spaced
Break lines	8		Free hand line for short breaks
	9		Ruled line and free hand zigzag for long breaks
Adjacent parts and alternate positions	10		Broken line made up of long dashes
Ditto lines	11		Indication of repeated detail. Short double dashes evenly spaced

FIG. 4. These line conventions correspond with ASA practice and are widely used.

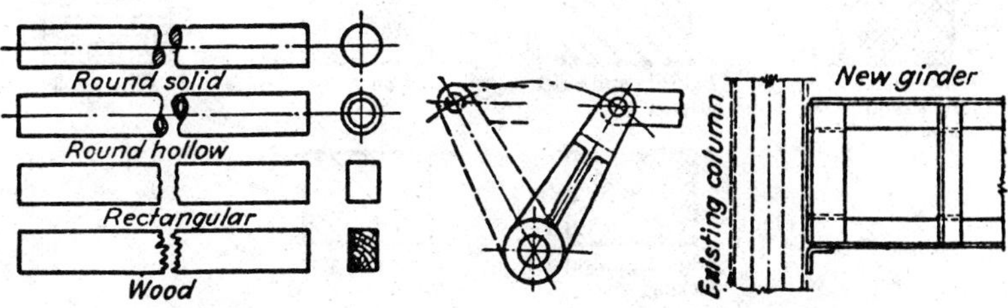

FIG. 5. *Left*, methods of showing how pieces are broken to conserve space. *Center*, an alternate position of the lever is shown by dash lines. *Right*, new additions to a machine or structure are shown in solid lines, the previously existing portion in dash lines.

Rules for Dimensioning

Dimensioning is the art of imparting definition to a pictorial representation (Fig. 9). A detail drawing, when it is dimensioned intelligently, should represent the part as it is to be received by the inspection department, regardless of its source of manufacture. It should usually deal cautiously with the method of manufacture,

ADJACENT PARTS
DO NOT SECTION STANDARD PARTS, SHAFTS AND LIKE PARTS.

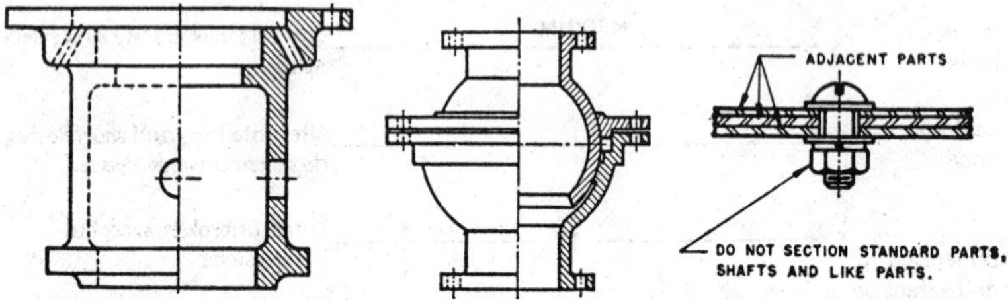

FIG. 6. Hidden lines are used in a detail half section (*left*), but omitted in a half-sectioned assembly (*center*). Conventional parts are not crosshatched in section views (*right*).

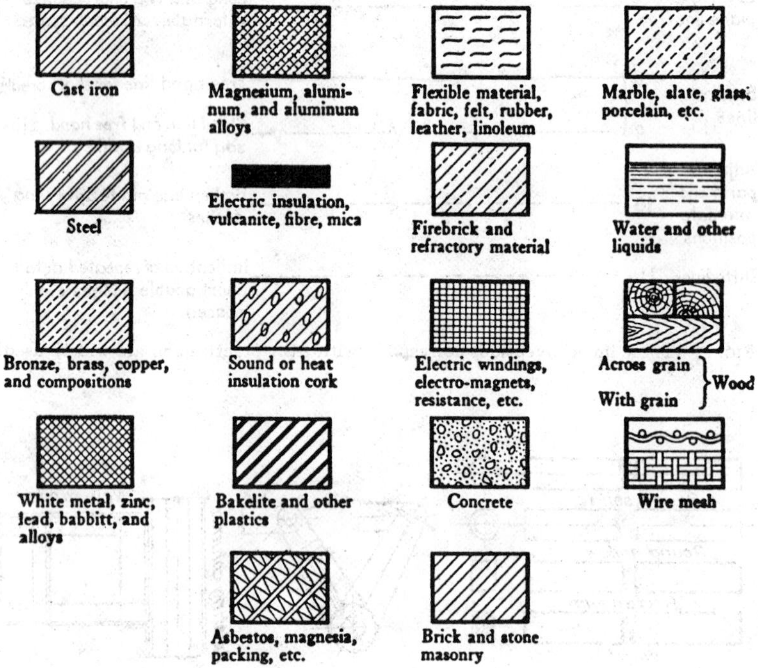

Cast iron
Magnesium, aluminum, and aluminum alloys
Flexible material, fabric, felt, rubber, leather, linoleum
Marble, slate, glass, porcelain, etc.
Steel
Electric insulation, vulcanite, fibre, mica
Firebrick and refractory material
Water and other liquids
Bronze, brass, copper, and compositions
Sound or heat insulation cork
Electric windings, electro-magnets, resistance, etc.
Across grain
With grain
Wood
White metal, zinc, lead, babbitt, and alloys
Bakelite and other plastics
Concrete
Wire mesh
Asbestos, magnesia, packing, etc.
Brick and stone masonry

FIG. 7. Symbols for materials are often used in sectioning assemblies. However, the material must generally have a specific composition and often a heat-treatment, which are best taken care of by notes. In such cases, sectioning for cast iron may be sufficient.

specifying the result to be obtained rather than the method of obtaining it, in order to permit the maximum latitude in optional methods of manufacture.

The following general rules form a basis for intelligent application of dimensions:

1. Plan the location of dimensions in an orderly, uncrowded arrangement.
2. Dimension the piece rather than the picture.
3. Be sure that the dimensions and their relation to one another express the engineering intent clearly.
4. Use sufficient dimensions to define accurately the size of every portion of the represented part.
5. Leave nothing to judgment of the user.
6. Select a method of dimensioning which will insure interchangeability of parts from multiple sources without rejection of usable material.
7. Every drawing should be dimensioned so that it will stand up in any court of law as a clear, uncontroversial statement of all that is required.

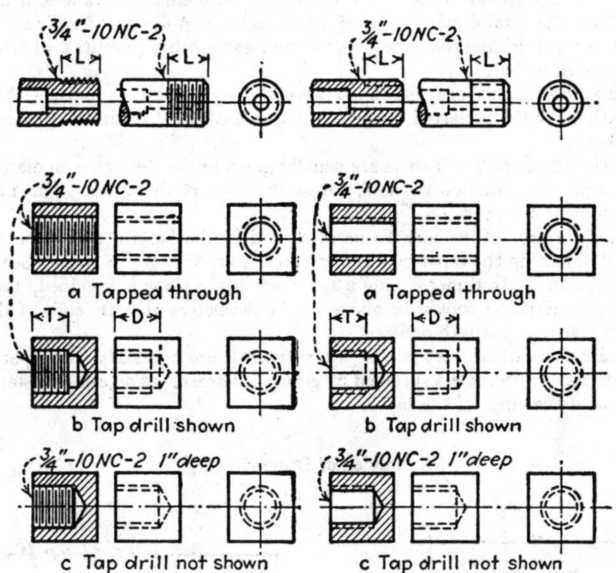

FIG. 8. Screw-thread symbols of the regular type may be used (*left*), but the simplified symbols (*right*) save a great deal of time.

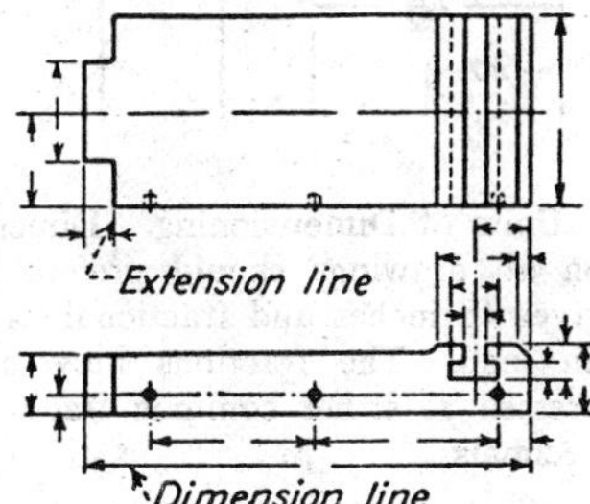

FIG. 9. Proper application of dimension lines is essential to easy reading of a drawing.

8. Keep the dimensions outside the outline of the part, wherever possible.

9. Avoid duplication of dimensions, because a duplicate dimension may be overlooked when changes are made.

10. Where gaging is necessary, specify clearly the extent and relation of gage inspection required and the amount of variation which the gage may allow.

11. Avoid crossing dimension lines with witness lines or leaders to notes.

12. Dimensions should be given a liberal tolerance wherever possible.

Dimensions of parts that can be measured or that can be produced with sufficient accuracy by using an ordinary scale should be written in units and common fractions. Parts requiring greater accuracy should be dimensioned in decimal fractions.

Wire, tubing walls, sheet metal, standard structural sections, etc., should be described by commercial designation followed by dimension in decimals.

Dimensions up to and including 72 in. should preferably be expressed in inches; and those greater than this length, in feet and inches.

Where dimensions call for accurate machining with small tolerances, it is recommended that the total dimension be given in inches and decimal fractions.

In automotive, locomotive, sheet metal, and some other practices all dimensions are specified in inches.

Limiting dimensions are shown in several ways, such as 8.627 – 8.623; or 8.625 ±0.002; or the preferred dimension with a unilateral tolerance, plus or minus, but not both.

Dimensioning Tapers. Tapers are usually given in the difference in diameter per foot of length, measured on the axis, not on the slope of the taper. Three methods of measurement are used.

Standard Tapers. Give one diameter or width, the length, and insert note on drawing designating the taper by number taken from ASA B5.10-1943. See Sec. 35.

Special Tapers. In dimensioning a taper when the slope is specified, the length and only one diameter should be given, or the diameters at both ends of the taper should be given, and length omitted.

In certain cases where very precise measurements are necessary, the taper surface, either external or internal, is specified by giving a diameter at a certain distance from a surface and the slope of the taper.

Examples of Dimensioning

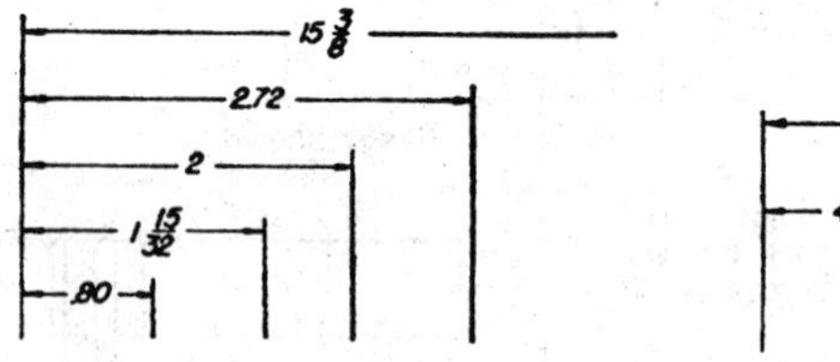

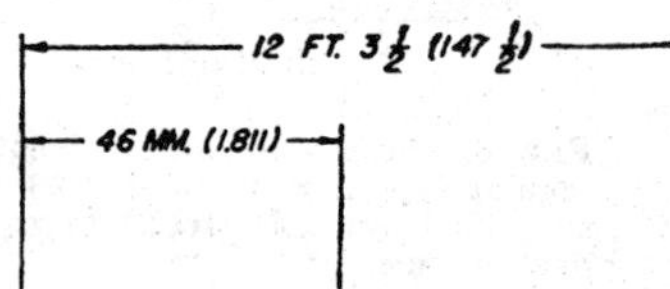

Units of Dimensioning. Dimensions on all drawings should preferably be given in inches and fractional parts of an inch. The fractions may be expressed as either common fractions or decimals.

Where peculiarities of the product require dimensions to be given in feet or in metric increments, only that portion of the dimension not expressed in inches should be designated as feet, millimeters, etc. The equivalent in inches may be shown.

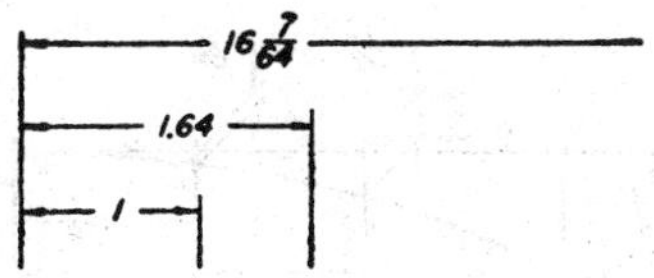

The term "inches" or its symbol is not used in connection with linear dimensions.

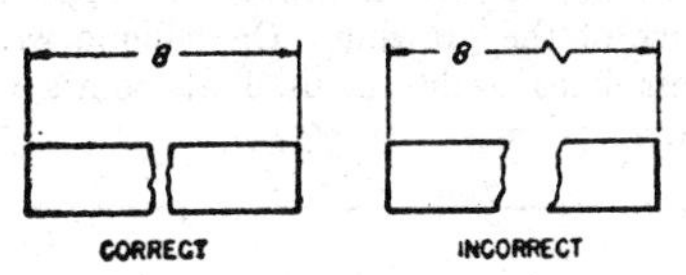

Dimension Lines. Dimension lines are not broken when dimensioning a part which is broken.

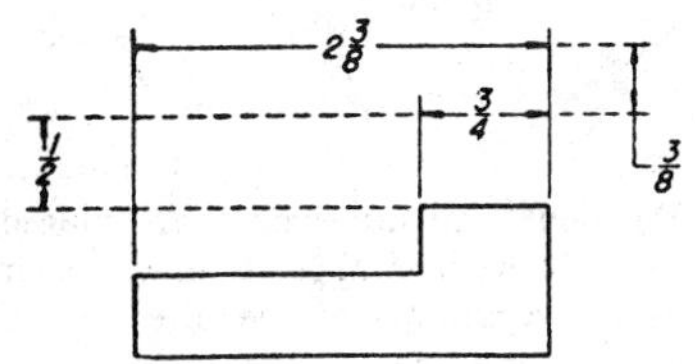

Dimension lines are spaced a reasonable distance from the outline of the part to present a good appearance, approximately $\frac{1}{2}$ in. They are spaced uniformly relative to one another—preferably not less than $\frac{3}{8}$ in. apart.

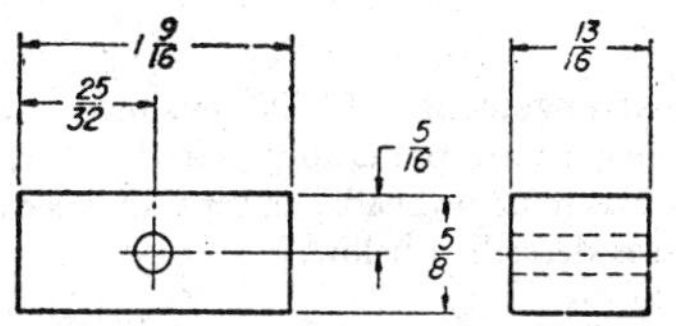

Dimensions are placed outside the outline of the part and between the views wherever possible. Where center lines are used as witness lines, they are shown as such beyond the outline of the part.

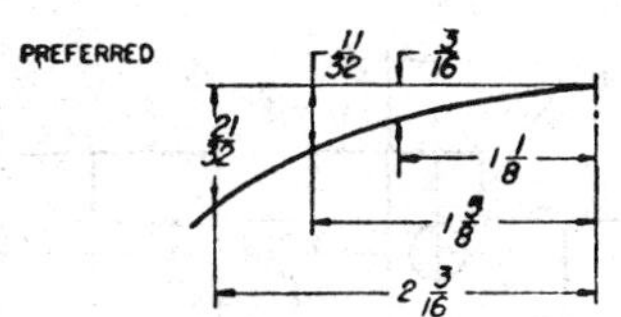

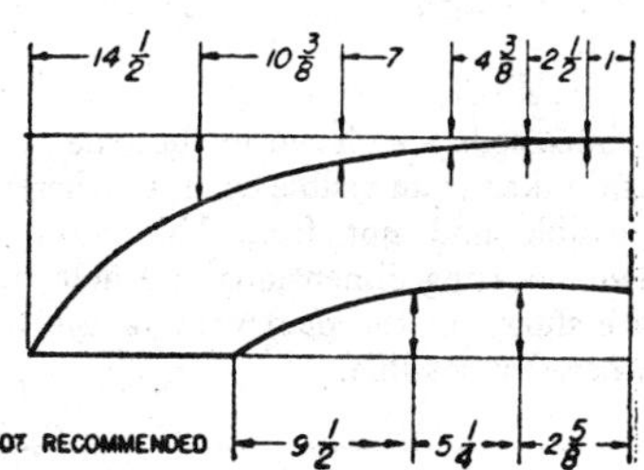

Where a number of ordinates and abscissas are used to dimension a curved outline or a series of holes it is desirable, in the interest of clarity, to extend each dimension line to the base line as marked "preferred," rather than to use methods such as those shown below, which may result in confusion because of lack of general understanding of their meaning.

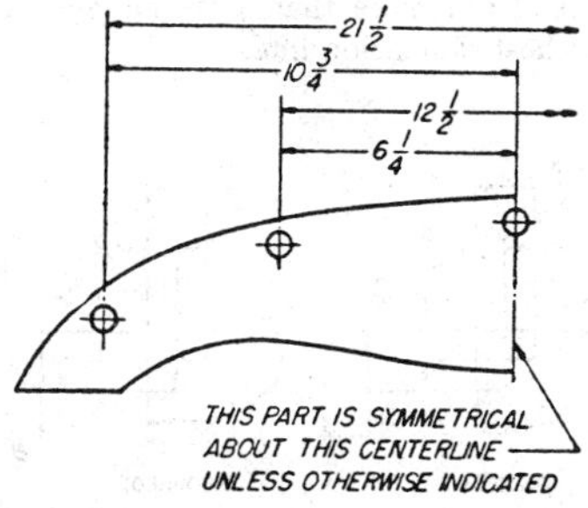

On drawings of symmetrical parts where only one half of the part is pictured, double arrows extending beyond the center line are used to indicate symmetry in combination with a note stating that the part is symmetrical about the center line unless otherwise indicated.

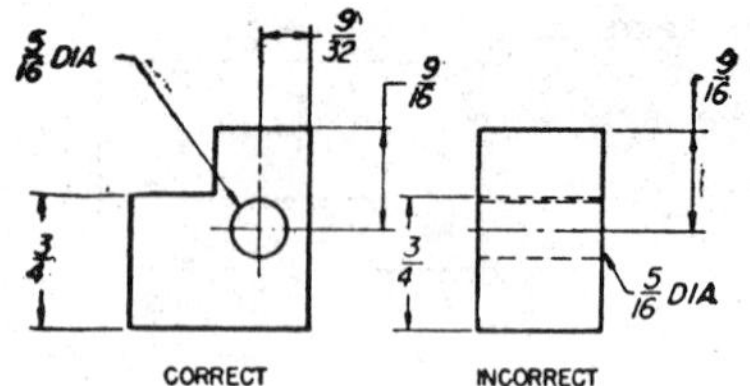

Dimensions are shown in the true views and taken from visible outlines wherever possible and not from hidden edges. Hole-locating dimensions and hole sizes are shown in the plan view of the hole wherever possible.

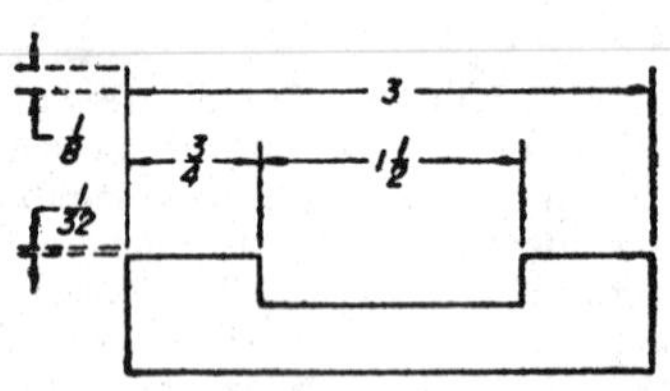

Witness Lines. Witness lines begin $\frac{1}{32}$ in. from the outline of the part and extend not more than $\frac{1}{8}$ in. beyond the farthest dimension line.

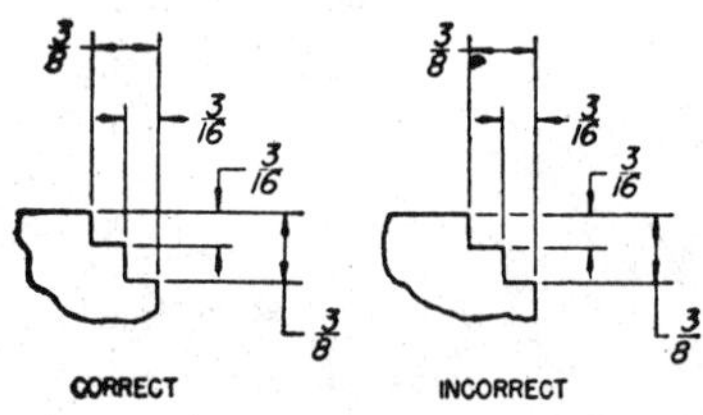

Crossing of witness lines is usually avoided. When crossing is necessary, the witness lines are not broken at the point of crossing as shown at right.

An exception is shown below. Witness lines may be broken when necessary to clarify the extent of dimension lines.

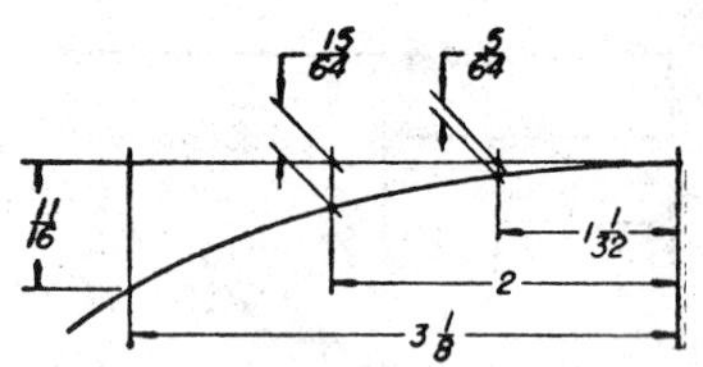

Oblique witness lines may be used when this is advantageous to avoid crowding or to improve the appearance of the drawing. On oblique witness lines a dot is used at points of intersection.

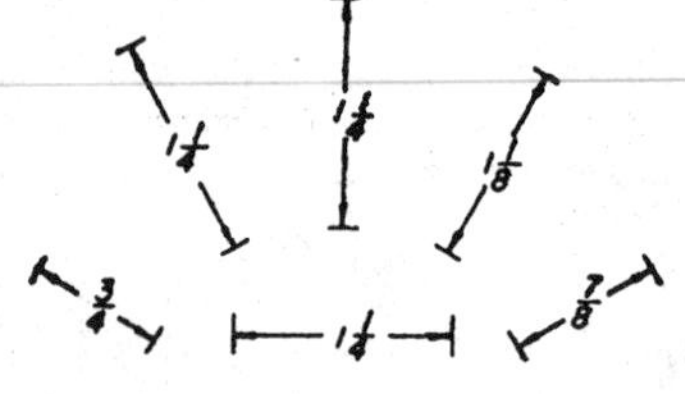

Position. All dimensions are placed to read in a horizontal position, wherever possible, regardless of the direction of the dimension line.

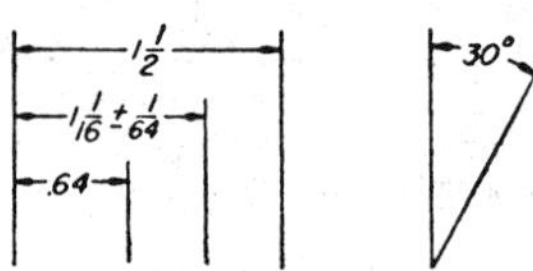

Arrangement of Dimensions. Dimensions are preferably placed midway between arrowheads, except in special cases described below.

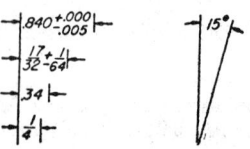

Where the nature or magnitude of the dimension makes it impractical to show dimensions in the manner above, it is permissible to place the dimension between the witness lines with the dimension lines and arrowheads outside of the witness lines.

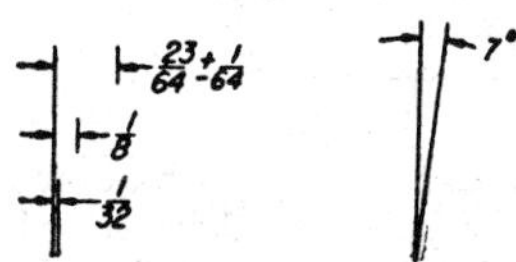

Where it is impractical to use either of the arrangements shown above, the dimension, its dimension lines, and arrowheads may be placed outside of the witness lines.

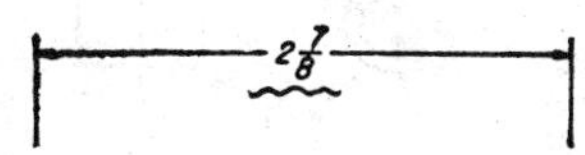

Out-of-scale Dimensions. Original detail drawings are never issued with dimensions out-of-scale to any appreciable extent. Where changes make it impractical to show the new dimension to scale, the dimension must be underlined with a wavy line. "NTS" is never used to indicate out-of-scale dimensions.

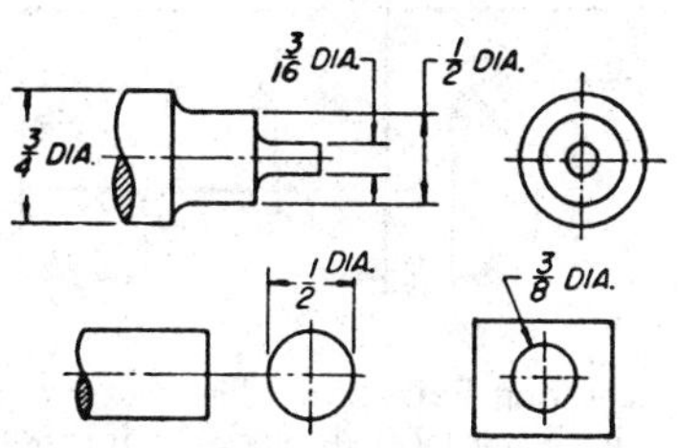

Diameters. Dimensions indicating the diameter of circles are followed by the abbreviation "dia."

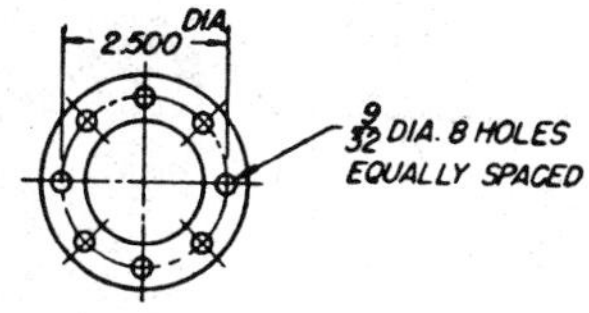

Holes on a Bolt Circle. Holes which are equally spaced around a bolt circle are dimensioned by a note and a bolt-circle diameter as shown.

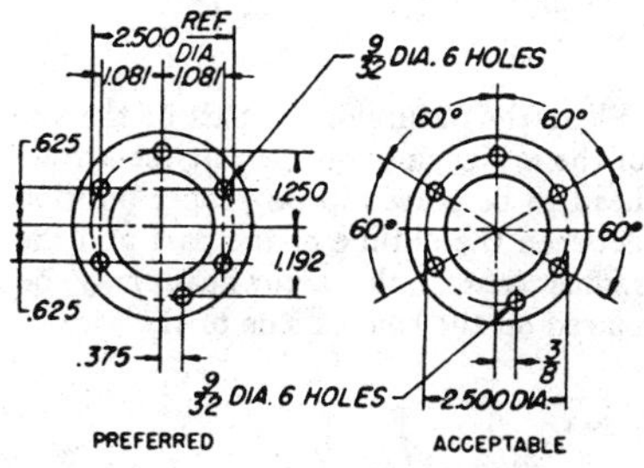

Where the spacing of holes around a bolt circle is other than equal, it is preferable to locate the holes either by ordinates and abscissas or angularly, rather than by descriptive notes which may tend to confuse.

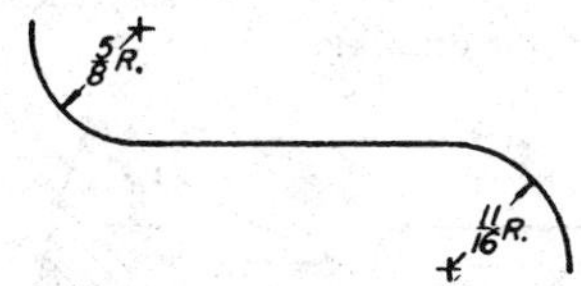

Radiuses. In dimensioning radiuses it is desirable that the leader pass through the center of the radius with the dimension, always followed by "R.," placed between the arrowhead and the center.

When the recommended practice shown above is impossible because of the size of the radius or the nature of its dimension, it is permissible to extend the

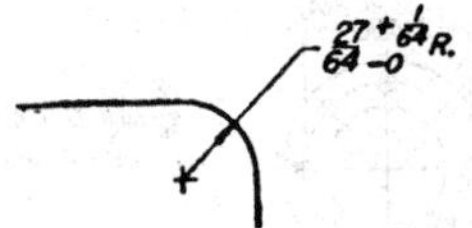

leader beyond the arrow and place the dimension on the convex side of the arc.

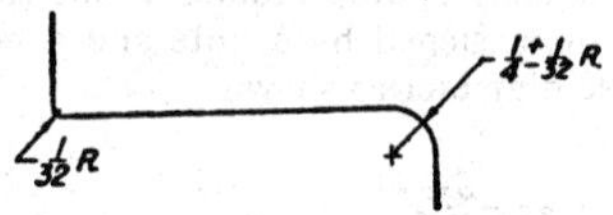

Where the nature of the part or the size of the radius does not permit the arrowhead to be placed in its proper position between the outline of the part and the radius center, the arrowhead may be placed on the convex side of the arc.

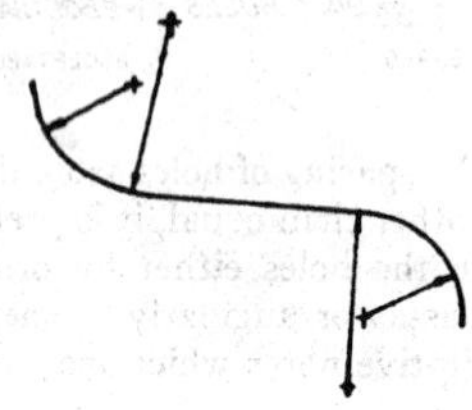

The centers of radiuses may be indicated by a small cross where this tends to clarify the drawing.

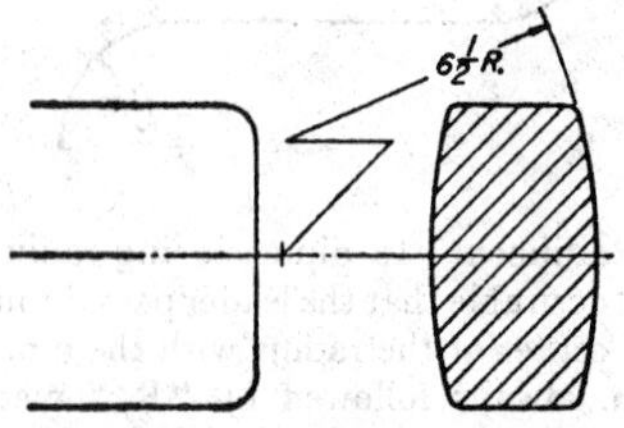

Where the center of the arc of a given radius lies outside the limits of the drawing, or within the boundaries of another view or other data, the leader may be broken. Only one break should be used, and that portion of the leader which ends in the arrowhead should point to the actual center of the radius.

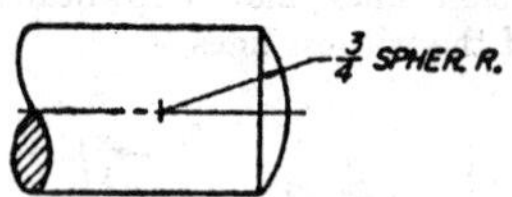

Spherical Radius. Whenever a portion of a sphere is dimensioned by a radius the dimension is followed by the term "spher. R."

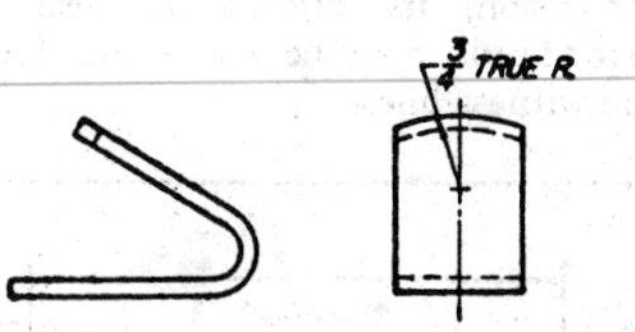

True Radius. In dimensioning a radius in a plane not perpendicular to the plane of projection, such as a radius projected from an inclined surface, the term "true R." may be used as a matter of convenience to avoid showing an auxiliary view.

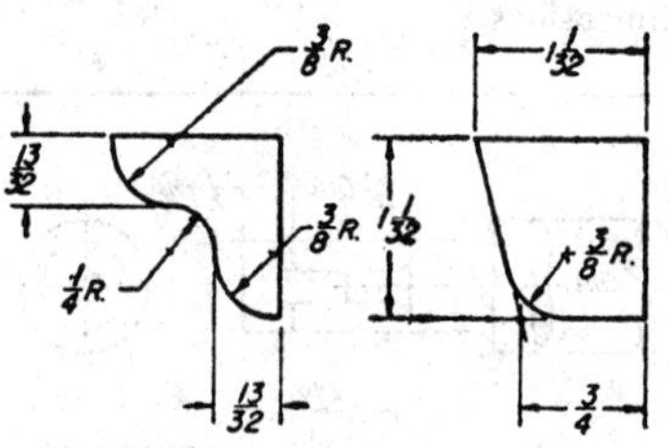

Locating Radiuses. The center of a radius is not located, especially if it falls outside of the outline of the part. It is preferable to locate the intersecting points of its tangents and merely dimension the radius.

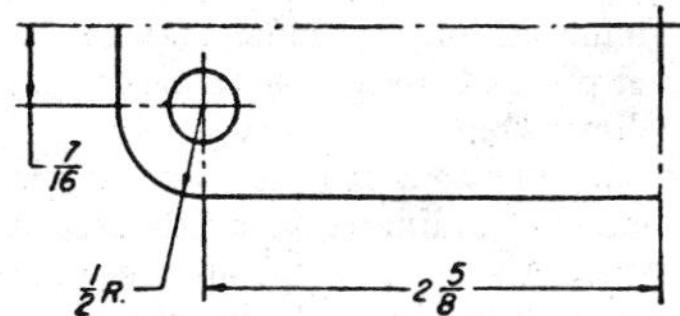

In cases where the center of a radius falls on the center of a hole or on some other center line which has been dimensioned, no additional location for the radius is needed.

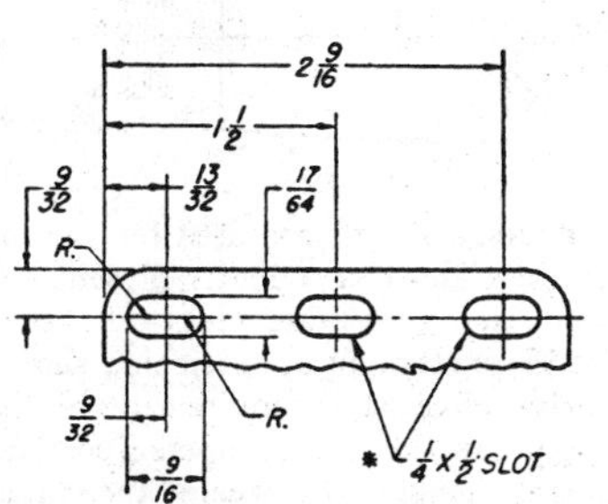

* Slots in sheet metal made by standard punches are specified as shown.

Slots. Slots of regular shape for the purpose of compensating for inaccuracies of manufacture or to provide for adjustment are dimensioned for size by over-all length and width dimensions, and for location by dimensions to their center lines.

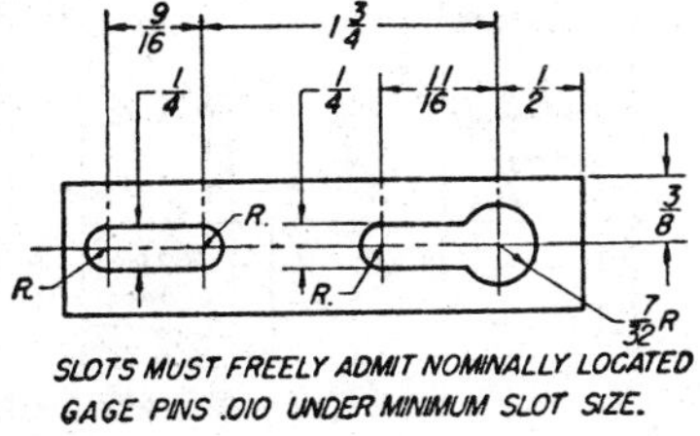

Slots which are intended to perform a mechanical function and whose size and location are subject to gage inspection are treated as two partial holes separated by space and dimensioned as such.

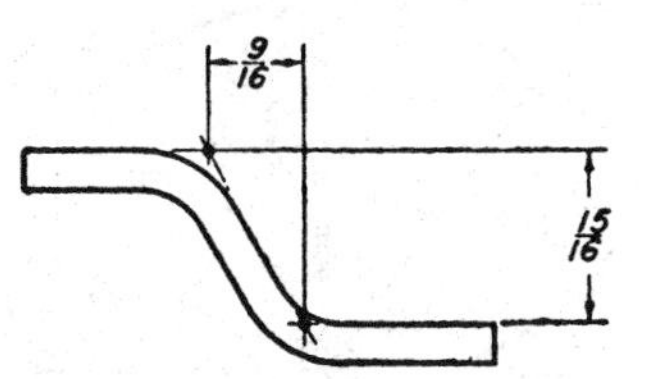

Offsets. Offsets are preferably dimensioned from the points of intersection of the tangents and to the same side of the material.

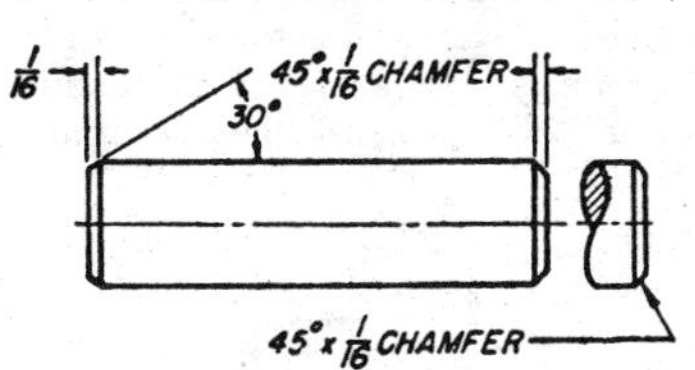

Chamfers. When a chamfer is specified by a note as "45° × $\frac{1}{16}$ chamfer" it is understood that the linear dimension is taken parallel to the axis of the part and not along the angular face of the chamfer. Chamfers other than 45° are specified by dimensions applied to the appropriate view rather than by a note. The purpose of this practice is to avoid misinterpretation.

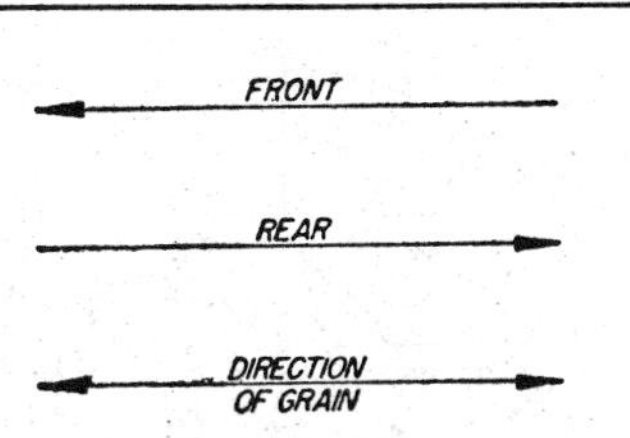

Direction. To indicate front, rear, direction of grain, or the direction for special views, use a simple heavy arrow.

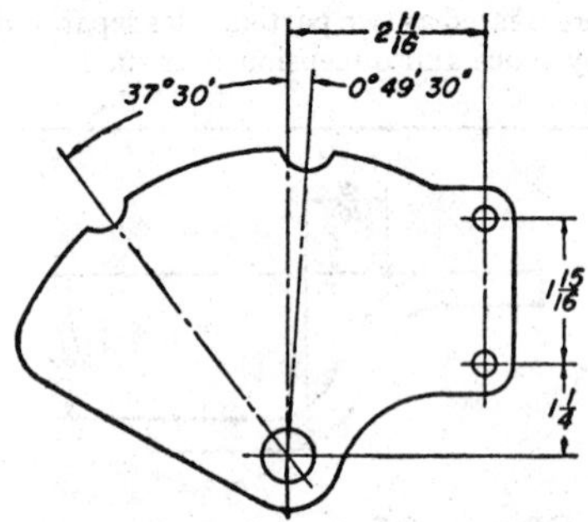

Angles. Angular dimensions are avoided wherever possible. Where linear dimensioning will not suffice, angles are expressed in degrees, minutes, and seconds—never in fractions of degrees or decimals. On angles of less than one degree "0°" precedes the specification of minutes and seconds. Dashes are never placed between the terms of angular dimensions.

Angular relations are preferably dimensioned by ordinates and abscissa to eliminate the need for calculations in the shop, and to insure better control of tolerances.

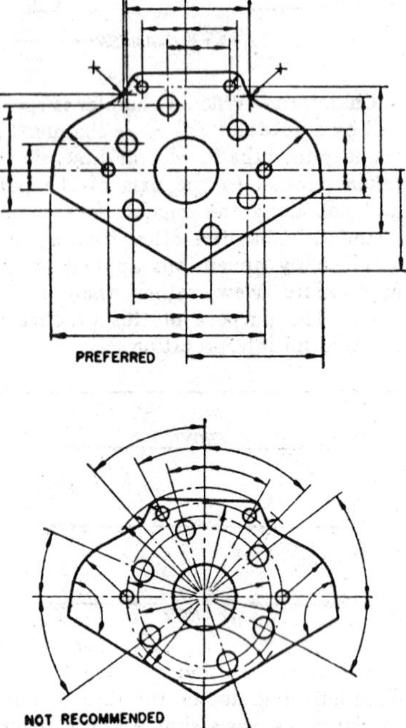

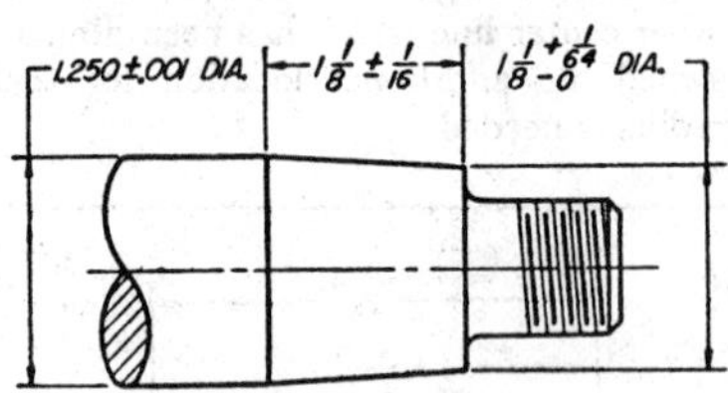

Tapers. Tapers are used for various purposes, and consequently present different design requirements. In each case the method of dimensioning should clearly convey the requirement of the design. In the case of tapered portions of machined parts where the taper serves no purpose other than to provide a gradual reduction of section between two functional surfaces the diameter at each end of the taper is dimensioned with the maximum permissible tolerance and a length is specified either as "approximate" or with a generous tolerance.

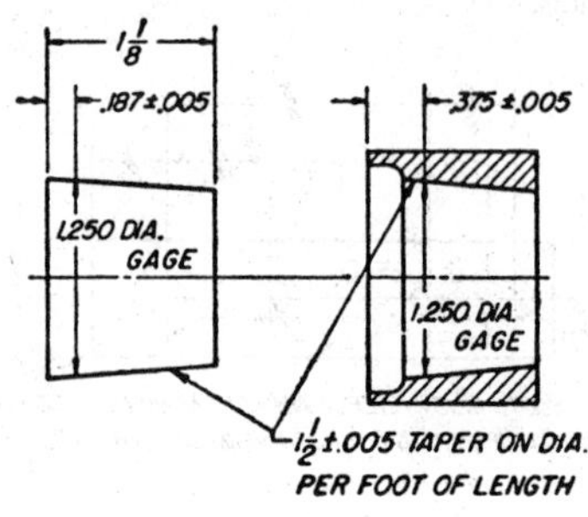

In the case of tapers which engage one another either permanently or intermittently, the mating parts are dimensioned by specifying the taper per foot with a tolerance and the diameter without tolerance located by a dimension and tolerance parallel to the axis.

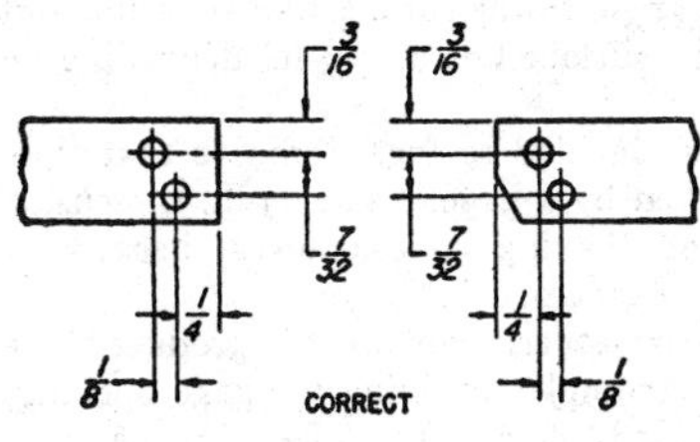

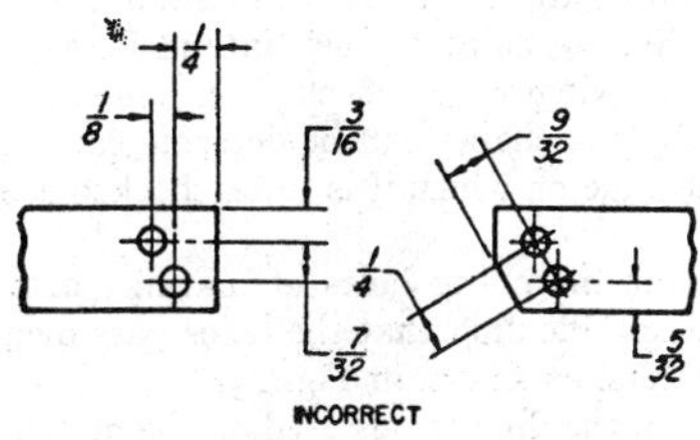

Mating Parts. Mating surfaces, holes, etc., on related parts are dimensioned according to the same pattern to facilitate fabrication and to insure the desired fit of the parts.

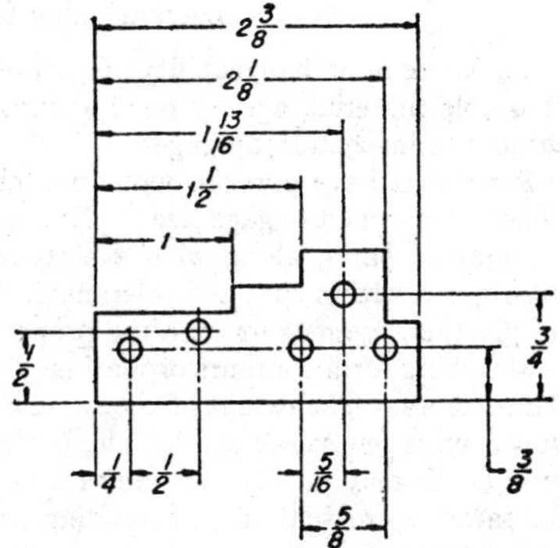

Dimensioning from Baseline. Generally, surfaces, groups of holes, etc., are located from a common baseline. This practice clearly indicates the requirements from a functional standpoint, and avoids an unnecessary accumulation of tolerances.

Tubing Bends. Avoid compound bends. Second and third bends from left could be compound, but are approximated by true radii at *G* and *H*. Hence, bends are true only in plane of bend—bend *G* in plane of points *J-K-L* and bend *H* in plane of points *K-L-M*.

Tubing is dimensioned to center line, at both bend radiuses and bend location. Exceptions: If radius fits an adjoining part, it is dimensioned to outside or inside of bend. If bends are to be made on a single die, it is easier to specify inside radius.

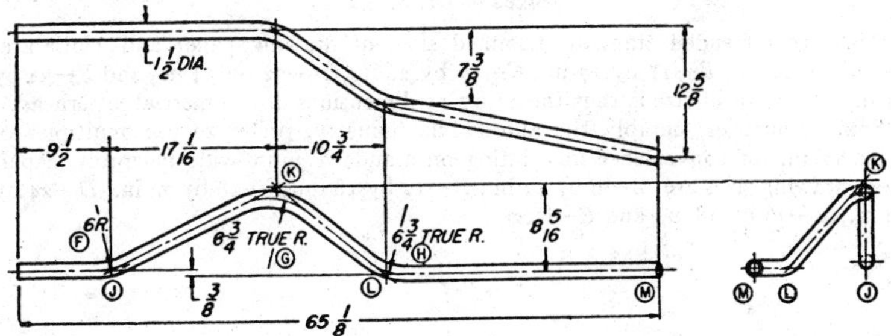

Dimensioning for Gage Inspection

To insure interchangeability of parts from multiple sources and to avoid rejection of usable material, specify on drawings exactly the extent to which the parts are subject to inspection by gages.

Parts that have portions which match with mating parts should have those portions dimensioned to gage sizes. Nominal dimensions should be specified for mating portions of parts, along with a statement that they must conform to nominally sized gages within specified tolerances. Nominal dimensions are used because they supply the gage makers with the proper gage sizes.

Matching form contours of mating parts are signaled with slanting lines drawn to the surfaces. The extent of the signaled area is dimensioned. Signaling should be shown in as few views as possible, preferably in the view which best covers its full extent. It may be shown at any convenient angle to the surface and spaced about the same as crosshatching, extending from the surface line ⅛ to ¼ in. depending on size of the part.

In the channel-sectioned part (Fig. 10) the signaled surfaces indicate that they match with mating parts and should be checked by a gage. The 2½-in. dimension which connects them is a nominal dimension (having no tolerances) because it denotes the size of the gage, which is made to exact size.

The tolerance notation indicates that the part is acceptable if it is produced in a size range which extends from maximum size (completely filling the gage opening) to minimum size (loose in the gage to the specified tolerance). A feeler of the tolerance thickness, as shown in Fig. 11, may be used to determine the minimum size.

Sheet-metal parts which lack sufficient rigidity to maintain their intended shape when unsupported by gages are indicated on drawings to be checked on nominally sized gages, because that is the only way their true shapes can be determined. A typical sheet-metal part, its tolerance, and the gage on which it is to be checked are shown in Fig. 12.

When dimensioning the distance between the matching holes of mating parts, nominal dimensions should be used. Write a specification that the holes pass over nominally located gage pins of specified diameters, as shown in Fig. 13.

Angular and linear tolerances are omitted from the dimensions used on the matching portions of mating parts, because they are included in the amount of permissible deviation that must be specified in connection with nominal dimensions. For instance, a part having portions which match with other parts, as shown in Fig. 15, may require gage-inspection operations as shown in Fig. 14. From these illustrations, it is evident that tolerances are included in the "tolerance" distances between the part and the gage.

Sizes of Drawings

ASA recommended standard trimmed sizes of drawing paper and cloth are: *A*—8½ by 11 in., *B*—11 by 17 in., *C*—17 by 22 in., *D*—22 by 34 in., and *E*—34 by 44 in. Basic sheet size is thus the 8½ by 11 dimensions of commercial letterhead.

Some industries, notably the automobile industry, prefer to use multiples of 9 by 12 in. for convenience in printing on standard 36-in.-wide blueprint paper. Here drawing sizes are: *A*—9 by 12 in., *B*—12 by 18 in., *C*—18 by 24 in., *D*—24 by 36 in., *E*—36 by 48 in., and *R*—larger.

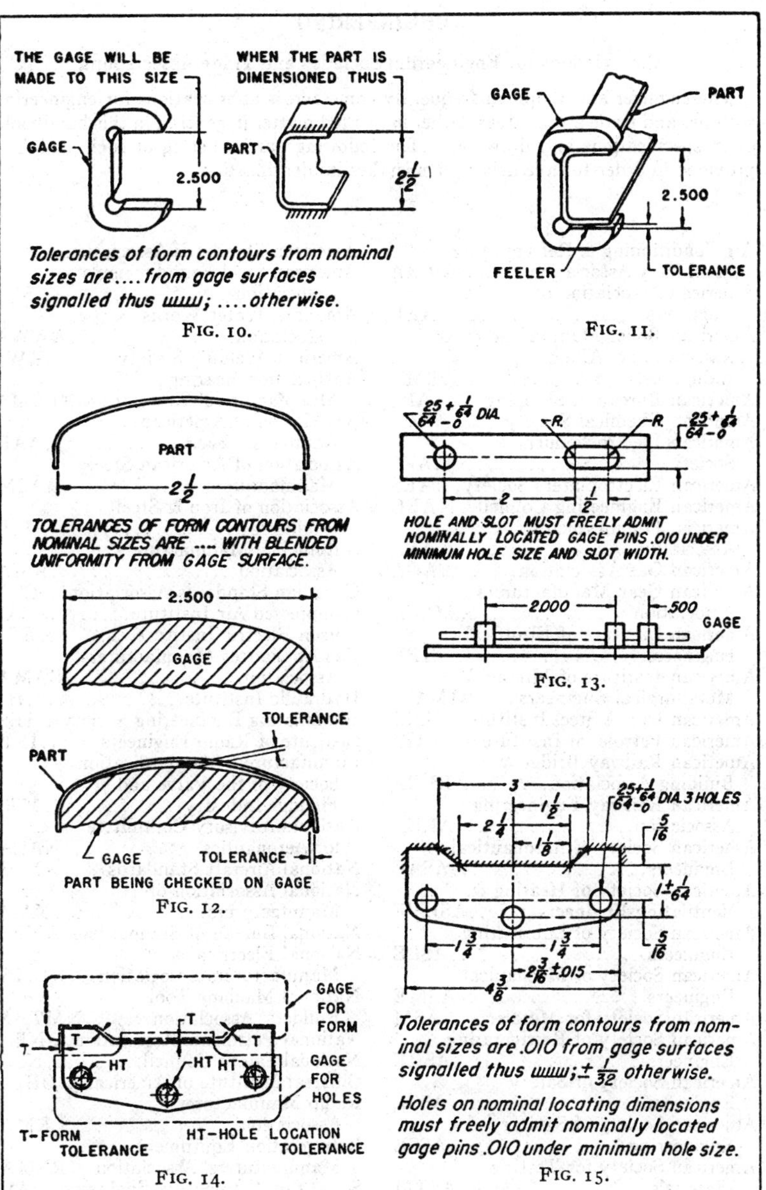
THE GAGE WILL BE MADE TO THIS SIZE
WHEN THE PART IS DIMENSIONED THUS
GAGE
PART
2.500
2 1/2
Tolerances of form contours from nominal sizes are.... from gage surfaces signalled thus ; .... otherwise.
GAGE
PART
2.500
FEELER
TOLERANCE
PART
2 1/2
TOLERANCES OF FORM CONTOURS FROM NOMINAL SIZES ARE .... WITH BLENDED UNIFORMITY FROM GAGE SURFACE.
2.500
GAGE
TOLERANCE
PART
GAGE
TOLERANCE
PART BEING CHECKED ON GAGE
25/64 +1/64 −0 DIA
R.
R.
25/64 +1/64 −0
2
1/2
HOLE AND SLOT MUST FREELY ADMIT NOMINALLY LOCATED GAGE PINS .010 UNDER MINIMUM HOLE SIZE AND SLOT WIDTH.
2.000
.500
GAGE
GAGE FOR FORM
T
T
T
T
HT
HT
HT
GAGE FOR HOLES
T-FORM TOLERANCE
HT-HOLE LOCATION TOLERANCE
3
1 1/2
2 1/4
1 1/8
25/64 +1/64 −0 DIA. 3 HOLES
5/16
1 ± 1/64
5/16
1 3/4
1 3/4
2 3/16 ±.015
4 3/8
Tolerances of form contours from nominal sizes are .010 from gage surfaces signalled thus ; ± 1/32 otherwise.
Holes on nominal locating dimensions must freely admit nominally located gage pins .010 under minimum hole size.

FIG. 10. FIG. 11. FIG. 12. FIG. 13. FIG. 14. FIG. 15.

FIGS. 10 to 15. Methods of dimensioning parts for gaging.

ENGINEERING

Abbreviations for Engineering Societies and Trade Associations

The engineer and shopman frequently come across abbreviations for engineering societies and trade associations, either in printed matter in general, in this handbook, or in specifications and drawings. The following partial listing of such bodies is provided in order to save delays or mistakes in identification.

Air Conditioning & Refrigerating Machinery Association......ACRMA
American Association of Engineers.....................AAE
American Boiler Manufacturers' Association & Affiliated Industries...................ABMA
American Bureau of Shipping.....ABS
American Chemical Society.......ACS
American Electrochemical Society......................AES
American Electroplaters Society...AES
American Engineering Council....AEC
American Foundrymen's Association..................AFA
American Gas Association........AGA
American Gear Manufacturers' Association.................AGMA
American Institute of Electrical Engineers...................AIEE
American Institute of Mining & Metallurgical Engineers.....AIMME
American Iron & Steel Institute...AISI
American Petroleum Institute.....API
American Railway Bridge & Building Association........ARBBA
American Railway Engineering Association..................AREA
American Society of Aeronautical Engineers..................ASAE
American Society of Heating & Ventilating Engineers.......ASHVE
American Society of Lubricating Engineers...................ASLE
American Society of Mechanical Engineers...................ASME
American Society for Metals......ASM
American Society of Refrigerating Engineers..................ASRE
American Society of Safety Engineers...................ASSE
American Society of Sanitary Engineering..................ASSE
American Society for Testing Materials...................ASTM
American Society of Tool Engineers..................ASTE
American Standards Association...ASA
American Steel Foundrymen's Association..................ASFA
American Water Works Association.................AWWA
American Welding Society........AWS
Anti-friction Bearing Manufacturers' Association.AFBMA
Association of American Railroads....................AAR
Association of American Steel Manufacturers...............AASM
Association of Iron & Steel Engineers...................AISE
Automobile Manufacturers' Association..................AMA
Canadian Standards Association...CSA
Compressed Air Institute.........CAI
Edison Electric Institute..........EEI
Gas Appliances Manufacturers' Association.................GAMA
Hydraulic Institute................HI
Illuminating Engineering Society...IES
Institute of Radio Engineers......IRE
Manufacturers Standardization Society of the Valve and Fittings Industry..............MSS
National Advisory Committee for Aeronautics..............NACA
National Aircraft Standards.......NAS
National Association of Manufacturers...............NAM
National Bureau of Standards.....NBS
National Electrical Manufacturers Association...NEMA
National Machine Tool Builders' Association.......NMTBA
National Petroleum Association...NPA
National Safety Council..........NSC
Oil Heat Institute of America....OHIA
Radio Manufacturers' Association..................RMA
Refrigeration Equipment Manufacturers' Association...REMA
Society of Automotive Engineers..SAE
Society of Industrial Engineers.....SIE
Underwriters' Laboratories, Inc.....UL

Standards and the Engineer

By specifying standard parts, the engineer saves the labor of designing many details, and he is more likely to create an economical and serviceable product. Approximately 1100 standards have been formulated by 110 national technical societies and trade associations affiliated with the American Standards. These standards cover a broad range of mechanical, electrical, and consumer's goods items.

By selecting standard components, the engineer achieves:

1. Speedy delivery of material.
2. Reduced product costs, because suppliers can price standard parts on basis of large output.
3. Reduced inventory.
4. Less chance of failure, because standard parts have had ample field testing, and quality is likely to be uniform.

The editors have drawn liberally upon American Standards, but it is manifestly impossible to reproduce all standards of interest in their entirety. Also, standards are revised periodically and new ones appear between handbook revisions. Therefore, those who have need of standards are advised to request the list of American Standards from the American Standards Association, 70 East 45th St., New York 17, N.Y.

Pointers for Lower Cost Designs

Instructions issued by the General Electric Co. to its designers contain in brief form many worth-while suggestions for reducing the cost of parts and assemblies. Periodic reference to the following matter will aid the designer in making an economical choice of material and manufacturing method:

Castings

1. Eliminate dry-sand (baked-sand) cores.
2. Minimize depth to obtain flatter castings.
3. Use minimum weight consistent with sufficient thickness to cast without chilling.
4. Choose simple forms.
5. Symmetrical forms produce uniform shrinkage.
6. Apply liberal radiuses—no sharp corners.
7. If surfaces are to be accurate with relation to each other, they should be in the same part of the pattern, if possible.
8. Locate parting lines so that they will not affect looks and utility, and need not be ground smooth.
9. Specify multiple patterns instead of single ones.
10. Metal patterns are preferable to wood.
11. Permanent molds are preferable to metal patterns.

Moldings

1. Eliminate inserts from parts.
2. Design molds with smallest number of parts.
3. Use simple shapes.
4. Locate flash lines so that the flash does not need to be filed and polished.
5. Minimize the weight.

Punchings

1. Punched parts are cheaper than molded, cast, machined, or fabricated parts.
2. Choose "nestable" punchings to economize on material.
3. Holes requiring accurate relation to each other should be made by the same die.
4. Design to use coil stock.
5. Design punchings to have minimum sheared length and maximum die strength with fewest die moves.

Formed Parts

1. Drawn parts instead of spun, welded, or forged parts.
2. Shallow draws if possible.
3. Liberal radiuses on corners.
4. Bent parts instead of drawn.
5. Parts formed of strip or wire instead of punched from sheet.

Pointers for Lower Cost Designs *Continued*

Fabricated Parts

1. Self-tapping screws instead of standard screws.
2. Drive pins or screws instead of standard screws.
3. Rivets instead of screws.
4. Hollow rivets instead of solid rivets.
5. Spot or projection welding instead of riveting.
6. Welding instead of brazing or soldering.
7. Use die castings or molded parts instead of fabricated construction requiring several parts.

Machined Parts

1. Use rotary machining processes instead of shaping methods.
2. Use automatic or semiautomatic machining instead of hand-operated.
3. Reduce the number of shoulders.
4. Omit finishes where possible.
5. Use rough finish when satisfactory.
6. Dimension drawings from same point as used by factory in measuring and inspecting.
7. Use centerless grinding instead of between-center grinding.
8. Avoid tapers and formed contours.
9. Allow a radius or undercut at shoulders.

Screw-machine Parts

1. Eliminate second operation.
2. Use cold-rolled stock.
3. Design for header instead of screw machine.
4. Use rolled threads instead of cut threads.

Welded Parts

1. Fabricated construction instead of castings or forgings.
2. Minimum sizes of welds.
3. Welds made in flat position rather than vertical or overhead.
4. Eliminate chamfering edges before welding.
5. Use "burn-outs" (torch-cut contours) instead of machined contours.
6. Lay out parts to cut to best advantage from standard rectangular plates and avoid scrap.
7. Use intermittent instead of continuous weld.
8. Design for circular or straight-line welding to use automatic machines.

Treatments and Finishes

1. Reduce baking time to minimum.
2. Use air drying instead of baking.
3. Use fewer or thinner coats.
4. Eliminate treatments and finishes entirely.

Assemblies

1. Make assemblies simple.
2. Make assemblies progressive.
3. Make only one assembly and eliminate trial assemblies.
4. Make component parts *right* in the first place so that fitting and adjusting will not be required in assembly.

This means that drawings must be correct, with proper tolerances, and that parts are made according to drawing.

Cams

Basic Types of Cams. Modifications and uses for basic types of cams are shown in Fig. 16. Edward Rahn, chief engineer of the Rowbottom Machine Co., Inc., has this to say about the several types:

Fig. 16—1. Flat-plate Cam. Essentially a displacement cam. With it, movement can be made from one point to another along any desired profile. Often used in place of taper attachments on lathes for form turning. Some have been built in sections up to 15 ft long for turning the outside profile on gun barrels. Such cams can be made on either milling machines or profiling machines.

FIG. 16—2. BARREL CAM. Sometimes called a cylindrical cam. The follower moves in a direction parallel to the cam axis, and lever movement is reciprocating. As with other types of cam, the base curve can be varied to give any desired movement. Internal as well as external barrel cams are practical. A limitation: internal cams less than 11 in. dia are difficult to make on cam millers because of clearance restrictions.

FIG. 16—3. INDEX CAM. Within limits, such cams can be designed for any desired acceleration, deceleration, and dwell period. A relatively short period for acceleration can be allowed on high-speed cams such as those used on zipper-making equipment on which indexing occurs 1200 to 1500 times per minute. Cams of this sort can also be designed with four or more index stations. Grooves can be either tapered or straight.

FIG. 16—4. SIDE CAM. Essentially a barrel cam having only one side. Can be designed for any type of motion, depending on requirements and speed of operation. Spring- or weight-loaded followers of either the pointed or roller type can be used. Either vertical or horizontal mounting is permissible. Cutting of the profile is usually done on a shaper or a cam miller equipped with a small-diameter cutter.

FIG. 16—5. NONUNIFORM FACE CAM. Sometimes called a disk cam. Follower can be either a roller, hexagon, or pointed bar. Profile can be derived from a straight line, modified straight line, harmonic, parabolic, or nonuniform base curve. Generally, the shock imposed by a cam designed on a straight-line base curve is undesirable.

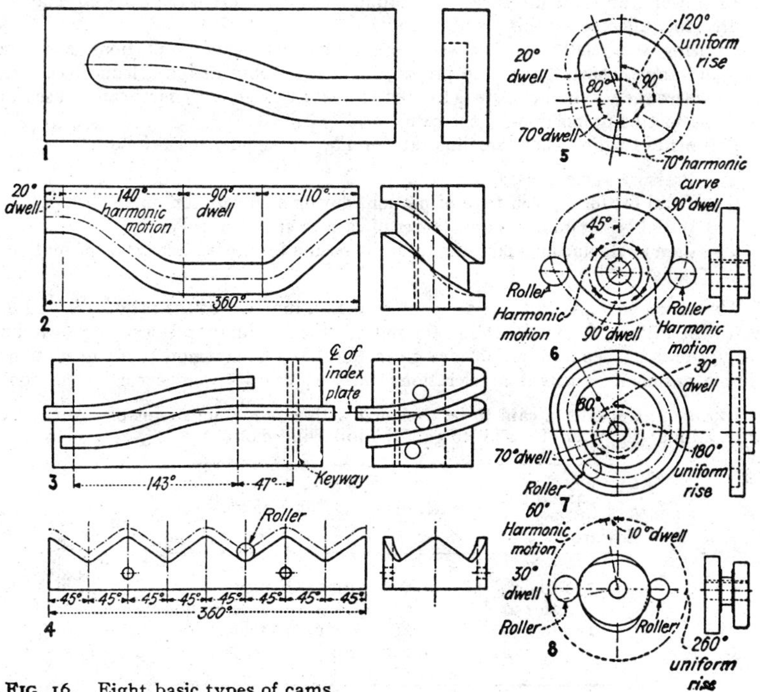

FIG. 16. Eight basic types of cams.

Follower usually is weight-loaded, although spring, hydraulic, or pneumatic loading is satisfactory.

FIG. 16—6. SINGLE-FACE CAM WITH TWO FOLLOWERS. Similar in action to a box or double-face cam, except that flexibility is less than that for the latter type. Cam action for feed and return motions must be the same to prevent looseness. Used in place of box cams or double-face cams to conserve space, and instead of single-face cams to provide more positive movement.

FIG. 16—7. BOX CAM. Gives positive movement in two directions. A profile can be based on any desired base curve, as with face cams, but a cam miller is needed to cut it; whereas with face cams, a hand saw and disk grinder could conceivably be used. No spring, pneumatic, or hydraulic loading is needed for the followers. This type of cam requires more material than for a face cam but is no more expensive to mill.

FIG. 16—8. DOUBLE-FACE CAM. Similar to single-face cam, except that it provides positive straight-line movement in two directions. The supporting fork for the rollers can be mounted separately or between the faces. If the fork fulcrum is extended beyond the pivot point, the cam can be used for oscillatory movement. With this cam, the return stroke on a machine can be run faster than the feed stroke. Cost is more than that for a box cam.

Pressure Angle in Cam Design. An important factor in design of radial-disk cams is the pressure angle between the cam surface and the radial roller or knife-edge follower. Variations in this angle affect the transverse forces which act on the follower. In some instances, the torque required to drive the camshaft may be appreciably affected by this pressure angle.

Utilizing the relationship between the geometrical cam proportions and the pressure angle, charts have been prepared which simplify design calculations. In cases where the maximum angle is limited to a given value, the charts can be used to find the specific proportions of the cam.

Pertinent cam dimensions are illustrated in Fig. 17, and defined as follows:

α = Pressure angle of cam, deg
L = Total lift for a given type of motion, during cam rotation, in.
R = Initial base radius of cam, center of cam to center of roller, in.
β = Cam angle during which the total lift occurs for a particular type of motion, radians

The curves in Figs. 18 to 21 are expressed as a function of the ratio L/R and β. With the ratio L/R and the total cam angle β, the maximum pressure angle can be readily found. Separate graphs are given for the most common types of motion: simple harmonic, constant acceleration, constant velocity, and cycloidal motion.

EXAMPLE: Consider a cam to be designed with a minimum radius of 2.0 in. and with a roller diameter of 1.0 in. to give the following motion:

Cam angle, deg	Lift, in.	Type of motion
0 to 60	0.5	Simple harmonic
60 to 150	0.75	Constant acceleration
150 to 250	−1.25	Constant velocity
250 to 360	0.0	Dwell

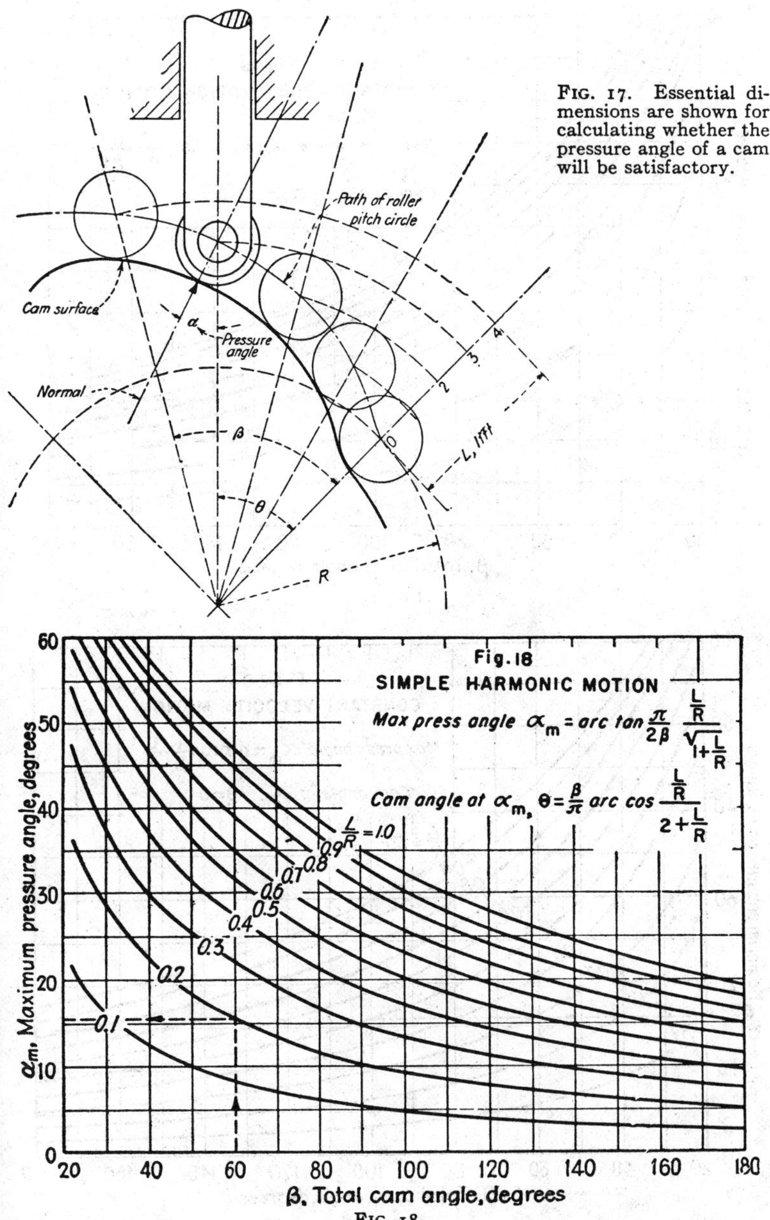

FIG. 17. Essential dimensions are shown for calculating whether the pressure angle of a cam will be satisfactory.

FIG. 18.

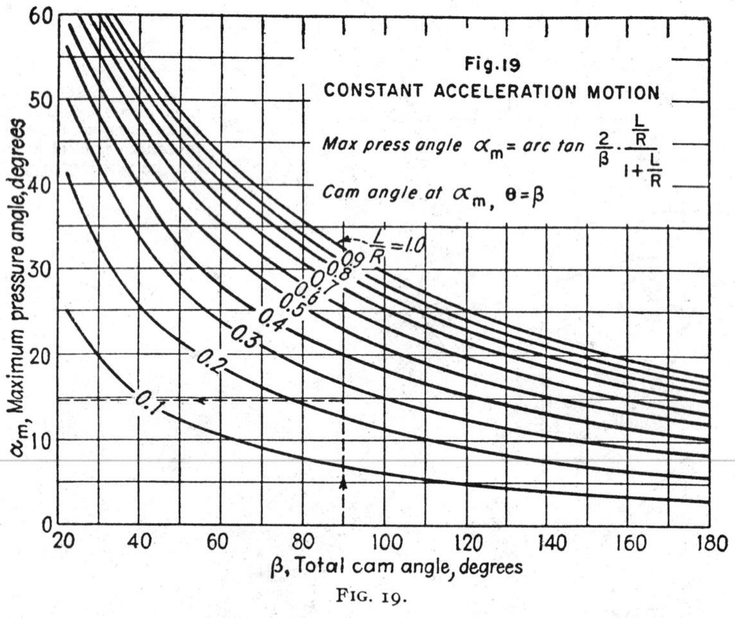
Fig.19
CONSTANT ACCELERATION MOTION
Max press angle α_m = arc tan $\frac{2}{\beta} \cdot \frac{\frac{L}{R}}{1+\frac{L}{R}}$
Cam angle at α_m, $\theta = \beta$
$\frac{L}{R}$=1.0
0.9
0.8
0.7
0.6
0.5
0.4
0.3
0.2
0.1
α_m, Maximum pressure angle, degrees
β, Total cam angle, degrees

FIG. 19.

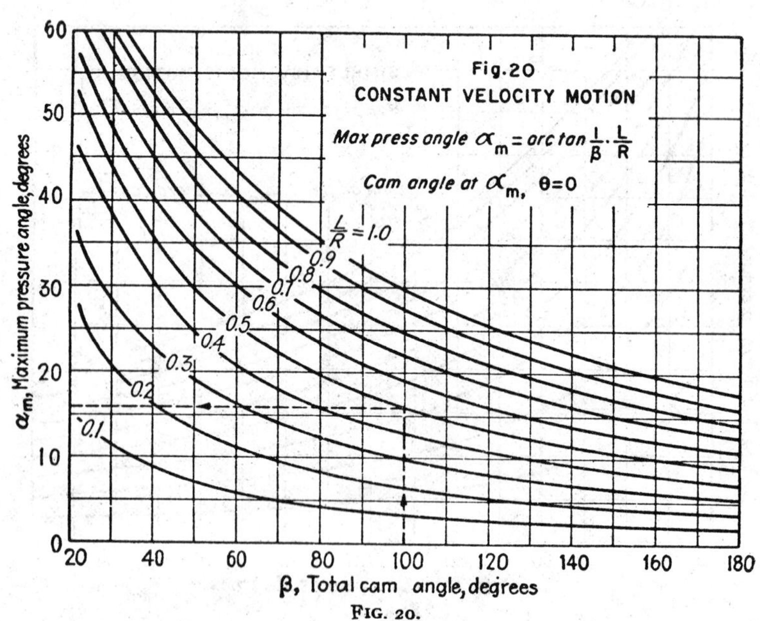
Fig.20
CONSTANT VELOCITY MOTION
Max press angle α_m = arc tan $\frac{1}{\beta} \cdot \frac{L}{R}$
Cam angle at α_m, $\theta = 0$
$\frac{L}{R}$=1.0
0.9
0.8
0.7
0.6
0.5
0.4
0.3
0.2
0.1
α_m, Maximum pressure angle, degrees
β, Total cam angle, degrees

FIG. 20.

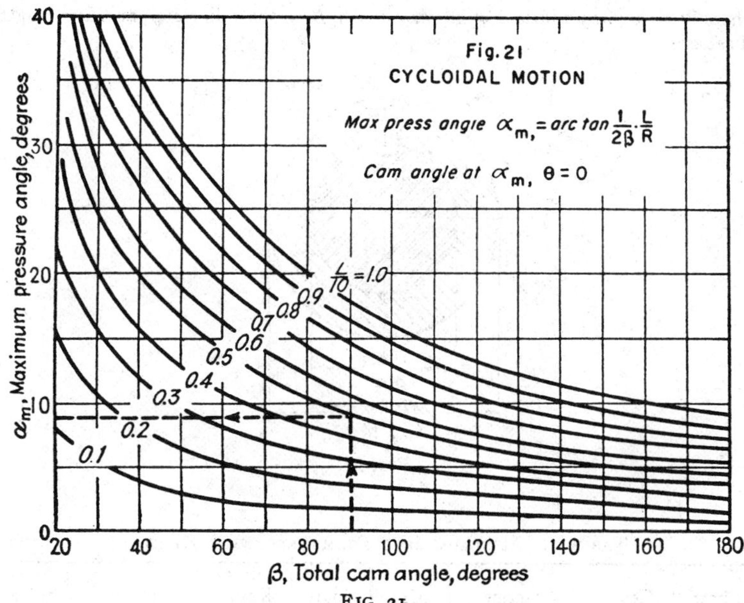

FIG. 21.

For the first motion: L equals 0.5 in., β is 60°, and the initial radius R is a minimum at 2.0 + 0.50 or 2.5 in.; thus the ratio L/R is 0.2. From Fig. 18, the maximum pressure angle α_m is 15.5°.

For the second motion: β is 90°, the minimum radius R becomes 2.5 + 0.5 or 3.0 in., and L/R is 0.25. From Fig. 19, α_m is equal to 14.5°.

For the third motion (downward) the initial base radius R is actually located at the end of motion and must be calculated from the cam dimensions. This special consideration is necessary only where the displacement is downward since these graphs have been based on a minimum base radius. For convenience, the cam may be considered to be rotating in a reverse direction for the constant velocity motion. Thus, R equals 2.0 + 0.5 or 2.5 in. and the ratio L/R is 0.5. From Fig. 20 with β equal to 100°, α_m is found to be 16.0°.

Hub Diameters

Values of minimum hub diameters (Fig. 22) are given in Table 1 for cams, gears, sprockets, and pulleys fixed to shafts from $\frac{1}{2}$ to 3 in. dia by a standard square key and setscrew over the key. Basis for the table is the formula

$$\text{Min hub dia} = \left[\sqrt{\left(\frac{D}{2}\right)^2 - \left(\frac{W}{2}\right)^2} + d + L \right] 2$$

where W = keyway width, d = keyway depth, D = shaft diameter, and L = length of setscrew.

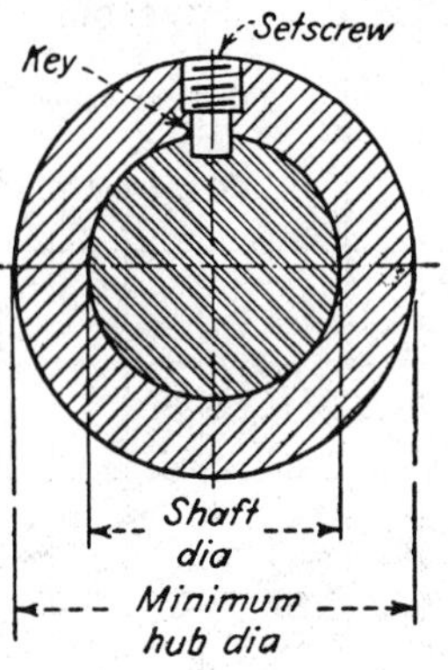

FIG. 22.

TABLE 1. MINIMUM HUB DIAMETERS WITH KEYWAY

Shaft Dia, In.	Keyway, In.	Setscrew Size and Length	Min Hub Dia, In.	Shaft Dia, In.	Keyway, In.	Setscrew Size and Length	Min Hub Dia, In.
1/2	1/8 × 1/16	1/4-20 × 1/4	1 3/32	1 13/16	1/2 × 1/4	9/16-12 × 9/16	3 11/32
9/16	1/8 × 1/16	1/4-20 × 1/4	1 5/32	1 7/8	1/2 × 1/4	9/16-12 × 9/16	3 13/32
5/8	3/16 × 3/32	5/16-18 × 5/16	1 13/32	1 15/16	1/2 × 1/4	9/16-12 × 9/16	3 1/2
11/16	3/16 × 3/32	5/16-18 × 5/16	1 15/32	2	1/2 × 1/4	9/16-12 × 9/16	3 9/16
3/4	3/16 × 3/32	5/16-18 × 5/16	1 17/32	2 1/16	1/2 × 1/4	9/16-12 × 9/16	3 5/8
13/16	3/16 × 3/32	5/16-18 × 5/16	1 19/32	2 1/8	1/2 × 1/4	9/16-12 × 9/16	3 11/16
7/8	3/16 × 3/32	5/16-18 × 5/16	1 21/32	2 3/16	1/2 × 1/4	9/16-12 × 9/16	3 3/4
15/16	1/4 × 1/8	3/8-16 × 3/8	1 29/32	2 1/4	1/2 × 1/4	9/16-12 × 9/16	3 13/16
1	1/4 × 1/8	3/8-16 × 3/8	1 31/32	2 5/16	5/8 × 5/16	5/8-11 × 5/8	4 3/32
1 1/16	1/4 × 1/8	3/8-16 × 3/8	2 1/32	2 3/8	5/8 × 5/16	5/8-11 × 5/8	4 5/32
1 1/8	1/4 × 1/8	3/8-16 × 3/8	2 3/32	2 7/16	5/8 × 5/16	5/8-11 × 5/8	4 7/32
1 3/16	1/4 × 1/8	3/8-16 × 3/8	2 5/32	2 1/2	5/8 × 5/16	5/8-11 × 5/8	4 9/32
1 1/4	1/4 × 1/8	3/8-16 × 3/8	2 7/32	2 9/16	5/8 × 5/16	5/8-11 × 5/8	4 11/32
1 5/16	5/16 × 5/32	7/16-14 × 7/16	2 15/32	2 5/8	5/8 × 5/16	5/8-11 × 5/8	4 13/32
1 3/8	5/16 × 5/32	7/16-14 × 7/16	2 17/32	2 11/16	5/8 × 5/16	5/8-11 × 5/8	4 15/32
1 7/16	3/8 × 3/16	1/2-13 × 1/2	2 3/4	2 3/4	5/8 × 5/16	5/8-11 × 5/8	4 17/32
1 1/2	3/8 × 3/16	1/2-13 × 1/2	2 13/16	2 13/16	3/4 × 3/8	3/4-10 × 3/4	4 31/32
1 9/16	3/8 × 3/16	1/2-13 × 1/2	2 7/8	2 7/8	3/4 × 3/8	3/4-10 × 3/4	5 1/32
1 5/8	3/8 × 3/16	1/2-13 × 1/2	2 15/16	2 15/16	3/4 × 3/8	3/4-10 × 3/4	5 3/32
1 11/16	3/8 × 3/16	1/2-13 × 1/2	3	3	3/4 × 3/8	3/4-10 × 3/4	5 5/32
1 3/4	3/8 × 3/16	1/2-13 × 1/2	3 1/16				

Four-bar Linkages

All mechanisms can be broken down into equivalent four-bar linkages. They can be thought of as the basic mechanism and are useful in many mechanical applications. Twenty-four linkages are shown in Fig. 23. Descriptions of these mechanisms are:

FIG. 23—1. FOUR-BAR LINKAGE. Two cranks, a connecting rod, and a line between the fixed centers of the cranks make up the basic four-bar linkage. Cranks can rotate if A is smaller than B or C or D. Link motion can be predicted.

FIG. 23—2. CRANK AND ROCKER. Following relations must hold for operation: $A + B + C > D$; $A + D + B > C$; $A + C - B < D$; and $C - A + B > D$.

FIG. 23—3. FOUR-BAR LINK WITH SLIDING MEMBER. One crank replaced by circular slot with effective crank distance of B.

FIG. 23—4. PARALLEL-CRANK FOUR-BAR. Both cranks of the parallel-crank four-bar linkage always turn at the same angular speed but they have two positions where the crank cannot be effective. They are used on locomotive drivers.

FIG. 23—5. DOUBLE PARALLEL CRANK. This mechanism avoids dead-center position by having two sets of cranks at 90° advancement. Connecting rods are always parallel. Sometimes used on driving wheels of locomotives.

FIG. 23—6. PARALLEL CRANKS. Steam-control linkage assures equal valve openings.

FIG. 23—7. NONPARALLEL EQUAL CRANK. The centrodes are formed as gears for passing dead center and can replace ellipticals.

FIG. 23—8. SLOW-MOTION LINK. As crank A is rotated upward, it imparts motion to crank B. When A reaches dead-center position, the angular velocity of crank B decreases to zero. This mechanism is used on the Corliss valve.

FIG. 23—9. TRAPEZOIDAL LINKAGE. This linkage is not used for complete rotation but can be used for special control. Inside moves through larger angle than outside with normals intersecting on extension of axle in cars.

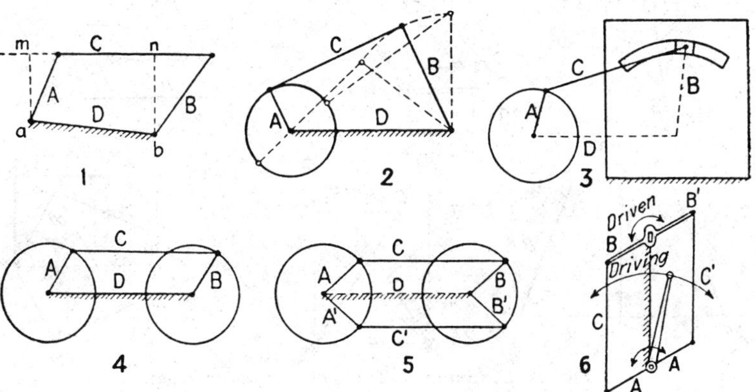

FIG. 23. Types of four-bar linkages.

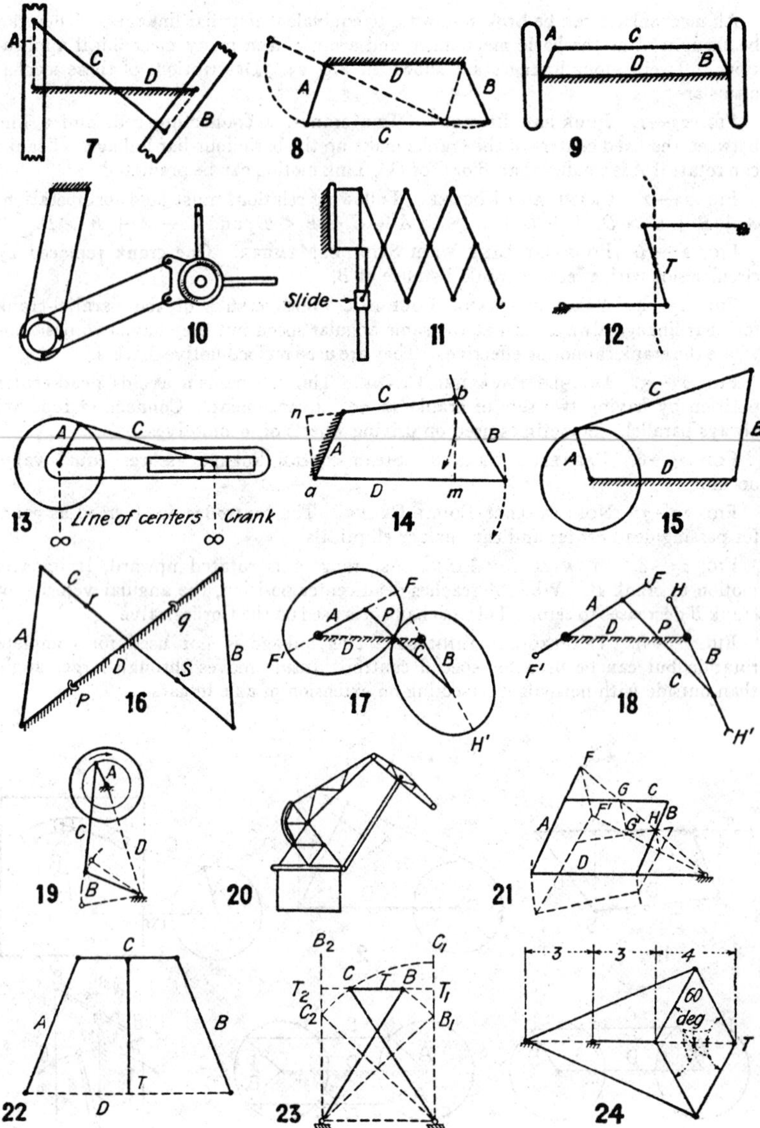

FIG. 23 (*Continued*). Types of four-bar linkages.

FIG. 23—10. DOUBLE PARALLEL-CRANK MECHANISM. This mechanism forms the basis for the universal drafting machine.

FIG. 23—11. ISOSCELES DRAG LINKS. "Lazy-tong" device made of several isosceles links; used for movable lamp support.

FIG. 23—12. WATTS'S STRAIGHT-LINE MECHANISM. Point T describes line perpendicular to parallel position of cranks.

FIG. 23—13. STRAIGHT SLIDING LINK. This is the form in which a slide is usually used to replace a link. The line of centers and the crank B are both of infinite length.

FIG. 23—14. DRAG LINK. This linkage used as the drive for slotter machines. For complete rotation: $B > A + D - C$ and $B < D + C - A$.

FIG. 23—15. ROTATING-CRANK MECHANISM. This linkage is frequently used to change a rotary motion to swinging movement.

FIG. 23—16. NONPARALLEL EQUAL CRANK. If crank A has uniform angular speed, B will vary.

FIG. 23—17. ELLIPTICAL GEARS. They produce same motion as nonparallel equal cranks.

FIG. 23—18. NONPARALLEL EQUAL CRANK. Same as first but with crossover points on link ends.

FIG. 23—19. TREADLE DRIVE. This four-bar linkage is used in driving grinding wheels and sewing machines.

FIG. 23—20. DOUBLE-LEVER MECHANISM. Slewing crane can move load in horizontal direction by using D-shaped portion of top curve.

FIG. 23—21. PANTOGRAPH. The pantograph is a parallelogram in which lines through F, G, and H must always intersect at a common point.

FIG. 23—22. ROBERTS'S STRAIGHT-LINE MECHANISM. The lengths of cranks A and B should not be less than $0.6D$; C is one-half D.

FIG. 23—23. TCHEBICHEFF'S MECHANISM. Links made in proportion: $AB = CD = 20$, $AD = 16$, $BC = 8$.

FIG. 23—24. PEUCELLIER'S CELL. When proportioned as shown, the tracing point T forms a straight line perpendicular to the axis.

Mechanisms for Specific Motions

Simple mechanisms that produce various types of motions have been laid out by Sigmund Rappaport[1] of the Ford Instrument Co. These devices are comprised of components that are easily procured on the open market or manufactured. Five mechanisms are illustrated in Fig. 24, and are described thus:

STRAIGHT-LINE MOTION

FIG. 24—1. No linkages or guides are used in this modified hypocyclic drive which is relatively small in relation to the length of its stroke. The sun gear of pitch diameter D is stationary. The drive shaft, which turns the T-shaped arm, is concentric with this gear. The idler and planet gears, the latter having a pitch diameter of $D/2$, rotate freely on pivots. Pitch diameter of the idler is of no geometrical significance, although this gear reverses the rotation of the planet gear, thus producing true hypocyclic motion with ordinary spur gears only. Such an arrangement occupies only about half as much space as does an equivalent mechanism containing an internal gear. Center distance R is the sum of $D/2$, $D/4$, and an arbitrary distance d, determined by a particular application. Points A and B on the driven link, which is fixed to the planet, describe straight-line paths through a

[1] *Product Eng.*, January, 1951.

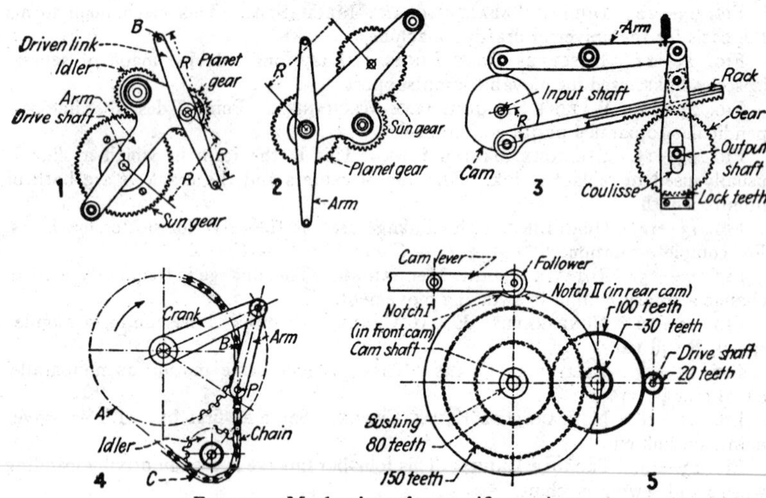

FIG. 24. Mechanisms for specific motions.

stroke of $4R$. All points between A and B trace ellipses, while the line AB envelopes an astroid.

PARALLEL MOTION

FIG. 24—2. A slight modification of the mechanism in Fig. 1 will produce another type of useful motion. If the planet gear has the same diameter as that of the sum gear, the arm will remain parallel to itself throughout the complete cycle. All points on the arm will thereby describe circles of radius R.

INTERMITTENT MOTION

FIG. 24—3. An operating cycle of 180° motion and 180° dwell is produced by this mechanism. The input shaft drives the rack which is engaged with the output shaft gear during half the cycle. When the rack engages, the lock teeth at the lower end of the coulisse are disengaged and, conversely, when the rack is disengaged, the coulisse teeth are engaged, thereby locking the output shaft positively. The change-over points occur at the dead-center positions so that the motion of the gear is continuously and positively governed. By varying R and the diameter of the gear, the number of revolutions made by the output shaft during the operating half of the cycle can be varied to suit requirements.

INTERMITTENT MOTION

FIG. 24—4. This mechanism can be adapted to produce a stop, a variable speed without stop, or a variable speed with momentary reverse motion. Uniformly rotating input shaft drives the chain around the sprocket and idler, the arm serving as a link between the chain and the end of the output shaft crank. The sprocket drive must be in the ratio $N:n$ with the cycle of the machine, where n is the number of teeth on the sprocket and N the number of links in the chain. When point P

travels around the sprocket from point A to position B, the crank rotates uniformly. Between B and C, P decelerates; between C and A it accelerates; and at C there is a momentary dwell. By changing the size and position of the idler, or the lengths of the arm and crank, a variety of motions can be obtained. If in the sketch, the length of the crank is shortened, a brief reverse period will occur in the vicinity of C; if the crank is lengthened, the output velocity will vary between a maximum and minimum without reaching zero.

Fast Cam-follower Motion

Fig. 24—5. Fast cam action every n cycles when n is a relatively large number can be obtained with this manifold cam and gear mechanism. A single notched cam geared $1/n$ to a shaft turning once a cycle moves relatively slowly under the follower. The double notched-cam arrangement shown is designed to operate the lever once in 100 cycles, imparting to it a rapid movement. One of the two identical cams and the 150-tooth gear are keyed to the bushing which turns freely around the camshaft. The latter carries the second cam and the 80-tooth gear. The 30- and 100-tooth gears are integral, while the 20-tooth gear is attached to the one-cycle drive shaft. One of the cams turns in the ratio of 20:80 or 1:4; the other in the ratio 20:100 times

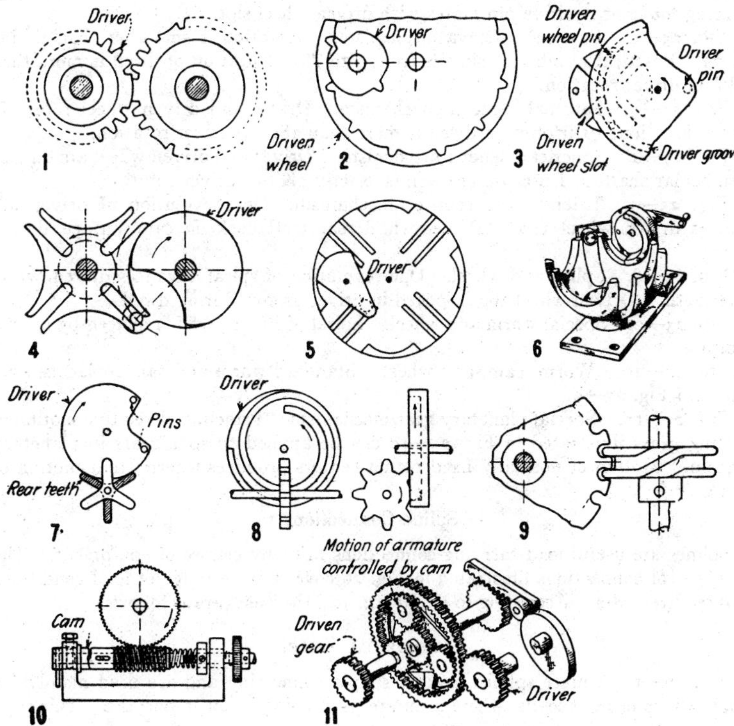

Fig. 25. Eleven mechanisms for intermittent rotary motion.

30:150, or 1:25. The notches therefore coincide once every 100 cycles (4 × 25). Lever movement is the equivalent of a cam turning in a ratio of 1:4 in relation to the drive shaft. To obtain fast cam action, *n* must be broken down into prime factors. For example, if 100 were factored into 5 and 20, the notches would coincide after every 20 cycles.

Mechanisms for Intermittent Rotary Motion

Eleven ways are given in Fig. 25 for converting uniform angular motion to intermittent angular motion. The conversion of a uniform rotation to an intermittent rotation need not be performed in a single step. Two ways in which this can be carried out in two steps are:

1. Conversion of uniform rotation to reciprocating motion followed by conversion of reciprocating motion to intermittent motion.

2. Conversion of uniform rotation to angular oscillation followed by conversion of angular oscillation to intermittent rotation.

Details of the several methods are:

FIG. 25—1. External timing gears.

FIG. 25—2. Internal timing gears.

FIG. 25—3. External Geneva mechanism. Difference from mechanism of Fig. 4 lies in method of locking. Driver grooves lock driven-wheel pins during dwell. During movement, drive pin mates with driven-wheel slot.

FIG. 25—4. External Geneva mechanism. Operation is smoother than timing gears. Practical number of slots is from 3 to 18. Duration of dwell is more than 180° of driver rotation.

FIG. 25—5. Internal Geneva mechanism. Driver and driven wheel rotate in same direction. Duration of dwell is more than 180° of driver rotation.

FIG. 25—6. Spherical Geneva mechanism. Driver and driven wheel are on perpendicular shafts. Duration of dwell is exactly 180° of driver rotation.

FIG. 25—7. Intermittent counter mechanism. One revolution of driver advances driven wheel 120°. Driven-wheel rear teeth lock on cam surface during dwell.

FIG. 25—8. Spiral and wheel. One revolution of spiral advances driven wheel one tooth. Driven-wheel tooth locked in driver groove during dwell.

FIG. 25—9. Special worm and wheel. Spiral of Fig. 25—8 is replaced by special worm.

FIG. 25—10. Worm, cam, and wheel. Standard worm and cam replace special worm of Fig. 25—9.

FIG. 25—11. Special planetary gear mechanism. Principle of relative motion of mating gears illustrated in Fig. 25—10 can be applied to spur gears in planetary system. Motion of normally fixed planet centers produces intermittent motion of sun gear.

Spline Connections

Splines are useful load-carrying connections in many classes of equipment. The 10 types of connections illustrated in Fig. 26 cover a wide range of load conditions and requirements. According to W. Heath, Jr.,[1] their uses are as follows:

CYLINDRICAL TYPES

FIG. 26—1. Square splines make a simple connection and are used mainly for applications of light loads, where accurate positioning is not important. This type

[1] *Product Eng.*, March, 1951.

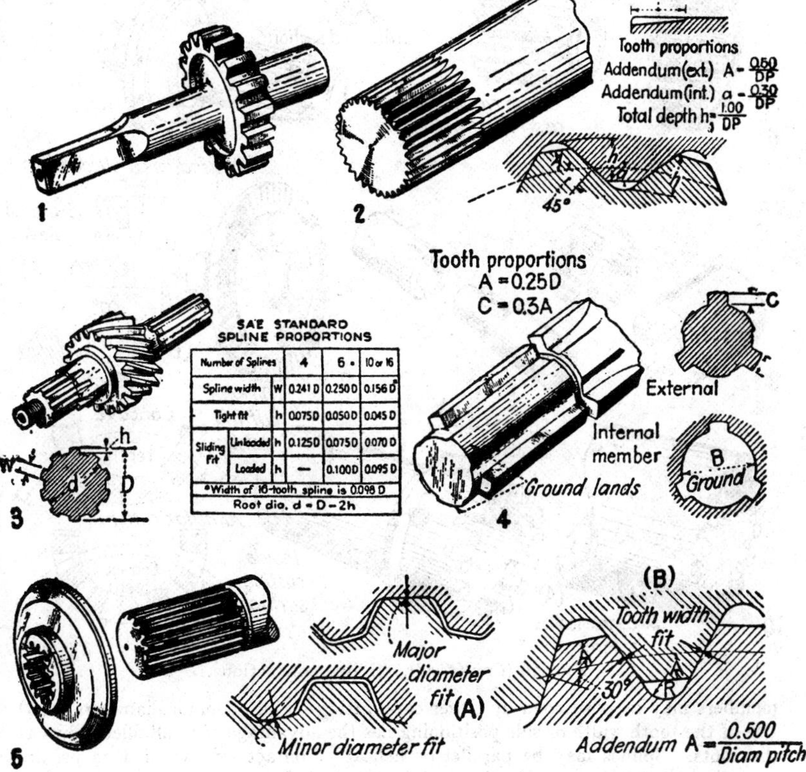

SAE STANDARD SPLINE PROPORTIONS

Number of Splines			4	6 *	10 or 16
Spline width		W	0.241 D	0.250 D	0.156 D*
Tight fit		h	0.075 D	0.050 D	0.045 D
Sliding Fit	Unloaded	h	0.125 D	0.075 D	0.070 D
	Loaded	h	—	0.100 D	0.095 D
*Width of 16-tooth spline is 0.098 D					
Root dia. d = D − 2h					

FIG. 26. Spline connections.

is commonly used on machine tools; a cap screw is necessary to hold the enveloping member.

FIG. 26—2. Serrations of small size are used mostly for applications of light loads. Forcing this shaft into a hole of softer material makes an inexpensive connection. Originally straight-sided and limited to small pitches, 45° serrations have been standardized (SAE) with large pitches up to 10 in. dia. For tight fits, serrations are tapered.

FIG. 26—3. Straight-sided splines have been widely used in the automotive field. Such splines are often used for sliding members. The sharp corner at the root limits the torque capacity to low pressures about 1000 psi on the spline projected area. For different applications, tooth height is altered as shown in table.

FIG. 26—4. Machine-tool spline has a wide gap between splines to permit accurate cylindrical grinding of the lands—for precise positioning. Internal parts can be readily ground to a close fit with the lands of the external member.

FIG. 26—5. Involute-form splines are used where high loads are to be transmitted. Tooth proportions are based on a 30° stub-tooth form. (*A*) Splined

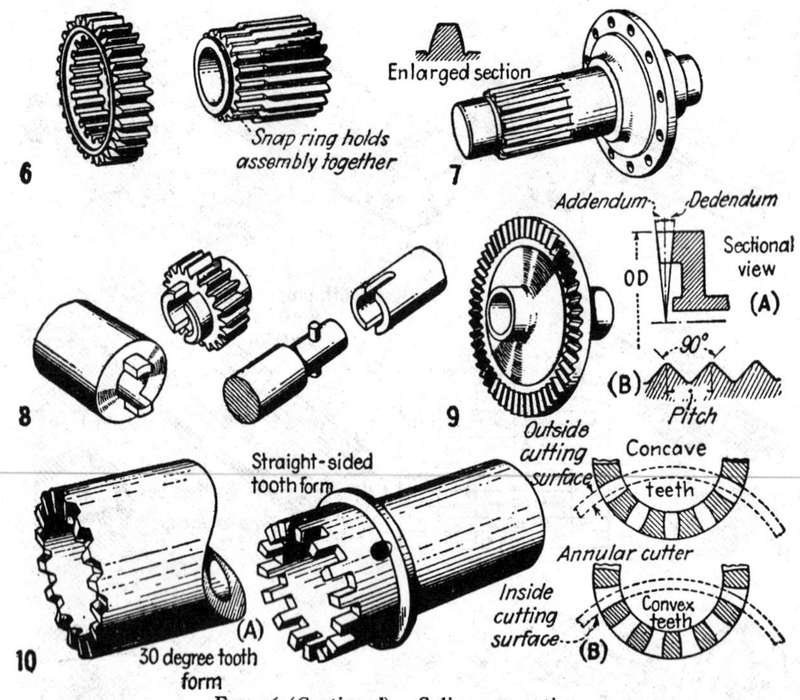

FIG. 26 (*Continued*). Spline connections.

members may be positioned by either close-fitting major or minor diameters. (*B*) Use of the tooth width or side positioning has the advantage of a full fillet radius at the roots. Splines may be parallel or helical. Contact stresses of 4000 psi are used for accurate hardened splines. Diametral pitch is the ratio of teeth to the pitch diameter.

FIG. 26—6. Special involute splines are made by using gear-tooth proportions. With full-depth teeth, greater contact area is possible. A compound pinion is shown made by cropping the smaller pinion teeth and internally splining the larger pinion.

FIG. 26—7. Taper-root splines are for drives which require positive positioning. This method holds mating parts securely. With a 30° involute stub tooth, this type is stronger than parallel-root splines and can be hobbed with a range of tapers.

FACE TYPES

FIG. 26—8. Milled slots in hubs or shafts make an inexpensive connection. This type is limited to moderate loads and requires a locking device to maintain positive engagement. Pin-and-sleeve method is used for light torques and where accurate positioning is not required.

FIG. 26—9. Radial serrations by milling or shaping the teeth make a simple connection. (*A*) Tooth proportions decrease radially. (*B*) Teeth may be straight-sided (castellated) or inclined; a 90° angle is common.

FIG. 26—10. Curvic coupling teeth are machined by a face-mill type of cutter. When hardened parts are used which require accurate positioning, the teeth can be ground. (*A*) This process produces teeth with uniform depth and can be cut at any pressure angle, although 30° is most common. (*B*) Because of the cutting action, the shape of the teeth will be concave (hourglass) on one member and convex on the other member to fit each other intimately.

Sorting and Feeding Mechanisms

The design of suitable sorting and feeding mechanisms is always a problem. Often a design is required, and there is no guide to a successful device. However, William Schwartz, consulting engineer, has found that the sorters and feeders illustrated in Figs. 27 to 37 worked well where basic requirements had been properly evaluated. See Table 2 for selection guide.

Drum Sorter for Flanged Work. A slowly revolving drum (Fig. 27), driven by ring gear and supported on idlers, has stationary part-feed troughs at ends. One line of holes around the drum allows a stream of parts to drop onto a slide at exit point. An external guard keeps parts from falling through elsewhere, and an internal

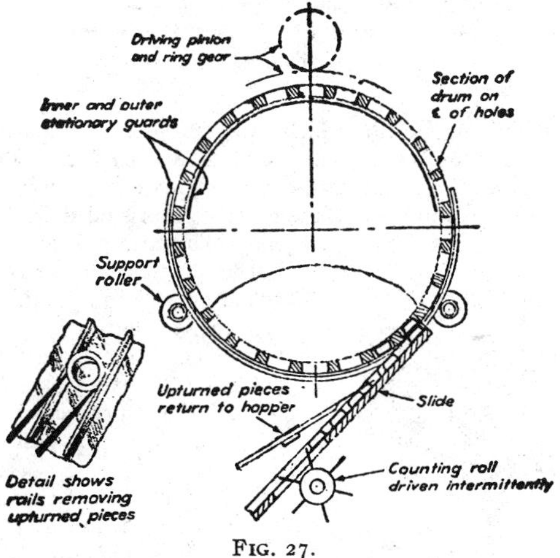

FIG. 27.

guard lining upper half keeps parts not discharged from falling back. Drum is slightly thicker than piece height.

A pair of slides projects into chute to pick up an upturned flange but clears those turned down, thus removing improperly oriented pieces. Rejected parts fall into tote box at slide end, or into hopper equipped with a return conveyor.

A counting roll holds parts on slide until needed. Slide will fill back to the drum. Parts carried to this point by drum will pass over last piece on slide and remain in holes until entry to slide is clear.

Wiper and Belt Sort Unequal-leg Clips. A clip angle with unequal legs is sorted with the short leg up and delivered to a chute by the mechanism shown in Fig. 28. Parallel wiper bars *K-K* move between end rails *B* and *C*, positioned by the rods *L-L*, and driven by the crank *N*. As the bars reach the right end of travel, projecting

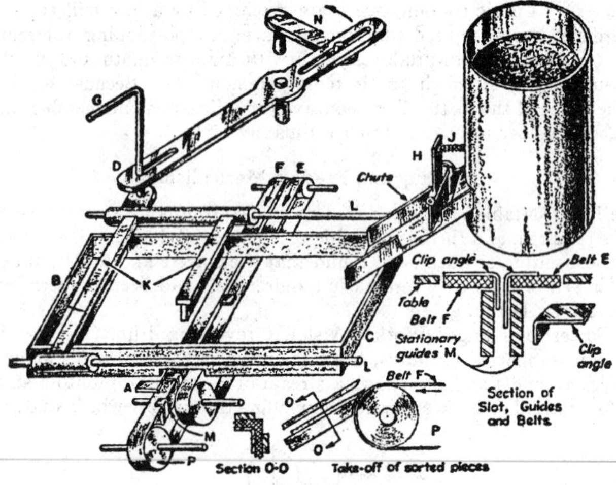

FIG. 28.

bar G strikes the pivoting gate H, which swings to pass a few parts down the chute from the hopper. Timing is such that they fall between K-K. The spring J keeps the gate closed unless struck, and feed can be regulated by lowering or raising G.

As the wiper bars move to the left, the mass of parts is piled against the right-hand one. This arm pushes them over the slot in the table between belts E and F. Some will drop one flange into the slot. These are carried outward by the belts, and restrained from falling through by the guides M-M. Clearance between the bottom of the wiper blades and the table is such that the wipers pass over parts falling into the slot.

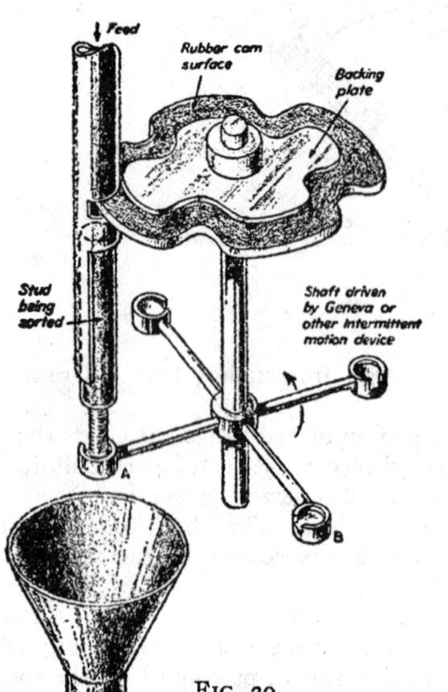

FIG. 29.

If only one line of clips is needed, belt E dumps its load into a tote box below the delivery end pulley. Belt F also carries parts, but half will have their long leg lying across its top surface. Block A is positioned just above belt F and projects inward just far enough to be struck by the long leg of the clip but not by the short one. Thus, any clips which are improperly oriented are knocked off. Take-off from the belt to a chute is shown in section O-O, where a knife-edge angle will pick clips off the belt.

Cupped Arms Feed Parts Big End Down. Work like shoulder studs is oriented and fed big end down by the device sketched in Fig. 29. The parts drop continuously into the funnel; there are no rejects. By synchronization of the shaft drive with other equipment, the discharge is timed or counted.

The drive shaft mounts a spider and a rubber-edged, four-lobe cam. While a cam lobe projects into a slot cut into the feed tube, the workpiece above cannot drop. But when the cutout in the cam coincides with the slot, the piece is free to fall onto a spider-arm cup. Timing is adjusted so that the piece hits the cup just after it is aligned with the tube center line. Each spider-arm cup has an OD corresponding to the shoulder diameter on the piece. Cup ID is slightly smaller than the stud diameter. If the shoulder is down, it rests on top of the cup. After the dwell period, the cup is jerked out from under the piece, which falls big end down into the funnel.

If the stud end is down, the piece is engaged by the cup. Indexing tends to carry the stud end along with the spider, and the big end slides down the cutaway portion of the feed tube. When the big end clears the feed tube, it falls into the funnel.

Drum-belt Feeding, Sorting Hopper. Items that nest, tangle, or have wires attached can be sorted to a specific end forward. In the case illustrated in Fig. 30, terminal clips are wanted in the orientation shown at *D*.

The drum is end-fed from a fixed trough and is lined with staggered angle brackets having small lips. The brackets are just large enough to hold one piece, though in operation others will cling. As the terminal clips ride into the upper right quadrant of sketch II, most parts not securely resting on the brackets will fall.

The belt is so placed that parts tipping out of the brackets above will fall onto it;

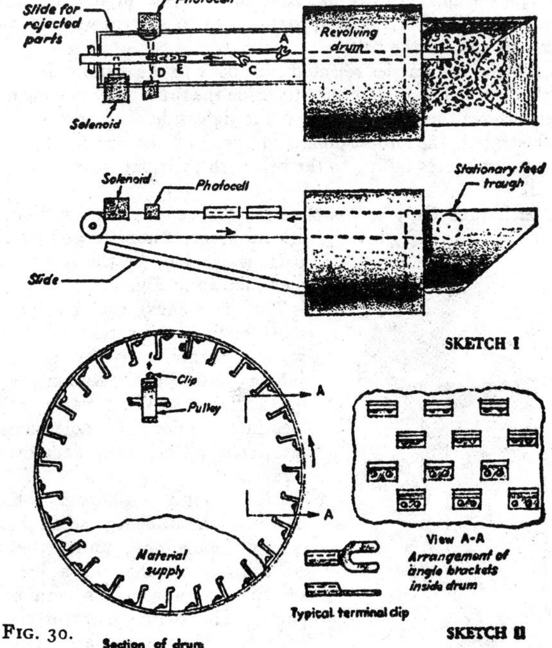

FIG. 30.

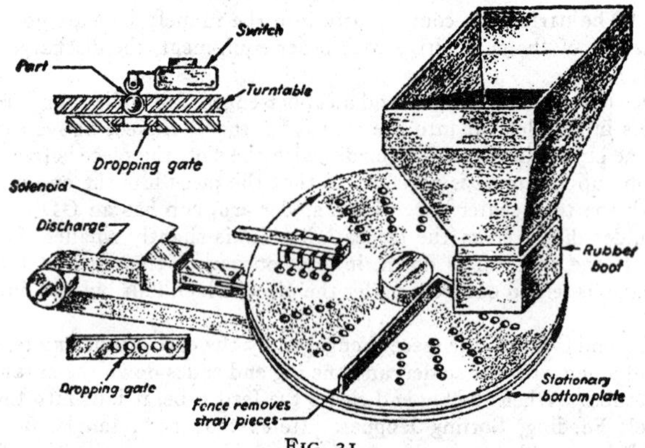

FIG. 31.

many will of course bounce off if the drop is too great or the belt is too resilient. The belt carries the clips past the fixed guides *A* and *C* which either line them up with the belt or push them off.

The photocell beam is interrupted by the high part of the clip (the wire ferrule). By a suitable time-delay circuit corresponding to the time of belt travel from cell to solenoid, the solenoid is correctly energized. The solenoid plunger extends and retracts quickly; if the clip is turned like *D* on sketch I, the plunger hits nothing, but if it is turned opposite like *E*, the forked part of the clip is in the way and the solenoid plunger knocks the clip off the belt.

A trough below the belt carries rejected clips back to the drum. If this feature is unnecessary, the lips on the angle brackets inside the drum may be omitted, and the belt located just above the center and near the right side of the drum. In the case of the clips illustrated, the proportions and speed of the drum and belt may be adjusted so that only pieces falling to the belt with their flat sides down will remain on it.

Turntable Discharges Only a Complete Parts Group. A convenient means of selecting only complete groups of identical parts is provided by the turntable arrangement shown in Fig. 31.

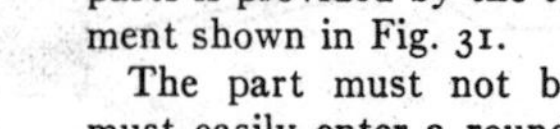

The part must not be too thin, and must easily enter a round hole. Delivery to the turntable is by a hopper with rubber outlet boot. Radial lines of holes in turntable pick up parts. Pieces that lock into the edge of a hole but do not fall into it are pushed into place by a fence or scraped off.

After passing the hopper, a line of holes moves to the discharge sector, where there is a dropping gate under the turntable. At this sector, there is a line of microswitches, with rollers set to contact the parts. The switches are of the normally open circuit type and are wired in series.

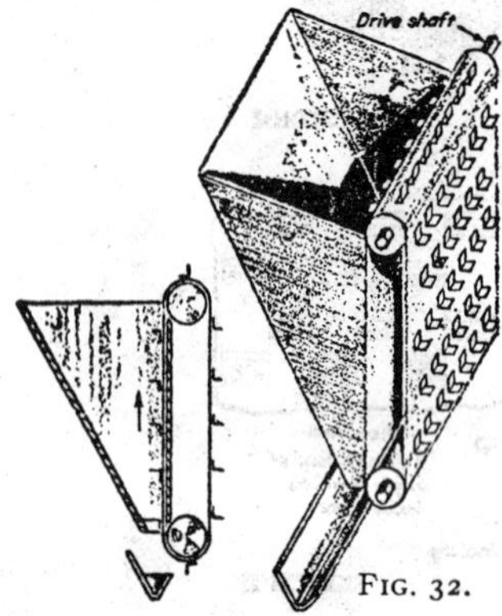

FIG. 32.

TABLE 2. SELECTION GUIDE FOR SORTERS AND FEEDERS

Materials and Shapes	27	28	29	30	31	32	33	34	35	36	37
Round and flat	X				X	X	X		X	X	X
Long cylinders			X	X		X	X	X		X	
Cubes				X				X	X	X	
Irregular shapes		X		X		X					
Symmetrical parts having projections			X	X		X	X	X	X		
Spheres	X				X				X		
Fragile or brittle	X		X				X			X	
Soft	X			X	X	X	X	X	X	X	
Nesting or tangling		X		X		X		X			
Delivery rate over 100 parts per min	X	X		X	X	X	X	X			
Delivery rate under 100 parts per min			X				X		X	X	X
Multiple-stream delivery	X				X			X			

Best applications of designs illustrated: The selections noted above can be only a rough guide to designers. Where several mechanisms could handle a given type the simplest designs have been noted; special considerations must be weighed in all cases.

If the entire row of holes is filled with parts, the circuit is completed. Now, a solenoid is energized to pull out the dropping gate so that the complete group of parts can fall onto delivery belt or into a tube or chute.

Cleated Belt Prevents Clogging of Hopper. Objects having a major dimension more than twice the minor dimension will clog unless the discharge opening of a stationary hopper is made several times their length. One positive anticlogging device is the belt illustrated in Fig. 32. Cleats staggered in arrangement are wide enough so that the entire length of the discharge opening is frequently cleared, but too small and narrow to support a piece lifted out of the top of the material heap. The hopper may drop parts to a belt below or be tilted so that the discharge opening is parallel to a sloping chute. When quarter full, this type of hopper increases its rate of discharge; otherwise the flow is relatively constant.

Blank Feeder Accepts Nonuniform Parts. In this device (Fig. 33), applicable only to round flat objects, the tube is stationary. The funnel reciprocates, driving the tube end through the material heap; at each downward stroke one or more parts will be passed through the slot in the top of the tube. Lubricant on the OD of the tube

FIG. 33.

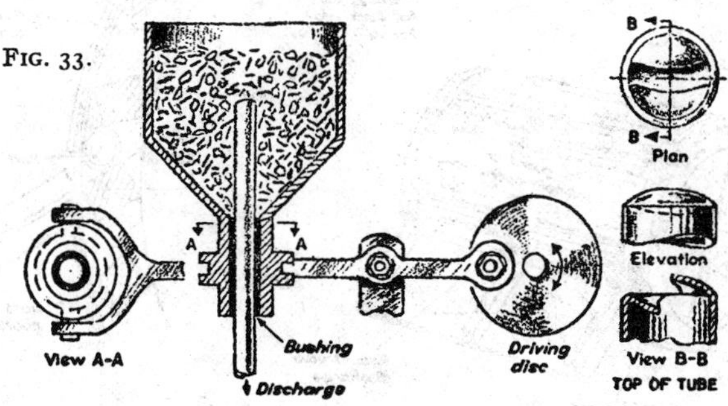

will dirty and contaminate the pieces handled. If this is objectionable, the bearing bushing must be dry. Parts which are too heavy will batter the end of the tube and burrs will reduce the number fed into it. Clearance between the OD of the piece and the ID of the tube may be sufficient so the design can be applied to parts not too uniform in size.

Bar-sided Drum Hopper Discharges onto Belt. The design in Fig. 34 can be used in the random-discharge version (shown in sketch I) for parts which tangle, as pieces not discharged are thrown back into the heap at the bottom. Sketch II shows the arrangement of the bars; and the outer guide ends at the discharge slide or belt.

Sketch III shows how the design can be adapted to discharge more than one stream of material by dividing the openings with separators, but parts which tangle would clog the openings. Cylindrical objects can be discharged at a timed rate or synchronized with other mechanism according to the operation of the gate solenoid.

Looking at sketch II upside down, the bottom pair of bars at the center line will be seen to form a funnel shape between surfaces *A* and *B*, which facilitates the parts dropping onto surface *C*, on which they ride in the lower half of the drum. An inner guide may be fitted into the upper half of the drum, in which case the spaces between the bars will remain filled until each slot passes the discharge point and finds the gate open.

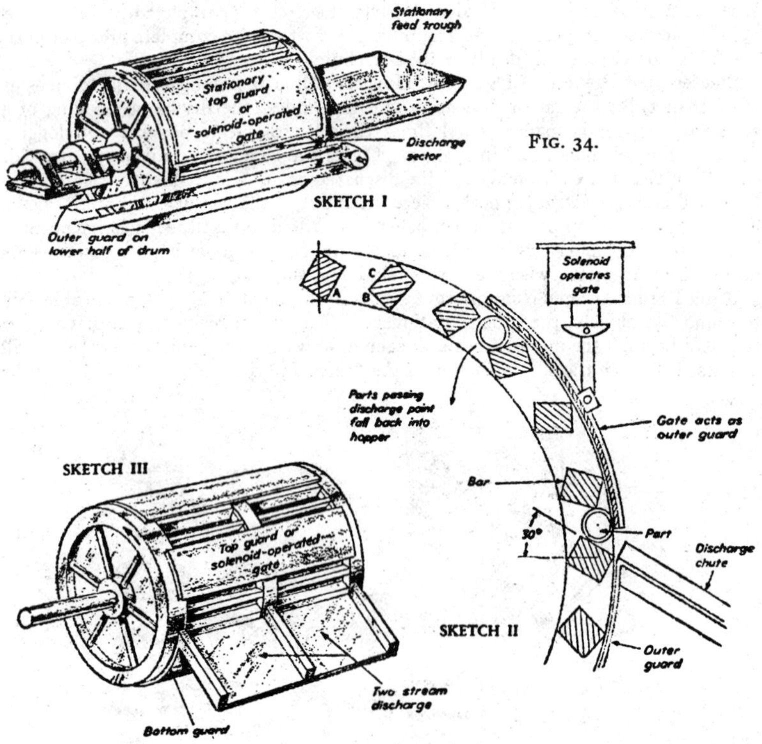

FIG. 34.

Drum-type Hopper Feeds at Rapid Rate. One of the oldest and most popular feeding and sorting devices, this commercially available hopper (Fig. 35) has been adapted to a variety of symmetrically shaped parts. The wedge-shaped blocks fastened to the rotating ring are spaced to permit the part to fall between them when they pass under the supply stored inside the cover. The baffle plate inside the rotating ring is stationary, being attached to the fixed central shaft which also supports the discharge track. This baffle plate keeps the 50-caliber bullet cores illustrated from falling inward until they are carried up to the end of the track.

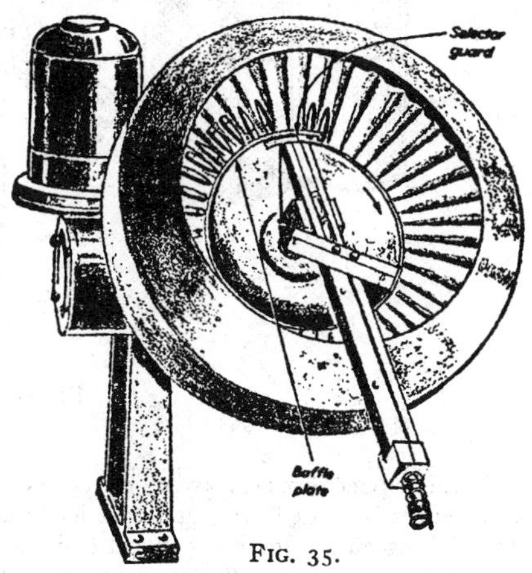

Fig. 35.

The selector guard attached to the baffle plate allows bullet cores oriented point first to pass into the track but rejects those having their blunt end down. For bulky parts which must be fed at a high rate, this hopper must be equipped with an auxiliary device to dump batches of parts inside the cover every few minutes, since the cover cannot hold a very large supply.

This hopper is usually driven by a constant-speed motor, a feed-limiting device or escapement being used to regulate the discharge. Feeds as high as 300 parts per min can be obtained, but light plastic or sheet-metal objects must be discharged at a much lower rate.

Squirrel-cage Hopper Sorts Rubber Parts. Rubber parts are difficult to sort because of their high coefficient of friction and resilience. Rubber casters (Fig. 36) pressed into the bottom of small electrical appliances could not be handled in any device dropping them even a short distance, and their center of gravity is such that orientation by movement would be difficult. However, lying in the outer cover of this hopper, they are tumbled until each piece happens to orient itself with the shank down. A slot passes under them, and they drop into it when other pieces permit.

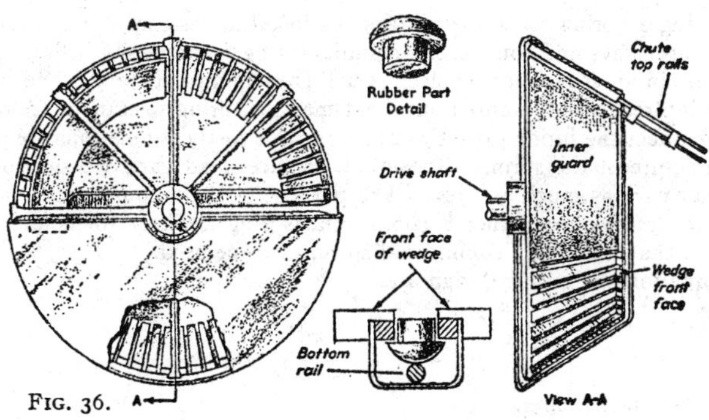

Fig. 36.

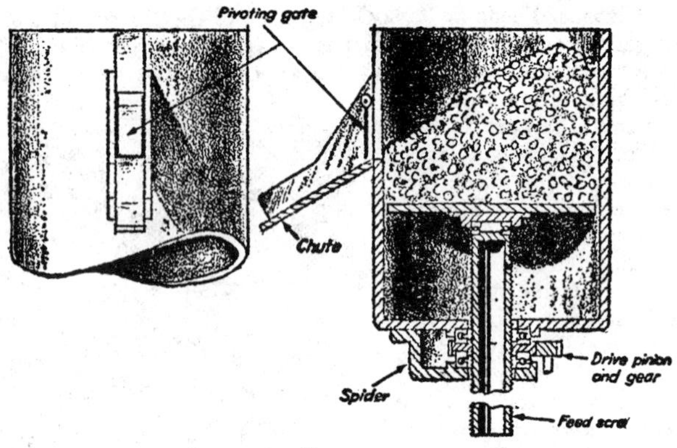

FIG. 37.

As the hopper turns clockwise, the caster lying with its shank in the slot between two wedges is carried into the upper left quadrant, where the inner guard keeps it from falling out. The inner guard ends at top center, where the chute picks up whatever pieces are in each slot and removes them. The weight of the piece is taken by point contact with the bottom rail, the top rails being stabilizers.

This type of chute is best for rubber and other high-friction material because it is open for clearing jams, has little friction resistance to movement, and is easy to fabricate. It will handle cubes, long cylinders, and other shapes and is especially good for material that has a high coefficient of friction.

Piston Hopper Proves Best for Fragile Items. Materials too fragile to be tumbled or agitated can be fed from the hopper shown in Fig. 37. The chute angle should be approximately equal to the angle of repose of the material. The drive pinion may be driven by a variable-speed motor and gear train, or the screw feed replaced by a hydraulic cylinder. A surprisingly constant flow of material can be maintained, but additional regulation by adjustment of the pivoting gate is difficult unless accompanied by control of the piston feed.

Helical-spring Design

In selecting a spring for a specific application, the designer must consider both design requirements and commercial manufacturing tolerances of helical springs, in order to obtain an economical choice. Carl Thumin and Hubert A. Reister of the ITE Circuit Breaker Co. pointed out[1] that manufacturing tolerances are ordinarily assumed to affect the spring force by $\pm 10\%$, or else costly stipulations are placed on the manufacture of the spring. In actual practice, load variation on commercial springs may vary as much as 27%. The purpose of the accompanying tables and charts is to determine whether a commercial spring is likely to handle the load properly, so that less costly commercial springs can be used.

Basic equations for spring design are

$$P = \frac{F}{K} = \frac{\pi d^3 S}{8KD} \tag{1}$$

[1] *Product Eng.*, March, 1951.

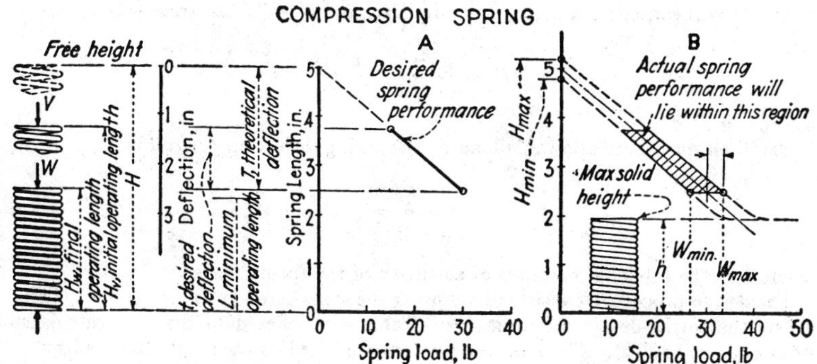

FIG. 38. Ultimate performance of springs should be compared with initial requirements. At *A* is the theoretical load-deflection curve for a specific application, but *B* shows that the actual performance depends upon manufacturing tolerances. In compression springs, the final operating length is limited by minimum operating length and not by solid height.

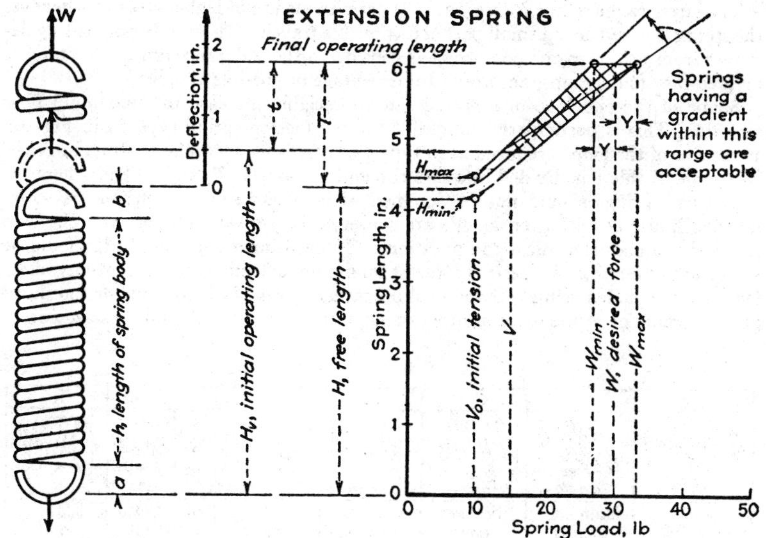

FIG. 39. Deviation in manufactured springs is determined by variations in maximum load and free length. Spring characteristics in the shaded region are considered satisfactory.

where P = maximum allowable load, F = nominal spring load, S = allowable fiber stress, D = mean coil diameter, d = wire size, and K = Wahl stress factor, or

$$K = \frac{4c - 1}{4c - 4} + \frac{0.615}{c}$$

where $c = D/d$.

Based on maximum allowable load P, the spring deflection *per coil* is

$$f = \frac{8PD^3}{Gd^4} \tag{2}$$

where G is the torsional modulus of elasticity of the spring material.

The design procedure consists of following the steps listed in Tables 3 to 6. Table 3 lists the spring-design requirements. Table 4 supplies data that are calculated from Eqs. (1) and (2). Since these equations are based on use of oil-tempered spring wire, correction factors from Table 7 must be applied when using other materials. In Table 5 are outlined computations to find the length of a tension or compression spring. Table 6 tabulates the manufacturing tolerances. Here Table 10 is useful for finding the load tolerance factor Y.

When materials other than oil-tempered spring wire are used, the materials conversion factors in Table 7 are used to find the true values of P and f.

For most efficient design, the actual stress range and the number of loading cycles during which a spring is expected to operate should be considered. These variables are included in Table 8 to establish a correction factor C (Table 3) which is used to obtain the modified spring force $P_1 = W/C$.

The stress range factor R (see Table 8) permits the use of higher stresses whenever the spring is flexed only a small part of its possible travel. The number of load cycles ranges from 100 to 10,000,000, which is considered the endurance limit. Factor R is divided into four groups, according to percentage of possible travel.

Shape of the end loop of a spring can appreciably increase the maximum fiber stresses. This is particularly noticeable for the tension-spring type T_1 in Fig. 40. If the size of the loop is reduced as in type T_2, there is no appreciable stress increase. Therefore, Table 8 is divided into two categories, namely, T_1 and all the others.

Tension springs should not be designed without some initial tension. Recommended limits of initial tension V_0 are conservatively given in Table 5. The wire size and coil diameter affect computations. In choosing a value of V_0, it should be somewhat lower than V (desired force at minimum deflection) to prevent any effect from error in coil spacing. If the final operating length H_w is too long or too short, another value of P and D or another value of initial tension should be selected.

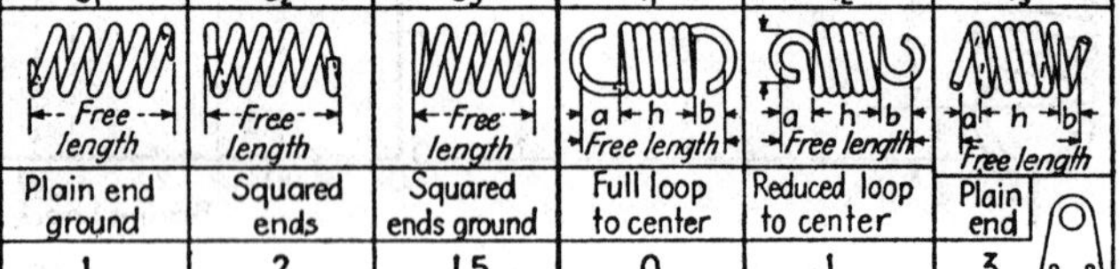

Type	C_1	C_2	C_3	T_1	T_2	T_3
Delineation of spring	Free length	Free length	Free length	a h b Free length	a h b Free length	a h b Free length
End designation	Plain end ground	Squared ends	Squared ends ground	Full loop to center	Reduced loop to center	Plain end
e (minimum)	1	2	1.5	0	1	3

Fig. 40. Spring configuration affects the value e, and in turn the maximum solid height.

Table 3. Spring-design Requirements

W = desired force at maximum deflection, lb
V = desired force at minimum deflection, lb
t = desired deflection between V and W, in.
T = theoretical deflection from zero to W, in. = $(W \times t)/(W - V)$
R = stress range = $(W - V)/W$
A = no. of operations for life of device
C = correction factors from Table 8 and Fig. 40
P_1 = modified force = W/C, lb

Table 4. Spring Data

Refer to Charts A or B, Fig. 38

P = spring load, lb [Eq. (1)]
f = spring deflection per coil, in. [Eq. (2)]
OD = Outside dia of spring, in.
d = Dia of wire, in.
B = dia tolerance, in. (Table 11)
f_1 = deflection per turn at W load = $(f \times W)/P$, in.
N = No. of turns = $T \times f_1$ (to 3 significant places)
For tension springs, parallel ends: N is to be either a full or half turn.
J = spring gradient = $P/(N \times f)$ to 3 significant places, lb per in.

Table 5. Spring Calculations

h = max solid height = $d(N + e)$. See Fig. 40 for e
L = min operating length, may vary from $(h + 0.125T)$ to $(h + 0.4T)$ in.
H = free length = $L + T$ (to nearest $\frac{1}{16}$ in.), in.
H_v = initial operating length = $H - (V/J)$, in.
H_w = final operating length = $H - (W/J)$, in. Check: $H_v - H_w = t$, in.
h = length of spring body = $d(N + 1 + e)$, in.
a = dimension of left-hand loop, in. See Fig. 40
b = dimension of right-hand loop, in. See Fig. 40
H = free length = $(h + a + b)$, in.
V_0 = initial tension, lb, may vary between $400d^2$ and $1250d^2$
$H_w = H + (W - V_0)/J$, in.
$H_v = H + (V - V_0)/J$, in. Check: $H_w - H_v = t$, in.

Table 6. Tolerances in Commercial Springs

OD_{max} = OD + B, in.
ID_{min} = OD − $2d$ − B in.
Coils per inch = N/H (for compression springs) = N/h (tension springs)
m = free length tolerance factor, Table 9
M = free length tolerance = $(m \times h)$ tension = $(m \times H)$ compression, in.
H_{max} = max free length = $H + M$, in.
H_{min} = min free length = $H - M$, in.
Z = load tolerance index = M/T
Y = load tolerance (from Table 10), or $(E \times W - V_0)$
For compression springs, $V_0 = O$, lb
$W_{max} = W + Y$, lb
$W_{min} = W - Y$, lb

Table 7. Materials Conversion Factor

When Using Same Wire Diam of	Multiply Values By	
	P	f
Tempered alloy steel	1.15	1.15
17-7 PH stainless steel	1	1
18-8 stainless steel	0.79	0.91
Beryllium copper	0.75	1.25
Monel metal	0.56	0.72

Table 9. Approximate Free Length Tolerance Factor

Coils per In	*m*
5	0.04
10	0.05
15	0.055
20	0.06

Table 10. Load Tolerance Factors

Load Tolerance Index	Load Tolerance Factor Y	
	Compression Springs	Tension Springs
0.01	0.085	0.085
0.02	0.090	0.09
0.03	0.090	0.095
0.04	0.095	0.10
0.05	0.10	0.11
0.06	0.105	0.12
0.07	0.11	0.13
0.08	0.115	0.14
0.09	0.125	0.155
0.10	0.135	0.165
0.11	0.140	0.175
0.12	0.15	0.19
0.13	0.16	0.20
0.14	0.165	0.215
0.15	0.175	0.23
0.16	0.18	0.24
0.17	0.19	0.25
0.18	0.20	0.27

Table 8. Correction Factor C Based on Fatigue Life and Stress Range

A	Stress Range R							
Expected Life (Cycles)	Spring Type T_1				For All Other Types			
	0.25	0.50	0.75	1.00	0.25	0.50	0.75	1.00
100	0.75	0.75	0.75	0.75	1.00	1.00	1.00	1.00
1,000	0.71	0.68	0.64	0.63	0.95	0.91	0.86	0.81
5,000	0.68	0.64	0.58	0.55	0.90	0.82	0.77	0.73
10,000	0.66	0.59	0.55	0.51	0.88	0.79	0.73	0.68
20,000	0.64	0.56	0.52	0.47	0.86	0.75	0.69	0.63
100,000	0.62	0.51	0.44	0.39	0.83	0.68	0.59	0.52
500,000	0.61	0.47	0.39	0.33	0.82	0.63	0.52	0.44
1,000,000	0.61	0.46	0.38	0.31	0.81	0.61	0.51	0.41
10,000,000	0.60	0.45	0.36	0.30	0.80	0.60	0.50	0.40

For use with all ferrous steels and beryllium copper. Divide these values by 2 for phosphor bronze and monel. Spring types are shown in Fig. 40.

Table 11. Outside Diameter Tolerance *B* for Spring Wire

Oil-tempered Wire		Music Wire	
OD	Tolerance B, in.	OD	Tolerance B, In.
$\frac{1}{8}$–$\frac{3}{16}$	0.002	0.025–$\frac{1}{8}$	0.002
$\frac{13}{64}$–$\frac{15}{64}$	0.003	$\frac{9}{64}$–$\frac{13}{64}$	0.003
$\frac{1}{4}$–$\frac{9}{32}$	0.005	$\frac{7}{32}$–$\frac{15}{64}$	0.004
$\frac{5}{16}$	0.006	$\frac{1}{4}$–$\frac{5}{16}$	0.005
$\frac{11}{32}$–$\frac{3}{8}$	0.007		
		$\frac{11}{32}$	0.006
$\frac{13}{32}$–$\frac{7}{16}$	0.008		
		$\frac{3}{8}$	0.007
$\frac{1}{2}$–$\frac{9}{16}$	0.009	$\frac{13}{32}$	0.008
$\frac{5}{8}$	0.012	$\frac{7}{16}$–$\frac{15}{32}$	0.009
$\frac{11}{16}$–$\frac{3}{4}$	0.014	$\frac{1}{2}$–$\frac{17}{32}$	0.010
		$\frac{9}{16}$	0.012
		$\frac{19}{32}$–$\frac{5}{8}$	0.013

Table 12. Allowable Fiber Stress *S* for Wire Diameter

Oil-tempered Wire			Music Wire	
Gage	d Decimal Equivalent	S Stress	d	S
22	0.0286	131,500	0.005	170,000
20	0.0348	129,000	0.010	165,000
18	0.0475	125,500		
16	0.0625	121,500	0.020	155,500
14	0.080	117,000	0.040	146,500
12	0.1055	112,000	0.062	140,300
$\frac{1}{8}$	0.125	109,000	0.100	129,000
$\frac{3}{16}$	0.1875	101,000	0.120	126,000
$\frac{7}{32}$	0.2187	98,000	0.140	120,600
$\frac{1}{4}$	0.250	96,000	0.160	115,700
$\frac{3}{8}$	0.375	90,500	0.180	112,500
$\frac{1}{2}$	0.500	86,500	0.200	110,000

For other materials use these values of S, since the spring load P is modified by a conversion factor from Table 7.

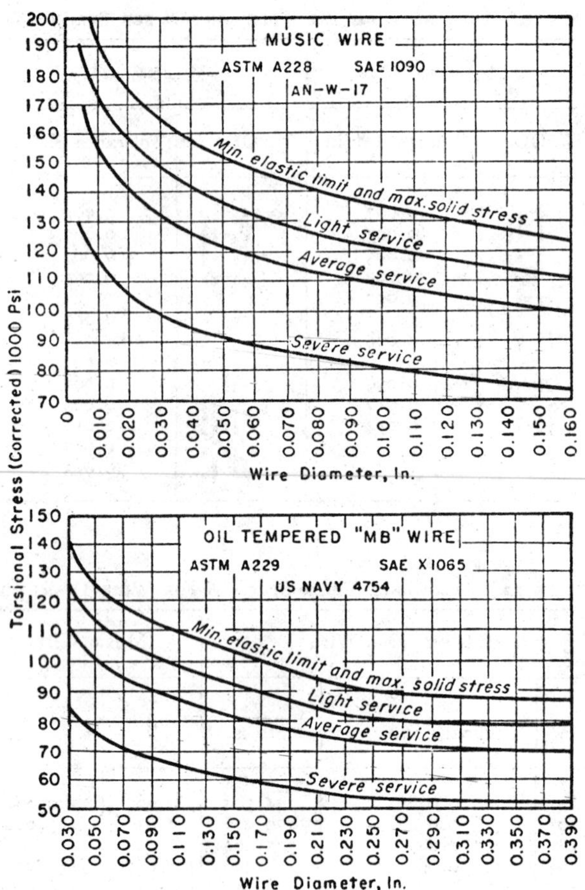

FIG. 41. Allowable working stresses for compression springs, based on severity of service.

Compression Springs. The spring designer is frequently at a loss to pick a suitable design stress for compression springs subjected to various kinds of service. Figure 41 gives allowable working stresses based on fatigue life. Music wire and oil-tempered "MB" steel spring wire are used for the bulk of springs, the latter especially for precision springs. In Fig. 41, the "light-service category" covers springs under static loads and those subject to small deflections with low stress ranges. The number of reversals is between 1000 and 10,000. The "average-service" group includes most springs in general use, characterized by normal frequency of deflection without shock loading and up to 100,000 stress reversals. The "severe-service" category refers to rapid deflections over long periods of time, as in automobile valve springs, and a minimum of 1,000,000 cycles.

Section 40

SURFACE FINISH

SECTION 40

SURFACE FINISH

Specification of surface finish is nominally the function of the designer, but the shop must be prepared to produce the desired surface finish and to inspect workpieces with various devices. "Surface finish" or "surface quality" is a condition, not a dimension. For many years, drawings bore finish marks like "f," "ff," and "fg," but these give rise to confusion and arguments when attempts are made to get people in the same organization, and in various plants, to work to the same degree of finish.

In recent years definitive specifications for surface finish have given way to definite standards and methods of measurement. Standards published by the American Standards Association cover the definitions of surface quality (ASA B46.1-1947) and for two types of physical specimens of surface roughness and lay (ASA B46.2-1952). There are no standards for suitable surface roughness for a given application, because this condition must be determined for each case by service tests.

Surface finish produced by various machining methods will vary. Each type of cutting tool or machining operation leaves individual markings. Shape of the cutting edge, its smoothness, type of tool, and its method of use all affect the smoothness and character of the finish produced. Irregularities in surface texture (Fig. 1) are deviations from the geometrically ideal form. "Roughness" is considered to be irregularities spaced less than $\frac{1}{32}$ in., "waviness" those irregularities spaced greater than $\frac{1}{32}$ in. When measuring roughness on a surface having a "regular" form, such as a plane, a cylinder or a cone, a quantitative description using an average height dimension (Fig. 2) of the scratches can be obtained in a number of ways from a profile curve made with a tracer instrument. Height dimensions h_{av} and h_{rms} usually do not differ more than the allowable error in a stylus-type tracer instrument.

Units of Measurement. Surface roughness is measured in microinches (millionths of an inch) and may be written "15 rms," meaning a root-mean-square average

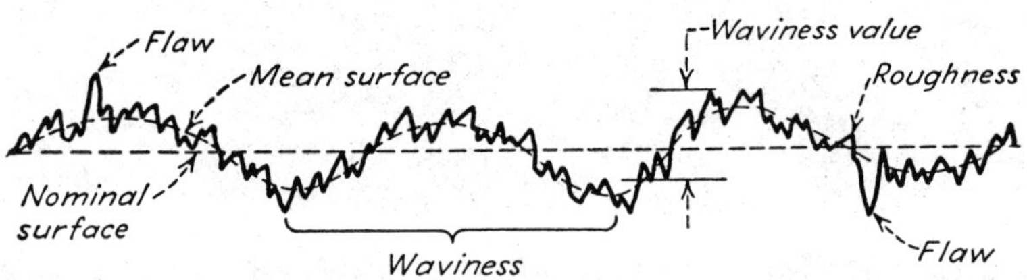

FIG. 1. Roughness is measured in millionths of an inch; waviness in decimal parts of an inch.

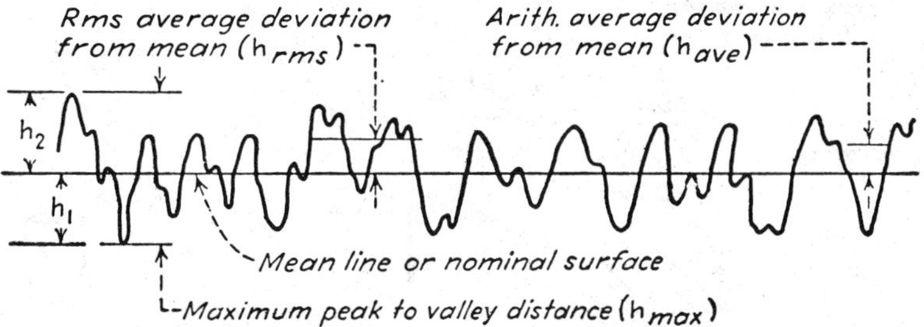

FIG. 2. By tracing the profile of a surface with a suitable instrument, the average height of "peaks" above the nominal surface can be determined.

height of 15 millionths. It may also be written as "15 mμin." or "15 μin." Waviness is measured in decimal parts of an inch, nominally in thousandths.

Roughness Standards. Prior to American Standard B46.1-1947, a number of standardizing groups and individual companies set up their own standards for surface-finish control and specification. Most of these standards parallel the new American Standard.

Most important of these standards are:

1. The "Aeronautical Standard on Surface Finish" (No. AS-107A) issued by the Society of Automotive Engineers principally for control of finishes on aircraft engine parts.

2. The "National Aircraft Standard on Surface Roughness Designation" (No. NAS-30) issued by the National Aircraft Standards Committee for control of machined surfaces on airframe parts.

3. The addition of a new section on "Surface Roughness—Designation and Measurements" to Ordnance Standard No. NAVORD OSTD 4 by the Navy Department's Bureau of Ordnance.

4. The preparation of Ordnance Standard No. URAX6 by the Army Ordnance Department to interpret previously established surface-finish code symbols for drawings in terms of microinch rms roughness values as well as to supersede those symbols as new drawings are prepared.

The first three of the above standards are basically the same as the American Standard. There is this one essential difference: roughness values in the American Standard can be stated in any one of four specified terms, while the first three of the above standards specify that roughness shall be measured in terms of average rms deviation from the mean surface.

Definitions. *Italics* in following list of definitions are from ASA B46.1-1947. Explanatory notes are from other sources.

Surface. *The surface of an object is the boundary which separates that object from another substance. Its shape and extent are usually defined by a drawing or descriptive specification.*

Nominal Surface. This term is sometimes used to specify a geometrically perfect two-dimensional boundary of separation whose shape and extent are defined by specification or drawing dimensions, and are the surface that would result if the peaks were leveled off to fill the valleys.

Surface Qualities. The physical characteristics of a surface may be described as roughness, waviness, size and shape of flaws, and lay. The surface profile, microstructure, bearing area, and luster may also be referred to.

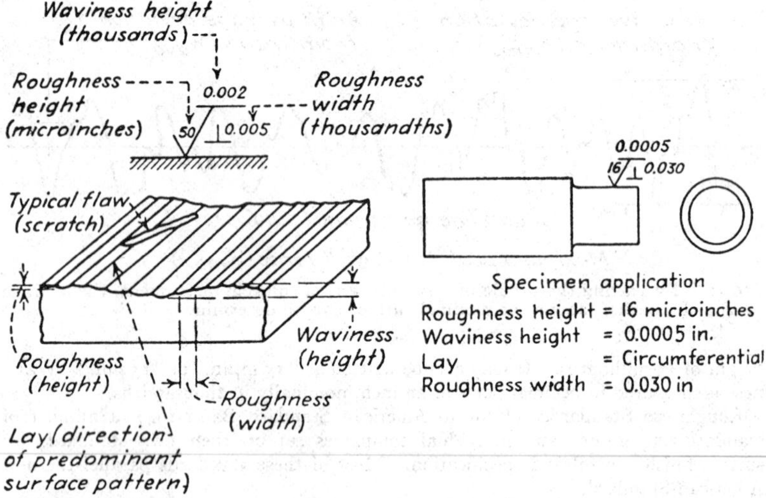

= Parallel to the boundary line of the surface indicated by the symbol.

⊥ Perpendicular to the boundary line of the surface indicated by the symbol.

X Angular in both directions to the boundary line of the surface indicated by the symbol.

M Multidirectional.

C Approximately circular relative to the center of the surface indicated by the symbol.

R Approximately radial relative to the center of the surface indicated by the symbol.

= 0.005 The numerical value in inches of the width of spacing of roughness is added to the right of the directional indication of the lay symbol as shown.

FIG. 3. Terminology of American Standard ASA B46.1-1947. *Top left*—standard symbol for drawings to indicate roughness, waviness, and lay requirements. *Left center*—graphical representation of principal terms. *Upper right*—method of showing surface-finish requirements on typical example. *Bottom*—symbols indicating direction of lay.

Profile. *The contour of any specified section through a surface.*

Roughness. *Relatively finely spaced surface irregularities. On surfaces produced by machining and abrading operations the irregularities produced by the cutting action of tool edges and abrasive grains and by the feed of the machine tool are roughness. Roughness may be considered as being superimposed on a "wavy" surface.* Roughness itself is not considered to affect the trueness of a surface as shown by a dimensional checking gage. Usually, roughness deviations are measured perpendicular to the nominal surface.

Waviness. *The surface irregularities which are of greater spacing than the roughness. On machined surfaces such irregularities may result from machine or work deflections, vibrations, etc. Irregularities of similar geometry may occur due to warping, strains, or other causes.*

Flaws. *Irregularities which occur at one place, or at relatively infrequent intervals in the surface, for example, a scratch, ridge, hole, peak, crack, or check.*

Lay. *The direction of the predominant surface pattern.* Normally determined by the production method used. A turned or ground surface has a definite direction of lay, whereas a lapped surface has not and is therefore considered multidirectional.

Roughness Width Rating. *The maximum permissible width of repetitive units of the dominant surface pattern. It may be specified in inches adjacent to the lay symbol. Irregularities having widths up to and including the maximum specified or, when no dimension is specified, up to and including the width of the irregularities due to machine feed shall be the basis for the roughness height specification.*

Waviness Width Rating. *Waviness widths may be specified directly in inches.*

Height Rating. *The height of the roughness or waviness shall be specified in one of the following terms:*

Maximum peak-to-valley height (h_{max}).
Average peak-to-valley height.
Average arithmetical deviation from mean surface (h_{av}).
Average rms deviation from mean surface (h_{rms}).

NOTE: *A general note should be included in all specifications and drawings indicating which type of height rating is intended.*

Lay Specification. *The lay of a surface shall be specified by the lay symbol indicating direction of the visible surface marks.* Some specifications call for the lay symbol to be inscribed in a half square.

Recommended Roughness and Waviness. American Standard B46.1-1947 tabulates the following recommended numbers for roughness and waviness. The use of one number to specify height or width of irregularities indicates the maximum allowable value. Any lesser degree will be satisfactory. When two numbers are used they specify the maximum and minimum permissible values.

ROUGHNESS HEIGHT VALUES, μIN.

¼	5	20	80	320
½	6	25	100	400
1	8	32	125	500
2	10	40	160	600
3	13	50	200	800
4	16	63	250	1000

WAVINESS HEIGHT VALUES, IN.

0.00002	0.00008	0.0003	0.001	0.005	0.015
0.00003	0.0001	0.0005	0.002	0.008	0.020
0.00005	0.0002	0.0008	0.003	0.010	

Measurement of Surface Roughness. There are numerous methods for measuring roughness of a surface. An exhaustive treatise, written by Ben C. Brosheer, appeared in *American Machinist*, Sept. 9 and 23, 1948, under the title "How Smooth Is Smooth." The three most practical methods appear to be:

TRACER-TYPE INSTRUMENTS. Of these, the Profilometer automatically produces the rms average (arithmetical average may be optional). This instrument, designed for shop use, takes readings with a hand-held stylus head or a motor-driven tracer.

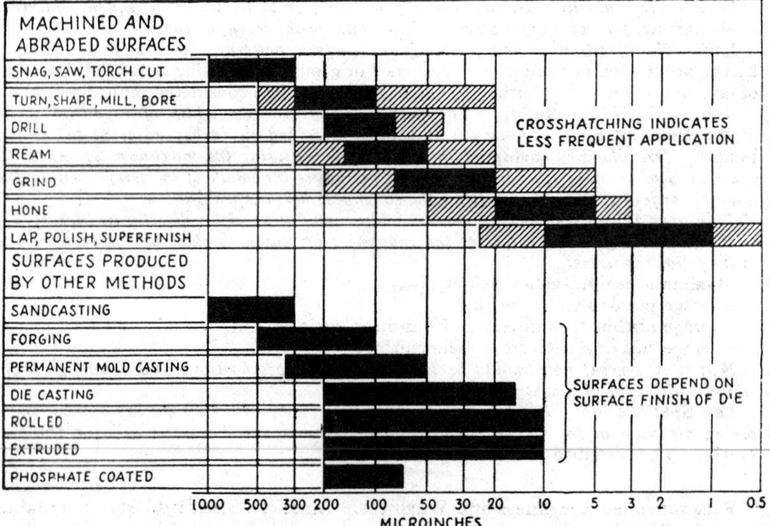

Fig. 4. Various factors affect the surface finish produced in a given machining operation. Charted values (*SAE Automotive Drafting Standards*) should not be considered hard and fast limits. In automotive practice, surfaces requiring control are usually confined to 50 μin. or less.

The Brush Surface Analyzer will inspect most surface irregularities normally encountered in machining operations. The instrument is limited normally to laboratory operations on small parts.

In general, the Profilometer will handle roughness widths up to $\frac{1}{32}$ in.; the Brush instrument, up to 0.025 in.

Commercial Surface Specimens. A number of companies make sets of machined specimen blocks for purposes of surface-quality control. Surfaces machined by different methods have somewhat different qualities, even when they show the same average roughness height value when measured with a stylus tracer instrument. Hence these sets often contain replicas of surfaces of several roughness value produced by turning, milling, grinding, honing, and lapping.

Translucent-film replicas of the surface of a workpiece may be made and studied for distribution and orientation of surface irregularities.

Precision Reference Specimens. By cooperative development, General Motors and Chrysler devised a method for making reproducible physical standards of surface roughness. These five blocks—20, 32, 50, 80, and 125 μin.—allow the calibration of stylus tracer instruments. In practice, nearly all machined surfaces are so variable in roughness over a $\frac{1}{8}$-in. trace that it is impossible to judge the roughness value accurately. In any event, the precision reference specimens make possible calibration of instruments so that readings taken at any time or place can be compared directly with measurements taken at any other time or place.

Part 9

MACHINE-TOOL STANDARDS

Section 41

STANDARD MACHINE-TOOL ELEMENTS

SECTION 41

STANDARD MACHINE-TOOL ELEMENTS

ORIGIN OF STANDARDS

Sectional Committee B5 of the American Society of Mechanical Engineers is responsible for that organization's work on the standardization of small tools and machine-tool elements. Technical committees appointed by the sectional committee have developed a number of standards on machine-tool elements with the help of representatives of groups like the Society of Automotive Engineers, National Machine Tool Builders Association, and the Metal Cutting Institute. The essential reference details of these standards are reproduced here, with the exception that lathe toolposts will be found under the section on Turning and Boring.

MACHINE TAPERS

A machine taper provides a separable connection for tool, center, workholder, or arbor with its mating part, such as a spindle socket or spindle nose. The purpose of a machine taper is to maintain accurate alignment of the two parts and to permit ready substitution of other tools, centers, or arbors. Two types of machine tapers—self-holding and steep—are embodied in American Standard ASA B5.10-1943.

Self-holding Tapers. Twenty-two numbered tapers of the self-holding variety have been combined in the ASA data from three original commercial series: Brown & Sharpe, Morse, and $\frac{3}{4}$ in. per ft (see Table 1). The eleven smallest self-holding tapers will not ordinarily loosen of themselves, if properly seated or not worn, and therefore they have a tang drive held in by friction. If more positive drive is required, Morse tapers from No. 3 to No. 7 may have a tang drive with the shank held in with key. Tapers 200 to 1200, inclusive, have a measure of self-holding power, but it is considered advisable to use a key drive with shank held in by key, or a key drive with shank held in with drawbolt.

Dimensional details of self-holding tapers with the four types of drive mentioned are given in Tables 3 to 6.

Steep Tapers. These self-releasing tapers (Table 2) have a taper of $3\frac{1}{2}$ in. per ft. The taper fit between the shank and socket acts only to maintain alignment. Hence, a positive locking device is required. Steep tapers are used for the spindle noses and arbors of milling machines and on type L lathe spindle noses.

Nonstandard Tapers. For many years Brown & Sharpe tapers from Nos. 1 to 18 (approximately $\frac{1}{2}$ in. per ft) have been used in industry, and the first three of these have been incorporated in the ASA self-holding series.

Jarno tapers were widely used in sizes from Nos. 2 to 20. The taper was 0.600 in. per ft. The number designated the small diameter in tenths of an inch.

Morse tapers range in eight sizes from No. 0 to 7 and have a taper of approximately $\frac{5}{8}$ in. per ft.

TABLE 1. SELF-HOLDING TAPERS—BASIC DIMENSIONS

All Dimensions Given in Inches

No. of Taper	Taper per Ft	Dia at Gage Line *A*	Origin of Series
0.239	0.50200	0.23922	Brown & Sharpe taper series
0.299	0.50200	0.29968	
0.375	0.50200	0.37525	
1	0.59858	0.47500	Morse taper series
2	0.59941	0.70000	
3	0.60235	0.93800	
4	0.62326	1.23100	
4½	0.62400	1.50000	
5	0.63151	1.74800	
6*	0.62565	2.49400	
7*	0.62400	3.27000	
200	0.750	2.000	¾ in. per ft taper series
250	0.750	2.500	
300	0.750	3.000	
350	0.750	3.500	
400	0.750	4.000	
450	0.750	4.500	
500	0.750	5.000	
600	0.750	6.000	
800	0.750	8.000	
1000	0.750	10.000	
1200	0.750	12.000	

* These sizes are continued in the Tang Drive series for the present to meet special needs.

TABLE 2. DIMENSIONS OF STEEP MACHINE TAPERS

All Dimensions Given in Inches

No. of Taper	Taper per Ft*	Dia at Gage Line		Length along Axis	
5	3.500	½	0.500	11/16	0.6875
10	**3.500**	**5/8**	**0.625**	**7/8**	**0.8750**
15	3.500	¾	0.750	1 1/16	1.0625
20	**3.500**	**7/8**	**0.875**	**1 5/16**	**1.3125**
25	3.500	1	1.000	1 9/16	1.5625
30	**3.500**	**1 1/4**	**1.250**	**1 7/8**	**1.8750**
35	3.500	1½	1.500	2¼	2.2500
40	**3.500**	**1 3/4**	**1.750**	**2 11/16**	**2.6875**
45	3.500	2¼	2.250	3 5/16	3.3125
50	**3.500**	**2 3/4**	**2.750**	**4**	**4.0000**
55	3.500	3½	3.500	5 3/16	5.1875
60	**3.500**	**4 1/4**	**4.500**	**6 3/8**	**6.3750**

The tapers numbered 10, 20, 30, 40, 50, and 60 that are printed in heavy-faced type are designated as the "Preferred Series." The tapers numbered 5, 15, 25, 35, 45, and 55 that are printed in light-faced type are designated as the "Intermediate Series."

* This taper corresponds to an included angle of 16° 35′ 33.4″.

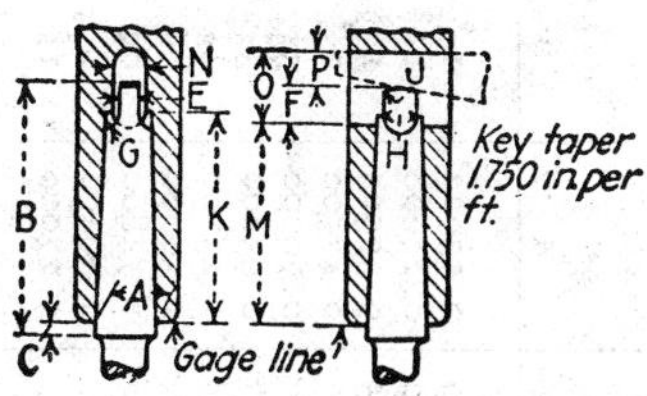

FIG. 1. Application of self-holding taper. Tang drive with shank retained by friction. For dimensions see Table 3.

TABLE 3. SELF-HOLDING TAPERS—TANG DRIVE WITH SHANK RETAINED BY FRICTION (SEE FIG. 1)
All Dimensions Given in Inches

	Dia at Gage Line*	Shank		Tang					Socket		Knockout Keyway		
		Over-all Length	Ex-posed Length	Thick-ness	Length	Radius of Mill	Dia	Radius	Depth	Gage Line to Key-way	Width	Length	Shank End to Back of Keyway
	A	B	C	E	F	G	H	J	K	M	N	O	P
0.239	0.23922	$1\frac{9}{32}$	$\frac{3}{32}$	0.125	$\frac{3}{16}$	$\frac{3}{16}$	$\frac{11}{64}$	$\frac{1}{32}$	$1\frac{1}{16}$	$\frac{15}{16}$	0.141	$\frac{3}{8}$	$\frac{1}{8}$
0.299	0.29968	$1\frac{19}{32}$	$\frac{3}{32}$	0.1562	$\frac{1}{4}$	$\frac{3}{16}$	$\frac{7}{32}$	$\frac{1}{32}$	$1\frac{5}{16}$	$1\frac{11}{64}$	0.172	$\frac{1}{2}$	$\frac{11}{64}$
0.375	0.37525	$1\frac{31}{32}$	$\frac{3}{32}$	0.1875	$\frac{5}{16}$	$\frac{3}{16}$	$\frac{9}{32}$	$\frac{3}{64}$	$1\frac{5}{8}$	$1\frac{15}{32}$	0.203	$\frac{5}{8}$	$\frac{7}{32}$
1	0.47500	$2\frac{9}{16}$	$\frac{1}{8}$	0.2031	$\frac{3}{8}$	$\frac{3}{16}$	$\frac{11}{32}$	$\frac{3}{64}$	$2\frac{3}{16}$	$2\frac{1}{16}$	0.213	$\frac{3}{4}$	$\frac{3}{8}$
2	0.70000	$3\frac{1}{8}$	$\frac{3}{16}$	0.250	$\frac{7}{16}$	$\frac{1}{4}$	$\frac{17}{32}$	$\frac{1}{16}$	$2\frac{21}{32}$	$2\frac{1}{2}$	0.260	$\frac{7}{8}$	$\frac{7}{16}$
3	0.93800	$3\frac{7}{8}$	$\frac{3}{16}$	0.3125	$\frac{9}{16}$	$\frac{9}{32}$	$\frac{23}{32}$	$\frac{5}{64}$	$3\frac{5}{16}$	$3\frac{1}{16}$	0.322	$1\frac{3}{16}$	$\frac{9}{16}$
4	1.23100	$4\frac{7}{8}$	$\frac{1}{4}$	0.4687	$\frac{5}{8}$	$\frac{5}{16}$	$\frac{31}{32}$	$\frac{3}{32}$	$4\frac{3}{16}$	$3\frac{7}{8}$	0.478	$1\frac{1}{4}$	$\frac{1}{2}$
$4\frac{1}{2}$	1.50000	$5\frac{3}{8}$	$\frac{1}{4}$	0.5625	$\frac{11}{16}$	$\frac{3}{8}$	$1\frac{13}{64}$	$\frac{1}{8}$	$4\frac{5}{8}$	$4\frac{5}{16}$	0.573	$1\frac{3}{8}$	$\frac{9}{16}$
5	1.74800	$6\frac{1}{8}$	$\frac{1}{4}$	0.625	$\frac{3}{4}$	$\frac{3}{8}$	$1\frac{13}{32}$	$\frac{1}{8}$	$5\frac{5}{16}$	$4\frac{15}{16}$	0.635	$1\frac{1}{2}$	$\frac{9}{16}$
6†	2.49400	$8\frac{9}{16}$	$\frac{5}{16}$	0.750	$1\frac{1}{8}$	$\frac{1}{2}$	2	$\frac{5}{32}$	$7\frac{13}{32}$	7	0.760	$1\frac{3}{4}$	$\frac{1}{2}$
7†	3.27000	$11\frac{5}{8}$	$\frac{3}{8}$	1.125	$1\frac{3}{8}$	$\frac{3}{4}$	$2\frac{5}{8}$	$\frac{3}{16}$	$10\frac{5}{32}$	$9\frac{1}{2}$	1.135	$2\frac{5}{8}$	$\frac{7}{8}$

* See Table 7 for plug gage dimensions.
† These sizes are continued in the Tang Drive series for the present to meet special needs.

TOLERANCES

For diameter of shank at gage line A:
All sizes, +0.002 − 0.000
For diameter of hole at gage line A:
All sizes, +0.000 − 0.002
For thickness of tang E:
Up to and including No. 5, +0.000 − 0.006
Larger than No. 5, +0.000 − 0.008
For width of knockout keyway N:
Up to and including No. 5, +0.006 − 0.000
Larger than No. 5, +0.008 − 0.000

For concentricity of tang E with center line of taper:
Up to and including No. 5, 0.0035 (indicator reading)
Larger than No. 5, 0.005 (indicator reading)
For concentricity of knockout keyway N with center line of taper:
Up to and including No. 5, 0.0035 (indicator reading)
Larger than No. 5, 0.005 (indicator reading)
Tolerances on fractional dimensions, ±0.010 unless otherwise specified.

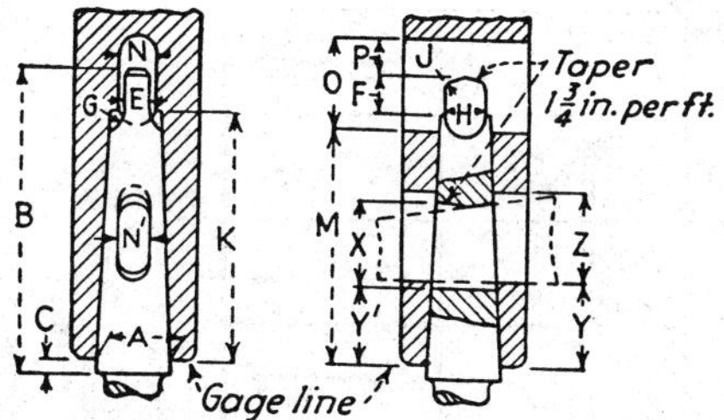

FIG. 2. Self-holding taper—tang drive with shank retained by key. For dimensions see Table 4.

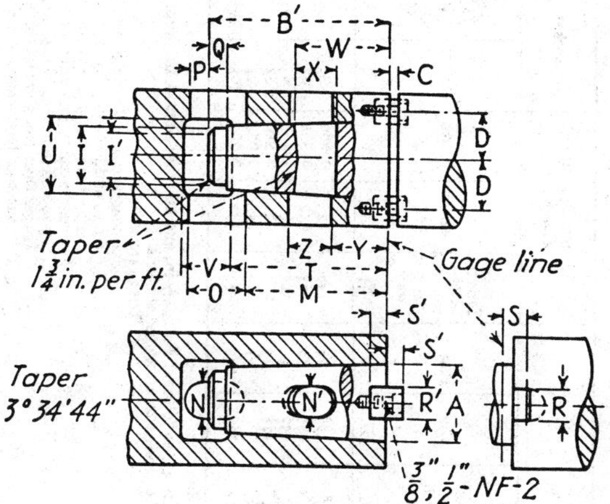

FIG. 3. Self-holding taper—key drive with shank retained by key. For dimensions see Table 5.

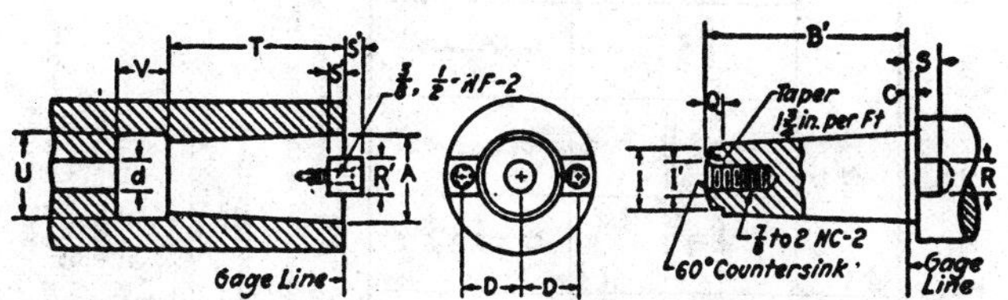

FIG. 4. Self-holding taper—key drive with shank retained by draw bolt. For dimensions see Table 6.

TABLE 4. SELF-HOLDING TAPERS—TANG DRIVE WITH SHANK RETAINED BY KEY (SEE FIG. 2)

All Dimensions Given in Inches

No. of Taper	Dia at Gage Line*	Shank		Tang					Holdback Keyway—Shank†			Socket		Knockout Keyway†			Holdback Keyway—Socket†		
		Over-all Length	Ex-posed Length	Thick-ness	Length	Radius of Mill	Dia	Radius	Gage Line to Bot-tom of Key-way	Length	Width	Depth	Gage Line to Key-way	Width	Length	Shank End to Back of Key-way	Gage Line to Front of Key-way	Length	Width
	A	B	C	E	F	G	H	J	Y′	X	N′	K	M	N	O	P	Y	Z	N′
3	0.93800	3 7/8	3/16	5/16	9/16	9/32	23/32	3/64	1 1/32	1 1/8	0.260	3 1/4	3 1/16	0.322	1 3/16	9/16	1 1/8	1 3/16	0.260
4	1.23100	4 7/8	1/4	15/32	5/8	5/16	31/32	3/32	1 13/32	1 3/16	0.385	4 1/8	3 7/8	0.478	1 1/4	1/2	1 1/2	1 1/4	0.385
4½	1.50000	5 3/8	1/4	9/16	11/16	3/8	1 13/64	1/8	1 21/32	1 1/4	0.447	4 9/16	4 5/16	0.573	1 3/8	9/16	1 13/16	1 3/8	0.447
5	1.74800	6 1/8	1/4	5/8	3/4	3/8	1 13/32	1/8	2	1 3/8	0.510	5 1/4	4 13/16	0.635	1 1/2	9/16	2 1/8	1 1/2	0.510
6‡	2.49400	8 9/16	5/16	3/4	1 1/8	1/2	2	5/32	2 1/8	1 3/8	0.635	7 3/8	7	0.760	1 3/4	1/2	2 1/4	1 3/4	0.635
7‡	3.27000	11 5/8	3/8	1 1/8	1 3/8	3/4	2 5/8	3/16	2 1/2	1 11/16	0.760	10 1/8	9 1/2	1.135	2 3/8	7/8	2 3/8	1 13/16	0.760

* See Table 7 for plug gage dimensions.
† Corners at entrance side of keyways shall be chamfered at 45° as follows: No. 3, 3/64 in. and all other sizes 1/16 in. deep.
‡ These sizes are continued in the Tang Drive series for the present to meet special needs.

TOLERANCES

For diameter of shank at gage line *A*:
All sizes, +0.002 − 0.000
For diameter of hole at gage line *A*:
All sizes, +0.000 − 0.002
For thickness of tang *E*:
Up to and including No. 5, +0.000 − 0.006
Larger than No. 5, +0.000 − 0.008
For width of knockout keyway *N*:
Up to and including No. 5, +0.006 − 0.000
Larger than No. 5, +0.008 − 0.000

For concentricity of tang *E* with center line of taper:
Up to and including No. 5, 0.0035 (indicator reading)
Larger than No. 5, 0.005 (indicator reading)
For concentricity knockout keyway *N* with center line of taper:
Up to and including No. 5, 0.0035 (indicator reading)
Larger than No. 5, 0.005 (indicator reading)
Tolerances on fractional dimensions, ÷0.010 unless otherwise specified.

TABLE 5. SELF-HOLDING TAPERS—KEY DRIVE WITH SHANK RETAINED BY KEY (SEE FIG. 3)
All Dimensions Given in Inches

No. of Taper	Dia at Gage Line*	Shank					Drive Keyway		Drive Key		Holdback Keyway in Shank†		
		Length from Gage Line	Exposed Length	Length of Relief	Dia of Flat	Dia of Relief	Width	Depth	Center Line to Center of Screw	Fine Thread Screw (NF)	Gage Line to Back of Keyway	Length	Width
	A	B'	C	Q	I'	I	R	S	D		W	X	N'
200	2.000	$5\frac{1}{8}$	Min 0.003 max 0.067 for all sizes	$\frac{1}{4}$	$1\frac{1}{8}$	$1\frac{5}{8}$	1.005	$\frac{9}{16}$	$1\frac{13}{32}$	$\frac{3}{8}$	$3\frac{7}{16}$	$1\frac{9}{16}$	0.656
250	2.500	$5\frac{7}{8}$		$\frac{1}{4}$	$1\frac{1}{8}$	$2\frac{1}{16}$	1.005	$\frac{9}{16}$	$1\frac{21}{32}$	$\frac{3}{8}$	$3\frac{11}{16}$	$1\frac{9}{16}$	0.781
300	3.000	$6\frac{5}{8}$		$\frac{1}{4}$	$1\frac{3}{8}$	$2\frac{1}{2}$	2.005	$\frac{9}{16}$	$2\frac{1}{4}$	$\frac{3}{8}$	$4\frac{1}{16}$	$1\frac{9}{16}$	1.031
350	3.500	$7\frac{7}{16}$		$\frac{5}{16}$	2	$2\frac{15}{16}$	2.005	$\frac{9}{16}$	$2\frac{1}{2}$	$\frac{3}{8}$	$4\frac{7}{8}$	2	1.031
400	4.000	$8\frac{3}{16}$		$\frac{5}{16}$	$2\frac{3}{8}$	$3\frac{5}{16}$	2.005	$\frac{9}{16}$	$2\frac{3}{4}$	$\frac{3}{8}$	$5\frac{5}{16}$	$2\frac{1}{4}$	1.031
450	4.500	9		$\frac{3}{8}$	$2\frac{3}{8}$	$3\frac{13}{16}$	3.005	$\frac{13}{16}$	3	$\frac{1}{2}$	$5\frac{7}{8}$	$2\frac{7}{16}$	1.031
500	5.000	$9\frac{3}{4}$		$\frac{3}{8}$	$2\frac{1}{2}$	$4\frac{1}{4}$	3.005	$\frac{13}{16}$	$3\frac{1}{4}$	$\frac{1}{2}$	$6\frac{7}{16}$	$2\frac{5}{8}$	1.031
600	6.000	$11\frac{5}{16}$		$\frac{7}{16}$	$2\frac{5}{8}$	$5\frac{3}{16}$	3.005	$\frac{13}{16}$	$3\frac{3}{4}$	$\frac{1}{2}$	$7\frac{7}{16}$	3	1.281
800	8.000	$14\frac{3}{8}$		$\frac{1}{2}$	$3\frac{1}{2}$	7	4.010	$1\frac{1}{16}$	$4\frac{3}{4}$	$\frac{1}{2}$	$9\frac{9}{16}$	4	1.781
1000	10.000	$17\frac{7}{16}$		$\frac{5}{8}$	$4\frac{1}{2}$	$8\frac{3}{4}$	4.010	$1\frac{1}{16}$			$11\frac{1}{2}$	$4\frac{3}{4}$	2.031
1200	12.000	$20\frac{1}{2}$		$\frac{3}{4}$	$5\frac{3}{8}$	$10\frac{1}{2}$	4.010	$1\frac{1}{16}$			$13\frac{3}{4}$	$5\frac{3}{4}$	2.531

TABLE 5. SELF-HOLDING TAPERS—KEY DRIVE WITH SHANK RETAINED BY KEY (SEE FIG. 3) (*Continued*)

No. of Taper	Socket						Knockout Keyway†			Holdback Keyway† in Socket		
	Drive Keyway		Gage Line to Front of Relief	Dia Relief	Depth Relief	Gage Line to Keyway	Width	Length	Shank End to Back of Keyway	Gage Line to Front of Keyway	Length	Width
	Width	Depth										
	R′	*S′*	*T*	*U*	*V*	*M*	*N*	*O*	*P*	*Y*	*Z*	*N′*
200	1.000	$\frac{1}{2}$	$4\frac{3}{4}$	$1\frac{13}{16}$	1	$4\frac{1}{2}$	0.656	$1\frac{9}{16}$	$\frac{15}{16}$	2	$1\frac{11}{16}$	0.656
250	1.000	$\frac{1}{2}$	$5\frac{1}{2}$	$2\frac{1}{4}$	1	$5\frac{3}{16}$	0.781	$1\frac{15}{16}$	$1\frac{1}{4}$	$2\frac{1}{4}$	$1\frac{11}{16}$	0.781
300	2.000	$\frac{1}{2}$	$6\frac{1}{4}$	$2\frac{3}{4}$	1	$5\frac{15}{16}$	1.031	$2\frac{3}{16}$	$1\frac{1}{2}$	$2\frac{5}{8}$	$1\frac{11}{16}$	1.031
350	2.000	$\frac{1}{2}$	$6\frac{15}{16}$	$3\frac{3}{16}$	$1\frac{1}{4}$	$6\frac{3}{4}$	1.031	$2\frac{3}{16}$	$1\frac{1}{2}$	3	$2\frac{1}{8}$	1.031
400	2.000	$\frac{1}{2}$	$7\frac{11}{16}$	$3\frac{5}{8}$	$1\frac{1}{4}$	$7\frac{1}{2}$	1.031	$2\frac{3}{16}$	$1\frac{1}{2}$	$3\frac{1}{4}$	$2\frac{3}{8}$	1.031
450	3.000	$\frac{3}{4}$	$8\frac{3}{8}$	$4\frac{3}{16}$	$1\frac{1}{2}$	8	1.031	$2\frac{3}{4}$	$1\frac{3}{4}$	$3\frac{5}{8}$	$2\frac{9}{16}$	1.031
500	3.000	$\frac{3}{4}$	$9\frac{1}{8}$	$4\frac{5}{8}$	$1\frac{1}{2}$	$8\frac{3}{4}$	1.031	$2\frac{3}{4}$	$1\frac{3}{4}$	4	$2\frac{3}{4}$	1.031
600	3.000	$\frac{3}{4}$	$10\frac{9}{16}$	$5\frac{1}{2}$	$1\frac{3}{4}$	$10\frac{1}{8}$	1.281	$3\frac{1}{4}$	$2\frac{1}{16}$	$4\frac{5}{8}$	$3\frac{1}{4}$	1.281
800	4.000	1	$13\frac{1}{2}$	$7\frac{7}{8}$	2	$12\frac{7}{8}$	1.781	$4\frac{1}{4}$	$2\frac{3}{4}$	$5\frac{3}{4}$	$4\frac{1}{4}$	1.781
1000	4.000	1	$16\frac{5}{16}$	$9\frac{3}{16}$	$2\frac{1}{2}$	$15\frac{3}{4}$	2.031	5	$3\frac{5}{16}$	7	5	2.031
1200	4.000	1	19	11	3	$18\frac{1}{2}$	2.531	6	4	$8\frac{1}{4}$	6	2.531

* See Table 7 for plug gage dimensions.

† Corners at entrance side of keyways shall be chamfered at 45°, as follows: Nos. 200 to 350, inclusive, $\frac{1}{16}$ in. deep; Nos. 400 to 600, inclusive, $\frac{3}{32}$ in. deep; Nos. 800 to 1200, inclusive, $\frac{1}{8}$ in. deep.

TOLERANCES

For diameter of shank at gage line *A*:
  All sizes, +0.002 − 0.000
For diameter of hole at gage line *A*:
  All sizes, +0.000 − 0.002
For width of knockout keyway *N*:
  +0.008 − 0.000
For width of drive keyway *R′*:
  In socket, +0.000 − 0.001

For width of drive keyway *R*:
  In shank, +0.010 − 0.000
For concentricity of knockout keyway *N*:
  With center line of spindle, 0.007
For concentricity of keyway *R′*:
  With center line of spindle, 0.002
Tolerances on fractional dimensions, ± 0.010 unless otherwise specified.

TABLE 6. SELF-HOLDING TAPERS—KEY DRIVE WITH SHANK RETAINED BY DRAW BOLT (SEE FIG. 4)
All Dimensions Given in Inches

No. of Taper	Dia at Gage Line	Shank											Drive Key		Socket					
		Length from Gage Line	Exposed Length	Screw Thread Holdback Rod		Drive Keyway		Dia of Flat	Depth of 60° Center	Length of Relief	Dia of Relief	Center Line to Center of Screw	Fine Thread Series Screw (NF)	Gage Line to Front of Relief	Dia of Relief	Depth Relief	Dia Draw Bolt Hole	Drive Keyway	
				Dia (NC)	Depth of Thread	Width	Depth											Width	Depth
	A	B'	C			R	S	I'		Q	I	D		T	U	V	d	R'	S'
200	2.000	$5\frac{1}{8}$		$\frac{7}{8}$	$1\frac{3}{4}$	1.005	$\frac{9}{16}$	$1\frac{1}{8}$	$\frac{1}{8}$	$\frac{1}{4}$	$1\frac{5}{8}$	$1\frac{13}{32}$	$\frac{3}{8}$	$4\frac{3}{4}$	$1\frac{13}{16}$	1	1	1.000	$\frac{1}{2}$
250	2.500	$5\frac{7}{8}$		$\frac{7}{8}$	$1\frac{3}{4}$	1.005	$\frac{9}{16}$	$1\frac{1}{8}$	$\frac{1}{8}$	$\frac{1}{4}$	$2\frac{1}{16}$	$1\frac{21}{32}$	$\frac{3}{8}$	$5\frac{1}{2}$	$2\frac{1}{4}$	1	1	1.000	$\frac{1}{2}$
300	3.000	$6\frac{5}{8}$	Min	1	2	2.005	$\frac{9}{16}$	$1\frac{3}{8}$	$\frac{3}{16}$	$\frac{1}{4}$	$2\frac{1}{2}$	$2\frac{1}{4}$	$\frac{3}{8}$	$6\frac{1}{4}$	$2\frac{3}{4}$	1	$1\frac{1}{8}$	2.000	$\frac{1}{2}$
350	3.500	$7\frac{7}{16}$	0.003	1	2	2.005	$\frac{9}{16}$	2	$\frac{3}{16}$	$\frac{5}{16}$	$2\frac{15}{16}$	$2\frac{1}{2}$	$\frac{3}{8}$	$6\frac{15}{16}$	$3\frac{3}{16}$	$1\frac{1}{4}$	$1\frac{1}{8}$	2.000	$\frac{1}{2}$
400	4.000	$8\frac{3}{16}$	max	$1\frac{1}{4}$	$2\frac{1}{2}$	2.005	$\frac{9}{16}$	$2\frac{3}{8}$	$\frac{1}{4}$	$\frac{5}{16}$	$3\frac{5}{16}$	$2\frac{3}{4}$	$\frac{3}{8}$	$7\frac{11}{16}$	$3\frac{5}{8}$	$1\frac{1}{4}$	$1\frac{3}{8}$	2.000	$\frac{1}{2}$
450	4.500	9	0.067	$1\frac{1}{4}$	$2\frac{1}{2}$	3.005	$\frac{13}{16}$	$2\frac{3}{8}$	$\frac{1}{4}$	$\frac{3}{8}$	$3\frac{13}{16}$	3	$\frac{1}{2}$	$8\frac{3}{8}$	$4\frac{3}{16}$	$1\frac{1}{2}$	$1\frac{3}{8}$	3.000	$\frac{3}{4}$
500	5.000	$9\frac{3}{4}$	for	$1\frac{1}{4}$	$2\frac{1}{2}$	3.005	$\frac{13}{16}$	$2\frac{1}{2}$	$\frac{3}{8}$	$\frac{3}{8}$	$4\frac{1}{4}$	$3\frac{1}{4}$	$\frac{1}{2}$	$9\frac{1}{8}$	$4\frac{5}{8}$	$1\frac{1}{2}$	$1\frac{3}{8}$	3.000	$\frac{3}{4}$
600	6.000	$11\frac{5}{16}$	all	$1\frac{1}{2}$	3	3.005	$\frac{13}{16}$	$2\frac{5}{8}$	$\frac{3}{8}$	$\frac{7}{16}$	$5\frac{3}{16}$	$3\frac{3}{4}$	$\frac{1}{2}$	$10\frac{9}{16}$	$5\frac{1}{2}$	$1\frac{3}{4}$	$1\frac{5}{8}$	3.000	$\frac{3}{4}$
800	8.000	$14\frac{3}{8}$	sizes	$1\frac{1}{2}$	3	4.010	$1\frac{1}{16}$	$3\frac{1}{2}$	$\frac{1}{2}$	$\frac{1}{2}$	7	$4\frac{3}{4}$	$\frac{1}{2}$	$13\frac{1}{2}$	$7\frac{3}{8}$	2	$1\frac{5}{8}$	4.000	1
1000	10.000	$17\frac{7}{16}$		2	4	4.010	$1\frac{1}{16}$	$4\frac{1}{2}$	$\frac{1}{2}$	$\frac{5}{8}$	$8\frac{3}{4}$			$16\frac{5}{16}$	$9\frac{3}{16}$	$2\frac{1}{2}$	$2\frac{1}{4}$	4.000	1
1200	12.000	$20\frac{1}{2}$		2	4	4.010	$1\frac{1}{16}$	$5\frac{3}{8}$	$\frac{1}{2}$	$\frac{3}{4}$	$10\frac{1}{2}$			19	11	3	$2\frac{1}{4}$	4.000	1

See Table 7 for plug gage dimensions.

For diameter of hole at gage line A:
All sizes, +0.000 − 0.002
For diameter of shank at gage line A:
All sizes, +0.002 − 0.000
For width of drive keyway R':
In socket, +0.000 − 0.001
For width of drive keyway R:
In shank, +0.010 − 0.000

TOLERANCES
For concentricity of drive keyway R:
With center line of spindle, 0.003
For concentricity of drive keyway R':
With center line of spindle, 0.001
Tolerances of fractional dimensions, ±0.010 unless otherwise specified.

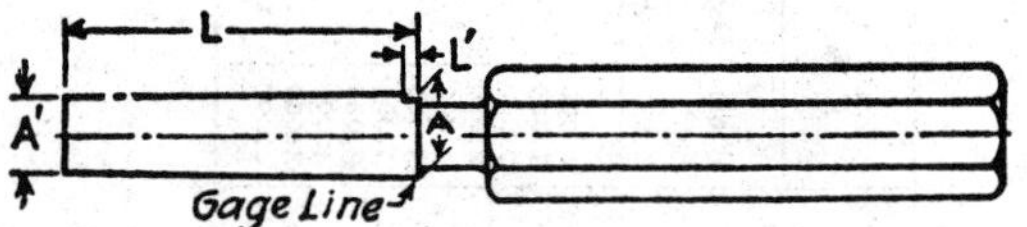

TABLE 7. PLUG GAGES FOR SELF-HOLDING TAPERS

No. of Taper	Taper† per Ft	Dia*·† at Gage Line A	Dia*·‡ at Small End A'	Length Gage Line to Small End L	Depth of Gaging Notch L'
0.239	0.50200	0.23922	0.20000	15/16	0.048
0.299	0.50200	0.29968	0.25000	1 3/16	0.048
0.375	0.50200	0.37525	0.31250	1 1/2	0.048
1	0.59858	0.47500	0.36900	2 1/8	0.040
2	0.59941	0.70000	0.57200	2 9/16	0.040
3	0.60235	0.93800	0.77800	3 3/16	0.040
4	0.62326	1.23100	1.02000	4 1/16	0.038
4 1/2	0.62400	1.50000	1.26600	4 1/2	0.038
5	0.63151	1.74800	1.47500	5 3/16	0.038
6§	0.62565	2.49400	2.11600	7 1/4	0.038
7§	0.62400	3.27000	2.75000	10	0.038
200	0.750	2.000	1.703	4 3/4	0.032
250	0.750	2.500	2.156	5 1/2	0.032
300	0.750	3.000	2.609	6 1/4	0.032
350	0.750	3.500	3.063	7	0.032
400	0.750	4.000	3.516	7 3/4	0.032
450	0.750	4.500	3.969	8 1/2	0.032
500	0.750	5.000	4.422	9 1/4	0.032
600	0.750	6.000	5.328	10 3/4	0.032
800	0.750	8.000	7.141	13 3/4	0.032
1,000	0.750	10.000	8.953	16 3/4	0.032
1,200	0.750	12.000	10.766	19 3/4	0.032

* Gage tolerance for diameters A and A':
Sizes 0.239 to No. 3, inclusive, +0.0001 − 0.0000.
Sizes No. 4 to No. 300, inclusive, +0.00015 − 0.0000.
Sizes No. 350 to No. 1200, inclusive, +0.0002 − 0.0000.
† Taper per foot and diameter at gage line basic dimensions.
‡ Dimension calculated for reference only.
§ These sizes are continued in the Tongue Drive series for the present to meet special needs.

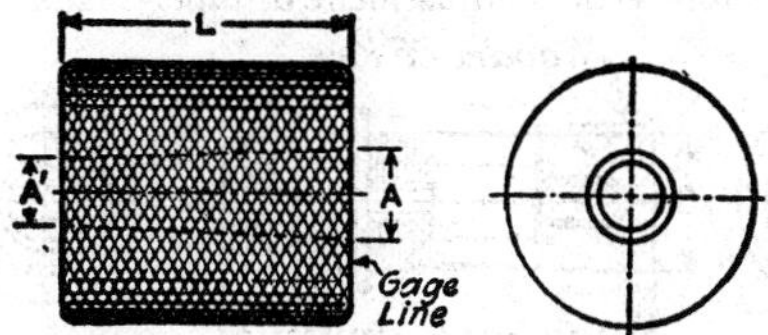

TABLE 8. RING GAGES FOR SELF-HOLDING TAPERS

No. of Taper	Taper per Ft†	Dia at Gage Line*,† A	Dia at Small End*,‡ A'	Length L
0.239	0.50200	0.23922	0.20000	15/16
0.299	0.50200	0.29968	0.25000	1 3/16
0.375	0.50200	0.37525	0.31250	1 1/2
1	0.59858	0.47500	0.36900	2 1/8
2	0.59941	0.70000	0.57200	2 9/16
3	0.60235	0.93800	0.77800	3 3/16
4	0.62326	1.23100	1.02000	4 1/16
4 1/2	0.62400	1.50000	1.26600	4 1/2
5	0.63151	1.74800	1.47500	5 3/16
6§	0.62565	2.49400	2.11600	7 1/4
7§	0.62400	3.2700	2.75000	10
200	0.750	2.0000	1.703	4 3/4
250	0.750	2.5000	2.156	5 1/2
300	0.750	3.0000	2.609	6 1/4
350	0.750	3.5000	3.063	7
400	0.750	4.0000	3.516	7 3/4
450	0.750	4.500	3.969	8 1/2
500	0.750	5.0000	4.422	9 1/4
600	0.750	6.0000	5.328	10 3/4
800	0.750	8.0000	7.141	13 3/4
1,000	0.750	10.0000	8.953	16 3/4
1,200	0.750	12.0000	10.766	19 3/4

* Gage tolerance for diameters A and A':
Sizes to No. 3 inclusive, −0.0001 + 0.0000.
Sizes No. 4 to No. 300 inclusive, −0.00015 + 0.0000.
Sizes No. 350 to 1200 inclusive, −0.0002 + 0.0000.

† Taper per foot and diameter at gage line basic dimensions.

‡ Dimensions calculated for reference only.

§ These sizes are continued in the Tongue Drive series for the present to meet special needs.

TABLE 9. ADAPTERS FOR MULTIPLE-SPINDLE DRILLHEADS

Used for vertical adjustment of taper-shank tools

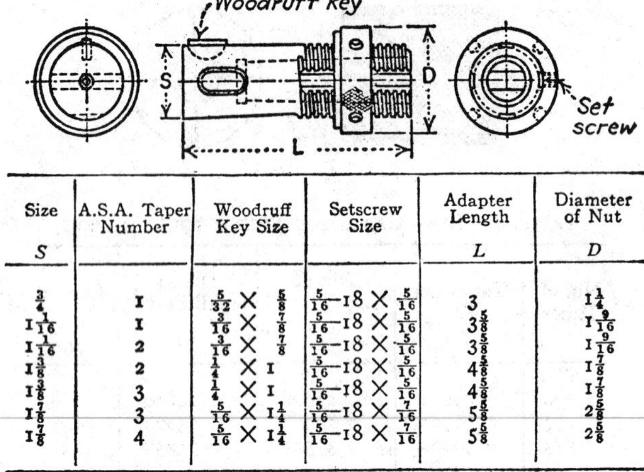

Size S	A.S.A. Taper Number	Woodruff Key Size	Setscrew Size	Adapter Length L	Diameter of Nut D
$\frac{3}{4}$	1	$\frac{5}{32} \times \frac{5}{8}$	$\frac{5}{16}$–18 $\times \frac{5}{16}$	3	$1\frac{1}{4}$
$1\frac{1}{16}$	1	$\frac{3}{16} \times \frac{7}{8}$	$\frac{5}{16}$–18 $\times \frac{5}{16}$	$3\frac{5}{8}$	$1\frac{9}{16}$
$1\frac{1}{16}$	2	$\frac{3}{16} \times \frac{7}{8}$	$\frac{5}{16}$–18 $\times \frac{5}{16}$	$3\frac{5}{8}$	$1\frac{9}{16}$
$1\frac{3}{8}$	2	$\frac{1}{4} \times 1$	$\frac{5}{16}$–18 $\times \frac{5}{16}$	$4\frac{5}{8}$	$1\frac{7}{8}$
$1\frac{3}{8}$	3	$\frac{1}{4} \times 1$	$\frac{5}{16}$–18 $\times \frac{5}{16}$	$4\frac{5}{8}$	$1\frac{7}{8}$
$1\frac{7}{8}$	3	$\frac{5}{16} \times 1\frac{1}{4}$	$\frac{5}{16}$–18 $\times \frac{7}{16}$	$5\frac{5}{8}$	$2\frac{5}{8}$
$1\frac{7}{8}$	4	$\frac{5}{16} \times 1\frac{1}{4}$	$\frac{5}{16}$–18 $\times \frac{7}{16}$	$5\frac{5}{8}$	$2\frac{5}{8}$

LATHE SPINDLE NOSES

American Standard B5.9–1948 lists four types of spindle noses—A, B, D, and L—for toolroom lathes, turret lathes, automatic lathes, and sundry other machine tools. As compared with the 1936 standard, the revision introduces the type L spindle nose with taper-key drive as an alternate for type D on engine lathes and extends the range, upward and downward, for types A, B, and D. Essential dimensions are tabulated for correlation of proper chucks and faceplates and the development of special workholding fixtures, if found needed.

The distinctions between the four types of spindle noses may be summarized thus:

Type A1. Has two rows of tapped holes and a driving button. The outer row of holes is used to attach faceplates, fixtures, and other chucks by means of cap screws; the inner row of holes is used to attach certain sizes of scroll chucks.

Type A2. The inner row of holes is omitted.

Type B1. Same as A1, except outer holes are clearance holes for bolts or studs passing through the flange.

Type B2. Same as B1 except inner row holes are omitted.

Type D1. Equipped to receive cam-lock studs on back of mating faceplate or chuck.

Type L. Has a long steep taper for locating and centering chucks and faceplates, a key to drive, and a flanged nut to hold the chuck or faceplate.

Spindle Noses Recommended for Different Types of Lathes

Bench Lathes. The 2-in. type D1 spindle nose is recommended for bench lathes of 6 to 8 in., inclusive, and the 3-in. type D1 or the type L00 spindle nose is recommended for 9- and 10-in. bench lathes.

Toolroom Lathes. The type D1 or the type L spindle noses are recommended as alternate standards for toolroom lathes. It is recommended that the size of the nose selected for each size of toolroom lathe be as follows:

Toolroom Lathes	*Noses*
10 in.	4-in. type D1 or type L0
12 to 16 in., inclusive	6-in. type D1 or type L1
Above 16 to 20 in., inclusive	6-in. type D1 or type L2

In each case the size of the toolroom lathe given is the nominal catalog size. Actual swing over the bed and carriage extensions has been established by American engine-lathe builders as being 2½ in. more than the catalog size specified.

Engine Lathes. Type D1 and type L spindle noses are recommended as alternate standards.

Engine Lathes	*Noses*
12 to 16 in., inclusive	6-in. type D1 or type L1
Above 16 to 20 in., inclusive	8-in. type D1 or type L2
Above 20 to 25 in., inclusive	8-in. type D1 or type L3
Above 25 to 32 in., inclusive	11-in. type D1 or type L3

Turret Lathes. The type A1 spindle nose is recommended for use on turret lathes, except that each manufacturer of turret lathes may use the type D1 nose.

Spindle Bores	*Noses*
Up to 1¾ in.	5 in. or smaller
Over 1¾ to 2 7/16 in., inclusive	6 in.
Over 2 7/16 to 3⅜ in., inclusive	8 in.
Over 3⅜ to 5⅜ in., inclusive	11 in.
Over 5⅜ to 8⅜ in., inclusive	15 in.
Over 8⅜ to 13 in., inclusive	20 in.
Over 13 to 21½ in., inclusive	28 in.

Single-spindle Automatic Lathes. The type A1 nose is recommended for single-spindle automatic lathes, size of nose depending upon the size of the heavy-duty chuck the machine normally takes, as follows:

Machines Taking	*Noses*
6-in. chucks or smaller	5 in. or smaller
8-in. chucks	6 in.
10- and 12-in. chucks	8 in.
15- and 18-in. chucks	11 in.

Other Applications. These spindle noses may be used on multiple-spindle automatic lathes, internal grinders, thread grinders, and hobbing machines.

For multiple-spindle automatic lathes, the type A1 or type A2 spindle nose is recommended. The size of the nose depends on the size of the hole in the spindle, or the size of heavy-duty chuck the machine normally takes. The size of the nose recommended for multiple-spindle chucking machines is as follows:

Machines Taking	*Noses*
4½-in. chucks	4 in.
6-in. chucks	5 in.
8-in. chucks	6 in.
10- and 12-in. chucks	8 in.
15- and 18-in. chucks	11 in.

Larger sizes of lathes of any type may be provided with any of the large spindle noses of type A1, A2, B1, B2, or D1.

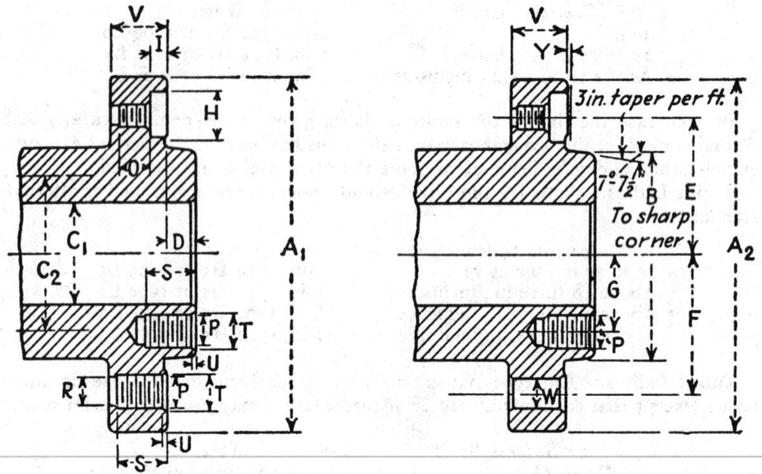

TYPE A SPINDLES

Type A1 is exactly as shown with tapped holes in both inner and outer bolt circles

Type A2 is same as shown except omit holes in inner bolt circle

TYPE B SPINDLES

Type B1 is exactly as shown with clearance holes in outer bolt circle and tapped holes in inner bolt circle

Type B2 is same as shown except omit tapped holes in inner bolt circle

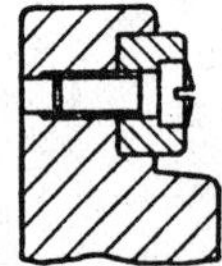

Section of spindle showing projection of driving button

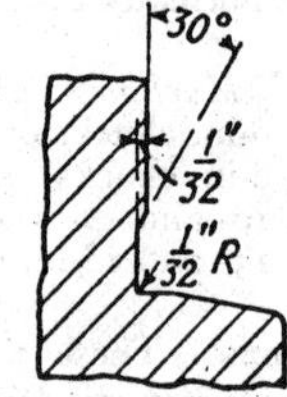

Enlarged view of undercut

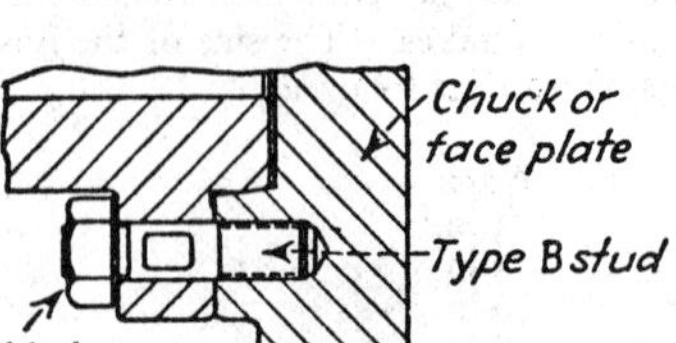

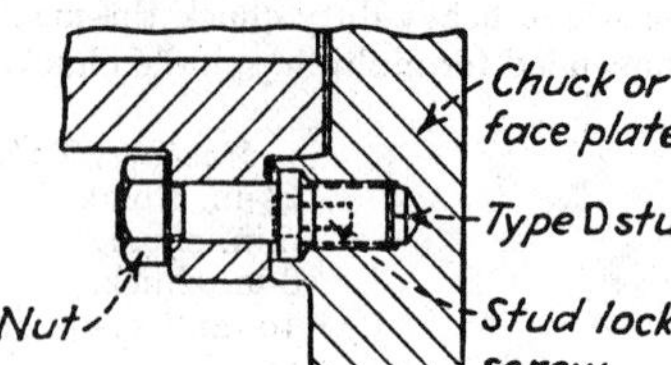

Methods of attaching chucks or face plates to type B spindles

FIG. 5. Types A and B lathe spindle noses and details of how chucks and faceplates are attached.

TABLE 10. DIMENSIONS OF TYPES A AND B LATHE SPINDLE NOSES
American Standard ASA B5.9–1948

Spindle Name	Dia	4-in. Nose	5-in. Nose	6-in. Nose	8-in. Nose	11-in. Nose	15-in. Nose	20-in. Nose	28-in. Nose
Dia of type A spindles	A_1	4.25	5.25	6.5	8.25	11.0	15.0	20.5	28.5
Dia of type B spindles	A_2	4.25	5.25	6.5	8.25	11.0	15.0	20.5	28.5
Pilot dia	B	2.5005 + 0.0005	3.2505 + 0.0005	4.1880 + 0.0005	5.50075 + 0.0005	7.75075 + 0.0005	11.251 + 0.001	16.251 + 0.001	23.001 + 0.001
Max hole in type $A1$ spindles	C_1	....	1.75	2.4375	3.375	5.375	8.375	13.0	19.0
Max hole in type $A2$ spindles	C_2	2.0	2.625	3.4375	4.625	6.75	10.125	15.0	21.5
Length of pilot	D	0.500 − 0.002	0.5625 − 0.002	0.625 − 0.002	0.6875 − 0.002	0.750 − 0.002	0.8125 − 0.002	0.875 − 0.002	1.0 − 0.002
Location of drive button	E	1.375 ± 0.003	2.0625 ± 0.003	2.625 ± 0.003	3.375 ± 0.003	4.625 ± 0.006	6.5 ± 0.006	9.125 ± 0.006	12.75 ± 0.006
Radius of outer bolt circle	F	1.625 ± 0.006	2.0625 ± 0.006	2.625 ± 0.006	3.375 ± 0.006	4.625 ± 0.008	6.5 ± 0.008	9.125 ± 0.008	12.75 ± 0.008
Radius of inner bolt circle	G	....	1.2187 ± 0.006	1.625 ± 0.006	2.1875 ± 0.006	3.25 ± 0.008	4.875 ± 0.008	7.25 ± 0.008	10.4375 ± 0.008
Hole for drive button	H	0.5625 + 0.002	0.625 ± 0.002	0.750 + 0.002	0.9375 + 0.002	1.125 + 0.002	1.375 + 0.002	1.625 + 0.002	2.000 + 0.002
Depth of counterbore	I	0.1875	0.25	0.3125	0.375	0.4375	0.5	0.625	0.8125
Thread	M	$\frac{3}{8}$-16NC-3	$\frac{5}{16}$-18NC-3	$\frac{5}{16}$-18NC-3	$\frac{3}{8}$-16NC-3	$\frac{3}{8}$-16NC-3	$\frac{1}{2}$-13NC-3	$\frac{1}{2}$-13NC-3	$\frac{1}{2}$-13NC-3
Full depth of thread	O	0.5625	0.5	0.5625	0.625	0.625	0.875	0.875	0.875
Thread	P	$\frac{7}{16}$-14NC-3	$\frac{7}{16}$-14NC-3	$\frac{1}{2}$-13NC-3	$\frac{5}{8}$-11NC-3	$\frac{3}{4}$-10NC-3	$\frac{7}{8}$-9NC-3	1-8NC-3	$1\frac{1}{8}$-7NC-3
Full depth of thread	S	0.75	0.75	0.875	1.0625	1.25	1.4375	1.625	2.0
Min width of flange	V	0.75	0.875	1.0	1.125	1.375	1.625	1.875	2.25
Bolt hole, type B spindle	W	0.4687	0.4687	0.5312	0.6562	0.7968	0.9218	1.0468	1.2968

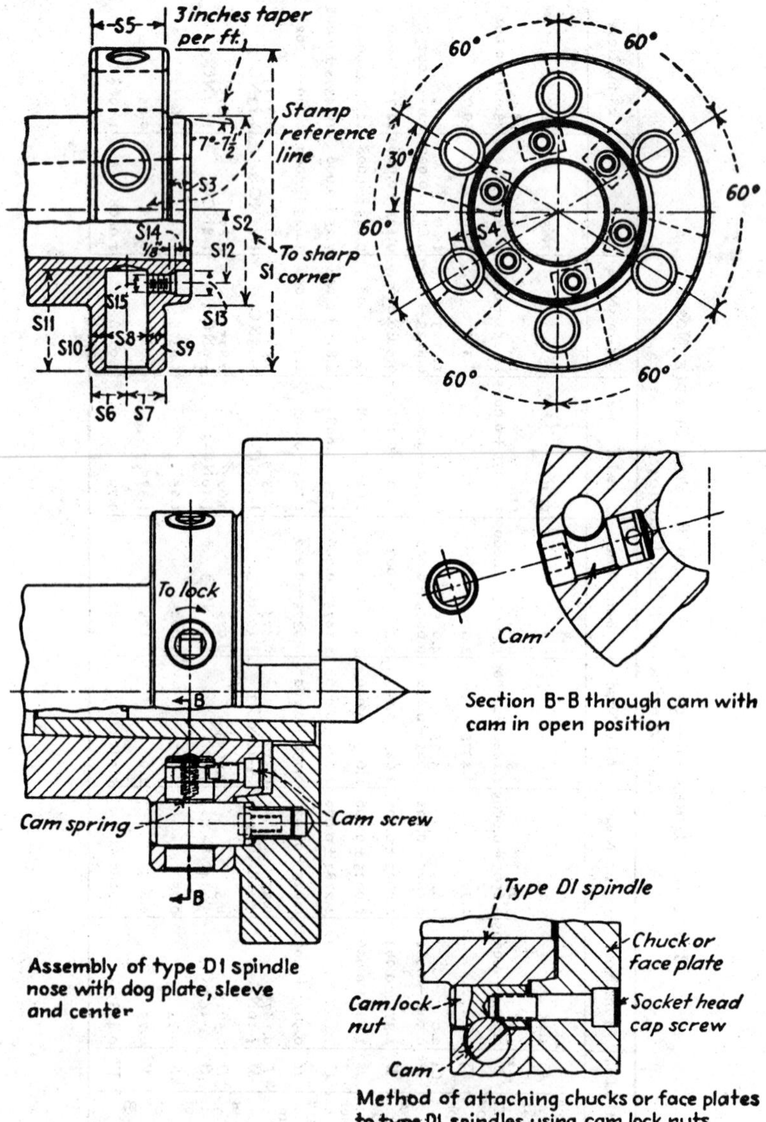

Assembly of type D1 spindle nose with dog plate, sleeve and center

Method of attaching chucks or face plates to type D1 spindles using cam lock nuts

FIG. 6. Type D1 spindle nose, also assembly with dog plate, sleeve, and center.

TABLE 11. ESSENTIAL DIMENSIONS OF TYPE D1 SPINDLE NOSES

Spindle Name	Dimension	5-in. Nose	6-in. Nose	8-in. Nose	11-in. Nose	15-in. Nose	20-in. Nose
Dia of spindle	S_1	5.75	7.125	8.875	11.75	15.875	21.5
Pilot dia	S_2	3.2505 +0.0005	4.1880 +0.0005	5.50075 +0.0005	7.75075 +0.0005	11.251 +0.001	16.251 +0.001
Length of pilot	S_3	0.5	0.5625	0.625	0.6875	0.75	0.8125
Radius of holes	S_4	2.0625 ±0.003	2.625 ±0.003	3.375 ±0.003	4.625 ±0.003	6.5 ±0.003	9.125 ±0.003
Min width of flange	S_5	1.5	1.75	2.0	2.375	2.75	3.25
Hole dia for cam	S_8	0.857 +0.002	1.000 +0.002	1.125 +0.002	1.250 +0.002	1.375 +0.002	1.625 +0.00:
Hole depth for cam	S_{11}	1.875 +0.016	2.250 +0.016	2.531 +0.016	2.9375 +0.016	3.3125 +0.016	3.6875 +0.016
Radius of cam screw	S_{12}	1.250 ±0.004	1.625 ±0.004	2.250 ±0.004	3.375 ±0.004	5.0625 ±0.004	7.5 ±0.004
Dia of counterbore	S_{13}	0.4531	0.5781	0.5781	0.5781	0.5781	0.5781
Depth of counterbore	S_{14}	0.3437	0.4062	0.4062	0.4062	0.4062	0.4062
Thread	S_{15}	$\frac{5}{16}$-18NC-3	$\frac{3}{8}$-16NC-3	$\frac{3}{8}$-16NC-3	$\frac{3}{8}$-16NC-3	$\frac{3}{8}$-16NC-3	$\frac{3}{8}$-16NC-

TABLE 12. DIMENSIONS OF SLEEVES FOR TYPE D1 SPINDLE NOSES

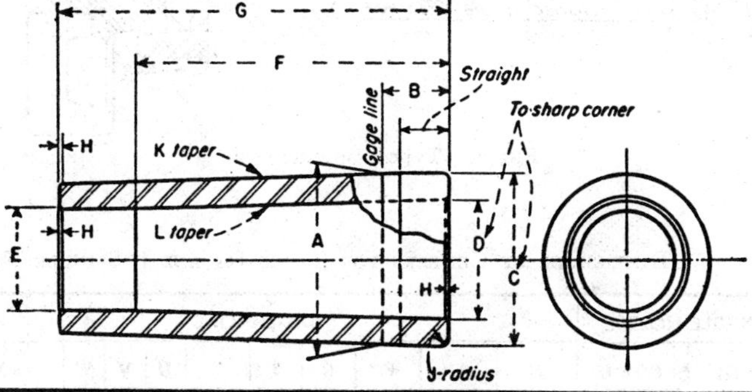

NOSE	A	B	C	D	E	F	G	H	J	K	L
2 in.	.938	$\frac{11}{16}$	−.005 .953	−.002 .475	+.005 .375	2	$3\frac{7}{8}$	$\frac{1}{32}$	$\frac{1}{32}$	#3 Am. std. taper	#1 Am. std. taper
3 in.	1.500	$\frac{1}{2}$	−.005 1.515	−.002 .700	+.005 .575	$2\frac{7}{16}$	5	$\frac{1}{16}$	$\frac{1}{16}$	#4½ Am. std. taper	#2 Am. std. taper
4 in. small sleeve	1.500	$\frac{11}{16}$	−.005 1.515	−.002 .938	+.005 .785	3	$5\frac{3}{16}$	$\frac{1}{16}$	$\frac{1}{16}$	#4½ Am. std. taper	#3 Am. std. taper
4 in. large sleeve	1.748	$\frac{11}{16}$	−.005 1.765	−.002 .938	+.005 .785	3	$5\frac{7}{8}$	$\frac{1}{16}$	$\frac{1}{16}$	#5 Am. std. taper	#3 Am. std. taper
5 in.	1.748	$\frac{7}{8}$	−.005 1.765	−.002 .938	+.005 .785	3	$6\frac{1}{16}$	$\frac{1}{16}$	$\frac{1}{16}$	#5 Am. std. taper	#3 Am. std. taper
6 in. small sleeve	1.748	$1\frac{1}{16}$	−.005 1.765	−.002 1.231	+.005 1.035	$3\frac{3}{4}$	$6\frac{1}{4}$	$\frac{1}{16}$	$\frac{1}{16}$	#5 Am. std. taper	#4 Am. std. taper
6 in. Large sleeve	2.000	$1\frac{1}{16}$	−.005 2.020	−.002 1.231	+.005 1.035	$3\frac{3}{4}$	$5\frac{13}{16}$	$\frac{1}{16}$	$\frac{1}{16}$	#200 Am. std. taper	#4 Am. std. taper

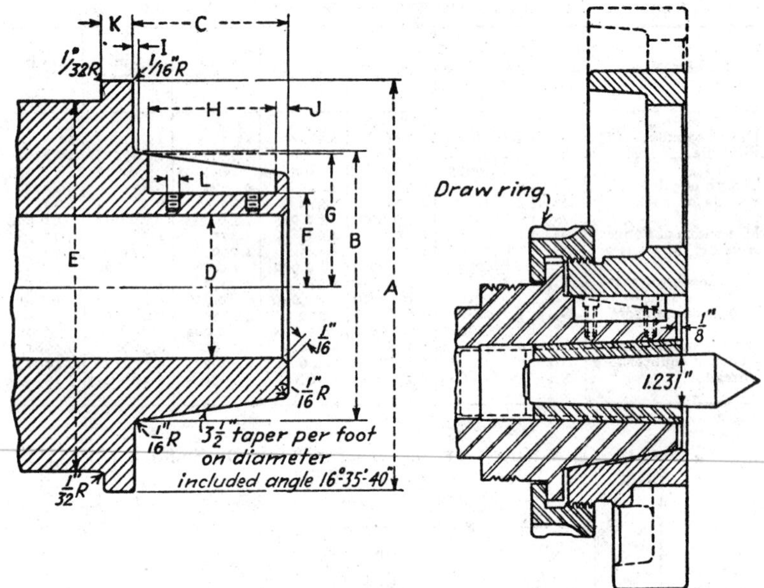

Fig. 7. Type L spindle nose.

Table 13. Dimensions of Centers for Types D1 and L Spindle Noses

Spindle nose		Dimensions									
Type D1	Type L	N	O	P	R	S	T	U	V	W	X
2 in.	....	.475	.493	27/64	23/64	2 1/16	2 29/32	5/16	1/64	1/64	#1 Am. std. taper
3 in.	....	.700	.728	5/8	9/16	2 1/2	3 13/16	7/16	1/32	1/64	#2 Am. std. taper
4 in.	LOO	.938	.982	27/32	7/8	3 1/16	4 29/32	5/8	1/32	1/32	#3 Am. std. taper
5 in.	LO	.938	.982	27/32	7/8	3 1/16	4 29/32	5/8	1/32	1/32	#3 Am. std. taper
6 in.	L1	1.231	1.276	1 3/32	7/8	3 7/8	5 31/32	3/4	1/16	1/32	#4 Am. std. taper
8 in.	L2	1.748	1.814	1 17/32	1 1/4	4 7/8	7 29/32	1 1/8	1/16	1/16	#5 Am. std. taper
11 in.	L3	2.494	2.559	2 5/32	1 1/4	6 7/8	10 17/32	1 3/4	1/16	3/32	#6 Am. std. taper

W = Maximum flat on point of nose.

TABLE 14. DIMENSIONS OF TYPE L SPINDLE NOSES

Spindle Name	Dimension	Loo Nose	Lo Nose	L1 Nose	L2 Nose	L3 Nose
Dia of flange	A	3.5	4.125	5.75	7.375	10.0
Dia of large end of pilot	B	2.75+0.002	3.25+0.002	4.125+0.002	5.25+0.002	6.5+0.002
Length of pilot	C	2.0	2.375	2.875	3.375	3.875
Max hole	D	1.5	2.0	2.25	3.0	3.875
Dia for draw-nut fit	E	2.870−0.005	3.495−0.005	4.745−0.005	6.37−0.005	8.995−0.005
Distance from center to bottom of keyseat	F	0.995−0.005	1.245−0.005	1.4325−0.005	1.87−0.005	2.245−0.005
Distance from center to top of key	G	1.37−0.007	1.62−0.007	2.0575−0.007	2.62−0.007	3.245−0.007
Length of keyseat	H	1.5+0.010	1.75+0.010	2.375+0.010	2.875+0.010	3.25+0.010
Width of flat	I	0.125	0.125	0.125	0.125	0.125
End of nose to end of keyseat	J	0.1875	0.25	0.25	0.25	0.25
Width of flange	K	0.2812	0.3125	0.4375	0.625	0.75
Thread	L	10-32NF-3	10-32NF-3	¼-20NC-3	5/16-18NC-3	5/16-18NC-3

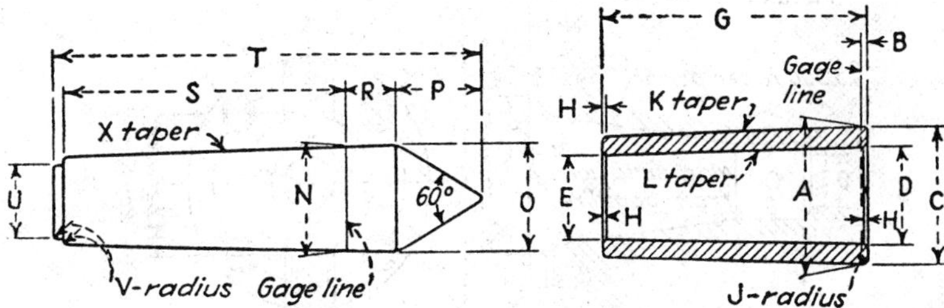

FIG. 8. Center for Types D1 and L spindle noses.

FIG. 9. Sleeve for Type L spindle nose.

TABLE 15. DIMENSIONS OF SLEEVES FOR TYPE L SPINDLE NOSES

SPINDLE NOSE	A	B	C	D	E	G	H	J	K	L
LOO	1.500	1/16	1.5032	—.002 .938	.7874	3	1/16	1/32	#4½ Am. std. taper	#3 Am. std. taper
LO	1.748	1/16	1.7513	—.002 .938	.7874	3	1/16	1/32	#5 Am. std. taper	#3 Am. std. taper
LI small sleeve	1.748	1/8	1.7546	—.002 1.231	1.0364	3¾	1/16	1/16	#5 Am. std. taper	#4 Am. std. taper
LI large sleeve	2.000	1/8	2.0078	—.002 1.231	1.0364	3¾	1/16	1/16	#200 Am. std. taper	#4 Am. std. taper
L2	2.500	1/8	2.5078	—.002 1.748	1.4980	4¾	1/16	1/16	#250 Am. std. taper	#5 Am. std. taper
L3	3.500	1/8	3.5078	—.002 2.494	2.1421	6¾	1/16	1/16	#350 Am. std. taper	#6 Am. std. taper

TABLE 16. STANDARD SPINDLE NOSES FOR MILLING MACHINES

	No. 30 (1 1/4 In.)	No. 40 (1 3/4 In.)	No. 50 (2 3/4 In.)	No. 60 (4 1/4 In.)
A	1 1/4	1 3/4	2 3/4	4 1/4
B	2.7493	3.4993	5.0618	8.7180
	2.7488	3.4988	5.0613	8.7175
C	0.692	1.005	1.568	2.381
	0.685	0.997	1.559	2.371
D	21/32	21/32	1 1/16	1 3/8
E	1/2	5/8	3/4	1 1/2
F	0.6255	0.6255	1.0006	1.000
	0.6252	0.6252	1.0002	0.999
G	5/16	5/16	1/2	1/2
H	1.315	1.819	2.819	4.819
	1.285	1.807	2.807	4.807
J	2.130	2.630	4.005	7.005
	2.120	2.620	3.955	6.995
K	3/8-16	1/2-13	5/8-11	3/4-10
L	2 7/8	3 7/8	5 1/2	8 5/8
M	5/8	13/16	1	1 1/4

TABLE 17. DIMENSIONS OF ARBORS FOR MILLING MACHINES

	No. 30 (1 1/4 In.)	No. 40 (1 3/4 In.)	No. 50 (2 3/4 In.)	No. 60 (4 1/4 In.)
N	1 1/4	1 3/4	2 3/4	4 1/4
O	27/64	17/32	7/8	1 7/64
P	41/64	15/16	1 1/2	2 9/32
Q	1/2-13	5/8-11	1-8	1 1/4-7
R	0.675	0.987	1.549	2.361
	0.673	0.985	1.547	2.359
S	13/16	1	1	1 3/4
T	1	1 1/8	1 3/4	2 1/4
U	2	2 15/16	3 1/2	4 1/4
V	2 3/4	3 3/4	5 1/8	8 5/16
W	1/16	1/16	1/8	1/8
X	0.640	0.890	1.390	2.400
	0.625	0.875	1.375	2.390
Y	0.630	0.630	1.008	1.008
	0.640	0.640	1.018	1.018

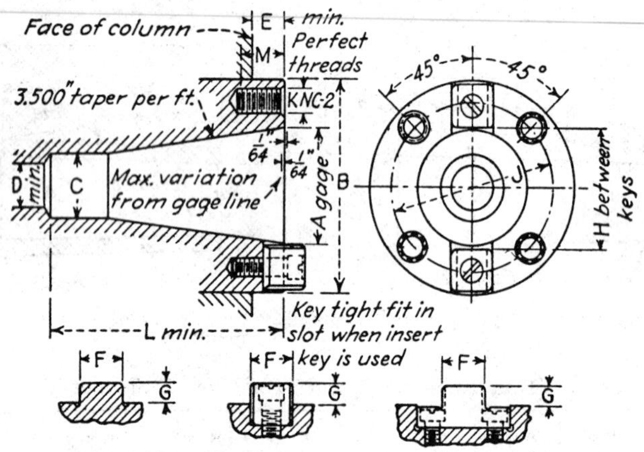

Standard spindle nose.

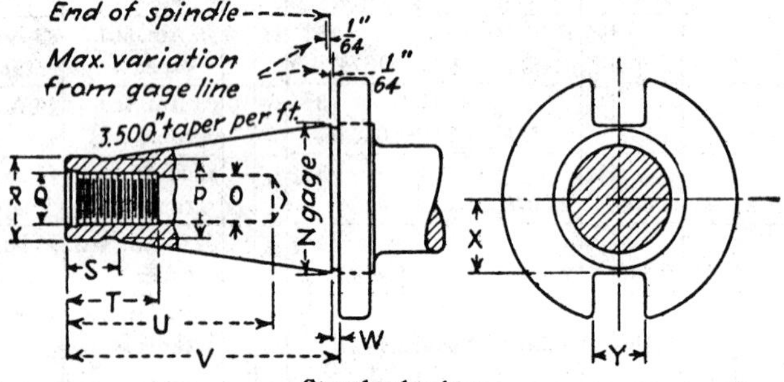

Standard arbor.

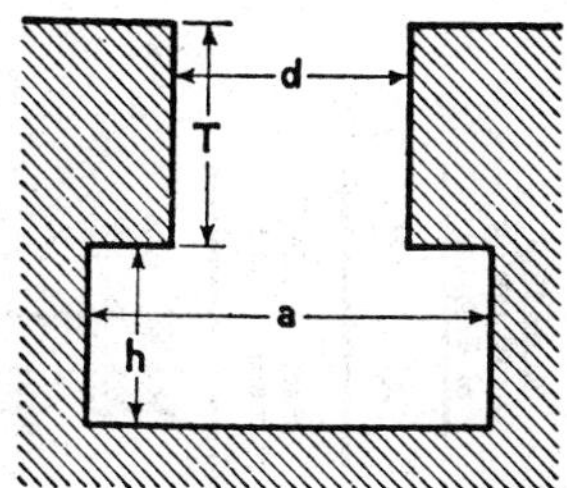

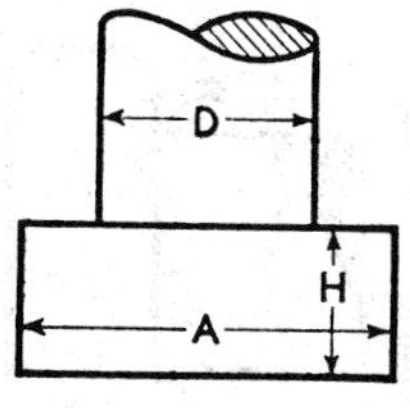

TABLE 18. DIMENSIONS OF T-SLOTS

Diameter of T-bolt	Width of Throat (d)	Depth of Throat (T)		Head Space Dimensions and Tolerances					
				Width (a)			Depth (h)		
		Maximum	Minimum	Maximum (Basic)	Tolerance (Minus)	Minimum	Maximum (Basic)	Tolerance (Minus)	Minimum
1/4	9/32	3/8	1/8	9/16	0.063	1/2	15/64	0.031	13/64
5/16	11/32	7/16	5/32	21/32	0.063	19/32	17/64	0.031	15/64
3/8	7/16	9/16	7/32	25/32	0.063	23/32	21/64	0.031	19/64
1/2	9/16	11/16	5/16	31/32	0.063	29/32	25/64	0.031	23/64
5/8	11/16	7/8	7/16	1 1/4	0.063	1 3/16	31/64	0.031	29/64
3/4	13/16	1 1/16	9/16	1 15/32	0.094	1 3/8	5/8	0.031	19/32
1	1 1/16	1 1/4	3/4	1 27/32	0.094	1 3/4	53/64	0.047	25/32
1 1/4	1 5/16	1 9/16	1	2 7/32	0.094	2 1/8	1 3/32	0.063	1 1/32
1 1/2	1 9/16	1 15/16	1 1/4	2 21/32	0.094	2 9/16	1 11/32	0.063	1 9/32

TABLE 19. DIMENSIONS OF T-BOLTS

Diameter of T-bolt (D)	Threads per Inch	Bolt Head Dimensions and Tolerances						
		Width across Flats (A)			Width across Corners	Height (H)		
		Maximum (Basic)	Tolerance (Minus)	Minimum		Maximum (Basic)	Tolerance (Minus)	Minimum
1/4	20	15/32	0.031	7/16	0.663	5/32	0.016	9/64
5/16	18	9/16	0.031	17/32	0.796	3/16	0.016	11/64
3/8	16	11/16	0.031	21/32	0.972	1/4	0.016	15/64
1/2	13	7/8	0.031	27/32	1.238	5/16	0.016	19/64
5/8	11	1 1/8	0.031	1 3/32	1.591	13/32	0.016	25/64
3/4	10	1 5/16	0.031	1 9/32	1.856	17/32	0.031	1/2
1	8	1 11/16	0.031	1 21/32	2.387	11/16	0.031	21/32
1 1/4	7	2 1/16	0.031	2 1/32	2.917	15/16	0.031	29/32
1 1/2	6	2 1/2	0.031	2 15/32	3.536	1 3/16	0.031	1 5/32

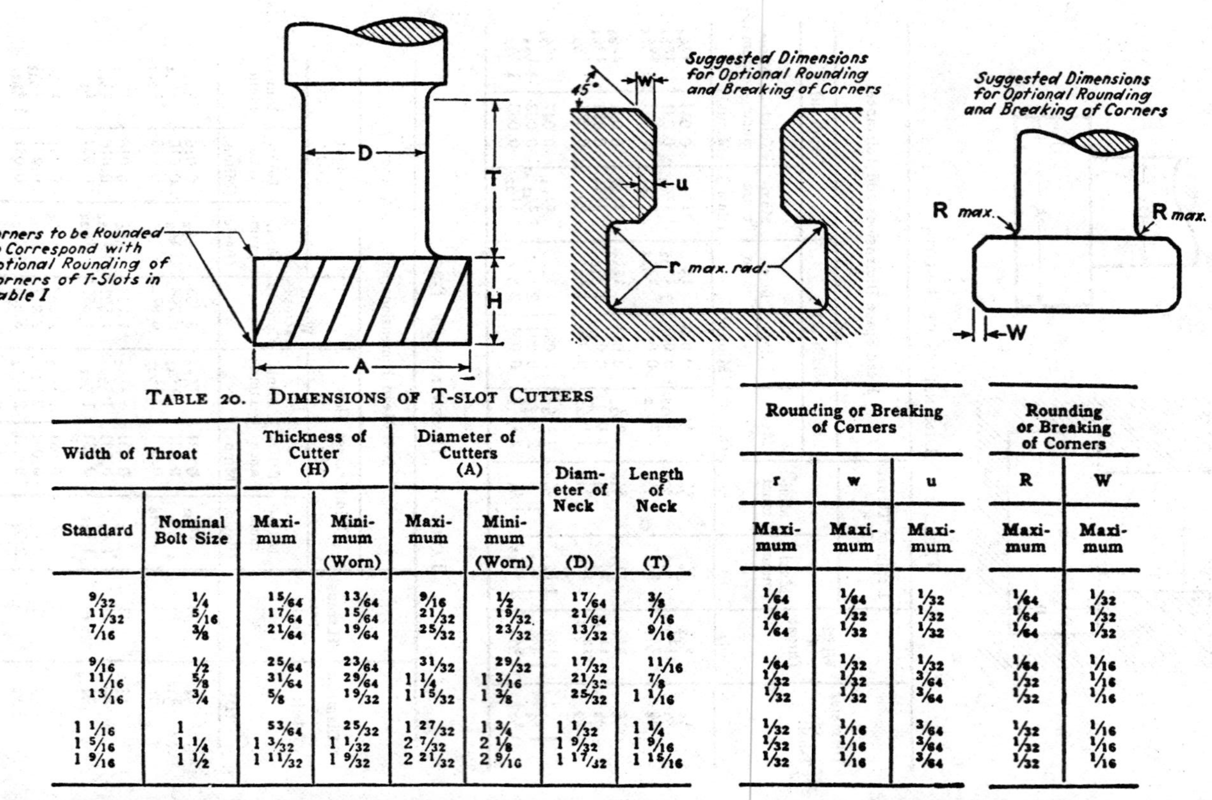

TABLE 20. DIMENSIONS OF T-SLOT CUTTERS

Width of Throat		Thickness of Cutter (H)		Diameter of Cutters (A)		Diameter of Neck	Length of Neck	Rounding or Breaking of Corners			Rounding or Breaking of Corners	
								r	w	u	R	W
Standard	Nominal Bolt Size	Maximum	Minimum (Worn)	Maximum	Minimum (Worn)	(D)	(T)	Maximum	Maximum	Maximum	Maximum	Maximum
9/32	1/4	15/64	13/64	9/16	1/2	17/64	3/8	1/64	1/64	1/32	1/64	1/32
11/32	5/16	17/64	15/64	21/32	19/32	21/64	7/16	1/64	1/32	1/32	1/64	1/32
7/16	3/8	21/64	19/64	25/32	23/32	13/32	9/16	1/64	1/32	1/32	1/64	1/32
9/16	1/2	25/64	23/64	31/32	29/32	17/32	11/16	1/64	1/32	1/32	1/64	1/16
11/16	5/8	31/64	29/64	1 1/4	1 3/16	21/32	7/8	1/32	1/32	3/64	1/32	1/16
13/16	3/4	5/8	19/32	1 15/32	1 3/8	25/32	1 1/16	1/32	1/32	3/64	1/32	1/16
1 1/16	1	53/64	25/32	1 27/32	1 3/4	1 1/32	1 1/4	1/32	1/16	3/64	1/32	1/16
1 5/16	1 1/4	1 3/32	1 1/32	2 7/32	2 1/8	1 9/32	1 9/16	1/32	1/16	3/64	1/32	1/16
1 9/16	1 1/2	1 11/32	1 9/32	2 21/32	2 9/16	1 17/32	1 15/16	1/32	1/16	3/64	1/32	1/16

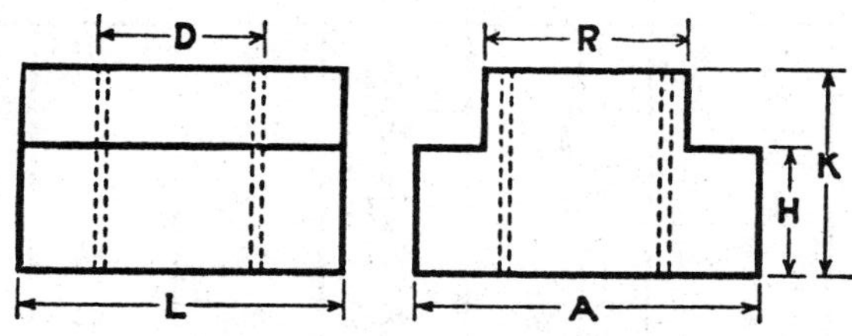

TABLE 21. DIMENSIONS OF T-NUTS

Tap for Stud (D)		Width of Throat T-slot	Width of Tongue (R)			Width of Nut (A)			Height of Nut (H)			Total Thickness, including Tongue (K)	Length of Nut (L)
Diameter	Threads per Inch		Maximum (Basic)	Tolerance (Minus)	Minimum	Maximum (Basic)	Tolerance (Minus)	Minimum	Maximum (Basic)	Tolerance (Minus)	Minimum		
1/4	20	11/32	0.330	0.010	0.320	9/16	0.031	17/32	3/16	0.016	11/64	9/32	9/16
5/16	18	7/16	0.418	0.010	0.408	11/16	0.031	21/32	1/4	0.016	15/64	3/8	11/16
3/8	16	9/16	0.543	0.010	0.533	7/8	0.031	27/32	5/16	0.016	19/64	17/32	7/8
1/2	13	11/16	0.668	0.010	0.658	1 1/8	0.031	1 3/32	13/32	0.016	25/64	5/8	1 1/8
5/8	11	13/16	0.783	0.010	0.773	1 5/16	0.031	1 9/32	17/32	0.031	1/2	25/32	1 5/16
3/4	10	1 1/16	1.033	0.015	1.018	1 11/16	0.031	1 21/32	11/16	0.031	21/32	1	1 11/16
1	8	1 5/16	1.273	0.015	1.258	2 1/16	0.031	2 1/32	15/16	0.031	29/32	1 5/16	2 1/16
1 1/4	7	1 9/16	1.523	0.015	1.508	2 1/2	0.031	2 15/32	1 3/16	0.031	1 5/32	1 5/8	2 1/2

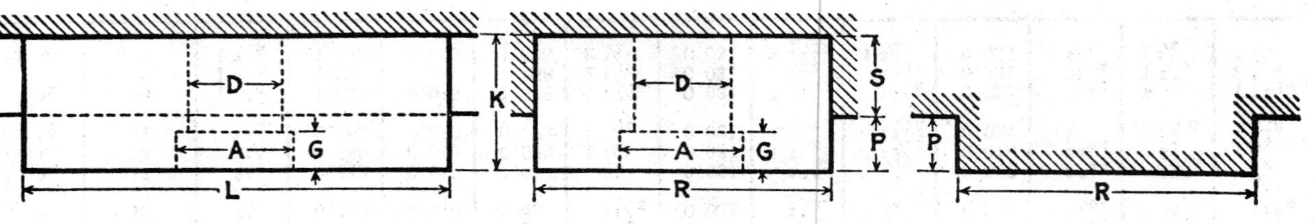

TABLE 22. INSERTED OR SOLID PLAIN TONGUES AND TONGUE SEATS FOR SINGLE-WIDTH T-SLOTS

Diameter of T-bolt	Tongue Dimensions: Width (R)	Tongue Dimensions: Length (L)	Tongue Dimensions: Projection (P)	Depth of Seat (S)	Total Thickness (K)	Screw Dimensions: Diameter of Screw (D)	Screw Dimensions: Number of Screw	Screw Dimensions: Threads per Inch	Screw Dimensions: Diameter of Head (A)	Screw Dimensions: Thickness of Head (G)
1/4	9/32	3/8	3/32	1/8	7/32	0.125	5	40	0.196	0.081
5/16	11/32	15/32	1/8	5/32	9/32	0.164	8	32	0.260	0.107
3/8	7/16	9/16	1/8	3/16	5/16	0.190	10	24	0.303	0.124
1/2	9/16	3/4	1/8	7/32	11/32	1/4	......	20	0.375	0.130
5/8	11/16	15/16	1/8	1/4	3/8	1/4	......	20	0.375	0.130
3/4	13/16	1 1/8	5/32	9/32	7/16	5/16	......	18	0.438	0.150
1	1 1/16	1 1/2	7/32	11/32	9/16	3/8	......	16	0.500	0.170
1 1/4	1 5/16	1 7/8	1/4	3/8	5/8	3/8	......	16	0.500	0.170
1 1/2	1 9/16	2 1/4	5/16	7/16	3/4	1/2	......	13	0.625	0.210

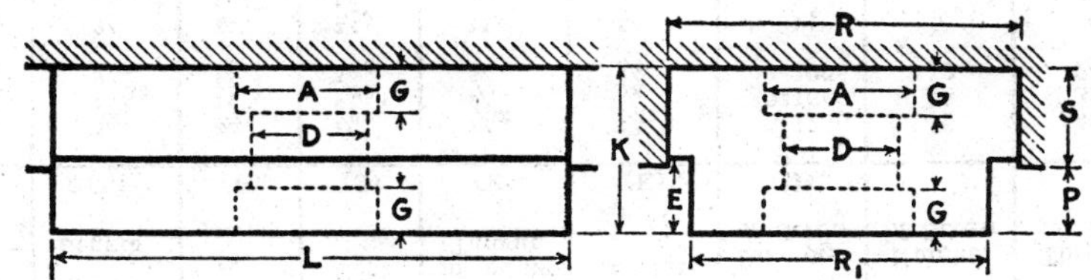

TABLE 23. REVERSIBLE TONGUES AND TONGUE SEATS FOR SLOTS FOR TWO SIZES OF T-BOLTS

Diameter of T-bolt		Tongue Dimensions						Depth of Seat	Total Thickness, including Tongue	Height of Shoulder	Thickness of Land	Screw Dimensions				
		Width				Length	Projection									
Small	Large	(R_1)	(R)	(R_3)	(R_2)	(L)	(P)	(S)	(K)	(E)	(C)	Diameter of Screw (D)	Number of Screw	Threads per Inch	Diameter of Head (A)	Thickness of Head (G)
1/4	5/16	9/32	11/32	5/16	1/4	15/32	1/8	3/16	5/16	1/8	1/16	0.164	8	32	0.260	0.107
5/16	3/8	11/32	7/16	3/8	5/16	9/16	1/8	7/32	11/32	9/64	1/16	0.190	10	24	0.303	0.124
3/8	1/2	7/16	9/16	1/2	3/8	3/4	1/8	1/4	3/8	5/32	1/16	0.250	......	20	0.375	0.130
1/2	5/8	9/16	11/16	5/8	1/2	15/16	1/8	9/32	13/32	5/32	3/32	1/4	......	20	0.375	0.130
5/8	3/4	11/16	13/16	3/4	5/8	1 1/8	5/32	5/16	15/32	3/16	3/32	5/16	......	18	0.438	0.150
3/4	1	13/16	1 1/16	1	3/4	1 1/2	7/32	3/8	19/32	1/4	3/32	3/8	......	16	0.500	0.170
1	1 1/4	1 1/16	1 5/16	1 1/4	1	1 7/8	1/4	7/16	11/16	9/32	1/8	3/8	......	16	0.500	0.170
1 1/4	1 1/2	1 5/16	1 9/16	1 1/2	1 1/4	2 1/4	5/16	1/2	13/16	11/32	1/8	1/2	......	13	0.625	0.210

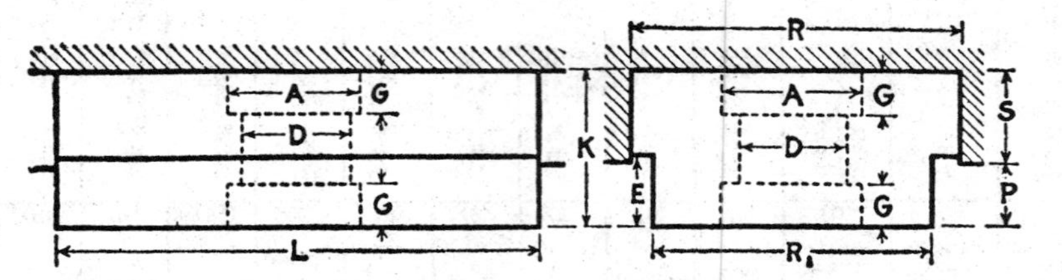

TABLE 24. REVERSIBLE TONGUES AND TONGUE SEATS FOR T-SLOTS OF TWO WIDTHS USING THE SAME SIZE OF T-BOLT

Diameter of T-bolt	Tongue Dimensions				Depth of Seat	Total Thickness, including Tongue	Height of Shoulder	Screw Dimensions				
	Width		Length	Projection				Diameter of Screw	Number of Screw	Threads per Inch	Diameter of Head	Thickness of Head
	(R_1)	(R)	(L)	(P)	(S)	(K)	(E)	(D)			(A)	(G)
1/4	1/4	9/32	3/8	3/32	1/8	7/32	3/32	0.125	5	40	0.196	0.081
5/16	5/16	11/32	15/32	1/8	5/32	9/32	1/8	0.164	8	32	0.260	0.107
3/8	3/8	7/16	9/16	1/8	3/16	5/16	9/64	0.190	10	24	0.303	0.124
1/2	1/2	9/16	3/4	1/8	7/32	11/32	5/32	1/4	......	20	0.375	0.130
5/8	5/8	11/16	15/16	1/8	1/4	3/8	5/32	1/4	......	20	0.375	0.130
3/4	3/4	13/16	1 1/8	5/32	9/32	7/16	3/16	5/16	......	18	0.438	0.150
1	1	1 1/16	1 1/2	7/32	11/32	9/16	1/4	3/8	......	16	0.500	0.170
1 1/4	1 1/4	1 5/16	1 7/8	1/4	3/8	5/8	9/32	3/8	......	16	0.500	0.170
1 1/2	1 1/2	1 9/16	2 1/4	5/16	7/16	3/4	11/32	1/2	......	13	0.625	0.210

Section 42

MACHINE-TOOL INSPECTION

SECTION 42

MACHINE-TOOL INSPECTION

ENGINE-LATHE ACCURACY

The following standards of accuracy for toolroom and engine lathes describe the methods for testing 25 matters and the tolerances that apply.[1]

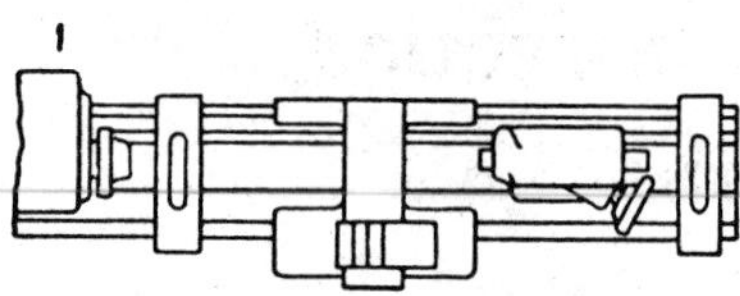

Bed level-transverse direction

1. When using level, both readings shall be:

Toolroom lathes—0.0005 in. in 12 in.
Engine lathes:
12- to 18-in. sizes—0.0005 in. in 12 in.
20- to 36-in. sizes—0.001 in. in 12 in.

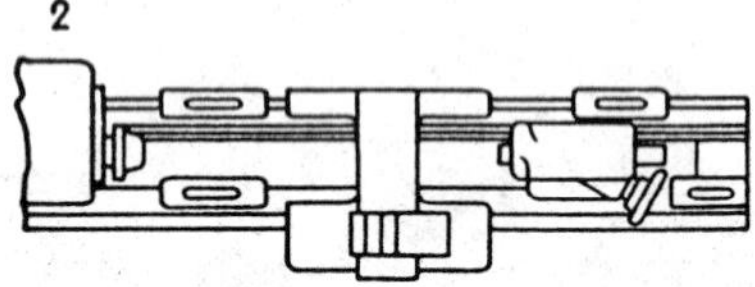

Bed level-longitudinal direction

2. When using precision level, maximum reading:

Toolroom lathes—0.0005 in. in 12 in.
Engine lathes:
12- to 18-in. sizes—0.0005 in. in 12 in.
20- to 36-in. sizes—0.001 in. in 12 in.

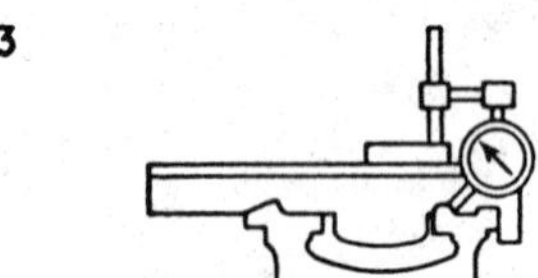

Tailstock way alignment

3. Maximum reading along bed:

Toolroom lathes—0.0005 in. in 48 in.
Engine lathes:
12- to 18-in. sizes—0.00075 in. in 48 in.
20- to 36-in. sizes—0.001 in. in 48 in.

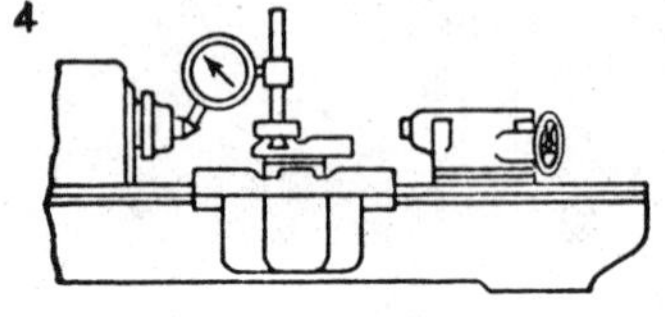

Spindle center runout

4. Total indicator reading:

Toolroom lathes—0 to 0.0004 in.
Engine lathes:
12- to 18-in. sizes—0 to 0.0005 in.
20- to 36-in. sizes—0 to 0.00075 in.

[1] Source: American Standard B5.16–1941.

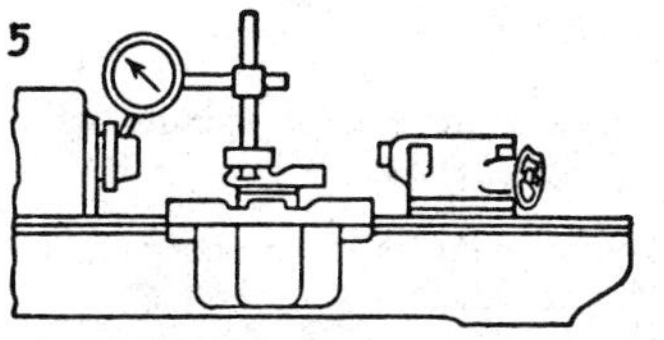

Spindle nose runout

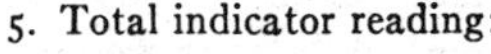

5. Total indicator reading:

Toolroom lathes—0 to 0.0003 in.
Engine lathes:
12- to 18-in. sizes—0 to 0.0004 in.
20- to 36-in. sizes—0 to 0.0006 in.

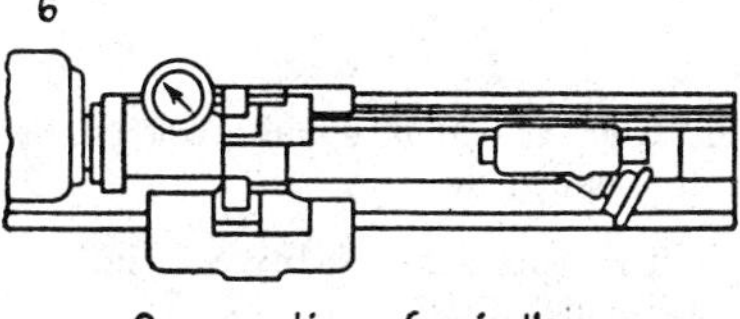

Cam action of spindle

6. Total indicator reading with indicator on rear side of test plate:

Toolroom lathes—0 to 0.0003 in.
Engine lathes:
12- to 18-in. sizes—0 to 0.0005 in.
20- to 35-in. sizes—0 to 0.00075 in.

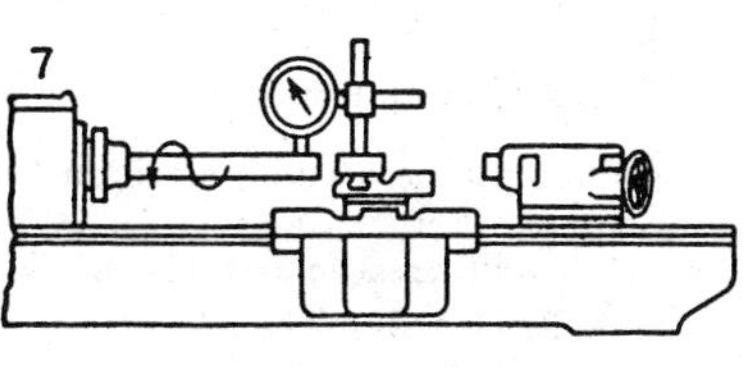

Spindle taper runout

7. Total indicator reading at end of:

Lathes	12-in. Test Bar	Spindle Nose
Toolroom......	0 to 0.0006	0.0003
Engine:		
12- to 18-in..	0 to 0.0008	0.0004
20- to 36-in..	0 to 0.00125	0.0006

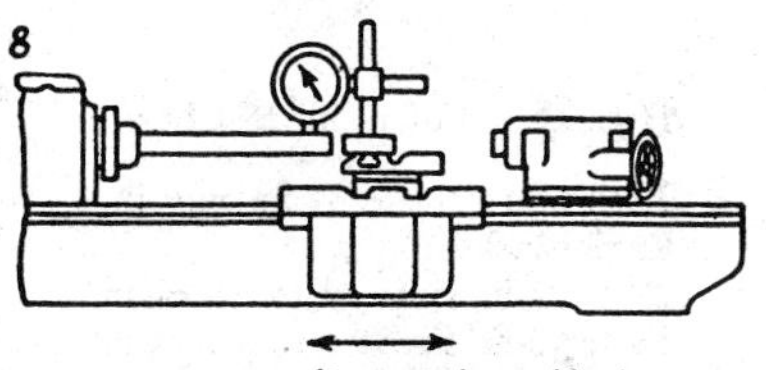

Headstock alignment-vertical

8. High at end of 12-in. test bar:

Toolroom lathes—0 to 0.0005 in.
Engine lathes:
12- to 18-in. sizes—0 to 0.001 in.
20- to 36-in. sizes—0 to 0.001 in.

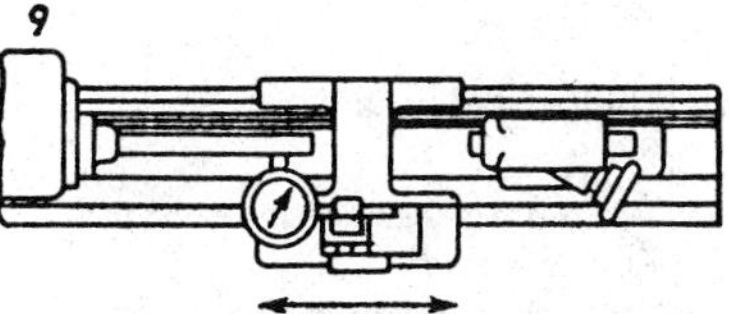

Headstock alignment horizontal

9. At end of 12-in. test bar:

Toolroom lathes—0 to ±0.0003 in.
Engine lathes:
12- to 18-in. sizes—0 to ±0.0005 in.
20- to 36-in. sizes—0 to ±0.0008 in.

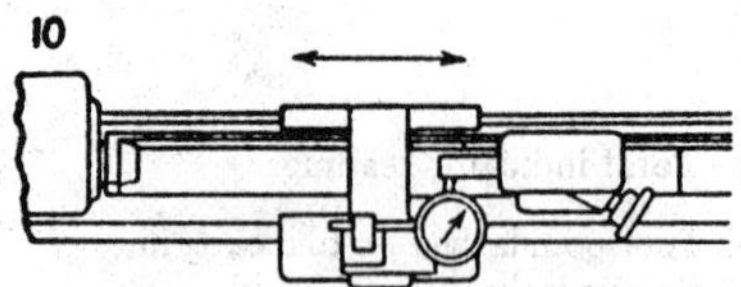

Tailstock spindle alignment-horizontal

10. Forward at end of spindle when fully extended:

Toolroom lathes—0 to 0.0005 in.
Engine lathes:
12- to 36-in. sizes—0 to 0.0005 in.

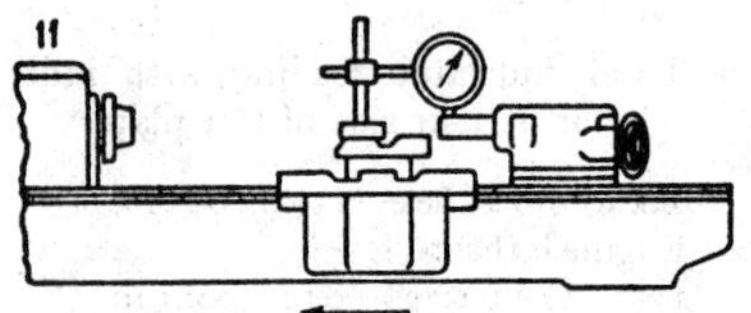

Tailstock spindle alignment-vertical

11. High at end of spindle when fully extended:

Toolroom lathes—0 to 0.0005 in.
Engine lathes:
12- to 36-in. sizes—0 to 0.0005 in.

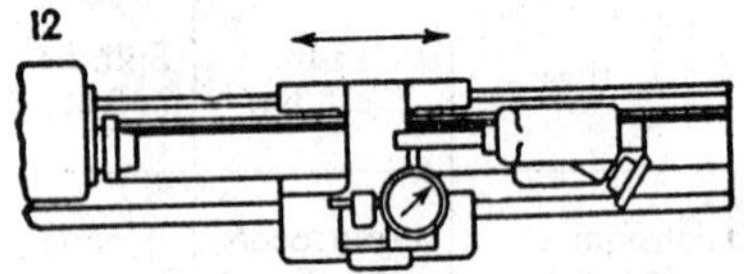

Tailstock taper alignment-horizontal

12. At end of 12-in. test bar:

Toolroom lathes—0 to ±0.0005 in.
Engine lathes:
12- to 18-in. sizes—0 to ±0.0008 in.
20- to 36-in. sizes—0 to ±0.0015 in.

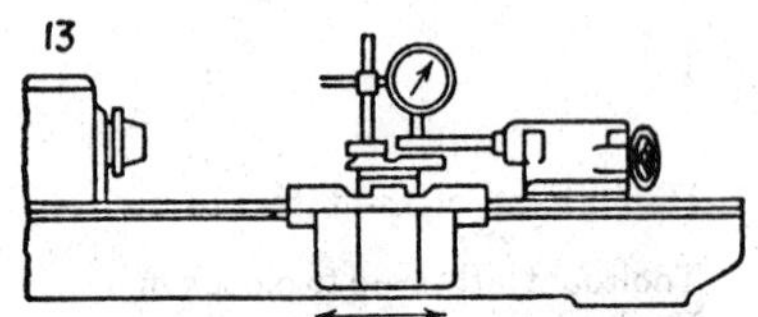

Tailstock taper alignment-vertical

13. High at end of 12-in. test bar:

Toolroom lathes—0 to 0.0005 in.
Engine lathes:
12- to 18-in. sizes—0 to 0.0008 in.
20- to 36-in. sizes—0 to 0.0015 in.

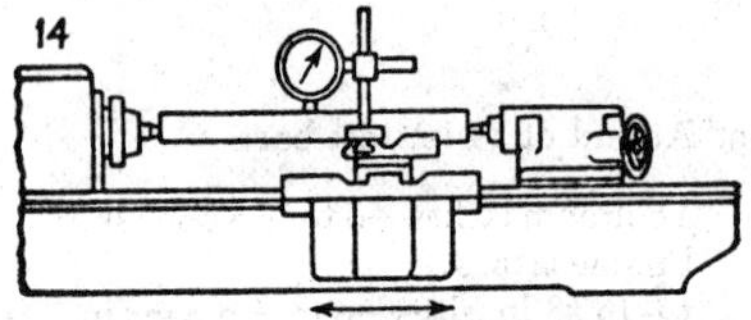

Vertical alignment of head & tail center

14. High at tailstock:

Toolroom lathes—0 to 0.0008 in.
Engine lathes:
12- to 18-in. sizes—0 to 0.001 in.
20- to 36-in. sizes—0 to 0.0015 in.

15

Readings for base length bed taken with lead screw stationary; add .001 inch for each additional 4 feet of bed length

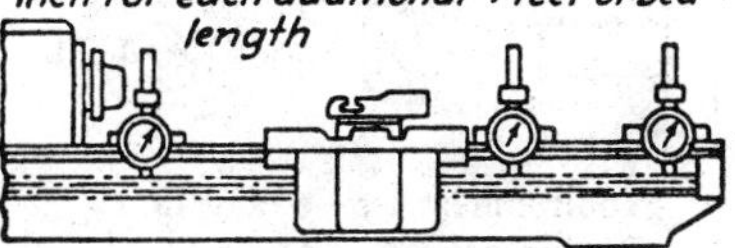

Lead screw alignment

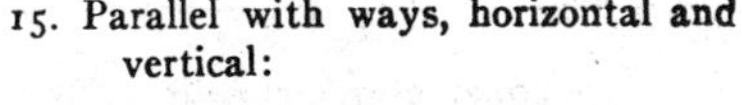

15. Parallel with ways, horizontal and vertical:

Toolroom and to 18-in. engine—0 to 0.004 in.
20- to 36-in. engine—0 to 0.006 in.

16

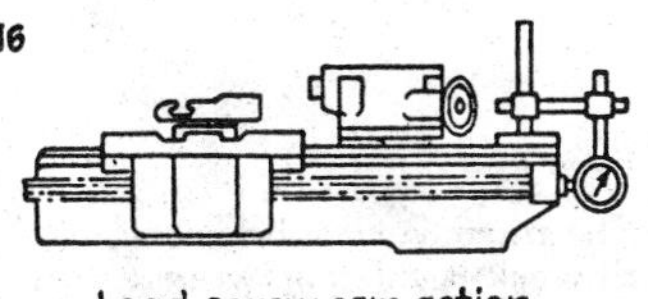

Lead screw cam action

16.

Toolroom lathes—0.0003 in. max
Engine lathes:
12- to 18-in.—0.0004 in. max
20- to 36-in.—0.0005 in. max

17

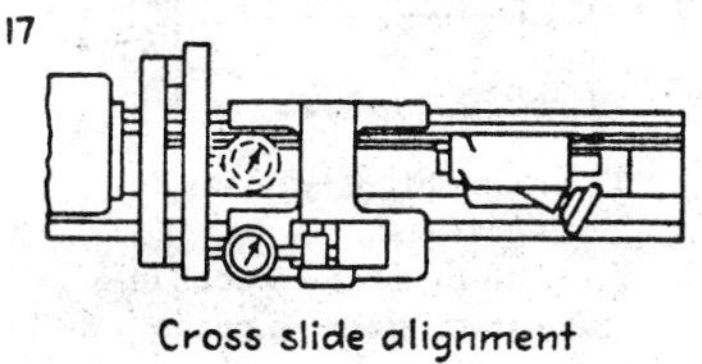

Cross slide alignment

17. To face hollow or concave on 12-in. dia

Toolroom lathes—0 to 0.005 in.
Engine lathes:
12- to 18-in.—0 to 0.001 in.
20- to 36-in.—0 to 0.001 in.

18

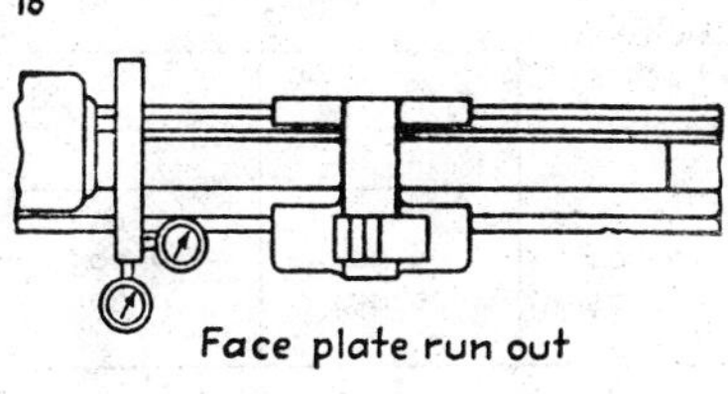

Face plate run out

18.

Lathes	On Dia, Zero to	On Face, Zero to
Toolroom.	0.0005	0.001
Engine:		
12- to 18-in. .	0.001	0.0015
20- to 36-in. .	0.0015	0.002

19

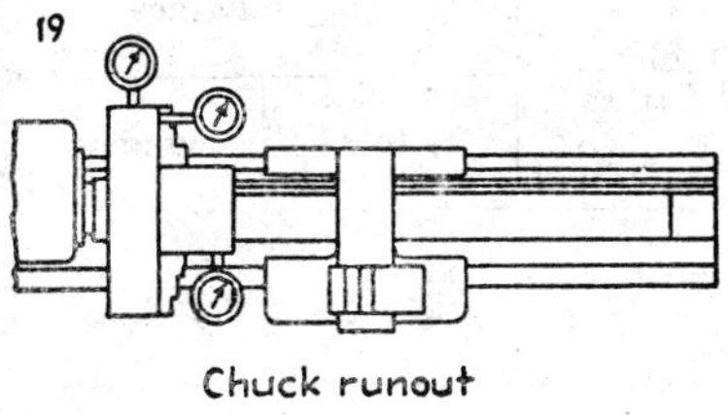

Chuck runout

19. Toolroom and engine lathes to 18-in.—face and periphery, 0.003; face of steps, 0.003; bar test 3 in. from end of jaw (bar dia same as hole), 0.003. Same items 0.004 for 20- to 36-in. lathes.

20

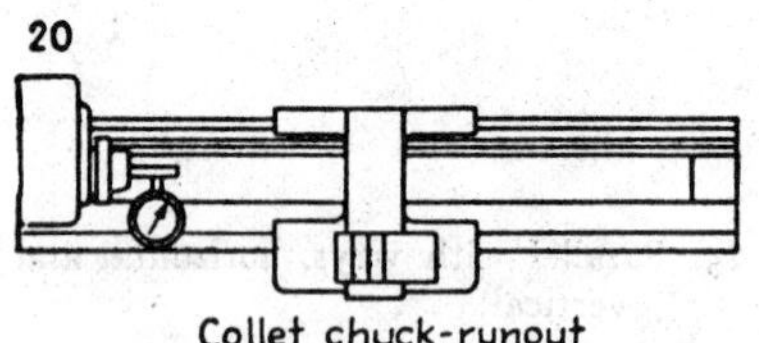

At 1 in. from spindle, 0 to 0.001 in. in all sizes.

21 Lathe must turn round with work mounted in chuck

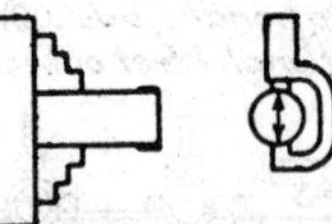

Toolroom lathes—0.0003 in.
Engine lathes:
12- to 18-in. sizes—0.0004 in.
20- to 36-in. sizes—0.0008 in.

22 Lathe must turn cylindrical with work mounted in chuck

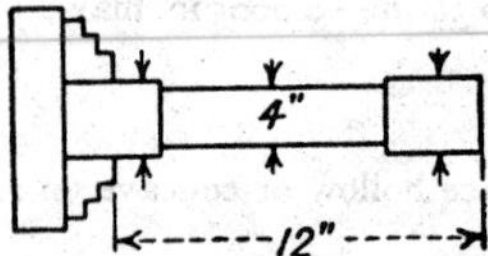

Toolroom lathes—0.0008 in.
Engine lathes:
12- to 18-in. sizes—0.0015 in.
20- to 36-in. sizes—0.002 in.

23 Lathe must turn cylindrical with work mounted between centers

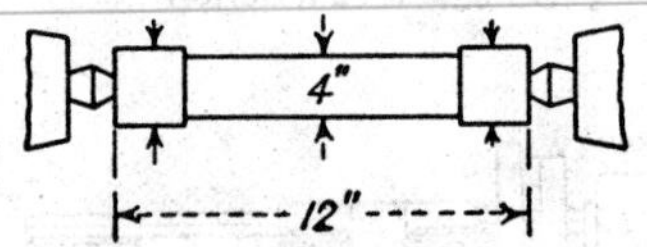

Toolroom lathes—0.0004 in.
Engine lathes:
12- to 18-in. sizes—0.0008 in.
20- to 36-in. sizes—0.001 in.

Backlash	Toolroom Lathes	Engine Lathes	
		12 to 18	20 to 36
On cross-feed screw	0.004	0.004	0.005
On compound-rest screw	0.004	0.004	0.005

Lead Screw	Toolroom Lathes	Engine Lathes	
		12 to 18	20 to 36
Lead per ft	±0.001	±0.0015	±0.002
Lead in any 4 in	±0.0004	±0.0005	±0.0007

Part 10

POWER-TRANSMISSION EQUIPMENT

Section 43

CHAIN, V-BELTS, AND MOTORS

SECTION 43

CHAIN, V-BELTS, AND MOTORS

TYPES OF CHAIN

Three types of chain are commonly found in the shop: roller, silent, and block. The first two are classified as power-transmission chain. Block chain is sometimes used for power transmission at slow speeds or light loads, but its major use is in conveyor applications. For transmission of power at average to high-speed conditions, it is essential that silent or roller chains be used.

Silent-chain drives have been standardized in pitches from 3/8 to 2 in. and tables of horsepower capacity have been established per inch of width, when running at small-sprocket speeds up to 6000 rpm in the case of 3/8-in.-pitch chain and up to 900 rpm for 2-in.-pitch chain.

Roller chain consists of a series of alternately assembled roller links and pin links, in which the pins articulate inside bushings, and the rollers are free to turn

TABLE 1. GENERAL ROLLER-CHAIN DIMENSIONS, INCHES

Standard Series							Heavy Series
Standard Chain No.	Pitch P	Max Roller Dia D_r	Width W	Pin Dia D_p	Link Plate Thickness LPT	Measuring Load, Lb	Link Plate Thickness LPT
25	1/4	0.130*	1/8	0.0905	0.030	18	
35	3/8	0.200*	3/16	0.141	0.050	18	
40	1/2	5/16	5/16	0.156	0.060	31	
50	5/8	0.400	3/8	0.200	0.080	49	
60	3/4	15/32	1/2	0.234	0.094	70	0.125
80	1	5/8	5/8	0.312	0.125	125	0.156
100	1 1/4	3/4	3/4	0.375	0.156	195	0.187
120	1 1/2	7/8	1	0.437	0.187	281	0.219
140	1 3/4	1	1	0.500	0.219	383	0.250
160	2	1 1/8	1 1/4	0.562	0.250	500	0.281
180	2 1/4	1 13/32	1 13/32	0.687	0.281	633	0.375
200	2 1/2	1 9/16	1 1/2	0.781	0.312	781	0.375
240	3	1 7/8	1 7/8	0.937	0.375	1125	0.500

Max pin dia = nominal pin dia + 0.0005.
Min hole in bushing = nominal pin dia + 0.0015.
Max width of roller link = nominal width of chain + 2.12 nominal link plate thickness.
Min distance between pin link plates = max width of roller link + 0.002.

* Without rollers.

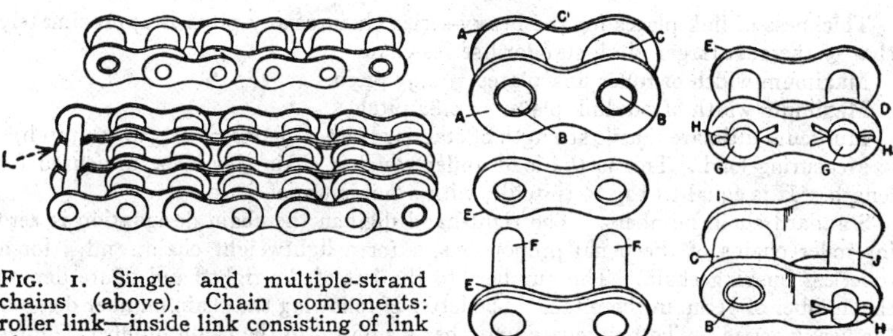

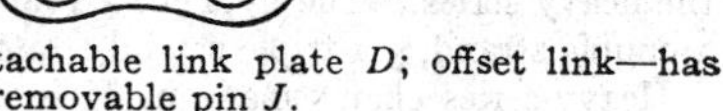

FIG. 1. Single and multiple-strand chains (above). Chain components: roller link—inside link consisting of link plates A, bushings B, and rollers C; pin link—two plates E assembled to pins F; connecting link—distinguished by detachable link plate D; offset link—has removable pin J.

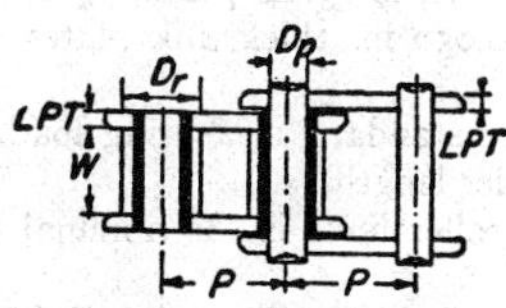

FIG. 2. Chain assembly, where P = pitch; D_r = roller dia; LPT = link plate thickness; W = chain width between link plates; D_p = pin dia between pin link plates.

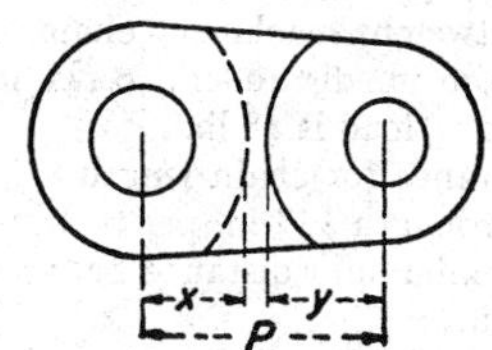

FIG. 3. Offset link plate.

on the bushings. Pins and bushings are press-fitted in their respective side plates. Standard chain numbers range from 25 to 240, in pitches from ¼ to 3. Horsepower-rating tables are based on a 15,000-hr life expectancy under a service factor of 1.

By revision of the tooth profiles of chain sprockets, these may now be used interchangeably with chain made by various manufacturers. Furthermore, data have been published that now make it possible for the machinery manufacturer to cut his own sprockets for original equipment or replacement purposes.

ROLLER CHAIN

Roller chain is a series of alternately assembled roller links and pin links in which the pins articulate inside the bushings and the rollers are free to turn on the bushings. Pins and bushings are press-fitted in their respective side plates. Chain may be single-strand having one row of roller links, or multiple-strand having more than one row of roller links and in which center plates L (Fig. 1) are located between the strands of roller links.

Chain Proportions and Designations (Table 1). Roller dia is approximately ⅝ pitch.

Chain width. The width is defined as the minimum distance between the roller link plates. This width is the nearest binary fraction to ⅝ pitches.

Pin dia is approximately 5⁄16 pitch or ½ of the roller dia.

Thickness of link plates for the of standard series chains = 0.125 × pitch.

Thickness of link plates for the heavy-series chain of any pitch is approximately that of the next larger pitch standard-series chain.

Maximum width of roller link plates = 0.95 pitch.

Maximum width of pin link plates = 0.82 pitch.

Minimum ultimate tensile strength of standard-series chains = $12{,}500 \times (\text{pitch})^2$.

Measuring load. This is the load under which a chain should be measured for length. It is equal to $125 \times (\text{pitch})^2$, with a minimum of 18 lb.

Standard chain numbers. The right-hand digit in the chain designation is zero for roller chains of the usual proportions, 1 for a lightweight chain, and 5 for a rollerless bushing chain. The numbers to the left of the right-hand figure denote the number of ⅛ in. in the pitch. The letter H following the chain number denotes the heavy series. The hyphenated number 2 suffixed to the chain number denotes a double strand, 3 a triple strand, 4 a quadruple strand chain, and so forth.

Heavy-series chains made in ¾ in. and larger pitches differ from the standard series in thickness of link plates. Their value is only in the acceptance of higher tensile or jerk loads during operation at lower speeds.

Lightweight machinery chain, designated as No. 41, is ½ in. pitch; ¼ in. wide; has 0.306 in. dia rollers; 0.141 in. pin dia; and 0.050 in. thick link plates. The measuring load is 18 lb.

Tolerance for chain length. New chains, under standard measuring load, may be over length 1/64 in. per ft, but must not be under length.

The minimum distance between link plates of roller links is the nominal width of chain.

Standard offset links are made to accommodate chains having roller link plates with a maximum width, after being beveled, equal to 95% of the pitch, and pin link plates with a maximum width, after being beveled, equal to 82% of the pitch. Therefore the standard minimum values of x and y (Fig. 3) are

$$x\ (\text{min}) = 0.41P + 0.008$$
$$y\ (\text{min}) = 0.475P + 0.008$$

where P = chain pitch.

Selection of Chain. For each pitch of chain there is a maximum sprocket rpm (Table 2) which should not be exceeded without consulting the chain manufacturer. Because of this, it is well to begin the selection of the drive with consideration of the fastest rotating shaft of the drive. Next, refer to Table 2 on maximum rpm and select the maximum pitch chain which will operate at the speed of the fastest turning shaft. In the selection it is always advisable to have at least 17 teeth in the smaller sprocket and preferably 24.

Check the sprocket for accommodation of the shaft and keyway. The bore should not exceed approximately seven-tenths of the maximum hub diameter.

The life of a roller chain drive can be affected adversely by any one of the following causes:

1. Small sprockets operating at high speed.
2. Insufficient and improper lubrication.
3. Inaccurate sprockets.
4. Misalignment of sprockets.
5. Overtensioning of chain during adjustment of centers or adjustment of idler sprocket.

Chain Lengths for Two-wheel Drives. A number of formulas have been developed to obtain the length of chain on two-wheel drives, but usually they involve more labor than that of making a careful layout to scale, or double size, and stepping

TABLE 2. RECOMMENDED MAXIMUM RPM OF SPROCKETS FOR AMERICAN STANDARD ROLLER CHAIN

Chain No..	25	35	41	40	50	60	80	100	120	140	160	180	200	240
Pitch......	$\frac{1}{4}$	$\frac{3}{8}$	$\frac{1}{2}$	$\frac{1}{2}$	$\frac{5}{8}$	$\frac{3}{4}$	1	$1\frac{1}{4}$	$1\frac{1}{2}$	$1\frac{3}{4}$	2	$2\frac{1}{4}$	$2\frac{1}{2}$	3
Teeth														
11	4310	2260	1020	1690	1220	920	580	415	325	235	200	165	145	110
12	4960	2590	1170	1940	1400	1050	670	475	375	270	230	190	165	125
13	5540	2900	1310	2180	1570	1180	750	535	415	305	260	215	185	140
14	6070	3170	1430	2380	1720	1290	820	585	455	335	280	235	205	155
15	6530	3420	1540	2560	1850	1390	880	630	490	360	305	255	220	165
16	6940	3630	1630	2720	1960	1480	935	670	520	380	325	270	235	175
17	7290	3810	1720	2860	2060	1550	985	700	550	400	340	285	245	185
18	7590	3970	1790	2980	2150	1610	1020	730	570	415	355	295	255	195
19	7840	4100	1850	3080	2220	1670	1060	755	590	430	365	305	265	200
20	8050	4210	1890	3160	2280	1720	1090	775	605	440	375	315	270	205
21	8230	4300	1940	3230	2330	1750	1110	790	620	450	385	320	280	210
22	8370	4380	1970	3290	2370	1780	1130	805	630	460	390	325	280	215
23	8480	4430	2000	3330	2400	1800	1150	815	640	465	395	330	285	215
24	8560	4480	2020	3360	2420	1820	1160	825	645	470	400	330	290	220
25	8610	4510	2030	3380	2440	1830	1160	830	650	475	400	335	290	220
30	8580	4490	2020	3370	2430	1830	1160	825	645	470	400	335	290	220
35	8200	4290	1930	3220	2320	1740	1110	790	615	450	380	320	275	210
40	7580	3970	1780	2970	2140	1610	1020	730	570	415	355	295	255	195
45	6820	3570	1600	2670	1930	1450	920	655	515	375	320	265	230	175
50	5950	3110	1400	2330	1680	1270	805	575	450	325	275	230	200	150
55	5010	2620	1180	1970	1420	1070	675	480	375	275	235	195	170	125
60	4020	2100	950	1580	1140	860	545	390	305	220	185	155	135	100

off with accurately spaced dividers. In any case, it is recommended that such layouts be given careful study, to provide proper ratios, ample adjustment, etc., and that they be submitted to the chain manufacturer for recheck and approval.

The following method may be used for accurately determining the chain lengths of two-wheel drives (after finding the approximate center distance, Fig. 4):

Subtract diameter of smaller sprocket from diameter of larger sprocket, and divide the difference by twice the center distance (in inches). Find in column *A*, Table 3, the value next larger than this result, to use as an index figure. The corresponding values (in same line) from columns *B*, *C*, and *D* should be substituted in the following formula:

Length of chain equals the sum of value *B* multiplied by center distance plus value *C* multiplied by number of teeth in smaller wheel, and by pitch, plus value *D* multiplied by number of teeth in larger wheel, and by pitch.

Horsepower Ratings of Single-strand Chain. The horsepower ratings (Table 4) have been established on a life expectancy of approximately 15,000 hr under optimum drive conditions and a service factor of 1. Principles and service factors subsequently outlined must be adhered to in making selections from the rating tables.

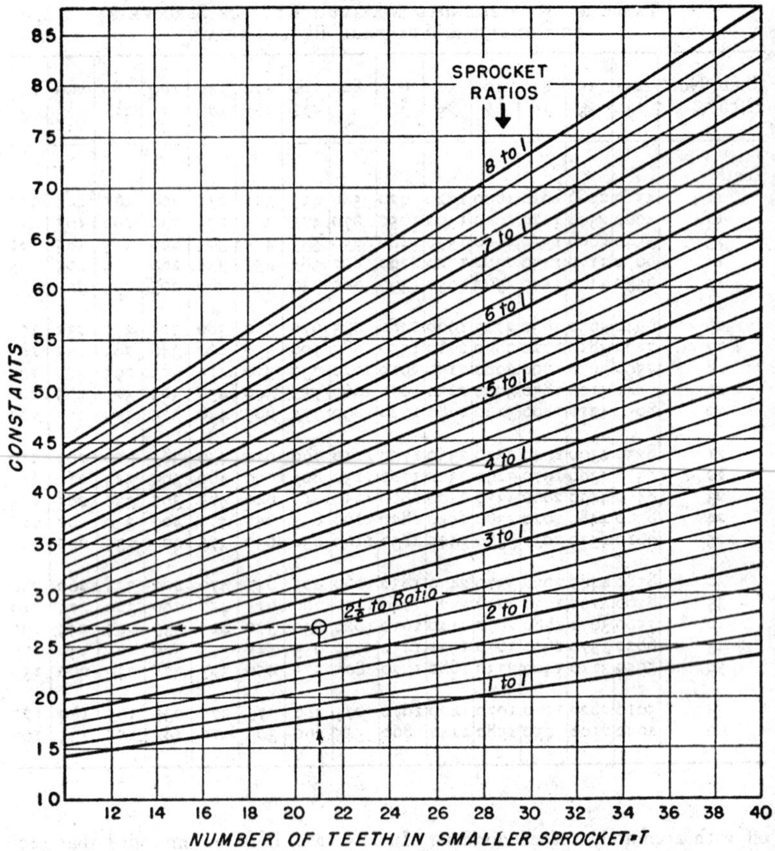

FIG. 4. Desirable chain centers are found as follows: Assume $2\frac{1}{2}:1$ ratio and 21 teeth in small sprocket. From 21 at bottom of chart move upward to $2\frac{1}{2}:1$ ratio, then left to obtain constant = 27. If chain pitch = 1 in., chain centers 27×1 = 27 in. If pitch = $\frac{1}{2}$ in., chain centers = $27 \times \frac{1}{2} = 13\frac{1}{2}$ in. The constants give approximate values of chain centers.

Failure to do so may result in unsatisfactory service from abnormal joint wear, accelerated chain elongation, and serious damage to sprockets.

Service factors as given in Table 5 should be used in connection with the horsepower ratings, the load being multiplied by the factor to obtain the required chain capacity.

Lubrication of roller chains is essential for effectively minimizing metal-to-metal contact of pin-bushing joints of the chain. Oil should be applied to link plate edges as explained under type I method of lubrication subsequently, since access to pin-bushing clearances is possible only through clearances between pin link plates and roller link plates of single chains, and in addition, between center link plates

TABLE 3. VALUES FOR LENGTH OF ROLLER CHAIN FOR TWO-WHEEL DRIVES

A	*B*	*C*	*D*	*A*	*B*	*C*	*D*
0.00000	2.0000	0.5000	0.5000	0.19509	1.9616	0.4375	0.5625
0.00436	2.0000	0.4986	0.5014	0.19937	1.9599	0.4361	0.5639
0.00873	1.9999	0.4972	0.5028	0.20364	1.9581	0.4347	0.5653
0.01309	1.9998	0.4958	0.5042	0.20791	1.9563	0.4333	0.5667
0.01745	1.9997	0.4944	0.5056	0.21218	1.9545	0.4319	0.5681
0.02181	1.9995	0.4931	0.5069	0.21644	1.9526	0.4306	0.5694
0.02618	1.9993	0.4917	0.5083	0.22070	1.9507	0.4292	0.5708
0.03054	1.9991	0.4903	0.5097	0.22495	1.9487	0.4278	0.5722
0.03490	1.9988	0.4889	0.5111	0.22920	1.9468	0.4264	0.5736
0.03926	1.9985	0.4875	0.5125	0.23345	1.9447	0.4250	0.5750
0.04362	1.9981	0.4861	0.5139	0.23769	1.9427	0.4236	0.5764
0.04798	1.9977	0.4847	0.5153	0.24192	1.9406	0.4222	0.5778
0.06234	1.9972	0.4833	0.5167	0.24615	1.9385	0.4208	0.5792
0.05669	1.9968	0.4819	0.5181	0.25039	1.9363	0.4194	0.5806
0.06105	1.9963	0.4806	0.5194	0.25460	1.9341	0.4181	0.5819
0.06540	1.9957	0.4792	0.5208	0.25882	1.9319	0.4167	0.5833
0.06976	1.9951	0.4778	0.5222	0.26303	1.9296	0.4153	0.5847
0.07411	1.9945	0.4763	0.5237	0.26724	1.9273	0.4139	0.5861
0.07846	1.9938	0.4750	0.5250	0.27144	1.9249	0.4125	0.5875
0.08281	1.9931	0.4736	0.5264	0.27564	1.9225	0.4111	0.5889
0.08716	1.9924	0.4722	0.5278	0.27983	1.9201	0.4097	0.5903
0.09150	1.9916	0.4708	0.5292	0.28402	1.9176	0.4083	0.5917
0.09585	1.9908	0.4694	0.5306	0.28820	1.9151	0.4070	0.5930
0.10019	1.9899	0.4681	0.5319	0.29237	1.9126	0.4056	0.5944
0.10453	1.9890	0.4667	0.5333	0.29654	1.9100	0.4042	0.5958
0.10887	1.9881	0.4653	0.5347	0.30071	1.9074	0.4028	0.5972
0.11320	1.9871	0.4639	0.5361	0.30486	1.9048	0.4014	0.5986
0.11754	1.9861	0.4625	0.5375	0.30902	1.9021	0.4000	0.6000
0.12187	1.9851	0.4611	0.5389	0.31316	1.8994	0.3986	0.6014
0.12620	1.9840	0.4597	0.5403	0.31730	1.8966	0.3972	0.6028
0.13053	1.9829	0.4583	0.5417	0.32144	1.8939	0.3958	0.6042
0.13485	1.9817	0.4569	0.5431	0.32557	1.8910	0.3944	0.6056
0.13917	1.9805	0.4555	0.5445	0.32969	1.8882	0.3931	0.6069
0.14349	1.9793	0.4542	0.5458	0.33381	1.8853	0.3917	0.6083
0.14781	1.9780	0.4528	0.5472	0.33792	1.8824	0.3903	0.6097
0.15212	1.9767	0.4514	0.5486	0.34202	1.8794	0.3889	0.6111
0.15643	1.9754	0.4500	0.5500	0.34612	1.8764	0.3875	0.6125
0.16074	1.9740	0.4486	0.5514	0.35021	1.8733	0.3861	0.6139
0.16505	1.9726	0.4472	0.5528	0.35429	1.8703	0.3847	0.6153
0.16935	1.9711	0.4458	0.5542	0.35837	1.8672	0.3833	0.6167
0.17365	1.9696	0.4444	0.5556	0.36244	1.8640	0.3819	0.6181
0.17794	1.9681	0.4430	0.5570	0.36650	1.8608	0.3806	0.6194
0.18224	1.9665	0.4416	0.5583	0.37056	1.8576	0.3792	0.6208
0.18652	1.9649	0.4403	0.5597	0.37461	1.8544	0.3778	0.6222
0.19081	1.9633	0.4389	0.5611	0.37865	1.8511	0.3764	0.6236

TABLE 3. VALUES FOR LENGTH OF ROLLER CHAIN FOR TWO-WHEEL DRIVES (*Continued*)

A	*B*	*C*	*D*	*A*	*B*	*C*	*D*
0.38268	1.8478	0.3750	0.6250	0.55557	1.6629	0.3125	0.6875
0.38671	1.8444	0.3736	0.6264	0.55919	1.6581	0.3111	0.6889
0.39073	1.8410	0.3722	0.6278	0.56280	1.6532	0.3097	0.6903
0.39474	1.8376	0.3708	0.6292	0.56641	1.6483	0.3083	0.6917
0.39875	1.8341	0.3694	0.6306	0.57000	1.6433	0.3069	0.6931
0.40275	1.8306	0.3681	0.6319	0.57358	1.6383	0.3055	0.6945
0.40674	1.8271	0.3667	0.6333	0.57715	1.6333	0.3042	0.6958
0.41072	1.8235	0.3653	0.6347	0.58070	1.6282	0.3028	0.6972
0.41469	1.8199	0.3639	0.6361	0.58425	1.6231	0.3014	0.6986
0.41866	1.8163	0.3625	0.6375	0.58779	1.6180	0.3000	0.7000
0.42262	1.8126	0.3611	0.6389	0.59131	1.6129	0.2986	0.7014
0.42657	1.8089	0.3597	0.6403	0.59482	1.6077	0.2972	0.7028
0.43051	1.8052	0.3583	0.6417	0.59832	1.6025	0.2958	0.7042
0.43445	1.8014	0.3569	0.6431	0.60182	1.5973	0.2945	0.7055
0.43837	1.7976	0.3556	0.6444	0.60529	1.5920	0.2932	0.7068
0.44229	1.7937	0.3542	0.6458	0.60876	1.5867	0.2917	0.7083
0.44620	1.7899	0.3528	0.6472	0.61222	1.5814	0.2903	0.7097
0.45010	1.7860	0.3514	0.6486	0.61566	1.5760	0.2889	0.7111
0.45399	1.7820	0.3500	0.6500	0.61909	1.5706	0.2875	0.7125
0.45787	1.7780	0.3486	0.6514	0.62251	1.5652	0.2861	0.7139
0.46175	1.7740	0.3472	0.6528	0.62592	1.5598	0.2847	0.7153
0.46561	1.7700	0.3458	0.6542	0.62932	1.5543	0.2833	0.7167
0.46947	1.7659	0.3445	0.6555	0.63271	1.5488	0.2819	0.7181
0.47332	1.7618	0.3431	0.6569	0.63608	1.5432	0.2805	0.7195
0.47716	1.7576	0.3417	0.6583	0.63944	1.5377	0.2792	0.7208
0.48099	1.7535	0.3403	0.6597	0.64279	1.5321	0.2778	0.7222
0.48481	1.7492	0.3389	0.6611	0.64612	1.5265	0.2764	0.7236
0.48862	1.7450	0.3375	0.6625	0.64945	1.5208	0.2750	0.7250
0.49242	1.7407	0.3361	0.6639	0.65276	1.5151	0.2736	0.7264
0.49622	1.7364	0.3347	0.6653	0.65606	1.5094	0.2722	0.7278
0.50000	1.7321	0.3333	0.6667	0.65935	1.5037	0.2708	0.7292
0.50377	1.7277	0.3320	0.6680	0.66262	1.4979	0.2694	0.7306
0.50754	1.7233	0.3305	0.6695	0.66588	1.4921	0.2680	0.7320
0.51129	1.7188	0.3292	0.6708	0.66913	1.4863	0.2667	0.7333
0.51504	1.7143	0.3278	0.6722	0.67237	1.4804	0.2653	0.7347
0.51877	1.7098	0.3264	0.6736	0.67559	1.4746	0.2639	0.7361
0.52250	1.7053	0.3250	0.6750	0.67880	1.4686	0.2625	0.7375
0.52621	1.7007	0.3236	0.6764	0.68200	1.4627	0.2611	0.7389
0.52992	1.6961	0.3222	0.6778	0.68518	1.4567	0.2603	0.7397
0.53361	1.6915	0.3208	0.6792	0.68835	1.4507	0.2583	0.7417
0.53730	1.6868	0.3194	0.6806	0.69151	1.4447	0.2569	0.7431
0.54097	1.6821	0.3180	0.6820	0.69466	1.4387	0.2556	0.7444
0.54464	1.6773	0.3166	0.6834	0.69779	1.4326	0.2542	0.7458
0.54829	1.6726	0.3153	0.6847	0.70091	1.4265	0.2528	0.7472
0.55194	1.6678	0.3139	0.6861	0.70401	1.4204	0.2514	0.7486
				0.70711	1.4142	0.2500	0.7500

TABLE 4. HORSEPOWER RATINGS FOR SINGLE-STRAND ROLLER CHAIN

Small Sprocket Teeth	Rpm, Small Sprocket									
	¼-in. Pitch—No. 25									
	100	400	800	1200	1600	2000	2400	3000	5000	6000
11	0.051	0.16	0.29	0.40	0.47	0.53	0.57	0.61		
13	0.61	0.20	0.37	0.50	0.61	0.70	0.76	0.84	0.90	
16	0.076	0.26	0.48	0.65	0.80	0.92	1.02	1.13	1.33	1.29
20	0.095	0.33	0.60	0.82	1.02	1.18	1.31	1.49	1.76	1.78
25	0.118	0.41	0.75	1.03	1.27	1.46	1.63	1.84	2.20	2.22
40	0.187	0.63	1.13	1.55	1.89	2.15	2.36	2.63	2.84	2.65
60	0.274	0.92	1.60	2.10	2.48	2.80	3.04	3.16		
	⅜-in. Pitch—No. 35									
	50	200	400	800	1200	1600	2000	2400	3200	4000
11	0.087	0.303	0.54	0.89	1.15	1.31	1.40			
13	0.105	0.369	0.66	1.13	1.48	1.73	1.91	2.02		
16	0.130	0.463	0.84	1.46	1.93	2.30	2.58	2.79	2.98	
20	0.164	0.585	1.07	1.86	2.49	2.97	3.37	3.65	4.01	4.06
25	0.204	0.730	1.33	2.31	3.09	3.71	4.19	4.54	4.98	5.04
40	0.322	1.14	2.05	3.50	4.59	5.39	5.96	6.32	6.58	
60	0.474	1.64	2.89	4.76	6.00	6.78	7.14			
	½-in. Pitch, Lightweight Machinery Chain—No. 41									
	50	200	400	600	800	1000	1200	1400	1600	1800
11	0.121	0.412	0.714	0.954	1.13	1.28				
13	0.146	0.505	0.888	1.20	1.46	1.67	1.85			
16	0.182	0.636	1.13	1.55	1.90	2.20	2.45	2.66	2.84	
20	0.228	0.804	1.44	1.97	2.44	2.83	3.17	3.46	3.70	3.89
25	0.285	1.00	1.80	2.47	3.04	3.54	3.96	4.31	4.64	4.85
40	0.449	1.56	2.75	3.73	4.54	5.20	5.76	6.19	6.52	
60	0.657	2.22	3.83	5.07	6.03					
	¾-in. Pitch—No. 60									
	50	150	300	500	700	900	1100	1300	1500	1700
11	0.66	1.70	2.93	4.12	4.93	5.41				
13	0.79	2.09	3.65	5.27	6.46	7.32	7.88			
16	0.99	2.64	4.66	6.82	8.52	9.80	10.77	11.46		
20	1.25	3.34	5.93	8.74	10.97	12.74	14.08	15.10	15.78	16.14
25	1.56	4.17	7.41	10.89	13.67	15.86	17.53	18.79	19.62	20.06
40	2.45	6.46	11.31	16.33	20.08	22.84	24.64	25.33	26.10	
60	3.58	9.21	15.66	21.80	25.81					
	1¼-in. Pitch—No. 100									
	10	50	100	150	200	300	400	500	600	700
11	0.663	2.88	5.19	7.10	8.78	11.4	13.6			
13	0.792	3.52	6.35	8.82	10.9	14.6	17.6	19.6		
16	0.987	4.37	8.07	11.2	14.1	18.9	23.2	26.1	28.7	
20	1.24	5.52	10.1	14.3	17.9	24.3	29.8	33.9	37.3	40.2
25	1.54	6.90	12.7	17.8	22.4	30.3	36.8	42.2	46.4	50.0
40	2.44	10.8	19.6	27.3	34.0	44.3	54.2	61.0	66.3	70.8
60	3.61	15.6	27.8	38.1	46.8	60.5				
	1¾-in. Pitch—No. 140									
	10	50	100	150	200	250	300	350	400	400
11	1.79	7.55	13.3	17.8	21.5					
13	2.13	9.19	16.3	22.3	27.3	31.6	35.2			
16	2.66	11.6	20.8	28.6	35.4	41.3	46.4	50.7		
20	3.34	14.6	26.4	36.4	45.3	53.0	59.8	65.6	70.7	
25	4.17	18.2	32.9	45.5	56.5	66.1	74.5	81.7	88.0	93.4
40	6.58	28.3	50.6	69.1	84.9	98.2	109.1	118.2	126.7	
60	9.67	40.7	70.9	94.8	113.9					
Lubrication:	Type I		Type II		Type III					

TABLE 5. SERVICE FACTORS FOR ROLLER CHAIN

Type of Load	Conditions of Service	
	10-hr Day	24-hr Day
Uniform load, average conditions	1.0	1.2
Moderate shock, abnormal conditions	1.2	1.4
Heavy shock, abnormal conditions	1.4	1.7

and roller link plates of multiple chains. Oil applied on center line of rollers cannot reach pin-bushing joints and therefore cannot retard chain wear elongation.

A good grade of mineral oil, without additives, of medium or light consistency, free-flowing at the prevailing temperature, should be used.

Drives should be protected against dirt and moisture. The method of lubrication, which is influenced by the speed of the chain and the amount of power transmitted, should be in accordance with one of the following:

Type I. Drip (4 to 10 drops per min), shallow bath, or manual with oil applied frequently with a brush or spout can to upper edges of all link plates in the lower span of chain.[1]

Type II. Rapid drip (20 drops per min minimum) or continuous with shallow bath, disk, or slinger.[1]

Type III. Continuous, with disk, slinger, or circulating pump.

The choice of method or type of lubrication should in general be guided by the following:

Chain Speed	*Method of Lubrication*
Up to 600 fpm	Type I
600 to 1500 fpm	Type II
Above 1500 fpm	Type III

Heavy oils and greases are not recommended for lubrication of roller chains except under unusual conditions of service because they are generally too stiff to enter and fill the small clearances between the chain parts.

SILENT-CHAIN DRIVES

ASA B29.2-1950

A silent or inverted-tooth chain consists of a series of links of the shape shown in Fig. 5 and assembled over pins to form the desired width. As the chain pitch lengthens from wear, the links adjust themselves to a larger pitch circle. Such chains can be run in either direction, quietly and smoothly, at speeds up to 4500 fpm, if sprockets with suitable numbers of teeth are used.

Design. In designing a silent-chain drive, matters to be taken into consideration are nature of the load, horsepower to be transmitted, and speed of the driving and driven units.

Horsepower ratings can be listed only for a uniform rate of work, where there is

[1] A circulating pump may be required when transmitted horsepower is substantial and center distance is short.

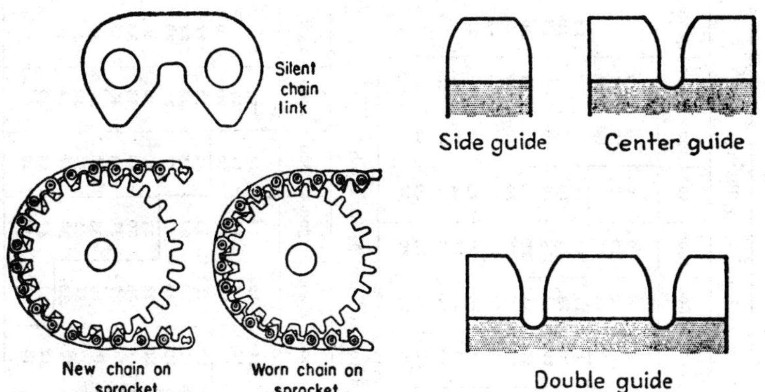

FIG. 5. Silent chain adjusts itself to wear.

FIG. 6. Profiles of silent-chain sprockets.

relatively little shock or load variation throughout a single revolution of the driven sprocket. The tabulated values (Table 6) are based upon life expectancy of approximately 20,000 hr under optimum drive conditions and a service factor of 1. Before the design is established, the nominal or given horsepower should be multiplied by a service factor (Table 7) to provide the necessary extra capacity to cope with the expected severity of operation. Engineering judgment must be applied here in connection with the source of power, the nature of the load, and the resulting effects of inertia, strain, and shock. Failure to do so may result in unsatisfactory service from abnormal joint wear, accelerated chain elongation, and serious damage to the sprockets.

Operating characteristics of all types of equipment, considering their source of power, cannot be given for space reasons. The following relationships of types of machine driven and source of power are provided for a general guide (it being recognized that some sources of power will not ordinarily be used with the cited machines, but that the combination of characteristics of machine and power source may conceivably be extended to other situations).

PITCH OF CHAIN. Select the pitch of chain according to Table 8. The higher shaft speed is the limiting factor and indicates the largest pitch chain that can be used. Also, the shaft diameter should be checked against the largest recommended bore in the sprocket selected. In general, the largest pitch gives the best drive, but limitations of sprocket diameters and the need for quiet operation often dictate smaller pitches. For the quietest drive, choose the smallest pitch that will carry the load. Large-pitch chains should be used for slow speeds and large power.

CHAIN WIDTH. Determine chain width on the basis of horsepower per inch of chain width, as given in Table 6. Thus chain width equals the nominal horsepower times the service factor divided by the horsepower capacity per inch of chain width. It is possible to make two or more sound selections, but strive for a pitch and sprocket size that will allow a chain width less than eight times the pitch.

CHAIN LENGTH. The length of chain is

$$L = 2C + \frac{(N + n)}{2} + \frac{[(N - n)^2/2\pi]}{C}$$

TABLE 6. HORSEPOWER CAPACITY OF SILENT CHAIN, PER INCH OF WIDTH

⅜-in. Pitch, Small-sprocket Rpm													Teeth, Small Sprocket	Small-sprocket Rpm, ½-in. Pitch										
100	500	1000	1200	1500	1800	2000	2500	3000	3500	4000	5000	6000		100	500	700	1000	1200	1800	2000	2500	3000	3500	4000
0.37	1.7	3.7	3.9	4.2	5.2	5.5	6.3	6.8	7.0	7.0	....	....	17*	0.66	3	4	5	6	8	9	9	9	9	
0.42	2.0	3.8	4.3	5.2	5.9	6.3	7.3	7.9	8.3	8.4	7.8	....	19*	0.74	3	4	6	7	9	10	11	11	11	
0.46	2.2	4.1	4.8	5.8	6.6	7.2	8.3	9.1	9.6	9.9	9.5	8.2	21	0.81	4	5	7	8	11	11	12	13	13	
0.50	2.4	4.5	5.3	6.4	7.4	8.0	9.2	10.0	11.0	11.0	11.0	9.5	23	0.89	4	6	8	9	12	13	14	15	15	14
0.55	2.6	4.9	5.8	7.0	8.1	8.8	10.0	11.0	12.0	12.0	12.0	11.0	25	0.97	4	6	8	10	13	14	16	17	17	16
0.59	2.8	5.4	6.3	7.6	8.8	9.5	11.0	12.0	13.0	14.0	14.0	13.0	27	1.0	5	7	9	10	14	15	17	19	19	18
0.64	3.0	5.8	6.8	8.2	9.5	10.3	12.0	13.0	14.0	15.0	15.0	14.0	29	1.1	5	7	10	11	15	17	19	20	20	20
0.68	3.3	6.2	7.3	8.8	10.0	11.0	13.0	14.0	15.0	16.0	16.0	15.0	31	1.2	6	8	10	12	17	18	20	22	22	22
0.72	3.5	6.6	7.8	9.4	11.0	12.0	14.0	15.0	17.0	17.0	17.0	16.0	33	1.3	6	8	11	13	18	19	22	23	24	23
0.77	3.7	7.0	8.3	10.0	12.0	13.0	15.0	16.0	18.0	18.0	18.0	17.0	35	1.4	6	9	12	14	19	20	23	25	25	24
0.82	3.9	7.3	8.7	11.0	12.0	13.0	16.0	17.0	19.0	19.0	19.0	....	37	1.5	7	9	13	15	20	21	24	26	26	
0.9	4.2	8.1	9.5	12.0	13.0	14.0	17.0	19.0	20.0	21.0	21.0	....	40	1.6	7	10	14	16	22	23	26	28	28	
1.0	4.8	9.1	10.0	13.0	15.0	16.0	19.0	21.0	22.0	23.0	....	....	45	2	8	11	15	18	24	24	29	31		
1.1	5.3	10.0	12.0	14.0	16.0	18.0	21.0	23.0	24.0	....	....	....	50	2	9	12	17	20	27	29	32			

⅝-in. Pitch, Small-sprocket Rpm										Teeth, Small Sprocket	Small-sprocket Rpm, ¾-in. Pitch								
100	500	700	1000	1200	1800	2000	2500	3000	3500		100	500	700	1000	1200	1500	1800	2000	2500
1.0	5	6	8	9	11	12	11	..	..	17*	1.5	6.5	8.6	11	12	13	14	14	
1.1	5	7	10	11	13	14	14	..	..	19*	1.6	7.4	10.0	12	14	16	17	17	
1.3	6	8	10	12	15	16	16	16	..	21	1.8	8.0	11.0	14	16	18	19	20	19
1.4	6	9	12	13	17	18	19	18	..	23	2.0	9.0	12.0	16	18	20	22	22	22
1.5	7	9	13	15	19	20	21	21	19	25	2.2	10.0	13.0	17	20	23	25	25	24
1.6	8	10	14	16	21	22	23	23	21	27	2.3	11.0	14.0	19	22	25	27	28	28
1.7	8	11	15	17	22	24	25	25	23	29	2.5	12.0	16.0	21	24	27	29	30	30
1.9	9	12	16	18	24	25	27	27	25	31	2.7	12.0	17.0	22	25	29	32	33	33
2.0	9	13	17	20	26	27	29	29	27	33	2.9	13.0	18.0	24	27	31	34	35	35
2.1	10	13	18	21	27	29	31	31	28	35	3.0	14.0	19.0	25	29	33	36	37	37
2.2	10	14	19	22	29	31	34	33	..	37	3.2	15.0	20.0	27	31	35	38	39	39
2.4	11	15	21	24	31	33	35	..	..	40	3.5	16.0	22.0	29	33	38	41	42	42
2.7	13	17	23	27	35	37	..	..	..	45	3.9	18.0	24.0	32	37	42	45	46	
3.0	14	19	26	30	38	40	..	..	..	50	4.3	20.0	27.0	36	41	46	49		

1-in. Pitch, Small-sprocket Rpm											Teeth, Small Sprocket	Small-sprocket Rpm, 1¼-in. Pitch										
100	200	300	400	500	700	1000	1200	1500	1800	2000		100	200	300	400	500	600	700	800	1000	1200	1500
3.0	5	7	9	11	14	17	18	..	..	..	17*											
3.0	6	8	10	12	16	20	21	22	..	..	19*	4.5	8	12	16	19	21	23	25	27	28	
3.0	6	9	12	14	18	23	25	26	26	..	21	5.0	9	14	18	21	24	26	29	32	33	
3.0	7	10	13	15	20	25	28	30	30	..	23	5.5	10	15	19	23	27	29	32	36	37	37
4.0	7	11	14	17	22	28	31	33	33	33	25	6.0	11	16	21	25	29	32	35	40	42	42
4.0	8	12	15	19	24	31	34	37	37	36	27	6.4	12	18	23	28	32	35	39	43	46	46
4.0	9	13	16	20	26	33	37	40	41	40	29	6.9	13	19	25	30	34	38	42	47	50	51
5.0	9	13	18	22	28	36	40	43	44	43	31	7.4	14	21	27	32	37	41	45	51	54	55
5.0	10	14	19	23	30	39	43	47	47	46	33	7.9	15	22	28	34	39	44	48	55	58	59
5.0	10	15	20	24	32	41	45	49	50	49	35	8.4	16	23	30	36	42	47	51	58	62	62
5.4	11	16	21	26	34	43	48	52	53	..	37	9.0	17	24	32	38	44	50	54	61	65	
6.0	12	18	23	28	36	47	52	56	..	..	40	9.6	19	27	35	42	48	54	59	66	70	
7.0	13	20	25	31	41	52	57	61	..	..	45	10.7	21	30	39	47	54	60	65	73		
8.0	15	22	28	34	45	57	62	..	..	..	50	12.0	23	34	43	52	59	66	72	80		

1½-in. Pitch, Small-sprocket Rpm											Teeth, Small Sprocket	Small-sprocket Rpm, 2-in. Pitch								
100	200	300	400	500	600	700	800	900	1000	1200		100	200	300	400	500	600	700	800	900
6.4	12	17	22	25	28	31	32	33	34	..	19*	11.0	21	29	35	40	43	45		
7.0	13	19	24	29	32	35	37	39	39	..	21	12.5	23	32	40	42	50	52		
8.0	15	21	27	32	36	39	42	44	45	44	23	13.5	26	36	44	51	56	59	60	
8.0	16	23	30	35	40	44	47	49	52	51	25	14.0	28	39	49	56	62	66	68	68
9.0	18	25	32	38	43	48	51	54	56	56	27	16.0	30	43	53	62	68	73	75	75
10.0	19	27	35	41	47	52	56	59	60	61	29	17.0	33	46	58	67	74	79	82	82
11.0	20	29	37	44	51	56	60	63	65	66	31	18.0	35	50	62	72	80	85	88	88
11.0	22	31	40	47	54	60	64	68	70	71	33	20.0	37	53	66	77	85	91	94	94
12.0	23	33	42	50	57	63	68	72	74	75	35	21.0	40	57	70	82	91	97	100	100
13.0	24	35	47	53	61	67	72	77	79	..	37	22.0	42	60	74	88	99	102	105	
14.0	26	38	53	58	66	72	78	84	..	..	40	24.0	46	65	81	94	103	110	113	
15.0	30	43	54	65	74	81	86	90	..	..	45	27.0	51	72	90	105	115	121		
17.0	33	47	60	71	81	89	94	..	..	..	50	30.0	57	80	100	115	125			

Lubrication: group to left of first zigzag line, use bath or splash, oil cup or brush; central group, use disk or circulating pump; right-hand group, consult manufacturer.

* For best results, smaller sprocket should have at least 21 teeth.

where L = chain length in pitches; C = center distance in pitches; N = number of teeth in large sprocket; n = number of teeth in small sprocket, and $2\pi = 6.2832$.

Center Distance. Minimum center distance between sprockets is the sum of the sprocket diameters under normal conditions. Long centers are practical on drives where provision is made for adjustment for chain wear or stretch.

Speed Ratios. The speed ratio is the basis of selection for driving and driven sprockets. The smaller sprocket is usually selected first on the following basis:

Speed ratio	1:1	2:1	3:1	4:1	5:1
Small-sprocket teeth	25–35	19–29	19–25	19–23	19–21

Table 7. Service Factors for Silent Chains

Operating Characteristics	10 Hr per Day	24 Hr per Day (Extra Long Life)
Boring mill, lathe, or drill press with motor of equal rating	1.0	1.3
Drop hammer, grinder, or miller with motor of equal rating	1.1	1.4
Punch press or shear with motor of equal rating	1.2	1.5
Drop hammer driven by oversize motor or drill press by gas engine	1.3	1.6
Punch press driven by oversize motor or lathe by gasoline engine	1.4	1.7
Drop hammer or punch press driven by gasoline engine	1.5	1.8
Tumbling barrel driven by motor of equal rating	1.6	1.9
Tumbling barrel driven by oversize motor	1.7	2.0

Table 8. Silent-chain Speeds for Various Pitches

Pitch, In.	Pitches per Ft	Chain Width, In.	Weight per Ft 1 In. Wide	Small-sprocket Driver		
				Normal Rpm	Max Rpm*	Min Teeth
3/8	32	3/8– 4	0.75	4000	6000	19
1/2	24	1/2– 7	1.00	2600	4000	19
5/8	19.2	1– 8	1.25	2300	3500	19
3/4	16	1–10	1.50	1650	2500	21
1	12	1 1/2–14	2.00	1300	2000	21
1 1/4	9.6	2 1/2–20	2.50	1000	1500	23
1 1/2	8	3–24	3.00	800	1200	23
2	6	4–30	4.00	600	900	25

* Under favorable conditions these values may be exceeded, and under unfavorable conditions they must be decreased.

Ratios as high as 8:1 are possible, but the use of sprockets with less than 17 or more than 120 teeth is to be avoided in the interest of long life and efficient operation. Table 9 gives desirable sprocket combinations for various ratios.

CHAIN SPEED. Chain speed in fpm is given by

$$S = \text{No. of teeth} \times \text{rpm} \times \text{pitch (in.)} \div 12$$

Lubrication. Wherever possible silent chains should be operated in an oiltight casing with provision for lubrication. When the chain speed is less than 1500 fpm, a drip or sight-feed oiler can be used. The oil should be applied to the inner surface of the chain at the rate of about one drop per minute.

When the chain speed is from 1500 to 3000 fpm, bath lubrication is recommended. Oil level in the casing is maintained so that the chain is immersed, while running, to a depth of approximately ½ pitch.

Viscosity of the oil used depends upon chain speed and temperature conditions. Straight mineral oil of SAE 30 viscosity is generally employed, but SAE 50 or 60 can be used on very slow moving chains. Avoid greases as they cannot penetrate the "joints" of the chain at ordinary temperatures.

Offset Links. Avoid use of offset links or chain lengths of an uneven number of pitches.

TABLE 9. SPROCKET COMBINATIONS FOR VARIOUS RATIOS

Ratio	Pinion	Sprocket	Ratio	Pinion	Sprocket	Ratio	Pinion	Sprocket
1.00	25	25	1.81	21	38	3.80	25	95
1.09	23	25	2.00	19	38	4.00	19	76
1.10	21	23	2.28	25	57	4.13	23	95
1.11	19	21	2.48	23	57	4.52	21	95
1.19	21	25	2.71	21	57	4.56	25	114
1.21	19	23	3.00	19	57	4.95	23	114
1.32	19	25	3.04	25	76	5.00	19	95
1.52	25	38	3.30	23	76	5.43	21	114
1.65	23	38	3.62	21	76			

TABLE 10. OVER-PIN DIAMETER TOLERANCES FOR SILENT-CHAIN SPROCKETS*
All tolerances are negative. Tolerance = $(0.004 + 0.001P\sqrt{N})$

Pitch	Up to 15	16–24	25–35	36–48	49–63	64–80	81–99	100–120	121–143	144 up
⅜	0.005	0.005	0.005	0.006	0.006	0.007	0.007	0.007	0.008	0.008
½	0.005	0.006	0.006	0.007	0.007	0.008	0.008	0.009	0.009	0.010
⅝	0.006	0.006	0.007	0.008	0.009	0.010	0.010	0.010	0.011	0.012
¾	0.006	0.007	0.008	0.009	0.010	0.011	0.011	0.012	0.013	0.014
1	0.007	0.008	0.009	0.010	0.011	0.012	0.013	0.014	0.015	0.016
1¼	0.008	0.009	0.010	0.011	0.013	0.014	0.015	0.017	0.018	0.019
1½	0.008	0.010	0.011	0.013	0.014	0.016	0.017	0.019	0.020	0.022
2	0.010	0.012	0.014	0.016	0.018	0.020	0.022	0.024	0.026	0.028

* Tolerances for OD of sprocket with square-top teeth = +0.000, $-0.050P$. Tolerance for maximum eccentricity of pitch dia with respect to the bore = $0.001PD$, but not less than 0.006 or more than 0.032 in.

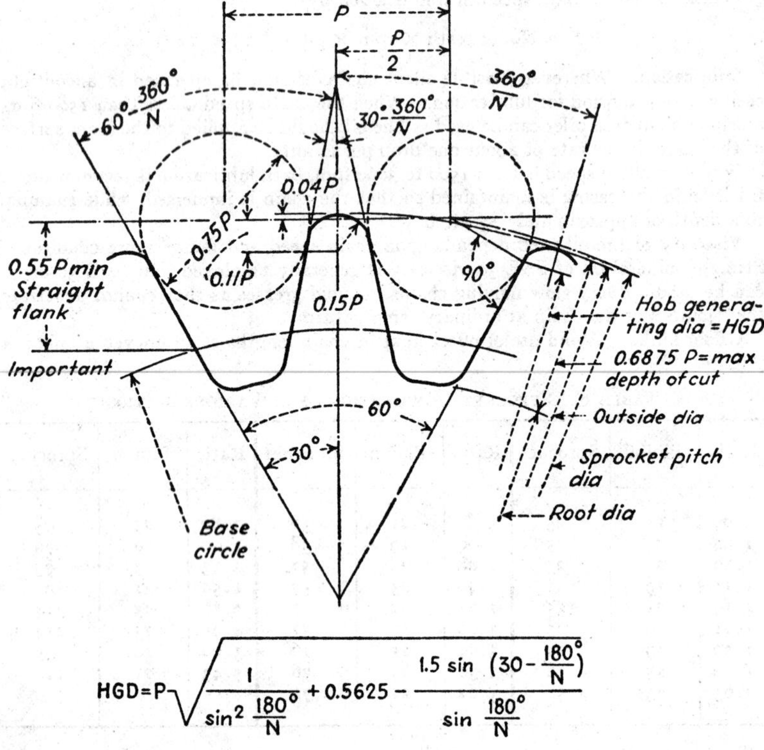

$$HGD = P\sqrt{\frac{1}{\sin^2 \frac{180°}{N}} + 0.5625 - \frac{1.5 \sin\left(30 - \frac{180°}{N}\right)}{\sin \frac{180°}{N}}}$$

FIG. 7. Hob-generating criteria for silent-chain sprockets. When the sprockets are cut, observe the over-pin diameters in Table 11.

FORMULAS FOR MAXIMUM HUB DIAMETERS

For hobbed teeth:

$$MHD = P(\cot 180°/N - 1.33)$$

For straddle-cut teeth:

$$MHD = P(\cot 180°/N - 1.25)$$

Elements of Silent-chain Sprockets

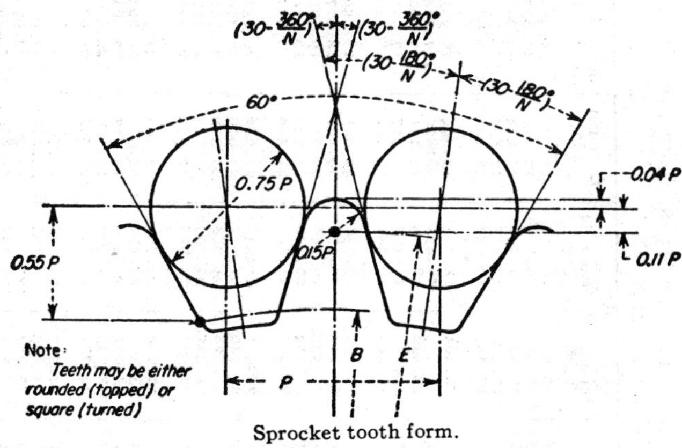

Sprocket tooth form.

P = chain pitch
PD = pitch dia
OD = outside dia
D_p = gage-pin dia
E = dia to center of topping curve
B = dia to base of working face
OPD = over-pin dia
G = max guide-groove dia
N = No. of teeth

Sprocket diameters.

Formulas for Sprockets

$$PD = P \div \sin 180°/N$$
$$D_p = 0.625P$$
$$E = P(\cot 180°/N - 0.22)$$
$$B = P\sqrt{1.5152 + (\cot 180°/N - 1.1)^2}$$

For even teeth:

$$OPD = PD - 0.125P \csc (30 - 180/N)° + 0.625P$$

For odd teeth:

$$OPD = \cos 90°/N[PD - 0.125P \csc (30 - 180/N)°] + 0.625P$$

For rounded teeth:

$$OD = P(\cot 180°/N + 0.08)$$

For square teeth:

$$OD = 2\sqrt{X^2 + L^2 - 2XL \cos \alpha}$$

where $X = Y \cos \alpha - \sqrt{(0.15P)^2 - (Y \sin \alpha)^2}$
$Y = P(0.500 - 0.375 \sec \alpha) \cot \alpha + 0.11P$
$L = Y + E/2$
$\alpha = (30 - 360/N)°$
$G \text{ (max)} = P(\cot 180°/N - 1.16)$

TABLE 11. STANDARD DIAMETERS OF SPROCKETS FOR SILENT CHAIN, 1-IN. PITCH, 0.625-IN. GAGE PIN

For other pitches—3/8 in. and larger—multiply these values by pitch

No. Teeth	Pitch Dia	OD Rounded* Teeth	OD Squared† Teeth	Over-pin Dia	Max Guide Groove Dia	No. Teeth	Pitch Dia	OD Rounded* Teeth	OD Squared† Teeth	Over-pin Dia	Max Guide Groove Dia
17	5.442	5.429	5.298	5.669	4.189	52	16.562	16.612	16.556	16.907	15.372
18	5.759	5.751	5.623	6.018	4.511	53	16.880	16.930	16.876	17.218	15.690
19	6.076	6.072	5.947	6.324	4.832	54	17.198	17.249	17.196	17.544	16.009
20	6.393	6.393	6.271	6.669	5.153	55	17.517	17.568	17.515	17.857	16.328
21	6.710	6.714	6.595	6.974	5.474	56	17.835	17.887	17.834	18.183	16.647
22	7.027	7.036	6.919	7.315	5.796	57	18.153	18.205	18.154	18.494	16.965
23	7.344	7.356	7.243	7.621	6.116	58	18.471	18.524	18.473	18.820	17.284
24	7.661	7.675	7.568	7.960	6.435	59	18.789	18.842	18.793	19.131	17.602
25	7.979	7.996	7.890	8.266	6.756	60	19.107	19.161	19.112	19.457	17.921
26	8.296	8.315	8.213	8.602	7.075	61	19.426	19.480	19.431	19.769	18.240
27	8.614	8.636	8.536	8.909	7.396	62	19.744	19.799	19.750	20.095	18.559
28	8.932	8.956	8.859	9.244	7.716	63	20.062	20.117	20.070	20.407	18.877
29	9.249	9.275	9.181	9.551	8.035	64	20.380	20.435	20.388	20.731	19.195
30	9.567	9.595	9.504	9.884	8.355	65	20.698	20.754	20.708	21.044	19.514
31	9.885	9.913	9.828	10.192	8.673	66	21.016	21.072	21.027	21.368	19.832
32	10.202	10.233	10.150	10.524	8.993	67	21.335	21.391	21.346	21.682	20.151
33	10.520	10.553	10.471	10.833	9.313	68	21.653	21.710	21.665	22.006	20.470
34	10.838	10.872	10.793	11.164	9.632	69	21.971	22.028	21.984	22.319	20.788
35	11.156	11.191	11.115	11.472	9.951	70	22.289	22.347	22.303	22.643	21.107
36	11.474	11.510	11.437	11.803	10.270	71	22.607	22.665	22.622	22.955	21.425
37	11.792	11.829	11.757	12.112	10.589	72	22.926	22.984	22.941	23.280	21.744
38	12.110	12.149	12.077	12.442	10.909	73	23.244	23.302	23.259	23.593	22.062
39	12.428	12.468	12.397	12.751	11.228	74	23.562	23.621	23.578	23.917	22.381
40	12.746	12.787	12.717	13.080	11.547	75	23.880	23.939	23.897	24.230	22.699
41	13.064	13.106	13.037	13.390	11.866	76	24.198	24.257	24.216	24.553	23.017
42	13.382	13.425	13.357	13.718	12.185	77	24.517	24.577	24.535	24.868	23.337
43	13.700	12.743	13.677	14.028	12.503	78	24.835	24.895	24.853	25.191	23.655
44	14.018	14.062	13.997	14.356	12.822	79	25.153	25.213	25.172	25.504	23.973
45	14.336	14.381	14.317	14.667	13.141	80	25.471	25.531	25.491	25.828	24.291
46	14.654	14.700	14.637	14.994	13.460	81	25.790	25.851	25.809	26.141	24.611
47	14.972	15.018	14.957	15.305	13.778	82	26.108	26.160	26.128	26.465	24.929
48	15.290	15.337	15.277	15.632	14.097	83	26.426	26.487	26.447	26.778	25.247
49	15.608	15.656	15.597	15.943	14.416	84	26.744	26.805	26.766	27.101	25.565
50	15.926	15.975	15.917	16.270	14.735	85	27.063	27.125	27.084	27.415	25.885
51	16.244	16.293	16.236	16.581	15.053	86	27.381	27.443	27.403	27.739	26.203

No. Teeth	Pitch Dia	OD Rounded* Teeth	OD Squared† Teeth	Over-pin Dia	Max Guide Groove Dia	No. Teeth	Pitch Dia	OD Rounded* Teeth	OD Squared† Teeth	Over-pin Dia	Max Guide Groove Dia
87	27.699	27.761	27.722	28.052	26.521	122	38.837	38.904	38.872	39.200	37.664
88	28.017	28.079	28.040	28.375	26.839	123	39.156	39.223	39.190	39.516	37.983
89	28.335	28.397	28.359	28.689	27.157	124	39.475	39.542	39.508	39.839	38.302
90	28.654	28.716	28.678	29.013	27.476	125	39.794	39.861	39.827	40.154	38.621
91	28.972	29.035	28.997	29.327	27.795	126	40.112	40.180	40.145	40.476	38.940
92	29.290	29.353	29.315	29.649	28.113	127	40.430	40.497	40.464	40.790	39.257
93	29.608	29.671	29.634	29.963	28.431	128	40.748	40.816	40.782	41.112	39.576
94	29.926	29.989	29.953	30.285	28.749	129	41.066	41.134	41.100	41.427	39.894
95	30.245	30.308	30.271	30.001	29.068	130	41.384	41.452	41.419	41.748	40.212
96	30.563	30.627	30.590	30.923	29.387	131	41.702	41.770	41.738	42.063	40.530
97	30.881	30.945	30.909	31.237	29.705	132	42.020	42.088	42.056	42.384	40.848
98	31.199	31.263	31.228	31.559	30.023	133	42.338	42.406	42.374	42.699	41.166
99	31.518	31.582	31.546	31.874	30.342	134	42.656	42.724	42.693	43.020	41.484
100	31.836	31.900	31.865	32.196	30.660	135	42.975	43.043	43.011	43.336	41.803
101	32.154	32.218	32.183	32.511	30.978	136	43.293	43.362	43.329	43.657	42.122
102	32.473	32.537	32.502	32.834	31.297	137	43.611	43.679	43.647	43.972	42.439
103	32.791	32.856	32.820	33.148	31.616	138	43.930	43.998	43.966	44.295	42.758
104	33.109	33.174	33.139	33.470	31.934	139	44.249	44.317	44.284	44.611	43.077
105	33.427	33.492	33.457	33.784	32.252	140	44.567	44.636	44.603	44.932	43.396
106	33.746	33.811	33.776	34.107	32.571	141	44.885	44.954	44.922	45.247	43.714
107	34.064	34.129	34.094	34.422	32.889	142	45.203	45.271	45.240	45.568	44.031
108	34.382	34.447	34.413	34.744	33.207	143	45.521	45.590	45.558	45.883	44.350
109	34.701	34.767	34.731	35.059	33.527	144	45.840	45.909	45.877	46.205	44.669
110	35.019	35.084	35.050	35.381	33.844	145	46.158	46.227	46.195	46.520	44.987
111	35.337	35.403	35.368	35.695	34.163	146	46.477	46.546	46.514	46.842	45.306
112	35.655	35.721	35.687	36.017	34.481	147	46.796	46.865	46.832	47.159	45.625
113	35.974	36.040	36.005	36.333	34.800	148	47.114	47.183	47.151	47.479	45.943
114	36.292	36.358	36.324	36.654	35.118	149	47.432	47.501	47.469	47.795	46.261
115	36.610	36.676	36.642	36.969	35.436	150	47.750	47.819	47.787	48.116	46.579
116	36.929	36.995	36.961	37.292	35.755						
117	37.247	37.313	37.279	37.606	36.073						
118	37.565	37.632	37.598	37.928	36.392						
119	37.883	37.950	37.916	38.243	36.710						
120	38.201	38.268	38.235	38.564	37.028						
121	38.519	38.586	38.553	38.879	37.346						

* Blank diameters shall be 0.020 larger than finished diameters tabulated above.
† Diameters given are maximum; all tolerances must be negative.

TABLE 12. GENERATING DATA FOR HOBBING SILENT-CHAIN SPROCKETS

Chain Pitch	Hob No.	Basic No. Teeth	Tooth Range of Hob	Generating Dia	Chain Pitch	Hob No.	Basic No. Teeth	Tooth Range of Hob	Generating Dia
SC3 = 0.375	1	20	17–23	2.311	SC8 = 1.000	1	20	17–23	6.163
	2	28	24–32	3.247		2	28	24–32	8.659
	3	38	33–43	4.428		3	38	33–43	11.809
	4	51	44–58	5.971		4	51	44–58	15.924
	5	69	59–79	8.114		5	69	59–79	21.636
	6	95	80–110	11.212		6	95	80–110	29.899
	7	130	111–150	15.385		7	130	111–150	41.026
SC4 = 0.500	1	20	17–23	3.082	SC10 = 1.250	1	20	17–23	7.704
	2	28	24–32	4.329		2	28	24–32	10.823
	3	38	33–43	5.904		3	38	33–43	14.761
	4	51	44–58	7.962		4	51	44–58	19.905
	5	69	59–79	10.818		5	69	59–79	27.045
	6	95	80–110	14.950		6	95	80–110	37.374
	7	130	111–150	20.513		7	130	111–150	51.283
SC5 = 0.625	1	20	17–23	3.852	SC12 = 1.500	1	20	17–23	9.245
	2	28	24–32	5.412		2	28	24–32	12.988
	3	38	33–43	7.381		3	38	33–43	17.713
	4	51	44–58	9.952		4	51	44–58	23.886
	5	69	59–79	13.522		5	69	59–79	32.454
	6	95	80–110	18.687		6	95	80–110	44.849
	7	130	111–150	25.641		7	130	111–150	61.539
SC6 = 0.750	1	20	17–23	4.623	SC16 = 2.000	1	20	17–23	12.327
	2	28	24–32	6.494		2	28	24–32	17.317
	3	38	33–43	8.857		3	38	33–43	23.618
	4	51	44–58	11.943		4	51	44–58	31.848
	5	69	59–79	16.227		5	69	59–79	43.272
	6	95	80–110	22.424		6	95	80–110	59.798
	7	130	111–150	30.770		7	130	111–150	82.052

HYDRAULIC-CYLINDER APPLICATION[1]

Shopmen can apply hydraulic cylinders in many ways: to speed up operations, to power fixtures, to build special equipment, and to rejuvenate old machine tools. Accompanying tables and formulas, with a sample calculation, will assist the practical man to use hydraulics.

Cylinders. It is impractical and uneconomical to make hydraulic cylinders, considering the high state of development and reliability of commercial products.

The double-acting cylinder is used for pressure up to 1500 psi, which is seldom exceeded in machine-shop applications. Various forms of attachments are available for the piston rod. Bores are standardized. Almost any stroke is available.

Tabular Data. To determine the pipe size suitable for a particular hydraulic system, refer to Table 13. Friction pressure drops are given at various oil velocities and the flows for various pipe sizes, plus loss through fittings. Table 14 gives the output forces of various sizes of cylinders, both push and pull.

Formulas. The following formulas are sufficient for calculations.

$$\text{Cylinder force } F = PA$$

where F = force, lb; P = unit pressure, psi; and A = piston area, sq in.

$$\text{Oil delivery gpm} = \frac{AV}{231}$$

where A = piston area, sq in.; and V = piston velocity, ipm.

$$\text{Oil velocity in pipes } V = \frac{0.32 \text{ gpm}}{A}$$

where V = oil velocity, fps; and A = internal pipe area, sq in.

$$\text{Theoretical horsepower } H_t = 0.000583P \times \text{gpm}$$

where H_t = approx input horsepower; P = operating pressure, psi; and gpm = pump delivery.

$$\text{Actual horsepower } H_a = H_t \times \frac{100}{E}$$

where E = efficiency of hydraulic pump and system, %.

Sample Problem. Assume that, to modernize a planer-table drive, the hydraulic cylinder must exert a pressure of 3000 lb on the cutting tool, permits a maximum cutting speed of 100 fpm, and operates at 1000 psi. On the cutting stroke, the equivalent pipe length is 12 ft and a four-way valve is used for control. To be found are: (1) size of cylinder, (2) required pump delivery, (3) pipe size, (4) return speed of table, (5) motor horsepower if pump efficiency is 90% and cylinder efficiency is 95%.

Solution: Cylinder area $A = F/P = 3000 \div 1000 = 3$ sq in. For a 2-in. bore, $A = 3.14$ sq in.

Pump delivery gpm $= AV/231 = 3.14 \times 100 \times 12 \div 231 = 16.31$.

Pipe size is determined from Table 13, knowing gpm. Oil flow of 16.6 gpm at 10 fps requires ¾-in. pipe.

Pressure drop for 12 ft of pipe (Table 13) equals $12/10 \times 3.70 = 4.44$ psi. Average pressure drop through a four-way valve is 30 psi. Hence, total pressure drop is 34.44 psi. Pump operating pressure equals $1000 + 34.44 = 1034.44$ psi.

[1] Woodrow L. Wroble, hydraulic engineer, Wroble Engineering Co.

TABLE 13. OIL FLOW AND FRICTION PRESSURE DROP FOR 10 FT OF PIPE*

Pipe Size	ID	Area, Sq In.	Gpm	Drop, Psi	Gpm	Drop, Psi	Gpm	Drop, Psi	Gpm	Drop, Psi	Gpm	Drop, Psi	Gpm	Drop, Psi
½	0.622	0.304	4.7	1.57	9.4	5.85	14.1	12.15	18.6	20.65	23.5	31.30	28.2	43.4
¾	0.824	0.533	8.3	1.17	16.6	3.70	24.9	7.10	33.2	15.20	41.5	23.00	49.8	31.7
1	1.049	0.864	13.5	0.90	26.9	3.23	40.4	6.73	53.8	11.55	67.3	17.25	80.8	24.4
1¼	1.380	1.495	23.3	0.64	46.5	2.31	69.8	4.88	93.0	7.55	116.3	12.40	139.6	17.4
1½	1.610	2.036	31.7	0.54	63.4	1.81	95.1	4.04	126.8	6.91	158.5	10.42	190.2	14.8
2	2.067	3.355	52.3	0.47	104.5	1.69	156.8	3.60	209.0	6.09	261.3	9.27	313.6	11.1

Pressure Drop in Fittings (Equivalent Pipe Ft)

Pipe Size	Elbow	Tee	Gate Valve	Globe Valve	Valves 3 and 4 Way
½	1.5	3.3	0.34	6.2	12 to 60
¾	2.2	4.6	0.46	9.4	20 to 100
1	2.8	5.7	0.60	12.4	25 to 125
1¼	3.7	7.8	0.81	15.6	30 to 150
1½	4.4	9.2	0.92	18.9	40 to 200
2	5.5	12.0	1.20	23.5	50 to 250

* Miller Motor Co.

Pressure drop shown is for ordinary wrought-iron pipe. For smooth new wrought-iron pipe, multiply the values shown by 0.7. For very smooth straight tubing, multiply the values shown by 0.54. The pressure drop is the same regardless of operating pressure. Avoid large pressure drops in low-pressure systems. Oil flows through large pipes at high velocity (up to 30 fps) with small pressure loss.

TABLE 14. PISTON AREAS, FLUID DISPLACEMENTS, AND OUTPUT FORCES FOR CYLINDERS WITH STANDARD PISTON ROD

Bore Size, In.	Piston Areas, Sq In.		Displacement, Cu In. per In. Stroke		250 Psi		500 Psi		1000 Psi		1500 Psi	
	Major Area	Minor Area	Head End	Rod End	Push	Pull	Push	Pull	Push	Pull	Push	Pull
1	0.78	0.58	0.78	0.58	195	145	390	290	780	580	1,170	870
1 9/16	1.91	1.47	1.91	1.47	477	368	955	735	1,910	1,470	2,865	2,205
2	3.14	2.54	3.14	2.54	785	635	1,570	1,270	3,140	2,540	4,710	3,810
2½	4.91	3.91	4.91	3.91	1,228	978	2,455	1,955	4,910	3,910	7,365	5,865
3	7.06	5.84	7.06	5.84	1,765	1,460	3,530	2,920	7,060	5,840	10,590	8,760
3½	9.62	7.86	9.62	7.86	2,405	1,965	4,810	3,930	9,620	7,860	14,430	11,790
4	12.56	10.80	12.56	10.80	3,140	2,700	6,280	5,400	12,560	10,800	18,840	16,200
4½	15.90	13.50	15.90	13.50	3,975	3,375	7,950	6,750	15,900	13,500	23,850	20,250
5	19.63	16.49	19.63	16.49	4,908	4,123	9,815	8,245	19,630	16,490	29,445	24,735
6	28.27	24.30	28.27	24.30	7,068	6,075	14,135	12,150	28,270	24,300	42,405	36,450
7	38.48	33.58	38.48	33.58	9,620	8,395	19,240	16,790	38,480	33,580	57,720	50,370
8	50.26	43.20	50.26	43.20	12,565	10,800	25,130	21,600	50,260	43,200	75,390	64,800

Return speed of the planer table is found by getting the area of the piston at the rod end, from Table 14. This value is 2.54 in. Hence, return speed $V = \text{gpm} \times 231/A = 16.31 \times 231 \div 2.54 = 1483$ ipm, or 124 fpm.

Pump horsepower $H_t = 0.000583 \times 1034.44 \times 16.31 \div 0.90 \times 0.95 = 11.4$ hp. Inasmuch as the driving motor is not under continuous load, a 10-hp electric motor will serve to drive the pump.

AC MOTORS[1]

New Frame Sizes. In 1953 the National Electrical Manufacturers Association issued revised dimensions of ac motors from 1 to 30 hp, the popular range of sizes for metalworking equipment. This revision, the first change in over-all mounting dimensions since 1929, was made possible by better insulating materials, silicon steels, lubrication, cooling methods, etc. In general, the motors are smaller than the older frame sizes but have larger shafts to transmit the higher horsepower rating per frame.

Transition bases are available to make the new NEMA motor fit in the old mounting dimensions. Thus, plants having an inventory of old and new motors should have little difficulty with making replacements on production machinery. However, coupling sizes on the new motors are larger.

Table 15 indicates the horsepower and speed for single-phase and three-phase motors and shows the old and the new NEMA frame size for standard motors. As in the past, motors of special design may not line up with the same frame-size rating chart. Mounting dimensions listed by NEMA are also shown for the old and the new ratings (Tables 16 and 17).

New Applications. The reduced weight and size of the new motors makes them more adaptable for application on new-equipment designs. Motor mounting space is always at a premium. Many times an intermediate drive must be employed, because of inability to apply a direct drive in the available space. The new motors are significantly smaller and have a liberal factor of safety in respect to temperature rise.

Types of Motor Construction. Motors can be divided into two general construction categories: (1) motors having special electrical features, and (2) motors having special mechanical features.

1. Electrical features available in the past will be continued with the new NEMA motors:

Normal-starting-torque motor, NEMA design B.

High-starting-torque motor, NEMA design C.

High-slip, high-starting motor for punch-press operation, NEMA design D.

Torque motors.

Multispeed motors.

Motors designed to operate on the various frequencies of power companies throughout the world.

Motors designed to operate on special voltages.

Motors designed for single-phase and three-phase operation.

2. Motors can be classified under mechanical features as follows:

Drip-proof motors.

Totally inclosed fan-cooled motors.

Motors for vertical operation.

Motors for end-shield mounting with the C-face or D-flange.

Motors with class B insulation for high-temperature operation.

Motors with special shafts to meet special mechanical requirements.

[1] C. G. Grant, Small Integral Motor Department, General Electric Co.

Table 15. New vs. Old NEMA Frame Numbers for Integral-horsepower AC Motors

Polyphase Motors, Designs A and B Horizontal and Vertical, 60 Cycles, 550 Volts and Less, General-purpose, Squirrel-cage

Open Type, 40 C

Hp	3600 Rpm		1800 Rpm		1200 Rpm		900 Rpm		720 Rpm		600 Rpm	
	Old	New	Old	New	Old	New	Old	New	Old	New	Old	New
½	....	....	....	....	....	....	204	182				
¾	....	....	....	....	203	182	224	184				
1	....	....	203	182	204	184	225	213	254	213	254	215
1½	203	182	204	184	224	184	254	213	254	215	284	254U
2	204	184	224	184	225	213	254	215	284	254U	324	256U
3	224	184	225	213	254	215	284	254U	324	256U	326	286U
5	225	213	254	215	284	254U	324	256U	326	286U	364	324U
7½	254	215	284	254U	324	256U	326	284U	364	324U	365	326U
10	284	254U	324	256U	326	284U	364	286U	365	326U		
15	324	256U	326	284U	364	324U	365	326U				
20	326	284U	364	286U	365	326U						
25	364S	286U	364	324U								
30	364S	324S	365	326U								
40	365S	326S										

Totally Inclosed, Fan-cooled Type, 55 C

Hp	3600 Rpm		1800 Rpm		1200 Rpm		900 Rpm		720 Rpm		600 Rpm	
	Old	New	Old	New	Old	New	Old	New	Old	New	Old	New
½	....	....	....	....	....	....	204*	182				
¾	....	....	....	....	203*	182	224*	184				
1	....	....	203*	182	204*	184	225*	213	254	213	254	215
1½	203*	182	204*	184	224*	184	254*	213	254	215	284	254U
2	204*	184	224*	184	225*	213	254*	215	284	254U	324	256U
3	224	184	225*	213	254*	215	284*	254U	324	256U	326	286U
5	225	213	254	215	284	254U	324	256U	326	286U	365	324U
7½	254	215	284	254U	324	256U	326	284U	365	324U	†	326U
10	284	254U	324	256U	326	284U	364	286U	†	326U		
15	324	256U	326	284U	364	324U	365	326U				
20	326	286U	364	286U	365	326U						
25	365S	326S	365	326U								

* Totally inclosed, nonventilated.
† No previously assigned standard.

Polyphase Motors, Design C, Horizontal and Vertical, 60 Cycles, 550 Volts and Less, General-purpose, Squirrel-cage

Open Type, 40 C

Hp	1800 Rpm		1200 Rpm		900 Rpm	
	Old	New	Old	New	Old	New
3	...	....	254	215	284	254U
5	254	215	284	254U	324	256U
7½	284	254U	324	256U	326	284U
10	324	256U	326	284U	364	286U
15	326	284U	364	324U	365	326U
20	364	286U	365	326U		
25	364	324U				
30	365	326U				

Totally Inclosed Fan-cooled Type, 55 C

Hp	1800 Rpm		1200 Rpm		900 Rpm	
	Old	New	Old	New	Old	New
3	...	....	254	215	284	254U
5	254	215	284	254U	324	256U
7½	284	254U	324	256U	326	284U
10	324	256U	326	284U	364	286U
15	326	284U	364	324U		
20	364	286U				
25	365	326U				

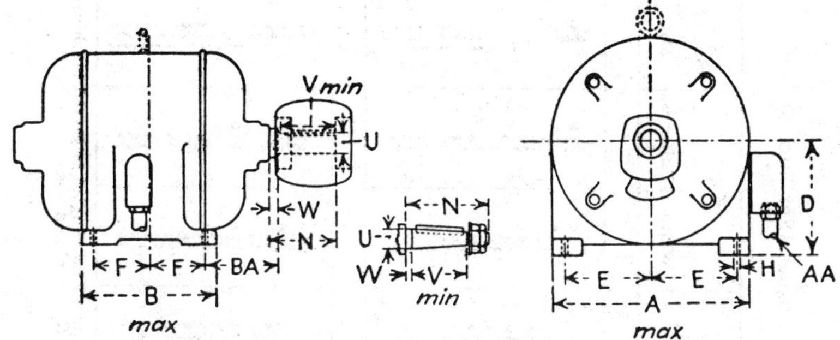

FIG. 8. Mounting dimensions of foot-mounted induction motors.

Motor Selection. Both open and totally inclosed motors are used in metal-working plants. Motors on machine tools are subject to chips and metal dust, splashing coolant, etc., and may be hidden away in machine compartments. The generally preferred type of motor, especially in production shops, is the totally inclosed fan-cooled type.

Machines like presses often operate under a widely fluctuating load during the cycle. The high-starting-torque, high-slip motor has been designed for this application.[1]

Conveyors and hoists create other motor-selection problems. For hoist applications, the high-torque squirrel-cage motor is recommended.

If only a case of outright replacement is involved, the nameplate rating is sufficient, because nameplate ratings of both the old and new motors are exactly the same.

For new designs of equipment or new plant installations, the motor application chart in Table 18 will be of some service. Tables of starting torques, inrush, and operating currents are not given in this handbook, because the developments in motors are being made so rapidly that the motor manufacturer or his local application engineer should be consulted. In fact, a good many motors have been successfully applied by rules of thumb and experience, but on the whole it is safer and better practice to get expert assistance from application engineers of motor manufacturers. If a designer is faced with selection of motors that will be used in quantity, he can often get special motors that are tailored to his needs for approximately the same cost, and maybe even a lower cost, than standard motors.

DC MOTORS

Easily adjusted speed is the primary advantage of the dc motor. However, until recently the dc motor has been greatly overshadowed by the ac motor because of power-distribution losses, cost, and maintenance of earlier types of controls.

Developments of comparatively recent origin are again focusing attention on the dc drive. For presses, it has been demonstrated that control over the number of strokes per minute and/or drawing speed enables the shop to pick the best speed for a given operation. Work characteristics and die life are thereby improved.

DC drives are now of two fundamental types:

1. In the constant-potential system, a dc motor with starter and rheostat speed

[1] Ed. note: Motor application on presses is a highly controversial subject.

TABLE 16. DIMENSIONS FOR SINGLE AND POLYPHASE, INDUCTION, FOOT-MOUNTED MOTORS (SEE FIG. 8)
All Dimensions in Inches

Frame No.	A Max	B Max	D*	E	F	BA	H	N-W	U	V Min	Key		AA Size of Conduit Min
											Width	Thickness	
Old-standard Frame Sizes													
203	10	7 1/2	5	4	2 3/4	3 1/8	13/32	2 1/4	3/4	2	3/16	3/16	3/4
204	10	8 1/2	5	4	3 1/4	3 1/8	13/32	2 1/4	3/4	2	3/16	3/16	3/4
224	11	8 3/4	5 1/2	4 1/2	3 3/8	3 1/2	13/32	3	1	2 3/4	1/4	1/4	3/4
225	11	9 1/2	5 1/2	4 1/2	3 3/4	3 1/2	13/32	3	1	2 3/4	1/4	1/4	3/4
254	12 1/2	10 3/4	6 1/4	5	4 1/8	4 1/4	17/32	3 3/8	1 1/8	3 1/8	1/4	1/4	1
284	14	12 1/2	7	5 1/2	4 3/4	4 3/4	17/32	3 3/4	1 1/4	3 1/2	1/4	1/4	1
324	16	14	8	6 1/4	5 1/4	5 1/4	21/32	4 7/8	1 5/8	4 5/8	3/8	3/8	1 1/4
326	16	15 1/2	8	6 1/4	6	5 1/4	21/32	4 7/8	1 5/8	4 5/8	3/8	3/8	1 1/4
364	18	15 1/4	9	7	5 5/8	5 7/8	21/32	5 5/8	1 7/8	5 3/8	1/2	1/2	1 1/2
364S	18	15 1/4	9	7	5 5/8	5 7/8	21/32	3 1/4	1 5/8	3	3/8	3/8	1 1/2
365	18	16 1/4	9	7	6 1/8	5 7/8	21/32	5 5/8	1 7/8	5 3/8	1/2	1/2	1 1/2
365S	18	16 1/4	9	7	6 1/8	5 7/8	21/32	3 1/4	1 5/8	3	3/8	3/8	1 1/2
1953 Standard Frame Sizes													
182	9	6 1/2	4 1/2	3 3/4	2 1/4	2 3/4	13/32	2 1/4	7/8	2	3/16	3/16	3/4
184	9	7 1/2	4 1/2	3 3/4	2 3/4	2 3/4	13/32	2 1/4	7/8	2	3/16	3/16	3/4
213	10 1/2	7 1/2	5 1/4	4 1/4	2 3/4	3 1/2	13/32	3	1 1/8	2 3/4	1/4	1/4	3/4
215	10 1/2	9	5 1/4	4 1/4	3 1/2	3 1/2	13/32	3	1 1/8	2 3/4	1/4	1/4	3/4
254U	12 1/2	10 3/4	6 1/4	5	4 1/8	4 1/4	17/32	3 3/4	1 3/8	3 1/2	5/16	5/16	1
256U	12 1/2	12 1/2	6 1/4	5	5	4 1/4	17/32	3 3/4	1 3/8	3 1/2	5/16	5/16	1
284U	14	12 1/2	7	5 1/2	4 3/4	4 3/4	17/32	4 7/8	1 5/8	4 5/8	3/8	3/8	1 1/4
286U	14	14	7	5 1/2	5 1/2	4 3/4	17/32	4 7/8	1 5/8	4 5/8	3/8	3/8	1 1/4
324U	16	14	8	6 1/4	5 1/4	5 1/4	21/32	5 5/8	1 7/8	5 3/8	1/2	1/2	1 1/2
324S	16	14	8	6 1/4	5 1/4	5 1/4	21/32	3 1/4	1 5/8	3	3/8	3/8	1 1/2
326U	16	15 1/2	8	6 1/4	6	5 1/4	21/32	5 5/8	1 7/8	5 3/8	1/2	1/2	1 1/2
326S	16	15 1/2	8	6 1/4	6	5 1/4	21/32	3 1/4	1 5/8	3	3/8	3/8	1 1/2

* Dimension *D* will never be greater than the above values, but it may be less, so that shims are usually required for coupled or geared machines. When the exact dimension is required, shims up to 1/32 in. may be necessary on frame sizes whose dimension *D* is 8 in. and less.

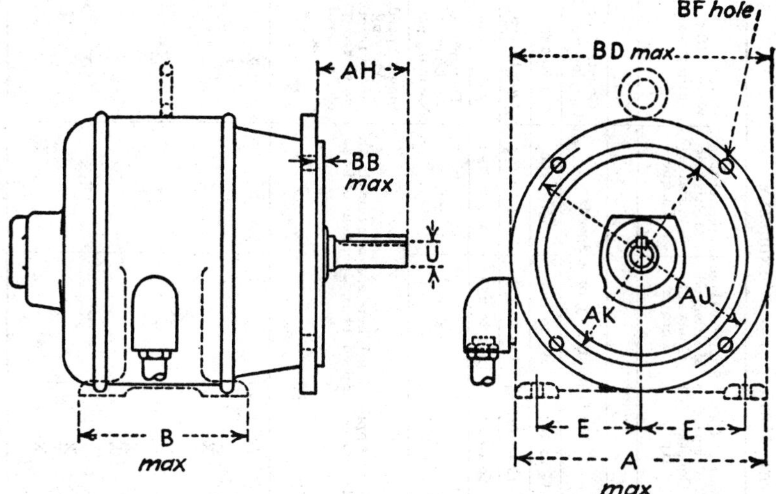

Fig. 9. Mounting dimensions of flange-mounted Type D motors, ac and dc, the type much favored in machine-tool application.

Table 17. Dimensions for Type D Flange-mounted, Single-phase and Polyphase, Squirrel-cage Induction Motors (See Fig. 9)

Frame No.	*AJ*	*AK*	*BD* (Max)	*AH*	*BB* (Max)
182D and 184D	10	9	11	$2\frac{1}{4}$	$\frac{1}{4}$
213D and 215D	10	9	11	3	$\frac{1}{4}$
254UD and 256UD	$12\frac{1}{2}$	11	14	$3\frac{3}{4}$	$\frac{1}{4}$
284UD and 286UD	$12\frac{1}{2}$	11	14	$4\frac{7}{8}$	$\frac{1}{4}$
324UD and 326UD	16	14	18	$5\frac{5}{8}$	$\frac{1}{4}$
324SD and 326SD	16	14	18	$3\frac{1}{4}$	$\frac{1}{4}$

Frame No.	*U*	Key		Clearance Hole *BF*	
		Width	Thickness	Size	Number
182D and 184D	$\frac{7}{8}$	$\frac{3}{16}$	$\frac{3}{16}$	$\frac{17}{32}$	4
213D and 215D	$1\frac{1}{8}$	$\frac{1}{4}$	$\frac{1}{4}$	$\frac{17}{32}$	4
254UD and 256UD	$1\frac{3}{8}$	$\frac{5}{16}$	$\frac{5}{16}$	$\frac{13}{16}$	4
284UD and 286UD	$1\frac{5}{8}$	$\frac{3}{8}$	$\frac{3}{8}$	$\frac{13}{16}$	4
324UD and 326UD	$1\frac{7}{8}$	$\frac{1}{2}$	$\frac{1}{2}$	$\frac{13}{16}$	4
324SD and 326SD	$1\frac{5}{8}$	$\frac{3}{8}$	$\frac{3}{8}$	$\frac{13}{16}$	4

TABLE 18. POLYPHASE MOTOR APPLICATION CHART FOR THE METALWORKING INDUSTRY*

Type of Motor	Standard Squirrel Cage Normal Torque, Normal Starting Current	Squirrel Cage High Torque, Low Starting Current	Squirrel Cage Punch Press High Torque, High Slip	Squirrel Cage Multispeed Constant Hp	Squirrel Cage Multispeed Constant Torque	Squirrel Cage Multispeed Variable Torque	Wound Rotor Slip Ring
NEMA Design	B	C	D				
Speed Classification	Constant Speed	Constant Speed	Constant or Multispeed	Multispeed, 2, 3, or 4 Speeds			Constant or Varying Speed
Range of Hp Ratings	¼ to 150	3 to 100	1 to 100	½ to 100	½ to 100	½ to 100	1 to 150
Starting Torque, % of Full Load	125 to 175	200 to 250	350	125 to 175	125 to 175	125 to 175	200 to 250
Pull-out Torque, % of Full Load	200 to 250	175 to 225	300 to 400	200 to 250	200 to 250	200 to 250	200 to 250
Speed, % Slip	2 to 5	3 to 5	5 to 13	2 to 5	2 to 5	2 to 5	Dependent on Load at Normal Speed
General Remarks, Characteristics	General Purpose Wide Application	Heavy Starting Loads, Simple Control	High Slip High Torque Suitable for Intermittent Load with Flywheel	Wide Application Where 2, 3, or 4 Speeds Are Desirable. Speed Is Independent of Load.			Single Speed or Varying Speed from 50 to 100% of Full Load Speed
Typical Applications							
Bending rolls	X	X	..	..	..	..	X
Bending brakes	..	..	X	..	..	..	X
Boring mills	X	..	..	X			
Broaching machines	X	..	..	X			
Burring machines	X	..	..	..	X		
Chuckers	X	..	..	..	X		
Conveyors, loaded start	..	X	..	..	X	..	X
Cranes, hoists, elevators	..	X	X	..	..	..	X
Drilling machines, radial, multiple	X	..	..	X			
Forging, forming machines	..	..	X	X			
Gear reducers (depending on application)	X	X	X	..	X	X	X
Gear cutters	X	..	..	X			
Grinders, buffers, polishers	X	..	..	X	X		
Hobbers	X	..	..	X			
Honers	X	..	..	..	X		

Type of Motor	Standard Squirrel Cage Normal Torque, Normal Starting Current	Squirrel Cage High Torque, Low Starting Current	Squirrel Cage Punch Press High Torque, High Slip	Squirrel Cage Multispeed Constant Hp	Squirrel Cage Multispeed Constant Torque	Squirrel Cage Multispeed Variable Torque	Wound Rotor Slip Ring
NEMA Design	B	C	D				
Speed Classification	Constant Speed	Constant Speed	Constant or Multispeed	Multispeed, 2, 3, or 4 Speeds			Constant or Varying Speed
Range of Hp Ratings	¼ to 150	3 to 100	1 to 100	½ to 100	½ to 100	½ to 100	1 to 150
Starting Torque, % of Full Load	125 to 175	200 to 250	350	125 to 175	125 to 175	125 to 175	200 to 250
Pull-out Torque, % of Full Load	200 to 250	175 to 225	300 to 400	200 to 250	200 to 250	200 to 250	200 to 250
Speed, % Slip	2 to 5	3 to 5	5 to 13	2 to 5	2 to 5	2 to 5	Dependent on Load at Normal Speed
General Remarks, Characteristics	General Purpose Wide Application	Heavy Starting Loads, Simple Control	High Slip High Torque Suitable for Intermittent Load with Flywheel	Wide Application Where 2, 3, or 4 Speeds Are Desirable. Speed Is Independent of Load.			Single Speed or Varying Speed from 50 to 100% of Full Load Speed
Typical Applications							
Keyseaters	X	..	..	..	X		
Lathes	X	..	..	X			
Line shafts (starting unloaded)	X						
Milling machines	X	..		X			
Punch presses, shears, and hammers	..	..	X	..	X	..	X
Shapers, slotters	X	..	..	..	X		
Shears	..	..	X	..	X	..	X
Swagers	X	X	X	..	X		
Superfinishers, lappers	X	..	..	..	X		
Threading, tapping machines	X	..	..	..	X		
Upsetters headers	..	..	X	..	X		
Welders	X	..	..	..	X		

Motors shown are electrical types. Each electrical type can be obtained in a variety of mechanical or physical modifications. All types of inclosures (open, drip-proof, totally inclosed fan-cooled, or explosion proof) are available in any electrical type.

* General Electric Company.

control is connected to a dc bus or to a large motor-generator set that supplies a number of machines. The rheostat varies the voltage to the motor-field windings and hence the motor speed.

2. In the adjustable potential system, the plant's ac power is converted to dc of the needed voltage by either a motor-generator set or an electronic rectifier. Relays are provided for stopping and starting the dc motors.

It appears that the second type of dc drive will become increasingly popular. Packaged units up to 200 hp are available.

There are two fundamental uses for dc drives on machine tools: (1) infinitely variable spindle speeds over a broad range, and (2) infinitely variable feeds over a broad range.

By putting a dc drive on a lathe spindle, for example, it is possible to face large surfaces, such as impeller wheels, at constant work surface speed. Theoretically, depending upon ratio of maximum to minimum diameter, a 50% decrease in machining time is possible when the work is progressively speeded up as the tool approaches center.

Simplicity of control also adapts dc drives to contour-turning lathes and turret lathes. In the latter case, the control box is preset to give the desired work rpm each time the turret faces are indexed.

In respect to machine feeds, the fundamental problems in designing a dc drive are: (1) a suitable speed range, (2) speed regulation, and (3) adequate low-speed torque.

One answer to these requirements is the "packaged" device that provides infinite feeds over a wide speed range.

V-BELT DRIVES[1]

Selection of a V-belt drive should be based on the nature of the load and the type of driving unit. The service factors in Table 19 correlate the type of load with the various types of driving units. To obtain the horsepower capacity of the proper V-belt drive, multiply the rated horsepower (nameplate rating) of the driving unit by the service factor.

If the arc of contact on the small sheave is less than 180°, divide the horsepower obtained above by the correction factor in Table 20.

$$\text{Arc of contact} = 180 - \frac{60(D - d)}{C}$$

where D = dia of large sheave or flat pulley, d = dia of small sheave, and C = center distance. A V-flat belt is composed of a small sheave and a large flat pulley.

V-belt dimensions are shown in Fig. 10. To find the belt cross-section, use the chart (Fig. 11). If the point found falls on the dividing line between two sizes, investigate the two possible choices.

[1] Multiple V-belt Drive and Mechanical Power Transmission Association, 1951.

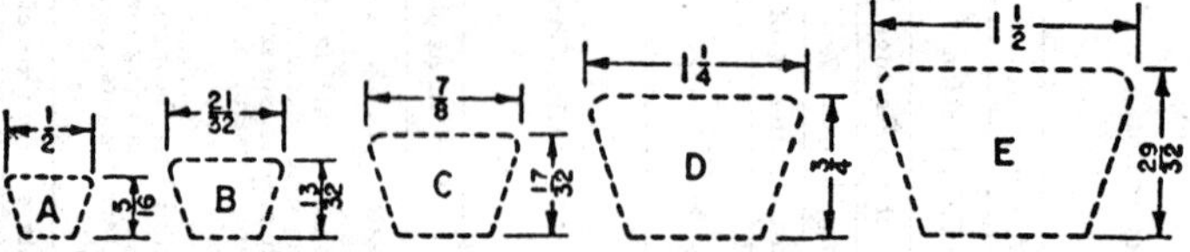

Fig. 10. Nominal cross-sections and dimensions of V-belts.

TABLE 19. SERVICE FACTORS FOR V-BELT DRIVES ACCORDING TO LOAD

Applications	AC								DC Motors	
	Squirrel Cage				Synchronous		Single Phase			
	Normal Torque Line Start	Normal Torque Compensator Start	High Torque	Wound Rotor (Slip Ring)	Normal Torque	High Torque	Repulsion and Split-phase	Capacitor	Shunt-wound	Compound-wound
Compressors:										
Centrifugal	1.2	1.2	...	1.4	1.4	...	...	...	1.2	
Rotary	1.2	1.2	...	1.4	1.4	...	1.2	1.2	1.2	
Reciprocating:										
3 or more cylinders	1.2	1.2	...	1.4	1.4	...	...	...	1.2	
1 or 2 cylinders	1.4	1.4	...	1.5	1.5	...	...	...	1.2	
Conveyors:										
Apron	...	1.4	1.6	...	...	...	...	...	1.4	
Belt (light package)	...	1.0	1.1	...	...	...	...	...	1.0	
Oven	...	1.0	1.1	...	...	...	...	...	1.0	
Screw	...	1.6	1.8	...	...	...	...	...	1.6	
Elevator	...	1.4	1.6	...	...	...	...	...	1.4	
Fans and Blowers:										
Centrifugal	1.2	1.2	...	1.4	...	...	...	...	1.2	
Propeller	1.4	1.4	2.0	1.6	...	2.0	...	...	1.4	
Induced draft	1.2	1.2	...	1.4	...	...	...	...	1.4	
Positive blowers	1.6	1.6	...	2.0	2.0	2.0				
Exhausters	1.2	1.2	...	1.4	...	...	...	...	1.4	
Machine Tools:										
Grinders	1.2	...	...	1.4	...	...	1.2	1.0	1.2	1.2
Boring mills	1.2	...	...	1.4	...	...	...	...	1.2	1.2
Lathes	1.0	...	...	1.2	...	...	1.0	1.0	1.0	1.0
Milling machines	1.2	...	...	1.4	...	...	...	...	1.2	1.2
Screw machines	1.0	...	...	1.0	...	...	1.0	1.0	1.0	1.0
Cam cutters	1.0	...	...	1.0	...	...	...	...	1.0	1.0
Planers	1.2	...	...	1.4	...	...	1.2	1.0	1.2	1.2
Shapers	1.0	...	...	1.0	...	...	1.0	1.0	1.0	1.0
Drill press	1.0	...	...	1.0	...	...	1.0	1.0	1.0	1.0
Drop hammers	1.0	...	...	1.0	...	...	1.0	1.0	1.0	1.0
Shears	1.2	...	...	1.4	...	...	1.2	1.2	1.2	1.2
Pumps:										
Centrifugal	1.2	1.2	1.4	1.4	...	...	1.2	1.2		
Gear	1.2	1.2	1.4	1.4	...	...	1.2	1.2		
Rotary	1.2	1.2	1.4	1.4	...	...	1.2	1.2	1.2	
Reciprocating:	1.2	1.2	...	1.4	1.4	1.6				
3 or more cylinders	1.4	1.4	...	1.6	1.6	1.8				

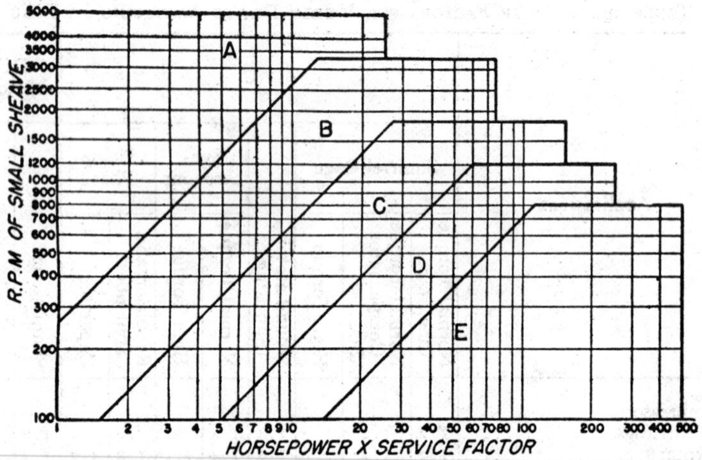

FIG. 11. Chart for determining the size of a V-belt, considering rpm of small sheave and horsepower to be transmitted.

Horsepower ratings for V-belts can be calculated:

Cross-section	Hp Formula	Max d
A	$\text{Hp} = \left(1.589 - \dfrac{2.702}{d} - 0.0146S^2\right) S$	5
B	$\text{Hp} = \left(2.822 - \dfrac{7.725}{d} - 0.0251S^2\right) S$	7
C	$\text{Hp} = \left(5.882 - \dfrac{26.971}{d} - 0.0397S^2\right) S$	12
D	$\text{Hp} = \left(12.628 - \dfrac{96.991}{d} - 0.0815S^2\right) S$	17
E	$\text{Hp} = \left(19.238 - \dfrac{223.04}{d} - 0.1233S^2\right) S$	28

S = belt speed, fpm, divided by 1000.

TABLE 20. CORRECTION FACTORS FOR CONTACT LESS THAN 180°

Arc of Contact	V-V	V-flat	Arc of contact	V-V	V-flat
180	1.00	0.75	130	0.86	0.86
170	0.98	0.77	120	0.82	0.82
160	0.95	0.80	110	0.78	0.78
150	0.92	0.82	100	0.74	0.74
140	0.89	0.84	90	0.69	0.69

TABLE 21. STANDARD V-BELT PITCH LENGTHS

Standard Nominal Length	Standard Pitch Lengths, In.					Permissible Deviations from Standard Pitch Length, In.		Matching Limits fo One Set, In.
	A	B	C	D	E			
26	27.3	.....	.....	.....	.....	+0.7	−0.3	0.10
31	32.3	.....	.....	.....	.....	+0.7	−0.3	0.10
33	34.3	.....	.....	.....	.....	+0.8	−0.4	0.20
35	36.3	36.8	.....	.....	.....	+0.8	−0.4	0.20
38	39.3	39.8	.....	.....	.....	+0.8	−0.4	0.20
42	43.3	43.8	.....	.....	.....	+0.8	−0.4	0.20
46	47.3	47.8	.....	.....	.....	+0.8	−0.4	0.20
48	49.3	49.8	.....	.....	.....	+0.9	−0.5	0.20
51	52.3	52.8	53.9	.....	.....	+0.9	−0.5	0.20
53	54.3	54.8	.....	.....	.....	+0.9	−0.5	0.20
55	56.3	56.8	.....	.....	.....	+0.9	−0.5	0.20
60	61.3	61.8	62.9	.....	.....	+0.9	−0.5	0.20
62	63.3	63.8	.....	.....	.....	+0.9	−0.5	0.20
64	65.3	65.8	.....	.....	.....	+0.9	−0.5	0.20
66	67.3	67.8	.....	.....	.....	+0.9	−0.5	0.20
68	69.3	69.8	70.9	.....	.....	+0.9	−0.5	0.20
71	72.3	72.8	.....	.....	.....	+0.9	−0.5	0.20
75	76.3	76.8	77.9	.....	.....	+0.9	−0.5	0.20
78	79.3	79.8	.....	.....	.....	+1.0	−0.5	0.30
80	81.3	.....	.....	.....	.....	+1.0	−0.5	0.30
81	.....	82.8	83.9	.....	.....	+1.0	−0.5	0.30
83	.....	84.8	.....	.....	.....	+1.0	−0.5	0.30
85	86.3	86.8	87.9	.....	.....	+1.0	−0.5	0.30
90	91.3	91.8	92.9	.....	.....	+1.0	−0.5	0.30
96	97.3	.....	98.9	.....	.....	+1.0	−0.5	0.30
97	.....	98.8	.....	.....	.....	+1.0	−0.5	0.30
105	106.3	106.8	107.9	.....	.....	+1.1	−0.5	0.40
112	113.3	113.8	114.9	.....	.....	+1.1	−0.5	0.40
120	121.3	121.8	122.9	123.3	.....	+1.2	−0.5	0.40
128	129.3	129.8	130.9	131.3	.....	+1.3	−0.6	0.40
136	.....	137.8	138.9	.....	.....	+1.3	−0.6	0.40
144	.....	145.8	146.9	147.3	.....	+1.4	−0.6	0.40
158	.....	159.8	160.9	161.3	.....	+1.5	−0.6	0.40
162	.....	.....	164.9	165.3	.....	+1.6	−0.6	0.40
173	.....	174.8	175.9	176.3	.....	+1.7	−0.7	0.50
180	.....	181.8	182.9	183.3	184.5	+1.7	−0.7	0.50
195	.....	196.8	197.9	198.3	199.5	+1.8	−0.8	0.50
210	.....	211.8	212.9	213.3	214.5	+2.0	−0.8	0.50
240	.....	240.3	240.9	240.8	241.0	+2.2	−0.9	0.50
270	.....	270.3	270.9	270.8	271.0	+2.4	−1.0	0.50
300	.....	300.3	300.9	300.8	301.0	+2.5	−1.2	0.60
330	.....	.....	330.9	330.8	331.0	+2.5	−1.2	0.60
360	.....	.....	360.9	360.8	361.0	+2.5	−1.2	0.60
390	.....	.....	390.9	390.8	391.0	+3.0	−1.5	0.70
420	.....	.....	420.9	420.8	421.0	+3.5	−2.0	0.80
480	.....	.....	.....	480.8	481.0	+4.0	−2.5	0.90
540	.....	.....	.....	540.8	541.0	+4.5	−3.0	1.10
600	.....	.....	.....	600.8	601.0	+5.0	−3.5	1.30
660	.....	.....	.....	660.8	661.0	+6.0	−4.0	1.50

Center distance and belt length

$$L = 2C + 1.57(D + d) + \frac{(D - d)^2}{4C}$$

$$C = \frac{b + \sqrt{b^2 - 32(D - d)^2}}{16}$$

where $b = 4L - 6.28(D + d)$, D = pitch dia of large sheave, d = pitch dia of small sheave, L = pitch length of belt, and C = center distance, all in inches.

Installation. After calculating a center distance from a standard pitch length (Table 21) make provision to move the centers closer together by the amount shown in Table 22, in order to facilitate installing belts without injury.

Take-up. The centers should be adjustable over the calculated distance by an amount, as given in Table 23, because of possible belt stretch and wear of the grooves.

TABLE 22. MINIMUM ALLOWANCE BELOW STANDARD CENTER DISTANCE FOR APPLICATION OF BELTS

Dimensions in Inches

Nominal Lengths*	*A*	B	C	D	E
26 to 38	3/4	1			
38 to 60	3/4	1	1½		
60 to 90	3/4	1¼	1½		
90 to 120	1	1¼	1½		
120 to 158	1	1¼	1½	2	
158 to 195	...	1¼	2	2	2½
195 to 240	...	1½	2	2	2½
240 to 270	...	...	2	2½	2½
270 to 330	...	...	2	2½	3
330 to 420	...	...	2	2½	3
420 and over	...	...	...	3	3½

* In each group the range is to, but not including, the second length.

TABLE 23. MINIMUM ALLOWANCE ABOVE STANDARD CENTER DISTANCE FOR STRETCH AND WEAR

Dimensions in Inches

*Nominal Lengths**	*All Sections*
26 to 38	1
38 to 60	1½
60 to 90	2
90 to 120	2½
120 to 158	3
158 to 195	3½
195 to 240	4
240 to 270	4½
270 to 330	5
330 to 420	6
420 and over	1.5% of belt length

* In each group the range is to, but not including, the second length.

V-BELT SHEAVES

Outside rims or faces of sheaves are grooved to work in conjunction with standard multiple V-belts. Sheaves may be cast iron, steel, or other materials. All grooves are to be smooth and the edges rounded (Fig. 12).

$$\text{Face width} = S(N - 1) + 2E$$

where N = number of grooves.

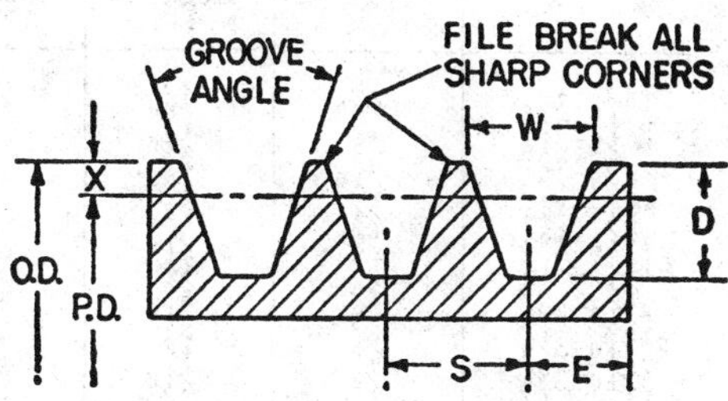

FIG. 12. Proportions of V-belt sheaves.

TABLE 24. DIMENSIONS OF SHEAVE GROOVES (SEE FIG. 12)

Belt	Min Recommended Pitch Dia	Pitch Dia	Groove Angle, Deg	Standard Groove					Deep Groove				
				W	*D*	*X*	*S*	*E*	*W*	*D*	*X*	*S*	*E*
A	3.0	2.6 to 5.4	34	0.494	0.490	0.125	5/8	3/8	0.589	0.645	0.280	3/4	7/16
		Over 5.4	38	0.504					0.611				
B	5.4	4.6 to 7.0	34	0.637	0.580	0.175	3/4	1/2	0.747	0.760	0.355	7/8	9/16
		Over 7.0	38	0.650					0.774				
C	9.0	7.0 to 7.99	34	0.879	0.780	0.200	1	11/16	1.066	1.085	0.505	1 1/4	13/16
		8.0 to 12.0	36	0.887					1.085				
		Over 12.0	38	0.895					1.105				
D	13.0	12.0 to 12.99	34	1.259	1.050	0.300	1 7/16	7/8	1.513	1.465	0.715	1 3/4	1 1/16
		13.0 to 17.0	36	1.271					1.541				
		Over 17.0	38	1.283					1.569				
E	21.0	18.0 to 24.0	36	1.527	1.300	0.400	1 3/4	1 1/8	1.816	1.745	0.845	2 1/16	1 5/16
		Over 24.0	38	1.542					1.849				

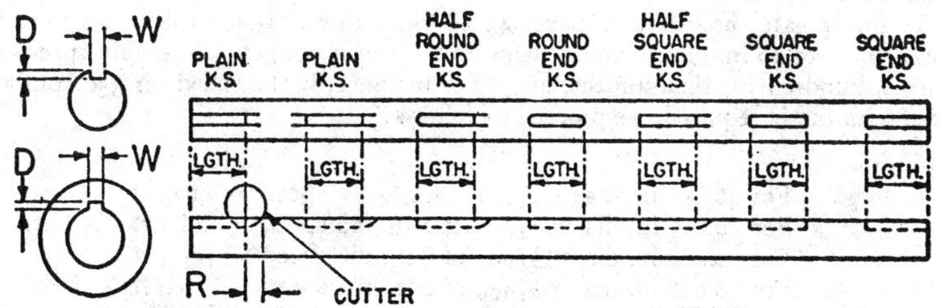

FIG. 13. Standard keyseats for V-belt drives (see Table 25).

TABLE 25. STANDARD KEYSEATS FOR V-BELT DRIVES (SEE FIG. 13)

Shaft Size	*W* Width +0.002 In. −0.000 In.	*D** Depth		*R* Max Cutter Runout	Max Dia of Cutter
		Regular	Shallow		
5/16–7/16	3/32	3/64	...	1/2	3 1/4
1/2–9/16	1/8	1/16	...	9/16	3 1/4
5/8–7/8	3/16	3/32	...	11/16	3 1/4
15/16–1 1/4	1/4	1/8	...	13/16	4
1 5/16–1 3/4	3/8	3/16	...	1 1/16	5
1 13/16–2 1/4	1/2	1/4	1/8	1 3/16	5
2 5/16–2 3/4	5/8	5/16	3/16	1 5/16	5
2 13/16–3 1/4	3/4	3/8	3/16	1 9/16	5 1/2
3 5/16–3 3/4	7/8	7/16	1/4	1 11/16	5 1/2
3 13/16–4 1/2	1	1/2	1/4	1 3/4	5 1/2
4 9/16–5 1/2	1 1/4	5/8	1/4	1 15/16	5 1/2
5 9/16–6 1/2	1 1/2	3/4	1/4	2 1/8	5 1/2
6 9/16–7 1/2	1 3/4	3/4	1/4	2 1/8	5 1/2
7 9/16–9	2	3/4	3/8	2 1/8	5 1/2
9 1/16–11	2 1/2	7/8	3/8	2 5/16	6
11 1/16–13	3	1	3/8	2 7/16	6

Shaft Keyseat. Always make straight, never taper. Use regular depth in shaft, even though shallow depth is used in hub.

Hub Keyseat. Make straight, unless a taper keyseat is specified. Make regular depth, except when shallow depth is specified.

Taper Keyseat. Use 1/8 in. per ft. Depth at large end equals *D*.

* Tolerance on Depth:
Shaft keyseat: +0.010 −0.000 in.
Straight keyseat in hub: +0.010 −0.000 in. (preferably +0.010 in.).
Taper keyseat in hub: +0.000 −0.010 in. (preferably −0.010 in.).

Tables of Circumferential Speeds

The tables on pages 43-38 to 43-39, which give circumferential speeds, can be used for obtaining gear and belt speeds and the speed of revolving parts of high-speed machinery.

For diameters greater than those given in the tables, the speeds can be obtained by adding together the speeds for two diameters whose sum equals that of the diameter for which we require the speed.

To interpolate the tables on pages 43-38 to 43-39 we can use the values given for speed for 1 to 10 in. dia, dividing them by 10, 100, 1,000, etc., to obtain speeds for tenths, hundredths, thousandths, etc. For instance, if the speed for 550 rpm and 46.186 in. dia is required, we proceed as follows:

For 46 in. dia	....	speed =	6,623 ft
For 0.1 in. dia = 1/10	of 1 in. dia	speed =	14.4 ft
For 0.08 in. dia = 1/100	of 8 in. dia	speed =	11.5 ft
For 0.006 in. dia = 1/1000	of 6 in. dia	speed =	0.9 ft
For 46.186 in. dia	....	speed =	6,650 ft

SPEEDS OF PULLEYS AND GEARS

The fact that the circumference of a pulley or gear is always 3.1416 or $3\frac{1}{7}$ times the diameter makes it easy to figure speeds by considering only the diameter of both driver and driven pulleys. Belting from one 6-in. pulley to another gives the same speed to both; but if the driving pulley is 16 in. and the driven pulley only 4 in. it is clear that the small pulley will turn 4 times for every turn of the large pulley. If this is reversed and the small pulley is the driver, the large pulley will make only one turn for every four of the small pulley. The same rule applies to gears if the *pitch diameter*, and not the outside diameter, is taken. The following rules have been arranged for convenience in finding any desired information about pulley or gear speeds.

Having	To Find	Rule
Dia driving pulley Dia driven pulley Speed of driving pulley	Speed of driven pulley	Multiply dia driving pulley by its speed and divide by dia driven pulley
Dia driving pulley Speed of driving pulley Speed of driven pulley	Dia driven pulley	Multiply dia of driving pulley by its speed and divide by speed of driven pulley
Dia driving pulley Dia driven pulley Speed of driven pulley	Speed of driving pulley	Multiply dia of driven pulley by its speed and divide by dia driving pulley
Dia driven pulley Speed of driven pulley Speed of driving pulley	Dia driving pulley	Multiply dia driven pulley by its speed and divide by speed of driving pulley

The rules given above apply equally well to a number of pulley belts together or to a train of gears if *all* the driving and *all* the driven pulley diameters and speeds are grouped together.

Power and Heat Equivalents

1 horsepower = 33,000 ft-lb per min = 550 ft-lb per sec
= 2545 Btu per hr = 746 watts

1 kilowatt = 1000 watts = 1.34 hp = 44236 ft-lb per min
= 737 ft-lb per sec = 3415 Btu per hr

$$\text{Steam engine hp} = \frac{PLAN}{33{,}000}$$

where P = mean effective pressure per sq in.; L = length of a double stroke, ft; A = area of piston, sq in.; and N = rpm.

$$\text{SAE hp} = D^2 \times \frac{N}{2.5}$$

where D = cylinder bore, in.; N = number of cylinders.

Table 26. Circumferential Speeds in Feet per Minute

Dia, In.	Rpm										
	50	100	150	200	250	300	350	400	450	500	550
1	13	26	39	52	65	79	92	105	118	131	144
2	26	52	79	105	131	157	183	209	236	262	288
3	39	79	118	157	196	236	275	314	353	393	432
4	52	105	157	209	262	314	367	419	471	523	576
5	65	131	196	262	328	393	458	524	589	654	720
6	79	157	236	314	393	471	550	628	707	785	863
7	92	183	275	367	458	550	641	733	825	916	1008
8	105	209	314	419	524	628	733	838	942	1047	1152
9	118	236	353	471	589	707	825	942	1060	1178	1296
10	131	262	393	524	655	785	916	1047	1178	1309	1440
11	144	288	432	576	720	864	1008	1152	1296	1440	1584
12	157	314	471	628	785	943	1100	1257	1414	1571	1728
13	170	340	511	681	851	1021	1191	1361	1532	1701	1872
14	183	367	550	733	916	1100	1283	1466	1649	1832	2016
15	196	393	589	785	982	1178	1375	1571	1767	1963	2160
16	209	419	628	838	1047	1257	1466	1675	1885	2094	2304
17	223	445	668	890	1113	1335	1558	1780	2003	2225	2442
18	236	471	707	943	1178	1414	1649	1885	2121	2356	2592
19	249	497	746	995	1244	1492	1741	1990	2238	2487	2730
20	262	524	785	1047	1309	1571	1833	2094	2356	2618	2880
21	275	550	825	1100	1374	1649	1924	2199	2474	2749	3024
22	288	576	864	1152	1440	1728	2016	2304	2592	2880	3168
23	301	602	903	1204	1505	1806	2107	2409	2710	3011	3312
24	314	628	943	1257	1571	1885	2199	2513	2827	3142	3456
25	327	655	982	1309	1636	1963	2291	2618	2945	3273	3600
26	340	681	1021	1361	1702	2042	2382	2723	3063	3403	3744
27	353	707	1060	1414	1767	2121	2474	2827	3181	3534	3888
28	367	733	1100	1466	1837	2199	2566	2932	3299	3665	4032
29	380	759	1139	1518	1898	2278	2657	3037	3417	3796	4176
30	393	785	1178	1571	1964	2356	2749	3142	3534	3927	4320
31	406	812	1217	1623	2029	2435	2840	3246	3652	4058	4464
32	419	838	1257	1676	2094	2513	2932	3351	3770	4189	4608
33	432	864	1296	1728	2160	2592	3024	3456	3888	4320	4752
34	445	890	1335	1780	2225	2670	3115	3560	4006	4451	4896
35	458	916	1375	1833	2291	2749	3206	3665	4123	4581	5040
36	471	943	1414	1885	2356	2827	3299	3770	4241	4712	5184
37	484	969	1453	1937	2422	2906	3390	3875	4359	4843	5328
38	497	995	1492	1990	2487	2985	3482	3979	4477	4974	5472
39	511	1021	1532	2042	2553	3063	3573	4084	4595	5105	5616
40	524	1047	1571	2094	2618	3142	3665	4189	4712	5236	5760
41	537	1073	1610	2147	2683	3220	3757	4294	4831	5367	5904
42	550	1100	1649	2199	2749	3299	3848	4398	4948	5498	6048
43	563	1126	1689	2251	2814	3377	3940	4503	5066	5629	6192
44	576	1152	1728	2304	2880	3456	4032	4608	5184	5760	6336
45	589	1178	1767	2356	2945	3534	4123	4712	5301	5891	6480
46	602	1204	1806	2408	3011	3613	4215	4817	5419	6021	6623
47	615	1231	1846	2461	3076	3692	4307	4922	5537	6152	6768
48	628	1257	1885	2513	3142	3770	4398	5027	5655	6283	6912
49	641	1283	1924	2566	3207	3849	4490	5131	5773	6414	7056
50	655	1309	1963	2618	3273	3927	4581	5236	5891	6545	7200

TABLE 27. CIRCUMFERENTIAL SPEEDS IN FEET PER MINUTE

Dia, In.			Rpm									
(2,400 … 4,400 rpm)			2,400	2,600	2,800	3,000	3,200	3,400	3,600	3,800	4,000	4,400
	(1,200 … 2,200 rpm)		1,200	1,300	1,400	1,500	1,600	1,700	1,800	1,900	2,000	2,200
		(600 … 1,100 rpm)	600	650	700	750	800	850	900	950	1,000	1,100
¼	½	1	157	170	183	196	209	223	236	249	262	288
½	1	2	314	340	367	393	419	445	471	497	524	576
¾	1½	3	471	510	550	589	628	668	707	746	785	863
1	2	4	628	681	733	785	838	890	942	995	1,047	1,152
1¼	2½	5	785	851	916	982	1,047	1,113	1,178	1,244	1,309	1,440
1½	3	6	942	1,021	1,100	1,178	1,257	1,335	1,414	1,492	1,571	1,728
1¾	3½	7	1,100	1,191	1,283	1,375	1,466	1,558	1,649	1,741	1,832	2,016
2	4	8	1,257	1,361	1,466	1,571	1,675	1,780	1,885	1,990	2,094	2,304
2¼	4½	9	1,414	1,531	1,649	1,767	1,885	2,003	2,121	2,238	2,356	2,592
2½	5	10	1,571	1,702	1,833	1,964	2,094	2,225	2,356	2,487	2,618	2,880
2¾	5½	11	1,728	1,872	2,016	2,160	2,304	2,448	2,592	2,736	2,880	3,168
3	6	12	1,885	2,042	2,199	2,356	2,513	2,670	2,827	2,984	3,143	3,456
3¼	6½	13	2,042	2,212	2,382	2,552	2,723	2,893	3,063	3,233	3,403	3,744
3½	7	14	2,199	2,382	2,566	2,749	2,932	3,115	3,299	3,482	3,665	4,032
3¾	7½	15	2,356	2,552	2,749	2,945	3,142	3,338	3,534	3,731	3,927	4,320
4	8	16	2,513	2,723	2,932	3,142	3,351	3,560	3,770	3,979	4,189	4,608
4¼	8½	17	2,670	2,893	3,115	3,338	3,560	3,783	4,006	4,228	4,451	4,896
4½	9	18	2,827	3,063	3,299	3,534	3,770	4,006	4,241	4,477	4,712	5,184
4¾	9½	19	2,985	3,233	3,482	3,731	3,979	4,228	4,477	4,725	4,974	5,472
5	10	20	3,142	3,403	3,665	3,927	4,189	4,451	4,712	4,974	5,236	5,760
5¼	10½	21	3,299	3,573	3,848	4,123	4,398	4,673	4,948	5,223	5,498	6,048
5½	11	22	3,456	3,744	4,032	4,320	4,608	4,896	5,184	5,472	5,760	6,336
5¾	11½	23	3,613	3,914	4,215	4,516	4,817	5,118	5,419	5,720	6,021	6,623
6	12	24	3,770	4,084	4,398	4,712	5,027	5,341	5,655	5,969	6,283	6,912
6¼	12½	25	3,927	4,254	4,581	4,909	5,236	5,563	5,891	6,218	6,545	7,200
6½	13	26	4,084	4,424	4,764	5,105	5,445	5,786	6,126	6,466	6,807	7,487
6¾	13½	27	4,241	4,594	4,948	5,301	5,655	6,008	6,362	6,715	7,069	7,775
7	14	28	4,398	4,764	5,131	5,498	5,864	6,231	6,597	6,963	7,330	8,063
7¼	14½	29	4,555	4,935	5,314	5,694	6,074	6,453	6,833	7,213	7,592	8,351
7½	15	30	4,712	5,105	5,498	5,890	6,283	6,676	7,069	7,461	7,854	8,639
7¾	15½	31	4,870	5,275	5,681	6,086	6,493	6,988	7,304	7,710	8,116	8,927
8	16	32	5,027	5,445	5,864	6,283	6,702	7,121	7,540	7,959	8,378	9,215
8¼	16½	33	5,184	5,615	6,048	6,479	6,912	7,343	7,775	8,207	8,640	9,503
8½	17	34	5,341	5,785	6,231	6,676	7,121	7,566	8,011	8,456	8,901	9,791
8¾	17½	35	5,498	5,956	6,414	6,872	7,330	7,789	8,247	8,705	9,163	10,079
9	18	36	5,655	6,126	6,597	7,069	7,540	8,011	8,482	8,954	9,425	10,367
9¼	18½	37	5,812	6,296	6,781	7,265	7,749	8,234	8,718	9,202	9,687	10,655
9½	19	38	5,969	6,466	6,964	7,461	7,959	8,456	8,954	9,451	9,948	10,943
9¾	19½	39	6,126	6,637	7,147	7,658	8,168	8,679	9,189	9,700	10,210	
10	20	40	6,283	6,807	7,330	7,854	8,378	8,901	9,425	9,948	10,472	
10¼	20½	41	6,440	6,977	7,514	8,050	8,587	9,124	9,660	10,197	10,734	
10½	21	42	6,597	7,147	7,697	8,247	8,797	9,346	9,896	10,446	10,996	
10¾	21½	43	6,754	7,317	7,880	8,443	9,006	9,569	10,131	10,695		
11	22	44	6,912	7,487	8,063	8,639	9,215	9,791	10,367	10,943		
11¼	22½	45	7,069	7,658	8,247	8,836	9,425	10,014	10,603			
11½	23	46	7,226	7,828	8,430	9,032	9,634	10,236	10,839			
11¾	23½	47	7,383	7,998	8,613	9,228	9,844	10,459				
12	24	48	7,540	8,168	8,797	9,425	10,053	10,681				
12¼	24½	49	7,697	8,338	8,980	9,621	10,263	10,903				
12½	25	50	7,854	8,508	9,163	9,818	10,472					
12¾	25½	51	8,011	8,679	9,346	10,014	10,681					
13	26	52	8,168	8,849	9,529	10,210	10,891					
13¼	26½	53	8,325	9,019	9,712	10,407						
13½	27	54	8,482	9,189	9,896	10,603						
13¾	27½	55	8,639	9,359	10,079	10,799						
14	28	56	8,797	9,530	10,263	10,996						
14¼	28½	57	8,954	9,700	10,446							
14½	29	58	9,111	9,870	10,629							
14¾	29½	59	9,268	10,040	10,812							
15	30	60	9,425	10,210	10,996							

Dia, In.			Rpm	
(2,400 … 2,600 rpm)			2,400	2,600
	(1,200 … 1,300 rpm)		1,200	1,300
		(600 … 650 rpm)	600	650
15¼	30½	61	9,582	10,380
15½	31	62	9,739	10,551
15¾	31½	63	9,896	10,721
16	32	64	10,053	10,891
16¼	32½	65	10,210	
16½	33	66	10,367	
16¾	33½	67	10,524	
17¾	34	68	10,681	
17¼	34½	69	10,839	

TABLE 28. COLD-FINISHED SHAFTING, STANDARD DIAMETERS AND LENGTHS

Adopted by American Standards Association, October, 1943
Diameters and Tolerances

Transmission Shafting Sizes, In.	Machinery Shafting Sizes, In.	Tolerances* on Dia (−), In.	Transmission Shafting Sizes, In.	Machinery Shafting Sizes, In.	Tolerances* on Dia (−), In.
	1/2	0.002	2 3/16	2 3/16	0.004
	9/16	0.002		2 1/4	0.004
	5/8	0.002		2 5/16	0.004
				2 3/8	0.004
	11/16	0.002			
	3/4	0.002	2 7/16	2 7/16	0.004
	13/16	0.002		2 1/2	0.004
	7/8	0.002		2 5/8	0.004
				2 3/4	0.004
15/16	15/16	0.002		2 7/8	0.004
	1	0.002			
	1 1/16	0.003	2 15/16	3	0.004
	1 1/8	0.003		3 1/8	0.004
				3 1/4	0.004
1 3/16	1 3/16	0.003		3 3/8	0.004
	1 1/4	0.003			
	1 5/16	0.003	3 7/16	3 1/2	0.004
	1 3/8	0.003		3 5/8	0.004
				3 3/4	0.004
1 7/16	1 7/16	0.003		3 7/8	0.004
	1 1/2	0.003			
	1 9/16	0.003	3 15/16	4	0.004
	1 5/8	0.003		4 1/4	0.005
			4 7/16	4 1/2	0.005
1 11/16	1 11/16	0.003		4 3/4	0.005
	1 3/4	0.003			
	1 13/16	0.003	4 15/16	5	0.005
	1 7/8	0.003		5 1/4	0.005
1 15/16	1 15/16	0.003	5 7/16	5 1/2	0.005
	2	0.003		5 3/4	0.005
	2 1/16	0.004			
	2 1/8	0.004	5 15/16	6	0.005

* These tolerances are *negative* and represent the maximum allowable variation *below* the exact nominal size. For instance, the maximum dia of the 1 15/16-in. shaft is 1.938 in. and its minimum allowable dia is 1.935 in.

Standard stock lengths for cold-finished shafting shall be 16, 20, and 24 ft.

Part 11

MATHEMATICS AND TABLES

Section 44

MATHEMATICS

SECTION 44

MATHEMATICS

Prime-number Fractions. The decimal equivalent of prime-number fractions is found with the table below. Example: A worm has 5 threads per inch, 7 DP, and 2 in. pitch dia. What is the thread angle? Tangent thread angle = tpi ÷ DP × pitch dia = $\frac{5}{7} \times \frac{1}{2}$ = 0.7143 (from table) × $\frac{1}{2}$ = 0.35715. From tangents table, the angle = 19° 39″.

DECIMAL EQUIVALENTS OF PRIME-NUMBER FRACTIONS
Denominators are prime numbers only

Numerators (Prime Numbers only)	97	89	83	79	73	71	67	61	59	53	47	43
1	.0103	.0112	.0120	.0126	.0137	.0141	.0149	.0164	.0169	.0189	.0213	.0233
3	.0309	.0337	.0361	.0380	.0411	.0423	.0448	.0492	.0508	.0566	.0638	.0698
5	.0515	.0562	.0602	.0633	.0685	.0704	.0746	.0820	.0847	.0943	.1064	1163
7	.0722	.0787	.0843	.0886	.0959	.0986	.1045	.1148	.1186	.1321	.1489	.1628
11	.1134	.1236	.1325	.1392	1507	.1549	.1642	.1803	1864	.2075	.2340	.2558
13	.1340	.1461	.1566	.1646	.1781	.1831	.1940	.2131	.2203	.2453	.2766	.3023
17	.1753	.1910	.2048	.2152	.2329	.2394	.2537	.2787	.2881	.3208	.3617	.3953
19	.1959	.2135	.2289	.2405	.2603	.2676	.2836	.3115	.3220	.3585	.4043	.4419
23	.2371	.2584	.2771	.2911	.3151	.3299	.3433	.3770	.3808	.4340	.4894	.5349
29	.2990	.3258	.3494	.3671	.3973	.4085	.4328	.4754	.4915	5472	.6170	.6744
31	.3196	.3483	.3735	.3924	.4247	.4366	.4627	.5082	.5254	5849	6596	.7209
37	.3814	.4157	.4458	.4684	.5068	.5211	.5522	.6066	.6271	.6981	.7872	.8605
41	.4227	.4607	.4940	.5190	.5616	.5775	.6119	.6721	.6949	7736	.8723	.9535
43	.4433	.4831	.5181	.5443	.5890	.6056	.6418	.7049	.7288	.8113	9149	
47	.4845	.5281	.5663	5949	.6438	.6620	7015	.7705	.7966	.8868		
53	.5464	.5955	.6386	.6709	7260	7465	.7910	.8689	.8983			
59	.6082	.6629	.7108	.7468	.8082	.8310	.8806	.9672				
61	.6289	.6854	7349	.7722	.8356	.8592	.9104					
67	.6907	.7528	.8072	.8481	.9178	.9437						
71	.7320	7978	.8554	8987	.9726							
73	.7526	.8202	.8795	.9241								
79	.8144	.8876	.9518									
83	.8557	.9326										
89	.9175											

Denominators (Prime Numbers Only)

Numerators (Prime Numbers only)	41	37	31	29	23	19	17	13	11	7	5	3
1	.0244	.0270	.0323	.0345	.0435	.0526	.0588	.0769	.0909	1429	.2000	.3333
3	.0732	.0811	.0968	.1034	.1304	.1579	.1765	.2308	2727	.4286	.6000	
5	.1220	.1351	.1613	.1724	.2174	.2632	.2941	.3846	.4545	.7143		
7	.1707	.1892	.2258	.2414	.3043	.3684	.4118	5385	.6364			
11	.2683	.2973	.3548	.3793	.4783	5789	.6471	.8462				
13	.3171	.3514	.4194	.4483	.5652	.6842	.7647					
17	.4146	.4595	.5484	.5862	.7391	.8947						
19	.4634	.5135	.6129	.6552	.8261							
23	5610	6212	.7419	.7931								
29	.7073	7838	.9355									
31	.7561	.8378										
37	.9024											

Only those common fractions having prime numbers for both the numerator and denominator are given in table Others can be found by simple multiplication or division.

Decimal Equivalents of Common Fractions

From 1/64 to 1 by 64ths

1/64 = .015625	17/64 = .265625	33/64 = .515625	49/64 = .765625
1/32 = .03125	9/32 = .28125	17/32 = .53125	25/32 = .78125
3/64 = .046875	19/64 = .296875	35/64 = .546875	51/64 = .796875
1/16 = .0625	5/16 = .3125	9/16 = .5625	13/16 = .8125
5/64 = .078125	21/64 = .328125	37/64 = .578125	53/64 = .828125
3/32 = .09375	11/32 = .34375	19/32 = .59375	27/32 = .84375
7/64 = .109375	23/64 = .359375	39/64 = .609375	55/64 = .859375
1/8 = .125	3/8 = .375	5/8 = .625	7/8 = .875
9/64 = .140625	25/64 = .390625	41/64 = .640625	57/64 = .890625
5/32 = .15625	13/32 = .40625	21/32 = .65625	29/32 = .90625
11/64 = .171875	27/64 = .421875	43/64 = .671875	59/64 = .921875
3/16 = .1875	7/16 = .4375	11/16 = .6875	15/16 = .9375
13/64 = .203125	29/64 = .453125	45/64 = .703125	61/64 = .953125
7/32 = .21875	15/32 = .46875	23/32 = .71875	31/32 = .96875
15/64 = .234375	31/64 = .484375	47/64 = .734375	63/64 = .984375
1/4 = .25	1/2 = .50	3/4 = .75	

How to Square Fractions

By Carl P. Nachod

Rule I. The square of a number ending in $\frac{1}{2}$ is the product of the next smaller and the next larger integers plus $\frac{1}{4}$.

$$(N\tfrac{1}{2})^2 = (N + 1)(N) + \tfrac{1}{4}$$

Example: Let $N = 8$. Then $(8\frac{1}{2})^2 = 9 \times 8 + \frac{1}{4} = 72\frac{1}{4}$

Proof: $(a - b)(a + b) = a^2 - b^2$

$$a^2 = (a - b)(a + b) + b^2$$

Let $a = 8\frac{1}{2}$ and $b = \frac{1}{2}$

$$(8\tfrac{1}{2})^2 = (8\tfrac{1}{2} - \tfrac{1}{2})(8\tfrac{1}{2} + \tfrac{1}{2}) + (\tfrac{1}{2})^2 = 8 \times 9 + \tfrac{1}{4} = 72\tfrac{1}{4}$$

Rule II. The square of a number ending in $\frac{1}{4}$ or $\frac{3}{4}$ is the product of number minus $\frac{1}{4}$ and the number plus $\frac{1}{4}$, all plus $\frac{1}{16}$.

$$(N\tfrac{1}{4})^2 = (N\tfrac{1}{4} - \tfrac{1}{4})(N\tfrac{1}{4} + \tfrac{1}{4}) + \tfrac{1}{16}$$
$$(N\tfrac{3}{4})^2 = (N\tfrac{3}{4} - \tfrac{1}{4})(N\tfrac{3}{4} + \tfrac{1}{4}) + \tfrac{1}{16}$$

Examples:

$$(4\tfrac{1}{4})^2 = 4 \times 4\tfrac{1}{2} + \tfrac{1}{16} = 18\tfrac{1}{16}$$
$$(4\tfrac{3}{4})^2 = 4\tfrac{1}{2} \times 5 + \tfrac{1}{16} = 22\tfrac{9}{16}$$

Rule III. The square of any two-digit integer is the product of the next smaller multiple of ten and the number plus this difference, all plus the square of the difference. To put it another way, the square of *any* number may be broken down into the product of the next smaller *convenient* number and a number larger than the original by the same amount plus the square of this amount.

EXAMPLES:

$$31^2 = 30 \times 32 + 1 = 960 + 1 = 961$$
$$34^2 = 30 \times 38 + 16 = 1140 + 16 = 1156$$
$$35^2 = 30 \times 40 + 25 = 1225$$
$$87^2 = 84 \times 90 + 9 = 7569$$
$$5\tfrac{1}{8}^2 = 5 \times 5\tfrac{1}{4} + \tfrac{1}{64} = 25\tfrac{7}{64}$$
$$5\tfrac{3}{8}^2 = 5 \times 5\tfrac{3}{4} + \tfrac{9}{64} = 28 + \tfrac{3}{4} + \tfrac{9}{64} = 28 - \tfrac{7}{64} = 28\tfrac{57}{64}$$

POWERS AND ROOTS

Mathematical formulas often incorporate numbers or symbols written as 10^2, 10^3, a^2, a^3, etc. The small figure at the upper right of the base figure is called the *exponent* and denotes the number of times that the base is to be multiplied by itself. The second power of a number, 10^2 for example, is called the square of the number; the third power of a number like 10^3 is called the cube.

Roots of numbers or symbols are written as $\sqrt{10}$, $\sqrt[3]{10}$, $\sqrt{a}$, $\sqrt[3]{a}$, etc. The square root is indicated by the symbol $\sqrt{\ }$ without the index number 2 at the upper left, and means that the number is to be separated into two equal factors. EXAMPLE: $\sqrt{16} = 4$, because $4 \times 4 = 16$.

Cube root of a number is one of the three equal factors into which a number divides. EXAMPLE: $\sqrt[3]{64} = 4$, because $4 \times 4 \times 4 = 64$.

Extraction of Square Root. The arithmetical method for extracting the square root of a large number such as 556,516 follows:

```
        55' 65  16'.|746 root
        49
 144¦   665
    ¦   576
1486¦   8916
    ¦   8916
```

SOLUTION: Starting at the decimal point, point off the number into periods of two numbers each. The number of periods established (in this case three) equals the number of figures in the root.

Find the greatest whole number squared that will go into the first period 55. This number is 7. Enter 7 as the first figure in the root, and subtract its square from 55.

Bring down the next period and annex to the remainder 6, making 665 the first remainder. Write twice the root, or 14, alongside of 665. It will be seen that 14 (the trial divisor) goes four times into the first two figures of the remainder 665. Thus, annex 4 to the trial divisor, making the true divisor 144. Put a 4 in the root. Multiply 144 by 4 and subtract from 665.

Bring down the last period 16 and annex to the remainder 89. The second trial divisor is $74 \times 2 = 148$. Write this figure to the left of 8916. It will be seen that 148 goes six times in the first three figures of the remainder. Place a 6 in the root and also to the right of the trial divisor 148. Divide 1486 into 8916, and find that goes exactly 6 times.

Cube Root. The extraction of cube root is a cumbersome operation by arithmetic and is seldom required. Hence the method is not shown.

Logarithms. The extraction of roots and many other calculations are facilitated by the use of logarithms. Although the use of logarithms is not difficult, the occasions requiring them in much design work are not frequent.

Functions of Fractions

(From 1/64 to 1/2 by 64ths)

Decimal Equivalents, Squares, Square Roots, Cubes, Cube Roots, and Circumferences and Areas of Circles

Fraction	Decimal Equivalent	Square	Square Root	Cube	Cube Root	Circumference of Circle	Area of Circle
1/64	.015625	.000244	.1250	.000003815	.2500	.04909	.000192
1/32	.03125	.0009765	.1768	.00003052	.3150	.09818	.000767
3/64	.046875	.002197	.2165	.000103	.3606	.1473	.001726
1/16	.0625	.003906	.2500	.0002442	.3968	.1963	.003068
5/64	.078125	.006104	.2795	.0004768	.4275	.2455	.004794
3/32	.09375	.008789	.3062	.0008240	.4543	.2945	.006903
7/64	.109375	.01196	.3307	.001308	.4782	.3436	.009396
1/8	.1250	.01563	.3535	.001953	.5000	.3927	.01228
9/64	.140625	.01978	.3750	.002781	.5200	.4438	.01553
5/32	.15625	.02441	.3953	.003815	.5386	.4909	.01916
11/64	.171875	.02954	.4161	.005078	.5560	.5400	.02321
3/16	.1875	.03516	.4330	.006592	.5724	.5890	.02761
13/64	.203125	.04126	.4507	.008381	.5878	.6381	.03241
7/32	.21875	.04786	.4677	.01047	.6025	.6872	.03758
15/64	.234375	.05493	.4841	.01287	.6166	.7363	.04314
1/4	.2500	.0625	.5000	.01562	.6300	.7854	.04909
17/64	.265625	.07056	.5154	.01874	.6428	.8345	.05541
9/32	.28125	.07910	.5303	.02225	.6552	.8836	.06213
19/64	.296875	.08813	.5449	.02616	.6671	.9327	.06922
5/16	.3125	.09766	.5590	.03052	.6786	.9817	.07670
21/64	.328125	.1077	.5728	.03533	.6897	1.031	.08456
11/32	.34375	.1182	.5863	.04062	.7005	1.080	.09281
23/64	.359375	.12913	.5995	.04641	.7110	1.129	.1014
3/8	.3750	.1406	.6124	.05273	.7211	1.178	.1104
25/64	.390625	.1526	.6250	.05960	.7310	1.227	.1226
13/32	.40625	.1650	.6374	.06705	.7406	1.276	.1296
27/64	.421875	.17800	.6495	.07508	.7500	1.325	.1398
7/16	.4375	.1914	.6614	.08374	.7592	1.374	.1503
29/64	.453125	.2053	.6732	.09304	.7681	1.424	.1613
15/32	.46875	.2197	.6847	.1030	.7768	1.473	.1726
31/64	.484375	.2346	.6960	.1136	.7853	1.522	.1843
1/2	.5000	.2500	.7071	.1250	.7937	1.571	.1963

Functions of Fractions (*Continued*)

(From 33/64 to 1 by 64ths)

Fraction	Decimal Equivalent	Square	Square Root	Cube	Cube Root	Circumference of Circle	Area of Circle
33/64	.515625	.2659	.7181	.1371	.8019	1.620	.2088
17/32	.53125	.2822	.7289	.1499	.8099	1.669	.2217
35/64	.546875	.2991	.7395	.1636	.8178	1.718	.2349
9/16	.5625	.3164	.7500	.1780	.8255	1.767	.2485
37/64	.578125	.3342	.7603	.1932	.8331	1.816	.2625
19/32	.59375	.3525	.7706	.2093	.8405	1.865	.2769
39/64	.609375	.3713	.7806	.2263	.8478	1.914	.2916
5/8	.6250	.3906	.7906	.2441	.8550	1.963	.3068
41/64	.640625	.4104	.8004	.2629	.8621	2.013	.3223
21/32	.65625	.4307	.8101	.2826	.8690	2.062	.3382
43/64	.671875	.4514	.8197	.3033	.8758	2.111	.3545
11/16	.6875	.4727	.8292	.3250	.8826	2.160	.3712
45/64	.703125	.4944	.8385	.3476	.8892	2.209	.3883
23/32	.71875	.5166	.8478	.3713	.8958	2.258	.4057
47/64	.734375	.5393	.8569	.3961	.9022	2.307	.4236
3/4	.7500	.5625	.8660	.4219	.9086	2.356	.4418
49/64	.765625	.5862	.8750	.4488	.9148	2.405	.4604
25/32	.78125	.6104	.8839	.4768	.9210	2.454	.4794
51/64	.796875	.6350	.8927	.5060	.9271	2.503	.4987
13/16	.8125	.6602	.9014	.5364	.9331	2.553	.5185
53/64	.828125	.6858	.9100	.5679	.9391	2.602	.5386
27/32	.84375	.7119	.9186	.6007	.9449	2.651	.5592
55/64	.859375	.7385	.9270	.6347	.9507	2.700	.5801
7/8	.8750	.7656	.9354	.6699	.9565	2.749	.6013
57/64	.890625	.7932	.9437	.7064	.9621	2.798	.6230
29/32	.90625	.8213	.9520	.7443	.9677	2.847	.6450
59/64	.921875	.8499	.9601	.7835	.9732	2.896	.6675
15/16	.9375	.8789	.9682	.8240	.9787	2.945	.6903
61/64	.953125	.9084	.9763	.8659	.9841	2.994	.7135
31/32	.96875	.9385	.9843	.9091	.9895	3.043	.7371
63/64	.984375	.9690	.9922	.9539	.9948	3.093	.7610
1	1	1	1	1	1	3.1416	.7854

Functions of Whole Numbers*

Number	Square	Cube	Square root	Cube root	Number = diameter	
					Circumference	Area
1	1	1	1.0000	1.0000	3.142	0.7854
2	4	8	1.4142	1.2599	6.283	3.1416
3	9	27	1.7321	1.4422	9.425	7.0686
4	16	64	2.0000	1.5874	12.566	12.5664
5	25	125	2.2361	1.7100	15.708	19.6350
6	36	216	2.4495	1.8171	18.850	28.2743
7	49	343	2.6458	1.9129	21.991	38.4845
8	64	512	2.8284	2.0000	25.133	50.2655
9	81	729	3.0000	2.0801	28.274	63.6173
10	100	1000	3.1623	2.1544	31.416	78.5398
11	121	1331	3.3166	2.2240	34.558	95.0332
12	144	1728	3.4641	2.2894	37.699	113.097
13	169	2197	3.6056	2.3513	40.841	132.732
14	196	2744	3.7417	2.4101	43.982	153.938
15	225	3375	3.8730	2.4662	47.124	176.715
16	256	4096	4.0000	2.5198	50.265	201.062
17	289	4913	4.1231	2.5713	53.407	226.980
18	324	5832	4.2426	2.6207	56.549	254.469
19	361	6859	4.3589	2.6684	59.690	283.529
20	400	8000	4.4721	2.7144	62.832	314.159
21	441	9261	4.5826	2.7589	65.973	346.361
22	484	10648	4.6904	2.8020	69.115	380.133
23	529	12167	4.7958	2.8439	72.257	415.476
24	576	13824	4.8990	2.8845	75.398	452.389
25	625	15625	5.0000	2.9240	78.540	490.874
26	676	17576	5.0990	2.9625	81.681	530.929
27	729	19683	5.1962	3.0000	84.823	572.555
28	784	21952	5.2915	3.0366	87.965	615.752
29	841	24389	5.3852	3.0723	91.106	660.520
30	900	27000	5.4772	3.1072	94.248	706.858
31	961	29791	5.5678	3.1414	97.389	754.768
32	1024	32768	5.6569	3.1748	100.531	804.248
33	1089	35937	5.7446	3.2075	103.673	855.299
34	1156	39304	5.8310	3.2396	106.814	907.920
35	1225	42875	5.9161	3.2711	109.956	962.113
36	1296	46656	6.0000	3.3019	113.097	1017.88
37	1369	50653	6.0828	3.3322	116.239	1075.21
38	1444	54872	6.1644	3.3620	119.381	1134.11
39	1521	59319	6.2450	3.3912	122.522	1194.59
40	1600	64000	6.3246	3.4200	125.66	1256.64
41	1681	68921	6.4031	3.4482	128.81	1320.25
42	1764	74088	6.4807	3.4760	131.95	1385.44
43	1849	79507	6.5574	3.5034	135.09	1452.20
44	1936	85184	6.6332	3.5303	138.23	1520.53
45	2025	91125	6.7082	3.5569	141.37	1590.43
46	2116	97336	6.7823	3.5830	144.51	1661.90
47	2209	103823	6.8557	3.6088	147.65	1734.94
48	2304	110592	6.9282	3.6342	150.80	1809.56
49	2401	117649	7.0000	3.6593	153.94	1885.74

* Courtesy of Bethlehem Steel Company.

Number	Square	Cube	Square root	Cube root	Number = diameter	
					Circumference	Area
50	2500	125000	7.0711	3.6840	157.08	1963.50
51	2601	132651	7.1414	3.7084	160.22	2042.82
52	2704	140608	7.2111	3.7325	163.36	2123.72
53	2809	148877	7.2801	3.7563	166.50	2206.18
54	2916	157464	7.3485	3.7798	169.65	2290.22
55	3025	166375	7.4162	3.8030	172.79	2375.83
56	3136	175616	7.4833	3.8259	175.93	2463.01
57	3249	185193	7.5498	3.8485	179.07	2551.76
58	3364	195112	7.6158	3.8709	182.21	2642.08
59	3481	205379	7.6811	3.8930	185.35	2733.97
60	3600	216000	7.7460	3.9149	188.50	2827.43
61	3721	226981	7.8102	3.9365	191.64	2922.47
62	3844	238328	7.8740	3.9579	194.78	3019.07
63	3969	250047	7.9373	3.9791	197.92	3117.25
64	4096	262144	8.0000	4.0000	201.06	3216.99
65	4225	274625	8.0623	4.0207	204.20	3318.31
66	4356	287496	8.1240	4.0412	207.35	3421.19
67	4489	300763	8.1854	4.0615	210.49	3525.65
68	4624	314432	8.2462	4.0817	213.63	3631.68
69	4761	328509	8.3066	4.1016	216.77	3739.28
70	4900	343000	8.3666	4.1213	219.91	3848.45
71	5041	357911	8.4261	4.1408	223.05	3959.19
72	5184	373248	8.4853	4.1602	226.19	4071.50
73	5329	389017	8.5440	4.1793	229.34	4185.39
74	5476	405224	8.6023	4.1983	232.48	4300.84
75	5625	421875	8.6603	4.2172	235.62	4417.86
76	5776	438976	8.7178	4.2358	238.76	4536.46
77	5929	456533	8.7750	4.2543	241.90	4656.63
78	6084	474552	8.8318	4.2727	245.04	4778.36
79	6241	493039	8.8882	4.2908	248.19	4901.67
80	6400	512000	8.9443	4.3089	251.33	5026.55
81	6561	531441	9.0000	4.3267	254.47	5153.00
82	6724	551368	9.0554	4.3445	257.61	5281.02
83	6889	571787	9.1104	4.3621	260.75	5410.61
84	7056	592704	9.1652	4.3795	263.89	5541.77
85	7225	614125	9.2195	4.3968	267.04	5674.50
86	7396	636056	9.2736	4.4140	270.18	5808.80
87	7569	658503	9.3274	4.4310	273.32	5944.68
88	7744	681472	9.3808	4.4480	276.46	6082.12
89	7921	704969	9.4340	4.4647	279.60	6221.14
90	8100	729000	9.4868	4.4814	282.74	6361.73
91	8281	753571	9.5394	4.4979	285.88	6503.88
92	8464	778688	9.5917	4.5144	289.03	6647.61
93	8649	804357	9.6437	4.5307	292.17	6792.91
94	8836	830584	9.6954	4.5468	295.31	6939.78
95	9025	857375	9.7468	4.5629	298.45	7088.22
96	9216	884736	9.7980	4.5789	301.59	7238.23
97	9409	912673	9.8489	4.5947	304.73	7389.81
98	9604	941192	9.8995	4.6104	307.88	7542.96
99	9801	970299	9.9499	4.6261	311.02	7697.69

Number	Square	Cube	Square root	Cube root	Number = diameter	
					Circumference	Area
100	10000	1000000	10.0000	4.6416	314.16	7853.98
101	10201	1030301	10.0499	4.6570	317.30	8011.85
102	10404	1061208	10.0995	4.6723	320.44	8171.28
103	10609	1092727	10.1489	4.6875	323.58	8332.29
104	10816	1124864	10.1980	4.7027	326.73	8494.87
105	11025	1157625	10.2470	4.7177	329.87	8659.01
106	11236	1191016	10.2956	4.7326	333.01	8824.73
107	11449	1225043	10.3441	4.7475	336.15	8992.02
108	11664	1259712	10.3923	4.7622	339.29	9160.88
109	11881	1295029	10.4403	4.7769	342.43	9331.32
110	12100	1331000	10.4881	4.7914	345.58	9503.32
111	12321	1367631	10.5357	4.8059	348.72	9676.89
112	12544	1404928	10.5830	4.8203	351.86	9852.03
113	12769	1442897	10.6301	4.8346	355.00	10028.7
114	12996	1481544	10.6771	4.8488	358.14	10207.0
115	13225	1520875	10.7238	4.8629	361.28	10386.9
116	13456	1560896	10.7703	4.8770	364.42	10568.3
117	13689	1601613	10.8167	4.8910	367.57	10751.3
118	13924	1643032	10.8628	4.9049	370.71	10935.9
119	14161	1685159	10.9087	4.9187	373.85	11122.0
120	14400	1728000	10.9545	4.9324	376.99	11309.7
121	14641	1771561	11.0000	4.9461	380.13	11499.0
122	14884	1815848	11.0454	4.9597	383.27	11689.9
123	15129	1860867	11.0905	4.9732	386.42	11882.3
124	15376	1906624	11.1355	4.9866	389.56	12076.3
125	15625	1953125	11.1803	5.0000	392.70	12271.8
126	15876	2000376	11.2250	5.0133	395.84	12469.0
127	16129	2048383	11.2694	5.0265	398.98	12667.7
128	16384	2097152	11.3137	5.0397	402.12	12868.0
129	16641	2146689	11.3578	5.0528	405.27	13069.8
130	16900	2197000	11.4018	5.0658	408.41	13273.2
131	17161	2248091	11.4455	5.0788	411.55	13478.2
132	17424	2299968	11.4891	5.0916	414.69	13684.8
133	17689	2352637	11.5326	5.1045	417.83	13892.9
134	17956	2406104	11.5758	5.1172	420.97	14102.6
135	18225	2460375	11.6190	5.1299	424.12	14313.9
136	18496	2515456	11.6619	5.1426	427.26	14526.7
137	18769	2571353	11.7047	5.1551	430.40	14741.1
138	19044	2628072	11.7473	5.1676	433.54	14957.1
139	19321	2685619	11.7898	5.1801	436.68	15174.7
140	19600	2744000	11.8322	5.1925	439.82	15393.8
141	19881	2803221	11.8743	5.2048	442.96	15614.5
142	20164	2863288	11.9164	5.2171	446.11	15836.8
143	20449	2924207	11.9583	5.2293	449.25	16060.6
144	20736	2985984	12.0000	5.2415	452.39	16286.0
145	21025	3048625	12.0416	5.2536	455.53	16513.0
146	21316	3112136	12.0830	5.2656	458.67	16741.5
147	21609	3176523	12.1244	5.2776	461.81	16971.7
148	21904	3241792	12.1655	5.2896	464.96	17203.4
149	22201	3307949	12.2066	5.3015	468.10	17436.6

Number	Square	Cube	Square root	Cube root	Number = diameter	
					Circumference	Area
150	22500	3375000	12.2474	5.3133	471.24	17671.5
151	22801	3442951	12.2882	5.3251	474.38	17907.9
152	23104	3511808	12.3288	5.3368	477.52	18145.8
153	23409	3581577	12.3693	5.3485	480.66	18385.4
154	23716	3652264	12.4097	5.3601	483.81	18626.5
155	24025	3723875	12.4499	5.3717	486.95	18869.2
156	24336	3796416	12.4900	5.3832	490.09	19113.4
157	24649	3869893	12.5300	5.3947	493.23	19359.3
158	24964	3944312	12.5698	5.4061	496.37	19606.7
159	25281	4019679	12.6095	5.4175	499.51	19855.7
160	25600	4096000	12.6491	5.4288	502.65	20106.2
161	25921	4173281	12.6886	5.4401	505.80	20358.3
162	26244	4251528	12.7279	5.4514	508.94	20612.0
163	26569	4330747	12.7671	5.4626	512.08	20867.2
164	26896	4410944	12.8062	5.4737	515.22	21124.1
165	27225	4492125	12.8452	5.4848	518.36	21382.5
166	27556	4574296	12.8841	5.4959	521.50	21642.4
167	27889	4657463	12.9228	5.5069	524.65	21904.0
168	28224	4741632	12.9615	5.5178	527.79	22167.1
169	28561	4826809	13.0000	5.5288	530.93	22431.8
170	28900	4913000	13.0384	5.5397	534.07	22698.0
171	29241	5000211	13.0767	5.5505	537.21	22965.8
172	29584	5088448	13.1149	5.5613	540.35	23235.2
173	29929	5177717	13.1529	5.5721	543.50	23506.2
174	30276	5268024	13.1909	5.5828	546.64	23778.7
175	30625	5359375	13.2288	5.5934	549.78	24052.8
176	30976	5451776	13.2665	5.6041	552.92	24328.5
177	31329	5545233	13.3041	5.6147	556.06	24605.7
178	31684	5639752	13.3417	5.6252	559.20	24884.6
179	32041	5735339	13.3791	5.6357	562.35	25164.9
180	32400	5832000	13.4164	5.6462	565.49	25446.9
181	32761	5929741	13.4536	5.6567	568.63	25730.4
182	33124	6028568	13.4907	5.6671	571.77	26015.5
183	33489	6128487	13.5277	5.6774	574.91	26302.2
184	33856	6229504	13.5647	5.6877	578.05	26590.4
185	34225	6331625	13.6015	5.6980	581.19	26880.3
186	34596	6434856	13.6382	5.7083	584.34	27171.6
187	34969	6539203	13.6748	5.7185	587.48	27464.6
188	35344	6644672	13.7113	5.7287	590.62	27759.1
189	35721	6751269	13.7477	5.7388	593.76	28055.2
190	36100	6859000	13.7840	5.7489	596.90	28352.9
191	36481	6967871	13.8203	5.7590	600.04	28652.1
192	36864	7077888	13.8564	5.7690	603.19	28952.9
193	37249	7189057	13.8924	5.7790	606.33	29255.3
194	37636	7301384	13.9284	5.7890	609.47	29559.2
195	38025	7414875	13.9642	5.7989	612.61	29864.8
196	38416	7529536	14.0000	5.8088	615.75	30171.9
197	38809	7645373	14.0357	5.8186	618.89	30480.5
198	39204	7762392	14.0712	5.8285	622.04	30790.7
199	39601	7880599	14.1067	5.8383	625.18	31102.6

Number	Square	Cube	Square root	Cube root	Number = diameter: Circumference	Number = diameter: Area
200	40000	8000000	14.1421	5.8480	628.32	31415.6
201	40401	8120601	14.1774	5.8578	631.46	31730 9
202	40804	8242408	14.2127	5.8675	634.60	32047.4
203	41209	8365427	14.2478	5.8771	637.74	32365.5
204	41616	8489664	14.2829	5.8868	640.88	32685.1
205	42025	8615125	14.3178	5.8964	644.03	33006.4
206	42436	8741816	14.3527	5.9059	647.17	33329.2
207	42849	8869743	14.3875	5.9155	650.31	33653.5
208	43264	8998912	14.4222	5.9250	653.45	33979.5
209	43681	9129329	14.4568	5.9345	656.59	34307.0
210	44100	9261000	14.4914	5.9439	659.73	34636.1
211	44521	9393931	14.5258	5.9533	662.88	34966.7
212	44944	9528128	14.5602	5.9627	666.02	35298.9
213	45369	9663597	14.5945	5.9721	669.16	35632.7
214	45796	9800344	14.6287	5.9814	672.30	35968.1
215	46225	9938375	14.6629	5.9907	675.44	36305.0
216	46656	10077696	14.6969	6.0000	678.58	36643.5
217	47089	10218313	14.7309	6.0092	681.73	36983.6
218	47524	10360232	14.7648	6.0185	684.87	37325.3
219	47961	10503459	14.7986	6.0277	688.01	37668.5
220	48400	10648000	14.8324	6.0368	691.15	38013.3
221	48841	10793861	14.8661	6.0459	694.29	38359.6
222	49284	10941048	14.8997	6.0550	697.43	38707.6
223	49729	11089567	14.9332	6.0641	700.58	39057.1
224	50176	11239424	14.9666	6.0732	703.72	39408.1
225	50625	11390625	15.0000	6.0822	706.86	39760.8
226	51076	11543176	15.0333	6.0912	710.00	40115.0
227	51529	11697083	15.0665	6.1002	713.14	40470.8
228	51984	11852352	15.0997	6.1091	716.28	40828.1
229	52441	12008989	15.1327	6.1180	719.42	41187.1
230	52900	12167000	15.1658	6.1269	722.57	41547.6
231	53361	12326391	15.1987	6.1358	725.71	41909.6
232	53824	12487168	15.2315	6.1446	728.85	42273.3
233	54289	12649337	15.2643	6.1534	731.99	42638.5
234	54756	12812904	15.2971	6.1622	735.13	43005.3
235	55225	12977875	15.3297	6.1710	738.27	43373.6
236	55696	13144256	15.3623	6.1797	741.42	43743.5
237	56169	13312053	15.3948	6.1885	744.56	44115.0
238	56644	13481272	15.4272	6.1972	747.70	44488.1
239	57121	13651919	15.4596	6.2058	750.84	44862.7
240	57600	13824000	15.4919	6.2145	753.98	45238.9
241	58081	13997521	15.5242	6.2231	757.12	45616.7
242	58564	14172488	15.5563	6.2317	760.27	45996.1
243	59049	14348907	15.5885	6.2403	763.41	46377.0
244	59536	14526784	15.6205	6.2488	766.55	46759.5
245	60025	14706125	15.6525	6.2573	769.69	47143.5
246	60516	14886936	15.6844	6.2658	772.83	47529.2
247	61009	15069223	15.7162	6.2743	775.97	47916.4
248	61504	15252992	15.7480	6.2828	779.11	48305.1
249	62001	15438249	15.7797	6.2912	782.26	48695.5

Number	Square	Cube	Square root	Cube root	Number = diameter	
					Circumference	Area
250	62500	15625000	15.8114	6.2996	785.40	49087.4
251	63001	15813251	15.8430	6.3080	788.54	49480.9
252	63504	16003008	15.8745	6.3164	791.68	49875.9
253	64009	16194277	15.9060	6.3247	794.82	50272.6
254	64516	16387064	15.9374	6.3330	797.96	50670.7
255	65025	16581375	15.9687	6.3413	801.11	51070.5
256	65536	16777216	16.0000	6.3496	804.25	51471.9
257	66049	16974593	16.0312	6.3579	807.39	51874.8
258	66564	17173512	16.0624	6.3661	810.53	52279.2
259	67081	17373979	16.0935	6.3743	813.67	52685.3
260	67600	17576000	16.1245	6.3825	816.81	53092.9
261	68121	17779581	16.1555	6.3907	819.96	53502.1
262	68644	17984728	16.1864	6.3988	823.10	53912.9
263	69169	18191447	16.2173	6.4070	826.24	54325.2
264	69696	18399744	16.2481	6.4151	829.38	54739.1
265	70225	18609625	16.2788	6.4232	832.52	55154.6
266	70756	18821096	16.3095	6.4312	835.66	55571.6
267	71289	19034163	16.3401	6.4393	838.81	55990.2
268	71824	19248832	16.3707	6.4473	841.95	56410.4
269	72361	19465109	16.4012	6.4553	845.09	56832.2
270	72900	19683000	16.4317	6.4633	848.23	57255.5
271	73441	19902511	16.4621	6.4713	851.37	57680.4
272	73984	20123648	16.4924	6.4792	854.51	58106.9
273	74529	20346417	16.5227	6.4872	857.65	58534.9
274	75076	20570824	16.5529	6.4951	860.80	58964.6
275	75625	20796875	16.5831	6.5030	863.94	59395.7
276	76176	21024576	16.6132	6.5108	867.08	59828.5
277	76729	21253933	16.6433	6.5187	870.22	60262.8
278	77284	21484952	16.6733	6.5265	873.36	60698.7
279	77841	21717639	16.7033	6.5343	876.50	61136.2
280	78400	21952000	16.7332	6.5421	879.65	61575.2
281	78961	22188041	16.7631	6.5499	882.79	62015.8
282	79524	22425768	16.7929	6.5577	885.93	62458.0
283	80089	22665187	16.8226	6.5654	889.07	62901.8
284	80656	22906304	16.8523	6.5731	892.21	63347.1
285	81225	23149125	16.8819	6.5808	895.35	63794.0
286	81796	23393656	16.9115	6.5885	898.50	64242.4
287	82369	23639903	16.9411	6.5962	901.64	64692.5
288	82944	23887872	16.9706	6.6039	904.78	65144.1
289	83521	24137569	17.0000	6.6115	907.92	65597.2
290	84100	24389000	17.0294	6.6191	911.06	66052.0
291	84681	24642171	17.0587	6.6267	914.20	66508.3
292	85264	24897088	17.0880	6.6343	917.35	66966.2
293	85849	25153757	17.1172	6.6419	920.49	67425.6
294	86436	25412184	17.1464	6.6494	923.63	67886.7
295	87025	25672375	17.1756	6.6569	926.77	68349.3
296	87616	25934336	17.2047	6.6644	929.91	68813.4
297	88209	26198073	17.2337	6.6719	933.05	69279.2
298	88804	26463592	17.2627	6.6794	936.19	69746.5
299	89401	26730899	17.2916	6.6869	939.34	70215.4

Number	Square	Cube	Square root	Cube root	Number = diameter	
					Circumference	Area
300	90000	27000000	17.3205	6.6943	942.48	70685.8
301	90601	27270901	17.3494	6.7018	945.62	71157.9
302	91204	27543608	17.3781	6.7092	948.76	71631.5
303	91809	27818127	17.4069	6.7166	951.90	72106.6
304	92416	28094464	17.4356	6.7240	955.04	72583.4
305	93025	28372625	17.4642	6.7313	958.19	73061.7
306	93636	28652616	17.4929	6.7387	961.33	73541.5
307	94249	28934443	17.5214	6.7460	964.47	74023.0
308	94864	29218112	17.5499	6.7533	967.61	74506.0
309	95481	29503629	17.5784	6.7606	970.75	74990.6
310	96100	29791000	17.6068	6.7679	973.89	75476.8
311	96721	30080231	17.6352	6.7752	977.04	75964.5
312	97344	30371328	17.6635	6.7824	980.18	76453.8
313	97969	30664297	17.6918	6.7897	983.32	76944.7
314	98596	30959144	17.7200	6.7969	986.46	77437.1
315	99225	31255875	17.7482	6.8041	989.60	77931.1
316	99856	31554496	17.7764	6.8113	992.74	78426.7
317	100489	31855013	17.8045	6.8185	995.88	78923.9
318	101124	32157432	17.8326	6.8256	999.03	79422.6
319	101761	32461759	17.8606	6.8328	1002.2	79922.9
320	102400	32768000	17.8885	6.8399	1005.3	80424.8
321	103041	33076161	17.9165	6.8470	1008.5	80928.2
322	103684	33386248	17.9444	6.8541	1011.6	81433.2
323	104329	33698267	17.9722	6.8612	1014.7	81939.8
324	104976	34012224	18.0000	6.8683	1017.9	82448.0
325	105625	34328125	18.0278	6.8753	1021.0	82957.7
326	106276	34645976	18.0555	6.8824	1024.2	83469.0
327	106929	34965783	18.0831	6.8894	1027.3	83981.8
328	107584	35287552	18.1108	6.8964	1030.4	84496.3
329	108241	35611289	18.1384	6.9034	1033.6	85012.3
330	108900	35937000	18.1659	6.9104	1036.7	85529.9
331	109561	36264691	18.1934	6.9174	1039.9	86049.0
332	110224	36594368	18.2209	6.9244	1043.0	86569.7
333	110889	36926037	18.2483	6.9313	1046.2	87092.0
334	111556	37259704	18.2757	6.9382	1049.3	87615.9
335	112225	37595375	18.3030	6.9451	1052.4	88141.3
336	112896	37933056	18.3303	6.9521	1055.6	88668.3
337	113569	38272753	18.3576	6.9589	1058.7	89196.9
338	114244	38614472	18.3848	6.9658	1061.9	89727.0
339	114921	38958219	18.4120	6.9727	1065.0	90258.7
340	115600	39304000	18.4391	6.9795	1068.1	90792.0
341	116281	39651821	18.4662	6.9864	1071.3	91326.9
342	116964	40001688	18.4932	6.9932	1074.4	91863.3
343	117649	40353607	18.5203	7.0000	1077.6	92401.3
344	118336	40707584	18.5472	7.0068	1080.7	92940.9
345	119025	41063625	18.5742	7.0136	1083.8	93482.0
346	119716	41421736	18.6011	7.0203	1087.0	94024.7
347	120409	41781923	18.6279	7.0271	1090.1	94569.0
348	121104	42144192	18.6548	7.0338	1093.3	95114.9
349	121801	42508549	18.6815	7.0406	1096.4	95662.3

Number	Square	Cube	Square root	Cube root	Number = diameter	
					Circumference	Area
350	122500	42875000	18.7083	7.0473	1099.6	96211.3
351	123201	43243551	18.7350	7.0540	1102.7	96761.8
352	123904	43614208	18.7617	7.0607	1105.8	97314.0
353	124609	43986977	18.7883	7.0674	1109.0	97867.7
354	125316	44361864	18.8149	7.0740	1112.1	98423.0
355	126025	44738875	18.8414	7.0807	1115.3	98979.8
356	126736	45118016	18.8680	7.0873	1118.4	99538 2
357	127449	45499293	18.8944	7.0940	1121.5	100098
358	128164	45882712	18.9209	7.1006	1124.7	100660
359	128881	46268279	18.9473	7.1072	1127.8	101223
360	129600	46656000	18.9737	7.1138	1131.0	101788
361	130321	47045881	19.0000	7.1204	1134.1	102354
362	131044	47437928	19.0263	7.1269	1137.3	103922
363	131769	47832147	19.0526	7.1335	1140.4	103491
364	132496	48228544	19.0788	7.1400	1143.5	104062
365	133225	48627125	19.1050	7.1466	1146.7	104635
366	133956	49027896	19.1311	7.1531	1149.8	105209
367	134689	49430863	19.1572	7.1596	1153.0	105784
368	135424	49836032	19.1833	7.1661	1156.1	106362
369	136161	50243409	19.2094	7.1726	1159.2	106941
370	136900	50653000	19.2354	7.1791	1162.4	107521
371	137641	51064811	19.2614	7.1855	1165.5	108103
372	138384	51478848	19.2873	7.1920	1168.7	108687
373	139129	51895117	19.3132	7.1984	1171.8	109272
374	139876	52313624	19.3391	7.2048	1175.0	109858
375	140625	52734375	19.3649	7.2112	1178.1	110447
376	141376	53157376	19.3907	7.2177	1181.2	111036
377	142129	53582633	19.4165	7.2240	1184.4	111628
378	142884	54010152	19.4422	7.2304	1187.5	112221
379	143641	54439939	19.4679	7.2368	1190.7	112815
380	144400	54872000	19.4936	7.2432	1193.8	113411
381	145161	55306341	19.5192	7.2495	1196.9	114009
382	145924	55742968	19.5448	7.2558	1200.1	114608
383	146689	56181887	19.5704	7.2622	1203.2	115209
384	147456	56623104	19.5959	7.2685	1206.4	115812
385	148225	57066625	19.6214	7.2748	1209.5	116416
386	148996	57512456	19.6469	7.2811	1212.7	117021
387	149769	57960603	19.6723	7.2874	1215.8	117628
388	150544	58411072	19.6977	7.2936	1218.9	118237
389	151321	58863869	19.7231	7.2999	1222.1	118847
390	152100	59319000	19.7484	7.3061	1225.2	119459
391	152881	59776471	19.7737	7.3124	1228.4	120072
392	153664	60236288	19.7990	7.3186	1231.5	120687
393	154449	60698457	19.8242	7.3248	1234.6	121304
394	155236	61162984	19.8494	7.3310	1237.8	121922
395	156025	61629875	19.8746	7.3372	1240.9	122542
396	156816	62099136	19.8997	7.3434	1244.1	123163
397	157609	62570773	19.9249	7.3496	1247.2	123786
398	158404	63044792	19.9499	7.3558	1250.4	124410
399	159201	63521199	19.9750	7.3619	1253.5	125036

Number	Square	Cube	Square root	Cube root	Number = diameter	
					Circumference	Area
400	160000	64000000	20.0000	7.3681	1256.6	125664
401	160801	64481201	20.0250	7.3742	1259.8	126293
402	161604	64964808	20.0499	7.3803	1262.9	126923
403	162409	65450827	20.0749	7.3864	1266.1	127556
404	163216	65939264	20.0998	7.3925	1269.2	128190
405	164025	66430125	20.1246	7.3986	1272.3	128825
406	164836	66923416	20.1494	7.4047	1275.5	129462
407	165649	67419143	20.1742	7.4108	1278.6	130100
408	166464	67917312	20.1990	7.4169	1281.8	130741
409	167281	68417929	20.2237	7.4229	1284.9	131382
410	168100	68921000	20.2485	7.4290	1288.1	132025
411	168921	69426531	20.1731	7.4350	1291.2	132670
412	169744	69934528	20.2978	7.4410	1294.3	133317
413	170569	70444997	20.3224	7.4470	1297.5	133965
414	171396	70957944	20.3470	7.4530	1300.6	134614
415	172225	71473375	20.3715	7.4590	1303.8	135265
416	173056	71991296	20.3961	7.4650	1306.9	135918
417	173889	72511713	20.4206	7.4710	1310.0	136572
418	174724	73034632	20.4450	7.4770	1313.2	137228
419	175561	73560059	20.4695	7.4829	1316.2	137885
420	176400	74088000	20.4939	7.4889	1319.5	138544
421	177241	74618461	20.5183	7.4948	1322.6	139205
422	178084	75151448	20.5436	7.5007	1325.8	139867
423	178929	75686967	20.5670	7.5067	1328.9	140531
424	179776	76225024	20.5913	7.5126	1332.0	141196
425	180625	76765625	20.6155	7.5185	1335.2	141863
426	181476	77308776	20.6398	7.5244	1338.3	142531
427	182329	77854483	20.6640	7.5302	1341.5	143201
428	183184	78402752	20.6882	7.5361	1344.6	143872
429	184041	78953589	20.7123	7.5420	1347.7	144545
430	184900	79507000	20.7364	7.5478	1350.9	145220
431	185761	80062991	20.7605	7.5537	1354.0	145896
432	186624	80621568	20.7846	7.5595	1357.2	146574
433	187489	81182737	20.8087	7.5654	1360.3	147254
434	188356	81746504	20.8327	7.5712	1363.5	147934
435	189225	82312875	20.8567	7.5770	1366.6	148617
436	190096	82881856	20.8806	7.5828	1369.7	149301
437	190969	83453453	20.9045	7.5886	1372.9	149987
438	191844	84027672	20.9284	7.5944	1376.0	150674
439	192721	84604519	20.9523	7.6001	1379.2	151363
440	193600	85184000	20.9762	7.6059	1382.3	152053
441	194481	85766121	21.0000	7.6117	1385.4	152745
442	195364	86350888	21.0238	7.6174	1388.6	153439
443	196249	86938307	21.0476	7.6232	1391.7	154134
444	197136	87528384	21.0713	7.6289	1394.9	154830
445	198025	88121125	21.0950	7.6346	1398.0	155528
446	198916	88716536	21.1187	7.6403	1401.2	156228
447	199809	89314623	21.1424	7.6460	1404.3	156930
448	200704	89915392	21.1660	7.6517	1407.4	157633
449	201601	90518849	21.1896	7.6574	1410.6	158337

Number	Square	Cube	Square root	Cube root	Number = diameter	
					Circumference	Area
450	202500	91125000	21.2132	7.6631	1413.7	159043
451	203401	91733851	21.2368	7.6688	1416.9	159751
452	204304	92345408	21.2603	7.6744	1420.0	160460
453	205209	92959677	21.2838	7.6801	1423.1	161171
454	206116	93576664	21.3073	7.6857	1426.3	161883
455	207025	94196375	21.3307	7.6914	1429.4	162597
456	207936	94818816	21.3542	7.6970	1432.6	163313
457	208849	95443993	21.3776	7.7026	1435.7	164030
458	209764	96071912	21.4009	7.7082	1438.8	164748
459	210681	96702579	21.4243	7.7138	1442.0	165468
460	211600	97336000	21.4476	7.7194	1445.1	166190
461	212521	97972181	21.4709	7.7250	1448.3	166914
462	213444	98611128	21.4942	7.7306	1451.4	167639
463	214369	99252847	21.5174	7.7362	1454.6	168365
464	215296	99897344	21.5407	7.7418	1457.7	169093
465	216225	100544625	21.5639	7.7473	1460.8	169823
466	217156	101194696	21.5870	7.7529	1464.0	170554
467	218089	101847563	21.6102	7.7584	1467.1	171287
468	219024	102503232	21.6333	7.7639	1470.3	172021
469	219961	103161709	21.6564	7.7695	1473.4	172757
470	220900	103823000	21.6795	7.7750	1476.5	173494
471	221841	104487111	21.7025	7.7805	1479.7	174234
472	222784	105154048	21.7256	7.7860	1482.8	174974
473	223729	105823817	21.7486	7.7915	1486.0	175716
474	224676	106496424	21.7715	7.7970	1489.1	176460
475	225625	107171875	21.7945	7.8025	1492.3	177205
476	226576	107850176	21.8174	7.8079	1495.4	177952
477	227529	108531333	21.8403	7.8134	1498.5	178701
478	228484	109215352	21.8632	7.8188	1501.7	179451
479	229441	109902239	21.8861	7.8243	1504.8	180203
480	230400	110592000	21.9089	7.8297	1508.0	180956
481	231361	111284641	21.9317	7.8352	1511.1	181711
482	232324	111980168	21.9545	7.8406	1514.2	182467
483	233289	112678587	21.9773	7.8460	1517.4	183225
484	234256	113379904	22.0000	7.8514	1520.5	183984
485	235225	114084125	22.0227	7.8568	1523.7	184745
486	236196	114791256	22.0454	7.8622	1526.8	185508
487	237169	115501303	22.0681	7.8676	1530.0	186272
488	238144	116214272	22.0907	7.8730	1533.1	187038
489	239121	116930169	22.1133	7.8784	1536.2	187805
490	240100	117649000	22.1359	7.8837	1539.4	188574
491	241081	118370771	22.1585	7.8891	1542.5	189345
492	242064	119095488	22.1811	7.8944	1545.7	190117
493	243049	119823157	22.2036	7.8998	1548.8	190890
494	244036	120553784	22.2261	7.9051	1551.9	191665
495	245025	121287375	22.2486	7.9105	1555.1	192442
496	246016	122023936	22.2711	7.9158	1558.2	193221
497	247009	122763473	22.2935	7.9211	1561.4	194000
498	248004	123505992	22.3159	7.9264	1564.5	194782
499	249001	124251499	22.3383	7.9317	1567.7	195565

Number	Square	Cube	Square root	Cube root	Number = diameter	
					Circumference	Area
500	250000	125000000	22.3607	7.9370	1570.8	196350
501	251001	125751501	22.3830	7.9423	1573.9	197136
502	252004	126506008	22.4054	7.9476	1577.1	197923
503	253009	127263527	22.4277	7.9528	1580.2	198713
504	254016	128024064	22.4499	7.9581	1583.4	199504
505	255025	128787625	22.4722	7.9634	1586.5	200296
506	256036	129554216	22.4944	7.9686	1589.6	201090
507	257049	130323843	22.5167	7.9739	1592.8	201886
508	258064	131096512	22.5389	7.9791	1595.9	202683
509	259081	131872229	22.5610	7.9843	1599.1	203482
510	260100	132651000	22.5832	7.9896	1602.2	204282
511	261121	133432831	22.6053	7.9948	1605.4	205084
512	262144	134217728	22.6274	8.0000	1608.5	205887
513	263169	135005697	22.6495	8.0052	1611.6	206692
514	264196	135796744	22.6716	8.0104	1614.8	207499
515	265225	136590875	22.6936	8.0156	1617.9	208307
516	266256	137388096	22.7156	8.0208	1621.1	209117
517	267289	138188413	22.7376	8.0260	1624.2	209928
518	268324	138991832	22.7596	8.0311	1627.3	210741
519	269361	139798359	22.7816	8.0363	1630.5	211556
520	270400	140608000	22.8035	8.0415	1633.6	212372
521	271441	141420761	22.8254	8.0466	1636.8	213189
522	272484	142236648	22.8473	8.0517	1639.9	214008
523	273529	143055667	22.8692	8.0569	1643.1	214829
524	274576	143877824	22.8910	8.0620	1646.2	215651
525	275625	144703125	22.9129	8.0671	1649.3	216475
526	276676	145531576	22.9347	8.0723	1652.5	217301
527	277729	146363183	22.9565	8.0774	1655.6	218128
528	278784	147197952	22.9783	8.0825	1658.8	218956
529	279841	148035889	23.0000	8.0876	1661.9	219787
530	280900	148877000	23.0217	8.0927	1665.0	220618
531	281961	149721291	23.0434	8.0978	1668.2	221452
532	283024	150568768	23.0651	8.1028	1671.3	222287
533	284089	151419437	23.0868	8.1079	1674.5	223123
534	285156	152273304	23.1084	8.1130	1677.6	223961
535	286225	153130375	23.1301	8.1180	1680.8	224801
536	287296	153990656	23.1517	8.1231	1683.9	225642
537	288369	154854153	23.1733	8.1281	1687.0	226484
538	289444	155720872	23.1948	8.1332	1690.2	227329
539	290521	156590819	23.2164	8.1382	1693.3	228175
540	291600	157464000	23.2379	8.1433	1696.5	229022
541	292681	158340421	23.2594	8.1483	1699.6	229871
542	293764	159220088	23.2809	8.1533	1702.7	230722
543	294849	160103007	23.3024	8.1583	1705.9	231574
544	295936	160989184	23.3238	8.1633	1709.0	232428
545	297025	161878625	23.3452	8.1683	1712.2	233283
546	298116	162771336	23.3666	8.1733	1715.3	234140
547	299209	163667323	23.3880	8.1783	1718.5	234998
548	300304	164566592	23.4094	8.1833	1721.6	235858
549	301401	165469149	23.4307	8.1882	1724.7	236720

Number	Square	Cube	Square root	Cube root	Number = diameter	
					Circumference	Area
550	302500	166375000	23.4521	8.1932	1727.9	237583
551	303601	167284151	23.4734	8.1982	1731.0	238448
552	304704	168196608	23.4947	8.2031	1734.2	239314
553	305809	169112377	23.5160	8.2081	1737.3	240182
554	306916	170031464	23.5372	8.2130	1740.4	241051
555	308025	170953875	23.5584	8.2180	1743.6	241922
556	309136	171879616	23.5797	8.2229	1746.7	242795
557	310249	172808693	23.6008	8.2278	1749.9	243669
558	311364	173741112	23.6220	8.2327	1753.0	244545
559	312481	174676879	23.6432	8.2377	1756.2	245422
560	313600	175616000	23.6643	8.2426	1759.3	246301
561	314721	176558481	23.6854	8.2475	1762.4	247181
562	315844	177504328	23.7065	8.2524	1765.6	248063
563	316969	178453547	23.7276	8.2573	1768.7	248947
564	318096	179406144	23.7487	8.2621	1771.9	249832
565	319225	180362125	23.7697	8.2670	1775.0	250719
566	320356	181321496	23.7908	8.2719	1778.1	251607
567	321489	182284263	23.8118	8.2768	1781.3	252497
568	322624	183250432	23.8328	8.2816	1784.4	253388
569	323761	184220009	23.8537	8.2865	1787.6	254281
570	324900	185193000	23.8747	8.2913	1790.7	255176
571	326041	186169411	23.8956	8.2962	1793.8	256072
572	327184	187149248	23.9165	8.3010	1797.0	256970
573	328329	188132517	23.9374	8.3059	1800.1	257869
574	329476	189119224	23.9583	8.3107	1803.3	258770
575	330625	190109375	23.9792	8.3155	1806.4	259672
576	331776	191102976	24.0000	8.3203	1809.6	260576
577	332929	192100033	24.0208	8.3251	1812.7	261482
578	334084	193100552	24.0416	8.3300	1815.8	262389
579	335241	194104539	24.0624	8.3348	1819.0	263298
580	336400	195112000	24.0832	8.3396	1822.1	264208
581	337561	196122941	24.1039	8.3443	1825.3	265120
582	338724	197137368	24.1247	8.3491	1828.4	266033
583	339889	198155287	24.1454	8.3539	1831.5	266948
584	341056	199176704	24.1661	8.3587	1834.7	267865
585	342225	200201625	24.1868	8.3634	1837.8	268783
586	343396	201230056	24.2074	8.3682	1841.0	269703
587	344569	202262003	24.2281	8.3730	1844.1	270624
588	345744	203297472	24.2487	8.3777	1847.3	271547
589	346921	204336469	24.2693	8.3825	1850.4	272471
590	348100	205379000	24.2899	8.3872	1853.5	273397
591	349281	206425071	24.3105	8.3919	1856.7	274325
592	350464	207474688	24.3311	8.3967	1859.8	275254
593	351649	208527857	24.3516	8.4014	1863.0	276184
594	352836	209584584	24.3721	8.4061	1866.1	277117
595	354025	210644875	24.3926	8.4108	1869.2	278051
596	355216	211708736	24.4131	8.4155	1872.4	278986
597	356409	212776173	24.4336	8.4202	1875.5	279923
598	357604	213847192	24.4540	8.4249	1878.7	280862
599	358801	214921799	24.4745	8.4296	1881.8	281802

Number	Square	Cube	Square root	Cube root	Number = diameter	
					Circumference	Area
600	360000	216000000	24.4949	8.4343	1885.0	282743
601	361201	217081801	24.5153	8.4390	1888.1	283687
602	362404	218167208	24.5357	8.4437	1891.2	284631
603	363609	219256227	24.5561	8.4484	1894.4	285578
604	364816	220348864	24.5764	8.4530	1897.5	286526
605	366025	221445125	24.5967	8.4577	1900.7	287475
606	367236	222545016	24.6171	8.4623	1903.8	288426
607	368449	223648543	24.6374	8.4670	1906.9	289379
608	369664	224755712	24.6577	8.4716	1910.1	290333
609	370881	225866529	24.6779	8.4763	1913.2	291289
610	372100	226981000	24.6982	8.4809	1916.4	292247
611	373321	228099131	24.7184	8.4856	1919.5	293206
612	374544	229220928	24.7386	8.4902	1922.7	294166
613	375769	230346397	24.7588	8.4948	1925.8	295128
614	376996	231475544	24.7790	8.4994	1928.9	296092
615	378225	232608375	24.7992	8.5040	1932.1	297057
616	379456	233744896	24.8193	8.5086	1935.2	298024
617	380689	234885113	24.8395	8.5132	1938.4	298992
618	381924	236029032	24.8596	8.5178	1941.5	299962
619	383161	237176659	24.8797	8.5224	1944.6	300934
620	384400	238328000	24.8998	8.5270	1947.8	301907
621	385641	239483061	24.9199	8.5316	1950.9	302882
622	386884	240641848	24.9399	8.5362	1954.1	303858
623	388129	241804367	24.9600	8.5408	1957.2	304836
624	389376	242970624	24.9800	8.5453	1960.4	305815
625	390625	244140625	25.0000	8.5499	1963.5	306796
626	391876	245314376	25.0200	8.5544	1966.6	307779
627	393129	246491883	25.0400	8.5590	1969.8	308763
628	394384	247673152	25.0599	8.5635	1972.9	309748
629	395641	248858189	25.0799	8.5681	1976.1	310736
630	396900	250047000	25.0998	8.5726	1979.2	311725
631	398161	251239591	25.1197	8.5772	1982.3	312715
632	399424	252435968	25.1396	8.5817	1985.5	313707
633	400689	253636137	25.1595	8.5862	1988.6	314700
634	401956	254840104	25.1794	8.5907	1991.8	315696
635	403225	256047875	25.1992	8.5952	1994.9	316692
636	404496	257259456	25.2190	8.5997	1998.1	317690
637	405769	258474853	25.2389	8.6043	2001.2	318690
638	407044	259694072	25.2587	8.6088	2004.3	319692
639	408321	260917119	25.2784	8.6132	2007.5	320695
640	409600	262144000	25.2982	8.6177	2010.6	321699
641	410881	263374721	25.3180	8.6222	2013.8	322705
642	412164	264609288	25.3377	8.6267	2016.9	323713
643	413449	265847707	25.3574	8.6312	2020.0	324722
644	414736	267089984	25.3772	8.6357	2023.2	325733
645	416025	268336125	25.3969	8.6401	2026.3	326745
646	417316	269586136	25.4165	8.6446	2029.5	327759
647	418609	270840023	25.4362	8.6490	2032.6	328775
648	419904	272097792	25.4558	8.6535	2035.8	329792
649	421201	273359449	25.4755	8.6579	2038.9	330810

Number	Square	Cube	Square root	Cube root	Number = diameter	
					Circumference	Area
650	422500	274625000	25.4951	8.6624	2042.0	331831
651	423801	275894451	25.5147	8.6668	2045.2	332853
652	425104	277167808	25.5343	8.6713	2048.3	333876
653	426409	278445077	25.5539	8.6757	2051.5	334901
654	427716	279726264	25.5734	8.6801	2054.6	335927
655	429025	281011375	25.5930	8.6845	2057.7	336955
656	430336	282300416	25.6125	8.6890	2060.9	337985
657	431649	283593393	25.6320	8.6934	2064.0	339016
658	432964	284890312	25.6515	8.6978	2067.2	340049
659	434281	286191179	25.6710	8.7022	2070.3	341083
660	435600	287496000	25.6905	8.7066	2073.5	342119
661	436921	288804781	25.7099	8.7110	2076.6	343157
662	438244	290117528	25.7294	8.7154	2079.7	344196
663	439569	291434247	25.7488	8.7198	2082.9	345237
664	440896	292754944	25.7682	8.7241	2086.0	346279
665	442225	294079625	25.7876	8.7285	2089.2	347323
666	443556	295408296	25.8070	8.7329	2092.3	348368
667	444889	296740963	25.8263	8.7373	2095.4	349415
668	446224	298077632	25.8457	8.7416	2098.6	350464
669	447561	299418309	25.8650	8.7460	2101.7	351514
670	448900	300763000	25.8844	8.7503	2104.9	352565
671	450241	302111711	25.9037	8.7547	2108.0	353618
672	451584	303464448	25.9230	8.7590	2111.2	354673
673	452929	304821217	25.9422	8.7634	2114.3	355730
674	454276	306182024	25.9615	8.7677	2117.4	356788
675	455625	307546875	25.9808	8.7721	2120.6	357847
676	456976	308915776	26.0000	8.7764	2123.7	358908
677	458329	310288733	26.0192	8.7807	2126.9	359971
678	459684	311665752	26.0384	8.7850	2130.0	361035
679	461041	313046839	26.0576	8.7893	2133.1	362101
680	462400	314432000	26.0768	8.7937	2136.3	363168
681	463761	315821241	26.0960	8.7980	2139.4	364237
682	465124	317214568	26.1151	8.8023	2142.6	365308
683	466489	318611987	26.1343	8.8066	2145.7	366380
684	467856	320013504	26.1534	8.8109	2148.8	367453
685	469225	321419125	26.1725	8.8152	2152.0	368528
686	470596	322828856	26.1916	8.8194	2155.1	369605
687	471969	324242703	26.2107	8.8237	2158.3	370684
688	473344	325660672	26.2298	8.8280	2161.4	371764
689	474721	327082769	26.2488	8.8323	2164.6	372845
690	476100	328509000	26.2679	8.8366	2167.7	373928
691	477481	329939371	26.2869	8.8408	2170.8	375013
692	478864	331373888	26.3059	8.8451	2174.0	376099
693	480249	332812557	26.3249	8.8493	2177.1	377187
694	481636	334255384	26.3439	8.8536	2180.3	378276
695	483025	335702375	26.3629	8.8578	2183.4	379367
696	484416	337153536	26.3818	8.8621	2186.5	380459
697	485809	338608873	26.4008	8.8663	2189.7	381553
698	487204	340068392	26.4197	8.8706	2192.8	382649
699	488601	341532099	26.4386	8.8748	2196.0	383746

Number	Square	Cube	Square root	Cube root	Number = diameter	
					Circumference	Area
700	490000	343000000	26.4575	8.8790	2199.1	384845
701	491401	344472101	26.4764	8.8833	2202.3	385945
702	492804	345948408	26.4953	8.8875	2205.4	387047
703	494209	347428927	26.5141	8.8917	2208.5	388151
704	495616	348913664	26.5330	8.8959	2211.7	389256
705	497025	350402625	26.5518	8.9001	2214.8	390363
706	498436	351895816	26.5707	8.9043	2218.0	391471
707	499849	353393243	26.5895	8.9085	2221.1	392580
708	501264	354894912	26.6083	8.9127	2224.2	393692
709	502681	356400829	26.6271	8.9169	2227.4	394805
710	504100	357911000	26.6458	8.9211	2230.5	395919
711	505521	359425431	26.6646	8.9253	2233.7	397035
712	506944	360944128	26.6833	8.9295	2236.8	398153
713	508369	362467097	26.7021	8.9337	2240.0	399272
714	509796	363994344	26.7208	8.9378	2243.1	400393
715	511225	365525875	26.7395	8.9420	2246.2	401515
716	512656	367061696	26.7582	8.9462	2249.4	402639
717	514089	368601813	26.7769	8.9503	2252.5	403765
718	515524	370146232	26.7955	8.9545	2255.7	404892
719	516961	371694959	26.8142	8.9587	2258.8	406020
720	518400	373248000	26.8328	8.9628	2261.9	407150
721	519841	374805361	26.8514	8.9670	2265.1	408282
722	521284	376367048	26.8701	8.9711	2268.2	409415
723	522729	377933067	26.8887	8.9752	2271.4	410550
724	524176	379503424	26.9072	8.9794	2274.5	411687
725	525625	381078125	26.9258	8.9835	2277.7	412825
726	527076	382657176	26.9444	8.9876	2280.8	413965
727	528529	384240583	26.9629	8.9918	2283.9	415106
728	529984	385828352	26.9815	8.9959	2287.1	416248
729	531441	387420489	27.0000	9.0000	2290.2	417393
730	532900	389017000	27.0185	9.0041	2293.4	418539
731	534361	390617891	27.0370	9.0082	2296.5	419686
732	535824	392223168	27.0555	9.0123	2299.6	420835
733	537289	393832837	27.0740	9.0164	2302.8	421986
734	538756	395446904	27.0924	9.0205	2305.9	423138
735	540225	397065375	27.1109	9.0246	2309.1	424292
736	541696	398688256	27.1293	9.0287	2312.2	425447
737	543169	400315553	27.1477	9.0328	2315.4	426604
738	544644	401947272	27.1662	9.0369	2318.5	427762
739	546121	403583419	27.1846	9.0410	2321.6	428922
740	547600	405224000	27.2029	9.0450	2324.8	430084
741	549081	406869021	27.2213	9.0491	2327.9	431247
742	550564	408518488	27.2397	9.0532	2331.1	432412
743	552049	410172407	27.2580	9.0572	2334.2	433578
744	553536	411830784	27.2764	9.0613	2337.3	434746
745	555025	413493625	27.2947	9.0654	2340.5	435916
746	556516	415160936	27.3130	9.0694	2343.6	437087
747	558009	416832723	27.3313	9.0735	2346.8	438259
748	559504	418508992	27.3496	9.0775	2349.9	439433
749	561001	420189749	27.3679	9.0816	2353.1	440609

Number	Square	Cube	Square root	Cube root	Number = diameter	
					Circumference	Area
750	562500	421875000	27.3861	9.0856	2356.2	441786
751	564001	423564751	27.4044	9.0896	2359.3	442965
752	565504	425259008	27.4226	9.0937	2362.5	444146
753	567009	426957777	27.4408	9.0977	2365.6	445328
754	568516	428661064	27.4591	9.1017	2368.8	446511
755	570025	430368875	27.4773	9.1057	2371.9	447697
756	571536	432081216	27.4955	9.1098	2375.0	448883
757	573049	433798093	27.5136	9.1138	2378.2	450072
758	574564	435519512	27.5318	9.1178	2381.3	451262
759	576081	437245479	27.5500	9.1218	2384.5	452453
760	577600	438976000	27.5681	9.1258	2387.6	453646
761	579121	440711081	27.5862	9.1298	2390.8	454841
762	580644	442450728	27.6043	9.1338	2393.9	456037
763	582169	444194947	27.6225	9.1378	2397.0	457234
764	583696	445943744	27.6405	9.1418	2400.2	458434
765	585225	447697125	27.6586	9.1458	2403.3	459635
766	586756	449455096	27.6767	9.1498	2406.5	460837
767	588289	451217663	27.6948	9.1537	2409.6	462041
768	589824	452984832	27.7128	9.1577	2412.7	463247
769	591361	454756609	27.7308	9.1617	2415.9	464454
770	592900	456533000	27.7489	9.1657	2419.0	465663
771	594441	458314011	27.7669	9.1696	2422.2	466873
772	595984	460099648	27.7849	9.1736	2425.3	468085
773	597529	461889917	27.8029	9.1775	2428.5	469298
774	599076	463684824	27.8209	9.1815	2431.6	470513
775	600625	465484375	27.8388	9.1855	2434.7	471730
776	602176	467288576	27.8568	9.1894	2437.9	472948
777	603729	469097433	27.8747	9.1933	2441.0	474168
778	605284	470910952	27.8927	9.1973	2444.2	475389
779	606841	472729139	27.9106	9.2012	2447.3	476612
780	608400	474552000	27.9285	9.2052	2450.4	477836
781	609961	476379541	27.9464	9.2091	2453.6	479062
782	611524	478211768	27.9643	9.2130	2456.7	480290
783	613089	480048687	27.9821	9.2170	2459.9	481519
784	614656	481890304	28.0000	9.2209	2463.0	482750
785	616225	483736625	28.0179	9.2248	2466.2	483982
786	617796	485587656	28.0357	9.2287	2469.3	485216
787	619369	487443403	28.0535	9.2326	2472.4	486451
788	620944	489303872	28.0713	9.2365	2475.6	487688
789	622521	491169069	28.0891	9.2404	2478.7	488927
790	624100	493039000	28.1069	9.2443	2481.9	490167
791	625681	494913671	28.1247	9.2482	2485.0	491409
792	627264	496793088	28.1425	9.2521	2488.1	492652
793	628849	498677257	28.1603	9.2560	2491.3	493897
794	630436	500566184	28.1780	9.2599	2494.4	495143
795	632025	502459875	28.1957	9.2638	2497.6	496391
796	633616	504358336	28.2135	9.2677	2500.7	497641
797	635209	506261573	28.2312	9.2716	2503.8	498892
798	636804	508169592	28.2489	9.2754	2507.0	500145
799	638401	510082399	28.2666	9.2793	2510.1	501399

Number	Square	Cube	Square root	Cube root	Number = diameter	
					Circumference	Area
800	640000	512000000	28.2843	9.2832	2513.3	502655
801	641601	513922401	28.3019	9.2870	2516.4	503912
802	643204	515849608	28.3196	9.2909	2519.6	505171
803	644809	517781627	28.3373	9.2948	2522.7	506432
804	646416	519718464	28.3549	9.2986	2525.8	507694
805	648025	521660125	28.3725	9.3025	2529.0	508958
806	649636	523606616	28.3901	9.3063	2532.1	510223
807	651249	525557943	28.4077	9.3102	2535.3	511490
808	652864	527514112	28.4253	9.3140	2538.4	512758
809	654481	529475129	28.4429	9.3179	2541.5	514028
810	656100	531441000	28.4605	9.3217	2544.7	515300
811	657721	533411731	28.4781	9.3255	2547.8	516573
812	659344	535387328	28.4956	9.3294	2551.0	517848
813	660969	537367797	28.5132	9.3332	2554.1	519124
814	662596	539353144	28.5307	9.3370	2557.3	520402
815	664225	541343375	28.5482	9.3408	2560.4	521681
816	665856	543338496	28.5657	9.3447	2563.5	522962
817	667489	545338513	28.5832	9.3485	2566.7	524245
818	669124	547343432	28.6007	9.3523	2569.8	525529
819	670761	549353259	28.6182	9.3561	2573.0	526814
820	672400	551368000	28.6356	9.3599	2576.1	528102
821	674041	553387661	28.6531	9.3637	2579.2	529391
822	675684	555412248	28.6705	9.3675	2582.4	530681
823	677329	557441767	28.6880	9.3713	2585.5	531973
824	678976	559476224	28.7054	9.3751	2588.7	533267
825	680625	561515625	28.7228	9.3789	2591.8	534562
826	682276	563559976	28.7402	9.3827	2595.0	535858
827	683929	565609283	28.7576	9.3865	2598.1	537157
828	685584	567663552	28.7750	9.3902	2601.2	538456
829	687241	569722789	28.7924	9.3940	2604.4	539758
830	688900	571787000	28.8097	9.3978	2607.5	541061
831	690561	573856191	28.8271	9.4016	2610.7	542365
832	692224	575930368	28.8444	9.4053	2613.8	543671
833	693889	578009537	28.8617	9.4091	2616.9	544979
834	695556	580093704	28.8791	9.4129	2620.1	546288
835	697225	582182875	28.8964	9.4166	2623.2	547599
836	698896	584277056	28.9137	9.4204	2626.4	548912
837	700569	586376253	28.9310	9.4241	2629.5	550226
838	702244	588480472	28.9482	9.4279	2632.7	551541
839	703921	590589719	28.9655	9.4316	2635.8	552858
840	705600	592704000	28.9828	9.4354	2638.9	554177
841	707281	594823321	29.0000	9.4391	2642.1	555497
842	708964	596947688	29.0172	9.4429	2645.2	556819
843	710649	599077107	29.0345	9.4466	2648.4	558142
844	712336	601211584	29.0517	9.4503	2651.5	559467
845	714025	603351125	29.0689	9.4541	2654.6	560794
846	715716	605495736	29.0861	9.4578	2657.8	562122
847	717409	607645423	29.1033	9.4615	2660.9	563452
848	719104	609800192	29.1204	9.4652	2664.1	564783
849	720801	611960049	29.1376	9.4690	2667.2	566116

Number	Square	Cube	Square root	Cube root	Number = diameter	
					Circumference	Area
850	722500	614125000	29.1548	9.4727	2670.4	567450
851	724201	616295051	29.1719	9.4764	2673.5	568786
852	725904	618470208	29.1890	9.4801	2676.6	570124
853	727609	620650477	29.2062	9.4838	2679.8	571463
854	729316	622835864	29.2233	9.4875	2682.9	572803
855	731025	625026375	29.2404	9.4912	2686.1	574146
856	732736	627222016	29.2575	9.4949	2689.2	575490
857	734449	629422793	29.2746	9.4986	2692.3	576835
858	736164	631628712	29.2916	9.5023	2695.5	578182
859	737881	633839779	29.3087	9.5060	2698.6	579530
860	739600	636056000	29.3258	9.5097	2701.8	580880
861	741321	638277381	29.3428	9.5134	2704.9	582232
862	743044	640503928	29.3598	9.5171	2708.1	583585
863	744769	642735647	29.3769	9.5207	2711.2	584940
864	746496	644972544	29.3939	9.5244	2714.3	586297
865	748225	647214625	29.4109	9.5281	2717.5	587655
866	749956	649461896	29.4279	9.5317	2720.6	589014
867	751689	651714363	29.4449	9.5354	2723.8	590375
868	753424	653972032	29.4618	9.5391	2726.9	591738
869	755161	656234909	29.4788	9.5427	2730.0	593102
870	756900	658503000	29.4958	9.5464	2733.2	594468
871	758641	660776311	29.5127	9.5501	2736.3	595835
872	760384	663054848	29.5296	9.5537	2739.5	597204
873	762129	665338617	29.5466	9.5574	2742.6	598575
874	763876	667627624	29.5635	9.5610	2745.8	599947
875	765625	669921875	29.5804	9.5647	2748.9	601320
876	767376	672221376	29.5973	9.5683	2752.0	602696
877	769129	674526133	29.6142	9.5719	2755.2	604073
878	770884	676836152	29.6311	9.5756	2758.3	605451
879	772641	679151439	29.6479	9.5792	2761.5	606831
880	774400	681472000	29.6648	9.5828	2764.6	608212
881	776161	683797841	29.6816	9.5865	2767.7	609595
882	777924	686128968	29.6985	9.5901	2770.9	610980
883	779689	688465387	29.7153	9.5937	2774.0	612366
884	781456	690807104	29.7321	9.5973	2777.2	613754
885	783225	693154125	29.7489	9.6010	2780.3	615143
886	784996	695506456	29.7658	9.6046	2783.5	616534
887	786769	697864103	29.7825	9.6082	2786.6	617927
888	788544	700227072	29.7993	9.6118	2789.7	619321
889	790321	702595369	29.8161	9.6154	2792.9	620717
890	792100	704969000	29.8329	9.6190	2796.0	622114
891	793881	707347971	29.8496	9.6226	2799.2	623513
892	795664	709732288	29.8664	9.6262	2802.3	624913
893	797449	712121957	29.8831	9.6298	2805.4	626315
894	799236	714516984	29.8998	9.6334	2808.6	627718
895	801025	716917375	29.9166	9.6370	2811.7	629124
896	802816	719323136	29.9333	9.6406	2814.9	630530
897	804609	721734273	29.9500	9.6442	2818.0	631938
898	806404	724150792	29.9666	9.6477	2821.2	633348
899	808201	726572699	29.9833	9.6513	2824.3	634760

Number	Square	Cube	Square root	Cube root	Number = diameter	
					Circumference	Area
900	810000	729000000	30.0000	9.6549	2827.4	636173
901	811801	731432701	30.0167	9.6585	2830.6	637587
902	813604	733870808	30.0333	9.6620	2833.7	639003
903	815409	736314327	30.0500	9.6656	2836.9	640421
904	817216	738763264	30.0666	9.6692	2840.0	641840
905	819025	741217625	30.0832	9.6727	2843.1	643261
906	820836	743677416	30.0998	9.6763	2846.3	644683
907	822649	746142643	30.1164	9.6799	2849.4	646107
908	824464	748613312	30.1330	9.6834	2852.6	647533
909	826281	751089429	30.1496	9.6870	2855.7	648960
910	828100	753571000	30.1662	9.6905	2858.8	650388
911	829921	756058031	30.1828	9.6941	2862.0	651818
912	831744	758550528	30.1993	9.6976	2865.1	653250
913	833569	761048497	30.2159	9.7012	2868.3	654684
914	835396	763551944	30.2324	9.7047	2871.4	656118
915	837225	766060875	30.2490	9.7082	2874.6	657555
916	839056	768575296	30.2655	9.7118	2877.7	658993
917	840889	771095213	30.2820	9.7153	2880.8	660433
918	842724	773620632	30.2985	9.7188	2884.0	661874
919	844561	776151559	30.3150	9.7224	2887.1	663317
920	846400	778688000	30.3315	9.7259	2890.3	664761
921	848241	781229961	30.3480	9.7294	2893.4	666207
922	850084	783777448	30.3645	9.7329	2896.5	667654
923	851929	786330467	30.3809	9.7364	2899.7	669103
924	853776	788889024	30.3974	9.7400	2902.8	670554
925	855625	791453125	30.4138	9.7435	2906.0	672006
926	857476	794022776	30.4302	9.7470	2902.1	673460
927	859329	796597983	30.4467	9.7505	2912.3	674915
928	861184	799178752	30.4631	9.7540	2915.4	676372
929	863041	801765089	30.4795	9.7575	2918.5	677831
930	864900	804357000	30.4959	9.7610	2921.7	679291
931	866761	806954491	30.5123	9.7645	2924.8	680752
932	868624	809557568	30.5287	9.7680	2928.0	682216
933	870489	812166237	30.5450	9.7715	2931.1	683680
934	872356	814780504	30.5614	9.7750	2934.2	685147
935	874225	817400375	30.5778	9.7785	2937.4	686615
936	876096	820025856	30.5941	9.7819	2940.5	688084
937	877969	822656953	30.6105	9.7854	2943.7	689555
938	879844	825293672	30.6268	9.7889	2946.8	691028
939	881721	827936019	30.6431	9.7924	2950.0	692502
940	883600	830584000	30.6594	9.7959	2953.1	693978
941	885481	833237621	30.6757	9.7993	2956.2	695455
942	887364	835896888	30.6920	9.8028	2959.4	696934
943	889249	838561807	30.7083	9.8063	2962.5	698415
944	891136	841232384	30.7246	9.8097	2965.7	699897
945	893025	843908625	30.7409	9.8132	2968.8	701380
946	894916	846590536	30.7571	9.8167	2971.9	702865
947	896809	849278123	30.7734	9.8201	2975.1	704352
948	898704	851971392	30.7896	9.8236	2978.2	705840
949	900601	854670349	30.8058	9.8270	2981.4	707330

Number	Square	Cube	Square root	Cube root	Number = diameter	
					Circumference	Area
950	902500	857375000	30.8221	9.8305	2984.5	708822
951	904401	860085351	30.8383	9.8339	2987.7	710315
952	906304	862801408	30.8545	9.8374	2990.8	711809
953	908209	865523177	30.8707	9.8408	2993.9	713306
954	910116	868250664	30.8869	9.8443	2997.1	714803
955	912025	870983875	30.9031	9.8477	3000.2	716303
956	913936	873722816	30.9192	9.8511	3003.4	717804
957	915849	876467493	30.9354	9.8546	3006.5	719306
958	917764	879217912	30.9516	9.8580	3009.6	720810
959	919681	881974079	30.9677	9.8614	3012.8	722316
960	921600	884736000	30.9839	9.8648	3015.9	723823
961	923521	887503681	31.0000	9.8683	3019.1	725332
962	925444	890277128	31.0161	9.8717	3022.2	726842
963	927369	893056347	31.0322	9.8751	3025.4	728354
964	929296	895841344	31.0483	9.8785	3028.5	729867
965	931225	898632125	31.0644	9.8819	3031.6	731382
966	933156	901428696	31.0805	9.8854	3034.8	732899
967	935089	904231063	31.0966	9.8888	3037.9	734417
968	937024	907039232	31.1127	9.8922	3041.1	735937
969	938961	909853209	31.1288	9.8956	3044.2	737458
970	940900	912673000	31.1448	9.8990	3047.3	738981
971	942841	915498611	31.1609	9.9024	3050.5	740506
972	944784	918330048	31.1769	9.9058	3053.6	742032
973	946729	921167317	31.1929	9.9092	3056.8	743559
974	948676	924010424	31.2090	9.9126	3059.9	745088
975	950625	926859375	31.2250	9.9160	3063.1	746619
976	962576	929714176	31.2410	9.9194	3066.2	748151
977	954529	932574833	31.2570	9.9227	3069.3	749685
978	956484	935441352	31.2730	9.9261	3072.5	751221
979	958441	938313739	31.2890	9.9295	3075.6	752758
980	960400	941192000	31.3050	9.9329	3078.8	754296
981	962361	944076141	31.3209	9.9363	3081.9	755837
982	964324	946966168	31.3369	9.9396	3085.0	757378
983	966289	949862087	31.3528	9.9430	3088.2	758922
984	968256	952763904	31.3688	9.9464	3091.3	760466
985	970225	955671625	31.3847	9.9497	3094.5	762013
986	972196	958585256	31.4006	9.9531	3097.6	763561
987	974169	961504803	31 4166	9.9565	3100.8	765111
988	976144	964430272	31.4325	9.9598	3103.9	766662
989	978121	967361669	31.4484	9.9632	3107.0	768214
990	980100	970299000	31.4643	9.9666	3110.2	769769
991	982081	973242271	31.4802	9.9699	3113.3	771325
992	984064	976191488	31.4960	9.9733	3116.5	772882
993	986049	979146657	31.5119	9.9766	3119.6	774441
994	988036	982107784	31.5278	9.9800	3122.7	776002
995	990025	985074875	31.5436	9.9833	3125.9	777564
996	992016	988047936	31.5595	9.9866	3129.0	779128
997	994009	991026973	31.5753	9.9900	3132.2	780693
998	996004	994011992	31.5911	9.9933	3135.3	782260
999	998001	997002999	31.6070	9.9967	3138.5	783828

Mensuration of Plane Figures*

Figure	Name and Definition	Mensuration Formulas
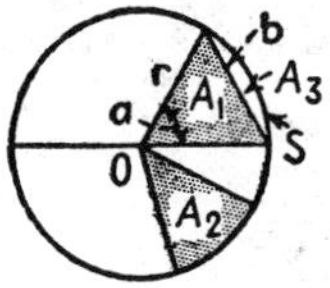	*Circle:* A plane closed curve every point of which is distant r from a fixed point O inside the curve. *Circular segment* $= A_3$ *Circular sector* $= A_2$ Sector-segment $= A_1 = A_2 - A_3$	$d = 2r$ $c = \pi d = 2\pi r$ $A = \pi r^2$ = area of circle $s = r\alpha$ where α is measured in radians $b = 2r \sin\left(\frac{\alpha}{2}\right) = 2r \sin\left(\frac{90s}{\pi r}\right)^\circ$ $A_1 = \frac{1}{2} br \cos\left(\frac{\alpha}{2}\right)$ $A_2 = \frac{\alpha r^2}{2} = \frac{sr}{2}$ $A_3 = A_2 - A_1$
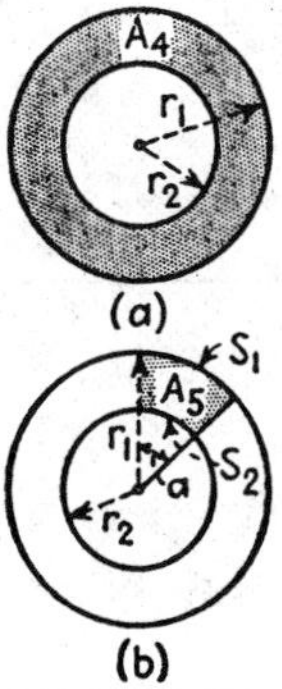 (a) (b)	*Annulus* $= A_4 =$ area between two concentric circles (i.e., circles having the same center). *Portion of an annulus* $= A_5$	$A_4 = \pi(r_1^2 - r_2^2)$ $A_5 = \frac{\alpha}{2}(r_1^2 - r_2^2)$ where α is measured in radians $= \frac{\alpha}{2}(r_1 - r_2)(r_1 + r_2)$ $= \left(\frac{r_1 - r_2}{2}\right)(s_1 + s_2)$
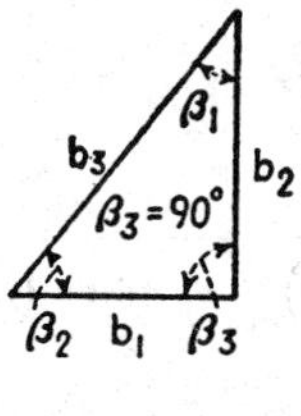	*Right triangle:* A closed three-sided plane figure two of whose sides are perpendicular.	$c = b_1 + b_2 + b_3, \quad b_3 = \sqrt{b_1^2 + b_2^2}$ $A = \frac{1}{2}b_1b_2 = \frac{1}{2}b_1^2 \tan\beta_2 = \frac{1}{2}b_2 \tan\beta_1$ $= \frac{1}{2}b_1b_3 \sin\beta_2 = \frac{1}{2}b_2b_3 \sin\beta_1$ $\sin\beta_2 = \frac{b_2}{b_3}, \quad \cos\beta_2 = \frac{b_1}{b_3}, \quad \tan\beta_2 = \frac{b_2}{b_1}$ $\sin\beta_1 = \frac{b_1}{b_3}, \quad \cos\beta_1 = \frac{b_2}{b_3}, \quad \tan\beta_1 = \frac{b_1}{b_2}$ $\beta_1 + \beta_2 + \beta_3 = 180^\circ = \pi$ radians $\beta_1 + \beta_2 = 90^\circ = \frac{\pi}{2}$ radians
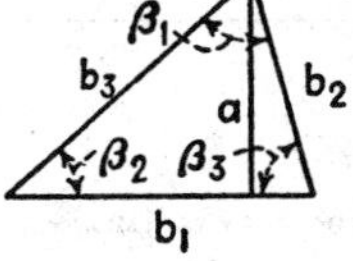	*Any triangle:* Any closed three-sided plane figure.	$c = b_1 + b_2 + b_3, \quad \frac{c}{2} = \frac{b_1 + b_2 + b_3}{2} = k$ $A = \frac{1}{2}ab_1 = \frac{1}{2}b_1b_3 \sin\beta_2 = \frac{1}{2}b_1b_2 \sin\beta_3$ $a = \frac{2}{b_1}\sqrt{k(k - b_1)(k - b_2)(k - b_3)}$ $\frac{\sin\beta_1}{b_1} = \frac{\sin\beta_2}{b_2} = \frac{\sin\beta_3}{b_3}$ $b_1^2 = b_2^2 + b_3^2 - 2b_2b_3 \cos\beta_1$ $\tan\left(\frac{\beta_1}{2}\right) = \sqrt{\frac{(k - b_2)(k - b_3)}{k(k - b_1)}}$ $\beta_1 + \beta_2 + \beta_3 = 180^\circ = \pi$ radians

* Reprinted from O'Rourke, "General Engineering Handbook," McGraw-Hill Book Company, Inc., 1940, by permission.

Mensuration of Plane Figures (*Continued*)

Figure	Name and Definition	Mensuration Formulas
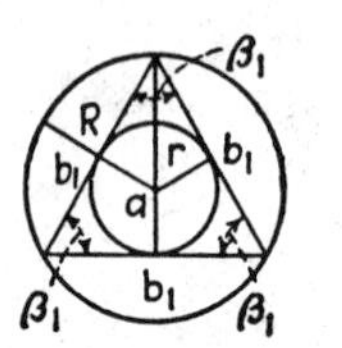	*Equilateral triangle:* A triangle whose sides are equal.	$c = 3b_1, \quad \frac{c}{2} = k = \frac{3b_1}{2}$ $A = \frac{1}{2} ab_1 = \frac{1}{2} b_1^2 \sin \beta_1 = b_1^2 \frac{\sqrt{3}}{4}$ $a = \frac{b_1 \sqrt{3}}{2}$ $r = \frac{b_1}{2\sqrt{3}}, \quad R = 2r = \frac{b_1}{\sqrt{3}}$ $3\beta_1 = 180°; \quad \therefore \beta_1 = 60°$
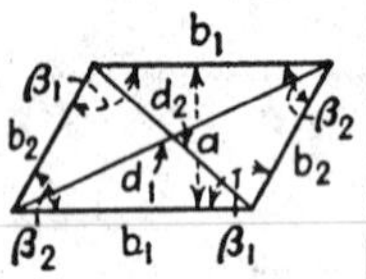	*Parallelogram:* A four-sided plane figure (a quadrilateral) the opposite sides of which are parallel and equal.	$c = 2(b_1 + b_2)$ $d_1 = \sqrt{b_1^2 + b_2^2 - 2b_1b_2 \cos \beta_1}$ $d_2 = \sqrt{b_1^2 + b_2^2 - 2b_1b_2 \cos \beta_2}$ $a = b_2 \sin \beta_2$ $A = b_1a = b_1b_2 \sin \beta_2$ $(\beta_1 + \beta_2) = 180° = \pi$ radians
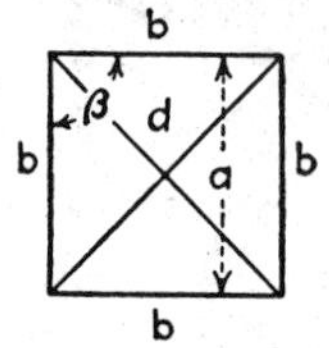	*Square:* A parallelogram whose sides are equal and perpendicular.	$c = 4b$ $d = b\sqrt{2}$ $a = b$ $A = b^2$ $\beta = 90°$
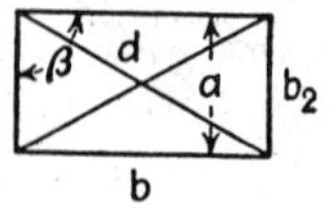	*Rectangle:* A parallelogram whose sides are perpendicular but unequal.	$c = 2(b_1 + b_2) \quad d = \sqrt{b_1^2 + b_2^2}$ $a = b_2 \quad A = b_1b_2 \quad \beta = 90°$
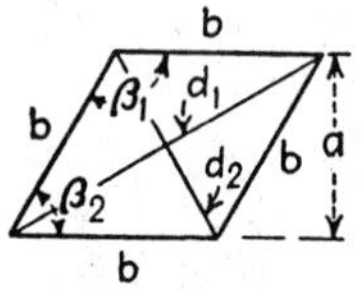	*Rhombus:* A parallelogram whose sides are equal but not perpendicular.	$c = 4b, \quad d_1 = b \sin \left(\frac{\beta_1}{2}\right), \quad d_2 = b \sin \left(\frac{\beta_2}{2}\right)$ $a = b \sin \beta_2$ $A = ba = b^2 \sin \beta_2$ $\beta_1 + \beta_2 = 180° = \pi$ radians
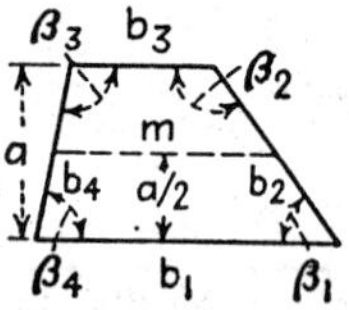	*Trapezoid:* A quadrilateral only two sides of which are parallel.	$c = b_1 + b_2 + b_3 + b_4$ $m = \frac{(b_1 + b_3)}{2}$ $a = b_2 \sin \beta_1 = b_4 \sin \beta_4$ $A = ma.$ $\beta_1 + \beta_2 + \beta_3 + \beta_4 = 360° = 2\pi$ radians

MENSURATION OF PLANE FIGURES (*Continued*)

Figure	Name and Definition	Mensuration Formulas
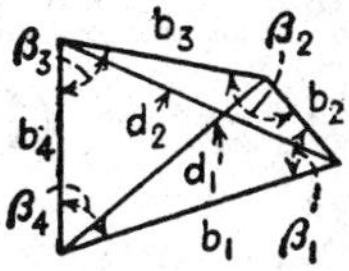	*Trapezium:* A quadrilateral no two sides of which are parallel.	$c = b_1 + b_2 + b_3 + b_4$ A = sum of areas of the two triangles formed by either one of the diagonals and the four sides $= (b_1 b_2 \sin \beta_1 + b_3 b_4 \sin \beta_3) \div 2$ $\beta_1 + \beta_2 + \beta_3 + \beta_4 = 360° = 2\pi$ radians
	*Regular Polygon of $n > 4$ sides:* An n-sided plane figure all of whose sides are equal.	$c = nb$ $\alpha = \frac{2\pi}{n}$ radians = central angle $\beta = (n - 2)\frac{\pi}{n}$ radians = vertex angle A = area of n isoceles triangles $= \left(\frac{n}{2}\right) Rb \cos\left(\frac{\alpha}{2}\right)$
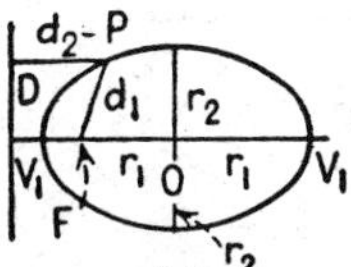 (a) 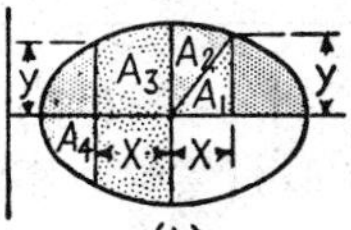 (b)	*Ellipse:* A closed plane curve each point of which satisfies the relation $\frac{d_1}{d_2} = e < 1$. d_1 = distance from a fixed point F called the *focus* d_2 = distance from a fixed line D called the *directrix* e = eccentricity V_1 and V_2 are the vertices $2r_1$ and $2r_2$ are the major and minor axes, respectively O is a point midway between v_1 and v_2 and is the center of the ellipse	$c = r_1\left[4 + 1.1\left(\frac{r_2}{r_1}\right) + 1.2\left(\frac{r_2}{r_1}\right)^2\right]$ approximately *Note:* The error in the value of c is always less than 2 % and even less than 1 % if $\frac{r_2}{r_1} > \frac{1}{3}$. $A = \pi r_1 r_2$ $A_1 = \left(\frac{r_1 r_2}{2}\right)\cos^{-1}\left(\frac{x}{r_1}\right)$ $A_2 = \left(\frac{r_1 r_2}{2}\right)\cos^{-1}\left(\frac{x}{r_1}\right)$ $A_3 = xy + r_1 r_2 \sin^{-1}\left(\frac{x}{r_1}\right)$ $A_4 = -xy + r_1 r_2 \cos^{-1}\left(\frac{x}{r_1}\right)$
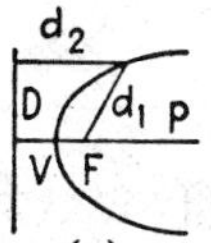 (a) 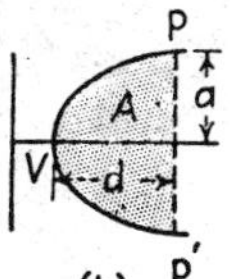 (b)	*Parabola:* An open plane curve each point of which satisfies the relation $\frac{d_1}{d_2} = 1$, where d_1, d_2, V, and F have the same meaning as for ellipse (see above).	c = length of arc pVp', $n = \frac{2d}{a}$ $c = a\left[\sqrt{n^2+1} + \left(\frac{1}{n}\right)\ln\left(\sqrt{n^2+1} + n\right)\right]$ $= a\left[\sqrt{n^2+1} + \left(\frac{1}{n}\right)\sinh^{-1} n\right]$ $A = \frac{4ad}{3}$

Mensuration of Plane Figures (*Continued*)

Figure	Name and Definition	Mensuration Formulas
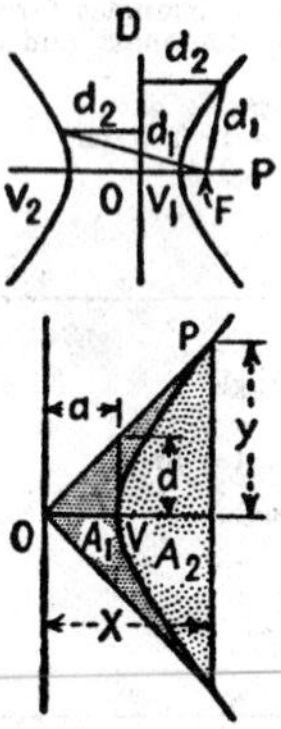	*Hyperbola:* An open plane curve each point of which satisfies the relation $\frac{d_1}{d_2} = e > 1$, where $d_1, d_2, D, e, F, O, V_1$, and V_2 have the same meaning as for ellipse. The line pF, which is perpendicular to D, is called the axis of the hyperbola.	c = length of arc pVp'. (Note: This yields to an expression involving elliptic functions and will not be given here.) $A_1 = ad \ln\left[\left(\frac{x}{a}\right) + \left(\frac{y}{d}\right)\right] = ad \cosh^{-1}\left(\frac{x}{a}\right)$ $A_2 = xy - ad \ln\left(\frac{x}{a} + \frac{y}{d}\right)$ $= xy - ad \cosh^{-1}\left(\frac{x}{a}\right)$
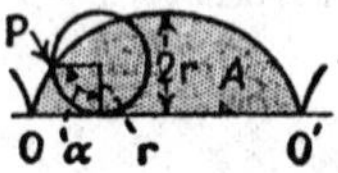	*Cycloid:* A plane curve traced by a fixed point p on a circle of radius r, as that circle rolls on a straight line.	c = length of arc $opo' = 8r$ s = length of arc $op = 4r\left[1 - \cos\left(\frac{\alpha}{2}\right)\right]$ $A = 3\pi r^2$
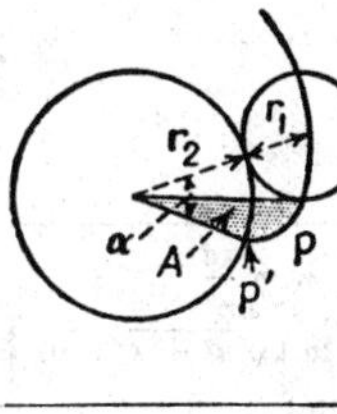	*Epicycloid:* A plane curve traced by a fixed point p on a circle of radius r_1 as that circle rolls on another circle of radius r_2.	s = length of arc pp' $= \left(\frac{4r_1}{r_2}\right)(r_1 + r_2)\left[1 - \cos\left(\frac{r_2\alpha}{2r_1}\right)\right]$ $A = \left(\frac{r_1}{2r_2}\right)(r_1 + r_2)(2r_1 + r_2)\left(\frac{r_2\alpha}{r_1} - \sin\frac{r_2\alpha}{r_1}\right)$
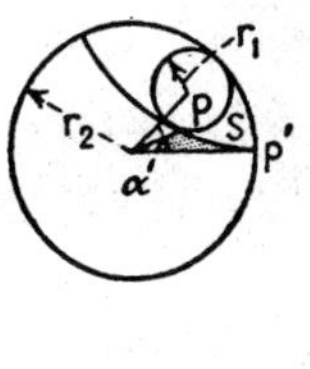	*Hypocycloid:* A plane curve traced by a fixed point p, on a circle of radius r_1, as that circle rolls inside another circle of radius $r_2 > r_1$.	s = length of arc pp' $= \left(\frac{4r_1}{r_2}\right)(r_2 - r_1)\left[1 - \cos\left(\frac{r_2\alpha}{2r_1}\right)\right]$ $A = \left(\frac{r_1}{2r_2}\right)(r_2 - r_1)(r_2 - 2r_1)$ $\left[\frac{r_2\alpha}{r_1} - \sin\left(\frac{r_2\alpha}{r_1}\right)\right]$

Mensuration of Plane Figures (*Continued*)

Figure	Name and Definition	Mensuration Formulas
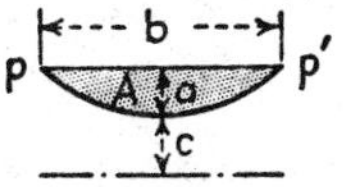	*Catenary:* A plane curve made by a cable which hangs freely between two points of support p, p'.	s = length of arc $pp' = 2c \sinh\left(\frac{b}{2c}\right) \cong b\left(1 + \frac{8a^2}{3b^2}\right)$ $A = b(a + c) - c^2 \sinh\left(\frac{b}{2c}\right)$
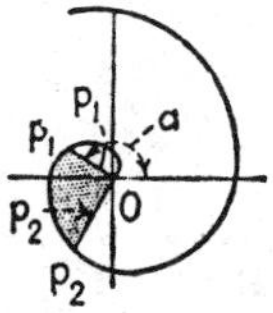	*Spiral of Archimedes:* A plane curve traced by the end point p of a line ρ which revolves about a fixed point O such that $\rho = a\alpha$.	s = length of arc op_1 $= \left(\frac{a}{2}\right)[\alpha\sqrt{1 + \alpha^2} + \ln(\alpha + \sqrt{1 + \alpha^2})]$, where α is expressed in radians $A = \frac{\rho_2^3 - \rho_1^3}{6a}$
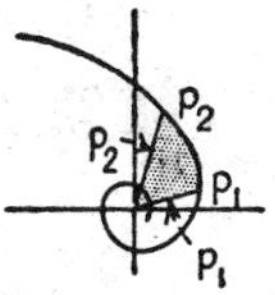	*Hyperbolic spiral:* A plane curve traced by the end point p of a line ρ which revolves about a fixed point O such that $\rho\alpha = a$.	s = length of arc p_1p_2 $= \sqrt{p_2^2 + a^2} - \sqrt{p_1^2 + a^2} + a \ln \left[\frac{(\sqrt{\rho_2^2 + a^2} + \rho_2 - a)(\sqrt{\rho_1^2 + a^2} + \rho_1 + a)}{(\sqrt{\rho_2^2 + a^2} + \rho_2 + a)(\sqrt{\rho_1^2 + a^2} + \rho_1 - a)}\right]$ $A = \frac{a}{2}(\rho_2 - \rho_1)$
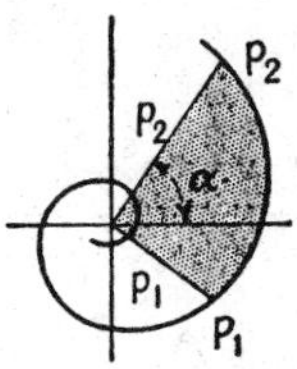	*Logarithmic spiral:* A plane curve traced by the end point p of line ρ which revolves about a fixed point O such that $\rho = a^{\alpha a}$.	s = length of arc $p_1p_2 = \left(\frac{\rho_2 - \rho_1}{a}\right)\sqrt{a^2 + 1}$ $A = \frac{1}{4a}(\rho_2^2 - \rho_1^2)$
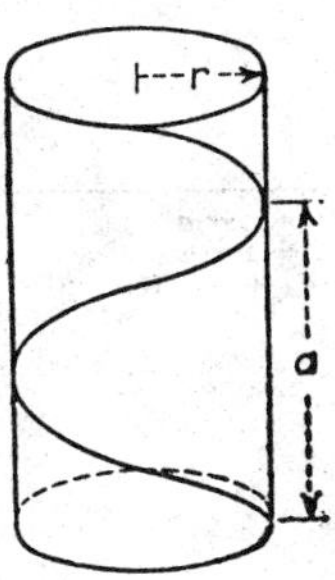	*Helix:* Draw a straight line on a paper and wrap the paper around a cylinder. The curve formed by that line is a helix. a is the pitch of the helix, the radius r of the cylinder is that of the helix, and n is the number of turns.	s = total length of helix $= n\sqrt{(2\pi r)^2 + a^2}$

Mensuration of Solid Figures

Figure	Name and Definition	Mensuration Formulas
	*Sphere:* A solid figure bounded by a surface all points of which are a distance r from a given point O called the *center.* r is the radius of the sphere.	$A = 4\pi r^2$ $V = \frac{4}{3}\pi r^3$
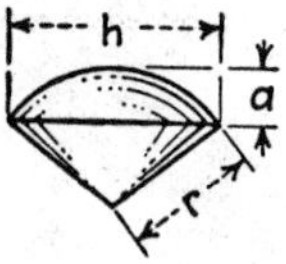	*Spherical sector:* The portion of a sphere shown in figure.	$A = \frac{1}{2}\pi r(4a + h)$ $V = \frac{2\pi r^2 a}{3}$
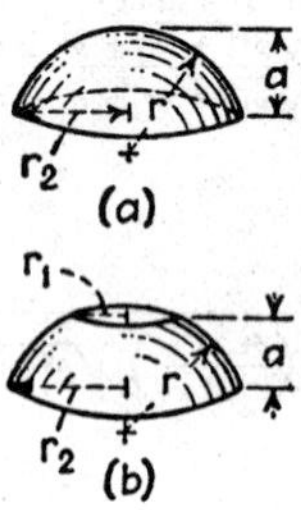	*Spherical segment:* The portion of a sphere shown in Fig. *a* or Fig. *b*.	*Fig. a:* $A = 2\pi ra$, where $a = r - \sqrt{r^2 - r_2^2}$ $V = \frac{1}{6}\pi a(3r_2^2 + a^2)$ *Fig. b:* $A = 2\pi ra$, where $a = \sqrt{r^2 - r_1^2} - \sqrt{r^2 - r_2^2}$ $V = \frac{1}{6}\pi a(3r_1^2 + 3r_2^2 + a^2)$
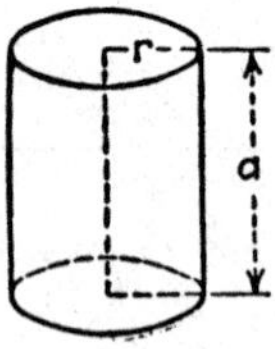	*Right circular cylinder:* A solid figure traced by a rectangle of height a and width r rotated about the side a as an axis.	$A = 2\pi ra$ $V = \pi r^2 a$
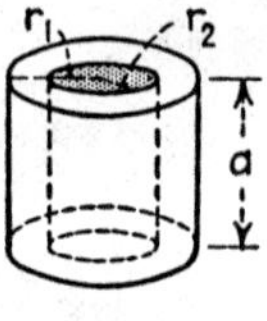	*Pipe:* A hollow cylinder.	A_1 = external lateral area = $2\pi r_1 a$ A_2 = internal lateral area = $2\pi r_2 a$ A_t = total area = $A_1 + A_2 + 2\pi(r_1^2 - r_2^2)$ $V = \pi(r_1^2 - r_2^2)a$

MENSURATION OF SOLID FIGURES (*Continued*)

Figure	Name and Definition	Mensuration Formulas
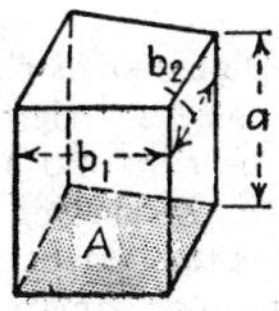	*Prism:* A solid whose bases are similar or equal polygons and whose faces are parallelograms.	Let $b_1, b_2 \cdots b_n$ = width of lateral edges a = height of any edge A_l = lateral area $= a(b_1 + b_2 + \cdots + b_n)$ $V = Aa$, where A = area of base (see page 44-28)
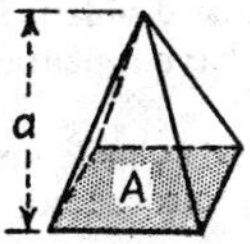	*Pyramid:* A figure whose base is a polygon and whose sides are triangles.	A_l = lateral area = sum of area of triangular sides (see page 44-27) $V = \frac{Aa}{3}$, where A = area of base
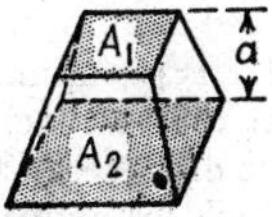	*Truncated pyramid:* The part left of a pyramid which is cut off by a plane parallel to the base.	A_l = lateral area = sum of areas of the trapezoidal sides (see page 44-28) $V = \frac{a}{3}(A_1 + A_2 + \sqrt{A_1A_2})$, where A_1 = area of truncated section and A_2 = area of base
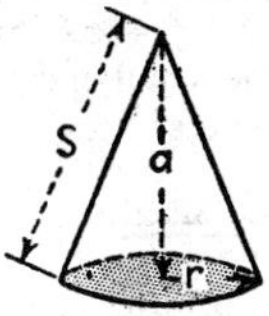	*Right circular cone:* A solid figure generated by a right triangle of base r and altitude a which is rotated about its altitude as an axis.	A_l = lateral area $= \pi rs = \pi r\sqrt{r^2 + a^2}$ A_t = total area $= A_l + \pi r^2$ $V = \frac{\pi r^2 a}{3}$
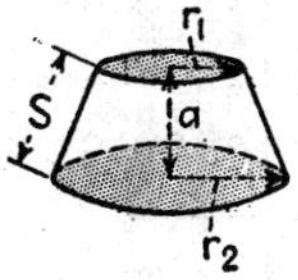	*Frustum of a right circular cone:* The part of a cone left after the cone is cut by a plane parallel to its base.	A_l = lateral area $= \pi s(r_1 + r_2) = \pi(r_1 + r_2)\sqrt{a^2 + (r_2 - r_1)^2}$ A_t = total area $= A_l + \pi(r_1^2 + r_2^2)$ $V = \frac{\pi a(r_1^2 + r_2^2 + r_1r_2)}{3}$
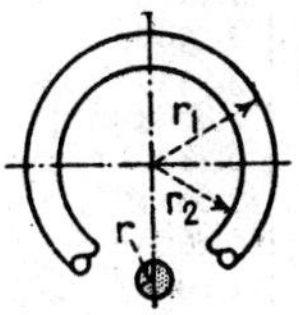	*Torus (circular section):* A solid of revolution obtained by revolving a circle about an axis outside the circle but parallel to the plane of the circle.	$A_l = \pi^2(r_1 + r_2)(r_1 - r_2)$ $V = \left(\frac{\pi^2}{4}\right)(r_1 + r_2)(r_1 - r_2)^2$

LAYING OUT REGULAR POLYGONS

The cut-and-try method is to draw a circle and space it off, but it saves time to know what spacing to use or how large a circle to draw to get a figure of the right size. Suppose we wish to lay out any regular figure, such as a pentagon or five-sided figure, having sides $1\frac{1}{2}$ in. long.

How to Use Table. Looking in the third column, we find "Diameter of circle that will just inclose it," and opposite "pentagon" we find 1.7012 as the circle that will just inclose a pentagon having a side equal to 1. This may be 1 in. or 1 anything else, so, as we are dealing in inches, we call it inches. As the side of the pentagon is to be $1\frac{1}{2}$ in., we multiply 1.7012 by $1\frac{1}{2}$ and get 2.5518 as the diameter of circle to draw, and take half of this, or the radius 1.2759 ($1\frac{9}{32}$ in. approximately), in the compass to draw the circle. Then with $1\frac{1}{2}$ in. in the dividers, we space around the circle, and if the work has been carefully done, it will just divide it into five equal parts. Connect these points by straight lines, and you have a pentagon with sides $1\frac{1}{2}$ in. long.

If the pentagon is to go inside a circle of given diameter, say 2 in., look under column 5, which gives "Length of side where diameter of inclosing circle equals 1," and find 0.5878. Multiply by 2 as this is for a 2-in. circle, and the side will be $2 \times 0.5878 = 1.1756$ or $1\frac{11}{64}$ approximately.

Assume that it is necessary to have a triangular end on a round shaft, how large must the shaft be to give a triangle 1.5 in. on a side?

Look in the table under column 3, and opposite triangle, find 1.1546, meaning that, where the side of a triangle is 1, the diameter of a circle that will just inclose it is 1.1546. As the side is 1.5, we have $1.5 \times 1.1546 = 1.7318$, the diameter of the shaft required. If the corners need not be sharp, a shaft 1.625 would be ample.

Reversing this to find the size of a bearing that can be turned on a triangular bar of this size, look in column 4, which gives the largest circle that will go inside a triangle with a side equal to 1. This gives 0.5774. Multiplying 0.5774 by $1.5 = 0.8661$.

A square taper reamer is to be used which must ream 1 in. at the small end and 1.5 at the back. What size must this be across the flats at both places?

Number of Sides	Name of Figure	Diameter of Circle that will just enclose when side is 1	Diameter of circle that will just go inside when side is 1	Length of side where diameter of enclosure circle equals 1	Length of side where inside circle equals 1	Angle formed by lines drawn from center to corners	Angle formed by outer sides of figures	To find Area of Figure multiply side by itself and by number in this column
3	Triangle ..	1.1546	.5774	.866	1.732	120°	60°	.4330
4	Square	1.4142	1.	.7071	1.	90	90	1.
5	Pentagon ..	1.7012	1.3764	.5878	.7265	72	108	1.7204
6	Hexagon ..	2.	1.732	.5	.5774	60	120	2.5980
7	Heptagon .	2.3048	2.0766	.4338	.4815	51°-26′	128 $\frac{4}{7}$	3.6339
8	Octagon...	2.6132	2.4142	.3827	.4142	45	135	4.8284
9	Nonagon ..	2.9238	2.7474	.342	.3639	40	140	6.1818
10	Decagon ..	3.236	3.0776	.309	.3247	36	144	7.6942
11	Undecagon	3.5494	3.4056	.2817	.2936	32°-43	147 $\frac{3}{11}$	9.3656
12	Dodecagon	3.8638	3.732	.2588	.2679	30	150	11.1961

Under column 5 find 0.7071 as the length of the side of a square when the diameter of the inclosing circle is 1, so that this will be the side of the small end of the reamer, and 1.5 × 0.7071 = 1.0606, the side of the reamer at the large end.

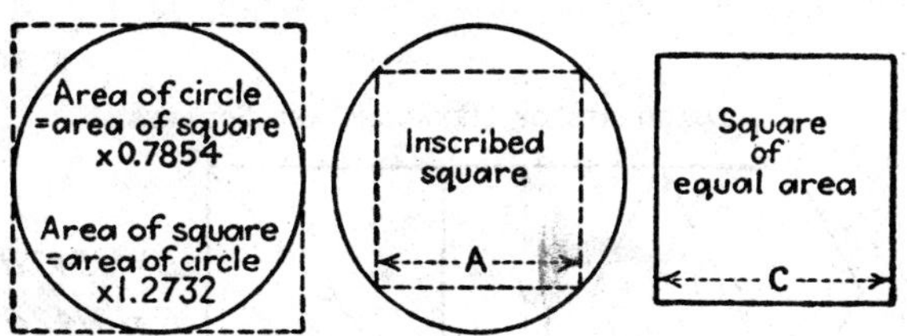

RULES FOR FINDING DIMENSIONS OF CIRCLES AND SQUARES

To Find	Having Given	Rule
Circumference...	Diameter	Multiply diameter by 3.1416 or divide diameter by 0.3183
Circumference...	Area	Divide area by 0.07958 and find square root of quotient
Circumference...	Side of an inscribed square *A*	Multiply side *A* by 4.443
Circumference...	Side of square of equal area *C*	Multiply side *C* by 3.545
Diameter........	Circumference	Multiply circumference by 0.3183 or divide circumference by 3.1416
Diameter........	Area	Divide area by 0.7854 and find square root of quotient
Diameter........	Side of an inscribed square *A*	Divide side *A* by 0.7071 or multiply side *A* by 1.4142
Diameter........	Side of square of equal area *C*	Multiply side *C* by 1.1284 or divide side *C* by 0.8862
Radius..........	Circumference	Multiply circumference by 0.15915 or divide circumference by 6.28318
Area............	Circumference	Multiply the square of the circumference by 0.07958
Area............	Diameter	Multiply the square of the diameter by 0.7854
Area............	Radius	Multiply the square of the radius by 3.1416
Area............	Circumference and diameter	Multiply the circumference by one-quarter the diameter
Side of an inscribed square *A*............	Diameter	Multiply diameter by 0.7071
Side of an inscribed square *A*	Circumference	Multiply circumference by 0.2251 or divide circumference by 4.4428
Side of a square of equal area *C*	Diameter	Multiply diameter by 0.8862 or divide diameter by 1.1284
Side of a square of equal area *C*	Circumference	Multiply circumference by 0.2821 or divide circumference by 3.545

Properties of Figures

Diagonals of Hexagons and Squares

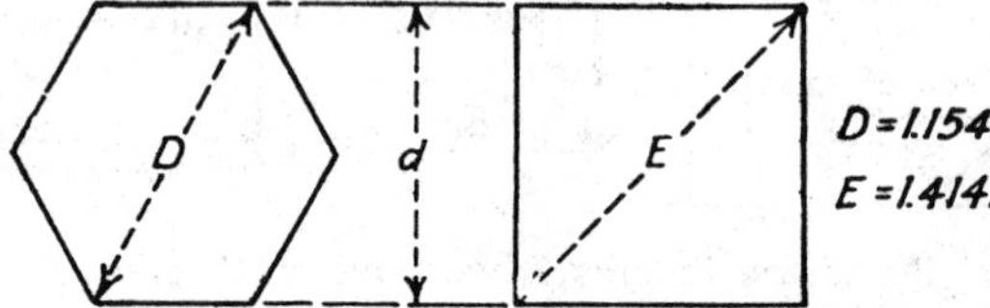

d	D	E	d	D	E	d	D	E
1/4	0.2886	0.3535	1 1/4	1.4434	1.7677	2 5/16	2.6702	3.2703
9/32	0.3247	0.3977	1 9/32	1.4794	1.8119	2 3/8	2.7424	3.3587
5/16	0.3608	0.4419	1 5/16	1.5155	1.8561	2 7/16	2.8145	3.4471
11/32	0.3968	0.4861	1 11/32	1.5516	1.9003	2 1/2	2.8867	3.5355
3/8	0.4329	0.5303	1 3/8	1.5877	1.9445	2 9/16	2.9583	3.6239
13/32	0.4690	0.5745	1 13/32	1.6238	1.9887	2 5/8	3.0311	3.7123
7/16	0.5051	0.6187	1 7/16	1.6598	2.0329	2 11/16	3.1032	3.8007
15/32	0.5412	0.6629	1 15/32	1.6959	2.0771	2 3/4	3.1754	3.8891
1/2	0.5773	0.7071	1 1/2	1.7320	2.1213	2 13/16	3.2476	3.9794
17/32	0.6133	0.7513	1 17/32	1.7681	2.1655	2 7/8	3.3197	4.0658
9/16	0.6494	0.7955	1 9/16	1.8042	2.2097	2 15/16	3.3919	4.1542
19/32	0.6855	0.8397	1 19/32	1.8403	2.2539	3	3.4641	4.2426
5/8	0.7216	0.8839	1 5/8	1.8764	2.2981	3 1/16	3.5362	4.3310
21/32	0.7576	0.9281	1 21/32	1.9124	2.3423	3 1/8	3.6084	4.4194
11/16	0.7937	0.9723	1 11/16	1.9485	2.3865	3 3/16	3.6806	4.5078
23/32	0.8298	1.0164	1 23/32	1.9846	2.4306	3 1/4	3.7527	4.5962
3/4	0.8659	1.0606	1 3/4	2.0207	2.4708	3 5/16	3.8249	4.6846
25/32	0.9020	1.1048	1 25/32	2.0568	2.5190	3 3/8	3.8971	4.7729
13/16	0.9380	1.1490	1 13/16	2.0929	2.5632	3 7/16	3.9692	4.8613
27/32	0.9741	1.1932	1 27/32	2.1289	2.6074	3 1/2	4.0414	4.9497
7/8	1.0102	1.2374	1 7/8	2.1650	2.6516	3 9/16	4.1136	5.0381
29/32	1.0463	1.2816	1 29/32	2.2011	2.6958	3 5/8	4.1857	5.1265
15/16	1.0824	1.3258	1 15/16	2.2372	2.7400	3 11/16	4.2579	5.2149
31/32	1.1184	1.3700	1 31/32	2.2733	2.7842	3 3/4	4.3301	5.3033
1	1.1547	1.4142	2	2.3094	2.8284	3 13/16	4.4023	5.3917
1 1/32	1.1907	1.4584	2 1/32	2.3453	2.8726	3 7/8	4.4744	5.4801
1 1/16	1.2268	1.5026	2 1/16	2.3815	2.9168	3 15/16	4.5466	5.5684
1 3/32	1.2629	1.5468	2 3/32	2.4176	2.9610	4	4.6188	5.6568
1 1/8	1.2990	1.5910	2 1/8	2.4537	3.0052	4 1/8	4.7631	5.8336
1 5/32	1.3351	1.6352	2 5/32	2.4898	3.0494	4 1/4	4.9074	6.0104
1 3/16	1.3712	1.6793	2 3/16	2.5259	3.0936	4 3/8	5.0518	6.1872
1 7/32	1.4073	1.7235	2 1/4	2.5981	3.1820	4 1/2	5.1961	6.3639

Areas and Circumferences of Circles

D = 1/32 to 9 7/8

Dia	Area	Circum	Dia	Area	Circum	Dia	Area	Circum
1/32	0.00077	0.098175	2	3.1416	6.28319	5	19.635	15.7080
3/64	0.00173	0.147262	1/16	3.3410	6.47953	1/16	20.129	15.9043
1/16	0.00307	0.196350	1/8	3.5466	6.67588	1/8	20.629	16.1007
3/32	0.00690	0.294524	3/16	3.7583	6.87223	3/16	21.135	16.2970
1/8	0.01227	0.392699	1/4	3.9761	7.06858	1/4	21.648	16.4934
5/32	0.01917	0.490874	5/16	4.2000	7.26493	5/16	22.166	16.6897
3/16	0.02761	0.589049	3/8	4.4301	7.46128	3/8	22.691	16.8861
7/32	0.03758	0.687223	7/16	4.6664	7.65763	7/16	23.221	17.0824
1/4	0.04909	0.785398	1/2	4.9087	7.85398	1/2	23.758	17.2788
9/32	0.06213	0.883573	9/16	5.1572	8.05033	9/16	24.301	17.4751
5/16	0.07670	0.981748	5/8	5.4119	8.24668	5/8	24.850	17.6715
11/32	0.09281	1.07992	11/16	5.6727	8.44303	11/16	25.406	17.8678
3/8	0.11045	1.17810	3/4	5.9396	8.63938	3/4	25.967	18.0642
13/32	0.12962	1.27627	13/16	6.2126	8.83573	13/16	26.535	18.2605
7/16	0.15033	1.37445	7/8	6.4918	9.03208	7/8	27.109	18.4569
15/32	0.17257	1.47262	15/16	6.7771	9.22843	15/16	27.688	18.6532
1/2	0.19635	1.57080	3	7.0686	9.42478	6	28.274	18.8496
17/32	0.22166	1.66897	1/16	7.3662	9.62113	1/8	29.465	19.2423
9/16	0.24850	1.76715	1/8	7.6699	9.81748	1/4	30.680	19.6350
19/32	0.27688	1.86532	3/16	7.9798	10.0138	3/8	31.919	20.0277
5/8	0.30680	1.96350	1/4	8.2958	10.2102	1/2	33.183	20.4204
21/32	0.33824	2.06167	5/16	8.6179	10.4065	5/8	34.472	20.8131
11/16	0.37122	2.15984	3/8	8.9462	10.6029	3/4	35.785	21.2058
23/32	0.40574	2.25802	7/16	9.2806	10.7992	7/8	37.122	21.5984
3/4	0.44179	2.35619	1/2	9.6211	10.9956	7	38.485	21.9911
25/32	0.47937	2.45437	9/16	9.9678	11.1919	1/8	39.871	22.3838
13/16	0.51849	2.55254	5/8	10.321	11.3883	1/4	41.282	22.7765
27/32	0.55914	2.65072	11/16	10.680	11.5846	3/8	42.718	23.1692
7/8	0.60132	2.74889	3/4	11.045	11.7810	1/2	44.179	23.5619
29/32	0.64504	2.84707	13/16	11.416	11.9773	5/8	45.664	23.9546
15/16	0.69029	2.94524	7/8	11.793	12.1737	3/4	47.173	24.3473
13/32	0.73708	3.04342	15/16	12.177	12.3700	7/8	48.707	24.7400
1	0.78540	3.14159	4	12.566	12.5664	8	50.265	25.1327
1/16	0.88664	3.33794	1/16	12.962	12.7627	1/8	51.849	25.5255
1/8	0.99402	3.53429	1/8	13.364	12.9591	1/4	53.456	25.9181
3/16	1.1075	3.73064	3/16	13.772	13.1554	3/8	55.088	26.3108
1/4	1.2272	3.92699	1/4	14.186	13.3518	1/2	56.745	26.7035
5/16	1.3530	4.12334	5/16	14.607	13.5481	5/8	58.426	27.0962
3/8	1.4849	4.31969	3/8	15.033	13.7445	3/4	60.132	27.4889
7/16	1.6230	4.51604	7/16	15.466	13.9408	7/8	61.862	27.8816
1/2	1.7671	4.71239	1/2	15.904	14.1372	9	63.617	28.2743
9/16	1.9175	4.90874	9/16	16.349	14.3335	1/8	65.397	28.6670
5/8	2.0739	5.10509	5/8	16.800	14.5299	1/4	67.201	29.0597
11/16	2.2365	5.30144	11/16	17.257	14.7262	3/8	69.029	29.4524
3/4	2.4053	5.49779	3/4	17.721	14.9226	1/2	70.882	29.8451
13/16	2.5802	5.69414	13/16	18.190	15.1189	5/8	72.760	30.2378
7/8	2.7612	5.89049	7/8	18.665	15.3153	3/4	74.662	30.6305
15/16	2.9483	6.08684	15/16	19.147	15.5116	7/8	76.589	31.0232

AREAS AND CIRCUMFERENCES OF CIRCLES (*Continued*)

D = 10 to 27 7/8

Dia	Area	Circum	Dia	Area	Circum	Dia	Area	Circum
10	78.540	31.4159	16	201.06	50.2655	22	380.13	69.1150
1/8	80.516	31.8086	1/8	204.22	50.6582	1/8	384.46	69.5077
1/4	82.516	32.2013	1/4	207.39	51.0509	1/4	388.82	69.9004
3/8	84.541	32.5940	3/8	210.60	51.4436	3/8	393.20	70.2931
1/2	86.590	32.9867	1/2	213.82	51.8363	1/2	397.61	70.6858
5/8	88.664	33.3794	5/8	217.08	52.2290	5/8	402.04	71.0785
3/4	90.763	33.7721	3/4	220.35	52.6217	3/4	406.49	71.4712
7/8	92.886	34.1648	7/8	223.65	53.0144	7/8	410.97	71.8639
11	95.033	34.5575	17	226.98	53.4071	23	415.48	72.2566
1/8	97.205	34.9502	1/8	230.33	53.7998	1/8	420.00	72.6493
1/4	99.402	35.3429	1/4	233.71	54.1925	1/4	424.56	73.0420
3/8	101.62	35.7356	3/8	237.10	54.5852	3/8	429.13	73.4347
1/2	103.87	36.1283	1/2	240.53	54.9779	1/2	433.74	73.8274
5/8	106.14	36.5210	5/8	243.98	55.3706	5/8	438.36	74.2201
3/4	108.43	36.9137	3/4	247.45	55.7633	3/4	443.01	74.6128
7/8	110.75	37.3064	7/8	250.95	56.1560	7/8	447.69	75.0055
12	113.10	37.6991	18	254.47	56.5487	24	452.39	75.3982
1/8	115.47	38.0918	1/8	258.02	56.9414	1/8	457.11	75.7909
1/4	117.86	38.4845	1/4	261.59	57.3341	1/4	461.86	76.1836
3/8	120.28	38.8772	3/8	265.18	57.7268	3/8	466.64	76.5765
1/2	122.72	39.2699	1/2	268.80	58.1195	1/2	471.44	76.9690
5/8	125.19	39.6626	5/8	272.45	58.5122	5/8	476.26	77.3617
3/4	127.68	40.0553	3/4	276.12	58.9049	3/4	481.11	77.7544
7/8	130.19	40.4480	7/8	279.81	59.2976	7/8	485.98	78.1471
13	132.73	40.8407	19	283.53	59.6903	25	490.87	78.5398
1/8	135.30	41.2334	1/8	287.27	60.0830	1/8	495.79	78.9325
1/4	137.89	41.6261	1/4	291.04	60.4757	1/4	500.74	79.3252
3/8	140.50	42.0188	3/8	294.83	60.8684	3/8	505.71	79.7179
1/2	143.14	42.4115	1/2	298.65	61.2611	1/2	510.71	80.1105
5/8	145.80	42.8042	5/8	302.49	61.6538	5/8	515.72	80.5033
3/4	148.49	43.1969	3/4	306.35	62.0465	3/4	520.77	80.8960
7/8	151.20	43.5896	7/8	310.24	62.4392	7/8	525.84	81.2887
14	153.94	43.9823	20	314.16	62.8319	26	530.93	81.6814
1/8	156.70	44.3750	1/8	318.10	63.2246	1/8	536.05	82.0741
1/4	159.48	44.7677	1/4	322.06	63.6173	1/4	541.19	82.4668
3/8	162.30	45.1604	3/8	326.05	64.0100	3/8	546.35	82.8595
1/2	165.13	45.5531	1/2	330.06	64.4026	1/2	551.55	83.2522
5/8	167.99	45.9458	5/8	334.10	64.7953	5/8	556.76	83.6449
3/4	170.87	46.3385	3/4	338.16	65.1880	3/4	562.00	84.0376
7/8	173.78	46.7312	7/8	342.25	65.5807	7/8	567.27	84.4303
15	176.71	47.1239	21	346.36	65.9734	27	572.56	84.8230
1/8	179.67	47.5166	1/8	350.50	66.3661	1/8	577.87	85.2157
1/4	182.65	47.9093	1/4	354.66	66.7588	1/4	583.21	85.6084
3/8	185.66	48.3020	3/8	358.84	67.1515	3/8	588.57	86.0011
1/2	188.69	48.6947	1/2	363.05	67.5442	1/2	593.96	86.3938
5/8	191.75	49.0874	5/8	367.28	67.9369	5/8	599.37	86.7865
3/4	194.83	49.4801	3/4	371.54	68.3296	3/4	604.81	87.1792
7/8	197.93	49.8728	7/8	375.83	68.7223	7/8	610.27	87.5719

Areas and Circumferences of Circles (*Continued*)

D = 28 to 45⅞

Dia	Area	Circum	Dia	Area	Circum	Dia	Area	Circum
28	615.75	87.9646	34	907.92	106.814	40	1256.6	125.664
⅛	621.26	88.3573	⅛	914.61	107.207	⅛	1264.5	126.056
¼	626.80	88.7500	¼	921.32	107.600	¼	1272.4	126.449
⅜	632.36	89.1427	⅜	928.06	107.992	⅜	1280.3	126.842
½	637.94	89.5354	½	934.82	108.385	½	1288.2	127.235
⅝	643.55	89.9281	⅝	941.61	108.788	⅝	1296.2	127.627
¾	649.18	90.3208	¾	948.42	109.170	¾	1304.2	128.020
⅞	656.84	90.7135	⅞	955.25	109.563	⅞	1312.2	128.413
29	660.52	91.1062	35	962.11	109.956	41	1320.3	128.805
⅛	666.23	91.4989	⅛	969.00	110.348	⅛	1328.3	129.198
¼	671.96	91.8916	¼	975.91	110.741	¼	1336.4	129.591
⅜	677.71	92.2843	⅜	982.84	111.134	⅜	1344.5	129.993
½	683.49	92.6770	½	989.80	111.527	½	1352.7	130.376
⅝	689.30	93.0697	⅝	996.78	111.919	⅝	1360.8	130.769
¾	695.13	93.4624	¾	1003.8	112.312	¾	1369.0	131.161
⅞	700.98	93.8551	⅞	1010.8	112.705	⅞	1377.2	131.554
30	706.86	94.2478	36	1017.9	113.097	42	1385.4	131.947
⅛	712.76	94.6405	⅛	1025.0	113.490	⅛	1393.7	132.340
¼	718.69	95.0332	¼	1032.1	113.883	¼	1402.0	132.732
⅜	724.64	95.4259	⅜	1039.2	114.275	⅜	1410.3	133.125
½	730.62	95.8186	½	1046.3	114.668	½	1418.6	133.518
⅝	736.62	96.2113	⅝	1053.5	115.001	⅝	1427.0	133.910
¾	742.64	96.6040	¾	1060.7	115.454	¾	1435.4	134.303
⅞	748.69	96.9967	⅞	1068.0	115.846	⅞	1443.8	134.696
31	754.77	97.3894	37	1075.2	116.239	43	1452.2	135.088
⅛	760.87	97.7821	⅛	1082.5	116.632	⅛	1460.7	135.481
¼	766.99	98.1748	¼	1089.8	117.024	¼	1469.1	135.874
⅜	773.14	98.5675	⅜	1097.1	117.417	⅜	1477.6	136.267
½	779.31	98.9602	½	1104.5	117.810	½	1486.2	136.659
⅝	785.51	99.3529	⅝	1111.8	118.202	⅝	1494.7	137.052
¾	791.73	99.7456	¾	1119.2	118.596	¾	1503.3	137.445
⅞	797.98	100.138	⅞	1126.7	118.988	⅞	1511.9	137.837
32	804.25	100.531	38	1134.1	119.381	44	1520.5	138.230
⅛	810.54	100.924	⅛	1141.6	119.773	⅛	1529.2	138.623
¼	816.86	101.316	¼	1149.1	120.166	¼	1537.9	139.015
⅜	823.21	101.709	⅜	1156.6	120.559	⅜	1546.6	139.408
½	829.58	102.102	½	1164.2	120.951	½	1555.3	139.801
⅝	835.97	102.494	⅝	1171.7	121.344	⅝	1564.0	140.194
¾	842.39	102.887	¾	1179.3	121.737	¾	1572.8	140.586
⅞	848.83	103.280	⅞	1186.9	122.129	⅞	1581.6	140.979
33	855.30	103.673	39	1194.6	122.522	45	1590.4	141.372
⅛	861.79	104.065	⅛	1202.3	122.915	⅛	1599.3	141.764
¼	868.31	104.458	¼	1210.0	123.308	¼	1608.2	142.157
⅜	874.85	104.851	⅜	1217.7	123.700	⅜	1617.0	142.550
½	881.41	105.243	½	1225.4	124.093	½	1626.0	142.942
⅝	888.00	105.636	⅝	1233.2	124.486	⅝	1634.9	143.335
¾	894.62	106.029	¾	1241.0	124.878	¾	1643.9	143.728
⅞	901.26	106.421	⅞	1248.8	125.271	⅞	1652.9	144.121

Areas and Circumferences of Circles (*Continued*)

D = 46 to 63 7/8

Dia	Area	Circum	Dia	Area	Circum	Dia	Area	Circum
46	1661.9	144.513	52	2123.7	163.363	58	2642.1	182.212
1/8	1670.9	144.906	1/8	2133.9	163.756	1/8	2653.5	182.605
1/4	1680.0	145.299	1/4	2144.2	164.148	1/4	2664.9	182.998
3/8	1689.1	145.691	3/8	2154.5	164.541	3/8	2676.4	183.390
1/2	1698.2	146.084	1/2	2164.8	164.934	1/2	2687.8	183.783
5/8	1707.4	146.477	5/8	2175.1	165.326	5/8	2699.3	184.176
3/4	1716.5	146.869	3/4	2185.4	165.719	3/4	2710.9	184.569
7/8	1725.7	147.262	7/8	2195.8	166.112	7/8	2722.4	184.961
47	1734.9	147.655	53	2206.2	166.504	59	2734.0	185.354
1/8	1744.2	148.048	1/8	2216.6	166.897	1/8	2745.6	185.747
1/4	1753.5	148.440	1/4	2227.0	167.290	1/4	2757.2	186.139
3/8	1762.7	148.833	3/8	2237.5	167.683	3/8	2768.8	186.532
1/2	1772.1	149.226	1/2	2248.0	168.075	1/2	2780.5	186.925
5/8	1781.4	149.618	5/8	2258.5	168.468	5/8	2792.2	187.317
3/4	1790.8	150.011	3/4	2269.1	168.861	3/4	2803.9	187.710
7/8	1800.1	150.404	7/8	2279.6	169.253	7/8	2815.7	188.103
48	1809.6	150.796	54	2290.2	169.646	60	2827.4	188.496
1/8	1819.0	151.189	1/8	2300.8	170.039	1/8	2839.2	188.888
1/4	1828.5	151.582	1/4	2311.5	170.431	1/4	2851.0	189.281
3/8	1837.9	151.975	3/8	2322.1	170.824	3/8	2862.9	189.674
1/2	1847.5	152.367	1/2	2332.8	171.217	1/2	2874.8	190.066
5/8	1857.0	152.760	5/8	2343.5	171.609	5/8	2886.6	100.459
3/4	1866.5	153.153	3/4	2354.3	172.002	3/4	2898.6	190.852
7/8	1876.1	153.544	7/8	2365.0	172.395	7/8	2910.5	191.244
49	1885.7	153.938	55	2375.8	172.788	61	2922.5	191.637
1/8	1895.4	154.331	1/8	2386.6	173.180	1/8	2934.5	192.030
1/4	1905.0	154.723	1/4	2397.5	173.573	1/4	2946.5	192.423
3/8	1914.7	155.116	3/8	2408.3	173.066	3/8	2958.5	192.815
1/2	1924.2	155.509	1/2	2419.2	174.358	1/2	2970.6	193.208
5/8	1934.2	155.904	5/8	2430.1	174.751	5/8	2982.7	193.601
3/4	1943.9	156.294	3/4	2441.1	175.144	3/4	2994.8	193.993
7/8	1953.7	156.687	7/8	2452.0	175.536	7/8	3006.9	194.386
50	1963.5	157.080	56	2463.0	175.929	62	3019.1	194.779
1/8	1973.3	157.472	1/8	2474.0	176.322	1/8	3031.3	195.171
1/4	1983.2	157.865	1/4	2485.0	176.715	1/4	3043.5	195.564
3/8	1993.1	158.258	3/8	2496.1	177.107	3/8	3055.7	195.957
1/2	2003.0	158.650	1/2	2507.2	177.500	1/2	3068.0	196.350
5/8	2012.9	159.043	5/8	2518.3	177.893	5/8	3080.3	196.742
3/4	2022.8	159.436	3/4	2529.4	178.285	3/4	3092.6	197.135
7/8	2032.8	159.829	7/8	2540.6	178.678	7/8	3104.9	197.528
51	2042.8	160.221	57	2551.8	179.071	63	3117.2	197.920
1/8	2052.8	160.614	1/8	2563.0	179.463	1/8	3129.6	198.313
1/4	2062.9	161.007	1/4	2574.2	179.856	1/4	3142.0	198.706
3/8	2073.0	161.399	3/8	2585.4	180.249	3/8	3154.5	199.098
1/2	2083.1	161.792	1/2	2596.7	180.642	1/2	3166.9	199.491
5/8	2093.2	162.185	5/8	2608.0	181.034	5/8	3179.4	199.884
3/4	2103.3	162.577	3/4	2619.4	181.427	3/4	3191.9	200.277
7/8	2113.5	162.970	7/8	2630.7	181.820	7/8	3204.4	200.660

Areas and Circumferences of Circles (*Continued*)

D = 64 to 81 7/8

Dia	Area	Circum	Dia	Area	Circum	Dia	Area	Circum
64	3217.0	201.062	70	3848.5	219.911	76	4536.5	238.761
1/8	3229.6	201.455	1/8	3862.2	220.304	1/8	4551.4	239.154
1/4	3242.2	201.847	1/4	3876.0	220.697	1/4	4566.4	239.546
3/8	3254.8	202.240	3/8	3889.8	221.090	3/8	4581.3	239.939
1/2	3267.5	202.633	1/2	3903.6	221.482	1/2	4596.3	240.332
5/8	3280.1	203.025	5/8	3917.5	221.875	5/8	4611.4	240.725
3/4	3292.8	203.418	3/4	3931.4	222.268	3/4	4626.4	241.117
7/8	3305.6	203.811	7/8	3945.3	222.660	7/8	4641.5	241.510
65	3318.3	204.204	71	3959.2	223.053	77	4656.6	241.903
1/8	3331.1	204.596	1/8	3973.1	223.446	1/8	4671.8	242.295
1/4	3343.9	204.989	1/4	3987.1	223.838	1/4	4686.9	242.688
3/8	3356.7	205.382	3/8	4001.1	224.231	3/8	4702.1	243.081
1/2	3369.6	205.774	1/2	4015.2	224.624	1/2	4717.3	243.473
5/8	3382.4	206.167	5/8	4029.2	225.017	5/8	4732.5	243.866
3/4	3395.3	206.560	3/4	4043.3	225.409	3/4	4747.8	244.259
7/8	3408.2	206.952	7/8	4057.4	225.802	7/8	4763.1	244.652
66	3421.2	207.345	72	4071.5	226.195	78	4778.4	245.044
1/8	3434.3	207.738	1/8	4085.7	226.587	1/8	4793.7	245.437
1/4	3447.2	208.131	1/4	4099.8	226.930	1/4	4809.0	245.830
3/8	3460.2	208.523	3/8	4114.0	227.373	3/8	4824.4	246.222
1/2	3473.2	208.916	1/2	4128.2	227.765	1/2	4839.8	246.615
5/8	3486.3	209.309	5/8	4142.5	228.158	5/8	4855.2	247.008
3/4	3409.4	209.701	3/4	4156.8	228.551	3/4	4870.7	247.400
7/8	3512.5	210.094	7/8	4171.1	228.944	7/8	4886.2	247.793
67	3525.7	210.487	73	4185.4	229.336	79	4901.7	248.186
1/8	3538.8	210.879	1/8	4199.7	229.729	1/8	4917.2	248.579
1/4	3552.0	211.272	1/4	4214.1	230.122	1/4	4932.7	248.971
3/8	3565.2	211.665	3/8	4228.5	230.514	3/8	4948.3	249.364
1/2	3578.5	212.058	1/2	4242.9	230.907	1/2	4963.9	249.757
5/8	3591.7	212.450	5/8	4257.4	231.300	5/8	4979.5	250.149
3/4	3605.0	212.843	3/4	4271.8	231.692	3/4	4995.2	250.542
7/8	3618.3	213.236	7/8	4286.3	232.085	7/8	5010.9	250.935
68	3631.7	213.628	74	4300.8	232.478	80	5026.5	251.327
1/8	3645.0	214.021	1/8	4315.4	232.871	1/8	5042.3	251.720
1/4	3658.4	214.414	1/4	4329.9	233.263	1/4	5058.0	252.113
3/8	3671.8	214.806	3/8	4344.5	233.656	3/8	5073.8	252.506
1/2	3685.3	215.199	1/2	4359.2	234.049	1/2	5089.6	252.898
5/8	3698.7	215.592	5/8	4373.8	234.441	5/8	5105.4	253.291
3/4	3712.2	215.984	3/4	4388.5	234.834	3/4	5121.2	253.684
7/8	3725.7	216.337	7/8	4403.1	235.227	7/8	5137.1	254.076
69	3739.3	216.770	75	4417.9	235.619	81	5153.0	254.469
1/8	3752.8	217.163	1/8	4432.6	236.012	1/8	5168.9	254.862
1/4	3766.4	217.555	1/4	4447.4	236.405	1/4	5184.9	255.254
3/8	3780.0	217.948	3/8	4462.2	236.798	3/8	5200.8	255.647
1/2	3793.7	218.341	1/2	4477.0	237.190	1/2	5216.8	256.040
5/8	3807.3	218.733	5/8	4491.8	237.583	5/8	5232.8	256.433
3/4	3821.0	219.126	3/4	4506.7	237.976	3/4	5248.9	256.825
7/8	3834.7	219.519	7/8	4521.5	238.368	7/8	5264.9	257.218

AREAS AND CIRCUMFERENCES OF CIRCLES (*Continued*)

D = 82 to 99⅞

Dia	Area	Circum	Dia	Area	Circum	Dia	Area	Circum
82	5281.0	257.611	88	6082.1	276.460	94	6939.8	295.310
⅛	5297.1	258.003	⅛	6099.4	276.853	⅛	6958.2	295.702
¼	5313.3	258.396	¼	6116.7	277.846	¼	6976.7	296.095
⅜	5329.4	258.789	⅜	6134.1	277.638	⅜	6995.3	296.488
½	5345.6	259.181	½	6151.4	278.031	½	7013.8	296.881
⅝	5361.8	259.574	⅝	6168.8	278.424	⅝	7032.4	297.273
¾	5378.1	259.967	¾	6186.2	278.816	¾	7051.0	297.666
⅞	5394.3	260.359	⅞	6203.7	279.209	⅞	7069.6	298.059
83	5410.6	260.752	89	6221.1	279.602	95	7088.2	298.451
⅛	5426.9	261.145	⅛	6238.6	279.994	⅛	7106.9	298.844
¼	5443.3	261.538	¼	6256.1	280.387	¼	7125.6	299.237
⅜	5459.6	261.930	⅜	6273.7	280.780	⅜	7144.3	299.629
½	5476.0	262.323	½	6291.2	281.173	½	7163.0	300.022
⅝	5492.4	262.716	⅝	6308.8	281.565	⅝	7181.8	300.415
¾	5508.8	263.103	¾	6326.4	281.958	¾	7200.6	300.807
⅞	5525.3	263.501	⅞	6344.1	282.351	⅞	7219.4	301.200
84	5541.8	263.894	90	6361.7	282.743	96	7238.2	301.593
⅛	5558.3	264.286	⅛	6379.4	283.136	⅛	7257.1	301.986
¼	5574.8	264.679	¼	6397.1	283.529	¼	7276.0	302.378
⅜	5591.4	265.072	⅜	6414.9	283.921	⅜	7294.9	302.771
½	5607.9	265.465	½	6432.6	284.314	½	7313.8	303.164
⅝	5624.5	265.857	⅝	6450.4	284.707	⅝	7332.8	303.556
¾	5641.2	266.250	¾	6468.2	285.100	¾	7351.8	303.949
⅞	5657.8	266.643	⅞	6486.0	285.492	⅞	7370.8	304.342
85	5674.5	267.035	91	6503.9	285.885	97	7389.8	304.734
⅛	5691.2	267.428	⅛	6521.8	286.278	⅛	7408.9	305.127
¼	5707.9	267.821	¼	6539.7	286.670	¼	7428.0	305.520
⅜	5724.7	268.213	⅜	6557.6	287.063	⅜	7447.1	305.913
½	5741.5	268.606	½	6575.5	287.456	½	7466.2	306.305
⅝	5758.3	268.999	⅝	6593.5	287.848	⅝	7485.3	306.698
¾	5775.1	269.392	¾	6611.5	288.241	¾	7504.5	307.091
⅞	5791.9	269.784	⅞	6629.6	288.634	⅞	7523.7	307.483
86	5808.8	270.177	92	6647.6	289.027	98	7543.0	307.876
⅛	5825.7	270.570	⅛	6665.7	289.419	⅛	7562.2	308.269
¼	5842.6	270.962	¼	6683.8	289.812	¼	7581.5	308.661
⅜	5859.6	271.355	⅜	6701.9	290.205	⅜	7600.8	309.064
½	5876.5	271.748	½	6720.1	290.597	½	7620.1	309.447
⅝	5893.5	272.140	⅝	6738.2	290.990	⅝	7639.5	309.840
¾	5910.6	272.533	¾	6756.4	291.383	¾	7658.9	310.232
⅞	5927.6	272.926	⅞	6774.7	291.775	⅞	7678.3	310.625
87	5944.7	273.319	93	6792.9	292.168	99	7697.7	311.018
⅛	5961.8	273.711	⅛	6811.2	292.561	⅛	7717.1	311.410
¼	5978.9	274.104	¼	6829.5	292.954	¼	7736.6	311.803
⅜	5996.0	274.497	⅜	6847.8	293.346	⅜	7756.1	312.196
½	6013.2	274.889	½	6866.1	293.739	½	7775.6	312.588
⅝	6030.4	275.282	⅝	6884.5	294.132	⅝	7795.2	312.981
¾	6047.6	275.675	¾	6902.9	294.524	¾	7814.8	313.374
⅞	6064.9	276.067	⅞	6921.3	294.917	⅞	7834.4	313.767

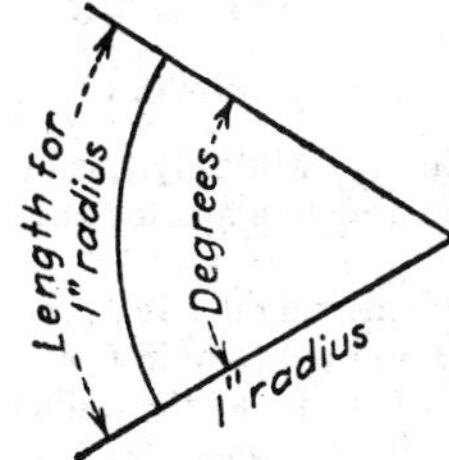

Lengths of Arcs

The table gives the lengths of circular arcs to the radius of one, for angles from 1° to 180°. The lengths for minutes of arcs are given at the right.

To find the length of a circular arc with radius of 1 in. and angle of 45° 20′. Opposite 45° find 0.7854, and opposite 20′ find 0.0058. Adding these gives 0.7912 in. as the length of arc. If the radius is 2 in., multiply the lengths in the table by 2.

Lengths of Circular Arcs to Radius of 1

Degree	Length	Degree	Length	Degree	Length	Degree	Length	Min.	Length	Min.	Length
0	0.0000	45	0.7854	90	1.5708	135	2.3562	0	0.0000	45	0.0131
1	0.0175	46	0.8029	91	1.5882	136	2.3736	1	0.0003	46	0.0134
2	0.0349	47	0.8203	92	1.6057	137	2.3911	2	0.0006	47	0.0137
3	0.0524	48	0.8378	93	1.6232	138	2.4086	3	0.0009	48	0.0140
4	0.0698	49	0.8552	94	1.6406	139	2.4260	4	0.0012	49	0.0143
5	0.0873	50	0.8728	95	1.6581	140	2.4435	5	0.0015	50	0.0145
6	0.1047	51	0.8901	96	1.6755	141	2.4609	6	0.0017	51	0.0148
7	0.1222	52	0.9076	97	1.6930	142	2.4784	7	0.0020	52	0.0151
8	0.1396	53	0.9250	98	1.7104	143	2.4958	8	0.0023	53	0.0154
9	0.1571	54	0.9425	99	1.7279	144	2.5133	9	0.0026	54	0.0157
10	0.1745	55	0.9599	100	1.7453	145	2.5307	10	0.0029	55	0.0160
11	0.1920	56	0.9774	101	1.7628	146	2.5482	11	0.0032	56	0.0163
12	0.2094	57	0.9948	102	1.7802	147	2.5656	12	0.0035	57	0.0166
13	0.2269	58	1.0123	103	1.7977	148	2.5831	13	0.0038	58	0.0169
14	0.2443	59	1.0297	104	1.8151	149	2.6005	14	0.0041	59	0.0172
15	0.2618	60	1.0472	105	1.8326	150	2.6180	15	0.0044	60	0.0175
16	0.2793	61	1.0647	106	1.8500	151	2.6354	16	0.0047		
17	0.2967	62	1.0821	107	1.8675	152	2.6529	17	0.0050		
18	0.3142	63	1.0996	108	1.8850	153	2.6704	18	0.0052		
19	0.3316	64	1.1170	109	1.9024	154	2.6878	19	0.0055		
20	0.3491	65	1.1345	110	1.9199	155	2.7052	20	0.0058		
21	0.3665	66	1.1519	111	1.9373	156	2.7227	21	0.0061		
22	0.3840	67	1.1694	112	1.9548	157	2.7402	22	0.0064		
23	0.4014	68	1.1868	113	1.9722	158	2.7576	23	0.0067		
24	0.4189	69	1.2043	114	1.9897	159	2.7751	24	0.0070		
25	0.4363	70	1.2217	115	2.0071	160	2.7925	25	0.0073		
26	0.4538	71	1.2392	116	2.0246	161	2.8100	26	0.0076		
27	0.4712	72	1.2566	117	2.0420	162	2.8274	27	0.0079		
28	0.4887	73	1.2741	118	2.0595	163	2.8449	28	0.0081		
29	0.5061	74	1.2915	119	2.0769	164	2.8623	29	0.0084		
30	0.5236	75	1.3090	120	2.0944	165	2.8798	30	0.0087		
31	0.5411	76	1.3265	121	2.1118	166	2.8972	31	0.0090		
32	0.5585	77	1.3439	122	2.1293	167	2.9147	32	0.0093		
33	0.5760	78	1.3614	123	2.1468	168	2.9322	33	0.0096		
34	0.5934	79	1.3788	124	2.1642	169	2.9496	34	0.0099		
35	0.6109	80	1.3963	125	2.1817	170	2.9671	35	0.0102		
36	0.6283	81	1.4137	126	2.1991	171	2.9845	36	0.0105		
37	0.6458	82	1.4312	127	2.2166	172	3.0020	37	0.0108		
38	0.6632	83	1.4486	128	2.2340	173	3.0194	38	0.0111		
39	0.6807	84	1.4661	129	2.2515	174	3.0369	39	0.0113		
40	0.6981	85	1.4835	130	2.2690	175	3.0543	40	0.0116		
41	0.7156	86	1.5010	131	2.2864	176	3.0718	41	0.0119		
42	0.7330	87	1.5184	132	2.3038	177	3.0892	42	0.0122		
43	0.7505	88	1.5359	133	2.3212	178	3.1067	43	0.0125		
44	0.7679	89	1.5533	134	2.3387	179	3.1241	44	0.0128		

Table for Finding Radiuses of Segments

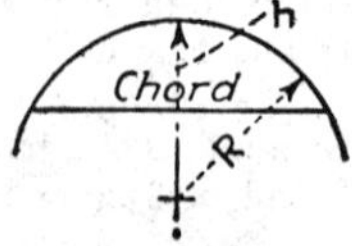

Measure width and height of segment to find ratio $\frac{h}{C}$ or $\frac{\text{height}}{\text{chord}}$. Then find quotient in Col. 1 and opposite in Col. 2 find the chord C for segment of radius 1. Multiply this by the width of actual segment for the correct radius for that segment. Thus:

If the width of chord C is 3 in. and the height 1 in., then $\frac{1}{3}$ equals 0.333. Find this in Col. 1 and opposite in Col. 2 read 0.5419. Multiply by 3, giving 1.6257 in. as the radius required for the segment.

$\frac{h}{C}$	Constant	$\frac{h}{C}$	Constant	$\frac{h}{C}$	Constant	$\frac{h}{C}$	Constant	$\frac{h}{C}$	Constant
0.010	12.500	0.045	2.800	0 080	1.602	0.114	1 1534	0 149	0 9134
0 011	11.362	0 046	2.740	0 081	1.583	0.115	1.1445	0.150	0.9083
0.012	10.418	0.047	2.681	0 082	1.565	0.116	1.1362	0 150	0 9083
0.013	9.611	0.048	2 628	0 083	1.548	0 117	1 1269	0 151	0 9033
0.014	8.846	0.049	2.575	0 084	1.530	0.118	1 1182	0 152	0 8984
0.015	8.341	0.050	2.525	0 085	1.513	0.119	1.1099	0 153	0.8935
0.016	7.821	0.051	2.476	0.086	1.496	0 120	1 1017	0 154	0.8887
0.017	7.361	0.052	2.430	0 087	1.481	0.121	1.0934	0 155	0.8839
0.018	6.838	0.053	3.385	0 088	1 465	0.122	1.0856	0 156	0.8794
0.019	6.581	0.054	2.342	0 089	1.449	0 123	1 1078	0.157	0 8747
0.020	6.260	0.055	2.300	0 090	1.434	0 124	1 0700	0.158	0 8701
0.021	5.962	0.056	2.260	0 091	1.419	0 125	1.1063	0 159	0 8657
0.022	5.693	0.057	2.221	0.092	1.404	0 126	1.0551	0.160	0 8613
0.023	5.446	0.058	2.184	0.093	1.390	0.127	1.0477	0 161	0 8569
0.024	5.231	0.059	2.148	0 094	1.377	0 128	1.0406	0 162	0 8526
0.025	5.012	0.060	2.114	0 095	1.363	0 129	1.0335	0 163	0 8483
0.026	4.812	0 061	2.080	0 096	1.350	0 130	1.0265	0 164	0 8441
0.027	4.643	0.062	2.047	0 097	1.337	0 131	1.0186	0 165	0 8401
0.028	4.478	0.063	2 015	0.098	1.324	0 132	1 0130	0 166	0.8360
0.029	4.325	0.064	1.965	0 099	1.312	0 133	1 0062	0 167	0.8320
0.030	4.181	0.065	1.955	0 100	1 300	0 134	0.9998	0 168	0.8280
0.031	4.048	0.066	1.945			0 135	0 9935	0.169	0.8241
0.032	3.922	0.067	1.899	0.101	1.2871	0 136	0 9871	0 170	0.8203
0.033	3.804	0.068	1.870	0.102	1 2763	0 137	0 9809	0.171	0.8165
0.034	3.693	0.069	1.846	0.103	1.2651	0 138	0 9748	0.172	0.8127
0.035	3.589	0.070	1.821	0.104	1.2539	0.139	0 9688	0 173	0.8091
0.036	3.490	0.071	1.796	0.105	1.2429	0.140	0 9629	0.174	0.8054
0.037	3.400	0.072	1.772	0.106	1.2323	0 141	0.9570	0.175	0.8018
0.038	3.308	0.073	1.749	0.107	1.2217	0 142	0.9513	0.176	0.7983
0.039	3.224	0.074	1.726	0 108	1 2114	0 143	0.9457	0.177	0.7947
0.040	3.130	0.075	1.704	0 109	1.2013	0.144	0.9401	0 178	0.7912
0.041	3.069	0.076	1.683	0.110	1.1903	0 145	0.9345	0.179	0.7877
0.042	2 997	0.077	1.662	0 111	1.1815	0 146	0 9292	0.180	0.7844
0.043	2.928	0.078	1.641	0.112	1.1720	0 147	0.9238	0 181	0.7811
0.044	2.863	0 079	1.602	0.113	1.1628	0 148	0.9185	0 182	0.7778

Radiuses of Segments (*Continued*)

$\frac{h}{C}$	Constant	$\frac{h}{C}$	Constant	$\frac{h}{C}$	Constant	$\frac{h}{C}$	Constant	$\frac{h}{C}$	Constant
0.183	0.7746	0.232	0.6548	0.282	0.5843	0.332	0.5425	0.382	0.5182
0.184	0.7713	0.233	0.6529	0.283	0.5832	0.333	0.5419	0.383	0.5179
0.185	0.7682	0.234	0.6514	0.284	0.5821	0.334	0.5412	0.384	0.5175
0.186	0.7651	0.235	0.6494	0.285	0.5811	0.335	0.5406	0.385	0.5172
0.187	0.7629	0.236	0.6477	0.286	0.5801	0.336	0.5400	0.386	0.5168
0.188	0.7590	0.237	0.6459	0.287	0.5790	0.337	0.5394	0.387	0.5165
0.189	0.7559	0.238	0.6441	0.288	0.5780	0.338	0.5387	0.388	0.5162
0.190	0.7529	0.239	0.6425	0.289	0.5770	0.339	0.5382	0.389	0.5158
0.191	0.7500	0.240	0.6408	0.290	0.5760	0.340	0.5376	0.390	0.5155
0.192	0.7470	0.241	0.6392	0.291	0.5751	0.341	0.5371	0.391	0.5152
0.193	0.7441	0.242	0.6375	0.292	0.5741	0.342	0.5365	0.392	0.5149
0.194	0.7413	0.243	0.6359	0.293	0.5731	0.343	0.5459	0.393	0.5146
0.195	0.7386	0.244	0.6343	0.294	0.5721	0.344	0.5354	0.394	0.5143
0.196	0.7357	0.245	0.6327	0.295	0.5711	0.345	0.5348	0.395	0.5139
0.197	0.7330	0.246	0.6311	0.296	0.5703	0.346	0.5334	0.396	0.5137
0.198	0.7307	0.247	0.6296	0.297	0.5694	0.347	0.5337	0.397	0.5134
0.199	0.7276	0.248	0.6280	0.298	0.5685	0.348	0.5332	0.398	0.5131
0.200	0.7250	0.249	0.6265	0.299	0.5676	0.349	0.5327	0.399	0.5128
		0.250	0.6250	0.300	0.5666	0.350	0.5321	0.400	0.5125
0.201	0.7224	0.251	0.6235	0.301	0.5658	0.351	0.5316	0.401	0.5122
0.202	0.7199	0.252	0.6220	0.302	0.5649	0.352	0.5311	0.402	0.5119
0.203	0.7173	0.253	0.6205	0.303	0.5641	0.353	0.5306	0.403	0.5117
0.204	0.7147	0.254	0.6290	0.304	0.5632	0.354	0.5302	0.404	0.5114
0.205	0.7123	0.255	0.6177	0.305	0.5623	0.355	0.5296	0.405	0.5111
0.206	0.7098	0.256	0.6163	0.306	0.5615	0.356	0.5291	0.406	0.5109
0.207	0.7074	0.257	0.6150	0.307	0.5607	0.357	0.5287	0.407	0.5106
0.208	0.7050	0.258	0.6135	0.308	0.5599	0.358	0.5282	0.408	0.5104
0.209	0.7026	0.259	0.6122	0.309	0.5590	0.359	0.5277	0.409	0.5101
0.210	0.7003	0.260	0.6108	0.310	0.5582	0.360	0.5272	0.410	0.5099
0.211	0.6978	0.261	0.6094	0.311	0.5575	0.361	0.5268	0.411	0.5096
0.212	0.6957	0.262	0.6081	0.312	0.5566	0.362	0.5263	0.412	0.5094
0.213	0.6933	0.263	0.6066	0.313	0.5558	0.363	0.5258	0.413	0.5092
0.214	0.6911	0.264	0.6054	0.314	0.5551	0.364	0.5254	0.414	0.5089
0.215	0.6889	0.265	0.6042	0.315	0.5543	0.365	0.5249	0.415	0.5087
0.216	0.6867	0.266	0.6029	0.316	0.5536	0.366	0.5245	0.416	0.5085
0.217	0.6845	0.267	0.6016	0.317	0.5528	0.367	0.5241	0.417	0.5082
0.218	0.6820	0.268	0.6004	0.318	0.5521	0.368	0.5237	0.418	0.5080
0.219	0.6803	0.269	0.5992	0.319	0.5513	0.369	0.5233	0.419	0.5078
0.220	0.6782	0.270	0.5980	0.320	0.5506	0.370	0.5228	0.420	0.5076
0.221	0.6761	0.271	0.5968	0.321	0.5509	0.371	0.5224	0.421	0.5074
0.222	0.6740	0.272	0.5955	0.322	0.5492	0.372	0.5220	0.422	0.5072
0.223	0.6720	0.273	0.5944	0.323	0.5485	0.373	0.5216	0.423	0.5070
0.224	0.6700	0.274	0.5933	0.324	0.5478	0.374	0.5212	0.424	0.5068
0.225	0.6680	0.275	0.5920	0.325	0.5471	0.375	0.5208	0.425	0.5066
0.226	0.6660	0.276	0.5909	0.326	0.5463	0.376	0.5205	0.426	0.5064
0.227	0.6641	0.277	0.5898	0.327	0.5457	0.377	0.5201	0.427	0.5062
0.228	0.6623	0.278	0.5886	0.328	0.5451	0.378	0.5197	0.428	0.5061
0.229	0.6603	0.279	0.5875	0.329	0.5444	0.379	0.5192	0.429	0.5059
0.230	0.6585	0.280	0.5864	0.330	0.5438	0.380	0.5189	0.430	0.5057
0.231	0.6566	0.281	0.5853	0.331	0.5431	0.381	0.5186	0.431	0.5055

RADIUSES OF SEGMENTS (*Continued*)

$\frac{h}{C}$	Constant	$\frac{h}{C}$	Constant	$\frac{h}{C}$	Constant	$\frac{h}{C}$	Constant	$\frac{h}{C}$	Constant
0.432	0.5054	0.447	0.5031	0.461	0.50164	0.476	0.50061	0.491	0 50009
0.433	0.5052	0.448	0.5030	0.462	0.50156	0.477	0.50055	0.492	0 50007
0.434	0.5050	0.449	0.5029	0.463	0.50148	0.478	0.50050	0.493	0.50005
0.435	0.5048	0.450	0.5028	0.464	0.50140	0.479	0.50048	0.494	0.50004
0.436	0.5047	0.450	0.50278	0.465	0.40132	0.480	0.50042	0.495	0.50003
0.437	0.5045	0.451	0.50266	0.466	0.50124	0.481	0.50037	0.496	0.50002
0.438	0.5044	0.452	0.50255	0.467	0.50116	0.482	0.50033	0.497	0.50001
0.439	0.5043	0.453	0.50244	0.468	0.50109	0.483	0.50029	0.498	0.50000
0.440	0.5042	0.454	0.50233	0.469	0.50102	0.484	0.50026	0.499	0.50000
0.441	0.5039	0.455	0.50223	0.470	0.50096	0.485	0.50023	0.500	0.50000
0.442	0.5038	0.456	0.50213	0.471	0.50089	0.486	0.50022		
0.443	0.5037	0.457	0.50203	0.472	0.50083	0.487	0.50021		
0.444	0.5035	0.458	0.50193	0.473	0.50077	0.488	0.50018		
0.445	0.5034	0.459	0.50181	0.474	0.50071	0.489	0.50015		
0.446	0.5033	0.460	0.50174	0.475	0.50066	0 490	0.50012		

Areas of Segments

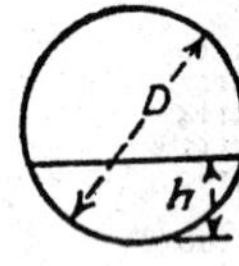

Areas of segments are given in percentage of area of a circle, according to percentage of h to D. Example: $h = 1$, $D = 4$. Then $h/D \times 100 = 25$ per cent. Opposite 25 in column "Per Cent Diameter" read 19.54 under column "Per Cent Area." Area of segment is $0.1954 \times 12.566 = 2.455$, where 12.566 is area of circle with diameter D (see page 44-37).

Showing Relation between Percentages of Diameter and Percentages of Area; Diameter Percentages Taken in Even Advances of 1 Per Cent

Per Cent Diameter	Per Cent Area	Per Cent Diameter	Per Cent Area	Per Cent Diameter	Per Cent Area	Per Cent Diameter	Per Cent Area	Per Cent Diameter	Per Cent Area
1	0.170	21	15 28	41	38 60	61	63 90	81	86 77
2	0.477	22	16 30	42	39 88	62	65 14	82	87.76
3	0 873	23	17 37	43	41 11	63	66 37	83	88.74
4	1 337	24	18 45	44	42 37	64	67 59	84	89.68
5	1 869	25	19 54	45	43 64	65	68 82	85	90.60
6	2 443	26	20 66	46	44 90	66	70 03	86	91.50
7	3.077	27	21 78	47	46 17	67	71 24	87	92.36
8	3.746	28	22 91	48	47 44	68	72 43	88	93 20
9	4 457	29	24 05	49	48 72	69	73 60	89	94.02
10	5.203	30	25 23	50	50	70	74 77	90	94.80
11	5 983	31	26 40	51	51 28	71	75 95	91	95 56
12	6.795	32	27 57	52	52 56	72	77 09	92	96.25
13	7 638	33	28 76	53	53 83	73	78 22	93	96.92
14	8 502	34	29 97	54	55 10	74	79 34	94	97.61
15	9.403	35	31 18	55	56 36	75	80 46	95	98.13
16	10 32	36	32 41	56	57 63	76	81 55	96	98.66
17	11.26	37	33 63	57	58 89	77	82 63	97	99.13
18	12 24	38	34 86	58	60 12	78	83 70	98	99 52
19	13 28	39	36 10	59	61 40	79	84 74	99	99 83
20	14.23	40	37 35	60	62 65	80	85 77	100	100 00

Volume of Spherical Segments

h = height of segment R = radius of sphere

Determine $\frac{h}{R}$ from data at hand

To find volume of segment:

Method A: Multiply volume of sphere by number opposite under Relative Volume

Method B: Multiply R^3 by number opposite $\frac{h}{R}$ under Volume, R = Unity

$\frac{h}{R}$	Relative Volume	Volume, R = Unity	$\frac{h}{R}$	Relative Volume	Volume, R = Unity
0.01	0.00007475	0.00031311	0.51	0.16191225	0.678216
0.02	0.000298	0.00124826	0.52	0.167648	0.702243
0.03	0.00066825	0.00279916	0.53	0.17345575	0.726571
0.04	0.001184	0.00495953	0.54	0.179334	0.751193
0.05	0.00184375	0.00772308	0.55	0.18528125	0.776104
0.06	0.002646	0.01108355	0.56	0.191296	0.801300
0.07	0.00358925	0.0150346	0.57	0.19737675	0.826771
0.08	0.004672	0.0195700	0.58	0.203522	0.852251
0.09	0.00589275	0.0246835	0.59	0.20973025	0.878515
0.10	0.00725	0.0303688	0.60	0.216	0.904779
0.11	0.00874225	0.0366195	0.61	0.22232975	0.931293
0.12	0.010368	0.0434293	0.62	0.228718	0.958053
0.13	0.01212575	0.0507923	0.63	0.23516225	0.985050
0.14	0.014014	0.0587017	0.64	0.241664	1.011581
0.15	0.01603125	0.0671502	0.65	0.24821875	1.039738
0.16	0.018176	0.0761354	0.66	0.254826	1.067414
0.17	0.02044675	0.0856772	0.67	0.26148425	1.095305
0.18	0.022842	0.0956804	0.68	0.268192	1.123400
0.19	0.02536025	0.1062288	0.69	0.27494775	1.151700
0.20	0.028	0.1172863	0.70	0.28175	1.180193
0.21	0.03075975	0.128846	0.71	0.28859725	1.208875
0.22	0.033638	0.140903	0.72	0.295488	1.237738
0.23	0.03663325	0.153449	0.73	0.30242075	1.266780
0.24	0.039744	0.166480	0.74	0.309394	1.295985
0.25	0.04296875	0.1177987	0.75	0.31640625	1.325363
0.26	0.046306	0.193967	0.76	0.323456	1.354890
0.27	0.04975425	0.208410	0.77	0.33054175	1.384570
0.28	0.053312	0.223313	0.78	0.337662	1.414397
0.29	0.05697775	0.238667	0.79	0.34481525	1.444359
0.30	0.06075	0.254469	0.80	0.352	1.474459
0.31	0.06462725	0.270710	0.81	0.35921475	1.504678
0.32	0.068608	0.287385	0.82	0.366458	1.535011
0.33	0.07269075	0.304486	0.83	0.37372825	1.565471
0.34	0.076874	0.322009	0.84	0.381024	1.596030
0.35	0.08115625	0.399947	0.85	0.38834375	1.626697
0.36	0.085536	0.358293	0.86	0.395686	1.657446
0.37	0.09001175	0.377040	0.87	0.40304925	1.688289
0.38	0.094582	0.396184	0.88	0.410432	1.719214
0.39	0.09924525	0.415718	0.89	0.41783275	1.750210
0.40	0.104	0.435634	0.90	0.42525	1.781283
0.41	0.10884475	0.445928	0.91	0.4326825	1.812417
0.42	0.113778	0.476593	0.92	0.440128	1.843606
0.43	0.11879825	0.407621	0.93	0.44758475	1.874845
0.44	0.123904	0.519008	0.94	0.455054	1.906127
0.45	0.12909375	0.540747	0.95	0.46253125	1.937445
0.46	0.134366	0.562832	0.96	0.470016	1.968798
0.47	0.13971925	0.585256	0.97	0.477750675	2.000175
0.48	0.145152	0.608011	0.98	0.485002	2.031570
0.49	0.15066275	0.631096	0.99	0.49250025	2.062982
0.50	0.15625	0.654498	1.00	0.5	2.094397

Finding Circle Diameters without the Center

It sometimes happens in measuring up a machine that we need to know the radius of curves when the center is not accessible. Three such cases are shown in Figs. 1, 2, and 3, the first two being a machine and the last a broken pulley. In Fig. 1, the rule is short enough to go in the curve, but in Fig. 2, it has one end touching and the other across the sides. It makes no difference which is used so long as the distances are measured correctly, the short distance, or *versed sine*, being taken at the exact center of the chord and at right angles to it.

If the chord is 6 in., as in Fig. 1, and the height $1\frac{1}{4}$ in., we have:

$$D = \frac{(\frac{1}{2}\text{ chord})^2 + \text{height}^2}{\text{height}} = \frac{3^2 + 1\frac{1}{4}^2}{1\frac{1}{4}} = \frac{9 + 1\frac{9}{16}}{1\frac{1}{4}} = \frac{10\frac{9}{16}}{1\frac{1}{4}} = 8.45 \text{ in.}$$

Or, as in Fig. 2, call the chord $10\frac{3}{4}$ in. and the height $1\frac{3}{8}$ in., then the figures are:

$$D = \frac{5\frac{3}{8}^2 + 1\frac{3}{8}^2}{1\frac{3}{8}} = 22.4 \text{ in.}$$

In Fig. 3 we have a piece of broken pulley, and we find the chord B to be 24 in. and the height A to be 2 in. Thus

$$D = \frac{12^2 + 2^2}{2} = \frac{144 + 4}{2} = \frac{148}{2} = 74$$

Flat-square Method. The method described is approximate only, because the measurement of the height will usually fall in decimals that cannot be read by scale, and a very small error in the height will be greatly multiplied in the resulting diameter. For example: with a chord of $10\frac{3}{4}$ in. and a height of $1\frac{3}{8}$ in., as in Fig. 2, an error of 0.01 in. in the height will make a difference of about $\frac{1}{4}$ in. in the diameter.

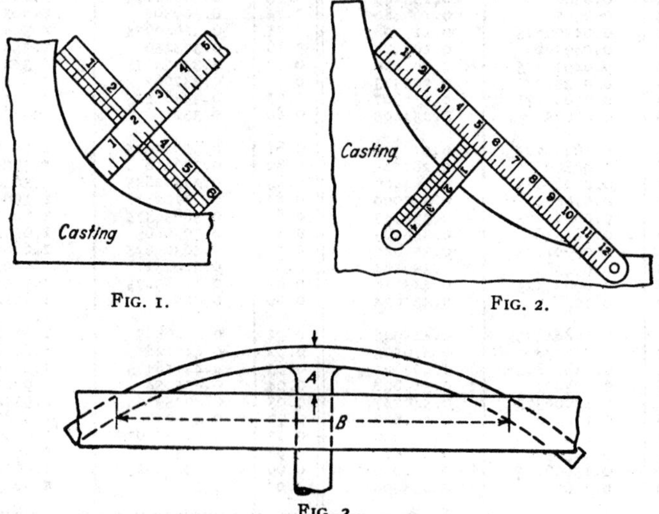

FIG. 1. FIG. 2.

FIG. 3.

Then, too, it is difficult to measure the chord accurately with a scale. For greater accuracy, the height should be measured by a depth micrometer.

For outside measurements, a better method is to use a flat square, as in Fig. 4. In this case, the height H from the surface of the work to the inside apex of the square multiplied by 4.8286 equals the diameter of the work. When great accuracy is required, as in sizing gear segments, an instrument having legs at an included angle of 60°, as in Fig. 5, should be used. With such an instrument, the height H is always equal to one-half the diameter of the work, so that, to find the diameter, the height must be multiplied by two. Either of these instruments can have a micrometer height gage attached for greater accuracy in reading the height, or the height can be measured by an inside micrometer having a pointed anvil.

With an instrument having legs at any included angle, the cosecant of half the included angle is always equal to the distance between the center of the work and the inner apex of the angle. Thus, a constant by which to multiply the height H to find the diameter D can be found by subtracting 1 from the cosecant of half the included angle and dividing 2 by the remainder, thus:

$$\frac{2}{\text{cosecant angle } A - 1}$$

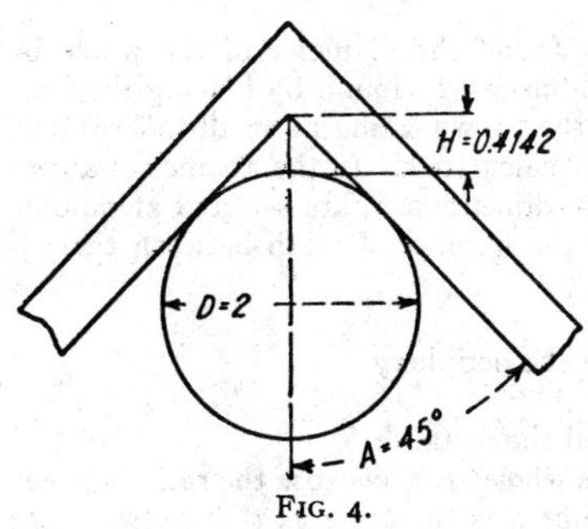

FIG. 4.

EXAMPLE: In an instrument having legs at an included angle of 90°, as in Fig. 4, the angle A is 45°, the cosecant of which is 1.4142.

Then

$$D = \frac{2}{1.4142 - 1} \times 0.4142 = 2$$

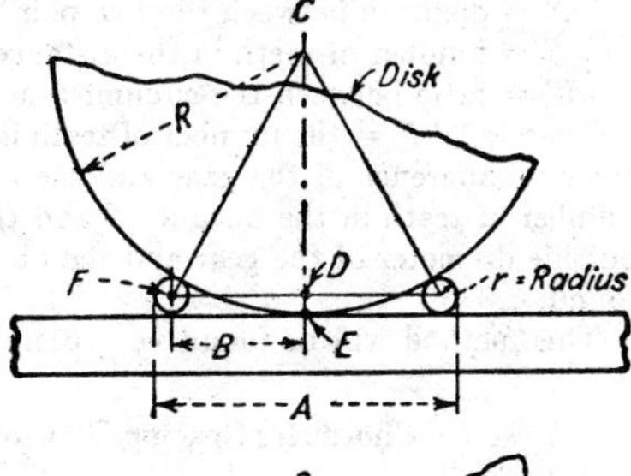

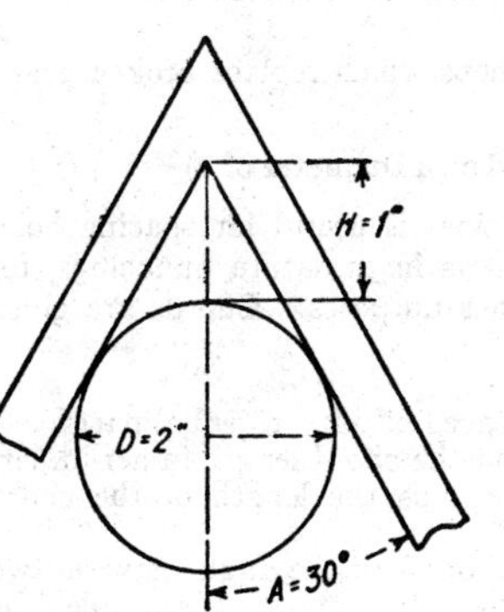

FIG. 5.

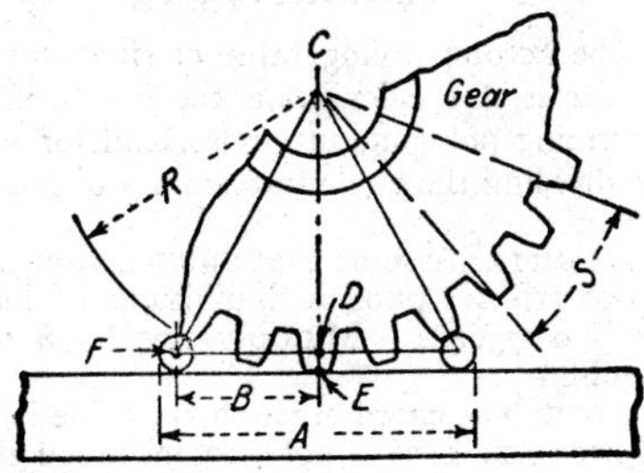

FIG. 6.

In an instrument having legs at an included angle of 60°, as in Fig. 5, the angle A is 30°, the cosecant of which is 2. If the height H is 1 in., then $\frac{2}{2-1} = 2$, and the height (1 in.) multiplied by 2 = 2 in., which is the diameter D of the work.

When using such instruments for very accurate work, it is always well to calibrate them by using a gage or a disk of known diameter and as near the size of the work as possible.

Another Method. Another method is to stand the fragment of either pulley or gear on a flat plate, as in Fig. 6. Place two pieces of drill rod on the plate in contact with the face of the fragment, as shown, and fasten them to the plate with a few drops of solder. Remove the fragment and measure the distance across the pieces of drill rod with a micrometer.

Let A = micrometer measurement across the pieces of drill rod
r = radius of the drill rod
R = radius of the whole disk or gear

Then

$$B = \frac{A - 2r}{2} \quad \text{and} \quad R = \frac{B^2}{4r}$$

If the fragment is that of a gear, having found the diameter of the whole by multiplying the radius by 2, the diametral pitch can be found by placing the fragment on such a gear chart as can be found in the Brown & Sharpe small-tool catalog. Having these two items, all other data can be calculated. In the absence of a gear chart, two points, such as are included in the dimension S, are selected at random and are measured with a flexible scale, and the number of teeth between them is counted.

Let H = circumference of the gear
S = distance between the two points on the periphery
F = number of teeth in the distance S
W = ratio between the circumference and the distance S

Then $W \times F$ = the number of teeth in the whole gear, because the ratio between the circumference of the gear and the distance S is the same as that between the number of teeth in the distance S and the number of teeth in the gear. With the outside diameter of the gear and the number of teeth known, all other data can be found.

This method will be found very useful in job shops which replace broken gears.

Chords for Spacing Off Circles (Based on a Diameter of 1)

The accompanying table of sides, angles, and sines is useful for spacing holes in circles, for calculating the chord of spacing slots in armature punchings, for inscribing polygons in circles, and for various other purposes. Chords are given for dividing the circle into from 3 to 300 parts.

EXAMPLE: Assume that an 8-in. circle is to be spaced off into 18 equal parts.

SOLUTION: Opposite 18 divisions in the table, find the chord length in a 1-in. circle = 0.1736481. Multiply this by 8 to get 1.3892 as the length of the chord required.

NOTE: The chord given in the table is the sine of one-half the angle between two adjacent sides of a polygon inscribed in the circle. In the above example, the included angle = 360 ÷ 18 = 20°, and the sine of 10° = 0.1736+.

Chordal Data for Inscribed Polygons
3 to 100 Sides

No. Sides	Angle Deg. Min. Sec.	Sine	No. Sides	Angle Deg. Min. Sec.	Sine
3	60	.8660254	52	3–27–41.53	.0603784
4	45	.7071067	53	3–23–46.41	.0592405
5	36	.5877852	54	3–20	.0581448
6	30	.5000000	55	3–16–21.81	.0570887
7	25–42–51.42	.4338828	56	3–12–51.42	.0560704
8	22–30	.3826834	57	3– 9–28.42	.0550877
9	20–	.3420201	58	3– 6–12.41	.0541388
10	18–	.3090170	59	3– 3– 3.05	.0532221
11	16–21–49.09	.2817325	60	3–	.0523360
12	15–	.2588190	61	2–57– 2.95	.0514787
13	13–50–46.15	.2393157	62	2–54–11.61	.0506491
14	12–51–25.71	.2225208	63	2–51–25.71	.0498458
15	12	.2079116	64	2–48– 45	.0490676
16	11–15	.1950903	65	2–46– 9.23	.0483133
17	10–35–17.64	.1837405	66	2–43–38.18	.0475819
18	10–	.1736481	67	2–41–11.64	.0468722
19	9–28–25.26	.1645945	68	2–38–49.41	.0461834
20	9–	.1564344	69	2–36–31.30	.0455145
21	8–34–17.14	.1490422	70	2–34–17.14	.0448648
22	8–10–54.54	.1423148	71	2–32– 6.76	.0442333
23	7–49–33.91	.1361666	72	2–30	.0436194
24	7–30–	.1305262	73	2–27–56.71	.0430222
25	7–12–	.1253332	74	2–25–56.75	.0424411
26	6–55–23.07	.1205366	75	2–24–	.0418757
27	6–40	.1160929	76	2–22– 6.31	.0413249
28	6–25–42.85	.1119644	77	2–20–15.58	.0407885
29	6–12–24.82	.1081189	78	2–18–27.69	.0402659
30	6–	.1045284	79	2–16–42.53	.0397565
31	5–48–23.22	.1011683	80	2–15–	.0392598
32	5–37–30	.0980171	81	2–13–20	.0387753
33	5–27–16.36	.0950560	82	2–11–42.45	.0383027
34	5–17–38.82	.0922683	83	2–10– 7.22	.0378414
35	5– 8–34.28	.0896392	84	2– 8–34.28	.0373911
36	5–	.0871557	85	2– 7– 3.54	.0369515
37	4–51–53.51	.0848058	86	2– 5–34.88	.0365220
38	4–44–12.63	.0825793	87	2– 4– 8.27	.0361023
39	4–36–55.38	.0804665	88	2– 2–43.63	.0356923
40	4–30–	.0784591	89	2– 1–20.89	.0352914
41	4–23–24.87	.0765492	90	2–	.0348995
42	4–17– 8.57	.0747301	91	1–58–40.87	.0345160
43	4–11– 9.76	.0729952	92	1–57–23.47	.0341410
44	4– 5–27.27	.0713391	93	1–56– 7.74	.0337741
45	4	.0697565	94	1–54–53.61	.0334149
46	3–54–46.95	.0682423	95	1–53–41.05	.0330633
47	3–49–47.23	.0667926	96	1–52–30.	.0327190
48	3–45–	.0654031	97	1–51–20.41	.0323818
49	3–40–24.49	.0640702	98	1–50–12.24	.0320515
50	3–36–	.0627905	99	1–49– 5.45	.0317279
51	3–31–45.88	.0615609	100	1–48–	.0314107

CHORDAL DATA FOR INSCRIBED POLYGONS (*Continued*)

101 to 200 Sides

No. Sides	Angle Deg. Min. Sec.	Sine	No. Sides	Angle Deg. Min. Sec.	Sine
101	1–46–55.84	.0310998	151	1–11–31.39	.0208037
102	1–45–52.94	.0307950	152	1–11– 3.15	.0206668
103	1–44–51.26	.0304961	153	1–10–35.29	.0205318
104	1–43–50.76	.0302029	154	1–10– 7.79	.0203985
105	1–42–51.42	.0299154	155	1– 9–40.64	.0202669
106	1–41–53.20	.0296332	156	1– 9–13.84	.0201370
107	1–40–56.07	.0293564	157	1– 8–47.38	.0200087
108	1–40–	.0290847	158	1– 8–21.26	.0198821
109	1–39– 4.95	.0288179	159	1– 7–55.47	.0197571
110	1–38–10.90	.0285560	160	1– 7–30	.0196336
111	1–37–17.83	.0282488	161	1– 7– 4.84	.0195117
112	1–36–25.71	.0280462	162	1– 6–40	.0193913
113	1–35–34.51	.0277981	163	1– 6–15.46	.0192723
114	1–34–44.21	.0275543	164	1– 5–51.21	.0191548
115	1–33–54.78	.0273147	165	1– 5–27.27	.0190387
116	1–33– 6.20	.0270793	166	1– 5– 3.61	.0189241
117	1–32–18.46	.0268479	167	1– 4–40.23	.0188107
118	1–31–31.52	.0266204	168	1– 4–17.14	.0186988
119	1–30–45.38	.0263968	169	1– 3–54.31	.0185881
120	1–30–	.0261769	170	1– 3–31.76	.0184788
121	1–29–15.37	.0259606	171	1– 3– 9.47	.0183708
122	1–28–31.47	.0257478	172	1– 2–47.44	.0182640
123	1–27–48.29	.0255386	173	1– 2–25.66	.0181584
124	1–27– 5.80	.0253326	174	1– 2– 4.13	.0180541
125	1–26–24	.0251300	175	1– 1–42.85	.0179509
126	1–25–42.85	.0249306	176	1– 1–21.81	.0178489
127	1–25– 2.36	.0247344	177	1– 1– 1.01	.0177481
128	1–24–22.50	.0245412	178	1– 0–40.44	.0176484
129	1–23–43.25	.0243509	179	1– 0–20.11	.0175498
130	1–23– 4.61	.0241637	180	1– –	.0174524
131	1–22–26.56	.0239793	181	–59–40.11	.0173559
132	1–21–49.09	.0237976	182	–59–20.43	.0172605
133	1–21–12.18	.0236188	183	–59– 0.98	.0171663
134	1–20–35.82	.0234425	184	–58–41.73	.0170730
135	1–20–	.0232689	185	–58–22.70	.0169807
136	1–19–24.70	.0230978	186	–58– 3.87	.0168894
137	1–18–49.92	.0229292	187	–57–45.24	.0167991
138	1–18–15.65	.0227631	188	–57–26.30	.0167097
139	1–17–41.87	.0225994	189	–57– 8.57	.0166214
140	1–17– 8.57	.0224380	190	–56–50.52	.0165339
141	1–16–35.74	.0222789	191	–56–32.67	.0164473
142	1–16– 3.38	.0221220	192	–56–15	.0163617
143	1–15–31.46	.0219673	193	–55–57.51	.0162769
144	1–15–	.0218148	194	–55–40.20	.0161930
145	1–14–28.96	.0216644	195	–55–23.07	.0161100
146	1–13–58.35	.0215160	196	–55– 6.12	.0160278
147	1–13–28.16	.0213697	197	–54–49.34	.0159464
148	1–12–58.37	.0212253	198	–54–32.72	.0158659
149	1–12–28.99	.0210829	199	–54–16.28	.0157862
150	1–12–	0209424	200	–54–	.0157073

CHORDAL DATA FOR INSCRIBED POLYGONS (*Continued*)
201 to 300 Sides

No. Sides	Angle Min. Sec.	Sine	No. Sides	Angle Min. Sec.	Sine
201	53–43.88	.0156244	251	43– 1.67	.0125160
202	53–27.92	.0155518	252	42–51.43	.0124663
203	53–12.12	.0154752	253	42–41.26	.0124171
204	52–56.47	.0153993	254	42–31.18	.0123682
205	52–40.97	.0153242	255	42–21.18	.0123197
206	52–25.63	.0152498	256	42–11.25	.0122715
207	52–10.44	.0151764	257	42– 1.40	.0122238
208	51–55.38	.0151033	258	41–51.63	.0121764
209	51–40.48	.0150310	259	41–41.93	.0121294
210	51–25.71	.0149595	260	41–32.31	.0120827
211	51–11.09	.0148886	261	41–22.76	.0120364
212	50–56.60	.0148183	262	41–13.28	.0119905
213	50–42.25	.0147487	263	41– 3.88	.0119449
214	50–28.04	.0146798	264	40–54.54	.0118997
215	50–13.96	.0146115	265	40–45.28	.0118548
216	50–	.0145439	266	40–36.09	.0118102
217	49–46.17	.0144769	267	40–26.96	.0117660
218	49–32.48	.0144104	268	40–17.91	.0117221
219	49–18.91	.0143446	269	40– 8.93	.0116786
220	49– 5.46	.0142794	270	40–	.0116353
221	48–52.13	.0142148	271	39–51.14	.0115923
222	48–38.92	.0141508	272	39–42.35	.0115497
223	48–25.83	.0140874	273	39–33.63	.0115074
224	48–12.86	.0140245	274	39–24.96	.0114654
225	48–	.0139622	275	39–16.36	.0114237
226	47–47.26	.0139004	276	39– 7.83	.0113823
227	47–34.63	.0138392	277	38–59.35	.0113412
228	47–22.11	.0137785	278	38–50.94	.0113004
229	47– 9.69	.0137183	279	38–42.58	.0112599
230	46–57.39	.0136587	280	38–34.28	.0112197
231	46–45.19	.0135995	281	38–26.05	.0111798
232	46–33.10	.0135409	282	38–17.87	.0111401
233	46–21.11	.0134828	283	38– 9.75	.0111008
234	46– 9.23	.0134252	284	38– 1.69	.0110617
235	45–57.45	.0133681	285	37–53.68	.0110229
236	45–45.76	.0133115	286	17–45.73	.0109844
237	45–34.18	.0132553	287	37–37.84	.0109461
238	45–22.69	.0131996	288	37–30	.0109081
239	45–11.29	.0131444	289	37–22.21	.0108704
240	45–	.0130896	290	37–14.48	.0108329
241	44–48.80	.0130353	291	37– 6.80	.0107957
242	44–37.68	.0129814	292	36–59.18	.0107587
243	44–26.67	.0129280	293	36–51.60	.0107220
244	44–15.74	.0128750	294	36–44.08	.0106855
245	44– 4.90	.0128225	295	36–36.61	.0106493
246	43–54.15	.0127704	296	36–29.19	.0106133
247	43–43.48	.0127187	297	36–21.82	.0105776
248	43–32.40	.0126674	298	36–14.50	.0105421
249	43–22.41	.0126165	299	36– 7.22	.0105068
250	43–12	.0125661	300	36–	.0104718

TRIGONOMETRY

Trigonometry deals with the relations of angles and sides of triangles and is one of the most useful branches of mathematics to the shopman and engineer. The word "relations" is used advisedly, because in any given triangle there are always fixed proportions between the three angles and the three sides. Thus, all the sides and angles can be determined if the given data include: one angle and two sides, or two angles and one side, or all three sides.

Functions of Angles. Figures 7 and 8 show the classical representation of the functions of angles for a right triangle ABC. As radius AB moves counterclockwise from coincidence with AE to AH, the value of the function, or ratio, BC/AB (the sine of A) will change in value from 0 to 1. And at the same time all other functions of A change. But note this: If the triangle ABC is proportionately expanded or contracted in any amount, and angle A remains unchanged, the other functions of A remain unchanged. Then, if A is 30° and $AB = 1$, $BC = \frac{1}{2}$, and if $AB = 4$, $BC = 2$, because the sine of 30° is always 0.5000.

Any problem dealing with triangles can be solved by use of the formulas given and reference to the table of natural trignometric functions beginning on page 44–66. For example, if the sine of an angle is found by calculation to be 0.30071, reference to the table will disclose that the angle is 17°30′. The same applies to cosines, tangents, cotangents, secants, and cosecants, whichever function is most convenient to use.

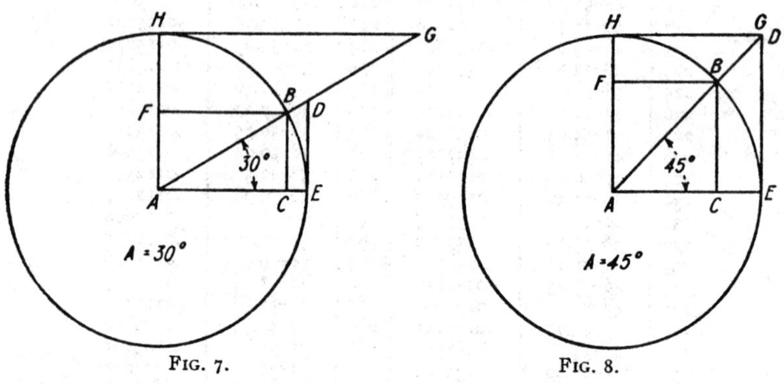

FIG. 7. FIG. 8.

$$BC = \text{sine } A = \frac{\text{side opposite}}{\text{hypotenuse}} \qquad AD = \text{secant } A = \frac{\text{hypotenuse}}{\text{adjacent side}}$$

$$AC = \text{cosine } A = \frac{\text{side adjacent}}{\text{hypotenuse}} \qquad AG = \text{cosecant } A = \frac{\text{hypotenuse}}{\text{opposite side}}$$

$$DE = \text{tangent } A = \frac{\text{side opposite}}{\text{side adjacent}} \qquad CE = \text{versed sine } A = AB - AC \div AB$$

$$GH = \text{cotangent } A = \frac{\text{side adjacent}}{\text{side opposite}} \qquad FH = \text{coversed sine } A = AB - BC \div AB$$

$$\text{Hypotenuse} = \frac{\text{side opposite}}{\sin A} = \frac{\text{side adjacent}}{\text{cosine } A}$$

NOTE: sine $B = AC \div AB$, cosine $B = BC \div AB$, tangent $B = AC \div BC$, etc.

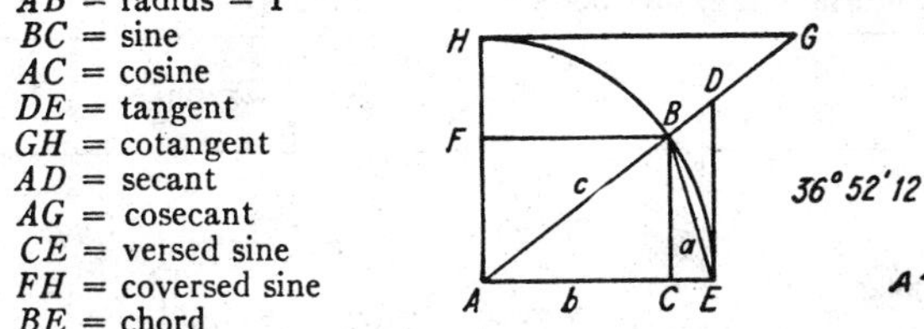

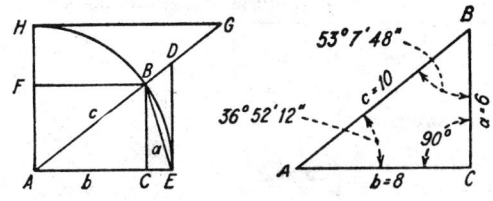

FORMULAS FOR RIGHT TRIANGLES

To find	Knowing	Formula	Example
Angle A	Angles B and C	$C - B$	$90° - 53°7'48'' = 36°52'12''$
Angle B	Angles A and C	$C - A$	$90° - 36°52'12'' = 53°7'48''$
Angle C	Angles A and B	$A + B$	$36°52'12'' + 53°7'48'' = 90°$
Sine A	Sides a and c	$\frac{a}{c}$	$\frac{6}{10} = 0.60000$ $= \sin 36°52'12''$
	Cosecant A	$\frac{1}{\text{cosec } A}$	$\frac{1}{1.66666} = 0.60000$ $= \sin 36°52'12''$
	Tangent A and cosine A	$\tan A \times \cos A$	0.75000×0.80000 $= 0.60000 = \sin 36°52'12''$
	Tangent A and secant A	$\frac{\tan A}{\sec A}$	$\frac{0.75000}{1.25000} = 0.60000$ $= \sin 36°52'12''$
	Cosine A and cotangent A	$\frac{\cos A}{\cot A}$	$\frac{0.80000}{1.33333} = 0.60000$ $= \sin 36°52'12''$
	Cosine A	$\sqrt{1 - \cos A^2}$	$\sqrt{0.36} = 0.60000$ $= \sin 36°52'12''$
Cosine A	Sides b and c	$\frac{b}{c}$	$\frac{8}{10} = 0.80000$ $= \cos 36°52'12''$
	Secant A	$\frac{1}{\text{Sec } A}$	$\frac{1}{1.25000} = 0.80000$ $= \cos 36°52'12''$
	Cotangent A and sine A	$\cot A \times \sin A$	1.33333×0.60000 $= 0.80000$ $= \cos 36°52'12''$
	Sine A and tangent A	$\frac{\sin A}{\tan A}$	$\frac{0.60000}{0.75000} = 0.80000$ $= \cos 36°52'12''$
	Cotangent A and cosecant A	$\frac{\cot A}{\text{cosec } A}$	$\frac{1.33333}{1.66666} = 0.80000$ $= \cos 36°52'12''$
	Sine A	$\sqrt{1 - \sin A^2}$	$\sqrt{0.64} = 0.80000$ $= \cos 36°52'12''$

FORMULAS FOR RIGHT TRIANGLES (*Continued*)

To find	Knowing	Formula	Example
Tangent A	Sides a and b	$\frac{a}{b}$	$\frac{6}{8} = 0.75000 = \tan 36°52'12''$
	Cotangent A	$\frac{1}{\cot A}$	$\frac{1}{1.33333} = 0.75000 = \tan 36°52'12''$
	Secant A and sine A	$\sec A \times \sin A$	$1.25000 \times 0.60000 = 0.75000 = \tan 36°52'12''$
	Sine A and cosine A	$\frac{\sin A}{\cos A}$	$\frac{0.60000}{0.80000} = 0.75000 = \tan 36°52'12''$
	Secant A and cosecant A	$\frac{\sec A}{\operatorname{cosec} A}$	$\frac{1.25000}{1.66666} = 0.75000 = \tan 36°52'12''$
	Secant A	$\sqrt{\sec A^2 - 1}$	$\sqrt{0.5625} = 0.75000 = \tan 36°52'12''$
Cotangent A	Sides b and a	$\frac{b}{a}$	$\frac{8}{6} = 1.33333 = \cot 36°52'12''$
	Tangent A	$\frac{1}{\tan A}$	$\frac{1}{0.75000} = 1.33333 = \cot 36°52'12''$
	Cosecant A and cosine A	$\operatorname{cosec} A \times \cos A$	$1.66666 \times 0.80000 = 1.33333 = \cot 36°52'12''$
	Cosine A and sine A	$\frac{\cos A}{\sin A}$	$\frac{0.80000}{0.60000} = 1.33333 = \cot 36°52'12''$
	Cosecant A and secant A	$\frac{\operatorname{cosec} A}{\sec A}$	$\frac{1.66666}{1.25000} = 1.33333 = \cot 36°52'12''$
	Cosecant A	$\sqrt{\operatorname{cosec} A^2 - 1}$	$\sqrt{1.77777} = 1.33333 = \cot 36°52'12''$
Secant A	Sides c and b	$\frac{c}{b}$	$\frac{10}{8} = 1.25000 = \sec 36°52'12''$
	Cosine A	$\frac{1}{\cos A}$	$\frac{1}{0.80000} = 1.25000 = \sec 36°52'12''$
	Cosecant A and tangent A	$\operatorname{cosec} A \times \tan A$	$1.66666 \times 0.75000 = 1.25000 = \sec 36°52'12''$
	Tangent A and sine A	$\frac{\tan A}{\sin A}$	$\frac{0.75000}{0.60000} = 1.25000 = \sec 36°52'12''$
	Cosecant A and cotangent A	$\frac{\operatorname{cosec} A}{\cot A}$	$\frac{1.66666}{1.33333} = 1.25000 = \sec 36°52'12''$
	Tangent A	$\sqrt{\tan A^2 + 1}$	$\sqrt{1.5625} = 1.25000 = \sec 36°52°12''$

FORMULAS FOR RIGHT TRIANGLES (*Continued*)

To find	Knowing	Formula	Example
Cosecant A	Sides c and a	$\frac{c}{a}$	$\frac{10}{6} = 1.66666 = \text{cosec } 36°52'12''$
	Sine A	$\frac{1}{0.60000}$	$\frac{1}{0.60000} = 1.66666 = \text{cosec } 36°52'12''$
	Secant A and cotangent A	$\sec A \times \cot A$	$1.25000 \times 1.33333 = 1.66666 = \text{cosec } 36°52'12''$
	Secant A and tangent A	$\frac{\sec A}{\tan A}$	$\frac{1.25000}{0.75000} = 1.66666 = \text{cosec } 36°52'12''$
	Cotangent A and cosine A	$\frac{\cot A}{\cos A}$	$\frac{1.33333}{0.80000} = 1.66666 = \text{cosec } 36°52'12''$
	Cotangent A	$\sqrt{\cot A^2 + 1}$	$\sqrt{2.77777} = 1.66666 = \text{cosec } 36°52'12''$
Side a	Sides c and b	$\sqrt{c^2 - b^2}$	$\sqrt{36} = 6 = \text{side } a$
	Side c and sine A	$c \times \sin A$	$10 \times 0.60000 = 6 = \text{side } a$
	Side b and tangent A	$b \times \tan A$	$8 \times 0.75000 = 6 = \text{side } a$
	Side b and cotangent A	$\frac{b}{\cot A}$	$\frac{8}{1.3333} = 6 = \text{side } a$
	side c and cosecant A	$\frac{c}{\text{cosec } A}$	$\frac{10}{1.66666} = 6 = \text{side } a$
Side b	Sides c and a	$\sqrt{c^2 - a^2}$	$\sqrt{64} = 8 = \text{side } b$
	Side c and cosine A	$c \times \cos A$	$10 \times 0.80000 = 8 = \text{side } b$
	Side a and cotangent A	$a \times \cot A$	$6 \times 1.33333 = 8 = \text{side } b$
	Side a and tangent A	$\frac{a}{\tan A}$	$\frac{6}{0.7500} = 8 = \text{side } b$
	Side c and secant A	$\frac{c}{\sec A}$	$\frac{10}{1.25000} = 8 = \text{side } b$
Side c	Sides a and b	$\sqrt{a^2 + b^2}$	$\sqrt{6^2 + 8^2} = \sqrt{36 + 64} = \sqrt{100} = 10 = \text{side } c$
	Side b and secant A	$b \times \sec A$	$8 \times 1.25000 = 10 = \text{side } c$
	Side a and cosecant A	$a \times \text{cosec } A$	$6 \times 1.66666 = 10 = \text{side } c$
	Side a and sine A	$\frac{a}{\sin A}$	$\frac{6}{0.60000} = 10 = \text{side } c$
	Side b and cosine A	$\frac{b}{\cos A}$	$\frac{8}{0.80000} = 10 = \text{side } c$
	Sides c and b and versed sine A	$\frac{c - b}{\text{vers } A}$	$\frac{2}{0.20000} = 10 = \text{side } c$

FORMULAS FOR RIGHT TRIANGLES (*Continued*)

To find	Knowing	Formula	Example
Side c (*continued*)	Sides c and a and coversed sine A	$\frac{c-a}{\text{covers } A}$	$\frac{4}{0.40000} = 10 =$ side c
Chord A	Sine $\frac{1}{2}A$	$2c \sin \frac{A}{2}$	$\sin 18°26'6'' = 0.31623$; $0.31623 \times 2 \times 10 = 6.3246$ = chord $36°52'12''$
Versed sine A	Sides c and b	$\frac{c-b}{c}$	$\frac{2}{10} = 0.20000$ = versin $36°52'12''$
	Cosine A	$1 - \cos A$	$1 - 0.80000 = 0.20000$ = versin $36°52'12''$
Coversed sine A	Sides c and a	$\frac{c-a}{c}$	$\frac{4}{10} = 0.40000$ = covers $36°52'12''$
	Sine A	$1 - \sin A$	$1 - 0.60000 = 0.40000$ = covers $36°52'12''$

Abbreviations and Symbols

sin = sine; cos = cosine; tan = tangent; cot = cotangent; sec = secant; csc = cosecant; and ∠ = angle

Signs of Functions. The values of trigonometric functions change from plus to minus, and vice versa, according to the four quadrants of the circle. These quadrants are: first quadrant—upper right; second quadrant—upper left; third quadrant—lower left, and fourth quadrant—lower right. Algebraic signs of the functions as used in formulas are:

	Quadrants			
	First	Second	Third	Fourth
Sine and cosecant	+	+	−	−
Cosine and secant	+	−	−	+
Tangent and cotangent	+	−	+	−

Oblique

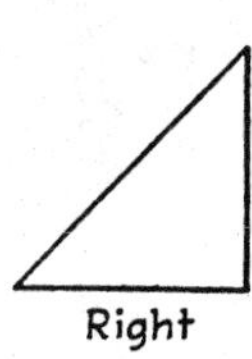

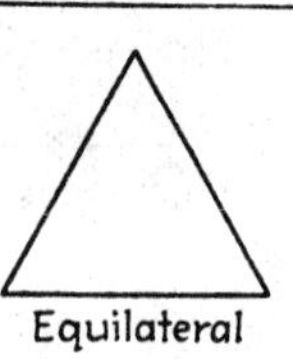

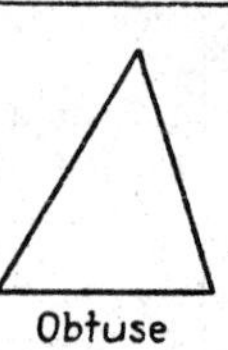

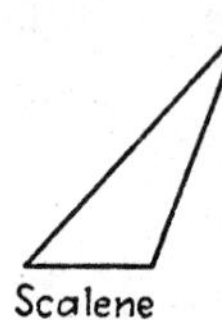

FIG. 9. Types of triangles.

Formulas for Oblique Triangles (Figs. 10 and 11)

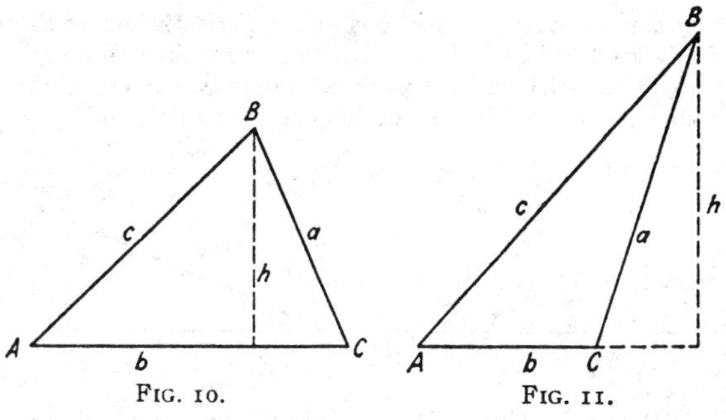

FIG. 10. FIG. 11.

From law of sines: *the sides are proportional to sines of opposite angles.*

$$a = \frac{b \sin A}{\sin B} = \frac{c \sin A}{\sin C} \qquad c = \frac{a \sin C}{\sin A} = \frac{b \sin C}{\sin B}$$

$$b = \frac{a \sin B}{\sin A} = \frac{c \sin B}{\sin C} \qquad h = c \sin A = a \sin C$$

If a, b, and B are given: $A = \sin^{-1} \dfrac{a \sin B}{b}$, where the expression $\sin^{-1}$ is read "the angle whose sine is."

From law of cosines: *the square of any side equals the sum of the squares of the other two sides minus twice the product of these two sides multiplied by the cosine of the angle between them.*

$$a^2 = b^2 + c^2 - 2bc \cos A$$
$$b^2 = a^2 + c^2 - 2ac \cos B$$
$$c^2 = a^2 + b^2 - 2ab \cos C$$

Obviously, $\cos A = \dfrac{b^2 + c^2 - a^2}{2bc}$, etc.

Integral Right-angled Triangles

The erection of a perpendicular by the construction of a triangle whose sides are respectively 3, 4, and 5 units in length is a familiar device. The table gives a greater range of choice in the proportions of the triangle employed. The table is a list of all integral, or whole-number, right-angled triangles wherein the least side does not exceed 20.

Height	Base	Hypotenuse	Height	Base	Hypotenuse	Height	Base	Hypotenuse
3	4	5	12	16	20	17	144	145
5	12	13	12	35	37	18	24	30
6	8	10	13	84	85	18	80	82
7	24	25	14	48	50	19	180	181
8	15	17	15	20	25	20	21	29
9	12	15	15	36	39	20	48	52
9	40	41	15	112	113	20	99	101
10	24	26	16	30	34			
11	60	61	16	63	65			

Laying Out Square Corners

It sometimes happens that we wish to lay out a perfectly square corner and have no square of any kind handy. Here is a way that requires nothing but a scale or rule, or even a straight stick without any graduations whatever will do. Using this stick, draw a line as *AC* (in left-hand accompanying figure), and at one end of this

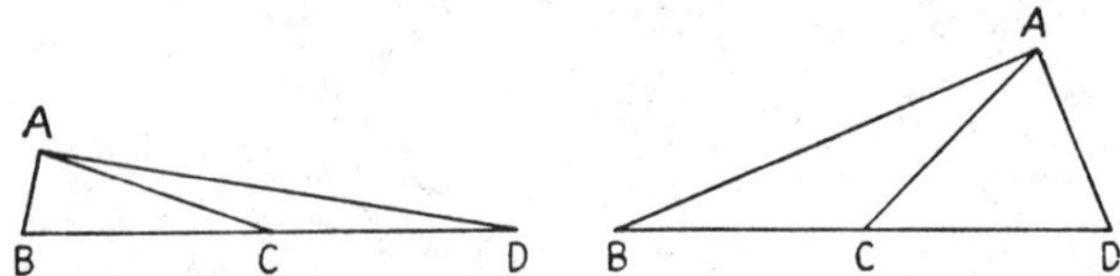

draw the line *BD* at any angle. This line must be straight, twice as long as *AC* and of equal length each side of the point *C*. Then, if you join points *DAB*, you have an exact right angle or square corner.

The right-hand accompanying figure is simply another example of this, in which the line *AC* has been drawn at a very different angle to show that it works in any position. Joining the ends *DAB* as before also gives an exact right angle.

Another Method. Another method is by what is known as the "6, 8, and 10 rule." This means that if a triangle has sides in the ratio of 6, 8, and 10, the angle is 90°. Lay down a line 6 units long, either inches, feet, or yards. Lay off another line 8 units long as nearly right angles as possible. Measure across the ends of the two lines and adjust until this distance is 10 units, which makes it a right angle. These distances may be 3, 4, and 5 or 12, 16, and 20, or any combination in this ratio. It is largely used in laying out large corners.

Use of Chords to Construct Angles

To construct any angle from the table of chords, page 44-61: Let the required angle be 36°38′; the nearest angles in the table are 36°30′ and 36°40′, and the chords are respectively 0.6263 and 0.6291, the difference 0.0028 corresponding to an angular difference of 10′. To find the amount which must be added to 0.6263 (the chord corresponding to 36°30′) in order to obtain the chord for a 36°38′ arc, multiply 0.0028 by $\frac{8}{10} = 0.00224$. $0.6263 + 0.00224 = 0.62854$. Then, if the radius is 1 in. and the angle 36°38′, the chord will be 0.62854 in.

In laying out an angle, as in the accompanying illustration, a base line *AB* can be drawn, say 10 in. long; then with a radius *AB* and center *A*, arc *BC* can be struck. Multiply chord 0.62854 in. by 10 giving 6.2854 in., as the radius of an arc to be struck from center *B* and cutting arc *BC* at *C*. Through point *C* draw a line *AC* and the angle *BAC* will equai 36°38′.

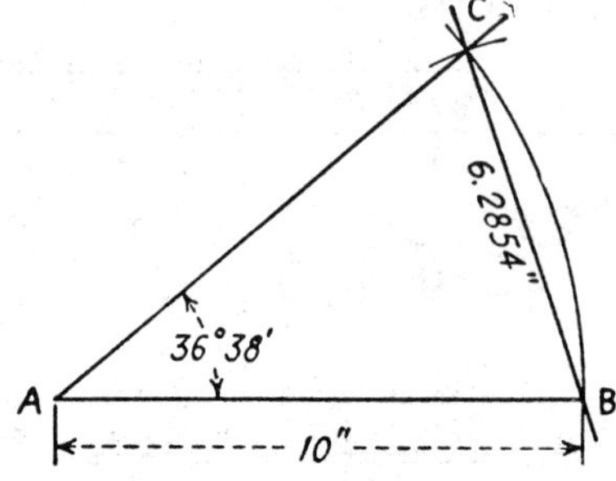

Where the angle, required is in even degrees or sixths of degrees (as 10°20′, etc.), the corresponding chord may be taken directly from the table. A 10:1 layout is particularly convenient.

Chords Subtended by Angles of 0° to 45°

Tabulated quantities = twice the sine of one-half the angle

Deg	0′	10′	20′	30′	40′	50′	60′
0	.0000	.0029	.0058	.0087	.0116	.0145	.0174
1	.0174	.0204	.0233	.0262	.0291	.0320	.0349
2	.0349	.0378	.0407	.0436	.0465	.0494	.0523
3	.0523	.0553	.0582	.0611	.0640	.0669	.0698
4	.0698	.0727	.0756	.0785	.0814	.0843	.0872
5	.0872	.0901	.0930	.0959	.0988	.1017	.1047
6	.1047	.1076	.1105	.1134	.1163	.1192	.1221
7	.1221	.1250	.1279	.1305	.1337	.1366	.1395
8	.1395	.1424	.1453	.1482	.1511	.1540	.1569
9	.1569	.1598	.1627	.1656	.1685	.1714	.1743
10	.1743	.1772	.1801	.1830	.1859	.1888	.1917
11	.1917	.1946	.1975	.2004	.2033	.2062	.2090
12	.2090	.2119	.2148	.2177	.2206	.2235	.2264
13	.2264	.2293	.2322	.2351	.2380	.2409	.2437
14	.2437	.2466	.2495	.2524	.2553	.2582	.2610
15	.2610	.2639	.2668	.2697	.2726	.2755	.2783
16	.2783	.2812	.2841	.2870	.2899	.2927	.2956
17	.2956	.2985	.3014	.3042	.3071	.3100	.3129
18	.3129	.3157	.3186	.3215	.3243	.3272	.3301
19	.3301	.3330	.3358	.3387	.3416	.3444	.3473
20	.3473	.3502	.3530	.3559	.3587	.3616	.3645
21	.3645	.3673	.3702	.3730	.3759	.3788	.3816
22	.3816	.3845	.3873	.3902	.3930	.3959	.3987
23	.3987	.4016	.4044	.4073	.4101	.4130	.4158
24	.4158	.4187	.4215	.4243	.4272	.4300	.4329
25	.4329	.4357	.4385	.4414	.4442	.4471	.4499
26	.4499	.4527	.4556	.4584	.4612	.4641	.4669
27	.4669	.4697	.4725	.4754	.4782	.4810	.4838
28	.4838	.4867	.4895	.4923	.4951	.4979	.5008
29	.5008	.5036	.5064	.5092	.5120	.5148	.5176
30	.5176	.5204	.5232	.5261	.5289	.5317	.5345
31	.5345	.5373	.5401	.5429	.5457	.5485	.5513
32	.5513	.5541	.5569	.5596	.5624	.5652	.5680
33	.5680	.5708	.5736	.5764	.5792	.5820	.5847
34	.5847	.5875	.5903	.5931	.5959	.5986	.6014
35	.6014	.6042	.6069	.6097	.6125	.6153	.6180
36	.6180	.6208	.6236	.6263	.6291	.6318	.6346
37	.6346	.6374	.6401	.6429	.6456	.6484	.6511
38	.6511	.6539	.6566	.6594	.6621	.6649	.6676
39	.6676	.6703	.6731	.6758	.6786	.6813	.6840
40	.6840	.6868	.6895	.6922	.6950	.6977	.7004
41	.7004	.7031	.7059	.7086	.7113	.7140	.7167
42	.7167	.7194	.7222	.7249	.7276	.7303	.7330
43	.7330	.7357	.7384	.7411	.7438	.7465	.7492
44	.7492	.7519	.7546	.7573	.7600	.7627	.7654
45	.7654	.7680	.7707	.7734	.7761	.7788	.7815

Chords Subtended by Angles of 46° to 90°

Tabulated values = twice the sine of one-half the angle

Deg	0′	10′	20′	30′	40′	50′	60′
46	.7815	.7841	.7868	.7895	.7921	.7948	.7975
47	.7975	.8001	.8028	.8055	.8081	.8108	.8135
48	.8135	.8161	.8188	.8214	.8241	.8267	.8294
49	.8294	.8320	.8347	.8373	.8400	.8426	.8452
50	.8452	.8479	.8505	.8531	.8558	.8584	.8610
51	.8610	.8636	.8663	.8689	.8715	.8741	.8767
52	.8767	.8793	.8820	.8846	.8872	.8898	.8924
53	.8924	.8950	.8976	.9002	.9028	.9054	.9080
54	.9080	.9106	.9132	.9157	.9183	.9209	.9235
55	.9235	.9261	.9286	.9312	.9338	.9364	.9389
56	.9389	.9415	.9441	.9466	.9492	.9518	.9543
57	.9543	.9569	.9594	.9620	.9645	.9671	.9696
58	.9696	.9722	.9747	.9772	.9798	.9823	.9848
59	.9848	.9874	.9899	.9924	.9949	.9975	1.0000
60	1.0000	1.0025	1.0050	1.0075	1.0100	1.0126	1.0151
61	1.0151	1.0176	1.0201	1.0226	1.0251	1.0276	1.0301
62	1.0301	1.0326	1.0350	1.0375	1.0400	1.0425	1.0450
63	1.0450	1.0475	1.0500	1.0524	1.0550	1.0574	1.0598
64	1.0598	1.0623	1.0648	1.0672	1.0697	1.0721	1.0746
65	1.0746	1.0770	1.0795	1.0819	1.0844	1.0868	1.0893
66	1.0893	1.0917	1.0941	1.0966	1.0990	1.1014	1.1039
67	1.1039	1.1063	1.1087	1.1111	1.1135	1.1159	1.1184
68	1.1184	1.1208	1.1232	1.1256	1.1280	1.1304	1.1328
69	1.1328	1.1352	1.1376	1.1400	1.1424	1.1448	1.1471
70	1.1471	1.1495	1.1519	1.1543	1.1567	1.1590	1.1614
71	1.1614	1.1638	1.1661	1.1685	1.1708	1.1732	1.1756
72	1.1756	1.1780	1.1803	1.1826	1.1850	1.1873	1.1896
73	1.1896	1.1920	1.1943	1.1966	1.1990	1.2013	1.2036
74	1.2036	1.2059	1.2083	1.2106	1.2129	1.2152	1.2175
75	1.2175	1.2198	1.2221	1.2244	1.2267	1.2290	1.2313
76	1.2313	1.2336	1.2360	1.2382	1.2405	1.2427	1.2450
77	1.2450	1.2473	1.2496	1.2518	1.2541	1.2564	1.2586
78	1.2586	1.2609	1.2631	1.2654	1.2677	1.2699	1.2721
79	1.2721	1.2744	1.2766	1.2789	1.2811	1.2833	1.2856
80	1.2856	1.2878	1.2900	1.2922	1.2945	1.2967	1.2989
81	1.2989	1.3011	1.3033	1.3055	1.3077	1.3099	1.3121
82	1.3121	1.3143	1.3165	1.3187	1.3209	1.3231	1.3252
83	1.3252	1.3274	1.3296	1.3318	1.3340	1.3361	1.3383
84	1.3383	1.3404	1.3426	1.3447	1.3469	1.3490	1.3512
85	1.3512	1.3533	1.3555	1.3576	1.3597	1.3619	1.3640
86	1.3640	1.3661	1.3682	1.3704	1.3725	1.3746	1.3767
87	1.3767	1.3788	1.3809	1.3830	1.3851	1.3872	1.3893
88	1.3893	1.3914	1.3935	1.3956	1.3977	1.3997	1.4018
89	1.4018	1.4039	1.4060	1.4080	1.4101	1.4121	1.4142
90	1.4142						

Solution of Trigonometric Problems

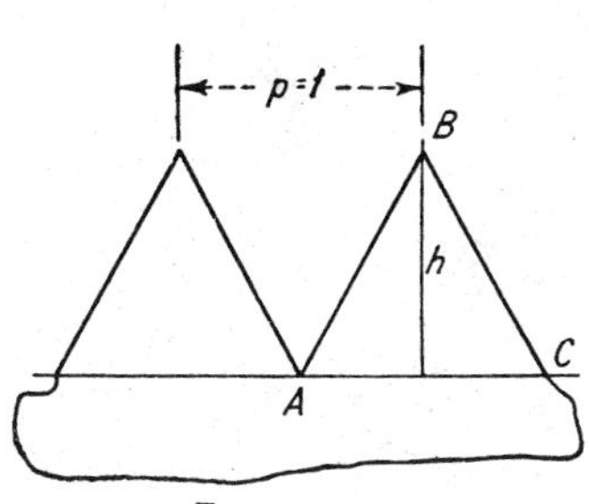

FIG. 12.

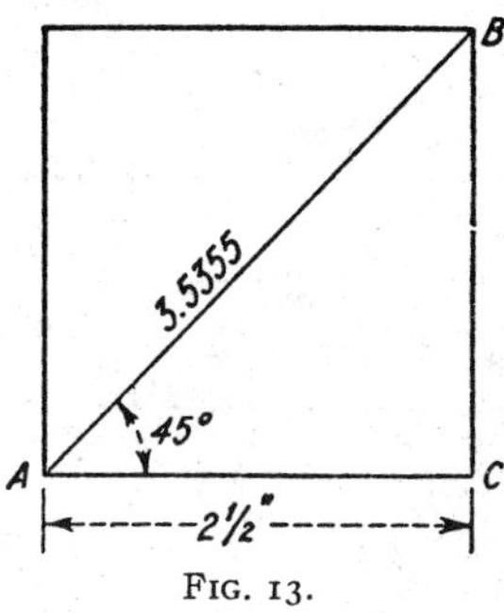

FIG. 13.

EXAMPLE 1: What is the depth of a sharp 60° thread? See Fig. 12.
Let pitch $p = 1$ in.
Then $AC = p = 1$.
And h bisects B.
Hence $h = \cos \frac{1}{2}B = \cos 30° = 0.866$.
EXAMPLE 2: What is the length of the diagonal of a 2½-in. square? See Fig. 13.
Diagonal AB bisects the square.
Hence $A = 45°$.
$AB = \sec 45° \times AC = 1.4142 \times 2.5 = 3.5355$ in.
Or $AB =$ side opposite $\div \sin 45° = 2.5 \div 0.7171 = 3.5355$ in.

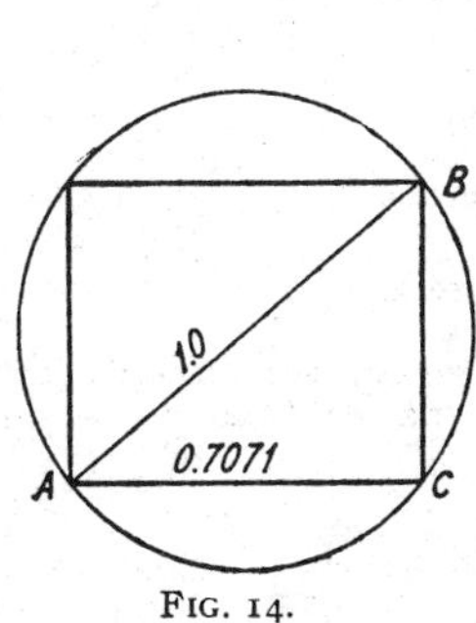

FIG. 14.

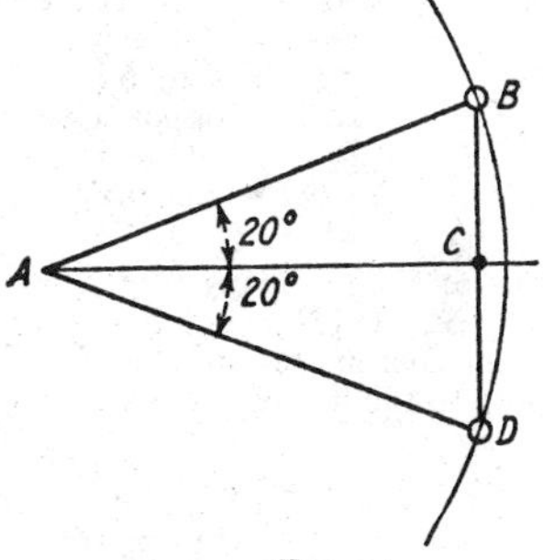

FIG. 15.

EXAMPLE 3: Find the largest square that can be milled on the end of a 1-in. round bar. See Fig. 14.
Let AB = dia of circle = hypotenuse of triangle ABC.
Then $AC = \cos 45° \times AB = 0.7071 \times 1 = 0.7071$ in.
EXAMPLE 4: What is the length of chord required to space off nine boltholes in a 14-in. circle? See Fig. 15.
$360° \div 9 = 40°$ per division.
Chord $BD = BC + CD$.
AC bisects angle A. Hence $BC = CD$.
$AB = 14 \div 2 = 7$ in.
$\sin \frac{1}{2}A = BC \div AB$. Then $BC = \sin 20° \times 7 = 0.342 \times 7 = 2.394$.
$BD = 2BC = 2 \times 2.394 = 4.788$ in.

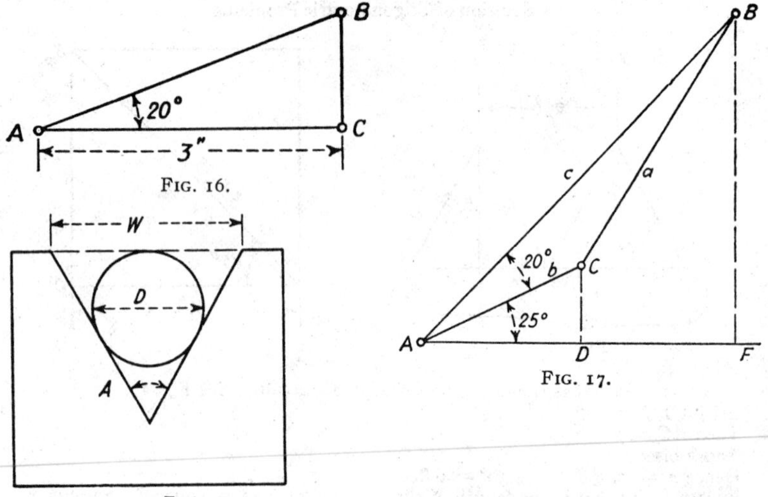

Fig. 16.

Fig. 17.

Fig. 18.

Example 5: A jig drawing shows three holes A, B, and C (Fig. 16) and gives the dimension AC = 3 in. and angle A = 20°. What are the lengths of AB and BC?

$$
\begin{aligned}
\sec A &= \text{hypotenuse} \div \text{adjacent side} = AB \div 3 \\
1.0642 &= AB \div 3 \\
AB &= 3.1926 \text{ in.} \\
\tan A &= \text{opposite side} \div \text{adjacent side} \\
0.36397 &= BC \div 3 \\
BC &= 1.09191 \text{ (or 1.092 approx)}
\end{aligned}
$$

Example 6: A jig drawing (Fig. 17) shows three holes.
Angle $A = 20°$, $\angle CAD = 25°$, $b = 4$, and $c = 12$.
What are the coordinates of holes B and C, and what is the dimension a?
According to the law of cosines for an oblique triangle ABC:

$$
\begin{aligned}
a &= \sqrt{b^2 + c^2 - 2bc \cos A} \\
&= \sqrt{16 + 144 - 2 \times 4 \times 0.9397} \\
&= \sqrt{69.79} = 8.345 \\
AE &= BE = c \sin 45^\circ = 12 \times 0.7171 = 8.605 \\
AD &= 4 \cos 25^\circ = 4 \times 0.9063 = 3.6252 \\
CD &= 4 \sin 25^\circ = 4 \times 0.4226 = 1.6904
\end{aligned}
$$

Checking Angles with Plugs

A plug is often used to check correctness of the width W in the angular opening of a templet or tool. Both W and angle A (Fig. 18) are correct when the correct size of plug is tangent with the opening indicated by the dash line. Plug diameter is given by the formula

$$D = W \tan \left(45^\circ - \frac{A}{4}\right)$$

Right Triangles. Plugs can also be used to check templets which have corners cut off or in which the body is notched with a 90° angle, as shown in Figs. 19 to 20. Here $D = a + b - c$. For example, in Figs. 20 and 27, $A = 30°$ and $b = 1.3$. What is the diameter of plug that will contact a, b, and c?

$$a = b \tan A = 1.30 \times 0.57735 = 0.75056$$
$$c = b \sec A = 1.3 \times 1.1547 = 1.5011$$

Then $$D = 0.75056 + 1.3 - 1.5011 = 0.5495$$

When only A and b are known, as often given on drawings:

$$D = 2b \div (\cot \tfrac{1}{2}A + 1)$$

Then for the above example

$$D = 2 \times 1.3 \div (\cot 15° + 1)$$
$$= 2.6 \div (3.73205 + 1)$$
$$= 0.5495$$

Angles Other Than Right Angles. The formula for the diameter of the plug D to fill the angle B in Figs. 27 and 28 and contact all three sides of an oblique triangle ABC is

$$D = 2W \div (\cot \tfrac{1}{2}A + \cot \tfrac{1}{2}B)$$

EXAMPLE: $A = 40°$, $B = 68°$, $W = 2$. Then

$C = 180° - (40° + 68°) = 72°$ $\quad D = 4 \div (\cot 20° + \cot 36°) = 0.9698$

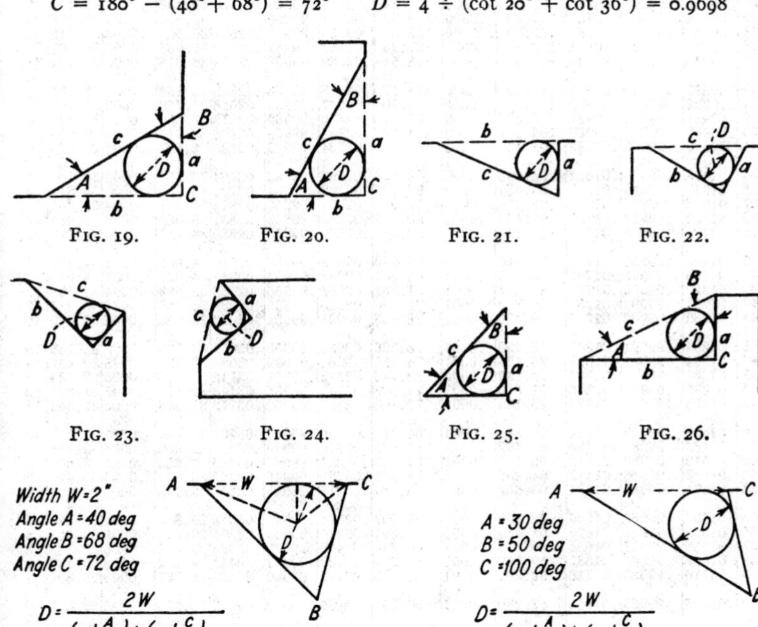

NATURAL TRIGONOMETRIC FUNCTIONS

	0°						1°						
′	sin	cos	tan	cot	sec	cosec	sin	cos	tan	cot	sec	cosec	′
0	.00000	1.0000	.00000	Infinite	1.0000	Infinite	.01745	.99985	.01745	57.290	1.0001	57.299	60
1	.00029	.0000	.00029	3437.7	.0000	3437.7	.01774	.99984	.01775	56.350	.0001	56.359	59
2	.00058	.0000	.00058	1718.9	.0000	1718.9	.01803	.99984	.01804	55.441	.0001	55.450	58
3	.00087	.0000	.00087	1145.9	.0000	1145.9	.01832	.99983	.01833	54.561	.0002	54.570	57
4	.00116	.0000	.00116	859.44	.0000	859.44	.01861	.99983	.01862	53.708	.0002	53.718	56
5	.00145	1.0000	.00145	687.55	.0000	687.55	.01891	.99982	.01891	52.882	.0002	52.891	55
6	.00174	.0000	.00174	572.96	.0000	572.96	.01920	.99981	.01920	52.081	.0002	52.090	54
7	.00204	.0000	.00204	491.11	.0000	491.11	.01949	.99981	.01949	51.303	.0002	51.313	53
8	.00233	.0000	.00233	429.72	.0000	429.72	.01978	.99980	.01978	50.548	.0002	50.558	52
9	.00262	.0000	.00262	381.97	.0000	381.97	.02007	.99980	.02007	49.816	.0002	49.826	51
10	.00291	0.99999	.00291	343.77	.0000	343.77	.02036	.99979	.02036	49.104	.0002	49.114	50
11	.00320	.99999	.00320	312.52	.0000	312.52	.02065	.99979	.02066	48.412	.0002	48.422	49
12	.00349	.99999	.00349	286.48	.0000	286.48	.02094	.99978	.02095	47.739	.0002	47.750	48
13	.00378	.99999	.00378	264.44	.0000	264.44	.02123	.99977	.02124	47.085	.0002	47.096	47
14	.00407	.99999	.00407	245.55	.0000	245.55	.02152	.99977	.02153	46.449	.0002	46.460	46
15	.00436	.99999	.00436	229.18	.0000	229.18	.02181	.99976	.02182	45.829	.0002	45.840	45
16	.00465	.99999	.00465	214.86	.0000	214.86	.02210	.99975	.02211	45.226	.0002	45.237	44
17	.00494	.99999	.00494	202.22	.0000	202.22	.02240	.99975	.02240	44.638	.0002	44.650	43
18	.00524	.99999	.00524	190.98	.0000	190.99	.02269	.99974	.02269	44.066	.0002	44.077	42
19	.00553	.99998	.00553	180.93	.0000	180.93	.02298	.99974	.02298	43.508	.0003	43.520	41
20	.00582	.99998	.00582	171.88	.0000	171.89	.02326	.99973	.02327	42.964	.0003	42.976	40
21	.00611	.99998	.00611	163.70	.0000	163.70	.02356	.99972	.02357	42.433	.0003	42.445	39
22	.00640	.99998	.00640	156.26	.0000	156.26	.02385	.99971	.02386	41.916	.0003	41.928	38
23	.00669	.99998	.00669	149.46	.0000	149.47	.02414	.99971	.02415	41.410	.0003	41.423	37
24	.00698	.99997	.00698	143.24	.0000	143.24	.02443	.99970	.02444	40.917	.0003	40.930	36
25	.00727	.99997	.00727	137.51	.0000	137.51	.02472	.99969	.02473	40.436	.0003	40.448	35
26	.00756	.99997	.00756	132.22	.0000	132.22	.02501	.99969	.02502	39.965	.0003	39.978	34
27	.00785	.99997	.00785	127.32	.0000	127.32	.02530	.99968	.02531	39.506	.0003	39.518	33
28	.00814	.99997	.00814	122.77	.0000	122.78	.02559	.99967	.02560	39.057	.0003	39.069	32
29	.00843	.99996	.00844	118.54	.0000	118.54	.02589	.99966	.02589	38.618	.0003	38.631	31
30	.00873	.99996	.00873	114.59	.0000	114.59	.02618	.99966	.02618	38.188	.0003	38.201	30
31	.00902	.99996	.00902	110.89	.0000	110.90	.02647	.99965	.02648	37.769	.0003	37.782	29
32	.00931	.99996	.00931	107.43	.0000	107.43	.02676	.99964	.02677	37.358	.0003	37.371	28
33	.00960	.99995	.00960	104.17	.0000	104.17	.02705	.99963	.02706	36.956	.0004	36.969	27
34	.00989	.99995	.00989	101.11	.0000	101.11	.02734	.99963	.02735	36.563	.0004	36.576	26
35	.01018	.99995	.01018	98.218	.0000	98.223	.02763	.99962	.02764	36.177	.0004	36.191	25
36	.01047	.99994	.01047	95.489	.0000	95.495	.02792	.99961	.02793	35.800	.0004	35.814	24
37	.01076	.99994	.01076	92.908	.0000	92.914	.02821	.99960	.02822	35.431	.0004	35.445	23
38	.01105	.99994	.01105	90.463	.0001	90.469	.02850	.99959	.02851	35.069	.0004	35.084	22
39	.01134	.99993	.01134	88.143	.0001	88.149	.02879	.99958	.02880	34.715	.0004	34.729	21
40	.01163	.99993	.01164	85.940	.0001	85.946	.02908	.99958	.02910	34.368	.0004	34.382	20
41	.01193	.99993	.01193	83.843	.0001	83.849	.02937	.99957	.02939	34.027	.0004	34.042	19
42	.01222	.99992	.01222	81.847	.0001	81.853	.02967	.99956	.02968	33.693	.0004	33.708	18
43	.01251	.99992	.01251	79.943	.0001	79.950	.02996	.99955	.02997	33.366	.0004	33.381	17
44	.01280	.99992	.01280	78.126	.0001	78.133	.03025	.99954	.03026	33.045	.0004	33.060	16
45	.01309	.99991	.01309	76.390	.0001	76.396	.03054	.99953	.03055	32.730	.0005	32.745	15
46	.01338	.99991	.01338	74.729	.0001	74.736	.03083	.99952	.03084	32.421	.0005	32.437	14
47	.01367	.99991	.01367	73.139	.0001	73.146	.03112	.99951	.03113	32.118	.0005	32.134	13
48	.01396	.99990	.01396	71.615	.0001	71.622	.03141	.99951	.03143	31.820	.0005	31.836	12
49	.01425	.99990	.01425	70.153	.0001	70.160	.03170	.99950	.03172	31.528	.0005	31.544	11
50	.01454	.99989	.01454	68.750	.0001	68.757	.03199	.99949	.03201	31.241	.0005	31.257	10
51	.01483	.99989	.01484	67.402	.0001	67.409	.03228	.99948	.03230	30.960	.0005	30.976	9
52	.01512	.99988	.01513	66.105	.0001	66.113	.03257	.99947	.03259	30.683	.0005	30.699	8
53	.01542	.99988	.01542	64.858	.0001	64.866	.03286	.99946	.03288	30.411	.0005	30.428	7
54	.01571	.99988	.01571	63.657	.0001	63.664	.03315	.99945	.03317	30.145	.0005	30.161	6
55	.01600	.99987	.01600	62.499	.0001	62.507	.03344	.99944	.03346	29.882	.0005	29.899	5
56	.01629	.99987	.01629	61.383	.0001	61.391	.03374	.99943	.03375	29.624	.0006	29.641	4
57	.01658	.99987	.01658	60.306	.0001	60.314	.03403	.99942	.03405	29.371	.0006	29.388	3
58	.01687	.99986	.01687	59.266	.0001	59.274	.03432	.99941	.03434	29.122	.0006	29.139	2
59	.01716	.99985	.01716	58.261	.0001	58.270	.03461	.99940	.03463	28.877	.0006	28.894	1
60	.01745	.99985	.01745	57.290	.0001	57.299	.03490	.99939	.03492	28.636	.0006	28.654	0
′	cos	sin	cot	tan	cosec	sec	cos	sin	cot	tan	cosec	sec	′
	89°						88°						

* Courtesy of Bethlehem Steel Company.

	2°						**3°**						
′	sin	cos	tan	cot	sec	cosec	sin	cos	tan	cot	sec	cosec	′
0	.03490	.99939	.03492	28.636	1.0006	28.654	.05234	.99863	.05241	19.081	1.0014	19.107	**60**
1	.03519	.99938	.03521	28.399	.0006	28.417	.05263	.99861	.05270	18.975	.0014	19.002	**59**
2	.03548	.99937	.03550	28.166	.0006	28.184	.05292	.99860	.05299	18.871	.0014	18.897	**58**
3	.03577	.99936	.03579	27.937	.0006	27.955	.05321	.99858	.05328	18.768	.0014	18.794	**57**
4	.03606	.99935	.03608	27.712	.0006	27.730	.05350	.99857	.05357	18.665	.0014	18.692	**56**
5	.03635	.99934	.03638	27.490	.0007	27.508	.05379	.99855	.05387	18.564	.0014	18.591	**55**
6	.03664	.99933	.03667	27.271	.0007	27.290	.05408	.99854	.05416	18.464	.0015	18.491	**54**
7	.03693	.99932	.03696	27.056	.0007	27.075	.05437	.99852	.05445	18.365	.0015	18.393	**53**
8	.03722	.99931	.03725	26.845	.0007	26.864	.05466	.99850	.05474	18.268	.0015	18.295	**52**
9	.03751	.99930	.03754	26.637	.0007	26.655	.05495	.99849	.05503	18.171	.0015	18.198	**51**
10	.03781	.99928	.03783	26.432	.0007	26.450	.05524	.99847	.05532	18.075	.0015	18.103	**50**
11	.03810	.99927	.03812	26.230	.0007	26.249	.05553	.99846	.05562	17.980	.0015	18.008	**49**
12	.03839	.99926	.03842	26.031	.0007	26.050	.05582	.99844	.05591	17.886	.0016	17.914	**48**
13	.03868	.99925	.03871	25.835	.0007	25.854	.05611	.99842	.05620	17.793	.0016	17.821	**47**
14	.03897	.99924	.03900	25.642	.0008	25.661	.05640	.99841	.05649	17.701	.0016	17.730	**46**
15	.03926	.99923	.03929	25.452	.0008	25.471	.05669	.99839	.05678	17.610	.0016	17.639	**45**
16	.03955	.99922	.03958	25.264	.0008	25.284	.05698	.99837	.05707	17.520	.0016	17.549	**44**
17	.03984	.99921	.03987	25.080	.0008	25.100	.05727	.99836	.05737	17.431	.0016	17.460	**43**
18	.04013	.99919	.04016	24.898	.0008	24.918	.05756	.99834	.05766	17.343	.0017	17.372	**42**
19	.04042	.99918	.04045	24.718	.0008	24.739	.05785	.99832	.05795	17.256	.0017	17.285	**41**
20	.04071	.99917	.04075	24.542	.0008	24.562	.05814	.99831	.05824	17.169	.0017	17.198	**40**
21	.04100	.99916	.04104	24.367	.0008	24.388	.05843	.99829	.05853	17.084	.0017	17.113	**39**
22	.04129	.99915	.04133	24.196	.0008	24.216	.05872	.99827	.05883	16.999	.0017	17.028	**38**
23	.04158	.99913	.04162	24.026	.0009	24.047	.05902	.99826	.05912	16.915	.0017	16.944	**37**
24	.04187	.99912	.04191	23.859	.0009	23.880	.05931	.99824	.05941	16.832	.0018	16.861	**36**
25	.04217	.99911	.04220	23.694	.0009	23.716	.05960	.99822	.05970	16.750	.0018	16.779	**35**
26	.04246	.99910	.04249	23.532	.0009	23.553	.05989	.99820	.05999	16.668	.0018	16.698	**34**
27	.04275	.99908	.04279	23.372	.0009	23.393	.06018	.99819	.06029	16.587	.0018	16.617	**33**
28	.04304	.99907	.04308	23.214	.0009	23.235	.06047	.99817	.06058	16.507	.0018	16.538	**32**
29	.04333	.99906	.04337	23.058	.0009	23.079	.06076	.99815	.06087	16.428	.0018	16.459	**31**
30	.04362	.99905	.04366	22.904	.0009	22.925	.06105	.99813	.06116	16.350	.0019	16.380	**30**
31	.04391	.99903	.04395	22.752	.0010	22.774	.06134	.99812	.06145	16.272	.0019	16.303	**29**
32	.04420	.99902	.04424	22.602	.0010	22.624	.06163	.99810	.06175	16.195	.0019	16.226	**28**
33	.04449	.99901	.04453	22.454	.0010	22.476	.06192	.99808	.06204	16.119	.0019	16.150	**27**
34	.04478	.99900	.04483	22.308	.0010	22.330	.06221	.99806	.06233	16.043	.0019	16.075	**26**
35	.04507	.99898	.04512	22.164	.0010	22.186	.06250	.99804	.06262	15.969	.0019	16.000	**25**
36	.04536	.99897	.04541	22.022	.0010	22.044	.06279	.99803	.06291	15.894	.0020	15.926	**24**
37	.04565	.99896	.04570	21.881	.0010	21.904	.06308	.99801	.06321	15.821	.0020	15.853	**23**
38	.04594	.99894	.04599	21.742	.0010	21.765	.06337	.99799	.06350	15.748	.0020	15.780	**22**
39	.04623	.99893	.04628	21.606	.0011	21.629	.06366	.99797	.06379	15.676	.0020	15.708	**21**
40	.04652	.99892	.04657	21.470	.0011	21.494	.06395	.99795	.06408	15.605	.0020	15.637	**20**
41	.04681	.99890	.04687	21.337	.0011	21.360	.06424	.99793	.06437	15.534	.0021	15.566	**19**
42	.04711	.99889	.04716	21.205	.0011	21.228	.06453	.99791	.06467	15.464	.0021	15.496	**18**
43	.04740	.99888	.04745	21.075	.0011	21.098	.06482	.99790	.06496	15.394	.0021	15.427	**17**
44	.04769	.99886	.94774	20.946	.0011	20.970	.06511	.99788	.06525	15.325	.0021	15.358	**16**
45	.04798	.99885	.04803	20.819	.0011	20.843	.06540	.99786	.06554	15.257	.0021	15.290	**15**
46	.04827	.99883	.04832	20.693	.0012	20.717	.06569	.99784	.06583	15.189	.0022	15.222	**14**
47	.04856	.99882	.04862	20.569	.0012	20.593	.06598	.99782	.06613	15.122	.0022	15.155	**13**
48	.04885	.99881	.04891	20.446	.0012	20.471	.06627	.99780	.06642	15.056	.0022	15.089	**12**
49	.04914	.99879	.04920	20.325	.0012	20.350	.06656	.99778	.06671	14.990	.0022	15.023	**11**
50	.04943	.99878	.04949	20.205	.0012	20.230	.06685	.99776	.06700	14.924	.0022	14.958	**10**
51	.04972	.99876	.04978	20.087	.0012	20.112	.06714	.99774	.06730	14.860	.0023	14.893	**9**
52	.05001	.99875	.05007	19.970	.0012	19.995	.06743	.99772	.06759	14.795	.0023	14.829	**8**
53	.05030	.99873	.05037	19.854	.0013	19.880	.06772	.99770	.06788	14.732	.0023	14.765	**7**
54	.05059	.99872	.05066	19.740	.0013	19.766	.06801	.99768	.06817	14.668	.0023	14.702	**6**
55	.05088	.99870	.05095	19.627	.0013	19.653	.06830	.99766	.06846	14.606	.0023	14.640	**5**
56	.05117	.99869	.05124	19.515	.0013	19.541	.06859	.99764	.06876	14.544	.0024	14.578	**4**
57	.05146	.99867	.05153	19.405	.0013	19.431	.06888	.99762	.06905	14.482	.0024	14.517	**3**
58	.05175	.99866	.05182	19.296	.0013	19.322	.06918	.99760	.06934	14.421	.0024	14.456	**2**
59	.05204	.99864	.05212	19.188	.0013	19.214	.06947	.99758	.06963	14.361	.0024	14.395	**1**
60	.05234	.99863	.05241	19.081	.0014	19.107	.06976	.99756	06993	14.301	.0024	14.335	**0**
′	cos	sin	cot	tan	cosec	sec	cos	sin	cot	tan	cosec	sec	′
	87°						**86°**						

	4°						5°						
′	sin	cos	tan	cot	sec	cosec	sin	cos	tan	cot	sec	cosec	′
0	.06976	.99756	.06993	14.301	1.0024	14.335	.08715	.99619	.08749	11.430	1.0038	11.474	60
1	.07005	.99754	.07022	14.241	.0025	14.276	.08744	.99617	.08778	11.392	.0038	11.436	59
2	.07034	.99752	.07051	14.182	.0025	14.217	.08773	.99614	.08807	11.354	.0039	11.398	58
3	.07063	.99750	.07080	14.123	.0025	14.159	.08802	.99612	.08837	11.316	.0039	11.360	57
4	.07092	.99748	.07110	14.065	.0025	14.101	.08831	.99609	.08866	11.279	.0039	11.323	56
5	.07121	.99746	.07139	14.008	.0025	14.043	.08860	.99607	.08895	11.242	.0039	11.286	55
6	.07150	.99744	.07168	13.951	.0026	13.986	.08889	.99604	.08925	11.205	.0040	11.249	54
7	.07179	.99742	.07197	13.894	.0026	13.930	.08918	.99601	.08954	11.168	.0040	11.213	53
8	.07208	.99740	.07226	13.838	.0026	13.874	.08947	.99599	.08983	11.132	.0040	11.176	52
9	.07237	.99738	.07256	13.782	.0026	13.818	.08976	.99596	.09013	11,095	.0040	11.140	51
10	.07266	.99736	.07285	13.727	.0026	13.763	.09005	.99594	.09042	11.059	.0041	11.104	50
11	.07295	.99733	.07314	13.672	.0027	13.708	.09034	.99591	.09071	11.024	.0041	11.069	49
12	.07324	.99731	.07343	13.617	.0027	13.654	.09063	.99588	.09101	10.988	.0041	11.033	48
13	.07353	.99729	.07373	13.563	.0027	13.600	.09092	.99586	.09130	10.953	.0041	10.998	47
14	.07382	.99727	.07402	13.510	.0027	13.547	.09121	.99583	.09159	10.918	.0042	10.963	46
15	.07411	.99725	.07431	13.457	.0027	13.494	.09150	.99580	.09189	10.883	.0042	10.929	45
16	.07440	.99723	.07460	13.404	.0028	13.441	.09179	.99578	.09218	10.848	.0042	10.894	44
17	.07469	.99721	.07490	13.351	.0028	13.389	.09208	.99575	.09247	10.814	.0043	10.860	43
18	.07498	.99718	.07519	13.299	.0028	13.337	.09237	.99572	.09277	10.780	.0043	10.826	42
19	.07527	.99716	.07548	13.248	.0028	13.286	.09266	.99570	.09306	10.746	.0043	10.792	41
20	.07556	.99714	.07577	13.917	.0029	13.235	.09295	.99567	.09335	10.712	.0043	10.758	40
21	.07585	.99712	.07607	13.146	.0029	13.184	.09324	.99564	.09365	10.678	.0044	10.725	39
22	.07614	.99710	.07636	13.096	.0029	13.134	.09353	.99562	.09394	10.645	.0044	10.692	38
23	.07643	.99707	.07665	13.046	.0029	13.084	.09382	.99559	.09423	10.612	.0044	10.659	37
24	.07672	.99705	.07694	12.996	.0029	13.034	.09411	.99556	.09453	10.579	.0044	10.626	36
25	.07701	.99703	.07724	12.947	.0030	12.985	.09440	.99553	.09482	10.546	.0045	10.593	35
26	.07730	.99701	.07753	12.898	.0030	12.937	.09469	.99551	.09511	10.514	.0045	10.561	34
27	.07759	.99698	.07782	12.849	.0030	12.888	.09498	.99548	.09541	10.481	.0045	10.529	33
28	.07788	.99696	.07812	12.801	.0030	12.840	.09527	.99545	.09570	10.449	.0046	10.497	32
29	.07817	.99694	.07841	12.754	.0031	12.793	.09556	.99542	.09599	10.417	.0046	10.465	31
30	.07846	.99692	.07870	12.706	.0031	12.745	.09584	.99540	.09629	10.385	.0046	10.433	30
31	.07875	.99689	.07899	12.659	.0031	12.698	.09613	.99537	.09658	10.354	.0046	10.402	29
32	.07904	.99687	.07929	12.612	.0031	12.652	.09642	.99534	.09688	10.322	.0047	10.371	28
33	.07933	.99685	.07958	12.566	.0032	12.606	.09671	.99531	.09717	10.291	.0047	10.340	27
34	.07962	.99682	.07987	12.520	.0032	12.560	.09700	.99528	.09746	10.260	.0047	10.309	26
35	.07991	.99680	.08016	12.474	.0032	12.514	.09729	.99525	.09776	10.229	.0048	10.278	25
36	.08020	.99678	.08046	12.429	.0032	12.469	.09758	.99523	.09805	10.199	.0048	10.248	24
37	.08049	.99675	.08075	12.384	.0032	12.424	.09787	.99520	.09834	10.168	.0048	10.217	23
38	.08078	.99673	.08104	12.339	.0033	12.379	.09816	.99517	.09864	10.138	.0048	10.187	22
39	.08107	.99671	.08134	12.295	.0033	12.335	.09845	.99514	.09893	10.108	.0049	10.157	21
40	.08136	.99668	.08163	12.250	.0033	12.291	.09874	.99511	.09922	10.078	.0049	10.127	20
41	.08165	.99666	.08192	12.207	.0033	12.248	.09903	.99508	.09952	10.048	.0049	10.098	19
42	.08194	.99664	.08221	12.163	.0034	12.204	.09932	.99505	.09981	10.019	.0050	10.068	18
43	.08223	.99661	.08251	12.120	.0034	12.161	.09961	.99503	.10011	9.9893	.0050	10.039	17
44	.08252	.99659	.08280	12.077	.0034	12.118	.09990	.99500	.10040	9.9601	.0050	10.010	16
45	.08281	.99656	.08309	12.035	.0034	12.076	.10019	.99497	.10069	9.9310	.0050	9.9812	15
46	.08310	.99654	.08339	11.992	.0035	12.034	.10048	.99494	.10099	9.9021	.0051	9.9525	14
47	.08339	.99652	.08368	11.950	.0035	11.992	.10077	.99491	.10128	9.8734	.0051	9.9239	13
48	.08368	.99649	.08397	11.909	.0035	11.950	.10106	.99488	.10158	9.8448	.0051	9.8955	12
49	.08397	.99647	.08426	11.867	.0035	11.909	.10134	.99485	.10187	9.8164	.0052	9.8672	11
50	.08426	.99644	.08456	11.826	.0036	11.868	.10163	.99482	.10216	9.7882	.0052	9.8391	10
51	.08455	.99642	.08485	11.785	.0036	11.828	.10192	.99479	.10246	9.7601	.0052	9.8112	9
52	.08484	.99639	.08514	11.745	.0036	11.787	.10221	.99476	.10275	9.7322	.0053	9.7834	8
53	.08513	.99637	.08544	11.704	.0036	11.747	.10250	.99473	.10305	9.7044	.0053	9.7558	7
54	.08542	.99634	.08573	11.664	.0037	11.707	.10279	.99470	.10334	9.6768	.0053	9.7283	6
55	.08571	.99632	.08602	11.625	.0037	11.668	.10308	.99467	.10363	9.6493	.0053	9.7010	5
56	.08600	.99629	.08632	11.585	.0037	11.628	.10337	.99464	.10393	9.6220	.0054	9.6739	4
57	.08629	.99627	.08661	11.546	.0037	11.589	.10366	.99461	.10422	9.5949	.0054	9.6469	3
58	.08658	.99624	.08690	11.507	.0038	11.550	.10395	.99458	.10452	9.5679	.0054	9.6200	2
59	.08687	.99622	.08719	11.468	.0038	11.512	.10424	.99455	.10481	9.5411	.0055	9.5933	1
60	.08715	.99619	.08749	11.430	.0038	11.474	.10453	.99452	.10510	9.5144	.0055	9.5668	0
′	cos	sin	cot	tan	cosec	sec	cos	sin	cot	tan	cosec	sec	′
	85°						84°						

	6°						7°						
′	sin	cos	tan	cot	sec	cosec	sin	cos	tan	cot	sec	cosec	′
0	.10453	.99452	.10510	9.5144	1.0055	9.5668	.12187	.99255	.12278	8.1443	1.0075	8.2055	60
1	.10482	.99449	.10540	.4878	.0055	.5404	.12216	.99251	.12308	.1248	.0075	.1861	59
2	.10511	.99446	.10569	.4614	.0056	.5141	.12245	.99247	.12337	.1053	.0076	.1668	58
3	.10540	.99443	.10599	.4351	.0056	.4880	.12273	.99244	.12367	.0860	.0076	.1476	57
4	.10568	.99440	.10628	.4090	.0056	.4620	.12302	.99240	.12396	.0667	.0076	.1285	56
5	.10597	.99437	.10657	9.3831	.0057	9.4362	.12331	.99237	.12426	8.0476	.0077	8.1094	55
6	.10626	.99434	.10687	.3572	.0057	.4105	.12360	.99233	.12456	.0285	.0077	.0905	54
7	.10655	.99431	.10716	.3315	.0057	.3850	.12389	.99229	.12485	.0095	.0078	.0717	53
8	.10684	.99428	.10746	.3060	.0057	.3596	.12418	.99226	.12515	7.9906	.0078	.0529	52
9	.10713	.99424	.10775	.2806	.0058	.3343	.12447	.99222	.12544	.9717	.0078	.0342	51
10	.10742	.99421	.10805	9.2553	.0058	9.3092	.12476	.99219	.12574	7.9530	.0079	8.0156	50
11	.10771	.99418	.10834	.2302	.0058	.2842	.12504	.99215	.12603	.9344	.0079	7.9971	49
12	.10800	.99415	.10863	.2051	.0059	.2593	.12533	.99211	.12633	.9158	.0079	.9787	48
13	.10829	.99412	.10893	.1803	.0059	.2346	.12562	.99208	.12662	.8973	.0080	.9604	47
14	.10858	.99409	.10922	.1555	.0059	.2100	.12591	.99204	.12692	.8789	.0080	.9421	46
15	.10887	.99406	.10952	9.1309	.0060	9.1855	.12620	.99200	.12722	7.8606	.0080	7.9240	45
16	.10916	.99402	.10981	.1064	.0060	.1612	.12649	.99197	.12751	.8424	.0081	.9059	44
17	.10944	.99399	.11011	.0821	.0060	.1370	.12678	.99193	.12781	.8243	.0081	.8879	43
18	.10973	.99396	.11040	.0579	.0061	.1129	.12706	.99189	.12810	.8062	.0082	.8700	42
19	.11002	.99393	.11069	.0338	.0061	.0890	.12735	.99186	.12840	.7882	.0082	.8522	41
20	.11031	.99390	.11099	9.0098	.0061	9.0651	.12764	.99182	.12869	7.7703	.0082	7.8344	40
21	.11060	.99386	.11128	8.9860	.0062	.0414	.12793	.99178	.12899	.7525	.0083	.8168	39
22	.11089	.99383	.11158	.9623	.0062	.0179	.12822	.99174	.12928	.7348	.0083	.7992	38
23	.11118	.99380	.11187	.9387	.0062	8.9944	.12851	.99171	.12958	.7171	.0084	.7817	37
24	.11147	.99377	.11217	.9152	.0063	.9711	.12879	.99167	.12988	.6996	.0084	.7642	36
25	.11176	.99373	.11246	8.8918	.0063	8.9479	.12908	.99163	.13017	7.6821	.0084	7.7469	35
26	.11205	.99370	.11276	.8686	.0063	.9248	.12937	.99160	.13047	.6646	.0085	.7296	34
27	.11234	.99367	.11305	.8455	.0064	.9018	.12966	.99156	.13076	.6473	.0085	.7124	33
28	.11262	.99364	.11335	.8225	.0064	.8790	.12995	.99152	.13106	.6300	.0085	.6953	32
29	.11291	.99360	.11364	.7996	.0064	.8563	.13024	.99148	.13136	.6129	.0086	.6783	31
30	.11320	.99357	.11393	8.7769	.0065	8.8337	.13053	.99144	.13165	7.5957	.0086	7.6613	30
31	.11349	.99354	.11423	.7542	.0065	.8112	.13081	.99141	.13195	.5787	.0087	.6444	29
32	.11378	.99350	.11452	.7317	.0065	.7888	.13110	.99137	.13224	.5617	.0087	.6276	28
33	.11407	.99347	.11482	.7093	.0066	.7665	.13139	.99133	.13254	.5449	.0087	.6108	27
34	.11436	.99344	.11511	.6870	.0066	.7444	.13168	.99129	.13284	.5280	.0088	.5942	26
35	.11465	.99341	.11541	8.6648	.0066	8.7223	.13197	.99125	.13313	7.5113	.0088	7.5776	25
36	.11494	.99337	.11570	.6427	.0067	.7004	.13226	.99121	.13343	.4946	.0089	.5611	24
37	.11523	.99334	.11600	.6208	.0067	.6786	.13254	.99118	.13372	.4780	.0089	.5446	23
38	.11551	.99330	.11629	.5989	.0067	.6569	.13283	.99114	.13402	.4615	.0089	.5282	22
39	.11580	.99327	.11659	.5772	.0068	.6353	.13312	.99110	.13432	.4451	.0090	.5119	21
40	.11609	.99324	.11688	8.5555	.0068	8.6138	.13341	.99106	.13461	7.4287	.0090	7.4957	20
41	.11638	.99320	.11718	.5340	.0068	.5924	.13370	.99102	.13491	.4124	.0090	.4795	19
42	.11667	.99317	.11747	.5126	.0069	.5711	.13399	.99098	.13520	.3961	.0091	.4634	18
43	.11696	.99314	.11777	.4913	.0069	.5499	.13427	.99094	.13550	.3800	.0091	.4474	17
44	.11725	.99310	.11806	.4701	.0069	.5289	.13456	.99090	.13580	.3639	.0092	.4315	16
45	.11754	.99307	.11836	8.4489	.0070	8.5079	.13485	.99086	.13609	7.3479	.0092	7.4156	15
46	.11783	.99303	.11865	.4279	.0070	.4871	.13514	.99083	.13639	.3319	.0092	.3998	14
47	.11811	.99300	.11895	.4070	.0070	.4663	.13543	.99079	.13669	.3160	.0093	.3840	13
48	.11840	.99296	.11924	.3862	.0071	.4457	.13571	.99075	.13698	.3002	.0093	.3683	12
49	.11869	.99293	.11954	.3655	.0071	.4251	.13600	.99071	.13728	.2844	.0094	.3527	11
50	.11898	.99290	.11983	8.3449	.0071	8.4046	.13629	.99067	.13757	7.2687	.0094	7.3372	10
51	.11927	.99286	.12013	.3244	.0072	.3843	.13658	.99063	.13787	.2531	.0094	.3217	9
52	.11956	.99283	.12042	.3040	.0072	.3640	.13687	.99059	.13817	.2375	.0095	.3063	8
53	.11985	.99279	.12072	.2837	.0073	.3439	.13716	.99055	.13846	.2220	.0095	.2909	7
54	.12014	.99276	.12101	.2635	.0073	.3238	.13744	.99051	.13876	.2066	.0096	.2757	6
55	.12042	.99272	.12131	8.2434	.0073	8.3039	.13773	.99047	.13906	7.1912	.0096	7.2604	5
56	.12071	.99269	.12160	.2234	.0074	.2840	.13802	.99043	.13935	.1759	.0097	.2453	4
57	.12100	.99265	.12190	.2035	.0074	.2642	.13831	.99039	.13965	.1607	.0097	.2302	3
58	.12129	.99262	.12219	.1837	.0074	.2446	.13860	.99035	.13995	.1455	.0097	.2152	2
59	.12158	.99258	.12249	.1640	.0075	.2250	.13888	.99031	.14024	.1304	.0098	.2002	1
60	.12187	.99255	.12278	8.1443	.0075	8.2055	.13917	.99027	.14054	7.1154	.0098	7.1853	0
′	cos	sin	cot	tan	cosec	sec	cos	sin	cot	tan	cosec	sec	′
	83°						82°						

	8°						9°						
′	sin	cos	tan	cot	sec	cosec	sin	cos	tan	cot	sec	cosec	′
0	.13917	.99027	.14054	7.1154	1.0098	7.1853	.15643	.98769	.15838	6.3137	1.0215	6.3924	60
1	.13946	.99023	.14084	.1004	.0099	.1704	.15672	.98764	.15868	.3019	.0125	.3807	59
2	.13975	.99019	.14113	.0854	.0099	.1557	.15701	.98760	.15898	.2901	.0125	.3690	58
3	.14004	.99015	.14143	.0706	.0099	.1409	.15730	.98755	.15928	.2783	.0126	.3574	57
4	.14032	.99010	.14173	.0558	.0100	.1263	.15758	.98750	.15958	.2665	.0126	.3458	56
5	.14061	.99006	.14202	7.0410	.0100	7.1117	.15787	.98746	.15987	6.2548	.0127	6.3343	55
6	.14090	.99002	.14232	.0264	.0101	.0972	.15816	.98741	.16017	.2432	.0217	.3228	54
7	.14119	.98998	.14262	.0117	.0101	.0827	.15844	.98737	.16047	.2316	.0128	.3113	53
8	.14148	.98994	.14291	6.9972	.0102	.0683	.15873	.98732	.16077	.2200	.0128	.2999	52
9	.14176	.98990	.14321	.9827	.0102	.0539	.15902	.98727	.16107	.2085	.0129	.2885	51
10	.14205	.98986	.14351	6.9682	.0102	7.0396	.15931	.98723	.16137	6.1970	.0129	6.2772	50
11	.14234	.98982	.14380	.9538	.0103	.0254	.15959	.98718	.16167	.1856	.0130	.2659	49
12	.14263	.98978	.14410	.9395	.0103	.0112	.15988	.98714	.16196	.1742	.0130	.2546	48
13	.14292	.98973	.14440	.9252	.0104	6.9971	.16017	.98709	.16226	.1628	.0131	.2434	47
14	.14320	.98969	.14470	.9110	.0104	.9830	.16045	.98704	.16256	.1515	.0131	.2322	46
15	.14349	.98965	.14499	6.8969	.0104	6.9690	.16074	.98700	.16286	6.1402	.0132	6.2211	45
16	.14378	.98961	.14529	.8828	.0105	.9550	.16103	.98695	.16316	.1290	.0132	.2100	44
17	.14407	.98957	.14559	.8687	.0105	.9411	.16132	.98690	.16346	.1178	.0133	.1990	43
18	.14436	.98952	.14588	.8547	.0106	.9273	.16160	.98685	.16376	.1066	.0133	.1880	42
19	.14464	.98948	.14618	.8408	.0106	.9135	.16189	.98681	.16405	.0955	.0134	.1770	41
20	.14493	.98944	.14648	6.8269	.0107	6.8998	.16218	.98676	.16435	6.0844	.0134	6.1661	40
21	.14522	.98940	.14677	.8131	.0107	.8861	.16246	.98671	.16465	.0734	.0135	.1552	39
22	.14551	.98936	.14707	.7993	.0107	.8725	.16275	.98667	.16495	.0624	.0135	.1443	38
23	.14579	.98931	.14737	.7856	.0108	.8589	.16304	.98662	.16525	.0514	.0136	.1335	37
24	.14608	.98927	.14767	.7720	.0108	.8454	.16333	.98657	.16555	.0405	.0136	.1227	36
25	.14637	.98923	.14796	6.7584	.0109	6.8320	.16361	.98652	.16585	6.0296	.0136	6.1120	35
26	.14666	.98919	.14826	.7448	.0109	.8185	.16390	.98648	.16615	.0188	.0137	.1013	34
27	.14695	.98914	.14856	.7313	.0110	.8052	.16419	.98643	.16644	.0080	.0137	.0906	33
28	.14723	.98910	.14886	.7179	.0110	.7919	.16447	.98638	.16674	5.9972	.0138	.0800	32
29	.14752	.98906	.14915	.7045	.0111	.7787	.16476	.98633	.16704	.9865	.0138	.0694	31
30	.14781	.98901	.14945	6.6911	.0111	6.7655	.16505	.98628	.16734	5.9758	.0139	6.0588	30
31	.14810	.98897	.14975	.6779	.0111	.7523	.16533	.98624	.16764	.9651	.9139	.0483	29
32	.14838	.98893	.15004	.6646	.0112	.7392	.16562	.98619	.16794	.9545	.9140	.0379	28
33	.14867	.98889	.15034	.6514	.0112	.7262	.16591	.98614	.16824	.9439	.0140	.0274	27
34	.14896	.98884	.15064	.6383	.0113	.7132	.16619	.98609	.16854	.9333	.0141	.0170	26
35	.14925	.98880	.15094	6.6252	.0113	6.7003	.16648	.98604	.16884	5.9228	.0141	6.0066	25
36	.14953	.98876	.15123	.6122	.0114	.6874	.16677	.98600	.16914	.9123	.0142	5.9963	24
37	.14982	.98871	.15153	.5992	.0114	.6745	.16705	.98595	.16944	.9019	.0142	.9860	23
38	.15011	.98867	.15183	.5863	.0115	.6617	.16734	.98590	.16973	.8915	.0143	.9758	22
39	.15050	.98862	.15213	.5734	.0115	.6490	.16763	.98585	.17003	.8811	.0143	.9655	21
40	.15068	.98858	.15243	6.5605	.0115	6.6363	.16791	.98580	.17033	5.8708	.0144	5.9554	20
41	.15097	.98854	.15272	.5478	.0116	.6237	.16820	.98575	.17063	.8605	.0144	.9452	19
42	.15126	.98849	.15302	.5350	.0116	.6111	.16849	.98570	.17093	.8502	.0145	.9351	18
43	.15155	.98845	.15332	.5223	.0117	.5985	.16878	.98565	.17123	.8400	.0145	.9250	17
44	.15183	.98840	.15362	.5097	.0117	.5860	.16906	.98560	.17153	.8298	.0146	.9150	16
45	.15212	.98836	.15391	6.4971	.0118	6.5736	.16935	.98556	.17183	5.8196	.0146	5.9049	15
46	.15241	.98832	.15421	.4845	.0118	.5612	.16964	.98551	.17213	.8095	.0147	.8950	14
47	.15270	.98827	.15451	.4720	.0119	.5488	.16992	.98546	.17243	.7994	.0147	.8850	13
48	.15298	.98823	.15481	.4596	.0119	.5365	.17021	.98541	.17273	.7894	.0148	.8751	12
49	.15328	.98818	.15511	.4472	.0119	.5243	.17050	.98536	.17303	.7794	.0148	.8652	11
50	.15356	.98814	.15540	6.4348	.0120	6.5121	.17078	.98531	.17333	5.7694	.0149	5.8554	10
51	.15385	.98809	.15570	.4225	.0120	.4999	.17107	.98526	.17363	.7594	.0150	.8456	9
52	.15413	.98805	.15600	.4103	.0121	.4878	.17136	.98521	.17393	.7495	.0150	.8358	8
53	.15442	.98800	.15630	.3980	.0121	.4757	.17164	.98516	.17423	.7396	.0151	.8261	7
54	.15471	.98796	.15659	.3859	.0122	.4637	.17193	.98511	.17453	.7297	.0151	.8163	6
55	.15500	.98791	.15689	6.3737	.0122	6.4517	.17221	.98506	.17483	5.7199	.0152	5.8067	5
56	.15528	.98787	.15719	.3616	.0123	.4398	.17250	.98501	.17513	.7101	.0152	.7970	4
57	.15557	.98782	.15749	.3496	.0123	.4279	.17279	.98496	.17543	.7004	.0153	.7874	3
58	.15586	.98778	.15779	.3376	.0124	.4160	.17307	.98491	.17573	.6906	.0153	.7778	2
59	.15615	.98773	.15809	.3257	.0124	.4042	.17336	.98486	.17603	.6809	.0154	.7683	1
60	.15643	.98769	.15838	6.3137	.0125	6.3924	.17365	.98481	.17633	5.6713	.0154	5.7588	0
′	cos	sin	cot	tan	cosec	sec	cos	sin	cot	tan	cosec	sec	′
	81°						80°						

	10°						11°						
′	sin	cos	tan	cot	sec	cosec	sin	cos	tan	cot	sec	cosec	′
0	.17365	.98481	.17633	5.6713	1.0154	5.7588	.19081	.98163	.19438	5.1445	1.0187	5.2408	60
1	.17393	.98476	.17663	.6616	.0155	.7493	.19109	.98157	.19468	.1366	.0188	.2330	59
2	.17422	.98471	.17693	.6520	.0155	.7398	.19138	.98152	.19498	.1286	.0188	.2252	58
3	.17451	.98465	.17723	.6425	.0156	.7304	.19166	.98146	.19529	.1207	.0189	.2174	57
4	.17479	.98460	.17753	.6329	.0156	.7210	.19195	.98140	.19559	.1128	.0189	.2097	56
5	.17508	.98455	.17783	5.6234	.0157	5.7117	.19224	.98135	.19589	5.1049	.0190	5.2019	55
6	.17537	.98450	.17813	.6140	.0157	.7023	.19252	.98129	.19619	.0970	.0191	.1942	54
7	.17565	.98445	.17843	.6045	.0158	.6930	.19281	.98124	.19649	.0892	.0191	.1865	53
8	.17594	.98440	.17873	.5951	.0158	.6838	.19309	.98118	.19680	.0814	.0192	.1788	52
9	.17622	.98435	.17903	.5857	.0159	.6745	.19338	.98112	.19710	.0736	.0192	.1712	51
10	.17651	.98430	.17933	5.5764	.0159	5.6653	.19366	.98107	.19740	5.0658	.0193	5.1636	50
11	.17680	.98425	.17963	.5670	.0160	.6561	.19395	.98101	.19770	.0581	.0193	.1560	49
12	.17708	.98419	.17993	.5578	.0160	.6470	.19423	.98095	.19800	.0504	.0194	.1484	48
13	.17737	.98414	.18023	.5485	.0161	.6379	.19452	.98090	.19831	.0427	.0195	.1409	47
14	.17766	.98409	.18053	.5393	.0162	.6288	.19480	.98084	.19861	.0350	.0195	.1333	46
15	.17794	.98404	.18083	5.5301	.0162	5.6197	.19509	.98078	.19891	5.0273	.0196	5.1258	45
16	.17823	.98399	.18113	.5209	.0163	.6107	.19537	.98073	.19921	.0197	.0196	.1183	44
17	.17852	.98394	.18143	.5117	.0163	.6017	.19566	.98067	.19952	.0121	.0197	.1109	43
18	.17880	.98388	.18173	.5026	.0164	.5928	.19595	.98061	.19982	.0045	.0198	.1034	42
19	.17909	.98383	.18203	.4936	.0164	.5838	.19623	.98056	.20012	4.9969	.0198	.0960	41
20	.17937	.98378	.18233	5.4845	.0165	5.5749	.19652	.98050	.20042	4.9894	.0199	5.0886	40
21	.17966	.98373	.18263	.4755	.0165	.5660	.19680	.98044	.20073	.9819	.0199	.0812	39
22	.17995	.98368	.18293	.4665	.0166	.5572	.19709	.98039	.20103	.9744	.0200	.0739	38
23	.18023	.98362	.18323	.4575	.0166	.5484	.19737	.98033	.20133	.9669	.0201	.0666	37
24	.18052	.98357	.18353	.4486	.0167	.5396	.19766	.98027	.20163	.9594	.0201	.0593	36
25	.18080	.98352	.18383	5.4396	.0167	5.5308	.19794	.98021	.20194	4.9520	.0202	5.0520	35
26	.18109	.98347	.18413	.4308	.0168	.5221	.19823	.98016	.20224	.9446	.0202	.0447	34
27	.18138	.98341	.18444	.4219	.0169	.5134	.19851	.98010	.20254	.9372	.0203	.0375	33
28	.18166	.98336	.18474	.4131	.0169	.5047	.19880	.98004	.20285	.9298	.0204	.0302	32
29	.18195	.98331	.18504	.4043	.0170	.4960	.19908	.97998	.20315	.9225	.0204	.0230	31
30	.18223	.98325	.18534	5.3955	.0170	5.4874	.19937	.97992	.20345	4.9151	.0205	5.0158	30
31	.18252	.98320	.18564	.3868	.0171	.4788	.19965	.97987	.20375	.9078	.0205	.0087	29
32	.18281	.98315	.18594	.3780	.0171	.4702	.19994	.97981	.20406	.9006	.0206	.0015	28
33	.18309	.98309	.18624	.3694	.0172	.4617	.20022	.97975	.20436	.8933	.0207	4.9944	27
34	.18338	.98304	.18654	.3607	.0172	.4532	.20051	.97969	.20466	.8860	.0207	.9873	26
35	.18366	.98299	.18684	5.3521	.0173	5.4447	.20079	.97963	.20497	4.8788	.0208	4.9802	25
36	.18395	.98293	.18714	.3434	.0174	.4362	.20108	.97957	.20527	.8716	.0208	.9732	24
37	.18424	.98288	.18745	.3349	.0174	.4278	.20136	.97952	.20557	.8644	.0209	.9661	23
38	.18452	.98283	.18775	.3263	.0175	.4194	.20165	.97946	.20588	.8573	.0210	.9591	22
39	.18481	.98277	.18805	.3178	.0175	.4110	.20193	.97940	.20618	.8501	.0210	.9521	21
40	.18509	.98272	.18835	5.3093	.0176	5.4026	.20222	.97934	.20648	4.8430	.0211	4.9452	20
41	.18538	.98267	.18865	.3008	.0176	.3943	.20250	.97928	.20679	.8359	.0211	.9382	19
42	.18567	.98261	.18895	.2923	.0177	.3860	.20279	.97922	.20709	.8288	.0212	.9313	18
43	.18595	.98256	.18925	.2839	.0177	.3777	.20307	.97916	.20739	.8217	.0213	.9243	17
44	.18624	.98250	.18955	.2755	.0178	.3695	.20336	.97910	.20770	.8147	.0213	.9175	16
45	.18652	.98245	.18985	5.2671	.0179	5.3612	.20364	.97904	.20800	4.8077	.0214	4.9106	15
46	.18681	.98240	.19016	.2588	.0179	.3530	.20393	.97899	.20830	.8007	.0215	.9037	14
47	.18709	.98234	.19046	.2505	.0180	.3449	.20421	.97893	.20861	.7937	.0215	.8969	13
48	.18738	.98229	.19076	.2422	.0180	.3367	.20450	.97887	.20891	.7867	.0216	.8901	12
49	.18767	.98223	.19106	.2339	.0181	.3286	.20478	.97881	.20921	.7798	.0216	.8833	11
50	.18795	.98218	.19136	5.2257	.0181	5.3205	.20506	.97875	.20952	4.7728	.0217	4.8765	10
51	.18824	.98212	.19166	.2174	.0182	.3124	.20535	.97869	.20982	.7659	.0218	.8967	9
52	.18852	.98207	.19197	.2092	.0182	.3044	.20563	.97863	.21012	.7591	.0218	.8630	8
53	.18881	.98201	.19227	.2011	.0183	.2963	.20592	.97857	.21043	.7522	.0219	.8563	7
54	.18909	.98196	.19257	.1929	.0184	.2883	.20620	.97851	.21073	.7453	.0220	.8496	6
55	.18938	.98190	.19287	5.1848	.0184	5.2803	.20649	.97845	.21104	4.7385	.0220	4.8429	5
56	.18967	.98185	.19317	.1767	.0185	.2724	.20677	.97839	.21134	.7317	.0221	.8362	4
57	.18995	.98179	.19347	.1686	.0185	.2645	.20706	.97833	.21164	.7249	.0221	.8296	3
58	.19024	.98174	.19378	.1606	.0186	.2566	.20734	.97827	.21195	.7181	.0222	.8229	2
59	.19052	.98168	.19408	.1525	.0186	.2487	.20763	.97821	.21225	.7114	.0223	.8163	1
60	.19081	.98163	.19438	5.1445	.0187	5.2408	.20791	.97815	.21256	4.7046	.0223	4.8097	0
′	cos	sin	cot	tan	cosec	sec	cos	sin	cot	tan	cosec	sec	′
	79°						78°						

	12°						13°						
′	sin	cos	tan	cot	sec	cosec	sin	cos	tan	cot	sec	cosec	′
0	.20791	.97815	.21256	4.7046	1.0223	4.8097	.22495	.97437	.23087	4.3315	1.0263	4.4454	60
1	.20820	.97809	.21286	.6979	.0224	.8032	.22523	.97430	.23117	.3257	.0264	.4398	59
2	.20848	.97803	.21316	.6912	.0225	.7966	.22552	.97424	.23148	.3200	.0264	.4342	58
3	.20876	.97797	.21347	.6845	.0225	.7901	.22580	.97417	.23179	.3143	.0265	.4287	57
4	.20905	.97790	.21377	.6778	.0226	.7835	.22608	.97411	.23209	.3086	.0266	.4231	56
5	.20933	.97784	.21408	4.6712	.0226	4.7770	.22637	.97404	.23240	4.3029	.0266	4.4176	55
6	.20962	.97778	.21438	.6646	.0227	.7706	.22665	.97398	.23270	.2972	.0267	.4121	54
7	.20990	.97772	.21468	.6580	.0228	.7641	.22693	.97391	.23301	.2916	.0268	.4065	53
8	.21019	.97766	.21499	.6514	.0228	.7576	.22722	.97384	.23332	.2859	.0268	.4011	52
9	.21047	.97760	.21529	.6448	.0229	.7512	.22750	.97378	.23363	.2803	.0269	.3956	51
10	.21076	.97754	.21560	4.6382	.0230	4.7448	.22778	.97371	.23393	4.2747	.0270	4.3901	50
11	.21104	.97748	.21590	.6317	.0230	.7384	.22807	.97364	.23424	.2691	.0271	.3847	49
12	.21132	.97741	.21621	.6252	.0231	.7320	.22835	.97358	.23455	.2635	.0271	.3792	48
13	.21161	.97735	.21651	.6187	.0232	.7257	.22863	.97351	.23485	.2579	.0272	.3738	47
14	.21189	.97729	.21682	.6122	.0232	.7193	.22892	.97344	.23516	.2524	.0273	.3684	46
15	.21218	.97723	.21712	4.6057	.0233	4.7130	.22920	.97338	.23547	4.2468	.0273	4.3630	45
16	.21246	.97717	.21742	.5993	.0234	.7067	.22948	.97331	.23577	.2413	.0274	.3576	44
17	.21275	.97711	.21773	.5928	.0234	.7004	.22977	.97324	.23608	.2358	.0275	.3522	43
18	.21303	.97704	.21803	.5864	.0235	.6942	.23005	.97318	.23639	.2303	.0276	.3469	42
19	.21331	.97698	.21834	.5800	.0235	.6879	.23033	.97311	.23670	.2248	.0276	.3415	41
20	.21360	.97692	.21864	4.5736	.0236	4.6817	.23061	.97304	.23700	4.2193	.0277	4.3362	40
21	.21388	.97686	.21895	.5673	.0237	.6754	.23090	.97298	.23731	.2139	.0278	.3309	39
22	.21417	.97680	.21925	.5609	.0237	.6692	.23118	.97291	.23762	.2084	.0278	.3256	38
23	.21445	.97673	.21956	.5546	.0238	.6631	.23146	.97284	.23793	.2030	.0279	.3203	37
24	.21473	.97667	.21986	.5483	.0239	.6569	.23175	.97277	.23823	.1976	.0280	.3150	36
25	.21502	.97661	.22017	4.5420	.0239	4.6507	.23202	.97271	.23854	4.1921	.0280	4.3098	35
26	.21530	.97655	.22047	.5357	.0240	.6446	.23231	.97264	.23885	.1867	.0281	.3045	34
27	.21559	.97648	.22078	.5294	.0241	.6385	.23260	.97257	.23916	.1814	.0282	.2993	33
28	.21587	.97642	.22108	.5232	.0241	.6324	.23288	.97250	.23946	.1760	.0283	.2941	32
29	.21615	.97636	.22139	.5169	.0242	.6263	.23316	.97244	.23977	.1706	.0283	.2838	31
30	.21644	.97630	.22169	4.5107	.0243	4.6201	.23344	.97237	.24008	4.1653	.0284	4.2836	30
31	.21672	.97623	.22200	.5045	.0243	.6142	.23373	.97230	.24039	.1600	.0285	.2785	29
32	.21701	.97617	.22230	.4983	.0244	.6081	.23401	.97223	.24069	.1546	.0285	.2733	28
33	.21729	.97611	.22261	.4921	.0245	.6021	.23429	.97216	.24100	.1493	.0286	.2681	27
34	.21757	.97604	.22291	.4860	.0245	.5961	.23458	.97210	.24131	.1440	.0287	.2630	26
35	.21786	.97598	.22322	4.4799	.0246	4.5901	.23486	.97203	.24162	4.1388	.0288	4.2579	25
36	.21814	.97592	.22353	.4737	.0247	.5841	.23514	.97196	.24192	.1335	.0288	.2527	24
37	.21843	.97575	.22383	.4676	.0247	.5782	.23542	.97189	.24223	.1282	.0289	.2476	23
38	.21871	.97579	.22414	.4615	.0248	.5722	.23571	.97182	.24254	.1230	.0290	.2425	22
39	.21899	.97573	.22444	.4555	.0249	.5663	.23599	.97175	.24285	.1178	.0291	.2375	21
40	.21928	.97566	.22475	4.4494	.0249	4.5604	.23627	.97169	.24316	4.1126	.0291	4.2324	20
41	.21956	.97560	.22505	.4434	.0250	.5545	.23655	.97162	.24346	.1073	.0292	.2273	19
42	.21985	.97553	.22536	.4373	.0251	.5486	.23684	.97155	.24377	.1022	.0293	.2223	18
43	.22013	.97547	.22566	.4313	.0251	.5428	.23712	.97148	.24408	.0970	.0293	.2173	17
44	.22041	.97541	.22597	.4253	.0252	.5369	.23740	.97141	.24439	.0918	.0294	.2122	16
45	.22070	.97534	.22628	4.4194	.0253	4.5311	.23768	.97134	.24470	4.0867	.0295	4.2072	15
46	.22098	.97528	.22658	.4134	.0253	.5253	.23797	.97127	.24501	.0815	.0296	.2022	14
47	.22126	.97521	.22689	.4074	.0254	.5195	.23825	.97120	.24531	.0764	.0296	.1972	13
48	.22155	.97515	.22719	.4015	.0255	.5137	.23853	.97113	.24562	.0713	.0297	.1923	12
49	.22183	.97508	.22750	.3956	.0255	.5079	.23881	.97106	.24593	.0662	.0298	.1873	11
50	.22211	.97502	.22781	4.3897	.0256	4.5021	.23910	.97099	.24624	4.0611	.0299	4.1824	10
51	.22240	.97495	.22811	.3838	.0257	.4964	.23938	.97092	.24655	.0560	.0299	.1774	9
52	.22268	.97489	.22842	.3779	.0257	.4907	.23966	.97086	.24686	.0509	.0300	.1725	8
53	.22297	.97483	.22872	.3721	.0258	.4850	.23994	.97079	.24717	.0458	.0301	.1676	7
54	.22325	.97476	.22903	.3662	.0259	.4793	.24023	.97072	.24747	.0408	.0302	.1627	6
55	.22353	.97470	.22934	4.3604	.0260	4.4736	.24051	.97065	.24778	4.0358	.0302	4.1578	5
56	.22382	.97463	.22964	.3546	.0260	.4679	.24079	.97058	.24809	.0307	.0303	.1529	4
57	.22410	.97457	.22995	.3488	.0261	.4623	.24107	.97051	.24840	.0257	.0304	.1481	3
58	.22438	.97450	.23025	.3430	.0262	.4566	.24136	.97044	.24871	.0207	.0305	.1432	2
59	.22467	.97443	.23056	.3372	.0262	.4510	.24164	.97037	.24902	.0157	.0305	.1384	1
60	.22495	.97437	.23087	4.3315	.0263	4.4454	.24192	.97029	.24933	4.0108	.0306	4.1336	0
′	cos	sin	cot	tan	cosec	sec	cos	sin	cot	tan	cosec	sec	′
	77°						76°						

	14°						15°						
′	sin	cos	tan	cot	sec	cosec	sin	cos	tan	cot	sec	cosec	′
0	.24192	.97029	.24933	4.0108	1.0306	4.1336	.25882	.96592	.26795	3.7320	1.0353	3.8637	60
1	.24220	.97022	.24964	.0058	.0307	.1287	.25910	.96585	.26826	.7277	.0353	.8595	59
2	.24249	.97015	.24995	.0009	.0308	.1239	.25938	.96577	.26857	.7234	.0354	.8553	58
3	.24277	.97008	.25025	3.9959	.0308	.1191	.25966	.96570	.26888	.7191	.0355	.8512	57
4	.24305	.97001	.25056	.9910	.0309	.1144	.25994	.96562	.26920	.7147	.0356	.8470	56
5	.24333	.96994	.25087	3.9861	.0310	4.1096	.26022	.96555	.26951	3.7104	.0357	3.8428	55
6	.24361	.96987	.25118	.9812	.0311	.1048	.26050	.96547	.26982	.7062	.0358	.8387	54
7	.24390	.96980	.25149	.9763	.0311	.1001	.26078	.96540	.27013	.7019	.0358	.8346	53
8	.24418	.96973	.25180	.9714	.0312	.0953	.26107	.96532	.27044	.6976	.0359	.8304	52
9	.24446	.96966	.25211	.9665	.0313	.0906	.26135	.96524	.27076	.6933	.0360	.8263	51
10	.24474	.96959	.25242	3.9616	.0314	4.0859	.26163	.96517	.27107	3.6891	.0361	3.8222	50
11	.24502	.96952	.25273	.9568	.0314	.0812	.26191	.96509	.27138	.6848	.0362	.8181	49
12	.24531	.96944	.25304	.9520	.0315	.0765	.26219	.96502	.27169	.6806	.0362	.8140	48
13	.24559	.96937	.25335	.9471	.0316	.0718	.26247	.96494	.27201	.6764	.0363	.8100	47
14	.24587	.96930	.25366	.9423	.0317	.0672	.26275	.96486	.27232	.6722	.0364	.8059	46
15	.24615	.96923	.25397	3.9375	.0317	4.0625	.26303	.96479	.27263	3.6679	.0365	3.8018	45
16	.24643	.96916	.25428	.9327	.0318	.0579	.26331	.96471	.27294	.6637	.0366	.7978	44
17	.24672	.96909	.25459	.9279	.0319	.0532	.26359	.96463	.27326	.6596	.0367	.7937	43
18	.24700	.96901	.25490	.9231	.0320	.0486	.26387	.96456	.27357	.6554	.0367	.7897	42
19	.24728	.96894	.25521	.9184	.0320	.0440	.26415	.96448	.27388	.6512	.0368	.7857	41
20	.24756	.96887	.25552	3.9136	.0321	4.0394	.26443	.96440	.27419	3.6470	.0369	3.7816	40
21	.24784	.96880	.25583	.9089	.0322	.0348	.26471	.96433	.27451	.6429	.0370	.7776	39
22	.24813	.96873	.25614	.9042	.0323	.0302	.26499	.96425	.27482	.6387	.0371	.7736	38
23	.24841	.96865	.25645	.8994	.0323	.0256	.26527	.96417	.27513	.6346	.0371	.7697	37
24	.24869	.96858	.25676	.8947	.0324	.0211	.26556	.96409	.27544	.6305	.0372	.7657	36
25	.24897	.96851	.25707	3.8900	.0325	4.0165	.26584	.96402	.27576	3.6263	.0373	3.7617	35
26	.24925	.96844	.25738	.8853	.0326	.0120	.26612	.96394	.27607	.6222	.0374	.7577	34
27	.24953	.96836	.25769	.8807	.0327	.0074	.26640	.96386	.27638	.6181	.0375	.7538	33
28	.24982	.96829	.25800	.8760	.0327	.0029	.26668	.96378	.27670	.6140	.0376	.7498	32
29	.25010	.96822	.25831	.8713	.0328	3.9984	.26696	.96371	.27701	.6100	.0376	.7459	31
30	.25038	.96815	.25862	3.8667	.0329	3.9939	.26724	.96363	.27732	3.6059	.0377	3.7420	30
31	.25066	.96807	.25893	.8621	.0330	.9894	.26752	.96355	.27764	.6018	.0378	.7380	29
32	.25094	.96800	.25924	.8574	.0330	.9850	.26780	.96347	.27795	.5977	.0379	.7341	28
33	.25122	.96793	.25955	.8528	.0331	.9805	.26808	.96340	.27826	.5937	.0380	.7302	27
34	.25151	.96785	.25986	.8482	.0332	.9760	.26836	.96332	.27858	.5896	.0381	.7263	26
35	.25179	.96778	.26017	3.8436	.0333	3.9716	.26864	.96324	.27889	3.5856	.0382	3.7224	25
36	.25207	.96771	.26048	.8390	.0334	.9672	.26892	.96316	.27920	.5816	.0382	.7186	24
37	.25235	.96763	.26079	.8345	.0334	.9627	.26920	.96308	.27952	.5776	.0383	.7147	23
38	.25263	.96756	.26110	.8299	.0335	.9583	.26948	.96301	.27983	.5736	.0384	.7108	22
39	.25291	.96749	.26141	.8254	.0336	.9539	.26976	.96293	.28014	.5696	.0385	.7070	21
40	.25319	.96741	.26172	3.8208	.0337	3.9495	.27004	.96285	.28046	3.5656	.0386	3.7031	20
41	.25348	.96734	.26203	.8163	.0338	.9451	.27032	.96277	.28077	.5616	.0387	.6993	19
42	.25376	.96727	.26234	.8118	.0338	.9408	.27060	.96269	.28109	.5576	.0387	.6955	18
43	.25404	.96719	.26266	.8073	.0339	.9364	.27088	.96261	.28140	.5536	.0388	.6917	17
44	.25432	.96712	.26297	.8027	.0340	.9320	.27116	.96253	.28171	.5497	.0389	.6878	16
45	.25460	.96704	.26328	3.7983	.0341	3.9277	.27144	.96245	.28203	3.5457	.0390	3.6840	15
46	.25488	.96697	.26359	.7938	.0341	.9234	.27172	.96238	.28234	.5418	.0391	.6802	14
47	.25516	.96690	.26390	.7893	.0342	.9190	.27200	.96230	.28266	.5378	.0392	.6765	13
48	.25544	.96682	.26421	.7848	.0343	.9147	.27228	.96222	.28297	.5339	.0393	.6727	12
49	.25573	.96675	.26452	.7804	.0344	.9104	.27256	.96214	.28328	.5300	.0393	.6689	11
50	.25601	.96667	.26483	3.7759	.0345	3.9061	.27284	.96206	.28360	3.5261	.0394	3.6651	10
51	.25629	.96660	.26514	.7715	.0345	.9018	.27312	.96198	.28391	.5222	.0395	.6614	9
52	.25657	.96652	.26546	.7671	.0346	.8976	.27340	.96190	.28423	.5183	.0396	.6576	8
53	.25685	.96645	.26577	.7627	.0347	.8933	.27368	.96182	.28454	.5144	.0397	.6539	7
54	.25713	.96638	.26608	.7583	.0348	.8890	.27396	.96174	.28486	.5105	.0398	.6502	6
55	.25741	.96630	.26639	3.7539	.0349	3.8848	.27424	.96166	.28517	3.5066	.0399	3.6464	5
56	.25769	.96623	.26670	.7495	.0349	.8805	.27452	.96158	.28549	.5028	.0399	.6427	4
57	.25798	.96615	.26701	.7451	.0350	.8763	.27480	.96150	.28580	.4989	.0400	.6390	3
58	.25826	.96608	.26732	.7407	.0351	.8721	.27508	.96142	.28611	.4951	.0401	.6353	2
59	.25854	.96600	.26764	.7364	.0352	.8679	.27536	.96134	.28643	.4912	.0402	.6316	1
60	.25882	.96592	.26795	3.7320	.0353	3.8637	.27564	.96126	.28674	3.4874	.0403	3.6279	0
′	cos	sin	cot	tan	cosec	sec	cos	sin	cot	tan	cosec	sec	′
	75°						74°						

′	16° sin	16° cos	16° tan	16° cot	16° sec	16° cosec	17° sin	17° cos	17° tan	17° cot	17° sec	17° cosec	′
0	.27564	.96126	.28674	3.4874	1.0403	3.6279	.29237	.95630	.30573	3.2708	1.0457	3.4203	60
1	.27592	.96118	.28706	.4836	.0404	.6243	.29265	.95622	.30605	.2674	.0458	.4170	59
2	.27620	.96110	.28737	.4798	.0405	.6206	.29293	.95613	.30637	.2640	.0459	.4138	58
3	.27648	.96102	.28769	.4760	.0406	.6169	.29321	.95605	.30668	.2607	.0460	.4106	57
4	.27675	.96094	.28800	.4722	.0406	.6133	.29348	.95596	.30700	.2573	.0461	.4073	56
5	.27703	.96086	.28832	3.4684	.0407	3.6096	.29376	.95588	.30732	3.2539	.0461	3.4041	55
6	.27731	.96078	.28863	.4646	.0408	.6060	.29404	.95579	.30764	.2505	.0462	.4009	54
7	.27759	.96070	.28895	.4608	.0409	.6024	.29432	.95571	.30796	.2472	.0463	.3977	43
8	.27787	.96062	.28926	.4570	.0410	.5987	.29460	.95562	.30828	.2438	.0464	.3945	52
9	.27815	.96054	.28958	.4533	.0411	.5951	.29487	.95554	.30859	.2405	.0465	.3913	51
10	.27843	.96045	.28990	3.4495	.0412	3.5915	.29515	.95545	.30891	3.2371	.0466	3.3881	50
11	.27871	.96037	.29021	.4458	.0413	.5879	.29543	.95536	.30923	.2338	.0467	.3849	49
12	.27899	.96029	.29053	.4420	.0413	.5843	.29571	.95528	.30955	.2305	.0468	.3817	48
13	.27927	.96021	.29084	.4383	.0414	.5807	.29598	.95519	.30987	.2271	.0469	.3785	47
14	.27955	.96013	.29116	.4346	.0415	.5772	.29626	.95511	.31019	.2238	.0470	.3754	46
15	.27983	.96005	.29147	3.4308	.0416	3.5736	.29654	.95502	.31051	3.2205	.0471	3.3722	45
16	.28011	.95997	.29179	.4271	.0417	.5700	.29682	.95493	.31083	.2172	.0472	.3690	44
17	.28039	.95989	.29210	.4234	.0418	.5665	.29710	.95485	.31115	.2139	.0473	.3659	43
18	.28067	.95980	.29242	.4197	.0419	.5629	.29737	.95476	.31146	.2106	.0474	.3627	42
19	.28094	.95972	.29274	.4160	.0420	.5594	.29765	.95467	.31178	.2073	.0475	.3596	41
20	.28122	.95964	.29305	3.4124	.0420	3.5559	.29793	.95459	.31210	3.2041	.0476	3.3565	40
21	.28150	.95956	.29337	.4087	.0421	.5523	.29821	.95450	.31242	.2008	.0477	.3534	39
22	.28178	.95948	.29368	.4050	.0422	.5488	.29848	.95441	.31274	.1975	.0478	.3502	38
23	.28206	.95940	.29400	.4014	.0423	.5453	.29876	.95433	.31306	.1942	.0478	.3471	37
24	.28234	.95931	.29432	.3977	.0424	.5418	.29904	.95424	.31338	.1910	.0479	.3440	36
25	.28262	.95923	.29463	3.3941	.0425	3.5383	.29932	.95415	.31370	3.1877	.0480	3.3409	35
26	.28290	.95915	.29495	.3904	.0426	.5348	.29959	.95407	.31402	.1845	.0481	.3378	34
27	.28318	.95907	.29526	.3868	.0427	.5313	.29987	.95398	.31434	.1813	.0482	.3347	33
28	.28346	.95898	.29558	.3832	.0428	.5279	.30015	.95389	.31466	.1780	.0483	.3316	32
29	.28374	.95890	.29590	.3795	.0428	.5244	.30043	.95380	.31498	.1748	.0484	.3286	31
30	.28401	.95882	.29621	3.3759	.0429	3.5209	.30070	.95372	.31530	3.1716	.0485	3.3255	30
31	.28429	.95874	.29653	.3723	.0430	.5175	.30098	.95363	.31562	.1684	.0486	.3224	29
32	.28457	.95865	.29685	.3687	.0431	.5140	.30126	.95354	.31594	.1652	.0487	.3194	28
33	.28485	.95857	.29716	.3651	.0432	.5106	.30154	.95345	.31626	.1620	.0488	.3163	27
34	.28513	.95849	.29748	.3616	.0433	.5072	.30181	.95337	.31658	.1588	.0489	.3133	26
35	.28541	.95840	.29780	3.3580	.0434	3.5037	.30209	.95328	.31690	3.1556	.0490	3.3102	25
36	.28569	.95832	.29811	.3544	.0435	.5003	.30237	.95319	.31722	.1524	.0491	.3072	24
37	.28597	.95824	.29843	.3509	.0436	.4969	.30265	.95310	.31754	.1492	.0492	.3042	23
38	.28624	.95816	.29875	.3473	.0437	.4935	.30292	.95301	.31786	.1460	.0493	.3011	22
39	.28652	.95807	.29906	.3438	.0438	.4901	.30320	.95293	.31818	.1429	.0494	.2981	21
40	.28680	.95799	.29938	3.3402	.0438	3.4867	.30348	.95284	.31850	3.1397	.0495	3.2951	20
41	.28708	.95791	.29970	.3367	.0439	.4833	.30375	.95275	.31882	.1366	.0496	.2921	19
42	.28736	.95782	.30001	.3332	.0440	.4799	.30403	.95266	.31914	.1334	.0497	.2891	18
43	.28764	.95774	.30033	.3296	.0441	.4766	.30431	.95257	.31946	.1303	.0498	.2861	17
44	.28792	.95765	.30065	.3261	.0442	.4732	.30459	.95248	.31978	.1271	.0499	.2831	16
45	.28820	.95757	.30096	3.3226	.0443	3.4698	.30486	.95239	.32010	3.1240	.0500	3.2801	15
46	.28847	.95749	.30128	.3191	.0444	.4665	.30514	.95231	.32042	.1209	.0501	.2772	14
47	.28875	.95740	.30160	.3156	.0445	.4632	.30542	.95222	.32074	.1177	.0502	.2742	13
48	.28903	.95732	.30192	.3121	.0446	.4598	.30569	.95213	.32106	.1146	.0503	.2712	12
49	.28931	.95723	.30223	.3087	.0447	.4565	.30597	.95204	.32138	.1115	.0504	.2683	11
50	.28959	.95715	.30255	3.3052	.0448	3.4532	.30625	.95195	.32171	3.1084	.0505	3.2653	10
51	.28987	.95707	.30287	.3017	.0448	.4498	.30653	.95186	.32203	.1053	.0506	.2624	9
52	.29014	.95698	.30319	.2983	.0449	.4465	.30680	.95177	.32235	.1022	.0507	.2594	8
53	.29042	.95690	.30350	.2948	.0450	.4432	.30708	.95168	.32267	.0991	.0508	.2565	7
54	.29070	.95681	.30382	.2914	.0451	.4399	.30736	.95159	.32299	.0960	.0509	.2535	6
55	.29098	.95673	.30414	3.2879	.0452	3.4366	.30763	.95150	.32331	3.0930	.0510	3.2506	5
56	.29126	.95664	.30446	.2845	.0453	.4334	.30791	.95141	.32363	.0899	.0511	.2477	4
57	.29154	.95656	.30478	.2811	.0454	.4301	.30819	.95132	.32395	.0868	.0512	.2448	3
58	.29181	.95647	.30509	.2777	.0455	.4268	.30846	.95124	.32428	.0838	.0513	.2419	2
59	.29209	.95639	.30541	.2742	.0456	.4236	.30874	.95115	.32460	.0807	.0514	.2390	1
60	.29237	.95630	.30573	3.2708	.0457	3.4203	.30902	.95106	.32492	3.0777	.0515	3.2361	0
′	73° cos	73° sin	73° cot	73° tan	73° cosec	73° sec	72° cos	72° sin	72° cot	72° tan	72° cosec	72° sec	′

′	18°						19°						′
′	sin	cos	tan	cot	sec	cosec	sin	cos	tan	cot	sec	cosec	′
0	.30902	.95106	.32492	3.0777	1.0515	3.2361	.32557	.94552	.34433	2.9042	1.0576	3.0715	60
1	.30929	.95097	.32524	.0746	.0516	.2332	.32584	.94542	.34465	.9015	.0577	.0690	59
2	.30957	.95088	.32556	.0716	.0517	.2303	.32612	.94533	.34498	.8987	.0578	.0664	58
3	.30985	.95079	.32588	.0686	.0518	.2274	.32639	.94523	.34530	.8960	.0579	.0638	57
4	.31012	.95070	.32621	.0655	.0519	.2245	.32667	.94514	.34563	.8933	.0580	.0612	56
5	.31040	.95061	.32653	3.0625	.0520	3.2216	.32694	.94504	.34595	2.8905	.0581	3.0586	55
6	.31068	.95051	.32685	.0595	.0521	.2188	.32722	.94495	.34628	.8878	.0582	.0561	54
7	.31095	.95042	.32717	.0565	.0522	.2159	.32749	.94485	.34661	.8851	.0584	.0535	53
8	.31123	.95033	.32749	.0535	.0523	.2131	.32777	.94476	.34693	.8824	.0585	.0509	52
9	.31150	.95024	.32782	.0505	.0524	.2102	.32804	.94466	.34726	.8797	.0586	.0484	51
10	.31178	.95015	.32814	3.0475	.0525	3.2074	.32832	.94457	.34758	2.8770	.0587	3.0458	50
11	.31206	.95006	.32846	.0445	.0526	.2045	.32859	.94447	.34791	.8743	.0588	.0433	49
12	.31233	.94997	.32878	.0415	.0527	.2017	.32887	.94438	.34824	.8716	.0589	.0407	48
13	.31261	.94988	.32910	.0385	.0528	.1989	.32914	.94428	.34856	.8689	.0590	.0382	47
14	.31289	.94979	.32943	.0356	.0529	.1960	.32942	.94418	.34889	.8662	.0591	.0357	46
15	.31316	.94970	.32975	3.0326	.0530	3.1932	.32969	.94409	.34921	2.8636	.0592	3.0331	45
16	.31344	.94961	.33007	.0296	.0531	.1904	.32996	.94399	.34954	.8609	.0593	.0306	44
17	.31372	.94952	.33039	.0267	.0532	.1876	.33024	.94390	.34987	.8582	.0594	.0281	43
18	.31399	.94942	.33072	.0237	.0533	.1848	.33051	.94380	.35019	.8555	.0595	.0256	42
19	.31427	.94933	.33104	.0208	.0534	.1820	.33079	.94370	.35052	.8529	.0596	.0231	41
20	.31454	.94924	.33136	3.0178	.0535	3.1792	.33106	.94361	.35085	2.8502	.0598	3.0206	40
21	.31482	.94915	.33169	.0149	.0536	.1764	.33134	.94351	.35117	.8476	.0599	.0181	39
22	.31510	.94906	.33201	.0120	.0537	.1736	.33161	.94341	.35150	.8449	.0600	.0156	38
23	.31537	.94897	.33233	.0090	.0528	.1708	.33189	.94332	.35183	.8423	.0601	.0131	37
24	.31565	.94888	.33265	.0061	.0539	.1681	.33216	.94322	.35215	.8396	.0602	.0106	36
25	.31592	.94878	.33298	3.0032	.0540	3.1653	.33243	.94313	.35248	2.8370	.0603	3.0081	35
26	.31620	.94869	.33330	.0003	.0541	.1625	.33271	.94303	.35281	.8344	.0604	.0056	34
27	.31648	.94860	.33362	2.9974	.0542	.1598	.33298	.94293	.35314	.8318	.0605	.0031	33
28	.31675	.94851	.33395	.9945	.0543	.1570	.33326	.94283	.35346	.8291	.0606	.0007	32
29	.31703	.94841	.33427	.9916	.0544	.1543	.33353	.94274	.35379	.8265	.0607	2.9982	31
30	.31730	.94832	.33459	2.9887	.0545	3.1515	.33381	.94264	.35412	2.8239	.0608	2.9957	30
31	.31758	.94823	.33492	.9858	.0546	.1488	.33408	.94254	.35445	.8213	.0609	.9933	29
32	.31786	.94814	.33524	.9829	.0547	.1461	.33435	.94245	.35477	.8187	.0611	.9908	28
33	.31813	.94805	.33557	.9800	.0548	.1433	.33463	.94235	.35510	.8161	.0612	.9884	27
34	.31841	.94795	.33589	.9772	.0549	.1406	.33490	.94225	.35543	.8135	.0613	.9859	26
35	.31868	.94786	.33621	2.9743	.0550	3.1379	.33518	.94215	.35576	2.8109	.0614	2.9835	25
36	.31896	.94777	.33654	.9714	.0551	.1352	.33545	.94206	.35608	.8083	.0615	.9810	24
37	.31923	.94767	.33686	.9686	.0552	.1325	.33572	.94196	.35641	.8057	.0616	.9786	23
38	.31951	.94758	.33718	.9657	.0553	.1298	.33600	.94186	.35674	.8032	.0617	.9762	22
39	.31978	.94749	.33751	.9629	.0554	.1271	.33627	.94176	.35707	.8006	.0618	.9738	21
40	.32006	.94740	.33783	2.9600	.0555	3.1244	.33655	.94167	.35739	2.7980	.0619	2.9713	20
41	.32034	.94730	.33816	.9572	.0556	.1217	.33682	.94157	.35772	.7954	.0620	.9689	19
42	.32061	.94721	.33848	.9544	.0557	.1190	.33709	.94147	.35805	.7929	.0622	.9665	18
43	.32089	.94712	.33880	.9515	.0558	.1163	.33737	.94137	.35838	.7903	.0623	.9641	17
44	.32116	.94702	.33913	.9487	.0559	.1137	.33764	.94127	.35871	.7878	.0624	.9617	16
45	.32144	.94693	.33945	2.9459	.0560	3.1110	.33792	.94118	.35904	2.7852	.0625	2.9593	15
46	.32171	.94684	.33978	.9431	.0561	.1083	.33819	.94108	.35936	.7827	.0626	.9569	14
47	.32199	.94674	.34010	.9403	.0562	.1057	.33846	.94098	.35969	.7801	.0627	.9545	13
48	.32226	.94665	.34043	.9375	.0563	.1030	.33874	.94088	.36002	.7776	.0628	.9521	12
49	.32254	.94655	.34075	.9347	.0565	.1004	.33901	.94078	.36035	.7751	.0629	.9497	11
50	.32282	.94646	.34108	2.9319	.0566	3.0977	.33928	.94068	.36068	2.7725	.0630	2.9474	10
51	.32309	.94637	.34140	.9291	.0567	.0951	.33956	.94058	.36101	.7700	.0632	.9450	9
52	.32337	.94627	.34173	.9263	.0568	.0925	.33983	.94049	.36134	.7675	.0633	.9426	8
53	.32364	.94618	.34205	.9235	.0569	.0898	.34011	.94039	.36167	.7650	.0634	.9402	7
54	.32392	.94608	.34238	.9208	.0570	.0872	.34038	.94029	.36199	.7625	.0635	.9379	6
55	.32419	.94599	.34270	2.9180	.0571	3.0846	.34065	.94019	.36232	2.7600	.0636	2.9355	5
56	.32447	.94590	.34303	.9152	.0572	.0820	.34093	.94009	.36265	.7575	.0637	.9332	4
57	.32474	.94580	.34335	.9125	.0573	.0793	.34120	.93999	.36298	.7550	.0638	.9308	3
58	.32502	.94571	.34368	.9097	.0574	.0767	.34147	.93989	.36331	.7525	.0639	.9285	2
59	.32529	.94561	.34400	.9069	.0575	.0741	.34175	.93979	.36364	.7500	.0641	.9261	1
60	.32557	.94552	.34433	2.9042	.0576	3.0715	.34202	.93969	.36397	2.7475	.0642	2.9238	0
′	cos	sin	cot	tan	cosec	sec	cos	sin	cot	tan	cosec	sec	′
	71°						70°						

	20°						21°						
′	sin	cos	tan	cot	sec	cosec	sin	cos	tan	cot	sec	cosec	′
0	.34202	.93969	.36397	2.7475	1.0642	2.9238	.35837	.93358	.38386	2.6051	1.0711	2.7904	60
1	.34229	.93959	.36430	.7450	.0643	.9215	.35864	.93348	.38420	.6028	.0713	.7883	59
2	.34257	.93949	.36463	.7425	.0644	.9191	.35891	.93337	.38453	.6006	.0714	.7862	58
3	.34284	.93939	.36496	.7400	.0645	.9168	.35918	.93327	.38486	.5983	.0715	.7841	57
4	.34311	.93929	.36529	.7376	.0646	.9145	.35945	.93316	.38520	.5960	.0716	.7820	56
5	.34339	.93919	.36562	2.7351	.0647	2.9122	.35972	.93306	.38553	2.5938	.0717	2.7799	55
6	.34366	.93909	.36595	.7326	.0648	.9098	.36000	.93295	.38587	.5916	.0719	.7778	54
7	.34393	.93899	.36628	.7302	.0650	.9075	.36027	.93285	.38620	.5893	.0720	.7757	53
8	.34421	.93889	.36661	.7277	.0651	.9052	.36054	.93274	.38654	.5871	.0721	.7736	52
9	.34448	.93879	.36694	.7252	.0652	.9029	.36081	.93264	.38687	.5848	.0722	.7715	51
10	.34475	.93869	.36727	2.7228	.0653	2.9006	.36108	.93253	.38720	2.5826	.0723	2.7694	50
11	.34502	.93859	.36760	.7204	.0654	.8983	.36135	.93243	.38754	.5804	.0725	.7674	49
12	.34530	.93849	.36793	.7179	.0655	.8960	.36162	.93232	.38787	.5781	.0726	.7653	48
13	.34557	.93839	.36826	.7155	.0656	.8937	.36189	.93222	.38821	.5759	.0727	.7632	47
14	.34584	.93829	.36859	.7130	.0658	.8915	.36217	.93211	.38854	.5737	.0728	.7611	46
15	.34612	.93819	.36892	2.7106	.0659	2.8892	.36244	.93201	.38888	2.5715	.0729	2.7591	45
16	.34639	.93809	.36925	.7082	.0660	.8869	.36271	.93190	.38921	.5693	.0731	.7570	44
17	.34666	.93799	.36958	.7058	.0661	.8846	.36298	.93180	.38955	.5671	.0732	.7550	43
18	.34693	.93789	.36991	.7033	.0662	.8824	.36325	.93169	.38988	.5649	.0733	.7529	42
19	.34721	.93779	.37024	.7009	.0663	.8801	.36352	.93158	.39022	.5627	.0734	.7509	41
20	.34748	.93769	.37057	2.6985	.0664	2.8778	.36379	.93148	.39055	2.5605	.0736	2.7488	40
21	.34775	.93758	.37090	.6961	.0666	.8756	.36406	.93137	.39089	.5583	.0737	.7468	39
22	.34803	.93748	.37123	.6937	.0667	.8733	.36433	.93127	.39122	.5561	.0738	.7447	38
23	.34830	.93738	.37156	.6913	.0668	.8711	.36460	.93116	.39156	.5539	.0739	.7427	37
24	.34857	.93728	.37190	.6889	.0669	.8688	.36488	.93105	.39189	.5517	.0740	.7406	36
25	.34884	.93718	.37223	2.6865	.0670	2.8666	.36515	.93095	.39223	2.5495	.0742	2.7386	35
26	.34912	.93708	.37256	.6841	.0671	.8644	.36542	.93084	.39257	.5473	.0743	.7366	34
27	.34939	.93698	.37289	.6817	.0673	.8621	.36569	.93074	.39290	.5451	.0744	.7346	33
28	.34966	.93687	.37322	.6794	.0674	.8599	.36596	.93063	.39324	.5430	.0745	.7325	32
29	.34993	.93677	.37355	.6770	.0675	.8577	.36623	.93052	.39357	.5408	.0747	.7305	31
30	.35021	.93667	.37388	2.6746	.0676	2.8554	.36650	.93042	.39391	2.5386	.0748	2.7285	30
31	.35048	.93657	.37422	.6722	.0677	.8532	.36677	.93031	.39425	.5365	.0749	.7265	29
32	.35075	.93647	.37455	.6699	.0678	.8510	.36704	.93020	.39458	.5343	.0750	.7245	28
33	.35102	.93637	.37488	.6675	.0679	.8488	.36731	.93010	.39492	.5322	.0751	.7225	27
34	.35130	.93626	.37521	.6652	.0681	.8466	.36758	.92999	.39525	.5300	.0753	.7205	26
35	.35157	.93616	.37554	2.6628	.0682	2.8444	.36785	.92988	.39559	2.5278	.0754	2.7185	25
36	.35184	.93606	.37587	.6604	.0683	.8422	.36812	.92978	.39593	.5257	.0755	.7165	24
37	.35211	.93596	.37621	.6581	.0684	.8400	.36839	.92967	.39626	.5236	.0756	.7145	23
38	.35239	.93585	.37654	.6558	.0685	.8378	.36866	.92956	.39660	.5214	.0758	.7125	22
39	.35266	.93575	.37687	.6534	.0686	.8356	.36893	.92945	.39694	.5193	.0759	.7105	21
40	.35293	.93565	.37720	2.6511	.0688	2.8334	.36921	.92935	.39727	2.5171	.0760	2.7085	20
41	.35320	.93555	.37754	.6487	.0689	.8312	.36948	.92924	.39761	.5150	.0761	.7065	19
42	.35347	.93544	.37787	.6464	.0690	.8290	.36975	.92913	.39795	.5129	.0763	.7045	18
43	.35375	.93534	.37820	.6441	.0691	.8269	.37002	.92902	.39828	.5108	.0764	.7026	17
44	.35402	.93524	.37853	.6418	.0692	.8247	.37029	.92892	.39862	.5086	.0765	.7006	16
45	.35429	.93513	.37887	2.6394	.0694	2.8225	.37056	.92881	.39896	2.5065	.0766	2.6986	15
46	.35456	.93503	.37920	.6371	.0695	.8204	.37083	.92870	.39930	.5044	.0768	.6967	14
47	.35483	.93493	.37953	.6348	.0696	.8182	.37110	.92859	.39963	.5023	.0769	.6947	13
48	.35511	.93482	.37986	.6325	.0697	.8160	.37137	.92848	.39997	.5002	.0770	.6927	12
49	.35538	.93472	.38020	.6302	.0698	.8139	.37164	.92838	.40031	.4981	.0771	.6908	11
50	.35565	.93462	.38053	2.6279	.0699	2.8117	.37191	.92827	.40065	2.4960	.0773	2.6888	10
51	.35592	.93451	.38086	.6256	.0701	.8096	.37218	.92816	.40098	.4939	.0774	.6869	9
52	.35619	.93441	.38120	.6233	.0702	.8074	.37245	.92805	.40132	.4918	.0775	.6849	8
53	.35647	.93431	.38153	.6210	.0703	.8053	.37272	.92794	.40166	.4897	.0776	.6830	7
54	.35674	.93420	.38186	.6187	.0704	.8032	.37299	.92784	.40200	.4876	.0778	.6810	6
55	.35701	.93410	.38220	2.6164	.0705	2.8010	.37326	.92773	.40233	2.4855	.0779	2.6791	5
56	.35728	.93400	.38253	.6142	.0707	.7989	.37353	.92762	.40267	.4834	.0780	.6772	4
57	.35755	.93389	.38286	.6119	.0708	.7968	.37380	.92751	.40301	.4813	.0781	.6752	3
58	.35782	.93379	.38320	.6096	.0709	.7947	.37407	.92740	.40335	.4792	.0783	.6733	2
59	.35810	.93368	.38353	.6073	.0710	.7925	.37434	.92729	.40369	.4772	.0784	.6714	1
60	.35837	.93358	.38386	2.6051	.0711	2.7904	.37461	.92718	.40403	2.4751	.0785	2.6695	0
′	cos	sin	cot	tan	cosec	sec	cos	sin	cot	tan	cosec	sec	′
	69°						68°						

	22°						23°						
′	sin	cos	tan	cot	sec	cosec	sin	cos	tan	cot	sec	cosec	′
0	.37461	.92718	.40403	2.4751	1.0785	2.6695	.39073	.92050	.42447	2.3558	1.0864	2.5593	60
1	.37488	.92707	.40436	.4730	.0787	.6675	.39100	.92039	.42482	.3539	.0865	.5575	59
2	.37514	.92696	.40470	.4709	.0788	.6656	.39126	.92028	.42516	.3520	.0866	.5558	58
3	.37541	.92686	.40504	.4689	.0789	.6637	.39153	.92016	.42550	.3501	.0868	.5540	57
4	.37568	.92675	.40538	.4668	.0790	.6618	.39180	.92005	.42585	.3482	.0869	.5523	56
5	.37595	.92664	.40572	2.4647	.0792	2.6599	.39207	.91993	.42619	2.3463	.0870	2.5506	55
6	.37622	.92653	.40606	.4627	.0793	.6580	.39234	.91982	.42654	.3445	.0872	.5488	54
7	.37649	.92642	.40640	.4606	.0794	.6561	.39260	.91971	.42688	.3426	.0873	.5471	53
8	.37676	.92631	.40673	.4586	.0795	.6542	.39287	.91959	.42722	.3407	.0874	.5453	52
9	.37703	.92620	.40707	.4565	.0797	.6523	.39314	.91948	.42757	.3388	.0876	.5436	51
10	.37730	.92609	.40741	2.4545	.0798	2.6504	.39341	.91936	.42791	2.3369	.0877	2.5419	50
11	.37757	.92598	.40775	.4525	.0799	.6485	.39367	.91925	.42826	.3350	.0878	.5402	49
12	.37784	.92587	.40809	.4504	.0801	.6466	.39394	.91913	.42860	.3332	.0880	.5384	48
13	.37811	.92576	.40843	.4484	.0802	.6447	.39421	.91902	.42894	.3313	.0881	.5367	47
14	.37838	.92565	.40877	.4463	.0803	.6428	.39448	.91891	.42929	.3294	.0882	.5350	46
15	.37865	.92554	.40911	2.4443	.0804	2.6410	.39474	.91879	.42963	2.3276	.0884	2.5333	45
16	.37892	.92543	.40945	.4423	.0806	.6391	.39501	.91868	.42998	.3257	.0885	.5316	44
17	.37919	.92532	.40979	.4403	.0807	.6372	.39528	.91856	.43032	.3238	.0886	.5299	43
18	.37946	.92521	.41013	.4382	.0808	.6353	.39554	.91845	.43067	.3220	.0888	.5281	42
19	.37972	.92510	.41047	.4362	.0810	.6335	.39581	.91833	.43101	.3201	.0889	.5264	41
20	.37999	.92499	.41081	2.4342	.0811	2.6316	.39608	.91822	.43136	2.3183	.0891	2.5247	40
21	.38026	.92488	.41115	.4322	.0812	.6297	.39635	.91810	.43170	.3164	.0892	.5230	39
22	.38053	.92477	.41149	.4302	.0813	.6279	.39661	.91798	.43205	.3145	.0893	.5213	38
23	.38080	.92466	.41183	.4282	.0815	.6260	.39688	.91787	.43239	.3127	.0895	.5196	37
24	.38107	.92455	.41217	.4262	.0816	.6242	.39715	.91775	.43274	.3109	.0896	.5179	36
25	.38134	.92443	.41251	2.4242	.0817	2.6223	.39741	.91764	.43308	2.3090	.0897	2.5163	35
26	.38161	.92432	.41285	.4222	.0819	.6205	.39768	.91752	.43343	.3072	.0899	.5146	34
27	.38188	.92421	.41319	.4202	.0820	.6186	.39795	.91741	.43377	.3053	.0900	.5129	33
28	.38214	.92410	.41353	.4182	.0821	.6168	.39821	.91729	.43412	.3035	.0902	.5112	32
29	.38241	.92399	.41387	.4162	.0823	.6150	.39848	.91718	.43447	.3017	.0903	.5095	31
30	.38268	.92388	.41421	2.4142	.0824	2.6131	.39875	.91706	.43481	2.2998	.0904	2.5078	30
31	.38295	.92377	.41455	.4122	.0825	.6113	.39901	.91694	.43516	.2980	.0906	.5062	29
32	.38322	.92366	.41489	.4102	.0826	.6095	.39928	.91683	.43550	.2962	.0907	.5045	28
33	.38349	.92354	.41524	.4083	.0828	.6076	.39955	.91671	.43585	.2944	.0908	.5028	27
34	.38376	.92343	.41558	.4063	.0829	.6058	.39981	.91659	.43620	.2925	.0910	.5011	26
35	.38403	.92332	.41592	2.4043	.0830	2.6040	.40008	.91648	.43654	2.2907	.0911	2.4995	25
36	.38429	.92321	.41626	.4023	.0832	.6022	.40035	.91636	.43689	.2889	.0913	.4978	24
37	.38456	.92310	.41660	.4004	.0833	.6003	.40061	.91625	.43723	.2871	.0914	.4961	23
38	.38483	.92299	.41694	.3984	.0834	.5985	.40088	.91613	.43758	.2853	.0915	.4945	22
39	.38510	.92287	.41728	.3964	.0836	.5967	.40115	.91601	.43793	.2835	.0917	.4928	21
40	.38537	.92276	.41762	2.3945	.0837	2.5949	.40141	.91590	.43827	2.2817	.0918	2.4912	20
41	.38564	.92265	.41797	.3925	.0838	.5931	.40168	.91578	.43862	.2799	.0920	.4895	19
42	.38591	.92254	.41831	.3906	.0840	.5913	.40195	.91566	.43897	.2781	.0921	.4879	18
43	.38617	.92242	.41865	.3886	.0841	.5895	.40221	.91554	.43932	.2763	.0922	.4862	17
44	.38644	.92231	.41899	.3867	.0842	.5877	.40248	.91543	.43966	.2745	.0924	.4846	16
45	.38671	.92220	.41933	2.3847	.0844	2.5958	.40205	.91531	.44001	2.2727	.0925	2.4829	15
46	.38698	.92209	.41968	.3828	.0845	.5841	.40301	.91519	.44036	.2709	.0927	.4813	14
47	.38725	.92197	.42002	.3808	.0846	.5823	.40328	.91508	.44070	.2691	.0928	.4797	13
48	.38751	.92186	.42036	.3789	.0847	.5805	.40354	.91496	.44105	.2673	.0929	.4780	12
49	.38778	.92105	.42070	.3770	.0849	.5787	.40381	.91484	.44140	.2655	.0931	.4764	11
50	.38805	.92164	.42105	2.3750	.0850	2.5770	.40408	.91472	.44175	2.2637	.0932	2.4748	10
51	.38832	.92152	.42139	.3731	.0851	.5752	.40434	.91461	.44209	.2619	.0934	.4731	9
52	.38859	.92141	.42173	.3712	.0853	.5734	.40461	.91449	.44244	.2602	.0935	.4715	8
53	.38886	.92130	.42207	.3692	.0854	.5716	.40487	.91437	.44279	.2584	.0936	.4699	7
54	.38912	.92118	.42242	.3673	.0855	.5699	.40514	.91425	.44314	.2566	.0938	.4683	6
55	.38939	.92107	.42276	2.3654	.0857	2.5681	.40541	.91414	.44349	2.2548	.0939	2.4666	5
56	.38966	.92096	.42310	.3635	.0858	.5663	.40567	.91402	.44383	.2531	.0941	.4650	4
57	.38993	.92084	.42344	.3616	.0859	.5646	.40594	.91390	.44418	.2513	.0942	.4634	3
58	.39019	.92073	.42379	.3597	.0861	.5628	.40620	.91378	.44453	.2495	.0943	.4618	2
59	.39046	.92062	.42413	.3577	.0862	.5610	.40647	.91366	.44488	.2478	.0945	.4602	1
60	.39073	.92050	.42447	2.3558	.0864	2.5593	.40674	.91354	.44523	2.2460	.0946	2.4586	0
′	cos	sin	cot	tan	cosec	sec	cos	sin	cot	tan	cosec	sec	′
	67°						66°						

′	24° sin	cos	tan	cot	sec	cosec	25° sin	cos	tan	cot	sec	cosec	′
0	.40674	.91354	.44523	2.2460	1.0946	2.4586	.42262	.90631	.46631	2.1445	1.1034	2.3662	60
1	.40700	.91343	.44558	.2443	.0948	.4570	.42288	.90618	.46666	.1429	.1035	.3647	59
2	.40727	.91331	.44593	.2425	.0949	.4554	.42314	.90606	.46702	.1412	.1037	.3632	58
3	.40753	.91319	.44627	.2408	.0951	.4538	.42341	.90594	.46737	.1396	.1038	.3618	57
4	.40780	.91307	.44662	.2390	.0952	.4522	.42367	.90581	.46772	.1380	.1040	.3603	56
5	.40806	.91295	.44697	2.2373	.0953	2.4506	.42394	.90569	.46808	2.1364	.1041	2.3588	55
6	.40833	.91283	.44732	.2355	.0955	.4490	.42420	.90557	.46843	.1348	.1043	.3574	54
7	.40860	.91271	.44767	.2338	.0956	.4474	.42446	.90544	.46879	.1331	.1044	.3559	53
8	.40886	.91260	.44802	.2320	.0958	.4458	.42473	.90532	.46914	.1315	.1046	.3544	52
9	.40913	.91248	.44837	.2303	.0959	.4442	.42499	.90520	.46950	.1299	.1047	.3530	51
10	.40939	.91236	.44872	2.2286	.0961	2.4426	.42525	.90507	.46985	2.1283	.1049	2.3515	50
11	.40966	.91224	.44907	.2268	.0962	.4418	.42552	.90495	.47021	.1267	.1050	.3501	49
12	.40992	.91212	.44942	.2251	.0963	.4395	.42578	.90483	.47056	.1251	.1052	.3486	48
13	.41019	.91200	.44977	.2234	.0965	.4379	.42604	.90470	.47092	.1235	.1053	.3472	47
14	.41045	.91188	.45012	.2216	.0966	.4363	.42630	.90458	.47127	.1219	.1055	.3457	46
15	.41072	.91176	.45047	2.2199	.0968	2.4347	.42657	.90445	.47163	2.1203	.1056	2.3443	45
16	.41098	.91164	.45082	.2182	.0969	.4332	.42683	.90433	.47199	.1187	.1058	.3428	44
17	.41125	.91152	.45117	.2165	.0971	.4316	.42709	.90421	.47234	.1171	.1059	.3414	43
18	.41151	.91140	.45152	.2147	.0972	.4300	.42736	.90408	.47270	.1155	.1061	.3399	42
19	.41178	.91128	.45187	.2130	.0973	.4285	.42762	.90396	.47305	.1139	.1062	.3385	41
20	.41204	.91116	.45222	2.2113	.0975	2.4269	.42788	.90383	.47341	2.1123	.1064	2.3371	40
21	.41231	.91104	.45257	.2096	.0976	.4254	.42815	.90371	.47376	.1107	.1065	.3356	39
22	.41257	.91092	.45292	.2079	.0978	.4238	.42841	.90358	.47412	.1092	.1067	.3342	38
23	.41284	.91080	.45327	.2062	.0979	.4222	.42867	.90346	.47448	.1076	.1068	.3328	37
24	.41310	.91068	.45362	.2045	.0981	.4207	.42893	.90333	.47483	.1060	.1070	.3313	36
25	.41337	.91056	.45397	2.2028	.0982	2.4191	42920	.90321	.47519	2.1044	.1072	2.3299	35
26	.41363	.91044	.45432	.2011	.0984	.4176	.42946	.90308	.47555	.1028	.1073	.3285	34
27	.41390	.91032	.45467	.1994	.0985	.4160	.42972	.90296	.47590	.1013	.1075	.3271	33
28	.41416	.91020	.45502	.1977	.0986	.4145	.42998	.90283	.47626	.0997	.1076	.3256	32
29	.41443	.91008	.45537	.1960	.0988	.4130	.43025	.90271	.47662	.0981	.1078	.3242	31
30	.41469	.90996	.45573	2.1943	.0989	2.4114	.43051	.90258	.47697	2.0965	.1079	2.3228	30
31	.41496	.90984	.45608	.1926	.0991	.4099	.43077	.90246	.47733	.0950	.1081	.3214	29
32	.41522	.90972	.45643	.1909	.0992	.4083	.43104	.90233	.47769	.0934	.1082	.3200	28
33	.41549	.90960	.45678	.1892	.0994	.4068	.43130	.90221	.47805	.0918	.1084	.3186	27
34	.41575	.90948	.45713	.1875	.0995	.4053	.43156	.90208	.47840	.0903	.1085	.3172	26
35	.41602	.90936	.45748	2.1859	.0997	2.4037	.43182	.90196	.47876	2.0887	.1087	2.3158	25
36	.41628	.90924	.45783	.1842	.0998	.4022	.43208	.90183	.47912	.0872	.1088	.3143	24
37	.41654	.90911	.45819	.1825	.1000	.4007	.43235	.90171	.47948	.0856	.1090	.3129	23
38	.41681	.90899	.45854	.1808	.1001	.3992	.43261	.90158	.47983	.0840	.1092	.3115	22
39	.41707	.90887	.45889	.1792	.1003	.3976	.43287	.90145	.48019	.0825	.1093	.3101	21
40	.41734	.90875	.45924	2.1775	.1004	2.3961	.43313	.90133	.48055	2.0809	.1095	2.3087	20
41	.41760	.90863	.45960	.1758	.1005	.3946	.43340	.90120	.48091	.0794	.1096	.3073	19
42	.41787	.90851	.45995	.1741	.1007	.3931	.43366	.90108	.48127	.0778	.1098	.3059	18
43	.41813	.90839	.46030	.1725	.1008	.3916	.43392	.90095	.48162	.0763	.1099	.3046	17
44	.41839	.90826	.46065	.1708	.1010	.3901	.43418	.90082	.48198	.0747	.1101	.3032	16
45	.41866	.90814	.46101	2.1692	.1011	2.3886	.43444	.90070	.48234	2.0732	.1102	2.3018	15
46	.41892	.90812	.46136	.1675	.1013	.3871	.43471	.90057	.48270	.0717	.1104	.3004	14
47	.41919	.90790	.46171	.1658	.1014	.3856	.43497	.90044	.48306	.0701	.1106	.2990	13
48	.41945	.90778	.46206	.1642	.1016	.3841	.43523	.90032	.48342	.0686	.1107	.2976	12
49	.41972	.90765	.46242	.1625	.1017	.3826	.43549	.90019	.48378	.0671	.1109	.2962	11
50	.41998	.90753	.46277	2.1609	.1019	2.3811	.43575	.90006	.48414	2.0655	.1110	2.2949	10
51	.42024	.90741	.46312	.1592	.1020	.3796	.43602	.89994	.48449	.0640	.1112	.2935	9
52	.42051	.90729	.46348	.1576	.1022	.3781	.43628	.89981	.48485	.0625	.1113	.2921	8
53	.42077	.90717	.46383	.1559	.1023	.3766	.43654	.89968	.48521	.0609	.1115	.2907	7
54	.42103	.90704	.46418	.1543	.1025	.3751	.43680	.89956	.48557	.0594	.1116	.2894	6
55	.42130	.90692	.46454	2.1527	.1026	2.3736	.43706	.89943	.48593	2.0579	.1118	2.2880	5
56	.42156	.90680	.46489	.1510	.1028	.3721	.43732	.89930	.48629	.0564	.1120	.2866	4
57	.42183	.90668	.46524	.1494	.1029	.3706	.43759	.89918	.48665	.0548	.1121	.2853	3
58	.42209	.90655	.46560	.1478	.1031	.3691	.43785	.89905	.48701	.0533	.1123	.2839	2
59	.42235	.90643	.46595	.1461	.1032	.3677	.43811	.89892	.48737	.0518	.1124	.2825	1
60	.42262	.90631	.46631	2.1445	.1034	2.3662	.43837	.89879	.48773	2.0503	.1126	2.2812	0
′	cos	sin	cot	tan	cosec	sec	cos	sin	cot	tan	cosec	sec	′
	65°						64°						

	26°						27°						
′	sin	cos	tan	cot	sec	cosec	sin	cos	tan	cot	sec	cosec	′
0	.43837	.89879	.48773	2.0503	1.1126	2.2312	.45399	.89101	.50952	1.9626	1.1223	2.2027	60
1	.43863	.89867	.48809	.0488	.1127	.2798	.45425	.89087	.50989	.9612	.1225	.2014	59
2	.43889	.89854	.48845	.0473	.1129	.2784	.45451	.89074	.51026	.9598	.1226	.2002	58
3	.43915	.89841	.48881	.0458	.1131	.2771	.45477	.89061	.51062	.9584	.1228	.1989	57
4	.43942	.89828	.48917	.0443	.1132	.2757	.45503	.89048	.51099	.9570	.1230	.1977	56
5	.43968	.89815	.48953	2.0427	.1134	2.2744	.45528	.89034	.51136	1.9556	.1231	2.1964	55
6	.43994	.89803	.48989	.0412	.1135	.2730	.45554	.89021	.51172	.9542	.1233	.1952	54
7	.44020	.89790	.49025	.0397	.1137	.2717	.45580	.89008	.51209	.9528	.1235	.1939	53
8	.44046	.89777	.49062	.0382	.1139	.2703	.45606	.88995	.51246	.9514	.1237	.1927	52
9	.44072	.89764	.49098	.0367	.1140	.2690	.45632	.88981	.51283	.9500	.1238	.1914	51
10	.44098	.89751	.49134	2.0352	.1142	2.2676	.45658	.88968	.51319	1.9486	.1240	2.1902	50
11	.44124	.89739	.49170	.0338	.1143	.2663	.45684	.88955	.51356	.9472	.1242	.1889	49
12	.44150	.89726	.49206	.0323	.1145	.2650	.45710	.88942	.51393	.9458	.1243	.1877	48
13	.44177	.89713	.49242	.0308	.1147	.2636	.45736	.88928	.51430	.9444	.1245	.1865	47
14	.44203	.89700	.49278	.0293	.1148	.2623	.45761	.88915	.51466	.9430	.1247	.1852	46
15	.44229	.89687	.49314	2.0278	.1150	2.2610	.45787	.88902	.51503	1.9416	.1248	2.1840	45
16	.44255	.89674	.49351	.0263	.1151	.2596	.45813	.88888	.51540	.9402	.1250	.1828	44
17	.44281	.89661	.49387	.0248	.1153	.2583	.45839	.88875	.51577	.9388	.1252	.1815	43
18	.44307	.89649	.49423	.0233	.1155	.2570	.45865	.88862	.51614	.9375	.1253	.1803	42
19	.44333	.89636	.49459	.0219	.1156	.2556	.45891	.88848	.51651	.9361	.1255	.1791	41
20	.44359	.89623	.49495	2.0204	.1158	2.2543	.45917	.88835	.51687	1.9347	.1257	2.1778	40
21	.44385	.89610	.49532	.0189	.1159	.2530	.45942	.88822	.51724	.9333	.1258	.1766	39
22	.44411	.89597	.49568	.0174	.1161	.2517	.45968	.88808	.51761	.9319	.1260	.1754	38
23	.44437	.89584	.49604	.0159	.1163	.2503	.45994	.88795	.51798	.9306	.1262	.1742	37
24	.44463	.89571	.49640	.0145	.1164	.2490	.46020	.88781	.51835	.9292	.1264	.1730	36
25	.44489	.89558	.49677	2.0130	.1166	2.2477	.46046	.88768	.51872	1.9278	.1265	2.1717	35
26	.44516	.89545	.49713	.0115	.1167	.2464	.46072	.88755	.51909	.9264	.1267	.1705	34
27	.44542	.89532	.49749	.0101	.1169	.2451	.46097	.88741	.51946	.9251	.1269	.1693	33
28	.44568	.89519	.49785	.0086	.1171	.2438	.46123	.88728	.51983	.9237	.1270	.1681	32
29	.44594	.89506	.49822	.0071	.1172	.2425	.46149	.88714	.52020	.9223	.1272	.1669	31
30	.44620	.89493	.49858	2.0057	.1174	2.2411	.46175	.88701	.52057	1.9210	.1274	2.1657	30
31	.44646	.89480	.49894	.0042	.1176	.2398	.46201	.88688	.52094	.9196	.1275	.1645	29
32	.44672	.89467	.49931	.0028	.1177	.2385	.46226	.88674	.52131	.9182	.1277	.1633	28
33	.44698	.89454	.49967	.0013	.1179	.2372	.46252	.88661	.52168	.9169	.1279	.1620	27
34	.44724	.89441	.50003	1.9998	.1180	.2359	.46278	.88647	.52205	.9155	.1281	.1608	26
35	.44750	.89428	.50040	1.9984	.1182	2.2346	.46304	.88634	.52242	1.9142	.1282	2.1596	25
36	.44776	.89415	.50076	.9969	.1184	.2333	.46330	.88620	.52279	.9128	.1284	.1584	24
37	.44802	.89402	.50113	.9955	.1185	.2320	.46355	.88607	.52316	.9115	.1286	.1572	23
38	.44828	.89389	.50149	.9940	.1187	.2307	.46381	.88593	.52353	.9101	.1287	.1560	22
39	.44854	.89376	.50185	.9926	.1189	.2294	.46407	.88580	.52390	.9088	.1289	.1548	21
40	.44880	.89363	.50222	1.9912	.1190	2.2282	.46433	.88566	.52427	1.9074	.1291	2.1536	20
41	.44906	.89350	.50258	.9897	.1192	.2269	.46458	.88553	.52464	.9061	.1293	.1525	19
42	.44932	.89337	.50295	.9883	.1193	.2256	.46484	.88539	.52501	.9047	.1294	.1513	18
43	.44958	.89324	.50331	.9868	.1195	.2243	.46510	.88526	.52538	.9034	.1296	.1501	17
44	.44984	.89311	.50368	.9854	.1197	.2230	.46536	.88512	.52575	.9020	.1298	.1489	16
45	.45010	.89298	.50404	1.9840	.1198	2.2217	.46561	.88499	.52612	1.9007	.1299	2.1477	15
46	.45036	.89285	.50441	.9825	.1200	.2204	.46587	.88485	.52650	.8993	.1301	.1465	14
47	.45062	.89272	.50477	.9811	.1202	.2192	.46613	.88472	.52687	.8980	.1303	.1453	13
48	.45088	.89258	.50514	.9797	.1203	.2179	.46639	.88458	.52724	.8967	.1305	.1441	12
49	.45114	.89245	.50550	.9782	.1205	.2166	.46664	.88444	.52761	.8953	.1306	.1430	11
50	.45140	.89232	.50587	1.9768	.1207	2.2153	.46690	.88431	.52798	1.8940	.1308	2.1418	10
51	.45166	.89219	.50623	.9754	.1208	.2141	.46716	.88417	.52836	.8927	.1310	.1406	9
52	.45191	.89206	.50660	.9739	.1210	.2128	.46741	.88404	.52873	.8913	.1312	.1394	8
53	.45217	.89193	.50696	.9725	.1212	.2115	.46767	.88390	.52910	.8900	.1313	.1382	7
54	.45243	.89180	.50733	.9711	.1213	.2103	.46793	.88376	.52947	.8887	.1315	.1371	6
55	.45269	.89166	.50769	1.9697	.1215	2.2090	.46819	.88363	.52984	1.8873	.1317	2.1359	5
56	.45295	.89153	.50806	.9683	.1217	.2077	.46844	.88349	.53022	.8860	.1319	.1347	4
57	.45321	.89140	.50843	.9668	.1218	.2065	.46870	.88336	.53059	.8847	.1320	.1335	3
58	.45347	.89127	.50879	.9654	.1220	.2052	.46896	.88322	.53096	.8834	.1322	.1324	2
59	.45373	.89114	.50916	.9640	.1222	.2039	.46921	.88308	.53134	.8820	.1324	.1312	1
60	.45399	.89101	.50952	1.9626	.1223	2.2027	.46947	.88295	.53171	1.8807	.1326	2.1300	0
′	cos	sin	cot	tan	cosec	sec	cos	sin	cot	tan	cosec	sec	′
	63°						62°						

	28°						29°						
′	sin	cos	tan	cot	sec	cosec	sin	cos	tan	cot	sec	cosec	
0	.46947	.88295	.53171	1.8807	1.1326	2.1300	.48481	.87462	.55431	1.8040	1.1433	2.0627	60
1	.46973	.88281	.53208	.8794	.1327	.1289	.48506	.87448	.55469	.8028	.1435	.0616	59
2	.46998	.88267	.53245	.8781	.1329	.1277	.48532	.87434	.55507	.8016	.1437	.0605	58
3	.47024	.88254	.53283	.8768	.1331	.1266	.48557	.87420	.55545	.8003	.1439	.0594	57
4	.47050	.88240	.53320	.8754	.1333	.1254	.48583	.87405	.55583	.7991	.1441	.0583	56
5	.47075	.88226	.53358	1.8741	.1334	2.1242	.48608	.87391	.55621	1.7979	.1443	2.0573	55
6	.47101	.88213	.53395	.8728	.1336	.1231	.48633	.87377	.55659	.7966	.1445	.0562	54
7	.47127	.88199	.53432	.8715	.1338	.1219	.48659	.87363	.55697	.7954	.1446	.0551	53
8	.47152	.88185	.53470	.8702	.1340	.1208	.48684	.87349	.55735	.7942	.1448	.0540	52
9	.47178	.88171	.53507	.8689	.1341	.1196	.48710	.87335	.55774	.7930	.1450	.0530	51
10	.47204	.88158	.53545	1.8676	.1343	2.1185	.48735	.87320	.55812	1.7917	.1452	2.0519	50
11	.47229	.88144	.53582	.8663	.1345	.1173	.48760	.87306	.55850	.7905	.1454	.0508	49
12	.47255	.88130	.53619	.8650	.1347	.1162	.48786	.87292	.55888	.7893	.1456	.0498	48
13	.47281	.88117	.53657	.8637	.1349	.1150	.48811	.87278	.55926	.7881	.1458	.0487	47
14	.47306	.88103	.53694	.8624	.1350	.1139	.48837	.87264	.55964	.7868	.1459	.0476	46
15	.47332	.88089	.53732	1.8611	.1352	2.1127	.48862	.87250	.56003	1.7856	.1461	2.0466	45
16	.47357	.88075	.53769	.8598	.1354	.1116	.48887	.87235	.56041	.7844	.1463	.0455	44
17	.47383	.88061	.53807	.8585	.1356	.1104	.48913	.87221	.56079	.7832	.1465	.0444	43
18	.47409	.88048	.53844	.8572	.1357	.1093	.48938	.87207	.56117	.7820	.1467	.0434	42
19	.47434	.88034	.53882	.8559	.1359	.1082	.48964	.87193	.56156	.7808	.1469	.0423	41
20	.47460	.88020	.53919	1.8546	.1361	2.1070	.48989	.87178	.56194	1.7795	.1471	2.0413	40
21	.47486	.88006	.53957	.8533	.1363	.1059	.49014	.87164	.56232	.7783	.1473	.0402	39
22	.47511	.87992	.53995	.8520	.1365	.1048	.49040	.87150	.56270	.7771	.1474	.0392	38
23	.47537	.87979	.54032	.8507	.1366	.1036	.49065	.87136	.56309	.7759	.1476	.0381	37
24	.47562	.87965	.54070	.8495	.1368	.1025	.49090	.87121	.56347	.7747	.1478	.0370	36
25	.47588	.87951	.54107	1.8482	.1370	2.1014	.49116	.87107	.56385	1.7735	.1480	2.0360	35
26	.47613	.87937	.54145	.8469	.1372	.1002	.49141	.87093	.56424	.7723	.1482	.0349	34
27	.47639	.87923	.54183	.8456	.1373	.0991	.49166	.87078	.56462	.7711	.1484	.0339	33
28	.47665	.87909	.54220	.8443	.1375	.0980	.49192	.87064	.56500	.7699	.1486	.0329	32
29	.47690	.87895	.54258	.8430	.1377	.0969	.49217	.87050	.56539	.7687	.1488	.0318	31
30	.47716	.87882	.54295	1.8418	.1379	2.0957	.49242	.87035	.56577	1.7675	.1489	2.0308	30
31	.47741	.87868	.54333	.8405	.1381	.0946	.49268	.87021	.56616	.7663	.1491	.0297	29
32	.47767	.87854	.54371	.8392	.1382	.0935	.49293	.87007	.56654	.7651	.1493	.0287	28
33	.47792	.87840	.54409	.8379	.1384	.0924	.49318	.86992	.56692	.7639	.1495	.0276	27
34	.47818	.87826	.54446	.8367	.1386	.0912	.49343	.86978	.56731	.7627	.1497	.0266	26
35	.47844	.87812	.54484	1.8354	.1388	2.0901	.49369	.86964	.56769	1.7615	.1499	2.0256	25
36	.47869	.87798	.54522	.8341	.1390	.0890	.49394	.86949	.56808	.7603	.1501	.0245	24
37	.47895	.87784	.54559	.8329	.1391	.0879	.49419	.86935	.56846	.7591	.1503	.0235	23
38	.47920	.87770	.54597	.8316	.1393	.0868	.49445	.86921	.56885	.7579	.1505	.0224	22
39	.47946	.87756	.54635	.8303	.1395	.0857	.49470	.86906	.56923	.7567	.1507	.0214	21
40	.47971	.87742	.54673	1.8291	.1397	2.0846	.49495	.86892	.56962	1.7555	.1508	2.0204	20
41	.47997	.87728	.54711	.8278	.1399	.0835	.49521	.86877	.57000	.7544	.1510	.0194	19
42	.48022	.87715	.54748	.8265	.1401	.0824	.49546	.86863	.57039	.7532	.1512	.0183	18
43	.48048	.87701	.54786	.8253	.1402	.0812	.49571	.86849	.57077	.7520	.1514	.0173	17
44	.48073	.87687	.54824	.8240	.1404	.0801	.49596	.86834	.57116	.7508	.1516	.0163	16
45	.48099	.87673	.54862	1.8227	.1406	2.0790	.49622	.86820	.57155	1.7496	.1518	2.0152	15
46	.48124	.87659	.54900	.8215	.1408	.0779	.49647	.86805	.57193	.7484	.1520	.0142	14
47	.48150	.87645	.54937	.8202	.1410	.0768	.49672	.86791	.57232	.7473	.1522	.0132	13
48	.48175	.87631	.54975	.8190	.1411	.0757	.49697	.86776	.57270	.7461	.1524	.0122	12
49	.48201	.87617	.55013	.8177	.1413	.0746	.49723	.86762	.57309	.7449	.1526	.0111	11
50	.48226	.87603	.55051	1.8165	.1415	2.0735	.49748	.86748	.57348	1.7437	.1528	2.0101	10
51	.48252	.87588	.55089	.8152	.1417	.0725	.49773	.86733	.57386	.7426	.1530	.0091	9
52	.48277	.87574	.55127	.8140	.1419	.0714	.49798	.86719	.57425	.7414	.1531	.0081	8
53	.48303	.87560	.55165	.8127	.1421	.0703	.49823	.86704	.57464	.7402	.1533	.0071	7
54	.48328	.87546	.55203	.8115	.1422	.0692	.49849	.86690	.57502	.7390	.1535	.0061	6
55	.48354	.87532	.55241	1.8102	.1424	2.0681	.49874	.86675	.57541	1.7379	.1537	2.0050	5
56	.48379	.87518	.55279	.8090	.1426	.0670	.49899	.86661	.57580	.7367	.1539	.0040	4
57	.48405	.87504	.55317	.8078	.1428	.0659	.49924	.86646	.57619	.7355	.1541	.0030	3
58	.48430	.87490	.55355	.8065	.1430	.0648	.49950	.86632	.57657	.7344	.1543	.0020	2
59	.48455	.87476	.55393	.8053	.1432	.0637	.49975	.86617	.57696	.7332	.1545	.0010	1
60	.48481	.87462	.55431	1.8040	.1433	2.0627	.50000	.86603	.57735	1.7320	.1547	2.0000	0
′	cos	sin	cot	tan	cosec	sec	cos	sin	cot	tan	cosec	sec	′
	61°						60°						

	30°						31°						
′	sin	cos	tan	cot	sec	cosec	sin	cos	tan	cot	sec	cosec	′
0	.50000	.86603	.57735	1.7320	1.1547	2.0000	.51504	.85717	.60086	1.6643	1.1666	1.9416	60
1	.50025	.86588	.57774	.7309	.1549	1.9990	.51529	.85702	.60126	.6632	.1668	.9407	59
2	.50050	.86573	.57813	.7297	.1551	.9980	.51554	.85687	.60165	.6621	.1670	.9397	58
3	.50075	.86559	.57851	.7286	.1553	.9970	.51578	.85672	.60205	.6610	.1672	.9388	57
4	.50101	.86544	.57890	.7274	.1555	.9960	.51603	.85657	.60244	.6599	.1674	.9378	56
5	.50126	.86530	.57929	1.7262	.1557	1.9950	.51628	.85642	.60284	1.6588	.1676	1.9369	55
6	.50151	.86515	.57968	.7251	.1559	.9940	.51653	.85627	.60324	.6577	.1678	.9360	54
7	.50176	.86500	.58007	.7239	.1561	.9930	.51678	.85612	.60363	.6566	.1681	.9350	53
8	.50201	.86486	.58046	.7228	.1562	.9920	.51703	.85597	.60403	.6555	.1683	.9341	52
9	.50226	.86471	.58085	.7216	.1564	.9910	.51728	.85582	.60443	.6544	.1685	.9332	51
10	.50252	.86457	.58123	1.7205	.1566	1.9900	.51753	.85566	.60483	1.6534	.1687	1.9322	50
11	.50277	.86442	.58162	.7193	.1568	.9890	.51778	.85551	.60522	.6523	.1689	.9313	49
12	.50302	.86427	.58201	.7182	.1570	.9880	.51803	.85536	.60562	.6512	.1691	.9304	48
13	.50327	.86413	.58240	.7170	.1572	.9870	.51827	.85521	.60602	.6501	.1693	.9295	47
14	.50352	.86398	.58279	.7159	.1574	.9860	.51852	.85506	.60642	.6490	.1695	.9285	46
15	.50377	.86383	.58318	1.7147	.1576	1.9850	.51877	.85491	.60681	1.6479	.1697	1.9276	45
16	.50402	.86369	.58357	.7136	.1578	.9840	.51902	.85476	.60721	.6469	.1699	.9267	44
17	.50428	.86354	.58396	.7124	.1580	.9830	.51927	.85461	.60761	.6458	.1701	.9258	43
18	.50453	.86339	.58435	.7113	.1582	.9820	.51952	.85446	.60801	.6447	.1703	.9248	42
19	.50478	.86325	.58474	.7101	.1584	.9811	.51977	.85431	.60841	.6436	.1705	.9239	41
20	.50503	.86310	.58513	1.7090	.1586	1.9801	.52002	.85416	.60881	1.6425	.1707	1.9230	40
21	.50528	.86295	.58552	.7079	.1588	.9791	.52026	.85400	.60920	.6415	.1709	.9221	39
22	.50553	.86281	.58591	.7067	.1590	.9781	.52051	.85385	.60960	.6404	.1712	.9212	38
23	.50578	.86266	.58630	.7056	.1592	.9771	.52076	.85370	.61000	.6393	.1714	.9203	37
24	.50603	.86251	.58670	.7044	.1594	.9761	.52101	.85355	.61040	.6383	.1716	.9193	36
25	.50628	.86237	.58709	1.7033	.1596	1.9752	.52126	.85340	.61080	1.6372	.1718	1.9184	35
26	.50653	.86222	.58748	.7022	.1598	.9742	.52151	.85325	.61120	.6361	.1720	.9175	34
27	.50679	.86207	.58787	.7010	.1600	.9732	.52175	.85309	.61160	.6350	.1722	.9166	33
28	.50704	.86192	.58826	.6999	.1602	.9722	.52200	.85294	.61200	.6340	.1724	.9157	32
29	.50729	.86178	.58865	.6988	.1604	.9713	.52225	.85279	.61240	.6329	.1726	.9148	31
30	.50754	.86163	.58904	1.6977	.1606	1.9703	.52250	.85264	.61280	1.6318	.1728	1.9139	30
31	.50779	.86148	.58944	.6965	.1608	.9693	.52275	.85249	.61320	.6308	.1730	.9130	29
32	.50804	.86133	.58983	.6954	.1610	.9683	.52299	.85234	.61360	.6297	.1732	.9121	28
33	.50829	.86118	.59022	.6943	.1612	.9674	.52324	.85218	.61400	.6286	.1734	.9112	27
34	.50854	.86104	.59061	.6931	.1614	.9664	.52349	.85203	.61440	.6276	.1737	.9102	26
35	.50879	.86089	.59100	1.6920	.1616	1.9654	.52374	.85188	.61480	1.6265	.1739	1.9093	25
36	.50904	.86074	.59140	.6909	.1618	.9645	.52398	.85173	.61520	.6255	.1741	.9084	24
37	.50929	.86059	.59179	.6898	.1620	.9635	.52423	.85157	.61560	.6244	.1743	.9075	23
38	.50954	.86044	.59218	.6887	.1622	.9625	.52448	.85142	.61601	.6233	.1745	.9066	22
39	.50979	.86030	.59258	.6875	.1624	.9616	.52473	.85127	.61641	.6223	.1747	.9057	21
40	.51004	.86015	.59297	1.6864	.1626	1.9606	.52498	.85112	.61681	1.6212	.1749	1.9048	20
41	.51029	.86000	.59336	.6853	.1628	.9596	.52522	.85096	.61721	.6202	.1751	.9039	19
42	.51054	.85985	.59376	.6842	.1630	.9587	.52547	.85081	.61761	.6191	.1753	.9030	18
43	.51079	.85970	.59415	.6831	.1632	.9577	.52572	.85066	.61801	.6181	.1756	.9021	17
44	.51104	.85955	.59454	.6820	.1634	.9568	.52597	.85050	.61842	.6170	.1758	.9013	16
45	.51129	.85941	.59494	1.6808	.1636	1.9558	.52621	.85735	.61882	1.6160	.1760	1.9004	15
46	.51154	.85926	.59533	.6797	.1638	.9549	.52646	.85020	.61922	.6149	.1762	.8995	14
47	.51179	.85911	.59572	.6786	.1640	.9539	.52671	.85004	.61962	.6139	.1764	.8986	13
48	.51204	.85896	.59612	.6775	.1642	.9530	.52695	.84989	.62003	.6128	.1766	.8977	12
49	.51229	.85881	.59651	.6764	.1644	.9520	.52720	.84974	.62043	.6118	.1768	.8968	11
50	.51254	.85866	.59691	1.6753	.1646	1.9510	.52745	.84959	.62083	1.6107	.1770	1.8959	10
51	.51279	.85851	.59730	.6742	.1648	.9501	.52770	.84943	.62123	.6097	.1772	.8950	9
52	.51304	.85836	.59770	.6731	.1650	.9491	.52794	.84928	.62164	.6086	.1775	.8941	8
53	.51329	.85821	.59809	.6720	.1652	.9482	.52819	.84912	.62204	.6076	.1777	.8932	7
54	.51354	.85806	.59849	.6709	.1654	.9473	.52844	.84897	.62244	.6066	.1779	.8924	6
55	.51379	.85791	.59888	1.6698	.1656	1.9463	.52868	.84882	.62285	1.6055	.1781	1.8915	5
56	.51404	.85777	.59928	.6687	.1658	.9454	.52893	.84866	.62325	.6045	.1783	.8906	4
57	.51429	.85762	.59967	.6676	.1660	.9444	.52918	.84851	.62366	.6034	.1785	.8897	3
58	.51454	.85747	.60007	.6665	.1662	.9435	.52942	.84836	.62406	.6024	.1787	.8888	2
59	.51479	.85732	.60046	.6654	.1664	.9425	.52967	.84820	.62446	.6014	.1790	.8879	1
60	.51504	.85717	.60086	1.6643	.1666	1.9416	.52992	.84805	.62487	1.6003	.1792	1.8871	0
′	cos	sin	cot	tan	cosec	sec	cos	sin	cot	tan	cosec	sec	′
	59°						58°						

	32°						33°						
′	sin	cos	tan	cot	sec	cosec	sin	cos	tan	cot	sec	cosec	′
0	.52992	.84805	.62487	1.6003	1.1792	1.8871	.54464	.83867	.64941	1.5399	1.1924	1.8361	60
1	.53016	.84789	.62527	.5993	.1794	.8862	.54488	.83851	.64982	.5389	.1926	.8352	59
2	.53041	.84774	.62568	.5983	.1796	.8853	.54513	.83835	.65023	.5379	.1928	.8344	58
3	.53066	.84758	.62608	.5972	.1798	.8844	.54537	.83819	.65065	.5369	.1930	.8336	57
4	.53090	.84743	.62649	.5962	.1800	.8836	.54561	.83804	.65106	.5359	.1933	.8328	56
5	.53115	.84728	.62689	1.5952	.1802	1.8827	.54586	.83788	.65148	1.5350	.1935	1.8320	55
6	.53140	.84712	.62730	.5941	.1805	.8818	.54610	.83772	.65189	.5340	.1937	.8311	54
7	.53164	.84697	.62770	.5931	.1807	.8809	.54634	.83756	.65231	.5330	.1939	.8303	53
8	.53189	.84681	.62811	.5921	.1809	.8801	.54659	.83740	.65272	.5320	.1942	.8295	52
9	.53214	.84666	.62851	.5910	.1811	.8792	.54683	.83724	.65314	.5311	.1944	.8287	51
10	.53238	.84650	.62892	1.5900	.1813	1.8783	.54708	.83708	.65355	1.5301	.1946	1.8279	50
11	.53263	.84635	.62933	.5890	.1815	.8775	.54732	.83692	.65397	.5291	.1948	.8271	49
12	.53288	.84619	.62973	.5880	.1818	.8766	.54756	.83676	.65438	.5282	.1951	.8263	48
13	.53312	.84604	.63014	.5869	.1820	.8757	.54781	.83660	.65480	.5272	.1953	.8255	47
14	.53337	.84588	.63055	.5859	.1822	.8749	.54805	.83644	.65521	.5262	.1955	.8246	46
15	.53361	.84573	.63095	1.5849	.1824	1.8740	.54829	.83629	.65563	1.5252	.1958	1.8238	45
16	.53386	.84557	.63136	.5839	.1826	.8731	.54854	.83613	.65604	.5234	.1960	.8230	44
17	.53411	.84542	.63177	.5829	.1828	.8723	.54878	.83597	.65646	.5233	.1962	.8222	43
18	.53435	.84526	.63217	.5818	.1831	.8714	.54902	.83581	.65688	.5223	.1964	.8214	42
19	.53460	.84511	.63258	.5808	.1833	.8706	.54926	.83565	.65729	.5214	.1967	.8206	41
20	.53484	.84495	.63299	1.5798	.1835	1.8697	.54951	.83549	.65771	1.5204	.1969	1.8198	40
21	.53509	.84479	.63339	.5788	.1837	.8688	.54975	.83533	.65813	.5195	.1971	.8190	39
22	.53533	.84464	.63380	.5778	.1839	.8680	.54999	.83517	.65854	.5185	.1974	.8182	38
23	.53558	.84448	.63421	.5768	.1841	.8671	.55024	.83501	.65896	.5175	.1976	.8174	37
24	.53583	.84433	.63462	.5757	.1844	.8663	.55048	.83485	.65938	.5166	.1978	.8166	36
25	.53607	.84417	.63503	1.5747	.1846	1.8654	.55002	.83469	.65980	1.5156	.1980	1.8158	35
26	.53632	.84402	.63543	.5737	.1848	.8646	.55090	.83453	.66021	.5147	.1983	.8150	34
27	.53656	.84386	.63584	.5727	.1850	.8637	.55121	.83437	.66063	.5137	.1985	.8142	33
28	.53681	.84370	.63625	.5717	.1852	.8629	.55145	.83421	.66105	.5127	.1987	.8134	32
29	.53705	.84355	.63666	.5707	.1855	.8620	.55169	.83405	.66147	.5118	.1990	.8126	31
30	.53730	.84339	.63707	1.5697	.1857	1.8611	.55194	.83388	.66188	1.5108	.1992	1.8118	30
31	.53754	.84323	.63748	.5687	.1859	.8603	.55218	.83372	.66230	.5099	.1994	.8110	29
32	.53779	.84308	.63789	.5677	.1861	.8595	.55242	.83356	.66272	.5089	.1997	.8102	28
33	.53803	.84292	.63830	.5667	.1863	.8586	.55266	.83340	.66314	.5080	.1999	.8094	27
34	.53828	.84276	.63871	.5657	.1866	.8578	.55291	.83324	.66356	.5070	.2001	.8086	26
35	.53852	.84261	.63912	1.5646	.1868	1.8569	.55315	.83308	.66398	1.5061	.2004	1.8078	25
36	.53877	.84245	.63953	.5636	.1870	.8561	.55339	.83292	.66440	.5051	.2006	.8070	24
37	.53901	.84229	.63994	.5626	.1872	.8552	.55363	.83276	.66482	.5042	.2008	.8062	23
38	.53926	.84214	.64035	.5616	.1874	.8544	.55388	.83260	.66524	.5032	.2010	.8054	22
39	.53950	.84198	.64076	.5606	.1877	.8535	.55412	.83244	.66566	.5023	.2013	.8047	21
40	.53975	.84182	.64117	1.5596	.1879	1.8527	.55436	.83228	.66608	1.5013	.2015	1.8039	20
41	.53999	.84167	.64158	.5586	.1881	.8519	.55460	.83211	.66650	.5004	.2017	.8031	19
42	.54024	.84151	.64199	.5577	.1883	.8510	.55484	.83195	.66692	.4994	.2020	.8023	18
43	.54048	.84135	.64240	.5567	.1886	.8502	.55509	.83179	.66734	.4985	.2022	.8015	17
44	.54073	.84120	.64281	.5557	.1888	.8493	.55533	.83163	.66776	.4975	.2024	.8007	16
45	.54097	.84104	.64322	1.5547	.1890	1.8485	.55557	.83147	.66818	1.4966	.2027	1.7999	15
46	.54122	.84088	.64363	.5537	.1892	.8477	.55581	.83131	.66860	.4957	.2029	.7992	14
47	.54146	.84072	.64404	.5527	.1894	.8468	.55605	.83115	.66902	.4947	.2031	.7984	13
48	.54171	.84057	.64446	.5517	.1897	.8460	.55629	.83098	.66944	.4938	.2034	.7976	12
49	.54195	.84041	.64487	.5507	.1899	.8452	.55654	.83082	.66986	.4928	.2036	.7968	11
50	.54220	.84025	.64528	1.5497	.1901	1.8443	.55678	.83066	.67028	1.4919	.2039	1.7960	10
51	.54244	.84009	.64569	.5487	.1903	.8435	.55702	.83050	.67071	.4910	.2041	.7953	9
52	.54268	.83993	.64610	.5477	.1906	.8427	.55726	.83034	.67113	.4900	.2043	.7945	8
53	.54293	.83978	.64652	.5467	.1908	.8418	.55750	.83017	.67155	.4891	.2046	.7937	7
54	.54317	.83962	.64693	.5458	.1910	.8410	.55774	.83001	.67197	.4881	.2048	.7929	6
55	.54342	.83946	.64734	1.5448	.1912	1.8402	.55799	.82985	.67239	1.4872	.2050	1.7921	5
56	.54366	.83930	.64775	.5438	.1915	.8394	.55823	.82969	.67282	.4863	.2053	.7914	4
57	.54391	.83914	.64817	.5428	.1917	.8385	.55847	.82952	.67324	.4853	.2055	.7906	3
58	.54415	.83899	.64858	.5418	.1919	.8377	.55871	.82936	.67366	.4844	.2057	.7898	2
59	.54439	.83883	.64899	.5408	.1921	.8369	.55895	.82920	.67408	.4835	.2060	.7891	1
60	.54464	.83867	.64941	1.5399	.1922	1.8361	.55919	.82904	.67451	1.4826	.2062	1.7883	0
′	cos	sin	cot	tan	cosec	sec	cos	sin	cot	tan	cosec	sec	′
	57°						56°						

′	34° sin	cos	tan	cot	sec	cosec	35° sin	cos	tan	cot	sec	cosec	′
0	.55919	.82904	.67451	1.4826	1.2062	1.7883	.57358	.81915	.70021	1.4281	1.2208	1.7434	60
1	.55943	.82887	.67493	.4816	.2064	.7875	.57381	.81898	.70064	.4273	.2210	.7427	59
2	.55967	.82871	.67535	.4807	.2067	.7867	.57405	.81882	.70107	.4264	.2213	.7420	58
3	.55992	.82855	.67578	.4798	.2069	.7860	.57429	.81865	.70151	.4255	.2215	.7413	57
4	.56016	.82839	.67620	.4788	.2072	.7852	.57453	.81848	.70194	.4246	.2218	.7405	56
5	.56040	.82822	.67663	1.4779	.2074	1.7844	.57477	.81832	.70238	1.4237	.2220	1.7398	55
6	.56064	.82806	.67705	.4770	.2076	.7837	.57500	.81815	.70281	.4228	.2223	.7391	54
7	.56088	.82790	.67747	.4761	.2079	.7829	.57524	.81798	.70325	.4220	.2225	.7384	53
8	.56112	.82773	.67790	.4751	.2081	.7821	.57548	.81781	.70368	.4211	.2228	.7377	52
9	.56136	.82757	.67832	.4742	.2083	.7814	.57572	.81765	.70412	.4202	.2230	.7369	51
10	.56160	.82741	.67875	1.4733	.2086	1.7806	.57596	.81748	.70455	1.4193	.2233	1.7362	50
11	.56184	.82724	.67917	.4724	.2088	.7798	.57619	.81731	.70499	.4185	.2235	.7355	49
12	.56208	.82708	.67960	.4714	.2091	.7791	.57643	.81714	.70542	.4176	.2238	.7348	48
13	.56232	.82692	.68002	.4705	.2093	.7783	.57667	.81698	.70586	.4167	.2240	.7341	47
14	.56256	.82675	.68045	.4696	.2095	.7776	.57691	.81681	.70629	.4158	.2243	.7334	46
15	.56280	.82659	.68087	1.4687	.2098	1.7768	.57714	.81664	.70673	1.4150	.2245	1.7327	45
16	.56304	.82643	.68130	.4678	.2100	.7760	.57738	.81647	.70717	.4141	.2248	.7319	44
17	.56328	.82626	.68173	.4669	.2103	.7753	.57762	.81630	.70760	.4132	.2250	.7312	43
18	.56353	.82610	.68215	.4659	.2105	.7745	.57786	.81614	.70804	.4123	.2253	.7305	42
19	.56377	.82593	.68258	.4650	.2107	.7738	.57809	.81597	.70848	.4115	.2255	.7298	41
20	.56401	.82577	.68301	1.4641	.2110	1.7730	.57833	.81580	.70891	1.4106	.2258	1.7291	40
21	.56425	.82561	.68343	.4632	.2112	.7723	.57857	.81563	.70935	.4097	.2260	.7284	39
22	.56449	.82544	.68386	.4623	.2115	.7715	.57881	.81546	.70979	.4089	.2263	.7277	38
23	.56473	.82528	.68429	.4614	.2117	.7708	.57904	.81530	.71022	.4080	.2265	.7270	37
24	.56497	.82511	.68471	.4605	.2119	.7700	.57928	.81513	.71066	.4071	.2268	.7263	36
25	.56521	.82495	.68514	1.4595	.2122	1.7693	.57952	.81496	.71110	1.4063	.2270	1.7256	35
26	.56545	.82478	.68557	.4586	.2124	.7685	.57975	.81479	.71154	.4054	.2273	.7249	34
27	.56569	.82462	.68600	.4577	.2127	.7678	.57999	.81462	.71198	.4045	.2276	.7242	33
28	.56593	.82445	.68642	.4568	.2129	.7670	.58023	.81445	.71241	.4037	.2278	.7234	32
29	.56617	.82429	.68685	.4559	.2132	.7663	.58047	.81428	.71285	.4028	.2281	.7227	31
30	.56641	.82413	.68728	1.4550	.2134	1.7655	.58070	.81411	.71329	1.4019	.2283	1.7220	30
31	.56664	.82396	.68771	.4541	.2136	.7648	.58094	.81395	.71373	.4011	.2286	.7213	29
32	.56688	.82380	.68814	.4532	.2139	.7640	.58118	.81378	.71417	.4002	.2288	.7206	28
33	.56712	.82363	.68857	.4523	.2141	.7633	.58141	.81361	.71461	.3994	.2291	.7199	27
34	.56736	.82347	.68899	.4514	.2144	.7625	.58165	.81344	.71505	.3985	.2293	.7192	26
35	.56760	.82330	.68942	1.4505	.2146	1.7618	.58189	.81327	.71549	1.3976	.2296	1.7185	25
36	.56784	.82314	.68985	.4496	.2149	.7610	.58212	.81310	.71593	.3968	.2298	.7178	24
37	.56808	.82297	.69028	.4487	.2151	.7603	.58236	.81293	.71637	.3959	.2301	.7171	23
38	.56832	.82280	.69071	.4478	.2153	.7596	.58259	.81276	.71681	.3951	.2304	.7164	22
39	.56856	.82264	.69114	.4469	.2156	.7588	.58283	.81259	.71725	.3942	.2306	.7157	21
40	.56880	.82247	.69157	1.4460	.2158	1.7581	.58307	.81242	.71769	1.3933	.2309	1.7151	20
41	.56904	.82231	.69200	.4451	.2161	.7573	.58330	.81225	.71813	.3925	.2311	.7144	19
42	.56928	.82214	.69243	.4442	.2163	.7566	.58354	.81208	.71857	.3916	.2314	.7137	18
43	.56952	.82198	.69286	.4433	.2166	.7559	.58378	.81191	.71901	.3908	.2316	.7130	17
44	.56976	.82181	.69329	.4424	.2168	.7551	.58401	.81174	.71945	.3899	.2319	.7123	16
45	.57000	.82165	.69372	1.4415	.2171	1.7544	.58425	.81157	.71990	1.3891	.2322	1.7116	15
46	.57023	.82148	.69415	.4406	.2173	.7537	.58448	.81140	.72034	.3882	.2324	.7109	14
47	.57047	.82131	.69459	.4397	.2175	.7529	.58472	.81123	.72078	.3874	.2327	.7102	13
48	.57071	.82115	.69502	.4388	.2178	.7522	.58496	.81106	.72122	.3865	.2329	.7095	12
49	.57095	.82098	.69545	.4379	.2180	.7514	.58519	.81089	.72166	.3857	.2332	.7088	11
50	.57119	.82082	.69588	1.4370	.2183	1.7507	.58543	.81072	.72211	1.3848	.2335	1.7081	10
51	.57143	.82065	.69631	.4361	.2185	.7500	.58566	.81055	.72255	.3840	.2337	.7075	9
52	.57167	.82048	.69674	.4352	.2188	.7493	.58590	.81038	.72299	.3831	.2340	.7068	8
53	.57191	.82032	.69718	.4343	.2190	.7485	.58614	.81021	.72344	.3823	.2342	.7061	7
54	.57214	.82015	.69761	.4335	.2193	.7478	.58637	.81004	.72388	.3814	.2345	.7054	6
55	.57238	.81998	.69804	1.4326	.2195	1.7471	.58661	.80987	.72432	1.3806	.2348	1.7047	5
56	.57262	.81982	.69847	.4317	.2198	.7463	.58684	.80970	.72477	.3797	.2350	.7040	4
57	.57286	.81965	.69891	.4308	.2200	.7456	.58708	.80953	.72521	.3789	.2353	.7033	3
58	.57310	.81948	.69934	.4299	.2203	.7449	.58731	.80936	.72565	.3781	.2355	.7027	2
59	.57334	.81932	.69977	.4290	.2205	.7442	.58755	.80919	.72610	.3772	.2358	.7020	1
60	.57358	.81915	.70021	1.4281	.2208	1.7434	.58778	.80902	.72654	1.3764	.2361	1.7013	0
′	cos	sin	cot	tan	cosec	sec	cos	sin	cot	tan	cosec	sec	′
	55°						54°						

	36°						37°						
′	sin	cos	tan	cot	sec	cosec	sin	cos	tan	cot	sec	cosec	′
0	.58778	.80902	.72654	1.3764	1.2361	1.7013	.60181	.79863	.75355	1.3270	1.2521	1.6616	60
1	.58802	.80885	.72699	.3755	.2363	.7006	.60205	.79846	.75401	.3262	.2524	.6610	59
2	.58825	.80867	.72743	.3747	.2366	.6999	.60228	.79828	.75447	.3254	.2527	.6603	58
3	.58849	.80850	.72788	.3738	.2368	.6993	.60251	.79811	.75492	.3246	.2530	.6597	57
4	.58873	.80833	.72832	.3730	.2371	.6986	.60274	.79793	.75538	.3238	.2532	.6591	56
5	.58896	.80816	.72877	1.3722	.2374	1.6979	.60298	.79776	.75584	1.3230	.2535	1.6584	55
6	.58920	.80799	.72921	.3713	.2376	.6972	.60320	.79758	.75629	.3222	.2538	.6578	54
7	.58943	.80782	.72966	.3705	.2379	.6965	.60344	.79741	.75675	.3214	.2541	.6572	53
8	.58967	.80765	.73010	.3697	.2382	.6959	.60367	.79723	.75721	.3206	.2543	.6565	52
9	.58990	.80747	.73055	.3688	.2384	.6952	.60390	.79706	.75767	.3198	.2546	.6559	51
10	.59014	.80730	.73100	1.3680	.2387	1.6945	.60413	.79688	.75812	1.3190	.2549	1.6552	50
11	.59037	.80713	.73144	.3672	.2389	.6938	.60437	.79670	.75858	.3182	.2552	.6546	49
12	.59060	.80696	.73189	.3663	.2392	.6932	.60460	.79653	.75904	.3174	.2554	.6540	48
13	.59084	.80679	.73234	.3655	.2395	.6925	.60483	.79635	.75950	.3166	.2557	.6533	47
14	.59107	.80662	.73278	.3647	.2397	.6918	.60506	.79618	.75996	.3159	.2560	.6527	46
15	.59131	.80644	.73323	1.3638	.2400	1.6912	.60529	.79600	.76042	1.3151	.2563	1.6521	45
16	.59154	.80627	.73368	.3630	.2403	.6905	.60552	.79582	.76088	.3143	.2565	.6514	44
17	.59178	.80610	.73412	.3622	.2405	.6898	.60576	.79565	.76134	.3135	.2568	.6508	43
18	.59201	.80593	.73457	.3613	.2408	.6891	.60599	.79547	.76179	.3127	.2571	.6502	42
19	.59225	.80576	.73502	.3605	.2411	.6885	.60622	.79530	.76225	.3119	.2574	.6496	41
20	.59248	.80558	.73547	1.3597	.2413	1.6878	.60645	.79512	.76271	1.3111	.2577	1.6489	40
21	.59272	.80541	.73592	.3588	.2416	.6871	.60668	.79494	.76317	.3103	.2579	.6483	39
22	.59295	.80524	.73637	.3580	.2419	.6865	.60691	.79477	.76364	.3095	.2582	.6477	38
23	.59318	.80507	.73681	.3572	.2421	.6858	.60714	.79459	.76410	.3087	.2585	.6470	37
24	.59342	.80489	.73726	.3564	.2424	.6851	.60737	.79441	.76456	.3079	.2588	.6464	36
25	.59365	.80472	.73771	1.3555	.2427	1.6845	.60761	.79424	.76502	1.3071	.2591	1.6458	35
26	.59389	.80455	.73816	.3547	.2429	.6838	.60784	.79406	.76548	.3064	.2593	.6452	34
27	.59412	.80437	.73861	.3539	.2432	.6831	.60807	.79388	.76594	.3056	.2596	.6445	33
28	.59435	.80420	.73906	.3531	.2435	.6825	.60830	.79371	.76640	.3048	.2599	.6439	32
29	.59459	.80403	.73951	.3522	.2437	.6818	.60853	.79353	.76686	.3040	.2602	.6433	31
30	.59482	.80386	.73996	1.3514	.2440	1.6812	.60876	.79335	.76733	1.3032	.2605	1.6427	30
31	.59506	.80368	.74041	.3506	.2443	.6805	.60899	.79318	.76779	.3024	.2607	.6420	29
32	.59529	.80351	.74086	.3498	.2445	.6798	.60922	.79300	.76825	.3016	.2610	.6414	28
33	.59552	.80334	.74131	.3489	.2448	.6792	.60945	.79282	.76871	.3009	.2613	.6408	27
34	.59576	.80316	.74176	.3481	.2451	.6785	.60968	.79264	.76918	.3001	.2616	.6402	26
35	.59599	.80299	.74221	1.3473	.2453	1.6779	.60991	.79247	.76964	1.2993	.2619	1.6396	25
36	.59622	.80282	.74266	.3465	.2456	.6772	.61014	.79229	.77010	.2985	.2622	.6389	24
37	.59646	.80264	.74312	.3457	.2459	.6766	.61037	.79211	.77057	.2977	.2624	.6383	23
38	.59669	.80247	.74357	.3449	.2461	.6759	.61061	.79193	.77103	.2970	.2627	.6377	22
39	.59692	.80230	.74402	.3440	.2464	.6752	.61084	.79176	.77149	.2962	.2630	.6371	21
40	.59716	.80212	.74447	1.3432	.2467	1.6746	.61107	.79158	.77196	1.2954	.2633	1.6365	20
41	.59739	.80195	.74492	.3424	.2470	.6739	.61130	.79140	.77242	.2946	.2636	.6359	19
42	.59762	.80177	.74538	.3416	.2472	.6733	.61153	.79122	.77289	.2938	.2639	.6352	18
43	.59786	.80160	.74583	.3408	.2475	.6726	.61176	.79104	.77335	.2931	.2641	.6346	17
44	.59809	.80143	.74628	.3400	.2478	.6720	.61199	.79087	.77382	.2923	.2644	.6340	16
45	.59832	.80125	.74673	1.3392	.2480	1.6713	.61222	.79069	.77428	1.2915	.2647	1.6334	15
46	.59856	.80108	.74019	.3383	.2483	.6707	.61245	.79051	.77475	.2907	.2650	.6328	14
47	.59879	.80090	.74064	.3375	.2486	.6700	.61268	.79033	.77521	.2900	.2653	.6322	13
48	.59902	.80073	.74809	.3367	.2488	.6694	.61290	.79015	.77568	.2892	.2656	.6316	12
49	.59926	.80056	.74855	.3359	.2491	.6687	.61314	.78998	.77614	.2884	.2659	.6309	11
50	.59949	.80038	.74900	1.3351	.2494	1.6681	.61337	.78980	.77661	1.2876	.2661	1.6303	10
51	.59972	.80021	.74946	.3343	.2497	.6674	.61360	.78962	.77708	.2869	.2664	.6297	9
52	.59995	.80003	.74991	.3335	.2499	.6668	.61383	.78944	.77754	.2861	.2667	.6291	8
53	.60019	.79986	.75037	.3327	.2502	.6661	.61405	.78926	.77801	.2853	.2670	.6285	7
54	.60042	.79968	.75082	.3319	.2505	.6655	.61428	.78908	.77848	.2845	.2673	.6279	6
55	.60065	.79951	.75128	1.3311	.2508	1.6648	.61451	.78890	.77895	1.2838	.2676	1.6273	5
56	.60088	.79933	.75173	.3303	.2510	.6642	.61474	.78873	.77941	.2830	.2679	.6267	4
57	.60112	.79916	.75219	.3294	.2513	.6636	.61497	.78855	.77988	.2822	.2681	.6261	3
58	.60135	.79898	.75264	.3286	.2516	.6629	.61520	.78837	.78035	.2815	.2684	.6255	2
59	.60158	.79881	.75310	.3278	.2519	.6623	.61543	.78819	.78082	.2807	.2687	.6249	1
60	.60181	.79863	.75355	1.3270	.2521	1.6616	.61566	.78801	.78128	1.2799	.2690	1.6243	0
′	cos	sin	cot	tan	cosec	sec	cos	sin	cot	tan	cosec	sec	′
	53°						52°						

	38°						39°						
'	sin	cos	tan	cot	sec	cosec	sin	cos	tan	cot	sec	cosec	'
0	.61566	.78801	.78128	1.2799	1.2690	1.6243	.62932	.77715	.80978	1.2349	1.2867	1.5890	60
1	.61589	.78783	.78175	.2792	.2693	.6237	.62955	.77696	.81026	.2342	.2871	.5884	59
2	.61612	.78765	.78222	.2784	.2696	.6231	.62977	.77678	.81075	.2334	.2874	.5879	58
3	.61635	.78747	.78269	.2776	.2699	.6224	.63000	.77660	.81123	.2327	.2877	.5873	57
4	.61658	.78729	.78316	.2769	.2702	.6218	.63022	.77641	.81171	.2320	.2880	.5867	56
5	.61681	.78711	.78363	1.2761	.2705	1.6212	.63045	.77623	.81219	1.2312	.2883	1.5862	55
6	.61703	.78693	.78410	.2753	.2707	.6206	.63067	.77605	.81268	.2305	.2886	.5856	54
7	.61726	.78675	.78457	.2746	.2710	.6200	.63090	.77586	.81316	.2297	.2889	.5850	53
8	.61749	.78657	.78504	.2738	.2713	.6194	.63113	.77568	.81364	.2290	.2892	.5854	52
9	.61772	.78640	.78551	.2730	.2716	.6188	.63135	.77549	.81413	.2283	.2895	.5839	51
10	.61795	.78622	.78598	1.2723	.2719	1.6182	.63158	.77531	.81461	1.2276	.2898	1.5833	50
11	.61818	.78604	.78645	.2715	.2722	.6176	.63180	.77513	.81509	.2268	.2901	.5828	49
12	.61841	.78586	.78692	.2708	.2725	.6170	.63203	.77494	.81558	.2261	.2904	.5822	48
13	.61864	.78568	.78739	.2700	.2728	.6164	.63225	.77476	.81606	.2254	.2907	.5816	47
14	.61886	.78550	.78786	.2692	.2731	.6159	.63248	.77458	.81655	.2247	.2910	.5811	46
15	.61909	.78532	.78834	1.2685	.2734	1.6153	.63270	.77439	.81703	1.2239	.2913	1.5805	45
16	.61932	.78514	.78881	.2677	.2737	.6147	.63293	.77421	.81752	.2232	.2916	.5799	44
17	.61955	.78496	.78928	.2670	.2739	.6141	.63315	.77402	.81800	.2225	.2919	.5794	43
18	.61978	.78478	.78975	.2662	.2742	.6135	.63338	.77384	.81849	.2218	.2922	.5788	42
19	.62001	.78460	.79022	.2655	.2745	.6129	.63360	.77365	.81898	.2210	.2926	.5783	41
20	.62023	.78441	.79070	1.2647	.2748	1.6123	.63383	.77347	.81946	1.2203	.2929	1.5777	40
21	.62046	.78423	.79117	.2639	.2751	.6117	.63405	.77329	.81995	.2196	.2932	.5771	39
22	.62069	.78405	.79164	.2632	.2754	.6111	.63428	.77310	.82043	.2189	.2935	.5766	38
23	.62092	.78387	.79212	.2624	.2757	.6105	.63450	.77292	.82092	.2181	.2938	.5760	37
24	.62115	.78369	.79259	.2617	.2760	.6099	.63473	.77273	.82141	.2174	.2941	.5755	36
25	.62137	.78351	.79306	1.2609	.2763	1.6093	.63495	.77255	.82190	1.2167	.2944	1.5749	35
26	.62160	.78333	.79354	.2602	.2766	.6087	.63518	.77236	.82238	.2160	.2947	.5743	34
27	.62183	.78315	.79401	.2594	.2769	.6081	.63540	.77218	.82287	.2152	.2950	.5738	33
28	.62206	.78297	.79449	.2587	.2772	.6077	.63563	.77199	.82336	.2145	.2953	.5732	32
29	.62229	.78279	.79496	.2579	.2775	.6070	.63585	.77181	.82385	.2138	.2956	.5727	31
30	.62251	.78261	.79543	1.2572	.2778	1.6064	.63608	.77162	.82434	1.2131	.2960	1.5721	30
31	.62274	.78243	.79591	.2564	.2781	.6058	.63630	.77144	.82482	.2124	.2963	.5716	29
32	.62297	.78224	.79639	.2557	.2784	.6052	.63653	.77125	.82531	.2117	.2966	.5710	28
33	.62320	.78206	.79686	.2549	.2787	.6046	.63675	.77107	.82580	.2109	.2969	.5705	27
34	.62342	.78188	.79734	.2542	.2790	.6040	.63697	.77088	.82629	.2102	.2972	.5699	26
35	.62365	.78170	.79781	1.2534	.2793	1.6034	.63720	.77070	.82678	1.2095	.2975	1.5694	25
36	.62388	.78152	.79829	.2527	.2795	.6029	.63742	.77051	.82727	.2088	.2978	.5688	24
37	.62411	.78134	.79876	.2519	.2798	.6023	.63765	.77033	.82776	.2081	.2981	.5683	23
38	.62433	.78116	.79924	.2512	.2801	.6017	.63787	.77014	.82825	.2074	.2985	.5677	22
39	.62456	.78097	.79972	.2504	.2804	.6011	.63810	.76996	.82874	.2066	.2988	.5672	21
40	.62479	.78079	.80020	1.2497	.2807	1.6005	.63832	.76977	.82923	1.2059	.2991	1.5666	20
41	.62501	.78061	.80067	.2489	.2810	.6000	.63854	.76958	.82972	.2052	.2994	.5661	19
42	.62524	.78043	.80115	.2482	.2813	.5994	.63877	.76940	.83022	.2045	.2997	.5655	18
43	.62547	.78025	.80163	.2475	.2816	.5988	.63899	.76921	.83071	.2038	.3000	.5650	17
44	.62570	.78007	.80211	.2467	.2819	.5982	.63921	.76903	.83120	.2031	.3003	.5644	16
45	.62592	.77988	.80258	1.2460	.2822	1.5976	.63944	.76884	.83169	1.2024	.3006	1.5639	15
46	.62615	.77970	.80306	.2452	.2825	.5971	.63966	.76865	.83218	.2016	.3010	.5633	14
47	.62638	.77952	.80354	.2445	.2828	.5965	.63989	.76847	.83267	.2009	.3013	.5628	13
48	.62660	.77934	.80402	.2437	.2831	.5959	.64011	.76828	.83317	.2002	.3016	.5622	12
49	.62683	.77915	.80450	.2430	.2834	.5953	.64033	.76810	.83366	.1995	.3019	.5617	11
50	.62706	.77897	.80498	1.2423	.2837	1.5947	.64056	.76791	.83415	1.1988	.3022	1.5611	10
51	.62728	.77879	.80546	.2415	.2840	.5942	.64078	.76772	.83465	.1981	.3025	.5606	9
52	.62751	.77861	.80594	.2408	.2843	.5936	.64100	.76754	.83514	.1974	.3029	.5600	8
53	.62774	.77842	.80642	.2400	.2846	.5930	.64123	.76735	.83563	.1967	.3032	.5595	7
54	.62796	.77824	.80690	.2393	.2849	.5924	.64145	.76716	.83613	.1960	.3035	.5590	6
55	.62819	.77806	.80738	1.2386	.2852	1.5919	.64160	.76698	.83662	1.1953	.3038	1.5584	5
56	.62841	.77788	.80786	.2378	.2855	.5913	.64189	.76679	.83712	.1946	.3041	.5579	4
57	.62864	.77769	.80834	.2371	.2858	.5907	.64212	.76660	.83761	.1939	.3044	.5573	3
58	.62887	.77751	.80882	.2364	.2861	.5901	.64234	.76642	.83811	.1932	.3048	.5568	2
59	.62909	.77733	.80930	.2356	.2864	.5896	.64256	.76623	.83860	.1924	.3051	.5563	1
60	.62932	.77715	.80978	1.2349	.2867	1.5890	.64279	.76604	.83910	1.1917	.3054	1.5557	0
'	cos	sin	cot	tan	cosec	sec	cos	sin	cot	tan	cosec	sec	'
	51°						50°						

′	40° sin	cos	tan	cot	sec	cosec	41° sin	cos	tan	cot	sec	cosec	′
0	.64279	.76604	.83910	1.1917	1.3054	1.5557	.65606	.75471	.86929	1.1504	1.3250	1.5242	60
1	.64301	.76586	.83959	.1910	.3057	.5552	.65628	.75452	.86980	.1497	.3253	.5237	59
2	.64323	.76567	.84009	.1903	.3060	.5546	.65650	.75433	.87031	.1490	.3257	.5232	58
3	.64345	.76548	.84059	.1896	.3064	.5541	.65672	.75414	.87082	.1483	.3260	.5227	57
4	.64368	.76530	.84108	.1889	.3067	.5536	.65694	.75394	.87133	.1477	.3263	.5222	56
5	.64390	.76511	.84158	1.1882	.3070	1.5530	.65716	.75375	.87184	1.1470	.3267	1.5217	55
6	.64412	.76492	.84208	.1875	.3073	.5525	.65037	.75356	.87235	.1463	.3270	.5212	54
7	.64435	.76473	.84257	.1868	.3076	.5520	.65759	.75337	.87287	.1456	.3274	.5207	53
8	.64457	.76455	.84307	.1861	.3080	.5514	.65781	.75318	.87338	.1450	.3277	.5202	52
9	.64479	.76436	.84357	.1854	.3083	.5509	.65803	.75299	.87389	.1443	.3280	.5197	51
10	.64501	.76417	.84407	1.1847	.3086	1.5503	.65285	.75280	.87441	1.1436	.3284	1.5192	50
11	.64523	.76398	.84457	.1840	.3089	.5498	.65847	.75261	.87492	.1430	.3287	.5187	49
12	.64546	.76380	.84506	.1833	.3092	.5493	.65869	.75241	.87543	.1423	.3290	.5182	48
13	.64568	.76361	.84556	.1826	.3096	.5487	.65891	.75222	.87595	.1416	.3294	.5177	47
14	.64590	.76342	.84606	.1819	.3099	.5482	.65913	.75203	.87646	.1409	.3297	.5171	46
15	.64612	.76323	.84656	1.1812	.3102	1.5477	.65934	.75184	.87698	1.1403	.3301	1.5166	45
16	.64635	.76304	.84706	.1805	.3105	.5471	.65956	.75165	.87749	.1396	.3304	.5161	44
17	.64657	.76286	.84056	.1798	.3109	.5466	.65978	.75146	.87801	.1389	.3307	.5156	43
18	.64679	.76267	.84806	.1791	.3112	.5461	.66000	.75126	.87852	.1383	.3311	.5151	42
19	.64701	.76248	.84856	.1785	.3115	.5456	.66022	.75107	.87904	.1376	.3314	.5146	41
20	.64723	.76229	.84906	1.1778	.3118	1.5450	.66044	.75088	.87955	1.1369	.3318	1.5141	40
21	.64745	.76210	.84956	.1771	.3121	.5445	.66066	.75069	.88007	.1363	.3321	.5136	39
22	.64768	.76191	.85006	.1764	.3125	.5440	.66087	.75049	.88058	.1356	.3324	.5131	38
23	.64790	.76173	.85056	.1757	.3128	.5434	.66109	.75030	.88110	.1349	.3328	.5126	37
24	.64812	.76154	.85107	.1750	.3131	.5429	.66131	.75011	.88162	.1343	.3331	.5121	36
25	.64834	.76135	.85157	1.1743	.3134	1.5424	.66153	.74992	.88213	1.1336	.3335	1.5116	35
26	.64856	.76116	.85207	.1736	.3138	.5419	.66175	.74973	.88265	.1329	.3338	.5111	34
27	.64878	.76097	.85257	.1729	.3141	.5413	.66197	.74953	.88317	.1323	.3342	.5106	33
28	.64900	.76078	.85307	.1722	.3144	.5408	.66218	.74934	.88369	.1316	.3345	.5101	32
29	.64923	.76059	.85358	.1715	.3148	.5403	.66240	.74915	.88421	.1309	.3348	.5096	31
30	.64945	.76041	.85408	1.1708	.3151	1.5398	.66262	.74895	.88472	1.1303	.3352	1.5092	30
31	.64967	.76022	.85458	.1702	.3154	.5392	.66284	.74876	.88524	.1296	.3355	.5087	29
32	.64989	.76003	.85509	.1695	.3157	.5387	.66305	.74857	.88576	.1290	.3359	.5082	28
33	.65011	.75984	.85559	.1688	.3161	.5382	.66327	.74838	.88628	.1283	.3362	.5077	27
34	.65033	.75965	.85609	.1681	.3164	.5377	.66349	.74818	.88680	.1276	.3366	.5072	26
35	.65055	.75946	.85660	1.1674	.3167	1.5371	.66371	.74799	.88732	1.1270	.3369	1.5067	25
36	.65077	.75927	.85710	.1667	.3170	.5366	.66393	.74780	.88784	.1263	.3372	.5062	24
37	.65100	.75908	.85761	.1660	.3174	.5361	.66414	.74760	.88836	.1257	.3376	.5057	23
38	.65121	.75889	.85811	.1653	.3177	.5356	.66436	.74741	.88888	.1250	.3379	.5052	22
39	.65144	.75870	.85862	.1647	.3180	.5351	.66458	.74722	.88940	.1243	.3383	.5047	21
40	.65166	.75851	.85912	1.1640	.3184	1.5345	.66479	.74702	.88992	1.1237	.3386	1.5042	20
41	.65188	.75832	.85963	.1633	.3187	.5340	.66501	.74683	.89044	.1230	.3390	.5037	19
42	.65210	.75813	.86013	.1626	.3190	.5335	.66523	.74664	.89097	.1224	.3393	.5032	18
43	.65232	.75794	.86064	.1619	.3193	.5330	.66545	.74644	.89149	.1217	.3397	.5027	17
44	.65254	.75775	.86115	.1612	.3197	.5325	.66566	.74625	.89201	.1211	.3400	.5022	16
45	.65276	.75756	.86165	1.1605	.3200	1.5319	.66588	.74606	.89253	1.1204	1.3404	1.5018	15
46	.65298	.75737	.86216	.1599	.3203	.5314	.66610	.74586	.89306	.1197	.3407	.5013	14
47	.65320	.75718	.86267	.1592	.3207	.5309	.66631	.74567	.89358	.1191	.3411	.5008	13
48	.65342	.75700	.86318	.1585	.3210	.5304	.66653	.74548	.89410	.1184	.3414	.5003	12
49	.65364	.75680	.86368	.1578	.3213	.5299	.66675	.74528	.89463	.1178	.3418	.4998	11
50	.65386	.75661	.86419	1.1571	.3217	1.5294	.66697	.74509	.89515	1.1171	.3421	1.4993	10
51	.65408	.75642	.86470	.1565	.3220	.5289	.66718	.74489	.89567	.1165	.3425	.4988	9
52	.65430	.75623	.86521	.1558	.3223	.5283	.66740	.74470	.89620	.1158	.3428	.4983	8
53	.65452	.75604	.86572	.1551	.3227	.5278	.66762	.74450	.89672	.1152	.3432	.4979	7
54	.65474	.75585	.86623	.1544	.3230	.5273	.66783	.74431	.89725	.1145	.3435	.4974	6
55	.65496	.75566	.86674	1.1537	.3233	1.5268	.66805	.74412	.89777	1.1139	.3439	1.4969	5
56	.65518	.75547	.86725	.1531	.3237	.5263	.66826	.74392	.89830	.1132	.3442	.4964	4
57	.65540	.75528	.86775	.1524	.3240	.5258	.66848	.74373	.89882	.1126	.3446	.4959	3
58	.65562	.75509	.86826	.1517	.3243	.5253	.66870	.74353	.89935	.1119	.3449	.4954	2
59	.65584	.75490	.86878	.1510	.3247	.5248	.66891	.74334	.89988	.1113	.3453	.4949	1
60	.65606	.75401	.86929	1.1504	.3250	1.5242	.66913	.74314	.90040	1.1106	.3456	1.4945	0
′	cos	sin	cot	tan	cosec	sec	cos	sin	cot	tan	cosec	sec	′
	49°						48°						

	42°						43°						
′	sin	cos	tan	cot	sec	cosec	sin	cos	tan	cot	sec	cosec	′
0	.66913	.74314	.90040	1.1106	1.3456	1.4945	.68200	.73135	.93251	1.0724	1.3673	1.4663	60
1	.66935	.74295	.90093	.1100	.3460	.4940	.68221	.73115	.93306	.0717	.3677	.4658	59
2	.66956	.74276	.90146	.1093	.3463	.4935	.68242	.73096	.93360	.0711	.3681	.4654	58
3	.66978	.74256	.90198	.1086	.3467	.4930	.68264	.73076	.93415	.0705	.3684	.4649	57
4	.66999	.74236	.90251	.1080	.3470	.4925	.68285	.73056	.93469	.0699	.3688	.4644	56
5	.67021	.74217	.90304	1.1074	.3474	1.4921	.68306	.73036	.93524	1.0692	.3692	1.4640	55
6	.67043	.74197	.90357	.1067	.3477	.4916	.68327	.73016	.93578	.0686	.3695	.4635	54
7	.67064	.74178	.90410	.1061	.3481	.4911	.68349	.72996	.93633	.0680	.3699	.4631	53
8	.67086	.74158	.90463	.1054	.3485	.4906	.68370	.72976	.93687	.0674	.3703	.4626	52
9	.67107	.74139	.90515	.1048	.3488	.4901	.68391	.72956	.93742	.0667	.3707	.4622	51
10	.67129	.74119	.90568	1.1041	.3492	1.4897	.68412	.72937	.93797	1.0661	.3710	1.4617	50
11	.67150	.74100	.90621	.1035	.3495	.4892	.68433	.72917	.93851	.0655	.3714	.4613	49
12	.67172	.74080	.90674	.1028	.3499	.4887	.68455	.72897	.93906	.0649	.3718	.4608	48
13	.67194	.74061	.90727	.1022	.3502	.4882	.68476	.72877	.93961	.0643	.3722	.4604	47
14	.67215	.74041	.90780	.1015	.3506	.4877	.68497	.72857	.94016	.0636	.3725	.4599	46
15	.67237	.74022	.90834	1.1009	.3509	1.4873	.68518	.72837	.94071	1.0630	.3729	1.4595	45
16	.67258	.74002	.90887	.1003	.3513	.4868	.68539	.72817	.94125	.0624	.3733	.4590	44
17	.67280	.73983	.90940	.0996	.3517	.4863	.68561	.72797	.94180	.0618	.3737	.4586	43
18	.67301	.73963	.90993	.0990	.3520	.4858	.68582	.72777	.94235	.0612	.3740	.4581	42
19	.67323	.73943	.91046	.0983	.3524	.4854	.68603	.72757	.94290	.0605	.3744	.4577	41
20	.67344	.73924	.91099	1.0977	.3527	1.4849	.68624	.72737	.94345	1.0599	.3748	1.4572	40
21	.67366	.73904	.91153	.0971	.3531	.4844	.68645	.72717	.94400	.0593	.3752	.4568	39
22	.67387	.73885	.91206	.0964	.3534	.4839	.68666	.72697	.94455	.0587	.3756	.4563	38
23	.67409	.73865	.91259	.0958	.3538	.4835	.68688	.72677	.94510	.0581	.3759	.4559	37
24	.67430	.73845	.91312	.0951	.3542	.4830	.68709	.72657	.94565	.0575	.3763	.4554	36
25	.67452	.73826	.91366	1.0945	.3545	1.4825	.68730	.72637	.94620	1.0568	.3767	1.4550	35
26	.67473	.73806	.91419	.0939	.3549	.4821	.68751	.72617	.94675	.0562	.3771	.4545	34
27	.67495	.73787	.91473	.0932	.3552	.4816	.68772	.72597	.94731	.0556	.3774	.4541	33
28	.67516	.73767	.91526	.0926	.3556	.4811	.68793	.72577	.94786	.0550	.3778	.4536	32
29	.67537	.73747	.91580	.0919	.3560	.4806	.68814	.72557	.94841	.0544	.3782	.4532	31
30	.67559	.73728	.91633	1.0913	.3563	1.4802	.68835	.72537	.94896	1.0538	.3786	1.4527	30
31	.67580	.73708	.91687	.0907	.3567	.4797	.68856	.72517	.94952	.0532	.3790	.4523	29
32	.67602	.73688	.91740	.0900	.3571	.4792	.68878	.72497	.95007	.0525	.3794	.4518	28
33	.67623	.73669	.91794	.0894	.3574	.4788	.68899	.72477	.95062	.0519	.3797	.4514	27
34	.67645	.73649	.91847	.0888	.3578	.4783	.68920	.72457	.95118	.0513	.3801	.4510	26
35	.67666	.73629	.91901	1.0881	.3581	1.4778	.68941	.72437	.95173	1.0507	.3805	1.4505	25
36	.67638	.73610	.91955	.0875	.3585	.4774	.68962	.72417	.95229	.0501	.3809	.4501	24
37	.67709	.73590	.92008	.0868	.3589	.4769	.68983	.72397	.95284	.0495	.3813	.4496	23
38	.67730	.73570	.92062	.0862	.3592	.4764	.69004	.72377	.95340	.0489	.3816	.4492	22
39	.67752	.73551	.92116	.0856	.3596	.4760	.69025	.72357	.95395	.0483	.3820	.4487	21
40	.67773	.73531	.92170	1.0849	.3600	1.4755	.69046	.72337	.95451	1.0476	.3824	1.4483	20
41	.67794	.73511	.92223	.0843	.3603	.4750	.69067	.72317	.95506	.0470	.3828	.4479	19
42	.67816	.73491	.92277	.0837	.3607	.4746	.69088	.72297	.95562	.0464	.3832	.4474	18
43	.67837	.73472	.92331	.0830	.3611	.4741	.69109	.72277	.95618	.0458	.3836	.4470	17
44	.67859	.73452	.92385	.0824	.3614	.4736	.69130	.72256	.95673	.0452	.3839	.4465	16
45	.67880	.73432	.92439	1.0818	.3618	1.4732	.69151	.72236	.95729	1.0446	.3843	1.4461	15
46	.67901	.73412	.92493	.0812	.3622	.4727	.69172	.72216	.95785	.0440	.3847	.4457	14
47	.67923	.73393	.92547	.0805	.3625	.4723	.69193	.72196	.95841	.0434	.3851	.4452	13
48	.67944	.73373	.92601	.0799	.3629	.4718	.69214	.72176	.95896	.0428	.3855	.4448	12
49	.67965	.73353	.92655	.0793	.3633	.4713	.69235	.72156	.95952	.0422	.3859	.4443	11
50	.67987	.73333	.92709	1.0786	.3636	1.4709	.69256	.72136	.96008	1.0416	.3863	1.4439	10
51	.68008	.73314	.92763	.0780	.3640	.4704	.69277	.72115	.96064	.0410	.3867	.4435	9
52	.68029	.73294	.92817	.0774	.3644	.4699	.69298	.72095	.96120	.0404	.3870	.4430	8
53	.68051	.73274	.92871	.0767	.3647	.4695	.69319	.72075	.96176	.0397	.3874	.4426	7
54	.68072	.73254	.92926	.0761	.3651	.4690	.69340	.72055	.96232	.0391	.3878	.4422	6
55	.68093	.73234	.92980	1.0755	.3655	1.4686	.69361	.72035	.96288	1.0385	.3882	1.4417	5
56	.68115	.73215	.93034	.0749	.3658	.4681	.69382	.72015	.96344	.0379	.3886	.4413	4
57	.68136	.73195	.93088	.0742	.3662	.4676	.69403	.71994	.96400	.0373	.3890	.4408	3
58	.68157	.73175	.93143	.0736	.3666	.4672	.69424	.71974	.96456	.0367	.3894	.4404	2
59	.68178	.73155	.93197	.0730	.3669	.4667	.69445	.71954	.96513	.0361	.3898	.4400	1
60	.68200	.73135	.93251	1.0724	.3673	1.4663	.69466	.71934	.96569	1.0355	.3902	1.4395	0
′	cos	sin	cot	tan	cosec	sec	cos	sin	cot	tan	cosec	sec	′
	47°						46°						

44°

′	sin	cos	tan	cot	sec	cosec	′
0	.69466	.71934	.96569	1.0355	1.3902	1.4395	60
1	.69487	.71914	.96625	.0349	.3905	.4391	59
2	.69508	.71893	.96681	.0343	.3909	.4387	58
3	.69528	.71873	.96738	.0337	.3913	.4382	57
4	.69549	.71853	.96794	.0331	.3917	.4378	56
5	.69570	.71833	.96850	1.0325	.3921	1.4374	55
6	.69591	.71813	.96907	.0319	.3925	.4370	54
7	.69612	.71792	.96963	.0313	.3929	.4365	53
8	.69633	.71772	.97020	.0307	.3933	.4361	52
9	.69654	.71752	.97076	.0301	.3937	.4357	51
10	.69675	.71732	.97133	1.0295	.3941	1.4352	50
11	.69696	.71711	.97189	.0289	.3945	.4348	49
12	.69716	.71691	.97246	.0283	.3949	.4344	48
13	.69737	.71671	.97302	.0277	.3953	.4339	47
14	.69758	.71650	.97359	.0271	.3957	.4335	46
15	.69779	.71630	.97416	1.0265	.3960	1.4331	45
16	.69800	.71610	.97472	.0259	.3964	.4327	44
17	.69821	.71589	.97529	.0253	.3968	.4322	43
18	.69841	.71569	.97586	.0247	.3972	.4318	42
19	.69862	.71549	.97643	.0241	.3976	.4314	41
20	.69883	.71529	.97700	1.0235	.3980	1.4310	40
21	.69904	.71508	.97756	.0229	.3984	.4305	39
22	.69925	.71488	.97813	.0223	.3988	.4301	38
23	.69945	.71468	.97870	.0218	.3992	.4297	37
24	.69966	.71447	.97927	.0212	.3996	.4292	36
25	.69987	.71427	.97984	1.0206	.4000	1.4288	35
26	.70008	.71406	.98041	.0200	.4004	.4284	34
27	.70029	.71386	.98098	.0194	.4008	.4280	33
28	.70049	.71366	.98155	.0188	.4012	.4276	32
29	.70070	.71345	.98212	.0182	.4016	.4271	31
30	.70091	.71325	.98270	1.0176	.4020	1.4267	30
31	.70112	.71305	.98327	.0170	.4024	.4263	29
32	.70132	.71284	.98384	.0164	.4028	.4259	28
33	.70153	.71264	.98441	.0158	.4032	.4254	27
34	.70174	.71243	.98499	.0152	.4036	.4250	26
35	.70194	.71223	.98556	1.0146	.4040	1.4246	25
36	.70215	.71203	.98613	.0141	.4044	.4242	24
37	.70236	.71182	.98671	.0135	.4048	.4238	23
38	.70257	.71162	.98728	.0129	.4052	.4233	22
39	.70277	.71141	.98786	.0123	.4056	.4229	21
40	.70298	.71121	.98843	1.0117	.4060	1.4225	20
41	.70319	.71100	.98901	.0111	.4065	.4221	19
42	.70339	.71080	.98958	.0105	.4069	.4217	18
43	.70360	.71059	.99016	.0099	.4073	.4212	17
44	.70381	.71039	.99073	.0093	.4077	.4208	16
45	.70401	.71018	.99131	1.0088	.4081	1.4204	15
46	.70422	.70998	.99189	.0082	.4085	.4200	14
47	.70443	.70977	.99246	.0076	.4089	.4196	13
48	.70463	.70957	.99304	.0070	.4093	.4192	12
49	.70484	.70936	.99362	.0064	.4097	.4188	11
50	.70505	.70916	.99420	1.0058	.4101	1.4183	10
51	.70525	.70895	.99478	.0052	.4105	.4179	9
52	.70546	.70875	.99536	.0047	.4109	.4175	8
53	.70566	.70854	.99593	.0041	.4113	.4171	7
54	.70587	.70834	.99651	.0035	.4117	.4167	6
55	.70608	.70813	.99709	1.0029	.4122	1.4163	5
56	.70628	.70793	.99767	.0023	.4126	.4159	4
57	.70649	.70772	.99826	.0017	.4130	.4154	3
58	.70669	.70752	.99884	.0012	.4134	.4150	2
59	.70690	.70731	.99942	.0006	.4138	.4146	1
60	.70711	.70711	1.00000	1.0000	.4142	1.4142	0
′	cos	sin	cot	tan	cosec	sec	′

45°

Section 45

REFERENCE TABLES

SECTION 45

REFERENCE TABLES

COMMON WEIGHTS AND MEASURES

Measures of Length

12 inches = 1 foot   3 feet = 1 yard
$5\frac{1}{2}$ yards = 1 rod   40 rods = 1 furlong
8 furlongs = 1 mile

Equivalent Measures

Inches		Feet		Yards		Rods		Furlongs		Mile
36	=	3	=	1						
198	=	16.5	=	5.5	=	1				
7920	=	660	=	220	=	40	=	1		
63,360	=	5280	=	1760	=	320	=	8	=	1

Square Measure

144 square inches = 1 square foot
9 square feet = 1 square yard
$30\frac{1}{4}$ square yards = 1 square rod
160 square rods = 1 acre
640 acres = 1 square mile
1 acre = 43,560 square feet = 208.7 feet on side
10 acres = a square 660 feet on each side

Equivalent Measure

Sq Mi		A		Sq Rd		Sq Yd		Sq Ft		Sq In
1	=	640	=	102,400	=	3,097,600	=	27,878,400	=	4,014,489,600
		1	=	160	=	4840	=	43,560	=	6,272,640

Cubic Measure

1728 cubic inches = 1 cubic foot   128 cubic feet = 1 cord
27 cubic feet = 1 cubic yard $24\frac{3}{4}$ cubic feet = 1 perch
1 cu yd = 27 cu ft = 46,656 cu in.

Weight—Avoirdupois

437.5 grains = 1 ounce   100 pounds = 1 hundredweight
16 ounces = 1 pound   2000 pounds = 1 ton
2240 pounds = 1 long ton
1 ton = 20 cwt = 2000 lb = 32,000 oz = 14,000,000 gr

Weight—Troy

24 grains = 1 pennyweight   20 dwt = 1 ounce
12 ounces = 1 pound
1 lb = 12 oz = 240 dwt = 5760 gr

Dry Measure

2 pints = 1 quart   8 quarts = 1 peck
4 pecks = 1 bushel
1 bu = 4 pk = 32 qt = 64 pt
U.S. bushel = 2150.42 cu in. British = 2218.19 cu in.

Liquid Measure

4 gills = 1 pint   4 quarts = 1 gallon
2 pints = 1 quart $31\frac{1}{2}$ gallons = 1 barrel
2 barrels or 63 gal = 1 hogshead
1 hhd = 2 bbl = 63 gal = 252 qt = 504 pt = 2016 gi

The U.S. gallon contains 231 cu in. = 0.134 cu ft.
One cubic foot = 7.481 gallons.
One cubic foot weighs 62.425 lb at 39.2 F.
One gallon weighs 8.345 lb. British Imperial gallon weighs 10 lb.
For rough calculations 1 cu ft is called $7\frac{1}{2}$ gallons and 1 gallon as $8\frac{1}{3}$ lb.

Angles or Arcs

60 seconds = 1 minute   90 degrees = 1 right angle or quadrant
60 minutes = 1 degree   360 degrees = 1 circle
1 circle = 360° = 21,600′ = 1,296,000″

1 minute of arc on the earth's surface is 1 nautical mile = 1.15 times a land mile or 6080 feet.

Measures of Pressure

1 psi = 2.0416 in. mercury at 62 F
1 psi = 27.71 in. water at 62 F
1 atmosphere = 14.7 psi
1 atmosphere = 33.947 ft water at 62 F
1 atmosphere = 30.011 in. mercury at 62 F
1 in. water 62 F = 6.236 psf

Water Conversion Factors

U.S. gallons	×	8.33	= pounds
U.S. gallons	×	0.13368	= cubic feet
U.S. gallons	×	231	= cubic inches
U.S. gallons	×	0.83	= English gallons
U.S. gallons	×	3.78	= liters
English gallons (Imperial)	×	10	= pounds
English gallons (Imperial)	×	0.16	= cubic feet
English gallons (Imperial)	×	277.274	= cubic inches
English gallons (Imperial)	×	1.2	= U.S. gallons
English gallons (Imperial)	×	4.537	= liters
Cubic inches of water (39.1 F)	×	0.036125	= pounds
Cubic inches of water (39.1 F)	×	0.004329	= U.S. gallons
Cubic inches of water (39.1 F)	×	0.003607	= English gallons
Cubic inches of water (39.1 F)	×	0.576384	= ounces
Cubic feet (of water) (39.1 F)	×	62.425	= pounds
Cubic feet (of water) (39.1 F)	×	7.48	= U.S. gallons
Cubic feet (of water) (39.1 F)	×	6.232	= English gallons
Cubic feet (of water) (39.1 F)	×	0.028	= tons
Pounds of water	×	27.72	= cubic inches
Pounds of water	×	0.01602	= cubic feet
Pounds of water	×	0.12	= U.S. gallons
Pounds of water	×	0.10	= English gallons

THE METRIC SYSTEM

The metric system is based on the meter, which was designed to be one ten-millionth (1/10,000,000) part of the earth's meridian quadrant, through Dunkirk and Formentera. Later investigations, however, have shown that the meter exceeds one ten-millionth part by almost one part in 6400. The value of the meter, as authorized by the United States government, is 39.37 in. The metric system was legalized by the United States government in 1866.

The three principal units are the meter, the unit of length, the liter, the unit of capacity, and the gram, the unit of weight. Multiples of these are obtained by prefixing the Greek words: deka (10), hekto (100), and kilo (1000). Divisions are obtained by prefixing the Latin words: deci($\frac{1}{10}$), centi ($\frac{1}{100}$), and milli ($\frac{1}{1000}$). Abbreviations of the multiples begin with a capital letter, and of the divisions with a small letter, as in the following tables:

Measures of Length

10 millimeters (mm)	=	1 centimeter (cm)
10 centimeters	=	1 decimeter (dm)
10 decimeters	=	1 meter (m)
10 meters	=	1 dekameter (dm)
10 dekameters	=	1 hektometer (hm)
10 hektometers	=	1 kilometer (km)

Measures of Surface (not Land)

100 square millimeters (mm^2)	=	1 square centimeter (cm^2)
100 square centimeters	=	1 square decimeter (dm^2)
100 square decimeters	=	1 square meter (m^2)

Measures of Volume

1000 cubic millimeters (mm^3)	=	1 cubic centimeter (cm^3)
1000 cubic centimeters	=	1 cubic decimeter (dm^3)
1000 cubic decimeters	=	1 cubic meter (m^3)

Measures of Capacity

10 milliliters (ml)	=	1 centiliter (cl)
10 centiliters	=	1 deciliter (dl)
10 deciliters	=	1 liter (l)
10 liters	=	1 dekaliter (dl)
10 dekaliters	=	1 hektoliter (hl)
10 hektoliters	=	1 kiloliter (kl)

Note: The liter is equal to the volume occupied by 1 cubic decimeter.

Measures of Weight

10 milligrams (mg)	=	1 centigram (cg)
10 centigrams	=	1 decigram (dg)
10 decigrams	=	1 gram (g)
10 grams	=	1 dekagram (dg)
10 dekagrams	=	1 hektogram (hg)
10 hektograms	=	1 kilogram (kg)
1000 kilograms	=	1 ton (T)

Note: The gram is the weight of 1 cubic centimeter of pure distilled water at a temperature of 39.2 F; the kilogram is the weight of 1 liter of water; the ton is the weight of 1 cubic meter of water.

Metric and English Conversion Tables

Measures of Length

1 meter = 39.37 inches; 3.28083 feet; 1.0936 yards

1 centimeter = 0.3937 inch

1 millimeter = 0.03937 inch, or $\frac{1}{25}$ inch nearly

1 kilometer = 0.62137 mile

1 foot = 0.3048 meter

1 inch = 2.54 centimeters; 25.4 millimeters

Measures of Surface

1 square meter = 10.764 square feet; 1.196 square yards

1 square centimeter = 0.155 square inch

1 square millimeter = 0.00155 square inch

1 square yard = 0.836 square meter

1 square foot = 0.0929 square meter

1 square inch = 6.452 square centimeters; 645.2 square millimeters

Measures of Volume and Capacity

1 cubic meter = { 35.314 cubic feet; 1.308 cubic yards; 264.2 gallons (231 cubic inches) }
1 cubic decimeter = { 61.023 cubic inches; 0.0353 cubic foot }
1 cubic centimeter = 0.061 cubic inch
1 liter = { 1 cubic decimeter; 61.023 cubic inches; 0.0353 cubic foot; 1.0567 quarts (U.S.); 0.2642 gallon (U.S.); 2.202 lb of water at 62 F }

1 cubic yard = 0.7645 cubic meter
1 cubic ft = { 0.02832 cubic meter; 28.317 cubic decimeters; 28.317 liters }
1 cubic inch = 16.387 cubic centimeters
1 gallon (British) = 4.543 liters
1 gallon (U.S.) = 3.785 liters

Measures of Weight

1 gram = 15.432 grains
1 kilogram = 2.2046 pounds
1 metric ton = { 0.9842 ton of 2240 pounds; 19.68 hundred weight; 2204.6 pounds }

1 grain = 0.0648 gram
1 ounce avoirdupois = 28.35 grams
1 pound = 0.4536 kilogram
1 ton of 2240 pounds = { 1.016 metric tons; 1016 kilograms }

Miscellaneous Conversion Factors

1 kilogram per meter = 0.6720 pound per foot
1 gram per square millimeter = 1.422 pounds per square inch
1 kilogram per square meter = 0.2048 pound per square foot
1 kilogram per cubic meter = 0.0624 pound per cubic foot
1 pound per foot = 1.488 kilograms per meter
1 pound per square foot = 4.882 kilograms per square meter
1 pound per cubic foot = 16.02 kilograms per cubic meter
1 calorie (French thermal unit) = 3.968 Btu (British thermal unit)

Inches to Millimeters

Inches	Millimeters	Inches	Millimeters	Inches	Millimeters	Inches	Millimeters
1	25.4	26	660.4	51	1295.4	76	1930.4
2	50.8	27	685.8	52	1320.8	77	1955.8
3	76.2	28	711.2	53	1346.2	78	1981.2
4	101.6	29	736.6	54	1371.6	79	2006.6
5	127.0	30	762.0	55	1397.0	80	2032.0
6	152.4	31	787.4	56	1422.4	81	2057.4
7	177.8	32	812.8	57	1447.8	82	2082.8
8	203.2	33	838.2	58	1473.2	83	2108.2
9	228.6	34	863.6	59	1498.6	84	2133.6
10	254.0	35	889.0	60	1524.0	85	2159.0
11	279.4	36	914.4	61	1549.4	86	2184.4
12	304.8	37	939.8	62	1574.8	87	2209.8
13	330.2	38	965.2	63	1600.2	88	2235.2
14	355.6	39	990.6	64	1625.6	89	2260.6
15	381.0	40	1016.0	65	1651.0	90	2286.0
16	406.4	41	1041.4	66	1676.4	91	2311.4
17	431.8	42	1066.8	67	1701.8	92	2336.8
18	457.2	43	1092.2	68	1727.2	93	2362.2
19	482.6	44	1117.6	69	1752.6	94	2387.6
20	508.0	45	1143.0	70	1778.0	95	2413.0
21	533.4	46	1168.4	71	1803.4	96	2438.4
22	558.8	47	1193.8	72	1828.8	97	2463.8
23	584.2	48	1219.2	73	1854.2	98	2489.2
24	609.6	49	1244.6	74	1879.6	99	2514.6
25	635.0	50	1270.0	75	1905.0	100	2540.0

Millimeters to Inches

Millimeters	Inches	Millimeters	Inches	Millimeters	Inches	Millimeters	Inches
1	0.039370	26	1.023622	51	2.007874	76	2.992126
2	0.078740	27	1.062992	52	2.047244	77	3.031496
3	0.118110	28	1.102362	53	2.086614	78	3.070866
4	0.157480	29	1.141732	54	2.125984	79	3.110236
5	0.196850	30	1.181102	55	2.165354	80	3.149606
6	0.236220	31	1.220472	56	2.204724	81	3.188976
7	0.275591	32	1.259843	57	2.244094	82	3.228346
8	0.314961	33	1.299213	58	2.283465	83	3.267717
9	0.354331	34	1.338583	59	2.322835	84	3.307087
10	0.393701	35	1.377953	60	2.362205	85	3.346457
11	0.433071	36	1.417323	61	2.401575	86	3.385827
12	0.472441	37	1.456693	62	2.440945	87	3.425197
13	0.511811	38	1.496063	63	2.480315	88	3.464567
14	0.551181	39	1.535433	64	2.519685	89	3.503937
15	0.590551	40	1.574803	65	2.559055	90	3.543307
16	0.629921	41	1.614173	66	2.598425	91	3.582677
17	0.669291	42	1.653543	67	2.637795	92	3.622047
18	0.708661	43	1.692913	68	2.677165	93	3.661417
19	0.748031	44	1.732283	69	2.716535	94	3.700787
20	0.787402	45	1.771654	70	2.755906	95	3.740157
21	0.826772	46	1.811024	71	2.795276	96	3.779528
22	0.866142	47	1.850394	72	2.834646	97	3.818898
23	0.905512	48	1.889764	73	2.874016	98	3.858268
24	0.944882	49	1.929134	74	2.913386	99	3.897638
25	0.984252	50	1.968504	75	2.952756	100	3.937008

Microns and Microinches

One micron = 1/1,000,000 meter = 1/25,000 inch
One microinch = 1/1,000,000 inch

Units of Power

1 horsepower = { 33,000 foot-pounds per minute; 746 watts }

1 watt (unit of electrical power) = { 0.00134 horsepower; 44.24 foot-pounds per minute }

1 kilowatt = { 1000 watts; 1.34 horsepower; 44,240 foot-pounds per minute }

Conversion of Fractions and Decimals of an Inch to Millimeters

Fraction of Inch	Decimal of Inch	Milli-meters	Fraction of Inch	Decimal of Inch	Milli-meters
1/64	0.015625	0.3968	33/64	0.515625	13.0966
1/32	0.03125	0.7937	17/32	0.53125	13.4934
3/64	0.046875	1.1906	35/64	0.546875	13.8903
1/16	0.0625	1.5875	9/16	0.5625	14.2872
5/64	0.078125	1.9843	37/64	0.578125	14.6841
3/32	0.09375	2.3812	19/32	0.59375	15.0809
7/64	0.109375	2.7780	39/64	0.609375	15.4778
1/8	0.125	3.1749	5/8	0.625	15.8747
9/64	0.140625	3.5718	41/64	0.640625	16.2715
5/32	0.15625	3.9686	21/32	0.65625	16.6684
11/64	0.171875	4.3655	43/64	0.671875	17.0653
3/16	0.1875	4.7624	11/16	0.6875	17.4621
13/64	0.203125	5.1592	45/64	0.703125	17.8590
7/32	0.21875	5.5561	23/32	0.71875	18.2559
15/64	0.234375	5.9530	47/64	0.734375	18.6527
1/4	0.25	6.3498	3/4	0.75	19.0496
17/64	0.265625	6.7467	49/64	0.765625	19.4465
9/32	0.28125	7.1436	25/32	0.78125	19.8433
19/64	0.296875	7.5404	51/64	0.796875	20.2402
5/16	0.3125	7.9373	13/16	0.8125	20.6371
21/64	0.328125	8.3342	53/64	0.828125	21.0339
11/32	0.34375	8.7310	27/32	0.843750	21.4308
23/64	0.359375	9.1279	55/64	0.859375	21.8277
3/8	0.375	9.5248	7/8	0.875	22.2245
25/64	0.390625	9.9216	57/64	0.890625	22.6214
13/32	0.40625	10.3185	29/32	0.90625	23.0183
27/64	0.421875	10.7154	59/64	0.921875	23.4151
7/16	0.4375	11.1122	15/16	0.9375	23.8120
29/64	0.453125	11.5091	61/64	0.953125	24.2089
15/32	0.46875	11.9060	31/32	0.96875	24.6057
31/64	0.484375	12.3029	63/64	0.984375	25.0026
1/2	0.5	12.6997	1	1.0	25.3995

Ten-thousandths of an Inch to Millimeters

INCH	MM	INCH	MM	INCH	MM	INCH	MM	INCH	MM
.0010	.02540	.0050	.12700	.0090	.22860	.0130	.33020	.0170	.43180
.0011	.02794	.0051	.12954	.0091	.23114	.0131	.33274	.0171	.43434
.0012	.03048	.0052	.13208	.0092	.23368	.0132	.33528	.0172	.43688
.0013	.03302	.0053	.13462	.0093	.23622	.0133	.33782	.0173	.43942
.0014	.03556	.0054	.13716	.0094	.23876	.0134	.34036	.0174	.44196
.0015	.03810	.0055	.13970	.0095	.24130	.0135	.34290	.0175	.44450
.0016	.04064	.0056	.14224	.0096	.24384	.0136	.34544	.0176	.44704
.0017	.04318	.0057	.14478	.0097	.24638	.0137	.34798	.0177	.44958
.0018	.04572	.0058	.14732	.0098	.24892	.0138	.35052	.0178	.45212
.0019	.04826	.0059	.14986	.0099	.25146	.0139	.35306	.0179	.45466
.0020	.05080	.0060	.15240	.0100	.25400	.0140	.35560	.0180	.45720
.0021	.05334	.0061	.15494	.0101	.25654	.0141	.35814	.0181	.45974
.0022	.05588	.0062	.15748	.0102	.25908	.0142	.36068	.0182	.46228
.0023	.05842	.0063	.16002	.0103	.26162	.0143	.36322	.0183	.46482
.0024	.06096	.0064	.16256	.0104	.26416	.0144	.36576	.0184	.46736
.0025	.06350	.0065	.16510	.0105	.26670	.0145	.36830	.0185	.46990
.0026	.06604	.0066	.16764	.0106	.26924	.0146	.37084	.0186	.47244
.0027	.06858	.0067	.17018	.0107	.27178	.0147	.37338	.0187	.47498
.0028	.07112	.0068	.17272	.0108	.27432	.0148	.37592	.0188	.47752
.0029	.07366	.0069	.17526	.0109	.27686	.0149	.37846	.0189	.48006
.0030	.07620	.0070	.17780	.0110	.27940	.0150	.38100	.0190	.48260
.0031	.07874	.0071	.18034	.0111	.28194	.0151	.38354	.0191	.48514
.0032	.08128	.0072	.18288	.0112	.28448	.0152	.38608	.0192	.48768
.0033	.08282	.0073	.18542	.0113	.28702	.0153	.38862	.0193	.49022
.0034	.08636	.0074	.18796	.0114	.28956	.0154	.39116	.0194	.49276
.0035	.08890	.0075	.19050	.0115	.29210	.0155	.39370	.0195	.49530
.0036	.09144	.0076	.19304	.0116	.29464	.0156	.39624	.0196	.49784
.0037	.09398	.0077	.19558	.0117	.29718	.0157	.39878	.0197	.50038
.0038	.09652	.0078	.19812	.0118	.29972	.0158	.40132	.0198	.50292
.0039	.09906	.0079	.20066	.0119	.30226	.0159	.40386	.0199	.50546
.0040	.10160	.0080	.20320	.0120	.30480	.0160	.40640	.0200	.50800
.0041	.10414	.0081	.20574	.0121	.30734	.0161	.40894	.0201	.51054
.0042	.10668	.0082	.20828	.0122	.30988	.0162	.41148	.0202	.51308
.0043	.10922	.0083	.21082	.0123	.31242	.0163	.41402	.0203	.51562
.0044	.11176	.0084	.21336	.0124	.31496	.0164	.41656	.0204	.51816
.0045	.11430	.0085	.21590	.0125	.31750	.0165	.41910	.0205	.52070
.0046	.11684	.0086	.21844	.0126	.32004	.0166	.42164	.0206	.52324
.0047	.11938	.0087	.22098	.0127	.32258	.0167	.42418	.0207	.52578
.0048	.12192	.0088	.22352	.0128	.32512	.0168	.42672	.0208	.52832
.0049	.12446	.0089	.22606	.0129	.32766	.0169	.42926	.0209	.53086

TEN-THOUSANDTHS OF AN INCH TO MILLIMETERS (*Continued*)

INCH	MM	INCH	MM	INCH	MM	INCH	MM	INCH	MM
.0210 -	.53340	.0250 -	.63500	.0290 -	.73660	.0330 -	.83820	.0370 -	.93980
.0211 -	.53594	.0251 -	.63754	.0291 -	.73914	.0331 -	.84074	.0371 -	.94234
.0212 -	.53848	.0252 -	.64008	.0292 -	.74168	.0332 -	.84328	.0372 -	.94488
.0213 -	.54102	.0253 -	.64262	.0293 -	.74422	.0333 -	.84582	.0373 -	.94742
.0214 -	.54356	.0254 -	.64516	.0294 -	.74676	.0334 -	.84836	.0374 -	.94996
.0215 -	.54610	.0255 -	.64770	.0295 -	.74930	.0335 -	.85090	.0375 -	.95250
.0216 -	.54864	.0256 -	.65024	0296 -	.75184	.0336 -	.85344	.0376 -	.95504
.0217 -	.55118	.0257 -	.65278	.0297 -	.75438	.0337 -	.85598	.0377 -	.95758
.0218 -	.55372	.0258 -	.65532	.0298 -	.75692	.0338 -	.85852	.0378 -	.96012
.0219 -	.55626	.0259 -	.65786	.0299 -	.75946	.0339 -	.86106	.0379 -	.96266
.0220 -	.55880	.0260 -	.66040	.0300 -	.76200	.0340 -	.86360	.0380 -	.96520
.0221 -	.56134	.0261 -	.66294	.0301 -	.76454	.0341 -	.86614	.0381 -	.96774
.0222 -	.56388	.0262 -	.66548	.0302 -	.76708	.0342 -	.86868	.0382 -	.97028
.0223 -	.56642	.0263 -	.66802	.0303 -	.76962	.0343 -	.87122	.0383 -	.97282
.0224 -	.56896	.0264 -	.67056	.0304 -	.77216	.0344 -	.87376	.0384 -	.97536
.0225 -	.57150	.0265 -	.67310	.0305 -	.77470	.0345 -	.87630	.0385 -	.97790
.0226 -	.57404	.0266 -	.67564	.0306 -	.77724	.0346 -	.87884	.0386 -	.98044
.0227 -	.57658	.0267 -	.67818	.0307 -	.77978	.0347 -	.88138	.0387 -	.98298
.0228 -	.57912	.0268 -	.68072	.0308 -	.78232	.0348 -	.88392	.0388 -	.98552
.0229 -	.58166	.0269 -	.68326	.0309 -	.78486	.0349 -	.88646	.0389 -	.98806
.0230 -	.58420	.0270 -	.68580	.0310 -	.78740	.0350 -	.88900	.0390 -	.99060
.0231 -	.58674	.0271 -	.68834	.0311 -	.78994	.0351 -	.89154	.0391 -	.99314
.0232 -	.58928	.0272 -	.69088	.0312 -	.79248	.0352 -	.89408	.0392 -	.99568
.0233 -	.59182	.0273 -	.69342	.0313 -	.79502	.0353 -	.89662	.0393 -	.99822
.0234 -	.59436	.0274 -	.69596	.0314 -	.79756	.0354 -	.89916	.0394 -	1.00076
.0235 -	.59690	.0275 -	.69850	.0315 -	.80010	.0355 -	.90170	.0395 -	1.00330
.0236 -	.59944	.0276 -	.70104	.0316 -	.80264	.0356 -	.90424	.0396 -	1.00584
.0237 -	.60198	.0277 -	.70358	.0317 -	.80518	.0357 -	.90678	.0397 -	1.00838
.0238 -	.60452	.0278 -	.70612	.0318 -	.80722	.0358 -	.90932	.0398 -	1.01092
.0239 -	.60706	.0279 -	.70866	.0319 -	.81026	.0359 -	.91186	.0399 -	1.01346
.0240 -	.60960	.0280 -	.71120	.0320 -	.81280	.0360 -	.91440	.0400 -	1.01600
.0241 -	.61214	.0281 -	.71374	.0321 -	.81534	.0361 -	.91694	.0401 -	1.01854
.0242 -	.61468	.0282 -	.71628	.0322 -	.81788	.0362 -	.91948	.0402 -	1.02108
.0243 -	.61722	.0283 -	.71882	.0323 -	.82042	.0363 -	.92202	.0403 -	1.02362
.0244 -	.61976	.0284 -	.72136	.0324 -	.82296	.0364 -	.92456	.0404 -	1.02616
.0245 -	.62230	.0285 -	.72390	.0325 -	.82550	.0365 -	.92710	.0405 -	1.02870
.0246 -	.62484	.0286 -	.72644	.0326 -	.82804	.0366 -	.92964	.0406 -	1.03124
.0247 -	.62738	.0287 -	.72898	.0327 -	.83058	.0367 -	.93218	.0407 -	1.03378
.0248 -	.62992	.0288 -	.73152	.0328 -	.83312	.0368 -	.93472	.0408 -	1.03632
.0249 -	.63246	.0289 -	.73406	0329 -	.83566	.0369 -	.93726	.0409 -	1.03886

SQUARE IN. TO SQUARE CM.

sq. in.	sq. cm.	sq. in.	sq. cm.	sq. in.	sq. cm.	sq. in.	sq. cm.
1/4	1.613	7	45.16	45	290.3	200	1290
1/2	3.226	8	51.61	50	322.6	250	1613
3/4	4.839	9	58.06	55	354.8	300	1935
1	6.452	10	64.52	60	387.1	350	2258
1 1/2	9.677	12	77.42	65	419.4	400	2581
2	12.90	14	90.32	70	451.6	450	2903
2 1/2	16.13	16	103.2	75	483.9	500	3226
3	19.35	18	116.1	80	516.1	600	3871
3 1/2	22.58	20	129.0	85	548.4	700	4516
4	25.81	25	161.3	90	580.6	800	5161
4 1/2	29.03	30	193.5	95	612.9	900	5806
5	32.26	35	225.8	100	645.2	1000	6452
6	38.71	40	258.1	150	967.7		

1 sq. in. = 6.45160 sq. cm.

SQUARE CM. TO SQUARE IN.

sq. cm.	sq. in.	sq. cm.	sq. in.	sq. cm.	sq. in.	sq. cm.	sq. in.
1	0.155	18	2.790	80	12.40	600	93.00
2	0.310	20	3.100	85	13.18	700	108.5
3	0.465	25	3.875	90	13.95	800	124.0
4	0.620	30	4.650	95	14.73	900	139.5
5	0.775	35	5.425	100	15.50	1000	155.0
6	0.930	40	6.200	150	23.25	1500	232.5
7	1.085	45	6.975	200	31.00	2000	310.0
8	1.240	50	7.750	250	38.75	2500	387.5
9	1.395	55	8.525	300	46.50	3000	465.0
10	1.550	60	9.300	350	54.25	3500	542.5
12	1.860	65	10.08	400	62.00	4000	620.0
14	2.170	70	10.85	450	69.75	5000	775.0
16	2.480	75	11.63	500	77.50		

1 sq. cm. = 0.155000 sq. in.

GALLONS* TO LITERS

gal.	liters	gal.	liters	gal.	liters	gal.	liters
1	3.785	10	37.85	90	340.7	300	1136
2	7.571	20	75.71	100	378.5	400	1514
3	11.36	30	113.6	120	454.2	500	1893
4	15.14	40	151.4	140	529.9	600	2271
5	18.93	50	189.3	160	605.7	700	2650
6	22.71	60	227.1	180	681.4	800	3028
7	26.50	70	265.0	200	757.1	900	3407
8	30.28	80	302.8	250	946.3	1000	3785
9	34.07						

1 liter = 0.2641775 gal.*

LITERS TO GALLONS*

liters	gal.	liters	gal.	liters	gal.	liters	gal.
1	0.264	10	2.642	250	66.04	1000	264.2
2	0.528	20	5.284	300	79.25	2000	528.4
3	0.793	30	7.925	400	105.7	3000	792.5
4	1.057	40	10.57	500	132.1	4000	1057
5	1.321	50	13.21	600	158.5	5000	1321
6	1.585	100	26.42	700	184.9	6000	1585
7	1.849	150	39.63	800	211.3	8000	2113
8	2.113	200	52.84	900	237.8	10000	2642
9	2.378						

1 gal.* = 3.785334 liter.

SPECIFIC GRAVITIES AND WEIGHTS OF NONMETALLIC SUBSTANCES*

Substance	Sp gr	Weight, lb per cu ft
Bituminous substances		
Coal, anthracite	1.4–1.7	97
Coal, bituminous	1.2–1.5	84
Coal, coke	1.0–1.4	75
Petroleum, benzine	0.73–0.75	46
Petroleum, gasoline	0.66–0.69	42
Tar, bituminous	1.20	75
Coal and coke, piled		
Coal, anthracite	....	47–58
Coal, bituminous, lignite	....	40–54
Coal, coke	....	23–32
Timber, U.S. seasoned		
Ash, white-red	0.62–0.65	40
Cedar, white-red	0.32–0.38	22
Chestnut	0.66	41
Cypress	0.48	30
Fir, Douglas spruce	0.51	32
Fir, eastern	0.40	25
Elm, white	0.72	45
Hemlock	0.42–0.52	29
Hickory	0.74–0.84	49
Locust	0.73	46
Maple, hard	0.68	43
Maple, white	0.53	33
Oak, chestnut	0.86	54
Oak, live	0.95	59
Oak, red, black	0.65	41
Oak, white	0.74	46
Pine, Oregon	0.51	32
Pine, red	0.48	30
Pine, white	0.41	26
Pine, yellow, longleaf	0.70	44
Pine, yellow, shortleaf	0.61	38
Poplar	0.48	30
Redwood, California	0.42	26
Spruce, white, black	0.40–0.46	27
Walnut, black	0.61	38
Walnut, white	0.41	26
Moisture contents:		
Seasoned timber, 15 to 20%		
Green timber, up to 50%		

SPECIFIC GRAVITIES AND WEIGHTS OF NONMETALLIC SUBSTANCES* (*Continued*)

Substance	Sp gr	Weight, lb per cu ft
Concrete masonry		
Cement, stone, sand	2.2–2.4	144
Cement, slag, etc	1.9–2.3	130
Cement, cinder, etc	1.5–1.7	100
Building material		
Ashes, cinders	....	40–45
Cement, portland, loose	....	90
Cement, portland, set	2.7–3.2	183
Lime, gypsum, loose	....	53–64
Mortar, set	1.4–1.9	103
Gases, air = 1		
Air, 0 C, 760 mm	1.0	0.08071
Ammonia	0.5920	0.0478
Carbon dioxide	1.5291	0.1234
Carbon monoxide	0.9673	0.0781
Gas, illuminating	0.35–0.45	0.028–0.036
Gas, natural	0.47–0.48	0.033–0.039
Hydrogen	0.0693	0.00559
Nitrogen	0.9714	0.0784
Oxygen	1.1056	0.0892
Liquids		
Alcohol, 100%	0.79	49
Acids, muriatic 40%	1.20	75
Acids, nitric 91%	1.50	94
Acids, sulfuric 87%	1.80	112
Lye, soda 66%	1.70	106
Oils, mineral, lubricants	0.90–0.93	57
Water, 4 C, max density	1.0	62.428
Water, 100 C	0.9584	59.830
Water, ice	0.88–0.92	56
Water, snow, fresh fallen	0.125	8
Water, sea water	1.02–1.03	64

* Baldwin Southwark Div., Baldwin Locomotive Works.

Index

Subjects are followed by double numbers to indicate each page reference. The bold-face number indicates the section, the light-face number is the page number in that section.

T

U

V

W

Z